215/04.09.09
20 071 070 4X
AF762225
UNIVERSITY OF NORTHUMBRIA
AT NEWCASTLE LIBRARY
WITHDRAWN

ENCYCLOPEDIA OF QUATERNARY SCIENCE

ENCYCLOPEDIA OF QUATERNARY SCIENCE

Editor-in-Chief

SCOTT A. ELIAS

ELSEVIER

Amsterdam • Boston • Heidelberg • London • New York • Oxford
Paris • San Diego • San Francisco • Singapore • Sydney • Tokyo

The Boulevard, Langford Lane, Kidlington, Oxford OX5 1GB, UK
Radarweg 29, PO Box 211, 1000 AE Amsterdam, The Netherlands

First edition 2007

Copyright © 2007 Elsevier B.V. All rights reserved

The following articles are US government works in the public domain and are not subject to copyright:
ICE CORE RECORDS/Thermal Diffusion Paleotemperature Records
LOESS DEPOSITS, ORIGINS AND PROPERTIES
LOESS RECORDS/North America
PALEOCEANOGRAPHY
PALEOCEANOGRAPHY, RECORDS/Late Pleistocene North Atlantic
PALEOCEANOGRAPHY, RECORDS/Postglacial North Atlantic
PALEOCLIMATE RECONSTRUCTION/Pliocene Environments
PALEOSOLS AND WIND-BLOWN SEDIMENTS/Overview
POLLEN RECORDS, LATE PLEISTOCENE/Western North America

No part of this publication may be reproduced, stored in a retrieval system or transmitted in any form or by any means electronic, mechanical, photocopying, recording or otherwise without the prior written permission of the publisher

Permissions may be sought directly from Elsevier's Science & Technology Rights Department in Oxford, UK: phone (+44) (0) 1865 843830; fax (+44) (0) 1865 853333; email: permissions@elsevier.com. Alternatively you can submit your request online by visiting the Elsevier web site at http://elsevier.com/locate/permissions, and selecting *Obtaining permission to use Elsevier material*

Notice
No responsibility is assumed by the publisher for any injury and/or damage to persons or property as a matter of products liability, negligence or otherwise, or from any use or operation of any methods, products, instructions or ideas contained in the material herein. Because of rapid advances in the medical sciences, in particular, independent verification of diagnoses and drug dosages should be made

Library of Congress Catalog Number: 2006929723

British Library Cataloguing in Publication Data
A catalogue record for this book is available from the British Library

ISBN-13: 978-0-444-51919-1
ISBN-10: 0-444-51919-X

For information on all Elsevier publications visit our website at books.elsevier.com

Printed and bound in Spain

06 07 08 09 10 10 9 8 7 6 5 4 3 2 1

Working together to grow libraries in developing countries
www.elsevier.com | www.bookaid.org | www.sabre.org
ELSEVIER BOOK AID International Sabre Foundation

NORTHUMBRIA UNIVERSITY
LIBRARY
1447279
ITEM No. 20071 0704X
CLASS No. 551. 7903 ENC

NORTHUMBRIA UNIVERSITY
LIBRARY
1447279
ITEM No. 20071 0704X
CLASS No. 551. 7903 ENC

DEDICATION

We dedicate this encyclopedia to the memory of Professor Sir Nicholas Shackleton, or 'Nick' as he preferred to be called, who died 24 January 2006. He needs no introduction to Quaternary scientists, but his contributions to the geosciences, and particularly to palaeoceanography, were truly outstanding. Nick was born in 1937; his father was an eminent geologist and his mother was a professional musician. Both of these fields were to play important roles in his life. Nick studied at Cambridge University, and went on to join the faculty there. Early in 1961, Harry Godwin encouraged Nick to analyze oxygen isotopes taken from fossil Foraminifera from deep sea sediments for his doctoral research. Nick was awarded a PhD in 1967 for his thesis entitled 'The measurement of palaeotemperatures in the Quaternary Era.' This was a ground-breaking piece of research, demonstrating that the dominant cause of changes in oxygen isotope composition in these marine fossils was not temperature, as had previously been thought, but the isotopic composition of the ocean water. From then on, Nick remained at the forefront of research in the field, through the continued development and refinement of the techniques required for high-resolution isotope stratigraphy. These methods were used to obtain the first high resolution isotope records showing the history of ice-sheet fluctuations during the Quaternary. Perhaps his best-known achievement was the establishment of a time scale for the isotope fluctuations that had been defined (Marine Isotope Stages). Shackleton and Neil Opdyke obtained a core from the eastern Pacific which extended beyond the Brunhes-Matuyama magnetic reversal (0.78 Ma). Their work demonstrated that global ice volume could be determined throughout the intervals of Pleistocene glaciation. This classic study (Shackleton and Opdyke, 1973) on marine core V28-238 demonstrated that the isotopic signature of fossils in deep sea sediments, even from tropical waters, represents a global record, not just local conditions. This set the stage for many subsequent advances paleoceanography that have revolutionized Quaternary science.

The carbon isotopic composition of the oceans is also affected by storage on land. In 1983, Nick Shackleton demonstrated that atmospheric carbon dioxide concentrations had changed through the last glacial cycle, in parallel with those found in the Vostok ice-core record from Antarctica. By comparing the timing of the records from ice cores and the deep-ocean sediments, it was possible to show that ice volume (as shown by the oxygen-isotope record) lags behind the carbon dioxide variations. Therefore Nick concluded that variations in the volume of ice sheets do not drive changes in atmospheric carbon dioxide concentration. Rather, changes in carbon dioxide concentration appear to play a significant role in the climatic changes associated with interglacial to glacial transitions. The recognition that changes in atmospheric carbon dioxide concentration play such a central role in the Earth's climatic variation was a profoundly important observation, the important implications for future global warming scenarios. In later years, Nick's work continued with the use of various proxy climatic indicators (isotopes, lithology, faunal counts, magnetic susceptibility) to investigate the relationship between orbital forcing and climatic change. One of the results was the development of high-resolution "astronomically tuned" time scales for geological sequences. The new chronologies have brought about re-evaluations of the supposed ages of crucial boundary events such as the Matuyama-Brunhes geomagnetic polarity reversal (Shackleton et al. 1990).

Nick was elected to the Royal Society in 1985 and knighted in 1998 in recognition of his services to Earth Sciences. His many awards and medals include the Crafoord Prize (1995), the Milankovitch Medal (1999), the Wollaston Medal (1996), the Vetlesen Prize (2004) and, most recently, the Blue Planet Prize (2005). In 2000 he was made a Foreign Associate of the U.S. Academy of Sciences and he was elected President of the International Union for Quaternary Research (INQUA) for 1999–2003. In addition to his remarkable achievements in climate history and palaeoceanography, he was a devoted clarinet player and student of the instrument. He built an unrivalled personal collection of woodwind instruments during his lifetime.

Nick Shackleton's career exemplifies the kind of multidisciplinary science that makes Quaternary research both fascinating to undertake and important for society. Our current knowledge of the complexity and rapidity of climate change, and therefore of environmental change during recent earth history, rests upon the concepts established by Nick and his co-workers. Sir Isaac Newton has famously said, "If I have seen further it is by standing on the shoulders of Giants." It is fair to say that Quaternary scientists have been standing on Nick Shackleton's shoulders for the last 40 years.

References cited:

Shackleton, N. J. & Opdyke, N. 1973: Oxygen isotope and palaeomagnetic stratigraphy of the equatorial Pacific core V28-238: oxygen isotope temperatures and ice volumes on a 10^5 and 10^6 year scale. Quaternary Research 3, 39–55.

Shackleton, N. J., Berger, A. & Peltier, W. R. 1990: An alternative astronomical calibration of the lower Pleistocene timescale based on ODP site 677. Transactions of the Royal Society of Edinburgh: Earth Sciences 81, 251–261.

EDITORIAL ADVISORY BOARD

EDITOR-IN-CHIEF

Scott A. Elias
Royal Holloway, University of London, Surrey, UK

ASSOCIATE EDITORS

David M Anderson
NOAA Paleoclimatology Program, Boulder, CO, USA

Richard Bradshaw
University of Liverpool, Liverpool, UK

Ed Brook
Oregon State University, Corvallis, OR, USA

Marianne Douglas
University of Toronto, Toronto, ON, Canada

Mary E Edwards
University of Southampton, Southampton, UK

Jürgen Ehlers
Universitat Hamburg, Hamburg, Germany

David J A Evans
University of Durham, Durham, UK

Clive Gamble
Royal Holloway, University of London, Surrey, UK

Philip L. Gibbard
University of Cambridge, Cambridge, UK

AJT Jull
University of Arizona, Tucson, AZ, USA

Cary Mock
University of South Carolina, Columbia, SC, USA

Dan Muhs
U.S. Geological Survey, Denver, CO, USA

Brad Pillans
Australian National University, ACT, Australia

Danielle Schreve
Royal Holloway, University of London, Surrey, UK

Henry Schwarcz
McMaster University, Hamilton, ON, Canada

Ian Shennan
Durham University, Durham, UK

Torbjörn E. Törnqvist
Tulane University, New Orleans, USA

SECTION EDITORS

Hilary Birks
University of Bergen, Bergen, Norway and Bjerknes Centre for Climate Research, Norway

John Gosse
Dalhousie University, Halifax, NS, Canada

Charles Harris
Cardiff University, Cardiff, UK

Konrad Hughen
Woods Hole Oceanographic Institution, Woods Hole, MA, USA

Vivienne Jones
University College, London, UK

Olav Lian
University College of the Fraser Valley, King Road, BC, Canada

Gifford Miller
University of Colorado, Boulder, CO, USA

Dan Smith
University of Victoria, Victoria, BC, Canada

FOREWORD

At these times, when we are all aware of climate and environmental change, it has never been more important to seek a perspective of the scale and rates at which the characteristics of the natural environment have developed and changed in response to natural forces. These changes concern all aspects of the Earth upon which we live: the land-surface, soils, rivers, waves and glaciers, plants, animals, oceans and climate, as well as the changes from within the earth: earthquakes and volcanic eruptions. We now know that over the period during which humans have been on the Earth, all of these processes, and more, have changed frequently, often rapidly and at different scales in different places. The subject that has developed to investigate and understand these patterns and changes is Quaternary Science.

By definition, Quaternary Science is multi-disciplinary, spanning the geography, biology, physics and chemistry of the Earth. By nature of the complexity of changes that occur across space and through time, Quaternary Science is a complex subject requiring an understanding of: 1) present and past processes, 2) a wide range of methods for determining past processes, climates and environments, 3) the timescale at which these past processes occurred and 4) the methods of modelling these past climates and environments in order to understand what caused the changes in the past. Only with this knowledge will we be able us to attempt to predict the changes that may occur in the future. Truly, there are few subjects that are more complex and challenging and more relevant to society at the present day.

The *Encyclopedia of Quaternary Science* seeks to provide an authoritative and comprehensive coverage of this subject. It is the latest in the series of major reference works, published by Elsevier, on such topics as Geology, Ocean Sciences and Atmospheric Sciences. The encyclopedia will consist of four volumes and is edited by Scott Elias of Royal Holloway, University of London, UK. Briefly, it will aim to provide the definitive work on the nature and causes of changes in the natural environment of the Earth over the last c. 2.6 million years. The encyclopedia will be produced in both hard copy as four books and as an electronic reference source on the Elsevier website ScienceDirect (http://www.sciencedirect.com).

The *Encyclopedia of Quaternary Science* covers the major topics of the subject, each edited by an authority within the field. For instance dating methods is edited by Tim Jull, Quaternary stratigraphy by Brad Pillans, paleoclimate by Cary Mock, palaeoceanography by David Anderson, ice core records by Ed Brook, Quaternary glaciations by David Evans and Quaternary vertebrates by Danielle Schreve, amongst many others. Each of the editors have called on specialists within their particular field, who have provided within the volume template of c. 4,000 words, a brief, up-to-date description of the topic and an explanation that is in accord with current thinking. Colour photographs and diagrams are used liberally to elaborate the text and all the subjects are supported by a carefully selected list of references. When all the references are considered together they provide an un-rivalled bibliography of Quaternary Science.

Behind all of the content material, the Editor has attempted to emphasize the inter-connectedness of Quaternary science, and he seeks to stress the multidisciplinary nature of the science and the immense strength of the multiproxy approach to Quaternary palaeoenvironmental reconstruction. He sees this as the hallmark of the discipline, and draws attention to the wide range of skills required to understand and contribute to Quaternary science, maybe in some cases as an individual, but in most cases as a member of a multi-disciplinary (and usually multi-national) team.

He has attempted to make the volume accessible and of value to all levels of the science, from the university undergraduate through to the academic and consulting professional. This connectedness is seen as the *raison d'etre* of the encyclopedia, and the example is drawn that the Palaeolithic archaeologist can only make sense of prehistory if there is adequate knowledge of the environment in which the early humans lived. This environmental evidence is provided by the landforms and drainage (geomorphology), sediments (sedimentology), soils (soil science and geochemistry), vegetation and animals (palaeobotany and palaeozoology respectively) and climate (palaeoclimatology, geochemistry, biological climatic proxies). Likewise the palaeoceanographer can only understand the past ocean currents, ocean temperatures, heat transport, biota and sea-level with reference to the taxonomy and environmental requirements of organisms that lived in and around the ocean (foraminifera and coral), the stable isotope geochemistry of the organisms that lived in the ocean, the sea-bed sediment properties and the coastal landforms.

In both of these examples, and in all others, Quaternary scientists must collaborate with those who can provide age estimates of the past environmental and climatic events, and those able to correlate and test the validity of like events in different places, or in some cases, dissimilar events that occurred at similar times in different places.

In addition there are those who evaluate the influence of human activity upon the natural environment. Some see local changes to vegetation, animal populations, and soil erosion that appear with the occurrence of humans in the landscape, others see the presence of humans as causing incremental changes, that over time are of sufficient importance to actually change the global climate. Cleary, without an understanding of how the complex natural environment works it would be impossible to evaluate human influence, and our chances of continuing to live successfully upon planet Earth would be diminished.

This major reference work gives access to all the various topics within the multi-facetted sphere of Quaternary Science. It does so in a way that is authoritative and accessible.

James Rose
Royal Holloway, University of London,
Surrey, UK

INTRODUCTION

It has been my privilege in developing this encyclopedia to work with more than 400 leading Quaternary scientists. This team of writers and editors represent 28 countries in Europe, Asia, Africa, the Americas, Australia and New Zealand. For the past three years, I have had my finger on the pulse of Quaternary science, and this branch of science is truly pulsating! Information now comes from an incredible variety of disciplines: geochemistry, numerical modelling, history, vulcanology, paleobiology, nuclear physics, stratigraphy, sedimentology, climatology, anthropology, archaeology, glacial geology, soil science, ice-stream modelling – the list is staggering. This highly disparate group of people are bound together by one common thread: the desire to know the history of the planet during the last 2.6 million years – the time of the ice ages. For Quaternary scientists, this is a pressing need, not an idle curiosity. Any doubts about this statement can be easily dispelled by a consideration of the lengths to which many of them go to gather the necessary data. Some of them have worked for months in sub-zero temperatures on top of very high mountains or near the center of polar ice sheets, collecting ice cores. Others have spent many weeks on some of the roughest seas in the world, drilling deep-sea sediment cores. Often the work is more mundane. An oxygen isotope curve for a lengthy marine sediment core represents thousands of hours of patiently picking tiny fossils from layer after layer of sediments, in order to obtain sufficient numbers of calcium-carbonate shells to yield samples for isotopic analysis. A map showing proposed ice limits from the last glaciation represents thousands of hours of field mapping of glacial features by dozens of people. Why do all of these people devote their lives to this pursuit of knowledge? Does it really matter so much? The answer becomes clear when you step back and examine the topic of Quaternary science in its proper context.

The world we inhabit has largely been shaped by the events of the Quaternary. All the biological communities that exist today are the end product of a long series of species associations that came together in the past, largely driven by climatic change during the Pleistocene. We cannot properly understand the functioning of modern ecosystems without a solid knowledge of their history, any more than we can understand the plot of a long novel by reading just the final page. We are also living in a time of alarming climate changes. Even though the pace and intensity of some of these changes has not been seen in historical times, there were many rapid, large-scale climatic shifts in the Pleistocene. The best way to predict the effects of global warming on the planet's climate and ecosystems is to look at the history of similarly intense, rapid changes in the prehistoric past. The interval that is most relevant to today is the most recent geologic period: the Quaternary. As human populations rise exponentially, increasing numbers of people are exposed to geologic hazards, such as earthquakes (and attendant tsunamis) and volcanic eruptions. These are short-lived events that take place only rarely in any one region. The interval between major events, such as volcanic eruptions, may be centuries or millennia. How do we come to grips with predicting the future likelihood of such erratic phenomena? Again, the answers come from piecing together the ancient history of such events, over many thousands of years.

The Quaternary has been the time when our own species came of age. The beginning of the Quaternary, roughly 2.6 million years ago, was about the time when the earliest member of our genus (*Homo*) first appeared in Africa. Pleistocene environments shaped the course of human evolution, culminating in anatomically modern *Homo sapiens* spreading from Africa throughout most of the world during the last glaciation. Even though human beings largely shape their own environments today, for the vast majority of our species' history, it has been the environment that has shaped us. Our direct ancestors' adaptations to environmental change are deeply ingrained in our genes. Thus an understanding of the environmental conditions that shaped our species is critical to our understanding of humanity.

Quaternary science is a rapidly changing field, and the articles that appear in this encyclopedia reflect this. New dating techniques, such as cosmogenic nuclide dating, are revolutionizing our understanding of many earth surface processes. The ability to analyze increasingly smaller samples for radiocarbon and stable isotopes of oxygen and hydrogen means that we are gaining a level of precision in the reconstruction of past events that was unheard of, just a few years ago. Stable isotope studies of air bubbles trapped in ice cores from Greenland and Antarctica have given Quaternary scientists an entirely new perspective on the rapidity and intensity of climatic change during the last glacial cycle and beyond. Likewise, the discovery of long sequences of annually laminated sediments in both marine and freshwater environments has provided a great leap forward in our ability to resolve the timing of environmental changes in non-polar regions. The ability to extract and analyze ancient DNA sequences from Pleistocene fossils (both plants and animals) is revolutionizing the field of paleobiology. We are beginning to be able to trace the genetic lineages of a number of different organisms, from beetles to bison. In short, these are very exciting times to be a Quaternary scientist! While it is virtually impossible for any Quaternary researcher or student to keep abreast of all the new discoveries in this multi-faceted science, this encyclopedia can be a great help. The articles contained here represent the state-of-the-art in a huge variety of topics, and they offer the opportunity to dig deeper into their respective subjects by providing full citations of the most pertinent literature available. I invite you to come and explore the Quaternary Period in the pages that follow. It is a fascinating story.

SCOTT A. ELIAS

GUIDE TO USE OF THE ENCYCLOPEDIA

Structure of the Encyclopedia

The material in this Encyclopedia is organised into two broad sections. At the start of Volume 1 are four **Introductory Articles**. The introductory articles serve as a broad introduction to the field. It is intended that these articles should serve as the first port of call for readers, to enable them to embark on any of the main entries without the need to consult other material beforehand.

Following the Introductory Articles, the main body of the Encyclopedia is arranged as a series of entries in alphabetical order. These entries fill the remainder of Volume 1 and all of the subsequent volumes (2–4).

To help you realize the full potential of the material in the Encyclopedia we have provided four features to help you find the topic of your choice: a contents list by subject, an alphabetical contents list, cross-references, and a full subject index.

1. Contents list by subject

Your first point of reference will probably be the contents list by subject. This list appears at the front of each volume, and groups the entries under subject headings describing the broad themes of quaternary science. This will enable the reader to make quick connections between entries and to locate the entry of interest. Under each main section heading, you will find several subject areas and under each subject area is a list of those entries that cover aspects of that subject, together with the volume and page numbers on which these entries may be found.

2. Alphabetical contents list

The alphabetical contents list, which also appears at the front of each volume, lists the entries in the order in which they appear in the Encyclopedia. This list provides both the volume number and the page number of each entry. On the opening page of an entry a contents list is provided so that the full details of any articles within the entry are immediately available.

You will find "dummy entries" where obvious synonyms exist for entries or where we have grouped together related topics. Dummy entries appear in both the contents list and the body of the text.

Example

If you were attempting to locate material on peat studies via the alphabetical contents list:

PEAT STUDIES *see* PLANT MACROFOSSIL METHODS AND STUDIES: Mire and Peat Macros; Holocene Peatland Records

The dummy entry directs you to two other entries in which peat studies are covered. At the appropriate locations in the contents list, the volume and page numbers for these entries are given.

If you were trying to locate the material by browsing through the text and you had looked up Peat Studies, then the following information would be provided in the dummy entry:

Peat studies *see* **Plant Macrofossil Methods and Studies:** Mire and Peat Macros; Holocene Peatland Records

3. Cross-references

All of the entries in the Encyclopedia have been extensively cross-referenced. The cross-references, which appear at the end of an entry, serve three different functions:

i. To indicate if a topic is discussed in greater detail elsewhere
ii. To draw the reader's attention to parallel discussions in other entries
iii. To indicate material that broadens the discussion

Example
The following list of cross-references appears at the end of the entry GLACIAL LANDFORMS, ICE SHEETS/ Growth and Decay.

See also: **Introduction**/History of Quaternary Science. **Glacial Climates**/Thermohaline Circulation. **Glaciation, Causes**/Milankovitch Theory and Paleoclimate. **Glaciations**/Overview. **Ice Core Records**/Correlations Between Greenland and Antarctica. **Paleoceanography. Paleoclimate Modeling**/Last Glacial Maximum GCMs. **Paleoclimate Reconstruction**/Sub-Milankovitch (DO/ Heinrich) Events. **Sea Level Studies**/Overview.

Here you will find examples of all three functions of the cross-reference list: a topic discussed in greater detail elsewhere (e.g. **Paleoclimate Modeling**: Last Glacial Maximum GCMs), parallel discussion in other entries (e.g. **Glacial Climates**: Thermohaline Circulation) and reference to entries that broaden the discussion (e.g. **Glaciations**: Overview).

4. Index

The index will provide you with the page number where the material is located. The index entries differentiate between material that is a whole entry, is part of an entry, or is data presented in a figure or table. Detailed notes are provided on the opening page of the index.

5. Contributors

A full list of contributors appears at the beginning of each volume.

LIST OF CONTRIBUTORS

L Abbazzi
University of Florence, Firenze, Italy

S T Abbott
Southern Cross University, NSW, Australia

J S Aber
University of Minnesota, MN, USA

F Abrantes
Instituto Nacional de Engenharia, Amadora, Portugal

B Alloway
GNS Science, Lower Hutt, New Zealand

B Ammann
Institute of Plant Sciences, Bern, Switzerland

D M Anderson
University of Colorado, CO, USA

A A Andreev
Alfred-Wegener-Institut für Polar- und Meerforschung, Potsdam, Germany

J T Andrews
University of Colorado, CO, USA

P Antoine
Paris, France

D Antoniades
Université Laval, QC, Canada

A C Ashworth
North Dakota State University, ND, USA

A Aslan
Mesa State College, CO, USA

V Astakhov
St. Petersburg University, St. Petersburg, Russia

D E Atkinson
University of Alaska Fairbanks, AK, USA

W Austin
University of St. Andrews, St. Andrews, UK

G Bailey
University of York, York, UK

R G Baker
University of Iowa, Iowa, USA

G Balco
University of Minnesota, MN, USA

C K Ballantyne
University of St. Andrews, St. Andrews, UK

K E Barber
University of Southampton, Southampton, UK

D Barboni
Université Paul Cézanne, Aix-en-Provence, France

R Barendregt
University of Minnesota, MN, USA

P A Barker
Lancaster University, Lancaster, UK

S Barker
Lamont-Doherty Earth Observatory of Columbia University, NY, USA

I Barnes
Rayal Holloway, University of London, Ergham, UK

P J Bartlein
University of Oregon, OR, USA

F C Bassinot
LSCE, Gif-sur-Yvette, France

H Behling
Georg-August-University of Göttingen, Göttingen, Germany

C J Bell
The University of Texas at Austin, Texas, USA

D I Benn
University of St Andrews, St Andrews, UK

O Bennike
Geological Survey of Denmark and Greenland, Copenhagen, Denmark

M J Bentley
Durham University, Durham, UK

A Berger
Université Catholique de Louvain, Louvain-la-Neuve, Belgium

S M Bernasconi
Geological Institute, Zurich, Switzerland

J C Berrio
University of Leicester, Leicester, UK

E A Bettis III
University of Iowa, IA, USA

P R Bierman
University of Minnesota, MN, USA

N H Bigelow
University of Alaska Fairbanks, AK, USA

R Bindschadler
NASA Goddard Space Flight Center, MD, USA

M I Bird
University of St. Andrews, St. Andrews, UK

H H Birks
University of Bergen, Bergen, Norway

H J B Birks
University of Bergen, Bergen, Norway

L C Bishop
Liverpool John Moores University, Liverpool, UK

S Björck
Geobiosphere Science Centre, Lund, Sweden

M D Blum
Louisiana State University, LA, USA

H Bocherens
Universität Tübingen, Tübingen, Germany

R K Booth
Lehigh University, PA, USA

K V Boyle
University of Cambridge, Cambridge, UK

R Bradshaw
University of Liverpool, Liverpool, UK

D R Braun
Rutgers University, NJ, USA

S Brewer
Université Paul Cézanne, Aix-en-Provence, France

J Brigham-Grette
University of Massachusetts, MA, USA

A J Broccoli
Rutgers University, NJ, USA

E J Brook
Oregon State University, OR, USA

S J Brooks
The Natural History Museum, London, UK

V Brovkin
Potsdam Institute for Climate Impact Research (PIK), Potsdam, Germany

A Brunelle
University of Utah, UT, USA

S L Buchardt
University of Copenhagen, Copenhagen, Denmark

M J Bunting
University of Hull, Hull, UK

C R Burn
Carleton University, ON, Canada

G S Burr
University of Arizona, AZ, USA

N G Cameron
University College London, London, UK

C Carcaillet
Centre of Bio-Archaeology and Ecology, Montpellier, France

S J Carpenter
The University of Iowa, IA, USA

J L Carrivick
University of Leeds, Leeds, UK

J A Carter
Victoria University of Wellington, Wellington, New Zealand

R M Carter
James Cook University, Townsville, Australia

T Cerling
University of Utah, UT, USA

Y Chaimanee
Bureau of Geological Survey, Bangkok, Thailand

J Chappellaz
Laboratoire de Glaciologie et Géophysique de l'Environnement, St. Martin d'Hères, France

B M Chase
University of Oxford, Oxford, UK

R Cheddadi
Universite Montpellier II, Montpellier, France

M Chenoweth
Elkridge, MD, USA

B C Christner
Montana State University, MT, USA

S J Clarke
Charles Sturt University, New South Wales, Australia

A E Close
University of Washington, WA, USA

C Ó Cofaigh
University of Durham, Durham, UK

P M Colgan
Grand Valley State University, MI, USA

E Colhoun
University of Newcastle, Newcastle, Australia

G T Cook
Scottish Universities Environmental Research Centre, East Kilbride, UK

G R Coope
University of London, Egham, UK

J A G Cooper
University of Ulster, Coleraine, UK

A Coronato
Centro Austral de Investigaciones Científicas, Ushuaia, Argentina

E Cortijo
CEA-CNRS, Gif sur Yvette, France

R Cosgrove
La Trobe University, Melbourne, Australia

K M Cuffey
University of California, Berkeley, USA

W Davies
University of Southampton, Southampton, UK

M E Davis
The Ohio State University, Columbus, USA

J-L de Beaulieu
Centre National de la Recherche Scientifique, Aix-En-Provence, France

T de Garidel-Thoron
CEREGE-CNRS, Aix-Marseille Université, France

A de Vernal
Université du Québec à Montréal, Montréal, Canada

J de Vos
Nationaal Natuurhistorisch Museum, Leiden, The Netherlands

M Delfino
University of Florence, Firenze, Italy

R Dennell
University of Sheffield, Sheffield, UK

E Derbyshire
Royal Holloway, University of London, Surrey, London

A C Dieffenbacher-Krall
University of Maine, ME, USA

A E Dodonov
Russian Academy of Sciences, Moscow, Russia

J R Dodson
Brunel University, London, UK

M S V Douglas
University of Toronto, ON, Canada

H J Dowsett
U S Geological Survey, VA, USA

D G Drucker
Allée de l'Université, Nanterre, France

A S Dyke
Geological Survey of Canada, Ottawa, Canada

M E Edwards
University of Southampton, Southampton, UK

R J Edwards
University of Dublin, Dublin, Ireland

H Eggermont
Ghent University, Ghent, Belgium

J Ehlers
Geologisches Landesamt, Hamburg, Germany

S A Elias
Royal Holloway, University of London, Surrey, UK

J M Erickson
St. Lawrence University, NY, USA

D J A Evans
Durham University, Durham, UK

I S Evans
Durham University, Durham, UK

N Eyles
University of Toronto, ON, Canada

M P Ferretti
University of Florence, Firenze, Italy

B Finney
University of Alaska, AK, USA

J-A Flores
Universidad de Salamanca, Salamanca, Spain

R J Flower
University College London, London, UK

C M Foreman
Montana State University, MT, USA

R Francois
The University of British Columbia, BC, Canada

G Frédéric
University Paul Cezanne (Aix-Marseilles III), Aix-en-Provence cedex, FRANCE

S C Fritz
University of Nebraska, NE, USA

G Frühbeck
Universitaria de Navarra, Pamplona, Spain

H Gärtner
Swiss Federal Research Institute WSL, Birmensdorf, Switzerland

M-J Gaillard
University of Kalmar, Kalmar, Sweden

C Gamble
Royal Holloway, University of London, Surrey, UK

W R Gehrels
University of Plymouth, Plymouth, UK

Philip L. Gibbard
University of Cambridge, Cambridge, UK

T Giesecke
University of Liverpool, Liverpool, UK

M T P Gilbert
University of Copenhagen, Copenhagen, Demmark

L Giosan
Woods Hole Oceanographic Institution, MA, USA

J Gómez-Ambrosi
Universitaria de Navarra, Pamplona, Spain

S Gonzalez
Liverpool John Moores University, Liverpool, UK

S L Goodbred Jr.
Vanderbilt University, TN, USA

J C Gosse
Dalhousie University, Nova Scotia, Canada

A M Grachev
Oregon State University, OR, USA

D E Granger
Purdue University, IN, USA

J O Grimalt
Institute of Chemical and Environmental Research (CSIC), Barcelona, Spain

E C Grimm
Illinois State Museum, IL, USA

R Grün
Australian National University, Canberra, Australia

T P Guilderson
University of California, CA, USA

J Guiot
Université Paul Cézanne, Aix-en-Provence, France

J Hall
Glendale Community College, CA, USA

C R Harington
Canadian Museum of Nature, ON, Canada

C Harris
Cardiff University, Cardiff, UK

J W K Harris
Rutgers University, NJ, USA

S P Harrison
University of Bristol, Bristol, UK

C Hatté
University of Arizona, Arizona, USA

M Hayashi
Hoshizaki Green Foundation, Izumo, Japan

O Heiri
Ultrecht University, Ultrecht, The Netherlands

S R Hemming
Lamont-Doherty Earth Observatory, NY, USA

S Hicks
University of Oulu, Oulu, Finland

J F Hiemstra
University of Wales, Swansea, UK

B Hönisch
University of Bremen, MARUM, Germany

H Hooghiemstra
University of Amsterdam, Amsterdam, The Netherlands

B P Horton
University of Pennsylvania, PA, USA

K A Hughen
Woods Hole Oceanographic Institution, MA, USA

P D Hughes
University of Manchester, Manchester, UK

C S Hvidberg
University of Copenhagen, Copenhagen, Denmark

Ó Ingólfsson
University of Iceland, Reykjavík, Iceland

S Ivy-Ochs
Institute of Particle Physics, Zürich, Switzerland

C S Jackson
The University of Texas at Austin, Austin, USA

S T Jackson
University of Wyoming, WY, USA

P M Jacobs
University of Wisconsin-Whitewater, WI, USA

S Jacomet
Basel University, Basel, Switzerland

C E Jennings
University of Minnesota, MN, USA

S J Johnsen
Niels Bohr Institute, University of Copenhagen, Denmark

R T Jones
University of Exeter, Exeter, UK

V Jones
University College London, London, UK

J T Jordan
University of Exeter, Exeter, UK

J Jouzel
Laboratoire des Sciences du Climat et de l'Environnement, Gif-sur-Yvette, France

A J Timothy Jull
University of Arizona, AZ, USA

A Kääb
University of Oslo, Oslo, Norway

X Ke
University of Regina, Saskatchewan, Canada

A E Kehew
Western Michigan University, MI, USA

J Keltner
Pennsylvania State University, PA, USA

R A Kemp
University of London, Surrey, UK

P Kershaw
Monash University, Vic, Australia

M W Kerwin
University of Denver, CO, USA

T Kiefer
PAGES, Bern, Switzerland

F Kienast
Alfred Wegener Institute for Polar and Marine Research, Potsdam, Germany

F Kober
Institute of Geology, Zürich, Switzerland

N Koç
Norwegian Polar Institute, Tromso, Norway

I Koizumi
Atsubetsu-kita, Sapporo, Japan

A Korhola
University of Helsinki, Helsinki, Finland

A Koutavas
The City University of New York, New York, USA

A L Kozlowski
Susquehanna University, Selinsgrove, PA, USA

K J Kreutz
University of Maine, ME, USA

S Kuzmina
University of London, Surrey, UK

L Labeyrie
Université Versailles St. Quentin and CEA-CNRS, Gif sur Yvette, France

D Lal
Scripps Institution of Oceanography, CA, USA

N Lancaster
Desert Research Institute, NV, USA

G Larsen
University of Iceland, Reykjavik, Iceland

D Lazarus
Humboldt Universität, Berlin, Germany

M Lazorek
University of Toronto, ON, Canada

M J Leng
University of Nottingham, Nottingham, UK

D Lewis
University of Victoria, Victoria, Canada

A-M Lézine
Centre National de la Recherche Scientifique, Gif-sur-Yvette Cedex, France

O B Lian
University College of the Fraser Valley, British Columbia, Canada

N Lifton
Lamont-Doherty Earth Observatory, NY, USA

C S Lingle
University of Alaska, Fairbanks, USA

B Linsley
University at Albany, NY, USA

K Liu
Louisiana State University, LA, USA

N Loader
University of Wales – Swansea, Swansea, UK

J F Lopez
Department of Environmental Chemistry, Barcelona, Spain

M L Lord
Western Carolina University, NC, USA

P Loubere
Northern Illinois University, IL, USA

M-F Loutre
Université Catholique de Louvain, Louvain-la-Neuve, Belgium

D J Lowe
University of Waikato, Hamilton, New Zealand

J J Lowe
University of London, Egham, UK

A V Lozhkin
Russian Academy of Sciences, Magadan, Russia

B H Luckman
University of Western Ontario, Ontario, Canada

J Lundqvist
Stockholm University, Stockholm, Sweden

G M MacDonald
University of California – Los Angeles, CA, USA

A W Mackay
University College London, London, UK

J Magee
The Australian National University, ACT, Australia

M Magny
Centre National de la Recherche Scientifique, Besançon, France

J Mangerud
University of Bergen, Bergen, Norway

M E Mann
Pennsylvania State University, PA, USA

T M Marchitto
University of Colorado, CO, USA

A Markova
Institute of Geography, Russian Academy of Sciences, Moscow, Russia

M Marra
University of Canterbury, Christchurch, New Zealand

J A Mason
University of Wisconsin-Madison, WI, USA

J Massaferro
Universidad del Comahue, Bariloche, Argentina

V Masson-Delmotte
Laboratoire des Sciences du Climat et de l'Environnement, Gif-sur-Yvette, France

M D Mastrandrea
Stanford University, CA, USA

J A Matthews
University of Wales – Swansea, Swansea, UK

D Mauquoy
University of Aberdeen, Aberdeen, UK

D McCarroll
University of Wales – Swansea, Swansea, UK

D P McCarthy
Brock University, ON, Canada

J R McConnell
Desert Research Institute, NO, USA

L McFadden
University of New Mexico, NM, USA

S McGowan
University of Nottingham, Nottingham, UK

J A McKenzie
Geological Institute, Zurich, Switzerland

J S McLachlan
University of California – Davis, CA, USA

J McNabb
University of Southampton, Southampton, UK

J I Mead
Northern Arizona University, AZ, USA

M E Meadows
University of Cape Town, Rondebosch, South Africa

D J Merritts
Franklin and Marshall College, PA, USA

S Metcalfe
University of Nottingham, Nottingham, UK

D M Mickelson
University of Wisconsin-Madison, Madison, USA

G H Miller
University of Colorado, CO, USA

G Milne
Durham University, Durham, UK

T Minichillo
University of Washington, WA, USA

C J Mock
University of South Carolina, Columbia, USA

M Montoya
Universidad Complutense de Madrid, Madrid, Spain

D R Muhs
U.S. Geological Survey, CO, USA

S Mulitza
University of Bremen, Bremen, Germany

C V Murray-Wallace
University of Wollongong, New South Wales, Australia

J Murton
University of Sussex, Brighton, UK

N D Naeser
National Center United States Geological Survey, VA, USA

P D Naidu
National Institute of Oceanography, Goa, India

A R Nelson
US Geological Survey, CO, USA

F Nelson
University of Delaware, DE, USA

A Nesje
University of Bergen and Bjerknes Centre for Climate Research, Bergen, Norway

C O'Cofaigh
Durham University, Durham, UK

B V Odgaard
University of Aarhus, Aarhus, Denmark

J D Ortiz
Kent State University, OH, USA

L A Owen
University of Cincinnati, OH, USA

K Pahnke
Massachusetts Institute of Technology, MA, USA

D K Pak
University of California – Santa Barbara, CA, USA

A S Palmer
Massey University, Palmerston North, New Zealand

M Pante
Rutgers University, NJ, USA

A Paul
University of Bremen, Bremen, Germany

M D Petraglia
University of Cambridge, Cambridge, UK

J Pike
Cardiff University, Cardiff, UK

B Pillans
The Australian National University, ACT, Australia

P A Pirazzoli
CNRS – Laboratoire de Géographie Physique, Paris, France

R B Platt Jr
SUNY College of Environmental Science and Forestry, NY, USA

P Ponel
Mediterranean Institute for Ecology and Paleoecology, France

R Z Poore
Center for Coastal and Watershed Studies, FL, USA

N Porch
Monash University, VIC, Australia

S C Porter
University of Washington, WA, USA

G J Prideaux
Western Australian Museum, Perth, WA, Australia

J C Priscu
Montana State University, MT, USA

A Puzachenko
Institute of Geography Russian Academy of Sciences, Moscow, Russia

J Quade
University of Arizona, AZ, USA

T M Quinn
University of South Florida, FL, USA

J Rabassa
Centro Austral de Investigaciones Científicas, Ushuaia, Argentina

T Radi
Université du Québec à Montréal, Québec, Canada

S Rahmstorf
Potsdam Institute for Climate Impact Research, Potsdam, Germany

M Rautio
Université Laval, Québec, QC, Canada

B Rea
University of Aberdeen, Aberdeen, UK

D K Rea
The University of Michigan, MI, USA

N Reeh
Technical University of Denmark, Lyngby, Denmark

P J Reimer
Queen's University Belfast, Belfast, UK

R W Reimer
Queen's University Belfast, Belfast, UK

H M Roberts
University of Wales, Aberystwyth, UK

A Rochon
Université du Québec à Rimouski, Québec, Canada

L Rook
University of Florence, Firenze, Italy

Y Rosenthal
Rutgers University, NJ, USA

N Ross
Cardiff University, Cardiff, UK

D-D Rousseau
Université Montpellier, Montpellier, France

C W Rovey II
University of Minnesota, MN, USA

M Roy
University of Minnesota, MN, USA

G Royston-Bishop
University of Bristol, Bristol, UK

M Rundgren
Lund University, Lund, Sweden

A J Russell
University of Newcastle, Newcastle, UK

J P Sachs
Massachusetts Institute of Technology, MA, USA

L Sadori
Università di Roma "La Sapienza," Rome, Italy

R Sardella
Università di Roma "La Sapienza," Rome, Italy

J M Schaefer
Lamont-Doherty Earth Observatory, NY, USA

S A Schellenberg
San Diego State University, CA, USA

S H Schneider
Stanford University, CA, USA

B R Schöne
University of Frankfurt, Frankfurt, Germany

D Schreve
Royal Holloway, University of London, Surrey, UK

J Schwander
University of Bern, Bern, Switzerland

H Schwarcz
McMaster University, Hamilton, Canada

E M Scott
University of Glasgow, Glasgow, UK

H Seppä
University of Helsinki, Helsinki, Finland

P Shane
The University of Auckland, Auckland, New Zealand

I Shennan
Durham University, Durham, UK

A Sher
Russian Academy of Sciences, Moscow, Russia

N I Shiklomanov
University of Delaware, DE, USA

D T Shindell
NASA Goddard Institute for Space Studies and Columbia University, NY, USA

J Shulmeister
University of Newcastle, Newcastle, Australia

B Shuman
University of Minnesota, MN, USA

M J Siegert
University of Edinburgh, Edinburgh, UK

F J Sierro
Universidad de Salamanca, Salamanca, Spain

H Simola
University of Joensuu, Joensuu, Finland

M J Singer
University of California, CA, USA

L Skinner
University of Cambridge, Cambridge, UK

J L Smellie
British Antarctic Survey, Cambridge, UK

D Smith
University of Victoria, Victoria, Canada

R Smittenberg
Massachusetts Institute of Technology, MA, USA

B Stauffer
University of Bern, Bern, Switzerland

T E Steele
Max Planck Institute for Evolutionary Anthropology, Leipzig, Germany

C E Stickley
Norwegian Polar Institute, Tromsø, Norway

S Sugita
University of Minnesota, MN, USA

J Sun
Chinese Academy of Sciences, Beijing, China

J I Svendsen
University of Bergen, Bergen, Norway

A Svensson
University of Copenhagen, Copenhagen, Denmark

P E Tarasov
Free University of Berlin, Berlin, Germany

K Taylor
University of Nevada, NV, USA

I Tegen
Leibniz Institute for Tropospheric Research, Leipzig, Germany

L G Thompson
Ohio State University, OH, USA

R S Thompson
US Geological Survey, CO, USA

W G Thompson
Woods Hole Oceanographic Institution, MA, USA

L H Thorleifson
University of Minnesota, MN, USA

W Tinner
Institute of Plant Sciences, Bern, Switzerland

T E Törnqvist
Tulane University, LA, USA

R H Towner
The University of Arizona, AZ, USA

A Turner
Liverpool John Moores University, Liverpool, UK

R M Twyman
Market Weighton, York, UK

C Tzedakis
University of Leeds, Leeds, UK

P Tzedakis
University of Aberdeen, Aberdeen

M Ubilla
INGEPA, Montevideo, Uruguay

A-P Valérie
Université Paul Cézanne, Aix en Provence Cedex 04, France

S van der Kaars
Monash University, Vic, Australia

J van der Plicht
University of Groningen, Groningen, The Netherlands

B Van Geel
Universiteit van Amsterdam, Amsterdam, The Netherlands

T van Kolfschoten
Leiden University, Leiden, The Netherlands

J Vandenberghe
Vrije Universiteit, Amsterdam, The Netherlands

M Vandergoes
University of Maine, ME, USA

J Vassiljev
Tallinn University of Technology, Tallinn, Estonia

K L Verosub
University of California, CA, USA

D Verschuren
Ghent University, Ghent, Belgium

B M Vinther
University of Copenhagen, Denmark

E D Waddington
University of Washington, Seattle, USA

C P Wake
University of New Hampshire, NH, USA

A Walanus
Rzeszów University, Rzeszów, Poland

I R Walker
University of British Columbia Okanagan, British Columbia, Canada

M J C Walker
University of Wales, Lampeter, UK

K Wasylikowa
Polish Academy of Sciences, Kraków, Poland

H J T Weerts
TNO Built Environment and Geosciences/Geological Survey of The Netherlands, Utrecht, The Netherlands

G Wefer
University of Bremen, Bremen, Germany

W E Westerhoff
Netherlands Institute of Applied Geoscience TNO - National Geological Survey, Utrecht, The Netherlands

J A Westgate
University of Toronto, ON, Canada

C Whitlock
Montana State University, MT, USA

P E Wigand
University of Nevada, NV, USA

D A Willard
US Geological Survey, VA, USA

E Willerslev
University of Copenhagen, Copenhagen, Demmark

J W Williams
University of Wisconsin, WI, USA

P Wilson
University of Ulster, Northern Ireland, UK

C Winguth
University of Wisconsin, Madison, USA

A G Wintle
University of Wales, Aberystwyth, UK

S A Wolfe
Natural Resources Canada, ON, Canada

C D Woodroffe
University of Wollongong, Wollongong, Australia

G Yu
Chinese Academy of Sciences, Nanjing, People's Republic of China

M A Zárate
Universidad Nacional de la Pampa, La Pampa, Argentina

G D Zazula
Simon Fraser University, British Columbia, Canada

Z Zhang
Massachusetts Institute of Technology, MA, USA

B Zolitschka
University of Bremen, Bremen, Germany

Y Zong
University of Durham, Durham, UK

CONTENTS LIST BY SUBJECT

Location references refer to the volume number and page number (separated by a colon).

INTRODUCTION AND HISTORY OF SCIENCE

PALEOCLIMATOLOGY

Overview and Approaches

Paleoclimate Reconstructions

Climate Forcing Mechanisms

Climate Modelling

DATING TECHNIQUES

Radiometric Methods

Geomagnetic Excursions and Secular Variations in the Magnetic Field

Biological Methods

Annual Layer Methods

Chemical Methods

QUATERNARY STRATIGRAPHY

QUATERNARY GLACIATION

Glacial Landforms

Cold Regions Landforms

FLUVIAL ENVIRONMENTS

HISTORY OF QUATERNARY GLACIATIONS

SEA LEVEL STUDIES

PALEOSOLS AND WIND-BLOWN SEDIMENTS

Paleosol Research

Dune Studies

Loess Studies

LAKE LEVEL STUDIES

PALEOBOTANY

Overview

Pollen Studies

Charred Particle Analysis

Quaternary Pollen Records

Phytoliths

Ancient Plant DNA

Plant Macrofossil Studies

Plant Macrofossil Records

Diatom Studies

PALEOLIMNOLOGY

VERTEBRATE STUDIES

INSECT FOSSIL STUDIES

Beetle Studies

Chironomid Studies

Oribatid mite studies

PALEOCEANOGRAPHY

Paleoceanographic Studies

Paleoceanographic Records

ICE CORE STUDIES

Ice Core Studies

Ice Core Records

STABLE ISOTOPE RESEARCH - CARBONATES

HUMANS IN THE QUATERNARY

CONTENTS

VOLUME 1

F

G

VOLUME 2

G

H

I

K

L

P

VOLUME 3

P

VOLUME 4

P

Q

R

S

V

W

Y

Introductory Articles

INTRODUCTION

Contents

Introduction: Societal Relevance of Quaternary Research

S A Elias, Royal Holloway, University of London, Surrey, UK

© 2007 Elsevier B.V. All rights reserved.

Introduction

The great 19th-century geologist and natural historian, Charles Lyell (1830), is generally credited as being the author of the concept of Uniformitarianism (*see* History of Quaternary Science). Simply stated, this concept says that "the present is the key to the past." In other words, we may interpret the ancient history of the planet through our understanding of modern-day processes. This concept represented a major step forward in scientific thought, greatly influencing contemporary scientists, such as Charles Darwin (*see* History of Quaternary Science). However, in this article I hope to make it clear that the reverse of Lyell's concept is equally true: the past is the key to the present. In other words, a knowledge of the past is key to our understanding of the present world, especially in the multiple disciplines known as the environmental sciences. No matter which aspect of the environment one considers, be it the atmosphere, the oceans, the cryosphere, the lithosphere, or the biosphere, it is impossible to fully comprehend modern processes and conditions without a knowledge of the past. To try to understand present-day environments without a knowledge of their Quaternary history would be like trying to understand the plot of a long novel by reading only the last page.

Climate History

The Quaternary period saw a large number of climatic oscillations (*see* Introduction) on a scale that was probably greater than at any other time in the last 60 million years (Bradley, 1999). Thus, a study of Quaternary climate change provides us with an understanding of climatic variation on a much larger scale than has been recorded in historic records of the last few centuries. An understanding of the magnitudes and rates of climate change during the Quaternary period is necessary to develop our comprehension of modern climate. Much of climate modeling is based on data derived from reconstructions of ancient climates. These models require long-term data, and these are only available from ancient climate records, reconstructed from various kinds of proxies, such as geochemical records from polar ice cores, and terrestrial and oceanic fossil records.

The predicted amplitude of global warming in the 21st century varies greatly from one scientific study to another, but in the absence of any mitigating policies, Wigley and Raper (2001) predicted with a 90% probability that by the end of the 21st century, global mean temperatures will rise between 1.7 to 4.9°C. This level of warming is certainly unprecedented within recorded history, but it falls within the boundaries of more than one previous interglacial period, including the last interglacial (ca. 120,000–115,000 years ago), when mean temperatures were substantially warmer than they are today in many parts of the world. Evidence from the Vostok Ice Core in Antarctica (*see* Antarctic Stable Isotopes) indicates that at the beginning of the interglacial, between about 130,000 and 127,000 years ago, regional temperatures climbed as much as 3°C higher than today (Kukla *et al.*, 2002) (**Fig. 1**). As Velichko

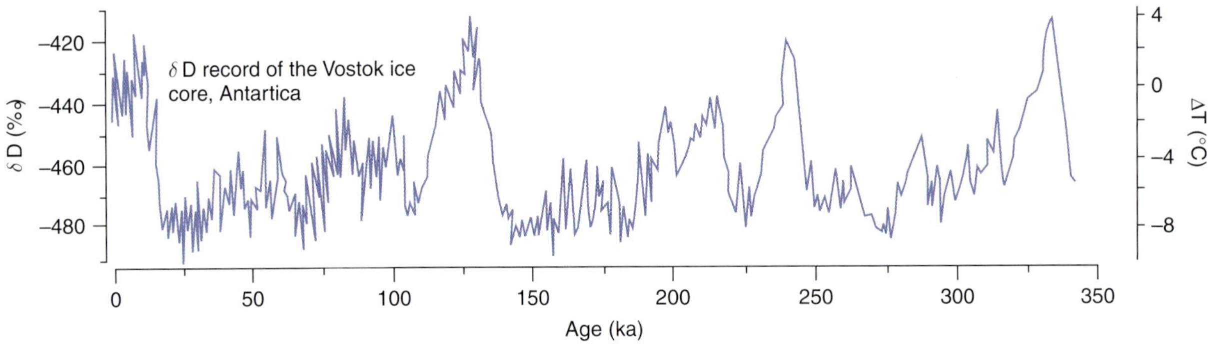

Figure 1 Deuterium isotope record from the Vostok ice core, Antarctica, showing large-scale fluctuations in temperature during the last four glacial–interglacial cycles. After Petit *et al.* (1999).

et al. (1993) observed, there is abundant proxy evidence of large-scale climatic change during the Quaternary, especially in the northern high latitudes. During the long series of glacial–interglacial cycles, high latitude temperatures appear to have fluctuated by about the same amount as that projected for the next century (*see* Paleoclimate Relevance to Global Warming).

Predictions of future temperature changes, based on orbital forcing and increased greenhouse gas concentrations in the atmosphere (Loutre and Berger, 2003), suggest that anthropogenic (human-induced) effects may force global climate into an interglacial mode that may persist for 30,000–40,000 years, not unlike an interglacial period that occurred about 405,000–340,000 years ago.

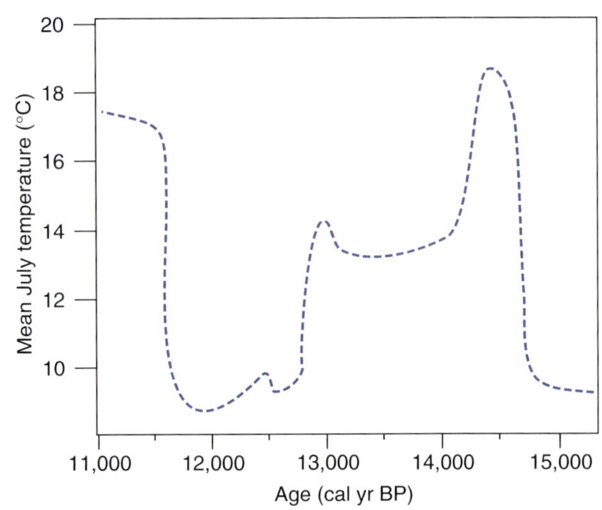

Figure 2 Reconstructed mean July temperature change during the interval 15,000–11,000 years ago in Britain, based on insect fossil analysis. Data from Lowe *et al.* (1995).

Pace of Climate Change

Predicted rates of global warming in the coming decades are much faster than the rates of temperature change observed in recent centuries. For instance, Vinnikov and Grody (2003) analyzed upper atmospheric (troposphere) temperatures for the interval 1978 to 2002, and from these data they inferred a future global warming trend of 0.22 to 0.26 °C per decade. Has the Earth previously experienced temperature changes as rapid as this? Fossil beetle evidence from Britain indicates that even greater rates of warming have occurred in the past. For instance, at the end of the last glaciation, the temperature change during the transition from the glacial climates of the Younger Dryas interval to the post-glacial climates at the beginning of the current interglacial was extremely rapid and intense. The beetle evidence (*see* Late Pleistocene of Europe) suggests that mean summer temperatures warmed on the order of 0.35 °C per decade during this transition (Lowe *et al.*, 1995) (**Fig. 2**). This climate change led to wholesale changes in the flora and fauna of many regions. We cannot hope to predict the kinds of changes that may take place in the modern biota without a knowledge of how past biotic communities responded to large-scale climate change.

Ocean History

By studying the Quaternary fossil record from deep-sea sediments, paleoceanographers have come to understand the role of oceans in the transfer of heat from the tropics to the high latitudes, via ocean currents (*see* Paleoceanography). Broecker's (1987) model of the oceanic thermohaline conveyor belt demonstrated how this system functions, both past and present (**Fig. 3**). This thermohaline conveyor belt acts as a major conduit for the exchange of thermal energy, nutrients, and dissolved oxygen between the shallow and deep oceans of the world. There is little doubt that the conveyor belt's capability to transfer heat in large volumes and across vast distances profoundly influences global climate. There is some

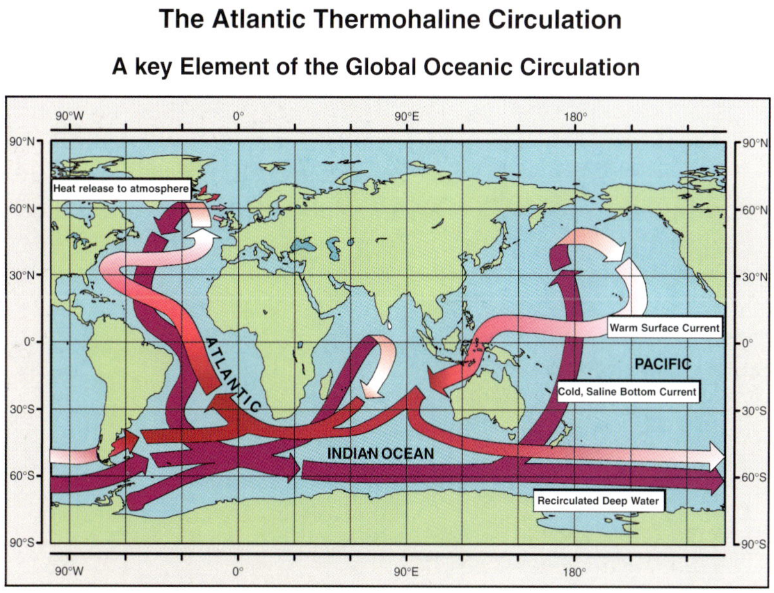

Figure 3 Schematic diagram of the global ocean circulation pathways, the 'conveyer' belt (after Broecker W, modified by Maier-Reimer E).

evidence that the current climatic warming, especially in the high latitudes, may be weakening the thermohaline circulation (Häkkinen and Rhines, 2004). Current oceanographic research (Curry and Mauritzen, 2005) suggests that the northern regions of the North Atlantic Ocean have been diluted by increasing amounts of freshwater since the 1960s. If the conveyor belt were to shut down, climate models indicate that much of Europe would be 5–10 °C colder, the equatorial regions would be 4–5 °C warmer, and Greenland would be as much as 16 °C colder than it is today. Rainfall patterns would radically change, and the atmosphere would become dustier (Broecker, 1999). Again, without the paleontological data from Quaternary marine sediments, we would have little or no understanding of these phenomena.

Sea Level Change

There is a growing body of evidence that the current level of global warming is going to bring about substantial rises in sea level, as polar ice caps melt. The Intergovernmental Panel on Climate Change (IPCC) (2001) has estimated that sea level will rise 50 cm above current levels by the year 2100 (**Fig. 4**). This is a substantially greater rise in sea level than occurred in the last century. The ecological and human impacts of rising oceans would be substantial, including increased flooding, coastal erosion, salination of aquifers, and loss of coastal agricultural land and living space (IPCC, 2001). The loss of living space is particularly critical, given that 50% of

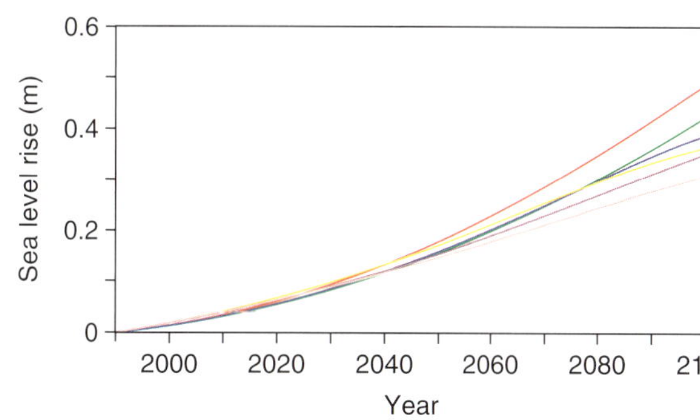

Figure 4 Predicted global sea-level rise during the 21st century, based on six different models. Data from IPCC, (2001).

world's population lives within 6 km of the sea, and that 14 of the 15 largest cities are coastal.

Again, one must refer to the Quaternary record to find sea levels that were as high, or higher than the predicted levels in the coming century. Reconstructions of global mean sea level during the last interglacial period (*see* Eustatic Sea-Level Changes, Glacial-Interglacial Cycles) place it 6–7 m above modern mean sea level at about 125,000 years ago (Chapell and Shackleton, 1986; Bard *et al.*, 1993). By studying the effects that these higher sea levels had on ancient coastal landscapes, we are able to develop more useful models of what may happen in the future.

Coastal Deposition and Erosion

As sea levels rise and storm intensity increases with global warming, many of the world's coastlines are being eroded (**Fig. 5**). The impacts on society are relatively large and costly. For instance, a recent report by the Heinz Center for Science, Economics and the Environment states that in the United States, approximately 1,500 homes will be lost to coastal erosion each year for the next several decades, at an annual cost to coastal landowners of approximately $530 million (Heinz Center, 2000).

Of course, coastal erosion has been going on for many thousands of years, and has been documented by marine geologists studying Quaternary near-shore deposits. What is happening today is just the most recent episode in an ongoing Holocene marine transgression. The erection of coastal defenses has halted the process, at least temporarily, but ultimately the phenomenon is inexorable. Quaternary geologists have been studying such processes as the supply and sink of coastal sediments, interactions of the retreating shoreface with older deposits, and long-term changes in the direction of shoreline changes (*see* High Energy Coasts Sedimentary Indicators) (McNinch, 2004). They have found that variations in the rates and directions of shoreline change are strongly controlled by changes in ocean wave and tidal energy (van der Molen and van Dijck, 2000). These are highly complex problems that are difficult to model, but data from Quaternary studies are essential to the process.

Environmental Change in Oceanic Ecosystems

The world's continental shelves are highly productive marine ecosystems (Lohrenz *et al.*, 2002). However, we do not fully understand the role of continental shelves in the global climate system. The science of paleoceanography is providing the information needed to gain a better understanding of this role. Specifically, we need to know more about changes in continental shelf productivity during times of different sea levels, the storage of nutrients on continental shelves, and how changes in sea level affect the total sea surface area. As with the other problems discussed here, these are not just 'academic' issues. For instance, they affect the biological productivity and diversity of the oceans, both of which have direct implications for the world's fisheries.

Figure 5 Coastal erosion. Photo from Peter French.

Cryosphere History

The waxing and waning of ice sheets has been the dominant environmental phenomenon of the Quaternary, either directly or indirectly affecting every region of the globe. In the current climate, polar ice caps are thinning, mountain glaciers are retreating in most regions, and many are in danger of disappearing altogether. Notable among these are the glaciers and snowfields that are perched on the tops of high mountains near the Equator (*see* Africa). A recent study of Mt Kilimanjaro's summit in Africa (Irion, 2001) showed a 33% loss of ice since 1989 and an 82% decline since 1912. At the current rate of melting, this ice cap will be gone within a few years (**Fig. 6**). Similarly, the massive Quelccaya ice cap in the Peruvian Andes is melting (*see* South America). The ice cap shrank from 56 km^2 in 1976 to just 44 km^2 in 2001, and it may disappear within the next 20 years.

The Antarctic and Greenland ice sheets together hold 33 million km^3 of ice, representing enough water to raise global sea level by 70 m (Rignot and Thomas,

Figure 6 Aerial photograph of the top of Mt Kilimanjaro, showing the remnants of the ice cap. Photo courtesy of Byrd Polar Center, Ohio State University.

Figure 7 Photograph of Pine Island Glacier taken by Tom Kellogg onboard the U.S. Coast Guard icebreaker *Glacier*, 1985, in Pine Island Bay.

Table 1 1991 soil degradation statistics by region (after Oldeman *et al.*, 1991)

Region	*Human-induced soil degradation (million ha)*	*Total land surface (million ha)*	*Percent of land with soil degradation*
Africa	494	2,966	17
Asia	748	4,256	18
South America	243	1,768	14
Central America	63	306	21
North America	95	1,885	5
Europe	219	950	23
Australasia	103	882	12
World	1,964	13,013	15

2002). Current research by several different research groups indicates that the Greenland Ice Sheet is thinning, especially near the coasts (*see* Dynamics of the Greenland Ice Sheet). The Antarctic story is more complicated. It appears that the West Antarctic Ice Sheet is thinning (**Fig.** 7) (*see* Dynamics of the West Antarctic Ice Sheet) overall, but it is thickening in some regions and thinning in others (Rignot and Thomas, 2002). Large sectors of ice in southeast Greenland, West Antarctica, and the Antarctic Peninsula are changing rapidly, but glaciologists do not yet understand the processes involved. Modeling is a key element in developing that understanding, and the best sources of data for such models come from paleoenvironmental research. It is clear that just as global climate affects polar ice sheets, so also ice sheets affect global climate. As Clark *et al.* (1999) have said, given the importance of ice sheets in the climate system, establishing the factors that control their evolution and behavior is necessary to understanding their influence on climate over the long term. This applies equally to questions about the future stability of the polar ice sheets.

Lithosphere History

The study of geomorphology, or the processes that act to shape the surface of the Earth, is rooted firmly in Quaternary science. Understanding the Earth's geomorphic history is the key to understanding current geomorphological problems. A knowledge of contemporary Earth surface processes is critical for the study of the hazards and sustainability of the landscape.

Soil Studies

Modern soils in the rich agricultural regions of the middle latitudes of the Northern Hemisphere owe much of their formation to Pleistocene glaciations. Glaciers and ice sheets pulverized layers of bedrock and organic matter that fell within the ice boundaries, creating the mixtures of sand, silt, and clay that became the best suited soils in which to grow crops. This is due to their moisture retention characteristics and nutrient storage ability. These soils can be considered part of the world's Pleistocene heritage (*see* Overview)

These rich agricultural soils are the product of thousands of years of glacial and postglacial environments. They are durable but still susceptible to degradation through over-use and poor land management practices. Soils in other regions of the world vary in their ability to sustain agriculture. By one estimate (Oldeman, *et al.*, 1991), about 15% of the Earth's land surface has been seriously degraded by human activities (**Table 1**). Perhaps 25% more of the Earth's soils are at risk of serious degradation (World Resources Institute, 1990). The loss of agricultural land through erosion is estimated at 6 or 7 million hectares (ha) per year, with an additional annual loss of 1.5 million ha as a result of waterlogging, salinization, and alkalinization (Brundtland *et al.*, 1987). Thus the ever-expanding human population of the world is essentially using up the vital resource of arable soil at an alarming rate. It took many thousands of years for these soils to develop, but only a few centuries to degrade.

Geologic Hazards

Our understanding of most geologic hazards has been greatly enhanced by Quaternary studies. The reason we must probe into the past to understand these phenomena is that most of them are highly sporadic and largely unpredictable, despite the best efforts of geoscientists. Such catastrophes as large volcanic eruptions, earthquakes and tsunamis, take place only rarely in any given region. These phenomena are discussed in a document published by the United Nations Environment Program (2005). The interval between events may be hundreds or even thousands of years – well beyond the span of recorded history. The Quaternary record affords us insights into the timing and intensity of past catastrophic events.

Volcanic Eruptions

Sitting astride the boundaries of the world's tectonic plates, volcanoes are a major hazard to human life in many parts of the world. Volcanic eruptions have claimed more than a quarter of a million lives since the destruction of Pompeii in AD 79 (**Table 2**). The principal hazard is not lava, which usually flows sufficiently slowly to allow all but the closest inhabitants to the volcano to escape. Volcanic ash, or tephra, is the principal hazard. It is ejected in far larger quantities than lava in most eruptions (**Fig. 8**), and can travel thousands of kilometers, often circling the globe in a plume that is visible from space (**Fig. 9**). In rare cases, ignimbrites are ejected during volcanic eruptions. These are pyroclastic rocks consisting of glass shards, crystals, and lithic fragments. Ignimbrites are formed by the deposition of hot, rapidly expanding, turbulent magmatic gases. This happened in the 1902 eruption of Mt Pelée in the Caribbean island of Martinique, killing all but two people on the island. When volcanic ash and rock mix with water, they may form a lahar, which sweeps rapidly down the mountain into surrounding valleys. Volcanic tephra layers deposited in Quaternary sediments are the chief means by which scientists reconstruct the history of prehistoric eruptions (*see* Tephrochronology). We can say with absolute certainty that some of the volcanic eruptions that took place in the Pleistocene were far greater, and affected larger areas, than anything in recorded history. For instance, 600,000 years ago, the central part of what is now Yellowstone National Park, Wyoming, exploded in an enormous volcanic eruption The eruption spewed out nearly 1,000 km^3 of debris. What is now the park's central portion then

Figure 8 Photograph of volcanic eruption, Kanaga Volcano, Aleutian Islands, Alaska, 1994. Photo courtesy of the U.S. Geological Survey.

Figure 9 Eruption of Rabaul volcano, Papua New Guinea, 1994, as seen from the space shuttle. Photo courtesy of NASA.

Table 2 Major historic volcanic eruptions

Locality	*Date*	*Estimated human death toll*
Pompeii and Herculaneum, Italy (Vesuvius)	AD 79	Thousands (uncounted)
Iceland (Laki volcano)	AD 1783	9350
Kyushu, Japan	AD 1793	14,300 (triggered avalanche and tsunami)
Sumbawa, Indonesia (Tambora volcano)	AD 1815	92,000 (10,000 directly; 82,000 through starvation and disease)
Central Ecuador (Mt Cotopaxi)	AD 1877	1,000
Krakatau, Indonesia	AD 1883	36,000
St Vincent, West Indies (Soufriére volcano)	AD 1902	1,680
Martinique, West Indies (Mt. Pelée)	AD 1902	40,000
Northern Columbia (Nevada del Ruiz)	AD 1985	25,000
Total loss of life		>250,000

collapsed, forming a 45- by 75-km caldera (or basin). Tephra from this eruption is known to have blanketed more than 5 million km^2 of western and central North America (Smith and Siegel, 2000). If an eruption on this scale were to happen today, it would ecologically devastate large regions, and cause climatic cooling for several years afterwards because of the screening of sunlight by volcanic ash and gases in the atmosphere. To put this Yellowstone eruption into context, the largest volcanic eruption in historic times was from Tambora, in Indonesia, in 1815. That eruption is estimated to have ejected about 150 km^3 of ash into the atmosphere (Oppenheimer, 2003). This is less than one-third the amount estimated for the Yellowstone eruption. Anomalously cold weather hit the northeastern United States, maritime provinces of Canada, and Europe the following year. The year of 1816 came to be known as the 'Year without a summer' in these regions.

The Yellowstone volcanic eruption story is a cautionary tale. Far greater eruptions have taken place in prehistory than have happened in historic times. Our knowledge of the prehistoric eruptions has come through the analysis of ancient volcanic deposits. Tephrochronology, the study of volcanic ash deposits, has helped scientists determine the source of ash deposits (through chemical finger printing). By piecing together regional histories of volcanic activity, Quaternary scientists have been able to reconstruct the size and timing of ancient volcanic eruptions. By working with modern vulcanologists, we have begun the difficult task of predicting volcanic eruptions. The science of volcanic eruption prediction is far from exact, but it has enabled the evacuation of some threatened regions before major eruptions took place. For instance, in 1991, vulcanologists were able to warn the people living near Mt Pinatubo in the Philippines, a few days before devastating eruptions took place. In total, 58,000 people were evacuated from a 30-km radius around the volcano (Wolfe and Hoblitt, 1996).

Earthquakes and Tsunamis

There is no doubt that earthquakes and the resulting tsunamis (**Fig. 10**) have killed more people in historic times than any other geologic hazard. As shown in **Table 3**, major earthquakes in the past 1,200 years are known to have killed more than 4 million people. This total includes more than 2 million people in just the 20th century. As human populations rise, and the number of poorly-built multi-story buildings increases, these numbers are likely to grow larger. But, like volcanic eruptions, earthquakes are relatively rare events in any one region, and we have yet to develop any meaningful ways of predicting them. Geoscientists may be able to predict that there will be a major earthquake somewhere along a given fault system, but as of now they can only say that it will probably happen within the next decade or the next century. Some of the

Figure 10 Satellite image of a tsunami striking the coast of Sri Lanka following the massive earthquake in the Indian Ocean, 29 December 2004. The image shows the backwash of the tsunami. Photo copyright DigitalGlobe.

Table 3 Major historic earthquakes (>10,000 casualties)

Locality	*Date*	*Estimated human death toll*
Damghan, Iran	AD 857	200,000
Ardabil, Iran	AD 893	150,000
Aleppo, Syria	AD 1138	230,000
Chihli, China	AD 1290	100,000
Shaanxi province, China	AD 1556	830,000
Shemakha, Caucasia	AD 1667	80,000
Sicily, Italy	AD 1693	60,000
Tabriz, Iran	AD 1727	77,000
Lisbon, Portugal	AD 1755	70,000
Calabria, Italy	AD 1783	50,000
Messina, Sicily	AD 1908	70,000–100,000
Avezzano, Italy	AD 1915	29,980
Gansu province, China	AD 1920	200,000
near Tokyo, Japan	AD 1923	>140,000
Xining, China	AD 1927	200,000
Gansu, China	AD 1932	70,000
Quetta, Pakistan	AD 1935	30,000–60,000
Northern Turkey	AD 1939	100,000
Ashgabat, Turkmenistan	AD 1948	110,000
Assam, India	AD 1950	20,000–30,000
Peru	AD 1970	66,000
Tangshan, China	AD 1976	655,000
Tabas, Iran	AD 1978	25,000
Mexico City	AD 1985	25,000
Armenia	AD 1988	25,000
Northwest Iran	AD 1990	>50,000
Izmit, Turkey	AD 1999	>17,000
Bhuj, India	AD 2001	>20,000
Bam, Iran	AD 2003	>30,000
Sumatra, Indonesia (marine earthquake and tsunami)	AD 2004	>226,000
Total loss of life from major historic earthquakes		>4,000,000

most useful data in the study of earthquakes comes from Quaternary science, where ancient earthquakes have left behind clear indicators in buried sediments. For instance, Fumai *et al.* (1993) were able to reconstruct a 100-year recurrence interval for the San Andreas Fault zone, 70 km northeast of Los Angeles, based on sedimentary evidence. Earthquakes caused debris-flow deposits, ruptured, tilted, and folded bedding planes in the sediments.

The timing of an earthquake in the year AD 1700 along the coast of Washington state, USA was able to be accurately dated through analysis of tree-rings (*see* Dendrochronology) Yamaguchi *et al.* (1997) studied Japanese historical evidence for a large tsunami of previously unknown origin, and were able to pin down the source of the earthquake that caused it. They proposed that a magnitude 9 earthquake in the Cascadia subduction zone of the Pacific Northwest region of North America was the trigger for the tsunami that occurred on 26 January 1700. Tree-ring records from the central Cascadia region (**Fig. 11**) support this theory. When the earthquake struck, much of the coast between southern British Columbia and northern California was abruptly lowered, submerging some coastal forests in more than a meter of tidewater. Yamaguchi *et al.* studied the annual growth rings of some Sitka spruce trees that survived this tidal submergence, and the rings in some trees show changes in width and anatomy consistent with disturbance (tilting, increased flooding, or both) in AD 1700 and for a few years afterward.

Tsunamis may also occur in tectonically stable areas, as shown by the Storegga Slide that struck the coasts of Scotland and Norway, about 7,000 yr BP (Dawson *et al.*, 1988).

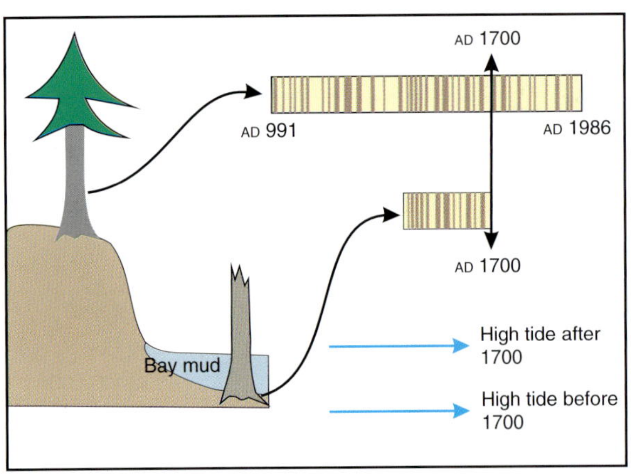

Figure 11 Tree-ring dating of the 1700 Cascadia earthquake. The dead tree on the right was drowned in salt water due to subsidence during the earthquake. The outermost tree-rings from a number of dead snags in regional salt marshes correspond to the year AD 1700 in tree-ring sequences from living (upland) trees in the region.

Biosphere History

Our knowledge of how ecosystems function has been greatly aided by Quaternary paleontological studies. In the absence of such studies, modern ecologists would have little or no idea about the longevity of ecosystems and their resistance to change in the face of large-scale environmental variations.

Longevity, Resistance and Resilience of Ecosystems

Studies of plant and animal remains from the Quaternary have shown that the current ecosystems are not composed of species that are inherently the best suited to their environments. Rather, modern ecosystems represent the latest reshuffling of species, and the mixture of species in biological communities at any given time has been shaped by many forces. In the middle and high latitudes, the catalyst for changes in ecosystems has been large-scale climatic fluctuations of the Pleistocene, accompanied by the waxing and waning of continental ice sheets and mountain glaciers (*see* Evidence of Glacier and Ice Sheet Extent). There have been hundreds of these large-scale fluctuations in climate during the last 2.4 million years. Amazingly, some groups of organisms (notably beetles) have remained intact throughout this interval, shifting their ranges in response to climatic change (*see* Overview). Other groups, such as mammals, have undergone waves of speciation during the Quaternary (*see* Vertebrate Overview). This includes, of course, the hominid lineage that gave rise to our species (*see* Overview). Towards the end of the last glaciation, a large number of megafaunal mammals (species with adult weight greater than 40 kg) became extinct. During the middle of the last glaciation, more than 150 genera of megafauna were alive on the planet. By 10,000 years ago, that number was reduced to just over 50 genera (*see* Late Pleistocene Megafaunal Extinctions). Thus approximately two-thirds of the large mammal gr-oups became extinct during this interval (**Fig. 12**). This extinction event has been the subject of hot scientific debate for the last 40 years.

Pleistocene Megafaunal Extinctions

The nature of the debate over the megafaunal mammal extinctions at the end of the Pleistocene rests in determining its cause. Paleontological evidence has shown that previous glacial-to-interglacial transitions in the Pleistocene did not bring about wholesale extinctions of species. This evidence has led some scientists to conclude that human predation was what dealt the final blow to the Pleistocene megafauna, at the end of the last ice age (Martin, 1989; Flannery, 1994). Others believe that changing

Figure 12 Megafaunal mammals that became extinct towards the end of the last glaciation. A) *Macrauchenia* (South American browser – no living relatives); B) *Megaloceros* (Giant deer); C) *Eremotherium* (Ground sloth); D) *Mammuthus* (Woolly mammoth); E) *Megatherium* (Giant ground sloth); F) *Procoptodon* (Giant short-faced kangaroo); G) *Coelodonta* (Woolly rhinoceros); H) *Smilidon* (sabre-tooth cat). Drawings courtesy of Corel Corp.

environments brought about this extinction, and express the view that humans had little or nothing to do with it (Grayson and Meltzer, 2003). A third group of scientists holds a middle view, that perhaps people played a part in megafaunal extinction, but the degree of human impact has yet to be determined (Barnowsky *et al.*, 2004).

Modern Extinction Rates

We may never find the 'smoking gun' for the end-of-Pleistocene extinction event, but the fossil record demonstrates that such large-scale extinction events have been rare in Earth's history. There have been at least five well-documented mass extinctions in the planet's history. Some, such as the dinosaur extinction, 65 million years ago, are thought to have been caused by meteor impacts. The causes of other extinction events remain unknown. However, the causes of the modern extinction of species are all too obvious. As human populations swell and human land use alters once-pristine ecosystems, the biodiversity of the planet has been plummeting. For the past 300 years, recorded extinctions for some groups of organisms have shown rates of extinction at least several hundred times the rate expected on the basis of the geological record (Dirzo and Raven, 2003).

Human 'Coming of Age' in the Quaternary

As we have seen, the study of Quaternary environments and organisms forms the basis for gauging much of what is happening in the modern world. Quaternary science provides the necessary information for assessing potential impacts of future global warming, recurrence rates for geologic hazards, modern species extinction rates, and a host of other topics of vital interest to humanity. Our own species, *Homo sapiens*, first appeared only about 160,000 years ago (Clark *et al.*, 2003). Although modern societies are quite capable of shaping their immediate environment to suit their needs, for more than 90% of human history, the environment has been shaping us. Our species, and its immediate ancestors, were all products of the Pleistocene, first shaped by environmental changes in Africa, followed by Eurasia, finally colonizing most of the globe within the last 12,000 years (since the end of the last glaciation) (*see* Human Migrations During the Late Pleistocene). *Homo sapiens* means 'wise man.' Will our species live up to its scientific name?

See also: **Introduction**: History of Quaternary Science. **Archaeological Records**: Overview. **Beetle Records**: Overview; Late Pleistocene of Europe. **Dendrochronology**. **Glacial Landforms, Ice Sheets**: Evidence of Glacier and Ice Sheet Extent. **Ice Core Records**: Africa; South America; Antarctic Stable Isotopes. **Ice Cores**: Dynamics of the Greenland Ice Sheet; Dynamics of the West Antarctic Ice Sheet. **Paleoceanography**; **Paleoclimate Relevance to Global Warming**. **Paleoclimate**: Introduction. **Paleosols and Wind-Blown Sediments**: Overview. **Quaternary Stratigraphy**: Tephrochronology. **Sea Level Studies**: High Energy Coasts Sedimentary Indicators. **SEA Level Studies**: Eustatic Sea-Level Changes, Glacial-Interglacial Cycles. **Vertebrate Overview**. **Vertebrate Records**: Late Pleistocene Megafaunal Extinctions.

References

Bard, E., Stuiver, M., and Shackleton, N. (1993). How accurate are our chronologies of the Past? In *Global Changes in the Perspective of the Past* (J. A. Eddy and H. Oeschger, Eds.), pp. 109–120. Wiley, New York.

Barnowsky, A. D., Koch, P. L., Feranec, R. S., Wing, S. L., and Shabel, A. B. (2004). Assessing the causes of Late Pleistocene extinctions on the continents. *Science* **306**, 70–75.

Broecker, W. S. (1987). The biggest chill. *Natural History* **96**, 74–82.

Broecker, W. (1999). What if the conveyor were to shut down? *GSA Today* **9**, 1–6.

Brundtland, G. H., and Khalid, M., *et al.* (1987). *Our common future. Report of World Commission on Environment and Development presented to the chairman of Intergovernmental Intersessional Preparatory Committee, UNEP Governing Council.* Oxford University Press, Oxford.

Chappell, J., and Shackleton, N. J. (1986). Oxygen isotopes and sea level. *Nature* **324**, 137–140.

Clark, J. D., Beyene, Y., Wolde-Gabriel, G., *et al.* (2003). Stratigraphic, chronological and behavioural contexts of Pleistocene *Homo sapiens* from Middle Awash, Ethiopia. *Nature* **423**, 747–752.

Clark, P. U., Alley, R. B., and Pollard, D. (1999). Northern Hemisphere Ice-Sheet Influences on Global Climate Change. *Science* **286,** 1104–1111.
Curry, R., and Mauritzen, C. (2005). Dilution of the Northern North Atlantic Ocean in recent decades. *Science* **308,** 1772–1774.
Dawson, A. G., Long, D., and Smith, D. E. (1988). The Storegga slides: Evidence from Eastern Scotland for a possible Tsunami. *Marine Geology* **82,** 271–276.
Dirzo, R., and Raven, P. H. (2003). Global state of biodiversity and loss. *Annual Review of Environment and Resources* **28,** 137–167.
Flannery, T. F. (1994). *The Future Eaters. An ecological history of Australasian lands and people.* Reed New Holland, Sydney.
Fumai, T. E., Pezzopane, S. K., Weldon, R. J., and Schwartz, D. P. (1993). A 100-year average recurrence interval for the San Andreas Fault at Wrightwood, California. *Science* **259,** 199–203.
Grayson, D. K., and Meltzer, D. J. (2003). A requiem for North American overkill. *Journal of Archaeological Science* **30,** 585–593.
Häkkinen, S., and Rhines, P. B. (2004). Decline of subpolar North Atlantic circulation during the 1990s. *Science* **304,** 555–559.
Heinz Center (2000). Evaluation of erosion hazards, a collaborative research project of the H John Heinz III Center for Science, Economics and the Environment. Website at http://www.heinzcenter.org/.
Intergovernmental Panel on Climate Change (2001). *IPCC Third Assessment Report: Climate Change 2001.* IPCC, Geneva.
Kukla, G. J., Bender, M. L., deBeaulieu, J.-L., *et al.* (2002). Last interglacial climates. *Quaternary Research* **58,** 2–13.
Lohrenz, S. E., Redalje, D. G., Verity, P. G., Flagg, C. N., and Matulewski, K. V. (2002). Primary production on the continental shelf off Cape Hatteras, North Carolina. *Deep Sea Research Part II: Topical Studies in Oceanography* **49,** 4479–4509.
Loutre, M. F., and Berger, A. (2003). Marine Isotope Stage 11 as an analogue for the present interglacial. *Global and Planetary Change* **36,** 209–217.
Lowe, J. J., Coope, G. R., Sheldrick, C., Harkness, D. D., and Walker, M. J. C. (1995). Direct comparison of UK temperatures and Greenland snow accumulation rates, 15 000–12 000 years ago. *Journal of Quaternary Science* **10,** 175–180.
Lyell, C. (1830). *Principles of Geology.* John Murray, London.
McNinch, J. E. (2004). Geologic control in the nearshore: shore-oblique sandbars and shoreline erosional hotspots, Mid-Atlantic Bight, USA. *Marine Geology* **211,** 121–141.
Oldeman, L. R., Hakkeling, R. T. A., and Sombroek, W. G. (1991). *World map of the status of human-induced soil degradation: an explanatory note. Second Revised Edition.* International Soil Reference and Information Centre, The Netherlands, Wageningen.
Oppenheimer, C. (2003). Climatic, environmental and human consequences of the largest known historic eruption: Tambora volcano (Indonesia) 1815. *Progress in Physical Geography* **27,** 230–259.
Petit, J. R., Jouzel, J., Raynaud, D., Barkov, N. I., Barnolam, J.-M., Basile, I., Bender, M., Chappellaz, J., Davis, M., Delayque, G., Delmotte, M., Kotlyakov, V. M., Legrand, M., Lipenkov, V. Y., Lorius, C., Pépin, L., Ritz, C., Saltzman, E., and Stievenard, M. (1999). Climate and atmospheric history of the past 420,000 years from the Vostok ice core, Antarctica. *Nature* **399,** 429–436.
Rignot, E., and Thomas, R. H. (2002). Mass balance of polar ice sheets. *Science* **297,** 1502–1506.
Smith, R. B., and Siegel, L. (2000). *Windows into the Earth: the geologic story of Yellowstone and Grand Teton National Park.* Oxford University Press, New York.
United Nations Environment Program (2005). *One Planet Many People: Atlas of Our Changing Environment.* pp. 336.
van der Molen, J., and van Dijck, B. (2000). The evolution of the Dutch and Belgian coasts and the role of sand supply from the North Sea. *Global and Planetary Change* **27,** 223–244.
Velichko, A. A., Borlsova, O. K., Zelikson, E. M., Faure, H., Adams, J. M., Branchu, P., and Faure-Denard, L. (1993). Greenhouse warming and the Eurasian biota: are there any lessons from the past? *Global and Planetary Change* **7,** 51–67.
Vinnikov, K. Y., and Grody, N. C. (2003). Global warming trend of mean tropospheric temperature observed by satellites. *Science* **302,** 269–272.
Wigley, T. M. L., and Raper, S. C. B. (2001). Interpretation of high projections for global-mean warming. *Science* **293,** 451–454.
Wolfe, E. W., and Hoblitt, R. P. (1996). Overview of the Eruptions. In *Fire and Mud, Eruptions and Lahars of Mount Pinatubo, Philippines,* pp. 1–5 (C. G. Newhall and S. Punongbayan, Eds.). University of Washington Press, Seattle.
World Resources Institute (1990). *World Resources 1990–1991: A guide to the global environment. World Resources Institute in collaboration with the United Nations Environment Programme and the United Nations Development Programme.* Oxford University Press, Oxford.
Yamaguchi, D., Atwater, B. F., Bunker, D. E., Benson, B. E., and Reid, M. S. (1997). Tree-ring dating the 1700 Cascadia earthquake. *Nature* **389,** 922–923.

Introduction: History of Quaternary Science

S A Elias, Royal Holloway, University of London, Surrey, UK

© 2007 Elsevier B.V. All rights reserved.

Introduction

The Quaternary sciences represent the systematic study of the Quaternary, or most recent geologic period. This period is generally characterized by a series of glaciations, or ice ages, interspersed with relatively warm, interglacial intervals, such as the current interglacial, the Holocene. The study of Quaternary environments began in the late eighteenth century. Quaternary geology and paleontology came of age in the nineteenth century, and other important aspects of Quaternary science, such as paleoceanography (*see* Paleoceanography), paleoecology, and paleoclimatology (*see* Introduction), developed to a much greater extent in the twentieth century. As with many branches of science, the pioneers in Quaternary studies had to work hard to overcome many widely held, erroneous ideas from previous generations of scholars.

At the beginning of the nineteenth century, science itself was rapidly changing. Up until that time, university professors and other scholars who performed scientific research were mostly generalists who dabbled in many different fields. They looked upon themselves as natural historians, studying the workings of the natural world, as their whimsy led them. The early nineteenth century saw the beginnings of specialization in science. As the level of scientific knowledge was rapidly increasing, it was no longer possible for individual scholars to keep abreast of all the new discoveries. People began to devote their time and energy to one or just a few lines of research. This new, focused style of scientific study brought great leaps forward for science as a whole, and for Quaternary science, in particular, as we shall see, below.

Establishing the Geologic Framework

The term 'Quaternary' was coined by an Italian mining engineer, Giovanni Arduino (1714–95). He distinguished four orders of strata comprising all of Earth's history: Primary, Secondary, Tertiary, and Quaternary (Schneer (1969), p.10). Arduino (**Fig. 1**) distinguished four separate stages or 'orders' which he recognized on the basis of very large strata arranged one above the other.

These four 'orders' were expressed regionally in Italy, as the Atesine Alps, the Alpine foothills, the sub-Alpine hills, and the Po River plain, respectively. The term 'Quaternary' apparently was not used again until the French geologist Desnoyers (1829) used it to differentiate Tertiary from Younger strata in the Paris basin. It was redefined by another Frenchman Reboul (1833) to include all strata containing extant flora and fauna.

The Quaternary period, as we now know it, is divided into two epochs: the Pleistocene and the Holocene (*see* Overview). The history of these terms

Figure 1 Giovanni Arduino (1714–95).

Figure 2 Charles Lyell (1797–1875).

is likewise complicated. The term 'Pleistocene' was coined by Scottish geologist, Charles Lyell (**Fig. 2**) in 1839, to replace his previous term 'Newer Pliocene' (1833).

Lyell defined the Pleistocene as the 'most recent' geologic era, and further specified that Pleistocene rocks and sediments are characterized by containing more than 90% fossil mollusks that are recognized as living species. As glacial theory began to take shape (see below), Forbes (1846) redefined the Pleistocene as equivalent to the 'Glacial Epoch.' Then Hörnes (1853) introduced the term Neogene to include Lyell's Miocene and Pliocene. In response, Lyell (1873) specified that the term Pleistocene should be used 'as strictly synonymous with post-Pliocene.' In the same publication, Lyell also separated the Pleistocene (glacial) from the 'Recent' (postglacial) time. The term 'Recent' was later replaced by the term 'Holocene' by Gervais (1867–69).

Thus, by the end of the nineteenth century, the stratigraphic nomenclature of the Quaternary period was firmly established (*see* Overview). However, no one knew when the Tertiary ended and the Quaternary began. In geology, it is standard procedure to designate a type locality that typifies such boundaries between major stratigraphic units. The 18th International Geological Congress (London, 1948) resolved to select a type locality for the Pliocene–Pleistocene (Tertiary–Quaternary) boundary. After three decades of deliberations, the Vrica section in Calabria, southern Italy, was finally selected. Hence the Plio-Pleistocene boundary was established at this site, where the boundary falls at ca. 1.64 Ma (Aguirre and Pasini, 1985; Bassett, 1985). Hilgen (1991) calibrated this age, based on an orbital forcing chronology, to an age of 1.81 Ma.

These age designations were only made possible through the invention of radiometric dating methods, which came about in the latter half of the twentieth century (see below).

The Discovery of Pleistocene Mammals

The threads of research that eventually led to modern Quaternary science came from a variety of disciplines, and were driven by scientific observations in a number of fields. One of these was the field of vertebrate paleontology (*see* Vertebrate Overview). As with many branches of science, pivotal discoveries often launch major new lines of research. One such discovery was made at a Pleistocene site in Kentucky, called Big Bone Lick. The site lies on a tributary of the Ohio River, about 30 km southwest of Cincinnati Ohio. It was the first major New World fossil locality known to Europeans. Baron Charles de Lougueuil, the commander of a French military expedition, may have been the first European to visit the site in 1739. He collected some mastodon fossils that were later studied by the French naturalists, Daubenton, Buffon, and Cuvier. Cuvier (1825) published a description of the Big Bone Lick mastodon remains. Inspired by this and other Pleistocene fossil discoveries, Cuvier developed his theory of global cooling that led to the extinction of these strange beasts.

In 1807, at the behest of Thomas Jefferson, William Clark conducted a major collecting expedition at Big Bone Lick that yielded about 300 specimens, most of which can still be found either at the National Museum of Natural History in Paris or at the Academy of Natural Sciences in Philadelphia. Thus, the fossils from this one site helped to launch Pleistocene vertebrate paleontology on two continents. The discovery of mastodon and other large Pleistocene mammal remains at this site sparked the imagination of scientists and politicians alike. In 1803, the United States purchased the Louisiana Territory from France. This territory included more than 2 million sq. km of land extending from the Mississippi River to the Rocky Mountains. When President Thomas Jefferson sent Meriwether Lewis and William Clark to explore and map this new American territory, he expected that they might find living specimens of mastodon and other large Pleistocene mammals, roaming the uncharted wilderness of the West. Jefferson was an avid naturalist, and took great interest in the fossil bones from Big Bone Lick.

Based on discoveries such as these, the field of vertebrate paleontology was starting to take shape during the late eighteenth and early nineteenth centuries. As discussed above, one of the most important leaders in this newly emerging field was the French scientist, Georges Cuvier (**Fig. 3**). At the start of the nineteenth century, Cuvier was a professor of animal anatomy at the Musée National d'Histoire Naturelle (National Museum of Natural History) in Paris.

Figure 3 Georges Cuvier (1769–1832).

An opponent of the theory of evolution, Cuvier's most important contribution to science was the establishment of extinction of ancient species, based on fossil records. Until the nineteenth century, most philosophers and natural historians rejected the idea that some species had died out, and that new species had evolved over time. Most Europeans held to a strict, literal interpretation of the Bible which dictated that the Earth was created in just 6 days, only a few thousand years ago. But the fossil record that was just beginning to be unearthed by a handful of paleontologists began to challenge this view.

Although Cuvier remained a Creationist, the fossils he was describing were re-shaping his views on the nature of that creation. Cuvier believed in the great antiquity of the Earth, and held that more-or-less modern conditions had been in existence for most of Earth's history. However, in order to explain the extinction of species Cuvier had seen in the fossil record, he invoked periodic 'revolutions' in Earth's history. Each 'revolution' was a natural event that had brought about the extinction of a number of species. Unlike others of his time (notably the Reverend William Buckland, who invoked the Biblical Flood), Cuvier did not equate these 'revolutions' with Biblical or historical events.

Cuvier considered that the last great 'revolution,' the one that brought about the extinction of such spectacular animals as mammoths and mastodons, might have been a severe and sudden cooling of the planet. Louis Agassiz (**Fig. 4**) took this idea and developed it further, into the concept of a 'Great Ice Age.'

Figure 4 Louis Agassiz (1807–73).

Agassiz was a Swiss naturalist who started his career as Cuvier's assistant. Agassiz thought that mammoths and other extinct mammals must have been adapted to a tropical climate. Here is how he described their demise in the face of the oncoming ice age:

> The gigantic quadrupeds, the Mastodons, Elephants, Tigers, Lions, Hyenas, Bears, whose remains are found in Europe from its southern promontories to the northernmost limits of Siberia and Scandinavia· · ·may indeed be said to have possessed the earth in those days. But their reign was over. A sudden intense winter, that was also to last for ages, fell upon our globe; it spread over the very countries where these tropical animals had their homes, and so suddenly did it come upon them that they were embalmed beneath masses of snow and ice, without time even for the decay which follows death.
>
> (Agassiz (1866), p. 208).

The Discovery of Pleistocene Glaciations

Agassiz's theory of the Great Ice Age was first presented to the Swiss Society of Natural Sciences in Neuchâtel in 1837. This was an ideal setting in which to convince geologists and natural historians, Agassiz could demonstrate the effects of glacial ice in the landscapes of the Alps (*see* Vertebrate Overview). He pointed to large boulders that had been transported by ice (glacial erratics), piles of rocks moved by glacial ice (glacial moraines), and scratched surface lines in bedrock, made by the passage of glacial ice and debris. Agassiz published his theory in the books *Étude sur les glaciers*, in 1840, and *Système glaciare*, in 1847. These books summarized his findings from Europe. He later found even more evidence of glaciation in North America. Agassiz's theory was initially rejected by many leading geologists, who still held to the idea that the transportation of surficial sediments was mainly due to the effects of the Biblical Flood. Agassiz's ideas on the glaciation eventually won the day, but his ideas about the nature of the Pleistocene megafauna turned out to be largely nonsensical. Far from being tropically adapted animals, the mammoths, mastodons, and other Ice Age mammals of Europe were adapted to the very same glacial environments to which Agassiz had ascribed their demise. Most of these animals died out during the transition to *warm* climate at the *end* of the last glaciation, not at its beginning.

Evidence for glaciation had been seen by some of Agassiz's contemporaries in other parts of Europe. For instance, Esmark noted the existence of glacial deposits in Norway, Bernhardi found evidence for glaciation in Germany, and de Venetz and Charpentier found evidence for the advance of glacial ice far beyond the limits of modern Alpine glaciers in Switzerland (*see* Vertebrate Overview). Agassiz himself traveled to Britain and North America and argued that surficial deposits that had previously been considered flood deposits should be reclassified as glacial.

Convinced by Agassiz's ice age theory, field geologists of the middle and late nineteenth century began looking for evidence to help reconstruct the actual history of glacial events. Agassiz had proposed a single, massive glacial event in which ice sheets covered much of the middle latitudes, as well as the high latitudes of the Earth. Evidence started accumulating that pointed to multiple glaciations, separated by warm periods. By the 1850s, evidence was pointing toward at least two major glaciations in Europe. By 1877, James Geikie (**Fig. 5**) had developed the concept of four or five large glaciations during the Pleistocene, based on stratigraphic evidence.

Evidence from North America made it clear that the last glaciation had not been the largest one, because it had not entirely destroyed the evidence for earlier, larger glaciations (*see* Late Quaternary in North America). Geologists coined the terms 'Nebraskan,' 'Kansan,' 'Illinoian,' and 'Wisconsinan,' to describe a sequence of four glacial epochs in North America. These were separated by three warm, or interglacial periods (the Aftonian, Yarmouthian, and Sangamon), based on the presence of ancient soils buried between glacial deposits.

Pioneering work on establishing the European glacial sequence was carried out by Albrecht Penck and Eduard Brückner (**Fig. 6**), who identified four glaciations, the Günz, Mindel, Riss, and Würm (*see* Late Quaternary in Lowland Asia).

Figure 5 James Geikie (1839–1915).

Figure 6 Eduard Brückner (1862–1927) and Albrecht Penck (1858–1945).

These glaciations were named after four rivers in southern Germany. Penck and Brückner's (1909) work was based on the identification of the stratigraphic sequence of river terraces in the valleys of the northern Alps (**Fig. 7**). In many parts of the world, diligent field studies in the last century have failed to find evidence for more than four glaciations on land.

The ways in which these glaciations were recognized varied from one part of the world to another. In Europe, only the ice advances that reached farther south than younger ones were recognized as separate glaciations. The traces of any intermediate ice expansion were essentially overridden and destroyed by subsequent larger glacial advances. American glaciations were originally defined as times when the ice sheets extended south to the American Midwest. Interglacials were the times when the Midwest region was ice free. The classical North American Pleistocene subdivision is one of long interglacials and short glacials, whereas the North European system recognizes short interglacials and long glacials (Kukla, 2005). Penck and Brückner's Alpine glaciation scheme was the most widely used system of classification in the twentieth century for the correlation of Pleistocene deposits between continents (Flint, 1971) (*see* Overview).

Development of Theories on the Causes of Glaciation

As we have seen, by the late nineteenth century, the geologic evidence for repeated, large-scale glaciations of the globe was firmly established. The causes of glaciation, however, remained a mystery. Geikie's geologic evidence from Scotland showed that warm intervals had developed between glaciations. While the relative length of glacial and interglacial periods remained unknown, it was becoming clear that large-scale climatic oscillations had taken place over many thousands of years of Earth's recent history. Various suggestions were put forward to explain these cycles. Changes in carbon dioxide levels were proposed, as well as changes in solar intensity.

Croll's Orbital Theory

One of the earliest theories on the cause of glacial/interglacial cycles was developed by the Scottish scientist, James Croll (**Fig. 8**). Croll had little formal education, but he was a voracious reader.

In 1859, his pursuit of knowledge led him to enter academia 'through the back door,' by becoming a janitor at the museum at Anderson's Institution in Glasgow. Once there, he began developing a theory about the causes of glaciation. He began writing letters to Charles Lyell, discussing his ideas on the connections between glaciation and variations in the Earth's orbit. Lyell was suitably impressed with Croll's scholarship, and helped him obtain a clerical position at the Geological Survey of Scotland in 1867.

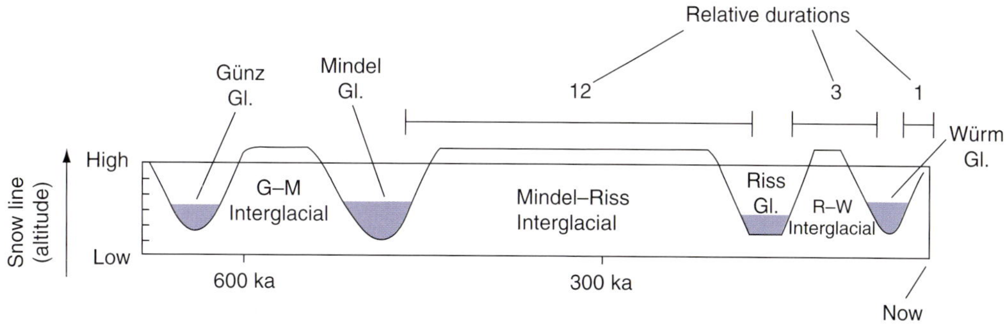

Figure 7 Diagram of European ice ages, their relative durations and the relative snow line during each. Note: During glacial periods snow fell at lower altitudes than during interglacial periods. Gl – Glacial; Intergl. – Interglacial. From Penck A and Brückner E (1909) *Die Alpen im Eiszeitalter.* Leipzig: Tachnitz.

Figure 8 James Croll (1821–90). Photo by J. C. Irons, 1896.

It was here that Croll was encouraged by Archibald Geikie to further develop his theory. Charles Darwin was also a regular correspondent with Croll, and both scientists benefited from this exchange of ideas.

Croll started publishing his theories in 1867, and his major contributions include *Climate and Time, in their Geological Relations* (1875) and *Climate and Cosmology* (1885).

In 1846, French astronomer Urbain Le Verrier published formulas that allow the calculation of changes in the shape of a planet's orbit and its axial precession. In 1864, Croll used these formulas to plot changes in the shape of Earth's orbit (called orbital eccentricity) over the past 3 Myr. He found that a pattern of high eccentricity had persisted for hundreds of thousands of years, followed by a pattern of low eccentricity, as is the case today. The more elliptical the orbit, the greater the difference in incoming solar radiation (insolation) between the different seasons of the year. Croll realized the importance of calculating the seasonality of insolation, which is one of his major contributions to the science of paleoclimatology. Changes in Earth's orbit that act to prolong the winter season cause greater amounts of snow to accumulate in the high latitudes (**Fig. 9**).

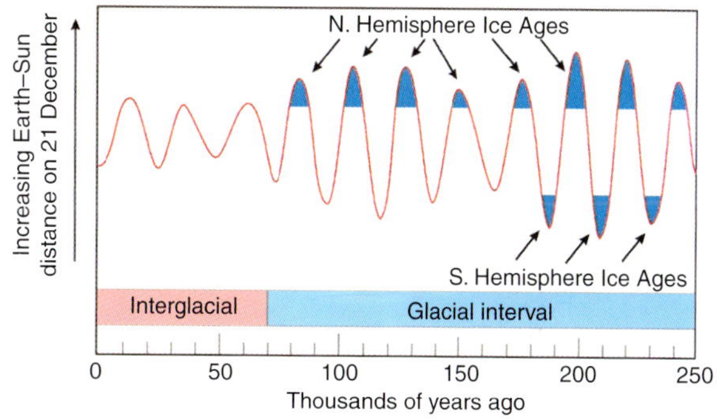

Figure 9 Diagram illustrating Croll's (1887) explanation of ice ages, based on changes in Earth's orbit around the Sun.

The extra snow cover reflects more solar energy back out into space, thereby amplifying the orbital effects. Croll argued that this amplification is what triggers the growth of ice sheets.

Croll's theory introduced important new concepts in the field of climatology. Subsequent research has shown that Croll's theory is insufficient to explain the global pattern of Pleistocene glaciations, and Croll's chronology of glaciations has been shown to be in error. Specifically, Croll's scheme made the last ice age much older than was inferred from the geologic evidence of Geikie and others. Ultimately, Croll failed to convince most of his contemporaries, and his ideas remained largely ignored by other researchers until the 1940s.

The Milankovitch Theory

Milutin Milankovitch (**Fig. 10**) was a Serbian mathematician who specialized in astronomy and geophysics.

Figure 10 Portrait of Milutin Milankovitch (1879–1958) by Paja Jovanovic, 1943. Courtesy of the Serbian Academy of Sciences and Arts.

In 1909 he became a member of the faculty in applied mathematics at the University of Belgrade.

Imprisoned by the Austro-Hungarian Army in the First World War, he recommenced work on his mathematical theory of climate change in 1920, completing this work in 1941. Milankovitch built his theory from previous work done by J.A. Adhemar and James Croll. In 1842 Adhemar explained glacial climate using only precession. Milankovitch used Croll's work to help him develop a mathematical model of climate change. This model incorporates the cyclical variations in three elements of Earth's orbit around the Sun: eccentricity, obliquity, and precession. Using these three orbital factors, Milankovitch developed a comprehensive mathematical model that calculated latitudinal differences in insolation and the corresponding surface temperatures during the last 600 kyr (*see* Milankovitch Theory and Paleoclimate, and Introduction) (**Fig. 11**).

The next step in Milankovitch's work was an attempt to correlate the orbital variations with glacial/interglacial cycles. Milankovitch worked on the assumption that radiation changes in some latitudes and seasons are key to triggering glaciation and deglaciation. Working with German Climatologist Vladimir Koppen, he chose the summer insolation values at 65° N as the critical latitude and season. Their reasoning was that the continental ice sheets grew near this latitude, and that cooler summers might reduce

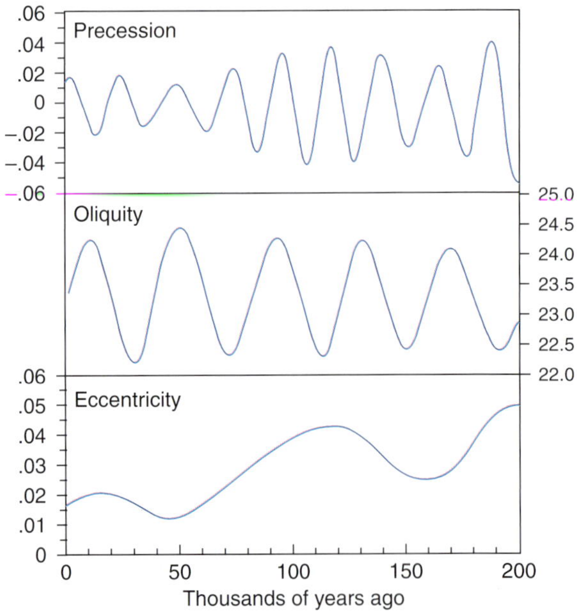

Figure 11 Orbital variations predicted by the Milankovitch theory. From Berger A and Loutre M F (1991) Insolation values for the climate of the last 10 million years. *Quaternary Science Reviews* 10: 297–317.

summer snowmelt, leading to a buildup of snow pack, and eventually to the growth of ice sheets.

Sadly, Milankovitch's theory was largely ignored for decades. However, in 1976, Hays *et al.* published a study of deep-sea sediment cores and found that Milankovitch's predictions matched their own interpretations of the timing and intensity of climate change during the last 450 kyr (*see* Paleoceanography). Specifically, they found that major variations in climate were closely associated with changes in the eccentricity, obliquity, and precession of Earth's orbit.

The Invention of Dating Methods

Without a means of obtaining an absolute age for events in the Quaternary, there would have been no way to test the validity of Milankovitch's orbital variation theory. Until the latter half of the twentieth century, Quaternary scientists lacked the tools to obtain such absolute ages, and could only infer the ages of events through relatively dating techniques. In other words, they could sometimes establish the 'sequence' of events, for instance, by determining the relative stratigraphic position of various kinds of fossils. But they could not tell whether a given sequence of events took place 50 or 150 ka, unless they were dealing with long sequences of sedimentary layers that accumulated in recognizable, annual layers (a very rare phenomenon).

Uranium-Series Dating

Radiometric dating methods were developed in the twentieth century, and have revolutionized Quaternary science. In 1902, physicists Ernest Rutherford and Frederick Soddy had discovered that radioactive elements broke down into other elements in a definite sequence or series, through the process of nuclear fission. The possibility of using this radioactivity as a means of measuring geologic time was first discussed by Rutherford in 1904. In 1906, Rutherford began calculating the rate of radioactive decay of uranium. This decay process (uranium decaying to lead) has since been discovered to go through multiple steps, with intermediate daughter products. It is now possible to use various uranium-series decay processes to derive age estimates for uranium-bearing fossils and sediments, back many millions of years (*see* Paleoceanography).

Radiocarbon Dating

Perhaps the most important breakthrough in the absolute dating of Quaternary fossils and sediments was the invention of radiometric dating methods, especially radiocarbon dating. In 1940, American physicists Martin Kamen and Sam Ruben discovered the long-lived radioactive carbon isotope, carbon-14. Kamen used ^{14}C as a tracer in biological systems. Kamen found that some of the nitrogen in the atmosphere was turned into carbon-14 when bombarded with cosmic rays. The existence of ^{14}C had been postulated since 1934, but it had never been directly observed nor characterized. Kamen succeeded in preparing ^{14}C in sufficient amounts to determine its half-life (5700 yr), that is, the amount of time it takes for half of a sample of ^{14}C to break down to the stable ^{14}N isotope of nitrogen (*see* Conventional Method).

Building on Kamen's discoveries, in 1947 American chemist Willard Libby (**Fig. 12**) determined that plants absorb traces of ^{14}C during their uptake of carbon in photosynthesis. At death, the plant would stop absorbing carbon, and the ^{14}C it contained would decay at its usual rate without being replaced. By measuring the concentration of ^{14}C left in the remains of a plant, Libby (1952) discovered that it was possible to calculate the amount of time since the plant had died. In addition, it was found that the same concentrations of ^{14}C occur in the tissues of animals as in plants, since animals either directly or indirectly ingest the carbon from plant tissues as their food. Given that it is possible to measure the concentration of remaining ^{14}C back to nine or ten half-lives, it has thus become possible to obtain absolute age estimates of fossil specimens (both plant and animal), back to about 45–50 kyr.

Figure 12 Photograph of Willard F. Libby, inventor of the radiocarbon dating method. Photo courtesy of Geoscience Analytical Inc.

For his work on carbon-14 dating, Libby received the Nobel prize in chemistry in 1960.

Conclusions

Other articles in this Encyclopedia will highlight the state of the art in the above-mentioned fields of Quaternary stratigraphy, vertebrate paleontology, Pleistocene glaciology, paleoclimatology, and dating methods. As with all branches of science, the current generation of researchers has built on the foundations of people such as Agassiz, Lyell, Cuvier, Milankovitch, and Libby. We owe these pioneers an enormous debt of gratitude. Many of these people worked in relative obscurity during their own lifetimes, and their theories were openly ridiculed by their contemporaries. Many survived major political upheavals and wars in the rapidly changing world of the nineteenth and twentieth centuries. The unifying themes of their lives are their intellectual curiosity, their diligence and perseverance, and their breadth of vision. May the same be said of twenty-first century Quaternary scientists, by future generations.

See also: **Glaciation, Causes**: Milankovitch Theory and Paleoclimate. **Glaciations**: Overview; Late Quaternary in Lowland Asia; Late Quaternary in North America. **Introduction**: History of Recent Major Projects. **Paleoceanography**. **Paleoclimate**: Introduction. **Quaternary Stratigraphy**: Overview. **Radiocarbon Dating**: Conventional Method. **Vertebrate Overview**.

References

Agassiz, L. (1840). *Etudes Sur Les Glaciers.* Jent and Gassmann, Neuchâtel.

Agassiz, L. (1866). *Structure of Animal Life: Six Lectures Delivered at the Brooklyn Academy of Music in January and February, 1866.* Charles Scribner and Co, New York.

Aguirre, E., and Pasini, G. (1985). The Pliocene–Pleistocene boundary. *Episodes* **8**, 116–120.

Bassett, M. G. (1985). Towards a "common language" in stratigraphy. *Episodes* **8**, 87–92.

Berger, A., and Loutre, M. F. (1991). Insolation values for the climate of the last 10 million years. *Quaternary Science Reviews* **10**, 297–317.

Croll, J. (1867a). On the eccentricity of the Earth's orbit, and its physical relations to the glacial epoch. *Philosophical Magazine* **33**, 119–131.

Croll, J. (1867b). On the change in the obliquity of the ecliptic, its influence on the climate of the polar regions and on the level of the sea. *Philosophical Magazine* **33**, 426–445.

Croll, J. (1875). *Climate and Time in their Geological Relations.* E. Stanford, London.

Cuvier, G. (1825). *Recherches Sur Les Ossemens Fossiles: Où l'on Rétablit Les Charactères de Plusieurs Animaux Dont Les Révolutions du Globe ont Détruit Les Espèces.* G. Dufour et E. d'Ocagne, Paris.

Desnoyers, J. (1829). Observations sur un ensemble de dépôts marins plus recents que les terrains tertiaries du bassin de la Seine, et constituant une formation geologique distincte; precedees d'une aperçu de la non-simulaneite des bassins tertiares. *Annals Sciences Naturelles, Paris* **16**, 171–214, 402–491.

Flint, R. F. (1971). *Glacial and Quaternary Geology.* Wiley, New York.

Forbes, E. (1846). On the connection between the distribution of existing fauna and flora of the British Isles, and the geological changes which have affected their area, especially during the epoch of the Northern Drift. *Great Britain Geological Survey Memoir* **1**, 336–342.

Gervais, P. (1867–69). *Zoologie et paleontology Générales. Nouvelles recherches sur les animaux vertétebrés et fossils (two volumes).* Bertrand, Paris.

Hays, J. D., Imbrie, J., and Shackleton, N. J. (1976). Variations in the Earth's Orbit: Pacemaker of the Ice Ages. *Science* **194**, 1121–1132.

Hilgen, F. J. (1991). Astronomical calibration of Gauss to Matuyama sapropels in the Mediterranean and implication for the geomagnetic polarity time scale. *Earth and Planetary Science Letters* **104**, 226–244.

Hörnes, M. (1853). Mittheilung an Prof. Bronn gerichtet. Wien, 3. Okt., 1853. In *Neues Jahrbuch Mineralogie Geologie Geognosie und Petrefaktenkunde,* pp. 806–810.

Kukla, G. (2005). Saalian supercycle, Mindel/Riss interglacial and Milankovitch's dating. *Quaternary Science Reviews* **24**, 1573–1583.

Libby, W. F. (1952). *Radiocarbon Dating.* University of Chicago Press, Chicago.

Lyell, C. (1833). *Principles of Geology, being an Attempt to Explain the Former Changes of the Earth's Surface by Reference to Causes Now in Operation,* vol. 3. John Murray, London.

Lyell, C. (1839). *Eléments de Geologie.* Pitois-Levrault, Paris.

Lyell, C. (1873). *The Geological Evidence of the Antiquity of Man,* 4th edn. John Murray, London.

Milankovitch, M. (1941). Kanon der Erdbestrahlung und seine Anwendung auf das Eiszeitenproblem. *Académie Royale Serbe Editions Speciales Section des Sciences Mathématiques et Naturelles* **133**.

Penck, A., and Brückner, E. (1909). *Die Alpen im Eiszeitalter.* Tachnitz, Leipzig.

Reboul, H. (1833). *Géologie de la période Quaternaire et Introduction a L'histoire Ancienne.* F.G. Levrault, Paris.

Schneer, C. J. (1969). *Toward a History of Geology.* MA: MIT Press, Cambridge.

Introduction: History of Dating Methods

A G Wintle, University of Wales, Aberystwyth, UK

© 2007 Elsevier B.V. All rights reserved.

The first methods for dating Quaternary materials resulted from the discovery of radioactivity 120 years ago. In addition, other dating methods have been developed that rely on other time-dependent changes that occur in natural materials. The different methods can be used to obtain the age of formation of either organic or inorganic materials.

Early Developments

Numerical dating methods based on natural radioactive phenomena were initially developed as a means of determining the age of the Earth (Dalrymple, 1991). Arguments on this topic raged in the second half of the 19th century, a time when the effects of recent glaciation were also being debated. However, it was not until the discovery of radioactivity in the 1880s that consideration was given to radioactive isotopes for use as natural clocks. As early as 1906, Rutherford suggested that ages could be obtained based on the production of helium by the decay of uranium in rocks. The first calculation of the age of the Earth based on the amount of radium in the Earth's crust was made by Russell in 1921. However, it was not until the 1930s, when Nier brought together the understanding of natural isotopes in the uranium and thorium decay chains and the construction of the first mass spectrometers, that it became possible to measure a range of isotopes. Also at this time, the potential of several different decay series that could be used for dating was proposed, based on a better understanding of the atomic structure of elements in the periodic table. The relationship between these early radiometric dates and the evolution of the geological timescale has been covered in a history of the work of Arthur Holmes (Lewis, 2000).

Radiocarbon Dating

The speed at which new geochronological tools based on radioactivity became relevant to Quaternary geology can be seen by examining the prefaces to the 1958 edition of Zeuner's book *Dating the Past.* In 1946, when the first edition was published, geochronology was very much based on the study of superposition of geological deposits and botanical and zoological successions. By the time of the third edition in 1952, theoretical and practical approaches to radiocarbon (^{14}C) dating had been put forward by Libby and the first radiocarbon laboratory was established. Soon after, lists of radiocarbon dates started to appear, although at this time radiocarbon dating was mainly applied to archeological materials, particularly from Upper Paleolithic sites. Another important result of the application of radiocarbon dating was in providing a chronology for the development of farming in Europe and the Near East. Libby was at the heart of radiocarbon dating development for three decades, producing the first book on the topic in 1952 and receiving the Nobel Prize in chemistry in 1960 for his work on radiocarbon dating. A comprehensive history of radiocarbon dating, and one that demonstrates the role played by Libby, can be found in Chapter 6 of the book by Taylor (1987) and in the edited volume by Taylor *et al.* (1992).

The early measurement of beta particles released from solid carbon was rapidly replaced by the use of more efficient counting systems using gases or liquids made from the organic material being used for dating. Both methods enable dating back to 60 ka, although the dating range is usually limited to approximately 40 ka due to sample contamination. In the 1950s, gas counting using carbon dioxide, methane, or acetylene was introduced and continued to be used for many decades. These gases (CO_2, CH_4, and C_2H_2) contained molecules with ^{14}C in their structure. Preparing the sample as a gas had the additional advantage that it was possible to enrich the concentration of ^{14}C relative to ^{12}C and ^{13}C in part of the gas volume and thus extend the potential age range back to 75 ka (Grootes *et al.*, 1975). This process was known as isotopic enrichment.

In the 1960s, liquid scintillation counting using benzene (C_6H_6) as the carrier of the ^{14}C was developed. Benzene was easy to prepare directly from acetylene. As a liquid, it had two advantages: It was possible to mix the sample and the organic scintillator chemicals, and it had a far greater density than acetylene and thus took up a smaller volume in the counting equipment. The latter property enabled a reduction in background count rate, and by the 1970s automated liquid scintillation measurement systems were being set up in laboratories worldwide.

The next major breakthrough in measurement techniques came in the 1970s, when methods were put forward in which it was not necessary to measure the beta particles released as the ^{14}C atoms decayed. The objective was to measure the ratio of ^{14}C to the more stable isotope of carbon (^{12}C) using a mass spectrometer. The first results were achieved using a particle accelerator known as a cyclotron, which had the advantage of removing all interfering particles except ^{14}N. In other studies, electrostatic accelerators were used in which molecules with similar charge-to-mass ratios as ^{14}C needed to be removed before being allowed into the mass spectrometer. Destruction of these molecules was accomplished by ionizing the sample and accelerating the ions, prior to passing them through a foil. Subsequently, small accelerators were built and the measurement technique of accelerator mass spectrometry (AMS) was born (Hedges, 1981). Being able to measure the atoms directly and not have to wait for them to decay speeded up measurement time. It also led to the dating of very small samples containing only 1 or 2 mg of elemental carbon. This led to a revolution in applications. A significant development has also been made in sample preparation using acid–base–wet oxidation and target production for AMS measurements. This has resulted in finite ^{14}C ages older than 50 ka being obtained for charcoal.

The development of radiocarbon dating was not without minor setbacks. Early tests of the dating technique were carried out on material from the tombs of the ancient Egyptian kings. In the 1960s, extensive sets of measurements were carried out on tree rings for which calendar ages could be determined by counting the rings. This method of dating is known as dendrochronology; the first tree-ring laboratory was set up in Arizona in 1937. Radiocarbon ages were obtained on rings from continuous and overlapping sequences cut from dead and living examples of the bristlecone pine, a tree that grows high up in the White Mountains of eastern California (**Fig. 1**). When the radiocarbon ages

Figure 1 Long-lived bristlecone pines growing in the White Mountains, California, used to provide a tree-ring record for radiocarbon calibration. Photo by Mike Walker.

were compared with the calendar dates from the counted tree-ring chronology, a significant deviation was observed for samples older than 3,000 years old. This deviation was the result of long-term changes in the production rate of ^{14}C in the upper atmosphere.

By 1986, calibration curves from many laboratories were brought together and a combined calibration curve was published. Use of bristlecone pines and fossil trees from Ireland and Germany resulted in a continuous calibration curve (IntCal98) for radiocarbon ages (Stuiver and Van der Plicht, 1998) back to 10,299 years BP. A new replacement calibration curve (IntCal04) has been constructed back to 26,000 years BP using paired radiocarbon and uranium series dates on corals and foraminifera (Reimer *et al.*, 2004).

Potassium–Argon and Argon–Argon Dating

The potential of dating using the decay of ^{40}K to ^{40}Ar was first reported in 1948. The K-Ar technique was developed in the 1950s for determining the age of rocks (Dalrymple and Lanphere, 1969); the first studies were mainly on pre-Quaternary rocks because of the difficulty in measuring the small amounts of argon that have accumulated in the younger rocks. The $^{40}Ar/^{39}Ar$ dating method, in which it is not necessary to measure absolute abundances of either argon or potassium in the sample, was developed in 1966. In the first study, it was noted that the new procedure would result in more precise ages, would allow measurements on very small samples, and could provide information on the thermal history of crystalline material if step heating was used. In the early 1970s, such experiments were carried out on a range of terrestrial rocks, with comparisons being made with conventional K-Ar dating.

An important application of both techniques was the dating of lava flows (**Fig. 2**), particularly since they contained evidence of the direction of the Earth's magnetic field at the time that they cooled. During the Quaternary period, the Earth's magnetic field has changed direction several times and the record of these changes is also found in sedimentary rocks (e.g., deep-sea cores and loess). At approximately the beginning of the Quaternary, the field changed from being in the direction that it is today to being totally reversed, and then at approximately 780 ka it changed back again (Jacobs, 1994). These periods of different direction are referred to as chrons and are named after scientists who have studied magnetic behavior. The change from the Gauss normal chron to the Matuyama reversed chron was dated to 2.47 Ma, and the change from the Matuyama chron to the Brunhes normal chron was first dated by the potassium–argon method to 730 ka. Most recently, an age of 775.6 ± 1.9 ka has been obtained using the $^{40}Ar/^{39}Ar$ method. Within the Matuyama reversed chron, there were a number of times when the field direction was normal for a few thousand years; these subchrons, named after places where they were first recorded, occurred at approximately 0.9 Ma (Jaramillo), 1.11 Ma (Cobb Mountain), 1.7 Ma (Olduvai), and 2.1 Ma (Reunion). These reversals have been the subject of extensive K-Ar and $^{40}Ar/^{39}Ar$ dating studies during the past 30 years (Jacobs, 1994). This record of reversals has been found in the 195 m of loess that overlies red clay in the Chinese Loess Plateau. This has allowed the transfer of an independent age scale onto the loess and the demonstration of the coincidence of the loess/red clay transition and the beginning of the Quaternary.

Figure 2 A lava flow on Etna that would provide a record of magnetic field direction and that could be dated by the potassium–argon method.

This geomagnetic polarity timescale has also been used, often in conjunction with other dating techniques, to obtain ages for sediments containing skeletal remains of early hominids in Africa. In the Turkana Basin of East Africa, the paleomagnetic history recorded in the sediments has been used along with K-Ar and $^{40}Ar/^{39}Ar$ dating to provide a framework for the archeological material found therein. In exposed sections, layers of volcanic ash (tephra), known as tuff, can be seen, relating to explosive eruptions of volcanoes hundreds of kilometers to the north. The first K-Ar age was for the KBS Tuff, found above the remains of an important skull of *Homo habilis*. However, the age was severely overestimated as a result of the incorporation of detrital minerals during fluvial reworking of the ash after it had been laid down. During the next 20 years, this tuff and approximately 20 others have been dated (and redated) as technological advances have been made (McDougall, 1990; McDougall and Harrison, 1999). The KBS tuff (named after the graduate student who had found stone tools within the ash layer) and the controversy generated by the dating of it and other tuffs were the subject of two chapters in Lewin's book (1987) on human origins in Africa. In the decade that followed the announcement of the first K-Ar date on the KBS tuff, a number of new measurements were carried out, concentrating on the separation of crystals of anorthoclase (a plagioclase feldspar found in the pumice clasts). These were used for both conventional K-Ar dating and $^{40}Ar/^{39}Ar$ dating using step heating, a process that would identify the presence of partially heated grains. By 1981, an age of 1.88 ± 0.03 Ma was accepted for the KBS tuff.

At approximately the same time, it was demonstrated that laser light focused onto a small spot on the surface of a grain could release argon. By the end of the 1980s, it became possible to obtain $^{40}Ar/^{39}Ar$ ages on individual grains of potassium-rich feldspars of Quaternary age, with the first results for samples of pumice and other volcanic material from a site in Kenya being published in 1990. This method became known as single crystal laser fusion (SCLF) dating (Walter, 1997). In the next few years, this method was employed on crystals from other hominid sites in Kenya and in Ethiopia. The earlier dates that had been obtained for the Turkana Basin by conventional K-Ar and $^{40}Ar/^{39}Ar$ techniques were confirmed by the new SCLF technique. On a more recent timescale, SCLF methods were used for sanidine crystals from the Laacher See Tephra in the Eifel region of Germany. Combining the ages for 16 individual crystals, an age of 12.7 ± 0.8 ka was obtained for the eruption.

The ability to use $^{40}Ar/^{39}Ar$ for Holocene volcanic samples has been demonstrated using subsamples made up of a few tens of sanidine crystals, feldspars with very high potassium contents. These crystals, up to approximately 1 mm in diameter, were extracted from a pumice clast from the city of Pompeii, buried by the eruption of Vesuvius in 79 AD. The age of 1925 ± 94 years before 1997 AD was obtained not by laser fusion but by using incremental heating of the grains, thus taking methods of analysis to an even higher level. A review of the principles and methodology of K-Ar and $^{40}Ar/^{39}Ar$ dating has been published by Renne (2000).

Fission Track Dating

Explosive volcanic eruptions not only produce potassium-rich feldspars but also produce volcanic glass and small zircon and apatite grains. These materials are suitable for yet another dating method, fission track dating (Westgate *et al.*, 1997). Fission track dating is based on fission of the uranium isotope ^{238}U, a process that has a half-life of 8.2×10^{15} years. The use of fission tracks for dating was born out of the study of fissile material during World War II. Fission is the name given to the process in which an atom spontaneously breaks apart with a tremendous release of energy. In the case of ^{238}U, two nuclei (fission fragments) are formed. They travel through the crystal away from the site of the parent ^{238}U atom in opposite directions, creating a damage trail that is a few micrometers long and approximately one-hundredth of a micrometer in diameter. This damage trail can only be seen as a fission track when the mineral face containing the track is exposed to a chemical etchant that attacks the damage zone along the length of the track. The number of tracks that cross the polished face of the mineral is proportional to the age and ^{238}U content. To obtain an age, this track density is compared with that induced by irradiating the mineral with neutrons, a process that causes the less abundant ^{235}U atoms to undergo fission. The tracks caused in this way are etched and measured as for the spontaneous tracks formed in nature by ^{238}U.

Because the tracks can be removed by heating (annealing), fission track measurements have also been used to determine the time since some thermal event subsequent to the formation of the mineral grains (thermochronology). The historical development of fission track measurement methods was discussed by Fleischer *et al.* (1975), and a more recent account of fission track dating over the whole of geological time was provided by Wagner and van den Haute (1992). This book also includes discussion of how two different values for the fission half-life were used for many years; this resulted in ages up to 20% different, depending on which half-life was used in the age equation. This problem was resolved by using standard minerals of known age (as obtained by $^{40}Ar/^{39}Ar$ dating).

For Quaternary volcanic events, dating has been limited to the use of zircon and apatite grains and volcanic glass. Zircon is the mineral of choice because of its high uranium content; zircon grains will thus have proportionally more tracks than volcanic glass for the same time period. However, very few zircon grains are usually found, perhaps as little as one grain in 1 kg of volcanic sediment (Dumitru, 2000). One of the earliest successes in fission track dating of zircon was the establishment of different ages for a number of volcanic ashes derived from the Quaternary volcanoes in Yellowstone National Park. Two geochemically different examples of the Pearlette Ash, previously used as a time-marker horizon, were found to have ages of 1.9 ± 0.1 and 0.6 ± 0.1 Ma, in agreement with K-Ar ages for two ashes closer to the source.

Even when present, zircon grains tend to be small and thus quite a few are needed to make a reliable measurement. The counting of several grains has the advantage of being able to identify contaminant grains. In the case of a volcanic ash bed in Ethiopia, it was found that the sample contained two populations of zircon grains (Walter, 1989). The ash was dated to 2.3 ± 0.5 Ma by dating the younger grains, but it was shown to contain grains with an age of 13.1 ± 3.2 Ma. If the different track densities had not been noticed, an age of 4.0 ± 0.7 Ma would have resulted.

Because of the scarcity of zircon grains, volcanic glass is usually preferred for Quaternary deposits. Early ages on glass were found to be too young compared with zircon ages from the same tephra. This was the result of the more rapid annealing of tracks in the case of volcanic glass. To overcome this problem, it was suggested that both irradiated and unirradiated aliquots should be heated to different temperatures for fixed periods of time prior to etching. When a temperature is reached for which the measured track densities are in a fixed ratio, this ratio can be used to obtain the age of the glass.

A more widely used approach has been developed using a lower temperature and longer times. This approach is known as isothermal plateau fission track (ITPFT) dating. More recently, a relationship between the average diameter of the fission tracks and the track density has been found; both are reduced by thermal annealing. ITPFT dating of volcanic glass has proved invaluable for dating ash layers found in the loess deposits of Alaska (**Fig. 3**), the Yukon, and New Zealand (Westgate *et al.*, 1997). A major result of the application of fission track dating of volcanic glass was the dating of the Toba tephra from western India, which, together with geochemical analysis of all known occurrences, firmly establishes that it has a single age, namely 75 ka.

Figure 3 Volcanic ash (Old Crow tephra) from Alaska that has been dated by the fission track method. The ash is overlain and underlain by loess that has been dated using luminescence techniques. Photo by Helen Roberts.

Luminescence Dating

Another method that measures the effect of nuclear processes in minerals is based on the luminescence properties of quartz and feldspars. Unlike fission track dating, it is the effect of ionizing radiation resulting from the decay of ^{238}U, ^{235}U, ^{232}Th, and ^{40}K in the minerals and their surroundings that results in the age-dependent signal. Unlike most of the other dating methods, luminescence dating techniques do not provide the age of formation of the quartz and feldspar grains to which they are applied. Instead, they determine the point in time at which the grains were either heated or exposed to light. For sedimentary grains, determining the time since the grains were last exposed to light provides the depositional age. This has allowed direct determination of rates of sediment accretion.

Luminescence dating was born out of the need to measure the amounts of ionizing radiation in a laboratory handling radioactivity, or in the environment or as a result of nuclear explosions, and the ability to make the measurements provided by the development in the 1940s of a sensitive light-measuring device known as a photomultiplier tube. The earliest use of thermoluminescence (TL) for dating both archeological and geological materials was in 1953, but it was not until the 1960s that it was developed as a useful tool. The TL is released from irradiated minerals when they are heated from room temperature to 450°C in an oxygen-free environment. TL measurements on mineral grains extracted from pottery, which would have had their signal removed by heating during manufacture, were first made in 1964. For the next two decades, TL dating was confined to heated materials, including burnt flint and lava flows. The technical aspects of TL dating are fully covered by Aitken (1985). However, as early as the mid-1960s, TL ages for unheated sediments had been reported in the Russian literature, but they remained unnoticed in the West until the late 1970s when the dating of mineral grains from a deep-sea core was reported. During the next 15 years, TL dating was applied to a wide range of terrestrial deposits ranging from coastal dunes to loess.

The next major breakthrough was in the use of an optically stimulated luminescence (OSL) signal (Huntley *et al.*, 1985). The use of an optically stimulated signal was seen as clearly more appropriate for measuring a signal whose resetting mechanism was exposure to light. Indeed, a light-insensitive TL signal remained when sedimentary grains were exposed to light, either from the Sun or from laboratory light sources. OSL techniques became more widely used when small halogen bulbs, and then light-emitting diodes (LEDs) (**Fig. 4**), replaced the expensive argon ion laser used as the stimulation source.

A peculiar optical stimulation resonance for feldspars, but not quartz, was found in the infrared (IR) region of the spectrum. The availability of IR-emitting diodes soon led to the dating of feldspars using infrared stimulated luminescence (IRSL). This signal is rapidly zeroed by sunlight, thus making it suitable for dating sediments. The ease of zeroing the IRSL in a relatively short time using such diodes led to the development in 1991 of measurement procedures that could be applied to single portions (aliquots) of K-feldspar grains. This permitted multiple measurements of the equivalent dose (the measure of the past radiation exposure that is proportional to age) to be made.

The stability, power, and wavelength availability of LEDs has led to a rapid growth in methodological developments in the past 10 years (Wintle, 1997; Aitken, 1998). LEDs producing intense blue-green light became available, and a procedure was developed for quartz based on the use of a single aliquot.

Figure 4 A luminescence dating laboratory showing blue light used for optically stimulated luminescence dating of sand grains. Photo by Risø National Laboratory, Denmark.

The single-aliquot regenerative dose procedure is now the basis of most OSL dating applications. IRSL, or OSL, dating of feldspars is less widely applied since for many feldspars the luminescence signals are unstable over geological time (a phenomenon known as anomalous fading). Overcoming this problem for feldspars and getting around the early saturation of the OSL signal for quartz are the current challenges. It should be noted that because of early saturation of the OSL signal found for all quartz, older sediments that are known to be made up of well-bleached grains (e.g., coastal dune ridges) have continued to be dated using a light-sensitive TL signal from quartz.

Also, within the past decade, it has become possible to use a small laser to produce a beam of green light focused onto a 20-μm-diameter spot to obtain OSL from single quartz, or feldspar, grains. This has enabled the measurement of several thousand grains to be carried out and, depending on the brightness of the individual grains, the equivalent dose to be determined for a large number of them (Duller, 2004). Whether these values of equivalent dose form a single population or are scattered enables conclusions to be drawn about whether the grains making up a sediment unit were all well exposed to light at deposition. This is particularly important when sediments of glacial origin are to be dated or when postdepositional reworking is a possibility.

Electron Spin Resonance Dating

Another mineral property that provides a measure of past radiation dose is electron spin resonance (ESR). ESR measurements are made by placing a small aliquot of sample in a magnetic field and observing the amount of microwave energy that is absorbed. This relates to the trapped electrons in the crystalline structure and is proportional to the age since formation (Grün, 1997). The idea of measuring trapped electrons by their behavior in a magnetic field was first put forward in 1936. However, it was not until 1975 that it was used to date a speleothem. In this case, it was used to determine the time that had elapsed since the formation of the calcite that makes up the stalagmite or stalagtite. In the 1980s, there was a considerable amount of investigation of ESR as a dating method for a wide range of Quaternary materials, with the first use for dating corals from raised beaches to establish a terrestrial record of high sea levels. Also in the 1980s, ESR was explored as a method of dating minerals affected by the pressure produced at active faults. It has also been demonstrated that there are particular ESR signals in quartz grains that can be zeroed by sunlight and this would enable sediments to be dated in a way that is analogous to OSL dating. The first measurements of these signals in single grains are also encouraging. Single grain ESR measurements on quartz grains of volcanic origin have also provided an age for the Toba eruption of 74 ka, albeit with a large amount of scatter.

The potential of using the signal from tooth enamel was investigated in 1987 and by the early 1990s ESR dating had been applied at a number of sites containing evidence of modern humans. Tooth enamel is made up of hydroxyapatite, a material that provides sufficient ESR sensitivity and remains unaltered during burial. However, the dentine and cementum that make up the rest of the tooth absorb large amounts of uranium from the groundwater that moves through the surrounding sediment, and this has led to the need to model the dose rate through time. Until recently, the upper and lower limits for ESR ages have been calculated using the 'early' or 'linear' uptake models. Such calculations were used in the dating of animal teeth from the Paleolithic site of Pech de l'Azé in France (Grün and Stringer, 1991). Unfortunately, for samples around 150 ka, there is a 40-ka range of values as a result of the uncertainty as to when the uranium was absorbed by the dentine attached to the teeth being dated. In 1993, newly developed thermal ionization mass spectrometers were used for measuring U and Th isotopes when present in low concentrations in very small samples. For teeth beyond the usual range of U-series dating, some assumptions still need to be made about when the uptake of uranium occurred, but the ages obtained with the two extreme hypotheses are much closer, for example, resulting in a concordant age of approximately 400 ka for two teeth from the Hoxne interglacial site. More recent developments in U-series methods using laser ablation to provide atoms for the mass spectrometer have enabled even greater precision, with the isotope contents able to be obtained spatially across a thin section cut from a tooth (**Fig. 5**).

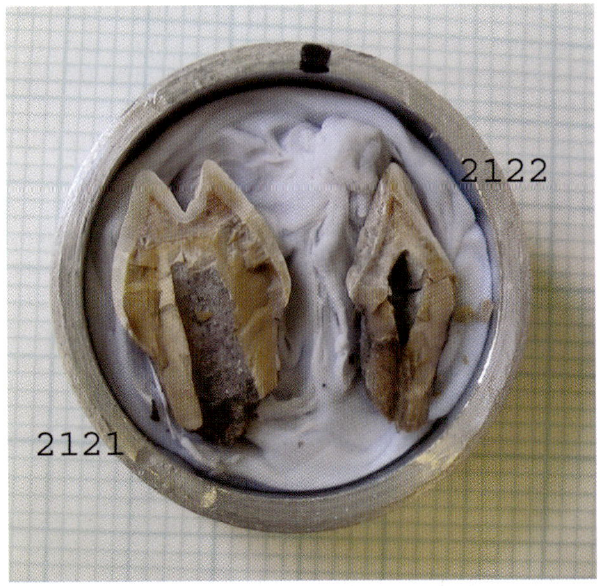

Figure 5 Sections of teeth from extinct kangaroos that have been used for combined electron spin resonance and MC-ICP-MS uranium series dating. Photo by Rainer Grün.

Figure 6 Cross section of a stalagtite showing growth rings that have been dated using the MC-ICP-MS uranium series dating method. Photo by Miryam Bar-Matthews.

Uranium Series Methods

The radioactive decay processes of the uranium (^{238}U and ^{235}U) and thorium (^{232}Th) decay chains were investigated in 1938, but at that time the main interest was in the lead isotopes that are at the ends of each decay chain (^{206}Pb, ^{207}Pb, and ^{208}Pb, respectively). In between these lead isotopes and the parent isotopes are a series of daughter isotopes of various elements, each daughter being radioactive. Each of these daughter isotopes will have different chemical properties, with solubility being the most important. ^{230}Th is a daughter product in the ^{238}U decay chain. Thorium (in the form of either ^{230}Th or ^{232}Th) is much less soluble in water than uranium and as a result is not found in groundwater. Thus, stalagmites and stalagtites (**Fig. 6**) formed in caves will take ^{238}U (and its daughter product ^{234}U) into the calcite structure as they are formed. ^{230}Th is then produced as the uranium isotopes decay. This in-growth of ^{230}Th increases with time and provides a chronological tool for such materials (Schwarcz, 1997).

The earliest applications of ^{230}Th measurements were made in the 1950s and 1960s and applied to lacustrine carbonates, marine sediments and corals, as well as to cave calcite deposits. Until the 1980s, measurements were made by dissolving samples, depositing the isotopes onto metal disks, and then counting the alpha particles released. The isotopic source of each alpha particle could be recognized by its characteristic energy, a technique known as alpha spectrometry. This approach enabled ages to be obtained back to approximately 350 ka, but it was time-consuming, particularly for older samples, which required several days of counting. Also, relatively large samples were required to obtain high enough isotope concentrations for counting levels to be sufficiently above background, and this led inevitably to the incorporation of inappropriate material (e.g., recrystallized or reworked material).

In the 1980s, these problems were overcome by the introduction of mass-spectrometric techniques. Using a mass spectrometer, the ^{238}U, ^{234}U, and ^{230}Th atoms could be measured directly, without waiting for their decay. This meant that small samples could be used, thus reducing the possibility of contamination, and counting statistics were improved so that very small errors could be obtained (e.g., a 2σ error of ± 1 ka for a sample with an age of 125 ka). The procedure was termed thermal ionization mass spectrometry (TIMS). Because of its high precision, TIMS has been applied to a wide range of material, including speleothem (Richards and Dorale, 2003), marine and lacustrine carbonates (Edwards *et al.*, 2003), and materials such as calcitic flowstones under cave art and ostrich eggshell as well as flowstones that seal off sediments containing other evidence related to early human activity (Pike and Pettitt, 2003). Multicollector inductively coupled-plasma mass spectrometry (MC-ICP-MS) is being developed as an even more precise tool for the *in situ* dating of growth rings in stalagtites, such as those shown in **Figure 6** (Eggins *et al.*, 2005).

Dating the Past 100 Years

^{210}Pb is another isotope further down the ^{238}U decay chain. It has been used for dating younger material because of its very short lifetime. It was introduced as a dating method in 1963 and is used for dating sediments from the past 100 years. Although several approaches have been developed (Noller, 2000), the current rate of supply model is most widely used, a method that requires the integrated activity of ^{210}Pb to be calculated for the deposit. Counting methods are employed following sample digestion, but it has been suggested that ^{210}Pb could be measured directly using MC-ICP-MS, as used for the longer lived isotopes in the ^{238}U decay chain. An artificial isotope, ^{137}Cs, has also been used as a marker horizon in even younger sediments.

Cosmogenic Nuclide Dating

During the 1990s, a suite of new techniques for dating appeared in the Quaternary literature (Gosse and Phillips, 2001). These methods are based on the production of isotopes in rocks at the Earth's surface and are collectively known as cosmogenic nuclide dating. Although it had been proposed by Davis and Schaeffer as far back as 1955 that cosmic rays would interact with atoms in minerals at the Earth's surface, it was not until the mid-1980s that this was explored. At that time, AMS measurement of cosmogenically produced ^{14}C (produced in the atmosphere) became possible. Using similar AMS techniques, it was possible to measure the small amounts of radioactive isotopes (^{10}Be, ^{26}Al, and ^{36}Cl) produced in rocks (Zreda and Phillips, 2000). In addition, the stable cosmogenic isotopes (^{3}He and ^{21}Ne) could be measured by mass spectrometry (Niedermann, 2002). The importance of these techniques is that they are able to provide direct information on the exposure history of both rocks and sediments. Thus, for the first time it was possible to obtain rates for geomorphic processes in the Quaternary.

Two types of application were developed. First, the exposure age of a rock surface can be measured directly because the number of atoms will be proportional to time, with due account being taken of both the production rate and radioactive decay. Second, when rock surfaces are removed by geomorphic processes, it is possible to use two isotopes with different lifetimes and obtain information on long-term landscape evolution. Some of the first applications were to glacial deposits, for which the fresh rock surfaces are exposed to cosmic rays as the result of erosion by glaciers (**Fig. 7**), and to the new rock surfaces provided by meteorite impacts. However, any process that has the potential to move previously buried rocks to the surface in a relatively short time can be investigated using cosmogenic isotopes. Such studies have been used to quantify rates of bedrock weathering (Gosse and Phillips, 2001; Cockburn and Summerfield, 2004).

Amino Acid Racemization

During the past 40 years, several other dating methods have been developed that use changes that occur as a result of chemical processes in the environment. Of these, the most widely applicable is amino acid racemization (AAR), which was first put forward as a dating technique in 1969. Amino acids are the basic chemical units from which proteins are

Figure 7 Glacial landscape in Patagonia showing granitic erratic for which an exposure age has been obtained using cosmogenic nuclides (^{10}Be). Photo by Neil Glasser.

formed and several different types (e.g., alanine and isoleucine) are found in organic materials (Hare *et al.*, 1997).

In the natural environment, these amino acids exist in two molecular forms, known as isomers. However, in living materials, they occur only in one form (L-type), and after death they convert to the other form (D-type) with the passage of time until equilibrium is reached. This process is racemization (or epimerization for some amino acids). Racemization occurs more rapidly when the environmental temperature is higher (e.g., during interglacials). As with any process occurring in biological material, there are additional complications relating to structure and these have led to some early problems.

One of the earliest applications of AAR dating was to bones at a native American site in California. However, due to the development of AMS ^{14}C dating, the early (apparently pre-Clovis) ages were found to be incorrect. Since then, no further dating of bones has been attempted. Instead, the main applications of AAR have been to mollusks, both marine and nonmarine, with isoleucine being the amino acid of choice (Wehmiller and Miller, 2000). Although a study has criticized use of the AAR technique for producing absolute ages, D/L ratios have been useful for providing correlations and relative stratigraphies over areas where there is little temperature variation. A more secure approach would be to use more than one amino acid, an approach currently being tested. New technological developments include the use of multiple collector ion gas chromatography and reverse phase chromatography and the analysis of free (rather than bound within the protein) amino acids. These have permitted the measurement of very small samples, such as a single foraminifera or ostracod valve (**Fig. 8**).

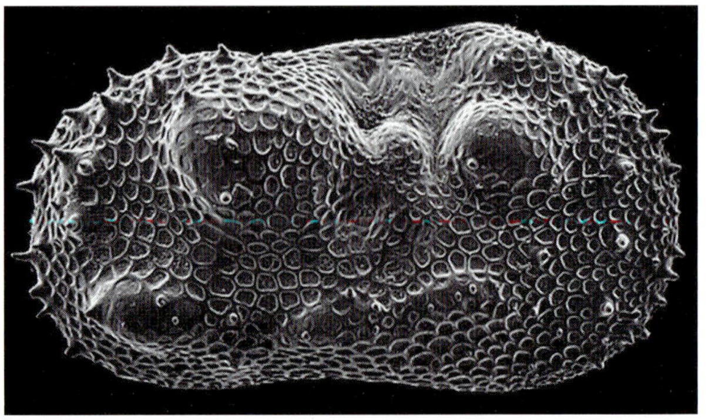

Figure 8 Ostracod valve suitable for amino acid racemization analysis: 0.8-mm-long SEM image of a freshwater cypridoidean ostracod found in British Pleistocene interglacial deposits of Marine Isotope Stage 9. Image by John Whittaker.

Another application of AAR has been to ostrich and other ratite bird eggshells. These have the advantage of having a mineral structure that is far less susceptible to diagenetic alteration during burial. Fossil eggshells are often found in association with archeological material and have provided chronological information at sites in southern Africa and Australia.

Summary

Besides the methods outlined in this brief overview, other time-dependent changes can be used to establish age equivalence, although they do not give rise to numerical ages. These and other methods are well described in Chapter 7 of Walker (2005). In addition, there are edited volumes covering many of the techniques (Taylor and Aitken, 1997; Noller *et al.*, 2000), as well as review papers dealing with one or more of the techniques discussed here (Wintle, 1996).

See also: **Cosmogenic Nuclide Dating**: Cosmic Ray Interactions in Minerals; Exposure Geochronology; Landscape Evolution. **Dating Techniques**. **Dendrochronology**. **Fission-Track Dating**. **Geomagnetic Excursions and Secular Variations**. **K/Ar and Ar/Ar Dating**. **Luminescence Dating**: Thermoluminescence; Optically-Stimulated Luminescence; Electron Spin Resonance Dating. **Radiocarbon Dating**: Conventional Method; AMS Method; Variations in Atmospheric ^{14}C; Calibration. **U-SERIES Dating**.

References

Aitken, M. J. (1985). *Thermoluminescence Dating*. Academic Press, London.

Aitken, M. J. (1998). *An Introduction to Optical Dating*. Oxford University Press, Oxford.

Cockburn, H. A. P., and Summerfield, M. A. (2004). Geomorphological applications of cosmogenic isotope analysis. *Progress in Physical Geography* **28**, 1–42.

Dalrymple, G. B. (1991). *The Age of the Earth*. Stanford University Press, Stanford, CA.

Dalrymple, G. B., and Lanphere, M. A. (1969). *Potassium–Argon Dating*. Freeman, San Francisco.

Davis, R., and Schaeffer, O. A. (1955). Chlorine-36 in nature. *Annals of the New York Academy of Science* **62**, 105–122.

Duller, G. A. T. (2004). Luminescence dating of Quaternary sediments: Recent advances. *Journal of Quaternary Science* **19**, 183–192.

Dumitru, T. A. (2000). Fission-track geochronology. In *Quaternary Geochronology: Methods and Applications* (J. S. Noller, J. M. Sowers and W. R. Lettis, Eds.), pp. 131–155. American Geophysical Union, Washington, DC.

Edwards, R. L., Gallup, C. D., and Cheng, H. (2003). Uranium-series dating of marine and lacustrine deposits. *Reviews in Mineralogy and Geochemistry* **52**, 363–405.

Eggins, S., Grün, R., McCulloch, M. T., *et al.* (2005). In situ U-series dating by laser-ablation multi-collector ICPMS: New prospects for Quaternary geochronology. *Quaternary Science Reviews* **24**, 2523–2538.

Fleischer, R. L., Price, P. B., and Walker, R. M. (1975). *Nuclear Tracks in Solids*. University of California Press, Berkeley.

Gosse, J. C., and Phillips, F. M. (2001). Terrestrial in situ cosmogenic nuclides: Theory and applications. *Quaternary Science Reviews* **20**, 1475–1560.

Grootes, P. M., Mook, W. G., Vogel, J. C., *et al.* (1975). Enrichment of radiocarbon for dating samples up to 75,000 years. *Zeitschrift für Naturforschung A* **30**, 1–14.

Grün, R. (1997). Electron spin resonance dating. In *Chronometric Dating in Archaeology* (R. E. Taylor and M. J. Aitken, Eds.), pp. 217–260. Plenum, New York.

Grün, R., and Stringer, C. B. (1991). Electron spin resonance dating and the evolution of modern humans. *Archaeometry* **33**, 153–199.

Hare, P. E., Von Endt, D. W., and Kokis, J. E. (1997). Protein and amino acid diagenesis dating. In *Chronometric Dating in Archaeology* (R. E. Taylor and M. J. Aitken, Eds.). Plenum, New York.

Hedges, R. E. M. (1981). Radiocarbon dating with an accelerator: Review and preview. *Archaeometry* **23**, 3–18.

Huntley, D. J., Godfrey-Smith, D. I., and Thewalt, M. L. W. (1985). Optical dating of sediments. *Nature* **313**, 105–107.

Jacobs, J. A. (1994). *Reversals of the Earth's Magnetic Field*. Cambridge University Press, Cambridge, UK.

Lewin, R. (1987). *Bones of Contention*. Simon & Schuster, New York.

Lewis, C. (2000). *The Dating Game*. Cambridge University Press, Cambridge, UK.

Libby, W. L. (1952). *Radiocarbon Dating*. University of Chicago Press, Chicago.

McDougall, I. (1990). Potassium–argon dating in archaeology. *Science Progress* **74**, 15–30.

McDougall, I., and Harrison, T. M. (1999). *Geochronology and Thermochronology by the $^{40}Ar/^{39}Ar$ Method*. Oxford University Press, New York.

Niedermann, S. (2002). Cosmic-ray-produced noble gases in terrestrial rocks: Dating tools for surface processes. *Reviews in Mineralogy and Geochemistry* **47**, 731–784.

Noller, J. S. (2000). Lead-210 geochronology. In *Quaternary Geochronology: Methods and Applications* (J. S. Noller, J. M. Sowers and W. R. Lettis, Eds.), pp. 115–120. American Geophysical Union, Washington, DC.

Noller, J. S., Sowers, J. M., and Lettis, W. R. (2000). *Quaternary Geochronology: Methods and Applications*. American Geophysical Union, Washington DC.

Pike, A. W. G., and Pettitt, P. B. (2003). U-series dating and human evolution. *Reviews in Mineralogy and Geochemistry* **52**, 607–630.

Reimer, P. J., Baillie, M. G. L., Bard, E., *et al.* (2004). IntCal04 terrestrial radiocarbon age calibration, 0–26 cal kyr BP. *Radiocarbon* **46**, 1029–1058.

Renne, P. R. (2000). K-Ar and 40Ar/39Ar dating. In *Quaternary Geochronology: Methods and Applications* (J. S. Noller, J. M. Sowers and W. R. Lettis, Eds.), pp. 77–100. American Geophysical Union, Washington, DC.

Richards, D. A., and Dorale, J. A. (2003). Uranium-series chronology and environmental applications of speleothems. *Reviews in Mineralogy and Geochemistry* **52**, 407–460.

Schwarcz, H. P. (1997). Uranium series dating. In *Chronometric Dating in Archaeology* (R. E. Taylor and M. J. Aitken, Eds.), pp. 159–182. Plenum, New York.

Stuiver, M., and Van der Plicht, J. (1988). IntCal98 calibration issue. *Radiocarbon* **40**, 1041–1164.

Taylor, R. E. (1987). *Radiocarbon Dating*. Academic Press, London.

Taylor, R. E., and Aitken, M. J. (1997). Chronometric dating in archaeology. In. Plenum, New York.

Taylor, R. E., Long, A., and Kra, R. S. (1992). *Radiocarbon after Four Decades: An Interdisciplinary Perspective*. Springer-Verlag, New York.

Wagner, G. A., and van den Haute, P. (1992). *Fission Track Dating. Dordrecht*. Kluwer, The Netherlands.

Walker, M. (2005). *Quaternary Dating Methods*. Wiley, Chichester, UK.

Walter, R. C. (1989). Application and limitation of fission-track geochronology to Quaternary tephras. *Quaternary International* **1**, 35–46.

Walter, R. C. (1997). Potassium–argon/argon–argon dating methods. In *Chronometric Dating in Archaeology* (R. E. Taylor and M. J. Aitken, Eds.), pp. 97–126. Plenum, New York.

Wehmiller, J. F., and Miller, G. F. (2000). Aminostratigraphic dating methods in Quaternary geology. In *Quaternary Geochronology: Methods and Applications* (J. S. Noller, J. M. Sowers and W. R. Lettis, Eds.), pp. 187–222. American Geophysical Union, Washington, DC.

Westgate, J., Sandhu, A., and Shane, P. (1997). Fission-track dating. In *Chronometric Dating in Archaeology* (R. E. Taylor and M. J. Aitken, Eds.), pp. 127–158. Plenum, New York.

Wintle, A. G. (1996). Archaeologically-relevant dating techniques for the next century. *Journal of Archaeological Science* **23**, 123–138.

Wintle, A. G. (1997). Luminescence dating: Procedures and protocols. *Radiation Measurements* **27**, 769–817.

Zeuner, F. E. (1958). *Dating the Past*. Methuen, London.

Zreda, M. G., and Phillips, F. M. (2000). Cosmogenic nuclide buildup in surficial materials. In *Quaternary Geochronology: Methods and Applications* (J. S. Noller, J. M. Sowers and W. R. Lettis, Eds.), pp. 61–76. American Geophysical Union, Washington, DC.

Introduction: History of Recent Major Projects

J J Lowe, University of London, Egham, UK
M J C Walker, University of Wales, Lampeter, UK
S C Porter, University of Washington, Seattle, WA, USA

© 2007 Elsevier B.V. All rights reserved.

Introduction

Quaternary Science is an interdisciplinary subject concerned with the history of environmental changes on Earth during the past 2.6 Myr. It is the goal of Quaternary scientists to interpret the geological record from this period in order to understand the key processes (physical, chemical, biological, atmospheric, human-induced) that trigger and modulate environmental change at all geographic and temporal scales. Given this very wide scope, it is not surprising that the technical and intellectual tools required to achieve this goal have been drawn from almost every scientific discipline. The accurate dating of many Quaternary events, for example, only became possible after major developments in atomic physics during the first half of the twentieth century; recent discoveries in evolutionary biology and biomolecular analysis have revolutionized our interpretation of the Quaternary fossil record; engineering breakthroughs just after the Second World War enabled sediment cores to be recovered from the deepest parts of the oceans and polar ice caps, the analysis of which revealed for the first time the highly sensitive nature of the Earth's climate; and space exploration has enabled processes at the Earth's surface to be measured and monitored independently.

Applications of these various approaches to the interpretation of Quaternary geological records over the course of the past 80 years or so has significantly changed the way in which we now interpret recent Earth history (Porter, 1999). The analytical tools available for studying the records grow ever more sophisticated and diverse, and past events can now be reconstructed at a much higher temporal resolution than was possible only a few decades ago. What emerges from these analyses is a highly resolved record of global climate change during glacial and interglacial episodes, with shifts in temperature of 10 °C or more sometimes occurring on millennial or, in some cases, centennial timescales.

Over the past 2.6 Myr, the Earth acquired the combination of physical characteristics (e.g., shape and disposition of continents and mountain belts, ocean circulation systems, atmospheric gas content, and major vegetation biomes) that we can observe today, and which differ markedly from those of earlier eras. The Quaternary geological record of this time interval has revealed just how frequently, and occasionally spasmodically, the global climate system can change. It provides the basis for testing outputs of Global Circulation Models (GCMs) (*see* Quaternary Environments), numerical models that simulate the workings of the present global climate system. It was also during the last 2.6 Myr that anatomically modern humans evolved, diffused globally, and became increasingly influential in modifying the Earth's surface and atmosphere.

Quaternary scientists integrate information obtained from the natural science disciplines, such as climatology, ecology, geology, and oceanography, as well as from subjects usually associated with the humanities, such as archaeology and anthropology. This reflects the belief, long held by Quaternary scientists, that a multidisciplinary approach is essential for understanding global environmental changes, how their effects might be controlled or mitigated, and how humans might learn to cope with environmental stress – knowledge which could well have a bearing on long-term human survival. This brief review considers the primary methodological developments that underpin current thinking about ways in which climatic fluctuations regulate Earth surface processes and conditions. We begin with the publication of the Astronomical Theory, which represented the first systematic attempt to explain the rhythmic nature of Quaternary climate change. Coincidentally, the formulation of the Astronomical Theory broadly coincided with the establishment of INQUA, the *International Union for Quaternary Research* (Porter, 1999). INQUA held its first international congress in 1928 and since then its diverse, multidisciplinary membership has been consistently at the cutting edge of research into global climate change.

Astronomical Rhythm of Climate Change

It had long been suspected that the Earth has experienced periodic changes in climate. The first tentative suggestions about these matters were made in the later years of the nineteenth and early years of the twentieth centuries, although in those times the geological evidence available for testing the emerging ideas was limited and equivocal. Since then, geological archives have become more robust and diverse. Rhythmical change is clearly reflected, for example, in the oscillatory nature of the stable oxygen isotope signal in deep-marine sediments and polar ice cores (*see* Oxygen Isotope Stratigraphy of the Oceans and Antarctic Stable Isotopes), in the regular occurrence of well-developed soil units (paleosols) within the thick accumulations of wind-blown dust (loess) in central China (*see* China), and in an increasing number of lake sediment records which suggest that the margins of forests in Europe and in high mountain regions, such as the Andes, regularly migrated backwards and forwards across considerable tracts of land (*see* West-Central-Europe and Latin America). These are manifestations of a global response to some rhythmical forcing agent, now widely believed to be variations in the Earth's orbit and axis. This is the 'Astronomical' or 'Milankovitch theory' (*see* Milankovitch Theory and Paleoclimate), named after the Serbian geophysicist, Milutin Milankovitch who, in a series of papers between 1920 and 1941, proposed that cyclical variations in the amount of solar radiation received at the surface of the Earth could induce major changes in global climates. The principal components of the theory are the precession of the equinoxes (apparent movement of the seasons around the sun) with a periodicity of ca. 19–23 ka, the obliquity of the ecliptic (variations in the tilt of the Earth's axis) with a periodicity of ca. 41 kya, and the eccentricity of the orbit (changes in the shape of the Earth's orbit) with a periodicity of ca. 100 kya.

Developing a convincing means of testing this theory has been one of the notable achievements of Earth science in the twentieth century. This was achieved initially through the detailed analysis of stable oxygen isotope variations in marine microorganisms (foraminifera) obtained from deep-ocean sediments (**Fig. 1**; *see* Oxygen Isotope Stratigraphy of the Oceans). The ratio of the lighter oxygen isotope (^{16}O) to the heavier isotope (^{18}O) is mainly controlled by the volume of global ice cover and hence is an index of global climatic change. Deviations towards heavier ratios (downward swings in the isotope curve) reflect times of increased land–ice cover ('glaciations'), and upward shifts represent warmer periods ('interglaciations') when it was significantly reduced.

This evidence offers a deep insight into how the Earth responds to the external stimulus of insolation (solar radiation), how the incoming radiation signal may influence processes operating at the surface of the Earth, and how the behavior of the Earth's climate at the 10^5yr scale may be predictable over the long term. Some of the more important features of the Earth's recent climate include the following:

(1) As **Figure 1** implies, cyclic change in global climate is the hallmark of the Quaternary. There is no evidence in the records of extended periods of climatic stability; instead, the Earth's climate system appears to have been constantly in transition throughout the past 2.6 Myr.

(2) The frequency of the climate cycle changed during the Quaternary. Milankovitch's 'precession cycle' appears to have been predominant during the late Pliocene, between ca. 2.8 and 2.6 Ma; the 'obliquity signal' dominated in the period 2.6 to 0.8 ka, and the 'eccentricity signal' is characteristic of the last 800 ka (Raymo and Ruddiman, 1992).

(3) Whereas the calculated values for variations in insolation vary symmetrically over time (Berger *et al.*, 1999), measured oscillations in oceanic stable isotope

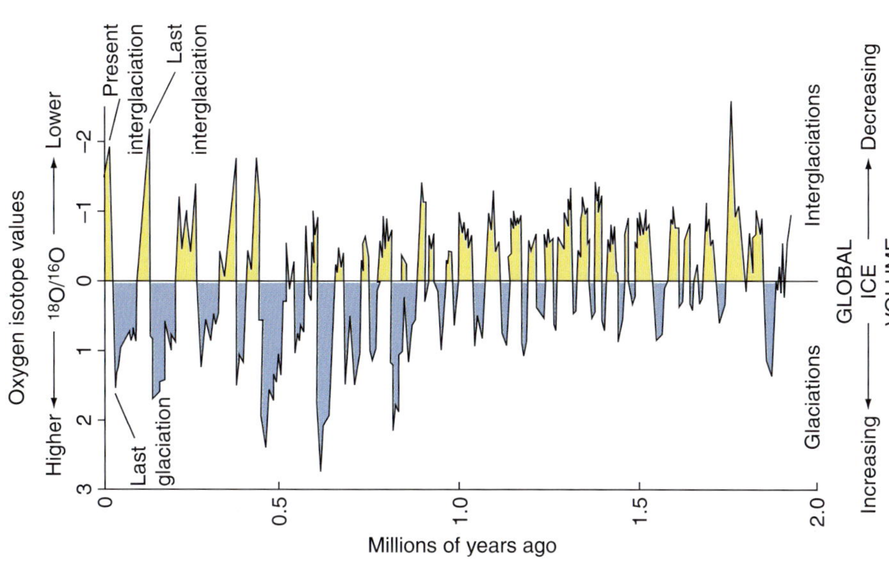

Figure 1 Stable oxygen isotope variations in marine cores spanning the last 2 Myr. The values are shown as deviations from a long-term mean. Variations in the marine isotope ratio reflect variations in the volume of global land ice, and hence the curve is primarily a 'paleoglaciation' curve, and only indirectly a paleoclimatic curve. Interglacial stages are shown in yellow, and glacial stages in blue. See Lowe and Walker (1997) and Lowe (2001) for further details.

ratios are nonsymmetrical (**Fig. 1**). Rates of warming at the start of interglacial stages are usually more rapid than the more gradual cooling trends leading to glacial stages. The implication is that climate response to astronomical variations is not linear, and that the radiation signal is modulated in some way by processes operating at the Earth's surface.

(4) **Figure 1** also shows how the amplitude of climate oscillation increased during the Quaternary, particularly within the last 800 ka or so, when the eccentricity cycle was predominant. Interglaciations became generally warmer during the later part of the Quaternary, while glaciations appear to have been characterized by colder extremes than during earlier times.

(5) During later Quaternary time, interglaciations have been relatively short. Periods during which conditions were as warm as the present appear to last, on average, about 15–17 ka, before a decline begins that is the precursor to a new glaciation.

Global climatic signals at orbital frequencies have been detected and confirmed in other contexts, most notably in Antarctic and Greenland ice cores (EPICA Members, 2004; Masson-Delmotte *et al.*, 2005), and in Chinese loess–paleosol sequences (Sun *et al.*, 2006). It seems evident that astronomical cycles will continue to exert a strong influence on the Earth's climate and that the present interglacial period will, in due course, be succeeded by the next glaciation. Indeed, some believe that the next glaciation may have already begun (e.g., Ruddiman (2003, 2005)).

Short-Term ('Sub-Milankovitch') Climatic Variations

Climatic perturbations that occur over much shorter frequencies cannot be explained by the astronomical cycles alone. Short-term perturbations in the global climate system are most clearly reflected in cores obtained from the polar ice sheets (Alley, 2000). During the 1960s, several deep boreholes were drilled in Greenland and Antarctica, while a number of smaller ice masses in other parts of the world were also investigated (*see* History of Research, Greenland and Antarctica). These pioneering studies showed that important paleoenvironmental information could be obtained from the analysis of stable isotope ratios, gas content, chemicals, and other impurities in ice that had accumulated over many thousands of years. Subsequent work in Greenland in the 1990s, most notably the GRIP and GISP-2 projects (Alley, 2000), and more recently, the NGRIP programme (2003), built on the previous research findings and provided penetrating new insights into recent global environmental change over the last glacial–interglacial cycle (NorthGRIP Members, 2004). In Antarctica, lower rates of snow accumulation mean that the ice-core sequences are generally stratigraphically less well-resolved than in Greenland, but there the record extends over a much longer time. The EPICA core, which is still being drilled, now extends back ca. 800 ka and may eventually reach 1 Myr (EPICA Community Members, 2004). The polar ice cores also contain a history of long-term stable isotope variations that generally match those obtained from marine cores that span the same time intervals. This indicates that the oceans and ice sheets are closely linked ('coupled') in both physical state and behavior. Mathematical models explain the exchange of stable isotopes between the oceans, the atmosphere, and the cryosphere, which are the dominant global reservoirs constituting the hydrosphere (*see* Greenland Stable Isotopes; Antarctic Stable Isotopes, Jouzel *et al.* (1987)). The stable isotope and other proxy records reveal patterns of climate change at both Milankovitch and sub-Milankovitch (short-term) frequencies. Millennial and submillennial climate variations are clearly recorded in the Greenland ice-core records. The central sector of the Greenland ice sheet is approximately 1,500 m thick and provides a continuous environmental sequence extending over more than 100 ka. The precise period represented is uncertain due to deformation and thinning of the basal ice layers, but conservative estimates place the minimum age of the basal ice at ca. 123 ka (*see* Chronologies; NGRIP Members, 2004). Distinctive annual layers occur in the upper levels of the core sequence, which means that past environmental changes can be resolved at a very high resolution, far higher than is normally achievable with marine- or lake-sediment records. An irregular series of peaks and troughs in the isotope signal (referred to as Dansgaard–Oeschger oscillations, after two leading ice-core scientists, Willi Dansgaard and Hans Oeschger) is replicated almost exactly between the GISP-2 and GRIP cores, and is equally well matched in the more northerly NGRIP core (**Fig. 2**). The remarkable degree of conformity between these three records suggests that the ice sheet as a whole responded cyclically to hemispherical climatic forcing, the shifts in the stable isotope record reflecting a combination of changes in ambient temperatures, moisture source and delivery, and wind strength and direction. Notable features of this high-resolution paleoclimatic record include the following:

(1) Modern isotope–temperature relationships suggest that the amplitude of isotopic variations represented in the Greenland ice-core records reflects shifts of up to 15 °C in ambient temperature during the last cold stage.

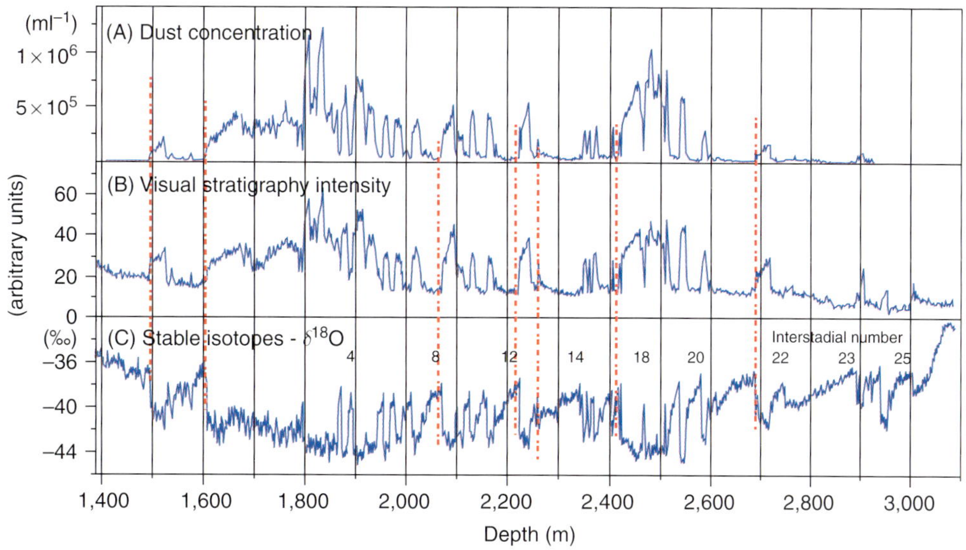

Figure 2 Stratigraphic variations in the NorthGRIP core spanning the period approximately 9 ka (1,400 m depth) to 120 ka (3,200 m depth). The lowest curve shows variations in stable oxygen isotope ratios, the key proxy that reflects temperature variations over central Greenland. The middle panel shows variations in visible stratigraphy (ice color and transmissivity), while the top panel shows variations in dust concentration in the ice. The red dotted lines emphasize the anti-phase relationships between oxygen isotope variations and dust concentration. Svensson *et al.* (2005). Numbers in the lower panel indicate sequence order of abrupt warming events (interstadials).

(2) During the last cold stage (115–11.5 ka), as many as 27 distinct warming events (interstades) occurred, some with maximum temperatures that approached, but did not quite reach, modern temperature values over central Greenland.

(3) The climatic oscillations throughout the last cold stage, as reflected in the stable isotope records, typically range between 500 and 2000 yr in duration, and the isotopic profiles show characteristic asymmetric shapes reflecting abrupt warmings that are followed by more gradual cooling trends.

(4) In the upper layers of the ice cores, the duration of the transitions from cold stadial intervals to maximum warmth can be estimated by ice-layer counting; remarkably, the evidence indicates that some of these marked climatic shifts have occurred within a few decades (i.e., within a modern human lifetime), while the most recent major warming, marking the start of the current interglacial period at ca. 11.7 ka, may have occurred over an even shorter interval.

Collectively, the ice-core data suggest a considerable degree of instability in Greenland climate during the last cold stage, with oscillations between stadial and interstadial modes, a process that Taylor *et al.* (1993) likened to the operation of a 'flickering switch.' Prior to the publication of these records, climatic shifts of this magnitude and over such restricted timescales were not considered feasible. Two questions arose: (1) Were these climatic oscillations confined to Greenland? (2) What mechanism might cause the signal to flicker in this way? Subsequent research has shown that the ice-core climatic record for the last cold stage is not unique to Greenland, for similar climate histories have been discovered in other areas of the world. The answer to question 2 lies in the deep oceans where changes in ocean circulation appear to be the principal forcing factor in the operation of the climatic switch.

The Role of the Oceans

Shortly before the publication of the GISP-2 and GRIP ice-core results, Heinrich (1988) drew attention to a series of coarse sediment layers (now termed 'Heinrich Layers') in cores obtained from the floor of the North Atlantic (*see* Sub-Milankovitch (DO/Heinrich) Events). These are believed to be the products of glacier-calving around the margins of the northern ice sheets, which resulted in massive 'armadas' of icebergs in the North Atlantic circulating in a counterclockwise flow (gyre) and releasing sediment (ice-rafted debris) as they melted (Hemming, 2004). Eight such "Heinrich events" occurred between 70 and 10 ka, indicating periodic and large-scale instability of the ice-sheet margins during times of maximal expansion. A close relationship was detected between episodes of coarse sediment influx into the North Atlantic, abundance of cold-adapted marine microfossils, and isotope variations in the GRIP record (Bond and Lotti, 1995; **Fig. 3**).

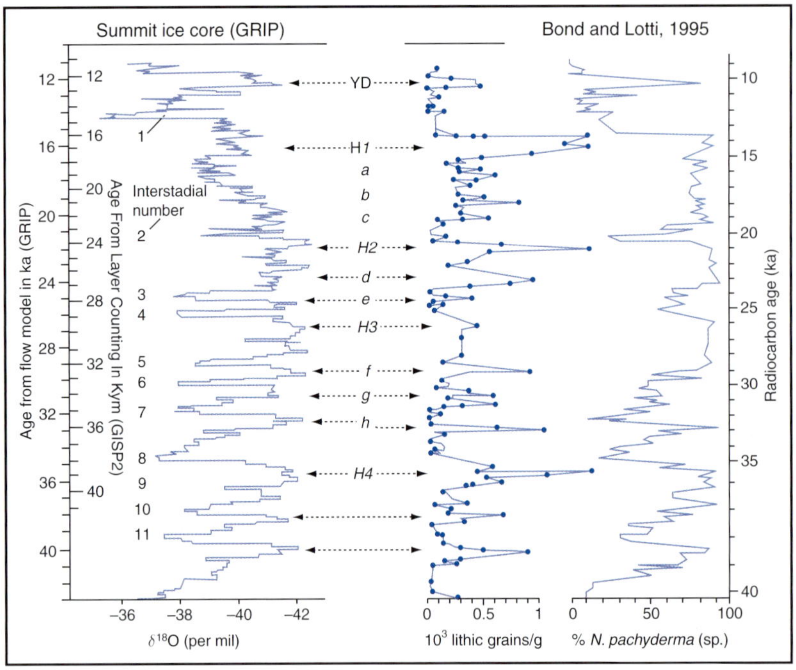

Figure 3 Variations in concentration of coarse mineral grains (lithic particles) and in abundance of the cold-adapted plankton species, *N. pachyderma* in sediment cores recovered from the floor of the North Atlantic, compared with the stable oxygen isotope record from the GRIP ice-core record. Bond and Lotti (1995).

Hughen *et al.* (1996) presented compelling evidence for an even closer match between sediment variations in the tropical Atlantic and the GISP-2 record (**Fig. 4**). Since then, geological archives from places as widely scattered as China, the Arabian Sea, and the sub-Antarctic have yielded proxy climate data that closely resemble oceanographic and climatic changes reflected in the North Atlantic records (e.g., Hemming (2004), Sachs and Anderson (2005), and Porter and Zhou (2006)).

Several important points arise from this work that have a bearing on our interpretation of the glacial record and of the global climate sequence:

(1) It is now clear that the history of glaciation is far more complex than was previously realized. Traditional models of glacier/ice-sheet advance and retreat have been based on painstaking investigation of continental records, but the evidence on land is partial simply because more recent glacier advances tend to bury or destroy the evidence of earlier glacier activity. The glacial–geological evidence from the last glaciation, for example, does not provide evidence of all the numerous stadial–interstadial oscillations that are reflected in the ice-core records.

(2) The close correspondence between the isotope signal from Greenland and that from the sedimentary record in the North Atlantic suggests that the ice sheet and oceans were responding to the same external stimulus during the course of the last cold stage.

(3) The often-abrupt climatic oscillations of the last cold stage were not confined to Greenland and the North Atlantic, but appear to have occurred in other parts of the world as well.

(4) It is frequently assumed, although it has not yet been tested satisfactorily, that these abrupt climatic changes occurred more-or-less simultaneously on a global scale, and that major shifts in climate somehow operate contemporaneously.

A number of explanations have been put forward to account for the periodic instability of the ice sheets. These include self-regulating processes that came into play as the ice sheets became large enough to influence their immediate environment, by increased melting at the base, by extending into deeper water where marginal calving would be accelerated, or by some complex interplay between these factors and concomitant changes in global sea level (*see* Growth and Decay; Glacimarine Sediments and Ice-Rafted Debris). It has also been suggested that the timing of the onset of the Greenland interstadial events during the last cold stage approximates a 1.47 kyr climate cycle and hence ice-sheet instability could, in part, be a response to a more regular forcing mechanism such as solar variations (e.g., Schulz (2002)). Others, however, favor a different explanation, namely changes in the rate or pattern of ocean circulation.

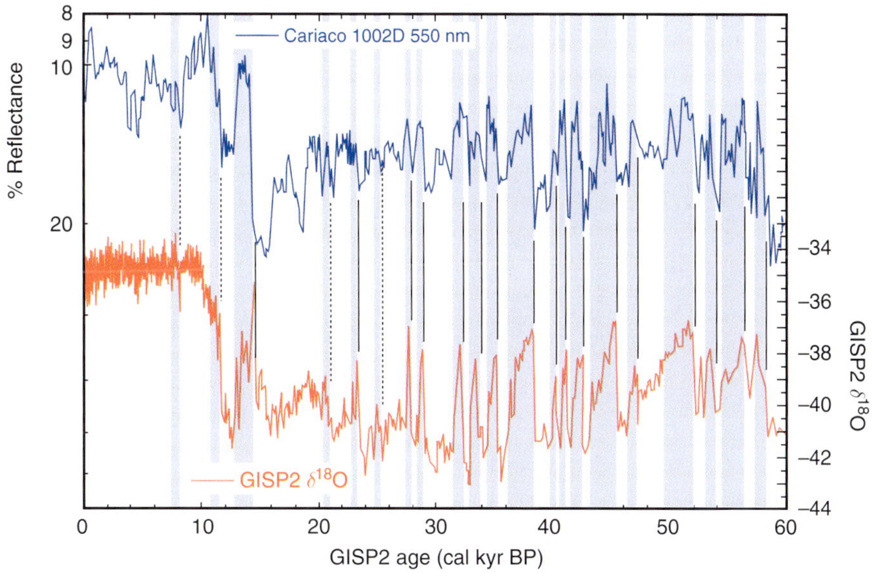

Figure 4 Comparison between variations in sediment optical properties (light reflectance) of a laminated sequence of marine sediments in the Cariaco Basin, north of Venezuela, and the GISP2 stable oxygen isotope record for the past 90 ka. Gray-shaded vertical bars represent Greenland interstadial events. Hughen (1996).

One of the most persuasive advocates of the latter hypothesis is Broecker (1994, 1997), who cites in its support the considerable body of paleoceanographic evidence which shows that the oceans, and the North Atlantic in particular, experienced marked changes in salinity, density, and rate of water movement during the last cold stage. These must, in Broecker's view, have had a major impact on the speed and pattern of flow of the thermohaline circulation, the network of surface and deep-water currents that circulates ocean water around the globe (Rahmstorf, 2000). The present arterial flows, such as the Gulf Stream, are maintained by thermal and salinity stratification which leads to warm salty surface water being transported northward across the North Atlantic where it cools, sinks, and forms a return southward-flowing deep-water current. This circulatory system, or 'conveyor,' keeps NW Europe warmer than other regions in the northern hemisphere at comparable latitudes, and also delivers moisture to the Greenland ice sheet and to the smaller ice bodies in Iceland, Spitsbergen, and Scandinavia. According to this hypothesis, rapid melting of ice sheets would have delivered enormous volumes of cold freshwater onto the ocean surface, and since this would be lighter (less dense), it would form a cap over the ocean, preventing sinking, and thereby reducing the pressure gradient that drives the conveyor. This, in turn. would lead to a marked cooling in the North Atlantic and NW Europe, and a reduction in moisture supply to the Greenland ice sheet.

This explanatory mechanism has been widely favored, partly because it is based on the physical parameters that govern ocean-water movement, and partly because numerical modeling of the North Atlantic shows how sensitive the three-dimensional circulation pattern is to changes in surface water fluxes (Stocker and Wright, 1991; Rahmstorf, 1995). In addition, quantitative estimates of the volumes of water released to the North Atlantic during ice-sheet wastage at the end of the last glaciation appear to provide adequate fluxes for major disruptions of the ocean circulation system (Teller *et al.*, 2002; Clark *et al.*, 2003). It also offers a plausible explanation for why ventilation of the deep ocean (exchange of waters, nutrients, and oxygen with surface waters) was periodically interrupted throughout the last cold stage (Bard, 2002). Finally, it might explain why sudden changes in climate in the North Atlantic were also experienced in other parts of the world.

What cannot yet be adequately explained by these inferred oceanographic changes, however, is the trigger mechanism for ice-sheet instability. In seeking a resolution to this problem, it may prove to be a mistake to focus solely on the ice sheets and the oceans, for there is now a growing body of evidence for the operation of other factors in controlling abrupt climate change.

Global Feedback Mechanisms

The ice-core records constitute an unparalleled archive for showing how the different components of the global environmental system interconnect. The gas bubbles locked in the ice, for example, provide an independent record of variations in atmospheric gas content (*see* CO_2 Studies; Methane Studies), which enables a direct comparison to be made with the stable isotope records that reflect ambient temperature. The evidence reveals a very close correlation between atmospheric levels of carbon dioxide (CO_2), deuterium, and global temperature (**Fig. 5**), a relationship that has stimulated wide public debate about the continued rise in greenhouse gases in the atmosphere, and the implications for future global climate (IPCC, 2001).

There is also a close relationship between isotopic variations and the dust content of polar ice, at both glacial–interglacial (Delmas, 1992) and stadial–interstadial scales (Svensson *et al.*, 2005). **Figure 2** shows how peaks in the record of dust flux to the Greenland ice sheet during the last cold stage coincided with stadial intervals (lighter isotope ratios), while abrupt warmings at the start of the interstades coincided with marked reductions in dust content. Clearly, the atmosphere was dustier during cold periods. This may partly reflect increased transport of wind-borne dust under a more vigorous atmospheric circulation regime from arid and semi-arid areas such as western China. It may also be partly explained by a reduction in stabilizing vegetation cover in such regions, and partly by wind erosion of exposed seafloor during times of reduced sea level.

Many other processes and environmental fluxes contributed to, and were affected by, abrupt climatic change, but the cases of CO_2 and dust serve to illustrate how difficult it is to determine with present knowledge which agents and processes were the key initiators of change. The problems are threefold: (1) The majority of geological records cannot yet be dated with a precision sufficient to determine the order in which different components of the Earth system change. (2) The linkages frequently are mutually dependent. For example, a change in climate might lead to a change in the global carbon cycle, but the latter could equally result from the operation of nonclimatic factors, for example, from an increase in ocean biomass due to a change in nutrient supply. One consequence of this might be an increase in the rate of extraction of CO_2 from the atmosphere, which then becomes locked in the oceans (a process referred to as 'draw-down' of CO_2) and which might, in turn, impact on climate. (3) The same changes may sometimes be induced by the operation of opposing or quite different processes. The amount of dust entrained in the atmosphere could, for example, increase with a shift to colder conditions through a lowering of sea level and exposure of silty areas to wind erosion, or by an increase in the transport of dust to polar regions from arid and semi-arid regions. A similar effect could result from a warming trend that led to a reduction in effective precipitation in some regions and the expansion of arid and semi-arid zones.

Understanding the complex chain of feedback mechanisms that constitutes the global environmental system is one of the most challenging areas of Quaternary science, but given the problems of

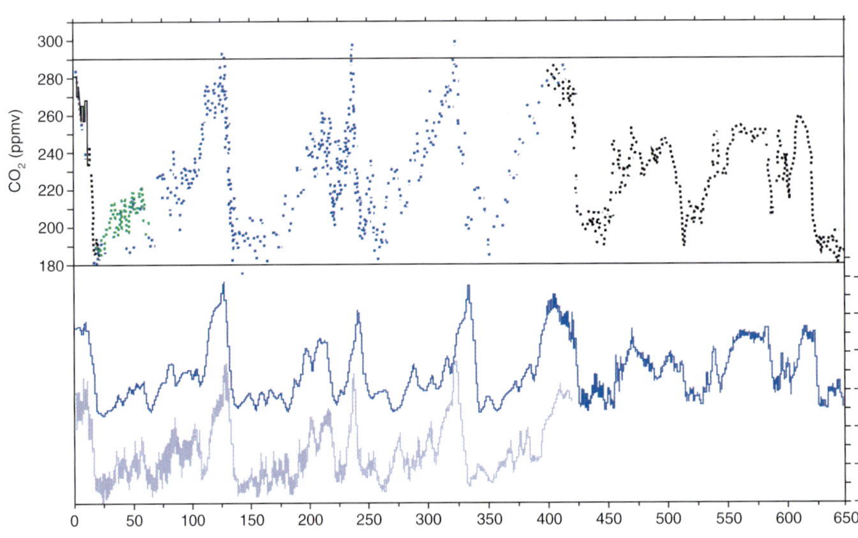

Figure 5 Composite records of variations in atmospheric CO_2 concentrations (upper panel) and deuterium ratios (a palaeotemperature index – lower panel) in gas bubbles in Antarctic ice cores spanning the lost 650 kyr. The upper panel shows combined data from the Dome C (black dots), Vostok (blue) and Taylor Dome (green) ice cores, while the lower panel shows deuterium records for Dome C (upper, dark blue curve) and Vostok (grey-blue). Siegenthaler *et al.* (2005).

interpretation referred to above, such an understanding will be difficult to achieve. However detailed and diverse Quaternary paleoenvironmental records may be, they do not, in themselves, constitute tests of causation between multiple competing variables. What is needed is some basis for experimentation, where key variables can be controlled while others are deliberately varied (and, perhaps, exaggerated) in order to isolate those that are the most influential. This takes us into another area where considerable progress has been made in Quaternary science in recent years: numerical modeling.

Numerical Environmental Modeling

Over the past several decades, increasingly sophisticated numerical models have been developed that can simulate a range of natural processes operating at the surface of the Earth (Schellnhuber, 1999). They are based on mathematical descriptions of the behavior of, and interactions between, specified variables, each component or interaction being mathematically prescribed to obey the laws of physics. Some models are simple, being constrained to focus on a particular process or on the interaction between selected variables within a limited geographical area, as, for example, a hydrological model of water-balance changes in a river catchment (Kutzbach, 1980). More complex models, however, are designed to represent the operation of Earth systems at a global scale. The best known and most sophisticated are general circulation models (GCMs), which simulate the operation of the global climate system (*see* Quaternary Environments; McGuffie and Henderson-Sellers, 2001). The capabilities of numerical models have increased dramatically within the last few decades following major advances in computer power and speed, software design, and the tools used to measure and monitor the natural fluxes between global reservoirs. Increased national and international cooperation in model design and intercomparison has also been a significant factor.

Quaternary scientists have long appreciated the potential of numerical models for providing analogs of the Earth system under different climate states, and for helping to clarify the key mechanisms that drive environmental change. Important integrated research programs employing numerical modeling have included simulations of the conditions in the ocean at the Last Glacial Maximum (CLIMAP Project Members, (1976, 1981)), and an analysis of the effects of insolation variations and concomitant changes in surface conditions on global climate over the past 18 ka (Kutzbach and Guetter, 1986; COHMAP Members, 1988). These were important milestones in a developing collaboration between Earth scientists who generate the paleodata and the numerical modeling community. The relationship has been reciprocal with, on the one hand, Quaternary scientists learning from the modelers about the types of input data that are needed to generate the best model results, while on the other, the modelers require snapshots of past environmental conditions (based on paleoenvironmental reconstructions) from the Quaternary science community to test model outputs (van Andel, 2002). The latter are especially important because models are simplifications of the natural world with built-in assumptions. Hence, model outputs invariably have large uncertainties. Nowhere is this more apparent (or relevant) than in the current debate centering around the wide divergence in predictions of the nature and magnitude of future climate change generated by different GCMs (e.g., Allen *et al.* (2000)).

One possible means of testing whether a GCM captures realistic climate situations, therefore, is to use the model to 'predict' an outcome for which an independent estimate already exists: this can be done by generating model evaluations of climate states for times in the recent geological past. In this procedure, a time period is selected for which the global boundary conditions are reasonably well known, and for which a detailed paleorecord can be compiled. The paleodata provide the basis for a reconstruction of the climatic conditions that may have prevailed at that time, and the model can be run to generate a stable climate state when forced by the prescribed (past) boundary conditions (e.g., Wright *et al.* (1993)). The two outcomes can then be compared. One problem, of course, is that any significant differences between the results would be difficult to reconcile, for there is no independent means of determining the extent to which either, or both, are in error. Nevertheless, model data comparison is widely practiced (e.g., Paleoclimate Modeling Intercomparison Project (PMIP) - http://www-lsce.cea.fr/pmip/), for there is a widespread view that improvements in both model design and in the amount and quality of the paleodata being generated will ultimately lead to greater convergence between the two outputs.

Numerical modeling plays an increasingly important role in contemporary Quaternary science because it permits speculative questions to be posed with some basis for constraining the answers provided. The balance between variables and inputs can be deliberately altered within a model in order to evaluate the nature, magnitude, and timing of environmental response. This approach has a range of applications. One example is the recent use of a

GCM, forced by the insolation cycle alone, to examine future climate trends (Hargreaves and Annan, 2002). This particular model suggests that the Earth will experience a significant cooling step, followed by a slower cooling trend that will culminate in the next glacial maximum in about 60 ka. It also predicts that any 'global warming' effect caused by anthropogenic pollution will arrest this trend for about 30 yr, but is unlikely to be large enough to prevent long-term cooling thereafter. A second example predicts the mass balance changes of the polar ice sheets and the subsequent effects on global sea level (Vaughan and Spouge, 2002), a topic of considerable concern given that a large percentage of the world's population lives within a few meters of present sea level. A third is the use of a three-dimensional ice-sheet model to examine the hypothesis that periodic large-scale oscillations of the North American Laurentide ice sheet may have been responsible for Heinrich Events (Papa *et al.*, 2006).

Although numerical modeling has provided important new insights into the workings of the global climate system, even the most sophisticated of models have limitations. Some of the main areas of concern are the following:

(1) Models invariably (and inevitably) oversimplify environmental complexity, for they cannot represent every component and synergistic interaction that occurs in nature. New discoveries continue to show how small-scale processes, such as feedback mechanisms involving cloud formation, can have far bigger impacts on climate (and also model output) than might have been anticipated (Williams *et al.*, 2006).

(2) Because numerical models can only handle a finite number of calculations, all the global processes and interactions represented in the models must be scaled down in order to be manageable. This inevitably leads to a 'scaling-up' problem when it comes to interpreting the output: this is the reverse logic problem of knowing whether the scaled-down, almost artificial microcosm, created in the model truly reflects the real operation of the system at the global scale.

(3) Interactions between global subsystems (the atmosphere, hydrosphere, cryosphere, biosphere, geosphere) are invariably nonlinear. Hence, whereas it may be possible to predict responses within a simplified subsystem when only a few variables are involved, the greater the number of variables, the greater the complexity in terms of links between different subsystems. This almost inevitably means that the outcomes are less predictable (Rind, 1999; Clark *et al.*, 1999).

(4) Only very small changes in one or more parameters can have significant effects on model output. Running the same model twice using exactly the same parameters and boundary conditions can generate quite different outputs. Therefore, the outcomes may be unpredictable (Sivakumar, 2004), which may help explain why GCMs often tend to produce quite different predictions of future climatic scenarios.

Paradoxically, however, if an element of chaotic behavior is involved in all natural processes, then the only viable way forward in the search for an understanding of the global climate system, and predicting its behavior, is through numerical modeling. If this challenge is to be met, future generations of numerical models must become even more sophisticated. They will have to incorporate a much wider range of variables and linkages to reflect more realistically the complexity of the Earth system.

Climate and Humans

Concern over the modern climate-warming trend and the extent to which it may have been generated by human activity has stimulated widespread debate. The degree of concern over this issue is reflected in the tenor of the policy statements drawn up by almost every national government, as well as the increasing acceptance of the need for agreed protocols to curb greenhouse gas emissions (*see* Introduction). By far the most compelling evidence has been provided by Antarctic ice cores, which contain a detailed record of atmospheric gas ratios stretching back over the past 650 ka. **Figure 6** shows the record of variations in concentration of atmospheric CO_2 and CH_4 over the past 400 ka obtained from the Vostok ice core (Petit *et al.*, 1999). These data, together with the longer records from a second Antarctic borehole, Dome Concordia (**Fig. 5**), indicate that natural atmospheric levels have normally been well below 300 ppm for CO_2 and 800 ppb for CH_4. The concentration of these greenhouse gases in the modern atmosphere is significantly above these levels (double in the case of CH_4), with the evidence pointing to a steady rise in concentrations from the time of the Industrial Revolution. However, it is the abrupt increase over the course of the second half of the twentieth century that gives the greatest cause for concern. Although the causal connections between atmospheric gas concentration and global temperature change continues to be a matter of dispute (e.g., Mann *et al*, (1998) and McIntyre and McKiltrick (2003)), there is a widespread acceptance that the late twentieth century warming detected in a range of proxy and instrumental records is anomalous

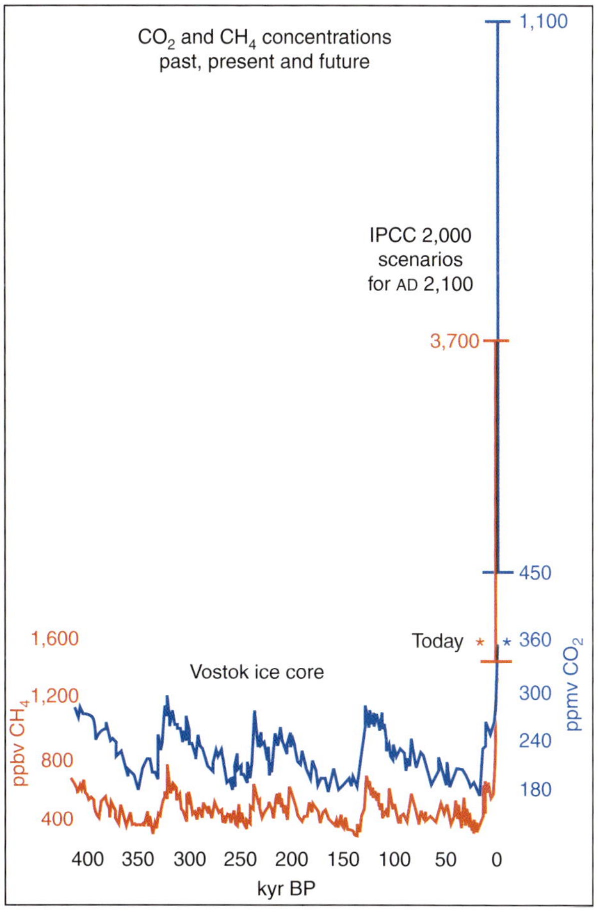

Figure 6 Record of CO_2 (blue curve) and CH_4 (red) variations in the Vostok (Antarctic) ice core over the past 400 ka and projected future levels based on IPCC estimates. PAGES web resources: http://www.pages.unibe.ch/about/index.html

(Osborn and Briffa, 2006), and that this is inextricably linked to increased atmospheric concentrations of greenhouse gases.

This view is reinforced by predictions, based on GCMs, of the likely atmospheric concentrations of these gases in AD 2100, should current fossil fuel consumption be allowed to continue unchecked. These values (**Fig. 6**) are far in excess of any recorded in the recent geological record. Should these predictions prove to be anywhere near reliable, then it is difficult to imagine that global climate will remain unaffected. Whether this will prove to be as detrimental to humans as is frequently predicted is difficult to assess, for there are no analogs in the Quaternary geological record for a world with climatic boundary conditions comparable to those of the present time having such high concentrations of atmospheric greenhouse gases.

The history of human impact on climate may, however, predate the Industrial Revolution, for it has recently been suggested that anthropogenically-enhanced emissions of greenhouse gases (notably CO_2 and CH_4) could have begun as early as 8 ka as a result of the spread of deliberate burning of vegetation and the adoption of animal husbandry, cultivation, and other farming practices (Ruddiman, (2003, 2005)). Ruddiman's analysis of the natural rhythm of climate variations during the Quaternary leads him to conclude that if it were not for the present abnormally high levels of atmospheric CO_2 and CH_4, global climate would be ca. 2 °C cooler than it is today, and the world would be about a third of the way towards full glacial conditions (Ruddiman *et al.*, 2005). Whether or not these conclusions are valid, they are examples of the types of question that are now being posed about the Earth's past and future climatic history, and it is in the rich archive of the Quaternary geological record where answers to such questions perhaps can be found best.

See also: **Glacial Climates**: Thermohaline Circulation. **Glacial Landforms, Ice Sheets**: Growth and Decay. **Glacial Landforms, Sediments**: Glacimarine Sediments and Ice-Rafted Debris. **Glaciation, Causes**: Milankovitch Theory and Paleoclimate. **Ice-Core Methods**: Chronologies; CO_2 Studies; Methane Studies. **Ice-Core Records**: Antarctic Stable Isotopes; Greenland Stable Isotopes. **Ice Cores**: History of Research, Greenland and Antarctica. **Lake-Level Studies**: Latin America; West-Central-Europe. **Loess Records**: China. **Paleoceanography, Physical and Chemical Proxies**: Oxygen Isotope Stratigraphy of the Oceans. **Paleoclimate Modeling**: Data-Model Comparisons; Quaternary Environments. **Paleoclimate Reconstruction**: Sub-Milankovitch (DO/Heinrich) Events. **Paleoclimate**: Introduction.

References

Allen, M. R., Stott, P. A., Mitchell, J. F. B., Schnur, R., and Delworth, T. L. (2000). Quantifying the uncertainty in forecasts of anthropogenic climate change. *Nature* **407**, 617–620.

Alley, R. B. (2000). *The Two-Mile Time Machine: Ice Cores, Abrupt Climate Change and Our Future*. Princeton University Press, Princeton.

Bard, E. (2002). Climate shock: Abrupt changes over millennial time scales. *Physics Today*, December 2002, 32–38.

Berger, A., Li, X. S., and Loutre, M. F. (1999). Modelling Northern Hemisphere ice volume over the last 3 Ma. *Quaternary Science Reviews* **18**, 1–11.

Bond, G. C., and Lotti, R. (1995). Iceberg discharges into the North Atlantic on millennial timescales during the last glaciation. *Science* **267**, 1005–1010.

Broecker, W. S. (1994). An unstable superconveyor. *Nature* **367**, 414–415.

Broecker, W. S. (1997). Thermohaline circulation, the Achilles Heel of our climate system: Will man-made CO_2 upset the current balance? *Science* **278**, 1582–1588.

Clark, P. U., Alley, R. B., and Pollard, D. (1999). Northern Hemisphere ice sheet influences on global climate change. *Science* **286,** 1104–1111.

Clark, G., Leverington, D., Teller, J., and Dyke, A. (2003). Superlakes, megafloods, and abrupt climatic change. *Science* **301,** 922–923.

CLIMAP Project Members (1976). The surface of ice-age Earth. *Science* **191,** 1131–1137.

CLIMAP Project Members (1981). Seasonal reconstructions of the Earth's surface at the Last Glacial Maximum. *Geological Society of America Map and Chart Series* MC-36, 1–18.

COHMAP Members (1988). Climatic changes of the last 18,000 years: Observations and model simulations. *Science* **241,** 1043–1052.

Delmas, J. (1992). Environmental information from ice cores. *Reviews of Geophysics* **30,** 1–21.

EPICA Community Members (2004). Eight glacial cycles from an Antarctic ice core. *Nature* **429,** 623–628.

Hargreaves, J., and Annan, J. (2002). Assimilation of paleo-data in a simple Earth system model. *Climate Dynamics* **19,** 371–381.

Heinrich, H. (1988). Origin and consequences of cyclic ice rafting in the northeast Atlantic Ocean during the past 130,000 years. *Quaternary Research* **29,** 143–152.

Hemming, S. R. (2004). Heinrich events: Massive Late Pleistocene detritus layers of the North Atlantic and their global climate imprint. *Review Geophysics* **42**(No. 1): RG1005.

Hughen, K. A., Overpeck, J. T., Peterson, L. C., and Trumbore, S. (1996). Rapid climate changes in the tropical Atlantic region during the last deglaciation. *Nature* **380,** 51–54.

Intergovernmental Panel on Climate Change (2001). *Climate Change 2001: The Scientific Basis.* Cambridge University Press, Cambridge.

Jouzel, J., Russell, G. L., Suozzo, R. J., Koster, R. D., White, J. W. C., and Broecker, W. S. (1987). Simulations of the HDO and H218O atmospheric cycles using the NASA/GISS general circulation model: The seasonal cycle for present-day conditions. *Journal of Geophysical Research* **92,** 14739–14760.

Kutzbach, J. (1980). Estimates of past climate at palaeolake Chad, North Africa, based on a hydrological and energy balance model. *Quaternary Research* **14,** 210–223.

Kutzbach, J., and Guetter, P. J. (1986). The influence of changing orbital parameters and surface boundary conditions on climate simulations for the past 18,000 years. *Journal of Atmospheric Sciences* **43,** 1726–1759.

Lowe, J. J. (2001). Quaternary geochronological frameworks. In *Handbook of Archaeological Sciences* (D. R. Brothwell and A. M. Pollard, Eds.), pp. 9–21. Wiley, Chichester.

Lowe, J. J., and Walker, M. J. C. (1997). *Reconstructing Quaternary Environments*, 2nd edn, Harlow, Addison, Wesley, Longman Ltd.

Mann, M. E., Bradley, R. S., and Hughes, M. K. (1998). Global-scale temperature patterns and climate forcing over the past six centuries. *Nature* **392,** 779–787.

Masson-Delmotte, V., Jouzel, J., Landais, A., *et al.* (2005). GRIP deuterium excess reveals rapid and orbital-scale changes in Greenland moisture origin. *Science* **309,** 118–121.

McGuffie, K., and Henderson-Sellers, A. (2001). Forty years of numerical climate modelling. *International Journal of Climatology* **21,** 1067–1109.

McIntyre, S., and McKitrick, R. (2003). Corrections to the Mann *et al.* (1998) proxy data base and Northern Hemispheric average temperature series. *Energy and Environment* **14,** 751–771.

Milankovitch, M. (1930). Mathematische klimalehre und astronomie theorie der klimaschwankungen. In Handbuch der Klimatologie (W. Köppen and R. Geiger, Eds.) vol.I(A), 1–176. Gebrüder Borntraeger, Berlin.

North Greenland Ice-Core Project (NorthGRIP) Members (2004). High resolution climate record of the Northern Hemisphere reaching into the last glacial interglacial period. *Nature* **431,** 147–151.

Osborn, T. J., and Briffa, K. R. (2006). The spatial extent of 20th-century warmth in the context of the past 1200 years. *Science* **311,** 841–845.

Papa, B. D., Mysak, L. A., and Wang, Z. (2006). Intermittent ice sheet discharge events in northeastern North America during the last glacial period. *Climate Dynamics* **26,** 201–216.

Petit, J. R., Jouzel, J., Raynaud, D., *et al.* (1999). Climate and atmospheric history of the past 420,000 years from the Vostok ice core, Antarctica. *Nature* **399,** 429–436.

Porter, S. C. (1999). INQUA and Quaternary Science at the millennium: A personal retrospective. *Quaternary International* **62,** 111–117.

Porter, S. C., and Zhou, W. J. (2006). Synchronism of Holocene East Asian monsoon variations and North Atlantic drift-ice tracers. *Quaternary Research* **65,** 443–449.

Rahmstorf, S. (1995). Bifurcation of the Atlantic thermohaline circulation in response to changes in the hydrological cycle. *Nature* **378,** 145–149.

Rahmstorf, S. (2000). The thermohaline ocean circulation: A system with dangerous thresholds? *Climatic Change* **46,** 247–256.

Rahmstorf, S. (2002). Ocean circulation and climate during the past 120,000 years. *Nature* **419,** 207–214.

Raymo, M. E., and Ruddiman, W. F. (1992). Tectonic forcing of late Cenozoic climate. *Nature* **359,** 117–124.

Rind, D. (1999). Complexity and climate. *Science* **284,** 105–107.

Ruddiman, W. F. (2003). The anthropogenic Greenhouse Era began thousands of years ago. *Climatic Change* **61,** 261–293.

Ruddiman, W. F. (2005). *Plows, Plagues and Petroleum: How Humans Took Control of Climate.* Princeton University Press, Princeton.

Ruddiman, W. F., Vavrus, S. J., and Kutzbach, J. E. (2005). A test of the overdue-glaciation hypothesis. *Quaternary Science Reviews* **24,** 1–10.

Sachs, J. P., and Anderson, R. F. (2005). Increased productivity in the sub-Antarctic ocean during Heinrich Events. *Nature* **434,** 1118–1121.

Schellnhuber, H. J. (1999). Earth System' analysis and the second Copernican revolution. *Nature* **402,** C19–C23.

Schulz, M. (2002). On the 1470-year pacing of Dansgaard-Oeschger warm events. *Paleoceanography* **17,** 1014–1029.

Siegenthaler, U., Stocker, T. F., Monnin, E., *et al.* (2005). Stable carbon cycle-climate relationship during the Late pleistocene. *Science* **310,** 1313–1317.

Sivakumar, B. (2004). Chaos theory in geophysics: Past, present and future. *Chaos, Solitons and Fractals* **19,** 441–462.

Stocker, T. F., and Wright, D. G. (1991). Rapid transitions of the ocean's deep circulation induced by changes in surface water fluxes. *Nature* **351,** 729–732.

Sun, Y., Clemens, S. C., An, Z., and Yu, Z. (2006). Astronomical timescale and palaeoclimatic implications of stacked 3.6 –Myr monsoon records from the Chinese Loess Plateau. *Quaternary Science Reviews* **25,** 33–48.

Svensson, A., Nielsen, S. W., Kipfstuhl, S., *et al.* (2005). Visual stratigraphy of the North Greenland Ice Core Project (NorthGRIP) ice core during the last glacial period. *Journal of Geophysical Research* **110**(D02), 108.

Taylor, K. C., Lamorey, G. W., Doyle, G. A., *et al.* (1993). The 'flickering switch' of Late Pleistocene climate change. *Nature* **361,** 432–436.

Teller, J. T., Leverington, D. W., and Mann, J. D. (2002). Freshwater outbursts to the oceans from glacial Lake Agassiz and their role in climate change during the last deglaciation. *Quaternary Science Reviews* **21,** 879–887.

Van Andel, T. H. (2002). The climate and landscape of the middle part of the Weichselian glaciation in Europe: The stage 3 project. *Quaternary Research* **57**, 2–8.
Vaughan, D. G., and Spouge, J. R. (2002). Risk estimation of collapse of the west Antarctic ice sheet. *Climatic Change* **52**, 65–91.
Williams, K. D., Ringer, M. A., Senior, C. A., *et al.* (2006). Evaluation of a component of the cloud response to climate change in an intercomparison of climate models. *Climate Dynamics* **26**, 145–165.
Wright, H. E., Jr, Kutzbach, J. E., Webb, T., III., Ruddiman, W. F., Street-Perrott, A. F. and Bartlein, P. J. (Eds.) (1993). *Global Climates Since the Last Glacial Maximum*. University of Minnesota Press, Minneapolis.

Relevant Website

www.lsce.cea.fr/pmip – Paleoclimate Modeling Intercomparison Project (PMIP).

A

Alkenone Studies *see* **Paleoceanography, Biological Proxies**: Alkenone Paleothermometry from Coccoliths

Allostratigraphy *see* **Quaternary Stratigraphy**: Morphostratigraphy-Allostratigraphy

AMINO-ACID DATING

G H Miller, University of Colorado, CO, USA
S J Clarke, Charles Sturt University, New South Wales, Australia

© 2007 Elsevier B.V. All rights reserved.

Amino acids are simple organic molecules that are the fundamental building blocks of all proteins. Although there are a large number of possible amino acids, only approximately 20 are common constituents of proteins. Amino acids are assembled in long chains, in which they are linked by peptide bonds. Except for glycine, one of the simplest forms, all protein amino acids have at least one asymmetric carbon atom or center of symmetry. Such molecules are considered chiral molecules (derived from the Greek word for 'hand'). Chiral molecules exhibit 'handedness' in that their mirror images are not superimposable (**Fig. 1**). Consequently, they can exist in at least two different configurations or enantiomers.

Chiral forms of the same amino acid were originally identified by the way in which they rotated plane-polarized light passed through a solution containing the pure enantiomer. Enantiomers rotating the light to the left are designated L-amino acids (levorotatory), whereas those enantiomers rotating the light to the right are D-amino acid (dextrorotatory). For reasons that remain debated, life is left-handed, and almost all living organisms utilize exclusively L-amino acids to build protein. However, once biological constraints are removed upon the death of an organism, diagenetic reactions begin to degrade proteinaceous residues into their constituent amino acids, which spontaneously interconvert to their D-configurations in a regular and predictable manner. The process whereby an L-amino acid interconverts to its D-configuration is known as racemization.

Philip Abelson, working at the Carnegie Institution of Washington's Geophysical Laboratory, was the first to recognize that amino acids could survive for millions of years at ambient Earth conditions, and that their unique initial status of 100% left-handed configuration offered the potential of using amino acid racemization as a geological clock or dating technique (Abelson, 1954). The first application of amino acid racemization as a dating tool was by Ed Hare, working with Abelson in the late 1960s (Hare and Abelson, 1968; Hare and Mitterer, 1967). Early reviews of the application of amino acid racemization to geological dating are provided by Schroeder and Bada (1976) and Williams and Smith (1977). Recent reviews include Johnson and Miller (1997), Kaufman and Miller (1992), Miller and Brigham-Grette (1989), Murray-Wallace and Kimber (1993), Wehmiller (1990), Wehmiller and Miller (2000), as well as two edited volumes focusing on the chemistry of amino acids in geological environments (Goodfriend *et al.*, 2000; Hare *et al.*, 1980).

Factors Influencing the State of Biomolecules over Geologic Time

Amino acid geochronology belongs to the chemical family of dating methods. Chemical

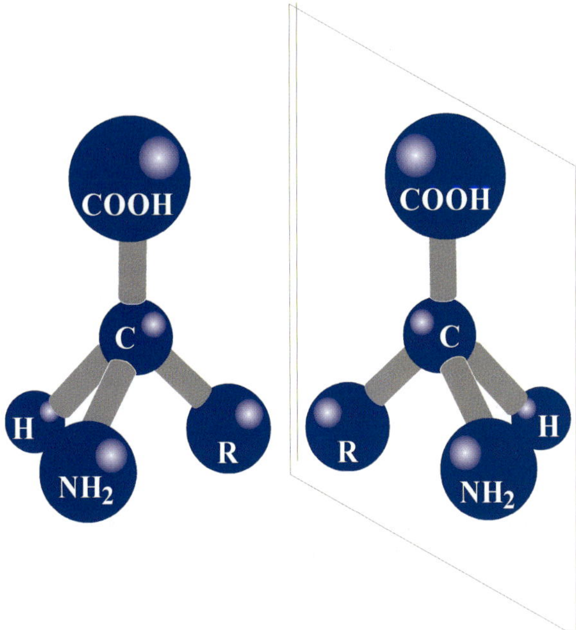

Figure 1 The structure of a generic amino acid and the mirror-image relation of its two stereoisomers or enantiomers.

methods differ from radioactive dating techniques in that their reaction rate depends on one or more environmental parameters, whereas radioactive decay remains constant regardless of environmental conditions. The rates at which proteins are degraded and their constituent amino acids altered, the processes fundamental to amino acid geochronology, are primarily dependent on the temperature the samples have experienced. Consequently, the extent of any particular chemical alteration of amino acids depends on the time elapsed since the death of the organism, the integrated thermal history, environmental pH, and the vital effects unique to each taxon.

Preservation of Amino Acids

The idea that amino acids might be preserved in biominerals over long periods of geological time was first proposed by Abelson (1954), who reported that the amino acids most stable at high temperatures in the laboratory were also present in fossils, the oldest of which was a 360 million-year-old Devonian fish. Hare and Abelson (1968) were able to separate L-isoleucine from its epimer, D-alloisoleucine, which showed a correlation between the ratio of these epimers and time. Subsequently, Hare and Mitterer (1969) demonstrated that changes in the amino acid composition of mollusk shells over geological time could be induced in the laboratory by heating, implying that the diagenesis of amino acids took place systematically. Supported by this observation, these authors were the first to estimate an age for a fossil based on the extent of amino acid racemization.

Amino acids are preserved in the geological record if they fail to enter the biotic nitrogen cycle. The nitrogen of polypeptides represents the major pool of organic nitrogen readily utilized in the biosphere. Without biological attack, the simplest amino acids are stable at ambient Earth surface temperatures for hundreds of millions of years. A common environment for the preservation of amino acids is the carbonate skeleton of invertebrates such as mollusks and foraminifera. Organic material is incorporated into carbonate exoskeletons either between the mineral crystals (intercrystalline), where it usually has a role in determining the form and shape of the mineral phase during precipitation, or within crystals (intracrystalline), where it may serve to provide resiliency to the mineral phase.

Diagenesis refers to reactions that degrade organic molecules and their constituent amino acids during their residence in geological environments following the death of the host organism or its disuse of the molecules (e.g., avian eggshell proteins). Ultimately, the original complex proteins are degraded into simple hydrocarbons and nitrogen compounds. The organic residues present in fossils are not proteins in a biochemical sense but are commonly referred to as indigenous protein residues, reflecting their biological origin. In nearly all situations, when a biomineral enters the geological environment, all of its amino acids are in the L-configuration; this defines the starting point of the geological clock.

Amino Acid Racemization

Racemization is the chemical reaction that interconverts an amino acid into its alternative enantiomer or mirror-image form (**Fig. 1**). The rate at which racemization proceeds varies between amino acids according to the ability of R groups to stabilize a carbanion intermediate. The carbanion intermediate is the molecule formed when the hydrogen side chain is abstracted from an amino acid. In an ideal system, upon readdition of the hydrogen atom to the carbanion intermediate, there is an equal probability of L- and D-amino acid formation, and the forward and reverse reaction rates are equal. Under these conditions, D/L progresses from values near zero in modern tissues until D/L reaches unity at racemic equilibrium. A consequence of these conditions is the slowing of the racemization rate as the extent of

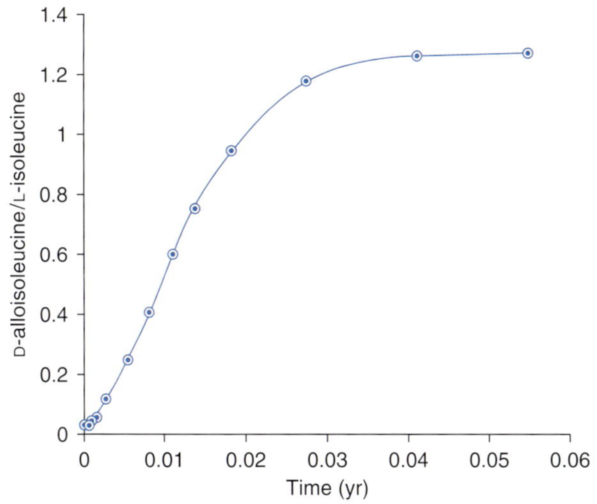

Figure 2 Characteristic trajectory of increasing A/I with increasing isothermal heating time of a modern biomineral, in this case a modern eggshell of the Australian emu heated at 143°C. Modified from Miller GH, Hart CP, Roark EB, and Johnson BJ (2000) Isoleucine epimerization in eggshells of the flightless Australian birds, *Genyornis* and *Dromaius.* In: Goodfriend G, Collins M, Fogel M, Macko S, and Wehmiller J (Eds.), *Perspectives in Amino Acid and Protein Geochemistry*, pp. 161–181. New York: Oxford University Press.

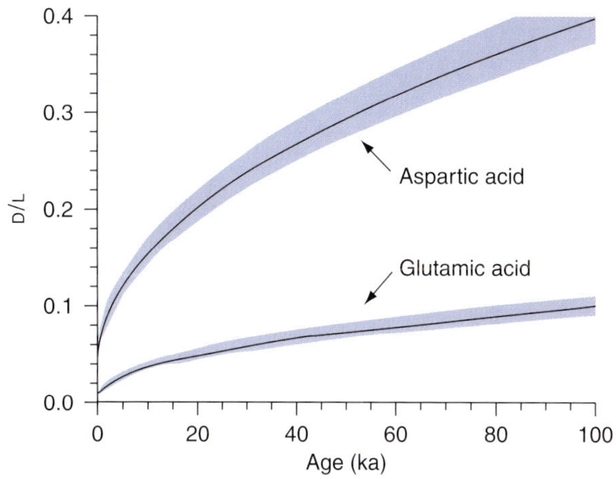

Figure 3 Relation of D/L for aspartic acid (Asp) and glutamic acid (Glu) with respect to calibrated age in the ostracode *Candona* from deep lakes in the western United States. Gray envelopes denote $\pm 1\sigma$ uncertainty. From Kaufman DS (2003b) Dating deep-lake sediments by using amino acid racemization in fossil ostracodes. *Geology* 31: 1049–1052.

racemization increases—the characteristic curvilinear trend produced when D/L ratios are plotted against linear time (**Fig. 2**).

Not all amino acids are suitable for geochronology. Glycine, the simplest and one of the most common amino acids, is not a chiral molecule. Other simple amino acids, such as alanine, are created in fossils by the decomposition of other more complex amino acids. Thermodynamically unstable amino acids, such as serine and threonine, have half-lives too short for most geological applications. The relatively slow-racemizing amino acids isoleucine, leucine, valine, and glutamic acid are widely used in geochronology, as is the relatively fast-racemizing aspartic acid (**Fig. 3**).

The amino acids commonly utilized for geochronological applications, with the exception of isoleucine, have a single chiral carbon atom and can exist in either their D- or L- forms, which have identical physical and chemical properties except in their interaction with other chiral substances. Isoleucine has two chiral carbon atoms and can racemize about either or both carbon atoms (**Fig. 4**). In nature, L-isoleucine racemizes almost exclusively about the alpha carbon, producing a new molecule, D-alloisoleucine ('allo' meaning other). Because D-alloisoleucine is not a mirror image of L-isoleucine, it has different chemical and physical characteristics, making it easier to separate in the laboratory. The reaction by which L-isoleucine converts to D-alloisoleucine is known as epimerization, and the extent of isoleucine epimerization is abbreviated A/I or aIle/Ile. The equilibrium ratio of D-alloisoleucine to L-isoleucine of approximately 1.30 indicates the reverse reaction (D-aIle formation) is 30% more likely than the forward reaction (L-Ile formation; Williams and Smith, 1977).

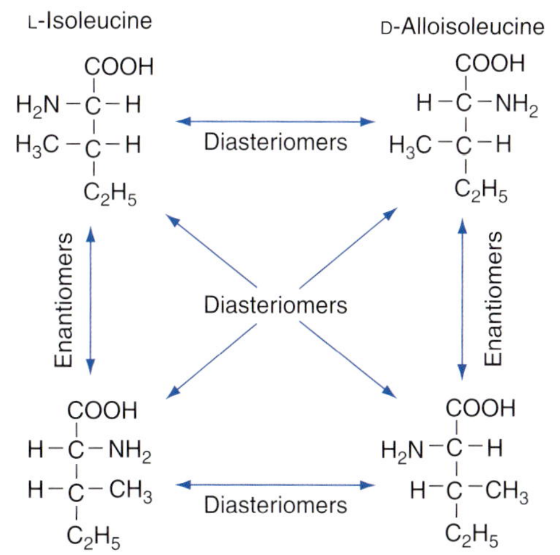

Figure 4 Possible racemization pathways for the protein amino acid L-isoleucine. In nature, nearly all of the L-isoleucine epimerizes to D-alloisoleucine.

Temperature

Rates of amino acid racemization and most other diagenetic reactions, such as hydrolysis, deamination, decarboxylation, and condensation, are highly sensitive to temperature. The temperature dependency of these reactions is described by the Arrhenius equation:

$$k_1 = A\, e^{(-E_a/RT)} \tag{1}$$

or

$$\ln(k_1) = \ln(A) - E_a/RT \tag{2}$$

where k_1 is the forward rate constant; A is a constant, often referred to as the Arrhenius factor; E_a is the activation energy; R is the universal gas constant; and T is the integrated thermal history T_{eff} in Kelvin.

The Arrhenius equation is derived empirically as the best fit to observed changes in reaction rates with respect to temperature. This is usually achieved by heating modern samples isothermally for specific time intervals in the presence of water vapor and determining the temperature-specific rate constant (k_1) for each temperature. In practice, this can be accomplished for temperatures between 80 and 160°C. Lower temperature rate constants are usually derived from radiocarbon-dated samples in sites of near-constant temperature. Equation 2 indicates that plotting ln (k_1) against the reciprocal of its absolute temperature will define a straight line, with the activation energy directly proportional to the slope of the line and the Arrhenius factor defined by the intercept (**Fig.** 5).

The rate constant, k_1, is related to time (t, measured in years) by the integrated rate equation (Bada and Schroeder, 1972):

$$k_1 = \ln[(1 + D/L)/(1 - K'D/L)] \times [(1 + K')t]^{-1} + c \tag{3}$$

where $K' = k_2/k_1$; c is a constant to account for laboratory-induced racemization: $c = \ln[(1 + D/L_0)/(1 - K'D/L_0)]$; and D/L_0 represents the extent of racemization measured in a modern specimen (e.g., live at time of collection, equivalent to time zero, or the start of the geologic clock). For most amino acids the forward and reverse rate constants, k_1 and k_2, are equal; therefore, $K' = 1$. For the interconversion of L-Ile to D-aIle, $k_1/k_2 = 1.3$, and $K' = 0.77$.

The activation energies for racemization of most amino acids preserved in carbonate fossils are between 28 and 30 kcal mol^{-1}. This range implies that the reaction rate approximately doubles for every 4°C increase in burial temperature at typical ambient temperatures (−10 to +30°C). The effect of temperature on reaction rate is readily apparent in a plot of racemization levels for last interglacial 'moderate' racemization rate mollusks from the European Arctic to the Mediterranean (**Fig.** 6).

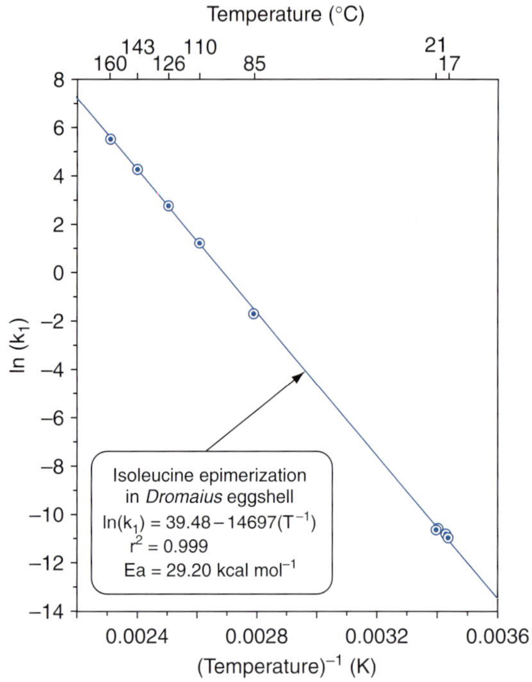

Figure 5 Arrhenius plot for isoleucine epimerization in eggshells of *Dromaius*, the Australian emu, based on isothermal heating of modern eggshell at 160, 143, 126, 110, and 85°C and radiocarbon-dated early Holocene specimens between 21 and 17°C. From Miller GH, Hart CP, Roark EB, and Johnson BJ (2000) Isoleucine epimerization in eggshells of the flightless Australian birds, *Genyornis* and *Dromaius*. In: Goodfriend G, Collins M, Fogel M, Macko S, and Wehmiller J (eds.), *Perspectives in Amino Acid and Protein Geochemistry*, pp. 161–181. New York: Oxford University Press. Reproduced with permission.

The exponential dependence of reaction rate on temperature (Eq. 1) places significant constraints on optimal geological contexts for samples. The T_{eff} experienced by a buried biomineral is strongly influenced by the magnitude of the temperature variations during the postdepositional interval. Consider a simple case. A fossil spends half of its time buried at a temperature of 0°C and half at 20°C. The arithmetic mean temperature is 10°C, but the extent of racemization exhibited by the fossil would be the same as if it had been stored at 16.7°C for the entire time, dramatically higher than the arithmetic mean temperature (**Fig.** 7). For this reason, racemization temperature is defined as the effective diagenetic temperature (T_{eff}). T_{eff} is the equivalent temperature required to explain the extent of observed racemization if the sample had

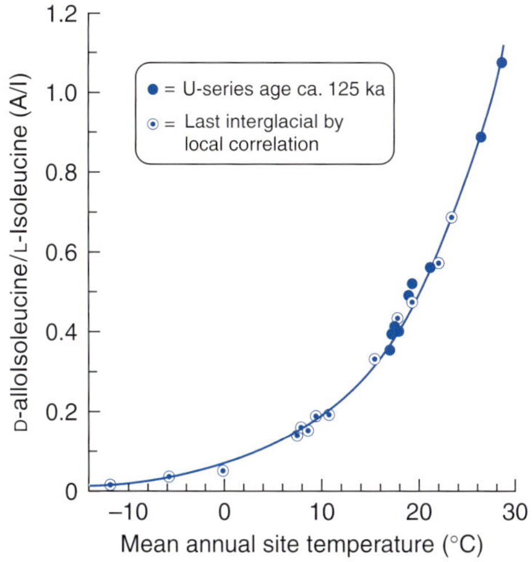

Figure 6 D-alloisoleucine/L-isoleucine (A/I) in moderate racemization rate mollusks from last interglacial (ca. 125 ka) sites across western Europe, ranging from Svalbard and Arctic Russia to the Mediterranean Basin, plotted against current mean annual site temperature. Last interglacial sites are dated by U/Th on corals or correlated to the last interglacial on the basis of diagnostic marine faunal elements. These data document the close correlation to and exponential dependency of racemization rate to site temperature. Unpublished data from G. H. Miller and P. J. Hearty, University of Colorado.

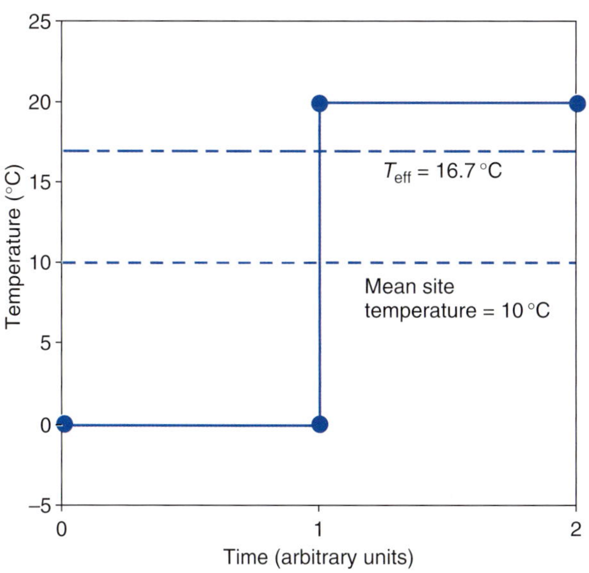

Figure 7 A simplified illustration of the impact of variable burial history. A fossil is buried for equal time at 20 and 0°C. The arithmetic mean temperature is 10°C. However, because of the exponential dependency of racemization rate on temperature, the fossil will exhibit the same amount of racemization as a sample maintained for the same total time at a much higher temperature (16.7°C). This is the effective diagenetic temperature (T_{eff}).

experienced a constant postburial temperature. It is always equal to or higher than the arithmetic mean temperature.

Sites with high diurnal or seasonal temperature fluctuations will have T_{eff} well above the arithmetic mean temperature. In most environments, samples only need to be buried a few centimeters to escape the diurnal heating cycle, but the seasonal cycle in continental climates, although rapidly damped, can extend to a depth of several meters. Reactions monitored at a continental site in southern New Mexico exhibit a seasonal temperature amplitude over 20°C in the upper few centimeters, resulting in an T_{eff} more than 10°C above the mean annual air temperature (MAT) (**Fig. 8**). The amplitude of the annual temperature cycle decreases rapidly so that by 2 m depth it is <6°C, at which point the difference between T_{eff} and MAT is negligible (<1°C; **Fig. 8**).

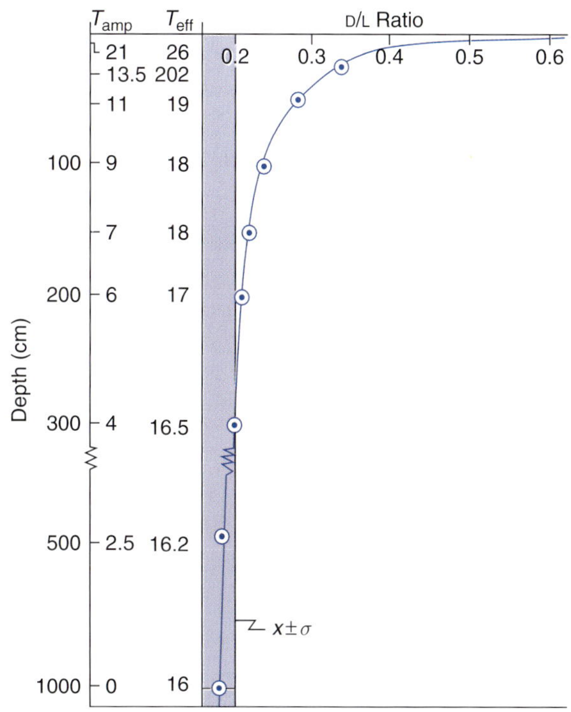

Figure 8 The difference between effective diagenetic temperature (T_{eff}) and mean annual air temperature (MAT; mean, 16.8°C) with depth based on multiyear instrumentation of a strongly continental site in southern Arizona. The relatively high amplitude of the annual temperature cycle near the surface (T_{amp} = 21°C) results in an T_{eff} 9°C above the MAT. The predicted A/I in a gastropod held at specific depths for 14 kyr is plotted on the right. By a depth of approximately 1 m, the difference between predicted and expected A/I is within analytical uncertainties, but at shallow burial depths the predicted A/I deviates sharply from that expected based on the site MAT. Modified from Miller GH and Brigham-Grette J (1989) Amino acid geochronology: Resolution and precision in carbonate fossils. *Quaternary International* 1: 111–128.

Sampling strategies must recognize the importance of avoiding high-amplitude temperature fluctuations. In exposed continental sites, collecting sites should have been buried 2 m or more for most of their postdepositional history. Shaded sites and sites from more maritime regions need not be so deeply buried.

Environmental pH

Racemization is known to be strongly base catalyzed, and biominerals heated in solutions above pH 9 racemize much more quickly than do the same biominerals heated at neutral pH. Carbonate fossils have the advantage of buffering ambient pH and maintaining a local environmental pH of approximately 8. Samples from strongly basic environments should be avoided.

Taxonomic Effects

Position in a polypeptide chain The negative correlation between the molecular weight of polypeptides and the extent of racemization among their amino acids is a product of the relationship between the racemization rate of an amino acid and its location within the polypeptide. For example, the extent of isoleucine epimerization in various molecular fractions of three *Genyornis* eggshells is always lowest in the highest molecular weight fraction, whereas the low-molecular-weight and free amino acid fractions have the highest A/I values (**Fig. 9**; Kaufman and Miller, 1995). Laboratory experiments indicate that amino acids located at interior positions within peptide chains feature slow rates of racemization, whereas amino acids located at terminal positions and incorporated in diketopiperazines racemize most rapidly. The D/L of free amino acids is commonly high but the reaction rate is low when the molecules exist in this state. This potential incongruity is explained by the preferential release of relatively highly racemized amino acids from terminal positions into the free pool. Relatively intact polypeptides have a higher ratio of interior to terminal amino acids than smaller polypeptides; consequently, high-molecular-weight separations feature a lower extent of racemization.

The differences in racemization rates produces two pools of amino acids with different average D/L that are readily separated: 'free' and 'total acid hydrolysate.' Normally, peptide bonds are hydrolyzed prior to amino acid analysis by heating the decalcified residue (typically 22 h at 110°C) in a strong acid (6 N HCl) to yield the total acid hydrolysate fraction. Alternatively, the unhydrolyzed residue, or free fraction, is dominated by naturally hydrolyzed amino acids. Due to the more advanced level of racemization in this pool, D/L in the free fraction may provide a more effective means of separating intervals of time than is attainable with D/L in the total acid hydrolysate.

Optimizing sample sites within a fossil Many fossils, including mollusks and eggshells, have different structural layers. Usually, each structural layer

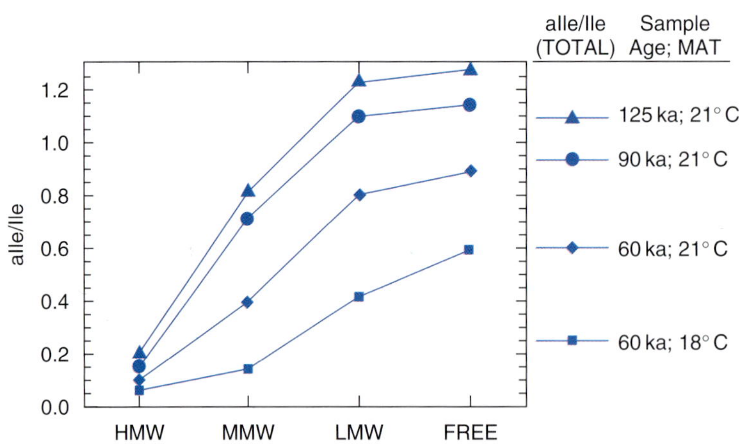

Figure 9 Extent of isoleucine epimerization (aIle/Ile) measured in four molecular weight fractions of four late Pleistocene eggshells of the extinct giant Australian bird, *Genyornis*. AIle/Ile was measured in high-molecular-weight (HMW), moderate-molecular-weight (MMW), low-molecular-weight (LMW), and free (FREE) fractions. In all four samples, aIle/Ile increased with decreasing molecular weight. Symbols on the right are aIle/Ile in the total acid hydrolysate. Modified from Kaufman DK and Miller GH (1995) Isoleucine epimerization and amino acid composition in molecular-weight separations of Pleistocene Genyornis eggshell. *Geochimica et Cosmochimica Acta* 59: 2757–2765.

contains a different suite of amino acids. Because the racemization rate for any specific amino acid depends to some extent on the nature of its bonds with adjacent amino acids in its peptide chain, racemization rates commonly vary in different structural layers. Consequently, sampling should be restricted to a single layer.

Tests of mollusk A/I variability indicate values may differ by up to 40% in different parts of even relatively simple bivalves (**Fig. 10**, top). In bivalve shells, most of the protein residues reside along intercrystalline growth, where they act to nucleate incremental shell growth. Leaching of amino acids and small peptides is enhanced where these growth bands intersect the surface of the shell bands (**Fig. 10**, bottom). Consequently, optimal sampling sites for AAR studies are located at sites 'A' or 'B' in **Figure 10** (top).

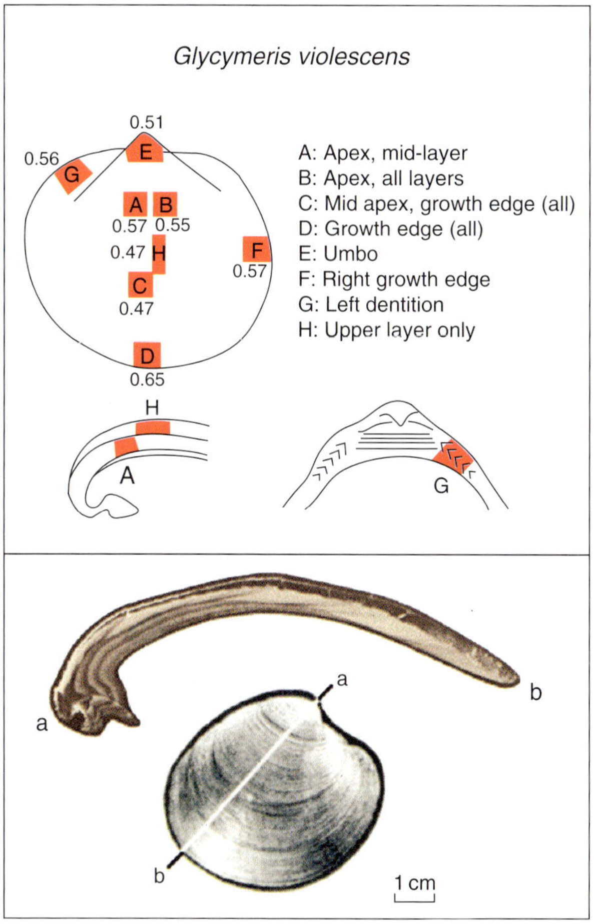

Figure 10 (Top) Typical intrashell variability in A/I—in this case, in a last interglacial age marine taxodont *Glycymeris* from the Mediterranean. (Bottom) A cross section through a marine pelecypod, showing how growth bands extending to the surface can serve as pathways that allow leaching of low-molecular-weight organic compounds. The optimal sample site in both examples is an inner layer, near the umbo region ('A' and 'B' in the top panel), where the shell is dense and there are fewer pathways for leaching.

Other vital effects Differences up to a factor of two in racemization rates according to taxonomy have been noted for mollusks, foraminifera, and avian eggshells. Among mollusks, there are significant differences between genera but broadly comparable reaction rates among species of the same genus. The reason for the taxonomic dependency is not precisely known, but because they are accompanied by differences in amino acid composition, likely explanations are differences either in the amino acid composition of the precursor proteins or in the relative abundance of the numerous proteins involved in the secretion of these carbonate structures. For example, Goodfriend *et al.* (1992) observed high concentrations and very rapid racemization of Asx residues in the coral *Porites australiensis*. Because peptide bonds at Asx sites are prone to hydrolysis, the authors interpreted these observations as cause and effect. Where multiple species are recovered in association, intergeneric differences in racemization rates provide an internal check on the reliability of D/L values. Similarly, because amino acids feature characteristic rates of racemization, a comparison of the D/L values of different amino acids within a sample provides an additional means of assessing the veracity of data.

Almost all biominerals contain a relative abundance of specific amino acids that are unique at the generic level. This characteristic signature can be used as a taxonomic tool and provides a test for impurities in most geological samples.

Leaching

The physical transport of amino acids out of mineral matrices by an aqueous phase, commonly water, is termed 'leaching'. The rate of leaching is controlled by the size of the mobile molecules and their sorption to the fossil matrix, the diameter and complexity of the diffusion pathway, and temperature (Collins and Riley, 2000). Molecules most susceptible to leaching are the low-molecular-weight solutes (i.e., free amino acids, small peptides, and diketopiperazines) in which protein residues are situated in pathways that lead to the exterior of the biomineral. Because low-molecular-weight residues tend to be more racemized, leaching decreases the overall level of racemization. Hot, alternately wet and dry environments are most conducive to leaching.

Protein residues preserved in intracrystalline sites are least susceptible to leaching. Strategies to isolate this subset of organic residues have been developed to minimize the potential effects of leaching and improve the correlation between D/L and fossil age.

Separation and Detection of D- and L-Amino Acids

The separation, detection, and quantification of the D- and L-amino acid enantiomers and diastereomers are accomplished by either gas or high-pressure liquid chromatography. L-Isoleucine can be separated from its D-alloisoleucine pair by conventional ion-exchange high-pressure liquid chromatography, which provides a quantitative assessment of most other protein amino acids and allows an independent assessment of sample quality. Most other D-/Lamino acid pairs are separated by gas chromatography or reverse-phase liquid chromatography using a chiral stationary or mobile phase to separate the enantiomers (**Fig. 11**). An interlaboratory standard (Wehmiller, 1984) provides an independent check on instrumental accuracy.

Applications

Amino acid racemization has potential application in the Quaternary sciences wherever amino acid residues are preserved over geological time and questions of time or temperature remain. To utilize amino acid racemization as a dating tool requires a medium that not only preserves indigenous protein amino acids from bacterial attack and leaching but also excludes contamination by other amino acids. The more tightly the amino acids are integrated into the biominerals, the closer the medium will be to a closed system. Commonly utilized fossils include marine bivalves, land and aquatic snail shells, eggshells, foraminifera, and ostracodes. Most are abundant in the geologic record and preserve amino acids well over geologically significant timescales. Other materials that have been analyzed, some with mixed success, include fish otoliths, woodrat middens, wood, bones and teeth, corals, and the carbonate spores of charophyte oogonia. Although most applications utilize amino acids derived from known biological sources, racemization studies in biogenic carbonate breccia (e.g., 'whole rock' eolianites), speleothems, and paleosols have proven successful.

The time interval over which racemization is effective is determined by the reaction rate: it is longer in cold climates (low rate) than in tropical climates (high rate). At room temperature (22°C), conversion from an optically pure L-amino acid (D/L = 0) to a racemic mixture lacking optical activity (D/L = 1.0 for most amino acids) requires several tens of thousands of years for most amino acids. At 160°C, a racemic mixture is reached within 10–20 h, whereas in Arctic regions (−10°C) racemic equilibrium requires 1–2 million years. The useful time range for dating under different mean annual temperatures for isoleucine epimerization (valine, leucine, and glutamic acid racemization have similar time ranges) is illustrated in **Figure 12**.

Aspartic acid racemizes approximately an order of magnitude faster than other amino acids and is

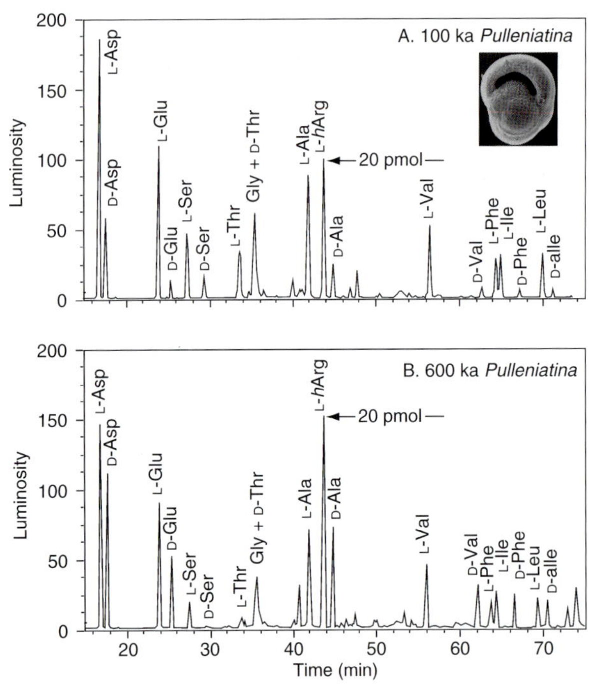

Figure 11 Reverse-phase liquid chromatography chromatograms of single individuals of the large planktonic foraminifer *Pulleniatina* from marine cores recovered off northeastern Australia, with approximate ages of 100 ka (A) and 600 ka (B). Modified from Hearty PJ, O'Leary MJ, Kaufman DS, Page M, and Bright J (2004) Amino acid geochronology of individual foraminifer (*Pulleniatina obliquiloculata*) tests, north Queensland margin, Australia: A new approach to correlating and dating Quaternary tropical marine sediment cores. *Paleoceanography* 19: PA4022.

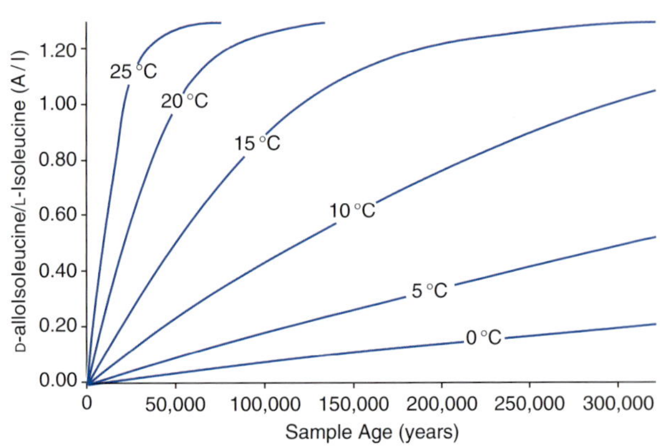

Figure 12 Relation between the extent of isoleucine epimerization (A/I) and sample age for T_{eff} between 0 and 25°C showing the strong dependence of epimerization rate on temperature. Lines are derived from kinetics in eggshells of the Australian emu (*Dromaius*) shown in Miller *et al.* (2000), but the relation is generally applicable to most carbonate fossils.

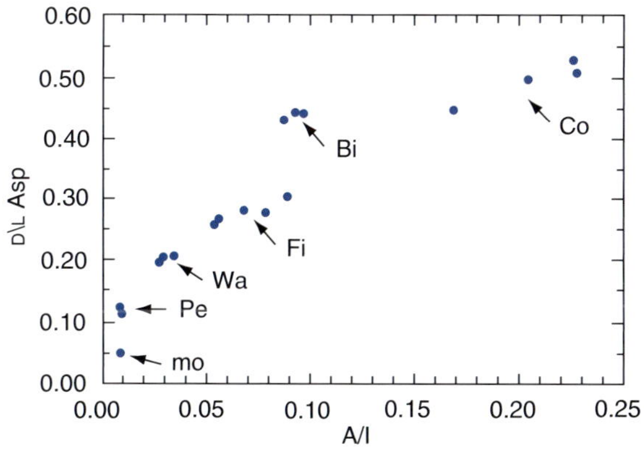

Figure 13 Fast-racemizing amino acids provide superior resolution for cold or very young samples. D/L aspartic acid and D-alloisoleucine/L-isoleucine (A/I) were measured in the marine bivalve *Hiatella arctica* collected from raised marine terraces along the north coast of Alaska (mean annual temperature, ca. −12°C). For the younger shorelines, D/L Asp provides a clearer separation of the sites, but this difference decreases for the older shorelines. Aminozones: mo, modern; Pe, Pelukian shoreline (last interglaciation); Wa, Wainwrightian shoreline; Fi, Fishcreekian shoreline; Bi, Bigbendian shoreline; and Co, Colvillian shoreline (probably late Pliocene). Modified from Goodfriend GA, Brigham-Grette J, and Miller GH (1996) Enhanced age resolution of the marine Quaternary record in the Arctic using aspartic acid racemization dating of bivalve shells. *Quaternary Research* 45: 176–187.

frequently used for very young materials or extremely cold sites (**Fig. 3**). In tropical settings, the rapid racemization of Asx residues provides age resolution for very young samples, whereas the slower epimerization of isoleucine is effective at distinguishing events that are thousands of years apart (**Figs. 3, 13** and **14**).

Aminostratigraphy

Amino acid racemization can be applied as a geological dating tool in a relative and absolute sense. As a relative dating tool, the technique is known as aminostratigraphy. Aminozones, a chemostratigraphic unit based on clustering in the extent of amino acid racemization in replicate fossils, are the fundamental units of aminostratigraphy (**Fig. 13**). The premise is that over a limited geographic range and elevation, samples experienced a similar thermal history (MAT, ±1°C; temperature amplitude at collection site, ≤6°C). Over this limited region, samples with similar D/L are of the same age, those sites with higher D/L are older, and those with lower D/L are younger (**Fig. 13**). Aminostratigraphy does not require an assessment of racemization kinetics or estimates of past thermal histories, and it is the least ambiguous application of amino acid geochronology.

Aminostratigraphy has been applied to mollusks excavated from raised marine terraces in coastal regions throughout the world. One example is a series of uplifted interglacial terraces along the Peruvian coast from which bivalve mollusks exhibit a regular increase in D/L leucine in higher (older) terraces, with corresponding absolute ages derived from electron spin resonance (ESR) dating (**Fig. 15**).

Absolute Age and Paleothermometry

Once the temperature-dependence of racemization in a particular biomineral has been derived, the equation describing racemization contains three unknowns: D/L (the extent of the reaction, which can be measured in the laboratory); t, the time since

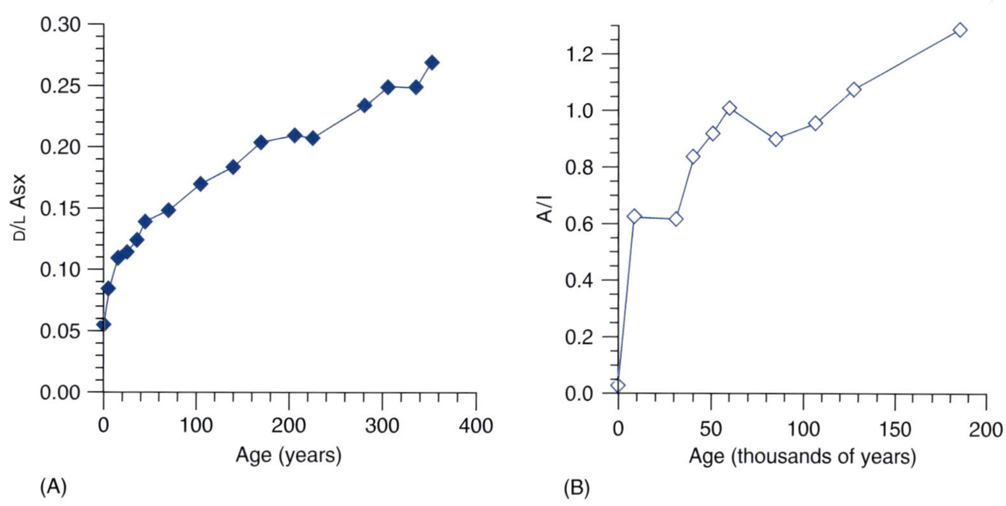

Figure 14 Examples of racemization applied to two very different timescales. D/L Asx in very young corals (A; data from Goodfriend GA, Hare PE, and Druffel RM (1992) Aspartic acid racemization and protein diagenesis in corals over the last 350 years. *Geochimica et Cosmochimica Acta* 56: 3847–3850) and A/I in the giant clam *Tridacna* from the Huon Peninsula (B; data from Hearty PJ and Aharon P (1988) Amino acid chronostratigraphy of late Pleistocene coral reef sites: Huon Peninsula, New Guinea and the Great Barrier Reef, Australia. *Geology* 16: 579–583). Note timescales differ by three orders of magnitude.

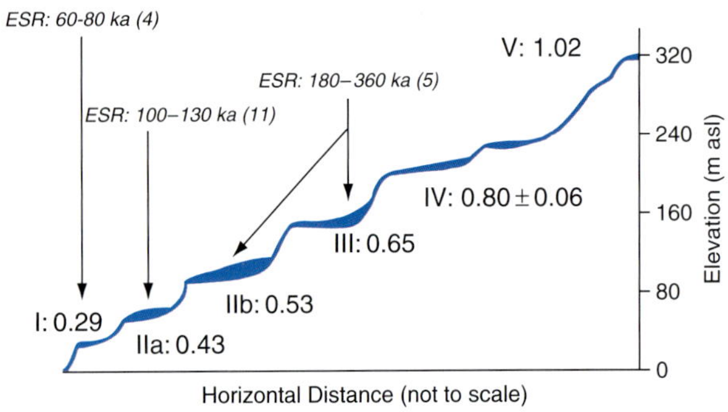

Figure 15 Schematic cross-section through a flight of raised marine terraces on the coast of Peru with aminozones (in roman numerals) defined on the extent of leucine racemization. ESR dates provide additional chronological control. The highest terraces are of early Quaternary age. Modified from Wehmiller JF (1993) Applications of organic geochemistry for Quaternary research: Aminostratigraphy and aminochronology. In: Engel M and Macko S (Eds.), *Organic Geochemistry*, pp. 755–783. New York: Plenum.

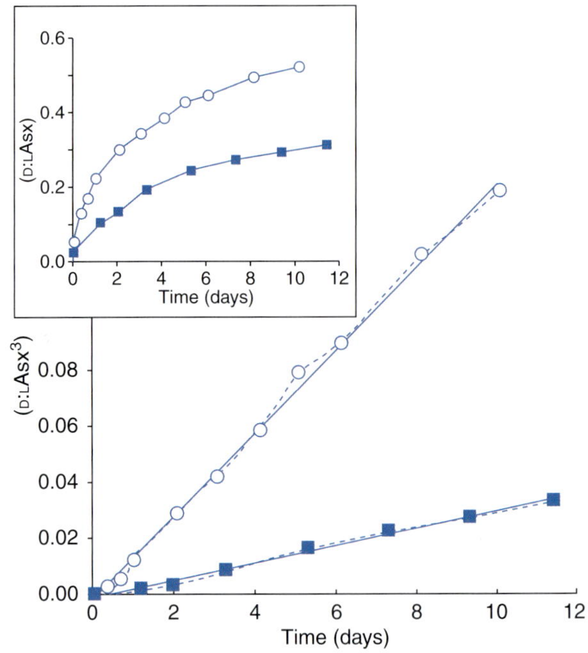

Figure 16 Cubic transformation of D/L Asx to linearize the original kinetics (inset), from which the slope is proportional to sample age. Original data from Goodfriend (1992) (land snails heated at 100°C, open circles) and Brinton and Bada (1995) (free asparagines heated at 100°C). From Collins MJ, Waite ER, and van Duin ACT (1999) Predicting protein decomposition, the case of aspartic acid racemization kinetics. *Royal Society of London Philosophical Transactions, Series B, Biological Sciences* 354: 51–64. Modified with permission.

death; and T, the integrated thermal history. Theoretically, if either time (t) or effective diagenetic temperature (T) is known, the other can be predicted (from Eqs. 1 and 3):

$$\mathrm{A}e^{(-E_a/RT)} = \ln[(1+D/L)/(1-K'D/L)] \times [(1+K')t]^{-1} \quad (4)$$

In theory, deriving absolute ages from the extent of amino acid racemization requires only an estimate (±1°C) of the T_{eff}. For samples of late Holocene age, this can be reasonably estimated from the instrumental record, but secure estimates for the time-dependent changes in mean annual temperatures across glacial/interglacial cycles are rarely possible. Bada and Protsch (1973) presented an alternative approach in which an appropriate in situ value for k_1 is determined from the D/L ratio of a specimen of known age. This 'calibrated' rate constant may be used to determine the age of nearby specimens from their D/L value, assuming they have experienced a comparable thermal history. For Quaternary samples, a reasonable assumption is that a calibration site of at least 100 kyr has a T_{eff} that is a reasonable estimate for most of the Quaternary.

Age estimates for many fossils can be complicated by kinetic uncertainties. An inherent assumption is that racemization follows simple, reversible first-order kinetics. Although this assumption is approximated during the early stages of racemization, most mollusk and foraminifera shells deviate from predicted linear kinetics in the later stages of diagenesis. This has led several researchers to utilize empirically derived models (**Fig. 16**) that more closely approximates the observed change in D/L with time.

In an evaluation of uncertainties, McCoy (1987a) showed that predicting paleotemperatures is inherently more precise than predicting ages. Unlike most biological proxies that record the instantaneous temperature at which an organism lived, racemization-derived paleotemperatures reflect the integrated thermal history (T_{eff}) since the organism died. In the case of deeply buried samples, this equates to the integrated mean annual temperature. If samples with independently derived ages are available from a limited region, then the evolution of temperature across the region can be calculated. For example, the Pleistocene/Holocene temperature change in semiarid Australia was quantified from A/I in emu eggshells, each of which had also been dated by ^{14}C (**Fig. 17**). Not all samples were deeply buried; hence, some samples exhibit more advanced levels of racemization than expected, but the change in slope between samples with the lowest A/I dated <16 ka and those >16 ka is unambiguous and reflects at least 7°C warming from the cold last glacial maximum to the warmth of the present interglacial.

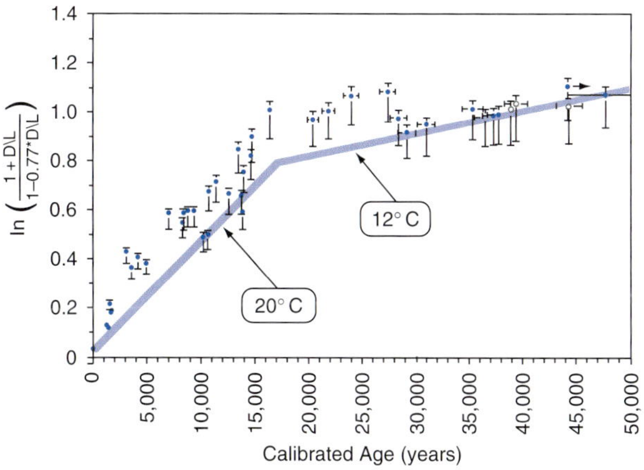

Figure 17 A/I (in its linearized form) plotted against corresponding calibrated radiocarbon ages for 43 samples from central Australia. Asymmetric vertical error bars incorporate the possibility of shallow burial. Horizontal error bars reflect $\pm 1\sigma$ dating errors where the uncertainty exceeds the diameter of the plotted point. Regression lines, which are directly proportional to temperature, go through the lowest A/I. The lower slope for samples older than ca. 16 ka indicates much lower Pleistocene temperatures in central Australia than during the Holocene. Modified from Miller GH, Magee JW, and Jull AJT (1997) Low-latitude glacial cooling in the Southern Hemisphere from amino acids in emu eggshells. *Nature* 385: 241–244.

Conclusions

Amino acid geochronology offers a simple but powerful relative dating tool (aminostratigraphy) that provides temporal constraints on biominerals beyond the range of radiocarbon dating. In all but the hottest regions, this includes all or most of the middle and late Quaternary, a time window not readily dated by many other methods. When independent calibration is available, the extent of racemization can be used to reconstruct past thermal histories with a precision of $\pm 1°C$. Also, with suitable calibration the extent of racemization can be converted to reliable absolute ages. Although the measured extent of racemization depends heavily on temperature, the racemization reaction does not involve the loss of reactants; hence, the measurement statistics for old samples are similar to those for young samples. For the most stable amino acids, the concentration of the summed D- and L-forms at racemic equilibrium is of a similar magnitude to the concentration of the original L-amino acid, providing there has been no diffusional loss from the biomineral matrix. The fast-racemizing Asx can be used to assess differences in age over a few decades to centuries in optimal settings.

Not all biominerals are optimal for amino acid geochronology. Those that approximate closed system behavior with respect to indigenous amino acids provide optimal samples. This is usually met in samples for which the amino acids are sequestered within the crystal matrix of a fossil.

See also: **Introduction**: History of Dating Methods. **Dating Techniques**. **Quaternary Stratigraphy**: Overview.

References

Abelson, P. H. (1954). Organic constituents of fossils. *Carnegie Institution of Washington Yearbook* **53**, 97–101.

Bada, J. L., and Protsch, R. (1973). Racemization reaction of aspartic acid and its use in dating fossil bones. *Proceeding of the National Academy of Sciences USA* **70**, 1331–1334.

Bada, J. L., and Schroeder, R. A. (1972). Racemization of isoleucine in calcareous marine sediments: Kinetics and mechanism. *Earth and Planetary Science Letters* **15**, 1–11.

Brinton, K. L., and Bada, J. L. (1995). Aspartic acid racemization and protein diagenesis in corals over the last 350 years—Comment. *Geochimica et Cosmochimica Acta* **59**, 415–416.

Collins, M. J., Waite, E. R., and van Duin, A. C. T. (1999). Predicting protein decomposition, the case of aspartic acid racemization kinetics. *Royal Society of London Philosophical Transactions, Series B, Biological Sciences* **354**, 51–64.

Goodfriend, G. A., and Mitterer, R. M. (1988). Late Quaternary land snails from the north coast of Jamaica: Local extinctions and climatic change. *Palaeogeography, Palaeoclimatology, Palaeoecology* **63**, 293–311.

Goodfriend, G. A., Collins, M. J., Fogel, M. L., Macko, S. A., and Wehmiller, J. F. (2000). *Perspectives in Amino Acid and Protein Geochemistry*. Oxford University Press, New York.

Hare, P. E., and Abelson, P. H. (1968). Racemization of amino acids in fossil shells. *Carnegie Institution Washington Year Book* **66**, 526–528.

Hare, P. E., and Mitterer, R. M. (1967). Nonprotein amino acids in fossil shells. *Carnegie Institution Washington Yearbook* **65**, 362–364.

Hare, P. E., and Mitterer, R. M. (1968). Laboratory simulation of amino acid diagenesis in fossils. *Carnegie Inst. Wash. Yrbk* 205–208.

Hare, P. E., and Mitterer, R. M. (1969). Laboratory simulation of amino acid diagenesis in fossils. *Carnegie Institution Washington Yearbook* 205–208.

Hare, P. E., Hoering, T. C., and King, K. (eds.) (1980). *Biogeochemistry of Amino Acids*. Wiley & Sons, John New York.

Hearty, P. J., and Aharon, P. (1988). Amino acid chronostratigraphy of Late Pleistocene coral reef sites: Huon Peninsula, New Guinea and the Great Barrier Reef, Australia. *Geology* **16**, 579–583.

Johnson, B. J., and Miller, G. H. (1997). Archaeological applications of amino acid racemization. *Archaeometry* **39**, 265–287.

Kaufman, D. K., and Miller, G. H. (1995). Isoleucine epimerization and amino acid composition in molecular-weight separations of Pleistocene Genyornis eggshell. *Geochimica et Cosmochimica Acta* **59**, 2757–2765.

Kaufman, D. S., and Miller, G. H. (1992). Overview of amino acid geochronology. *Comparative Biochemistry and Physiology* **102B**, 199–204.

McCoy, W. D. (1987a). The precision of amino acid geochronology and paleothermometry. *Quaternary Science Reviews* **6**, 43–54.

Miller, G. H., and Brigham-Grette, J. (1989). Amino acid geochronology: Resolution and precision in carbonate fossils. *Quaternary International* **1**, 111–128.

Schroeder, R. A., and Bada, J. L. (1976). A review of the geochemical applications of the amino acid racemization reaction. *Earth Science Reviews* **12**, 347–391.
Wehmiller, J. F. (1990). Amino acid racemization: Applications in chemical taxonomy and chronostratigraphy of Quaternary fossils. In *Skeletal Biomineralization: Patterns, Processes, and Evolutionary Trends* (J. Carter, Ed.), pp. 583–608. Van Nostrand Reinhold, New York.
Wehmiller, J. F. (1993). Applications of organic geochemistry for Quaternary research: Aminostratigraphy and aminochronology. In *Organic Geochemistry* (M. Engel and S. Macko, Eds.), pp. 755–783. Plenum, New York.
Wehmiller, J. F., and Miller, G. H. (2000). Aminostratigraphic dating methods in Quaternary geology. In *Quaternary Geochronology: Methods and Applications* (J. Noller, J. Sowers and W. Lettis, Eds.), pp. 187–222. American Geophysical Union, Washington, DC.
Williams, K. M., and Smith, G. G. (1977). A critical evaluation of the application of amino acid racemization to geochronology and geothermometry. *Origins of Life* **8**, 91–144.

Anatomically Modern Humans *see* **Archaeological Records**: Global Expansion 300,000-8000 years ago, Africa; Global Expansion 300,000-8000 years ago, Asia; Global Expansion 300,000-8000 years ago, Australia; Global Expansion 300,000-8000 years ago, Americas; Postglacial Adaptations

ARCHAEOLOGICAL RECORDS

Contents

Overview

C Gamble, Royal Holloway, University of London, Surrey, UK

© 2007 Elsevier B.V. All rights reserved.

Introduction

The study of hominids, hominins, and humans (**Table 1**) in the Quaternary period is an interdisciplinary endeavour that combines the expertise of the biological, physical, human, and natural sciences (**Fig. 1**). Scientific advances in this collective enterprise have been most marked in the past 50 years. They can be summarized by the appearance of a new subdiscipline, paleoanthropology, that emerged in the 1970s. The title includes, at a minimum, the following specialists: archeologists, physical anthropologists, molecular geneticists, geochronologists, and paleoecologists. Paleoanthropologists are not only interested in investigating well-dated sequences that contain environmental and hominin data but also in using these archives to study the behavior of our earliest ancestors. They work in the wider framework of evolutionary models and principles. Their primary interests lie in using the current wealth of paleoenvironmental data to understand changing adaptations in the various hominin lineages, and using the continuous records of climate change, particularly those from the ocean and ice-core archives, to examine whether a forcing mechanism existed that explains both anatomical and behavioral evolution. More recently, the field has developed an interest in the biogeography of hominins and in particular the timing of major dispersals and their explanation (Straus and Bar-Yosef, 2001).

The papers in this section address these and other themes in paleoanthropology on a geographical basis. The evidence is presented and assessed and

Table 1 Three important terms in paleoanthropology. Hominin replaced hominid in the 1990s when the genetic classification of extant apes and humans caused a rethink of their ancestral relations. The use of the term, however, is not yet ubiquitous

Human	All living people and their recent Pleistocene ancestors
Hominin	The above, and all the Pliocene and Pleistocene fossil ancestors (genera and species) of humans
Hominid	The above, and the African great apes: chimpanzee, bonobo, and gorilla, and their fossil ancestors.

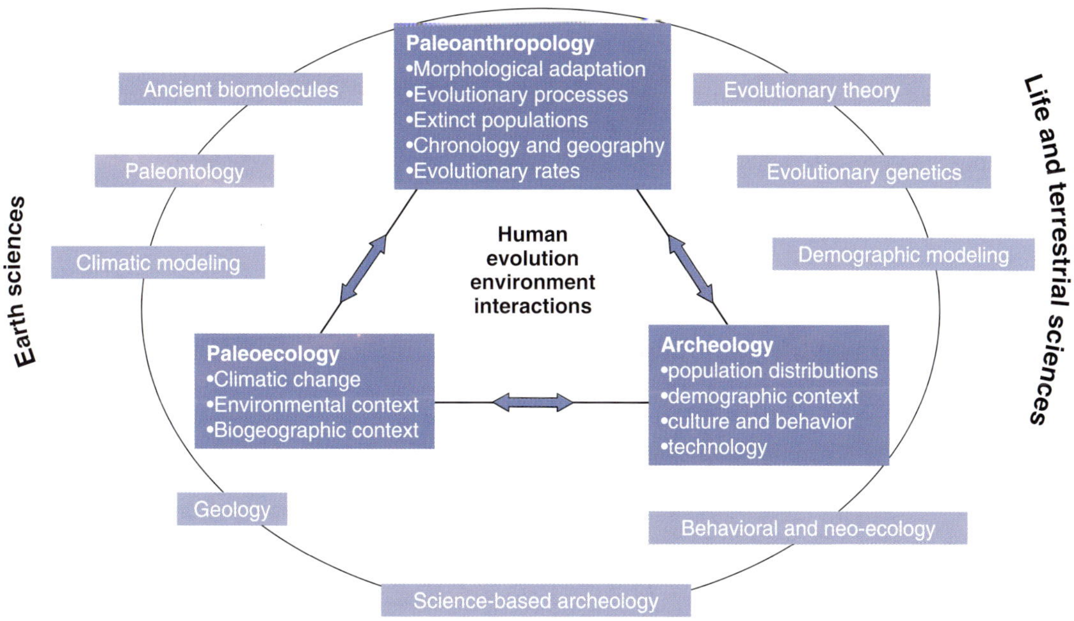

Figure 1 The interdisciplinary relationships involved in paleoanthropology. Foley (nd) with permission. Prepared for the NERC, Environmental Factors and Chronology in Human Evolution and Dispersal programme 2003.

a flavor of the common but distinctive traditions of regional and national research is provided. The date of 300 ka has been used to organize the archeological evidence. This corresponds to the beginning of MIS8 in the marine record. This cold stage does not in itself mark an abrupt change either in hominin archeology or anatomy, but after this date significant technical, cultural, and social changes occurred cumulatively, and especially during the Upper Pleistocene (MIS5-2).

The Transition to Paleoanthropology

The development of paleoanthropology within Quaternary science can be illustrated by some selected texts. In 1958, Frederick Zeuner published the fourth edition of his highly influential synthesis, *Dating the Past: An Introduction to Geochronology* (Zeuner, 1958). A year later, he followed it with *The Pleistocene Period: Its Climate, Chronology and Faunal Successions* (Zeuner, 1959). The two volumes provide a benchmark with which to assess the current work of paleoanthropologists.

Zeuner's syntheses were dominated by the Pleistocene sequences of Europe that had been established in southern Germany 50 years before by Penck and Brückner. In 1959, research had barely begun in Africa, Asia, and Australia, and the number of reliable sequences was very small. America did not possess a deep Pleistocene record for hominins although the glacial sequences had been established.

Zeuner's discussion of the archeology was dominated by European typology and taxonomy, particularly in *Dating the Past*. Europe also loomed large in his discussion of the fossil record. There are, for example, very few references to the Australopithecines in either book even though they had been named 30 years before in South Africa. The state of play in 1959 is summarized in **Table 2**.

However, 1959 was a significant year for paleoanthropology outside of Europe. During their annual field season at Olduvai Gorge, Tanzania, Mary and Louis Leakey discovered a robust Australopithecine (*Paranthropus (Zinjanthropus) boisei*) in their excavations. The development of potassium/argon dating techniques and their

Table 2 The Pleistocene framework for the study of hominins in 1959 (Zeuner 1959: 219). Zeuner was following the four ice age model of Penck and Brückner. The table reveals the paucity of fossil finds, mostly from Europe, and the classification of what there was into a very few species. The age estimates were based on a number of sources including the astronomical theory of the Serbian astrophysicist Milutin Milankovitch who died in 1958

Upper Pleistocene	Back to about 180 kyr BP	Last glaciation (1) and last interglacial	*Homo sapiens* *Homo neanderthalensis*
Middle Pleistocene	About 180 to 425 kyr BP	Penultimate glaciation (2) and penultimate interglacial	*Homo cf. sapiens*
Lower Pleistocene	About 425 to 600 kyr BP	Antepenultimate glaciation (3), antepenultimate interglacial, and early glaciation (4)	*Homo (Pithecanthropus) erectus* and *pekinensis Homo heidelbergensis*

BP = Before Present.

application soon after (Leakey *et al.*, 1961) to the volcanic tuffs that segment the long Olduvai sequence achieved a breakthrough that the Europeans could not match. The hominin 'living floor,' as Leakey (1971) referred to it, and where first *Zinjanthropus* and later *Homo habilis* (Leakey *et al.*, 1964) had been found, was now dated in Bed I to more than 1.7 Myr old, placing the finds in the Pliocene. Moreover, the association of the fossil hominins with rich faunal remains and stone tools provided an opportunity to conduct a paleoanthropological investigation into the capabilities and adaptive behavior of these remote ancestors. Discussions of these data then ranged around the existence of central place foraging (Isaac, 1978), the contribution of hunting and scavenging to the diet (Binford, 1981; Bunn, 1981), and the importance of using modern analogs to distinguish hominin behavior from that of other accumulators and modifiers of bone and stone such as hyenas and rivers (Behrensmeyer and Hill, 1980; Brain, 1981).

Following this lead similar issues were tackled in Europe. The adaptations of hominins to their environment were studied by means of their stone technology as well as by using evolutionary ecology to analyze the decisions they had taken about where to stay and which parts of the landscape they repeatedly visited. A pioneering project in this emerging paleoanthropology took place in the Ambrona valley, Spain, led by Clark Howell and Emiliano Aguirre. While issues of dating and comparative stone tool typology were uppermost at their excavations of the Lower Paleolithic sites of Torralba and Ambrona (Howell, 1966), their aim was also to turn those bones and stones into information about the process of human evolution. As at Olduvai, the questions dealt with adaptations to the food quest: how did they get their food, what did they eat, and also did they eat well?

The excavations impressed because of their scale (Santonja, 2005) and the association of stone tools with elephant bones. Clark Howell, in his popular book *Early Man* (Howell, 1965), presented this association as evidence for big-game hunting by *Homo erectus* using Acheulean handaxes. The conclusions emphasized their human capacity to plan, cooperate, share, and instruct.

These capacities were summarized by Butzer (1982) who used his geological work to reconstruct the wider landscape to which the hominins were adapting. In his study, the two sites, Torralba and Ambrona, formed a hub in a network of movements that radiated over large distances as hominins adjusted to seasonal fluctuations in their food supply (**Fig. 2**). Hominins would come together to hunt when game was migrating and then disperse in small groups across a much larger territory. An important factor in these seasonal movements was the high elevation (over 1,000 m) for the sites suggesting that winters were harsh and that prey species would migrate in search of food. The ability to kill big game and to adjust population to the seasons all indicated modern capabilities for these handaxe makers.

Subsequently, several of these claims have been disputed and revised (Villa, *et al.*, 2005), and the debate about the ability of early hominins either to hunt or to scavenge has seesawed with new and more precise evidence (Dominguez-Rodrigo, 2002; Stiner, 1994). However, whatever the outcome, and currently the hunting hypothesis commands the majority view (Gamble, 1999), these debates illustrate the questions that paleoanthropologists came to ask about Quaternary hominins. At the heart of these questions is the time depth of human rather than hominin behavior in the Quaternary and it is to these issues that the author will devote the rest of this contribution.

Adaptive Radiations and Climate Forcing

While paleoanthropologists work within the framework of evolutionary science, their approach, it has been claimed (Foley (2001), p. 5), is largely untouched by advances in evolutionary theory. New data and new methods in the last 50 years, and in particular the ongoing development of absolute chronologies,

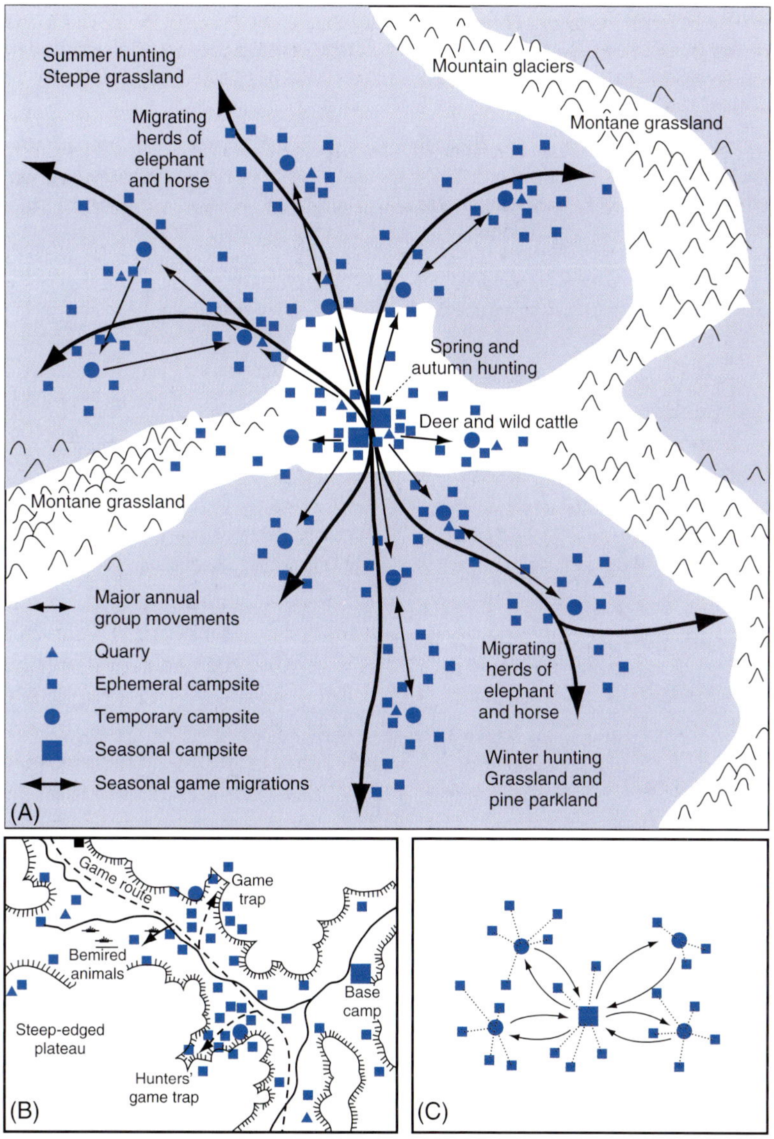

Figure 2 A reconstruction of the movements of Acheulean hunters on the Spanish meseta (after Butzer (1982)). (A) intercept hunting during the spring and autumn game migrations. (B) The Ambrona valley used as a game trap. (C) Seasonal movement between camps of varying size provides a model of the archeological evidence.

has led to the development of more complicated models of hominin ancestry. For example, one of the most significant findings in Bed I at Olduvai Gorge was the presence of two contemporary hominin genera, *Paranthropus* and *Homo*. The wealth of fossil hominins that have since been discovered in the Hadar region of Ethiopia, Koobi Fora in Kenya, and throughout the African Rift Valley, as well as the limestone caves of South Africa (Klein, 1999; Stringer and Andrews, 2005), have confirmed this pattern. At any one moment, there were several species present at a subcontinental scale and often within the same local geographical region. This is well illustrated at Koobi Fora where an African example of *H. erectus* was found in the same geological horizon, dated to 1.3 Ma, as a robust *Australopithecus* (Leakey and Walker, 1976).

These finds changed the basic model of hominin evolution. No longer was there a single species evolving in well-recognized anatomical stages toward modern humans. Instead, at any one time, there were several contenders, like branches on a bush, and the task of the paleoanthropologist was to decide, after a review of all the evidence, which one was indeed the ancestor.

The problem of deciding the path of human ancestry is illustrated by stone tools. For example, which of the two hominin genera at Koobi Fora (see above) had made the stone tools that were excavated from the same geological deposits? Furthermore, the observations of tool making and usage by chimpanzees at Gombe in the 1960s has now been extended to New World monkeys who use stones to dig for roots in Brazil. Such widespread technological competence has demanded a rewrite of the earlier certainties about technology and the evolution of the human brain. This is an ongoing debate currently concerned with the competence of the small-brained hominin *Homo floresiensis* from Indonesia to make the stone tools that have been excavated from Lateglacial levels in the Liang Bua cave (Morwood *et al.* 2004).

In order to inject a more theoretical perspective into paleoanthropology, Robert Foley (2002, figure 3) has recently proposed seven adaptive radiations in hominid evolution (**Table 3**; **Fig. 3**). This model not only responds to the empirical evidence of many hominin species but also establishes a behavioral basis to the pattern of change that is evident in both the archeology and anatomy.

Adaptive radiations, according to Foley, are based on the process of diversification that stems from dispersal, and the adaptive basis of these dispersals, rather than on the process of speciation. In his framework, dispersal is the mark of evolutionary success. It is an indication that speciation has occurred due to a combination of three factors: allopatry (or geographical isolation), local adaptation, and genetic drift.

The six hominin adaptive radiations are each characterized by a large number of species, and we can predict that an even larger number await discovery. The route to modern humans will not therefore be traced in a simple lineal fashion from one fossil species to another. Rather it will be through major adaptive

Table 3 Seven hominid adaptive radiations. The timescale is approximate

Suggested timescale	*Adaptive radiation*	*Adaptive pattern*	*Grade*	*Representative taxa*
Last 100 ka	7	Aquatic resources	Anatomically modern humans	*H. sapiens*
500 ka	6	Projectiles ?Fire	Larger-brained *Homo*	*H. neanderthalensis H. heidelbergensis*
1 Ma	5	Carnivory	*Homo*	*H. ergaster*
2 Ma	4		Earliest *Homo*	*H. rudolfensis*
3 Ma	3	Megadonty	*Paranthropus*	*P. boisei*
4 Ma	2	Bipedalism	*Australopithecus*	*A. afarensis*
5 Ma	1	African apes		

After Foley, R. A. (2002) Adaptive radiations and dispersals in hominin evolutionary ecology. *Evolutionary Anthropology* **11**, 32–37.

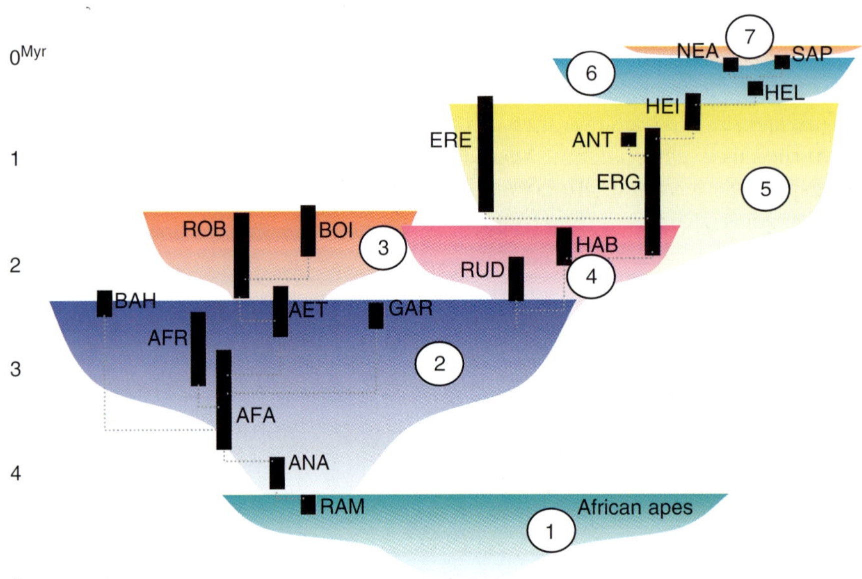

Figure 3 The adaptive radiations of the hominins (after Foley (2002), figure 3). The radiations are described in **Table 3** Key – *Ardipithecus*: RAM = *ramidus*. *Australopithecus*: ANA = *anamensis*; AFA = *afarensis*; AFR = *africanus*; BAH = *bahrelghazali*; AET = *aethiopicus*; GAR = *garhi*. *Paranthropus*: ROB = *robustus*; BOI = *boisei*. *Homo*: RUD = *rudolfensis*; HAB = *habilis*; ERG = *ergaster*, ERE = *erectus*; ANT = *antecessor*, HEI = *heidelbergensis*; HEL = *helmei*; NEA = *neanderthalensis*; SAP = *sapiens*.

changes that involved a suite of regional hominin forms that shared the major adaptive response of that radiation, such as carnivory or a composite weapon technology. The task of the paleoanthropologist, as reviewed by Potts (1998), is therefore to examine the evidence for significant territorial expansion and longer, sustained occupancy in habitats that experienced variable climatic selection.

One example is the turnover pulse model (Vrba, 1996) that draws an explicit link between the timing of speciation among a range of African mammals, including hominins in the Plio-Pleistocene, and the possible forcing effect of climate change (deMenocal, 2003). In particular, attention has been drawn to an explosion of new species between 2 and 2.4 Ma that is interpreted by Vrba as the impact of climate change on low-latitude forests and the opening up of new savannah habitats. This change would have favored a radiation of bovid species as wet and dry conditions alternated more rapidly. In Vrba's view, such turnovers are associated with ecological generalists who can adjust their feeding strategies to changing conditions in the environment. Once established, the possibility emerges for a rapid, pulse-like radiation into new niches, so producing a wealth of new taxa. Therefore, as soon as the environment has become 'saturated,' specialist adaptations with a different set of ecological strategies emerge (Gamble, 1993, pp. 87–89).

The argument is persuasive but a statistical examination (Foley, 1994) has shown that species appearance is not causally linked to climatic change as indicated in the marine cores. However, the same statistical study does show that species extinction is more likely to be climatically controlled (Foley (1994), p. 285). Foley's observation is important in establishing the nature of the link between climate change and hominin evolution on the large scale. The implication is that habitats where specialist feeders might normally be found are lost more quickly than they are gained. And if climate change during the Quaternary controls extinction, then we can see that a major biogeographical pattern in hominin evolution has been the repeated contraction and expansion of their geographical ranges. As Dennell (2003, p. 421) has pointed out, the question of 'how successful' these dispersals might have been has received far less attention that the 'when' and 'how often.' It is expected that within radiations 4–6 (**Table 3**), these contractions were frequent and short lived, especially in the northern areas of the Old World where climate exerted a more controlling effect through the seasonal distribution of food resources.

From Dispersals to Diasporas and the Legacy of the Pleistocene

The first six adaptive radiations (**Table 3**) all took place within the confines of the Old World (**Table 4**). Moreover, radiations 2–6, characterized by bipedalism, megadonty (large teeth), carnivory, and projectiles only achieved a restricted distribution within the Old World. For example, Siberia was not settled until the end of the Upper Pleistocene. This pattern helps to explain the late colonization of the Americas. Furthermore, even though stepping-stone islands existed to facilitate dispersal to a large, low-latitude island such as Madagascar, this did not occur during the Pleistocene. It was only adaptive radiation 7, marked by the use of aquatic resources, that saw a global distribution of hominins – a process, as the Australian and American evidence shows, that started no earlier than 60 kyr BP and probably more recently in all the three continents.

The legacy of the Quaternary viewed from a hominin perspective is therefore a single species globally distributed (**Table 4**). The process by which the Old World and subsequently the globe came to be occupied by hominins and humans is now described as a

Table 4 Aspects of the two adaptive radiations of *Homo*

Adaptive radiations	*1–6*	*7*
Global habitats occupied	25%	100%
Duration	2–2.5 Ma	300–100 ka
Biological diversity	Multiple species, genera	Geographical populations, races
Technological skills	Generic and transferable	Place specific
Social networks	Local	Continental
Release from social proximity	Limited	Achieved
Climate change response	Retreat to refuge	Track from refuge
Representative taxa in *Homo*	*erectus/ergaster/heidelbergensis/ helmei/neanderthalensis*	*sapiens*

After Gamble, C. S., Davies, W., Pettitt, P., and Richards, M. (2004). Climate change and evolving human diversity in Europe during the last glacial. *Philosophical Transactions of the Royal Society Biological Sciences* **359**, 243–254.

series of first dispersals (Gamble, 1993) and then in adaptive radiation 7, diasporas (LL Cavalli-Sforza and F Cavalli-Sforza, 1995). The length of time it took to colonize the Earth and so realize the Quaternary legacy has been put down to cognitive limitations, a lack of transport technology (especially boats), and environmental barriers (particularly oceans) (Gamble (1993), chapters 2 and 3).

Three examples question these views.

1. The earlier dispersals of 2 Ma BP are characterized as multiple and frequent. Africa is regarded as the source and Asia/Europe as the recipient of population. However, this view has been questioned for the earliest dispersals. Robin Dennell (Dennell and Roebroeks, 2005) has pointed to evidence from Asia which suggests that one of paleoanthropology's most basic paradigms – dispersal by early *Homo* Out of Africa – needs rethinking. The Sahara was not the barrier to movement as often presented so that an earlier hominid dispersal could have resulted in the origin of hominins, and in particular the genus *Homo*, outside Africa with subsequent dispersal back 'into' that continent. This scenario is by no means proven but serves as an alternative hypothesis to the Out of Africa 1 model for the origin of the genus *Homo*. Consequently, Dennell and Roebroeks (2005, p. 1101) argue that what paleoanthropologists need are spatial units other than modern political boundaries in order to investigate such dispersals and their internal movements. They propose 'Savannahstan' as one such unit to model adaptations and dispersals since it links the Pliocene grasslands from west Africa to northern China.

2. The multiple dispersals of hominins is therefore a parallel to the multiple species that are now expected, apart from the present day, at any one time in the fossil record. The implications of such a pattern have been discussed by Marta Lahr and Robert Foley (Foley and Lahr, 1997) in the context of the origins of modern humans. Rather than a single diaspora of *H. sapiens*, the Out of Africa 2 model, they argue for repeated dispersals after 300 ka BP and that these included the much later diasporas into Australia and the Americas. The multiple dispersal model combines genetic, anatomical, and archeological data.

3. The diasporas of the last 60 kyr can be studied in greater detail as a result of enhanced preservation, greater chronological precision, and the use of molecular genetics to produce phylogeographies. The articles in this Encyclopedia on Australia and the Americas indicate the fine-grained control that is possible while a combination of genetic, archeological, and radiocarbon evidence is allowing cross-checking on issues of timing and direction in the recolonization of formerly glaciated and periglacial areas of northern Europe (Gamble *et al.*, 2004; Straus *et al.*, 1996). Comparison of the radiocarbon dating of these Lateglacial diaspora and settlement patterns with the GRIP ice-core chronology of Greenland (**Fig. 4**) reveals that another of paleoanthropology's basic paradigms – that climate warming encourages recolonization – is not supported, since populations left the southern glacial refuges in GS-2a, a time of extreme cold. Following a rise in population numbers during GI-1 (Bocquet-Appel and Demars, 2000), the return to cold conditions in GS-1, also known as the Younger Dryas, did lead to a reduction in settlement numbers confirming, as we saw above, that Quaternary climates controlled extinction (population loss) rather than speciation (population gain) (Gamble *et al.*, 2004).

Social Rather than Environmental Change

The move from dispersals to diaspora is a further example of the continuing debate among paleoanthropologists of where to draw the hominin to human line. Cognitive, technological, and environmental reasons do not seem sufficient to account for the limited Old World distribution of all hominins prior to the late diasporas of modern humans. The novelty of these later diasporas, as opposed to the earlier dispersals, was the release from social proximity which they achieved. This release characterizes a human as compared to a primate community (Rodseth *et al.*, 1991). In other words, human social life was extended to include participants who were not always living together and who might be physically absent from their immediate social networks for lengthy periods (Gamble, 1999). This pattern of social extension is a hallmark of humans that does not appear to have been shared in such a dramatic way with other hominins. The evidence that their social lives were local (**Table 4**) comes from the limited distances that raw materials were transported (Féblot-Augustins, 1997). These were usually less than 5 km and rarely more than 80 km from source (Gamble (1999), Table 3.4). By comparison, diasporic humans exchanged shells and stone sometimes over distances of 1,000 km and regularly over 100 km. This extension in the transfer and exchange of raw materials is supported by the ocean crossing to Australia and the rapid peopling of its hyperarid Pleistocene interior (Smith, 2005), that involved very low population densities and physical separation. But such colonization did not lead to a breakdown in social life as must have been the case when earlier hominins faced such socially challenging environments. With the final,

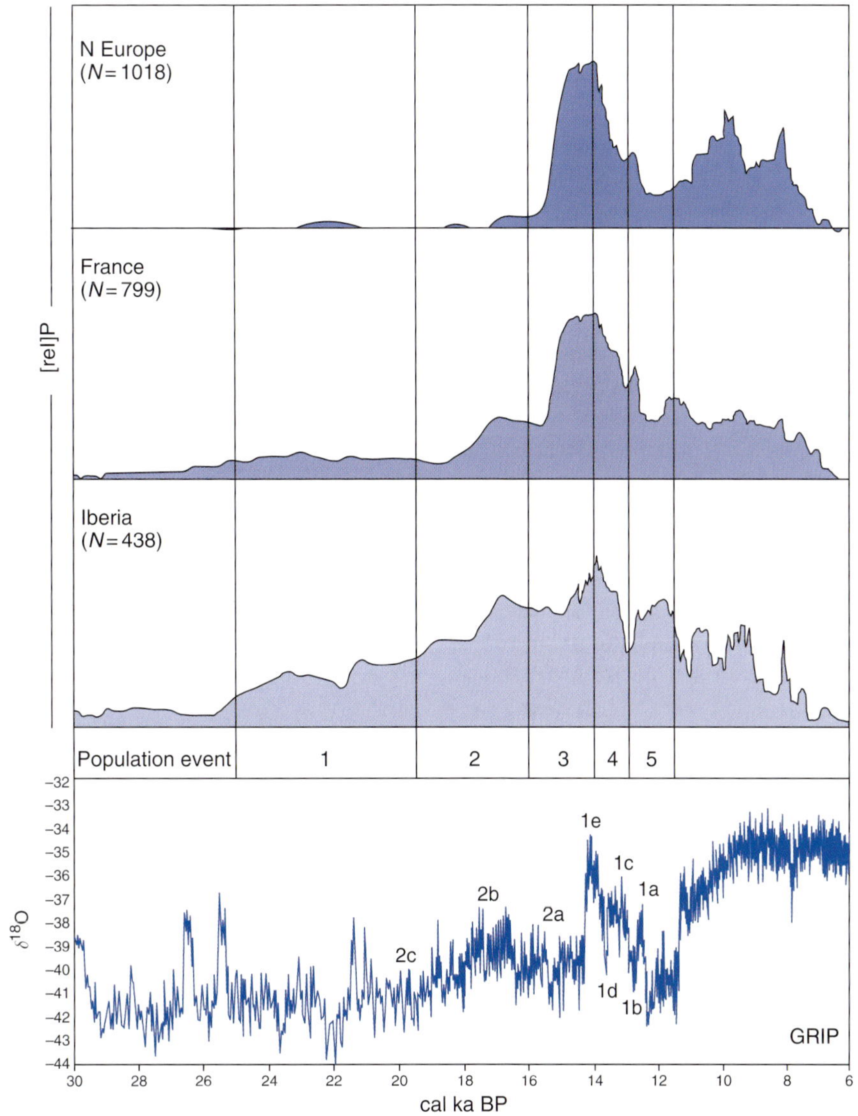

Figure 4 The pattern of recolonization of northern Europe from a southern refuge during the Last Glacial Maximum. The frequeny curves show the distribution of radiocarbon dates and their increase points to the timing of population dispersal (after Gamble *et al.* (2004), figure 2). The radiocarbon determinations are calibrated using CALPAL (Weninger and Jöris, 2000) and presented by three major regions. Five population events associated with the dispersal are indicated. The distribution of radiocarbon dates points to the major dispersal, population event 3, preceding the warming of the Greenland interstadial. N = number of calibrated radiocarbon determinations in the frequency curve. Stratotype boundaries in GRIP ice-core years are as follows: GI-2 = 21.8–21.2; GS-2 = 21.2–14.7; GI-1 = 14.7–12.7; GS-1 = 12.7–11.5 kyr BP.

aquatic (7) radiation, the emphasis now shifted to the construction of society over longer timescales and distances. This release from proximity was achieved by connecting people and places in an extended social network.

Technological Modes and the Evolution of a Social Brain

Stone tools have been the main source for studying the significance of encephalization in hominin evolution (Schick and Toth, 1993; Wynn, 1993). Changes in Paleolithic traditions (**Table 5**) were once seen as a demonstration of progressive evolution. Each stage in the Lower–Middle–Upper (Zeuner, 1958) sequences was linked to a fossil hominin. However, as we have seen above, such simple correlations (**Table 2**) are no longer possible and the criteria which define such classification as the African Earlier–Middle–Later Stone Age are not as hard and fast as once thought. For example, the advent of radiometric dating and the investigation of long stratified sequences, as at Klasies River Mouth, South Africa (Klein, 1999), have shown that stone blades, once considered a

Table 5 Technological modes in the Paleolithic

	Dominant lithic technologies	*Conventional divisions in Europe*	*Conventional divisions in Africa*
Mode 5	Microlithic components of composite artifacts	Mesolithic	Later Stone Age
Mode 4	Punch-struck blades with steep retouch	Upper Paleolithic	
Mode 3	Flake tools from prepared cores	Middle Paleolithic	Middle Stone Age
Mode 2	Bifacially flaked handaxes	Lower Paleolithic	Earlier Stone Age
Mode 1	Chopper tools and flakes		

Clark offered the modes as a homotaxial sequence which at least in Africa and, at the time he was writing, in Eurasia followed the progression from mode 1 to mode 5. However, although he was writing a world prehistory, Clark never claimed his modes were universal stages (1969, p. 30). Instead, they were his attempt to escape from the historical thorn bushes of what to call prehistoric hunters and gatherers in different continents. For example, they certainly were neither Paleolithic nor Stone Age in Australia or the Americas. Reproduced from Clark JGD (1969) *World Prehistory a New Outline*, 2nd edn., p. 31. Cambridge: Cambridge University Press.

diagnostic type fossil of the European Upper Paleolithic and the African Later Stone Age, have a much greater ancestry. Blades are any intentionally struck stone blank that is at least twice as long as it wide. With such a definition, it is not surprising that blades frequently appear as elements in Middle Paleolithic and Middle Stone Age assemblages and the age of their appearance has been extended from some 40 kyr BP to over 300 kyr BP (Bar-Yosef and Kuhn, 1999).

An alternative scheme, proposed by Clark (1969), recognized five technological modes (**Table 5**). His approach has been widely followed as studies of Quaternary hominins and their technology became a truly worldwide endeavour and where the European Paleolithic and African Stone Age terms appeared anachronistic. However, one problem with Clark's modes is that they unduly privilege stone tools, and all other aspects of technology and material culture are forgotten. These would include wooden spears, hearths, huts, art, and items of display, all of which have a varied and complex history.

Clark's five technological modes do however highlight two aspects of hominin technology during the Quaternary. First, it was very static, and second, it did not follow the ethnographic pattern found among today's hunters and gatherers, where the local environment exerts a selection pressure on the shape, form, and complexity of technology. For example, mode 2, typified by large bifaces such as Acheulean handaxes, is first found in east Africa at 1.5 Myr BP and persists for more than a million years (Klein, 1999). Moreover, these bifaces are found from the equator to the northern latitudes of Britain and in environments with very different plant and animal resources, shorter growing seasons, and a range of adaptive challenges. As a result, the technological skills of the early hominins can be characterized as generic (**Table 4**). They could be transferred to environments as different as the tropical savannahs and the northern grasslands. A very different picture is presented by mode 4 technologies that, after the 60 ka diaspora, compare well in their diversity and complexity with those of contemporary hunters and gatherers and where the skills are specific to local conditions and places. However, the picture is further complicated by the evidence from Australia where mode 3 continued to the present day. Technology does not therefore correlate with particular hominin fossils and their assumed levels of intelligence and problem solving. In addition, the role of environmental selection on innovation and adoption of new techniques is restricted.

Technology, especially the appearance of art and ornament, has been used as a guide to major cognitive and behavioral changes, including language and symbolic thought (Mithen, 1996). However, an examination of hominin encephalization (**Fig. 5**) by Aiello and Dunbar (1993) revealed that the most significant increase in brain size occurred half a million years ago by which time hominin brains were more than twice the size, proportionately, than any of the earlier Australopithecines. Using a comparative sample of living primates, they also showed how early hominin brain sizes implied much larger social groupings. They calculated that group size had risen at this time to at least 120, whereas the largest among primates is 60. Dunbar (2003) has since expanded this observation into a social brain hypothesis: our social lives drove encephalization. In explanation, bigger brains have proportionately larger amounts of neocortex, the thinking part of the brain, that also manages relationships and interactions between group members. By 500 kyr BP, hominin group size was now so much larger than those of primates that the traditional means of maintaining and organizing groups through physical grooming was no longer an option. Dunbar's suggestion is that language appeared at that time to facilitate interaction and organize larger social groupings. However, such an age is well before the flourescence of art and ornament that appears in Europe some 40 ka. The social brain hypothesis focuses attention on how exactly technology, and material culture more generally,

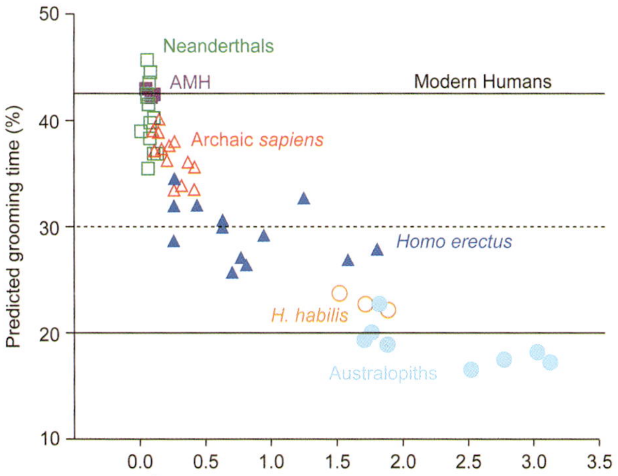

Figure 5 The implications of increases in brain size during 3.5 Ma of hominin evolution (after Dunbar (2003)). At 500 ka, there is a marked encephalization. According to the social brain hypothesis, this resulted in an increase in group sizes and the amount of time that would have been needed daily to undertake physical grooming to maintain social ties. The 20% (lower solid line) of the day spent in social activities such as grooming represents a significant limit for today's great apes since they have to spend much of the day traveling and looking for food. This threshold would have been similar for the small-brained Australopithecines. The 30% of daily activity spent on grooming (dotted line) represents an upper limit beyond which other mechanisms of social integration such as language are needed. The primate-style social model was therefore exceeded at least 500 ka. AMH = anatomically modern human.

contributed to those tipping points in hominin evolution such as language.

The Human Revolution: Gradual or Abrupt?

My final theme, the human revolution, became one of the central concerns of paleoanthropologists in the 1980s (Mellars and Stringer, 1989). This was not the first time that the idea of a human revolution had been applied to the fossil and archeological record. What marked this application out was the claim for recency. As Paul Mellars has shown (Mellars, 2005), the conjunction in Europe of innovation in technology and material culture coincides with the replacement of the native Neanderthals by incoming modern humans *H. sapiens* (Stringer and Gamble, 1993). The model of replacement was initially strongly contested by an ancient origin for modern humans, often described as multiregional evolution (Wolpoff, 1989). Its supporters argue that through a process of convergent evolution the founding populations of Out of Africa 1 in the late Pliocene evolved into the distinctive geographical populations, or races, that are encountered today. The model of a recent African origin for *H. sapiens* is also known as Out of Africa 2. The evidence now available from dating both the anatomical (McDougall *et al.*, 2005) and genetic changes (Cann *et al.*, 1987; Krings, 1997) indicates conclusively that modern humans originated in Africa.

The current interest in the human revolution model centers on its timing. The Upper Paleolithic revolution of Europe has been questioned by paleoanthropologists working in Africa (Wadley, 2001) and characterized as 'the revolution that wasn't' (McBrearty and Brooks, 2000). Excavations at Blombos Cave, South Africa, provide evidence that shell ornaments, engraved ochres, bone tools, and culturally specific projectile points were all present in excess of 80 kyr BP (Henshilwood and Marean, 2003). Confirmation comes from Israel, based on the deliberate burial of modern people with items that can be interpreted as intentional grave goods more than 100 kyr BP (Grün *et al.*, 2005).

The issue that the human revolution therefore addresses is explaining why such a long delay occurred between the appearance of anatomically modern people in Africa and the suite of cultural changes that led eventually to the Australian and American diasporas. The African record is indeed more complex than once thought and of much greater antiquity. The claims (d'Errico *et al.*, 1998) are also not convincing that Europe's Neanderthals produced their own independent cultural revolution that just happened to coincide with the arrival of modern humans from Africa (Mellars, 2005). As a result, attention has focused on the effect of major environmental events such as the eruption at 71 kyr BP of Toba on Sumatra (Ambrose, 1998) and its impact through a prolonged volcanic winter on human populations that favored a subsequent African diaspora.

Revolution, with its accent on abrupt, dramatic change, is not an appropriate term at Quaternary timescales. Neither is the alternative tenable, namely, a gradual addition of innovation in culture and society. Much of the argument revolves around the identification of what constitutes art, symbolic behavior, language, and hence the modern mind (Mithen, 1996) among archeological data. One person's evidence for art is another's scratched pebble (Marshack, 1972). These are difficult issues not helped by small samples, widely distributed through the Old World.

However, what is apparent is that the major African diaspora that led to modern humans peopling Australia, Europe, and later the Americas depended upon the social ability to extend society across time as well as space. This occurred with, as was the case in Europe, or without, as in America, either much art and ornament or mode 4 technologies, as the Australian record shows. Nobody disputes that these diasporas involved modern humans as defined genetically, anatomically, and socially. This cultural diversity suggests that paleoanthro-pology can no longer rely on universal processes to construct an understanding of the timing and sequence of changes among hominins. While the evidence from the paleoecological archives is vital to such discussions, the trend toward the detailed history of social and cultural development in selected regions is returning to favor (Straus and Bar-Yosef, 2001).

Future Prospects

These historical accounts will be assisted by a number of recent breakthroughs in Quaternary science, promising an equally productive and exciting next 50 years. Among many, three with the potential to transform the subject can be singled out:

1. adding to the continued precision of dating through such techniques as microtephras permitting independent, long-distance correlation that bring the study of contemporaneity down to below a single human generation within any single MIS;
2. tracing the phylogenies of fossils by analyzing their ancient DNA; and
3. using the direct, measures contained in stable isotopes to characterize hominin diet, and as a proxy for the provenance, and dispersal of hominins.

All of these, and many more techniques, will allow paleoanthropologists to use archeological evidence in new ways. They will be able to analyze the material culture of our earliest ancestors as historical information in the unfolding story of human evolution during the Quaternary, and provide an examination of how deep-seated, or superficial, our current differences might be.

See also: **Archaeological Records**: 2.7 MYR-300,000 years ago in Africa; 2.7 MYR-300,000 years ago in Asia; 1.9 MYR-300,000 years ago in Europe; Global Expansion 300,000-8000 years ago, Africa; Global Expansion 300,000-8000 years ago, Asia; Global Expansion 300,000-8000 years ago, Australia; Global Expansion 300,000-8000 years ago, Americas; Neanderthal Demise; Postglacial Adaptations. **Vertebrate Studies**: Interactions with Hominids.

References

Aiello, L., and Dunbar, R. (1993). Neocortex size, group size and the evolution of language. *Current Anthropology* **34**, 184–193.

Ambrose, S. H. (1998). Late Pleistocene human population bottlenecks, volcanic winter, and differentiation of modern humans. *Journal of Human Evolution* **34**, 623–651.

Bar-Yosef, O., and Kuhn, S. (1999). The big deal about blades: Laminar technologies and human evolution. *American Anthropologist* **101**, 322–338.

Behrensmeyer, A. K., and Hill, A. P. (1980). *Fossils in the Making.* University of Chicago Press, Chicago.

Binford, L. R. (1981). *Bones: Ancient Men and Modern Myths.* Academic Press, New York.

Bocquet-Appel, J.-P., and Demars, P.-Y. (2000). Population kinetics in the Upper Palaeolithic in western Europe. *Journal of Archaeological Science* **27**, 551–570.

Brain, C. K. (1981). *The Hunters or the Hunted?* Chicago University Press, Chicago.

Bunn, H. T. (1981). Archaeological evidence for meat-eating by Plio-Pleistocene hominids from Koobi Fora and Olduvai Gorge. *Nature* **291**, 574–577.

Butzer, K. W. (1982). *Archaeology as Human Ecology* Cambridge University Press, Cambridge.

Cann, R., Stoneking, M., and Wilson, A. (1987). Mitochondrial DNA and human evolution. *Nature* **325**, 31–36.

Cavalli-Sforza, L. L., and Cavalli-Sforza, F. (1995). *The Great Human Diasporas: The History of Diversity and Evolution.* Perseus Books, Reading, MA.

Clark, J. G. D. (1969). . Cambridge University Press, Cambridge.

d'Errico, F., Zilhao, J., Julien, M., Baffier, D., and Pelegrin, J. (1998). Neanderthal acculturation in Western Europe?: A critical review of the evidence and its interpretation. *Current Anthropology* **39 (supplement 1)**, S1–S44.

deMenocal, P. B. (2003). African climate and faunal evolution during the Pliocene–Pleistocene. *Earth and Planetary Science Letters* **220**, 3–4.

Dennell, R. (2003). Dispersal and colonisation, long and short chronologies: How continuous is the Early Pleistocene record for hominids outside East Africa. *Journal of human evolution* **45**, 421–440.

Dennell, R., and Roebroeks, W. (2005). An Asian perspective on early human dispersal from Africa. *Nature* **438**, 1099–1104.

Dominguez-Rodrigo, M. (2002). Hunting and scavenging by early humans: The state of the debate. *Journal of World Prehistory* **16**, 1–54.

Dunbar, R. I. M. (2003). The social brain: Mind, language, and society in evolutionary perspective. *Annual Review of Anthropology* **32**, 163–181.

Féblot-Augustins, J. (1997). *La circulation des Matières Premières au Paléolithique*, vol. 75. ERAUL, Liège.

Foley, R., and Lahr, M. M. (1997). Mode 3 technologies and the evolution of modern humans. *Cambridge Archaeological Journal* **7**, 3–36.

Foley, R. A. (1994). Speciation, extinction and climatic change in hominid evolution. *Journal of Human Evolution* **26**, 275–289.

Foley, R. A. (2001). In the shadow of the modern synthesis? Alternative perspectives on the last fifty years of palaeoanthropology. *Evolutionary Anthropology* **10**, 5–15.

Foley, R. A. (2002). Adaptive radiations and dispersals in hominin evolutionary ecology. *Evolutionary Anthropology* **11**, 32–37.

Gamble, C. S. (1993). *Timewalkers: The Prehistory of Global Colonization.* Harvard University Press, Cambridge, MA.

Gamble, C. S. (1999). *The Palaeolithic societies of Europe.* Cambridge University Press, Cambridge.

Gamble, C. S., Davies, W., Pettitt, P., and Richards, M. (2004). Climate change and evolving human diversity in Europe during the last glacial. *Philosophical Transactions of the Royal Society Biological Sciences* **359**, 243–254.

Grün, R., Stringer, C., McDermott, F., *et al.* (2005). U-series and ESR analyses of bones and teeth relating to the human burials from Skhul. *Journal of Human Evolution* **49**, 316–334.

Henshilwood, C. S., and Marean, C. W. (2003). The origin of modern human behaviour: Critique of the models and their test implications. *Current Anthropology* **44**, 627–651.

Howell, F. C. (1965). *Early man.* Time Life Books, London.

Howell, F. C. (1966). Observations of the earlier phases of the European Lower Palaeolithic. *American Anthropologist* **68,** 88–201.

Isaac, G. (1978). The food sharing behaviour of proto-human hominids. *Scientific American* **238,** 90–108.

Klein, R. G. (1999). *The Human Career: Human Biological and Cultural Origins*, 2nd edn. University of Chicago Press, Chicago.

Krings, M. (1997). Neanderthal DNA sequences and the origins of modern humans. *Cell* **90,** 19–30.

Leakey, L. S. B., Evernden, J. F., and Curtis, G. H. (1961). Age of Bed I, Olduvai Gorge, Tanganyika. *Nature* **191,** 478–479.

Leakey, L. S. B., Tobias, P. V., and Napier, J. R. (1964). A new species of the genus Homo from Olduvai Gorge. *Nature* **202,** 308–312.

Leakey, M. D. (1971). *Olduvai Gorge: Excavations in Beds I and II 1960–1963.* Cambridge University Press, Cambridge.

Leakey, R. E. F., and Walker, A. (1976). *Australopithecus, Homo erectus* and the single species hypothesis. *Nature* **261,** 572–574.

Marshack, A. (1972). *The Roots of Civilization.* McGraw-Hill, New York.

McBrearty, S., and Brooks, A. S. (2000). The revolution that wasn't: A new interpretation of the origin of modern humans. *Journal of Human Evolution* **39,** 453–563.

McDougall, I., Brown, F. H., and Fleagle, J. G. (2005). Stratigraphic placement and age of modern humans from Kibish, Ethiopia. *Nature* **433,** 733–736.

Mellars, P. (2005). The impossible coincidence: A single-species model for the origins of modern human behaviour in Europe. *Evolutionary Anthropology* **14,** 12–27.

Mellars, P. A., and Stringer, C. (Eds.) (1989). *The Human Revolution: Behavioural and Biological Perspectives on the Origins of Modern Humans* Edinburgh University Press, Edinburgh.

Mithen, S. (1996). *The Prehistory of the Mind.* Thames and Hudson, London.

Morwood, M. J., Soejono, R. P., Roberts, R. G., *et al.* (2004). Archaeology and age of a new hominin from Flores in eastern Indonesia. *Nature* **431,** 1087–1091.

Potts, R. (1998). Environmental hypotheses of hominin evolution. *Yearbook of Phyisical Anthropology* **41,** 93–136.

Rodseth, L., Wrangham, R. W., Harrigan, A., and Smuts, B. B. (1991). The human community as a primate society. *Current Anthropology* **32,** 221–254.

Santonja, M. (Ed.) (2005). *Esperando el diluvio: Ambrona y Torralba hace 400,000 anos.* Museo Arqueologico Regional, Madrid.

Schick, K. D., and Toth, N. (1993). *Making Silent Stones Speak: Human Evolution and the Dawn of Technology.* Simon and Schuster, New York.

Smith, M. A. (2005). Moving into the southern deserts: An archaeology of dispersal and colonisation. In *23°S: Archaeology and Environmental History of the Southern Deserts* (M. A. Smith and P. Hesse, Eds.) pp. 92–107. National Museum of Australia Press, Canberra.

Stiner, M. C. (1994). *Honor Among Thieves: A Zooarchaeological Study of Neandertal Ecology.* Princeton University Press, New Jersey.

Straus, L. G., and Bar-Yosef, O. (Eds.) (2001). *Out of Africa in the Pleistocene. Quaternary International* 75.

Straus, L. G., Eriksen, B. V., Erlandson, J. M. and Yesner, D. R. (Eds.) (1996). *Humans at the End of the Ice Age: The Archaeology of the Pleistocene–Holocene Transition* Plenum, New York.

Stringer, C., and Andrews, P. (2005). *The Complete World of Human Evolution.* Thames and Hudson, London.

Stringer, C., and Gamble, C. (1993). *In Search of the Neanderthals: Solving the Puzzle of Human Origins.* Thames and Hudson, London.

Villa, P., Soto, E., Santonja, M., *et al.* (2005). New data from Ambrona: Closing the hunting debate. *Quaternary International* **126–128,** 223–250.

Vrba, E. S. (1996). Climate, heterochrony, and human evolution. *Journal of Anthropological Research* **52,** 1–28.

Wadley, L. (2001). What is cultural modernity? A general view and a South African perspective from Rose Cottage Cave. *Cambridge Archaeological Journal* **11,** 201–221.

Weninger B and Jöris O (2000). "CALPAL", 1.10.00 edition. Köln: Radiocarbon Laboratory Institut Fur Urgeschichte Und Fruhgeschichte Weyertal 125 D-50923, http://www.calpal.de.

Wolpoff, M. A. (1989). Multiregional evolution: The fossil alternative to Eden. In *The Human Revolution: Behavioural and Biological Perspectves on the Origins of Modern Humans* (P. Mellars and C. Stringer, Eds.) pp. 62–109. Edinburgh University Press, Edinburgh.

Wynn, T. (1993). Two developments in the mind of early. *Homo. Journal of Anthropological Archaeology* **12,** 299–322.

Zeuner, F. E. (1958). *Dating the Past: An Introduction to Geochronology,* 4th edn. Methuen, London.

Zeuner, F. E. (1959). *The Pleistocene Period: Its Climate, Chronology and Faunal Successions.* Hutchinson, London.

2.7 MYR-300,000 years ago in Africa

J W K Harris, D R Braun and M Pante, Rutgers University, NJ, USA

© 2007 Elsevier B.V. All rights reserved.

Introduction

The continent of Africa preserves the longest record of archeological manifestations of hominin behavior beginning with the discoveries in the Kada Gona region of Ethiopia (Semaw *et al.*, 1997). These artifacts, dating to 2.5 Ma, are Oldowan core and flake industries with associated hominin-modified bone. This overview focuses entirely on Africa during the long formative period of human prehistory known as the Early Stone Age (ESA, Oldowan and Acheulean Industries) spanning from about 2.5 Ma to 200–300 ka. All of the assemblages discussed in this section predate technological innovations such as point production, hafting, blade production, and core-reduction methods such as the Levallois technique. These later techniques characterize the Middle and Later Stone Age and will not be part of this review.

The ESA of Africa is divisible into the Oldowan Industry (Industrial Complex) or the Acheulean Industry (Industrial Complex). The Oldowan tradition of craftsmanship is characterized by simple fist-sized cores and whole flakes produced by hand-to-hand hard hammer percussion flaking, where the cobble or nodule from which flakes are to be removed is held in one hand and a dynamic blow is delivered with a stone hammer from the opposing hand. Instances of the bipolar technique, where the cobble is placed on a hard surface or anvil and a flake or detached piece is removed by initiating a wedging fracture with a hard hammer, also exist at some sites. Fixed design core forms are usually lacking and there is minimal secondary retouch on whole flakes. More formalized tools that appear in some industries appear to be the result of raw material shapes and the properties of these stones. Assemblages attributed to the Oldowan Industry *sensu stricto* range in age from 2.5 to 1.8/1.7 Ma. The Oldowan was first described by Mary Leakey from assemblages found at Olduvai Bed I (Leakey, 1971). These industries fall into the mode I technique described by Clark (1969). The transition to the beginnings of the Acheulean Industry dates to between 1.7 and 1.5 Myr and is marked by highly variable assemblages that contain examples of larger cores, larger flakes, rare instances of crude bifaces as well as the previous suite of artifacts usually associated with the Oldowan Industrial Complex. Industries in this time range have collectively been called the Developed Oldowan (see **Fig. 1**).

While there is no total agreement on the actual definition of this term, technological studies of these assemblages have referred to these industries as early Acheulean (Ludwig and Harris, 1998).

The earliest well-documented and unequivocally Acheulean assemblages are found at Konso-Gardula (Ethiopia), FxJj 63 (Kenya), and EF-HR (Olduvai, Tanzania) dating to approximately 1.5 Ma. These assemblages are characterized by large bifacially flaked hand-axes and cleavers made on much larger flakes than is evident in the Oldowan. These flakes are struck from heavier and larger cores. Smaller but nonetheless significant retouched flake tool forms constitute a conspicuous part of the Acheulean stone tool assemblages. Through the course of the Acheulean Industry, there appears to be a general trend toward thinner cores produced through techniques of bifacial thinning. Bifacial thinning is often conducted using a soft hammer, usually a wood or bone billet. These assemblages are often associated

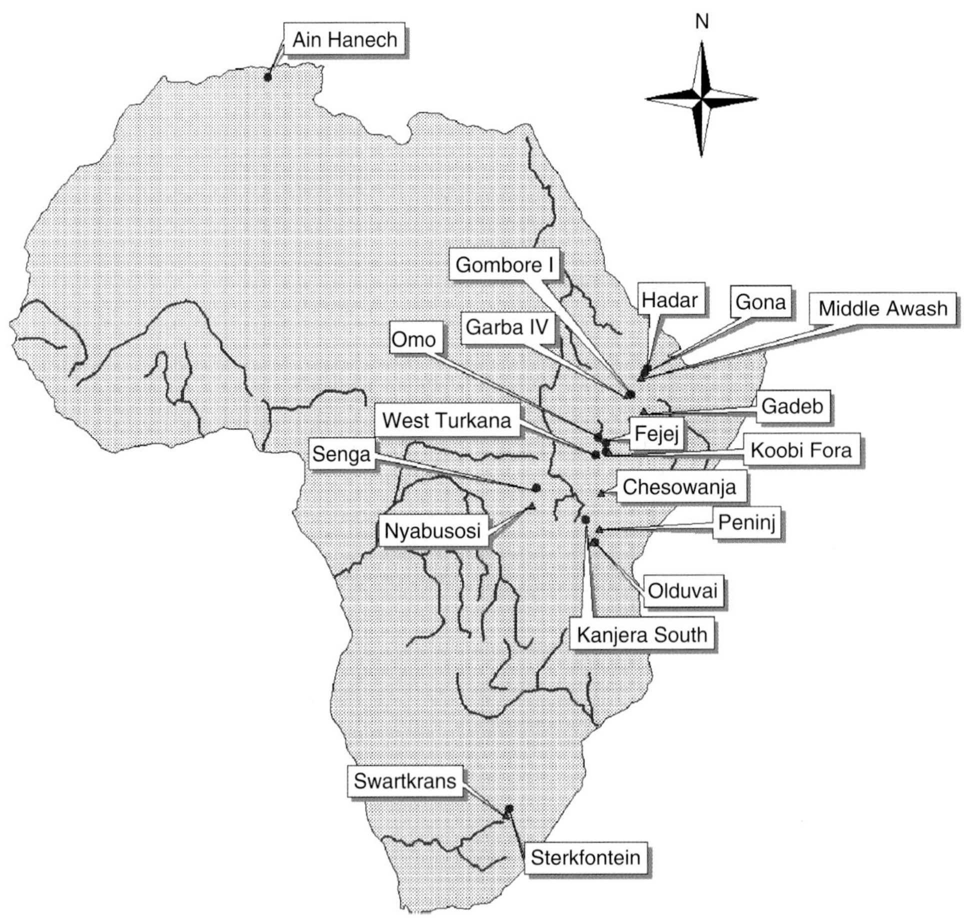

Figure 1 Site localities attributed to the Oldowan Industrial Complex on the African Continent. These include sites that have been collectively called Developed Oldowan and Oldowan.

with the onset of the Levallois core technique of core preparation. Sites from the Acheulean industry are found in a much broader geographical range than the previous Developed Oldowan sites (see **Fig. 2**). The Acheulean Industry is succeeded in the geological record by stone assemblages attributed to the Middle Stone Age (MSA) between 200 and 300 ka with the earliest MSA. dating to around 280 ka in the Kapthurin Formation, Kenya (Tryon and McBrearty, 2002). The Acheulean would fall into the mode II of the schema proposed by Clark (1969).

One of the most important events in human evolution was the dietary shift toward meat-eating that, for the first time, put our earliest stone tool-using ancestors in competition with large carnivores that remained dangerous predators. In previous ESA studies, some interpretations of hominin subsistence strategies have relied heavily upon the stratigraphic association of fragmentary animal bones and stone artifacts and overlooked the range of taphonomic processes that may have contributed to the accumulations of artifacts and bones excavated from archeological sites. Archeologists have since realized that agents other than hominins, including carnivores, disease, droughts, and water flow, may contribute to the formation of an archeological site. As a result, identifying the relative contribution of hominins to the accumulation of archeofaunas has become the focus of zooarcheological investigations, requiring a theoretically grounded body of research directed at identifying the traces left by various taphonomic agencies (Binford, 1981). Bone-surface modifications

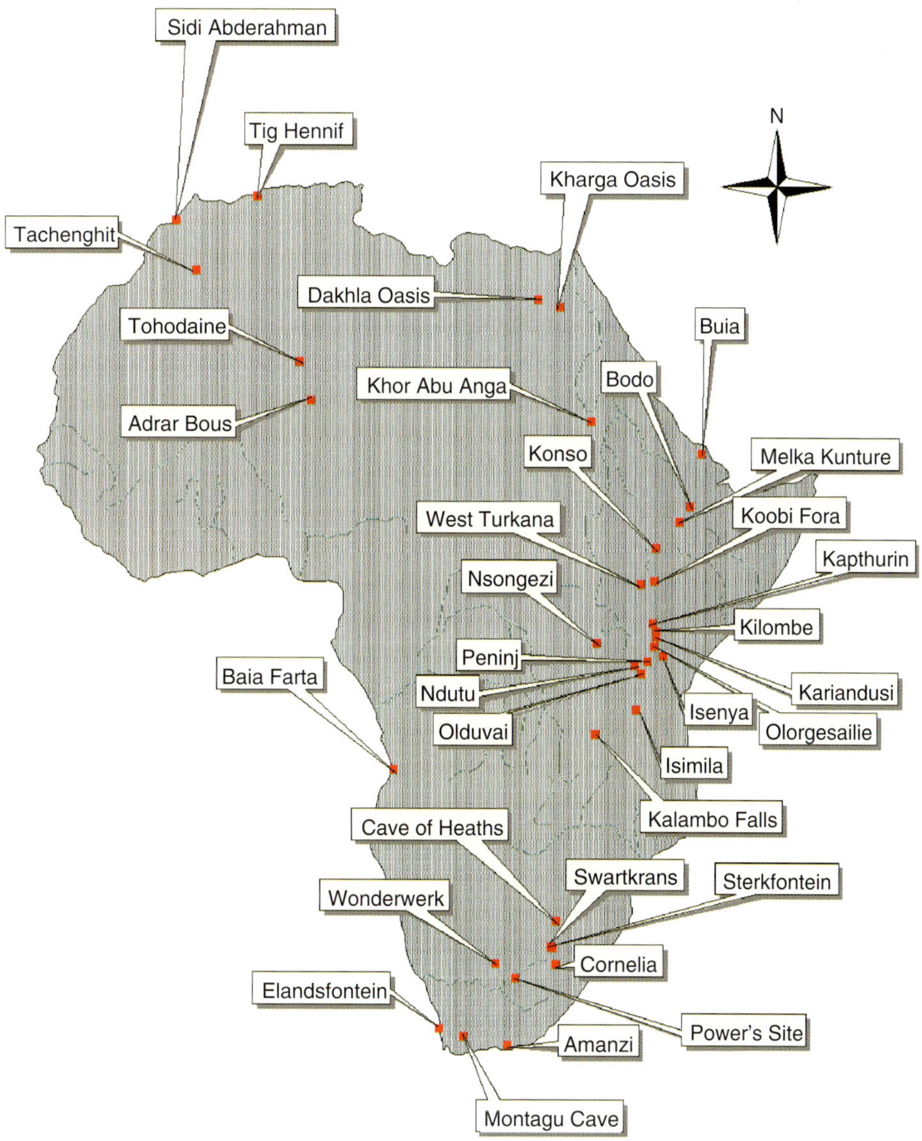

Figure 2 Site localities attributed to the Acheulan Industrial Complex on the African Continent.

including stone tool cut marks, hammerstone percussion marks, and carnivore tooth marks have proved to be the most reliable indicators of hominin and carnivore involvement in the accumulation of bone assemblages and since the first descriptions of cut marks by Bunn (1981) archeologists are no longer dependent on the association of artifacts and bones to infer hominin involvement in site formation. Therefore, two types of bone assemblages will be discussed here, those that exhibit evidence of hominin-induced bone-surface modifications and those that do not.

Oldowan

The discovery of artifacts in the Kada Gona in sediments that are securely dated to 2.5 Myr sets the base of the archeological record (Semaw *et al.*, 1997). It is very likely that the earliest tool use has not been found in the record because sites are too diffuse or scattered (Panger *et al.*, 2002). However, current archeological data support the evidence of 2.6 Myr as the onset of technological evolution. Oldowan assemblages that represent the technological system of the earliest tool makers are confined to time periods between 2.6 and 1.5 Myr (see **Fig. 3**).

The majority of these sites are confined to the eastern rift valley of Ethiopia, Kenya, and Tanzania, although sites can be found as far north as Algeria (Ain Hanech) and as far south as South Africa (Sterkfontein) (see **Fig. 1**). Further fieldwork in northeast (e.g., Eritrea) and Northwest Africa (e.g., Morocco) could begin to fill in the current gaps in the geographical range of Oldowan industries. A notable exception to the Africa-centered distribution of Oldowan sites is the remarkable discovery of Oldowan-like artifacts in the Republic of Georgia at the site of Dmanisi (Gabunia *et al.*, 2001). New

Figure 3 Oldowan core, initially classified by M.D. Leakey (1971) as a chopper in her terminology.

research in eastern and central Asia suggests hominins began encroaching on these areas in the Early Pleistocene.

The classic African Oldowan assemblages are found in sediments that are dated to between 2.6 and 1.8 Myr. The earliest sites are found in the lower Awash River Valley of Ethiopia at three major Oldowan research areas. The oldest of these three localities are found in the Gona region in drainages feeding into the modern Awash River (Semaw *et al.*, 1997). These sites were located near large cobble conglomerates, likely on the floodplains of a paleo-Awash river. These sites also preserve some of the earliest evidence of stone tool use in the form of cut marked bones at a few localities in the Dana Aouli and Ounda Gona drainage system (Dominguez-Rodrigo *et al.*, 2005). Sites in the adjacent Hadar region also preserve numerous evidence of stone-flaking activity (Kimbel *et al.*, 1996). Further south in the Middle Awash research area, evidence of stone tool use occurs only in the form of hominin-modified bone at the site of Bouri (de Heinzelin *et al.*, 1999). The lack of stone artifacts at this site has been suggested to be the product of early hominin 'curation' of stone used for the manufacture of tools. Similar patterns of artifact transport have been suggested for sites in East Turkana where hominin-modified bones are found without associated tools (Harris *et al.*, 2002).

The Turkana Basin in northern Kenya arguably contains the most productive deposits for studying the earliest archeology. The oldest sites are found in the lower Omo valley in the Shungura Formation. The assemblages are dominated by quartz raw material and are often suggested to be the product of expedient bipolar reduction of pebbles. However, re-analysis suggests that a relatively high degree of planning went into the hard hammer percussion of small pebbles in some of these early Omo assemblages (de la Torre, 2004). This interpretation of the Omo assemblages is particularly interesting given the fact that sites of a similar age in the Nachukui Formation on the West Side of Lake Turkana show evidence of markedly variable skill levels in artifact production. The Lokalalei 2c site shows evidence of well-defined rules in stone tool production and core reduction (Delagnes and Roche, 2005). The appearance of stone artifacts is presently only known from the KBS Member (1.89–1.65 Myr) in the Koobi Fora Formation on the eastern shores of Lake Turkana (Isaac, 1997). These sites are dominated by freehand percussion of basaltic materials (Isaac, 1997). However, further north on the east side of Lake Turkana at the site of Fejej (de Lumley and Beyene, 2004), bipolar reduction of small quartz pebbles is the most common form of core reduction. This dichotomy in stone tool production is considered to be caused by the differences in available raw material in the different parts of the basin and attests to the limited mobility of the earliest tool makers (Rogers *et al.*, 1994).

However, new evidence from the site of Kanjera South in western Kenya suggests the majority of the raw material used for artifact manufacture was transported to this site (Plummer, 2004). It is quite likely that the technological system found at the Kanjera sites is unique among Oldowan localities. Transport and selection of specific raw materials seems to be a very tightly constrained part of the technological organization at Kanjera (Braun *et al.*, 2005). Similarly, transport of materials at Olduvai Gorge seems to follow some specific patterns about the use and manufacture of tools. Recent landscape archeology studies in the Olduvai basin have suggested that hominins practiced selective transport over long distances but movement of raw materials over shorter distances did not seem to abide by the usual relationship between distance from stone source and attrition of a specific raw material in an assemblage.

Although there is general agreement that Oldowan hominins could consistently flake stone, there is considerable disagreement over the relative skill level of the first toolmakers. Research at the Gona sites has suggested that even at the very inception of stone tool technology, Oldowan hominins had already mastered stone fracture mechanics (Semaw *et al.*, 1997). However, the evidence from West Turkana shows marked variation in the ability of early hominins to produce flakes and reduce cores in a systematic way (Delagnes and Roche, 2005).

Faunal assemblages associated with Oldowan artifact assemblages are dispersed throughout Africa with the earliest of these occurrences in East Africa. Hominin-induced modifications have only been identified from a few of the Oldowan artifact-bone concentrations that have been excavated. However, the absence of hominin-induced bone modifications is not always an indicator of the lack of hominin involvement in the formation of a bone assemblage. In fact, it may be the result of poor bone-surface preservation that obscures bone-surface modifications due to the simple fact that few bone assemblages have been subjected to analyses concerned with the identification of bone-surface modifications. Localities that have produced Oldowan bone assemblages that have been identified as archeofaunas based on their spatial association with artifacts include Hadar, Middle Awash, Omo (Shungura Formation), and Melka Kunture in Ethiopia; Upper Semliki Valley in the Democratic Republic of Congo; Ain Hanech and El Kherba in Algeria; and Sterkfontein in South Africa (Plummer, 2004). Localities that have produced Oldowan bone assemblages that exhibit evidence of hominin-induced modifications include Gona and Bouri in Ethiopia; Kanjera South, West Turkana, and Koobi Fora in Kenya; and Olduvai Gorge in Tanzania. These sites have the potential to provide insights into the subsistence capabilities of hominins and the interactions between hominins and carnivores who were competing for animal carcasses.

Sites that preserve bone-surface modifications range in composition from a few cut-marked bones found on the surface to thousands of bones that preserve modifications induced by both hominins and carnivores. The earliest examples of cut-marked bones come from the finds at Gona and Bouri in Ethiopia and date between 2.5–2.6 Myr. These sites are important in that they represent the beginning of a dietary shift toward meat eating that coincides with the earliest evidence of stone tool manufacture. However, due to the extremely limited number of bones preserving hominin-induced modifications at these sites, it is difficult to determine the extent of hominin carnivory in this remote time period. As a result our conception of early tool-using hominins as hunters or scavengers is still open to interpretation.

The majority of attention from studies interested in interpreting early hominin food acquisition has been focused on the bone assemblages excavated from the FLK 22 site, Olduvai Gorge, Tanzania and the co-occurrence of hominin- and carnivore-induced modifications that are associated with carcass consumption. Some authors assert that the abundance of cut marks on bones from Plio–Pleistocene archeological sites indicates that hominins had primary access to flesh (Dominguez-Rodrigo, *et al.*, 2002), whereas others argue that cut marks result only from the removal of scraps of flesh after large cats had primary access to carcasses (Blumenschine, 1995). While a consensus has still not been reached on the food acquisition and consumption practices of Oldowan hominins at FLK 22, archeologists generally agree that by 1.8 Myr hominins were depending more on nutrients obtained from carcasses than their predecessors at 2.6 Myr. These findings have extensive implications for the behavior of *Homo erectus*, a species that appears in the fossil record at about this time. It is possible that a major reason for the success and geographic radiation of *H. erectus* is the incorporation of larger amounts of high-quality protein (meat) in the diet.

Transitional Industries: Developed Oldowan–Early Acheulean

Assemblages of artifacts and bones found between 1.7 and 1.5 Myr are often described as Developed Oldowan industries. Many of these sites appear to have a mixture of features associated with both mode I and mode II industries. These 'transitional' assemblages include the variability of the earliest Oldowan assemblages but are also precursors of the imposition of form on core morphology. This pattern is manifested in the repeated use of particular forms or core-reduction techniques. The clearest example of this appears in the Karari Industry on the east side of Lake Turkana (Harris and Isaac, 1976) with the introduction of larger 'scraper' forms that dominate the Oldowan assemblages after 1.6 Myr. Similar patterns can be found in sites such as Peninj, Chesowanja, and Gadeb (see **Fig. 4**).

The appearance of these 'transitional' industries that are usually grouped with the Oldowan but show technological similarities with mode II industries suggests a continuity in the Paleolithic record which may argue against previously suggested models of episodic change in 'cultural evolution' (Semaw *et al.*, 1997). These transitional industries are particularly interesting because at about 1.8 Myr (or possibly earlier), fossil remains attributed to *Homo ergaster* (or early African *H. erectus*) began to appear in sediments in East Africa. Well-preserved cranial and postcranial remains from Lake Turkana basin, most notably the Nariokotomie skeleton (KNM-ER 15000), reflects major changes in the anatomy and physiology of hominins at this time. The concurrent change in the archeological record represented by transitional (Developed Oldowan and Early Acheulean) assemblages may record a shift in behavioral complexity of this new hominin form (see **Fig. 5**).

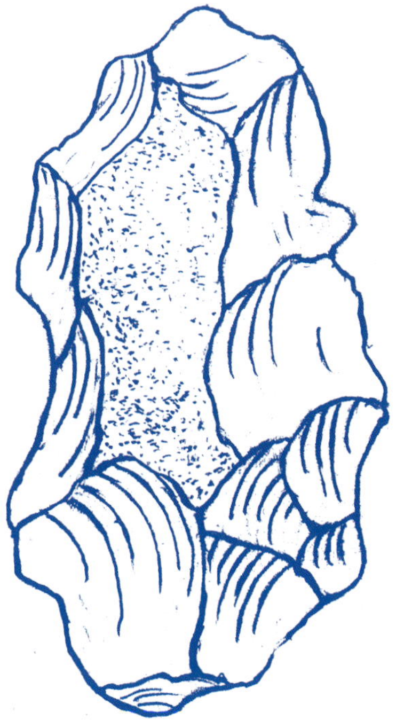

Figure 4 Developed Oldowan or Karari Industry core, initially classified as a Karari core scrapper Harris and Isaac (1976).

Behavioral traces from this time period began to be found in habitats that were not previously exploited by early hominins, such as highland plateaus (Gadeb, Melka Kunture) and possibly drier savannah habitats, on the floor of the East African Rift Valley at sites such as Koobi Fora and Olduvai. Although it is difficult to make a strict link between these transitional assemblages and the appearance of *H. ergaster*, certain aspects of the biology of a larger hominin, such as secondary altriciality and larger home range sizes, suggest links between behavior and biology. These anatomical changes complement the archeological record, which suggests increased efficiency in raw material procurement and use in these transitional industries. This is consistent with possibly complex foraging strategies, which may have been linked to an economic organization which included a highly variable tool-kit adapted to more variable ecological settings.

Acheulean

The Acheulean Industry is often described as mode II industry. **Figure 2** displays the localities of Acheulean sites distributed over large parts of the African continent. All of them include some of the characteristic hand-axes that mark the transition to mode II industry. Often referred to as large cutting tools, these larger implements are the hallmark of the Acheulean Industry (see **Fig. 6**).

Acheulean sites are now known to exist in the African interior in particular, in more open, elevated, drier parts of sedimentary basins (e.g., Olduvai Bed III, Koobi Fora-Chari Member), and on the high plateaus of the flanks of the east African rift valley at high altitude (e.g., Gadeb, Isenya). Acheulean sites are found in a wide variety of settings including marine beach localities in northwest Africa (e.g., Sidi Abderrahman) and along the eastern and western coastlines of the southern part of the continent (Baia Foura and Amanzi). The distribution of sites does not necessarily indicate the penetration of the tropical rainforest in the watersheds and main drainages of the Congo River basin or the more forested regions of West Africa. However, the presence of Acheulean toolmakers is indicated in the more closed habitats of east Africa such as the site of Nsongezi,

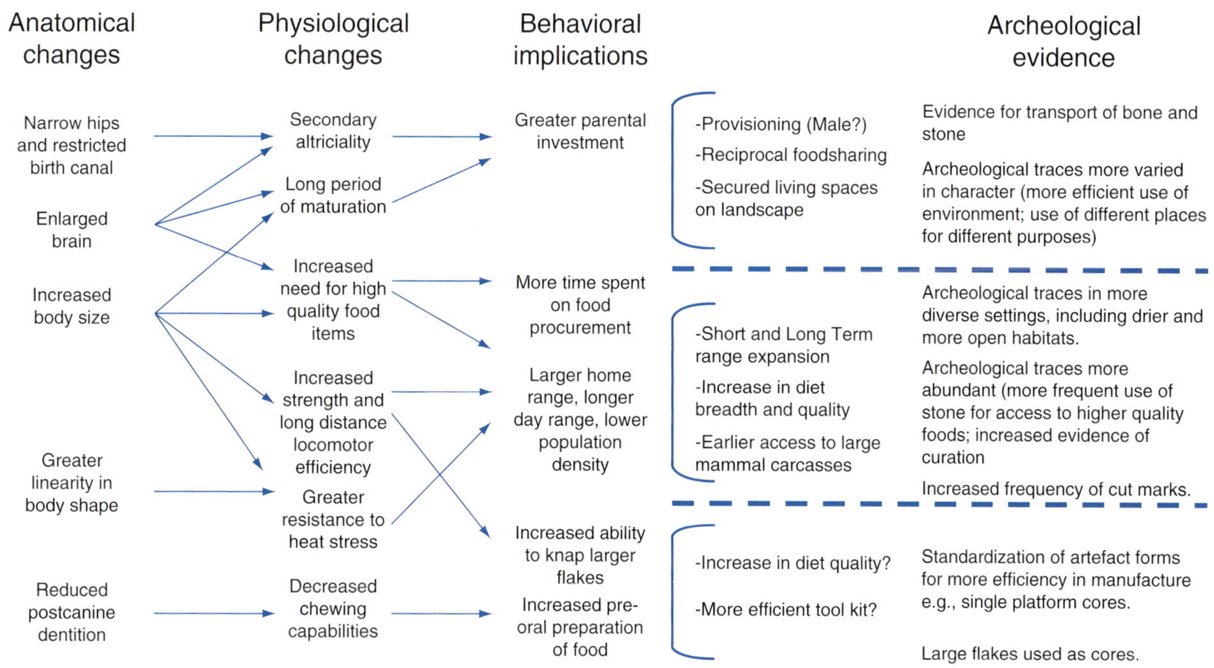

Figure 5 Important shifts in behavioral complexity of early African *Homo erectus or Homo ergaster* associated with Developed Oldowan and Early Acheulean archaeological assemblages (modified after Rogers, Harris, and Fiebel, 1994).

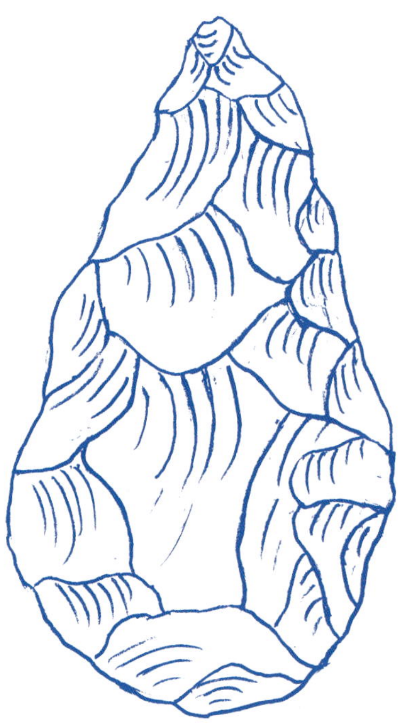

Figure 6 Acheulean biface or handaxe, which may have had multiple uses, including as a core for the production of flakes.

Uganda. Sites located in this setting commonly have thick and broadly flaked bifaces, often attributed to the Sangoan tradition. Acheulean sites are also found in and around the margins of the Sahara desert. During the Pleistocene, these more arid regions may have undergone dramatic oscillations, which would have been host to less harsh steppe environments (Isaac, 1982).

In addition to the diversity of topographic and ecological settings, Acheulean sites are known from open air and cave settings. In most instances, Acheulean sites are characteristically found near water sources (spring deposits or fluvial systems). Overall, the presence of these sites in these varied habitats suggests that *H. ergaster* and subsequently *H. heidelbergensis* exploited a diversity of habitats. It can be inferred that the increased dietary requirements of these more encephalized hominins forced increased home and day ranges, which led to an ever-increasing overall geographic range resulting in a pan-African distribution. The perils of exploiting open habitats would have been ameliorated by the larger body size of *H. erectus* as well as the possible use of fire. The behavioral package displayed by *H. erectus* (and later *H. heidelbergensis*) would have allowed this hominin to survive in a niche that would have required direct competition with the Pleistocene carnivore guild. Controversial evidence for the control of fire exists as early as 1.5 Myr at Koobi Fora, Chesowanja, and Gadeb. It has yet to be proved whether fire was an adaptive tool to enhance access to high-quality food items in more open grasslands as well as high-altitude settings subject to freezing overnight temperatures. Yet by 300–400 ka, evidence at Kalambo Falls of charred logs, ash, and fire-reddened soil indicates unequivocal use of fire by hominins by the Middle Pleistocene (Clark and Harris, 1985). Furthermore, ash deposits at the Acheulean sites of Cave of Hearths and Montagu Cave in southern Africa indicate that this was definitely part of Middle Pleistocene hominin adaptations. The substantial changes in the body size of Middle Pleistocene hominins and concurrent encephalization would have required a shift in diet that facilitated increased nutritional intake (Milton, 1987). The exploitation of larger mammal carcasses may have provided a source for this increased nutritional intake. Therefore, assessing changes in the subsistence capabilities between *H. ergaster* and their Oldowan hominin ancestors is essential to determining the significance of meat eating in Plio–Pleistocene hominin evolution. Despite the obvious importance of Acheulean archeofaunas, very few of these assemblages have enjoyed the full suite of taphonomic analyses. Acheulean sites that have produced extensive faunal assemblages include Ternifine, Sidi Abderrahman, and Bir Tarfawi in north Africa; Bodo, Melka Kunture, Gadeb, Karari (Koobi Fora), Konso, Kapthurin, Kanjera North, Kariandusi, Kilombe, Olorgesailie, Isenya, Lainyamok, Olduvai (Bed III-IV), Ndutu, and Peninj in east Africa; and Kalambo Falls, Wonderwerk, Swartkrans, Sterkfontein, Amanzi, and Elandsfontein in southern Africa. Despite an abundance of sites, meaningful interpretations of Acheulean hominin behavior are limited because several of these sites are from secondary contexts. Winnowed or transported assemblages often do not have the integrity required to make meaningful behavioral interpretations. Subsequently, many of these assemblages are overlooked in discussions of Acheulean paleoenvironmental context or the dietary adaptations of Acheulean hominins. Recent zooarcheological investigations of Acheulean assemblages are beginning to produce interesting results. New rounds of data analysis, including sites such as Swartkrans (South Africa) and Peninj (Tanzania), are intermediary sites that are associated with 'Developed Oldowan' and Acheulean lithic assemblages. The association between the faunal remains and lithic assemblages at Swartkrans is still tenuous. Yet, the faunal assemblages at Peninj in Tanzania demonstrate clear differences between the ecological contexts of accumulations associated with Oldowan tools and those associated with Acheulean bifaces (Dominguez-Rodrigo *et al.*, 2001). At Peninj, Oldowan-like tools are found near the basin margin and are associated with butchery activities. In contrast, the riverine deposits further from the Natron basin axis are found in fluvial contexts and are devoid of archeofaunal evidence. Phytolith analysis of Acheulean hand-axes at Peninj suggests these bifaces were being used as wood-working tools. The subsequent inference is that the behavioral repertoire represented at these sites also included wooden spears, which subsequently eroded away. At all the sites found in Peninj, hominins are believed to have acquired early access to carcasses based on the anatomical location of cut marks on long bone fragments (Domingo-Rodriguez *et al.*, 2002). This may indicate that Acheulean hominins were hunting or confrontationally scavenging to obtain meat, rather than passively scavenging flesh scraps and marrow from carnivore kills as their Oldowan hominin ancestors are believed to have done.

Excavations at other Acheulean sites, including Olorgesailie and Olduvai Gorge, have recovered Acheulean bifaces in association with extensive faunal remains (Leakey and Roe, 1994; Potts *et al.*, 1999). Olorgesailie preserves an association between the bones of 57 giant gelada baboons and over 4,000 stone artifacts, which, along with breakage patterns, lead some researchers to infer that hominins, were hunting baboons (Shipman *et al.*, 1981). Taphonomic analyses of this assemblage and other assemblages are in progress and will presumably reveal the diversity of hominin dietary adaptations during the Middle Pleistocene. Of particular interest is whether the behavioral patterns seen at Peninj are a local response to the ecological contexts of the Natron basin or are indicative of an overall Acheulean pattern. The Middle Awash Valley of Ethiopia has preserved an extensive record of the Acheulean, dating between 1.0 Ma and 0.1 Myr (Clark and Schick, 2000). The artifact-bone concentrations are found in both primary and secondary contexts with several localities preserving bifaces in association with hominin-modified bone. The Herto Member (400–100 ka) hippo butchery sites in the Bouri Formation hold special interest in this regard. Here, bifaces are found in association with several hippo butchery sites; they exhibit evidence of hominin-induced modification to bone including a hippo cranium with large chop marks, suggesting bifaces may have been used to butcher large animal carcasses.

The incorporation of plant foods into the diet of *H. erectus* cannot be underestimated, even though plant remains are sparse in the archeological record. Edible plant foods such as fruits, nuts, berries, roots, tubers, and seeds may have been important 'fall-back foods' in times of environmental stress. Two Acheulean sites, Amanzi Springs and Kalambo Falls, in southern Africa have evidence of preserved plant materials due to the waterlogged conditions of the sediments. At both sites, woods and seeds are preserved. At Kalambo Falls, evidence of fruit underscores the potential of these food items. These discoveries at Acheulean sites underscore the importance of plant foods in the overall diet of *H. erectus*.

The presence of bone tools in Early Pleistocene archeological sites in South Africa and East Africa contributes to our understanding of the diversity of tools incorporated into early hominin's foraging practices. A recent study of the collection of bone tools from Member 1–3 (1.8–1.0 Myr) at Swartkrans (South Africa) shows that, in some instances, there was an intentional grinding to modify the polished pointed surface of bone tools. It is believed that this practice increased the digging efficiency of these tools, particularly in opening termite mounds (Backwell and D'Errico, 2005). The clear time investment in tool manufacture for termite extraction emphasizes the importance of insect foods as a source of protein in the Acheulean diet. The bone tools from Olduvai Gorge in East Africa (Leakey and Roe, 1994) were deliberately flaked into mostly irregular shapes. Some of the more heavy-duty hammer and anvil tools may have been used in large mammal carcass-processing (Backwell and D'Errico, 2005). The geographical differences may be a reflection of regional adaptations to foraging for variable food resources. This would be taken to reflect the flexibility of *H. erectus* to exploit a great diversity of food resources that might have been only seasonally available in different ecological settings. One interpretation of the variation in bone tool production strategies is that different hominin taxa (*Paranthropus robustus* in South Africa and *Australopithecus boisei* in East Africa) created these different industries.

Highly variable assemblages including more advanced core preparation and flake removal techniques attributed to the Levallois technique also characterize the final stages of the Acheulean. In addition, numerous Middle Pleistocene assemblages are associated with more elongated thick core axes, pick-like bifaces and a variety of small retouched or utilized flakes generally designated as Sangoan. The Sangoan tool assemblages could be viewed as a transitional industry as they continue the tradition of bifacial core axes and pointed forms. These industries could represent early manifestations that are present in younger industries that characterize the MSA.

Conclusion

A major unresolved issue in the ESA is whether the beginnings of Oldowan tool manufacture and use occurs abruptly in a punctuated period of cultural evolution or the origin of tool use has been a gradual event. An alternative hypothesis would be that stone tools were part of a behavioral repertoire of earlier Pliocene hominins that have yet to be recovered in the archeological record. Based upon the current evidence in the Kada Gona region, the beginnings of stone tool manufacture could be interpreted as an important adaptive threshold or innovation occurring abruptly, perhaps in response to changing environmental conditions brought about by a major global cooling event between 2.8 and 2.4 Myr. This time interval also coincides with an adaptive radiation of the Homininae. It is possible that a more efficient processing of sources of high-quality protein facilitated this adaptive radiation. The incorporation of high-quality protein (presumably meat and marrow) and other nutritious food sources into the diet of one hominin species (most likely early *Homo* or by multiple hominins including the australopithecines and paranthropines) surely changed regional ecological dynamics in the Early Pleistocene.

Clearly by 1.5 Myr (if not earlier), there does seem to be a correspondence between anatomical change and technological elaboration characterizing the Developed Oldowan and early Acheulean toolkits. The increasingly widespread nature, geographically and ecologically, of Acheulean sites may be linked to the emergence of a bigger-bodied hominin (early African *H. erectus* or *H. ergaster*). Through the course of the time period associated with Acheulean assemblages, large bifacially flaked cutting tools appear to become more refined and the end of the Acheulean is associated with prepared core technologies.

The growing complexity of the toolkit and evidence of a wide-ranging hominid inhabiting more diverse locations on the landscape may indicate the increasing intelligence, mental mapping, and growing social complexity in Acheulean lifeways. A possible indication of interpersonal group conflict comes from cut marks found on the human cranium recovered at the site of Bodo, in the Middle Awash region of Ethiopia. This specimen, attributed to *H. heidelbergensis* is associated with a large occurrence of Acheulean tools. The extent of the hominin modification found on this specimen suggests it was intentionally defleshed. This rare Bodo example may be one of the earliest recorded instances of cannibalism. This evidence may indicate that symbolic behavior, more characteristic of anatomically modern humans and their associated MSA technologies, had its beginnings in the late Acheulean. It is possible that many of the features Paleolithic archeologists often associate with the development of modern human behavior had their roots in the deep time of the ESA.

See also: **Archaeological Records**: Overview; 2.7 MYR-300,000 years ago in Asia; 1.9 MYR-300,000 years ago in Europe; Global Expansion 300,000-8000 years ago, Africa; Global Expansion 300,000-8000 years ago, Asia; Global Expansion 300,000-8000 years ago, Australia; Global Expansion 300,000-8000 years ago, Americas; Neanderthal Demise; Postglacial Adaptations. **Vertebrate Studies**: Interactions with Hominids.

References

Backwell, L., and D'Errico, F. (2005). The origin of bone tool technology and the identification of early hominid cultural traditions. In *From Tools to Symbols* (F. D'Errico and L. Backwell, Eds.), pp. 238–275. Witwatersrand University Press, Johannesburg.

Binford, L. R. (1981). *Bones, Ancient Men and Modern Myths.* Academic Press, New York.

Blumenschine, R. J. (1995). Percussion marks, tooth marks and experimental determinations of the timing of hominid and carnivore access to long bones at FLK Zinjanthropus, Olduvai Gorge, Tanzania. *Journal of Human Evolution* **29,** 21–51.

Braun, D. R., Plummer, T., Ferraro, J. V., *et al.* (2005). *Oldowan technology at Kanjera South, Kenya: The context of technological diversity.* PaleoAnthropology.

Bunn, H. T. (1981). Archaeological evidence for meat-eating by Plio-Pleistocene hominids from Koobi Fora and Olduvai Gorge. *Nature* **291,** 547–577.

Clark, G. (1969). *World Prehistory: A New Outline.* Cambridge University Press, Cambridge.

Clark, J. D., and Harris, J. W. K. (1985). Fire and its roles in early hominid lifeways. *African Archaeological Review* **3,** 3–27.

Clark, J. D., and Schick, K. (2000). Acheulean archaeology of the eastern middle awash. In *The Acheulean and the Plio-Pleistocene Deposits of the Middle Awash Valley, Ehtiopia, Vol. 104: Annales Sciences Geologiques* (J. de Heinzelin, J. D. Clark, K. Schick and W. H. Gilbert, Eds.). Musee Royal De L'Afrique Centrale, Tervuren, Belgique.

de Heinzelin, J., Clark, D., White, T., *et al.* (1999). Environment and behavior of 2.5-million-year-old Bouri hominids. *Science* **284,** 625–629.

Delagnes, A., and Roche, H. (2005). Late Pliocene hominid knapping skills: The case of Lokalalei 2C, West Turkana, Kenya. *Journal of Human Evolution* **48,** 435–472.

De la Torre, I. (2004). Omo revisited: Evaluating the technological skill of Pliocene hominids. *Current Anthropology* **45,** 439–465.

deLumley, H., and Beyene, Y. (2004). *Les Sites Prehistoriques de la Region de Fejej, Sud-Omo, Ethiopie, dans leur Contexte Stratigraphique et Paleontologique.* EditionsRecherche sur les Civilisations, Paris.

Dominguez-Rodrigo, M. (1997). Meat-eating by early hominids at the FLK 22 Zinjanthropus site, Olduvai Gorge (Tanzania): An experimental approach using cut-mark data. *Journal of Human Evolution* **33,** 669–690.

Dominguez-Rodrigo, M., de la Torre, I., de Luque, L., Alcala, L., Mora, R., Serrallonga, J., and Medina, V. (2002). The ST site complex at Peninj, West Lake Natron, Tanzania: Implications for early hominid behavioural models. *Journal of Archaeological Science* **29,** 639–665.

Dominguez-Rodrigo, M., Pickering, T. R., Semaw, S., and Rogers, M. J. (2005). Cutmarked bones from Pliocene archaeological sites at Gona, afar, Ethiopia: Implications for the function of the world's oldest stone tools. *Journal of Human Evolution* **48,** 109–121.

Dominguez-Rodrigo, M., Serrallonga, J., Jaun-Tresserras, J., Alcala, L., and Luque, L. (2001). Woodworking activities by early humans: A plant residue analysis on Acheulean stone tools from Peninj (Tanzania). *Journal of Human Evolution* **40,** 289–299.

Gabunia, L., Anton, S. C., Lordkipanidze, D., Vekua, A., Justus, A., and Swisher, C. C. (2001). Dmanisi and dispersal. *Evolutionary Anthropology* **10,** 158–170.

Harris, J. W. K., Braun, D., McCoy, J. T., Pobiner, B. L., and Rogers, M. (2002). New directions and preliminary results from a landscape approach to the study of archaeological traces for the behavior of Plio-Pleistocene hominids at Koobi Fora. *Journal of Human Evolution.* **42,** A15.

Harris, J. W. K., and Isaac, G. L. L. (1976). The Karari Industry: Early Pleistocene archaeological evidence from the terrain east of Lake Turkana, Kenya. *Nature* **262,** 102–107.

Isaac, G. (1982). The earliest archaeological traces. In *Cambridge History of Africa* (J. D. Clark, Ed.) vol. I, pp. 157–247. Cambridge University Press, Cambridge.

Isaac, G. Ll. (1997). *Koobi Fora Research Project: Plio-Pleistocene Archaeology.* Clarendon Press, Oxford.

Kimbel, W. H., Walter, R. C., Johanson, D. D., *et al.* (1996). Late Pliocene *Homo* and Oldowan tools from the Hadar formation (Kada Hadar Member), Ethiopia. *Journal of Human Evolution* **31,** 549–561.

Leakey, M. D. (1971). *Olduvai Gorge: Excavations in Beds I and II, 1960–1963.* Cambridge University Press, Cambridge.

Leakey, M. D., and Roe, D. (Eds.) (1994). In *Olduvai Gorge Excavations in Beds III, IV, and the Masek Beds 1968–1971* Chapters 2–7, Cambridge University Press, Cambridge.

Ludwig, B., and Harris, J. W. K. (1998). Towards a technological reassessment of East African Plio-Pleistocene lithic assemblages. In *Early Human Behavior in Global Context* (M. D. Petraglia and R. Korrisettar, Eds.), pp. 84–107. Routledge, London and New York.

Milton, K. (1987). Primate diets and gut morphology: Implications for hominid evolution. In *Food and Evolution Toward a Theory of Human Food Habits,* pp. 93–115. Temple University Press, Philadelphia.

Panger, M. A., Brooks, A., Richmond, B., and Wood, B. (2002). Older than the Oldowan? Rethinking the emergence of hominin tool use. *Evolutionary Anthropology* **11,** 235–245.

Plummer, T. (2004). Flaked stones and old bones: Biological and cultural evolution at the dawn of technology. *Yearbook of Physical Anthropology* **47,** 118–164.

Potts, R., Behrensmeyer, A. K., and Ditchfield, P. (1999). Paleolandscape variation and Early Pleistocene hominid activities: Members 1 and 7, Olorgesailie formation, Kenya. *Journal of Human Evolution* **37,** 747–788.

Rogers, M. J., Harris, J. W. K., and Feibel, C. S. (1994). Changing patterns of land use by Plio-Pleistocene hominids in the Lake Turkana basin. *Journal of Human Evolution* **27,** 139–158.

Semaw, S., Renne, P., Harris, J. W. K., *et al.* (1997). 2.5 Million-year-old stone tools from Gona, Ethiopia. *Nature* **385,** 333–336.

Shipman, P., Bosler, W., and Davis, K. L. (1981). Butchering of giant geladas at an Acheulian site. *Current Anthropology* **22,** 257–264.

Tryon, C. A., and McBrearty, S. (2002). Tephrostratigraphy and the Acheulian to Middle Stone Age transition in the Kapthurin, Kenya. *Journal of Human Evolution* **42,** 211–235.

2.7 MYR-300,000 years ago in Asia

R Dennell, University of Sheffield, Sheffield, UK

© 2007 Elsevier B.V. All rights reserved.

Introduction

At the beginning of the twenty-first century, we know as little about early human evolution in Asia as we knew about that of East Africa in the early 1960s (when *Homo habilis* and *Zinjanthropus* were first discovered), and compared to Europe and much of Africa, there is still a chronic shortage of well-dated sites >300 ka, especially those with associated faunal, fossil hominin, and environmental evidence. Consequently, most statements about Asia's earliest inhabitants are highly vulnerable to new discoveries. Relevant examples from the last decade are the wholly unexpected discoveries of very primitive and small-brained hominins dating from 1.75 Ma in Dmanisi, Georgia, the diminutive Late Pleistocene hominins from Flores, and the re-dating of the earliest Javan hominins, and the earliest archeological evidence from China to >1.5 Ma. The first issue to consider is when hominins first inhabited Asia.

Early Pleistocene Hominins in Asia

The history of hominins in Asia before 300 ka is often summarized as the outcome of a dispersal event from Africa in the Early Pleistocene. However, very little is known about the late Pliocene faunal record of much of Asia before 1.7 Ma, and it remains unclear when hominins first left Africa. Because australopithecines were already living in Chad, 2,500 km west of the Rift Valley, 3.0–3.5 Ma, some may already have been in Southwest and other parts of Asia by this time. Hominins might also have left Africa shortly after the flaking of stone to make tools became routine ca. 2.6 Ma in East Africa (Dennell and Roebroeks, 2005). Indications of when hominins might have left Africa are provided by the sapropel record of the East Mediterranean (Kroon *et al.*, 1998), and the loess/paleosol record of North China (Huayu Lu *et al.* (1999); see **Fig. 1**). Sapropels are organic-rich sediments associated with greater inputs of fresh water into the sea; those in the East Mediterranean resulted largely from increased influx from the Nile, presumably during periods of higher rainfall. The Chinese loess and paleosol sequence indicates alternating cool and dry glacial periods, and warmer and moister interglacial ones. Both records indicate several 'windows of opportunity' before 1.7 Ma, during which rainfall and temperatures, and hence biological productivity, were higher, and conditions more favorable for hominins to disperse. In intervening cooler and drier periods, deserts (particularly those between northeast Africa and Southwest Asia) would have encroached onto grasslands, and reduced opportunities for animals such as hominins to move freely between Africa and Asia. As can be seen, hominins had several opportunities to migrate out of (and possibly into) Africa before and after 1.7 Ma.

At present, the earliest skeletal evidence for hominins in Asia dates from ca. 1.75 Ma from Dmanisi, Georgia and ca. 1.6 Ma from Java (see **Fig. 2** and **Table 1**).

Early Pleistocene Hominin Remains from Java and Dmanisi

The hominin remains from Java are classified as *H. erectus sensu stricto*, and were first discovered there in 1891. The child's cranium from Mojokerto has been dated by Ar^{40} to 1.81 ± 0.04 Ma, but doubts persist over its provenance and age; the earliest hominins from Sangiran date from ca. 1.6 Ma. These early Javan hominins are usually regarded as descended from East African populations of *H. erectus* (sometimes classified as *H. ergaster*), which date from ca. 1.9 to 1.7 Ma (Antón, 2002). The Dmanisi hominins have been precisely dated by paleomagnetism and Ar^{40} to ca. 1.75 Ma. They are remarkably small-brained and were probably 1.5 m or less in height. Apart from one metatarsal fragment, information is not yet available on their body proportions and bipedal competence. The Dmanisi hominins have proved difficult to classify. One mandible has been assigned to a new taxon, *H. georgicus*. The crania are very primitive, and show many resemblances to *H. habilis*. The most recent taxonomic assessment is that they are a very primitive form of *H. erectus sensu lato*, and may be ancestral to both early East African populations of *H. erectus* (or *H. ergaster*) as well as the Javan *H. erectus sensu stricto*. That is to say, *H. erectus* may have originated in Asia, and then migrated back into Africa, as well as eastwards across southern Asia to Java and possibly north China. Until more is known of the fossil record of Southwest Asia in the late Pliocene, this possibility cannot be excluded (Rightmire *et al.*, 2006). The most significant aspect of the Dmanisi evidence is that hominins did not need large brains or bodies in order to venture out of Africa; this clearly implies that earlier, small-bodied and small-brained hominins might also have done the same.

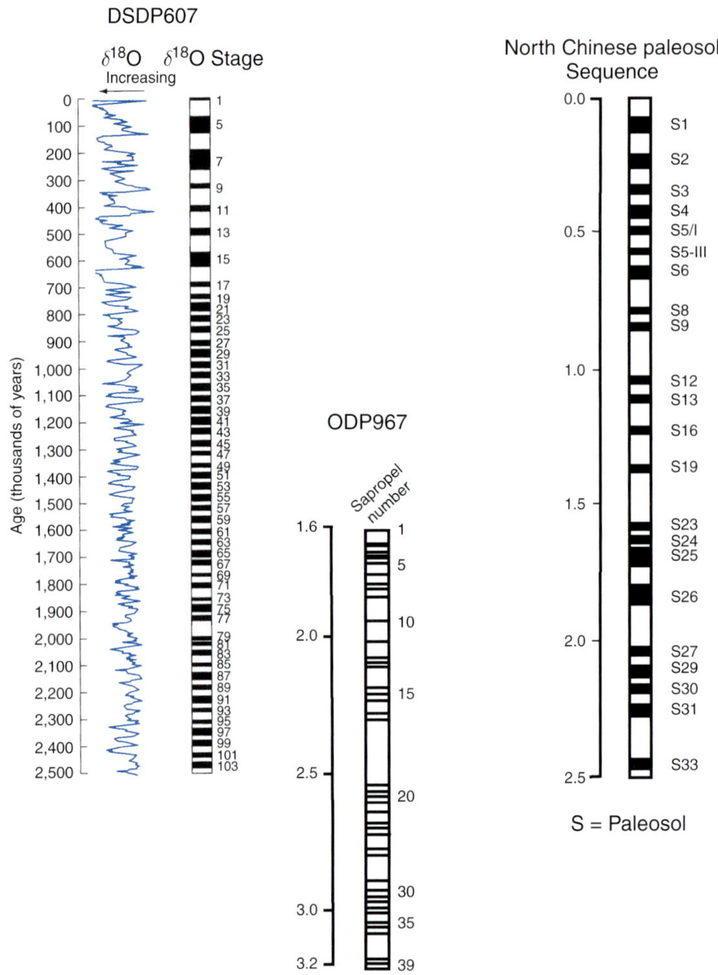

Figure 1 Climatic change and hominin dispersals: the oxygen isotope, East Mediterranean sapropel and Chinese loess records. As can be seen, there were numerous opportunities for hominins to disperse out of, into, and within Asia during the late Pliocene and early Pleistocene. Wilson *et al.* (2000), figure 4.15; Kroon *et al.* (1998), figure 3; and Hyayu Lu *et al.*, (1999), figures 2 and 3.

Early Pleistocene Lithic Technology in Asia

The main evidence for the stone technology used by early *H. erectus* in Asia comes from Dmanisi (de Lumley *et al.* (2005); see **Fig. 3**), and sites such as Majuangou and Xaiochangliang in the Nihewan Basin, North China, that are dated by paleomagnetism to ca. 1.66 and 1.36 Ma, respectively (Zhu *et al.* (2003, 2004); see **Fig. 4**). These assemblages are broadly similar to Oldowan ones from East Africa that date from 2.6 Ma (*see* 2.7 MYR-300,000 years ago in Africa), and comprise flakes and the cores from which they were struck. The cores are simple in that they were not shaped prior to flaking, and no attempts were made to detach flakes of particular sizes or shapes. The main types recognized are choppers (cores with a few flakes struck from opposing angles), knives and scrapers (i.e., flakes with a sharp edge suitable for cutting or scraping). The worked stone was obtained locally: basalt in the case of Dmanisi, and low-quality chert at Majuangou, Nihewan Basin, and most was used and discarded within 5 km of its source. The artifacts from Dmanisi are regarded as very simple (a conclusion in keeping with the primitive nature of their makers), and have been called pre-Oldowan on technological grounds because of the absence of retouched tools.

One of the most informative Early Pleistocene sites in Asia is 'Ubeidiya, Israel, which comprises a long lacustrine sequence dated faunistically to ca. 1.4–1.0 Ma (Bar-Yosef, 1998). Stone tool assemblages were mainly found in or on swamp deposits and former lake beaches, as well as in fluvial conglomerates. Assemblages that lacked bifaces are described as Developed Oldowan, and those with bifaces, or cores that have been flaked on both sides to make an ovate or pointed tool and to remove most of the cortex, are

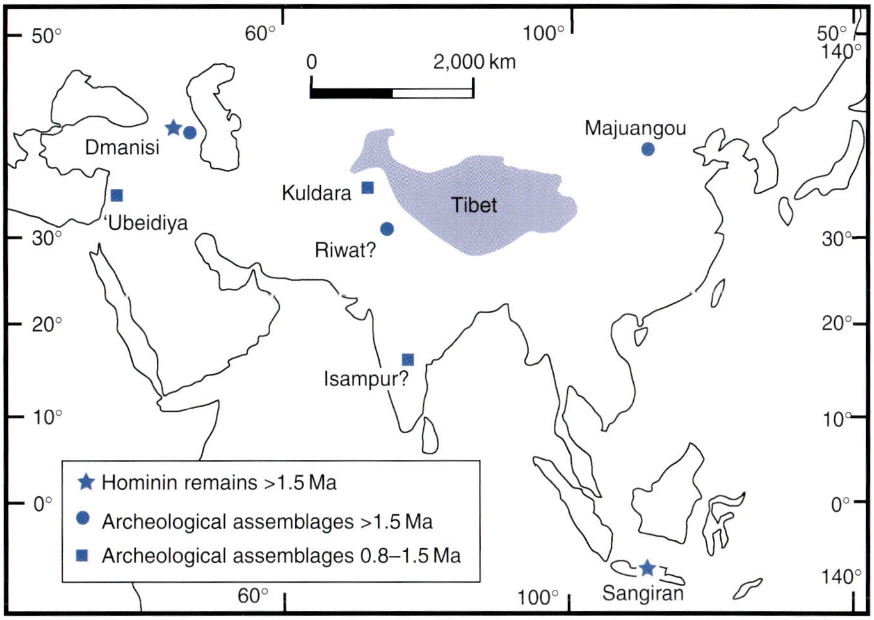

Figure 2 Early Pleistocene hominin and archeological sites in Asia.

Table 1 Cranial capacities of specimens of early Pleistocene *Homo erectus* specimens from Dmansi and Java, and middle Pleistocene ones from Zhoukoudian, China. Those from Dmanisi are ca. 1.75 Ma, and those from Java are between 1.0 and 1.6 Ma, except for Mojokerto, which may be slightly older

Origin	*Specimen*	*Description*	*Cranial capacity (cc^3)*
Dmanisi, Georgia	D2280	Adult braincase, possibly male	775
	D2282/D211	Partial cranium of a young adult	650–660
	D2700/D2735	Complete skull of small subadult	600
	D3444/D3900	Edentulous old cranium	625
Java, Indonesia	Mojokerto	Child's cranium	800[a]
	Trinil 2		940
	Sangiran 2	Calotte	813
	Sangiran 4	Calvarium	908
	Sangiran 10	Calotte	855
	Sangiran 12	Calotte	1,059
	Sangiran 17	Calvarium	1,004
	Sangiran IX	Calotte	845-
	Sangiran	Average of Sangiran 2, 4,10,12,17, IX and Trinil 2	918
Zhoukoudian, China			
	Zhoukoudian II	Middle Pleistocene, ca. 500 ka	1,030
	Zhoukoudian III	Middle Pleistocene, ca. 500 ka	915
	Zhoukoudian V	Middle Pleistocene, ca. 500 ka	1,140
	Zhoukoudian VI	Middle Pleistocene, ca. 500 ka	850
	Zhoukoudian X	Middle Pleistocene, ca. 500 ka	1,225
	Zhoukoudian XI	Middle Pleistocene, ca. 500 ka	1,015
	Zhoukoudian XII	Middle Pleistocene, ca. 500 ka	1,030
Average of crania II,III,V,VI,X–XII			1,029

[a]Estimated adult size.
Rightmire *et al.* (2005) for Dmanisi, and Antón (2002): Table 1, for Java and Zhoukoudian.

classified as early Acheulean (see **Figs. 5** and **6**). It remains unclear whether these were made by different hominins, or are part of the same lithic technocomplex.

Early Pleistocene Hominin Subsistence and Settlement

Because animal bones can be associated with stone tools for a variety of reasons (such as stream action

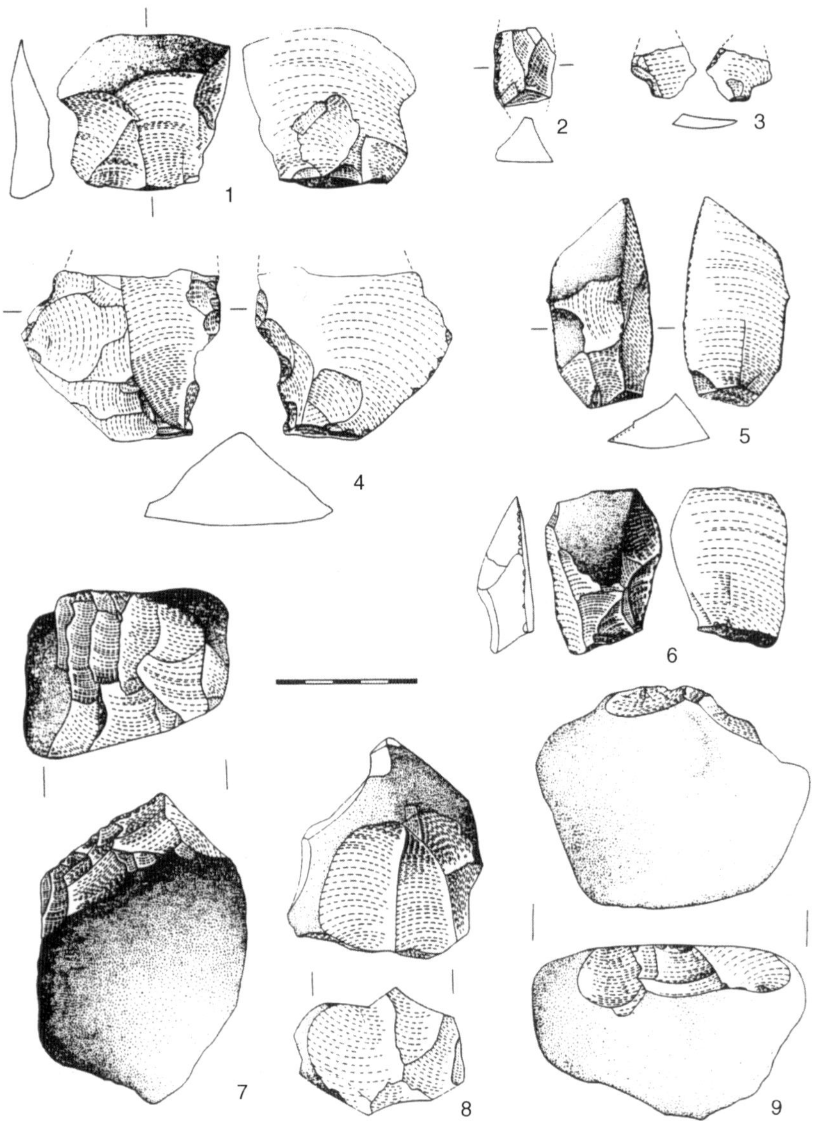

Figure 3 Stone artifacts from Dmanisi. Bar-Yosef (1998), figure 8.4.

mixing the two together, or hominins discarding stone tools in an area previously used by a carnivore), the main evidence for the subsistence activities of early Asian hominins depends primarily upon the discovery of bone fragments with cut marks resulting from hominins butchering carcasses with stone tools. (As plant remains are preserved only very exceptionally, we know almost nothing about their use of plant foods.) At present, little is known about the hunting or scavenging activities of Early Pleistocene Asian hominins. There is no evidence for the use of fire, artificial shelters, or projectiles for hunting. Recent faunal evidence from 'Ubeidiya has been interpreted as showing the butchery and hunting of medium-sized game, although the evidence of hominin involvement in either activity is very slight (Gaudzinski, 2004). The nonhominin faunal remains at Dmanisi are still being studied; those from Majuangou were very fragmented, but percussion marks, indicative of marrow extraction, were found on some deer- and horse-sized bones. Inferences from contemporaneous sites in East Africa suggest that Early Pleistocene hominins did not hunt, but obtained most of their meat through scavenging, either passively (i.e., taking whatever a carnivore had left) or actively, by capturing carcasses from carnivores as they were feeding (Domínguez-Rodrigo, 2002). Either possibility underlines the importance of predators in the daily life of early *H. erectus* – either to avoid being eaten, or to obtain meat in a world where competition for it was fierce.

The likely size of the home ranges of early *H. erectus* has been estimated from their inferred

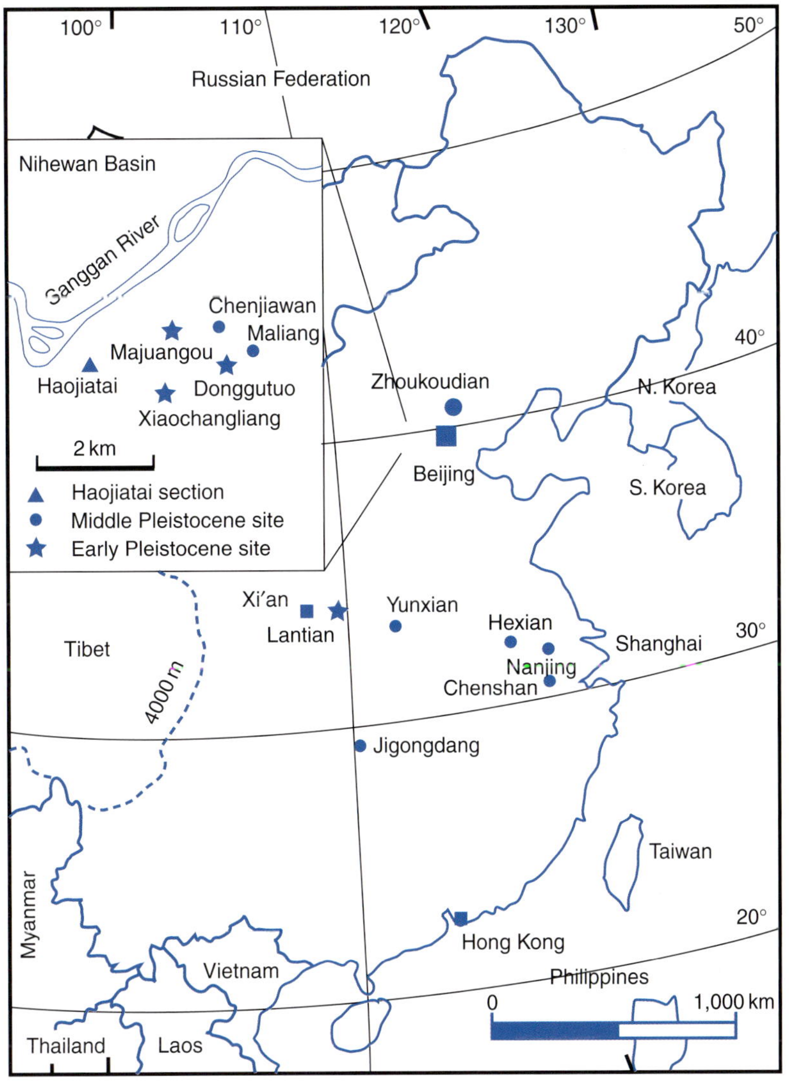

Figure 4 Early and middle Pleistocene sites in China.

height and weight as only 330–430 ha per individual (Antón and Swisher, 2004), or an operating radius of 5.1–5.6 km for a group of 25, which is probably the minimum viable group size. They were thus constrained to environments where the basic necessities of water, foods from animal and plant resources, stone for making tools, and shelter from carnivores could be found within a small area. Small lake basins, such as Dmanisi, those in North China, or 'Ubeidiya appear to have been the preferred hominin location in Asia; such basins are also ones where the chances of finding their remains are also high (*see* Asia).

Although hominins are evidenced in Israel, North China, Java and possibly Pakistan before 1.6 Ma, and are also recorded in Central Asia by 0.9 Ma (Ranov, 1995), it is probable that populations were frequently discontinuous geographically and temporally (see **Fig.** 7), particularly because climatic fluctuations between glacials and interglacials every 41 ka after 2.5 Ma (*see* Pliocene Environments) would have greatly influenced the extent of areas suitable for hominins. Many dispersal events are likely to have been short term and unsuccessful in establishing populations that were permanently resident (Dennell, 2003). Out of Africa 1 – the first movement of hominins from Africa – was probably a series of dispersal events that were not necessarily one-way from Africa to Asia.

Middle Pleistocene Hominin Archives in Asia, 800–300 ka

In the Middle Pleistocene, climatic alternations between warm and cold intervals were more pronounced than before, and cold phases lasted much longer, on a 100 ka cycle instead of 41 ka

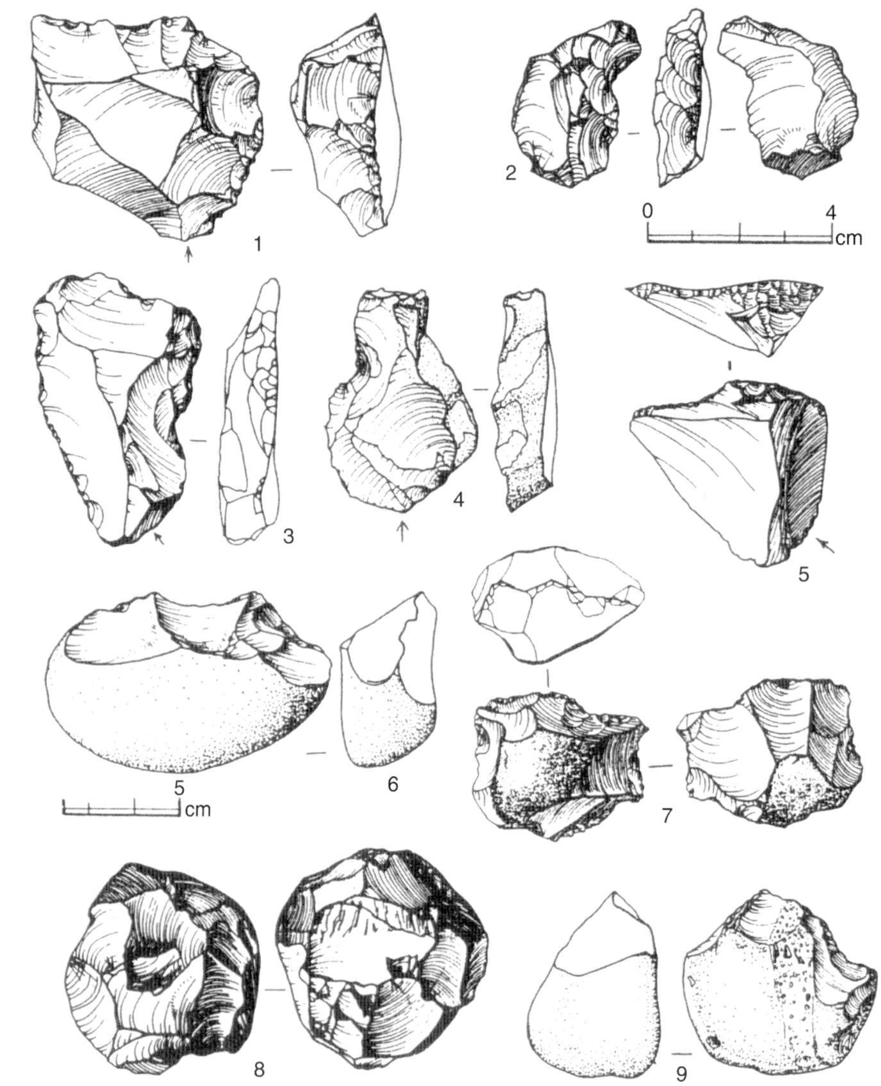

Figure 5 Selected retouched flakes including a scraper, and also core-choppers and a spheroid from 'Ubeidiya. Bar-Yosef (1998), figure 8.9.

(*see* Mid-Quaternary in Eurasia). Sea level changes were more dramatic, with much of Indonesia joined to mainland Southeast Asia and an extensive coastal shelf exposed off China. Major faunal changes also occurred during what is often referred to as the Galerian Faunal Turnover, which led to the emergence of a modern type of mammalian fauna across Siberia, South Asia, and western Asia (*see* Mid-Pleistocene of Southern Asia). The Middle Pleistocene also witnessed substantial tectonic activity, especially along the forefront of the Himalayas and Karakorum, and the Qinling Range that now divides North from South China (*see* Tectonic Uplift-Continental Configurations). Hominin regional records thus need to be considered against a background of climatic, faunal, and even tectonic changes (Dennell, 2004).

The Middle Pleistocene evidence for hominins in Asia is much more extensive than that for the Early Pleistocene (see **Fig. 8**), but is still largely confined to latitudes south of 40 °N, although they may have up dispersed to 50 °N in Kazakhstan, where, as yet undated, Acheulean bifaces have been found (Vishnyatsky, 1999). There are several Middle Pleistocene regional records from Asia, of which the best are from the Levant (**Fig. 9**), India (**Fig. 11**), and Tajikistan (**Fig. 13**).

The Levant

There are two 'flagship' sites in this region, Gesher Benot Ya'aqov, Israel (Goren-Inbar *et al.*, 2002, 2004), and Latamne, Syria (Bar-Yosef, 1998). The former lies in the Dead Sea Valley. Its deposits are

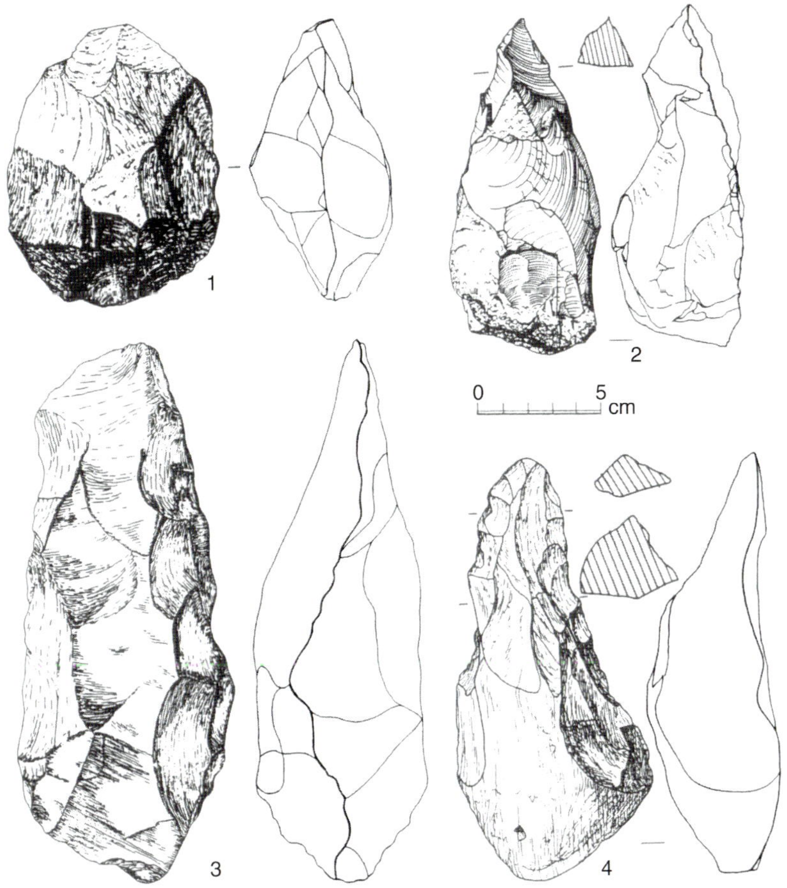

Figure 6 Acheulean bifaces (1,3,4) and a trihedral (2) from 'Ubeidiya. Bar-Yosef (1998), figure 8.8.

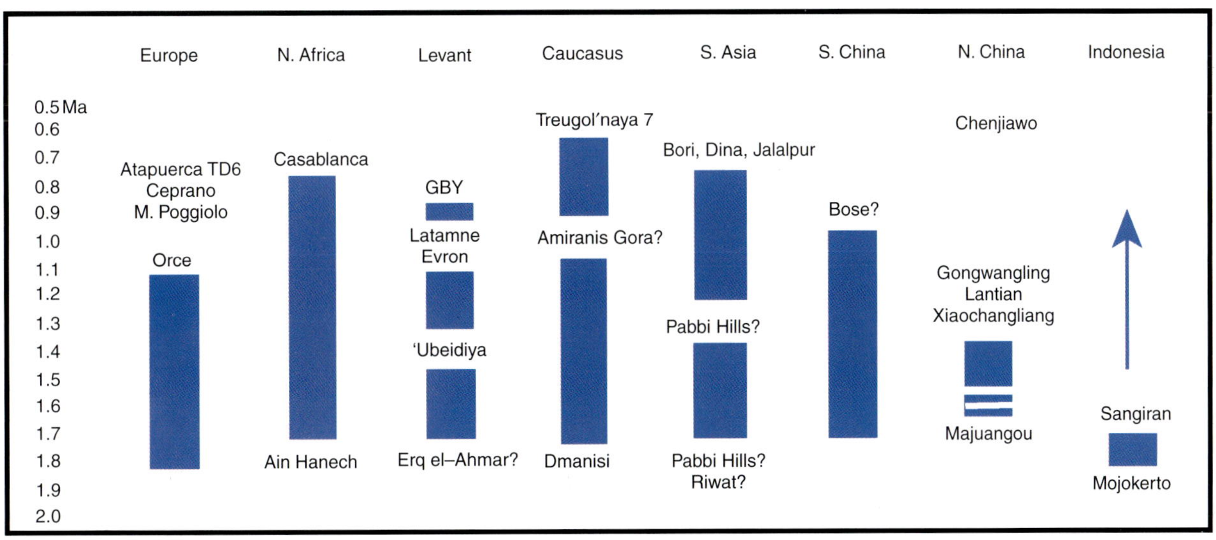

Figure 7 Discontinuities in hominin occupation in the early and middle Pleistocene of Asia. Dennell (2004), figure 4.

strongly tilted former river- and lake-shore sediments, dated to ca. 800 ka by paleomagnetism. The site contains Acheulean assemblages containing large numbers of basalt cleavers, often made by the Kombewa technique (i.e., from flakes with two opposing bulbs of percussion). The choice of basalt as a raw material, the use of the Kombewa technique, and production of cleavers strongly imply the

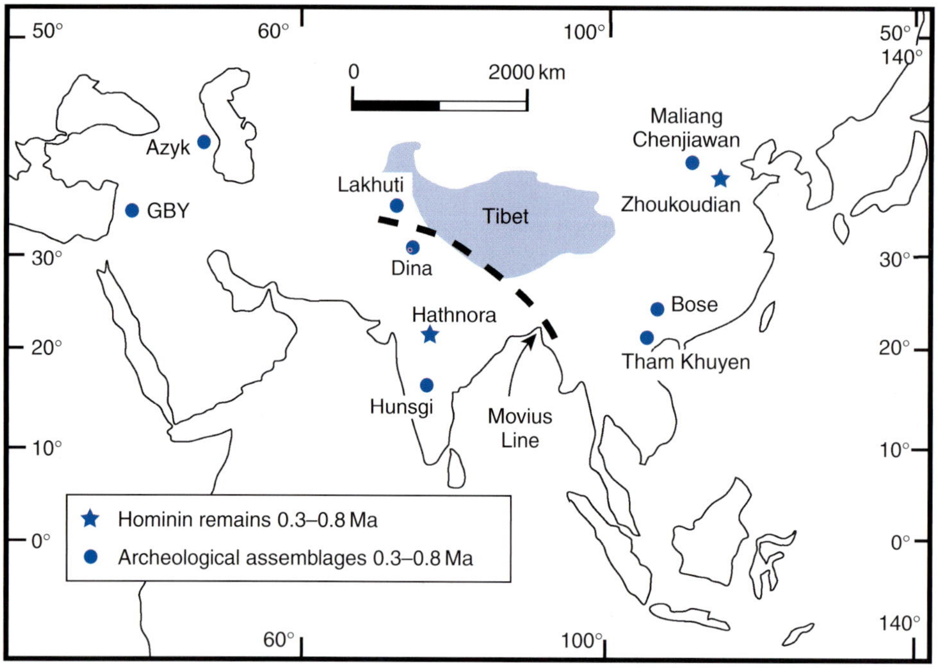

Figure 8 Middle Pleistocene sites in Asia.

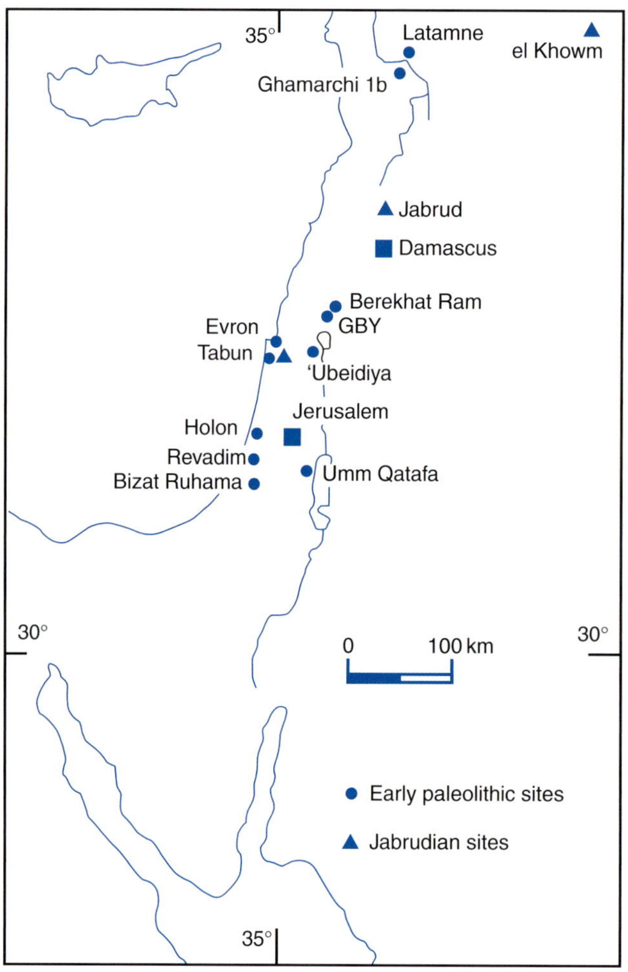

Figure 9 Principal archeological sites in the Levant.

dispersal of hominins or ideas from Africa, where all these traits occur earlier. In one area of the site, the remains of an elephant skull and a small tool assemblage were found, which has been interpreted as evidence of hunting, rather than the natural death of an elephant in an area where hominins also discarded stone tools. As the site is waterlogged, there is also evidence of plant remains, including charcoal, seed, and nut fragments. According to the excavators, Gesher Benot Ya'aqov contains the earliest evidence for the controlled use of fire, for processing nuts (a highly nutritious resource), and for gathering other plant foods. Details are not yet available on the faunal remains.

Latamne lies on the banks of the Orontes, and is an example of a 'living site' (see **Fig. 10**). It is probably younger than Gesher Benot Ya'aqov, and is dated to the mid-Middle Pleistocene. An area ca. 60 m^{-2} was excavated, within which several hundreds of stone artifacts were found; there was no evidence of structures or fire, and faunal remains were not preserved. Latamne is usually interpreted as a living site, where a group camped while hunting game. Alternatives are that it is a palimpsest of numerous short visits, or a site used seasonally by part of a group.

Another important Levantine Lower Paleolithic site is Berekhat Ram (dated as between 230 and 780 ka) that contains an Acheulean assemblage made from basalt, and an intriguing 'figurine' that is a small piece of volcanic tuff that was intentionally modified. Whether or not this is the earliest example

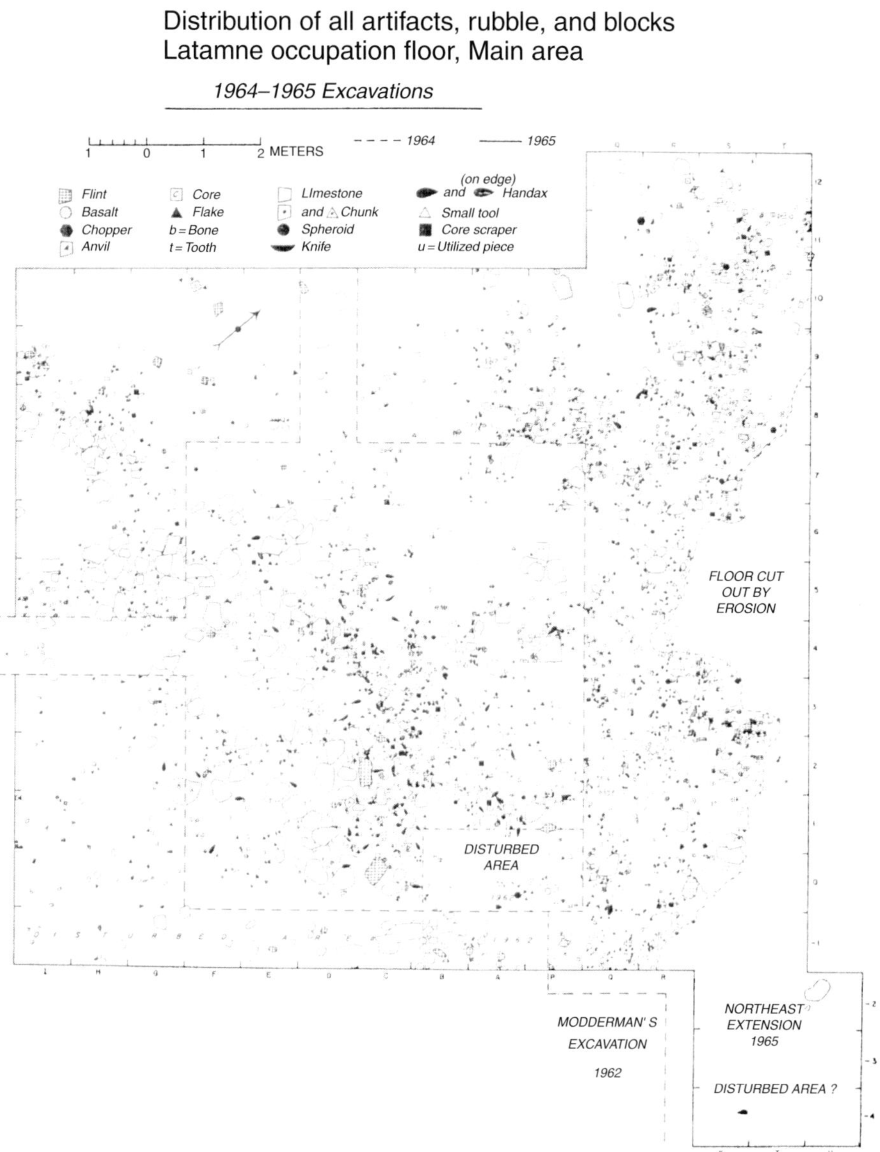

Figure 10 Plan of the 'living floor' at Latamne, Syria. Clark (1969), figure 10.

of representational art, it is clearly nonutilitarian, and indicates at least the capacity for some degree of symbolism. Other Levantine sites that are >800 ka are Evron and Bizat Ruhama, the Acheulean at Holon, Revadim, Umm Qatafa, the basal layers (F and G) of Tabun Cave, and Ghamarchi 1b are younger but probably >300 ka (Bar-Yosef, 1998).

Although the Acheulean is the most conspicuous part of the Levantine Early Palaeolithic, there are Lower Paleolithic sites such as Bizat Ruhama that lack bifaces and contain only flakes and small cores. The final part of the Levantine Early Paleolithic also contains a distinct regional variant known as the Yabrudian. This is found over Syria, Lebanon, and northern Israel, and comprises a non-Levallois flake industry with a distinctive convergent (*déjeté*) side scraper. Yabrudian assemblages in layer E of Tabun Cave date from ca. 350 to 250 ka, after which they are replaced by Middle Paleolithic, Levallois-Mousterian assemblages (*see* Global Expansion 300,000-8000 years ago, Asia). Nothing is known of the subsistence and social strategies associated with each of these traditions.

India

It is not clear when the Indian subcontinent was first colonized. The only Early Pleistocene evidence is a small stone tool assemblage in a secondary context from Riwat, Pakistan, dated to $\geq$1.9 Ma, and a quarry site at Isampur, Karnataka, for which an ESR date of 1.2 Ma has recently been obtained (Paddayya *et al.*, 2002), but which requires confirmation. At present, the earliest archeological evidence from the subcontinent dates to the Early Middle Pleistocene, as evidenced by Acheulean hand axes from Dinar and Jalalpur, Pakistan, and Bori, India (Petraglia, 1998) (**Fig. 11**).

The Indian Lower Paleolithic includes several sites and thousands of bifaces and cleavers, but

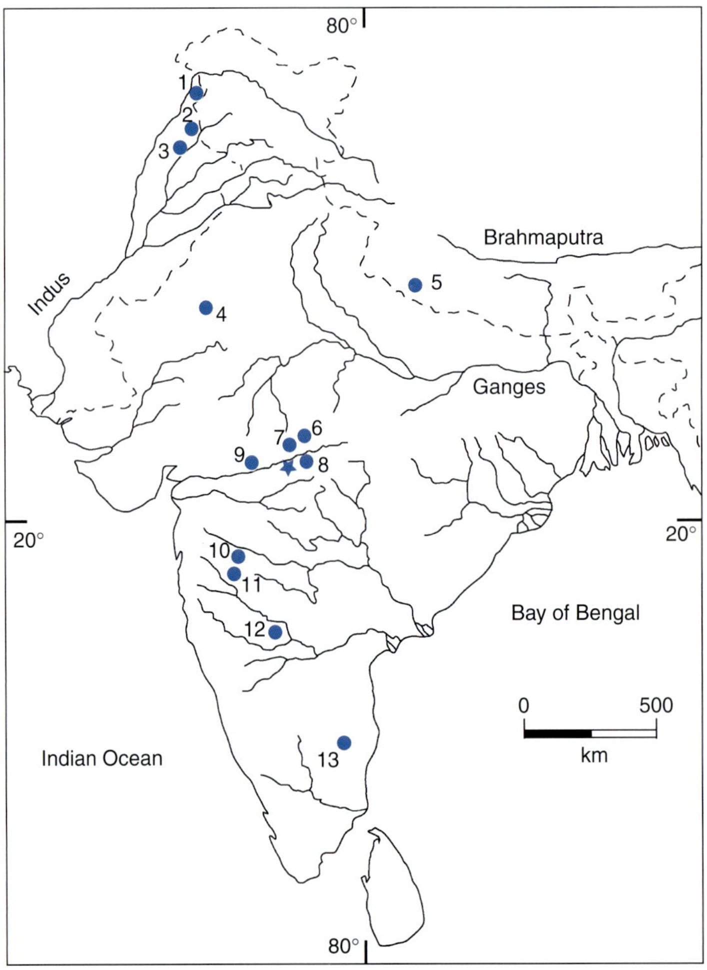

Figure 11 Principal archeological sites in India. 1, Riwat; 2, Dina; 3, Jalalpur, Singhi Talav; 5, Dang Valley; 6, Raisen Complex; 7, Bhimbetka; 8, Adamgarh; 9, Durkadi; 10, Chirki-Nevasa; 11, Kukdi; 12, Hunsgi-Baichbal Complex; 13, Attirampakkam. The star represents Hathnora.

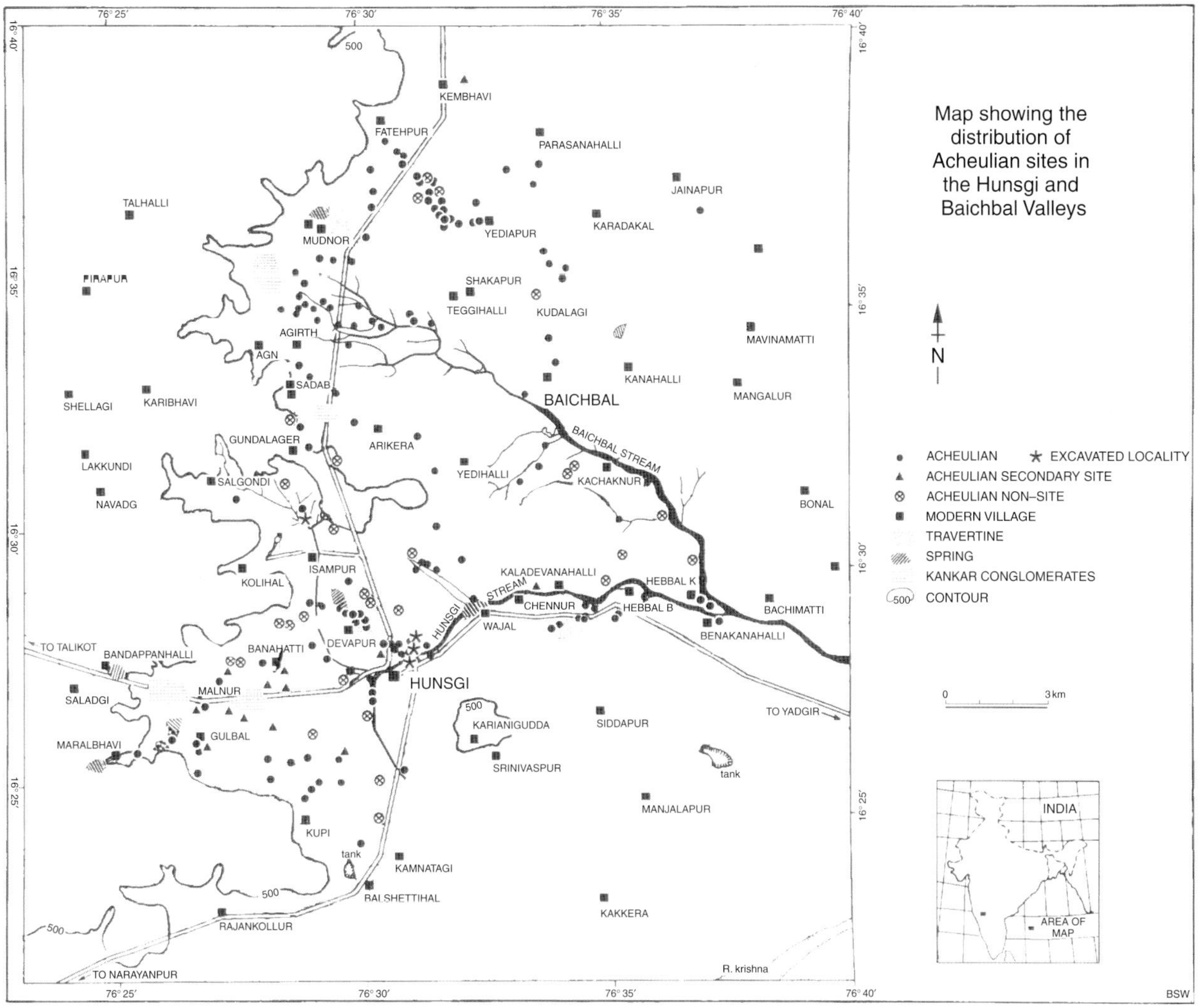

Figure 12 Acheulean sites in the Hunsgi-Baichbal Valleys, India. Paddayya (2001), figure 16.1.

unfortunately few sites and sequences are dated, and faunal remains are very rarely preserved (Petraglia, 1998). Apart from a poorly dated but probably Late Middle Pleistocene cranial fragment from Hathnora in the Narmada Valley (Cameron *et al.*, 2004), there are no hominin remains from India >30 ka. Nevertheless, there is a long cave sequence for rock shelter FIII-23 at Bhimbetka from the Acheulean to the Mesolithic that appears to show an unbroken record of continuity in lithic traditions. Open-air sites are known from Raisen, where more than 90 Early Paleolithic sites were recorded. Best known of all is the remarkable, world-class record from Hunsgi-Baichbal, Karnataka, that contains >200 Acheulean sites (see **Fig. 12**), and a settlement pattern that appears to have been bipolar, with dry season aggregation camps near water holes, and smaller wet season camps from which plant foods and small game were obtained (Paddayya, 2001). Such is the quality of information from sites such as Isampur that the actions of the hominins that quarried rock to make cleavers can be monitored in detail (Petraglia *et al.*, 2005).

Tajikistan

The loess and paleosol sequence (*see* Central Asia) of southern Tajikistan, between the Pamirs and Hindu Kush, is another Asian region with an impressive archeological record (see **Fig. 13**). Sites are always found in interglacial pedocomplexes, a fact that probably reflects the inhospitable nature of the area in glacial periods, as well as the scarcity of stone at times of active loess deposition. The oldest site found so far is Kuldara, in pedcocomplex 10–11, and dated by paleomagnetism and correlation with the marine isotope record to 855–950 ka (see **Fig. 14**). The stone tools were made from small (typically <5 cm long) chert and metamorphic pebbles, from which small

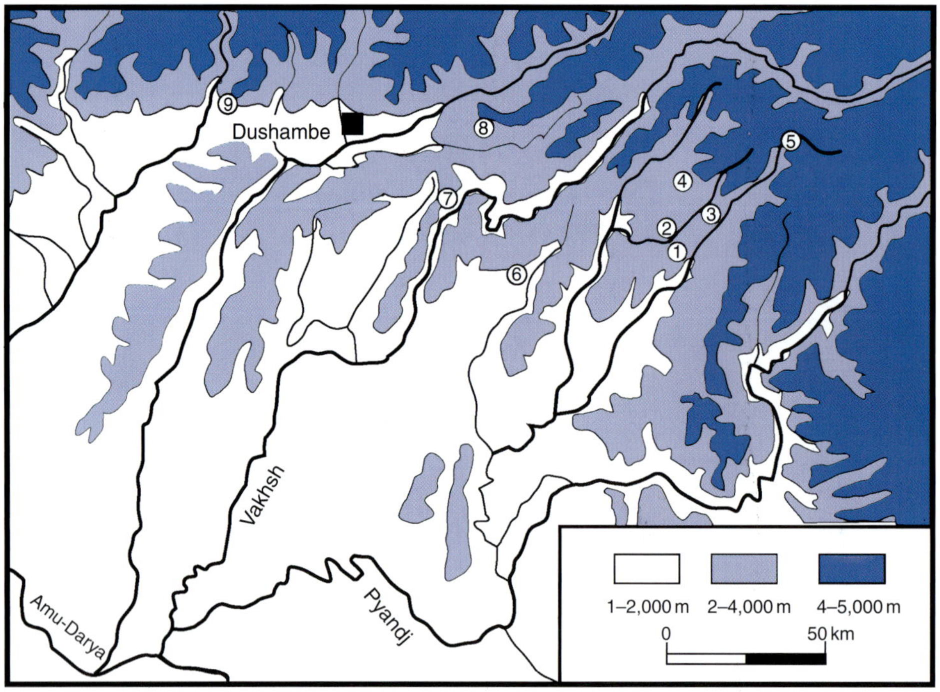

Figure 13 Principal archeological sites in Tajikistan. Paleolithic sites: 1, Kuldara; 2, Obi-Mazar/Lakhuti; 3, Khanako; 4, Chashmanigar; 5, Shugnou; 6, Ogizkuchlik; 7, Karatau; 8, Karamaidan; 9, Khudji. Ranov and Dodonov (2003), figure 1., pp. 133–147.

flakes were obtained (see **Fig. 15**). There was no standardization of tool types, minimal retouch, or core preparation. Given the size of the raw material, it is unsurprising that hand axes were absent. Later sites, also from interglacial pedocomplexes, are known from Karatau, Lakhuti, and Obi-Mazar, and extend the Tajik record into the Upper Pleistocene (Ranov, 1995).

The Caucasus

After Dmanisi, dated at 1.75 Ma, there is no clear evidence of hominins in this region until the Middle Pleistocene, a fact that may indicate an absence of relevant deposits, lack of fieldwork, or genuine absence. In the Middle Pleistocene, several caves were formed that were subsequently used for denning by cave bear (the remains of which virtually dominate most faunal sequences) and occasionally by hominins. Examples are Azyk in Azerbaijan, the caves of Kudaro I, III, and Tcona in Georgia, and Treugol'naya in southern Russia. All except Treugol'naya have Acheulean assemblages containing some hand axes and a few cleavers. Layer 7 at Treugol'naya has been dated by TL to 583 ± 25 ka; the basal assemblage from Azyk may be slightly older, and the other caves rather younger (Ljubin and Bosinski, 1995).

China and Southeast Asia

In eastern Asia, simple (mode 1) core and flake assemblages continued into the Middle Pleistocene, and the type of Acheulean bifaces and cleavers that are so widespread in western and southern Asia were seldom if ever made. Good examples are the sites of Chenjiawan (970–900 ka) and Maliang ($\leq$730 ka) in the Nihewan Basin (Chen Shen and Wei Qi (2004); see also **Fig. 4**). Here, analysis of refits, use(wear analysis, and stone reduction sequences show deliberate selection of high-quality stone, and the progressive reduction of cores by centripetal flaking (see **Figs. 16** and **17**). Although these assemblages can easily be dismissed as 'simple,' their makers nonetheless exhibited considerable skill in their use of flaking techniques and sequences when making tools for cutting, scraping, and processing meat. The best-known site in northern China is the fissure of Zhoukoudian (formerly Choukoutien) near Beijing, where hundreds of fragments of *H. erectus sensu stricto* were discovered before 1939. The 20 m deep section of locality I, where the main discoveries were made, is dated to ca. 500 ka. It is primarily a record of the giant hyanea, *Pachycrocuta brevirostris*, which accumulated almost all the other bones, including those of *H. erectus*. Hominins occasionally used the cave (hence their artifacts) but there is no evidence that they hunted (Boaz *et al.*, 2000).

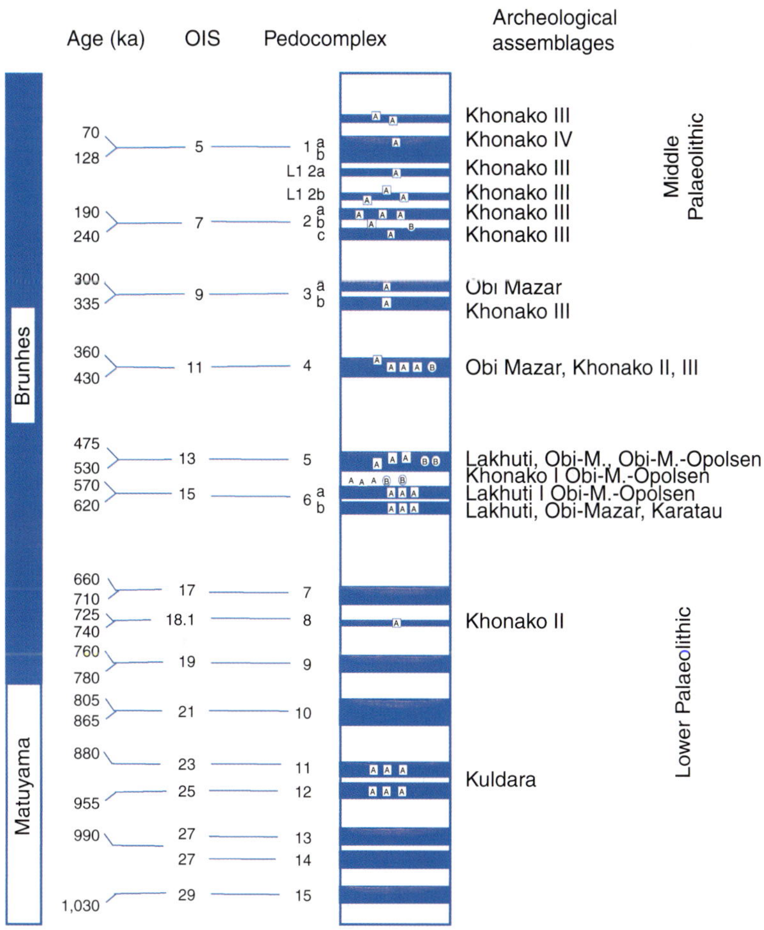

Figure 14 The stratigraphic context and age of sites in the 'loess Paleolithic' of Tajikistan. Dodonov (2002), tables 9 and 14; Ranov and Dodonov (2003), figure 10.

The same type of core and flake assemblages are found in South China and Southeast Asia, but the archeological record is less clear. Sites such as Chenshan and Jigongshan, and hominin remains from Hexian, Nanjing, and Yunxian show that the Yangste River south of the Qinling Range was occupied from the Early Middle Pleistocene (Wang Youping, 2001). Less is known about the Middle Pleistocene record of Indo-China and the southern part of China, which were tropical or subtropical, and had a distinctive *Stegodon-Auluripoda* fauna (*see* Mid-Pleistocene of Southern Asia). One of the earliest examples is Bose, South China, where large bifaces have been found, some of which look like Acheulean bifaces; these have been dated to ca. 800 ka based on their association with tektites that were part of the fallout from the Australasian Tektite Shower that occurred at that time (Yamei Hou *et al.*, 2000). Another important site is Tham Khuyen, northern Vietnam, where teeth of *H. erectus* dated at 475 ± 125 ka have been found with those of *Gigantopithecus*, the largest primate that ever lived (Ciochon *et al.*, 1996). The most important archeological evidence from Indonesia comes from the island of Flores, where artifacts ca. 800 ka have been found (Morwood *et al.*, 1998); even at times of low sea level, this island can be reached only by crossing open water (though not necessarily by boat) (Smith, 2001).

At a continental scale, there are two issues concerning the Middle Pleistocene hominin record of Asia that deserve attention.

The Movius Line

One of the most durable aspects of the Asian Paleolithic is the distinction first recognized by Movius (1949) between the Acheulean (mode 2) biface assemblages of Southwest and South Asia, and simpler core and flake (mode 1) assemblages of Southeast Asia and China. (The northern boundary of the Movius Line is less distinct, as the Tajik 'loess Paleolithic' is a simple core and flake industry, but bifaces (currently undated) are known from Iran and Kazakhstan.) The earliest Acheulean assemblages

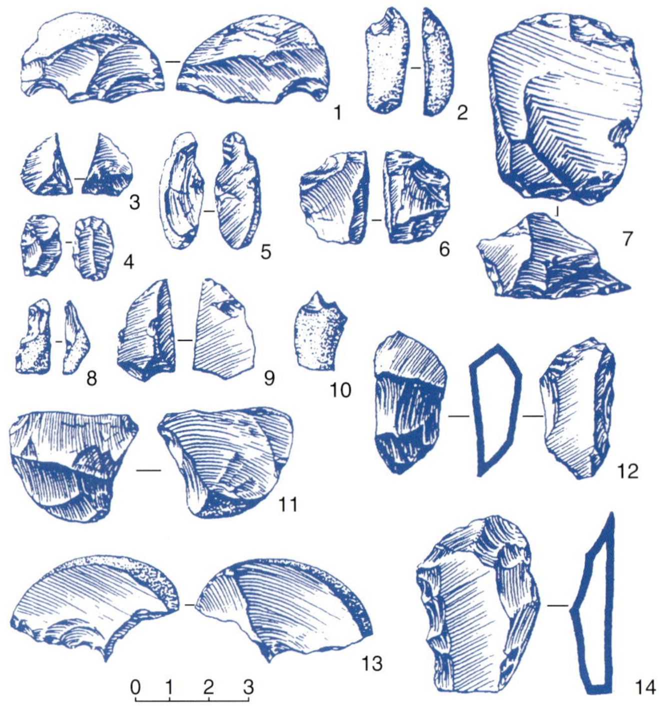

Figure 15 Stone tools from Kuldara, Tajikistan. Ranov (1995), Figure 3.

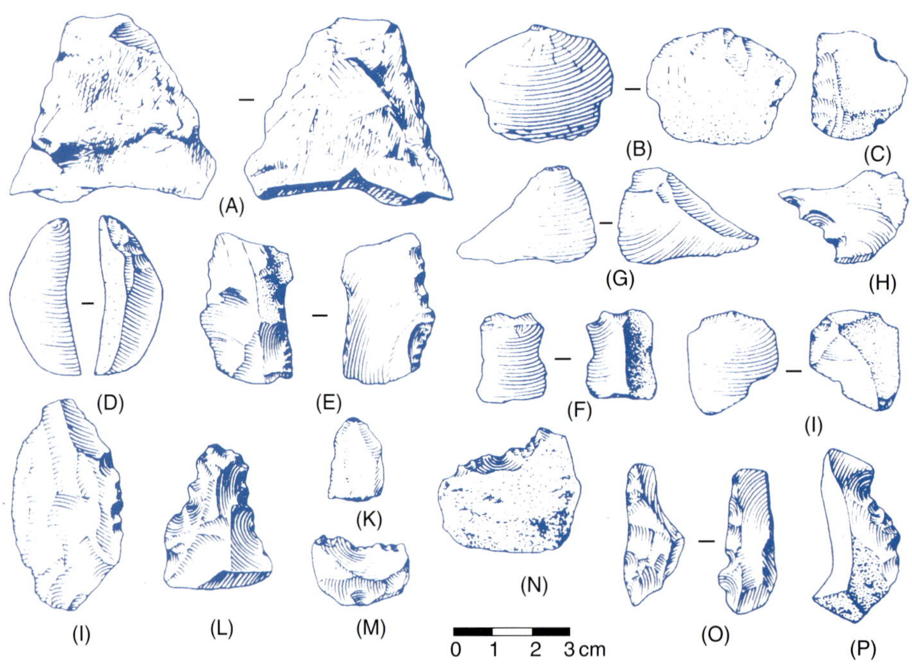

Figure 16 Stone artifacts from Chenjiawan, Nihewan basin, China. C,E,H,I,M, side-scrapers; P, notch; A,B,F,G,I, flake.

(marked mainly by bifaces and cleavers) are found in East Africa, ca. 1.6 Ma, and the earliest known in Asia are dated to ca. 1.4 Ma, from 'Ubeidiya, Israel. (The Acheulean of the Arabian peninsula and Southwest Asia outside the Levant is undated, and may prove to be similar in age, or older.) In South Asia, Acheulean hand axes were definitely in use by 600–700 ka, and possibly as early (as at Isampur) by 1.2 Ma. Although Acheulean-like bifaces are known from Bose in South China, Vietnam and possibly also

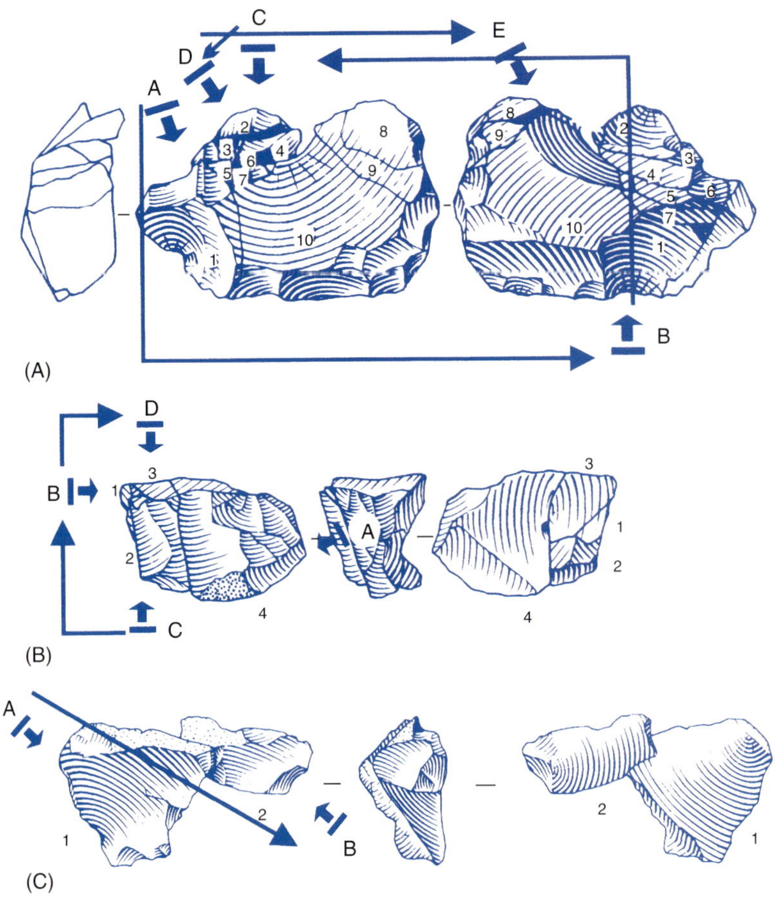

Figure 17 Refitting stone sets from Chenjiawan, Nihewan basin, China. (A) Refitting-set #1-1986; (B) Refitting-set #3-1992; (C) Refitting-set #1-1992. Chen Shen and Wei Qi (2004), figures 2 and 3.

Korea, they are very rare east of the Ganges, compared with the thousands known from India and the Levant. In South and Southwest Asia, the Acheulean is remarkably homogenous (see **Fig. 18**). Because it is earlier in Africa, its presence in western Asia is widely regarded as the result of the diffusion of ideas or hominins. If so, it seems that neither succeeded in entering China via a corridor north of the Tibetan Plateau, or the more forested regions of Southeast Asia.

Several reasons have been suggested for this division (Schick, 1994). One suggestion is that hand axes and cleavers were not needed because bamboo was widely available, and highly versatile as a raw material. Others are that the complex rules for making hand axes were not viable in the forested parts of Southeast Asia; or that hominins beyond the Movius Line did not require an overdesigned way of obtaining a cutting or chopping edge. Another reason may be related to the size of foraging groups; if these were small, there would have been fewer hands to carry bulky items such as bifaces as well as any food that had been obtained.

Middle Pleistocene Hominins: Speciation Events and Isolation?

In eastern Asia, *H. erectus sensu stricto* was the main type of hominin throughout the Middle Pleistocene (see **Table 1**). By ca. 500 ka, its cranial capacity (one of our few direct indicators of cognitive abilities and body size) had increased to within the modern human range of ca. 1,300 cc, and thus it would have had a greater range of skills and abilities with which to survive. The same pattern is probably true of western Asia, although its Middle Pleistocene fossil hominin record is almost nonexistent. Western Asia may have been inhabited by a different type of hominin from *H. erectus*. The poorly dated Narmada skull cap has been identified as *H. heidelbergensis*, a type better known from Middle Pleistocene Europe, and possibly sub-Saharan Africa. Southwest Asia *c.* 1 Ma may also have been occupied by *H. cepraniensis*, which is defined by specimens from Ceprano, Italy, and Bouri, Ethiopia. Because *H. erectus* is the only type of Middle Pleistocene hominin known from China and Indonesia, it is possible that the

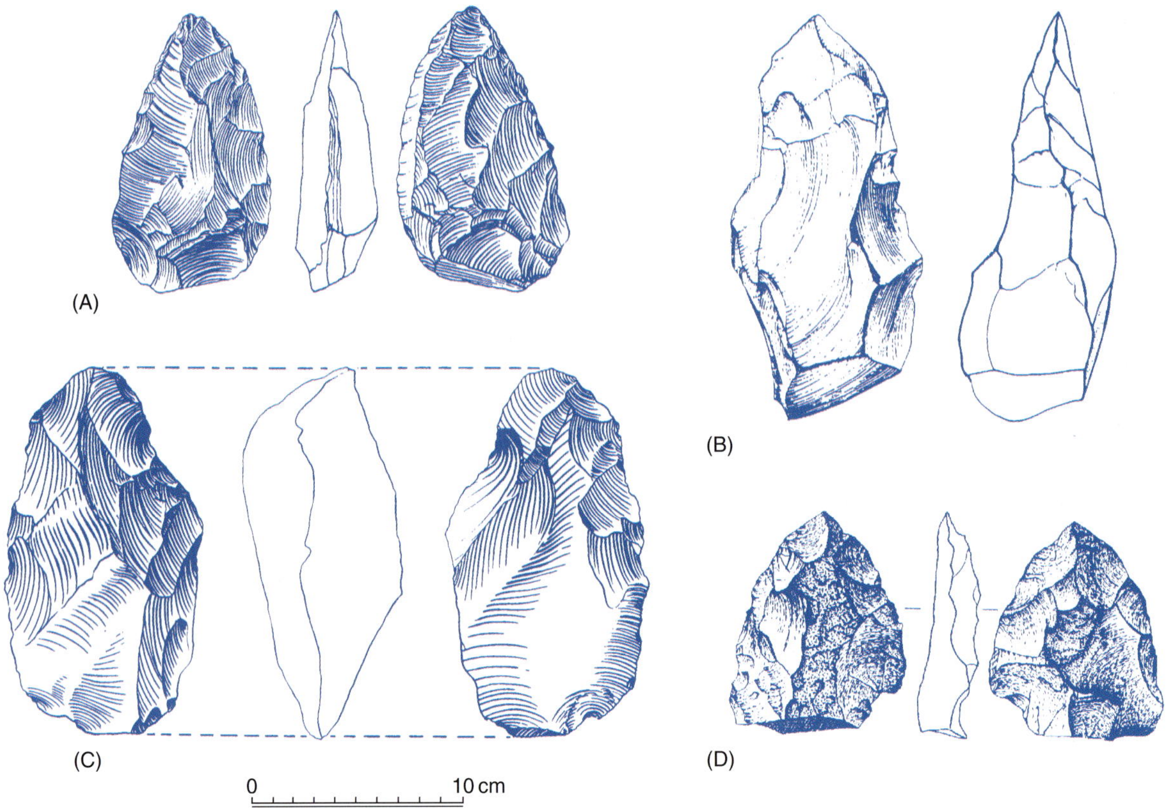

Figure 18 Acheulean bifaces from: (A) Cona cave, Georgia; (B) Evron, Israel; (C) Isampur India; and (D) the Krasnovodsk Plateau, Kazakhstan. Note that biface (D) was broken at the proximal end. Sources: Ljubin and Bosinski (1995), figure 24; Bar-Yosef (1998), figure 8.10; Paddayya (2001), figure 16.12; and Vishnyatsky (1999), figure 2.

increased severity and duration of cold, glacial periods in the Middle Pleistocene resulted in hominin populations in eastern Asia becoming more isolated than in the Early Pleistocene (*see* Mid-Quaternary in Eurasia, Overview). Sustained isolation may be a major factor in explaining why the Acheulean (and Later, the Middle Paleolithic (*see* 2.7 MYR-300,000 years ago in Asia) never reached China and Southeast Asia.

See also: **Archaeological Records**: Overview; 2.7 MYR-300,000 years ago in Africa; Global Expansion 300,000-8000 years ago, Asia. **Glaciation, Causes**: Tectonic Uplift-Continental Configurations. **Glaciations**: Mid-Quaternary in Eurasia. **Lake Level Studies**: Asia. **Loess Records**: Central Asia. **Paleoclimate Reconstruction**: Pliocene Environments. **Vertebrate Records**: Mid-Pleistocene of Southern Asia.

References

Antón, S. C. (2002). Evolutionary significance of cranial variation in Asian *Homo erectus*. *American Journal of Physical Anthropology* **118**, 301–323.

Antón, S., and Swisher, C. C., III (2004). Early dispersals of *Homo* from Africa. *Annual Review of Anthropology* **33**, 271–296.

Bar-Yosef (1998). Earlycolonizations and cultural continuities in the Lower Palaeolithic of western Eurasia. In *Early Human Behavior in Global Context: The Rise and Diversity of the Lower Palaeolithic Record* (M. Petraglia and R. Korisettar, Eds.), pp. 221–279. Routledge, London and New York.

Boaz, N. T., Ciochon, R. L., Xu Qinqi, and Liu Jinji (2000). Large mammalian carnivores as a taphonomic factor in the bone accumulation at Zhoukoudian. *Acta Anthropology Sinica* (supplement), **19**, 224–234.

Cameron, D., Patnaik, R., and Sahni, A. (2004). The phylogenetic significance of the Middle Pleistocene Narmada cranium from Central India. *International Journal of Osteoarchaelogy* **14**, 419–447.

Chen Shen, and Wei Qi (2004). Lithic technological variability of the Middle Pleistocene in the eastern Nihewan Basin, northern China. *Asian Perspectives* **43**(2), 281–301.

Ciochon, R., Long, V. T., Larick, R., *et al.* (1996). Dated co-occurrence of *Homo erectus* and *Gigantopithecus* from Tham Khuyen Cave, Vietnam. *Proceedings of the National Academy of Sciences of the United States of America* **93**(7), 3016–3020.

Clark, J. D. (1969). The Middle Acheulean site at Latamne, Northern Syria (Second paper). *Quaternaria* **10**, 1–60.

Dennell, R. W. (2003). Dispersal and colonisation, long and short chronologies: How continuous is the Early Pleistocene record

for hominids outside East Africa?. *Journal of Human Evolution* **45**, 421–440.
Dennell, R. W. (2004). Hominid dispersals and Asian biogeography during the Lower and Early Middle Pleistocene, ca. 2.0 – 0.5 Mya. *Asian Perspectives* **43**(2), 205–226.
Dennell, R. W., and Roebroeks, W. (2005). Out of Africa: An Asian perspective on early human dispersal from Africa. *Nature* **438**, 1099–1104.
de Lumley, H., de Nioradzé, M., Barsky, D., *et al.* (2005). Les industries lithiques préoldowayennes du début du Pléistocène inférieur du site de Dmanisi en Géorgie. *L'Anthropologie* **109**(1), 1–182.
Domínguez-Rodrigo, M. (2002). Hunting and scavenging by early humans: The state of the debate. *Journal World Prehistory* **16**(1), 1–54.
Dodonov, A. (2002). *Quaternary of Middle Asia: Stratigraphy, Correlation and Paleogeography*. Geos. (In Russian), Moscow.
Gaudzinski, S. (2004). Early hominind subsistence in the Levant: Taphonomic studies of the Plio–Pleistocene 'Ubeidya Formation (Israel). In *Human Paleoecology in the Levantine Corridor* (N. Goren-Inbar and J. D. Speth, Eds.), pp. 75–88. Oxbow Books.
Goren-Inbar, N., Sharon, G., Melamed, Y., and Kislev, M. (2002). Nuts, nut-cracking, and pitted stones at Gesher Benot Ya'aqov, Israel. *Proceedings of the National Academy of Sciences of the United States of America* **99**(4), 2455–2460.
Goren-Inbar, N., Alperson, N., Kislev, M. E., *et al.* (2004). Evidence of hominin control of fire at Gesher Benot Ya'aqov, Israel. *Science* **30**, 725–727.
Huayu Lu, Liu, X., Zjhang, F., An, Z., and Dodson, J. (1999). Astronomical calibration of loess-paleosol deposits at Luochan, central Chinese Loess Plateau. *Palaeogeography, Palaeoclimatology, Palaeoecology* **154**, 237–246.
Kroon, D., Alexander, I., Little, M., *et al.* (1998). Oxygen isotope and sapropel stratigraphy in the eastern Mediterranean during the last 3.2 million years. *Proceedings of the Ocean Drilling Program, ScientificResults* **160**, 181–189.
Ljubin, V. P., and Bosinski, G. (1995). The earliest occupation of the Caucasus region. In *The Earliest Occupation of Europe* (W. Roebroeks and T. van Kolfschoten, Eds.), pp. 207–253. University of Leiden Press.
Morwood, M. J., O'Sullivan, P. B., Aziz, F., and Raza, A. (1998). Fission-track ages of stone tools and fossils on the east Indonesian island of Flores. *Nature* **392**, 173–176.
Movius, H. (1949). The lower Palaeolithic cultures of southern and eastern Asia. *Transaction of the American Philosophical Society* **38**(4), 329–420.
Paddayya, K. (2001). The Acheulean Culture Project of the Hunsgi and Baichbal Valleys, peninsular India. In *Human Roots: Africa and Asia in the Middle Pleistocene* (L. Barham and K. Robson-Brown, Eds.), pp. 235–258. Western Academic and Specialist Press.
Paddayya, K., Blackwell, B. A. B., Jhaldiyal, R., *et al.* (2002). Recent findings on the Acheulean of the Hunsgi and Baichbal Valleys, Karnataka, with special reference to the Isampur excavation and its dating. *Current Science* **83**(5), 641–647.
Petraglia, M. D. (1998). The Lower Palaeolithic of India and its bearing on the Asian record. In *Early Human Behavior in Global Context: The Rise and Diversity of the Lower Palaeolithic Record* (M. Petraglia and R. Korisettar, Eds.), pp. 343–390. Routledge.
Petraglia, M. D., Shipton, C., and Paddayya, K. (2005). Life and mind in the Acheulean: A case study from India. In *The Hominid Individual in Context: Archaeological Investigations of Lower and Middle Palaeolithic Landscapes, Locales and Artefacts* (C. Gamble and M. Porr, Eds.), pp. 197–219. Routledge.
Ranov, V. A. (1995). The 'Loessic Palaeolithic' in South Tadjikistan, Central Asia: Its industries, chronology and correlation. *Quaternary Science Review* **14**, 731–745.
Ranov, V. A., and Dodonov, A. E. (2003). Small instruments of the Lower Palaeolithic site Kuldara and their geoarchaeological meaning. In *British Archaeological Reports (International Series) 1115: Lower Palaeolithic Small Tools in Europe and Asia* (J. M. Burdukiewicz and A. Ronen, Eds.) 133–147.
Rightmire, G. P., Lordkipanidze, D., and Vekua, A. (2006). Anatomical descriptions, comparative studies and evolutionary significance of the hominin skulls from Dmanisi, Republic of Georgia. *Journal of Human Evolution* **50**, 115–141.
Schick, K. D. (1994). The Movius Line reconsidered: perspectives on the Earlier Paleolithic of Eastern Asia. In *Integrative Paths to the Past: Paleoanthropological Advances in Honor of F.Clark Howell* (R. Corruchini and S. Ciochon, Eds.), pp. 569–594. Prentice Hall.
Smith, J. M. B. (2001). Did early hominids cross sea gaps on natural rafts? In Faunal and Floral Migration and Evolution in SE Asia-Australia. (J. M. B. Smith, M. Morwood, and I. Davidson, Eds.), pp. 409–416. Swets & Zeitlinger, Lisse, Netherlands.
Vishnyatsky, L. B. (1999). The paleolithic of Central Asia. *Journal of World Prehistory* **13**(1), 69–122.
Wang Youping (2001). The Middle Pleistocene archaeology of the valley of the Yangtze River. In .*Human Roots: Africa and Asia in the Middle Pleistocene* (L. Barham and K. Robson-Brown, Eds.), pp. 149–157. Western Academic and Specialist Press.
Wilson, R. C. L., Drury, S. A., and Chapman, J. L. (2000). *The Great Ice Age*. Routledge.
Yamei Hou, Potts, R., Baoyin, Y., Zhengtang, G., Deino, A., and Wei, W. (2000). Mid-Pleistocene Acheulean-like stone technology of the Bose Basin, South China. *Science* **287**, 1622–1626.
Zhu, R., Zhinsheng, A.n., Potts, R., and Hoffman, K. A. (2003). Magnetostratigraphy of early humans in China. *Earth-Science Reviews* **61**, 341–359.
Zhu, R. X., Potts, R., Xie, F., *et al.* (2004). New evidence on the earliest human presence at high northern latitudes in northeast Asia. *Nature* **431**, 559–562.

1.9 MYR-300,000 years ago in Europe

J McNabb, University of Southampton, Southampton, UK

© 2007 Elsevier B.V. All rights reserved.

Introduction

The Middle Pleistocene transition (MPT) represents the period of time between ca. 942 and 641 kyr when the Earth's orbit shifted from a more circular to a more elliptical path around the Sun. After this time, the amplitude and duration of Northern

Hemisphere glaciations changed markedly when compared to those before the transition (see **Figure 1**; Lisiecki, 2005; Lisiecki and Raymo, 2005). Against this climatic backdrop, the hominin colonization and settlement of southern and northern Europe occurred. The MPT makes a useful framework through which this process event can be viewed.

The Pattern of Hominin Colonization of Western Europe

Prior to the MPT

The earliest occupants of Western Europe appear to have arrived either from Africa or from Eurasia. In the former case, they may have represented *Homo ergaster*, or some other small brained species of African *Homo*, moving up the African Rift Valley and its Levantine extension. At the so-called Gates of Europe, the site of Dmanisi, in the Caucasus, represents the earliest Eurasian hominin site with stone tools so far discovered outside of Africa. The site is dated to ca. 1.7 Myr, and the stone tool assemblage was one of cores and flakes – see **Table 1**. Some of the hominins may have been the prey of hyaena, as a number of the fossils and artifacts were recovered from the infill of their dens. Rightmire and colleagues (Rightmire *et al.*, 2006) suggest there may be two hominin populations at this site, *H. ergaster* and a large hominin known from a single jaw, *H. georgicus*. Whether this latter was an African hominin, a purely Eurasian species, or a Eurasian descendant of an earlier African migrant remains to be determined. We do not know which hominin was responsible for the manufacture of the stone tools.

South of the Caucasus there are a number of early sites. In the Erq-el-Ahmar Formation, in the Jordan Valley, Israel, sediments are dated to the Olduvai magnetic event at 1.96–1.78 Myr (Ron and Levi, 2001). Other sediments, some 1.5 km away, but in the same formation, contain cores and flakes. Some researchers are not convinced there are artefacts. Ubeidiya, also in the Jordan Valley, is dated to ca. 1.4 Myr (Bar-Yosef and Goren-Inbar, 1993). Here a series of handax and non-handax assemblages alternate with each other. A non-handax assemblage is present at Bizat Ruhama on the southern coastal plain of Israel at about 1.0 Myr.

Another possible crossing point for hominins into western Europe is from the North African coast via the Straits of Gibraltar. Core- and flake-using hominins, presumably *H. ergaster*, were certainly present in North Africa, at sites such as Ain Hanech, at 1.8 Myr (Sahnouni *et al.*, 2002). The earliest sites in western Europe are the Barranco León 5 and Fuente Nueva 3 localities in the Guadix–Baza basin (previously the Orce sites; Gibert *et al.* (2001)) in Spain. These have been redated on paleomagnetic, but mostly on faunal, grounds to slightly younger than Dmanisi, perhaps ca. 1.6 Myr in age (Gibert *et al.* (2006), and references therein). These sites contain core-and-flake assemblages. Also in Spain is a small collection of cores and flakes from levels E9 to E14 at the Sima del Elefante, in the Sierra Atapuerca. They are certainly older than the Brunhes/Matuyama boundary at ca. 780 kyr and may be close to ca. 1.0 Myr as indicated by microfauna (Parés *et al.*, 2006). Contemporary with this, or perhaps slightly younger, are a series of sites at Ceprano, Italy. Small collections of cores and flakes have been recovered and dated between 980 kyr and 1.0 Myr.

Whichever species is ultimately shown to be the original colonizer of Western Europe, current evidence does not support an extensive initial migration. Dennell (2003) has argued that first appearances may be deceptive; hominin occupation may represent, in these cases, brief incursions into new areas. Bar-Yosef and Belfer-Cohen have noted that a successful migrant does not always make a successful colonist (Bar-Yosef and Belfer-Cohen, 2001). The paucity of early sites before the MPT in Western Europe, compared to the stronger hominin signal to the east of the region, remains an enigma (see Dennell (2003) and McNabb (2005) for reviews and references therein). Do the Spanish/Italian data imply a limited, or ultimately unsuccessful, one-off colonization event over the Gibraltar Straits? We do not know.

Across the MPT

Once again a hominin presence on the eastern margins of western Europe is attested to by sites in the Caucasus and in the Levantine Rift Valley. Gesher Benot Ya'akov (GBY) is a handax site on the banks of the Jordan River dated to ca. 780 kyr (Goren-Inbar *et al.*, 2000), and notable for early evidence of the use of fire. The strong similarity between its handax and cleaver component, and those found at contemporary African Acheulean sites, has led some researchers to suggest GBY was a stopping-off point for African handax-making hominins migrating up the rift valley. At the Evron Quarry, on the coastal plain of western Galilee, in this case situated away from the rift valley, there are hand axes (found out of context) which date anywhere between the Brunhes–Matuyama boundary and possibly 1.0 Myr (best estimates suggest the earlier end of the range). To the north there is hominin occupation in the Caucasus at sites such as Amiranis-Gora (Akhalkalaki), an open-air site with a core-and-flake assemblage dated on faunal grounds to the Late-Early Pleistocene/earlier-

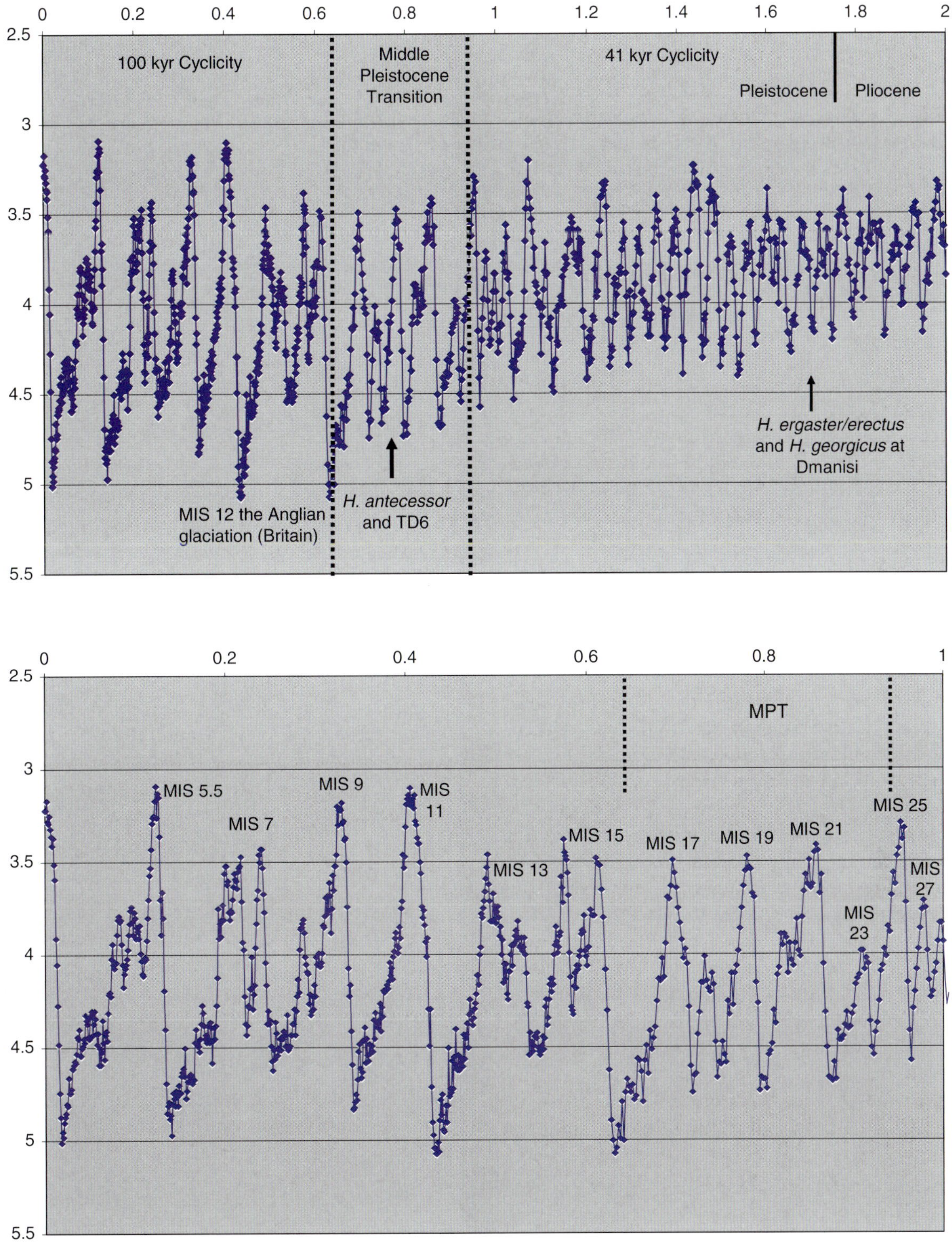

Figure 1 Marine Isotope Stages (MIS) and climate curve from combined data stack (LR04) of 57 different records from a global distribution. Data from Lisiecki 2005, and Lisiecki and Raymo 2005. Data used with permission. Horizontal scale is time measured in millions of years (myr). Vertical scale is delta ^{18}oxygen record of benthic foraminifera. Upper diagram shows overall climate change from Dmanisi (1.7 myr) until the Anglian glaciation (MIS 12) of Britain, herein taken as the boundary between the earlier Middle Pleistocene and the later Middle Pleistocene – see **table 1**. The upper diagram makes the point that after the shift to orbital eccentricity (period of change in orbit = Middle Pleistocene Transition - MPT), glaciations are colder, and interglacial peaks are warmer. In the lower diagram the climate curve is depicted for the last 1.0 myr. MIS stages have been added.

Table 1 Selected stone tool assemblages, and trends in stone tool assemblage types from Europe between 1.0 myar (MIS 28) and the advent of Levallois/prepared core technology (PCT) marking the beginning of the Middle Palaeolithic in Europe. Key to table: NHA = non-handaxe assemblages; A = Acheulean (handaxe) assemblages; BG = Boxgrove; HL = High Lodge; Ar = Arago; StA = St Acheul; SH = Sima de los Huesos; Cep = Ceprano; MP = Monte Poggiolo; TD6 = Atapuerca Trinchera Dolina 6; TD 4 = Atapuerca Trinchera Dolina 4; S. de Elefante = Atapuerca Sima del Elefante; B = Bilzingsleben; V = Vértesszöllös; Sc = Schöningen; K-H = Kärlich H; K-G = Kärlich G; M1 = Miesenheim 1; T = Treugol'naya Cave; GBY = Gesher Benot Ya'aqov; A-G = Amiranis-Gora (Akhalkalaki). Dotted lines indicate period boundaries and their continuation. Double headed black arrows indicate potential date range for individual sites. Single headed black arrows indicate continuation over time. For sources see text citations and references

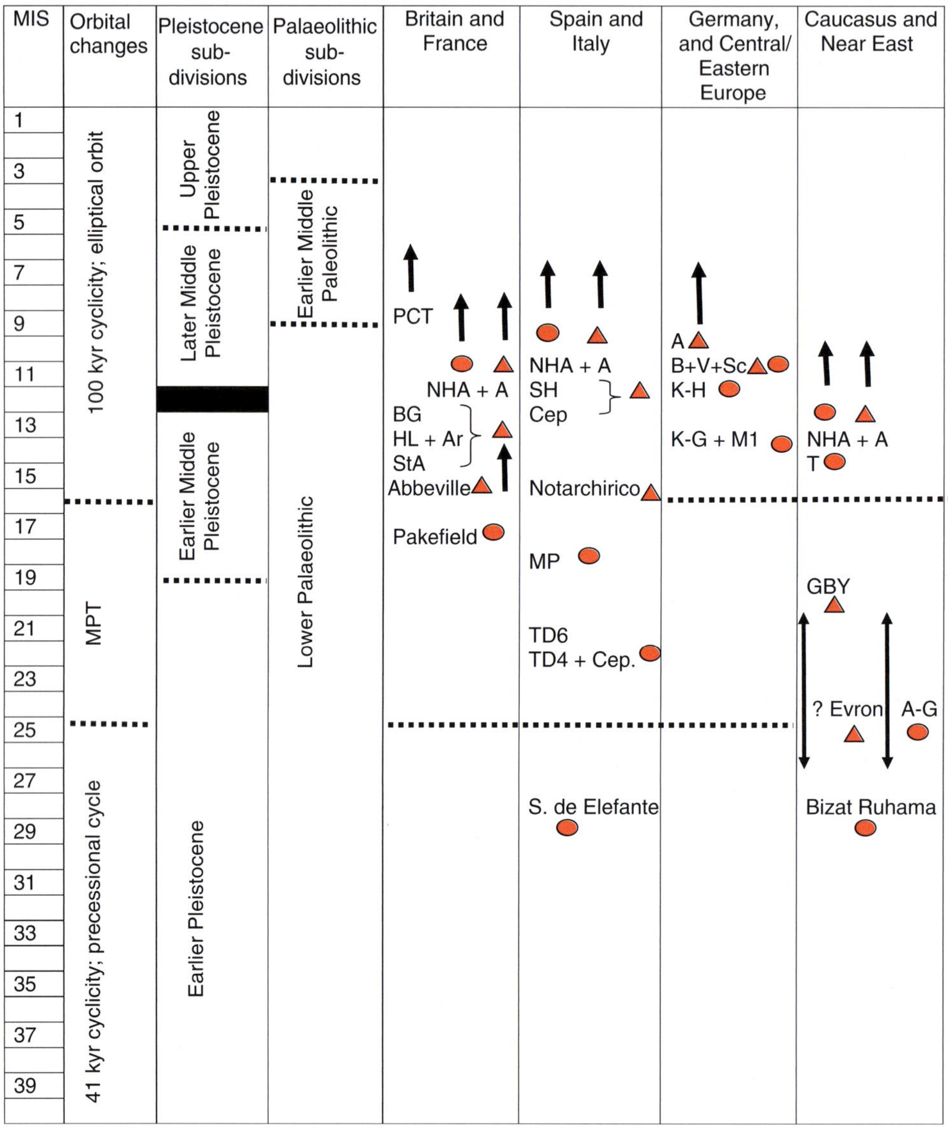

Middle Pleistocene transition (see **Table 1**; see papers in Lordkipanidze *et al.* (2000)).

There are three, and possibly four, occurrences in southern western Europe, where archeological sites dating to the MPT are known: two in Spain and two in Italy. The two Spanish examples represent different layers at the same site, the Trinchera Dolina in the Sierra Atapuerca. The lower of the two levels, TD4, represents a small collection of cores and flakes, made on local quartzite, and dated to MIS

22. The higher layer, TD6, is a more significant occurrence (Bermúdez de Castro *et al.*, 2004). Over 300 stone tools, but no hand axes, have been discovered. Many of them have been retouched to form scrapers or denticulates. Associated with these artifacts are specimens of a new hominin type, *H. antecessor*. Just as with the Dmanisi hominins, this may be an African migrant, the descendant of one, or a purely Eurasian species, we do not as yet know. To date, no examples of this species have been found outside of southern Europe. Cut marks on the *H. antecessor* bones suggest that this species was a cannibal. Two Italian sites are dated to the MPT. At Ceprano, already noted to have core-and-flake assemblages near the MPT boundary, a skull cap was discovered higher up in the valley sequence than the level of the archeological sites (Manzi, 2004). Unaccompanied by stone tools, the skull cap may represent a portion of the cranium of an adult *H. antecessor* (Manzi, 2004). Monte Poggiolo, in Italy, is suggested to date from this period; however, uncertainties concerning its radiometric dating, and the clear association of the artifacts with securely dated deposits, have left doubts hanging over this substantial early core-and-flake assemblage (Villa, 2001).

To the north of the Alps, only one site has been securely dated to the MPT. This is the site of Pakefield, Suffolk, England (Parfitt *et al.*, 2005). Flakes and cores have been discovered in sediments of the Cromer Forest-bed Formation. Sediments from this section of the formation date to at least ca. 700 kyr (MIS 17). The site clearly demonstrates that earlier phases of the earliest occupation of Europe were not confined to the south of Europe. Associated fauna suggest a warm Mediterranean-like climate.

After the MPT

One thing is clear from **Table 1**: it is only after the shift to the 100-kyr cycle that hand axes appear in the archeological record of western Europe with any frequency. From this time on, the only hominin present in Europe is *H. heidelbergensis*, whose evolutionary trajectory ultimately led to the Neanderthals. It is tempting to associate the appearance of *H. heidelbergensis* and hand axes with the shift to the 100-kyr periodicity. Was the climatic change sufficient to open up previously closed environmental niches and so initiate the movement of animals, and the hominins that preyed on them, into new habitats? Did *H. heidelbergensis* migrate into Europe in one wave, or was it a series of migrations? How was it that this species seemed better able to survive European mosaic environments than *H. antecessor*? Within its suite of adaptive capabilities, what allowed it to cope with European seasonality, and to overcome the difficulties that accompanied overwintering in European environments? At present, we cannot answer these questions. See Roebroeks (2001) for a summary and review.

Throughout western, central, and eastern Europe, assemblages with and without handaxes have been found across a variety of locally variable open-air and cave sites. At present, there is no reason to doubt these were all made by groups of *H. heidelbergensis*. The handax sites are all subsumed under the label Acheulean. The non-handax sites are given a series of names reflecting either regionally significant cultural patterns, or local variations on the non-handax assemblage type. It is a moot point whether these non-handax sites represent distinct cultural phenomena, the result of extended lineages of intergenerational learning, or whether they represent locally variable responses by hominins who normally made hand axes. For a good review of the Europe-wide variability in Middle Pleistocene stone tool assemblages, see Roebroeks and van Kolfschoten (1995).

The frequency of stone tool assemblages in western, central, and eastern Europe, which date to the Later Middle Pleistocene (post-MIS 12 glaciation), suggest an increase in population, though the poor degree of survival of older sites should be factored into such interpretations. From the point of view of the stone tool assemblages, however, there is little change in the character of what was made until the wide-scale adoption of Levallois/prepared core technology (PCT), possibly at the end of MIS 9. This date of ca.300 kyr, and the adoption of PCT, is taken to be the end of the Lower Paleolithic and the beginning of the Middle Paleolithic – see **Table 1**. In some areas, the beginning of the Middle Paleolithic may have been earlier. Many scholars see a link between the handax phenomenon and the rise of Levallois. French archeologists in particular can track nascent tendencies toward Levallois-like PCT throughout this period (Orgnac 3 (Ardèche) and Baume Bonne (Quinson, Alpes de Haute-Provence; see papers in Tuffreau (1996).Whether this implies geographically distinct multiple local inventions, or alternatively the widespread diffusion of ideas or Levallois-using hominin groups, remains unclear.

Social Construction

Archeology reveals very few clues that relate to the construction of Middle Pleistocene society, by which I mean demonstrating the nature of relationships between individuals, and between individuals and larger groups at a series of hierarchical levels (e.g., personal core group, hunting coalition, family group, whole group, etc.).

Currently, we can only speculate on the nature of relationships between individuals. Artifact analysis may shed some light on these relationships, given the assumption that social signals are encoded within the making, character, and use of material culture. A number of scholars have argued that very large and symmetrical hand axes have little practical use. Rather, they represent social signaling devices aimed at attracting mates (Kohn and Mithen 1999) – in effect the hominin males' substitute for the peacock's tail; as such, they would represent an individual fitness indicator. Much work remains to be done on this aspect of material culture studies.

Gamble (1999) has argued that hominins during the Lower Paleolithic lived in small groups, moving around familiar territories from which they rarely, if ever, strayed. Their knowledge of that territory was acquired through movement along well-known paths and trails. This world, and the hominins' comprehension of it, was dominated by the local and the familiar. Because of a limited linguistic capacity, relations were maintained with individuals via visual signals and the familiarity of often-repeated socially sanctioned gestures and activities. The local group represented the totality of the social world, and contact with other groups in other territories would have been discouraged. Gamble uses the transport of raw stone materials, and finished artifacts whose rock type can be traced to their source, as evidence of the extent of group territories.

Essential to identifying any form of social construction is the presence of language and of the ability to correctly interpret the intentions and actions of others. Dunbar (2004) has argued that *H. heidelbergensis* at this time was capable of limited grammatical language, and had sufficient cognitive abilities (see below) to empathize to some degree with other individuals. Such abilities would greatly empower socially cooperative action. How much of this would have encouraged a sense of individuality and personality within and between group members is a subject that will engage much future research. See papers in Gamble and Porr (2005).

Subsistence Behavior

The question of how *H. antecessor* and *H. heidelbergensis* acquired their food, and what they ate, is intimately bound with the character of social relations between individuals, sub-groups, and between the sexes. These relationships would have influenced the pattern of hunting and gathering adopted. It may well be that socially and geographically distant groups had very different subsistence practices. There are no *a priori* grounds why a single species-wide pattern need have prevailed everywhere.

At a site in Schöningen, Germany, dated to MIS 11, a series of wooden spears, made of spruce wood, has been excavated in a context strongly suggestive of their use in deliberate hunting. A herd of horses is interpreted to have been mired along the margins of an ancient lake and then killed. Deliberate fires were set along the shoreline, possibly as part of the hunt, possibly for carcass processing and cooking/eating of the meat. Interestingly, this site contained no handaxes. Another wooden spear point of the same date was recovered from the core-and-flake site at Clacton-on-Sea, Essex, England; this one was made of yew wood. Spears are taken to be definitive hunting equipment, whereas stone tools, whether they be the handaxes of the Acheulean, or the sharp flakes and retouched flake tools of the non-handax assemblages, were hand-held carcass-processing equipment, used on the ground (among other things – see below).

At Boxgrove, an MIS 13 site in Sussex, England, hunting is posited on the basis of the successful defense of the carcass of a large species of extinct horse. The extensive cut marks from stone tools left on the horses' bones indicate that the hominins were in no hurry to leave, and they controlled the carcass long enough to extensively butcher it. Rhinoceros bones from this site may show similar evidence. The basis of this assertion is the primacy of stone tool cut marks on the bones. Here a tool has slipped during the process of filleting the meat and removing sinew, cutting into the bone by accident. These marks are overlain by the teeth marks of carnivores, who must then have only gained access to the carcass after the hominins had finished with it. Whether deliberate hunting was involved at Boxgrove, or whether a party of hominins came upon an injured or sick animal that they were able to dispatch with relatively little effort, remains to be proven. The horse butchery was accomplished with the aid of hand axes. Both Boxgrove and Schöningen very clearly imply sufficient social cooperation between individuals acting as a group in order to secure and then defend the carcass against other predators and scavengers. See papers in Gamble and Porr (2005) for more details, and Gamble (1999) for an overview.

By the time of *H. heidelbergensis*, and *H. antecessor*, genus *Homo* was adapted to a predominantly meat-based diet. Only from such a highly protein-rich food source would hominins have obtained sufficient energy resources to survive winter conditions when edible plant resources would have been at a minimum. But the question of how early *Homo* acquired meat remains frustratingly difficult to resolve; was this animal a predator-hunter, or a scavenger, or something in between? The archeological record is contradictory and fragmented. At the two Spanish sites of Aridos I and II, to the south of Madrid, there is ample evidence of hominins interacting with elephant carcasses. In the former site, bifaces and a small collection of Levallois pieces were found scattered among the bones of the animal. Butchery is strongly implied. In the latter site, the case is similar, except that no Levallois pieces were found. Both assemblages would class as Acheulean. Unfortunately, it was not possible to discern how the elephants died. Two other Spanish sites, in the mountain heartland of Spain, Torralba and Ambrona, were for many years considered good evidence of big game hunting – elephants driven into swampy ground and subsequently butchered. More recent studies set the two sites against a background of occasional visits by hominins to an area not intensively inhabited, thus explaining the low frequency of cut marks and stone tools from the sites. While it can be proved that hominins butchered occasional animals in the vicinity, proactive big-game hunting can not be clearly demonstrated.

As important as meat itself is a supply of fats, which aid in the breaking down of meat proteins. A diet of animal tissue represents the only effective overwintering solution for hominins surviving the cold and resource-scarce winter months and the early spring. Late winter/early spring may have been a time of dietary stress for hominins. Although animals would still have been present, their fat reserves would have been severely depleted over the winter. For hominins, a surfeit of lean meat would have represented a source of poor nutrition without the fat content to break the meat proteins down. Seasonal dietary stress may have been a frequent part of Middle Pleistocene hominin life.

While it is an assumption that edible plant material would have formed an important contribution to the diet, just what was eaten, and when, remains unknown. Presumably, the majority of edible plant foodstuffs would be a seasonal, late spring/summer resource. Fish, marine mammals, and birds, may also have contributed to the diet, but as yet there are no substantive data to support the idea that these resources were exploited frequently.

Tool Behavior

Most tractable rock types are capable of being knapped, and were used to manufacture stone tools in the Lower Paleolithic. In regions dominated by fine-grained quartzites, and by siliceous rocks such as flint, these appear to have been extensively used. Knapping involved either the use of a hard hammer stone, usually a cobble, for working cores to make flakes, and roughing out blanks for hand axes, and/or the use of a soft hammer to thin and shape already-worked rough-outs for hand axes. Soft hammers may have been made of antler, bone, wood, or even softer rock types.

Although there are regional variations on what was made throughout Europe during the Lower Paleolithic, for the most part these represent variations on a small number of basic tool themes. These were handaxes, and their characteristic debitage, thinning flakes; cores knapped for flakes; the un-retouched flakes which could be used as sharp cutting knives; and a variety of retouched tools. The latter were flakes or pieces of shattered raw material whose edges were trimmed by direct percussion along a length of one or more edges. Commonly this trimming was applied to make a scraper, but other retouched tool types were equally as common: notches, denticulates, flaked flakes, etc. As noted above, stone tools were a hand-held processing technology. They were used for carcass dismemberment and processing; usually only the hand axes were removed from a butchery site for further use. The cores, flakes, and retouched tools were left behind. It is only during the Middle Paleolithic that deliberate stone weapon tips begin to appear in the archeological record.

Materials other than stone were utilized, probably extensively, but the evidence for organic materials is very sparse, a reflection of chance preservation. Shaped wooden spears from Schöningen and Clacton-on-Sea have already been discussed. Schöningen also preserved lengths of wood that have been split, and the discoverers interpreted these as composite tools, the wooden handles into which worked stone items were inserted (Thieme, 2005). Other wooden tools were interpreted as a throwing stick and even a charred cooking stick, used to hang meat over the fire. At a number of sites there are a small number of hand axes that were flaked on fragments of thick bone. They may represent a regionally specific behavioral response to shortages of suitable raw materials since they almost all come from Italy. There is an excellent German example from Bilzingsleben. From Boxgrove, a series of soft hammers (Pitts and Roberts, 1997) has been discovered in antler and in bone. They show extensive wear from use and were clearly used as knapping tools for some time before being abandoned or lost.

The hominin use of fire, its early discovery and the nature of its subsequent usage, has always been surrounded by controversy. But sites like

Schöningen and Beeches Pit, Suffolk, England clearly demonstrate the controlled use of fire. Gowlett (2006) argues that hominins may have used fire, and been able to sustain it, but may not have been able to kindle it. As such, it was more of an ecological tool.

What the Archeology can tell us about Cognition

The material culture of *H. antecessor* is confined to cores and flakes, but the more varied repertoire of *H. heidelbergensis* allows some insights into the cognitive capabilities of this species. To begin with the ability to retain a complicated sequence of thoughts/ actions in the memory, and act them out, is implied by the process of cooperative behavior (see above). This is also implied in the ability to conceive of composite tools (see above), especially since they comprise elements of two different media (stone/ organic). The enchainment of separate actions and distinct concepts, as part of a single activity, is further implied in the use of two different hammer types (hard and soft) to make many Acheulean handaxes. The evidence of curated soft hammers from Boxgrove underscores this and implies the capacity for forward planning (Hallos, 2005).

But the signals are contradictory. The curated soft hammers from Boxgrove may have been abandoned when they were worn out, but the spears from Schöningen could certainly have been used again. While some hand axes appear to have been removed from kill sites for later use, a great many perfectly serviceable examples are scattered round the landscape, seemingly abandoned for no particular reason. Was *H. heidelbergensis*' memory/cognitive capacity sufficient only for the immediate task? Or is the paucity of information merely blurring a situationally variable behavioral pattern? Apart from the bone handaxes cited above, and despite the survival of bones on a great number of Middle Pleistocene sites, no other bone tools have ever been found. It is worth noting that *H. heidelbergensis* made only handaxes out of bone – no other tool types appear to have been fashioned. Does this actually imply a rather limited cognitive capacity, or alternatively a lack of cognitive flexibility in developing novel solutions for familiar problems, and adapting old practices to new situations?

The Ideational Realm

We have no direct clues as to what early *Homo* in Europe, or anywhere else for that matter, thought about its world, its place within it, and what its relationship to the other world might be – if hominins thought about it all. There are no clear-cut examples of deliberate burials in the Middle Pleistocene of Europe. At the Sima de los Huesos, in the Sierra Atapuerca, Spain, some 28, or possibly more, *H. heidelbergensis* individuals were thrown down a deep vertical shaft at the back of the cave. The excavators interpret this as deliberate interment, and cite the presence of a single handax found in and among the remains as an example of an offering left among the dead (Bermúdez de Castro *et al.*, 2004). But the hominin bones were found among carnivore remains, and the lower chamber may have been a bear's den with a lower entrance as yet undiscovered. This implies that the hominins' remains were brought in as prey. Even if they were deliberately thrown down the shaft, it may represent nothing more than the disposal of corpses. The case for deliberate interment, and any associated ritualistic qualities, remains to be proved.

There are no examples of Lower Paleolithic artwork in Europe. There are two examples of hominin figurines, one from Berekhat Ram in Israel, and one from outside the area at Tan-Tan in Morocco, associated with Acheulean handaxes (Bednarik, 2003). But in both cases, the resemblance to a human-like figure is questionable, and their interpretation should remain in a suspense account until validated by unambiguous examples.

A sense of the esthetic is often claimed for hand axes which are exceptionally large and which show a clear effort to produce bilateral symmetry. This is a possibility, and these artifacts may have had a distinct social resonance (see above). In other cases, hand axes which have natural holes in the center, around which the knapper has carefully flaked a working edge along the margins of the nodule or blank, might imply a sense of visual appreciation, or even humor. Hand axes with fossils in the center of the blank or nodule are also sometimes suggested to have been purposefully knapped to emphasize the decorative character of the fossil. But like hand axes with a hole in the center, these are so few and far between that gifting their knappers with a clear sense of their overall esthetic value is difficult. They may well have failed to recognize fossils as having any form of visually pleasing quality, and a hole in the center of a knappable piece of flint may have been nothing more than that.

Conclusion

The archeological record from the earliest occupation of Europe until ca. 300 kyr and the introduction of PCT is a rich and varied one. The European data provide key insights into the variability present in the adaptation and lifestyles of early species of *Homo*. This record is every bit as rich as that found in other parts of the Old World. Future research directives will need to focus on the nature of social behaviour in the Middle Pleistocene, and especially the significance of material culture is its social context. Other critical research agendas include the date and character of the earliest occupation of Europe, the question of the relationship between *Homo antecessar* and *Homo heidelbergensis*, and the explanation of non-handaxe assemblages in Europe.

Glossary

Acheulean Name given to an assemblage, or assemblage type, characterized by the presence of handaxes or thinning flakes. First identified at St. Acheul, Abbeville, France.

Assemblage A collection of artifacts, usually stone tools, but can refer to any preserved material, that is united by a common feature or set of related features. This commonality may be the presence or absence of diagnostic tools. It may, on the other hand, be a particular spatial or temporal boundary that marks the collection of artifacts as discreet. For example, a collection of stone tools from a site/layer which lack a particular type of stone tool might be grouped together and characterized by that absence; alternatively, the layer, or restriction to a part of the site, might impart distinctiveness or discreetness.

Assemblage Type A group of assemblages that share a common feature or features.

Composite Tools Tools made up of two or more different components, often of two different substances. For example, a wooden or antler shaft, with a stone tip.

Cores A nodule or river cobble of rock that has been shaped by knapping. Flakes have been removed by direct percussion.

Denticulate A particular type of stone tool usually made on a flake. The edge is retouched to form a ragged and uneven edge when seen in profile. Often described as saw-toothed.

Direct Percussion The process of striking a block of raw material directly with a percussor or hammer (hard or soft).

Fitness Indicator A physical attribute of an animal that stands proxy for a statement on the physical health, genetic quality, social standing, and desirability as a mate of an individual (phenotype).

Flaked Flake A particular type of stone tool made on a flake. This is a flake that has had another flake removed from any edge or off either upper or lower face of the flake. Sometimes the concavity left by the detachment was used, sometimes the spall was used, sometimes both were.

Flakes The portions of raw material that are detached from a core during knapping by direct percussion.

Formation A single sedimentary layer, or group of related sedimentary/rock strata having a common feature that unites them. For example, a river terrace may be thought of as a formation, it may be made of one layer of gravel, but it is more likely to be made of a number of different gravel layers interspersed with other layers of sand, silt, or clay.

HandAxe A stone tool made by shaping a block of raw material, cobble, or big flake. The tool is bifacial – knapped on both upper and lower faces, and the whole is thinned and shaped to a number of different outline shapes. The earliest are about 1.7 Myr from Africa, and they continued to be in use until the end of the Middle Paleolithic. The process of thinning the handax produces a characteristic type of flake – the thinning flake.

HandAx Assemblage An assemblage of stone tools from one layer at a site, as from a number of related layers which contains handaxes or thinning flakes, or both.

Hard Hammer or Hammer Stone A pebble or cobble of rock used in knapping by direct percussion, usually cores, but also in the early stages of roughing out a handax.

Hominin A term replacing the older word hominid. It has been adopted because of the genetic work which demonstrates that humans, their ancestors, and related genera, are closer genetically to chimpanzees. But gorillas separated from this group of related genera a long time ago. The old use of the word hominid included gorillas with the human family on the basis that their skeletons showed marked similarities.

Knapping The name given to the process of controlled breakage of a block of raw material. The process may be employed to shape a block of raw material into a tool. Alternatively, it may be to use the flakes that come off during the process of knapping to make tools from.

Levallois/Prepared Core Technology (PCT) A way of knapping a block of raw material that carefully shapes and prepares one surface of that block. The resulting core, with its carefully prepared surface, can be used to detach specific types/shapes of stone tool, depending on the particular type of surface preparation applied to the core. The first appearance of Levallois is taken to mark the beginning of the Middle Paleolithic. It may occur at different times in different regions.

Microfauna Small mammals, such as small rodents.

Middle Pleistocene Transition (MPT) The period of time it took for the Earth's orbit to shift from a more circular path around the Sun to a more elliptical one. Prior to the MPT, the glacial–interglacial cycle was dominated by the precessional cycle. The rotation of the Earth around its axis wobbles like a spinning top. The 41-kyr precessional cycle is the length of time it takes to complete one rotational wobble. After the MPT, the orbital pattern was more elliptical, and at this time the amplitude and wavelength of the glacial–interglacial cycle changed. The cycle is now more closely tuned to a 100-kyr pattern. But whether the orbital eccentricity has been the major factor in causing this change remains unclear.

Non-handAx Assemblage A site, or group of related layers from a site layer from a site, where the stone tool assemblage lacks hand axes or the evidence of their manufacture – the thinning flakes.

Percussor A hammer such as a hard hammer stone or a soft hammer such as one made out of bone, antler, or wood.

Notch A particular kind of stone tool usually made on a flake. A concavity is made on the edge of a flake by small contiguous retouch removals.

Retouch or Retouched Tools The process of shaping an edge, usually a flake, in order to make a hand-held tool on a flake. Retouch is usually confined to the flake's edge, and only shapes or sculpts the edge.

Scraper A particular type of stone tool usually made on a flake. A length of flake's edge is shaped by retouch to make a long continuous retouched edge with a smooth profile.

Soft Hammer Usually made of wood, bone, or antler. These are used in direct percussion in the Lower Paleolithic for the manufacture of handaxes. They are used in thinning and shaping a handax as they take off much less volume of surface raw material during the thinning process than a hammer stone. It is possible that soft rock types can also be used as soft hammers.

See also: **Archaeological Records**: Overview; 2.7 MYR-300,000 years ago in Asia; Global Expansion 300,000-8000 years ago, Africa; Global Expansion 300,000-8000 years ago, Asia; Global Expansion 300,000-8000 years ago, Australia; Global Expansion 300,000-8000 years ago, Americas; Neanderthal Demise; Postglacial Adaptations. **Vertebrate Studies**: Interactions with Hominids.

References

Bar-Yosef, O., and Belfer-Cohen, A. (2001). From Africa to Eurasia – Early dispersals. *Quaternary International* **75**, 19–28.

Bar-Yosef, O., and Goren-Inbar, N. (1993). *The lithic assemblages of 'Ubeidiya: A Lower Palaeolithic Site in the Jordan Valley.* The Hebrew University of Jerusalem, Jerusalem.

Bednarik, R. G. (2003). A figurine from the African Acheulian. *Current Anthropolgy* **44**, 405–413.

Bermúdez de Castro, J. M., Martinón-Torres, M., Carbonell, E., *et al.* (2004). The Atapuerca sites and their contribution to the knowledge of Human Evolution in Europe. *Evolutionary Anthropology* **13**, 25–41.

Dennell, R. (2003). Dispersal and colonisation, long and short chronologies: How continuous is the Early Pleistocene record for hominids outside East Africa. *Journal of Human Evolution* **45**, 421–440.

Dunbar, R. (2004). *The Human Story.* Faber and Faber, London.

Gamble, C. (1999). *The Palaeolithic Societies of Europe.* Cambridge University Press.

Gamble, C., and Porr, M. (2005). *The Hominid Individual in Context.* Routledge, Abingdon.

Gibert, J., Gibert, L., Ferràndez-Canyadell, C., Iglesias, A., and González, F. (2001). Venta Micena, Barranco León-5 and Fuentenueva-3: Three archaeological sites in the Early Pleistocene deposits of Orce, South-East Spain. In *A Very Remote Period Indeed, Papers on the Palaeolithic Presented to Derek Roe* (S. Milliken and J. Cook, Eds.), pp. 144–152. Oxbow Books, Oxford.

Gibert, L., Scott, G., and Ferràndez-Canadell, C. (2006). Evaluation of the Olduvai Subchron in the Orce ravine (SE Spain): Implications for Plio-Pleistocene mammal biostratigraphy and the age of the Orce archaeological sites. *Quaternary Science Reviews* **25**, 507–525.

Goren-Inbar, N., Feibel, C. S., Verosub, K. L., *et al.* (2000). Pleistocene milestones on the Out of Africa corridor at Gesher Benot Ya'aqov, Israel. *Science* **289**, 944–946.

Gowlett, J. A. J. (2006). The early settlement of Northern Europe: Fire history in the context of climate change and the social brain. *Comptes Rendus Palevol* **5**, 299–310.

Hallos, J. (2005). "15 Minutes of Fame": Exploring the temporal dimension of Middle Pleistocene lithic technology. *Journal of Human Evolution* **49**, 155–179.

Kohn, M., and Mithen, S. (1999). Handaxes: products of sexual selection? *Antiquity* **73**, 518–526.

Lisiecki, L. (2005). http://www.lorraine-lisiecki.com

Lisiecki, L., and Raymo, M. E. (2005). A Pliocene-Pleistocene stack of 57 globally distributed benthic delta^{18}oxygen records. *Paleoceanography* **20**, 1–17, (PA1003. doi10.1029/2004PA001071).

Lordkipanidze, D., Bar-Yosef, O., and Otte, M. (2000) Early humans at the gates of Europe. *Proceedings of the First International Symposium Dmanisi,* Tbilisi (Georgia) September 1998. Etudes et Recherches Archéologiques de l'Université de Liège (E.R.A.U.L.) 92.

Manzi, G. (2004). Human evolution at the Matuyama-Brunhes Boundary. *Evolutionary Anthropology* **13**, 11–24.

McNabb, J. (2005). Hominins and the Early-Middle Pleistocene transition: Evolution, culture and climate in Africa and Europe. In *Early-Middle Pleistocene Transitions: The Land-Ocean Evidence* (M. J. Head and P. L. Gibbard, Eds.), pp. 287–304. Geological Society, Special Publications 247, The Geological Society.

Parés, J. M., Pérez-González, A., Rosas, A., Benito, A., *et al.* (2006). Matuyama-age tools from the Sima del Elefante site, Atapuerca (northern Spain). *Journal of Human Evolution* **50**, 163–169.

Parfitt, S. A., Barengregt, R. W., Breda, M., *et al.* (2005). The earliest record of human activity in Northern Europe. *Nature* **438**, 1008–1012.

Pitts, M., and Roberts, M. (1997). *A Fairweather Eden.* Century Books.

Rightmire, G. P., Lordkipanidze, D., and Vekua, A. (2006). Anatomical description, comparative studies and evolutionary significance of the hominin skulls from Dmanisi, Republic of Georgia. *Journal of Human Evolution* **50**, 115–141.

Roebroeks, W. (2001). Hominid behaviour and the earliest occupation of Europe: An exploration. *Journal of Human Evolution* **41**, 437–461.

Roebroeks, W., and van Kolfschoten, T. (1995). The earliest occupation of Europe. *Proceedings of the European Science Foundation workshop at Tautavel (France), 1993.* Analecta Praehistorica Leidensia 27. University of Leiden, Leiden.

Ron, H., and Levi, S. (2001). When did hominids first leave Africa? New high- resolution paleomagnetic evidence from the Erk El-Ahmar Formation, Israel. *Geology* **29**, 887–890.

Sahnouni, M., Hadjouis, D., Made, Jvd., *et al.* (2002). Further research at the Oldowan site of Ain Hanech, North-Eastern Algeria. *Journal of Human Evolution* **43**, 925–937.

Thieme, H. (2005). The Lower Palaeolithic art of hunting. In *The Hominid Individual in Context* (C. Gamble and M. Porr, Eds.), pp. 115–132. Routledge.

Tuffreau, A. (1996). L'Acheuléen dans l'ouest de l'Europe. Centre d'Etudes et de Recherches Préhistoriques Université des Sciences et Technologies de Lille. Publicatons du CERP, p. 4.

Villa, P. (2001). Early Italy and the colonization of Western Europe. *Quaternary International* **75**, 113–130.

Global Expansion 300,000-8000 years ago, Africa

A E Close and T Minichillo, University of Washington, WA, USA

© 2007 Elsevier B.V. All rights reserved.

Models of the Geographical Origin of Modern Humans

There were, until recently, two major opposing models for the origin of our species, the out-of-Africa and the multiregional model. Of the two, the latter is now largely abandoned except by a small group of adherents. However, in light of their continued existence in the literature, a summary of the multiregional model is appropriate.

The Multiregional Model

The multiregional model currently holds that *Homo sapiens* evolved in Africa, but, by means of gene flow and hybridization, modern peoples in regions of the Old World retain genetic contributions from earlier hominids: specifically from *Homo erectus* in East Asia and *Homo neanderthalensis* in Europe. The model has two forms. The strong form holds that *H. sapiens* has been the only species of *Homo* since at least 1800 ka, and that the modern form evolved in several places at once, with earlier local populations making substantial genetic contributions to their modern counterparts. The weak form holds that a genetic contribution of some kind was made by local 'archaic' populations to the earliest local modern ones. All proponents of the model have shifted to increasingly weaker forms as new evidence has made the stronger forms increasingly unlikely (Pearson, 2004). Since it is agreed that most (if not all) of the genetic origin of *H. sapiens* is African, the debate has become one of statistically remote possibilities and is essentially over.

The Out-of-Africa Model

The out-of-Africa model, also referred to as the out-of-Africa-2 model (since there was at least one earlier *Homo* exodus from Africa), the African Eve model, or the replacement model, holds that *H. sapiens* evolved only in Africa and then migrated to all other parts of the globe with very little or no genetic contribution from other hominid species, which were rapidly replaced.

Evidence for the Geographical Origin of Modern Humans

Three forms of evidence have been brought to bear on each of these two major models: morphological, genetic, and archeological. Together, they support only the out-of-Africa model for the origin of our species.

Skeletal Morphological Evidence

Supporters of the out-of-Africa model note continuities in the skeletal morphology of African populations that occur nowhere else. They believe that the earliest modern people in Europe were morphologically more African than later Europeans, and that any morphological similarities to preceding archaic *Homo* populations outside Africa are the result of convergent evolution under similar environmental stresses or of retained traits from a very ancient common ancestor.

The earliest skeletal materials assigned to *H. sapiens* are all African. The oldest of all *H. sapiens* are the Omo finds from near Lake Turkana, Ethiopia, which have recently been dated to 196 ka (McDougall *et al.*, 2005) (**Fig. 1**). The second oldest *H. sapiens* material comes from near Herto, Ethiopia, and has been dated to between 154 and 160 ka (White *et al.*, 2003) (**Fig. 1**). Specimens from southern African Middle Stone Age (MSA) sites identified as *H. sapiens* include material from Border Cave, Klasies River Mouth (Singer and Wymer, 1982), Die Kelders, and Pinnacle Point 13B (**Fig. 1**). The recent application of multiple dating methods has led to a consensus that at least some of the Klasies and Border Cave material is older than 100 kyr and the rest of the Klasies material is older than 60 kyr.

In northwestern Africa, Jebel Irhoud (Morocco) yielded several well-preserved fossils, which are almost certainly older than OIS 4 (Hublin, 1992, pp. 186–187), and electron spin resonance (ESR) dates place them between 190 and 90 ka. There are no human fossils of comparable age in northeastern Africa, but modern humans at Skhul and Qafzeh (Israel) are dated to 90–100 ka, are associated with African faunal elements, and have African body proportions. Hublin (1992) considers them to be very similar to those from Jebel Irhoud. They represent a temporary expansion by moderns out of northeastern Africa.

General features such as cranial capacity, tooth size and morphology, and stature, when they can be measured, fall within the range of modern (Holocene) African populations. However, the sample of early *H. sapiens* is still small, so that it is difficult to assess population-scale questions of variability and change. The variability shown has been interpreted both as

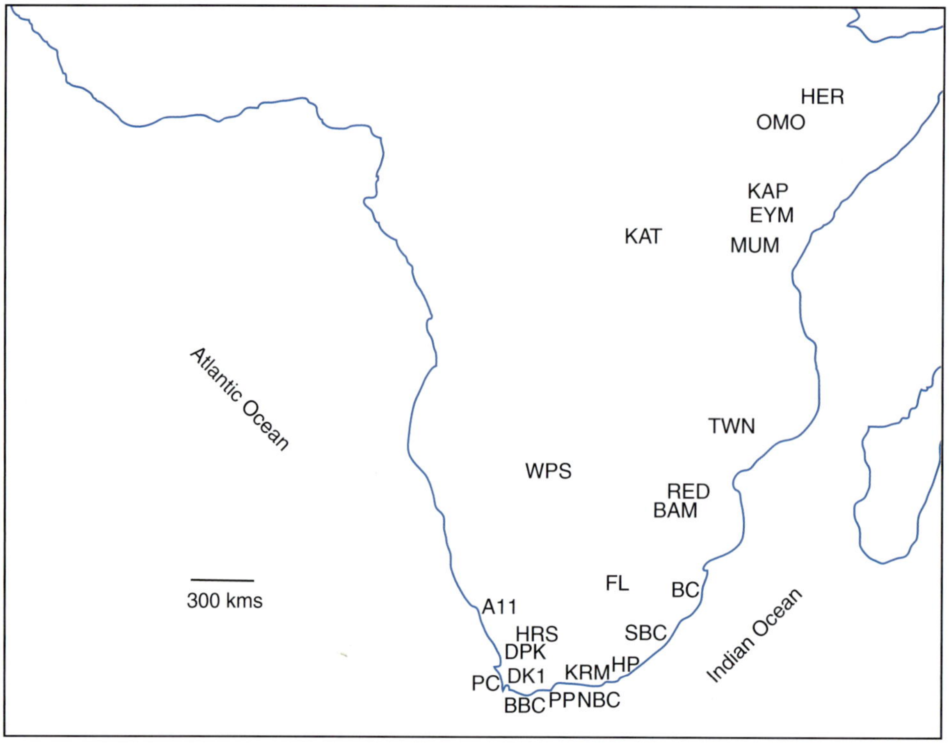

Figure 1 Middle Stone Age sites in sub-Saharan Africa. Key: A11 = Apollo 11, BAM = Bambata, BBC = Blombos Cave, BC = Border Cave, DK1 = Die Kelders, DPK = Diepkloof, HP = Howieson's Poort, EYM = Enkapune ya Muto, FL = Florisbad, HER = Herto, HRS = Hollow Rock Shelter, KAP = Kapthurin, KAT = Katanda, KRM = Klasies River Mouth; MUM = Mumba, NBC = Nelson Bay Cave, OMO = Omo, PC = Peers Cave, PP = Pinnacle Point, RED = Redcliff, SBC = Sibudu Cave, TWN = Twin Rivers, WPS = White Paintings Shelter. McBrearty and Brooks (2000), Mitchell (2002), and Minichillo (2005).

exactly what would be expected from the founding modern population, and also as having no discernible relationship to living populations.

Genetic Evidence

The genetic evidence strongly supports a single African origin for all living people. This was first applied to modern human origins in a germinal study (Cann *et al.*, 1987) which compared the mitochondrial DNA (mtDNA) from living populations. Cann *et al.* found that living Africans have the greatest diversity and, thus, the greatest age, and they estimated that all living people are descended from a related group of African women who lived between 200 and 100 ka.

The study was initially attacked as being flawed in methodology. However, since then, many unflawed studies of living populations (Pearson, 2004) have reached conclusions very similar to the first study: that our species has a single origin in Africa.

Harpending *et al.* (1998) utilized both microsatellites on the Y chromosome and mtDNA and found that our ancestors underwent a rapid population expansion about 100 ka. They concluded that the population expansion took place after some bottleneck event, that it occurred in Africa, and that it is completely incompatible with any version of the multiregional model.

Archeological Evidence from Southern Africa

Skeletal and genetic evidence refers to human biology or anatomy. Archeological evidence refers to behavior: a profoundly vague concept that includes social relations, customs, technology, language, intellectual capacity, and many other aspects. Within evolutionary biology, it is generally held that if anatomical and behavioral changes do not coincide, then it is the behavioral changes that occur first. In the study of modern humans, however, there has been uncritical acceptance of the idea that anatomical change preceded behavioral. This idea has persisted for over three decades, perhaps because it allows the European record some relevance. However, in the words of Lewin (1998, p. 113), 'the argument must be recognized as special pleading with no empirical basis.'

MSA studies in sub-Saharan Africa have recently undergone a major paradigm shift, involving reinterpretation of previously known sites and new research goals. For convenience, we name the earlier paradigm that prevailed in MSA studies from the late

1970s until recently, the Klasies paradigm, and the newer one, which has emerged since the late 1990s, the Blombos paradigm.

The Klasies paradigm The MSA cave sites at the Klasies River Mouth (KRM) in southernmost Africa are crucial to research on modern human origins (**Fig. 1**). Excavated in 1966–68, the MSA deposits there were massive and provided tens of thousands of lithic and faunal artifacts from gross-scale, stratigraphic units. There was also evidence that the oldest deposits were least 120 kyr old (Singer and Wymer, 1982). An important pattern in the lithic assemblage was what appeared to be long periods (some 80 kyr altogether) without much technological change and near-absence of more modern artifacts, such as fully bifacial points and bone tools (**Fig. 2**). Indeed, the last paragraph of the report includes the dismal statement, 'Most significant is the correlation of such a long period of occupation... with little if anything to show any change in subsistence and economy' (Singer and Wymer, 1982, p. 210).

Figure 2 Unifacial point (top left), Still Bay bifacial points and worked bone point from Blombos Sands and Kleinjongensfontein, near Still Bay, South Africa. The Heese collection at IZIKO: South African Museum. Bar is 1 cm. Minichillo, 2005.

On the basis of the Klasies sequence, Singer and Wymer outlined the MSA sequence as shown in **Table 1**, with the Howiesons Poort as the only (very brief) break in the monotony. The Howiesons Poort substage, dating to around 60 ka, includes small-backed blades and true-backed bladelets (**Fig. 3**) that we now know were hafted as parts of composite tools (**Fig. 4**) (Minichillo, 2005). Both backing of lithics and the creation of composite tools are considered to be advanced technologies. The increase in mineral pigment use and color variability during this period has been interpreted as increasingly symbolic behavior (**Fig. 5**).

In 1981, when Volman found that the Klasies sequence did not fit that of Border Cave, he devised the sequence shown in the second column of **Table 1**. Before long, this proved inadequate as well.

The fauna from Klasies has also played a critical role in discussions of the modernity of MSA people. One such discussion was engendered by Binford's (1984) interpretation, which focused on the relative abundance of parts of animals. He proposed that the MSA people were primarily scavengers and had poor organizational and planning skills. This interpretation has been dismissed on the grounds that the materials he analyzed lacked long bone shaft fragments due to excavation and analytical bias, and because there is now direct evidence for MSA hunting of large bovids.

Other patterning in the faunal assemblages of the MSA continues to be used to construct models of human behavior. Klein (1999) compared the patterns of faunal exploitation in local MSA and LSA sites to argue forcefully that MSA people were inefficient

Table 1 Proposed substages of the Middle Stone Age in southern Africa

Singer and Wymer (1982)	*Volman (1981)*	*Wurz (2002)*	*Minichillo (2005)*	*Sites/Layers*	*Age estimates (years ago)*
MSA III and IV	MSA 3/4	Post-Howiesons Poort	Post-Howiesons Poort	KRM (Upper), Rose Cottage Cave, Sibudu, Ysterfontein?	55–35,000
Howiesons Poort	Howiesons Poort	Howiesons Poort	Howiesons Poort	KRM (HP), Howiesons Poort shelter (S&H)	65–55,000
MSAII	MSA 2b		Die Kelders	Die Kelders, Diepkloof? Howiesons Poort shelter (D&D)?	70–60,000
		Still Bay	Still Bay	Blombos Cave (M1), Dale Rose Parlour	79–69,000
			Blombos	Blombos Cave (M2)	?
		Mossel Bay	Mossel Bay	KRM (SAS)	72–68,000 (too young??)
MSA I	MSA 2a MSA 1	Klasies	Klasies	KRM (LBS) Border Cave	>120–105,000 >125,000

Adapted from Wurz (2002) and Minichillo (2005).

Figure 3 Backed segments and crescents for the Howiesons Poort layers at Klasies River Mouth, the Singer and Wymer collection at IZIKO: South African Museum. Bar is 1 cm.

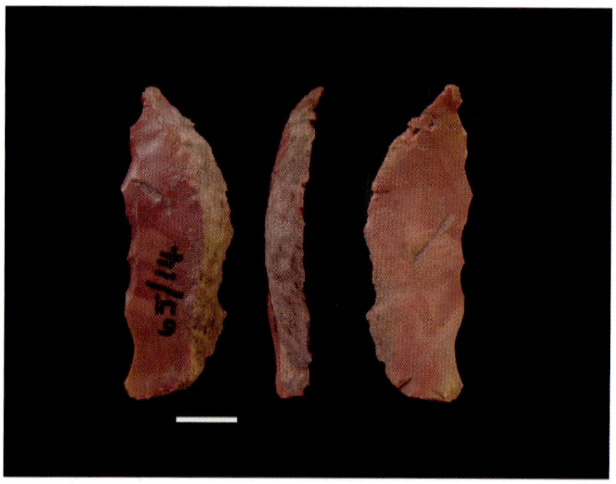

Figure 4 Backed blade with adhesive residue from hafting as a wood-scraper from the Howieson's Poort shelter, near Grahamstown, South Africa. The Deacon and Deacon collection, Albany Museum. Bar is 1 cm. Minichillo, 2005.

Figure 5 Ochre crayons from the Howieson's Poort shelter, near Grahamstown, South Africa. The Deacon and Deacon collection, Albany Museum. Bar is 1 cm.

hunters compared with subsequent LSA people. This argument has two parts: first, that fewer species were exploited during the MSA, and, second, that aggressive species, such as adult cape buffalo and bush-pig, were avoided. Klein went on to link this directly to his idea that MSA people lacked some modern intellectual capacities until 50–40 ka However, a statistical analysis of the same faunal data found that the faunal patterns are poor markers of behavioral modernity, good markers of climatic change at the Pleistocene–Holocene transition, and are better explained by sample sizes and the application of foraging theory models than by changes in human cognition (Minichillo, 2005).

Within the Klasies paradigm, all models of MSA behavior had to take into account the apparent stasis observed there.

The Blombos paradigm Blombos Cave (**Fig. 1**) was excavated using extremely fine-scaled recovery and stratigraphic techniques, and the MSA has precise age-estimates based on single-grain optically stimulated luminescence. The sequence is quite different from that at Klasies, and Wurz (2002) has thus proposed a new scheme for organizing the southern African MSA (**Table 1**, column 3).

The MSA at Blombos is sealed below a dune dating to 65–69 ka. The uppermost MSA deposits, M1, have yielded bifacial points of the Still Bay type and are sometimes referred to as the Still Bay layers. These have been dated to 74 ± 5 ka (Henshilwood *et al.*, 2004). Mineral pigments abound, and some engraved pieces of ochre are seen as the earliest expression of symbolic thought. There are also perforated marine shells interpreted as a necklace and, thus, the oldest known personal adornment.

The underlying M2 is best known for the many pieces of worked bone. Careful excavation revealed that the abundant bifacial points in M1 grade into the abundant bone points in M2; that is, two types of artifacts vary inversely. This is quite different from the transitions at Klasies where there were hiatus in occupation between major stratigraphic units.

Recent work at Klasies confirmed the general chronology of the site, but not the specific interpretations. New dates, the lack of climatic change, and low density of shells suggest that the main stratigraphic unit (the SAS Member) was not deposited over tens of millennia, but over a very few millennia or perhaps even centuries (Minichillo, 2005). This removes the appearance of stasis; the stability within the SAS Member still represents stylistic unity (Wurz, 2002; Wurz *et al.*, 2003), but for a much briefer period. Rapid deposition has also been suggested for the large MSA deposit at Die Kelders, and technological change within a few millennia is clearly shown at Blombos.

In sum, the MSA archeological record in southern Africa shows several profound technological changes occurring in periods of no more than a few millennia (**Table 1**); Sophisticated and composite artifacts, fine bone-work, use of color and incised patterns, and perhaps even jewelry — these are surely modern behaviors.

The Expansion of Modern Humans

Since there are the remains of early modern humans from the southernmost Cape to Morocco, we know that they expanded throughout Africa. When modern humans began to expand beyond Africa, it was almost certainly from northern Africa, into either Iberia or the Levant. We do not know enough about Arabia to assess the probability of a route through the Horn of Africa. We will outline the archeological evidence from northern Africa up to 8 ka and then discuss what this reflects about expansion.

Archeological Evidence from Northern Africa

The great Sahara Desert is the largest and driest desert in the world; very little of it receives more than 25 mm precipitation per annum and some parts receive none at all. At some periods in the past, the Sahara has seen significant rainfall and human occupation extending from the modern sahel to the Mediterranean. During other periods, it has expanded northwards to meet the sea itself, as it does today. Rainfall has always been the most important factor in North African prehistory.

Middle Paleolithic or Middle Stone Age As in southern Africa, the middle Paleolithic (or Middle Stone Age) of northern Africa probably began by 300 –250 ka. It includes two major units, the Mousterian and the Aterian, which overlapped considerably in time. The Mousterian predominates in northeastern Africa and the Aterian in northwestern, but the Mousterian is known as far west as Morocco and the Aterian as far east as the Nile Valley. Within the Nile Valley, other variants have been defined, such as the Nubian complex and the Taramsan, but these are undated. In fact, the only long Middle Paleolithic sequence is from the Haua Fteah, Libya, which was excavated in the early 1950s and is therefore essentially undated.

In artifacts, the Mousterian is very similar to Charentian and Denticulate Mousterian as known in Europe and southwestern Asia, and has a much higher frequency of retouching than is found in southern Africa. Bifacial foliates occur in the Nubian Complex in the Nile Valley and the Red Sea Hills and also in the Aterian to the west. The Aterian is best known for the presence of tanged tools (not merely tanged points), which are generally taken to indicate hafting. Wurz *et al.* (2005) concluded that there is the same kind of variability in the MSA technology from Klasies as there is in the Middle Paleolithic of the Nile Valley. However, the latter remains undated and, at this level of technology, convergence is not surprising.

During a warm period about 100–90 ka, the northern African moderns moved briefly into the Levant, as did several other African animals. This colonization failed and the Africans returned south when the climate became colder. The successful and permanent occupation beyond Africa was closer to 50 ka, or later (Hublin, 2000).

There is very little information about when the middle Paleolithic ended. In the Nile Valley, Vermeersch (2000, table 14.1) estimates that the middle Paleolithic ended about 80–70 ka. The Sahara could be inhabited only during periods of increased rainfall, and the last wet period before the final Pleistocene lies far beyond the range of radiocarbon dating. Thus, the end of the middle Paleolithic could be dated only when techniques were developed that extend beyond 40 ka (luminescence, ESR, uranium-series and others). Age estimates for the last wet phase are available for at least seven regions of the Sahara. They range for a low estimate of ≥60 ka in the Tadrart Acacus in central Libya to high estimates of ≥80 ka in the southcentral Sahara and the region of the Great Eastern Erg. The middle Paleolithic in the Sahara must therefore all be older than 65–70 kyr (**Fig. 6**).

There are few reliable absolute dates for the end of the middle Paleolithic. The date of 45 ka at the Haua Fteah is minimal. At Jebel el Gharbi, western Libya, two sites have radiocarbon dates of about 44–43 ka (Garcea, 2004, pp. 33–34). Dating the end of the Middle Paleolithic is particularly contentious in the Maghreb, where the Mousterian sometimes underlies the Aterian. Recent critical compilations of dates for the Aterian (Cremaschi *et al.*, 1998, table I; Bouzouggar *et al.*, 2002, table 7; Close, in press, table I) indicate that it may not have lasted much (if at all) beyond about 40 ka.

Upper/Late Paleolithic or Late Stone Age The earliest Upper Paleolithic (in the European and southwest Asian sense) in northern Africa is the Dabban in Libya. It dates to about 40 ka in Cyrenaica (McBurney (1967, p. 326, 1975, p. 419)) and to about 30 ka in Jebel el Gharbi (Garcea (2004, pp. 33–34)). It is unknown in the Maghreb, where there is no archeological record between ≥40 ka and the first bladelet industry (Iberomaurusian) not long before 20 ka (**Fig. 7**).

The Dabban persisted at Jebel el Gharbi until no later than about 25 ka (Garcea 2004, pp. 33–34).

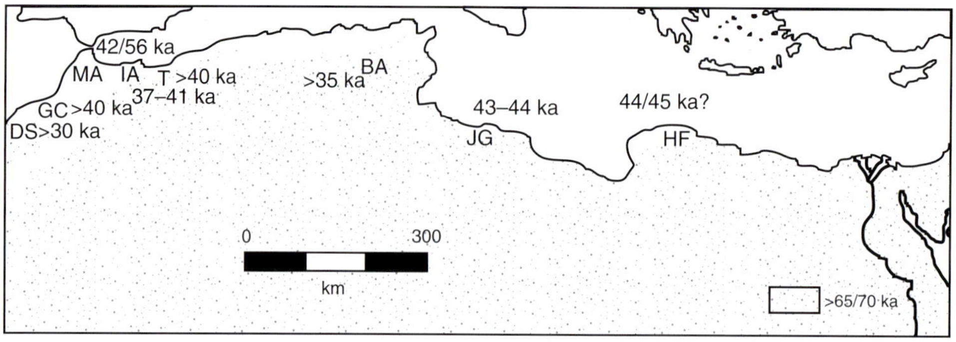

Figure 6 Dates (in thousands of years ago) for the end of the middle Paleolithic in northern Africa. Key: DS = Dar-es-Soltane, GC = Grotte des Contrebandiers (Témara), MA = Mugharet el 'Aliya, IA = Ifri n'Ammar, T = Taforalt (Grotte des Pigeons), BA = Bir el Ater, JG = Jebel el Gharbi, HF = Haua Fteah.

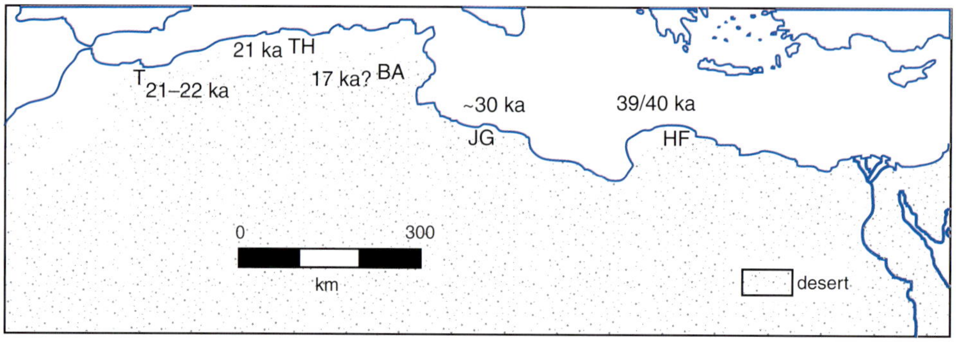

Figure 7 Dates (in thousands of years ago) for the beginning of the Upper or late Paleolithic in northern Africa. Key: T = Taforalt (Grotte des Pigeons), TH = Tamar Hat, BA = Bir Oum Ali, JG = Jebel el Gharbi, HF = Haua Fteah.

At the Haua Fteah, the only two dates that definitely refer to the Dabban are about 33.1 and 28.5 ka (McBurney 1967, table III.1). The Dabban may have disappeared from both Cyrenaica and Jebel el Gharbi at about the same time, and both areas then remained unoccupied for several millennia.

In the Nile Valley, chert mines date to 35–30 ka followed by a hiatus until about 25 ka (Vermeersch (1992, pp. 121–122)).

At about 20 ka, there was a profound change in technology, to one directed almost exclusively toward the production of bladelets, some of which were then backed. These appeared in Nubia and Upper Egypt at about 20 ka in Cyrenaica by 19 ka, and in the Maghreb before 20 ka at Tamar Hat and Taforalt (Close and Wendorf, 1989) (**Fig. 7**). At the Haua Fteah, the change from the Dabban was so great McBurney (1967) suggested an actual change of population; this would, indeed, be the case if Cyrenaica had been unoccupied for several millennia. The Iberomaurusian certainly represents a new population in the Maghreb, which had been unoccupied for some 20 kyr. The Sahara remained uninhabitable. The origin of the bladelet-makers remains unclear but may have been in sub-Saharan Africa.

Epipaleolithic and Neolithic About 12 ka, the monsoon-belt moved northward and rain came again to the Sahara. For the next 4 kyr, all of northern Africa was occupied by descendants of earlier local populations, as far as we know. In the Nile Valley, along the coast and in the Maghreb, freshwater was relatively copious. There were, in addition, the rich resources of the Nile itself and of the Mediterranean. In these regions, therefore, people were able to continue a gathering–hunting–fishing way of life until after 8 ka.

In the desert, however, particularly in what is now the southern Sahara, water was much less abundant – probably less than 20 mm per annum except in mountainous areas – and subsistence was more precarious. People thus ensured their survival as foragers by also keeping domestic animals. Cattle were domesticated in the southern Sahara before 8 ka, but their herders remained essentially gatherer-hunters (Close, 1992). Farming did not become an important factor in northern Africa until the arrival of the southwest Asian domesticated plants and animals, some time after 8 ka.

Routes Out of and into Northern Africa

The archeological record suggests that between 300 and about 20 ka, northern Africa was sometimes occupied by modern humans and, with the possible exception of the Nile Valley, was sometimes not occupied by anyone. Periods of occupation could be separated by millennia, especially in northwestern Africa. Boats are more recent, so any contact would have to be overland into the Levant, or, perhaps, across the very narrow Strait of Gibraltar.

Although the Strait of Gibraltar is only about 14 km wide at its narrowest point, it is also extremely turbulent and the safest crossing is west of 6° W and ≥100 km long. Nonetheless, it is possible to see across the strait today, and, during periods of lowest sea levels, a series of rocky islands would have been exposed, reducing the maximum water crossing to about 4 km (Bouzouggar *et al.* (2002, figure 23)). The earliest (nonmodern) humans from Iberia almost certainly crossed Gibraltar (Bermudez de Castro *et al.*, 1997), so it is difficult to think that modern humans could not have crossed in the Upper Pleistocene. However, those earliest humans moved into empty territory with no direct competition, whereas until after 30 ka, moderns would have found the hominid niche in Iberia occupied by Neanderthals and their predecessors. In this case, moderns would have been archeologically visible only if there was a massive invasion of (Hublin 2000, p. 170).

Similarly, only a massive invasion would result in early moderns having a detectable genetic effect in Iberia (Hublin (2000, pp. 170–171)). Today, living Iberians share a common ancestry dating to the Upper Paleolithic, and they are genetically much closer to other West European populations than they are to North Africans. This is the genetic pattern of living populations, so that failure to cross the strait has nothing to do with modernity. Genetic comparison of living people in northern Africa, Europe, and the Levant shows that the oldest European group and the main group in northern Africa (including the Berbers) shared a common Levantine ancestor until about 50 ka, suggesting that Europe and northern Africa underwent independent colonization by moderns from the Levant. Moderns initially spread from Africa into the Levant by around 50 ka and some of their descendants returned to recolonize northwestern Africa (at least) after it had been empty of people for ≥20 kyr (40—20 ka).

The Insularity of the Coast of Northern Africa

Modern humans evolved in Africa some 200 ka. Since then and unexpectedly, much of northern Africa has been empty of people for tens of millennia. The discrepancy may lie in our view of northern Africa.

The Arab invaders of the seventh century AD named northwestern Africa *el Jezirat el Maghreb* – the Island of the West (Vaufrey 1955, p. 9) – because it was completely isolated between the Mediterranean and the desert. This isolation is true of almost the entire coast of northern Africa and its prehistory becomes more understandable when it is viewed as a series of islands (Shaw (2003, p. 98)): the Nile Valley, Jebel el Akhdar, Jebel el Gharbi, and the Maghreb (**Fig. 8**). The islands have permanent sources of drinking water and are separated from each other by hundreds of kilometers of coastline, where the Sahara meets the Mediterranean and drinking water is rare to lacking.

During wetter phases, the islands were easily accessible from the rest of northern Africa and most of what is now desert was occupied. During phases of aridity, contact between the islands is very unlikely and each would become isolated. By 40 ka, human occupation had ceased, certainly in the Maghreb, probably in Jebel el Gharbi and perhaps in

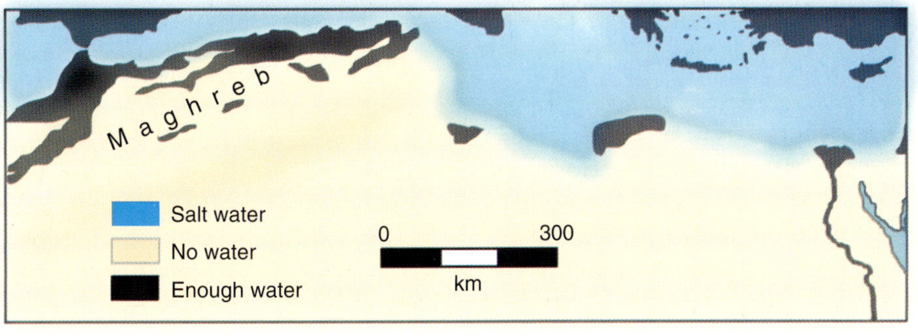

Figure 8 Northern Africa seen as a chain of islands. The boundary between the desert and the sea is deliberately blurred because its precise location is not crucial, and the islands remain insular no matter what the sea level.

Cyrenaica. We conclude that, in each area, modern humans may have become locally extinct.

The presence of the Mediterranean and the absence of boats meant that any travel between the islands of northern Africa had to be overland. Historically, such journeys have depended upon the use of domestic animals. There were no domestic animals on the coast (although they existed in the southern Sahara) until after 8 ka and people had to carry everything they would need themselves, especially water.

Nevertheless, inter-island journeys were sometimes made successfully and islands could be recolonized for periods of millennia. Thus, about 40 ka, the makers of the Dabban reached Cyrenaica from the east and remained there until well after 30 ka. Their descendants (presumably) can be seen as far west as Jebel el Gharbi till about 30 ka; the area was occupied until about 25 ka. But they seem not to have gone farther west into the Maghreb. The final colonization of the islands began somewhat more than 20 ka with the makers of the backed-bladelet industries; modern human beings have occupied all of the islands ever since then.

See also: **Archaeological Records**: Overview; 2.7 MYR-300,000 years ago in Asia; 1.9 MYR-300,000 years ago in Europe; Global Expansion 300,000-8000 years ago, Asia; Global Expansion 300,000-8000 years ago, Australia; Global Expansion 300,000-8000 years ago, Americas; Neanderthal Demise; Postglacial Adaptations. **Vertebrate Studies**: Interactions with Hominids.

References

Bermúdez de Castro, J. M., Arsuaga, J. L., Carbonell, E., Rosas, A., Martínez, I., and Mosquera, M. (1997). A hominid from the Lower Pleistocene of Atapuerca, Spain: Possible ancestor to Neanderthals and modern humans. *Science* **276**, 1392–1395.

Binford, L. R. (1984). *Faunal Remains from Klasies River Mouth*. Academic Press.

Bouzouggar, A., Kozlowski, J. K., and Otte, M. (2002). Étude des ensembles lithiques atériens de la grotte d'El Aliya à Tanger (Maroc). *L'Anthropologie* **106**, 207–248.

Cann, R. L., Stoneking, M., and Wilson, A. C. (1987). Mitochondrial DNA and human evolution. *Nature* **325**, 31–36.

Close, A. E. (1992). Holocene occupation of the eastern Sahara. In *New Light on the Northeast African Past: Current Prehistoric Research* (F. Klees and R. Kuper, Eds.), pp. 154–183. Heinrich-Barth-Institut.

Close, A. E., and Wendorf, F. (1989). North Africa at 18 000 BP. In *The World at 18 000 BP. Volume 2: The Low Latitudes* (C. Gamble and O. Soffer, Eds.), pp. 41–57. Unwin Hyman.

Close, A. E., (in press) On the absence of a Middle–Upper Paleolithic transition in Mediterranean Northwest Africa. In *The Mediterranean from 50,000 to 25,000 BP: Turning Points and New Directions*. Camps M and Szmidt C (eds.) Tempus Reparatum (BAR).

Cremaschi, M., Di Lernia, S., and Garcea, E. A. A. (1998). Some insights on the Aterian in the Libyan Sahara: Chronology, environment and archaeology. *African Archaeological Review* **15**, 261–286.

Garcea, E. A. A. (2004). Crossing deserts and avoiding seas: Aterian North-African–European relations. *Journal of Anthropological Research* **60**, 27–53.

Harpending, H. C., Batzer, M. A., Gurven, M., Jorde, L. B., Rogers, A. R., and Sherry, S. T. (1998). Genetic traces of ancient demography. *Proceedings of the National Academy of Sciences of the United States of America* **95**, 1961–1967.

Henshilwood, C., d'Errico, F., Vanhaeren, M., van Niekerk, K., and Jacobs, Z. (2004). Middle Stone Age shell beads from South Africa. *Science* **304**, 404.

Hublin, J.-J. (1992). Recent human evolution in northwestern Africa. *Philosophical Transactions of the Royal Society of London B* **337**, 185–191.

Hublin, J.-J. (2000). Modern–nonmodern hominid interactions: A Mediterranean perspective. In *The Geography of Neanderthals and Modern Humans in Europe and the Greater Mediterranean* (O. Bar-Yosef and D. Pilbeam, Eds.), pp. 157–182. Peabody Museum of Archaeology and Ethnology.

Klein, R. G. (1999). *The Human Career: Human Biological and Cultural Origins*, 2nd edn. University of Chicago Press.

Lewin, R. (1998). *The Origin of Modern Humans*. Scientific American Library.

McDougall, I., Brown, F. H., and Fleagle, J. G. (2005). Stratigraphic placement and age of the modern humans from Kibish, Ethiopia. *Nature* **433**, 733–736.

McBurney, C. B. M. (1967). *The Haua Fteah (Cyrenaica) and the Stone Age of the South–East Mediterranean*. Cambridge University Press.

McBurney, C. B. M. (1975). Current status of the Lower and Middle Paleolithic in the entire region from the Levant through North Africa. In *Problems in Prehistory: North Africa and the Levant* (F. Wendorf and A. E. Marks, Eds.), pp. 411–426. Southern Methodist University Press.

Minichillo T (2005) *Middle Stone Age Lithic Study, South Africa: An Examination of Modern Human Origins*. PhD Dissertation, University of Washington, Seattle.

Pearson, O. M. (2004). Has the combination of genetic and fossil evidence solved the riddle of modern human origins? *Evolutionary Anthropology* **13**, 145–159.

Shaw, B. D. (2003). A peculiar island: Maghrib and Mediterranean. *Mediterranean Historical Review* **18**, 93–125.

Singer, R., and Wymer, J. (1982). *The Middle Stone Age at Klasies River Mouth in South Africa*. University of Chicago Press.

Vaufrey, R. (1955). *Préhistoire de l'Afrique. Tome I: Le Maghreb*. Masson.

Vermeersch, P. M. (1992). The Upper and Late Palaeolithic of northern and eastern Africa. In *New Light on the Northeast African Past: Current Prehistoric Research* (F. Klees and R. Kuper,Eds.), pp. 99–153. Heinrich-Barth-Institut.

Vermeersch, P. M. (Ed.) (2000). *Palaeolithic Living Sites in Upper and Middle Egypt*. Leuven University Press.

Volman, T. P. (1981) *The Middle Stone Age in the Southern Cape*. PhD Dissertation, University of Chicago.

White, T. D., Asfaw, B., DeGusta, D., *et al.* (2003). Pleistocene *Homo sapiens* from Middle Awash, Ethiopia. *Nature* **423**, 742–747.

Wurz, S. (2002). Variability in the Middle Stone Age lithic sequence, 115,000–60,000 years ago at Klasies River, South Africa. *Journal of Archaelogical Science* **29**, 1001–1015.

Wurz, S., le Roux, N. J., Gardner, S., and Deacon, H. J. (2003). Discriminating between the end products of the earlier Middle

Stone Age sub-stages at Klasies River using biplot methodology. *Journal of Archaelogical Science* 30, 1107–1126.

Wurz, S., Van Peer, P., le Roux, N. J., Gardner, S., and Deacon, H. J. (2005). Continental patterns in stone tools: A technological and biplot-based comparison of Early Late Pleistocene assemblages from northern and southern Africa. *African Archaelogical Review* 22, 1–24.

Global Expansion 300,000-8000 years ago, Asia

M D Petraglia, University of Cambridge, Cambridge, UK
R Dennell, University of Sheffield, Sheffield, UK

© 2007 Elsevier B.V. All rights reserved.

Transitions in Lithic Production after 300 ka

In western Asia (including the Caucasus), changes in lithic production resulted in the type of Mousterian industries seen in Europe, whereas changes were more muted in southern Asia, and resulted in a non-Mousterian Middle Paleolithic. In China and the Far East, changes were even more subtle, and stone tool assemblages continued without major industrial innovations. Outside the Levant, details are still sketchy as to when and how these Late Middle Pleistocene transitions occurred.

The Lower to Middle Paleolithic transition in the Levant occurred ca. 250 ka (Shea, 2003) (**Table 1**). The immediate precursor of the Levantine Mousterian is the Yabrudian industry, which dates from 350 ka, and is the earliest regional lithic variant in Asia. It is found in Syria, Lebanon, and northern Israel, and features steeply retouched scrapers and relatively thick flakes (**Fig. 1**); the Levallois technique is absent, and bifaces are rare or absent. The principal Middle Paleolithic industry of the Levant is Mousterian. Its most conspicuous attribute is the use of recurrent Levallois core reduction strategies to produce triangular and subtriangular flakes. The earliest Mousterian lithic assemblages are distinctive, with high values for Levallois and laminar indices. The transition from hand-held handaxes in the Lower Paleolithic to the use of hafted tools (i.e., stone points or blades mounted on wooden shafts) in the Middle Paleolithic may be considered a profound technological shift, as later hominids employed stone projectile weaponry to obtain game (e.g., Shea (2003)). Early Mousterian industries in the Levant range from ca. 250 to 128 ka, thus including a transition from warm and humid conditions during oxygen isotope stage (OIS) 7 to colder and more arid conditions in OIS 6. It is not clear which hominin(s) made the earliest Levantine Middle Paleolithic tools because both Neanderthals and early modern humans in the Levant are associated with the same types of Middle Paleolithic assemblages.

The Lower to the Middle Paleolithic transition in South Asia is poorly dated, though there is good stratigraphic evidence to indicate that the transition from large bifacial industries of the Acheulean to prepared core and flake industries of the Middle Paleolithic occurred by ca. 150 ka (Misra, 2001; Petraglia *et al.*, 2003) (**Fig. 2**). Acheulean industries in India are thought to be associated with *Homo heidelbergensis*, based upon the identification of a cranial fragment from the Narmada river valley (Cameron *et al.*, 2004). The hominin(s) that made the stone tools associated with the Indian Middle Paleolithic remain unknown.

The Early Middle Paleolithic in Central Asia is extremely poorly dated, but those from the last glaciation are Mousterian, or have Mousterian affinities. The cave of Teschik-Tasch, Uzbekistan, provides clear evidence that Neanderthals inhabited this region before 40 ka. The only well-dated Early Middle Paleolithic assemblages are from Tajikistan; these date from OISs 5 and 7, and show an increase in Levallois flaking. Their makers are unknown, but usually assumed to be Neanderthal.

Although a Middle Paleolithic has been recognized in China on the basis of stone tools classifiable as

Table 1 Periodization of the Levantine Middle Paleolithic (after Shea 2003: Table V)

Period	*Dates (ka)*	*Marine OIS and Levantine climate*	*Hominins*	*Lithic industries*
Late Lower Paleolithic	350–250	OIS 8–7. Cold, then warmer	*Homo* sp.	Late Acheulean and Acheulo-Yabrudian
Early Middle Paleolithic	250–128	OIS 7–6. Warm, then colder	*Homo* sp.	Phase 1 Levantine Mousterian
Mid Middle Paleolithic	128–71	OIS 5. Initially warm and hyperarid but growing colder, more humid	Neandertals and early modern humans	Phase 2 Levantine Mousterian
Late Middle Paleolithic	71 to <47	OIS 4-early 3. Cold and dry	Neandertals	Phase 3 Levantine Mousterian
Initial Upper Paleolithic	47–32	Mid-late OIS 3. Cold and dry	Modern humans	Early Upper Paleolithic (Emiran and Ahmarian), possible late Levantine Mousterian

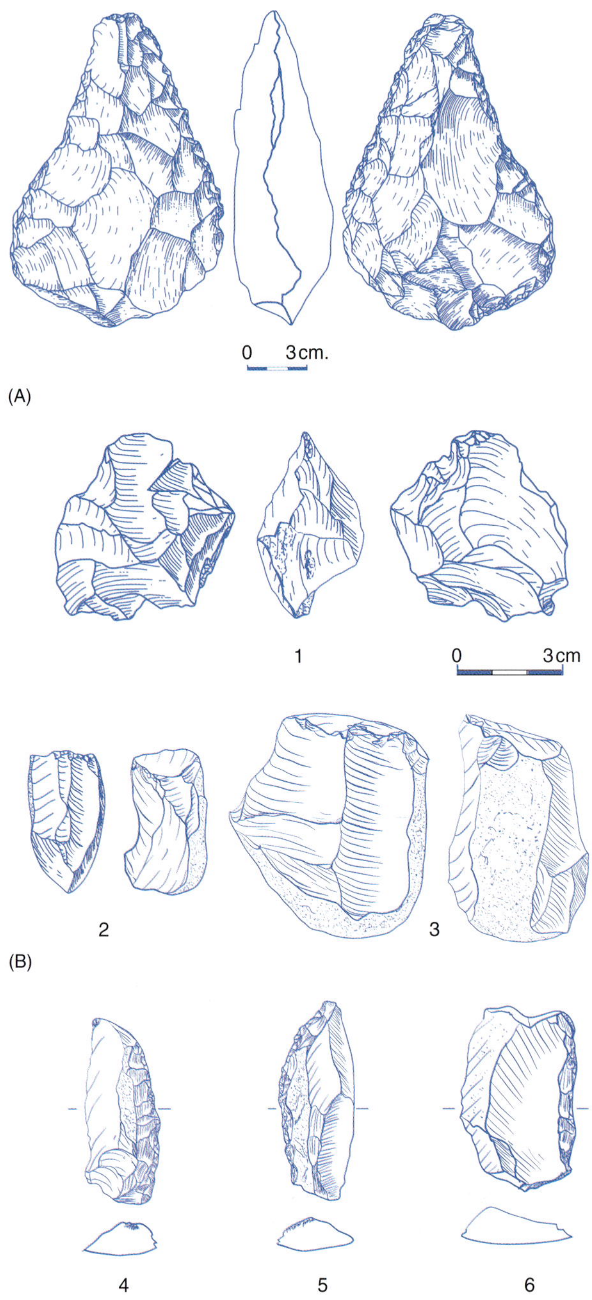

Figure 1 Transitional Acheulo-Yabrudian assemblages from Qesem Cave, Israel, dated to about 300–200 ka. (A) Handax; (B) radial core (1), cores on nodules (2, 3); (C) side-scrapers (4–6). Adapted from Gopher A, Barkai R, Shimelmitz R, *et al.* (2005) Qesem Cave: An Amudian site in Central Israel. *Journal of the Israel Prehistoric Society* 35: 69–92.

scrapers, points, and choppers, these assemblages are so similar to earlier ones that it is doubtful whether it is useful to regard them as Middle Paleolithic rather than a later variant of the Early Paleolithic (Gao and Norton, 2002). Additionally, very few Chinese sites are sufficiently well-dated and -documented to allow detailed analyses of the changes that occurred between 300 and 50 ka. The best-documented Late Middle Pleistocene site in China is currently the cave of Panxian Dadong (ca. 100–250 ka) in southern China (Miller-Antonio *et al.*, 2004); although lithic reduction sequences were complex, they were not dependent upon the type of prepared-core technologies so common in western Asia.

The Emergence of the Upper Paleolithic and the Expansion of Modern Humans

The earliest specimens of anatomically modern humans (AMH) in Northeast Africa date from ca. 190 ka, and are assumed to have originated in that continent. By ca. 25–30 ka, AMH is the only type of humanity in mainland Eurasia, and its expansion at the expense of indigenous populations (encapsulated by the 'Out of Africa 2' model) was a complex, lengthy, and still very poorly understood process. The main archeological change that occurs across much of Asia as part of this process is the replacement of Middle Paleolithic assemblages by Upper Paleolithic ones. The relationship of these changes to the expansion of modern humans is not clear-cut: in the Levant, for example, the Later Middle Paleolithic was made by both Neanderthals and modern humans, and early Upper Paleolithic assemblages are far from being unambiguous indicators of modern humans. Additionally, the quality of regional records of this process generally declines from west to east across Eurasia.

The Levant

The Israeli caves at Mount Carmel have yielded extensive fossil material of Neanderthals and anatomically modern humans (Shea, 2003). The cave sites of Tabun, Kebara, and Amud have yielded Neanderthals, whereas Qafzeh and Skhul contain anatomically modern humans, of which the earliest date to ca. 100 ka. This date is widely regarded as indicating the time when modern humans first entered Asia from Africa; however, given that there is no hominin skeletal data in western Eurasia between 100 and 300 ka, an earlier date cannot be excluded. Dated specimens of Neanderthals indicate that they were present in the Levant at and probably between 120 and 60 ka. Although the dating evidence implies that moderns and Neanderthals coexisted in the same region, it is of course possible that they never met, as different populations could have expanded during particular climatic phases, the Neanderthals occupying the region during colder periods, and the moderns during warmer episodes. The presence of both Neanderthals and modern humans in the region does

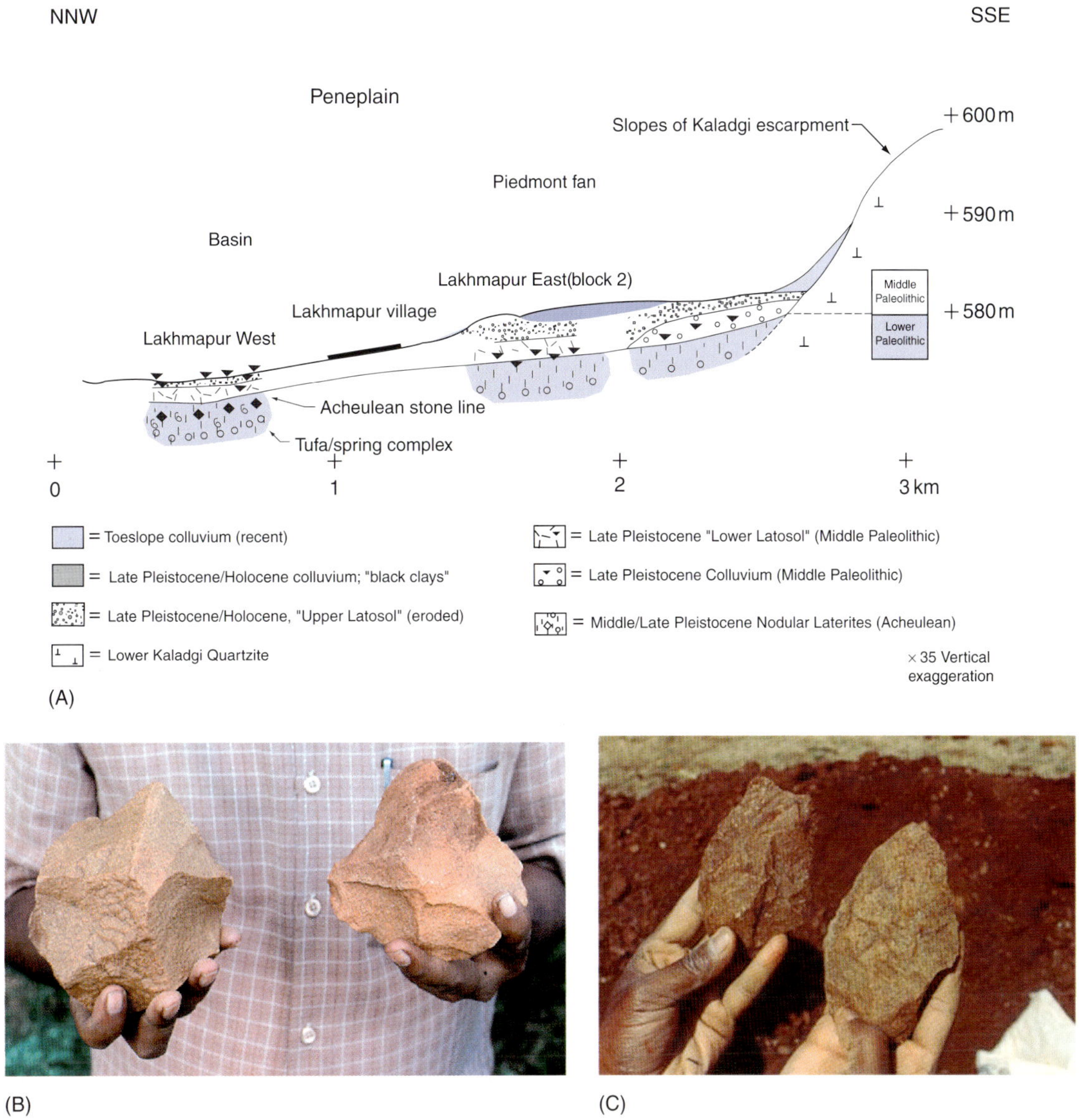

Figure 2 (A) Landscape relations between transitional Lower and Middle Paleolithic assemblages from Lakhmapur, Malaprabha Valley, India. (B) Diminutive handaxes and cleavers are associated with (C) prepared cores in early Middle Paleolithic assemblages. (A) Adapted from Petraglia MD, Schuldenrein J, and Korisettar R (2003) Landscapes, activity, and the Acheulean to Middle Paleolithic transition in the Kaladgi Basin, India. *Journal of Eurasian Prehistory* 1(2): 3–24.

however support the single-origin model of human evolution, as Neanderthals clearly became extinct in the Later Pleistocene (Stringer, 2002).

Significant technological changes occur at ca. 47–45 ka in the Levant, and early Upper Paleolithic industries occur in Ksar Akil (Lebanon) and Boker Tachtit (Israel) at around 45–38 ka. Two later Upper Paleolithic assemblages, known as the Ahmarian and the Aurignacian, are dated to ca. 38–30 ka. The Ahmarian is characterized by a blade and bladelet industry whereas the Aurignacian consists of a flaked-based industry with lamellar retouch on carinated scrapers and burins. The latter type of assemblage may have originated in Central Asia or in western Iran, and is generally regarded as the earliest unequivocal Upper Paleolithic tradition in Europe. Though these earliest Levantine Upper Paleolithic assemblages are not associated with human fossils, it is generally assumed that they were produced by modern humans (Bar-Yosef, 2002).

Anatolia and the Taurus-Zagros Mountains

Beyond the Eastern Mediterranean of the Levant, Neanderthals lived in the Taurus-Zagros Mountain zone, the Caucasus, and Central Asia as far east as Teschik Tasch, Uzbekistan. The Middle Paleolithic industries of this region are differentiated from those of the Levant as they often have low Levallois indices, little centripetal core preparation, and few retouched tools. One of the Neanderthals from Shanidar Cave, Iraq, has signs of disease and numerous injuries, perhaps indicating that this individual was cared for by others. The Baradostian industry, dating to ca. 40–35 ka, directly overlies Mousterian assemblages in cave sites of Iran and Iraq. The Baradostian is characterized by high percentages of burins, and rare carinate scrapers and busked burins, reminiscent of European and Levantine Aurignacian assemblages. The Mousterian assemblages in Karain Cave of western Turkey contain high frequencies of retouched and resharpened tools, including side-scrapers, points, and convergent scrapers of nonlocal materials (Kuhn, 2002). The assemblages are primarily discoid, though Levallois technology appears in at least one stratigraphic layer. The Karain assemblages are thought to have affinities with the Zagros Mousterian or the Middle Paleolithic of the Balkans and southeastern Europe, rather than the Levant.

Upper Paleolithic sites are generally rare in elevated zones of the Zagros, the Iranian plateau, and the Anatolian plateau, leading some investigators to wonder if gaps in the archeological records indicated abandonment during OISs 2 and 3. Few have been intensively studied and/or dated, and the Middle to Upper Paleolithic transition in this region is still poorly understood. Üçağızlı Cave is an early Upper Paleolithic sequence in Turkey that has yielded a combination of typical Middle and Upper Paleolithic tools that have been dated to ca. 40 ka (Kuhn, 2002) (**Fig. 3**). The forms include end-scapers, burins, and retouched blades, many pieces classifiable as Levallois flakes, blades, and points. The upper layers contain assemblages that are similar to early Ahmarian ones, including a prismatic blade technology dating to between 33 and 28 ka. Bone tools are common and the assemblages contain shell ornaments. Epipaleolithic or late Upper Paleolithic assemblages with microlithic-backed pieces have been reported throughout the region. The number of sites and the diversity of ecological niches exploited suggest localized specializations involving hunting of particular species.

South Asia

Further eastward, South Asia contains numerous Middle Paleolithic sites, but these are not associated with any human fossils (Misra, 2001; James and

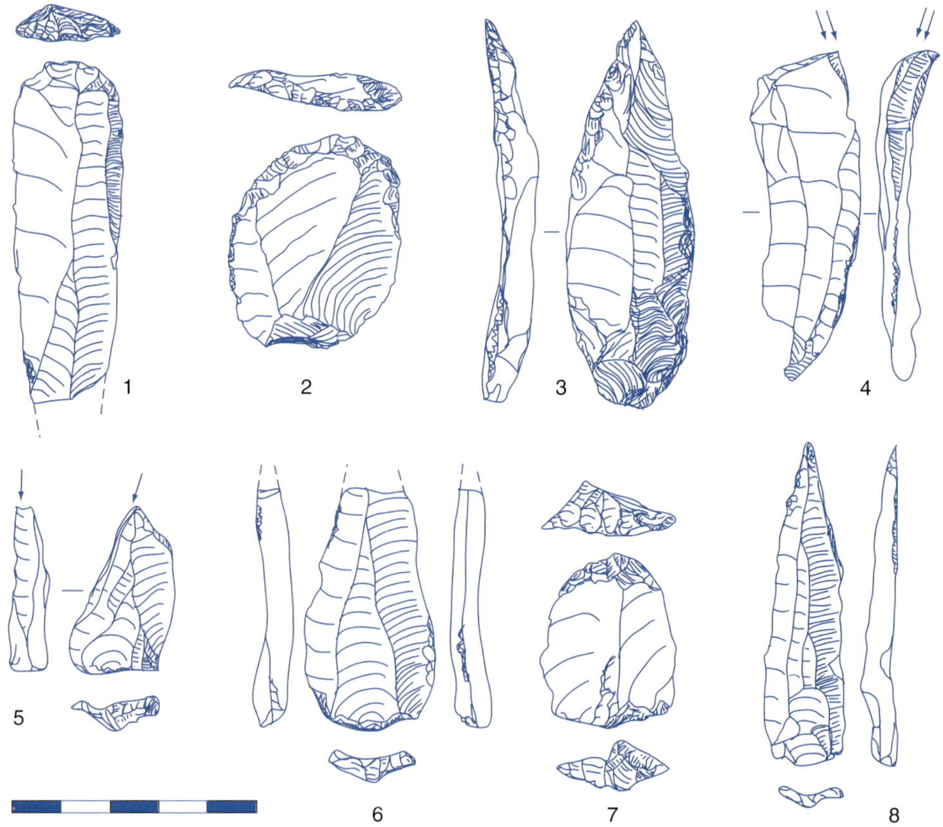

Figure 3 Early Upper Paleolithic artifacts from Üçağızlı Cave, Turkey. Ahmarian (1–4) and initial Upper Paleolithic (5–8) tools. Adapted from Kuhn SL (2004) Upper Paleolithic raw material economies at Üçağızlı cave, Turkey. *Journal of Anthropological Archaeology* 23: 431–448, figure 3.

Petraglia, 2005). During the Middle Paleolithic, there does not appear to be a single or favored technique of core preparation throughout the Indian subcontinent, though Levallois and discoidal techniques are most often described. As in Southwest Asia outside the Levant, there is urgent need for more well-dated sequences between 70 and 25 ka. From approximately 45 ka, increases in blade production and variability in assemblage composition characterize the South Asian record, with assemblages divided into flake-blade, blade-based, and blade and burin industries (Murty, 1979). Microlithic industries are as old as 28.5 ka, indicating that they were part of a diverse late Paleolithic package (**Fig. 4**). In sites such as Patne, geometric microlithic technology developed from an industry characterized by backed blades and burins (Sali, 1989) (**Fig. 5**). With the exception of the geometric microliths, the standardization of retouched artifact forms is not comparable to those seen in the Upper Paleolithic of Europe or Western Asia. There does not appear to be a sudden shift to 'classic' prismatic cores at the onset of the late Paleolithic, but rather, there is contemporaneity between flake- and blade-based technology, with the introduction of microlithic industries by 28.5 ka. There is some evidence for deliberate construction of shelters by about 45 ka and some evidence for symbolism after 28.5 ka, though the evidence is rare (James and Petraglia, 2005).

Eastern Asia

Late Middle Pleistocene hominins such as the Dali and Jinnuishan specimens have been classified as archaic *Homo sapiens* (Wu and Poirier, 1995), late *H. erectus* (Thorne and Wolpoff, 1992) or even as *H. heidelbergensis* (Rightmire, 2001), on the grounds that they are intermediate (at least chronologically) between *H. erectus* and *H. sapiens*. Discussion is often obfuscated by lack of agreement over what constitutes an 'archaic' or 'modern' feature, or suite of features. Further evidence is clearly needed to clarify what was probably a complicated set of biogeographic and phylogenetic relationships in eastern Asia (Lahr, 1996).

Late Paleolithic Far Eastern lithic artifact assemblages vary in reduction techniques, and include

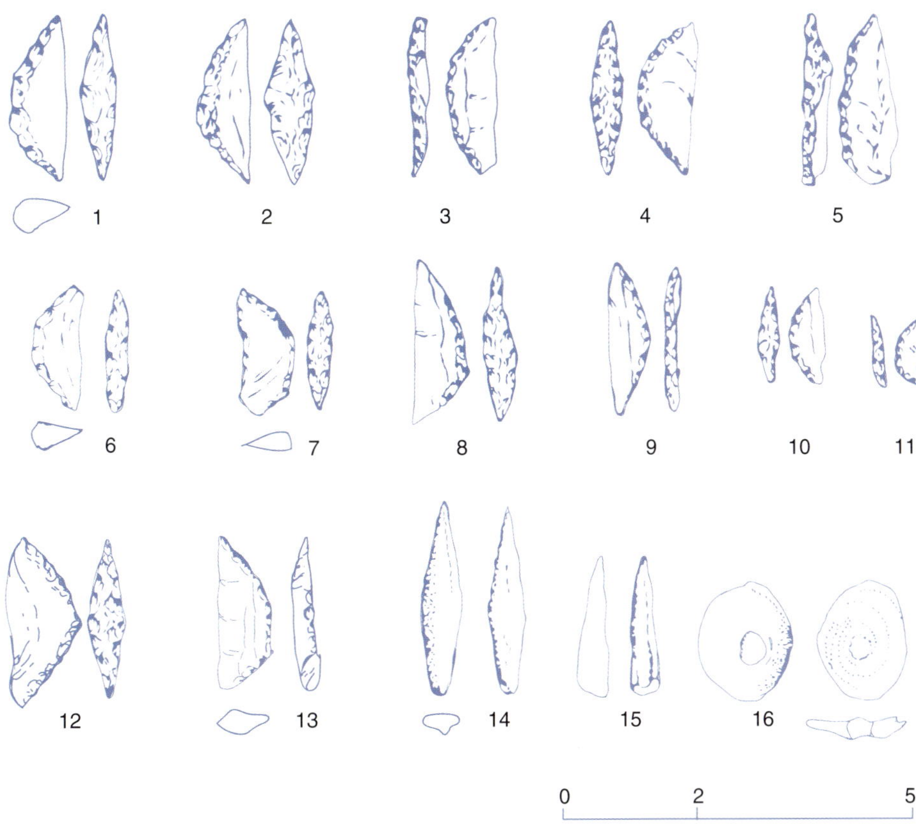

Figure 4 Microlithic artifacts from Sri Lanka. Geometric microliths (1–13), bone points (14, 15) and bead (16). Adapted from Deraniyagala SU (1992) *The Prehistory of Sri Lanka: An Ecological Perspective*. Colombo: Department of the Archaeological Survey, Government of Sri Lanka, figure 59.

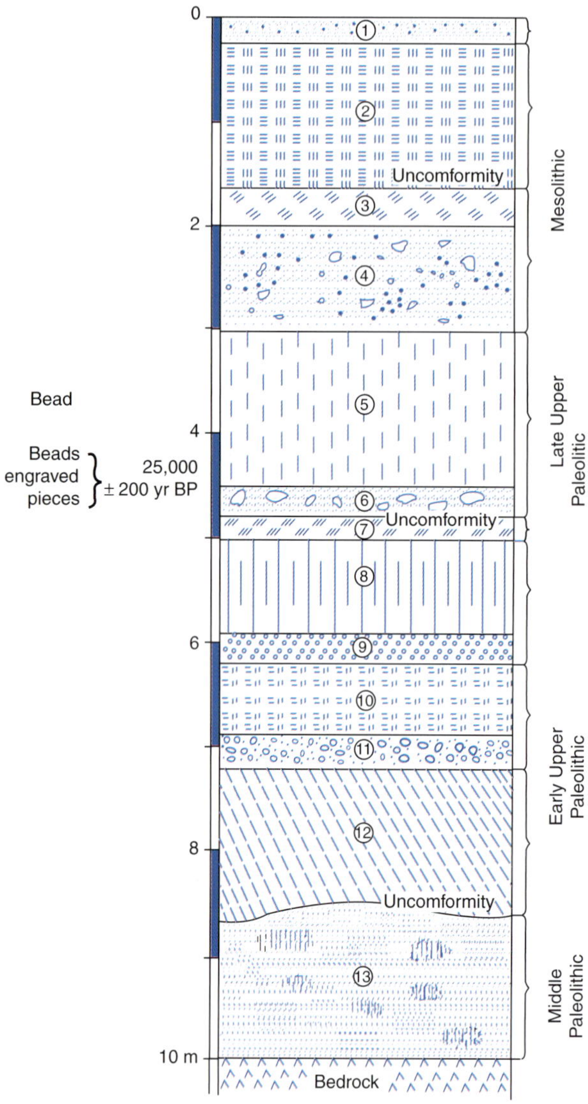

Figure 5 Late Pleistocene stratigraphic profile of Patne, India, showing Middle to Mesolithic sequence and location of symbolic artifacts and radiocarbon date. Adapted from Sali SA (1989) *The Upper Palaeolithic and Mesolithic Cultures of Maharashtra.* Pune: Deccan College Post Graduate and Research Institute, figure 11.

bipolar flakes, Levallois blades, and microliths (Jia and Huang, 1985; Brantingham *et al.*, 2001) (**Fig. 6**). The Zhoukoudian Upper Cave, which is associated with modern humans, contains an assortment of elaborate artifacts, including bone needles and harpoons, perforated marine shells, stone beads, and grave goods in burials. The association of modern humans with symbolic artifacts may indicate their spread, and the replacement of archaic hominins. Microlithic artifacts, perhaps as old as 28 ka, occur in a variety of ecological settings, from deserts to lakeside settings, indicating a wide range of adaptations.

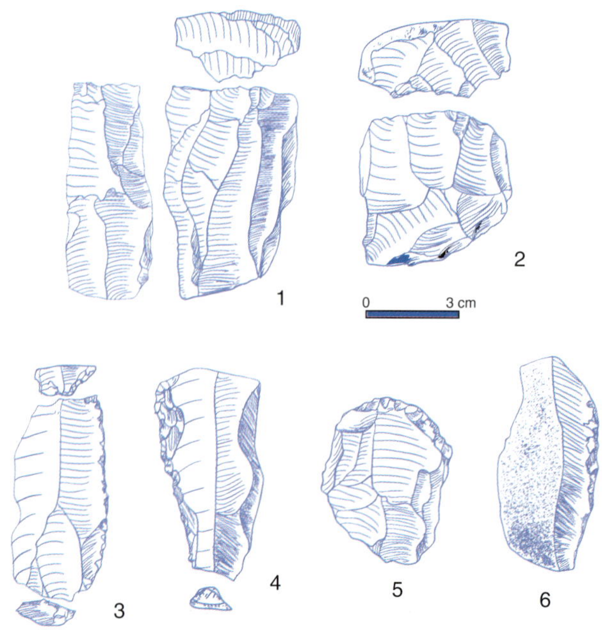

Figure 6 Initial Upper Paleolithic assemblages from Shuidonggou, China (adapted from Brantingham *et al.* (2001), figure 5). Flat-faced 'Levallois' cores (1–2); Levallois blades (3–4); retouched flake-scraper (5); side-scraper (6).

The Replacement of Indigenous Populations of Neanderthals, *H. erectus* and *H. floresiensis* by Modern Humans

The prevailing view among many geneticists and paleoanthropologists is that all living humans in Asia today are descended from a relatively small population which originally arose in Africa (e.g., Ingram *et al.* (2000)). Based on slight and generally ambiguous fossil and archeological evidence, it has been argued that modern humans reached the Levant by 100 ka, and later, other populations traveled (perhaps along the Indian Ocean rim) to reach Australasia by 60 ka (Lahr and Foley, 1994; Field and Lahr, 2006) (**Fig. 7**). An alternative scenario is that the expansion of the modern human genotype occurred without large-scale migrations of people (Eswaran *et al.*, 2005). Whichever model is preferred, modern humans replaced Neanderthals in Southwest and Central Asia, indigenous populations (type(s) unknown in South Asia), populations of *H. erectus* in Eastern Asia, and ultimately, *H. floresiensis* in the island of Flores, Indonesia.

The appearance of anatomically modern humans in Africa is accompanied by a variety of economic, social, and symbolic traits that are often associated with modern human culture, although the materialization of such traits is temporally and geographically patchy over the last 250 ka (McBrearty and Brooks,

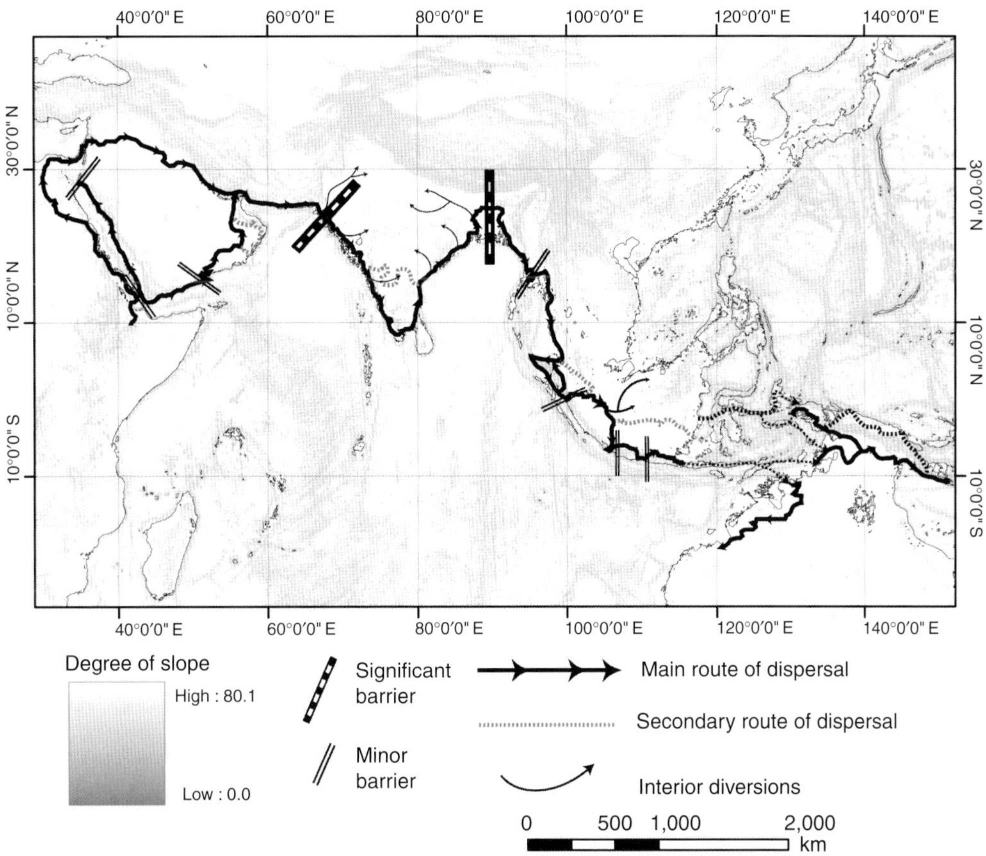

Figure 7 Potential routes of anatomically modern humans during OIS 4, showing routes and barriers. Adapted from Field JS and Lahr MM (2006) Assessment of the southern dispersal: GIS-based analyses of potential routes at oxygen isotopic stage 4. *Journal of World Prehistory*, figure 14.

2000), and few of these traits are both widespread and exclusive to modern humans.

The period between 45 and 30 ka in the Levant is marked by the extinction of the Neanderthals and the expansion of modern humans equipped with Upper Paleolithic adaptations (Shea, 2003). Neanderthals throughout western and Central Eurasia appear to have been well-adapted to temperate woodlands and cold steppes, with adaptations geared toward meat acquisition. Economic intensification in later Neanderthal adaptations in the Levant are sindicated by multiseasonal site occupations, a broadening of the subsistence base, and production of spear points to procure prey (Shea, 2003). Though Neanderthals may have had complex social and economic behaviors, small behavioral and adaptive advantages on the part of modern humans may have eventually led to their replacement by modern humans.

In addition to the expansion of modern humans into the Levant, populations may have used the Bab al Mandab strait across the southern end of the Red Sea for expansions during OIS 5, as is exemplified by the presence of Middle Paleolithic sites in southern Arabia with Northeast African affinities (Petraglia and Alsharekh, 2003; Rose, 2004). Though it is clear that modern humans reached Australia by 45–42 ka and possibly earlier (Bowler *et al.*, 2003; O'Connell and Allen, 2004), the spread of modern humans across Asia has not been documented from skeletal evidence or archeological assemblages, and remains at present a hypothesis based largely on modern genetic evidence. Y-chromosome and MtDNA data suggest the colonization of South Asia by modern humans originating in Africa, sometime between 73 and 55 ka (Kivisild *et al.*, 2003). The genetic data suggest a single, early migration was responsible for the initial settlement toward South Asia and toward the east (Thangaraj *et al.*, 2005). It has been suggested that modern movements into South Asia may have originally been accompanied by a Middle Paleolithic technology, presenting problems in discriminating such assemblages from those being used by resident populations of archaic hominins (James and Petraglia, 2005). Though modern humans were probably present in South Asia sometime after 70 ka, evidence for modern human behavior develops slowly and such signs are temporally and spatially patchy. The development of new cultural innovations in the Late Paleolithic of South Asia may be related, in part, to fluctuations in the environment and demographic changes, indicating that it may be difficult to discern movements of modern humans. Modern humans probably reached Eastern Asia by 40–30 ka, as implied by a more elaborate archeological record and signs of symbolism, and the skull of a modern human at Niah Cave, Borneo, dated at ca. 43 ka (Barker, 2002). Though encounters between archaic and modern humans are unclear across most of Asia, the replacement of indigenous archaic populations was probably completed by 18 ka, with the extinction of hominins such as *H. floresiensis* (Brown *et al.*, 2004) in remote island settings. The last appearance of *H. erectus* remains uncertain. In Java, the Ngandong assemblage of *H. erectus* has been dated to 28–54 ka (Swisher *et al.*, 1996), implying they overlapped with modern humans: others argue, however, on paleontological grounds that the Ngandong assemblage is Late Middle Pleistocene and thus >125 ka (see e.g., Storm *et al.* (2005)).

The Colonization of New Territories after 125 ka

Before the last interglacial, hominin populations inhabited the same parts of Asia as had been colonized in the Middle Pleistocene, that is, up to ca. 40–45° N, with perhaps short forays further north when climatic conditions permitted. As with so much of early Asian prehistory, more dates are needed to indicate when and how the expansion of humans across the whole of Asia occurred. Neanderthals in Central Asia appear to have been the first to expand northward into southern Siberia up to 55° N, probably in the early part of OIS 4 in the last glaciation, and possibly earlier. This process of colonization accelerated after 40 ka, and was most likely accomplished by modern humans, which began to colonize areas >55° N after 26 ka (Goebel, 1999). Another region that appears to have been colonized by modern humans rather than *H. erectus* is Borneo, which would have formed part of the Sunda Continental Shelf in the last glaciation; here, the earliest evidence from Niah Cave dates from ca. 43 ka. Finally, the Tibetan Plateau, 5 million sq km in extent and averaging 5,000 m a.s.l., was uninhabited until the late glacial, and was one of the last parts of Eurasia to be settled (Brantingham *et al.*, 2001).

Settling and Sedentism: Developments after the Last Glacial Maximum

Major changes in subsistence occurred after the Last Glacial Maximum, ca. 16–18 ka, and resulted in the emergence of agricultural societies across much of Asia by 8 kyr BP (Mithen, 2003). The background to this process was the ending of the last ice age and onset of modern climatic conditions. In Southeast Asia, the most conspicuous environmental change was the flooding of the Sunda Shelf, which resulted in the loss of >1.5 million sq km of land, or three-quarters the present day area of mainland Southeast Asia (Voris, 2000). The archeology of foraging populations in the Levant between 20 and 14 ka is relatively well known (Bar-Yosef, 1998). The Kebaran is characterized by obliquely truncated blades, bladelets, and microliths, with the addition of groundstone implements, perhaps used to process nuts and grasses. The Kebaran is thought to reflect seasonal mobility by hunting and gathering groups between highland and lowland settings. Ohalo II, on the shore of the Sea of Galilee of Israel, is a spectacular waterlogged site dated to ca. 20 ka (Nadel, 2002). A community geared toward fishing and hunting-gathering left behind brush huts, hearths, hundreds of thousands of charred seeds/fruits, and fish and animal bones (**Fig. 8**). During a humid interval at around 14.5–13 ka, an expansion of settlement occurs into the interior and highland zones of the Negev, Sinai, and southern Jordan. Climatic improvement appears to be responsible for population expansions into the steppes and desertic areas.

The date of 11 kyr BP is usually given as the beginning of food production based on the domestication of relatively few wild plant and animal species (Diamond and Bellwood, 2003). Until the end of the Pleistocene, all people in Asia lived as hunter-gatherers. The introduction of food production led to major changes in subsistence and settlement practices in some populations, whereas in other populations it led to territorial marginalization or to even to extinction. Some foraging populations continued to live alongside Neolithic peoples, sometimes engaging in trade and exchange with settled groups.

The emergence and subsequent expansion of Natufian populations in the Levant is dated to about 13–10 ka (Bar-Yosef, 1998) (**Fig. 9**). Some large sites, such as Ain Mallaha, hint at prolonged occupations based on substantial semi-subterranean dwellings and multiple superimposed occupation floors. Stone tool assemblages include numerous crescentic microliths and bone, and horncore tools include harpoons, fishhooks, spear and arrow points, awls, and needles. Large groundstone querns are often times found and may have functioned in the processing of nuts or

Figure 8 Ohalo II excavations, showing (A) hut remains and (B) human skeleton. Photographs by D. Nadel.

cereals. The availability of cereal grasses and semisedentary lifestyles were likely major contributing factors in leading to agriculture. The Pre-Pottery Neolithic witnessed an increase in the number and diversity of sites, and changes in settlement patterns may be partly a reflection of climatic changes occurring between 11 and 10 ka (Bar-Yosef, 1998). Abu Hureyra, located in Syria and occupied from about 11.5 to 7 ka, is a good example of changes in lifestyle and economy that occurred over this period (Moore *et al.*, 2000). The first period of habitation, before agricultural development, included a settlement that appears to represent year-round occupation. The populations gathered a variety of wild seeds, including cereals, and hunted migrating gazelle populations. After about 10 ka, Abu Hureyra was inhabited by permanently settled communities who had cultivated and processed a variety of domesticated seeds, such as oats, barley, chickpeas, emmer, and lintels. Such changes in Pre-Pottery Neolithic communities were precursors to more significant technological, social, and economic changes that were to take place throughout Western Asia.

By 13–8 ka, many parts of Asia were filled with complex hunting and gathering communities. Siberia, which was largely uninhabited during the Last Glacial Maximum, was rapidly recolonized up to the Arctic Ocean by 12.5 ka (Goebel, 1999). After 12.5 ka, Jomon sites in Japan appear to show production of

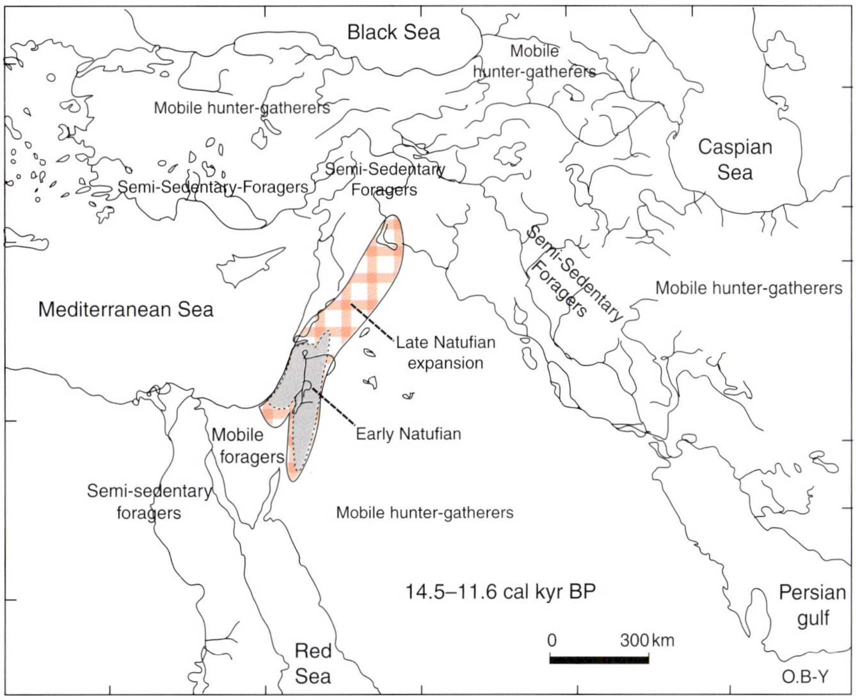

Figure 9 Map of initial Natufian origin and later expansions. Adapted from Bar-Yosef O (1998) The Natufian culture in the Levant, threshold to the origins of agriculture. *Evolutionary Anthropology* 6: 159–177, figure 1.

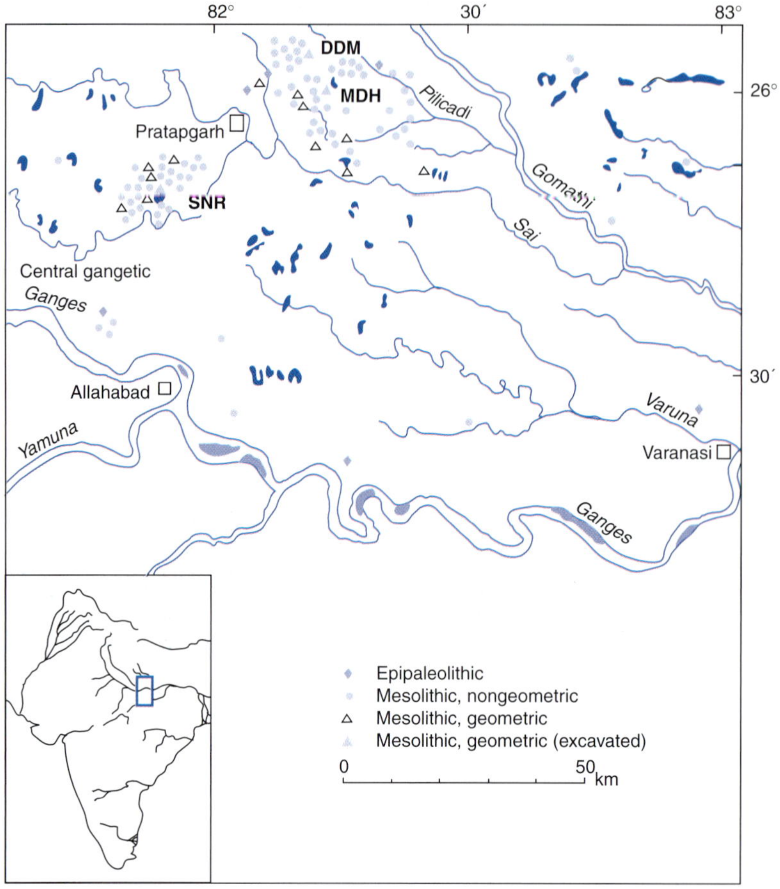

Figure 10 Map showing Epipaleolithic and Mesolithic sites of the central Ganga Plains. Adapted from Chattopadhyaya UC (1996) Settlement pattern and the spatial organization of subsistence and mortuary practices in the Mesolithic Ganges Valley, north-central India. *World Archaeology* 27(3): 461–476, figure 1.

pottery and sedentary lifestyles based on fishing, with supplementation of cereal grains by around 9 ka. Mesolithic sites are well represented in the Indian subcontinent, leading archeologists to infer that the large number of sites relates to marked growth in human populations (Misra, 2001). The increased food supply available in the Mesolithic is thought to have led to a reduction in mobility, as reflected in the large size of sites, evidence for huts and communal hearths, the substantial appearance of habitation deposits, and the presence of cemeteries, particularly in the Ganga Valley (Chattopadhyaya, 1996) (**Fig. 10**).

Recent archeological evidence suggests two distinct centers of early rice cultivation in Asia (Fuller, 2006). In China, it is widely accepted that rice cultivation was underway in the Middle Yangzi, and adjacent South China by 8 ka, and in India, rice cultivation may date back to as early as ca. 9 ka. While further research is needed in South Asia, the recent evidence indicates at the very least that foragers were exploiting wild rice from about 9 ka. In northwestern South Asia, the dominant crops from the time of earliest evidence derived from the Southwest Asian Neolithic founder crops. Representatives of this crop package spread to western Pakistan by the time of Neolithic Mehrgarh after 9 ka (Jarrige *et al.*, 2006), and Central Asia by ca. 8 ka. Such subsistence and settlement shifts set into motion fundamental changes in the demography, health, and lifestyles of populations inhabiting Asia.

See also: **Archaeological Records**: Overview; 2.7 MYR-300,000 years ago in Asia; 1.9 MYR-300,000 years ago in Europe; Global Expansion 300,000-8000 years ago, Africa; Global Expansion 300,000-8000 years ago, Australia; Global Expansion 300,000-8000 years ago, Americas; Neanderthal Demise; Postglacial Adaptations. **Vertebrate Studies**: Interactions with Hominids.

References

Barker, G. W. W. (2002). Prehistoric foragers and farmers in South-east Asia: Renewed investigations at Niah Cave, Sarawak. *Proceedings of the Prehistoric Society* **68**, 147–164.

Bar-Yosef, O. (1998). The Natufian culture in the Levant, threshold to the origins of agriculture. *Evolutionary Anthropology* **6,** 159–177.

Bar-Yosef, O. (2002). The Upper Paleolithic revolution. *Annual Reviews of Anthropology* **31,** 363–393.

Bowler, J. M., Johstone, H., Olley, J., *et al.* (2003). New ages for human occupation and climate change at Lake Mungo, Australia. *Nature* **421,** 837–840.

Brantingham, P. J., Krivoshapkin, A. I., Jinzeng, L., and Tserendagva, Y. (2001a). The initial Upper Paleolithic in Northeast Asia. *Current Anthropology* **22,** 735–747.

Brantingham, P. J., Olsen, J. W., and Schaller, G. B. (2001b). Lithic assemblages from the Chang Tang region, northern Tibet. *Antiquity* **75,** 319–327.

Brown, P., Sutikna, T., Morwood, M. J., *et al.* (2004). A new small bodied hominin from the Late Pleistocene of Flores, Indonesia. *Nature* **431,** 1055–1061.

Cameron, D., Patnaik, R., and Sahni, A. (2004). The phylogenetic significance of the Middle Pleistocene Narmada hominin cranium from central India. *International Journal of Osteoarchaeology* **14,** 419–447.

Chattopadhyaya, U. C. (1996). Settlement pattern and the spatial organization of subsistence and mortuary practices in the Mesolithic Ganges Valley, north-central India. *World Archaeology* **27**(3), 461–476.

Deraniyagala, S. U. (1992). *The Prehistory of Sri Lanka: An Ecological Perspective.* Department of the Archaeological Survey, Government of Sri Lanka, Colombo.

Diamond, J., and Bellwood, P. (2003). Farmers and their languages: The first expansions. *Science* **300,** 597–603.

Eswaran, V., Harpending, H., and Rogers, A. R. (2005). Genomics refutes an exclusively African origin of humans. *Journal of Human Evolution* **49,** 1–18.

Field, J. S., and Lahr, M. M. (2006). Assessment of the southern dispersal: GIS-based analyses of potential routes at oxygen isotopic stage 4. *Journal of World Prehistory.*

Fuller, D. (2006). Non-human genetics, agricultural origins and historical linguistics in South Asia. In *The Evolution and History of Human Populations in South Asia: Inter-disciplinary Studies in archaeology, biological anthropology, linguistics and genetics* (M. D. Petraglia and B. Allchin, Eds.), (in press), Springer, The Netherlands.

Gao, X., and Norton, C. (2002). A critique of the Chinese "Middle Palaeolithic". *Antiquity* **76,** 397–412.

Goebel, T. (1999). Pleistocene human colonization of Siberia and peopling of the Americas: An ecological approach. *Evolutionary Anthropology* **8**(6), 208–227.

Gopher, A., Barkai, R., Shimelmitz, R., *et al.* (2005). Qesem Cave: An Amudian site in Central Israel. *Journal of the Israel Prehistoric Society* **35,** 69–92.

Ingman, M., Kaessmann, H., Paabo, S., and Gyllensten, D. (2000). Mitochondrial genome variation and the origin of modern humans. *Nature* **408,** 708–713.

James, H. V. A., and Petraglia, M. D. (2005). Modern human origins and the evolution of behavior in the Later Pleistocene record of South Asia. *Current Anthropology* **46,** S3–S27.

Jarrige, J.-F., Jarrige, C., and Quivron, G. (2006). Mehrgarh Neolithic: The updated sequence. In *South Asian Archaeology 2001* (C. Jarrige and V. Lefèvre Eds.), pp. 129–142. ADPF Éditions Recherche sur les Civilisations, Paris.

Jia, L., and Huang, W. (1985). The Late Palaeolithic of China. In *Palaeoanthropology and Palaeolithic Archaeology in the People's Republic of China* (R. Wu and J. W. Olsen, Eds.), pp. 211–223. Academic Press, London.

Kivisild, T., Rootsi, S., Metspalu, M., *et al.* (2003). The genetic heritage of the earliest settlers persists both in Indian tribal and caste populations. *American Journal of Human Genetics* **72,** 313332.

Kuhn, S. L. (2002). Paleolithic archaeology in Turkey. *Evolutionary Anthropology* **11,** 198–210.

Kuhn, S. L. (2004). Upper Paleolithic raw material economies at Üçağızlı cave, Turkey. *Journal of Anthropological Archaeology* **23,** 431–448.

Lahr, M. M. (1996). *The Evolution of Modern Human Diversity.* Cambridge University Press, Cambridge.

Lahr, M. M., and Foley, R. (1994). Multiple dispersals and modern human origins. *Evolutionary Anthropology* **3,** 48–60.

McBrearty, S., and Brooks, A. (2000). The revolution that wasn't: A new interpretation of the origin of modern human behaviour. *Journal of Human Evolution* **39,** 453–563.

Miller-Antonio, S., Schepartz, L., Karkanas, P., Hou Yamei, Huang Weiwen, and Bakken, D. (2004). Lithic raw material use at the Late Middle Pleistocene site of Panxian Dadong. *Asian Perspectives* **43**(2), 314–332.

Misra, V. N. (2001). Prehistoric human colonization of India. *Journal of Bioscience* **26**(4), 491–531.

Mithen, S. J. (2003). *After the Ice: A Human Global History.* Weidenfeld and Nicolson, London.

Moore, A. M. T., Hillman, G. C., and Legge, A. J. (2000). *Village on the Euphrates.* Oxford University Press, Oxford.

Murty, M. L. K. (1979). Recent research on the Upper Palaeolithic Phase in India. *Journal of Field Archaeology* **6,** 301–320.

Nadel, D. (Ed.) (2002). *Ohalo II – a 23,000 Year-Old Fisher-Hunter-Gatherers' Camp on the Shore of the Sea of Galilee.* Hecht Museum, Haifa.

O'Connell, J. F., and Allen, J. (2004). Dating the colonization of Sahul (Pleistocene Australia-New Guinea): A review of recent research. *Journal of Archaeological Science* **31,** 835–853.

Petraglia, M. D., and Alsharekh, A. (2003). The Middle Palaeolithic of Arabia: Implications for modern human origins, behaviour and dispersals. *Antiquity* 77(298), 671–684.

Petraglia, M. D., Schuldenrein, J., and Korisettar, R. (2003). Landscapes, activity, and the Acheulean to Middle Paleolithic transition in the Kaladgi Basin, India. *Journal of Eurasian Prehistory* **1**(2), 3–24.

Rightmire, G. P. (2001). Comparison of Middle Pleistocene hominids from Africa and Asia. In *Human Roots: Africa and Asia in the Middle Pleistocene* (L. Barham and K. Robson-Brown, Eds.), pp. 123–135. Western Academic and Specialist Press, Bristol.

Rose, J. I. (2004). New evidence for the expansion of an Upper Pleistocene population out of East Africa, from the site of Station One, northern Sudan. *Cambridge Archaeological Journal* **14**(2), 205–216.

Sali, S. A. (1989). *The Upper Palaeolithic and Mesolithic Cultures of Maharashtra.* Deccan College Post Graduate and Research Institute, Pune.

Shea, J. J. (2003). The Middle Paleolithic of the East Mediterranean Levant. *Journal of World Prehistory* **17,** 313–394.

Storm, P., Aziz, F., de Vos, J., Kosasih, D., Baskoro, S., Ngaliman, and van den Hoek Ostende, L. W. (2005). Late Pleistocene *Homo sapiens* in a tropical rainforest fauna in East Java. *Journal of Human Evolution* **49,** 536–545.

Stringer, C. (2002). Modern human origins: Progress and prospects. *Philosophical Transactions of the Royal Society, London B* **357,** 563–579.

Swisher, G. C., Rink, W. J., Anton, S. C., *et al.* (1996). Latest *Homo erectus* of Java: Potential contemporaneity with *Homo sapiens* in Southeast Asia. *Science* **274,** 1870–1874.

Thangaraj, K., Chaubey, G., Kivisild, T., *et al.* (2005). Reconstructing the origin of Andaman Islanders. *Science* **308,** 996.

Thorne, A. G., and Wolpoff, M. H. (1992). The multiregional evolution of humans. *Scientific American* **266**(4), 76–83.
Voris, H. (2000). Maps of Pleistocene sea levels in Southeast Asia: Shorelines, river systems and time durations. *Journal of Biogeography* **27**, 1153–1167.
Wu, X. Z., and Poirier, F. E. (1995). *Human Evolution in China*. Oxford University Press, New York.

Global Expansion 300,000-8000 years ago, Australia

R Cosgrove, La Trobe University, Melbourne, Australia

© 2007 Elsevier B.V. All rights reserved.

Introduction

The human colonization of the Sahul landmass, the continent that joined New Guinea, Australia, and Tasmania after dry land was exposed due to lowered sea level, was a significant event in our species global expansion. It heralded the first sea crossing by peoples whose ancestors spread out of Africa perhaps 70,000 years ago (ya). It would have involved the construction of water craft sophisticated enough to make voyages of up to 100 km from Timor and lesser distances through the Indonesian archipelago. Clearly, organizational skills and language must have been present to make such trips. Crossing this water gap brought people into a world that, in some respects, would have been familiar while in others unique. Many plants would have been recognizable especially rainforest varieties, as well as marine life but most animals encountered would have been alien to them. On passing through the biogeographical transitional zone of Wallacea and across Lydekkers Line, which forms the faunal boundary of the Australian region, people entered a land of giant marsupials, egg-laying mammals, giant flightless birds, and enormous reptiles (**Fig. 1**).

Unlike the Eurasian fauna human groups would have co-existed with, there were no large carnivorous pack animals such as hyenas or wolves. Large cave bears and felids were absent as were smaller carnivores like the fox (until European settlement), wolverine, and lynx. Only three marsupial carnivores were present, the thylacine (*Thylacinus cynocephallus*), Tasmanian devil (*Sarcophilus harrisii*), and the marsupial lion (*Thylacoleo carnifex*), the former two surviving across the continent until about 3,000 years before present (BP) and in Tasmania into modern times (**Fig. 2**). Indeed Sahul, like the Americas, also had no earlier hominins. This is despite the recent Indonesian discoveries of *Homo erectus* and their 'Hobbit' descendents, *H. floresiensis*, living between 800,000 and 12,000 ya just 100 km from Sahul's coastline (Morwood *et al.*, 2004). As yet, there is no convincing evidence that premodern humans had colonized Sahul (O'Connell and Allen, 2004). The people that entered the continent were behaviorally modern humans.

Paleoenvironments

Research on the past climate of Sahul (Pleistocene New Guinea–Australia–Tasmania) has identified major changes in sea levels and terrestrial environments. These have had a significant influence on the ecological structure of this continent, affecting the distribution of potential food and the ability of humans to make a living off the land, especially around the coasts, mountains, and arid interior. Human responses were not uniform however and different behavioral patterns are documented in archeological sites from different regions. These climatic changes were driven by large-scale Milankovitch Cycles that changed the distribution of incoming solar radiation permitting the buildup of ice at the poles. Vast amounts of water were frozen and locked up as glaciers, reducing the sea levels by at least 100 m across the globe. Research on deep-sea cores and uplifted coal terraces on the Huon Peninsula, Papua New Guinea, suggest that sea levels have fluctuated over the last 100,000 years or more. The last glacial cycle began about 130,000 ya with a series of sea-level rises and falls trending toward lower levels until about 18,000 ya after which the sea rose, stabilizing about 6,000 years BP (Lambeck and Chappell, 2001).

Previously, it had been thought that the last glacial was a spike of much colder temperatures and lower precipitation between 20,000 and 18,000 ya. New data suggest that it was more active over a longer period with increased magnitude and frequency from about 30,000 ya (Lambeck *et al.*, 2002). The climatic cycles appear to have oscillated rapidly approaching the Last Glacial Maximum (LGM). These changes would have had implications for humans over the medium term with influences on the distribution of prey animals and plant foods. More significantly for people, embedded within these large environmental fluctuations were smaller, more dramatic El Niño-Southern Oscillation (ENSO) events that began at least 45,000 ya (Kershaw *et al.*, 2003). During this period there was an increase in fire-tolerant vegetation and a steady decline in precipitation that in turn increased periods of dust circulation blowing west out over the Indian Ocean and east into the Tasman Sea (Hesse and McTainsh, 2003; Shulmeister *et al.*,

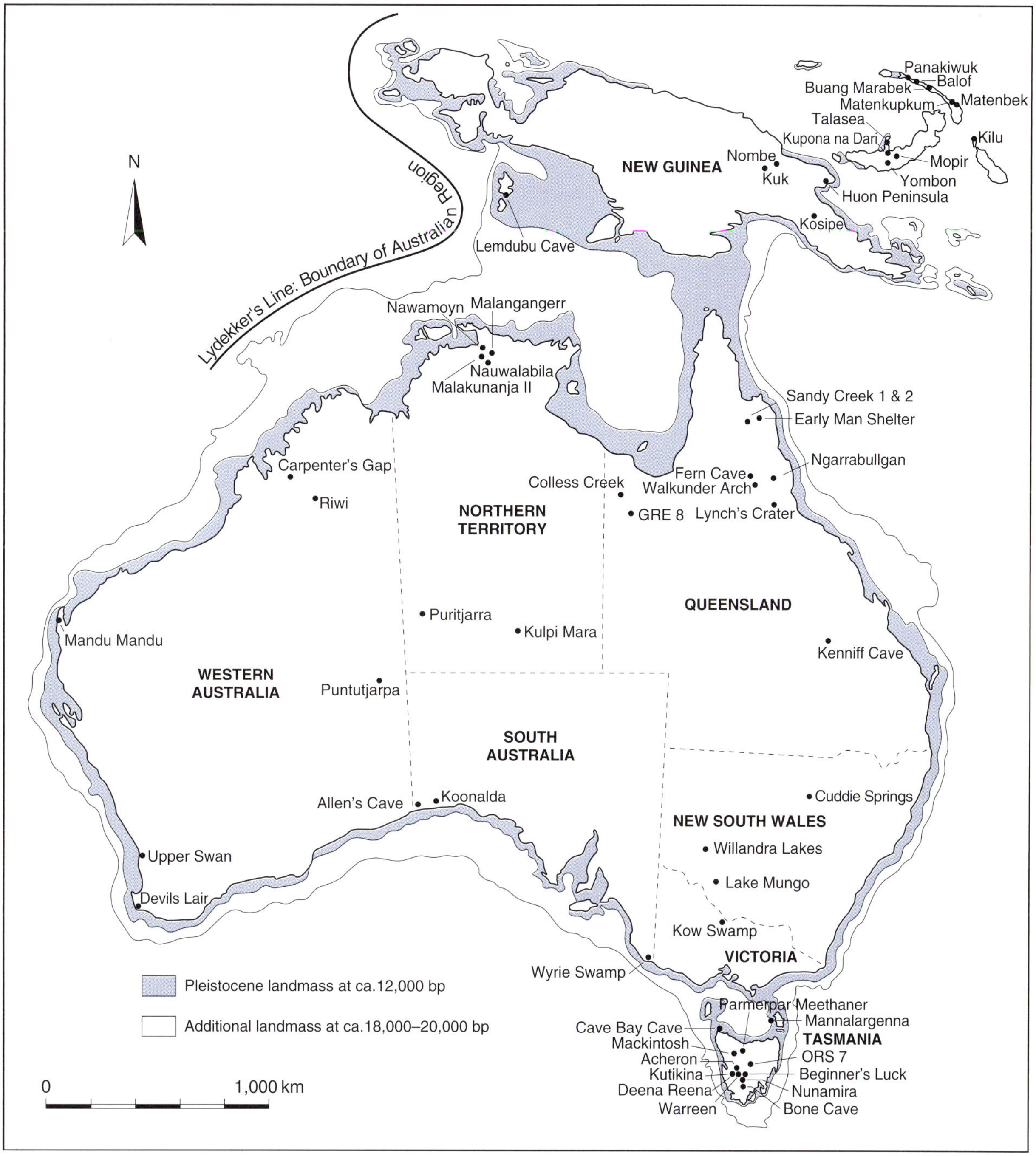

Figure 1 Selected late Pleistocene sites from New Guinea, Australia, and Tasmania. During this time the continent expanded to over 11.5 million square kilometers when sea levels dropped to 130 m below present day levels.

2004). Although some researchers have suggested wind speeds were at least 8 km/hour greater during the LGM than present day, dust particle analysis would indicate that it was more arid than windier.

Terrestrial evidence from the Willandra Lakes system in southeastern Australia suggests alternating wet and dry periods recorded in the lakeside lunettes as sand and clay layers, respectively (Bowler, 1998). These support evidence from the marine chronology for moister conditions before 43,000 years BP when the lake system was full. A sharp but short-lived decline in water levels was recorded around 40,000 years BP. It is at this time that Australia's oldest human remains were recorded, Mungo 3. From this period the lake levels fluctuated with less severe conditions represented by alternating clay and gravel

Figure 2 The Tasmanian Devil (*Sarcophilus harrisi*) remains Australia's sole marsupial carnivore. Unlike *Thylacinus cynocephalus* and *Thylacoleo carnifex* it is primarily a scavenger (photograph by R. Cosgrove).

Figure 3 Lynch's Crater on the Atherton Tablelands, north Queensland has a pollen record stretching back at least 130,000 years. The extinct volcano contains a long continuous record of peat accumulation which records the change from rainforest to fire-adapted species like *Eucalyptus* around 30,000 years BP until 8,000 years BP when rainforest again invaded the landscape (photograph by R. Cosgrove).

deposits. Freshwater shell midden and aquatic fauna are abundant in human sites around the lakes at this time. Fishing with gill nets has been argued based on the remains of fish otoliths in the middens. After 16,000 years BP there was a steady decline in human occupation as the lakes dried up completely.

In the tropical north, pollen cores from Lynch's Crater on the Atherton Tableland have provided detailed sediment records back to at least 130,000 years BP (Kershaw, 1994; **Fig. 3**).

From then until about 78,000 years BP rainforest angiosperms were dominant with some minor dry sclerophyll woodland species. Increasing amounts of *Eucalyptus* pollen, particularly around 35,000 years BP indicates further drying in northern Australia concomitant with changes in southern Australia. Increases in charcoal around 45,000 years BP have been argued to represent human occupation of the Tableland (Turney *et al.*, 2001) although recent archeological research suggests the earliest dates for humans here is around 7,500 years BP (Cosgrove, 2005). A major period of glacial advance and global cooling after 30,000 years BP saw sclerophyll vegetation become dominant on the Atherton Tablelands, all but replacing rainforest species (Kershaw, 1994). Some tropical rainforest refugia remained that sustained endemic species like the tree kangaroo and cassowary into the Holocene. It was only after 8,000 years BP that rainforests began to expand again after increases in rainfall during the climatic optimum.

The arid center of Australia gradually become drier, particularly after about 20,000 years BP with increased dust and lower temperatures enhanced by reduced vegetation cover and a trend towards grass species (Field *et al.*, 2003; Hesse and McTainsh, 2003; Pack *et al.*, 2003). At this time, the complete drying of lake systems around the arid margins by 16,000 years BP led to reactivation of dune and lunette building activity (Bowler, 1998). Other environmental changes to Australia's Pleistocene ecosystems have been said to be a direct result of human firing regimes (Miller *et al.*, 2005) which led to the extinction of *Genyornis newtonii*, a large flightless bird in central Australia 50,000 years BP. Results from other studies on the edge of the arid zone show this bird coexisted with humans until at least 30,000 years BP (Trueman *et al.*, 2005).

New Guinea and Tasmania were finally cut off from Australia by rising seas toward the end of the LGM around 9,000 and 14,000 years BP, respectively.

Chronology of Human Arrivals

There has been continuing debate about the timing of human entry into Australia. Several challenges have confronted archeologists when addressing this question. The first and most contentious is the accuracy of the range of dating techniques, the dynamics of site formation processes, and taphonomy. The protagonists fall into two main groups; those advocating settlement 40,000–45,000 years BP and those arguing for occupation before 60,000 years BP. There have been claims of occupation before 100,000 years BP but so far direct archeological evidence for this has been lacking.

Much of the debate hinges on demonstrating the clear and unequivocal relationships between dates, artifacts, and sediments (O'Connell and Allen, 1998, 2004). Because the timing of human occupation intersects with the present known limits of radiocarbon, it has been necessary for geochronologists to refine dating methods and to understand archeologists' concerns regarding site-formation processes. Developments in accelerator mass spectroscopy (AMS) and luminescence dating have provided a better chronological framework to the problem but have exposed weaknesses in the application of the results to archeological records. Concerns have been raised about the taphonomic effects of termites, contamination by older sediments of younger archeological deposits, movement of stone tools within sandy deposits, and the inadequate bleaching of the quartz grains to reset the luminescence clock (**Fig. 4**).

O'Connell and Allen (2004) list eight sites that they argue have unequivocal archeological evidence for earliest human occupation between 40,000 and 45,000 years BP. Geographically these sites are located across the continent, from the northern tropics (e.g., Riwi, Huon Peninsula, and Carpenter's Gap), to the temperate southern and arid regions of Devils Lair, Allen's Cave and Lake Mungo. More controversial is the claim for artifacts in the oldest Lower Mungo sedimentary unit dated to about 50,000 ya (Bowler, 1998). Other researchers remain unconvinced since no refitting of artifacts from this level has been undertaken (O'Connell and Allen, 2004). In higher levels, stone flakes have been displaced from their cores by up to 60 cm, implying similar mechanisms for reworking the sediments and their contents in the lower layers. If the post-50,000 years BP chronological patterns hold, it suggests that a wide variety of environments were settled relatively swiftly, within 5,000 years of initial landfall. O'Connell and Allen dismiss the two sites claimed to be 50,000 years BP of Malakunanja and Nauwalabila as having unconvincing associations of stratigraphy, chronology, and artifacts (**Fig. 5**).

By the time the dry Bass Plain had been exposed (40,000 years BP), people took the opportunity to cross into what was then the toe of Sahul, now Tasmania. Here evidence based on radiocarbon chronology suggests Tasmania was first settled 35,000 years BP. The two earliest dates come from Warreen Cave and Parmerpar Meethaner in the southwest and north of Tasmania, respectively (Cosgrove, 1999; **Figs. 6** and **7**).

In Melanesia, humans had settled in the Bismarck Archipelago and the Solomon Islands by at least 40,000 ya at Buang Marabek and 29,000 ya at Kilu (Leavesley *et al.*, 2002). By 20,000 ya people were using boats to move possums and obsidian across water gaps as a means of subsistence and technological adaptations to new lands (Allen *et al.*, 1989). There is also data to show humans had begun utilizing tropical rainforest on New Britain at the open site of Yombon by 35,000 years BP (Pavildes, 2004). The spread and relative speed of settlement across Sahul suggests that environmental barriers did not restrain early modern human movement. They appear to have taken the opportunities to expand their range and adapt to the many and varied ecosystems of the Sahul land mass. These achievements point to an established social and economic system, probably already in place before they crossed the sea separating Southeast Asia and Sahul (**Figs. 8** and **9**).

Figure 4 Dating the earliest human arrivals in Sahul has been contentious. Recent use of optically stimulated luminescence (OSL) techniques has pushed back the ages to between 45,000 and 55,000 ya. Problems arise through contamination and the inadequate resetting of the luminescence clock. Termite activity can be a source of this because they continually circulate sand grains above and below ground. A large termite mound is located within the overhang of this rock shelter and when it collapses may contribute to this problem (photograph by J. Allen).

Figure 5 The site of Nauwalabila is claimed to be at least 50,000 years old (photograph by J. Allen).

Figure 6 Parmerpar Meethaner is one of the oldest Tasmanian sites dating back to at least 34,000 years BP. It is located in the Forth River valley in northern Tasmania (photograph by R. Cosgrove).

Figure 7 The archeological section of Parmerpar Meethaner reveals a series of layers containing stone artifacts, bone, and charcoal. The upper 30 cm is Holocene dating to between 3,500 years BP and modern times. From the middle section to the base of the trench ages range from 18,000 to 34,000 years BP. A date of 40,000 years BP in the lowest section does not contain any human artifacts (photograph by R. Cosgrove).

Figure 8 Matemkupkum on New Ireland dates back at least 35,000 years. It contains the oldest evidence of sea fishing anywhere in the world (photograph by R. Frank).

Figure 9 Yombon has a series of volcanic ash layers seen here as alternating dark and light bands. Dr Christina Pavlides indicates the lowest late Pleistocene layer dated to 35,570 years BP. Over 20 chert stone tools were found here. The raw material for these artifacts was quarried locally while the obsidian found in the layers above was imported from the other side of the island (photograph by C. Pavildes).

Physical Anthropology

Much of what we know about the physical anthropology of the earliest Australians has come from southeast Australia. Kow Swamp and Lake Mungo have provided much of this evidence although there are widely differing interpretations of these human fossil remains. The debate largely stems from the two competing models of the origin and global spread of humans known as the Out-of-Africa and Regional Continuity (Mulvaney and Kamminga, 1999). The former asserts that anatomically and behaviorally modern humans moved out of Africa perhaps 100,000 ya replacing an earlier migration of hominid forms represented by robust *H. erectus* and *H. ergaster* that settled in Eurasia and Southeast Asia. Some archeologists question the chronology although arguments have been mounted to show anatomically modern humans did not establish beyond the Levant after reaching there 70,000 years BP. It is only after 45,000 years BP that there is unequivocal data to show behaviorally modern humans reaching Southeast Asia and Sahul. Others propose an earlier movement of modern humans out of Africa earlier, around 70,000 years BP along the southern coasts of Eurasia into Southeast Asia, and then perhaps into Sahul. However there is no reliable archeological evidence to suggest a human landfall of this antiquity apart from problematic dating of the Mungo 3 skeleton at around 60,000 years BP (Bowler and Magee, 2000; Gillespie and Roberts, 2000; Thorne and Curnoe, 2000; **Fig. 10**).

Put simply, the Regional Continuity model proposes that there was only one migration out of Africa by *H. erectus/ergaster* over 1 million ya whose ancestors evolved *in situ* into regional populations of later modern humans. Physical anthropologists who argue for this model stress change through time from robust to gracile morphology through gene flow between geographically separate populations. However, significant gaps in the fossil record and the appearance of similar cranial morphologies on different continents 10,000 km apart continue to

Figure 10 Lake Mungo and the Walls of China. Here severe gully erosion can have a profound influence on the location of artifacts and bones. Although human settlement is well dated to about 45,000 years BP, there is still some disquiet about the integrity of some of the lowest artifacts found below the burial site of the Mungo 3 skeleton (photograph J. Allen).

hamper the testing of this model (Stringer, 2002). In addition, the new finds from Flores and their late dates suggest that *H. erectus* did not evolve into later modern humans but remained a separate species until only 12,000 ya in the form of *H floresiensis* (Morwood *et al.*, 2004). Its discovery on the island of Flores poses more questions than answers and reflects the complex nature of the human evolutionary process.

The Continuity Model asserts that the earliest Australian arrivals were essentially robust, represented by the incomplete WLH50 skull from the Willandra Lakes, although others argue that its form is the result of pathology, large size, or deformation. The model posits two waves of migration into Australia; an early one represented by WLH50 from Java and a later one represented by more gracile morphology from China. The proponents argue that the present day Aboriginal populations are the result of a mixture of the robust and gracile groups during the Holocene. A second competing view sees the presence of the two morphologies as the result of skeletal variations and sexual dimorphism within a single population. It has also been argued that cultural isolation and artificial head binding also influenced the morphological characteristics found in the Kow Swamp population (Brown, 2005; Pardoe, 1995).

Both models imply an increasing gracilization over time, where robust forms are followed by gracile skeletal morphology, particularly of the skull. The interesting aspect of the Australian data is that the earliest skeletal remains at Lake Mungo are gracile. These date to around 40,000–45,000 years BP. Mungo 3 and Mungo 1 contrast with the robust remains from Kow Swamp, Cohuna, and the Willandra Lakes hominid WLH50. These latter specimens are all dated younger than 20,000 years BP, and in the case of the best preserved at Kow Swamp, date only between 13,000 and 9,500 years BP. That the Mungo fossils predate the robust group by a considerable time implies a significant degree of diversity and variation in Aboriginal populations in Sahul. The evidence does not support the notion of separate ancestral lineages (Stringer, 2002). It suggests a more complex situation involving multiple waves of people entering Sahul, affected morphologically by the combined effects of a harsh glacial climatic, generally low population numbers, and cultural and/or geographic isolation. Only one thing is clear, that the samples are small and represent individuals rather than populations. The study of a more representative sample of pre-Holocene skeletal remains will bring a better understanding of the past population dynamics of Sahul.

Impacts of Humans on Environments

Two issues dominate the debates on human influences on past environments, the extinction of Sahul's megafauna and the effect of fire on the biota (Bowman, 1998; Field, 2004; Roberts *et al.*, 2001; Wroe *et al.*, 2004). More than 50 species of animals became extinct during the Late Pleistocene, and while many weighed less than 40 kg, a significant proportion weighed over 100 kg. The most well-known include *Diprotodon*, *Sthenurus*, and the large flightless bird, *Genyornis*. Arguments over the causes and timing of their extinction have been intense. The explanations include the over-kill hypothesis or 'blitzkrieg' whereby humans, shortly upon arrival, hunted these large beasts to extinction, climatic effects of increasing aridity, ecosystem modification, or a combination of climate and human hunting (**Figs. 11** and **12**).

Evidence for the swift demise of megafauna due to overhunting has come from systematic dating of sediments using luminescence techniques. Age estimates on some 12 taxa suggest extinction between 40,000 and 50,000 ya. The dates come from sediments associated with articulated megafauna bones, not from archeological sites. The results have been questioned by a number of researchers who are critical of the selection of naturally deposited semi-articulated skeletons. They argue that to truly understand the mechanism and age of the extinctions, human sites associated with megafauna must be studied (Wroe *et al.*, 2004).

Figure 11 The skull of a *Diprotodon optatum*. Note the large upper and lower incisors as well as the bony structure on top of the maxilla thought to be for the attachment of powerful muscles. These large herbivores weighed more than 1,000 kg. They were probably browsers preferring the open, arid lands, using their sharp incisors to crop hard-stemmed vegetation (photograph by R. Cosgrove).

Figure 12 The postcranial skeleton of *Diprotodon* suggests a very heavily built animal with long proximal and short distal limbs. The feet were relatively small and probably had heavy padding (photograph by R. Cosgrove).

Other geochronologists have dated eggshell from the extinct *Genyornis* and extant emu (*Dromaius*) and found isotope variation in the two species from about 140,000 ya (Miller *et al.*, 2005). Their work included the analysis of thousands of eggshell pieces of each species from central Australia. They argue that *Genyornis* became extinct about 45,000 years BP under a regime of altered ecological conditions as shown in the changed isotope values beginning about 50,000 years BP. They dismiss climate as a cause and propose human alternation to natural vegetation patterns by fire as an explanation. Drought adapted vegetation, including fertile grasses, gave way to fire adapted desert scrub. Because isotope values in the eggshell demonstrate *Genyornis* was apparently restricted to a narrow range of grasses, it is suggested that it could not adapt to the fire-prone desert vegetation and died out. Other workers reject the stated impact of Aboriginal firing on the Pleistocene landscape, arguing that it has been overemphasized (Bowman, 1998). They see the relationship between past Aboriginal firing practices and the resultant landscape as a highly complex ecological interaction. To understand it requires an apposite temporal resolution of the ecological and archeological record.

Other data contradict the view that there was widespread extinction of *Genyornis* by 50,000 years BP (Trueman *et al.*, 2005). Evidence from the archeological site of Cuddie Springs suggests that *Genyornis* survived for another 15,000 years in coexistence with humans. In addition, the site contains thousands of megafaunal bones in layers below the lowest stone tools showing a diminishing species diversity in the absence of humans through time. The implication is

Figure 13 Cuddie Springs is situated in the centre of a shallow depression. Many thousands of megafauna bones have been excavated in association with artifacts (photograph by R. Cosgrove).

Figure 14 Excavation at Cuddies Springs revealed the long bones of *Genyornis* lying next to stone implements (arrowed). The lowest human occupation is dated to 35,000 years BP (photograph by R. Cosgrove).

that megafauna may have already been in decline due to sustained long-term landscape changes driven by climatic factors (**Figs. 13** and **14**).

The most common species at Cuddie Springs are *Diprotodon optatum* and *Genyornis newtonii* and these are clearly associated with artifacts sandwiched between two hardened pavements dated to between 30,000 and 35,000 years BP (Field, 2004). The most parsimonious explanation for the data is that these species retreated to the better-watered edges of the expanding arid zone, leaving the less specialized emus to survive in the drier interior where they live at present day. *Genyornis*'s prolonged co-existence with humans at Cuddie Springs can be explained this way without recourse to the overdramatization of the human-caused extinction theory. Nevertheless they appear to have finally succumbed to a drying continent and human predation (**Figs. 15** and **16**).

Figure 15 An artistic impression of *Diprotodon optatum*. It probably stood about 1.8 m at the shoulder and had an overall length of 3.4 m (illustration A. Musser and Australian Museum).

Figure 16 *Genyornis newtonii* is a distant relative of the emu but was larger. It stood over 2 m and had large robust feet (illustration A. Musser and Australian Museum).

Ultimately the question of megafauna extinction and human influence is undeniably linked to a determination of the arrival time of people on the Sahul continent. Until this is resolved the cause of extinctions remains problematic. In Tasmania there is a reasonably good understanding of the human's entry time of about 35,000 years BP. Although megafauna were present in Tasmania, none have been recorded in over 600,000 bones excavated from seven Late Pleistocene human cave sites (Cosgrove and Allen, 2001). All are extant animal species. It strongly suggests that megafauna were already locally extinct before humans arrived (**Figs. 17** and **18**).

Figure 17 Bone Cave in southwest Tasmania has human occupation extending back at least 29,000 years. It contains two main occupation period; from 29,000 years BP to about 23,000 years BP then a hiatus until 16,000 years BP when occupation intensified (photograph by R. Cosgrove).

Figure 18 Southwest Tasmanian sites are characterized by there extreme richness of stone tools and bones. It is not uncommon to find 250,000 bones and 35,000 stone tools in less than a cubic meter of deposit (photograph by R. Cosgrove).

Continent-wide explanations are therefore unlikely to address this issue in the short term. The expectation is that these will be much more complex and focused at local or regional scales. Additional work needs to be done in identifying causal mechanisms and the effects. Ecological modeling and the identification of other archeological sites to establish regional variability may go some way in solving this vexatious question.

Technology and Economy

Late Pleistocene Sahul stone technology has been characterized as conservative and for the most part unchanging. In the past, these notions have been used to portray the early colonists as technologically backward, without regional variability. Recent advances in fieldwork, method, and theory have largely debunked this view, recognizing much more behavioral variability across Sahul. It is clear that the first settlers must have had reliable sea-going rafts or canoes. A level of organizational capacity and language must have also been present to carry out these voyages. Obviously the organic material necessary for the construction of these boats has long since decayed although sea faring continued throughout the Bismarck Archipelago during the Late Pleistocene as demonstrated by the presence of traceable stone artifact raw material. Significant quantities of obsidian have been recorded in the Late Pleistocene sites on New Britain and New Ireland for example. Recent research at Kupona na Dari indicates that it was used soon after human arrival about 40,000 years BP (Torrence *et al.*, 2004). Obsidian was transported by boat up to 32 km and in the case of sites on New Ireland at Matenbek, up to 300 km from sources at Mopir and Talasea by 20,000 ya (**Fig. 19**).

Coupled with this movement is evidence for the economic structuring by human colonizers. It has been argued that the early settlers to the Bismarks were highly mobile, utilizing resource patches, cleared with 'waisted' axes and fire as a means of breaking the tropical canopy. Before 20,000 years BP water crossings were made into previously uninhabited areas

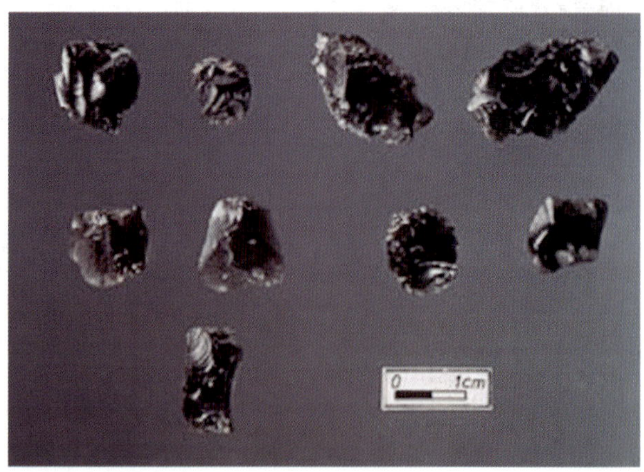

Figure 19 Small obsidian artifacts found in the sites on New Ireland. These implements have been worked down to a small size, indicative of raw-material rationing (photograph by R. Frank).

where resource levels and interior geography were unknown. Significant marine distances were crossed where the next landfall could not be seen before leaving their origin point, as was the case of Kilu, Solomon Islands 29,000 years BP. Exploration included a degree of deliberate voyaging to locate stone quarries that meant groups were mobile, exploiting a wide range of resources from both the coast and inland. After 20,000 years BP there were transfers of animals as a risk-reducing device, raw materials, and increased sailing to more remote islands.

The earliest and most typologically distinct tools are ground-edge axes that have been found across northern Australia and New Guinea. They only appear in the archeological record in southern Australia after 5,000 years BP. 'Waisted' and ground-edge hatches are dated to over 40,000 years on the Huon Peninsula and 26,000 years at the New Guinea highlands site of Kosipe. Ground-edge hatchets were recovered from the sites of Malangangerr and Nawamoyn in the Northern Territory that date to 23,000 years BP. Archeologists have suggested that the typological links indicate a degree of cultural connectedness and perhaps a common origin of human groups operating at this time. They were probably used in forest clearance to create small, disturbed forest patches to encourage useful and productive plant foods (**Figs. 20** and **21**).

Although typological schemes have been put forward in the past to describe the flaked stone technology of Sahul, several objections have been raised that counter the usefulness of typological groupings (Hiscock and Attenbrow, 2003). Indeed with the recent discovery of microliths dated to 15,000 years BP at GRE 8 and Walkunder Arch in Queensland aged 13,000 years BP, their usefulness as chronological markers of mid-Holocene technological change is no longer viable (**Fig. 22**).

The appearance of grindstones perhaps 30,000 years BP also suggests that early tools for seed-grinding exploitation were instrumental in opening

Figure 20 This 23,000-year-old ground-edge hatchet was found at the northern Australian site of Malangangerr. They are made from porphyritic dolerite. It is one of the oldest such implements in the world (photograph by J. Allen).

Figure 21 Two main phases of occupation were found at Malangangerr; an early Pleistocene and a later Estuarine phase containing shell midden and smaller-sized artifacts (photograph by J. Allen).

Figure 22 Walkunder Arch is a large limestone cave that contains a series of fine soil lenses dating to the Holocene period. In the darker reddish layers small backed microliths were recovered indicating that these implements have a late Pleistocene origin (photograph by R. Cosgrove).

Figure 23 Bennett's wallaby is the most common prey animal in the Tasmanian ice sites. Studies of the annuli in their teeth show that the majority were hunted during the autumn and late spring, with many killed during winter. This was the harshest time of the year but people were able to use the valleys in a planned and systematic manner that indicates organization skill and forward planning (photograph by R. Cosgrove).

up opportunities for the settlement of challenging environments such as deserts. The use of organic technology has also been recorded at Wyrie Swamp where barbed spears and boomerangs dated to at least 10,000 years BP have been preserved. In Tasmania and southwest Western Australia bone points dating back 30,000 years BP have been identified. These were probably used as awls, needles, and spear points (Mulvaney and Kamminga, 1999). Given these developments the classification of the Late Pleistocene stone technology under the rubric, Core Tool and Scraper Tradition is now considered redundant.

In Tasmania, humans were exploiting wallabies as the major prey animals. Hunters used sites in the valleys primarily between autumn and early spring, capturing the macropods for their fat, fur, and sinews. Body-part analysis of the butchered carcasses suggests selection of the lower limb bones processed for marrow. The pattern of exploitation suggests optimal foraging of Bennett's wallaby. They were an excellent prey choice as they took little effort to obtain, energy returns were higher for the effort spent in their search, and killing and processing took place in the patches of grassland on which they fed (Cosgrove and Pike-Tay, 2004; **Fig. 23**).

Abbreviations and Glossary

AMS	accelerator mass spectroscopy
artifact	any object shaped and made by humans
BP	radiocarbon years before the present, conventionally measured from 1950
ENSO	El Niño-Southern Oscillation
obsidian	a natural glass produced from erupting volcanoes. Molten rock ejected from the crater solidifies extremely rapidly prohibiting the growth of crystals and forming glass
LGM	Last Glacial Maximum
luminescence dating	a method of dating inorganic material like clay and sediments. Electrons built up and trapped in the crystal lattice structure of quartz grains are artificially stimulated by light that releases them. The amount of energy given off is an indication of how long these sediments have been buried
taphonomy	study of the laws of burial, particularly transformation of organics from the biosphere to the lithosphere
ya	years ago

See also: **Archaeological Records**: Overview; 2.7 MYR-300,000 years ago in Asia; 1.9 MYR-300,000 years ago in Europe; Global Expansion 300,000-8000 years ago, Africa; Global Expansion 300,000-8000 years ago, Asia; Global Expansion 300,000-8000 years ago, Americas; Neanderthal Demise; Postglacial Adaptations. **Vertebrate Records**: Mid-Pleistocene of Africa. **Vertebrate Studies**: Interactions with Hominids.

References

Allen, J., Gosden, C., and White, J. P. (1989). Human Pleistocene adaptations in the tropical island Pacific: recent evidence from New Island, a Greater Australian outlier. *Antiquity* **63** 548–561.

Bowler, J. (1998). Willandra Lakes revisited: environmental framework for human occupation. *Archaeology in Oceania* 33.

Bowler, J., and Magee, J. W. (2000). Redating Australia's oldest human remains: a sceptic's view. *Journal of Human Evolution* 38.

Bowman, D. M. J. S. (1998). The impact of Aboriginal landscape burning on the Australian biota. *New Phytologist* **140** 385–410.

Brown P (2005) Peter Brown's Australian and Asian palaeoanthropology. http://www-personal.une.edu.au/~pbrown3/palaeo.html (lasted accessed 3 June 2006).

Cosgrove, R. (1999). Forty-two degrees south: the archaeology of Late Pleistocene Tasmania. *Journal of World Prehistory* **13** 357–402.

Cosgrove, R. (2005). Coping with noxious nuts. *Nature Australia* 28 46–53.

Cosgrove, R., and Allen, J. (2001). Prey choice and Hunting Strategies in the Late Pleistocene: Evidence from Southwest Tasmania. In *Histories of Old Ages: Essays in Honour of Rhys Jones* (A. Anderson, S. O'Connor and I. Lilley, Eds.). Coombs Academic Publishing, Australian National University, Canberra.

Cosgrove, R. and Pike-Tay, A. (2004). The Middle Palaeolithic and Late Pleistocene Tasmania hunting behaviour: A reconsideration of the attributes of modern human behaviour. *International Journal of Osteoarchaeology* **14**, 321–332.

Field, J. (2004). Australian Late Pleistocene faunal extinctions and the archaeological record: a view from Cuddie Springs. In *Archaeology from Australia* (T. Murray, Ed.). Australian Scholarly Publishing, Melbourne.

Field, J. H., Dodson, J. R., and Prosser, I. P. (2003). A Late Pleistocene vegetation history from the Australian semi-arid zone. *Quaternary Science Reviews* **21**, 1023–1037.

Gillespie, R., and Roberts, R. (2000). On the reliability of age estimates for human remains at Lake Mungo. *Journal of Human Evolution* **38**, 727–732.

Hesse, P. P., and McTainsh, G. H. (2003). Australian dust deposits: modern processes and the Quaternary record. *Quaternary Science Reviews* **22**, 2007–2035.

Hiscock, P., and Attenbrow, V. (2003). Early Australian implement variation: a reduction model. *Journal of Archaeological Science* **30**, 239–249.

Kershaw, P. (1994). Pleistocene vegetation of the humid tropics of northeastern Queensland, Australia. *Palaeogeography, Palaeoclimatology, Palaeoecology* **109**, 399–412.

Kershaw, P., van der Kaars, S., and Moss, P. (2003). Late Quaternary Milankovich-scale climatic change and variability and its impact on monsoonal Australasia. *Marine Geology* **201**, 81–95.

Lambeck, K., and Chappell, J. (2001). Sea level change through the last glacial cycle. *Science* **292**, 672–686.

Lambeck, K., Esat, T. M., and Potter, E. K. (2002). Links between climate and sea levels for the past three million years. *Nature* **419**, 199–206.

Leavesley, M. G., Bird, M. I., Fifield, L. K., *et al.* (2002). Buang Merabak: early evidence for human occupation in the Bismark Archipelago, Papua New Guinea. *Australian Archaeology* **54**, 55–57.

Miller, G. H., Fogel, M. L., Magee, J. W., Gagan, M. K., Clarke, S. J., and Johnson, B. J. (2005). Ecosystem Collapse in Pleistocene Australia and a Human Role in Megafaunal Extinction. *Science* **309**, 287–290.

Morwood, M. J., Soejono, R. P., Roberts, R. G., *et al.* (2004). Archaeology and age of a new hominin from Flores in eastern Indonesia. *Nature* **431**, 1087–1091.

Mulvaney, J., and Kamminga, J. (1999). *Prehistory of Australia.* Allen and Unwin, Sydney.

O'Connell, J. F., and Allen, J. (1998). When did humans first arrive in Greater Australia and why is it important to know? *Evolutionary Anthropology* **6**, 132–146.

O'Connell, J. F., and Allen, J. (2004). Dating the colonization of Sahul (Pleistocene Australia-New Guinea): a review of recent research. *Journal of Archaeological Science* **31**, 835–853.

Pack, S. M., Miller, G. H., Fogel, M. L., and Spooner, N. A. (2003). Carbon isotopic evidence for increased aridity in northwestern Australia through the Quaternary. *Quaternary Science Reviews* **22**, 629–643.

Pardoe, C. (1995). Riverine, biological and cultural evolution in southeastern Australia. *Antiquity* **69**, 696–713.

Pavildes, C. (2004). From Misisil Cave to Eliva Hamlet: rediscovering the Pleistocene in interior west New Britain. *Records of the Australian Museum* **29**, 97–108.

Roberts, R., Flannery, T., Ayliffe, L. K., *et al.* (2001). New ages for the last Australian megafauna: continent-wide extinction about 46,000 years ago. *Science* **292**, 1888–1892.

Shulmeister, J., Goodwin, I., Renwick, J., *et al.* (2004). The Southern Hemisphere westerlies in the Australasian sector over the last glacial cycle: a synthesis. *Quaternary International* **118–119**, 23–53.

Stringer, C. (2002). Modern human origins: progress and prospects. *Philosophical Transactions of the Royal Society London, Series B* **357**, 563–579.

Thorne, A., and Curnoe, D. (2000). Sex and significance of Lake Mungo 3: reply to Brown "Australian Pleistocene variation and the sex of Lake Mungo 3. *Journal of Human Evolution* **39**, 587–600.

Torrence, R., Neall, V., Dolman, T., *et al.* (2004). Pleistocene colonisation of the Bismark Archipelago: new evidence from West New Britain. *Archaeology in Oceania* **39**, 101–130.

Trueman, C., Field, J. H., Dortch, J., Charles, B., and Wroe, S. (2005). Prolonged coexistence of humans and megafauna in Pleistocene Australia. *Proceedings of the National Academy of Sciences* **102**, 8381–8385.

Turney, C. S. M., Kershaw, A. P., Moss, P., *et al.* (2001). Redating the onset of burning at Lynch's Crater (North Queensland): Implications for human settlement in Australia. *Journal of Quaternary Science* **16**, 767–771.

Wroe, S., Field, J., Fullagar, R., and Jermin, L. S. (2004). Megafaunal extinction in the Late Quaternary and the global overkill hypothesis. *Alcheringa* **28**, 291–332.

Global Expansion 300,000-8000 years ago, Americas

S Gonzalez, Liverpool John Moores University, Liverpool, UK

© 2007 Elsevier B.V. All rights reserved.

Introduction

Anatomically modern humans (*Homo sapiens*) moved from their origins in Africa, migrating across the different continents until finally reaching the Americas. A number of issues about this final part of the journey are still hotly debated by scientists today. These include questions such as from where are the First Americans, when exactly did this happen and who were they? The answers to these questions are of vital importance in human evolution because they will contribute a wealth of information about the general processes of human migration and dispersal across the planet, and in particular about the timing of technological capabilities of humans advancing in unknown continents and their relationship with the environment.

There are two main hypothesis to try to explain when the early settlement of the Americas happened:

1. **The Clovis First Model** (Late Entry Model)
2. **The Pre-Clovis** (Early Entry Model)

The Clovis First Model

At the height of the Last Glacial Maximum (18,000 years ago), with large amounts of water stored in the Ice Caps, the global sea level dropped by an estimated maximum of 125 m. This caused large areas of previously submerged continental shelf to become dry land, creating a series of land bridges. One of these is the area that is now known as the Bering Sea, which created a land bridge that connected Northeast Asia with western Alaska. The area known as Beringia (**Fig. 1**), was a flat, dry, and cold landscape covered mainly by grassland, shrubs, and small trees. Humans and animals could have used Beringia to walk from Siberia to Alaska and vice versa. It is known that this land bridge existed from 25,000 to 10,000 years ago.

The traditional view is that this migration was from Siberia no more than 12,000 RCYBP, by a group of humans that were hunting megafauna (mainly mammoths), using sophisticated fluted lithics known as 'Clovis points', named after the locality where they were first discovered in 1932 (**Figs. 1** and **2A**) (Haynes, 1980; 1987). They were highly mobile and spread rapidly from Alaska in the north to Patagonia in the south of the continent. Their rapid expansion was closely associated with the major extinction of several large animal species (known with the general term of megafauna, such as mammoth, camel, horse, sabre-tooth cat, mastodon) broadly about 10,000 years ago, during the transition between the Pleistocene and the Holocene.

There is no question that humans were in America at least at Clovis times, at around 11,200 years ago. It is not known if they were the first Americans due to the lack of a lithic point in Alaska as a predecessor to Clovis points and the theoretical demands of explaining how and why Clovis groups apparently migrated so rapidly through the Americas (Meltzer, 1995). There is also the fact that the technology and lifestyle represented in the earliest sites in South America, (e.g., Monte Verde, Chile, and Pedra Pintada, Brazil), are very different from the technology and patterns of space use characteristic of Clovis people in the north of the continent (Roosevelt *et al.*, 2002; Miotti, 2004). In South America, the characteristic lithics found in association with the earliest sites are different in shape to Clovis and they include Fishtail points and El Jobo points (**Fig. 2**). The Paleoindian site of Turrialba, in Costa Rica (**Fig. 1**) is important because it is the only site in the Americas where both Clovis points and Fishtail points are found together. However, because they were surface finds there are no dates for the site (Snarskis, 1979).

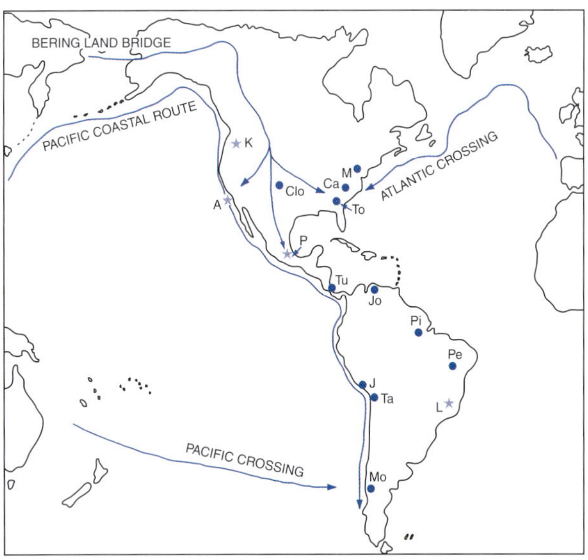

Figure 1 Map showing the main migration routes proposed to enter the Americas together with the location of some important early sites across the continent: Black stars = Early American remains: K = Kennewick Man, USA; A = Arlington Spring femora, USA; P = Peñon Woman III, Mexico, L = Lapa Vermelha IV (Luzia), Brazil. Black circles = Early American Sites: M = Meadowcroft rockshelter; Clo = Clovis; Ca = Cactus Hill; To = Topper; Turrialba, Costa Rica; Jo = El Jobo, Venezuela; Pi = Piedra Pintada, Brazil, Pe = Pedra Furada, Brazil; J = Quebrada Jaguay, Peru; Ta = Quebrada Tacahuay, Peru; Mo = Monte Verde, Chile.

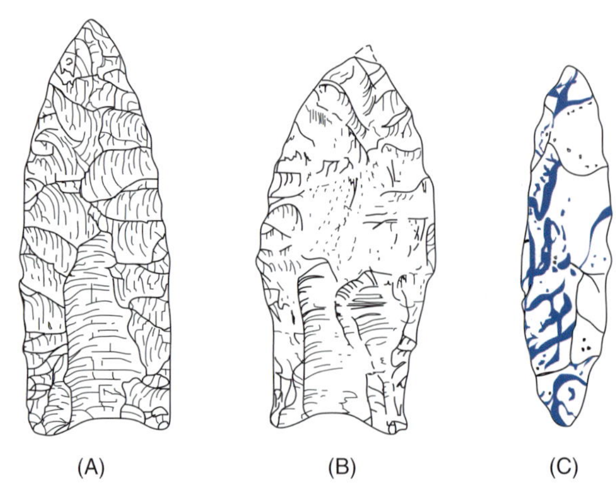

Figure 2 Example of early lithic points found in the Americas: (A) Clovis-type fluted points associated with mammoth kill sites, found mainly in the southwest and east of the USA reaching until Central America and dated between 11,200 to 10,900 RCYBP; (B) Fishtail-type fluted points found mainly in South and Central America, dated between 11,000 to 10,500 RCYBP; (C) El Jobo-type bipoint from Venezuela associated with mastodon kill sites, with ages from 10,000 to more than 11,000 RCYBP (redrawn from Bryan, 1999).

The Early Entry Model

Another hypothesis supports the view that human settlement happened much earlier, perhaps between 20,000 to 50,000 years ago (Bryan, 1999). Proposed sites which have been suggested include Meadowcroft, Rockshelter in Pennsylvania, USA (Adovasio, *et al.*, 1999); the Cactus Hill site in Virginia, USA (McAvoy and McAvoy, 1997); the Topper Site in South Carolina (Goodyear, 1999); Valsequillo, in Central Mexico (Irving-Williams, 1972; González *et al*, 2005); Pedra Furada in Brazil (Guidon and Delibrias, 1986) and Monte Verde in Chile (Dillehay, 1989; 1997). The main problems with the Early Entry Model are dating questions and the acceptance of the cultural origin of proposed artifacts or suggested archaeological evidence (e.g. hearths, charcoal) (Meltzer, 1995).

Proposed Migration Routes into the Americas

There are several migration routes that have been proposed as entry to the Americas (**Fig. 1**):

The Bering Land Bridge Route

This traditional model suggests the hypothesis that humans came from Siberia using the Beringia land bridge during low sea levels, crossing through the Middle of the American Continent using an ice free corridor, following and hunting megafauna until reaching Patagonia in the south in a period of about 1,000 years (Haynes, 1987; Fiedel and Haynes, 2004).

The Pacific Coastal Route

This model has been put forward on the basis of geological and linguistic evidence (Fladmark, 1979; Gruhn 1994; Dixon, 1999), together with computer models of patterns of genetic diversity in Amerindian populations. A coastal migration route is now gaining more acceptance with newly emerging evidence, suggesting that people with boats moved along the Pacific coast, hopping between islands, until reaching Alaska and Northwest Canada and continuing south until finally reaching Peru and Chile by 12,500 years ago, or perhaps much earlier. Archaeological evidence in Australia, Melanesia, and Japan indicate that boats were in use from 25–40,000 years ago. Several early sites along the West Coast of America (Canada, California, Peru, Ecuador, and Chile) have been dated between 10–12,000 years ago (Erlandson, 2002). The main problem is that many potential Pleistocene early sites are now submerged due to modern high sea levels, making archaeological investigation very difficult in coastal areas.

The Atlantic Crossing Route

From the shorelines of the Iberian Peninsula, humans traveled bordering along the Arctic Ice cap in the Atlantic Ocean until finally reaching the East Coast of America. Apparently there are a lot of similarities between lithic points of the Solutrean cultures of the Iberian Peninsula and the characteristic Clovis points of North America. (Jelinek, 1971; Stanford and Bradley, 2002). The two cultures also share bone shaping techniques, pebble-decorating artistry, and the tradition of burying stone tools in caches filled with ochre.

The Pacific Crossing Route

Humans reached America by crossing the Pacific Ocean using watercraft technology, using the islands in the Pacific as stepping stones. It is possible that this was a very recent migration route.

It is likely that all of these routes have been used in the past to reach the Americas at different times but current research is trying to establish a robust timeline to explain when they happened and by which population groups.

Late Pleistocene–Holocene Megafaunal Extinctions

Large-scale extinction of megafauna generally coincided with the arrival of modern humans in different parts of world, such as Australia or Tasmania. One such large extinction event happened in America during the Pleistocene–Holocene transition about 10,000 years ago. There are several causes that have been proposed to try to explain such rapid extinctions:

1. **Overkill:** humans swept across America from Siberia to Patagonia in a 'killing front' that moved perhaps 100 miles in a decade. They were hunting megafauna (larger animals) with their Clovis points (Martin, 1967, 1984; Fiedel and Haynes, 2004).
2. **Hyperdisease** spread across the continent, after the entrance of humans and their dogs into the Americas, because the new species of animals they encountered had no immunity against the diseases they carried with them (MacPhee and Marx, 1997).
3. **Rapid climatic changes** during the transition between the Pleistocene–Holocene caused large changes in temperatures and vegetation types, which had profound negative effects on the fauna,

particularly the larger animals (megafauna), such as mammoths, camels, horses, glyptodonts, and sabre-toothed cats, until they finally became extinct (Graham, 2001).

The Latest dates obtained on megafauna bones found in America in association with lithic tools or other evidence of human presence (i.e., worked bone and burned bones in hearths) suggest that there was a major influx of people around 11,000 years ago. However, this process does not rule out the presence of small bands of humans at an earlier date.

Linguistic Evidence

Studies of the native languages of the Americas have shown that they are very diverse, representing nearly 200 distinct families. Early claims that these languages descend from only three separate linguistic stocks, Amerind, Na-Dene, and Eskimo Aleut, suggesting three separate migrations (Greenberg, 1986), that occurred only during the last 12,000 years, are regarded today by most specialists as extremely unlikely. New proposals have explored deep structural affinities among American Indian languages with Circum-Pacific Old World languages. Most linguists believe that the large number and variety of Native American languages needs a long period of language diversification. Nichols (1990) estimates that language diversification in the Americas began as early as 35,000 years ago, and suggests a coastal route of migration into the New World, rather than an interior route, because the centers of greatest language diversity are located along the Pacific Coast of North America and the northwest corner of South America. Studying the linguistic history of the Americas is a complex problem that will be under investigation for years to come.

Dating and Craniometric Evidence on Human Skeletons

There are very few humans in the Americas that have been dated and are in the Late Pleistocene (older than 10,000 RCYBP) (See **Table 1**). There seems to be many problems with rapid mineralization and contamination of the bones that makes accurate dating difficult. Some of the oldest directly radiocarbon dated human remains in the Americas include the Arlington Springs femora from the Channel Islands of California with an age of 10,960 ± 80 RCYBP and the Anzick infant skull fragments from Montana with reported ages of 10,820 ± 100 and 11,040 ± 40 RCYBP. However, due to their fragmentary state they do not give us information about the face or type of individual. Peñon Woman III, a semi-complete skeleton found by chance on an island in the Middle of the Basin of Mexico, is one of the most complete skeletons, with a well-preserved skull, with a radiocarbon AMS date on collagen of 10,755 ± 75 RCYBP (González *et al.*, 2003) (**Fig. 3**). In South America the oldest human so far known is the skeleton known as Lapa Vermelha IV or Luzia from Brazil, with ages between 10,200 ± 220 to 11,680 ± 500. In this case, it is the layer in which the skeleton was found that has been dated using charcoal (Neves *et al.*, 1998).

Table 1 The oldest dated humans from the Americas (Late Pleistocene or older than 10,000 years old), modified after Haynes (2002)

Name of Site	*Date (RCYBP) and Laboratory Number*	*Comments*
1. Anzick, Montana, USA	8,620 to 10,500; redated to 11,550 ± 60 (CAMS-35912)	Infant, cranial and postcranial fragments with ochre and artifacts
2. Arlington Springs, Santa Rosa Island, California, USA	10,960 ± 80 (CAMS-16810) on collagen	Only femora found
3. Buhl Woman, Idaho, USA	10,675 ± 95 (Beta-4055/ETH-7729) on collagen	Skeleton, female
4. Fishbone Cave, Nevada, USA	10,900 ± 300; 11,200 ± 250 (L-245) on bark	Dates on bark wrapping the burial
5. Lapa Vermelha IV (Luzia), Brazil	10,200 ± 220 to 11,680 ± 500 (GIF-3726) on charcoal	Cranium and postcranial elements, adult female
6. Marmes, Washington, USA	10,130 ± 300 (W-2218) on charcoal; 9,840 ± 300 on charcoal; 9,820 ± 300 (W-2209) on shell	Cranial fragments
7. Pampa de Fósiles, Peru	10,200 ± 180 (GIF-3781)	Adult and adolescent
8. Peñon Woman III, Mexico City, Mexico	10,755 ± 75 (OxA-10112) on collagen	Semicomplete female skeleton, aged 25 years
9. Tlapacoya I Man, Mexico City, Mexico	10,200 ± 65 (OxA-10225)	Calvarium, male aged 30–35 years
10. Warm Mineral Springs, Florida, USA	10,620 ± 190 (GaK-3998)	Postcranial fragments

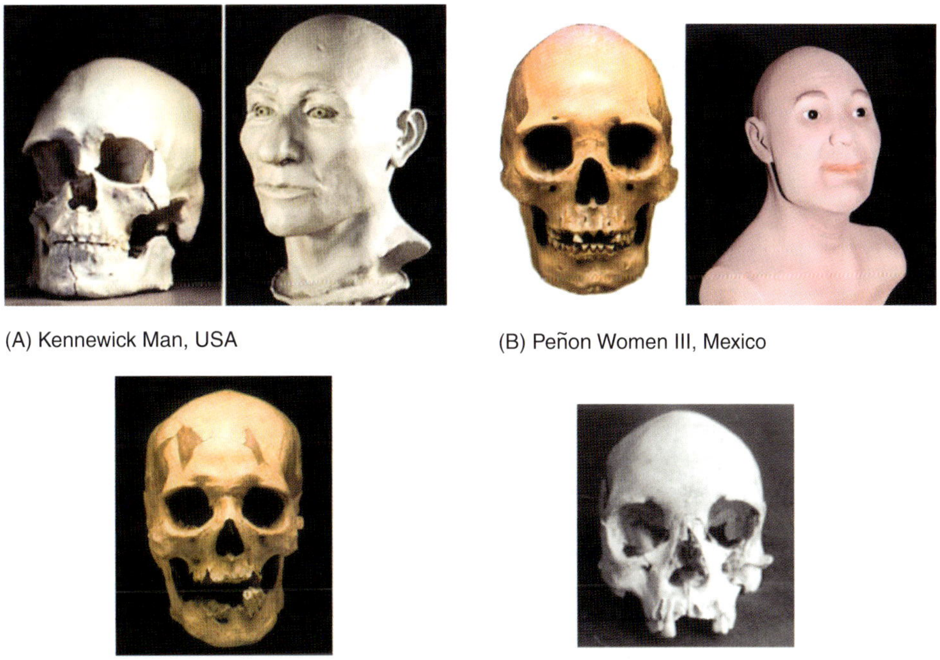

(A) Kennewick Man, USA

(B) Peñon Women III, Mexico

(C) Texcal Man, Mexico

(D) Luzia, Brazil

Figure 3 Examples of the wide variation of skull morphology for some of the earliest Americans: (A) Kennewick Man, USA with an age of 8,410 RCYBP with a skull that has been interpreted as 'Caucasoid' or with Ainu-like traits; (B) Peñon Woman, Mexico with an age of 10,755 RCYBP showing a long and narrow (dolichocephalic) skull frequently found in the late Pleistocene humans from America; (C) Texcal Man, Mexico with an age of 7,480 RCYBP showing the typical 'mongoloid' shape, found frequently in Modern Native Americans, with rounded (brachycephalic) skulls; (D) Lapa Vermelha IV (Luzia), Brazil with an age range between 10–11,000 RCYBP with a skull that has been interpreted to have negroid or Melanesian affinities.

Studies of the variation of skull shape (craniometry) in human skeletons from the Americas have shown biological differences between the Paleoindians (11,000 to 7,000 years ago), the Archaic populations dating to less than 7,000 years ago, and the Modern Native Americans (Jantz and Owsley, 2001; Steele and Powell, 2002; González-Jose *et al.*, 2005). Paleoindians in general show a wider range of variation in their craniofacial characteristics compared with later Native Americans (**Fig.** 3), showing long and narrow cranial vaults, with short and relatively gracile faces. In contrast, modern Native Americans show broad faces, prominent cheek bones and round cranial vaults. These differences suggest two possibilities:

1. that the Paleoindians and subsequent Amerindian populations may have arisen from two temporally distinct migrations that originated in different parts of Asia; or
2. that these craniofacial differences resulted from the effects of genetic drift and adaptation over the past 10,000 years (Powell and Neves, 1999). Many Paleoindian crania, including Kennewick Man, an 8,500 years old human skeleton, seem to bear resemblance to those of ancient Eurasian/East Asian populations (Chatters, 2000). However, it is important to stress that the sample of human skeletons from the Americas with a Late Pleistocene and early Holocene age is still relatively small and many more need to be found before final conclusions can be made on their biological and ethnic affinities.

Genetic evidence

The use of genetic markers such as mitochondrial DNA (mtDNA) has emerged as a powerful tool to study the age, number, and timing of migrations into the Americas. It is known that mtDNA is passed down the maternal line, evolving rapidly by mutation. A haplogroup is a collection of closely related haplotypes that all share one or more mutations in common as compared to other haplogroups. In the Americas so far, four main haplogroups, mainly of Asiatic origin, have been identified from the study of Modern Native Americans: A, B, C, and D (Wallace *et al.*, 1985; Wallace and Torroni, 1992), together with lower percentages of haplogroups such as the X, which is more commonly associated with Europeans (Brown *et al.*, 1998). Stanford and Bradley (2002) have interpreted the haplogroup X in America as a remnant of the Atlantic Migration Route of Solutreans from Spain about 16,500 RCYBP, based upon similar stone flaking technologies and tool shapes. Attempts to apply molecular clocks to the problem of the initial peopling of the Americas have generated estimates as early as 14,000–40,000 years ago (Torroni *et al.*, 1994; Schurr, 2004). If these estimates are correct, the Clovis First Model cannot be correct. However, it is necessary to stress that the genetic makeup of humans in America today is not necessarily the same as from humans living 11,000 years ago, with the presence of perhaps some extinct haplogroups that are not represented in modern Native American populations. Another possibility is the problem of admixture resulting from New World populations, possibly migrating back into Siberia and Northeast Asia. If this happened, Modern native residents in Alaska and Siberia may be descendants of more recent migrations across the Bering Strait or from migrations back into the area from within the interior of the continent. For these reasons attempts to obtain ancient DNA information from Paleoindian and Archaic populations has been a priority in recent years. However, due in part to problems of preservation and contamination with modern DNA, no reliable ancient DNA has been recovered so far from the Late Pleistocene humans from America. Further work with mtDNA, nuclear DNA (which is more representative of the entire genome) and Y-chromosome (the male related counterpart of mtDNA), will allow better estimates of the genetic similarities between Europe, Asia, and America to help to determine when they would have shared a common ancestor.

Summary

Strong archaeological evidence indicates that humans arrived into the Americas long before 12,000 RCYBP and that the Clovis people were not the first group to migrate into the New World. New information derived from linguistics, craniometric, and genetic evidence strongly suggest that the first Americans were very diverse from an ethnic point of view. However, the study of many more early human skeletons is required to construct a real picture.

See also: **Archaeological Records**: Overview; Postglacial Adaptations. **Vertebrate Studies**: Interactions with Hominids.

References

Adovasio, J. M., Pedler, D. R., Donahue, J., and Stuckenrath, R. (1999). No vestige of a beginning nor prospect for an end: Two decades of debate on Meadowcroft Rockshelter. In *Ice Age People of North America: Environments, origins and adaptations of the first Americans* (R. Bonnichsen and K. L. Turnmire, Eds.), pp. 416–431. Oregon State University Press, Corvallis.

Brown, M. D., Hosseini, S. H., Torroni, A., Bandelt, H. J., Allen, J. C., Schurr, T. G., Scozzari, R., Cruciani, F., and Wallace, D. C. (1998). mtDNA haplogroup X: An ancient link between Europe/Western Asia and North America? *American Journal of Human Genetics* **63(6)**, 1852–1861.

Bryan, A. L. (1999). El Poblamiento Originario. In Historia General de America Latina, Volumen I: Las sociedades originarias. Editorial Trotta, Ediciones UNESCO, pp. 41–68.

Chatters, J. (2000). The recovery and first analysis of an early Holocene human skeleton from Kennewick, Washington. *American Antiquity* **65**, 291–316.

Dillehay, T. D. (1989). *Monte Verde: A Late Pleistocene Settlement in Chile.* Smithsonian Institution Press, London and Washington.

Dillehay, T. D. (1997). *Monte Verde: A Late Pleistocene Settlement in Chile Vol. 2: The Archaeological Context.* Smithsonian Institution Press, Washington, D.C.

Dixon, E. J. (1999). *Bones Boats and Bison: Archaeology of the First Colonization of Western North America.*, p. 332. The University of New Mexico Press, Albuquerque, NM.

Erlandson, J. M. (2002). Anatomically modern humans, maritime voyaging, and the Pleistocene colonization of the New World. In *The First Americans, the Pleistocene colonization of the New World* (N. G. Jablonski, Ed.). Memoirs of the California Academy of Sciences 27, San Francisco.

Fiedel, S., and Haynes, G. (2004). A premature burial: comments on Grayson and Meltzer's 'Requiem for overkill'. *Journal of Archaeological Sciences* **31**, 121–131.

Fladmark, K. R. (1979). Routes: Alternative Migration corridors for Early Man in North America. *American Antiquity* **44**, 55–69.

González-José, R., Neves, W., Lahr, M. M., González, S., Pucciarelli, H., Martinez, M. H., and Correal, G. (2005). Late Pleistocene/Holocene craniofacial morphology in Mesoamerican Paleoindians: Implications for the peopling of the New World. *American Journal of Physical Anthropology (In press).*

González, S., Jiménez-López, J. C., Hedges, R., Huddart, D., Ohman, J. C., Turner, A., and Pompay Padilla, J. A. (2003). Earliest humans in the Americas: New evidence from Mexico. *Journal of Human Evolution* **44**, 379–387.

González, S., Huddart, D., Bennett, M. R., Schwenninger, J. L., Higham, T., Bronk-Ramsey, C., Grün, R., González-Huesca, A., Kelley, S., Pike, A., and Nathan, R. (2005). Human Footprints in Central México older than 40,000 years ago. *Quaternary Sciences Reviews (In press).*

Goodyear, A. C., III (1999). The Early Holocene occupation of the Southeastern United States: a geoarchaeological summary. In *Ice Age Peoples of North America* (R. Bonnichsen and K. L. Turnmire, Eds.), pp. 432–481. Center for the Study of the First Americans, Corvallis.

Graham, R. W. (2001). Late Quaternary Biogeography and Extinction of Proboscideans in North America. Proceedings of the 1st International Congress. In *The World of Elephants* (G. Cavarreta, P. Giola, M. Mussi and M. R. Palombo, Eds.) Roma 16–20 October, 707–709.

Greenberg, J. H., Turner, C. G., and Zegura, S. L. (1986). The settlement of the Americas: A comparison of the linguistic, dental and genetic evidence. *Current Anthropology* **27**, 477–497.

Gruhn, R. (1994). The Pacific Coast Route of Initial Entry: An Overview. In *Methods and Theory for Investigating the Peopling of the Americas* (R. Bonnichsen and G. D. Steele,

Eds.), pp. 249–256. Center for the Study of the First Americans Oregon State University, Corvallis.
Guidon, N., and Delibrias, G. (1986). Carbon-14 dates point to man in the Americas 32,000 years ago. *Nature* **321**, 769–771.
Haynes, C. V., Jr. (1980). The Clovis Culture. *Canadian Journal of Anthropology* **1**(1), 115–121.
Haynes, V.ance, Jr. (1987). Clovis Origins Update. *The Kiva* **52**(2), 83–93.
Irving-Williams, C. (1972). Summary of archaeological evidence from the Valsequillo region, Puebla, Mexico. In *Cultural continuity in Mesoamerica* (D. Browman, Ed.), pp. 7–22. Aldine, Chicago.
Jantz, R. L., and Owsley, D. W. (2001). Variation among early North American crania. *American Journal of Physical Anthropology* **114**, 146–155.
Jelinek, A. J. (1971). Early Man in the New World: A technological perspective. *Arctic Anthropology* **8**, 15–21.
MacPhee, R. D. E., and Marx, P. A. (1997). The 40,000 year plague: humans, hyperdisease and first-contact extinctions. In *Natural Change and Human Impact in Madagascar* (S. Goodman and B. Patterson, Eds.) pp. 169–217.
Martin, P. S. (1967). Prehistoric overkill. In *Pleistocene Extinctions: The Search for a Cause* (P. S. Martin and H. E. Wright, Eds.), pp. 75–120. Yale University Press, New Haven.
Martin, P. S. (1984). Prehistoric extinctions: The global model. In *Quaternary Extinctions: A Prehistoric Revolution* (P. S. Martin and R. G. Klein, Eds.), pp. 354–403. University of Arizona Press, Tucson.
McAvoy, J. M., and McAvoy, L. D. (1997). *Archaeological investigations of Site 40SX02, Cactus Hill, Sussex Country, Virginia*. Richmond. Va. Dep. Hist. Resourc., Res. Rep. Ser. 8
Meltzer, D. J. (1995). Clocking the First Americans. *Annual Review of Anthropology* **24**, 21–45.
Miotti, L. L. (2004). Quandary: The Clovis Phenomenon, the First Americans, and the View from Pa tagonia. *New Perspectives on the First Americans* 31–36.
Neves, W. A., Powell, J., Prous, A., and Ozolin, E. G. (1998). Lapa Vermelha IV, Hominid 1: Morphological affinities of the earliest known American. *American Journal of Physical Anthropology Supplement* **26**, 169.
Nichols, J. (1990). Linguistic diversity and the first settlement of the New World. *Language* **66**(3), 475–521.
Roosevelt, A. C., Douglas, J., and Brown, L. (2002). The migrations and adaptations of the first Americans: Clovis and pre-Clovis viewed from South America. In *The First Americans: The Pleistocene of the New World* (N. G. Jablonski, Ed.), pp. 159–235. California University Press, San Francisco California.
Schurr, T. G. (2004). The Peopling of the New World: Perspectives from Molecular Anthropology. *Annual Reviews in Anthropology* **33**, 551–583.
Snarskis, M. J. (1979). Turrialba: A Paleo-Indian Quarry and Workshop Site in Eastern Costa Rica. *American Antiquity* **44**(1), 125–138.
Steele, D. G., and Powell, J. F. (2002). Facing the Past: A View of the North American Human Fossil Record. In *The First Americans: The Pleistocene of the New World* (N. G. Jablonski, Ed.), pp. 93–122. California University Press, San Francisco California.
Stanford, D., and Bradley, B. (2002). Ocean trails and Prairie Paths? Thoughts about Clovis origins. In *The First Americans: The Pleistocene of the New World* (N. G. Jablonski, Ed.), pp. 255–271. California University Press, San Francisco California.
Torroni, A., Neel, J. V., Barrantes, R., Schurr, T. G., and Wallace, D. C. (1994). Mitochondrial DNA 'clock' for the Amerinds and its implications for timing their entry into North America. *Proceedings of the National Academy of Sciences* **91**, 1158–1162.
Wallace, D. C., Garrison, K., and Knowler, W. (1985). Dramatic founder effects in Amerindian mitochondrial DNAs. *American Journal Physical Anthropology* **68**, 149–155.
Wallace, D. C., and Torroni, A. (1992). American Indian prehistory as written in the mitochondrial DNA: A review. *Human Biology* **64**, 403–416.

Neanderthal Demise

W Davies, University of Southampton, Southampton, UK

© 2007 Elsevier B.V. All rights reserved.

Introduction

The extinction of the Neanderthals (*Homo neanderthalensis*), and the arrival of modern humans (*Homo sapiens*) in Europe between 45 and 30 thousand years ago ('kyr BP') is one of the most hotly debated topics in Paleolithic archeology and paleoanthropology. Why? What is the evolutionary and behavioral importance of this transition?

Most researchers in this period of human evolution incline more toward a single-origin model for the evolution of modern humans, with a subsequent expansion out of Africa into Asia, Greater Australasia, and Europe, ultimately replacing the Neanderthals in the latter subcontinent (e.g., Mellars (1989), Stringer (1989), and d'Errico *et al.* (1998)). However, a substantial minority of researchers believe the multiregional origin hypothesis to be more appropriate, with modern humans evolving independently in many regions of the Old World, yet sufficiently connected to each other to facilitate the exchange of genes, allowing *Homo sapiens* to develop at a global scale as a single species for the last two million years (e.g., Wolpoff and Caspari (1997)). The models have distinctive temporal and spatial scales: single-origin models tend to think in terms of speciation locations (with subsequent dispersals) and events, while multiregional ones think in terms of global population interactions and broad (sympatric) speciation processes. Both models can conceive of population extinctions and replacements, although these are more geographically restricted if one is a multiregionalist (Wolpoff and Caspari, 1997). Because this author has difficulties in believing that multiregionalism can operate at the large spatio-temporal scales proposed by its adherents, the single-origin ('Out-of-Africa 2') model will be given stronger preference for much of the discussion in this article.

Species Definitions

Neanderthals

Although specimens had been discovered in 1830 (Engis, Belgium) and 1848 (Gibraltar), it was not until a third one was recovered from the Neander valley ('Neanderthal') in Germany in 1856, that species-specific traits were identified (King, 1864) and a new hominid species was classified, indeed the first human species to be distinguished since the publication of Darwin's *On the Origin of Species* in 1859. Neanderthals were notable for their stocky build, but it is their cranial characteristics that are the most diagnostic: a protruding mid-facial region, emphasized by the lack of a distinct chin, and thick, well-developed brow-ridges over the eyes, which grade backward into a weakly developed forehead. Seen from above, it is as if someone has taken the modern human skull shape we are used to, and stretched it longitudinally. Nevertheless, the cranial capacity (brain size) of Neanderthals is not notably different from that of modern humans, averaging some 1,400 cc. Perhaps because of our simplistic views of what characterized the climatic conditions of ice ages, it has often been claimed that Neanderthals were adapted to exceptionally cold ('hyperarctic') conditions (e.g., Holliday (1997)). However, if Neanderthals preferred cold conditions, then how can we explain their presence in the Near East or Iberia: two regions not known for their climatic affinity with the Arctic (see Aiello and Wheeler (2003))?

Two enlightenment-derived perceptions of Neanderthals have tended to dominate, both crystallized from mythology: men-as-monsters and men-as-gods. An example of the former is to see Neanderthals as something akin to Shakespeare's Caliban: a distillation of all that is monstrous in humans; an example of the latter is surely the concept of the 'Noble Savage.' Presumably owing to their 'coarse' physical features, Neanderthals have tended to be seen as more monstrous than noble, but recently the pendulum has begun to swing the other way (e.g., d'Errico *et al.* (1998)). It will be seen, however, that such Cartesian dualities are difficult to test or falsify against evidence that is largely ambiguous.

The origins and extent of Neanderthal symbolic behavior is one of the most problematic debates in Paleolithic archeology, and is dependent upon the vagaries and chronological precision of the extant archeological record. It was frequently thought that Neanderthals were incapable of complex spoken language, not just on the grounds of mental ability but also of physical capability. The claims of, among others, Lieberman (1989), that the configuration and morphology of the Neanderthal vocal tract ensured that they could not vocalize like ourselves, have been largely discredited by the morphological similarity of the hyoid bone from a Near Eastern Neanderthal (at Kebara, Israel) to those of modern humans (Arensburg, 1989).

Modern Humans

Homo sapiens are one of the most poorly defined species in the animal kingdom, with relatively few diagnostic physiological features. In comparison to Neanderthals, we have a much flatter mid-facial region, with a pronounced chin and steep forehead (with weak brow-ridges). The elongated body shapes of many modern humans, today and in the past, are congruent with our proposed origins in the tropics (of Africa). Behaviorally, modern humans have been defined by the universal use of complex spoken language, with a strong capacity for abstract and metaphorical thought (e.g., Mellars (1989)). In recent decades, this 'symbolism' has leapt to the forefront of the debate on the origins of modern humans, with positions on the assembly of the 'modern human behavioral package' being dichotomized as either 'long chronology' (the gradual accretion of modern characteristics) or 'short chronology' (a sudden explosion of unprecedented modern behavior) (e.g., McBrearty and Brooks (2000) vs. Klein (1995)).

Neanderthal and modern human technologies **Tables 1** and **2** give brief summaries of the archeological groupings we shall consider, and their skeletal associations (where known). Some terms, especially in **Table 1**, need further explanation, and the archeological groupings, henceforth called technocomplexes, need to be set into their spatial and temporal contexts. A technocomplex (following Clarke (1968)) is a group of cultures that share a range of common characteristics, but differ – perhaps by region or time period – in the presence/absence of specific artifact types. The term 'culture,' while widely used, is seldom defined, and carries much interpretative baggage. Hence, technocomplex will be preferred here; in any case, its broad spatio-temporal scaling probably better matches most of the archeological groupings we are discussing.

The technocomplexes listed in **Tables 1** and **2** are not exhaustive, but represent a fair selection of those thought to be significant, and/or associated with human remains. They can be grouped into two larger conceptual units, defined largely by how the stone tools were made. Middle Paleolithic assemblages

('Mousterian', if they contain Mousterian points) are generally flake based, that is, their 'blanks' for tool production are generally wide and relatively short, and were frequently obtained by a knapping technique called 'Levallois.' Classic Levallois entails the careful preparation of a flat stone nodule by first regularizing its entire margin through the removal of overlapping flakes from one (normally the upper) surface, making the core ovoid in shape. The flat surfaces created by these peripheral flake scars are then used as a platform to remove overlapping flakes centripetally from the whole upper surface of the core, creating a 'tortoise-shell' gently domed surface that will yield an oval, gently domed flake when struck from one end. Control of the convexity and regularity of this surface is critical. Variants of this Levallois technique exist, producing triangular flakes ('Levallois points') and blades (blanks whose lengths are at least twice their widths): the different shapes are created by changes in the orientation of the upper

Table 1 Characteristics of selected western Eurasian archeological technocomplexes at the transition between Neanderthals and modern humans

		Technocomplexes	Diagnostic characteristics
Middle Paleolithic	SW Asia	Tabun C-type Mousterian	Squat, thin triangular flakes (Levallois points) are common.
		Tabun B-type Mousterian	Succeeds the Tabun-C variant, and has higher quantities of more elongate Levallois points and retouched tools.
	East Europe	Caucasian Mousterian	Several variants are present: 'para-Micoquian' in the north Caucasus. Typical Mousterian (Levallois *and* non-Levallois, sometimes with bifaces), (Levallois) Denticulate Mousterian on the NE Black Sea coast, and a Charentian Mousterian site with bifaces as well as sidescrapers.
		Crimean Mousterian	Several variants: the 'Ak-Kaya ('para-Micoquian') tradition,' with bifacial (foliate) knives and points, sidescrapers and some Charentian characteristics (e.g., slug-like highly retouched stone tools called 'limaces'), also encompasses the Kiik-Koba and Staroselian variants. Typical Mousterian assembages (see below) are concentrated in the nearby Dnestr basin, and they are founded upon Levallois technology.
		East European Micoquian	Perhaps better called *Keilmessergruppen*, named after the characteristic large, bifacially retouched knife-like tool with an acute cutting edge on one margin. Also contains elongate bifaces with asymmetric cross sections (*Halbkeile*), small pointed bifaces (*Faüstel*) and foliate bifaces (*Faustkeilblätter*). Many of these bifacial forms were made on flat stone nodules; the Levallois technique is sometimes present.
	West Europe	Typical Mousterian	Relatively undifferentiated from other western European Mousterian variants, except that bifaces and backed knives are absent or very scarce. Contains sidescrapers (some of Quina type), Mousterian points, some denticulates and notches.
		Mousterian of Acheulean Tradition ("MTA")	Two phases: (1) MTA-A, with triangular and pointed ovate ('cordiform') bifaces and scarce backed knives, then (2) MTA-B, with less common, exclusively cordiform bifaces and numerous backed knives (some of which anticipate the Châtelperron knife in their morphology). Both phases also contain Mousterian points, sidescrapers, notches and denticulates.
		Charentian Mousterian	Two variants, in roughly chronological order: (1) Ferrassie type, with presence of the Levallois knapping technique and numerous Mousterian points, then (2) Quina type, whose prevalent 'salami-slice' knapping method ('slicing off' flakes from a sausage-shaped stone nodule) was primarily used to make thick, distinctively retouched sidescrapers. Both variants are dominated by sidescrapers (some with bifacially worked convex edges), with fewer notches, denticulates, burins, and endscrapers.
		Denticulate Mousterian	Slightly more distinctive than the typical Mousterian, with high (≤80%) levels of notches and denticulates, generally ≤10% sidescrapers, scarce endscrapers, borers, and burins; other Mousterian tools are (near-)absent.

(Continued)

Table 1 (*Continued*)

Upper Paleolithic	SW Asia	Emiran	Mousterian-like assemblages, characterized by Levallois points carrying bifacial thinning on their bulbed (proximal) ends.
		Baradostian	Production of bladelets from carinated cores, and their transformation into Dufour bladelets and points resembling Font-Yves ones has led some to call it 'Zagros Aurignacian' (Olszewski and Dibble, 1994), but other elements, such as backed blades and 'retouched rods' are not so easy to fit within the Aurignacian.
		Ahmarian	Blade/bladelet dominated assemblages, with diagnostic 'chamfered' pieces, el-Wad points, backed blades/bladelets. Surviving symbolic activity is rare, and bone-working is limited.
	East Europe	Bohunician	Contains a mixture of Middle and Upper Paleolithic elements: sidescrapers, Mousterian points, and a few bifacial leafpoints in association with endscrapers (some Aurignacian-type) and burins. Both (Upper Paleolithic) blade reduction and Levallois techniques were applied to the same cores.
		Szeletian	Defined by bifacial leafpoints, with endscrapers (a few Aurignacian type), sidescrapers, Mousterian points and pointed retouched blades; more flake orientated than the Bohunician.
		Jerzmanowician	Defined by laminar leafpoints retouched mainly on the bulbed end of one (the ventral) surface. Splintered pieces and some Mousterian tool types (no sidescrapers) also present.
	West Europe	Lincombian	Characterized by 'blade-points' (a type of unifacial leafpoint: cf. Jerzmanowician), with some bifacial points. Otherwise very heterogeneous (mixed assemblages?), with Aurignacian tools sometimes present, e.g., *busqué* burins.
		Châtelperronian	Defined by the backed Châtelperron knife/point; associated with Mousterian-type tools (sidescrapers, etc.). A mixture of flake and blade-orientated strategies is present. Beads, pendants and some bone-working found in some northerly sites.
		Uluzzian	Defined by generally scarce small backed lunates, mostly on flake fragments, although apparently on blades in Greece; also Mousterian tools, splintered pieces, endscrapers, and some bone-working ("awls").
		Aurignacian	Characterized by osseous projectiles, notably the split-based antler point, but also simple-based/lozangic/Mladeč and biconical ones; lithic tools include Dufour bladelets, Font-Yves/Krems/el-Wad points, Vachons and *busqué* burins, nosed and carinated scrapers, and marginally retouched Aurignacian blades. Beads/pendants are common; some anthropomorphic and zoomorphic figurines.
		Gravettian	Defined by the Gravette point, associated variously with Noailles and Bassaler burins, *fléchettes*, Font-Robert points and Kostenki points. Prismatic blade(let) technologies are strongly present. Symbolic activity is especially characterized by the female 'Venus' figurines.

Allsworth-Jones, 1986; Bar-Yosef, 1993; Bietti, 1997; Bordes, 1968, 1977; Bosinski, 1967; Campbell, 1980; Chabai, 1996; Cohen and Stepanchuk, 1999; Davies, 2001; Garrod, 1955; Gilead, 1991; Hole and Flannery, 1967; Hublin *et al.*, 1996; Jacobi, 1990; Kozlowski, 1990, 2000; Otte, 1981; Richter, 1997; Svoboda *et al.*, 1996; Veil *et al.*, 1994.

surface's preparatory removals rather than in the convexity of the surface. Upper Paleolithic assemblages, conversely, frequently used blades produced from prismatic – often conical – cores: not only are these blades at least twice as long as they are wide, but they also carry parallel scars of blades previously removed from the core, whose scar edges form longitudinal ridges on their upper ('dorsal') surfaces, typically giving them a trapezoidal cross section. Bladelets are small blades, generally less than 12 mm wide, and sometimes attaining lengths of 50–60 mm (Hassan, 1972).

Having generalized that Middle Paleolithic (flake-based) technologies are associated with Neanderthals, and Upper Paleolithic (blade-based) ones with modern humans, we should be aware that there are important exceptions. In the Levant of southwest Asia, modern human remains are associated with 'Tabūn C'-type Mousterian assemblages at sites such as Qafzeh and Skhūl at about 100–

Table 2 Skeletal associations of archeological technocomplexes in western Eurasia

<table>
<tr><th></th><th></th><th>Homo neanderthalensis</th><th>Homo sapiens</th><th>Unknown (Homo sp.)</th></tr>
<tr><td rowspan="2">Middle Paleolithic</td><td rowspan="2"></td><td>European Middle Paleolithic and Mousterian</td><td></td><td></td></tr>
<tr><td colspan="2">SW Asian Middle Paleolithic and Mousterian</td><td></td></tr>
<tr><td rowspan="2">Upper Paleolithic</td><td>SW Asia</td><td></td><td>Ahmarian</td><td>Emiran
Baradostian</td></tr>
<tr><td>Europe</td><td>Szeletian?
Châtelperronian</td><td>Aurignacian[a]
Gravettian</td><td>Bohunician
Uluzzian
Jerzmanowician
Lincombian</td></tr>
</table>

[a]Aurignacian in the broad sense, including the 'Bachokirian' of eastern Europe.

90 cal kyr BP. Unlike thermoluminescence (TL), optically stimulated luminescence (OSL), uranium series, and electron-spin resonance (ESR), radiocarbon dates need to be converted into calendar years (normally designated as 'cal BP'). Only recently, however, have calibration curves been applied to radiocarbon dates in our study period (van Andel *et al.* (2003), etc.), and readers should be aware that there is not yet any universally accepted calibration curve, and probably never will be, as such curves tend to be revised regularly to accommodate new data. The calibrated radiocarbon dates given here (using the CalPal 2005 default curve; Weninger and Jöris (2005) – http://www.calpal.de) should be treated broadly and with a degree of caution, but they have the advantage of giving more accurate age estimates than uncalibrated radiocarbon dates, and they allow comparison with the other, aforementioned dating techniques. There appears to be little in their behavioral repertoires to distinguish them from the contemporary and subsequent Neanderthals in that area (Bar-Yosef (1993); Mellars (1989): p. 354). At about the same time in northern Europe, Neanderthals appear to be making blades in exactly the same way as would become the norm in the Upper Paleolithic, before apparently reverting to more 'typical' flake-based technologies (e.g., Révillion and Tuffreau (1994)). Therefore, our debate should be less about technological capacity than about preferred technological modes, and about the frequencies of particular core-reduction strategies employed to produce tool blanks.

In most parts of western Eurasia, the Middle Paleolithic technocomplexes listed in **Table 1** have disappeared by ca. 35 cal kyr BP, although some persisted, particularly in southern Iberia, Croatia, the Crimea, and the Caucasus, until perhaps 30 cal kyr BP (e.g., van Andel *et al.* (2003)). Outside these apparently isolated 'pockets' of survival, it becomes much more difficult to identify surviving Neanderthals. The Châtelperronian of France (and perhaps northern Iberia) is the only Upper Paleolithic technocomplex attributed to Neanderthals with any confidence, lasting between ca. 43 and 30 cal kyr BP. In particular, this association is based upon the finds from two sites in northern central France: Roche à Pierrot (St.-Césaire) and Grotte du Renne (Arcy-sur-Cure) (e.g., Hublin *et al.* (1996)). Thus, toward the end of their existence we can see Neanderthals' behavior changing rapidly in some environments, and utilizing, for apparently the first time, bone and ivory to produce not just tools (awls, etc.) but also elements of personal adornment (beads, pendants, etc.), apparently providing evidence for the application of social meaning to particular artifacts by this species. The big debate in archeology at the moment is whether this apparent increase in symbolic activity among some Neanderthals was the result of socially internal developmental trajectories, or the result of contact and social interactions with incoming modern human groups – with the behavior of both groups being changed (e.g., d'Errico *et al.* (1998)). Perhaps the deteriorating conditions in the northern part of the Neanderthal range in France required a change in their social and symbolic behavior: beads, etc., might have been needed to reinforce links between groups in adverse conditions. Certainly, there is little evidence of identifiable symbolic behavior in Neanderthals prior to the Châtelperronian; there are very few isolated and unique finds from some Middle Paleolithic sites, but nothing that can be compared stylistically at the intersite scale. Paradoxically, it seems that Neanderthals were adopting explicitly symbolic behavior (and thus hinting to many archeologists at the capacity for modern human behavior) just at the time of their extinction (Mellars *et al.*, 1999).

The other European early Upper Paleolithic technocomplexes that are often assumed to have been made by Neanderthals, by analogy with the 'transitional' characteristics of the Châtelperronian, are actually blank slates on which archeologists can impose their prejudices. The Uluzzian, Szeletian, Jerzmanowician-Lincombian, and Bohunician have not yet yielded any (diagnostic) human remains, although this has not stopped them being attributed to Neanderthals (e.g., d'Errico *et al.* (1998)). The Levantine Emiran, perhaps dated to about 50 cal kyr BP (or earlier), appears to have developed into the Ahmarian *in situ* within the Levant (Marks and Volkman, 1983); the latter technocomplex was probably associated with modern humans, if the skeletal remains from the Lebanese site of Ksar 'Akil are representative and reliable (Ohnuma and Bergman, 1990). The Emiran has also been connected with the Bohunician of central Europe – they share a similar use of Levallois technology – and it has been postulated that both might have been made by modern humans, who dispersed from the Levant to Europe at around 45 cal kyr BP (e.g., Tostevin (2003)). At present, it is difficult to test this hypothesis, although the Bohunician does seem to represent a technological break with the local Middle Paleolithic of central Europe. This is different from the western European situation, where the Châtelperronian shares several characteristics with the MTA-B (**Table 1**), for example, backed knives/points, and probably was derived from it.

The pan-European Aurignacian and Gravettian technocomplexes have been extensively studied, and are presumed to be 'true' Upper Paleolithic entities. In addition to the working of bone, antler, and ivory (not just to produce awls), the use of bladelets in both was extensive, and is not really seen in preceding (Neanderthal) European technocomplexes (Bar-Yosef and Kuhn, 1999). Bladelets permit flexibility, in that they can be arrayed in a shaft in different combinations to perform different tasks; any component unit ('microlith') can be replaced easily if it breaks.

It has often been argued that the Aurignacian was carried out of the Levant into eastern Europe, and then continued to move westward (e.g., Mellars (1989)). However, the earliest Aurignacian-type assemblages appear to be eastern European, dated at about 43 cal kyr BP, and slightly later Aurignacian assemblages have been found further east in the Levant, as well as further west in Germany, Franco-Cantabria, and Belgium. At face value, the relative lateness and restricted spatial distribution of Levantine Aurignacian sites might appear to contradict hypotheses arguing for a spread of modern humans out of Africa into Europe via the Levant, but it can easily be argued that what we know now as the 'Aurignacian' developed as people had to respond to different environmental conditions and spatial distributions of raw materials and resources. The Levantine Aurignacian might thus represent a back-migration of the European Aurignacian into the Near East. The Aurignacian seems to have been made by modern humans, as far as we can judge, although in contrast to the preceding Middle Paleolithic and succeeding Gravettian technocomplexes, it has tended to yield fragmentary associated human remains (e.g., Henry-Gambier *et al.* (2004)). Perhaps the desire to break up bodies was related to the apparent adaptations to high mobility shown by the Aurignacian. The Gravettian can be more confidently ascribed to modern humans, as whole human bodies were interred in identifiably symbolic burials. Some of the earliest modern human remains in Europe have no associated archeology (**Table 3**): they are certainly contemporary with the Aurignacian, although we cannot be certain that they themselves were its authors.

How and why did Neanderthals become extinct? We shall explore Neanderthal responses to changing climate in this section, which of course means we must consider elements of chronology and how they exploited food and other resources. Intercomparability of dating techniques is important, as it allows us to compare our results with the Greenland ice-core climate records (given in cal BP). Taking the dated sites we have at face value, it would appear that Neanderthals were most successful between 60 and 40 cal kyr BP, which was the stable warm phase of (marine) oxygen isotope stage (OIS) 3, immediately after the cold maximum of OIS-4 (72–60 cal kyr BP) (van Andel *et al.*, 2003). After 40 cal kyr BP, the climate began to cool, and by 30 cal kyr BP Neanderthals appear to be extinct.

Naturally, the dating methods we use, and how we apply them, are crucial if we are to be confident in the chronology we have (*see* Dating Techniques; AMS Method; U-Series Dating; Thermoluminescence; Electron Spin Resonance Dating). Radiocarbon is still the main dating technique for our study period – owing to reasons of its affordability, the prevalence of

Table 3 Directly dated late Neanderthal remains and those of early modern humans in Europe between ca. 50 and 30 cal kyr BP

	Site/specimen	Lab. No.	Date (uncal BP)	Date (cal BP)	Comments
Neanderthal	Mezmaiskaya: layer 3d	Ua-14512	29,195 ± 965	33,855 ± 1,318	A late Neanderthal from the Caucasus mountains.
	Vindija Cave: layer G1	OxA-8296	29,080 ± 400	34,180 ± 640	Both C14 dates (OxA-) are much younger than the 2 uranium series ones.
		**	51,000 ± 8,000	51,000 ± 8,000	
		**	46,000 ± 7,000	46,000 ± 7,000	
		OxA-8295	28,020 ± 360	32,560 ± 728	
	Feldhöfer-kirche (Neandertal)	[ETH-??]	39,240 ± 670	43,468 ± 594	Both dates are from the Neanderthal type-specimen(s).
		[ETH-??]	39,900 ± 620	43,806 ± 564	
	Banyoles: lake travertine (5m depth) around mandible	**	45,000 ± 4,000	45,000 ± 4,000	A striking difference between the very young C14 date (UCLA-) and the uranium series one.
		UCLA-930	17,600 ± 1,000	21,098 ± 1,151	
Modern human	Pestera cu Oase: mandible	OxA-11711	>35,200	N/A	This site appears to have the earliest-known modern human in Europe; no associated archeology however.
		GrA-22810	34,290 + 970/–870	39,823 ± 1,301	
	Oblazowa: layer VIII: phalange	OxA-4586	31,000 ± 550	36,110 ± 496	
	Mladeč: ulna & teeth	VERA-2736	26,330 ± 170	30,997 ± 130	Some associated (Aurignacian) artefacts, but the dates are quite variable: the ulna (VERA-2736) and the 'brown-colored collagen fraction' (VERA-3075) are notably younger than the other two dates (on teeth).
		VERA-3073	31,190 + 400/–390	36,263 ± 381	
		VERA-3074	31,320 + 410/–390	36,367 ± 382	
		VERA-3075	30,680 + 380/–360	35,873 ± 385	
		VERA-3076A	31,500 + 420/–400	36,512 ± 408	
		VERA-3076B	27,370 ± 230	31,646 ± 290	
	Grotte de Cussac, loc.1 : rib	Beta-156643	25,120 ± 120	30,103 ± 193	A Gravettian, not Aurignacian, age.
	Kent's Cavern: maxilla	OxA-1621	30,900 ± 900	36,065 ± 851	The earliest modern human in western Europe?
	Paviland (Goat's Hole)	OxA-1815	26,350 ± 550	30,896 ± 339	A Gravettian, not Aurignacian, age from a complete burial.
		OxA-8025	25,840 ± 280	30,684 ± 232	

Trinkaus E *et al.* (2003) *Proceedings of the National Academy of Sciences* 100: 11231–11236; Wild EM *et al.* (2005) *Nature* 435: 332–335; other references in Davies W (2001) Stage Three Project Archaeological Database: http://www.esc.cam.ac.uk/oistage3/details/homepage.html

organic remains at many archeological sites, and its availability over more than 55 years – even though its reliability decreases beyond 30 kyr BP and only a few laboratories produce reliable age estimates beyond 40 kyr BP. 50 kyr BP is essentially the limit of the method. Other dating methods, such as TL, OSL, uranium series, and ESR, are needed to extend our (reliable) chronology further back in time, certainly beyond 30 kyr BP.

If we believe that replacement of Neanderthals by modern humans occurred in Eurasia, then we have recourse to two competing hypotheses: either Neanderthals were outcompeted by modern humans (what one might call 'confrontational replacement'), or they had already become extinct before modern humans arrived in an area ('successional replacement'). To test these models, archeologists have traditionally used two lines of chronological/associative evidence: the relative stratigraphic positions of assemblages attributed to both human species, and/or absolute dating of the latter. While relative stratigraphic positions can give us an idea of succession, or perhaps indicate contemporaneity if interstratification can be proved, they give no real sense of chronology: without absolute dating methods, we do not know the duration of each depostional (or erosional) phase represented by a geological layer. In addition, it is by no means evident that a geological layer is equivalent to a phase of human occupation at a given site. It is instead obvious that in many cases a geological layer contains the remnants of more than one occupation episode; taphonomic disturbance between levels is not necessarily the only explanation for wide age estimate ranges for a given layer. Obtaining direct dates on anthropogenic material, ideally diagnostic artifacts, is obviously the way to identify such occupational phases, but has only recently begun to be practiced.

Current evidence does not tend to support arguments of interstratification, with the claimed occurrences at the French sites of Roc de Combe and Le Piage being now generally attributed to taphonomic disturbance and redeposition of sediments (see d'Errico *et al.* (1998)). Nevertheless, a recent claim has been made for interstratification of Châtelperronian (assumed to be Neanderthal) and Aurignacian (modern human) levels at the French site of Châtelperron (see Mellars *et al.* (1999), p. 346). Notwithstanding the stratigraphic evidence, current absolute dates for both Neanderthal and modern human technocomplexes do overlap temporally at the regional scale, especially in western Europe (France and Iberia), between ca. 43–33 kyr BP (ca. 47–38 cal kyr BP).

Stratigraphic information, then, would suggest that Neanderthals almost always preceded modern humans at sites where technocomplexes from both species are found, while the absolute ages we have (including those from sites with technocomplexes attributed to just one species) suggest a broader contemporaneity. Both lines of evidence are actually telling us subtly different things: the stratigraphic evidence is telling us about the chronology of occupation in a site, and, when taken at the regional scale, is telling us something about how groups of each human species used their landscape and its resources. The absolute ages are telling us about when sites were occupied, and by incorporating ages from sites with, say, only Aurignacian remains (and none attributed to Neanderthals), we can plot the distributions of Neanderthals over time and space: at certain periods, were some areas 'Neanderthal' or 'modern human?' Any sites with interstratified Neanderthal and modern human archeological assemblages would thus represent shifting boundaries between the two species.

The desire of d'Errico *et al.* (1998) to see Neanderthals never overlap with modern humans, based on site stratigraphic successions, favors independent invention by Neanderthals of blade technology and bone working without any influence (known as 'acculturation') by incoming modern humans. Having found new solutions to (new) problems, they then suffer the misfortune of local extinction. Acculturation is very much in the thoughts of those who favor a contemporaneous existence for both species prior to Neanderthal extinction, with social connections between Neanderthal and modern human groups catalyzing behavioral changes, either equally or unequally distributed between each species. At present, our data are not good enough to identify any possible acculturation in either direction between modern humans and Neanderthals, but a useful start to addressing this problem would be made if the bones of contention (diagnostic artifacts) were themselves dated, and the chronologies then compared directly.

Archeologists like to identify boundaries in the Paleolithic world, and a major one that has arisen in the literature over the last 15 years has been the so-called 'Ebro frontier' (e.g., d'Errico *et al.* (1998)). It is claimed that modern humans were either unable or unwilling to disperse into the parts of Iberia south of the river Ebro until about 30 cal kyr BP; until that time, the region remained a Neanderthal bastion. It is of course doubtful that a river could form a boundary to a species that had already dispersed over huge distances, and even reached Australia, and the frontier's proponents thus claim that it was ecological factors (notably the structure of habitats, and the resources available) that held the marauding moderns at bay. When the Mousterian sites in Iberia are plotted on the map, it is evident that the sites we have are concentrated around the margins of this peninsula: the central, upland Meseta region is more or less uninhabited, and perhaps it formed a 'buffer zone' between the largely modern human occupied north, and the late Neanderthals in the southerly coastal regions.

The oddity is that these last Neanderthals in southern Iberia were relatively conservative in behavioral terms: they did not make Châtelperronian assemblages, and continued with the Mousterian. Perhaps they could afford to behave in this fashion because environmental change in this region was lesser than elsewhere in western Europe; the tightly packed mosaic of different habitats ensuring that Neanderthal groups did not have to range too widely to obtain the resources they needed to survive. Modern humans, in contrast, tended to range more widely, certainly in terms of acquiring stone raw materials for their tools, and to make greater use of these 'exotic' raw materials; they also obtained marine shells over long distances for use in personal adornment (*see* Mellars (1989), p. 361). However, in terms of animal species exploited, there are fewer clear-cut differences between the two human species: both tended to hunt medium-to-large mammals (reindeer, red deer, wild boar, horse, mammoth), although at some sites we can see that modern humans exploited an impressive range of animals, ranging from mammoths at one end of the scale to hares, fish, and birds at the other. Only in the last two decades or so have archeologists begun to ask how Neanderthals and modern humans exploited the same food species, and whether their exploitation strategies show any clear differences (Stewart, 2004). It has been observed that Neanderthal skeletons show high levels of trauma and injury: probably their hunting methods led to frequent close-contact encounters with their prey, but in the absence of comparable early modern human remains, we cannot say if these injuries were unique to Neanderthals (Berger and Trinkaus, 1995). Nevertheless, we are still a long way from identifying seasonal differences in the occupation periods at sites: where Neanderthals and modern humans were at particular times of year is not a question we can yet answer.

As fine-grained mosaics of different habitats became increasingly rare in Europe after 37 cal kyr BP, replaced by swathes of more homogeneous environments, Neanderthals would have found it more difficult to obtain the resources they needed and to move to surviving refuges. Fragmentation into small pockets of suitable habitats would render them much more susceptible to extinction, especially with modern humans 'playing cuckoo' in the interstitial areas they once inhabited. Whether or not modern humans had a significant behavioral advantage (they at least managed to survive the last cold maximum of the ice age at ca. 21 cal kyr BP), they could have administered the fatal blow to Neanderthals simply because they were not restricted to Europe: waves of modern humans coming in to Europe could always replenish groups that had become extinct, whereas Neanderthals could never call upon reinforcements from outside Europe.

We stand today at the threshold of a new dating revolution in the transition from Neanderthals to modern humans: being able to date diagnostic organic artifacts, for example, the bone industries from some Châtelperronian sites, will allow us to state with more confidence whether they can be attributed to Neanderthal authorship or not. If they are not of the same age, then we can break the presumed association. Similarly, we can apply the same dating techniques to the skeletal remains and artifacts presumed to belong to the earliest modern humans: AMS radiocarbon dating of small (=0.3 g) organic samples (diagnostic pieces) and TL dating of heated (diagnostic) stone tools. The direct dating of human remains and particular classes of tools will allow archeologists to compare like with like across regions, and to assess the validity and spatio-temporal coherence of our presumed technocomplexes, rather than quibble over the quality of association on a site-by-site basis. Only when we have placed our chronology on a firmer footing can the transition from Neanderthals to modern humans be reassessed.

See also: **Dating Techniques**. **Luminescence Dating**: Thermoluminescence; Electron Spin Resonance Dating. **Radiocarbon Dating**: AMS Method. **U-Series Dating**.

References

Aiello, L. C., and Wheeler, P. (2003). Neanderthal thermoregulation and the glacial climate. In *Neanderthals and Modern Humans in the European Landscape During the Last Glaciation* (T. H. van Andel and W. Davies, Eds.), pp. 147–166. McDonald Institute Monographs, Cambridge.

Allsworth-Jones, P. (1986). *The Szeletian and the Transition from Middle to Upper Palaeolithic in Central Europe.* Clarendon, Oxford.

Arensburg, B. (1989). New skeletal evidence concerning the anatomy of Middle Palaeolithic populations in the Middle East. In *The Human Revolution: Behavioural and Biological Perspectives in the Origins of Modern Humans* (P. Mellars and C. Stringer, Eds.), pp. 165–171. Edinburgh University Press, Edinburgh.

Bar-Yosef, O. (1993). The Role of Western Asia in Modern Human Origins. In *The Origin of Modern Humans and the Impact of Chronometric Dating* (M. J. Aitken, C. B. Stringer and P. A. Mellars, Eds.), pp. 132–147. Princeton University Press, Princeton.

Bar-Yosef, O., and Kuhn, S. L. (1999). The big deal about blades: Laminar technologies and human evolution. *American Anthropologist* **101**, 322–338.

Berger, T. D., and Trinkaus, E. (1995). Patterns of trauma among the Neandertals. *Journal of Archaeological Science* **22**, 841–852.

Bietti, A. (1997). The transition to anatomically modern humans: The case of peninsular Italy. In *Conceptual Issues in Modern Human Origins Research* (G. A. Clark and C. M. Willermet, Eds.), pp. 132–147. Aldine de Gruyter, New York.

Bordes, F. (1968). *The Old Stone Age*. Weidenfeld and Nicolson (World University Library), London.

Bordes, F. (1977). Time and space limits in the Mousterian. In *Stone Tools as Cultural Markers* (R. V. S. Wright, Ed.), pp. 37–39. Australian Institute of Aboriginal Studies, Canberra.

Bosinski, G. (1967). *Die Mittelpaläolithischen Funde im Westlichen Mitteleuropa*. Fundamenta, Reihe A, Bd. 4, Köln.

Campbell, J. B. (1980). Le problème des subdivisions du Paléolithique supérieur britannique dans son cadre européen. *Bulletin de la Société Royale Belge d'Anthropologie et de Préhistoire* **91**, 39–77.

Chabai, V. P. (1996). Kabazi-II in the context of the Crimean Middle Paleolithic. *Préhistoire Européenne* **9**, 31–48.

Clarke, D. L. (1968). *Analytical Archaeology*. Methuen, London.

Cohen, V. Y., and Stepanchuk, V. N. (1999). Late Middle and Early Upper Paleolithic evidence from the East European Plain and Caucasus: A new look at variability, interactions, and transitions. *Journal of World Prehistory* **13**, 265–319.

Davies, W. (2001). A very model of a modern human industry: New perspectives on the origins and spread of the Aurignacian in Europe. *Proceedings of the Prehistoric Society* **67**, 195–217.

d'Errico, F., Zilhão, J., Julien, M., Baffier, D., and Pélegrin, J. (1998). Neanderthal acculturation in western Europe? A critical review of the evidence and its interpretation. *Current Anthropology* **39**(supplement), S1–S44.

Garrod, D. A. E. (1955). The Mugharet el Emireh in lower Galilee: Type station of the Emiran industry. *Journal of the Royal Anthropological Institute* **85**, 141–162.

Gilead, I. (1991). The Upper Paleolithic period in the Levant. *Journal of World Prehistory* **5**, 105–154.

Hassan, F. A. (1972). Toward a definition of "Lames," "Lamelles," and "Microlamelles. *Libyca* **20**, 163–267.

Henry-Gambier, D., Maureille, B., and White, R. (2004). Vestiges humains des niveaux de l'aurignacien ancient du site de Brassempouy (Landes). *Bulletins et Mémoires de la Société d'Anthropologie de Paris (n.s.)* **16**, 49–87.

Hole, F., and Flannery, K. V. (1967). The Prehistory of Southwestern Iran: A preliminary report. *Proceedings of the Prehistoric Society* **33**, 147–206.

Holliday, T. W. (1997). Postcranial evidence of cold adaptation in European Neandertals. *American Journal of Physical Anthropology* **104**, 245–258.

Hublin, J.-J., Spoor, F., Braun, M., Zonneveld, F., and Condemi, S. (1996). A late Neanderthal associated with Upper Palaeolithic artefacts. *Nature* **381**, 224–226.

Jacobi, R. M. (1990). Leaf-points and the British early Upper Palaeolithic. In *Feuilles de Pierre: Les Industries à Pointes Foliacées du Paléolithique Supérieur Européen* (J. K. Kozlowski, Ed.), (Actes du Colloque de Cracovie, 1989), pp. 271–289. Études et Recherches Archéologiques de l'Université de Liège, 42, Liège.

King, W. (1864). The reputed fossil man of the Neanderthal. *Quarterly Journal of Science* **1**, 88–97.

Klein, R. G. (1995) Anatomy Behavior, and Modern Human origins. *Journal of word prehistory* **9**, 167–198.

Kozlowski, J. K. (1990). Certains aspects techno-morphologiques des pointes foliacées de la fin du Paléolithique moyen et du début du Paléolithique supérieur en Europe centrale. In *Paléolithique moyen récent et paléolithique superieur ancien en Europe. Ruptures et transitions: Examen critique des documents archéologiques* (C. Farizy, Ed.), pp. 125–133. Mémoires d'Île-de-France, 3, Nemours.

Kozlowski, J. K. (2000). Châtelperronien, Uluzzien et quoi plus à l'est?. *Anthropologie* **38**, 249–259.

Lieberman, P. (1989). The origins of some aspects of human language and cognition. In *The Human Revolution: Behavioural and Biological Perspectives in the Origin of Modern Humans* (P. Mellars and C. Stringer, Eds.), pp. 391–414. Edinburgh University Press, Edinburgh.

Marks, A. E., and Volkman, P. W. (1983). Changing core reduction strategies: A technological shift from the Upper Paleolithic in the Southern Levant. In *The Mousterian Legacy: Human Biocultural Change in the Upper Pleistocene* (E. Trinkaus, Ed.), pp. 13–33. British Archaeological Reports, International Series, 164, Oxford.

Mellars, P. (1989). Major issues in the emergence of modern humans. *Current Anthropology* **30**, 349–385.

Mellars, P. A., Otte, M., and Straus, L. G. (1999). The Neanderthal problem continued [with reply from Zilhão J and d'Errico F]. *Current Anthropology* **40**, 341–364.

Ohnuma, K., and Bergman, C. A. (1990). A technological analysis of the Upper Palaeolithic levels (XXV–XXVI) of Ksar Akil, Lebanon. In *The Emergence of Modern Humans: An Archaeological Perspective* (P. Mellars, Ed.), pp. 91–138. Edinburgh University Press, Edinburgh.

Olszewski, D. I., and Dibble, H. L. (1994). The Zagros Aurignacian. *Current Anthropology* **35**, 68–75.

Otte, M. (1981). *Le Gravettien en Europe Centrale*.,

Stringer, C. B. (1989). The origin of early modern humans: A comparison of the European and non–European evidence. In *The Human Revolution: Behavioural and Biological Perspectives in the Origin of Modern Humans* (P. Mellars and C. Stringer, Eds.), pp. 232–244. Edinburgh University Press, Edinburgh.

Svoboda, J., Ložek, V., and Vlček, E. (1996). *Hunters between East and West: The Paleolithic of Moravia*. Plenum, New York.

Trinkaus, E., Mowovan, O., Milota, S., *et al.* (2003). An early modern human from the peştera ca Oase, Romania. *Proceedings of the National Academy of Sciences* **100**, 11,231–11,236.

Tostevin, G. B. (2003). A quest for antecedents: A comparison of the Terminal Middle Palaeolithic and early Upper Palaeolithic of the Levant. In *More than Meets the Eye: Studies on Upper Palaeolithic diversity in the Near East* (A. N. Goring-Morris and A. Belfer-Cohen, Eds.), pp. 54–67. Oxbow Monographs, Oxford.

van Andel, T. H., Davies, W., and Weninger, B. (2003). The human presence in Europe during the last glacial period. I: Human migrations and the changing climate. In *Neanderthals and Modern Humans in the European Landscape During the Last Glaciation* (T. H. van Andel and W. Davies, Eds.), pp. 31–56. McDonald Institute Monographs, Cambridge.

Veil, S., Breest, K., Höfle, H.-C., *et al.* (1994). Ein mittelpaläolithischer Fundplatz aus der Weichsel-Kaltzeit bei Lichtenberg, Ldkr. Lüchow-Dannenberg: Zwischenbericht über die archäologischen und geowissenschaftlichen Untersuchungen 1987–1992. *Germania* **72**, 1–66.

Wild, E. M., Teschler-Nicola, M., Kutschera, W., Steies, P., Trinkans, E., and Wanck, B. (2005). Direct dating of Early upper Palaeolithic human remains from Mladeč. *Nature* **435**, 332–335.

Wolpoff, M., and Caspari, R. (1997). *Race and Human Evolution*. Simon and Schuster, New York.

Relevant Websites

Davies, W. (2001b). Stage Three Project Archaeologal Database http://www.esc.cam.ac.uk/oistage3/Details/Homepage.html

Weninger Jöris, O. (2005). Cologne Radiocarbon Calibration & Paleoclimate Research Package http://www.calpal.de

Postglacial Adaptations

G Bailey, University of York, York, UK

© 2007 Elsevier B.V. All rights reserved.

World Overview

By archeological convention, the Postglacial is defined as the period beginning with the final melting of the Scandinavian ice sheet at about 10 radiocarbon kyr BP. This date also marks the conventional beginning of the Mesolithic period, seen as the continuation of a Paleolithic hunter-gatherer way of life adapted to the changed environmental conditions of the Postglacial – in particular, the change from the glacial steppe with its migrating herds of large mammals, to heavily forested environments with fewer resources. This is a view shaped by European data and intellectual history, in which the Mesolithic has traditionally been viewed as a period of cultural and demographic decline or stasis, marking time until Neolithic farmers from the Near East entered Europe, some 8 radiocarbon kyr BP, replacing the hunter-gatherer way of life across the continent over the following three millennia.

Archeological investigations now demonstrate that this notion of progressive evolution through uniform stages of development is greatly oversimplified. So far from being a period of cultural stagnation, these 'Mesolithic' millennia that preceded the development of full-scale village farming are now seen as representing a period of radical change and innovation, while Mesolithic achievements in their turn were built on developments, already well underway, in the closing millennia of the Last Glacial period (Mithen, 2003).

From an archeological point of view, there are good reasons to extend the period of 'Postglacial' adaptations back to the beginning of the warming trend initiated at about 13 radiocarbon kyr BP with the Late Glacial interstadials (*see* Late Glacial Events in Northwest Europe), and perhaps back to 16 radiocarbon kyr BP, when sea levels began their sustained rise from the low point of the Last Glacial Maximum (*see* Last Glacial Maximum GCMs). As sea level rose to reach its present position about 6 radiocarbon kyr BP, progressive inundation of the continental shelf removed extensive areas of lowland territory, breached intercontinental land connections, brought existing hinterlands within reach of milder 'oceanic' climates, and culminated in the creation of entirely new coastal landscapes. The impact of sea-level rise on paleoecology and paleogeography would have differed in different areas, but the effects would have been experienced worldwide. They were especially marked in regions with shallow continental shelves, such as northwest Europe, where large areas of hunting territory were lost, only to be replaced by the creation of shallow inshore waters, indented coastlines, and offshore islands and archipelagos, with more productive and more easily accessible supplies of fish, sea mammals, and intertidal mollusks.

It is during this Late Glacial period of changing environmental conditions, and especially between 13 and 12 radiocarbon kyr BP, that we see the first stages of population expansion into new territory exposed by glacial retreat in northern Eurasia, the first unequivocal evidence for the expansion of human populations into the Americas (Dillehay, 2000), the extensive use of microblade technologies with miniature bladelets and points blunted along one edge to provide inserts for hafted hunting weapons throughout large areas of the inhabited world, and the development of pottery in China, the Russian Far East, and Japan (Habu, 2004, Kuzmin *et al.*, 2004).

Regions strongly affected by glaciation saw the rapid spread of populations into deglaciated territory, and the development of coastal and inland economies with significant dependence on aquatic resources along the shorelines of newly created lakes and coastlines throughout the northerly regions of Eurasia and North America. In subsequent millennia, subsistence economies throughout the world diversified into new niches exploiting a wider range of animal and plant food resources, including the first evidence for sustained and intensive use of marine resources on many of the world's coastlines. In semi-arid environments at lower latitudes, stands of seed-bearing plants began to expand as climatic conditions improved, creating the basis for intensive harvesting of plant foods. In the Near East, this process was underway from as early as 12 radiocarbon kyr BP, ultimately leading to agriculture based on cereal cultivation and domestic animals.

In temperate latitudes, the same climatic trends favored the expansion of woodland cover and reimmigration of deciduous trees, resulting in a complex and rapidly shifting geographical mosaic of vegetation and faunal patterns, in which the large herds of steppe animals such as reindeer, horse, and bison were replaced by woodland species of deer and wild boar. At the same time, these changes in vegetation brought with them new opportunities for food – nut-bearing shrubs and trees such as hazel, oak, and chestnut, and edible roots, leafy greens, and fruits. Hunting parties found new opportunities for seeking out prey such as chamois and ibex on the higher slopes of mountains above the tree line made accessible by deglaciation. Other resources such as birds and small mammals played a larger role, facilitating greater residential stability, larger communities of people, and smaller territories.

High-latitude regions with sub-Arctic conditions were first occupied from at least 10 radiocarbon kyr BP on the coastlines of Norway and Alaska, and the final retreat of the North American ice sheets and the opening up of tundra environments in northern Canada and Greenland was closely followed by the expansion of 'paleoeskimo' populations into the high Artic regions of North America between 5 and 4 radiocarbon kyr BP. The occupation of this new niche was accompanied by exploitation of sea mammals alongside hunting of animals on land, and by dramatic and rapid expansions of human range. In Norway, 2,000 km of coastline was colonized in a matter of generations (Bjerck, in press). In North America, 4,800 km of new territory from the Alaskan border to northeast Greenland was opened up in a few centuries (Fiedel, 1987).

Europe

The European landmass, broadly defined, provides an appropriate microcosm of these global trends (**Fig. 1**). With a wide range of environments from the Eastern Mediterranean to the sub-Arctic shores of northern Scandinavia, and a long history of Mesolithic research stimulated by regular

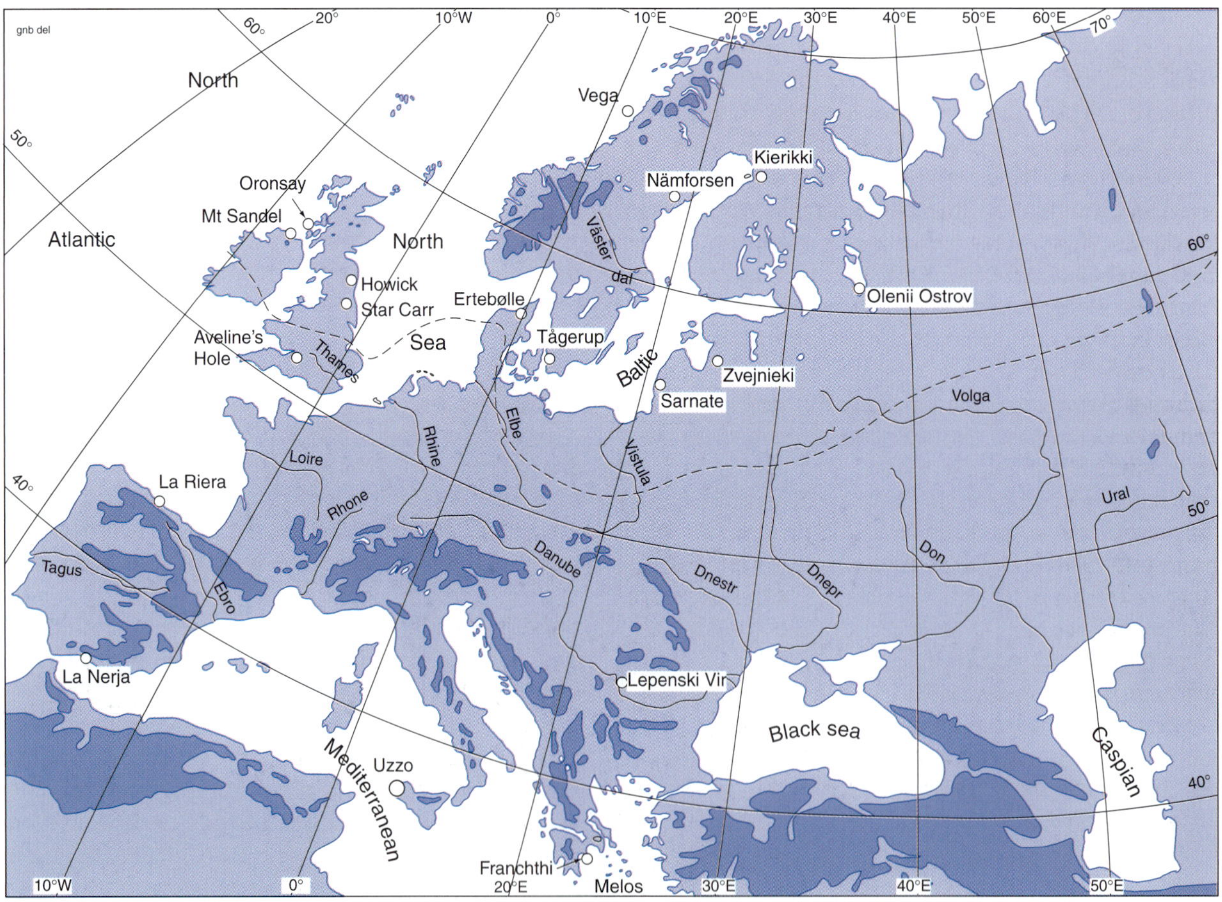

Figure 1 Map of Europe showing principal topographic features and sites (see text). Shaded areas indicate land above about 1,000 m, and the dashed line indicates the maximum extent of the Scandinavian ice sheet at the glacial maximum.

international meetings and periodic overviews (most recently Bailey and Spikins (in press), Larsson *et al.* (2003)), it provides an unparalleled range and richness of archeological material. We can subdivide developments into three broadly defined periods: (1) Late Glacial, ca. 13–10 radiocarbon kyr BP, characterized by expansion into newly available territory; (2) Early Mesolithic, ca. 10–8 radiocarbon kyr BP, with consolidation of settlement and adaptation of hunting and gathering to new vegetational conditions, and a renewed phase of expansion into high-latitude maritime environments; (3) Late Mesolithic, ca. 8–5 radiocarbon kyr BP, diversification of settlement and economy and greater regional differentiation, development of sedentary villages in productive aquatic environments on coastlines, rivers, and lakes, the husbandry of naturally occurring plant and animal resources, including evidence of environmental modification, and interaction with farmers and selective adoption of agricultural practices. These chronological boundaries vary according to region, but nevertheless appear to coincide with major climatic changes, particularly in more northerly regions most sensitive to the effects of deglaciation.

Glacial Prelude

At the height of the Last Glacial, cold and aridity resulted in environments dominated by open steppe or forest-steppe vegetation that extended from the periglacial margins of the ice sheets in the north to the shores of the Mediterranean. Trees were confined to localized stands of coniferous trees in sheltered localities, whereas deciduous species were restricted to pockets of warmer and wetter conditions in refugia at intermediate altitude in the Pyrenees and the circum-Mediterranean mountain ranges. Human populations were largely dependent on migratory herd animals such as reindeer, horse, and bison in colder regions and wild cattle, red deer, gazelle, and the steppe ass in the south and southeast. Woolly mammoth were abundant on the more extensive steppes of eastern Europe and were an important source of meat, bone for artifacts and fuel, and ivory for decorative items. Plant foods were scarce nearly everywhere, though probably exploited where available.

Human populations were distributed in small and mobile bands with large annual territories and social contacts with neighboring bands over long distances, signaled in the archeological record by the movement of exotic materials such as marine shells and rare and fine-grained flint many hundreds of kilometers from their sources. The most favorable environments, the largest concentrations of archeological sites, and the highest population densities were in southwest France and northern Spain in the west and the riverine basins of eastern Europe and the Ukraine in the east. These were also the centers of cultural innovation and of social aggregation and ritual, associated in the west with the flowering of Franco-Cantabrian rock art, and in the east with the production of a rich variety of portable art and the construction of dwelling structures built from mammoth bones (*see* Interactions with Hominids).

Late Glacial Expansion

It was from these centers that human populations first expanded into newly accessible territory as the ice sheets retreated and climate ameliorated. Many of the animal species of former significance moved their ranges northward, as with reindeer and horse, or became extinct, as with mammoth and bison, and were replaced by forest or forest-edge species such as elk, red deer, roe deer, cattle, and wild boar, according to regional and local circumstances.

From about 13 radiocarbon kyr BP onward, settlements associated with hunting of reindeer and, later, horse, cattle, and elk, became established across the north European plain in a broad swathe from the British lowlands in the west to the Crimea in the east (Dolukhanov, in press, Tolan-Smith, in press, Zvelebil, in press). Stone industries were characterized by tanged points for hafting on the ends of spear or arrow shafts, and by 'knife' blades with a curved and blunted edge. The southern end of the North Sea was still dry land at this period and the recovery of an antler harpoon dredged up from the Leman and Ower Bank and dated to 11.7 radiocarbon kyr BP shows that this territory was used from an early period (Tolan-Smith and Bonsall, 1999). Variations in the proportion of these artifacts and minor differences in morphology of the key artifact types have given rise to a variety of cultural labels, Swiderian in Poland, Hamburgian, Bromme, and Ahrensburgian in northern Germany and Denmark, and Federmesser and Creswellian in the Low Countries and Britain. Whether these represent stylistic markers of distinct social groupings or functional adaptations to different types of weapons and prey species is unclear. The smaller and lighter tanged points of the Ahrensburgian, associated with the final cold phase of the Younger Dryas period between 11 and 10 radiocarbon kyr BP (*see* Younger Dryas Oscillation, Global Evidence) could well have been used as arrow points, since actual bows have been recovered from this period. At any rate, the similarities across this vast territory are as impressive as any differences, with relatively small settlements, specialized toolkits of stone and antler, and a considerable degree of mobility. Importation of raw materials for making artifacts from favored sources over distances of 150 km or more reflects both the extended lines of communication in small mobile populations engaged in the early stages of colonizing new territory and lack of familiarity with less suitable materials available nearer to hand or the absence of any need to use them.

The Early Mesolithic

Technologically, there is no obvious break with the preceding period, although there is a greater diversity of raw materials, techniques of stone working, and tool types. Microlithic stone tools made from bladelet segments proliferated into new geometric forms, particularly rectangular pieces, triangles, and crescents. Many of these were most probably used to provide replacements for arrow tips, and the bow and arrow were widely in use throughout Europe during this period. Some may also have been used as replaceable components in other multipart tools such as knives with bone or wooden handles. There was also a general diversification of artifact types made from stone, bone, antler, and wood throughout the Mesolithic period, including macrolithic tools of worked antler, flaked flint, or other stone, and polished slate, used for working wood, digging the ground, or processing sea-mammal carcasses and fish. Harpoons of bone or antler and bone fishhooks are evidence of new technologies for exploiting marine and aquatic resources. Wooden artifacts include dugouts, paddles, fish weirs, spear and arrow shafts, and timber-framed houses. Artificial dwelling structures are more widely in evidence and range from lightly built tent-like structures to substantial stone-built constructions, depending on the available raw materials, climatic conditions, the function of the structure, the nature of the settlement, and the size and permanence of the resident population.

In areas of the north European plain, first occupied in the preceding Late Glacial, there was a consolidation of settlement across this extensive lowland region, a trend already apparent toward the end of the Late Glacial, but one that became more marked with the onset of the Preboreal period after 10 radiocarbon kyr BP (*see* The Last Interglacial). The archeological record indicates more sites, greater use of local raw materials, a wider range of game animals, and a marked preference in site locations for lake-edge settings, with evidence for fishing and fowling as well as hunting. Star Carr in Britain, dated at about 9.5 radiocarbon kyr BP is an early example of this pattern, with broad-blade industries typical of the early Mesolithic period, flaked stone axes, barbed harpoons of red deer antler, antler mattocks, and faunal remains of wild cattle, red deer, roe deer, elk, and boar, though fish are absent here (Mellars and Dark, 1998). The presence of a wooden paddle indicates the use of water transport, most probably dugouts. Similar sites are well documented in Denmark in association with the Maglemosean culture (Blankholm in press), and similar groupings are found in Poland, southern Sweden, the eastern Baltic and Russia (Dolukhanov, in press, Zvelebil, in press). At the very end of this period, about 7.8 radiocarbon kyr BP, the Arctic island of Zhokov in the East Siberian Sea was first visited when still accessible from the Siberian mainland, with a fauna dominated by reindeer and polar bear (Pitul'ko, 1995). Nuts, greens, roots, seeds, and fruits became more widely available and other resources such as birds and small mammals created more varied and diverse paleodiets, but in all cases seasonality evidence suggests seasonal occupations and relatively small social units of one or a few families, implying a large degree of residential mobility.

Maritime adaptations The earliest use of marine resources is obscured by the lowered sea levels that persisted throughout the glacial period and the removal or submergence of relevant evidence before sea level approximated its present position about 6 radiocarbon kyr BP. This problem of differential visibility of evidence is especially acute on the shallower and broader stretches of continental shelf around the coastlines of the North Sea basin and northern France, but even on the steeply shelving coastlines of southern Europe, the shorelines of the glacial maximum would have been at least 5 km away, far enough to take most marine resources out of reach of sites located on the present-day coastline. Underwater survey and excavation are beginning to reveal new evidence, and it seems plausible to argue that late Paleolithic hunters were hunting seals and taking other marine resources along these now submerged coastlines from an early period (Fischer 1995, 1996; Flemming 2004).

Quantities of marine mollusk shells appear in the coastal cave deposits of northern Spain, such as La Riera and El Juyo, from as far back as the glacial maximum, with evidence for increased quantities in the Late Glacial period (Straus, in press). Cueva de la Nerja in southeast Spain contains evidence of marine mollusks, fish, seal, and birds, associated with an early stage of the Late Glacial dating between about

14 and 12 radiocarbon kyr BP (Morales *et al.* 1998). Given that shorelines were still at some distance from the present coastline, this is a significant indication of interest in marine resources and most probably only the visible tip of a more extensive pattern of marine exploitation (Bailey and Craighead, 2003). As shorelines approached their present position, thick shell middens with many millions of discarded mollusk shells begin to appear. The earliest group of such sites in the Iberian Peninsula is the Asturian shell middens, accumulated in rockshelter mouths on the north Spanish coast between about 9 and 6 radiocarbon kyr BP. Here, as with the larger and more numerous shell middens that abound during the Late Mesolithic, hunting and gathering of resources on land continued alongside the collection of mollusks and other marine resources.

In the eastern Mediterranean, the occupants of the Franchthi Cave on the Argolid Peninsula of southern Greece were importing obsidian from the island of Melos 11 radiocarbon kyr BP, which would have required a sea journey of 100 km by the direct route, or a series of sea crossings of at least 20 km by a more circuitous route. However, apart from a few limpet shells, evidence for the exploitation of marine resources does not appear in the Franchthi deposits until somewhat later in the sequence, most probably because the coastline was still too far away (Pluciennik in press).

A notable exception to this general picture of submergence and loss of coastal archeology is the Norwegian coast, which has undergone substantial isostatic uplift following the removal of the Scandinavian ice sheet, so that 13-kyr-old shorelines formed when the eustatic level of the sea was still about 50 m below the present are now elevated above present sea level along much of the Norwegian coastline.

Despite the opportunities for preservation of early coastal sites and the fact that the Norwegian coastline was ice free from an early period, there is no convincing evidence of human occupation, apart from stray and ambiguous finds, until about 10 radiocarbon kyr BP, when numerous sites appear all along the Norwegian coastline in association with locally named Komsa and Fosna cultures characterized by large numbers of flake adzes and tanged points of Ahrensburgian type. Preservation of organic materials is almost nonexistent, so that there are few indications of the animals exploited, but the location of the sites (some of them on offshore islands), the nature of the resources available, and the need for substantial intake of animal fats for human survival in these northerly regions indicate that marine resources and especially sea mammals must have been a major source of subsistence, while the flake adzes would have been well suited for scraping blubber from seal carcasses (Bjerck, in press). The sites on the three large islands of Vega, some 20 km offshore, provide a good example. These were occupied from the earliest period, and by about 8.5 radiocarbon kyr BP comprised an interrelated network of some 12 settlements scattered across the islands. The largest site of Åsgarden was associated with the best natural harbor and had evidence of solid circular dwelling structures with foundations partially excavated into the subsoil ('pit dwellings'), stone or turf walls, and a wide range of artifacts suggesting a residential base. The other sites were smaller with a narrower range of tool types indicating shorter-lived occupations for specialist activities.

The absence of earlier sites on the Norwegian coast may be due to the lack of technology or skills to make the seaworthy boats that would have been essential for sea travel in the region. Or it may be that the resource options available for human survival were too narrow and too risky to encourage settlement until the retreat of the Scandinavian ice sheet and the expansion of the Gulf Stream created more productive conditions on land and at sea.

At about the same time, during the Preboreal period, the British Isles also witnessed a new phase of expansion northward and westward beyond the lowland regions of England, with a number of new settlements appearing around the coastlines of Scotland and in the Western Isles, dating from about 9.5 radiocarbon kyr BP (Tolan-Smith, in press). Ireland, which had never had a passable land connection to mainland Britain, was also occupied for the first time. As in Norway, the appearance of the Gulf Stream would have been a significant attraction to the new settlement. Seaworthy boats would have been essential for navigating the Western Isles and especially for the crossing to Ireland, regarded as one of the most treacherous sea crossings in the British Isles. Sturdy dwelling structures would also have been essential in the harsher winter climates of the north. Faunal preservation on many of these sites is poor, but Mt. Sandel in Northern Ireland has preserved evidence of salmon fishing and boar hunting, postholes indicating substantial dwelling structures and pits that could have been used for food storage. The coastal site of Howick on the North Sea coast of Northumberland in northern England belongs to this phase of expansion, and has yielded traces of a substantial timber-framed dwelling and evidence of sealing and boar hunting.

The Late Mesolithic

During this period, sea levels stabilized in most parts of Europe at about their present position, resulting in the development of estuarine and inshore mudflats, and climate became slightly warmer and wetter – the Climatic Optimum (*see* Overview) – with temperatures 2–3 °C higher on the average than today's. These changes would have had a number of generally beneficial effects on the productivity of coastlines, lakes, and rivers, and during this period we see the proliferation of large numbers of sites and settlements in these settings, many of them associated with permanent villages, food storage, cemeteries, and indications of social ranking. More specialized sites for hunting or gathering of plant foods associated with mobile settlement patterns continued in the less productive regions of Europe, or as outlying locations visited by hunting parties from residential bases elsewhere. Pollen evidence of woodland clearance in southern Sweden, southern Finland, and eastern Latvia (Zvelebil, in press), and in upland regions of Britain (Tolan-Smith, in press) is associated with deliberate clearance to improve the productivity of food-bearing shrubs and plants, either for direct human consumption or to concentrate animal resources. Freshwater or saltwater fish were a major source of protein, the domestic dog was widely in evidence, and pigs were commonly hunted, and may have been husbanded without necessarily involving biological domestication, the combination of fish, pigs, and nuts creating a powerful 'package' of resources capable of sustaining large communities.

The technological characteristics of this period represent a continuation or elaboration of existing trends. Among the geometric microliths, trapezes and rhomboid-shaped pieces replaced the earlier shapes across extensive territories. Pottery also came into regular use in some areas. The earliest examples occur between 9 and 7.3 radiocarbon kyr BP in European Russia and ceramic containers were widely used from 6.5 radiocarbon kyr BP in the Baltic and southern Scandinavia in late Mesolithic contexts.

Shell mounds Coastlines are associated with shell mounds of varying size and number, the largest and best known grouping being the Ertebølle in Denmark, dated between about 6.5 and 5 radiocarbon kyr BP, the largest mound, as at the type site of Ertebølle itself, being several hundred meters long and several meters thick. These sites were built up slowly over many centuries of repeated use and discard of oyster shells and other refuse, and over 400 shell mounds have been recorded. The larger sites were residential bases with a full range of artifacts including microlithic arrow barbs and flaked axes, an antler and bone industry, and pottery jars and bowls. Food remains include sea and land mammals, fish, and fowl (Blankholm, in press). Marine resources were most probably the major source of protein, among which fish rather than shellfish were the major food item, a conclusion reinforced by stable-isotope analyses of human bones recovered from these sites. Coastal sites without mollusk shells are just as common and probably played the same role in the settlement system, occurring on shorelines lacking local supplies of oysters. Some of the coastal settlements of this period have been submerged in shallow water because of isostatic submergence, resulting in spectacular conditions of organic preservation, notably at the underwater settlement of Tybrind Vig, which produced a dugout, a paddle richly incised with geometric designs, and large numbers of wooden stakes associated with a landing stage and a fish weir. Many other submerged sites have been discovered in Danish waters, including other examples of fish weirs, suggesting a major investment in communal facilities.

Other well-studied concentrations of shell mounds are those on the small island of Oronsay in the Western Islands of Scotland, used between about 6 and 5.5 radiocarbon kyr BP for seal hunting, fishing, and the collection of limpets – perhaps for fish bait (Tolan-Smith, in press), and the Portuguese shell mounds on the Muge and Sado tributaries of the Tagus estuary (Straus, in press). Around the Mediterranean coastline, fishing and shellgathering were combined with gathering of plant foods and hunting of mammals, supporting year-round occupations in some cases, notably at Franchthi Cave in Greece and the Uzzo Cave in Sicily, while the offshore islands of Mallorca, Corsica, Sardinia, and Cyprus were occupied or visited (Pluciennik, in press). The apparent 'explosion' in the numbers of these coastal sites may be misleading, given that earlier coastlines are now mostly submerged in much deeper water.

Lake and river settlements Elsewhere in Europe, large settlements have been recorded on the edges of lakes and rivers. Notable examples are Tågerup in Sweden (8–6.5 radiocarbon kyr BP), with timber dwellings, wooden jetties for boats, and basketry used in trapping and carrying fish. On the Latvian coast, Sārnate (7–4 radiocarbon kyr BP) has wooden houses with evidence of seal hunting, fishing, and large-scale processing of water chestnuts. In N. Finland, Kierikki (4.8–3.8 radiocarbon kyr BP) on the estuary of the River Li revealed more than 300 house floors of pit dwellings, very large pottery jars used for storing seal oil, a large stone structure 50 m × 30 m suggesting a communal or ritual purpose, and evidence of long-distance trade in flint, greenstone, amber, and copper (Zvelebil, in press). In the Iron Gates region on the middle reaches of the Danube River is a group of sites dated between about 8 and 7 radiocarbon kyr BP, such as Lepenski Vir, Vlasac, and Schela Cladovei, with trapezoidal dwelling structures and faunal remains including deer, dogs, fish, freshwater mollusks, and land snails (Bonsall, in press). Fish include the migratory sturgeon, which can grow to an individual weight of 150 kg, and stable isotope analyses of human bones indicate a major dependence on fish protein.

Burials and rock art Burials are common, either in intimate association with domestic dwellings or as separately demarcated cemetery areas. The earliest example is at Aveline's Hole, a cave in the Cheddar Gorge of southwest England, with a date of ca. 9.1 radiocarbon kyr BP, with more than 70 human burials, interpreted as the cemetery of a larger social group (Tolan-Smith, in press). Olenii Ostrov on an island in Lake Onega in Karelia, dated at about 7.5 radiocarbon kyr BP, had over 300 burials. At Zvejnieki in Latvia, 315 burials have been excavated indicating repeated use between 8.2 and 4.2 radiocarbon yr BP. Grave goods and the position of the skeletons indicate a variety of burial rites and evidence of social ranking based on age, gender, personal wealth, clan membership, and the presence of shamans (Zvelebil, in press). The Portuguese shell mounds of the Muge River are also estimated to have contained at least 300 burials (Straus, in press). In Denmark, human remains are rare in the Ertebølle shell mounds but 18 burials were found at the coastal settlement of Vedbaek. At Skateholm in southern Sweden, also belonging to the Ertebølle culture, a total of 85 burials were recovered including evidence of a wooden mortuary structure (Blankholm, 2006). There are many instances of violent deaths, suggesting conflict associated with increased competition or territoriality.

Rock art is commonly regarded as absent in the Mesolithic period, an absence taken as evidence of cultural impoverishment compared with the Upper Paleolithic. However, rock art is in fact quite common, as are decorative artifacts, but concentrated in the new centers of population growth located in different regions from their Upper Paleolithic predecessors, providing additional insights into ritual and symbolism. Well-dated rock art is present in northern Norway by about 9 kyr BP with naturalistic representations of elk, reindeer, bear, whale, and seal. Many hundreds of sites have been recorded in Norway and Sweden with images of anthropomorphs, deer, boats, sea mammals, bear, water birds, fish, reptiles, tracks, hunting and fishing implements, and abstract designs. One of the largest such concentrations is the site of Nämforsen (5.5–3.5 radiocarbon kyr BP) in Swedish Norland, which was a major focus of ritual and seasonal aggregation for settlements in the surrounding region. At the other end of Europe is the Levantine rock art of eastern Spain, with schematic and naturalistic representations including hunting scenes and humans, and the Addaura Cave in Sicily (Pluciennik, in press).

Conclusion

The major driving force behind these postglacial adaptations was the changing environmental conditions associated with deglaciation and climatic amelioration, which created new territories for human settlement, new supplies of food and raw materials for subsistence, new technological demands for tools, shelter, and transport, and new opportunities for population growth. These changes resulted in major reconfigurations of social geography, with expansion of human populations into new territory and the growth of new centers of population density, often associated with permanent 'villages' repeatedly occupied on a year-round basis by large communities, or with aggregation sites that formed a seasonal focus for more widely dispersed populations. The more ecologically productive environments capable of supporting higher population densities were often associated with a rich ritual and symbolic life and evidence of social differentiation. The contrasts often drawn with the Upper Paleolithic period thus reflect changes in the geographical distribution of population centers and the occupation of ecologically productive regions that were unavailable during the Last Glacial period, rather than an overall advance in social or economic complexity. It was these closing millennia of the hunting, fishing, and gathering way of life, in their turn, that witnessed the emergence and synthesis of the technological, economic, and social capabilities on which the later developments of agriculture, urban civilization, and long-distance trade were founded.

See also: **Archaeological Records**: Overview. **Glaciations**: Late Glacial Events in Northwest Europe. **Paleoclimate Modeling**: The Last Interglacial; Last Glacial Maximum GCMs. **Paleoclimate Reconstruction**: Younger Dryas Oscillation, Global Evidence. **Vertebrate Studies**: Interactions with Hominids.

References

Bailey, G., and Craighead, A. (2003). Late Pleistocene and Early Holocene coastal palaeoeconomies: A reconsideration of the molluscan evidence from Northern Spain. *Geoarchaeology: An International Journal* **18(2)**, 175–204.

Bailey, G., and Spikins, P. (Eds.) (in press) *Mesolithic Europe.* Cambridge: Cambridge University Press.

Bjerck, H., (in press) Norwegian Mesolithic trends: A review. In *Mesolithic Europe* (G. Bailey and P. Spikins, Eds.). Cambridge University Press, Cambridge.

Blankholm, HP., (in press) Southern Scandinavia. In *Mesolithic Europe* (G. Bailey and P. Spikins, Eds.). Cambridge University Press, Cambridge.

Bonsall, C., (in press) The Mesolithic of the iron gates. In *Mesolithic Europe* (G. Bailey and P. Spikins, Eds.). Cambridge University Press, Cambridge.

Dillehay, T. D. (2000). *The Settlement of the Americas. A New Prehistory.* Basic Books, New York.

Dolukhanov, P., (in press) The Mesolithic of European Russia, Byelorussia and the Ukraine. In *Mesolithic Europe* (G. Bailey and P. Spikins, Eds.) Cambridge University Press, Cambridge.

Fiedel, S. J. (1987). *Prehistory of the Americas.* Cambridge University Press, Cambridge.

Fischer, A. (Ed.) (1995). *Man and Sea in the Mesolithic: Coastal Settlement Above and Below Present Sea Level.* Oxbow, Oxford.

Fischer, A. (1996). At the border of human habitat: The late Palaeolithic and early Mesolithic in Scandinavia. In *The Earliest Settlement of Scandinavia and Its Relationship with Neighbouring Areas, Acta Archaeologica Lundensia, Series In 8, 24* (L. Larsson, Ed.), pp. 157–176. Almquist and Wiksell, Stockholm.

Flemming, N. (Ed.) (2004). *Submarine Prehistoric Archaeology of the North Sea: Research Priorities and Collaboration with Industry.* CBA Research Report 141, London.

Habu, J. (2004). *Ancient Jomon Japan.* Cambridge University Press, Cambridge.

Kuzmin, Y. V., Jull, A. J. T., Burr, G. S., and O'Malley, J. M. (2004). The timing of pottery origins in the Russian Far East: ^{14}C chronology of the earliest Neolithic complexes. In *Radiocarbon and Archaeology: Proceedings of the Fourth International Symposium, Oxford 2002, Monograph 62* (T. Higham, C. Bronk-Ramsey and C. Owen, Eds.), pp. 153–159. Oxford University School of Archaeology, Oxford.

Larsson, L., Kindgren, H., Knutsson, K., Loeffler, D. and Åkerlund, A. (Eds.) (2003). *Mesolithic on the move: papers presented at the Sixth International Conference on the Mesolithic in Europe, Stockholm 2000.* Oxford, Oxbow.

Mellars, P., and Dark, P. (1998). *Star Carr in Context.* McDonald Institute for Archaeological Research, Cambridge.

Mithen, S. (2003). *After the Ice.* Weidenfeld and Nicolson, London.

Morales, A., Roselló, E., and Hernández, F. (1998). Late upper Palaeolithic subsistence strategies in southern Iberia: Tardiglacial faunas from Cueva de Nerja (Málaga, Spain). *European Journal of Archaeology* **1**(1), 9–50.

Pitul'ko, V. (1995). High arctic Mesolithic culture: Man and Environment. In *Man and Sea in the Mesolithic: Coastal Settlement Above and Below Present Sea Level* (A. Fischer, Ed.), pp. 351–359. Oxbow, Oxford.

Pluciennik, M., (in press) The coastal Mesolithic of the European Mediterranean. In *Mesolithic Europe* (G. Bailey and P. Spikins, Eds.). Cambridge University Press, Cambridge.

Straus, L.G., (in press) The Mesolithic of Atlantic Iberia. In *Mesolithic Europe* (G. Bailey and P. Spikins, Eds.). Cambridge University Press, Cambridge.

Tolan-Smith, C. (in press) Mesolithic Britain. In *Mesolithic Europe* (G. Bailey and P. Spikins, Eds.). Cambridge University Press, Cambridge.

Tolan-Smith, C., and Bonsall, C. (1999). Stone Age studies in the British Isles: The impact of accelerator dating. In *14C et Archéologie. Actes du 3ème congrès international (Lyon, 6–10 avril 1998)* (J. Evin, C. Oberlin, J.-P. Daugas and J.-F. Salles, Eds.), pp. 249–257. Mémoires de la Société Préhistorique Franéaise 26, 1999 et Supplément 1999 de la Revue d'Archéometrie, Paris.

Vigne, J.-D., and Desse-Berset, N. (1995). The exploitation of animal resources in the Mediterranean Islands during the Pre-Neolithic: The example of Corsica. In *Man and Sea in the Mesolithic: Coastal Settlement Above and Below Present Sea Level* (A. Fischer, Ed.), pp. 309–318. Oxbow, Oxford.

Zvelebil, M., (in press) Innovative hunter-gatherers: The Mesolithic in the Baltic. In *Mesolithic Europe* (G. Bailey and P. Spikins Eds.). Cambridge University Press, Cambridge.

B

BEETLE RECORDS

Contents

Overview

S A Elias, Royal Holloway, University of London, Surrey, UK

© 2007 Elsevier B.V. All rights reserved.

The study of fossil beetles (Coleoptera) from Quaternary deposits began in earnest in the 1950s in Britain and has since spread to many areas of the world, most recently to Australia, New Zealand, and Japan. Fossil beetle research has led to many exciting breakthroughs in our understanding of the pace and intensity of climate change in terrestrial landscapes. In many ways, beetles are the ideal proxy for tracking terrestrial environmental change.

The Perfect Proxy?

If one were to design the ideal source of biological proxy for the reconstruction of Quaternary terrestrial environments, certain aspects of the nature of this ideal organism would have to conform to the following guidelines:

1. It must be sensitive to environmental change. Organisms that are able to withstand decades or centuries of adverse climate *in situ* before eventually dying out would be of far less use than organisms that simply could not survive such adversity.

2. There should be great species diversity within the group of organisms to be studied. Greater diversity yields more precise and diverse information on paleoenvironments. Taxonomic groups with only a few dozen species necessarily offer less precise information about past environments than groups with many thousands of species. Each species has its own ecological niche, so by studying the fossil record of a group that represents thousands of ecological niches, one is able to piece together a broad spectrum of information about past environments (climate, soils, water quality, and vegetation).

3. It should be highly mobile and thereby able to respond rapidly to environmental change by rapidly shifting its distribution into new areas where the environment is suitable. This quality ensures little or no problem with 'migration lag', wherein an organism requires decades, centuries, or even millennia to become established in regions with potentially suitable environments.

4. It should have an abundant, well-preserved fossil record. This allows regional reconstructions to be built up for given time intervals based on multiple study sites containing assemblages of the fossil type in question. If the fossil remains of an organism are too scarce or too poorly preserved to be identified, then their fossil record will contribute little to regional paleoenvironmental reconstructions.

5. It should be readily identifiable to the species level. Even if a group of organisms satisfy the first four criteria, their fossil record will be of limited value if their remains cannot be specifically identified. There is often a great deal of ecological variation among the species in a given genus, so a generic identification provides little that is useful to paleoenvironmental reconstruction.

6. It should have many species with narrow ecological or climatic tolerances. Again, this harkens back to the issue of specificity. The more specific the environmental requirements of the organism, the more specific the reconstruction that can be made, based on its presence in the fossil record. This criterion works together with criterion No. 2. If taxonomic groups with a great deal of species diversity also contain many species with very narrow ecological or climatic tolerances, then a complex mosaic image of the paleoenvironment can be constructed.

7. None of the previous criteria are of much use unless the taxonomic group has remained extant throughout the study period one is interested in. If the group of organisms has gone extinct, no matter how diverse, abundant, or well preserved, then we are left making educated guesses about the ecological or climatic requirements of the ancestral forms. The modern ecological and climatic data available for the surviving species in such a group of organisms can never be fully applied to extinct species.

8. Finally, it is a great help to the paleoecologist if the proxy group of organisms is well-known to science. In other words, if a group of organisms has been well studied in recent times, then we have at our disposal a great body of data, built up through many generations of scientific study, concerning the modern ranges and ecological requirements of the species.

The reason why beetles (Coleoptera) are so useful as proxy indicators of past terrestrial environments is that they fulfill all these requirements. Beetles are the most diverse group of organisms on Earth, with more than 1 million species known to science. A large proportion of beetle species are known to be quite sensitive to environmental change and to shift their distributions across continents in order to become established in regions of suitable environment (Elias, 1994). Beetles have a well-preserved, abundant fossil record in many regions of the world. Beetle exoskeletons are reinforced with chitin, a highly durable compound that resists decay in water-logged and other anerobic sedimentary environments. That chitinous exoskeleton is covered with taxonomically useful characters on the main body parts, including the head capsule, thoracic shield (pronotum), and wing covers (elytra). In many study regions, paleoentomologists are able to specifically identify up to 50% of these fossil beetle sclerites; in some well-studied regions, this percentage is closer to 75%. Quaternary beetle faunal assemblages frequently contain hundreds of identified species, more than any other group of organisms preserved in Quaternary deposits.

Many beetle species are stenothermic, adapted to a narrow range of temperatures (**Fig. 1**). Predatory and scavenging families of beetles comprise less than half the total number of families, but their species' numbers are often proportionally high in fossil assemblages. These are the two groups that receive the most attention in paleoclimatic reconstructions. The rationale behind this decision is that predators and scavengers are not tied to particular species of host plants and thus are able to become established most rapidly in new regions in response to changing climates. For instance, the Pleistocene fossil record indicates that these groups have colonized newly deglaciated landscapes within a few years of ice-margin retreat (Coope, 1977). Perhaps because of beetle mobility in the face of changing environments, the fossil evidence strongly indicates that nearly all species found in Quaternary assemblages remain extant today. It appears likely that beetle populations never remained genetically isolated for sufficient lengths of time in the past 2.6 million years to develop new species. The constant mixing of gene pools through shifting distributions, in response to 50+ glacial/interglacial cycles, prevented this from happening (Coope, 1978). Finally, beetles are some of the best studied groups of insects, with modern ecological requirements and ranges well understood for many species living in the temperate and high-latitude regions (where the vast majority of fossil investigations have taken place).

Where Fossil Beetles Are Found

Fossil beetle assemblages have been recovered from a wide variety of sedimentary environments, especially anoxic water-lain sediments that concentrate the remains in layers of organic detritus. Lacustrine (lake and pond) sediments have yielded abundant, diverse assemblages of fossil beetles, especially in deposits from the littoral zone and where a stream enters a lake or pond (**Fig. 2**). Such deltaic deposits can yield astonishing numbers of fossil beetle specimens. Fluvial sediments also yield fossil beetles. The most productive type of fluvial deposit is an accumulation of organic detritus, laid down in secondary channel bends, backflows, and pools between riffles (**Fig. 3**). Bogs, fens, and mires can also yield good accumulations of fossil beetles, especially in locations representing the edge of such bodies, where there is increased input from the upland beetle fauna (**Fig. 4**).

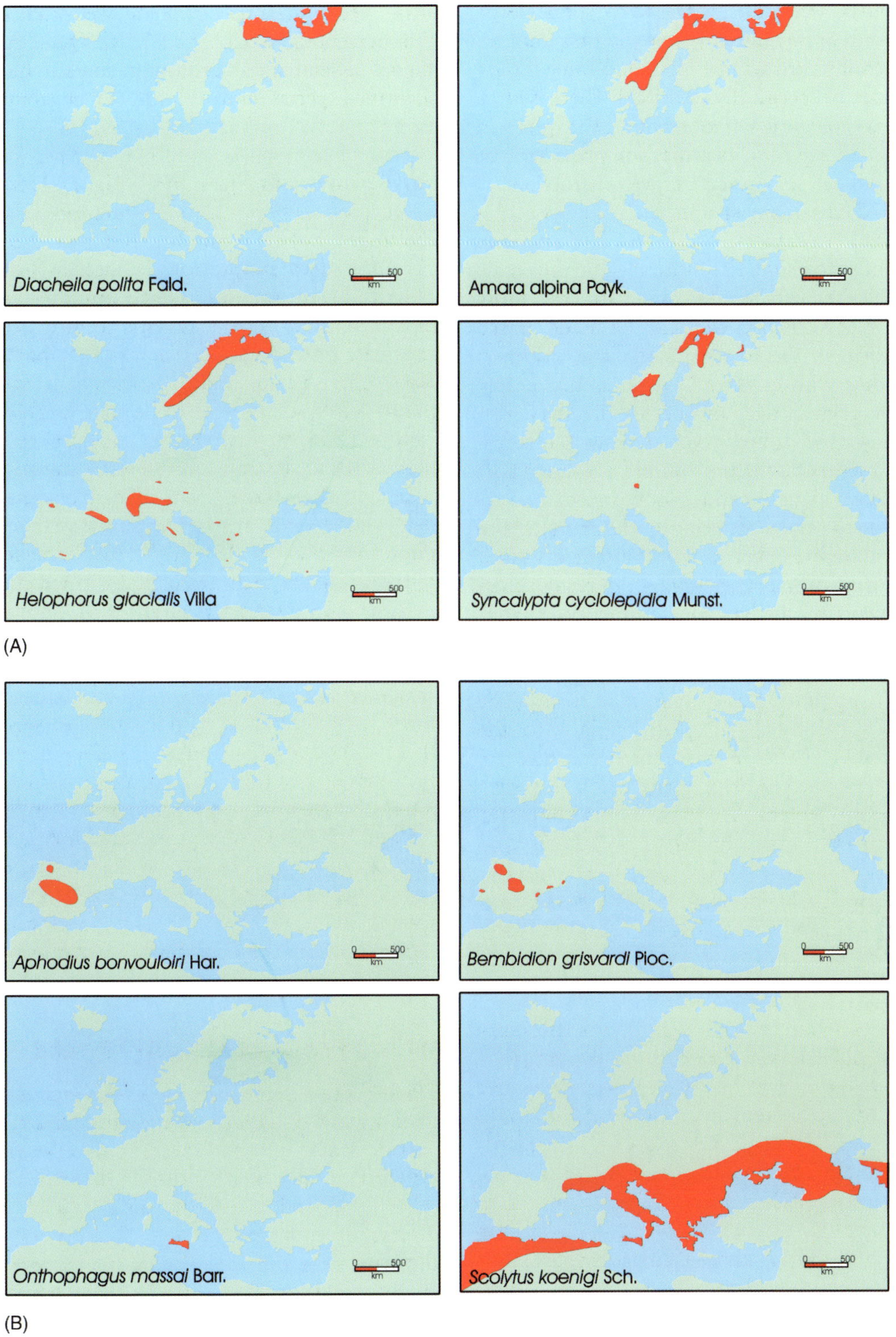

Figure 1 Modern European distributions of stenothermic beetle species found in British Pleistocene deposits. (A) Cold-adapted species; (B) warm-adapted species.

One of the more unusual types of deposit-yielding beetle remains are the middens of certain rodents, including packrats or woodrats (*Neotoma*), the native group of North American rats. Their middens (**Fig. 5**), intermittently wet and sticky from rat urine and feces, serve as traps for beetles entering the shallow caves and rock shelters of desert and semiarid regions where these rats thrive (Elias, 1990). Unlike

Figure 2 Deltaic deposit exposed at Lake Isabelle, Colorado. The layers of organic detritus yielded thousands of fossil beetle specimens. Photo by S. A. Elias.

Figure 3 Lense of organic detritus exposed after a flood on the Roaring River, Rocky Mountain National Park, Colorado. This late Holocene deposit contained a rich, diverse beetle fauna.

Figure 4 Holocene peat from a high mountain bog at La Poudre Pass, Colorado, exposed along an irrigation canal.

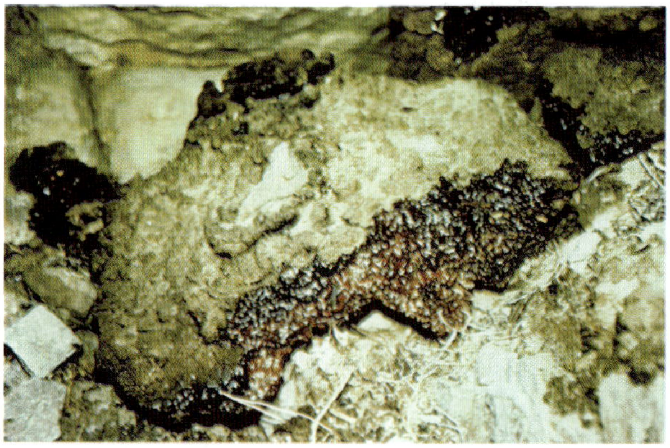

Figure 5 Packrat (*Neotoma*) midden sample from a rock shelter at Emery Falls, Grand Canyon, Arizona. Photo by Thomas Van Devender.

the waterlogged sediments of lakes, rivers, and bogs, packrat middens preserve beetle (and other arthropod) exoskeletons in extremely dry conditions, sometimes mummifying their soft tissues.

Methods of Sampling and Extraction

When sampling the kinds of sediments discussed previously, the aim is to obtain a sufficient quantity of material to yield at least a liter of organic detritus. The details of sampling and extraction procedures are discussed in Elias (1994). In lacustrine sediments that are only 3–5% organic detritus by volume, this requires the extraction of many kilograms of sediment, followed by wet screening to isolate the organic fraction.

Generally, the quantity of sediment obtainable by piston coring of lake and pond sediments is insufficient to yield adequate amounts of detritus. Although some success has been achieved by taking multiple, large-diameter cores, this is not often practical. The best results are obtained by sampling exposures of sediments, either natural or man-made. Natural exposures include cutbanks of streams and ocean and lake bluffs. Man-made exposures include gravel and clay pits, irrigation ditches, trenches, and building sites.

The sequence of steps in extraction and mounting of insect fossils is summarized in **Figure 6**. Fossil insect extraction is relatively safe, cheap, and easy. The only lengthy process that may be involved is the pretreatment of samples to disaggregate the organic detritus from inorganic matrix (i.e., calcareous sediments and clays) or to soften and then disperse felted peats or lignite. Processing procedures are largely a matter of personal preference and availability of equipment, and some workers use a slightly different procedure than the one discussed later. For instance, Russian workers sometimes prefer to skip the kerosene flotation step discussed later and sort through entire samples of washed organic detritus in order to maximize the recovery of such heavy-bodied groups as weevils (Curculionidae) that sometimes do not rise to the top during flotation. Japanese workers studying peat deposits prefer to split the individual layers of peat and examine each layer under a microscope.

Once disaggregated organic detritus is obtained, the next step is to wet screen the sample in a 300 μm sieve. This process removes fine particles, such as silt, that may fill the concavities of rounded insect sclerites. Once the detritus has been screened, the residual material is placed, still damp, in a large bowl with a spout, or a rectangular dishpan, and processed by kerosene flotation to isolate and concentrate insect fossils. Kerosene or other lightweight oil is added to cover the sample and gently worked into the sample by hand for several minutes. The oil adheres to the insect sclerites but not to plant detritus. The remaining kerosene is decanted from the bowl and cold water is vigorously added to the oily detritus. In most samples, nearly all of the insect sclerites will rise to the top and float at the oil–water interface. Within 15–60 minutes, most plant residue sinks to the bottom of the bowl, and the now-concentrated insects may be decanted onto a screen to be washed gently in detergent and dehydrated in 95% ethanol before microscopic sorting. Small samples should be sorted completely rather than processed with kerosene. Also, samples that may need to be submitted for radiocarbon dating should not be exposed to kerosene.

Specimen sorting is done under low-power (10×) binocular microscope in alcohol or water. Specimens may be stored in vials of ethanol, or more robust specimens may be mounted with gum tragacanth (a water-soluble glue) onto micropaleontological slides. Identification of fossil specimens is usually done through comparison with modern identified specimens (often from museum collections). This is another potentially lengthy process.

Exoskeletal Parts Preserved in Quaternary Deposits

The chitinous remains of beetles often preserve extremely well in Quaternary sediments and can be matched exactly with modern specimens. The principal sclerites that are used in the identification of Quaternary beetles are the head, pronotum, and elytra (**Fig. 7**). The male genitalia (aedeagi) are likewise reinforced with chitin and are sometimes recovered from compressed groups of abdominal segments found in fossil assemblages. The aedeagus is often the most diagnostic sclerite in the identification of species, especially in genera that contain dozens or hundreds of species.

Useful Characters

The process of identification begins with the shape and size of the main exoskeletal parts. Often, colors derived from pigments are preserved, and color patterns can be quite useful in the identification of groups that have spots, stripes, and other noticeable color variations on the elytra or pronota. Metallic coloration preserves extremely well because it is based on surface microsculpture rather than pigments. The depth, shape, and distribution of such surface features as striations, punctures, and microsculpture are all potentially diagnostic characters, as are the presence and placement of scales and setae in some groups. As discussed previously, the size and shape of the aedeagus are two of the most diagnostic characters for beetles, although these are not as commonly found as heads, pronota, and elytra.

Basic Assumptions for Use in Paleoecology

There are several basic assumptions made in the reconstruction of paleoenvironments based on fossil beetle assemblages. These assumptions fall within the framework of uniformitarianism in the sense that we assume that ancient beetle populations had the same environmental tolerances and behavior patterns as their modern counterparts.

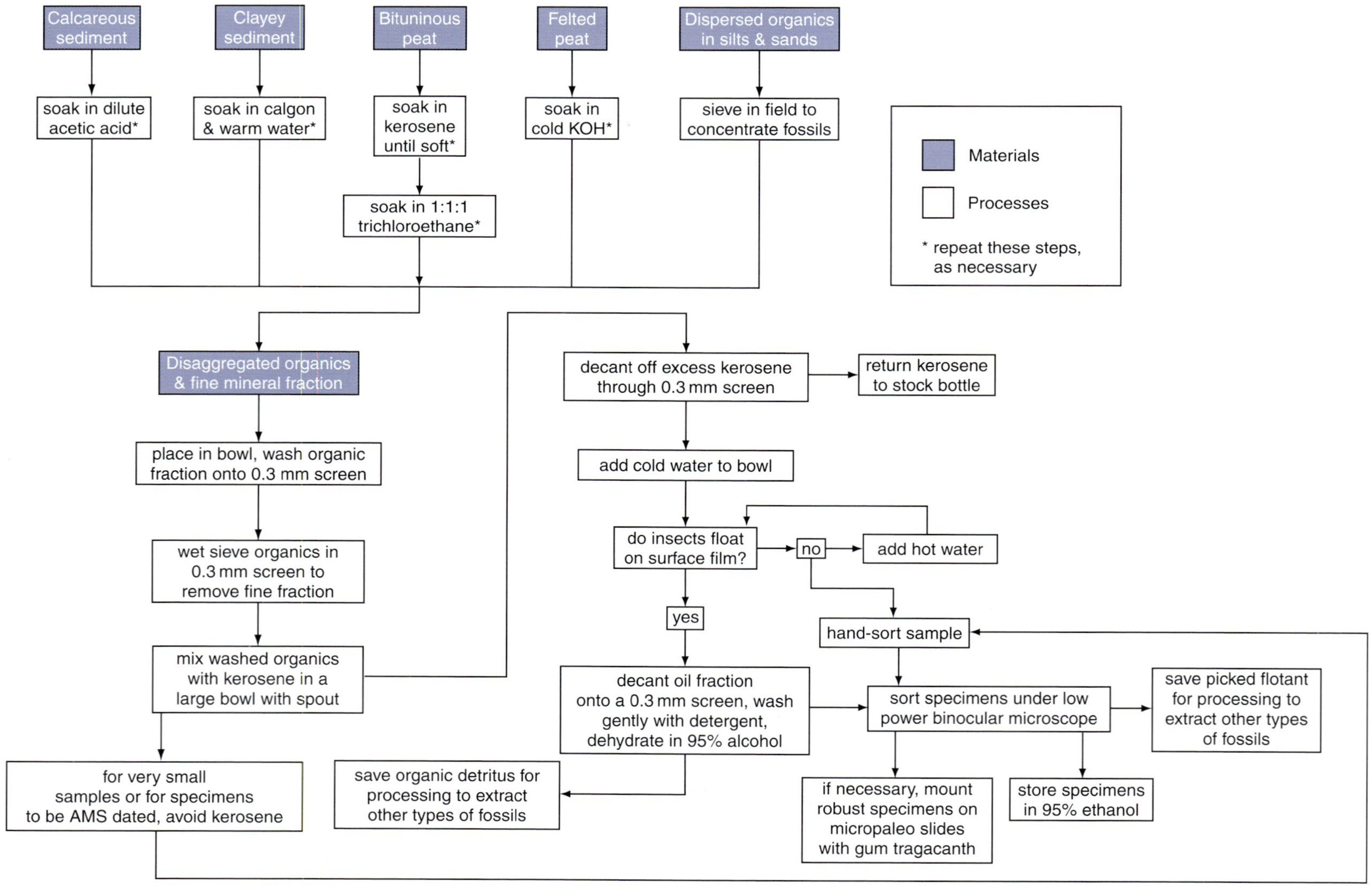

Figure 6 Flow chart of fossil beetle sample processing for kerosene flotation. Adapted from Elias (1994).

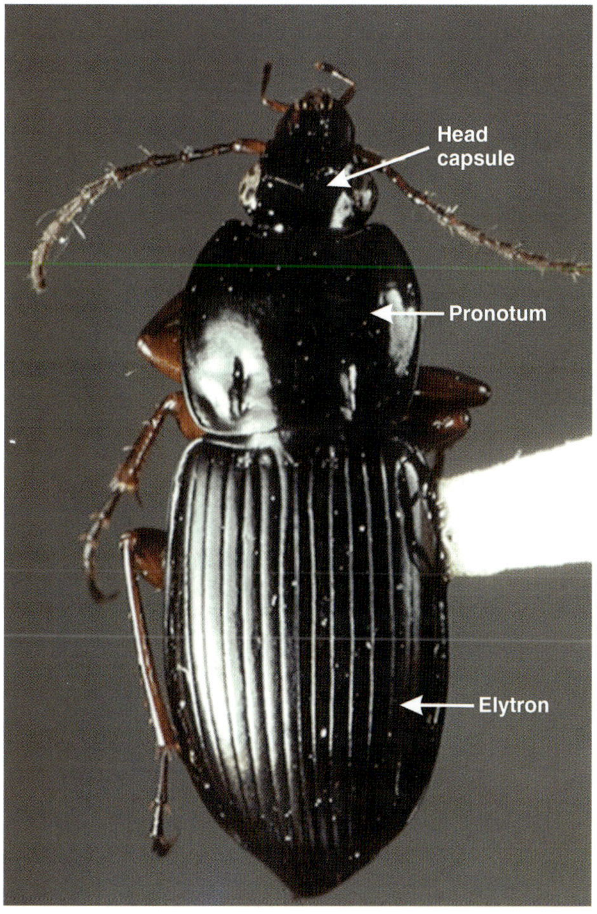

Figure 7 The ground beetle, *Pterostichus leconteianus*, with principal body parts found in fossil assemblages identified.

Species Constancy throughout the Quaternary

There are three lines of evidence that support the assumption that the beetle species found in Quaternary deposits are the same species that live today. The first is that the morphology of their exoskeletons has remained constant. In other words, we can find no differences between the fossil specimens and their modern counterparts that fall outside of the boundaries of known variation within the modern species. The youngest beetle fossils that exhibit morphological variations outside the bounds of modern species come from fossil assemblages that date to the Pliocene/Pleistocene boundary, approximately 2.6 million years ago. However, there are many reliably identified beetle species from Pliocene assemblages that exactly match modern species. The second line of evidence for Quaternary species constancy comes from the stability of shape and size of fossil aedeagae, as discussed previously.

The third line of evidence deals with the issue of constancy of physiology, as expressed in environmental tolerances. This issue can only be dealt with indirectly because the exoskeletal evidence does not speak to it. However, what has been repeatedly observed in fossil beetle assemblages from sites throughout the world and ranging in age from early to late Quaternary times is that the species found together in fossil assemblages remain ecologically compatible today, even if they no longer live in the same geographic regions. For instance, a fossil assemblage associated with Pleistocene periglacial environments may contain a mixture of species found today in arctic and alpine habitats in various areas of the Northern Hemisphere. Conversely, an interglacial fauna may be found to contain large numbers of warm-adapted beetle species that today are found in a variety of temperate or subtropical habitats in various regions. In order for this persistence of ecological compatibility to occur, one of two scenarios must be invoked. Either whole suites of species have evolved unidirectionally during the Quaternary so that their past environmental tolerances have changed together, or their environmental tolerances have not changed at all. The latter scenario seems by far the likeliest because evolutionary theory indicates that each species evolves separately from others in response to its own unique set of selective pressures (Elias, 1994).

Usefulness of Beetles as Paleoclimate Proxies

Insect ecologists support the idea that insect abundance and diversity may be controlled by biotic factors (e.g., predators and competition) in the central part of a species' range, but that climatic factors probably limit populations toward the edges of these ranges. Oxygen isotope records from deep-sea sediments and polar ice cores have shown that large-scale, climatic change was practically a constant feature of Pleistocene environments; it would appear that the shifting of insect distributions in response to these changes would have placed a large proportion of populations in this 'abiotic factor' zone of their respective ranges during much of the Pleistocene.

Thermal Tolerances of Beetles

The modern distribution patterns of many beetle species in the temperate and high latitudes have been shown to coincide closely with climatic zones. The distribution of individual species reflects temperature regime, especially summer warmth and degree of seasonality (i.e., annual temperature range). Stenothermic species with limited thermal tolerances are found in most fossil assemblages (**Fig. 1**). The physiology of stenothermic species is closely tied to their thermal requirements. Experiments on metabolic rates of ground beetles (Carabidae) (Thiele, 1977) showed that beetles' optimal metabolisms correspond with their thermal tolerance zones. This phenomenon is strengthened during the process of cold hardening, which takes place in late summer or fall. In preparation for winter cold, ground beetles in northern regions become progressively cold adapted (able to function at cold temperatures). Under these conditions, the beetles' metabolic rates peak at low temperatures, and they may become paralyzed or die if exposed to the temperatures they experience in mid-summer.

Mutual Climatic Range Analysis of Fossil Assemblages

The mutual climatic range (MCR) method of paleoclimate reconstruction is based on the concept that the proxy organisms (in this case, beetles) live within the bounds of a set of climatic parameters, as defined by the climatic conditions within their geographic range. Thus, beetles live in a quantifiable piece of climate space. In the MCR method, the first step is to develop a climate envelope for the species of predators and scavengers that are found in the fossil assemblage (**Fig. 8**). The complete method is described by Atkinson *et al.* (1986) for European studies and by Elias *et al.* (1996) for North American studies. These envelopes are typically defined on the basis of the mean July and mean January temperatures of the locations where the species are known to occur today. The climatic parameters of each of these localities are plotted on diagrams of mean July temperature (T_{max}) vs the difference between the mean July and mean January temperature (T_{range}). Once the individual species' climate envelopes are created, the envelopes for the species found together in a fossil assemblage are put together to find the area of overlap in climate space. This area is the mutual climatic range of the faunal assemblage. In the example shown in **Figure 9**, the climate envelopes of four cold-adapted species are combined to produce the MCR, shown in bright yellow.

European and North American MCR studies have focused on species that are predators or scavengers. Plant-feeding species have been avoided because of the possibility that the migration rate of the host plants of these beetles may lag behind climatic change. For instance, in recently deglaciated landscapes, the process of ecological succession from bare mineral soil to mature forest stands may take many centuries to achieve. Predators and scavengers are able to exploit such landscapes as part of the earliest pioneering biological communities, as soon as the climate is suitable (Elias, 1994).

Modern tests of the reliability of the MCR method have been performed for the European and North American beetle faunas. These tests compared predicted with observed temperatures for a broad spectrum of sites using the MCR of species found today at localities to predict modern temperatures of those localities and comparing these predictions with modern climatological data for the same sites. For 35 sites in North America, a linear regression of actual vs predicted T_{max} values yielded an r^2 value of 0.94. A regression of observed vs predicted T_{min} values yielded an r^2 value of 0.82. The standard deviation for T_{max} was 0.7° C and for T_{min} was 10° C. Similarly for Europe, a linear regression of actual vs predicted T_{max} values yielded an r^2 value of 0.88, and a regression of observed vs predicted T_{min} values yielded an r^2 value of 0.88. The standard deviation for T_{max} was 0.83° C, and for T_{min} it was 2.42° C. Beetles living in regions with cold winters are poor indicators of winter air temperatures because they seek shelter from exposure to winter air through various means (e.g., burrowing in mud, leaf litter, or under bark or stones).

The MCR method has allowed paleoentomologists to produce quantified paleotemperature estimates for both summer and winter seasons for hundreds of fossil assemblages ranging in age from late Tertiary (Elias and Matthews, 2002) to the late Holocene (Lavoie and Arseneault, 2001). Regional summaries of MCR reconstructions have been prepared for northwest Europe (Coope, 1987; Coope *et al.*, 1998), eastern and central North America (Elias *et al.*, 1996), the Rocky Mountain region (Elias, 1996), and Alaska and the Yukon Territory of Canada (Elias, 2001). The principal findings in both Europe and North America are that late Pleistocene climatic changes were rapid and often of large scale. Interglacial climates were at least 3° C warmer than today in some regions, but during the Last Glacial Maximum (LGM), T_{max} was depressed by 8–10° C in most temperate regions. The LGM summer temperatures of the Arctic regions were only depressed by 3–5° C in Beringia. T_{min} was depressed by as much as 25° C in some mid-latitude regions, but T_{min} was within 1 or 2° C of modern levels in most regions of eastern Beringia.

In New Zealand, collections of the modern beetle fauna are not sufficiently complete to allow species' climate envelopes to be constructed. Marra and colleagues have developed an alternative method, called the maximum likelihood envelope, that has allowed them to successfully reconstruct late Pleistocene climates. Another regional variation on the MCR method has been developed by Russian scientists working on northeastern Siberian fossil

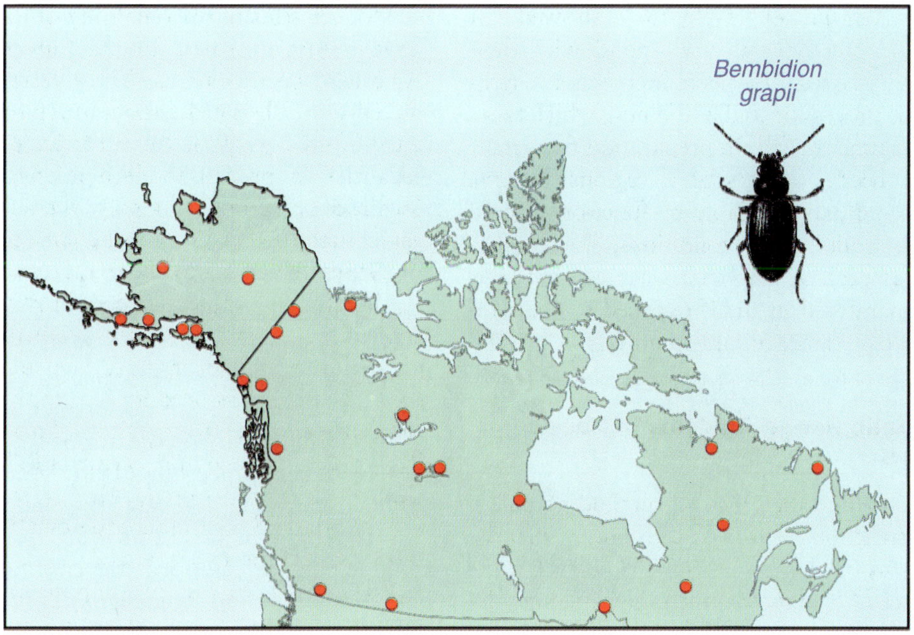

(A)

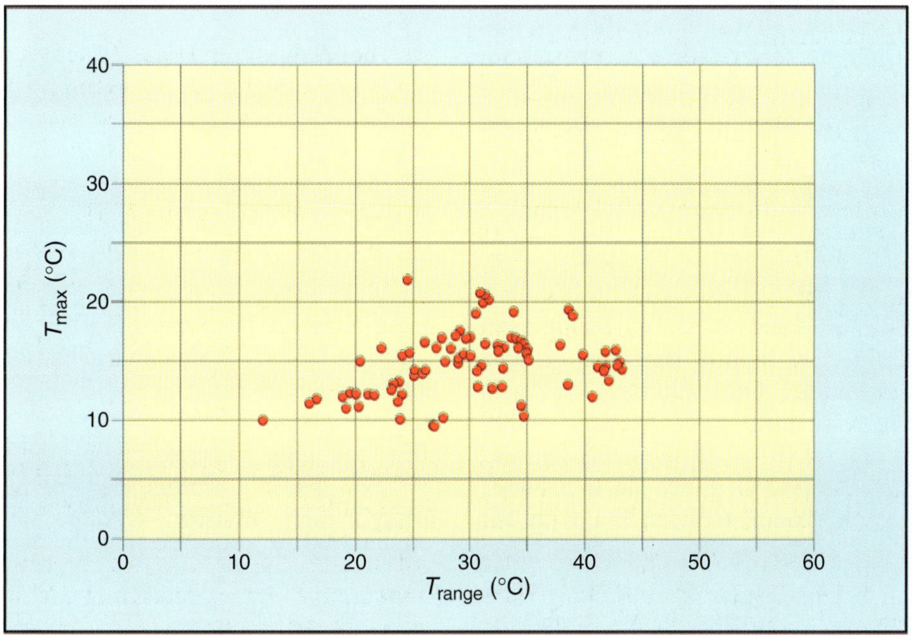

(B)

Figure 8 (A) Modern North American distribution of the ground beetle, *Bembidion grapii*. (B) Species climate envelope for *B. grapii*, based on climatic parameters (T_{max} and T_{range}) of collecting localities shown in A.

beetle assemblages. They argue that certain plant-feeding beetle species should be included in MCR reconstructions because they feed on plants such as sedges and mosses that remained ubiquitous in Siberia throughout the Pleistocene. Hence, there should be no problem with plant migration lag in these circumstances.

Multifaceted Paleoenvironmental Reconstructions

The order Coleoptera is the most diverse group of organisms on Earth, with a known modern fauna exceeding 1 million species. This number is roughly four times that of all the flowering plant species

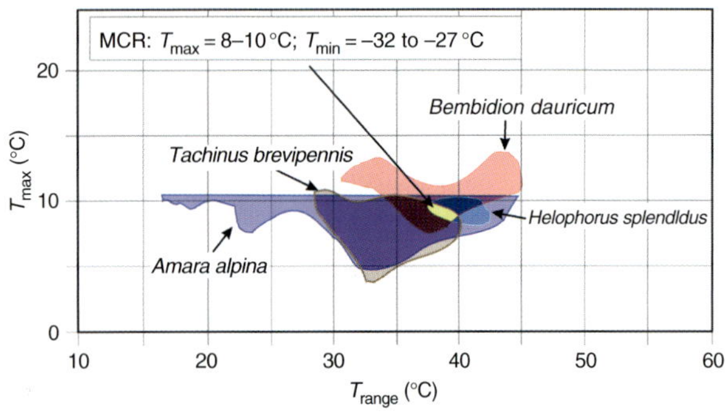

Figure 9 An example of the application of the MCR method showing the overlap of the climate envelopes of four species of cold-adapted beetles. In nearly all cases, far more species are involved in MCR reconstructions. The number is limited here for the sake of visual clarity.

(angiosperms) and more than 200 times the number of all mammal species. This incredible diversity carries through to Quaternary beetle faunal assemblages, which often contain between 100 and 200 species from dozens of families. Although predators and scavengers provide reliable information on past climates, plant-feeding species serve as proxies for the reconstruction of past plant communities and can provide such information as the relative maturity and health of forest stands and the presence of downed, rotting trees and associated fungi. In some cases, fossil beetle evidence may provide 'smoking gun' evidence for an environmental change. Such was the case when fossil remains of the elm bark beetle, *Scolytus scolytus*, were found in British sediments that date to the time of the Elm Decline in the mid-Holocene. The decline may have been due to Dutch elm disease, which is transmitted by the beetle (Girling, 1988).

Lacustrine aquatic beetles provide useful information on the temperature, trophic status, substrates, and size of lakes and ponds. Stream-dwelling beetles provide information on the size and velocity of streams, as well as temperature, trophic status, substrates, and degree of water clarity.

As in other macrofossil studies, the 'story' obtained from fossil beetle assemblages is a local one and needs substantiating with numerous regional replicates to produce a valid paleoenvironmental synthesis. However, because of the ubiquity of fossil beetle preservation in water-lain deposits, such regional syntheses are relatively easy (if time-consuming) to perform.

Other Types of Studies

In addition to straightforward paleoenvironmental reconstructions, Quaternary beetle assemblages have also been used to shed light on various aspects of biological and environmental history. Beetle species longevity, combined with our ability to identify the species of many fossil specimens, allows us greater understanding of how these animals have dealt with the vicissitudes of Pleistocene environmental change.

Zoogeography

One application of fossil beetle data has been for zoogeographical studies. By plotting the shifts of beetle distributions through time and space, researchers have been able to reconstruct at least partial zoogeographic histories of various species. These studies have reinforced the theory that beetles have shifted their ranges repeatedly, and sometimes on very large spatial scales, in order to find regions of suitable climate. Such was the case for the tiny rove beetle, *Micropeplus hoogendorni*. This species was initially thought to have become extinct in the Pliocene or early Pleistocene because it was only known from a 5.7 million-year-old deposit in Alaska. However, Coope found mid-Pleistocene specimens of this beetle in Britain, and it has been found to be synonymous with a Siberian species, *M. dokuchaevi* (Elias, 1994).

The history of comings and goings of various groups of beetles across the Bering Land Bridge have been discussed by Elias *et al.* (2000), and the late Pleistocene and Holocene zoogeographic history of Chihuahuan Desert beetles was analyzed by Elias (1992). The fossil evidence used to develop these histories frequently disproves zoogeographic theories based solely on modern distributions (Elias, 1994). Fossil beetle evidence documents the importance of glacial refugia in maintaining high-latitude faunas through glacial periods. Beringia was the most important refugium of the cold-adapted fauna in the Arctic. At the other end of the world, the beetle evidence suggests that ice-free regions of the Chilean coast formed a refuge for the cold-adapted fauna of South America.

Environmental Archeology

There are two types of fossil beetle studies done in aid of environmental archeology. The first is the reconstruction of natural environments associated with archeological sites. This kind of work has mainly been done in North America, where prehistoric human populations were mainly hunter–gatherers who had relatively light impacts on the landscape. With a few notable exceptions, these peoples built no permanent buildings, so they did not create any anthropogenic environments that attract warmth-loving insects incapable of surviving winter climate outdoors. This kind of research has also been applied to Paleolithic sites in Britain, where fossil beetle assemblages have shed light on mid- and late Pleistocene environments that were experienced by *Homo erectus* and *Homo neanderthalensis*.

The other type of environmental archeological studies done with beetle (and other arthropod) remains concerns the reconstruction of anthropogenic environments—the houses, barns, and other man-made buildings where warmth-loving insects live as uninvited guests. This type of research is most common in Europe, where large-scale human disturbance of landscapes began by at least the beginning of the Bronze Age. In fact, one of the difficulties of mid- and late Holocene beetle studies in Europe is the inability to tease apart natural environmental change from anthropogenic change.

See also: **Beetle Records**: Late Tertiary and Early Quaternary; Middle Pleistocene of Europe; Late Pleistocene of Japan; Late Pleistocene of South America; Late Pleistocene of Northern Asia.

References

Atkinson, T. C., Briffa, K. R., Coope, G. R., Joachim, M., and Perry, D. (1986). Climatic calibration of coleopteran data. In *Handbook of Holocene Palaeoecology and Palaeohydrology* (B. E. Berglund, Ed.)Wiley, New York.

Coope, G. R. (1977). Fossil Coleopteran assemblages as sensitive indicators of climatic changes during the Devensian (Last) cold stage. *Philosophical Transactions of the Royal Society of London, Series B* **280**, 313–340.

Coope, G. R. (1978). Constancy of insect species versus inconstancy of Quaternary environments. In *Diversity of Insect Faunas Symposium No. 9* (L. A. Mound and N. Waloff, Eds.), pp. 176–187. Royal Entomological Society of London, London.

Coope, G. R. (1987). Fossil beetle assemblages as evidence for sudden and intense climatic change in the British Isles during the last 45,000 years. In *Abrupt Climate Change* (W. H. Berger and L. D. Labeyrie, Eds.), pp. 147–150. Reidel, Dordrecht, The Netherlands.

Coope, G. R., Lemdahl, G., Lowe, J. J., and Walkling, A. (1998). Temperature gradients in northern Europe during the last glacial–Holocene transition (14-9 ^{14}C kyr BP) interpreted from coleopteran assemblages. *Journal of Quaternary Science* **13**, 419–433.

Elias, S. A. (1990). Observations on the taphonomy of late Quaternary insect fossil remains in packrat middens of the Chihuahuan Desert. *Palaios* **5**, 356–363.

Elias, S. A. (1992). Late Quaternary zoogeography of the Chihuahuan Desert insect fauna, based on fossil records from packrat middens. *Journal of Biogeography* **19**, 285–297.

Elias, S. A. (1994). *Quaternary Insects and Their Environments.* Smithsonian Institution Press, Washington, DC.

Elias, S. A. (1996). Late Pinedale and Holocene seasonal temperatures reconstructed from fossil beetle assemblages in the Rocky Mountains. *Quaternary Research* **46**, 311–318.

Elias, S. A. (2001). Mutual climatic range reconstructions of seasonal temperatures based on late Pleistocene fossil beetle assemblages in eastern Beringia. *Quaternary Science Reviews* **20**, 77–91.

Elias, S. A., and Matthews, J. V., Jr. (2002). Arctic North American seasonal temperatures in the Pliocene and early Pleistocene, based on mutual climatic range analysis of fossil beetle assemblages. *Canadian Journal of Earth Sciences* **39**, 911–920.

Elias, S. A., Anderson, K. H., and Andrews, J. T. (1996). Late Wisconsin climate in northeastern USA and southeastern Canada, reconstructed from fossil beetle assemblages. *Journal of Quaternary Science* **11**, 417–421.

Elias, S. A., Berman, D., and Alfimov, A. (2000). Late Pleistocene beetle faunas of Beringia: Where east met west. *Journal of Biogeography* **27**, 1349–1363.

Girling, M. A. (1988). The bark beetle *Scolytus scolytus* (Fabricius) and the possible role of elm disease in the early Neolithic. In Archaeology and the Flora of the British Isles, (M. Jones, Ed.) Oxford University Committee for Archaeology Monograph Vol. 14, pp. 34–38. Oxford University Press, Oxford.

Lavoie, C., and Arseneault, D. (2001). Late Holocene climate of the James Bay area, Québec, Canada, reconstructed using fossil beetles. *Arctic, Antarctic, and Alpine Research* **33**, 13–18.

Thiele, H. U. (1977). *Carabid Beetles in Their Environments.* Springer-Verlag, New York.

Late Tertiary and Early Quaternary

S A Elias, Royal Holloway, University of London, Surrey, UK

© 2007 Elsevier B.V. All rights reserved.

During the past 30 years, Quaternary insect paleontologists working in the northern high latitudes have had the opportunity to study extremely rare deposits of fossil insects that date back millions of years. In some cases, these fossil assemblages represent late Tertiary environments that preceded the earliest glaciations of the Quaternary. Other assemblages represent early Quaternary environments, when glacial–interglacial cycles were beginning. These fossil assemblages afford us rare glimpses into the history of the Arctic and sub-Arctic regions during periods when regional environments were startlingly different from today. In many cases, there are no modern analogs for the fossil assemblages. The challenge for insect paleontologists has been to make ecological sense of these fossil assemblages to reconstruct the environments in which they lived. It has been difficult, painstaking work that in some cases required decades of research to bring to completion.

These fossil insect faunas bridge a paleontological gap between the actual exoskeletal remains preserved in the unconsolidated sediments of the Pleistocene and the mineral replacements and trace fossils preserved in deposits of bedrock dating from the Tertiary back to the Paleozoic. What makes these late Tertiary and early Quaternary fossils unique is that they are millions of years old yet they are essentially the same kind of exoskeletal remains found in much younger (Pleistocene and Holocene) sediments. In other words, these fossils are not mineralized replacements of insect exoskeletons but, rather, the actual exoskeletons preserved in frozen deposits since their death. When removed from their frozen sedimentary matrix, these fossils have the same freshness of appearance as much younger specimens from other areas of the world. To the insect paleontologist studying these specimens under the microscope, it appears as if these insect specimens have spent up to several million years in a state of suspended animation. Together with frozen plant remains and vertebrate animal bones from these deposits, the Arctic insect fossils preserved from late Tertiary and early Quaternary deposits provide a treasure trove of information on the deep history of the Arctic regions.

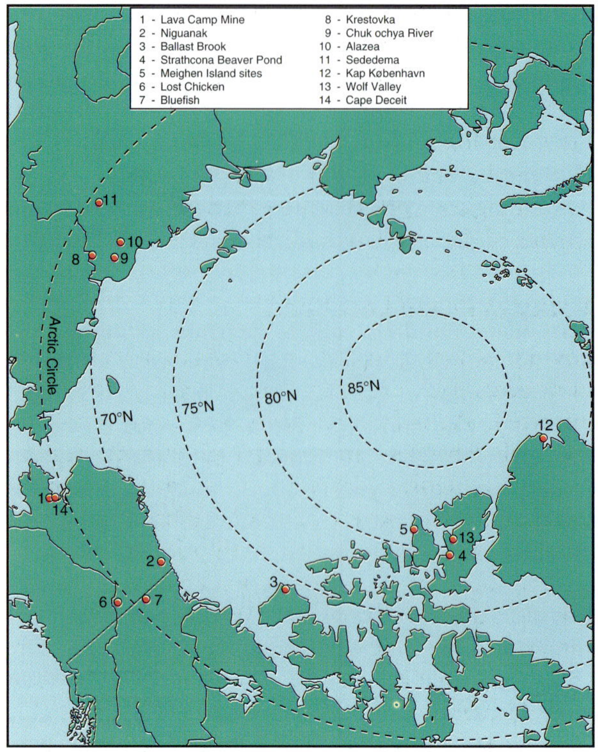

Figure 1 Map of the Arctic showing fossil sites discussed in text.

History of Late Tertiary/Early Quaternary Insect Fossil Studies

The existence of these fossil deposits was unknown until quite recently. Paleontological investigation of remote Arctic and sub-Arctic regions only occurred in the latter half of the 20th century. These studies require lengthy field expeditions that are logistically difficult and expensive. The possibility of fieldwork in such regions is generally limited to a brief summer season. The research has generally been carried out under the auspices of governmental organizations, such as the geological surveys of Canada and Denmark and the academy of sciences of the former Soviet Union.

One of the most difficult aspects of this research has been establishing the chronologies of the fossil assemblages. Although radiometric dating has been used to establish the age of a few samples (notably the Lava Camp assemblages from Alaska), most of the assemblages discussed here have only been dated on the basis of site stratigraphy (**Fig. 1**). These assemblages necessarily have only approximate age estimates. The ages discussed here represent the most up-to-date interpretations of the site investigators, but these chronologies are very likely to change in the future as more geologic research is done in the study regions.

Preservation of Late Tertiary/Early Quaternary Insect Fossils

Exoskeletal remains of insects from the late Tertiary and early Quaternary have been found almost exclusively in Arctic and sub-Arctic regions. Chitin, the main structural constituent of insect exoskeletons, is highly resistant to decay, but eventually it decomposes, especially when exposed to air. The long-term preservation of chitinous exoskeletons in the high latitudes is greatly facilitated by permanently frozen ground or permafrost.

Importance of Permafrost

Permafrost is key to the preservation of insect exoskeletons in the Arctic and sub-Arctic for several reasons. First, it greatly retards the pace of chemical reactions and bacterial decomposition, allowing chitin to remain virtually unaltered for as much as several million years. Second, permafrost protects insect exoskeletal remains from the physical abrasion that often takes place as sediments are reworked by wind, water, and gravity. Layers of frozen sediments tend to remain intact or are moved only as large blocks in response to thermokarst action (the melting of frozen sediments, leading to erosion and slumping). Thus, the frozen sediments of the North have entombed insect exoskeletons and have protected them from the physical and chemical degradation that breaks down chitin in other types of environments.

One topic currently being explored is the recovery of frozen DNA samples from insect fossils preserved in frozen sediments. It has been demonstrated that frozen sediments in Siberia, Alaska, and Canada can yield well-preserved fossil DNA from plants and mammals (Shapiro *et al.*, 2004; Willerslev *et al.*, 2003). It would appear likely that the frozen soft tissues of fossil insects will also yield ancient DNA. If it can be obtained, this DNA will be able to help entomologists work out the genetic history of insect species of the past few million years.

Rarity of Fossil Localities

Fossil sites containing insect remains more than 1 million years old are extremely rare, as might be expected. Permafrost environments only occur in the high latitudes, and most high-latitude regions were repeatedly glaciated in the Quaternary. Glaciers and ice sheets have obliterated underlying (late Tertiary and Quaternary) deposits, stripping the land surface down to bedrock in many arctic regions, leaving a mantle of reworked debris as their margins retreated. Thus, one of the main reasons why unconsolidated, organic-rich deposits of late Tertiary and early Quaternary age are rare is that they could only ever occur in regions that were not subject to repeated Quaternary glaciations. However, there were some regions in the Arctic that remained unglaciated through much if not all of the Pleistocene. Chief among these was the region known as Beringia, which included the unglaciated lowlands of northeastern Russia, Alaska, and the Yukon Territory, linked together by the Bering Land Bridge. This land bridge formed during glaciations as the growth of continental ice sheets trapped huge quantities of water, causing the global sea level to drop. The continental shelf regions of the Bering and Chukchi seas are relatively shallow (most of the shelf regions are less than 90 m below modern sea level), so these shelves became dry land during glaciations, connecting northeastern Siberia with western Alaska.

Types of Insects Preserved in Frozen Sediments

By far the most abundant and diverse group of insects preserved in these deposits are beetles (Coleoptera). The beetle families most often found in these ancient assemblages are ground beetles (Carabidae), water beetles (Dytiscidae and Hydrophilidae), rove beetles (Staphylinidae), leaf beetles (Chrysomelidae), and weevils (Curculionidae). Various kinds of flies (Diptera) have been found, especially non-biting midges (Chironomidae). Wasps and ants (Hymenoptera) are also commonly found. Finally, several families of oribatid mites (Arachnida: Oribatida) are common in these fossil assemblages. Oribatids are soil-dwelling mites that are the chief agents of organic decomposition in Arctic soils.

The fossil insect faunas of late Tertiary and early Quaternary age from North America and Greenland comprise 437 taxa of insects and arachnids. Of these, 180 have been identified to the species level (Matthews and Telka, 1997). Only beetle faunas (not other insects or arachnids) of this age have been published from northeastern Siberia by Kiselev (1981) and Kuzmina (1989). These include 139 beetle taxa, of which 114 species have been identified. Surprisingly, there are only 10 species in common between the Siberian and North American/ Greenland faunal lists for these ages of fossil assemblages.

Research in Alaska and Canada

John Matthews, Geological Survey of Canada, has studied a number of fossil insect assemblages of late Tertiary and early Quaternary age in Alaska and northern Canada, and his work has undoubtedly made the greatest contribution to our knowledge of insect faunas of this age. His work on this topic spans 25 years and 19 publications. Nine faunal assemblages dating from approximately 5.7 to 1.8 million years ago (Mya) have thus far been analyzed (**Table 1**). The sites range from just south of the Arctic Circle at the Lost Chicken site to north of 80° latitude at the Meighen Island and Wolf Valley sites.

The series of fossil insect assemblages (**Fig. 2**) from these sites document very warm, generally maritime late Tertiary climates. The insect faunas of this age are quite mixed in terms of their modern ranges. Some of these species live today only in Asia; others live today in the temperate regions of eastern North America. As might be expected for such warm-adapted insect faunas, there is considerable taxonomic diversity in these assemblages. The level of diversity declined during the onset of climatic cooling associated with Tertiary–Quaternary transition (Matthews and Telka, 1997). The late Tertiary faunas of the Arctic regions of North America lived in coniferous forests that were likewise more diverse than the modern boreal forests of Canada and Alaska. The ancient forests had some taxa that are less tolerant of cold climate, such as hemlocks and elderberry.

By approximately 2.5–2 Mya, there is reliable evidence for climatic cooling from both the insect and the plant fossil records. Nevertheless, temperatures remained well above their modern levels in the high Arctic. Sites such as Wolf Valley, Ellesmere Island (2.1–1.7 Mya) and Cape Deceit, Alaska (1.8 Mya) contain mixtures of arctic tundra and coniferous

Table 1 Late Tertiary and Early Quaternary Insect Fossil Sites

Site	*Age*	*Insect Fauna*	*Environmental Reconstruction*	*References*
1. Lava Camp Mine, Alaska	5.7 ± 0.2 Mya (late Miocene)	83 insect and arachnid taxa, including several extinct species; fauna indicative of coniferous forest.	Warm, maritime climate; fossil flora includes species found today in the forests of the Pacific Northwest region.	Hopkins *et al.* (1971), White *et al.* (1997, 1999), Elias and Matthews (2002)
2. Niguanak, Alaska	Early Pliocene?	47 insect and arachnid taxa, including extinct species, *Diacheila matthewsi* Fauna not yet fully analyzed but has some Asian affinities.	Plant fossils indicate coniferous forest, including hemlocks; climate was far warmer and less continental than today.	Matthews and Telka (1997), Elias and Matthews (2002)
3. Ballast Brook, Banks Island, NWT, Canada	5–3 Mya	39 insect and arachnid taxa, indicative of boreal environment	Coniferous forest, similar to modern forests in the sub-Arctic regions of Canada. There are indications that the site was near northern tree line at the time of fossil deposition. Climate was substantially warmer than today.	Fyles *et al.* (1994), Elias and Matthews (2002)
4. Strathcona Beaver Peat, Ellesmere Island, NWT, Canada	> 3.3 Mya	86 insect and arachnid taxa, including both arctic and boreal elements; some species with Asian affinities	Fauna and flora indicative of northern tree line setting; climate substantially warmer than today	Matthews and Telka (1997), Elias and Matthews (2002)
5. Meighen Island sites, NWT, Canada	3 Mya	198 insect and arachnid taxa, including both boreal and arctic tundra species. Faunal affinities with eastern North America and Asia	Flora and fauna indicative of northern tree line environment; climate substantially warmer than today	Matthews and Telka (1997), Elias and Matthews (2002)
6. Lost Chicken, Alaska	3 Mya	83 insect and arachnid taxa, including blind weevil, *Otibazo*, found today in Japan; boreal fauna with several extinct species	Flora and fauna indicative of coniferous forest; several warmth-loving plant species; climate with warmer winters than today	Matthews and Telka (1997), Elias and Matthews (2002)
7. Bluefish, Yukon Territory	< 3 Mya (late Pliocene)	26 insect and arachnid taxa, including *Notiophilus aeneus*, now confined to eastern North America, and extinct species, *Helophorus meighensis*	Bluefish plant fossil assemblages include species indicative of substantially warmer, less continental climate than exists in the Yukon today; MCR results indicate T_{max} up to 3°C warmer than today	Matthews and Ovenden (1990), Matthews and Telka (1997), Elias and Matthews (2002)
8. Krestovka, NE Siberia, Russia	Late Pliocene (Kutuyakh Fm.)	22 beetle taxa; mixture of boreal and arctic fauna; some elements of steppe tundra fauna already present	Larch forest tundra landscape with steppe patches	Kiselev (1981)

	Late Pliocene to early Quaternary (Olyorian Suite)	51 beetle taxa; most found today in arctic tundra regions; some taxa with steppe affinities	Steppe–tundra landscape with patches of shrub tundra and coniferous woodlands	Kiselev (1981), Kuzmina (1989)
9. Chukochya River, NE Siberia, Russia	Late Pliocene to early Quaternary (Olyorian Suite)	80 beetle species; most found today in arctic tundra regions; some taxa with steppe affinities	Steppe–tundra landscape with patches of shrub tundra and coniferous woodlands	Kiselev (1981), Kuzmina (1989)
10. Alazea, NE Siberia, Russia	Late Pliocene to early Quaternary (Olyorian Suite)	56 beetle species; most found today in arctic tundra regions; some taxa with steppe affinities	Steppe–tundra landscape with patches of shrub tundra and coniferous woodlands	Kiselev (1981), Kuzmina (1989)
11. Sededema, NE Siberia, Russia	Late Pliocene to early Quaternary (Olyorian Suite)	55 beetle species; most found today in arctic tundra regions; some taxa with steppe affinities	Steppe–tundra landscape with patches of shrub tundra and coniferous woodlands	Kiselev (1981), Kuzmina (1989)
12. Kap København, Greenland	2.5–2 Mya (early Quaternary)	154 insect and arachnid taxa, indicative of northern tree line environments	Probably deposited after first glaciation; less faunal diversity than Meighen Island faunas, indicating climatic cooling; climates still substantially warmer than today. T_{max} about 8°C warmer than today; T_{min} about 10°C warmer than today	Böcher (1995), Elias and Matthews (2002)
13. Wolf Valley, Ellesmere Island, NWT, Canada	2–1.7 Mya	21 insect and arachnid taxa, indicative of northern tree line environments	Plant and insect fossils suggest northern tree line near site at time of deposition; climate cooler than that of Beaufort Formation faunas; T_{max} 5–7°C warmer than today	Matthews and Fyles (2000), Elias and Matthews (2002)
14. Cape Deceit, Alaska	1.8 Mya	86 insect and arachnid taxa, indicative of northern tree line environments; some elements of tundra fauna seen for first time in eastern Berngian region	The vegetation record suggests that temperatures were slightly warmer than they are today MCR results indicate T_{max} about 2°C warmer than today; T_{min} similar to modern winter temperatures.	Matthews (1974), McDougall (1995), Matthews and Telka (1997), Elias and Matthews (2002)

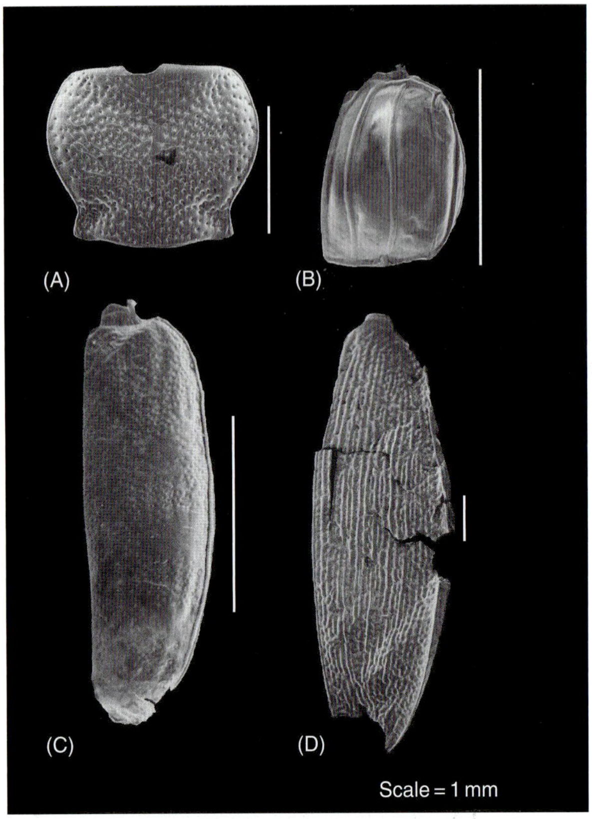

Figure 2 Scanning electron micrographs of fossil beetle specimens from Canada and Alaska. (A) Pronotum of *Diacheila matthewsi* from the Beaufort Formation, Meighen Island. (B) Elytron of extinct species of *Kalissus* from the Beaufort Formation, Meighen Island. (C) Elytron of extinct species of *Asaphidion* from the Lost Chicken site, Alaska. (D) Elytron of extinct species of *Carabus* from the Beaufort Formation, Meighen Island. Photos by Alice Telka and John Matthews, Geological Survey of Canada.

Figure 3 Fossil sites in northeastern Siberia. (A) The study site exposure on the Alazea River. (B) Exposure of Olyorian Suite Formation sediments at the Krestovka site. Photos by Andrei Sher.

forest insect fauna and flora. The Cape Deceit fauna includes some elements that typify the mesic tundra environments that existed throughout much of the late Pleistocene, including a wide variety of ground beetle species in the *Cryobius* group of the genus *Pterostichus*.

Research in Siberia

The Siberian faunal assemblages of late Tertiary and early Quaternary age have been described from four sites (**Fig. 1, Table 1**). The sites are along the banks of large rivers (**Fig. 3**). In contrast to the fossil record of the North American Arctic, the Kutuyakh Formation Beds of late Pliocene age in northeastern Siberia already have a well-developed cold-adapted insect fauna. These assemblages include species of *Pterostichus* (*Cryobius*) and *Blethisa catenaria* that live on mesic to wet tundra and forest–tundra, as well as woodland species, especially weevils associated with conifers (Kiselev, 1981). Faunal assemblages from stratigraphic horizons between the Kutuyakh and Olyorian Suite Formations contain mostly tundra species, including species of *Pterostichus* (*Cryobius*), as well as the more dry-adapted arctic ground beetle, *Amara alpina*. However, elements of the taiga fauna persist in these faunas, especially tree-feeding weevils. Some steppe-associated beetles made their first appearance in northern Siberian faunal assemblages of this age (late Pliocene). The combined faunal and floral evidence suggests a larch forest–tundra landscape with patches of steppe.

The Olyorian Suite is now thought to represent the Tertiary–Quaternary transition. Deposits of this age are very widespread in northeastern Siberia, especially in the Kolyma lowland region. The dominant feature of these beetle assemblages (**Figs. 4A–D**) is that they are virtually identical in composition to the subsequent regional faunas from middle and late Pleistocene times. The only difference recognized thus far (Kuzmina, 1989) is that the steppe component of Olyorian Suite faunas is less important

Figure 4 Light microscope photographs of fossil beetle specimens from Olyorian faunal assemblages at the Alazea site, northeastern Siberia (A–D) and the Kap København formation in Greenland (E–H): (A) pronotum of *Thanatophilus dispar*, (B) pronotum of *Upis ceramboides*, (C) elytron of *Carabus odoratus*, (D) elytron of *Chrysolina brunnicornis*, (E) elytron of *Elaphrus tuberculatus*, (F) elytron of *Scolytus piceae*, (G) elytron of *Grypus equiseti*, (H) pronotum of *Diacheila polita* from Kap København, and (I) pronotum of *Diacheila matthewsi*. Scale bars = 1 mm.

than it is in most of the middle and late Pleistocene insect assemblages. The typical Olyorian insect faunas consist mostly of tundra species, with dry tundra insects dominating over mesic and wet tundra taxa.

Research in Greenland

Only one Greenland site has been studied that contains insect fossil assemblages in the age range being considered here. Most of Greenland has been covered by ice during the many glaciations of the Quaternary, making the fossil site all the more remarkable. Kap København (**Fig. 5**) is the northernmost site studied for insect fossils of this age. It is situated near the northern tip of Greenland (**Fig. 1**, No. 12). The fossil insect assemblages (**Figs. 4E–H**) have been described by

Figure 5 Kap København. (Top) View of the modern polar desert landscape. (Bottom) View of the early Quaternary fossil beds. Photographs by Jens Böcher.

Jens Böcher (1995). The age of the deposits is estimated between 2.5 and 2 Mya, at approximately the boundary between the Pliocene and the Pleistocene. At that time, it is clear that northern high-latitude climates were remarkably warm. The Kap København flora and insect fauna are indicative of northern tree line environments. It appears that the Kap København fossil beds were deposited soon after the first glacial period, near the end of the Pliocene. The Meighen Island fauna from northern Canada is more diverse and contains more warm-adapted species than the Kap København fauna. The former fauna is thought to be 500,000–700,000 years older than the latter (Matthews and Telka, 1997).

Potential Importance of Research

The research outlined previously is important to a number of disciplines. It has helped define paleoclimate reconstructions for regions that formerly had little or no reconstructions for the critical climatic transition from the relatively warm climates of the Pliocene to the oscillating glacial and interglacial climates of the Quaternary. The wealth of paleoenvironmental data extracted from the fossil insect assemblages has also added greatly to our understanding of soils, moisture conditions, and vegetation cover of the Arctic and sub-Arctic regions during the late Pliocene and early Quaternary. The fossil insect evidence also provides important evidence concerning the longevity of many insect species, especially beetles. Finally, the combination of plant and insect fossil assemblages has given us important new insights into the evolution, functioning, and longevity of northern biological communities.

Insects as Paleoclimate Indicators

The fossil insect research described previously has demonstrated the morphological constancy of exoskeletal features in beetles and some other insects, over several million years for some species. This alone would seem to be a strong indicator that the species in question have remained constant throughout this time interval. However, insect exoskeletons provide little or no data on the internal physiology or ecological requirements of these organisms. In order to test these aspects of species' constancy, another, indirect method had to be employed. Fortunately, we can examine the ecological requirements of apparently extant species in fossil assemblages through the analysis of their modern counterparts. The modern distributions of species found in fossil assemblages fall within certain climatic parameters. When these parameters (e.g., mean temperature of the warmest or the coldest month of the year) are plotted for all the known modern collecting localities of a species, it is possible to generate a 'climate envelope' for that species (Atkinson *et al.*, 1986). These climate envelopes vary from broad to narrow, depending on the range of thermal tolerances of the species. Elias and Matthews (2002) analyzed the climatic compatibility of the late Tertiary and early Pleistocene fossil assemblages from Greenland, Canada, and Alaska, and in all cases the species in the fossil assemblages were found to be climatically compatible. Taking this process a step further, the climate envelopes of the species in the fossil assemblages overlapped to yield a mutual climatic range (MCR), a set of climatic parameters within which all the species in the assemblage can live.

Longevity of Insect Species

The MCR study indicates that the species in question have not evolved new sets of climatic tolerances since the late Tertiary. Dozens of species were used in the MCR reconstructions. If this kind of physiological evolution had taken place, then it would be extremely unlikely that the climate envelopes based on the modern distribution of these beetles would overlap to produce an MCR.

Given the fact that the modern representatives of the species in the fossil assemblages remain climatically compatible with each other, there has either been no evolution of their thermal tolerances since the late Tertiary or whole suites of species have evolved new tolerances in concert with each other. The latter scenario appears extremely unlikely because evolutionary theory holds that each species evolves independently from others.

However, as Matthews and Telka (1997) noted, the late Tertiary beetle faunas from the Arctic also contain 'clear examples of extinct species.' It appears likely that most of these species were relatively warm adapted, and that they failed to become established in more southerly regions during the climatic cooling that signaled the beginning of glaciations in the Quaternary. The fossil evidence demonstrates that the evolutionary 'clock' of beetles has not stopped; it is just moving extremely slowly. Some late Tertiary species apparently were precursors to modern species that evolved during the Quaternary. For instance, Matthews and Telka discussed the possible evolutionary history of a ground beetle in the genus *Asaphidion*, an extinct species closely related to the contemporary subarctic carabid *Asaphidion yukonense*. A fossil elytron from the Lost Chicken site (**Fig. 2C**) clearly shows bare (impunctate) patches that characterize the modern species, but the patches are less well developed and elytral microsculpture better developed than in the extant species. Fossils that are similar to *A. yukonense* have been found in other late Tertiary assemblages from the Arctic, but some have even more primitive elytral sculpture than the Lost Chicken specimens. These fossils could represent various stages in the evolution of the modern species. Matthews (1976) also described late Tertiary fossil specimens of the water scavenger beetle genus, *Helophorus*, that appear to be precursors to the modern species, *H. tuberculatus*. The fossils, from Lava Camp Mine and Meighen Island, exhibit more primitive elytral sculpture. Matthews traced the possible evolution of these fossil types into the extant species. Matthews (1970) also discussed the possible evolution of the more primitive rove beetles, *Micropeplus hoogendorni* and *M. hopkinsi*, leading to the modern species, *M. cribratus* and *M. punctatus*. Matthews assumed that the two former species had become extinct in the Quaternary, but *M. hoogendorni* has subsequently been found to be extant. Modern specimens were collected from Siberia by the

Russian entomologist, Rjabukhin. Unaware of the fossil evidence from Alaska, he described the Siberian specimens as a new species, *M. dokuchaevi* (Elias, 1994).

Insects as Paleoecological Indicators

Some of the environments reconstructed on the basis of these insect faunal assemblages have no modern analogs; in fact, most of them cannot be found today. This does not mean that the insect evidence is contradictory; it simply means that some of the ancient environments in which they lived have ceased to exist. This is especially true for the Arctic Pliocene environments. From 3–4 Mya, and up to parts of the early Quaternary at Kap København, coniferous forests were growing at very high latitudes. These regions are polar deserts today, with a minimum of tundra vegetation interspersed with areas of bare rock, sand, and gravel. Paleoclimatologists have yet to develop paleoclimate models that can satisfactorily explain the existence of very warm climates in polar regions during the transition from late Tertiary to Quaternary times, but the fossil data (plant and animal) clearly demonstrate its existence.

Origins and Longevity of Biological Communities

The fossil beetle evidence, in agreement with the fossil plant evidence from the same or similar deposits, demonstrates that biological communities are ephemeral associations of species. Certain functional types of communities may persist through great lengths of time (e.g., coniferous forests or arctic tundra), but the constituent species within these communities shift with great regularity in response to environmental changes. The late Tertiary and early Quaternary biota of the Arctic constituted boreal-style coniferous forests, but these were unlike any forests that exist today in any particular region.

One of the dominant trends in this sequence of Arctic faunas is the gradual depletion of the number of species through time. The earlier, warm-adapted faunas were relatively rich in species. The later, cold-adapted faunas contain fewer species. This makes sense from an ecological standpoint because few insects are adapted to life in extremely cold climates. Since they do not maintain their own internal temperature through metabolism, as do mammals, insects are at a distinct disadvantage in cold climates. The modern statistics clearly reveal this pattern. Danks (1981) listed just 1,800 named species of arthropods living today in the North American Arctic. The number of insect species living today in North America as a whole exceeds 90,000 species (Borror and White, 1970).

Mutual Climatic Range Reconstructions

As discussed previously, Elias and Matthews (2002) employed the predaceous and scavenging beetle species found in the late Tertiary and early Quaternary fossil assemblages to produce their MCR estimates (for a discussion of MCR methods, see Elias, 1994). These beetles are in various families, including ground beetles (Trachypachidae and Carabidae), predaceous diving beetles (Dytiscidae), water-scavenger beetles (Hydrophilidae), rove beetles (Staphylinidae), carrion beetles (Silphidae), and dung beetles (Scarabaeidae). Eleven fossil assemblages associated with this time interval contained sufficient numbers of identified species to yield MCR estimates. Elias and Matthews found that average winter temperatures (T_{min}) in the Arctic during the latest Miocene and Pliocene were substantially warmer than they are today. The MCR estimates agree with the evidence from paleobotanical data that Arctic Pliocene climates were far less continental than they are today. Several Pliocene-age assemblages from the high arctic yielded mean summer temperature (T_{max}) estimates 9 or 10° C warmer than modern values at the sites. This is the same degree of warming required to allow coniferous forests to grow in the high Arctic. By 3 Mya, a cooling trend is marked in both the paleobotanical and fossil beetle evidence from Alaska. All assemblages dating between 5.7 and 2 Mya yielded T_{max} values between approximately 12 and 14° C, regardless of location. Thus, the insect fossil data support the theory that there was far less latitudinal gradation in temperatures during the late Pliocene than there is today. These results agree well with those of a study of global environments during the mid-Pliocene (Dowsett *et al.*, 1999), which showed greatly reduced continental ice volume, greatly reduced sea ice with the Arctic being seasonally ice free, and an expansion of evergreen forests to the margins of the Arctic Ocean. The MCR reconstructions suggest that regional climatic cooling (especially winter temperatures) began by at least 2 Mya.

The fossil insect history of the Arctic provides fascinating glimpses into a world that ceased to exist approximately 2 Mya – a world in which a substantially warmer Arctic was home to coniferous forests, even at 80° N, near the northernmost point of land in the Northern Hemisphere. The ecosystems of this warm Arctic zone collapsed as the ice ages began in earnest, but, surprisingly, most of the insect fauna did not die out. The many species went their separate ways, some into eastern North America, some to unglaciated regions of Beringia, and others to other areas of Asia, such as Japan. Even blind, flightless beetle species managed to migrate into new regions. Each species has its own unique history, its own remarkable story of survival in the face of shifting climates. Before the fossil data were published, few biologists would have guessed that 90% of the Arctic fauna of the Pliocene has remained extant to the present day. Thus, the trend of beetles shifting their ranges in response to climate change, first noted for Pleistocene fossil assemblages (Coope, 1978), began far earlier than we might have expected.

See also: **Beetle Records**: Overview; Late Pleistocene of Northern Asia. **Glaciations**: Late Pleistocene Events in Beringia. **Paleobotany**: Ancient Plant DNA. **Vertebrate Studies**: Ancient DNA.

References

Atkinson, T. C., Briffa, K. R., Coope, G. R., Joachim, M., and Perry, D. (1986). Climatic calibration of coleopteran data. In *Handbook of Holocene Palaeoecology and Palaeohydrology* (B. E. Berglund, Ed.), pp. 851–858. Wiley, New York.

Böcher, J. (1995). Paleoentomology of the Kap København Formation, a Plio-Pleistocene sequence in Peary Land, North Greenland. *Medd. Grønland, Geoscience* 33, 1–82.

Borror, D. J., and White, R. E. (1970). *A Field Guide to the Insects of America North of Mexico*. Houghton Mifflin, Boston.

Coope, G. R. (1978). Constancy of insect species versus inconstancy of Quaternary environments. In *Diversity of Insect Faunas, Symposium 9* (L. A. Mound, and N. Waloff, Eds.), pp. 176–187. Royal Entomological Society of London, London.

Danks, H. V. (Ed.) (1981). *Arctic Arthropods*. Entomological Society of Canada, Ottawa, Ontario, Canada.

Dowsett, H. J., Barron, J. A., Poore, R. Z., *et al.* (1999). Middle Pliocene Paleoenvironmental Reconstruction: PRISM2. U.S. Geological Survey Open File Report, 99-535.

Elias, S. A. (1994). *Quaternary Insects and Their Environments*. Smithsonian Institution Press, Washington, DC.

Elias, S. A., and Matthews, J. V., Jr. (2002). Arctic North American seasonal temperatures in the Pliocene and early Pleistocene, based on mutual climatic range analysis of fossil beetle assemblages. *Canadian Journal of Earth Science* **39**, 911–920.

Fyles, J. G., Hills, L. V., Matthews, J. V., Jr. *et al.* (1994). Ballast Brook and Beaufort Formations (late Tertiary) on Northern Banks Island, Arctic Canada. *Quaternary International* **22/23**, 141–172.

Hopkins, D. M., Matthews, J. V., Wolfe, J. A., and Silberman, M. L. (1971). A Pliocene flora and insect fauna from the Bering Strait region. *Palaeogeography, Palaeoclimatology, Palaeoecology* **9**, 211–231.

Kiselev, S. V. (1981). *Late Cenozoic Coleoptera of Northeastern Siberia*. U.S.S.R. Academy of Sciences, Paleontological Institute, Izdatel'stvo, Moscow, Nauka Press. [In Russian]

Kuzmina, S. (1989). Late Pleistocene insects from the Alazea River (Kolyma lowland). *Bulletin MOIP, Geological Series* **64**, 42–55. [In Russian]

Matthews, J. V., Jr. (1970). Two new species of *Micropeplus* from the Pliocene of western Alaska, with remarks on the evolution of Micropeplinae (Coleoptera: Staphylinidae). *Canadian Journal of Zoology* **48**, 779–788.

Matthews, J. V., Jr. (1974). Quaternary environments at Cape Deceit (Seward Peninsula, Alaska): Evolution of a tundra ecosystem. *Geological Society of America Bulletin* **85**, 1353–1384.

Matthews, J. V., Jr. (1976). Evolution of the subgenus *Cyphelophorus* (genus *Helophorus*, Hydrophilidae): Description of two new fossil species and discussion of *Helophorus tuberculatus* Gyll. *Canadian Journal of Zoology* **54**, 653–673.

Matthews, J. V., Jr., and Fyles, J. G. (2000). Late Tertiary plant and arthropod fossils from the High Terrace Sediments on the Fosheim Peninsula of Ellesmere Island (Northwest Territories, District of Franklin). *Geological Survey of Canada Bulletin* **529**, 295–317.

Matthews, J. V., Jr., and Ovenden, L. E. (1990). Late Tertiary plant macrofossils from localities in arctic/subarctic North America: A review of the data. *Arctic* **43**, 364–392.

Matthews, J. V., Jr., and Telka, A. (1997). Insect fossils from the Yukon. In *Insects of the Yukon* (H. V. Danks, and J. A. Downes, Eds.), pp. 911–962. Biological Survey of Canada (Terrestrial Arthropods), Ottawa, Ontario, Canada.

McDougall, K. (1995). Age of the Fishcreekian Transgression. *Palaios* **10**, 199–220.

Shapiro, B., Drummond, A. J., Rambaut, A. *et al.* (2004). Rise and fall of the Beringian steppe bison. *Science* **306**, 1561–1565.

White, J. M., Ager, T. A., Adam, D. P. *et al.* (1997). An 18 million year record of vegetation and climate change in northwestern Canada and Alaska: Tectonic and global climatic correlates. *Palaeogeography, Palaeoclimatology, Palaeoecology* **130**, 293–306.

White, J. M., Ager, T. A., Adam, D. P. *et al.* (1999). Neogene and Quaternary quantitative palynostratigraphy and paleoclimatology from sections in Yukon and adjacent Northwest Territories and Alaska. *Geological Survey of Canada Bulletin* **543**, 1–30.

Willerslev, E., Hansen, A. J., Binladen, J. *et al.* (2003). Diverse plant and animal genetic records from Holocene and Pleistocene sediments. *Science* **300**, 791–794.

Middle Pleistocene of Europe

G R Coope, University of London, Egham, UK

© 2007 Elsevier B.V. All rights reserved.

Introduction

Middle Pleistocene insect assemblages have been described from organic deposits at many sites in southern and central England, but only a few insect faunas have been investigated in recent years from Western Europe. All these faunas are made up almost entirely of species that are still extant, even though their modern occurrences may be far from the places where they are found as fossils. On the assumption that this morphological stability is accompanied by a similar degree of physiological constancy, the occurrences of particular species in a fossil assemblage may be used to interpret past environments and climates. Altogether, several hundred beetle species have been recognized from these deposits, which have enabled a detailed sequence of environmental and climatic changes to be reconstructed.

In recent years the complexity of the Quaternary stratigraphy of Europe has been partially resolved by the application of a notional framework based on the Marine Isotope Stages (MIS). In these, global fluctuations in climate, oscillating between cold and warm periods, can be recognized. It has been possible to correlate, albeit provisionally, the climatic events recorded in terrestrial sequences to the various stages of the MIS sequence. In this article, the base of the middle Pleistocene is taken as the Brunhes/ Matuyama geomagnetic boundary at about MIS 19 and its upper limit as the base of MIS 5e, that is, the start of the Last (Ipswichian, Eemian, Wisconsinan) Interglacial. Wherever possible the beetle assemblages will be allocated to the various MIS events. This means that several of the earlier descriptions of beetle assemblages, which may be of middle Pleistocene age, will not be mentioned here because of the difficulty in relating these to modern Quaternary stratigraphy. However, most of these early papers have been included in Buckland and Coope (1991).

Middle Pleistocene Coleopteran assemblages

It is convenient to subdivide these middle Pleistocene Coleopteran faunas into two groups:

(1) those that date from before the extensive Anglian (Saalian) Glaciation; and
(2) those which post-date this major glaciation.

Pre-Anglian middle Pleistocene beetle localities

In the past, several organic deposits at the foot of the Norfolk cliffs have been grouped together as the 'The Cromer Forest Bed'. These are now known to represent a complex sequence of interglacials that cannot confidently be attributed to any particular MIS events. The deposits are highly fossiliferous and include abundant insect remains, clearly indicating different climates. All these deposits predate the Anglian (Saalian) Glaciation (MIS 12). The occurrence of numerous temporary exposures of organic deposits in active gravel workings elsewhere means that the understanding of their biostratigraphy is rapidly developing. A brief summary of the more important sites will be given here, but it must be borne in mind that others are currently under investigation and will be published in the near future.

Beeston The earliest Coleopteran assemblage from this period was obtained from deposits at Beeston, north Norfolk and can be correlated with the Pre-Pastonian. Because they rest directly on the chalk, it is difficult to give them a precise age. They are probably equivalent to MIS 18 or MIS 20 and may correlate to the Tiglian deposits of The Netherlands. They are included here because the insect fauna is the earliest Pleistocene assemblage from the British Isles (Coope, 2000). This fauna consists of 15 taxa of which 11 could be determined to species. All are of extant species with the exception of a species of the water beetle, *Agabus*, that has not been identified yet. The paleotemperature estimates were equivalent to present-day northern Britain.

Broomfield Details of the stratigraphy and paleontology of this site are given in Gibbard *et al.* (1996). At the moment, it is not possible to correlate these deposits precisely with any MIS event. Altogether 69 beetle taxa were recovered from the organic horizon of which 52 could be identified to species. The assemblage included a large number of species characteristic of running water and marginal riverine habitats. The discovery here of beetle species with southern European distributions at the present day indicates that the mean July temperature was at least 19° C.

Pakefield This site has long been known for its large vertebrate fossils. An account of the rich beetle fauna has been recently published (Parfitt *et al.*, 2005). The most likely correlation of these deposits is with MIS 17. Here, as at Broomfield, the fauna was characterized by numerous riverine and marsh species, indicating that this deposit originated in a mature river with both shallows and pools in which there were submerged trunks of decomposing oak trees. The beetles indicate the presence of dried-out carcasses that were partially consumed by maggots. The beetle assemblage included a number of southern European species, suggesting that the climate was warmer than present. Quantified estimates of the paleoclimate, using the Mutual Climatic Range method (MCR) on this Coleopteran assemblage (Atkinson *et al.*, 1987), indicate that the mean July temperature lay between 17° C and 23° C and that the mean January/February temperature lay between −6° C and 4° C.

Sugworth A Coleopteran assemblage has been described from preglacial channel deposit at Sugworth near Oxford (Osborne, 1979). Again, this fauna indicates that the deposit was largely riverine and also included several exclusively southern

European species. The presence at both Sugworth and Pakefield, but nowhere else, of the exotic species *Oxytelus* (=*Anotylus*) *opacus*, a southeastern European species with a distribution centered on the Danube basin, suggests that both sites may be correlated. MCR estimates on this beetle fauna indicate that the mean July temperature lay between 17° C and 22° C and the mean January/February temperature lay between −6° C and 4° C.

West Runton Freshwater Bed In 1995, a partial mammoth skeleton was excavated at West Runton near Cromer, north Norfolk. Samples of the associated sediment were analyzed for beetle remains (Coope, 2000). The upper parts of the sedimentary sequence were devoid of any insect fossils, probably due to their removal by oxidation. All beetle species in this assemblage can be found living in Britain at the present day. They indicate a reedy marsh with any open water overgrown with the duckweed *Lemna*. The climate was probably no different from that of Norfolk at the present day. Of particular interest is the scarcity of dung-beetles in this assemblage, suggesting that the mammoth carcass may have been transported *post mortem* into its present location.

Happisburgh An organic deposit is exposed from time to time below high tide level at Happisburgh. It predates the earliest evidence of glaciation in Britain. The insect fauna from the organic bed indicates stagnant water surrounded by a reedy marsh. The fauna includes several species that have relatively northern distributions at the present day. MCR estimates on the beetle assemblage indicate that the mean July temperature lay between 12° C and 15° C and the mean January/February temperature lay between −11° C and −3° C (Coope, 2006).

Ardleigh An interglacial beetle fauna, indicative of cool conditions equivalent to those at Happisburgh, has been described from middle Pleistocene deposits at Ardleigh, Essex (Gibbard *et al.* in prep.). Of particular interest is the presence directly above the interglacial deposits of a sediment lens (Ardleigh J) interbedded within gravels that yielded beetle species, which today have exclusively Arctic geographical ranges. One species, *Helophorus arcticus,* lives today in northern North America and extreme eastern Siberia. MCR estimates based on this assemblage indicates that the mean July temperature was below 10° C and mean January/February temperatures were probably −15° or lower. These deposits may represent an early climatic stage of the Anglian (Saalian) Glaciation.

High Lodge This site near Mildenhall, Suffolk, is famous for lower Paleolithic flint implements. Beetle assemblages have been recovered from the organic silts, yielding a fauna of 47 taxa of which 32 could be identified to species (Coope, 1992). The local environment was dominated by reedy swamp conditions. MCR estimates indicate that the mean July temperature lay between 15° C and 16° C and the mean January/February temperature lay between −4° C and 1° C.

Waverley Wood This site is situated south of Coventry, Warwickshire. The organic deposits were located in a series of channels directly underneath glacial deposits now attributed to the Anglian (Saalian) Glaciation. A large beetle assemblage was obtained from one of the lowermost of these channels (Shotton *et al.*, 1993). The beetles indicate that sedimentation commenced in running water, which later developed into a pond surrounded by a marsh. The presence of *Agonum quadripunctatus* is important because it is attracted by forest fires and it is thus significant that charcoal fragments occurred throughout the deposit. The climatic history during the channel filling is complex (Shotton *et al.*, 1993, Figure 9). At the base of Channel 2, MCR estimates indicate that the mean July temperature lay between 15° C and 18° C and the mean January/February temperature lay between −13° C and 2° C. However, at the top of Channel 2, the mean July temperatures have deteriorated to between 8° C and 11° C with mean January/February values to between −25° C and −10° C. At the base of the succeeding Channel 3, temperatures had returned to the same cool temperate conditions that prevailed at the base of Channel 2. A similar intense cold snap in an otherwise temperate episode is indicated by the beetle fauna from the MIS 11 interglacial.

Pools Farm Pit A similar beetle assemblage to that at Waverley Wood Pit has been recorded from a nearby gravel pit at Pools Farm near Brandon, Warwickshire (Maddy *et al.*, 1994).

Mathon This site was in the old gravel workings at The Brays Pit, near Mathon, Herefordshire, in the western English Midlands. The beetle assemblage was obtained from organic sediments at the base of the pit, underlying glacial deposits that in turn were cut into by a channel infilled with interglacial deposits attributed, on palynological grounds, to MIS 11. The beetle fauna from the basal sediments was dominated by running water species and others that depend on coniferous trees (Coope *et al.*, 2002). The climatic implications of this fauna are interesting in that one of the most abundant of these running water species, *Dupophilus brevis,* is not found today anywhere in northern regions but is confined to southwestern Europe, where it lives in streams that are relatively cool in summer.

Correlation between Happisburgh, Waverley Wood, Pools Farm, and Mathon In the absence of any evolutionary change amongst beetles and little evidence of global extinctions during the Quaternary, it is not possible to use the classical methodologies for paleontological correlation. It is always possible for similar assemblages to come together at different times whenever similar environments arise. Thus, gross faunal similarities do not necessarily indicate contemporaneity. However, the presence of exotic species carries more stratigraphical weight. The more unexpected the fossil occurrence of a species, the more likely it is to have stratigraphic value. The faunas from the above four sites have several exotic species in common (Coope, 2005). Perhaps the most unexpected of these is the small staphylinid species *Micropeplus hoogendorni*, a distinctive species originally described from Pliocene deposits of Alaska that were at least 5.7 million years old (Matthews, 1970). The species was believed by Matthews to be extinct but has now been recognized as the extant Siberian species *Micropeplus dokuchaevi.* It is interesting to note the presence in Britain of this exotic species at all four of these sites where it must have been fairly abundant, but so far it has been found nowhere else as a Pleistocene fossil. The presence of unexpected species, such as the above example, strongly suggests that all four deposits are of the same age.

Early middle Pleistocene Coleopteran faunas and lower Paleolithic Archaeology Several of the insect faunas mentioned above are associated with evidence of early human occupation of Britain, usually in the form of biface hand axes and/or characteristically struck flakes. At some sites there were also bones that had apparently been deliberately broken open, probably to extract the marrow. The beetle faunas provide intimate information about the living conditions at these times.

The fossiliferous deposits at Pakefield contain flint flakes that are most likely the earliest evidence of human presence in northern Europe (Parfitt *et al.*, 2005). The beetles indicate that the people lived beside a large meandering river, which occasionally inundated its floodplain. Alder and willow trees grew alongside the river and oak trees on the better drained ground. The climate at that time was substantially warmer than it is in Suffolk today (Parfitt *et al.*, 2005).

The sites at Happisburgh, High Lodge, and Waverley Wood yielded beautifully made hand-axes. The associated beetle assemblages indicate open marshland with nearby woodlands made up of both deciduous and coniferous trees. The climate at the time was rather cooler than today.

Post-Anglian middle Pleistocene beetle assemblages

The ice sheets of the Anglian (Saalian) glacial stage covered most of the British Isles and much of northwest Europe. This glaciation is relatively easy to recognize and most likely attributable to MIS 12. It is convenient to use this glacial event to subdivide the middle Pleistocene insect faunas. Most of these later insect assemblages have been obtained from interglacial deposits, though there are also a few faunas of cold-adapted species. Only the more important sites will be discussed here.

Interglacials correlated with MIS 11 Large Coleopteran assemblages have been recorded from five sites in Britain that can be attributed to MIS 11.

Hoxne, Suffolk This site is near the village of Hoxne in an abandoned brick-pit that is famous for beautifully made flint hand axes. The deposit is largely lake clay. Recent extensive excavations have been made to investigate the archaeology, vertebrate, and invertebrate paleontology, and also the paleobotany. Insect fossils have been studied from the early cold episode that preceded the interglacial, which may be interpreted as the final stage of the Anglian Glaciation. Beetles from a small sample of an organic bed at the base of the sequence included such cold-adapted species as *Diacheila polita, Bembidion hasti, Oreodytes alpinus, Tachinus caelatus*, and *Holoboreaphilus nordenskioeldi,* indicating that the climate was arctic and continental (Coope, 1993). However, the fauna is not simple, and includes some relatively thermophilous species. This suggests that the harsh climate may have already started to ameliorate, indicating the start of the MIS 11 interglacial.

The beetle fauna from the overlying interglacial deposits indicates still water surrounded by lush, marshy, meadow-like habitats in a mature forest of various deciduous trees, many of which were dead and decomposing. Recent investigations of the beetle faunas (Coope, unpublished) show that, after this temperate phase there was a brief return to cold-continental conditions with exclusively Arctic and Siberian species, such as *Helophorus obscurellus* and *Holoboreaphilus nordenskioeldi.*

Quinton, West Midlands Interglacial lake deposits were encountered in the construction of the M5 motorway, 10 km south of Birmingham. On palynological grounds, this deposit has been equated with those from Hoxne. The insect fauna was sparse in the lower part of the sequence but indicated temperate conditions somewhat warmer than those today. In the upper part, where the insect fauna was richer, they showed an episode of sudden and intense climatic deterioration with the appearance of exclusively arctic/continental species such as *Diacheila arctica, Helophorus obscurellus,* and *Holoboreaphilus nordenskioeldi.* Immediately after this cold episode, a cool temperate beetle fauna returned. This cannot be due to a re-deposition of fossils eroded out from the earlier temperate sediments, since they are a different suite of species (Coope and Kenward, in prep.). This sudden and intense cold snap has much in common with the climatic fluctuation recognized during the later part of the interglacial at Hoxne (see above).

Nechells, Birmingham, West Midlands A sequence of beetle assemblages has been described from borehole samples through interglacial lake deposits correlated with MIS 11; the Hoxnian interglacial (Shotton and Osborne, 1965). Unfortunately, only the first part of the interglacial was represented at Nechells. An interesting member of this assemblage is the southern European bark beetle *Platypus oxyurus*, which lives exclusively on the silver fir, *Abies.*

Woodston Beds, Peterborough, Cambridgeshire Quaternary deposits have been exposed for many years as channel-like deposits in the overburden being cleared from the Oxford clay used for brick making (Horton *et al.*, 1992). The insect assemblage was typically interglacial and dominated by running water species as well as indications of still or slowly flowing water, suggesting a mature river meandering across a floodplain. The likely age of these interglacial deposits is MIS 11, but again only the earlier part of the interglacial was represented.

Frog Hall Pit, Warwickshire Details of the stratigraphy and paleontology of the organic deposits at Frog Hall Pit have been described by Keen *et al.* (1997). The beetle assemblage indicates a river meandering across a marshy floodplain flanked by oak woodland. The climate was temperate, with MCR estimates of the paleoclimate indicating that mean July temperatures lay between 16° C and 18° C and mean January/February temperatures lay between −8° C and 12° C. Precipitation was sufficient to maintain a year-round flow in the river. Unfortunately, only the early part of the interglacial was represented at Frog Hall.

Interglacial beetle assemblages correlated with MIS 9
Only three insect faunas can be attributed with confidence to this period.

Hackney, North London A sequence of organic silts were exposed during building work at the Nightingale Estate. They yielded the only large beetle fauna that can be attributed to MIS 9. It comprised 254 taxa of which 181 could be identified to species (Green *et al.*, 2005). These provide a complex mosaic picture of the local environment. There was a mature meandering river with shallow rapid stretches and deeper pools. It is possible that some of the beetle species in this assemblage inhabited oxbow lakes on the floodplain. Nearby were marshes and damp meadow-like habitats frequented by large herbivorous mammals. The adjacent woodland consisted of both coniferous and deciduous trees. MCR estimates of the thermal climate gave mean July temperatures of about 18° C or somewhat warmer. Mean January temperatures were between −4° C and 1° C. Here again adequate precipitation must have been available to keep the river flowing throughout the year.

Barling, Essex During a multidiciplinary investigation, insect fossils were obtained from organic sediments infilling a channel in a gravel pit on the north side of the Thames estuary (Bridgland *et al.*, 2001). The beetle assemblage comprised 29 taxa of which 15 could be identified to species. Here again, the Coleoptera indicate a riverine and swampy grassland grazed by large herbivorous mammals. Abundant Paleolithic implements have been found in the immediate vicinity. MCR estimates based on the beetle fauna indicate that mean July temperatures lay between 17° C and 26° C and that mean January/February temperatures lay between −11° and 13° C.

Cudmore Grove, Mersea Island, Essex Organic sediments occupying a broad channel-like depression are exposed at the foot of the sea cliff. A rich beetle assemblage was obtained from near the base of the sequence (Coope, unpublished data), comprising 52 taxa of which 37 could be named to species. Here, the beetles indicate slow moving or stationary water with reedy marshes and grasslands, where large mammals were feeding. The climate was warmer than that of the present day. MCR estimates show that mean July temperatures lay between 16° C and 22° C and mean January/February temperatures lay between −7° C and 4° C.

Interglacials correlated with MIS 7 MIS 7 is climatically complex and in the near future insect faunas will be of value in its subdivision. Four large insect faunas are known that can be attributed to this period.

Stanton Harcourt, Oxfordshire An account of the stratigraphy and paleontology of this site have been given by Briggs *et al.* (1985). The fossiliferous deposits were in a channel cut into the Oxford clay. A single 10 kg sample yielded 122 beetle taxa of which 88 could be identified to species. This fauna indicates a moderately large river with riffles and pools, with bare clay banks in places and lush marshes alongside. This habitat attracted many large herbivorous mammals. Drier ground nearby was thinly vegetated with heathers and oaks. MCR estimates of the climate indicated mean July temperatures between 16° C and 18° C, with mean January/February temperatures close to 0° C. Numerous Paleolithic implements were found at this site.

Marsworth, Buckinghamshire The complicated stratigraphy and paleontology of this site has been described by Murton *et al.* (2001). A beetle fauna from the Lower Channel comprised 92 taxa of which 57 could be named to species. There is a curious rarity of water beetles and a complete absence of the species indicative of running water, suggesting that when the sediment was laid down the area had become a marsh surrounded by drier sandy soil. MCR estimates indicate that the mean July temperatures lay between 15° C and 17° C and mean January/February temperatures lay between −9° C and 1° C.

Stoke Goldington, Bedfordshire The stratigraphy and paleontology of this site has been described by Green *et al.* (1996). A beetle assemblage of 91 taxa, of which 60 could be identified to species, was obtained from the deposits in the 'lower channel'. This assemblage indicated very slow moving or stationary water with much swamp vegetation, notably the sweet grass *Glyceria* that is attractive to herbivorous mammals, notably mammoths, whose remains were also found at the site. The beetle assemblage indicates similar climatic conditions to those at Marsworth.

Upper Strensham, Worcestershire Organic deposits were encountered in a temporary exposure together with numerous mammoth bones (De Rouffgnac *et al.*, 1995). The associated beetle assemblage comprised 119 taxa of which 79 could be identified to species. They indicate that the sediment accumulated in a pond surrounded by meadow-like habitats. In spite of the apparent rarity of trees, the beetles were entirely temperate, suggesting that the mean July temperature was about 16° C and mean January/February temperatures were about 0° C.

Aveley Temporary exposures near Aveley, Essex, have attracted attention because of the occurrence of mammalian bones. These deposits are now generally accepted as being of MIS 7 age. Associated organic sediments have yielded large insect faunas (Schreve *et al.* in prep.). They indicate an extensive marsh with adjacent grassland and oak woodland. The climate was temperate, with MCR estimations based on the beetle fauna indicating mean July temperatures between 17° C and 18° C and mean January/February temperatures between −2° C and 0° C.

Cold-stage correlated with MIS 6

Balderton Sand and Gravel Terrace deposits of the river Trent near Lincoln have been equated with MIS 6 and are associated with a cold stage mammalian fauna. Associated organic sediments have yielded a beetle assemblage that included numerous exclusively Arctic/continental species (Coope and Taylor, 1991). These included *Diacheila polita, D. arctica, Helophorus obscurellus, H. sibiricus, H. jacutus, Boreaphilus henningianus*, and *Tachinus caelatus*. The local environment was reminiscent of tundra. MCR estimates based on the beetle fauna indicated that mean July temperatures lay between 10° C and 13° C and mean January/February temperatures lay between −25° C and −12° C.

Prospect for middle Pleistocene Coleopteran studies in Western Europe

Two important middle Pleistocene Coleopteran assemblages have been recently described from Western Europe. First, from interglacial deposits at Dingé, in northwest France, a large beetle assemblage has been described as part of a multidisciplinary study (Andrieu *et al.*, 1997). This fauna indicative of a warm temperate climate, is believed to date from MIS 7 or 9. Secondly, an extensive insect assemblage has been described as part of a multidisciplinary study from interglacial deposits at La Côte in the western French Alps that probably date from MIS 11 (Field *et al.*, 2000). The wealth of paleoecological data emerging from such investigations of Coleopteran assemblages points to new opportunities throughout Western Europe for reconstruction of past local environments and for giving quantified estimates of past climatic conditions.

Summary

The sequence of Coleopteran assemblages outlined above provides an ecological picture that is consistent with the other physical and biological environmental indicators. Thus, the fossil beetle assemblages show constancy in their environmental requirements as well as stability in the morphology of their component species. By using the MCR method, it has been possible to quantify climatic parameters for both interglacial and glacial events. In particular, some of the middle Pleistocene interglacial faunas included southern European species, indicating that the climate was substantially warmer than today. During the cold periods, the presence of exotic cold adapted beetle species in Britain showed that the climate was of arctic and extreme continental severity. In stratigraphical correlation, the presence of unexpected exotic species in assemblages may provide better indicators of similarity of age than their gross faunal (or floral) similarity.

See also: **Beetle Records**: Overview; Late Pleistocene of Europe; Postglacial Europe.

References

Andrieu, V., Field, M. H., Ponel, P., Guiot, P., de Beaulieu, J.-L., Reille, and Morzadec-Kerfourn (1997). Middle Pleistocene temperate deposits at Dingé, Ille-et-Vilaine, northwest France. *Journal of Quaternary Science* **12**(4), 309–331.

Atkinson, T. C., Briffa, K. R., and Coope, G. R. (1987). Seasonal temperatures in Britain during the past 22,000 years, reconstructed using beetle remains. *Nature* (London) **325**, 587–592.

Bridgland, D. R., Preece, R. C., Roe, H. M., Tipping, R. M., Coope, G. R., Field, M. H., Robinson, J. E., Schreve, D. C., and Crowe, K. (2001). Middle Pleistocene interglacial deposits at Barling, Essex, England: evidence for a longer chronology for the Thames terrace sequence. *Journal of Quaternary Science* **16**, 813–840.

Briggs, D. J., Coope, G. R., and Gilbertson, D. D. (1985). *The Chronology and Environmental Framework of Early Man in the Upper Thames Valley*. Vol. 137, pp. 1–176. British Archaeological Reports, Oxford.

Buckland, P. C., and Coope, G. R. (1991). *A Bibliography and Literature Review of Quaternary Entomology*. J Collis Publications, University of Sheffield, Sheffield.

Coope, G. R., and Taylor, B. J. (1991). Coleoptera from the Balderton Sand and Gravel. (A. Brandon and M. G. Sumbler, Eds.)The Balderton Sand and Gravel; pre-Ipswichian cold stage fluvial deposits near Lincoln, England. *Journal of Quaternary Science* **6**, 130–133.

Coope, G. R. (1992). The High Lodge insect fauna. In *High Lodge, Excavations* (N. M. Ashton, J. Cook, S. G. Lewis and J. Rose, Eds.), by Sieveking G de G, 1962–8, and Cook J, 1998, pp. 117–119. British Museum Press, London.

Coope, G. R. (1993). Late-Glacial (Anglian) and Late-Temperate (Hoxnian) Coleoptera. In *The Lower Palaeolithic Site at Hoxne, England* (R. Singer, B. G. Gladfelter and J. J. Wymer, Eds.), pp. 156–162. University of Chicago Press, Chicago.

Coope, G. R. (2000). Coleoptera from Beeston and West Runton, Norfolk. In *The Quaternary of Norfolk & Suffolk: Field Guide* (S. G. Lewis, C. A. Whiteman and R. C. Preece, Eds.), pp. 73–75. Quaternary Research Association, London.

Coope, G. R. (2006). Insect Faunas associated with Palaeolithic industries from five sites of pre-Anglian age in Central England, *Quaternary Science Reviews*.

Coope, G. R., and Kenward, H. K., (in prep) Evidence for a short but intense cold interlude during the late-temperate stage of the Hoxnian Interglacial (MIS11) based on Coleopteran assemblages from Quinton, West Midlands, UK. *Quaternary Science Reviews*.

Coope, G. R., Field, M. H., Gibbard, P. L., Greenwood, M., and Richards, A. E. (2002). Palaeontology and biostratigraphy of Middle Pleistocene river sediment in the Mathon Member, at Mathon, Herefordshire, England. *Proceedings of the Geologists Association* **113**, 237–258.

De Rouffignac, C., Bowen, D. Q., Coope, G. R., Keen, D. H., Lister, A. M., Maddy, D., Robinson, J. E., Sykes, G. A., and Walker, M. J. C. (1995). Late Middle Pleistocene interglacial deposits at Upper Strensham, Worcestershire, England. *Journal of Quaternary Science* **19**, 15–31.

Field, M. H., Beaulieu, J.-L.de, Guiot, J., and Ponel, P. (2000). Middle Pleistocene deposits at La Côte, Val- de-Lans, Isère Department, France: plant macrofossils, palynology and fossil insect investigations. *Palaeogeography, Palaeoclimatology, Palaeoecology* **159**, 53–83.

Gibbard, P. L., Aalto, M. M., Coope, G. R., Currant, A. P., McGlade, J. M., Pegler, S. M., Preece, R. C., Turner, C., Whiteman, C. A., and Wrayton, R. C. (1996).*The Early Middle Pleistocene in Europe* (C. Turner, Ed.), pp. 83–119. Balkema, Rotterdam.

Gibbard, P. L., Aalto, M. M., Allen, P., Bridgland, D. R., Coope, G. R., Pettit, M. E., Turner, C., Whiteman, C. A., and Wrayton, R. C. (*in prep.*) Middle Pleistocene fluvial deposits at Ardleigh, Essex, England and their stratigraphical significance. *Journal of Quaternary Science*.

Green, C. P., Coope, G. R., Jones, R. L., Keen, D. H., Bowen, D. Q., Currant, A. P., Holyoak, D. T., Ivanovich, M., Robinson, J. E., Rogerson, R. J., and Young, R. C. (1996). Pleistocene deposits at Stoke Goldington, in the valley of the Great Ouse, UK. *Journal of Quaternary Science* **11**, 59–87.

Green, C. P., Branch, N. P., Coope, G. R., Field, M. H., Keen, D. H., Wells, J. M., Schwenningen, J.-L., Preece, R. C., Schreve, D. C., Canti, G., and Gleed-Owen, C. P. (2005). Marine Isotope Stage 9 environments of fluvial deposits at Hackney, North London, UK. *Quaternary Science Reviews*.

Horton, A. (1989). Quinton. In *West Midlands Field Guide* (D. H. Keen, Ed.), p. 69. Quaternary Research Association, Cambridge.

Horton, A., Keen, D. H., Field, M. H., Robinson, J. E., Coope, G. R., Currant, A. P., Graham, D. K., Green, C. P., and Philipps, L. M. (1992). The Hoxnian Interglacial deposits at Woodston, Peterborough. *Philosophical Transactions of the Royal Society of London* **338**, 131–164.

Keen, D. H., Coope, G. R., Jones, R. L., Field, M. H., Griffiths, H. I., Lewis, S. G., and Bowen, D. Q. (1997). Middle Pleistocene deposits at Frog Hall Pit, Stretton on Dunsmore, Warwickshire, East Midlands, and their implications for the age of the type Wolstonian. *Journal of Quaternary Science* **12–3**, 183–208.

Maddy, D., Coope, G. R., Gibbard, P. L., Green, C. P., and Lewis, S. G. (1994). Reappraisal of Middle Pleistocene fluvial deposits near Brandon, Warwickshire and their significance for the Wolston glacial sequence. *Journal of the Geological Society* **151**, 221–233.

Matthews, J. V., Jr. (1970). Two new species of *Micropeplus* from the Pliocene of western Alaska with remarks on the evolution of Micropepliinae (Coleoptera, Staphylinidae). *Canadian Journal of Zoology* **48**, 779–788.

Murton, J. B., Baker, A., Bowen, D. Q., Castledine, C. J., Coope, G. R., Currant, A. P., Evans, J. G., Field, M. H., Green, C. P., Hatton, J., Ito, M., Jones, R. L., Keen, D. H., Kerney, M. P., McEwan, R., McGregor, D. F. M., Parish, D., Robinson, J. E., Schreve, D. C., and Smart, P. L. (2001). A late Middle Pleistocene temperate- periglacial temperate sequence (Oxygen Isotope Stages 7–5e) near Marsworth, Buckingham, UK. *Quaternary Science Reviews* **20**, 1787–1825.

Osborne, P. J. (1979). The Insect Fauna of the organic deposits at Sugworth and its environmental and stratigraphical implications. *Philosophical Transactions of the Royal Society of London* **289B**, 119–133.

Parfitt, S. A., Barenbregt, R. W., Breda, M., Candy, I., Collins, M. J., Coope, G. R., Durbridge, P., Field, M. H., Lee, J. R., Lister, A. M., Mutch, R., Penkman, K. E. H., Preece, R. C., Rose, J., Stringer, C. B., Symmonds, R., Whitaker, J. E. P., Wymer, J. J., and Stuart, A. J. (2005). The earliest humans in Northern Europs: artefacts from the Cromer Forest Formation at Pakefield. *Nature (London)* **438**, 1008–1012.

Schreve, D. C., Bridgland, D. R., Allen, P., Keen, D. H., White, M. J., Blackford, J. J., Coope, G. R., Field, M. H., Juby, C., Kemp, R. A., Gleed-Owen, C. P., Cooper, J. H., and Green, F. M. L. (*in prep.*) Late Middle Pleistocene River Thames terrace deposits at Aveley, Essex, UK: a multiproxy framework for the penultimate (MIS 7) interglacial. *Quaternary Science Reviews*.

Shotton, F. W., and Osborne, P. J. (1965). The fauna of the Hoxnian Interglacial deposits of Nechells, Birmingham. *Philosophical Transactions of the Royal Society of London* **248B**, 353–378.

Shotton, F. W., Keen, D. H., Coope, G. R., Currant, A. P., Gibbard, P. L., Aalto, M. M., Pegler, S. M., and Robinson, J. E. (1993). The Middle Pleistocene deposits at Waverley Wood Pit, Warwickshire, England. *Journal of Quaternary Science* 8(4), 293–325.

Late Pleistocene of Australia

N Porch, Monash University, VIC, Australia

© 2007 Elsevier B.V. All rights reserved.

Introduction

General Significance for Quaternary Entomology

Limited evidence from the Australian early Pleistocene suggests that Northern Hemisphere precepts of Quaternary entomology that include species constancy and a lack of extinction through the Quaternary (Coope, 1995) do not necessarily apply to the Australian record. In Australia there are almost certainly significant levels of extinction between the early and late Pleistocene paralleling, unsurprisingly, the plant record for this period (Jordan, 1997). Fossil beetles from the early Pleistocene site of Stony Creek Basin include a number of species from well-studied groups that do not match any modern material. It is, of course, possible that some or all of these taxa will be discovered to be extant, but the high levels of extinction in the plant record from this site (Sniderman *et al.*, 2004) certainly suggest extinction of beetles and other biota is likely to have occurred. Data from the late Quaternary of Australia, however, suggest little or no extinction and no difference in the nature of the species present. This provides the necessary context for paleoclimatic reconstruction using the extant environmental requirements of existing species.

Australia's Beetle Fauna

In contrast to most areas from which Quaternary beetle records have been examined, the beetle fauna of Australia is large – recent estimates suggest as many as 80,000–100,000 species (Yeates *et al.*, 2003) or more (Hammond, 1994), although only a fraction of these (20–30%) have been formally described (Lawrence and Britton, 1994). This has presented special problems for Quaternary beetle analysis in Australia. In particular, there are relatively few published ecological and distributional data for most taxa, necessitating extensive research into the distribution and ecological requirements of taxa recovered from fossil records. Fortunately, although the entire beetle fauna is relatively poorly known taxonomically and ecologically, in many groups, including those often common in the fossil record, the situation is substantially better.

Paleoclimatic Reconstruction Methods

Appropriate Methods for the Australian Region

Several factors set Australia apart from areas where the majority of Quaternary beetle research has been undertaken. First, winter temperatures are less significant in determining the distribution of taxa. Second, variability in precipitation regimes is a major determinant of the distribution of many taxa. This is especially true, for example, for taxa that have soil or litter dwelling life stages, those associated with fungi, and taxa that live in rotting wood. For these taxa, and taxa from desert regions, the presence of suitable moisture regimes often has greater significance in limiting their distribution than temperature regimes. Finally, there are fewer distribution records for most taxa with which to define their climatic requirements. For these reasons, the mutual climatic range (MCR) method as initially conceived for reconstructing seasonal temperature regimes using beetle assemblages (Atkinson *et al.*, 1986) is inappropriate for Australia. The fact that Australian and regional climatic fluctuations are dominated by greater variability in moisture rather than in temperature regimes (Turney *et al.*, 2006) means methods that ignore the influence of changes in the amount and seasonality of precipitation, like MCR, are liable to provide a limited picture of Quaternary climate change in this region (although see Elias (1997)).

Climate Profiles for Australian Taxa

BIOCLIM is a computer program (currently included in ANUCLIM 5.1) used to derive a range of parameter-based climate data for point or grid data from precipitation and temperature surfaces of a particular region (Houlder *et al.*, 2001; Nix, 1986; contrary to the statement of Marra *et al.* (2004)) BIOCLIM is not a 'method' of paleoclimatic reconstruction. For each taxon, geocoded locality data (latitude, longitude, altitude) form the basic BIOCLIM input. BIOCLIM output includes files describing the climate at each point at which the taxon was found (bio files) and a summary of the parameter statistics for the species – the taxon profile (pro). In the Australian context, this includes 35 parameters describing temperature, precipitation, radiation, and moisture indices that are derived from standardized climate surfaces that span the continent. **Table 1** lists the temperature and precipitation parameters produced by the current version of BIOCLIM. Radiation and moisture index parameters are not considered further. **Figure 1** illustrates the spatial distribution of two temperature and two precipitation parameters across Australia.

A total of more than 780 bioclimatic profiles for taxa that have either been found in the fossil record, or taxa that are considered likely to be found in the fossil record, have been produced for the Australian region. The profiles are dominated by aquatic and riparian taxa that are usually diverse and abundant in the fossil record, dung beetles that are consistently present in small numbers, and a selection of other taxa that are associated with a range of habitats, especially wet forest and rainforest litter. These profiles form the core data set with which the utility of beetles for reconstruction of past climates was assessed. **Figure 2** shows the distribution of sites contributing at least one record to the bioclimatic profiles of the taxa examined.

Table 1 Temperature and precipitation parameters estimated by BIOCLIM. Parameters in bold are those considered further for their paleoclimatic utility on the basis of their potential ecological significance and predictability using beetle assemblage data

Parameters	*Abbreviations*
Temperature Parameters	
1. Annual mean temperature	***Tmat***
2. Mean diurnal range	
3. Isothermality (Parameter 2/Parameter 7)	
4. Temperature seasonality (coefficient of variation)	
5. Maximum temperature of warmest month	***TmaxM***
6. Minimum temperature of coldest month	***TminM***
7. Temperature annual range (5–6)	***Trange***
8. Mean temperature of wettest quarter	
9. Mean temperature of driest quarter	
10. Mean temperature of warmest quarter	***TwarmQ***
11. Mean temperature of coldest quarter	***TcoldQ***
Precipitation parameters	
12. Annual mean precipitation	***Pmap***
13. Precipitation of wettest month	
14. Precipitation of driest month	
15. Precipitation seasonality (C of V)	***Pseas***
16. Precipitation of wettest quarter	
17. Precipitation of driest quarter	
18. Precipitation of warmest quarter	***PwarmQ***
19. Precipitation of coldest quarter	***Pcold***

Assessment of BIOCLIM Parameters for Paleoclimatic Reconstruction

Not all bioclimatic parameters produced by BIOCLIM are biologically significant. There are a several ways to determine the relative utility of the parameters. These included examination of the ecology of the taxa being studied, especially the significance of seasonal temperature and moisture regimes in determining activity periods, analyses of regional variability of parameters in the modern environmental space and, perhaps most usefully, the ability to precisely reconstruct modern climates using the climatic requirements of species in extant assemblages. The results of modern reconstructions must, however, be carefully framed within an understanding of present and past environmental space when applied to the Quaternary record.

When there are sufficient records, usually more than 50 for stenotopic taxa and over 100 for eurytopic taxa, cumulative frequency plots (produced in BIOCLIM) or histograms (see, e.g., **Fig. 3**) reflect the relatively normal responses of the beetle taxa to mean temperature parameters. This is significant for two reasons. First, the close approximation to a normal distribution has significance for estimating errors for poorly known taxa; the assumption that the response takes a different form would potentially significantly underestimate the actual range of an

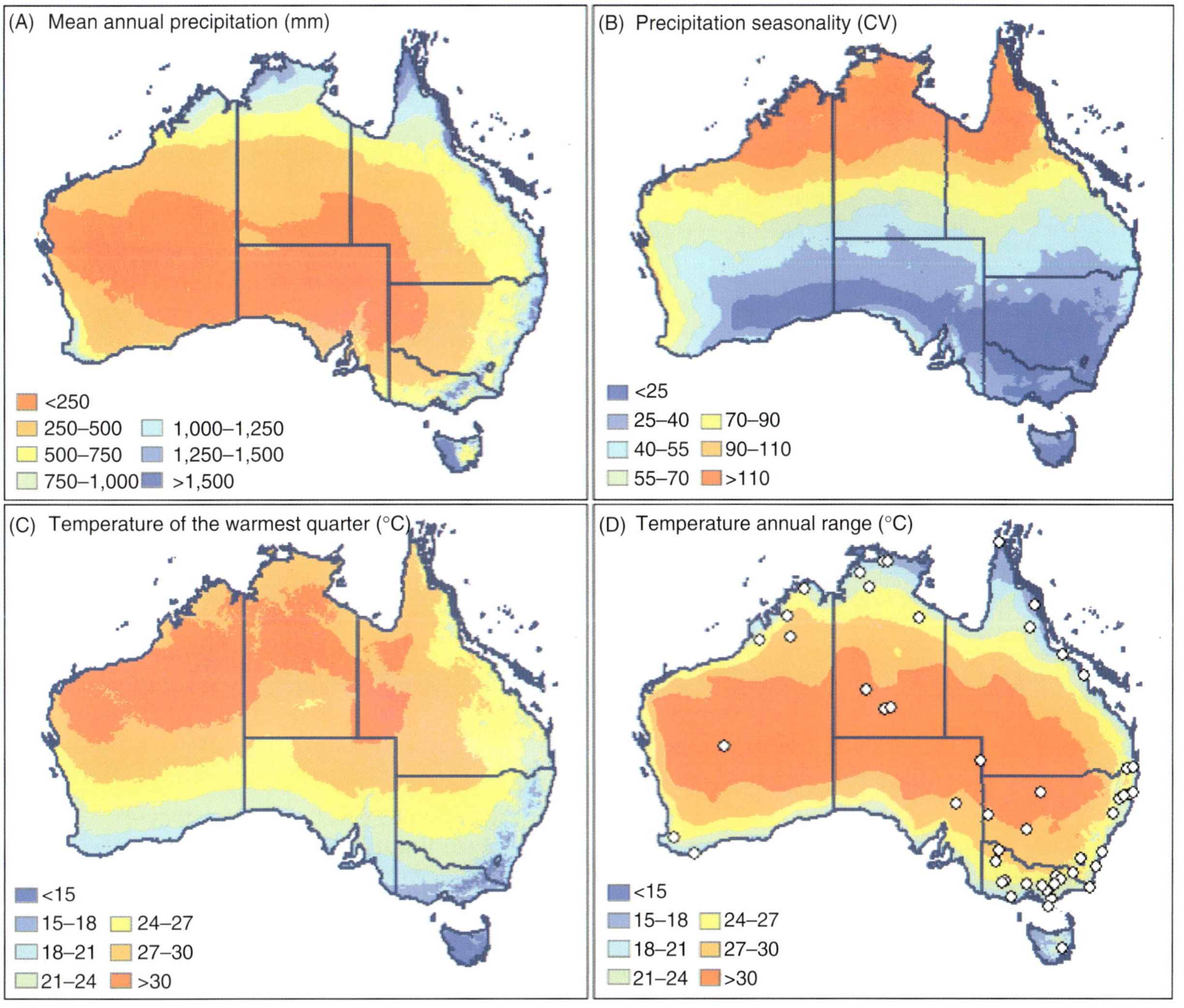

Figure 1 Regional climate summaries of selected precipitation and temperature parameters derived from BIOCLIM analysis. (A) *Mean annual precipitation*; (B) *Precipitation seasonality*; (C) *Temperature of the warmest quarter*; (D) *Temperature annual range*. The locations of modern test sites, referred to in the text, are shown as white circles in (D). Note, in particular, the strong correlation of areas with low *mean annual precipitation* (< 500 mm) and high *temperature annual range*. Also note the correlation of cool summer climates (*temperature of the warmest quarter*) and higher levels of rainfall (*mean annual precipitation*) in southern, particularly southeastern Australia.

infrequently collected taxon. Second, it means that maximum and minimum values are equally suitable for estimation. In contrast, precipitation values are usually strongly skewed by higher values that occur infrequently in bioclimatic profiles, largely because areas of high precipitation are relatively rare in environmental space (see **Fig. 1A**). The example shown in **Figure 3**, for the click beetle *Agrypnus caliginosus*, is typical. Subsequently, unless a taxon is extremely well collected (hundreds of records), it is unlikely that its upper precipitation limits have been sampled.

Error Estimation

The relatively small number of distribution records for most species recorded in the Australian fossil record has meant that their climatic ranges are generally underestimated. Marra *et al.* (2004) attempted to resolve this problem in New Zealand by using error estimates derived from maximum likelihood analyses. For Australia an alternative method has been used that relies on the fact that a number of taxa have multiple bioclimatic profiles: new profiles were produced for taxa when a significant amount of new distribution data were accessed (sometimes as much as four times the original data volume). Comparison of the relationship between the original profile (*raw*) data and the data-enhanced profile for a range of taxa allowed the development of equations for estimated error for temperature and precipitation parameters (*estimated parameter ranges*). These equations are based on the range per record value for each parameter and for each taxon. The climatic range of a taxon becomes better

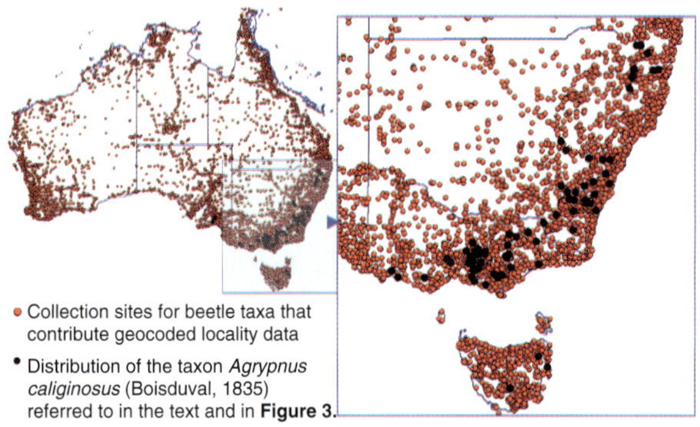

Figure 2 The distribution of collection localities that contribute at least one data point to the bioclimatic profiles of the 780 taxa used to examine the modern relationships between actual (modern) and predicted values (see text and **Figs. 5** and **7**) and for interpretation of the paleoclimatic significance of the fossil assemblages. The inset shows the sampling density in the region of southeastern Australia that has thus far been the focus of Quaternary beetle analyses and the distribution of *Agrypnus caliginosus* (Elateridae) (see **Fig. 3**).

known as the range per record value diminishes: for temperature parameters, a value of less than 0.1° C is considered well known. For precipitation parameters, errors were constructed only for minimum values and using the 0–25th percentile spread of values to exclude the influence of the skew to high values with increasing data volumes.

Coexistence Methods for Paleoclimatic Reconstruction

Overview Coexistence (or coincidence) methods were used for paleoclimatic reconstruction (Kershaw and Nix, 1988; Mosbrugger and Utescher, 1997). Coexistence methods utilize the same premise as MCR; however, they consider single parameters negating some of the issues noted above, particularly the more limited data sets and the lack of climate analogs. For each BIOCLIM parameter, the climatic estimates for modern and fossil assemblages include a range defined by the overlap of the taxa in the assemblage; the median of this range is the best estimate. **Figure 4**, for example, illustrates the *TwarmQ* raw data ranges for an assemblage of water beetles from northern Australia and the derivation of the estimate for the assemblage. Using this method, estimation of modern climate at a range of localities across Australia (locations shown in **Fig. 1D**) was undertaken to assess the precision and accuracy of beetle-based reconstructions. To provide a context for understanding the nature of the method and its limitations, indicative results are shown in **Figs. 5–7** and briefly discussed below. Estimation of paleoclimates has used an identical method based on the taxa in fossil assemblages that have bioclimatic profiles in the modern data set. In the following discussion and for Pleistocene assemblages, results are based on estimated parameter range data rather than raw data values.

Temperature parameters The results of actual/predicted tests for four BIOCLIM temperature parameters are shown in **Figure 5**. The beetle-based predictions for the parameters *Tmat* and *TwarmQ* are the best performing in terms of the accuracy and precision in relation to the modern (actual) value. Increasing numbers of taxa in a modern assemblage significantly increase the precision of the

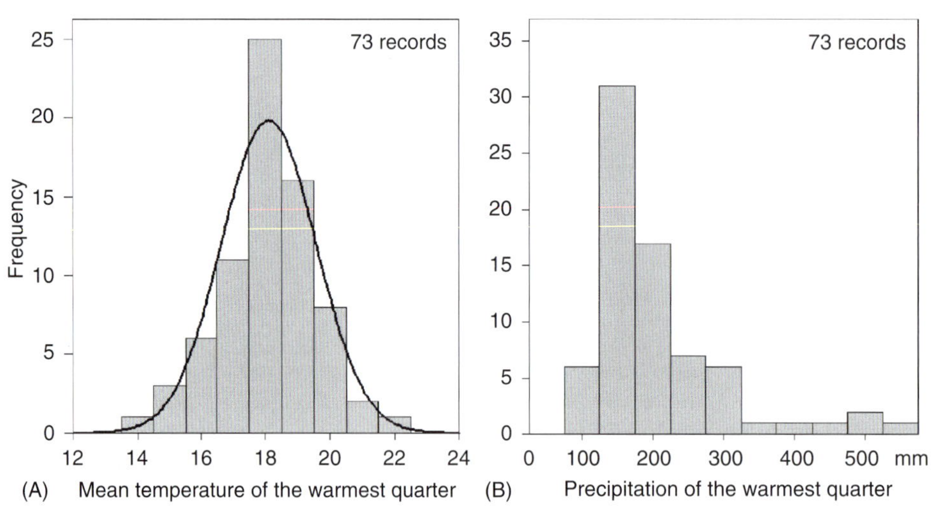

Figure 3 Histograms illustrating the contrasting nature of the response of the click beetle *Agrypnus caliginosus* (Elateridae) to (A) temperature and (B) precipitation parameters. The relatively normal response to mean temperature values is typical of taxa with large numbers of collection localities. The strongly skewed response to mean precipitation values is similarly typical, however, in this case only for taxa that inhabit mesic environments; in arid and semi-arid environments upper precipitation limits are significant, resulting less skewed responses. The distribution of this species is shown in the inset of **Figure 2**.

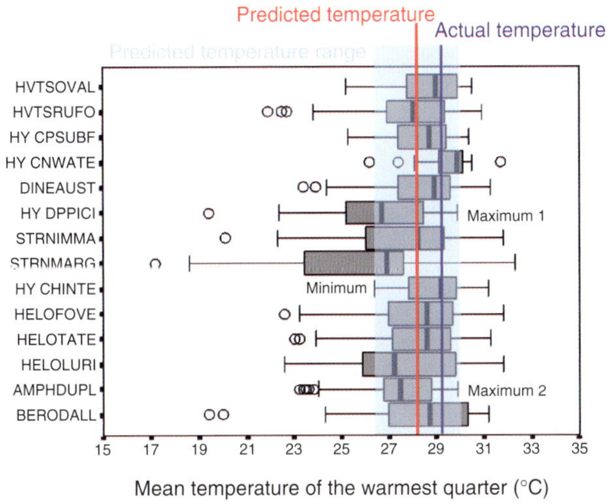

Figure 4 An assemblage of 14 aquatic beetles (Dytiscidae and Hydrophilidae – codes listed on left) from a site in northern Australia showing the derivation of the predicted temperature range (envelope) and the best estimate (predicted temperature). Data for each taxon are displayed as box-plots which allows examination of the structure of the data underlying the predictions. In this case a single taxon (labeled *minimum*) determines the lower (cooler) bound and two taxa (labeled *maximum*) determine the upper (warmer) bound of the predicted range envelope. In this example, the *raw* data values are utilized rather *estimated parameter range* values (see text)

estimate – when there are more than 20 taxa a precision of ±1° C is routinely attained for *TwarmQ* (**Fig.** 6). Winter temperatures (*TcoldQ*) are reconstructed with similar accuracy but with less precision. It is likely that this is related to the significant correlation in modern Australian environmental space of summer and winter temperatures rather than the unrecognized biological significance of winter temperatures in Australia. In line with expectations, based on the lack of strong seasonality in Australian environmental space and the lack of significance of winter temperatures, *Trange* is one of the worst performing BIOCLIM temperature parameters, with low precision and accuracy.

Precipitation parameters The results of actual/predicted tests for four BIOCLIM temperature parameters are shown in **Figure** 7. The best performing precipitation parameter is *PwarmQ*. Although the median for some predictions significantly exceeds the actual modern value, the minimum of the estimated range differs little, suggesting that the precipitation estimates are best considered minimum values. This reflects the skewed nature of the precipitation response (see above) for many taxa and the possible lack of ecological significance of precipitation maxima. For precipitation parameters, for these reasons, only the minimum values for each parameter are estimated.

Late Pleistocene Records

Yarra Creek, King Island, Tasmania – MIS 5

Sediments exposed in coastal cliffs immediately south of the Yarra Creek outlet into the City of Melbourne Bay on King Island (location shown on **Fig.** 8) have yielded the only marine isotope stage (MIS) 5 beetle assemblage from Australia thus far. Unfortunately, chronological evidence, based on thermoluminescence data, does not allow the age of the biota to be determined with more precision. The beetle assemblages are associated with a micro- and macroflora that is characteristic of cool temperate rainforest – a formation that does not occur on King Island today and almost certainly did not occur at any time during the Holocene.

The Yarra Creek insect assemblage was deposited by a moderate to fast flowing stream, which was small, oxygen rich, and stony in some places. The domination of the fauna by saproxylic and rainforest/wet sclerophyll forest litter-inhabiting beetles indicates that the stream was flowing through dense forest with deep leaf litter and abundant decaying wood. There are no identified taxa that definitively indicate the composition of the forest flora; however, the abundance of several taxa commonly (but not exclusively) associated with cool temperate rainforest suggests this formation.

The assemblage has no modern analog and is composed of essentially four elements. The first is taxa that are widespread across mesic southeastern Australia including Tasmania (most of the assemblage). The second contains taxa that are restricted to Tasmania but do not reach King Island (Lucanidae: *Ceratognathus westwoodi*). The third contains taxa that are widespread in mainland southeastern Australia and reach King Island at their southernmost expression (Prostomidae: *Prostomis cornuta*, Carabidae: *Mecyclothorax* '*cordicollis*'). The last group taxa are restricted to the mainland and do not, on the basis of current knowledge, reach King Island or indeed anywhere in Tasmania (Tenebrionidae: *Archaeoglenes australis* Lawrence).

The beetle-based reconstructions of the climate at the time of the deposition of the assemblage suggest that the MIS 5 King Island climate was substantially different from that of the present day in several ways. Summer rainfall was significantly greater than today (*PwarmQ*: modern 137 mm, predicted >176 mm). Temperature seasonality was enhanced, resulting in hotter summers in terms of maximum temperatures

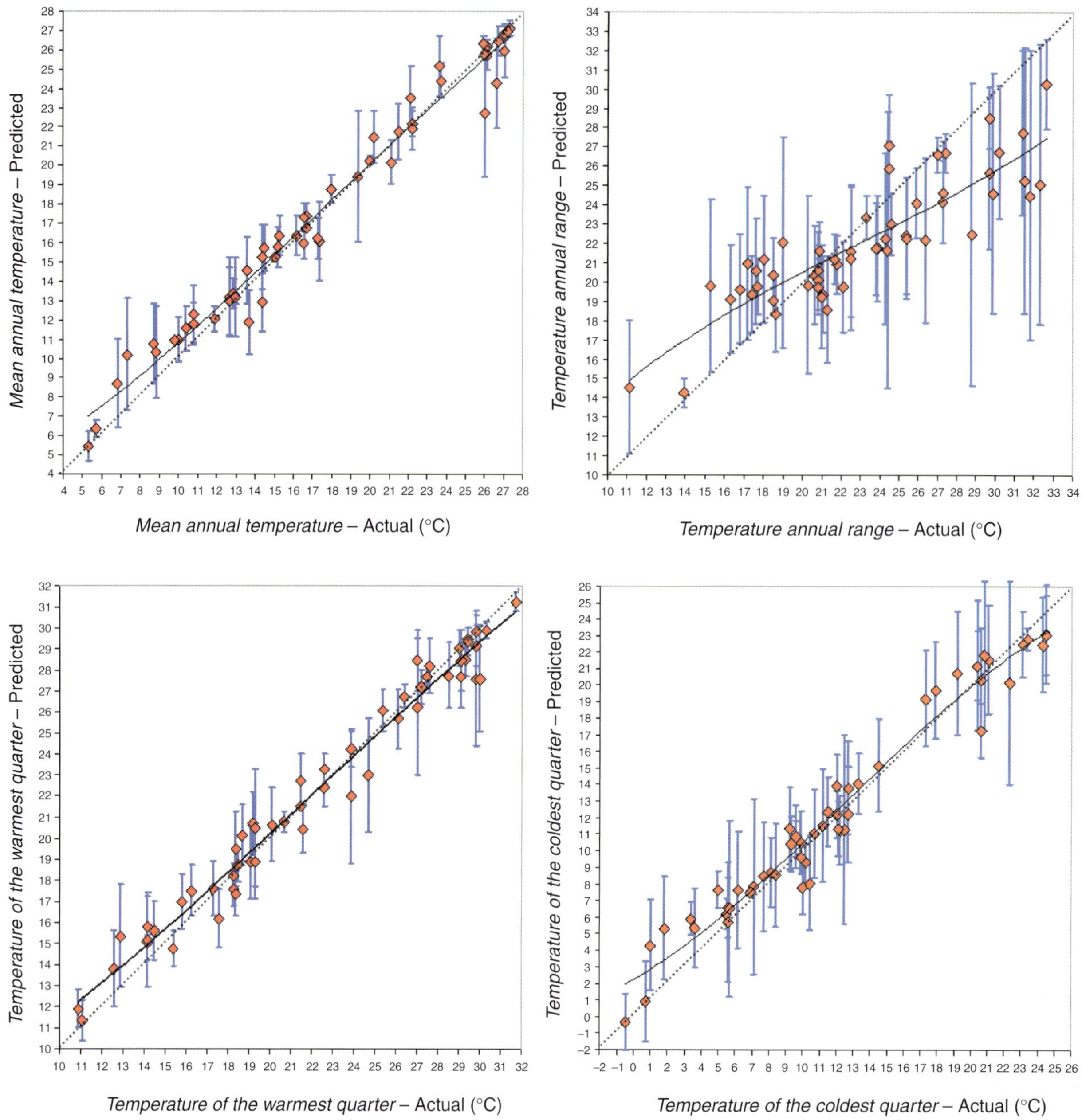

Figure 5 *Actual* (modern) vs. *predicted* values for modern beetle assemblage across Australia for four BIOCLIM temperature parameters. Locations of the assemblages are shown in **Figure 1D** and the relationship between number of taxa in the assemblage and the precision of the prediction shown in **Figure 6**. The diamond represents the median value of the predicted range which is shown as a bar. The dashed line is the line of equivalence and the curved line the best fitting cubic regression.

(*TmaxM*), even though the summer mean was similar (*TwarmQ*: modern 16.0° C, predicted 15.6° C), and the winters were colder (*TcoldQ*: modern 10.1° C, predicted 6.2° C; *TminM*: 7.2° C, predicted 2.1° C).

The fossil plant record is entirely consistent with this reconstruction – indeed several taxa would require more summer rainfall to occur within the Yarra Creek catchment. Within southeastern Australia there are few firmly dated MIS 5 records with which to compare the results, and none with published quantified climatic data. Harle *et al.* (2004, p. 125) suggest that expansion of cool and warm temperate rainforest elements during a period interpreted as MIS 5e at Lake Wangoom in western Victoria reflects increased effective precipitation and potentially warmer environments (**Figs. 9** and **10**).

Spring Creek, Western Victoria – Last Glacial?

The Spring Creek locality, western Victoria (**Fig. 8**), is a megafaunal site, initially argued to provide

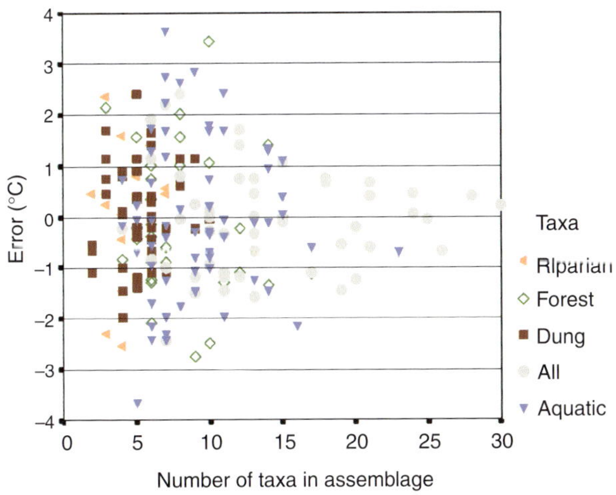

Figure 6 The relationship between the number of taxa in an assemblage used for climatic reconstruction and the precision (error) of the estimate (median value of the estimate range) – in this case for the BIOCLIM parameter *temperature of the warmest quarter*. Results are shown for a range of ecologically based assemblages including riparian, forest, dung, and aquatic taxa. The category *All* refers to assemblages that contain taxa from two or more ecological categories. The narrow climatic ranges of many riparian and dung beetles means fewer taxa are required to give a precise climate estimate.

evidence for late survival, to the LGM, of a number of extinct taxa (Flannery and Gott, 1984). Fossil bone has been recovered, and still erodes from, the edge of a thin exposure of heterogeneous gravely silt that is overlain by a bed of variable dark, clayey silts. Flannery and Gott (1984) presented details of the vertebrate fauna and associated plant remains including a short list of pollen types recorded from the deposit, and reported a single conventional radiocarbon determination of 19,800 ± 390 (Teledyne 1-11,018). More recently, four AMS radiocarbon determinations were obtained on bone (White and Flannery, 1995); the results were interpreted to indicate contamination of the bone samples and they proposed that the true age of the Spring Creek fauna was greater than 35 kyr. To examine the age of the sedimentary context, a series of AMS determinations on plant macrofossils, insect chitin, and pollen preparations were obtained. The determinations on all macrofossils are either background or >40 kyr BP, whereas the pollen preparation determinations are finite, but close to background. This suggests contamination by modern carbon and implies that the age of the fauna is greater than 40 kyr BP.

Insect assemblages were derived from two localities: sediments containing megafaunal remains and a previously unrecognized sequence of bedded fluvial silt that is a downstream equivalent of the megafaunal bearing sediments. Like the macrofossil plant and pollen assemblages, the insect faunas from all samples are essentially identical, suggesting rapid accumulation. The fauna is dominated by aquatic insects, especially Trichoptera (caddisflies) and a number of aquatic beetle taxa that collectively indicate deposition in clean, flowing, oxygen-rich water over a sandy or gravel bottom (e.g., Elmidae: *Austrolimnius* spp., *Simsonia* sp., ecnomid and hydropsychid Trichoptera).

The relatively small terrestrial fauna contains no taxa that are indicative of a wooded environment, but does include a range of taxa that are indicative of open grassy or heathy habitats, either in riparian microhabitats (Carabidae: *Bembidion blackburni*) or elsewhere (Carabidae: *Scopodes* spp., Anthribidae: *Eucioides suturalis*). The only taxon that could potentially be associated with trees (*Casuarina* or *Allocasuarina*), the buprestid *Germarica lilliputana,* can also be found in shrubs of the latter genus, macrofossils of which are abundant in the sediments. The dung beetle fauna, which contains three aphodiine species (*Ataenius basiceps*, *Proctophanes sculptus*, and *Podotenus* sp.), is commonly associated with marsupial dung in open habitats beside water (*A. basiceps*) or in a range of environments (the other two taxa). The interpretation of a small stream flowing through a treeless environment is supported by both the plant macrofossil and pollen record in which arboreal taxa are consistently <5% of the dryland pollen sum.

The predicted ranges for all temperature parameters overlap the modern values for the site. Median values suggest slightly more seasonal climates with summers around 0.5° C warmer than the present, and winters approximately 1.0° C colder than the present, although given the overlap in the predicted ranges it is impossible to know whether these differences are meaningful. Precipitation reconstructions suggest a similar precipitation regime to the present. There is, however, an indication, based on the taxa present in the dung beetle fauna, that summer precipitation was fractionally greater than the present (*PwarmQ*: modern 108 mm, predicted >116 mm).

The reconstruction of a climate similar to the present presents an interesting conundrum in terms of the nature of the pollen record, which is essentially "glacial" in its character. Glacial age pollen records are characterized by their domination by pollen of Poaceae and Asteraceae relative to trees (*see* Australia and New Zealand) and are usually interpreted to represent colder and/or drier climates. This will be discussed further in the conclusion.

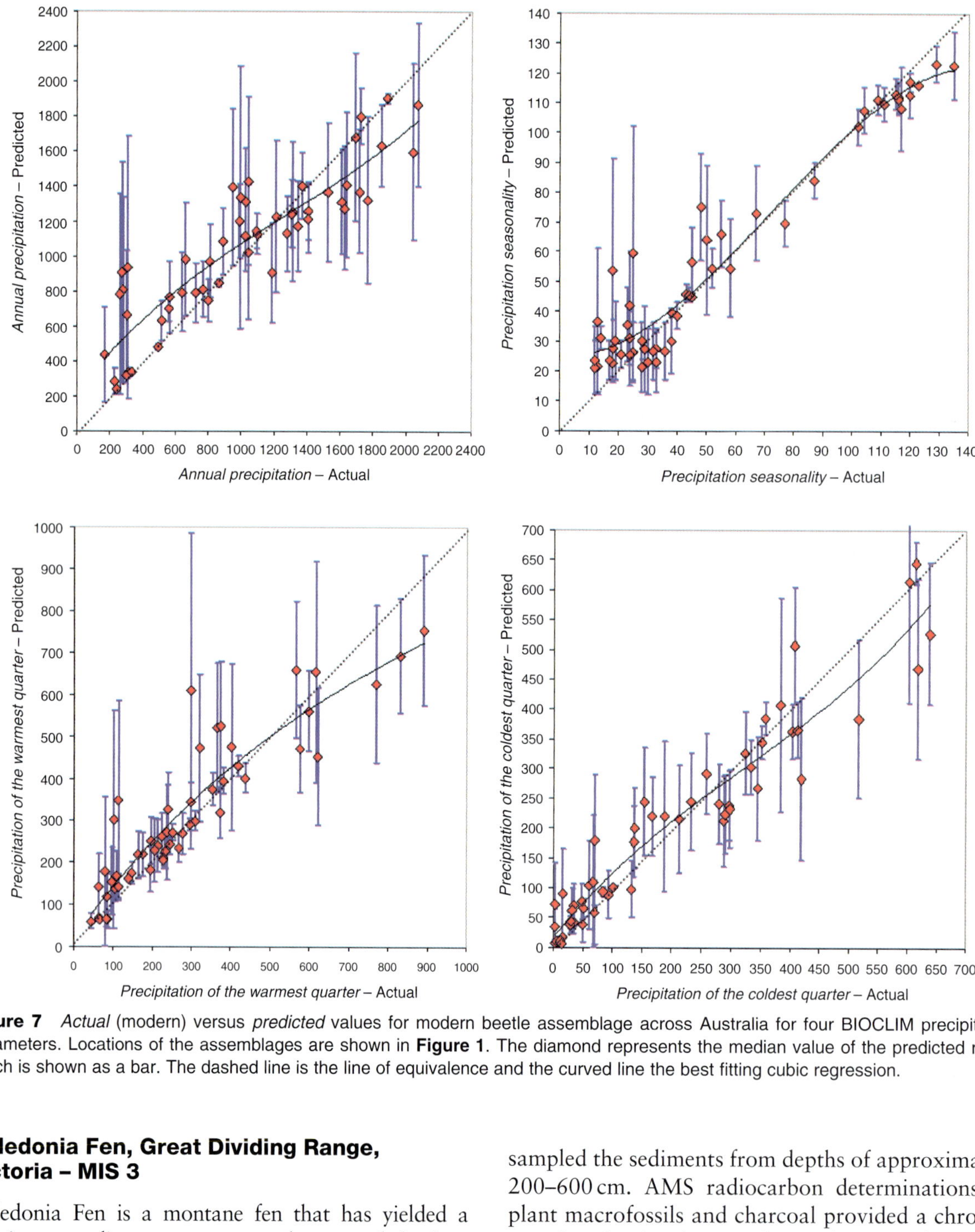

Figure 7 *Actual* (modern) versus *predicted* values for modern beetle assemblage across Australia for four BIOCLIM precipitation parameters. Locations of the assemblages are shown in **Figure 1**. The diamond represents the median value of the predicted range which is shown as a bar. The dashed line is the line of equivalence and the curved line the best fitting cubic regression.

Caledonia Fen, Great Dividing Range, Victoria - MIS 3

Caledonia Fen is a montane fen that has yielded a continuous sedimentary sequence that extends back to MIS 6 (*see* Australia and New Zealand for more details; **Fig. 8**). The pollen record is dominated by grasses and daisies indicating the local and regional presence of steppe-grassland for the majority of the record for times other than for the Holocene, MIS 5e, and a short period at the beginning of MIS 3 when forest composed primarily of *Eucalyptus* was established.

A series of small beetle assemblages were recovered from three large diameter cores (80 mm) which sampled the sediments from depths of approximately 200–600 cm. AMS radiocarbon determinations on plant macrofossils and charcoal provided a chronology for the beetle samples, indicating they were deposited from approximately 50 to 25 radiocarbon kyr BP. Most of the core samples yielded a limited fauna of a few species, but several samples contained larger faunas that indicate subalpine rather than alpine conditions.

Most significantly, from the base of the sampled sequence to approximately 32–35 radiocarbon kyr BP, specimens of the hydraenid *subgenus Hygrotympanogaster* and the elmid *Austrolimnius* were

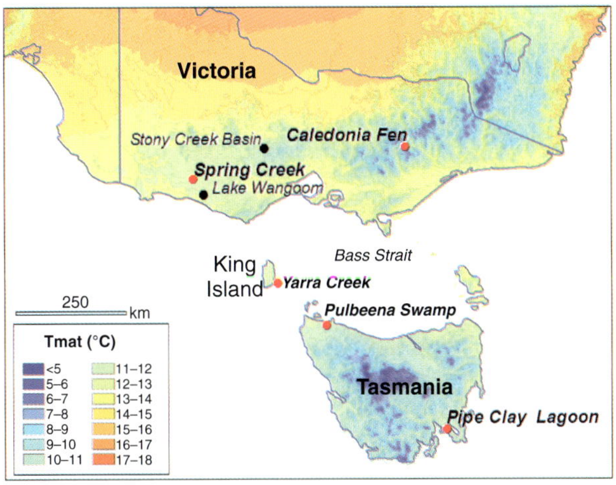

Figure 8 Locations of late Pleistocene beetle faunas (red) and other sites (black) discussed in the text shown in relation to *mean annual temperature*.

commonly present. *Hygrotympanogaster* are found exclusively in hygropetric habitats either in the spray zone of waterfalls or in the splash zone of rocky streams, whereas *Austrolimnius* are restricted to sandy and gravely microhabitats in flowing streams. The presence of these taxa indicates that through this period the fen was fed by a perennial stream (there is no inflowing stream presently), implying that effective precipitation was higher than it is today.

Pulbeena Swamp, Northwestern Tasmania – MIS 3

Pulbeena Swamp is a large, drained swamp in northwest Tasmania (**Fig. 8**). The deposits are at 30 m elevation and consist of bedded late Quaternary peat, biochemically precipitated marls, shelly marls, and peaty marls. A 5 m section exposed during the late 1970s has been analyzed for its stratigraphic and pollen sequences (Colhoun *et al.*, 1982). The strongly humified nature of the MIS 2 peat means that insect fossils are extremely rare or very poorly preserved. Less humified, marly peat, that dates to the latter part of MIS 3 and earlier, do contain small, mainly terrestrial, beetle faunas. The youngest peaty layer that contains well-preserved beetles dates to the latter

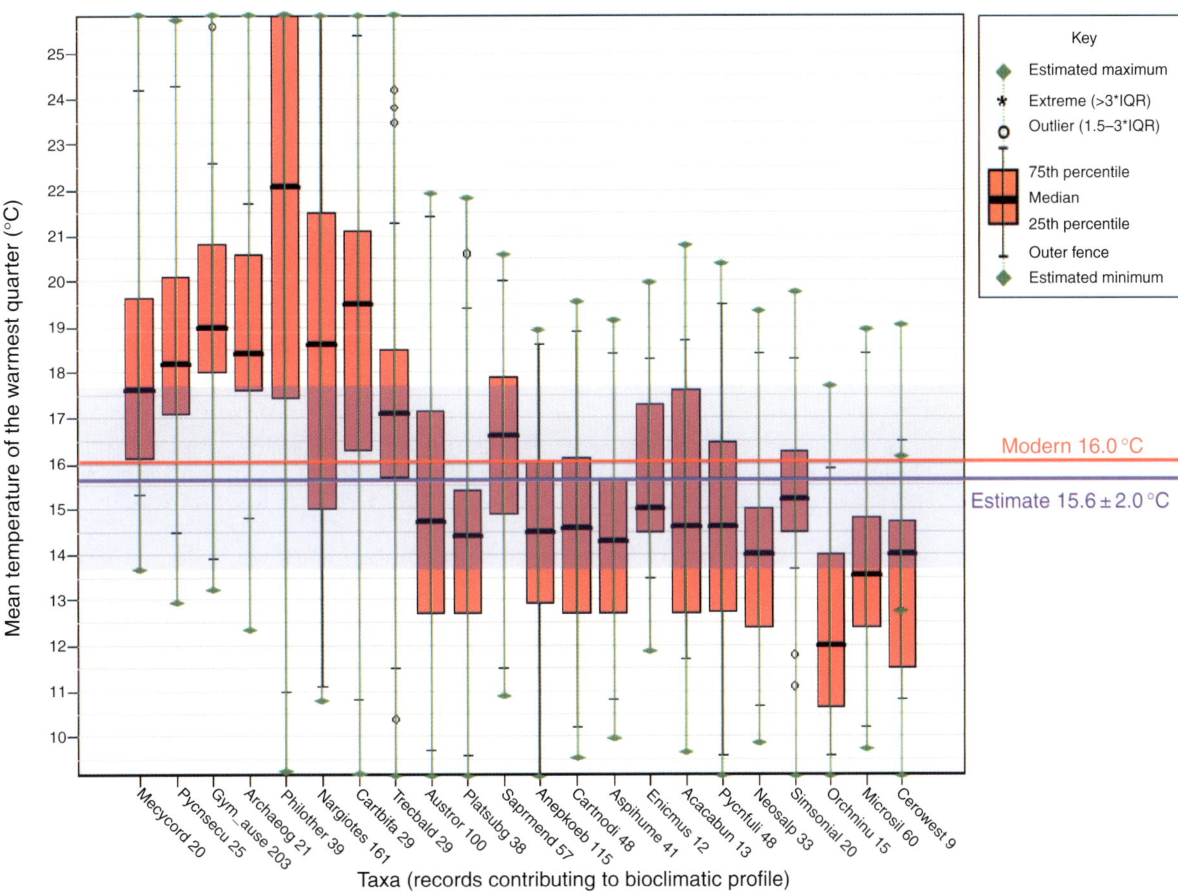

Figure 9 *Mean Temperature of the Warmest Quarter* (*TwarmQ*) box-plots with estimated 'error' ranges for 22 beetle taxa from the Yarra Creek assemblage. Based on the *estimated parameter ranges* (using 'errors') the value for the assemblage of 15.6° C ± 2.0° C, overlaps the modern value of 16.0° C. *Raw* data values (inner bars or outliers) narrow this range to 15.45° C ± 0.45° C on the basis of the ranges of two species, both of which are, however, known from 20 or fewer records.

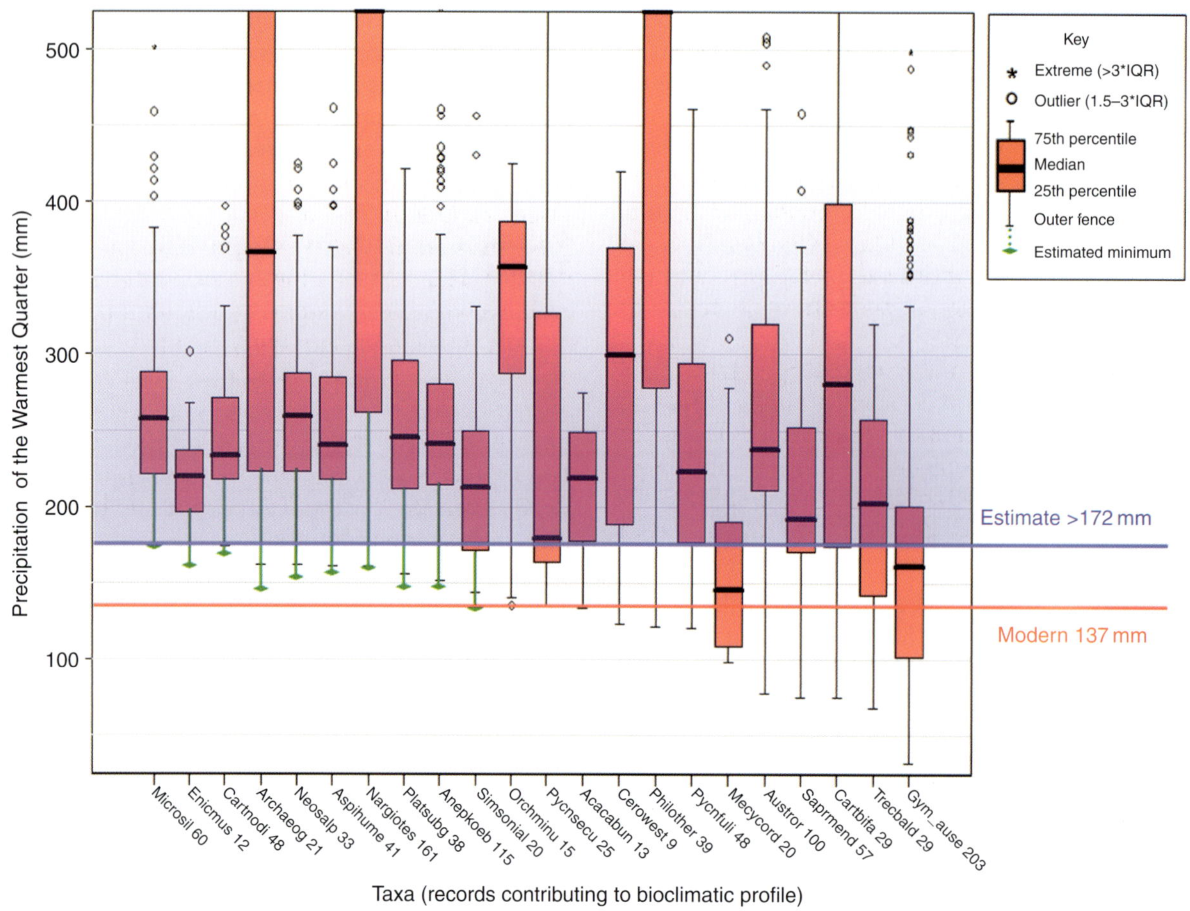

Figure 10 *Precipitation of the Warmest Quarter* (*PwarmQ*) box-plots with estimate minima 'error' ranges for 22 beetle taxa from the Yarra Creek assemblage. *PwarmQ* is estimated to be >172 mm. Nine taxa in the assemblage have *PwarmQ* minima greater than the modern value of 137 mm when using the *estimated parameter range* values increasing to 11 of 22 using the *raw* data.

part of MIS 3 (28–30 radiocarbon kyr BP), on the basis of a series of conventional and AMS determinations.

The fauna of this layer is indicative of deposition in a riparian swamp forest. The diversity of forest/scrub litter inhabiting Staphylinidae (including many pselaphinae) and cryptorhynchine weevils is typical of closed habitats while the presence of carabid taxa like *Bembidion blackburni* and *Pseudoceneus* sp. is indicative of more open riparian habitats. The presence of a range of taxa, including several anobiids and the wet forest/rainforest carabid *Promecoderus viridiaeneus*, indicate a closed canopy forest, probably of the scrub taxa *Melaleuca* and/or *Leptospermum* rather than *Eucalyptus* or cool temperate rainforest species, on the basis of the pollen record. Unfortunately, only a few of the specifically identified taxa have bioclimatic profiles and all but one have broad climatic tolerances. *Promecoderus viridiaeneus*, however, is informative suggesting that *Pmap* and *PwarmQ* were at least as high as the present and that summer temperatures were lower than the present, possibly by only 0.5° C, potentially by up to a maximum of 6° C.

Pipe Clay Lagoon, Eastern Tasmania – MIS 3/2 Boundary

The sedimentary sequence referred to as the Pipe Clay Lagoon site was described by Colhoun (1977) on the basis of a section of inorganic and organic sediments exposed on the eastern side of Pipe Clay Lagoon, a marine embayment in eastern Tasmania (**Fig. 8**). Colhoun (1977) described the sedimentary sequence and its pollen record, which was dated using radiocarbon to the period spanning the transition from MIS 3 to MIS 2. It is notable as one of the first glacial-age pollen records from southern Australia, particularly because it preserves, in a lowland rain-shadow area (< 600 mm annual precipitation), organic sediments deposited just prior to the height of the LGM.

The insect assemblages described here are derived from the uppermost of two organic sand beds.

Colhoun (1977, p. 8) listed two radiocarbon determinations from the same sample from this layer, 19,840 ± 280 yr BP (SUA153) and a more intensively pre-treated 21,920 ± 330 yr BP (SUA153/2). A single AMS determination on the upper organic sand was obtained on *Neopaxia* (Portulacaceae) seeds to attempt to resolve the original chronological issues. The result, 21,070 ± 100 (OZF799), 25,279 ± 356 cal yr BP, generally confirms the published chronology for the upper organic sand, placing it firmly at the MIS 3–MIS 2 transition.

The insect fauna is dominated by aquatic and riparian taxa. The aquatic taxa belong primarily to species that are common in a wide range of aquatic habitats. The fauna suggests deposition in shallow (<1 m, probably <30 cm) standing, peaty water. The riparian fauna includes a range of taxa common at the margins of standing water in southeastern Australia. Also present is the carabid *Cyphotrechodes gibbipennis*, which is today restricted to dense accumulations of litter beside water, especially *Sphagnum* and other riparian mosses in high rainfall areas (raw data: >970 mm *Pmap*). The small terrestrial fauna includes two aphodiine dung beetles: *Ataenius basiceps*, which today attains a toehold in the warm lowlands of northeastern Tasmania, and *Podotenus erosus*, which is apparently restricted to moist habitats in central and northern Tasmania.

The beetle-based climatic reconstruction for the Pipe Clay Lagoon assemblage suggests that the climate was similar or slightly warmer than today in terms of summer temperatures (*TwarmQ*: modern 15.5° C, predicted 16.6° C; *TmaxM*: modern 20.9° C, predicted 23.6° C), cooler than today in terms of winter temperatures (*TcoldQ*: modern 8.2° C, predicted 7.3° C; *TminM*: modern 3.8° C, predicted 1.7° C) and substantially wetter than present (*Pmap*: modern 653 mm, predicted >934 mm) including over the summer season (*PwarmQ*: modern 121 mm, predicted >141 mm). Increase in summer temperatures and decrease in winter temperatures would be expected to occur with the increased continentality of MIS 3/2; the magnitude of the expected changes in this regard is difficult to determine.

Conclusions

The results described above indicate that in the last glacial assemblages there can be significant discrepancies between the vegetation inferred paleoclimate and the paleoclimate reconstructed on the basis of beetle assemblages. For the Australian record there are several potential explanations for these discrepancies: the lack of taxonomic variability in the southeastern Australian lowland pollen record means it is paleoclimatically uninformative; the beetle-based estimates are unreliable; or finally, and most probably, that the differences reflect unresolvable discrepancies in the nature of environmental space that existed during glacials (see Jackson and Overpeck (2000) for a discussion).

Kershaw *et al.* (2004) suggest influences such as extreme winter events, lower carbon dioxide levels, and, unconvincingly, altered precipitation seasonality (all implying changes in the nature of past environmental spaces) could potentially explain the lack of trees in 'glacial' pollen assemblages. The two glacial beetle records that provide significant evidence in this regard, Pipe Clay Lagoon and Spring Creek, clearly indicate that typical 'glacial' pollen assemblages, dominated by grasses and daisies, can be associated with climates that had, at least during summer, temperatures and rainfall much the same as, or even wetter than, today. There is clearly much potential for Quaternary beetle research to provide definitive data on the nature of past climates in the Australian region.

See also: **Beetle Records**: Overview; Late Pleistocene of New Zealand. **Pollen Records, Late Pleistocene**: Australia and New Zealand.

References

Atkinson, T. C., Briffa, K. R., Coope, G. R., Joachim, M., and Perry, D. (1986). Climatic calibration of coleopteran data. In *Handbook of Holocene Palaeoecology and Palaeohydrology* (B. E. Berglund, Ed.), pp. 851–858. Wiley, New York.

Colhoun, E. A. (1977). A sequence of late Quaternary deposits at Pipe Clay Lagoon, southeastern Tasmania. *Papers and Proceedings of the Royal Society of Tasmania* **111**, 1–12.

Colhoun, E. A., van de Geer, G., and Mook, W. G. (1982). Stratigraphy, pollen analysis and paleoclimatic interpretation of Pulbeena Swamp, northwestern Tasmania. *Quaternary Research* **18**, 108–126.

Coope, G. R. (1995). Insect faunas in ice age environments: Why so little extinction? In *Extinction Rates* (J. H. Lawton and R. M. May, Eds.), pp. 55–74. Oxford University Press, Oxford.

Elias, S. A. (1997). The mutual climatic range method of palaeoclimate reconstruction based on insect fossils: New applications and interhemispheric comparisons. *Quaternary Science Reviews* **16**, 1217–1225.

Flannery, T. F., and Gott, B. (1984). The Spring Creek locality, southwestern Victoria, a late surviving megafaunal assemblage. *Australian Zoologist* **22**, 385–422.

Hammond, P. M. (1994). Practical approaches to the estimation of the extent of biodiversity in speciose groups. *Philosophical Transactions of the Royal Society of London B* **345**, 119–136.

Harle, K. J., Kershaw, A. P., and Clayton, E. (2004). Patterns of vegetation change in southwest Victoria (Australia) over the last two glacial/interglacial cycles: Evidence from Lake Wangoom. *Proceedings of the Royal Society of Victoria* **116**, 105–137.

Houlder, D., Hutchinson, M., Nix, H., and McMahon, J. (2001). *ANUCLIM 5.1. User's Guide*. Centre for Resource and Environmental Studies, Canberra.

Jackson, S. T., and Overpeck, J. T. (2000). Responses of plant populations and communities to environmental changes of the late Quaternary. *Paleobiology* **26**, 194–220.

Jordan, G. J. (1997). Evidence of Pleistocene plant extinction and diversity from Regatta Point, western Tasmania, Australia. *Botanical Journal of the Linnean Society* **123**, 45–71.

Kershaw, A. P., and Nix, H. A. (1988). Quantitative palaeoclimatic estimates from pollen data using bioclimatic profiles of extant taxa. *Journal of Biogeography* **15**, 589–602.

Kershaw, A. P., Tibby, J., Penny, D., *et al.* (2004). Lastest Pleistocene and Holocene vegetation and environmental history of the Western Plains of Victoria, Australia. *Proceedings of the Royal Society of Victoria* **116**, 139–161.

Lawrence, J. F., and Britton, E. B. (1994). *Australian Beetles*. Melbourne University Press, Melbourne.

Marra, M. J., Smith, E. G. C., Shculmeister, K., and Leschen, R. (2004). Late Quaternary climate change in the Awatere Valley, South Island, New Zealand using a sine model with a maximum likelihood envelope on fossil beetle data. *Quaternary Science Reviews* **23**, 1637–1650.

Mosbrugger, V., and Utescher, T. (1997). The coexistence approach – A method for quantitative reconstructions of tertiary terrestrial palaeoclimate data using plant fossils. *Palaeogeography, Palaeoclimatology, Palaeoecology* **134**, 61–86.

Nix, H. A. (1986). A biogeographic analysis of Australian elapid snakes. In *Snakes: Atlas of Elapid Snakes of Australia* (R. Longmore, Ed.), pp. 4–15. Australian Government Publishing Service, Canberra.

Sniderman, J. M. K., O'Sullivan, P. B., Hollis, J. D., and Kershaw, A. P. (2004). Late Pliocene vegetation and climate change at Stony Creek Basin, Western Uplands of Victoria, Australia. *Proceedings of the Royal Society of Victoria* **116**, 77–92.

Turney, C. S. M., Kershaw, A. P., Lowe, J. J., *et al.* (2006). Climatic variability in the southwest Pacific during the Last Termination (20–10 kyr BP). *Quaternary Science Reviews* **25**, 886–903.

White, J. P., and Flannery, T. (1995). Late Pleistocene fauna at Spring Creek, Victoria: A re-evaluation. *Australian Archaeology* **40**, 13–17.

Yeates, D. K., Harvey, M. S., and Austin, A. D. (2003). New estimates for terrestrial arthropod species-richness in Australia. *Records of the Australian Museum Monograph Series* **7**, 231–241.

Late Pleistocene of Europe

M-J Gaillard, University of Kalmar, Kalmar, Sweden
G R Coope, University of London, Egham, UK

© 2007 Elsevier B.V. All rights reserved.

Introduction

In this article we summarize the response of Coleoptera (beetles) to Upper Pleistocene climatic changes. This should cast light upon our understanding of the nature of these changes and thus predict the ways in which other animal and plant species might respond to any future climatic events. By using the changes in the range of beetle species during the Quaternary, it has been possible to quantify thermal climatic conditions and to estimate both the amounts and also the rates of change involved. Furthermore, it has been possible to map the regional differences in thermal climates within northwestern Europe.

The beetle assemblages described here include those from the last (Eemian, Ipswichian) interglacial, the last (Weichselian, Devensian) glaciation and the transition to the present (Holocene) interglacial. Most of the examples will be drawn from sites in the British Isles and northern Europe because it is there that these faunas have been most intensively investigated. The beetle record shows that the climate oscillated rapidly between temperate-oceanic and Arctic-continental conditions. Where possible, we have correlated with marine isotope stages (MIS). In the account that follows paleotemperature estimates are given using the mutual climatic range (MCR) method (Atkinson *et al.* (1987) cited in Middle Pleistocene of Europe), where T_{max} indicates the mean temperature of the warmest month (July) and T_{min} indicates the mean temperature of the coldest months (January and February).

Upper Pleistocene Beetle Assemblages

Eemian Interglacial (MIS 5e)

Beetle assemblages of this age are known from seven localities in southeast England (Coope, 2000). These faunas include a number of relatively southern European species (Coope, 1990). Quantitative estimates of the thermal climate of these sites (**Fig. 1**) gave the following figures: T_{max} 18 to 24° C, T_{min} −6 to 6° C. It is likely that the actual mean July temperatures may have been about 21° C and mean January/February temperatures may have been about 4° C. Thus, the mean temperature of the warmest month was of the order of 3° C warmer than it is in southern England today. Winter figures are more uncertain but may not have been much different from those of the present day. Precipitation at that time was sufficient to maintain a vigorous flow in the rivers throughout the year. The termination of the Eemian Interglacial is difficult to investigate because it has often been eroded away, but a borehole at Elsing in the valley of the river Wensum, about 22 km west of Norwich, penetrated Eemian deposits immediately overlain by sediments with arthropod faunas indicative of rapidly alternating climates ranging from temperate to Arctic conditions (Taylor and Coope, 1985).

At the Oerel site, in northwest Germany (Behre *et al.*, 2005), beetle assemblages were recovered

Figure 1 Map showing Eemian to full-glacial Weichselian sites where fossil beetle assemblages were studied: (1) early Weichselian sites in northern Sweden, (2) Dimlington, (3) Chelford, (4) Elsing near Norwich, (5) Latton, (6) Cassington near Oxford, (7) South Kensington, (8) Isleworth, (9) Peelo, (10) Oerel, (11) Gross Todtshorn, (12) Belchatów, (13) La Grande Pile, (14) Gossau, (15) full-glacial site near Verona.

from sediments representing the late part of the interglacial. They were divided in two different faunal units, in which the older is rather species rich, and indicative of temperate conditions compared with the younger sparser fauna which lacked thermophiles. MCR estimates based on the older assemblages are: T_{max} 14 to 24° C, T_{min} −11 to 6° C.

Further southwest, at the site La Grande Pile, Vosges, France, beetle remains were analyzed from ten samples assigned Eemian deposits (Ponel, 1995). The beetle assemblages were divided into an early and a late faunal unit. The early unit is characterized by an abundance of beetles dependent on deciduous trees and thermophilous predators and scavengers. MCR estimates based on this assemblage are: T_{max} 16 to 22° C, T_{min} 0 to 10° C. The faunal unit corresponding to the later part of the interglacial includes species feeding on both deciduous and coniferous trees. A number of species indicate a mild and humid climate: T_{max} 17 to 26° C, T_{min} −6 to 12° C.

Herning Stadial (MIS 5d)

The insect assemblages from the Oerel and Gross Todtshorn sites in Germany (Walkling, 1997), attributed to this period, are made up of only a few taxa. At Oerel, a few species confined to a marshy environment were identified. At Gross Todtshorn typical heath taxa were recovered. However, the faunas were too small to permit MCR reconstructions to be made. At La Grande Pile one sample included a number of species living on open ground, such as open grassland. However, the MCR estimates do not indicate arctic conditions during this time: T_{max} 15 to 19° C, T_{min} −15 to 9° C.

Brørup Interstadial (MIS 5c)

Coleoptera and other insect remains were found and identified from deposits of Brørup age (Peräpohjola) at ten different sites in northernmost Sweden (Lemdahl, 1997). All the recorded taxa give a very consistent picture of a xeric to mesic tundra environment. A number of the identified species are not found within Fennoscandia today, but in northern Russia and Mongolia. MCR estimates indicate cold, harsh conditions: T_{max} 9 to 11° C, T_{min} −20 to −9° C. Beetle faunas characteristic of mixed conifer and birch woodland are known from at least two sites of this age in the English Midlands. The most extensively studied is from Chelford, Cheshire (Coope, 1959). MCR analysis of this fauna gives the following figures: T_{max} 15 to 18° C, T_{min} −11 to 1° C. It is likely that the actual figure for T_{max} was closer to 15° C. Thus, the summer temperatures were similar to those in central England today but the winters were substantially colder.

Results from the Oerel site present two contrasting faunas from this interstadial. The earliest part, of low diversity, included northern species. This was followed by a striking increase in number of species, with terrestrial and aquatic beetles that today are found only in temperate regions where they inhabit fens and coniferous woodlands. Similar patterns are found in the results from Gross Todtshorn. MCR estimates from Oerel are as follows: T_{max} 12 to 19° C, T_{min} −21 to 6° C. At La Grande Pile, the assemblage from this time is characterized by a rich fauna of tree-dependent species living on both deciduous and coniferous trees. Climate conditions were likely less temperate than during the Eemian: T_{max} 13 to 24° C, T_{min} −13 to 10° C.

Rederstal Stadial (MIS 5b) and Odderade Interstadial (MIS 5a)

Insect assemblages dating to the Odderade Interstadial (Tärendö) were recovered from four Swedish sites (Lemdahl, 1997). They are characterized by Arctic tundra faunas of low diversity. The climatic conditions seem to have been even more severe than those of the Brørup: T_{max} 10 to 11° C, T_{min} −20 to −17° C. There are no beetle faunas from Britain that can be unequivocally attributed to MIS 5b and 5a.

The Rederstall assemblages from Oerel and Gross Todtshorn are made up of only a few beetle taxa and indicate the presence of shrub tundra. T_{max} estimates from Gross Todtshorn are around 11° C. The Odderade Interstadial is well represented at Oerel where two different insect assemblages reflect two different climate regimes during the period. The first part of the interstadial is characterized by a thermophilus marsh fauna (T_{max} 13 to 25° C, T_{min} −8 to 10° C). During the latter part, there is a rather rapid change in which cold stenotherm species replace the temperate fauna (T_{max} 13 to 14° C, T_{min} −14 to −10° C). At Gröbern in eastern Germany, the beetle succession showed large-scale post-Eemian climatic oscillations that probably belong to this period. These involved changes of at least 7° C in mean July temperatures (Walkling and Coope, 1996). Beetle assemblages dating to the Rederstall from La Grande Pile show a fall in the number of tree-dependent taxa and a cold-adapted species was also recorded (T_{max} 13 to 20° C, T_{min} −16 to 8° C). During the Odderade, the number of tree-dependent beetles increased again. The climate was probably slightly cooler than during the Brørup (T_{max} 12 to 14° C, T_{min} 0 to 6° C).

Middle Weichselian Stadials and Interstadials (MIS 4)

This period was climatically complex. There are at least two separate Middle Weichselian Interstadial beetle faunas known from central and southern England, separated from one another by a severely cold interval. Both of the temperate episodes postdate MIS 5 (Coope, 2000b) and yield pollen indicating a treeless landscape. Beetle assemblages are known from three sites in England that date from the earlier of these two Weichselian Interstadials, namely, from Isleworth, (Coope and Angus, 1975), Cassington (Maddy *et al.*, 1998), and Latton (Lewis *et al.*, 2006). The beetle assemblages from these sites are rich and diverse and are made up wholly of temperate/oceanic species. Average MCR analyses based on these beetle faunas from these three sites gave the following estimates: T_{max} 17 to 18° C, T_{min} −4 to 4° C. At Cassington, these temperate interstadial deposits are overlain by gravels in the middle of which organic silts filled a small paleochannel with a fauna made up of exclusively Arctic and continental species. MCR analysis of the beetle assemblage yielded the following estimates: T_{max} 7 to 11° C, T_{min} −30 to −10° C. This highly characteristic cold-adapted beetle assemblage was also found at South Kensington, London, providing a faunal link between the two sites (Coope *et al.*, 1997). The level containing this Arctic fauna was overlain almost immediately by organic silty gravels which yielded a rich interstadial fauna composed entirely of temperate/oceanic beetle species even though the pollen was almost totally devoid of evidence for trees. MCR calculations of this temperate interlude gave the following estimates: T_{max} 16 to 18° C, T_{min} −6 to −1° C. This temperate interstadial period follows so closely on the arctic bed, both of them occurring within the same small channel deposit, that it seems likely that the climatic warming took place very rapidly. However, in the absence of reliable dates, the actual rate of climatic change cannot be determined.

Middle Weichselian (MIS 3)

Numerous organic deposits are known from Britain which have yielded beetle assemblages that clearly can be attributed to the Middle Weichselian on stratigraphical grounds (Coope, 1987a, b), but it is not yet possible to allocate them to their precise equivalent positions in the Greenland ice-cores chronology. Several sites that have been radiocarbon-dated to about 42 kyr BP have yielded beetle assemblages that include relatively southern species (e.g., Coope *et al.* (1961)). Because of the unreliability of radiocarbon dates of this age, it is not possible at the moment to be sure if these warmer episodes correlate with any particular Weichselian Interstadial. The beetle assemblages from many deposits that apparently date from 40 to 20 kyr BP are remarkably consistent and include numerous species with exclusively Arctic and Asiatic distributions today (Coope, 1995b). Many of these exotic species are represented by large numbers of individuals. MCR estimates have been made on over 20 of these faunas (Coope, 1987b) and an approximate average may be summarized as follows: T_{max} 9 to 11° C, T_{min} −25 to −10° C.

A Middle Weichselian insect fauna from Poland included a remarkable group of high Arctic and Asiatic species (Kasse *et al.*, 1998). MCR estimations based on this assemblage gave the following estimates: T_{max} 8 to 12° C, T_{min} −27 to −20° C. A similar Middle Weichselian insect assemblage is known from Peelo in The Netherlands where many exclusively Arctic/Asiatic species were present (Coope, 1969). These species indicate a similar climatic regime to that discussed above. Insect remains were found in deposits referring to the Oerel interstadial both at the Oerel site and at Gross Todtshorn. The beetle assemblages are composed of northern Pale-Arctic species, which indicate a treeless environment and cold climate (T_{max} 12 to 13° C, T_{min} −9 to −2° C). At the Oerel site, insect remains were also recovered from a peat layer referred to the younger Glinde Interstadial. The beetle assemblage is rather similar in composition to those from the earlier interstadial. Climate conditions were also likely very similar (T_{max} 9 to 14° C, T_{min} −17 to −2° C).

In Switzerland two insect assemblages of Middle Weichselian age have been recorded from Gossau (Jost-Stauffer *et al.*, 2005). They include several exclusively arctic and Asiatic species that do not occur anywhere in central Europe at the present day. MCR estimates gave the following figures: T_{max} 8 to 13° C, T_{min} −21 to −7° C. Middle Weichselian beetle assemblages from La Grande Pile are characterized by the presence of almost entirely arctic and continental species. Obligate tree-dependent species were not recorded. Climate was cold in general through the period (T_{max} around 12° C, T_{min} −10° C or below).

Last Glacial Maximum (MIS 2)

Insect assemblages from the Last Glacial Maximum are rare. A small beetle assemblage, obtained from Dimlington, East Yorkshire, has been radiocarbon-dated to 18.5 kyr BP (Penny *et al.*, 1969). Given the smallness of this insect fauna, only provisional MCR estimates are possible: T_{max} 9 to 11° C, T_{min} −27 to −7° C. These figures do not necessarily represent the thermal climate at the glacial maximum as a whole, since from the sedimentary context it would seem that the single fossiliferous horizon was probably atypically warm. Beetle assemblages from La Grande Pile, dating to this period, indicate very cold and harsh conditions with T_{max} mostly around 10° C and T_{min} well below −10° C. Fossil remain of beetles, the oribatid mites, which are dating to around 18.8 kyr BP, were recovered from a site close to Verona, northeastern Italy (Foddai and Minelli, 1994). They indicate climate conditions colder and wetter than today. T_{max} may have been 8–9° C lower than today.

Lateglacial Period

This period falls well within the radiocarbon dating techniques. Dates are cited here in radiocarbon years BP and calibrated calender years are shown in parentheses as cal yr BP. Correlations are made with the Greenland ice-core record (Walker *et al.*, 1999). A considerable number of studies have been carried out on material from sites (**Fig. 2**) scattered over Europe (cf. Coope *et al.* (1998)). Only a representitive selection is discussed here.

Lateglacial Stadial (GS 2) Faunas that date from the stadial (GS 2) immediately before the Lateglacial Interstadial, namely, from just before 13 kyr BP (ca. 14.7 cal kyr BP) are rather rare. In Britain, the most stratigraphically secure faunal assemblage of this age is from the North Wales coast at Glanllynnau, where a sequence of insect-bearing deposits spans almost the whole of the early part of the Lateglacial Interstadial (Coope and Brophy (1972) cited in Coope *et al.* (1998)). The stadial deposits yielded arctic and Siberian species. MCR calculations on this assemblage gave the following estimates: T_{max} 10 to 11° C, T_{min} −26 to −23° C. Only two beetle assemblages of this age are known from northern Europe. Both of these faunas are from Kullaberg, southwestern Sweden, and can be dated with confidence to this period (Lemdahl (1988) cited in Coope *et al.* (1998)). Although these include only a few species, they resemble those from Britain. Similar faunas are also known from Poland, and from the Swiss Plateau (Elias and

Figure 2 Map showing Lateglacial sites where fossil beetle assemblages were studied.

Wilkinson 1983; Gaillard and Lemdahl 1994; Coope and Elias 2000; Hadorn *et al.* 2000). MCR estimates gave the following figures: T_{max} 10 to 13° C, T_{min} −17 to 0° C).

Lateglacial Interstadial (GI 1) Beetle faunas from the Lateglacial Interstadial (GI 1) are widespread across the whole of northwestern Europe. In Britain, important faunas have been described from Glanllynnau, North Wales (Coope and Brophy (1972) cited in Coope *et al.* (1998)), St Bees, Cumbria (Coope and Joachim (1980) cited in Coope *et al.* (1998)), Gransmoor, East Yorkshire (Walker *et al.* (1993) cited in Coope *et al.* (1998)), Hollywell Coombe, Kent (Coope (1998) cited in Postglacial Europe), and from Llanilid, South Wales (Walker *et al.*, 2003). Faunas from the early part of the GI 1e interstadial (the Bølling phase), from about 12.8 kyr BP (14.7 cal kyr BP) or slightly earlier, include a wide variety of relatively southern species. At this time the vegetation was sparse and made up of pioneer species. The only trees that were present were birches and willows confined to the damp valley bottoms, and the landscape of northern Europe as a whole was largely without a woodland cover in spite of the fact that the climate was warm enough to have supported mixed oak forest to have developed.

Average MCR estimates from these sites give approximate figures: T_{max} 17 to 18° C, T_{min} −7 to −1° C. The initial warming involved a rapid rise in mean July temperatures of at least 7° C and a corresponding rise in winter temperatures substantially greater than this. The rate at which the mean July temperature rose was about 1° C per decade and even higher than this for the mean annual temperature (Coope and Brophy (1972) cited in Coope *et al.* (1998)). The beetle reconstruction that yielded this unexpectedly high rate of climatic change has subsequently been supported by evidence from the Greenland Ice Cores which shows a dramatic climatic amelioration at about this time. This episode of rapid climate warming is not indicated by beetle assemblages from southern Sweden (Lemdahl (1988) cited in Coope *et al.* (1998)) or from the coast of western Norway. During this period the Scandinavian beetle assemblages are of an arctic or boreal character. However, faunas from Poland and The Netherlands (Geel *et al.* (1989) cited in Coope *et al.* (1998)) indicate temperatures similar to those of the British Isles. Moreover, during about the same time, a warm-demanding beetle fauna turns up at the

Swiss Plateau (Elias and Wilkinson 1983; Gaillard and Lemdahl 1994; Coope and Elias 2000; Hadorn *et al.*, 2000). Similar faunal changes are traced in southern France (Ponel and Coope (1990) cited in Postglacial Europe; Ponel *et al.* (2001)).

Shortly before 12 kyr BP there was a sudden climatic cooling. T_{max} figures suddenly fell to around 15° C and never recovered again during the rest of the interstadial (i.e., the whole of the Allerød phase was cooler than the Bølling), though there appears to have been a sequence of relatively minor climatic oscillations, well illustrated by the climatic curves from Gransmoor (Walker *et al.* (1993) cited in Coope *et al.* (1998)) and Llanilid (Walker *et al.* (2003), and in summary by Lowe and Walker (1997) cited in Coope *et al.* (1998)). The frequency of these oscillations makes it difficult to give useful MCR estimates for the latter-half of the Lateglacial Interstadial. In contrast to the British Isles, the beetle assemblages from southern Sweden show that the highest temperatures in southern Sweden were reached during the latter part of the Lateglacial Interstadial. MCR estimates were, however, similar to British figures at this time. In Norway arctic faunas are recorded for the Allerød phase (Birks *et al.* (1993), Lemdahl (2000a); cited in Coope *et al.* (1998)). In Poland, species with a boreal distribution appear during this period (Lemdahl (1991a) cited in Coope *et al.* (1998)), which also seems to be the case in southern France. These beetle assemblages show that during the Lateglacial Interstadial (GI 1) there were distinct regional differences in climate and steep temperature gradients across much of northern Europe (Coope and Lemdahl (1995) cited in Coope *et al.* (1998)).

Lateglacial Stadial GS 1 The beginning of the GS 1 (Younger Dryas) cold period was probably about 11 kyr BP (12.9 cal kyr BP). Many beetle assemblages are known from the Younger Dryas period (e.g., Coope and Joachim (1980) cited in Coope *et al.* (1998), Coope (1998) cited in Postglacial Europe, and Walker *et al.* (1993, 2003)). These assemblages include a large number of exclusively arctic species, many of which were previously encountered in the assemblages from full-glacial times, although the exclusively Asiatic species are rare in Younger Dryas faunas. An average of MCR estimates based on faunas from this period indicates the following figures: T_{max} 9 to 11° C, T_{min} −10 to −20° C. MCR reconstructions show that, in contrast to the Lateglacial Interstadial, relatively similar climatic conditions seem to have prevailed in most of northern Europe at this time. High arctic beetle faunas colonized eastern Finland in recently deglaciated areas (Bondestam *et al.*, 1994). Although Younger Dryas beetle assemblages are very rare from the Swiss Plateau, they suggest cold, arctic conditions (Lemdahl (2000b) cited in Postglacial Europe). Many sites that include the transition from the Lateglacial to the Holocene show signs of a hiatus, making it difficult to estimate the rate of change involved. However all the available beetle coleopteran evidence suggests that the change was abrupt and intense and probably similar in magnitude to the one at the start of the Lateglacial Interstadial (e.g., Ashworth (1973), Lemdahl (1991b) cited in Coope *et al.* (1998) and Lemdahl (2000a)). There is no evidence of any gradual warming.

See also: **Beetle Records**: Overview; Middle Pleistocene of Europe; Postglacial Europe.

References

Bondestam, K., Vasari, A., Vasari, Y., Lemdahl, G., and Eskonen, K. (1994). Younger Dryas and Preboreal in Salpausselkä foreland, Finnish Karelia. *Dissertationes Botaniceae* **234**, 161–206.

Behre, K.-E., Hölzer, A., and Lemdahl, G. (2005). Botanical macro-remains and insects from the Eemian and Weichselian site at Oerel (NW-Germany) and their evidence for the history of climate. *Vegetation History and Archaeobotany* **14**, 31–53.

Coope, G. R. (1959). A Late Pleistocene insect fauna from Chelford, Cheshire. *Proceedings of the Royal Society of London B* **151**, 70–86.

Coope, G. R. (1968). An insect fauna from Mid-Weichselian deposits at Brandon, Warwickshire. *Philosophical Transactions of the Royal Society of London B* **254**, 425–456.

Coope, G. R. (1969). Insect remains from Mid Weichselian deposits at Peelo, the Netherlands. *Mededelingen Rijks Geologische Dienst* **20**, 79–83.

Coope, G. R. (1970). Interpretations of quaternary insect fossils. *Annual Review of Entomology* **15**, 97–120.

Coope, G. R. (1987a). The response of late quaternary insect communities to sudden climatic changes. In *Organisation of Communities – Past and Present* (J. H. R. Gee and P. S. Giller, Eds.), pp. 421–438. Blackwell, Oxford.

Coope, G. R. (1987b). Fossil beetle assemblages as evidence of sudden and intense climatic changes in the British Isles during the last 45,000 years. In *Abrupt Climatic Change, Environment and Implications* (W. H. Berger and L. D. Labeyrie, Eds.), pp. 147–150. Reidel, Dordrecht.

Coope, G. R. (1990). The invasion of Northern Europe during the Pleistocene by Mediterranean species of Coleoptera. In *Biological Invasions in Europe and the Mediterranean Basin* (F. di Castri, A. J. Hansen and M. Debussche, Eds.), pp. 203–215. Kluwer, Dordrecht.

Coope, G. R. (1995a). Insect faunas in ice age environments: Why so little extinction? In *Extinction Rates* (J. H. Lawton and R. M. May, Eds.), pp. 55–74. Oxford University Press, Oxford.

Coope, G. R. (1995b). The effects of Quaternary climatic changes on insect populations: Lessons from the past. In *Insects in a Changing Environment* (R. Harrington and N. E. Stork, Eds.), pp. 29–48. Academic Press, London.

Coope, G. R. (2000). The climatic significance of coleopteran assemblages from the Eemian deposits of southern England. *Geologie en Mijnbouw* **79**, 257–267.

Coope, G. R. (2000b). Middle Devensian (Weichselian) coleopteran assemblages from Earith, Cambridgeshire (UK) and their bearing on the interpretation of "Full glacial" floras and faunas. *Journal of Quaternary Science* **15**, 779–788.

Coope, G. R. (2002). Changes in the thermal climate in northwestern Europe during marine oxygen isotope stage 3, estimated from fossil insect assemblages. *Quaternary Research* **57**, 401–408.

Coope, G. R., and Angus, R. B. (1975). An ecological study of a temperate interlude in the middle of the last glaciation, based on fossil Coleoptera from Isleworth, Middlesex. *Journal of Animal Ecology* **44**, 365–391.

Coope, G. R., and Elias, S. A. (2000). The environment of Upper Palaeolithic (Magdalenian and Azilian) hunters at Hauterive-Champréveyres, Neuchâtel, Switzerland, interpreted from coleopteran remains. *Journal of Quaternary Science* **15**, 157–175.

Coope, G. R., Gibbard, P. L., Hall, A. R., Preece, R. C., Robinson, J. E., and Sutcliffe, A. J. (1997). Climatic and environmental reconstructions based on fossil assemblages from Middle Devensian (Weichselian) deposits of the River Thames at South Kensington, Central London UK. *Quaternary Science Reviews* **6**, 1163–1195.

Coope, G. R., and Lemdahl, G. (1995). Regional differences in the Lateglacial climate of northern Europe based on coleopteran analysis. *Journal of Quaternary Science* **10**, 391–395.

Coope, G. R., Lemdahl, G., Lowe, J. J., and Walkling, A. P. (1998). Temperature gradients in northern Europe during the last glacial–interglacial transition (14–9 ^{14}C kyr BP.) interpreted from coleopteran assemblages. *Journal of Quaternary Science* **13**, 419–433.

Coope, G. R., Shotton, F. W., and Strachan, I. (1961). A late Pleistocene fauna and flora from Upton Warren, Worcestershire. *Philosophical Transactions of the Royal Society of London B* **244**, 379–421.

Elias, S. A., and Wilkinson, B. (1983). Lateglacial insect fossil assemblages from Lobsigensee (Swiss Plateau). Studies in the late quaternary of Lobsigensee 3. *Revue de Paléobiologie* **2**, 189–204.

Foddai, D., and Minelli, A. (1994). Fossil arthropods from a full-glacial site in northeastern Italy. *Quaternary Research* **41**, 336–342.

Gaillard, M.-J., and Lemdahl, G. (1994). Lateglacial insect assemblages from Grand-Marais, South-western Switzerland – Climatic implications and comparison with pollen and plant macrofossil data. *Dissertationes Botaniceae* **234**, 287–308.

Hadorn, P., Thew, N., Coope, G. R., Lemdahl, G., Hajdas, I., and Bonani, G. (2000). A Late-Glacial and early Holocene environment and climate history for the Neuchâtel region (CH). In *Équilibres et Ruptures dans les Ecosystems Durant les 20 Derniers Millénaires en Europe de l'Ouest, Actes du Colloque International de Besanç on, Septembre 2000* (H. Richard and A. Vignot, Eds.), pp. 75–90. Presses Universitaires Franc-Comtoises, Besançon.

Jost-tauffer, M., Coope, G. R., and Schlüchter, C. (2005). Environmental and climatic reconstructions during marine oxygen isotope stage 3 from Gossau, Swiss Midlands, based on coleopteran assemblages. *Boreas* **34**, 53–60.

Kasse, C., Huitzer, A. S., Krzyszkowski, D., Bohncke, S. J. P., and Coope, G. R. (1998). Weichselian Late-Pleniglacial and Late-Glacial depositional environment, Coleoptera and periglacial climate records from central Poland (Belchatów). *Journal of Quaternary Science* **13**, 455–469.

Lemdahl, G. (1997). Early Weichselian insect faunas from northern Sweden: Climatic and environmental implications. *Arctic and Alpine Research* **29**, 63–74.

Lemdahl, G. (2000a). Late-glacial and early-Holocene Coleoptera assemblages as indicators of local environment and climate at Kråkenes Lake, western Norway. *Journal of Paleolimnology* **23**, 57–66.

Lewis, S. G., Maddy, D., and Buckingham, C. (2006). Pleistocene fluvial sediments, palaeontology and archaeology of the upper River Thames at Latton, Wiltshire, England. *Journal of Quaternary Science* **21**, 181–205.

Maddy, D., Lewis, S. G., Scaife, R. G., *et al.* (1998). The Upper Pleistocene deposits at Cassington, near Oxford, England. *Journal of Quaternary Science* **13**, 205–231.

Penny, L. F., Coope, G. R., and Catt, J. A. (1969). Age and insect fauna of the Dimlington silts, East Yorkshire. *Nature* **224**, 65–67.

Ponel, P. (1995). Rissian, Eemian and Würmian Coleoptera assemblages from La Grande Pile (Vosges, France). *Palaeogeography, Palaeoclimatology, Palaeoecology* **114**, 1–41.

Ponel, P., Andrieu-Ponel, V., Parchoux, F., Juhasz, I., and de Beaulieu, J.-L. (2001). Late-Glacial and Holocene high-altitude environmental changes in Vallée des Merveilles (Alpes Maritimes, France): Insect evidence. *Journal of Quaternary Science* **16**, 795–812.

Taylor, B. J., and Coope, G. R. (1985). Arthropods in the quaternary of East Anglia – Their role as indices of local palaeoenvironments and regional palaeoclimates. *Modern Geology* **9**, 159–185.

Walker, M. J. C., Coope, G. R., and Lowe, J. J. (1993). The Devensian (Weichselian) Lateglacial Palaeoenvironmental Record from Gransmoor, East Yorkshire, England. *Quaternary Science Reviews* **12**, 659–680.

Walker, M. J. C., Björck, S., Lowe, J. J., *et al.* (1999). Isotopic "events" in the GRIP ice-core: A stratotype for the late Pleistocene. *Quaternary Science Reviews* **18**, 1143–1150.

Walker, M. J. C., Coope, G. R., Sheldrick, C., *et al.* (2003). Devensian lateglacial environmental changes in Britain: A multi-proxy environmental record from Llanilid, South Wales, UK. *Quaternary Science Reviews* **22**, 475–520.

Walking, A. P. (1997). Käferkundliche Untersuchungen an weiselzeitlichen Ablagerungen der Bohrung Gross Todtshorn (Kr. Harburg; Niedersachsen). *Schriftenreihe derDeutschen Geologischen Gesellschaft* **4**, 87–102.

Walkling, A. P., and Coope, G. R. (1996). Climatic reconstructions from the Eemian/early Weichselian transition in Central Europe based on the coleopteran record from Göbern, Germany. *Boreas* **25**, 145–159.

Late Pleistocene of Japan

M Hayashi, Hoshizaki Green Foundation, Izumo, Japan

© 2007 Elsevier B.V. All rights reserved.

Introduction

The study of Japanese Quaternary insects began in 1978 at the Tategahana Site of the Nojiri-ko Site Group, Nagano Prefecture. This study was carried out by the Fossil Insect Research Group for Nojiri-ko Excavation (FIRGNE). Tategahana is a Paleolithic site, yielding bone artifacts and large-mammal fossils, formed on the shore of Lake Nojiri

(Japanese name, Nojiri-ko) during the last glaciation. In total, the research group has so far identified 85 species of beetles from the site (see discussion of the Nojiri-ko insect fauna, below). This research group also studied the methodology of Quaternary Entomology and published a Japanese handbook on fossil insects (FIRGNE, 1988).

In Japan, Quaternary peat beds have yielded abundant fossil beetles, especially ground beetles (Carabidae) and donaciine leaf beetles (Donaciinae). Last glacial peat beds are known from many Japanese regions but only a few studies on fossil beetles have so far been performed. However, Japanese Quaternary entomologists have recorded 115 species of beetles from 16 sites (**Tables 1, 2**). These beetle records are reviewed here.

Fossil Sites

Sixteen fossil sites are shown in **Table 1** and **Fig. 1**. Most sites are on Honshu, the main island of Japan, and only one site is on Kyushu, in western Japan (**Fig. 1**). The late Pleistocene fossil beds are mainly discovered in archeological excavations, terrace deposits, and lake deposits.

The Owatari II Site is on the lower terrace of Waga River, Yuda-machi, Iwate Prefecture. The terrace deposit is intercalated with three peat beds, identified as the "1st" peat (*ca.* 14–12 ka), the "2nd" peat (*ca.* 23–19 ka), and the "3rd" peat (*ca.* 25–24 ka), respectively (Mori, 1995). Twenty-eight taxa of beetles from these deposits have been recorded by Mori (1995). These include *Elaphrus japonicus* (Carabidae), *Ilybius poppiusi* (Dytiscidae), *Dineutus orientalis* (Gyrinidae), *Enochrus* sp. (Hydrophilidae), *Plesiophthalmus* sp. (Tenebrionidae), and *Plateumaris sericea* (Chrsysomelidae). Mori recognized two unidentified species of the genus *Agabus* (Dytiscidae) as being extinct species in Japan.

The Tomizawa Site is on Sendai Plain, Sendai City, Miyagi Prefecture. Late Pleistocene beetles were found from the '27th' to '25th' beds of the site. Mori and Itoh (1992) recorded 43 taxa of beetles from the beds, such as *Pterostichus* sp. (Carabidae), *Noterus japonicus* (Noteridae), *Ilybius poppiusi* (Dytiscidae), *Laccobius bedeli* (Hydrophilidae), *Geotrupes auratus* (Scarabaeidae), *Donacia splendes* (Chrysomelidae), and *Limnobaris* sp. (Curculionidae).

The Maebashi Peat Formation (**Fig. 2**) is found in the Maebashi Terrace, Takasaki City and Maebashi City, Gunma Prefecture. The formation is characterized by peat beds intercalated with silt, sand, gravel beds, and volcanic ash layers. Hayashi (1996, 2005) reported fossil beetles from peat beds

Table 1 Late Pleistocene fossil sites in Japan

No.	*Site*	*Strata*	*Age*	*References*
	Iwate Pref.			
1	Owatari II Site, Yuda-machi	3rd to 1st Peat Beds	*ca.* 25-12 ka	Mori, 1995
	Miyagi Pref.			
2	Tomizawa Site, Sendai City	27-25th Beds	*ca.* 24-19 ka	Mori and Itoh, 1992
	Gunma Pref.			
3	Sojya, Maebashi City	Maebashi Peat Bed	*ca.* 14-13 ka	Hayashi, 1996
4	Shimowada, Takasaki City	Maebashi Peat Bed	*ca.* 24-13 ka	Hayashi, 2005
5	Taira, Annaka City	Tsukumogawa Peat Bed	*ca.* 25-24 ka	Hayashi and Shimadu, 2005
6	Nukui, Agatsuma-machi	Nukui Peat Bed	*ca.* 25-24 ka	Hayashi, 1999
	Saitama Pref.			
7	Taguro, Tamagawa-mura	Taguro Peat Bed	Last glacial age	Nirei and Hayashi, 1998
	Nagano Pref.			
8	Ikejirigawa Site, Shinano-machi	Nojiri-ko Formation	*ca.* 50-10 ka	FIRGNE, 1987, 1990
9a	Nakamachi Site, Shinano-machi	Nojiri-ko Formation	*ca.* 50-10 ka	FIRGNE, 1993b, 1996b, 2000b
9b	Nakamachi Site, Shinano-machi	Kannoki Formation	*ca.* 60 ka	FIRGNE, 1993b, 1996b, 2000b
9c	Nakamachi Site, Shinano-machi	Biwajimaoki Peat F.	*ca.* 70 ka	FIRGNE, 1993b, 1996b, 2000b
10	Tategahana Site, Shinano-machi	Nojiri-ko Formation	*ca.* 50-10 ka	FIRGNE, 1980, 1984, 1987, 1990, 1993a, 1996a, 2000a, 2003
11	Ono, Tatsuno-machi	Ono Peat Bed	*ca.* 70-60 ka	Hayashi, 1998
	Niigata Pref.			
12	Tarukuchi Site, Asahi-mura	Terrace deposit	*ca.* 25-24 ka	Hayashi and Miyatake, 1996
13	Garahagi Site, Asahi-mura	Terrace deposit	*ca.* 40-30 ka	Hayashi and Miyatake, 1996
14	Kujiranami, Kashiwazaki City	Yasuda Formation	*ca.* 100 ka	Kashiwazaki Naumann's Elephant Research Group, 1991
	Gifu Pref.			
15	Miyanomae Site, Miyagawa-mura	Terrace deposit	*ca.* 14-12 ka	Mori *et al.* 1997
	Kagoshima Pref.			
16	Shimogose, Yoshimatsu-cho	Mizozono Formation	unknown	Hayashi *et al.* 2002

Table 2 Late Pleistocene beetle records in Japan. Site number refers to **Table 1**.

Scientific name/Fossil site	1	2	3	4	5	6	7	8	9a	9b	9c	10	11	12	13	14	15	16
Carabidae																		
Calosoma inquisitor (Linnaeus)												+						
Carabus albrechti Morawitz												+						
Carabus granulatus Linne								+	+		+	+						
Carabus insulicola Chaudoir								+				+				+		
Carabus vanvolxemi Putzeys								+			+			+				
Damaster blaptoides Kollar						+						+						
Elaphrus japonicus S.Ueno	+						+	+			+				+			
Scarites terricola pacificus Bates																+		
Trigonognatha aurescens Bates														+				
Lesticus magnus (Motschulsky)																+		
Pterostichus leptis Bates																	+	
Pterostichus longinquus Bates											+							
Pterostichus microcephalus (Motschulsky)											+							
Pterostichus planicollis (Motschulsky)									+							+		
Pterosticus prolongatus Morawitz											+							
Colpodes japonicus (Motschulsky)											+							
Epomis nigricans (Wiedemann)																+		
Chlaenius circumdatus Brulle																+		
Chlaenius gebleri Ganglbauer										+	+							
Chlaenius pallipes Gebler											+							
Calleida lepida Redtenbacher											+							
Dytiscidae																		
Platambus pictipennis (Sharp)									+			+						
Agabus conspicuus Sharp							+		+									
Agabus japonicus Sharp				+							+		+					
Agabus miyamotoi Nakane														+				
Agabus optatus Sharp												+						
Ilybius apicalis Sharp								+	+									
Ilybius poppiusi Zaitsev	+	+		+														
Ilybius weymarni Balfour-Browne			+															
Rhantus erraticus Sharp											+							
Dytiscus marginalis czerskii Zaitzev												+						
Dytiscus sharpi Wehncke										+								
Cybister brevis Aube								+			+							
Cybister japonicus Sharp																+		
Hydrophilidae																		
Coelostoma orbiculare (Fabricius)	+			+	+			+	+		+	+			+	+		+
Coelostoma stultum (Walker)		+						+				+						
Pachysternum haemorrhoum Motschulsky												+						
Anacaena asahinai M.Sato	+								+									
Enochrus japonicus (Sharp)											+					+		
Hydrophilus acuminatus Motschulsky								+			+	+				+		
Sternolophus rufipes (Fabricius)												+				+		
Hydrochara affinis (Sharp)								+				+						
Hydrochara libera (Sharp)				+	+			+			+	+						+
Regimbartia attenuata (Fabricius)	+																	
Gyrinidae																		
Dineutus orientalis (Modeer)	+							+				+						
Histeridae																		
Atholus duodecimstriatus (Gyllenhal)											+							
Hister concolor Lewis												+						
Hister simplicisternus Lewis																		+
Margarinotus niponicus (Lewis)											+							
Silphidae																		
Nicrophorus vespilloides (Herbst)											+							
Oiceoptoma thoracicum (Linne)												+						
Silpha longicornis Portevin								+				+		+				
Eusilpha japonica (Motschulsky)												+						+
Phosphuga atrata (Linnaeus)												+						

(*Continued*)

Table 2 (*Continued*)

Scientific name/Fossil site	1	2	3	4	5	6	7	8	9a	9b	9c	10	11	12	13	14	15	16
Staphylinidae																		
Paederus parallelus Weise												+						
Scaphidiidae																		
Scaphidium rufopygum Lewis												+						
Lucanidae																		
Nipponodorcus rubrofemoratus (Snellen van Vollenhoven)												+						
Platycerus acuticollis Y. Kurosawa								+										
Geotrupidae																		
Geotrupes auratus Motschulsky								+			+	+			+			
Scarabaeidae																		
Copris pecuarius Lewis												+						
Copris tripartitus Waterhouse																+		
Caccobius jessoensis Harold												+						
Caccobius nikkoensis Lewis											+							
Onthophagus lenzii Harold											+							+
Aphodius brachysomus Solsky								+				+						
Aphodius breviusculus (Motschulsky)												+						
Aphodius elegans Allibert												+						
Aphodius igai Nakane												+						
Aphodius quadratus Reiche								+			+							
Aphodius rectus											+	+						
Aphodius rufipes (Linne)											+	+						
Aphodius yamato Nakane												+						
Phyllopertha diversa Waterhouse		+																
Mimela holosericea (Fabricius)		+																
Mimela testaceipes Motschulsky												+						+
Mimela cuprea Hope												+						
Anomala octiescostata (Burmeister)																+		
Rhomborrhina japonica Hope																	+	
Rhomborrhina unicolor Motschulsky											+	+				+		
Eucetonia roelofsi (Harold)										+		+						
Byrrhidae																		
Cytilus sericeus (Forster)										+	+		+					
Elateridae																		
Selatosomus puncticollis (Motschulsky)											+							
Lagriidae																		
Luprops cribrifrons Marseul																	+	
Trogossitidae																		
Leperina squamulosa (Gebler)												+						
Coccinellidae																		
Harmonia axyridis (Pallas)												+						
Cerambycidae																		
Pterolophia caudata Bates												+						
Spondylis buprestoides Linnaeus																	+	
Chrysomelidae																		
Plateumaris constricticollis (Jacoby)		+	+	+	+		+	+	+		+	+			+			
Plateumaris sericea (Linne)	+						+				+		+	+				
Plateumaris shirahatai Kimoto		+																
Donacia flemora Goecke																		+
Donacia japana Chujo et Goecke								+			+							+
Donacia lenzi (Schonfeldt)									+		+					+		+
Donacia ozensis Nakane											+	+				+		
Donacia provostii Fairmaire																+		
Donacia sparganii Ahrens								+			+							
Donacia splendens Jacobson				+	+		+	+	+		+	+			+			
Donacia vulgaris Zschach																+		
Basilepta fulvipes (Motschulsky)																+		
Chrysolina aurichalcea (Mannerheim)												+		+		+		
Chrysolina exanthematica (Wiedemann)				+														
Gastrolina depressa Baly								+										

(*Continued*)

Table 2 (*Continued*)

Scientific name/Fossil site	1	2	3	4	5	6	7	8	9a	9b	9c	10	11	12	13	14	15	16
Gastrolina peltoidea (Geblar)												+						
Linaeidea aenea (Linne)											+	+						
Gonioctena japonica Chujo et Kimoto												+						
Cneorane elegans Baly																	+	
Liroetis coeruleipennis Weise		+																
Agelastica coerulea Baly												+						
Attelabidae																		
Byctiscus puberulus (Motschulsky)	+											+						
Byctiscus venustus (Pascoe)												+						
Euops punctatostriatus (Motschulsky)												+						
Curculionidae																		
Limnobaris japonica Yoshihara et Morimoto					+						+							
Hylobitelus pinastri (Gyllenhal)	+																	
Dyscerus roelofsi (Harold)								+				+						
Niphades variegatus (Roelofs)	+																	

Figure 1 Late Pleistocene fossil sites in Japan. Site numbers match those in **Table 1**.

Figure 2 The Maebashi Peat Formation in Shimowada, Takasaki City, Gunma Pref. © 2005, Photograph, M. Hayashi.

at two sites, Sojya in Maebashi City and Shimowada in Takasaki City. The base of the peat at the Shimowada site is older than that of the Sojya site. The age of the Shimowada peat is *ca.* 2.0 ka (Nirei and Hayashi, 2004). Hayashi (1996) recorded 10 taxa from the Sojya Site and Hayashi (2005) recorded 14 taxa from the Shimowada Site.

The Tsukumogawa Peat Bed is in the lower terrace deposit of the Tsukumogawa River, Annaka City, Gunma Prefecture. The Aira-Tn tephra, dated *ca.* 25–24 ka, overlies the peat bed. Hayashi and Shimadu (2005) recorded 12 taxa of beetles from the bed, such as *Chlaenius* sp. (Carabidae), *Agabus* sp. *Ilybius* sp., *Hydrochara libera* (Hydrophilidae), and *Donacia splendens*.

The Nukui Peat Bed is found in Nukui, Agastuma-machi, Gunma Prefecture. The peat bed is intercalated with the Aira-Tn tephra. *Damaster blaptoides* (Carabidae) and *Plateumaris* sp. (Chrysomelidae) were found in this bed.

The Taguro Peat Bed is distributed in Taguro, Tamagawa-mura, Saitama Prefecture. A radiocarbon date from the wood was more than 38,020 yr BP (Gak-19303). Nirei and Hayashi (1998) recorded six taxa of beetles from this deposit: *Elaphrus japonicus*, *Tachyura* sp., *Agabus conspicuus*, *Plateumaris constricticollis*, *Plateumaris sericea*, and *Donacia splendens*.

Figure 3 The Tategahana Site in Shinano-machi, Nagano Pref. At 12th Excavation in 1993. © 2005, Photograph, FIRGNE.

The Nojiri-ko Site Group in Shinano-machi, Nagano Prefecture includes Tategahana, Nakamachi, Ikejirigawa, and other sites (**Fig. 3**). Fossil beetles were found from three of the sites. The Upper Pleistocene in the site groups is divided into three strata: the Biwajima-oki Peat Formation (*ca.* 100–70 ka), the Kannoki Formation (*ca.* 60 ka) and the Nojiri-ko Formation (*ca.* 50–10 ka), respectively. The Nojiri-ko Excavation Research Group has excavated the Nojiri-ko Formation from these three sites. The Biwajima-oki Peat Formation and the Kannoki Formation have been excavated at the Nakamachi Site. The last glacial beetles from the group of sites have been reported by FIRGNE (1980, 1984, 1987, 1990, 1993a, b, 1996a, b, 2000a, b, 2003).

The Ono Peat bed is located in Ono, Tatsuno-machi, Nagano Prefecture. Its age is *ca.* 60 ka. Hayashi (1998) recorded eight species of beetles from the bed, including well-preserved fossils of *Agabus japonicus* (Dytiscidae) and *Plateumaris sericea*.

The Miomote Site Group is on river terraces of the Miomote River, Asahi-mura Niigata Prefecture. This group includes Tarukuchi (**Fig. 4**), Garahagi, and other sites. Two peat beds were found from the Tarukuchi Site. The upper peat bed overlies the Aira-Tn tephra. These peat beds yielded *Carabus vanvolxemi* (Carabidae), *Agabus miyamotoi* (Dytiscidae), *Silpha longicornis* (Silphidae), *Plateumaris sericea* and some undetermined beetles (Hayashi and Miyatake, 1996). A radiocarbon age from the peat bed of the Garahagi Site was more than 35,900 yr BP (Gak-18168). Hayashi and Miyatake (1996) recorded nine taxa of beetles from the bed, including *Elaphrus japonicus*, *Agabus* sp., *Coelostoma orbiculare*, *Donacia splendens*, and *Plateumaris constricticollis*.

The Yasuda Formation is a deposit from the last interglacial (MIS 5e) in Kashiwazaki City, Niigata Prefecture. Abundant fossil beetles comprising 24 taxa were obtained from the formation during the excavation of the Naumann's Elephant (*Palaeoloxodon naumanni*) in 1986 (Kashiwazaki Naumann's Elephant Research Group, 1991). Beetle faunas from the last interglacial are poorly studied in Japan. The fauna is dominated by water beetles, ground beetles, plant-feeding beetles, and dung beetles. Most species are still found today at the fossil site and in the surrounding regions, but a dung beetle, *Copris tripartitus* is no longer living on the island of Honshu. The ground beetles include *Carabus insulicola*, *Scarites terricola pacificus*, *Lesticus magnus*, *Pterostichus planicollis*, *Epomis nigricans* and *Chlaenius circumdatus*. The plant-feeding beetles include four donaciine beetles of *Donacia ozensis*, *Donacia lenzi*, *Donacia provostii* and *Donacia*

Figure 4 (A), The Tarukuchi Site in Asahi-mura, Niigata Pref.; (B), peat bed in the site. © 2005, Photograph, M. Hayashi.

vulgaris. The paleoenvironments of the Yasuda Formation have been reconstructed as ranging from forest edges to grassland, surrounding an area of still water with emergent and floating vegetation.

The Miyanomae Site is on a terrace of the Miyagawa River, Miyagawa-mura, Gifu Prefecture. Twenty-four taxa of beetles were found from the late Paleolithic horizon, radiocarbon dated between 14–12 ka (Mori *et al.*, 1997).

The Mizozono Formation (**Fig. 5**) is a late Pleistocene deposit in Yoshimatsu-cho, Kagoshima Prefecture, but its exact age is unknown. Hayashi *et al.* (2002) recorded 36 taxa of beetles, such as *Hister simplicisternus* (Histeridae), *Eusilpha japonica* (Silphidae), and *Onthophagus lenzii* (Scarabaeidae), and *Donacia flemora* (Chrysomelidae) from this site.

Fossil Records

The fossil beetles identified from the various Upper Pleistocene fossil assemblages in Japan are shown in **Table 2**.

Figure 5 The Mizozono Formation in Shimogose, Yoshimatsu-cho, Kagoshima Pref. © 2005, Photograph, E. Kitabayashi.

Carabidae

Abundant ground beetle fossils are typically found in Japanese peat beds, but their specific identification has been difficult for Japanese Quaternary entomologists. Therefore, most fossils have been identified only to the subfamily or tribe level. Twenty-one species are recorded from ten sites (**Table 2**): six species of Carabinae, one species of Elaphrinae, one species of Scaritidae, eight species of Pterostichinae, four species of Callistinae, and one species of Lebiineae. *Elaphrus japonicus* (**Fig. 6**) is recorded from five sites and *Carabus granulatus* (**Fig. 7**) is recorded from four sites. Today these species inhabit marshes.

Dytiscidae

Thirteen species of predaceous diving beetles are recorded from eleven Japanese sites (**Table 2**). These comprise nine species of Colymbetinae and four species of Dytiscinae. *Agabus japonicus* (**Fig. 8**) and *Ilybius poppiusi* (**Fig. 9**) are each known from three sites. Two species of *Dytiscus* are recorded from the Nojiri-ko Site Group: *D. marginalis czerskii* (**Fig. 10**) from the Nojiri-ko Formation in the Tategahana, and *D. sharpi* (**Fig. 11**) from the Kannoki Formation in the Nakamachi Site.

Hydrophilidae

Ten species of water scavenger beetles have been recorded from ten sites (**Table 2**): three species of Sphaeridiinae, two species of Hydrobiinae, four species of Hydrophilinae, and one species of Chaetritrinae. *Coelostoma orbiculare* (**Fig. 12**) is known from nine sites and *Hydrochara libera* is known from six sites.

Figure 6 *Elaphrus japonicus*, mesosternum and right elytron, from the Nojiri-ko Formation in the Ikejirigawa Site. © 2005, Photograph, FIRGNE. Scale bar = 1 mm.

Figure 7 *Carabus granulatus*, left elytron, from the Nojiri-ko Formation in the Tategahana Site. © 2005, Photograph, FIRGNE. Scale bar = 1 mm.

Figure 8 *Agabus japonicus*, elytra and sternum, from the Ono Peat Bed. © 2005, Photograph, M. Hayashi. Scale bar = 1 mm.

Figure 9 *Ilybius poppiusi*, left elytron, from the Maebashi Peat Formation. © 2005, Photograph, M Hayashi. Scale bar = 1 mm.

Most of these beetles are aquatic species but *Pachysternum haemorrhoum* is a dung-feeder.

Gyrinidae

One species of whirligig beetle, *Dineutus orientalis*, is recorded from the Owatari II, Ikejirigawa, and Tategahana Sites. Undetermined specimens of *Gyrinus* spp. are known from several sites.

Histeridae

Four species of hister beetles are recorded from three sites (**Table 2**): *Atholus duodecimstriatus* (**Fig. 13**) and *Margarinotus niponicus* from the Nakamachi Site, *Hister concolor* (**Fig. 14**) from the Tategahana Site, and *Hister simplicisternus* from Shimogose, Kagoshima Prefecture.

Figure 10 *Dytiscus marginalis czerskii*, right elytron, from the Nojiri-ko Formation in the Tategahana Site. © 2005, Photograph, FIRGNE. Scale bar = 1 mm.

Figure 11 *Dytiscus sharpi*, head, from the Kannoki Formation in the Nakamachi Site. © 2005, Photograph, FIRGNE. Scale bar = 1 mm.

Silphidae

Five species of carrion beetles are recorded from five sites: one species of Nicrophorini and four species of Silphini. *Silpha longicornis* (**Fig. 15**) is known from three sites. This species is currently found in the cool-temperate zone forests in Japan.

Figure 12 *Coelostoma orbiculare*, left elytron, from the Nojiri-ko Formation in the Ikejirigawa Site. © 2005, Photograph, FIRGNE. Scale bar = 1 mm.

Figure 13 *Atholus duodecimstriatus*, right elytron, from the Biwajima-oki Peat Formation the Nakamachi Site. © 2005, Photograph, FIRGNE. Scale bar = 1 mm.

Staphylinidae

Indetermined fossils of the rove beetle family are commonly found from the Pleistocene peat beds but one species, *Paederus parallelus* is recorded from the Tategahana Site.

Figure 14 *Hister concolor*, right elytron, from the Nojiri-ko Formation in the Tategahana Site. © 2005, Photograph, FIRGNE. Scale bar = 1 mm.

Figure 16 *Scaphidium rufopygum*, left elytron, from the Nojiri-ko Formation in the Tategahana Site. © 2005, Photograph, FIRGNE. Scale bar = 1 mm.

Figure 15 *Silpha longicornis*, pronotum, from the Nojiri-ko Formation in the Ikejirigawa Site. © 2005, Photograph, FIRGNE. Scale bar = 1 mm.

Figure 17 *Platycerus acuticollis*, pronotum, from the Nojiri-ko Formation in the Ikejirigawa Site. © 2005, Photograph, FIRGNE. Scale bar = 1 mm.

Scaphidiidae

Fossils of this family (shining fungus beetles) are poorly known in Japan. One species, *Scaphidium rufopygum* (**Fig. 16**) is recorded from the Tategahana Site.

Lucanidae

Two species of stag beetles are recorded (**Table 2**): *Nipponodorcus rubrofemoratus* from the Tategahana Site and *Platycerus acuticollis* (**Fig. 17**) from the Ikejirigawa Site. *P. acuticollis* currently lives in the cool-temperate zone forest in Japan.

Geotrupidae

One species of earth-boring dung beetle, *Geotrupes auratus* (**Fig. 18**), is recorded from four sites. It is found today in broad-leaved forests of mountain regions, where it feeds on animal dung.

Figure 18 *Geotrupes auratus*, protibia, from the Nojiri-ko Formation in the Ikejirigawa Site. © 2005, Photograph, FIRGNE. Scale bar = 1 mm.

Scarabaeidae

Twenty-one scarab beetle species are recorded from seven sites (**Table 2**): five species of Scarabaeinae, eight species of Aphodiinae, five species of Rutelinae, three species of Cetoniinae. Thirteen species of Scarabaeinae and Aphodiinae are dung-feeders and the other species of subfamilies are plant-feeders.

Byrrhidae

One species of pill beetle, *Cytilus sericeus* (**Fig. 19**), is recorded from the Nakamachi Site and Ono, Nagano Prefecture.

Elateridae

Undetermined fossils of click beetles are commonly found from the Pleistocene peat beds. One species, *Selatosomus puncticollis* is recorded from the Nakamachi Site.

Lagriidae

Fossils of the long-jointed beetle family are poorly known in Japan. One species, *Luprops cribrifrons* is recorded from the Miyanomae Site.

Figure 19 *Cytilus sericeus*, right elytron, from the Biwajima-oki Peat Formation in the Nakamachi Site. © 2005, Photograph, FIRGNE. Scale bar = 1 mm.

Trogossitidae

Fossils of the bark-gnawing beetle family are poorly known in Japan. One species, *Leperina squamulosa* (**Fig. 20**) is recorded from the Tategahana Site.

Figure 20 *Leperina squamulosa*, pronotum, from the Nojiri-ko Formation in the Tategahana Site. © 2005, Photograph, FIRGNE. Scale bar = 1 mm.

Coccinellidae

Fossils of the ladybird family are poorly known in Japan. One species, *Harmonia axyridis* is recorded from the Tategahana Site.

Cerambycidae

Two species of long-horn beetles are recorded, *Pterolophia caudata* (**Fig. 21**) from the Tategahana Site and *Spondylis buprestoides* from the Miyanomae Site.

Chrysomelidae

Twenty-one species of leaf beetles are recorded from fifteen sites (**Table 2**): eleven species of Donaciinae, one species of Eumolpinae, six species of Chrysomelinae, and three species of Galerucinae. *Plateumaris constricticollis* (**Fig. 22**) is known from nine sites and *Donacia splendens* (**Fig. 23**) is known from seven sites. These two species of Donaciinae are commonly found from the Pleistocene peat beds in Japan.

Attelabidae

Three species of leaf-rolling beetles are recorded from two sites (**Table 2**): *Byctiscus puberulus* from the Owatari II Site and the Tategahana Site, *Byctiscus*

Figure 21 *Pterolophia caudata*, right elytron, from the Nojiri-ko Formation in the Tategahana Site. © 2005, Photograph, FIRGNE. Scale bar = 1 mm.

Figure 22 *Plateumaris constricticollis*, pronotum, from the Nojiri-ko Formation in the Ikejirigawa Site. © 2005, Photograph, FIRGNE. Scale bar = 1 mm.

Figure 23 *Donacia splendens*, pronotum and elytra, from the Biwajima-oki Peat Formation in the Nakamachi Site. © 2005, Photograph, FIRGNE. Scale bar = 1 mm.

venustus and *Euops punctatostriatus* from the Tategahana Site.

Curculionidae

Four weevil species are recorded from five sites (**Table 2**): one species of Baridinae and three species of Hylobiinae. *Limnobaris japonica* (**Fig. 24**) is known from two sites that lives in marsh and eats leaves of *Carex* spp.

Figure 24 *Limnobaris japonica*, head and pronotum, from the Biwajima-oki Peat Formation in the Nakamachi Site. © 2005, Photograph, FIRGNE. Scale bar = 1 mm.

Figure 26 *Hydrophilus acuminatus*, head, from the Nojiri-ko Formation in the Tategahana Site. © 2005, Photograph, FIRGNE. Scale bar = 1 mm.

Nojiri-Ko Insect Fauna

The Nojiri-ko insect fauna comprises the last glacial insects from the Nojiri-ko Site Group, including 85 species of beetles from the Nojiri-ko, the Kannoki, and the Biwajima-oki Peat Formations. The fauna is dominated by water beetles, ground beetles, plant-feeding beetles, and dung beetles. Most species are found today in the cool-temperate zone of Japan.

The aquatic beetles include Dytiscidae, Hydrophilidae and Gyrinidae. Species of *Agabus*, *Ilybius*, *Rhantus*, *Dytiscus*, *Cybister* (**Fig.** 25), *Coelostoma*, *Anacaena*, *Enochrus*, *Hydrophilus* (**Fig.** 26), *Sternolophus*, *Hydrochara*, *Regimbartia*, and *Dineutus* live in areas of still water with emergent and floating vegetation. *Platambus pictipennis* (**Fig.** 27) lives in streams. This species is found in the Nojiri-ko Formation at the Tategahana Site.

Figure 25 *Cybister brevis*, head, from the Biwajima-oki Peat Formation in the Nakamachi Site. © 2005, Photograph, FIRGNE. Scale bar = 1 mm.

Figure 27 *Platambus pictipennis*, left elytron, from the Nojiri-ko Formation in the Tategahana Site. © 2005, Photograph, FIRGNE. Scale bar = 1 mm.

The terrestrial beetles include Carabidae, Histeridae and Silphidae. Most of them live on the forest floor and/or its edge. Five species live in wet ground with vegetation, including *Carabus granulatus* (**Fig.** 7), *Elaphrus japonicus* (**Fig.** 6), *Pterostichus prolongatus* (**Fig.** 28), *Epomis nigricans*, and *Chlaenius gebleri* (**Fig.** 29). Two species, *Calosoma inquisitor* (**Fig.** 30) and *Calleida lepida* live in trees, not on the ground.

Figure 28 *Pterosticus prolongatus*, pronotum, from the Biwajima-oki Peat Formation in the Nakamachi Site. © 2005, Photograph, FIRGNE. Scale bar = 1 mm.

Figure 29 *Chlaenius gebleri*, pronotum, from the Biwajima-oki Peat Formation in the Nakamachi Site. © 2005, Photograph, FIRGNE. Scale bar = 1 mm.

The plant-feeding beetles include Lucanidae, Scarabaeidae (the genera *Mimela*, *Anomala*, *Rhomborrhina* and *Eucetonia*), Cerambycidae, Chrysomelidae, Attelabidae and Curculionidae. Most of these species live in forests and/or the forest edge. Species of *Plateumaris*, *Donacia*, and *Limnobaris* are tied to aquatic host-plants, such as Cyperaceae, Potamogetonaceae, Sparganiaceae, Typhaceae, and Nymphaeaceae.

The dung-feeding beetles include *Pachysternum haemorrhoum* (**Fig. 31**) (Hydrophilidae) and dung beetles (Scarabaeidae) in the genera *Copris* (**Fig. 32**), *Caccobius*, *Onthophagus* and *Aphodius* (**Fig. 33**). Two extinct mammals, Naumann's elephant (*Palaeoloxodon naumanni*) and Yabe's giant fallow deer (*Sinomegaceros yabei*) have been excavated from the Nojiri-ko Formation in the Tategahana Site.

Figure 30 *Calosoma inquisitor*, elytron, from the Nojiri-ko Formation in the Tategahana Site. © 2005, Photograph, FIRGNE. Scale bar = 1 mm.

Figure 31 *Pachysternum haemorrhoum*, left elytron, from the Nojiri-ko Formation in the Tategahana Site. © 2005, Photograph, FIRGNE. Scale bar = 1 mm.

Figure 32 *Copris pecuarius*, pronotum, from the Nojiri-ko Formation in the Tategahana Site. © 2005, Photograph, FIRGNE. Scale bar = 1 mm.

Figure 33 *Aphodius brachysomus*, left elytron, from the Nojiri-ko Formation in the Tategahana Site. © 2005, Photograph, FIRGNE. Scale bar = 1 mm.

The dung beetles from the Tategahana Site may have fed on the dung of these mammals.

Biogeography

The Japanese Islands are surrounded by sea, therefore one might expect many endemic species to be found on the islands. However, during the Pleistocene, these islands were connected to continental Asia during intervals of lowered sea level. Modern entomologists have speculated that many recent species immigrated to the Japanese Islands during the last glaciation (e.g. Ishikawa, 1988). However, late Pleistocene fossil records indicate that the recent fauna was formed before the last glaciation.

Distributional ranges of beetles have shifted in response to environmental changes in geologic time. The transition from glacial to interglacial intervals exerted a particularly strong influence on these shifts. For example, *Copris tripartitus*, recorded from the Yasuda Formation in Niigata Prefecture, is found today on Tsushima Island, Chejido Island, Korea, and China. The fossil record indicates that *C. tripartitus* lived on Honshu during the last interglaciation, and died out there during the last glacial age. On the other hand, two dytiscid beetles, *Ilybius poppiusi* (**Fig. 9**) and *Ilybius weymarni* (**Fig. 34**), were recorded from several fossil assemblages from the last glaciation on eastern Honshu. These are currently found on Hokkaido and in northern Asia. The fossil records indicate that they had been living on Honshu during the last glaciation and died out there during the Holocene. *Cytilus sericeus* (**Fig. 19**) was also recorded from several last glacial deposits

Figure 34 *Ilybius weymarni*, left elytron, from the Maebashi Peat Bed. © 2005, Photograph, M. Hayashi. Scale bar = 1 mm.

on eastern Honshu. It is found today on Hokkaido and across Asia to Europe. The distribution of these species may have been influenced by the climatic change of the Pleistocene-Holocene transition.

Future of Japanese Quaternary Entomology

Japanese Quaternary entomology will continue to supply vital information on the evolution of Japanese beetle fauna. The framework of speciation in the islands differs from that in continental regions. The formation of land bridges between island and continent or island and island are the most important element of the regional zoogeography. In the island arcs, the formation of land-bridges was caused by a combination of lowered sea level and/or uplifting of the sea floor. Japanese evolutionary scientists try to reconstruct biogeographic histories based on molecular phylogenetic trees and paleogeographic maps. However, they lack useable timescales for their theories, because the land bridge was repeatedly formed during geologic time. The fossil beetle record from Quaternary deposits provides concrete evidence that species were living at the fossil locality in times past. Therefore, evolutionary scientists are seeking information from Quaternary entomologists about the fossil records. To date, the Japanese Quaternary entomologists have insufficient data to meet these requests. However, our studies will continue, and new Quaternary entomologists will be trained.

See also: **Beetle Records**: Overview; Late Pleistocene of Northern Asia.

References

Fossil Insect Research Group for Nojiri-ko Excavation (1980). Fossil insects from the Nojiri-ko Formation. *The Memoirs of the geological Society of Japan* **19**, 147–159.

Fossil Insect Research Group for Nojiri-ko Excavation (1984). Fossil insects obtained from the Nojiri-ko Excavations in 1978 to 1982. *Monograph of the Association for the Geological Collaboration in Japan* **27**, 137–156.

Fossil Insect Research Group for Nojiri-ko Excavation (1987). Fossil insects obtained from the Nojiri-ko Excavations in the 9th Nojiri-ko Excavation and the 4th Hill Site Excavation. *Monograph of the Association for the Geological Collaboration in Japan* **32**, 117–136.

Fossil Insect Research Group for Nojiri-ko Excavation (1988). *Handbook for Fossil Insects* New Science Co. Tokyo, p. 129.

Fossil Insect Research Group for Nojiri-ko Excavation (1990). Fossil insects obtained from the Nojiri-ko Formation during the 10th Nojiri-ko and the 5th Hill Site Excavations. *Monograph of the Association for the Geological Collaboration in Japan* **37**, 93–110.

Fossil Insect Research Group for Nojiri-ko Excavation (1993a). Insect fossils obtained from the Nojiri-ko Formation during the 11th Nojiri-ko Excavation. *Bulletin of the Nojiri-ko Museum* **4**, 69–81.

Fossil Insect Research Group for Nojiri-ko Excavation (1993b). Insect fossils obtained from the Nojiri-ko Formation during the 6th Hill Site Excavation. *Bulletin of the Nojiri-ko Museum* **4**, 203–214.

Fossil Insect Research Group for Nojiri-ko Excavation (1996a). Insect fossils obtained from the Nojiri-ko Formation during the 12th Nojiri-ko Excavation. *Bulletin of the Nojiri-ko Museum* **4**, 71–80.

Fossil Insect Research Group for Nojiri-ko Excavation (1996b). Insect fossils obtained from the Nojiri-ko Formation during the 7th Nojiri-ko Hill Site Excavation. *Bulletin of the Nojiri-ko Museum* **4**, 199–206.

Fossil Insect Research Group for Nojiri-ko Excavation (2000a). Insect fossils obtained from the Nojiri-ko Formation during the 13th Nojiri-ko Excavation. *Bulletin of the Nojiri-ko Museum* **8**, 53–58.

Fossil Insect Research Group for Nojiri-ko Excavation (2000b). Insect fossils obtained from the Upper Pleistocene during the 8th Nojiri-ko Hill Site Excavation. *Bulletin of the Nojiri-ko Museum* **8**, 139–147.

Fossil Insect Research Group for Nojiri-ko Excavation (2003). Insect fossils obtained from the Nojiri-ko Formation during the 14th Nojiri-ko Excavation. *Bulletin of the Nojiri-ko Museum* **11**, 83–88.

Hayashi, M. (1996). Fossils of a Dytiscid beetle, Ilybius weymarni, from Late Pleistocene Maebashi Peat in Maebashi City, central Japan. *Daiyonki-kenkyu* **35**, 305–312.

Hayashi, M. (1998). Upper Pleistocene fossil beetles from the Ono Peat Bed in Tatsuno-machi, Nagano Prefecture, Japan. *Bulletin of the Mizunami Fossil Museum* **25**, 117–125.

Hayashi, M. (1999). Fossil donaciine beetles (Coleoptera: Chrysomelidae: Donaciinae) from the Pleistocene around the Haruna Volcano, Gunma Prefecture, central Japan. *Bulletin of the Mizunami Fossil Museum* **26**, 149–162.

Hayashi, M. (2005). Fossil Insects and reconstructed paleoenviromnets of the Upper Pleistocene Maebashi Peat in Takasaki, Gunma Prefecture, Central Japan. *Bulletin of the Gunma Museum of Natural History* **9**, 93–99.

Hayashi, M., and Shimadu, Y. (2005). Stratigraphy and fossil insects of the Upper Pleistocene Tsukumogawa Peat Bed in Annaka, Gunma Prefecture, Japan. *Bulletin of the Gunma Museum of Natural History* **9**, 101–107.

Hayashi, M., and Miyatake, Y. (1996). Last Glacial insect fossils from the Oku Miomote site group, Niigata Prefecture, Japan. *Chikyu Kagaku* **50**, 188–193.

Hayashi, M., Yahiro, K., and Kitabayashi, E. (2002). Late Pleistocene insects from the Mizozono Formation in Yoshimatsu-cho, Kagoshima Prefecture, Japan. *Bulletin of the Mizunami Fossil Museum* **29**, 161–168.

Ishikawa, R. (1988). Origin and speciation of the Japanese Caraninae. In: Sato M (ed.) *The Japanese Coleoptera: its origin and speciation*, pp. 23–32. Tokai: Tokai University Press.

Kashiwazaki Naumann's Elephant Research Group (1991). Palaeoloxodon naumanni (Makiyama) found in the upper Pleistocene, Yasuda Formation in the Kashiwazaki City, Niigata Prefecture, Central Japan, and its palaeoenvironment. *Chikyu-Kagaku* **45**, 161–176.

Mori, Y. (1995). Fossil insects and the paleoenviromnets. In. Reports on Owatari II Site Excavation, pp. 351–370. Iwate: Archaeological Research Center, Iwate.

Mori, Y., and Itoh, T. (1992). Paleoenvironments of the Tomizawa Site reconstructed by fossil insects and diatoms. In *Reports on Tomizawa Site (Paleolithic)*, pp. 126–141. Broad of Education, Sendao City.

Mori, Y., Nakamura, T., and Yoshii, R. (1997). Insect and macroplant fossils found from the Miyanomae Site. *Abstract of the Japan Association for Quaternary Research* **27**, 130–131.

Nirei, T., and Hayashi, M. (1998). Fossil pollen and insect assemblage from Late Pleistocene peat deposits at Taguro, Tamagawa Village, western margin of Kanto Plain, Central Japan. *Bulletin of the Saitama Museum of Natural History* **16**, 15–21.

Nirei, T., and Hayashi, M. (2004). Fossil pollen assemblages of the Late Pleistocene Maebashi Peat Bed and palaeoclimatic oscillation at Takasaki, Gunma, central Japan. *The Natural Environmental Science Research* **17**, 43–49.

Late Pleistocene of South America

A C Ashworth, North Dakota State University, ND, USA

© 2007 Elsevier B.V. All rights reserved.

Background

Studies of fossil beetle assemblages in South America have been directed at coming to a better understanding of how beetles respond to climate change and to the use of that information for interpreting paleoclimate. One of the primary challenges of the research has been to demonstrate that the results would be sufficiently robust for other researchers in archeology, plant paleoecology, glacial geology, and paleoclimatology to use with confidence in their studies. Of particular interest has been the question of whether a climatic reversal occurred in southern South America that was of similar magnitude and timing to the Younger Dryas interval defined from the North Atlantic region. The sites from which Pleistocene fossil assemblages have been studied in South America are located primarily in the Chilean lake region and the Chilean channels, between 40 and 50° S (**Fig. 1**). A single fossil assemblage of early to mid-Holocene age has also been examined from the Falkland Islands in the Atlantic Ocean between 51 and 52° S (Buckland and Hammond, 1997).

The Landscape of Southern Chile

In the Región de Los Lagos (lake region), the terrain is mountainous with the Cordillera de la Costa (coastal ranges) separated from the stratovolcanoes of the Cordillera de los Andes (Andes) by a wide north–south trending lowland, the Valle Longitudinal (central valley). Dense forests once cloaked the entire landscape of the lower elevations

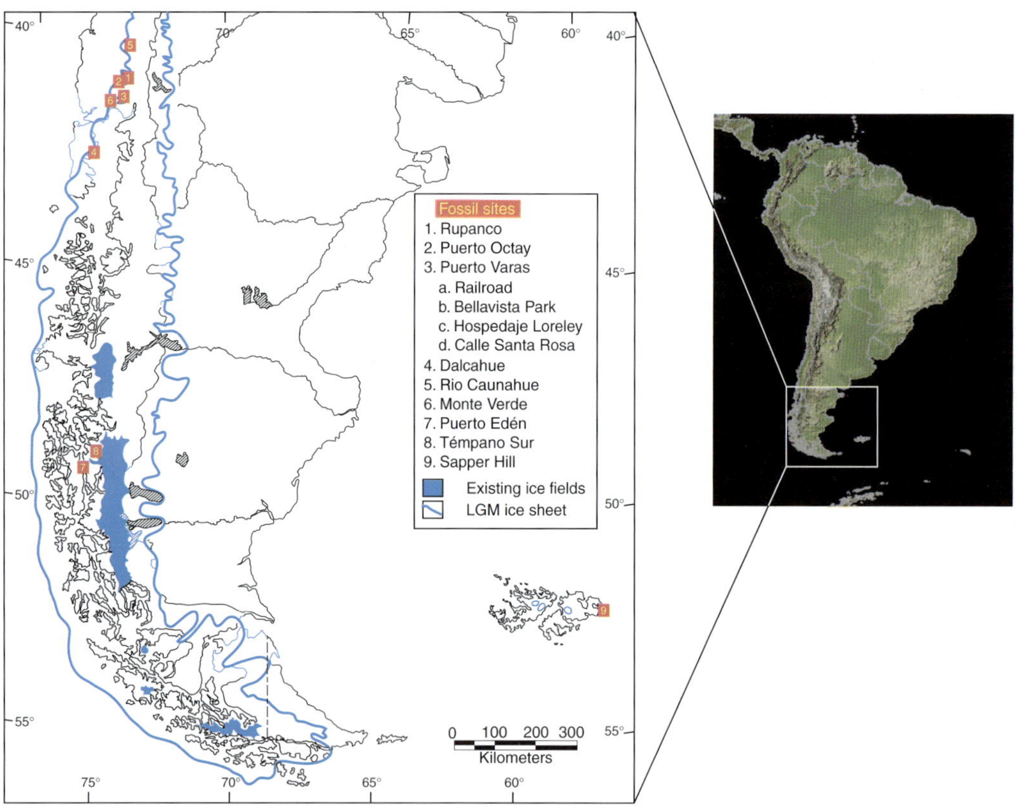

Figure 1 Location of fossil beetle assemblages from southern South America. The representation of the Patagonian ice sheet during the last glaciation is from Hollin and Schilling (1981).

Figure 2 Habitats in the Chilean lake region: (A) The lower Valdivian rain forest in the lake region near Aguas Calientes, Parque Nacional Puyehue; (B) the Andean tundra and krummholz at tree line east of Antillanca, Parque Nacional Puyehue. Volcán Casablanca is in the background. Photographs by A. C. Ashworth.

Figure 3 Habitats in the Chilean channels: (A) The Glaciar Pio XI in the Chilean channels; (B) dense rainforest of *Nothofagus betuloides* in the Fiordo Bernardo. (C) Magellanic moorland in the coastal mountains west of Puerto Edén. Photographs by A. C. Ashworth.

but have mostly been cleared for agriculture. Today, extensive tracts of native Valdivian and North Patagonian rain forests survive only at higher elevations in the lake region (**Fig. 2A**). Above tree line, the forests are replaced by an Andean tundra (**Fig. 2B**) that extends to the fell fields, and permanent snow and ice capping the higher volcanic peaks. Further south in the Chilean channels, the landscape is more fragmented. Island archipelagoes, crisscrossed by deep glacially excavated fjords, extend to the tip of the continent. The permanent snow line is lower, and outlet glaciers of the North and South Patagonian ice fields calve icebergs into the fjords (**Fig. 3A**). The lowland vegetation is a dense cold-temperate rainforest (**Fig. 3B**) which, at higher elevations, is replaced by the saturated soils and blanket bogs of the Magellanic moorland (**Fig. 3C**). The climate in southern South America, from the mid- to the high latitudes, is dominated by a westerly circulation stronger in the south than in the north. Whitlock *et al.* (2001) discuss the relationship between climate, vegetation, and the insect fauna.

The Existing Beetle Fauna

The beetle fauna of the forests, moorlands, and steppes of southern South America is unusually rich in endemic taxa. Endemism implies ancient evolutionary relationships and Patagonia is the home of many surviving lineages of ancient clades. The Migadopini, for example, are an exclusively Southern Hemisphere tribe of Carabidae (ground beetles) that have a typically Gondwanan distribution. They occur throughout southern South America, including the Falkland Islands, Tasmania and southeastern Australia, and New Zealand and the Auckland Islands. Similar restricted geographic distributions occur within several families of beetles in the southern South American fauna, notably the Protocujidae, Chalcodryidae, Nemonychidae, and Belidae. These groups are either monotypic or represented by a small number of genera that are associated with *Nothofagus* forest. The taxa in many of these families evolved during the late Cretaceous when southern South America was still connected to Antarctica. The connection was severed by seafloor spreading to form the Drake Passage between 34 and 30 Ma. The recent discovery of fossils of listroderine weevils in Neogene deposits, 500 km from the South Pole, confirms the evolutionary ties that South America had with Antarctica (Ashworth and Kuschel, 2003). Listroderine weevils are presently widely distributed in the forests, steppes, and moorlands of southern South America.

Joseph Banks, naturalist on the HMS Endeavour, Captain James Cook's first voyage of exploration, collected beetles from southern South America in 1769. One of those specimens, the ground beetle *Ceroglossus suturalis*, was described by Johann Fabricius in 1775. In the nineteenth century, Charles Darwin and other naturalists, who followed, sent their specimens back to Europe to be described by specialists in the natural history museums in London and Paris. Darwin was responsible for collecting one of the most distinctive beetles of the Chilean rain forest, the lucanid *Chiasognathus grantii* Stephens. The male is characterized by mandibles that are almost twice as long as its body (**Fig. 4A**) giving rise to its common name in southern Chile of 'ciervo volante' or 'flying deer.' In *The Descent of Man*, Darwin wrote 'But the mandibles were not strong enough so as to cause actual pain.' The specimens are preserved in the Natural History Museum, London, where the male is known as 'the beetle that bit Darwin.' The nineteenth century naturalist who made the greatest contribution to the knowledge of the Chilean fauna was Claudio Gay. He made extensive collections of the fauna during travels between 1828 and 1842, and those specimens are described and illustrated in color in his 1849 and 1851 volumes of *La Historia Física y Política de Chile* (**Fig. 4A**). More recently, Elizabeth Arias (2000) has described the most distinctive species of the fauna in *Coleópteros de Chile*.

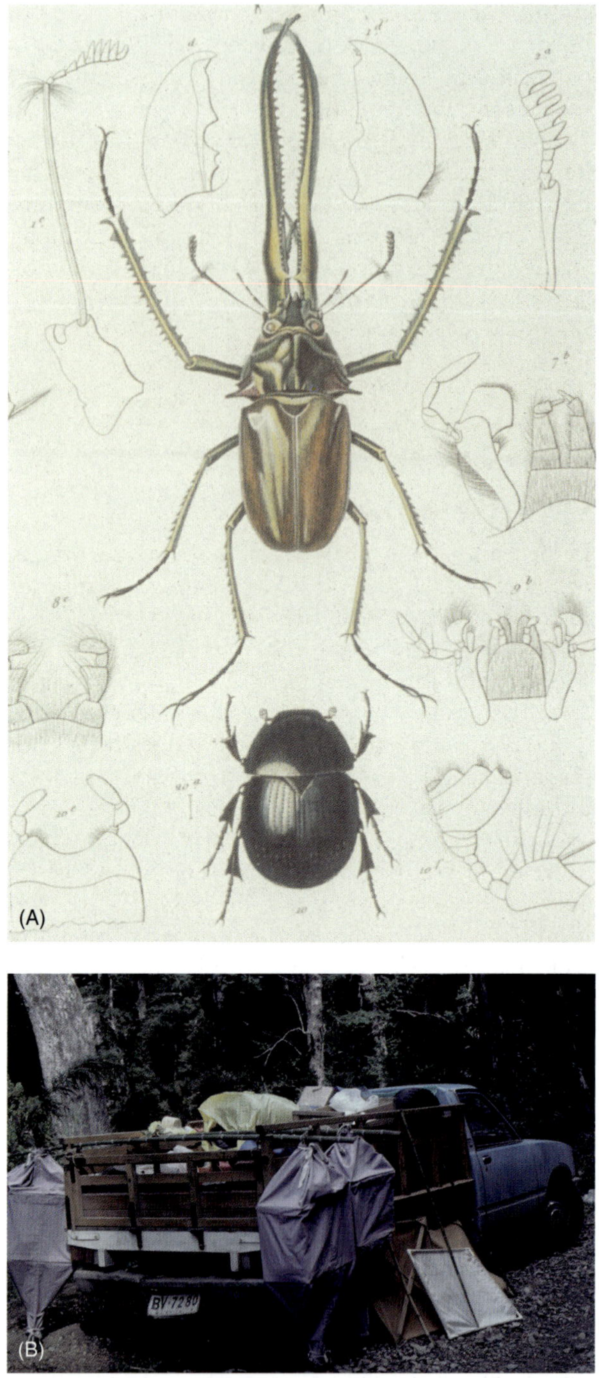

Figure 4 Modern beetle fauna: (A) Enlargement of one of the plates from Gay's nineteenth-century *Historia Física y Política de Chile*. The beetle with the large mandibles is the distinctive Valdivian rain forest lucanid *Chiasognathus grantii*; (B) studies of the modern beetle fauna were essential to establishing the base for paleoclimatic interpretation. Photograph by A. C. Ashworth.

There are numerous excellent taxonomic treatments of the existing beetle fauna which are invaluable as aids in the identification of Pleistocene fossils. There are few faunistic and ecological studies, however, on which to base paleoenvironmental and paleoclimatic interpretations. Ashworth and Hoganson (1987) made extensive collections in the Parque Nacional Puyehue and other undisturbed habitats in the Chilean lake region (**Fig. 4B**). They described a diverse fauna of 462 species from habitats collected systematically along an elevational transect from sea level to 1,200 m. Their ordination analysis showed changes in beetle communities along the transect which generally correlated with changes in vegetation.

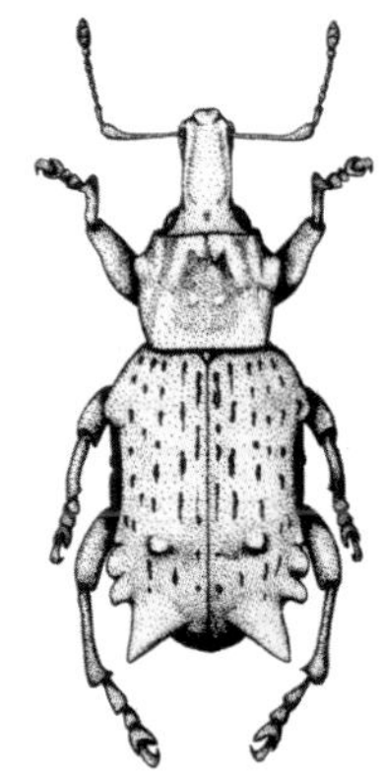

Figure 5 The weevil *Germainiellus dentipennis* (Morrone, 1993) whose exoskeletal remains are the most abundant in fossil assemblages dating from 24 to 14.000 yr BP. Reproduced from Morrone JJ (1993) Revisión systemática de un nuero género de Rhytirrhinini (Coleoptera, Curculionid), con un análisis biogeográphico del domino subantártico. Boletin de la socieded de Biológia de Conceptión, Chile, 64: 121–145.

Fossil Studies

Chilean Lake Region

In the Chilean lake region, fossil assemblages ranging in age from 24.000 to 4.525 kyr BP have been examined. Assemblages from sites at Puerto Octay, Puerto Varas (Park and Railroad), Rio Caunahue, and Monte Verde have been described and their significance discussed in a series of papers (Hoganson *et al.*, 1989; Hoganson and Ashworth, 1992; Ashworth and Hoganson, 1993). Additionally, unpublished fossil assemblages have been studied by Ashworth from sites at Rupanco, Dalcahue, and Puerto Varas (Loreley and Calle Santa Rosa) (**Fig. 1**). All of the sites, with the exception of Monte Verde, are from locations in the lowlands of the central valley that were glaciated during the Last Glacial Maximum. The Monte Verde site is located west of the terminal moraines of the last glaciation and is associated with a drainage that developed on glacial outwash.

The oldest fossil beetle assemblages to have been examined are from peat deposits near the west end of Lago Rupanco (**Fig. 1**). Fossils were extracted from an *in situ* peat that is overlain by gravel, till, and volcanic deposits (Mercer, 1976; Denton *et al.*, 1999). The assemblages range in age from 19.500 to 24.000 yr BP and are representative of the beetle fauna of the lake region immediately prior to the Last Glacial Maximum. The assemblages are the least diverse of any examined from the lake region consisting of only 9–10 species. Exceptionally well-preserved skeletal parts of the listroderine weevil *Germainiellus dentipennis* (Germain) (**Fig. 5**), including heads, pronota, elytra, sternites, leg segments, and genitalia, make up 90% of the fossils.

Similar assemblages dominated by *G. dentipennis* occur within the lake region until about 14.000 to 14.500 yr BP. The youngest of these assemblages, dated at 14.3400 yr BP, is from near Dalcahue on La Isla Grande de Chiloé (Chiloé). The peaty deposits are overlain by outwash gravels and till of the final major glacial advance in the mid-latitudes of South America (**Fig. 6**). Included in the 22 taxa from this assemblage is the weevil *Paulsenius carinicollis* (Blanchard) and the ground beetle *Cascellius septentrionalis* Roig-Juñent that today both inhabit higher elevations in the Andes. Villagrán (1990) made the observation that Magellanic moorland plants, which currently inhabit relict patches at 700 m elevation in the coastal ranges of Chiloé, grew near sea level during the last glaciation Villagrán (1990). Using an adiabatic lapse rate of 0.65° C/100 m, Ashworth and Hoganson (1993) estimated that mean summer temperature during the glacial interval was 4–5° C lower than today based on the species composition of fossil beetle assemblages. Denton *et al.* (1999) estimated that snow and tree line were lowered by 1,000 m between 29.400 and 14.400 yr BP representing mean summer temperatures 6–8° C lower than today.

Fossil assemblages postdating 14.000 yr BP are fundamentally different in their composition. Not only are they considerably more diverse but also contain higher percentages of taxa that are obligate forest inhabitants. Most of the assemblages that have been examined are from exposed cut bank sections on the Rio Caunahue, a large river that originates in the Andes and enters Lago Ranco on the east side. Rhythmically laminated deposits containing abundant organics and volcanic ashes were deposited in a narrow arm of the lake immediately following deglaciation (**Fig. 7A**). The basal sediments are inorganic clays representing the deposition of rock flour. The 7-m sequence, with a basal age of 13.900 yr BP (Ashworth and Hoganson, 1993), contains exceptionally well-preserved fossils (**Fig. 7B**). The species

Figure 6 (A) At Dalcahue, Isla Grande de Chiloé, organic sediments (brown) contain the youngest fossil beetle assemblages representing the glacial beetle fauna. Fossil wood at the top of the section (by the knife) has an age of 14.34 kyr BP; (B) organic sediments accumulated in a small depression on the surface of a weathered till. The surface was buried by outwash gravel (top of the section) during the last glacial advance in the lake region. Scale = cm divisions. Photographs by A. C. Ashworth.

compositions of 30 intervals pre-dating 10.000 yr BP were reported in Hoganson and Ashworth (1992); the number of taxa identified was 154, of which 41–48% consisted of obligate forest taxa. All of the taxa currently inhabit the forests of the lake region, mostly the Valdivian rain forest.

The only other fossil beetle assemblage post-dating 14.000 yr BP that has been examined from the lake region is from the archeological site at Monte Verde, near Puerto Montt (**Fig. 8**). At Monte Verde, humans are believed to have constructed primitive shelters and hunted mastodon. Fossil beetles were retrieved from horizons considered to be contemporaneous with the human habitation (MV-6) dated between 13.565 and 11.790 yr BP, and also from a peat bed (MV-5) that sealed the cultural horizon and is dated between 11.81 and 10.860 yr BP (Dillehay and Pino, 1989). Statistical comparison of the 95 taxa occurring in MV-6 and MV-5 demonstrated that the assemblages were similar to one another (Hoganson *et al.*, 1989).

Using cluster analysis, the species compositions of the fossil assemblages were compared to modern communities from 41 locations within the Chilean lake region. The Monte Verde assemblage was most similar to the communities from slightly disturbed coastal forests and from the Valdivian Rain Forest at about 500 m elevation (Hoganson *et al.*, 1989). The interpretation that the paleoclimate at Monte Verde was essentially similar to that of today during the human occupation is consistent with interpretations from similar-aged fossil beetle assemblages at the Rio Caunahue site.

One of the intriguing puzzles of the Monte Verde beetle assemblage is that no synanthropic beetles were identified. In Europe, fossil beetle assemblages from archeological sites ranging in age from Mesolithic to historical times show evidence of

Figure 7 (A) The Rio Caunahue assemblages are from laminated silts interbedded with volcanic ashes. The bench on which the person is standing has an age of 12.81 kyr BP and the prominent white volcanic ash above his head has an age of about 10.8 kyr BP. (B) The fused fossil elytra (wing cases) of a specimen of the ground beetle *Ceroglossus*, that is from the horizon dated at 10 kyr BP. (C) A modern specimen of *Ceroglossus* from the Valdivian rain forest. Scale bars = 5 mm. Photographs by A. C. Ashworth.

Figure 8 The Monte Verde archeological excavation. An excavated hearth-like structure is visible in the foreground. The peat MV-5 which contains abundant fossil beetles and is dated between 11.81 and 10.86 kyr BP is visible at the base of the cut on the far wall of the excavation. Photograph by A. C. Ashworth.

modification, especially those associated with agricultural. Why there would be no synanthropic insects at Monte Verde is unknown, but it is possible that hunter-gatherer groups did not disturb the natural environment to such an extent that it caused an ecological response in the beetle fauna. A similar observation was made for early Postglacial beetle faunas associated with human occupation in North America (Elias, 1994).

To examine longer-term trends represented by the fossil beetle assemblages, the total taxa and the ratio of taxa dependent on trees are examined for 1 kyr intervals between 24.000 and 9.000 yr BP (**Fig. 9A**). The beetle fauna from 24.000 yr BP to the time of the last glacial advance about 14.000 yr BP was composed of species of wet moorland habitats. The assemblages are characterized by low diversity and low numbers of taxa dependent on shrubs and trees. Between 14.000 and 13.000 yr BP the diversity is low but the percentage of arboreal taxa is higher. Between 13.000 and 12.000 yr BP, the assemblages are at least 5 times as diverse as those of the glacial interval. The changes indicate that by 13.000 yr BP, moorland habitats, which had existed in the central valley for more than 10.000 yr, had been replaced by forested habitats similar to those that exist in the region today.

The numerical data are also used to test two null hypotheses (**Fig. 9B**). The first is that no relationship exists between total taxa and ratio of arboreal taxa. This hypothesis was tested using linear regression ($\alpha = 0.05$). The null hypothesis was shown not to be true, and the relationship between total taxa and

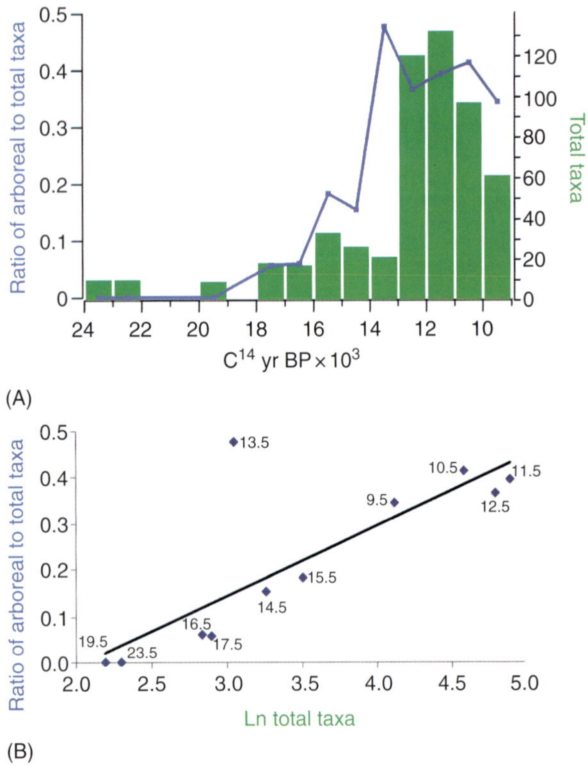

Figure 9 Longer-term trends for the interval 24–9 kyr BP represented by a compilation of data from all of the fossil assemblages: (A) plots of the total number of taxa (bar graph) and the ratio of obligate arboreal taxa (line graph); (B) the relationship between total and ratio of arboreal taxa. The *x*-axis is natural log transformed. The labels shown are mid-ranges for samples, each of which represents a 1 kyr interval of time. The line represents a significant linear regression. The data point at 13.5 (14–13 kyr BP) is a significant outlier representing the time of major change between moorland and forest assemblages.

ratio of arboreal taxa was found to be highly significant ($P = 0.001$, $r^2 = 0.62$). The interval from 14.000 to 13.000 yr BP has a studentized residual value of 11.5 making it very different from all other samples. The difference is explained because the sample represents most of the change in the transition between moorland and forested biomes.

The second null hypothesis is that no significant difference exists between the ratios of arboreal taxa for two subsets of samples between 24.000–15.000 yr BP and 13.000–9.000 yr BP. The sample representing the 13.000–14.000 yr BP interval was not included in the test because it is a statistical outlier. The null hypothesis was tested using a two-sample T-test ($\alpha = 0.05$). The result was that the null hypothesis was shown not to be true. The ratios of arboreal taxa to total taxa for the glacial interval 24.000–15.000 yr BP are significantly different than for the postglacial interval between 13.000 and 9.000 yr BP ($t = 9.842$, $df = 8.5$, $P < 0.001$).

Chilean Channels

Fossil beetle assemblages have been examined from two locations in the Chilean channels. The Puerto Edén site on the Canal Messier is from an excavation in a bog (**Fig. 10**), and the Témpano Sur site is from an outcrop in the wall of an abandoned meltwater channel about 1 km from the margin of the Glaciar Témpano, an outlet glacier of the South Patagonian ice field (**Fig. 11**). At both sites, pollen and fossil beetles were examined from the same samples (Ashworth and Markgraf, 1989; Ashworth *et al.*, 1991). As the glacier at Puerto Edén receded westward to the coastal mountains, peat began to accumulate at about 13.000 yr BP in a shallow depression between morainal ridges. Throughout the peat

Figure 10 The bog at Puerto Edén is located between morainic arcs deposited by a glacier that retreated into the coastal ranges. Glacial ice also filled the Paso del Indio, the marine channel that leads northward to the Messier Canal. Photograph by A. C. Ashworth.

Figure 11 The fossil site at Témpano Sur is located behind the bedrock ridge in the foreground. It is located in an abandoned meltwater channel that has an outlet to the fjord at the extreme right of the photograph. The floating ice is from melting of the Glaciar Témpano about 1 km to the left of the image. Photograph by A. C. Ashworth.

profile there is excellent agreement between the interpretations based on fossil beetles and those based on pollen (Ashworth *et al.*, 1991). Immediately following deglaciation, the colonizing plants and beetles were those of open ground communities, similar to those inhabiting active glacial outwash plains about 45 km east of the site. From 13 to about 9.500 yr BP the biota was that of a *Nothofagus* woodland which persisted until 5.500 yr BP, when it was replaced by a rainforest-moorland mosaic similar to that existing in the region today. Sometime between 5.500 and 3.000 yr BP, the biota changed dramatically on the moraines surrounding the bog, and the bog itself dried up. After 3.000 yr BP, a biota similar to that which had existed between 9.500 and 5.500 yr BP returned. Puerto Edén has a precipitation of about 3,000 mm yr^{-1}. For the bog to completely dry up requires a very significant climatic change. Markgraf *et al.* (2003) reported changes in the sedimentology, paleontology, and geomorphology of Lago Cardiel (49° S) on the Patagonian steppe, east of Puerto Edén indicating severe droughts between 6 and 5,000 kyr BP. They point out that widespread mid-Holocene aridity has been reported in records from the Patagonian steppe to the North American plains, but for which, as yet, there is no compelling climatological explanation.

At the head of the Témpano fjord, interbedded sands and peats infilled a bedrock-floored channel that was later eroded by meltwater. The basal peat is dated between 11.180 and 10.130 yr BP. Fossil beetle and pollen assemblages from the peat indicate a fauna and flora very similar to that inhabiting stable substrates within the area today. The implication is that the South Patagonian ice field during the Younger Dryas interval was no larger than it is today and could even have been smaller (Ashworth and Markgraf, 1989).

Biogeographic Considerations

In the montane and island-fragmented landscape of southern Chile, the onset of the last phase of the glaciation is dated at 29.400 yr BP (Denton, *et al.*, 1999). Nothing is known of the beetle fauna that existed before the final phase of glaciation; however, it is reasonable to assume that it was a forest fauna based on available pollen analyses. Further, it is reasonable to speculate that as the climate cooled, species became regionally extinct. It is assumed that these species survived in more northerly locations and were available for recolonization when the climate warmed between 14.000 and 13.000 yr BP. The fauna that survived glaciation in the moorland habitat of the lake region has no modern analog. It consisted of: (1) species that had dispersed downslope from habitats high in the coastal ranges and the Andes and (2) species that were 'climatic generalists' that could survive the lower temperatures. There is no evidence that cold-adapted species from more southerly latitudes migrated to the lake region.

The absence of evidence for latitudinal movement of beetle taxa in response to climate change during the last glaciation makes southern South America very different than either Europe or North America where species were displaced more than 1,000 km from their existing distributions (Ashworth, 2001). The prevailing hypothesis is that the biota south of Chiloé was completely extirpated during the last glaciation and consequently would not have been available to colonize more northerly regions. The hypothesis has serious flaws (see below), but even if pockets of biota did survive to the south, northward dispersal would have been blocked by barriers imposed by marine inlets and large glaciers.

Reconstructions of ice cover for southern South America during the Last Glacial Maximum portray an ice sheet that was continuous from latitude 43° S in the lake region to 56° S in Tierra del Fuego (Hollin and Schilling, 1981) (**Fig. 1**). At the northern end, parts of Chiloé, the coastal ranges, and the central valley were ice free. South of Chiloé, however, ice is shown as covering the coastal mountains and extending to the Pacific Ocean. An equilibrium line altitude lowering of 900 m was used by Hubbard *et al.* (2005) to model the central regions of the Patagonian ice sheet between 45° and 48° S. Their model shows that by 21.000 yr BP ice covered the Taitaó Peninsula extending to the coastal shelf, comparable to the Hollin and Schilling (1981) reconstruction. Using climatic forcing modified from the Vostok ice core, their model shows an ice sheet that slowly retreated until 14.500 yr BP, after which it collapsed rapidly reaching its present-day configuration by 11.000 yr BP after briefly stabilizing during the Antarctic Cold Reversal.

The effect of glaciation is considered to have been devastating to the flora and fauna, resulting in the total extirpation of species south of Chiloé. Two pieces of evidence, however, suggest that the extent of glaciation and extirpation of the biota may have been overestimated. The first is that there are several species of large- to moderate-sized flightless beetles in the basal layer of the bog at Puerto Edén with an age of about 13.000 yr BP. The fossils include the ground beetles *Ceroglossus suturalis* F., *Creobius eydouxi* Guérin, *Cascellius gravesii* Curtis, and *C. septentrionalis* Roig-Juñnent, and the weevils, *Aegorhinus kuscheli* Elgueta and *Germainiellus rugipennis* (Blanchard). For flightless species to be among the earliest colonizers of the bog implies that they had to have survived glaciation locally, either in the coastal mountains or more probably along the coast. The second piece of evidence supporting survival in local refugia is the high genetic diversity measured in isolated populations of the tree species *Pilgerodendron uvifera* (D. Don) Florin (Premoli *et al.*, 2002). If these populations were Holocene immigrants derived by long-distance dispersal from northern refugia, then it would be expected that the severity of the genetic bottleneck they have been through would have resulted in the populations having much lower genetic diversity.

The Younger Dryas Conundrum

One of the enduring problems associated with paleoclimatic interpretation in the Chilean lake region is whether the pattern of climatic changes at the end of the last glaciation was identical to that of the North Atlantic region. In particular, the problems focused on the existence of a Younger Dryas equivalent between 11.000 and 10.000 yr BP. Currently, the question involves whether there is a climatic cooling, the Huelmo-Mascardi cold reversal (Hajdas *et al.*, 2003), between 12.400 and 10.000 yr BP. The initial disagreement centered on several interpretations based on pollen studies for a climatic reversal of 5–10° C precisely correlated with the Younger Dryas (Heusser, 1974). The studies were considered to be strong evidence for the Younger Dryas being a global event. Ashworth and Hoganson (1993), however, argued that there was no evidence for any significant amount of cooling indicated by fossil beetle assemblages after 14.000 yr BP precipitating a disagreement that continues today.

In southern South America, outside of the Chilean lake region, the evidence for climatic cooling coeval with the Younger Dryas is equally problematical. Ariztegui *et al.* (1997) reported indirect evidence for a Younger Dryas glacial advance on Monte Tronadór, based on studies of sediments, diatoms, and pollen from Lago Mascardi. Redating of that event has demonstrated that the proposed cooling includes the Younger Dryas but pre-dates it by 550 yr (Hajdas *et al.*, 2003). Earlier, pollen studies in the region showed that there had been no response of vegetation to climate change between 11.000 and 10.000 yr BP (Markgraf, 1984). In pollen records from the Chonos (44° S) and Taitaó peninsulas (46°), the vegetation shows a unidirectional shift from heath to forest, with essentially the modern forest developed by 12.400 yr BP (Bennett *et al.*, 2000; Haberle and Bennett, 2004). No evidence was found for cooling during either the Younger Dryas or the earlier Antarctic Cold Reversal. However, one of the possible explanations for changes in chironomid (midge) fossil assemblages from a lake studied by Bennett *et al.* (2000) was for climatic cooling during the Younger Dryas (Massaferro and Brooks, 2002). Further south in the Chilean Channels (49° S) combined analyses of fossil beetle assemblages and pollen also showed no evidence of a Younger Dryas climatic cooling (Ashworth *et al.*, 1991). At the Témpano glacier, Ashworth and Markgraf (1989) reported that the North Patagonian ice field was within its present boundaries during the Younger Dryas. In Tierra del Fuego, Heusser and Rabassa (1987) reported a glacial and vegetational response consistent with cooling during the Younger Dryas. From the same region, Markgraf (1993) reported the occurrence of charcoal during the Lateglacial and preferred to interpret the pollen in terms of local environmental instability rather than climatic cooling.

Moreno *et al.*, (1999) reported a climatic cooling of 2–3° C between 12.200 and 9.800 yr BP in the Chilean lake region based mostly on the occurrence of the pollen of *Podocarpus nubigena* Lindl., a conifer which occurs in the North Patagonian and Valdivian rain forests. They discussed how the proposed cooling produced only minor changes in forest structure but still considered climate change to be the probable cause rather than disturbance from a nonclimatic cause. Ashworth and Hoganson (1987) were unable to discriminate statistically between the modern beetle communities of the North Patagonian and upper Valdivian rain forest which, in terms of paleoclimatic interpretation, would translate to ±1° C. The difference between interpretations based on pollen and those on fossil beetle assemblages is now within ±1° C or what many paleoecologists would consider noise within the records and irresolvable between climatic and nonclimatic causes.

The situation now is very different than when disagreements between interpretations were initially stated. Then, the pollen evidence was unequivocally for a climatic cooling of similar magnitude and timing to the Younger Dryas in the North Atlantic. Currently, even proponents for climate change argue that it is subtle and that it begins 500–1,200 yr before the Younger Dryas. Moreno (1997), Moreno *et al.* (1999), and Hajdas *et al.* (2003) still consider the cooling to be part of a globally synchronous event but Ashworth and Hoganson (1993), Markgraf (1993), and Bennett *et al.* (2000) find no evidence to support climatic reversals at the time of the Younger Dryas (11.000–10.000 yr BP) or during the Antarctic Cold Reversal (14.500–12.900 yr BP).

See also: **Beetle Records**: Overview; Late Tertiary and Early Quaternary. **Pollen Records, Late Pleistocene**: South America.

References

Arias, E. T. (2000). *Coleópteros de Chile*. Fototeknika, Santiago de Chile.

Ariztegui, D., Bianchi, M. M., Massaferro, J., Lafargue, E., and Niessan, F. (1997). Interhemispheric synchrony of late-glacial climatic instability as recorded in proglacial Lake Mascardi, Argentina. *Journal of Quaternary Science* **12**, 133–138.

Ashworth, A. C. (2001). Perspectives on Quaternary Beetles and Climate Change. In *American Association of Petroleum Geologists Studies in Geology, Geological Perspectives of Global Climate Change*, (L. Gerhard, W. Harrison and B. Hanson, Eds.) Vol. 47, pp. 153–168. American Association of Petroleum Geologists, Tulsa, Oklahoma.

Ashworth, A. C., and Hoganson, J. W. (1987). Coleoptera bioassociations along an elevational gradient in the lake region of southern Chile and comments on the postglacial development of the fauna. *Annals of the Entomological Society of America* **80**, 865–895.

Ashworth, A. C., and Hoganson, J. W. (1993). Magnitude and rapidity of the climate change marking the end of the Pleistocene in the mid-latitudes of South America. *Palaeogeography, Palaeoclimatology, Palaeoecology* **101**, 263–270.

Ashworth, A. C., and Kuschel, G. (2003). Fossil weevils (Coleoptera:Curculionidae) from latitude 85° S Antarctica. *Palaeogeography, Palaeoclimatology, Palaeoecology*, **191**, 191–202.

Ashworth, A. C., and Markgraf, V. (1989). Climate of the Chilean channels between 11,000 and 10,000 yr BP based on fossil beetle and pollen analyses. *Revista Chilena de Historia Natural* **62**, 61–74.

Ashworth, A. C., Markgraf, V., and Villagrán, C. (1991). Late Quaternary climatic history of the Chilean channels based on fossil pollen and beetle analysis, with an analysis of the modern vegetation and pollen rain. *Journal of Quaternary Science* **6**, 279–291.

Bennett, K. D., Haberle, S. G., and Lumley, S. H. (2000). The last Glacial–Holocene transition in southern Chile. *Science* **290**, 325–328.

Buckland, P. C., and Hammond, P. M. (1997). The origins of the biota of the Falkland Islands and South Georgia. In *Quaternary Proceedings 5: Studies in Quaternary Entomology–An Inordinate Fondness for Insects* (A. C. Ashworth, P. C. Buckland and J. P. Sadler, Eds.), pp. 59–66. Wiley, Chichester.

Denton, G. H., Lowell, T. V., Heusser, C. J., *et al.* (1999). Geomorphology, stratigraphy, and radiocarbon chronology of Llanquihue drift in the area of the southern Lake District, Seno Reloncavi, and Isla Grandé de Chilo., Chile. *Geografiska Annaler* **A81**, 167–229.

Dillehay, T. D., and Pino, M. (1989). Stratigraphy and chronology. In *Monte Verde – A Late Pleistocene Settlement in Chile* (T. D. Dillehay, Ed.), pp. 133–145. Smithsonian Institution Press, Washington.

Elias, S. A. (1994). *Quaternary insects and their environments.* Smithsonian Institution Press, Washington.

Haberle, S. G., and Bennett, K. D. (2004). Postglacial formation and dynamics of North Patagonian rainforest in the Chonos Archipelago, southern Chile. *Quaternary Science Reviews* **23**, 2433–2452.

Hajdas, I., Bonani, G., Moreno, P. I., and Ariztegui, D. (2003). Precise radiocarbon dating of late-glacial cooling in mid-latitude South America. *Quaternary Research* **59**, 70–78.

Heusser, C. J. (1974). Vegetation and climate of the southern Chilean lake district during and since the last interglaciation. *Quaternary Research* **4**, 290–315.

Heusser, C. J., and Rabassa, J. (1987). Cold climatic episode of Younger Dryas in South America. *Nature* **328**, 609–611.

Hoganson, J., Gunderson, M., and Ashworth, A. (1989). Fossil-beetle analysis. In *Monte Verde – A late Pleistocene settlement in Chile* (T. D. Dillehay, Ed.), pp. 211–226. Smithsonian Institution Press, Washington.

Hoganson, J. W., and Ashworth, A. C. (1992). Fossil beetle evidence for climatic change 18,000–10,000 years BP in South-Central Chile. *Quaternary Research* **37**, 101–116.

Hollin, J. T., and Schilling, D. H. (1981). Late-Wisconsin-Weichselian mountain glaciers and small ice caps. In *The Last Great Ice Sheets* (G. H. Denton and T. J. Hughes, Eds.), pp. 179–206. Wiley, New York.

Hubbard, A., Hein, A. S., Kaplan, M. R., Hulton, N. R. J., and Glasser, N. (2005). A modelling reconstruction of the late glacial maximum ice sheet and its deglaciation in the vicinity of the Northern Patagonian Icefield, South America. *Geografiska Annaler* **A87**, 375–391.

Markgraf, V. (1984). Late Pleistocene and Holocene vegetation history of temperate Argentina: Lago Morenito, Barailoche. *Dissertationes Botanicae* **72**, 235–254.

Markgraf, V. (1993). Younger Dryas in southernmost South America – An update. *Quaternary Science Reviews* **12**, 351–355.

Markgraf, V., Bradbury, J. P., Schwalb, A., *et al.* (2003). Holocene palaeoclimates of southern Patagonia: Limnological and environmental history of Lago Cardiel, Argentina. *The Holocene* **13**, 597–607.

Massaferro, J., and Brooks, S. J. (2002). Response of chironomids to late Quaternary environmental change in the Taitaó peninsula, southern Chile. *Journal of Quaternary Science* **17**, 101–111.

Mercer, J. H. (1976). Glacial history of southernmost South America. *Quaternary Research* **6**, 125–166.

Moreno, P. I. (1997). Vegetation and climate near Lago Llanquihue in the Chilean Lake district between 20,200 and 9500 14C yr BP. *Journal of Quaternary Science* **12**, 485–500.

Moreno, P. I., Lowell, T. V., Jacobson, G. L., jr, and Denton, G. H. (1999). Abrupt vegetation and climate changes during the last glacial maximum and last termination in the Chilean Lake District: A case study from Canal de la Puntilla (41° S). *Geografiska Annaler* **A81**, 285–311.

Morrone, J. J. (1993). Revisión sistemática de un nuevo género de Rhytirrhinini (Coleoptera, Curculionidae), con un análisis biogeográphico del domino Subantártico. *Boletín de la Sociedad de Biológia de Concepción, Chile* **64**, 121–145.

Premoli, A. C., Souto, C. P., Rovere, A. E., Allnut, T. R., and Newton, A. (2002). Patterns of isozyme variation as indicators of biogeographic history in *Pilgerodendron uviferum* (D. Don) Florin. *Diversity and Distributions* **8**, 57–66.

Villagrán, C. (1990). Glacial climates and their effects on the history of the vegetation of Chile: A synthesis based on palynological evidence from Isla de Chiloé. *Review of Palaeobotany and Palynology* **65**, 17–24.

Whitlock, C., Bartlein, P. J., Markgraf, V., and Ashworth, A. C. (2001). The midlatidudes of North and South America during the last glacial maximum and early Holocene: Similar paleoclimatic sequences despite differing large-scale controls. In *Interhemispheric Climate Linkages* (V. Markgraf, Ed.), pp. 391–416. Cambridge University Press, Cambridge.

Late Pleistocene of North America

S A Elias, Royal Holloway, University of London, Surrey, UK

© 2007 Elsevier B.V. All rights reserved.

Beetles are the most diverse group of organisms on Earth, and their exoskeletal remains preserve well in water-logged sediments of Pleistocene age. Studies of late Pleistocene fossil beetle assemblages have been carried out throughout many regions of North America in recent decades. This article reports on the major findings of these studies in terms of paleoenvironmental reconstructions and shifting distribution patterns (biogeography) of the species found in the fossil assemblages, all of which remain extant today.

The paleotemperature estimates reported here are derived from the mutual climatic range (MCR) of paleoclimate analysis, as discussed in the overview article for this section. For the sites in eastern and central North America, the MCR method was used to provide estimates of mean temperatures of the warmest (T_{max}) and coldest (T_{min}) months of the year. T_{max} and T_{min} estimates are discussed in terms of departures from the modern T_{max} and T_{min} of the study sites in question.

Research in Eastern North America

Pleistocene insect fossil studies began in earnest in the 1970s, with the most intensively studied region of North America being the central and eastern United States and southeastern Canada. Several fossil beetle assemblages in eastern North America are thought to represent Marine Isotope Stage (MIS) 5e, the last interglaciation. However, the ages of these fossil assemblages are tentative. One of the richest beetle faunas in this group is from Innerkip, Ontario (**Fig. 1**, No. 1). This fauna includes several warm-adapted species, and MCR analysis of the fauna (**Table 1**) yielded an estimate of T_{max} that is essentially the same as modern parameters. The fossil assemblage from the Pointe-Fortune site on the southern Ontario–Quebec border (**Fig. 1**, No. 2) is also thought to derive from MIS 5 (Anderson *et al.*, 1990). This fauna contains boreal species that indicate colder than modern climate (**Table 1**), so perhaps it reflects a late MIS 5 environment. Likewise,

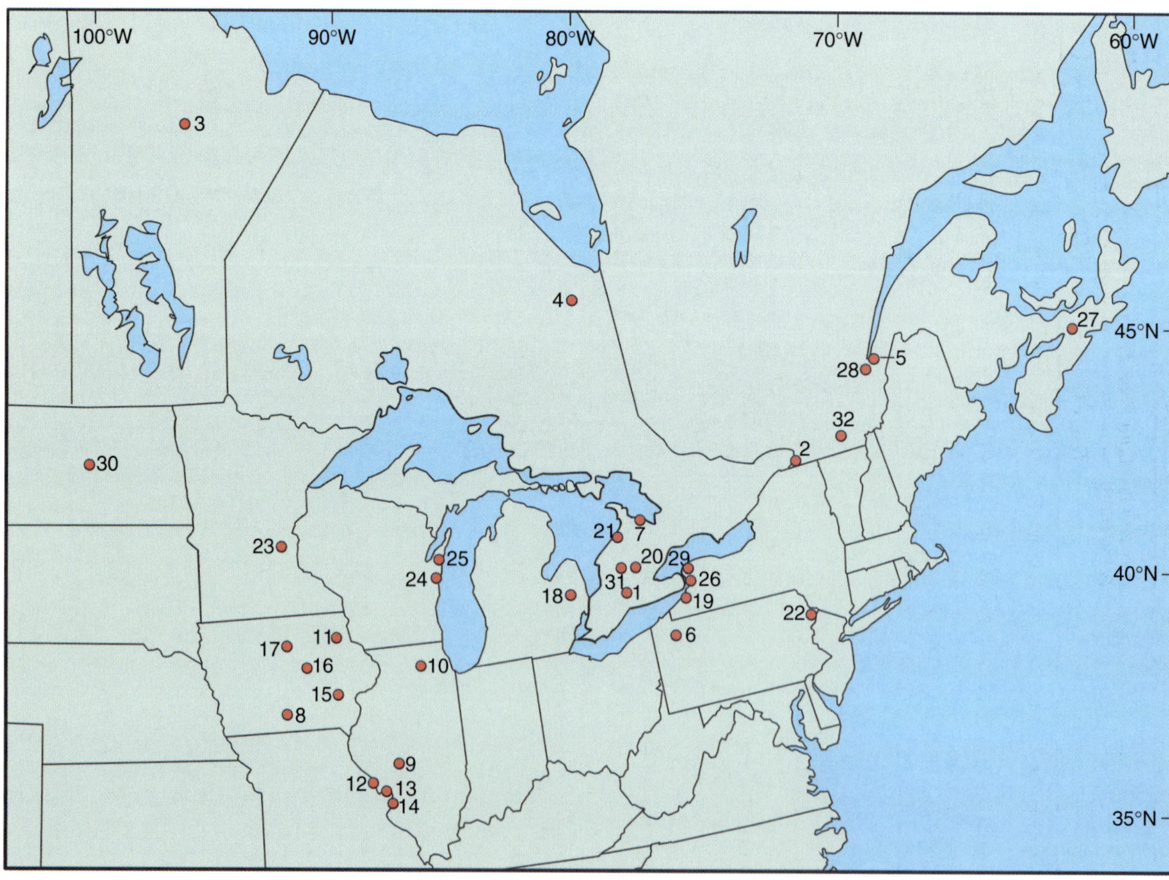

Figure 1 Map of eastern and central North America showing locations of fossil sites discussed in text. Site numbers match those in **Table 1**.

Table 1 Summary of fossil beetle assemblage data from sites in central and eastern North America

Site	Age (cal yr BP × 1000)	Late Pleistocene		Modern		Change in Temperature		References[a]
		T_{max} (°C)	T_{min} (°C)	T_{max} (°C)	T_{min} (°C)	July ΔT (°C)	January ΔT (°C)	
1. Innerkip, Ont.	MIS 5e	16–19	−16 to −8	20.4	−6.3	−4.4 to −1.4	−9.7 to −1.7	Filny and Morgan (1987), Churcher *et al.* (1990)
2. Pointe Fortune, Que	MIS 5	14–17	−22.5 to −6	19.3	−9.8	−5.3 to −2.3	−12.7 to +3.8	Anderson *et al.* (1990)
3. Henday, Man.	MIS 5	8.5–10	−32.5 to −27.5	15.3	−25.8	−6.8 to −5.3	−6.7 to −1.7	Nielsen *et al.* (1986)
4. Owl Creek, Ont.	MIS 5	13–15	−24 to −19	17.4	−17.5	−4.4 to −2.4	−6.5 to −1.5	Mott and DiLabio (1990)
5. Chaudiere Valley, Que	>52	10–10.5	−30 to −28	18	−13	−8 to −7.5	−17 to −10	Matthews and Mott (1987), Elias (1999)
6. Titusville, PA #6	46.5	12.75–13.5	−19.3 to −17.5	20.3	−5.1	−7.4 to −6.8	−14.2 to −12.4	Cong *et al.* (1996), Elias (1999)
6. Titusville, PA #5	45.7	15–15.5	−24 to −23	20.3	−5.1	−5.3 to −4.8	−18.9 to −17.9	Cong *et al.* (1996), Elias (1999)
6. Titusville, PA #4	45	18.5–19.3	−10.5 to −8.25	20.3	−5.1	−1.8 to −1	−5.4 to −3.2	Cong *et al.* (1996), Elias (1999)
6. Titusville, PA #3	44.3	13.5–14.5	−30.5 to −27.5	20.3	−5.1	−6.8 to −5.8	−25.4 to −22.4	Cong *et al.* (1996), Elias (1999)
6. Titusville, PA #2	43.7	12–13	−31 to −28	20.3	−5.1	−8.3 to −7.3	−25.9 to −17.5	Cong *et al.* (1996), Elias (1999)
6. Titusville, PA #1	43.2	11.8–12.5	−31.3 to −22.5	20.3	−5.1	−8.5 to −7.8	−26.2 to −17.4	Cong *et al.* (1996), Elias (1999)
7. Clarksburg, Ont.	>40.6	10–12	−31 to −17	19.1	−8.9	−9.1 to −7.1	−22.1 to −8.1	Warner *et al.* (1988), Elias (1999)
8. St. Charles, IA	40.1	17–20.5	−16.5 to −6.5	23.8	−6.9	−6.8 to −3.3	−9.6 to +0.4	Baker *et al.* (1991), Elias (1999)
9. Athens, IL	29.9–26.7	13–15	−19.5 to −13	24.6	−4.1	−11.6 to −9.6	−15.4 to −8.9	Morgan (1987), Elias (1999)
10. Wedron, IL	25.8	12–12.5	−30 to −26.5	23.6	−5.9	−11.6 to −11.1	−24.1 to −20.6	Garry *et al.* (1990), Elias (1999)
11. Elkader, IA	24.5	13–15	−27 to −22	22.6	−9.2	−9.6 to −7.6	−17.8 to −12.8	Schwert (1992), Elias (1999)
12. Salt River, MO	23.3	14.5–16	−22.3 to −15.5	22	−4.4	−7.5 to −6	−17.9 to −11.1	Schwert *et al.* (1997), Elias (1999)
13. Salt Spring Hollow, IL	22.5	12.8–14.8	−27.3 to −19.3	24.6	−3.5	−11.8 to −9.8	−23.8 to −15.8	Schwert *et al.* (1997), Elias (1999)
14. Bonfils Quarry, MO	20.8	14–16.3	−26 to −17	26.6	−1.5	−12.6 to −10.3	−24.5 to −15.5	Schwert *et al.* (1997), Elias (1999)
15. Conklin Quarry, IA	20.7	11.5–12.5	−29 to −27	22.3	−9	−10.8 to −9.8	−20 to −18	Baker *et al.* (1986), Elias *et al.* (1996a)
16. Saylorville, IA	19.4	12–16	−30 to −10	22.7	−9.1	−10.7 to −6.7	−20.9 to −0.9	Schwert (1992), Elias *et al.* (1996a)
17. Ft. Dodge, IA/I	18	12–13	−26 to −20	23.1	−9.4	−11.9 to −10.9	−16.6 to −10.6	Schwert (1992), Elias *et al.* (1996a)
17. Ft. Dodge, IA/II	17.3	11.5–12.5	−28.5 to −19.5	23.1	−9.4	−11.4 to −12.4	−19.1 to −10.1	Schwert (1992), Elias *et al.* (1996a)
18. Weaver Drain, MI	17	12.5–16.5	−29.5 to −6.5	21.1	−6.2	−8.6 to −4.6	−23.3 to −0.3	Morgan *et al.* (1981), Elias *et al.* (1996a)
19. Winter Gulf, NY/W08	15.4	17.7–20.5	−10 to −4.5	20.4	−5.2	−2.9 to +0.1	−4.8 to +0.7	Schwert and Morgan (1980), Elias *et al.* (1996a)
20. Gage St., Ont./G8	15.0–14.5	11.5–13.5	−25.5 to −11.5	19.9	−7.5	−8.4 to −6.4	−18 to −4	Schwert *et al.* (1985), Elias *et al.* (1996a)

(*Continued*)

Table 1 (*Continued*)

Site	*Age (cal yr BP × 1000)*	*Late Pleistocene*		*Modern*		*Change in Temperature*		*References*[a]
		T_{max} (°C)	T_{min} (°C)	T_{max} (°C)	T_{min} (°C)	*July* ΔT (°C)	*January* ΔT (°C)	
19. Winter Gulf, NY/W06	14.8	16.5–21	−11.5 to −4	20.4	−5.2	−3.9 to +0.6	−8.6 to +1.2	Schwert and Morgan (1980), Elias *et al.* (1996a)
21. Rostock, Ont.	14.7	12–15	−28 to −20	19.4	−7.8	−7 to −4.4	−20.2 to −12.2	Pilny and Morgan (1987), Elias *et al.* (1996a)
22. Newton, PA	14.6	10–15	−31 to −9	19.7	−5.6	−9.7 to −4.7	−25.4 to −3.4	Barnowsky *et al.* (1988), Elias *et al.* (1996a)
23. Norwood, MN/Peat	14.5	16.5–21.5	−17.5 to −3	22.1	−12.3	−5.6 to −0.6	−5.2 to +9.3	Ashworth *et al.* (1981), Elias *et al.* (1996a)
19. Winter Gulf, NY/W04	14.5	15–21.5	−14 to 0	20.4	−5.2	−5.4 to +1.1	−9.8 to +5.2	Schwert and Morgan (1980), Elias *et al.* (1996a)
20. Gage St., Ont./G7	14.5–12.8	20–21	−5.5 to −2.5	19.9	−7.5	0 to +1.1	+2 to +5	Schwert *et al.* (1985), Elias *et al.* (1996a)
23. Norwood, MN/M.S.	13.8	15–20	−25 to −12	22.1	−12.3	−7.1 to −2.1	−12.7 to −0.3	Ashworth *et al.* (1981), Elias *et al.* (1996a)
21. Rostock, Ont/65–75	13.6	15–20.5	−22.5 to −4.5	19.4	−7.8	−4 to +0.6	−14.7 to +3.3	Pilny and Morgan (1987), Elias *et al.* (1996a)
24. Two Creeks, WI	13.6	12.5–13.5	−16.5 to −11.5	21.1	−7.3	−8.8 to −7.8	−10.2 to −4.2	Morgan and Morgan (1979), Elias *et al.* (1996a)
25. Kewaunee, WI	13.5	12.5–17.5	−28 to −5	20.8	−7.8	−8.3 to −3.3	−20.2 to +2.8	Garry *et al.* (1990), Elias *et al.* (1996a)
23. Norwood, MN/U.S.	13.4	17–18	−11 to −7	22.1	−12.3	−5.1 to −4.1	+1.3 to +5.3	Ashworth *et al.* (1981), Elias *et al.* (1996a)
21. Rostock, Ont/55–65	13.3	17–21	−15.5 to −2.5	19.4	−7.8	−2.4 to +1.6	−7.7 to +5.3	Pilny and Morgan (1987), Elias *et al.* (1996a)
27. Brookside, NS	13	12.5–20	−31.5 to 5	18.4	−7.4	−5.9 to +1.6	−24.1 to +12.4	Mott *et al.* (1986)
28. St. Eugene, Que	12.9	10–12.5	−31 to −18.5	17.2	−13.1	−7.2 to −4.7	−17.9 to −5.4	Mott *et al.* (1981)
29. Lockport Gulf, NY L01	12.8	17.5–21.5	−14 to −3	21.5	−4.8	−4 to 0	−9.2 to +1.8	Miller and Morgan (1982)
30. Johns Lake, ND	12.7	13.5–15.5	−26.5 to −20	20	−17	−6.5 to −4.5	−6.5 to 0	Ashworth and Schwert (1992)
31. Eighteen Mile R., Ont.	12.4	14–15	−16.5 to −10	19.8	−5.8	−5.8 to −4.8	−10.7 to −4.2	Ashworth (1977)
29. Lockport Gulf, NY L02	12	12.5–17.5	−25 to −5	21.5	−4.8	−9 to −4	−20.2 to +0.2	Miller and Morgan (1982)
32. St. Hilaire, Que	11.7	16.5–17.5	−15 to −10	20.7	−10.2	−4.2 to −3.2	−4.8 to +0.2	Mott *et al.* (1981)
26. Nichols Brook, NY-C	11.5–10.1	14.5–18	−21 to −12.5	18.1	−10	−3.5 to −0.1	−11 to −2.5	Fritz *et al.* (1987)

[a]References available from the QBib Web site, www.bugs2000.org/qbib.html

faunal assemblages from the Henday site in northern Manitoba (**Fig. 1**, No. 3) and the Owl Creek site in northeastern Ontario (**Fig. 1**, No. 4) indicate climates that were colder than today (**Table 1**), based on the presence of Arctic ground beetles (Carabidae).

Assemblages from MIS 4 and 3 have been described from several localities in the Great Lakes region, as summarized by Elias (1994) and Ashworth (2004). MIS 4 assemblages from southeastern Canada contain tundra and tree-line beetle faunas. Faunal assemblages containing the same species also occur in regional MIS 2 assemblages, so it appears that tundra and tree-line species inhabited the margins of the Laurentide ice sheet continually during Wisconsinan time.

A series of insect faunas indicate changing conditions during the last (Wisconsin) glaciation. To date, late Pleistocene beetle assemblages have been published from 26 sites in this region (**Table 1**). These range in age from >52,000 calibrated years before present (>52 kya) to 11.5 kya. The mid-Wisconsin interstadial, MIS 3, is thought to have lasted from approximately 65–25 ka. Isotopic data from Greenland ice cores and North Atlantic oceanic cores show multiple, abrupt climate changes during marine isotope stage 3 (Grootes *et al.*, 1993). Events within the first half of this interval cannot be dated by the radiocarbon method, and organic deposits that are stratigraphically correlated to the mid-Wisconsin interval are poorly dated in many regions of North America. The analysis of fossil beetle assemblages has progressed sufficiently in North America to allow reconstructions of the timing and intensity of MIS 3 climatic change (Elias, 1999).

Paleoenvironments of Central and Eastern North America during the Last Glaciation

The oldest MIS 3 beetle fauna considered here comes from the Chaudière valley, southeastern Quebec (**Fig. 1**, No. 5). The fauna is associated with a mid-Wisconsin cold interval. Radiocarbon ages of this formation include a date of >54 ka. The sediments that yielded the fossils were underlain by a glacial till and overlain by late Wisconsin till. The beetle fauna is indicative of arctic tundra environments and includes many cold-adapted species. T_{max} was depressed by approximately 6° C, and T_{min} was approximately 13° C colder than today (**Table 1**).

The clearest indication of the strength of interstadial warming in eastern North America comes from the series of beetle faunal assemblages from Titusville, Pennsylvania (**Fig. 1**, No. 6). The six fossil assemblages were deposited between approximately 46 and 43 kya. Within this interval, the Titusville faunas reflect climatic amerlioration, followed by cooling. The oldest fauna (46.5 kya) indicates cooling of both T_{max} and T_{min} to approximately the same level as at the Chaudière valley site (**Table 1**). At the height of regional warming (ca. 45.7 kya), T_{max} was only 2° C less than modern levels, and T_{min} was approximately 6° C colder than modern. This warm interval occurred at approximately the same time as the Upper Warren Interstadial in Britain and the warm sea-surface temperature interval between Heinrich events 4 and 5 (Elias, 1999). By the end of the interstadial event (43.2 kya), regional climates had cooled to even lower levels than previously. Thus, in the space of approximately 3,000 years, regional climates oscillated from subarctic to boreal and back again.

A beetle assemblage dated 38.5 kya from St. Charles, Iowa (**Fig. 1**, No. 8), reflects prairie or savanna environments with patches of conifer–hardwood forest (Baker *et al.*, 1991). The St. Charles beetle fauna was indicative of somewhat milder climate than during full glacial conditions but well below the levels of warming suggested by the interstadial fauna from Titusville. Later in the MIS 3 interval (32.9–28.9 kya), fossil assemblages from Illinois and Iowa indicate closed spruce forests. The sketchy picture that emerges for the early and mid-Wisconsinan is one of spatial and temporal heterogeneity in which populations of plants and insects were responding more dynamically than at any time during the Holocene (Ashworth, 2004).

By approximately 30 kya, the beetle evidence suggests that central North American climates had cooled to full-glacial levels (**Table 1**). T_{max} was approximately 10° C colder than today at sites such as Athens and Wedron in Illinois (**Fig. 1**, Nos. 9 and 10). Regional faunas contained mixtures of boreal and arctic species, depending on latitude. Southerly sites, such as Salt Spring Hollow, Illinois (**Fig. 1**, No. 13), contained boreal beetle assemblages just before and during the Last Glacial Maximum (LGM). More northerly sites, such as Conklin Quarry, Iowa (**Fig. 1**, No. 15), contained beetles with Arctic and subArctic affinities.

In the mid-western region, Arctic beetles replaced the boreal forest fauna along the southern margin of the Laurentide ice sheet at approximately 25.8 kya. Colonization was probably from populations that dispersed southward in front of the growing ice sheet and from populations that dispersed westward and eastward from montane refugia in the Appalachian and Rocky Mountains, respectively (Schwert and Ashworth, 1988). From 25.8–17.7 kya, Arctic and subArctic species inhabited a discontinuous tundra zone along the margin of the ice sheet. Fossil assemblages typical of this time have been reported from Iowa to New York (Ashworth, 2004). At the end of the glaciation, these cold-adapted faunas were forced to migrate in order to survive. Some species found refuge in high mountains, such as the Rocky Mountains in the west and the tops of the Appalachian Mountains in the east. Other species died out in the middle latitudes of the continent, and only their Beringian populations in ice-free regions of northwestern North America survived beyond the end of the Pleistocene.

The earliest sign of warming following the LGM comes from Weaver Drain, Michigan (**Fig. 1**, No. 18). This site was ice proximal at 17 kya, but MCR analysis (**Table 1**) shows that T_{max} had started to rise above the levels reconstructed from a 17.3-ka assemblage at Fort Dodge, Iowa (**Fig. 1**, No. 17). Ice proximal sites showed the effects of local cooling until the ice margin retreated and pockets of stranded dead ice melted, after approximately 12.6 kya.

By approximately 15.4 kya, substantial warming had occurred at Winter Gulf, New York (**Fig. 1**, No. 19). T_{max} values were less than 3° C cooler than modern, and T_{min} values were less than 5° C cooler than modern. This degree of warming was not seen in the oldest assemblages from Gage St., Ontario (**Fig. 1**, No. 20). However, during the interval 14.5–12.8 kya, T_{max} and T_{min} at Gage St. rose above modern levels (**Table 1**). This was also the case at Nichols Brook, New York (**Fig. 1**, No. 26) and Rostock, Ontario (**Fig. 1**, No. 21) between 13.3 and 11.5 kya.

Between approximately 12.9 and 11.5 kya, a third set of sites (St. Eugene and St. Hilaire, Quebec, Eighteen Mile River, Ontario, and Lockport Gulf, New York) were situated near large bodies of glacial meltwater (proglacial lakes and the Champlain Sea). The large bodies of chilled water adjacent to these sites strongly affected local climates (Ashworth, 2004). After 11.5 kya the ice sheet margin had retreated well north of the study sites, and all the fossil assemblages reflect T_{max} values near modern parameters (**Table 1**).

Beetle MCR reconstructions from northwest Europe show a sudden rise in seasonal temperatures by approximately 15.9 kya, followed by a general cooling through the late glacial interstadial, before dramatic cooling at 12.9 kya (the Younger Dryas chronozone), followed by a sudden rise in temperatures at approximately 11.5 kya (Coope and Lemdahl, 1995). The eastern North American results show almost a plateau of summer temperatures from 13–10 ka. The vegetation history, as reconstructed from fossil pollen assemblages, also suggests no climatic reversals during this interval (Grimm and Jacobson, 2004). The lack of oscillations in the North American assemblages may be due to the effects of continental climate, whereas late glacial climates in northwest Europe were strongly affected by shifts in North Atlantic surface water (Coope and Lemdahl, 1995). The only insect evidence for cooling between 12.9 and 11.5 kya in North America comes from the maritime provinces of Canada (Walker, *et al.*, 1991) and from Arctic Alaska.

Research in Western North America

Late Pleistocene beetle assemblages have also been studied from a number of sites in western North America (**Table 2**). Insect fossils of this age from the Rocky Mountain region are discussed by Elias (1990, 1991). Full glacial age faunas, such as the assemblage dated 17.7 kya from Lamb Spring, Colorado (**Fig. 2**, No. 1), indicate that T_{max} was 10–11° C colder than today. The earliest evidence for the start of postglacial warming is a 16.9-kya assemblage at the Mary Jane site in Colorado (**Fig. 2**, No. 2). Rapid warming occurred after 13.3 kya, with summer temperatures becoming warmer than present by 11.5 kya at False Cougar Cave, Montana (**Fig. 2**, No. 4). Many of the cold-adapted species in the Late Pleistocene Rocky Mountain fossil assemblages have since retreated upslope to the Alpine tundra, but not all species were able to do this. For instance, the cold-adapted rove beetle, *Holoboreaphilus nordenskioeldi*, was found in a late glacial age assemblage from Marias Pass, Montana (**Fig. 2**, No. 3). This species apparently died out in the midlatitudes at the end of the last glaciation. It is found today only in Arctic North America (Elias, 1991).

In the past 20 years, insect fossils from arid regions of the southwestern United States and northern Mexico have received intensive study. These studies are based on fossil remains extracted from packrat (*Neotoma*) middens. Packrat middens are caches of objects, including edible plants, cactus spines, insect and vertebrate remains, small pebbles and feces, brought to the rat's den site for a variety of reasons, including food, curiosity, and den protection; then they are cemented into black tarry masses by packrat urine. Late Pleistocene insect fossils have been studied from the Great Basin, but the most intensively studied region is the Chihuahuan Desert (**Fig. 3**). Most packrat midden beetle assemblages from the Sonoran Desert date to the Holocene.

Late Pleistocene faunas from the arid southwest were mixtures of temperate and desert species not seen in any one region today. Since the end of the Pleistocene, some of these species have become established in different desert regions (Elias, 1992). Others now live outside the desert zone, in temperate

Table 2 Summary of fossil beetle assemblage data from sites in western North America

Site	*Age (cal yr BP × 1000)*	*Late Quaternary*		*Modern*		*Change in Temperature*		*References*[a]
		T_{max} (°C)	T_{min} (°C)	T_{max} (°C)	T_{min} (°C)	July ΔT (°C)	January ΔT (°C)	
1. Lamb Spring, CO	17.7	10–11	−31 to −27	21.4	−1.3	−11.4 to −10.4	−29.7 to −25.7	Elias and Toolin, 1990; Elias, 1996
2. Mary Jane, CO	16.2	9.8–10.2	−29.3 to −27.6	13.4	−8.6	−3.6 to −3.2	−20.7 to −19	Short and Elias, 1987; Elias, 1996
2. Mary Jane, CO	15.4	10–10.2	−29.1 to −27.6	13.4	−8.6	−3.4 to −3.2	−20.5 to −19	Short and Elias, 1987; Elias, 1996
3. Marias Pass, MT	14.2	11–14	−30 to −16	13.9	−9	−2.9 to +0.1	−21 to −7	Elias, 1988, 1996
4. False Cougar Cave, MT	11.5	15.5–17.7	−4.5 to +0.25	13.5	−10.1	+2 to +4.25	+5.6 to +10.4	Elias, 1990, 1996
Chihuahuan Desert and Great Basin packrat midden faunas								
5. S. Organ Mountains, NM	>46.3–41.2	NA	NA	NA	NA	NA	NA	Elias and Van Devender (1992)
6. Hueco Mountains, TX	44.9–12.4	NA	NA	NA	NA	NA	NA	Elias and Van Devender (1992)
7. Kaetan Cave, AZ	35.8–17.5	NA	NA	NA	NA	NA	NA	Elias *et al.* (1992)
8. Maravillas Canyon, TX	32.7–13.8	NA	NA	NA	NA	NA	NA	Elias and Van Devender (1990)
9. Salt Creek Canyon, UT	32.4	NA	NA	NA	NA	NA	NA	Elias *et al.* (1992)
10. Baby Vulture Den, TX	31–17.5	NA	NA	NA	NA	NA	NA	Elias and Van Devender (1990)
11. Bida Cave, AZ	29.2–19.5	NA	NA	NA	NA	NA	NA	Elias *et al.* (1992)
12. Tunnel View, TX	28.9–12.3	NA	NA	NA	NA	NA	NA	Elias and Van Devender (1990)
13. Owl Roost, AZ	25.7	NA	NA	NA	NA	NA	NA	Elias *et al.* (1992)
14. Sacramento Mountains, NM	22.1–12.4	NA	NA	NA	NA	NA	NA	Elias and Van Devender (1992)
15. Bennett Ranch, TX	21.9–12.9	NA	NA	NA	NA	NA	NA	Elias and Van Devender (1992)
16. Streeruwitz Hills, TX	21.7–12.7	NA	NA	NA	NA	NA	NA	Elias and Van Devender (1992)
17. Ernst Tinaja, TX	20.6–13.3	NA	NA	NA	NA	NA	NA	Elias and Van Devender (1990)
18. Shafter, TX	19.2–13.3	NA	NA	NA	NA	NA	NA	Elias and Van Devender (1992)
19. Terlingua, TX	18.3	NA	NA	NA	NA	NA	NA	Elias and Van Devender (1990)
20. Bechan Cave, UT	16.2–13.4	NA	NA	NA	NA	NA	NA	Elias *et al.* (1992)
21. Puerto de Ventanillas, Coahuila, Mexico	16.2–15.2	NA	NA	NA	NA	NA	NA	Elias *et al.* (1993)
22. Cañon de la Fragua, Coahuila, Mexico	16.1–14.5	NA	NA	NA	NA	NA	NA	Elias *et al.* (1993)
23. Guadalupe Mountains, TX	15.9–13.8	NA	NA	NA	NA	NA	NA	Elias and Van Devender (1992)
24. Bonneville Estates, UT	13.6	17.5–19	NA	23.9	NA	−6.4 to −4.9	NA	Rhode (2000)
25. Guadalupe Mountains, NM	12.4–11.6	NA	NA	NA	NA	NA	NA	Elias and Van Devender (1992)
Pacific regional sites								
26. Kalaloch, WA	24.1–20.5	14–17	−19 to −11	18.9	5.2	−4.9 to −1.9	−24.2 to −16.2	Cong and Ashworth (1986)
27. Mary Hill, BC	22.6–22.5	11–18	−29 to −4	18.1	4.8	−7.1 to −0.1	−24.2 to −8.8	Miller *et al.* (1985)
28. Port Moody, BC	22.1	11–18	−29 to −4	18.1	4.8	−7.1 to −0.1	−24.2 to −8.8	Miller *et al.* (1985)
29. Discovery Park, WA	18	15–17	−16 to −10	18.1	4.8	−3.1 to −1.1	−24.8 to −14.8	Nelson and Coope (1982)
30. McKittrick, CA	14.3	NA	NA	NA	NA	NA	NA	Miller (1983)

[a]References available from the QBib Web site, www.bugs2000.org/qbib.html

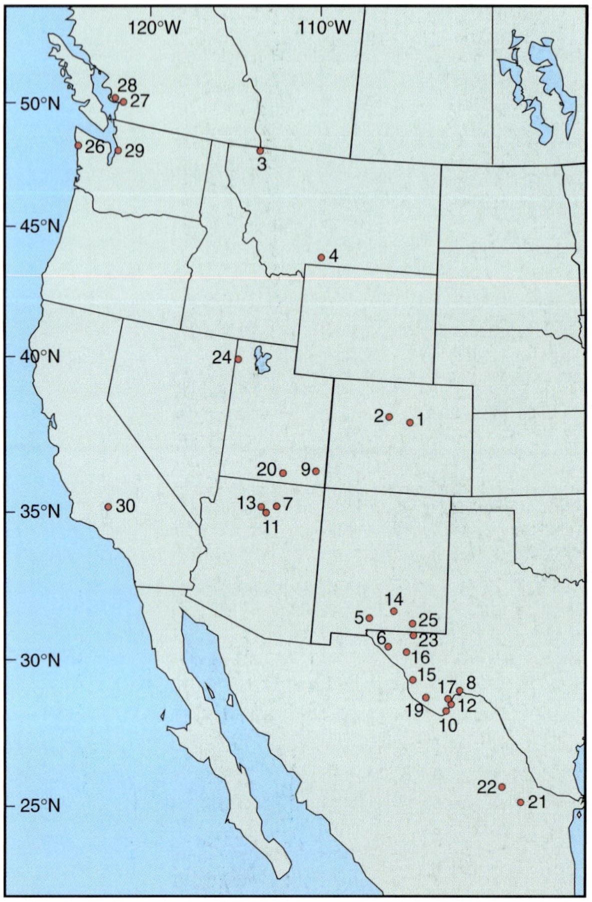

Figure 2 Map of western North America showing locations of fossil sites discussed in text. Site numbers match those in **Table 2**.

regions. The fossil insect record indicates that even sedentary, flightless beetles (e.g., the heavy-bodied weevils) have undergone marked distributional shifts in the American southwest within the space of a few centuries. Moreover, even highly specialized cave dwellers have somehow managed to move from one cave system to another in response to changes in late Wisconsin and Holocene environments.

Late Pleistocene beetle assemblages from the southern Chihuahuan Desert of Mexico (**Fig. 2**, Nos. 21 and 22) contain mixtures of desert and temperate zone species (Elias *et al.*, 1995). Midden assemblages from locations farther north in the Chihuahuan Desert are generally separated into glacial age faunas with temperate zone affinities and postglacial faunas with desert zone affinities. Faunal assemblages with no modern analog are indicative of the late Quaternary environments unlike any that exist today. This conclusion is also borne out by the plant macrofossil record.

The insect faunas from the Big Bend region of southwest Texas (**Fig. 2**, Nos. 8, 10, 12, 17, and 19) suggest that there was greater effective moisture from 35–14 kya (Elias and Van Devender, 1990). During the late Wisconsin (MIS 2), many temperate grassland species lived in the Big Bend region. After 14 kya, most of these species were replaced by either desert species or more cosmopolitan taxa. The faunal change suggests a climatic shift from cool, moist conditions of full glacial times to hotter, drier conditions during the late glacial and early Holocene.

In the northern Chihuahuan Desert, full glacial (26.5–21.6 kya) faunal assemblages (**Fig. 2**, Nos. 5, 6, 14–16, 23, and 24) suggest widespread coniferous woodland at elevations as low as 1,200–1,400 m above see level (Elias and Van Devender, 1992). These woodland environments persisted until 12.9 kya, but the insect data suggest considerable open ground, with grasses at least locally important at the midden sites. The grassland nature of the arthropod fauna was also suggested in the regional vertebrate record. The transition from the temperate Wisconsin fauna to the more xeric postglacial fauna started by 14.8 kya. The timing of this faunal change was essentially synchronous throughout the region. A major difference between the Big Bend and northern Chihuahuan Desert scenarios is the nature of this faunal change. In the Big Bend region, the transition was characterized by the disappearance from the record of all but one of the temperate insect species by approximately 14 kya. However, the xeric-adapted fauna did not appear in the Big Bend records until approximately 8.3 kya. In the northern Chihuahuan Desert assemblages, the xeric species first appeared at 14.8 kya, and several of the temperate grassland species from the Wisconsin interval persisted well into the Holocene. This mixture of xeric and temperate elements makes ecological sense because these northern faunas were living close to the edge of the Chihuahuan Desert. The gradual shifting of northern desert boundaries in postglacial times probably created many marginal habitats for temperate species in ecotones between grassland and desert scrub.

Work on late Quaternary insect faunas from the Colorado Plateau region (**Fig. 2**, Nos 7, 9, 11, 13, and 20) suggests that late Wisconsin climatic conditions there were cooler and moister than present, and that the plateau supported a mosaic of grassland and shrub communities without modern analog (Elias *et al.*, 1992).

MCR analysis of arid southwestern faunas (Elias, 1998) suggests that effective moisture exerts greater control on desert beetle distributions than does temperature. The author's MCR analysis of late Pleistocene beetle assemblages from the Mojave Desert (Elias, unpublished data) indicates that T_{max} was 7–8° C cooler than it is today from 38–36 kya,

Figure 3 Scanning electron micrographs of late Pleistocene fossil beetle specimens from the Chihuahuan Desert. (A) Elytron of *Onthophagus lecontei*; (B) elytron of *Pachybrachis mitis*; (C) elytron of *Piosoma setosum*; (D) head, pronotum, and elytra of *Rhadine longicolle*; (E) intact exoskeleton of *Ptinus*; (F) head of *Onthophagus cochisus*; (G) head of *Sapotes longipillis*; (H) elytron of *Hypoccacus estriatus*; (I) head of *Rhagodera costata*.

and mean annual precipitation (MAP) was 3.5–4.4 times greater than it is today. In the eastern Great Basin, the Bonneville Estates site yielded a beetle fauna dated 13.6 kya. Several of the species in this fauna are found today in the Pacific Northwest region. MCR analysis of this fauna indicated that T_{max} was 5–6° C cooler than today, and MAP was 525–675 mm, compared to modern MAP at the site of approximately 190–210 mm. These results agree well with paleoclimate reconstructions based on ancient pluvial lake levels (Benson, 2004) and regional paleobotany (Thompson *et al.*, 2004).

Miller (1983) discussed insect fossil assemblages from California asphalt deposits at McKittrick (**Fig. 2**, No. 29) and at Rancho La Brea in Los Angeles. The Rancho La Brea assemblages remain undated; the late Pleistocene McKittrick fauna dates to approximately 14.3 kya. Unlike the faunas of most other regions discussed here, the McKittrick and Rancho La Brea assemblages reflect conditions similar to those in southern California today.

Nelson and Coope (1982) described a diverse insect fauna dating from 18 kya from Seattle, Washington (**Fig. 2**, No. 28). Although the pollen spectra associated with this assemblage suggest conditions substantially colder than present, the insects are characteristic of the modern Puget lowland. Nelson and Coope suggest that the discrepancy between the flora and insect fauna may be due to increased climatic continentality just before the last glacial advance. MCR analysis of this assemblage (**Table 2**) suggests that T_{max} was 1–3° C colder than modern at the site, and that T_{min} was considerably colder than modern.

Cong and Ashworth (1996) described a beetle fauna from the Kalaloch site in northwest Washington (**Fig. 2**, No. 25). This fauna dates from 24.1–20.5 kya, and MCR analysis indicates that summer temperatures were only slightly cooler than today (**Table 2**). Most of this fauna can be found today in the lowlands of the Pacific Northwest, but a few species are restricted today to the northern boreal forest regions (Ashworth, 2004).

In southern British Columbia, Miller *et al.* (1985) studied insect fossils from 22.6–22.1 kya at Mary Hill and Port Moody (**Fig. 2**, Nos. 26 and 27). These assemblages represent an open coniferous forest floor community, developed in cool, dry climatic conditions between two advances of the Cordilleran ice sheet.

Throughout western North America, both temperature and precipitation regimes changed dramatically throughout the late Pleistocene. These climatic changes brought about wholesale changes in regional ecosystems, as reflected in the fossil beetle record. Unlike eastern North America, the western regions are topographically diverse, with dozens of mountain ranges interspersed with valleys, canyons, and flatlands. Cold-adapted beetles dominated many late Pleistocene faunal assemblages, expanding their ranges downslope and southward in response to climatic cooling, then retreating upslope and northward in response to warming.

Research in Eastern Beringia

During the late Pleistocene, eastern Beringia comprised the unglaciated regions of Alaska and the Yukon Territory. This region was an important refuge for arctic biota, including insects. Because of preservation in permafrost, the fossil beetle record of this region extends back into the late Tertiary. The nature of Pleistocene environments in eastern Beringia is a topic of considerable debate. Some patterns concerning Beringian paleoenvironments are beginning to emerge, but our reconstructions are far from complete. When I reported on the state of Pleistocene insect fossil research in this region in 1994, I noted that 'an enormous amount of work is yet to be done.' Although our knowledge has increased in the past decade, the statement remains equally true today. The following discussion deals with published fossil data from 28 sites (**Table 3**, **Fig. 4**). The evidence has been reviewed most recently by Elias (2000, 2001) and Ashworth (2004).

Based on their stratigraphic position, a number of fossil beetle assemblages can be confidently placed in MIS 6 (**Table 3**), including assemblages from the Old Crow site (**Fig. 4**, No. 1), Chi'jee's Bluff (**Fig. 4**, No. 2), and Kulukbuk Bluffs (**Fig. 4**, No. 3). These faunas reflect climates with depressed summer temperatures, but some sites yielded MCR estimates of T_{min} that are near modern levels. Based on stratigraphic relationships with sediments containing the Old Crow tephra (dated 140 kya), several beetle assemblages from Alaska and the Yukon indicate that prior to the onset of full interglacial conditions in MIS 5e, temperatures were beginning to warm. This evidence comes from several sites on the Noatak River in northwestern Alaska (**Fig. 4**, Nos. 4–6) and the Nuyakuk site in southwestern Alaska (**Fig. 4**, No. 7). Full interglacial warming peaked at levels that varied from region to region. In southwestern Alaska, the height of MIS 5e amelioration was as much as 3.5° C warmer than modern at the Nuyakuk site. At the NK-37 site in northwestern Alaska, T_{max} climbed to as much as 4.5° C above modern levels. Further east, summer temperatures were probably closer to modern levels. The best constrained estimate of average winter temperatures during MIS 5e comes from Chi'Jee's Bluff. This estimate suggests that T_{min} was approximately 4–7° C warmer than today, even though average summer temperatures at this site may have been slightly cooler than modern. Boreal insects and paleobotanical evidence combine to document the presence of coniferous forests in much of eastern Beringia during MIS 5e.

Beetle faunas associated with younger intervals of MIS 5 have yielded MCR estimates showing climatic cooling, with T_{max} levels declining 2–3°C below modern parameters (**Table 3**). Two faunas thought to derive from MIS 4 indicate that summer temperatures continued to cool. The MIS 4 faunal assemblage from Hungry Creek, Yukon (**Fig. 4**, No. 9), yielded a T_{max} estimate 6–9° C cooler than modern.

The long MIS 3 interstadial complex in eastern Berngia is represented by 25 faunal assemblages from 18 different sites. Warming intervals are indicated by faunas from the Titaluk River in northern Alaska (**Fig. 4**, No. 12) (46.4, 36, and 33.6 kya), from Cape Deceit on the Seward Peninsula (**Fig. 4**, No. 14) (42.2 kya), and from Kulukbuk Bluffs in southwestern Alaska (35 kya). The strongest indication of interstadial warming comes from the Titaluk River fauna dated 33.6 kya. This fauna yielded a T_{max}

Table 3 Summary of fossil beetle assemblage data from sites in eastern Beringia

Site	*Age (cal yr BP × 1000)*	*Late Quaternary*		*Modern*		*Change in Temperature*		*References*[a]
		T_{max} (°C)	T_{min} (°C)	T_{max} (°C)	T_{min} (°C)	July ΔT (°C)	Jan ΔT (°C)	
1. Old Crow CRH15 78–91	MIS 6	8.75–10	−29 to −31.75	16	−29	−7.25 to −6	−2.75 to 0	Morlan and Matthews (1983)
2. Chi'jee's Bluff Stn 1A, 4	MIS 6	12.5–13.25	−20.5 to −18.5	16	−29	−3.5 to −2.75	+8.5 to +10.5	Matthews *et al.* (1990a)
3. Kulukbuk Bluff 25A	MIS 6	7.25–9.25	−32.75 to −28.75	14.8	−22.6	−7.55 to −5.55	−10.15 to −6.15	Elias (1992a)
4. Noatak NK-26 Unit 4	140	3.75–10.5	−17.5 to −36.25	11	−25.8	−7.25 to −0.5	−10.45 to +8.3	Elias *et al.* (1998a)
5. Noatak NK29a unit 3, lower	140	5–9	−29 to −21.5	11	−29	−6.0 to −2.0	0 to +7.5	Elias *et al.* (1998a)
2. Chi'jee's Bluff Stn 4, I&J	140	3.5–9.5	−36 to −21	16	−29	−12.5 to −5.5	−7.0 to +8.0	Matthews *et al.* (1990a)
2. Chi'jee's Bluff 87-2	MIS 6-5 transition	11.75–12.5	−28 to −27.5	16	−29	−4.25 to −3.5	+1.0 to +1.5	Matthews *et al.* (1990a)
5. Noatak 29A unit 3, upper	MIS 6-5 transition	12–12.5	−29 to −27	11	−25.8	+1.0 to +1.5	−3.2 to −1.2	Elias *et al.* (1998a)
6. Noatak 37 MEE 0.13–0.22	MIS 6-5 transition	9–11	−25.5 to −22	11	−25.8	−2.0 to 0	+0.3 to +3.8	Edwards *et al.* (2003)
7. Nuyakuk 13-1	MIS 6-5 transition	12.5–13	−29.25 to −28	12.7	−9.7*	−0.2 to +0.3	NA	Elias and Short (1992)
6. Noatak 37 MEE2 140–153	MIS 5e	12.5–15	−28.5 to −19	11	−25.8	+1.5 to +4	−2.7 to +6.8	Edwards *et al.* (2002)
6. Noatak 37 MEE2 0.97–1.06	MIS 5e	12–16	−27 to −16	11	−25.8	+1 to +5	−1.2 to +9.8	Edwards *et al.* (2002)
8. Eva Creek 3-1A	MIS 5e	12.5–13	−22.5 to −20	16.9	−23.4	−4.5 to −3.9	+0.9 to +3.4	Matthews (1968)
3. Kulukbuk Bluff 25Z	MIS 5e	11.75–12	−22.5 to −20.5	14.8	−22.6	−3.05 to −2.8	+0.1 to +2.1	Elias (1992a)
2. Chi'jee's Bluff Stn 4, N-A	MIS 5e	12–13.25	−23 to −20.25	16	−29	−4.0 to −2.75	+6.0 to +8.75	Matthews *et al.* (1990a)
2. Chi'jee's Bluff Stn 4, N-B	MIS 5e	14.5–15.5	−25.25 to −21.5	16	−29	−1.5 to −0.5	+3.75 to +7.5	Matthews *et al.* (1990a)
7. Nuyakuk 13-2	MIS 5e	13–16.25	−30 to −21.75	12.7	−9.7*	+0.3 to +3.55	NA	Elias and Short (1992)
7. Nuyakuk 13-4	Late MIS 5	12.5–12.75	−12 to −11.25	12.7	−9.7*	−0.2 to +0.05	NA	Elias and Short (1992)
7. Nuyakuk 13-7	Late MIS 5	10.5–12	−20 to −19.5	12.7	−9.7*	−2.2 to −0.7	NA	Elias and Short (1992)
6. Noatak 37 94-25	Late MIS 5	7.5–9	−26.5 to −22.5	11	−25.8	−3.5 to −2	−0.7 to +3.3	Elias (1997)
9. Hungry Creek 52&53	Early MIS 4	6.5–10.25	−34.5 to −24.75	16	−29	−9.5 to −5.75	−5.5 to +4.25	Hughes *et al.* (1981)
10. Igushik 9-M-2	Early MIS 4	9–10	−30 to −23	12.7	−9.7*	−3.7 to −2.7	NA	Lea *et al.* (1991)
7. Chi'jee's Bluff Sample A	52 K	8.25–10	−32.75 to −27.25	16	−29	−7.25 to −6	−3.75 to +1.75	Matthews and Telka (1997)
9. Hungry Creek 27	Early MIS 3	11.5–12.5	−30 to −28	16	−29	−4.5 to −3.5	−1.0 to +1.0	Hughes *et al.* (1981)
10. Igushik 1-4	Early MIS 3	8.5–10	−30.5 to −22	12.7	−9.7*	−4.2 to −2.7	NA	Lea *et al.* (1991)
10. Igushik 4-M-1	Early MIS 3	12.5–15.5	−28 to −19	12.7	−9.7*	−0.2 to +2.8	NA	Lea *et al.* (1991)
11. Kuskokwim KBR/24	Early MIS 3	11–12.75	−24.25 to −19.75	14.8	−22.6	−3.8 to −2.05	−1.65 to +2.85	Elias (1992a)
11. Kulukbuk Bluff Unit C, A	Early MIS 3	9.75–11.5	−31.25 to −26.5	14.8	−22.6	−5.05 to −3.3	−8.65 to −3.9	Elias (1992a)
9. Hungry Creek 35	Mid-MIS 3	10–12.5	−19.5 to −13	16	−29	−6 to −3.5	+9.5 to +16.0	Hughes *et al.* (1981)
1. Old Crow 77-51	MIS 3	14.5–15.5	−28.5 to −25	16	−29.0	−1.5 to −0.5	+0.5 to +4.0	Morlan and Matthews (1983)
12. Titaluk River 17A	46.4	9.5–10	−31 to −28	10.3	−25.2	−0.8 to −0.3	−5.8 to −2.8	Elias (1997)
13. Cutler River	46.3	8–9.5	−26 to −22	11	−25.8	−3.0 to −1.5	−0.2 to +3.8	Elias (1997)
14. Cape Deceit Peat 5	43.3	6–10.5	−35.5 to −18	12.1	−20.4	−6.1 to −1.6	−15.1 to +2.4	Matthews (1968)
15. Upper Porcupine River	42.2	9.5–10	−29 to −27.5	16	−29	−6.5 to −6.0	0 to +1.5	Matthews and Telka (1997)
14. Cape Deceit S-5	42.2	11–11.75	−20.25 to −19.75	12.1	−20.4	−1.1 to −0.35	+0.15 to +0.65	Matthews (1968)
16. Blue Babe Site	41.8	6–10.5	−32.5 to −18.5	16	−24.4	−10 to −5.5	−8.1 to +5.9	Guthrie (1990)
8. Eva Creek 3-1A	38.5	11–13.5	−24 to −17	16.9	−23.4	−5.9 to −3.4	−0.6 to +6.4	Matthews (1968)
7. Chi'jee's Bluff Stn 1, B	37.8	10–11	−27.5 to −23	16	−29	−6.0 to −5.0	+1.5 to +6.0	Matthews *et al.* (1990a)
17. Isabella Basin IS-69	36.6	12–14.25	−32 to −22.25	16.9	−23.4	−4.9 to −2.65	−8.6 to +1.15	Matthews (1974)
12. Titaluk River 7D15	36	8–9.5	−31.5 to −25.5	10.3	−25.2	−2.3 to −0.8	−6.3 to −0.3	Nelson and Carter (1987)

(*Continued*)

Table 3 (*Continued*)

Site	*Age (cal yr BP × 1000)*	*Late Quaternary*		*Modern*		*Change in Temperature*		*References*[a]
		T_{max} (°C)	T_{min} (°C)	T_{max} (°C)	T_{min} (°C)	July ΔT (°C)	Jan ΔT (°C)	
8. Eva Creek 3-3B	35.2	8.5–9.5	−31.5 to −26	16.9	−23.4	−8.4 to −7.4	−8.1 to −2.6	Matthews (1968)
11. Kulukbuk Bluffs Unit C, B	35	12.25–13	−30.75 to −28	14.8	−27.6	−2.55 to −1.8	−3.15 to −0.4	Elias (1992a)
12. Titaluk River 20A	33.6	10.75–12.25	−20 to −19.25	10.3	−25.2	+0.45 to +1.95	+5.2 to +5.95	Nelson and Carter (1987)
18. Mayo Village Unit 1	31.5	9–9.75	−29.75 to −28	15.2	−29	−6.2 to −5.45	−0.75 to +1.0	Matthews *et al.* (1990b)
19. Baldwin Peninsula	31.5	8.5–9.5	−31 to −28	12.1	−18.3*	−3.6 to −2.6	NA	Hopkins *et al.* (1976)
20. Rock River	30.1	9.5–10.5	−28 to −22.5	16	−29	−6.5 to −5.5	+1.0 to +6.5	Matthews and Telka (1997)
9. Hungry Creek 49	26.1	11–12.25	−29.5 to −25.5	16	−29	−5.0 to −3.75	−0.5 to +3.5	Hughes *et al.* (1981)
10. Bering Shelf 78-15	24.9	5.5–10.5	−34.5 to −18.75	10.8	−15.6*	−5.3 to −0.3	NA	Elias *et al.* (1996, 1998b)
21. Bluefish	24.1	8.5–9.5	−31.5 to −28	16	−27.8	−7.5 to −6.5	−3.7 to −0.2	Matthews and Telka (1997)
22. Bering Land Bridge Park	21.7	5.5–9.5	−34 to −21	8.4	−18.3*	−2.5 to +1.1	NA	Elias (1997)
14. Cape Deceit S-1	ca. 21.6	9.5–10.5	−25 to −22	16	−20.4*	−6.5 to −5.5	NA	Matthews (1968)
23. Colorado Creek	18.2	11.25–12.5	−21.25 to −18.25	14.8	−22.6	−3.55 to −2.3	+1.35 to +4.35	Elias (1992b)
24. Clarence Lagoon	15.3	9.25–15.25	−35.25 to −19.25	7.5	−26	+1.75 to +7.75	−9.25 to +6.75	Matthews (1975)
25. Flounder Flat 3-M-8	15	12.5–13	−30.5 to −28	12.7	−9.4*	−0.2 to +0.3	NA	Elias (1992b)
26. Bering Shelf 76–121	13.4	11.5–13	−28.5 to −23.5	10.8	−15.6*	+0.7 to +2.2	NA	Elias *et al.* (1996, 1998b)
6. Noatak NK35-95AHA-60	13.2	15–16	−26 to −20.5	11	−25.8	+4.0 to +5.0	−0.2 to +5.3	Elias (1997)
27. Chukchi Shelf 85-69	12.9	9.25–10.25	−32.25 to −27.25	4	−27.7*	+5.25 to +6.25	NA	Elias *et al.* (1996, 1998b)
6. Noatak NK35-94AHA-24	12.3	12.5–13.5	−22 to −18	11	−25.8	+1.5 to +2.5	+3.8 to +7.8	Elias (1997)
6. Noatak NK32A-94AHA-17	12.3	10–10.5	−30 to −28.5	11	−25.8	−1.0 to −0.5	−4.2 to −2.7	Elias (1997)
28. Eagle River	11.5	12–13.5	−31.25 to −27	14.9	−30.5	−2.9 to −1.4	−0.75 to +3.5	Matthews and Telka (1997)

[a]References available from the QBib Web site, www.bugs2000.org/qbib.html
*Coastal sites with unreliable reconstructions of TMIN. See text for discussion.

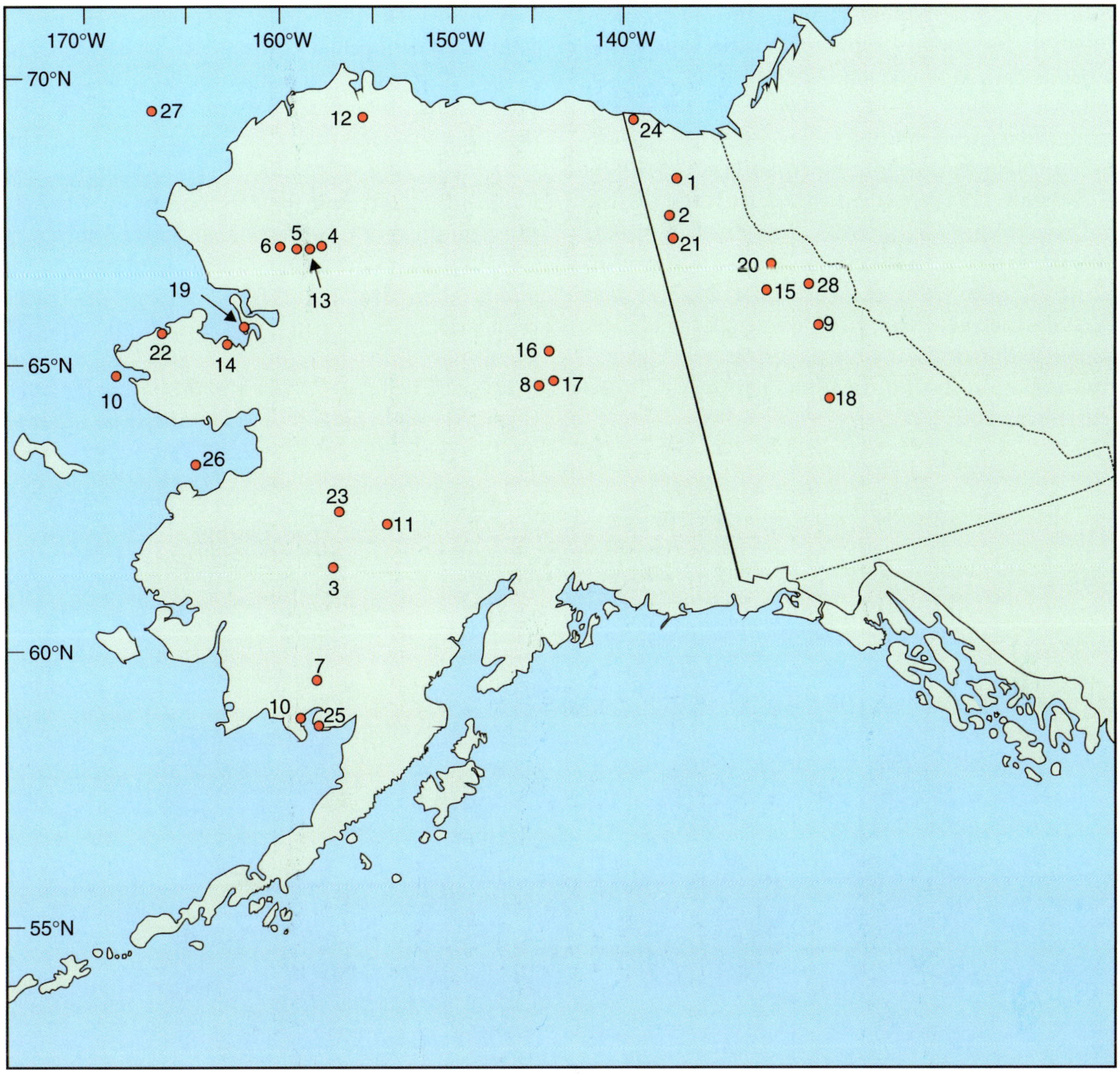

Figure 4 Map of northwestern North America showing locations of fossil sites discussed in text. Site numbers match those in **Table 3**.

estimate 0.5–2° C warmer than modern. The other faunas discussed previously yielded T_{max} estimates that were 0.5–2° C cooler than modern. Interestingly, a fauna dated 31.5 kya from Mayo Village, Yukon (**Fig. 4**, No. 18), indicates that regional T_{max} had fallen to 5–6° C colder than modern levels. Likewise, a fauna dated 35.2 kya from Eva Creek, interior Alaska (**Fig. 4**, No. 8), indicated T_{max} levels 7–8° C colder than modern. Thus, within the space of 2,000 years, temperatures appear to have oscillated dramatically in eastern Beringia. The beetle faunas that yielded the indications of interstadial warming are composed of species found today in open-ground habitats within the boreal forest. They do not rely on the presence of trees. The paleobotanical evidence suggests that most if not all of eastern Beringia remained open-ground tundra or steppe–tundra throughout MIS 3.

During the LGM, the beetle evidence indicates that T_{max} was depressed in eastern Beringia, but that T_{min} was within 1–2°C of modern levels in most regions. Elias *et al.* (1999) discussed difficulties in reconstructing T_{min} from faunal assemblages sampled from coastal sites. First, the ancient coastline of Beringia was removed from its current position by more than 1,000 km in some regions. Therefore, the modern T_{min} of a coastal site, greatly influenced by maritime climate, bears little relation to the ancient climate of the same site when it was hundreds of kilometers inland because of lowered sea level. Also, even the modern T_{min} of coastal sites is a poor reflection of winter temperatures at these sites because of the incursion of Arctic air masses during the winter season. These periods of very low temperature are sufficiently long, and sufficiently frequent, to virtually eliminate the less cold-resistant beetle fauna of coastal Alaska.

Following the LGM, beetle faunas from throughout eastern Beringia are indicative of climatic amelioration, beginning by at least 15.3 kya. This warming is especially noteworthy in the Arctic zone, where T_{max} rose above modern levels during the late glacial interval. The warming indicated by faunas from the subArctic zone is less marked during this interval. There is also evidence, again from the Arctic assemblages, for a brief climatic oscillation during the Younger Dryas chronozone (Elias, 2000).

Two principal ecosystems appear to have dominated eastern Beringia during late Pleistocene glacial periods: steppe–tundra and mesic tundra. Steppe–tundra, or 'mammoth steppe' as Guthrie (1990) described it, was rich in grasses and sagebrush (*Artemisia*). It supported a diverse, abundant megafauna of grazers and their predators. In Alaska and the Yukon, this ecosystem also had a characteristic beetle fauna, dominated by the ground beetles *Amara alpina* (**Fig. 5H**) and several species of *Harpalus* (**Fig. 5D**) pill beetles (Byrrhidae) in the genus *Morychus* (**Fig. 5I**), dung beetles in the genus *Aphodius*, and the weevils *Lepidophorus lineaticollis* (**Figs 5J** and **5K**) and *Connatichela artemisiae*

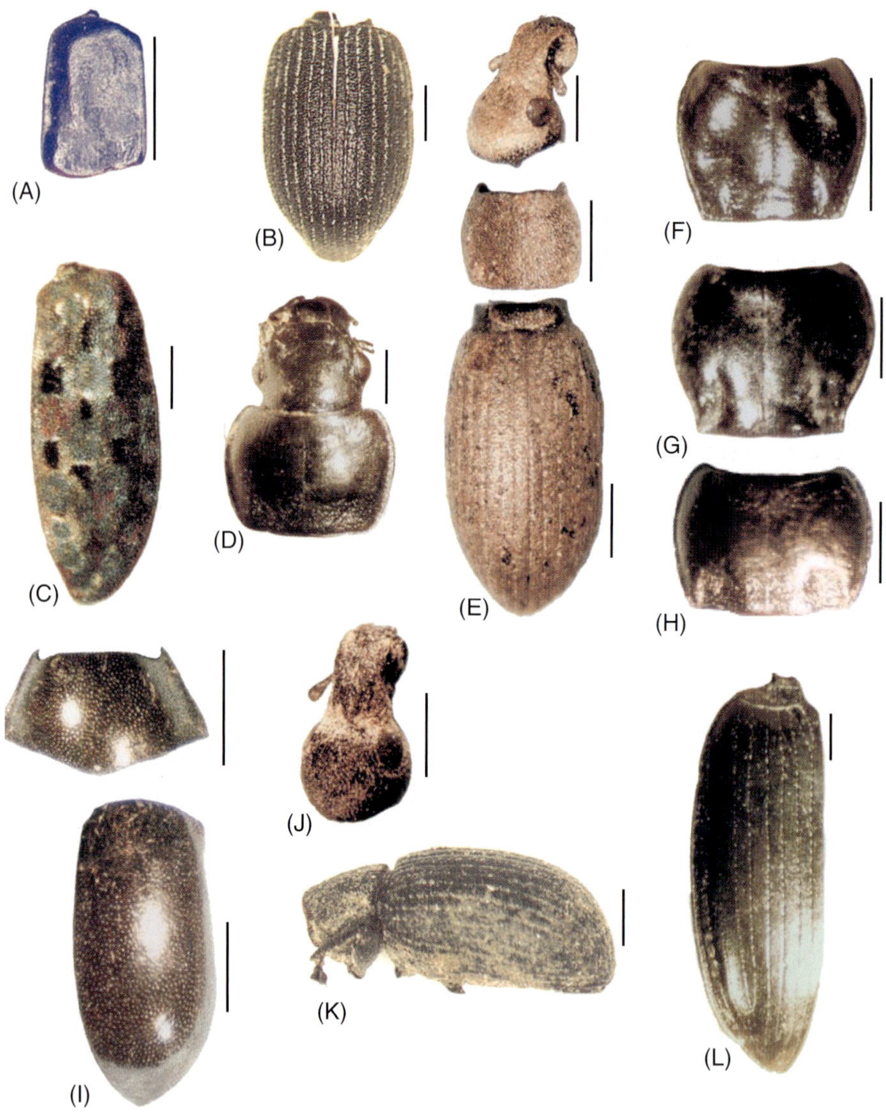

Figure 5 Light microscope photographs of late Pleistocene fossil beetle specimens from eastern Beringia. (A) Left elytron of *Tachinus brevipennis*; (B) elytra of *Vitavitus thulius*; (C) right elytron of *Elaphrus parviceps*; (D) head and pronotum of *Harpalus amputatus*; (E) head, pronotum, and elytra of *Connatichela artemisiae*; (F) pronotum of *Pterostichus brevicornis*; (G) pronotum of *Pterostichus pinguedineus*; (H) pronotum of *Amara alpina*; (I) pronotum and right elytron of *Morychus* sp.; (J) head of *Lepidophorus linneaticollis*; (K) pronotum and elytron of *L. linneaticollis*; (L) left elytron of *Pterostichus nearcticus*. Scale bars = 1 mm.

(**Fig. 5E**). This steppe–tundra fauna, first recognized by Matthews (1983), is commonly found in fossil assemblages throughout the Yukon and interior Alaska. However, in other regions of Alaska, a different suite of beetle taxa dominate late Pleistocene assemblages. This group includes the *Cryobius* group of ground beetles in the genus *Pterostichus* (**Figs 5F** and **5G**), several other ground beetles associated with mesic and moist habitats, rove beetles in the subfamily Omaliinae, and weevils that feed on dwarf birch and dwarf willow. This mesic tundra fauna has been found in fossil assemblages throughout southwestern and northwestern Alaska, as well as in samples taken from the ancient Bering Land Bridge that connected Alaska to Siberia during intervals of lowered sea level (Elias *et al.*, 2000). The land bridge may have served as a biological filter, keeping dry-adapted species, such as the woolly rhinoceros, from entering North America (Guthrie, 2001).

Conclusions

Late Pleistocene beetle assemblages from sites throughout North America have documented large-scale, rapid climatic changes. During MIS 3, beetle assemblages have demonstrated oscillating interstadial warming and cooling events that coincide with climatic patterns inferred from oxygen isotope records in Greenland ice cores. The pace and severity of such changes have often gone undetected in paleoenvironmental reconstructions based on fossil pollen, although new analytical methods are making breakthroughs (Grimm and Jacobson, 2004). As has been noted in the history of the European beetle fauna (Coope, 1978), North American beetles appear to have responded to the vagaries of Pleistocene climates by shifting their distributions in order to stay within climatic conditions to which they are best adapted. Thus, the Pleistocene beetle fauna of North America remains extant today, albeit living in different regions. Each new climatic regime of the Pleistocene brought new associations of species together in a given region. Many of the species in given types of associations remained constant through long stretches of time, but different species were added or subtracted from the fauna of any specific region as ecosystems waxed and waned.

See also: **Beetle Records**: Overview; Late Tertiary and Early Quaternary; Late Pleistocene of Europe; Postglacial North America. **Ice Core Records**: Greenland Stable Isotopes. **Paleoceanography, Physical and Chemical Proxies**: Oxygen Isotope Stratigraphy of the Oceans. **Plant Macrofossil Methods and Studies**: Rodent Middens.

References

Anderson, T. W., Matthews, J. V., Jr., Mott, R. J., and Richard, S. H. (1990). The Sangamonian Pointe-Fortune site, Ontario–Quebec border. *Géographie Physique et Quaternaire* **44**, 271–287.

Ashworth, A. C. (2004). Quaternary Coleoptera of the United States and Canada. In *The Quaternary Period in the United States* (A. R. Gillespie, S. C. Porter, and B. F. Atwater, Eds.), pp. 505–517. Elsevier, New York.

Baker, R. G., Schwert, D. P., Bettis, E. A., III, Kemmis, T. J., Horton, D. G., and Semken, H. A. (1991). Mid-Wisconsinan stratigraphy and paleoenvironments at the St. Charles site in south-central Iowa. *Geological Society of America Bulletin* **103**, 210–220.

Benson, L. (2004). Western lakes. In *The Quaternary Period in the United States* (A. R. Gillespie, S. C. Porter, and B. F. Atwater, Eds.), pp. 185–204. Elsevier, New York.

Cong, S., and Ashworth, A. C. (1996). Palaeoenvironmental interpretation of middle and late Wisconsinan fossil coleopteran assemblages from western Olympic Peninsula, Washington, USA. *Journal of Quaternary Science* **11**, 345–356.

Coope, G. R. (1978). Constancy of insect species versus inconstancy of Quaternary environments. In *Diversity of Insect Faunas* (L. A. Mound and N. Waloff, Eds.), Symposium No. 9, pp. 176–187. Royal Entomological Society of LondonLondon.

Coope, G. R., and Lemdahl, G. (1995). Regional differences in the late glacial climate of northern Europe based on coleopteran analysis. *Journal of Quaternary Science* **10**, 391–395.

Elias, S. A. (1990). The timing and intensity of environmental changes during the Paleoindian period in western North America: Evidence from the fossil insect record. In *Megafauna and Man* (L. D. Agenbroad, J. I. Mead and L. W. Nelson, Eds.), pp. 11–14. Northern Arizona Press, Flagstaff.

Elias, S. A. (1991). Insects and climate change: Fossil evidence from the Rocky Mountains. *BioScience* **41**, 552–559.

Elias, S. A. (1992). Late Quaternary zoogeography of the Chihuahuan Desert insect fauna, based on fossil records from packrat middens. *Journal of Biogeography* **19**, 285–297.

Elias, S. A. (1994). *Quaternary Insects and Their Environments.* Smithsonian Institution Press, Washington, DC.

Elias, S. A. (1998). The mutual climatic range method of paleoclimate reconstruction based on insect fossils: New applications and interhemispheric comparisons. *Quaternary Science Reviews* **16**, 1217–1225.

Elias, S. A. (1999). Mid-Wisconsin seasonal temperatures reconstructed from fossil beetle assemblages in eastern North America: Comparisons with other proxy records from the Northern Hemisphere. *Journal of Quaternary Science* **14**, 255–262.

Elias, S. A. (2000). Late Pleistocene climates of Beringia, based on fossil beetle analysis. *Quaternary Research* **53**, 229–235.

Elias, S. A. (2001). Mutual climatic range reconstructions of seasonal temperatures based on late Pleistocene fossil beetle assemblages in eastern Beringia. *Quaternary Science Reviews* **20**, 77–91.

Elias, S. A., and Van Devender, T. R. (1990). Fossil insect evidence for late Quaternary climatic change in the Big Bend region, Chihuahuan Desert, Texas. *Quaternary Research* **34**, 249–261.

Elias, S. A., and Van Devender, T. R. (1992). Insect fossil evidence of late Quaternary environments in the northern Chihuahuan Desert of Texas and New Mexico: Comparisons with the paleobotanical record. *Southwestern Nature* **37**, 101–116.

Elias, S. A., Mead, J. I., and Agenbroad, L. D. (1992). Late Quaternary anthropods from the Colorado Plateau, Arizona and Utah. *Great Basin Naturalist* **52**, 59–67.

Elias, S. A., Van Devender, T. R., and De Baca, R. (1995). Insect fossil evidence of late glacial and Holocene environments in the

Bolson de Mapimi, Chihuahuan Desert, Mexico: Comparisons with the paleobotanical record. *Palaios* **10**, 454–464.
Elias, S. A., Andrews, J. T., and Anderson, K. H. (1999). New insights on the climatic constraints on the beetle fauna of coastal Alaska derived from the mutual climatic range method of paleoclimate reconstruction. *Arctic, Antarctic, and Alpine Research* **31**, 94–98.
Elias, S. A., Berman, D., and Alfimov, A. (2000). Late Pleistocene beetle faunas of Beringia: Where east met west. *Journal of Biogeography* **27**, 1349–1363.
Grimm, E. C., and Jacobson, G. L., Jr. (2004). Late Quaternary vegetation history of the eastern United States. In *The Quaternary Period in the United States* (A. R. Gillespie, S. C. Porter, and B. F. Atwater, Eds.), pp. 381–402. Elsevier, New York.
Grootes, P. M., Stuiver, M., White, J. W. C., Johnsen, S., and Jouzel, J. (1993). Comparison of oxygen isotope records from the GISP2 and GRIP Greenland ice cores. *Nature* **366**, 552–554.
Guthrie, R. D. (1990). *Frozen Fauna of the Mammoth Steppe. The Story of Blue Babe.* University of Chicago Press, Chicago.
Guthrie, R. D. (2001). Origins and causes of the mammoth steppe: A story of cloud cover, woolly mammoth tooth pits, buckles, and inside-out Beringia. *Quaternary Science Reviews* **20**, 549–574.
Matthews, J. V., Jr. (1983). A method for comparison of northern fossil insect assemblages. *Géographie Physique et Quaternaire* **37**, 297–306.
Miller, R. F., Morgan, A. V., and Hicock, S. R. (1985). Pre-Vashon fossil Coleoptera of Fraser age from the Fraser Lowland, British Columbia. *Canadian Journal of Earth Science* **22**, 498–505.
Miller, S. E. (1983). Late Quaternary insects of Rancho La Brea and McKittrick, California. *Quaternary Research* **20**, 90–104.
Nelson, R. E., and Coope, G. R. (1982). A late Pleistocene insect fauna from Seattle, Washington. American Quaternary Association seventh biennial meeting, program and abstracts, 146.
Schwert, D. P., and Ashworth, A. C. (1988). Late Quaternary history of the northern beetle fauna of North America: A synthesis of fossil distributional evidence. *Memoirs of the Entomological Society of Canada* **144**, 93–107.
Thompson, R. S., Shafer, S. L., Strickland, L. E., Van de Water, P. K., and Anderson, K. H. (2004). Quaternary vegetation and climate change in the western United States: Developments, perspectives, and prospects. In *The Quaternary Period in the United States* (A. R. Gillespie, S. C. Porter, and B. F. Atwater, Eds.), pp. 403–426. Elsevier, New York.
Walker, I. R., Mott, R. J., and Smol, J. P. (1991). Allerød-Younger Dryas lake temperatures from midge fossils in Atlantic Canada. *Science* **253**, 1010–1012.

Late Pleistocene of New Zealand

M Marra, University of Canterbury, Christchurch, New Zealand

© 2007 Elsevier B.V. All rights reserved.

Introduction

Fossil beetle reconstructions are relatively new in New Zealand (NZ) with the first published work dating only to 2000. Before 2000, terrestrial paleoenvironmental reconstructions relied mainly on pollen records, which defined regional vegetation patterns for the late Quaternary and from which broad climate inferences were drawn. Quantitative estimates for past climate remained less well defined because there is no clear relationship between temperature and vegetation. Hence, there was a need for additional proxies and particularly for proxies capable of quantifiable paleoclimatic reconstruction.

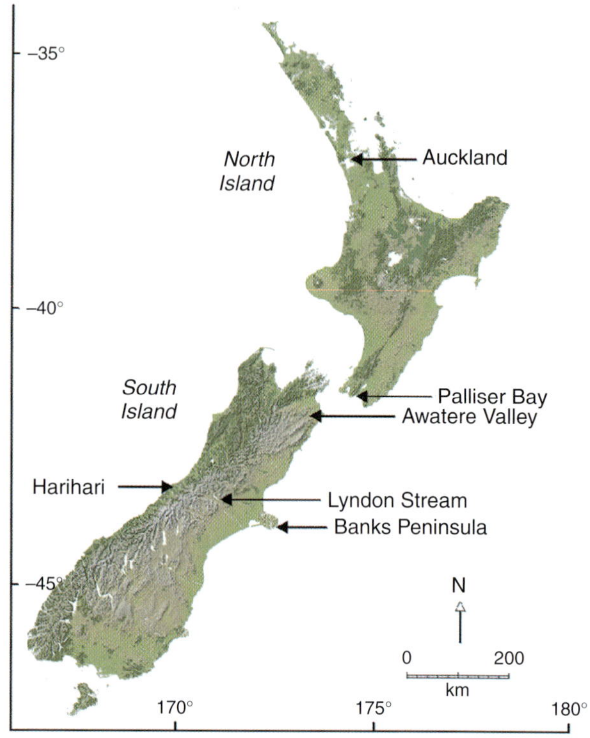

Figure 1 Map of New Zealand showing location of research sites mentioned in the text. (map courtesy of Geographx)

To date, late Pleistocene beetle assemblages have been recorded from five NZ sites (**Fig. 1**) ranging in age from marine isotope stage (MIS) 7 to MIS 2. The late Pleistocene literature to date includes the development of a beetle-based climate estimation model and its application to two Last Glacial Maximum (LGM) sites. The paleoecology is described for each site, detailing ecosystems at a local scale and at community and species levels. The following is a review of this literature prefaced by a summary of present knowledge of the modern beetle fauna, and a brief description of the NZ physical environment, identifying factors that affect beetle distribution.

Background

Modern Fauna

Knowledge of the modern fauna underpins both fossil identification and the ecological controls on distribution that are the basis for climate modeling. The NZ beetle fauna is relatively rich: ca. 5,223 native and 356 introduced species have been described, comprising ca. 1,100 genera in 82 families (Klimaszewski and Watt, 1997). The actual number of species is unknown but is certain to be significantly higher (Leschen *et al.*, 2003). The Lynfield survey (Kuschel, 1990), regarded as the most comprehensive survey so far in NZ, identified ca. 1,000 species. As this is thought to be about 8% of the total NZ beetle fauna, it suggests the total number is around 10,000–10,500 species (Klimaszewski and Watt, 1997). Consequently, the 5,223 species held in NZ collections represent about half the total NZ beetle fauna, but several groups need review. Endemism is high at most levels, with one endemic family, several endemic subfamilies, high but unquantified endemism of genera and about 90% of species are endemic. A summary of the dominant families and the number of described species in each of these families is listed in **Table 1**.

NZ Setting

The NZ geography imposes many barriers on the dispersal of insects and the tracking of ecosystems across the landscape in response to climate change. It is an island archipelago focused on two main islands (North Island, South Island), and has a land area of ca. 270,550 km^2. South Island is dominated by an axial mountain range, the Southern Alps, extending about 500 km along the western side of the island, with peaks reaching over 3,000 m. During the LGM, the Southern Alps developed a piedmont icecap that extended the length of the Alps. On the western side, glaciers advanced to the coast at least south of Harihari (**Fig. 1**), while to the east they extended beyond the range fronts into inland basins and coastal outwash fans. A barrier to dispersal during interglacials is the separation of the North and South islands by the 32 km wide Cook Strait. However, during glaciations when the sea level is lower, a land bridge exists. North Island dispersal barriers and impacts on distribution include a series of north/south trending axial ranges, active volcanism, and a high elevation central plateau.

Table 1 List of dominant eight families in NZ fauna and number of described species in each family

Family in order of dominance	*Number of described NZ species*
Curculionidae (incl. Scolydidae)	1,542
Staphylinidae (incl. Pselaphidae)	1,021
Carabidae	445
Scymaenidae	202
Colydiidae	196
Cerambycidae	188
Chrysomelidae	153
Scarabaeidae	144

Klimaszewski and Watt, 1997.

Late Pleistocene Beetle Distributions

Shifting Distributions

The Last Interglaciation (MIS 5) The late Pleistocene fossil beetle records show that beetle assemblages in some locations were very different from those found there today. The best example is a site in Palliser Bay on the southern coast of North Island (**Fig. 1**) (Marra, 2003a; Shulmeister *et al.*, 2000). The site dates from the Last Interglaciation, MIS 5. Two fossil beetle assemblages totaling 63 taxa were identified from two organic silt beds occurring in a sequence of fluvial and lacustrine sediments (**Fig. 2**).

Both assemblages are similar in that they include ecological successions from lake margins to forest, but they comprise different taxa. Of the two fossiliferous layers, the lower contained beetles that are found today some 600–700 km north of the study site. This fauna represents a significant southwards displacement compared to modern distributions, despite the physical barrier of a central North Island volcanic plateau. This is a minimum southward dispersal because Cook Strait would have impeded further movement to the south. No MIS 5 sites have been found on South Island to test whether these beetles did cross the Strait.

The penultimate interglaciation (MIS 7) Although the modern NZ beetle fauna is not well-known in general, the fauna of a few regions is better documented. As noted, the Lynfield Survey (Kuschel, 1990) of

Figure 2 The MIS 5 Palliser Bay research site. The arrows indicate the organic beds sampled for fossil beetles. Horizontal distance between sampled beds is ca. 7 m.

the Auckland Region was carried out during the 1980s. This survey provided the background against which a fossil assemblage of 98 beetle taxa extracted from an interglacial buried forest unit (**Fig. 3**) identified at late MIS 7 could be interpreted (Marra *et al.*, 2006b). All but two of the fossil taxa occur in the local modern fauna. Based on the Lynfield Survey, the fossil fauna represents 48% of families, 20% of genera, and 13% of species in the local modern native fauna (**Fig. 4**).

The assemblage shows that the same fauna has effectively reassembled in the same place across two glacial cycles, changes in vegetation, and nearly 200 kyr. The beetles, supported by plant macrofossils and pollen, identify the vegetation and climate as being similar to that of the present day. The reassembly of the beetle fauna two glacial cycles later is probably a consequence of a similar environment during warm stages and the lack of topographical barriers to reoccupation of the area after glaciations.

Figure 3 The MIS 7 paleo-forest unit sampled for beetles. The paleo-forest was destroyed and buried by several meters of volcanic deposits. Arrow indicates the sample site.

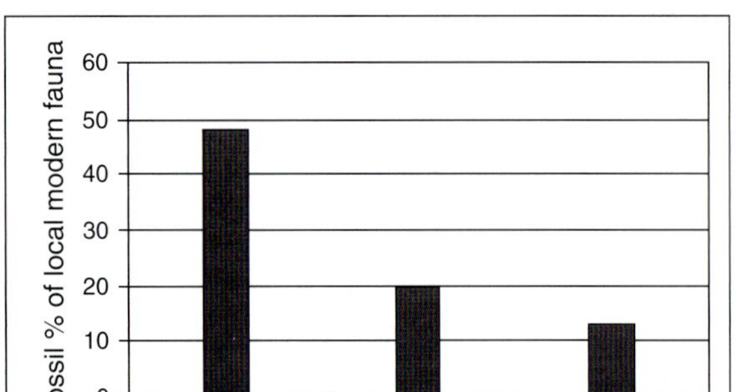

Figure 4 The comparison between Auckland site fossil taxa and modern Auckland fauna showing the representativeness of modern families, genera, and species in the fossil assemblage.

The penultimate glaciation (MIS 6) The modern beetle fauna of Banks Peninsula (**Fig. 1**) is also well-known from several published natural histories and inventories (e.g., Johns (1986)). These inventories provide a basis for comparison with a fossil assemblage from Gebbies Valley, Banks Peninsula.

Nineteen beetle fossils extracted from core material are reported in Marra, (2003b). The fossil assemblage comprised montane forest taxa in a coastal setting. Based on the modern distributions and upper altitudinal limits of the fossil fauna, Marra, (2003b) interpreted the climate as interstadial rather than pleniglacial, and by using a simple environmental lapse ($0.55 \pm 0.05°$ C per 100 m), estimated the lowering of the present-day montane zone by ca. 450 m to a sea-level position. This equates to a temperature depression of around 2–3° C.

Although the reconstruction of this site was restricted by the limited amount of material available from the core, the beetle fossils were abundant and provided additional information of the paleoenvironment. Originally the site was described from pollen evidence as a back barrier, freshwater setting (Shulmeister *et al.*, 1999; Soons *et al.*, 2002). However, the beetle fossils indicate that the paleoenvironment was estuarine (Marra, 2003b). The pollen-based description of the surrounding vegetation is described as low forest-shrubs (Soons *et al.*, 2002). The fossil beetle *Rhicnobelus metallicus* (Belidae), that is host specific to *Podocarpus totara* and *P. hallii*, provided confirmation of this forest and identified the forest type. Both *P. totara* and *P. hallii* grow on Banks Peninsula but as *P. hallii* dominates the higher elevations, it was probably *P. hallii* present at this lowland site during the cooler MIS 6 as part of the low forest described from the pollen.

The MIS 6 fossil fauna is a very different beetle fauna than the modern Banks Peninsula fauna. Of the 19 beetle species identified, only three of the fossil species (*R. metallicus* (Belidae), *Eucossonus setiger* (Curculionidae), *Cecyropa modesta* (Curculionidae)) are found in the modern Banks Peninsula fauna (although undetermined carabids and species of *Nestrius* (Curculionidae) and *Bledius* (Staphylinidae) could also be present). Species of *Epichorius* (Byrrhidae), *Spinditeles* (Anobiidae), and *Mandalotus Baeosomus*, *Microcryptorhynchus*, or *Scelodolichus* (Curculionidae) in the fossil assemblage are not part of the modern Banks Peninsula fauna. The different faunal composition of MIS 6 is a result of the montane environmental conditions at the site at that time.

The Last Glacial Maximum (MIS 2) The LGM beetles at Nina Stream in the Awatere Valley (**Fig. 1**) in northeastern South Island have been reported in

Figure 5 The Awatere Valley site stratigraphy. Inset indicates the lower three organic silts sampled for fossil insects.

Marra and Leschen (2004). The Awatere Valley flows northeast along the northern side of the Inland Kaikoura Range (2,880 m). At the site, the Nina Stream is transected by the Awatere fault, one of the series of active faults that continue the NZ plate boundary into North Island. Movement on the fault impedes stream drainage, which presents stratigraphically as a sequence of fan gravels intersected by organic silts (**Fig. 5**). Beetle fossils were extracted from the lower three organic beds. The calibrated radiocarbon ages for the bottom organic sample of 24 cal kyr BP and 22–23 cal kyr for the top sample place the deposits within the LGM. The site is presently 200 m above sea level.

The LGM Awatere Valley beetle fauna includes sub-Alpine beetle taxa. Two of these are ground beetles (Carabidae), *Notagonum feredayi* and *Bembidion hokitikense*, which are at present found near water in tussock grasslands. Other sub-Alpine species include *Peristoreus fusconotatus* (Curculionidae) that lives on *Olearia* (a genus of tree daisies), and a species of *Lyperopais* (Curculionidae) that so far has been collected only above the tree line. *Acritelater reversus* (Elateridae) and *Aleochara hammondi* (Staphylinidae) also both inhabit sub-Alpine tussock grasslands. The fauna suggests an altitudinal shift in the vegetation of several hundred meters from higher elevations to the lowland river plain, presumably in response to the cold climate conditions.

Quantifying Past Climates

Geographic Overlap Method

The Last Interglaciation (MIS 5) Based on the hypothesis that modern distributions indicate the preferred climate for an assemblage of beetles, these past beetle distributions provide a basis for quantifying climatic conditions in the past. The Palliser Bay fossil fauna described above provides the basis of a rudimentary climate reconstruction using a geographic overlap method (Marra, 2003a). By mapping the modern distributions of the fossil fauna, which, as stated earlier, is now up to 700 km further north, the climate was estimated from the overlapping distributions of modern fauna (**Fig. 6**). The temperature of the overlapping region was then compared with the temperature of the study site to give the estimated paleotemperature.

The temperature for this MIS 5 site was estimated at 1.6–2.5° C warmer in the summer (January) and 2.3–3.2° C warmer in the winter (July) than present day. These estimates agree with local and regional reconstructions of warmer, wetter conditions at the thermal maximum of the Last Interglaciation. From this, the lower sample was assigned to the thermal peak of the Last Interglaciation, substage 5e.

Although this method was used in the early years of Quaternary entomology and in some instances is still the preferred approach, its application is limited because when distribution is governed by a climatic variable, a species may be found over a wide geographic area but at the same time have a very narrow climatic range. This is particularly evident in montane or Alpine species that may be distributed geographically over much of the NZ mountains, but only at certain elevations, so the true geographic range is overestimated. A better approach is to convert the geographic range of each taxon into a climate range.

Method for Climate Quantification

Method development A maximum likelihood envelope (MLE) method has been developed (Marra *et al.*, 2004) that is suitable for quantifying past climates from beetle fossils in NZ. MLE was developed to deal with problems associated with the incomplete knowledge of the NZ fauna. Incomplete knowledge limits robust climatic reconstruction because if the observed distribution of species is poorly known, the actual distribution of the ecological tolerances may be underestimated.

The MLE method is based on the premise that beetle distributions are climate driven and that the distribution of each taxon is controlled by finite physiological requirements. It, therefore, uses a sine

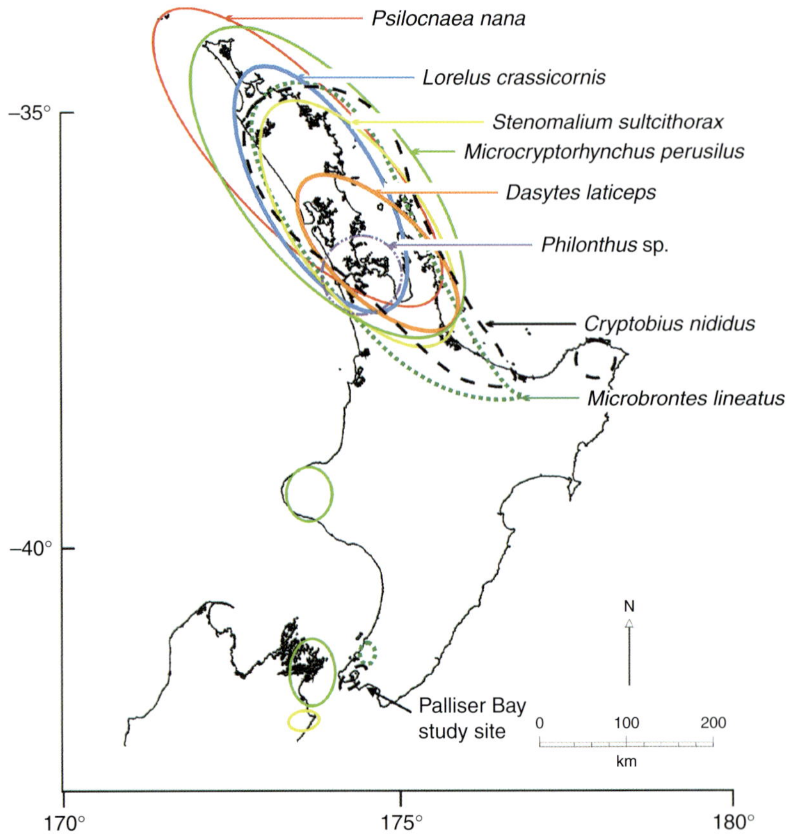

Figure 6 The modern day distribution of species as they relate to the fossil site at Palliser Bay. Marra (2003a).

distribution model rather then a Gaussian distribution model because a Gaussian model does not prescribe absolute limits to distributions needed for the present inside-absent outside approach intrinsic to beetle climate modeling (e.g., mutual climate range (Atkinson *et al.*, 1987)). Maximum likelihood (ML) best-high and best-low values are determined and fitted to the sine distributions to allow for the uncertainty that the samples are realistic representations of the actual climatic distributions. **Figure** 7 illustrates how the estimated maximum likelihood endpoints effectively increase the distribution of a taxon. The endpoints define a 95% confidence distribution.

Climate variables for reconstruction The climate reconstructions are based on mean February and mean minimum July temperatures. Although MCR is based on summer temperature and seasonality, seasonality is much more subdued in NZ than in continental locations, so it is unlikely that this is a major climatic consideration. Anecdotal evidence suggests that length of summer and extreme winter temperatures may drive beetle distributions, hence mean February and minimum July temperatures.

These climate variables were tested in Marra *et al.* (2004) by pooling the distribution data points of all the species. Each datum (x_i) was transformed to

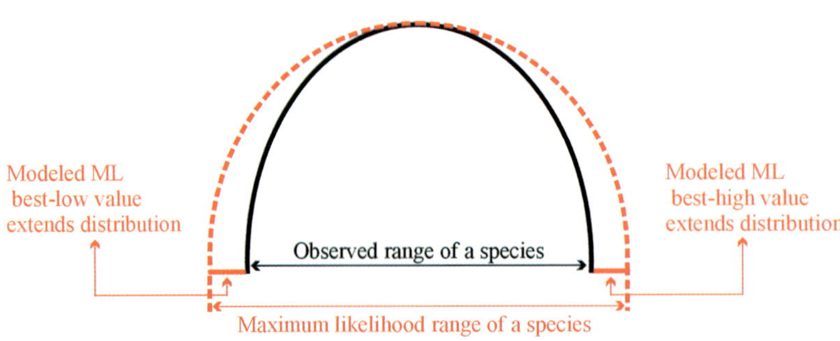

Figure 7 Maximum likelihood best-high and best-low endpoints attached to a sine distribution model, effectively increasing the climate range of a species.

$(x_i - a)/(b - a)$ to give each species' actual distribution and relative position on a standard range (0,1) of all the data points (all the distribution points of all the species) (**Fig. 8**).

The test showed that the sine model fit the July minimum temperature rather well, but the February mean temperature pooled data were significantly positively skewed. As the same skewness is not apparent in the July minimum temperature data as expected, it suggests that, aside from temperature, some other climate variable was affecting beetles significantly in summer but not in winter. Tests to determine the nature of this climate factor are underway with initial results (unpublished data) suggesting that available moisture during the summer is a factor governing beetle distribution.

LGM Climate Reconstruction

Lyndon Stream Using MLE, Marra *et al.* (2006a) estimated the LGM mean temperature for February and the mean minimum July temperature at the Lyndon Stream in the eastern foothills of the Southern Alps (alt. ca. 700 m) (**Fig. 1**). During the LGM, the site was very close to the major Rakaia glacier. The site comprises lake beds bounded by LGM proximal outwash, confining the lake within the LGM chron (**Fig. 9**). The site was first examined by Soons and Burrows (1978) who dated the paleolake to between 24.84–26.1 cal kyr BP (base of the section) and 21.98–23.28 cal kyr BP (top of the section). Using pollen and plant macrofossils, they described an environment during the LGM that was nearly identical to modern conditions.

Marra *et al.* (2005) used nine fossil beetle taxa in their LGM reconstruction of mean February (summer) and mean minimum temperatures for July (the coldest month).

The February temperature reconstruction is illustrated in **Figure 10**. The broken line is the modern temperature at the site, and horizontal bars are the temperature ranges for the taxa. The shadings at the end of each bar are the ML best-high and best-low statistical endpoints. The envelopes (yellow or shaded) are the mutual temperature overlaps of all the taxa, which are defined at the lower limit by the species with the highest minimum, and the upper limit by species with the lowest maximum. The yellow shows the overlap of the known temperature ranges, whereas the shaded envelope is the MLE estimate that has statistically extended the known ranges. The estimates indicate that the mean temperature at the site at LGM was between 0.5° C warmer and 1.9° C cooler in February than at the present day. The modeled mean minimum winter temperature was similarly close to

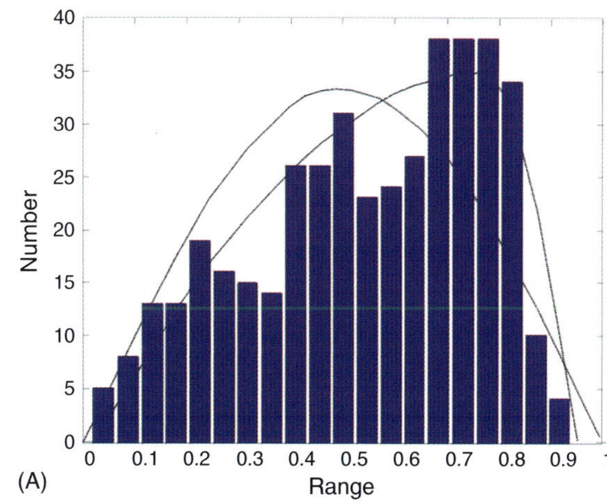

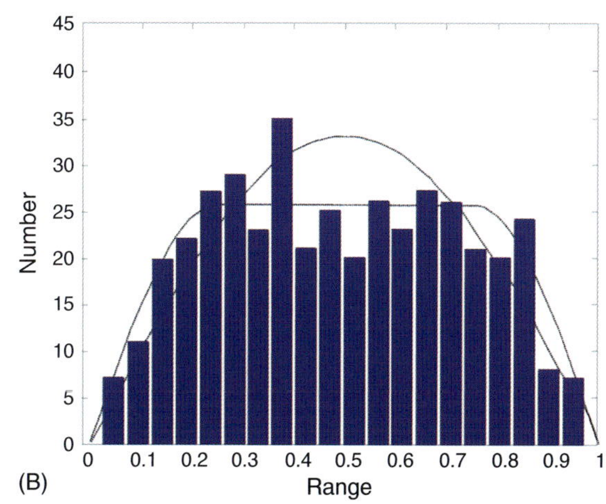

Figure 8 Graphs showing the fit of sine models to the climate variables for: (A) mean February temperature and (B) mean July minimum temperature. (A) Pooled mean February data and best-fitting sine model and bi-sine model and (B) pooled mean July minimum temperature data and two sine models: sine and sine-tapered boxcar. From Marra *et al.* (2004).

Figure 9 The Lyndon Stream site lake beds bounded by LGM proximal outwash.

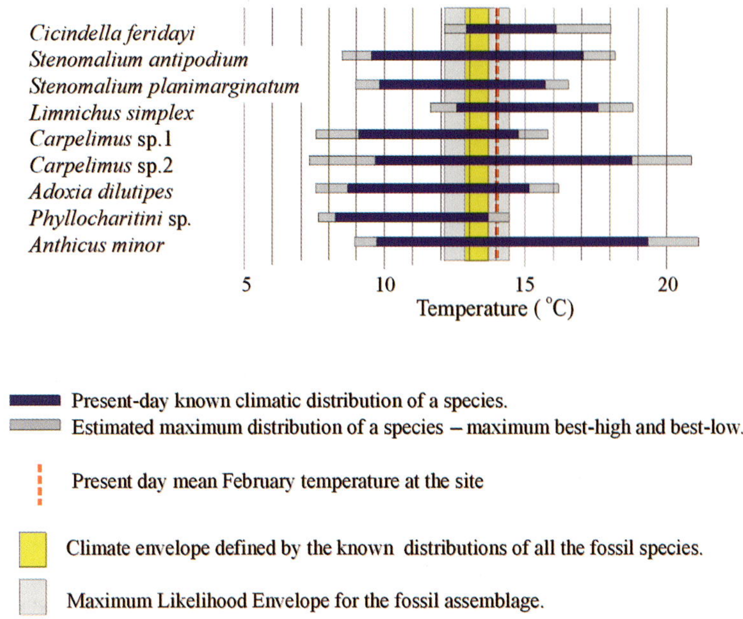

Figure 10 Application of MLE to reconstruct the mean February temperature for the LGM at Lyndon Stream showing the difference between the present day temperature at the site to the estimated temperature at the LGM. Marra *et al.* (2006a).

present values, being between 1° C warmer and 2.2° C cooler than present day.

Even at the cooler extremes of these modeled estimates, the climate during the LGM appears to have been milder than the traditional view, for which the temperature depression has been estimated at 4–7° C (e.g., Porter (1975), Soons (1979), and Shulmeister *et al.* (2001)). However, the predictions from beetle environmental requirements fit remarkably well with the reconstructions from plant macrofossils and are consistent with new evidence of relatively mild conditions (Hormes *et al.*, 2003) during at least some periods within the LGM chron in NZ.

Awatere Valley In contrast, the MLE climate reconstruction for the Awatere Valley ca. 300 km to the north, reported in Marra *et al.* (2004) shows a very much cooler climate during the same time (24.65–22.05 cal kyr BP) during LGM. At this site, summer (February) mean temperature was ca. 3.5–4° C cooler, and July (winter) mean daily minimum temperature was ca. 4–5° C cooler than present-day temperatures. The maximum likelihood estimates broadened the reconstructed temperature ranges to 2.5–5° C cooler for February temperatures and 3.5–6.0° C cooler for mean minimum daily temperature of the coldest month (July). These estimates are more consistent with the traditional cooler LGM model of 4–7° C cooler.

The conflict between these sites of apparently similar age needs to be resolved. It is most likely to be a result of poorly constrained ages. Tighter age control may reveal that these sites are close but not identical in age, and that the different environmental estimates reflect rapid climate change from cold glacial to warm interstadial such as described above at the Lyndon Stream site. Another possibility is that the Lyndon Stream site plots warmer because it is sheltered from southerly and cold easterly winds.

Reconstructions of rates of climate change have not yet been attempted. The ability to determine rates of climates change and to identify short-term climatic oscillations is the real strength of fossil beetle analyses and these should be investigated. There is also no literature on the postglacial warming, a time-frame covering a period of rapid climate change about which there are much conflicting data.

Late Pleistocene Beetles and Paleoecology

Paleoecology of the LGM

Awatere Valley Late Pleistocene beetle research has highlighted the importance of scale in reconstructions (e.g., Marra and Leschen (2004)). It is evident that pollen provides regional-scaled vegetation information, but the beetles provide local-scale insight. This is clearly evident in eastern South Island beetle data. The pollen indicates that the vegetation for this region was predominantly grassland from 26 to 12 kyr BP (Moar, 1980). Low values of tree pollen

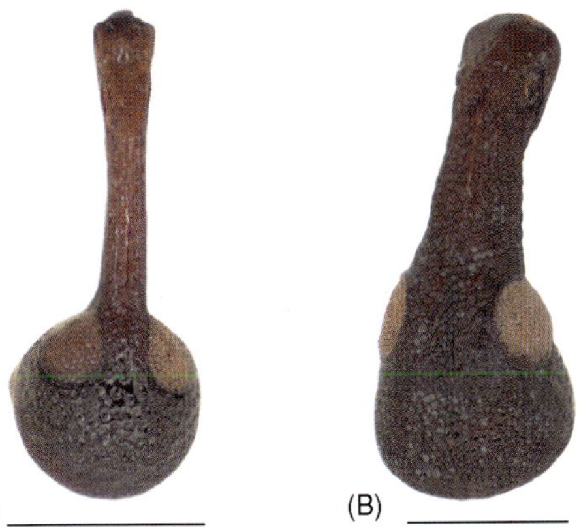

Figure 11 (A) The head of *Hypotagea lewisi* (Curculionidae). (B) Head of *Rhopalomerus tenuirostris* (Curculionidae). (Scale bars = 1 mm.) Images by B. Rhode of Landcare Research. (a) Marra and Leschen (2004).

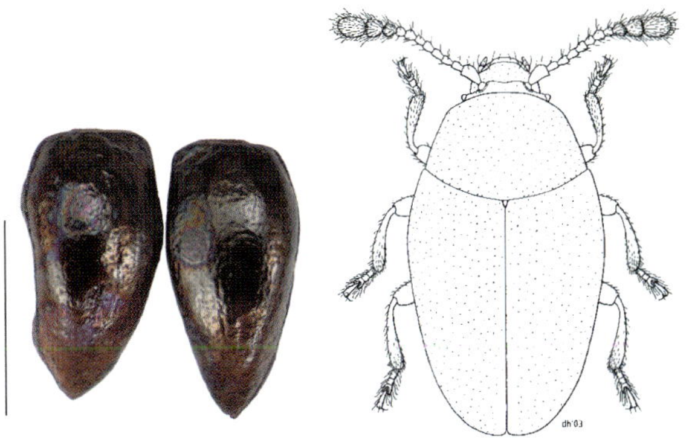

Figure 12 Illustration of new species of Phyllocharitini (Chrysomelidae) and the fossil wingcases. The illustration is by D.W. Helmore and courtesy of Landcare Research, NZ. Image by B. Rhode of Landcare Research. (Scale bar = 1 mm).

indicated that trees were present throughout this period, although the actual source of the pollen remained a mystery. The beetle fossils from the Awatere Valley precisely locate one of these LGM forest patches.

The fossil beetle assemblage is a complex of montane sub-Alpine tussock (discussed earlier), forest, and aquatic taxa. The most significant discovery was of forest-dwelling beetles, particularly the weevil *Hypotagea lewisi* (**Fig. 11A**) which is host specific on living *Nothofagus* trees, both the larvae and adults feeding on *Nothofagus* parenchyma tissue.

The fossil of *Hypotagea lewisi* pinpoints a stand of *Nothofagus* forest in Marlborough at the LGM, the first confirmed eastern forest refugium for the LGM. No pollen was found at the site, but recent additional (unpublished) data include additional beetle taxa associated with *Nothofagus* plus fossil leaves of *Nothofagus solandri var. cliffortioides*, both of which confirm the original interpretation of this forest refugium.

The Lyndon Stream Similarly, although the data are not as conclusive as those from the Awatere Valley, the Lyndon Stream LGM fossil beetle fauna also contains at least two forest taxa. The pollen and plant macrofossils at the site indicate that the local vegetation at the LGM consisted of grasses and shrubs. However, the beetles suggest a more diverse ecosystem at that time. The 25 fossil beetle taxa identified from Lyndon Stream represent a rich and diverse ecosystem at the paleo-Lyndon lake at some period during the LGM (Marra *et al.*, 2006a). The beetles are typical of those found high-country Canterbury lakes, which is consistent with the inferred interstadial climate discussed above, though it must be noted that the modern landscape is partially deforested by human activity. Although most of the fossil beetles are aquatic and riparian taxa, included in the assemblage is a species of *Rhopalomerus tenuirostris* (Curculionidae) (**Fig. 11B**). Adults of *R. tenuirostris* feed on pollen, but the larvae inhabit very rotten wood, typically on the forest floor. In addition, the fauna included a species of Phyllocharitini (Chrysomelidae) (**Fig. 12**). This is a newly discovered taxon that is found in wet forests. Although not conclusive, the presence of these two taxa supports forest habitat at the site.

Peculiarities of NZ Fossil Beetles

Where Are All the Fossil Carabids?

A striking peculiarity of the NZ late Pleistocene fossil beetle faunas is the absence or very low representation of ground beetles compared with fossil assemblages in other parts of the world. The poor representation of these taxa in the NZ fossil assemblages is surprising because the modern beetle fauna includes some 650 species (450 described) of carabid (Larochelle and Lariviere, 2001). Weevils regularly dominate fossil assemblages, which correlates well with their abundance in the modern fauna (**Table 1**), so the apparent absence of carabid fossils is mysterious.

Marra (2003a) presented two possible explanations. First, there are no carabids in NZ that are specifically associated with swamp or peat environments, which is in stark contrast with North America where several genera are specific to swampy environments. Second, information from collectors indicates that fecundity and biomass (both eggs and adults) of NZ forest carabids are very low. Pitfall trapping yields of carabids in the Northern Hemisphere are one to two orders of magnitude larger than they are in NZ, and life histories on *Plocamosthetus planiusculus* and *Megadromus turgidiceps* indicate that both species produce ca. six eggs, about 75% fewer than the number produced by *Megadromus* species in North America (Cartellieri and Lövei, 2000).

There is one exception: carabid fossils occur in abundance in fossil petrel burrows. The site where this was discovered had previously been investigated for avian fossils and is described in Worthy (1998) and dated as late Holocene by Holdaway *et al.* (2002). Hundreds of large carabid fossils comprising several species (including species of *Megadromus*) were recovered from less than a kilogram of petrel-nest detritus (unpublished data). Hundreds of fossil snails were also present in the sample, which may have been the main attraction for the beetles. In addition, the beetles indicate an open or forest margin at that time, which is a habitat where modern carabids are found to be more abundant than they are in closed forest habitats.

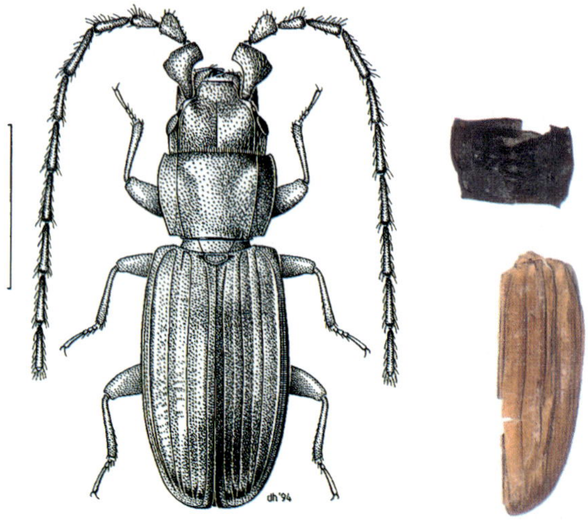

Figure 13 Illustration of *Microbrontes lineatus* (Laemophloeidae) and the prothorax and wingcase of this species. The illustration is by D. W. Helmore and courtesy of Landcare Research, NZ. Image by B. Rhode of Landcare Research. (Scale bar = 1 mm).

Extinction/Local Extinction

The beetle fossil evidence suggests that some species were more abundant in the past. For example, the North Island species *Microbrontes lineatus* (Laemophloeidae) (**Fig. 13**) is very rare in modern collections yet it has been found as a fossil in two sites, one of which is in the South Island. A fossil latridiid (**Fig. 14A**), probably indicating a subalpine habitat, is abundant in LGM samples. Although all fossil elements have been collected providing good detail for its identification, it has no known modern equivalent.

A small number of distinct and well-preserved fossils remain unidentified. With the uncertainties over the modern fauna, it is difficult to say whether the unidentifiable fossils belong to extinct species or are taxa that have yet to be collected. A well-preserved prothorax fossil that is clearly Limnichidae from the Last Interglacial remains a mystery: the prothorax has a distinctive oblique ridge cutting across each posterior pronotal angle (**Fig. 14B**), a feature that is not found on any of the modern limnichids in New Zealand or Australia. Similarly, four distinctively ornamented weevil elytra (**Fig. 14C**) from an MIS 6 site in the eastern South Island are also unknown in the modern fauna.

Extinctions of NZ beetles are well attested for the time since the arrival of predators within the past 2 kyr and settlement ca. 750 yr ago and the subsequent destruction of forest and shrubland habitats. For example, the large weevil *Ectopsis ferrugalis* occurs as fossils in South Island assemblages but it is not found today in the South Island. Its recent

Figure 14 (A) Head, prothorax and wingcases of an unknown species of Latridiidae. (B) Prothroax of an unknown species/genus of Limnichidae. (C) Wingcase of an unknown species of Cryptorhynchini (Curculionidae). (D) Wingcases of two species of *Baeosomus*. Wingcase on the right shows the elytral tubercles that are typical of lowland species of *Baeosomus*. Images by B. Rhode of Landcare Research. (Scale bars = 1 mm).

absence has been attributed to introduced predators (Kuschel and Worthy, 1996). This species is important as a fossil because it has a specific host tree and its larvae have specific tree-size burrowing requirements, which not only identify the host tree present but its size as well. In some cases, the fossil record is the only evidence of some taxa. For example, the largest species of Ulodidae is an extinct species of 24.5 mm. It was found in cave deposits and has been described from the fossil evidence (Leschen and Rhode, 2002). Its extinction is also attributed to introduced predators.

Emerging Indicators

With so few fossil beetle faunas known in NZ at present, no single species or group of species has yet emerged that occur regularly enough (such as are present in the fossil faunas of North America and Britain) to be used as definitive indicators of past climate change. This problem may resolve itself as more fossil faunas are assessed. However, at present taxa such as *Baeosomus* that are regular components of fossil assemblages do provide some information. *Baeosomus* is a genus of moss weevils that occur from lowland to alpine environments, but only the lowland species have uneven elytral tubercles (**Fig. 14D**). Therefore, in lowland sites, *Baeosomus* lacking tubercles suggests downslope movement in response to cooler climate.

See also: **Beetle Records**: Late Pleistocene of Australia; Late Pleistocene of South America. **Glaciations**: Late Pleistocene of the SW Pacific Region. **Phytoliths**. **Pollen Records, Late Pleistocene**: Australia and New Zealand.

References

Atkinson, T. C., Briffa, K. R., and Coope, G. R. (1987). Seasonal temperatures in Britain during the last 22,000 years, reconstruction using beetle remains. *Nature* **325**, 587–592.

Cartellieri, M., and Lövei, G. L. (2000). Seasonal dynamics and reproductive phenology of *Plocamostethus planiusculus* (White) and *Megadromus turgidiceps* (Broun), two endemic New Zealand ground Beetles. In *Natural History and Applied Ecology of Carabid Beetle* (P. Brandmayr, G. L. Lövei, T. Zetto-Brandmayr, A. Casale and A. Vigna Taglianti, Eds.), pp. 179–184. Pensoft, Sofia.

Holdaway, R. N., Roberts, R. G., Beavan Athfield, N. R., Olley, J. M., and Worthy, T. H. (2002). Optical dating of quartz sediments and accelerator massspectrometry 14C dating of bone gelatin and moa eggshell: A comparison of age estimates for non-archaeological deposits in New Zealand. *Journal of the Royal Society of New Zealand* **32**, 463–505.

Hormes, A., Preusser, F., Denton, G., *et al.* (2003). Radiocarbon and luminescence dating of overbank deposits in outwash sediments of the Last Glacial Maximum in north Westland, New Zealand. *New Zealand Journal of Geology and Geophysics* **46**, 95–106.

Johns, P. M. (1986). Arthropods of Banks Peninsula Reserves. *Report to the Commissioner of Crown Lands, Christchurch.* Zoology Department, University of Canterbury, Christchurch, New Zealand.

Klimaszewski, J., and Watt, J. C. (1997). Coleoptera: Family-group review and keys to identification. *Fauna of New Zealand 37*. Manaaki Whenua Press, New Zealand.

Kuschel, G. (1990). Beetles in a suburban environment: A New Zealand case study. *DSIR Plant Protection report No. 3*, DSIR Plant Protection, Mt Albert Research Centre, Auckland, New Zealand.

Kuschel, G., and Worthy, T. H. (1996). Past distribution of large weevils (Coleoptera: Curculionidae) in the South Island of New Zealand, based on Holocene fossil remains. *New Zealand Entomologist* **19**, 15–21.

Larochelle, A., and Lariviere, M.-C. (2001). Carabidae (Insecta: Coleoptera): Catalogue. *Fauna of New Zealand* **43**, p. 285 Manaaki Whenua Press, Lincoln.

Leschen, R. A. B., Lawrence, J. F., Kuschel, G., Thorpe, S., and Wang, Q. (2003). Coleoptera general of New Zealand. *New Zealand Entomologist* **26**, 15–28.

Leschen, R. A. B., and Rhode, B. E. (2002). A new genus and species of large extinct Ulodidae (Coleoptera) from New Zealand. *New Zealand Entomologist* **25**, 57–64.

Marra, M. J. (2003a). A last interglacial beetle fauna from New Zealand. *Quaternary Research* **59**, 122–131.

Marra, M. J. (2003b). A description and interpretation of a fossil beetles assemblage from marine isotope stage 6 from Banks Peninsula, New Zealand. *New Zealand Journal of Geology and Geophysics* **46**, 523–528.

Marra, M. J., and Leschen, R. A. B. (2004). Late Quaternary paleoecology from fossil beetle communities in the Awatere Valley, South Island, New Zealand. *Journal of Biogeography* **31**, 571–586.

Marra, M. J., Shulmeister, J., and Smith, E. G. C. (2006a). Reconstructing temperature during the Last Glacial Maximum from Lyndon Stream, South Island, New Zealand using beetle fossils and maximum likelihood envelopes. *Quaternary Science Reviews* **25**, 1841–1849.

Marra, M. J., Smith, E. G. C., Shulmeister, J., and Leschen, R. (2004). Late Quaternary climate change in the Awatere Valley, South Island, New Zealand using a sine model with a maximum likelihood envelope on fossil beetle data. *Quaternary Science Reviews* **23**, 1637–1650.

Marra, M. J., Alloway, B. V., and Newnham, R. M. (2006b). Paleoenvironmental reconstruction of a well-preserved Stage 7 forest sequence catastrophically buried by basaltic eruptive deposits, northern New Zealand. *Quaternary Science Review* In Press.

Moar, N. T. (1980). Late Otiran and early Aranuian grassland in central South Island. *New Zealand Journal of Ecology* **3**, 4–12.

Porter, S. C. (1975). Equilibrium – line altitudes of late Quaternary glaciers in the Southern Alps, New Zealand. *Quaternary Research* **5**, 27–47.

Shulmeister, J., Barrett, P., and Marra, M. (2000). Modern and ancient sedimentary environments of Southern Wairarapa. Field trip guide GSNZ conference. *Geological Society of New Zealand Miscellaneous Publication* **108B**, 31–47.

Shulmeister, J., Okuda, M., Carter, J. A., *et al.* (2001). A long late-Quaternary record from Lake Poukawa, Hawke's Bay, New Zealand. *Palaegeography, Palaeoclimatology, Palaeoecology* **176**, 81–107.

Shulmeister, J., Soons, J. M., Berger, G. W., *et al.* (1999). Environmental and sea-level changes on Banks Peninsula

(Canterbury, New Zealand) through three glaciation-interglaciation cycles. *Palaegeography, Palaeoclimatology, Palaeoecology* **152**, 101–127.
Soons, J. M. (1979). Late Quaternary environments in the central South Island of New Zealand. *New Zealand Geographer* **35**, 16–23.
Soons, J. M., and Burrows, C. J. (1978). Dates for Otiran deposits, including plant microfossils and macrofossils, from Rakaia Valley. *New Zealand Journal of Geology and Geophysics* **21**, 607–615.
Soons, J. M., Moar, N., Shulmeister, J., Wilson, H. D., and Carter, J. A. (2002). Quaternary vegetation and climate changes on Banks Peninsula, South Island, New Zealand. *Global and Planetary Change* **33**, 301–314.
Worthy, T. H. (1998). Quaternary fossil faunas of Otago, South Island, New Zealand. *Journal of the Royal Society of New Zealand* **28**, 412–521.

Late Pleistocene of Northern Asia

A Sher[1] and S Kuzmina[1,2], [1]Russian Academy of Sciences, Moscow, Russia, [2]University of London, Surrey, UK

© 2007 Elsevier B.V. All rights reserved.

Late Pleistocene Beetle Records from Northern Asia

Preservation, Abundance and Collecting Fossil Beetles in the Siberian Arctic

Insect fossils are common in various types of Quaternary deposits in Northern Asia. They are most abundant and well preserved in silt or sandy silt with plant detritus. Two main factors – the predominantly fine-grained composition of sediment matrix and ubiquitous permafrost – allow excellent preservation of the late Pleistocene insect fossils in this region (**Figs. 1** and **2**). Massive peat layers in the Arctic usually contain few insect remains, but thin-bedded peat intercalating with silt may contain rich insect faunas. The coarser sediment, such as sand with fine gravel deposited in rapidly running water, usually preserves only the most durable insect remains (**Fig. 3**).

The average abundance of insect fossils in the sediment is not very high (**Table 1**). This means that in order to obtain a statistically representative sample with a minimum number of 100 individuals, we must process 50 to 60 kg of moist sediment with an average content of fossils, and sometimes up to 200 kg of sediment with a poor content of fossils. Taking into account the difficulties and expense of working in remote arctic regions, we usually spend several weeks at one site and collect 20 to 30 samples there. The enormous cost of transportation forces us to reduce the weight of samples to be shipped, so we screen our samples in the field (**Figs. 4–6**).

The identification of insect fossils in Northern Asia is difficult because of insufficient knowledge of both taxonomy and distribution of modern species. That is why we always collect modern insects during our field season in these remote regions. This only marginally advances the cause of regional beetle taxonomy, however. Many groups are still poorly studied.

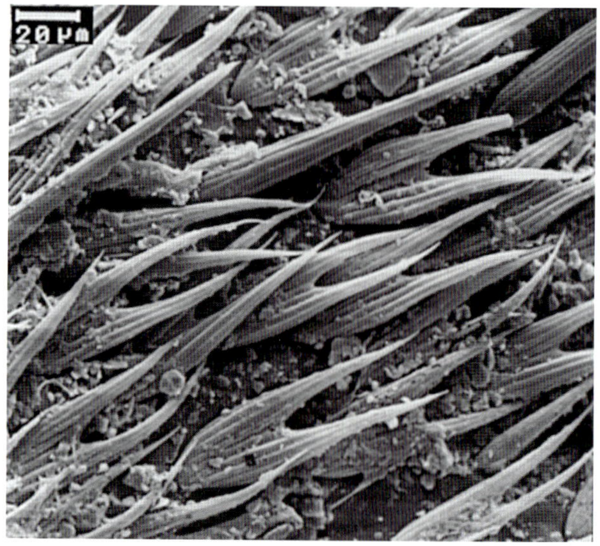

Figure 1 Fossil scales on the surface of the weevil, *Hypera ornata*, Bykovsky Peninsula, Late Pleistocene. The condition of fossil chitin in frozen silts is usually as good as in modern beetles. Even such delicate and fragile elements as scales, hairs, and antennae are sometimes preserved.

Figure 2 Fossil head, pronotum, and leg of a dung beetle *Aphodius* sp., from the Duvanny Yar site, Late Pleistocene. Excellent preservation is common for beetle remains recovered from buried ground squirrel nests, where some specimens preserve legs in their original position.

Figure 3 Fossil head capsules of weevils (above and right) and a predaceous diving beetle (below, left) from a sandy–gravel unit, Keremesit River.

Localities of Fossil Beetles in Northern Asia

The first studies of Pleistocene insects in Northern Asia took place in the 1960s (Grushevskiy and Medvedev, 1962). These studies were associated with famous discoveries of complete mammal carcasses in permafrost or rich accumulations of mammoth bones, the so called 'mammoth cemeteries' (Medvedev and Voronova, 1977). The regular sampling of Quaternary sediments for insect fossils in northeastern Siberia and their systematic research was started in the 1970s by Sergey Kiselyov (1981). Currently, Svetlana Kuzmina continues Kiselyov's work in that region, and Evgeny Zinovyev studies Quaternary insects in northwestern Siberia.

More than 30 years of regular paleoentomological research in Northern Asia has resulted in a large number of studied sites and analyzed assemblages, but their distribution over such a vast territory is uneven and discontinuous. For example, Zinovyev (1997) studied about 36 sites in northwestern Siberia, most of them of late Pleistocene age. However, further east, the huge region between the mouth of the Ob' River and almost to the Lena River Delta is virtually unstudied. From the Lena Delta to Chukotka (northeastern Siberia) the number of sites is much higher, but they are unevenly distributed (**Fig. 7**). More than 200 late Pleistocene beetle assemblages have been studied from 44 localities in this region (**Table 2**). The most important are discussed below.

Geological Background of the Late Pleistocene Beetle Record in Northern Asia

Northern Asia from west to east Even if we restrict our consideration of Northern Asia to the areas north of the Arctic Circle, it still remains a huge region comprising very different physiographic provinces with contrasting geological histories. While northwestern Siberia and the Taimyr Peninsula were strongly affected by the advances of the Barents-Kara Ice Sheet (Svendsen *et al.*, 2004), no ice sheets existed east of the Taimyr. Only some high mountain ridges received enough moisture to build valley glaciers. Marine sediments exposed above present sea level are also found mainly in the glacially-affected areas. Only at the easternmost part of Asia, close to the Bering Strait (Chukotka Peninsula), were both glaciations (mountain and piedmont types) and marine transgressions more significant in the Pleistocene. Thus, the huge territory of Northern Asia between the Taimyr and Chukotka (i.e., between about 112 and 175°E) has never experienced major glaciations. Instead, this area was a broad arena for permafrost and periglacial

Table 1 Concentration of insect fossils in the Late Pleistocene frozen silts. The Lena Delta, Late Weichselian, Mamontouy Khayata section

Sample No	*Weight of the screened wet sediment, kg*	*Number of insect fossils extracted*	*MNI (minimum number of individuals)*	*The number of fossils per kilo*	*The number of insect individuals per kilo*	*Weight of matrix to be screened to get 100 insect individuals, kg*
MKh99-18	9	52	30	6.1	3.5	28
MKh99-16	20	32	25	1.6	1.2	82
MKh99-14	20	10	9	0.5	0.4	227
MKh99-12	9	17	12	2.0	1.4	71
MKh99-22	17	160	88	9.4	5.2	19
MKh99-6	26	109	56	4.3	2.2	46
MKh99-5a	9	42	23	4.9	2.7	37
MKh99-5o	14	111	47	8.2	3.5	29
MKh99-5s	14	146	57	10.7	4.2	24
MKh99-3a	7	85	43	12.5	6.3	16
Average				6.0	3.1	62

NOTE. The volume of screened sediments in these samples is small, as they were screened by geologists. However, the weight of samples was measured, and this case can be used to estimate concentration of fossils.

Figure 4 We screen sediments in the field, using a large hand-made box with 0.5 mm mesh sieve, letting river, lake, or pond water in and out through the sieve only. The screening is aimed at removing most, but not all of the mineral component, so some amount of silt remains in the box, helping preserve the shape of many fossils smaller than 0.5 mm that might otherwise be lost. These specimens usually float on top of the silt and remain in the sample.

Figure 5 The screen residues are dried as much as the weather allows, at first openly, later in fabric bags.

Figure 6 Quite often, silt sticks to the fossils, or fills the insides of pronota, head capsules, or joined elytra. This limits the effectiveness of the laboratory flotation method. Instead, we separate dry residuals into the size fractions through regular soil sieves, then pick the insect and other fossils by hand, under binocular microscope for smaller fractions.

processes. At present, the environment of the vast lowlands bordering the East Siberian and Laptev Seas is strongly influenced by the Arctic Ocean, but in the late Pleistocene these flatlands had a much more continental climate, as they extended much further north during long periods when the shallow continental shelf was above sea level.

What is Yedoma? The most remarkable and widespread Pleistocene sedimentary unit on these flatlands is ice-rich silt and sandy silt with thick ice wedges (**Fig. 8**). It usually builds the main surface of the lowland (up to 60–100 m above the rivers) or its remaining hills when it is destroyed by thermokarst and erosion. Local people call these hills 'Yedoma', and geologists used that name for this kind of sediment (**Fig. 9**). The high content of ice in Yedoma sediments (up to 80%) (**Figs. 10 and 11**) explains another common name for this formation – Ice Complex. Some important features of Yedoma, or Ice Complex, are shown in **Figures 12–14**.

The same kind of ice-rich silt builds not only thick sequences of high Yedoma, but also lower terraces (**Fig. 15**). It is also found on the slopes of rock hills and fills the valleys of small creeks in the highlands. In other words, a similar kind of sediment can occur in many different depositional environments. That is

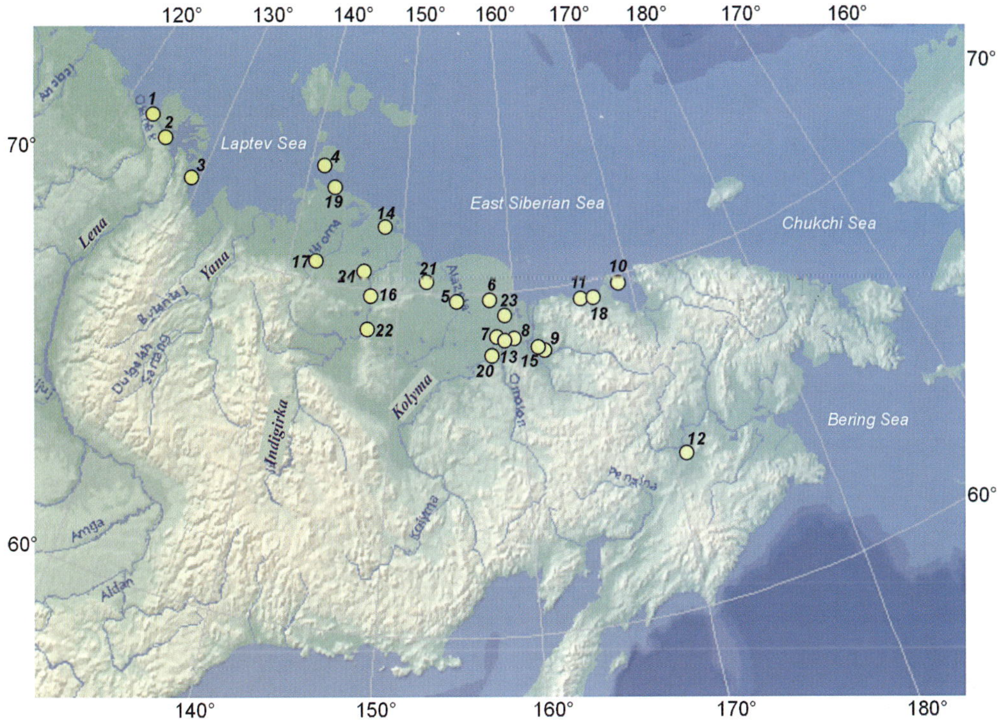

Figure 7 Map of Late Pleistocene fossil beetle localities in north-eastern Siberia. 1 – 3 Lena Delta region: 1 – Olenyok Channel, Nagym, 2 – Byor-Khaya, 3 – Bykovsky Peninsula, 4 – Bolshoy Lyakhovsky Island, 5 – Alazeya, 6 – Chukochya, 7 – Alyoshkina Zaimka, 8 – Duvanny Yar, 9 – Stanchikovsky Yar, 10 – Ayon Island, 11 – Milkera, 12 – Main River, Ledovy Obryv (Ice Bluff), 13 – Omolon, 15 – Krasivoye, 16 – Shamanovo, 17 – Khroma River, 18 – Primorsky, 19 – Oyagosskiy Yar, 20 – Krestovka, 21 – Khomus-Yuryakh, 22 – Sypnoy Yar, 23 – Yakutskoe Lake, 24 – Achchagyy-Allaikha.

Table 2 The distribution of studied insect assemblages of the Late Pleistocene age in North–eastern Siberia

Region	*Number of studied sections*	*Number of insect samples with MNI >30*
Lena Delta	3	63
Indigirka Lowland	10	28
New Siberian Islands	4	11
Kolyma Lowland	16	50
Medvezhyi Islands	2	4
West Chukotka (Chaun)	3	16
Ayon Island	4	38
South Chukotka (Main R.)	2	7
TOTAL	44	217

Figure 8 One of the famous exposures of Late Pleistocene icy silts – Vorontsovskiy Yar on the Indigirka River. The top of this bluff, which is about 40 m high, lies about 90 m above the river level.

why earlier heated debates about the origin of Yedoma silts (fluvial, eolian, eluvial-colluvial, etc.) gradually gave way to the understanding that the ice-rich Yedoma silts are of polygenetic origin. They include sediments of different facies, which appear similar due to the homogeneity of the original material generated by cryogenic weathering during the Pleistocene and due to the activity of permafrost processes that universally affected sedimentation in various environments.

In closer proximity to mountain areas, active denudation yielded mostly coarser sediments, especially sands. These are the so-called Sand Yedoma sections known in various regions of northeastern Siberia. Usually, but not always, they include much smaller

Figure 9 The Duvannyy Yar site, in the lower course of the Kolyma River, is a cut bank exposure of the so-called Omolon-Anyuy Yedoma – the high eroded surface between the Omolon and Anyuy Rivers – large tributaries of the Kolyma.

Figure 10 Mamontovyy Khayata Cliff, Bykovsky Peninsula, eastern side of the Lena R. Delta. Extensive ice bodies such as this, which in earlier times were interpreted as buried glaciers, in fact represent polygonal systems of syngenetic (growing simultaneously with continued sedimentation) ice wedges.

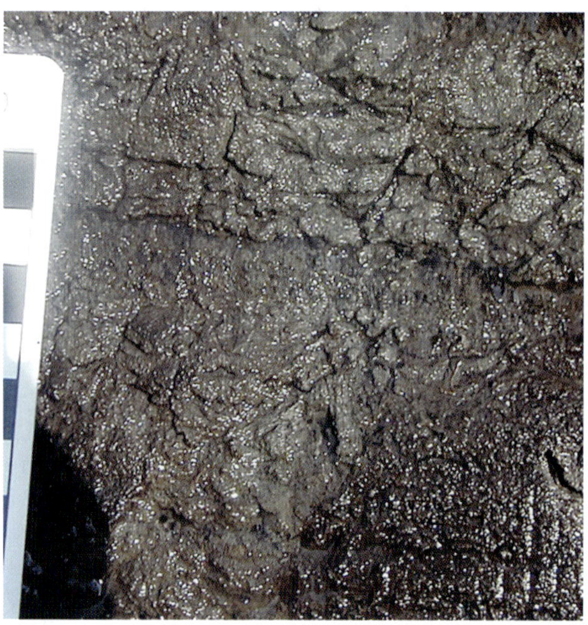

Figure 11 The ground pillars between the ice wedges also contain lots of ice, but as small layers and microscopic lenses segregated when the silt was freezing.

Figure 12 One of the typical features of Yedoma silts is the abundance of very thin plant roots, most likely belonging to grasses. They occur *in situ* and usually form complicated branching systems, permeating frozen silt.

Figure 13 Some frozen silt layers include woody roots of small shrubs.

ice wedges (**Figs. 16–17**). A similar situation can occur along large rivers, building sand terraces with scattered ice wedges.

Thus Yedoma is not a facies, but a type of sedimentation reflecting a specific climate and environment, and in this sense the synonym 'Ice Complex' seems more correct. Interestingly, similar sedimentation took place in northwestern Siberia, when the Barents-Kara ice sheet retreated north in the Late Weichselian (Svendsen *et al.*, 2004). The widest distribution of

Figure 14 The Yedoma silts are very rich in various kinds of fossils – plants, freshwater shells, and insects. The most famous remains are the large number of mammal bones and complete or partial frozen carcasses of Pleistocene mammals.

Figure 15 The same kind of ice-rich silt builds not only thick sequences of high Yedoma, but also terrace-like bodies of lower elevation (Low Yedoma, 15–25 m above the river), more clearly positioned in river valleys (Omolon River, Kolyma River basin).

the Ice Complex sediments and their overall similarity make their stratigraphic characterization difficult.

Other types of sections Numerous sections of 'alasses' – thermokarst depressions formed by the thawing of the Ice Complex – are another widespread element of regional geology. The majority of them were formed at the Pleistocene–Holocene transition, and are composed of reworked Yedoma silt or sand with low ice content (**Fig. 18**) (*see* Ice Wedges and Ice Wedge Casts). They may include a lacustrine unit of former thermokarst lakes and are overlain by a more or less thick peat layer, usually of Early Holocene age. Earlier thermokarst events are recorded in buried alas units. They have a similar composition, but are overlain by an Ice Complex unit, and so mark periods of regional, or at least local, degradation of

Figure 16 Impressive sand sequences up to 60 m high are usually exposed where large rivers run in low mountain areas. The best known examples are Sypnoy Yar on the Indigirka River and in the city of Verkhoyansk area (Yana and Adycha River basins). A coarser sediment (sands) limited the scale of frost cracks, and the ice wedges in these fluvial sequences are much smaller than in classical Yedoma. They are restricted to the lenses of silty sand (floodplain and ox-bow deposits).

Figure 17 The Ice Bluff section, Main River, southern Chukotka. Sand sections such as this, with a higher contents of silt, allowed the continuous growth of ice wedges.

Figure 18 In the thermokarst (alas) sections the Ice Complex sediments have thawed through, i.e. reworked in place. As a result, the ice content is much lower here, as former ice wedges have been replaced by ice-wedge casts, or pseudomorphs.

permafrost in the past. In several sites where ancient alas sediments underlie the Weichselian Ice Complex, and overlie supposedly Middle Pleistocene units, they probably represent the Last Interglacial (Eemian). Only one such case has been documented by direct dating (IRSL, Krest-Yuryakh at Bolshoy Lyakhovsky Island, Andreev *et al.*, 2004). If a thermokarst lake reappeared several times in one place, then composite sections formed, consisting of a few superimposed alas complexes (**Fig. 19**).

The late Pleistocene stratigraphy in Northeastern Siberia The stratigraphic subdivision of Ice Complex (Yedoma) sequences is not easy because of the high homogeneity of the sediments. It is mostly based on radiocarbon dating, which allows us to reliably distinguish the late Weichselian units, but we encounter problems with the Middle Weichselian

Figure 19 One such reworked section, Molotkovsky Kamen' at the Malyy Anyuy River, had been selected as the regional type of Middle Weichselian Interstadial (Molotkov Horizon) with alternating warmings and coolings. Subsequent re-dating of the buried peat layers in this section, however, showed much earlier ^{14}C dates, so the age of the Molotkov Horizon should be revised (Sher et al., 2005).

(Sher and Plakht, 1988). Ice Complex sections are referred to the Early Weichselian if they have an infinite ^{14}C age and are overlain by the dated middle to late Weichselian sediments. Strictly speaking, such arguments are not fully conclusive. There are Ice Complex sections dated from the late Middle Pleistocene (Kaplina *et al.*, 1980), or even earlier. On the other hand, the Ice Complex units with 'finite' ^{14}C dates in the range 35 to 45 ka do not necessarily belong to the middle Weichselian, as there were many cases when such units turned out to be much older (Sher *et al.*, 2005).

Table 3 summarizes the geological position and the age of the late Pleistocene sites from which the insect assemblages have been studied.

Peculiar Features of Fossil Beetle assemblages in Northeastern Asia

General composition The species composition of the Pleistocene beetle fauna of northeastern Asia is strange, compared with the other arctic regions:

(1) It has higher diversity and abundance of plant-feeding insects, mostly weevils (Curculionidae) and leaf beetles (Chrysomelidae). Even among ground beetles (Carabidae), mostly a predatory family, one of the commonly found species is a plant-feeder *Curtonotus alpinus* in northeastern Siberia (although this has not been observed in its populations from northern Europe or arctic North America).

(2) Fossil assemblages usually include many species that currently live in tundra and forest tundra, but the diversity of some groups, such as litter inhabitants (rove beetles), carrion beetles, and fungus beetles, is notably lower than in modern tundra communities. Also, the diversity of water and riparian species in the fossil assemblages is less than today in the Arctic.

(3) Along with tundra species, most Pleistocene assemblages show significant numbers of steppe, meadow-steppe, and other xerophilic insects. Especially notable is the abundance of the pill beetle *Morychus viridis*, an inhabitant of cold steppe-like biotopes in northeastern Siberia today (see below). This species makes up as much as 90% of some late Pleistocene insect assemblages.

(4) Xylophages and other dendrophilic species are very rare in the late Pleistocene insect assemblages in this region, so no true forest environment can be reconstructed here for any part of the late Pleistocene.

The above features characterize many late Pleistocene insect assemblages in northeastern Asia, which combine species presently inhabiting a variety of different regions and ecosystems. For example,

Table 3 Geological position and the age of the Late Pleistocene sites in Northeastern Siberia, from which the insect assemblages have been studied

The section type	*Sediment*	*Geocryology*	*Sections & Map No, Fig*	*Age range*	*Reference*
High Yedoma section (40–60 m and more above the river or sea level)	Mostly silt with sandy silt	One penetrating system of large ice wedges	Mamontovyy Khayata (**Fig. 7**, No 3, **Fig. 10**)	EW (?) - MW-LW (>50–12 ka)	Sher *et al.*, 2005
			Duvannyy Yar (**Fig. 7**, No 8, **Fig. 9**)	Mostly EW (?) - MW?. Probably LW at the top (>50–18 ka)	Sher and Plakht, 1988
		Multiple (2–3) overlying systems of ice wedges	Khomus-Yuryakh, Loc. 83 (**Fig. 7**, No 21)	Late middle Pleistocene	Sher, unpublished
High Sand section (>30, up to 60 m and more above the river or sea level)	Sand, silty sand	Relatively small ice wedges are restricted to more silty lenses.	Sypnoy Yar (**Fig. 7**, No 22, **Fig. 16**)	LP undivided (>45,000 yr BP in the lower third of the section)	Kaplina, 1981
			Utatgyr Suite, Ayon Island (**Fig. 7**, No 10)	LP undivided	Svitoch, 1980; Kiselyov, 1994
		One penetrating system of large ice wedges	Ledovyy Obryv (Ice Bluff) (**Fig. 7**, No 12, **Fig. 17**)	MW-LW (42–15,000)	Anderson and Lozhkin, 2002; Sher, unpublished
Topping Yedoma - Ice Complex up to 20–25 m thick on the top of composite sections	Mostly silt with sandy silt	One penetrating system of large ice wedges	Buor-Khaya (**Fig. 7**. No 2)	EW-LW (>50–17 ka)	Schirrmeister *et al.*, 2003
			Nagym (**Fig. 7**, No 1)	MW ? (44–45 ka)	Schirrmeister *et al.*, 2003
			Zimovye R. Mouth, Bol. Lyakhovsky Island (**Fig. 7**, No 4)	EW-MW (100–28 ka)	Schirrmeister *et al.*, 2004
			Yakutskoe Lake (**Fig. 7**, No 23)	LW ?	Kiselyov, 1994
			Khroma, Alazeya, Chukochya, Krestovka (**Fig. 7**, No 17, 5, 6, 20)	EW-MW? (Alazea, the date of 38 ka in the lower part of the unit)	Kaplina *et al.*, 1981
			Achchagyy-Allaikha (**Fig. 7**, No 24)	Late middle Pleistocene	Kaplina *et al.*, 1980
			Oyagosskiy Yar (**Fig. 7**, No 19)	Late middle Pleistocene - LW	Nikol'skiy and Basilyan, 2003
Low Yedoma terraces (20–30 m high above the river level)	Mostly silt with sandy silt	One penetrating system of large ice wedges	Omolon I (**Fig. 7**, No 13, **Fig. 15**)	LW (24–15 ka)	Sher, unpublished
			Krasivoye Bluff (**Fig. 7**, No 15)	LW (18,700 yr BP)	Kiselyov, 1994
			Shamanovo (**Fig. 7**, No 16)	MW-LW (33 ka and younger)	Kaplina *et al.*, 1980

(*Continued*)

Table 3 (*Continued*)

The section type	*Sediment*	*Geocryology*	*Sections & Map No, Fig*	*Age range*	*Reference*
Low Sand terraces (20–25 m high above the river level)	Sand, silty sand	Relatively small ice wedges are restricted to more silty lenses.	Alyoshkina Zaimka (**Fig. 7**, No 7)	LW (17.2–13.2 ka)	Sher *et al.*, 1979; Alfimov *et al.*, 2003
Multilayer alasses	Silt, peat	Ice-wedge pseudomorphs	Primorskiy, coast near Rauchua R. mouth (**Fig. 7**, No 18)	>28,000 yr BP in the lower part of the sequence, 24,800 yr BP in the middle part, 10.4 ka at the top	Kiselyov, 1994
Buried alass complexes	Silt, peat	Ice-wedge pseudomorphs	Khroma, Kyl-Bastakh Layers (**Fig. 7**, No 17)	LI ? (overiles mid-Pleistocene Khroma Suite, overlain by Ice Complex)	Kaplina *et al.*, 1983
			Alazeya, Kuobakh Unit (**Fig. 7**, No 5)	LI ? (overiles mid-Pleistocene (?) Maastakh Suite, overlain by Ice Complex)	Kaplina *et al.*, 1981
			Zimovye R. Mouth, Bol. Lyakhovsky Island (**Fig. 7**, No 4)	LI ? (overiles mid-Pleistocene (?) Kuchchuguy Suite, overlain by Ice Complex)	Andreev *et al.*, 2004
Low terraces		16 m terrace	Mil'kera (**Fig. 7**, No 11)	MW (32,800) ?	Kiselyov, 1994

NOTES

Abbreviations: LI – Last Interglacial, LP – Late Pleistocene, EW – Early Weichselian, MW – Middle Weichselian. LW – Late Weichselian

Sources:

Alfimov, A.V., Berman, D.I., Sher, A.V., 2003. Tundra-steppe insect assemblages and reconstruction of Late Pleistocene climate in the lower reaches of the Kolyma River. Zoologicheskiy Zhurnal, 82, 2, 281–300 (In Rissian).

Anderson, P.M., Kotov, A.N., Lozhkin, A.V., and Trumpe, M.A. 2002. Palynological and radiocarbon data from Late Quaternary deposits of Chukotka. In: Anderson, P.M., Lozhkin, A.V., Eds., Late Quaternary Vegetation and Climate of Siberia and Russian Far East (Palynological and Radiocarbon Database. Magadan: NESC FEB RAS, 35–79.

Kaplina, T.N. 1981 History of frozen sequences of northern Yakutia in the late Cenozoic. In: Dubikov G.I., Baulin V.V., Eds., Istoriya razvitiya mnogoletnemerzlykh porod Evrazii., M., Nauka, 1981, 153–181 (In Russian)

Kaplina, T.N., Lakhtina, O.V., Rybakova, N.O. 1981 Cenozoic deposits in the middle course of the Alazeya R. (Kolyma Lowland). Izvestiya Academii Nauk SSSR. Ser. Geologicheskaya, 8, pp. 51–63 (In Russian)

Kaplina, T.N., Ovander, M.G., Lozhkin, A.V., Zhigulyovtseva, S.N., Rybakova, N.O., Nikitin, V.P., Kiselyov, S.V., Bakhtin, A.V., Pirumova, L.G., Virina, E.I. 1983 Quaternary deposits in the middle current of the Khroma River (Yana-Indigirka Lowland). In: Shilo N.A., Ed., Stratigrafiya i paleogeografiya pozdnego kaynozoya Vostoka SSSR., Magadan, SV KNII DVNTs AN SSSR, 1983, 80–95 (In Russian)

Nikolsky, P.A. and Basilyan, A.E. 2003 Svyatoy Nos Cape – the key section of the Pleistocene in the North of Yana-Indigirka Lowland. In: The natural history of the Russian East Arctic in the Pleistocene and Holocene. GEOS Publishers, Moscow, 5–13. (In Russian)

Schirrmeister, L., Andreev, A., Grosse, G., Meyer, H. 2004 Saalian to Holocene Paleoenvironmental History Documented in Permafrost Sequences of Arctic Siberia (New Siberian Archipelago, Bolshoy Lyakhovsky Island) AGU Fall Meeting, 13–17 December 2004 San Francisco, California C13B-0286

Schirrmeister, L., Kunitsky, V. V., Grosse, G., Schwamborn, G., Andreev, A. A., Meyer, H., Kuznetsova, T., Bobrov, A., Oezen, D. 2003. Late Quaternary history of the accumulation plain north of the Chekanovsky Ridge (Lena Delta, Russia) – a multidisciplinary approach, Polar Geography, 27 (4), 277–319.

Svitoch, A.A. 1980. The description of the lates deposits [of the Ayon Island]. In Kaplin, P.A. (ed.). The latest deposits and paleogeography of the Pleistocene in Chukotka. Moscow, Nauka, 180–202. (In Russian).

such arctic species as the ground beetle *Pterostichus costatus*, the leaf beetles *Chrysolina tolli* and *Ch. subsulcata*, and the willow weevil *Isochnus arcticus*, are often found in one fossil assemblage with such steppe leaf beetles as *Chrysolina perforata*, and the weevils *Stephanocleonus eruditus*, *S. fossulatus*, and *S. incertus*, which now live in southern Siberian and Mongolian steppe. On the other hand, these assemblages sometimes include a few species of insects whose current distribution is restricted mostly to the taiga zone, such as the boreal ant species *Camponotus herculeanus* and *Formica gagatoides*, the weevils *Hylobius piceus* and *Pissodes irroratus*, and others.

Table 4 lists the most common species of late Pleistocene beetles found in more than 40% of all late Pleistocene (LP) assemblages and, additionally, a few less abundant, but rather indicative species. **Figure 20** shows the fossils of some beetle species found in the LP sediments in Northeastern Siberia.

Non-analog assemblages. What is tundra-steppe? The fossil insect assemblages of the kind discussed above have no modern analog. They resemble non-analog Pleistocene mammal communities, in which, for example, saiga antelope, a modern inhabitant of dry steppe and semi-desert, existed together with arctic fox and musk ox. It has been shown that such assemblages are not a random mixture of fossils, but a natural combination of species related to a peculiar type of Pleistocene environment (Vangengeim, 1977; Sher, 1988, 1990). In Russia, this environment and associated plant and animal communities are traditionally called 'tundra-steppe' after Tugarinov (1929). In Europe, tundra-steppe communities were related mostly to the periglacial zone south of the ice sheets, but in Siberia, under a more continental climate, they were distributed much more widely and existed more permanently (Vangengeim and Ravsky, 1965). Kiselyov (1981) has shown that insect assemblages of the tundra-steppe type were very common in northeastern Siberia during the Pleistocene. Moreover, he demonstrated that such insect communities developed in Siberia as early as the late Pliocene. Steppe elements were especially abundant in tundra faunas during relatively cold (glacial) periods, but they survived even during warmer (interglacial) times (Kiselyov, 1981, 1994). The significant destruction of tundra-steppe communities occurred only at the Pleistocene-Holocene boundary, when the dry Pleistocene grassland mostly disappeared, giving way to wet tundra and bogs. This brought about the extinction of the most remarkable large mammalian herbivores, such as mammoth and woolly rhino, and the dramatic shrinkage of ranges of others, including horses, bison, and saiga antelope. However, the plant communities of the tundra-steppe type persisted in relatively small relic habitats (Yurtsev, 1981). The ranges of steppe insects also shrank, but many of them survived in relic steppe habitats. Such habitats are widespread in the inner mountains of northeastern Siberia that have an extremely continental climate. Many thermophilic steppe species found refugia there on south-facing slopes (Berman, et al., 2001). Less thermophilic steppe beetles were able to survive in the warmest localities in the Arctic, these habitats being especially abundant on Wrangel Island (Berman, 1986; Khruleva, 2004). The persistence of these beetle species – relics of the Pleistocene tundra-steppe – is one of the most remarkable features of the present insect fauna of northeastern Siberia.

Comparison of Insect Assemblages of Different Ages using Ecological Grouping

The high frequency of occurrence of a restricted number of species in fossil assemblages (**Table 4**), and the permanent presence of insects associated with tundra, steppe and other xerophilic habitats makes the species composition of various assemblages from the Pleistocene of northeastern Siberia rather monotonous. They differ from each other mostly in the relative abundance of fossils of certain taxa. That is why Kiselyov (Sher *et al.*, 1979; Kiselyov, 1981) used a quantitative approach to trace faunal and environmental changes through time. He classified all species into a number of ecological groups according to their preferred modern habitat, then he counted the total number of individuals belonging to all species in each group. It was much easier to compare assemblages from different horizons or even sites by the percentage of ecological groups. We use the same method with minor variations (**Table 5**).

For large sections with many insect samples, we use percentage diagrams for better presentation. Recently, we suggested the establishment of Insect Faunal Zones in such sections, based on the ecological structure of assemblages. This zonation scheme is similar in some ways to pollen zones (Sher *et al.*, 2005). In **Figure 21** you can see how the composition of insect assemblages changed in the Lena Delta during the last 50 ka (this sequence will be considered in more detail below).

The Late Pleistocene Beetle Faunas of Certain Periods and Areas

Below we describe the most important Late Pleistocene beetle assemblages, starting with the

Table 4 Characteristics of the most common and some indicative beetle species in the Late Pleistocene assemblages in Northeastern Siberia

Family and species	*% of all LP assemblages, where the species is found*	*Description*	*Taxonomic notes*
Ground beetles, Carabidae			
Pterostichus (Cryobius) spp.	78	Most *Cryobius* are common in modern tundra and northern taiga and prefer boggy and wet habitats. *P. (C.) nigripalpis* Popp. is abundant today on Wrangel Island on dry patches with xeropilic vegetation (O. Khruleva, personal comm.).	Of numerous species in this subgenus we can identify fossils of *P. (C.) ventricosus* Esch., *P. (C.) pinguedineus* Esch., *P. (C.) brevicornis* (Kirby) and probably *P. (C.) nigripalpis* Popp.
Curtonotus alpinus (Payk)	92	Very widespread species from taiga to arctic tundra; prefers relatively dry and warm patches; feed on grasses. Hardly distinguishable fossils of a close species, *C. bokori*, can be present in some assemblages, but *C. alpinus* is much more common.	In Europe and America this species is usually called *Amara alpina* Payk. In Russia, the subgenus *Curtonotus*, to which this species belongs, was raised to the generic level (Kryzhanovsky 1965, Kryzhanovsky *et al*, 1995).
Poecilus (Derus) nearcticus Lth.	49	Modern specimens of *P. nearcticus* were collected on dry sandy slopes at a few sites in the Siberian Arctic and Alaska. In the Pleistocene, it was much more common, though not very abundant. It is a typical relict of the tundra-steppe beetle complex.	
Pterostichus (Tundraphilus) sublaevis Sahlb.	35	This beetle lives now in tundra and prefers dry and warm habitats, but it is relatively rare. It was more widespread and abundant during the Pleistocene.	
Small carrion and round fungus beetles, Leiodidae			
Cholevinus sibiricus (Jean.)	47	This is a rare species in modern tundra, preferring wet habitats. It is probably necro-, copro-, or/and detritophagous. *Ch. sibiricus* was more important in the Pleistocene tundra-steppe communities (Perkovsky, Kuzmina, 2001). Despite rather large populations of grazing mammals in tundra-steppe ecosystem, the revealed diversity of necro- and coprophagous beetles is very low. *Ch. sibiricus* might use this kind of food resources.	Prior to 1999, this species was called *Cryocatops poppiusi* Jean., family Catopidae (Perkovsky, 1999)
Rove beetles, Staphylinidae			
Tachinus arcticus Motsch. and *T. brevipennis* Sahlb.	45	Both species live in wet tundra habitats, but *T. arcticus* is more abundant in the north, while *T. brevipennis* – in the south of the zone, and is probably more thermophilic.	Earlier these fossils were referred to *T. apterus* Maekl (Kiselyov, 1981). Later, the identification was changed to *T. arcticus*. Now we suspect that it can be a mixture of fossils of two different species.

(Continued)

Table 4 (*Continued*)

Family and species	*% of all LP assemblages, where the species is found*	*Description*	*Taxonomic notes*
Dung beetles, Scarabaeidae			
Aphodius sp.	44	This dung beetle occurs in many Pleistocene sites and is clearly associated with tundra-steppe assemblages. Its well-preserved fossils have been found in buried ground squirrel nests.	Probably an extinct species
Pill beetles, Byrrhidae			
Morychus viridis Kuzm. et Kor.	86	This beetle is a real symbol of the Pleistocene biota in Northeastern Siberia. Found in the majority of fossil assemblages, often in very high numbers, it is clearly associated with thermophilic steppe. At present, it lives in cold steppe-like habitats in the Kolyma Highlands, Chukotka, and Wrangel Island, called 'hemicryophytic steppe' after Yurtsev (1981). Modern habitats of *M. viridis* have extremely contrasting microclimatic conditions – very hot and dry in summer, and snowless and very cold in winter (Berman *et al.*, 2001).	Before this species was described (Kuzmina and Korotyaev, 1987), its fossils were referred to *Chrysobyrrhulus rutilans* Motsh. (Medvedev, Voronova, 1977; Kiselyov 1981).
Leaf beetles, Chrysomelidae			
Chrysolina tolli Jac.	25	This is one of the most cold-resistant species of beetles in the world. *Ch. tolli* was found in polar desert on Novaya Zemlya, Severnaya Zemlya and Novosibirsk Islands. In the tundra zone, it occupies different habitats, and feeds on Crucifera. *Ch. tolli* is found in many Pleistocene assemblages, though in low numbers, and is most common in northern sites.	The fossils were referred to *Ch cavigera* Sahlb. until the recent revision by Bienkowski (1999), who showed that the modern arctic populations belong to *Ch. tolli*, while *Ch cavigera* is restricted to the Kamchatka Peninsula and the Okhotsk Sea coast
Weevils, Curculionidae			
Sitona borealis Kor.	46	This weevil feeds on legumes. Currently it lives mostly in meadow-steppe habitats in the taiga zone, but it also occurs on dry warm patches in tundra. It was more common in the Pleistocene than it is now.	Formerly named as *S. ovipennis* Hochh. (Kiselyov, 1981), then *S. ovipennis borealis* Kor., this weevil is now considered as a separate species *S. borealis* Kor. (Korotyaev, 1996).
Stephanocleonus eruditus Faust	45	The main modern range of this weevil is Southern Siberia and Mongolia (mountain steppe). It is the most common steppe element in the Pleistocene assemblages. Having been very widely distributed in the Pleistocene, it survived in scattered relic steppe communities in the Kolyma-Indigirka Highlands (Berman *et al.*, 2001). This species is a good indicator of relatively high summer temperature (Berman and Alfimov, 1993).	

(*Continued*)

Table 4 (*Continued*)

Family and species	*% of all LP assemblages, where the species is found*	*Description*	*Taxonomic notes*
Lepyrus nordenskioeldi Faust	46	This weevil lives on willow, rarely on birch, and is common in tundra, taiga and relic steppe modern environments, but prefers drier habitats with sandy soil. Thus, being very often a single representative of shrub insect group in the Pleistocene assemblages, *L. nordenskjoeldi* can be considered as an indicator of relatively dry environment, although we count this species among the forest-tundra species.	
Hypera ornata (Cap.)	48	The distribution and ecology of this weevil *are* similar to those of *Sitona borealis* (see above).	Former name *Phytonomus ornatus* Cap.
Isochnus arcticus (Kor.)	44	This species lives on willows, but its distribution is restricted to the typical tundra subzone. It is most abundant on Wrangel Island, but was also found on the Taimyr Peninsula, in western Chukotka and northern Alaska. It can be found in the Pleistocene assemblages of different age in small numbers, but became the dominant species during the Last Glacial Maximum at some (mostly northern) sites.	Former name *Rhynchaeus arcticus* Kor.

Sources
Alfimov, A.V., Berman, D.I., Sher, A.V., 2003. Tundra-steppe insect assemblages and reconstruction of Late Pleistocene climate in the lower reaches of the Kolyma River. Zoologicheskiy Zhurnal, 82, 2, 281–300 (In Rissian).
Berman, D.I., Alfimov, A.V., 1993. Microclimatic conditions of existence of steppe ecosystems in the Subarctic of north-east Asia. Bulletin of Moscow Society of Naturalists, Biology Section 98 (3) (In Rissian).
Bienkowski, A.O. 1999. A review of the species of the subgenus Chrysolina (Pleurosticha) Motschulsky (Insecta: Coleoptera, Chrysomelidae). Ent. Abhandl., 58, 10, p. 165–179.
Korotyaev, B.A. 1996. 165. Sitona Germ. In: Egorov, A.B., Zherichin, V.V., and Korotyaev, B.A. Sem. Curculionidae - Dolgonosiki. Key to the insects of the Russian Far East, Vol. 3, Coleoptera, Pt. 3, Supplement. Vladivostok, Dal'nauka: 499–503. (In Russian).
Kryzhanovsky, O.L., 1965. Fam. Carabidae – ground beetles. Guide book to the identification of the insects of European part of USSR. V.2. Moscow-Leningrad, Nauka, p. 29–77. (In Russian).
Kryzhanovsky, O.L., Belousov, I.A., Kabak, I.I., Kataev, B.M., Makarov, K.V., Shilenkov, V.G. 1995. A checklist of the ground-beetles of Russia and adjacent lands (Insecta, Coleoptera, Carabidae). Sofia - Moscow: Pensoft Publishers. 272 p.
Kuzmina, S.A., Korotyaev, B.A. (1987). New species of the pill beetle genus Morychus Er. (Coleoptera, Byrrhidae) from the North-East of the USSR. Entomol. Review, v. 66,N 2, p. 342–344 (in Russian).
Perkovsky, E.E. 1999. Species of Cholevinus (Coleoptera, Leiodidae, Cholevini) from North of Eastern Siberia and Chukotka. Vestnik zoologii, 33, 3, p. 66 (in Russian).
Perkovsky, E.E., Kuzmina, S.A. (2001). Beetles of the genus Cholevinus (Coleoptera, Leiodidae, Cholevinae) from the Nord-West of Asia from the Pleistocene to today. Vestnik Zoologii, pp. 31–38. (in Russian).

youngest, because they are the most reliably dated (**Table 3**).

Late Weichselian insect assemblages The fossil insect assemblages dated from about 24 to 12 ka were first studied in the Kolyma Lowland, which later proved to be almost the centre of continentality and formation of steppe habitats in the whole of Northern Asia during the Pleistocene. The most interesting of these sites is Alyoshkina Zaimka on the lower course of the Kolyma River (**Fig.** 7, No. 7, **Table 3**). This site yielded four rich insect faunas within the range of 17 to 13 ka (all dates in this article are uncalibrated radiocarbon ages), almost completely (up to 80–95%) dominated by thermophilic xerophiles, including true steppe species (10–20% of the whole assemblage), with a very

Figure 20 Late Pleistocene fossil beetles from Western Beringia: A – *Harpalus vittatus kiselevi* Kat. et Shil., right elytron; B – *Amara glacialis* Mnnh. – left elytron; C – *Dicheirotrichus mannerheimi* Sahlb., left elytron; D – *Pterostichus agonus* Horn., pronotum; E – *Cyrtoplastus irregularis* Rtt., right elytron; F – *Helophorus splendidus* Sahlb.right elytron; G – *Aphodius sp.*, body without head and pronotum; H – *Morychus viridis* Kuzm. et Kor., left elytron; I – *Chrysolina tolli* Jac., left elytron; J – *Sitona borealis* Kor., head; K – *Coniocleonus cinerascens* Hochh., head and pronotum; L – *Stephanocleonus eruditus* Faust, head; M – *Lepyrus nordenskioeldi* Faust, head and pronotum; N – *Isochnus arcticus* Kor., body without head and pronotum; O – *Mesotrichapion wrangelianum* Kor., left elytron; P – *Hypera diversipunctata* Schrank., right elytron. A–F, H–L, N – Lena Delta region; M, O, P – Chukotka. Light microscope photographs, scale bar equals 1 mm.

low percentage of tundra beetles (Kiselyov, 1981). The assemblages show high diversity and abundance of steppe and meadow steppe species (**Table 6**). Either meadow steppe beetles, or a pill beetle *Morychus viridis* dominate all samples. These faunas represent the most 'steppic' dominated type of non-analog tundra-steppe assemblages.

Scattered similar assemblages of the same age are known elsewhere, but mostly within the Kolyma Lowland. For example, they are found in the Omolon-I section on the Kolyma River (**Fig. 7**, No 13, **Fig. 15**, **Table 3**), Assemblages with more than 40% of meadow-steppe and steppe species are present here in the interval between 7.0 and 12 m depth

Table 5 Ecological groups of insects, used for quantitative comparison of fossil assemblages

Ecological group and symbol	*Group description*	*Main species*
Steppe (**st**)	The members of this group live today in the steppe zone, some in the mountain belt of southern Siberia and Mongolia. Some of them survive under special microclimatic conditions in the relic steppe patches in NE Asia. However, the presence of any of these species can be considered evidence of unusually high summer temperatures (Berman *et al.*, 2001).	Ground beetles *Harpalus pusillus* Motsch., *Cymindis arctica* Kryzh. et Em., leaf beetles *Chrysolina pertorata* Gebl., *Ch. brunnicornis bermani* Medv., *Galeruca interrupta circumdata* Duft. weevils *Stephanocleonus eruditus* Faust, *S. fossulatus* F-W. *S. incertus* T-M.
Meadow-steppe (**ms**)	This group consists of species common in dry (steppe-like) meadows in various ecosystems. They are more cold-tolerant than the previous group. They survive today even in the tundra zone, on the warmest localities.	Ground beetles *Harpalus vittatus kiselevi* Kat. et Shil, soft-winged beetle *Troglocollops arcticus* L. Medv., the leaf beetle *Chrysolina arctica* Medv., weevils *Phyllobius kolymensis* Kor. et Egorov, *Coniocleonus cinerascens* Hochh., *C. ferrugineus* Fahr., *C. astragali* T.-M.et Kor.
Insects of hemicryophytic steppe (**ss**)	Hemicryophytic steppe (Yurtsev, 1981) is a cold steppe-like environment, dominated by a few species of xerophilic sedges and mosses. It has extremely contrasting microclimatic conditions – very hot and dry in summer, and snowless and very cold in winter. It occurs on very dry, snowless hill tops in the taiga or, rarely in tundra zone; (Berman *et al.*, 2001). This environment is currently found in the Kolyma Highlands, Chukotka, and Wrangel Island. It is believed to have many features in common with some kinds of Pleistocene tundra-steppe.	Pill beetle *Morychus viridis* Kuzm. et Kor. is the only represenatative of this group
Xerophilic insects of various habitats (**ks**)	Insects of xerophilic habitats of tundra, taiga and steppe.	Ground beetle *Notiophilus aquaticus* L., dung beetle *Aphodius* sp
Insects of dry tundra habitats (**dt**),	The insects of this group usually occupy the warmest sites in the tundra zone, most commonly well-drained sites with diverse grasses and herbs. Most of them do not reach the Arctic tundra today, but they quite often occupy parts of the taiga zone, in suitable biotopes. Some species can also be found today on relict steppe patches.	Ground beetles *Bembidion dauricum* Motsch., *Poecilus (Derus) nearcticus* Lth., *Pterostichus (Petrophilus) abnormis* Sahlb., *P. (Tundraphilus) sublaevis* Sahlb., *Stereocerus haematopus* (Dej.), *Curtonotus alpinus* Payk., *Amara interstitialis* Dej., *A. glacialis* Mnnh., *Dicheirotrichus mannerheimi* Sahlb., leaf beetle *Chrysolina marginata borealis* Medv., weevils *Mesotrichapion wrangelianum* Kor., *Hemitrichapion tschernovi* T.-M., *Sitona borealis* Kor., *Hypera ornata* Cap., *H. diversipunctata* Schrank., *Vitavitus thulius* Kiss.
Insects of typical and arctic tundra (**tt**)	Most cold-resistant species from different habitats in the Arctic	Leaf beetles *Chrysolina tolli* Jac., *Ch. subsulcata* Mnnh., *Ch. bungei* Jac., weevil *Isochnus arcticus* Kor.
Insects of mesic tundra habitats (**mt**)	The insects of this group are now found in different, mostly wet and relatively cold habitats in tundra; some occur on boggy patches in forest-tundra.	Ground beetles *Carabus truncaticollis* Esch., *Blethisa catenaria* Brown. *Diacheila polita* Fald., *Pterostichus (Cryobius) ventricosus* Esch., *P. (Cryobius) pinguedineus* Esch., *P. (Cryobius) brevicornis* (Kirby), *P. (Lenapterus) vermiculosus* Men., *P. (Lenapterus) costatus* Men., *P. (Lenapterus) agonus* Horn., small carrion beetle *Cholevinus sibiricus* (Jean.), rove beetles *Olophrum consimile* Gyll., *Tachinus arcticus* Motsch., *T. brevipennis* Sahlb., leaf beetle *Chrysolina septentrionalis* Men.
Shrub (**sh**)	These beetles feed on shrubs, mostly willows, except dwarf shrubs of typical and Arctic tundra	Leaf beetles *Chrysomela blaisdelli* Van Dyke, *Phratora polaris* Schn., *Ph. vulgatissima* L., weevils *Lepyrus nordenskioeldi* Faust, *L. gemellus* Kirby, *Dorytomus rufulus amplipennis* Tourn., *Isochnus flagellum* Erics.

(*Continued*)

Table 5 (*Continued*)

Ecological group and symbol	*Group description*	*Main species*
Meadows (**me**),	Mostly in the forest zone, although some species can occur on tundra meadows	Pill beetles *Byrrhus fasciatus* Forst., *Cytilus sericeus* Forst., click beetle *Hypnoidus hyperboreus* Gyll., leaf beetles *Bromius obscurus* L., *Phaedon concinnus* Steph., *Ph. armoraciae* L., weevils *Phyllobius virideaeris* Laich., *Sitona lineellus* Bonsd.
Taiga (**ta**),	Tree-associated insects and insects whose range is restricted to the taiga zone.	Ground beetle *Pterostichus (Petrophilus) magus* Man., click beetle *Denticollis varians* Germ., weevils *Hylobius piceus* DeG., *Pissodes insignatus* Boh., *P. irroratus* Reitt., bark beetles *Ips cembrae* Heer, *Polygraphus* sp., ants *Leptothorax acervorum* Fabr., *Formica gagatoides* Ruzs., *Camponotus herculeanus* L.
Riparian (**ri**),	These insects live near water in several zones, but most of them are restricted to the taiga zone.	Ground beetles *Pelophila borealis* Payk., *Nebria frigida* Sahlb., *Elaphrus lapponicus* Gyll., *E. riparius* L., *Dyschiriodes nigricornis* Motsch., *Bembidion umiatense* Lindrt., *Agonum quinquepunctatum* Motsch., rove beetles *Stenus sp.*, lady-bird *Hippodamia arctica* Schneid., leaf beetles *Donacia sp., Hydrothassa hannoverana* F., weevils *Phytobius leucogaster* Marsh., *Tournotaris bimaculatus* F., bugs of Saldidae family.
Aquatic (**aq**),	These insects live in various tundra and taiga water bodies, mostly small ponds.	Predaceous diving beetles *Hydroporus acutangulus* Thoms., *Agabus moestus* (Curt.), *Colymbetes dolabratus* (Payk.), whirligig beetle *Gyrinus opacus* Salb. water scavenger beetles *Helophorus splendidus* Sahlb., *Hydrobius fuscipes* F., water bug *Sigara* sp.
Others (**oth**).	Insects of intrazonal habitats (such as plant litter, or fungi) and fossils unresolved to a species (of genera with ecologically various members)	Small carrion beetle *Cyrtoplastus irregularis* carrion beetle *Thanatophilus dispar* Hbst., rove beetles *Eucnecosum* sp., *Lathrobium* sp., *Quedius* sp., lady-bird *Coccinella* sp., minute brown savenger beetle *Corticaria* sp.

(ca. 23–18 ka) (Kiselyov *et al.*, 1987). Tundra-steppe assemblages with extremely high numbers of *M. viridis* (**Table** 7), but less diverse and abundant steppe species occur below and above this interval.

The *M. viridis*-dominated late Weichselian assemblages were distributed much more widely than just in the Kolyma Lowland. They are also known from the Yedoma sections in the Indigirka and Khroma valleys (**Table** 3), and much further north, in sites from the Lena Delta. In the Mamontovy Khayata (MKh) section (**Fig.** 7, No 3) they form a series of samples associated with the termination of the late Weichselian (IFZ 4) and dated about 14 to 12 ^{14}C ka (**Fig. 21**). *M. viridis* exclusively dominates four samples (50–60%). In the Lena Delta, these samples represent the highest content of *Morychus*, steppe and other xerophilic species during the whole Weichselian, so this stage has been recognized as the Late Glacial Event (LGE), a brief interval of higher summer temperatures and drier environments in this Arctic region (Sher *et al.*, 2005).

The LGM insect assemblages in the Lena Delta are strikingly different from those in the Kolyma Lowland. In the MKh section they are recognized as IFZ 3 (**Fig. 21**). Of nine samples within the range of 24 to 14 ka, six are strongly dominated by Arctic tundra insects (40–67%), mainly *Isochnus arcticus*. Xerophilic insects constitute only 1.5 to 3%, with a maximum of 5% of the fauna – the lowest values in the whole Weichselian. Even the ubiquitous *M. viridis* was encountered in only 4 of 9 samples (below 2%). True steppe insects were totally absent between ca. 23 and 20 ka and comprised just 1 to 3% of the faunas in the other samples. These peculiar assemblages (Arctic type of tundra-steppe) are interpreted as evidence of a very cold climate with dry summers (Sher *et al.*, 2005). This reconstruction is corroborated by the plant macrofossil assemblages from the same sites (Kienast *et al.*, 2005).

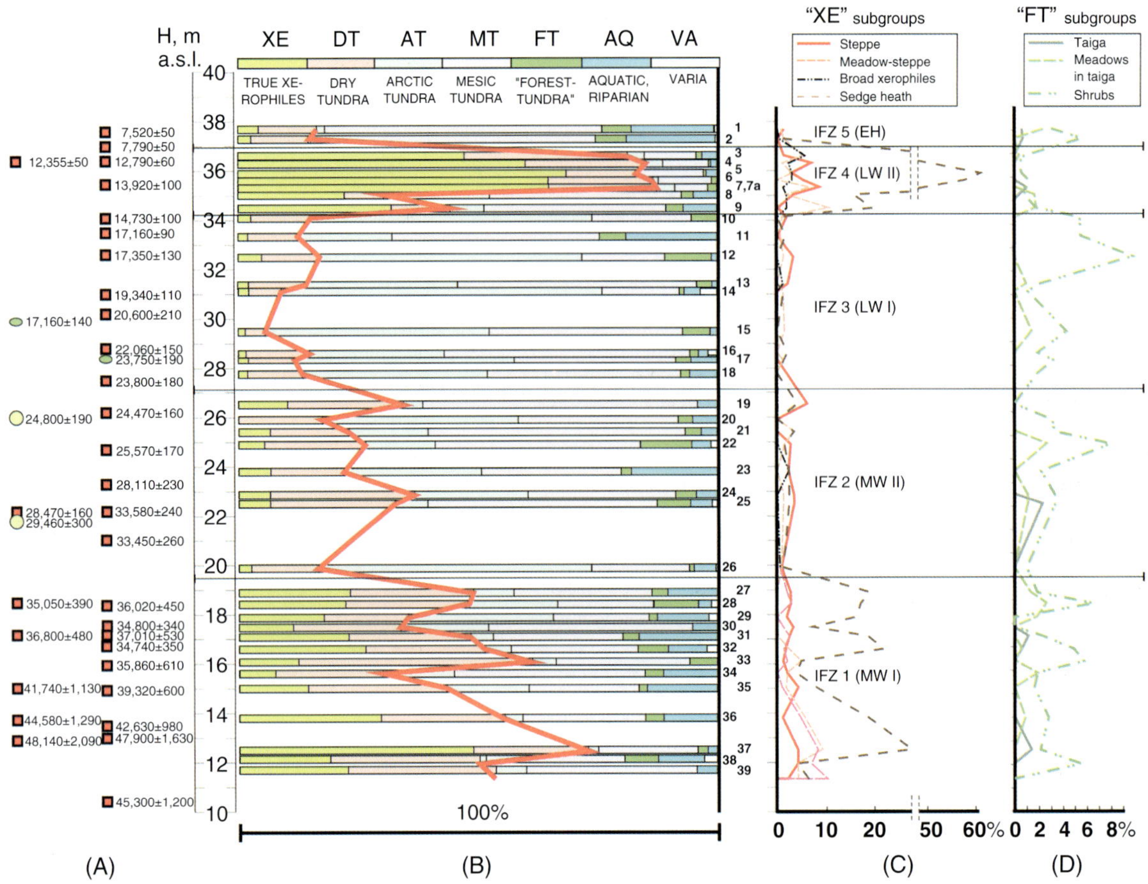

Figure 21 The sequence of fossil beetle assemblages in the Mamontovy Khayata key section, Bykovsky Peninsula, Lena Delta, ranging in age from greater than 50 ka to 8 ka (from Sher et al., 2005).

A similar assemblage is known from the Buor-Khaya exposure on the Central Lena Delta (from the upper part of the Ice Complex), where it is dated about 17 ka (Kuzmina, 2001). Outside the Lena Delta, the Arctic-type tundra-steppe fauna was found in a single, poor sample from the late Weichselian Ice Complex in the northern Yana-Indigirka Lowland (Volchya River, 71.7° N).

Thus, during the LGM, tundra-steppe insect faunas of the Arctic type existed only in the northernmost parts of the continent. Further south, they included increasingly thermoxerophilic and steppe elements. For example, at the Krasivoye site on the Malyy Anyuy River (68.3° N) (**Fig.** 7, No 15, **Table** 3), the Ice Complex assemblage dated about 19 ka was dominated by dry tundra (29%) and xerophilic beetles (53%), with sedge steppe, meadow steppe and true steppe components each comprising over 15%. Assemblages with an even more pronounced xerophilic component are common in sections (or their members) that are tentatively referred to the late Weichselian, all over northeastern Siberia. Many exposures in the Kolyma Lowland have yielded faunas with 65 to 90% xerophiles (alternatively dominated by sedge steppe or meadow steppe beetles) in the upper part of the Ice Complex (Duvanny Yar, Lakes Yakutskoye and others (**Fig.** 7, No 8, **Fig.** 9, **Table** 3). In the Indigirka Lowland, the late Weichselian (LW) sites on the Khroma, Indigirka (Suturuokha), and Keremesit Rivers are also dominated by xerophiles (60–80%), with *Morychus viridis* dominating the whole spectrum (30–60%). Similar kinds of insect assemblages (with xerophiles forming 50 to 95% of the whole spectrum) have been described from the upper part of the Ice Complex of Ayon Island (Utatgyr Suite, Northwestern Chukotka, ca. 69.5° N (**Fig.** 7, No 10, **Table** 3), but here they have an even stronger contribution of meadow-steppe or/and true steppe species (30–90%). In contrast to this, at the eastern edge of Asia, closer to the Pacific, the steppe component in the LGM faunas is much less pronounced, or absent altogether. In southern Chukotka, for example (the Ice Bluff section on the Main River, **Fig.** 7, No 12, **Table** 3), throughout the

Table 6 Occurrence of xerophilic beetles in the Alyoshkina Zaimka Late Weichselian sequence of samples

Family	*TAXON*	*Eco Code*	*% in the sample*	
			MIN	*MAX*
Carabidae	*Curtonotus fodinae* Mnnh.	st	0.1	1.7
Carabidae	*Harpalus amputatus amputatoides Mlynar*	ms	0.4	1.4
Carabidae	*Harpalus vittatus kiselevi* Kat. et Shil.	ms	0.4	3.9
Carabidae	*Harpalus pusillus* Motsch.	st	0.73	0.73
Carabidae	*Cymindis arctica* Kryzh. et Em.	st	0.7	3.9
Silphidae	*Aclypea* cf. *jacutica* (Rjab.)	ms	0.24	0.24
Scarabaeidae	*Aphodius* sp. nov. ?	ks	0.98	8.92
Byrrhidae	*Morychus viridis* Kuzm. et Kor.	ss	7.51	64.0
Melyridae	*Troglocollops* cf. *arcticus* L.Medv.	ms	0.7	5.9
Chrysomelidae	*Chrysolina aurichalcea* Mnnh.	ms	2.44	2.44
Chrysomelidae	*Chrysolina perforata* Gebl.	st	0.65	0.85
Chrysomelidae	*Chrysolina aeruginosa* Fald.	st	0.48	0.65
Chrysomelidae	*Chrysolina brunnicornis bermani* Medv.	st	1.45	4.89
Chrysomelidae	*Chrysolina rufilabris* Fald.	st	0.47	3.91
Chrysomelidae	*Colaphellus alpinus* Gebl.	ms	0.24	0.24
Chrysomelidae	*Galeruca daurica* Jeann.	st	0.49	0.49
Apionidae	*Loborhynchapion* cf. *amethystinum* Miller	ms	0.5	29.6
Curculionidae	*Phyllobius* cf. *kolymensis* Kor. et. Egorov	ms	5.1	29.1
Curculionidae	*Phyllobius (Angarophyllobius)* sp.	ms	0.72	0.72
Curculionidae	*Coniocleonus cinerascens* Hochh.	ms	0.97	0.97
Curculionidae	*Coniocleonus ferrugineus* Fahr.	ms	0.12	0.98
Curculionidae	*Coniocleonus astragali* T.-M. et Kor.	ms	0.22	0.22
Curculionidae	*Stephanocleonus eruditus* Faust	st	0.43	4.16
Curculionidae	*Stephanocleonus fossulatus* F.-W.	st	0.65	1.22
Curculionidae	*Stephanocleonus foveifrons* Chevr.	st	2.42	13.9
Curculionidae	*Stephanocleonus* cf. *anceps* Chevr.	st	0.22	0.22

Table 7 Change of percentage of xerophilic insects along the late Weichselian section Omolon-I

Estimated age, years BP	*Steppe species (st)*	*Morychus viridis (ss)*	*Meadow-steppe species (ms)*
15,000	6.44	77.65	8.14
17,000	5.49	83.65	3.64
19,000	16.13	41.40	26.88
21,000	1.73	49.14	41.29
23,000	4.29	39.64	41.79
24,000	11.45	76.72	5.34

interval from 25 to 15 ka, all groups of xerophilic insects rarely reach 30% (usually below 10%) and are mostly represented by *Morychus viridis*. On the other hand, arctic beetles are abundant in assemblages dated around 21 ka (up to 40%). This is probably indicative of a milder (Pacific) version of the cold-adapted (Arctic type) LGM faunas.

Middle Weichselian insect assemblages The middle Weichselian interval (from about 50 to 25 ka) was traditionally considered as a 'warmer climate interstadial' interval in Northern Asia (called 'Karginsky'), including several cycles of cooling (Kind, 1974). In many natural sections, where organic-rich deposits (usually peat) occurred under the Ice Complex, or inside it, or separating two 'Ice Complexes' (two members with their own system of polygonal ice wedges), these peats have been ^{14}C dated to 35 to 40 ka. The pollen spectra of these layers usually show a high percentage of arboreal (mostly shrub) pollen, in contrast to the completely grass-herb dominated spectra of the Ice Complex itself. In recent years, however, several important new discoveries have been made about this interval. First, repeated ^{14}C dating showed that most of the samples that yielded ^{14}C ages near 35 ka and older, were indeed much older and could actually represent ages beyond the ^{14}C age range (Sher and Plakht, 1988). Second, it was shown that the 'Karginsky' sections in the type area (Lower Yenissey) had been misinterpreted, and their dating was erroneous. In fact, they turned out to be much older, probably as old as the last interglacial (Astakhov, 2001). Third, new radiocarbon ages from the regional type sections of the 'Karginsky Interstadial' in northeastern Siberia also yielded much older dates (Sulerzhitsky, 1998). Finally, the multidisciplinary study of a continuous section through most of the late Weichselian in the Lena Delta, well-dated by ^{14}C, revealed a completely different regional history of this period (Andreev *et al.*, 2002; Schirrmeister *et al.*, 2002). The study of the detailed record of fossil insect assemblages through this sequence allowed us (Sher *et al.*, 2005) to suggest a very different reconstruction of middle Weichselian environments.

It has been shown that tundra-type insect assemblages persisted in the Lena Delta throughout the middle Weichselian interval, with a variable, but almost consistent presence of xerophilic species. Although today the northern limit of taiga (sparse larch forest) is not far south of the site, no fossil assemblages of the forest or even forest-tundra type have ever been found here. The contribution of xerophilic species was more noticeable in the first half of the middle Weichselian interval, especially in the assemblages ^{14}C dated about 45 to 48 ka, when they reached 30 to 50%. We cannot exclude the possibility, however, that these assemblages are actually early Weichselian in age. Taken together with the dry tundra species contribution, the percentage of relatively thermophilic forms gradually decreased from 48 to 34 ka (Insect Faunal Zone 1, **Fig. 21**). At the same time, the contribution of arctic species increased, and by 34 ka it almost reached LGM levels (more than 50%). The second half of the middle Weichselian (34–24 ka, IFZ 2 on **Fig. 21**) is marked by the lowest levels of xerophilic species (usually below 7%) and the predominance of mesic tundra insects, sometimes with a high contribution of arctic species. This sequence, the most complete in Northern Asia, is interpreted as a gradual decrease in summer temperatures towards the LGM, probably accompanied by some increased moisture. However, these moisture levels did not reach the levels observed in modern tundra environments. This indicates that during the middle Weichselian the climate was more continental than today in this coastal area. This conclusion is in agreement with various observations that sea level in the Eastern Arctic remained low at that time.

Several sites in more southern areas of northeastern Siberia yielded representative insect assemblages obtained from the Ice Complex and associated with late middle Weichselian ^{14}C dates. For example, in the Shamanovo Yedoma section on the Indigirka River (**Fig.** 7, No 16, **Table 3**) a rich insect sample came from the unit dated by *in situ* grass roots at 29,300 to 32,900 yr BP (Kaplina et al., 1980). The assemblage is dominated by *M. viridis* and other xerophilic insects (66%) and dry habitat species prevail in the tundra group (Kiselyov, 1994). On the Khroma River (**Fig.** 7, No 17, **Table 3**), the assemblage dated 28,800 yr BP (Kaplina et al., 1983) has the same percentage of xerophilic species, but *M. viridis* is much more abundant, while the contribution of steppe species is low.

Interesting assemblages, presumably associated with the middle Weichselian, were found in the Primorsky alas section in western Chukotka (**Fig.** 7, No 18, **Table 3**). The sampled unit contains tree birch wood, and is dated more than 27,870 yr BP (Kiselyov, 1994). The assemblages (**Table 7**) include unusually high percentages of forest-tundra species (7%), comprising such forest inhabitants as the rove beetles *Acidota* sp., *Arpedium* sp., *Xantholinus* sp., *Anthobium* sp., and *Bolitobius* sp., as well as the birch-associated weevil *Betulapion simile*. On the other hand, these assemblages contain high percentages of *M. viridis* and a few true steppe beetles. A peat layer from about 12 m higher in the section is dated at 24,800 ± 400 yr BP. Two insect assemblages from this unit are similar to the lower ones, but show even higher percentages of forest-tundra beetles (up to 12%) represented by the same species. The steppe, meadow-steppe beetles and *M. viridis* are still present, although in lower numbers.

However, in southern Chukotka, insect assemblages from the same age interval (34–24 ka) (Ice Bluff locality, **Fig.** 7, No 12, **Fig. 17**, **Table 3**) do not contain any forest beetles (Kiselyov, 1994). The percentage of xerophilic species is high (38%), mostly at the expense of dung beetles (*Aphodius* sp. nov. ?) and *M. viridis*. True steppe and meadow-steppe species are represented only by single specimens.

Early Weichselian insect assemblages Taking into account the dating problems, it is difficult to confidently assign any fossil insect assemblage to the early Weichselian interval. On the other hand, the lower parts of many thick Ice Complex sections may belong to this time interval. For example, in the key section of the Ice Complex of Duvannyy Yar (**Fig.** 7, No 8, **Fig.** 9, **Table** 3) a large number of infinite ^{14}C dates (>45 ka) in the lower part of the section (mixed with dates in the range of 33 to 45 ka) may suggest an early Weichselian age (Sher, 1991). The same is possible for the Oyagosskiy Yar assemblage (**Fig.** 7, No 19, **Table** 3) and the sections on the southern coast of Bol. Lyakhovsky Island. (**Fig.** 7, No 4, **Table** 3) In the Kolyma Lowland, the insect assemblages from the lower parts of Ice Complex sections are dominated by *M. viridis* with a small admixture of other xerophiles (usually below 10%) (Duvannyy Yar, Sher *et al.*, 1979). However, the northern localities, such as Oyagos, Lyakhovsky, or Nagym in the Lena Delta (**Fig.** 7, No 1, **Table** 3), are mostly dominated by tundra species, sometimes with a high percentage of Arctic tundra species or mesic tundra inhabitants. The total percentage of all xerophiles is usually below

15%, and the main role in this group is played by *M. viridis*, while true steppe species are either extremely rare or completely absent (Kuzmina, 2001).

Thus, we may conclude that the same regularities in spatial distribution of insect assemblages in the late Weichselian existed during early Weichselian time.

'Last Interglacial' insect assemblages The problems of recognition of Last Interglacial sediments were discussed above (cf. also Sher, 1991). Tentatively, we refer some parts of particular complex sections to this interval. These sediments overlie sediments of presumed middle Pleistocene age and they underlie the late Pleistocene Ice Complex. Also, they contain abundant organic remains. Such localities usually represent buried alas complexes (**Table 3**). The characteristics of insect assemblages forming three of them are shown in **Table 8**.

Table 8 The insect assemblages of presumably 'Last Interglacial' age in Northeastern Siberia

Unit	*Location*	*Sediment description*	*Insect assemblages*
Kyl-Bastakh Layers	Khroma River (**Fig. 7**, No. 17).	Organic-rich filling of ice-wedge pseudomorphs (the casts of former ice wedges), intruding the top of the Khroma Suite, presumably referred to the late middle Pleistocene, and overlain by the Yedoma Suite (Ice Complex).	Three fossil assemblages are dominated by tundra species (>30%), both mesic and xeric, and the percentage of forest-tundra species is probably the highest elsewhere in the regional record (about 10%). Among the latter, such forest zone species as *Phyllobius virideaeris* Laich. (a meadow weevil) and *Carabus maeander* Fisch. (a taiga ground beetle), shrub leaf beetles *Chrysomela lapponica* L. and *Phratora polaris* Schn. are recorded. At the same time, the thermophilic xerophiles comprise 15–20% of the whole fauna, and among them are not only the 'sedge-steppe' species *Morychus viridis* Kuzm. et Kor. and the dung beetles *Aphodius* sp. nov., but true steppe insects, such as the ground beetles *Harpalus pusillus* Motsch. and *Cymindis arctica* Kryzh. et Em., a leaf beetle *Chrysolina perforata* Gebl. and a weevil *Stephanocleonus foveifrons* Chevr. Remarkably, the assemblages include a few specimens of Arctic tundra leaf beetle *Chrysolina subsulcata* Mnnh. and an arctic weevil *Isochnus arcticus* Kor. They also contain a large numbert of hydrophilic beetles, which is commonly found in peat deposits.
Kuobakh Unit	Alazeya River (**Fig. 7**, No. 5)	A peat layer and organic-rich ice-wedge pseudomorphs are observed between the late middle Pleistocene Maastakh Suite and the Yedoma Ice Complex	Similarly to the Kyl-Bastakh, the assemblages combine a high percentage of hydrophilic and tundra species with a large number (up to 33%) of forest-tundra insects. The latter include mostly the ant *Formica (Serviformica) gagatoides* Ruzs. and various meadow species from the forest zone, such as a click beetle *Hypnoidus hyperboreus* Gyll., an ant-like beetle *Anthicus ater* Pz., a leaf beetle *Phaedon concinnus* Steph., and a weevil *Phyllobius virideaeris* Laich. Xerophiles comprise up to 8–10% of the faunal assemlages, but do not include steppe species
Krest-Yuryakh Unit	Bolshoy Lyakhovsky Island (**Fig. 7**, No. 4)	Two layers of organic-rich ice-wedge pseudomorphs	The unit is in the same stratigraphic position, but the locality is much further north, so it is dominated by mesic tundra species (35–45%) and dry tundra insects (20–30%). However, the Krest-Yuryakh assemblages include both forest-tundra species (2–5%) and up to 10–20% of thermophilic xerophiles, including a few occurrences of steppe species, such as the ground beetle *Cymindis arctica* Kryzh. et Em., and the weevils *Stephanocleonus eruditus* Faust and *S. fossulatus* F.-W.

The most remarkable character of all these assemblages is the extreme heterogeneity of their ecological composition, combining representatives from practically all ecological groups. Many species comprising those assemblages do not occur together today. At present their ranges are often separated by many hundreds of kilometers. This is especially evident for the northernmost assemblages (Bolshoy Lyakhovsky Island), where about 80% of identified insect species currently live much further south. This indicates the non-analog character of the assemblage, which certainly existed in a warmer climate than today, with higher summer and probably somewhat higher winter temperatures. According to the paleobotanical data (Andreev *et al.*, 2004), summer temperatures could have been 4 to 5° C higher than present. In general, the climate was more continental than today on Lyakhovsky Island, but less so than during the Weichselian interval. Compared with the modern environment of the island, the humidity was certainly lower in summer, but probably somewhat higher in winter. A thicker snow cover would have allowed the growth of tall shrubs at this high latitude.

Summary

This review demonstrates the large scope of the Pleistocene beetle research in Northeastern Siberia during the recent years. However, many gaps remain in our knowledge and some significant problems are still unsolved. Evidently, the characterized beetle faunas are peculiar, but we have a poor idea of how far south and west they were distributed. Very few fossil assemblages are known south of the Arctic Circle in this region, and none to the west of the Lena Delta. We also have serious problems with the dating of fossil assemblages older than 30 ka. We believe that future research should be concentrated on continuous sedimentary sequences, such as the Bykovsky section, and this research must be a part of broad multidisciplinary projects. Recent experience shows that successful projects of such kinds must be based on international collaboration.

Glossary

Dendrophilic – species that live and feed on trees.
Thermophilic – species that live in relatively warm climates.
Xerophilic – species that prefer dry and relatively warm habitats, often on open ground, with or without sparse shrubs and trees.
Xylophages – species that feed on wood.

See also: **Beetle Records**: Overview. **Periglacial Landforms**: Ice Wedges and Ice Wedge Casts.

References

Andreev, A. A., Grosse, G., Schirrmeister, L., Kuzmina, S. A., Novenko, E. Y.u, Bobrov, A. A., Tarasov, E., Ilyashuk, B. P., Kuznetsova, T. V., Krbetschek, M., Meyer, H., and Kunitsky, V. V. (2004). Late Saalian and Eemian palaeoenvironmental history of the Bol'shoy Lyakhovsky, Island Laptev Sea region. *Arctic Siberia. Boreas* **33**(4), 319–348.

Andreev, A. A., Schirrmeister, L., Siegert, C.h, Bobrov, A. A., Demske, D., Seiffert, D., and Hubberten, H.-W. (2002). Paleoenvironmental changes in Northeastern Siberia during the Late Quaternary – evidences frompollen records of the Bykovsky Peninsula. *Polarforschung* **70**, 13–25.

Astakhov, V. (2001). The stratigraphic framework for the Upper Pleistocene of the glaciated Russian Arctic: changing paradigms. *Global and Planetary Change* **31**(1–4), 283–295.

Berman, D. I. (1986). Invertebrate fauna and population in tundrasteppe communities of Wrangel Island. In pp. 146–160 (in Russian).

Berman, D. I., Alfimov, A. V., Mazhitova, G. G., Grishkan, I. B., and Yurtsev, B. A. (2001). *Cold steppe in north-eastern Asia. – Vladivostok, Dal'nauka,* p. 183. (in Russian).

Eremin, P. K. (1990). For systematic of the ground beetles of *Cryobius* Chd. subgenus (Coleoptera, Carabidae, *Pterostichus*) from USSR fauna. In *Fauna and ecology of ground beetles. Kishinev*, pp. 18–19. (In Russian).

Grushevskiy I. I, and Medvedev, L. N. (1962). Preliminary data on the application of coleopterological analysis to the study of Quaternary deposits of Northern Yakutiya. In *Collected articles on paleontology and biostratigraphy of Scientific Research Institute of Arctic Geology* (Shvedev, Ed.), (In Russian), pp. 38–72. Scientific Research Institute of Arctic Geology, Leningrad.

Kaplina, T. N., Sher, A. V., Giterman, R. E., Zazhigin, V. S., Kiselyov, S. V., Lozhkin, A. V., and Nikitin, V. P. (1980). Key section of Pleistocene deposits on the Allaikha river (lower reaches of the Indigirka). Bulletin of the Commission on Quaternary Period Research. *USSR Academy of Sciences* **50**, 73–95, (In Russian).

Kaplina, T. N., Shilova, G., and Pirumova, L. G. (1980). Shamanovo key section of Late Pleistocene and Holocene deposits on the Indigirka. *Izvestiya Academii Nauk SSSR. Ser. Geologicheskaya* **9**, 74–171, (In Russian).

Khruleva, O. A. (2004). Tundra-steppe leaf beetle *Chrysolina brunnicornis vrangeliani* (Coleoptera, Chrysomelidae): distribution, life history and habitats. In *New developments in the biology of Chrysomelidae* (P. PJolivet, J. A. JSantiago-Blay and M. Schmitt, Eds.), pp. 541–550. SPB Academic Publishing bv, The Hague.

Kienast, F., Schirrmeister, L., Siegert, C.h, and Tarasov, P. (2005). Palaeobotanical evidence for warm summers in the East Siberian Arctic during the last cold stage. *Quaternary Research* **63**(3), 283–300.

Kind, N. V. (1974). *Late Quaternary Geochronology According to Isotope Data (Geokhronologia Pozdniego Antropogena Po Izotopnym Dannym)* Moscow: Nauka. p. 255. (in Russian).

Kiselyov, S. V. (1981). *Late Cenozoic Coleoptera of North-East Siberia.* Nauka, Moscow. (In Russian).

Kiselyov SV (1994) Northern Eurasia environment in the Pleistocene and Holocene (according to Coleoptera researches). Thesis in Geology, p. 32 Moscow: MSU, (in Russian).

Kiselyov, S. V., Kolesnikov, S., and Rybakova, N. O. (1987). Climate of the vegetation period during the 'Ice Complex' deposits forming from the Omolon River Site). *Byulleten Moskovskogo obshchestva ispytatelei prirody* **62**, 113–119, (In Russian).

Kuzmina, S. A., and Korotyaev, B. A. (1987). A new species of a pill beetle Morychus Er. Coleoptera, Byrrhidae) from Northern-East of USSR. *Entomol. obozrenie* 66 **2**, 342–344, (in Russian).

Kuzmina, S. A. (2001) Quaternary Insects of the Coastal Lowlands of Yakutia. PhD (Candidate) Thesis in Biology. Paleontological Institute. Moscow: RAS. (In Russian).

Medvedev, L. N., and Voronova, N. N. (1977). Coleopterological analysis of geological sections of mammoth cemeteries in northern Yakutia. In *The mammoth fauna of the Russian Plain and Eastern Siberia* (O. Scarlato, Ed.), (In Russian), pp. 72–76. Nauka, Leningrad.

Schirrmeister, L., Siegert, C., Kuznetsova, T., Kuzmina, S., Andreev, A. A., Kienast, F., Meyer, H., and Bobrov, A. (2002). Paleoenvironmental and paleoclimatic records from permafrost deposits in the Arctic region of northern Siberia. *Quaternary International* **89**, 97–118.

Sher, A. V. (1988). Problems of ecology of Pleistocene mammals. In *Aktual'nye problemy morfologii i ekologii vysshikh pozvonochnykh*, pp. 432–464, Chast' II, Moskva: Izdatelstvo AN SSSR. (In Russian).

Sher, A. V. (1990). Actualism and disconformism in the studies of Pleistocene mammals ecology. *Journal of General Biology* **51**(2), 163–177.

Sher, A. V. (1991). Problems of the last interglacial in Arctic Siberia. *Quaternary International* **10–12**, 215–222.

Sher, A. V., Kaplina, T. N., Giterman, R. E., Lozhkin, A., Arkhangelov, A. A., Kiselyov, S. V., Kouznetsov, Yu. V., Virina, E. I., and Zazhigin, V. S., (1979) Late Cenozoic of the Kolyma Lowland: XIV Pacific Science Congress, Tour Guide XI, Khabarovsk August 1979, pp. 1–116. Academy of Sciences of the USSR: Moscow.

Sher, A. V., Kuzmina, S. A., Kuznetsova, T. V., and Sulerzhitsky, L. D. (2005). New insights into the Weichselian environment and climate of the Eastern-Siberian Artic, derived from fossil insects, plants, and mammals. *Quaternary Science Reviews* **24**(5–6), 533–569.

Sher, A. V., and Plakht, I. R. (1988). Radiocarbon dating and problems of the Pleistocene stratigraphy in lowlands of the Northeast USSR. Izvestiya AN SSSR, ser. Geol. 8, 17–31 (in Russian); In English. *International Geology Review* **30**(8), 853–867.

Sulerzhitsky, L. D. (1998). On the microbial contamination organic matter from permafrost observed during radiocarbon dating. *Kriosfera Zemli* (*Earth Cryosphere*) **2**(2), 76–80, (in Russian).

Svendsen, J. I., Alexanderson, H., Astakhov, V. I., Demidov, I., Dowdeswell, J. A., Funder, S., Gataullin, V., Henriksen, M., Hjort, C., Houmark-Nielsen, M., Hubberten, H. W., Ingólfsson, O., Jakobsson, M., Kjær, K. H., Larsen, E., Lokrantz, H., Lunkka, J. P., Lyså, A., Mangerud, J., Matioushkov, A., Murray, A., Möller, P., Niessen, F., Nikolskaya, O., Polyak, L., Saarnisto, M., Siegert, C., Siegert, M. J., Spielhagen, R. F., and Stein, R. (2004). Late Quaternary ice-sheet history of Northern Eurasia. *Quaternary Science Reviews* **23**, 1229–1271.

Tugarinov, A.Ya (1929). On the origin of arctic fauna. *Priroda* **7–8**, 653–680, (in Russian).

Vangengeim, E. A. (1977). *Paleontological foundation of Anthropogene stratigraphy in Northern Asia (by mammals)*, pp 1–172, Moscow: Nauka. (In Russian).

Vangengeim, E. A., and Ravsky, E. I. (1965). On the intra-continental type of natural zonality in Eurasia during the Quaternary Period (Anthropogene). In *The problems of Cenozoic stratigraphy* (V. V. Menner, Ed.), (In Russian), pp. 128–141. Nedra Publishers, Moscow.

Yurtsev, B. A. (1981). *Relict steppe vegetation complexes of Northern Asia (Problems of reconstruction of the cryoxerotic landscapes of Beringia)*. Nauka, Novosibirsk. (in Russian).

Zinoviev, E. V. (1997). Quaternary insects of West Siberian Plain. In *Successes of entomology on Ural*, pp. 153–157, Ekaterinburg. (In Russian).

Postglacial Europe

P Ponel, Mediterranean Institute for Ecology and Paleoecology, France

© 2007 Elsevier B.V. All rights reserved.

Insect analyses carried out from Holocene sequences are relatively rare in Britain and northern Europe, in contrast to late Pleistocene and Late Glacial sequences, which have been thoroughly investigated during the past 50 years. Therefore, the early to mid-Holocene history of the coleopteran fauna in western Europe relies for the most part on the analysis of scattered assemblages obtained from natural or seminatural deposits. Late Holocene studies, especially from the historical period, are much more abundant, but the vast majority of data concerning this period derive from analyses carried out in archeological sites from anthropogenic contexts and are dominated by synanthropic insects. These studies provide a wealth of data, mainly concerning the living conditions of man. However, they offer relatively restricted information on past environments. A further complication hampering the paleoenvironmental interpretation of postglacial beetle records is the growing influence of man on natural ecosystems from the mid-Holocene onwards through forest clearance, development of agriculture, pastoralism, and fragmentation of habitats. These landscape modifications make it difficult to disentangle the respective role of climate and human impact on coleopteran faunas.

Considering the wide range of subjects treated by Quaternary entomologists and archeoentomologists from postglacial beetle records, this article is necessarily focused on a selection of issues, such as changes in coleopteran faunal assemblages at the onset of the Holocene and their paleoclimate and paleoenvironmental implications, the impact of human activities on landscapes and Coleoptera (especially the effect of forest clearance on saproxylophagous species), and reconstruction of paleoenvironmental changes in Mediterranean mountains from continuous sedimentary sequences.

Climatic Context

The Quaternary is marked by a climatic cyclicity of cold and warm periods (glacial–interglacial) and by the appearance and evolution of humans. The postglacial or Holocene that started approximately 11,000 years ago is the last warm period of the Quaternary (the current interglacial) and can be considered an approximate replica of the penultimate interglacial, the Eemian (130,000–110,000 BP). The Holocene interglacial has been deeply affected by human impact on natural ecosystems, especially in its second half. The glaciation that immediately preceeded the Holocene is biologically important since glacial conditions that prevailed at that time (60,000–15,000 BP) eradicated most of temperate biota from central and northern Europe, including insects, which were relegated to temperate refugia (probably located in Mediterranean peninsulas). The recolonization of Europe by insects induced by the Holocene warming is the result of the northwards shift of biota from these southern 'reservoirs'. Far from being a steady process, the transition from the last glaciation to the current interglacial, approximately 15,000 and 10,000 BP, was marked by a complex series of sharp warm and cold events. This period of climatic instability, commonly named the Late Glacial, includes a first climatic warming between 15,000 and 13,000 BP (the Oldest Dryas), a predominantly warm period between 13,000 and 10,700 BP, the Late Glacial interstadial, and a short but drastic climatic cooling between 10,700 and 10,300 BP (the Younger Dryas). The Late Glacial is particularly well documented from beetle records (Coope, 1994), and one of the major achievements of Quaternary entomology is the demonstration of the amazing rapidity of the Late Glacial warming at 13,000 BP since mean July temperatures in Britain rose by approximately 1° C per decade (for a total of approximately 7° C), and corresponding winter temperatures rose by approximately 20° C (Atkinson *et al.*, 1987). Following the second episode of rapid warming (by 10,000 BP), temperate climate conditions prevailed during most of the Holocene.

Vegetation History

Deciduous forest (with hazel, oak, elm, and linden) spread throughout much of Europe by 10,000 BP. Based on paleobotanical reconstructions, a thermal optimum is identified between 8,000 and 6,000 BP, and it is at that time that an unprecedented event, the Neolithic 'revolution,' arose in the Near East and expanded quickly into Europe, inducing forest clearance and agriculture. By 5,000 BP, the 'Urwald,' or primitive forest, started to shrink in most of lowland Europe including Britain. 'Slash-and-burn' cultivation practices induced a deep change in plant communities, as indicated by pollen analysis: There was a decline of many forest taxa and a rise of heliophilous plants and many weeds (*Plantago*, and *Urtica*, plants associated with disturbance). The Iron Age (approximately 3,500–2,500 BP) corresponds with a time of marked human population growth, furthering the development of agricultural landscapes. By 2,500 BP, the intensification of agricultural disturbance on natural vegetation structures is superimposed with a climate change at the subboreal (warm/dry climate)–subatlantic (cool/wet climate) transition.

The Onset of the Holocene: Beetles at the Younger Dryas–Preboreal Transition

Shortly before the climatic and environmental changes that occurred in Europe during the early Holocene (preboreal interval), an Arctic/Alpine beetle fauna, including many species restricted today to very high latitudes in Arctic regions of Scandinavia and northern Russia, thrived in Britain and in most of Europe. This cold-adapted fauna is associated with the Younger Dryas (from 10,700 to approximately 10,300 BP), a short but intense cold event. Most of these species were already present at the end of the last glaciation, approximately 15,000 BP, but were temporarily eradicated from mid-latitudes by the warm episode of the Late Glacial interstadial. This extremely cold-adapted fauna includes many characteristic species, such as

Nebria nivalis	*Amara alpina*
Diacheila arctica	*Pterostichus middendorffi*
Elaphrus lapponicus	*Agonum consimile*
Dyschirius septentrionum	*Agonum sahlbergi*
Bembidion dauricum	*Helophorus fennicus*
Bembidion fellmanni	*Helophorus glacialis*
Bembidion hasti	*Helophorus obscurellus*
Bembidion hyperboraeum	*Arpedium brachypterum*
Bembidion lapponicum	*Boreaphilus henningianus*
Bembidion mckinleyi	*Olophrum boreale*

Some Siberian and eastern Asiatic species (*Pterostichus magus*, *Helophorus jacutus*, and *Tachinus jacuticus*) suggest a climate more continental than today. At the time of the Younger Dryas the landscape in most of Europe was comparable to an

Arctic tundra, and the average July temperature was close to 10° C.

At the close of the Late Glacial, by 9,500 BP, this very cold-adapted fauna was completely and rapidly eradicated. In the British Isles and certainly also in most of central and northern Europe, the representatives of the Arctic–Alpine fauna disappear suddenly, to be rapidly replaced by a thermophilous fauna, including various saproxylophagous species (beetles associated with mature old trees). The early Holocene sediments from the sedimentary sequence at Holywell Coombe (Coope, 1998) yielded such postglacial assemblages, with many rather thermophilous Coleoptera, such as

Clivina fossor	*Prosternon tessellatum*
Bembidion octomaculatum	*Denticollis linearis*
Bembidion biguttatum	*Melasis buprestoides*
Patrobus atrorufus	*Eucnemis capucina*
Pterostichus strenuus	*Dirhagus pygmaeus*
Pterostichus minor	*Anaglyptus mysticus*
Platynus assimilis	*Pogonochaerus hispidulus*
Licinus depressus	*Taphrorychus villifrons*
Megarthrus hemipterus	*Coenorhinus aeneovirens*
Olophrum piceum	*Rhyncolus elongatus*
Lesteva heeri	

This faunal change clearly suggests the rapid establishment of a climate marked by temperatures as high, or even higher, than those of the present day. Elsewhere in Europe, the same pattern is also recorded from a variety of sites scattered from Scandinavia (Lemdahl, 1997) to central France (Ponel and Coope, 1990), Switzerland (Lemdahl, 2000), and Italy (Ponel, 1997).

The very rich fauna from Lea Marston, analyzed by Osborne (1974) and dated to approximately 9,500 BP, is especially interesting since most of the beetles from the assemblages are rather thermophilous woodland species (such as *Calosoma inquisitor* and *Xylodrepa quadripunctata*, predators on tree-feeding caterpillars). This fauna is compatible with a mixed oak forest and therefore suggests a warm climate with summer temperatures on the order of 16 or 17° C, comparable with that of southern England or southern Sweden today. However, pollen evidence shows that the landscape was dominated at that time by willow and birch woodland, not by thermophilous trees. This time lag between the colonization of warm-adapted beetles during a period of very rapid climate amelioration and the northwards thermophilous tree expansion is analogous to that described by Coope and Brophy (1972) at the onset of the Late Glacial interstadial and should be attributed to the interaction of various factors, such as tree dispersal rate, location of refuges, soil maturation, and competition. A similar observation was made by Coope (1998) at Holywell Coombe: The lithological boundary between clayey Late Glacial 'cold' deposits and more organic postglacial 'warm' deposits is not synchronous with the faunal change since the first occurrences of thermophilous beetles are recorded earlier—that is, below the lithological boundary. Again, beetles seem to respond more rapidly than plants to the warming and appear in the area before the rise in organic sedimentation triggered by the expansion of vegetation.

Climatic Indications from Coleoptera during the Holocene

The synthesis by Mayewski *et al.* (2004) provides evidence for a highly variable climate during the Holocene and reveals at least six periods of rapid climate change at a global scale. Concerning beetles, however, as clearly stated by Coope (1986), "It is easy to demonstrate that many species of Coleoptera have had their geographical ranges drastically altered during this time, but it is much more difficult to sort out the contributions made by changes in climate from those imposed by human interference". (p. 712). The quantification of Holocene climate changes reconstructed from beetle data is therefore a relatively neglected issue despite the high number of sites investigated, especially in the British Isles. Computerized climatic reconstruction methods such as mutual climatic range have rarely been used.

Despite these limitations, several attempts to retrace Holocene climate changes from beetle records have been made in Britain (Osborne, 1976, 1997; Buckland and Wagner, 2001; Kenward, 2004). Some useful inferences may be drawn from beetles not associated with dead wood habitat since saproxylophagous taxa are more likely to respond to forest clearance by man than to climate changes. After the initial rapid rise in temperature 10,000 years ago, the following period is rather poorly known and beetle assemblages do not show any obvious evidence of significant climate fluctuations until approximately 4,830 BP. At that time, a climatic deterioration is suggested by the occurrence in the British Midlands (Kelly and Osborne, 1965) of several 'northern' Coleoptera (*Bembidion atroviolaceum*, *Sphaerites glabratus*, *Tropiphorus obtusus*, and *Barynotus squamosus*). This cool phase seems relatively short since Girling (1984) described an early Neolithic beetle assemblage (5,000 BP) suggesting temperatures slightly warmer than today. This fauna contained the thermophilous and continental *Oodes gracilis*.

Younger assemblages do not indicate thermal conditions very different from those of the present time. A marked change is recorded in Bronze Age assemblages (Osborne, 1982, 1988), with the appearance of an apparently warm fauna including several dung beetles (mainly *Onthophagus*) that today are extremely rare or even extinct in Britain. These assemblages probably suggest a thermal climate warmer than today during the Bronze Age at approximately 4,000–3,000 BP. The late Iron Age site of Fisherwick, at approximately 2,130 BP (Osborne, 1979), yielded fossils of the dung beetle *Heptaulacus testudinarius*, again a very southern species in Britain, and is interpreted as evidence for a warm climate. However, the climatic significance of these dung beetle occurrences remains a matter of debate since it is possible that pastures may have suffered from human perturbations, leading to the retraction or eradication of beetles associated with old grasslands (Buckland and Dinnin, 1993).

The vast majority of records for the historic period have been obtained from urban anthropogenic contexts such as heated buildings, allowing many thermophilous and synanthropic species to survive outside their natural range and blurring the climatic signal. Nevertheless, Kenward (2004) suggested that a careful interpretation of data concerning the nonsynanthropic elements included in insect assemblages from archeological occupation deposits may indeed provide useful paleoclimatic data.

Climatic reconstructions may be drawn from the comparison of past and modern ranges of *Heterogaster urticae*, a bug mainly dependent on *Urtica dioica*, a very common nettle present everywhere in Britain. The nettlebug is now restricted to southern England, but numerous fossil records from the 1st to the 16th century (i.e., before the Little Ice Age) in northern England suggest that in this region temperatures were probably 1 or 2° C higher than today (Kenward, 2004). The fossil occurrence in Roman sites of several other bugs (*Sehirus luctuosus*, *Syromastus rhombeus*, and *Aphanus rolandri*), and also beetles (*Airaphilus elongatus*, *Heptaulacus sus*, *H. testudinarius*, *Odacantha melanura*, *Plateumaris braccata*, and *Bagous czwalinai*), to the north of their modern range also suggests that temperatures during the 2nd through 4th centuries were 1 or 2° C higher than today (Kenward, 2004). This corresponds with the Roman climate optimum. However, Holocene distribution shifts of semiaquatic Coleoptera should be interpreted with caution in terms of climate changes since it was suggested by Andrieu-Ponel and Ponel (1999) that in the French region of Provence (and certainly elsewhere in Europe), the local extinction of fen species such as *Odacantha melanura* may have been a consequence of fen drainage and other human activities.

During the 9th through 11th centuries, there is also a wealth of data from Heteroptera and Coleoptera supporting the hypothesis of summer temperatures higher than today by 1 or 2° C (in agreement with the climate optimum around 1000 AD). In northern England, examples of this include the presence of such species as *Eurydema oleracea*, *Coreus marginatus*, *Ischnodemus sabuleti* (Heteroptera), *Odacantha melanura*, *Cryptolestes duplicatus*, *Phymatodes testaceus*, *Pediacus dermestoides*, *Anthicus antherinus*, *Anthicus bifasciatus*, *Acritus homoeopathicus*, *Apion difficile*, several *Platysthetus* species, and *Xestobium rufovillosum* (Coleoptera).

There are few beetle records dated to the later medieval and post-medieval period (Kenward, 2004), and therefore the effect of the Little Ice Age (approximately AD 1450–1880) on insect ranges is still poorly documented. A review of this topic by Buckland and Wagner (2001) emphasized the difficulty of extracting a climatic signal from beetle assemblages deposited in an environment increasingly perturbated by human activities and fragmentation of habitat, but it seems likely that some species were indeed affected by Little Ice Age cooling. These beetles became extinct in the northern parts of their range or even eradicated from Britain. This may be what happened to the beetles *Airaphilus elongatus* (Osborne, 1997), *Gyrinus colymbus* (Girling, 1984), *Aphodius quadrimaculatus* (Osborne, 1969), *Xestobium rufovillosum*, *Odacantha melanura*, and *Hydrophilus piceus* (Buckland and Dinnin, 1993) and the bug *Heterogaster urticae* (Kenward, 2004).

The present northward expansion of many insects may be interpreted as a consequence of a probable global warming that has taken place since the Little Ice Age. This has been particularly well documented for the Lepidoptera, but there is also evidence from many other insect orders, especially Heteroptera (Kenward, 2004). The discovery of several beetle species in southern France, well beyond their northern limit, may also be attributed to an increase in temperature. This is probably the case of the darkling beetles (Tenebrionidae) *Gonocephalum yelamosi* and *Pseudoseriscius pruinosus* and the ground beetles *Paracelia simplex* and *Syntomus fuscomaculatus* (Ponel and Jeanne, 1997).

Impact of Forest Clearance on Saproxylophagous and Grassland Coleoptera

The overwhelming majority of data concerning this issue have been obtained from British sites. The impact of forest clearance on Urwald insects (associated with old primeval forest) from the mid-Holocene onwards was perceived early in the history of Quaternary entomology, but more recent detailed studies have provided a very accurate picture of the profound changes that affected the beetle fauna in response to the progressive destruction by man of the natural oak and linden forest that dominated the British landscape from the early Holocene to approximately 5,000 BP. Of particular interest is the site of Thorne Moors, thoroughly analyzed by Buckland (1979). This Bronze Age site, dated to approximately 3,000 BP, yielded a very rich Urwald beetle assemblage. This fauna included many saproxylophagous species now extinct in Britain, such as *Rhysodes sulcatus*, *Isorhipis melasoides*, *Prostomis mandibularis*, *Mycetina cruciata*, and *Zimioma grossa*, or with an extremely limited distribution in Britain, such as *Dryophthorus corticalis* or *Eucnemis capucina*. According to Buckland and Dinnin (1993), at least 30 beetle taxa found in British Holocene deposits are now apparently extinct, and this list is likely to expand in the future with the analysis of more sites.

A similar pattern emerges from the study of continental sites, especially in France, since several records of Urwald beetles were obtained from seminatural archeological contexts dated to approximately 5,000 BP in the lacustrine village of Chalain (unpublished data). Examples include *Zimioma grossa*, *Lichenophanes varius*, *Bothrideres contractus*, and *Dryophthorus corticalis*. Unlike Britain, all these species are still present in France today; however, some of them (e.g., *Z. grossa*) are now relegated to old-growth mountain forests, especially in the Pyrénées. This is probably because the steep slopes of their last strongholds have not been disturbed by human action. Their inability to recolonize suitable habitats from such refuges may be a consequence of a low dispersal potential, but it is also possible that suitable old-growth forests are not available elsewhere in France.

Interestingly, the removal of the original forest cover did not affect only saproxylophagous beetles, but also other species not directly associated with woodland. This is probably the case for the water beetle *Agabus wasastjernae* found today in Scandinavian native pine forest, where it lives in acid pools (Buckland, 1979), and almost eradicated from Britain along with the coniferous cover.

Wholesale forest clearance during the Neolithic period not only resulted in the retraction and eradication of saproxylophagous beetles but also, as might be expected, lead to the parallel expansion of open environment and grassland beetles. This faunal change is well documented in many British sites, particularly from the Bronze Age, and beetle assemblages from that time are progressively dominated by species associated with dung, such as the representatives of the large coprophagous genera *Onthophagus* and *Aphodius* and the coprophilous rove beetles *Oxytelus* and *Platysthetus*. Many grass root feeders, such as *Adelocera murina* and *Phyllopertha horticola*, are also found in significant numbers. Such assemblages dominated by dung and grassland beetles suggest that grazing animals were abundant, in a predominantly open environment. Most of the studied sites show that beetle assemblages are almost totally devoid of tree-dependent species at that time and suggest an increasing representation of open environments in the British landscape, but it is probable that remnants of forests and extensive grasslands coexisted until the Late Bronze Age (Osborne, 1978). During the Roman period, the landscape in the British Isles was probably very similar to the present. In southern France, the analysis of beetle assemblages sampled near the Rhône valley suggests also that the landscape was already completely open 2,000 years ago and has remained so until now (Andrieu-Ponel *et al.*, 2000).

Other Consequences of Human Action

As indicated previously, there is increasing evidence concerning the role of man, rather than that of climate, on the retraction of the range of hygrophilous or semiaquatic beetle species. This phenomenon has been described from the ancient marsh of Tourves (southern France) in a Holocene sequence spanning the Neolithic transition. At approximately 4,310 BP, a marked change is recorded by the beetles as a rich marshy fauna is suddenly replaced by an impoverished fauna in which running water elements suggest a change in the hydrological regime; *Odacantha melanura* and *Cerapheles lateplagiatus* disappear from the assemblages. Both species are now restricted to Camargue, the last extensive marshy area in Provence, approximately 100 km westward (Andrieu-Ponel and Ponel, 1999). It is suggested that hydrologic changes were induced by drainage associated with the intensification of cultivation practices in a region densely populated by Neolithic people.

The disappearance of riffle beetles (Elmidae) from rivers in Britain also seems to be attributed to forest

Figure 1 A beetle assemblage containing several synanthropic species from a Medieval archeological site near the Vieux-Port of Marseille, France.

Figure 2 Large-diameter corer (ϕ 20 cm) used for sampling at La Restanque, French Pyrénées.

clearance. These running water beetles, adapted to highly oxygenated rivers, are found only on the gravelly bottom of streams, where they hang on the lower part of shingle by means of their oversized claws. It is likely that a secondary impact of the removal of the forest cover was an increase in soil erosion leading to a rise of alluviation processes in lowland areas, making the biotope unsuitable for Elmidae (Smith, 2000).

A further consequence of the development of agriculture is the range extension of opportunist beetle species now abundant in cultivated areas in Britain but unknown before Roman times. This is the case of the ground beetles *Pterostichus madidus* and *Zabrus tenebrioides*, but also of thermophilous species now very rarely found in the wild but usually associated with human habitations and livestock shelters, such as *Aglenus brunneus*, *Tenebroides mauritanicus*, *Oryzaephilus surinamensis*, and many others. The case of the dissemination of several species such as the grain weevil *Sitophilus granarius* is especially interesting since they are strictly associated with stored food products and are unable to live outside human habitations (**Fig. 1**). Their geographic origin remains obscure and it is likely that some of them originate from tropical regions, but this issue is still a matter of debate (Buckland, 1981).

Holocene History of the Beetle Fauna of Mediterranean Mountains

During the past 15 years, particular attention has been paid to the Holocene history of insect faunas in southern France and especially in the Mediterranean mountains (southern Alps and eastern Pyrénées). Coring of long continuous Holocene sequences was performed with either multiple coring techniques or with large-volume corers (**Fig. 2**), allowing sufficient quantities of sediment to be obtained.

The key site of La Restanque, located 1,620 m above sea level in the eastern part of the Pyrénées, provides an excellent example of climatic and paleoenvironmental reconstructions based on a succession of Late Glacial–Holocene insect assemblages. A 5-cm-resolution analysis of this 4-m profile yielded one of the most detailed analyses of the past 13,000 years. This record (Ponel *et al.*, 2004) provides original data concerning climatic and environmental changes during the Late Glacial–Holocene transition. This history documents changes in insect faunas, in conjunction with vegetational changes, and human impact on the environment (**Fig. 3**).

The lower part of the La Restanque insect sequence is characterized by an abundance of cold-adapted taxa, particularly of *Helophorus glacialis*, a beetle restricted to the margin of melting snow patches. This species lives at sea level in northern Scandinavia but well above 2,000 m in southern Europe. It is a good climate indicator since it requires mean July temperatures below 10° C to complete its life cycle. The dominance of this cold-adapted insect fauna clearly indicates that no forest cover was present at this time. This part of the sequence may be attributed to the Late Glacial interstadial since it is likely that at such high altitudes the Late Glacial warming did not reach a sufficient level to allow warm-adapted insects to survive there, but it allowed cold-adapted taxa to settle soon after the Würmian deglaciation. Comparison with pollen analyses carried out on the same sequence confirms this hypothesis.

The perceptible rise in cold-adapted beetles in the next faunal unit suggests that this unit corresponds to the Younger Dryas. Immediately after this cold interval, at the onset of the Holocene, the disappearance of cold-adapted beetle species was dramatic, signaling the

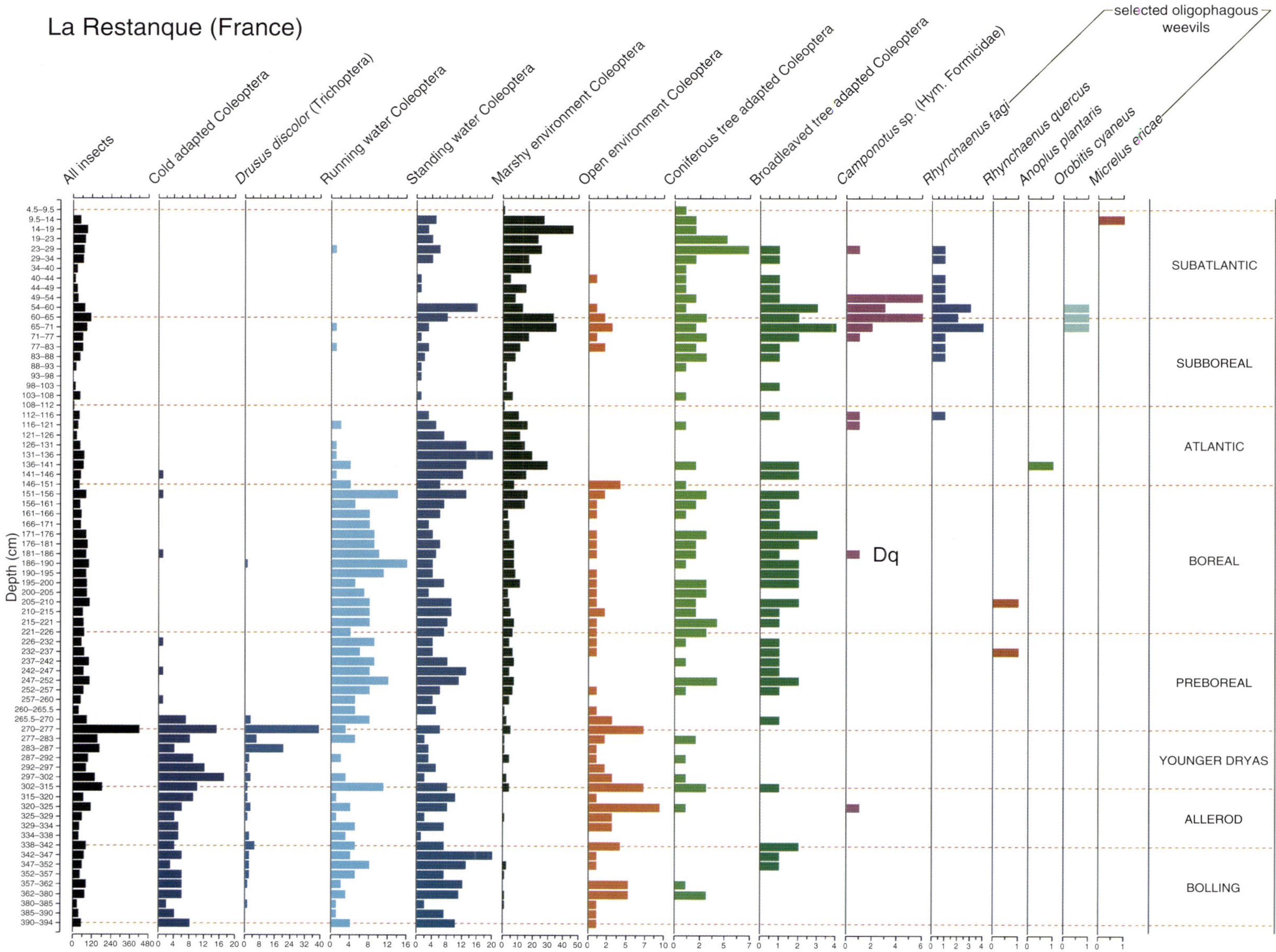

Figure 3 Succession of insect assemblages at La Restanque, French Pyrénées (1,620 m), compared with pollen chronozones. The diagram shows the minimum number of individuals. Dq, isolated occurrence of the thermophilous ant *Dolichoderus quadripunctatus*. Preferred host plants of a selection of monophagous or oligophagous weevils: *Rhynchaenus fagi* associated with beech (*Fagus sylvatica*), *Rhynchaenus quercus* associated with oak (*Quercus*), *Anoplus plantaris* associated with birch (*Betula*), *Orobitis cyaneus* associated with violet (*Viola*), *Micrelus ericae* associated with heather (*Calluna vulgaris* and *Erica tetralix*).

beginning of Holocene warming. This event is synchronous with a sharp peak of *Drusus discolor*, a caddisfly (Trichoptera) that lives today in cold and rapid mountain streams. This caddisfly is especially interesting since it suggests a short period of high-energy streams, certainly linked with large-scale snow melting induced by climatic warming. In its turn, this taxa disappears quickly, suggesting rapid environmental change, the disappearance of most of the snow cover, and a concurrent decrease in stream velocities.

This initial warming phase was followed by a period characterized by regular occurrences of arboreal beetles. Beetles associated with both broadleaved and coniferous trees were present, suggesting that forest cover was established in the vicinity of the site. This period corresponds with the boreal chronozone (9,000–8,000 BP). Surprisingly few forest-dependent terrestrial beetles (associated with shaded environment) were identified, suggesting that the forest cover was not dense. At this period, temperatures much higher than today are indicated by several taxa, including the arboreal ant *Dolichoderus quadripunctatus*. This is a predominantly lowland species that is rather thermophilous and unable to live today at such high altitude.

The next unit, between 8,000 and 4,700 BP, is marked by the disappearance of running water beetles and by the rise of standing water and hygrophilous marshy species. This period corresponds with the Atlantic chronozone. It is possible that these events could be related to changes in local hydrology, such as a shift from a fluvial regime to a lacustrine environment, associated with the infilling of the depression.

During the the subboreal (4,700–2,700 BP) and the Subatlantic (2,700 BP onward), the human impact on the fauna and the forest cover is suggested by a complex sequence of events, including a decrease in conifer-dependent beetles, a reappearance of open-environment beetles, and a peak of the carpenter ant, *Camponotus*. Most of the representatives of this genus are dependent on dead wood since they dig galleries in old stumps and trunks to establish their nests. This peak of *Camponotus* is likely attributed to the increased availability of dead wood after forest clearance (fires and tree felling) and not to climatic influence since no cold-adapted taxa are recorded at that time. Simultaneously, the beech forest became established nearby, as attested by the occurrence of the weevil *Rhynchaenus fagi*, which feeds exclusively on beech. The presence of another weevil, *Orobitis cyanea*, which feeds exclusively on *Viola* in shaded underwoods, suggests that a true forest environment prevailed at this period. At that time, it appears that the Restanque site was situated in an ecotone, at the boundary between a decimated conifer forest and an expanding beech forest. The uppermost samples of the sequence, corresponding to the end of the Subatlantic interval, are marked by an extremely impoverished insect fauna, partly for taphonomic reasons but certainly also because heathlands expanded, at the expense of forest. This is suggested by the isolated occurrence of the beetle *Micrelus ericae*, a weevil dependent on heaths (Ericaceae).

Despite some possible events of local significance, it is likely that the broad pattern of faunal successions described at La Restanque is valid at a regional scale and accurately reflects the progression of beetle assemblages and environments at high-altitude regions in the eastern part of the Pyrénées.

Conclusion

Beetles are recognized as excellent biological markers of ancient environments and they display a great sensitivity to climate change. The analysis of beetle fauna successions during the postglacial period in Europe provides ample evidence for climate and environmental changes at the onset of the Holocene and independent proxy data concerning the disruption of landscapes and vegetation under human activity (forest clearance, habitat fragmentation, agriculture, and pastoralism) from the Neolithic period onwards. However, a detailed reconstruction of subtle Holocene climate evolution from beetle data is still not available, precisely because human action and climate-induced faunal changes during this period are extremely difficult to disentangle. A special effort devoted to this problem should be made in the future.

See also: **Beetle Records**: Late Pleistocene of Europe; Postglacial North America.

References

Andrieu-Ponel, V., and Ponel, P. (1999). Human impact on Mediterranean wetland coleoptera: An historical perspective at Tourves (Var, France). *Biodiversity & Conservation* 8, 391–407.

Andrieu-Ponel, V., Ponel, P., Bruneton, H., and Leveau, P. (2000). Palaeoenvironments and cultural landscape of the last 2000 years reconstructed from pollen and coleopteran records in the Lower Rhône Valley, southern France. *The Holocene* 10, 341–355.

Atkinson, T. C., Briffa, K. R., and Coope, G. R. (1987). Seasonal temperatures in Britain during the past 22,000 years, reconstructed using beetles remains. *Nature* **325**, 587–592.

Buckland, P. C. (1979). *Thorne Moors: A Palaeoecological Study of a Bronze Age Site (a Contribution to the History of the British Insect Fauna)*, Publication No. 8. Department of

Geography Occasional, University of Birmingham, Birmingham, UK.
Buckland, P. C. (1981). The early dispersal of insect pests of stored products as indicated by archaeological records. *Journal of Stored Product Research* **17**, 1–12.
Buckland, P. C., and Dinnin, M. J. (1993). Holocene woodlands: The fossil insect evidence. In *Dead Wood Matters: The Ecology and Conservation of Saproxylic Invertebrates in Britain* (K. Kirby and C. M. Drake, Eds.), English Nature Science No. 7, pp. 6–20. English Nature, Peterborough, UK.
Buckland, P. C., and Wagner, P. E. (2001). Is there an insect signal from the 'Little Ice Age'? *Climatic Change* **48**, 137–149.
Coope, G. R. (1986). Coleoptera analysis. In *Handbook of Holocene Palaeoecology and Palaeohydrology* (B. E. Berglund, Ed.), pp. 703–713. Wiley, Chichester, UK.
Coope, G. R. (1994). The response of insect faunas to glacial–interglacial climatic fluctuations. *Philosophical Transactions of the Royal Society of London B* **344**, 19–26.
Coope, G. R. (1998). Insects. In *Late Quaternary Environment Change in Northwest Europe: Excavations at Holywell Coombe, Southeast England* (R. C. Preece and D. R. Bridgland, Eds.), pp. 213–233. Chapman & Hall, London.
Coope, G. R., and Brophy, J. A. (1972). Late Glacial environmental changes indicated by a coleopteran succession from North Wales. *Boreas* **1**, 97–142.
Girling, M. A. (1984). A Little Ice Age extinction of a water beetle from Britain. *Boreas* **13**, 1–4.
Kelly, M., and Osborne, P. J. (1965). Two faunas and floras from the alluvium at Shustoke, Warwickshire. *Proceedings of the Linnean Society of London* **176**, 37–65.
Kenward, H. (2004). Do insect remains from historic-period archaeological occupation sites track climate change in Northern England? *Environmental Archaeology* **9**, 47–59.
Lemdahl, G. (1997). Late Weichselian and early Holocene colonisation of beetle faunas in S. Sweden. *Quaternary Proceedings* **5**, 153–164.
Lemdahl, G. (2000). Lateglacial and early Holocene insect assemblages from sites at different altitudes in the Swiss Alps—Implications on climate and environment. *Palaeogeography, Palaeoclimatology, Palaeoecology* **159**, 293–312.
Mayewski, P. A., Rohling, E. E., Stager, J. C., *et al.* (2004). Holocene climate variability. *Quaternary Research* **62**, 243–255.
Osborne, P. J. (1969). An insect fauna of late Bronze Age date from Wilsford, Wiltshire. *Journal of Animal Ecology* **38**, 555–566.
Osborne, P. J. (1974). An insect assemblage of Early Flandrian age from Lea Marston, Warwickshire and its bearing on the contemporary climate and ecology. *Quaternary Research* **4**, 471–486.
Osborne, P. J. (1976). Evidence from the insects of climatic variations during the Flandrian period, a preliminary note. *World Archaeology* **8**, 150–158.
Osborne, P. J. (1978). Insect evidence for the effect of man on the lowland landscape. In *The Effects of Man on the Landscape: The Lowland Zone* (S. Limbrey and J. G. Evans, Eds.), Research Report No. 21, pp. 32–34. Council for British Archaeology, London.
Osborne, P. J. (1979). Insect remains. In *Fisherwick: The Reconstruction of an Iron Age Landscape* (C. Smith, Ed.), No. 61, pp. 189–193, 85–87. British Archaeological Reports, Oxford.
Osborne, P. J. (1982). Some British later prehistoric insect faunas and their climatic implications. In *Climatic Change in Later Prehistory* (A. Harding, Ed.), pp. 68–74. Edinburgh University PressEdinburgh, UK.
Osborne, P. J. (1988). A Late Bronze Age insect fauna from the River Avon, Warwickshire, England: Its implications for the terrestrial and fluvial environment and for climate. *Journal of Archaeological Science* **15**, 715–727.
Osborne, P. J. (1997). Insects, man and climate in the British Holocene. *Quaternary Proceedings* **5**, 193–198.
Ponel, P. (1997). Late Pleistocene coleopteran fossil assemblages in high altitude sites: A case study from Prato Spilla (Northern Italy). *Quaternary Proceedings* **5**, 207–218.
Ponel, P., and Coope, G. R. (1990). Lateglacial and Early Flandrian Coleoptera from La Taphanel, Massif Central, France: Climatic and ecological implications. *Journal of Quaternary Science* **5**, 235–249.
Ponel, P., and Jeanne, C. (1997). Un carabique nouveau pour la faune française: *Syntomus fuscomaculatus* (Motschoulsky, 1844) dans les Bouches-du-Rhône [Coleoptera Lebiidae]. *Nouvelle Revue d'Entomologie* **14**, 353–357.
Ponel, P., Andrieu-Ponel, V., Reille, M., and Guiter, F. (2004). A cross-comparison of insect and pollen data of the last 13 ka in the French Pyrénées, XI International Palynological Congress, Granada (Spain) pp. 4–9, July 2004. *Polen* **14**, 234.
Smith, D. N. (2000). Disappearance of elmid 'riffles beetles' from lowland river systems—The impact of alluviation. In *People as an Agent of Environmental Change* (R. A. Nicholson and T. P. O'Connor, Eds.), pp. 75–80. Oxbow Books, Oxford.

Postglacial North America

S A Elias, Royal Holloway, University of London, Surrey, UK

© 2007 Elsevier B.V. All rights reserved.

Introduction

The exoskeletal remains of beetles (Coleoptera) preserve well in waterlogged sediments of Pleistocene and Holocene age. Although the transition from the last glaciation to postglacial environments was a time-transgressive phenomenon, in most regions of North America this climatic transition took place between 14,000 and 11,550 years ago (12,000–10,000 ^{14}C yr BP). Most of the North American fossil beetle assemblages greater than 11,500 years in age are discussed in the article entitled 'Late Pleistocene Beetle Records from North America.' In this article, I discuss the late glacial fossil beetle record of just the Maritime region of Canada, as well as Holocene records from across the continent.

In recent decades, studies of postglacial fossil beetle assemblages have been carried out in many regions of North America. **Table 1** lists 24 postglacial faunal sites from eastern and central North America, and **Table 2** lists 20 postglacial sites from western North America. This article reports on the major findings of these studies, in terms of paleoenvironmental reconstructions and shifting distribution patterns (biogeography) of the species found in the fossil assemblages, all of which remain extant today. Some of the fossil assemblages discussed here formed part

Table 1 Summary of fossil beetle assemblage data from sites in central and eastern North America

Site	Age (cal yr BP × 1000)	Holocene		Modern		Change in Temperature		References[1]
		TMAX (°C)	TMIN (°C)	TMAX (°C)	TMIN (°C)	July ΔT	January ΔT	
1) Benacadie, Nova Scotia	13.73	9–13	−31 to −14	18.1	−5.5	−9.1 to −5.1	−25.5 to −8.5	Miller and Elias, 2000
2) Collins Pond, Nova Scotia/Lower	13.73	10–12.5	−20.5 to −11	16.3	−3.9	−6.3 to −3.8	−16.6 to −7.1	Miller and Elias, 2000
3) Todd Mountain, New Brunswick	13.42	10–13	−30.5 to −18	19.1	−9.1	−9.1 to −6.1	−21.4 to −8.9	Miller and Elias, 2000
4) Campbell, Nova Scotia	13.16	11.5–14.5	−27.5 to −8	18.6	−8	−7.1 to −4.1	−19.5 to 0	Miller and Elias, 2000
2) Collins Pond, Nova Scotia/Upper	12.82	11–14	−21.5 to −7.5	16.3	−3.9	−5.3 to −2.3	−17.4 to −3.6	Miller and Elias, 2000
5) Amaguadees, Nova Scotia	12.73	14.5–15.5	−28 to −23.5	18.1	−5.5	−3.6 to −2.6	−22.5 to −18	Miller and Elias, 2000
6) West Mabou, Nova Scotia	12.53	11–14	−18 to −5	18.6	−8	−7.6 to −4.6	−10 to +3	Miller and Elias, 2000
7) Rainy River, Ontario/Site E-13	12.24–11.75	13–15	−27 to −22	18.8	−16.2	−5.8 to −3.8	−10.8 to −5.8	Bajc *et al.* 2000
8) Wales, Ontario	12.1	N/A	N/A	20.5	−8.1	N/A	N/A	Miller *et al.* 1987
9) Rainy River, Ontario/Site RR54	11.68	14.5–15.5	−27 to −23.5	18.8	−16.2	−4.3 to −3.3	−10.8 to −7.3	Bajc *et al.* 2000
10) Mosbeck, Minnesota/Unit E	11.50	15–20	−21 to −5	20.2	−16.9	−5.2 to −0.2	−4.1 to +11.9	Ashworth *et al.* 1972
11) Lockport Gulf, New York/L03	11.45	17–20.5	−15.5 to −3.5	21.5	−4.8	−4.5 to −1	−10.7 to +0.3	Miller and Morgan, 1982
12) Seibold, North Dakota	11.08	17–21.5	−16 to −5	20.5	−14.9	−3.5 to +1	−0.6 to +9.9	Ashworth and Brophy, 1972
10) Mosbeck, Minnesota Unit D	10.91	16.5–20	−18 to −6.5	20.2	−16.9	−3.7 to −0.2	−1.1 to +10.4	Ashworth *et al.* 1972
11) Lockport Gulk, New York/LO4	10.87	15–22	−22.5 to −1	21.5	−4.8	−6.5 to +0.5	−17.7 to +4.3	Miller and Morgan, 1982
11) Lockport Gulk, NY/LO5	10.36	18.5–22	−19.5 to −10.5	21.5	−4.8	−3 to +0.5	−14.2 to −5.2	Miller and Morgan, 1982
10) Mosbeck, Minnesota Unit C	9.44	15–21	−24 to −1.5	20.2	−16.9	−5.2 to +0.8	−7.1 to +15.4	Ashworth *et al.* 1972
13) Saint Flavien, Quebec Unit 1	7.83	16.75–20	−12.25 to −7	18.7	−13	−1.95 to+1.3	+0.75 to +6	Lavoie *et al.* 1997a
14) Ennadai Lake, NWT Unit D	7.2–6.86	10–14.5	−28.5 to −23	13	−32	−3 to +1.5	+3.5 to +9	Elias, 1982a
14) Ennadai Lake, NWT Unit C	6.86–2.92	12–15	−25 to −19	13	−32	−1 to +2	+7 to +13	Elias, 1982
15) Boniface River, Quebec	6.66–1.91	11.8–14.5	N/A	9.4	−24.8	+2.4 to +5.1	N/A	Lavoie *et al.* 1997
16) Radisson, Quebec	5.84–0	14.5–17.5	N/A	13.4	−3.4	+1.1 to +4.1	N//A	Lavoie, 2001
17) Charlevoix, Quebec	5.25–0	N/A	N/A	19.2	−12.8	N/A	N/A	Lavoie, 2001
18) Au Sable River, Michigan	4.66–4.55	17–18	−14 to −12	20	−5	−3 to −2	−9 to −7	Morgan *et al.* 1985
13) St Flavien, Quebec Unit 2	4.49–3.55	15–20	−14 to −5	18.7	−13	−3.7 to +1.3	−1 to +9	Lavoie *et al.* 1997
19) Bongards, Minnesota	3.55–0	N/A	N/A	23.1	−11.2	N/A	N/A	Schwert and Ashworth, 1985
20) Roberts Creek, Iowa	3.19–0	N/A	N/A	22.4	−8.9	N/A	N/A	Schwert, 1996
13) St Flavien, Quebec/Unit 4	3.19–0	16–20	−23.5 to −5	18.7	−13	−2.7 to +1.3	−10.5 to +8	Lavoie *et al.* 1997
21) Umiakoviarusek, Labrador	2.21–0.89	12–13.5	−23 to −12	10.7	−18.5	+1.3 to +2.8	−5.5 to +6.5	Elias, 1982
22) Cahokia, Illinois	AD 1050	N/A	N/A	26.6	−1.5	N/A	N/A	Pauketat *et al.* 2002
23) Boston, Massachusetts	17th century	N/A	N/A	23.1	−1.9	N/A	N/A	Bain, 1998
24) Quebec City, Quebec	19th century	N/A	N/A	19.2	−12.8	N/A	N/A	Bain, 2001

[1]References available from the 'QBib' web site: http://www.bugs2000.org/qbib.html

Table 2 Summary of fossil beetle assemblage data from sites in western North America

Site	Age (cal yr BP×1000)	Holocene		Modern		Change in Temperature		References[1]
		TMAX (°C)	TMIN (°C)	TMAX (°C)	TMIN (°C)	July ΔT	January ΔT	
1) False Cougar Cave, Montana	11.59	15.5–17.75	−4.5 to +0.25	13.5	−10.1	+2 to +4.25	+5.6 to +10.4	Elias, 1996
2) Huntington Canyon, Utah	11.46	15–17	−6 to 0	16	−9.1	−1 to +1	+3.1 to +9.1	Elias, 1996
3) La Poudre Pass, Colorado	11.39	15–18	−17.5 to −7	11.3	−7.7	+3.7 to +6.7	−9.8 to +0.7	Elias, 1996
4) Lake Isabelle Delta, Colorado	10.12	11.75–14.5	−31.25 to −15	10.8	−8.2	+1 to +3.7	−23 to −6.8	Elias, 1996
3) La Poudre Pass, Colorado	9.89	13.5–16.5	−19.5 to −9	11.3	−7.7	+2.2 to +5.2	−13.6 to −1.3	Elias, 1996
4) Lake Isabelle Delta, Colorado	9.48	10.5–13	−23.5 to −16	10.8	−8.2	−0.3 to +2.2	−15.3 to −7.8	Elias, 1996
4) Lake Isabelle Delta, Colorado	8.70	11–13	−14 to −9	10.8	−8.2	+0.2 to +2.2	−5.8 to −0.8	Elias, 1996
4) Lake Isabelle Fen, Colorado	7.90	10.25–13	−14.75 to −7.5	10.8	−8.2	−0.6 to +2.2	−6.6 to +0.7	Elias, 1996
3) La Poudre Pass, Colorado	6.14	12.5–13.5	−12.5 to −11.5	10.8	−8.2	+1.7 to +2.7	−4.3 to −3.3	Elias, 1996
3) La Poudre Pass, Colorado	3.79	11.75–15	−21.25 to −14.5	11.3	−7.7	+0.4 to +3.7	−11.8 to −6.8	Elias, 1996
4) Lake Isabelle Fen, Colorado	3.17	10.25–13	−14.5 to −7.5	10.8	−8.2	−0.6 to +2.2	−6.3 to +0.7	Elias, 1996
5) Longs Peak Inn, Colorado	3.14	12–15.5	−26.5 to −14	14.5	−5	−2.5 to +1	−21.5 to −9	Elias, 1996
5) Longs Peak Inn, Colorado	2.82	13.5–15.5	−24 to −15	14.5	−5	−1 to +1	−19 to −10	Elias, 1996
6) Roaring River, Colorado	2.49	14.25–14.75	−18.25 to −16.5	14.3	−5.5	0 to +0.5	−12.75 to −11	Elias, 1996
7) Rat's Nest Cave, Alberta	1.97–0	N/A	N/A	16.2	−8.9	N/A	N/A	Bain *et al.* 1997
8) Mount Ida Bog, Colorado	0.85	10.25–12	−14.75 to −9	9.3	−11.5	+1 to +2.7	−3.25 to +2.5	Elias, 1996
5) Longs Peak Inn, Colorado	0.43	13.5–15.5	−24.5 to −15	14.5	−5	−1 to +1	−19.5 to −10	Elias, 1996

(*Continued*)

Table 2 (*Continued*)

Site	Age (cal yr BP×1000)	Holocene		Modern		Change in Temperature		References[1]
		TMAX (°C)	TMIN (°C)	TMAX (°C)	TMIN (°C)	July ΔT	January ΔT	
Chihuahuan Desert Packrat Midden Faunas								
9) Sacramento Mountains, NM	11.91–0.44	Not available						Elias and Van Devender, 1992
10) Tunnel View, BBNP, Texas	11.39–10.05	Not available						Elias and Van Devender, 1992
11) Hueco Mountains, Texas	11.18–2.60	Not available						Elias and Van Devender, 1992
12) Sierra de la Misericordia, Coahuila	10.57–0.720	Not available						Elias *et al.* 1995
13) Puerto de Ventanillas, Coahuila	9.75–.59	Not available						Elias *et al.* 1995
14) Maravillas Canyon Cave, Texas	9.22–.09	Not available						Elias and Van Devender, 1990
15) Cañon de la Fragua, Coahuila	6.52–.51	Not available						Elias *et al.* 1995
Great Basin Packrat Middens								
16) Glen Canyon region, Utah	9.69–1.42	Not available						Elias *et al.* 1992
17) Canyonlands National Park, Utah	7.82–4.24	Not available						Elias *et al.* 1992
Sonoran Desert Packrat Mideens								
18) Puerto Blanco Mountains, Arizona	12.40–0	Not available						Hall *et al.* 1990
19) Ajo Mountains, Arizona	11.47–1.09	Not available						Hall *et al.* 1989
20) Cataviña, Baja California	1.1	Not available						Clark and Sankey, 1999

[1]References available from the 'QBib' web site: http://www.bugs2000.org/qbib.html

of purely paleontological research projects. Others formed part of archeological research, as noted in the text.

The paleotemperature estimates discussed are derived from the Mutual Climatic Range (MCR) method of paleoclimate analysis, as discussed in the Overview for this section. For sites in eastern and central North America, the MCR method was used to provide estimates of mean temperatures of the warmest (TMAX) and coldest (TMIN) months of the year. TMAX and TMIN estimates are discussed in terms of departures from the modern TMAX and TMIN of the study sites in question.

Research in Eastern North America

Fossil beetle assemblages of postglacial age have been studied in eastern Canada, New England, along the Mississippi drainage in the mid-western United States, and in the Canadian Northwest Territory of Keewatin (**Fig. 1**). Many of these assemblages represent early postglacial environments (**Table 1**, sites 1–14). Only a few mid-Holocene assemblages have been described from these regions (**Table 1**, sites 15–18). Late Holocene assemblages, including three from archeological sites, span the interval 3.5 to 0.1 kya (**Table 1**, sites 19–24).

Late Pleistocene-Holocene Transition

The fossil beetle record from many parts of North America indicates a relatively smooth transition from glacial to interglacial climates, beginning about 17 kya (calibrated years ago × 1,000) (Elias *et al.*, 1996). Throughout most of North America, paleotemperature reconstructions based on beetle assemblage data do not show a climatic reversal during the late glacial interval (17–11 kya). The one prominent exception to this comes from the Maritime region of Canada. Miller and Elias (2000) reconstructed mean July temperatures (TMAX) and mean January temperatures (TMIN) from radiocarbon-dated fossil beetle assemblages from a suite of late-glacial sites in this region. These assemblages date from 14.9 kya to 12.5 kya. In the Maritime region, multiple lines of evidence, including stratigraphy, palynology, chironomid fossils, and beetles, suggest a climatic deterioration during the Younger Dryas chronozone. MCR estimates suggest several 'events' in the regional fossil beetle record that appear similar to climate events recorded in the GRIP ice core record, including the Younger Dryas cooling event from GI-1a to GS-1, beginning ca. 12,650 GRIP yr BP. The results of MCR-based temperature reconstruction suggests a drop in TMAX of about 5° C between 13 kya (at Brookside, Nova Scotia) and 12.5 kya (at West

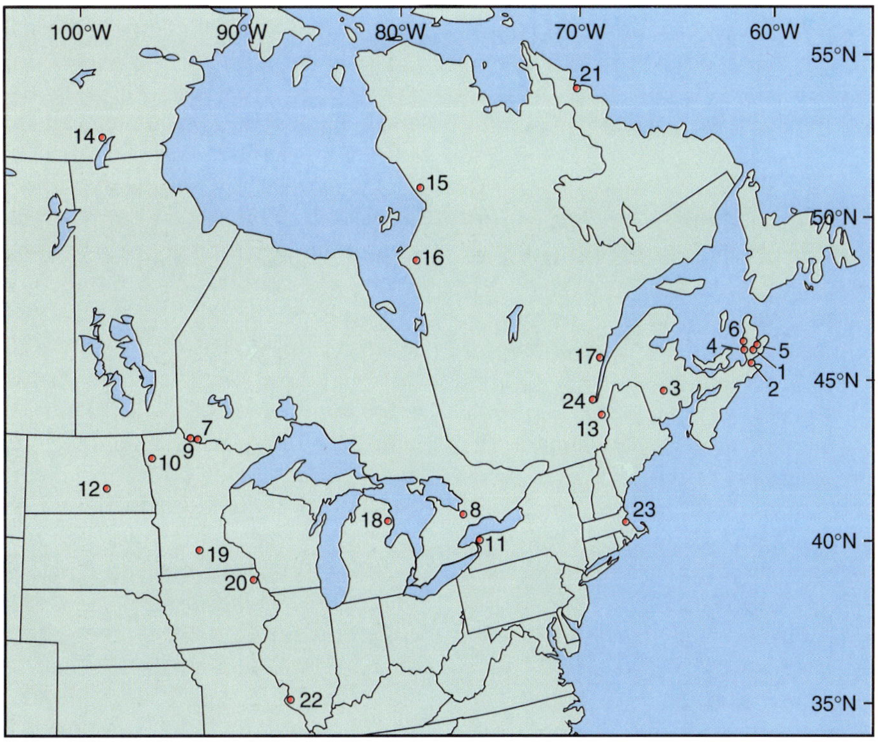

Figure 1 Map of eastern and central North America, showing location of fossil sites discussed in text. Site numbers match those in **Table 1**.

Mabou, Nova Scotia). These results compare well with the temperature reconstruction of summer surface waters based on chironomids (Levesque *et al.*, 1997; Lotter *et al.*, 1999). The fact that there is little fossil beetle evidence for a Younger Dryas cooling event elsewhere in North America lends support for the paleoceanographic model (Duplessy *et al.*, 1996) that links this cooling event with changes in circulation patterns in the North Atlantic. The only other North American region from which beetle records indicate a Younger Dryas cooling is arctic Alaska (Elias, 2000). However, few continuous series of beetle assemblages have been studied from this time interval in most regions of North America.

Most early postglacial beetle faunas from eastern and central North America are indicative of climatic regimes close to modern parameters (**Table 1**). The exceptions to this include sites that were located in close proximity to large bodies of glacial meltwater, such as Lockport Gulf, New York. Here, at 11.45 kya, TMAX was 1–4.5° C colder than today (Elias *et al.*, 1996). Climatic amelioration through the early Holocene is indicated by the mid-western beetle assemblages. Schwert and Ashworth (1988) discussed the discrepancy between early postglacial environments deduced from insect assemblages and paleobotanical reconstructions in the mid-western United States as follows. Pollen and plant macrofossils that accumulated in ice-marginal deposits just after deglaciation reflect tundra-like communities with sedges and *Dryas*, and are lacking trees and shrubs. However, the insect faunas are consistently different, being analogous to modern faunas from the middle of the boreal forest zone. At each study site, the discrepancy between the plant and insect evidence has been accounted for by invoking a substantial lag in the arrival and establishment of woody plants on deglaciated landscapes. As in central Europe, this lag may represent the time needed for ecological succession from open ground to herbaceous cover, shrub cover, and finally forest cover. The most recent paleobotanical reconstructions for the eastern and central United States (e.g., Grimm and Jacobson, 2004) suggest that such lags may have lasted for centuries, not for millennia. The paleoclimatic interpretations from the two disciplines (beetles and pollen) seem to be coming into closer agreement in recent years. For instance, a syntheses of palynological data (Grimm and Jacobson, 2004) concluded that "the climate of the Northeast was warmest and driest during the early Holocene··· from about 10,200–6,000 cal yr BP—a time of enhanced summer insolation." This conclusion agrees well with fossil beetle reconstructions.

Elias (1982), Lavoie *et al.* (1997), and Lavoie and Arsenault (2001) studied late Holocene beetle assemblages from sites near arctic tree line in northern Quebec and Labrador, Canada. Based on MCR analysis, Lavoie *et al.* (1997) found that summer temperatures in northern Quebec were 2.8–5.5° C higher than modern from 6.7–1.9 kya. Elias (1982) found evidence of climatic cooling in northeastern Labrador from 2.7–0.9 kya. However, Lavoie and Arsenault (2001) found no evidence of cooling in the James Bay region during the last 5.7 kya. This suggests either that climates varied in different parts of Quebec–Labrador during the late Holocene, or that the northern tree-line faunas were more sensitive to small-scale changes than faunas that lived nearer the center of the boreal forest.

Environmental Archeology Studies

One of the principal differences between European and North American postglacial environments has been the relatively strong human influence on the landscapes of Europe during at least the last 5,000 years, compared with the relatively weak human disturbance of many North American landscapes until at least late prehistoric times. The difficulties experienced by European researchers in disentangling the effects of human perturbation from the effects of climate change on postglacial beetle faunas (*see* Postglacial Europe) have not hindered North American workers to the same extent. The hunter-gatherer societies of the Americas did not clear forests and plant crops until the last few centuries.

Fossil beetle assemblages taken from archeological sites in eastern and central North America have documented physical environments and human living conditions at several sites. Fossil insects were studied from disposal pits at Cahokia, Illinois (Pauketat *et al.*, 2002). One of the aims of this study was to determine the timing of the disposal events represented by deposition of refuse in barrow pits. All but one of the studied deposits contained sizable quantities of insects active during the warm months between April and October. The fossil assemblages were dominated by beetles (58%), fly pupae (28%), and ants (14%). Some assemblages were indicative of rich mounds of discarded plant and animal wastes that were exposed for sufficient time, several days to several weeks, to allow insect colonization.

Bain (1998) studied beetle remains from a colonial-era archeological site in Boston, Massachusetts. The fossil assemblages were extracted from sediments that accumulated in a latrine. Bain identified 42 beetle species, including 22 European taxa, and one that was possibly of Caribbean origin. Five beetle species in the fossil assemblage normally attack wood, suggesting that woodworking took place nearby, such as a sawmill, a furniture shop, house building, or stacks of firewood. A large number of granary weevils (*Sitophilus granarius*) were found, indicating weevil infestation of stored grain or flour.

Bain's (2001) study of insects from the Îlot Hunt site in Quebec City, Canada indicates that the late nineteenth century was a time of great change in urban sanitary regulations and awareness of public health problems. Bain analyzed insect fossil samples from two latrines, a drain, and an abandoned well, representing both domestic and commercial establishments.

Research in Western North America

Postglacial sites in western North America include eight sites in the Rocky Mountain region (**Table 2**, sites 1–8) and twelve sites in the arid southwest (**Table 2**, sites 9–20). The Rocky Mountain sites represent organic deposits sampled from lake sediments, peat bogs, and one cave deposit (Rat's Nest Cave, Alberta). The southwestern sites are all woodrat middens.

Studies in the Rocky Mountains

In many ways, the Rocky Mountain region is ideally suited for Quaternary insect studies. The vertical relief of the mountain chain has provided regional biota with a wide variety of physical environments, expressed in the development of ecological communities ranging from grasslands in the eastern foothills zone through several types of montane and subalpine forest communities, to the alpine tundra at or near mountaintops. Furthermore, the ecotones, or boundaries, between these communities have undergone noticeable shifts in elevation in the late Quaternary, facilitating the study of environmental change, as a given site documents different biological communities through time.

Fossil insects from the Rocky Mountains have left us important records of past events and biotic responses. The atmospheric moisture caught by the Rockies makes this mountain chain wetter than the adjacent plains, and this in turn has led to the accumulation and preservation of organic deposits.

Rocky Mountain study sites (**Fig. 2**) span an elevational range from the foothills zone to the Alpine tundra. The general trend in regional faunas was that cold-adapted species lived at high elevations during the postglacial warm interval, but during the last glaciation, these beetles shifted their ranges downslope by 300–1,500 m (Short and Elias, 1987; Elias and Toolin, 1989).

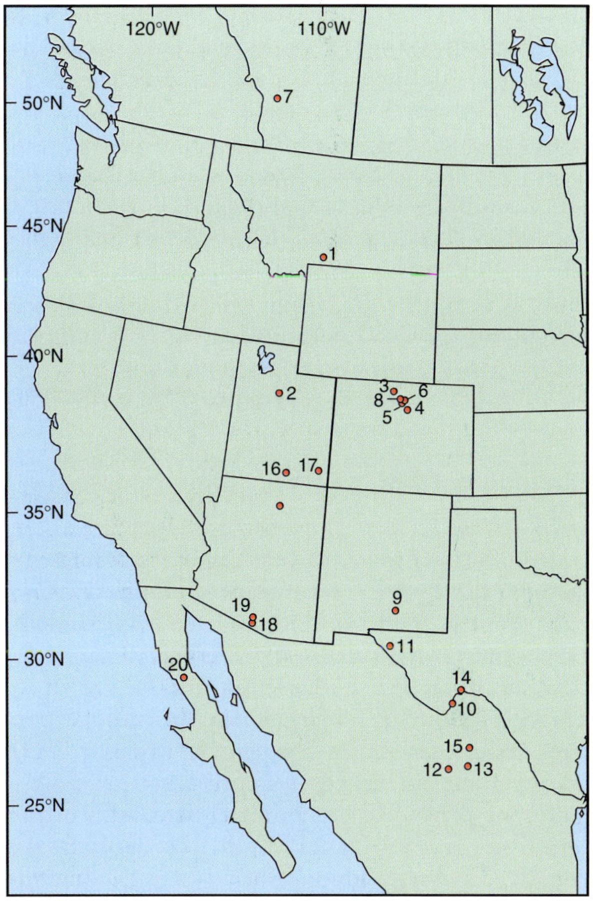

Figure 2 Map of western North America, showing location of fossil sites discussed in text. Site numbers match those in **Table 2**.

MCR analysis of fossil beetle assemblages from the Rockies indicates that following deglaciation, postglacial summer warming was rapid. This is indicated by an assemblage dating 16.2 kya, with TMAX only 2.1° C cooler than modern. By 11.6 kya, several assemblages indicate warmer than modern TMAX and TMIN values. Pollen evidence suggests that coniferous forests did not reach their modern elevations until about 10 kya, more than 1,000 years after climatic amelioration had reached modern levels, based on the beetle evidence. By that time, beetle MCR reconstructions indicate that both summer and winter temperatures were already beginning to decline from early postglacial maxima.

TMAX values remained above modern levels and TMIN values remained below modern levels until about 3.2 kya. These long-term trends were followed by a series of small-scale oscillations in the late Holocene. Pollen spectra from sites in the Colorado Front Range show substantial changes after 3.8 kya, interpreted as either a lowering of tree line, or a thinning of subalpine forests, or both (Short, 1985). Pine pollen also became less abundant in subalpine pollen spectra at this time. These changes were inferred to be a response by the vegetation to climatic cooling.

The site of Rat's Nest Cave in Alberta yielded organic detritus that accumulated at the bottom of a deep natural shaft inside the mouth of a cave (Bain *et al.*, 1997). The beetle faunas span the last 2,000 years, and their composition suggests environmental conditions similar to modern parameters at the site. This is in contrast to beetle assemblages from this interval from farther south, in the Central Rocky Mountains. The latter are indicative of a series of climatic oscillations during this time, with a relatively warm period about 1 kya, followed by a relatively cold period in the last 500 years.

Biogeography of the Rocky Mountain Beetle Fauna

Because of the persistently cold climates at high elevations, the Rockies have been a refuge for many cold-adapted species of both plants and animals, even during the warmest interglacial intervals. Species that shifted southward in front of glacial ice advances in the interior of North America had only two options for survival when the climate began warming and the ice receded at the end of the last glaciation. They had to migrate either northward or up the slopes of mountains, in search of suitably cold climate. Thus, cold-adapted beetles that lived on the plains south of Denver toward the end of the last glaciation have retreated to the Alpine tundra in the Colorado Front Range (Elias, 1986). In contrast to the Rocky Mountain fauna, cold-adapted species living in the central Great Plains region were not able to migrate north at the end of the last glaciation. At that time, climatic warming caused the extirpation (regional extinction) of much of the cold-adapted beetle fauna in this region, with small populations of Arctic and subArctic species surviving in Alpine habitats of the Cordillera and Appalachians, or in short-lived habitats associated with stagnant ice (Schwert and Ashworth, 1988).

There have been a number of regional extirpations from the Rocky Mountain beetle fauna during the last 17 kya, probably as a result of large-scale climatic change (**Fig. 3**). Six of the eight species extirpated before the Holocene have shifted their distributions to the north. These are all cold-adapted animals, either Arctic or boreo-Arctic species. The most extreme displacement of distribution is that of a rove beetle (Staphylinidae), *Holoboreaphilus nordenskioeldi*, which lives today in Arctic Canada, 3,000 km north of its late glacial locality at Marias Pass, Montana (**Fig. 2**). Two additional species made their last Rocky Mountain appearance in an assemblage from Marias Pass, dated at just over 13.3 kya. These are a ground beetle (Carabidae), *Bembidion sulcipenne hyperboroides*, and a carrion beetle (Silphidae), *Thanatophilus trituberculatus*. *B. sulcipenne hyperboroides* is a northwestern North American species, living today in northern British Columbia, the Yukon Territory, and Alaska. *T. trituberculatus* is a northern, cold-adapted carrion-feeder, which probably also feeds on other insects inhabiting carrion. Its modern range extends north to the Arctic in Canada and Alaska, but its southern limit does not reach the lower 48 states of the United States.

Two species make their last appearance in the Rocky Mountain fossil record at approximately 14.7 kya, from the Mary Jane site in Colorado. One of these is the water scavenger beetle

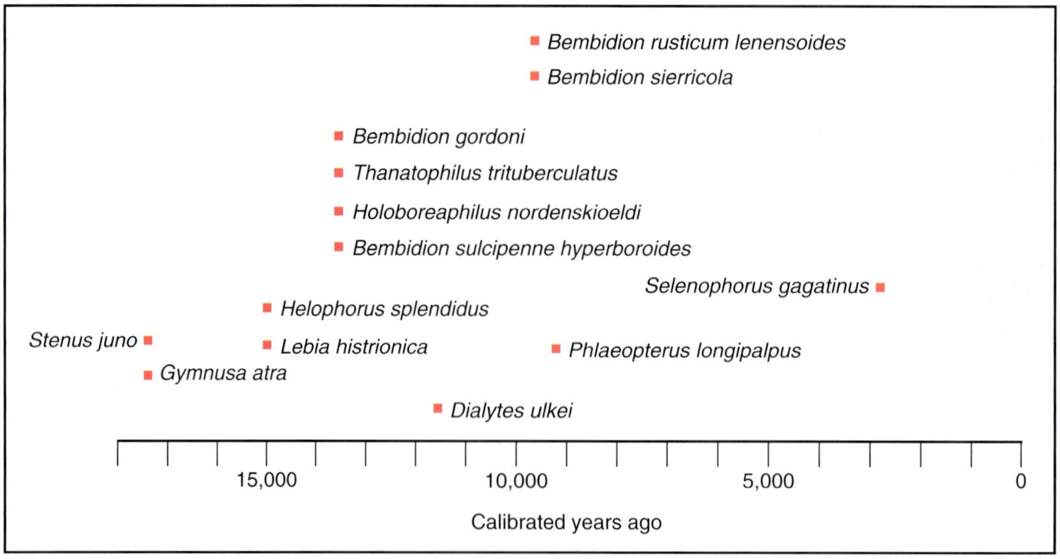

Figure 3 Last recorded fossil occurrence of various beetle species in Rocky Mountain region postglacial fossil assemblages. After Elias, 1981.

(Hydrophilidae), *Helophorus splendidus*. This beetle is found today in the northern Holarctic regions, including northeastern Siberia and the arctic coast of the Canadian Northwest Territories. It was found also in the 17.8 kya-assemblage from Lamb Spring, and is indicative of tundra conditions at this site.

The Huntington Canyon site, in the subAlpine zone of the Wasatch Mountains of Utah, yielded one species that has been regionally extirpated. The assemblage dates to 13.2 kya. *Bembidion gordoni* is a riparian ground beetle that lives on the banks of fast-running mountain streams. Today, however, its distribution is restricted to the mountains of the Pacific Northwest, from British Columbia to Oregon. Based solely on its modern distribution, a biogeographer might describe *B. gordoni* as endemic to the Pacific Northwest, but the Huntington fossil record demonstrates that it was more widespread in the recent past (i.e., 'recent' in terms of the longevity of the species).

Of the species extirpated during the Holocene, only one has shifted its distribution to the north. The riparian ground beetle, *Bembidion rusticum*, now lives only in British Columbia and northward, a distance of more than 2,600 km from Lake Emma, Colorado, where it was found as an early Holocene fossil. Lake Emma also yielded *Bembidion sierricola*, which is restricted today to the mountains of the Pacific Northwest, from California to southern British Columbia (Lindroth, 1963). The dung beetle (Scarabaeidae) *Dialytes ulkei*, was extirpated from the Rockies sometime after 9,600 yr BP. Its last occurrence in the regional fossil record was at La Poudre Pass, Colorado (**Fig. 2**). *D. ulkei* is an eastern North American species today, ranging from southern Quebec, southward in the Appalachian Mountain chain to South Carolina. Finally, the ground beetle, *Selenophorus gagatinus*, was last recorded from Colorado in a 2,400 yr BP assemblage from the Roaring River, in Rocky Mountain National Park (**Fig. 2**). Today *S. gagatinus* is an eastern North American species, ranging as far west as Indiana. In the late Quaternary interval, it was widespread in western North America, as it was also found in ancient packrat middens from the northern Chihuahuan Desert (**Fig. 4**), in assemblages ranging from 24 to 4.7 kya (Elias and Van Devender, 1990).

Based on the fossil evidence discussed above, it is clear that the current biological communities of the Rocky Mountain region developed only since the end of the last glaciation. They are the product of changing Quaternary environments. Changes in distribution have been large-scale and rapid. These changes demonstrate the great mobility of many beetle species, and their ability to move great distances in order to satisfy their ecological requirements in the face of changing climates. These changes also suggest that the current insect communities of the Rocky Mountain region are simply the latest reshuffling of species, and that species composition is probably in a more or less continuous state of flux (Elias, 1991).

Since the 1980s, insect fossils from packrat middens have received intensive study. Fossil insect research has been done in the Sonoran Desert and the Great Basin, but the most intensively studied region (191 midden insect fossil samples from 27 sites) is the Chihuahuan Desert (Elias and Van Devender, 1990, 1992; Elias *et al.*, 1995). The fossil insect assemblages from the southern Chihuahuan Desert in Mexico contain mixtures of desert- and temperate-zone species (**Fig. 4**) in almost every interval from the late glacial through to the late Holocene. Midden assemblages from locations farther north in the Chihuahuan Desert are generally separated into glacial-age faunas with temperate-zone affinities and postglacial faunas with desert-zone affinities. The 'no modern analog' faunal assemblages indicate that the late Quaternary environments in this region were unlike any that exist today. After 14 kya, most temperate-zone beetle species in the central and northern regions of the Chihuahuan Desert were replaced by either desert species or by more cosmopolitan taxa. The faunal change suggests a climatic shift from cool, moist conditions of Wisconsin glacial times to hotter, drier conditions in the early Holocene.

In the northern Chihuahuan Desert, woodland environments persisted until 13 kya, but the insect data suggest considerable open ground, with grasses at least locally important at the midden sites. The transition from the temperate (glacial) fauna to the more xeric postglacial fauna started by 14.7 kya. The timing of this faunal change was essentially synchronous throughout the northern Chihuahuan Desert. A major difference between the Big Bend and northern Chihuahuan Desert scenarios is the nature of this faunal change. In the Big Bend region, the transition was characterized by the disappearance of nearly all temperate species at about 14 kya. However, the xeric-adapted fauna did not appear in the Big Bend records until about 8.4 kya. In the northern Chihuahuan Desert assemblages, the xeric species first appeared at 14.7 kya, and several of the temperate grassland species that had lived there during the last glaciation persisted well into the Holocene. This mixture of xeric and temperate elements makes ecological sense, because these northern faunas were living close to the edge of the Chihuahuan Desert. The gradual shifting of northern desert boundaries in the

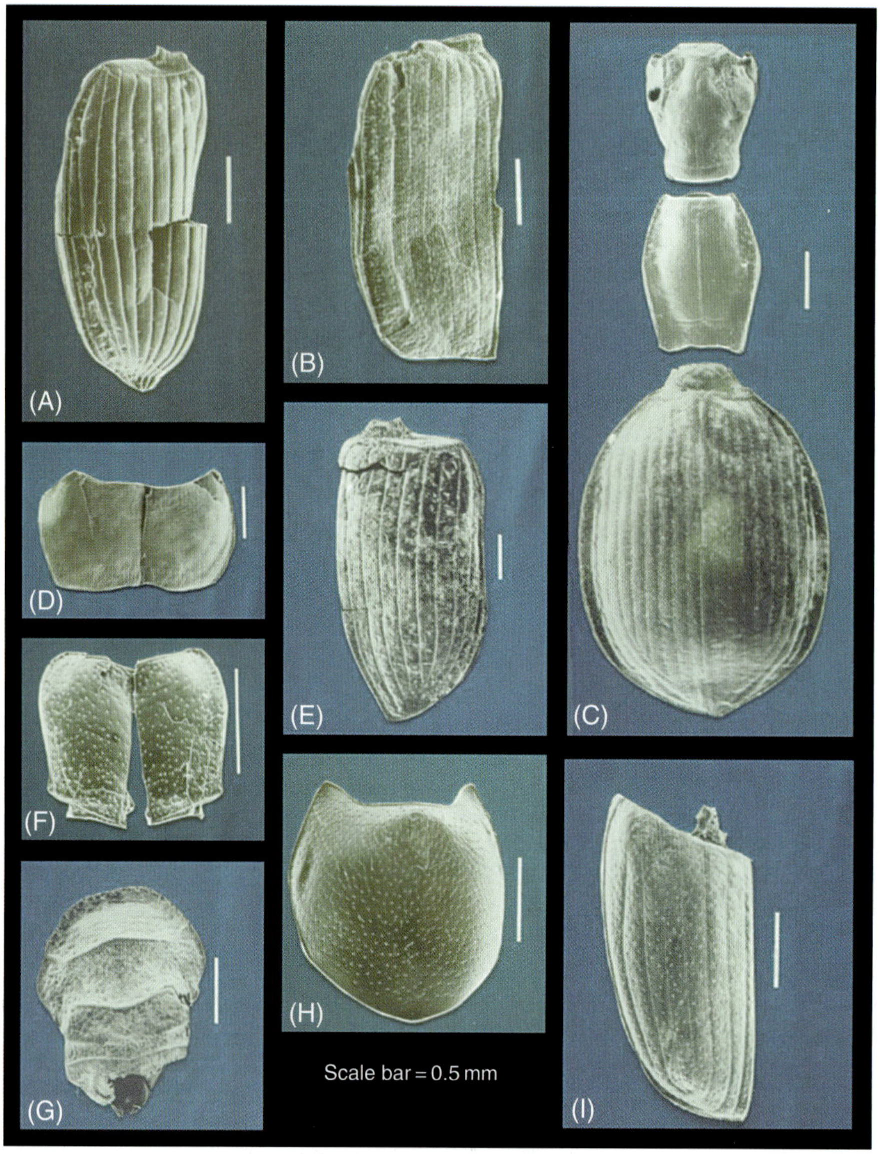

Figure 4 Scanning electron micrographs of fossil beetles from Chihuahuan Desert packrat midden assemblages. (A) left elytron of the ground beetle *Selenophorus gagatinus*; (B) left elytron of the ground beetle *Lebia lecontei*; (C) head, pronotum, and elytra of the ground beetle *Agonum (Rhadine) longicolle*; (D) pronotum of the ground beetle *Harpalus herbivagus*; (E) right elytron of the ground beetle *Piosoma setosum*; (F) pronotum of the ground beetle *Cymindis* cf. *borealis*; (G) head capsule of the dung beetle *Onthophagus brevifrons*; (H) pronotum of the dung beetle *Onthophagus cavernicollis*; (I) left elytron of the dung beetle *Onthophagus brevifrons*.

Holocene probably created many marginal habitats for temperate species. By about 8.4 kya, the dominance of xeric species indicates establishment of desert environments, including desert grasslands, throughout the Chihuahuan Desert region. After 2.6 kya, the last of the temperate species was replaced by species associated with desert-scrub communities.

Work on late Quaternary insect faunas from the Colorado Plateau region (Elias *et al.*, 1992) suggests that late Wisconsin climatic conditions were cooler and moister than present, and that the plateau supported a mosaic of grassland and shrub communities without modern analog. The temporal and spatial coverage of midden sites is relatively poor for this enormous region, especially in comparison with the work done on middens from the Chihuahuan Desert. However, it is possible to say that the transition from glacial to postglacial climatic regimes took place after 17 kya. Paleobotanical data suggest that major regional warming was well underway by 13 kya (Cole, 1990). The fossil insect data indicate that the most arid conditions in the Great Basin region occurred within the last 1,500 years.

Insect faunas have also been studied from several sites in the Sonoran Desert (Hall *et al.*, 1989, 1990; Clark and Sankey, 1999). Unlike the Chihuahuan

Desert insect fauna, the Sonoran Desert fauna indicates little change from the late Pleistocene through to modern times. All taxa found in the Sonoran middens probably live within a few kilometers of the midden sites today. The Sonoran arthropod fauna showed a marked increase in diversity during the late Holocene, as more subtropical plants and warmer climates became established about 4.5 kya. Many warm-adapted beetles probably dispersed into southern Arizona from Sonora, Mexico, during the last 4 kyr. The increase in diversity in fossil arthropod assemblages has been correlated with the frequency of winter freezes, and secondarily with the amount of summer precipitation. This late Holocene peak in species richness is in sharp contrast to the Chihuahuan Desert insect record, which showed the least number of species in late Holocene samples. If the Sonoran Desert insect fauna truly was stable through the late Quaternary, then this represents a significant difference between the Sonoran and Chihuahuan desert insect faunal histories.

On the whole, the results of North American postglacial beetle studies tend to agree well with reconstructions based on other proxy data, such as pollen and plant macrofossils. The only important discrepancies concern early postglacial environments. The fossil beetle records of all studied North American regions show unequivocally that at least summer temperatures reached modern levels very early in the postglacial period. Warm-adapted beetles were colonizing the recently deglaciated landscapes in the midwest and the Rockies, providing evidence of this warming, even though low herbaceous vegetation persisted in these regions for up to 1,000 years after deglaciation. Likewise, in the Chihuahuan Desert, the shift from cool-temperate to warm-xeric beetle faunas anticipated the major changes in regional vegetation (notably, the upslope shift in lower tree lines) by up to 1,000 years (Elias and Van Devender, 1992). The rapidity and magnitude of postglacial warming identified in the fossil beetle records from both North America and Europe has been verified by rapid changes seen in oxygen isotope ratios of Greenland ice cores (Dansgaard *et al.*, 1989).

See also: **Beetle Records**: Overview; Postglacial Europe.

References

Bain, A. L. (1998). A seventeenth century beetle fauna from colonial Boston. *Historical Archaeology* **32**, 38–48.

Bain, A. (2001). Archaeoentomological and Archaeoparasitological Reconstructions at Îlot Hunt (CeEt-110): New Perspectives in Historical Archaeology (1850–1900). *British Archaeological Reports* S973.

Bain, A. L., Morgan, A. V., Burns, J. A., and Morgan, A. (1997). The paleoentomology of Rat's Nest Cave, Grotto Mountain, Alberta, Canada. *Quaternary Proceeding* **5**, 23–34.

Clark, W. H., and Sankey, J. T. (1999). Late Holocene Sonoran Desert arthropod remains from a packrat midden, Cataviña, Baja California Norté, México. *Pan-Pacific Entomologist* **75**, 183–199.

Cole, K. L. (1990). Reconstruction of past desert vegetation along the Colorado River using packrat middens. *Palaeogeography, Palaeoclimatology, Palaeoecology* **76**, 349–366.

Dansgaard, W., White, J. W. C., and Johnson, S. J. (1989). The abrupt termination of the Younger Dryas climate event. *Nature* **339**, 532–534.

Duplessy, J. C., Labeyrie, L. D., and Paterne, M. (1996). North Atlantic sea surface conditions during the Younger Dryas cold event. In *Late Quaternary Palaeoceanography of the North Atlantic Margins* (J. T. Andrews, W. E. N. Austin, H. Bergsten, and A. E. Jennings, Eds.), *Geological Society Special Publication* **111**, 167–175.

Elias, S. A. (1982). Holocene insect fossil assemblages from northeastern Labrador. *Arctic Alpine Research* **14**, 311–319.

Elias, S. A. (1986). Fossil insect evidence for late Pleistocene paleoenvironments of the Lamb Spring site, Colorado. *Geoarchaeology* **1**, 381–386.

Elias, S. A. (1991). Insects and climate change: fossil evidence from the Rocky Mountains. *Bioscience* **41**, 552–559.

Elias, S. A. (2000). Late Pleistocene climates of Beringia, based on fossil beetle analysis. *Quaternary Research* **53**, 229–235.

Elias, S. A., Anderson, K., and Andrews, J. T. (1996). Late Wisconsin climate in the northeastern United States and southeastern Canada, reconstructed from fossil beetle assemblages. *Journal of Quaternary Science* **11**, 417–421.

Elias, S. A., Mead, J. I., and Agenbroad, L. D. (1992). Late Quaternary arthropods from the Colorado Plateau, Arizona and Utah. *Great Basin Naturalist* **52**, 59–67.

Elias, S. A., and Toolin, L. J. (1989). Accelerator dating of a mixed assemblage of late Pleistocene insect fossils from the Lamb Spring site, Colorado. *Quaternary Research* **33**, 122–126.

Elias, S. A., and Van Devender, T. R. (1990). Fossil insect evidence for Late Quaternary climatic change in the Big Bend region, Chihuahuan Desert, Texas. *Quaternary Research* **34**, 249–261.

Elias, S. A., and Van Devender, T. R. (1992). Insect fossil evidence of late Quaternary environments in the northern Chihuahuan Desert of Texas and New Mexico: comparisons with the paleobotanical record. *Southwest. Naturalist* **37**, 101–116.

Elias, S. A., Van Devender, T. R., and De Baca, R. (1995). Insect fossil evidence of Late Glacial and Holocene environments in the Bolson de Mapimi, Chihuahuan Desert, Mexico: comparisons with the paleobotanical record. *Palaios* **10**, 454–464.

Grimm, E. C., and Jacobson, G., Jr. (2004). Late-Quaternary vegetation history of the eastern United States. In *The Quaternary Period in the United States* (A. R. Gillespie, S. C. Porter and B. F. Atwater, Eds.), pp. 381–402. Elsevier, Amsterdam.

Hall, W. E., Olson, C. A., and Van Devender, T. R. (1989). Late Quaternary and modern arthropods from the Ajo Mountains of southwestern Arizona. *Pan-Pacific Entomologist* **65**, 322–347.

Hall, W. E., Olson, C. A., and Van Devender, T. R. (1990). Late Quaternary and modern arthropods from the Puerto Blanco Mountains, Organ Pipe Cactus National Monument, southwestern Arizona. In *Packrat Middens, the last 40,000 years of biotic change* (J. L. Betancourt, T. R. Van Devender, and P. S. Martin, Eds.), pp. 363–379. University of Arizona Press, Tucson.

Lavoie, C., Elias, S. A., and Payette, S. (1997). Holocene fossil beetles from a treeline peatland in subarctic Québec. *Canadian Journal of Zoolology* 75, 227–236.
Lavoie, C., and Arseneault, D. (2001). Late Holocene climate of the James Bay area, Québec, Canada, reconstructed using fossil beetles. *Arctic, Antarctic, and Alpine Research* 33, 13–18.
Levesque, A. J., Cwynar, L. C., and Walker, I. R. (1997). Exceptionally steep north-south gradients in lake temperatures during the last deglaciation. *Nature* 385, 423–426.
Lotter, A. F., Walker, I. R., Brooks, S. J., and Hofmann, W. (1999). An intercontinental comparison of chironomid palaeotemperature inference models: Europe vs North America. *Quaternary Science Reviews* 18, 717–735.
Miller, R. F., and Elias, S. A. (2000). Late-glacial climate in Maritime Canada, reconstructed from Mutual Climatic Range analysis of fossil Coleoptera. *Boreas* 29, 79–88.
Pauketat, T. R., Kelly, L. S., Fritz, G. J., Lopinot, N. H., Elias, S. A., and Hargrave, E. (2002). The residues of feasting and public ritual at early Cahokia. *American Antiquity* **67,** 52–65.
Schwert, D. P., and Ashworth, A. C. (1988). Late Quaternary history of the northern beetle fauna of North America: a synthesis of fossil distributional evidence. *Memoirs of the Entomological Society of Canada* **144,** 93–107.
Short, S. K. (1985). Palynology of Holocene sediments Colorado Front Range: vegetation and treeline changes in the subalpine forest. *American Association of Stratigraphic Palynologists Contribution Series* **16,** 7–30.
Short, S. K., and Elias, S. A. (1987). New pollen and beetle analysis at the Mary Jane site, Colorado: evidence for Late-Glacial tundra conditions. *Geological Society of American Bulletin* **98,** 540–548.

Beringia *see* **Archaeological Records**: Global Expansion 300,000-8000 years ago, Americas; **Beetle Records**: Late Pleistocene of North America; **Dune Fields**: High Latitudes; **Glaciations**: Late Pleistocene Events in Beringia; **Paleoceanography, Records**: Late Pleistocene North Pacific; **Plant Macrofossil Records**: Arctic North America; **Pollen Records, Late Pleistocene**: Northern North America; **Vertebrate Records**: Late Pleistocene of North America; Late Pleistocene of Northern Asia

Biogenic Carbonate Studies *see* **Carbonate Stable Isotopes**: Biogenic Studies

Bond Cycles *see* **Paleoclimate Reconstruction**: Sub-Milankovitch (DO/Heinrich) Events

C

CARBONATE STABLE ISOTOPES

Contents

Overview

H Schwarcz, McMaster University, Hamilton, Canada

© 2007 Elsevier B.V. All rights reserved.

Concepts in Stable Isotope Geochemistry

The light elements oxygen (O), hydrogen (H), carbon (C), and nitrogen (N) have two or more stable isotopes whose relative abundances vary slightly in nature. Since the early 1950s, geochemists have used isotopic variations in sedimentary materials to reconstruct past changes in climate. Emiliani (1955) first showed cyclical variations in the $^{18}O/^{16}O$ ratios of foraminifera from deep-sea sediments on a 10^5 year timescale. He inferred that these changes reflected cyclical advances and retreats of continental ice sheets causing coupled changes in both the temperature and $^{18}O/^{16}O$ ratio of seawater. While some of the best-known and most continuous isotopic records of Quaternary paleoclimate are arguably from deep-sea sediments (Shackleton and Opdyke, 1973) and continental ice cores, isotopic records from continental loci also provide important archives of past climate change.

The physical principles behind the use of isotope records can be summarized as follows. When a material is forming either as a result of chemical precipitation or biological activity, the isotope ratios of the constituent atoms can reflect the temperature of the depositional process. As well, the isotope ratio of the newly formed material reflects the ratio in the medium from which it is formed (usually water) and this in turn can be controlled by climate change. When crystals of calcium carbonate ($CaCO_3$; calcite or aragonite) precipitate at equilibrium, their $^{18}O/^{16}O$ ratio is offset from that of the aqueous medium by a fractionation factor α_{ct-w} that depends on temperature, generally decreasing (causing $^{18}O/^{16}O$ of calcite to increase) by $\sim$0.24‰/° C. Likewise, the $^{18}O/^{16}O$ and deuterium/hydrogen (D/H) ratios of meteoric precipitation increase with temperature (by $\sim$ 0.7‰/° C for $^{18}O/^{16}O$) except in the tropics where $^{18}O/^{16}O$ is a decreasing function of the amount of precipitation. Isotopes in surface waters derived from rain and snow-melt will likewise reflect changes in average temperature and rainfall. The $^{18}O/^{16}O$ ratios of minerals precipitated from water (including apatite in bones and teeth) reflect the ratio in environmental water.

Isotope ratios are reported using the delta (δ) notation. For example:

$$\delta^{13}C = \{[(^{13}C/^{12}C)_{sample}/(^{13}C/^{12}C)_{std}] - 1\} \times 1000$$

where 'std' is the standard: VPDB for $\delta^{13}C$ and $\delta^{18}O$ in calcite, VSMOW for $\delta^{18}O$ in water, and for δD (derived from $^2H/^1H \equiv D/H$).

There is a relative paucity of continuously deposited Quaternary sediments from which to extract stable isotopic records in the terrestrial realm compared to the marine record. Most terrestrial isotopic records for this period relate to specific, short-lived sedimentary loci or to collective records of disparate events such as soil profiles. The sections of the

Encyclopedia on this topic largely present methodological treatments of the use of isotopes in paleoclimate rather than summarizing detailed climatic records (except for the Holocene, for which quasi-continuous records are more common). The highlights of these sections are summarized below.

Speleothems

Some of the longest, most accurately dated, and continuous isotopic terrestrial isotopic records are taken from deposits of $CaCO_3$ formed in caves, principally stalagmites and horizontally laminated flowstone. These deposits form as a result of supersaturation of calcium-rich solutions dripping from the cave roof. When deposited deep inside poorly ventilated caves, they can be formed in O isotopic equilibrium with drip water and therefore have the potential to record paleotemperatures. Drip waters isotopically resemble average meteoric precipitation above the cave; therefore $\delta^{18}O$ of the calcite ($\delta^{18}O_{ct}$) depends on both $\delta^{18}O$ of precipitation (controlled by temperature and rainfall amount) as well as the fractionation between calcite and water, α_{ct-w}. The net effect can be either an increase or a decrease in $\delta^{18}O_{ct}$ with increasing atmospheric T (both are observed). Serial records drilled out of speleothems have been used to reconstruct the history of local and regional climate. Their precise U-series chronologies allow correlation with isotopic records in ice and ocean cores. Variation in $\delta^{13}C_{ct}$ records changes in organic activity above the cave including changes in amount of C_3 and C_4 plants.

Absolute temperatures of formation of speleothem calcite may be obtainable by analysis of fluid inclusions in the speleothem which represent trapped samples of original drip water. While the $^{18}O/^{16}O$ ratio may have changed due to post-trapping exchange, the D/H ratio of the included water can be used to reconstruct initial $\delta^{18}O$ using the meteoric water relationship $\delta D = 8\ \delta^{18}O + \delta_o$ where $\delta_o \sim 10‰$. Thus, speleothems have the potential to provide archives of surface temperatures over the last 0.4 million years (My).

Teeth and Bones

The teeth and bones of mammals consist of hydroxyapatite (HA) and collagen. They contain records of isotopes of O and C (in HA) and C and N (in collagen). The $^{18}O/^{16}O$ of HA is mainly determined by the $^{18}O/^{16}O$ of consumed water including leaf water (for herbivores). $\delta^{18}O$ of HA reflects temperature change as it affects $\delta^{18}O$ in precipitation, and also relative humidity (RH) which causes leaf water to increase in ^{18}O. Serial sections of enamel in single teeth record seasonal changes in temperature or rainfall amount, while HA in bone (which is continuously remodeled) records long-term average climate.

The C and N isotopic records of bone have also been shown to vary with climate. In arid-to-semiarid zones, $\delta^{15}N$ increases with decreasing rainfall (Heaton *et al.*, 1986) while $\delta^{13}C$ can increase as RH falls, due both to change in water balance and to increase in the proportion of C_4 plants in the diet.

Bones and teeth represent a highly discontinuous isotopic record, because they do not generally occur in long stratified sequences. An exception is at archeological sites in Eurasia and Africa where more or less continuous sequences can be assembled from faunal remains at these sites.

Biogenic Carbonates

The shells of mollusks and other $CaCO_3$-precipitating fauna are formed approximately in O isotopic equilibrium with the water in which they live. Therefore their $\delta^{18}O$ values reflect the temperature of deposition, partly because of the temperature-dependence of α_{ct-w} which decreases by 0.24‰/° C and also because the $\delta^{18}O$ of the meteoric source of most surficial waters increases by c. 0.7‰/° C. In larger lakes and rivers $\delta^{18}O$ of water is less variable and the effect of temperature on α_{ct-w} is dominant. There may be a *vital effect* leading to an offset of $\delta^{18}O$ from equilibrium, but usually the T-slope remains close to −0.24‰/° C. Seasonal changes in $\delta^{18}O_{ct}$ of shells are recorded in annual layers of the shells and can be used to reconstruct seasonality. $\delta^{13}C$ values of these shells mainly reflect the $\delta^{13}C$ of dissolved inorganic carbon (DIC) in the water mass. This in turn is influenced by photosynthetic activity (which raises $\delta^{13}C$ of DIC). The $\delta^{18}O_{ct}$ of terrestrial snail shells is controlled by the $\delta^{18}O$ of water in the leaves of the snail's diet, and mainly reflects changes in RH (Goodfriend *et al.*, 1989) while $\delta^{13}C$ reflects that of the leafy diet (Goodfriend, 1990).

The aragonitic otoliths of fish are also precipitated in O isotopic equilibrium while $\delta^{13}C$ reflects DIC of blood; the latter is controlled by metabolic activity which can change with age and also with the level of activity of the fish.

Long sequential isotopic records of fossil shells in terrestrial sequences are rare due to the lack of extensive exposures of lacustrine sediments, but these can be used effectively to monitor past climate change. Aragonitic ostracod tests have also been used for this purpose in proglacial lake sediments (Schwarcz and Eyles, 1991).

Lacustrine Sediments

In addition to the shells that are enclosed in these sediments, part or most of the sediment itself may consist of chemically or biochemically precipitated $CaCO_3$ in the form of marl. The $\delta^{18}O$ of these sediments (generally formed close to isotopic equilibrium) varies in response to temperature, but may also reflect RH in evaporative lakes. Carbon isotopes are controlled by $\delta^{13}C$ of DIC, and strongly reflect changes in photosynthetic activity in the lake.

Nonlacustrine Terrestrial Carbonates

$CaCO_3$ is deposited in some arid-region soils as a result of evaporation of Ca-rich water in the soil in the presence of biogenic carbon dioxide (CO_2) and bicarbonate (HCO_3). The $\delta^{18}O$ of these carbonates is strongly influenced by evaporative enrichment of the soil water in ^{18}O with the result that there is no simple relation between $\delta^{18}O_{ct}$ and temperature. A regional trend with a slope of +0.4‰/° C is observed for soils experiencing low evaporation, close to the value expected for the balance between the meteoric water and α_{ct-w} effects discussed above.

Diffusive loss of soil CO_2 to the atmosphere leads to a dependence of $\delta^{13}C$ in soil carbonates on partial pressure of CO_2 in the atmosphere. More important for Quaternary soils, $\delta^{13}C_{ct}$ reflects the balance between C_3 and C_4 plants growing on the soil. Because of the preference of C_4 plants for warm, dry climate, this allows us to use variations in $\delta^{13}C_{ct}$ in stacked sequences of paleosols as an archive of paleoclimate.

A unique terrestrial carbonate deposit, which could also be considered a speleothem, is the buildup of calcite on the walls of Devil's Hole, a warm spring in Nevada, USA (Winograd *et al.*, 1992). A well-dated record of secular variation in $\delta^{18}O$ resembles in many precise details the corresponding $\delta^{18}O$ record for deep-sea foraminifera. There is, however, a striking offset in the ages of terminations (interglacial–glacial transitions) between the U-series-dated Devil's Hole deposits and the astronomically dated marine record. This discrepancy remains unresolved.

Terrestrial Organic Materials

Plant remains are found in terrestrial sediments both as organic rich peaty deposits and as traces of plant materials in detrital sediments. Photosynthates (largely cellulose) are significantly depleted in ^{13}C with respect to atmospheric CO_2 as a result of multistage isotopic fractionations in leaves. The degree of depletion varies as a result of change in stomatal opening and is therefore an indicator of water availability. O and H isotopes reflect both the respective isotope ratios in meteoric water (and soil or ground water derived from it), as well as evaporative effects in leaves that enrich leaf-water in heavy isotopes. Thus, a combination of C, O, and H analyses of plant materials can provide a complex image of climatic change.

Serial samples of tree rings provide the most widely cited continuous isotopic records of plant materials. Peat deposits can provide sequences of organic matter mainly derived from mosses. Advances in mass spectrometry have opened up the possibility of isotopic analysis of pollen grains which form a minor component of many terrestrial sediments.

Conclusions

In general, stable isotope ratios in terrestrial materials are affected to approximately equal extents by two or more environmental parameters (including temperature, RH, rainfall, and $\delta^{18}O$ of meteoric or surface water). It is seldom possible to give a simple and unambiguous interpretation of a serial isotopic record in a sedimentary sequence. Nevertheless, the conflation of multiple isotopic indicators, sometimes from the same material (using different isotopes) can narrow down the possible interpretations assignable to a given series of wiggles on an isotopic graph. Our ability to analyze successively smaller samples opens up the possibility of extracting individual biomolecules for analysis, as well has higher spatial resolution in analyses of serial samples from single deposits such as speleothems or shells.

Abbreviations

$CaCO_3$	calcium carbonate
CO_2	carbon dioxide
δ	delta
D/H	deuterium/hydrogen
DIC	dissolved inorganic carbon
HA	hydroxyapatite
HCO_3	bicarbonate
My	million years
RH	relative humidity

See also: **Carbonate Stable Isotopes**: Speleothems; Terrestrial Teeth and Bones; Nonmarine Biogenic Carbonates; Terrestrial Organic Materials; Non-Lacustrine Terrestrial Studies. **Ice Core Records**: Chinese, Tibetan Mountains.

References

Emiliani, C. (1955). Pleistocene paleotemperatures. *Journal of Geology* **63**, 538–578.

Goodfriend, G. A. (1990). Rainfall in the Negev Desert during the Middle Holocene, based on ^{13}C of organic matter in land snail shells. *Quaternary Research* **34**, 186–197.

Goodfriend, G. A., Magaritz, M., and Gat, J. R. (1989). Stable isotope composition of land snail body water and its relation to environmental waters and shell carbonate. *Geochimica et Cosmochimica Acta* **53**, 3215–3221.

Heaton, J. H. E., Vogel, J. C., von la Chevallerie, G., and Collet, G. (1986). Climatic influence on the isotopic composition of bone nitrogen. *Nature* **322**, 822–823.

Schwarcz, H. P., and Eyles, N. (1991). Laurentide ice sheet extent inferred from stable isotopic composition (O,C) of ostracodes at Toronto, Canada. *Quaternary Research* **35**, 305–320.

Shackleton, N. J., and Opdyke, N. D. (1973). Oxygen isotope and paleomagnetic stratigraphy of equatorial Pacific core V28-238: oxygen isotope temperatures and ice volumes on a 10^5 and 10^6 year time scale. *Quaternary Research* **3**, 39–53.

Winograd, I. J., Coplen, T. B., Landwehr, J. M., *et al.* (1992). Continuous 500,000-year climate record from vein calcite in Devils Hole, Nevada. *Science* **258**, 255–260.

Speleothems

H Schwarcz, McMaster University, Hamilton, Canada

© 2007 Elsevier B.V. All rights reserved.

Speleothems are crystalline deposits of calcium carbonate ($CaCO_3$) formed in karstic caves as a result of precipitation from dilute aqueous solutions entering the cave, typically as drips from the roof. Karstic caves (q.v.) are formed by dissolution of soluble bedrock, usually a carbonate rock, by the action of meteoric water percolating from overlying soil or bedrock surfaces. This is partly due to organic activity in the soil that generates carbonic acid, which reacts with calcite or dolomite. Karstic dissolution forms cavities varying in size from less than 1 mm to passages many kilometers in length and tens of meters high (Ford and Williams, 1989). Karst can form in other kinds of rock (e.g., evaporitic sulfates), and speleothems can be formed of minerals other than $CaCO_3$. However, this article will deal only with Ca carbonates.

Speleothems can be seen forming today and we can infer that some have formed throughout the Quaternary epoch. They occur in many geometric forms, including stalagmites (which grow upward from the floor of caves beneath a drip site), stalactites (growing downward from the roof of a cave), and flowstone—sheet-like horizontally or inclined layers of calcite produced where water from one or more sources has spread as a film across a cave floor or wall (**Fig. 1**). When viewed in cross section (axially or longitudinally), many speleothems display a prominent laminated structure (**Fig. 2**). These preserve records of secular variation in the properties of the speleothem material that can be interpreted as changes in conditions of deposition in the cave and thus, indirectly, on the Earth's surface above. These properties include the following:

- Chronometric age and rate of growth
- Isotopic composition of the $CaCO_3$
- Abundance and isotopic composition of fluid inclusions of water trapped during growth
- Trace element concentrations in $CaCO_3$
- Abundance and composition of organic impurities (especially humic substances)
- Mineralogy (calcite, aragonite, etc.)

Here, we focus on the first three items.

Stable Isotope Systematics in Speleothems

Before percolating through the bedrock, surface water is exposed first to atmospheric CO_2 and then to soil gases enriched in biogenic CO_2. As water percolates through the bedrock overlying a cave, it can react extensively with the rock or may pass through with relatively little chemical interaction, depending on the nature of the openings through which it passes. Under conditions of diffuse flow through relatively tight fractures and joints, the water can approach chemical and isotopic equilibrium with the rock, whereas in larger, open channels entering caves, surface water may enter essentially unchanged. On entering the cave, drip waters can begin to precipitate $CaCO_3$ as a result of two discrete processes: evaporation, which increases the concentration (activity) of Ca ions until the solubility of $CaCO_3$ is exceeded, and outgassing of CO_2 from the water, which results in the slow precipitation of $CaCO_3$:

$$Ca^{2+} + 2HCO_3^- \rightarrow CaCO_3 + H_2O + CO_2 \quad (1)$$

$CaCO_3$ potentially contains a number of distinct isotopic signals, recorded as variations in the relative abundance of the isotopes of Ca, O, and C. Only the latter two have been extensively studied.

Oxygen Isotopes

Variations in the $^{18}O/^{16}O$ ratio in water and O-bearing minerals have been widely used as proxies for various environmental records, especially temperature. $CaCO_3$ can isotopically exchange with water:

$$1/3CaC^{18}O_3 + H_2{}^{16}O \leftrightarrow 1/3\ CaC^{16}O_3 + H_2{}^{18}O \quad (2)$$

Epstein *et al.* (1953) showed that the equilibrium constant for this reaction, $K_{ct-w} = (^{18}O/^{16}O)_{ct}/$

Figure 1 Typical forms of speleothems: (A) stalagmites (middle stalagmite is ~40 cm high); (B) stalactite, ~4 cm in diameter at top, flanked by 'soda straws'; (C) thin stalactites (soda straws), ~5 mm in diameter, showing drip water emerging; and (D) flowstone, ~1 m wide. Photos by P. Beddows.

Figure 2 Section through a stalagmite showing laminated structures (growth layers). Lighter zones are richer in fluid inclusions; note the detritus-rich layer at base. The color is due to the presence of humic matter in calcite.

$(^{18}O/^{16}O)_{H_2O}$, depended only on temperature. Therefore, if a sample of $CaCO_3$ had precipitated in isotopic equilibrium with water of known $^{18}O/^{16}O$ ratio, we could determine the temperature of precipitation. Hendy (1971) showed that under conditions widely encountered in deep caves, speleothems formed at oxygen isotopic equilibrium with drip waters. This happened if the mode of deposition was by relatively slow outgassing of CO_2 from the drip water, whereas evaporative deposition (or rapid outgassing) led to disequilibrium precipitation. That is, under the latter conditions the oxygen isotope fractionation was not equal to the equilibrium constant at the temperature of deposition, and deviation from equilibrium increased as drip water traversed the surface of the speleothem. Hendy showed that the calcite precipitated along this path became progressively enriched in ^{18}O. Therefore, a convenient test for equilibrium was to measure the $^{18}O/^{16}O$ ratio of calcite along several points spaced some centimeters apart on a single growth layer. For equilibrium deposits it was expected that $(^{18}O/^{16}O)_{ct}$ would remain constant, whereas disequilibrium growth layers would display (1) continuously varying $^{18}O/^{16}O$ and (2) a correlation between $^{18}O/^{16}O$ and $^{13}C/^{12}C$. The latter reflects the fact that rapid outgassing or evaporation were usually coupled with variations in the $^{13}C/^{12}C$ ratio of dissolved inorganic carbon (DIC) in the drip water. Speleothems satisfying the criteria of equilibrium deposition are invariably found deep inside long caves (>1 km) or in recently opened, formerly sealed cavities, where relative humidity was ~100% and where lack of significant movement of air resulted in high partial pressure of CO_2 with the result that outgassing of drip water was slow. Recent studies (Mickler *et al.*, 2004) suggest that kinetic effects may prevent the attainment of isotopic equilibrium between calcite and water. However, actively growing stalactites ('soda straws') from deep, high-humidity caves have been shown by other authors to yield isotopic temperatures similar to the temperature in the cave.

The discovery of this means of recognizing speleothems deposited at equilibrium opened up the possibility of using $(^{18}O/^{16}O)_{ct}$ variations in speleothems as a paleothermometer. Far from the entrances to most caves (>100–500 m), the temperature remains constant seasonally and approaches the mean annual surface temperature above the cave. Therefore, a paleotemperature record from an equilibrium speleothem would provide a record of past temperatures on the surface for the duration of growth of the speleothem.

For a given sample of calcite extracted from such an equilibrium deposit, the temperature is given by the paleotemperature equation (Epstein *et al.* 1953):

$$T(^{\circ}C) = 16.5 - 4.3(\delta^{18}O_{ct.} - \delta^{18}O_w) + 0.14(\delta^{18}O_{ct.} - \delta^{18}O_w)^2 \quad (3)$$

where $\delta^{18}O_x$ is defined as

$$\delta^{18}O = \{[(^{18}O/^{16}O)_x/(^{18}O/^{16}O)_{std}] - 1\} \times 1000 \quad (4)$$

where w is drip water, and std refers to one of the two international standards for $^{18}O/^{16}O$ ratios: Vienna Peedee Belemnite (VPDB) or Vienna Standard Mean Ocean Water. The units of $\delta^{18}O$ are ‰ (per mil). The precision of analyses of $\delta^{18}O$ is typically approximately ±0.1‰. Note, however, that in order to determine T, it is necessary to have values for $\delta^{18}O$ of both the calcite and the water from which it was precipitated.

$\delta^{18}O_{ct}$ is usually determined by reacting samples ≥100 μg with 100% phosphoric acid and analyzing the resultant CO_2 on a gas-source, isotope ratio-detecting mass spectrometer. Most studies of long-term variation in $\delta^{18}O_{ct}$ of speleothems have been done on stalagmites using axial sections (**Fig. 3**) oriented more or less in a diametric section of the stalagmite. Samples are drilled out of the polished surface of the section. High-resolution records (with micrometer spacing) have been obtained using computer-driven milling machines (Mangini *et al.*, 2005) and laser heating to decompose $CaCO_3$ to $CaO + CO_2$ (McDermott *et al.*, 2001). In most studies of

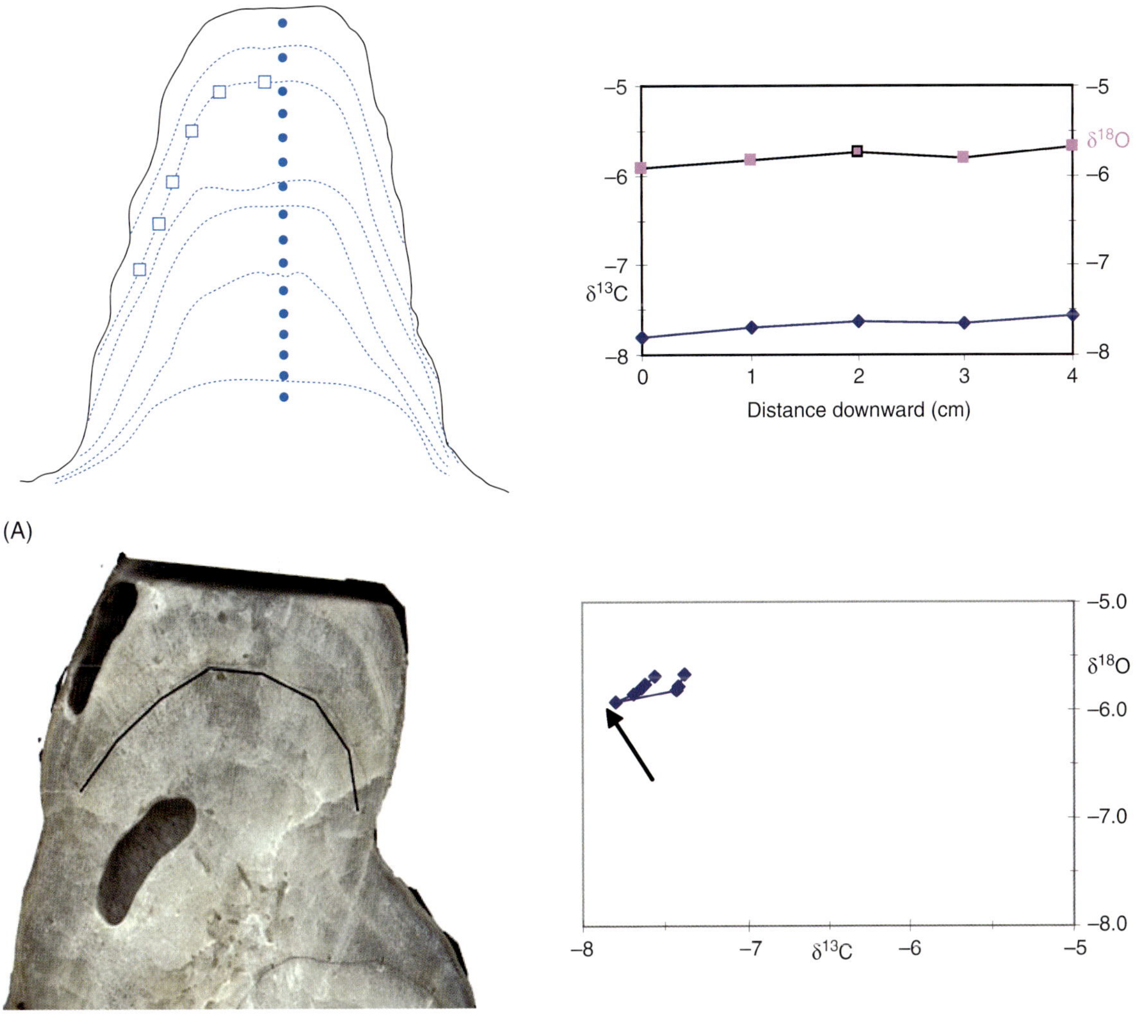

Figure 3 Analysis of layers: (A) Drawing of stalagmite showing sampling localities for Hendy test (squares) and climate records (circles); (B) Hendy test of a growth layer in a Holocene stalagmite from Marengo Cave, Indiana (R. Zhang, unpublished data). (Top graph) $\delta^{13}C$ and $\delta^{18}O$ along the left side of the growth layer are indicated by the curved line. (Bottom graph) Lack of correlation between $\delta^{13}C$ and $\delta^{18}O$; arrow marks the top of the growth layer.

secular variation in $\delta^{18}O_{ct}$, the authors use the so-called Hendy test to ensure that the deposits were formed in oxygen isotopic equilibrium, reporting $\delta^{18}O_{ct}$ and $\delta^{13}C_{ct}$ values for samples taken along single growth layers.

It has proven much more difficult to determine values for $\delta^{18}O_{W}$. Various approaches to this problem are discussed next.

Carbon Isotopes

The $^{13}C/^{12}C$ ratio of the carbonate in calcite is always measured at the same time as $\delta^{18}O_{ct}$ during analysis of CO_2 gas on isotope ratio mass spectrometers. Data are reported as $\delta^{13}C_{ct}$ with respect to the VPDB standard. The precision of analysis is $\pm 0.1‰$. Variations in $\delta^{13}C_{ct}$ are not significantly temperature dependent but are rather a consequence of various mixing processes during the formation of DIC.

Dating

Ages of speleothems have been obtained principally by ^{14}C and uranium (U) series dating. Electron spin resonance has been applied with limited success, possibly due to the short lifetime of trapped charges in calcite.

Radiocarbon

^{14}C is introduced into DIC and thus into speleothems through oxidation of plant-derived organic matter in the soil. Ages must be corrected for the presence of dead carbon, nonradioactive C introduced into DIC during dissolution of calcite in the soil zone. From Eq. (1), it appears that approximately half of the C atoms in HCO_3^- ions should be derived from the bedrock calcite, which in most circumstances contains no ^{14}C. Therefore, modern cave-deposited calcite should have a $^{14}C/^{12}C$ ratio approximately half that of modern carbon. The initial percentage of dead carbon (dcp) is in fact observed to range from ca. 5 to <40% (Genty *et al.*, 2001). Lower values are presumably a result of continued exchange between soil-derived CO_2 and recharge water as it percolates through air-filled passages on the way to the cave. The initial dcp cannot be assumed to be constant at a given site but may vary as a result of changing soil activity. Errors due to uncertainty in dcp decrease with age. Speleothems dated by U series have also been used to reveal large oscillations in atmospheric ^{14}C activity (Beck *et al.*, 2001).

U series

^{238}U decays to a series of short-lived daughter isotopes; after a long time (>2 My), the activity of each of these isotopes (in disintegrations per minute in a given sample) should equal the activity of the parent ^{238}U. However, the daughter isotopes can be geochemically separated from the parent U prior to deposition of calcite because uranium is soluble in groundwater, whereas daughter Th and Pa are insoluble. The age of calcite is determined from the extent of growth of the daughter isotopes thorium-230 (^{230}Th) and ^{234}U into secular equilibrium with ^{238}U and that of protactinium-231 (^{231}Pa) into equilibrium with ^{235}U (Richards and Dorale, 2003). The activity ratio $^{230}Th/^{234}U$ increases from zero at the time of deposition to a maximum value of 1 at infinite age. The ratio of ^{234}U to ^{238}U also varies because the radiogenic ^{234}U is a 'hot atom' whose chemical properties during weathering are different from those of its parent ^{238}U.

Speleothems are well suited to U-series dating because their calcite crystals are usually large and have little tendency to recrystallize after deposition. Speleothem calcite typically contains between 100 and 1,000 parts per billion (ppb) U, whereas aragonitic speleothem contains up to 100 ppm U. Some speleothems contain detritus, which is trapped in the calcite during flooding events in caves and which contaminates the samples with ^{230} Th so that the apparent age of the freshly deposited calcite is greater than zero. Analyses of such 'dirty' calcites can be obtained by use of isochrons, in which a nonradiogenic isotope, ^{232}Th, is used to estimate the amount of detritus (Bischoff and Fitzpatrick, 1991). Individual analyses can be corrected by using the $^{230}Th/^{232}Th$ ratio to estimate the amount of 'detrital' ^{230}Th present in the sample.

U-series isotopes are analyzed using thermal ionization mass spectrometry (TIMS; Edwards *et al.*, 2003) or multicollector inductively coupled plasma mass spectrometry. The precision of analyses of speleothems is better than 1% of the age. Older analyses done by counting of α particles were of lower precision (5–10% of age). The practical age limit for U-series dating is approximately 500,000 years. U-Pb dating by TIMS can be used for older speleothems, although the method requires moderately high U concentrations and analysis of coeval samples with varying U/Pb ratios (Richards *et al.*, 1998).

Paleoclimate Inferred from O Isotope Records in Speleothems

Eq. (1) shows that for speleothems formed at oxygen isotopic equilibrium, we can infer the temperature of deposition from $\delta^{18}O_{ct}$ and $\delta^{18}O_{ppt}$. In general, we can express a change $\delta^{18}O_{ct}$ as

$$\Delta\delta^{18}O_{ct} = [(d\varepsilon_{ct-w}/dT) + (d\delta^{18}O_{ppt}/dT)]\Delta T + \Delta\delta^{18}O_{sw} + \Delta\delta^{18}O_s$$

where sw is seawater: All meteoric water is derived from the ocean; thus, changes in $\delta^{18}O$ of seawater during the Pleistocene as a result of changes in ice volume would be reflected in $\delta^{18}O_{ppt}$. The last term is a correction to represent changes in the source of water as a result of secular shift in the average path of storm tracks. $\varepsilon_{ct-w} = (\alpha_{ct-w} - 1) \times 1{,}000$, and $d\varepsilon_{ct-w}/dT$ is the temperature (T) dependence of the calcite–water fractionation, approximately $-0.25‰/°$ C. The second term, $d\delta^{18}O_{ppt}/dT$, refers to the long-term (secular) dependence of $\delta^{18}O_{ppt}$ on temperature, which is not necessarily equivalent to short-term temperature dependence as observed, for example, in annual records at a single site. The latter T dependence is typically equal to approximately 0.7‰/° C (Rozanski *et al.*, 1993). If the secular T dependence were also ~0.7‰/° C, then the net T-dependent change in $\delta^{18}O_{ct}$ would be approximately 0.4‰/° C. In tropical and subtropical regions, $\delta^{18}O$ of drip waters can also change as a result of the so-called 'amount effect,' in which $\delta^{18}O_{ppt}$ decreases with increasing volume of individual rain events (Rozanski *et al.*, 1993). From the previous relationship, we would expect to find a positive correlation between temperature of deposition and $\delta^{18}O_{ct}$. We can test the nature of this correlation in a number of ways:

- Comparison of $\delta^{18}O_{ct}$ of modern (or interglacial) speleothems with glacial age samples
- Comparison of $\delta^{18}O_{ct}$ record with other coeval, dated paleoclimate records, such as $\delta^{18}O$ of foraminifera from deep-sea cores or ice cores or

temperatures obtained from Mg/Ca or alkenone ratios in foraminifera

- Association between a deposit and sedimentary or paleontological deposits of known climatic signature

In published studies of $\delta^{18}O_{ct}$ vs age, the correlation between inferred paleotemperature (i.e., paleoclimate, ranging between glacial (cold) and interglacial (warm)) is either positive or negative. We can express this fact in terms of the sign of $\gamma = d(\delta^{18}O_{ct})/dT$, which can be positive or negative. This represents the tendency of $\delta^{18}O_{ct}$ to rise or fall as local climate becomes warmer or cooler. In principle, we can also have sites with $\gamma = 0$, where the positive and negative climatic effects on $\delta^{18}O_{ct}$ exactly cancel out.

Determination of δ^{18}O of Paleowaters

In order to determine paleotemperatures from speleothems deposited at isotopic equilibrium, it is necessary to determine $\delta^{18}O_w$ of the drip water from which the speleothem formed (see Eq. 3). $\delta^{18}O$ of drip water is close to the annual average of $\delta^{18}O_{ppt}$, the $\delta^{18}O$ value of precipitation falling on land surface above the cave (Yonge *et al.*, 1985; Spötl *et al.*, 2005), but may be locally affected by factors such as surface evaporation and transpiration or selective recharge of seasonal precipitation (e.g., snow meltwater). Most papers reporting $\delta^{18}O_{ct}$ data do not explicitly attempt to evaluate variation in $\delta^{18}O_{ppt}$ but merely show comparisons between excursions in $\delta^{18}O_{ct}$ and other paleoclimate proxies as noted previously. However, at least three distinct approaches have been used to estimate isotopic composition of paleowaters:

- Assuming that secular variation in $\delta^{18}O_{ppt}$ is the same as annual (seasonal) variation (Dorale *et al.*, 1992). This assumes a particular value for $d(\delta^{18}O_{ppt})/dT$.
- Analysis of fluid inclusions in speleothem, which are presumed to represent the drip water present during growth (Schwarcz *et al.*, 1976). This water can be extracted by crushing the sample in vacuum, and $\delta^{18}O$ and δD can be determined (Dennis *et al.*, 2001). Although oxygen isotopes may have been exchanged postdepositionally with the calcite, the δD value of the water should be unchanged and can be used to reconstruct $\delta^{18}O_{ppt}$ from the meteoric water relationship $\delta D_{ppt} = 8\ \delta^{18}O_{ppt} + \delta_0$, where δ_0 is the so-called deuterium excess. For most continental regions of the world, $\delta_0 = 10‰$, but this value may have changed through the Pleistocene. For a detailed discussion of the use of fluid inclusions, see McDermott *et al.* (2006).
- Specific models for variation in $\delta^{18}O_{ppt}$, such as derivation of water vapor from seawater whose $\delta^{18}O$ varied in a known fashion (Frumkin *et al.*, 1999).

Application to Paleoclimate Studies

$\delta^{18}O_{ct}$

More than 100 papers report secular variation in $\delta^{18}O_{ct}$ of U-series dated speleothems deposited during the past 200,000 years. Most use $\delta^{18}O_{ct}$ as a proxy for paleoclimate, an approach that is analogous to the use of $\delta^{18}O$ analyses of foraminifera from deep-sea sediments or $\delta^{18}O$ (and δD) measurements of ice cores. Most studies identify features (excursions or trends) in the speleothem records that match coeval isotopic excursions in the better known proxy records as well as records of pollen, fauna, trace elements in sediments, etc. Thus, we can describe the use of most speleothem $\delta^{18}O_{ct}$ data as providing a formal (but not quantitative) proxy for Quaternary paleoclimate. Correlations with $\delta^{18}O_{ct}$ records from U-series dated speleothems can be used to give precise dates for critical climatic transitions in other proxy records; they also have been interpreted as indicating how local climate was coupled to global climate change.

An example of a formal proxy is the study of subaqueous calcite precipitated from thermal water in a fissure at Devil's Hole, Nevada (Winograd *et al.*, 1992). In all significant respects, this is identical to studies of speleothems, specifically in that it presents a detailed U-series dated sequence of analyses of $\delta^{18}O_{ct}$ for a laminated deposit formed at relatively low temperatures. The $\delta^{18}O_{ct}$ record (**Fig. 4**) shows signals corresponding to all the major climatic cycles observed in deep-sea cores from 500 to 70 ka. The timing of the $\delta^{18}O_{ct}$ cycles at Devil's Hole is distinctly discordant to that inferred from the Milankovitch theory (e.g., Termination II is approximately 15,000 years earlier than expected).

$\delta^{18}O_{ct}$ records from speleothems found in the Soreq Cave and other caves in Israel (Bar-Matthews and Ayalon, 2003; Frumkin *et al.*, 2000) show striking formal agreement between the $\delta^{18}O_{ct}$ record and data from ice cores and deep-sea sediments (**Fig. 5**). The correspondence persists over eight oxygen isotope stages, back to 250,000 BP. $\gamma < 0$ for these sites. McGarry *et al.* (2004) measured δD in fluid inclusions at Soreq. Noting that $\delta_0 = 20‰$ for Israel today, they used comparisons between alkenone-based sea-surface temperatures and $\delta^{18}O_{ct}$ values to infer secular change in δ_0.

The latter two studies made use of records that spanned glacial–interglacial transitions marked by

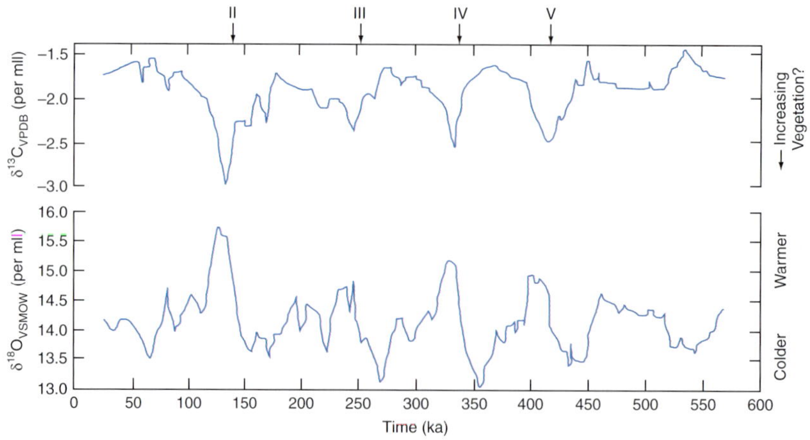

Figure 4 $\delta^{18}O$ record for Devil's Hole, Nevada. U-series dates are shown at top. From Winograd, I., Landwehr, J., Ludwig, K. R., Coplen, T., and Riggs, A. C. (1997). Duration and structure of the past four glaciations. *Quaternary Research* **48**, 141–154. Reproduced with permission from the National Speleological Society.

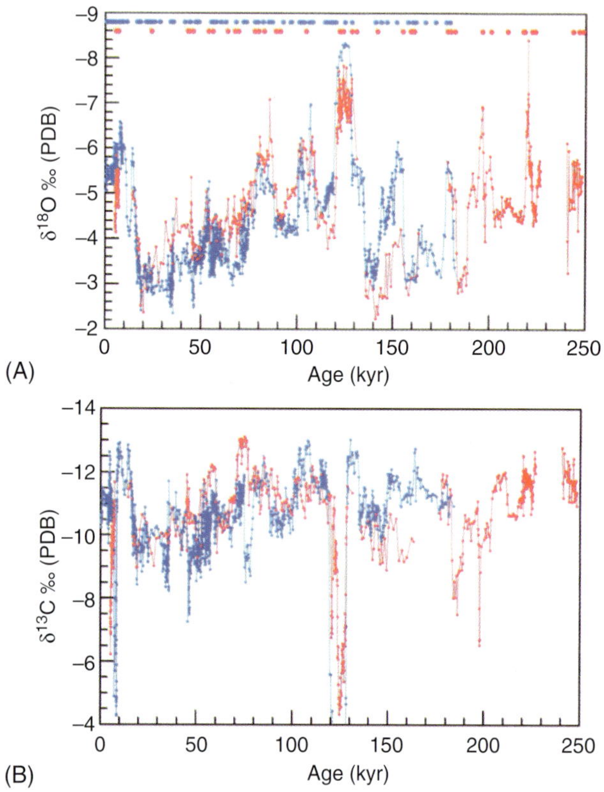

Figure 5 $\delta^{18}O$ (A) and $\delta^{13}C$ (B) records for speleothems from Soreq (blue) and Peqiin Cave (red), Israel. Circles in A show U-series ages. From Bar-Matthews, M., and Ayalon, A. (2003). Sea–land oxygen isotopic relationships from planktonic Foraminifera and speleothems in the eastern Mediterranean region and their implication for paleorainfall during interglacial intervals. *Geochimica et Cosmochimica Acta* **67**, 3181–3199. Reproduced with permission.

large shifts in $\delta^{18}O_{ct}$. Studies at other sites where the dated speleothem record did not cross a termination generally display smaller scale variations in $\delta^{18}O_{ct}$, which nevertheless have been correlated with shorter term cycles in the global paleoclimate record, such as Heinrich events and Dansgaard–Oeschger (D-O) cycles. Examples follow:

Hulu Cave, China: Several speleothems give concordant, partially overlapping records that span a 70,000-year period from ca. 80 to 10 ka (**Fig. 6**). Cycles with amplitudes of 0.5–1‰ ($\gamma < 0$) are correlated by these authors with coeval Heinrich events (cooler periods lasting a few kiloyears or less). The authors infer that these cycles record variations in monsoon activity (Wang *et al.*, 2001).

Oman: Speleothems from caves in Oman record climatic cycles on a time scale ranging from hundreds of years to 330 ka (Fleitmann *et al.*, 2004). $\delta^{18}O_{ct}$ in series of relatively short-lived stalagmites ranges over 8‰ ($\gamma > 0$?). These represent pluvial periods during which the source of moisture has shifted between a continental and monsoonal pattern, as shown by changing δD of fluid inclusions.

Austrian Alps: At the Kleegruben Cave, oscillations in $\delta^{18}O_{ct}$ appear to match D-O cycles seen in Greenland ice cores (Spötl and Mangini, 2002; **Fig. 7**) and were used to provide an improved chronology for these cycles; from the correlation with D-O cycles, we infer that $\gamma > 0$ at this site.

North America: At Crevice Cave, Iowa (Dorale *et al.*, 1998), the form of the $\delta^{18}O_{ct}$ record is correlated to that of the deep-sea record from 75 to 25 ka BP (with $\gamma > 0$). At Reed's Cave, South Dakota (Serefiddin *et al.*, 2003), oscillations in $\delta^{18}O_{ct}$ are matched to cycles in Greenland ice cores. Two speleothems from this site have opposite polarities of γ, which the authors attribute to differences in the seasonality of the supply of recharge water. Significant differences in $\delta^{18}O$ of coeval

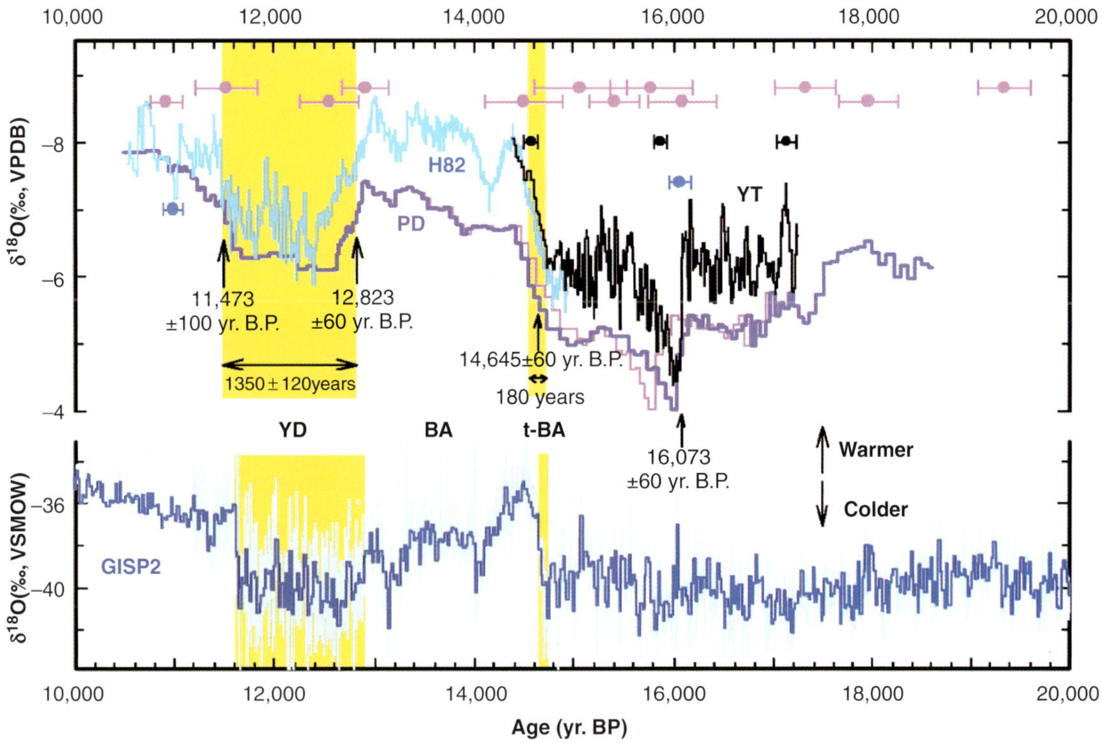

Figure 6 $\delta^{18}O$ of stalagmites from Hulu Cave, China (purple, black, and blue) and Greenland Ice (dark blue, 20-year averages; gray, 3-year averages) vs age. YD, Younger Dryas; BA, Bolling–Allerod; t-BA, transition into B. From Wang, Y. J., Cheng, H., Edwards, R. L., An, Z. S., Wu, J. Y., Shen, C. C., and Dorale, J. A. (2001). A high-resolution absolute-dated late Pleistocene monsoon record from Hulu Cave, China. *Science* **294**, 2345–2348. Reproduced with permission.

calcite deposits have been observed in speleothems from other caves.

Holocene records: Records of $\delta^{18}O_{ct}$ from Holocene speleothems have shed light on short-term, recent changes in climate in many regions. For example, Burns *et al.* (1998), who studied the Hoti Cave in Oman, identified pluvial periods in the early Holocene as well as the last interglacial. In Coldwater Cave, Iowa, early Holocene speleothems gave significant differences in $\delta^{18}O_{ct}$ between coeval deposits, attributable to evaporative effects on the recharge waters (Denniston *et al.*, 1999); secular changes in $\delta^{18}O_{ppt}$ are attributed to shifts in storm tracks over the site. Niggeman *et al.* (2003) showed a high-resolution $\delta^{18}O_{ct}$ record from a cave in south Germany, in which power spectrum analyses revealed the influence of solar cycles on $\delta^{18}O_{ppt}$.

The vast literature affords many other records of interest. For example, Bard *et al.* (2002) accounted for a drop in $\delta^{18}O_{ct}$ in an Italian speleothem approximately 170 ka BP as a result of the amount effect. This effect increased during a wet period identified by the occurrence of a sapropel layer in coeval sediments of the Mediterranean Sea (**Fig. 8**).

Paleotemperature Records

Relatively few studies have attempted to interpret variations in $\delta^{18}O_{ct}$ as paleotemperatures, even where Hendy tests gave positive results. Dorale *et al.* (1992) and van Beynen *et al.* (2004) assumed that the secular T dependence of $\delta^{18}O_{ppt}$ was equal or similar to the seasonal value and used that relationship to derive temperatures for speleothems in Iowa and New York state (**Fig. 9**). Although Dennis *et al.* (2001) analyzed fluid inclusions in speleothems, they did not use the data to calculate temperatures; Serefiddin *et al.* (2005) did this for three stalagmites and showed that coeval stalagmites with differing $\delta^{18}O_{ct}$ also differed in δD of inclusions so that the calculated temperatures of deposition were the same. Mangini *et al.* (2005) calibrated the relationship between $\delta^{18}O_{ct}$ and T by correlating historically reconstructed average annual temperature with $\delta^{18}O_{ct}$ values for speleothem SPA from an Alpine cave.

Carbon Isotopes

Values of $\delta^{13}C_{ct}$ directly reflect the isotopic composition of the drip water from which the speleothem

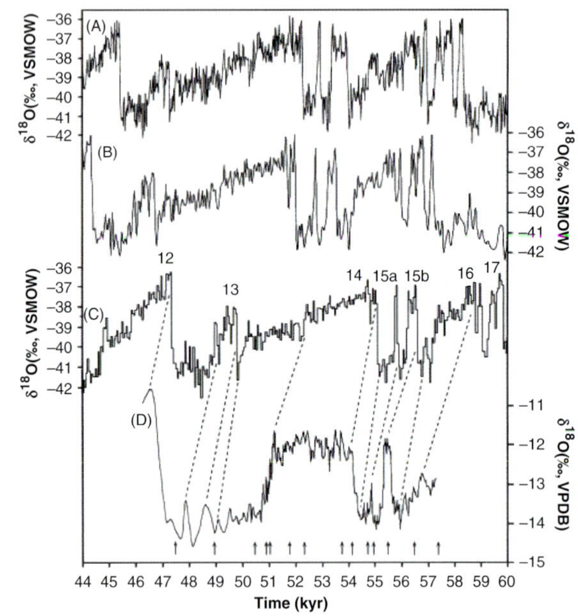

Figure 7 Comparison of oxygen isotope records of the Greenland ice cores and stalagmite SPA 49 from Kleegruben Cave, Austrian Alps: (A) GISP2 record; (B) GRIP, 1995 time scale; (C) GRIP, 2001 chronology; and (D) stalagmite SPA 49. Numbers refer to Greenland Interstadials. Upward arrows indicate positions of U-series samples. Dashed lines indicate major marker events used for cross-correlation. From Spötl, C., and Mangini, A. (2002). Stalagmite from the Austrian Alps reveals Dansgaard–Oeschger events during isotope stage 3; Implications for the absolute chronology of Greenland ice cores. *Earth and Planetary Science Letters* **203**, 507–518. Reproduced with permission.

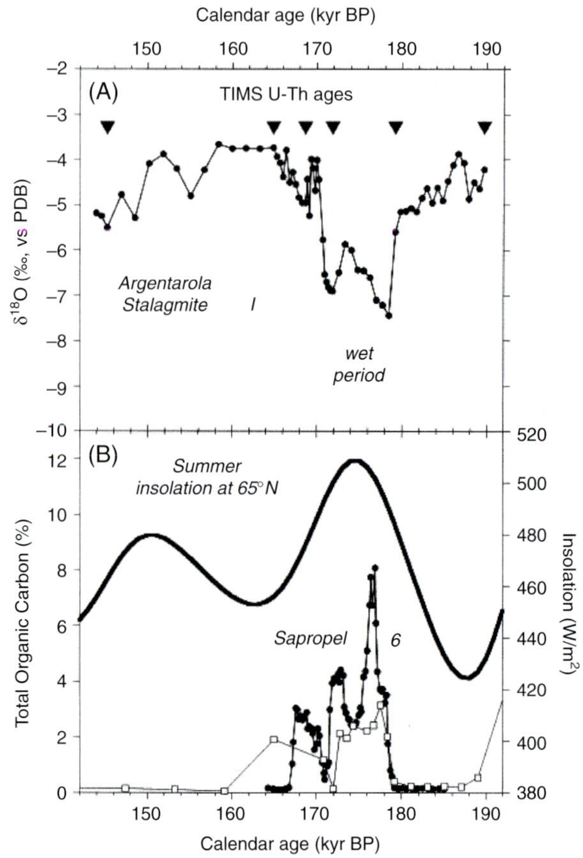

Figure 8 (A) $\delta^{18}O$ vs age for stalagmite from Argentarola Cave, Italy. Triangles: U-series dated points. (B) Upper curve: summer insolation at 65°N; lower curves: wt% total organic carbon in two nearby marine cores. From Bard, E., Delaygue, G., Rostek, F., Antonioli, F., Silenzi, S., and Schrag, D. P. (2002). Hydrological conditions over the western Mediterranean basin during the deposition of the cold sapropel 6 (ca. 175 kyr BP). *Earth and Planetary Science Letters* **202**, 481–494. Reproduced with permission.

grew because the temperature dependence of the carbon isotopic fractionation between calcite and DIC is relatively small. The $\delta^{13}C$ value of DIC depends critically on the sources of carbon: atmospheric CO_2 ($\delta^{13}C \sim -7‰$), carbonate derived from the bedrock ($\delta^{13}C \sim -5$ to 5‰), and oxidation of soil organic matter ($\delta^{13}C \sim -26$ to $-9‰$). Large variations in $\delta^{13}C_{ct}$ are observed on various timescales ranging from years (Frappier *et al.*, 2002) to tens of kiloyears. Variations in $\delta^{13}C_{ct}$ (the $\delta^{13}C$ value of the $CaCO_3$) can be caused by two distinct effects, both of which are potentially influenced by climate change:

- Change in proportion of C_3 to C_4 plants growing above the cave: $\delta^{13}C$ values of C_4 plants are 17‰ higher than those of C_3 plants; most C_4 plants are grasses; shifts from forest to grassland can cause an increase in $\delta^{13}C_{ct}$ (Dorale *et al.*, 1998).
- Change in the density of vegetation above the cave: Dissolution of limestone in the absence of soil cover occurs by reaction with atmospheric CO_2 ($\delta^{13}C \sim -7‰$), leading to higher $\delta^{13}C$ values of DIC in recharge water compared to that produced under C_3- and even C_4-based soils (Frumkin *et al.*, 2000).

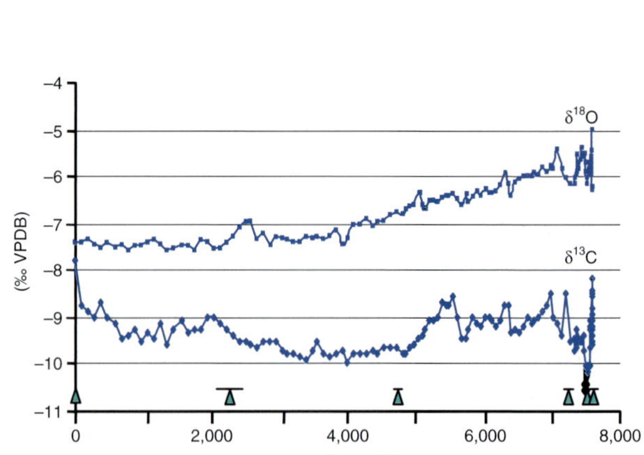

Figure 9 $\delta^{18}O$ and $\delta^{13}C$ record for stalagmite MF 1, McFail's Cave, New York state; gradient in $\delta^{18}O$ following hypsithermal at 7,500 BP represents temperature decrease of ca. 5 °C. From van Beynen, P. E., Schwarcz, H. P., Ford, D. C., and Morrison, J. (2004). Holocene climatic variability measured by speleothems in McFail's Cave, New York. *Journal of Karst and Cave Studies* **66**, 20–27. Reproduced with permission.

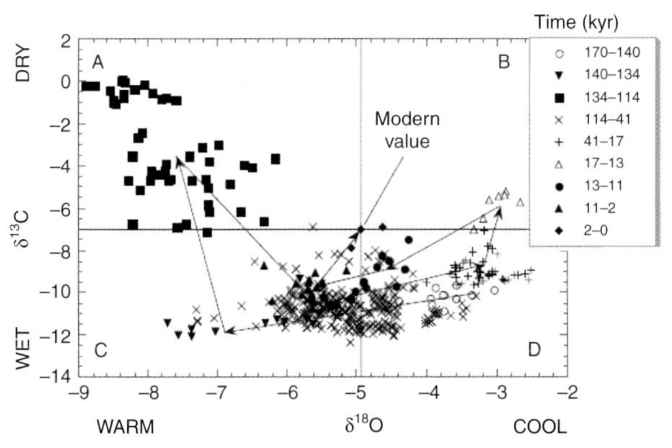

Figure 10 Isotopic and age data for stalagmite AF12 from Jerusalem, Israel, showing the relationship with climate. $\delta^{13}C$ increases with decreasing rainfall due to a decrease in the contribution of soil HCO_3^- to drip water. $\delta^{18}O$ increases with decreasing temperature (Frumkin *et al.*, 1999), inferred from correlation with marine record. From Frumkin, A., Ford, D. C., and Schwarcz, H. P. (2000). Paleoclimate and vegetation of the last glacial cycles in Jerusalem from a speleothem record. *Global Biogeochemical Cycles* **14**, 863–870. Reproduced with permission.

In addition, the extent of exchange of the recharge water with bedrock (limestone) may affect $\delta^{13}C$ of DIC of the emerging drip waters, depending on whether or not continued air exchange is possible in the passages (fissures and joints) through which the water flows (Hendy, 1971). Finally, during progressive outgassing of CO_2 into the cave atmosphere, both equilibrium and kinetic effects can affect the $\delta^{13}C$ of the precipitated calcite (Mickler *et al.*, 2004). These effects might be detectible as gradients in $\delta^{13}C_{ct}$ on growth layers.

Some studies have reported correlated shifts in $\delta^{18}O_{ct}$ and $\delta^{13}C_{ct}$ reflecting coupled changes in vegetation (e.g., C_3/C_4 ratios) and temperature (as reflected in $\delta^{18}O_{ct}$), whereas other studies show distinctly independent trends for each isotopic signal. Frumkin *et al.* (2000) showed that the trajectory of $\delta^{18}O_{ct}$ and $\delta^{13}C_{ct}$ defines simultaneous coupled changes in rainfall and temperature (**Fig. 10**).

Future Directions and Problems

Instrumental advances in the analysis of speleothems may allow further paleoclimatic insights. Higher resolution sampling of $\delta^{18}O_{ct}$ (and $\delta^{13}C_{ct}$) may be obtained using lasers or micromilling methods. Analyses of fluid inclusions for $\delta^{18}O$ and δD may provide better paleotemperatures, but these kinds of studies are challenged by problems in extraction and analysis of sub-microliter volumes of water. There may also be isotopic heterogeneity within the population of inclusions (Schwarcz and Yonge, 1983), partly revealed by infrared analysis (Fourier transform infrared spectroscopy) of calcite (unpublished data).

Future studies must take account of two concerns—one ethical and the other practical. Speleothems are admired for their natural beauty and should be disturbed as little as possible. Researchers must collect samples in a way that does not alter the appearance of the cave, either by using already broken stalagmites or working in caves that are in risk of destruction by erosion or human exploitation of the landscape. Second, we must be aware that, with the exception of actively growing speleothems, it is currently impossible to estimate the age of a speleothem *a priori.* Therefore, in general we cannot target a specific pre-Holocene period for study. As speleothem studies progress, previously analyzed, dated speleothems should be carefully archived for possible future research.

See also: **Carbonate Stable Isotopes**: Non-Lacustrine Terrestrial Studies. **Ice Core Records**: Correlations Between Greenland and Antarctica. **Radiocarbon Dating**: Calibration. **U-Series Dating**.

References

Bar-Matthews, M.., and Ayalon, A. (2003). Sea–land oxygen isotopic relationships from planktonic Foraminifera and speleothems in the eastern Mediterranean region and their implication for paleorainfall during interglacial intervals. *Geochimica et Cosmochimica Acta* **67**, 3181–3199.

Bard, E., Delaygue, G., Rostek, F., Antonioli, F., Silenzi, S., and Schrag, D. P. (2002). Hydrological conditions over the western Mediterranean basin during the deposition of the cold sapropel 6 (ca. 175 kyr BP). *Earth and Planetary Science Letters* **202**, 481–494.

Beck, J. W., Richards, D. A., Edwards, R. L., Silverman, B. W., Smart, P. L., Donahue, D. L., Hererra-Osterheld, S., Burr, G. S., Calsoyas, L., Jull, A. T., and Biddulph, D. (2001). Extremely large variations of atmospheric ^{14}C concentration during the last glacial period. *Science* **292**, 2453–2458.

Bischoff, J., and Fitzpatrick, J. (1991). U-series dating of impure carbonates; An isochron technique using total-sample dissolution. *Geochimica et Cosmochimica Acta* **55**, 543–554.

Burns, S. J., Matter, A., Frank, N., and Mangini, A. (1998). Speleothem-based paleoclimate record from northern Oman. *Geology* **26**, 499–502.

Dennis, P. F., Rowe, P. J., and Atkinson, T. C. (2001). The recovery and isotopic measurement of water from fluid inclusions in speleothems. *Geochimica et Cosmochimica Acta* **65**, 871–884.

Denniston, R. F., Gonzalez, L. A., Asmerom, Y., Baker, R. G., Reagan, M. K., and Bettis, E. A., 3rd (1999). Evidence for increased cool season moisture during the middle Holocene. *Geology* **27**, 815–818.

Dorale, J. A., Gonzalez, L. A., Reagan, M. K., Pickett, D. A., Murrell, M. T., and Baker, R. G. (1992). A high-resolution record of Holocene climate change in speleothem calcite from Cold Water Cave, northeast Iowa. *Science* **258**, 1626–1630.

Dorale, J. A., Edwards, R. L., Ito, E., and Gonzalez, L. A. (1998). Climate and vegetation history of the midcontinent from 75 to 25 ka: A speleothem record from Crevice Cave, Missouri, USA. *Science* **282**, 1871–1874.

Edwards, R. L., Gallup, C. D., and Cheng, H. (2003). Uranium-series dating of marine and lacustrine carbonates. *Reviews in Mineralogy and Geochemistry* **52**, 363–405.

Epstein, S., Buchsbaum, R., Lowenstam, H. A., and Urey, H. C. (1953). Revised carbonate-water isotopic temperature scale. *Bull. Geol. Soc. Amer* **64**, 1315–1326.

Fleitmann, D., Burns, S. J., Neff, U., Mudelsee, M., and Mangini, A. (2004). Palaeoclimatic interpretation of high-resolution oxygen isotope profiles derived from annually laminated speleothems from southern Oman. *Quaternary Science Reviews* **23**, 935–945.

Ford, D. C., and Williams, P. W. (1989). *Karst Geomorphology and Hydrology*. Unwin Hyman, London.

Frappier, A., Sahagian, D., Gonzalez, L. A., and Carpenter, S. J. (2002). El Nino events recorded by stalagmite carbon isotopes. *Science* **298**, 565.

Frumkin, A., Ford, D. C., and Schwarcz, H. P. (2000). Paleoclimate and vegetation of the last glacial cycles in Jerusalem from a speleothem record. *Global Biogeochemical Cycles* **14**, 863–870.

Frumkin, A., Ford, D. C., and Schwarcz, H. P. (1999). Continental paleoclimatic record of the last 170,000 years in Jerusalem. *Quaternary Research* **51**, 317–327.

Genty, D., Baker, A., Massault, M., Proctor, C., Gilmour, M., Pons-Branchu, E., and Hamelin, B. (2001). Dead carbon in stalagmites: Carbonate bedrock paleodissolution vs. ageing of soil organic matter. Implications for ^{13}C variations in speleothems. *Geochimica et Cosmochimica Acta* **65**, 3443–3457.

Hendy, C. H. (1971). The isotopic geochemistry of speleothems—The calculation of the effects of different modes of formation on the isotopic composition of speleothems and their applicability as palaeoclimatic indicators. *Geochimica et Cosmochimica Acta* **35**, 801–824.

Mangini, A., Spötl, C., and Verdes, P. (2005). Reconstruction of temperature in the Central Alps during the past 2000 yr from a $\delta^{18}O$ stalagmite record. *Earth and Planetary Science Letters* **235**, 741–751.

McDermott, F. C., Mattey, D. P., and Hawkesworth, C. J. (2001). Holocene climate variation insights from an exceptionally high-resolution speleothem $\delta^{18}O$ record from SW Ireland. *Science* **294**, 1328–1331.

McDermott, F., Schwarcz, H. P., and Rowe, P. (2006). Isotopes in speleothems. In *Developments in Paleoenvironmental Research, Vol. 10: Isotopes in Paleoenvironmental Research* (M. Leng, Ed.) Kluwer, Dordrecht, pp. 185–226.

McGarry, S., Bar-Matthews, M., Matthews, A., Vaks, A., Schilman, B., and Ayalon, A. (2004). Constraints on hydrological and paleotemperature variations in the eastern Mediterranean region in the last 140 ka given by the δD values of speleothem fluid inclusions. *Quaternary Science Reviews* **23**, 919–934.

Mickler, P. J., Banner, J. L., Stern, L., Asmerom, Y., Edwards, R. L., and Ito, E. (2004). Stable isotope variations in modern tropical speleothems; Evaluating equilibrium vs. kinetic. *Geochimica et Cosmochimica Acta* **68**, 4381–4393.

Niggemann, S., Mangini, A., Mudelsee, M., Richter, D. K., and Wurth, G. (2003). Sub-Milankovitch climatic cycles in Holocene stalagmites from Sauerland, Germany. *Earth and Planetary Science Letters* **216**, 539–547.

Richards, D. A., and Dorale, J. A. (2003). Uranium-series chronology and environmental applications of speleothems. *Reviews in Mineralogy and Geochemistry* **52**, 407–460.

Richards, D. A., Bottrell, S., Cliff, R. A., Stroehle, K., and Rowe, P. J. (1998). U-Pb dating of a speleothem of Quaternary age. *Geochimica et Cosmochimica Acta* **62**, 3683–3688.

Rozanski, K., Araguás-Araguás, L., and Gonfiantini, R. (1993). Isotopic patterns in modern global precipitation. In *Climate Change in Continental Isotopic Records* (P. K. Swart *et al.*, Eds.), Geophysical Monograph No. 78, pp. 1–36. American Geophysical Union, Washington, DC.

Schwarcz, H. P., and Yonge, C. (1983). Isotopic composition of paleowaters as inferred from speleothem and its fluid inclusions. In *Paleoclimates and Paleowaters: A Collection of Environmental Isotope Studies* (R. Gonfiantini, Ed.), Proceedings of Advisory Group Meeting STI/PUB/621, pp. 115–133. International Atomic Energy Agency, Vienna, Austria.

Schwarcz, H. P., Harmon, R. S., Thompson, P., and Ford, D. C. (1976). Stable isotope studies of fluid inclusions in speleothems and their paleoclimate significance. *Geochimica et Cosmochimica Acta* **40**, 657–665.

Serefiddin, F., Schwarcz, H. P., Ford, D. C., and Baldwin, S. (2003). Late Pleistocene paleoclimate in the Black Hills of South Dakota from oxygen isotope records in speleothems. *Palaeogeography, Palaeoclimatology, Palaeoecology* **203**, 1–17.

Serefiddin, F., Schwarcz, H., and Ford, D. (2005). Use of hydrogen isotope variations in speleothem fluid inclusions as an independent measure of paleoclimate. In *Isotopic and Elemental Tracers of Cenozoic Climate Change* (G. Mora and D. Surge, Eds.), Geological Society of America, Special Publication 395, p. 43–53.

Spötl, C., and Mangini, A. (2002). Stalagmite from the Austrian Alps reveals Dansgaard–Oeschger events during isotope stage 3; Implications for the absolute chronology of Greenland ice cores. *Earth and Planetary Science Letters* **203**, 507–518.

Spötl, C., Fairchild, I. J., and Tooth, A. (2005). Cave air control on dripwater geochemistry, Obir Caves (Austria): Implications for speleothem deposition in dynamically ventilated caves. *Geochimicia et Cosmochimica Acta* **69**, 2451–2468.

van Beynen, P. E., Schwarcz, H. P., Ford, D. C., and Morrison, J. (2004). Holocene climatic variability measured by speleothems in McFail's Cave, New York. *Journal of Karst and Cave Studies* **66**, 20–27.

Wang, Y. J., Cheng, H., Edwards, R. L., An, Z. S., Wu, J. Y., Shen, C. C., and Dorale, J. A. (2001). A high-resolution absolute-dated late Pleistocene monsoon record from Hulu Cave, China. *Science* **294**, 2345–2348.

Winograd, I. J., Coplen, T. B., Landwehr, J. M., Riggs, A. C., Ludwig, K. R., Szabo, B. J., Kolesar, P. T., and Revesz, K. M. (1992). Continuous 500,000-year climate record from vein calcite in Devils Hole, Nevada. *Science* **258**, 255–260.

Winograd, I., Landwehr, J., Ludwig, K. R., Coplen, T., and Riggs, A. C. (1997). Duration and structure of the past four glaciations. *Quaternary Research* **48**, 141–154.

Yonge, C., Ford, D. C., Gray, J., and Schwarcz, H. P. (1985). Stable isotope studies of cave seepage water. *Isotope Geoscience* **58**, 97–105.

Thermokarst Topography

C R Burn, Carleton University, ON, Canada

© 2007 Elsevier B.V. All rights reserved.

Introduction

Thermokarst terrain exhibits a pitted relief formed from the melting of ground ice. 'Thermokarst' is used widely to describe all processes and landforms that are associated with melting ground ice, but thermokarst, or thaw, lakes are probably the most well-known features (**Fig. 1**) (French, 1996). Thermokarst processes are generally only considered with respect to permafrost terrain, and soil

Figure 1 Thermokarst lakes near Mayo, Yukon Territory, Canada (see Burn and Smith (1990)). A, In July 1988; B, in July 1998, showing expansion that led to the merging of two lakes. Photographs taken by C. R. Burn.

displacement due to melting of seasonal frost is not included. Thermokarst topography and landforms are found throughout the permafrost regions, wherever ground ice is exposed or melted. Thermokarst features are uncommon in bedrock areas, but may occur locally in depressions where a sedimentary cover, including ground ice, has accumulated.

The geotechnical significance of permafrost derives from the presence of ice in the ground, and the terrain instability that follows thawing of ice-rich permafrost. The volume of water released as ice melts is commonly higher than the saturation water content of the soil, and so pore pressures rise. In flat terrain, consolidation of thawed soil leads to subsidence of the ground and an undulating topography, because the rate and extent of thawing is spatially variable. This variability stems primarily from the heterogeneous distribution of ground ice. In sloping terrain, high pore-water pressures lead to a reduction in effective stress and loss of soil strength so that slopes may fail. Landslides are common in permafrost terrain that has been burned by forest fire, because the resulting change in surface characteristics, not the heat of the fire, raises ground temperatures and leads to permafrost degradation (Burn, 1998). Thermokarst processes may be problematic for infrastructure and design of construction projects, which modify surface conditions and warm the ground. There are numerous examples of structures that have become unserviceable due to melting of ground ice beneath them (**Fig. 2**) (Crawford and Johnston, 1971; French, 1975). While the thermal effects of artificial surfaces can be modeled reliably (Andersland and Ladanyi, 1994), the ground ice distribution at a site is commonly less well-known and the source of most unforeseen incidents.

Figure 2 Two buildings in Dawson City, Yukon Territory, that are no longer in use because of thaw settlement beneath them. Photograph taken in July 1994 by C. R. Burn.

Permafrost and the Active Layer

Permafrost is ground that remains at or below 0° C for two or more years. Permafrost terrain consists of a seasonally thawed active layer underlain by perennially frozen ground. The base of the active layer is at the top of permafrost. Continental permafrost is found where the climate sustains frozen ground, with vast extents in northern regions of Canada, Alaska, and Siberia. Alpine or plateau permafrost occurs at high elevation, in areas where continental permafrost is not found. There are areas of permafrost in the major mountains of the world, and an extensive distribution in the Tibetan Plateau. In total, about 15% of the land surface of the Northern Hemisphere is underlain by permafrost (Zhang *et al.*, 2000).

Permafrost is classified by the proportion of the exposed land area of a region that contains perennially frozen ground. By definition, permafrost is considered continuous where it underlies more than 90% of the terrain. In these areas, the spatial mean ground temperature is well below 0° C, commonly >−5° C, and only a few locations have a microclimate that does not support permafrost. The amount of unfrozen ground increases in the discontinuous permafrost zones, so that the proportions of perennially frozen ground in the extensive and sporadic discontinuous permafrost zones are 90–50%, 50–10% respectively (Zhang *et al.*, 2000). Isolated patches of permafrost occur where less than 10% of the ground supports permafrost. The spatial mean ground temperature increases with the proportion of unfrozen ground. Thermokarst features occur in all permafrost regions, because the critical requirement is ground ice, while surface disturbances may raise ground temperatures above 0° C at any location if either permafrost is exposed to the atmosphere, or water accumulates to depths exceeding the winter ice thickness.

The thickness of the active layer depends on a variety of factors, principally the temperature and duration of the summer, thermal properties of near-surface materials, ice content of the soil, and the nature of heat transfer in the near surface (Williams and Smith, 1989). However, conditions during the preceding freezing season are also significant, particularly the average winter temperature and snow depth, which control the temperature of the active layer before thawing begins. The deepest active layers are found in coastal beaches and lake shores, where convection enhances the transfer of heat to the thawing front, and in bedrock where there is relatively little ice to melt each summer. Deep active layers are also common in well-drained dry ground, such as gravel. Relatively shallow active layers are found in peatlands, where evaporation within damp moss maintains cool ground temperatures and preserves permafrost. In the Northern Hemisphere, the southern limits of continental permafrost are found in peatlands.

Near-Surface Ground Ice

Characteristically, there is an ice-rich zone at the top of permafrost, just below the active layer. This layer may have particularly high ice contents in fine-grained sediments in moist environments (**Fig. 3**), but it has also been observed in bedrock. If the active layer deepens and permafrost thaws, then the near-surface ground ice will melt. The presence of this ice-rich horizon is the basis for thermokarst development, and the reason that permafrost terrain is considered sensitive to disturbance. Commonly, the ice occurs in subhorizontal lenses of centimeter thickness, segregating similarly sized beds of sediment (segregated ice lenses), or as bodies of almost pure 'massive' ice.

Figure 3 Ice-rich near-surface permafrost exposed in a slumped block at Pelly Island, western Arctic coast, Canada. The block is an ice-wedge polygon that was undercut by shoreline erosion, and split along the ice wedges. There is a black organic horizon at the top of permafrost, an ice vein has been truncated by the early Holocene thaw unconformity, and an ice wedge is exposed on the left side of the block. Determinations of the near-surface volumetric ice content of permafrost on the island ranged up to 58% (Burn, 1997). Photograph taken in Aug. 1993 by C.R. Burn.

The most common forms of massive ice are ice wedges that form from repeated infilling of thermal contraction cracks by snow melt. These linear bodies are V-shaped in cross section, are commonly capped by the active layer, and extend downward for about 5 m (**Fig. 3**). They form outline networks of tundra polygons, which are easily seen in lowlands, but are obscured on hill slopes by soil movement. Other massive icy bodies may be formed near the base of the active layer by injection of water during the autumn freezeback, due to pressures generated during freezing of coarse-grained sediments. In some locations, bodies of massive ice several meters thick may be buried glacier ice, remnant from the last glaciation, or masses formed during regional aggradation of permafrost.

Where the ice is distributed within the mineral soil, the ground is particularly ice rich in the upper 2 or 3 m of permafrost (**Fig. 3**). The ice-rich zone has been called the 'intermediate layer' of permafrost, because it may thaw episodically due to climate variation (Shur *et al.*, 2005). However, in some locations, these sediments were formerly within the active layer only during the Holocene climatic optimum, and have accumulated ice during the subsequent cooling and thinning of the active layer (Burn, 1997). Similarly ice-rich horizons are found just below the active layer in depositional environments, where the top of permafrost has aggraded with sedimentation or peat growth. The ice enrichment occurs due to water being drawn along the ground-temperature profile into freezing soil and permafrost at the end of summer and during freeze back of the active layer (Mackay, 1983).

Massive ice and individual ice lenses may be nearly 100% ice, although air bubbles are commonly distributed through the ice body. The ice-rich zone of segregated ice lenses is generally also of high ice content. In sandy materials, the volume of ice may be 30% greater than the saturated volume of thawed soil, but in fine-grained sediments, the super saturation may be greater (Burn, 1997).

Disturbances to Permafrost

The ground surface temperature, T_s, and thermal properties of ground materials control the heat flux through the active layer into permafrost. T_s is controlled by the surface energy balance, providing a solution for

$$Q^* = Q_H + Q_E + Q_G \quad [1]$$

where Q^* is the net radiation at a site, Q_H is the flux of sensible heat between the surface and the atmosphere, Q_E is the latent heat flux (evaporation or condensation) between the surface and the atmosphere, and Q_G is the ground heat flux between the surface and underlying soil. Changes to surface characteristics alter these fluxes. For instance, drying of the surface reduces Q_E, and a higher T_s is then required to increase Q_H and Q_G to maintain their equivalence with Q^*. In permafrost regions, changes in surface conditions characteristically lead to either deepening of the active layer, long-term degradation of permafrost, or both (Mackay, 1970). Deepening of the active layer is a result of surface warming in summer, but increases in snowfall or winter temperatures may also lead to an increase in thaw depth.

In addition to T_s, the thermal properties of surface materials are critical to the stability of permafrost, because they control the heat flux through the active layer. In a substantial portion of the permafrost region, especially in the discontinuous zones, perennially frozen ground is covered by a blanket of peat and mosses. This low-density organic layer has a low thermal conductivity and is commonly moist. As a result of its damp state, heat supplied from the surface is used in evaporation rather than warming the ground, thus maintaining cool conditions at depth. The organic layer is critical to the stability of the active layer, and disturbance of the surficial organic cover leads to thawing of near-surface permafrost (Mackay, 1970; Zoltai, 1993).

Forest fires create the most widespread surface disturbances in permafrost regions. Fires change several surface characteristics to influence the surface energy balance (**Fig. 4**).

First, Q^* increases at the ground surface following a reduction in shade and a decrease in surface albedo from charred debris. Second, the organic horizon may be incinerated, removing its effective insulation, and drying the surface. Third, the reduction in canopy cover and accumulation of deadfall increases snow depth, reducing heat loss from the ground in winter. Burn (1998) measured an increase in mean annual soil temperature, at 20-cm depth, of 1.25 °C between proximal burned and unburned sites in southern Yukon Territory, where an intense fire in 1958 had incinerated the surface organic horizon. The increase in ground temperature prompted long-term permafrost degradation at the burned site. A 25-yr record of active-layer development following fire at Inuvik, Northwest Territories, detected initial deepening of the active layer, but aggradation of permafrost as vegetation recovered (Mackay, 1995).

Figure 4 Change in surface conditions wrought by a forest fire in the boreal forest of central Yukon Territory, Canada, during July 1995. The Klondike Highway was used to control the fire. A, east of the highway where the forest was burned; Photographs by C. R. Burn. B, To west of the highway, with undisturbed forest.

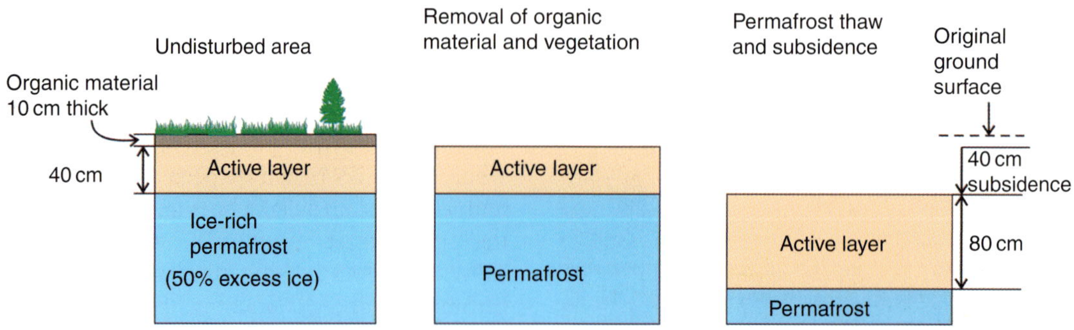

Figure 5 Thaw consolidation of permafrost terrain due to deepening of the active layer. The presence of excess ice in near-surface permafrost leads to lowering of the ground surface as the active layer deepens, and an absolute lowering of the permafrost table greater than the increase in active-layer depth. Diagram by S. Smith, Geological Survey of Canada. After Mackay JR (1970) Disturbances to the tundra and forest tundra environment of the western Arctic. *Canadian Geotechnical Journal* 7: 420–432.

Deepening of the active layer leads to subsidence of the ground where near-surface permafrost is ice rich (Mackay, 1970). **Figure 5** shows that active-layer deepening leads to lowering of the surface of permafrost by the sum of the increase in active layer thickness and ground subsidence. As the ice content of permafrost increases, for a given increase in active-layer depth, the surface subsidence increases. Where the degree of surface disturbance or near-surface ice content is spatially variable, subsidence will vary and thermokarst topography will develop. Mackay (1970) and French (1975) provide several examples of surface disturbances that led to ground subsidence.

Permafrost Degradation

Long-term degradation of permafrost will occur if the mean annual ground temperature rises above 0° C. Such long-term thaw is a function of the mean surface temperature, T, the volumetric latent heat in permafrost, L, and the thermal conductivity of thawed materials, λ. The Stefan solution has been successfully used under field conditions to approximate the relation between the depth of thaw, z, and the time, t, since thawing began (Burn and Smith, 1990; Burn, 1998, 2000):

$$z = \sqrt{(2\lambda Tt/L)} \qquad [2]$$

or,

$$z = b\sqrt{t} \qquad [3]$$

where,

$$b = \sqrt{(2\lambda T/L)} \qquad [4]$$

This model assumes that all heat entering the ground is used to thaw permafrost, rather than warm thawed soil above the permafrost or the permafrost itself, and that the surface temperature history involves a step change to the value that generates thawing. The assumptions are reasonable, because the latent heat of fusion is one to two orders of magnitude greater than the heat required to raise the same volume of soil materials by 1° C, and because natural changes in surface conditions require decades for recovery of vegetation to former conditions. The rate of permafrost degradation may be obtained by differentiating Eqn [3] with respect to time:

$$dz/dt = b/(2\sqrt{t}) \qquad [5]$$

Therefore, the rate of permafrost degradation declines with time since disturbance.

The Stefan solution is a one-dimensional analysis of permafrost degradation, and may be applied to the rate of thaw beneath a disturbance. However, at the edge of disturbances heat flow is both vertical and horizontal, in response to the temperature gradients across the boundary of the disturbance. In this case, the rate of horizontal expansion of the disturbance may be obtained from Eqn [5], with $t = 1$ (Burn and Smith, 1990). The relation may be applied when ground-ice conditions are spatially continuous. Commonly thermokarst disturbances are halted as local ground ice is exhausted.

Thermokarst Lakes

Ponding of water in depressions formed by melting of near-surface ground ice commonly leads to deepening of the active layer and degradation of permafrost. While the depth of a lake is less than one-half to two-thirds of the maximum winter-ice thickness, permafrost may be maintained beneath the lake, but once the pool exceeds this amount, the mean annual water-bottom temperature rises above 0° C, and permafrost is unsustainable beneath the pond (**Fig. 1**).

Figure 6 Tundra lake on Richards Island, western Arctic coast, Canada, showing the dark water of the deep central pool, and scars of retrogressive thaw slumps around the lake. Photograph taken in Aug. 1991 by C. R. Burn.

As a result, the pond deepens and expands. The rate of expansion depends on the ground ice content of near-surface permafrost. **Figure 1** shows actively expanding thermokarst lakes in central Yukon Territory, growing in ice-rich glaciolacustrine sediments (Burn and Smith, 1990). These lakes are enlarging at their shores at a rate of about $0.5\ \mathrm{m\,yr^{-1}}$.

There are numerous thaw lakes in permafrost regions because the frozen ground inhibits drainage. The Arctic coastal plains of northern Alaska, Siberia, and Canada's western Arctic contain thousands of lakes (**Fig. 6**), but most of these are thousands of years old (Rampton, 1988; Hinkel *et al.*, 2003). In many cases, the lakes have reached a steady state with the surrounding permafrost, so that the taliks of unfrozen ground beneath the lakes are of constant size (Burn, 2002). The taliks may be perched above permafrost for small lakes, but for large lakes, the talik may penetrate permafrost. The width of lakes where the taliks penetrate permafrost depends on the temperatures at the lake bottom and in the surrounding permafrost, and the geothermal gradient (Mackay, 1963). Many tundra lakes have shallow littoral terraces containing permafrost (**Fig. 6**), which may extend for over 100 m from the shore. The terraces shield permafrost around the lake from thaw, so these lakes are quasi-stable. Degradation of permafrost around these lakes may be initiated by terrestrial processes such as mass movements, rather than lacustrine effects.

The expansion of thermokarst lakes follows Eqn [2] (Burn and Smith, 1990). The rate of deepening depends on the ground-ice content, thermal conductivity of thawed lake-bottom sediments, and on the lake-bottom temperature. Lake-bottom temperature is sensitive to the extent of mixing in the water column. **Figure 7** indicates the annual temperature cycle in a small thermokarst lake in the boreal forest near Mayo, Yukon Territory, and a tundra lake near Canada's western Arctic coast. The tundra lake is well mixed in summer, so that warming of the lake bottom occurs rapidly. The forest surrounding the other lake acts as a wind break, and the lake is not mechanically mixed. Hence, warming at the lake bottom occurs by conduction,

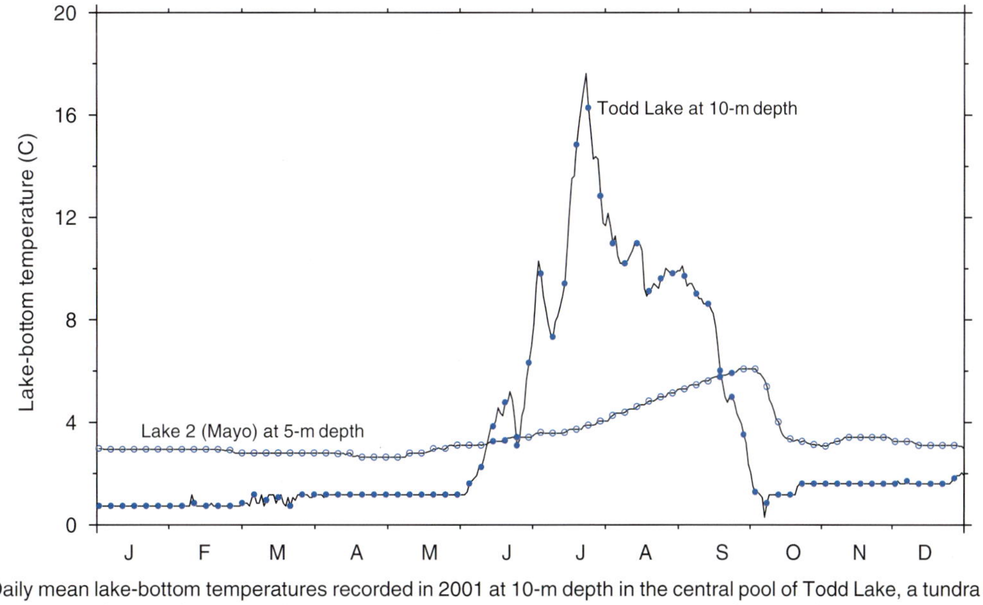

Figure 7 Daily mean lake-bottom temperatures recorded in 2001 at 10-m depth in the central pool of Todd Lake, a tundra lake near the western Arctic coast of Canada (Burn, 2002), and at 5-m depth in Lake 2 at Mayo (Burn and Smith, 1990). Todd Lake is well mixed by forced convection in summer, but Lake 2 is shielded from the wind and is poorly mixed. The series are plotted from daily data, but the symbols are plotted only every 5 days for clarity. The annual mean temperature at both of these depths is 3.5° C.

although cooling in fall is by density-induced free convection. A result is that the mean annual lake-bottom temperatures are similar at the sites. These data illustrate the independence of lake-bottom temperatures from climate, for the mean annual air temperature at Mayo is about −3° C, while at the coast it is about −10° C.

Thermokarst lakes may cease growing if the surrounding ground ice is exhausted, the basin becomes infilled by sedimentation or plant growth, or the lake drains. Drainage, or partial drainage, of thaw lakes is well documented, and commonly occurs if the active layer surrounding a lake deepens to below water level, so that once flow begins, it accelerates from a trickle to a torrent (Mackay, 1997). Drainage is rarely complete, and residual ponds are common in drained basins. Extensive depressions may be exposed by drainage of large lakes. These are best-known from Siberia, where they are called 'alasses.'

Retrogressive Thaw Slumps

Thaw slumps are the most active geomorphological features of permafrost terrain. Erosion of river banks, along the shores of lakes, or at the coast, may expose permafrost and initiate ablation of ice-rich ground. In river banks, high water during the spring flood may remove covering sediments and overhanging mats of peat, or undercut banks that subsequently collapse to reveal ice-rich ground. Along lake shores and at the coast, erosion during storms may similarly undercut ice-rich ground which is subsequently exposed (**Fig. 3**). Once exposed to air temperatures above 0° C, or to direct insolation, the ice-rich ground may thaw and turn to a mud slurry due to the high water content. This mud falls to the base of the exposure and flows down to the water body. The exposed permafrost commonly forms a steep headwall ranging from 20° to 90° (**Fig. 8A**), while the thawed materials accumulate in a more gentle foot slope of inclination between 3° and 15°, stretching from the foot of the headwall to the water body (**Figs. 8B** and **9**). Headwall ablation rates of up to $16\,m\,yr^{-1}$ have been measured, but more commonly the headwall retreat is about $10\,m\,yr^{-1}$ in a range of environments (Burn and Lewkowicz, 1990). Thaw slumps continue to develop until the headwall enters sediments of low ice content, as with the slump on **Figure 1**; the headwall is covered by an insulating layer of thawed debris, commonly an overhanging peat mat, or the terrain geometry allows the foot slope to rise and cover the permafrost. The longevity of thaw slumps therefore depends on ground-ice distribution and local topography. Many thaw slumps near the coast are rejuvenated by continuing coastal erosion, for instance, the slumps in **Figure 9** are within a larger stabilized area.

Direct thawing of permafrost occurs in the ablating headwall, where the gradient is too steep to retain thawed material on the surface. In contrast, on the foot slope, thawed material accumulates to shield the underlying permafrost. The disturbance of surface conditions from the previously undisturbed vegetation cover may be sufficient to cause long-term thawing of the permafrost beneath the foot slope. The foot slope commonly develops a deeper snow cover than the surrounding undisturbed ground because the rough accumulation of surface debris prevents drifting. In the boreal forest, trees are toppled by the thaw slump, so shading of the surface is less on the foot slope than in undisturbed ground. In the boreal

Figure 8 A retrogressive thaw slump near Mayo, central Yukon Territory, Canada. A, Ice-rich glaciolacustrine sediments exposed in the 10-m-high headwall (Jun. 1984); B, overview of the slump showing the headwall and footslope of about 350 m between the headwall and Stewart River (Aug. 1984). River erosion of ice-rich permafrost initiated the slump in 1949, and headwall retreat from the river at up to $16\,m\,yr^{-1}$. Photographs taken by C. R. Burn.

Figure 9 Retrogressive thaw slumps initiated by coastal erosion of Pelly Island, western Arctic Canada. Photograph taken in Aug. 1993 by C. R. Burn.

forest, such disturbance raises the mean ground temperature above 0° C, and permafrost is only reestablished after vegetation succession restores the original forest. An increase in mean annual ground temperature of about 4° C over undisturbed conditions has been measured at the thaw slump illustrated in **Figure 8** (Burn, 2000).

Active-Layer Detachments

In permafrost regions, failures commonly occur on low-angled slopes (McRoberts and Morgenstern, 1974). Characteristically, the failures occur during relatively warm summers or after forest fires (Harris and Lewkowicz, 2000). The failure involves detachment of the active layer from the underlying permafrost so that it may slide downslope (**Fig. 10**).

Such active-layer detachments have also been called 'skin flows' (McRoberts and Morgenstern, 1974) and 'active layer glides' (Mackay and Mathews, 1973). The detachment accumulates at the base of the slope as an involuted sheet, and may override and fan out there.

The development of active-layer detachments follows thawing of ice-rich ground at the bottom of, and below, the active layer. Water released by thawing excess ice is, by definition, of greater volume than the pore space in mineral soil, so the pore-water pressure increases. This reduces the effective stress at the thaw front, and may lead to failure on even gentle slopes (Harris and Lewkowicz, 2000). Slope stability may be maintained if the excess pore water drains as the thawed soil consolidates, reducing the pore pressure. The critical soil parameter is the thaw consolidation ratio, R (McRoberts and Morgenstern, 1974):

$$R = b/(2\sqrt{c_v}) \qquad [6]$$

where c_v is the coefficient of consolidation. The shear strength of the soil may decline over time due to slow deformation at the base of the active layer toward the end of summer, and reduce the excess pore pressures required for detachment (Harris and Lewkowicz, 2000). Active-layer detachments are common in fine-grained soil and diamicton, where the hydraulic conductivity is low and consolidation is slow, but they may occur in other materials if the rate of thaw is sufficiently high. They occur in clusters, both in space and time, due to their association with particular events and sedimentary materials.

Sedimentary Record

The sedimentary sequences in thermokarst basins contain primary deposits laid down during initial subsidence and secondary deposits transported into the basin. The primary deposits may contain disorganized remnant vegetation, particularly wood pieces,

Figure 10 Active-layer detachment failure on Pelly Island, western Arctic coast, Canada, August 1993. A, Aerial view of failure; B, failure surface showing slickensides. The slickensided surface of exposed ground indicates that the active layer has moved intact over permafrost. Photographs by C. R. Burn.

that grew before thermokarst initiation, and shells from pond fauna (Burn and Smith, 1990). In the boreal forest, trees either topple into thermokarst lakes or are gradually submerged (Burn and Smith, 1990), and such organic matter is well preserved in the anaerobic lake bottom. Primary deposits of thaw slumps are massive clastic sediments, due to thawing and consolidation of soil in the foot slope, containing dispersed organic accumulations of peat or woody material, representing debris that fell into the slump. The sedimentary structure left by active-layer detachment is convoluted, as a result of deformation of the active layer at the base of the slope.

Secondary deposits accumulate in thermokarst basins from mass wasting of the surrounding slopes (Murton, 1996). These follow a three-part sequence governed by the evolution of the basin. First, clastic sediments accumulate in the basin from thaw slumping or detachment around the edges of the basin or lake. Second, as ground ice around the basin is exhausted and as slope gradients decline, deposition shifts to sediments entrained around the lake edges and transported to the center in suspension. Third, after drainage, further slow mass wasting into the basin accompanies organic accumulation and eolian deposition. Thermokarst sequences may be observed in the headwalls of retrogressive thaw slumps (Burn and Smith, 1990), and, particularly, in coastal exposures (Murton, 1996).

Thermokarst and Climate Change

Thawing of permafrost is a consequence of climate warming (Jorgenson *et al.*, 2006), and deepening of the active layer increases emissions of methane and carbon dioxide due to increased respiration by soil microfauna (Christensen, 2004). The widespread climate warming in Arctic regions since the 1980s has been associated with thermokarst development in some areas (Payette *et al.*, 2004; Jorgenson *et al.*, 2006), and, in the past, periods of climate warming have also been periods of thermokarst lake initiation (Rampton, 1988; Burn and Smith, 1990). However, as discussed above, initiation of thermokarst features may occur on a site-specific basis, and may follow surface disturbance, irrespective of climate warming. For example, the lakes illustrated in **Figure 1** appear to have been initiated at various times during the last 800 yr (Burn and Smith, 1990). Furthermore, once a thermokarst basin is filled with water, the thermal regime of the lake bottom is disconnected from climatic conditions by the properties of the water, and thawing of ice-rich permafrost around the lake may proceed regardless of the climate (Mackay, 1978).

The prospect of thermokarst development accelerating during the twenty-first century has been a key factor stimulating research into permafrost conditions, especially with renewed interest in northern hydrocarbon development. Wholesale eradication of permafrost will take many millennia, but the upper layers of the ground appear to exhibit some response to climate warming already.

See also: **Ice Core Methods**: Borehole Temperature Records. **Periglacial Landforms**: Active Layer Processes; Ice Wedges and Ice Wedge Casts; Permafrost; Pingos and Pingo Scars; Slope Deposits and Forms.

References

Andersland, O. B., and Ladanyi, B. (1994). *An Introduction to Frozen Ground Engineering.*Chapman and Hall, New York.

Burn, C. R. (1997). Cryostratigraphy, paleogeography, and climate change during the early Holocene warm interval, western Arctic coast, Canada. *Canadian Journal of Earth Sciences* **34**, 912–925.

Burn, C. R. (1998). The response (1958–1997) of permafrost and near-surface ground temperatures to forest fire, Takhini River valley, southern Yukon Territory. *Canadian Journal of Earth Sciences* **35**, 184–199.

Burn, C. R. (2000). The thermal regime of a retrogressive thaw slump near Mayo, Yukon Territory. *Canadian Journal of Earth Sciences* **37**, 967–981.

Burn, C. R. (2002). Tundra lakes and permafrost, Richards Island, western Arctic coast, Canada. *Canadian Journal of Earth Sciences* **39**, 1281–1298.

Burn, C. R., and Lewkowicz, A. G. (1990). Retrogressive thaw slumps. *The Canadian Geographer* **34**, 273–276.

Burn, C. R., and Smith, M. W. (1990). Development of thermokarst lakes during the Holocene at sites near Mayo, Yukon Territory. *Permafrost and Periglacial Processes* **1**, 161–176.

Christensen, T. R., Johansson, T., Åkerman, H. J., *et al.* (2004). Thawing sub-arctic permafrost: Effects on vegetation and methane emissions. *Geophysical Research Letters* **31**, L04501, (doi:10.1029/2003GL018680).

Crawford, C. B., and Johnston, G. H. (1971). Construction on permafrost. *Canadian Geotechnical Journal* **8**, 236–251.

French, H. M. (1975). Man-induced thermokarst, Sachs Harbour airstrip, Banks Island, Northwest Territories. *Canadian Journal of Earth Sciences* **12**, 132–144.

French, H. M. (1996). *The Periglacial Environment*, 2nd edn. Longman, Harlow.

Harris, C., and Lewkowicz, A. G. (2000). An analysis of the stability of thawing slopes, Ellesmere Island, Nunavut, Canada. *Canadian Geotechnical Journal* **37**, 449–462.

Hinkel, K. M., Eisner, W. R., Bockheim, J. G., Nelson, F. E., Peyerson, K. M., and Dai, X. (2003). Spatial extent, age, and carbon stocks in drained thaw lake basins on the Barrow Peninsula, Alaska. *Arctic, Antarctic, and Alpine Research* **35**, 291–3000.

Jorgenson, M. T., Shur, Y. L., and Pullman, E. R. (2006). Abrupt increase in permafrost degradation in arctic Alaska. *Geophysical Research Letters* **33**, L04501, (doi:10.1029/2005GL024960).

Mackay, J. R. (1963). *The Mackenzie Delta Area, N.W.T. Geographical Branch Memoir 8.* Department of Mines and Technical Surveys, Ottawa.

Mackay, J. R. (1970). Disturbances to the tundra and forest tundra environment of the western Arctic. *Canadian Geotechnical Journal* 7, 420–432.
Mackay, J. R. (1978). Freshwater shelled invertebrate indicators of paleoclimate in northerwestern Canada during late glacial times: Discussion. *Canadian Journal of Earth Sciences* **15**, 461–462.
Mackay, J. R. (1983). Downward water movement into frozen ground, western arctic coast, Canada. *Canadian Journal of Earth Sciences* **20**, 120–134.
Mackay, J. R. (1995). Active layer changes (1968–1993) following the forest-tundra fire near Inuvik, N.W.T., Canada. *Arctic and Alpine Research* 27, 323–336.
Mackay, J. R. (1997). A full-scale field experiment (1978–1995) on the growth of permafrost by means of lake drainage, western Arctic coast: A discussion of the method and some results. *Canadian Journal of Earth Sciences* **34**, 17–33.
Mackay, J. R., and Mathews, W. H. (1973). Geomorphology and Quaternary history of the Mackenzie River Valley near Fort Good Hope, N.W.T., Canada. *Canadian Journal of Earth Sciences* **10**, 26–41.
McRoberts, E. C., and Morgenstern, N. R. (1974). The stability of thawing slopes. *Canadian Geotechnical Journal* **11**, 447–469.
Murton, J. B. (1996). Thermokarst-lake-basin sediments, Tuktoyaktuk Coastlands, western Arctic Canada. *Sedimentology* **43**, 737–760.
Payette, S., Delwaide, A., Caccianiga, M., and Beauchemin, M. (2004). Accelerated thawing of subarctic peatland permafrost over the last 50 years. *Geophysical Research Letters* **31**, L04501, (doi:10.1029/2004GL020358).
Rampton, V. N. (1988). Quaternary geology of the Tuktoyaktuk coastlands, Northwest Territories. *Geological Survey of Canada Memoir* **423**.
Shur, Y., Hinkel, K. M., and Nelson, F. E. (2005). The transient layer: Implications for geocryology and climate-change science. *Permafrost and Periglacial Processes* **16**, 5–17.
Williams, P. J., and Smith, M. W. (1989). *The Frozen Earth: Fundamentals of Geocryology*. Cambridge University Press, Cambridge.
Zhang, T., Heginbottom, J. A., Barry, R. G., and Brown, J. (2000). Further statistics on the distribution of permafrost and ground ice in the northern hemisphere. *Polar Geography* **24**, 126–131.
Zoltai, S. C. (1993). Cyclic development of permafrost in the peatlands of northwestern Alberta, Canada. *Arctic and Alpine Research* **25**, 240–246.

Terrestrial Teeth and Bones

H Bocherens, Universität Tübingen, Tübingen, Germany
D G Drucker, Allée de l'Université, Nanterre, France

© 2007 Elsevier B.V. All rights reserved.

Introduction

Since the 1970s, stable isotopes in terrestrial teeth and bones have been providing paleoenvironmental and paleobiological information on the Quaternary period. Indeed, the skeletal tissues of an animal through its diet and drinking water yield paleoenvironmental information in the form of isotopic ratios. This information is preserved in molecules and minerals found in fossil teeth and bones.

The most commonly used pairs of isotopes are $^{13}C/^{12}C$, $^{15}N/^{14}N$, and $^{18}O/^{16}O$. Isotopes correspond to different types of atoms for a given chemical element, meaning that two isotopes of the same element have the same number of protons, but differ in the number of neutrons. For instance, an atom of ^{12}C (carbon-12) has six protons and six neutrons in its core, while an atom of ^{13}C (carbon-13) has six protons and seven neutrons. Therefore, ^{13}C exhibits a higher atomic weight than ^{12}C; it is thus referred to as the heavy isotope (contrary to ^{12}C which is referred to as the light isotope). Nevertheless, both isotopes enter in the same molecules with the same types of chemical bonds. The relative proportion of ^{13}C and ^{12}C in a given molecule is not random but depends on the relative content of both isotopes in the source of carbon used to synthesize this molecule and on the differences of behavior between both isotopes during chemical reactions.

The relative abundances of different isotopes are much higher for light isotopes than for heavy ones (**Table 1**). The variations of isotopic abundances in natural samples are very small. They are measured using an isotopic ratio mass spectrometer, which separates and quantifies the number of heavy and light isotopes of a given element. In order to ensure accuracy, measurements are performed simultaneously on the test sample and on a standard, which allows corrections that make all results comparable to one another. Due to this measurement strategy, the results are expressed as relative abundances, known as 'delta' (δ) values, which are defined as follows: $\delta^E X = [(^E R_{\text{sample}} - {}^E R_{\text{standard}})/{}^E R_{\text{sample}}] \times 1000$ (‰), where $^E X$ stands for ^{13}C, ^{15}N, or ^{18}O, respectively. International reference standards have been established for $\delta^{13}C$ values (marine carbon PDB), for $\delta^{15}N$ values (atmospheric dinitrogen), and for $\delta^{18}O$ (an average mixture of oceanic waters SMOW).

Table 1 Stable isotopic compositions of carbon, nitrogen, and oxygen

Chemical element	*Atomic weight*	*Limits of observed terrestrial isotopic abundances (atom%)*		
Carbon	12	99.01	–	98.86
	13	1.15	–	0.99
Nitrogen	14	99.652	–	99.605
	15	0.395	–	0.348
Oxygen	16	99.7780	–	99.7510
	18	0.2084	–	0.1879

From Bocherens and Mariotti (2002).

Paleobiological Tracking by Isotopes in Teeth and Bones

Bones and teeth are composed of both organic and a mineral fractions which are synthesized during the lifetime of a vertebrate (**Fig. 1**). The organic fraction of bone and dentine is mainly formed of a protein, collagen, which contains carbon (around 40%) and nitrogen (around 15%), while the mineral fraction contains carbon and oxygen. Enamel is formed of a very small organic fraction devoid of collagen. In bone, dentine, and enamel, the mineral fraction is composed of a calcium phosphate (apatite) with many impurities, including 3–5% carbonate. The crystal size of apatite is much larger in enamel than in bone and dentine. The isotopic signatures of nitrogen are measured in the organic fraction, while oxygen isotopic signatures are measured in the mineral fraction, both in phosphate and carbonate. The isotopic signatures of carbon can be measured in the organic and mineral phases of bone and dentine, and in the mineral fraction of enamel.

Carbon

All the carbon of an organism comes from its dietary intake, in the form of proteins, carbohydrates, and lipids. Some of these nutrients are incorporated directly by the organism and sequestered in different tissues, while other molecules are synthesized by the organism from dietary nutrients. Due to these different biochemical characteristics and isotopic fractionations, the average carbon isotopic abundance of a vertebrate is close to that of its average diet, but those recorded in a given tissue or molecule exhibit specific differences (**Fig. 2**). The $\delta^{13}C$ values of

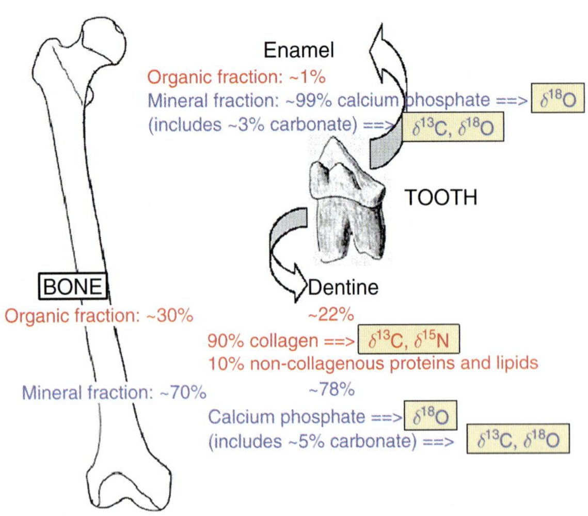

Figure 1 Summary of the composition of bones and teeth, with emphasis on the isotopic signatures that can be retrieved from bone, dentine, and enamel.

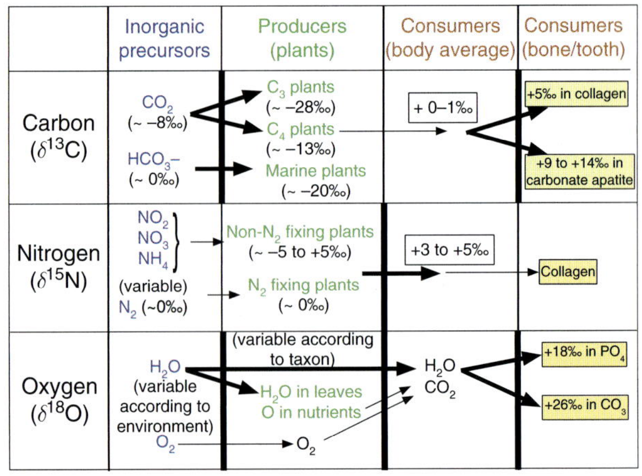

Figure 2 Summary of the isotopic fractionation factors associated with the transfer of carbon, nitrogen, and oxygen within food webs and the hydrological cycle, with emphasis on bone and tooth tissues. Bold arrows indicate steps where significant isotopic fractionation occurs.

collagen are typically 5‰ more positive than those of the average diet, while the $\delta^{13}C$ values of the carbonate fraction of bone and tooth mineral fraction is 9–14‰ more positive than that of the average diet (Cerling and Harris, 1999; Bocherens and Mariotti, 2002). Therefore, the $\delta^{13}C$ values of collagen and carbonate apatite are typically used to track the type of plant food consumed by herbivores and the type of plants at the base of the food web to which predators belong.

The $\delta^{13}C$ values discriminate between different types of plants, principally between marine and terrestrial plants. Among terrestrial plants, the carbon isotope signature varies between plants using the two main photosynthetic pathways, called C_3 and C_4 ('C_3 plants' and 'C_4 plants'). C_4 plants are absent or very limited in environments with a mild or cold growing season, as in Europe, northern latitudes, and high altitudes (Ehleringer *et al.*, 1997). When present, C_4 plants are grasses and forbs. In environments where all plants use the C_3 photosynthetic pathway, an isotopic distinction can be seen between plants growing under a closed canopy and those at the top of the canopy or growing in an open environment. Since herbivore teeth and bones record the $\delta^{13}C$ values of their plant food, it is possible to identify which kind of plant was consumed by a herbivore, and therefore the type of environment in which it lived (**Fig. 3**).

Nitrogen

Nitrogen is incorporated through dietary intake in the organic molecules of an organism's tissues. It is usually measured in the collagen preserved in fossil bones. Contrary to carbon, the isotopic signature of nitrogen is significantly enriched in vertebrate tissues

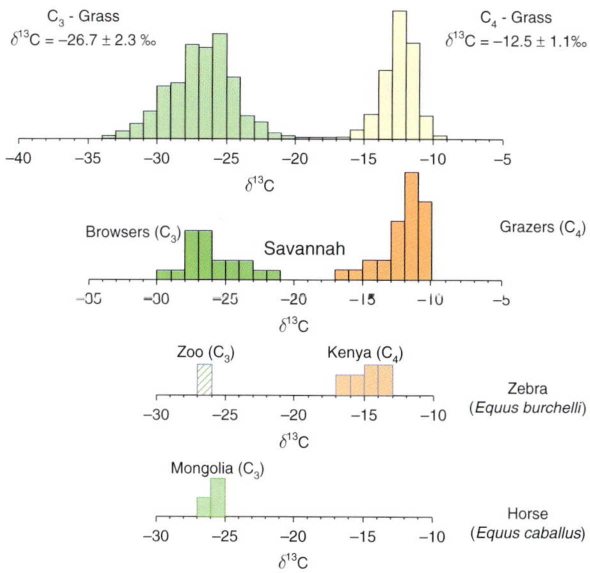

Figure 3 An example of the reconstruction of the carbon isotopic composition of the diet consumed by modern equids (horses and allies) based on their habitat. The $\delta^{13}C$ of diet have been calculated by adding 14 ‰ to the $\delta^{13}C$ values measured in carbonate hydroxylapatite (Cerling and Harris, 1999). The reconstructed diet of zebras (*Equus burchelli*) living in a Kenyan savanna exhibit $\delta^{13}C$ values similar to those of C_4 grasses, while those living in a zoo and fed with C_3 plants track this different diet, and exhibit $\delta^{13}C$ values similar to those of horses (*Equus caballus*) eating C_3 plants in a Mongolian grassland.

relative to its average diet, typically by 3–5 ‰ (**Fig. 2**, review in Bocherens and Mariotti (2002)). Therefore, the nitrogen isotopic signature of a given individual depends on the isotopic signature at the base of the food web to which it belongs (in the plants), and on the position of the species within the food web (herbivore or predator).

Oxygen

Oxygen isotopic signatures are measured in the mineral phase of enamel, dentine, or bone in the phosphate or carbonate fractions (Kohn and Cerling, 2002). For both fractions, the $\delta^{18}O$ values are related to those of environmental water through drinking water. However, there are complications due to the incorporation of water from food, and mixture with oxygen from respiration (**Fig. 2**). This leads to varying species-dependent relationships between $\delta^{18}O$ values in carbonate or phosphate and $\delta^{18}O$ values in environmental water taken by the organisms. However, both $\delta^{18}O$ values are linked by a clear relationship independent of the organism.

Variation in oxygen isotope ratios studied from terrestrial environments is ascribed to environmental temperature changes, with warmer weather resulting in more positive $\delta^{18}O$ values and cooler weather resulting in more negative $\delta^{18}O$ values in meteoric waters. However, the pattern is different in warm environments, where the temperature is higher than 20° C. In such contexts, the $\delta^{18}O$ values decrease when the amount of precipitation increases. Local parameters, such as evaporation in ponds or streams originating at high altitudes, may provide drinking water to terrestrial vertebrates with $\delta^{18}O$ values different from those of local precipitation, and thus complicate the interpretation of the oxygen isotopic composition of fossil teeth and bones.

Chronological Resolution of Skeletal Tissues

Different tissues record the isotopic signature of food or drinking water during the period of their synthesis. Typically, bone grows during the first stages of ontogeny, but in higher vertebrates such as mammals and birds, it continues to remodel and therefore incorporates newer carbon, nitrogen, and oxygen atoms. An adult mammal thus dies with its bone isotopic signature reflecting the last years of its lifetime. In contrast to this, mammalian dental tissues such as dentine and enamel form during a limited period of an individual's lifetime and, once formed, do not remodel. In an adult mammal, teeth exhibit isotopic signatures corresponding to the early years of the individual, depending on the chronology of tooth development for a given species. Cross sections of dentine and enamel retain their isotopic signature in a chronological order reflecting time of assimilation. Tooth enamel may thus record seasonal variations of $\delta^{18}O$ values (**Fig. 4**). The level of resolution in such tissues, especially tooth enamel, is still under debate since each volume of tissue forms over a significant amount of time (e.g., Kohn and Cerling (2002)).

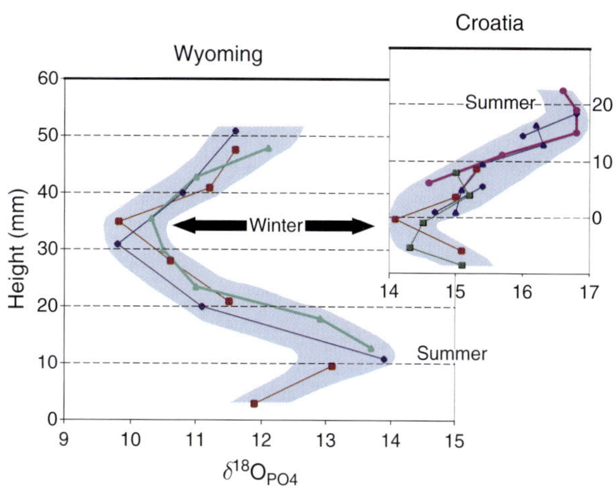

Figure 4 Variations of $\delta^{18}O$ in deer enamel reflect seasonal variations, in specimens from Wyoming and Croatia (Fricke *et al.*, 1998). The differences in amplitudes and absolute values of $\delta^{18}O$ values are due to different climatic regimen in both areas.

Physiological Effects (Suckling, Hibernation, Starvation Stress)

Some events in the life of a vertebrate can complicate the record of dietary isotopic signatures. In mammals, diet changes dramatically between the nursing and adult stages, following weaning. Maternal milk exhibits isotopic signatures different from those of the average adult diet, especially with more positive $\delta^{15}N$ values and more positive $\delta^{18}O$ values (Wright and Schwarcz, 1998). Therefore, suckling mammals exhibit isotopic differences relative to their adult counterparts, and tissues formed during this formative period of life, such as deciduous teeth and first molars, usually exhibit isotopic shifts relative to other teeth and bone tissues formed after weaning.

In hibernating mammals, such as bears, carbon and nitrogen isotopic signatures shift in the blood during the winter sleep and these shifts are recorded in tooth dentine (Bocherens, 2004). Starvation stress is also a phenomenon that may induce ^{15}N shifts in a vertebrate tissue (Ambrose, 1991).

Diagenetic Alteration

The chemical composition of bone and tooth is modified after an individual's death, through physical, chemical, and biochemical mechanisms called diagenesis. The intensity of diagenetic alteration depends not only on the time elapsed since death, but also on sedimentary conditions. High temperatures and humidity as well as high levels of microbial activity will increase the intensity of diagenetic transformation, increasing the chances of alteration of the biogenic isotopic signatures.

Organic Fraction

Collagen in fossil bones and dentine can survive for tens of thousands of years under favorable conditions. Cold and dry climatic conditions are more favorable than warm and humid ones, while cave deposits are more favorable than open air sites. Collagen preservation in fossil skeletal remains can be assessed by the nitrogen content of bulk sample as a proxy (e.g., Bocherens *et al.* (2005a)).

Fossil collagen can be purified through several methods. The reliability of the isotopic signatures of ancient collagen is assessed through its chemical similarity to collagen from fresh bone. Fossil organic extracts with atomic C:N ratios lower than 2.9 or higher than 3.6 are considered unreliable, as well as those containing less than 5% nitrogen (e.g., Ambrose (1990)).

Mineral Fraction

While it is almost always possible to measure carbon and oxygen signatures from fossil bones and teeth, the key question is to determine whether the measured values correspond to those recorded in the organism's tissues when it was alive. Evaluating the extent of diagenetic alteration can be done either using the pattern of isotopic variations observed in the fossil samples as compared to equivalent modern ones, or using indirect tracers of modifications, such as crystallinity indexes and uptake of trace elements (e.g., Kohn and Cerling (2002)). It is commonly assumed that the isotopic signatures of enamel are much more stable through time than those of dentine and bone, while the phosphate fraction is more stable than the carbonate fraction. However, the phosphate fraction can also be affected under special circumstances. The best approach is to assess the extent of diagenetic alteration in each studied site using as many different proxies as possible (e.g., Kohn *et al.* (1999)).

Reconstruction of Terrestrial Paleoenvironments

Tropical Environments

Paleoenvironments where C_3 and C_4 plants coexist are particularly favorable for isotopic tracking with carbon in teeth and bones. Indeed, C_4 plants increase in proportion as climatic conditions become drier and warmer, and C_3 plants as moisture increases. Since the difference between $\delta^{13}C$ values of these plants relative to C_3 plants is around 12‰, the $\delta^{13}C$ value recorded in the skeletal tissues of fossil vertebrates document these changes in the proportions of consumed plants. However, other factors such as the atmospheric pressure of CO_2 also play a role in this equation (Ehleringer *et al.*, 1997; Koch *et al.*, 2004). The main climatic parameter in these environments is the amount of annual rainfall, and nitrogen isotopic signatures of herbivores can be used to track past variations in aridity. Some examples are given below.

Development of C_4 biomes during the late Quaternary The open environments of the southern half of the North American continent are a place of competition between C_3 and C_4 herbaceous plants. Today there is a clear gradient of decreasing proportions of C_4 plants towards the North, with about 10% C_4 plants at 48° N. Carbon and oxygen isotopic studies of late Pleistocene herbivore tooth enamel from southwestern USA demonstrated the role of low atmospheric CO_2 in the expansion of C_4 plants during the Last Glacial Maximum (LGM), an expansion that could not be predicted based on temperature and precipitation changes alone (e.g., Koch *et al.* (2004)).

Studies based on late Quaternary herbivore tooth enamel showed that the proportions of C_3 and C_4 plants in the Northern Cape Province, South Africa, did not reflect the predictions of climatic models about winter and summer rainfall regimes (Lee-Thorp and Beaumont, 1995) and documented episodes of wetter and drier conditions since 16 ka (Smith *et al.*, 2002).

Aridity In tropical environments, aridity has been recognized as a factor influencing the δ^{15}N of herbivore collagen, with more positive δ^{15}N values under more arid conditions (e.g., Ambrose (1991) and Schwarcz *et al.* (1999)). This led to the possibility of quantifying past annual rainfall based on the δ^{15}N of herbivorous taxa, for instance macropods (kangaroos) in Australia (Gröcke *et al.*, 1997).

Decreasing humidity has an effect on bone because of increasing evapotranspiration in leaves leading to δ^{18}O increase in leaf water, which is recorded in bone phosphate (e.g., Ayliffe and Chivas (1990) and Luz *et al.* (1990)).

Temperate and Boreal Environments

In terrestrial environments, where all plants use the C_3 photosynthetic pathway, plant δ^{13}C values exhibit some differences related to environmental parameters (e.g., Heaton, 1999). In particular, plants growing under a closed canopy exhibit significantly more negative δ^{13}C values relative to those growing at the top of the canopy or in open environments. This carbon isotopic fractionation is transferred to herbivores consuming such plants in dense canopy forests. It is thus possible to track the changes of habitat of large herbivores at the beginning of the Holocene in Europe (**Fig.** 5; e.g., Drucker *et al.* (2003a))

Variations in δ^{15}N values in ungulate bone collagen seem to relate to changes in soil microbial activity during the climatic oscillations that occurred in Europe since 30 ka (Drucker *et al.*, 2003a, b; Richards and Hedges, 2003; Hedges *et al.*, 2004): low temperatures lead to decreased microbial activity and thus reduced nitrogen isotopic fractionation that is transferred in plants consumed by herbivores. Indeed, a decrease in δ^{15}N values is recorded during the cold peaks of the LGM and Younger Dryas, while an increase in δ^{15}N values is documented during the warming of the Bölling–Alleröd Interstadial and that of the Early Holocene (Preboreal and Boreal). These nitrogen isotopic excursions present geographic variations according to the intensity of the temperature changes.

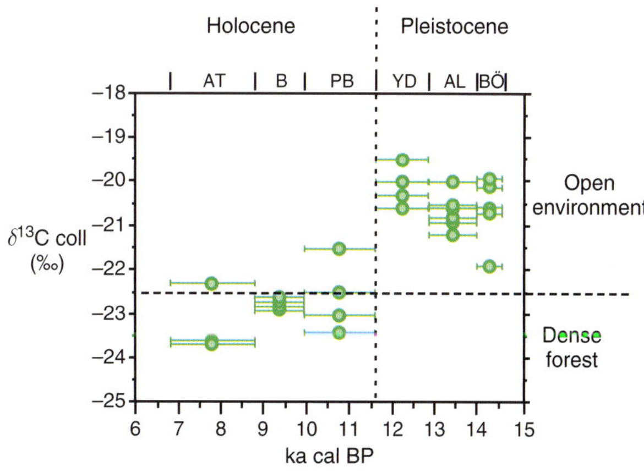

Figure 5 Evolution of collagen δ^{13}C values in red deer (*Cervus elaphus*) bones from French Jura (Drucker *et al.*, 2003a). The late Pleistocene samples exhibit δ^{13}C values indicative of open environments, while Holocene specimens present more negative δ^{13}C values, similar to those exhibited by modern large herbivores dwelling in a closed canopy forest. Abbreviation-keys for chronozones: AL = Alleröd, AT = Atlantic, B = Boreal, BÖ = Bölling, YD = Younger Dryas, PB = Preboreal.

Seasonality

Using carbon and oxygen isotopic variations in hypsodont tooth enamel, as in horse and bison, allowed the reconstruction of seasonal changes in consumed plant food and precipitation isotopic signatures, thus leading to a better understanding of past climatic regimen and events preceding the death of ancient animals (e.g., Gadbury *et al.* (2000), Kohn and Cerling (2002), Passey and Cerling (2002), and Higgins and MacFadden (2004)).

Reconstruction of the Paleobiology of Extinct Species

Using skeletal material from which collagen can be extracted, it is possible to reconstruct some aspects of animal diet using carbon and nitrogen isotopic signatures, because different dietary items exhibit different isotopic signatures, such as C_3 or C_4 plants, animal flesh, or food resources of marine origin. With material too old or too altered to yield collagen, the carbon isotopic signatures of tooth enamel can be used to test dietary hypotheses in contexts where different food webs start with plants exhibiting different δ^{13}C values, such as tropical environments where C_4 grasses coexist with C_3 tree leaves.

Subsistence Patterns of Ancient Hominids

Environmental and dietary changes are often linked to key stages of the evolution of Hominids. In some cases, the possible impact of such changes can be tested using the isotopic signature of fossil bones and teeth.

The importance of savanna for African Plio–Pleistocene hominids A hotly debated issue in the study of Quaternary terrestrial paleoenvironments is the role of savanna in the evolution of hominids in Africa. Indeed, modern African apes are restricted to forested environments, and savanna environments require special skills in order to be exploited by primates, such as a bipedal gait and strong social bonds. The $\delta^{13}C$ values of tooth enamel of fossil mammals, including hominids from South Africa, ranging in age from 3 to 1.7 million years old, indicate a significant C_4 component in the environment, as well as significant C_4 dietary inputs for hominids of the species *Australopithecus africanus*, *A. robustus*, and *Homo ergaster* (**Fig. 6**; Lee-Thorp *et al.*, 2003). This C_4 dietary input could be linked to the consumption of C_4 plant parts, but also to the consumption of animals feeding on C_4 plants, such as invertebrates, small vertebrates, or grazer meat. This supports the view that hominid evolution is linked to savanna, but also contradicts the hypothesis that different hominids used savanna resources in differing amounts.

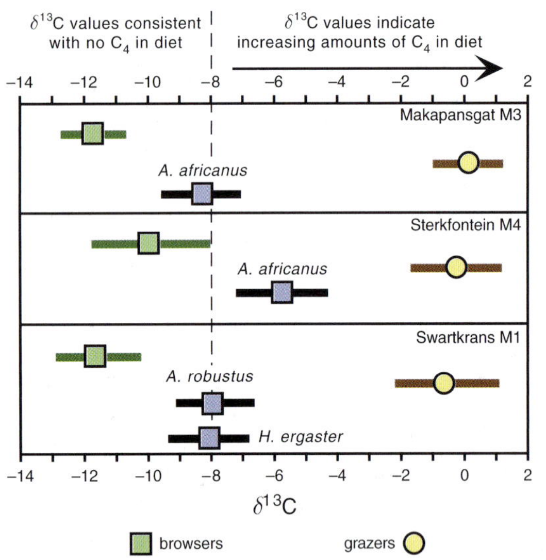

Figure 6 Reconstruction of the paleoenvironment and paleodiet of fossil hominids from South Africa, from Makapansgat Member 3 ($\geq$3 Ma), Sterfontein Member 4 ($\sim$2.4–2.6 Ma), and Swartkrans Member 1 ($\sim$1.7 Ma). The $\delta^{13}C$ values of browser and grazer enamel are similar to those of modern herbivores consuming C_3 plants (browsers) and C_4 plants (grazers), indicating preservation of the carbon isotopic signatures in the fossil material. The $\delta^{13}C$ values of *Australopithecus africanus*, *Australopithecus robustus*, and *Homo ergaster* show that a significant amount of food resources coming from C_4 environments (most likely savanna) were consumed by these three hominid species during the Pliocene and Pleistocene in South Africa (Lee-Thorp *et al.* (2003)).

The diet of Neanderthals and early anatomically modern humans During the Upper Pleistocene, Europe was populated by a distinctive hominid form, the Neanderthals. The demise of this hominid around 30 ka coincides with the appearance of anatomically modern humans in Europe, seemingly using a more sophisticated culture. Several Neanderthal fossils have yielded preserved collagen that could be analyzed for carbon- and nitrogen-stable isotopic compositions, and compared to those of Upper Paleolithic anatomically modern humans (e.g., Bocherens *et al.* (1999, 2005b) and Richards *et al.* (2000, 2001)). Comparing Neanderthals with predators such as hyenas suggests that Neanderthals consumed mainly large herbivore meat, including a large proportion of megaherbivores such as mammoth and woolly rhinoceros (Bocherens *et al.*, 2005b). It appears also that Neanderthals living under different environmental conditions and at different periods had similar diets. The last Neanderthals had diets similar to those of early anatomically modern humans (Drucker and Bocherens, 2004). This supports the hypothesis of dietary competition between Neanderthals and anatomically modern humans.

Paleodiet of Ancient Bears

Bears, as omnivorous carnivores, form an interesting group to study using the isotopic approach since their diet can be quite variable, and sometimes difficult to determine based on their morphological features. They are all the more interesting since they have an abundant fossil record, for instance for cave bears, and cover a dietary spectrum similar to that of humans. Detecting changes in bear diet may yield direct information on the availability of food resources relevant for human diet. Moreover, the paleogenetics of ancient bears is intensively studied and combining genetical and paleodietary data through the evolution of bear lineages provides direct evidence about their evolutionary biology. Using carbon- and nitrogen-stable isotopic signatures, cave bears (*Ursus spelaeus*) were shown to be essentially vegetarian animals (Bocherens *et al.*, 1994), while the extinct giant short-faced bear (*Arctodus simus*) from North America has been demonstrated to be a meat-eater (Barnes *et al.*, 2002). Interestingly, the diet of ancient brown bears living at the same time as extinct bears from specialized species shifts when the dietary competition stops with the extinction of their competitor. For instance, brown bears from North America were vegetarian when they coexisted with the carnivorous short-faced bears, but they became much more

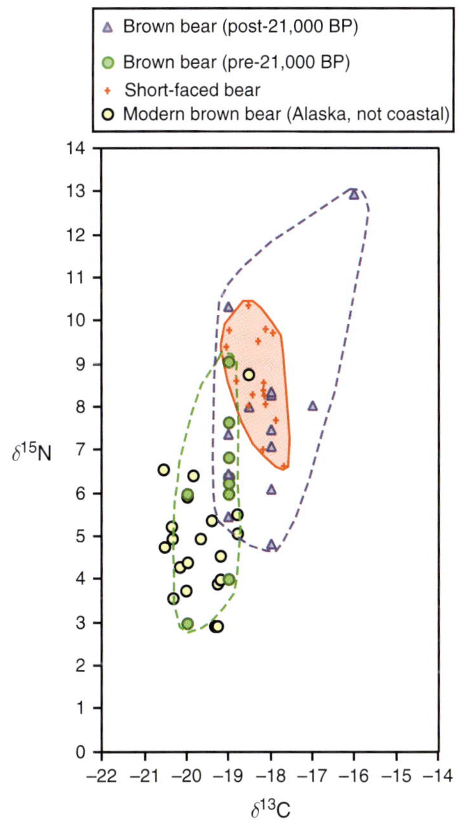

Figure 7 $\delta^{13}C$ and $\delta^{15}N$ values of brown bear (*Ursus arctos*) and short-faced bear (*Arctodus simus*) collagen from Alaska (Barnes *et al.*, 2002). These data show that short-faced bears were carnivorous, and that brown bears living when the carnivorous short-faced bears were still around, until 21 kyr BP, were mostly vegetarian, with carbon and nitrogen isotopic signatures similar to those of modern brown bears consuming mainly plant food. Brown bears younger than 21 kyr BP lived after the extinction of short-faced bears and had a more diverse diet, including meat. The extinction of carnivorous short-faced bears may have opened a new ecological niche for brown bears.

carnivorous after the extinction of these meat-eating bears (**Fig. 7**; Barnes *et al.*, 2002).

Paleodiet of Ancient Ungulates

Large herbivorous mammals often exhibit specialized diets, in connection with their tooth morphology and digestive physiology. The occurrence of some herbivorous mammals in Quaternary fossil and archaeological localities can be used to infer the paleoenvironments around the site. Isotopic analyses of fossil herbivores have sometimes yielded additional information about the ancient environments in which they used to dwell, and thus allows more precise reconstructions of the actual diet of extinct species. For instance, isotopic investigations of tooth enamel have shown that in southern North America, horses were mixed feeders consuming C_4 grasses and C_3 shrubs, while bisons were grazers, eating only C_4 grasses (Koch *et al.*, 2004). Horses have hypsodont (high-crowned) teeth and are traditionally thought to be specialized grazers. Another example of hypsodont ungulates with diverse diets are extinct camelids from North America (Feranec, 2003). Therefore the specialized morphology of some herbivore teeth is not always indicative of specialized diet, but rather indicates dietary flexibility.

Perspectives

Stable isotopes in terrestrial teeth and bones have already yielded invaluable information on Quaternary ecosystems, and this approach is expected to develop further during the coming years.

A better knowledge of isotopic variations in modern ecosystems and modern animals will most probably allow more accurate paleoenvironmental and paleodietary reconstructions to be performed, based on the stable isotopic signatures of teeth and bones of Quaternary terrestrial vertebrates. Our understanding of the fractionation factors in tissues prone to fossilization still needs improvement, based on controlled diet experiments of species closely related to the fossil taxa under study, and on investigations of large mammals in monitored wild contexts. Also the chronology of isotopic records in incrementally grown tissues, such as tooth enamel, needs more accurate quantification in the species found in Quaternary fossil assemblages.

The impact of diagenetic alteration on the isotopic signatures of teeth and bones needs to be better understood, especially when dealing with the mineral fraction of fossil vertebrates.

Collagen is currently used for its carbon and nitrogen isotopic signatures, but it could yield the isotopic signatures of additional chemical elements. For instance, the amount of deuterium, the stable heavy isotope of hydrogen, could be used to study paleoclimates and possible migrations, since it varies in a way similar to oxygen in relationship with temperatures (Cormie *et al.*, 1994; Bowen *et al.*, 2005). The isotopic signatures of sulfur in collagen may be used to improve paleodietary reconstructions and to provide identification of geographic origin (Richards *et al.*, 2003).

Improving biomolecular technologies will allow the analysis of isotopic signatures of other organic compounds, such as osteocalcin, single amino acids, cholesterol, and fatty acids in fossil bones and teeth (e.g., Corr *et al.* (2005) and Smith *et al.* (2005)). This is expected to lead to breakthroughs in retrieving paleobiological information at different timescales within a single individual, and in obtaining isotopic paleobiological signals in fossil material older than around 100 kyr.

See also: **Archaeological Records**: Neanderthal Demise. **Carbonate Stable Isotopes**: Speleothems; Lake Sediments. **Ice Core Methods**: Stable isotopes. **Ice Core Records**: Antarctic Stable Isotopes; Greenland Stable Isotopes. **Vertebrate Overview**. **Vertebrate Studies**: Ancient DNA.

References

Ambrose, S. H. (1990). Preparation and characterization of bone and tooth collagen for isotopic analysis. *Journal of Archaeological Science* **17,** 431–451.

Ambrose, S. H. (1991). Effects of diet, climate and physiology on nitrogen isotope abundances in terrestrial foodwebs. *Journal of Archaeological Science* **18,** 293–317.

Ayliffe, L. K., and Chivas, A. R. (1990). Oxygen isotope composition of the bone phosphate of Australian kangaroos: Potential as a palaeoenvironmental recorder. *Geochimica et Cosmochimica Acta* **54,** 2603–2609.

Barnes, I., Mattheus, P., Shapiro, B., Jensen, D., and Cooper, A. (2002). Dynamics of Pleistocene population extinctions in Beringian brown bears. *Science* **295,** 2267–2270.

Bocherens, H. (2004). Cave bear palaeoecology and stable isotopes: Checking the rules of the game. In: *Proceedings of the 9th International Cave Bear Conference. Cahiers scientifiques du Centre de Conservation et d'Etude des Collections* (M. Philippe, A. Argant and J. Argant, Eds.), Hors Série No 2, pp. 183–188. Muséum d'Histoire naturelle de Lyon, Lyon.

Bocherens, H., Fizet, M., and Mariotti, A. (1994). Diet, physiology and ecology of fossil mammals as inferred by stable carbon and nitrogen isotopes biogeochemistry: Implications for Pleistocene bears. *Palaeogeography, Palaeoclimatology, Palaeoecology* **107,** 213–225.

Bocherens, H., Billiou, D., Patou-Mathis, M., Otte, M., Bonjean, D., Toussaint, M., and Mariotti, A. (1999). Palaeoenvironmental and palaeodietary implications of isotopic biogeochemistry of late interglacial Neandertal and mammal bones in Scladina Cave (Belgium). *Journal of Archaeological Science* **26,** 599–607.

Bocherens, H., Drucker, D., Billiou, D., and Moussa, I. (2005). Une nouvelle approche pour évaluer l'état de conservation de l'os et du collagène pour les mesures isotopiques (datation au radiocarbone, isotopes stables du carbone et de l'azote). *L'Anthropologie* **109,** 557–567.

Bocherens, H., Drucker, D. G., Billiou, D., Patou-Mathis, M., and Vandermeersch, B. (2005). Isotopic evidence for diet and subsistence pattern of the Saint-Césaire I Neanderthal: Review and use of a multi-source mixing model. *Journal of Human Evolution* **49,** 71–87.

Bocherens, H., and Mariotti, A. (2002). Paléoenvironnements et paléoalimentations: Biogéochimie isotopique des vertébrés. In *Géologie de la Préhistoire* (J.-C. Miskowski, Ed.), pp. 1323–1344. GEOPRE, Presses Universitaires de Perpignan, Perpignan.

Bowen, G. J., Wassenaar, L. I., and Hobson, K. A. (2005). Global application of stable hydrogen and oxygen isotopes to wildlife forensics. *Oecologia* **143,** 337–348.

Cerling, T. E., and Harris, J. M. (1999). Carbon isotope fractionation between diet and bioapatite in ungulate mammals and implications for ecological and paleoecological studies. *Oecologia* **120,** 347–363.

Cormie, A. B., Schwarcz, H. P., and Gray, J. (1994). Relation between hydrogen isotopic ratios of bone collagen and rain. *Geochimica et Cosmochimica Acta* **58,** 377–391.

Corr, L. T., Sealy, J. C., Horton, M. C., and Evershed, R. P. (2005). A novel marine dietary indicator utilising compound-specific bone collagen amino acid $\delta^{13}C$ values of ancient humans. *Journal of Archaeological Science* **32,** 321–330.

Drucker, D. G., and Bocherens, H. (2004). Carbon and nitrogen stable isotopes as tracers of diet breadth evolution during Middle and Upper Palaeolithic in Europe. *International Journal of Osteoarchaeology* **14,** 162–177.

Drucker, D. G., Bocherens, H., and Billiou, D. (2003b). Evidence for shifting environmental conditions in Southwestern France from 33,000 to 15,000 years ago derived from carbon-13 and nitrogen-15 natural abundances in collagen of large herbivores. *Earth and Planetary Science Letters* **216,** 163–173.

Drucker, D., Bocherens, H., Bridault, A., and Billiou, D. (2003a). Carbon and nitrogen isotopic composition of Red Deer (*Cervus elaphus*) collagen as a tool for tracking palaeoenvironmental change during Lateglacial and Early Holocene in northern Jura (France). *Palaeogeography, Palaeoclimatology, Palaeoecology* **195,** 375–388.

Ehleringer, J. R., Cerling, T. E., and Helliker, B. R. (1997). C_4 photosynthesis, atmospheric CO_2, and climate. *Oecologia* **112,** 285–299.

Feranec, R. S. (2003). Stable isotopes, hypsodonty, and the paleodiet of *Hemiauchenia* (Mammalia, Camelidae): A morphological specialization creating ecological generalization. *Paleobiology* **29,** 230–242.

Fricke, H. C., Clyde, W. C., and O'Neil, J. R. (1998). Intra-tooth variations in $\delta^{18}O$ (PO_4) of mammalian tooth enamel as a record of seasonal variations in continental climate variables. *Geochimica et Cosmochimica Acta* **62,** 1839–1950.

Gadbury, C., Todd, L., Jahren, A. H., and Amundson, R. (2000). Spatial and temporal variations in the isotopic composition of bison tooth enamel from the Early Holocene Hudson-Meng Bone Bed, Nebraska. *Palaeogeography, Palaeoclimatology, Palaeoecology* **157,** 79–93.

Gröcke, D. R., Bocherens, H., and Mariotti, A. (1997). Annual rainfall and nitrogen-isotope correlation in Macropod collagen: Application as a paleoprecipitation indicator. *Earth and Planetary Science Letters* **153,** 279–285.

Heaton, T. H. E. (1999). Spatial, species, and temporal variations in the $^{13}C/^{12}C$ ratios of C_3 plants: Implications for palaeodiet studies. *Journal of Archaeological Science* **26,** 637–649.

Hedges, R. E. M., Stevens, R. E., and Richards, M. P. (2004). Bone as a stable isotope archive for local climatic information. *Quaternary Science Reviews* **23,** 959–965.

Higgins, P., and MacFadden, B. J. (2004). "Amount effect" recorded in oxygen isotopes of Late Glacial horse (*Equus*) and bison (*Bison*) teeth from the Sonoran and Chihuahuan deserts, southwestern United States. *Palaeogeography, Palaeoclimatology, Palaeoecology* **206,** 337–353.

Koch, P. L., Diffenbaugh, N. S., and Hoppe, K. A. (2004). The effects of late Quaternary climate and pCO_2 change on C_4 plant abundance in the south-central United States. *Palaeogeography, Palaeoclimatology, Palaeoecology* **207,** 331–357.

Kohn, M. J., and Cerling, T. E. (2002). Stable isotope compositions of biological apatite. In *Reviews in Mineralogy and Geochemistry, Vol. 48: Phosphates, Geochemical, Geobiological, and Materials Importance* (M. J. Kohn, J. Rakovan and J. M. Hughes, Eds.), pp. 455–488. Mineralogical Society of America, Washington DC.

Kohn, M. J., Schoeninger, M. J., and Barker, W. W. (1999). Altered states: Effects of diagenesis on fossil tooth chemistry. *Geochimica et Cosmochimica Acta* **63,** 2737–2747.

Lee-Thorp, J. A., and Beaumont, P. B. (1995). Vegetation and seasonality shifts during the Late Quaternary deduced from $^{13}C/^{12}C$ ratios of grazers at Equus Cave, South Africa. *Quaternary Research* **43,** 426–432.

Lee-Thorp, J. A., Sponheimer, M., and Van der Merwe, N. J. (2003). What do stable isotopes tell us about Hominid dietary and ecological niches in the Pliocene? *International Journal of Osteoarchaeology* **13**, 104–113.
Luz, B., Cormie, A., and Schwarcz, H. P. (1990). Oxygen isotope variations in phosphate of deer bones. *Geochimica et Cosmochimica Acta* **54**, 1723–1728.
Passey, B. H., and Cerling, T. E. (2002). Tooth enamel mineralization in ungulates: Implications for recovering a primary isotopic time series. *Geochimica et Cosmochimica Acta* **18**, 3225–3234.
Richards, M. P., and Hedges, R. E. M. (2003). Variations in bone collagen $\delta^{13}C$ and $\delta^{15}N$ values of fauna from Northwest Europe over the last 40 000 years. *Palaeogeography, Palaeoclimatology, Palaeoecology* **193**, 261–267.
Richards, M. P., Fuller, B. T., Sponheimer, M., Robinson, T., and Ayliffe, L. (2003). Sulphur isotopes in palaeodietary studies: A review and results from a controlled feeding experiment. *International Journal of Osteoarchaeology* **13**, 37–45.
Richards, M. P., Pettitt, P. B., Stiner, M. C., and Trinkaus, E. (2001). Stable isotope evidence for increasing dietary breadth in the European mid-Upper Paleolithic. *Proceedings of the National Academy of Science USA* **98**, 6528–6532.
Richards, M. P., Pettitt, P. B., Trinkaus, E., Smith, F. H., Paunovic, M., and Karavanic, I. (2000). Neandertal diet at Vindija and Neandertal predation: The evidence from stable isotopes. *Proceedings of the National Academy of Science USA* **97**, 7663–7666.
Schwarcz, H. P., Dupras, T. L., and Fairgrieve, S. I. (1999). ^{15}N enrichment in the Sahara: In search of a global relationship. *Journal of Archaeological Science* **26**, 629–636.
Smith, C. I., Craig, O. E., Prigodich, R. V., *et al.* (2005). Diagenesis and survival of osteocalcin in archaeological bone. *Journal of Archaeological Science* **32**, 105–113.
Smith, J. M., Lee-Thorp, J. A., and Sealy, J. C. (2002). Stable carbon and oxygen isotopic evidence for late Pleistocene to middle Holocene climatic fluctuations in the interior of southern Africa. *Journal of Quaternary Science* **17**, 683–695.
Wright, L. E., and Schwarcz, H. P. (1998). Stable carbon and oxygen isotopes in Human tooth enamel: Identifying breast-feeding and weaning in prehistory. *American Journal of Physical Anthropology* **106**, 1–18.

Nonmarine Biogenic Carbonates

S J Carpenter, The University of Iowa, IA, USA

© 2007 Elsevier B.V. All rights reserved.

Introduction

Biogenic carbonates – predominantly calcite and aragonite precipitated by plants and animals in nonmarine surface waters – are ubiquitous in Quaternary age deposits. Carbon and oxygen isotope ratios ($\delta^{13}C$ and $\delta^{18}O$ values) of these materials are routinely used as proxies for climate change. The ultimate utility of these data rests on scientists' ability to calibrate modern analogs, understand biological fractionations or 'vital effects', and use appropriate sampling methods. These materials have the potential for providing detailed data (sub-monthly resolution) for time periods that range from 1 to 50 years in length. This temporal resolution can be used in conjunction with longer term, yet lower temporal resolution, terrestrial isotope records (e.g., ice cores, stalagmites, soil profiles, and lake sediment cores) to examine seasonal variation in temperature, precipitation, vegetation, productivity, and water column stratification.

Equilibrium Precipitation

Isotope geochemists often seek to determine whether $CaCO_3$ is precipitated in isotopic equilibrium with ambient waters (i.e., the $\delta^{18}O_{water}$ and $\delta^{13}C_{DIC}$ values). By knowing this relation, it may be possible to determine the temperature or the isotopic composition of waters in which ancient carbonates formed. As with most estimates of this sort, assumptions must be made. Unfortunately, carbonate oxygen isotope thermometry requires knowledge of the $\delta^{18}O$ value of ambient waters. Conversely, estimates of $\delta^{18}O_{water}$ values require knowledge of the temperature at which the carbonate precipitated. To evaluate equilibrium precipitation, the results of numerous experimental and empirical studies must be considered.

Equilibrium $\delta^{18}O$ values of calcite are most accurately estimated using the $\delta^{18}O$ value of ambient water, water temperature (in kelvins), and a calcite (low-Mg calcite), the calcite-water fractionation factor (α), and an expression that relates these variables (Friedman and O'Neil, 1977; Kim and O'Neil, 1997).

$$\alpha(CaCO_3 water) = (1000 + \delta^{18}O_{CaCO3(SMOW)}) / (1000 + \delta^{18}O_{Water(SMOW)}) \quad [1]$$

$$10^3 \ln\alpha(\text{calcite-water}) = 2.78(10^6/T^2) - 2.89 \quad [2]$$
(Friedman and ONeil, 1977)

$$10^3 \ln\alpha(\text{calcite-water}) = 18.03(10^3/T) - 32.42 \quad [3]$$
(Kim and O'Neil, 1997)

These values can be corrected for the Mg-content of the calcite (+0.06‰ per mol% $MgCO_3$; Tarutani *et al.*, 1969).

To examine carbonate equilibrium $\delta^{13}C$ values, it is necessary to know the $\delta^{13}C$ value of water bicarbonate (HCO_3^-) or the dissolved inorganic carbon ($\delta^{13}C_{DIC}$). The calcite-HCO_3^- carbon isotope enrichment factor is +1.0‰ (and is constant from 10 to

40 °C). The aragonite-HCO_3^- enrichment factor is +2.7‰ throughout the same temperatures range (Romanek *et al.*, 1992). Assuming equilibrium precipitation, one can estimate ambient $\delta^{13}C_{DIC}$ values by subtracting 1.0 or 2.7‰ from measured calcite and aragonite values. Likewise, aragonite equilibrium $\delta^{13}C$ values can be estimated by adding the aragonite-calcite enrichment factor to measured calcite $\delta^{13}C$ values (+1.6‰ of Rubinson and Clayton, 1969; +1.7‰ of Romanek *et al.*, 1992).

The only empirically derived oxygen isotope fractionation factor for freshwater fish otoliths is:

$$10^3 \ln\alpha(\text{aragonite-water}) = 18.56(10^3/T) - 33.49 \quad \text{(Patterson } et\ al., 1993) \qquad [4]$$

The only empirically derived expression describing the oxygen isotope fractionation factor in Unionid aragonite is:

$$10^3 \ln\alpha(\text{aragonite-water}) = 2.559(10^6/T^2) + 0.715 \quad \text{(Dettman } et\ al., 1999) \qquad [5]$$

Equations 4 and 5 have slopes similar to that of Grossman and Ku (1986) for biogenic aragonite:

$$10^3 \ln\alpha(\text{aragonite-water}) = 18.07(10^3/T) - 31.079 \quad \text{(Grossman and Ku, 1986)} \qquad [6]$$

'Vital Effects' and Modern Calibration

The most significant impediment to effectively using freshwater biogenic carbonates as climate or environmental proxies is a generally poor understanding of biological fractionation of stable isotopes during formation. 'Vital effects' (Urey *et al.*, 1951) are in many ways a means of categorizing processes that are poorly understood. Numerous studies have examined "vital effects' in a variety of marine organisms (see these and citations within; corals, Adkins *et al.*, 2003; calcareous algae, Lee and Carpenter, 2001; brachiopods, Carpenter and Lohmann, 1995; foraminifera, Zeebe, 1999). Common among these studies is a less-than-satisfactory explanation of mechanisms and rates of carbonate precipitation, the incorporation of metabolic CO_2, and the positive covariance of $\delta^{13}C$ and $\delta^{18}O$ values in many biogenic carbonates. Far less work has been conducted on "vital effects" in freshwater biogenic carbonates.

More work is needed to examine the relation between the $\delta^{13}C$ and $\delta^{18}O$ value of ambient waters and their biogenic carbonates. These efforts require relatively long term monitoring of waters and related carbonates (a minimum of one year). In addition to monitoring the $\delta^{13}C_{DIC}$, $\delta^{18}O_{water}$, and temperature, measurements of pH, [HCO_3^-], overall stream/lake productivity, the $\delta^{13}C$ values of potential food sources are needed to examine the seasonal variation in the $\delta^{13}C_{DIC}$ and the potential physico-chemical and biological causes of 'vital effects' (e.g., Dettman *et al.*, 1999). In addition to environmental monitoring, detailed examination of the mechanism of calcification and physiological variation during the life history of individual organisms is needed.

Once such data are collected, empirical fractionation factors can be established for geologically significant species of plants and animals. A concerted effort by funding agencies is needed to establish these baseline data for modern organisms. Until these data are in hand, the utility of stable isotope data from archaeological and geological specimens will be in question.

Micro-sampling Techniques

The development of micro-sampling or micro-milling techniques (the work of K.C Lohmann and associates) to extract small amounts of powdered carbonate from ancient rocks, paleontological specimens, modern shells or otoliths, etc., together with the improved sensitivity of isotope ratio mass spectrometers (IRMS) has created a means of collecting very high temporal resolution data from such materials (Dettman and Lohmann, 1995; Wurster *et al.*, 1999; Frappier *et al.*, 2002). These automated micro-sampling devices are capable of routinely milling 15 to 20 μm-wide paths from biogenic carbonates – permitting sub-monthly sampling resolution in most specimens (**Fig. 1**). These devices employ low eccentricity dental hand pieces equipped with precision drill bits, precision X-Y-Z stages and actuators, microscopes equipped with digital imaging devices, and custom software and motion controllers. Currently, these devices can be purchased commercially from two manufacturers.

Micro-sampling of biogenic carbonates with well-defined growth increments (mollusk shells, otoliths) is critical in deciphering ontogenetic variation in these materials (**Fig. 2**). The importance of collecting sub-monthly resolution records from biogenic carbonates is the understanding of ontogenetic variations (associated with physiological changes such as sexual maturation and overall metabolic activity) and seasonal isotopic variation (induced by temperature and ambient conditions such as photosynthesis, respiration, and water column stratification). By collecting data at approximately bi-weekly intervals,

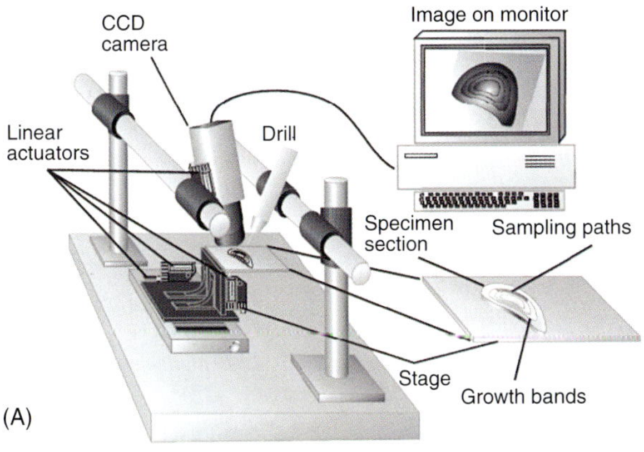

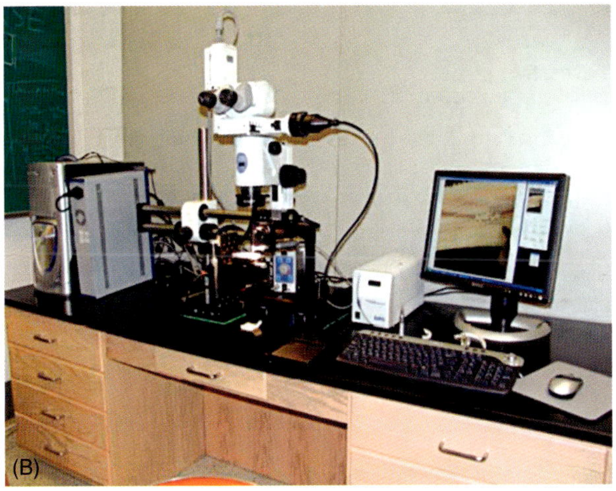

Figure 1 A) Schematic diagram of an automated micro-sampler using a video camera for image capture (From Patterson and Smith, 2001, after Wurster *et al.*, 1999). B) Photograph of the Carpenter Microsystems CM-2 Micro-sampling System at the University of North Carolina. This system is equipped with a precision X-Y-Z stage with linear actuators permitting two inches of travel, a horizontally mounted precision handpiece, 15:1 stereoscopic zoom microscope equipped with both transmitted light and an epifluorescence illumination systems and a 12 mega-pixel digital camera. These systems allow for real time/on-screen digitization of sampling paths.

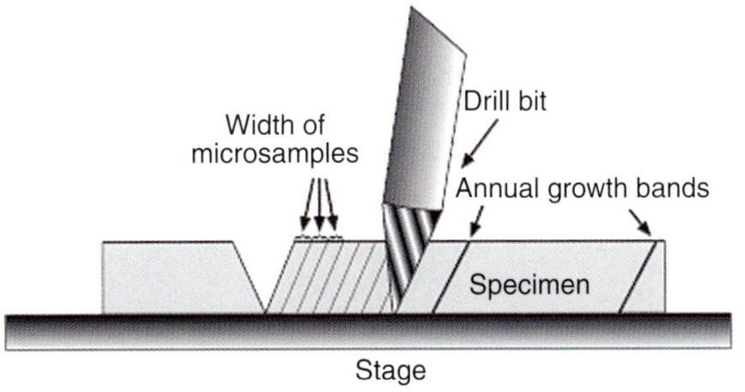

Figure 2 Close up schematic illustration of how individual micro-samples are milled from biogenic carbonate specimens using a rotating, faceted tungsten carbide drill bit. A first pass coarsely mills a trough parallel to the sample paths that allows the bit to shave small increments off one wall of the trough. These subsequent high-resolution paths are approximately 15 to 20 μm wide. Sample powders are manually removed from the specimen using a surgical scalpel. (From Patterson and Smith, 2002, after Wurster *et al.*, 1999).

one can examine short-term variation that would be overlooked by bulk analyses or lower-resolution sampling. Weekly to monthly sampling resolution is preferred as this is the temporal scale at which weather events occur (storms and surface water changes, cold front migrations, etc.) and to which the δ^{18}O and δ^{13}C values of biogenic carbonates react.

Where shell growth rates are sufficiently rapid, high-resolution micro-sampling of biogenic carbonates is not needed. In such cases, more routine, manual sampling can be conducted to collect the appropriate temporal resolution for the project. The non-linear growth of unionid shells, both during an annual cycle (rapid extension rates in the Summer and slow growth rates in the Winter) and through ontogeny (diminished growth rates with age) requires adjustments in sampling strategies.

Mid-latitude unionids from the Northern Hemisphere are thought to greately reduce or cease carbonate precipitation below ~12° C (e.g., Howard, 1921 (*Lampsilis luteola*); Negus, 1966 (*Anodonta anatine*); Dettman *et al.*, 1999). Both *Dreissena* sp. (Neumann *et al.*, 1993) and *Corbicula* sp. (Joy, 1985) display little growth below 10° C. Interestingly, *Lampsilis radiata siliquoidea* from the Mackenzie River, NWT, Canada precipitates carbonate in water temperatures below 12° C (Carpenter *et al.*, in prep.). The same species from more southerly latitudes in the United States (southeastern Michigan; e.g., Dettman *et al.*, 1999) does slow shell production at this temperature. The occurrence of non-winter growth bands (pseudoannual bands) in unionids (slowing of shell carbonate production) seems to correspond to abrupt changes in ambient conditions (decreases in water temperature and O_2 levels; Carpenter unpublished data). Clearly, cessation of shell growth is not fully understood and requires examination of genera or species that are geographically widespread.

Cessation of shell growth in unionids results in a lack of data corresponding to the late Fall through early Spring in mid-latitudes. This lack of growth diminishes the ability to examine annual temperature minima, the δ^{18}O value of winter precipitation, and the oxidation of organic matter in lakes and streams. However, a uniform, Winter "shut-down" temperature allows calculation of ambient water $\delta^{18}O$ values by fixing temperature (assuming a known fractionation factor; e.g., Dettman *et al.*, 1999). Given variation in shell growth rates, it's important to sample materials in a way that maximizes sampling efficiency with increased resolution during the Fall through Spring and perhaps a decreased resolution in the summer when shells are growing rapidly. This temporal "accordion effect" and the occurrence of Winter "shut downs" require special sampling strategies and data interpretations (see shell structure of pseudoannual and annual bands of Tevesz and Carter, 1980).

Many researchers have chosen to mill samples from the interior of unionid shells to avoid potential diagenetic alteration of the thin, outer prismatic layer of the shell. Although one should always assess diagenetic alteration, Quaternary age materials often do not exhibit extensive diagenetic alteration. In such cases, the outer surface of shells can be microsampled. One must use care to avoid drilling too deeply into the shell surface as these samples may sample carbonate that is older than that expressed at the shell surface. Dettman *et al.* (1999) found no difference in the $\delta^{18}O$ and $\delta^{13}C$ values of the prismatic and nacreous layers of a modern *Lampsilis radiata siliquoidea* shell (**Fig. 4**).

Unionid Bivalves and Freshwater Mollusks

Seasonal variation of $\delta^{13}C$ and $\delta^{18}O$ values in freshwater mollusk shells from temperate settings produces sigmoidal or sinusoidal ontogenetic trends. In theory, the carbon and oxygen isotope ratios of shell carbonate will be negatively correlated with lower $\delta^{18}O$ value and higher $\delta^{13}C$ values occurring during the summer months when temperatures and primary productivity are highest (**Fig. 3**). During the winter months (characterized by low temperatures and minimal productivity), $\delta^{18}O$ values reach their maximum and $\delta^{13}C$ values reach their minimum. The $\delta^{18}O_{water}$, temperature, and the $\delta^{13}C$ value of DIC (dissolved inorganic carbon or ΣCO_2) should play important roles in determining shell isotope ratios. There are several complicating factors that modify these theoretical parameters (e.g., specialized food sources, carbon cycling lag times, variable effects of chemical change on food-web dynamics, and the buffering capacity of the ambient water mass (the mass of carbon and oxygen upon which physico-chemical processes work)).

Lampsilis sp. – Huron River, Michigan

Dettman *et al.*'s (1999) examination of unionds from southeastern Michigan streams provided an

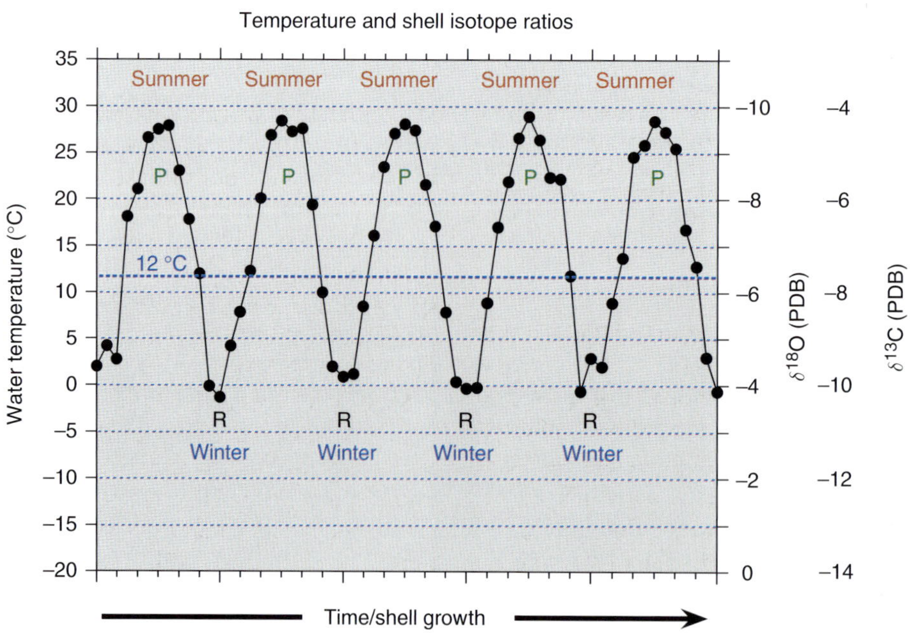

Figure 3 Idealized plot of unionid bivalve shell $\delta^{18}O$ and $\delta^{13}C$ values and water temperature. Sinusoidal trends represent both $\delta^{18}O$ and $\delta^{13}C$ values (scales for each at right). Winters (cold temperatures) are characterized by relatively high $\delta^{18}O$ values and low $\delta^{13}C$ values (due to oxidation or respiration (R) of plant material). Summers (warm temperatures) are characterized by relatively low $\delta^{18}O$ values and high $\delta^{13}C$ values (due to the removal of ^{12}C by photosynthesis (P) leaving DIC enriched in ^{13}C). Natural data will likely have seasonal changes in $\delta^{13}C$ values lagging behind $\delta^{18}O$ values. The 12 °C line represents the temperature at which shell carbonate may stop being precipitated. As a result, all data below this line may be missing in the unionid shell record. The resulting sinusoidal trends will likely have diminished seasonal amplitude and significant missing time associated with annual winter growth bands.

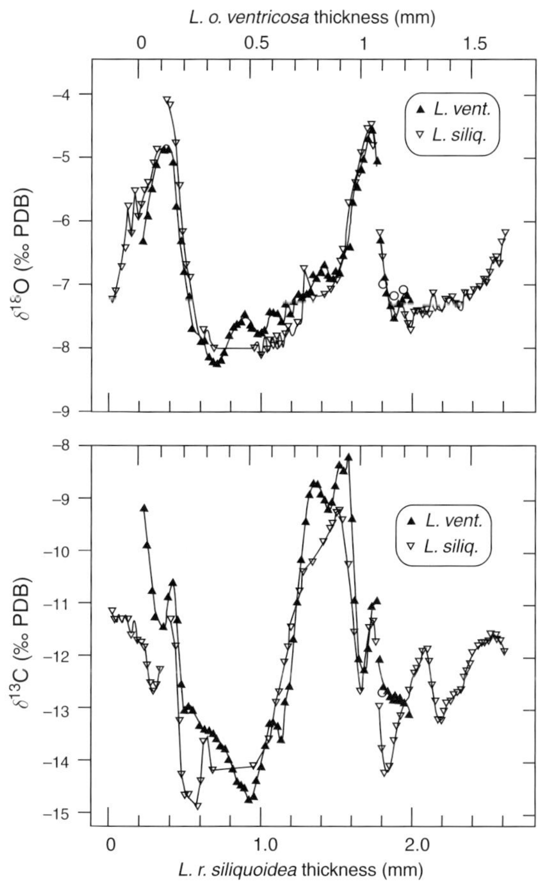

Figure 4 Shell $\delta^{18}O$ and $\delta^{13}C$ values of two unionids from the Huron River, SE Michigan (*Lampsilis ovata ventricosa* and *Lampsilis radiata siliquoidea*). Shell thickness is measured perpendicular to nacreous plates, beginning at an arbitrary point before the 1990–1991 hibernation mark. The isotope records are aligned at $\delta^{18}O$ maxima. *L. o. ventricosa* data have been stretched by 45% to compensate for its slower growth. Lines connecting points are separated at visible "hibernation" or "shut down" features. Open circles are prismatic layer measurements of *L. o. ventricosa*. From Dettman *et al.* (1999).

introduction to how high-resolution data can be extracted and interpreted. Their data from *Lampsilis* (Huron River) are worthy of mention (**Fig. 4**). There is a strong correlation between the $\delta^{18}O$ values collected from *L. o. ventricosa* and *L. r. siliquoidea*. Shell $\delta^{18}O$ values also are in isotopic equilibrium with ambient river water (**Eqn. 5, Figs. 4** and **5**). The timing of variation in shell $\delta^{13}C$ values also correlates, yet maximum and minimum $\delta^{13}C$ values vary significantly between co-occurring species (**Fig. 4**). Similarities between prismatic and nacreous layer isotope ratios confirm the utility of sampling the outer prismatic layer.

One can conclude that both species of *Lampsilis* are responding to similar environmental changes. The seasonal variation in $\delta^{13}C$ values is difficult to interepret in terms of measured variation in river temperature and $\delta^{13}C_{DIC}$ values and the relative timing of $\delta^{13}C$ value variation in the shells. It appears that there is a significant lag between stream productivity increases (marked by an increase in river $\delta^{13}C_{DIC}$ values) and shell $\delta^{13}C$ values.

Ricken *et al.* (2003) have found similar isotopic equilibrium precipitation in *Anodonta* and *Unio* from the Rhine River. This study examines the relation between isotopically distinct discharge from alpine and non-alpine catchments, using well documented surface water data. Their *Anodonta* specimens exhibit large decreases in $\delta^{13}C$ values during winter months (corresponding with high $\delta^{18}O$ values). Again, these values do not appear to be in isotopic equilibrium with ambient water DIC.

Pyganodon lacustris - Glovers Pond, New Jersey

The lacustrine unionid bivalve, *Pyganodon lacustris* occurs in *Chara*-rich sediment within the shallow water, carbonate bank (~1 m water depth) of Glovers Pond, NJ (**Figs. 6** and **7**). Glovers Pond is a temperate, dimictic lake, in northwestern New Jersey and is spring-fed (through Ordovician dolomite). Today, Glovers Pond is thermally and chemically stratified with a maximum depth of 9.5 meters and occupies a glacially modified and dammed fault valley currently being infilled concentrically by four types of sediment (peat, marl, calcareous-organic-rich silt, and gyttja, Erickson, 1968). The hypolimnion is undersaturated with respect to $CaCO_3$ which presently prohibits deposition of calcium carbonate in the deepest portions of the lake (Erickson, 1968). $CaCO_3$ (calcite) is deposited on a shallow shelf both on the surfaces of *Potamogeton* sp. and to a greater extent by blanketing growths of the green alga *Chara* sp. Marl and peat deposits form a platform that is prograding into the gyttja lithotope in the center of the lake basin (**Fig. 7**).

Shell $\delta^{18}O$ values of *Pyganodon lacustris* record predicted seasonal changes in oxygen isotope ratios associated with variation in ambient temperature, rainfall, and evaporation. Shell $\delta^{13}C$ values record seasonal changes in lake productivity and stratification/overturn events (**Fig. 6**). Warming of lake waters in the Spring yields a marked decrease in shell $\delta^{18}O$ values and the concomitant increase in productivity causes an increase in shell $\delta^{13}C$ values. Fall overturns (Oct.-Nov.) mix the cold, ^{13}C-depleted deep water (where organic matter is oxidized) with shallow waters. This produces a marked increase in shell $\delta^{18}O$ values and can produce as much as a 5‰ decrease in shell $\delta^{13}C$

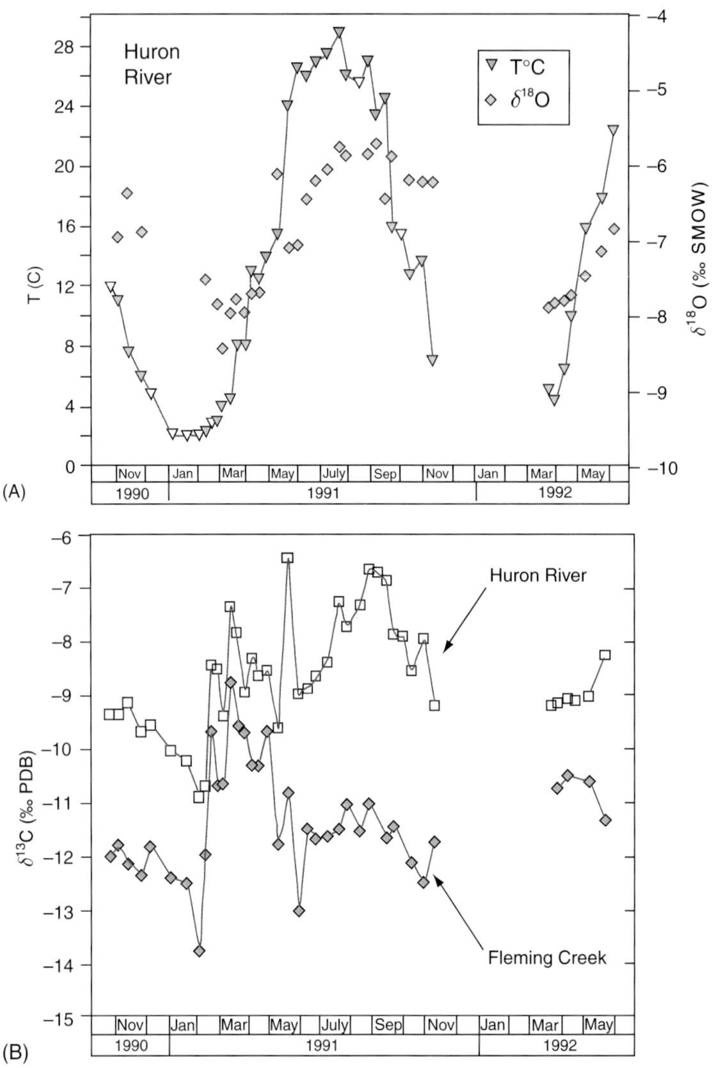

Figure 5 (A) Water temperatures and $\delta^{18}O$ values for the Huron River from October, 1990 to June, 1992. Filled triangles are temperatures that have associated water $\delta^{18}O$ measurements. (B) The $\delta^{13}C$ values of DIC from the Huron River and Fleming Creek, Michigan. These values correspond with shell data in Figure 4. From Dettman *et al.* (1999).

values. Following this overturn, shell $\delta^{13}C$ values rise briefly (due to continued productivity in the lake) and then decrease abruptly (~7‰) as low temperatures diminish productivity in the lake and organic matter is oxidized (returning ^{13}C-depleted carbon from plant material back to lake waters). As stated previously, winter "shut downs" represent an unconformity in the shell isotope record when water temperatures fall below ~12 °C. Spring overturns (Feb.–March) cannot be separated from the annual $\delta^{13}C$ minimum established by the cessation of productivity during the winter and the missing shell record. Return of lake productivity in the spring and summer produces a gradual increase in $\delta^{13}C$ values (by ~6‰) until this trend is modified by the next fall overturn.

Although these data suggest that lacustrine bivalves can provide a detailed seasonal record of temperature, stratification, and productivity in small lake basins, the $\delta^{13}C$ values of *Pyganodon lacustris* are lower than predicted equilibrium $\delta^{13}C$ values for shell aragonite. Glovers Pond $\delta^{13}C_{DIC}$ values are ~−6‰ (Quevedo, 2003). Therefore, equilibrium $\delta^{13}C$ values of shell aragonite should be ~−3.3‰. This finding, like those of other studies of unionids, is disappointing (e.g., Dettman *et al.*, 1999). However, unlike other studies, data from *Pyganodon lacustris* indicate that it is responding directly to processes at work in this lake. Its growth within the *Chara* mat ringing the lake margin may provide a food source that quickly responds to changes in photosynthesis and respiration in the lake.

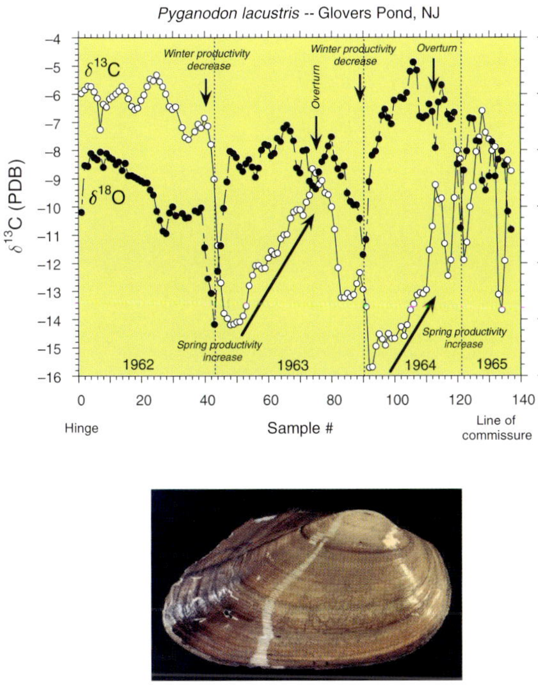

Figure 6 Plot of shell $\delta^{18}O$ and $\delta^{13}C$ values from *Pyganodon lacustris* collected from Glovers Pond, New Jersey. Shell height is 43 mm. Data are plotted relative to sequential sample number to avoid compression of data. Samples were collected from the exterior of the shell (note white sampling swath) using manual micro-sampling techniques (from Carpenter and Erickson, in preparation). Winters represent a missing time/isotopic record.

The work of Nichols and Garling (2000) and Christian *et al.* (2004) suggests that the food source of unionids is fine particulate organic matter that has a bacterial origin (not algal) and whose $\delta^{13}C$ and $\delta^{15}N$ values vary with season (summer FPOM has lower $\delta^{13}C$ values than fall FPOM). Further examination of unionid food sources and growth environments is critical in understanding the complicated record of shell $\delta^{13}C$ values. Among the carbon fluxes in lakes and rivers are: carbonate precipitation, formation and oxidation of organic matter (algae, bacteria, mosses, terrestrial plant material, etc.), methane generation (in lacustrine peats), and bicarbonate-rich ground water.

Corbicula fluminea - Indonesia

Corbicula fluminea and *Dreissena polymorpha* are two invasive species that recently have grown in number in the St. Lawrence Seaway, the Great Lakes, and the Mississippi River due to transport by ships that carry these species in ballast water (see Drake and Bossenbroek, 2004; Holeck *et al.*, 2004). As a result, these rather adaptive species have been studied extensively. Isotopic examination of their shells also has been conducted (Wurster and Patterson, 2001; Carpenter *et al.*, 2005; **Figs. 8** and **9**).

Corbicula sp. occurs naturally in streams and rivers in southeast Asia. A modern *Corbicula* sp. shell collected alive from a small stream in central Java has $\delta^{18}O$ and $\delta^{13}C$ values that reflect a monsoonal moisture pattern (rainy and dry seasons; **Fig. 8A**). The source of waters that feed these streams is rainfall on nearby volcanic peaks (~3,000 meters high). Shell $\delta^{13}C$ values range from −14.9 to −12.3‰ and reflect a riparian vegetation of C3 plants (trees and shrubs).

The Upper Sangiran Formation in Central Java, Indonesia (~1.66 to 1.5 Ma) contains abundant *Corbicula* sp. shells in a paludal to lacustrine setting. This locality is of importance as it has yielded skeletal remains of *Homo erectus*. The first occurrence of *H. erectus* in the Upper Sangiran coincides with wet conditions and soil organic matter and fossil grasses that have high $\delta^{13}C$ values (~−12‰; C4 Grasses). Grasses of this sort occur in poorly drained, wet conditions in southeast Asia (e.g., Johnston, 1996).

Corbicula sp. from this portion of the Upper Sangiran has $\delta^{13}C$ values that range from −7 to −1.8‰ (**Fig. 8B**). Carbonate soil nodules from this unit have similarly high $\delta^{13}C$ values. Both soil and molluscan carbonates have $\delta^{13}C$ values that are ~10 to 11‰ higher than associated plant organic matter. This offset is consistent with the carbon isotope fractionation between CO_2 and carbonate (Emrich *et al.*, 1970). Shell $\delta^{13}C$ values from modern *Corbicula* from this region are ~10‰ higher than their Pleistocene counterparts (**Fig. 8**). This suggests that the oxidation of plant material in soils and near streams is controlling the $\delta^{13}C$ value of shell carbonate (presumably via the $\delta^{13}C_{DIC}$). Given the tropical setting, cycling of carbon in these systems is likely very rapid, thus efficiently delivering CO_2 from organic matter decomposition to surface waters. This suggests that fossil mollusks from tropical regions may accurately record local vegetation.

In addition, the $\delta^{18}O$ and $\delta^{13}C$ values of the Pleistocene *Corbicula* display no similarities with the modern monsoonal moisture pattern, but instead exhibit a sinusoidal trend reminiscent of temperate bivalves. In addition, Pleistocene $\delta^{18}O$ values are significantly higher than analogous modern data (**Fig. 8**) suggesting either greater seasonal temperature variation and/or a markedly different source of precipitation that feeds local surface waters. Modern conditions are controlled by monsoonal rainfall with pronounced dry and rainy seasons. The dry season (May–Oct.) occurs when the

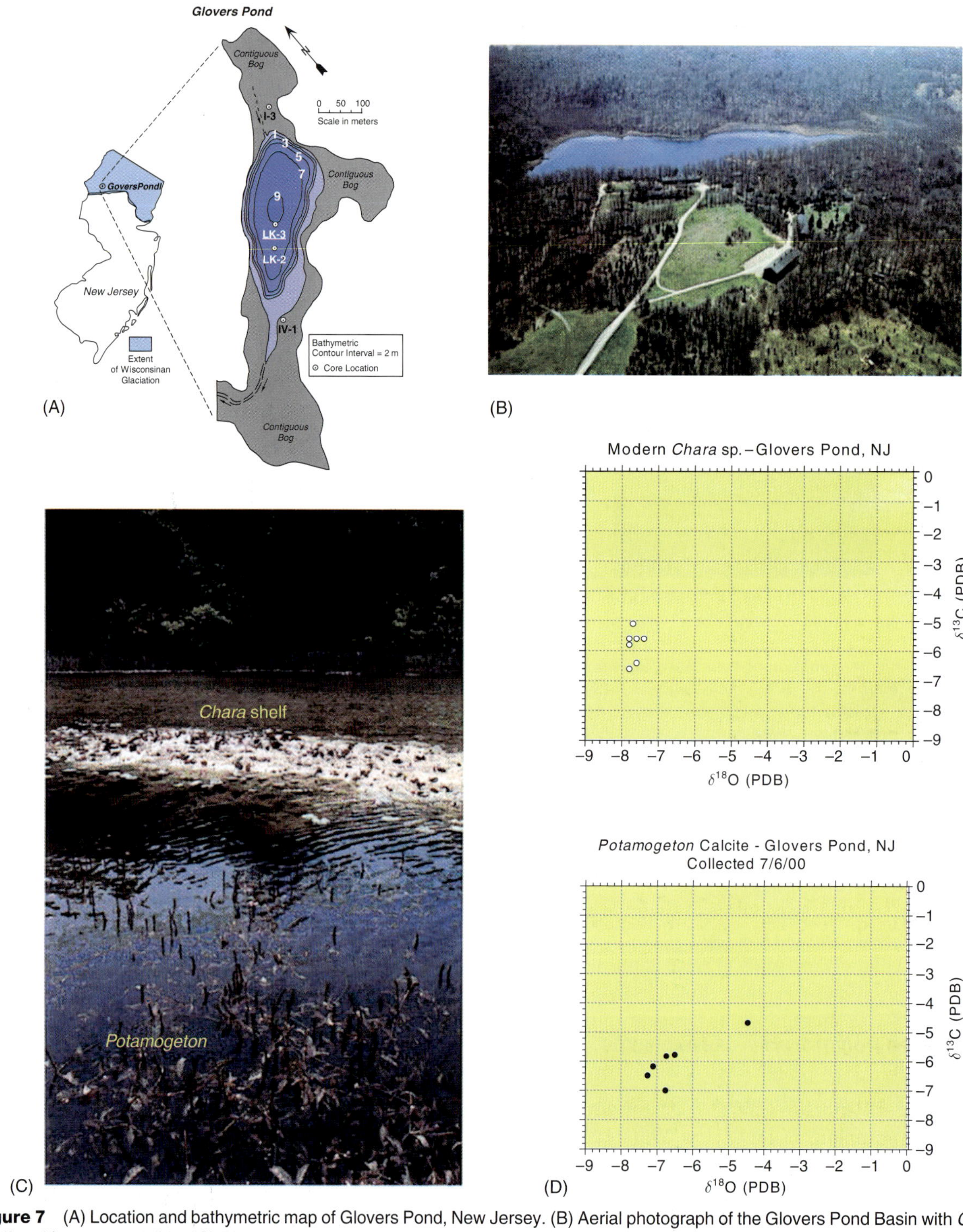

Figure 7 (A) Location and bathymetric map of Glovers Pond, New Jersey. (B) Aerial photograph of the Glovers Pond Basin with *Chara* shelf and bog areas visible. (C) Photograph of the margin of Glovers Pond with the well developed *Chara* shelf and actively calcifying *Potamogeton* growing on the shelf slope. (D) Plots of $\delta^{13}C$ and $\delta^{18}O$ values of modern *Chara* and *Potamogeton* carbonate. $\delta^{13}C_{DIC}$ is -6.5‰ and $\delta^{18}O_{water}$ is -4.8‰ (SMOW) following overturns. Precipitation of these carbonates under summer temperatures finds that these are precipitated in isotopic equilibrium with ambient waters.

ITCZ shifts to the north and dry air masses off Australia dominate. The rainy season (Nov. to April) occurs when the ITCZ shifts southward and humid air masses off SE Asia dominate. Paleosol and isotope data from the *Corbicula* in **Figure 8B** suggest that wet conditions persisted throughout the year without a dry season – perhaps during a glacial event.

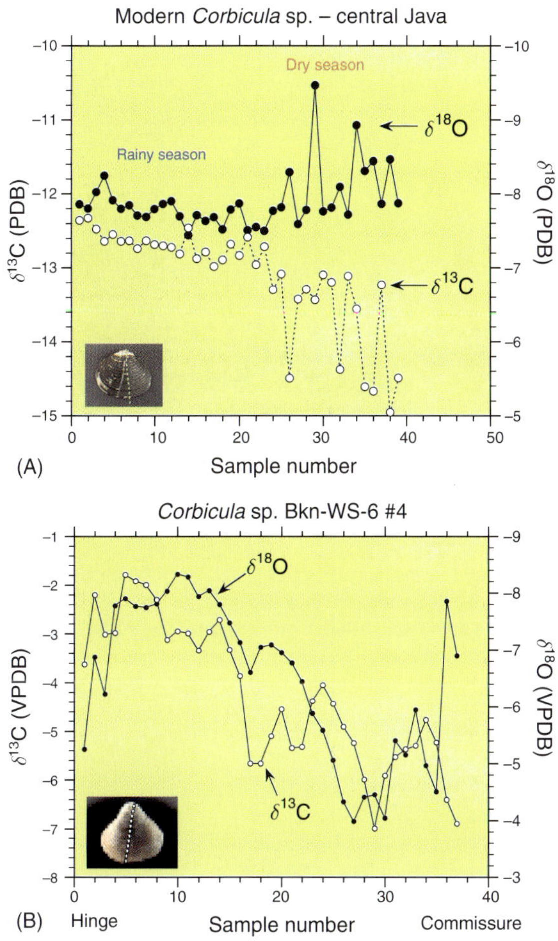

Figure 8 (A) Seasonal variation in $\delta^{13}C$ and $\delta^{18}O$ values in a modern *Corbicula* sp. shell from Central Java. This plot depicts the differences between rainy and dry seasons. (B) Seasonal variation in $\delta^{13}C$ and $\delta^{18}O$ values of a Pleistocene *Corbicula* sp. (Sangiran Fm) – note the higher $\delta^{13}C$ values and the presence of a sinusoidal trend – suggesting a lack of a monsoonal moisture pattern.

Dreissena polymorpha - Keuka Lake, New York

Wurster and Patterson (2001) have conducted high-resolution sampling of the Zebra Mussel (*Dreissena polymorpha*) from Keuka Lake, New York (**Figure 9**). Shell $\delta^{18}O$ values closely match the equilibrium aragonite $\delta^{18}O$ values predicted by using measured water $\delta^{18}O$ values, the fractionation factor of Dettman *et al.* (1999)(Equation 5) and both water and air temperatures. Micro-sampling yielded 277 carbonate samples over 14 mm of shell growth (206 were analyzed). While one might debate the utility of sampling at this density, this work indicates that sampling methods can begin to address short-term events (i.e., storms, floods) in molluscan shell records.

Calcareous Algae and Aquatic Plants

Perhaps the most important producer of biogenic carbonate formed in freshwater is calcareous green algae. The family *Characea* (often referred to as Charophytes and *Chara*) is composed of macrophytic green algae that grow in oligotrophic ponds and lakes with relatively high bicarbonate contents. Most species of *Chara* sp. occur in the upper few meters of the water column and under elevated pH and [Ca^{2+}] conditions, calcify heavily.

Chara, together with carbonate precipitation on aquatic plants such as *Potamogeton* are, to a large extent, responsible for the relatively rapid accumulation of carbonate sediments in marl lakes (**Fig. 7**). This type of lacustrine calcification occurs predominantly during the late Spring to early Fall in temperate latitudes (due to elevated temperatures, photosynthesis, pH, and [HCO_3^-]; McConnaughey *et al.*, 1994). Therefore, carbonates analyzed from lake sediment cores (marl lakes) must exhibit a strong bias toward Summer-time precipitation (e.g., Jones *et al.*, 1996). This requires an explanation of stable isotope ratios that is consistent with physico-chemical conditions within the higher range of seasonal temperature variation (from ~20 to 35 °C for temperate settings). Another factor in the interpretation of the $\delta^{18}O$ values of *Chara* calcite is its Mg content (e.g., Tarutani *et al.*, 1969; Anadón *et al.*, 2002).

Interpretation of $\delta^{13}C$ and $\delta^{18}O$ values collected from cores of lacustrine carbonate sediment requires an understanding of the calcification method and the degree to which it is precipitated in equilibrium with ambient waters. There is some debate over the degree to which Charaphyte carbonate is precipitated in isotopic equilibrium with ambient water (Huon and Mojon, 1993; Jones *et al.*, 1996; Coletta *et al.*, 2001). Both *Chara* sp. and *Potamogeton* sp. from Glovers Pond, New Jersey precipitate carbonate in carbon and oxygen isotopic equilibrium with ambient lake waters (Carpenter and Erickson, in prep.).

McConnaughey and Falk (1991) and McConnaughey (1991) describe possible calcification methods in *Chara* sp. where calcification occurs along alkaline bands on the exterior of the plant. Their data indicate a 1 : 1 ratio of calcification to photosynthesis. The extracellular precipitation of calcite by *Chara* as described by McConnaughey (1991) argues for minimal control of the calcification process by the plant and a greater likelihood of equilibrium precipitation (e.g., *Acetabularia* sp., Lee and Carpenter, 2001). If CO_2 hydroxylation plays a significant role in *Chara* calcification then one would anticipate a linear relation between $\delta^{13}C$ and $\delta^{18}O$ values of its carbonate (McConnaughey, 1989a, b).

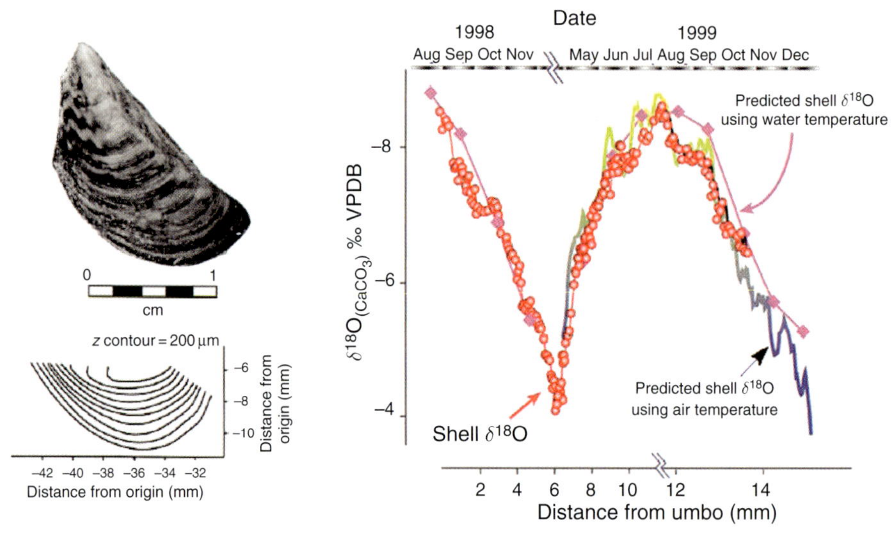

Figure 9 Example of high-resolution microsampling from freshwater bivalve *Dreissena polymorpha* from Keuka Lake, central New York (from Wurster and Patterson, 2001b). Digital image of *D. polymorpha* and illustration of a portion of the morphology of *D. polymorpha* determined using digital path coordinates; Comparison between predicted shell $\delta^{18}O$ values and measured shell $\delta^{18}O$ values for *D. polymorpha*. In order to satisfactorily compare the data and account for seasonal changes in the deposition of carbonate, different growth scales were used. A different growth scale was used from the umbo to the minimum shell $\delta^{18}O$ value and from the minimum shell $\delta^{18}O$ value to the last sample. In order to show a more detailed comparison, predicted shell $\delta^{18}O$ values determined using air temperatures from Rochester, New York and the fractionation factor of Dettman *et al.* (1999) (Eqn [5]) are also shown. Note that the scale was not changed for the date axis. However, Nov. through Mar. are omitted because predicted shell $\delta^{18}O$ values could not be determined for these months. Modified from Wurster and Patterson (2001b).

In contrast, $\delta^{13}C$ and $\delta^{18}O$ values collected from modern *Chara* exhibit minimal covariation in these values (Jones *et al.*, 1996; Carpenter and Erickson, in prep.). Positive, linear covariation of $\delta^{13}C$ and $\delta^{18}O$ values collected from carbonate-rich sediment cores (composed of predominantly Charaphytes) from temperate, marl lakes (Drummond *et al.*, 1995; Jones *et al.*, 1996) is inconsistent with data from modern *Chara*. This linear covariation has been ascribed to a variety of lake physico-chemical processes (stratification, CO_2 degassing, temperature variation, evaporation, seasonal changes in precipitation, etc.; Drummond *et al.*, 1995; Jones *et al.*, 1996). The time scales on which these data are formed (months for modern *Chara* and 10 to100 years for sediment samples) likely account for some of this difference. As lake basins evolve over thousands of years, climate, reservoir size, vegetation, and various other parameters change. It is likely that this variation will impact the $\delta^{13}C$ and $\delta^{18}O$ values of carbonate sediments. Clearly, more detailed examination of these processes is warranted.

Fish Otoliths

Otoliths (carbonate ear stones) of freshwater fish can be used to examine multi-year seasonal variation in $\delta^{18}O$ and $\delta^{13}C$ values. The well-defined

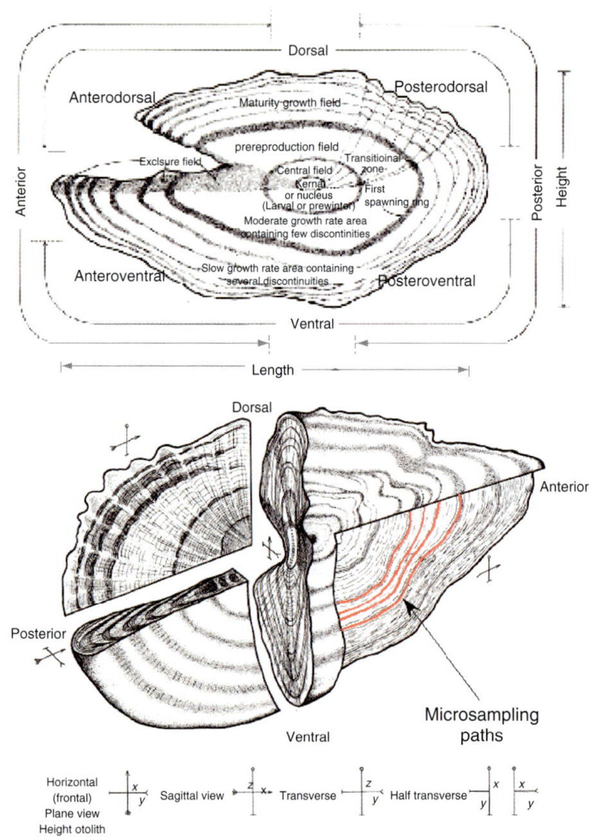

Figure 10 Schematic representation of a sagittal otolith depicting idealized growth bands (timelines). Microsampling paths follow these bands to extract intraotolith variation in $\delta^{13}C$ and $\delta^{18}O$ values. Modified from Pannella (1980).

sclerochronology of sagittal otoliths (sagittae) provides a record that can be exploited using microsampling techniques (**Fig. 10**). Typically aragonite, sagittae can be found in Quaternary age midden sites where fish were caught, eaten, and discarded (e.g., Wurster and Patterson, 2003). Preservation of otoliths is typically good in Quaternary deposits where materials have not been exposed to ground waters for prolonged periods of time. Stable isotope data have been collected from aragonite otoliths from Mesozoic age deposits (e.g., Patterson, 1999; Carpenter *et al.*, 2003).

Two studies of the stable isotope ratios of freshwater fish otoliths are worthy of note. The first is an analysis of fossil freshwater drum (*Aplodinotus grunniens*) otoliths from an archaeological site in northeastern Tennessee (**Fig. 11**; Wurster and Patterson, 2003). They examine intra-otolith variation in δ^{18}O and δ^{13}C values from specimens that span the last 5,500 years. This study contrasts data from high- and low-resolution sampling (**Fig. 11**). This comparison points out how high-resolution sampling provides significantly more information about fish life histories (**Fig. 11A–H**). Wurster and Patterson (2003) use the δ^{13}C and δ^{18}O values of otoliths from different periods in the Holocene to estimate changes in bioenergetics induced by climate (a combination of temperature, metabolic activity, and other factors).

Wurster *et al.* (2005) examine intra-otolith variation of δ^{13}C and δ^{18}O value in chinook salmon (*Oncorhynchus tshawytscha*) from Lake Ontario

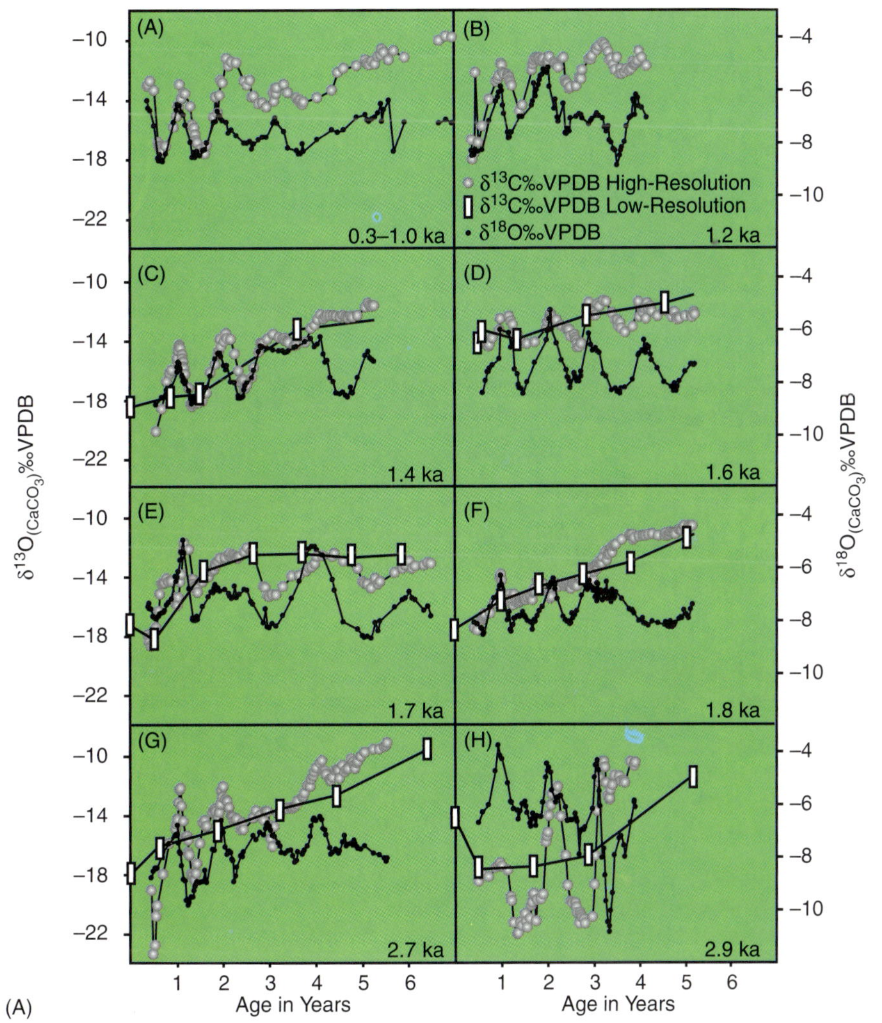

Figure 11 Otolith δ^{13}C and δ^{18}O values of the freshwater drum *Aplondinotus grunniens* from northeastern Tennessee (Wurster and Patterson, 2003). (A) Intraotolith variation in high-resolution δ^{13}C and δ^{18}O values of micromilled sagittae from time periods 0.3–2.9 ka. Samples were taken only from the first 8 yr of the sagittae or less. Each subplot represents an individual otolith. Each high-resolution sample represents an individual carbonate microsample. Low-resolution samples that approximate the average values of microsamples are plotted for comparison. Represented time periods are 0.3–1.0, 1.2, 1.4, 1.6, 1.7, 1.8, 2.7, and 2.9 ka. Specimens from A and B were destroyed before low-resolution sampling could be performed.

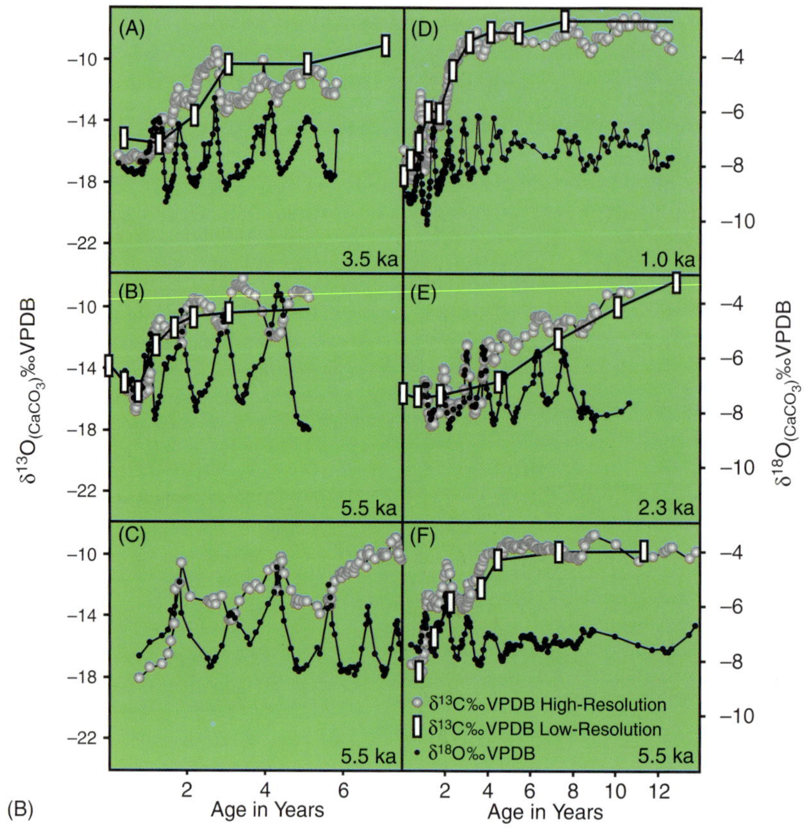

Figure 11 (B) Intra-otolith variation in high-resolution $\delta^{13}C$ and $\delta^{18}O$ values of micro-sampled sagittae from time periods 1.0 Ka, 2.3 Ka, 3.5 Ka, and 5.5 Ka. Each subplot represents an individual otolith. Each high-resolution sample represents an individual carbonate aliquot removed by micro-sampling. Low-resolution samples that approximate the average values of micro-sampled aliquots are plotted for comparison. Note different age axes on subplots. Subplots on the left side include samples only from the first eight years of the sagittae or less, whereas subplots on the right include samples from beyond the first eight years. The specimen from C was destroyed before low-resolution sampling could be performed. From Wurster and Patterson (2003).

(**Fig. 12**). They study thermal histories (otolith $\delta^{18}O$ thermometry), stress, and metabolism (otolith $\delta^{13}C$ values) in this migratory fish. During the summer these fish inhabit epilimnetic waters with temperatures of ~19–20 °C (**Fig. 12**). Chinook would approach but rarely exceed their reported upper incipient lethal limit of approximately 22 °C, which suggests that these fish were seeking

Discontinuous Records of Climate Change

The occurrence of freshwater biogenic carbonates during the Quaternary is quite variable. The most continuous carbonate records are of *Chara* sp. in lacustrine settings (i.e., lake sediment cores). Where rivers transport mollusks into lake basins, there is a potential to preserve more continuous records of mollusk shells. However, riverine deposits are by nature, discontinuous due to a variety of geological processes (e.g., meandering and channel migration, base level change and erosion, etc.). *In situ* riverine deposits containing mollusks and fish otoliths typically permit brief, episodic glimpses of seasonality and climate during periods of rapid deposition (e.g., flood events). Somewhat more continuous records of mollusk shells can be found in midden deposits. Where such deposits occur on flood plains, their continuous nature relies on the depositional and/or erosional qualities of the river and the extent to which humans or other organisms occupy a given site. Careful attention must be paid to the stratigraphic position, dating techniques used, position of paleosols, and bioturbation of midden deposits. In deposits where fire was used, researchers must use caution when analyzing charred biogenic carbonates.

Collection of long-term, continuous climate records from Quaternary age freshwater mussels and fish otoliths is not likely. Instead, isotopic records can be collected from relatively short intervals of time (typically a few thousand years) dictated by climatic conditions that are conducive to midden production. On a brighter note, well preserved specimens can yield a great deal of information regarding changes in seasonal temperature and water composition. Likewise, season of collection can be obtained in archaeological materials – a useful tool in understanding the timing of human habitation. For long-term climate interpretation, the best-case scenario is one in which well preserved materials occur episodically throughout a deposit where co-occurring dateable materials can provide precise limits on the timing of carbonate deposition (in particular, climatically important time intervals during the Late Pleistocene and Holocene).

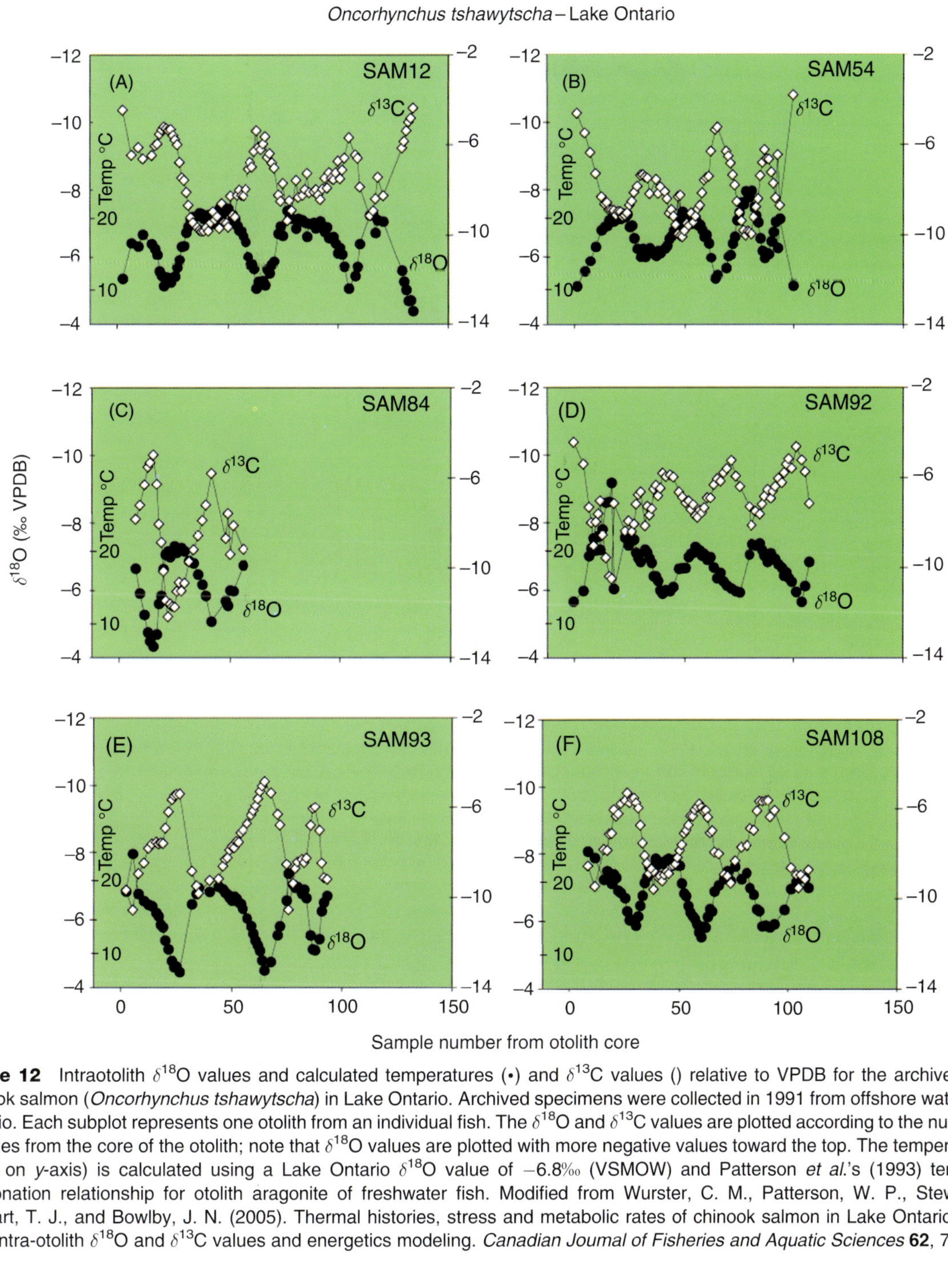

Figure 12 Intraotolith $\delta^{18}O$ values and calculated temperatures (•) and $\delta^{13}C$ values () relative to VPDB for the archived group of chinook salmon (*Oncorhynchus tshawytscha*) in Lake Ontario. Archived specimens were collected in 1991 from offshore waters of Lake Ontario. Each subplot represents one otolith from an individual fish. The $\delta^{18}O$ and $\delta^{13}C$ values are plotted according to the number of the samples from the core of the otolith; note that $\delta^{18}O$ values are plotted with more negative values toward the top. The temperature scale (inset on *y*-axis) is calculated using a Lake Ontario $\delta^{18}O$ value of −6.8‰ (VSMOW) and Patterson *et al.*'s (1993) temperature–fractionation relationship for otolith aragonite of freshwater fish. Modified from Wurster, C. M., Patterson, W. P., Stewart, D. J., Stewart, T. J., and Bowlby, J. N. (2005). Thermal histories, stress and metabolic rates of chinook salmon in Lake Ontario: Evidence from intra-otolith $\delta^{18}O$ and $\delta^{13}C$ values and energetics modeling. *Canadian Journal of Fisheries and Aquatic Sciences* **62**, 700–713.

water with temperatures as high as was tolerable while otolith growth occurred. These results contrast with expected midsummer temperatures for this typically cold-water fish. Bioenergetic simulations find that these salmon undergo significant stress at these temperatures. Increased catabolism and decreased gross conversion efficiency relative to nominal simulations where chinook salmon are modeled nearer their preferred temperature, confirms previous inferences that the chinook salmon population may be near the limits of sustainability (Wurster *et al.*, 2005).

Conclusions

Stable isotope data from freshwater biogenic carbonates have great potential for use in evaluating a variety of environmental conditions in the

Quaternary record. Among these are: temperature, $\delta^{18}O$ value of surface waters, vegetation, timing of seasonal surface water productivity changes, and lake stratification. To enhance the utility of these measurements several advances need to be made: 1) Additional calibration of geologically significant organisms, 2) Additional examination of the mechanisms responsible for biological fractionations or 'vital effects', and 3) Simultaneous analysis of carbonate minor element chemistry to develop an independent paleo-thermometer (e.g., Sr/Ca and Mg/Ca ratios). Elemental analyses come with their own caveats – such as calibration of modern materials, and understanding carbonate precipitation rates. Even with such advances, examination of multiple proxies is highly recommended (e.g., mollusks, otoliths, soil organic matter, stalagmites, lacustrine sediments (carbonate, organic matter, pollen, mites, ostracodes, etc.).

See also: **Carbonate Stable Isotopes**: Speleothems. **Ice Core Methods**: Overview; Methane Studies.

References

Adkins, J. F., Boyle, E. A., Curry, W. B., and Lutringer, A. (2003). Stable isotopes in deep-sea corals and a new mechanism for "vital effects". *Geochim. Cosmochim. Acta* **67**, 1129–1143.

Anadón, P., Utrilla, R., and Vazquez, A. (2002). Mineralogy and Sr-Mg geochemistry of charophyte carbonates: a new tool for paleolimnological research. *Earth and Planetary Science Letters* **197**, 205–214.

Carpenter, S. J., Erickson, J. M., and Holland, F. D. (2003). Migration of a Late Cretaceous fish. *Nature* **423**, 70–74.

Carpenter, S. J., and Lohmann, K. C. (1995). $\delta^{18}O$ and $\delta^{13}C$ values of modern brachiopod shells. *Geochim. Cosmochim. Acta* **59**, 3749–3764.

Carpenter, S.J. and Erickson, J.M. (in preparation). $\delta^{13}C$ and $\delta^{18}O$ values of shell carbonate from the freshwater mussel *Pyganodon lacustris*: temperature, stratification, and productivity records of Glovers Pond, New Jersey. *Palaeogeography, Palaeoclimatology, Palaeoecology*, 21 ms. pgs.

Carpenter, S.J., Weirich, F.H., Blesius, L.J., and Erickson, J.M. (in preparation). A record of discharge and turbidity in the Mackenzie River, NWT, Canada: a coordinated study of stable isotope data from *Lampsilis radiata siliquoidea* and remote sensing. *Geology, 12 ms. pgs.*

Carpenter, S. J., Bettis, E. A., III, Milius, A. K., and Ciochon, R. L. (2005). Carbon isotope ratios of tropical freshwater mollusk shells as indicators of ancient vegetation – reconstructing paleoenvironments at a Homo erectus locality in the Solo Basin, Central Java, Indonesia. *Geological Society of America, Abstracts with Program* 37(7), 363.

Christian, A. D., Smith, B. N., Berg, D. J., Smoot, J. C., and Findlay, R. H. (2004). Trophic position and potential food sources of 2 species of unionid bivalves (Mollusca: Unionidae) in 2 small Ohio streams. *Journal of the North American Benthological Society* **23**(1), 101–113.

Coletta, P., Pentecost, A., and Spiro, B. (2001). Stable isotopes in charophyte incrustations: relationships with climate and water chemistry. *Palaeogeography, Palaeoclimatology, Palaeoecology* **173**, 9–19.

Dettman, D. L., and Lohmann, K. C. (1995). Approaches to microsampling carbonates for stable isotope and minor element analysis: Physical separation of samples on a 20 micrometer scale. *Journal of Sedimentary Petrology* **v. A65**, 566–569.

Dettman, D. L., Reische, A. K., and Lohmann, K. C. (1999). Controls on the stable isotope composition of seasonal growth bands in aragonitic fresh-water bivalves (*Unionidae*). *Geochimica et Cosmochimica Acta* **v. 63**, 1049–1057.

Drake, J. M., and Bossenbroek, J. M. (2004). The potential distribution of Zebra Mussels in the United States. *BioScience* **54**, 931–941.

Drummond, C. N., Patterson, W. P., and Walker, J. C. G. (1995). Climate forcing of carbon-oxygen isotopic covariance in temperate region marl lakes. *Geology* **23**, 1031–1034.

Emrich, K., Ehhalt, D. H., and Vogel, J. C. (1970). Carbon isotope fractionation during the precipitation of calcium carbonate. *Earth and Planetary Science Letters* **8**, 363–371.

Erickson, J.M. (1968). The geologic and limnologic history of Glovers Pond, northwestern New Jersey, Unpublished M.S. Thesis, University of North Dakota, Grand Forks, North Dakota, p. 149

Frappier, A., Sahagian, D., González, L. A., and Carpenter, S. J. (2002). El Niño events recorded by stalagmite carbon isotopes. *Science* **298**, 565.

Grossman, E. L., and Ku, T.-L. (1986). Oxygen and carbon isotope fractionation in biogenic aragonite: Temperature effects. *Chem. Geol.* **59**, 59–74.

Holeck, K. T., Mills, E. L., Macissac, H. J., Dochoda, M. R., Colautti, R. I., and Ricciardi, A. (2004). Bridging Troubled Waters:Biological Invasions,Transoceanic Shipping, and the Laurentian Great Lakes. *BioScience* **54**, 919–929.

Howard, A. D. (1921). Experiments in the culture of freshwater mussels. *Bulletin of the U.S. Bureau of Fisheries* **38**, 63–90.

Huon, S., and Mojon, P.-O. (1994). Déséquilibre isotopique entre eau et parties calcifées de Charophytes actuels (Chara globularis): résultats préliminaries. *C.R. Acad. Sci. Paris série II* **318**, 205–210.

Johnston, W. H. (1996). The place of C4 grasses in temperate pastures in Australia. *New Zealand Journal of Agricultural Research* **39**, 527–540.

Jones, T. J., Fortier, S. M., Pentecost, A., and Collinson, M. E. (1996). Stable carbon and oxygen isotopic compositions of recent charophyte oosporangia and water from Malham Tarn, U.K.: Paleontological implications. *Biogeochemistry* **34**, 99–112.

Joy, J. E. (1985). A 40-week study on growth of the Asian Clam, Corbicula fluminea (Müller) in the Kanawha River, West Virginia. *Nautilus* **99**, 110–116.

Kim, S.-T., and O'Neil, J. R. (1997). Equilibrium and nonequilibrium oxygen isotope effects in synthetic carbonates. *Geochimica et Cosmochimica Acta* **61**, 3461–3476.

Lee, D., and Carpenter, S. J. (2001). Isotopic Disequilibrium in Marine Calcareous Algae. *Chemical Geology – Isotope Geoscience Section* **172**, 307–329.

McConnaughey, T. (1989a). ^{13}C and ^{18}O isotopic disequilibrium in biological carbonates. I: Patterns. *Geochimica et Cosmochimica Acta* **53**, 151–162.

McConnaughey, T. (1989b). ^{13}C and ^{18}O isotopic disequilibrium in biological carbonates. II: In vitro simulation of kinetic isotope effects. *Geochimica et Cosmochimica Acta* **53**, 163–171.

McConnaughey, T. (1991). Calcification in *Chara corallina*: CO_2 hydroxylation generates protons for bicarbonate assimilation. *Limol. Oceanogr.* **36**, 619–628.

McConnaughey, T., and Falk, R. H. (1991). Calcium-proton exchange during algal calcification. *Biol. Bull.* **180**, 185–195.

McConnaughey, T. A., LaBaugh, J. W., Rosenberry, D. O., Reddy, M. M., Schuster, P. F., and Carter, V. (1994). Carbon budget for a groundwater-fed lake: calcification supports summer photosynthesis. *Limnol. Oceanogr.* **39**, 1319–1332.

Negus, C. L. (1966). A quantitative study of growth and production of unionid mussels in the River Thames at Reading. *J. Anim. Ecolog.* **35**, 513–532.

Neumann, D., Borcherding, J., and Jantz, B. (1993). Growth and seasonal reproduction of *Dreissena polymorpha* in the Rhine River and adjacent waters. In *Zebra Mussels: Biology, Impacts, and Control* (T. F. Nalepa and D. W. Schloesser, Eds.), pp. 95–110. Lewis Publishers.

Nichols, S. J., and Garling, D. (2000). Food-web dynamics and trophic-level interactions in a multispecies community of freshwater unionids. *Canadian Journal of Zoology* **78**, 871–882.

Pannella, G. (1980). Growth patterns in fish sagittae. In *Skeletal Growth of Aquatic Organisms – Biological records of environmental Change* (D. C. Rhoads and R. A. Lutz, Eds), pp. 519–560.

Patterson, W. P. (1998). North American continental seasonality during the last millennium: high-resolution analysis of sagittal otoliths. *Palaeogeography, Palaeoclimatology, Palaeoecology* **138**, 271–303.

Patterson, W. P. (1999). Oldest isotopically characterized fish otoliths provide insight to Jurassic continental climate of Europe. *Geology* **27**, 199–202.

Patterson, W. P., and Smith, G. R. (2002). Uses of fishes in analysis of paleolimnology in Tracking Environmental Change Using Lake Sediments: Zoological Indicators. In *Paleolimnology* (J. P. Smol, H. J. B. Birks and W. M. Last, Eds.) , Special vol. 4, ch. 9, pp. 173–187.

Patterson, W. P., Smith, G. R., and Lohmann, K. C. (1993). Continental paleothermometry and seasonality using the isotopic composition of aragonitic otoliths of freshwater fishes. In *Amer. Geophys. Union Monogr. Continental Climate Change from Isotopic Indicators* (P. Swart, K. C. Lohmann, J. McKenzie and S. Savin, Eds.), pp. 191–202.

Quevedo, H.L. (2003). A high resolution stable isotope record of climate change from NW New Jersey, Unpublished Ph.D. Dissertation, University of Iowa, Iowa City, Iowa, p. 159.

Ricken, W., Steuber, T., Freitag, H., Hirschfeld, M., and Niedenzu, B. (2003). Recent and historical discharge of a large European river system – oxygen isotopic composition of river water and skeletal aragonite of Unionidae in the Rhine. *Palaeogeography, Palaeoclimatology, Palaeoecology* **193**, 73–86.

Romanek, C. S., Grossman, E. L., and Morse, J. W. (1992). Carbon isotopic fractionation in synthetic aragonite and calcite: Effects of temperature and precipitation rate. *Geochimica et Cosmochimica Acta* **56**, 419–430.

Rubinson, M., and Clayton, R. N. (1969). Carbon-13 fractionation between aragonite and calcite. *Geochimica et Cosmoschimica Acta* **33**, 997–1002.

Tarutani, T., Clayton, R. N., and Mayeda, T. K. (1969). The effect of polymorphism and magnesium substitution on oxygen isotope fractionation between calcium carbonate and water. *Geochimica et Cosmoschimica Acta* **33**, 987–996.

Tevesz, M. J. S., and Carter, J. G. (1980). Environmental relationships of shell form and structure of Unionacean bivalves. In *Skeletal Growth of Aquatic Organisms – Biological records of environmental Change* (D. C. Rhoads and R. A. Lutz, Eds.), pp. 295–322.

Urey, H. C., Lowenstam, H. A., Epstein, S., and McKinney, C. R. (1951). Measurement of paleotemperatures and temperatures of the upper Cretaceous of England, Denmark, and the southeastern United States. *Bull. Geol. Soc. Am.* **62**, 399–416.

Wurster, C. H., Patterson, W. P., and Cheatham, M. M. (1999). Advances in computer-based microsampling of biogenic carbonates. *Computers in Geosciences* **25**, 1155–1162.

Wurster, C. M., and Patterson, W. P. (2001). Stable oxygen and carbon isotope values recovered from lacustrine freshwater mollusks: paleoclimatic implications for sub-weekly temperature records. *Journal of Paleolimnology* **26**, 205–218.

Wurster, C. M., and Patterson, W. P. (2003). Late Holocene metabolic rate changes of freshwater drum (*Aplodinotus grunniens*): evidence from high-resolution sagittal otolith stable isotope ratios of carbon. *Paleobiology* **29**, 492–505.

Wurster, C. M., Patterson, W. P., Stewart, D. J., Stewart, T. J., and Bowlby, J. N. (2005). Thermal histories, stress, and metabolic rates of chinook salmon in Lake Ontario: evidence from intra-otolith $\delta^{18}O$ and $\delta^{13}C$ values and energetics modeling. *Canadian Journal of Fisheries and Aquatic Sciences* **62**, 700–713.

Zeebe, R. E. (1999). An explanation of the effect of seawater carbonate concentration on foraminiferal oxygen isotopes. *Geochimica et Cosmochimica Acta* **63**, 2001–2007.

Terrestrial Organic Materials

D McCarroll and N Loader, University of Wales – Swansea, Swansea, UK

© 2007 Elsevier B.V. All rights reserved.

Introduction

The elements carbon, hydrogen, and oxygen, which are the main constituents of terrestrial organic material, each have more than one stable (nonradioactive) isotope with nearly identical chemical properties but a difference in mass. By convention, the ratio of two stable isotopes is expressed relative to a standard using the delta notation in parts per mil (‰). In terrestrial organic material, the relevant isotope ratios are $\delta^{13}C$, $\delta^{18}O$, and $\delta^{2}H$ (or δD for deuterium). They are measured by first producing a suitable target gas, by either combustion or pyrolysis, and subsequent analysis by isotope ratio mass spectrometry (McCarroll and Loader, 2005). Since physical and biochemical processes are often sensitive to differences in mass, the ratios of the stable isotopes in organic material are not the same as those in the source (usually air or water). The degree of difference between the source and the organic product, or the total fractionation, may contain information not only on the source material, including the carbon isotopic ratios of atmospheric CO_2, and the isotopic ratios of source water, but also on the degree of change that occurred as a result of environmental influences on fractionation. The isotopic signatures of the sources and changes due to environmental influences both may contain paleoenvironmental information. The most important continuous archives of terrestrial organic material suitable for stable isotopic analyses are tree rings and peat profiles.

Stable Isotopes in Tree Rings

As natural archives of paleoclimate information, trees have the great advantage that, under many conditions, they produce annual rings. A few tree species can live for millennia, but even for shorter-lived species, counting, measuring, and cross-matching the rings of many overlapping generations allow very long, well-replicated, annually resolved, and precisely dated chronologies to be constructed. Variations in the width of the annual rings in these long chronologies provide one of the most commonly used high-resolution 'proxy' sources of paleoclimate information. Suitable trees are widespread, and sampling and measuring are relatively inexpensive. However, the time series of ring widths need to be analyzed very carefully because they contain trends that relate to the age of the trees, and removing these trends without degrading the climate information is challenging. It is particularly difficult to retain long-term, low-frequency climate changes. One attraction of stable isotopes in tree rings is that the time series probably does not contain long-term trends related to tree age, and may thus retain climate information at all temporal frequencies. Also, the growth of trees, and therefore the annual ring width, is influenced by a myriad of interconnected factors, and so is difficult to model mechanistically. The interpretation of ring-width series, therefore, generally relies on statistical inference and in particular on regression techniques. In contrast, the factors that control the fractionation of isotopes in trees are quite well understood and relatively simple in comparison, so that a mechanistic understanding of the processes involved, and of the ecophysiology of particular tree species, can be used to guide paleoenvironmental interpretations. Early attempts at 'stable isotope dendroclimatology' used blocks of wood covering several years and/or pooled the wood of several trees prior to analysis. However, recent advances in sample preparation and mass spectrometry have greatly accelerated the rate at which organic samples can be analyzed, and the use of stable isotopes in tree rings to reconstruct past climate is a rapidly expanding field of research (McCarroll and Loader, 2004, 2005).

Isotopic Fractionation by Trees

The main constituents of wood are cellulose, which forms the framework of cells, and lignin, which adds strength. These components contain carbon that is taken from CO_2 that enters the leaves via the stomata, whereas the oxygen and hydrogen are derived from water taken up through the roots. As a consequence of this composite structure, in ancient samples there is a potential for preferential decomposition of lignins or cellulose. To avoid these effects, a single compound is isolated from the wholewood prior to isotopic analysis. Typically, the fraction analyzed is the cellulose, as this signal is derived exclusively from the dated rings of the tree. However, studies with lignin have also been demonstrated to preserve climatic information at high (annual) resolution (Loader *et al.*, 2003).

The stable isotopes of carbon in CO_2 are fractionated first as the gas enters the stomata and again as it is taken up by the photosynthetic enzyme, in both cases leading to an increase in the relative abundance of the lighter isotope and therefore a reduction in $\delta^{13}C$ (more negative values). The first step occurs because lighter molecules diffuse through a gap more easily than heavier ones. As gas molecules bounce off each other, the lighter ones bounce furthest, so the fractionation due to diffusion is controlled by the difference in mass. This means that the fractionation of carbon as it diffuses into the stomatal chambers remains at about −4.4‰, almost irrespective of the environmental conditions or the size of the stomatal opening. It only increases under the rare conditions when the stomatal opening is so small that molecules bouncing off the sides, rather than each other, become important. The second fractionation, when the CO_2 is taken up by the photosynthetic enzyme, is similarly insensitive to environmental conditions, including temperature, and remains at about −27‰ (Farquhar *et al.*, 1982) (**Fig. 1**).

The insensitivity of the fractionations to environmental variables means that the isotopic ratios of the gas that enters the leaf is controlled by that of the ambient air, so that tree-ring $\delta^{13}C$ must be affected by changes in the isotopic composition of the atmosphere. The second fractionation, however, does not act on ambient air but on the CO_2 in air that is within the stomatal chambers. As CO_2 is removed and assimilated, the relative abundance of the heavier isotopes remaining in this intercellular CO_2 must increase, and as more of this enriched CO_2 is assimilated, the $\delta^{13}C$ of the resulting sugars must also increase, even though the fractionation remains at about −27‰.

The fractionation of carbon that occurs in tree leaves is not a direct response to environmental variables, but rather a measure of the relative rates at which CO_2 enters and is removed from the stomatal chambers. The controls are therefore stomatal conductance and photosynthetic rate. Where stomatal conductance is high relative to photosynthetic rate, the assimilates will have low $\delta^{13}C$ values, but where stomatal openings become restricted in response to moisture stress, or when photosynthetic rates are very high, the internal concentration of CO_2 will drop, becoming enriched in ^{13}C and yielding assimilates with higher $\delta^{13}C$

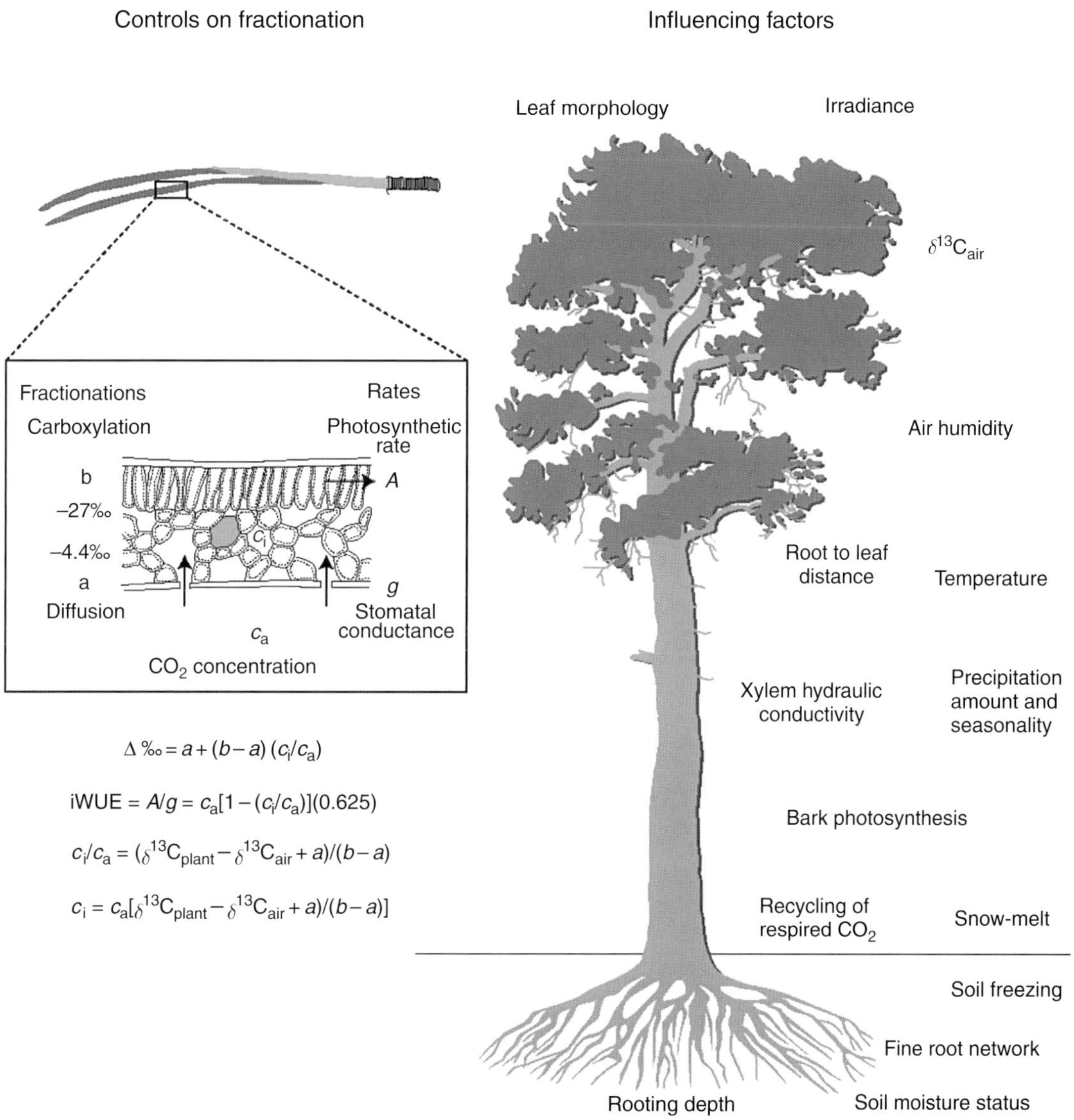

Figure 1 Diagram of a needle-leaf tree showing the main controls on fractionation of carbon isotopes and the environmental factors that influence them. Discrimination against ^{13}C (Δ) is expressed in per mil (‰) relative to a standard (Vienna Pee Dee Belemnite (VPDB)). Discrimination occurs during diffusion through the stomata (a) and again during carboxylation (b). Where the ambient concentration of CO_2 (c_a) is known, the key variable is the internal concentration of CO_2 (c_i), which is controlled by the balance between stomatal conductance (g) and photosynthetic rate (A). This relationship can be used to calculate intrinsic water use efficiency (iWUE). After McCarroll D and Loader NJ (2004) Stable isotopes in tree rings. *Quaternary Science Reviews* 23: 771–801.

values. The sugars produced in the leaf may be used directly to produce new wood, in which case, the climatic conditions that influenced the signal imparted in the leaf will be transferred to the wood cells produced in the same year, or they may be stored and remobilized later, leading to further fractionation and a lag between the climatic forcing and its expression in the isotopic composition of the tree ring (Helle and Schleser, 2004). This effect is most pronounced in the earlywood fraction of the tree ring; consequently, to avoid remobilized carbon, and thus maintain the annual resolution of the signal, many recent studies have used only the latewood portion of tree rings, which forms during the summer.

The ratios of the stable isotopes of oxygen and hydrogen in tree rings are more complex than those of carbon because they conflate two separate signals: a source water signal and an isotopic enrichment due to leaf transpiration. The isotopic ratios in precipitation are controlled by a number of factors (Darling *et al.*, 2005), including (but not always dominated by) temperature. However, to regard water-isotope ratios in tree rings as a simple paleothermometer is overly simplistic. The rainfall must first reach the tree roots, so the depth and structure of the root system of different species can mean that water of different ages is utilized. Wood formed in a given summer may, therefore, utilize water molecules from summer rainfall, precipitation from the previous winter (Robertson *et al.*, 2001), or over a much longer time period. There is no fractionation when roots take up the soil water, but on reaching the leaves, evaporation results in a preferential loss of the lighter isotopes. The controls are very similar to those that govern fractionation due to evaporation from a lake, though there is an additional fractionation as water vapor diffuses though the stomata and leaf boundary layer. Although there is a small temperature effect, the dominant controls are the isotopic composition of water in the atmosphere, relative to the source water, and the difference between the ambient and intercellular vapor pressures. Low air humidity thus leads to greater enrichment of the heavier isotopes of both oxygen and hydrogen. Backward movement of isotopically enriched water from the sites of evaporation is opposed by the supply of unenriched source water, so the isotopic values of total leaf water are lower than those at the sites of evaporation (Barbour *et al.*, 2001). There is a further fractionation when the water is incorporated in sucrose formed in the leaf, leading to a further enrichment of ^{18}O (+27‰) but a larger and less certain depletion of deuterium (−100‰ to −171‰) (**Fig. 2**).

The situation is further complicated because a proportion of the oxygen atoms in the sugars that form in the leaf are able to exchange with (unenriched) atoms from xylem (source) water moving up the stem at the site where the wood is formed. The amount of exchange is difficult to predict, but controls the relative importance of the source water and leaf enrichment signals. The potential for exchange is probably larger for lignin than for cellulose. When sugars are used to form wood cells, there is a large fractionation of hydrogen in the opposite direction and similar in magnitude to that which occurs when the sugar is formed (+144‰ to +166‰), and as with oxygen, there is potential for exchange with xylem water and thus for altering the balance of the two signals (McCarroll and Loader, 2004, 2005).

Interpreting Stable Isotopes from Tree Rings

Early work on stable isotopes in tree rings interpreted changes in both the carbon- and the water-isotope ratios in terms of changes in temperature, treating them as paleothermometers. Unfortunately, this is a gross oversimplification and in reality none of them is controlled in a simple way by a single climatic variable. Carbon-isotope ratios record the balance between stomatal conductance and photosynthetic rate, whereas the water isotopes take part of their signal from the isotopic ratios of the source water but then enrich it by evaporation in the leaf. Although each of these isotopes may, under certain circumstances, be strongly correlated with temperature, it is unlikely that temperature is actually the controlling factor. Temperature may correlate with carbon-isotopic ratios through covariance with changes in sunlight, which drives photosynthesis, or through covariance with relative humidity and rainfall, which control stomatal conductance. Temperature is only one of the factors that influence the isotopic ratios in source water, and although there is a small direct influence of temperature on fractionation at the sites of evaporation in the leaf, the dominant control is the difference in vapor pressure inside and outside of the leaf, which is strongly dependent on air relative humidity. Thus, although the correlation with temperature during the period when meteorological records are available can be used to reconstruct past temperatures using regression-based techniques, it is important to remember that correlation is not causation and that any change in one of the

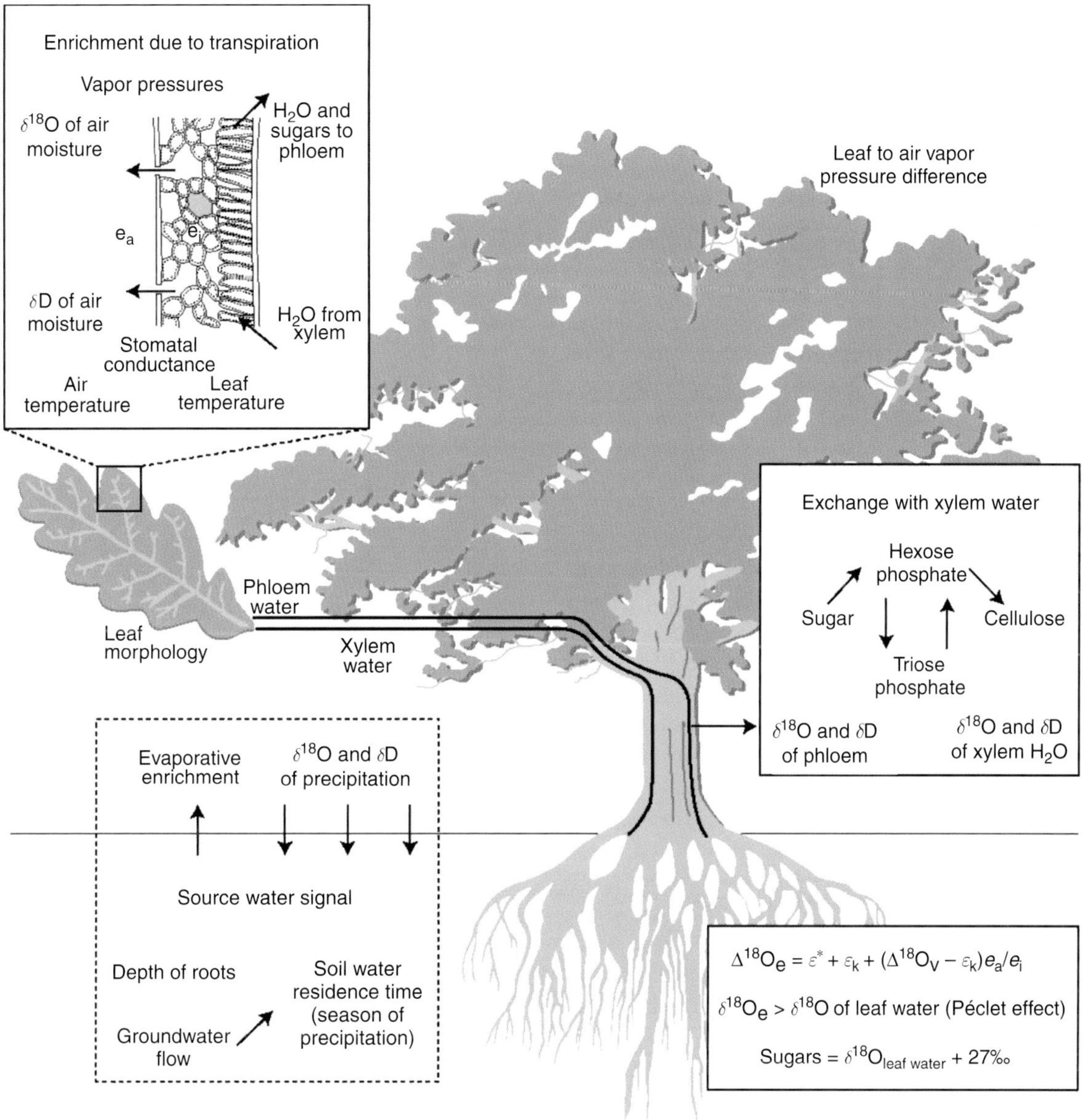

Figure 2 Diagram of a broad-leaved tree showing the main controls on the fractionation of oxygen isotopes and the environmental factors that influence them. There is no fractionation when source water is taken up by the roots. The level of enrichment of leaf water above source water, at the sites of evaporation ($\delta^{18}O_e$), depends upon the proportional depression of water vapor pressure by the heavier $H_2^{18}O$ ($\varepsilon*$), fractionation as water diffuses through the stomata and leaf boundary layer (ε_k), the isotopic composition of water vapor in the atmosphere, relative to source water ($\Delta^{18}O_v$), and the ambient and intercellular vapor pressures (e_a and e_i). The same equations control the fractionation of hydrogen isotopes, though the fractionation factors are different. After McCarroll D and Loader NJ (2004) Stable isotopes in tree rings. *Quaternary Science Reviews* 23: 771–801.

real controlling factors will change the isotopic ratios, irrespective of whether there was a change in temperature. Since the instrumental period coincides with the period of greatest anthropogenic disturbance, to both forests and the atmosphere, changes in the relationship between temperature and the controlling factors may well have occurred. Of course, this logic applies to most natural archives used as proxy sources of paleoclimate information, it is simply more apparent for tree ring stable isotopes. This is because trees provide absolute temporal control and a clearer (mechanistic) understanding of signal preservation than is often observed in other terrestrial paleorecords.

Rather than relying on the correlation between temperature and isotopic ratios to reconstruct paleotemperatures using 'black box' methods based on regression, it might be more enlightening to take advantage of the mechanistic understanding of fractionation and attempt to tease out the real environmental controls. This can be achieved in part by careful selection of trees and sites so that one control is likely to dominate (McCarroll and Loader, 2005). Shallow-rooted trees on sites with very freely draining soils, for example, are likely to be sensitive to the stomatal conductance control on carbon-isotope ratios and to respond rapidly to changes in source water-isotope ratios. Deeply rooted trees with access to a plentiful supply of groundwater, on the other hand, are likely to yield carbon-isotope ratios that are more responsive to changes in photosynthetic rate and water-isotope ratios, in which the source water signal is smoothed so that the evaporative enrichment signal is clearer. Perhaps even greater potential lies in combining different isotopes and other paleoclimate proxies that can be extracted from trees in a multiproxy approach that is guided by theory rather than purely by statistical inference (McCarroll *et al.*, 2003).

Application of isotopic analysis is not limited to annual increments. Through high-resolution sampling of individual rings, rhythmic trends and periodicities can be identified that can be related to plant physiological processes occurring within the growth cycle (Helle and Schleser, 2003). This approach can be taken forward further to tropical species that do not exhibit regular annual banding, yet where an apparent (possibly seasonal/annual) cyclicity may be identified in their isotopic composition.

Stable Isotopes in Peat Profiles

Peat deposits provide one of the best-known archives of paleoclimate information for the Holocene. Peat accumulation can be continuous and the acidic conditions are suitable for the preservation of a wide range of organic materials including leaves, other plant macrofossils, and pollen. The preserved macrofossils, or identifiable remains of the peat-forming plants, can be radiocarbon-dated to provide a chronology. Stable isotopic analysis of leaves provides a potential source of information on the stable isotope ratios in the vegetation surrounding the peat at different times (Rundgren *et al.*, 2000). More attention has been paid, however, to isotopic analysis of the mosses and other peat-forming plants.

Some mosses, such as *Sphagnum*, differ from the higher plants in that, although photosynthesis appears to follow the C3 pathway, they do not possess functioning guard cells around their stomata and hence do not regulate water loss. The dominant controls on the fractionation of carbon isotopes, therefore, are likely to be the isotope ratios of the source gas, photosynthetic rate, and the resistance to diffusion of CO_2. Resistance to diffusion is likely related to the thickness of the water layer, so that at higher moisture contents mosses would be expected to yield less negative values. This relationship with wetness is the opposite to that seen in the higher plants, including trees, where dry conditions lead to a reduction in stomatal conductance, a decline in internal concentration of CO_2, and thus to less negative $\delta^{13}C$ values. In the mosses, resistance to diffusion is equivalent to stomatal conductance, but a reduction in moisture content decreases resistance to diffusion, leading to more negative $\delta^{13}C$ values. This is supported by results from several species of *Sphagnum* from two peat bogs in Ireland, where plants growing on relatively dry hummocks yielded higher $\delta^{13}C$ values than those in wetter hollows (Price *et al.*, 1997).

As with other plants, the isotopic ratio of the source gas used by mosses must be reflected in the isotopic ratios of the photosynthetic products. However, whereas mature trees have access to well-mixed atmospheric gases, mosses on the surface of a peat mire may be in contact with significant amounts of recycled CO_2, which will produce relatively light $\delta^{13}C$ values. Price *et al.* (1997) have argued that this effect will be most marked where the mosses are submerged, and particularly where there is a low density of living plants in pools. For this reason, they argue that hummock-top species of *Sphagnum* are likely to be the most useful for stable carbon-isotopic analysis.

The differences in carbon-isotope fractionation between mosses and sedges in peat have been proposed as a relatively high-resolution method for reconstructing past changes in the concentration of atmospheric CO_2, perhaps linked to changes in thermohaline circulation (Figge and White, 1995). The efficacy of this technique is contested, however, since some of the differences could be due to the effects of changes in moisture regime on fractionation by the mosses.

The fractionation of the water isotopes in mosses should be the same as for higher plants, in that isotope ratios should reflect the isotopic values of the source water and isotopic enrichment due to evaporation. However, where the plants in question are fully submerged, evaporation does not take place and so the isotopic values in the plant tissue should mirror the isotopic values of the source water, with a consistent offset due to fractionation during photosynthesis. Oxygen isotope results from *Sphagnum* cellulose confirm that submerged species yield values that mirror those of the water, whereas emergent species give a wider range of values, reflecting differences in the amount of evaporation (Zanazzi and Mora, 2005). Stable hydrogen and oxygen-isotopic analysis of submerged moss species from peat profiles may provide a powerful proxy for past changes in the isotopic composition of the bog water.

Stable Isotope Analysis of Pollen

The physical (numerical) analysis of pollen assemblages preserved in terrestrial and marine sedimentary archives provides one of the most established and powerful tools for the study of paleoecology and vegetation dynamics throughout the Quaternary.

More recently, significant interest has developed into the (CHO) isotopic analysis of pollen grains. This approach has only been made feasible through the development of continuous-flow mass spectrometry methods, which require smaller sample sizes than traditional off-line preparation and dual inlet techniques. A series of pilot studies has emerged within the last decade, which serves to identify the potential of this approach for paleoecological research. Loader and Hemming (2004) provide a review of the current state-of-the-art and discuss the current limitations and research opportunities for this emerging proxy.

Like many terrestrial paleoarchives, pollen grains comprise a range of chemical components. Typically, a single grain may be considered to consist of three principal structural elements: an inner nucleus which contains the reproductive and genetic materials, a cellulose inner layer or 'intine,' and a β-carotenoid ester outer layer or 'exine,' which represents ca. 65–80% of the grain and together may be further bound with lipids or coated with a thin polysaccharide layer. There is still debate as to the exact chemical structure of the exine, although it is believed to approximate $C_{90}H_{150}O_{33}$. Although incredibly resistant, being of a composite origin, pollen is susceptible to differential diagenesis. Typically, it is the exine that is selected for isotopic analysis, as it is this component that preserves in its morphology, the key to the species/genus of plant from which it originated (**Fig. 3**).

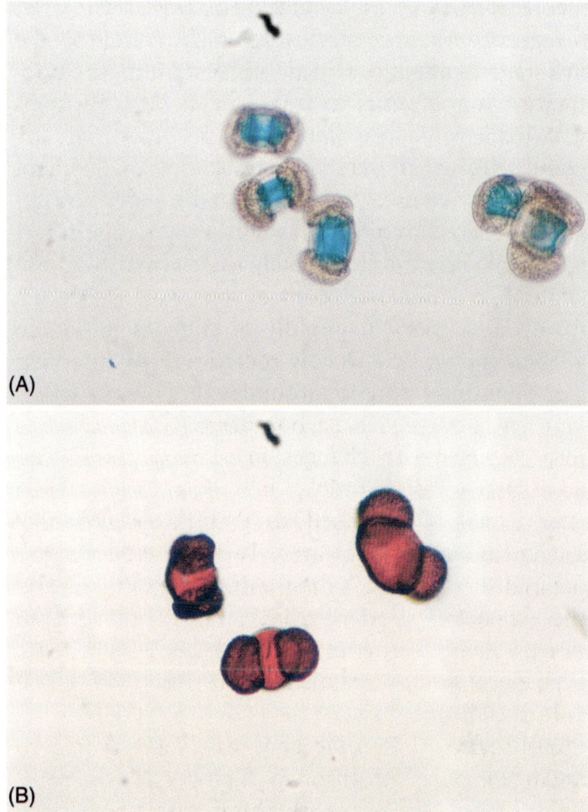

Figure 3 (A) Untreated *Pinus sylvestris* pollen composite-stained with safranin-fast green, highlighting the cellulose intine of the grains (green). Note: No cellulose is present in the saccae and the exine is only partially stained (pink) due to the presence of a polysaccharide layer. (B) Acid treated *Pinus sylvestris* pollen composite-stained with safranin-fast green. The strong safranin staining of the exine (pink) demonstrates removal of the outer polysaccharide layer. The cellulose intine has been completely removed. After Loader N. J. and Hemming D. L. (2000) Preparation of pollen for stable carbon isotope analyses. *Chemical Geology* **165**, 339–344.

Preparation of Pollen Samples

For conventional palynology, exines are purified from the sediment matrix through a suite of aggressive chemical processes. Isotopic analysis of raw and extracted modern pollen exines exhibit a species-specific isotopic offset of between ca. 2 and 4 per mil $\delta^{13}C$. Consequently, to avoid the effects of differential diagenesis, a chemical preparatory approach should be considered when preparing pollen extracts from the paleoarchive for isotopic analysis. Direct transfer of methodologies developed for conventional pollen preparation is likely to be inappropriate, as acetylation, for example, has been shown to contaminate isotopically the end product.

Carbon-isotopic analysis of modern pollen samples from different species have been shown to record different signals. It is therefore likely that in order to achieve a robust paleorecord, selection of a single species or genus may be required. At present, the greatest limitation to the wider application of this method lies in the extraction and purification of sufficient pollen grains for isotopic analysis. Sample concentration by sieving, dense media separation, and centrifugation, followed by manual picking, provide one of the most accessible methods currently available. Less time-consuming sample-preparation methods are urgently required. Application of fluid dynamic separation (split flow lateral transport thin separation (SPLITT)) methods or automated image recognition and micromanipulation exhibit the greatest potential for future advances in this field. Nevertheless, even using existing labor-intensive picking methods, it is possible to extract sufficient material for either radiocarbon dating or stable isotope analyses.

Large, easily extractable genera such as *Picea, Abies*, and, to a lesser extent, *Pinus* represent likely candidate species for paleoecological reconstruction based upon isotopic analysis (500 spruce grains ≈8 μg of carbon). Species differences are less important in applications where Poaceae pollen have been analyzed isotopically to establish the relative proportion of grassland community species which use either C3 or C4 photosynthetic mechanisms. Changes in the isotopic ratio of the bulk Poaecae signal reflect within it changes in (paleo)-grassland composition from which paleoclimate may be inferred (Amundson *et al.*, 1997). In all cases, representative sampling and thorough replication are essential, as is an understanding of pollen production, dispersal, and preservation. Interpretation of the isotopic data should also include consideration of the physical pollen record.

Environmental Signals

Unlike the formation of cellulose in tree rings, the imprinting of an environmental signal in pollen grains is still a subject of some debate. It is likely that many of the mechanisms described for tree rings are relevant to those responsible for pollen formation; however, the subsequent fate of the photoassimilates and the mechanisms and pathways they follow during their remobilization/use in the production of pollen remains largely unknown. Empirical studies provide some insight into these processes. Carbon isotopes are likely to reflect the composition of the photosynthates occurring during the formation of the pollen initials (year $t-1$) and the conditions prevailing during pollen maturation and photosynthesis prior to pollen fly (year t). Mechanistically, the oxygen- and hydrogen-isotope ratios of the photosynthates used to form the pollen should reflect the composition of the source water, with evaporative enrichment (and potentially subsequent re-exchange with source water). Initial investigations have broadly demonstrated this relationship for hydrogen, but it would appear that the signal preserved in the oxygen-isotope ratio of pollen exines is more complex and may reflect the relative degree to which the source water signal is transferred to the pollen grain or differences in biochemical fractionations occurring along the respective photosynthetic pathways. Further work is needed to evaluate and better quantify these processes.

See also: **Carbonate Stable Isotopes**: Terrestrial Organic Materials. **Dendrochronology**. **Dendroclimatology**. **Pollen Analysis, Principles**.

References

Amundson, R., Evett, R. R., Jahren, A. H., and Bartoleme, J. (1997). Stable carbon isotope composition of Poaceae pollen and its potential in paleovegetational reconstructions. *Review of Palaeobotany and Palynology* **99**, 17–24.

Barbour, M. M., Andrews, T. J., and Farquhar, G. D. (2001). Correlations between oxygen isotope ratios of wood constituents of *Quercus* and *Pinus* samples from around the world. *Australian Journal of Plant Physiology* **28**, 335–348.

Darling, W. G., Bath, A. H., Gibson, J. J., and Rozanski, K. (2005). Isotopes in water. In *Isotopes in Palaeoenvironmental Research* (M. J. Leng, Ed.), pp. 1–66. Springer, The Netherlands.

Farquhar, G. D., O'Leary, M. H., and Berry, J. A. (1982). On the relationship between carbon isotope discrimination and intercellular carbon dioxide concentration in leaves. *Australian Journal of Plant Physiology* **9**, 121–137.

Figge, R. A., and White, J. W. C. (1995). High resolution Holocene and late glacial atmospheric CO_2 record: Variability tied to changes in thermohaline circulation. *Global Biogeochemical Cycles* **9**, 391–403.

Helle, G., and Schleser, G. H. (2004). Beyond CO_2-fixation by Rubisco – An interpretation of C-13/C-12 variations in tree rings from novel intra-seasonal studies on broad-leaf trees. *Plant, Cell Environment* **27**, 367–380.

Loader, N. J., and Hemming, D. L. (2000). Preparation of pollen for stable carbon isotope analyses. *Chemical Geology* **165**, 339–344.

Loader, N. J., and Hemming, D. L. (2004). The stable isotope analysis of pollen as an indicator of terrestrial palaeoenvironmental change: A review of progress and recent developments. *Quaternary Science Reviews* **23**, 893–900.

Loader, N. J., Robertson, I., and McCarroll, D. (2003). Comparison of stable carbon isotope ratios in the whole wood, cellulose and lignin of oak tree rings. *Palaeogeography, Palaeoclimatology, Palaeoecology* **196**, 395–407.

McCarroll, D., Jalkanen, R., Hicks, S., *et al.* (2003). Multi-proxy dendroclimatology: A pilot study in northern Finland. *The Holocene* **13**, 829–838.

McCarroll, D., and Loader, N. J. (2004). Stable isotopes in tree rings. *Quaternary Science Reviews* **23**, 771–801.

McCarroll, D., and Loader, N. J. (2004). Isotopes in tree rings. In *Isotopes in Palaeoenvironmental Research* (M. J. Leng, Ed.), pp. 67–116. Springer, The Netherlands.

Price, G. D., McKenzie, J. E., Pilcher, J. R., and Hoper, S. T. (1997). Carbon-isotope variation in Sphagnum from hummocks-hollow complexes: Implications for Holocene climate reconstruction. *The Holocene* **7**, 229–233.

Robertson, I., Waterhouse, J. S., Barker, A. C., Carter, A. H. C., and Switsur, V. R. (2001). Oxygen isotope ratios of oak in east England: Implications for reconstructing the isotopic composition of precipitation. *Earth and Planetary Science Letters* **191**, 21–31.
Rundgren, M., Loader, N. J., and Beerling, D. J. (2000). Variations in the carbon isotope composition of late Holocene plant macrofossils: A comparison of whole-leaf and cellulose trends. *The Holocene* **10**(1), 149–154.
Zanazzi, A., and Mora, G. (2005). Palaeoclimate implications of the relationship between oxygen isotope ratios of moss cellulose and source water in wetlands of Lake Superior. *Chemical Geology* **222**, 281–291.

Non-Lacustrine Terrestrial Studies

J Quade, University of Arizona, AZ, USA
T Cerling, University of Utah, UT, USA

© 2007 Elsevier B.V. All rights reserved.

Introduction

The carbon- (expressed in ‰ as $\delta^{13}C$ (VPDB)) and oxygen-($\delta^{18}O$ (VPDB)) isotopic composition of surficial carbonate has been widely used in paleoclimatic and paleoenvironmental reconstruction during the Quaternary. In surficial carbonates we include: soil carbonates and spherosiderites, spring travertines, and a variety of early meteoric cements, usually formed in the saturated zone. Other important surficial meteoric carbonates covered by other articles in this Encyclopedia include lacustrine (*see* Lake Sediments) and speleothem (*see* Speleothems) carbonates. There is a large literature on this topic that cannot be comprehensively reviewed in this short article. Instead, we examine a few representative soil carbonate records of environmental change over the Neogene and Pleistocene, and groundwater carbonate records from Devils Hole, Nevada.

Formation of soil carbonates and groundwater carbonates can be viewed as expressions of a single groundwater flow continuum, summarized in **Figure 1**. The general equation for weathering (to the right) and calcite precipitation (to the left) in this continuum is:

$$CaCO_3(s) + CO_2(g) + H_2O = Ca^{++}(aq) + 2HCO_3^-(aq) \quad [1]$$

Here we use calcite as the parent material but we could just as easily have used most silicate minerals in its place. Calcite or silicate dissolution begins in the

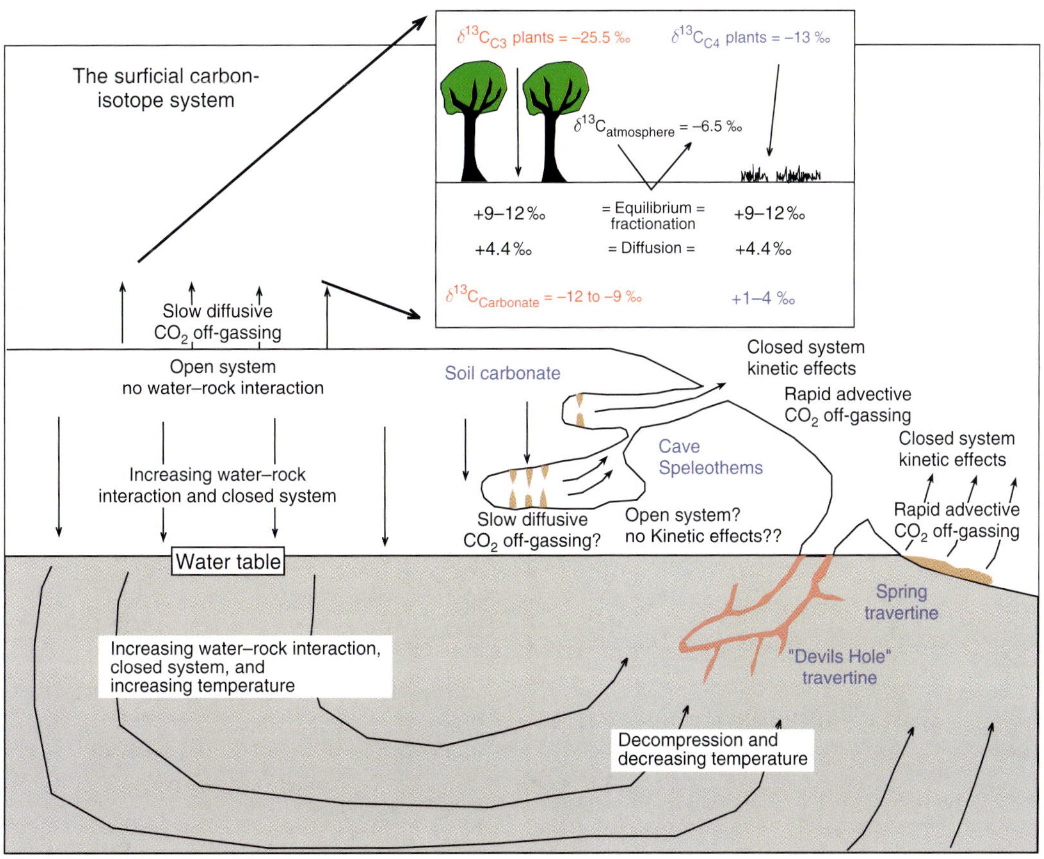

Figure 1 Summary diagram of formation processes for soil carbonates (inset), spring, and deep groundwater (Devils Hole) carbonates.

soil (vadose) zone in the presence of elevated soil pCO_2 (>350 ppmV) produced by plant respiration and decay. This dissolution process can continue below the vadose zone and into the saturated zone until significant levels of supersaturation with respect to calcite are reached, whereupon carbonate precipitates. In dry soil settings, supersaturation often develops prior to the next soil flushing event, due largely to soil dewatering by plants and by evaporation. In the shallow saturated zone, such in some caves and springs, calcite precipitation is driven by off-gassing of dissolved CO_2. In deeper systems, such as Devils Hole in southern Nevada, decompression during ascent of deeply circulated groundwater may induce carbonate formation.

Carbonate is among the most abundant of secondary minerals in the geologic record, and can form in a large range of burial depths and conditions. The use of the soil diffusion model (see below) as an interpretive framework for carbon-isotope values in soils assumes that the material being analyzed is a well-preserved soil carbonate. One should rely on a suite of features for accurate identification of soil carbonate over other secondary forms of calcite cementation. These criteria include (1) firm associations with other soil features (**Fig. 2**), (2) nodule form (**Figs. 2B** and **C**) and micritic texture (**Figs. 2I** and **J**), (3) reworking of nodules into contemporaneous channels (**Figs. 2H** and **I**), and (4) the absence of detrital carbonate contamination. (*see* Soil Morphology in Quaternary Studies).

The Devils Hole calcites in southern Nevada lie toward the distal end of a large (~12,000 km^2) flow system hosted largely by fractured carbonate rocks. Coarse-grained, laminated calcites coat fractures below the top of the water table (**Fig. 3**). Access to these fracture coatings is possible at Devils Hole, where collapse of the roof of one large fracture has exposed the aquifer and permitted divers to enter, as well as at various locations in nearby Death Valley, where recent tectonism has exhumed other fracture coatings (**Fig. 3**).

Figure 2 Diagnostic features of soil and paleosols. A = organic-rich A horizon, Bt = clay-rich argillic horizon, and Bk = calcic horizon. (A) Modern soil, Samos, Greece (hammer ~28 cm); (B) buried paleosol, late Miocene, Chinji Formation, Pakistan, (hammer ~66 cm); (C) soil carbonate nodules, late Miocene, Dhok Pathan Formation, Pakistan, (knife ~8 cm); (D) buried paleosol, late Pliocene, Pakistan (hammer ~28 cm); (E) buried organic-rich paleosol, Pliocene, Surai Khola, Nepal; (F) paleovertisol (hammer ~66 cm) with pseudo-anticlines and underlying airfall tuff; (G) detail of slickensides on ped (pen ~14 cm); (H) soil carbonate nodules reworked into contemporaneous channel lag (knife ~8 cm), middle Pleistocene, Busidima Formation, Gona, Ethiopia; (I) photomicrograph of soil nodules reworked into conglomerate and cemented by secondary spar, Pliocene, Pakistan; (J) photomicrograph of soil micrite with burrow infill and secondary spar cementing fractures, Pakistan, late Miocene.

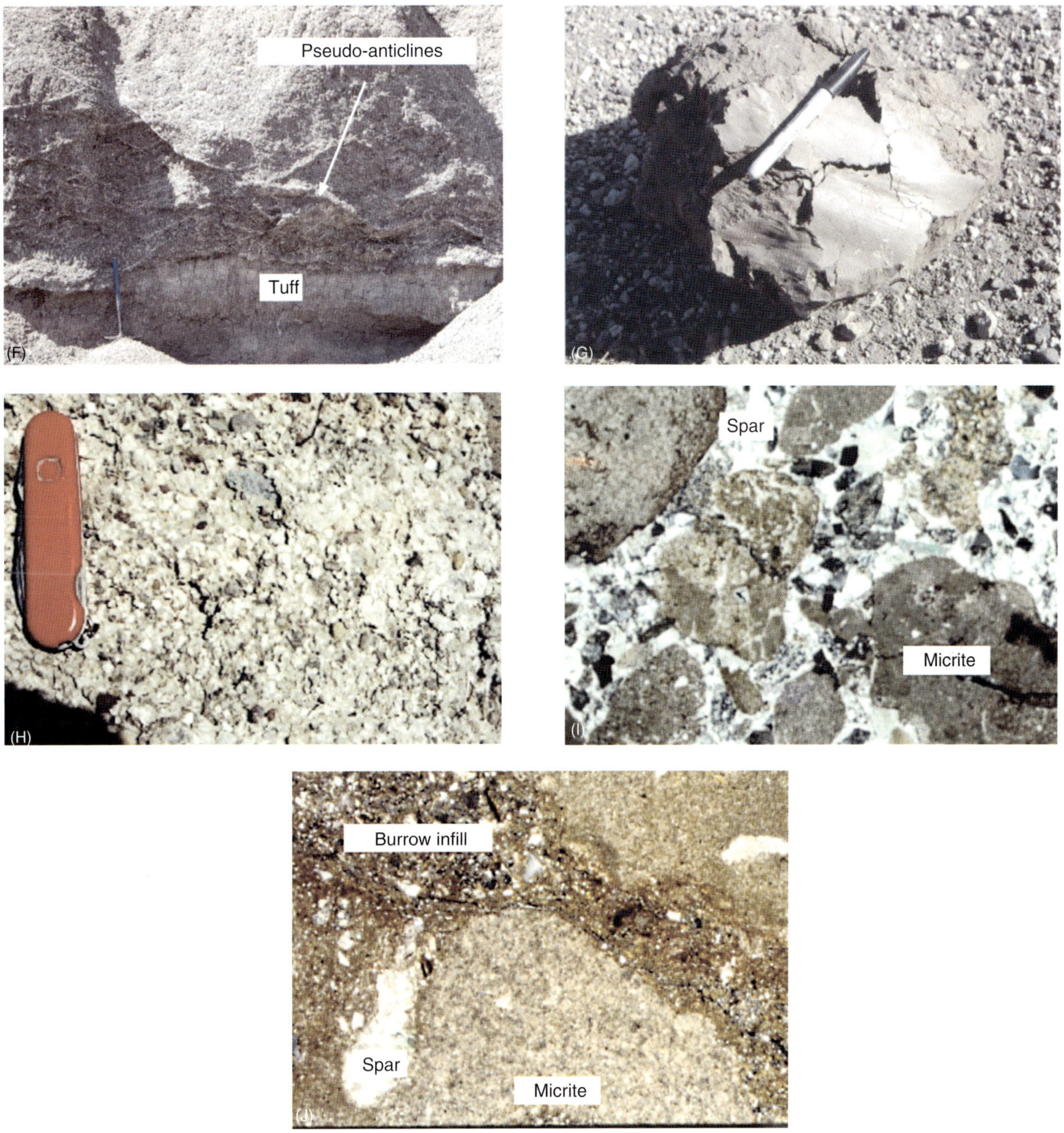

Figure 2 (*Continued*)

The Oxygen-Isotopic Composition of Surficial Meteoric Carbonates

Background

The oxygen-isotopic composition of surficial meteoric carbonate ($\delta^{18}O_{cc}$) (VPDB) is determined by both air and local vadose zone temperature, by the $\delta^{18}O$ value of rainfall ($\delta^{18}O_p$), and by soil evaporation.

Temperature effects Temperature can potentially influence $\delta^{18}O_{cc}$ values in two ways, through the influence of (1) soil temperature on fractionation between water and calcite during soil carbonate formation, and (2) air temperature on $\delta^{18}O_p$.

The temperature dependence of the fractionation factor (α_{c-w}) between calcite and water is:

$$\alpha_{c-w} = \exp\left(\frac{(18{,}030/T) - 32.42}{1{,}000}\right) \qquad [2]$$

Figure 3 Groundwater calcite, Death Valley, USA. (A) groundwater calcite veins cutting the Plio-Pleistocene Furnace Creek Formation; (B) core site of DH-11 (Devils Hole), photo by Ray Hoffman, courtesy of Alan Riggs (USGS); (C) vein calcite cutting the lower Furnace Creek Formation; (D) detail of laminated calcite fracture fill (pencil ~15 cm).

with temperature (T) in K (Kim and O'Neil, 1997). This equates to an approx. -0.22 to $-0.24‰\,C^{-1}$ change in $\delta^{18}O_{cc}$ values with soil temperature at constant $\delta^{18}O_p$. Temperature in the shallow vadose zone (2–10 m) approximates mean annual temperature (MAT), reducing potentially large seasonal temperature extremes. At depths of >100 m, the effects of local geothermal gradients (2–3° C per 100 m) will develop, as in the case of the mildly geothermal waters at Devils Hole.

The oxygen-isotope value of precipitation ($\delta^{18}O_p$) is correlated with air temperature at mid- to high latitudes. The slope of this dependence (the 'Dansgaard relationship') ranges between $+0.2$ and $0.9‰\,C^{-1}$ but averages $+0.6 \pm 0.1‰\,C^{-1}$ in most mid-latitude settings of soil carbonate formation.

These two influences of temperature on $\delta^{18}O_{cc}$ values are opposite in sign ($-0.23 \pm 0.1‰\,C^{-1}$ versus $+0.6 \pm 0.1‰\,C^{-1}$). They should combine to increase $\delta^{18}O_{cc}$ values by $+0.25$ to $+0.5‰\,C^{-1}$. In agreement with this prediction, in carbonates from modern soils with rainfall exceeding ~10 cm yr^{-1} (low evaporation), the observed slope is $+0.44‰\,C^{-1}$ ($r^2 = 0.56$) (**Fig. 4**).

The $+0.25$ to $+0.5‰\,C^{-1}$ dependence means that the large uncertainties in the $\delta^{18}O_p$ value produce major uncertainties in the reconstructed paleotemperature, but conversely, that large uncertainties in paleotemperature produce relatively small uncertainties in the reconstructed $\delta^{18}O_p$. This is why the $\delta^{18}O_{cc}$ values are poor paleothermometers but can be useful in constraining paleo-$\delta^{18}O_p$ values.

δ^{18}O composition of precipitation ($\delta^{18}O_p$) $\delta^{18}O_p$ is a strong predictor of $\delta^{18}O_{cc}$ values, especially in temperature soils where evaporation rates are minimized (**Fig. 5**). The oxygen value of precipitation is influenced by several factors, including air temperature, changes

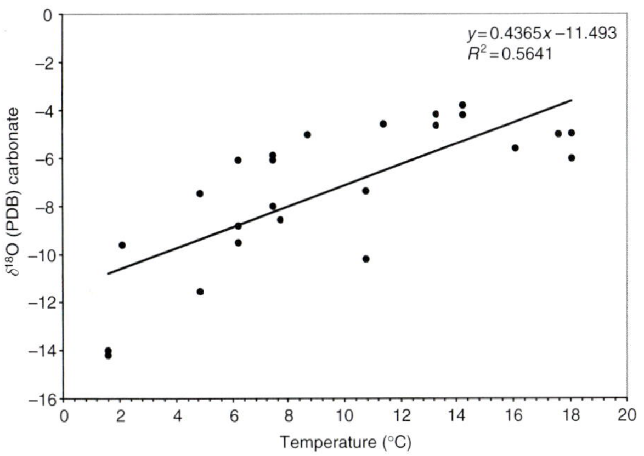

Figure 4 $\delta^{18}O_{cc}$ values from modern soils vs. mean annual temperature. The slope (+0.44 C/‰) of the best fit line matches the slope predicted by temperature effects of +0.25 °C/‰ to +0.5 C/‰.

in the $\delta^{18}O$ value of seawater ($\delta^{18}O_{seawater}$) and, at low latitudes, by changes in rainfall amount, referred to as the amount effect. The amount effect is the dominant control on $\delta^{18}O_p$ at low latitudes and is ~−1.5‰ per 10 cm rainfall globally (Rozanski *et al.*, 1993). This effect is not only visible in rainfall records from island stations, but also in monsoonal systems, where the $\delta^{18}O_p$ values during the summer rainy season are much lower than in the winter nonrainy season.

The ocean is the source of most atmospheric moisture, hence changes in the $\delta^{18}O_{seawater}$ values should translate directly into changes in $\delta^{18}O_p$ values. During glacial periods in the Quaternary, $\delta^{18}O_{seawater}$ values increase by ~1.0‰. At mid- to high latitudes, increases in $\delta^{18}O_{seawater}$ during glacial periods would dampen decreases in $\delta^{18}O_{cc}$ values due to lower temperatures.

Soil water evaporation When RH is high, $\delta^{18}O_{soilwater} = \delta^{18}O_p$, as discussed above. In sphaerosiderites and groundwater carbonates, the evaporation term is probably unimportant. In soils from dry regions, evaporation can strongly increase $\delta^{18}O_{soilwater}$ values with respect to $\delta^{18}O_p$ values. In general, evaporative enrichment in ^{18}O of water increases linearly with decreased relative humidity. Results from the Atacama Desert and the Death Valley portion of the Mojave Desert (**Fig. 5**) illustrate the strong effects of evaporation on $\delta^{18}O_{cc}$ values.

The Carbon-Isotopic Composition of Surficial Meteoric Carbonates

Organic processes tend to dominate the $\delta^{13}C$ value of surficial carbonates. In this article, we focus on the carbon isotopes in soil carbonates, since the $\delta^{13}C$ values of other types of carbonates, although of interest from a process point of view, do not contribute much to paleoenvironmental reconstruction. For example, spring travertines and many cave speleothems (Fantidas and Ehhalt, 1970; Hendy, 1971; reviewed in Quade (2003)) undergo kinetic enrichment in ^{13}C due to CO_2 off-gassing and evaporation, greatly

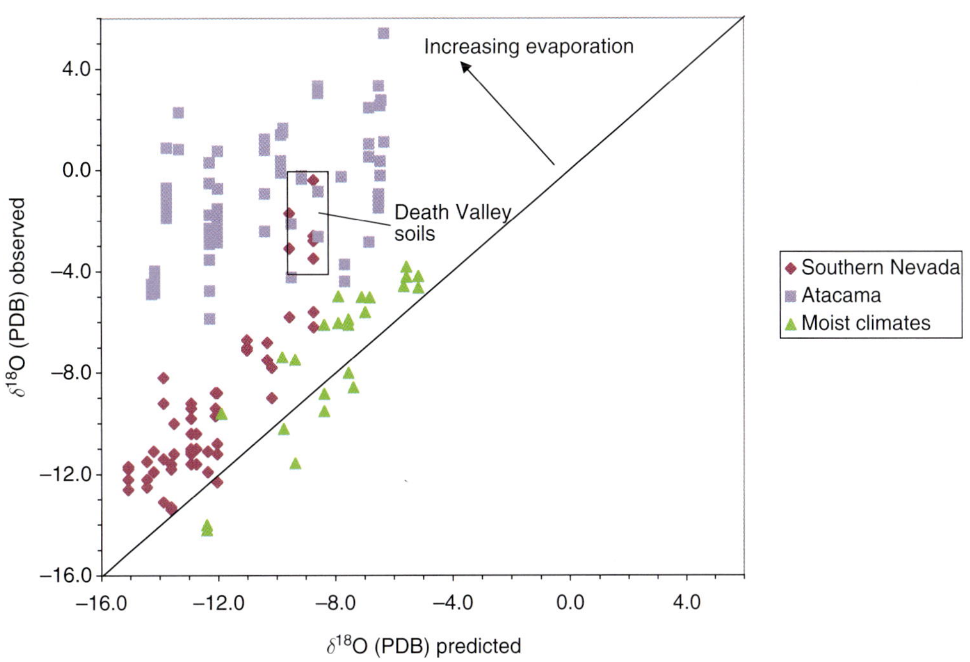

Figure 5 The $\delta^{18}O_{cc}$ (observed) (data from Cerling and Quade (1993), Quade *et al.* (1989b), and unpublished data) vs. predicted $\delta^{18}O_{cc}$ values based on local mean annual temperature and $\delta^{18}O_p$ of local rainfall.

reducing their usefulness in paleoclimatic reconstruction. The $\delta^{13}C$ value of sphaerosiderites (Ludvigson *et al.*, 2002) and other meteoric calcites can be strongly influenced by methanogenesis, especially in wet-logged, organic-rich settings. In other instances, water–limestone interactions can alter the primary (surficial) $\delta^{13}C$ value of dissolved bicarbonate, such as in speleothems (Baker *et al.*, 1997) and groundwater calcites (Devils Hole: Reviewed in Quade (2003)).

Soil Carbonate, Climate, and Vegetation

The $\delta^{13}C$ value of soil carbonate is largely determined by the proportion of C_3 and C_4 plants growing in the immediate vicinity. Globally, C_3 plants include virtually all trees, almost all shrubs, and those grasses growing under low-light and cool conditions. They average −27‰ in $\delta^{13}C$ (PDB). C_4 plants occur in 18 plant families; in terms of biomass they comprise mostly grasses and sedges. Of all the grass species, 21% are C_4, and compared to C_3 grasses, C_4 grasses tend to favor high-light and higher temperature conditions (i.e., warm growing seasons) (Sage, 1999). Globally, C_4 photosynthesis makes up *c.* 25% of terrestrial net primary productivity. CAM plants can be found in some ecosystems, but they are not abundant in many soil carbonate-forming settings.

C_3 plants dominate anywhere, forest or shrubs dominate over grass, irrespective of setting, and C_3 grasses dominate over C_4 grasses above mean monthly temperatures for the warmest month of ~22° C (Collatz *et al.*, 1998). The temperature threshold favoring C_4 grasses may have been higher in the past if pCO_2 was higher (Collatz *et al.*, 1998) and may have been lower during glacial episodes when pCO_2 was lower. Other C_4 plants include a few herbaceous species and saltbush, very common in high desert settings. C_4 grasses include several biochemical variants, termed NAD malic enzyme (NAD-ME), NADP malic enzyme (NADP-ME), and phosphenoenolpyruvate (PCK). All these variants have distinct $\delta^{13}C$ values. The proportion of NAD-ME and PCK variant grasses increases with decreasing rainfall and tends to dominate areas where most soil carbonate forms. $\delta^{13}C$ values of NAD-ME and PCK variant grasses average -13.8 ± 1.4‰ today, based on one compilation of data ($n = 105$) presented in Cerling *et al.* (2003) for East Africa. To reconstruct past vegetation, these isotopic averages must be increased by 1.5‰ due to fossil fuel burning (the Suess effect; Friedli *et al.* (1986)). As a result, prior to 1850, pure C_3 biomass in the drier areas is assumed to have averaged -25.5 ± 1.3‰ ($-27.0 + 1.5$‰) in $\delta^{13}C$ (VPDB) prior to 1850, and -11.8 ± 1.4‰ for pure C_4 plants.

Soil Carbonate Formation and the Soil Diffusion Model

Some early models (e.g., Salomans *et al.* (1978)) of soil carbonate formation invoked the simple stochiometry of Eqn [1] to suggest that half the C in solution as HCO_3^- derives from plant-derived CO_2 and half from dissolution of antecedent $CaCO_3$ such as marine limestone. We now know this view to be incorrect for soils because it fails to take into account isotopic equilibrium considerations in addition to chemical ones. Two key features of soils – abundant CO_2 produced by plants compared to the small amount of C in soil carbonate, and slow dewatering rates in soils – ensure that soil carbonate undergoes continuous, open-system exchange with a large, isotopically unchanging plant CO_2 reservoir. The pore space of soils contains high concentrations (up to 10,000 ppm V) of CO_2, the result of root respiration and tissue decay of plants. At deep soil levels, this plant CO_2 reservoir is much larger than any potential contribution from the atmosphere (270 ppm V) or from dissolution of limestone from parent material, typically added at the rate of ~0.005 mmol $m^{-2} h^{-1}$. Another critical feature of soils, particularly drier soils, is that they tend to dewater slowly in the days to weeks after a rainfall event. The long dewatering times (>day) of most soils greatly exceed the chemical rate constants of carbonate reactions (seconds to minutes) between carbon species (e.g., Dulinski and Rozanski (1990)). These considerations make it likely that carbon-isotopic equilibrium is maintained between carbon species during soil carbonate formation.

Plant-derived CO_2 – and hence the local proportions of C_3 to C_4 plants growing at the soil surface – can therefore be predicted to determine the $\delta^{13}C$ value ($\delta^{13}C_{cc}$) of carbonate found deep (>50 cm) in soils, where plant-derived soil CO_2 is more concentrated. Shallow in soils, plant-derived CO_2 mixes with atmospheric CO_2, and the proportion of C_3 to C_4 plants cannot be calculated from soil $\delta^{13}C_{cc}$ values. For this reason, it is essential to obtain soil carbonate at depth, generally >50 cm, as discussed below.

The one-dimensional soil diffusion model presented in Cerling and Quade (1993), and modified in Davidson (1995), provides the quantitative framework for interpretation of soil-carbonate-isotopic results. These studies show that at steady state, soil CO_2 concentration varies with depth according to:

$$C_s = \frac{k^2 \phi_s^0}{D_s}(1 - e^{-z/k}) + C_a \qquad [3]$$

where D_s = the diffusion coefficient of CO_2 in soils (in $cm^2 s^{-1}$); C_s = soil CO_2 concentration (in mmol cm^{-3}); C_a = atmospheric CO_2 concentration (in mmol cm^{-3}); k = the characteristic production depth (cm) of CO_2 in soils, typically 20–25 cm; z = depth (cm); $\phi_s^o = CO_2$ is the production rate at the soil surface ($z = 0$) (in mmol $cm^{-2}\,s^{-1}$); and assuming several boundary conditions, including $C_s = C_a$ at $z = 0$ and $dC/dz = 0$ at some specified depth L.

Determination of C_s contributes to estimation of the average $\delta^{13}C$ value of some mixture of C_3 and C_4 plants growing at a soil site ($\delta^{13}C_{plant}$), using the approximation presented in Davidson (1995) but solved in terms of $\delta^{13}C_{plant}$:

$$\delta^{13}C_{plant} = \frac{\delta^{13}C_s - 4.4 - \delta^{13}C_a R(z) + 4.4R(z)}{1.0044(1 - R(z))} \quad [4]$$

where $\delta^{13}C_s$ and $\delta^{13}C_a$ are the $\delta^{13}C$ values of soil CO_2 and air, respectively, and $R(z) = C_a/C_s$ at the depth of sampling z. This equation subsumes the effects of kinetic enrichment in ^{13}C of soil CO_2 during diffusion. This sets up a temperature-independent enrichment of 4.2–4.4‰ CO_2 in the soil with respect to the CO_2 flux leaving the soil (Davidson, 1995). This enrichment factor has been observed empirically (Cerling and Quade, 1993) and can be calculated using the Stefan–Maxwell equation applied to CO_2 diffusing through air. At isotopic equilibrium, soil carbonate will be enriched in ^{13}C with respect to soil CO_2 by 9–12‰, depending on temperature (25–0 C). The $\delta^{13}C_s$ values required to solve for $\delta^{13}C_{plant}$ in Eqn [4] can be obtained from the $\delta^{13}C$ value of soil carbonate ($\delta^{13}C_{cc}$):

$$\delta^{13}C_s = \frac{(\delta^{13}C_{cc} + 1{,}000)}{\left(\dfrac{11.98(\pm 0.13) - 0.12(\pm 0.1)T}{1{,}000} + 1\right)} - 1000 \quad [5]$$

where $11.98(\pm 0.13) - 0.12(\pm 0.1)T$ expresses the temperature-dependent (T = temperature in C) enrichment factor between calcite and gaseous CO_2 (Romanek *et al.*, 1992). Inspection of this equation shows that the fractionation factor is not strongly sensitive temperature, with a slope of ~0.12‰ C^{-1}.

Substitution of Eqn [5] into Eqn [4] allows $\delta^{13}C_{plant}$ to be calculated from $\delta^{13}C_{cc}$. The fraction C_4 biomass that once grew on the landscape can in turn be calculated from:

$$\text{fraction } C_4 \text{ biomass} = (\delta^{13}C_{plant} - \delta^{13}C_{C3}) / (\delta^{13}C_{C4} - \delta^{13}C_{C3}) \quad [6]$$

where $\delta^{13}C_{C3}$ and $\delta^{13}C_{C4}$ refer to the average, pre-industrial carbon isotopic values of C_3 and C_4 plants, respectively.

In summary, where soil respiration rates are high (>3–5 mmol $m^{-2}\,hr^{-1}$) and for low atmospheric CO_2 concentrations (< 500 ppmV), the isotopic composition of soil carbonate should differ from that of coexisting plant cover by about 13–16‰ (gray-shaded zone, **Fig. 6**). This difference ($\Delta_{cc-om} = \delta^{13}C_{cc} - \delta^{13}C_{om}$, where the subscript 'om' denotes soil organic matter) involves both equilibrium fractionation between $CaCO_3$ and CO_2 of 9–12‰, and kinetic diffusion effects of ~4.4‰ (**Fig. 1**). This difference is observed in modern soils with moderate to high respiration rates, verifying the general validity of the model (**Fig. 6**). Where respiration rates are lower, such as in deserts, Δ_{cc-om} > 13–16‰, as observed from a sampling of soils from the Atacama Desert (**Fig. 6**).

Long-Term Records of Climate Change from Nonlacustrine Calcites

Devils Hole

The long isotopic records from Devils Hole archive ~500 kyr of climate change in southern Nevada. The best-known groundwater calcite records in the region come from two cores obtained by scuba divers at Devils Hole. DH-2 spans 50–310 kyr BP and DH-11 50–550 kyr BP (**Fig. 7A**; Winograd *et al.*, 1992), based on multiple alpha and TIMS U-series dates from both cores. Neither record has any apparent depositional hiatuses. Dates from the cores lag climatic events by the time required for water to travel from the recharge area to Devils Hole. Thus, dates on laminated calcite (**Fig. 3**) from Devils Hole represent minimum ages (by < 10 kyr) for the recorded climate events.

$\delta^{18}O_{cc}$ (SMOW) values from Devils Hole fluctuate between +13.05 and +15.75‰ and display the same sawtooth pattern as marine benthic foram $\delta^{18}O_c$ records over the same timespan (**Fig. 7B**). Overall, the similarity in the form of the $\delta^{18}O_c$ curve from the Devils Hole record to marine (viewed as a proxy for ice volume changes) and ice records such as Vostok (viewed as a proxy for Antarctic air temperature) is striking.

How is the $\delta^{18}O_c$ record from Devils Hole being interpreted? The Devils Hole flow system is large (~12,000 km^2) and receives most of its recharge as winter/spring moisture in the Spring and Sheep mountains >30 km away. Mixing and additions of water from such a large recharge area dampen out short-term variations, but preserve them at a longer (≥1 yr) scale (Winograd *et al.*, 1992). The temperature of formation of the calcite has likely remained constant due to the deep circulation of water and the

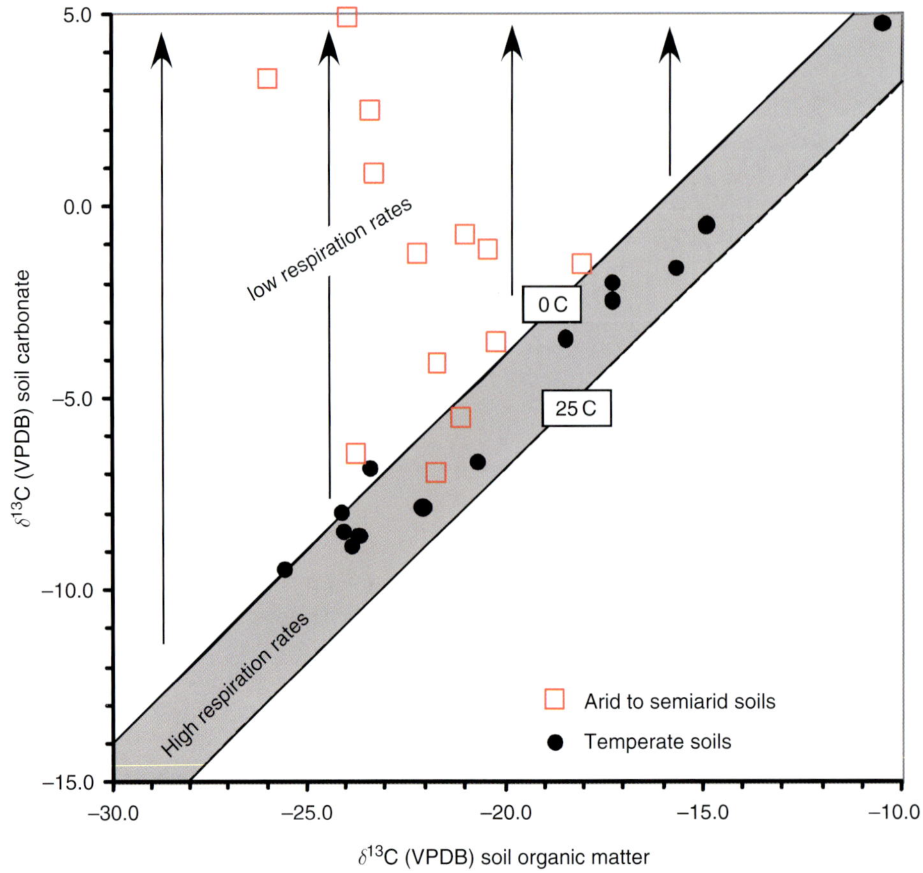

Figure 6 The $\delta^{13}C$ (VPDB) value of soil organic matter vs. that of coexisting soil carbonate. The shaded zone denotes Δ_{cc-om} ($= \delta^{13}C_{cc} - \delta^{13}C_{om}$) values of 13.4–16.4‰ predicted by the soil diffusion model of Cerling and Quade (1993), and observed in most modern soils with moderate to high respiration rates (solid circles). Most values in which $\Delta_{cc-om} > 13$–16‰ are produced in soils with low respiration rates, as exemplified by sampling of soils from the Atacama Desert (open squares).

large volume of the flow system. The temperature of the water (25–35° C) is low enough and the volume of water large enough that changes in primary $\delta^{18}O_p$ due to water–rock interaction are negligible. Thus, Winograd *et al.* (1992) have suggested that the main determinant of $\delta^{18}O_{cc}$ in the Devils Hole records is the value of $\delta^{18}O_p$ falling mainly in winter and spring on the recharge area to the Devils Hole. The $\delta^{18}O_p$ value is in turn determined by $\Delta\delta^{18}O_{seawater}$ and by changes in the average winter/spring temperature in cloud masses arriving over the recharge area.

The one feature of the Devils Hole record that has sparked great controversy is the dating of glacial terminations II and III at 140 ± 3 and $253 \pm 3 \times 10^3$ yr BP (2σ), respectively. These are minimum age estimates because of the time lag in the Devils Hole record mentioned before. In the orbitally tuned SPECMAP records, these same terminations date to 128 ± 3 and $244 \pm 3 \times 10^3$ yr BP, respectively (**Fig. 7C**). The offset of ≥ 9 kyr between age estimates for the terminations is important and challenges the widely held view that changes in the Earth's orbital parameters drive the ice ages. For example, the orbitally tuned date of $128 \pm 3 \times 10^3$ yr BP for termination II from SPECMAP coincides with a pronounced summer insolation maximum at $128 \pm 1 \times 10^3$ yr BP for the Northern Hemisphere (**Figs. 7B** and **C**), strongly supporting a causal link between maximum summer heating in the Northern Hemisphere and ice-sheet retreat. If, however, the Devils Hole chronology is correct, then the linkage between orbital forcing and waxing and waning of glaciation is not so straightforward.

The central question is whether the Devils Hole $\delta^{18}O_c$ record represents a regional versus more global record of temperature change. Winograd *et al.* (1992) have convincingly argued that local air temperature in the recharge area is the dominant control on the $\delta^{18}O_c$ value on the Devils Hole record. However, Herbert *et al.* (2001) contend that changes in air temperature in the region have not changed in phase with average global air temperature. The basis for this is their paleotemperature estimates

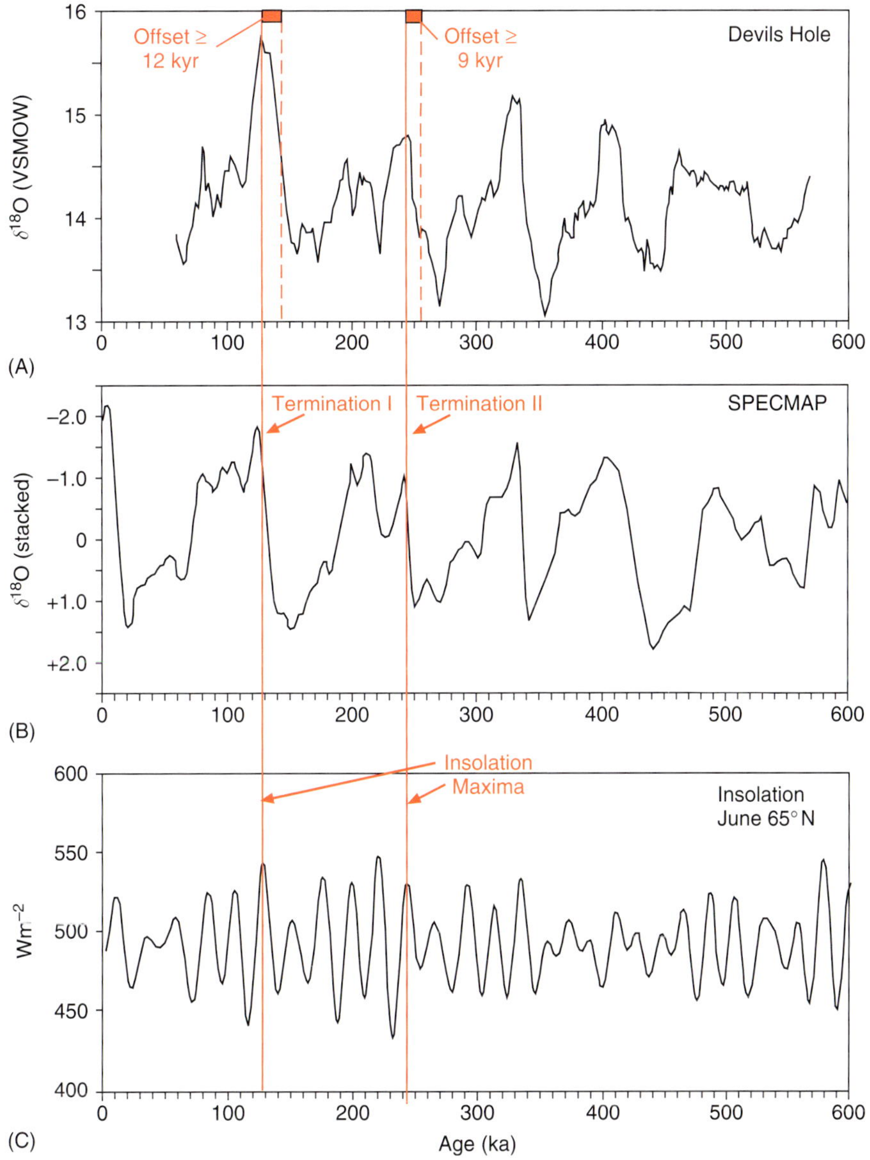

Figure 7 (A) δ^{18}O (SMOW) values of a calcite from Devils Hole (Winograd *et al.*, 1992); (B) stacked benthic forams from SPECMAP (Imbrie *et al.*, 1984); and (C) June insolation in W m^{-2} for 60 °N (Berger and Loutre, 1991).

spanning multiple glacial and interglacial cycles using unsaturated alkenones from marine sediments off the west coast of North America. The whole issue of how globally representative the Devils Hole record is remains unclear, and awaits, in part, development of well-dated groundwater calcite and speleothem records outside of southwestern North America.

The Global Neogene and Pleistocene Paleosol Record

A number of paleosol isotopic records spanning the Neogene have been developed over the past 15 yr that show dramatic environmental change globally. The shift in δ^{18}O values must be related to a global climate change, probably aridification. The shift in δ^{18}C values reflects a largely concurrent expansion in C_4 grasses at the expense of C_3 grasses, shrubs, and trees. Within the Pleistocene, however, change in C_4 biomass has been conservative globally except in East Africa, where grasslands expanded significantly, possibly influencing the evolution of the genus *Homo*.

δ^{18}O record Most of the oxygen-isotope records from soil carbonates show gradual (Argentina, Latorre *et al.* (1997); Greece/Turkey, Quade *et al.* (1994)) to sharp (Pakistan, Quade *et al.* (1989a); Nepal, Quade *et al.* (1995); Africa, Cerling (1992), Quade and Levin (in review)) increases in $\delta^{18}O_{cc}$ values starting in the late Miocene to early Pliocene (8–4 Ma) (**Fig. 8**). Only the North American (Fox and

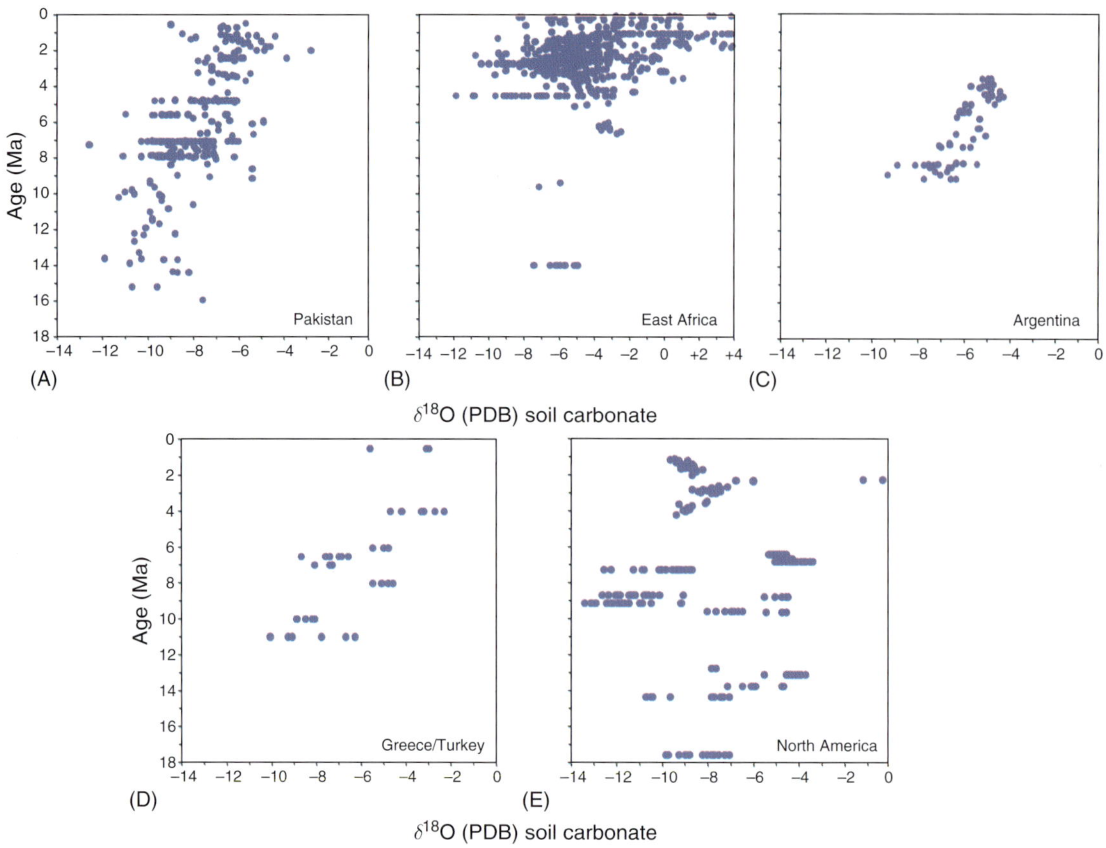

Figure 8 Global Neogene trends in $\delta^{18}O_{cc}$ values from (A) Pakistan (Quade *et al.*, 1989a and unpublished data); (B) East Africa (Quade and Levin, in review); (C) Argentina (Latorre *et al.*, 1997); (D) Greece and Turkey (Quade *et al.*, 1994); (E) North America (Fox and Koch, 2004; data courtesy of D. Fox).

Koch, 2004) soil carbonate record shows no increases over the late Neogene.

The largely global increase in $\delta^{18}O_{cc}$ values must reflect important changes in global climate. The shift is too large (3–5‰) and the wrong sign to be explained by temperature declines through the Neogene. As was discussed previously the slope of the theoretical and observed temperature–$\delta^{18}O_{cc}$ relationship in modern 'temperate' soil carbonate is +0.44‰ C^{-1}. The increase in $\delta^{18}O_{seawater}$ is the right sign but far too small (< 1.0‰) to explain the increases in $\delta^{18}O_{cc}$ values. At mid- to high latitudes, this leaves increased soil water evaporation as the dominant cause for increases in $\delta^{18}O_{cc}$ values. At lower latitudes, decreases in rainfall amount also contributed to the increase in $\delta^{18}O_{cc}$ values, as well as increased soil water evaporation (Quade and Levin, in review).

δ^{13}C record Most of the carbon-isotope records from soil carbonate show gradual (Argentina, Latorre *et al.* (1997)) to sharp (Pakistan, Quade *et al.* (1989a); Nepal, Quade *et al.* (1995); Africa, Cerling *et al.* (1992), Quade and Levin (in review)) increases in δ^{13}C values starting in the late Miocene (8–6 Ma) (**Fig. 9**). Others show more modest increases (North America, Fox and Koch (2004)), starting later, in the early Pliocene. One record – the only long record from a Mediterranean (winter rainfall-dominated) climate – shows no appreciable shift through the Neogene (Quade *et al.*, 1994) (**Fig. 9**).

Most of the increase in δ^{13}C values can be attributed to increases in C_4 grasses, although the contribution by lower respiration rates to higher δ^{13}C values has not been unequivocally excluded in some records (Latorre *et al.*, 1997; Fox and Koch, 2004). The lack of change in δ^{13}C values from the eastern Mediterranean record accords with the lack of C_4 plants in that region today. In some areas, the ecologic shift involved a change in life form: C_4 grasses replaced C_3 forest (e.g., Nepal, Quade *et al.* (1995)). In others, photosynthetic pathway but not life form changed: C_4 grasses apparently replaced a substantial fraction of C_3 grasses (e.g., North America: Fox and Koch (2004)). Cerling *et al.* (1997) observed similar increases over the late Neogene in the diet of large herbivores, attributed to increases in the dietary intake of C_4 grasses.

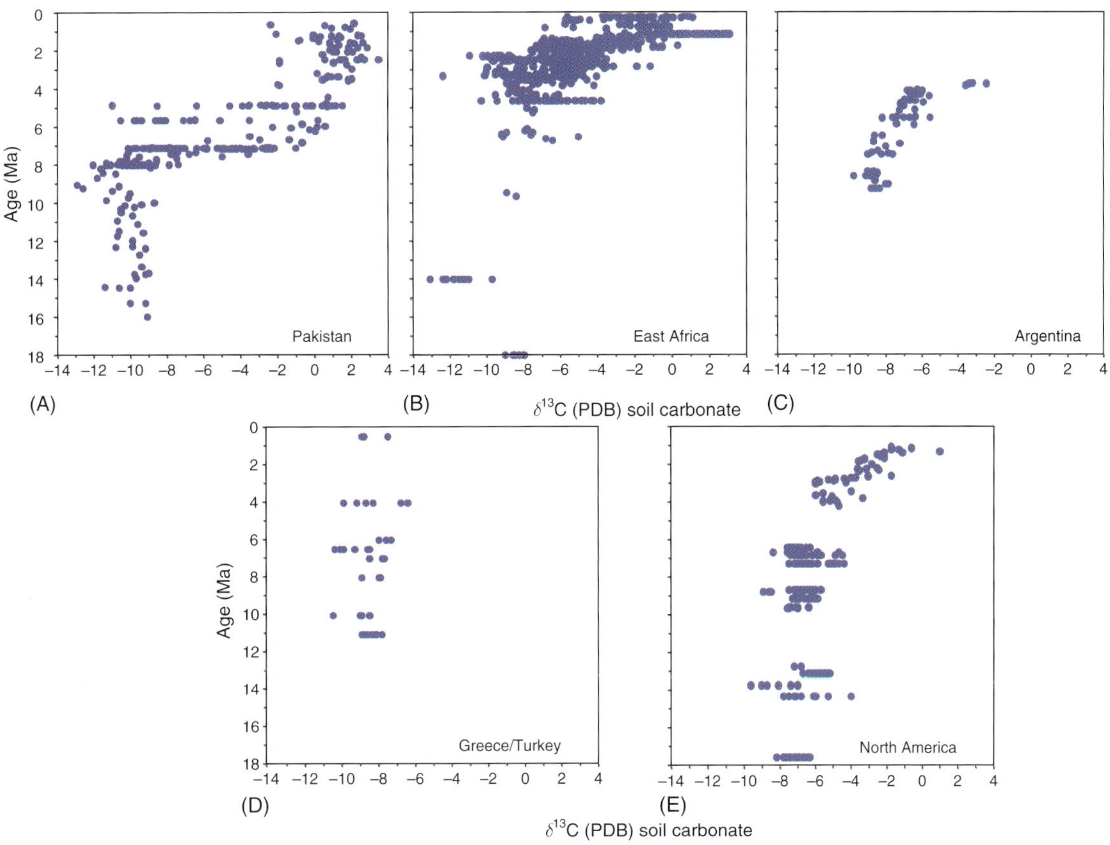

Figure 9 Global Neogene trends in $\delta^{13}C_{cc}$ values from (A) Pakistan (Quade *et al.*, 1989a, and unpublished data); (B) East Africa (Quade and Levin, in review); (C) Argentina (Latorre *et al.*, 1997); (D) Greece and Turkey (Quade *et al.*, 1994); and (E) North America (Fox and Koch, 2004; data courtesy of D. Fox).

The cause of this dramatic and largely global change in plant cover remains unclear. The C_4 expansion might be explained by a decline in pCO_2 globally, one of the environmental conditions that favors C_4 metabolism (Cerling *et al.*, 1997). However, there is no evidence of decline in pCO_2 over this period (Pagani *et al.*, 1999). Increased moisture stress, like pCO_2 decline, would also favor C_4 plant expansion. This explanation is supported by the clear increases over the same timespan in $\delta^{18}O_{cc}$ values, a trend we have already linked to aridification. The main weakness of this explanation is that it does not explain why C_4 metabolism did not expand earlier. Although more prevalent in the late Neogene, semiarid environments must have been present in the pre-late Miocene. C_4 plants are favored by fire, and an increase in fire frequency would support the expansion of C_4 biomass. This, too, is a hypothesis worth exploring.

The global carbon-isotope record from paleosols shows little change over the Pleistocene. A significant exception to this is the sharp increase in $\delta^{13}C_{cc}$ values observed from East Africa (**Fig. 9B**; summarized in Quade and Levin (in review)). At many study sites, $\delta^{13}C_{cc}$ values increase for the first time above −2‰ during the Pleistocene, denoting the appearance of savanna grasslands (>80% C_4 biomass) so characteristic of the region today. The pattern of expansion is uneven temporally and geographically, commencing as early as the late Miocene at Lothagam, but much later, in the 2.7–1.0 Ma timespan, at most other sites. Many researchers draw the possible connection between this expansion of savanna ecosystems and the development of the genus *Homo* in Africa starting sometime in the latest Pliocene (see review of Klein (1995)). The Pleistocene-age increase in $\delta^{13}C$ values in East Africa is embedded in a local, longer-term trend toward higher $\delta^{18}O$ values, again a pattern we attribute to aridification.

The impact on regional vegetation and climate during shifts from glaciations to interglaciations is also apparent in a number of isotopic records from paleosols from the Pleistocene. As just two examples, glacial-age soil carbonates from the Wind River Mountains of Wyoming, USA (Amundson *et al.*, 1996) yielded $\delta^{18}O_{cc}$ values significantly higher than modern soil carbonate values. This was interpreted to be the result of a combination of colder temperatures and the increased contribution of high $\delta^{18}O_{cc}$ waters from the Gulf of Mexico. In the Great Basin, western USA, $\delta^{13}C_{cc}$ values decreased during glaciations, as C_3-dominated vegetation at higher elevations today was displaced downward (Quade *et al.*, 1990). Recent advances in $^{230}Th/U$ dating of individual laminas in desert soil carbonates (e.g. Sharp *et al.*, 2003) should lead to much more extensive use of soil isotopic archives in shorter-term (10^4–10^5 yr) paleoclimate reconstruction.

See also: **Carbonate Stable Isotopes**: Lake Sediments; Speleothems. **Paleosols and Wind-Blown Sediments**: Soil Morphology in Quaternary Studies.

References

Amundson, R., Chadwick, O., Kendall, C., Wang, Y., and DeNiro, M. (1996). Isotopic evidence for shifts in atmospheric circulation during the Quaternary in mid-North America. *Geology* **24,** 23–26.

Baker, A., Ito, E., Smart, P. L., and McEwan, R. G. (1997). Elevated and variable values of ^{13}C in speleothems in a British cave system. *Chemical Geology* **136,** 263–270.

Berger, A., and Loutre, M. F. (1991). Insolation values for the climate of the last 10 million years. *Quaternary Science Reviews* **10**(4), 297–317.

Cerling, T. E. (1992). Development of grasslands and savannas in east Africa during the Neogene. *Palaeogeography, Palaeoclimatology, Palaeoecology* **97,** 241–247.

Cerling, T. E., Harris, J. M., MacFadden, B. J., *et al.* (1997). Global vegetation change through the Miocene/Pliocene boundary. *Nature* **389,** 153–157.

Cerling, T. E., Harris, J. M., and Passey, B. H. (2003). Diets of east African Bovidae based on stable isotope analysis. *Journal of Mammalogy* **84**(2), 456–470.

Cerling, T. E., and Quade, J. (1993). Stable carbon and oxygen isotopes in soil carbonates. In *Continental Indicators of Climate, Proceedings of Chapman Conference, Jackson Hole, Wyoming, American Geophysical Union Monograph* (P. Swart, J. A. McKenzie and K. C. Lohman, Eds.) **78,** pp. 217–231.

Collatz, G., Berry, J., and Clark, J. (1998). Effects of climate and atmospheric CO_2 partial pressure on the global distribution of C_4 grasses: Present, past, and future. *Oecologia* **114,** 441–454.

Davidson, G. R. (1995). The stable isotopic composition and measurement of carbon in soil CO_2. *Geochimica et Cosmochimica Acta* **59**(12), 2485–2489.

Dulinski, M., and Rozanski, K. (1990). Formation of $^{13}C/^{12}C$ ratios in speleothems: A semi-dynamic model. *Radiocarbon* **32,** 7–16.

Fantidas, J., and Ehhalt, D. H. (1970). Variations of the carbon and oxygen isotopic composition in stalagmites and stalactites; Evidence of non-equilibrium isotopic effects. *Earth and Planetary Science Letters* **10,** 136–144.

Fox, D., and Koch, P. L. (2004). Carbon and oxygen isotopic variability in Neogene paleosol carbonates: Constraints on the evolution of C_4 grasslands. *Palaeography, Palaeoclimatology, Palaeoecology* **207,** 305–329.

Freidli, H., Lotscher, H., Oeshger, H., Siegenthaler, U., and Stauffer, B. (1986). Ice core record of $^{13}C/^{12}C$ ratio of atmospheric CO_2 in the past two centuries. *Nature* **324,** 237–238.

Hendy, C. H. (1971). The isotope geochemistry of speleothems. I: The calculation of different modes of formation on the isotopic composition of speleothems and their applicability as paleoclimatic indicators. *Geochimica et Cosmochimica Acta* **35,** 801–824.

Herbert, T. D., Schuffert, J. D., Andreasen, D., *et al.* (2001). Collapse of the California current during glacial maxima linked to climate change in land. *Science* **293,** 71–76.

Imbrie, J., Hays, J. D., Martinson, D. G., *et al.* (1984). In *Milankovitch and Climate, Part 1* (A. L. Berger, J. Imbrie, J. Hays, G. Kukla and B. Saltzman, Eds.), pp. 269–305. Reidel, Dordrecht.

Kim, S., and O'Neil, J. R. (1997). Equlilibrium and nonequilibrium oxygen isotope effects in synthetic carbonates. *Geochimica et Cosmichimica Acta* **61**(16), 3461–3475.

Klein, R. (1995). *The Human Career.* University of Chicago Press, Chicago.

Latorre, C., Quade, J., and McIntosh, W. C. (1997). The expansion of C_4 grasses and global change in the late Miocene: Stable isotope evidence from the Americas. *Earth and Planetary Science Letters* **146,** 83–96.

Ludvigson G, A., Gonzalez, L. A., Metzger, R. A., *et al.* (2002). Meteroic spherosiderite lines and their use for paleohydrology and paleoclimatology. *Geology* **26,** 1039–1042.

Pagani, M., Freeman, K. H., and Arthur, M. A. (1999). Late Miocene atmospheric CO_2 concentrations and expansion of C_4 grasses. *Nature* **285,** 876–879.

Quade, J. (2003). Isotopic records from ground-water and cave speleothem calcite in North America. In *The Quaternary Period in the United States* (S. Gillespie, C. Porter and B. F. Atwater, Eds.), pp. 205–220. Elsevier Science, New York.

Quade, J., Cater, J. M. L., Ojha, T. P., Adam, J., and Harrison, T. M. (1995). Dramatic carbon and oxygen isotopic shift in paleosols from Nepal and late Miocene environmental change across the northern Indian sub-continent. *Geological Society of America Bulletin* **107,** 1381–1397.

Quade, J., Cerling, T. E., and Bowman, J. R. (1989a). Development of the Asian Monsoon revealed by marked ecological shift in the latest Miocene in northern Pakistan. *Nature* **342,** 163–166.

Quade, J., Cerling, T. E., and Bowman, J. R. (1989b). Systematic variation in the carbon and oxygen isotopic composition of Holocene soil carbonate along elevation transects in the southern Great Basin, USA. *Geological Society of America Bulletin* **101,** 464–475.

Quade, J., Cerling, T. E., and Bowman, J. R. (1990). Stable isotopic evidence for a pedogenic origin of carbonates in Trench 14 near Yucca Mountain, Nevada. *Science* **250,** 1549–1552.

Quade, J., Solounias, N., and Cerling, T. E. (1994). Stable isotopic evidence form paleosol carbonates and fossil teeth in Greece for forest or woodlands over the past 11 Ma. *Palaeography, Palaeoclimatology, Palaeoecology* **108,** 41–53.

Romanek, C. S., Grossman, E. T., and Morse, J. W. (1992). Carbon isotopic fractionation in synthetic aragonite and calcite: Effects of temperature and precipitation rate. *Geochimica et Cosmochimica Acta* **56,** 419–430.

Rozanski, K., Araguas-Araguas, L., and Gonfiantini, R. (1993). Isotopic patterns in modern global precipitation. In *Continental Indicators of Climate, Proceedings of Chapman Conference, Jackson Hole, Wyoming, American Geophysical Union Monograph 78* (P. Swart, J. A. Mckenzie and K. C. Lohman, Eds.) pp. 1–36.

Sage, R. F. (1999). Why C_4 photosynthesis? In *C_4 Plant Biology* (R. F. Sage and R. K. Monson, Eds.), pp. 3–16. Academic Press, San Diego.

Salomans, W., Goudie, A., and Mook, W. G. (1978). Isotopic composition of calcrete deposits from Europe, Africa and India. *Earth Surface Processes* 3, 43–57.

Sharp, W. D., Ludwig, K. R., Chadwick, O., Amundson, R., and Glaser, L. L. (2003). Dating fluvial terraces by ^{230}Th/U on pedogenic carbonate, Wind River Basin, Wyoming. *Quaternary Research* 59, 139–150.

Winograd, I. J., Coplen, T. B., Landwehr, J. M., *et al.* (1992). Continuous 500,000-year climate record from Devils Hole, Nevada. *Science* 258, 255–260.

Lake Sediments

S M Bernasconi and J A McKenzie, Geological Institute, Zurich, Switzerland

© 2007 Elsevier B.V. All rights reserved.

Introduction

Lakes are widely distributed across the continents and are found in climatic zones ranging from Antarctica to the tropics and from moist to relatively arid environments. Lacustrine sediments are important archives of climatic and environmental change. Their wide spatial distribution enables the study of the regional impact of global climate change in a variety of climatic regimes and with high geographic and temporal resolution. Lakes precipitating authigenic carbonate or containing carbonate in the form of biogenic shell material have been subjects of considerable research since the pioneering studies in the early 1970s (e.g., Stuiver (1970)). Because of the relative ease of analysis and the development of automated analytical systems, isotope studies of bulk lacustrine carbonates have become relatively common. However, the extraction of quantitative climatic and paleoenvironmental information from bulk carbonate records is not straightforward or simple. The main difficulties in interpreting oxygen-isotope records are related to three main factors: (1) bulk carbonates in lakes represent a mixture of particles of different origins; (2) the isotopic composition of the precipitated calcium carbonate is dependent on the water temperature and the isotopic composition of the lake water, and, thus, on the hydrological balance of a lake; and (3) biological activity in the lake, by influencing the carbonate chemistry, may influence the mechanisms, water depth, and timing of carbonate precipitation. In particular, if the hydrology of the lake is not well understood, a quantitative reconstruction of climate is prone to large errors.

Global Spatial and Age Distribution of Carbonate-Precipitating Lakes

Carbonate-precipitating modern lakes and paleolakes are found in all continents, at various latitudes and altitudes. The main control on the presence of carbonate in a lake is the composition of the bedrock. Significant calcium carbonate precipitation in the water column is observed only in lakes that have a large part of the catchment consisting of calcium-carbonate-rich rocks, which supply most of the calcium and carbonate ions through weathering reactions. Lakes within catchments dominated by silicate rocks are normally devoid of carbonates. Calcium carbonate supersaturation in the bottom waters and the sediments is also necessary to avoid postdepositional dissolution of calcite, which is promoted by CO_2 production related to organic matter decomposition.

The spatial distribution and age of lakes is dependent on their origin. Lakes of glacial origin are very common in formerly glaciated areas of Europe, Asia, and the Americas. Although some lakes may have originated in the Tertiary (e.g., the lakes of the insubric region in northern Italy and Switzerland), most of the sediments preserved, or accessible to sampling, in these lakes are younger than the Last Glacial Maximum, about 18 kyr BP. Older sediments were either removed by glacial erosion during the glacial advances or are compacted and present at sediment depths that are reachable only by deep drilling. Lakes of tectonic origin are globally less numerous than glacial lakes, but they are a prominent feature for their size and permanence, and have provided some of the longest records of continental climate. Lakes of tectonic origin include all of the major lakes in Africa; the Aral and Caspian seas, Lake Baikal and the Dead Sea (and its paleolake equivalent Lake Lisan) in Eurasia; Lake Titicaca in South America; the Great Salt Lake, Utah and its paleolake equivalent Lake Bonneville and Pyramid Lake, Nevada in North America. Crater lakes, which also have provided important long paleoclimatic records, are mostly devoid of carbonates, due to the chemistry of the waters.

Sampling of Lacustrine Sediments

Paleoclimatic studies on lacustrine sediments are mostly carried out on cores obtained by various platforms and coring devices ranging from hand-held gravity cores to complex drilling platforms. For a review of coring devices, see Glew *et al.* (2001). Sedimentation rates, and thus the length of cores required for long paleoclimatic records and the sampling resolution achievable, vary dramatically in lakes. In peri-Alpine lakes of glacial origin, sediment accumulation rates are in the range of 0.5–5 mm yr^{-1}. Typical lengths of cores are 1–2 m for simple hand-held gravity cores, which allow studies of the last few hundred to a few thousand years in very small lakes with low sedimentation rates. For the recovery of longer sediment sequences of up to 10–20 m, larger platform systems have to be used. Earlier Quaternary paleorecords have been also collected from paleolakes with drilling (Hsü and Kelts, 1984) or more rarely on outcrops.

Nature of Carbonate in Lacustrine Sediments

The carbonate fraction of lacustrine sediments can contain particles of multiple *in situ* origins: (1) an inorganically precipitated carbonate, (2) a biogenic fraction formed in the water column, and (3) a fraction which can be formed by diagenesis after the deposition of the sediments. In addition, because carbonate-precipitating, hard-water lakes are found in catchments composed, at least partially, of carbonate bedrocks, some fraction of the carbonate will be of detrital origin. Thus, for a quantitative interpretation of the isotopic signal, the origin of the carbonate particles has to be determined by X-ray diffractometry and by petrographic examination, such as with light or electron microscopy. Only the authigenic particles (**Fig. 1**) formed in the surface waters carry relevant isotopic information for paleoclimatic reconstruction, and, thus, they should be the only, or at least the dominant, fraction of the sediments used to obtain meaningful information from bulk carbonate isotope analyses.

Lake Albano, italy

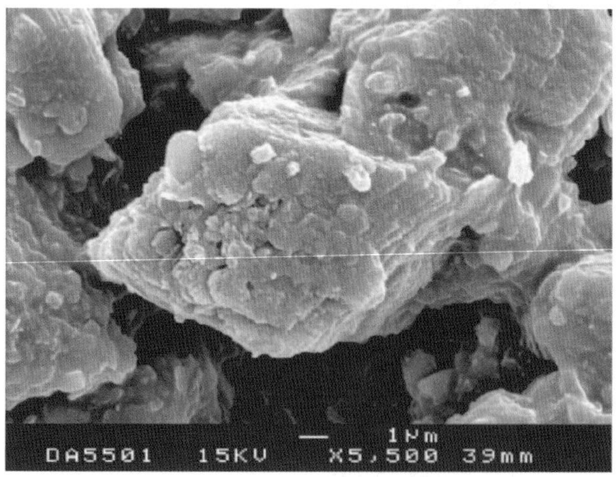

Lake Butrinti, Albania

Figure 1 Scanning electron photomicrographs of authigenic calcite crystals from Lake Albano, Italy and Lake Butrinti, Albania. Note the rhombohedral faces and the crystal-growth steps. Photos courtesy of D. Ariztegui, University of Geneva.

Carbonate Precipitation in Mid- to High-Latitude, Hard-Water Lakes

Carbonate precipitation in hydrologically open, hard-water lakes is induced by increases in pH and carbonate ion concentration related to the utilization of aqueous CO_2 by photosynthetic organisms, which increases calcium carbonate supersaturation (Kelts and Hsu, 1978). An active role of cyanobacterial picoplankton in carbonate precipitation has been postulated by Thompson *et al.* (1997) and Dittrich *et al.* (2004), and also has to be considered for the understanding of carbon- and oxygen-isotope geochemistry of carbonates. Carbonate precipitation in mid- to high-latitude lakes is commonly limited to the period between early spring and late autumn, when higher temperatures and light levels stimulate algal production. This period follows the annual recharging of the surface waters with essential nutrients as a result of the winter overturn of the water column. The isotopic composition of carbonate precipitate, therefore, predominantly reflects spring and summer conditions, whereas the sedimentation in the late fall and winter months consists predominantly of clastic material and organic matter. If the bottom waters of a lake are at least temporarily anoxic, the rhythmic sedimentation induced by the annual productivity cycle can be preserved in the sediment in the form of chemical varves of alternating white summer and dark winter layers (**Fig. 2**). The presence of an annual lamination allows the accurate dating of sedimentary sequences by varve counting. In addition, it facilitates the acquisition of paleoenvironmental records at annual or even at semi-annual resolution (e.g., Teranes *et al.* (1999b)). Even in this case, however, it has to be kept in mind that the residence time of the water in the catchment area and the lake itself is the limiting factor for the true temporal resolution of a sedimentary record.

The precipitation of calcium carbonate occurs close to the surface of the lake, where the biological activity is

Bernasconi & McKenzie

Figure 2 Photo of a gravity core collected form 130 m depth in Lake Zürich, Switzerland. For the last 100 yr, a clear annual lamination is visible in the sediments. The light layers are composed mainly of authigenic calcite formed in the water column in spring to late summer. The dark layers are rich in diatoms, organic matter, and clays, and represent autumn and winter sedimentation. The thick layers at 25–29 cm and 33–34 cm are fine-grained redeposited sediments derived from slope instabilities. The bottom of the core shows homogeneous carbonate-rich sediment deposited before significant eutrophication led to the preservation of the annual lamination.

most intense. The time in the season and the water depth, at which calcium carbonate precipitates, may vary from year to year, potentially creating spurious variations in the isotopic signatures of the carbonates. Large temperature gradients of up to 15° C in the top 20 m of the water column are common, and the temperature range at the lake surface between spring and summer can reach 10 to 15° C (e.g., Teranes *et al.* (1999a), Livingstone and Lotter (1999)). Because the precipitation of carbonate is dependent on biological activity, oxygen-isotope records from lakes, which have undergone strong changes in trophic level and productivity, have to be considered with caution. Possible effects of eutrophication on oxygen-isotope records have also been critically discussed, for example, by Gat and Lister (1995). Fronval *et al.* (1995) have shown that carbonate precipitation can occur in isotopic disequilibrium with the water under strongly eutrophic conditions, leading to a negative isotope shift in $\delta^{18}O$. Teranes *et al.* (1999a), on the other hand, have shown in the eutrophic Baldeggersee, Switzerland, that, although isotopic disequilibrium is observed during events of rapid calcite precipitation, the largest part of the calcite can precipitate close to equilibrium.

Carbonate Precipitation in Saline and Closed Basin Lakes

In hydrologically closed lakes and in arid regions, carbonates are precipitated mainly as a consequence of a negative water balance. In these lakes, the effects of biological activity are less important because the main driving force in carbonate precipitation is the increase in Ca^{2+} and CO_3^{2-} concentrations related to evaporation. In contrast to hard-water lakes where the mineralogy of the authigenic carbonates is dominantly low-magnesium calcite, a larger variety of carbonate minerals are found in saline lakes, including high-magnesium calcite, aragonite, and dolomite, as well as more soluble Na–Ca carbonates such as gaylussite and trona (see, e.g., Talbot (1990)). The mineralogy is related to the degree of salinity of the water and provides an independent way to estimate evaporation levels. For example, in Lake Van, Turkey, the Mg content of calcite increases with increasing salinity and can be used in conjunction with isotopes to constrain evaporation/precipitation ratios (Wick *et al.*, 2003). Isotope records from strongly evaporative soda lakes are relatively rare, due mainly to concerns related to recrystallization of the very soluble carbonates. However, there is potential to extract useful climatic information, even from such highly soluble carbonates (Bernasconi *et al.*, 1997).

Isotopic Composition of Lacustrine Carbonate

In paleoclimatic studies, the stratigraphic variation of the oxygen-isotope composition of carbonates is used to reconstruct changes in temperature and/or oxygen-isotope composition of the lake water. Both these parameters depend on local climate variations and on the hydrologic characteristics of the lake. Thus, before quantitative inferences on climate can be made, the main controls on the oxygen-isotopic composition of the lake water have to be evaluated, as each lake shows a different response to climatic forcing. When the hydrology of the lake is known, more precise climate information can be extracted from the oxygen-isotope record.

In contrast, the carbon-isotope composition of carbonates does not contain direct climatic information, mainly because carbon-isotope fractionation between dissolved inorganic carbon and calcite is only very weakly temperature dependent. The $\delta^{13}C$ value of carbonates mainly reflects changes in the inputs of dissolved inorganic carbon from the catchment area and exchange with the atmosphere. Moreover, it is strongly influenced by biological activity within the lake, leading to the development of surface-to-bottom water carbon-isotope gradient, similar to that found in the marine environment (McKenzie, 1985). Essentially, there is a transfer of carbon-12 from the surface to the deeper waters by a photosynthesis–respiration pathway, and the degree of this photosynthetic activity is reflected in the $\delta^{13}C$ value of the dissolved inorganic carbon and incorporated into the carbonate precipitate (**Fig. 3**). Thus, the carbon-isotope record archived in lacustrine carbonates documents changes in the paleoproductivity of lakes and, because of the intrinsinc link between the inorganic carbonate precipitate and the associated organic carbon, information about the dominant photosynthetic mechanism (CO_2 vs HCO_3^- utilization) can be interpreted (Hollander and McKenzie, 1991).

In addition, the carbon-isotope composition of the carbonates provides important information on changes in the catchment and within the lakes, which can be used to better interpret the oxygen-isotope signal. For example, the $\delta^{13}C$ and $\delta^{18}O$ values for lacustrine carbonates from many closed-basin lakes have been found to covary (Talbot (1990) and references therein). As this $\delta^{13}C$–$\delta^{18}O$ covariance is absent or weak in hydrological open lakes, it can be a useful indicator to interpret the open/closed history of a paleolake system. Li and Ku (1997) have evaluated the causal mechanism for the observed $\delta^{13}C$–$\delta^{18}O$ covariance in the closed-basin lake setting and attribute its occurrence and varying intensity to a number of factors, including size of lake, duration of hydrological closure, vapor exchange, lake productivity, and alkalinity. ($\delta^{18}O$ the oxygen isotope composition defined as $[((^{18}O/^{16}O_{sample})/(^{18}O/^{16}O_{standard})) - 1] \times 1000$.)

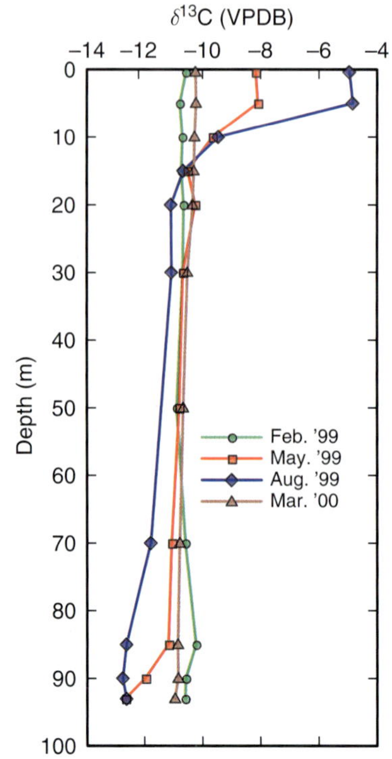

Figure 3 Carbon-isotope profiles in the water columns of Lake Lugano in different months. Note the constant composition in winter; after the water column has mixed during spring and summer, there is a progressive enrichment in ^{13}C in the photic zone due to the preferential assimilation of ^{12}C by the algae. In contrast, in the bottom waters, the oxidation of organic matter leads to a decease in $\delta^{13}C$ of the dissolved inorganic carbon. Data from Lehmann *et al.* (2004).

Oxygen-Isotope Composition of Lake Water and Its Relation to Climate and Local Hydrology

Lakes are dynamic systems linked to the hydrological cycle via surface and subsurface inflow and outflow and via direct precipitation and evaporation from the lake surface. The factors controlling the oxygen-isotope composition of lake waters are schematically summarized in **Figure 4**. Mathematically, the isotopic water balance of a lake can be expressed as:

$$V_{lake}\delta^{18}O_{lake} = \Sigma I\delta\ ^{18}O_I + P\delta^{18}O_P - \Sigma O\delta^{18}O_{lake} - E\delta^{18}O_E$$

where V is the volume of the lake, $\delta^{18}O$ is the oxygen isotopic composition of the various components, I is the inputs (rivers and surface runoff, groundwater input), P is direct precipitation on the lake, O is the output (surface and subsurface output), and E is the evaporation. Note that the surface and groundwater outputs have the same isotopic composition as the lake water. Evaporation is the only output component which imparts a fractionation on the isotopic composition of waters. Thus, this factor is the most important

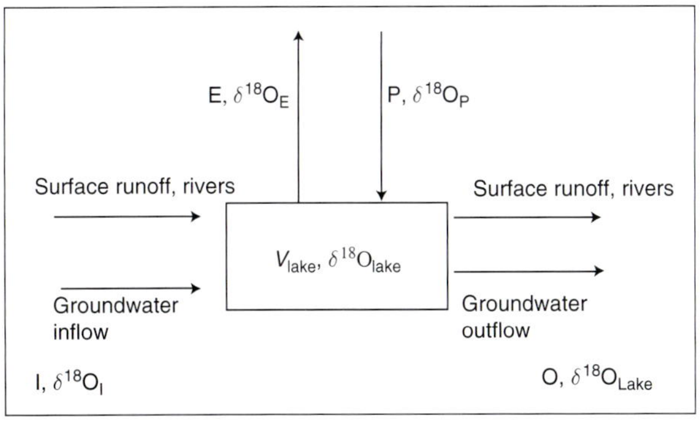

Figure 4 Schematic diagram of the main factors controlling the isotope composition of lake waters.

parameter modifying the oxygen-isotope composition of a lake besides the mixing of the various inputs. In hydrologically open basins, the $\delta^{18}O$ of lake water depends predominantly on the relative fluxes and $\delta^{18}O$ values of the inputs with more or less important modification induced by evaporation processes. In closed lake basins, where there is no surface or subsurface output ($O = 0$), the water balance of the lake, and thus its isotopic composition, are determined by the difference between evaporation and precipitation in the lake and the inputs in the catchment.

The isotopic composition of the lake water, as it is recorded by sedimentary carbonates formed in the surface waters, is additionally influenced by the pattern of vertical mixing of the lake. In monomictic lakes during the stratification period in the summer when most of the carbonate precipitates, the surface waters become enriched in ^{18}O due to the preferential removal of ^{16}O by evaporation. Thus, the carbonate will not necessarily reflect the bulk oxygen-isotope composition of the lake but will show a variably enriched composition which is dependent on evaporation rates (**Fig. 5**). This enrichment can be significant (up to a few ‰) and introduces an additional uncertainty factor in the interpretation of carbonate oxygen-isotope records.

The oxygen-isotopic composition of precipitation is dependent on air temperature (Dansgaard, 1964; Rozanski *et al.*, 1993), the source of moisture (i.e., large-scale atmospheric circulation), and the seasonal distribution of precipitation (Rozanski *et al.*, 1993). Dansgaard (1964) first described the temperature dependence of oxygen-isotope composition of precipitation. He noted a global relationship of 0.69‰ C^{-1} for mid- and high northern latitudes, with stronger variations depending on the location. The more recent compilation of Rozanski *et al.* (1993) provides an excellent updated analysis of the IAEA/GNIP database (International Atomic Energy Agency Global Network of Isotopes in Precipitation,

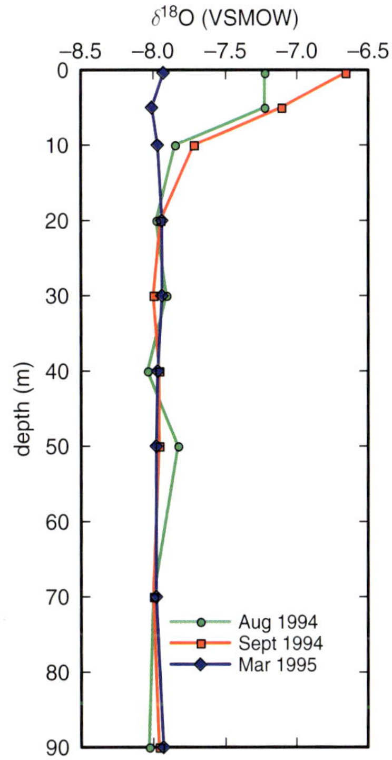

Figure 5 Oxygen-isotope composition of lake waters in a water-column profile from Lake Lugano, Switzerland, for three different months in 1994–95. Note the increase in the $\delta^{18}O$ values of up to 1.5 ‰ during summer stratification and the homogeneous composition of the water column in March after the lake has mixed.

http://isohis.iaea.org). This database is an irreplaceable tool for the calibration of the modern relationship at many locations around the world, which, in turn, can be used for the calibration and better interpretation of paleorecords of precipitation.

In a recent study, Kohn and Welker (2005) have reanalyzed data from four stations in the United States and concluded that the oxygen-isotope mean temperature correlation can be improved, showing a slope always close to 0.55‰ C^{-1}, if the isotope data are correlated with the average temperature on the days when precipitation occurs, rather than the overall average temperature. This effect can be easily understood if one considers that, in summer, the average temperature on the days with rain is systematically cooler than days without precipitation, whereas, in winter, the opposite effect is seen (Kohn and Welker, 2005). Thus, the $\delta^{18}O$ of precipitation most closely reflects a 'precipitation temperature' rather than the average yearly or monthly temperature. This adds an additional uncertainty in paleoclimatic reconstructions, because, unfortunately, these two temperatures are not well correlated. Changes in moisture source, in the contribution of re-evaporated moisture, and seasonal distribution of precipitation also need to be considered. Changes in moisture source or changes in re-evaporation along the trajectory of the air mass can change the local temperature/oxygen-isotope relationships. Air masses with different sources may have different isotope compositions due to different evaporation sources and rainout history (e.g., Clark and Fritz (1997), p. 53). An additional important factor is the seasonal distribution of precipitation. In summer, the $\delta^{18}O$ of precipitation is higher than in winter (e.g., Rozanski *et al.* (1993)). Thus, a switch from more dominant winter to summer precipitation can lead to changes in the $\delta^{18}O$ of lake water, which could be erroneously interpreted as a temperature change. The effect of changes in atmospheric circulation on lacustrine isotope records has been discussed and holds important potential for developing regional paleoclimate archives (e.g., McKenzie and Hollander (1993), Hodell *et al.* (1998), Smith and Hollander (1999), Teranes and McKenzie (2001), Kirby *et al.* (2002), Hammarlund *et al.* (2002)).

The above discussion illustrates that, because of the complex nature of the signal, it is virtually impossible to quantitatively reconstruct temperature variations or relative temperature changes without making some important assumptions. For this reason, oxygen-isotope studies on lacustrine carbonates should be coupled with other proxies to achieve a more precise climatic reconstruction. These may include palynological or micropaleontological studies, as well as other novel geochemical proxies, such as compound-specific hydrogen isotopes in organic matter (e.g., Sauer *et al.* (2001), Huang *et al.* (2004)). In addition, as suggested by Kohn and Welker (2005), air temperature reconstructions based on the reconstruction of the $\delta^{18}O$ of precipitation should be validated with modeling efforts conducted using global climate models (GCMs).

Oxygen-Isotope Temperature Calculations

Following the discussion of the factors controlling the oxygen-isotope composition of lake water, we next consider the factors controlling the oxygen-isotope composition of the associated carbonates. These are an indirect archive for the oxygen-isotope composition of precipitation and, as such, are an additional filter and source of noise, which has to be understood if quantitative reconstructions of climate change are to be obtained. In particular, the primary question that has to be asked when using oxygen-isotope data derived from bulk carbonate in lacustrine sediments is: do they record water temperature or rather the oxygen-isotope composition of the water?

The oxygen-isotope fractionation between water and calcite has been extensively studied, and a number of paleotemperature equations have been published. In a recent review, Leng and Marshall (2004) proposed the adoption of the following modified form of the Kim and O' Neil (1997) calibration:

$$T(\mathrm{C}) = 13.8 - 4.58\ (\delta_c - \delta_w) + 0.08(\delta_c - \delta_w)^2$$

where δ_c is the oxygen-isotope composition of the carbonate with respect to the VPDB standard and δ_w is the oxygen-isotope composition of the water with respect to the VSMOW standard (Coplen, 1994). (Standards are VSMOW (Vienna Standard Mean Ocean Water) for waters and VPDB (Vienna Pee Dee Belemnite) for carbonate minerals. Units are in per mill.) The examination shows that there are two unknowns, the temperature of precipitation and the $\delta^{18}O$ of the lake water. Thus, for paleoclimate reconstructions, assumptions are necessary to separate these two factors. In the case of tropical to subtropical lakes, even at the scale of glacial to interglacial cycles, the temperature variations at a location are relatively minor, on the order of a few degrees centigrade. Thus, the oxygen-isotope composition of the carbonate can be interpreted as predominantly a reflection of the $\delta^{18}O$ of the lake water. Neglecting possible temperature changes will introduce only a minor error in the estimates. In this case, the stratigraphic variations will be dependent on changes in the water balance, origin of air masses, and/or changes in the catchment area (Rosenmeier *et al.*, 2002). As discussed above, in hydrologically open mid- to high-latitude hard-water lakes, carbonate precipitation occurs mainly between April and October, as a consequence of biological activity and increasing temperatures. Thus, possible temperature effects have to be taken into consideration. As shown by Livingstone and Lotter (1999), lake-water temperature in the upper epilimnion closely correlates with mean air temperature; therefore, in this case, temperature influences the oxygen-isotope signal in carbonates. An interesting aspect is that increasing average air temperatures lead to an increase in the $\delta^{18}O$ of lake water due to the positive correlation between air temperature and $\delta^{18}O$ of precipitation which is on the order of 0.55‰°C^{-1} (Kohn and Welker, 2005). At the same time, increasing water temperatures leads to a decrease in oxygen-isotope composition of the calcite because the fractionation between water and calcite decreases with increasing temperature (Kim and O'Neil, 1997), at a rate of approximately 0.2‰°C^{-1}. Thus, the net effect of temperature increase on carbonate isotope composition is of the order of 0.35‰°C^{-1}. Unfortunately, other factors discussed below may also play a role in determining the oxygen-isotope composition of the carbonate, and only rarely a direct inference of average annual temperature can be made. This approximation has been confirmed in case studies, such as one for Baldeggersee, Switzerland, where, based on an isotope mass-balance model, the $\delta^{18}O_c$/temperature relationship was calculated to be 0.39‰° C^{-1} (Teranes and McKenzie, 2001).

Environmental Effects (e.g., Vital Effects, pH) on thc Oxygen-Isotope Composition of Carbonate

Calcite precipitation occurs in lakes only after relatively high levels of supersaturation have been reached (Kelts and Hsu, 1978). Thus, strictly speaking, carbonate precipitation is not an equilibrium process, and the possibility of isotopic disequilibrium has to be taken into consideration. Although most paleoclimatological studies tacitly assume that carbonate precipitation occurs in equilibrium with the water, this has rarely been demonstrated in water-column studies. Sediment trap studies in eutrophic lakes indicate that most of the carbonate does precipitate in isotopic equilibrium (Teranes *et al.*, 1999a; Fisher, 1996). However, it is not always possible to demonstrate this unambiguously, because the water depth at which carbonate precipitation occurs may vary with the trophic level of a lake or with the type of organisms responsible for inducing carbonate precipitation.

The mechanism and water depth of carbonate precipitation are important, because the location and timing in the annual cycle of carbonate precipitation influences its isotopic composition. This is a particularly important factor in lakes, as compared to the oceans, because the water column is characterized by strong vertical gradients in temperature and, in summer, also in oxygen-isotope composition. In addition, the annual productivity cycle can be variable and produce shifts in the timing of precipitation of carbonates, which can introduce spurious variations in their oxygen-isotope composition.

The pH of the lake water is also an important parameter to consider in the interpretation of carbonate records. Recently, studies pioneered by Zeebe (1999) have proposed a model of pH dependence of oxygen-isotope composition of carbonates, which can be used to explain effects previously attributed to disequilibrium, to kinetic fractionations, and/or to poorly understood 'vital effects.' These studies show that the observed decrease in oxygen-isotope composition with the concentration of CO_3^{2-}, first observed by Spero *et al.* (1996) and commonly considered a 'vital effect' or disequilibrium, can be explained by carbonate precipitation at different pH values. This model is based on the fact that dissolved HCO_3^- has an approximately 16‰ higher $\delta^{18}O$ than(CO_3^{2-} (at 19 C). Thus, if carbonate precipitating as calcite contains a mixture of bicarbonate and carbonate, precipitation at higher pH will lead to a progressively lower oxygen-isotope composition. This is caused by the higher abundance of CO_3^{2-} with increasing pH. This process is potentially particularly important in lakes because the pH of the surface waters can vary over 2–3 pH units throughout the year. It is also significant that vertical pH gradients arc observed in the water column. These gradients are particularly important during summer when primary productivity is at maximum in conjunction with stable thermal stratification in the mixolimnion. A pH effect may explain the decrease in $\delta^{18}O$ observed by Teranes *et al.* (1999b) in their sediment trap study of Lake Baldegg, Switzerland. A disturbance of the oxygen-isotope records of calcite may also be expected when a lake undergoes eutrophication. At higher levels of primary productivity, the pH in the surface waters can be expected to reach higher values. Thus, some of the negative $\delta^{18}O$ shifts observed in highly productive lakes may be due to higher average pH related to increased productivity rather than to changes in the $\delta^{18}O$ of the waters or to a shift to carbonate precipitation later in the summer, when water temperatures are higher.

Quaternary Lacustrine Isotopic Records

Quaternary marine oxygen-isotope records show a dominant pattern of change modulated by glacial–interglacial changes in ice volume and temperature, which can be found at most latitudes and longitudes. This clear and reproducible pattern of variation has been used to define robust chronology, the marine isotope stages, which can be used as an accurate dating tool. In contrast, lacustrine oxygen-isotope records are much more dominated by local effects. Therefore, it is not possible to establish a general lacustrine quaternary isotope stratigraphy equivalent to the marine one. Nevertheless, as discussed above, lacustrine isotope records, when compared with marine and other terrestrial archives, provide important climatic constraints. Depending on the type of lake, isotope composition of lacustrine carbonates may reflect changes in average air temperature, seasonal distribution of precipitation, evaporation/precipitation ratio (aridity), or, for example, changes in atmospheric circulation patterns.

Summary and Conclusions

In spite of the complexities in the interpretation of bulk lacustrine carbonate oxygen-isotope values, as discussed above, lacustrine oxygen-isotope records are arguably an invaluable source of paleoclimatic data. Recent years have seen major progress in our understanding of the hydrological effects on the oxygen-isotope composition of lakes and on the significance and problems related to the interpretation of bulk lacustrine isotope records. The quantitative interpretation of lacustrine oxygen-isotope records requires, first of all, a good understanding of the hydrology of the studied lake. Important parameters include a good characterization of the catchment area, which may not be easily determined in karstic areas, where significant underground sources and outflow may be present, and information on the residence time of the water in both the catchment and the lake itself, which controls the temporal resolution of the sedimentary record. In addition, the relationship between $\delta^{18}O$ of the precipitation in the catchment area and the lake water has to be understood and calibrated. Possible changes in the catchment area through time and their effects on the hydrological and isotopic balance must also be considered. Soil and vegetation development may change the residence time of water in the catchment and diversion of tributaries, by natural or anthropogenic causes, may change the size and characteristics of the catchment area and, hence, the isotope composition of the inputs.

Finally, the sediment has to be carefully examined to determine possible disturbances of the primary signal by detrital or authigenic carbonate components of variable origin, which may lead to spurious variations in $\delta^{18}O$ of the bulk carbonate. Possible effects due to changes in trophic state have to be considered an additional complication. This is especially important for lakes strongly affected by anthropogenic or natural eutrophication. Current developments include the inclusion of precipitation isotope data in GCMs, which can be used to evaluate the relative importance of temperature versus changes in atmospheric circulation on the oxygen-isotope records. In addition, the robustness of oxygen-isotope records needs to be validated with independent records, such as those derived from tree rings or compound-specific hydrogen isotopes. In conclusion, there are ongoing efforts to establish stable isotope maps of precipitation for different time slices, using a combination of terrestrial isotope records including lakes. These maps can be used to test and calibrate GCMs to improve our quantitative understanding of paleoclimatic variations.

See also: **Carbonate Stable Isotopes**: Terrestrial Organic Materials. **Diatom Methods**: $\delta^{18}O$ Records; Salinity Reconstructions from Continental Lakes. **Varved Lake Sediments**.

References

Bernasconi, S. M., Dobson, J., McKenzie, J. A., *et al.* (1997). Preliminary isotopic and palaeomagnetic evidence for Younger Dryas and Holocene climate evolution in NE Asia. *Terra Nova* **9**, 246–250.

Clark, I. D., and Fritz, P. (1997). *Environmental Isotopes in Hydrogeology*, p. 328. CRC Press, Boca Raton.

Coplen, T. B. (1994). Reporting of stable hydrogen, carbon, and oxygen isotopic abundances. *Pure and Applied Chemistry* **66**, 273–276.

Dansgaard, W. (1964). Stable isotopes in precipitation. *Tellus* **16**, 436–468.

Dittrich, M., Kurz, P., and Wehrli, B. (2004). The role of autotrophic picocyanobacteria in calcite precipitation in an oligotrophic lake. *Geomicrobiology Journal* **21**, 45–53.

Fisher (1996) *Isotopengecemische Untersuchungen im Wasser und in den Sedimenten des Soppensees (Kt. Luzer, Schweiz)*. PhD Dissertation ETH Zürich No. 11924.

Fronval, T., Jensen, N. B., and Buchardt, B. (1995). Oxygen isotope disequilibrium precipitation of calcite in Lake Arreso, Denmark. *Geology* **23**, 463–466.

Gat, J. R., and Lister, G. S. (1995). The "catchment effect" on the isotopic composition of lake waters; its importance in palaeolimnological interpretations. In Problems of stable isotopes in tree-rings, lake sediments and peat-bogs as climatic evidence for the Holocene (B. Frenzel, Ed.), vol. 15, pp. 1–16. Akademie der Wissenschaften und der Literatur, Stuttgart, Germany.

Glew, J. R., Smol, J. P., and Last, W. M. (2001). Sediment core collection and extrusion. In *Tracking environmental change using lake sediments, 1* (W. M. Last and J. P. Smol, Eds.), vol. 1, pp. 73–106. Dordrecht, Kluwer.

Hammarlund, D., Barnekov, L., Birks, H. J. B., Burchadt, B., and Edwards, T. W. D. (2002). Holocene changes in atmospheric circulation recorded in the oxygen-isotope stratigraphy of lacustrine carbonates from northern Sweden. *The Holocene* **12**, 339–351.

Hodell, D. A., Schelske, C. L., Fahnenstiel, G. L., and Robbins, L. L. (1998). Biologically induced calcite precipitation and its isotopic composition in Lake Ontario. *Limnology and Oceanography* **43**, 187–199.

Hollander, D. J., and McKenzie, J. A. (1991). CO_2 control on carbon-isotope fractionation during aqueous photosynthesis: A paleo-*p*-CO_2 barometer. *Geology* **19**, 929–932.

Hsü, K. J., and Kelts, K. R. (Eds.) (1984). *Quaternary Geology of Lake Zurich: An Interdisciplinary Investigation by Deep-Lake Drilling Contributions to Sedimentology No. 13, E,210p.* Schwizerbart's Verlagsbuchhandlung, Stuttgart.

Huang, Y., Shuman, B., Wang, Y., and Webb, T. (2004). Hydrogen isotope ratios of individual lipids in lake sediments as novel tracers of climatic and environmental change: A surface sediment test. *Journal of Paleolimnology* **31**, 363–375.

Kelts, K., and Hsu, K. J. (1978). Freshwater carbonate sedimentation. In *Lakes: Geology, Chemistry, Physics* (A. Lerman, Ed.), pp. 295–323. Springer, New York.

Kim, S. T., and O'Neil, J. R. (1997). Equilibrium and nonequilibrium oxygen isotope effects in synthetic carbonates. *Geochimica et Cosmochimica Acta* **61**, 3461–3475.

Kirby, M. E., Patterson, W. P., Mullins, H. T., and Burnett, A. W. (2002). Post-Younger Dryas climate interval linked to circumpolar vortex variability: Isotopic evidence from Fayetteville Green Lake, New York. *Climate Dynamics* **19**, 321–330.

Kohn, M. J., and Welker, J. M. (2005). On the temperature correlation of delta O-18 in modern precipitation. *Earth and Planetary Science Letters* **231**, 87–96.

Lehmann, M. F., Bernasconi, S. M., Barbieri, A., Simona, M., and McKenzie, J. A. (2004). Seasonal variation of the $\delta^{13}C$ and

$\delta^{15}N$ of particulate and dissolved carbon and nitrogen in lake Lugano: Constraints on biogeochemical cycling in a eutrophic Lake. *Limnology and Oceanography* **49**, 415–429.

Leng M, J., and Marshall, J. D. (2004). Palaeoclimate interpretation of stable isotope data from lake sediment archives. *Quaternary Science Reviews* **23**, 811–831.

Li, C., and Ku, T. L. (1997). $\delta^{13}C$–$\delta^{18}O$ covariance as a paleohydrological indicator for closed-basin lakes. *Palaeogeography, Palaeoclimatology, Palaeoecology* **133**, 69–80.

Livingstone, D. M., and Lotter, A. F. (1999). The relationship between air and water temperatures in lakes of the Swiss Plateau: A case study with palæolimnological implications. *Journal of Paleolimnology* **19**, 181–198.

Lotter A, F., Sturm, M., Teranes J, L., and Wehrli, B. (1997). Varve formation since 1885 and high-resolution varve analyses in hypereutrophic Baldeggersee (Switzerland). *Aquatic Sciences* **59**, 304–326.

McKenzie, J. A. (1985). Carbon isotopes and productivitiy in the lacustrine and marine environment. In *Chemical Processes in Lakes* (W. Stumm, Ed.), pp. 99–118. Wiley-Interscience, New York.

McKenzie, J. A., and Hollander, D. J. (1993). Oxygen isotope record in recent carbonate sediments from Lake Griefen, Switzerland (1750–1986): Application of continental isotopic indicator for evaluation of changes in climate and atmospheric circulation patterns. In *Geophysical Monographs, 78: Climate Change in Continental Isotopic Records* (P. K. Swart, K. C. Lohmann, J. A. McKenzie and S. Savin, Eds.), pp. 101–112. American Geophysical Union, Washington.

Rosenmeier, M. F., Hodell, D. A., Brenner, M.., Curtis, J. H., and Guilderson, T. P. A. (2002). A 4000-year lacustrine record of environmental change in the southern Maya lowlands, Peten, Guatemala. *Quaternary Research* **57**, 183–190.

Rosenmeier, M. F., Hodell, D. A., Brenner, M., *et al.* (2005). Influence of vegetation change on watershed hydrology: implications for paleoclimatic interpretation of lacustrine delta O-18 records. *Journal of Paleolimnology* **27**, 117–131.

Rozanski, K., Araguas-Araguas, L., and Gonfiantini, R. (1993). Isotopic patterns in modern precipitation. In *Geophysical Monographs, 78: Climate Change in Continental Isotopic Records* (P. K. Swart, K. C. Lohmann, J. A. McKenzie and S. Savin, Eds.), pp. 1–36. American Geophysical Union, Washington.

Sauer, P. E., Eglinton, T. I., Hayes, J. M., Schimmelmann, A., and Sessions, A. (2001). Compound-specific D/H ratios of lipid biomarkers from sediments as a proxy for environmental and climatic conditions. *Geochimica et Cosmochimica Acta* **65**, 213–222.

Smith, M. A., and Hollander, D. J. (1999). Historical linkage between atmospheric circulation patterns and the oxygen isotopic record of sedimentary carbonates from Lake Mendota, Wisconsin, USA. *Geology* **27**, 589–592.

Spero, H. J., Bijma, J., Lea, D. W., and Bemis, B. E. (1996). Effect of seawater carbonate concentration on foraminiferal carbon and oxygen isotopes. *Nature* **390**, 497–500.

Stuiver, M. (1970). Oxygen and carbon isotope ratios of freshwater carbonates as climatic indicators. *Journal of Geophysical Research* **75**, 5247–5257.

Talbot, M. R. (1990). A review of the paleohydrological interpretation of carbon and oxygen isotopic-ratios in primary lacustrine carbonates. *Chemical Geology* **80**, 261–279.

Teranes, J. L., and McKenzie, J. A. (2001). Lacustrine oxgen isotope record of 20th-century climate change in central Europe: Evaluation of climatic controls on oxygen isotopes in precipitation. *Journal of Paleolimnology* **26**, 131–146.

Teranes, J. L., McKenzie, J. A., Bernasconi, S. M., Lotter, A. F., and Sturm, M. (1999a). A study of oxygen isotopic fractionation during bio-induced calcite precipitation in eutrophic Baldeggersee, Switzerland. *Geochimica et Cosmochimica Acta* **63**, 1981–1990.

Teranes, J. L., McKenzie, J. A., Lotter, A. F., and Sturm, M. (1999b). Stable isotope response to lake eutrophication: Calibration of a high-resolution lacustrine sequence from Baldeggersee, Switzerland. *Limnology and Oceanography* **44**, 320–333.

Thompson, J. B., Schultze-Lam, S., Beveridge, T. J., and Des-Marais D, J. (1997). Whiting events: Biogenic origin due to the photosynthetic activity of cyanobacterial picoplankton. *Limnology and Oceanography* **42**, 133–141.

Wick, L., Lemcke, G., and Sturm, M. (2003). Evidence of Late glacial and Holocene climatic change and human impact in eastern Anatolia: High-resolution pollen, charcoal, isotopic and geochemical records from the laminated sediments of Lake Van, Turkey. *The Holocene* **13**, 665–675.

Zeebe, R. E. (1999). An explanation of the effect of seawater carbonate concentration on foraminiferal oxygen isotopes. *Geochimica et Cosmochimica Acta* **63**, 2001–2007.

Carbon-Dioxide, Atmospheric Concentrations *see* **Carbonate Stable Isotopes**: Overview; **Ice Core Methods**: CO_2 Studies; **Paleoceanography, Physical and Chemical Proxies**: Carbon Cycle Proxies; **Plant Macrofossil Methods and Studies**: CO_2 Reconstruction from Fossil Leaves

Cave Art *see* **Vertebrate Studies**: Interactions with Hominids

Charcoal Studies *see* **Paleobotany**: Charred Particle Analysis

CHIRONOMID OVERVIEW

I R Walker, University of British Columbia Okanagan, British Columbia, Canada

© 2007 Elsevier B.V. All rights reserved.

Introduction

The Chironomidae constitute a family of true flies (Diptera) prominent as living larvae in the bottom communities of almost all freshwater habitats. They are abundantly preserved in Quaternary lake sediments, and are widely used for paleoenvironmental reconstruction. The biology and applications of Chironomidae in Quaternary science have been extensively reviewed by Armitage *et al.* (1995), Frey (1976, 1988), Hofmann (1986, 1988), Porinchu and MacDonald (2003), and Walker (1987, 1993, 1995, 2001). This entry provides a summary of these, and other important and relevant works.

Adult Chironomidae seldom bear a proboscis, and although they and their phantom midge cousins (family Chaoboridae) are mosquito-like, neither family shares the biting habit of mosquitoes (family Culicidae). Consequently, chironomids are commonly referred to as non-biting midges.

Three related families of aquatic midges do bite as adults and are also preserved as head capsules in lake sediments: Ceratopogonidae ('no-see-ums' or biting midges), Culicidae (mosquitoes), and Simuliidae (black flies or buffalo gnats). Members of these families are vectors for many diseases of animals, including humans. Although Chaoboridae do not have a head capsule record, their mandibles are common fossils. Collectively, the preserved remains of Chironomidae, Chaoboridae, Culicidae, Ceratopogonidae, and Simuliidae are often referred to as fossil aquatic midges. The earliest known fossils of Chironomidae are of Triassic age.

The larvae, pupae and imagoes of all of these groups are important food items in the diets of fish. The imagoes are an important food for swallows and martins.

Biology

Chironomidae are amongst the most ubiquitous of insects, found on islands distributed throughout the world's oceans, and on all continents from the lowland tropics to the high arctic, Antarctic, and alpine. Chironomidae have a life cycle consisting of four stages: egg, larva, pupa, and imago.

Upon hatching, the first instar larva begins a period of growth, which eventually necessitates the shedding of the chitinous exoskeleton. Upon replacing the exoskeleton, the larva then continues growth in its second instar. Two further episodes of ecdysis (replacement of the exoskeleton) and growth follow, defining the third and fourth larval instars.

Although bearing a well developed, strongly sclerotized head capsule, the elongate soft-bodied larvae otherwise resemble maggots. As mature fourth instar larvae, they vary from 1 to 30 mm in length. Lacustrine chironomid larvae inhabit the uppermost sediments of both littoral and profundal (deepwater) environments, cling or burrow into aquatic plants, tunnel in moist wood, or live as parasites or commensal partners on other invertebrates. Similar habitats are occupied in streams, and chironomid taxa are known to inhabit littoral marine environments, water-logged soils, peats, and dung.

The first instar larva of lacustrine midges is generally planktonic, allowing for dispersal of larvae to suitable substrates. Larvae from habitats other than standing water, and the later instars of lacustrine midges, are largely sedentary and benthic (bottom dwelling). The first instar is typically brief, with each later instar usually of progressively longer duration. Although the first instar larvae may derive some nourishment from the egg yolk, they feed on a variety of materials including algae, organic detritus, aquatic plants, and other invertebrates. Some are predaceous, whereas most other chironomids combine algae and organic detritus to form their diet.

During the latter part of the fourth larval instar, the thoracic segments expand, achieving a pre-pupal condition. Following ecdysis, the chironomid enters the pupal stage in which morphology is re-organized to that of the imago (adult). Apart from the Tanypodinae, almost all pupae are sedentary. The pupal stage is brief, persisting for a few days at most. At maturity, the pupa moves to the water surface, allowing emergence of the winged adult imago.

Lasting for a few weeks only, the adult stage permits dispersal and reproduction. The duration of the midge life cycle varies greatly among taxa, and is also related to habitat. Long life cycles typify arctic taxa, occasionally extending to several years. In warmer climates, several generations are common in a single season.

Classification

Eleven subfamilies constitute the Chironomidae. One subfamily, the Telmatogetoninae, is almost completely restricted to marine environments. However, in Hawaii this family also inhabits fresh water. Three subfamilies are very rare, or known only from a small segment of the world. For example, the Chilenomyiinae, represented by a single species, is known only from southern Chile. The Buchonomyiinae includes three species native to Europe, Asia (Burma), and Central America (Costa Rica). The Usumbaromyiinae is a new subfamily recently described from Tanzania (Andersen and Sæther, 1994).

The subfamily Aphroteniinae, including four genera, is known only from South America, South Africa, and Australia. Such a distribution suggests that this primitive group originated before the Mesozoic disintegration of Gondwanaland. An extinct Aphroteniinae tribe, the Electroteniini, records this subfamily's former, more widespread Cretaceous distribution, extending into Siberia.

Three subfamilies, the Orthocladiinae, Tanypodinae, and Chironominae, are widely distributed and constitute the majority of taxa encountered in lake sediments. The remaining subfamilies, Podonominae, Diamesinae, and Prodiamesinae, are predominantly cold-stenothermous taxa inhabiting temperate and montane, to polar and alpine climatic regions.

Fossils and their Identification

Taphonomy

Given an appropriate sedimentary environment, remains of larvae, pupae, and adults (imagoes) may all be preserved. Imagoes are well preserved in Cretaceous amber, but in lake sediments their bodies are commonly disarticulated. Occasionally the hypopygium, a part of the male imago's genitalia, is preserved and allows species-level identifications. Thoracic horns of Tanypodinae pupae have rarely been found in lacustrine deposits, and the thin, delicate bodies of larvae and pupae are too fragile to normally preserve well. In contrast, Quaternary interglacial lake deposits typically contain 50 to 200 well-preserved larval head capsules per cubic centimeter. Thus, these well sclerotized head capsules are far more abundant in lakebeds than the remains of all other insects combined. They provide the basis for almost all Quaternary paleoenvironmental work involving the Chironomidae.

One head capsule is produced by each chironomid, during each larval instar. Thus, a chironomid surviving to the fourth larval instar will have produced four head capsules. A chironomid expiring prior to attaining its fourth instar will produce fewer head capsules. The head capsules subsequently preserved could include both those shed with exuviae during ecdysis, and the remains of dead larvae. However, there exist numerous reasons why the abundance of fossil midge remains might not reflect the abundance of the living larvae from which they were derived.

Quality of preservation may vary among taxa, among instars, and between head capsules derived from expired insects and those shed as exuviae. Bivoltine and multivoltine taxa produce more head capsules per season than univoltine taxa of the same abundance. Re-distribution of head capsules will influence the concentration and composition of fossil assemblages. Remains of stream Chironomidae are often deposited in lakes, and remains of littoral species may be transported offshore to deepwater sites. Erosion along lakes and streams facilitates the redeposition of semi-terrestrial species into lacustrine settings.

Iovino (1975) determined that first and second instars of *Chironomus attenuatus* usually dissolve the procuticle completely prior to ecdysis, leaving only a thin, non-chitinous, epicuticular exuviae. In contrast, third instar exuviae usually include chitin, and fourth instars of *Chironomus attenuatus* were chitinous almost without exception. Thus, exuviae of early instars are much less durable than both those of later instars and remains of expired larvae.

These experiments have only been conducted on *Chironomus attenuatus*. The results possibly apply to all chironomid species. This, in part, would explain the great under-representation of early instar head capsules in lake sediments. Early instar remains may also be lost through the use of coarse sieves during processing.

Tanypodinae head capsules are more delicate than those of other taxa, and may be poorly preserved. First instar head capsules of all chironomid taxa are normally uncommon to rare as fossils, but may dominate records preserved in the anoxic, deepwater regions of meromictic basins.

Fossil Morphology

Identification of Quaternary fossils relies principally on the morphology of the head capsule, although other larval body parts (e.g., antennae and various mouthparts) are useful when they are still attached to the head. The structure of the head capsule varies, but the greatest part of this variation may be illustrated with reference to fossils of the subfamilies Chironominae, Orthocladiinae, and Tanypodinae (**Fig. 1**).

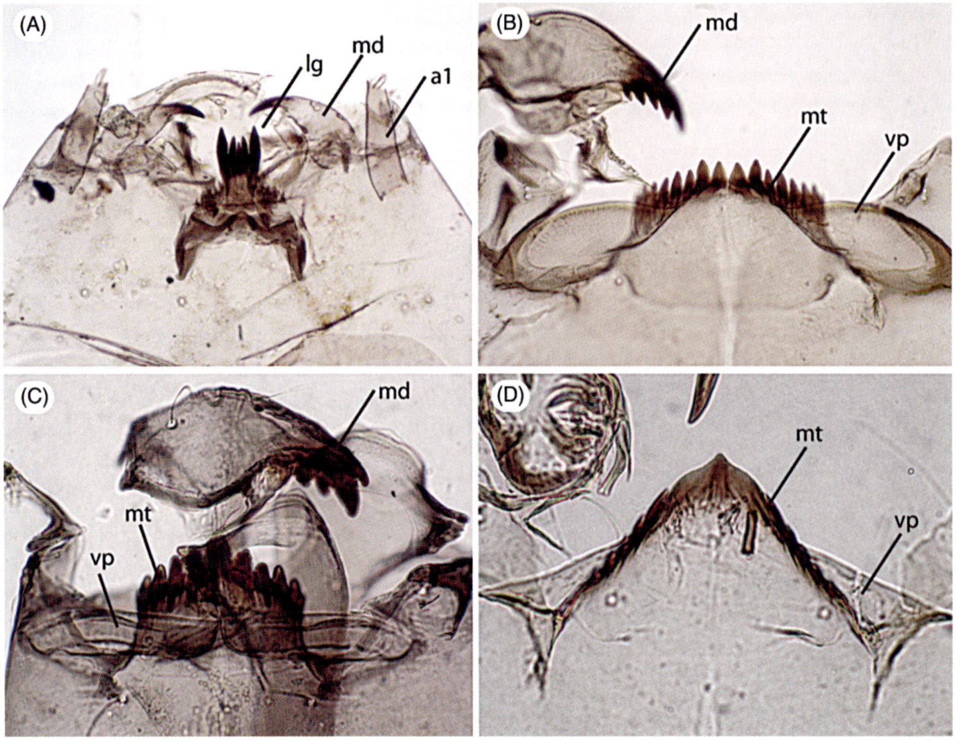

Figure 1 Representative fossils of Chironomidae, illustrating the principal structures used in their identification: (A) *Procladius* (subfamily Tanypodinae), (B) *Sergentia* (tribe Chironomini, subfamily Chironominae), (C) Tribe Tanytarsini (subfamily Chironominae), (D) *Psectrocladius* cf. *calcaratus* (Subfamily Orthocladiinae). Abbreviations: a1, 1st antennal segment; lg, ligula; md, mandible; mt, mentum; vp, ventromental plate.

The head capsules of Tanypodinae (**Fig. 1A**) are generally weakly pigmented and somewhat elongate. The retracted first segment of each antenna is often visible within fossil head capsules. The proportions (ratio of length to width) of the head capsule and first antennal segment can be important characters for fossil identification. A fork-shaped ligula is centrally located within the head capsule. The ligula may become separated from head capsules during burial or processing of fossil material. Variations in the shape, number, and color of ligula teeth also aid identification. The dorsomental teeth, lacking in the tribe Pentaneurini, are otherwise situated laterally adjacent to the ligula and will usually remain with the head capsule. In well preserved material, one or both mandibles may be retained. The arrangement of minute hairs (cephalic setae) on the fossil head capsule is also important for identification of larval fossils (Rieradevall and Brooks, 2001).

The principal features facilitating identification of fossil Chironominae (**Figs. 1B** and **C**) and Orthocladiinae (**Fig. 1D**) are the ventromental plates and the teeth of the mentum. The mentum is situated in an anterior median ventral position. The pair of ventromental plates are situated laterally, adjacent to the mentum on either side. In the Chironominae, the ventromental plates are very large and usually conspicuously striated. Although fan-shaped in the tribe Chironomini (**Fig. 1B**), laterally elongated, strap-shaped plates occur on members of the Tanytarsina (a subtribe of the tribe Tanytarsini) (**Fig. 1C**), and in the tribe Pseudochironomini.

In the Orthocladiinae (**Fig. 1D**), the ventromental plates are typically much less conspicuous, and never striated. In many Orthocladiinae, these plates are greatly reduced, vestigial structures. In all Chironominae and Orthocladiinae, the shape, number, and arrangement of the mental teeth and characteristics of the ventromental plates are critical diagnostic features. Most other structures are normally lost from the head capsule.

Although taxonomic keys for living larvae rarely permit identification of fossil larval head capsules, reference to associated illustrations and diagnoses for each genus usually permit classification. Fortunately, most of the characters employed by systematists are borne on the head capsule and consequently may be available for fossil work. Few keys have been designed specifically for fossil identifications.

Sample Preparation

Meticulous preparation, sorting, and identification of midge remains are requisites for analysis of past communities. Small lakes in forested watersheds usually contain sediments with a high organic content. In such lakes, 1 or 2 ml of wet sediment typically contains 50 to 200 head capsules, sufficient for analysis. Inorganic sediments, such as those frequently encountered in late glacial deposits, may yield much lower concentrations. However, head capsule concentrations vary greatly among sites.

Samples are usually deflocculated in 5 to 10% KOH prior to analysis. In calcareous sediments, an acid wash (10% HCl) may also be necessary. If excessive siliceous mineral matter is present, HF may be used to further concentrate and clean the head capsules. Following these chemical treatments, head capsules are generally separated from finer debris by means of sieving. Methodology varies, but a 100 μm or finer sieve will retain most head capsules and is recommended. Subsequently, the residue is back-washed from the sieve into a container and stored wet until the material is sorted. Unless sorted shortly after sieving, preservation in 99% ethanol is advisable. Sorting of wet residues is best accomplished in a Bogorov counting tray at 40 to 50X. As remains are located, each is transferred with forceps to coverslips, for mounting in an appropriate medium (e.g., Canada balsam, Euparal®, or Entellan®) and subsequent identification.

Ecology and Distribution

Early work on the ecology and distribution of Chironomidae was conducted by biological limnologists (e.g., August Thienemann, Lars Brundin, Ole Sæther) who sought to classify lakes principally with regard to biological productivity. They also emphasized the importance of Chironomidae as bio-indicators of water quality. The importance of water quality parameters, especially food quantity, food quality, and the oxygen concentration at the mud–water interface were considered of paramount importance in regulating the distributions of larval Chironomidae.

Large-bodied larvae, which contained hemoglobin (especially *Chironomus*), were noted as best adapted to the oxygen-deficient bottom waters typical of eutrophic lakes (nutrient-enriched lakes with high algal production). Hemoglobin facilitates the capture and storage of oxygen, whereas large body size allows the larvae to disrupt, and extend their bodies above, the most oxygen-deficient micro-layer at the mud–water interface. In contrast, small-bodied larvae, which lacked hemoglobin (several Tanytarsina and Orthocladiinae), were better adapted to low food availability in the well-oxygenated bottom waters of oligotrophic lakes (nutrient-poor lakes with low algal productivity). Because of these relationships among chironomids, water quality, and oxygen concentrations, extant chironomid communities are sampled extensively for environmental biomonitoring purposes. The Benthic Quality Index, for example, was developed as a means for environmental assessment in Nordic freshwater lakes (Wiederholm, 1980).

Detailed surveys of chironomid distributions in North America and Europe have revealed the importance of other environmental variables. Recent studies continue to recognize the importance of oxygen and biological productivity as controls on species distributions, but greater importance is commonly extended to climatic variables, especially summer air and water temperatures. The Chironomini and Tanypodinae are abundant and widely-distributed in temperate to tropical climates, but the subfamilies Orthocladiinae and Diamesinae, and some members of the subtribe Tanytarsina, are more abundant where colder, alpine and polar climates prevail. In addition to temperature and oxygen, salinity, pH, dissolved organic carbon, nutrient concentrations, and lake depth are considered important variables defining the distributions of individual chironomid species.

Quaternary Paleoecology

Lake ontogeny

Early chironomid paleoecological research was primarily directed to answering questions pertaining to lake ontogeny (i.e., those physical, chemical, and biological processes that influence how lakes intrinsically tend to develop through time). Einar Naumann (1919) speculated that lakes should gradually become less productive as a consequence of constant leaching of catchment soils. Many other limnologists, including Naumann's contemporary August Thienemann (1926), perceived eutrophication (a gradual increase in lake productivity) as the dominant, if not universal process. Whiteside (1983) traces this perception to early works published by Lindeman (1942) and Deevey (1955). Deevey emphasized that the gradual infilling of lakes, by reducing the volume of the hypolimnion (the cold-water region at the bottom of the lake) could generate an O_2-poor hypolimnion. Release of phosphorus, an important nutrient, from the sediments beneath the hypolimnion could then produce an increase in lake productivity. This process was dubbed morphometric eutrophication.

This appealing theory was broadly supported by early chironomid paleoecological studies, which suggested a shift from midge communities typical of well-oxygenated, nutrient-poor (oligotrophic) lakes, to those prevalent in the deep water of oxygen-deficient, nutrient-enriched (eutrophic) lakes. However, the theory was developed from a relatively small subset of the world's lakes (i.e., temperate lakes, formed during the relatively recent deglaciation of eastern North America and northwestern Europe), and was never tested more generally on lakes with different origins, or in other climatic regions.

These early investigators also failed to recognize or acknowledge the potential importance of climate in driving the observed changes in both midge fauna and lake productivity. Major changes in the midge faunas typically occurred at each site with the development of forests and rapid deglacial warming approximately 10,000 ^{14}C yr BP.

Today, these changes in productivity and midge communities are more likely to be interpreted as a consequence of climatic change, not as an intrinsic tendency of lakes to become more productive over time. The importance of climate as a driving force in lake productivity and midge community change is especially evident in late-glacial records, where increases and decreases in the apparent productivity of lakes and changes in midge faunas, coincide with independent evidence of climatic changes. The apparent productivity of lakes abruptly decreases and oligotrophic midge communities return, with the onset of colder conditions.

Human impact studies

Although the natural productivity of lakes and their midge faunas may be regulated more by climate than initially recognized, human activities have also had major impacts. Given the long history of environmental biomonitoring using living Chironomidae, there has also been long interest in using the fossils in paleoenvironmental assessments of human impacts. Some researchers have used mouthpart deformities as indicators of environmental stress (e.g., in the Rhine), but most results are derived from stratigraphic studies of midge community composition, as preserved in lake sediments.

Eutrophication Increased nutrient loading, for example, from urban and agricultural land use may dramatically increase the productivity of some lakes, to the point where algal blooms foul shorelines and their decomposition produces anaerobic conditions within the hypolimnion. In addition to this increased organic production from nutrient enrichment, external inputs of sewage and other organic wastes compound the de-oxygenation problem.

Since chironomid communities respond dramatically to such trophic alterations, their remains provide a detailed record of eutrophication events. Most studies (e.g., at Esthwaite Water and Lough Neagh (UK), Lake Washington (US), Wood Lake, and Bay of Quinte, Lake Ontario (Canada), Lake Mälaren and Lake Växjösjön (Sweden), Lake Vanajavesi (Finland)) have relied on qualitative interpretation of midge stratigraphies for impact assessment. In recent years, researchers have instead attempted to develop quantitative reconstructions of productivity related variables.

Most researchers would argue that hypolimnetic oxygen concentrations have the most important and most direct impact on living larvae; thus a quantitative model that infers paleo-oxygen conditions, should ideally be developed. Oxygen conditions vary continuously throughout the year, and also at different depths in the water column, so it is difficult to characterize a lake's oxygen level with a single oxygen measurement. Furthermore, the duration of anoxic conditions and the build-up of toxic H_2S in anoxic hypolimnia, are probably equally important variables. The relationship between midge communities and oxygen concentrations is even more tenuous in shallow lakes (lacking a hypolimnion) and in near-shore (littoral) habitats, than in deep, thermally stratified hypolimnia.

As an alternative to reconstructing oxygen levels, researchers have developed models to infer phosphorus or chlorophyll concentrations. The connection between inorganic nutrients, such as phosphorus, and their impacts on chironomid larvae is indirect, but nutrient concentrations are more easily measured and more widely used to characterize lake productivity. These midge-productivity models are now being used in North America and Europe to better quantify human impacts on the environment. Brodersen and Quinlan (in press) review recent pertinent literature.

Acidification During the 1980s, lake acidification and the long-range transport of air pollutants were amongst the most controversial and internationally important environmental issues. Much environmental research was devoted to tracing the paths of pollutants, and assessing their damage to ecosystems in both eastern North America and northern Europe. Little was known about the distributions of chironomids with respect to pH, alkalinity, or related environmental variables, but they were used as biomonitoring tools in several studies of lake acidification.

pH is a master variable in aquatic chemistry. Many other variables, for example, lake alkalinity, dissolved organic carbon, and metal concentrations, tend to co-vary with pH, thus it is difficult to separate the direct effects of pH from the effects of these covariables. At low pH, metals (e.g., Fe, Al) are more soluble, and some (e.g., Al) are known to be toxic to aquatic organisms. However, many acidic lakes are also humic lakes, containing high concentrations of dissolved organic acids.

The organic acids are derived from the incomplete decomposition of terrestrially-derived organic matter, especially from adjacent bogs. These organic acids are natural products, which may contribute to the acidity of lakes, but may also slow the rate of decomposition and bind to nutrients (e.g., P) and metals (e.g., Al), reducing the impact of these pollutants on aquatic life.

Chironomidae are well represented even in strongly-acidic waters (pH 3.5 to 5.5). *Chironomus*, *Phaenopsectra*, *Psectrocladius*, and *Zalutschia* are reported at greater relative abundance at low pH. Favored taxa tend to be large-bodied, having smaller surface areas relative to volume. It may be easier for large larvae to maintain their internal pH balance. Those midges abundant in acidified lakes include taxa reported to be more common in fish diets. Thus acidification, by eliminating fish, could also favor these taxa. Furthermore, the increased buffering capacity afforded by hemoglobin may also favor hemoglobin-bearing larvae.

Although several paleolimnological studies have documented changes in midge populations that accompanied lake acidification, these insects have never been used to quantitatively reconstruct pH. Although technically feasible, the chironomid-pH relationships have never been properly quantified. Once the pH optima and tolerances of chironomid taxa are quantified, they should be useful bio-indicators of lake acidification processes. This is an obvious area to target in future research.

Climate change

Paleotemperature Although researchers had previously pondered the possible climatic significance of midge stratigraphies, it was not until the late 1980s and early 1990s that midge–climate relationships were explored in earnest. At first, the possibility of using chironomids for paleoclimatic inference was highly controversial, stimulating a lively debate in Quaternary Research, the Journal of Paleolimnology, and the Canadian Journal of Fisheries and Aquatic Sciences

Key to resolving the issue was an international effort to map the distribution and abundance of chironomid taxa across long climatic gradients in both North America and Europe. Statistical analyses of these data repeatedly demonstrated that summer air and water temperatures were highly correlated with midge distributions.

Summer climate is now recognized as potentially the single most important variable regulating midge distributions. Fossil midges are also recognized as one of the most reliable indicators of past climatic change. Through the use of statistical inference procedures (e.g., weighted averaging, partial least squares, analog matching, and Bayesian methods) midge–climate models are routinely used in the quantitative reconstruction of late-glacial and Holocene paleotemperatures. Details of the midge–climate relationship, however, are still widely debated. At least in part, midges may be responding to climate indirectly via the influence of climate on lake productivity.

Surveys of midge distributions spanning the tree line commonly reveal distinct forest and tundra communities, with species turnover being most rapid in the vicinity of the tree line. This observation suggests that tree line communities may be highly sensitive to climate change, and these sensitive sites are targeted by paleoecologists for paleotemperature reconstructions. Conversely, these transects suggest that the fauna within the forest region is more uniform, and insensitive, so such insensitive sites tend to be ignored by paleoecologists. This bias is also likely to impact on midge-derived paleotemperature reconstructions. From the perspective of a researcher wishing to accurately reconstruct paleotemperatures, tree line sites may be overly sensitive, revealing paleotemperature signals larger than the changes that actually occurred, whereas forested sites may pose the opposite dilemma. This issue has implications for accurately reconstructing: 1) the temporal magnitude of paleoclimatic changes, and 2) spatial temperature gradients, for example, as they existed at the margin of ice sheets (Levesque *et al.*, 1997).

Brooks (in press), and Walker and Cwynar (in press) review recent progress in paleotemperature reconstruction using fossil midges.

Lake levels Statistical analyses have revealed a number of other environmental variables potentially critical for understanding midge distributions. Lake depth, for example, is often just as important as temperature. Since lake levels have often been used as indicators of hydrological balance, lake depth inferences might provide another useful tool for paleoclimatologists. During warm, dry intervals lake levels may drop. During cool, wet intervals lake levels will rise.

Midge–depth relationships are easily modeled for use in lake level reconstructions. Despite this opportunity, midges are seldom used for lake level reconstructions. Model errors indicate that satisfactory reconstructions will only be possible where the amplitude of lake level changes has been very large. Since the surface discharge from lakes can change enormously with small changes in lake level, inferences obtained from open basin lakes are unlikely to provide useful information. The magnitude of lake level changes in closed basin lakes is potentially much larger. These lakes are common in semi-arid regions and potentially contain more useful paleohydrological records.

Paleosalinity Although not extensively studied, chironomid communities are known to be salinity sensitive. This is never evident in freshwater regions, but saline lakes are common landscape features, for example, in Australia, east Africa, central Asia, and western North America. As in other aquatic organisms, the osmotic stress imposed by high salinities tends to dehydrate larval midges. Many species are intolerant, restricted entirely to fresh water, but a few (e.g., *Cricotopus ornatus* and *Tanypus nubifer*) thrive in highly saline lakes or in marine intertidal habitats.

Since hydrological balance (evaporation/precipitation) potentially regulates the salinity of these lakes, midges and other paleosalinity indicators (e.g., diatoms) may provide important paleoclimate proxy data from these sites. During warm, dry intervals salinity is expected to be high, and during cool, wet intervals salinity should be low. Midge paleosalinity reconstructions are now available for several sites in Africa and western North America. Recent evidence suggests that the paleohydrology of these lakes may be very complex. Paleosalinity reconstructions may be more difficult to interpret climatically than previously assumed.

In addition to their potential climatic significance, midge paleosalinity records may be useful for sea-level studies. Heinrichs and Walker (in press), and Verschuren and Eggermont (in press) have reviewed the potential of midge paleosalinity records for both paleohydrological and sea-level studies.

See also: **Chironomid Records**: Africa; Late Pleistocene of Europe; Postglacial Europe; Postglacial Southern Hemisphere. **Diatom Methods**: Salinity Reconstructions from Continental Lakes.

References

Andersen, T., and Sæther, O. A. (1994). *Usambaromyia nigrala* gen. n., sp. n., and Usambaromyiinae, a new subfamily among the Chironomidae (Diptera). *Aquatic Insects* **16**, 21–29.

Armitage, P. D., Cranston, P. S. and Pinder, L. C. V. (Eds.) (1995). *The Chironomidae: Biology and ecology of non-biting midges.* Chapman & Hall, London.

Brodersen, K. P., and Quinlan, R. in press. Midges as paleoindicators of lake productivity, eutrophication and hypolimnetic oxygen. *Quaternary Science Reviews.*

Brooks, S. J. in press. Fossil midges (Diptera: Chironomidae) as palaeoclimatic indicators for the Eurasian region. *Quaternary Science Reviews.*

Deevey, E. S., Jr. (1955). The obliteration of the hypolimnion. *Mem. Ist. Ital. Idrobiol., Suppl* **8**, 9–38.

Frey, D. G. (1976). Interpretation of Quaternary paleoecology from Cladocera and midges, and prognosis regarding usability of other organisms. *Canadian Journal of Zoology* **54**, 2208–2226.

Frey, D. G. (1988). Littoral and offshore communities of diatoms, cladocerans and dipterous larvae, and their interpretation in paleolimnology. *Journal of Paleolimnology* **1**, 179–191.

Heinrichs, M. L. in press. Fossil midges and palaeosalinity: potential as indicators of hydrological balance and sea level change. *Quaternary Science Reviews.*

Hofmann, W. (1986). Chironomid analysis. In *Handbook of Holocene Palaeoecology and Palaeohydrology* (B. E. Berglund, Ed.), pp. 715–727. John Wiley & Sons, Chichester.

Hofmann, W. (1988). The significance of chironomid analysis (Insecta: Diptera) for paleolimnological research. *Palaeogeography, Palaeoclimatology, Palaeoecology* **62**, 501–509.

Iovino, A. J. (1975) Extant Chironomid Larval Populations and the Representativeness and Nature of their Remains in Lake Sediments. Unpublished PhD. thesis, Indiana University, Bloomington.

Levesque, A. J., Cwynar, L. C., and Walker, I. R. (1997). Exceptionally steep north-south gradients in lake temperatures during the last deglaciation. *Nature* **385**, 423–426.

Lindeman, R. L. (1942). The trophic-dynamic aspect of ecology. *Ecology* **23**, 399–418.

Porinchu, D. F., and MacDonald, G. M. (2003). The use and application of freshwater midges (Chironomidae: Insecta: Diptera) in geographical research. *Progress in Physical Geography* **27**, 378–422.

Rieradevall, M., and Brooks, S. J. (2001). An identification guide to subfossil Tanypodinae larvae (Insecta: Diptera: Chironomidae) based on cephalic setation. *Journal of Paleolimnology* **25**, 81–99.

Thienemann, A. (1926). Der Nahrungskreislauf im Wasser. *Verhandlungen der Deutschen zoologischen Gesellschaft* **31**, 29–79.

Verschuren, D., and Eggermont, H. in press. Quaternary paleoecology of aquatic Diptera in tropical and southern hemisphere regions, with special reference to the Chironomidae. *Quaternary Science Reviews.*

Walker, I. (1995). Chironomids as indicators of past environmental change. In *The Chironomidae: Biology and ecology of non-biting midges* (P. D. Armitage, P. S. Cranston and L. C. V. Pinder, Eds.), pp. 405–422. Chapman & Hall, London.

Walker, I. R. (1987). Chironomidae (Diptera) in paleoecology. *Quaternary Science Reviews* **6**, 29–40.

Walker, I. R. (1993). Paleolimnological biomonitoring using freshwater benthic macroinvertebrates. In *Freshwater Biomonitoring and Benthic Macroinvertebrates* (D. M. Rosenberg and V. H. Resh, Eds.), pp. 306–343. Chapman & Hall, New York.

Walker, I. R. (2001). Midges: Chironomidae and related Diptera. In *Tracking Environmental Change Using Lake Sediments,* vol. 4: Zoological Indicators. *Developments in Paleoenvironmental Research* (J. P. Smol, H. J. B. Birks and W. M. Last, Eds.), pp. 43–66. Kluwer, Dordrecht.

Walker, I. R., and Cwynar, L. C. in press. Midges and palaeotemperature reconstruction – the North American experience. *Quaternary Science Reviews.*

Whiteside, M. C. (1983). The mythical concept of eutrophication. *Hydrobiologia* **103**, 107–111.

Wiederholm, T. (1980). Use of benthos in lake monitoring. *Journal of the Water Pollution Control Federation* **52**, 537–547.

CHIRONOMID RECORDS

Contents

Africa

H Eggermont and D Verschuren, Ghent University, Ghent, Belgium

© 2007 Elsevier B.V. All rights reserved.

Introduction

Building on the long-standing tradition to exploit chironomid larvas (Insecta: Diptera: Chironomidae) as biological indicators in contemporary water-quality monitoring of lakes and rivers (Brinkhurst, 1974; Rosenberg, 1993), paleolimnologists are making increasingly profitable use of their fossil remains preserved in lake sediments to reconstruct long-term environmental change affecting lakes and adjacent landscapes (Walker, 1995). In north-temperate Europe and North America, chironomid paleoecology (and particularly chironomid paleoclimatology) is now a mature research discipline (Walker, 2001), involving statistically robust inference models for quantitative reconstruction of past changes in temperature, salinity, lake trophic status, and oxygen regime (*see* Chironomid Overview, Postglacial Europe). In tropical regions including Africa, the growth of chironomid paleoecology as a rigorous, quantitative research discipline has been decidedly slower. The principal reasons for this geographical imbalance are incomplete taxonomic knowledge of tropical Chironomidae (and in particular their larval stage), the scarcity of autecological information that would confer bioindicator value to local taxa, and the logistic (sometimes also political) difficulties hampering the comprehensive ecological lake surveying needed to obtain this information. Although continuous high-resolution chironomid records spanning the entire postglacial period in Africa are thus far lacking, African chironomid paleoecology has made substantial progress over the past decade (Verschuren and Eggermont, 2006). In this article we provide an overview of its present status, dealing successively with chironomid records of past environmental change, advances in methodology, and taxonomy.

Chironomid Records of Past Environmental Change in Africa

Fossil Chironomidae have so far contributed to three late Quaternary climate reconstructions from tropical Africa. Two studies (Mees *et al.*, 1991; Verschuren *et al.*, 2000a) primarily exploited the varying salinity tolerance of individual species as indicator of past changes in the water balance of closed-basin lakes in, respectively, Sudan and Kenya. The third study (Verschuren *et al.*, 1996) used the relative abundance of (semi)terrestrial, lake, and stream taxa as indicator of past hydrological change in a high-altitude peat bog in Burundi. Mees *et al.* (1991) reconstructed Holocene lake-level and salinity changes at Malha crater lake in Sudan (15° N 26° E), one of the very few permanent waters in the eastern Sahara desert. The lithology and mineralogy of its sediment sequence indicated that six successive lake phases had occurred over the past 8,300 ^{14}C yr, separated by severe lowstands or complete desiccation (**Fig. 1**). The timing of these hydrological changes is poorly constrained, due to lack of high-quality radiocarbon dating targets, but their pattern is consistent with the regional history of millennial-scale fluctuations in rainfall and drought (Hoelzmann *et al.*, 2004). In the early to mid-Holocene, the lake was deep and meromictic, with surface waters evolving from fresh (as indicated by the presence of *Dicrotendipes fusconotatus*, *D. sudanicus*, and *Nilodorum brevibucca*) to mesosaline (indicated by presence of the salt-loving *Kiefferulus disparilis*). The authors interpreted this evolution as a reflection of decreasing rainfall and shrinking groundwater input, which both reduced new freshwater inflows and caused wind-driven mixing to bring increasing amounts of deep salty water to the surface. During the last ~4 kyr Malha crater lake was most often shallow and holomictic, and poly- to hypersaline. Chironomidae became locally extinct, and were replaced only by scattered occurrences of other salt-loving midges, such as the brine fly *Ephydra* (Ephydridae), the biting midge cf. *Culicoides* (Ceratopogonidae), and an unidentified

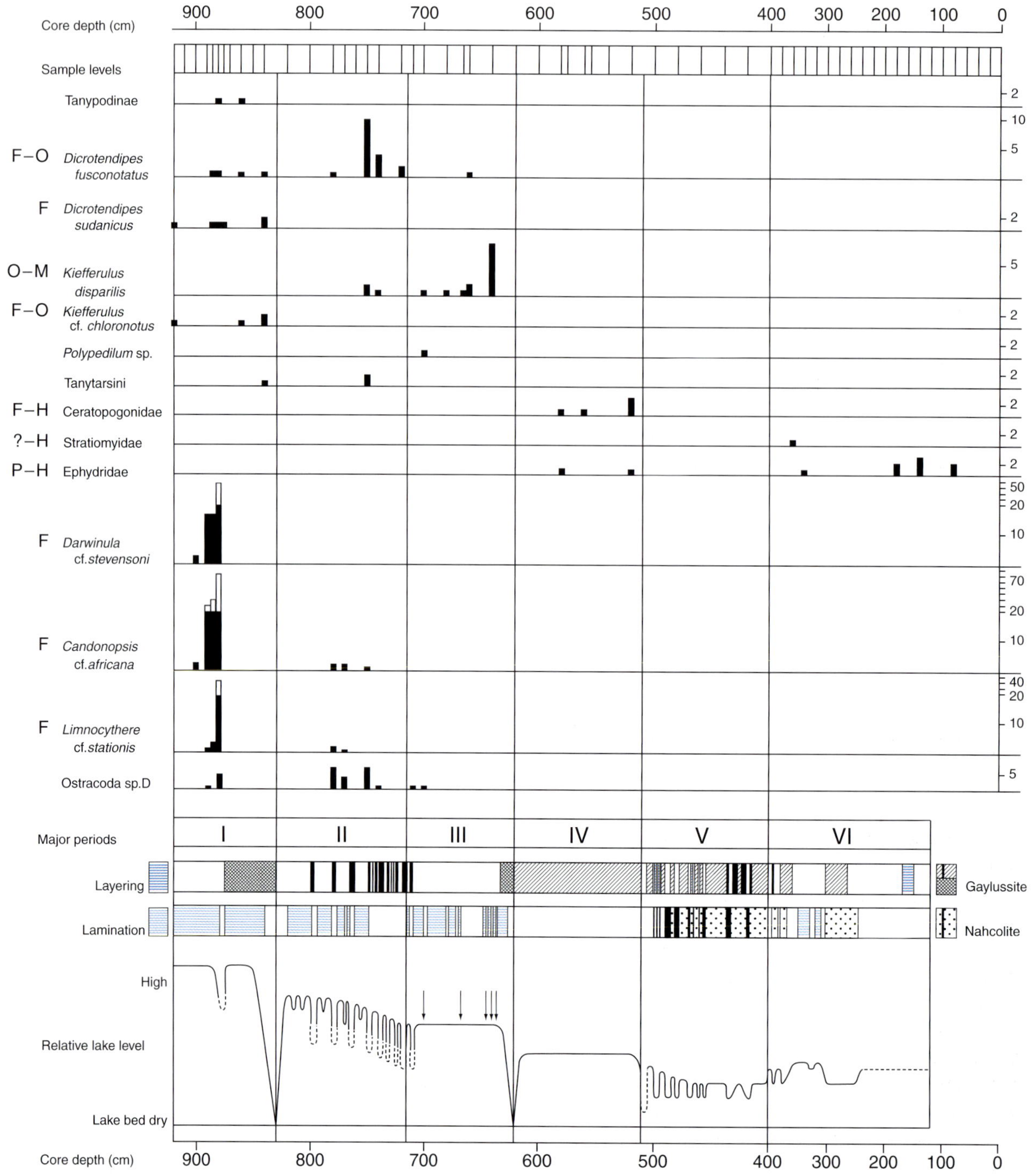

Figure 1 Distribution of invertebrate remains in the sediment core from Malha crater lake, with schematic sediment stratigraphy and evolution of lake level since 8,300 radiocarbon years ago. Core depths less than 400 cm scaled down to 1/2. Frequency of remains indicated as number recovered; frequencies above 20 scaled down to 1/5 (open bars). Codes preceding indicator species stand for freshwater (F), oligosaline (O), mesosaline (M), polysaline (P), and hypersaline (H). The reconstructed lake levels are relative; lake-level changes are indicated by stratigraphy and estimates of salinity in littoral and surface waters. Arrows indicate episodic events of turbulence affecting the profundal lake bottom. From Mees F, Verschuren D, Nijs R, and Dumont HJ (1991) Holocene evolution of the crater lake at Malha, northwest Sudan. *Journal of Paleolimnology* 5: 227–253.

soldier fly larva (Stratiomyidae). Clearly, the late Holocene aquatic environment in Malha Crater approached the limit of what African aquatic invertebrate communities can tolerate.

Verschuren *et al.* (1996) used fossil Chironomidae and water fleas (Cladocera) extracted from a peat deposit at Kashiru, Burundi (2,100 m a.s.l., 3° S 30° E) to describe the general late Quaternary evolution (last ~40 kyr) of this cold aquatic habitat in the mountains northeast of Lake Tanganyika. Peat sequences are the traditional source of African pollen records, but this is the only study so far which exploited them for chironomid paleoecology. Chironomid genera that are widely distributed in lowland African lakes (*Chironomus*, *Cladotanytarsus*, *Dicrotendipes*, *Kiefferulus*, *Nilodorum*, *Polypedilum*, *Procladius*) were absent or scarce in the fossil assemblages found at this site; consequently, the ecological literature on African lakes was of little use to their interpretation. Instead, the authors made inferences on past habitat conditions mostly by transferring Holarctic ecological typology (Merritt and Cummins, 1984), assuming that cold-adapted African representatives of Holarctic genera would have similar habitat preferences. Chironomid fossils were markedly concentrated in core sections consisting of decomposed or only slightly fibrous peat, which presumably had been deposited in peat hollows. Strongly fibrous peat (from hummocks) and sandy horizons (indicating an episode when the local environment too dynamic to allow peat accumulation) were typically devoid of fossils. Overall the fossil chironomid fauna at Kashiru indicated long-term persistence of a cold montane peat bog with a fragmented water surface through which small streams occasionally flowed. Stratigraphic trends in the species assemblages support pollen evidence for drier conditions in equatorial East Africa during the Last Glacial Maximum (ca. 25–15 ka) and a fairly humid climate before that time (Bonnefille *et al.*, 1990). *Chironomus formosipennis*, another *Chironomus* sp., and *Harnischia*, which all require mud bottoms and open standing water, are indeed limited to the period prior to 25 ka. Also concentrated in this section are the remains of *Chaetocladius*, which typically occur in swampy lakes, and *Rheocricotopus*, indicating some streamflow through the fen. Lowering of the water table during the Last Glacial Maximum is indicated by prevalence of taxa resembling Holarctic genera with semiterrestrial larvas, such as cf. *Pseudosmittia* and *Bryophaenocladius*. The bog surface did not desiccate completely however, as attested by the persistence of benthic water fleas (chydorid Cladocera) and a minority component of aquatic Chironomidae (*Polypedilum*, *Tanytarsus*, Tanypodinae indet.). The onset of the postglacial Pluvial period around 15 ka resulted in the disappearance of *Bryophaenocladius* and cf. *Pseudosmittia*, and a return to higher chironomid faunal diversity.

The best-resolved African climate-proxy record based in part on fossil chironomid data is that of Crescent Island Crater (1° S 36° E), a satellite basin of Lake Naivasha in the central Kenya Rift Valley (Kenya). This fossil record documents water-balance variations over the past 1.1 kyr with decade-scale time resolution (Verschuren *et al.*, 2000a). For this study, an exploratory salinity-inference model was developed (Verschuren *et al.*, 2004), see also below), using as calibration data set a compilation of published distribution data on 20 common chironomid species in 32 African surface waters (Verschuren, 1997). The resulting salinity reconstruction (**Fig. 2**) is fully consistent with a parallel reconstruction based on fossil diatom assemblages, and with a lake-level reconstruction based on changes in the lithological characteristics of the sediment sequence. The combined data show that equatorial East Africa was significantly drier than today during the Medieval Warm Period (~AD 1000–1270) and that fairly wet conditions during the Little Ice Age (~AD 1270–1850) were interrupted around AD 1380–1420, 1560–1620, and 1760–1840 by three decade-scale episodes of severe aridity. From ~AD 1670–1770, that is, the period which encompasses the Maunder Minimum in solar radiation, Crescent Island Crater stood higher than at any other time in the past millennium. The local chironomid fauna of that time was dominated by *Dicrotendipes septemmaculatus*, in line with the positive relationship between this species and lake depth in the historical (post-1880) portion of the sediment record. Presumably, this reflects expansion of its preferred habitat of submerged macrophyte beds during lake highstands. Superimposed on pronounced lake-level fluctuation during the past 1.1 kyr is a long-term freshening trend, beginning with the transition to a more frequently positive water balance (rainfall and river inflow exceeding evaporation and loss to groundwater) after AD 1270. In the chironomid data, this is reflected in the punctuated disappearance of salt-loving species (*Kiefferulus disparilis*, *Microchironomus deribae*) after each new highstand episode, to the benefit of a diverse freshwater assemblage including some salt-intolerant taxa (e.g., *Psectrocladius viridescens*).

The observation that recurrent drought is a natural characteristic of Kenya's semiarid Rift Valley climate also followed from a 200-yr paleoenvironmental reconstruction at Lake Oloidien, another satellite basin of Lake Naivasha (Verschuren *et al.*, 1999b). This study was intended to provide a long-term historical perspective on the problem of rising salinity during the late 1980s and early 1990s, and the possible role of rapidly increasing water extractions by the sprawling horticulture industry around Lake Naivasha. Fossil evidence for the historical

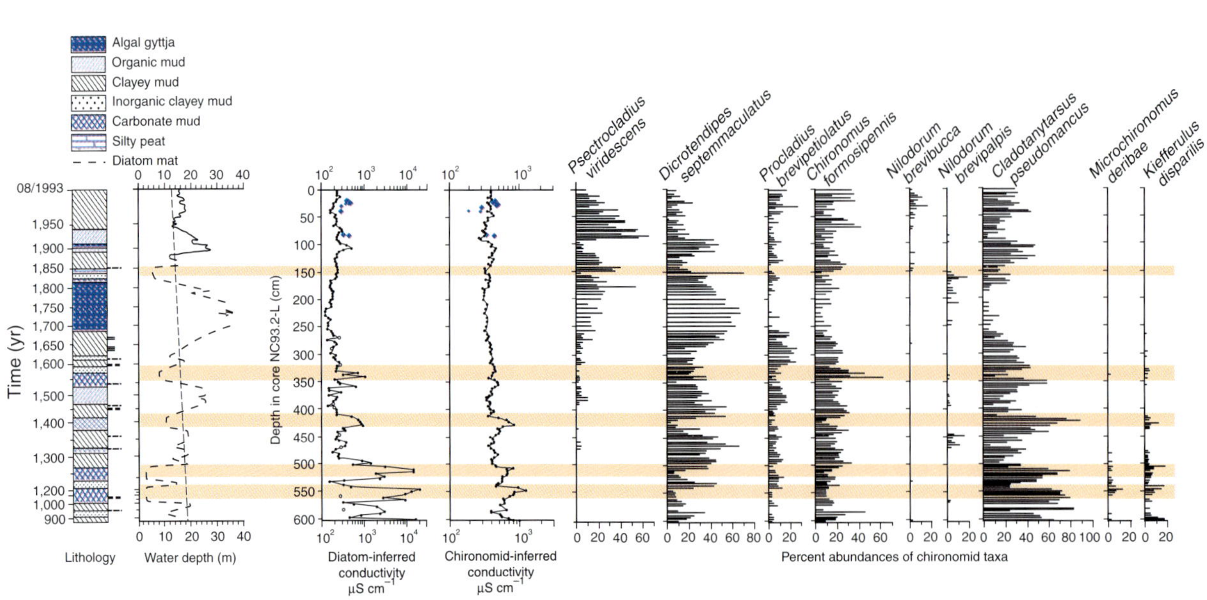

Figure 2 Sedimentological and biological evidence for past lake-depth and salinity fluctuations in the Crescent Island Crater basin of Lake Naivasha, with episodes of pronounced drought highlighted in orange. From left to right: lithology-inferred lake-level change over the past ~1,100 yr (solid and stippled line) relative to the elevation of the sill separating Crescent Island Crater from Naivasha (broken line); diatom-inferred conductivity reconstruction using the inference model of Gasse *et al.* (1995); chironomid-inferred conductivity reconstruction using the hybrid model of Verschuren *et al.* (2004); and stratigraphic distribution of selected fossil midge taxa with species sorted according to increasing conductivity optima. Modified from Verschuren D, Cumming BF, and Laird KR (2004) Quantitative reconstruction of past salinity variations in African lakes using fossil midges (Diptera: Chironomidae): Assessment of inference models in space and time. *Canadian Journal of Fisheries and Aquatic Sciences* 61: 986–998.

alternation of freshwater and salt-tolerant aquatic biota (including Chironomidae) throughout the nineteenth and twentieth centuries revealed that soda-lake conditions are a natural recurrent phenomenon in Lake Oloidien, due to major natural rainfall variability combined with the dependence of Lake Oloidien on episodes of confluence with Lake Naivasha to remain fresh. The recent salinity rise was temporarily halted in early 1998, thanks to a 2.6-m lake-level rise following high rainfall driven by El Niño. However, as commented upon by the authors, unchecked diversion of river inflows for cropland irrigation and direct water extractions for the horticulture industry will increasingly limit the ability of such periodic high-rainfall events to restore confluence with Lake Naivasha, and eventually result in Lake Oloidien becoming permanently saline.

With surface areas between 0.04 and 5.1 km^2, the lakes in the above-mentioned studies are sufficiently small and shallow for chironomid paleoecologists to assume a more or less complete spatial integration of chironomid death assemblages across the lake before permanent burial. Under those conditions, if no differential preservation among taxa occurs (*see* Chironomid Overview), relative abundances of chironomid species and ecological groups present in the fossil assemblages will reflect the fractional distribution of various types of bottom habitat at the time of deposition. These conditions are clearly not met in the African great lakes, such as Lake Victoria and Lake Tanganyika. These lake basins are so extensive that fossil assemblages must be taken to only represent habitat conditions close to their area of deposition at the time of their deposition, contrasting with other (distant) locations within the same lake. In these conditions, surface-sediment death assemblages deposited recently at various other locations and microhabitats within the lake can be used as modern reference data for comparison with fossil assemblages deposited at the master core location during certain times in the past. Verschuren *et al.* (2002) and Eggermont and Verschuren (2003) exploited this principle in their use of fossil Chironomidae as tracers of anthropogenic impact on the Lake Victoria and Lake Tanganyika ecosystems.

The study by Verschuren *et al.* (2002) addressed the question whether severe eutrophication and apparent degradation of Lake Victoria's ecological functions in recent decades resulted primarily from food-web alterations following the widely publicized population explosion of Nile perch (*Lates niloticus*) in the early 1980s, or alternatively from increased nutrient loading resulting from large-scale agricultural conversion of its drainage basin over the past half century. Fossil chironomid assemblages extracted from a mid-lake sediment core (**Fig. 3**) traced the evolution of deep-water oxygenation through time. To strengthen their inferences, the authors first calibrated the relationship

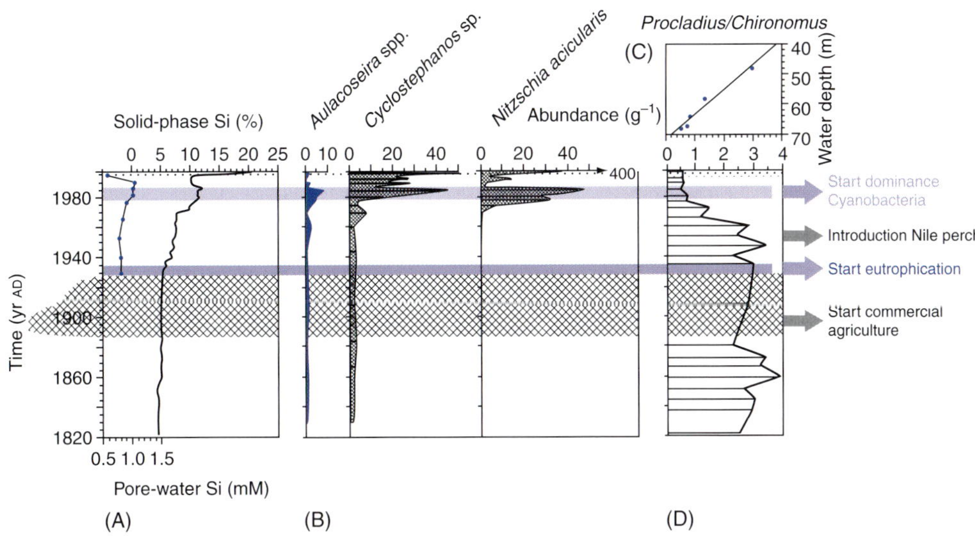

Figure 3 Evidence for lake-wide ecosystem change in Lake Victoria, East Africa. (A) Stratigraphy of biogenic silica (mostly fossil diatoms) and dissolved silica in pore waters; the cross-hatched zone reflects the inferred core hiatus corresponding to the unconformity in unsupported ^{210}Pb activity at 25 cm core depth. (B) Fossil diatom stratigraphy; peak concentrations in flocculent surface muds represent diatoms subject to further dissolution before permanent burial. (C) Abundance ratio of the anoxia-intolerant *Procladius brevipetiolatus* vs. the anoxia-tolerant *Chironomus* nr. *alluaudi* in mid-lake fossil assemblages compared with its modern-day gradient in relation to water depth as determined by the local duration of seasonal bottom anoxia. From Verschuren D, Johnson TC, Kling HJ, *et al.* (2002) History and timing of human impact on Lake Victoria, East Africa. *Proceedings of the Royal Society of London B* 269: 289–294.

between the fossil abundance ratio of *Procladius brevipetiolatus* to *Chironomus* nr. *alluaudi* (originally identified as *C. imicola*) and lake depth, or more specifically the duration of seasonal bottom anoxia, which increases with depth. They did so by analyzing recently deposited chironomid death assemblages in a transect of surface-sediment samples between 48 and 68 m water depth. The core analyses then revealed that the hypolimnion of Lake Victoria was adequately oxygenated year-round until about 1960. Seasonal depletion of deep-water oxygen started shortly after that time and expanded strongly in the 1970s, hence possibly contributing to the 1980s collapse of indigenous fish stocks by eliminating suitable habitat for deep-water cichlids. Concentration data on diatom silica in the same core profile (**Fig. 3**) further indicated that the primary (algal) productivity of Lake Victoria had already started to increase from the 1930s onward, that is, well before the introduction of Nile perch, and in proportion to historical agricultural production in the Lake Victoria region. The combined data thus suggested that the ongoing eutrophication of Lake Victoria, characterized by water quality loss and blooms of noxious algae, is mainly due to the excess nutrient loading resulting from anthropogenic landscape disturbance.

The study by Eggermont and Verschuren (2003) used fossil chironomid remains as model organisms to investigate the impact of sediment pollution (due to agricultural soil erosion) on the benthic invertebrate fauna of Lake Tanganyika. These authors did not look at stratigraphic changes in chironomid faunal composition through time, but compared the species composition of recently buried death assemblages in bottom habitat adjacent to pristine and disturbed watersheds. The analysis involved 36 surface-sediment samples between 53 and 189 m water depth, each representing an area of near-shore sublittoral habitat adjacent to different size classes ($<10\,km^2$, <30–$50\,km^2$, <50–$162\,km^2$; $>162\,km^2$) of either pristine or disturbed drainages. This data set was explored with uni- and multivariate techniques to evaluate relationships between species distribution and selected environmental variables. In this analysis, water depth was used as a proxy for the oxygen regime at the sampling site, and sediment composition (organic matter and carbonate content, fractions of coarse plant debris and sand) as a proxy for local substrate and food quality. The distance to shore (distance in m), the surface area of the adjacent drainage, and its land-use regime were used as proxies for the susceptibility of each particular site to excess sedimentation. The authors found that in each of the three smaller drainage size classes (of the largest size class, no pristine example remained in the study area), catchment disturbance had resulted in similar shifts in chironomid faunal composition (**Fig. 4**), which apparently reflected an expansion of coarse,

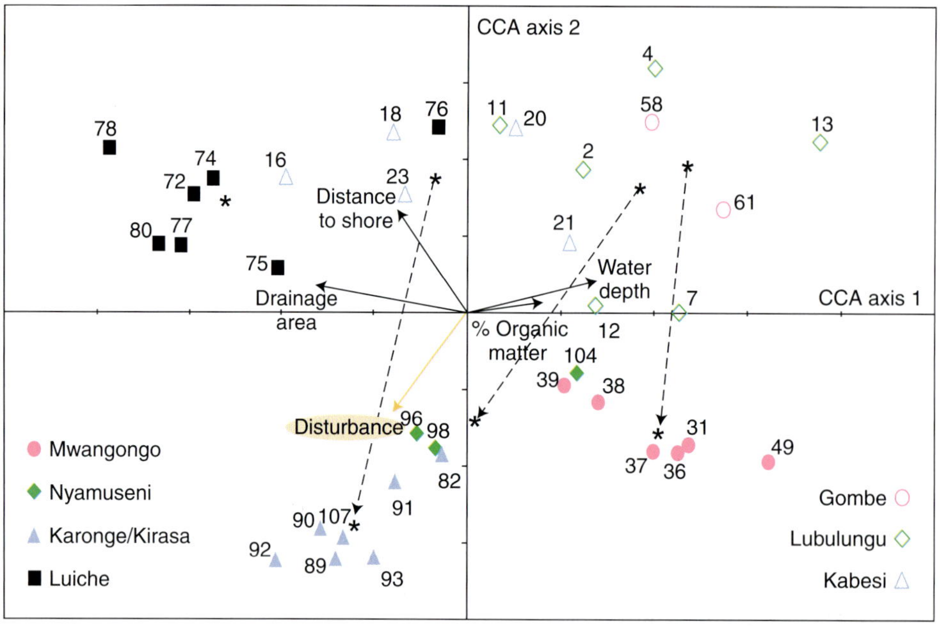

Figure 4 CCA ordination of surface-sediment chironomid faunal assemblages in 38 sublittoral bottom environments of northeastern Lake Tanganyika, East Africa, adjacent to pristine (open symbols) and disturbed (closed symbols) drainages ranging from very small (<10 km^2: circles) to very large (>162 km^2: squares). Excess clastic sedimentation due to human disturbance of adjacent stream drainages resulted in directional shifts in faunal composition (arrows), indicating habitat change equivalent to reductions in water depth and distance to shore. From Eggermont H and Verschuren D (2003) Impact of soil erosion in disturbed tributary drainages on the benthic invertebrate fauna of Lake Tanganyika, East Africa. *Biological Conservation* 113: 99–109.

low-organic substrates into deeper water, at the expense of the soft organic muds which normally characterize the deep-water environment. Because the area of habitable lake bottom in Lake Tanganyika is bounded by the permanent oxycline at ~100 m depth, these results indicated that sediment pollution is already negatively affecting oxygenated soft-bottom environments in Lake Tanganyika, and that intensifying soil erosion in its tributary drainages may eventually threaten the survival of endemic fish and mollusk species restricted to those same environments. This study established the value of chironomid death assemblages as biological indicators of ecosystem health on the scale at which human impact occurs, which is that of entire drainage basins.

Methodological Advances in African Chironomid Paleoecology

Qualitative Inferences Using the Indicator-Species Approach

Comprehensive lake surveying and collecting efforts in tropical Africa often face considerable logistical problems, sometimes compounded by political obstacles. Consequently, until recently no large reference data sets existed on the distribution of chironomid species across modern climate and ecotonal gradients (i.e., uniform collections of ecological reference data). This has obviously limited paleoecologists' opportunities for chironomid-based quantitative reconstruction of past environmental change. In addition, existing autecological information on local species is often anecdotal, and in contrast with other biological proxy indicators, such as diatoms, it can be used for paleoecological inferences only when both adult and larval morphology are known. As a result, workers have often been content to interpret stratigraphic change in fossil chironomid assemblages on the basis of traditional ecological typology and genus-level autecological data transferred from the Holarctic region. Prominent in this context is *Chironomus*, a genus of which most member species in both tropical and Holarctic regions are tolerant of poor oxygen conditions (as is often associated with anthropogenic nutrient enrichment and high primary productivity, but occurs naturally in many stratified tropical lakes). Another example is the subfamily Diamesinae, the members of which are largely restricted to cold-water environments, and therefore characteristic indicators of cold climatic episodes during the last deglaciation in north-temperate Europe (*see* Chironomid Overview, Postglacial Europe) and North America. Fossil Diamesinae have also been reported from south-temperate New Zealand, but not yet from Africa. However, the occurrence of at least three *Diamesa* species in Alpine lakes and streams on Africa's highest mountains suggests that they might become valuable tracers of tropical glacier advance and retreat.

It follows that, chironomid-based paleoclimate reconstruction in Africa has, to date, mostly employed the indicator-species approach. Studies on past salinity change in lowland lakes (Mees *et al.*, 1991; Verschuren *et al.*, 1999b) took advantage of a fair amount of pre-existing autecological information on the local species (Verschuren, 1997), while the study dealing with montane peat-bog fauna (Verschuren *et al.*, 1996) relied almost exclusively on ecological data from related Holarctic taxa.

Testing the Response of African Chironomidae to Climate Change

Verschuren *et al.* (1999a, 2000b) investigated the operative limits of African Chironomidae as biological climate proxies by tracing the signatures of their response to changes in lake level, water-column circulation, salinity, and the local distribution of various types of bottom substrate and microhabitat in dated sediment records linked to long time series of historical data. These studies employed the sediment records of the various satellite basins of Lake Naivasha in Kenya, which, because of their hydrological connections, have all experienced the same 120-yr-history of documented lake-level change. Instrumentally recorded lake depth was linked to the sediment record by ^{210}Pb dating, whereas past mixing regime, salinity, and benthic habitat conditions were reconstructed using lithological indicators, fossil diatoms, and plant macrofossils, respectively. In both studies, chironomid species-assemblage changes through time are entirely consistent with the available autecological data and known salinity-tolerance ranges. Salinity change and the temporal occurrence of littoral papyrus swamp together explained about half of the historical variation in benthic community composition, and their effects appear to be largely independent from each other. Most of all these studies illustrate the complexity of biological response to short-term (up to decade scale) water-balance fluctuations. More specifically, they indicate that the direct physiological problem of osmotic stress caused by salinity change affects African zoobenthos communities only at the extremes of the salinity gradient, and that a significant portion of the observed correlation between salinity and invertebrate community structure is in fact an indirect effect of diffuse covariance between salinity, lake level, and the distribution of various types of benthic habitat.

Development of Quantitative Inference Models

Accurate, quantitative reconstruction of lake and climate history based on fossil chironomid assemblages requires calibration of the relationship between chironomid species distribution and modern environmental conditions in an appropriate number of physically, chemically, and/or ecologically diverse lakes. Initially, lacking a traditional surface-sediment calibration data set, in which this relationship is determined using not the modern living communities but recent death assemblages of larval remains extracted from surface sediments, the first exploratory model for quantitative salinity inference was developed using published distribution data based on regional surveys of the living fauna (adults and/or larvas; Verschuren *et al.* (2004)). Using the computionally simple and straightforward method of weighted averaging, the authors produced inference models with a statistical performance comparable to that of weighted-averaging salinity-inference models based on African diatoms (Gasse *et al.*, 1995). For reconstruction of lake-level change in the Kenya Rift Valley, Verschuren *et al.* (2000a) then performed a hybrid procedure in which this inference model, calibrated with presence/absence distributional records, was applied to abundance-weighted fossil data. This ensured that not only the new appearance or complete disappearance of taxa but also changes in their relative abundance would generate a proxy climate signal. Because of the rather small number of calibration data (20 taxa recovered from 32 sites, or 147 species-conductivity combinations only), the model's statistical power is somewhat limited. Regression of inferred over observed conductivity values (in $\mu S\ cm^{-1}$) in the 32 reference lakes also suggested a threshold response of African chironomid communities at the transition between freshwater and salt-lake environments ($\sim$3,000 $\mu S\ cm^{-1}$), as opposed to the more continuous species replacement along the salinity gradient seen in diatoms (Gasse *et al.*, 1995). To evaluate whether this does indeed reflect ecological reality, or is an artifact of the exploratory inference model due to the use of presence–absence calibration data that perhaps do not adequately cover the regional salinity gradient, Eggermont *et al.* (in press) constructed a true paleoenvironmental calibration data set based on percent-abundance data of 65 chironomid morphotypes recovered from the surface sediments of 67 lakes in equatorial East Africa (Kenya, Uganda, Ethiopia, and Tanzania), and then applied various numerical techniques to this calibration data set to develop a series of salinity-inference models. These new models are evidently more robust statistically than the exploratory model developed previously, but the observed threshold response of African chironomid communities to increasing salinity remained manifestly present. To improve model sensitivity, the authors experimented with modified calibration data sets in which the regional salinity gradient was extended further (by adding dilute high-elevation lakes from Mt. Kenya and the Ruwenzori Mts.) or in which fresh and saline lakes were represented in equal numbers (to shift upward the salinity optimum of salt-tolerant taxa). However, neither experiment appreciably improved the inference models' statistical performance.

Validation of the new African chironomid-salinity inference models in the time domain, in part based on direct comparison of chironomid-inferred salinity trends with those based on fossil diatom assemblages, revealed significant variation among models in reconstructed salinity trends through time, due to their different sensitivity to the presence or relative abundance of certain key chironomid taxa. Models based on simple weighted averaging and the modern analog technique are inferior to more sophisticated models (such as maximum likelihood models) on the basis of error statistics. However, these simpler models produced the most trustworthy salinity reconstructions. The authors suggested that this occurred exactly because their step change of inferred conductivity near $3{,}000\,\mu S\,cm^{-1}$ mirrors the rapid climate-driven transitions between fresh and saline lake phases which characterize these shallow closed-basin lakes. These lakes persist in a semiarid tropical climate regime, thanks to substantial river or groundwater inflows. In a broader context, Eggermont *et al.* (in press) stressed that selection of the most appropriate transfer function for chironomid-based environmental reconstruction (or based on any other biological proxy indicator) should not only compare their statistical performance, but carefully consider whether models produce ecologically meaningful results with high signal content.

With appropriate documentation of the varied environmental controls on the distribution of African chironomid species, their fossil counterparts may also be developed as quantitative tracers for environmental variables other than salinity. Analysis of surface-sediment samples from high-mountain lakes revealed a clear gradient in chironomid species distribution which occurs in alpine lakes situated between 2,500 and 4,500 m altitude on Mt. Kenya and in the Ruwenzori Mts. of western Uganda (Eggermont and Verschuren). This points to the potential of African chironomids as a tropical paleothermometer, their main function in European and North American paleoclimate studies (*see* Postglacial Europe). The above-mentioned modest correlation between chironomid species composition and salinity in the broad range of African freshwaters (less than $3{,}000\,\mu S\,cm^{-1}$) allows for strong environmental control by other habitat characteristics, such as the interrelated variables of lake depth, mixing regime, and bottom oxygenation.

Taxonomy and Biogeography of African Chironomidae

Progress in the discrimination of subfossil morphotypes of African chironomid larvas (e.g., Eggermont and Verschuren (2004a–b; Eggermont *et al.*, 2005); **Fig. 5**) continues to increase the understanding of the taxonomic diversity of fossil assemblages. Judging from the comparison with morphological diversity within and between genera in the better-known Holarctic fauna, most larval morphotypes currently distinguished in African fossil assemblages must be functionally equivalent to biological species, or groups of closely related species. Maintaining a strong link to official chironomid taxonomy and systematics based on the morphology of adult midges ensures compatibility between ecological data derived from living and fossil assemblages, and justifies the use of chironomid fossils as biological proxy indicators. Given this level of taxonomic resolution, the African fossil material can also be exploited to address biogeographical issues about the living fauna, such as its patterns of alpha- and beta-diversity (taxonomic richness of lakes and regions: e.g., Eggermont *et al.* (2005)). It can also be exploited to determine the relative biological uniqueness of particular aquatic ecosystems in the context of biodiversity conservation. Most of all, in regions such as Africa, with only fragmentary taxonomic and ecological knowledge of individual chironomid species, taxon-discrimination criteria based on fossil material can also improve species discrimination in the living material. If this stimulates new applied ecological research on the benthic invertebrate communities of Africa's modern aquatic ecosystems, the knowledge gained may eventually feed back into improved ecological interpretation of fossil chironomid stratigraphy for the purpose of environmental reconstruction.

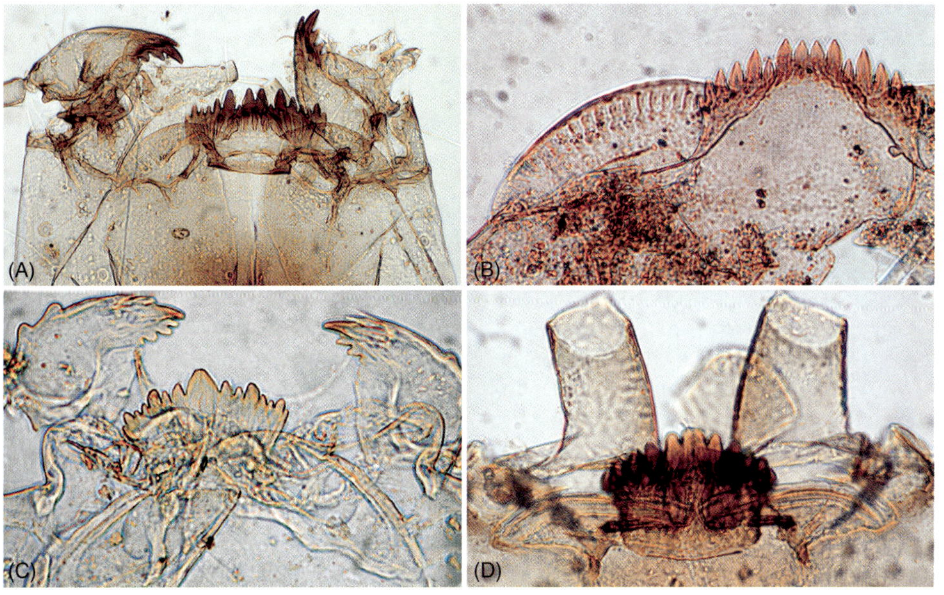

Figure 5 Subfossil remains of larval Chironomidae recovered from African lake sediments. (A) *Polypedilum* near *deletum* Goetghebuer; (B) *Conochironomus* type Gagouba; (C) Chironomini indet. type Tanganyika sp. 1.; (D) Tanytarsini indet. type Large Hall Tarn. Photos by H. Eggermont.

Glossary

Alpha-Diversity It refers to the diversity within a particular area or ecosystem, and is usually measured as the number of species (i.e., species richness) in that ecosystem.

Anoxia The absence of oxygen necessary for sustaining most life.

Autoecology The ecology of an individual species in its natural environment, without consideration of its interactions with other species.

Beta-Diversity The degree of species difference between habitats or ecosystems within a (biogeographical) region, usually measured as the amount of species turnover between sites.

Benthos Collectively, all organisms living on or in the bottom of a lake, stream, or other water body.

Calibration Data Set A set of species abundance data collected from modern sites along with relevant environmental data, often related to one another using a series of quantitative inference models. Calibration data sets form the basis for quantitative paleoecological reconstruction when used in conjunction with fossil assemblages. Used to generate inference models. See inference model.

Canonical Correspondence Analysis (CCA) A constrained ordination technique based on correspondence analysis, where the ordination axes are constrained to be linear combinations of the supplied environmental (predictor) variables.

Conductivity Here a measure of the ability of water to conduct an electric current due to the presence of dissolved solids, normally expressed in microsiemens cm^{-1} ($\mu S\ cm^{-1}$) or micromhos cm^{-1} ($\mu mhos\ cm^{-1}$).

Correspondence Analysis (CA) An ordination method that simultaneously ordinates species (variables) and sites (samples) with the aim to maximize correlation between sample and variable scores. Widely used in ecology because it assumes a unimodal response of species population abundance to underlying environmental gradients.

Direct Ordination An ordination in which the important gradients are known and measured. See ordination.

Ecotone The boundary or transition area between two or more habitats or communities.

Halobiont Living or growing in a salty environment.

Hypolimnion Lower water column in thermally-stratified lakes.

Holomictic Used for lakes which have at least one episode of complete circulation of the water column per year.

Hypersaline Having a total dissolved solid content (salinity) greater than about 40 ppt (parts per thousand) or $60{,}000\ \mu S\ cm^{-1}$.

Indirect Ordination An ordination in which gradients are unknown *a priori*, and are inferred from species composition data.

Indicator Species A species whose presence is a marker for a specific habitat condition or environmental value.

Inference Model A mathematical function that describes the relationship between biological species and environmental variables, and allows past values of an environmental variable to be inferred from the composition of a fossil assemblage.

Littoral Zone The zone in a lake or pond extending from the shore to the greatest depth at which rooted plants occur.

Maximum Likelihood A method of environmental reconstruction based on the idea that the relation between the abundance of a taxon and the environmental variable of interest can be modeled by an ecological response curve consisting of a systematic and a random (error) component. Together, the curves of all taxa determine what taxa composition is expected at a given conductivity value. This model of responses and their error structure is used to calculate the probability that a particular value would occur with a given assemblage. The value that gives the highest probability is the maximum likelihood estimate.

Meromictic Used for lakes which do not undergo complete mixing of the water column at least once each year.

Mesosaline Degree of salinity in brackish waters; water having a total dissolved solid content (salinity) of 5.0–18.0 ppt (parts per thousand) or 8,000–30,000 $\mu S\,cm^{-1}$.

Modern Analog Technique A method of environmental reconstruction based on matching fossil samples to modern surface sediment assemblages of similar taxonomic composition.

Morphotype In this context: group of fossil specimens which are considered to belong to the same species, group of species, or a higher unit based on shared morphological characters alone.

Oligosaline Having a total dissolved solid content (salinity) between 0.5 and 5 ppt (parts per thousand)

Ordination The collective term for multivariate techniques that arrange sites along axes on the basis of data on species composition; the usual objective of ordination is to help generate hypotheses about the relationship between the species composition at a site and the underlying environmental gradients.

Osmotic Stress Physiological stress caused when the salinity of an animal's environment changes drastically.

Oxycline A horizontal boundary layer in the water column, at which dissolved oxygen content changes sharply with depth.

Polysaline Water having a salinity of 18.0–30.0 ppt (parts per thousand) or 30,000–45,000 $\mu S\,cm^{-1}$.

Proxy Data In this context, data obtained from natural environmental archives and used as surrogates (proxies) for the past condition of another variable, such as past temperature or precipitation, that cannot be measured directly.

Profundal Pertaining to deep-water bottom habitats in lakes.

Weighted Averaging In environmental reconstruction, a numerical technique to derive inference models by estimating species optima from the calibration data set (i.e., regression), and then using these optima to estimate the past value of an environmental variable from a fossil assemblage (i.e., calibration). The fundamental assumption of the technique is that the weighted average of a species along a gradient reflects the optimal habitat conditions for this species. The species optima and tolerances are estimated from the weighted abundances and weighted standard deviations, respectively.

Weighted-Averaging Partial Least Squares An extension of weighted averaging and calibration where two or more components, that utilize the residual structure in the modern biological data, are used to improve the predictive power of the inference model.

Zoobenthos Animals that live in or on the bottom of a lake, stream, or other water body.

See also: **Chironomid Overview. Chironomid Records:** Postglacial Europe.

References

Bonnefille, R., Roeland, J. C., and Guiot, J. (1990). Temperature and rainfall estimates for the past 40,000 years in equatorial Africa. *Nature* **346**, 347–349.

Brinkhurst, R. O. (Ed.) (1974). *The benthos of lakes,* p. 190. The Macmillan Press Ltd., London.

Eggermont, H., and Verschuren, D. (2003). Impact of soil erosion in disturbed tributary drainages on the benthic invertebrate fauna of Lake Tanganyika, East Africa. *Biological Conservation* **113**, 99–109.

Eggermont, H., and Verschuren, D. (2004a). Subfossil Chironomidae from East Africa. 1: Tanypodinae and Orthocladiinae. *Journal of Paleolimnology* **32**, 383–412.

Eggermont, H., and Verschuren, D. (2004b). Subfossil Chironomidae from East Africa. 2: Chironomini and Tanytarsini. *Journal of Paleolimnology* **32**, 413–455.

Eggermont, H., Heiri, O., and Verschuren, D. (2006). Fossil Chironomidae (Insecta: Diptera) as quantitative indicators of past salinity in African lakes. *Quaternary Science Reviews* **25**, 1966–1994.

Eggermont, H., Verschuren, D., and Dumont, H. (2005). Taxonomic diversity and biogeography of Chironomidae (Insecta: Diptera) in lakes of tropical west Africa, using subfossil remains extracted from surface sediments. *Journal of Biogeography* **32**, 1063–1083.

Gasse, F., Juggins, S., and Khelifa, L. B. (1995). Diatom-based transfer functions for inferring hydrochemical characteristics of African paleolakes. *Palaeogeography, Palaeoclimatolology, Palaeoecology* **177**, 31–54.

Hoelzmann, P., Gasse, F., Dupont, L. M., Salzmann, U., Staubwasser, M., Leuschner, D. C., and Sirocko, F. (2004). Palaeoenvironmental changes in the arid and subarid belt (Sahara-Sahel-Arabian Peninsula) from 150 kyr to present. In *Past Climate Variability through Africa and Europe* (R. W. Batterbee, F. Gasse and C. E. Stickley, Eds.) vol. 6, pp. 219–256. Springer, Dordrecht.

Mees, F., Verschuren, D., Nijs, R., and Dumont, H. J. (1991). Holocene evolution of the crater lake at Malha, northwest Sudan. *Journal of Paleolimnology* **5**, 227–253.

Merritt, R. W., and Cummins, K. W. (Eds.) (1984). *An Introduction to the Aquatic Insects of North America*, p. 722. Kendall/Hunt Publishing Company, Dubuque.

Rosenberg, D. M. (Ed.) (1993). *Freshwater Biomonitoring and Benthic Macroinvertebrates*, p. 488. Chapman and Hall, New York.

Verschuren, D. (1997). Taxonomy and ecology of subfossil Chironomidae (Insecta: Diptera) from Rift Valley lakes in central Kenya. *Archiv für Hydrobiologie Beihefte* **107**, 467–512.

Verschuren, D., Cocquyt, C., Tibby, J., Roberts, N., and Leavitt, P. R. (1999a). Long-term dynamics of algal and invertebrate communities in a fluctuating tropical soda lake. *Limnology and Oceanography* **44**, 1216–1231.

Verschuren, D., Cumming, B. F., and Laird, K. R. (2004). Quantitative reconstruction of past salinity variations in African lakes using fossil midges (Diptera: Chironomidae): Assessment of inference models in space and time. *Canadian Journal of Fisheries and Aquatic Sciences* **61**, 986–998.

Verschuren, D., Dumont, H. J., and Armengol-Diaz, J. (1996). Utilisation de cladocères et chironomides fossiles pour réconstruire l'évolution hydrologique de leur habitat marécageux dans la tourbière de Kashiru (Burundi) depuis 40.000 ans BP. *Paleoecology of Africa and the Surrounding Islands* **24**, 109–120.

Verschuren, D., and Eggermont, H. (2006). Quaternary paleoecology of aquatic Diptera in tropical and Southern Hemisphere regions, with special reference to the Chironomidae. *Quaternary Science Reviews* **25**, 1926–1947.

Verschuren, D., Johnson, T. C., Kling, H. J., *et al.* (2002). History and timing of human impact on Lake Victoria, East Africa. *Proceedings of the Royal Society of London B* **269**, 289–294.

Verschuren, D., Laird, K. R., and Cumming, B. F. (2000a). Rainfall and drought in equatorial East Africa during the past 1100 years. *Nature* **403**, 410–414.

Verschuren, D., Tibby, J., Leavitt, P. R., and Roberts, C. N. (1999b). The environmental history of a climate-sensitive lake in the former 'White Highlands' of central Kenya. *Ambio* **28**, 494–501.

Verschuren, D., Tibby, J., Sabbe, K., and Roberts, C. N. (2000b). Effects of lake level, salinity and substrate on the invertebrate community of a fluctuating tropical lake. *Ecology* **81**, 164–182.

Walker, I. R. (1995). Chironomids as indicators of past environmental change. In *The Chironomidae: The Biology and Ecology of Non-Biting Midges* (P. Armitage, P. S. Cranston and L. V. C. Pinder, Eds.), pp. 405–422. Chapman and Hall, London.

Walker, I. R. (2001). Midges: Chironomidae and related Diptera. In *Tracking Environmental Change using Lake sediments: Zoological Indicators* (J. P. Smol, H. J. B. Birks and W. Last, Eds.) pp. 43–66. Kluwer, Dordrecht.

Late Pleistocene of Europe

S J Brooks, The Natural History Museum, London, UK

© 2007 Elsevier B.V. All rights reserved.

Introduction

Chironomids, nonbiting midges, are among the most versatile indicators of paleoenvironmental change. They have been used to reconstruct, both quantitatively and qualitatively, a variety of environmental variables (see other chapters). They have proved to be particularly useful as temperature proxies. Until the 1990s, paleoecologists used chironomids in qualitative studies of climate change. They assessed changes in chironomid assemblages, from those dominated by thermophilic taxa to those dominated by cold stenothermic taxa, as a reflection of relative changes in climate. However, with the advent of numerical techniques, which take the modern temperature optima of taxa and transfer these to fossil assemblages (see other chapters), it is now possible to use chironomids to reconstruct quantitatively past climatic change at high temporal resolution with an error of about $\pm$ 1.0° C. In this chapter, the use of chironomids as climate proxies for sequences during the late Pleistocene in Europe is reviewed. A beginning is made by reviewing papers in which chironomids were used qualitatively, and then an account of studies in which chironomids have been used quantitatively to estimate changes in mean July air temperature is presented. Finally, the results of these studies with those obtained from other climate proxies are compared and the relative performance of chironomids as climate proxies is assessed.

Chironomids as Qualitative Indicators of Climate Change in the Late Pleistocene

Chironomids have been used for 80 yr to investigate past environmental change. However, like Gams' (1927) pioneering work, the majority of these studies has focused on Holocene sequences. Nevertheless, among the earliest uses of chironomids in a paleoecological context was the work of Andersen (1938), who investigated a late glacial sequence from Denmark. However, it was not until the 1970s–1980s that the late glacial once more became the focus of paleoenvironmental studies using chironomids, when a series of papers were published by Wolfgang Hofmann.

Hofmann (1978) found a diverse late glacial chironomid fauna at Grosser Segeberger See in northern Germany. The Younger Dryas was dominated by cold stenothermic taxa of the *Tanytarsus lugens* community. During the Bølling, *Corynocera ambigua* comprised more than 40% of the chironomid fauna. Other taxa included *Microtendipes* and *Procladius*. In several other studies (see below), *Corynocera ambigua* has been found to rapidly dominate chironomid assemblages and just as rapidly decline after a relatively short period of time. Hofmann found similar trends in late glacial chironomid assemblages from Schöhsee (Hofmann, 1971). Insufficient data was available from the Allerød interval to draw any firm conclusions.

Lobsigensee, on the Swiss Plateau, was the focus of a study into the response of midge assemblages to climate change during the late glacial (Hofmann, 1983a). The Oldest Dryas was characterized by an assemblage dominated by *Sergentia coracina* and *Chironomus anthracinus*-type. Early in this period *Microtendipes*, *Paratanytarsus*, and *C. ambigua* were also abundant, but toward the end of the Oldest Dryas these taxa declined and were replaced by *Psectrocladius*, which coincided with the establishment of *Betula nana* in the catchment. These changes are consistent with climatic warming from relatively cold conditions (when the fauna included cold stenotherms such as *S. coracina*, *C. ambigua*, and *Microtendipes*, which is cool tolerant) to a temperate climate in which the thermophilic taxon *Psectrocladius* thrived. A major change in the fauna occurred at the onset of the Bølling when the fauna became dominated by just three thermophilic taxa, namely, *Dicrotendipes*, *Chironomus*, and *Psectrocladius*. In the Allerød, the fauna became even less diverse when 89% of the specimens were either *Chironomus* or *Tanytarsus* sp. C (Hofmann, 1971), but there was also a large increase in *Dasyhelea*, a ceratopogonid biting midge. This suggests that while temperate conditions continued to prevail, water levels in the lake also fell. The Lobsigensee core was taken in the littoral zone of the lake, making it particularly sensitive to lake-level change. Thus, changes in the chironomid fauna apparently indicated both changes from cold to temperate conditions and also a possible decline in precipitation, although changes in other factors (e.g., evaporation rates or stream inputs and outputs) could also be responsible. The core was sampled at relatively coarse intervals and so it was not possible to differentiate any response to relatively short-duration climatic oscillations.

In order to determine whether late glacial climate was driving changes in chironomid assemblages at other sites, Hofmann (1983b) analyzed sediments from a littoral core from Poolsee, a shallow lake in northern Germany. Once again he found that the chironomid fauna early in the late glacial was dominated by taxa such as *Micropsectra*, *Heterotrissocladius*, *Protanypus*, *Sergentia coracina*, and *Paracladius*, which are typical of cold, oligotrophic conditions. However, Hofmann argued that because these taxa, which occur in the profundal zone of temperate lakes, were occurring in the littoral zone of the lake, they must be indicative of a cold climate rather than trophic conditions. These taxa were replaced by thermophilic taxa as temperature warmed in postglacial times. Similar patterns emerged from studies at other sites. The early late glacial chironomid fauna at Meerfelder Maar, southern Germany, (Hofmann, 1984), was dominated by cold stenothermic taxa such as *Diamesa*, *Paracladopelma nigritula*, and *Micropsectra* but was later replaced by taxa associated with more thermophilic conditions such as *Psectrocladius* and *Paratanytarsus*.

In a study of cores from three lakes in the Eifel region of southern Germany and a lake from the Massif Central in central France, Hofmann (1990, 1991) found that Upper Pleniglacial sediments had relatively few chironomid remains but those present in the littoral zone of lakes (*Diamesa*, *Protanypus*, *Paracladopelma*, and *Micropsectra*) were all indicative of cold conditions. During the Middle–Lower Pleniglacial and early glacial, the increase in abundance and changes in dominance between *Micropsectra* and *Paracladius* indicated that climatic conditions may have been slightly warmer. During the Eemian Interglacial, increasing species diversity and the presence of thermophilic species, including *Glyptotendipes*, indicated that conditions had become considerably warmer.

At Lago Albano, central Italy, the response of chironomids to climatic change between 28 and 17 kyr BP was investigated as part of a multiproxy study (Manca *et al.*, 1996; Guilizzoni *et al.*, 2000). Between 28 and 26 kyr BP, chironomids were abundant and the fauna was dominated by *Paracladopelma* and *Micropsectra*. These taxa are indicative of cool, oligotrophic conditions. From 26 to 17 kyr BP, very few chironomids were recovered but most were *Micropsectra*. During the late glacial, numbers increased once more and the fauna became more diverse, including *Micropsectra*, but with additional temperate, littoral taxa such as *Dicrotendipes* and *Cladotanytarsus* (Manca *et al.*, 1996). Increases in the overall abundance and diversity of chironomids were probably in response to increases in temperature, whereas cooler, drier periods resulted in a decrease in lake productivity and thus chironomid abundance.

Brooks (2000) studied the late glacial–early Holocene chironomid assemblages of four lakes on an altitudinal transect in the Swiss Alps (Gerzensee, 603 m; Leysin, 1,230 m; Regenmoos, 1,260 m; Hérémence, 2,300 m). This was a high-resolution study, with samples analyzed at 0.5 cm intervals in some sections of the cores, in order to detect the chironomid response to low-amplitude, short-duration cold oscillations. Because the lakes were each at different altitudes, they supported quite different chironomid assemblages. However, if the chironomids were influenced by regional climate change, then a synchronous response would be expected in all the lakes and this could be disentangled from local responses to in-lake environmental change, which would not affect the chironomid assemblages of all the lakes at the same time. A detailed chronology was obtained by wiggle-matching high-resolution oxygen-isotope records from each lake against the Greenland ice core project (GRIP) oxygen-isotope record. Only Gerzensee and Leysin will be considered here since late glacial sediments were not well represented at the other two sites. A high-resolution record was obtained from Gerzensee where the sequence between 13,440 and 11,154 cal yr BP was analyzed at approximately 10–30 yr intervals (**Fig. 1**). The late Interstadial was dominated by a temperate taxon, *Parakiefferiella bathophila*, which accounted for 40–60% of the chironomid assemblage. However, during the Gerzensee Oscillation (13.2–12.8 cal kyr BP), there was a significant increase in the abundance of a second *Parakiefferiella* species, which later dominated the Younger Dryas assemblage. This taxon was referred to as *Parakiefferiella* type A, and is the same as *Parakiefferiella* sp. B of Walker *et al.* (2003). At the start of the Younger Dryas, *P. bathophila* decreased abruptly, and finally disappeared in the second half of this period. It was not until the early Holocene that *P. bathophila* returned and rapidly became a dominant part of the fauna once more. This taxon was apparently close to its thermal threshold at Gerzensee and

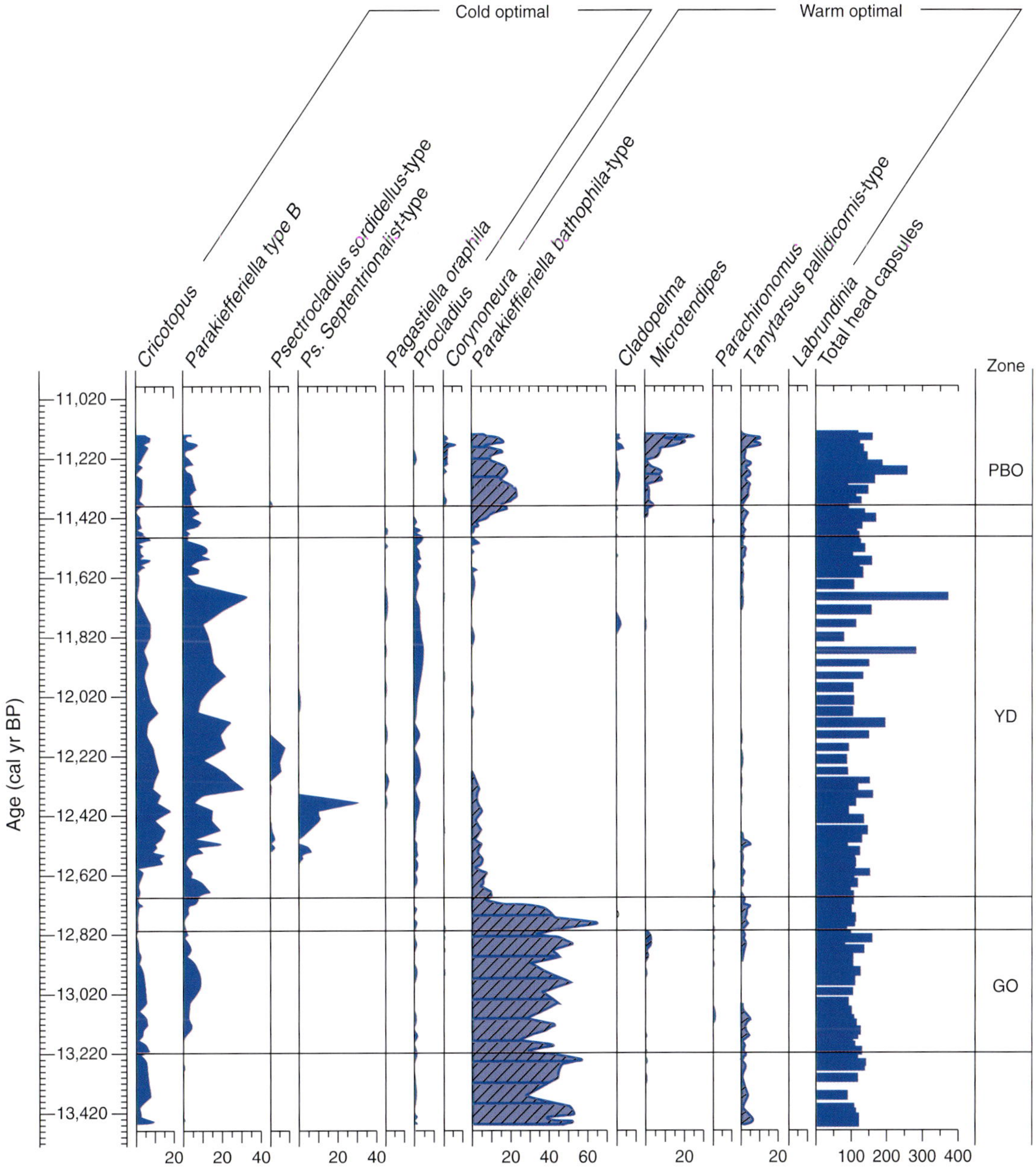

Figure 1 Response of Chironomidae to climate change around the time of the Younger Dryas at Gerzensee, Switzerland (Brooks, 2000).

therefore was a sensitive indicator of climatic conditions at this site.

At Leysin (**Fig. 2**), *Parakiefferiella bathophila* was absent. Presumably temperatures were too low at this altitude to support a population. *Parakiefferiella* type A once more became more abundant during the Gerzensee Oscillation and Younger Dryas, but at this site *Einfeldia* was apparently close to its thermal threshold and was only present in the fauna during the cold oscillations. This study clearly demonstrated that chironomids could be useful paleoclimatic indicators because they responded to shifts in regional climate. Synchronous responses to cool oscillations, which were independently identified from an oxygen-isotope signal, were apparent in each lake from different elements of the chironomid fauna. Unfortunately, despite the development of a quantitative chironomid–temperature inference model based on a modern training set of Swiss lakes (Heiri, 2001; Heiri *et al.*, 2003), it has not been possible to estimate temperatures from either of these two sites because *Parakiefferiella* type A was not represented in the training set. This taxon is present in southern

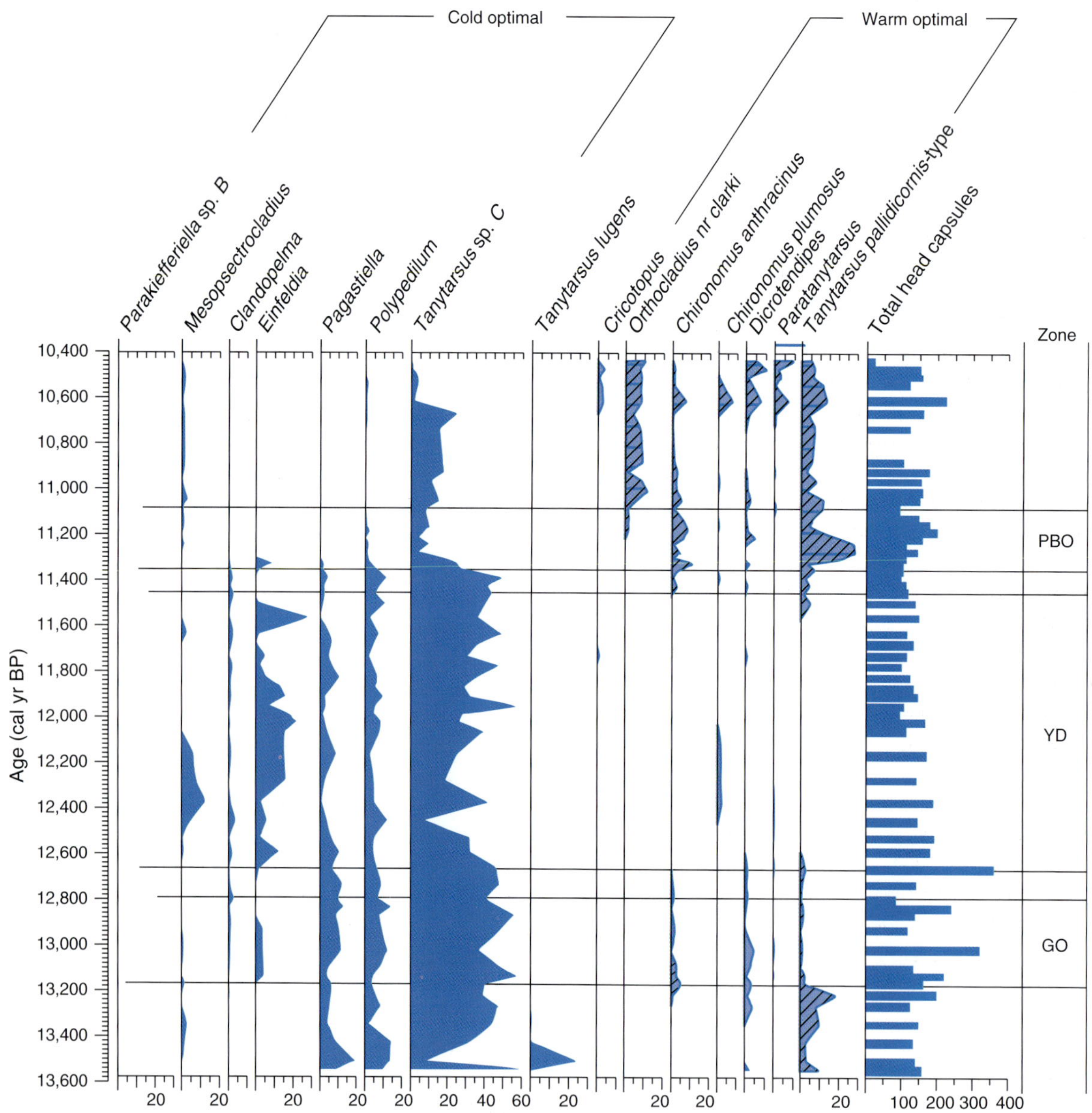

Figure 2 Response of Chironomidae to climate change around the time of the Younger Dryas at Leysin, Switzerland (Brooks, 2000).

Norway today but has apparently been eliminated from the Swiss fauna since the late glacial.

Gandouin and Franquet (2002) described a late glacial and Holocene chironomid fauna from Lac Long Inférieur in southeast France. The late glacial was not examined in detail but the fauna during this period was dominated by *T. lugens*-type and other cold stenotherms including *Paracladius* and *Diamesa*, which indicate that cool, oligotrophic conditions prevailed. A peak in the abundance of *Micropsectra* suggested a cold oscillation in the mid-Interstadial but the chronology of the sequence was not sufficiently resolved to determine to which of the Greenland Interstadial events this may be referable. A gradual decline in *T. lugens*-type and an increase in taxon diversity throughout the Interstadial may indicate an increase in temperature. However, this is contrary to most temperature reconstructions for this period. Another possible explanation for this result may be a decrease in lake level.

Chironomids as Quantitative Indicators of Climate Change in the Late Pleistocene

The development of quantitative chironomid–mean July air temperature inference models (see other chapters) has revolutionized paleoclimate research. Now it is possible to obtain high-resolution, low-error, summer air temperature estimates from lake sediment samples with the knowledge that the biological proxy has completed its development *in situ*. At the sampling resolution of most studies, the response of midge taxa to climate change shows little or no delay by virtue of the long-distance, wind-assisted, rapid dispersion of chironomid adults across the landscape. These chironomid-inferred temperature reconstructions can be compared with high-resolution oxygen-isotope records and quantitative reconstructions based on other proxies such as pollen and beetle assemblages to test their reliability.

The first late glacial chironomid-inferred summer temperature records from Europe were published as part of a multiproxy study of Kråkenes Lake, western Norway (Brooks and Birks, 2000a). The chironomid-inferred temperatures were derived from a modern Norwegian 44-lake training set. The training set covered a temperature range of 5.7–14.0° C but in fact only two lakes in the training set experienced mean July air temperatures of less than 8° C and so the chironomid-inferred temperatures estimates derived from this model were likely to overestimate the coldest temperatures of the Younger Dryas. Details of the chironomid stratigraphy are shown in **Figure 3**.

The early Interstadial chironomid fauna was not only dominated by cold stenothermic taxa such as Diamesinae, *Heterotrissocladius grimshawi*, *T. lugens*-type, *Micropsectra radialis*, *S. coracina*, and *Arctopelopia*, but also included more temperate taxa such as *Cricotopus*, *Psectrocladius sordidellus*-type, and *Microtendipes*. As the Interstadial continued to warm, the proportion of temperate taxa increased at the expense of the cold stenotherms and Diamesinae disappeared from the fauna altogether. The onset of the Younger Dryas caused a sudden and major change in the chironomid fauna. The temperate taxa and some of the cold stenothermic taxa, such as *Sergentia*, *T. lugens*-type, and *Arctopelopia*, were eliminated. *M. radialis* and *H. grimshawi* continued to dominate the fauna. Diamesinae returned but several cold stenothermic taxa appeared in the fauna for the first time, including *Heterotrissocladius subpilosus* and *Paracladius alpicola*. These faunistic changes indicated that the lake went from cool, mesotrophic to very cold and ultraoligotrophic conditions.

Temperatures of 9–10° C were inferred from the chironomid assemblage for the Younger Dryas, which were about 5° C higher than those inferred from plant macrofossils. This indicated that the range of temperatures covered by the chironomid training set was not sufficient to reconstruct reliable temperatures for the late glacial period. Therefore, an additional 65 lakes were added, which extended the range of temperatures in the modern training set to 3.5–15.6° C, and in addition provided a more even distribution of lakes across the temperature gradient (Brooks and Birks, 2001). Using the expanded training set, new temperature estimates for the late glacial and early Holocene were obtained that closely supported the temperatures indicated by the plant macrofossils (**Fig. 4**).

Gradually warming temperatures from about 9.5 to 10.5° C are indicated during the Interstadial, followed by a sudden drop of about 3.5° C over a 60 yr period at the start of the Younger Dryas. This fall is still not as great as that inferred from vegetational changes, which suggest that temperatures fell by as much as 6° C to reach 4–5° C at the start of the Younger Dryas (Birks, 1994). Chironomid-inferred temperatures began to increase before the end of the Younger Dryas, shortly after 12 cal kyr BP, and continued to rise rapidly into the early Holocene.

The 109-lake Norwegian training set was also used to reconstruct late glacial temperatures at Whitrig Bog in southeast Scotland (Brooks and Birks, 2000b, 2001). High-resolution analysis of the core meant that a highly detailed picture of climatic fluctuations emerged. Following the deglacial period when the chironomid assemblage was dominated by

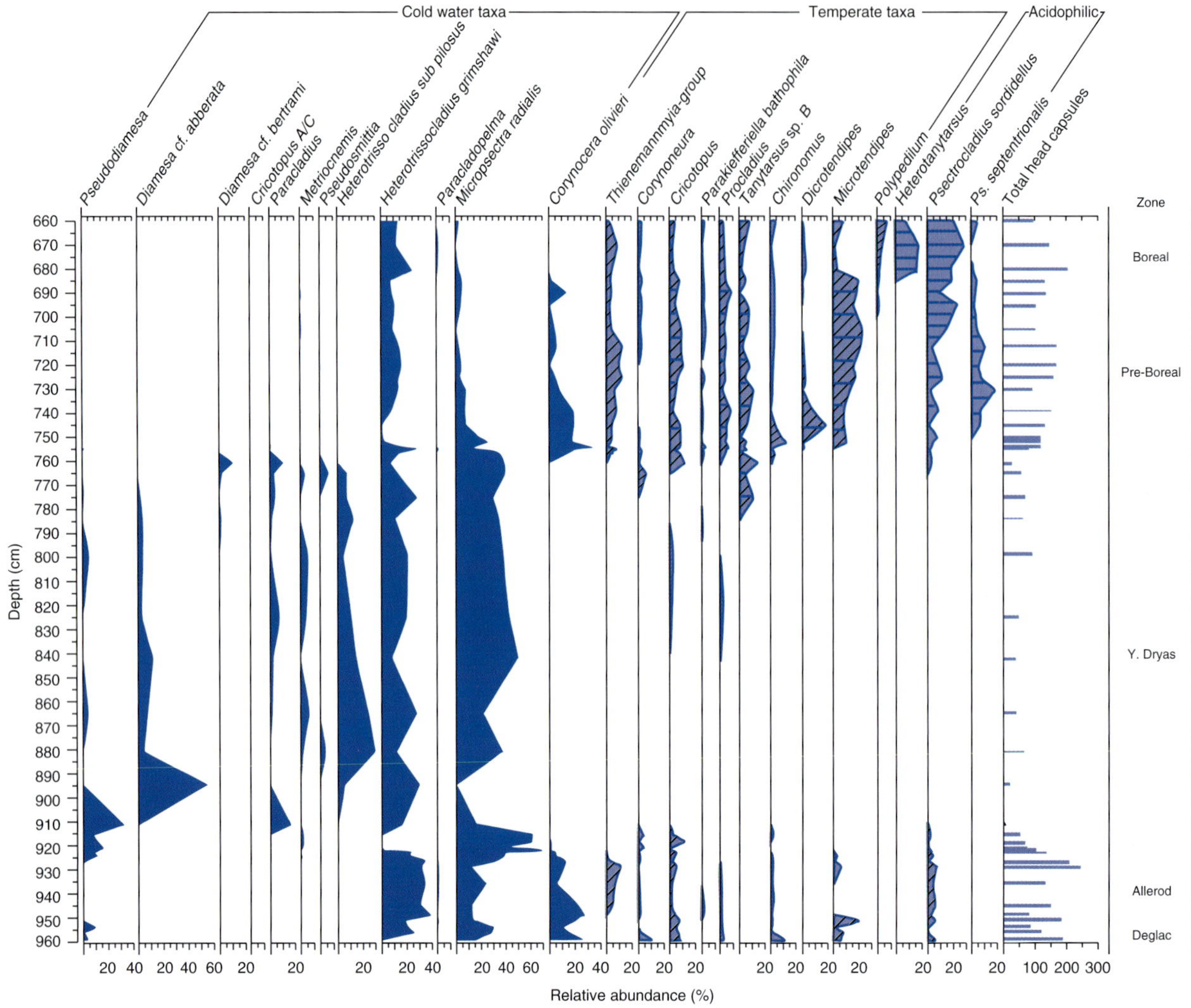

Figure 3 Response of Chironomidae to climate change during the late glacial and early Holocene at Kråkenes, Norway (Brooks, 1997).

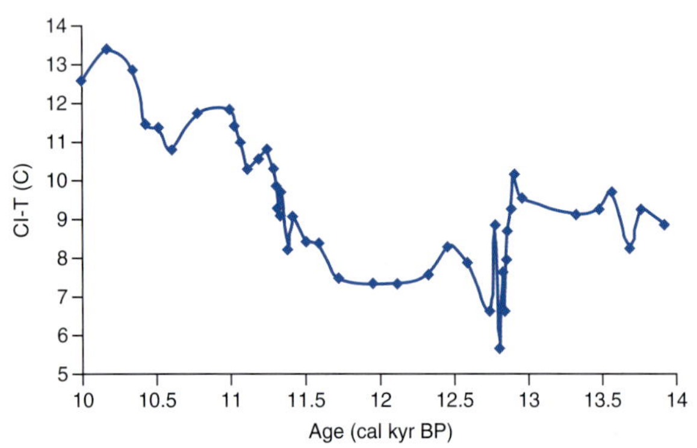

Figure 4 Chironomidae-inferred mean July air temperature (CI-T) reconstruction of the late glacial and early Holocene at Kråkenes, Norway (Brooks and Birks, 2001).

cold stenotherms, the Interstadial fauna was dominated for the most part by temperate taxa such as *Psectrocladius*, *Chironomus*, *Tanytarsus* spp., and Pentaneurini (Brooks *et al.*, 1997). However, there were two relatively short periods when there was a resurgence of cold-water taxa. In the earliest period, eight cold taxa were present and there were declines in most of the temperate taxa, including *Chironomus* (**Fig. 5**). During the second cool period, only three cold-water taxa were present and, although several temperate taxa declined, *Chironomus* did not decline in abundance. At the onset of the Younger Dryas, there was a sudden replacement of the entire temperate fauna with a cold-water fauna, dominated by *M. radialis*-type, which had not occurred in the lake before, and several other taxa which had previously only occurred at low abundance during the previous cold periods. These faunistic changes implied that the Younger Dryas was the coldest period in the late glacial, that temperatures during the early cool oscillation were a little less cold but that they were considerably colder than the second cool oscillation.

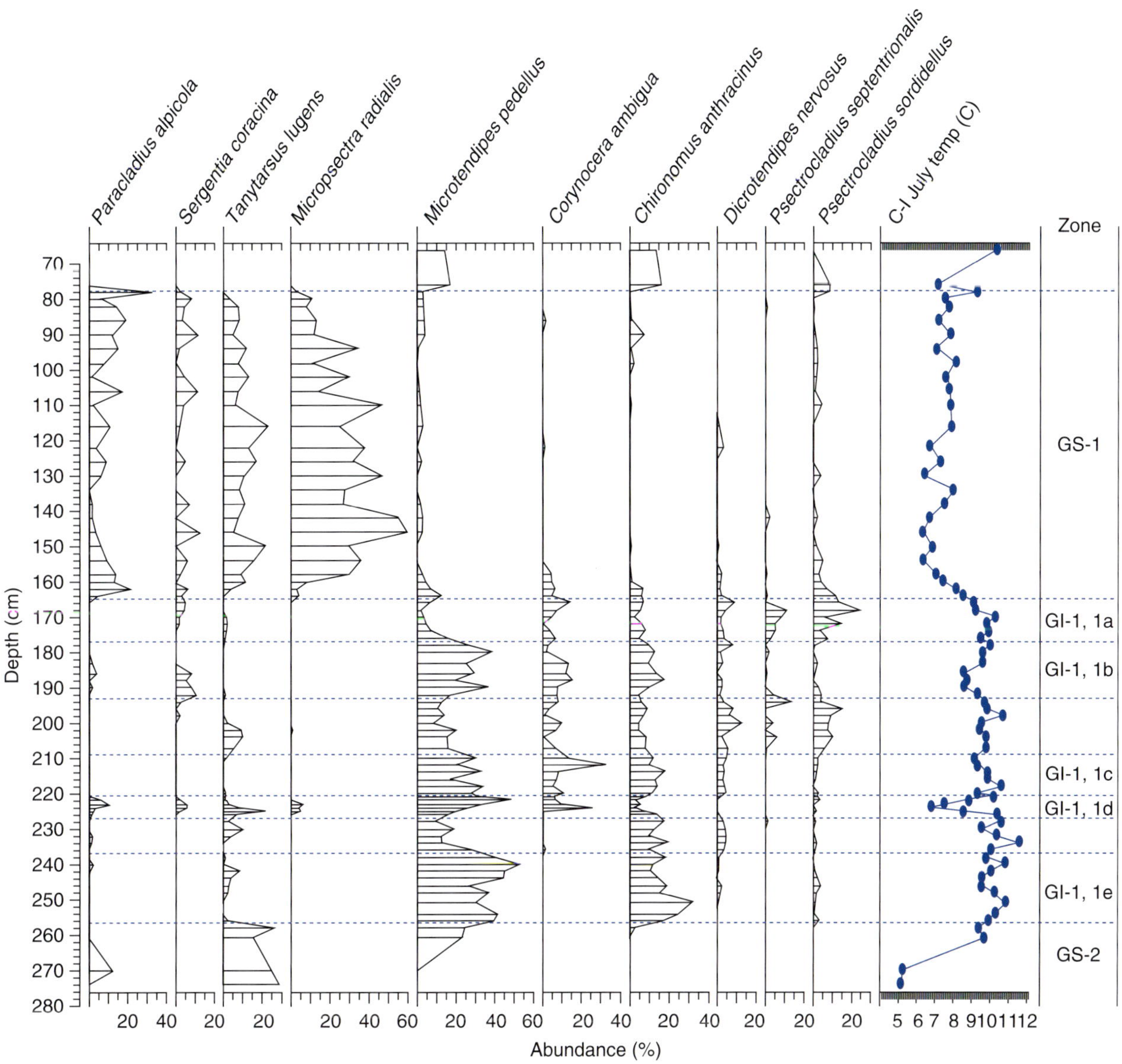

Figure 5 Response of Chironomidae to climate change during the late glacial and early Holocene at Whitrig Bog, southeast Scotland (Brooks and Birks, 2000b).

These qualitative inferences were confirmed when the 109-lake Norwegian transfer function was applied to the data. Interstadial temperatures gradually fell from ca. 11.5 to 11.0° C but were punctuated by two major cold oscillations. During the first oscillation, temperatures fell to ca. 8.0° C. This oscillation is referable to Greenland Interstadial event 1 d (GI-1d), which has been recorded in northern Switzerland as the Aegelsee Oscillation (Lotter *et al.*, 1992) and elsewhere in Europe (Björck *et al.*, 1998). The second oscillation, when temperatures fell to ca. 10.0° C, is referable to Greenland Interstadial event 1 b (GI-1b) and has also been recorded elsewhere in Europe as the Gerzensee or Amphi-Atlantic Oscillation (Lotter *et al.*, 1992; Levesque *et al.*, 1993; Bedford *et al.*, 2004). Early in the Younger Dryas, temperatures fell to ca. 7.5° C but gradually increased to ca. 8.5° C at the end of this period. These trends are closely reflected in the GRIP ice-core oxygen-isotope record (**Fig. 6**), in particular the gradual warming during the Younger Dryas, the two major cold oscillations in the Interstadial, together with the suggestion of two additional minor Interstadial oscillations just before and just after GI-1d. However, there are two obvious differences between the records. First, the cooling trend in Interstadial temperatures is more pronounced in the ice-core record than the Whitrig Bog record. Second, GI-1d is warmer than GI-1b in the GRIP record, whereas at Whitrig Bog the reverse is indicated. The chironomid-inferred temperature

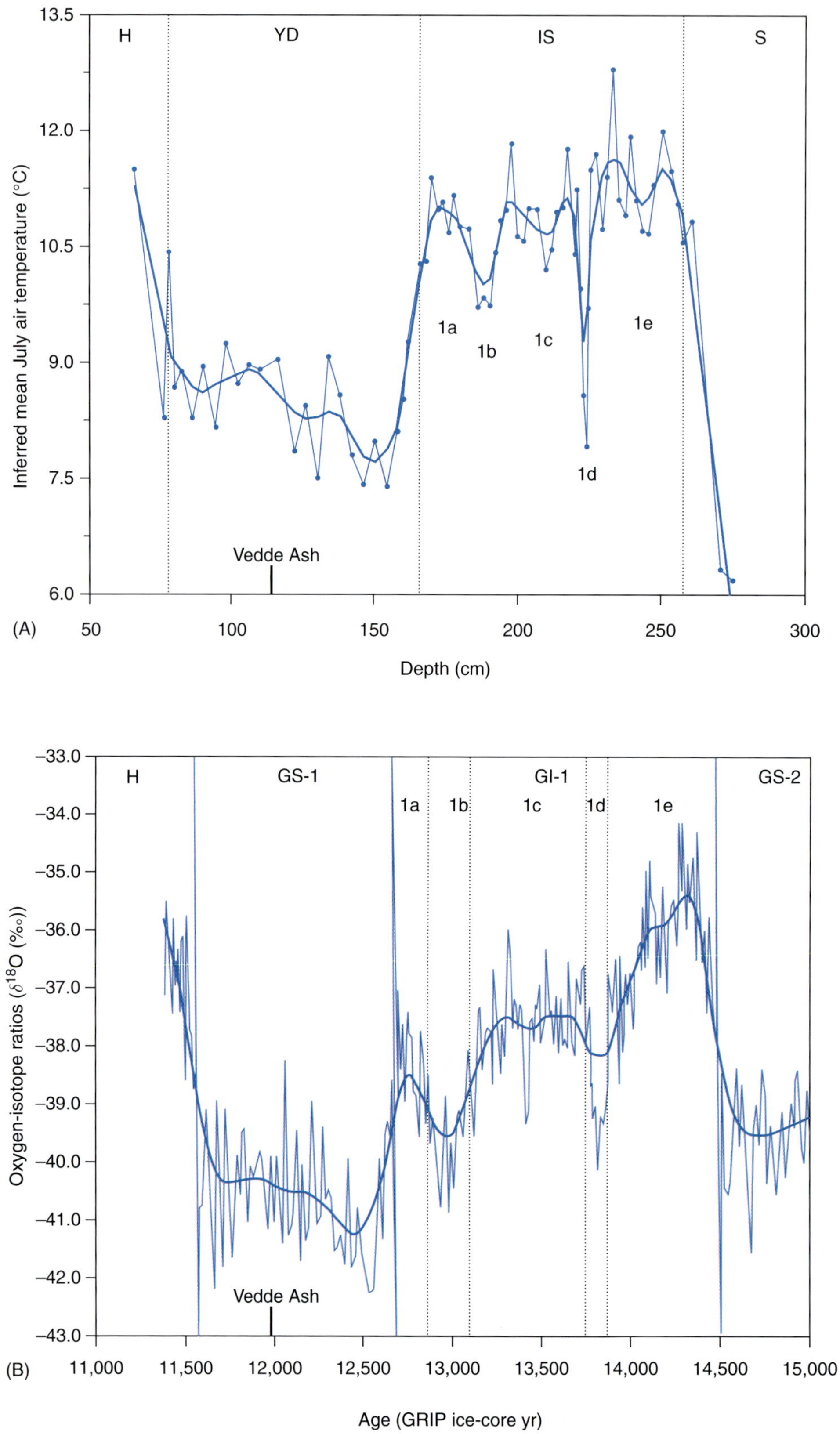

Figure 6 Comparison of chironomid-inferred mean July air temperatures for the late glacial at Whitrig Bog, southeast Scotland, and the oxygen-isotope record from the GRIP ice core (Johnsen *et al.*, 1992) (Brooks and Birks, 2000b).

estimates are closely corroborated by temperature estimates derived from plant macrofossils (Conolly, 1961; Sheldrick 1997; Birks, unpublished), although warmer than those derived from glacial data (Sissons, 1979; Sutherland, 1984; Ballantyne, 1989) for lowland Scotland. However, late glacial beetle-inferred estimates for southern Scotland are consistently about 2° C higher than the chironomid estimates (Atkinson *et al.*, 1987; Coope and Lrmdahl, 1995; Coope *et al.*, 1998). Reasons for this discrepancy are not obvious but may relate to differences in the numerical approaches used to derive temperatures.

A second British quantitative reconstruction of late glacial temperatures based on chironomids is available from Hawes Water in northwest England (Bedford *et al.*, 2004). The chironomid assemblage (**Fig.** 7) is similar to that of Whitrig Bog, even though Hawes Water is a hard-water lake whereas Whitrig Bog has soft water. The Interstadial is once more dominated by temperate taxa such as *Microtendipes*, *Chironomus*, *Psectrocladius*, and Pentaneurini. However, there are two periods during the Interstadial in which the cold stenotherm, *Sergentia*, increases in abundance. The Younger Dryas is dominated by a cold-water fauna very similar in composition to that at Whitrig Bog. The main difference between the two assemblages is that *C. ambigua* is most abundant during the Younger Dryas at Hawes Water, whereas at Whitrig Bog this taxon reaches maximum abundance during the Allerød. This may reflect in-lake differences between the sites since *C. ambigua* is often associated with charophyte beds (Brodersen and Lindegaard, 1999). The Norwegian training set was once more used to reconstruct chironomid-inferred temperatures. Interstadial temperatures gradually fell from ca. 13.0 to 12.0° C but were punctuated by two abrupt cold oscillations. The first, when temperatures fell by about 2° C to ca. 11.0° C, was more severe than the second when temperatures fell by about 1° C to ca. 11.5° C. Temperatures fell rapidly at the start of the Younger Dryas to ca. 7.5° C but then gradually increased to ca. 9.5° C at the end of this period. The chironomid-inferred temperature trend at Hawes Water therefore closely reflects that inferred at Whitrig Bog. The main difference is that temperatures are consistently warmer at Hawes Water than Whitrig Bog, but this probably reflects the more southerly location of the former site. An oxygen-isotope record is also available from Hawes Water (**Fig.** 8) and this closely reflects the chironomid-inferred trends. Notably, the Hawes Water isotope record, like the chironomid-inferred temperature records, indicates that the Aegelsee Oscillation was colder than the Gerzensee Oscillation and that there was only a gradual long-term cooling trend during the Interstadial. These trends are also similar to the oxygen-isotope record from Ammersee in southern Germany (von Grafenstein *et al.*, 1999) but differ from the GRIP record.

More late glacial studies from England, Scotland, and Norway are currently in progress by Brooks and coworkers and should be published over the next few years. These also show similar trends as those described above and further support the view that there are important and consistent differences in detail between the Greenland oxygen isotope records and chironomid-inferred temperature records and isotope records from Europe.

Following a detailed account by Millet *et al.* (2003) of the late glacial chironomid fauna at Lac Lautrey (Jura, eastern France), the Swiss chironomid-temperature inference model (Heiri, 2001; Heiri *et al.*, 2003) was used to reconstruct late glacial mean July air temperatures at this site by Heiri and Millet (2005). This high-resolution record (**Fig.** 9) begins at about 16 cal kyr BP when the fauna was dominated first by the cold stenotherm *Stictochironomus* and then by *C. ambigua* and temperatures were inferred at about 11–12° C (**Fig.** 10). There was an abrupt rise in temperature to 14.5° C at the beginning of the Bølling (14.6 cal kyr BP) when a diverse fauna of widely distributed and lowland taxa replaced the alpine and subalpine fauna. During the second half of the Bølling (14.2–13.7 cal kyr BP), temperatures continued to rise more gradually to reach 16.5° C as temperate taxa increased in abundance. These estimates are within the range of pollen-based reconstructions from central Europe (Renssen and Isarin, 2001). There was no apparent response to the Aegelsee cool fluctuation (at ca. 14 cal kyr BP); however, two centennial-scale cold oscillations of 1.5–2.0° C were indicated at 13.2 cal kyr BP and 12.9 cal kyr BP that appear to be synchronous with the beginning and end of the Gerzensee Oscillation. Interstadial temperatures peaked at about 18° C toward the end of the Allerød, which is warmer than present-day values from northern Switzerland and considerably warmer than late Allerød temperatures inferred from Cladocera and pollen assemblages (Lotter *et al.*, 2000).

Temperatures declined to 14° C at the start of the Younger Dryas but then gradually increased and were about 1° C higher by the end of the Younger Dryas, which reflects the trends in other European records (von Grafenstein *et al.*, 1999; Brooks and Birks, 2000b). When corrected for altitude, the Lac Lautrey Younger Dryas estimates agree closely with chironomid-inferred records from the northern Swiss Alps (Heiri *et al.*, 2003) but are warmer than pollen-inferred estimates (Lotter *et al.*, 2000). A warming of about 1.5° C was indicated at the Younger Dryas/Holocene transition, which is in close agreement with

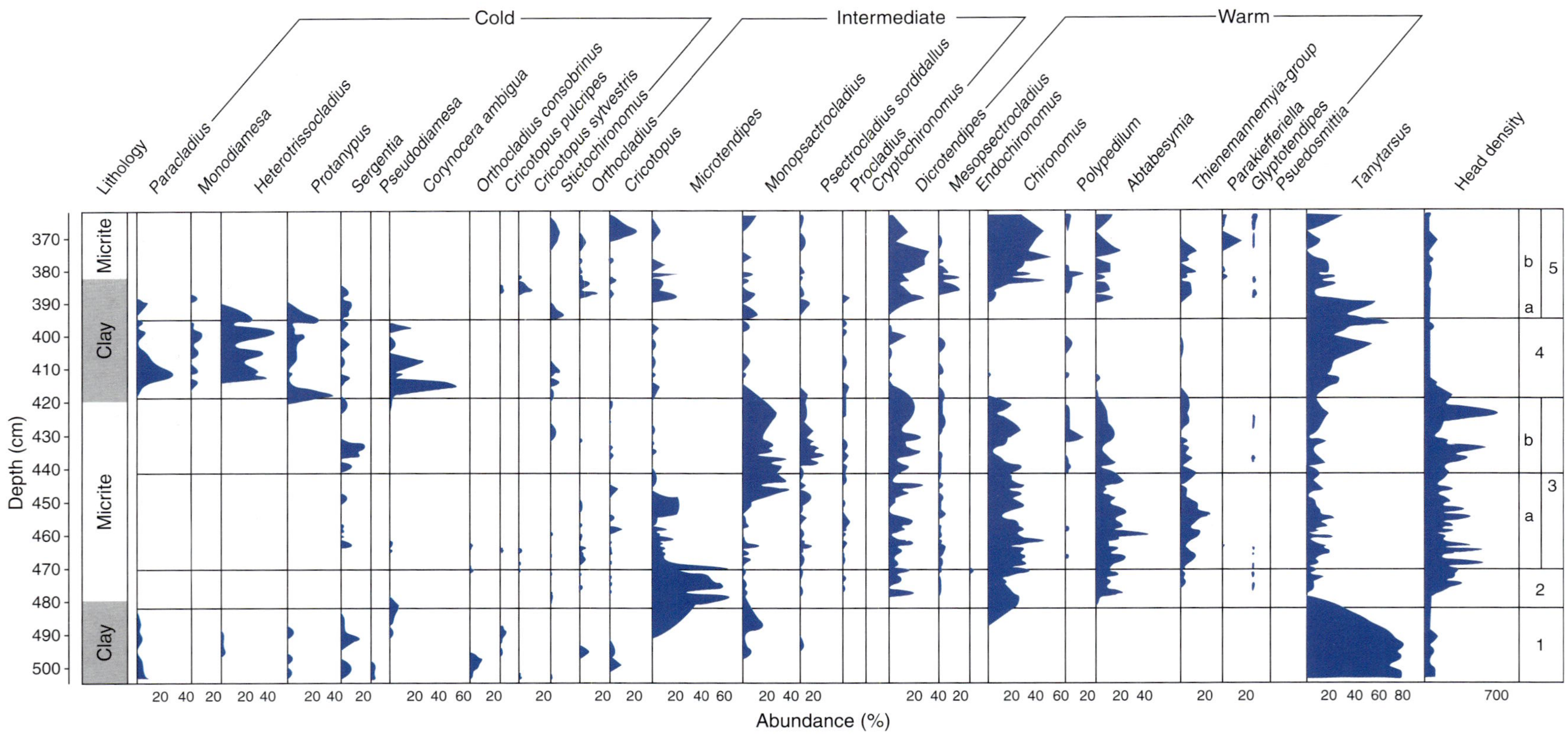

Figure 7 Response of Chironomidae to climate change during the late glacial and early Holocene at Hawes Water, northwest England (Bedford *et al.*, 2004).

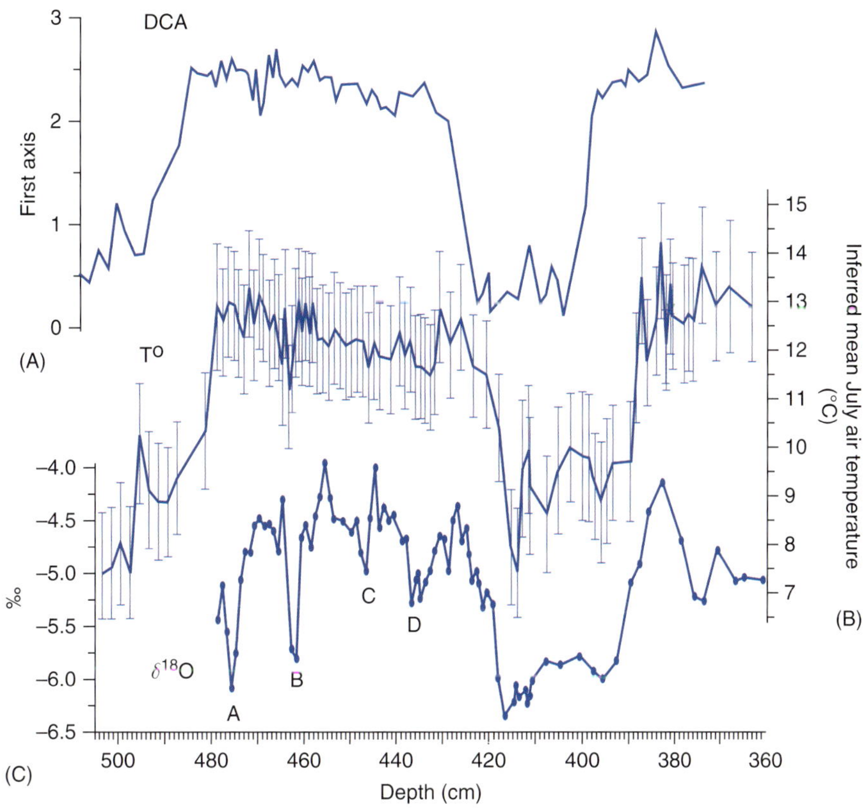

Figure 8 Comparison of chironomid-inferred mean July air temperatures for the late glacial at Hawes Water, northwest England, the DCA axis 1 sample scores and the oxygen-isotope record from Ammersee, southern Germany (von Grafenstein *et al.*, 1999) (Bedford *et al.*, 2004).

other proxy records from the region (Heiri *et al.*, 2003; Lotter *et al.*, 2000; Magny *et al.*, 2001). The Lac Lautry temperature record differs in detail from the GRIP oxygen-isotope record but shares similarities with the Ammersee isotope record from southern Germany (von Graffenstein *et al.*, 1999). Most notably, chironomid-inferred temperatures before the Bølling warming were cooler than those during the Younger Dryas, there is a two-step increase in early Bølling temperatures, the GI-1d event is not clearly reflected in the records, and there is not a strong temperature decline during the Interstadial.

Conclusions

At present, relatively few quantitative chironomid-inferred temperature reconstructions are available from the late glacial in Europe. However, it is clear from those that have been published that such sequences can provide valuable high-resolution, low-error records of climate change during this period. The chironomid-inferred temperature trends largely agree with those derived from oxygen-isotope records, which suggests that the chironomid records are reliable, and so provide quantitative details of short-lived, small-amplitude climate fluctuations as well as long-term trends. However, differences between the GRIP oxygen-isotope record and the isotope records from Hawes Water and Ammersee are also reflected in the chironomid-inferred temperature records from Whitrig Bog, Hawes Water, and Lac Lautrey, which more closely resemble the local isotope trends than GRIP. This suggests that climate change was not uniform throughout the North Atlantic region and central Europe and that it may be too simplistic to rely on the GRIP ice-core record to provide an 'event stratigraphy' for the whole region (Björck *et al.*, 1998). Similarly, late glacial chironomid-inferred temperature estimates are largely corroborated by estimates derived from other proxies. Where conflicts do exist, there are currently insufficient records to resolve them. Therefore, it is particularly important that more chironomid-derived climate records become available, which will help to underline consistencies and inconsistencies between them and other proxy records. This will be necessary to improve our understanding of the limitations of the different proxy records. In particular, discrepancies between chironomid- and beetle-inferred estimates need to be resolved. However, the current lack of same site quantitative reconstructions that include both beetles and chironomids impedes the resolution of this problem.

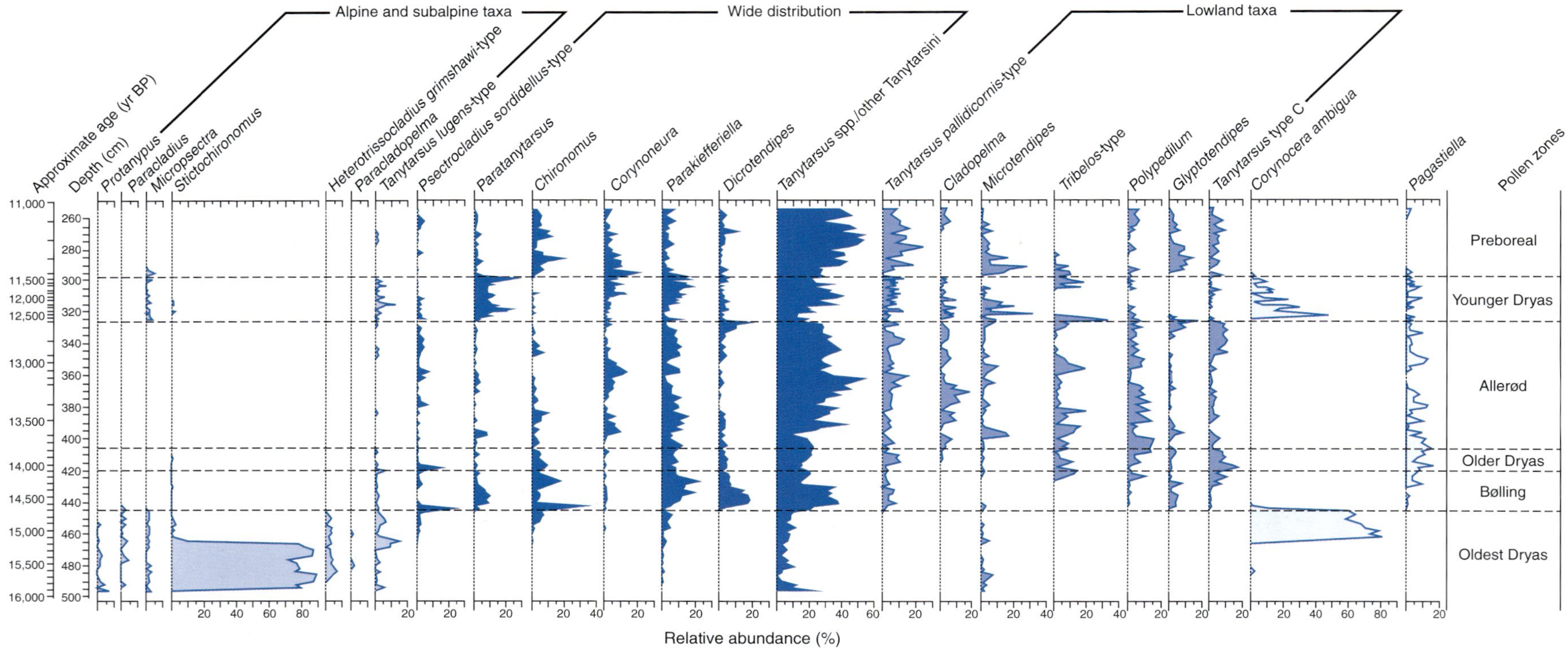

Figure 9 Response of Chironomidae to climate change during the late glacial and early Holocene at Lac Lautrey, Jura, eastern France (Heiri and Millett, 2005).

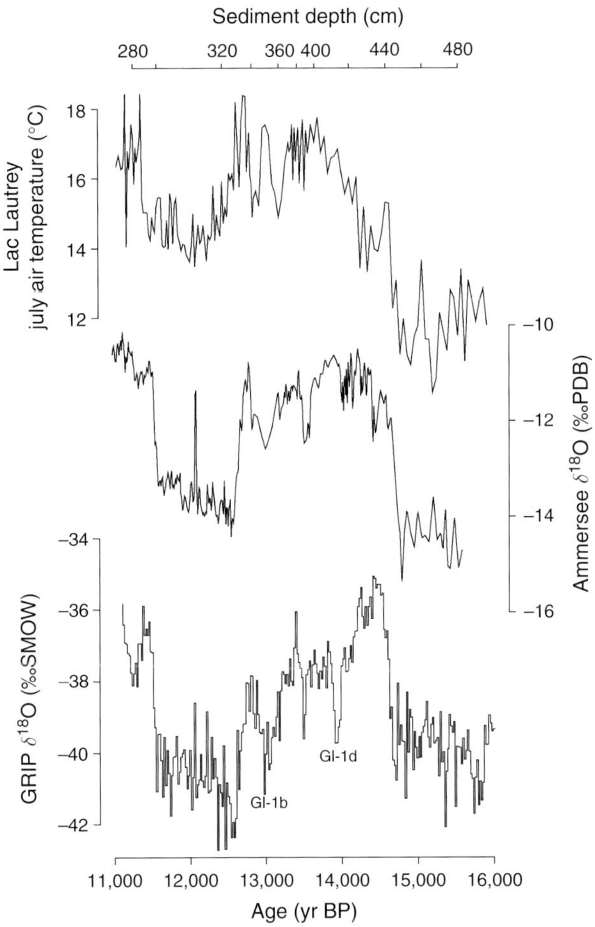

Figure 10 Comparison of chironomid-inferred mean July air temperatures for the late glacial at Lac Lautrey, Jura, eastern France, the oxygen-isotope record from Ammersee, southern Germany (von Grafenstein *et al.*, 1999), and the oxygen-isotope record from the GRIP ice core (Johnsen *et al.*, 2001) (Heiri and Millett, 2005).

See also: **Beetle Records**: Late Tertiary and Early Quaternary. **Chironomid Overview**. **Chironomid Records**: Africa; Postglacial Southern Hemisphere.

References

Andersen, F. S. (1938). Spätglaziale Chironomiden. *Meddelelser Dansk geologisk Forening* **9**, 320–326.

Atkinson, T. C., Briffa, K. R., and Coope, G. R. (1987). Seasonal temperatures in Britain during the past 22,000 years reconstructed using beetle remains. *Nature* **352**, 587–592.

Ballantyne, C. K. (1989). The Loch Lomond readvance on the Isle of Skye, Scotland: Glacier reconstruction and palaeoclimatic implications. *Journal of Quaternary Science* **4**, 95–108.

Bedford, A., Jones, R. T., Lang, B., Brooks, S. J., and Marshall, J. D. (2004). A late-glacial chironomid record from Hawes Water, N.W. England. *Journal of Quaternary Science* **19**, 281–290.

Birks, H. H. (1994). Late-glacial vegetational ecotones and climatic patterns in western Norway. *Vegetation History and Archaeobotany* **3**, 107–119.

Björck, S., Walker, M. J. C., Cwynar, L. C., *et al.* (1998). An event stratigraphy for the last termination in the north Atlantic region based on the Greenland ice-core record: A proposal by the INTIMATE group. *Journal of Quaternary Science* **13**, 283–292.

Brodersen, K. P., and Lindegaard, C. (1999). Mass occurrence and sporadic distribution of *Corynocera ambigua* Zetterstedt (Diptera, Chironomidae) in Danish lakes. Neo- and palaeolimnological records. *Journal of Paleolimnology* **22**, 41–52.

Brooks, S. J. (1997). The response of Chironomidae (Insecta: Diptera) assemblages to late-glacial climatic change in Kråkenes Lake, western Norway. *Quaternary Proceedings* **5**, 49–58.

Brooks, S. J. (2000). Late-glacial fossil midge stratigraphies (Insecta: Diptera: Chironomidae) from the Swiss Alps. *Palaeogeography, Palaeoclimatology, Palaeoecology* **159**, 261–279.

Brooks, S. J., and Birks, H. J. B. (2000a). Chironomid-inferred late-glacial and early-Holocene mean July air temperatures for Kråkenes Lake, western Norway. *Journal of Palaeolimnology* **23**, 77–89.

Brooks, S. J., and Birks, H. J. B. (2000b). Chironomid-inferred late-glacial air temperatures at Whitrig Bog, southeast Scotland. *Journal of Quaternary Science* **15**, 759–764.

Brooks, S. J., and Birks, H. J. B. (2001). Chironomid-inferred air temperatures from late-glacial and Holocene sites in north-west Europe: Progress and problems. *Quaternary Science Reviews* **20**, 1723–1741.

Brooks, S. J., Lowe, J. J., and Mayle, F. E. (1997). The Late Devensian lateglacial palaeoenvironmental record from Whitrig Bog, S.E. Scotland. 2. Chironomidae (Insecta: Diptera). *Boreas* **26**, 297–308.

Conolly, A. P. (1961). Some climatic and edaphic indications from the late-glacial flora. *Proceedings of the Linnean Society of London* **172**, 56–62.

Coope, G. R., and Lemdahl, G. (1995). Regional differences in the lateglacial climate of northern Europe based on coleopteran analysis. *Journal of Quaternary Science* **10**, 391–395.

Coope, G. R., Lemdahl, G., Lowe, J. J., and Walkling, A. (1998). Temperature gradients in northern Europe during the last glacial–interglacial transition (14–9 ^{14}C kyr BP) interpreted from coleopteran assemblages. *Journal of Quaternary Science* **13**, 419–433.

Gams, H. (1927). Die Gesschicte der Lunzer seen, moore und Wälder. *Internationale revue der gesamten Hydrobiologie und Hydrographie* **18**, 305–387.

Gandouin, E., and Franquet, E. (2002). Late Glacial and Holocene chironomid assemblages in 'Lac Long Inférieur' (southern France, 2090 m): Palaeoenvironmental and palaeoclimatic implications. *Journal of Paleolimnology* **28**, 317–328.

Guilizzoni, P., Marchetto, A., Lami, A., *et al.* (2000). Evidence for short-lived oscillations in the biological records from the sediments of Lago Albano (Central Italy) spanning the period ca. 28 to 17 k yr BP. *Journal of Paleolimnology* **23**, 117–127.

Heiri, O., (2001). *Holocene Paleolimnology of Swiss Mountain Lakes Reconstructed Using Subfossil Chironomid Remains: Past Climate and Prehistoric Human Impact on Lake Ecosystems*. PhD Thesis, University of Bern.

Heiri, O., and Millet, L. (2005). Reconstruction of Late Glacial summer temperatures from chironomid assemblages in Lac Lautrey (Jura, France). *Journal of Quaternary Science* **20**, 33–44.

Heiri, O., Lotter, A. F., Hausmann, S., and Kienst, F. (2003). A chironomid-based Holocene summer air temperature reconstruction from the Swiss Alps. *The Holocene* **13**, 477–484.

Hofmann, W. (1971). Die postglaziale Entwicklung der Chironomiden- und *Chaoborus*-Fauna (Dipt.) des Schöhsees. *Archiv für Hydrobiologie* (supplement 40): 1–74.

Hofmann, W. (1978). Analysis of animal microfossils from the Großer Segeberger See (F.R.G.). *Archiv für Hydrobiologie* **82**, 316–346.

Hofmann, W. (1983a). Stratigraphy of subfossil Chironomidae and Ceratopogonidae (Insecta: Diptera) in late glacial littoral sediments from Lobsigensee (Swiss Plateau). Studies in the Late Quaternary of Lobsigensee 4. *Revue de Paléobiologie* **2**, 205–209.

Hofmann, W. (1983b). Stratigraphy of Cladocera and Chironomidae in a core from a shallow North German lake. *Hydrobiologia* **103**, 235–239.
Hofmann, W. (1984). Stratigraphie subfossiler Cladocera (Crustacea) und Chironomidae (Diptera) in zwei Sedimentprofilen des Meerfelder Maares. *Courier Forschungs Institut Senckenberg* **65**, 67–80.
Hofmann, W. (1990). Weichselian chironomid and cladoceran assemblages from maar lakes. *Hydrobiologia* **214**, 207–211.
Hofmann, W. (1991). Stratigraphy of Chironomidae (Insecta: Diptera) and Cladocera (Crustacea) in Holocene and Wurm sediments from Lac du Bouchet (Haute Loire, France). *Documents du CERLAT* **2**, 363–386.
Johnsen, S. J., Dahl-Jensen, D., Gundestrup, N., Steffensen, J. P., Clausen, H. B., Miller, H., Masson-Delmotte, V., Sveinbjörnsdottir, A. E., and White, J. (2001). Oxygen isotope and palaeotemperature records from six Greenland ice core-stations: Camp Century, Dye-3, GRIP, GISP2, Renland and North GRIP. *Journal of Quaternary Science* **16**, 299–307.
Johnsen, S. J., Hammer, C. U., Iversen, P., Jouzel, J., Stauffer, N., and Steffensen, J. P. (1992). Irregular glacial interstadials recorded in a new Greenland ice core. *Nature* **359**, 311–331.
Levesque, A. J., Mayle, F. E., Walker, I. R., and Cwynar, L. C. (1993). The amphi-Atlantic oscillation: A proposed late-glacial climatic event. *Quaternary Science Reviews* **12**, 629–643.
Lotter, A. F., Birks, H. J. B., Eicher, U., Hofmann, W., Schwander, J., and Wick, L. (2000). Younger Dryas and Allerød summer temperatures at Gerzensee (Switzerland) inferred from fossil pollen and cladocrean assemblages. *Palaeogeography, Palaeoclimatology, Palaeoecology* **159**, 349–361.
Lotter, A. F., Eicher, U., Birks, H. J. B., and Siegenthaler, U. (1992). Late-glacial climatic oscillations as recorded in Swiss lake sediments. *Journal of Quaternary Science* **7**, 187–204.
Magny, M., Guiot, J., and Scoellammer, P. (2001). Quantitative reconstruction of Younger Dryas to mid-Holocene paleoclimates at Le Locle, Swiss Jura, using pollen and lake-level data. *Quaternary Research* **56**, 170–180.
Manca, M., Nocentini, A. M., Belis, C. A., Comoli, P., and Corbella, L. (1996). Invertebrate fossil remains as indicators of late Quaternary environmental changes in Latium crater lakes (L. Albano and L. Nemi). *Memorie dell'Istituto Italiano Idrobiologia* **55**, 149–176.
Millet, L., Verneaux, V., and Magny, M. (2003). Lateglacial paleoenvironmental reconstruction using subfossil chironomid assemblages from Lake Lautrey (Jura, France). *Archiv für Hydrobiologie* **156**, 405–429.
Renssen, H., and Isarin, R. F. B. (2001). The two major warming phases of the last deglaciation at ~14.7 and ~11.5 ka cal. BP in Europe: Climate reconstruction and AGCM experiments. *Global and Planetary Change* **30**, 117–153.
Sheldrick C (1997) *Plant Macrofossil Records from UK Lake Sediments Spanning the Last Glacial-Interglacial Transition, ca. 14–9^{14}C ka BP*. PhD thesis, University of London.
Sissons, J. B. (1979). Palaeoclimatic inferences from former glaciers in Scotland and the Lake District. *Nature* **278**, 518–521.
Sutherland, D. G. (1984). Modern glacier characteristics as a basis for inferring former climates with particular reference to the Loch Lomond stadial. *Quaternary Science Reviews* **3**, 291–309.
Von Graffenstein, U., Erlenkeuser, H., Brauer, A., Jouzel, J., and Johnsen, S. (1999). A mid-European decadal isotope climate record from 15,500 to 5000 years BP. *Science* **284**, 1654–1657.
Walker, I. R., Levesque, A. J., Pienitz, R., and Smol, J. P. (2003). Freshwater midges of the Yukon and adjacent Northwest Territories: A new tool for reconstructing Beringian paleoenvironments?. *Journal of the North American Benthological Society* **22**, 323–337.

Postglacial Europe

O Heiri, Ultrecht University, Ultrecht, The Netherlands

© 2007 Elsevier B.V. All rights reserved.

Introduction

The Chironomidae (nonbiting midges) are a family of the Diptera (true flies) whose larvae are abundant in most freshwater habitats. Individual lakes can contain more than 100 different chironomid species, many of which have a restricted distribution in respect to water chemistry parameters (e.g., nutrient condition, oxygen, pH) or temperature. The remains of chironomid larvae preserve well in the lake sediments and remain identifiable, usually to species-group or genus level. Therefore, past changes in the chironomid fauna of individual lakes can be reconstructed based on the fossil record (*see* Australia). The relationship between fossil chironomid 'types' (or taxa) and the lacustrine environment they originate from can be studied by analyzing the taxonomic composition of subfossil chironomid assemblages in the surface sediments of lakes. Commonly, the surficial 1–2 cm of sediment will be analyzed in each lake, usually originating from a sediment core taken from the lake center. These samples incorporate chironomid remains sedimented during the past years to a few decades at most and provide an integrated picture of the chironomid taxa inhabiting a lake. Surface sediment samples analyzed from lakes with different limnological conditions indicate that, even though the remains of many chironomids cannot be identified to species level, many of the taxa identifiable in fossil assemblages are indicative of the environment the larvae lived in (*see* Australia).

In Europe, analyses of fossil chironomid remains in Postglacial lake deposits have mainly focused on three research themes: (1) the natural development of lakes, (2) the effects of pollution on lake ecosystems and the reconstruction of past changes in water quality, and (3) the reconstruction of summer temperatures during the past ca. 11.5 kyr.

Reconstructions of Lake Ontogeny

Early applications of fossil chironomid analysis to lacustrine sediment records were mainly intended to reconstruct changes in lake nutrient and oxygen conditions related to the natural development of lake ecosystems. Since chironomids were considered a key indicator group for the classification of lakes according to their trophic state, fossil chironomid records were used to explore and test hypotheses related to lake ontogeny, such as the natural eutrophication of lakes, decreasing hypolimnetic oxygen conditions in lakes as a consequence of lake infilling, and the effects of natural acidification on lake ecosystems (see Walker (1987) for an extensive review of these topics). Although most of these studies originate from the early days of chironomid-based paleoenvironmental reconstruction, chironomids are still used occasionally to reconstruct the natural development of lakes. For example, Solem *et al.* (1997) analyzed chironomids to study the colonization and evolution of four Norwegian lakes after uplift and isolation of the lake basins from the sea.

Effects of Pollution on Lake Ecosystems and Reconstruction of Water Quality

Chironomids are one of the few benthic animal groups occurring in freshwater ecosystems that preserve well and remain identifiable in the fossil record. Furthermore, chironomids are known to be sensitive to changes in lake productivity, and the modern chironomid fauna has been used to classify lakes according to their nutrient conditions (e.g., Sæther (1979)). As a consequence, fossil chironomid records have been used to reconstruct and assess the effects of pollution on benthic macroinvertebrate assemblages in lakes. One commonly used approach is to analyze chironomids in the framework of a multiproxy study, usually including analyses of a number of other biotic and abiotic proxy indicators. Changes in chironomid assemblages are then examined in view of the ecology of the chironomid taxa found and in the context of the pollution history of the lake as inferred by the other proxy records analyzed in the same sediment core. In Europe, this approach has been widely used to study the response of chironomid assemblages to eutrophication and acidification (e.g., Brodin (1982), Heiri and Lotter (2003), Henrikson *et al.* (1982), Ilyashuk *et al.* (2003), Ilyashuk and Ilyashuk (2001), Itkonen *et al.* (1999), Kansanen (1985), Meriläinen and Hamina (1993), Räsänen *et al.* (1992), Schnell (1996), and Wiederholm and Eriksson (1979)). For example, Itkonen *et al.* (1999) used a combination of chironomid analysis and geochemical measurements to reconstruct the eutrophication history of Lake Päijänne, southern Finland. Chironomid assemblages in this deep lake (maximum depth 95 m) were dominated by *Heterotrissocladius subpilosus*, a species typical for oligotrophic lakes, and were comparatively stable during a large part of the past 8 kyr. However, in the uppermost part of the sediment sequence, accumulation rates of *H. subpilosus* remains were strongly reduced, and a number of other chironomid taxa such as *Sergentia coracina*, *Micropsectra* spp., and *Tanytarsus* spp. increased in abundance. These changes were dated to ca. AD 1970 and Itkonen *et al.* (1999) conclude that human impact during the past few decades had changed the ecosystem of Lake Päijänne to an extent never experienced before during the past 8 kyr. In a different subbasin of the same lake, Meriläinen and Hamina (1993) analyzed subfossil chironomid assemblages in a sediment core encompassing the past ca. 150 yr. Based on the identified chironomid taxa, these authors calculated the benthic quality index (BQI), a water quality index defined for live chironomid assemblages in deep stratified lakes (Wiederholm, 1980), in order to trace the local eutrophication history of the lake (**Fig. 1**). Meriläinen and Hamina's (1993) results indicate that the studied subbasin of Lake Päijänne experienced the strongest phase of eutrophication during the 1970s when a number of subfossil remains of *Chironomus anthracinus* were deposited in the sediments, and that the state of the lake has slightly improved since. Another example of the use of fossil chironomid records for the reconstruction of lake pollution is the study by Brodin and Gransberg (1993) that describes the effects of acidification on the chironomid fauna of Round Loch of Glenhead, a Scottish lake. Changes in chironomid assemblages during the past ca. 130 yr were compared with a diatom-based pH reconstruction, and the authors concluded that decreasing pH clearly affected the chironomid assemblages. However, the chironomids abundant before acidification were still present at a diatom-inferred pH below 5, and Brodin and Gransberg (1993) suggested that chironomids may be more resistant to acidification than other aquatic animal groups. Often the pollution history of a lake can be further constrained by using other sources, for example, published water-sampling series, or historical records. If the surface sediment layers can be adequately dated (e.g., using ^{137}Cs or ^{210}Pb dating), the downcore chironomid record can be directly compared with instrumental measurements, or records of past industrial pollution. In the case of the chironomid-based reconstruction from Lake Päijänne, Meriläinen and Hamina (1993) were able to show that, during the period of highest pollution input into the lake, the deterioration of the water quality as indicated by the BQI showed a close agreement with the biochemical oxygen demand (BOD) of discharge from pulp mills located close to the lake basin where the sediment core was taken. However, BQI values tended to lag changes in BOD loading by several years (Meriläinen and Hamina, 1993).

Recently attempts have been made to produce quantitative inference models (or transfer functions (*see* Chironomid Overview)) relating water-quality

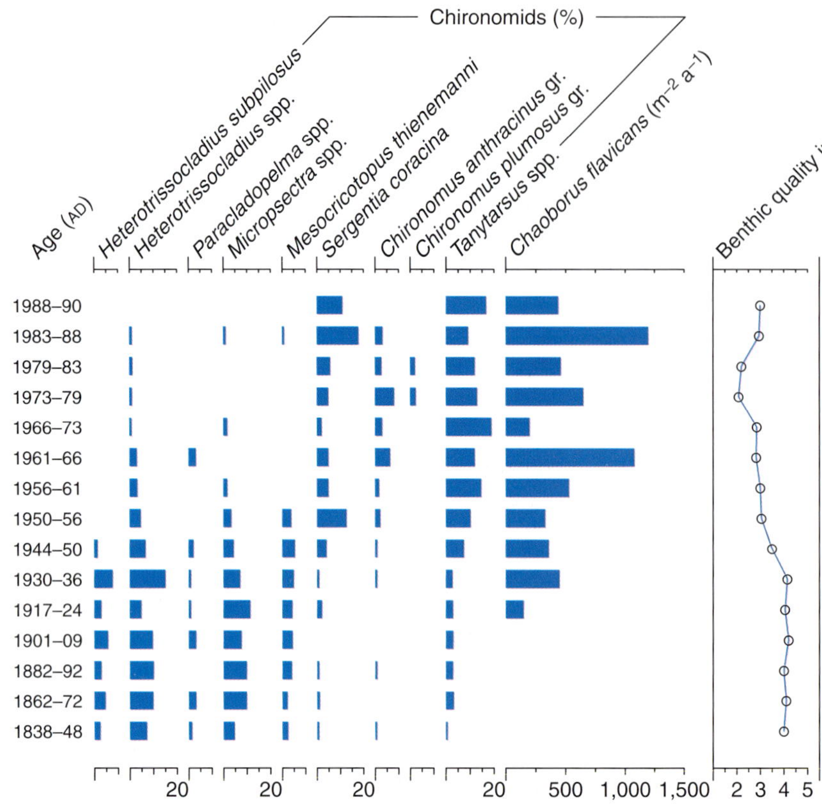

Figure 1 Abundance of selected fossil chironomid taxa (as % of total chironomid remains) and *Chaoborus flavicans* mandibles (as accumulation rates) in the uppermost sediments of Poronselkä basin, Lake Päijänne, south Finland. The benthic quality index (BQI) was calculated for individual samples based on the fossil chironomid assemblages. Data and BQI values from Meriläinen JJ and Hamina V (1993). Recent environmental history of a large, originally oligotrophic lake in Finland: A palaeolimnological study of chironomid remains. *Journal of Paleolimnology* 9: 129–140.

variables such as chlorophyll *a* or total phosphorus concentrations to subfossil chironomid assemblages analyzed in surface sediments of lakes. Initial down-core applications of these models have provided promising results (e.g., Brodersen *et al.* (2001) and Brooks *et al.* (2001)) and show that, within the prediction error of the applied models, chironomid-inferred water chemistry variables were consistent with diatom-inferred reconstructions. However, these transfer functions remain to be tested on additional sediment records and versus instrumental water-quality measurement series. In contrast with North America, where fossil chironomids have been used to quantitatively reconstruct past oxygen conditions in lakes a chironomid–oxygen transfer function is not yet available for Europe.

Chironomid-Based Summer Temperature Reconstruction

Fossil chironomids have received by far the most attention in recent years as quantitative paleotemperature indicators. At temperate and subarctic latitudes in Europe, the distribution of chironomid taxa in lake-surface sediments is closely related to summer air and water temperature. Hence, chironomid–summer temperature transfer functions can be developed based on these chironomid assemblages. At present, chironomid–summer air or water temperature transfer functions are available from different regions in Europe, including Finnish Lapland (e.g., Olander *et al.* (1999)), northern Sweden (e.g., Larocque *et al.* (2001)), Norway (e.g., Brooks and Birks (2001)), and the Swiss Alps (e.g., Heiri *et al.* (2003)). Chironomid-based temperature reconstruction has shown great potential for inferring past summer temperature during the major climatic fluctuations during the Lateglacial period (*see* Late Pleistocene of Europe). Due to a number of reasons, however, temperature reconstructions during the Postglacial period have proved to be more problematic. Available evidence indicates that temperature changes in Europe during the past c. 11.5 kyr were of the same order of magnitude as the prediction errors of chironomid–temperature transfer functions (ca. 1–1.5° C). Furthermore, during the Postglacial period, changes in chironomid-inferred temperatures are often driven by relatively minor changes in fossil chironomid assemblage composition. Therefore, these temperature signals are often difficult to detect in chironomid stratigraphies and difficult to separate from other influences on the chironomid fauna. For example, Brooks (2003) presents a chironomid-based summer temperature record from Holebudalen, southern Norway. In this reconstruction, a distinct cooling in inferred July air temperature of ca. 1.2° C is reconstructed at ca. 8.3 cal kyr BP (**Fig. 2**). This cooling most likely represents the 8.2 ka cooling event and is clearly expressed in the chironomid-based temperature reconstruction. However, the episode is accompanied by only comparatively minor changes in the chironomid fauna (**Fig. 2**). Brooks (2003) indicates that this decrease in inferred temperatures is largely driven by the absence of *Chironomus anthracinus* type from the fossil chironomid assemblages at ca. 8.3 cal kyr BP, a taxon at present restricted to comparatively warm Norwegian lakes.

A further confounding factor for Postglacial chironomid-based temperature reconstruction is that it is presently difficult to anticipate which Postglacial temperature development should be expected for different parts of Europe, and this complicates the assessment of whether individual records successfully reconstruct past summer temperature variability. This is illustrated by the situation in Swedish and Finnish Lapland where a number of Holocene temperature reconstructions have been developed using two different chironomid–temperature transfer functions (Korhola *et al.*, 2002; Larocque and Hall, 2004; Seppä *et al.*, 2002). All three reconstructions from northern Sweden (Larocque and Hall, 2004) and one record from Finnish Lapland (Seppä *et al.*, 2002) show a more or less continuous summer temperature decrease from 9 cal kyr BP to present. This would be expected for high northern latitudes during the mid- and late Holocene due to a decrease in summer insolation in this region related to orbital changes. The second chironomid-based summer temperature record from Finnish Lapland also indicates higher temperatures at ca. 9 cal kyr BP than at present (Korhola *et al.*, 2002). However, it also features periods of distinctly cooler inferred temperatures between ca. 6 and 4 kyr BP and, depending on the numerical method used for the transfer function, possibly another cooling episode at ca. 3–2 cal kyr BP. These coolings are not expressed as distinctly in the other chironomid-based temperature reconstructions from northern Fennoscandia, and it is difficult to decide whether these cold episodes in one of the five records are related to regional differences in Postglacial climate in northern Scandinavia, whether they are biases related to the use of different numerical methods and transfer functions, or whether they are the consequence of other factors than temperature influencing the chironomid assemblages. In southern Norway, Velle *et al.* (2005) applied a chironomid–July air temperature transfer function to six Postglacial chironomid profiles. These authors also report differences in the chironomid-inferred temperature development between the five different records and suggest that these discrepancies are either related to different microclimates at the five study sites, or related to a different sensitivity of the chironomid fauna of the individual lakes to changes in summer temperature. Velle *et al.* (2005) provide a detailed discussion of potential causes other than differences in microclimate for these discrepancies in the temperature records and recommend a multisite approach to chironomid-based temperature reconstruction for inferring millennial-scale temperature trends during the Holocene.

An alternative approach to evaluate downcore temperature reconstructions is the comparison of individual chironomid-based temperature records with independent reconstructions from the same geographical region that are based on other paleotemperature proxy indicators. For example, Heiri *et al.* (2004) present a chironomid-based July air temperature reconstruction for the early and mid-Holocene from the Swiss Alps which shows a close agreement with an independently dated pollen-based reconstruction of past tree-line altitude (**Fig. 3**). At present, tree-line altitude in the Alps is largely constrained by temperatures during the growing-season (*see* Treeline Studies). The close agreement between the chironomid-based July temperature record and the reconstruction of past tree-line altitude from two sites more than 60 km apart suggests that, in this case, the chironomid-based reconstruction successfully captured past centennial-scale changes in summer temperature.

As discussed in Velle *et al.* (2005) and Brooks (2003), changes in the lake catchment and in lake-water chemistry can influence the Postglacial development of chironomid assemblages and potentially lead to a bias in chironomid-inferred temperatures. For example, natural changes in catchment vegetation and soils can affect nutrient conditions in lakes. As discussed above, chironomid assemblages react sensitively to changes in lake trophic state, and it can therefore be expected that natural changes in lake primary productivity in the past may have affected fossil chironomid records. Similarly, vegetation changes, especially on acid bedrock, can affect pH and dissolved organic carbon (DOC) concentrations of lake water and thereby chironomid assemblages. For example, in Scandinavia a distinct chironomid assemblage tends to dominated humic (high-DOC) lakes, which typically feature a low

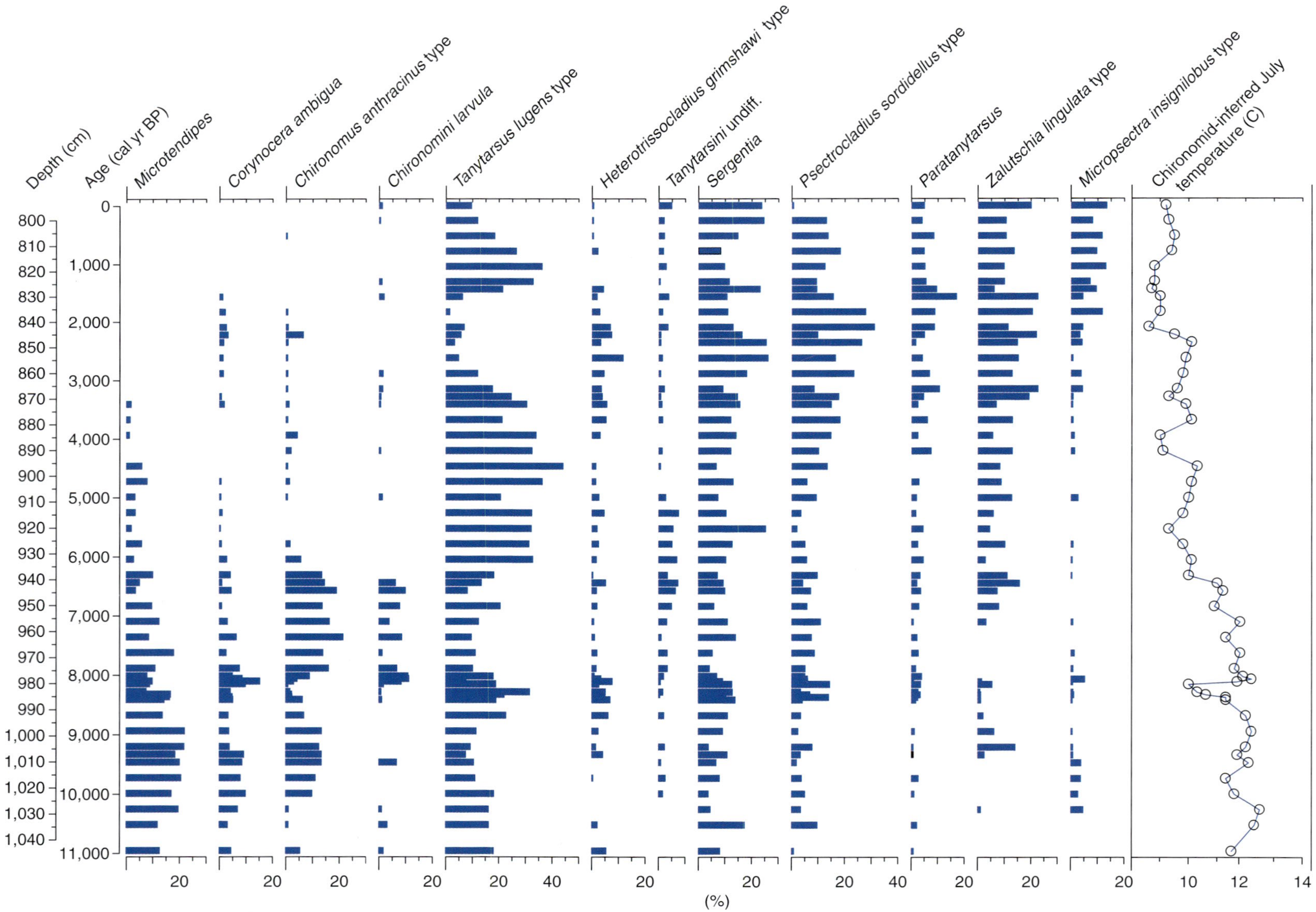

Figure 2 Chironomid assemblages in the Holocene sediments of Holebudalen, a small Alpine lake in western Norway, and chironomid-inferred July air temperatures (°C) for individual assemblages. Data from Brooks SJ (2003). Chironomid analysis to interpret and quantify Holocene climate change. In: Mackay A, Batterbee RW, Birks HJB, and Oldfield F (eds.) *Global change in the Holocene*, p. 328–341. London: Arnold; Velle G, Brooks SJ, Birks HJB, and Willassen E. (2005). Chironomids as a tool for inferring Holocene climate: An assessment based on six sites in southern Scandinavia. *Quaternary Science Reviews* 24: 1429–1462.

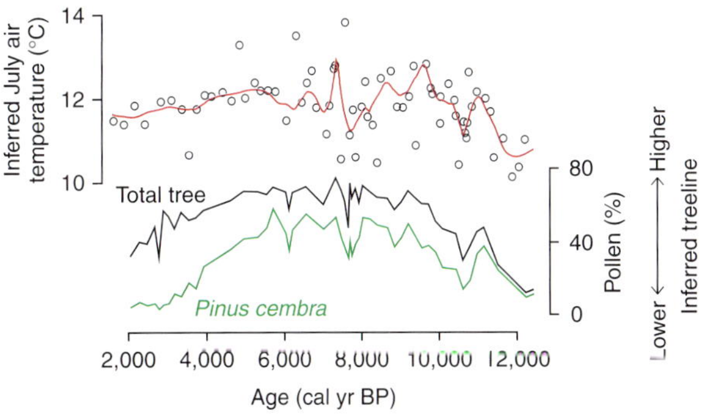

Figure 3 Chironomid-inferred temperatures and reconstruction of relative shifts in treeline altitude from the Swiss Alps. July air temperatures are reconstructed from chironomid assemblages in the sediments of Hinterburgsee, as small subalpine lake in the Northern Swiss Alps. The record is smoothed using a locally weighted regression (LOWESS) smoother. Relative changes in past tree-line altitude were reconstructed based on total tree and *Pinus cembra* pollen in the sediments of Gouillé Rion, a small lake situated at present-day tree line in the Central Swiss Alps. Data from Heiri O, Tinner W, and Lotter AF (2004). Evidence for cooler European summers during periods of changing melt-water flux to the North Atlantic. *Proceedings of the National Academy of Sciences of the United States of America* 101: 15285–15288.

productivity combined with low oxygen conditions in the bottom waters. In the Postglacial period human impact on lake ecosystems can further complicate the use of chironomids for paleotemperature reconstruction purposes. During the historical period, human land use and industrial activity have led to a widespread eutrophication of lakes in many parts of Europe. Furthermore, prehistoric agriculture and land-use changes may have influenced lacustrine ecosystems as well. Prehistoric anthropogenic effects on lakes were probably not as widespread as lake eutrophication during the historical period. However, evidence suggests that these effects may have been considerable locally. For example, Heiri and Lotter (2003) present a fossil chironomid record covering the past 9 kyr from the Bernese Alps, Switzerland. During three distinct phases, the earliest starting at ca. 4 cal kyr BP, the accumulation rate of the dominant lacustrine chironomid taxa was strongly reduced in the record, whereas most running water and littoral taxa showed no changes in accumulation rate. A comparison with other records produced within the framework of a multiproxy study indicated that these decreases in lacustrine chironomids occurred during phases for which geochemical measurements indicate reduced oxygen conditions in the lake. Furthermore, for all three of these phases, paleobotanical proxies showed clear evidence of increased human activity in the lake catchment, such as forest clear-cutting and pasturing. Combined, these records suggest that prehistoric human activity led to anoxic conditions during three distinct periods in the lake's past and reduced the abundance of or even eliminated some of the chironomid species living in the lake. When a chironomid–temperature transfer function was applied to this record, these episodes were accompanied by distinct coolings in the chironomid-inferred temperatures, and Heiri and Lotter (2003) concluded that these apparent coolings were most likely an artifact of prehistoric human impact on the lake ecosystem rather than a genuine temperature signal.

The examples discussed above indicate that postglacial chironomid-based paleotemperature records can be adversely affected by influences other than summer temperature on chironomid assemblages. Next to changes in water chemistry variables, such as nutrients, pH, DOC, and oxygen, these influences potentially include variations in lake level, stocking of lakes with fish, contamination of lake water or sediments by heavy metals or persistent organic pollutants, and, in some parts of Europe, changes in the salinity of standing waters. It is likely that many of these factors have influenced fossil chironomid records during the Lateglacial period as well. However, since postglacial temperature fluctuations in Europe were of lower amplitude and less distinct than variations during the Lateglacial, it is more difficult to separate the climatic signal from nonclimatic influences in chironomid profiles and in chironomid-based temperature reconstructions. It is therefore essential that, whenever possible, postglacial chironomid-based paleotemperature records are evaluated against other available reconstructions of within-lake conditions and of past climatic variability to assess the potential effects of nonclimatic influences on the reconstructions. Approaches and experimental schemes that have been successfully used to achieve this aim and to produce chironomid-based temperature records with a high reliability include single-site multiproxy reconstructions (see, e.g., examples given in Brooks (2003)), multisite chironomid-based reconstructions (e.g., Larocque and Hall (2004) and Velle *et al.* (2005)), or multiproxy, multisite reconstructions (e.g., Heiri *et al.* (2004)). The third of these options seems particularly promising as it allows the combination of the rather labor-intensive chironomid-based approach to temperature reconstruction with proxies that have been shown to be sensitive to past temperature change, but remain difficult to use on their own to quantitatively reconstruct postglacial temperature variability (e.g., pollen or oxygen isotopes in lake marl). The role of the chironomid-based reconstruction is then to quantify past summer temperature changes, whereas the additional nonchironomid proxies provide independent support for the chironomid-based inferences or point toward problematic parts of the record. In many combinations the different paleotemperature proxies would be affected by past temperature change via fundamentally different mechanisms. A close agreement between different paleotemperature records based on different proxies and from different localities in the same region would therefore increase the confidence that past temperatures rather than other environmental variables were the driving force behind chironomid-inferred changes in summer temperature.

Future Developments

In recent years most research on fossil chironomids has focused on their use as paleotemperature proxies. Not surprisingly most of these studies originate from regions for which local chironomid–temperature transfer functions are available (**Fig. 4**). Temperature transfer functions are most easily developed in regions with large temperature gradients in a geographically restricted area. Therefore, with the exception of a chironomid-based temperature inference model from Finland, all available transfer functions and many of the available postglacial temperature reconstructions in Europe originate from mountainous regions. Considering the promising results of the initial studies, the ongoing public interest in climate research, and the need for more information on natural climate variability during the preinstrumental period, it is expected that postglacial chironomid-based temperature reconstruction will continue to remain a focus of research in the near future. Next to methodological advancements such as the development of new numerical tools for transfer function development and an increase in the taxonomic resolution of fossil chironomid records, one of the main foci will most likely be the expansion of the spatial coverage of chironomid-based temperature records and the taxonomic harmonization and cross-evaluation of the existing calibration data sets and transfer functions. At present chironomid-based paleotemperature reconstructions covering significant parts of the Postglacial period have been published from Scandinavia, the British Isles, Iceland, the Alps, and the Kola Peninsula (**Fig. 4**). In contrast, chironomid-based temperature reconstructions for this period are still lacking from Eastern Europe, the Northwest European lowlands, France, Germany, and with the exception of a single record covering the past 200 yr, southern Europe.

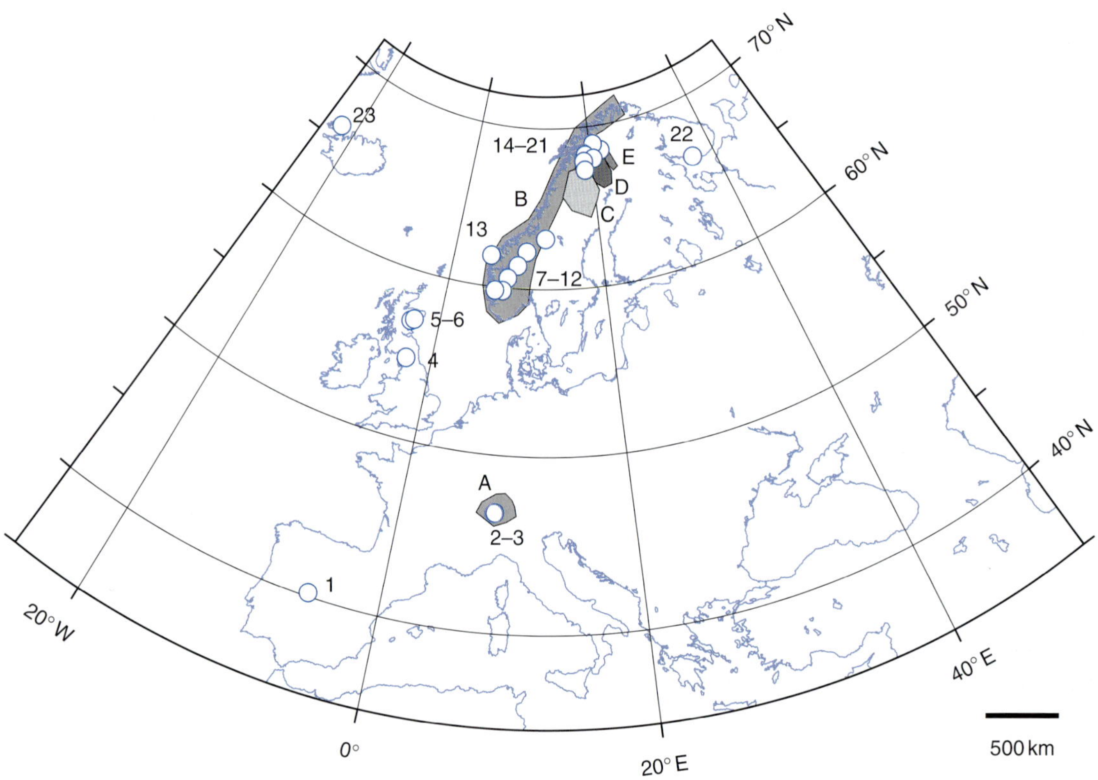

Figure 4 Surface-sediment transects used to produce chironomid–summer temperature transfer functions (shaded areas) and postglacial chironomid-based temperature reconstructions available from Europe (circles) (data compiled August 2005). A. Heiri *et al.* (2003), B. Brooks and Birks (2001), C. Rosén *et al.* (2001), D. Larocque *et al.* (2001), E. Olander *et al.* (1999), 1. Granados and Toro (2000), 2. Heiri *et al.* (2004), 3. Heiri and Lotter (2003), 4. Langdon *et al.* (2004), 5. Brooks and Birks (2001), 6. Dalton *et al.* (2005), 7–12. Velle *et al.* (2005), 13. Brooks and Birks (2001), 14. Rosén *et al.* (2001), 15. Rosén *et al.* (2003), 16. Heinrichs *et al.* (2005), 19. Larocque and Hall (2004), 20. Korhola *et al.* (2002), 21. Seppä *et al.* (2002), 22. Ilyashuk *et al.* (2005), 23. Caseldine *et al.* (2003).

Chironomid remains are highly resistant to microbial degradation and chemical alteration after their burial in the sediments. Recently, oxygen isotopes have been successfully measured on chironomid fossils to produce records of $\delta^{18}O$ in paleoprecipitation, and indirectly paleotemperature variability, from Baffin Island and Greenland (Wooller *et al.*, 2004). A widespread application of this method is still hampered by methodological problems, foremost the large number of chironomid remains needed for an individual measurement. However, if successfully replicated, this approach could provide an alternative to stable isotope (*see* Lake Sediments) measurements on other lacustrine sediment components such as lake marl and diatoms, and significantly expand the applicability and potential of fossil chironomids as paleoclimatic indicators.

Next to paleoclimate reconstruction, a second research field within paleolimnology that has received an increasing amount of attention is the reconstruction of preanthropogenic water-quality and ecosystem-baseline conditions for freshwater ecosystems. In Europe, this research has obtained further relevance due to recent developments in European legislature, since the ecological status of lakes and rivers is now defined in relation to the expected state of these ecosystems under minimal anthropogenic influence. Compared with other freshwater indicators such as diatoms, fossil chironomids have received comparatively little attention for reconstructing parameters reflecting past water-quality conditions. However, the examples mentioned above indicate that fossil chironomids have some potential as quantitative indicators of past nutrient conditions and productivity in lakes. Furthermore, chironomids are one of the few benthic macroinvertebrate groups living in profundal environments that preserve well and remain identifiable in most lacustrine sediments. There is, therefore, considerable potential for using chironomids to reconstruct the natural faunal assemblages in European lakes during the preanthropogenic period, and for following the North American example and using chironomids as quantitative indicators of past oxygen conditions in the bottom waters of lakes.

See also: **Carbonate Stable Isotopes**: Lake Sediments. **Chironomid Overview**. **Chironomid Records**: Late Pleistocene of Europe. **Lake Level Studies**: Australia. **Plant Macrofossil Methods and Studies**: Treeline Studies.

References

Brodersen, K. P., Anderson, N. J., and Odgaard, B. V. (2001). Long-term trends in the profundal chironomid-fauna in nitrogen-limited Lake Esrom, Denmark: A combined palaeolimnological/historical approach. *Archiv Für Hydrobiologie* **150**, 393–409.

Brodin, Y. (1982). Palaeoecological studies of the recent development of the lake Växjösjön IV. Interpretation of the eutrophication process through the analysis of subfossil chironomids. *Archiv Für Hydrobiologie* **93**, 313–326.

Brodin, Y. W., and Gransberg, M. (1993). Responses of insects, especially Chironomidae (Diptera), and mites to 130 years of acidification in a Scottish lake. *Hydrobiologia* **250**, 201–212.

Brooks, S. J. (2003). Chironomid analysis to interpret and quantify Holocene climate change. In *Global Change in the Holocene* (A. Mackay, R. W. Batterbee, H. J. B. Birks and F. Oldfield, Eds.), pp. 328–341. Arnold, London.

Brooks, S. J., Bennion, H., and Birks, H. J. B. (2001). Tracing lake trophic history with a chironomid-total phoshorus inference model. *Freshwater Biology* **46**, 513–533.

Brooks, S. J., and Birks, H. J. B. (2001). Chironomid-inferred air temperatures from Lateglacial and Holocene sites in north-west Europe: Progress and problems. *Quaternary Science Reviews* **20**, 1723–1741.

Caseldine, C., Geirsdottir, A., and Langdon, P. (2003). Efstadalsvatn–a multi-proxy study of a Holocene lacustrine sequence from NW Iceland. *Journal of Paleolimnology* **30**, 55–73.

Dalton, C., Birks, H. J. B., Brooks, S. J., Cameron, N. G., Evershed, R. P., Peglar, S. M., Scott, J. A., and Thompson, R. (2005). A multi-proxy study of lake-development in response to catchment changes during the Holocene at Lochnagar, north-east Scotland. *Palaeogeography, Palaeoclimatology, Palaeoecology* **221**, 175–201.

Granados, I., and Toro, M. (2000). Recent warming in a high mountain lake (Laguna Cimera, Central Spain) inferred by means of fossil chironomids. *Journal of Limnolology* **59**, 109–119.

Heinrichs, M. L., Peglar, S. M., Bigler, C., and Birks, H. J. B. (2005). A multi-proxy palaeoecological study of Alanen Laanijarvi, a boreal-forest lake in Swedish Lapland. *Boreas* **34**, 192–206.

Heiri, O., and Lotter, A. F. (2003). 9000 years of chironomid assemblage dynamics in an Alpine lake: Long-term trends, sensitivity to disturbance, and resilience of the fauna. *Journal of Paleolimnology* **30**, 273–289.

Heiri, O., Lotter, A. F., Hausmann, S., and Kienast, F. (2003). A chironomid-based Holocene summer air temperature reconstruction from the Swiss Alps. *Holocene* **13**, 477–484.

Heiri, O., Tinner, W., and Lotter, A. F. (2004). Evidence for cooler European summers during periods of changing meltwater flux to the North Atlantic. *Proceedings of the National Academy of Sciences of the United States of America* **101**, 15285–15288.

Henrikson, L., Olofson, J. B., and Oscarson, H. G. (1982). The impact of acidification on Chironomidae (Diptera) as indicated by subfossil stratification. *Hydrobiologia* **86**, 223–229.

Ilyashuk, B. P., and Ilyashuk, E. A. (2001). Response of alpine chironomid communities (Lake Chuna, Kola Peninsula, northwestern Russia) to atmospheric pollution. *Journal of Paleolimnology* **25**, 467–475.

Ilyashuk, B., Ilyashuk, E., and Dauvalter, V. (2003). Chironomid responses to long-term metal contamination: A paleolimnological study in two bays of Lake Imandra, Kola Peninsula, northern Russia. *Journal of Paleolimnology* **30**, 217–230.

Ilyashuk, E. A., Ilyashuk, B. P., Hammarlund, D., and Larocque, I. (2005). Holocene climatic and environmental changes inferred from midge records (Diptera: Chironomidae, Chaoboridae, Ceratopogonidae) at Lake Berkut, southern Kola Peninsula, Russia. *The Holocene* **15**, 897–914.

Itkonen, A., Marttila, V., Meriläinen, J. J., and Salonen, V.-P. (1999). 8000-year history of palaeoproductivity in a large boreal lake. *Journal of Paleolimnology* **21**, 271–294.

Kansanen, P. H. (1985). Assessment of pollution history from recent sediments in Lake Vanajavesi, southern Finland. II: Changes in the Chironomidae, Chaoboridae and Ceratopogonidae (Diptera) fauna. *Annales Zoologici Fennici* **22**, 57–90.

Korhola, A., Vasko, K., Toivonen, H., and Olander, H. (2002). Holocene temperature changes in northern Fennoscandia reconstructed from chironomids using Bayesian modeling. *Quaternary Science Reviews* **21**, 1841–1860.

Langdon, P., Barber, K. E., and Lomas-Clarke, S. H. (2004). Reconstructing climate and environmental change in northern England through chironomid and pollen analyses: evidence from Talkin Tarn, Cumbria. *Journal of Paleolimnology* **32**, 197–213.

Larocque, I., and Hall, R. I. (2004). Holocene temperature estimates and chironomid community composition in the Abisko Valley, northern Sweden. *Quaternary Science Reviews* **23**, 2453–2465.

Larocque, I., Hall, R. I., and Grahn, E. (2001). Chironomids as indicators of climate change: a 100-lake training set from a subarctic region of northern Sweden (Lapland). *Journal of Paleolimnology* **26**, 307–322.

Meriläinen, J. J., and Hamina, V. (1993). Recent environmental history of a large, originally oligotrophic lake in Finland: A palaeolimnological study of chironomid remains. *Journal of Paleolimnology* **9**, 129–140.

Olander, H., Birks, H. J. B., Korhola, A., and Blom, T. (1999). An expanded calibration model for inferring lakewater and air temperatures from fossil chironomid assemblages in northern Fennoscandia. *Holocene* **9**, 279–294.

Räsänen, M., Salonen, V.-P., Salo, J., Walls, M., and Sarvala, J. (1992). Recent history of sedimentation and biotic communities in lake Pyhajarvi, SW Finland. *Journal of Paleolimnology* **7**, 107–126.

Rosén, P., Segerström, U., Eriksson, L., Renberg, I., and Birks, H. J. B. (2001). Holocene climatic change reconstructed from diatoms, chironomids, pollen and near-infrared spectroscopy at an alpine lake (Sjuodjijaure) in northern Sweden. *The Holocene* **11**, 551–562.

Rosén, P., Segerström, U., Eriksson, L., and Renberg, I. (2003). Do diatom, chironomid, and pollen records consistently infer Holocene July air temperature? A comparison using sediment cores from four alpine lakes in northern Sweden. *Arctic, Antarctic and Alpine Research* **35**, 279–290.

Saether, O. A. (1979). Chironomid communities as water quality indicators. *Holarctic Ecology* **2**, 65–74.

Schnell, Ø. A. (1996). The chironomid (Diptera) communities in two sediment cores from Store Hovvatn, S. Norway, an acidified lake. *Annals of Limnology* **32**, 45–61.

Seppä, H., Nyman, M., Korhola, A., and Weckström, J. (2002). Changes of treelines and alpine vegetation in relation to postglacial climate dynamics in northern Fennoscandia based on pollen and chironomid records. *Journal of Quaternary Science* **17**, 287–301.

Solem, J. O., Solem, T., Aagard, K., and Hanssen, O. (1997). Colonization and evolution of lakes on the central Norwegian coast following deglaciation and land uplift 9500 to 7800 years B.P. *Journal of Paleolimnology* **18**, 269–281.

Velle, G., Brooks, S. J., Birks, H. J. B., and Willassen, E. (2005). Chironomids as a tool for inferring Holocene climate: An assessment based on six sites in southern Scandinavia. *Quaternary Science Reviews* **24**, 1429–1462.

Walker, I. R. (1987). Chironomidae (Diptera) in Paleolimnology. *Quaternary Science Reviews* **6**, 29–40.

Wiederholm, T. (1980). Chironomids as indicators of water quality in Swedish lakes. *Acta Universitatis Carolinae – Biologia* **1978**, 275–283.

Wiederholm, T., and Eriksson, L. (1979). Subfossil chironomids as evidence of eutrophication in Ekoln Bay, Central Sweden. *Hydrobiologia* **62**, 195–208.

Wooller, M. J., Francis, D., Fogel, M. L., Miller, U., Walker, I. R., and Wolfe, A. P. (2004). Quantitative paleotemperature estimates from $\delta^{18}O$ of chironomid head capsules preserved in arctic lake sediments. *Journal of Paleolimnology* **31**, 267–274.

Postglacial Southern Hemisphere

J Massaferro, Universidad del Comahue, Bariloche, Argentina
M Vandergoes, University of Maine, ME, USA

© 2007 Elsevier B.V. All rights reserved.

Introduction

The use of fossil chironomid remains as indicators of past environmental conditions is now a well-established approach (Walker (2001); *see* Chironomid Overview). From reconstructing trophic status to climate, chironomids have been deemed among the best indicators for their ability to provide quantifiable estimates of change. The use of chironomids in postglacial environmental reconstructions from Australia, New Zealand, and South America has not been extensive in the past, but in recent years has undergone a resurgence of interest by many workers following the success of chironomid-based reconstructions from Europe and North America. A recent review (Verschuren and Eggermont, 2006) provides a detailed account of chironomid paleoecology in the Southern Hemisphere to date. In this article, we review the published information about postglacial fossil chironomids from Australia, New Zealand, and South America and outline the directions that chironomid research has followed in these areas. We also discuss the importance of chironomid research in the Southern Hemisphere, identifying directions for future research.

Australasian–South American Fossil Chironomid Studies

Australia

Two paleoecological studies that primarily focus on chironomids as paleoindicators have been published from Australia. One study investigated changes in paleoproductivity, the other reconstructed paleoclimate. Paterson and Walker (1974) used the environmental range of tolerance of the chironomid *Tanytarsus barbitarsis* to investigate historical salinity changes in Lake Werowrap (38° 15′ S, 143° 30′ E) in western Victoria (**Fig. 1**). Based on experiments identifying the range of tolerance of *T. barbitarsis* to total dissolved solids (salinity), the authors concluded that freshwater conditions occurred in the early part of the lake's history. An increase in abundance of *T. barbitarsis* midway through the sequence signified a shift to more saline conditions. This condition persisted until present, with only minor episodes of salinity reduction during this time. The time span for the Lake

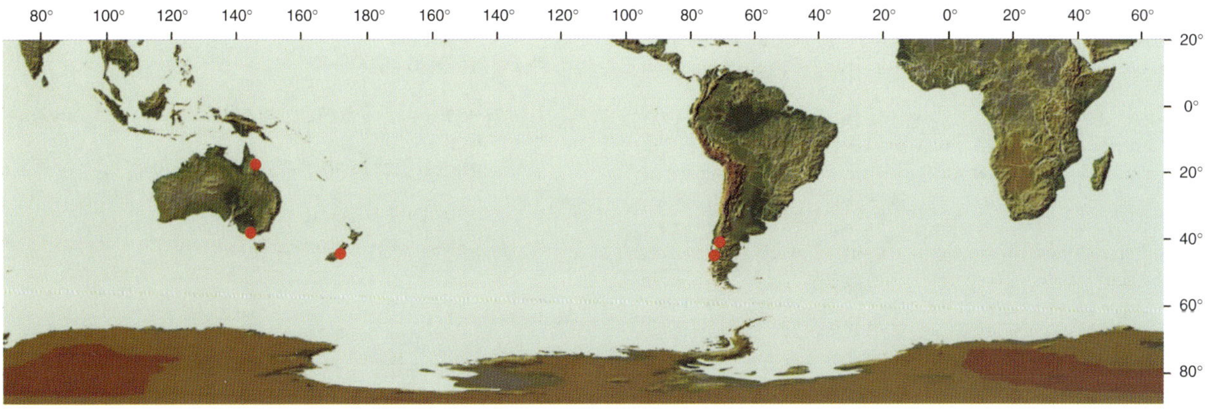

Figure 1 Main regions for postglacial chironomid studies in Australia, New Zealand, and South America.

Werowrap sequence remains unknown due to the lack of chronology, but it is likely to represent changes since the mid- to late Holocene.

The only study in the Australia to provide a quantitative reconstruction of Holocene climate using chironomids was carried out by Dimitriadis and Cranston (2001) from Lake Barrine (17° 15′ S, 145° 38′ E), on the Atherton Tablelands in tropical northeast Queensland. This study examined the relationship between the physical and chemical parameters of more than 68 water bodies and their modern chironomid fauna to assess the relative importance of climate and environmental variables on eastern Australian chironomids. They verified a significant relationship between climate and chironomids. There was a correlation (r^2) of 0.94 with mean air temperature of the warmest quarter and 0.59 with mean precipitation of the driest quarter. Through the use of a mutual climatic range method (*see* History of Quaternary Science), rather than the more commonly used weighted averaging method (*see* Late Pleistocene of Europe; Chironomid Overview), Holocene temperatures were reconstructed for Lake Barrine over the last 9 cal kyr. Temperature reconstructions indicated that annual temperatures were warmer by ~2.5° C from 9.01 to 6.89 cal kyr BP but may have been up to 6.6° C cooler in the summer months between 6.89 and 6.09 cal kyr BP (**Fig. 2**). Temperatures remained approximately 2.5° C warmer than the current mean from 6.09 cal kyr BP to the present day. Precipitation estimates indicate drier climates by approximately 500 mm of moisture per year from 6.89 to 6.09 and 5.130 to 2.650 cal kyr BP, but large ranges associated with these estimates make these reconstructions equivocal. Although estimates of 6.6° C temperature depressions seem extreme for periods during the Holocene, evidence from regional pollen records provide some support for the chironomid-based estimates and imply similar changes in temperature at these times during the Holocene. The overall results from this study are extremely promising and the authors note that these approaches will produce greater reliability as more information about the full distribution of Australian chironomids and their habitats becomes available.

New Zealand

Paleoecological studies involving chironomids from New Zealand are primarily qualitative investigations focusing on paleolimnological changes such as lake productivity or catchment and lake system development. Early work was carried out by Deevey (1955) from a swamp deposit sequence in the upper Pyramid Valley (42°58′S, 172° 35′ E), central South Island. In this study, Deevey (1955) used a range of terrestrial and aquatic macrofossils including cladocera, ostracods, mollusks, plant macrofossils, charophytes, green algae, pollen, diatoms, and chironomids to reconstruct lake history. The dynamics of these proxies were interpreted to indicate trophic changes and lake-level fluctuations during the past ca. 3.9 cal kyr. Interpretations from this study are generalized mainly due to a poor taxanomic resolution and ecological understanding of New Zealand freshwater biota at the time.

In two other studies, chakau (1986, 1990) examined chironomids from Holocene sequences in two lakes, Lakes Taylor (42°46′ S, 172°14′ E) and Grasmere (43° 05′ S, 171°45′ E), both located in the central South Island of New Zealand. In Lake Taylor (Shakau, 1986), the chironomid assemblage spanning the last ca. 3.2 cal kyr indicated a higher abundance of Tanytarsini in the older sediments and a shift in dominance from Tanytarsini to Chironomini after ca. 2.5 cal kyr BP. This shift was not statistically significant and therefore not considered to be a response to climate change or lake trophic status.

At Lake Grasmere (**Fig. 3**) (Shakau, 1990), changes in chironomid assemblages were used to reconstruct lake paleoenvironmetal history over the last 6.8 cal kyr (**Fig. 3B**). The sequence is dominated by *Corynocera*

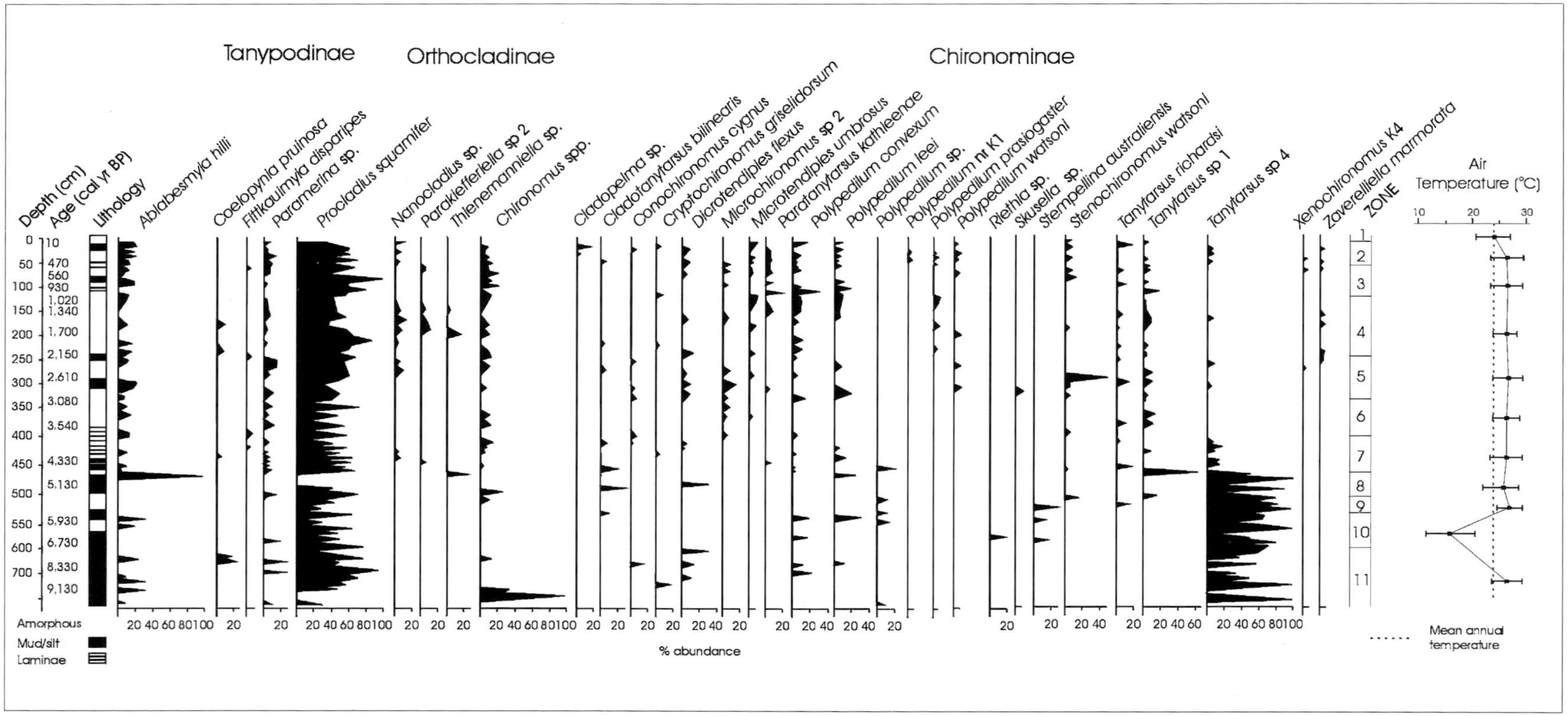

Figure 2 Holocene trends in chironomid abundance and reconstructed temperature estimates from Lake Barrine, Australia. Temperature error bars represent mean temperatures of the warmest and coldest quarter at 75 percentile (Dimitriadis and Cranston, 2001).

Figure 3A Lake Grasmere (center, 600 m a.s.l, 0.67 km^2, 12 m water depth) located in the foothills of the Southern Alps (background), New Zealand. The lake has formed by glacial retreat and alluvial aggradation since the end of the Last Glacial Maximum. Other lakes in the photo are lake's Sarah (right) and Pearson (left). GNS Science/*Photo: L. Homer.*

sp. but shows marked shifts in faunal abundance and composition characterized by increases in Chironomini, Orthocladinae, and Tanytarsini genera other than *Corynocera*. The observed changes were not interpreted to be driven by climate. Pollen analysis from the site likewise indicated the absence of any major vegetation changes that might be driven by climate change during this time. The dominance of *Corynocera* sp. and the relatively stable faunal assemblage suggest a general period of environmental stability over the past 6.8 kyr. Distinct changes in chironomid structure throughout this time were interpreted to be related to short-term changes in lake productivity, hydrological inflow, and rate and type of sedimentation. Changes in chironomid assemblage over the last 700–1,000 yr were explained as a response to human disturbance and forest clearance in the catchment area. Eastern end of Lago Mascardi (**Fig. 4**) (800 m a.s.l, 30 km^2 surface area, 200 m water deph) located ca. 15 km from the Mt Tronador ice cap (background) which feeds Lago Mascardi. Glacial ice from Mt Tronador filled this valley during the Last Glacial Maximum. Fossil core site located outside the photograph.

An important recent advance has been the development of the first chironomid inference models in New Zealand based on 37 lakes (Woodward and Shulmeister, in press). Temperature was identified as the most significant driver of chironomid distribution and produced a correlation (r^2_{jack}) of 0.77 and a prediction error (root mean squared error of prediction, $RMSEP_{jack}$) of 1.31° C. A less reliable model was produced for chlorophyll *a* with a correlation (r^2_{jack}) of 0.49 and a prediction error ($RMSEP_{jack}$) of 0.46 Log_{10} $\mu g\,l^{-1}$. These and other such studies (Dieffenbacher-Krall and Vandergoes, unpublished) provide a crucial cornerstone for the development of modern chironomid temperature calibration sets and the use of chironomids to quantify postglacial climatic change in New Zealand

In a recent paleolimnological study, Woodward and Shulmeister (in press) combine pollen, formaminifera, Trichoptera, chironomids, and algal abundance to reconstruct the development of coastal Lake Forsyth (43° 48′ S, 172° 44′ E), South Island, from a tidal bay to a barrier dammed lake. The main fluctuations recorded in chironomids (mainly *Chironomus* spp.) and other aquatic taxa during this time were interpreted as a response to changes in salinity, catchment deforestation, and eutrophication throughout the past 7 cal kyr.

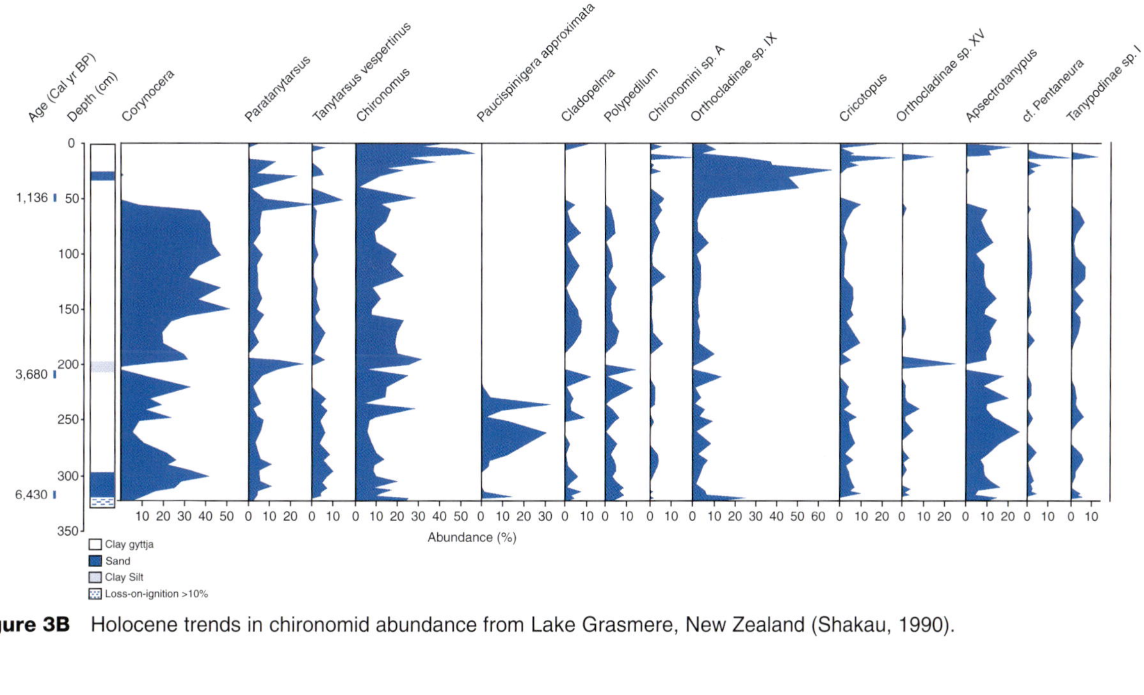

Figure 3B Holocene trends in chironomid abundance from Lake Grasmere, New Zealand (Shakau, 1990).

Figure 4 Eastern end of Lago Mascardi (800 m a.s.l, 30 km^2 surface area, 200 m water deph) located ca. 15 km from the Mt Tronador ice cap (background) which feeds Lago Mascardi. Glacial ice from Mt Tronador filled this valley during the Last Glacial Maximum. Fossil core site located outside the photograph.

This study indicates a significant relationship between the abundance of aquatic taxa and regional vegetation, and proposes catchment deforestation resulting from human disturbance as the main cause of biological change in the lake history. Increases in abundance and diversity of chironomids and other aquatic taxa after AD 1840 suggested an increase in sediment and nutrient input into the lake and a decrease in salinity as a result of deforestation. In the late 1800s, a significant peak in freshwater algae and the drop in abundance and diversity of chironomids and other aquatic invertebrates were interpreted as a period of lake eutrophication, followed by increasing salinity as a consequence of the opening of the lake to the Pacific Ocean. This study emphasizes the advantages of using multiproxy analyses to interpret environmental change.

South America

Pioneer chironomid work has been carried out in the Neotropics and three studies dealing with fossil midges have been published. Goulden (1966a, b) studied fossil midges from Laguna de Petenxil and another pond in Guatemala. Interpretations from both studies are vague mainly due to the low abundance and poor preservation of the remains and the lack of an accurate chronology. In Venezuela, Binford (1982) carried out a paleolimnological investigation in Lake Valencia using animal remains and some chemical and physical variables to reconstruct the last 12.5 kyr of lake history. Results of this work indicate that lake-level fluctuations and trophic changes occurred during the timespan of the core.

More recently, fossil chironomid investigations have been conducted in the southern Andes in Argentina and Chile (Ariztegui *et al.*, 1997; Bianchi *et al.*, 1997, 2000; Corley and Massaferro, 1998; Massaferro, 2000; Massaferro and Corley, 1998; Massaferro and Brooks, 2002; Massaferro *et al.*, 2004; Massaferro *et al.*, 2005). These studies were concentrated in two areas: in Northern Patagonia, Argentina, within the limits of the Nahuel Huapi National Park (41° S) and in Southern Chile on the Peninsula de Taitao and the Chonos Archipelago (46° S). These articles are reviewed based on their main focus, paleoclimate, and paleoproductivity reconstructions.

Paleoclimatic reconstructions Studies focusing mainly on Lateglacial climate reconstructions address evidence of Lateglacial Northern Hemisphere climate signals (i.e., Younger Dryas (YD) cooling) in South America. To date, results of investigations at four sites using fossil chironomids to reconstruct past climate during the Lateglacial period in Argentinean and Chilean Patagonia have been published.

In Ariztegui *et al.* (1997), changes in pollen, fossil chironomids, and geochemistry during the last 15 cal kyr at Lake Mascardi (41° 20′ S, 71° 34′ W) were interpreted as a response to a climate cooling coincident with the timing of the YD Northern Hemisphere cold event. Results showed that between 13.47 and 11.96 cal kyr BP, there was a decrease in the total pollen influx as a consequence of an increase in inorganic sedimentation followed by a sharp decrease in the hydrogen index. These results were interpreted as a response to cooling accompanied by an increase in subglacial erosion due to an advance of the Tronador ice cap that feeds the proglacial lake. The disappearance of the warm-adapted *Chironomus* and the decline in the total chironomid abundance at this time also suggested a climatic deterioration.

In a multiproxy paleolimnological study carried out in Lake El Trebol (41°04′ S, 71°30′ W), Bianchi *et al.* (1997) analyzed fossil chironomids, pollen, and geochemistry to reconstruct paleoenvironmental history. Changes in these proxies were interpreted to indicate climatic fluctuations during the last 14 cal kyr. The increase of fossil midges, organic matter, biogenic silica, and nutrients from ca. 12.42 cal kyr BP was interpreted as a change in lake productivity due to climatic amelioration. Subsequently, Bianchi *et al.* (1999) combined pollen, chironomids, rare earth elements, heavy metals, and fossil pigments to show the postglacial evolution of the landscape and biota in the surrounding area of Lake El Trebol, providing the basis for future accurate paleoenvironmental reconstructions. Based on changes in the proxies analyzed, three different periods were recognized: the Lateglacial, a transition, and the Holocene period. Chironomids showed a gradual replacement of littoral Orthocladinae fauna, abundant in the transition zone, by profundal Tanytarsini taxa, present in the Holocene, indicating more autochthonous lake production. Changes in pollen and paleopigments were consistent with changes in midge assemblages. These results suggested a climatic amelioration during the Lateglacial–Holocene transition. Climate interpretations were consistent with those observed at Lake Mascardi (Ariztegui *et al.*, 1997). Chironomid assemblages and biogeochemical parameters also indicated changes in lake productivity during Holocene times. Nevertheless, the interpretation of climate changes in both investigations remained speculative due to the lack of resolution and a precise chronology. Lake Grasmere (center, 600 m a.s.l, 0.67 km^2, 12 m water depth) located in the foothills of the Southern Alps (background), New Zealand. The lake has formed by glacial retreat and alluvial aggradation since the end of the Last Glacial Maximum. Other lakes in the photo are lake's Sarah (right) and Pearson (left). GNS Science/*Photo: L. Homer*.

The first high-resolution late Quaternary environmental reconstruction using chironomids in South America was published by Massaferro and Brooks (2002) (**Fig. 5**). Changes in the chironomid assemblages at Laguna Stibnite (46°25′ S, 74°24′ W) in the Taitao Peninsula in Chile suggested that the climate in southern Chile was drier and at its coolest during the YD chronozone. Between ca. 13 and 11.2 cal kyr BP, the cold-stenothermic Podonominae were consistently present in every sample reaching a peak abundance during this period. This suggests that lake waters were cool and oligotrophic during the YD. The low concentration of head capsules also reflects low lake productivity. The dynamics of the late Quaternary chironomid fauna of Lake Stibnite shows a remarkably cyclical pattern, which was interpreted as being linked to changes in precipitation. During the Holocene, the presence of a *Limnophyes*/*Chironomus*/*Phaenopsectra*/*Parachironomus* assemblage between 10.6 and 7.2 cal kyr BP and 2.5 and 1.5 cal kyr BP suggested dry periods of low lake levels. Within a regional context, results from beetles from Puerto Edén (Ashworth *et al.*, 1991; *see* Late Pleistocene of South America) located 300 km south of Taitao Peninsula also suggested changes in precipitation during the Holocene, although they were almost out-of-phase to those inferred at Laguna Stibnite. If these interpretations are both correct, changes in precipitation patterns may be a consequence of latitudinal shifts of the westerlies (Massaferro and Brooks, 2002).

However, the climatic interpretation that suggested cooling during the YD, is not in agreement with other investigations carried out in the lake. Pollen records from Lake Stibnite (McCulloch *et al.*, 2000) and other lakes in this region of Chile show no evidence of cooling during the Lateglacial period (e.g., Haberle and Bennet (2004)). In contrast, other sites in Patagonia show changes in pollen assemblages at the end of the Lateglacial that have been interpreted as a response to climatic cooling (e.g., Heusser *et al.* (1996) and Moreno (1997)). The inconsistencies between these interpretations have made the presence of the YD cooling signal in South America a controversial topic.

Recently, Massaferro *et al.* (2005) carried out a study on pollen and fossil chironomids in Laguna Facil (44°19′ S, 74°17′ W), located in the Chonos Archipelago in Chile. In this article, the authors argue that chironomids respond to local rather than regional environmental changes. Chironomid stratigraphical changes coincided closely but not synchronously with changes in the pollen and charcoal. The lag in the response of midges compared with the forest trees would suggest that the midge fauna responds to changes brought about by the migration and colonization of these trees in the lake catchment. This could be in response to changes in the structure of the forest canopy and amount of tree shade close to the shores of the lake, or to changes in water chemistry brought about by soil development. However, a change in the chironomid assemblage at ca. 7.2 cal kyr BP preceding the resurgence of *Pilgerodendron* suggests a response to regional climatic change.

Despite the similarities of fauna and the proximity between Lake Facil and Lake Stibnite, no response to a cooling event coinciding with the YD was recorded at Lake Facil. The explanation pointed to the more northerly location of Lake Facil than Laguna Stibnite which is less likely to have been influenced by any resurgence of Andean glaciers during the YD. Geographic variability in glacial activity would indicate that climatic conditions were possibly insufficiently intense and/or of insufficient duration to uniformly affect regional paleobiotic changes; therefore, a response during the YD may be recognizable at some sites but not at others.

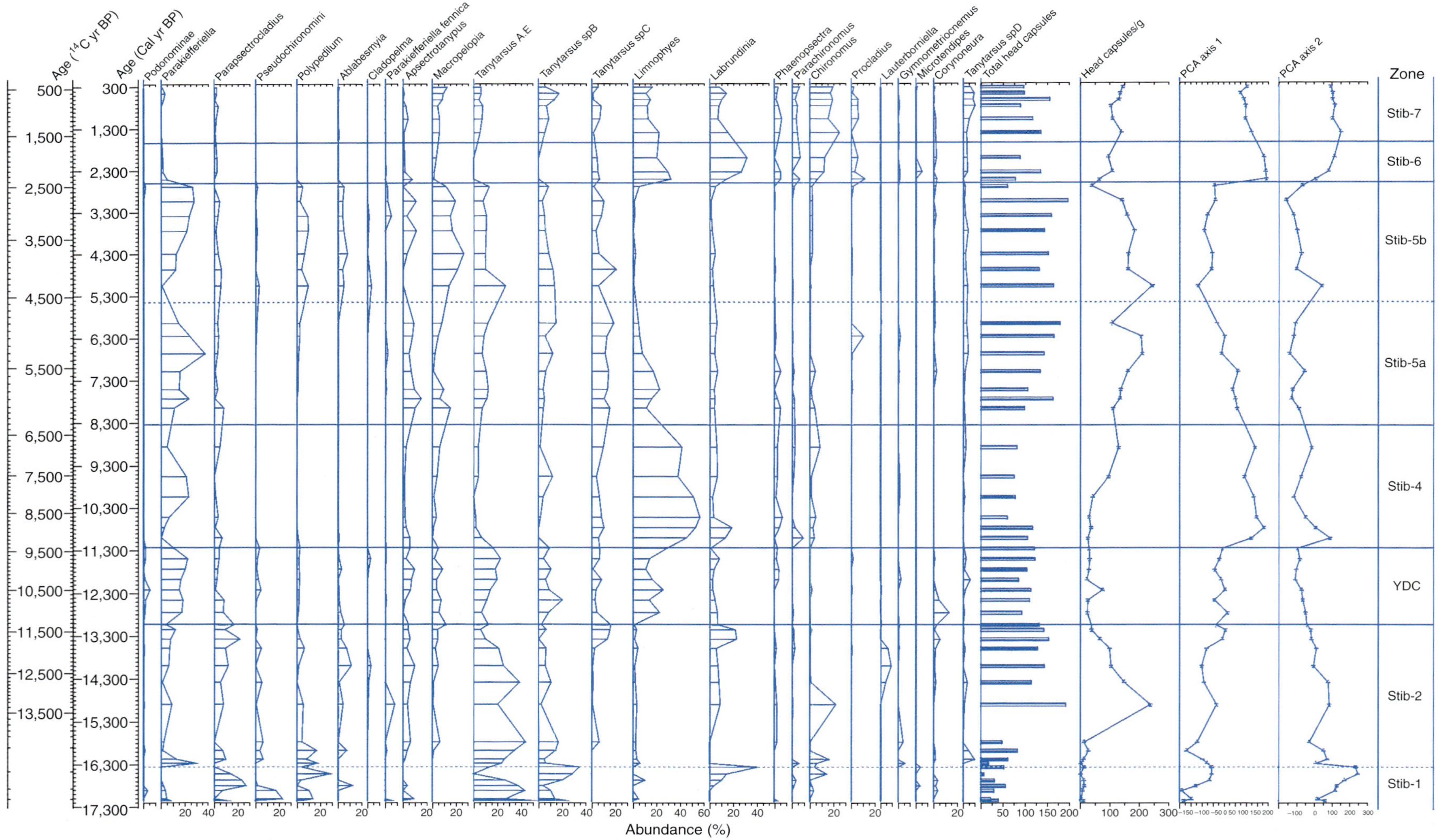

Figure 5 Postglacial trends in chironomid abundance from Lago Stibnite, Taitao Peninsula, Chile (Massaferro and Brooks, 2002).

Paleoproductivity reconstructions A number of studies from South America focus on the relationships between chironomids and lake productivity.

Massaferro and Corley (1998) studied the chironomid paleodiversity from Lake Mascardi. The study demonstrates how major environmental episodes such as volcanic eruptions, climatic cooling/warming, and changes in lake level occurring during the Lateglacial period affected midge diversity patterns. The authors suggest that long-term studies reconstructing the history of lake diversity provide important information for present-day conservation management. In conjunction with this study, Corley and Massaferro (1998) published the first fossil insect community dynamics study in Patagonia. Using a long-term apparent turnover, the authors show the extinction/immigration chironomid community patterns along a 10 m sediment core representing 15 kyr of lake history. This kind of work is important from an ecological perspective to facilitate the interpretation of current trends in diversity and succession patterns.

Subsequently, Massaferro (2000) published the complete chironomid assemblage from Lake Mascardi covering the last 15 kyr BP. In this study, changes in chironomid species abundance are compared to diversity indices and also to changes in total organic carbon (TOC). Sampling resolution was not enough to discern the response of the chironomids; however, changes in midge assemblages were interpreted to be related to temperature and also to changes in lake productivity. Massaferro *et al.* (2004) studied geochemical and chironomid records from a short core from Lake Morenito (41°03′ S, 71°31′ W), located near the city of San Carlos de Bariloche in Patagonia. Due to the expansion of the population in recent years, Lake Morenito and surrounding areas have experienced moderate human impact such as the deforestation, fish introduction, and building construction. This work shows the response of chironomids and geochemistry to natural and anthropogenic (post-European human) environmental disturbances during the last 200 yr. The main fluctuations recorded in chironomid assemblages during the time span of the core were interpreted as a response to human disturbances in the catchment area, demonstrating the value of chironomids in reconstructing paleoproductivity. It also demonstrates the utility of combining multiple proxies (chironomids, tephra and sediment analysis, organic matter, and chronology) in the interpretation of such changes.

Hemispheric and Interhemispheric Comparison

Although more high-resolution paleotemperature studies are warranted for the Southern Hemisphere, some tentative comparisons with other regions of the globe can be made from those that have already addressed the Lateglacial–Interglacial climate change in the mid-latitudes of the Southern Hemisphere. This topic has been extremely controversial in the past with some paleoclimate records indicating changes that are more coincident with cooling noted in Antarctic ice-core records, that is, Antarctic Cold Reversal (ACR), than with cooling noted from the North Atlantic region (associated with the YD chronozone). Other records have suggested gradual warming with no reversals in the Southern Hemisphere. Studies from lacustrine environments of South America can help resolve this debate.

Chironomid records suggest that a distinct climatic reversal similar to the YD cool phase of the North Atlantic region occurred in sites from Central Chile and Argentina. This mid-latitude cooling was recently identified as the Huelmo/Mascardi Cold Reversal (HMCR) (Hajdas *et al.*, 2003). Nevertheless, Chironomid assemblages also indicate that a cooling trend started earlier than the HMCR, at ca. 14.4 cal kyr BP, that corresponds with the timing of the ACR. Comparison of chironomid results with both EPICA Antarctica and GISP2 Greenland ice-core records indicates similar Lateglacial climatic events, but their full interpretation is limited by lack of information about taxonomy, development of a robust inference model, precise chronologies, and insufficient records spread across the Southern Hemisphere.

Current Chironomid Investigations

Two unpublished and independent chironomid studies from Lake Mascardi (Argentina) and Huelmo (Chile) are currently underway and appear to provide new evidence of climate change during the Lateglacial (Massaferro, unpublished). Recent work has been undertaken to explore and develop a statistical model to infer paleoenvironmental changes in two lakes in southern Chile (Gilchrist, unpublished PhD thesis). Studies are currently being carried out for New Zealand (Vandergoes and Dieffenbacher-Krall, unpublished; Woodward and Shulmeister (in press 2006)) and Tasmania (Cwynar, personal communication) to produce reliable data sets for quantitative reconstruction of past temperatures during the Lateglacial and Last Glacial Maximum using chironomids. Identification keys for the modern chironomid fauna are under development for New Zealand (Boothroyd, personal communication), and for the subfossil remains of New Zealand (Dieffenbacher-Krall, and Vandergoes, personal communication), Tasmania (Cwynar, personal communication), and South America (Massaferro, personal communication).

Future Research Directions

The fact that chironomids have the potential to provide detailed paleotemperature reconstructions is extremely important in the Southern Hemisphere where quantitative estimates of temperature change are limited. Combined with other proxies including pollen, diatoms, and isotopic measurements, chironomids can contribute substantially to multiproxy environmental reconstructions to produce more robust estimates of climate change. They can also be used to calibrate regional paleotemperature estimates derived from nonecological indicators such as snowline reconstructions and ice cores. The importance of attaining quantitative temperature estimates from landmasses within the Southern Hemisphere is critical to clarify climate relationships between the Northern and Southern Hemispheres, and therefore identify possible coupling mechanisms. Much progress has been made in the study of fossil midges from the Southern Hemisphere; however, further work is required in two main areas.

Modern Taxonomy and Ecology

Early studies by Edwards (1931) and Brundin (1966) were the first to recognize a circum-Antarctic chironomid faunal connection and make significant contributions to the knowledge of Southern Hemisphere chironomid taxonomy and biogeography. Since then, other researchers from New Zealand (e.g., Forsyth (1971), Sublette and Wirth (1980), and Boothroyd (2002)) and South America (e.g., Andersen and Contrera-Ramos (1999), Cranston (2000b), Fitkau and Reiss (1973), and Paggi (1998)) have made notable advances to furthering knowledge taxonomy and distribution of Southern Hemisphere chironomids while work by Cranston (2000a) has made an exceptional contribution to the identification and ecology of Australian larval fauna.

Even with these recent advances, knowledge of chironomid taxonomy of Australia, New Zealand, and South America is still limited. The number of species currently documented is considered conservative and many taxa remain undescribed (Boothroyd, personal communication; Cranston, 2000). A further problem is that the chironomid descriptions are based on the characteristics of the adult or pupae and the larval stages have largely been ignored. As a consequence, in New Zealand and South America, many of the existing taxa can only be identified to the generic level (**Fig. 6**) and identification keys for the larval stages of chironomids in these regions are scarce. Many chironomid taxa remain undescribed because of a lack of ecological investigations. The result is that the knowledge of ecological requirements, tolerances, and species distribution of these insects in Australia, New Zealand, and South America is incomplete. These factors hamper our understanding of the causes of change in modern chironomid assemblages and inhibit the use of fossil chironomid assemblages, which relies solely on the identification subfossil larval head capsules for environmental reconstructions.

There is a need for the improvement of taxonomic resolution of modern and subfossil chironomids. In the Southern Hemisphere, there is just one key for the identification of chironomid fauna from Australia (Cranston, 2000a), but similar keys have yet to be published for New Zealand or South America. For the first time, Massaferro and Brooks (2002) described specimens of the subfossil chironomid fauna from a Chilean lake and identified additional taxonomic groups that could be ecologically related to European taxa. However, more taxonomic and ecological work is needed in the Southern Hemisphere, especially regarding the distribution and habitat requirements of chironomids. A standardized taxonomic approach together with an improvement in taxonomic resolution will allow direct comparisons between chironomid investigations from Australia, New Zealand, South America, and Africa, providing an invaluable tool for biogeographical and paleolimnological research.

Chironomid-Temperature Inference Models

Quantitative estimates of paleotemperature can only be achieved through the development of chironomid-temperature inference models specific to the geographic region of the paleoclimate analysis. In the Northern Hemisphere, a number of studies have used the training set approach to determine how environmental variables, including summer temperature, are important in determining chironomid community composition. To date only one study from Australia (Dimitriadis and Cranston, 2001) has provided quantitative postglacial temperature estimates for the Southern Hemisphere. The recent development of inference models from New Zealand (Woodward and Shulmeister (in press 2006)) represents a significant advance in the use of chironomids as paleoenvironmental indicators. However, improvements to this initial calibration data set including a more refined taxonomy and the addition of more lakes at an even distribution along a wider temperature gradient are needed to better the performance of these models. At present, the lack of chironomid temperature inference models from South America is a serious impediment in providing robust quantifiable estimates of temperature from existing Lateglacial and Holocene chironomid records.

There is a strong need to further develop regional chironomid temperature training sets and inference models, especially for New Zealand and South America. Only by defining the temperature tolerance of chironomid species and developing regional inference

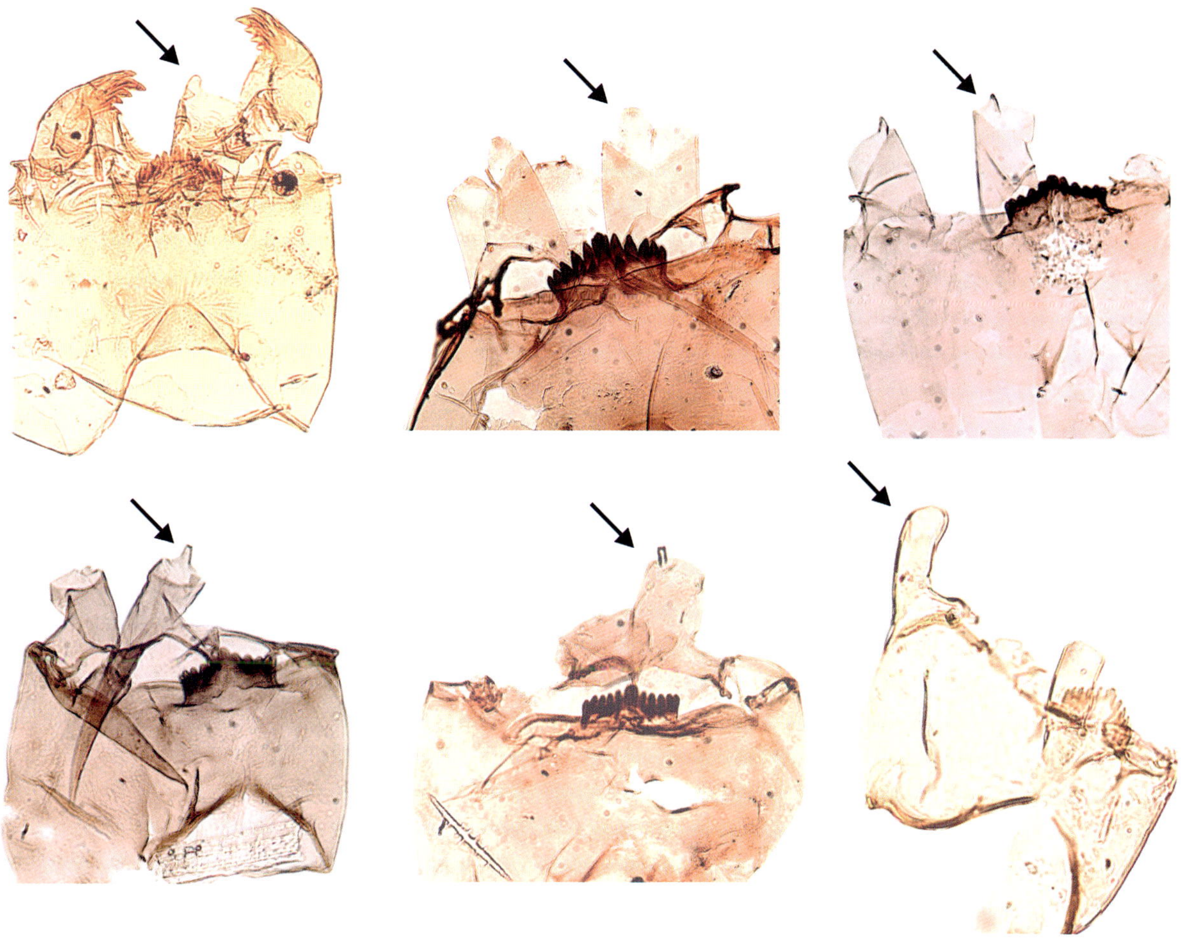

Figure 6 Head capsule remains of South American *Tanytarsus* sp. yet to be classified below generic level. Distinction of subgroups based on the differing characteristics of the spur on the antenna pedestal (arrows). Photographs by J. Massaferro.

models we will be able to provide robust quantifiable temperature reconstructions for existing and future chironomid records.

See also: **Chironomid Overview**. **Chironomid Records**: Late Pleistocene of Europe.

References

Andersen, T., and Contreras-Ramos, A. (1999). First record of Antillocladius Sæther from continental South America (Chironomidae, Orthocladiinae). *Acta Zoologica Academiae Scientiarum Hungaricae* **45**(2), 149–154.

Ariztegui, D., Bianchi, M. M., Masaferro, J., Lafargue, E., and Niessen, F. (1997). Interhemispheric synchrony of Late-glacial climatic instability as recorded in proglacial Lake Mascardi, Argentina. *Journal of Quaternary Science* **12**, 333–338.

Ashworth, A. C., Markgraf, V., and Villagran, C. (1991). Late Quaternary climatic history of the Chilean Channels based on fossil pollen and beetle analysis, with an analysis of the modern vegetation and pollen rain. *Journal of Quaternary Science* **6**, 279–291.

Bianchi, M. M., Massaferro, J., Roman Ross, G., and Lami, A. (2000). Geochemical and biological tracers of climate fluctuations during Late Pleistocene and Holocene in Lake El Trébol, Northern Patagonia, Argentina. *Journal of Paleolimnology* **22**, 137–148.

Bianchi, M. M., Massaferro, J., Roman Ross, G., del Valle, R., Tatur, A., and Amos, A. (1997). The Pleistocene-Holocene boundary from cores of Lake El Trebol, Patagonia Argentina. *Archiv für Limnologie* **26**, 805–808.

Binford, M. W. (1982). Ecological history of Lake Valencia, Venezuela: Interpretation of animal microfossils and some chemical, physical and geological features. *Ecological Monographs* **52**, 307–333.

Boothroyd, I. K. G. (2002). *Cricotopus* and *Paratricocladius* (Chironomidae: Insecta) in New Zealand, with description of *C. hollyfordensis* n.sp., and redescriptions of adult and immature stages of *C. zealandicus* and *P. plurisetrialis*. *New Zealand Journal of Marine and Freshwater Research* **36**, 775–788.

Brundin, L. (1966). Transantarctic relationships and their significance, as evidenced by chironomid midges, with a monograph of the subfamilies Podonominae and Aphroteniinae and the austral Heptagyiae. *Kungl Svenska Vetenskapsakademiens Handlingar* **11**, 1–472.

Corley, J. C., and Massaferro, J. I. (1998). Long term turnover of a fossil community of chironomids (Diptera) from Lake Mascardi (Patagonia, Argentina). *Journal of the Kansas Entomological Society* **71**, 407–413.

Cranston PS (2000a). An electronic guide to the Chironomidae of Australia. http://entomology.ucdavis.edu/chiropage/index.html.

Cranston, P. S. (2000b). *Parapsectrocladius*: A new genus of Orhocladiinae Chironomidae (Diptera) from Patagonia, the southern Andes. *Insect Systematics and Evolution* **31**(1), 103–120.

Deevy, E. S. (1955). Paleolimnology of the upper swamp deposit, Pyramid Valley. *Records of the Canterbury Museum* **6**, 291–344.

Dimitriadis, S., and Cranston, P. S. (2001). An Australian Holocene climate construction using Chironomidae from a tropical volcanic maar lake. *Palaeogeography, Palaeoclimatology, Palaeoecology* **176**, 109–131.

Edwards, F. W. (1931). Fascicle 5 – Chironomidae. In *Diptera of Patagonia and South Chile.*, vol. 2(5), pp. 233–331. Natural History Museum, London.

Fitkau, E. J., and Reiss, F. (1973). AmazonischeTanytarsini (Chironomidae Diptera)I: Die riopreto Gruppe der Gattung Tanytarsus. *Studies in Neotropical Fauna* **8**, 1–16.

Forsyth, D. J. (1971). Some New Zealand Chironomidae. *Journal of the Royal Society of New Zealand* **1**, 113–144.

Gilchrist, S. (2005). *Late Quaternary Paleoclimatic Reconstructions in Patagonia Using Chironomid Analysis.* PhD Thesis. NERC fellowship. University of Edinburgh, UK.

Goulden, C. E. (1966a). La Aguada de Santa Ana Vieja: An interpretive study of the cladoceran microfossils. *Archiv für Hydrobiologie* **62**, 373–404.

Goulden, C. E. (1966b). The animal microfossils of Laguna Petenxil, Guatemala. *Memoirs of the Connecticut Academy of Arts and Sciences* **17**, 84–120.

Haberle, S. G., and Bennet, K. D. (2004). Postglacial formation and dynamics of North Patagonian rainforest in the Chonos Archipelago, Southern Chile. *Quaternary Science Reviews* **23**, 2433–2452.

Hajdas, I., Bonani, G., Moreno, P., and Ariztegui, D. (2003). Precise radiocarbon dating of Late-Glacial cooling in mid-latitude South America. *Quaternary Research* **59**, 70–78.

Heusser, C. J. (1993). Late-glacial of southern South America. *Quaternary Science Reviews* **12**, 345–350.

Heusser, C. J., Lowell, T. V., Heusser, L. E., Hauser, A., Andersen, B. G., and Denton, G. H. (1996). Full-glacial–late-glacial palaeoclimate of the Southern Andes: Evidence from pollen, beetle, and glacial records. *Journal of Quaternary Science* **11**, 173–184.

Massaferro, J. (2000). Fossil assemblages in an oligotrophic lake of Patagonia (Lago Mascardi, Argentina). In: *Late 20th century research on Chironomidae* (Odwin Hoffrichter, Ed.) pp. 127–131. Shaker Verlag, Aachen, Germany.

Massaferro, J., and Brooks, S. J., (2002). The response of Chironomids to Late Quaternary environmental change in the Taitao Penninsula, Southern Chile. *Journal of Quaternary. Sciences* **17**(2), 101–111.

Massaferro, J., and Corley, J. (1998). Environmental disturbance and chironomid paleodiversity: 15 kyr BP of history at Lake Mascardi (Patagonia, Argentina). *Aquatic Conservation: Marine and Freshwater Ecosystem* **8**, 315–323.

Massaferro, J., Brooks, S. J., and Haberle, S. G. (2005). The dynamics of chironomid assemblages and vegetation during the Late Quaternary at Laguna Facil, Chonos Archipelago, southern Chile. *Quaternary Sciences Reviews* **24**, 2510–2522.

Massaferro, J., Ribeiro Guevara, S., Rizzo, A., and Arribere, M. (2004). Short term environmental changes in Lake Morenito, (41° S, Patagonia, Argentina) from the analysis of subfossil chironomids. *Aquatic Conservation: Marine and Freshwater Ecosystems* **14**, 123–134.

McCulloch, R. D., Bentley, M. J., Purves, R. S., *et al.* (2000). Climatic inferences from glacial and palaeoecological evidence at the last glacial termination, southern South America. *Journal of Quaternary Science* **15**, 409–417.

Moreno, P.I. (1997). Vegetation and climate near Lago Llanquihue in the Chilean lake district between 20,200 and 9,500 ^{14}C yr BP. *Journal of Quaternary Sciences* 12, 485–500.

Paggi, A. C. (1998). Chironomidae. In *Biodiversidad de Artrópodos argentinos. Una perspectiva biotaxonómica* (J. J. Morrone and y Coscarón, Eds.), Cap. 31, pp. 327–337. Ediciones Sur, La Plata.

Paterson, C. G., and Walker, K. F. (1974). Recent history of *Tanytarsus barbitarsus* Freeman (Diptera: Chironomidae) in the sediments of a shallow saline lake. *Australian Journal of Marine and Fresh Water Research* **25**, 151–165.

Schakau, B. L. (1986). Preliminary study of the development of the subfossil chironomid fauna (Diptera) of Lake Taylor, South Island, New Zealand, during the younger Holocene. *Hydrobiologia* **43**, 287–291.

Schakau, B. L. (1990). Stratigraphy of the fossil Chironomidae (Diptera) from Lake Grasmere, South Island, New Zealand, during the last 6000 years. *Hydrobiologia* **214**, 213–221.

Sublette, J. E., and Wirth, W. W. (1980). The Chironomidae and Ceraptapogonae (Diptera) of New Zealand's subantarctic islands. *New Zealand Journal of Zoology* **7**, 299–378.

Verschuren, D., and Eggermont, H. (2006). Quaternary paleoecology of aquatic Diptera in tropical and Southern Hemisphere regions, with special reference to the Chironomidae. *Quaternary Science Reviews*, 25, 15–16, 1926–1947.

Walker, I. R. (2001). Midges: Chironomidae and related diptera. In *Tracking Environmental Change Using Lake Sediments* (J. P. Smol, H. J. B. Birks and W. M. Last, Eds.), Vol. 4, Zoological Indicators. pp. 43–67. Kluwer Academic Publishers, Dordrecht.

Woodward, C. A., and Shulmeister, J. (2005). A Holocene record of human induced and natural environmental change from Lake Forsyth (Te Wairewa), New Zealand. *Journal of Paleolimnology*. **34(4)**, 481–501.

Woodward, C. A., and Shulmeister, J. (in press 2006) New Zealand chironomids as proxies for human-induced and natural change: Transfer functions for temperature and lake production (chlorohll a). *Journal of Paleolimnology* **36, 4(pp?).**

Cladocera Studies *see* **Paleolimnology**: Cladocera

Climate Change *see* **Paleoclimate**: Introduction; Time Scales of Climate Change; **Paleoclimate Modeling**: Data-Model Comparisons; Quaternary Environments; The Last Interglacial; Last Glacial Maximum GCMs; Paleo-ENSO; **Paleoclimate Reconstruction**: Approaches; Pliocene Environments; Paleodroughts and Society; Sub-Milankovitch (DO/Heinrich) Events; Paleotempestology; Younger Dryas Oscillation, Global Evidence; The Last Millennium; Historical Climatology; **Paleoclimate Relevance to Global Warming**

Climate Modeling, Quaternary *see* **Paleoclimate Modeling**: Quaternary Environments

Coccolithophore Studies *see* **Paleoceanography, Biological Proxies**: Alkenone Paleothermometry from Coccoliths; Coccoliths

Coleoptera Fossil Records *see* **Beetle Records**: Overview; Late Tertiary and Early Quaternary; Middle Pleistocene of Europe; Late Pleistocene of Australia; Late Pleistocene of Europe; Late Pleistocene of Japan; Late Pleistocene of South America; Late Pleistocene of North America; Late Pleistocene of New Zealand; Late Pleistocene of Northern Asia; Postglacial Europe; Postglacial North America

Coral Studies *see* **Paleoceanography, Biological Proxies**: Corals, Sclerosponges and Mollusks; **Sea Level Studies**: Coral Records

COSMOGENIC NUCLIDE DATING

Contents

Overview

J C Gosse, Dalhousie University, Nova Scotia, Canada

© 2007 Elsevier B.V. All rights reserved.

Introduction

The advent of atmospheric cosmogenic radiocarbon dating by Libby in the 1950s brought high precision ages of important geological and archeological events and solutions to long-standing marine and terrestrial Quaternary problems. Over the next three decades, methods were developed that could measure cosmogenic isotopes in minerals at the precision necessary for surface exposure dating. However, since the mid-1980s, there has been a comparable revolution in Quaternary science due to applications of terrestrial cosmogenic nuclides (TCN) produced *in situ*. Many previously intractable geochronology questions are being addressed, many new questions are being asked, and there has been an explosion of ideas related to rates of surface processes.

Synthesis of TCN applications

Recent reviews of applications of TCN methods are provided by Bierman and Nichols (2004), Cockburn and Summerfield (2004), Gosse and Phillips (2001), Muzikar *et al.* (2003), and Niedermann (2002). To date, TCN methods comprise six routinely-utilized isotope-mineral systems: ^{3}He, ^{21}Ne, ^{10}Be, ^{26}Al, ^{36}Cl, and ^{14}C. The former two are stable noble gases and the latter four are listed in order of decreasing half lives – 1.5 Ma (currently debated), 0.7 Ma, 0.3 Ma, and 5.7 ka. Some TCNs are calibrated for specific minerals due to:

(1) diffusion constraints (e.g., ^{3}He diffuses too quickly from quartz to make it useful as a geochronology tool, but is retained in minerals such as pyroxenes and olivine);
(2) production rates (^{36}Cl cannot be produced in pure quartz); and
(3) other issues, such as high native concentrations of the isotope or its comparative isotope for AMS measurement (for instance, measurements of ^{26}Al in an aluminosilicate are difficult).

So far, one isotope has been calibrated for measurement in whole rock samples (^{36}Cl), so virtually any rock type can be used for TCN research. The stability of the noble gases permits the measurements of the oldest surfaces on Earth (tens of millions of years). An advantage of the radionuclides is that their concentration is dependent on decay rate as well as exposure duration, offering an extra degree of freedom for calculating long exposure durations or avoiding difficulties associated with inheritance. For instance, the rapid decay of ^{14}C offers a useful geochronology tool to date glaciated surfaces (or sediment derived from them) that have not been deeply eroded. Those surfaces would retain some TCN concentration from exposure prior to ice cover. However, if the ice cover was of sufficient duration (say 40 ka), most of the ^{14}C would be decayed and its concentration measured today would reflect the timing of exposure since the surface was uncovered. The youngest and still experimental member of the TCN toolbox is the noble gas ^{38}Ar. However, other new isotope and mineral systems are being developed to provide more applications (*see* Methods).

Every TCN experiment depends on the past century (Gosse and Phillips, 2001) of efforts of theoretical and experimental physics, analytical chemistry, geochemistry, geomorphology, and geophysics. Sample selection must consider geometry effects and the angular dependence of cosmic radiation. Sample preparation requires planning to ensure the proper isotope-mineral system is selected based on lithology and mineral abundance. To achieve the desired precision, the analytical limits of the isotope mass spectrometry, exposure duration, site production rate and sample depth, mass of sample, number of samples per site, abundance of non-cosmogenic species of the same isotope, and complexity of exposure history, must all be considered. The analyses of these nuclides require high precision atomic-grade noble gas mass spectrometers or accelerator mass spectrometers. Besides physical improvements to both of these systems, spectrometry methods and target preparation procedures have improved significantly over the past decade (both techniques routinely analyze TCN samples with 2 to 4% 1δ precision). Finally, when a TCN method is applied, numerous calculations and sensitivity assessments are required before interpreting the concentration as an age or an erosion rate. Cosmic ray physics and empirical measurements must be relied upon to scale production rates from sea level high latitude to the sample site. Decisions must be made regarding which production rate to use, and whether to compensate for geomagnetic field effects, snow cover, bulk density variations with time, and other factors. The intensity of sample specific calculations and analyses is a significant departure from most Quaternary dating methods, which often only require a thorough consideration of stratigraphic context before the data can be used.

A major premise of all TCN methods is that the nuclides are produced at known rates. Production of cosmogenic isotopes in minerals exposed to secondary radiation depends on the nuclide and mineral system and other environmental factors. Secondary radiation is produced when primary galactic cosmic rays (primarily fast protons) interact with nuclei in the Earth's atmosphere or within the upper few meters of the lithosphere. TCN production rates vary spatially as a function of geomagnetic field effects (dipolar and non-dipolar), and atmospheric depth. TCN production also varies temporally, in a way similar to atmospheric radiocarbon production variation under stronger or weaker magnetic field intensities. Additionally, the concentration of TCNs in rock depends on erosion, movement, or burial of the surface. Production rates for most locations are probably known to within 15% at the 1δ confidence level. However, the uncertainty is not well defined due to the number of variables involved and the difficulty in establishing a highly accurate production rate for a given time and place. The principal objective of two concurrent major international research programs, CRONUS-Earth in North America and CRONUS-EU in Europe, is to improve knowledge of TCN production rates, so by 2010 the level of uncertainty should be substantially diminished.

TCN methods can be used to directly date rock or sediment landforms. By 2006, hundreds of lava flows and thousands of boulders on moraines have been dated directly with TCN measurements. Exposure of surfaces in hyper-arid regions, such as the Antarctic dry valleys and the Atacama Desert (Dunai *et al.*, 2005) record exposure histories since the Pliocene and Miocene. The upper time range of the method is limited by surface stability although for long (hundred million years) exposures, changes in the diffusion of the noble gases and the production of non-cosmogenic species can become the limiting factors.

Secondary cosmic radiation (primarily fast, epithermal, and thermal neutrons and muons) is attenuated with depth below the surface at a rate dependent on the particle attributes (type, charge, energy) and the bulk density of the rock or sediment. Production of TCN isotopes is routinely detected at depths exceeding 5 m. The attenuation of the cosmic ray secondaries with mass has many useful implications. For instance, the differences in the decrease in total production rate of ^{3}He and ^{38}Ar with atmospheric depth may lead to future uses of their ratio for paleo-altimetry (surface uplift) measurements. Furthermore, the predictable decrease in concentration with depth creates an important advantage for TCN over other geochronology methods. Because the concentration is sensitive to depth, the method can be used to determine directly the rate and style of erosion of landforms and landscapes. A third implication of the penetration of cosmic rays through mass is that the definition of 'surface' can include surface evolution, so that it is possible to yield a minimum exposure age on an eroding surface, although the original surface may have been changed or completely stripped.

TCN exposure methods have already made lasting impacts on society and science outside Quaternary geology. For instance, Partridge *et al.* (2003) have provided $^{26}Al/^{10}Be$ burial ages on sediment at Sterkfontein, South Africa, showing that associated hominid fossils were deposited in the Lower Pliocene. The capacity to date surfaces with millions of years of exposure has provided an important extension of the billion- to million-year timescale of rock exhumation history derived from high- and low-temperature thermochronology methods, such as apatite (U-Th)/He.

Preface

The purpose of the following four articles is to provide a review of the current state of knowledge concerning:

(1) the theory of cosmic ray interactions with minerals;
(2) recent developments in production systematics and analytical and geochemical methodologies;
(3) applications of the technique for geochronology; and
(4) applications of the technique for landscape evolution studies.

Lal (*see* Cosmic Ray Interactions in Minerals) provides a systematic review of the important interactions that produce TCN. The article describes the origin, composition, and energy spectra of cosmic rays, and their fate from origin to Earth's surface. Lal provides a summary of the founding physics theory and experimentation upon which the TCN technique is based, and discusses some of the shortcomings of knowledge that remain as important sources of systematic error in the methods. Schaefer and Lifton (*see* Methods) provide a thorough review of the recent progress that has been made in TCN production systematics and the isotopic procedures. They highlight one of many exciting strategies undertaken by the international CRONUS initiative to improve TCN techniques. Ivy-Ochs and Kober (*see* Exposure Geochronology) summarize a wide range of applications of TCN methods for dating surfaces and events, and include caveats and limitations to different approaches. Granger (*see* Landscape Evolution) demonstrates how the TCN methods offer more than geochronology, because their concentrations can be interpreted directly as erosion rates and burial histories to provide insights into landscape evolution that could otherwise not be resolved.

See also: **Cosmogenic Nuclide Dating**: Methods; Cosmic Ray Interactions in Minerals; Exposure Geochronology; Landscape Evolution.

References

Bierman, P., and Nichols, K. K. (2004). Rock to sediment – slope to sea with 10Be-rates of landscape change. *Annual Reviews of Earth Planetary Science* **32**, 215–255.

Cockburn, H. A. P., and Summerfield, M. A. (2004). Geomorphological applications of cosmogenic isotope analysis. *Progress in Physical Geography* **28**, 1–42.

Dunai, T. J., Gonzales Lopez, G. A., and Juez-Larre, J. (2005). Oligocene–Miocene age of aridity in the Atacama Desert revealed by exposure dating of erosion-sensitive landforms. *Geology* **33**, 321–324.

Gosse, J. C., and Phillips, F. M. (2001). Terrestrial *in situ* cosmogenic nuclides: theory and application. *Quaternary Science Reviews* **20**, 1475–1560.

Muzikar, P., Elmore, D., and Granger, D. (2003). Accelerator Mass Spectrometry in Geologic Research. *Geological Society of America Bulletin* **15**, 643–654.

Niedermann, S. (2002). Cosmic-ray-produced noble gases in terrestrial rocks: Dating tools for surface processes. In *Noble Gases in Geochemistry and Cosmochemistry: Reviews in Mineralogy and Geochemistry* (D. Porcelli, C. J. Ballentine and R. Wieler, Eds.) Vol. 47 pp. 731–784.

Partridge, T. C., Granger, D. E., Caffee, M. W., and Clarke, R. J. (2003). Lower Pliocene hominid remains from Sterkfontein. *Science* **300**, 607–612.

Methods

J M Schaefer and N Lifton, Lamont-Doherty Earth Observatory, NY, USA

© 2007 Elsevier B.V. All rights reserved.

Introduction

During the last 10–15 yr, now 50 yr after its initial discovery (Schaeffer and Davis, 1956), terrestrial cosmogenic nuclide (TCN) applications have had a revolutionary impact on various Earth scientific disciplines including Quaternary geology, geomorphology, sedimentology, volcanology, and paleoclimatology. The rapid progress has been fueled by recent methodological developments in TCN science, creating a quantitative foundation for studying a broad range of sediment processes on the Earth's surface. Simultaneously, it has become clear that the production systematics of TCNs on Earth are very complex and influenced by many parameters, known to varying degrees, all of which need to be taken into account for accurate interpretations in TCN applications (see Cosmic Ray Interactions in Minerals). At the beginning of the twenty-first century, TCN research is driven by the enormous potential of this tool in a rapidly growing number of applications in Earth science and beyond, but at the same time the effects of numerous uncertainties must be considered to avoid overinterpretation of TCN measurements.

A primary application of TCNs is the field of surface exposure dating (SED). The basic principle of SED is simple: secondary cosmic rays (mostly high-energy neutrons) interact with target atoms in rocks and sediments on the Earth's surface producing *in situ* TCNs (dominantly via spallation reactions). Several of these *in situ* TCNs can be measured in excess to the natural background. If the rate of TCN production per time is known, the exposure of the surface continuous, and the surface to sample is original, the measured cosmogenic nuclide concentrations can be directly translated into an exposure age. In reality, the application of TCN is more complex due to uncertainties in the quantitative knowledge of production rates and several natural processes modifying the surface with time (e.g., erosion) or interrupting the exposure (e.g., sediment/snow/glacier cover).

There are various recent review articles summarizing SED methods. The comprehensive paper by Gosse and Phillips (2001) presents the first complete overview of TCN science, including overall uncertainty discussions. Niedermann (2002) reviews in detail the cosmogenic noble gas method and presents an overview of modern applications. The reviews by Cerling and Craig (1994), Kurz and Brook, (1994), and Lal (1991) are already introductory classics, presenting the basic principles of SED and outlining the immense potential of the method. The rapid progress in the cosmogenic dating field is arguably best reflected in the breadth and depth of innovation reported since the recent reviews. This very recent progress is the focus of this short article. We will present and discuss the new findings in two different sections, nuclide specific news and production rate novelties.

TCN-Specific Progress

One of the major areas of recent progress is the broadening of the target minerals that can be used for TCN analyses and the extension of the time-scale over which TCN methods can be applied. We briefly discuss the recent developments for each TCN below.

^{3}He from Olivine, Pyroxene, Fe/Ti-Oxide Minerals, and Garnet

Cosmogenic ^{3}He is a widely used TCN, mostly due to its apparently well-calibrated production rate (Niedermann, 2002) and the wide dating time range from 1 kyr (e.g., Gayer *et al.* (2004)) to more than 10 Myr (Schäfer *et al.*, 1999). Most studies assume 100% spallation from high-energy nucleons as the production pathway for cosmogenic noble gases (i.e., assuming muogenic and thermal neutron production to be negligible). Cosmogenic helium is routinely analyzed from pyroxene and olivine (e.g., Kurz (1986) and Licciardi *et al.* (2001)). Some success in measuring cosmogenic helium from quartz had been reported earlier from cold and dry Antarctica (Brook and Kurz, 1993), but most studies have since found that quartz does not quantitatively retain cosmogenic ^{3}He (Niedermann (2002) and references therein). One of the challenges is to separate cosmogenic helium from the noncosmogenic helium trapped in the mineral (e.g., mantle helium), but several studies have convincingly shown that isotopic analysis of the helium released by cold crushing and subsequent analysis of the remaining helium released by heating allows accurate determination of the cosmogenic ^{3}He concentration (Kurz and Brook (1994), Niedermann (2002) and references therein).

^{3}He from Fe/Ti-oxide minerals and garnet Recent studies suggest that high-purity mineral separation can be challenging, particularly for the pyroxene method (Margerison *et al.*, 2005; Oberholzer, 2004). Extraction of cosmogenic ^{3}He from additional mineral phases would therefore be desirable.

Bryce and Farley (2002) presented cosmogenic ^{3}He data from magnetite at the Goldschmidt conference in Davos. Subsequently, Kober *et al.* (2005) presented a convincing cosmogenic ^{3}He data set from Fe/Ti-oxide minerals. Using the ^{21}Ne exposure ages from quartz from the same surfaces, they calibrated the production rate on five different samples (**Table 1**), resulting in much higher production rates from Fe/Ti-oxide minerals than previously suggested (Bryce and Farley, 2002; Margerison *et al.*, 2005). Kober *et al.* (2005) also modeled the ^{3}He production rates from the main target elements (i.e., O, Mg, Al, Si, Fe, and Ni; **Table 1**) and reported excellent agreement between experimentally and theoretically derived values. The main findings were the high production rate values from O, Mg, and Al, indicating that elemental analysis for each sample is a requirement for all cosmogenic ^{3}He studies. This has not been done routinely since cosmogenic ^{3}He was assumed to be produced at very similar rates from the main target elements given above.

Kober *et al.* (2005) also determined the degassing pattern of cosmogenic ^{3}He from Fe/Ti-oxide minerals, showing that almost the entire helium inventory is degassed between 800° C and 1,400° C. The production rate of cosmogenic ^{3}He in Fe/Ti-minerals remains unsettled for now, owing to these recent insights, and improvements are continuously developing.

Gayer *et al.* (2004) reported cosmogenic ^{3}He data from Himalayan garnets and compared those to ^{10}Be data from coexisting quartz. Their ^{3}He production rate calibration results show a trend toward higher ^{3}He/^{10}Be production ratio with increasing altitude, which they interpret as an altitude dependence of this production ratio. While certainly more studies are needed to answer the question of element-dependent scaling factors, this study introduces cosmogenic ^{3}He from garnet to the TCN family.

^{21}Ne from Quartz, Olivine, Pyroxene, and Sanidine

Cosmogenic neon is a more complex system but has specific potential compared to cosmogenic helium. The production rate of cosmogenic ^{21}Ne (^{21}Ne is used because it has the lowest isotopic abundance in air) is less well calibrated, primarily relying on the studies by Niedermann and Graf (Niedermann, 2000; Niedermann *et al.*, 1994). However, the useful age range is the largest of all TCNs, ranging from 10 kyr (Niedermann *et al.*, 1994) to more than 30 Myr (Dunai *et al.*, 2005). Traditionally, cosmogenic ^{21}Ne has been analyzed from quartz (Niedermann *et al.*, 1994), olivine (Poreda and Cerling, 1992), and pyroxene (e.g., Bruno *et al.* (1997)).

^{21}Ne from sanidine Kober *et al.* (2005) reported successful measurement of ^{21}Ne from sanidine (for production rate values, see **Table 1**) showing that cosmogenic ^{21}Ne is quantitatively retained in sanidine. A very interesting aspect of this work is the apparently pure two-component mixture of atmospheric and cosmogenic neon for all samples analyzed. If nucleogenic neon is less abundant or absent in sanidine, implying that the separation of cosmogenic neon from the neon fraction trapped in the mineral lattice (mostly nucleogenic neon) is not required, then cosmogenic neon in sanidine will allow young surfaces to be dated. For all studies, analysis of the elemental composition of the mineral separates is required due to large differences between elemental production rates (see **Table 1**). So far, several elemental production rates rely on modeling, which is an additional uncertainty. The analytical measurement of cosmogenic neon is challenging due to double-charged ions (e.g., $^{40}Ar^{++}$, CO_2^{++}) interfering with the single-charged neon isotopes in the collector of the mass spectrometer, which might explain the relatively small number of published cosmogenic neon studies.

^{38}Ar from Calcium

Cosmogenic ^{38}Ar is produced in rocks by fast neutron spallation (*see* Cosmic Ray Interactions in Minerals) of target nuclides, in particular K and Ca. Though the presence of cosmogenic argon in Ca-rich minerals has been demonstrated earlier (Renne *et al.*, 2001), it has proven difficult to establish its production rate, due to problems connected to ^{36}Ar production by ^{35}Cl neutron capture and different production rates from K and Ca. A strategy to circumvent these problems is to analyze essentially Cl- and K-free minerals, as done by Niedermann *et al.* (2005) using pyroxene, which is low in potassium (**Table 1**). Although this is a step forwards, this methodology is still in the developmental stage.

^{10}Be from Quartz, Calcite, and Sanidine

In situ ^{10}Be produced in quartz has become the workhorse of TCN applications in the 1990s, evidenced by a drastic increase in published ^{10}Be studies. This is mostly due to a well-constrained production rate. Stone (2000) showed that most production rate

Table 1 Recently published TCN production rates at the Earth's surface at SLHL

Path-target-nuclide	*Method*	*Production rate (atom $g^{-1}y^{-1}$)*	*References*	*Comment*
Helium				
S- (Fe/Ti-Oxide)-^{3}He[a]	Cross-calibrated from cosmogenic ^{21}Ne in coexisting quartz	120 ± 11	(Kober *et al.*, 2005)	Valid for present-day magnetic field
S-(Fe/Ti-Oxide)-^{3}He	Experimental	69-77	(Bryce and Farley, 2002)	
S- (Fe/Ti-Oxide)-^{3}He	Calculated	124 ± 24	(Kober *et al.*, 2005)	
S-(O)-^{3}He	Calculated	191		
S-(Mg)-^{3}He	Calculated	182		
S-(Al)-^{3}He	Calculated	160	(Kober *et al.*, 2005)	
S-(Si)-^{3}He	Calculated	129		
S-(Fe)-^{3}He	Calculated	95		
S-(Ni)-^{3}He	Calculated	91		
Neon				
S-(sanidine)-^{21}Ne	Cross-calibrated from cosmogenic ^{21}Ne in coexisting quartz	30.4 ± 5.4	(Kober *et al.*, 2005)	Valid for present-day magenetic field
S-(sanidine)-^{21}Ne	Calculated	28 ± 5.6	(Kober *et al.*, 2005)	Valid for present-day magenetic field
S-(Na)-^{21}Ne	Calculated	208		
S-(Mg)-^{21}Ne	Calculated	189		
S-(Al)-^{21}Ne	Calculated	60		
S-(Si)-^{21}Ne	Calculated	44	(Kober *et al.*, 2005)	
S-(Ca)-^{21}Ne	Calculated	17		
S-(Fe)-^{21}Ne	Calculated	1		
Argon				
s-(calcium)-^{38}Ar		~200	(Niedermann *et al.*, 2005)	
Beryllium				
S-(sanidine)-^{10}Be	Cross-calibrated from cosmogenic^{10}Be in coexisting quartz	4.16 ± 0.39	(Kober *et al.*, 2005)	Valid for present-day geomagnetic field; the value given here is based on the ^{10}Be production rate of 5.1 at g^{-1} yr^{-1}
S-(sanidine)-^{10}Be	Calculated	4.55 ± 0.91	(Kober *et al.*, 2005)	
S-(O)-^{10}Be	Calculated	6.5		
S-(N)-^{10}Be	Calculated	24.5		
S-(Mg)-^{10}Be	Calculated	1.9		
S-(Al)-^{10}Be	Calculated	1.6	(Kober *et al.*, 2005)	
S-(Si)-^{10}Be	Calculated	3.3		
S-(Ca)-^{10}Be	Calculated	1.1		
S-(Fe)-^{10}Be	Calculated	1.9		
S-(Ni)-^{10}Be	Calculated	1.5		
S-(carbonate)-^{10}Be	Cross-calibrated from cosmogenic^{36}Cl	37.9	(Braucher *et al.*, 2005)	
In situ ^{14}*C*				
S-(quartz)-^{14}C	17.4 ± 0.3 kyr	15.1 ± 0.2	(Lifton *et al.*, 2001; Pigati and Lifton, 2004), Pigati *et al.* (ms in prep.)	Assumes 100% spallation production
		16.3 ± 0.2		Assumes 83% spallation, 2% fast muon, 15% slow muon production
S-olivine-^{14}C	17.3 ± 0.4	13.6 ± 0.2	(Pigati and Lifton, 2004), Pigati *et al.* (ms in prep.)	Assumes 100% spallation production
		14.3 ± 0.2		Assumes 83% spallation, 2% fast-muon, 15% slow muon production
				Both values based on quartz production rate, assuming production rate normalization to SiO_2 of *c.* 14% for olivine

[a]'S' means spallation.

calibration experiments are consistent with each other (yielding a sea level and high geomagnetic latitude (SLHL) production rate of 5.1 at g^{-1} yr^{-1}) if about 97% of the ^{10}Be is produced by spallation from fast neutrons and only 3% are produced by muons. Other factors for its growing popularity are its large applicable age range for dating (10^1–10^6) and the high analytical precision of accelerator mass spectrometry (AMS) (routinely about 2–4% 1σ). Recent progress in isotope dilution and AMS techniques allow exposure age dating of surfaces significantly younger than 1 kyr (Stone *et al.*, 2003). This opens up new perspectives for applications in fields such as archeology. On the high end, surfaces up to 3–4 Myr can be dated with ^{10}Be.

^{10}Be from calcite Braucher *et al.* (2005) report a first attempt to measure *in situ* ^{10}Be from calcite. By comparing the *in situ* ^{10}Be to *in situ* ^{36}Cl concentrations from a depth profile in carbonate rocks, they find constant ^{10}Be/^{36}Cl ratios for all samples, suggesting that all ^{10}Be was produced *in situ*. They determined a six-fold higher ^{10}Be production rate from calcite than from quartz (see **Table 1**). In their study they also reported that simple rinsing in de-ionized water over 24 h removes the meteoric ^{10}Be completely, implying that meteoric ^{10}Be does not adsorb on carbonates. Simple *in situ* ^{10}Be measurements from carbonates would represent a significant advance, particularly interesting for ^{10}Be–^{36}Cl burial dating techniques (*see* Landscape Evolution). However, as the authors state themselves, their first attempt must be verified in future studies.

^{10}Be from sanidine Kober *et al.* (2005) measured the ^{10}Be from sanidine following similar chemical extraction procedures than routinely done for ^{10}Be from quartz. Again, the production is element dependent, that is, the elemental composition of each mineral separate has to be measured.

^{26}Al from Quartz

Cosmogenic ^{26}Al is produced from quartz at a rate approximately 6 times higher than that of ^{10}Be (Nishiizumi *et al.*, 1989). *In situ* ^{26}Al and ^{10}Be can be extracted in parallel from the same quartz separate without significant extra effort during the chemical processing, resulting in a large number of combined ^{26}Al and ^{10}Be studies. The ^{26}Al production rate is nearly as well-known as the ^{10}Be production rate. Its useful age range spans 10^3–10^6 yr. The analytical uncertainty of ^{26}Al by AMS is poorer than that for ^{10}Be, typically some 5–7%. Another methodological problem is the need for extremely pure, essentially Al-free quartz separates (only 0.5% of feldspar in the quartz separate compromises the ^{26}Al analysis). An important recent application of ^{26}Al with ^{10}Be is the burial dating technique discussed by Granger (*see* Landscape Evolution).

In situ ^{14}C from Quartz and Olivine

Over the last decade, the AMS group at University of Arizona, Tucson, has developed the *in situ* ^{14}C method. Lifton *et al.* (2001) reported successful routine measurements of *in situ* ^{14}C from quartz. The major challenge, the separation of the atmospheric ^{14}C from the *in situ* ^{14}C, is achieved by a combination of acid etching and stepwise combustion with a degassed oxidizing flux. The major advantage of this new tool is its dating range covering the Holocene, where traditional TCN methods became problematic (with the exception of ^{10}Be see ^{10}Be from quartz, calcite, and sanidine). The short half-life also offers exciting applications combining ^{10}Be and ^{14}C to date burial periods over the Holocene. The production rate of ^{14}C from quartz is about 16 atom g^{-1} a^{-1} (**Table 1**).

Cosmogenic ^{14}C is currently playing an important role in the Cosmic Ray Produced Nuclide Systematics on Earth Project (CRONUS-Earth Project) (see Outlook). After about five half-lives, the production and the decay of *in situ* radionuclides reach secular equilibrium. Therefore, surfaces older than 25 kyr can be used to estimate the local ^{14}C production rate. This is a significant advantage over other radionuclides, for which, because they do not attain equilibrium quickly, the influence of surface erosion is greater for the short window of exposure recorded by *in situ* ^{14}C. Cosmogenic ^{14}C will be measured along transects in altitude and latitude to constrain time-integrated scaling factors – a goal of the CRONUS initiative.

^{14}C from olivine Recently, Pigati *et al.* (manuscript in preparation) have demonstrated a method for extracting *in situ* ^{14}C from olivine separates, extending the *in situ* ^{14}C method toward mafic lithologies. They take a similar approach to that of Lifton *et al.* (2001), in that they use chemical pretreatments followed by stepped combustion with an oxidizing flux. However, recovery of *in situ* ^{14}C with this method is dependent on the Mg:Fe ratio of the olivine, apparently due to chemical interactions in the melted flux. A large empirical correction based on this ratio, however, brings the recovered *in situ* ^{14}C concentrations into agreement with *in situ* ^{14}C measurements in quartz samples from similar locations. As with several other methods described here, this is an emerging technique, which needs to be verified by future studies.

^{36}Cl from Whole Rock and Feldspar

Cosmogenic ^{36}Cl production in rocks is complex due to its multiple production pathways (fast neutron spallation, muonic interactions, and epithermal and thermal captures are the principal pathways for ^{36}Cl production). This is partially reflected in inconsistencies of published production rates, which can differ by up to a factor of 2 (Phillips *et al.*, 1996; Stone *et al.*, 1996; Swanson, 1996; Swanson and Caffee, 2001), although differences in chemical procedure have also contributed to the production rate disparities. The studies presented by Stone *et al.* (1996) are the only ones based on mineral separates to isolate the different pathways as much as possible. Phillips (personal communication) recently reported good agreement between his reevaluated ^{36}Cl production rates from Ca and rates for the same target by Stone *et al.* (1996). The ^{36}Cl tool has large potential due to the combination of using a relatively simple chemical extraction procedure on whole-rock samples and a wide age range allowing dating of surfaces as young as several hundred years up to about 1 Myr. However, systematic uncertainties remain high due to the complicated production rate systematics, and potential problems with water mobilization of ^{36}Cl by weathering processes. Analyzing mineral separates is overcoming the former problem.

Production of Cosmogenic Nuclides in Terrestrial Rocks

As mentioned above, the rapid increase in applications of TCN methods is paralleled by a better understanding of the complexity of TCN production physics. In this section, we summarize the new findings affecting TCN production systematics and their scaling with altitude and latitude, as well as recently published temporal corrections for TCN production rates, such as corrections for geomagnetic field variations or long-term solar modulations (see previous section by Lal).

Calculation of an absolute surface exposure age using TCNs at a given location on Earth requires the knowledge of the local TCN production rate. TCN production rates are estimated by (1) measuring the cosmogenic nuclide concentrations of an independently dated surface, (2) measuring the cosmogenic radionuclide concentration of an erosion-free surface that is old enough so that the radionuclide concentration approaches saturation/secular equilibrium (i.e., a steady state between decay and production), or (3) theoretical modeling of the cosmic ray particle pathways and interactions down to the Earth's surface. The latter approach currently serves to support the geological calibration experiments since the uncertainties of the particle path simulation models remain high. The 'saturation' approach is particularly promising for *in situ* ^{14}C (see Outlook). The majority of the calibration studies so far have used independently dated surfaces. However, most sites cluster at mid-latitudes and altitudes lower than 2,000 m. For exposure dating studies elsewhere, the local production rate value needs to be scaled to the different altitudes and latitudes. Currently, all locally derived production rates are by convention scaled to SLHL for comparison. The resulting SLHL production rate values of each study depend not only on the local production rate value but also on the scaling procedure used, that is, production rate experiments and scaling procedures are interlinked and SLHL production rate values can only be compared if the scaling from the calibration site to SLHL is done consistently. For exposure age applications it might be advantageous to select a production rate calibration study close to the investigation site to keep the uncertainty due to scaling relatively low.

Table 1 gives an overview of the recently published production rates for the TCNs ^{3}He, ^{21}Ne, ^{38}Ar, ^{14}C, and ^{10}Be. This table is meant to be a complementary addition to table 4 in the review by Gosse and Phillips (2001). Despite the numerous experiments (*see* Exposure Geochronology, Landscape Evolution), the production rate of TCNs in the lithosphere and its variation in time and space arguably represents the biggest remaining uncertainty of the method. To satisfy the urgent need to improve this, the large-scale US/EU CRONUS initiative was undertaken (see Outlook).

Production of *in situ* cosmogenic nuclides in the lithosphere is complex, since TCN production is a multiparameter function of the cosmic ray flux at the Earth's surface. Variations of this flux are consequences of (1) variations in the primary galactic cosmic ray flux, (2) solar modulation of the GCR flux, (3) changes in atmospheric shielding, and (4) variations in the intensity and orientation of the Earth's magnetic field. While the low-frequency variation (1) is of minor importance for the application of TCN on timescales of 10^{3}–10^{6} yr, the other variations have to be considered. This has led to an increasing number of new studies designed to characterize the various effects.

Geomagnetic Variation and TCN Production

There are several recent studies investigating the impact of geomagnetic variations in intensity and orientation on the TCN production on the Earth's surface (Desilets and Zreda, 2003; Dunai, 2000;

Gosse and Phillips, 2001; Masarik *et al.*, 2001; Pigati and Lifton, 2004), but the results of the individual studies remain controversial. For example, while Masarik *et al.* (2001) report maximum corrections for geomagnetic variation of 10% (equator, exposure time >50 kyr), Pigati and Lifton, (2004) report corrections of up to 30% for ^{10}Be production rates. This large difference is primarily due to the different scaling and normalization approaches. While Masarik *et al.* (2001) use the modern production rate at the geographic latitude at a given site to normalize integrated production rates, others (Gosse and Phillips, 2001; Pigati and Lifton, 2004) use the modern production rate value at the geomagnetic latitude. In addition, different data sets have been used to reconstruct the geomagnetic intensity (e.g., Yang *et al.* (2000) versus Guyodo and Valet (1999)). This controversy remains unresolved and is one of the focus areas of the CRONUS initiative.

Solar Modulation of TCN Production

Lifton *et al.* (2005) introduced correction factors for millennial-scale solar modulation of TCN production rates based on model sunspot numbers over the last 11.4 kyr. They find effects on spallogenic nucleon scaling factors for SLHL of up to 10%, supporting the view that long-term solar modulation must be included in secondary cosmic ray scaling procedures.

Changes in Atmospheric Shielding

Stone (2000) introduced the effects of regional air pressure regimes different from the standard atmosphere air pressure. The biggest changes are in Antarctica, where air pressure is about 25 hPa lower on average than elsewhere, resulting in up to 30% higher production rates. This paper also publishes a map of global correction factors relative to standard atmospheric values, which should be used in all TCN studies. The influence of variations in the local air pressure regime during glacial periods toward lower air pressure was recently suggested by Ackert *et al.* (2003) as an explanation for their relatively high ^{3}He production rates constrained from ^{39}Ar/^{40}Ar-dated lava some 70 kyr and 110 kyr in age.

Element-Dependent Scaling of TCN Production Rates

Desilets and Zreda (2003) reported shifts of the nucleon energy spectrum of TCN producing cosmic rays toward lower energies with increasing atmospheric depth based on neutron monitor data. Since different TCN producing reactions have distinct excitation functions, this would result in element-dependent scaling of these production reactions. Experimental support for this theory is presented by Gayer *et al.* (2004), suggesting evidence for an altitude dependence of the ^{3}He/^{10}Be production ratio; however, the cosmogenic ^{3}He data from the Himalayan garnets set underlying this study remains controversial. This fundamental problem of production rate determination is another core experiment of the CRONUS initiative.

Outlook

This article focuses on the most recent progress in the method of cosmogenic surface exposure over the last few years, stressing the rapid progress in this field. Cosmogenic nuclide surface exposure dating has fundamentally changed our perspectives on quantifying Earth surface processes. At the same time, the unresolved controversies about TCN production systematics introduced above present a danger for TCN users to push the method beyond its limits. To improve the situation and to create a solid quantitative understanding of the TCN method, an international multidisciplinary effort, the CRONUS initiative (http://www.physics.purdue.edu) officially started in early 2005. The overall goal of the 5-yr-long CRONUS project is to significantly improve the accuracy and reliability of TCN science while making the required production rate systematics more transparent. The synergistic framework of this project is also expected to accelerate methodological progress. It is the vision of CRONUS to lay the robust foundation for a new era of quantitative Earth surface process science with reliably defined uncertainties. Future progress in this field can expand TCN applications to disciplines such as archeology, while remaining a prime tool for modern Earth sciences research.

As one example, the use of *in situ* ^{14}C from surfaces in secular equilibrium should be mentioned here. After five half-lives (i.e., after about 25 kyr of exposure), the production of *in situ* ^{14}C is in equilibrium with its decay. Measuring the secular equilibrium ^{14}C concentration directly yields a local production rate. Within the CRONUS framework, Lifton is attempting to reconstruct time-integrated production rates by analyzing the *in situ* ^{14}C concentration from widely distributed saturated surfaces (typically with independent evidence of great antiquity, either from other TCN analyses or from geomorphic indicators) on latitude and altitude transects (**Fig. 1**). If successful, this will yield a network of local production rate calibration experiments for *in situ* ^{14}C and, in turn, production rate scaling factors. Compared to contemporary production rates derived

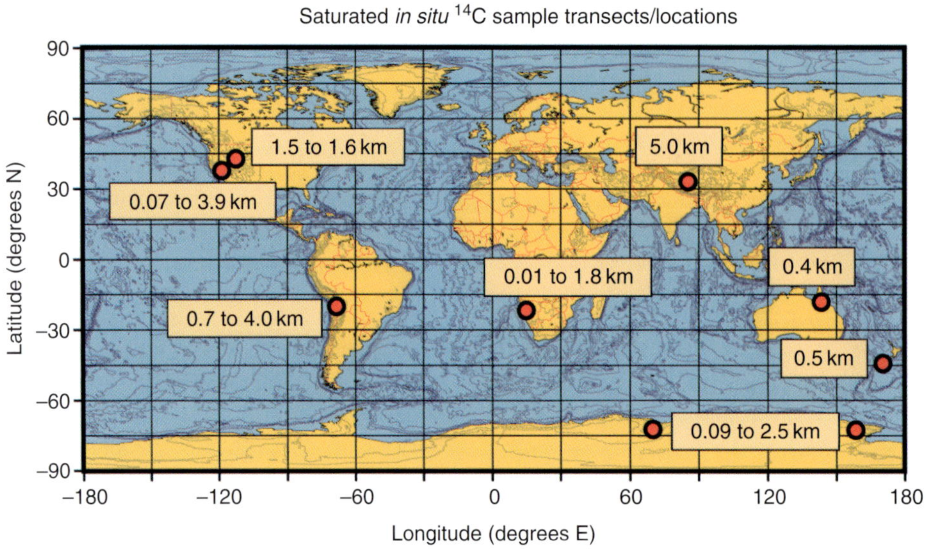

Figure 1 Map of the *in situ* ^{14}C survey within the CRONUS initiative.

from large targets exposed for short periods (1–2 yr), the impact of time-dependent fluctuations can be evaluated.

See also: **Cosmogenic Nuclide Dating**: Cosmic Ray Interactions in Minerals; Exposure Geochronology; Landscape Evolution.

References

Ackert, R. P., Singer, B. S., Guillou, H., Kaplan, M. R., and Kurz, M. D. (2003). Long-term cosmogenic 3He production rates from 40Ar/39Ar and K–Ar dated Patagonian lava flows at 47° S. *Earth and Planetary Science Letters* **210**(1–2), 119–136.

Braucher, R., Benedetti, L., Bourles, D., Brown, E. T., and Chardon, D. (2005). Use on *in-situ* produced Be−10 in carbonate-rich environments: A first attempt. *Geochimica et Cosmochimica Acta* **69**(6), 1473–1478.

Brook, E. J., and Kurz, M. D. (1993). Surface-exposure chronology using *in situ* cosmogenic 3He in antarctic quartz sandstone boulders. *Quaternary Research* **39**, 1–10.

Bruno, L. A., Baur, H., Graf, T., Schlüchter, C., Signer, P., and Wieler, R. (1997). Dating of Sirius group tillites in the antarctic dry valleys with cosmogenic 3He and 21Ne. *Earth and Planetary Science Letters* **147**, 37–54.

Bryce, J. G., and Farley, K. A. (2002). He-3 exposure dating of magnetite. *Goldschmidt Conference*, A108.

Cerling, T. E., and Craig, H. (1994). Geomorphology and *in-situ* cosmogenic isotopes. *Annual Reviews of Earth Planetary Science Letters* **22**, 273–317.

Desilets, D., and Zreda, M. (2003). Spatial and temporal distribution of secondary cosmic-ray nucleon intensities and applications to *in situ* cosmogenic dating. *Earth and Planetary Science Letters* **206**(1–2), 21–42.

Dunai, T. J. (2000). Scaling factors for production rates of *in-situ* produced cosmogenic nuclides: A critical re-evaluation. *Earth and Planetary Science Letters* **176**(1), 157–169.

Dunai, T. J., Gonzalez-Lopes, G. A., and Juez-Larre, J. (2005). Oligocene–Miocene age of aridity in the Atacama Desert revealed by exposure dating of erosion-sensitive landforms. *Geology* **33**(4), 321–324.

Gayer, E., Pik, R., Lave, J., France-Lanord, C., Bourles, D., and Marty, B. (2004). Cosmogenic He-3 in Himalayan garnets indicating an altitude dependence of the He-3/Be−10 production ratio. *Earth and Planetary Science Letters* **229**, 91–104.

Gosse, J. C., and Phillips, F. M. (2001). Terrestrial *in situ* cosmogenic nuclides: Theory and application. *Quaternary Science Reviews* **20**(14), 1475–1560.

Guyodo, Y., and Valet, J.-P. (1999). Global changes in intensity of the Earth's magnetic field during the past 800 kyr. *Nature* **399**, 249–252.

Kober, F., Ivy-Ochs, S., Leya, I., *et al.* (2005). *In situ* cosmogenic 10Be and 21Ne in sanidine and *in situ* cosmogenic 3He in Fe-Ti-oxide minerals. *Earth and Planetary Science Letters* **236**, 404–418.

Kurz, M. D. (1986). Cosmogenic helium in a terrestrial igneous rock. *Nature* **320**, 435–439.

Kurz, M. D., and Brook, E. J. (1994). Surface exposure dating with cosmogenic nuclides. In *Dating in Exposed and Surface Contexts* (C. Beck, Ed.)University of New Mexico Press.

Lal, D. (1991). Cosmic ray labeling of erosion surfaces: *In situ* nuclide production rates and erosion models. *Earth and Planetary Science Letters*, **104**, 424–439.

Licciardi, J. M., Clark, P. U., Brook, E. J., *et al.* (2001). Cosmogenic 3He and 10Be chronologies of the late Pinedale northern Yellowstone ice cap, Montana, USA. *Geology* **29**(12), 1095–1098.

Lifton, N. A., Bieber, J. W., Clem, J. M., *et al.* (2005). Addressing solar modulation and long-term uncertainties in scaling secondary cosmic rays for *in situ* cosmogenic nuclide applications. *Earth and Planetary Science Letters*, **239**, 140–161.

Lifton, N. A., Jull, A. J. T., and Quade, J. (2001). A new extraction technique and production rate estimate for *in situ* cosmogenic ^{14}C in quartz. *Geochimica et Cosmochimica Acta*, **65**(12), 1953–1969.

Margerison, H. R., Phillips, W. M., Stuart, F. M., and Sugden, D. E. (2005). Cosmogenic 3He concentrations in ancient flood deposits from the Coombs Hills, northern Dry Valleys, East Antarctica: Interpreting exposure ages and erosion rates. *Earth and Planetary Science Letters* **230**, 163–175.

Masarik, J., Frank, M., Schäfer, J. M., and Wieler, R. (2001). Correction of *in situ* cosmogenic nuclide production rates for geomagnetic field intensity variations during the past 800,000 years. *Geochimica et Cosmochimica Acta* **65**(17), 2995–3003.

Niedermann, S. (2000). The ^{21}Ne production rate in quartz revisited. *Earth and Planetary Science Letters* **183**, 361–364.

Niedermann, S. (2002). Cosmic-ray-produced noble gases in terrestrial rocks: Dating tools for surface processes. In *Noble Gases in Geochemistry and Cosmochemistry* (D. Porcelli, C. J. Ballentine and R. Wieler, Eds.), vol. 47 pp. 731–784. Mineralogical Society of America.

Niedermann, S., Graf, T., Kim, J. S., Kohl, C. P., Marti, K., and Nishiizumi, K. (1994). Cosmic-ray-produced ^{21}Ne in terrestrial quartz: The neon inventory of Sierra Nevada quartz separates. *Earth and Planetary Science Letters* **125**, 341–355.

Niedermann, S., Schaefer, J. M., Wieler, R., and Naumann, R. (2005). Experimental determination of the cosmogenic Ar production rate from Ca. *Eos, Transactions, American Geophysical Union* **86**(52) Fall Meeting Supplement, Abstract U33A-0003.

Nishiizumi, K., Winterer, E. L., Kohl, C. P., *et al.* (1989). Cosmic ray production rates of 10Be and 26Al in quartz from glacially polished rocks. *Journal of Geophysical Research* **94**(B12), 17907–17915.

Oberholzer P (2004) Reconstructing paleoclimate and landscape history in Antarctica and Tibet with cosmogenic nuclides. *Diss Nr. 15472, ETH Zuerich.*

Phillips, F. M., Zreda, M. G., Elmore, D., and Sharma, P. (1996). A reevaluation of cosmogenic 36Cl production rates in terrestrial rocks. *Geophysical Research Letters* **23**(9), 949–952.

Pigati, J. S., and Lifton, N. A. (2004). Geomagnetic effects on time-integrated cosmogenic nuclide production with emphasis on *in-situ* 14C and 10Be. *Earth and Planetary Science Letters* **226**, 193–205.

Pigati *et al.* (manuscript in prepration).

Poreda, R. J., and Cerling, T. E. (1992). Cosmogenic Neon in recent lavas from the western United-States. *Geophysical Research Letters* **19**(18), 1863–1866.

Renne, P. R., Farley, K. A., Becker, T. A., and Sharp, W. D. (2001). Terrestrial cosmogenic argon. *Earth and Planetary Science Letters* **188**(3–4), 435–440.

Schaeffer, O. A., and Davis, R. (1956). Chlorine-36 in nature. In: *Nuclear Processes in Geologic Settings*, pp. 172–180. National Academy of Science.

Schäfer, J. M., Ivy-Ochs, S., Wieler, R., *et al.* (1999). Cosmogenic noble gas studies in the oldest landscape on earth: Surface exposure ages of the Dry Valleys, Antarctica. *Earth and Planetary Science Letters* **167**, 215–226.

Stone, J. (2000). Air pressure and cosmogenic isotope production. *Journal of Geophysical Research* **105**(B10), 23753–23759.

Stone, J. O., Allan, G. L., Fifield, L. K., and Cresswell, R. G. (1996). Cosmogenic chlorine-36 from calcium spallation. *Geochimica et Cosmochimica Acta* **60**, 679–692.

Stone, J. O., Balco, G., Sugden, D. E., *et al.* (2003). Holocene deglaciation of Mary Bird Land, West Antarctica. *Science* **299**, 99–102.

Swanson, T. (1996). Determination of Cl-36 production rates form the deglaciation history of Whidbey and Fidalgo Islands, Washington. *Radiocarbon* **38**(1), 172.

Swanson, T. W., and Caffee, M. W. (2001). Determination of 36Cl production rates derived from the well-dated deglaciation surfaces of Whidbey and Fidalgo Islands, Washington. *Quaternary Research* **56**(3), 366–382.

Yang, S., Oday, H., and Shaw, J. (2000). Variations in the geomagnetic dipole moment over the last 12,000 years. *Geophysical Journal International* **140**, 158–162.

Relevant Website

http://www.physics.purdue.edu – Cosmic Ray Produced Nuclide Systematics on Earth Project (CRONUS-Earth Project).

Cosmic Ray Interactions in Minerals

D Lal, Scripps Institution of Oceanography, CA, USA

© 2007 Elsevier B.V. All rights reserved.

Introduction

Nuclear interactions occur everywhere in the universe since natural radiations are present everywhere. The types of interactions and their magnitude show a wide variety as would be expected. Nuclear interactions, particularly the inelastic variety, in which energy transfers occur between the incident particle and the target, have been increasingly used in the past few decades to study the history of evolution of planetary matter. Considering the topics covered in this encyclopedia, the emphasis is on nuclear interactions caused by cosmic radiation in the Earth's atmosphere and in the upper layers of the lithosphere, with a view to use interaction products as tracers for studying geomorphic processes. It must be pointed out here that this is a field which has now become accessible to quantitative modeling as a result of the recent advances that have made it possible to measure the ultralow concentrations of nuclear products arising from nuclear interactions in surficial and deep-seated materials (Lal, 1988a, 1991; Tuniz *et al.*, 1998; Gosse and Phillips, 2001; *see* Methods). Prior to this development, only sketchy information was available based on indirect assessments, such as coatings on surfaces and thermoluminescence.

To understand the principles and uncertainties of terrestrial *in situ* cosmogenic nuclide (TCN) methods, it is necessary to appreciate the theory and experimental basis for our understanding of primary cosmic ray interactions in the atmosphere and the subsequent interactions between cosmic ray secondaries and matter. Although the theoretical framework allows fairly accurate estimations of the temporal variations in the cosmic ray flux at the top of the atmosphere for hypothetical changes in the geomagnetic field and solar activity, and although the basic features of nuclear interactions are fairly well understood, it is indeed a difficult task to accurately estimate the total rates of nuclear disintegrations and particular nuclear interaction rates giving rise to specified nuclides at different locations in the atmosphere.

Cosmic rays attenuate as they interact with mass, at a rate referred to as an interaction mean free path. The Earth's atmosphere represents more than ten interaction mean free paths for the geometric nuclear interaction of nucleons (protons or neutrons) with air nuclei, about $90\,g\,cm^{-2}$. Within the lithosphere, the character of nuclear interacting particles changes dramatically, from primarily nucleons to slow and fast muons. Within the atmosphere, the total rate of nuclear disintegrations changes by a factor of 10^2 at the equator and 1.5×10^3 at the poles, between the top of the atmosphere and sea level (Lal and Peters, 1967). It is quite manifest that the atmosphere puts a heavy demand on any nuclear systematics used to treat the nuclear cascade development. Nuclear interactions are complex, with emission of different particles over a wide range of energies, and with a complex angular distribution (with respect to the incident particle). This requires parametrization of secondary particle emission characteristics as a function of energy for MeV to GeV energies. As an example, even if one makes an error of 10% in the assignment of the value of one of the relevant parameters, this would lead to an error of >100% in taking the nucleonic cascade to sea level (Lal, 2000). Even a 10% error in the assignment of geometric nuclear interaction cross section (the probability of interaction) produces the same effect. The propagation of the generating component of nucleons through the atmosphere depends critically on the inelasticity parameter, which fortunately does not show severe energy dependence. The observations of cosmic ray neutrons show that at low latitudes the secondary nucleonic component at atmospheric depths greater than $200\,g\,cm^{-2}$ is appreciably harder than at poles, which is not unexpected because of the much harder energy spectrum of the primary radiation incident at the equator (protons of $E > 14\,GeV$ and He nuclei above 6.6 GeV/n). The absorption mean free paths lie in the range of $220–160\,g\,cm^{-2}$ between the equator and the poles. A 10% uncertainty in the mean free paths corresponds to a 35% uncertainty in scaling factors between sea level at 60° and 3 km at equator.

There is a need for accurate estimates of production rates of nuclides in the atmosphere and at underground depths. Even if the total nuclear disintegration rates can be estimated, it is not generally feasible to make accurate estimations of nuclide production rates. One of the difficulties, in particular, is a lack of knowledge of partial cross sections for the formation of nuclides in different target elements as a function of neutron energy, since in the atmosphere, neutrons are the principal producers of nuclear interactions.

It is very difficult to depend entirely on theory to accomplish the tasks. One has to employ suitable 'process normalization' techniques, treating finite size chunks of matter as black boxes, without going into details of the interaction. In this treatment, as an example, one assumes that the more energetic component of nucleons is absorbed following certain observed features in high-energy interactions, whereby, after an interaction, the incident nucleon continues with an appreciable energy. The absorption of an incident nucleon (neutron or proton) in matter is then governed by simple absorption rules, again guided by experimental data. It further follows that within a few interaction mean free paths in the target (atmosphere or rock), there would exist certain relationships between the propagating nucleon components in different energy bands, with near-equilibrium between high- and low-energy nucleons. In fact, accepting this makes it possible to make very accurate predictions of fluxes of secondary nucleons in a target.

However, it must be mentioned that apparently some numerical codes are able to predict the buildup of nucleonic cascades in the atmosphere, and sometimes also predict the partial cross sections for the formation of different nuclides as a function of energy (Masarik and Reedy, 1995). However, the codes must also rest heavily on normalization techniques since data are simply not available to treat in any detail the processes which occur in nuclear interactions, as mentioned above.

A practical approach therefore is to consider the dynamics of nuclide production theoretically and to obtain absolute nuclide production rates wherever possible by exposing suitable targets to cosmic radiation at different altitudes and latitudes (Lal *et al.*, 1960), as was first shown in the case of radionuclides of half-lives <3 months. With the development of the accelerator mass spectrometry (AMS), it is now possible to extend this method to long half-life radionuclides (cf. Nishiizumi *et al.* (1996)). Currently utilized production rates of TCN have therefore been empirically based on measurements of the isotope for targets as short as a year to tens of thousands of years. The longer duration experiments make use of geological surfaces with exposure ages that have been determined independently (e.g., radiocarbon-dated lava).

Two approaches are currently being used in applications of the cosmogenic nuclides as tracers for production: one to determine the scaling factors for neutrons of different energies at different altitudes and latitudes, and the other to determine the absolute production rates of a nuclide in the atmosphere or in targets exposed in the atmosphere at a few sites. The latter is very necessary, partly because of lack of data on partial cross sections for formation of nuclides from different target elements. Additionally, semiempirical approaches have to be used to estimate nuclide production within rocks of different sizes exposed at different elevations in the atmosphere (Masarik and Wieler, 2003; Lal and Chen, 2005).

Cosmic Radiation: The Omnipresent Energetic Nuclear Radiation

Among the four principal radiations which are present everywhere in the universe, electromagnetic radiation, cosmic radiation, neutrinos and gravitational radiation, the former two occupy a large band in energy, >15 orders of magnitude. By this fact alone, they serve as windows to the universe to map the distribution of gases, dust, and different forms of matter including stars. Cosmic radiation, discovered in 1912 by Victor Hess by carrying his charged-particle detector instrument to an altitude of 17,500 feet in a gondola, consists of nuclei of elements in the periodic table with energies of a few electron volts to $>10^{20}$ eV. In cosmic ray physics, it is customary to measure nuclear particle energies in electron volts. Further, since one must often deal with nuclear matter moving with speeds comparable to the speed of light, it becomes necessary to use relativistic equations derived from Lorentz transformations. The magnitude of the electron volt and the relativistic relations between particle's mass, momentum, and energy are briefly outlined in Appendix 1.

As discussed below, most of the cosmic ray flux incident at the Earth lies in the region of a few thousand MeV. Since the binding energy of nucleons (neutrons or protons) in matter also lie in the range of few MeV, the cosmic ray particles are capable of fragmenting nuclei with which they interact. Indeed, they produce appreciable nuclear transformations in the Earth's atmosphere and even in the lithosphere down to great depths. This results in production of new forms of nuclei, some not initially present in the targets. In contrast, the electromagnetic radiation, which interacts very weakly with matter by virtue of its coulomb field, does not produce any appreciable changes in the nuclear composition of matter.

Appendix 1 Units Used in Cosmic Ray Physics

Since we are dealing with nuclear size particles, and since we generally deal with relativistic particles (velocities comparable to the speed of light, *c*), we work with a special unit system appropriate to particle masses and energies. The units and important relations between particle velocities, momentum, rigidity, and energy are defined below.

The unit of energy is electron volt, the kinetic energy gained by an electron when it is raised in potential by 1 volt (the energy gained by an He nucleus in being raised by 1 volt would be 2 eV). One electron volt equals 1.6×10^{-12} erg of energy (1 electron charge $= 4.8 \times 10^{-10}$ e.s.u. (1.6×10^{-20} e.m.u.) and 1 V = 1/300 e.s.u. (10^{8} e.m.u.)).

At relativistic energies, when particle velocities are comparable to speed of light, the following relations hold between momentum, *p*, and rigidity, *R* (*m* is the particle mass at rest, and *c* is the speed of light *in vacuo*):

$$E = mc^2\gamma \quad [A.1.1]$$

$$\gamma = 1/\sqrt{1-(v/c)^2} \quad [A.1.2]$$

$$p = mv\gamma; pc = mc^2(v/c)\gamma \quad [A.1.3]$$

$$R = pc/(Ze) \quad [A.1.4]$$

E is defined as the total energy (rest mass + kinetic energy, E_k). When *v c*, $\gamma \approx 1$, the total energy is close to its rest mass: m_0c^2. At Newtonian velocities, *v c*, the kinetic energy can be seen to be equal to $(1/2)mv^2$, from binomial expansion of Eqn [A.1.1].

The particle rigidity is so defined (Eqn [A.1.4]), based on the movement of a charged particle in magnetic field, balancing the centripetal force on the particle with the Lorentz force (Rossi and Olbert, 1970).The units of *E*, *p*, and *R* are electron volt (eV), eV/*c*, and volt (V), respectively.

Energy and momentum are related to each other by a very useful relation, as can be seen by using Eqns [A.1.1] and [A.1.3]:

$$E^2 = (E_k + mc^2)^2 = (pc)^2 + (mc^2)^2 \quad [A.1.5]$$

In this article, we discuss the salient features of cosmic radiation, their composition and energy spectra, development of the secondary cosmic ray beam in the atmosphere and the lithosphere, and production of nuclides in this process by the principal nuclear active particles. The incident cosmic ray flux at the top of the atmosphere is latitude dependent since it is strongly modulated by (1) the Earth's geomagnetic field and (2) solar plasma. These modulations show temporal variability because both geomagnetic field and solar activity are variable. We explicitly consider the nature of this variability.

Composition and Energy Spectrum of Cosmic Rays

'Cosmic rays' are composed of ordinary forms of matter accelerated by electromagnetic processes in certain flares in stars, and by shock processes in explosions of supernovas to relativistic energies. (The term cosmic radiation was originally adopted to reflect our lack of knowledge of their acceleration mechanisms and also the stellar sources primarily responsible for their acceleration. It remains today as the term of choice because of the multiplicity of its sources in the universe, from stars to supernovas.)

Although cosmic rays are accelerated in several astrophysical settings, the principal source of energy for galactic cosmic ray acceleration seems to be shock waves driven by the explosions of supernovas (see Higdon *et al.* (1998) and Ramaty *et al.* (1998), and references therein). Most (80–90%) of the galactic supernovas are type II and Ib/C (core collapses of massive O and B stars). Since the average supernova energy release is about 10^{50} erg, and their frequency is ~3 per century, the average cosmic ray power, 10^{41} erg s^{-1}, needed to maintain the level of 'galactic' cosmic rays is met by supernovas.

As would be expected, the cosmic ray beam is composed of fast nuclei up to relativistic energies, completely stripped of their electrons in the early stages of acceleration. Since hydrogen is the most abundant element in the universe, most of the flux is comprised of protons. Next in abundance are He, C, N, O, and heavier nuclei. Li, Be, B ions are rare in the universe. Their presence in the cosmic ray beam arises primarily due to nuclear interactions of heavier nuclei.

The wide range of cosmic ray energies requires that their energy spectra be described in terms of a power law function in kinetic energy, E_k, as differential or integral forms:

$$dn/dE_k = j(E_k) \qquad [1]$$

$$N(>E_k) = \int_{E_k} j(E_k)dE_k \qquad [2]$$

E_k is defined as kinetic energy of the particle. In the case of complex nuclei, such as He and heavier nuclei (C, N, O, Si, Fe, etc.), it is often defined as kinetic energy per nucleon. A typical form of $j(E_k)$ over a wide range of energies is: $a \times E_k^{-\Gamma}$, except at low energies, where it is of the form: $(a + E_k)^{-\Gamma}$, where a is a constant. At kinetic energies >3 GeV, the value of Γ is ~2.5. For further details, see Peters (1958), Simpson (1983, 1992), and Ziegler (1998).

Primary Galactic Cosmic Ray (GCR) Energy Spectra and its Modulation by Solar Plasma

The differential energy spectra of H, He, C, and Fe nuclei in the energy region 10–10^4 MeV per nucleon are shown in **Figure 1**, obtained using instruments flown in a large number of balloon flights at the top of the atmosphere (Simpson, 1992). The scatter in the

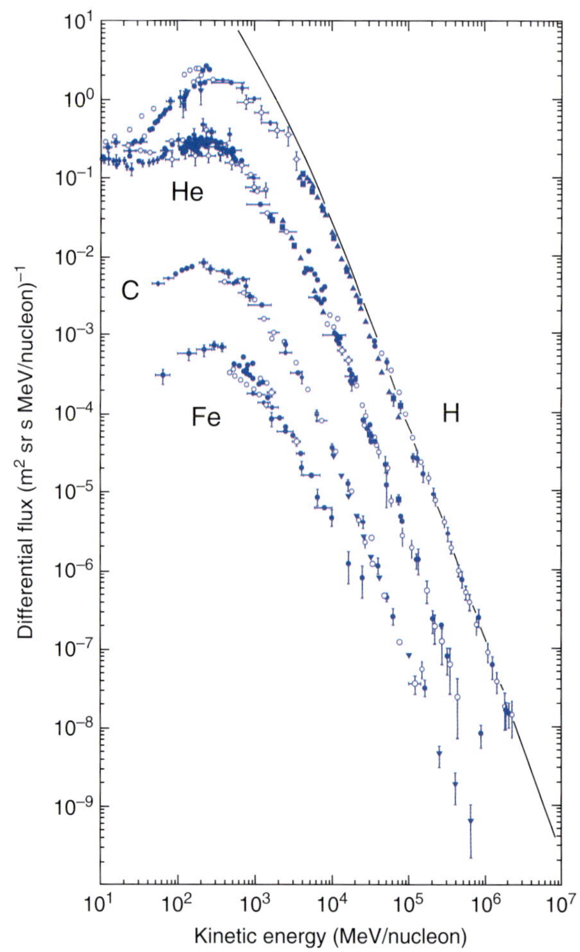

Figure 1 The energy spectra of elements in cosmic rays (from top: H, He, C and Fe) measured near Earth. The solid curve shows the hydrogen spectrum extrapolated for interstellar space, for no solar modulation. Figure reproduced from Simpson JA (1992) Cosmic radiation: Particle astrophysics in the heliosphere. *Annals of the New York Academy of Sciences* 655: 95–137.

measured data, taken at different periods, arises from varying modulation of the cosmic ray beam by the solar plasma. The primary GCR flux incident at the top of the Earth's atmosphere is anti-correlated with solar activity (with some phase lag). See Gleeson and Axford (1968) for a convection–diffusion theoretical framework for the transport of GCR particles through the heliosphere, and see Castagnoli and Lal (1980) and Papini *et al.* (1996) for the essential features of the experimental data on solar modulation of primary cosmic ray flux.

In terms of the solar modulation formulism (Gleeson and Axford, 1968), the effective change in the energy of a charged particle moving from a point P on the boundary along a dynamical trajectory to an interior point Q is $Ze\phi$, irrespective of its path to Q. If the measured kinetic energy of a charged particle at Q is E_k per nucleon, then it follows (Randall and Van Allen, 1986) that its kinetic energy per nucleon at P must have been $(E_k + Ze\phi)$. As shown by Randall and Van Allen (1986), this leads to the following simple equation relating differential fluxes at points P and Q:

$$\frac{j_Q(T)}{T(T+2mc^2)} = \frac{j_P(T+(Ze\phi/A))}{(T+(Ze\phi/A))(T+(Ze\phi/A)+2mc^2)} \quad [3]$$

In Eqn [3], A is the mass number, T is defined as the kinetic energy per nucleon, the total kinetic energy being AE_k. If the point P is taken at the heliosphere boundary, and the point Q at 1 AU, then the charged particle is modulated through a potential difference of ϕMeV per nucleon.

For more than four decades, the solar modulation of cosmic ray flux has been studied using satellites/spacecraft in orbit around the Earth, and more recently on deep space probes. The range of observed variations in the modulated near-Earth differential flux of GCR protons until the late 1970s is shown in **Figure 2** for protons, as described by different curves designated by the modulation parameter, ϕ (Randall and Van Allen, 1986; Castagnoli and Lal, 1980). The hypothetical curve, $\phi = 0$, corresponds to the predicted shape of the interstellar GCR proton flux, that is, the GCR flux outside the outer heliosphere. Information about near-Earth changes in flux have come largely from the charged-particle monitoring experiment (CPME) aboard the IMP8 satellite since 1973. For a recent summary of these data, reference is made to Simunac and Armstrong (2004).

Our knowledge of modulation of GCR through the outer heliosphere comes from deep space probes (Lockwood and Webber, 1995; Fuji and McDonald, 1997) which transmitted *in situ* cosmic ray data up to distances of more than 70 AU (Pioneers 10 and 11, Voyagers 1 and 2). In **Figure 3** are plotted the results of measurements of time variations in the differential particle fluxes of H ions of 130–225 MeV and He ions of 10–21 and 180–450 MeV/n kinetic energies (Fuji and McDonald, 1997), at distances of up to ~60 AU during 1972–95. In the same figure, the measured changes in the counting rates of the neutron monitor

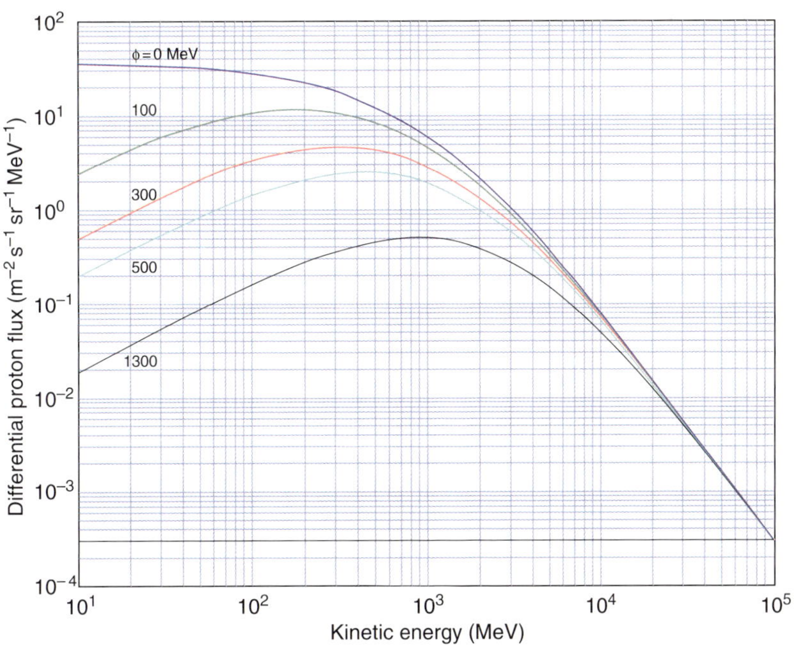

Figure 2 Estimated proton spectra for different solar modulation intensities as defined by ϕ and guided by observed proton spectra during different years (see text). Cosmic ray fluxes for different ϕ values are described well by eqn [3].

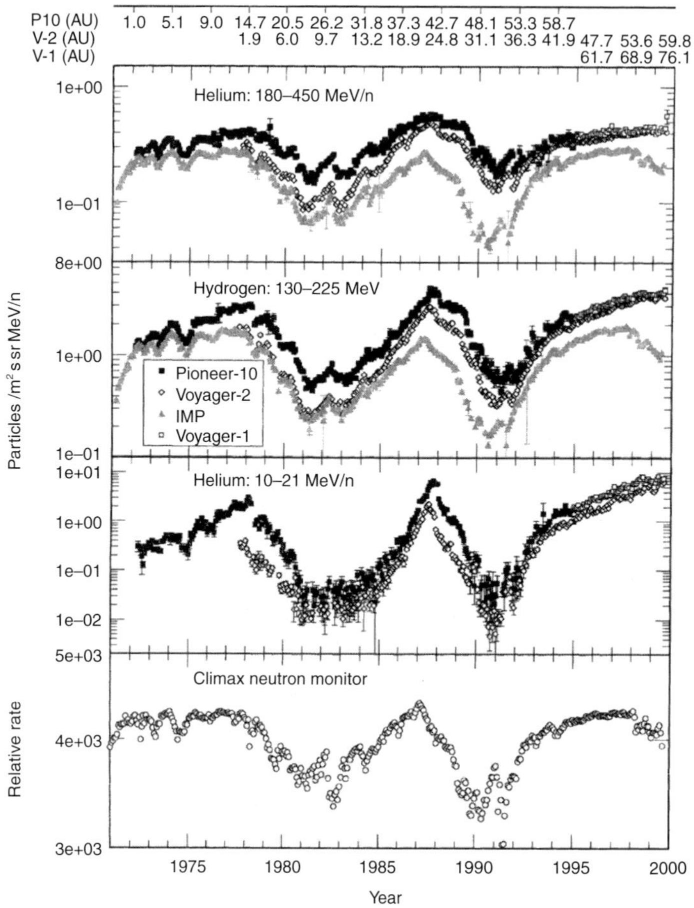

Figure 3 Time histories of the 26-day average galactic cosmic ray intensities of H and He, and of thermal neutrons measured at Climax. Figure reproduced from Fuji Z and McDonald FB (1997) Radial intensity gradients of galactic cosmic rays (1972–1995) in the heliosphere. *Journal of Geophysical Research* 102(A11): 24,201–24,208.

at Climax are also presented. Finally, **Figure 4** shows the data on energy spectra of protons measured with Pioneer 10 and Voyager 2 spacecrafts from 1978 to 1992 for GCR protons of >70 MeV kinetic energy and corresponding near Earth fluxes based on IMP satellite (Lockwood and Webber, 1995). The radial distances of Pioneer 10 and Voyager 2 were ~60 AU and 40 AU, respectively, in 1992.

Figures 2–4 show a consistent behavior of modulation of cosmic ray flux at radial distances of 1–60 AU space in the heliosphere, corresponding to modulation parameter, ϕ values ranging between ~350 and 1,300 MeV/n, inversely correlated with solar activity (often defined using the Zurich sunspot numbers). During sustained solar minimum periods, as during the Maunder Minimum, the value of ϕ may have been ~100 MeV/n (Castagnoli and Lal, 1980).

Solar Cosmic Ray (SCR) Energy Spectra

The energy spectra of energetic particles accelerated during certain solar flares are often represented as a power law (see Lal (1972, 1988b) and Aschwanden (2002)):

$$\mathrm{d}n/\mathrm{d}E_\mathrm{k} = \text{constant} \times E_\mathrm{k}^{-\Gamma} \qquad [4]$$

However, the value of Γ changes rapidly with energy in the interval (1 : 100) MeV/n; it usually lies between 2 and 4. A much better fit is the exponential rigidity spectrum:

$$\mathrm{d}n(R)/\mathrm{d}R = \text{constant}\,.\,\mathrm{e}^{(-R/R_0)} \qquad [5]$$

where R is the rigidity (Appendix 1) and R_0 is a constant for the time-averaged spectrum of the solar particles in a flare event. The value of R_0 ranges between 50 and 200 MV in most flares (Lal, 1988b).

The SCR energy spectra are very soft compared to the galactic cosmic ray spectra, as can be seen from **Figure 5** that presents the range of temporally observed variations in the GCR and SCR fluxes (Lal, 1988b). A quiet-time GCR spectrum in **Figure 5** implies that the spectrum was not affected by SCR protons at the time of measurement.

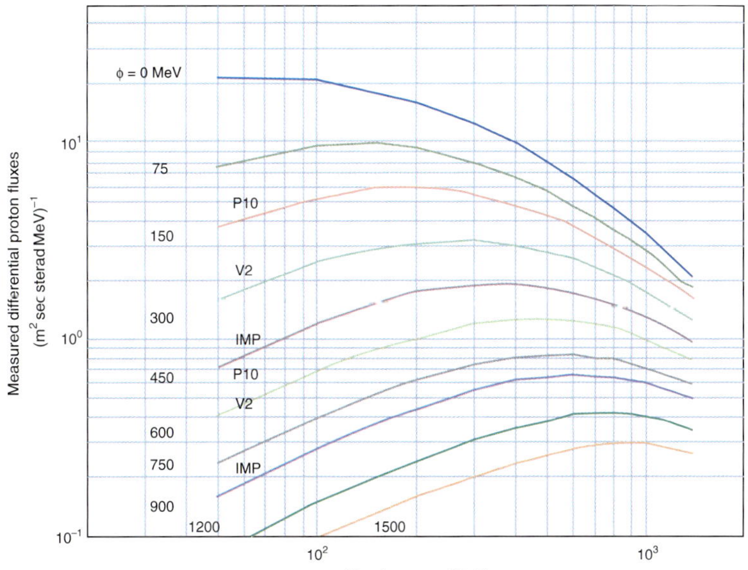

Figure 4 Observed cosmic ray fluxes of protons of kinetic energy >70 MeV by Pioneer 10 and Voyager 2 spacecrafts from 1978 to 1992 for GCR protons of 70 MeV kinetic energy, and near-Earth fluxes based on IMP satellite. The radial distances of Pioneer 10 and Voyager 2 were ~60 AU and 40 AU respectively in 1992. Figure redrawn from Lockwood JA and Webber WR (1995) An estimate of the location of modulation boundary for $E > 70$ MeV galactic cosmic rays using voyager and pioneer spacecraft data. *Ap. J.* 442: 852–860.

As can be seen from **Figure 5**, the energy of the bulk of SCR proton flux is >100 MeV. The range of a proton of 100 MeV in air is $<10\,g\,cm^{-2}$ (Ziegler, 2004), only about 10% of an interaction mean free path for its nuclear interaction (Appendix 2). Because of the geomagnetic cutoff (Appendix 3), SCR protons are only incident at geomagnetic latitudes >80°. Further they are lost by nuclear interactions in the upper atmosphere. Any neutrons produced by them would however be slowed down reaching greater depths in the atmosphere and would mostly give rise to ^{14}C by getting captured in ^{14}N nucleus.

In view of the above, the principal effect of SCR is to produce ^{14}C in the atmosphere. Even here, detailed calculations by Castagnoli and Lal (1980) show that the SCR production of ^{14}C is generally negligible compared to its production by GCR, but significant amounts could be produced in very large SCR events in some flares.

Development of the Secondary Particles of the Cosmic Radiation in the Earth's Atmosphere

A great deal is known about the basic processes which occur when energetic H, He, and heavier nuclei interact with matter, based on the early observations of nuclear interactions, and disintegrations of unstable particles as observed in photographic nuclear emulsions (Powell, 1959) and subsequent observations of high-energy particles accelerated up to ~30 GeV in particle accelerators. In **Figures 7** and **8**, we present examples of nuclear interactions of energetic protons and in **Figure 9**, an example of decay of an unstable particle (figures taken from Powell *et al.* (1959)). Several examples of first observations of interactions of multiply charged particles can be found in Bradt and Peters (1950).

A detailed description of the salient features of nuclear interactions is given by Lal and Peters (1967). We present here some of the important aspects of the propagation of the cosmic rays through the atmosphere.

The secondary cosmic ray phenomena are quite complex depending on the energy of the primary cosmic ray particle. Considering a group of interactions of an ensemble of cosmic ray primaries, the following is a good description of the secondary particles. The first-generation secondary particles are nucleons, pions, and a variety of unstable (elementary) particles. The nucleons and charged pions produce further nuclear interactions. Because of the low density of the atmosphere, most charged pions decay to muons before they interact. Neutral pions decay to photons which give rise to soft cascades (electrons and positrons). For properties of the elementary particles, see Review of Particle Physics (2004). We note here that the elementary particles are classified under the headings hadrons and leptons. Mesons (π and K) are bosons with spin 0, while nucleons, hyperons, leptons, and neutrinos are fermions with spin 1/2. Among leptons are electrons, positrons, muon, and

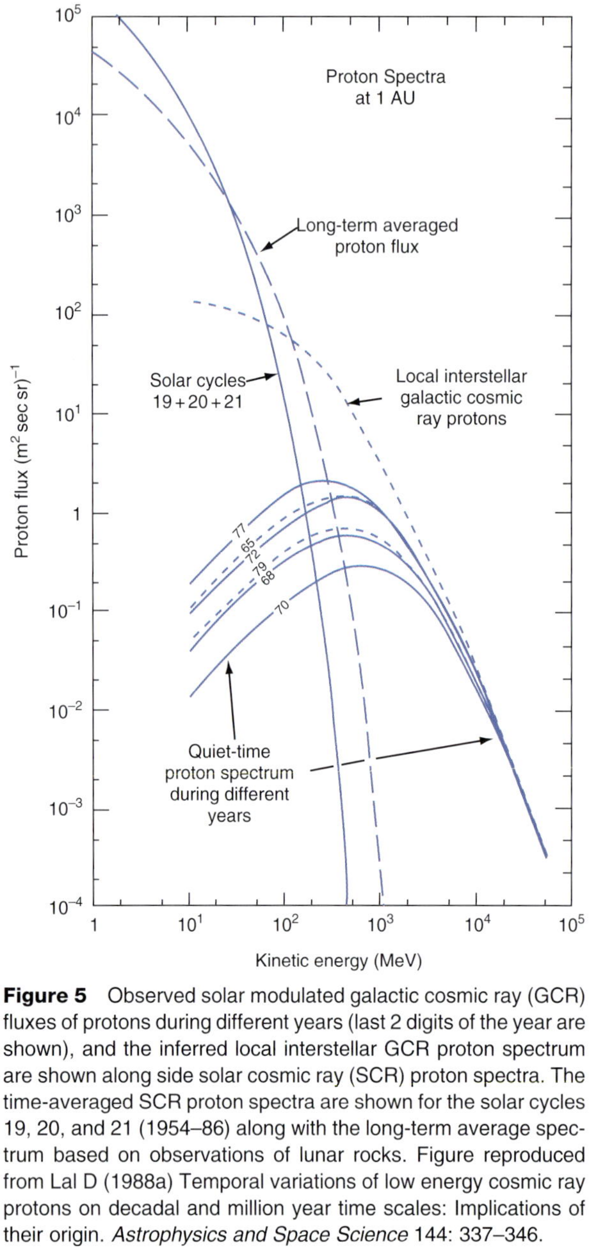

Figure 5 Observed solar modulated galactic cosmic ray (GCR) fluxes of protons during different years (last 2 digits of the year are shown), and the inferred local interstellar GCR proton spectrum are shown along side solar cosmic ray (SCR) proton spectra. The time-averaged SCR proton spectra are shown for the solar cycles 19, 20, and 21 (1954–86) along with the long-term average spectrum based on observations of lunar rocks. Figure reproduced from Lal D (1988a) Temporal variations of low energy cosmic ray protons on decadal and million year time scales: Implications of their origin. *Astrophysics and Space Science* 144: 337–346.

tauon. Note that muon is not called a meson because it is not a boson.

The fluxes of various cosmic ray secondary particles (nucleons, mesons, and electrons) change with depth in the atmosphere at geomagnetic latitude 45° and longitude 80° west (Peters, 1958; Lal and Peters, 1967) (**Fig. 10**). In this figure, the γ-ray flux is not shown; the photon : electron ratio is about unity at sea level and increases to a value of about 2 at 3500 m altitude. For approximate rates of nuclear disintegrations by nucleons and mesons, see Lal (1988a). It is seen from **Figure 10** that at atmospheric depths of ~150–200 g cm^{-2}, there occurs a transition in the fluxes of nucleons, mesons, and electrons when these fluxes build up to a maxima due to nuclear cascades. At depths exceeding 200 g cm^{-2}, nucleons in different energy regions come in to equilibrium, the nucleonic cascade being sustained by higher-energy nucleons of $E_k \geq 3$ GeV.

Very detailed information on the generation of the secondary component in nuclear interactions became available from observations of secondary particles in nuclear emulsions exposed at different altitudes and latitudes in the atmosphere. From measurements of grain densities of tracks in nuclear interactions, Camerini *et al.* (1949) deduced the energy spectra of secondary particles produced in these interactions. These data are very useful for estimating the development of secondary cosmic ray particles in the atmosphere, and nuclear disintegration rates (Powell *et al.*, 1959; Rossi, 1952; Lal, 1988a). From these data, Rossi (1952) has deduced the relative energy spectra of neutrons and protons at a depth of 700 g cm^{-2} in the atmosphere. This is in good agreement with the direct measurements of proton spectra, and the corresponding neutron fluxes based on the observed rates of nuclear disintegrations in nuclear emulsions by neutrons and protons (Lal, 1958; Lal and Peters, 1967). In **Figure 11**, we show the differential spectra of neutrons calculated in this way along with the measured proton spectrum at geomagnetic latitude 50° and atmospheric depth, 680 g cm^{-2}. The nuclear emulsion data provide the most direct means of obtaining neutron spectra in the atmosphere for kinetic energies in the MeV–GeV region; it is very difficult to obtain reliable neutron spectra using detectors, as was attempted by Hess *et al.* (1959).

It is seen from **Figure 11** that the fluxes of protons and neutrons are about the same at $E_k \geq 1{,}000$ MeV, but proton fluxes are progressively smaller at lower energies than those of neutrons. They are produced in about the same numbers in nuclear interactions. However, protons produce fewer interactions since lower-energy protons are brought to rest by ionization losses before they can interact. The ranges of a proton of 100, 500, and 1,000 MeV in air are ~10, 150, and 500 g cm^{-2}, respectively (Ziegler, 2004). The nuclear geometric and absorption mean free paths of nucleons are ~90 and 160 gm cm^{-2}, respectively (**Appendix 2**).

In the atmosphere, neutrons produce 90% of the nuclear interactions. The contribution of γ -radiation to nuclear interactions is about 5% at atmospheric depth of ~200 g cm^{-2} where its flux reaches a maximum (**Fig. 10**); however, it can be neglected when considering the atmosphere as a whole. The contribution of muons to nuclear interactions becomes important below the atmosphere. In the following, we present nuclear disintegration rates in the

Appendix 2 Definitions of Cosmic Ray Flux, and Geometric and Absorption Nuclear Interaction Rates

Crossing and Omnidirectional Fluxes

Unidirectional flux, $j(\theta, \phi)$, of cosmic ray particles arriving from any specified direction is defined as the number of particles per unit time, per unit solid angle, striking 1 cm^2 of a plane surface-oriented perpendicular to the incident direction; see **Figure 6** for definitions of θ (zenith angle) and ϕ (azimuthal angle) shown schematically around a sphere of radius, r. It then follows (since differential solid angle, $d\omega = \Delta A/r^2$) that the "total" number of particles striking a horizontal area, 1 cm^2 per unit time, F is given by:

$$F\ (cm^{-2}\ s^{-1}) = \iint_{\theta\ \phi} j(\theta, \phi)\ \cos(\theta)\ \sin(\theta)\ d\theta\ d\phi \qquad [A.2.1]$$

The $\cos(\theta)$ term in eqn [A.1.1] arises from the definition of 'flux.' It is often convenient to define an 'omnidirectional flux,' F_{omni}, the total number of particles striking an unit sphere of projected area 1 cm^2 from all directions, based on the definition of the unidirectional flux, $j(\theta,\phi)$:

$$F_{omni}\ (cm^{-2}\ sec^{-1}) = \iint_{\theta\ \phi} j(\theta, \phi)\ \sin(\theta)\ d\theta\ d\phi \qquad [A.2.2]$$

The secondary cosmic ray fast nucleon flux in the atmosphere at depths >200 g cm^{-2} shows a steep dependence with θ ($\cos^{2.3}(\theta)$) since particles incident at greater zenith angle traverse greater amount of matter in the atmosphere (cf. Conversi and Rothwell (1954)).

Nuclear Interaction Rates

The rate of nuclear reactions resulting in the production of a nuclide of interest in 1 g target, q (nuclei/g target/sec), by particles of kinetic energy, E, in the specified target atoms are given by the relation

$$q\ (\text{nuclei/g target/sec}) = F_{omni} n \sigma_i(E_k) \qquad [A.2.3]$$

$$= n \iiint_{\theta\ \phi E_k} j(\theta, \phi, E_k) \sigma_i(E_k) r\ \sin(\theta) r\ d\theta\ d\phi \qquad [A.2.4]$$

where n is the number of specified atoms in 1 g of the target, and $\sigma_i(E)$ is the kinetic energy dependent partial cross section for the formation of a nuclide. Note that in Eqn [A.2.4], the term $\cos(\theta)$ in the double integral of Eqn [A.2.1] is missing since particles incident at zenith angle, θ, have a longer path length in the unit target volume by a factor of $\sec(\theta)$. Here lies the utility of the concept of omnidirectional flux, F_{omni}.

Geometric Nuclear Interaction and Nuclear Absorption Cross Sections

For fast nucleons (neutrons or protons), the total geometric nuclear interaction cross section, σ_g, of a nucleus of mass number A is given by the following relation:

$$\sigma_g = \pi r^2 = \pi r_0^2 A^{2/3} \qquad [A.2.5]$$

where r is the effective nuclear radius and r_0 is the nucleon radius, taken to be 1.20×10^{-13} cm (Lang, 1980). Equation [A.2.5] assumes constant density of nuclear matter in complex nuclei. The geometric mean free path for nuclear interaction, λ, is defined as (see Eqn [A.1.3])

$$\lambda = 1/(n\ \sigma_g) \qquad [A.2.6]$$

The effective absorption cross section for fast nucleons is smaller than the geometric nuclear interaction cross section since the incident nucleon retains a large fraction of its initial kinetic energy (see text). If the elasticity, that is, the fraction of energy retained by the nucleon after interaction, is η, then it can be shown that for a power law nucleon differential spectrum of the form $a \times E_k^{-\Gamma}$((the emerging nucleons retain the same spectral form, and are attenuated with an absorption mean free path, Λ:

$$\Lambda = \lambda/(1 - \eta^{\Gamma-1}) \qquad [A.2.6a]$$

The corresponding effective absorption cross section, $\sigma_{a \text{ is}}$

$$\sigma_a = \sigma_g(1 - \eta^{\Gamma-1}) \qquad [A.2.7]$$

The above equations are valid for treatment of propagating flux of secondary high-energy nucleons in the atmosphere produced from primary protons of E_k3 GeV.

The calculated value of the geometric mean free path of nucleons in air is 90 g cm^{-2}. The corresponding values of the absorption mean free path of nucleons in air for $\eta = 0.50$, 0.55, and 0.60 are 155, 169, and 187 g cm^{-2}. The value of Γ has been taken to be 2.5 in these calculations.

Appendix 3 Modulation of Cosmic Ray Flux by the Geomagnetic Field; Cutoff Rigidity and Kinetic Energy

Before entering the Earth's atmosphere, primary cosmic ray particles being charged are influenced by the Earth's magnetic field. The present geomagnetic field is well represented by a dipole of strength, $M = 8.1 \times 10^{25}$ G cm^3, located 342 km from the Earth's geometric center, with its axis intersecting the Earth's surface at 78.50° N, 69.00° W and 78.50° S, 111.00° E (Sandstrom, 1965).The geomagnetic field acts as a rigidity analyzer. For any hypothetical direction of arrival, one can define a particle momentum, p_u, such that particles of this momentum arriving from infinity can reach the given point from all directions. Similarly, one can also define a rigidity value, p_c, so that particles of lower momentum cannot arrive the given point in a given direction. For particle rigidities between p_u and p_c, there are 'forbidden' and 'allowed' bands of momenta. Reference is made to Peters (1958), Sandstrom (1965), and Pomerantz (1971) for further details.

If one neglects the fact that the magnetic dipole is shifted from the Earth's geometric center, and also the fact that the Earth is not a perfect sphere, the minimum cutoff momentum, p_c, which a particle must have is given by

$$p_c = \frac{M}{r_e^2} \frac{\cos^4(\lambda)}{[1 + \sqrt{1 - \sin(\theta)\ \cos(\phi)\ \cos^2(\lambda)}\ ]^2} \qquad [A.3.1]$$

$$= 59.6 \frac{\cos^4(\lambda)}{[1 + \sqrt{1 - \sin(\theta)\ \cos(\theta)\ \cos^2(\lambda)}\ ]^2}\ \text{GeV}/c \qquad [A.3.2]$$

where r_e is the Earth's radius, θ and ϕ are respectively the zenith angle and azimuthal angles of the particle, and λ is the geomagnetic latitude. For vertical incidence, the corresponding cutoff rigidity, R_c (rigidity being defined as $R = p_c/Ze$; see Appendix 1), is given by:

$$R_c = 14.9 \cos^4(\lambda)\ \text{GV} \qquad [A.3.3]$$

The corresponding vertical cut-off kinetic energy, E_c (kinetic), of the particle can be estimated using relation [A.1.5]. The minimum kinetic energy of a vertically incident proton at geomagnetic latitudes of 0°, 10°, 20°, 30°, 40°, 50°, 60°, 70°, 80°, and 90° are respectively 14.0, 13.1, 10.7, 7.49, 4.28, 1.77, 0.384, 0.02, 0.0001, and 0 GeV.

The geomagnetic coordinates in the centered dipole approximation can be estimated from the corresponding geographical coordinates (cf. Hopper (1964) and Sandstrom (1965).

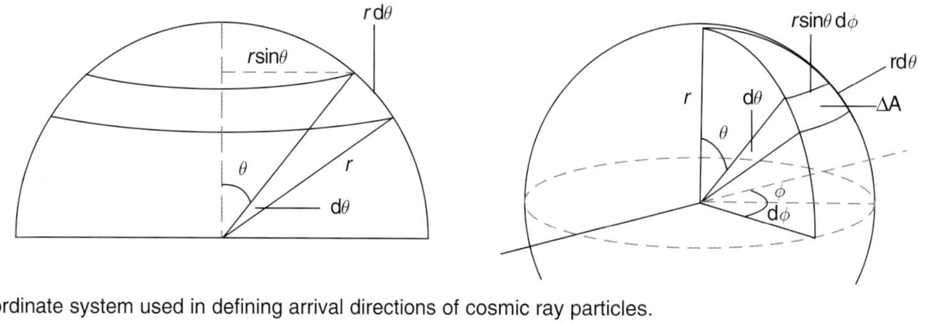

Figure 6 Coordinate system used in defining arrival directions of cosmic ray particles.

Figure 7 An example of a high-energy cosmic ray proton producing a nuclear interaction with a silver or bromine nucleus in the photographic nuclear emulsion. The incident proton track (shown by an arrow on the top) and other similarly appearing 'gray tracks' (five, marked S for secondary) are due to energetic particles (protons and unstable particles) of low ionization rate. Sixteen darker tracks, on the other hand, are due to slow protons and heavier ionization rates. Figure from Powell CF, Fowler PH, and Perkins DH (1959) *The Study of Elementary Particles by the Photographic Method*, 669p. New York: Pergamon.

Figure 8 Example of an energetic nuclear interaction produced by a cosmic ray proton (estimated to be of ~30 GeV kinetic energy) with a silver or bromine nucleus in the photographic nuclear emulsion. The incident proton track is shown by an arrow labeled P on the top. Nine tracks of energetic mesons (grey tracks) and 22 darker tracks of different ionization rates are seen. Figure from Powell CF, Fowler PH, and Perkins DH (1959) *The Study of Elementary Particles by the Photographic Method*, 669p. New York: Pergamon.

atmosphere and in the rocks and sediment of the upper tens of meters in the lithosphere.

Nuclear Disintegration Rates in the Atmosphere

The total rate of nuclear disintegrations in the atmosphere have been estimated, based on observed rates of nuclear disintegrations in nuclear emulsions and cloud chambers at different altitudes and latitudes, the measured low-energy neutron fluxes using neutron monitors located at several sites, and neutron fluxes measured in the atmosphere at different altitudes and latitudes using boron trifluoride counters flown by balloons (Lal and Peters, 1967). In the upper atmosphere ($<200\,\mathrm{gm\,cm^{-2}}$ depth), low-energy neutrons are not in equilibrium with fast nucleons. Furthermore, in this region, slow neutrons escape from the atmosphere. Even in the equilibrium region of the atmosphere ($>200\,\mathrm{g\,cm^{-2}}$ depth), only nucleons of $E_k < 400\,\mathrm{MeV}$ are in equilibrium with slow neutrons. Taking into account all these factors, the total rates of nuclear disintegrations produced in the atmosphere with energy release $\geq 40\,\mathrm{MeV}$ have been estimated (Lal and Peters, 1967; Lal, 1958, 1988c), as shown in **Figure 12**. These rates refer to a period of high solar activity since the neutron data come from observations during 1948–49.

The appreciable differences in the rate of nuclear interactions with latitude arise due to the modulation of cosmic ray flux by the Earth's geomagnetic field. As examples, the minimum kinetic energy which a vertically incident proton should have at geomagnetic latitudes of 0°, 30°, 60°, and 80° are, respectively, 14.0, 7.49, 0.384, and 0.0001 GeV (**Appendix 3**).

The effect of geomagnetic field changes on nuclear disintegration rates can be seen from **Figure 13**, which shows the global mean nuclear disintegration rate as a function of geomagnetic field strength (relative to present field) and solar modulation parameter, ϕ.

The minimum kinetic energy of a vertically incident proton at geomagnetic latitudes of 0°, 10°, 20°,

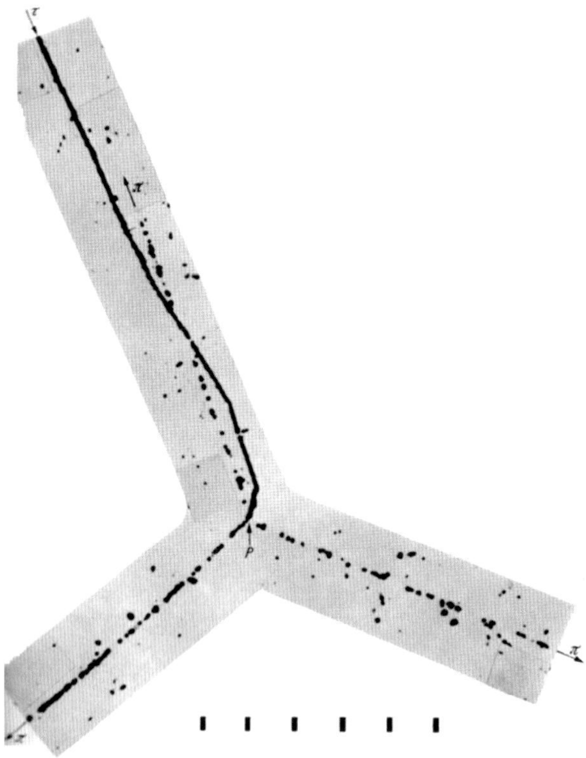

Figure 9 Example of decay of a τ-meson into three π-mesons in photographic nuclear emulsion. The darker wiggly track is due to τ-meson which comes to rest by ionization loss at the point P and decays giving rise to three π-meson tracks. Figure from Powell CF, Fowler PH, and Perkins DH (1959) *The Study of Elementary Particles by the Photographic Method*, p669. New York: Pergamon.

30°, 40°, 50°, 60°, 70°, 80°, and 90° are, respectively, 14.0, 13.1, 10.7, 7.49, 4.28, 1.77, 0.384, 0.02, 0.0001, and 0 GeV (Appendix 3). The differential primary fluxes at these energies are given in **Figure 2** for different states of solar modulation.

Temporal Variations in Cosmic Ray Nuclear Disintegration Rates

The incident cosmic ray flux at the top of the atmosphere is latitude dependent since it is strongly modulated by (1) the Earth's geomagnetic field and (2) solar plasma. Temporal variations in these parameters are therefore the two principal expected causes of changes in nuclear disintegration rates on the Earth. We do not expect to see any appreciable changes in the galactic cosmic ray intensity, except if a supernova explosion occurs nearby. This is supported by direct measurements of long-term constancy of cosmic radiation based on meteoritic evidence (Arnold *et al.*, 1961), and the long galactic confinement of cosmic rays in the galaxy (~20 Ma (Simpson and Connell, 1998)), combined with the fact that cosmic rays are replenished continually by supernovas whose frequency is quite high (~3 per century).

Temporal variations in nuclear disintegration rates due to changes in solar activity would depend strongly on altitude and latitude. Since the largest solar modulation effect is on the low energies, the temporal variations would be the greatest in the polar regions at high altitudes. As an example, the magnitude of change in cosmic ray intensity at different depths in the atmosphere between 1954 (a period of low solar activity) and 1958 (a period of unusually high solar activity) is shown in figure 15(B), page 579 of Lal and Peters (1967).

The effect of geomagnetic field changes on nuclear disintegration rates can be seen from **Figure 13**, which shows the global mean nuclear disintegration rate as a function of geomagnetic field strength (relative to present field) and solar modulation parameter, ϕ.

Nuclear Disintegration Rates in the Upper Lithosphere

As can be seen from **Figure 10**, the flux of fast neutrons decreases appreciably at sea level, decreasing with an absorption mean free path of $\sim 160\,\mathrm{gm\,cm^{-2}}$ (Lal and Peters, 1967; Gosse and Phillips, 2001), but the flux of fast mesons remains appreciable. In the atmosphere, most of the fast π-mesons decay before interacting, giving rise to fast muons. The latter component is absorbed slowly in the atmosphere; muons lose energy primarily by ionization, *bremsstrahlung* (braking radiation), and inelastic nuclear scattering. Loss of fast muons by nuclear interactions is not significant because they interact with an effective (weakly energy dependent) Coulombic (photonuclear) cross section of $\sim 3 \times 10^{-30}\,\mathrm{cm}^2$ per nucleon (cf. Wolfendale *et al.* (1972) and Heisinger *et al.* (2002a)). Compare this with geometric nuclear cross sections, $\sim 4.5 \times 10^{-26} A^{2/3}\,\mathrm{cm}^2$ (Appendix 2, Eqn [A.2.5]). Negative muon captures by nuclei result in nuclear disintegrations, with energy release of few tens of MeV at most. Energy release in nuclear interactions by fast muons is about an order of magnitude greater than in the capture of negative muons at rest.

Cosmic ray literature is extensive on cosmic rays underground (cf. Bhat and Ramana Murthy (1973) and Menon and Ramana Murthy (1967)). Cosmic ray muon intensities have been measured to undeground depths of ~3.7 km. In **Figure 14**, we present rates of nuclear disintegrations underground in standard rock (density = 3.09 g cm^{-3}; $Z^2/A = 6.3$), produced by neutrons and muons. For sources of data and the procedures used to determine the disintegration rates, reference is made to Lal (1987).

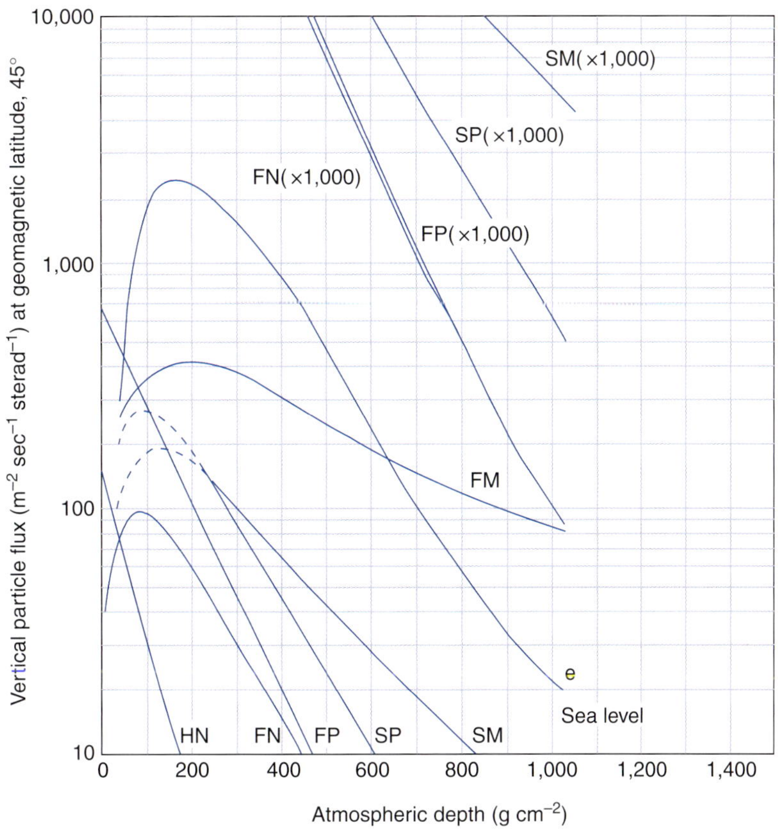

Figure 10 Composition of cosmic ray secondaries as a function of atmospheric depth at geomagnetic latitude 45° and longitude 80° west. The symbols in the figure are: HN (multiply charged (heavy) nuclei); FN (fast neutrons of $E_k \geq 3$ GeV); FP (fast protons of ≥ 3 GeV); SP (slow protons of $0.4 < E_k < 3$ GeV); FM (fast mesons of ≤ 220 MeV kinetic energy); SM (slow protons of $27 < E_k < 220$ MeV), and e ($E_k \geq 10$ MeV). Figure redrawn from Lal D and Peters B (1967) Cosmic ray produced radioactivity on the Earth. In: *Handbuch der Physik*, vol. 46(2), pp. 551–612. Berlin: Springer; see also Peters (1958).

Rates of Nuclear Interactions in Targets Exposed at Different Altitudes in the Atmosphere

Thus far we have presented rates of nuclear disintegrations in the atmosphere and in standard rock at underground depths. If a target is exposed in or at the bottom of the atmosphere, the rates of nuclear disintegrations in any mineral can be estimated. At altitudes within the atmosphere, nuclear interactions of fast muons are not important. Below, we discuss how nuclear disintegration rates for other particles can be estimated in minerals.

Fast Nucleons

The rates of nuclear disintegrations by fast nucleons in the target can be obtained by scaling the geometric cross section according to Eqn [A.2.5] in Appendix 2. For a given omnidirectional cosmic ray flux, the number of interactions per gram target is proportional to the product $n\sigma$ (Eqn [A.2.3]), (where n is the (number of target atoms per g) and σ is the geometric nuclear interaction cross section. If therefore the rate of nuclear interactions due to fast neutrons is known for air, the nuclear interaction rate in the target would be the product of nuclear reaction rate in air multiplied by a factor, f:

$$f = (A_{\text{target}}/\overline{A}_{\text{atmosphere}})^{-1/3} \qquad [6]$$

where $\overline{A}_{atmosphere}$ is the abundance weighted mean mass number of air, since n scales as A^{-1} and σ scales as $A^{2/3}$ (Eqn [A.2.5]).

Slow Negative Muons

In the atmosphere, the rate of stopping negative muons, S_{μ^-} is given by:

$$S_{\mu^-}(\text{g}^{-1}\text{s}^{-1}) = 5.1 \times 10^{-6}\, \text{e}^{(-X/\Lambda_a)} \qquad [7]$$

where X is the atmospheric depth (g cm^{-2}) and Λ_a is the absorption mean free path for negative muons (cf. Lal (1988)); the value of Λ_a is 247 g cm^{-2} for $160 < X < 1{,}030$ g cm^{-2}. In light nuclei, the rate of capture of negative muons by the nucleus depends on the target mass number. Negative muons are quickly

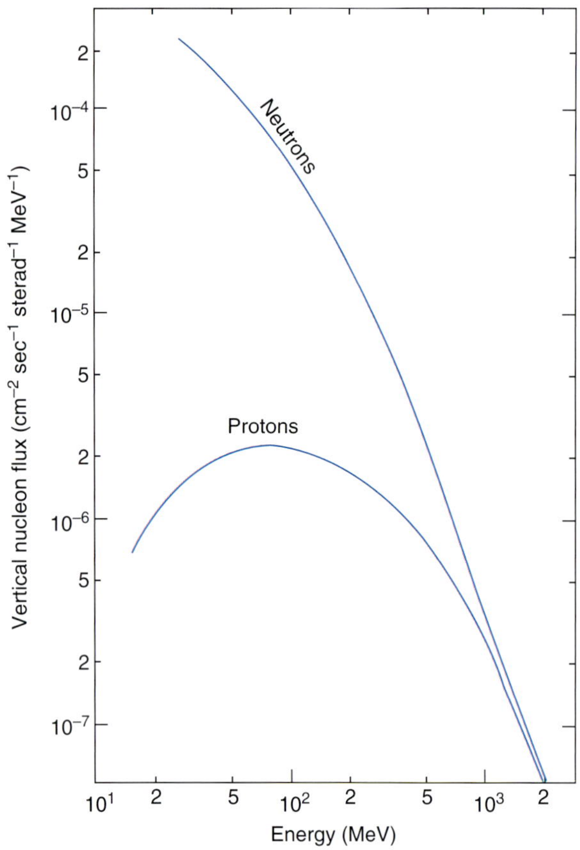

Figure 11 Differential kinetic energy spectra of protons (measured) and neutrons (estimated; see text) in the atmosphere at geomagnetic latitude 50° and depth 680 gm cm^{-2}. Figure taken from Lal D and Peters B (1967) Cosmic ray produced radioactivity on the Earth. In: *Handbuch der Physik*, vol. 46(2), pp. 551–612. Berlin: Springer.

captured in the K-shell of the atom but in light nuclei a fraction of the decay occurs before they are captured by the nucleus. For details, reference is made to **Figure 4** in Lal (1988b).

Thermal Neutrons

In the atmosphere, most of the thermal neutrons are captured by *N* since it has the greatest abundance and highest cross section among the atmospheric constituents. Rates of production of ^{14}C in the atmosphere during solar minimum and maximum periods observed during 1950–60 (Lingenfelter, 1963) are presented in **Figures 11** and **12** in Lal and Peters (1967). From these data, one can calculate the thermal neutron fluxes within the atmosphere.

The differences in the strength of interaction of these three components of secondary radiation have been utilized to help determine the exposure ages and the erosion rates of a wide range of landforms and landscapes over a wide timescale (see Nishizumi *et al.* (1993), Exposure Geochronology, and Landscape Evolution).

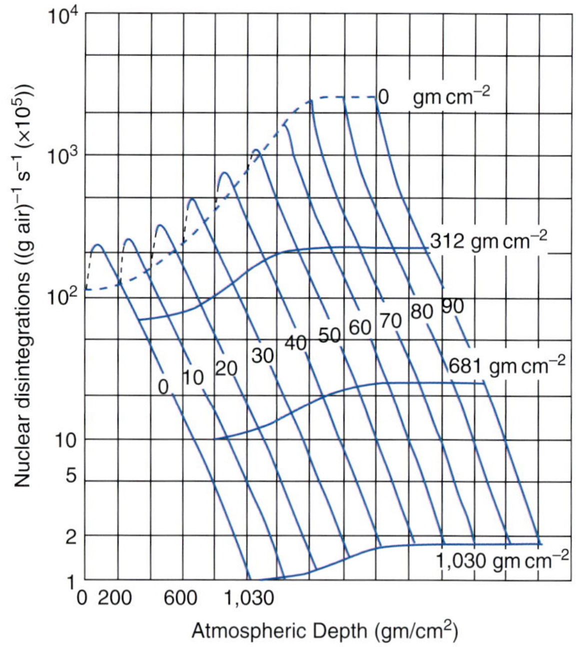

Figure 12 The total rate of nuclear disintegrations in the atmosphere, with energy release 40 MeV, is plotted as a function of altitude and geomagnetic latitude, in steps of 10°. Curves for different latitudes have been successively displaced along the abscissa by 1 km (based on Lal and Peters, 1967).

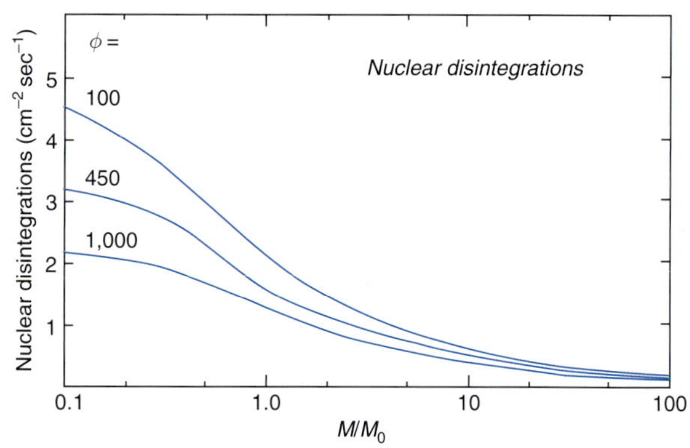

Figure 13 The global mean nuclear disintegration rate is plotted as a function of geomagnetic field strength, *M* (relative to present field, M_o) and solar modulation parameter $\phi = 100$, 450, and 1,000 MeV.

Rates of Production of Nuclides at Different Altitudes in the Atmosphere and at Depths in the Lithosphere

Detailed calculations of production rates of several nuclides are presented as a function of latitude and altitude by Lal and Peters (1967). The column-integrated global average production rates are also given by these authors. These estimates are based on three methods: (1) suitable (energy-dependent) normalization to nuclear disintegration rates in cases where

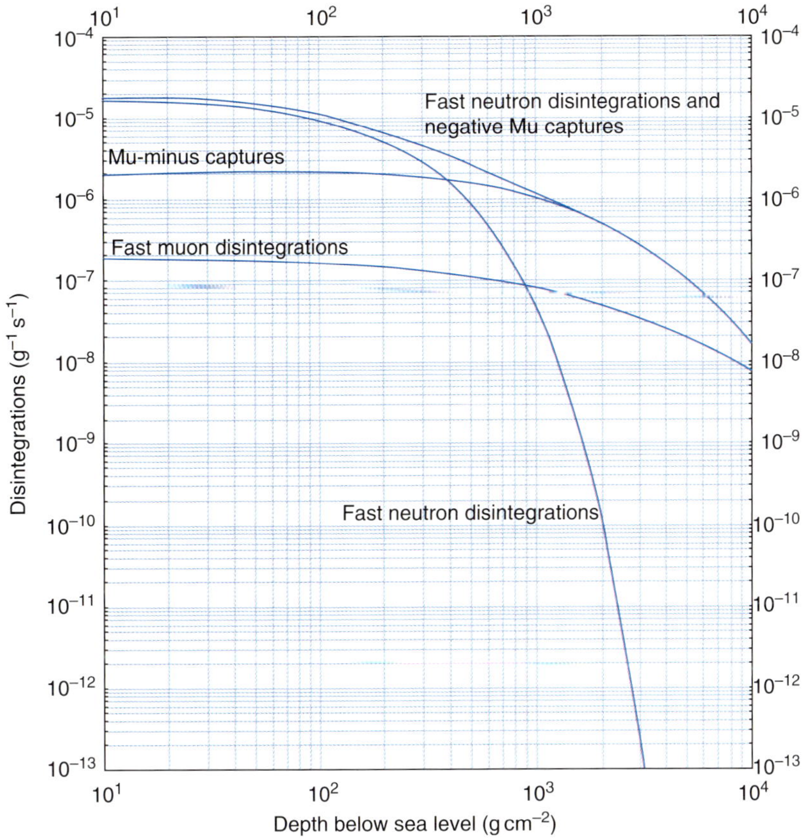

Figure 14 Nuclear disintegration rates due to fast neutrons and muons (fast muons and slow negative muons) in standard rock (density = 3.09 g cm^{-3}; $Z^2/A = 6.3$). Reference is made to Lal (1987) for sources of data and the procedures used to determine the disintegration rates. Depths given are below sea level (1,030 g cm^{-2}).

absolute production rates are available at some point(s) in the atmosphere; (2) based on the observed distribution of different sized nuclear disintegrations in nuclear emulsions, and the partitioning of isotopes in nuclear spallation at different energies; and (3) based on partial cross sections and the energy spectra of the nucleonic component. The first method is the most reliable; the second method fares much better than the third method, because it is closest to the nuclear reaction dynamics and because the relevant cross sections for neutrons, which produce most of the nuclides in the atmosphere, are generally few or are not available.

The latitudinal column production rates (cm^{-2} s^{-1}) of ^{14}C and ^{10}Be in the atmosphere are shown in **Figure 15** as a function of latitude for discrete values of the solar modulation parameter, ϕ (Lal, 1988c).

At underground depths, fluxes of slow and fast muons are presented by Heisinger *et al.* (2002a, b), along with a large number of cross sections for production of a large number of nuclides from different targets up to depths of 10^5 g cm^{-2}. This is the most comprehensive summary available to date, based on cosmic ray and cross-section data. These data allow fairly detailed calculations of the production rates of the frequently measured radionuclides, ^{10}Be, ^{14}C, and ^{26}Al, by fast muons, and slow negative muons. The production of nuclides by fast neutrons is important only at depths $<10^3$ g cm^{-2} (**Fig. 15**) (i.e., about 4 m of granite or 5 m of gravel).

The flux of thermal neutrons at sea level or near the atmosphere–lithosphere interface does not show a simple absorption. In fact, the flux is enhanced due to back-scattering neutrons produced in the crust/rock. Also, a depth transition occurs within the denser material in the neutron flux at depths of <50 g cm^{-2} (Phillips *et al.*, 2001), after which the thermal neutron flux decreases, following the absorption of the (parent) high-energy neutrons.

Rates of Production of Nuclides in Landforms Exposed at Different Altitudes in the Atmosphere and Lithosphere

The rate of production of a nuclide in a target is given by Eqn [A.2.4] in Appendix 2. Since the rates of nuclear disintegration are known in the atmosphere, one can derive the nuclide production rates in any

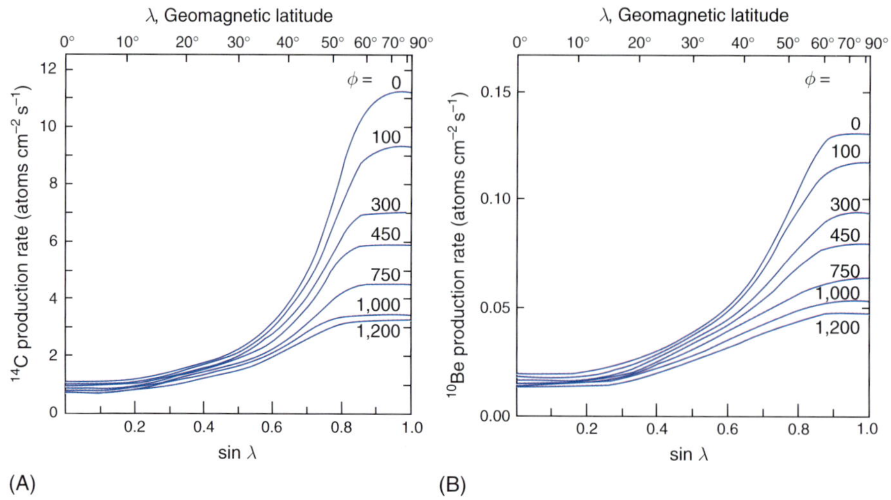

Figure 15 Total column production rates ($cm^{-2}\,s^{-1}$) of (A) ^{14}C and (B) ^{10}Be in the atmosphere are shown as a function of latitude for discrete values of the solar modulation parameter, ϕ.

target using any of the three methods discussed in the previous section. In large targets, such as boulders or tors, however, one has to take into account the absorption of the secondary cosmic ray beam.

The relative nuclide production rates at a point, P(x, y, z), within a body can then be obtained by integrating the following equation (Lal and Chen, 2005):

$$P(x,y,z) = \iint F(\theta, \text{surface})\sin(\theta)\, e^{-\rho \vec{r}\,(x,y,z,\theta,\phi)/\Lambda}\, d\theta\, d\phi \qquad [8]$$

where $\vec{r}\,(x,y,z,\theta,\phi)$ is the particular path length vector to reach the point (x, y, z) in the direction of incidence of the cosmic ray particle, θ, ϕ (see Appendix 2 for definitions of symbols). The parameter ρ signifies the mean density of the matrix and Λ is the mean absorption coefficient for nuclear interacting particles. A compilation of measured values of Λ can be found in Gosse and Phillips (2001). From studies of neutrons and nuclear interactions in the equilibrium region of the atmosphere ($\chi > 200\,g\,cm^{-2}$), the angular dependence of F in the atmosphere is known to be fairly well represented by

$$F(\theta, \text{surface}) = F(0, \text{surface})\cos^n(\theta) \qquad [9]$$

in terms of the vertical flux, F(0, surface) in the atmosphere, that is, the cosmic ray flux on the surface of the object. The value of n is $= 2.3 \pm 0.3$ (Conversi and Rothwell, 1954; Lal, 1988a). Equation [3] readily adapts itself to an object whose dimension is changing as a result of erosion (Lal and Chen, 2005).

The distribution of path length vectors within a target can easily be derived. This task has been carried out earlier in similar contexts (Lal, 1972; Bhattacharya *et al.*, 1973; Lal and Chen, 2005). For calculations of nuclide production rates in large boulders, see Masarik *et al.* (2000), Masarik and Wieler (2003), and Lal and Chen (2005). The latter authors have also considered relative nuclide production rates within rectangular bodies. They have concluded that since cosmic ray interactions are highly dependent on the geometry of exposure of the target, an examination of nuclide concentrations within a large target can be used to constrain the exposure history of objects, as their surfaces evolve due to weathering and erosion. This is a unique feature of the TCN 'exposure age dating', vis-É-vis other nuclear methods (Lal, 1991).

Conclusions

In the past five decades and beyond, a great deal has been learned about the composition and fluxes of cosmic ray particles on the Earth, as well as about their most significant origins in the galaxy. The primary cosmic ray beam is composed of the ordinary matter we see in the universe, mostly H nuclei, since this is most abundant in the cosmic matter. The incidence of the cosmic radiation on the Earth's atmosphere leads to a very complex cascade of nucleons, mesons, muons, and a variety of elementary particles and soft cascades composed of electrons and positrons. In this article, the discussion has been confined to the primary cosmic rays and their composition; the composition of the secondary particles and their inelastic nuclear interactions leading to changes in the nuclear composition of matter. Knowledge of the primary cosmic radiation and its temporal variation due to solar modulation, and the development of the secondary cosmic ray beam are essential for accurately deducing the production of new forms of matter in cosmic ray interactions in the atmosphere as well as in solids in the lithosphere.

One may wonder how one is able to deal practically with the above problem in spite of its complexity. The text shows that we have been fairly successful in this attempt, and no doubt further improvements will be made in future.

See also: **Cosmogenic Nuclide Dating**: Overview; Exposure Geochronology; Landscape Evolution.

References

Arnold, J. R., Honda, M., and Lal, D. (1961). Record of cosmic ray intensity in the meteorites. *Journal of Geophysical Research* **66,** 3519–3531.

Aschwanden, M. J. (2002). Particle acceleration and kinematics in solar flares. *Space Science Reviews* **101**, 1–227.

Bhat, P. N., and Ramana Murthy, P. V. (1973). Intensity measurements of low energy cosmic ray muons underground. *Journal of Physics A: Mathematical Nuclear and General Physics* **6,** 1960–1973.

Bhattacharya, S. K., Goswami, J. N., Gupta, S. K., and Lal, D. (1973). Cosmic ray effects induced in a rock exposed on the moon or in free space: Contrast in patterns for tracks and isotopes. *The Moon* **8,** 253–286.

Bradt, H. L., and Peters, B. (1950). The heavy nuclei of the primary cosmic radiation. *Physical Review* **77,** 54–75.

Camerini, U., Coor, T., Davies, J. H., *et al.* (1949). Nuclear transmutations produced by cosmic ray particles of great energy. *Philosophical Magazine* **40,** 1073.

Castagnoli, G., and Lal, D. (1980). Solar modulation effects in terrestrial production of carbon 14. Proceedings of the 10th International Radiocarbon Conference. *Radiocarbon* **22,** 133–158.

Connell, J. J., DuVernois, M. A., and Simpson, J. J. (1998). The cosmic-ray radioactive nuclide ^{36}Cl and its propagation in the galaxy. *The Astrophysical Journal* **509,** L97–L100.

Conversi, M., and Rothwell, P. (1954). Angular distributions in cosmic ray stars at 3500 meters. *Nuovo Cimento* **12,** 191–211.

Fuji, Z., and McDonald, F. B. (1997). Radial intensity gradients of galactic cosmic rays (1972–1995) in the heliosphere. *Journal of Geophysical Research* **102**(A11), 24,201–24,208.

Gleeson, L. J., and Axford, W. I. (1968). Solar modulation of galactic cosmic rays. *The Astrophysical Journal* **423,** 426–431, **154,** 1011–1026.

Gosse, J. C., and Phillips, F. M. (2001). Terrestrial *in situ* cosmogenic nuclides: Theory and application. *Quaternary Science Reviews* **20,** 1475–1560.

Heisinger, B., Lal, D., Jull, A. T. J., *et al.* (2002a). Production of selected cosmogenic radionuclides by muons. 1: Fast muons. *Earth and Planetary Science Letters* **200,** 345–355.

Heisinger, B., Lal, D., Jull, A. T. J., *et al.* (2002b). Production of selected cosmogenic radionuclides by muons. 2: Capture of negative muons. *Earth and Planetary Science Letters* **200,** 357–369.

Hess, W. N., Patterson, H. W., Wallace, R., and Chupp, E. L. (1959). Cosmic-Ray neutron energy spectrum. *Physical Reviews* **116,** 445–457.

Higdon, J. C., Lingenfelter, R. E., and Ramaty, R. (1998). Cosmic ray acceleration from supernova ejecta in superbubbles. *The Astrophysical Journal* **509,** L33–L36.

Hopper, V. D. (1964). p. 234. Prentice Hall, New Jersey.

Lal D (1958) *Investigations of Nuclear Interactions Produced by Cosmic Rays*, p. 90 PhD Thesis, University of Bombay.

Lal, D. (1972). Hard rock cosmic ray archaeology. *Space Science Reviews* **14,** 3–102.

Lal, D. (1987). Production of ^{3}He in terrestrial rocks. *Chemical Geology* **66,** 89–98.

Lal, D. (1988a). Temporal variations of low energy cosmic ray protons on decadal and million year time scales: Implications of their origin. *Astrophysics and Space Science* **144,** 337–346.

Lal, D. (1988b). *In-situ* Produced cosmogenic isotopes in terrestrial rocks. *Annual Review of Earth and Planetary Science Letters* **16,** 355–388.

Lal, D. (1988c). Theoretically expected variations in the terrestrial cosmic ray production rates of isotopes, Proceedings of the International School of Physics, 'Enrico Fermi' Course XCV, Varenna, June, 1985. In *Solar Terrestrial Relationships and the Earth Environment in the Last Millenia, Italian Physical Society, Bologna* (G. C. Castagnoli, Ed.), pp. 216–233. North Holland, Amsterdam.

Lal, D. (1991). Cosmic ray tagging of erosion surfaces: *In situ* production rates and erosion models. *Earth and Planetary Science Letters* **104,** 424–439.

Lal, D. (2000). Cosmogenic nuclide production rate systematics in terrestrial materials: Present knowledge, needs and future action for improvement. *Nuclear Instruments and Methods in Physics Research B* **172,** 772–781.

Lal, D., Arnold, J. R., and Honda, M. (1960). Cosmic ray production rates of Be^7 in Oxygen and P^{32}, P^{33}, S^{35} in Argon at mountain altitudes. *Physical Reviews* **118,** 1626–1632.

Lal, D., and Chen, J. (2005). Cosmic ray labeling of erosion surfaces. II: Special cases of exposure histories of boulders, soils and beach terraces. *Earth and Planetary Science Letters* **236**(3–4), 797–813.

Lal, D., and Peters, B. (1967). Cosmic ray produced radioactivity on the Earth. In *Handbuch der Physik* (K Sitte, Ed.), 551–612. Springer, Berlin.

Lang, K. R. (1980). *Astrophysical Formulae*, p. 783. Springer, Heidelberg.

Lingenfelter, R. E. (1963). Production of carbon-14 by cosmic ray neutrons. *Reviews of Geophysics* **1,** 35–55.

Lockwood, J. A., and Webber, W. R. (1995). An estimate of the location of modulation boundary for $E > 70$ MeV galactic cosmic rays using voyager and pioneer spacecraft data. *Astrophysical Journal* **442,** 852–860.

Lukasiak, A., Ferrando, P., McDonald, F. B., and Webber, W. R. (1994). The isotopic composition of cosmic ary beryllium and its implication for the cosmic ray age. *The Astrophysical Journal* **423,** 426–431.

Masarik, J., and Reedy, R. C. (1995). Terrestrial cosmogenic-nuclide production systematics calculated from numerical simulations. *Earth and Planetary Science Letters* **176,** 381–595.

Masarik, J., Kollar, D., and Vanya, S. (2000). Numerical simulation of *in situ* production of cosmogenic nuclides: Effects of irradiation geometry. *Nuclear Instruments and Methods B* **172,** 786–789.

Masarik, J., and Wieler, R. (2003). Production rates of cosmogenic nuclides in boulders. *Earth and Planetary Science Letters* **216,** 201–208.

Menon, M. G. K., and Ramana Murthy, P. V. (1967). Cosmic ray intensities deep underground. In *Progress in Elementary and Cosmic Ray Physics* (J. G. Wilson and S. A. Wothuysen, Eds.), vol. 9 pp. 163–243. Amsterdam, North Holland.

Muzikar, P., Elmore, D., and Granger, D. E. (2001). Accelerator mass spectrometry in geologic research. vol. 9, pp. 163–243. Amsterdam, North Holland.

Nishiizumi, K., Finkel, R. C., Klein, J., and Kohl, C. P. (1996). Cosmogenic production of ^{7}Be and ^{10}Be in water targets. *Journal of Geophysical Research* **101**, 22,225–22,232.

Nishiizumi, K., Kohl, C. P., Dorn, R., *et al.* (1993). Role of *in-situ* cosmogenic nuclides ^{10}Be and ^{26}Al in the study of diverse geomorphic processes. *Earth Surface Processes and Landforms* **18**, 407–425.

Nishiizumi, K., Winterer, E. L., Kohl, C. P., *et al.* (1989). Cosmic ray production rates of ^{10}Be and ^{26}Al in quartz from glacially polished rocks. *Journal of Geophysical Research* **94**(17), 907–916.

Papini, P., Grimani, C., and Stephens, S. A. (1996). An estimate of the secondary proton spectrum at small atmospheric depths. *Il Nuovo Cimento* **19C**(3), 367–388.

Peters, B. (1958). Cosmic rays. In *Handbook of Physics* (E. U. Condon and H. Odishaw, Eds.), ch. 12, pp. 201–244. McGraw-Hill, New York.

Phillips, F. M., Stone, W. D., and Fabryka-Martin, J. T. (2001). An improved approach to calculating low-energy cosmic-ray neutron fluxes near the land/ atmosphere interface. *Chemical Geology* **175**, 689–701.

Pomerantz, M. A. (1971). *Cosmic Rays* p. 186. Van Norstrand Reinhold, New York.

Powell, C. F., Fowler, P. H., and Perkins, D. H. (1959). *The Study of Elementary Particles by the Photographic Method* p. 669. Pergamon, New York.

Ramaty, R., Kozlovsky, B., and Lingenfelter, R. E. (1998). *Physics Today* **4**, 30–35.

Randall, B. A., and Van Allen, J. A. (1986). Heliocentric radius of the cosmic ray modulation boundary. *Geophysical Research Letters* **13**, 628–631.

Review of Particle Physics (2004). *Physics Letters B* **592**, 1–1109, (Edited by L. Alvarez-Gaume and 14 others).

Rossi, B. (1952). *High-nergy Particles* p. 561. Prentice Hall, Englewood Cliffs.

Rossi, B., and Olbert, S. (1970). *Introduction to the Physics of Space.*, p. 454. McGraw-Hill, New York.

Sandström, A. E. (1965). *Cosmic Ray Physics* p. 421. North Holland, Amsterdam.

Simpson, J. A. (1983). Elemental and isotopic composition of the galactic cosmic rays. *Annual Review of Nuclear Particle Science* **33**, 323–382.

Simpson, J. A. (1992). Cosmic radiation: Particle astrophysics in the heliosphere. *Annals of the New York Academy of Sciences* **655**, 95–137.

Simpson, J. A., and Connell, J. J. (1998). Cosmic ray 26Al and its decay in the galaxy. *The Astrophysical Journal* **497**, L85–L88.

Simunac, K. D. C., and Armstrong, T. P. (2004). Solar cycle variations in solar and interplanetary ions observed with Interplanetary Monitoring Platform. *Journal Geophysical Research* **109**, A10101.

Tuniz, C., Bird, J. R., Fink, D., and Herzog, G. F. (1998).., p. 371. CRC Press, Washington DC.

Wolfendale, A. W., Young, E. C. M., and Davis, R. (1972). Indirect observations of the photonuclear cross section above 20 GeV. *Nature Physical Science* **238**, 130–131.

Ziegler, J. F. (1998). Terrestrial cosmic ray intensities. *IBM Journal of Research and Development* **42**(1), 117–139.

Ziegler, J. F. (2004). SRIM 2003. *Nuclear Instruments and Methods* **1027–1036 219–220**, 1027–1036.

Exposure Geochronology

S Ivy-Ochs, Institute of Particle Physics, Zürich, Switzerland
F Kober, Institute of Geology, Zürich, Switzerland

© 2007 Elsevier B.V. All rights reserved.

Introduction

Thirty years ago it was impossible for most geologists and geomorphologists to think that one could date the surface of a rock or landform and obtain an age that is independent of the crystallization age of the rock, but shows instead how long that rock or landform surface has been exposed at the Earth's surface. Cosmogenic nuclides build up predictably with time within the lattices of minerals by reactions induced by cosmic ray secondaries. Measuring the concentrations of these nuclides allows calculation of how long a rock has been exposed. The surface of the Earth is constantly changing through the action of geological processes that work by the moving or wearing down of rocks. New landforms and new rock surfaces are formed as a result of both climatic events (e.g., moraines) and tectonic forces (e.g., fault surfaces). Prior to cosmogenic nuclide methods the ages of such events were estimated or dated indirectly through correlation with dated records or bracketing ages from underlying and overlying sediments were determined. With cosmogenic nuclide methods, it is now possible to date the rock surfaces themselves directly (Gosse and Phillips, 2001). With the obtained dates, chronologies are constructed. These can be compared with chronologies based on other dating methods, such as radiocarbon, luminescence techniques, U/Th methods, potassium–argon (K–Ar), or ^{40}Ar/^{39}Ar. But more information can be gleaned from the concentrations of cosmogenic nuclides measured in a rock surface than just an age. The dates themselves are used as a tool to glean fundamental information about the rates and styles of geomorphic processes.

In this chapter we describe the ways in which cosmogenic nuclides are used to date various landforms (**Fig. 1**). A number of review articles provide excellent overviews as well as comprehensive reference lists (Gosse and Phillips, 2001; Bierman *et al.*, 2002; Cockburn and Summerfield, 2004). In the following chapter, Granger describes the applications of TCN methods for landscape evolution.

Figure 1 Examples of exposure dated landforms and rock surfaces. (A) Boulder dated from the Kromer moraine (Austria) (Kerschner *et al.*, 2006), view is down valley, moraine ridge continues to the right. (B) Pahoehoe flow top from the ca. 90,000 year old (^{3}He) Bar Ten flow, Arizona (Fenton *et al.*, 2001). (C) Beryllium-10 dating of boulders indicates that this lobe of the Shephard Creek fan in Owens Valley, (California) is Holocene in age (M. Dühnforth unpublished data), view is up into the catchment. (D) Continuous sampling of the Kaparelli limestone fault surface (Greece) for ^{36}Cl exposure dating (Benedetti *et al.*, 2003). The ages for the fault surface range from 20 ky to recent.

Dating Rocks and Sediment with Cosmogenic Nuclides

Because cosmogenic nuclides build up predictably with time, measuring their concentrations allows calculation of how long a rock surface has been exposed to cosmic rays (Lal, 1991; Gosse and Phillips, 2001). In order to use the exposure age of a boulder surface to date a landform, the rock surface must have undergone single-stage (no pre-exposure), continuous (not covered) exposure in the same position (not shifted) since the landform formed. Thus limitations are imposed on the landforms that can be dated; they must have formed rapidly and be no longer active to interpret the concentration as a simple exposure age. However, because the TCN methods are sensitive to these environmental conditions, the concentration can be used to provide insights into their nature.

If a rock surface was exposed before coming to the position where it was sampled, it will contain inherited nuclide concentrations (pre-exposure). As shown in **Figure 2** the concentration of TCN is related to how deep into the bedrock the boulder originated and how long that bedrock surface was exposed.

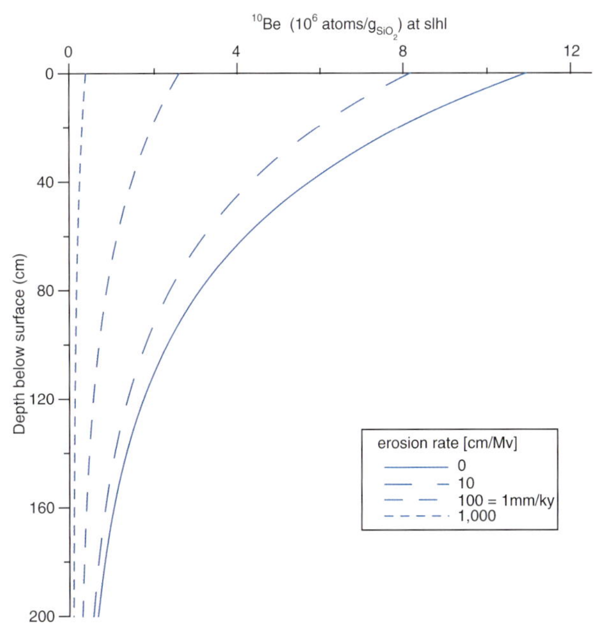

Figure 2 Depth profiles of saturation ^{10}Be concentrations in eroding bedrock surfaces (at sea level high latitude). Curve shape is dependent on exposure time and erosion rate. Other nuclides behave similarly. Curves constructed using production rates and depth-scaling formalism of Stone (2000) and Schaller *et al.* (2004).

The actual effect on the calculated age depends on the proportion of the final measured concentration that is due to the inherited nuclides. The inherited component will be proportionately greater in very young surfaces (Holocene).

Rock surfaces are constantly undergoing weathering (referred to here as erosion). As a rock surface weathers the outermost layers with the highest concentrations of cosmogenic nuclides are removed. This lowers the concentration and consequently the calculated age. As an aside, we note that because of the shape of depth profile of low-energy neutron production of chlorine-36 (^{36}Cl), some ^{36}Cl ages on eroding surfaces may be apparently 'older' (see Gosse and Phillips, 2001). For calculating exposure ages from a single nuclide concentration either an erosion rate of zero or a reasonable erosion rate for that rock surface is assumed.

During continuous exposure, radionuclide concentrations increase to the point where production and decay are balanced (saturation or secular equilibrium). Without erosion this is reached after four or five half-lives (**Fig. 3**). With erosion, saturation is reached earlier as the half-life is effectively shortened when the effect of erosion is included. The age range over which a surface can be dated is governed by the half-life of the nuclide and the erosion rate of the rock surface. Similarly, concentrations attained at depth are controlled by both exposure time and erosion rate (**Fig. 2**).

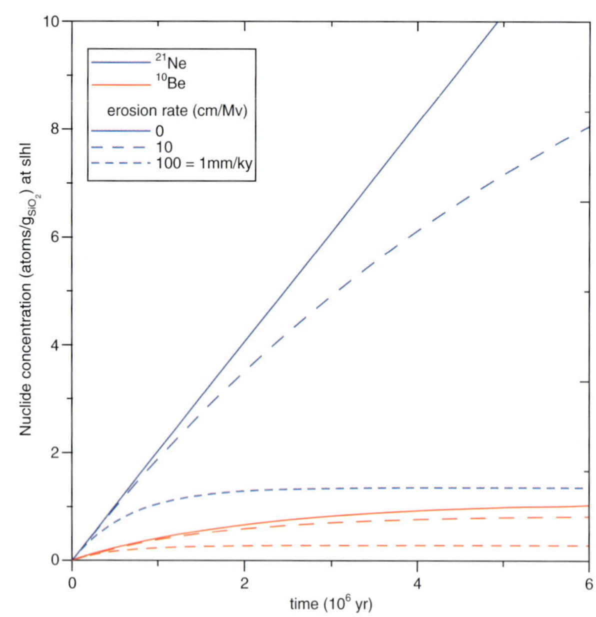

Figure 3 Increase in ^{10}Be and ^{21}Ne concentrations with time. As erosion rate increase saturation is reached earlier and at lower concentrations. Model parameters as in **Figure 2**, ^{21}Ne production rate after Niedermann (2002). In case of saturation, a minimum age or a maximum erosion rate can be derived.

In principle, when the concentrations of two nuclides from the same rock surface are measured, both time and erosion rate can be determined. An elegant way of portraying two nuclide data is the erosion-island plot (Lal, 1991; **Figs. 4A** and **B**). Certain caveats with respect to the erosion-island plot must be acknowledged. First, its utility is largely limited to settings characterized by very low erosion rates and thus old rock surfaces (ice coverage is an exception, see 'Glacial Landscapes' section), for example as in desert environments (see 'Old Slowly Weathering Landscapes' section). For young rock surfaces (exposed less than several hundred thousand years) or sites where erosion rates are higher than 1 m/million years (My; 1 mm/1,000 years (ky)) data points plot on the left-hand-side near the production ratio line. Second, the erosion-island area is relatively narrow with respect to the presently attainable measurement uncertainties. Nevertheless, the importance of the erosion-island plot lies in clear identification of data points that represent continuous simple or single-stage exposure (plot on or near the erosion-island) versus those that reflect complex-exposure by burial or spalling (plot below the aluminum (Al)/beryllium (Be) erosion-island and above the neon (Ne)/Be erosion-island). Third, there is no location on the diagram that can be reached by only one pathway that is a unique combination of exposure and erosion or burial. Data points for rocks or sediment that were exposed and then buried can arrive back inside the erosion-island after a sufficiently long period of (re)exposure. Nevertheless, the approach whereby the simplest explanation is the most likely (Occam's razor) can be applied. For points below the erosion-island, the minimum time to reach that point is made up of continuous exposure along the outer line followed by burial with complete shielding (zero production). All other possible scenarios (partial shielding or repeated burial and exposure) require more time (cf. Bierman *et al.*, 1999). As seen in **Figure 4B**, the use of ^{21}Ne (stable) in concert with ^{10}Be provides increased sensitivity for older, more slowly eroding surfaces.

The erosion-island plot also has another ingenious use. As a result of the difference in the half-lives of the two nuclides (^{26}Al and ^{10}Be), the length of time that sediments have been buried can be determined (Granger *et al.*, 1997). When sediment is completely shielded, the changes in concentrations are dependent only on the half-lives. Aluminum decays faster so the ratio decreases with time and the data points plot below the erosion island in the Al/Be erosion-island plot (**Fig. 4A**) and above in the Ne/Be plot (**Fig. 4B**). Transport and burial are assumed to be rapid.

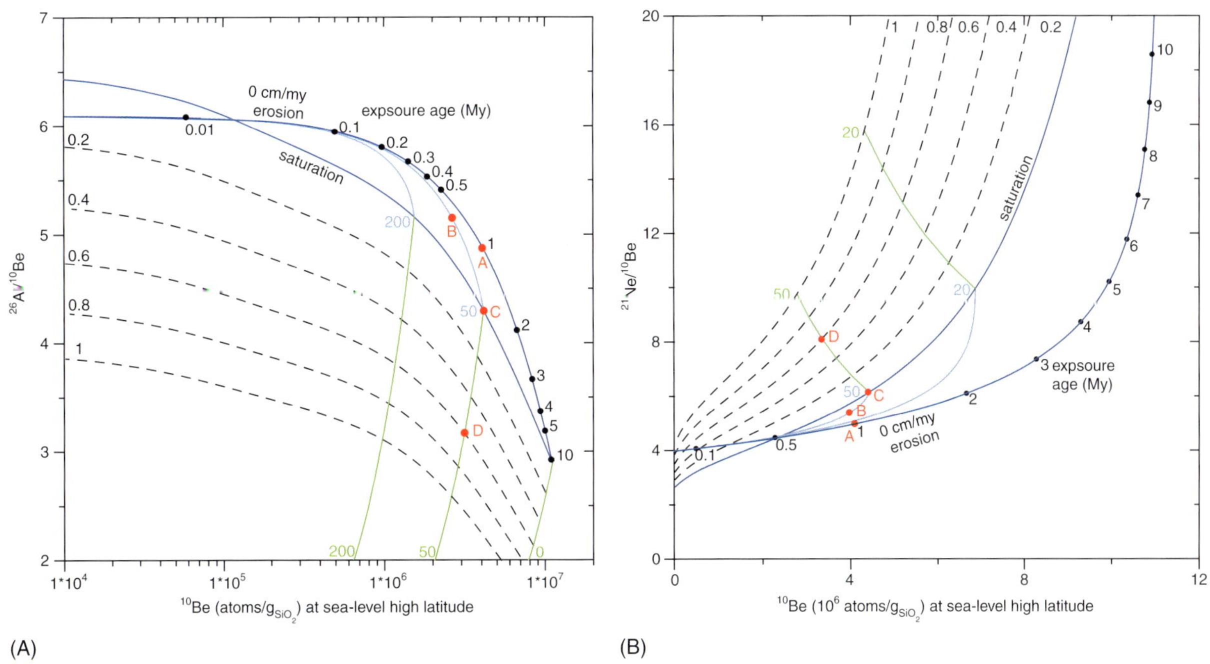

Figure 4 Erosion-island plots: $^{26}Al/^{10}Be$ versus ^{10}Be (A) and $^{21}Ne/^{10}Be$ versus ^{10}Be (B). With time nuclide concentrations increase and as ^{26}Al and ^{10}Be decay with different rates, the ratio decreases (Al–Be) or increases (Ne–Be). For a continuously exposed rock surface with no erosion and no coverage a data point would follow on the 'zero erosion' trajectory (sample point A). The lines inside the erosion-island are exposure trajectories for rocks continuously exposed but undergoing continuous erosion (sample point B). A data point on the saturation line indicates secular equilibrium of both nuclides with respect to a specific erosion rate (sample point C). The field between the zero erosion line and the saturation line is referred to as the erosion-island. Data points may plot in the field below (Al–Be) or above (Ne–Be) the erosion-island if the rock surface has been buried (e.g., by ice or sediment) and/or if thick slabs (tens of centimeters) have spalled off (sample point D). Dashed lines indicate time interval of completely-shielded burial in Mv. Figures constructed as outlined in **Figures 2** and **3**.

The Spectrum of Dated Landforms

Introduction

Within the time range of the late Pleistocene and Holocene, radiocarbon, luminescence methods and surface exposure dating with cosmogenic nuclides are broadly applied. Each dating method has advantages and disadvantages with respect to the material available and the relationship of the material dated to the event of interest. Surface exposure ages from hundreds of years to millions of years can be determined on rock surfaces and alluvial landforms. For landscapes formed in the last few hundred thousand years, single nuclide exposure ages allow the construction of a chronology. For older landscapes, two or more nuclides are needed to constrain erosion rates, reveal complex exposures, or discover step changes in erosion rates through time.

The minerals chosen depend on the rock type at the site and the nuclides to be combined. Beryllium-10, ^{26}Al, and ^{21}Ne are measured in quartz, while ^{3}He and ^{21}Ne are measured in olivine and pyroxene. Chlorine-36 can be measured in all rock types. However, new isotope-mineral systems are currently being developed (*see* Methods).

Glacial Landscapes

Glaciers are very sensitive indicators of climate change. A striking example is the nearly synchronous behavior of glaciers worldwide during the Little Ice Age. Moraines record past glacier advances. Prior to exposure dating, the timing of glacier variations was approximated based most often on correlation with local radiocarbon-dated records or with regional or hemispheric records, such as ice cores or marine sediment cores. However, fitting this into an established chronology meant that the information on past climate change contained in the glacial record itself was lost. The direct dating of boulders on moraines with cosmogenic nuclides is one of the most powerful tools for reconstructing the timing of glacier volume changes (Gosse, 2005). The timing of glacier variations with respect to regional or global climate events can be explored (e.g., Phillips *et al.*, 1996).

The time of stabilization of a moraine is determined by measuring the concentration of cosmogenic nuclides in the surface of a boulder on a moraine. The most coherent data sets stem from the dating of only the largest boulders (>1.5 m high and with large diameters) that are located in stable positions on the moraine (**Fig. 1A**). Depending on the moraine size and morphology, and the location of the boulders dated, the signal recorded is the point of abandonment by the glacier. For large (e.g., Last Glacial Maximum) glaciers an age progression from distal to proximal on the moraine may indicate the time period of construction of the moraine (Gosse *et al.*, 1995). Moraines from hundreds of years to hundreds of thousands of years old can be dated. Limiting factors are the weathering of the boulder surface and degradation of the moraine itself. Degradation of moraines and the associated exhumation and instability of boulders poses a severe limitation to accurate moraine dating (Putkonen and Swanson, 2003). Older moraines contain fewer boulders and the spread in boulder ages is often greater (e.g., Kaplan *et al.*, 2005). In only a few percent of cases has pre-exposure been identified in moraine boulders (Putkonen and Swanson, 2003).

Building chronologies in glaciated terrains need not be based solely on the dating of moraines. Dating of glacially modified bedrock surfaces (e.g., roches moutonnées) provides information about initial deglaciation of the valley bottom. However, results from bedrock surfaces should be used with caution. Often where excellent glacial polish is observed the rock surface has previously been covered by till and partially shielded, making it unsuitable for exposure dating. In addition, bedrock surfaces may contain inherited nuclides (cf. Fabel *et al.*, 2004). Indeed what can be seen on the one hand as a limitation (inherited nuclides) can be used on the other hand to map regions of cold- versus warm-based ice and fundamental information on past ice configurations and mechanisms of ice movement can be gleaned (e.g., Staiger *et al.*, 2005).

In a precocious application of burial dating with ^{26}Al and ^{10}Be, Balco *et al.* (2005) determined ages for soils on deeply buried till units. Based on lithostratigraphy, they were able to constrain the thickness and approximate ages of overburden and burial date ancient glaciations that were previously impossible to date.

Volcanic Terrains

Cosmogenic nuclides can be used to provide information about eruptive sequences (ages of lava and ignimbrite flows). In addition, the dating of volcanic features allows inferences to be drawn about fluvial incision rates and slip rates along faults (see also below). Formerly, in the time range of 45 thousand years ago (ka) or less, dating of lava flows by underlying or charcoal depended on fortuitous findings. With cosmogenic nuclides, lava flows that erupted hundreds to several hundred thousand years ago can be dated directly. The effective age range extends up to hundreds of thousands of years depending again on the rate of degradation of the flow surfaces. Exposure dating in mafic volcanic terrains has been carried out with the cosmogenic noble gases ^{3}He and/or ^{21}Ne (Niedermann, 2002; Cerling and Craig, 1994), because they are more suitable for analysis in mafic phenocrysts (olivine and pyroxene). Chlorine-36 is suitable for whole rock determinations on volcanic rocks, thus aphyric (phenocryst-free) lavas can be dated (Phillips, 2003). This also affords the opportunity for intercomparison between the eruptive (K–Ar or ^{40}Ar/^{39}Ar age) and the exposure age (e.g., Fenton *et al.*, 2001; Phillips, 2003). Lava flows barren of K-rich phases, and when excess Ar can be a problem, can be dated with cosmogenic nuclides.

To determine the eruptive age of the lava flow and develop a chronology for the flows in a volcanic field, original eruptive surfaces, such as spatter, frothy glassy texture of the cooling rinds on flows (**Fig. 1B**) or bombs, or tumuli ridges, must be present (Cerling and Craig, 1994; Fenton *et al.*, 2001; Phillips, 2003). Lavas of intermediate to silicic composition are chaotic and blocky making identification of original flow tops difficult. Regionally extensive weathering-resistant ignimbrite sequences with abundant quartz and sanidine have been exposure-dated to determine denudation rates in arid regions (Kober *et al.*, 2006).

Alluvial and Lacustrine Systems

Determining the age of fluvial and alluvial landforms (e.g., terraces and fans) provides fundamental information about depositional processes. For example, on composite landforms such as alluvial fans, boulders from sites across the fan lobe may reveal a time progression of deposition on the fan (**Fig. 1C**). By constructing a time series, it is possible to examine the contribution of the various forcing mechanisms, such as tectonic base-level or regional slope changes, climatically (discharge) induced or sediment supply dependent variations.

Effective dating of depositional landforms requires that they are a single unit and/or are deposited over a relatively short time interval (Hancock *et al.*, 1999). Boulder or amalgamated clast samples are collected from the highest, flattest part of the surface away from diffusing slopes. Two considerations are crucial: (i) the amount of inheritance in the clasts and

(ii) the degradation or aggradation rate of the top surface. These issues are addressed by measuring depth profiles that extend to several meters (Anderson *et al.*, 1996). The shape of the concentration versus depth curve depends on the exposure age, the erosion (or aggradation) rate, and the amount of inheritance (**Figs. 5C** and **D**). The inherited nuclides are produced before deposition, for example, during hill slope erosion and sediment transport. Where field relationships indicate that the top of the depositional surface has undergone little erosion (e.g., flat tread morphology as opposed to convex tops on terraces) or little aggradation (minimal silt layer below the top surface pavement) the shape of the depth curve reveals both age and inheritance (**Fig. 5A**). The lack of a decrease in nuclide concentration in upper several tens of centimeters may be due to mixing (Phillips *et al.*, 1998) (**Fig. 5B**). Depth-profile results indicating separate depositional units (**Fig. 5D**) are evaluated in light of field evidence.

It may be difficult to constrain the rate of surface modification. Measuring the concentrations from two nuclides in a depth profile can be used. In this way the difference of the half-lives (e.g., ^{26}Al and ^{10}Be) yields a decreasing ratio with depth, and the unit is effectively burial-dated (Wolkowinsky and Granger, 2004). Samples are taken at depths greater than several meters to ensure that they are completely shielded from production from the top of the landform.

The height and age of abandoned strath terraces (e.g., Burbank *et al.*, 1996) or fluvial terraces (e.g., Ward *et al.*, 2005) located along a given reach of a river provide information about fluvial incision rates and their variation with time. Fluvially sculpted channel walls may provide a more continuous record of incision than strath terraces (Schaller *et al.*, 2005). Sampled bedrock surfaces bear scars of fluvial erosion but should not have been covered. Boulders or amalgamated clast samples on debris-flow deposits on ancient straths are dated to determine incision rates (Fenton *et al.*, 2004). Another method to determine fluvial incision rates is to determine the burial age and the height of deeply buried sediments in caves (Granger *et al.*, 1997).

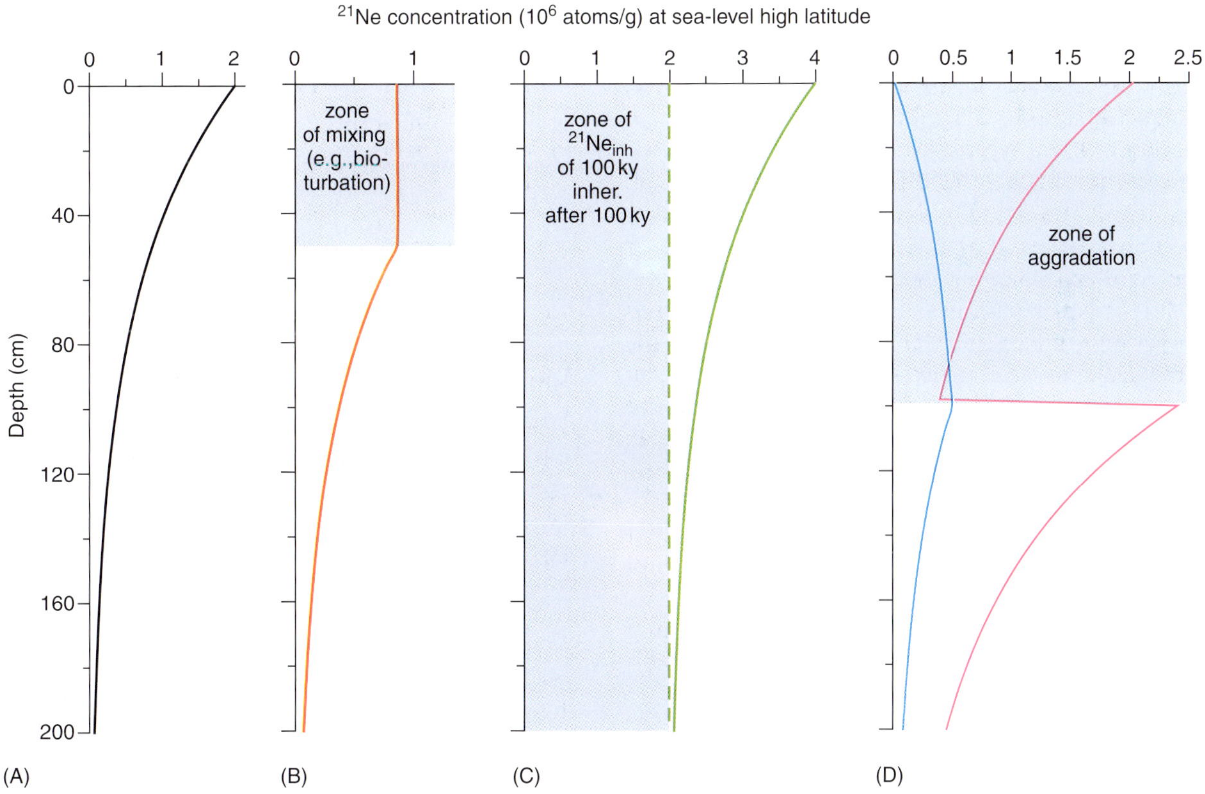

Figure 5 Depth profiles of ^{21}Ne reflecting various geomorphic processes. Depth profiles for alluvial landforms, erosion rate of the top of the surface is zero. (A) Depth profile showing nuclide build-up with time. (B) A flat profile in the upper few tens of centimeters, may indicate soil- or regolith production, bioturbation, or creep processes (e.g., Phillips *et al.*, 1998). (C) Inheritance or prior exposure is indicated by the shift to higher concentrations at several meters depth (compare to **Figure 5A**). (D) Two types of aggradation surfaces; the blue line shows continuous exposure of the first unit (the lower section) followed by aggradation on top of it at a rate of centimeters/1,000 years. The pink line portrays two separate single aggradation cycles each leaving one meter of sediment. Procedure and production rates used as outlined in **Figure 2** and **3**.

A time frame is required to access the climate signal contained in the record of high lake stands in arid regions. Wave-cut bedrock platforms (Lifton *et al.*, 2001), or loose clasts in ridge lines that mark ancient shorelines are dated. At Lake Lisan (Matmon *et al.*, 2003) problems with inheritance, pointing to intermediate storage of clasts, was recognized.

Rates of Tectonic Activity

Cosmogenic nuclides are used to determine rates of tectonic activity using two approaches: (i) indirect measure of slip rates along faults or growing folds by determining the ages of deformed or offset abandoned landforms and (ii) by dating bedrock (footwall) fault surfaces produced by rapid rupture events directly. Determined rates can be compared along various segments of a fault as well as through time.

Slip rates on strike-slip faults or rates of uplift on normal faults are determined by taking the amount of offset of the dated feature and its age. Offset alluvial fans, fluvial terraces, river channels, moraines, and lava flows have been dated for these determinations (e.g., Fenton *et al.*, 2001; Zehfuss *et al.*, 2001). Both inheritance in the sediment and the erosion (or aggradation) rate of the top of the surface must be constrained; depth profiles may be required.

With cosmogenic nuclides, the rock surfaces created during faulting are dated directly. The measurement of ^{36}Cl in limestone fault surfaces has been most widely applied. Mesoscopic (slickensides) and megascopic (broad undulating surfaces) features are often well preserved in limestones. Continuous sampling is done along an exposed bedrock fault face (**Fig. 1D**). Nuclide concentrations are plotted versus height above the base of the scarp whereby a pattern of steps in concentration that reflect periods of exposure separated by earthquake-related rupture events is revealed (e.g., Zreda and Noller, 1998; Mitchell *et al.*, 2001; Benedetti *et al.*, 2003). The recurrence interval of major earthquakes is constructed on the basis of this chronology.

Times Series of Mass Movements

Landsliding is an important process for the modification of valley slopes. The effect of high-magnitude, low-frequency events on landscape change can be evaluated when time series are constructed (Cockburn and Summerfield, 2004). By determining the timing of recurrence intervals we can understand processes of large-scale downslope movement of rock, and the interplay between various possible causes and triggering mechanisms. New rock surfaces suitable for exposure dating are created during large deep-seated mass movements. Each possible sampling site (head scarps, sliding planes, large boulders) has its own advantages and disadvantages with respect to exposure dating (Ivy-Ochs *et al.*, in press). In the breakaway zone postslide rock fall can be difficult to identify. Shallow sliding planes may yield ages that are too young due to coverage by snow and/or vegetation. Pre-exposure is not uncommon in rockslide boulders. A boulder may be derived from the outermost surface of the preslide bedrock as shown by the presence at Koefels (Austria) of glacially polished boulders on top of the slide debris (Ivy-Ochs *et al.*, 1998). Landslides are commonly characterized by multiple or progressive sliding. Each lobe must be dated. Numerous samples from different settings (boulders, sliding plane) may be necessary to constrain the age of a large landslide.

Ancient Landscapes

High concentrations of long-lived radionuclides (^{10}Be, ^{26}Al) or stable nuclides (^{3}He, ^{21}Ne) measured on bedrock or low-relief alluvial surfaces confirm the antiquity of landscapes in hyperarid regions of the planet. But are ages determined? Studies using two or more nuclides indicate that many of the ancient bedrock surfaces have been exposed continuously, but are clearly weathering. However, in most studies, data points plot within the erosion-island pointing to single stage exposure. Evidence of grusification and spalling (exfoliation, onionshell weathering) suggest significant erosion, including discontinuous erosion, during exposure. Kober *et al.*, 2006 pointed out that spalling of thin slabs may be the reason that data points for rock surfaces continuously exposed for millions of years do plot inside the erosion-island but do not plot on the saturation line (**Fig. 4**; cf. Bierman *et al.*, 1999). Nevertheless, the exposures are continuous and minimum ages are determined.

By assuming the nuclide concentrations have reached steady-state, erosion (denudation) rates are determined. Fundamental information about the age and mode of formation of ancient landscapes is gleaned. For example if the inselbergs in Australia are eroding in a self-same pattern, then they are the direct descendents of early Cenozoic or even Mesozoic landforms (Bierman and Caffee, 2002). Combined fission-track and consistently low cosmogenic nuclide-derived denudation rates across the Nambian escarpment indicate that great escarpments do not form by rapid (and ongoing) escarpment retreat, but must have formed early on after continental break-up (Cockburn *et al.*, 2000). High nuclide concentrations measured in rock surfaces in the Dry Valleys of Antarctica provided additional support for the premise that the East Antarctic Ice Sheet has been a stable feature since its inception (e.g., Schaefer *et al.*, 1999). Measured pre-Pleistocene apparent exposure ages and corresponding low denudation rates characterize the deserts of northern Chile, despite active uplift (Dunai *et al.*, 2005; Kober *et al.*, 2006; Nishiizumi *et al.*, 2005).

In old landscapes, the use of concentrations of cosmogenic nuclides to calculate erosion rates is dependent on the validity of the assumption that the concentration has reached steady state (saturation). But there are few ways to prove this. Landscape erosional steady state and nuclide steady state are not the same and need not operate on the same timescales. Depending on the coupling between perturbation and response time for the landscape, the geomorphic/geologic process and the nuclide system (one nuclide or several nuclides) may be in a transient state, and nonsaturation conditions may occur.

Archeological Contexts

In addition to quantifying the interaction of climate and tectonics in landscape evolution, cosmogenic nuclides are used to date the emergence and migration of early humans. Hominid fossils are indirectly dated with burial dating. Surfaces that were created during tool knapping are dated directly. Granger and coworkers used burial dating to determine that sand in lenses enclosing early hominid fossils in caves in South Africa were buried around four million years ago (Partridge *et al.*, 2003). Ivy-Ochs *et al.* (2001) measured ^{10}Be surface exposure ages for chert tools collected from limestone benches near Luxor, Egypt. Although the amount of inheritance is difficult to gauge, the exposure ages of several hundred thousand years are in good agreement with age estimates based on tool typology. Based on ^{10}Be concentrations, Verri *et al.* (2005) determined that chert tools in layers as old as 200–400 ky in the Tabun Cave in Israel, were made from nodules collected from surface pits or mines more than several meters in depth.

Summary and Outlook

The ability to use cosmogenic nuclides to determine how long rock surfaces have been exposed at the surface of the Earth provides an unrivaled tool for determining ages of landforms and therefore the rates of geomorphic processes. Rocks on landforms that have been exposed for hundreds of years can be dated with cosmogenic nuclides. As production rate and scaling formalities improve (see previous chapters by Lal and by Schaefer and Lifton), the precision of obtained ages improves. But the accuracy of the ages remains a question of geological uncertainties. Both rock surfaces and the landforms degrade with time. This imposes clear limitations on the time range and accuracy of dating. Similarly, the natural variability of samples depends on landform morphology and its age. Perhaps never before has the obtained age for a geological landform been so completely dependent on the choice of what to sample. Consequently, obtained exposure ages must be evaluated individually. The ages must conform with field relationships such as terrace or moraine stratigraphy and regional morphostratigraphic relationships; as well as independent age constraints for the same or correlative features. The dates are then used to construct a chronological framework and rates of geomorphic processes such as fault slip rates or the timing of glacier variations can be derived.

For millions of years old, bedrock landforms in arid regions minimum ages and more importantly, landform denudation rates are derived from cosmogenic nuclide concentrations. The unique opportunity of measuring different cosmogenic nuclides with different half-lives from individual surfaces allows us to decipher multistage exposures (with intervals of burial) and erosion rate histories. The addition of nuclides with shorter half-lives and additional production pathways (e.g., ^{14}C with the shortest half-life, low-energy production pathways of ^{36}Cl, respectively) provides more power to resolve and reveal erosion-controlled exposures or step changes in denudation rates over time.

Abbreviations

ky	thousand years
My	million years

See also: **Introduction**: History of Dating Methods. **Cosmogenic Nuclide Dating**: Overview; Methods; Cosmic Ray Interactions in Minerals. **Dating Techniques**; **U-Series Dating**.

Anderson, R. S., Repka, J. L., and Dick, G. S. (1996). Explicit treatment of inheritance in dating depositional surfaces using ^{10}Be and ^{26}Al. *Geology* **24**, 47–51.

Balco, G., Stone, J. O. H., and Mason, J. A. (2005). Numerical ages for Plio-Pleistocene glacial sediment sequences by ^{26}Al/^{10}Be dating of quartz in buried paleosols. *Earth and Planetary Science Letters* **232**, 179–191.

Benedetti, L., Finkel, R., King, G., *et al.* (2003). Motion on the Kaparelli Fault (Greece) prior to the 1981 earthquake sequence determined from ^{36}Cl cosmogenic dating. *Terra Nova* **15**, 118–224.

Bierman, P., and Caffee, M. (2002). Cosmogenic exposure and erosion history of Australian bedrock landforms. *Geological Society of America Bulletin* **114**, 787–803.

Bierman, P. R., Caffee, M. W., Davis, P. T., *et al.* (2002). Rates and timing of earth surface processes from in-situ produced cosmogenic ^{10}Be. In *Beryllium Reviews in Mineralogy and Geochemistry* (E. Grew, Ed.) 50, pp. 147–204.

Bierman, P. R., Marsella, K. A., Patterson, C., Davis, P. T., and Caffee, M. (1999). Mid-Pleistocene cosmogenic minimum-age limits for pre-Wisconsian glacial surfaces in southwestern Minnesota and southern Baffin Island: a multiple nuclide approach. *Geomorphology* **27**, 25–39.

Burbank, D. W., Leland, J., Fielding, E., *et al.* (1996). Bedrock incision, rock uplift and threshold hillslopes in the northwestern Himalayas. *Nature* **379**, 505–510.

Cerling, T. E., and Craig, H. (1994). Geomorphology and *in-situ* cosmogenic isotopes. *Annual Review of Earth and Planetary Sciences* **22**, 273–317.

Cockburn, H. A. P., Brown, R. W., Summerfield, M. A., and Seidl, M. A. (2000). Quantifying passive margin denudation and landscape development using a combined fission-track thermochronology and cosmogenic isotope analysis approach. *Earth and Planetary Science Letters* **179**, 429–435.

Cockburn, H. A. P., and Summerfield, M. A. (2004). Geomorphological applications of cosmogenic isotope analysis. *Progress in Physical Geography* **28**, 1–42.

Dunai, T. J., Gonzales Lopez, G. A., and Juez-Larre, J. (2005). Oligocene–Miocene age of aridity in the Atacama Desert revealed by exposure dating of erosion-sensitive landforms. *Geology* **33**, 321–324.

Fabel, D., Harbor, J., Dahms, D., *et al.* (2004). Spatial patterns of glacial erosion at a valley scale derived from terrestrial cosmogenic Be-10 and Al-26 concentrations in rock. *Annals of the Association of American Geographers* **94**, 241–255.

Fenton, C. R., Poreda, R., Nash, B. P., Webb, R. H., and Cerling, T. E. (2004). Geochemical discrimination of five Pleistocene lava-dam outburst-flood deposits, western Grand Canyon, Arizona. *Journal of Geology* **112**, 91–110.

Fenton, C. R., Webb, R. H., Pearthree, P. A., Cerling, T. E., and Rj, P. (2001). Displacement rates on the Toroweap and Hurricane faults: implications for Quaternary downcutting in the Grand Canyon, Arizona. *Geology* **29**, 1035–1038.

Gosse, J. C. (2005). The contributions of cosmogenic nuclides to unraveling alpine paleo-glacier histories. In *Global Change and Mountain Regions. A State of Knowledge Overview* (U. M. Huber, H. K. M. Bugmann and M. A. Reasoner, Eds.), 23, pp. 39–50. Springer, Dordrecht.

Gosse, J. C., and Phillips, F. M. (2001). Terrestrial *in situ* cosmogenic nuclides: theory and application. *Quaternary Science Reviews* **20**, 1475–1560.

Gosse, J. C., Klein, J., Evenson, E. B., Lawn, B., and Middleton, R. (1995). Be-10 dating of the duration and retreat of the last Pinedale glacial sequence. *Science* **268**, 1329–1333.

Granger, D. E., Kirchner, J. W., and Finkel, R. C. (1997). Quaternary downcutting rate of the New River, Virginia, measured from differential decay of cosmogenic ^{26}Al and ^{10}Be in cave-deposited alluvium. *Geology* **25**, 107–110.

Hancock, G. S., Anderson, R. S., Chadwick, O. A., and Finkel, R. C. (1999). Dating of fluvial terraces using ^{10}Be and ^{26}Al profiles, Wind River, Wyoming. *Geomorphology* **27**, 41–60.

Ivy-Ochs, S., Heuberger, H., Kubik, P. W., *et al.* (1998). The age of the Koefels event. Relative, ^{14}C and cosmogenic isotope dating of an early Holocene landslide in the Central Alps (Tyrol, Austria). *Zeitschrift für Gletscherkunde und Glazialgeologie* **134**, 57–70.

Ivy-Ochs, S., Wüst, R., Kubik, P. W., Müller-Beck, H., and Schlüchter, C. (2001). Can we use cosmogenic isotopes to date stone artifacts. *Radiocarbon* **43**, 759–764.

Ivy-Ochs, S., von Poschinger, A., Synal H.-A., and Maisch, M. (2006). Surface exposure dating of the Flims rockslide, Graubünden, Switzerland: *Geomorphology* (in press).

Kerschner, H., Hertl, A., Gross, G., Ivy-Ochs, S., and Kubik, P. W. (2006). Surface exposure dating of moraines in the Kromer valley (Silvretta Mountains, Austria) – evidence for glacial response to the 8.2 ka event in the Eastern Alps? *The Holocene* **16**, 7–15.

Kaplan, M. R., Douglass, D. C., Singer, B. S., and Caffee, M. W. (2005). Cosmogenic nuclide chronology of pre-last glacial maximum moraines at Lago Buenos Aires, 46 degrees S, Argentina. *Quaternary Research* **63**, 301–315.

Kober, F., Ivy-Ochs, S., Schlunegger, F., Baur, H., Kubik, P. W. and Wieler, R. (2006). Denudation rates and a topography-driven precipitation threshold in northern Chile: multiple cosmogenic nuclide data and sediment yield budgets: *Geomorphology* (in press).

Lal, D. (1991). Cosmic ray labelling of erosion surfaces: *in-situ* nuclide production rates and erosion models. *Earth and Planetary Science Letters* **104**, 424–439.

Lifton, N. A., Jull, A. J., and Quade, J. (2001). A new extraction technique and production rate estimate for in situ cosmogenic ^{14}C in quartz. *Geochimica et Cosmochimica Acta* **65**, 1935–1969.

Matmon, A., Crouvi, O., Enzel, Y., *et al.* (2003). Complex exposure histories of chert clasts in the late Pleistocene shorelines of Lake Lisan, southern Israel. *Earth Surface Processes and Landforms* **28**, 493–506.

Mitchell, S. G., Matmon, A., Bierman, P. R., Enzel, Y., Caffee, M., and Rizzo, D. (2001). Displacement history of a limestone normal fault scarp, northern Israel, from cosmogenic Cl-36. *Journal Of Geophysical Research* **B106**, 4247–4264.

Niedermann, S. (2002). Cosmic-ray-produced noble gases in terrestrial rocks: dating tools for surface processes. In *Noble Gases in Geochemistry and Cosmochemistry. Reviews in Mineralogy and Geochemistry* (D. Porcelli, C. J. Ballentine and R. Wieler, Eds.) 47 pp. 731–784.

Nishiizumi, K., Caffee, M., Finkel, R. C., Brimhall, G., and Mote, T. (2005). Remnants of a fossil alluvial fan landscape of Miocene age in the Atacama Desert of northern Chile using cosmogenic nuclide exposure age dating. *Earth and Planetary Science Letters* **237**, 499–507.

Partridge, T. C., Granger, D. E., Caffee, M. W., and Clarke, R. J. (2003). Lower Pliocene hominid remains from Sterkfontein. *Science* **300**, 607–612.

Phillips, F. M. (2003). Cosmogenic ^{36}Cl ages of Quaternary basalt flows in the Mojave Desert, California, USA. *Geomorphology* **53**, 199–208.

Phillips, F. M., Zreda, M. G., Benson, L. V., Plummer, M. A., Elmore, D., and Sharma, P. (1996). Chronology for fluctuations in late Pleistocene Sierra Nevada glaciers and lakes. *Science* **274**, 749–751.

Phillips, W. M., McDonald, E. V., Reneau, S. L., and Poths, J. (1998). Dating of soils and alluvium with cosmogenic ^{21}Ne depth profiles: case studies from the Pajarito Plateau, New Mexico, USA. *Earth and Planetary Science Letters* **160**, 209–223.

Putkonen, J., and Swanson, T. (2003). Accuracy of cosmogenic ages for moraines. *Quaternary Research* **59**, 255–261.

Schäfer, J. M., Ivy-Ochs, S., Wieler, R., *et al.* (1999). Cosmogenic noble gas studies in the oldest landscape on earth: surface exposure ages of the Dry Valleys, Antarctica. *Earth and Planetary Science Letters* **167**, 215–226.

Schaller, M., Hovius, N., Willet, S. D., Ivy-Ochs, S., Synal, H. A., and Chen, H. C. (2005). Fluvial bedrock incision in the active mountain belt of Taiwan from *in situ*-produced cosmogenic nuclides. *Earth Surface Processes and Landforms* **30**, 955–971.

Schaller, M., von Blanckenburg, F., Hovius, N., Veldkamp, A., van den Berg, M. W., and Kubik, P. W. (2004). Paleoerosion rates

from cosmogenic ^{10}Be in a 1.3 Ma terrace sequence: response of the River Meuse to changes in climate and rock uplift. *Journal of Geology* **112**, 127–144.
Staiger, J. K. W., Gosse, J. C., Johnson, J. V., *et al.* (2005). Quaternary relief generation by polythermal glacier ice. *Earth Surface Processes and Landforms* **30**, 1145–1159.
Stone, J. (2000). Air pressure and cosmogenic isotope production. *Journal of Geophysical Research* **105**, 23,753–23,759.
Verri, G., Barkai, R., Gopher, A., *et al.* (2005). Flint procurement strategies in the Late Lower Palaeolithic recorded by *in situ* produced cosmogenic Be-10 in Tabun and Qesem Caves (Israel). *Journal of Archaeological Science* **32**, 207–213.
Ward, D. J., Spotila, J. A., Hancock, G. S., and Galbraith, J. M. (2005). New constraints on the late Cenozoic incision of the New River, Virginia. *Geomorphology* **75**, 54–72.
Wolkowinsky, A. J., and Granger, D. E. (2004). Early Pleistocene incision of the San Juan River, Utah, dated with ^{26}Al and ^{10}Be. *Geology* **32**, 749–752.
Zehfuss, P. H., Bierman, P. R., Gillespie, A. R., Burke, R. M., and Caffee, M. W. (2001). Slip rates on the Fish Springs Fault, Owens Valley, California, deduced from cosmogenic ^{10}Be and ^{26}Al and soil development on fan surfaces. *Geological Society of America Bulletin* **113**, 241–255.
Zreda, M., and Noller, J. S. (1998). Ages of prehistoric earthquakes revealed by cosmogenic chlorine-36 in a bedrock fault scarp at Hebgen Lake. *Science* **282**, 1097–1099.

Landscape Evolution

D E Granger, Purdue University, IN, USA

© 2007 Elsevier B.V. All rights reserved.

Introduction

Landscapes evolve over time as material is eroded from bedrock, transported down hill slopes and river systems, and deposited in valleys or basins. River incision and aggradation transmit base-level change through a watershed, and help determine how landscapes respond to external factors such as tectonic uplift or climate change. To understand how landscapes change through time, it is important to characterize both erosion rates and valley incision rates. It is also important to establish the timing of major events in the landscape's history to understand its dynamic response to changes in tectonics, climate, or other factors. Cosmogenic nuclides have played a key role in quantifying both erosion rates and the timing of landscape evolution.

Until recently, measuring erosion rates has been very difficult. This is because eroding landscapes leave little evidence of their past. Erosion rates, after all, are a measure of how fast material has been removed from a landscape. Most of the eroded material is either transported away or dissolved. It is possible to estimate short-term erosion rates from sediment and solute loads in streams, but sediment loads are susceptible to rapid fluctuations and their measurements are often fraught with error. Moreover, modern sediment loads are strongly influenced by land use. Prior to cosmogenic nuclide methods, the issue of land use was so daunting that it was considered questionable that erosion rates could ever be measured reliably in large parts of the world.

Today, the application of cosmogenic nuclide methods allows us to circumvent the problems of land use and to measure erosion rates over timescales of thousands of years. This is because the cosmogenic nuclides present in rock or soil today were produced by secondary cosmic rays passing through the rock or sediment above them in the past (*see* Cosmic Ray Interactions in Minerals). Cosmogenic nuclide concentrations allow us to infer the presence of an overburden that is no longer there, and how quickly it disappeared. Moreover, because cosmogenic nuclides persist in mineral grains for millions of years, they allow us to determine paleo-erosion rates in the distant past and to date important events in landscape evolution (see previous section by Ivy-Ochs and Kober).

Cosmogenic Nuclide Buildup and Decay

Cosmogenic nuclides are such valuable monitors of surface processes because their production is rapidly attenuated within about a meter of the ground surface (Lal, 1991). Their buildup in a mineral grain thus indicates how much time that grain has spent within the production zone. Many different cosmogenic nuclides are produced in different mineral grains, depending on the mineral composition. Cosmogenic nuclides used in landscape evolution include stable nuclides such as ^{3}He and ^{21}Ne, as well as radioactive nuclides such as ^{10}Be, ^{14}C, ^{26}Al, and ^{36}Cl (Nishiizumi *et al.*, 1993; Cerling and Craig, 1994; Gosse and Phillips, 2001; Muzikar *et al.*, 2003). Each of these nuclides can be used in a slightly different way, taking advantage of their production rates and decay times. The great majority of landscape evolution studies use the nuclides ^{10}Be and ^{26}Al in quartz. This is because production rates of these nuclides have a simple depth dependence, and because quartz is a common mineral grain that is resistant to chemical and physical weathering, allowing it to persist over long periods of erosion and transport (Lal and Arnold, 1985).

Aluminum-26 and ^{10}Be are produced by two different types of secondary cosmic ray particles that result from a nuclear cascade through the atmosphere (Lal and Peters, 1967). The first type of particle is a fast nucleon (these are mainly neutrons but include some protons) which causes a nuclear spallation reaction within mineral grains. For example, in quartz (SiO_2), a nuclear collision can cause ^{28}Si to undergo spallation to the lighter nuclide ^{26}Al through the net loss of a proton and a neutron. Likewise, ^{16}O can be transformed to ^{10}Be through the net loss of three protons and three neutrons. Spallation reactions such as these occur very near the ground surface, because fast nucleons are rapidly attenuated as they pass through matter. The penetration length of a fast nucleon depends on the density of the material that it is passing through, and is approximately $160\,g\,cm^{-2}$. For example, in typical rock of density $2.6\,g\,cm^{-3}$, the nucleons travel 60 cm; in typical soil of density $1.5\,g\,cm^{-3}$ the nucleons travel over a meter.

The second type of particle that produces ^{26}Al and ^{10}Be is a muon. Muons are less reactive than nucleons, so they produce fewer cosmogenic nuclides at the ground surface, only 2–3% of total production (Brown *et al.*, 1995a); however, muons penetrate more deeply and dominate production at depths greater than a few meters. Although muons do not contribute much near the ground surface, they can account for up to 20% of the total cosmogenic nuclide inventory in an eroding terrain (Granger *et al.*, 2001). Muons are very important for dating sedimentary deposits and for determining paleo-erosion rates because they continue to produce ^{10}Be and ^{26}Al even when sediment is buried at depth. A proper interpretation of cosmogenic nuclide concentrations in a sedimentary deposit must account not only for the concentration inherited from the time of deposition, but also for postdepositional production by muons.

Determining Erosion Rates

To a close approximation, the depth dependence of ^{26}Al and ^{10}Be production can be described by the equation

$$P(z) = P_{\text{spallation}}(0)e^{-\rho z/\Lambda} + P_{\text{muon}}(0)e^{-\rho z/L} \quad [1]$$

where P is the production rate at depth z, the subscripts spallation and muon indicate production by the two separate mechanisms, ρ is density, and Λ ($= 160\,g\,cm^{-2}$) and L ($\sim 1{,}300\,g\,cm^{-2}$) are penetration lengths of fast nucleons and muons, respectively (Lal, 1991; Brown *et al.*, 1995a). Although the depth dependence of muons varies from this simple approximation at depths greater than about 5 m (Heisinger *et al.*, 2002), this equation will suffice for the purpose of discussion.

In an eroding landscape, mineral grains are exposed to an increasingly intense secondary cosmic ray flux as they are gradually exhumed from bedrock. In a slowly eroding landscape, mineral grains spend a lot of time near the surface and build up a large cosmogenic nuclide inventory; in contrast, in a rapidly eroding landscape the mineral grains approach the surface quickly and are eroded away with only a small cosmogenic nuclide inventory. More quantitatively, the cosmogenic nuclide concentration at a rock surface can be found by integrating Eqn [1] over time (Lal, 1991). For the case of long-term steady-state erosion, the concentration of a cosmogenic nuclide is given by

$$N = P_{\text{spallation}}/(1/\tau + \rho E/\Lambda) + P_{\text{muon}}/(1/\tau + \rho E/L) \quad [2]$$

where N is the concentration, E is the erosion rate and τ is the radioactive meanlife (a meanlife is equal to a half-life divided by the natural logarithm of 2). The meanlives of ^{26}Al and ^{10}Be are approximately 1.0 and 1.9 Myr , respectively.

It is important to realize that the timescale over which cosmogenic nuclides measure erosion rates is approximately the time taken to erode one fast nucleon penetration length (about 60 cm in rock; Lal, 1991). In most soil-mantled landscapes, erosion rates are averaged over a time that is comparable to that of soil formation. Cosmogenically derived erosion rates are usually averaged over a much longer timescale than that of agricultural land use, and so primarily reflect a pre-agriculture rate (Brown *et al.*, 1995b). The exception to this rule comes in rapidly eroding areas with a long history of land use, where several penetration lengths may have eroded away since the beginning of agriculture thousands of years ago. Cosmogenic nuclides not only circumvent the problem of land use in determining erosion rates, they also provide a useful tool for determining natural erosion rates as a baseline against which to ascertain the impacts of land use.

There are two different approaches to measuring erosion rates with cosmogenic nuclides. The first is to measure the erosion rate of a discrete site by analyzing material from the surface of eroding bedrock or sediment. This can be either an eroding outcrop, or material from bedrock covered by soil (**Fig. 1**). The latter case has been very important for understanding how erosion rates depend on soil cover.

The second approach is to measure erosion rates by analyzing sediment carried by streams. Streams in effect sample eroded material from the entire watershed, providing a natural averaging mechanism (**Fig. 2**). It can be shown that the average cosmogenic nuclide concentration in stream sediment reflects the average residence time of quartz in the watershed,

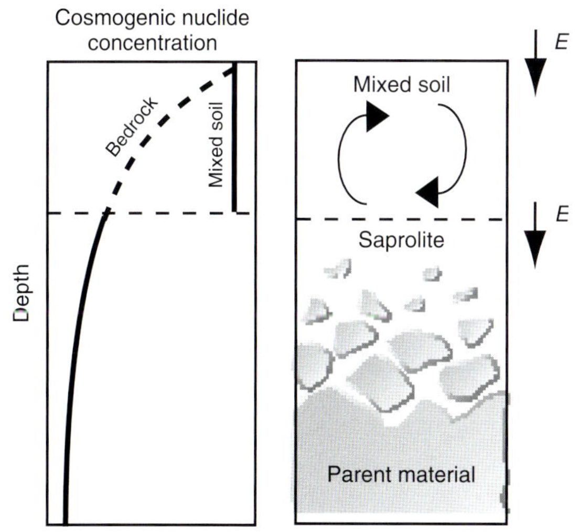

Figure 1 Cosmogenic nuclide concentrations in both bedrock (dashed line) and well-mixed soil (solid line) reach a value at the ground surface determined by the erosion rate (Eqn [2]). Erosion rates can be inferred from cosmogenic nuclide concentrations in samples collected at the surface of eroding rock or soil, as well as from saprolite collected beneath a soil cover, as long as soil depth is known.

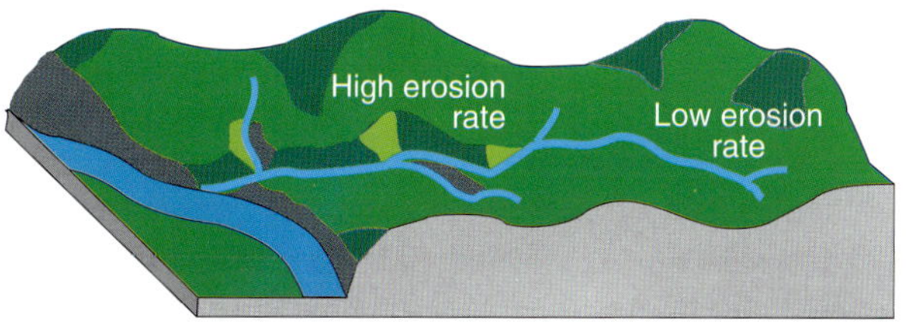

Figure 2 Cosmogenic nuclide concentrations in well-mixed stream sediment are determined by averaging a small amount of sediment from slowly-eroding areas with a large amount of sediment from rapidly eroding areas. Neglecting radioactive decay, the average cosmogenic nuclide concentration is approximately given by Eqn [2].

even in a watershed eroding at different rates in different places. This is because of a self-compensating tendency in sediment mixing. Areas that are eroding quickly produce a lot of sediment, but with a low cosmogenic nuclide concentration. Areas eroding slowly do not contribute much sediment, but that which they do produce contains a high cosmogenic nuclide concentration. These two effects balance so that the average cosmogenic nuclide concentration indicates the average erosion rate of the watershed, over the time interval required to erode about one penetration length, or 60 cm. There are many complications that can lead to errors. Care must be taken to ensure that the source of quartz in the watershed is known, and that stream sediment is indeed derived from the whole watershed, not just from a local landslide, for example. Nonetheless, repeated studies have shown that basin-averaged erosion rates can be accurately determined with this method (for recent reviews see Bierman and Nichols (2004) and von Blanckenburg (2005)).

Following are examples of each of these methods. The first example shows that bedrock erosion is highly dependent on the depth of soil cover. The second example shows that basin-averaged erosion rates are much more dependent on the long-term tectonic regime than on climate.

Example 1: Erosion Rates Beneath a Soil Cover—The Soil Production Function

It has long been postulated that the rate at which bedrock is converted to soil depends on soil depth. Erosional processes such as animal burrowing and tree throw are more vigorous under a thin soil cover, but decline for thick soils. Likewise, freeze–thaw and wet–dry cycles are more effective under a thin soil cover. The functional form by which soil production depends on soil depth, however, has remained debated. Cosmogenic nuclides provide a way to determine this relationship directly, by measuring erosion rates in a single landscape.

Heimsath *et al.* (1997, 2000, 2001) measured ^{10}Be concentrations in weathered bedrock beneath soil to determine the depth dependence of soil production rates. A successful experiment for determining the soil production function requires a landscape where both soil depths and erosion rates are likely to vary considerably. The landscapes they examined had thin soil on hill crests and ridges, and thicker soil in valley bottoms and hollows. All of their measurements were made on divergent hill slopes where it is most likely that soil is neither accumulating nor evacuating over time. It is important to realize that the rate of soil production they measured is not the rate of chemical weathering, but is instead the rate of mechanical disruption of preweathered bedrock into mobile regolith. For a steady-state soil depth, this is equivalent to the denudation rate.

Heimsath *et al.* found that soil production rates decrease exponentially with soil depth, reaching a maximum under very shallow soils. This implies that soil production and erosion is stabilized by a negative feedback, so that soil depths tend to vary smoothly across a landscape and regulate erosion rates. If soils become thin in a particular place, then bedrock is decomposed more quickly to produce a thicker soil. Likewise, thick soils inhibit erosion so that soils eventually thin.

The fact that Heimsath *et al.* found thin soils and fast erosion rates on hill crests and ridges indicates that the landscapes where they worked are not in steady state, but that their relief is decreasing through time as the hill crests lower faster than the surrounding hill slopes, at least locally. This is not necessarily the case in all landscapes, however. In areas with more uniform soil depths or with large areas of exposed bedrock, the relationship between erosion rates and soil depth is less apparent (Heimsath *et al.*, 2001; Wilkinson *et al.* 2005). Also, it should be realized that the rate of bedrock-weathering to produce saprolite may have a different soil depth dependence than the mechanical disruption of saprolite into soil. Indeed, exposed bedrock erodes far more slowly than soil-mantled saprolite in many terrains (Bierman, 1994; Heimsath *et al.*, 2000; Granger *et al.*, 2001).

Example 2: The Influence of Climate and Tectonics on Weathering and Erosion

Landscape-scale weathering and erosion rates depend on a wide variety of factors, including climatic parameters such as rainfall and temperature which have fluctuated dramatically through the Quaternary, as well as tectonic uplift rates, which vary at considerably slower rates. The importance of climate versus tectonics in determining regional erosion rates has been difficult to resolve using traditional methods. Many models have suggested that erosion rates may vary by orders of magnitude, for example, as rainfall increases from semiarid to humid environments. A suite of cosmogenic nuclide studies, however, has shown that although climate largely controls the degree of chemical alteration of rock as it is eroded, the net export of material from a watershed is closely matched to the long-term exhumation rate dictated by tectonics (see review by von Blanckenburg (2005)).

The influence of climate In a wide-ranging set of experiments, Riebe *et al.* analyzed ^{10}Be in sediment from nonglaciated granitic watersheds throughout a range of temperature and precipitation regimes to determine erosion rates (summarized by Riebe *et al.* (2003, 2004a)). They also inferred the degree of chemical weathering from the depletion of various elements relative to immobile zirconium.

Riebe *et al.* (2004a) found that although erosion rates were highly variable within a given climate regime, there is remarkably little variation in erosion rates between climate regimes. In other words, erosion rates show little dependence on climate. For example, in granitic sites throughout North America, extending over a mean annual temperature range from 2–25° C, and an annual precipitation gradient from 22–420 cm yr^{-1}, they found no apparent pattern in erosion rates.

Although erosion rates do not depend on climate, the degree of chemical weathering does. Tropical watersheds such as the Rio Icacos, Puerto Rico, show that 60% of the bedrock is chemically depleted during weathering, while more temperate watersheds in the Sierra Nevada show a chemical depletion of only 10–25%. Even on a local scale such as a single ridgeline, Riebe *et al.* (2004b) showed that the degree of chemical weathering varies with climate. Measurements at various altitudes in the Santa Rosa Mountains, Nevada, show that as vegetation decreases and snow cover increases, the degree of chemical depletion decreases from 20% to 0%, while the total erosion rate determined from ^{10}Be remains unchanged.

One implication of these studies is that the export of solutes from a watershed is generally limited by the supply of fresh minerals through physical erosion. A quickly eroding watershed in an arid environment supplies more solutes to the ocean than a slowly eroding watershed in a tropical environment (Riebe *et al.*, 2004a).

The influence of tectonics If erosion rates show such little dependence on climate, then what controls long-term erosion rates? A host of studies from throughout the world indicate that the long-term uplift and exhumation rate due to tectonics is the primary control of short-term erosion rates (von Blanckenburg, 2005).

Table 1 shows several examples of basin-averaged erosion rates compared to long-term exhumation

Table 1 Erosion rates determined from ^{10}Be in sediment compared to long-term erosion rates

Site	*^{10}Be erosion rate* (m/My)	*Long-term erosion rate* (m/My)
Himalaya (upper Ganges River)[a]	800–2,700	800–3,000
Sierra Nevada, USA[b]	15–60	25–60
Idaho Batholith, USA[c]	20–100	100
Sri Lanka[d]	5–11	4–15
Smoky Mtns., USA[e]	14–37	10–40
Blue Mtns., Australia[f]	10–20	<14
Bolivian Andes[g]	40–1,350 (avg. 420)	100–600

Data sources:
[a]Vance *et al.* (2003),
[b]Granger *et al.* (2001); Riebe *et al.* (2001),
[c]Kirchner *et al.* (2001),
[d]von Blanckenburg *et al.* (2004),
[e]Matmon *et al.* (2003),
[f]Wilkinson *et al.* (2005), and
[g]Safran *et al.* (2005).

rates determined from other methods such as thermochronometry or sediment budgets. Tectonically stable areas erode at only a few meters per million years, regardless of whether they are in a tropical region such as Sri Lanka, a temperate region such as Kentucky in the southeastern United States, or in an arid region such as central Australia or South Africa. In tectonically active areas such as the Himalaya or the Andes, erosion rates are kilometers per million years. In both cases, rates from cosmogenic and thermochronometric methods closely agree.

The match between cosmogenically derived erosion rates and long-term exhumation rates from thermochronometry indicates that landscapes tend to adjust themselves so that their erosion rate matches the regional uplift rate. These studies show that tectonics is far more important than climate in controlling erosion rates, at least in the absence of glaciation.

Measuring river incision rates River incision is intimately linked to basin-scale erosion. Rivers are the primary means for transporting sediment and solute from an eroding terrain. Rivers set the base level to which a watershed can erode. Because rivers can incise more quickly than hill slopes erode, they transmit base-level lowering signals to the basin's headwaters, thus regulating hill slope erosion in response to tectonic uplift and subsidence. Rivers also aggrade when they are choked by excess sediment from rapid hill slope erosion; thus, alluvial sediments often provide a record of fluctuations in erosion rates. Cosmogenic nuclides can be used to study the history of river incision and aggradation by dating alluvial sediments, dating terraces, and dating caves alongside the river.

Over timescales of tens to hundreds of thousands of years, cosmogenic nuclide exposure dating can be used to date bedrock surfaces or sediments on river terraces. Over longer timescales, however, surfaces are seldom preserved well enough for exposure dating. Yet it is important to understand how river incision varies through time over millions of years.

Cosmogenic nuclide burial dating A cosmogenic nuclide method that lends itself to dating sediments over million-year timescales is cosmogenic nuclide burial dating (Granger and Muzikar, 2001), which depends on radionuclide decay over time. If sediment containing cosmogenic nuclides is buried so that it is shielded from secondary cosmic radiation, then the inherited cosmogenic nuclides decay over time according

$$N(t) = N_{\mathrm{inh}}(0)e^{-t_{\mathrm{burial}}/\tau} \qquad [3]$$

where t_{burial} is the time since burial, and N_{inh} is the inherited cosmogenic nuclide concentration at the time of burial (Lal and Arnold, 1985). For a landscape eroding at steady state, N_{inh} can be determined using Eqn [2]. Burial dating works exceptionally well with the cosmogenic nuclide pair ^{26}Al–^{10}Be in quartz, because these nuclides are produced at a constant ratio, and they decay at different rates subsequent to burial. By solving Eqn [3] simultaneously for both ^{26}Al and ^{10}Be in the same sample of quartz, both the burial time and the paleo-erosion rate can be determined. The burial dating method is typically limited to the past 5 Myr, because ^{26}Al is usually undetectable beyond that time (Granger and Muzikar, 2001).

Burial dating requires that the sediment is shielded from secondary cosmic rays, including muons. The degree of shielding necessary to use this method depends on the age of the sediment being dated. For sediments older than a million years, burial depths of 10–20 m of sediment or rock are usually needed. Two situations that are appropriate for burial dating are thick alluvial terraces, where sediment packages are deposited in excess of 10 m, and caves, where sediment is deeply shielded underground.

Using caves to learn about river incision rates Caves provide a nearly perfect scenario for burial dating because sediment washed into the cave entrance is instantly shielded beneath tens of meters of solid rock. Because caves are formed by flowing water, they can also provide information about the water table at the time of sediment deposition. Quartz sediment can be washed into caves either by a surface river that passes near a pre-existing cave opening (**Fig. 3**), or by a sinking river that carries sediment underground as the cave is being formed. In either case, the burial date of quartz in the cave indicates the last time that water flowed through the passage. As a river cuts through bedrock, the cave is abandoned by flowing water. The fossil cave passage, together with its sediment, remains an indicator of when the river was at that particular elevation. In this way, caves are very similar to terraces; they both indicate past river levels and contain alluvial sediment. Caves sometimes offer an advantage over terraces, however, because the sediment within is shielded from weathering and erosion. Cave sediments can thus persist much longer than terrace sediments on the surface. Furthermore, burial in caves is instantaneous whereas burial under younger terrace sediment may, in some cases, have the disadvantage of only partial burial for an initial period.

Figure 3 River sediment deposited in caves such as this one above the New River, Virginia, can be dated with cosmogenic nuclides to infer long-term river incision rates.

Example: The Mammoth Cave System, Kentucky

Mammoth Cave is a multilevel cave along the Green River, Kentucky. It is the world's longest known cave system, with over 560 km of mapped passages. Multilevel caves such as the Mammoth Cave system form as the water table alternates between periods of stability and rapid incision. Each level of the cave represents a time of stability, with the highest passages being formed at the earliest time. Each successive level represents a phase of incision followed by stability. The sediments within each level of the cave thus represent the last time that the river occupied that particular passage.

The Mammoth Cave system formed south of the Laurentide ice sheet, but was nonetheless profoundly influenced as the ice sheet rearranged the drainage system of the Mississippi River. In particular, the Green River rapidly incised when the Ohio River, to the north, was formed along the edge of the Laurentide ice sheet.

Dates from the Mammoth Cave system show that the river has incised through 50 m of bedrock over the past 2.5 Myr . However, the data also show that river incision has been far from uniform over this time. Instead, it occurred in pulses associated with major climate changes. For example, the upper levels of the cave were filled with sediment and then abandoned at ~2.4 Ma, when the Laurentide ice sheet began to expand. Middle levels of the cave were abandoned at ~1.5 Ma, when the Ohio River formed along the ice-sheet margin. Lower levels of the cave were filled at ~800 ka, at the mid-Pleistocene Transition when the glacial periodicity changed from 41 kyr to 100 kyr. Other large caves throughout the region show a similar history of river incision, punctuated by these major climate transitions and episodic rearrangement of the Mississippi River system.

Additional Applications

There are many applications of cosmogenic nuclides in landscape evolution that have not been discussed here. For example, it is possible to use cosmogenic nuclides in buried sediments to infer not only the burial age, but also the paleoerosion rate. Data from Mammoth Cave, discussed above, show that local erosion rates remained very slow over the entire Quaternary. In particular, sandstone and conglomerate in the Mammoth Cave area eroded at 2–6 m per million years over the past 3–4 million years, despite major climate changes. However, paleo-erosion rates inferred for other watersheds around the world show stronger changes over this time (von Blanckenburg, 2005).

Another important application of cosmogenic nuclides is the discrimination of eroding versus noneroding areas beneath ice sheets and glaciers. It has long been suspected that cold-based glaciers, frozen to their beds, do not erode bedrock, while warm-based glaciers that slide over their beds are very effective at scouring bedrock. The degree of glacial erosion, though, has previously been difficult to measure. Cosmogenic nuclides provide a method for quantifying glacial erosion. This is because areas of bedrock that are eroded by less than a few meters retain cosmogenic nuclides generated prior to glaciation. Subsequent to deglaciation, these areas have an excess cosmogenic nuclide concentration relative to areas that are deeply scoured. This cosmogenic nuclide method has been used to demonstrate that significant areas of nonerosive ice existed in the Fennoscandian ice sheet (Fabel *et al.*, 2002) as well as in northern Canada (e.g., Staiger *et al.* (2005)).

Summary

The study of landscape evolution has historically been divided into two related but distinct approaches: denudation chronology and process geomorphology. Denudation chronology is concerned with establishing a sequence of events responsible for creating today's landscape. These events may be associated with, for example, tectonic uplift, climate change, or drainage capture. Process geomorphology, on the other hand, is less concerned with the sequence of events in the past, but instead focuses on the physical processes of how landscapes are changing today. Cosmogenic nuclides are contributing to both of these approaches. Measurements of erosion rates, soil production functions, and the influence of climate on erosion all increase our understanding of present-day processes that are shaping the landscape. River incision rates and paleoerosion rates, on the other hand, contribute more to understanding landscape history and denudation chronology. A fuller understanding of landscape evolution and the dynamic response of hill slopes to climate, tectonics, and land use change requires both these historical and mechanistic perspectives.

Glossary

Cosmogenic Nuclide A type of atom produced by a nuclear reaction with a secondary cosmic ray.

Nucleon A neutron or a proton; a type of secondary cosmic ray particle.

Spallation A type of nuclear reaction in which nucleons are lost from a nucleus in a collision.

Muon A type of secondary cosmic ray particle.

Radioactive Meanlife The time taken for a radioactive material to decay to $1/e$ of its abundance.

Saprolite Weathered bedrock that retains some original structure.

See also: **Cosmogenic Nuclide Dating**: Overview; Methods; Cosmic Ray Interactions in Minerals; Exposure Geochronology. **Dating Techniques**. **Fluvial Environments**: Terrace Sequences. **Glaciations**: Early Quaternary. **Periglacial Landforms, Rock Forms**: Rock Weathering. **Quaternary Stratigraphy**: Chronostratigraphy.

References

Bierman, P. R. (1994). Using *in situ* produced cosmogenic isotopes to estimate rates of landscape evolution: A review from the geomorphic perspective. *Journal of Geophysical Research* **99**, 13,885–13,896.

Bierman, P. R., and Nichols, K. K. (2004). Rock to sediment-slope to sea with ^{10}Be-rates of landscape change. *Annual Review of Earth and Planetary Science* **32**, 215–255.

Brown, E. T., Bourles, D. L., Colin, F., Raisbeck, G. M., Yiou, F., and Desgarceaux, S. (1995a). Evidence for muon-induced *in situ* production of ^{10}Be in near-surface rocks from the Congo. *Geophysical Research Letters* **22**, 703–706.

Brown, E. T., Stallard, R. F., Larsen, M. C., Raisbeck, G. M., and Yiou, F. (1995b). Denudation rates determined from the accumulation of *in situ*-produced ^{10}Be in the luquillo experimental forest, Puerto Rico. *Earth and Planetary Science Letters* **129**, 193–202.

Cerling, T. E., and Craig, H. (1994). Geomorphology and *in-situ* cosmogenic isotopes. *Annual Review of Earth and Planetary Science* **22**, 273–317.

Fabel, D., Stroeven, A. P., Harbor, J., Kleman, J., Elmore, D., and Fink, D. (2002). Landscape preservation under Fennoscandian ice sheets determined from *in situ* produced ^{10}Be and ^{26}Al. *Earth and Planetary Science Letters* **201**, 397–406.

Gosse, J. C., and Phillips, F. M. (2001). Terrestrial *in situ* cosmogenic nuclides: Theory and application. *Quaternary Science Reviews* **20**, 1475–1560.

Granger, D. E., and Muzikar, P. F. (2001). Dating sediment burial with cosmogenic nuclides: Theory, techniques, and limitations. *Earth and Planetary Science Letters* **188**, 269–281.

Granger, D. E., Riebe, C. S., Kirchner, J. W., and Finkel, R. C. (2001). Modulation of erosion on steep granitic slopes by boulder armoring, as revealed by cosmogenic ^{26}Al and ^{10}Be. *Earth and Planetary Science Letters* **186**, 269–281.

Heimsath, A. M., Chappell, J., Dietrich, W. E., Nishiizumi, K., and Finkel, R. C. (2000). Soil production on a retreating escarpment in southeastern Australia. *Geology* **28**, 787–790.

Heimsath, A. M., Dietrich, W. E., Nishiizumi, K., and Finkel, R. C. (1997). The soil production function and landscape equilibrium. *Nature* **388**, 358–361.

Heimsath, A. M., Dietrich, W. E., Nishiizumi, K., and Finkel, R. C. (2001). Stochastic processes of soil production and transport: Erosion rates, topographic variation and cosmogenic nuclides in the Oregon Coast Range. *Earth Surface Processes and Landforms* **26**, 531–552.

Heisinger, B., Lal, D., Jull, A. J. T., *et al.* (2002). Production of selected cosmogenic radionuclides by muons. 2: Capture of negative muons. *Earth and Planetary Science Letters* **200**, 357–369.

Kirchner, J. W., Finkel, R. C., Riebe, C. S., *et al.* (2001). Mountain erosion over 10 yr, 10 k.y., and 10 m.y. time scales. *Geology* **29**, 591–594.

Lal, D. (1991). Cosmic ray labeling of erosion surfaces: *In situ* nuclide production rates and erosion models. *Earth and Planetary Science Letters* **104**, 424–439.

Lal, D., and Arnold, J. R. (1985). Tracing quartz through the environment. *Proceedings of the Indian Academy of Sciences – Earth and Planetary Sciences* **94**, 1–5.

Lal, D., and Peters, B. (1967). Cosmic-ray produced activity on the earth. *Handbuch der Physik* **46**, 551–612.

Matmon, A., Bierman, P. R., Larsen, J., *et al.* (2003). Temporally and spatially uniform rates of erosion in the southern Appalachian Great Smoky Mountains. *Geology* **31**, 155–158.

Megahan, W. F. (2001). Mountain erosion over 10 yr, 10 k.y., and 10 m.y. time scales. *Geology* **29**, 591–594.

Muzikar, P., Elmore, D., and Granger, D. (2003). Accelerator mass spectrometry in geologic research. *Geological Society of America Bulletin* **15**, 643–654.

Nishiizumi, K., Kohl, C. P., Arnold, J. R., *et al.* (1993). Role of *in situ* cosmogenic nuclides ^{10}Be and ^{26}Al in the study of diverse geomorphic processes. *Earth Surface Processes and Landforms* **18**, 407–425.

Riebe, C. S., Kirchner, J. W., and Finkel, R. C. (2003). Long-term rates of chemical weathering and physical erosion from cosmogenic nuclides and geochemical mass balance. *Geochimica et Cosmochimica Acta* **67**, 4411–4427.

Riebe, C. S., Kirchner, J. W., and Finkel, R. C. (2004a). Erosional and climatic effects on long-term chemical weathering rates in granitic landscapes spanning diverse climate regimes. *Earth and Planetary Science Letters* **224**, 547–562.

Riebe, C. S., Kirchner, J. W., and Finkel, R. C. (2004b). Sharp decrease in long-term chemical weathering rates along an altitudinal transect. *Earth and Planetary Science Letters* **218**, 421–434.

Riebe, C. S., Kirchner, J. W., Granger, D. E., and Finkel, R. C. (2001). Strong tectonic and weak climatic control of long-term chemical weathering rates. *Geology* **29**, 511–514.

Safran, E. B., Bierman, P. R., Aalto, R., Dunne, T., Whipple, K. X., and Caffee, M. W. (2005). Erosion rates driven by channel network incision in the Bolivian Andes. *Earth Surface Processes and Landforms* **30**, 1007–1024.

Staiger, J. K. W., Gosse, J. C., Johnson, J. V., *et al.* (2005). Quaternary relief generation by polythermal glacier ice. *Earth Surface Processes and Landforms* **30**, 1145–1159.

Vance, D., Bickle, M., Ivy-Ochs, S., and Kubik, P. W. (2003). Erosion and exhumation in the Himalaya from cosmogenic isotope inventories of river sediments. *Earth and Planetary Science Letters* **206**, 273–288.

von Blanckenburg, F. (2005). The control mechanisms of erosion and weathering at basin scale from cosmogenic nuclides in river sediment. *Earth and Planetary Science Letters* **237**, 462–479.

von Blanckenburg, F., Hewawasam, T., and Kubik, P. W. (2004). Cosmogenic nuclide evidence for low weathering and denudation in the wet tropical highlands of Sri Lanka. *Journal of Geophysical Research* **109**, F03008, (doi:10.1029/2003JF000049).

Wilkinson, M. T., Chappell, J., Humphreys, G. S., Fifield, L. K., Smith, B., and Hesse, P. (2005). Soil production in heath and forest, Blue Mountains, Australia: Influence of lithology and paleoclimate. *Earth Surface Processes and Landforms* **30**, 923–934.

Cut-Marked Bone *see* **Vertebrate Studies**: Interactions with Hominids

D

Dansgaard-Oeschger Events *see* **Paleoclimate Reconstruction**: Paleodroughts and Society

DATING TECHNIQUES

A J Timothy Jull, University of Arizona, AZ, USA

© 2007 Elsevier B.V. All rights reserved.

Introduction

There are a number of important dating methods used for Quaternary studies. Many of these subjects are discussed in detail in the articles of this Encyclopedia. The purpose of all of these methods is to estimate the age of the material of interest. Sometimes, the need is for the age determination to be very precise; in other cases, relative age or approximate age may be sufficient. Hence, the purpose of this article is to summarize those methods and discuss briefly their advantages and deficiencies. **Figure 1** summarizes the age ranges covered by different dating methods.

The dating methods can be subdivided into three main categories:

1. radioactive nuclide methods (**Table 1**),
2. radiative dosimetry methods (**Table 2**), and
3. qualitative and comparative methods (**Table 2**).

Each of these methods discussed in separate sections is subdivided into the specific methods, and cross-references to the more detailed articles which deal with these methods are also given.

Radioactive Nuclides

The most important collection of methods for Quaternary dating involve radioactive decay, or in some cases the buildup of radioactive species. Since radioactive nuclides will increase or decrease with time, we can use the concentration of the specific nuclide to estimate the age of the material.

Radioactive species can be produced by two different kinds of nuclear processes: radioactive decay of a parent nuclide or production by nuclear reactions as a result of exposure to cosmic or other radiation.

Radioactive Decay

For example, in the simple case of the study of a nuclide present in its maximum concentration at the creation of the material, such as radiocarbon dating, the number of ^{14}C atoms will decay with time, according to the radioactive decay equation, originally shown by Ernest Rutherford (Dickin, 1995):

$$\mathrm{d}n/\mathrm{d}t = -\lambda N \qquad [1]$$

where the decay rate is proportional to the number of radioactive atoms present (N). The 'decay constant' λ defines the rate of decay. It is usual to refer to the 'half-life' of the radionuclide, the time for half of the remaining atoms to decay, which is related directly to the decay constant:

$$t_{1/2} = \frac{\ln 2}{\lambda} \qquad [2]$$

If Eqn [1] is integrated, within the limits $N = N_0$ at $t = 0$ and $N = 0$ at $t = \infty$, the result is a more common form:

$$\frac{N}{N_0} = \mathrm{e}^{-\lambda t} \qquad [3]$$

where N_0 is the number of atoms at time $t = 0$ and N is the number of atoms at any time t.

Ingrowth of a Radionuclide

Ingrowth of a radioactive species occurs when it is produced either from the decay of another nuclide, or

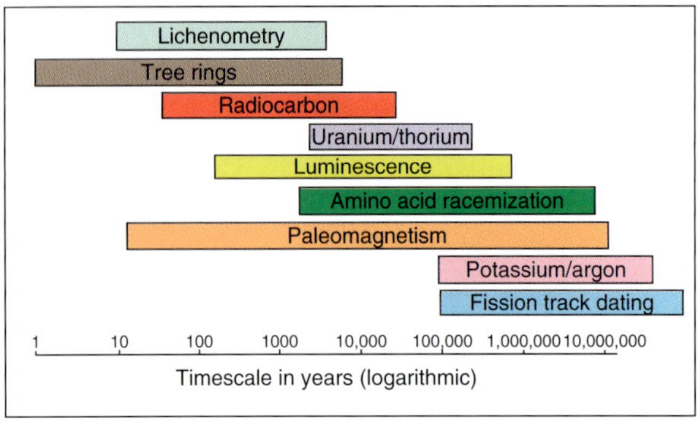

Figure 1 Timescales of different Quaternary geochronology methods. Courtesy of Scott Elias, Royal Holloway College, University of London.

directly in the sample. The latter example occurs often in the case of *in situ* cosmogenic nuclide studies (*see* Exposure Geochronology).

An example of a process where the ingrowth of a radionuclide is the ^{231}Th–^{235}U system. In this case, ^{231}Pa increases with time due to the decay of the parent nuclides ^{235}U, so that (Ku, 1968)

$$N_{231} = N_{235}(1 - e^{-\lambda t}) \qquad [4]$$

A more complex case occurs for ^{230} Th. In this case, ^{230}Th increases with time due to the decay of the parent nuclides ^{234}U, which is produced from ^{238}U decay (*see* U-Series Dating). Because the half-life of ^{234}U is very short compared to either the parent ^{238}U or daughter ^{230}Th, in principle it should drop out of the equation. This equation would be correct if the intermediate daughter product of ^{238}U, ^{234}U, was in secular equilibrium with the ^{238}U (Ku, 2000). Hence, the amount of ^{230}Th produced is a product of the decay of both the excess ^{234}U and the ^{238}U. The result is a more complex equation, described in detail in the article on U–Th dating (*see* U-Series Dating).

Radiocarbon Dating

As discussed in the articles on radiocarbon dating, this method is widely used for reconstructing the age of various kinds of carbon-containing materials. This involves a considerable amount of pretreatment chemistry to prepare the sample (**Fig. 2**). Here, organic or inorganic materials in equilibrium with the production of ^{14}C in the atmosphere and its removal into the oceans establish a consistent level of ^{14}C. When the animal or plant dies, it is removed from this quasi-equilibrium and so the level of ^{14}C decays according to Eqn [1]. There are a vast number of applications of ^{14}C, which can be measured either by counting the decays of the radioactive ^{14}C or by direct atom counting by accelerator mass spectrometry (AMS). Most measurements these days are made by AMS (see **Fig. 3**). More details of ^{14}C dating are given in the articles by Cook and van der Plicht (*see* Conventional Method), who discuss decay counting of ^{14}C, and Jull (*see* AMS Method), who discusses AMS. Other articles by van de Plicht (*see* Variations in Atmospheric ^{14}C) Burr (*see* Causes of Temporal Variations) discuss the causes of ^{14}C variations over time.

U-Series Methods

Studies of uranium-series nuclides (Wagner, 1995) rely on the extensive decay sequence in the uranium decay series. We can take some of these pairs such as ^{234}U–^{238}U, ^{230}Th–^{238}U, ^{231}Pa–^{235}U, and ^{226}Ra–^{230}Th as chronometers for measurements of time on various timescales, depending on the radioactive pair chosen (Ku, 2000). Problems can arise, such as where there is some initial ^{230}Th. Hence, calculations of the age must take into account if there is disequilibrium in the ^{238}U–^{234}U system (Cohen, 2005). U-series nuclides have a wide variety of applications to dating of sediments, soil horizons, peat, bones, corals, and carbonates of all kinds.

Table 1 Radioactive Quaternary dating methods

Isotopic system	*Parent nuclide or process*	*Half-life*	*Timescale*	*Material*	*Reference*
^{14}C	^{14}C(n,p) ^{14}N	5730 yr	0–50 ka	Organic material, carbonates	Taylor, 1987
^{230}Th	^{238}U and ^{234}U	75,200 yr	1000–500,000 yr	Carbonates	Schwarcz, 1989
^{210}Pb	U-series decay sequence	22.3 yr	0–150 yr	Fine-grained sediments	Olsson, 1986
U–He	U isotopes α-decay	Depends on parent nuclides	0–300,000 yr	Tephras	Kohn *et al.* 2000
^{137}Cs	U fission	30	After AD 1950	Fine-grained sediments	Pennington *et al.* 1973
^{40}K–^{40}Ar	^{40}K β-decay	1.25×10^9 yr	10^4–10^9 yr	K-bearing minerals, bones, tephras	McDougall and Harrison, 1988

Table 2 Nonradiometric Quaternary dating methods

Technique	*Process*	*Timescale*	*Material*	*References*
Fission-track dating	α-Particle tracks in minerals	10^5–10^9 yr	Tephras, glass	Westgate and Naeser, 1995
Amino acid racemization	Rate of equilibration of D- and L- isomers of amino acids	10–10^6 yr	Molluscs, ostracods, egg shells, some fine-grained sediments	Wehmiller and Miller, 2000
Thermoluminescence	Luminescence resulting from heating to rest electronic traps caused by radiation damage	100–10^6 yr	Fine-grained quartz or feldspar in sediments	Aiken, 1998
Optically stimulated luminescence	Luminescence resulting from optical stimulation to rest electronic traps caused by radiation damage	100–10^6 yr	Fine-grained quartz or feldspar in sediments	Stokes, 1999
Electron spin resonance	Intensity of ESR signal proportional to number of radiation-damage electron traps	100 yr to 50 kyr	Fine-grained quartz	Lee and Schwarcz, 2000
Varve chronologies	Annual varve deposits in sediment	Any age, up to 45 ka correlated with radiocarbon chronology	Fine-grained sediments	O'Sullivan, 1983, Wohlfarth *et al.* 1993 Hughen *et al.* 1996
Paleomagnetism	Changes in magnetic intensity and direction	10^2–10^8 yr	Fine-grained sediments and volcanics	Thompson and Oldfield, 1986; Verosub 1988
Tephrochronology	Volcanic ash layers	Varies	Tephras	Westgate and Naeser, 1995
Biostratigraphy	Correlation of similar macrofossils	Any age		

Figure 2 Sample preparation chemistry at the University of Arizona AMS Laboratory.

Figure 3 The power supply of the 3 MV accelerator mass spectrometer at the University of Arizona.

Other Dating Methods Based on U-Series Decay

Although sometimes listed as a separate method, in U–He dating, one observes the ingrowth of the He daughter product of U. This method is sometimes used for dating bones, mollusks, and corals (Wagner, 1995). Similarly, the U-daughter nuclide ^{210}Pb can be used for dating of sediments in the last ~200 yr (Wagner, 1995; Cohen, 2005).

K–Ar Dating

The K–Ar method and its companion the ^{40}Ar/^{39}Ar method are based on the radioactive decay of ^{40}K to

^{40}Ar (Renne, K/Ar and Ar/Ar Dating). ^{40}K undergoes a branching decay to both ^{40}Ar and ^{40}Ca, so it is important to revise the simple radioactive decay equation to take into account the decay via different processes. The ^{40}Ar–^{39}Ar process is based on the same idea as the K–Ar method, except that neutron activation is used to estimate the amount of ^{39}K by activation to ^{39}Ar. The ingrowth of ^{40}Ar is a variant of the ingrowth Eqn [4], since there is actually a branching decay to ^{40}Ca and ^{40}Ar from ^{40}K:

$$^{40}\mathrm{Ar} = {}^{40}\mathrm{Ar}_{\mathrm{initial}} + 0.105\,{}^{40}\mathrm{K}(1 - \mathrm{e}^{-\lambda t}) \qquad [5]$$

where t is the age and 0.105 is the branching ratio to ^{40}Ar. K–Ar dating has many applications, particularly during the Quaternary to the dating of bones and tephras.

Cosmogenic Radionuclides

Similar equations to those for ingrowth of a nuclide also apply to the case of the buildup of a cosmogenic radionuclide such as ^{10}Be:

$$N = \frac{P}{\lambda}(1 - \mathrm{e}^{-\lambda t}) \qquad [6]$$

Here, P is the production rate in atoms/yr and λ is the decay constant (in yr^{-1}). More complex scenarios can arise where the sample is buried and re-exposed. The application of cosmogenic nuclide for dating of Quaternary surfaces and deposits has increased dramatically in the past decade (Dickin, 1995; Zreda and Phillips, 2000), as presented elsewhere (*see* Exposure Geochronology).

In the case where the cosmogenically produced nuclide is stable, the number of atoms produced is simply $N = Pt$. There have been a number of studies of stable noble gas nuclides in various rock surfaces.

Postnuclear Tracers

The radionuclides produced as the result of anthropogenic nuclear activities can be used as tracers for recent processes. Nuclides such as ^{137}Cs, ^{90}Sr, and postbomb levels of ^{14}C can be used to date event horizons since AD 1950. These markers are often of great use in the study of recently deposited sediments and other materials. Recent sediment studies are often done in conjunction with ingrowth of ^{210}Pb from U-series decay (Cohen, 2005).

Radiative Dosimetry Methods

Radiative dosimetry methods are basically studies by which we investigate the buildup of radiation damage. This is usually measured by studying some luminescence phenomenon. Common methods are thermoluminescence, optically stimulated luminescence, and electron spin resonance. The techniques rely on the production of radiation damage in the substrate (usually a silicate mineral). When these solids are exposed to ionizing radiation, electron/hole pairs are formed and some of these become trapped in the solid.

The luminescence age appears simple enough, and is constrained by

$$\mathrm{Age} = \frac{\mathrm{ED}}{\Sigma\mathrm{D}}$$

where ED is the 'equivalent dose' (in units of grays, 1 Gy = 100 rad) and ΣD is the sum of annual doses from α, β, γ, and cosmic radiation (Forman, 2000). ED is determined in several different ways, including estimates of the chemical composition of radioactive sources in the sample, regeneration of an additional dose of the luminescence signal in the laboratory, and partial or total bleaching (i.e., removal of the luminescence signal by exposure to light).

The main problem with luminescence methods is the number of possible complications which can arise due to the fact that the original sediment was not totally bleached by sunlight before the accumulation of the dose under study, or that later partial bleaching occurs or there is an unexplained loss of the signal, called 'anomalous fading' (Wintle, 1973).

There are two main methods which use radiative dose: fission-track dating and luminescence methods, which include methods based on thermoluminescence and optically stimulated luminescence (Aiken, 1998). Another radiative dose method which can sometimes be used is the study of the buildup of electronic 'traps' by ESR (electron spin resonance) dating (Lee and Schwarcz, 1993). All of these methods have limitations due to the necessity of accurately estimating the levels of the source of the radiation, usually α-particles (from the decay of U) and sometimes γ-radiation (Wagner, 1995). These topics are all discussed in various articles of this Encyclopedia.

Amino Acid Racemization

The technique of amino acid racemization relies on the kinetics of changes in the symmetry of amino acid molecules (Miller and Brigham-Grette, 1989). Since amino acids can have two different optically active isomers, there is one biologically preferred form (usually the L-form). After initial deposition, these amino acids react over time to approach an equal mixture of D- and L- isomers, which is called a racemic mixture (**Fig. 4**). The decay rates of many of

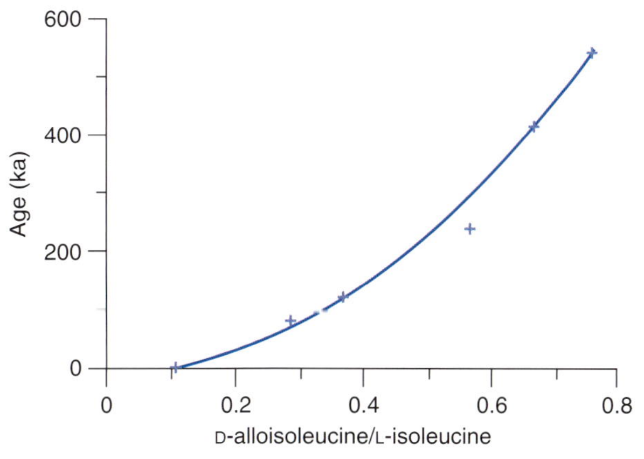

Figure 4 Exchange rates of amino acids at different temperatures. After Hearty PS and Kaufman DS (2000) Whole-rock aminostratigraphy and Quaternary sea-level history of the Bahamas. *Quaternary Research* 54: 163–173.

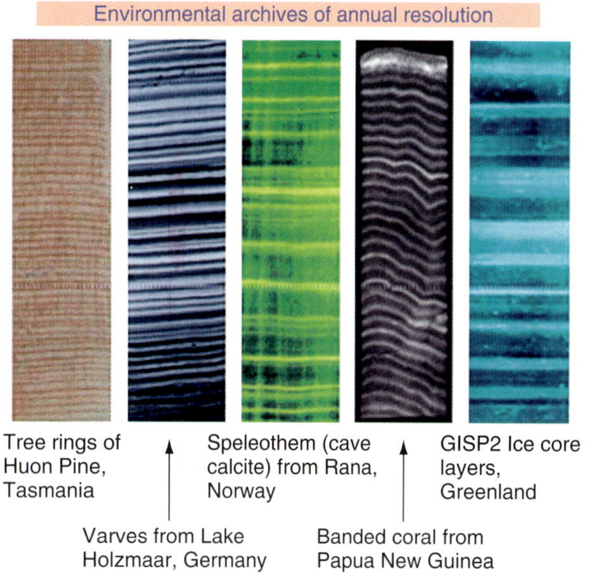

Figure 5 Examples of different Quaternary archives of annual resolution. Courtesy of Scott Elias, the Royal Holloway College, University of London, http://www.gg.rhul.ac.uk

these reactions are known, although they are sensitive to changes in temperature, as are nearly all chemical reactions. Hence, in order to calculate an age from an amino acid composition, it is important to understand the reaction rates and the dependence of the reaction rates on temperature and other parameters (Wehmiller and Miller, 2000). The amino acid composition is usually expressed as the D/L ratio of the amino acid of interest. If the goal is absolute dating, it is usually necessary to compare the rates of change with some independent dating method such as ^{14}C. The method can also be used in conjunction with independent age estimates to measure temperature (Miller *et al.*, 1997). Otherwise, serious errors can result, such as the very early age assignments of early humans in the New World, which were later shown to be in error by radiocarbon dating (Stafford *et al.*, 1984).

Qualitative and Comparative Methods

There are several methods used for dating of geological materials which rely on what are best described as relative or incremental techniques (**Fig. 5**). These methods rely on various observations to estimate the passage of time.

Lake varve chronology. Deposition of regular additions of sediment on a cyclical fashion is called rhythmites. These are often called 'varves' if annually deposited. Lake varve chronologies were originally proposed as a dating technique by the Swedish geologist De Geer. These chronologies have also been proposed as ways to cross-correlate to other chronologies (e.g., Wohlfahrt, 1994). In general, there is a degree of skepticism about some of these chronologies since it is difficult to establish *a priori* that the deposition is annual.

Marine varve chronology. Long varve chronologies have been established for marine sediments (e.g., Hughen, Varved Marine Sediments) and these chronologies have been incorporated into some radiocarbon calibration schemes (*see* History of Recent Major Projects). These varved sediments in principle have the same problems as lake sediments, but in the case of the Cariaco Basin chronology (Hughen *et al.*, 2004) there is good evidence that the annual deposition assumption can be taken as valid over certain parts of the timescale.

Lichenometry. The incremental growth of certain lichen species has been used to estimate the age of rock surfaces. The technique was pioneered by Bull (2000), who used measurements of the diameter of the largest lichens on a surfaces as a way of estimating the age of the surface. This approach appeared to give approximate ages for some surfaces which could be confirmed by other methods.

Dendrochronology. Trees produce annual growth rings and hence, the variation of the thickness of tree rings can be cross-correlated with established chronologies (**Fig. 6**). In cases where a piece of wood can be dendrochronologically dated, results can be precise to a few years.

Paleomagnetic dating. The remnant magnetization in magnetically susceptible materials in sediments and ceramics can sometimes be correlated with known geomagnetic fluctuations (Butler, 1992). Both the intensity of the geomagnetic field and its direction change with time (Cohen, 2005). The sequence of geomagnetic signals in a sediment sequence can then be used to estimate the age of sequence.

Figure 6 A bristlecone pine from the White Mountains of California. Courtesy of the University of Arizona Tree Ring Laboratory.

Correlation of isotopic signals. This is also an important method for Quaternary dating. The long record of $\delta^{18}O$ in ice cores and marine sediments is often used to cross-correlate between different marine and ice core records, and therefore establish definite time markers. An excellent example is the ability to cross-correlate dramatic changes in climate from glacial to interglacial, or a return to glacial-like conditions, such as the Younger Dryas event at about 13 ska (Markgraf, 2001), through the oxygen-isotope record (Mayewski *et al.*, 1997).

Correlation of changes in flora and pollen. The correlation of changes in flora and pollen can sometimes also be correlated between different sites in the same region, or over wider areas (Bradbury, 2001).

See also: **Introduction**: History of Recent Major Projects. **Cosmogenic Nuclide Dating**: Exposure Geochronology. **Radiocarbon Dating**: Conventional Method; AMS Method; Variations in Atmospheric ^{14}C; Causes of Temporal Variations. **U-Series Dating**.

References

Aiken, M. J. (1998). *Thermoluminescence Dating.* Academic Press, New York.

Bradbury, J. P. (2001). Full and late glacial lake records along the PEP1 transect: Their role in developing interhemispheric paleoclimate interactions. In *Interhemispheric Climate Linkages* (V. Markgraf, Ed.). Academic Press, San Diego.

Bull, W. B. (2000). Lichenometry: A new way of dating and locating prehistorical earthquakes. In *Quaternary Geochronology: Methods and Applications* (J. S. Noller *et al.*, Eds.), pp. 157–176. American Geophysical Union, Washington, DC.

Butler, R. (1992). *Paleomagnetism: Magnetic Domains to Geologic Terrains.* Blackwell, Boston.

Cohen, A. S. (2005). *Paleolimnology: The History and Evolution of Lake Systems.* Oxford University Press, Oxford.

Dickin, A. P. (1995). *Radiogenic Isotope Geology.* Cambridge University Press, Cambridge.

Forman, S. (2000). Thermoluminescence dating. In *Quaternary Geochronology: Methods and Applications* (J. S. Noller *et al.*, Eds.), pp. 157–176. American Geophysical Union, Washington, DC.

Hearty, P. S., and Kaufman, D. S. (2000). Whole-rock aminostratigraphy and Quaternary sea-level history of the Bahamas. *Quaternary Research* **54**, 163–173.

Hughen, K. A., Overpeck, J. T., Peterson, L. C., and Anderson, R. F. (1996). The nature of varved sedimentation in the Cariaco basin, Venezuela, and its palaeoclimatic significance. In *Palaeoclimatology and Palaeoceanography from Laminated Sediments* (A. E. S. Kemp, Ed.), (special publication no. 116), pp. 171–183. Geological Society, London.

Kohn, B. P., Farley, K. A., and PIllans, B. (2000). (U–Th)/He and fission track dating of the Pleistocene Rangitawa tephra, North Island, New Zealand: A comparative study. In *9th International Conference on Fission Track Dating and Thermochronology,* pp. 207–208, Melbourne, Australia: Geological Society of Australia, abstract series 58.

Ku, T. L. (2000). In *Quaternary Geochronology: Methods and Applications* (J. S. Noller *et al.*, Eds.), American Geophysical Union, Washington, DC.

Lee, H., and Schwarcz, H. (2000). Electron spin resonance dating of fault rocks. In *Quaternary Geochronology: Methods and Applications* (J. S. Noller *et al.*, Eds.), pp. 177–186. American Geophysical Union, Washington, DC.

Markgraf, V. (2001). *Interhemispheric Climate Linkages.* Academic Press, San Diego.

Mayewski, P. A., Meeker, L. D., Twickler, M. S., *et al.* (1997). Major features and forcing of high-latitude northern hemisphere atmospheric circulation using a 110,000-year-long glaciochemical series. *Journal of Geophysical Research-Oceans* **102**, 26,345–26,366.

McDougall, I. (1995). Potassium–argon dating in the Pleistocene. In *Dating Methods for Quaternary Deposits* (N. W. Rutter and N. R. Catto, Eds.), pp. 1–14. Geological Survey of Canada, Ottawa.

McDougall, I., and Harrison, T. M. (1988). *Geochronology and Thermochronology by the $^{40}Ar/^{39}Ar$ Method.* Oxford University Press, New York.

McSween, H. Y., Richardson, S. M., and Uhle, M. E. (2003). *Geochemistry: Pathways and Processes.* Columbia University Press, New York.

Miller, G. H., and Brigham-Grette, J. (1989). Amino acid geochronology – resolution and precision in carbonate fossils. *Quaternary International* **1**, 111–128.

Miller, G. H., Magee, J. W., and Jull, A. J. T. (1997). Subtropical southern hemisphere Pleistocene cooling deduced from racemization in emu eggshells from Australia. *Nature* **385**, 241–244.

Olsson, I. U. (1986). Radiometric dating. In *Handbook of Holocene Palaeoecology and Palaeohydrology* (B. E. Berglun, Ed.), pp. 273–312. Wiley, London.

O'Sullivan, P. E. (1983). Annually-laminated sediments and the study of Quaternary environmental changes – A review. *Quaternary Science Reviews* **1**, 245–313.

Pennington, W., Cambray, R. S., and Fisher, E. M. (1973). Observations on lake sediments using ^{137}Cs. *Nature* **242**, 324–326.

Schwarcz, H. (1989). Uranium series dating. *Quaternary International* **1**, 7–17.

Schwarcz H, Buhay W, and Grün R (1988) Electron spin resonance (ESR) dating of fault gouge. *US Geological Survey Open File Report 87-673*, pp. 50–64, Reston, UA.

Stafford, T. W., Jr., Jull, A. J. T., Zabel, T. H., *et al.* (1984). Holocene age of the Yuha burial: Direct radiocarbon determinations by accelerator mass spectrometry. *Nature* **308**, 446–447.

Stokes, S. (1999). Luminescence dating applications in geomorphological research. *Geomorphology* **29**, 153–171.

Taylor, R. E. (1987). *Radiocarbon Dating: An Archaeological Perspective*. Academic Press, New York.

Tuniz, C., Bird, J. R., Fink, D., and Herzog, G. F. (1998). *Accelerator Mass Spectrometry: Ultasensitive Analysis for Global Science*. CRC Press, Boca Raton, FL.

Wagner, G. A. (1995). *Age Determination of Young Rocks and Artifacts*. Springer, Berlin.

Wehmiller, J. F., and Miller, G. H. (2000). Aminostratigraphic dating methods in Quaternary geology. In *Quaternary Geochronology, Methods and Applications* (J. S. Noller *et al.*, Eds.), pp. 187–222. American Geophysical Union, Washington, DC.

Westgate, J. A., and Naeser, N. D. (1995). Tephrochronology and fission-track dating. In *Dating Methods for Quaternary Deposits* (N. W. Rutter and N. R. Catto, Eds.), pp. 15–28. Geological Survey of Canada, Ottawa.

Wintle, A. G. (1973). Anomalous fading of thermoluminescence in mineral samples. *Nature* **245**, 143–144.

Wohlfarth, B., Björck, S., Possnert, G., *et al.* (1993). AMS dating of Swedish varved clays of the last glacial/interglacial transition and potential difficulties of calibrating Late Weichselian "absolute" chronologies. *Boreas* **22**, 113–128.

Zreda, M., and Phillips, F. (2000). Cosmogenic nuclide buildup in surficial materials. In *Quaternary Geochronology, Methods and Applications* (J. S. Noller *et al.*, Eds.), pp. 61–76. American Geophysical Union, Washington, DC.

Relevant Website

http://www.gg.rhul.ac.uk – Royal Holloway: Department of Geography.

Dendroarchaeology *see* **Plant Macrofossil Methods and Studies**: Dendroarchaeology

DENDROCHRONOLOGY

D Smith and D Lewis, University of Victoria, Victoria, Canada

© 2007 Elsevier B.V. All rights reserved.

Dendrochronology

Dendrochronology is the science that deals with the dating and study of the annual growth layers, or tree rings, in woody trees and shrubs. In temperate climates, these layers of wood (tree rings) contain seasonal cell structures (earlywood and latewood) that signify one annual growth ring. When all the trees at a site are affected by a common environmental factor, such as climate, crossdating provides an accurate chronological record that can be used to date events or describe variations in environmental conditions. Due to the annual resolution possible throughout an entire tree-ring record, dendrochronological analyses provide both reliable and ubiquitous archives for paleoenvironmental reconstruction.

Origins and History

Although observations of tree-ring structures and related phenomena have a lengthy history, dendrochronology as a scientific discipline emerged relatively recently. As early as the fifteenth century, Leonardo da Vinci noted the annual nature of tree-rings, recognizing a relationship between tree ring widths and precipitation (Stallings, 1937). The first recorded application of dendrochronology occurred in 1737 when Du Hamel and De Buffon noted a prominent frost damaged tree ring in recently felled trees that was later recorded in both Sweden and Germany (Studhalter, 1955, 1956). By the mid-1800s, the ecological foundations of dendrochronology were established following Theodore Hartig's studies of wood structures and annual tree-ring development (Schweingruber, 1988). By the end of the nineteenth century, Hartig's son, Robert, had published numerous papers on the anatomy and ecology of tree-rings, and had used tree-rings to date hail, frost, and insect damage in trees (Schweingruber, 1988).

Astronomer Andrew Ellicott Douglass pioneered the science of dendrochronology in North America. In 1904, he began examining the annual rings of Ponderosa pine trees in the southwestern United States as a potential source for preinstrumental climate records. He soon established that narrow tree rings were formed in very dry years over large geographical areas, thereby confirming that the crossdating of tree-ring patterns between sites could be used as a chronological tool to identify the exact calendar year a ring was produced (Douglas, 1909). By 1914, Douglass had developed a 500-year crossdated pine chronology and established a positive correlation between ring width and precipitation of the preceding winter (Douglass, 1914). Douglass subsequently devoted the next 15 years to developing long tree-ring chronologies to date wood samples from prehistoric sites of the American southwest. His successful crossdating of a 585-year floating (undated) archeological tree-ring sequence provided precise dates of construction for numerous previously undated pueblo sites, bringing the science of dendrochronology to age as a dating tool (Dean, 1997).

Prior to pioneering the subfields of dendroclimatology and dendrohydrology, Edmund Schulman began working with Douglass to develop better methods of tree-ring analysis (Schulman, 1945). In his quest for improving the length and sensitivity of tree-ring records, Schulman also discovered the oldest living trees in the world, the over 4,000-year-old bristlecone pines of the California White Mountains (Schulman, 1954). This discovery was key to calibrating atmospheric ^{14}C records and set the foundation for precision radiocarbon dating methods (LaMarche and Harlan, 1973).

The basic premise of radiocarbon dating relies on the fixation of an unstable atmospheric carbon isotope (carbon-14) by organisms (e.g., trees) in the biosphere. As long as plants and animals are living and continue to take up air in their biophysical processes, their tissues absorb carbon-14. Once the organism dies, its intake of carbon atoms ceases and the amount of carbon-14 within their tissue begins to decrease logarithmically at a fixed rate, which can then be used to determine the time since death (Libby *et al.*, 1949). It was soon realized, however, that raw radiocarbon measurements were not equivalent to calendar dates, as the level of atmospheric carbon-14 has not been constant with time. Fortunately, long-lived trees, like the bristlecone pines studied by Sherman (1954), retain an annual measure of the amount of carbon-14 present in the atmosphere. By crossdating the tree-ring records from both living and dead trees, and measuring the amount of carbon-14 present in each tree ring, researchers have developed annually resolved time-series of radiocarbon concentrations, or radiocarbon calibration curves. The raw radiocarbon date of any sample within this timeframe can then be converted to a calendar date using the calibration curves (Damon *et al.*, 1974). These tree-ring based calibrations have successfully increased the accuracy of radiocarbon dating and presently provide annually-resolved carbon-14 records for the last 12,000 years (Friedrich *et al.*, 2005).

After the Second World War, German botanist Bruno Huber established an ecological approach to dendrochronology in Europe. His discovery that the year-to-year variation in tree-rings was not as pronounced in temperate tree species, as it was in those of semiarid zones, led to the development of a more quantitative approach to dendrochronology. Huber developed the statistical test of similarity, the Gleichlaeufigkeit (Huber, 1943; Eckstein and Pilcher, 1990) that remains an integral component of statistical crossdating in European laboratories.

In the early 1960s, Hal Fritts established the basic physiology of radial tree growth by employing a dendrograph (Haasis, 1933) to measure the daily growth of tree-rings. As detailed in his seminal publication *Tree Rings and Climate* (1976), Fritts was able to describe many of the complex connections between tree-rings and climate, and used these insights to provide annually-resolved climate reconstructions for eastern and western North America (Fritts, 1991). During the same time period, Polge (1966) applied X-ray densitometry to the study of tree rings and illustrated how latewood density provided higher resolution dendroclimatic insights than was possible from an analysis of the total annual ring-width increment (Lenz *et al.*, 1976).

In the last 50 years, dendrochronology has evolved from a discipline primarily applied to the dating of monuments and archaeological pueblos, to one with a variety of applied applications including studies of: the history and effects of pollution (Schweingruber, 1988); geobotanical dating (Winchester and Harrison, 2000); earthquakes (Jacoby *et al.*, 1997); snow avalanches (Smith *et al.*, 1994); glacier movement (Luckman and Villalba, 2001); volcanoes (Jones *et al.*, 1995); river flooding (St George and Nielson, 2003); and insect life cycles (Swetnam and Lynch, 1993). In addition to the continued application of the techniques first pioneered by Douglass, the introduction of new analytical techniques (i.e., X-radiography, X-ray densitometry, neutron activation analysis) are providing insights into climate-induced morphological and chemical changes within tree-rings and wood cells, and are being used to described trace-element and isotopes signatures (Kuniholm, 2001).

Biological Foundation of Dendrochronology

Tree-ring formation

Tree-ring dating is possible due to the physiological behavior of trees growing in locations with seasonal climates. In temperate climates, this annual growth appears as 'rings' that vary in width in response to fluctuating environmental variables (Fritts, 1976). Although there are differences in the anatomical structure of tree rings between angiosperms (deciduous) and gymnosperms (coniferous), both tree types are commonly used in dendrochronology.

Radial growth in trees is achieved by the division of vascular cambium cells that, in seasonal climates, are active for only part of the year. During winter, the cambium is dormant and cell production stops until the warmer and/or wetter days of spring reactivate cambium activity (Fritts, 1976). The xylem cells formed at the beginning of the season tend to be large and thinned-walled, and are commonly referred to as springwood or earlywood. As the growing season progresses and conditions become less favorable (e.g., hotter and drier), cambial activity slows and the xylem being produced has smaller cells with thicker walls, and is referred to as summer or latewood (Fritts, 1976). It is this marked difference between earlywood and latewood cells that make it possible to discern the annual growth rings (**Figs. 1** and **2**).

Although a single tree ring is normally formed every growing season, in some years a tree may fail to develop an annual ring (i.e., a missing ring) or the ring may be locally absent when viewed in cross-section. False rings (intra-annual growth bands) are sometimes produced if seasonal growing conditions temporarily deteriorate (Schulman, 1938; Fritts, 1976). Missing, locally absent, and false rings can usually be detected by cross-dating trees from the same plot or region. However, false rings are usually easily detectible, as the transition from one cell type to another is gradual, while the resumption of earlywood production after winter dormancy produces a distinct cell boundary.

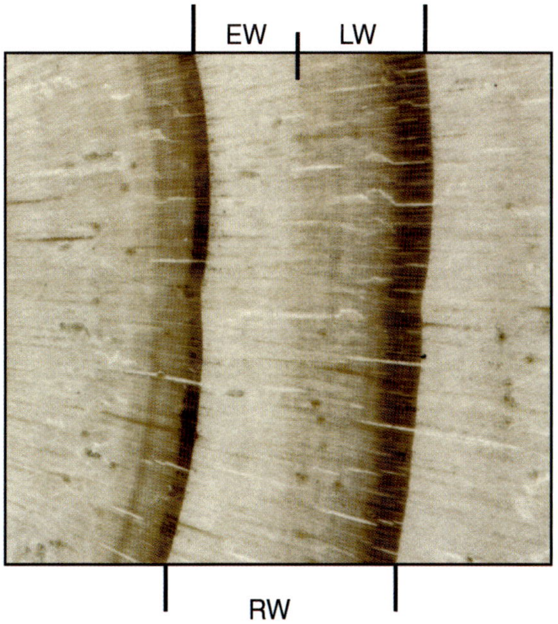

Figure 1 Cross-sectional view of tree rings scanned from a mountain hemlock (*Tsuga mertensiana*) tree. Highlighted are the annual ring-width (RW), latewood (LW), and earlywood (EW). © D. Lewis 2006.

Figure 2 A cross-section from a young sub-Alpine fir (*Abies lasiocarpa*) illustrating the annual concentric ring pattern around the stem of tree. © D. Lewis 2006.

Principles of Dendrochronology

Although different tree species and individual treesexhibit distinct growth characteristics, dendrochronological analysis of common patterns within their annual growth rings allows for comparative study. Seven widely accepted principles govern the application of dendrochronology. While some of these principles are specific to dendrochronology, others have been borrowed and modified from other disciplines.

The Uniformitarian Principle

The principle of uniformitarianism was originally proposed by James Hutton in 1785. Applied to dendrochronology, it states that the physical and biological processes that link contemporary environmental processes to current variations in radial tree growth existed in the past. This does not mean that past conditions were the same as present ones, but rather that the tree growth was influenced in the past in the same manner as in the present (Fritts, 1976). Hence, by understanding present climate tree-ring growth relationships, dendrochronologists are able to reconstruct paleoenvironmental records from tree-ring records. For example, **Figure 3** shows a reconstructed glacier mass balance record that was developed by calibrating historical tree-ring width data with glacier mass balance records. Based on the principle that growth-response conditions must have been similar in the past, a lengthy tree-ring width record was statistically developed as a proxy (or substitute) for actual mass balance anomalies prior to the historical record.

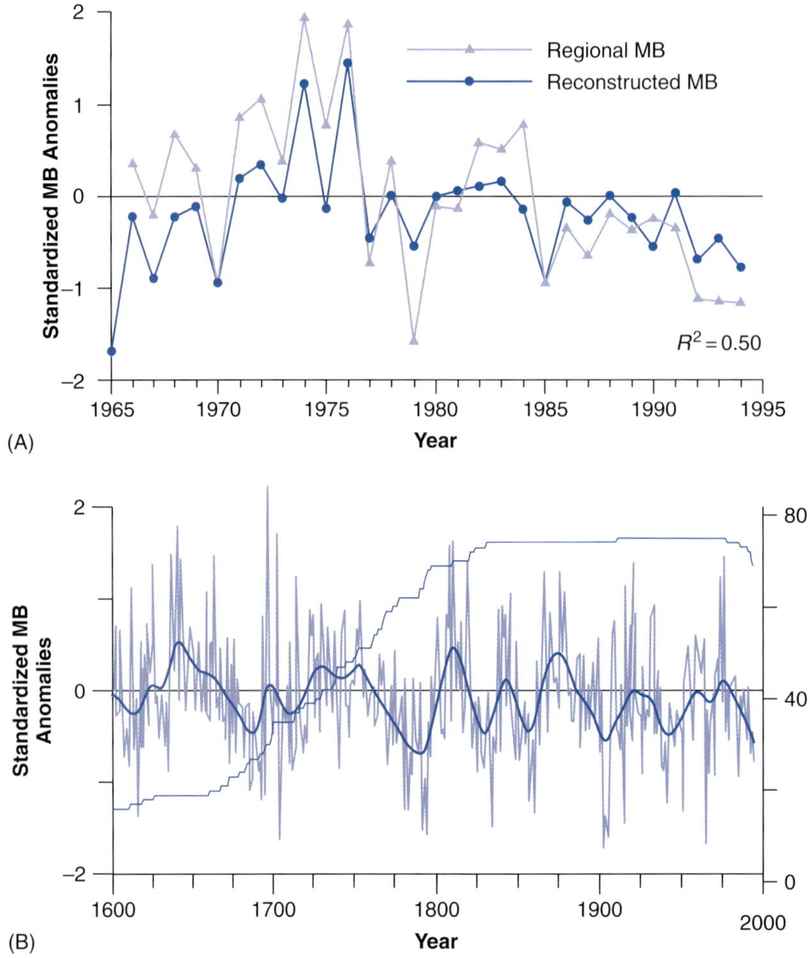

Figure 3 Reconstructed standardized mass balance anomalies for the period 1966 to 1994. The upper graph is the validation against the instrumental mass balance record. The lower graph (B) is the mass balance anomaly reconstruction back to 1600 AD. The thin gray line is the annual values and the thick black line is a 25-year spline. © D. Lewis 2006.

The Principle of Limiting Factors

This principle states that tree-ring growth cannot proceed faster than permitted by the most limiting factor (Fritts, 1976). For instance, in arid and semi-arid regions, a lack of precipitation often limits tree growth and the annual radial increment is a direct function of the amount of precipitation. Other limiting factors include air temperature and non-climatic parameters such as soil condition (e.g. well drained soil). This principle is important to dendrochronology because tree rings can be crossdated only if one or more of the environmental factors become significantly limiting, persist sufficiently long enough, and act over a wide enough geographical area to cause widespread ring width or density variation.

The Principle of Aggregate Tree Growth

This principle states that the individual tree-ring growth series of a tree can be broken down into an aggregate of environmental factors, both human and natural, that affect the patterns of tree-ring growth over time. A simplified linear model of a tree ring from Cook (1987) is given below:

$$R_t = A_t + C_t + D1_t + D2_t + E_t$$

In this example, the observed tree-ring width (R) in any one year (subscript $_t$) is a factor of: (A), the age-related growth trend due to normal physiological aging processes; (C) the climate related signal during that year; (D1) the pulse of a localized disturbance factor within the forest stand (e.g., blowdown, endogenous disturbance); (D2) the occurrence of a stand-wise disturbance factor (exogenous disturbance, e.g., fire, insect outbreak), and; (E) the unexplained year-to-year variability not accounted for by these other processes (Cook, 1987). Therefore, to maximize the desired component (signal), the other factors (noise) should be minimized. For example, to maximize the climate signal, age-related growth trends must be removed, and trees and sites selected to minimize the possibility of internal and external ecological processes affecting tree growth.

The Principle of Ecological Amplitude

This principle states that tree growth is particularly sensitive to environmental factors at the latitudinal, longitudinal, and elevation limits of the distributional range of the species (Fritts, 1976). For example, the growth of trees found growing at their altitudinal limit in mountainous settings is frequently discovered to be restricted by summer air temperatures, a relationship that has been used in some instances to develop millennial-long temperature reconstructions (Luckman and Wilson, 2005).

The Principle of Site Selection

Dendrochronologists recognize that the most useful locations for environmental reconstruction can often be identified based upon site characteristics. For example, trees found growing on bare rocky outcrops are often moisture-sensitive and their tree-ring records contain an historical record of drought conditions. Conversely, trees found growing in mesic (wet) sites usually contain complacent tree-ring records, as their annual ring growth is unlikely to be limited by a lack of moisture (precipitation) (**Fig. 4**).

The Principle of Crossdating

Probably the most important principle in dendrochronology, crossdating ensures that individual tree rings are assigned their exact year of formation. This is accomplished by identifying recognizable patterns of wide and narrow rings within a living tree-ring chronology. These tree-ring patterns develop in response to the sensitivity of trees to limiting factors or site condition changes. Crossdating of ring-width patterns in living trees allows for the identification of comparable patterns in undated wood samples; thereby providing a means to develop tree-ring chronologies that extended back in time beyond the age of living trees or to date an event. For instance, the date when a tree was killed by an advancing glacier can be determined by matching the tree-ring patterns of a glacially-sheared stump with those found within a chronology spanning the same period of time (**Fig. 5**).

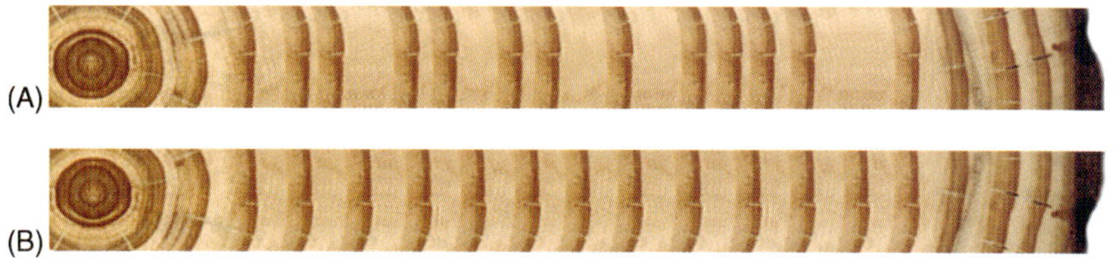

Figure 4 Illustration of effects of the Principles of Ecological Amplitude and Site Selection. Shown is the radial growth response of trees to limiting environmental factors under different site conditions. The top ring-width pattern (A) is from a tree found growing near its ecological limit and as a result of being is temperature and moisture stressed, it produces variable ring-widths sensitive. The sample below (B), is from a tree near the center of ecological range and not limited by environmental factors. As a result, the tree is not stressed and is producing complacent tree rings with little variability. © Arctic, Antarctic & Alpine Research 2004, V36, p. 602.

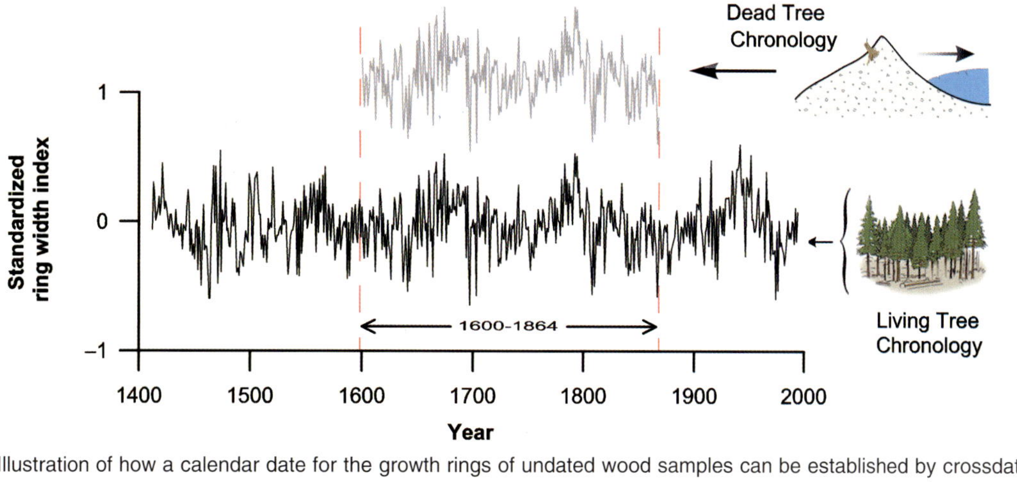

Figure 5 Illustration of how a calendar date for the growth rings of undated wood samples can be established by crossdating with a dated living chronology. By anchoring floating chronologies to the living chronology, it is possible to determine when a sample was killed.

The Principle of Replication

This principle states that an environmental signal of interest can be maximized, and the amount of 'noise' minimized by sampling more than one stem radius per tree, and more than one tree per site (Fritts, 1976). Examining and crossdating multiple samples from multiple trees, avoids the possibility that all collected samples may be missing a ring or could include a false ring. Obtaining more than one sample per tree reduces the amount of 'intra-tree variability', the amount of non-desirable environmental signal peculiar to a given tree. Obtaining numerous trees from one site, and perhaps several sites in a region, ensures that the amount of 'noise' (environmental factors not being studied, such as air pollution) is minimized (Cook, 1987).

Application of Dendrochronology

Tree-ring dating typically begins with the measurement of tree-ring widths or a density parameter, followed by the development of a master tree-ring chronology. The primary objective in developing tree-ring chronologies is to develop stationary time-series that are directly comparable and appropriate for statistical analysis (Fritts, 1976). Chronology development follows a three-step procedure:

(1) *Crossdating*: is used to check for measurement errors and false or missing rings. Although skeleton plots of tree-ring series are useful for visually identifying dating or measurement problems (Stokes and Smiley, 1963), the application of quality checking software programs provides a statistical measure of quality control (Holmes, 1983).

(2) *Standardization*: Although ring-width increments are assumed to change in response to climate variability, they also vary in size with tree age, stem height above the ground surface, in response to site-specific characteristics, and productivity (Fritts, 1976). For the purpose of climate reconstruction, it is essential to determine if the changes in ring width are associated with aging or disturbance factors, and to then remove these influences from the chronology. This correction is known as standardization, and the transformed values are referred to as ring-width indices (Fritts, 1976). Each set of tree-ring measurements requires a different standardization, depending on the statistical and morphological properties of the ring-width series (Cook and Kairiukstis, 1990).

(3) *Master chronology development*: The transformed time-series are combined into a single dimensionless ring-width index, or site chronology, that reflects the extrinsic (environmental) constraints on growth (Cook and Holmes, 1986). Crossdating provides the means to assign an absolute date to each ring in a tree-ring series of unknown age by matching the undated ring-width patterns to those of a known dated series (**Fig. 5**). If the entire ring sequence is present in the outermost perimeter of the undated sample (early-wood and latewood tissue), it is assumed that the sample was killed sometime between the end of one year's growth period and the beginning of another. Precise dating is possible when a sample has bark, as this infers that the last year of growth is present (Baillie, 1982).

Summary

Dendrochronology has proved to be an invaluable tool for dating events and for providing robust annually-resolved paleoenvironmental insights. The ongoing development of long tree-ring chronologies offers opportunities to develop paleoenvironmental reconstructions at local to hemispheric scales. The recent focus on developing tools and techniques suitable for discerning the climatic signal inherent in sub-annual cellular structures has the potential to provide reliable dating applications and studies of environmental variables that act on seasonal timescales.

See also: **Introduction**: Societal Relevance of Quaternary Research; History of Dating Methods. **Glacial Landforms, Tree Rings**: Dendrogeomorphology. **Plant Macrofossil Methods and Studies**: Dendroarchaeology.

References

Baillie, M. G. L. (1982). *Tree-Ring Dating and Archaeology*. Croom Helm, London.

Cook, E. R., and Holmes, R. L. (1986). Users Manual for Program ARSTAN. In *Tree-Ring Chronologies of Western North America: California, Eastern Oregon and Northern Great Basin* (R. L. Holmes, R. K. Adams and H. C. Fritts, Eds.), pp. 50–65. Laboratory of Tree-Ring Research, University of Arizona, Tucson AZ.

Cook, E., and Kairiukstis, L. (1990). *Methods of Dendrochronology*, 1st edn. Kluwer Academic.

Cook, E. R. (1987). The decomposition of tree-ring series for environmental series. *Tree-Ring Bulletin* **47**, 37–59.

Damon, P. E., Ferguson, C. W., Long, A., and Wallick, E. I. (1974). Dendrochronologic calibration of the radiocarbon time scale. *American Antiquity* **39**, 350–366.

Dean, J. S. (1997). Dendrochronology. In *Chronometric Dating in Archaeology* (R. E. Aitken and M. J. Taylor, Eds.), Vol. 2, pp. 31–64.

Douglass, A. E. (1909). Weather cycles in the growth of big trees. *Monthly Weather Review* **37**(5), 225–237.

Douglass, A. E. (1914). A method for estimating rainfall by the growth of trees. In *The Climatic Factor as Illustrated in Arid America, Publication 192* (E. Huntington, Ed.), pp. 101–121. Carnegie Institute of Washington, Washington, DC.

Douglass, A. E. (1919). Climatic Cycles and Tree Growth: Vol. I: A Study of the Annual Rings of Trees in Relation to Climate and

Solar Activity. *Carnegie Institute of Washington Publication* **289**, 1–127.
Eckstein, D., and Pilcher, J. R. (1990). Dendrochronology in western Europe. In *Methods of Dendrochronology: Applications in the Environmental Sciences* (E. R. Cook and L. A. Kairiukstis, Eds.), pp. 11–13. Kluwer Academic Publishers, Boston, MA.
Friedrich, M., Remmele, Kromer, B., Hofmann, J., Spurk, M., Kaiser, K., Orcel, C., and Kuppers, M. (2005). The 12,460-Year Hohenheim Oak and Pine tree-ring chronology from Central Europe-a unique annual record for radiocarbon calibration and paleoenvironment reconstructions. *Radiocarbon* **46**, 1111–1122.
Fritts, H. C. (1976). *Tree Rings and Climate*. Academic Press, New York.
Fritts, H. C. (1991). *Reconstructing large-scale Climate Patterns from Tree-Ring Data: A Diagnostic Analysis*. University of Arizona Press, Arizona.
Haasis, F. W. (1933). Shrinkage and expansion in woody cylinders of living trees. *American Journal of Botany* **20**(2), 85–91.
Holmes, R. L. (1983). Computer-assisted quality control in tree-ring dating and measurement. *Tree-Ring Bulletin* **43**, 69–75.
Huber, B. (1943). Über die Sicherheit jahrringchronologische Datierung. *Holz als Roh und Werkstoff* **6**(10/12), 263–268.
Jacoby, G. C., Bunker, D. E., and Benson, B. E. (1997). Tree-ring evidence for an A.D. 1700 Cascadia earthquake in Washington and northern Oregon. *Geology* **25**(11), 999–1002.
Jones, P. D., Briffa, K. R., and Schweingruber, F. H. (1995). Tree-ring evidence of the widespread effects of explosive volcanic eruptions. *Geophysical Research Letters* **22**(11), 1333–1336.
Kuniholm, P. I. (2001). Dendrochronology and other applications of tree-ring studies on archaeology. In *Handbook of Archaeological Sciences* (D. R. Brothwell and A. M. Pollard, Eds.), pp. 35–46. John Wiley, New York.
LaMarche, V. C., Jr., and Harlan, T. P. (1973). Accuracy of tree-ring dating of bristlecone pine for calibration of the radiocarbon time scale. *Journal of Geophysical Research* **78**(36), 8849–8858.
Lenz, O., Schär, E., and Schweingruber, F. H. (1976). Methodische Probleme bei der radiographisch-densitometrischen Bestimmung der Dichte und der Jahrringbreiten von Holz. *Holzforschung* **30**(4), 114–123.
Libby, W. F., Anderson, E. C., and Arnold, J. R. (1949). Age determination by radiocarbon content: world wide assay of natural radiocarbon. *Science* **109**, 227–228.
Luckman, B. H., and Villalba, R. (2001). Assessing the synchronicity of glacier fluctuations in the Western Cordillera of the Americas during the last millennium. In *Interhemispheric Climate Linkages* (Markgraf Ed.), Academic Press, San Diego.
Luckman, B. H., and Wilson, R. J. (2005). Summer temperatures in the Canadian Rockies during the last millennium: a revised record. *Climate Dynamics* **24**(2–3), 131–144.
Polge, H. (1966). Etablissement des courbes de variation de la densité du bois par exploration densitométrique de radiographies d'chantillons prélevés à la tariére sur des arbres vivants. Applications dans les domaines technologiques et physiologiques. *Annales des Sciences forestières* **23**, 1–206.
Schulman, E. (1938). Classification of false annual rings in Monterey pine. *Tree-Ring Bulletin* **4**(3), 4–7.
Schulman, E. (1945). Tree-ring hydrology of the Colorado Basin. In: Laboratory of Tree-Ring Research Bulletin 2. *University of Arizona Bulletin* **16**(4), 1–51.
Schulman, E. (1954). Longevity under adversity in conifers. *Science* **119**, 396–399.
Schweingruber, F. H. (1988). *Tree Rings: Basics and Applications of Dendrochronology*. D. Reidel, Dordrecht, The Netherlands.
Smith, D. J., McCarthy, D. P., and Luckman, B. H. (1994). Snow-avalanche impact pools in the Canadian Rocky Mountains. *Arctic and Alpine Research* **26**(2), 116–127.
Stallings, J. (1937). Some early papers on tree rings. *Tree-Ring Bulletin* **3**, 27–28.
St George, S., and Nielsen, E. (2003). Palaeoflood records for the Red River, Manitoba, Canada, derived from anatomical tree-ring signatures. *The Holocene* **13**(4), 547–555.
Stokes, M. A., and Smiley, T. L. (1968). *An Introduction to Tree-Ring Dating*, p. 73. University of Chicago Press, Chicago, IL.
Studhalter, R. A. (1955). Tree growth: I. Some historical chapters. *Botanical Review* **21**, 1–72.
Studhalter, R. A. (1956). Early history of crossdating. *Tree-Ring Bulletin* **21**, 1–4.
Swetnam, T. W. (1993). Fire history and climate change in giant sequoia groves. *Science* **262**, 885–889.
Swetnam, T. W., and Lynch, A. M. (1993). Multicentury, regional-scale patterns of western spruce budworm outbreaks. *Ecological Monographs* **63**(4), 399–424.
Winchester, V., and Harrison, S. (2000). Dendrochronology and lichenometry: Colonization, growth rates and dating of geomorphological events on the east side of the North Patagonian Icefield, Chile. *Geomorphology* **34**(3–4), 181–194.

DENDROCLIMATOLOGY

B H Luckman, University of Western Ontario, Ontario, Canada

© 2007 Elsevier B.V. All rights reserved.

Tree-ring series provide the most widely distributed and easily accessible archive of annually resolved proxy climate data. In regions with well-defined seasonal growth, the annual growth rings of trees provide both chronological control and a continuous time series of proxy environmental variables. The year-to-year variability of the physical (e.g., width and density) and chemical properties of these annual rings provides potential proxies for the environmental factors that influence tree growth. Dendroclimatology can be defined simply as 'the science that uses tree-rings to study present climate and reconstruct past climate' (Grissino-Meyer, n.d.), and during the past 30 years it has become a major tool in the reconstruction of climates of the past millennium in many areas of the world (Hughes, 2002).

Basic Principles and Methods

The process of matching patterns of tree-ring width between series (cross-dating) is one of the fundamentals of dendrochronology. It is used to verify the dating of tree-ring series and, in some cases, to demonstrate that the series are annual. For cross-dating to be successful, the ring parameters must vary synchronously among series, reflecting a common response to an external variable. Tree growth is influenced by many factors, but often only one is strongly and consistently limiting. Successful dendroclimatological studies target sampling sites where a single climate parameter is, or is expected to be, the limiting factor of growth and therefore the primary control of interannual variability in ring width. The presence and strength of the common signal between tree-ring series within a site is a measure of the degree of climate control of growth: Correlation between tree-ring series from different sites across a broad area almost inevitably reflects a common climate control because no other controlling factors vary systematically at these spatial scales. Therefore, successful cross-dating usually establishes the presence of a climate signal in tree-ring series.

Trees growing at their range limits are often the most sensitive to climate variability, whereas trees in mid-range or interior forest sites may show 'complacent' ring-width series with little interannual variability. Classically, latitudinal and altitudinal treeline sites are most sensitive to temperature variations, whereas lower treeline and/or forest border sites are moisture sensitive; however, there are exceptions. Bald cypress trees growing in swamps in the southeastern United States are sensitive to water level (and therefore precipitation), and growth at many treeline sites in arid mountains is sensitive to both precipitation and temperatures (e.g., Bristlecone pines (*Pinus aristata* and *P. longaeva*)). The key to dendroclimate research is careful site selection because the limiting factor to growth can also vary based on microsite considerations. Dendroclimate studies further assume that the growth-limiting factor at a site does not change over time and therefore contemporary tree-ring climate relationships can be used to reconstruct past climate conditions.

Annual tree-rings can be divided into earlywood (larger, thin-walled cells) and latewood (smaller cells with thicker walls).The earlywood–latewood transition is usually gradational, but the latewood–earlywood boundary is sharply defined and marks the end of the growth year. Traditionally, ring width is the primary variable utilized for dendroclimate work, but other variables, such as earlywood width or latewood width, have been successfully used for seasonal precipitation reconstructions (Stahle *et al.*, 2000; Meko and Baisan, 2001). Tree-ring densitometry was initially developed by Hubert Polge (1966) in France and applied to climate studies by Parker and Henoch (1971). Precise thin sections are cut from the cores, X-rayed, and the negatives scanned with a high-resolution densitometer to produce a continuous density trace orthogonal to the rings. Several ring-width and density variables can be obtained from these scans. Maximum latewood density (MXD) reflects the thickening of cell walls at the end of the growing season and therefore represents a narrower climatic window than more traditional ring-width parameters (Schweingruber *et al.*, 1978). MXD in northern conifers is very strongly related to summer half-year temperatures and has been used in several important temperature reconstructions.

Chronology Development

The fundamental data in dendroclimatology are tree-ring chronologies—average series of, for example, tree-ring widths from a stand or sample site of a single species growing under similar conditions. Individual chronologies usually consist of paired radii from at least 10–20 erect trees without growth anomalies. The age assigned to each measured ring is verified by the process of cross-dating series against each other and/or reference sequences (matching ring patterns either visually or by computer software) to ensure secure dating and to check for the presence of 'false' or locally absent rings. Prior to averaging, each tree-ring series is converted to indices by removing low-frequency age-related trends. This 'standardization' process is usually accomplished by dividing the measured ring parameter for each year by values derived from a theoretical curve (commonly a negative exponential or linear regression) that converts these data to a stationary series with a mean of 1 and similar variance throughout the series. This ensures that each sample contributes equally to the average series by removing differences in mean-ring widths between series (e.g., wider rings in relatively young trees, and narrower rings in older trees). Several approaches to standardization are available (Fritts, 1976) and are the subject of continuing discussion because standardization can modify the common (climate) signal retained within these chronologies. This is a significant concern in the construction of long chronologies from overlapping cross-dated segments of living and subfossil material, where standardization may remove climatically related low-frequency trends from tree-ring series. Studies indicate that the frequencies retained may be as little as one-third of the mean segment length in the chronology (the 'segment length curse'; Cook *et al.*, 1995). Techniques have been developed to try to retain this low-frequency signal (Briffa *et al.*, 2001: Esper *et al.*, 2003).

In any chronology, sample depth decreases further back in time because there are fewer older trees. Expressed population signal (EPS) is an objective measure of signal strength, based on intercore correlation and sample size, that can be used to identify those parts of a chronology (usually the oldest sectors) where signal strength falls below acceptable levels. Wigley *et al.* (1984) suggested that an EPS value of 0.85 is a reasonable threshold for adequate signal strength. The effective sample size for chronology development depends on the strength of the common signal between these series; for example, at sites where intercore correlation averages ca. 0.6, four or five trees may provide adequate signal strength, but more than 20 trees would be needed at sites with mean intercorrelation values of ca. 0.2 (Briffa and Jones, 1990).

Reconstruction of Climate Variables

Identification and capture of the climate signal in tree-ring series is accomplished through the process of sampling and chronology development. Isolation of the climate signal is best described by the conceptual linear aggregate model developed by Cook (1990) following Graybill (1982). In this model the variation in ring width can be decomposed into several components, namely $R_t = A_t + C_t + \delta D1_t + \delta D2_t + E_t$. In this expression, R_t is the ring-width series, A_t is the age-size-related trend of ring width, C_t is the climatically related environmental signal, $\delta D1_t$ and $\delta D2_t$ are local (D1) or standwide (D2) nonclimatic disturbances, and E_t is unexplained year-to-year variability (Cook, 1990). Ideally, the E term is random, minor, and minimized by averaging sufficient tree-ring series. The disturbance effects may be present or absent and can be avoided by examination of the individual series (D1), by careful site selection prior to sampling (D2), or by comparison of the record with that from adjacent undisturbed sites. The age-related term is removed by standardization. Averaging of the indexed tree-ring series emphasizes the common climate signal and conceptually the equation simplifies to $R_t \approx C_t$.

Identification of the climate signal in tree-ring series requires calibration against instrumental climate data that overlap part of the tree-ring series. These climate data should be spatially representative and homogeneous (i.e., high-quality series that are continuous, complete, and corrected for any changes in station location, instrumentation, measurement practices, etc.). They may be single station data or aggregated at appropriate spatial scales that vary with the parameter being investigated. Although proximal sites are ideal, proximity is more significant for precipitation than temperature data because of the greater spatial coherence of temperature fields (Jones and Thompson, 2003). Exploratory response function analyses are used to identify the strongest relationships between tree growth and a wide range of monthly, seasonal, or annual temperature or precipitation variables for periods up to 2 years prior to the growth year (Fritts, 1976). Subsequent analyses develop a transfer function model between the more significantly related climate variable (annual precipitation, summer temperatures, spring precipitation, etc.) and a series of candidate tree-ring-derived variables (ring width, ring width of prior year, maximum density, etc.). The tree-ring data are then entered into the model to reconstruct or hindcast the climate variable. Dendroclimate reconstructions are unique in that they are routinely verified against data withheld from the initial calibration, and error terms can be applied to these reconstructions (Cook and Kairiukstis, 1990). Ideally, if the calibration data set is long, then split-period verification is used. Calibration (transfer function development) is carried out using half of the data and the predicted values are compared ('verification') against the data withheld from calibration. Subsequently, the procedure is reversed and, if these results are satisfactory (i.e., the relationship is time stable) and pass a series of statistical tests, the entire joint data set is used to derive the transfer function used to reconstruct the climate variable for the entire chronology. Reconstructions are generally modeled linearly using regression-based procedures. For short calibration data sets, the 'leave out one' procedure may be used (Gordon, 1982). In addition to simple reconstructions for individual sites, extensive single- or multispecies chronology networks allow the reconstruction of temperature, precipitation, or atmospheric fields over large areas.

Historical Development

Although there are scattered European observations on tree-rings dating back several hundred years (Schweingruber, 1988), the 'father' of dendrochronology is generally considered to be the astronomer A. E. Douglass, who first developed the principles of cross-dating and built chronologies based on drought-sensitive trees from the American Southwest, ultimately using them to date archeological sites. In 1937, he established the Laboratory of Tree-Ring Research (LTRR) in Tucson, Arizona, which was the dominant center for dendrochronology in the Americas for the next 50 years. Edmund Schulman made significant early developments in dendroclimate research at Tucson during the 1930s–1950s, 'discovering' the ancient Bristlecone pines in the White Mountains that were used for the first calibration of ^{14}C dating. The oldest known pine was 4,862 years old and the oldest currently living tree is >4,765 years old. Schulman also carried out pioneer reconnaissance studies of trees in similar semiarid environments in Canada, Argentina, Chile, and Mexico. In Europe, Bruno Huber took up Douglass' ideas and with his students undertook studies of trees within the temperate mountains of Europe between the late 1930s and 1960s. Modern quantitative dendroclimatology, although drawing on these earlier studies, stems largely from the classic work of Hal Fritts (1976) at LTRR, who documented many of the basic principles of the science and led the application of computing techniques and statistics to tree-ring analysis that revolutionized the discipline. Much of the classic work from Tucson focused on ring-width studies of arid land conifers. However, in the past 25 years the research field has rapidly expanded and diversified with the development of new laboratories, techniques, and research centers, engendered in part by the climate change debate and the need for long, annually resolved proxy climate records (Hughes, 2002). Dendroclimatology is now applied to a broad range of species and environments that cover the globe from forest tundra at 72 °N in Siberia to scrubby rainforest at 56 °S at Cape Horn, and from sea level in Alaska to higher than 4,900 m on the Bolivian Altiplano.

Recognition of the increasing utility of tree-ring records for paleoclimate study and the expanding scales of investigation created the need for the archiving and exchange of data and the development of an easily accessible international database with standard protocols for format. The International Tree-Ring Data Bank was founded in 1974 and has been housed at the National Oceanic and Atmospheric Administration Paleoclimatology Data Center in Boulder, Colorado, since 1992. Initially based on the LTRR collections, it is now a truly global (if unevenly distributed) database that holds more than 2,000 chronologies from more than 150 species on six continents. These data are freely accessible on line at www.ncdc.noaa.gov/paleo/paleo.html, together with statistical programs for the analysis of tree-ring data from LTRR and other laboratories.

Applications of Dendroclimatology

Dendroclimatological techniques have been used to reconstruct a wide range of paleoclimate variables that have progressively expanded from local to regional and continental scales. In providing examples of these applications, the following sections proceed from relatively simple, single chronology examples to results from large-scale networks. For a more comprehensive review, see Hughes (2002).

Single Site Reconstructions

The earliest recognition of climate sensitivity in North American tree-rings came from the American Southwest, where Douglass first exploited the relationships between narrow rings and precipitation to infer major droughts based on tree-ring records from precipitation-sensitive trees. Drought-sensitive conifers show some of the strongest relationships with ring-width values (**Fig. 1**). Many reconstructions have been developed from such relationships to show seasonal or annual precipitation totals. In some cases, where there is more than one 'wet season,' seasonal precipitation amounts may be reconstructed separately, based on differences between the earlywood, latewood, or total ring chronologies. For example, latewood widths from Douglas fir (*Pseudotsuga menziesii*) in central Mexico are strongly correlated with the April–June rainfall that is critical for the rain-fed maize crop. Therefore, the reconstruction of maize yields in **Figure 2** is essentially a precipitation reconstruction in which the major droughts are verified by historical records (Therrell *et al.*, 2006). Drought magnitude and frequency can also be determined from reconstructions of climate indices such as the Palmer Drought Severity Index (PDSI).

Reconstructions of temperature from ring-width data are usually for summer season or annual values. Jacoby and Cook (1981) were the first to demonstrate strong linear relationships between spruce ring widths and June and July temperatures for a site in the central Yukon (65 °N). This record shows strong low-frequency variability and, together with 19 other treeline chronologies from North American, Scandinavian, and Russian treelines, provided one of the early reconstructions of a Northern Hemisphere temperature record (D'Arrigo and Jacoby, 1993). Significant summer temperature reconstructions have also been developed for northern conifers based on MXD data or combined ring width and MXD data.

Routinely recoverable tree-ring records in many areas of the world are less than 200–300 years long because heart rot, fire, or other disturbances limit the age of trees. Long, climate-sensitive chronologies are therefore of particular importance, and attention has been given to sites with long-lived species, well-preserved subfossil wood, or both. Many of these chronologies come from the range limits of species, particularly at Northern Hemisphere treelines with old slow-growing trees and/or good preservation of wood at dry or cold sites (Briffa, 2000). The greatest concentration of these long chronologies is in the

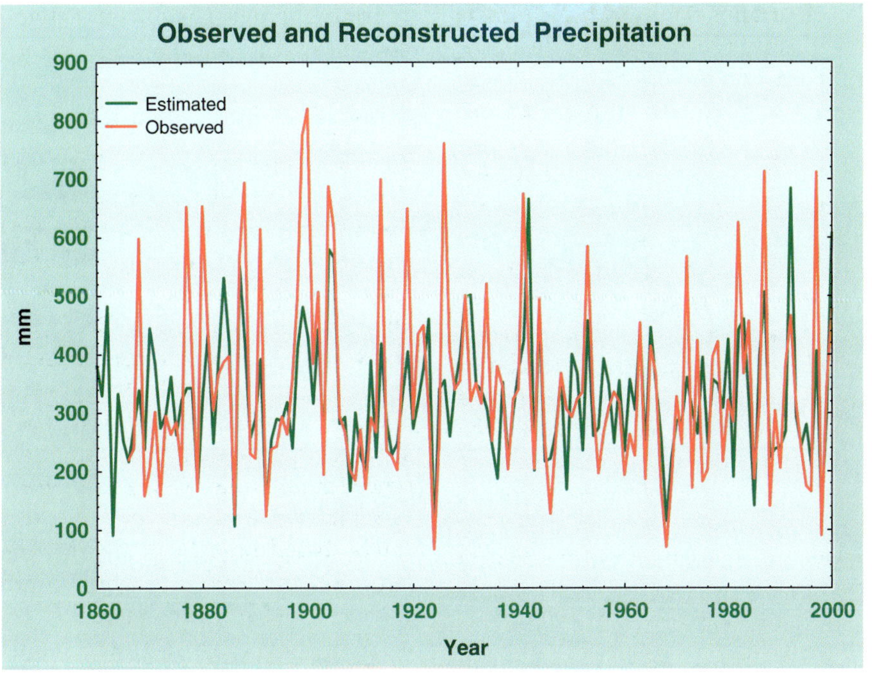

Figure 1 Actual and reconstructed precipitation for Santiago de Chile, 1866–2000, based on ring widths of *Austrocedrus chilensis* (J. Boninsegna, personal communication). The adjusted r^2 value is 0.54.

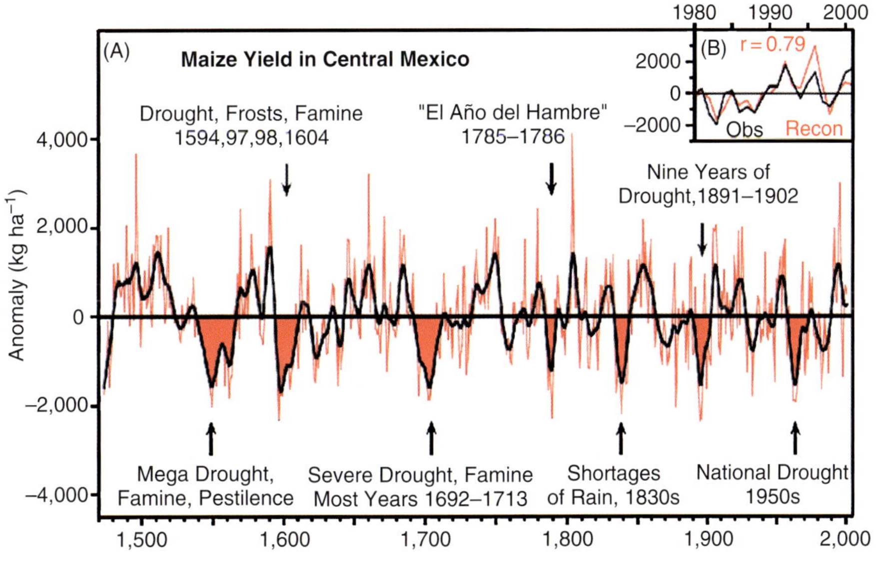

Figure 2 Reconstructed crop yield for central Mexico from 1480 to 2000 based on latewood ring width of Douglas fir growing at Cuauhtemoc la Frangua, Mexico. Maize cultivation is 'rain-fed' and the yield is strongly correlated with April–June rainfall. Dating for the major droughts is based on documentary records. After Therrell *et al.*, 2006. Tree-ring reconstructed maize yield in Central Mexico: 1474–2001. *Climatic Change*.

southwest United States (Hughes and Graumlich, 1996), but millennial chronologies are now developed in northern Europe (Briffa and Mathews, 2002), Canada (Luckman and Wilson, 2005), the Alps (Büntgen *et al.*, 2005), and Asia (Esper *et al.*, 2003; D'Arrigo *et al.*, 2001). These long chronologies, combined with other high-resolution proxy climate series, are primary inputs for several Northern Hemisphere annual or summer temperature reconstructions for the past millennium (Mann *et al.*, 1999; Esper *et al.*, 2002; Briffa *et al.*, 2004). Although these show a fairly similar pattern for the past 300–400 years (**Fig. 3**), there are differences in the earlier records related to the relatively sparse and

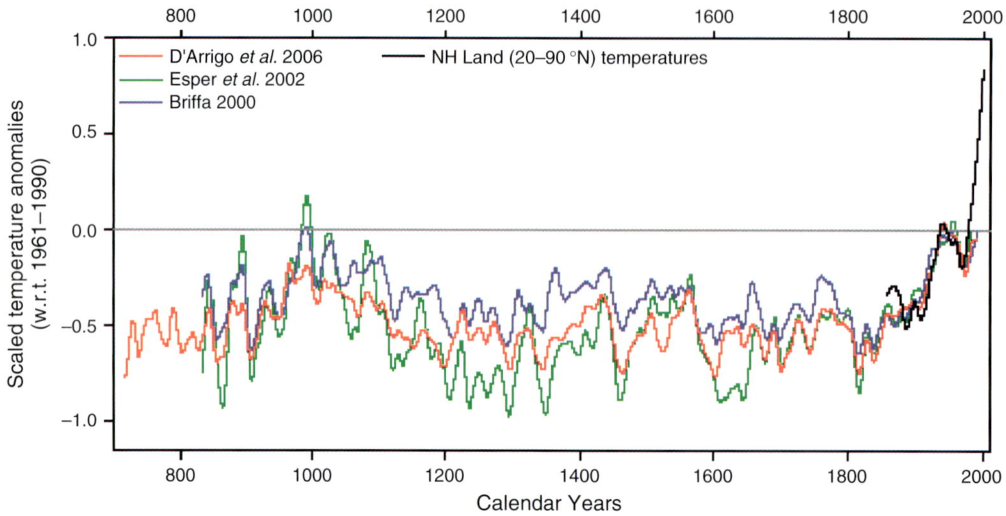

Figure 3 Comparison of three millennial-length Northern Hemisphere temperature reconstructions based exclusively on tree-ring data and standardized to emphasize low-frequency variance. The reconstructions are from Briffa (2000), Esper *et al.* (2002), and D'Arrigo, Wilson and Jacoby (2006) and are scaled to NH land temperatures (R. Wilson, personal communication).

unevenly distributed sample base and the methods of standardization used (Rutherford *et al.*, 2005). This situation should improve as increasingly more widely distributed long chronologies become available. However, it is clear from reconstructions of the past few hundred years that, although some variations may be globally synchronous (e.g., those forced by solar activity), there is considerably spatial variability at all timescales.

Currently, there are relatively few long chronologies (and related temperature reconstructions) from the Southern Hemisphere, in part because of the restricted land area. Millennial and multimillennial temperature reconstructions have been made from the Huon Pine in Tasmania (Cook *et al.*, 2000), silver pine (*Manoao colensoi*) in New Zealand (Cook *et al.*, 2002b), and *Fitzroya cupressoides* (Alerce) in Chile and Argentina (Lara and Villalba, 1993; Lara *et al.*, 2000).

Tree-Ring Networks

The widespread distribution of suitable sites for dendroclimatology has resulted in considerable effort to build single or multi-chronology networks to reconstruct spatial patterns of climate variability. The first explicit attempt was made by Harold Fritts (1976, 1991) and colleagues, who used a network of 65 moisture-sensitive ring-width chronologies to reconstruct spatial patterns of temperature, precipitation, and atmospheric pressure across western North America. Subsequent expansion of this network, incorporating results from many investigators, led to the development of a network of 155 PDSI reconstructions for a 2° latitude by 3° longitude grid, essentially covering the coterminous United States (Cook *et al.*, 1999). This been expanded to 286 gridpoints (2.5° × 2.5°) that cover most of North America from the Yukon to Yucatan (Cook *et al.*, 2004). These reconstructions provide a powerful tool for the analysis of hydroclimate variability across North America (**Figs. 4** and **5**) and exploration of the spatial extent and duration of drought, placing 20th-century records in a long-term context (Fye *et al.*, 2003). In the western United States, this new reconstruction has 68 grid squares with records that date back to 800 AD, showing the extent of significant droughts in the 12th and 13th centuries and even a major 8th-century drought that have had important impacts on the cultural history in pre-Hispanic America (Acuna-Soto *et al.*, 2005).

In a parallel endeavor, Fritz Schweingruber and colleagues built a circumpolar network of more than 387 MXD and ring-width chronologies that permit the development of hemispheric-scale warm-season

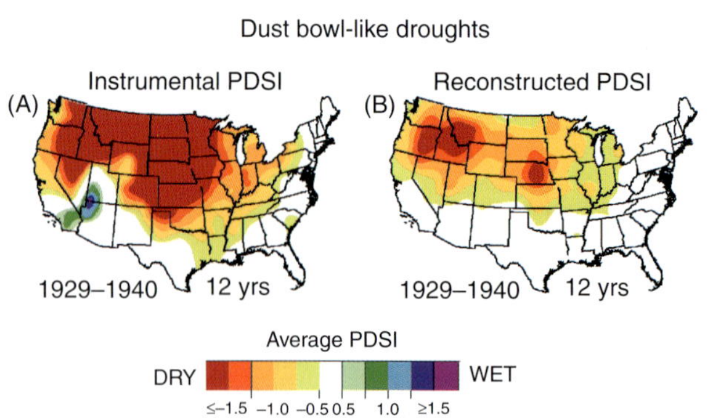

Figure 4 Comparison of instrumental and reconstructed PDSI data for the United States during the 'dustbowl' drought of 1929–1940 (Fye *et al.*, 2003).

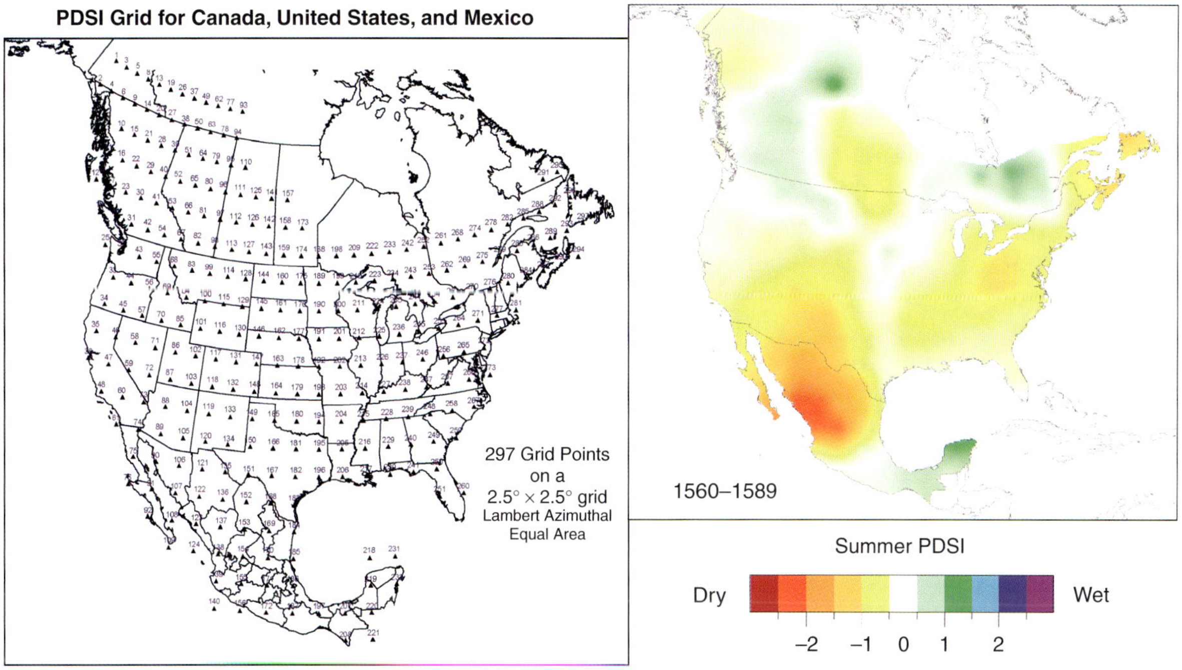

Figure 5 Coverage of the 2.5° × 2.5° gridded PDSI data set for North America based on tree-ring data. The diagram on the right shows the average reconstructed PDSI for the major drought between 1560 and 1589 (D. Stahle, personal communication; see Cook *et al.*, 2004).

temperature histories and the study of spatial patterns over the past 300 years. Assigning 303 chronologies from this network to 92 5° × 5° temperature grid boxes has yielded maps of circumpolar summer temperature patterns for each year since 1600 (**Fig. 6**) (Briffa *et al.*, 2002). The composite Northern Hemisphere (20–70 °N) series developed from these series clearly shows the global impact of major volcanic eruptions on summer temperatures during the past 600 years (**Fig. 7**). Cool and/or wet summers following major volcanic eruptions have poorly developed latewood and are characterized by 'light' rings in MXD chronologies. Decreased temperatures following major volcanic activity may also be inferred by the

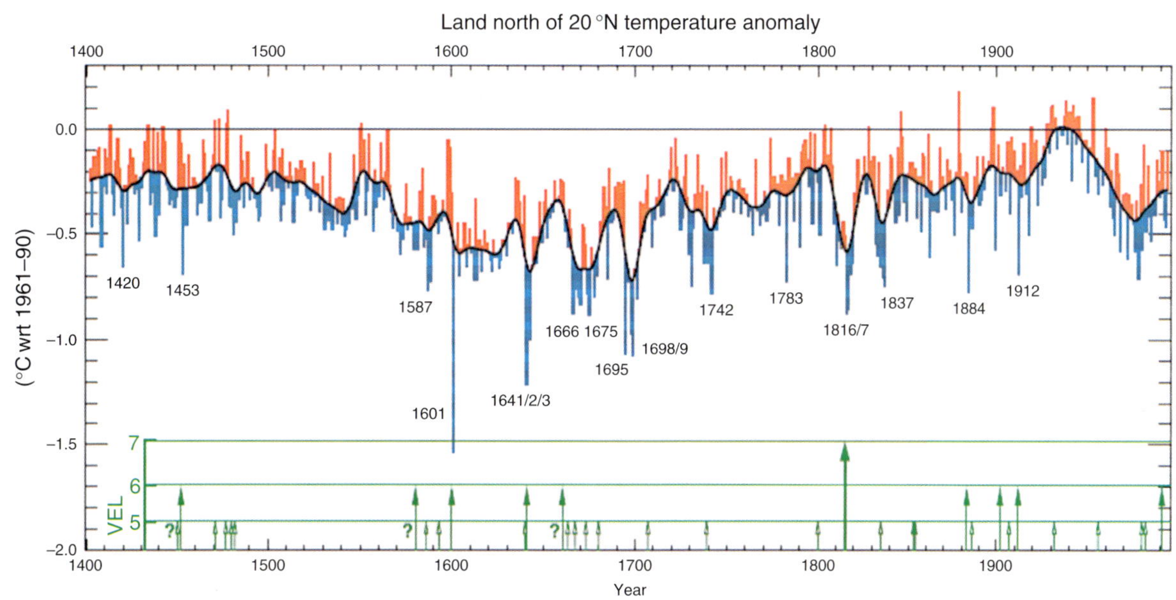

Figure 6 Estimated April–September temperature anomalies for the land area between 20 and 70 °N, smoothed with a 25-year filter and based on MXD. The lower diagram shows values of the Volcanic Explosivity Index. Eruptions with uncertain dating are indicated by question marks. Note the strong cooling trend reconstructed for the late 20th century, reflecting a loss of temperature signal in these data (Briffa *et al.*, 2004).

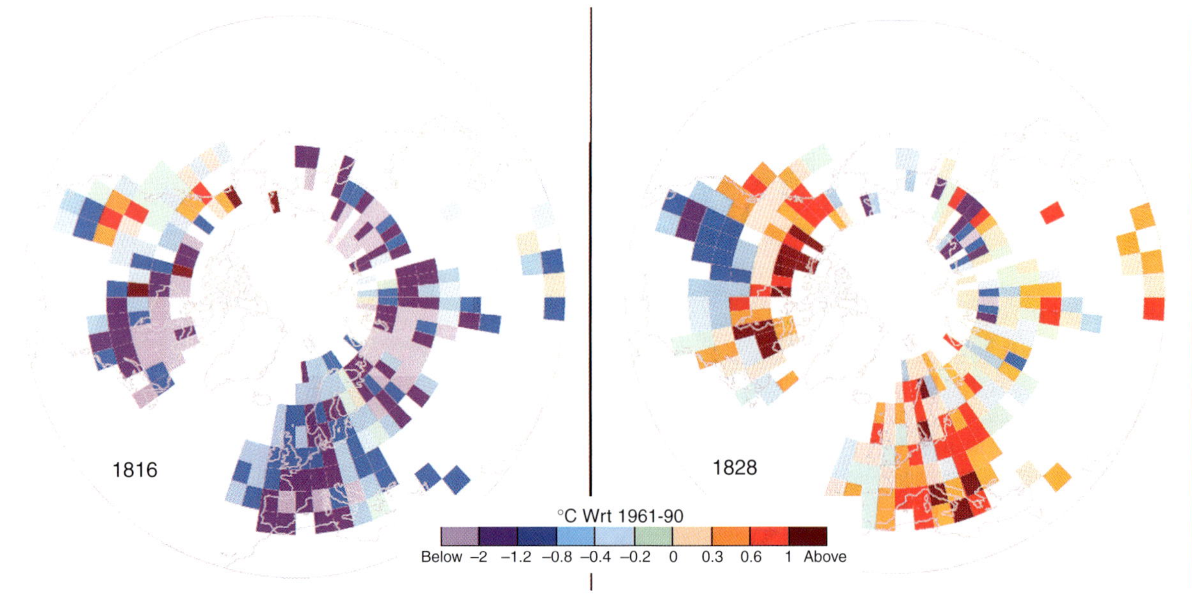

Figure 7 Mean April–September gridded temperature anomalies for 1816 and 1828 estimated from MXD tree-ring data. These are two of the most extreme years in the 19th century (1816 follows the Tambora eruption). Note the regional differences in temperatures (K. Briffa, personal communication; see Briffa *et al.*, 2002, 2004).

presence of 'frost rings' that show cell disruption due to freezing at certain high-elevation sites (Lamarche and Hirschboeck, 1984). The chronological precision of these tree-ring dates is comparable to documentary sources and superior to that available from ice cores or radiocarbon dating for events prior to the historic record, enabling tentative calendar dates to be ascribed to major volcanic or possibly cometary impact events in the past 2,000–3,000 years (Baillie, 1996).

Noncontiguous sets of tree-ring data have been assembled to examine spatial patterns and teleconnections associated with large-scale atmospheric phenomena. This is the basis for the field of synoptic dendroclimatology, which examines climate variability through reconstructing changes in the atmospheric circulation patterns that control temperatures and precipitation fields (Hirschboeck *et al.*, 1996) and attempts to reconstruct climate fields from limited proxy data (Evans *et al.*, 2001). The strong relationships between winter precipitation, the El Niño/Southern Oscillation (ENSO), and ring width in the southwest United States and adjacent Mexico first led to the recognition of a strong ENSO signal in many of these chronologies. Combination of these data with coral, ice core, and tree-ring data (including a teak chronology from Indonesia) allowed the reconstruction of the Southern Oscillation Index between 1707 and 1977 (Stahle *et al.*, 1998). A subsequent tree-ring-based reconstruction of the Niño3 Index, calibrated on sea-surface temperatures, extends back to 1408 (D'Arrigo *et al.*, 2005), and long precipitation-sensitive chronologies from *Austrocedrus* in Chile and Argentina allow the reconstruction of ENSO patterns back more than 1,500 years in this region (J. Boninsegna, 2003, personal communication). Chronologies from *Polylepis tarapacana* (queñoa) growing at over 4,500 m on the Bolivian Altiplano also show a strong correlation with ENSO (Villalba *et al.*, in press). Similar studies have reconstructed the Pacific Decadal Oscillation (PDO; Gedalof *et al.*, 2002), North Atlantic Oscillation (NAO; Cook *et al.*, 2002a), and other climate indices from tree-ring data. These large-scale investigations have also been important in the exploration of teleconnection patterns over time. For example, Villalba *et al.* (2001) demonstrated that the phase relations between tree-ring series from Patagonia and coastal Alaska show similar patterns at different frequencies during the past 300 years. These studies are of particular importance in documenting whether the decadal and longer scale variability seen in available instrumental records is characteristic of the long-term behavior of the climate system.

Climate-Related Reconstructions

As available databases and investigations diversify, dendroclimatological studies are expanding into several related fields. Temperature-sensitive chronologies have been used to reconstruct sea-surface temperatures (D'Arrigo *et al.*, 2005, in press). An obvious extension of work on moisture-sensitive sites has been the

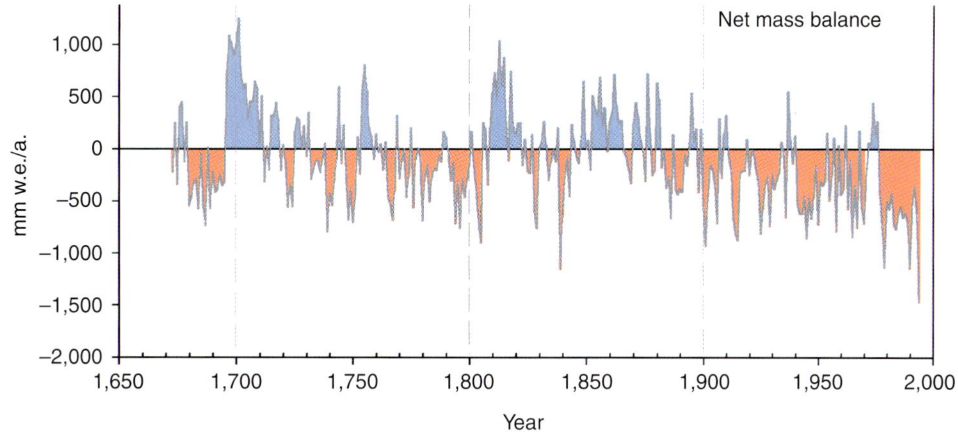

Figure 8 Reconstructed glacier mass balance for Peyto Glacier, Alberta, Canada, based on separate reconstructions of winter and summer balances using precipitation- and temperature-sensitive tree-ring chronologies (Watson and Luckman, 2004).

reconstruction of streamflow. In an early study, Stockton (1975) used 17 chronologies in the Colorado River basin to develop a long streamflow reconstruction, which demonstrated that the 1564–1961 mean flow was approximately $2.5 \times 10^9\,m^3$ less than the 1896–1960 mean and considerably less than the $22 \times 10^9\,m^3$ base period on which allocation of water rights was based in the 1922 Colorado compact between the United States and Mexico. In a warming world, more attention is being paid to water supply and streamflow reconstructions, and these types of calculations will become increasingly important in watershed planning (Meko and Woodhouse, 2005). Watson and Luckman (2004) used both temperature and precipitation-sensitive tree-ring series to reconstruct the mass balance of Peyto Glacier in Canada during the past 322 years (**Fig. 8**). Such studies are important in determining the relative roles of precipitation and temperature variability in the past and future behavior of these glaciers in response to climate change. Dendrochronology has also been used extensively to reconstruct forest fire history and examine the linkages between fire, drought, and other climatic factors that can provide a basis for forest and fire management (Swetnam and Baisan, 2003).

Investigations in the Tropics

One of major gaps in dendroclimatic studies concerns the tropics, where few trees have annual rings. In these areas, annual ring development is mainly found in regions with marked seasonality in precipitation (Worbes, 2002). Work has extended chronologies into the subtropics at high-elevation sites in the Americas to within 20° of the equator (Luckman and Boninsegna, 2004). Precipitation-sensitive chronologies up to 600 years long have been developed from *Polylepis tarapacana* on the Bolivian Altiplano between 18° and 22 °S (Argollo *et al*, 2004). Major new collections in Southeast Asia are beginning to yield excellent results (D'Arrigo *et al.*, 2006, in press), but much exploratory work is still needed. There is also promise that isotope dendroclimatology can yield proxy climate records from species that lack annual rings (Evans and Shragg, 2004).

Changing Signals

Dendroclimatic reconstructions, like many other paleoclimatic techniques, are based on uniformitarian assumptions. However, several authors have noted a decline in the temperature-related signal in both ring width (Jacoby and D'Arrigo, 1995) and MXD series from high-latitude Northern Hemisphere sites (Briffa *et al.*, 2004), and other studies report changes in signal relating to atmospheric loading effects (Wilson and Elling, 2004). A variety of causes have been proposed, including increased moisture sensitivity once summer temperatures pass a critical threshold (D'Arrigo *et al.*, 2004), possible ozone-related effects (Briffa *et al.*, 2004), or even chronology development techniques (Wilmking *et al.*, 2005). Although not universal, these effects need careful scrutiny and further study to resolve this potentially critical issue.

Conclusions

Dendroclimatological studies are providing new, annually resolved, and precisely dated proxy climate records for large areas of the Earth's land surface and parts of the adjacent oceans. These allow the reconstruction of climate history, trends, and patterns at an expanding range of temporal and spatial scales. The increasing availability of long chronologies from major continental areas can address climate variability at decadal to millennial timescales while retaining annual resolution. The increasing density and distribution of chronologies provide networks that can examine the spatial variability and teleconnection patterns associated with major modes of global climate variability, such as ENSO, PDO, and NAO, seen in the instrumental record. Because these networks of annually resolved data are at least double or triple the length of instrumental records (in many remote areas they may be the only records), they can be used to establish whether these patterns are time stable or how they have responded to past changes in forcing. In conjunction with other proxy climate data, they provide the best records to benchmark and understand the natural modes of climate variability at decade to century scales over much of the Earth's surface.

See also: **Dating Techniques**. **Paleoclimate**: Introduction. **Paleoclimate Modeling**: Paleo-ENSO. **Paleoclimate Reconstruction**: Approaches. **Radiocarbon Dating**: Plant Macrofossils.

References

Acuna-Soto, R., Therrell, M. D., Gomex Chavez, S., and Cleaveland, M. K. (2005). Drought, epidemic disease, and the fall of classic period cultures in Mesoamerica (AD 750–950). Hemorrhagic fevers as a cause of massive population loss. *Medical Hypotheses* **65**, 405–409.

Argollo, J., Soliz, C., and Villalba, R. (2004). Potencialidad dendrocronológica de *Polylepis tarapacana* en los Andes Centrales de Bolivia. *Ecología en Bolivia* **39**(1), 5–24.

Baillie, M. G. L. (1996). Extreme environmental events and the linking of tree-ring and ice-core records. In *Tree Rings, Environment and Humanity* (J. Dean, D. M. Meko and T. W. Swetnam, Eds.), pp. 703–711. Radiocarbon, Tucson, AZ.

Briffa, K. R. (2000). Annual variability in the Holocene: Interpreting the message of ancient trees. *Quaternary Science Reviews* **19**, 87–105.

Briffa, K. R., and Jones, P. D. (1990). Estimating the chronology signal strength. In *Methods of Dendrochronology: Applications in the Environmental Sciences* (E. R. Cook and L. A. Kairiukstis, Eds.), pp. 137–152. Kluwer, Dordrecht, The Netherlands.

Briffa, K. R., and Matthews, J. A. (2002). Analysis of dendrochronological variability and associated natural climates in Eurasia (Advance-10K) (Special issue). *The Holocene* **12**(6), 639–794.

Briffa, K. R., Osborn, T. J., Schweingruber, F. H., *et al.* (2001). Low-frequency temperature variations from a northern tree-ring network. *Journal of Geophysical Research* **106**, 2929–2941.

Briffa, K. R., Osborn, T. J., Schweingruber, F. H., *et al.* (2002). Tree-ring width and density data around the Northern Hemisphere: Part 2. Spatio-temporal variability and associated climate patterns. *The Holocene* **12**, 759–789.

Briffa, K. R., Osborn, T. J., and Schweingruber, F. H. (2004). Large-scale temperature inferences from tree-rings: A review. *Global and Planetary Change* **40**, 11–26.

Büntgen, U., Esper, J., Frank, D. C., Nicolussi, K., and Schmidhalter, M. (2005). A 1052-year tree-ring proxy for Alpine summer temperatures. *Climate Dynamics* **25**, 141–153.

Cook, E. R. (1990). A conceptual linear aggregate model for tree-rings. In *Methods of Dendrochronology: Applications in the Environmental Sciences* (E. R. Cook and L. A. Kairiukstis, Eds.), pp. 98–103. Kluwer, Dordrecht, The Netherlands.

Cook, E. R., and Briffa, K. R. (1990). A comparison of some tree-ring standardization methods. In *Methods of Dendrochronology: Applications in the Environmental Sciences* (E. R. Cook and L. A. Kairiukstis, Eds.), pp. 153–163. Kluwer, Dordrecht, The Netherlands.

Cook, E. R., and Kairiukstis, L. A. (Eds.) (1990). *Methods of Dendrochronology: Applications in the Environmental Sciences*, pp. 153–162. Kluwer, Dordrecht, The Netherlands.

Cook, E. R., Briffa, K. R., Meko, D. M., Graybill, D. M., and Funkhouser, G. (1995). The 'segment length curse' in long tree-ring chronology development for paleoclimate studies. *The Holocene* **5**, 229–237.

Cook, E. R., Meko, D. M., Stahle, D. W., and Cleaveland, M. K. (1999). Drought reconstructions for the continental United States. *Journal of Climate* **12**, 1145–1162.

Cook, E. R., Buckley, B. M., D'Arrigo, R. D., and Peterson, M. J. (2000). Warm-season temperatures since 1600 B.C. reconstructed from Tasmanian tree-rings and their relationship to large-scale sea surface temperature anomalies. *Climate Dynamics* **16**, 79–91.

Cook, E. R., D'Arrigo, R. D., and Mann, M. E. (2002a). A well-verified, multiproxy reconstruction of the winter North Atlantic Oscillation Index since A.D. 1400. *Journal of Climate* **15**, 1754–1764.

Cook, E. R., Palmer, I. G., and D'Arrigo, R. D. (2002b). Evidence for a 'medieval warm period' in a 1100 year tree-ring reconstruction of past Austral summer temperatures in New Zealand. *Geophysical Research Letters* **29**(14), 10.1029/2001GL014580.

Cook, E. R., Woodhouse, C. A., Eakin, C. M., Meko, D. M., and Stahle, D. W. (2004). Long-term aridity changes in the western United States. *Science* **306**, 1015–1018.

D'Arrigo, R. D., and Jacoby, G. C. (1993). Secular trends in high northern latitude temperature reconstructions based on tree-rings. *Climatic Change* **25**, 163–177.

D'Arrigo, R. D., Jacoby, G. C., Frank, D., *et al.* (2001). 1738 years of Mongolian temperature variability inferred from a tree-ring width chronology of Siberian pine. *Geophysical Research Letters* **28**, 543–546.

D'Arrigo, R. D., Kaufmann, R. K., Davi, N., *et al.* (2004). Thresholds for warming-induced growth decline at elevational treeline in Yukon Territory, Canada. *Global Biogeochemical Cycles*18, GB3021, doi: 10.1029/2004GB00249.

D'Arrigo, R., Cook, E. R., Wilson, R. J., Allan, R., and Mann, M. E. (2005). On the variability of ENSO over the past six centuries. *Geophysical Research Letters* **32**, L03711, doi: 10.1029/2004GL022055.

D'Arrigo, R., Wilson, R., Palmer, J., *et al.* (2006). Monsoon drought over Java, Indonesia during the past two centuries. *Geophysical Research Letters* 33, L04709, doi: 10.1029/2005GL025465.

D'Arrigo, R., Wilson, R., Palmer, J., *et al.* (in press). Reconstructed Indonesian warm pool SSTs from tree-rings and corals: Linkages to Asian monsoon drought and ENSO. *Paleoceanography.*

D'Arrigo, R., Wilson, R., and Jacoby, G. (2006). On the long-term context for late 20th century warming. *Journal of Geophysical Research*, vol. **III**, No. D3, D03103, 10.1029/2005JD006352.

Esper, J., Cook, E. R., and Schweingruber, F. H. (2002). Low-frequency signals in long tree-ring chronologies for reconstructing past temperature variability. *Science* **295**, 2250–2253.

Esper, J., Cook, E. R., Krusic, P. J., Peters, K., and Schweingruber, F. H. (2003a). Tests of the RCS method for preserving low-frequency variability in long tree-ring chronologies. *Tree Ring Research* **39**, 81–98.

Esper, J., Shiyatov, S. G., Mazepa, V. S., *et al.* (2003b). Temperature–sensitive Tien Shan tree-ring chronologies show multi-centennial growth trends. *Climate Dynamics* **21**, 699–706.

Evans, M. N., and Schrag, D. P. (2004). A stable isotope-based approach to tropical dendroclimatology. *Geochimica et Cosmochimica Acta* **68**, 3295–3305.

Evans, M. N., Kaplan, A., Cane, M. A., and Villalba, R. (2001). Globality and optimality in climate field reconstructions from proxy data. In *Interhemispheric Climate Linkages* (V. Markgraf, Ed.), pp. 53–72. Academic Press, New York.

Fritts, H. C. (1976). *Tree Rings and Climate.* Academic Press, New York.

Fritts, H. C. (1991). *Reconstructing Large-Scale Climatic Patterns from Tree-Ring Data.* University of Arizona Press, Tucson.

Fye, F. K., Stahle, D. W., and Cook, E. R. (2003). Paleoclimate analogs to 20th century moisture regimes across the US. *Bulletin of the American Meteorological Society* **84**, 901–909.

Gedalof, Z., Mantua, N. J., and Peterson, D. L. (2002). A multi-century perspective of variability in the Pacific Decadal Oscillation: New insights from tree-rings and coral. *Geophysical Research Letters* **29**(24), 2204.

Gordon, G. A. (1982). Verification of dendroclimatic reconstructions. In *Climate from Tree Rings* (M. K. Hughes, P. M. Kelly, J. P. Pilcher and V. C. LaMarche, Jr., Eds.), pp. 58–61. Cambridge University Press, Cambridge, UK.

Graybill, D. A. (1982). Chronology development and analysis. In *Climate from Tree Rings* (M. K. Hughes, P. M. Kelly, J. P. Pilcher and V. C. LaMarche, Jr., Eds.), pp. 21–28. Cambridge University Press, Cambridge, UK.

Grissino-Meyer, H. (n.d.). The ultimate tree-ring web site. http:web.utk.edu~grissino

Hirschboeck, K., Ni, F., Wood, M. L., and Woodhouse, C. A. (1996). Synoptic dendroclimatology: Overview and outlook. In *Tree Rings, Environment and Humanity* (J. Dean, D. M. Meko and T. W. Swetnam, Eds.), pp. 205–223. Radiocarbon, Tucson, AZ.

Hughes, M. K. (2002). Dendrochronology in climatology—The state of the art. *Dendrochronologia* **20**, 95–116.

Hughes, M. K., and Graumlich, L. J. (1996). Multimillennial dendroclimatic studies from the western United States. In *Climatic Variations and Forcing Mechanisms of the Last 2000 Years,* NATO ASI Series *1. Global Environmental Change* (P. D. Jones, R. S. Bradley and J. Jouzel, Eds.), 41, pp. 109–124.

Jacoby, G. C., and Cook, E. R. (1981). Past temperature variations inferred from a 400-year tree-ring chronology from Yukon Territory, Canada. *Arctic and Alpine Research* **13**(4), 409–418.

Jacoby, G. C., and D'Arrigo, R. D. (1995). Tree-ring width and density evidence of climatic and potentially forest change in Alaska. *Global Biogeochemical Cycles* **9**, 227–234.

Jones, P. D., and Mann, M. E. Climate over past millennia. *Reviews of Geophysics* 42, RG2002/2004 1–42.

Jones, P. D., and Thompson, R. (2003). Instrumental records. In *Global Change in the Holocene* (A. MacKay, R. Battarbee, J. Birks and F. Oldfield, Eds.), pp. 140–158. Hodder Arnold, London.

LaMarche, V. C., Jr., and Hirschboeck, K. K. (1984). Frost rings in trees as records of major volcanic eruptions. *Nature* **307**, 121–128.

Lara, A., and Villalba, R. (1993). A 3620 year temperature reconstruction from *Fitzroya cupressoides* in southern South America. *Science* **260**, 1104–1106.

Lara, A., Villalba, R., Aravena, J. C., Wolodarsky, A., and Neira, E. (2000). Desarrollo de una red de cronologías de Fitzoya *cupressoides* (Alerce) para Chile y Argentina. In *Dendrocronolgía en América Latina* (F. Roig, Ed.), pp. 217–244. EDIUNC, Universidad Nacional de Cuyo, Mendoza, Argentina.

Luckman, B. H., and Boninsegna, J. A., (2004). *Climate Variability from Tree Rings in Treeline Environments. Annual Report 2002–2003,* pp. 34–51. San Jose dos Campos, Brazil: Inter-American Institute for Global Change Research.

Luckman, B. H., and Wilson, R. J. S. (2005). Summer temperature in the Canadian Rockies during the last millennium—A revised record. *Climate Dynamics* **24**, 131–144.

Mann, M. E., Bradley, R. S., and Hughes, M. K. (1999). Northern Hemisphere temperatures during the past millennium: Inferences, uncertainties and limitations. *Geophysical Research Letters* **26**, 759–762.

Meko, D. M., and Baisan, C. H. (2001). Pilot study of latewood-width of conifers as an indicator of variability of summer rainfall in the North American monsoon region. *International Journal of Climatology* **21**, 697–708.

Meko, D. M., and Woodhouse, C. A., (2005). Tree-ring footprint of joint hydrologic drought in Sacramento and Upper Colorado River Basins, western USA. *Journal of Hydrology* **308**, 196–213.

Parker, M. L., and Henoch, W. E. S. (1971). The use of Engelmann spruce latewood density for dendrochronological analysis. *Canadian Journal of Forest Research* **1**, 90–98.

Polge, H. (1966). Etablissement des Courbes de Variation de la Densité du Bois par Exploration Densitométrique de Radiographies d' Échantillions Prelevès à la Tarière sur des Arbres Vivants. *Annales des Sciences Forestières* **23**, 1–206.

Rutherford, S., Mann, M. E., Osborn, T. J., *et al.* (2005). Proxy-based Northern Hemisphere surface temperature reconstructions: Sensitivity to method, predictor network, target season, and target domain. *Journal of Climate* **18**, 2308–2329.

Schweingruber, F. H. (1988). *Tree Rings.* Reidel, Dordrecht, The Netherlands.

Schweingruber, F. H., and Briffa, K. R. (1996). Tree-ring density networks for climate reconstruction. In *Climate Variations and Forcing Mechanisms of the Last 2000 Years* (P. D. Jones, R. S. Bradley and J. Jouzel, Eds.), pp. 43–66. Springer, Berlin.

Schweingruber, F. H., Fritts, H. C., Braeker, O. U., Drew, L. G., and Schaer, E. (1978). The x-ray technique as applied to dendroclimatology. *Tree-Ring Bulletin* **38**, 61–91.

Stahle, D. W., D'Arrigo, R. D., Krusic, P. J., *et al.* (1998). Experimental dendroclimatic reconstruction of the Southern Oscillation. *Bulletin of the American Meteorological Society* **79**, 2137–2152.

Stahle, D. W., Villanueva, J., Cleaveland, M. K., *et al.* (2000). Recent tree-ring research in Mexico. In *Dendrochronologia en América Latina* (F. Roig, Ed.), pp. 285–306. EDIUNC, Universidad Nacional de Cuyo, Mendoza, Argentina.

Stockton, C. W. (1975). *Long Term Streamflow Records Reconstructed from Tree Rings*, Paper No. 5. University of Arizona Press, Tucson.

Swetnam, T. W., and Baisan, C. H. (2003). Tree-ring reconstructions of fire and climate history in the Sierra Nevada and southwestern United States. In *Fire and Climate in Temperate Ecosystems of the Western Americas* (T. T. Veblen, W. L. Baker, G. Montenegro and T. W. Swetnam, Eds.), pp. 158–195. Springer-Verlag, New York.

Therrell, M. D., Stahle, D. W., Cleaveland, M. K., Villanueva-Diaz, J., and Cornejo Oviedo, E., (2006). Tree-ring reconstructed maize yield in Central Mexico: 1474–2001. *Climatic Change* **74**, 493–504.

Villalba, R., D'Arrigo, R. D., Cook, E. R., Jacoby, G. C., and Wiles, G. (2001). Decadal-scale climatic variability along the

extratropical western coast of the Americas: Evidence from tree-ring records. In *Interhemispheric Climate Linkages* (V. Markgraf, Ed.), pp. 155–172. Academic Press, New York.
Villalba, R., Lara, A., Boninsegna, J. A., *et al.* (2003). Large-scale temperature changes across the southern Andes: 20th-century variations in the context of the past 400 years. *Climatic Change* **59**, 177–232.
Villalba, R., Luckman, B. H., Boninsegna, J. A. *et al.* (in press). Dendroclimatology from regional to continental scales: Understanding local processes to reconstruct large-scale climatic variations across the Western Americas. In Hughes, M. K., Swetnam, T. W. and Diaz, H. F. (Eds.), Dendroclimatology: Progress and Prospects, Springer.
Watson, E., and Luckman, B. H. (2004). Tree-ring estimates of mass balance at Peyto Glacier for the last three centuries. *Quaternary Research* **62**, 9–18.
Wigley, T. M. L., Briffa, K. R., and Jones, P. D. (1984). On the average value of correlated time series with applications in dendroclimatology and hydrometeorology. *Journal of Climate and Applied Meteorology* **23**, 201–213.
Wilmking, M., D'Arrigo, R. D., Jacoby, G. C., and Juday, G. P. (2005). Increased temperature sensitivity and divergent growth trends in circumpolar boreal forests. *Geophysical Research Letters* **32**, L15715 doi: 10.1029/2005GL023331.
Wilson, R. J. S., and Elling, W. (2004). Temporal instabilities of tree-growth/climate response in the Lower Bavarian Forest Region: Implications for dendroclimatic reconstruction. *Trees: Structure and Function* **18**(1), 19–28.
Worbes, M. (2002). One hundred years of tree-ring research in the tropics—A brief history and an outline of future challenges. *Dendrochronologia* **20**, 217–231.

DIATOM INTRODUCTION

V Jones, University College London, London, UK

© 2007 Elsevier B.V. All rights reserved.

The Biology of Diatoms

The word 'diatom' derives from the Greek meaning cut in two. Diatoms (Bacillariophyta) are unicellular, eukaryotic (complex cells where the genetic material is in membrane-bound nuclei) microscopic organisms belonging to the Kingdom Protista. They are pigmented and usually appear golden-brown in color; they are photosynthetic although a few species are facultative or obligate heterotrophs (i.e., feeding on other organisms). The characteristic feature of the diatom is its cell wall which is almost always heavily impregnated with silica ($SiO_2 \cdot nH_2O$).

Globally, diatoms are very important as members of the planktonic food chain for large parts of the planet. They play a key role in the global carbon cycle. Diatom photosynthesis in the ocean yields as much as 40% of the organic carbon produced each year in the sea, and their role in global carbon cycling is predicted to be comparable to that of all the terrestrial rain forests combined, fixing about 20% of the carbon each year. Over geological time, diatoms may have influenced global climate by changing the flux of atmospheric carbon dioxide into the oceans.

Diatoms reflect a fundamentally different evolutionary history to the higher plants that dominate photosynthesis on land. They share biogeochemical features of both plants and animals. The marine centric diatom *Thalassiosira pseudonana* was chosen as the first eukaryotic marine phytoplankton species for whole-genome sequencing. The genome is relatively small at 34 mega base pairs (Armbrust *et al.*, 2004).

There is a well-documented fossil record of the diatoms from the middle Cretaceous with some records from the early Jurassic gap. There are a large number of Eocene/Miocene diatom fossil deposits situated around the world, which are mined commercially as diatomite; Flower deals with the Quaternary diatomites. Further details of diatom biology are given in Round *et al.* (1990) and van den Hoek *et al.* (1995).

The Structure of the Diatom Frustule

The silica cell wall of diatoms is box-like in structure, always made of two halves, called valves. The upper one (the epivalve or epitheca) is slightly larger than the lower valve (hypovalve or hypotheca; **Fig. 1**) and they are linked together with girdle bands. The valves are intricately sculptured and when organic material is removed by cleaning, the frustule splits in two. Most Quaternary applications are based on the observation of these fragments.

Diatoms are the most species-rich group of algae and there may be as many as 200,000 species. Although traditionally most diatom systematic studies have been based on valve features observed under the light microscope, recent developments through electron microscopy, the examination of living material, and molecular techniques have resulted in many nomenclatural changes, and consequently the taxonomy can be confusing.

Round *et al.* (1990) propose three classes of diatom: the Centrics, (Coscinodiscophyceae), pennates without raphe (Fragilariophyceae), and pennates

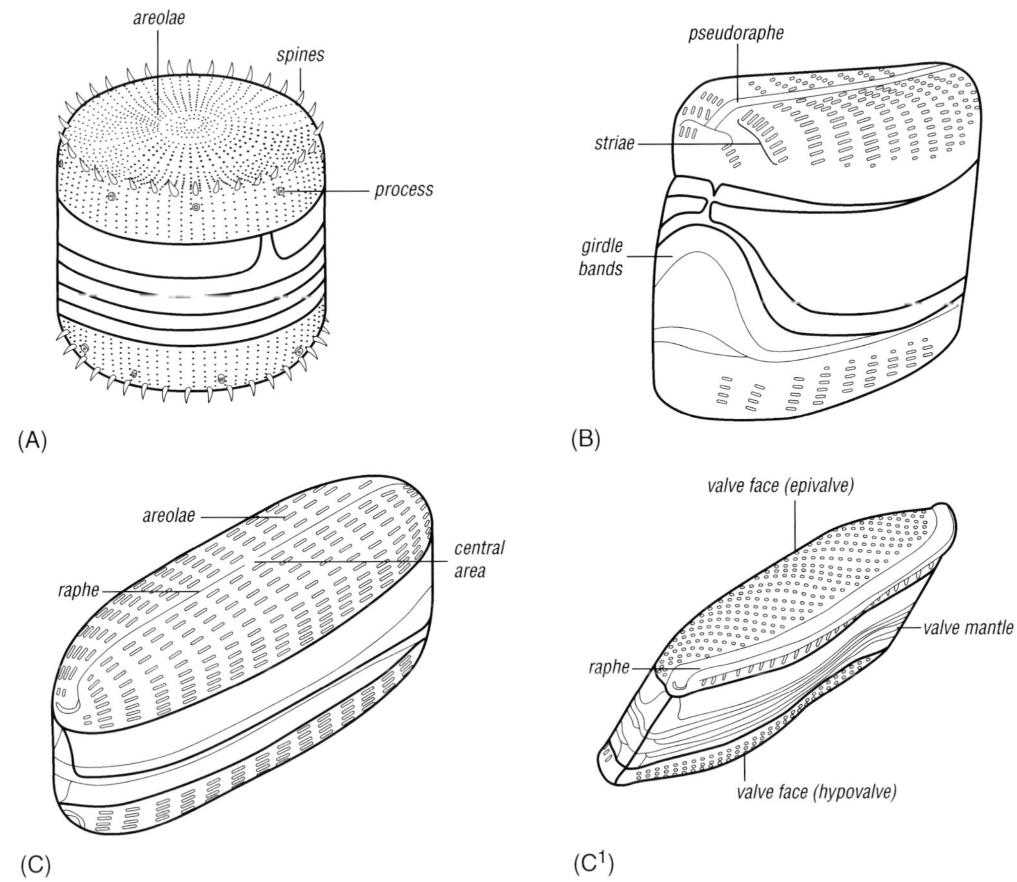

Figure 1 A schematic diagram illustrating the three main classes of diatom: (A) a typical centric form Coscinodiscophyceae, showing radial symmetry (e.g., *Stephanodiscus*) and pattern of areolae radiating from the center of the valve; (B) a typical pennate diatom without a raphe Fragilariophyceae (e.g., *Staurosira*) with areolae are usually arranged in rows or striae, in this case striae are parallel; (C) a typical pennate diatom with a raphe Bacillariophyceae (e.g., *Pinnularia*) with a naviculoid raphe running along the central axis of the valve which is divided into two separated by the central area; (C^1) a pennate diatom Bacillariophyceae (e.g., *Nitzschia*) with an eccentric raphe. Adapted from Round *et al.* (1990) with kind permission of Cambridge University Press.

with a raphe (Bacillariophyceae). These are shown in schematic form in **Figure 1**.

Features of the Valve

Most diatoms contain areolae (**Fig. 1**) that appear to be simple perforations through the cell wall, although internally they are occluded by a finely perforated velum. These areolae are usually arranged in rows of strias. The characteristics of these are used extensively for identification purposes, e.g., the spacing between strias (i.e., the strias density) and the pitch of the strias. The raphe is one of the most important features used to separate suborders or classes. It is an elongated fissure or pair of fissures through the valve wall. The naviculoid raphe (**Fig. 1C**) runs along the central axis of the valve and is divided in two, separated by the central nodule.

Diatom shape varies, usually being round or boat shape, but sometimes triangular, square, or elliptical. The cell wall is highly patterned with an array of pores, ribs, spines, and processes that can be used for identification purposes. Their spectacular forms have been appreciated since the invention of the microscope and have formed the subjects of many wonderful illustrations (**Fig. 5**).

Cells are solitary (**Fig. 4**) or united into colonies (**Fig. 3**) which may be linked by siliceous structures such as spines or bipolar elevations (e.g., *Eucampia*) or by mucilage (e.g., in tubes in such genera as *Nitzschia*).

Motility is limited to those pennate species that have a raphe, and occurs through secretion of mucopolysaccharide filaments which become attached to the solid substrate enabling the diatom to 'ooze' along a surface. A unique way of moving is displayed by the pennate diatom *Bacillaria paradoxa* (= *Bacillaria paxillifera*); this is a small colony in which individual cells slide back and forth over one another so the shape of the colony can change from a square to a linear form and back again.

The uptake and deposition of silica is one of the characteristic features of diatoms. This leads to rigidity of the cell walls and preservation as fossils. The characteristics of the silica cell wall are also used in their taxonomy. Leng and Barker give a more detailed account of the uptake of silica.

Lifecycles and Life Form

Diatoms are unusual in that vegetative cell division usually results in the size of the population becoming successively smaller and smaller. This is because the daughter cell is formed within the confines of the parent cell. In some species this can, over many generations, result in a considerable reduction in size which can lead to problems in their identification and classification as the full size ranges of most species are unknown. In centric species, the main features of the valve are unaltered but in pennate species the width often decreases less than the length (see **Fig. 2**). However, generally, size reduction is ignored by Quaternary scientists.

Size is restored after sexual reproduction, and diatoms show a range of different types of sex. The Centrics are oogamous (i.e., they have eggs and flagellate sperm) while the pennates have morphologically similar gametes, which are nonflagellate. In all cases, copulation takes place between the two haploid gametes producing a diploid zygote, the auxospore. The auxospore then swells and when it has enlarged fully, a new vegetative cell is formed within it. The complete lifecycle has rarely been observed in natural populations, but in some species at least, it seems to take a considerable time, for example, at least 6 years for *Nitzschia sigmoidea*, and 7–8 years (from varved sediments) for *Tabellaria fenestrata* (Mann 1998).

Diatom resting spores are dormant cells that are produced as a response to unfavorable conditions, mainly by centric diatoms. The spores contain stored energy in the form of photosynthetic products and tough, thickened cell walls. They sink to the bottom of the water column and can return to normal cell functioning if favorable conditions return. Some are morphologically distinct (e.g., *Chaetoceros* spp.) while others are not (e.g., *Aulocoseira subarctica*). In *Chaetoceros*, resting spore formation is associated with nutrient exhaustion in the water column, after mass multiplication of vegetative cells following an upwelling event. *Chaetoceros* resting spores, therefore, provide a record of paleobloom events in the ocean. In continental saline waters the very heavily silicified resting spores are often all that remain of a diatom assemblage after dissolution.

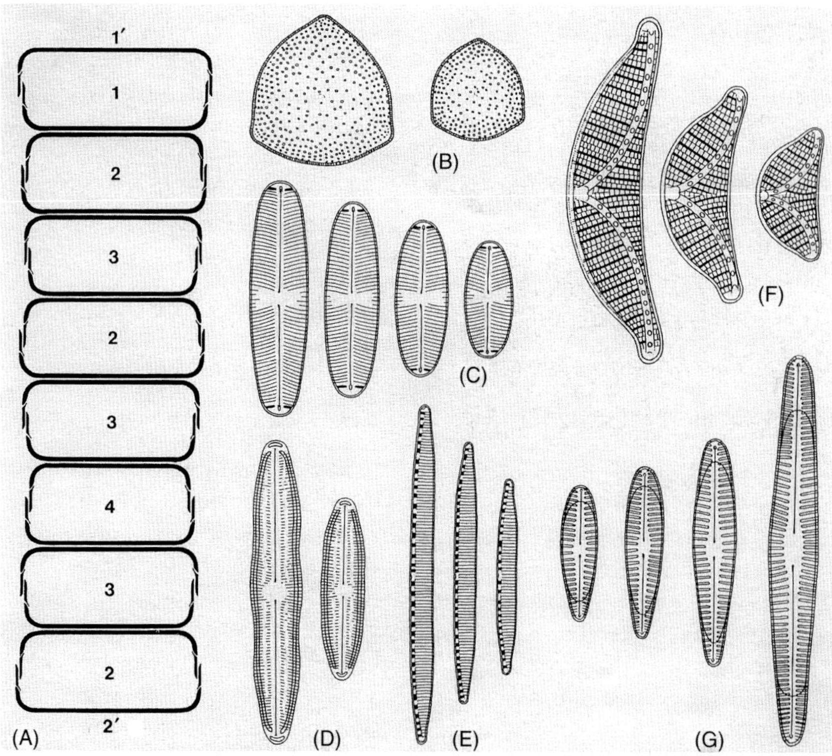

Figure 2 Diatom vegetative reproduction. Change in valve size following normal vegetative (mitotic)cell division 1′ and 2′ indicate the epivalve and hypovalve of the original cell: (B)–(G) examples of changes in valve shape and size following division (B) *Stictodiscus*, (C) *Sellaphora*, (D) *Brachysira*, (E) *Nitzschia*, (F) *Epithemia*, and (G) *Rhocosphenia*. From Round *et al.* (1990) reproduced with kind permission of Cambridge University Press.

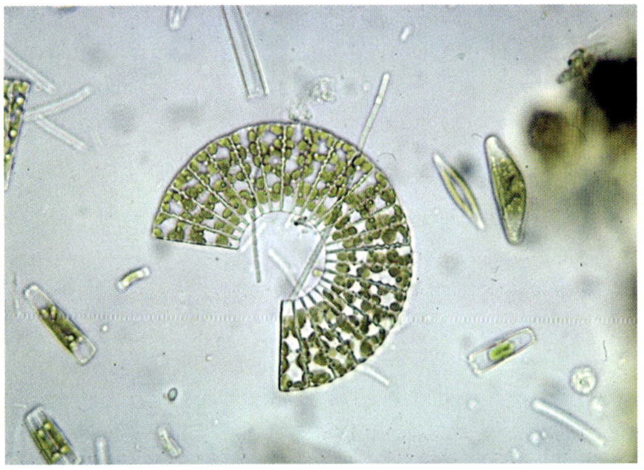

Figure 3 A light micrograph of a colony of *Meridion circulare,* a common stream diatom, showing the silica cell walls and chloroplasts (photograph S Spaulding).

Figure 4 A scanning electron micrograph of *Cyclostephanos tholiformis,* a typical centric diatom (photograph C Sayer).

Diatom Habitats

Diatoms are distributed in all waters except the hottest and most hypersaline and most samples obtained from aquatic habitats contain some cells.

Marine Habitats

Diatoms occur in all oceans from the poles to the tropics, but are most abundant in polar-to-temperate regions. These are mostly planktonic species. True planktonic species spend all of their lifecycle in open water, depending primarily on water currents for their buoyancy. Large blooms sometimes descend the water column *en masse* as mats.

As well as growing as plankton in open waters, they grow attached to various substrates in the coastal zone, e.g., as epiphytes on seaweeds or as epilithon on rocks. Diatoms also form biofilms on the surface of intertidal mudflats, the motile forms showing a remarkable tidal rhythm. On mudflats diatoms influence sediment stability by increasing sediment cohesion through the excretion of extracellular polysaccharides. By reducing erosive stress, diatoms promote the silt content of the sediment, and high densities of diatoms significantly reduce the frequency and magnitude of erosion events in some coastal areas.

In the coastal zone, Quaternary scientists have used diatom salinity preferences and life forms for environmental reconstruction (Cooper, 1999; Denys and de Wolf, 1999; Snoejis, 1999; Sullivan, 1999). Applications include sea-level reconstruction, studies of basin isolation, paleotidal reconstructions, studies of salt marsh development, inference of tsunami events, and biological indicators of pollution (e.g., eutrophication, radionuclide contamination). Zong and Horton (1999) modeled the relationship between diatom assemblages and tidal levels and developed a diatom-based tidal-level transfer function which was applied successfully to estimate paleotidal levels from fossil diatom data recorded in late Holocene coastal sequences.

A unique habitat is formed by the seasonal expansion and contraction of sea ice at the poles. Diatoms can grow attached to the underside of the ice as well as growing interstitially and attached to ice platelets. These different communities can be distinguished and have been used to trace the development of sea ice in the Quaternary.

Freshwater Habitats

Planktonic diatoms As in the oceans, planktonic diatoms are common in open water. In lakes, however, many forms are not truly planktonic as they spend some of their lifecycle resting on the sediment (meroplanktonic; e.g., some *Aulocoseira* spp.). However, planktonic species from both lakes and oceans are subject to two special influences: the availability of silicate and the tendency of diatoms to sink as they are as dense or denser than water. However, the photic zone is often turbulent due to winds, currents, and convection, therefore, many diatoms exhibit features such as a long cylindrical shape or long chains which slow their sinking

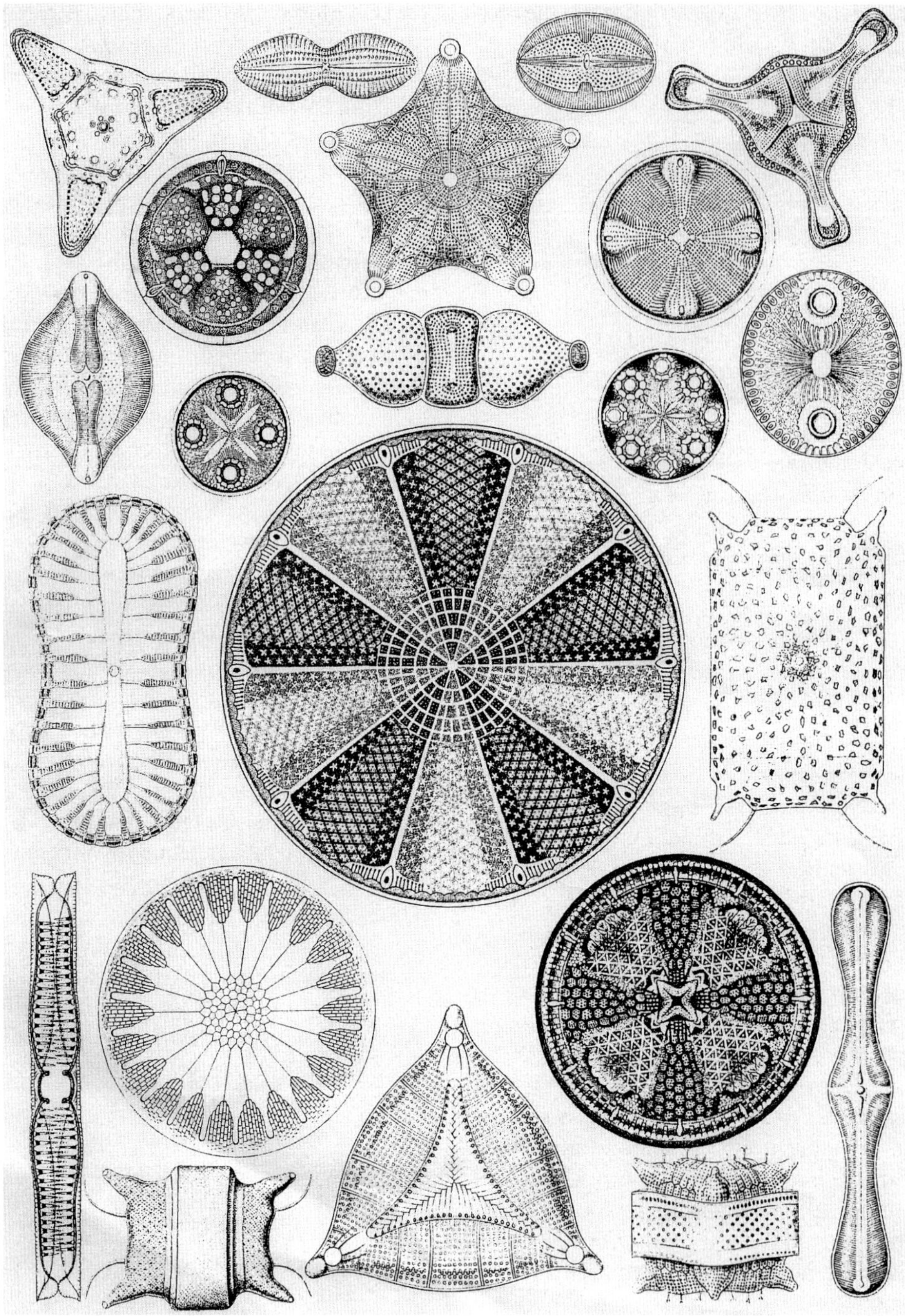

Figure 5 The variety of marine diatoms as illustrated by E Haeckel (1904; Artforms in Nature. Dover Press Pictorial Archive Series, Copyright-free).

velocity. In northern temperate lakes there is a well-established pattern of seasonal succession in the plankton with a spring bloom, a subsequent decline (as silica is depleted) and a second peak in autumn. The species are often distinct due mainly to nutrient differences between spring and autumn, and therefore this information has been used to interpret climate and weather patterns. In Elk Lake, Bradbury *et al.* (2002) interpreted changes in the ratio of *Stephanodiscus minutulus* and *Fragilaria crotonensis* as indicative of shifts between long, dry springs and warm, stormy summers. In addition, wavelet analysis suggested strong periodic correlations between individual diatom species and various sunspot cycles such as the 11-year Schwabe cycle (see also Mackay *et al.*, 2003).

In both marine and freshwater habitats the distinction between plankton and benthos is not absolute, as well as the meroplanktonic forms; species which have their true habitat in the benthos can often be found resuspended in the water column (tychoplanktonic).

Benthic diatoms Benthic diatoms are often very diverse and can live either free on or in the sediments, or attached to the substratum e.g., rocks (epilithon), sand grains (epipsammon), and plants (epiphyton). Diatoms living on soft sediments (epipelon) where the chances of burial are high are often motile species that are able to migrate up and down in the sediment matrix. Attached communities can be adnate (i.e., closely attached to the substrate; e.g., *Cocconeis*) or attached by mucilaginous stalks or pads (e.g., *Gomphonema*). Often diatoms can group themselves into 3-dimensional communities forming complex microscopic 'forests'. Diatoms can also form part of the diverse 3-dimensional microbial communities of filamentous algal mats which form in streams and lakes, in some cases forming perennial layers up to 30 cm thick.

Although diatoms rarely have very discrete habitats (there are often overlaps between species from different communities), habitat information is still useful for Quaternary scientists. For example, Sayer (2001) suggested that a recent increase in *Cocconeis placentula*, a very common epiphyte in Groby Pool was due to an increase in macrophytes in the lake over the last 50 years caused by increasing nutrient enrichment. In this study a shift from benthic *Fragilaria* communities to planktonic communities dominated by *Stephanodiscus parvus*, *Cyclostephanos tholiformis* (**Fig. 4**) and *Cyclostephanos dubius* was also seen as a consequence of nutrient enrichment. Although the ratio of benthic to planktonic species can give useful information about the changing trophic state of some lakes, there has been a general shift to the use of quantitative methods for reconstructing lake eutrophication.

Aerophilic diatoms Diatoms can also grow as subaerial forms (aerophilic) on damp rock faces, terrestrial soils, or even on the leaves of plants growing in damp environments. These include species such as *Hantzschia amphioxys*, *Diadesmis* spp., some *Pinnularia* spp., and large chain-forming *Orthoseira* spp.; some of these are obligate aerophiles (Johansen, 1999). When these are preserved in sediments they can provide useful information for environmental scientists, indicating drying out of a water body for example.

River diatoms In rivers, diatoms can grow attached to substrates, and in large rivers they can form a planktonic community. Some pennate species such as *Meridion circulare* (**Fig. 3**) and *Hannaea arcus* are characteristic of running waters while a characteristic centric species of running water is *Melosira varians*. Stream velocity is an important determinant of both species composition and the physical structure of communities.

Quaternary applications in lotic ecosystems are clearly limited, owing to the lack of continually accumulating sediments. However, undisturbed sediments can be found in reservoirs or deltas and in paleochannels and floodplain lakes in lowland river systems. The floodplains of relatively large, lowland rivers often have old channel systems (paleochannels) and associated oxbow cutoff lakes (or billabongs). Such systems are low energy and thus provide suitable sedimentary environments for paleoecological study. Thoms *et al.* (1999) determined the impact of agriculture and river regulation on water quality in billabongs of southeastern Australia using the fossil diatom record. Reavie and Smol (1997) developed reliable transfer functions for reconstructing past habitat conditions based on diatom habitat preferences (rock, *Cladophora*, macrophytes) in the St. Lawrence River, Canada. The models were applied to a fluvial sediment sequence and showed a marked increase in macrophyte-associated diatom taxa which the authors attributed to accelerated nutrient inputs. Ludlam *et al.* (1996) used a simple indicator for the relative contribution of running-water communities in an arctic lake. The Lotic Index showed a clear positive relationship to sedimentation rate as recorded in the varves, and therefore seem to be a good indicator of stream flow.

Biogeography and Endemism

Although very few diatoms are common in both marine and freshwater environments, there is often an emphasis on the cosmopolitan nature of diatoms. While some species have a global distribution, e.g., *Achnanthidium minutissimum*, others may not be as widespread as has been suggested. This is due to a number of factors, for example, the uncritical use of floras from other regions, and taxonomic clumping of closely related taxa. Certainly it seems that the numbers of diatom species have been underestimated and when careful taxonomic work is done several species that can not be attributed to published species are almost always identified for the first time.

Endemism describes diatoms that are restricted to a particular location. The best known examples of endemic diatom floras are island floras and long-lived or ancient freshwater lake systems such as those found in Lake Baikal, the African Rift Lakes, or Lake Ochrid (Macedonia). The endemic flora of Lake Baikal is discussed by Mackay.

Ecology

As well as the types of habitats available, there are other controls on the composition of the diatom flora. Physical controls include temperature, turbulence, and light while chemical controls include factors such as pH, nutrients, and salinity. Biological controls such as grazing and parasitism are also important. Ecology is dealt with in more detail by Korhola.

Collecting and Studying Live Diatoms

Live diatom material can be collected from different habitats using a variety of means; attached diatoms can simply be scraped off rocks or plants using a brush, from a known area if necessary. A lens tissue can be used to collect motile forms. Planktonic forms can be sampled with a bottle or plankton net. Diatoms can also be frozen *in situ* to preserve their 3-dimensional architecture (see **Fig. 6**). Normally though samples are preserved with a few drops of Lugol's iodine to retain the form and position of the chloroplasts. Even though most taxonomy is based on the cleaned frustule it is possible to identify living diatoms using cell and colony features (Cox, 1992). Although often ignored by Quaternary scientists the identification of live vs dead cells can be informative as to whether modern diatoms collected from a particular habitat are actually living there or whether they have been transported there after death. Diatomists working with lake sediments need to sample both living and fossil communities to develop transfer functions and to understand the relationship between source communities and fossil assemblages. Further details of field sampling and coring are given in Battarbee *et al.* (2001).

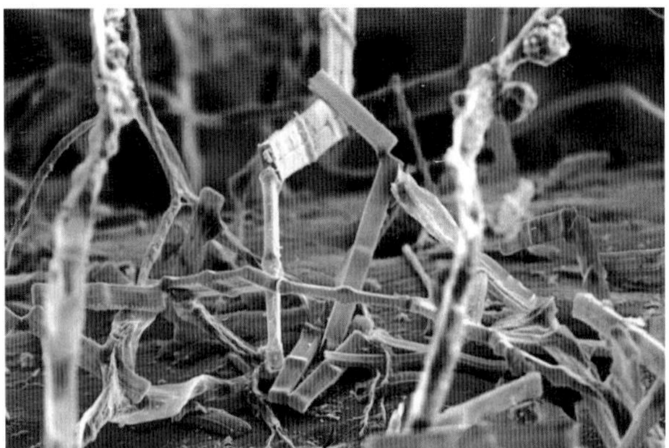

Figure 6 A colony of *Tabellaria quadriseptata* frozen *in situ* and examined by scanning electron microscopy (photograph R Flower).

Cleaning and Sample Preparation

In most Quaternary applications fossil diatoms need to be separated from the sedimentary matrix. The precise steps depend on the nature of the material and care needs to be taken not to lose or damage valves, for example, delicate diatoms can easily be broken by rapid centrifugation. **Figure** 7 illustrates the basic steps needed, but often the removal of organic matter by oxidation with either hydrogen peroxide, strong acids, or by incineration is all that is required. Once cleaned material is obtained, permanent slides can be made by mounting in a suitable medium with a high refractive index (>1.7) such as Naphrax (refractive index 1.73). Diatom concentrations can be calculated by adding a known quantity of markers such as divinylbenzene microspheres which are commercially available in the 5–10 µm size range. For further details on preparation techniques see Battarbee *et al.* (2001).

Identification

Most diatom assemblages are diverse and some expertise is needed for identification. Identification to the generic level is relatively straightforward although it requires some familiarity with terminology. Recent keys to genera include Barber and Haworth (1981), Kelly (2000) and Tomas (1996). Once familiar with the genera more specialized floras are needed (e.g., Krammer and Lange-Bertalot, 1986), ideally these should be generated from the same geographic region as the diatom sample. In addition, there are more specialized floras for specific ecosystems or specific genera which are available as books, CD-ROMs, and on the Internet.

Databases such as the European Diatom Database http://craticula.ncl.ac.uk/Eddis/jsp/ combine information on the distribution of diatoms in surface sediment datasets, including images, with the ability to perform online reconstructions.

Enumerating Diatoms

Most routine counts are based on the light microscope using oil immersion objectives and magnifications of ×750–×1,500. Brightfield, phase contrast, or differential interference contrast illumination can be used, and an eyepiece graticule is useful for measuring. Digital cameras with image capture software are increasingly being used to manipulate and store images.

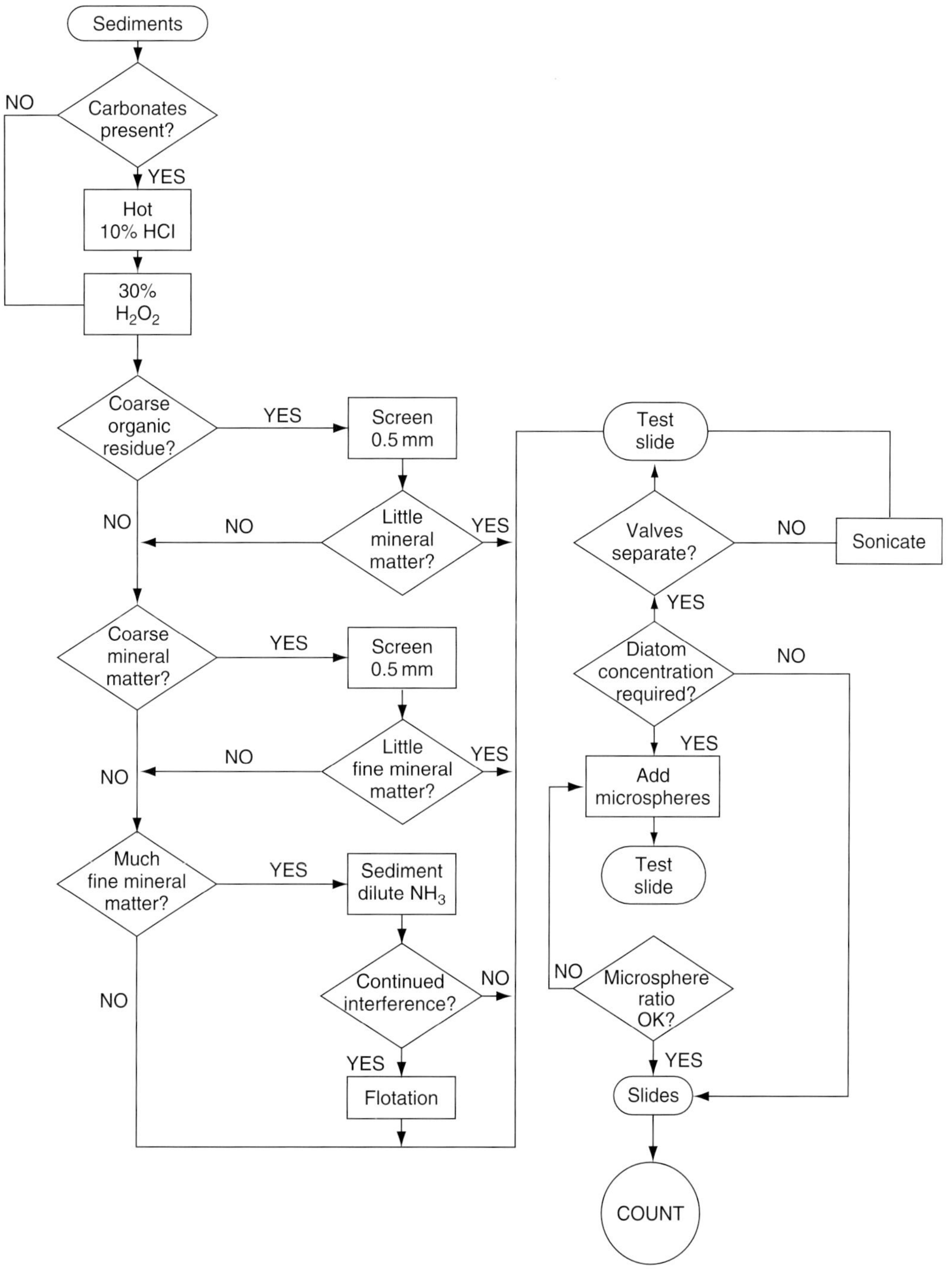

Figure 7 Flow chart for sample and slide preparation (from Battarbee *et al.*, 2001) with kind permission of Springer Science and Business Media.

Diatoms are commonly expressed as a percentage of a total sum or as a concentration value. Commonly the sum comprises all taxa but occasionally it is useful to exclude some groups, e.g., when these can be identified as clear contaminants from reworked material. Appropriate counting strategies need to be developed for individual sequences and to ensure a representative proportion of the coverslip is examined, and that broken fragments of valves are identified and enumerated. Concentration data can be expressed as number of valves per gram of sediment, or more usefully can be used to produce a diatom accumulation rate in valves per gram per year. As diatoms vary considerably in size this statistic may not be particularly meaningful unless it is converted to biovolume.

Preservation and Taphonomy

The preservation of diatoms in sediments ranges from excellent in diatomites (see Flower) to poor in some saline lakes (Fritz) and in extreme cases no diatoms are preserved in the sediment at all. Factors affecting preservation include salinity, temperature, alkalinity, and sediment accumulation rate as well as the silica content of the cell and the concentration gradient of dissolved silica between the sediment and overlying water. Differential dissolution can occur leading to bias in the composition of the assemblage to more robust forms.

Whether diatoms preserved in the sediment actually reflect the communities from which they are derived is a key question. Taphonomic processes include removal from the outflow of lakes, resuspension and reworking of older sediments, losses through grazing, bioturbation in the sediment, and dissolution. Some studies show a good agreement between the composition of live communities (e.g., Cameron 1995) and the sediment record while some studies show considerable differences between diatoms found in the water column and the sediment record. For example in Lake Baikal the majority of diatoms dissolve before reaching the sediment surface (see Mackay).

Quaternary Applications

Diatoms are numerous, diverse, frequently well-preserved, and often have well-known ecological preferences. Consequently there is a wide variety of applications in Quaternary science which are covered in the following sections.

See also: **Diatom Methods**: Data Interpretation; Quaternary Geological Records; Salinity Reconstructions from Continental Lakes. **Diatom Records**: Large Lakes; Marine Laminated Sequences.

References

Armbrust, E. V., Berges, J. A., Bowler, C., *et al.* (2004). The genome of the diatom *Thalassiosira pseudonana*: ecology, evolution, and metabolism. *Science* **306**, 79–86.

Barber, H. G. and Haworth, E. Y. (1981). *A Guide to the Morphology of the Diatom Frustule.*Freshwater Biological Association, Ambleside.

Battarbee, R. W., Jones, V. J., Flower, R. J., *et al.* (2001). Diatoms. In *Tracking Environmental Change Using Lake Sediments. Vol. 3: Terrestrial, Algal and Siliceous Indicators* (J. P. Smol, H. J. B. Birks, and W. M. Last, Eds.), pp. 155–202. Kluwer Academic, Dordrect.

Bradbury, P., Cumming, B., and Laird, K. (2002). A 1500-year record of climatic and environmental change in Elk Lake, Minnesota III: measures of past primary productivity. *Journal of Paleolimnology* **27**, 321–340.

Cameron, N. G. (1995). The representation of diatom communities by fossil assemblages in a small acid lake. *Journal of Paleolimnology* **14**, 185–223.

Cooper, S. R. (1999). Estuarine paleoenvironmental reconstructions using diatoms. In *The Diatoms: Applications for the Environmental and Earth Sciences* (E. F. Stoermer and J. P. Smol, Eds.), pp. 352–373. Cambridge University Press, Cambridge.

Cox, E. J. (1996). *Identification of Freshwater Diatoms from Live Material.* Chapman and Hall, London.

Denys, L. and de Wolf, H. (1999). Diatoms as indicator of coastal paleoenvironments and relative sea-level change. In *The Diatoms: Applications for the Environmental and Earth Sciences* (E. F. Stoermer and J. P. Smol, Eds.), pp. 277–297. Cambridge University Press, Cambridge.

Johansen, J. (1999). Diatoms of aerial habitats. In *The Diatoms: Applications for the Environmental and Earth Sciences* (E. F. Stoermer and J. P. Smol, Eds.), pp. 264–273. Cambridge University Press, Cambridge.

Kelly, M. (2000). Identification of common benthic diatoms in rivers. *Field Studies* **9**, 583–700.

Krammer, K. and Lange-Bertalot, H. (1986). *Bacillariophyceae.* Gustav Fisher Verlag, Stuttgart.

Ludlam, S. D., Feeney, S., and Douglas, M. S. V. (1996). Changes in the importance of lotic and littoral diatoms in a high arctic lake over the last 191 years. *Journal of Paleolimnology* **16**, 187–204.

Mackay, A. W., Jones, V. J., and Battarbee, R. W. (2003). Approaches to Holocene climate reconstruction using diatoms. In *Global Change in the Holocene* (A. W. Mackay, R. W. Battarbee, H. J. B. Birks, and F. Oldfield, Eds.), pp. 294–309. Arnold, London.

Mann, D. G. (1988). Why didn't Lund see sex in *Asterionella*? A discussion of the diatom life cycle in nature. In *Algae and the Aquatic Environment* (F. E. Round, Ed.), pp. 383–412. Biopress, Bristol.

Reavie, E. D., and Smol, J. P. (1997). Diatom-based model to infer past littoral habitat characteristics in the St. Lawrence River. *Journal of Great Lakes Research* **23**, 339–348.

Round, F. E., Crawford, R. M., and Mann, D. G. (1990). *The Diatoms. Biology and Morphology of the Genera.* Cambridge University Press, Cambridge.

Sayer, C. D. (2001). Problems with the application of diatom-total phosphorus transfer functions: examples from a shallow English lake. *Freshwater Biology* **46**, 743–757.

Snoeijs, P. (1999). Diatoms and environmental change in brackish waters. In *The Diatoms: Applications for the Environmental and Earth Sciences* (E. F. Stoermer and J. P. Smol, Eds.), pp. 298–333. Cambridge University Press, Cambridge.

Sullivan, M. J. (1999). Applied diatom studies in estuaries and shallow coastal environments. In *The Diatoms: Applications for the Environmental and Earth Sciences* (E. F. Stoermer and J. P. Smol, Eds.), pp. 334–351. Cambridge University Press, Cambridge.

Thoms, M. C., Ogden, R. W., and Reid, M. A. (1999). Establishing the condition of lowland floodplain rivers: a palaeo-ecological approach. *Freshwater Biology* **41**, 407–423.

Tomas, C. M. E. (1996). *Identifying Marine Diatoms and Dinoflagellates.* Academic Press, San Diego.

van den Hoek, C., Mann, D. G., and Jahns, H. M. (1995). *Algae: An Introduction to Phycology.* Cambridge University Press, Cambridge.

Zong, Y. and Horton, B. P. (1999). Diatom-based tidal-level transfer functions as an aid in reconstructing Quaternary history of sea-level movements in the UK. *Journal of Quaternary Science* **14**, 153–167.

DIATOM METHODS

Contents

δ^{18}O Records

M J Leng, University of Nottingham, Nottingham, UK
P A Barker, Lancaster University, Lancaster, UK

© 2007 Elsevier B.V. All rights reserved.

Introduction

Diatom silica is a form of biogenic opal ($SiO_2.nH_2O$) that contains oxygen isotopes of use in lacustrine and marine paleoenvironmental studies. The oxygen-isotope ratio ($^{18}O/^{16}O$) of diatom silica is expressed on the delta-scale (δ) in terms of per mille (or per mil) (‰):

$$\delta = \left[\left(R_{\text{sample}} / R_{\text{reference}} \right) - 1 \right] .1000‰ \quad [1]$$

where R is $^{18}O/^{16}O$, and 'reference' means the appropriate universally accepted reference material. The 'δ' for each element takes its name from the heavy isotope, thus δ^{18}O. For diatom oxygen the reference is VSMOW (Vienna Standard Mean Ocean Water, a specially prepared distilled seawater). There is no universally accepted standard material for diatom silica, although most laboratories analyze it alongside the NBS27 (quartz) and their own standard diatomites. VSMOW and NBS27 are held and distributed by the IAEA (International Atomic Energy Agency) in Vienna.

Diatoms form a shell or frustule, following a seasonal pattern defined by the variability of climate, nutrient supply, mixing regimes, and at high latitudes the period of ice cover. For example, in temperate regions, lakes tend to mix twice a year (spring and fall) when epilimnion and hypolimnion temperature and density are similar, thereby creating conditions where diatoms can exploit nutrient-rich waters, whereas the strong mixing would exclude other algae. In large tropical lakes, the majority of diatom productivity often takes place in the well-mixed dry season. In smaller polymictic lakes, seasonal productivity differences can also arise from nutrient and temperature variations.

Diatoms also dominate in nutrient-rich and turbulent marine waters, as found in coastal upwelling plumes, river plumes, ice edges, and regions of wind mixing. Seasons of thermal stratification can produce low light conditions that diatoms can take advantage of because of their ability to regulate buoyancy through chain formation and vertically migrating mats (Ragueneau, 2000). Therefore, the isotope signature acquired by diatoms will be skewed toward their major growing season and this will be specific to the lake or oceanic region under consideration.

Dissolution and Accumulation

Between 10% and 100% of diatom silica is recycled through dissolution within the water column but a fraction is incorporated into lake and marine sediments. Biogenic silica, comprising mainly diatoms with radiolaria locally important, has patchy distribution in ocean sediments (*see* North Atlantic and Arctic, Pacific, Antarctic Waters) although it is very common in lake sediments (Leng and Barker, 2006). In the marine environment, the most important location is the Southern Ocean where very high accumulation occurs (0.6–0.9 mol Si m^{-2} $year^{-1}$) (Ragueneau, 2000). The proportion of silica, derived from diatom productivity, that is incorporated into a sediment depends on a range of variables, for example, the amount of silica in solution, alkalinity, redox, and the presence of retardants such as Al and Fe. The organic membrane that covers the frustules also protects living diatoms and helps to adsorb metals but this decomposes on death and through bacterial activity. Without this protection rapid dissolution of the fresh frustule begins with the weakest and least dense parts of the frustule being lost first. The most important factor controlling the dissolution rate is the specific surface area of the fundamental building blocks of the frustule (silicon spheres and nanometer scale structures) and also the morphology of the whole diatom, which can lead to differential preservation of certain species.

Once the diatoms are incorporated into the sediment, the dissolution rate will tend to slow but diagenesis does not cease while pore waters are undersaturated with silica. The siliceous diatom frustules are highly adorned and dissolution will preferentially exploit structural ornamentation and lines of fracture (**Fig. 1**). Less obvious under optical microscopy is that the frustule itself is porous, possibly due to stacked spherical silicon building blocks, and that the pore spacing is generally lower with depth into the frustule. One consequence of this is that the densities of diatom frustules change with age as easily dissolved parts are lost and others are infilled by re-precipitated silica. Over long timescales amorphous diatom silica can be diagenetically altered to crystoballite. In extreme circumstances, zeolites and clay structures can be observed growing on diatom frustules from re-precipitated silica. Such diagenetic changes have important implications for the acquisition of oxygen-isotope data as discussed below (*see* discussion in Leng and Barker (2006)).

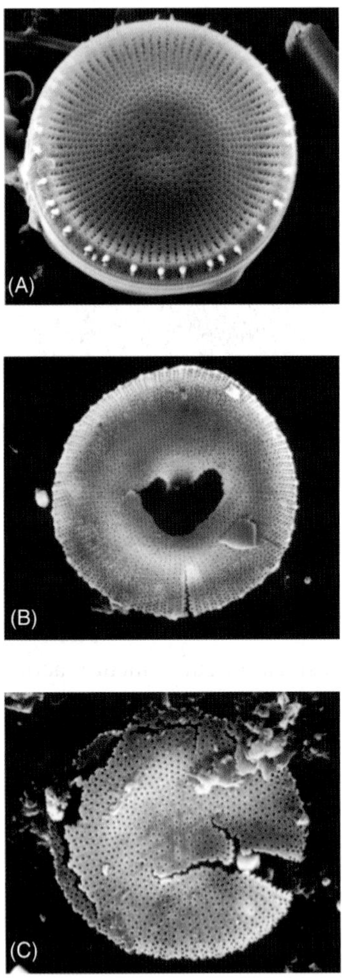

Figure 1 Once the diatoms are incorporated into the sediment the dissolution rate will tend to slow but does not cease while pore waters are undersaturated with silica, and further diagenesis occurs. (A) Diatom silica is highly adorned and (B) dissolution will preferentially exploit structural ornamentation and lines of fracture. (C) Further dissolution will enlarge structural holes and increase the overall porosity of the diatom.

Analytical Considerations

Analysis of the oxygen-isotope composition of diatom silica requires highly concentrated samples that are almost entirely made up of diatom frustules since the method generally used (i.e., fluorination techniques, e.g., *see* **Fig. 2**) will liberate oxygen from all the components in the sediment, for example, silt, clay, tephra, carbonates and organic matter. It is important to remove contaminants as δ^{18}O of the bulk sediment is directly proportional to the percentage of diatoms in the sediment (Leng and Barker, 2006). Published methods of removing contamination involve various stages of chemical attack (HCl, NaOH.Cl, H_2O_2, HNO_3), sieving, differential settling (**Figs. 3A** and **3B**) and the use of SPLITT (**Fig. 4**).

Frustules contain tetrahedrally bonded silica (Si–O–Si) surrounded by more hydrous material. The hydrous layer is freely exchangeable with any water the diatom silica comes into contact with and must be removed prior to oxygen-isotope measurement using oxidizing reagents and/or high temperatures. The distribution of oxygen in the frustule is more complex than simple layering and is related to the mode of formation of the frustule and its differential porosity. Biogenic opal has 20–30% water by weight that needs to be removed before stable values for δ^{18}O are reached. Aging also changes the degree of hydration of the diatom due to diagenesis and porosity changes. For example, Schmidt *et al.* (2001) found little relationship between diatom δ^{18}O from sediment traps and diatom δ^{18}O from the sediment itself.

The internal Si–O bond requires considerabe energy to break, and necessitates the use of an extremely powerful oxidizing reagent (i.e., a fluorine-based compound such as ClF_3 or BrF_5) or high temperatures. The difficulty of both the purification of the material as well as the extraction of formation water oxygen (described below) means that the analytical reproducibility of diatom silica oxygen-isotope ratios is generally quoted as 0.15–0.3%, although some studies suggest that this can be considerably higher (e.g., Schmidt *et al.* (2001)). The precision is significantly lower than for carbonate oxygen which has a precision generally in the range of ±0.05% for pure carbonate minerals to ±0.1% for authigenic carbonates within sediments.

To date there are three main methods that have been used to remove the hydrous layer from the biogenic silica and disassociate the silicon and

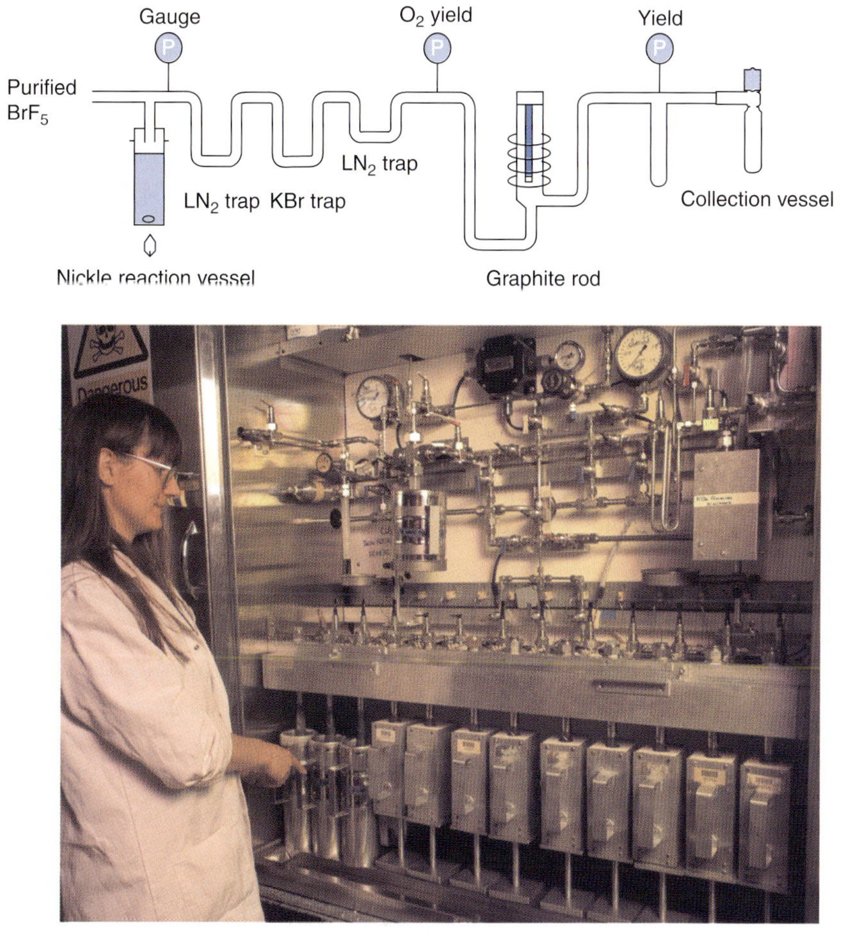

Figure 2 The fluorination line used at the NERC Isotope Geosciences Laboratory for the extraction of oxygen from diatom silica for isotope analysis was built by Mr P. B. Greenwood.

oxygen. These are vacuum dehydration, isotope exchange, and stepped fluorination techniques. The latter two involve fluorination techniques to liberate the oxygen from the diatom silica. The most recent dehydration technique (inductive high temperature carbon reduction, iHTR) is based on the reduction of silica by carbon at high temperatures (1,550° C for biogenic silica) (Lücke *et al.*, 2005). Further application of this methodology to sediment archive material is needed to assess its use in paleoclimate research in view of the identification of atomic diffusion at high temperatures. The controlled isotope exchange (CIE) method uses the fact that the surficial hydrous silica (SiOH) is reactive under much lower energy conditions than the inner tetrahedrally bound silica, which allows exchange of oxygen atoms of known isotope composition with the unstable hydrous material while leaving the skeletal silica intact (Labeyrie and Juillet, 1982). The method exchanges the oxygen of hydrous silica with oxygen of a known isotope value, and then through mass balance calculations determine the composition of the nonexchangeable silicate oxygen after fluorination. The third method, stepwise fluorination (SWF), involves reacting the silica with a fluorine compound in a series of steps and was first described by Haimson and Knauth (1983). The first oxygen fractions released are from the hydrous component of the biogenic silica leaving the structurally bound silica for subsequent reactions (Leng and Barker, 2006).

Diatom Oxygen Systematics

Changes in the oxygen-isotope composition of diatom silica can be used to infer changes in either the temperature or the oxygen-isotope composition of the water in which they grew (**Fig. 5**).

Temperature

A number of studies have determined the empirical relationship between temperature and the oxygen-isotope composition of carbonate minerals and the composition of the water from which they formed (Leng and Marshall, 2004). However, the

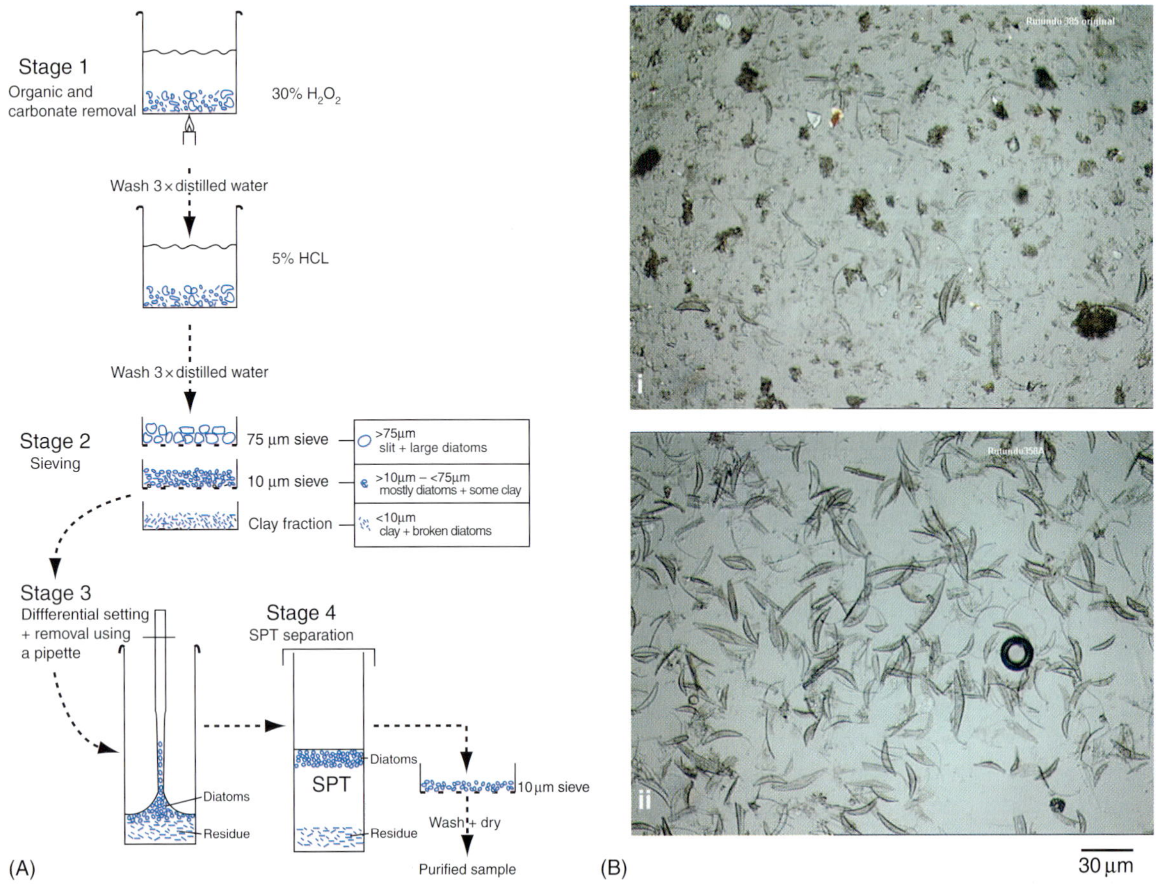

Figure 3 (A) Diagram showing the four stage cleaning method for concentrating diatom for oxygen-isotope analysis from lake sediments. Reproduced from Morley D.W., Leng M.J., Mackay A.W., Sloane H.J., Rioual P., and Battarbee R.W. (2004) Cleaning of lake sediment samples for diatom oxygen isotope analysis. *Journal of Palaeolimnology* 31: 391–401, with kind permission of Springer Science and Business Media. (B) Lake sediment samples from Lake Rutundu, Mount Kenya before and after cleaning, (i) shows the sample after standard diatom preparation using HCl and H_2O_2, (ii) shows the cleaned sample after a staged cleaning approach. Species diversity has been reduced by the cleaning procedure, leaving the sample dominated by large *Cymbella* and *Epithemia* spp.

temperature dependence of oxygen-isotope fractionation between diatom silica and water has not been rigorously derived, although the relationship has been estimated from analyses of diatoms from marine and freshwater sediments, coupled with estimates of the temperatures and isotope compositions of coexisting waters during silica formation (e.g., Labeyrie (1974)). The data from these calibration studies are limited, and are mainly based on bulk samples not individual diatom species (although Brandriss *et al.* (1998) provides a notable exception). The estimates of the average temperature dependence based on these bulk samples range from −0.2 to −0.5%/°C (Juillet-Leclerc and Labeyrie, 1987; Shemesh *et al.*, 1992). More recently, controlled diatom culturing experiments on marine (Brandriss *et al.*, 1998) and lacustrine diatoms (Moschen *et al.*, 2005) have shown that the diatom-temperature coefficient is around −0.2%/°C, although Schmidt *et al.* (1997), who analyzed the oxygen-isotope composition of diatom frustules collected live from the oceans, found no regular correlation between temperature and oxygen-isotope fractionation. These results led to the hypothesis that the temperature-dependent oxygen-isotope fractionation preserved in some biogenic opaline sediments may have been established during early diagenesis rather than acquired during growth. As yet, there is no evidence from paleoenvironmental studies to demonstrate a significant lag is introduced into the acquisition of the oxygen, especially that derived from the densest parts of the frustule analyzed after stepwise fluorination.

The diatom–water fractionation equation of Juillet-Leclerc and Labeyrie (1987) is:

$$T(^\circ\mathrm{C}) = 17.2–2.4(\delta^{18}\mathrm{O}_{\mathrm{diatom}} - \delta^{18}\mathrm{O}_{\mathrm{water}} - 40) - 0.2\,(\delta^{18}\mathrm{O}_{\mathrm{diatom}} - \delta^{18}\mathrm{O}_{\mathrm{water}} - 40)^2 \qquad [2]$$

Shemesh *et al.* (1992) assessed the silica–water temperature fractionation (Eqn. [2]) of Juillet-Leclerc and Labeyrie (1987), and reported that it

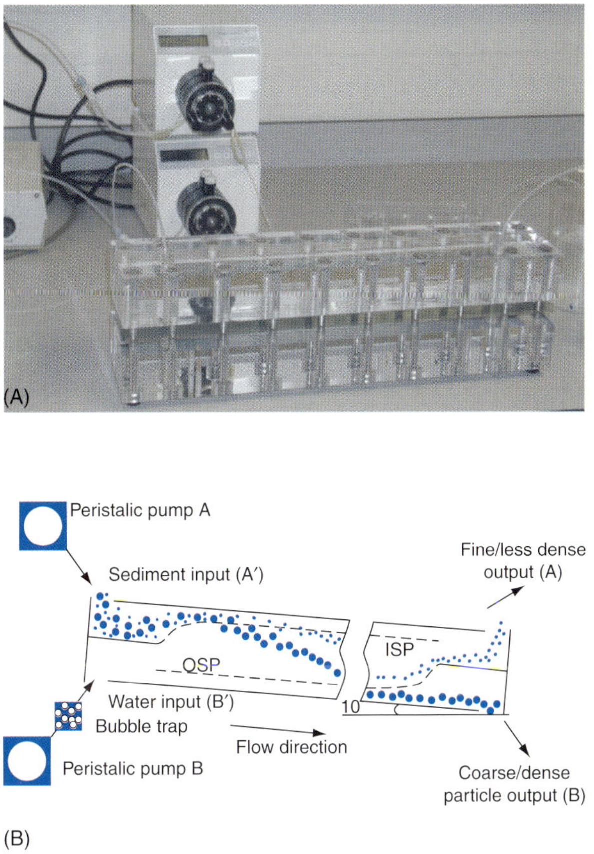

Figure 4 (A) An alternative approach to heavy liquid separation for cleaning diatom samples is gravitational split-flow thin fractionation (SPLITT) developed by J. C. Giddings at the University of Utah and first applied to the separation of diatoms at the University of Jülich (Schleser *et al.*, 2001; Rings *et al.*, 2004). (B) SPLITT works by introducing a sample (A′) into a thin channel where it meets a carrier fluid (B′, usually water for diatom samples) in a laminar flow. The inlet splitter plane (ISP) and outlet splitter plane (OSP) divide the flow. The velocity of the two flows is controlled to separate the sample into two streams: an upper A sample containing finer, less dense and more hydrodynamic particles passes through one outlet (A), while the remaining particles are collected through a lower outlet (B). The SPLITT cell is inclined to reduce deposition of sediment onto the splitter plate at low flows.

overestimates temperatures for polar oceanic regions and produced an alternative equation which has subsequently been used for many high-latitude sites:

$$T(\mathrm{C}) = 11.03\text{–}2.03(\delta^{18}\mathrm{O}_{\mathrm{diatom}} - \delta^{18}\mathrm{O}_{\mathrm{water}} - 40) \qquad [3]$$

In these equations δ^{18}O of the diatom silica and water are compared to the SMOW international standard.

Disequilibrium effects (**Fig.** 5) (commonly known as 'vital' effects in the case of biogenic carbonates) include a variety of rate effects and microenvironment-induced changes that cause the mineral to have an isotope composition that is different from that predicted purely by thermodynamics. Studies of oxygen-isotope ratios between different diatom species show only small differences, which are often within analytical error (see Leng and Marshall (2004)). Because lacustrine diatoms are physically difficult to separate into species, bulk samples are normally analyzed for $\delta^{18}\mathrm{O}_{\mathrm{diatom}}$. There are no studies which show consistent stratigraphical similarities between diatom species changes and $\delta^{18}\mathrm{O}_{\mathrm{diatom}}$ changes leading most authors to suggest that variations due to different vital effects are smaller than the overall analytical error (e.g., Leng *et al.* (2001)).

In the lacustrine environment the interpretation of the measured oxygen-isotope data from diatom silica in terms of paleotemperatures requires an understanding of two temperature effects that have opposing effects on the composition of diatom silica. There is the fractionation between diatom silica and water, described above, which has a gradient of between -0.5 and -0.2‰/°C. There is also the relationship between precipitation and temperature of approximately $+0.6$‰/°C (the so-called Dansgaard relationship $\delta p/\mathrm{d}T$; Dansgaard (1964)) at intermediate and high latitudes, and a worldwide range of between $+0.2$ and $+0.7$‰/°C (Dansgaard, 1964). Ideally, it

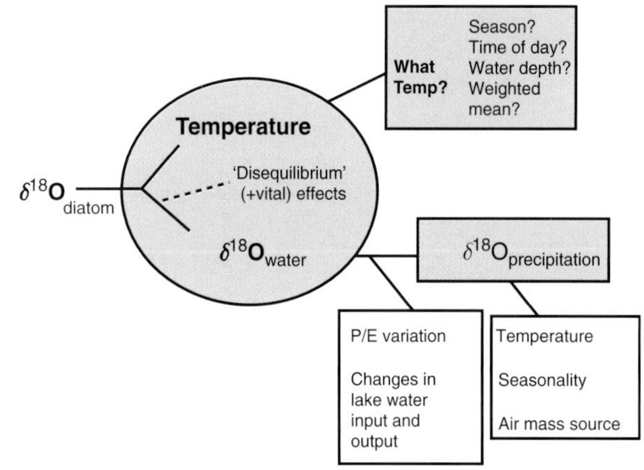

Figure 5 Controls on the oxygen-isotope composition of lacustrine diatom silica ($\delta^{18}O_{diatom}$). If diatom silica is precipitated in isotopic equilibrium, $\delta^{18}O_{diatom}$ depends entirely on temperature and the isotope composition of the lake water ($\delta^{18}O_{water}$). Disequilibrium effects, commonly known as 'vital effects' in biogenic precipitates, caused by local changes in micro-environment or rate of precipitation can induce systematic or nonsystematic offsets in the $\delta^{18}O_{diatom}$ signal. Interpretation of $\delta^{18}O_{diatom}$ signal in terms of temperature depends on a knowledge of the environment and time of year in which the diatom silica forms. For example at high-latitude sites diatoms typically grow in the summer months in the surface waters. Quantitative interpretation similarly demands an understanding of the local controls on the isotope composition of lake water ($\delta^{18}O_{lakewater}$). In some lakes there is a simple relationship to $\delta^{18}O_{precipitation}$ but in others the water composition is strongly influenced by processes such as evaporation within the catchment and within the lake itself. $\delta^{18}O_{precipitation}$ is increasingly being shown to be an important indicator of climate change: it typically changes with mean annual temperature but nonlinear responses to climate change are increasingly being recognized. See text for more discussion of this figure. Reproduced from Leng MJ and Barker PA (2006). Vol. 75, pp. 5–27. A review of the oxygen isotope composition of lacustrine diatom silica for palaeoclimate reconstruction. *Earth Science Reviews,* with permission from Elsevier.

is important to establish the relationship between the oxygen-isotope composition of precipitation (δp) and temperature at each site (see Leng *et al.* (2005a) and the IAEA-WMO GNIP database); it is worth noting here that a Dansgaard relationship is not common in coastal or monsoonal regions or where rainfall comes from two air masses.

For many lake records the diatom oxygen-isotope response to temperature will be dominated by the change in the isotope composition of precipitation and effectively 'damped' by the opposing effect of mineral-water fractionation. For diatoms the measured isotope composition will, therefore, covary with temperature – with an increase of ~0.1 to 0.4%/°C (cf. Leng and Marshall (2004)). This assumes that $\delta p/dT$ always changes according to the Dansgaard relationship. It is usually very difficult to derive quantitative paleotemperature/δp data with any certainty. It might be more realistic to use $\delta^{18}O_{diatom}$ in areas where there are likely to have been significant changes in the isotope composition of the lake water due to changes in the precipitation/evaporation balance or in the source of precipitation (see below). The change in $\delta^{18}O_{diatom}$ due to these factors is normally greater than temperature alone.

In the marine environment, diatom silica oxygen-isotope analysis can be used to investigate temperature variation in addition to foraminifera-based oxygen isotopes and in oceanic regions that are depleted in $CaCO_3$ and so do not preserve calcareous foraminifera. This is most prominent in high-latitude regions, such as the Southern Ocean, where foraminifera are not generally well preserved within the sedimentary record. However, where the two minerals co-occur, they can provide different information. For example, Shemesh *et al.* (1992) combined $\delta^{18}O_{diatom}$ with $\delta^{18}O_{foram}$ data, assuming the incorporation of oxygen into the foraminifera calcite shell and diatom frustule occurred in the same season and water depth, to calculate changes over time of paleo-$\delta^{18}O_{water}$ and paleotemperature in Southern Ocean surface waters. A major benefit of this approach is the ability to avoid using benthic $\delta^{18}O_{foram}$ records and their associated limitations to estimate the $\delta^{18}O_{Global\ Ice\ Volume}$ component present in both the $\delta^{18}O_{diatom}$ and planktonic $\delta^{18}O_{foram}$ signal (Maslin and Swann, 2005). Using this approach, a ca. 2.0° C SST increase was observed in the shift from glacial to Holocene conditions with a concordant $1.2 \pm 0.2‰$ decrease in $\delta^{18}O_{water}$. In another example of combined $\delta^{18}O_{foram}$ and $\delta^{18}O_{diatom}$ from the North West Pacific, the onset of major Northern Hemisphere Glaciation, ca. 2.75–2.73 Ma, was investigated. The foraminifera and diatoms are thought to have lived and precipitated their shells during different seasons, thus providing interseasonal information. The $\delta^{18}O$ record from spring calcifying planktonic foraminifera indicates a significant, 7.5° C, cooling of the surface waters at the onset of the Northern Hemisphere Glaciation, an autumnal blooming $\delta^{18}O_{diatom}$ species record consisting of *Coscinodiscus radiatus* and *Coscinodiscus marginatus* reveals a 4.6‰ decrease, equating to a significant freshening and warming of the surface waters (Haug *et al.*, 2005).

Changes in the Oxygen-Isotope Composition of the Host Water

Changes in the isotope composition of the lake water can be due to either evaporation or changes in the source of the rainfall. In closed (or terminal) lakes evaporation of $\delta^{18}O_{water}$ will usually be far greater than changes due to temperature or δp. $\delta^{18}O_{water}$

depends on the balance between the isotope composition of inputs (including the source and amount of precipitation, surface runoff, and groundwater inflow) and outputs (evaporation and groundwater loss). Unless there is significant groundwater seepage, most closed lakes will lose water primarily through evaporation, the rate of which is controlled by wind speed, temperature, and humidity. The phase change of evaporation results in light isotopes of oxygen (^{16}O) being preferentially evaporated from water bodies leaving water that is relatively enriched in the heavier isotope (^{18}O). Any effects of varying precipitation source or temperature on $\delta^{18}O$ are often small in comparison to evaporative concentration and measured $\delta^{18}O$ (and δD) values become elevated above those of ambient precipitation. In extreme circumstances, evaporation in the lake catchment area or from the surface of a lake can lead to significantly elevated $\delta^{18}O_{water}$ values (Leng *et al.*, 2005a).

The degree to which evaporation will increase $\delta^{18}O_{water}$ depends on the residence time of the lake (lake volume/throughput rate). Changes to a lake's residence time, caused by changes in basin hydrology or varying groundwater fluxes, will also influence the degree of enrichment, as will changes in the nature of catchment vegetation and soils. These factors have been considered important in the interpretation of a $\delta^{18}O_{diatom}$ record from Lake Tilo in East Africa which is interpreted as changes in aridity (Lamb *et al.* (2005), **Fig. 6**). They showed that high temperatures and a constant supply of solutes from hydrothermal springs enabled diatoms to grow throughout the year. The early Holocene lake level was close to the crater rim with a slow but progressive fall through the early Holocene. Major changes are recorded simultaneously by $\delta^{18}O_{diatom}$ and $\delta^{18}O_{calcite}$, although larger errors are associated with $\delta^{18}O_{diatom}$ and especially noteworthy is that the $\delta^{18}O_{diatom}$ data do not record the well-documented 7.8 kyr BP arid interval found in neighboring Ethiopian lakes.

Certain diatom records have been shown to be sensitive to other aspects of climate, such as amount

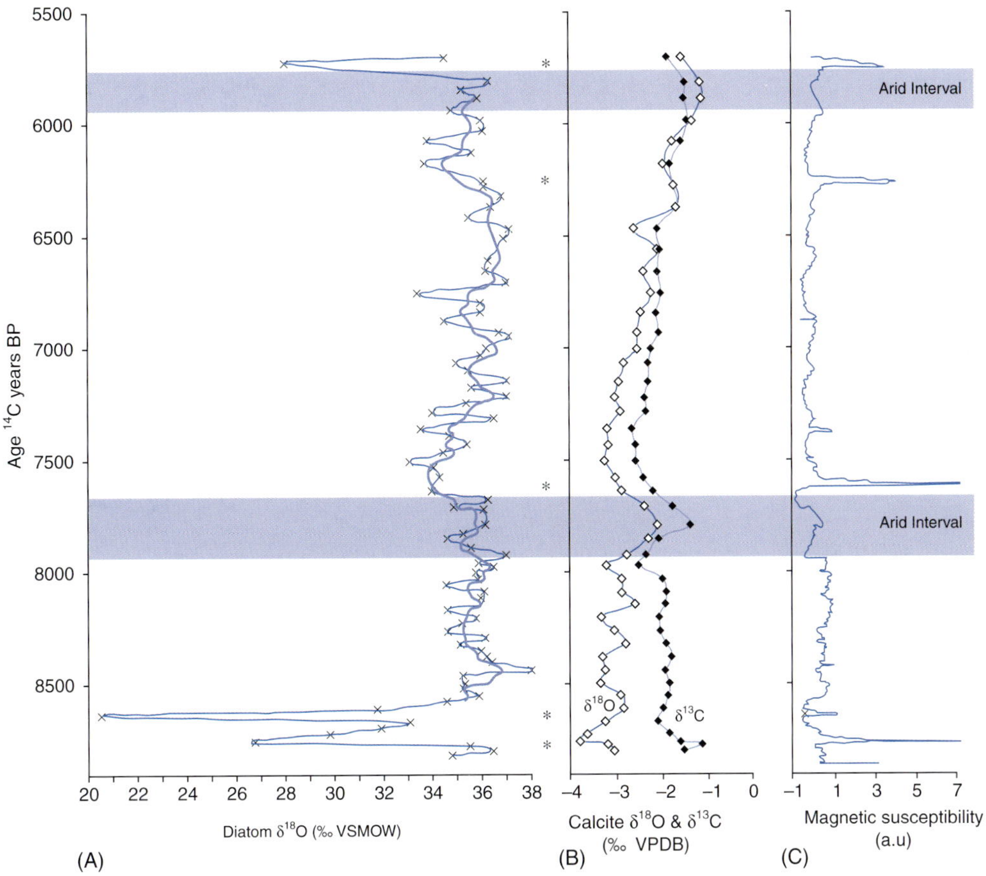

Figure 6 (A) $\delta^{18}O_{diatom}$ from Lake Tilo (8.8–5.5 kyr BP). Solid line is a three-point running mean and stars represent position of tephra deposits; (B) $\delta^{18}O_{calcite}$ and $\delta^{13}C_{calcite}$ from Lake Tilo (8.8–5.5 kyr BP); (C) magnetic susceptibility – peaks indicate high concentrations of tephra. Note: $\delta^{18}O/\delta^{13}C_{calcite}$ and $\delta^{18}O_{diatom}$ are on VPDB and VSMOW scales, respectively. The early Holocene arid intervals are indicated by gray shading. Reproduced from Lamb A.L., Leng M.J., Sloane H.J., and Telford R.J. (2005) A comparison of the paleoclimate signals from diatom oxygen isotope ratios and carbonate oxygen isotope ratios from a low latitude crater lake. *Palaeogeography, Palaeoclimatology, Palaeoecology* 223: 290–302, with permission from Elsevier.

or source of precipitation. For example, abrupt shifts of up to 18‰ in $\delta^{18}O_{diatom}$ have been found in a 14 kyr-long record from two Alpine lakes on Mount Kenya (Barker *et al.*, 2001), which cannot be entirely temperature related given the current knowledge of the diatom-temperature fractionation (**Fig.** 7). Instead, the variations have been interpreted as a moisture balance effect related to changes in Indian Ocean sea surface temperatures through the Holocene. Episodes of heavy convective precipitation in the $\delta^{18}O_{diatom}$ record correlate with enhanced alkenone-based sea surface temperature estimates.

There have been several studies of $\delta^{18}O_{diatom}$ from lakes in Northern Europe (see Leng and Barker (2006)). These lakes have similarities, in that many are open, through-flow systems with minimal evaporation. Changes in the stratigraphic record of $\delta^{18}O_{diatom}$ are interpreted as changes in the summer isotope composition of the lake water. In a proglacial lake covering the last 5 kyr in Northern Sweden, a $\delta^{18}O_{diatom}$ record combined with a sedimentary proxy for glacier fluctuations reflects changes in the isotope composition of summer precipitation (Rosqvist *et al.*, 2004). They found a general lowering trend of 3.5% with spikes (>1%) to lower values on a ca. 1 kyr cyclicity. Climate simultaneously sustained a positive glacier mass balance that caused the catchment glacier to advance. The authors suggest that a persistent change in the atmospheric circulation pattern could potentially have caused the changes in $\delta^{18}O_{diatom}$ because different air masses influence this area and these air masses have characteristic δ^{18}O signatures of their precipitation. They infer a steady increase but fluctuating dominance of colder and δ^{18}O-depleted air masses from the north/northeast. The $\delta^{18}O_{diatom}$ depletion and glacier events occur at times of relative ice-rafted-debris maxima in the North Atlantic, consistent with cold conditions and changes in surface wind directions (**Fig.** 8). The Northern European studies show some similarities in the diatom isotope records, in particular, a late-Holocene cooling, although there are regional differences, notably the relatively cold early Holocene on the Kola Peninsula is not seen in the records from Northern Sweden.

Changes in the source of precipitation have also been suggested as one reason for variations in $\delta^{18}O_{diatom}$ in a small, spring fed lake in the Central Mexican highlands (Leng *et al.*, 2005b). The lake has shallowed, and today is at one of its lowest lake levels, but the modern isotope data suggest that the modern water is not evaporated, indicating that hydrology or recharge rates, rather than evaporation is controlling lake level. Two main possibilities might have controlled variations in δ^{18}O in the past. Since Central Mexico gets rainfall that originates from the tropical Atlantic/Gulf of Mexico and also from both the tropical and extra-tropical Pacific (in summer and winter respectively), the authors postulate that changes in $\delta^{18}O_{precipitation}$ are connected to temperature and salinity variations in the Gulf of Mexico and/or changing contributions from these two different air masses (Gulf of Mexico vs. Pacific derived hurricanes).

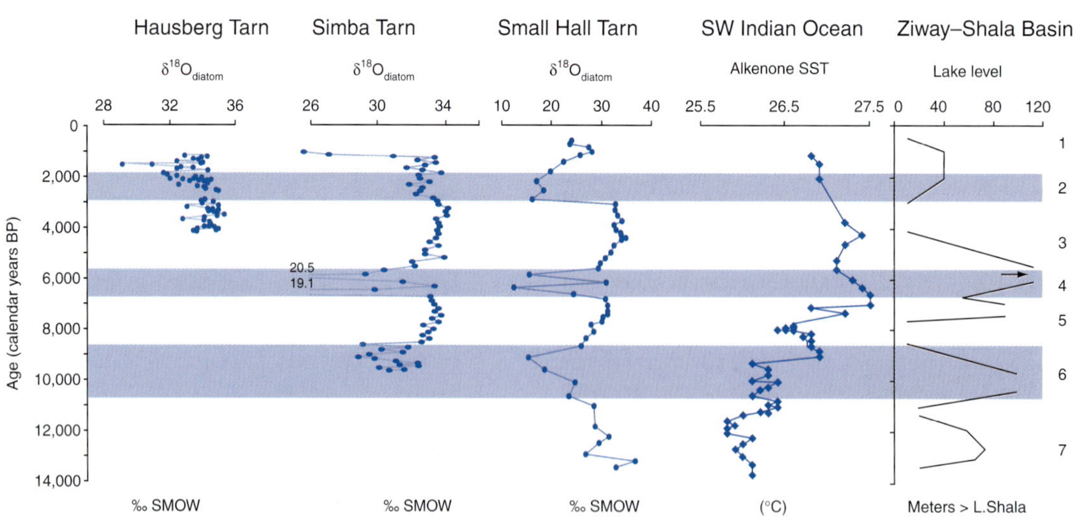

Figure 7 In three high-altitude, freshwater lakes on Mount Kenya the oxygen-isotope composition of biogenic silica is useful because the lake-water oxygen-isotope composition records lake moisture balance. The diatom oxygen-isotope data from Simba Tarn, Small Hall Tarn, and Hausberg Tarn (Mount Kenya) are compared to lake levels in the Ziway-Shala Basin, Ethiopia, derived from ^{14}C dated shorelines. Reprinted with permission from Barker, P.A., Street-Perrott, F.A., Leng, M.J, *et al.* (2001) A 14 ka oxygen isotope record from diatom silica in two alpine tarns on Mt Kenya. *Science* 292: 2307–2310. Copyright 2001 AAAS.

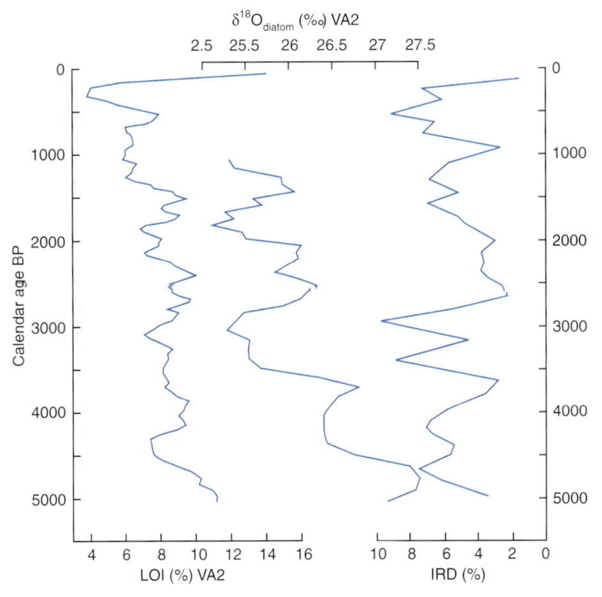

Figure 8 In a proglacial lake covering the last 5,000 yr in Northern Sweden, the oxygen-isotope record of diatom silica ($\delta^{18}O_{diatom}$), combined with a sedimentary proxy for glacier fluctuations, reflects changes in the isotope composition of summer precipitation. The $\delta^{18}O_{diatom}$ depletion and glacier events occur at times of relative ice-rafted-debris maxima in the North Atlantic. Reproduced from Rosqvist G, Jonsson C, Yam R, Karlen W, and Shemesh A (2004) Diatom oxygen isotopes in pro-glacial lake sediments from northern Sweden: a 5000 year record of atmospheric circulation. *Quaternary Science Reviews* 23: 851–860, with permission from Elsevier.

In the marine environment the most notable change in the host water occurs due to melt water influx during warmer periods. For example in studies in the Southern Ocean, $\delta^{18}O_{(diatom)}$ decreases are thought to be associated with glacial conditions through MIS 2 and the last ca. 430 kyr (Shemesh *et al.*, 1995, 2002). Relating these changes to an increasing temperature signal contradicts the numerous other proxy records which indicate the last glacial to have been much colder than present. The marked decreases in $\delta^{18}O_{diatom}$ are thought to be connected to meltwater influxes originating from Antarctica, releasing water containing an inherently lower value of δ^{18}O than seawater (Shemesh *et al.*, 1995).

There are only a few studies of oxygen-isotope ratios in lacustrine diatom silica that are older than the Holocene; these include interglacial investigations in France (Rioual *et al.*, 2001) and in Lake Baikal (Morley *et al.*, 2005).

Conclusions

Analysis of the oxygen-isotope composition of diatom silica requires samples that are almost pure diatomite since fluorination and other analytical techniques will liberate oxygen from all the components in the sediment. There is a generally acceptable protocol involving chemistry, sieving and settling techniques, and more recently laminar flow separation. However, all lakes require a dedicated procedure and every sample must be scrutinized with microscopy. Where sediments cannot be purified sufficiently, a semi-quantitative assessment of the diatom content can facilitate mass balance techniques. Diatom silica analysis requires the removal of the outer hydrous layer before a reaction to break the tetrahedrally bonded silica skeleton (Si–O–Si). To date there are three main methods (vacuum dehydration, isotope exchange and stepped fluorination techniques) to do this.

Diatom δ^{18}O can be an extremely useful indicator of paleoenvironmental change. Variation in $\delta^{18}O_{diatom}$ from sediments in areas where there are likely to have been large changes in the isotope composition of the host water due to changes in the precipitation/evaporation balance and the source of precipitation in lakes, or variations in melt water flux in the oceans, tend to be most successful. The change in $\delta^{18}O_{diatom}$ due to these factors is normally greater than temperature and methodological errors.

As with all isotope studies, the use of $\delta^{18}O_{diatom}$ in paleoenvironmental studies should include a study of the contemporary environment and the modern diatom silica. The presence of a systematic relationship between temperature, water isotope composition, and diatom isotope composition for the host water enables quantitative interpretation of the sediment record. It is particularly useful to know when and where the diatom silica 'bloom' in the modern environment, so details of diatom ecology (*see* Diatom Introduction) are important. This type of information will tell us whether the diatom record reflects specific seasons or year round conditions.

See also: **Diatom Introduction**. **Diatom Records**: Antarctic Waters; North Atlantic and Arctic; Pacific.

References

Barker, P. A., Street-Perrott, F. A., Leng, M. J., *et al.* (2001). A 14 ka oxygen isotope record from diatom silica in two alpine tarns on Mt Kenya. *Science* **292**, 2307–2310.

Brandriss, M. E., O'Neil, J. R., Edlund, M. B., and Stoermer, E. F. (1998). Oxygen isotope fractionation between diatomaceous silica and water. *Geochimica et Cosmochimica Acta* **62**, 1119–1125.

Dansgaard, W. (1964). Stable isotopes in precipitation. *Tellus* **16,** 436–468.

Epstein, S., Buchsbaum, R., Lowenstam, H. A., and Urey, H. C. (1953). Revised carbonate water isotopic temperature scale. *Geological Society of America Bulletin* **64,** 1315–1326.

Haimson, M., and Knauth, L. P. (1983). Stepwise fluorination – A useful approach for the isotopic analysis of hydrous minerals. *Geochimica et Cosmochimica Acta* **47,** 1589–1595.

Haug, G. H., Ganopolski, A., Sigmanm, D. M., *et al.* (2005). North Pacific seasonality and the glaciation of North America 2.7 million years ago. *Nature* **433,** 821–825.

Juillet-Leclerc, A., and Labeyrie, L. (1987). Temperature dependence of the oxygen isotopic fractionation between diatom silica and water. *Earth and Planetary Science Letters* **84,** 69–74.

Labeyrie, L. D. (1974). New approach to surface seawater palaeotemperatures using $^{18}O/^{16}O$ ratios in silica of diatom frustules. *Nature* **248,** 40–42.

Labeyrie, L. D., and Juillet, A. (1982). Oxygen isotopic exchangeability of diatom valve silica: interpretation and consequences for paleoclimatic studies. *Geochimica et Cosmochimica Acta* **46,** 967–975.

Lamb, A. L., Leng, M. J., Sloane, H. J., and Telford, R. J. (2005). A comparison of the paleoclimate signals from diatom oxygen isotope ratios and carbonate oxygen isotope ratios from a low latitude crater lake. *Palaeogeography, Palaeoclimatology, Palaeoecology* **223,** 290–302.

Leng, M. J., Barker, P., Greenwood, P., Roberts, N., and Reed, J. M. (2001). Oxygen isotope analysis of diatom silica and authigenic calcite from Lake Pinarbasi, Turkey. *Journal of Paleolimnology* **25,** 343–349.

Leng, M. J., and Marshall, J. D. (2004). Palaeoclimate interpretation of stable isotope data from lake sediment archives. *Quaternary Science Reviews* **23,** 811–831.

Leng, M. J., Lamb, A. L., Heaton, T. H. E., *et al.* (2005a). Isotopes in lake sediments. In *Isotopes in Palaeoenvironmental Research* (M. J. Leng, Ed.), pp. 147–184. Springer, Dordrecht.

Leng, M. J., Metcalfe, S. E., and Davies, S. J. (2005b). Investigating late Holocene climate variability in Central Mexico using carbon isotope ratios in organic materials and oxygen isotope ratios from diatom silica within lacustrine sediments. *Journal of Palaeolimnology* **34,** 413–431.

Leng, M. J., and Barker, P. A. (2006). A review of the oxygen isotope composition of lacustrine diatom silica for palaeoclimate reconstruction. *Earth Science Reviews* **75,** 5–27.

Lücke, A., Moschen, R., and Schleser, G. H. (2005). High temperature carbon reduction of silica: A novel approach for oxygen isotope analysis of biogenic opal. *Geochimica et Cosmochimica Acta* **69,** 1423–1433.

Maslin, M. A., and Swann, G. E. A., *et al.* (2005). Isotopes in marine sediments. In *Isotopes in Palaeoenvironmental Research* (M. J. Leng, Ed.), pp. 227–290. Springer, Dordrecht.

Morley, D. W., Leng, M. J., Mackay, A. W., Sloane, H. J., Rioual, P., and Battarbee, R. W. (2004). Cleaning of lake sediment samples for diatom oxygen isotope analysis. *Journal of Palaeolimnology* **31,** 391–401.

Morley, D. W., Leng, M. J., Mackay, A. W., and Sloane, H. J. (2005). Late glacial and Holocene atmospheric circulation change in the Lake Baikal region documented by oxygen isotopes from diatom. *Global and Planetary Change* **46,** 221–233.

Moschen, R., Lucke, A., and Schleser, G. H. (2005). Sensitivity of biogenic silica oxygen isotopes to changes in surface water temperature and palaeoclimatology. *Geophysical Research Letters* **32,** doi: 10.1029/2004GL022167.

Ragueneau, O., Treguer, P., Leynaert, A., *et al.* (2000). A review of the Si cycle in the modem ocean: Recent progress and missing gaps in the application of biogenic opal as a paleoproductivity proxy. *Global and Planetary Change* **26,** 317–365.

Rioual, P., Andrieu-Ponel, V., Battarbee, R. W., *et al.* (2001). High-resolution record of climate stability in France during the last interglacial period. *Nature* **413,** 293–296.

Rings, A., Lucke, A., and Schleser, G. H. (2004). A new method for the quantitative separation of diatom frustules from lake sediments. *Limnology and Oceanography: Methods* **2,** 25–34.

Rosqvist, G., Jonsson, C., Yam, R., Karlen, W., and Shemesh, A. (2004). *Diatom oxygen isotopes in pro-glacial lake sediments from northern Sweden: a 5000 year record of atmospheric circulation. Quaternary Science Reviews* **23,** 851–860.

Schleser, G. H., Lücke, A., Moschen, R., Rings, A., *et al.* (2001). Separation of diatoms from sediment and oxygen isotope extraction from their siliceous valves – A new approach. In *Terra Nostra, 2001/3. Schriften der Alfred-egener-tiftung (6th workshop of the European lake drilling programme, Potsdam.),* pp. 187–191.

Schmidt, M., Botz, R., Rickert, D., Bohrmann, G., Hall, S. R., and Mann, S. (2001). Oxygen isotopes of marine diatoms and relations to opal-A maturation. *Geochimica et Cosmochimica Acta* **65,** 201–211.

Schmidt, M., Botz, R., Stoffers, P., Anders, T., and Bohrmann, G. (1997). Oxygen isotopes in marine diatoms: A comparative study of analytical techniques and new results on the isotope composition of recent marine diatoms. *Geochimica et Cosmochimica Acta* **61,** 2275–2280.

Shemesh, A., Burckle, L. H., and Hays, J. D. (1995). Late Pleistocene oxygen isotope records of biogenic silica from the Atlantic sector of the Southern Ocean. *Paleoceanography* **10,** 179–196.

Shemesh, A., Charles, C. D., and Fairbanks, R. G. (1992). Oxygen isotopes in biogenic silica: Global changes in ocean temperature and isotopic composition. *Science* **256,** 1434–1436.

Shemesh, A., Hodell, D., Crosta, C., Kanfoush, S., Charles, C., and Guilderson, T. (2002). Sequence of events during the last deglaciation in Southern Ocean sediments and Antarctic ice cores. *Paleoceanography* **17,** 1056, doi: 10.1029/2000PA000599.

Data Interpretation

A Korhola, University of Helsinki, Helsinki, Finland

© 2007 Elsevier B.V. All rights reserved.

Introduction

Diatoms are single-celled, microscopic algae (Bacillariophyta) whose siliceous, glass-like cell walls preserve well in various sedimentary environments allowing reconstructions of past environments from subfossil diatom assemblages. They are a very successful group of organisms that can be found in virtually every aquatic habitat. Accordingly, diatom analysis is applicable across a wide range of aquatic ecosystem types, including freshwater, brackish, estuarine, and ocean, and includes lentic and lotic environments, wetlands and their associated damp, marginal, and littoral zones. This is why diatoms are one of the most widely used biological indicators in Quaternary paleoecology (Stoermer and Smol, 1999; *see* Overview).

Diatoms have played an important role in providing information on a wide spectrum of past environmental and ecological changes, such as climatic and lake-level changes, water pollution, eutrophication, and acidification, while their usefulness is also demonstrated in archeology, oil and gas exploration, and forensic science (Stoermer and Smol, 1999; Battarbee *et al.*, 2001; *see* Use in Archaeology). Most studies using fossil diatoms as indicators of environmental change have focused on the Lateglacial and Holocene sequences, although long diatom records encompassing several glacial/interglacial cycles are available from marine and freshwater sequences (e.g., *see* North Atlantic and Arctic; Large Lakes) and offer great potential for a very long-term perspective of environmental change. The standard method of inferring environmental conditions from fossil diatoms consists of an analysis of assemblage composition and consideration of the relevant autecological characteristics of the taxa present (indicator species). In recent decades, tremendous progress has been made in our methods of interpreting diatom stratigraphical data through powerful numerical techniques that allow surprisingly accurate quantitative reconstructions of environmental change to be made.

Why Are Diatoms Good Paleoindicators?

There are many reasons why diatoms are such excellent paleoindicators in Quaternary studies (Dixit *et al.*, 1992; Stoermer and Smol, 1999):

- Diatoms comprise a large proportion of total algal biomass in various aquatic systems and their assemblages are typically taxon rich. The large number of taxa enhances the ecological information gained, and hence increases the confidence of environmental inferences.
- They collectively show a broad range of tolerance along environmental and ecological gradients, with individual species having specific ecological and water quality requirements that are usually quantifiable.
- They have one of the shortest generation times of all biological indicators (~2 weeks), for which reason they reproduce and respond rapidly to environmental change.
- They are characterized by having an opaline, siliceous frustule that is resistant to degradation and hence is well preserved in different sedimentary environments.
- The taxonomy of diatoms is comprehensively documented, allowing species or even subspecies level identification of fossil diatom assemblages.

Analytic and Interpretative Procedures

Diatom Concentrations and Accumulation Rates

A prerequisite for the correct interpretation of stratigraphic diatom data is the careful collection, processing, and analysis of the initial diatom samples. Sampling and laboratory methods are discussed in detail by Battarbee *et al.* (2001). Diatom counts from sediments are most often expressed as percentage values, and paleoecologists generally base their environmental reconstructions on the assumption (that may not always be valid) that the percentage composition of diatoms in sediment assemblages accurately reflects the relative abundance of different taxa across the different living communities in the lake. Such proportional data lend themselves well to most numerical data treatment procedures and quantitative approaches. Estimation of diatom concentrations per unit weight (frustules per gram dry weight^{-1}) or unit volume (frustules cm^{-3}) can be used as a record of diatom paleoproductivity and possibly to indicate net primary productivity. If diatom concentrations are required, an easy method is to add a known quantity of external markers (e.g., microspheres) in cleaned sediment slurry (Battarbee *et al.*, 2001). Diatom concentration values *per se* have rather limited use in paleoecology, but they can be further converted to diatom accumulation rates (DAR; frustules per cm^2 per year, also referred to as diatom valve flux; **Fig. 1**). However, an excellent chronology is needed for this.

One has to bear in mind that different factors such as diatom preservation status (i.e., dissolution and diatom fragmentation during grazing or sample preparation) as well as potential variations in the rate of sediment accumulation, make interpretation of diatom concentration results more-or-less tentative. Also, because diatom cell sizes vary tremendously, simple concentrations may produce a misleading depiction of total diatom productivity. To minimize such biases, species-specific correction factors can be developed that allow the composition of the source populations to be reconstructed. Also, diatom concentrations can be converted into biovolumes per cm per year, using taxon-specific

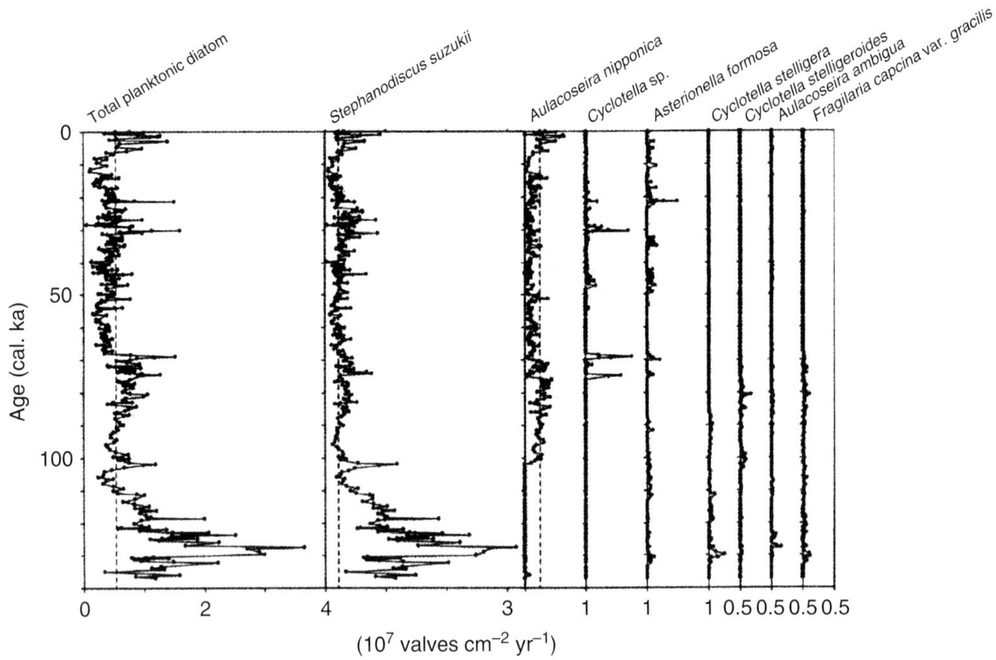

Figure 1 Records of total planktonic diatom flux (valves cm^{-2} yr^{-1}) and most important individual species fluxes from a 140 m long core taken from Lake Biwa, Shiga Prefecture, Japan, covering ca. 140 kyr. The dashed line indicates the present level of diatom flux for the total planktonic taxa and each individual taxa, respectively. Based on characteristics of past and modern diatom responses to possible climate variables, the authors interpreted changes in total planktonic diatom flux to reflect a combination of summer precipitation, winter temperature, and snowfall values. During the 140–101 ka interval, high values of *Stephanodiscus suzukii* productivity and the disappearance of *Aulacoseira nipponica* were interpreted to indicate weaker vertical mixing, possibly caused by increased temperatures and decreased snowfall levels in winter. During the 101–70 ka interval, the record showed levels near or above those observed in present times, indicating that winter water temperatures fell within the optimal range for *A. nipponica* to prosper. Low *Aulacoseira* fluxes during the 70–7 ka interval were interpreted to indicate weak winter vertical mixing and cold winters. From Kuwae M, Yoshikawa S, Tsugeki N, and Inouchi Y (2004) Reconstruction of a climate record for the past 140 kyr based on diatom valve flux data from Lake Biwa, Japan. *Journal of Paleolimnology* 32: 19–39.

measurements of whole-cell volumes. Surprisingly, the latter approach has rarely been implemented in fossil data, despite the fact that algal biovolumes have long been a standard and routine unit of measure in modern phytoplankton ecology. The concentration estimates can be further reinforced by multiple core and multiple indicator analyses. In conjunction with diatom studies, analysis of fossil pigments, organic carbon and chemical determination of biogenic silica may help reconstruct the biomass of diatom populations in a more reliable way. For example, Abrantes (2000) compared DAR records from several Atlantic upwelling sites with other independent productivity proxies such as silicoflagellates, radiolaria, total opal, and abundances and accumulation rates of organic carbon and benthic foraminifera, and suggested that maximum productivity of the past 200 kyr in the eastern Atlantic occurred at the Last Glacial Maximum (LGM), consistent with the models that attribute glacial atmospheric CO_2 decrease to higher low-latitude productivity.

Ecological and Habitat Classification – Qualitative Approaches

Despite the rapid development of quantitative reconstruction techniques during the past few decades, qualitative interpretations based on the relevant autecological characteristics of the individual taxa and the habitat and ecological classification of groups of taxa still form the basis for understanding and interpreting the stratigraphic diatom record. Many diatom species have clearly defined, and relatively narrow, preferences for particular habitats and water quality conditions. These species are indicative of several variables, including nutrient availability, acidity, light and thermal conditions, dissolved oxygen concentrations, and lake levels. As a result, a broad range of specific paleoenvironmental interpretations can be made on the basis of the life histories, habitats, and known ecological preferences of indicator species. Interpretation of diatom records is fortunate in this respect, because there is a considerable amount of tested species ecological data available. In particular, categorization of diatom taxa according to their acidity or salinity tolerances has been successfully used to interpret changes in surface water acidity, sea level, and hydrology of saline lakes (see, e.g., Palmer and Abbott (1986) and Stoermer and Smol (1999)).

Diatom species also have affinities for particular physical habitats for example there are open water (planktonic) and benthic forms. See Diatom Introduction for more details. In stratigraphic studies dealing with dynamic coastal environments, classification of diatoms according to habitat has also proved very useful since there are distinct communities associated with the different salt-marsh habitats (**Fig. 2**).

Qualitative interpretations based on expert knowledge have both advantages and disadvantages. Qualitative approaches may pursue a variety of theory-generating aims, including identifying potentially important variables or concepts, recognizing patterns and relationships, and generating coherent theories and hypotheses. However, qualitative studies rely heavily on subjective interpretation and expert knowledge. They do not typically generate answers but rather provide narrative accounts.

Toward More Objective Assessment

A great variety of classification and descriptive multivariate tools have been developed that can be used to more effectively manage the rich mosaic of ecological information inherent in diatom stratigraphic data. Typically, an important first step is the partitioning or zoning of the initial diatom diagram with the aim of identifying uniform segments in the data. A favored method for paleoecological data has been the constrained incremental sum-of-squares clustering (CONISS; Grimm, (1987)). However, it suffers from having no numerical basis for deciding on the number of reliable or statistically significant zones within a sequence. Bennett (1996) has resolved the problem by using the broken-stick and randomization models to assess the reliability of different zonations against a model of random distribution of zones within a sequence. For example, Korhola and Weckström (2004) applied Bennett's broken-stick model to core data from several lakes in Finnish Lapland to identify stratigraphic levels in which diatom floristic changes have been most distinct. Such analysis helped them elucidate the relative role of local versus regional environmental changes in explaining

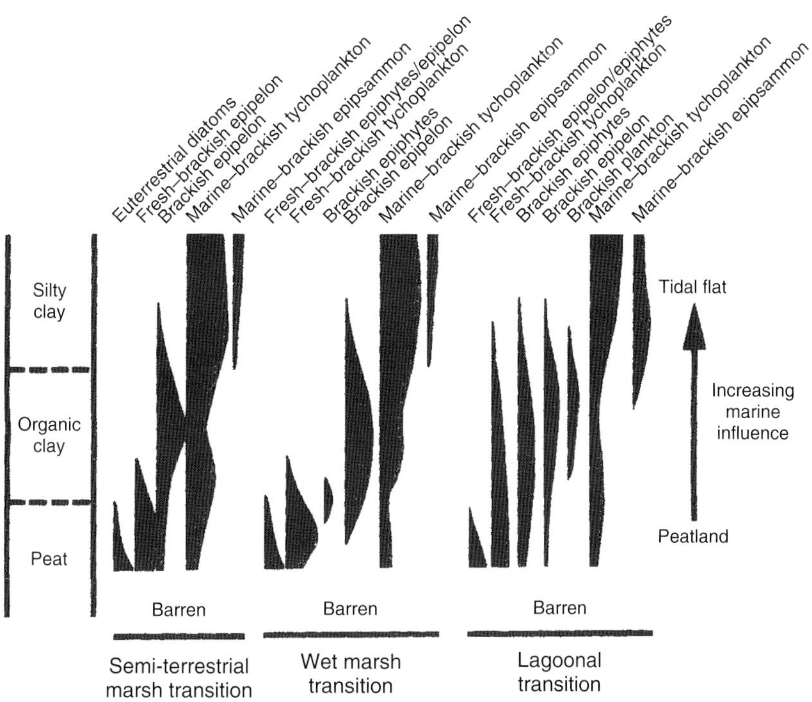

Figure 2 Idealized scheme of some possible diatom assemblage shifts occurring at the transgressive overlap of a peat with tidal deposits under different drainage conditions. Accuracy of paleotidal inference decreases to the right. Note that the relative representation of different groups will vary according to local conditions and input of external (allochtonous) material. From Denys L and De Wolf H (1999) Diatoms as indicators of coastal paleoenvironmental and relative sea-level change. In: Stoermer, E.F., and Smol, J.P(eds.) *The Diatoms: Applications for the Environmental and Earth Sciences*, pp. 277–297. Cambridge: Cambridge University Press.

the changes in fossil diatom assemblages. However, Bennett's method for determining the number of zones is based on the use of a single randomized sample as a comparison. Techniques that use a large set of randomized samples might provide more realistic means of making statistical evaluation of possible zonations.

Ordination techniques can be used to arrange species and/or samples along gradients. Ordination is especially useful for visualizing large and complex data sets in reduced dimensions that are ecologically easier to interpret. For example, stratigraphic sample scores derived from an indirect gradient analysis technique can be plotted chronologically, and when related together with species scores to ordination axes, used to determine the major ecological drivers of the temporal variation. Stager *et al.* (1997) used correspondence analysis (CA) to show that changes in the fossil diatom assemblages of a core from Lake Victoria occurred in response to climatic fluctuations during the Holocene. First, a CA ordination biplot of major diatom taxa distributed them into qualitative assemblage groups by quadrant (**Fig. 3A**). Based on available diatom habitat information, axis 1 was interpreted to approximate a gradient of water column stability and axis 2 to approximate a gradient of precipitation:evaporation (P:E) ratios or lake depth. When plotted stratigraphically, the sample scores of the two first CA axes were then able to be used as proxies of windiness, rainfall patterns, and monsoon variability (**Fig. 3B**). Large Lakes gives a more detailed account of the Lake Victoria diatom record and its use in paleoclimatic reconstruction.

Ordination procedures can also aid in the comparision and synthesis of multiple data sets. For example, Sorvari *et al.* (2002) summarized the diatom records from several short cores from Arctic Lapland using principal components analysis (PCA), and showed that the changes in diatom assemblages were significantly correlated with long-term variations in regional spring temperatures. Smol *et al.* (2005) used detrended canonical correspondence analysis (DCCA) to summarize the amount of species turnover in paleolimnological records (mostly diatoms) from 55 lakes in the circumpolar Arctic. They found widespread, parallel changes in species compositions that were likely to be driven by climate warming.

The multiple proxy indicator approach enables some independent variables representing different environmental processes to be identified. Variance decomposition (Borcard *et al.*, 1992) can then be effectively used to distinguish the independence and relative strength of different variables in explaining diatom species compositional changes through time. In each variance partitioning experiment, the diatoms are the response variable and other independent variables are used in different combinations as explanatory variables. The effect of time-dependent ecological and environmental processes can be further removed by introducing sample age as a covariable in all experiments. Finally, a variance decomposition experiment is performed with each of the predictor variables assigned as covariables to cancel out their effect on one another and assess redundancy proportions between variables. The explanatory strength of various variables (or their combinations) is assessed by testing the significance of the first ordination axis under different model conditions using a constrained Monte Carlo permutation test. For example, Barker *et al.* (2000) used partial redundancy analysis (RDA) to test whether the effect of tephra deposition 1.19 ka had a statistically significant impact on the diatom assemblages in Lake Massoko, a volcanic crater lake in southern Tanzania, and discriminated between floral changes in response to tephra and changes in climate (as derived from the pollen data), or those induced by physical catchment disturbance (as revealed by the magnetic properties of the sediment).

Given a modern reference data set (see below), canonical ordination can also be a useful visual aid in environmental reconstruction. Fossil diatom assemblages can be introduced in the ordination analysis of modern diatom data as supplementary (passive) samples and hence positioned into the ordination plane of modern training set samples on the basis of similarities in their biological composition. Changes in the diatom assemblages within a core can therefore be compared directly to the diatom-environment gradients within the modern data set to follow core trajectories. Engstrom *et al.* (2000) used this methodology effectively when studying chemical and biological trends during lake evolution in Glacier Bay, Alaska (**Fig. 4**). They summarized the limnological trends in lakes of varying age by projecting their fossil diatom assemblages onto the DCCA biplot of the modern samples. The core data of most lakes showed a net negative trajectory along DCCA axis 1, suggesting that the main trend of lake development with increasing age was a progressive loss of pH, alkalinity, and base cations, and a corresponding increase in dissolved organic carbon (DOC).

Diatom core data can be subjected to several numerical data manipulations to evaluate diatom community stability and diversity. For the former, rate of change analysis (Jacobson and Grimm,

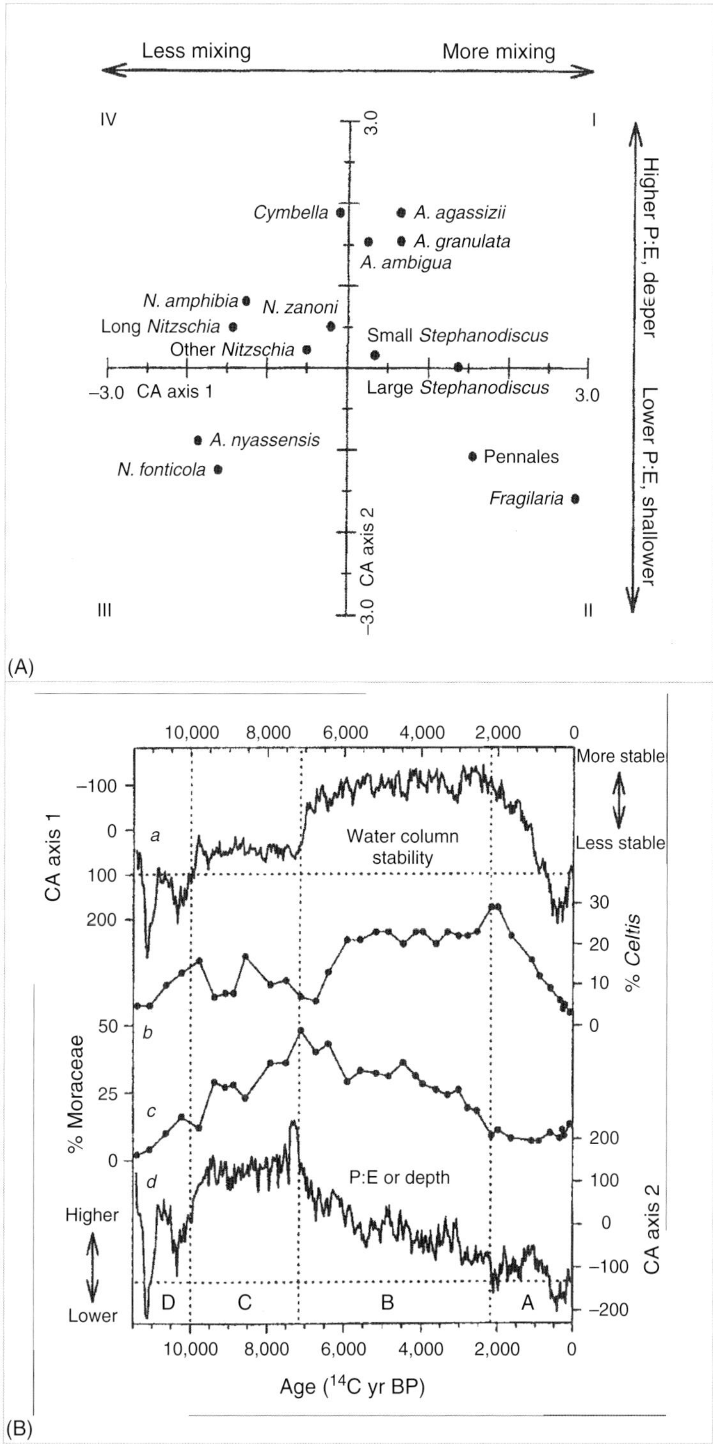

Figure 3 Interpretation of a diatom record of a core from Lake Victoria's Damba Channel. (A) CA biplot showing major diatom taxa of the core. (B) CA axis plots and pollen stratigraphy. Inverted CA axis 1 represents water column stability and resembles the *Celtis* (seasonally dry forest) pollen record. CA axis 2 represents P:E or lake depth, and resembles the Moraceae (moist forest) pollen record. Horizontal dotted lines mark present conditions. Vertical dotted lines delimit diatom zone boundaries. From Stager *et al.* (1997).

1986) has proved useful (**Figs. 5** and **9**). This technique simply calculates the dissimilarity between successive samples divided by the estimated period of deposition elapsed between the two, in order to produce measures of any assemblage changes per unit time. Chord distance is the dissimilarity metric most commonly applied to diatom data because of its statistical properties concerning the weighting of

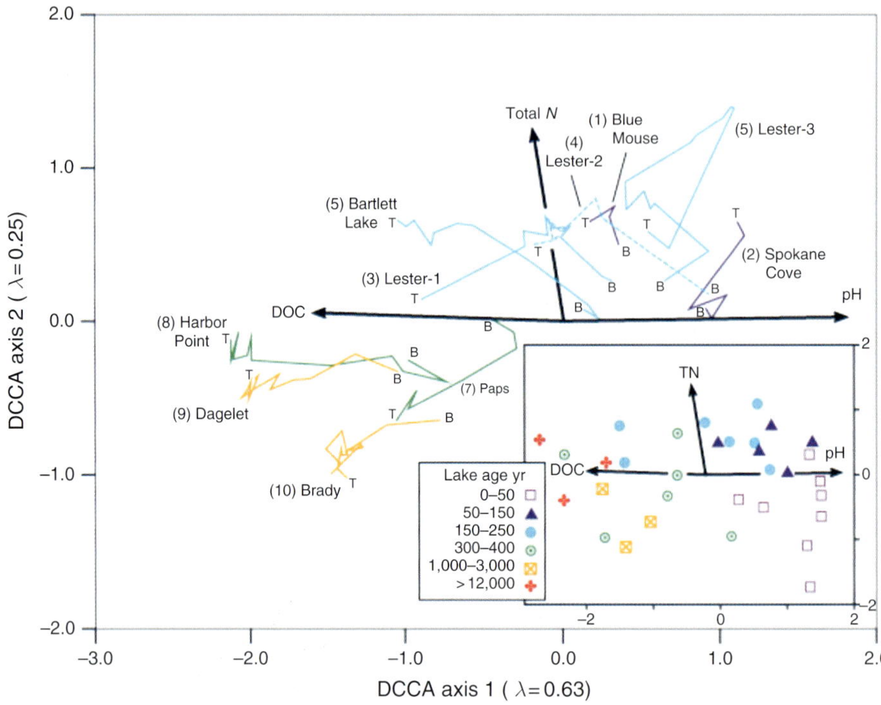

Figure 4 Detrended canonical correspondence analysis of modern and fossil diatom assemblages from the chronosequence of lakes in the Glacier Bay region of southeastern Alaska. In the inset, ordination of modern diatom assemblages (from surface sediments) and associated water-chemistry data are shown. The arrows depict the direction of increase and maximum variation of the measured environmental variables. Lakes plotting close to one another have similar diatom composition. In the main figure, the time-trajectories from 10 sediment cores (B, bottom; T, top) were projected onto the biplot by adding the core assemblages to the analysis as passive samples. From Engstrom *et al.* (2000).

species with low percentages. The rate of change method is critically dependent on the time standardization unit (TSU) used to standardize the estimated dissimilarity, so good chronological control is essential for this. As sediment accumulation rates are variable, sample intervals are unequally distributed through time. In order to correct this systematic bias, re-sampling by interpolation of new samples at even time intervals can be done. The resulting synthetic data set can then be subjected to smoothing and the chord distance coefficient calculated (Laird *et al.*, 1998). When analyzing large and complex data matrices, such as the multi-taxon diatom core data sets, often are, it is sometimes necessary to reduce the dimensionality in the data (e.g., by PCA) prior to estimating the rates of change (Barker *et al.*, 2000).

The assessment of past ecological diversity as deduced from the sedimentary record is a challenging task. Besides variables such as differential preservation of diatom fossils, one of the major issues in assessing past diversity is the choice of a sound measure that takes into account richness and equitability. The diversity of subfossil organisms cannot be investigated unless the count size of the samples is constant or standardized, because the number of identified taxa usually increases as the fossil count size increases. Rarefaction analysis (Heck *et al.*, 1975) has been shown to be very suitable for fossil diatom data (**Fig. 5**). Rarefaction uses a strategy of random selection without replacement, typically selecting a sample of the same size as the smallest count from the entire population. In this way realistic estimates of richness are produced without any bias associated with the variability of individual sample count sizes. However, when dealing with non-laminated sediments, differences in sediment accumulation rate need further to be taken into account when interpreting the diversity records.

Estimating Parameter Values Using Diatoms

Numerical modeling, with the aim of quantitative paleoenvironmental reconstruction from diatom assemblages, was first applied to reconstruct pH. The early applications were based on ecological species categories and pH indices, and later multiple linear regression models were developed to estimate pH according to the proportions of the established pH-preference groups (see Battarbee

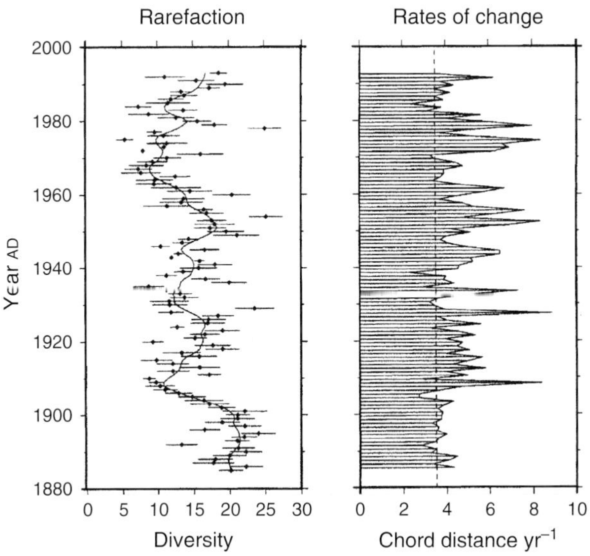

Figure 5 Diatom diversity as assessed by rarefaction analysis and rate of diatom assemblage change in the annually laminated sediment of Baldeggersee, Switzerland, between 1885 and 1993. The diatom diversity closely mirrors the different phases of nutrient enrichment in the lake with decreasing numbers of taxa during the eutrophication phases (cf. **Fig. 9**). The timing and direction of diatom assemblage changes also compare well with sedimentological and limnological changes (cf. **Fig. 9**). The error bars in the rarefaction diagram represent the 95% confidence interval, whereas the dashed line in the rates of change diagram indicates the approximate significance level (2.5th percentile). The line through the rarefaction estimates is based on a LOWESS smoother. From Lotter A F (1998) The recent eutrophication of Badeggersee (Switzerland) as assessed by fossil diatom assemblages. *The Holocene* 8: 395–405.

and Charles (1987) and Davis (1987) for extensive reviews). More recently, paleoecologists have cataloged the ecological optima and tolerances of diatom assemblages and have constructed transfer functions that allow environmental conditions to be inferred from fossil assemblage data (Birks *et al.*, 1990). Quantitative approaches using diatoms are now being applied to a wide array of environmental problems and questions, with many new methodologies currently being implemented.

In diatom palaeoecology, transfer functions simply describe the relationship between diatom assemblages and their environment with the ultimate goal to use the present-day ecology of diatoms to quantitatively infer past conditions. The basic approach is illustrated as a hypothetical example in **Figure 6**. The transfer function approach involves two steps: (1) the regression step in which the responses of taxa to the contemporary environment are modeled, and (2) a calibration step in which the environmental variable of interest is predicted from the fossil compositional data. Transfer functions are established using modern diatom assemblages and measured environmental variables from a calibration data set (also called a training set; **Fig. 6A**). These calibration sets involve sampling a range of sites (e.g., lakes, wetlands, estuaries, marine environments, usually >40 sites) for a suite of critically considered environmental variables, which are then related to the diatom assemblages preserved in the surface sediments (top 0.5–1 cm representing the last few years of sediment accumulation) of the systems concerned using various numerical procedures (**Fig. 6B**). The transfer function approach is based on the implicit assumption that the modern training sets cover the range of environmental conditions likely to be represented by the fossil assemblages. Hence, calibration data sets should ideally span all biologically relevant environmental gradients.

The mathematical methods for establishing a transfer function are commonly based on nonlinear regression techniques. The most popular approaches in diatom studies have been weighted averaging (WA; ter Braak and van Dam (1989)) and the WA partial least squares (WAPLS; ter Braak and Juggins (1993)) regression, because they perform well with noisy, species-rich, compositional data, with many zero values and some rogue observations (outliers). In both methods, the underlying response model is the Gaussian function, that is, each taxon has a bell shape (Gaussian) distribution along an environmental variable. From the Gaussian curve, the species optimum (the peak) and its tolerance (the standard deviation) can be estimated (**Fig. 7**).

In the WA methods, the estimate for the species optimum (u_k) is obtained as the average of the kth taxon in the calibration dataset (y_{ik}) weighted by the environmental variable (x_i) of interest:

$$\hat{u}_k = \frac{\sum_{i=1}^{n} y_{ik} x_i}{\sum_{i=1}^{n} y_{ik}} \quad [1]$$

WA can also be used to estimate tolerance values (**Fig. 7**) for diatoms and these values can be used to down-weight the importance of taxa with a wide tolerance relative to those with a narrow tolerance. **Figure 8** shows the response models and the estimated WA temperature optima and tolerances of the most frequent diatom genera and species for the Arctic Lapland temperature training set. These estimated optima and tolerances form the transfer function, which can then be used to infer past temperature from fossil diatom assemblages.

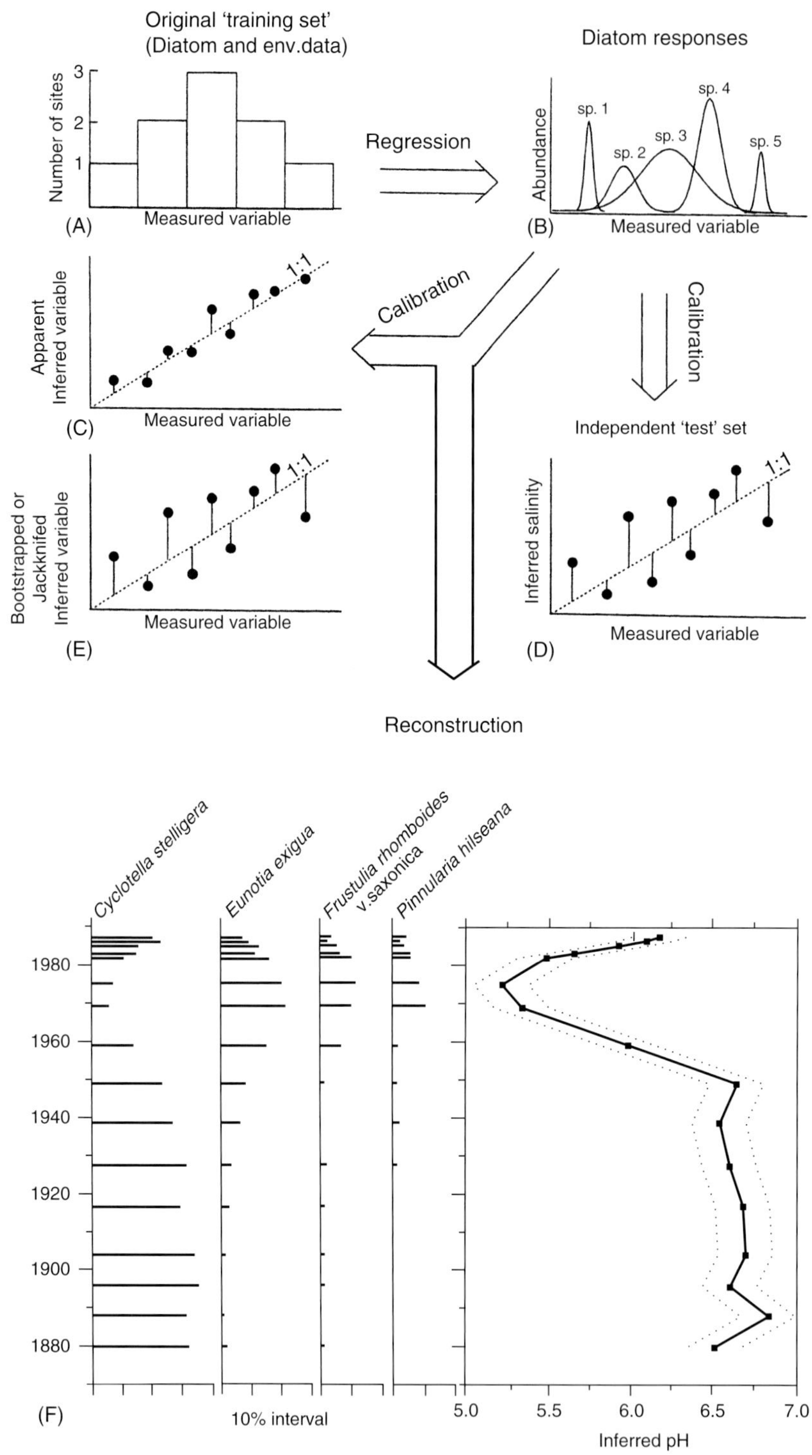

Figure 6 Simplified illustration of the typical steps involved in developing and assessing the predictive ability of a diatom-based calibration model and its application to fossil diatom data including: (A) selection of a modern training data set; (B) estimation of taxon responses based on their distribution in the training set (regression); (C) generation of a transfer function on the basis of the training set (calibration); (D) selection of an independent test set to assess the transfer function (calibration); (E) use of computer re-sampling techniques, such as jackknifing or bootstrapping, to assess the transfer function (calibration); and (F) reconstruction of parameter value from fossil diatom assemblages and estimation of sample-specific prediction errors by bootstrapping. Modified from Fritz *et al.* (1999) Fritz, S.C, Cumming, B.F., Gasse, F., and Laird, K.R (1997) Diatoms as indicators of hydrologic and climatic change in saline lakes. In: Stoermer EF and Smol JP (eds.) *The Diatoms: Applications for the Environmental and Earth Sciences*, pp. 41–72. Cambridge: Cambridge University Press.

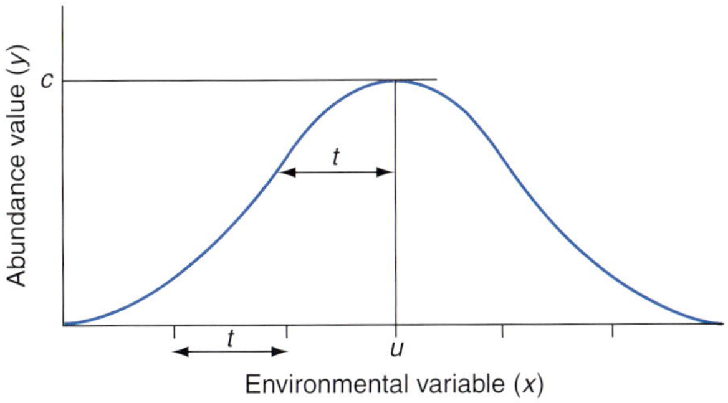

Figure 7 Gaussian response curve for the changing abundance of a species along an environmental gradient. t = tolerance; u = optimum; c = maximum. From ter Braak C. J. F. (1987) *Unimodal Models to Relate Species to Environment*, 152pp. Unpublished PhD Thesis, University of Wageningen.

In addition to WA, other routinely used calibration techniques include the maximum likelihood (ML), modern analog technique (MAT), and Gaussian logit regression (GLR) (reviewed in Birks (1995, 1998)). Recently, artificial neural networks (ANNs) have been introduced to diatom paleoecological research. However, Telford and Birks (2005) found that transfer functions based on a unimodal species–environment response are more robust to spatial autocorrelation of environmental variables in the training set (a serious problem for transfer functions) than MAT or ANN models, which are likely to overfit the data unless a spatially independent test set (see below) is available to help model selection. Of other novel techniques, Bayesian full probability models should be mentioned. Bayesian inference is shown to have many advantages over conventional frequency distribution methods in environmental reconstruction, including reduction of inherent systematic bias, more precise error estimation, the possibility of utilizing existing ecological and environmental information, and improved extrapolation capacity (Toivonen *et al.*, 2001; Vasko *et al.*, 2000; Korhola *et al.*, 2002).

Different modeling tools have their inherent philosophical and mathematical assumptions as well as limitations, so the prediction ability of the transfer models must be carefully evaluated. Ideally, independent data sets that are similar in size and environmental gradients would best suit this purpose (**Fig. 6D**), but are seldom available because they are laborious and expensive to collect. Instead, computer-intensive re-sampling techniques, such as jackknifing (leaving one-out substitution) and bootstrapping (creation of pseudoreplicate data sets by re-sampling), are used (**Fig. 6E**); these should provide more realistic assessments of the prediction ability of a transfer model than the pure or apparent error (**Fig. 6C**). The output of the reconstruction procedure will be a reconstructed parameter value for each core sample plus sample-specific prediction errors, commonly estimated by bootstrapping (**Fig. 6F**).

Evaluation of the Reconstruction

There are no means or 'skilled techniques' to prove that a reconstruction is correct. The most robust evaluation is to compare diatom-inferred values with measured long-term data series where these exist (**Figs. 9** and **10**). Close agreement between reconstructed and measured values increases confidence in the reconstruction whereas deviation between the two sets of data warrants caution. However, close agreement during the monitoring period does not necessarily imply accurate reconstructions for earlier periods, especially if assemblages have a very different composition. Conversely, divergence for the monitoring period does not necessarily imply that reconstructions for earlier periods are in error.

Another way to evaluate the robustness of a diatom-based reconstruction is to compare the inferred values with the ones reconstructed using other proxy sources (**Fig. 9**). Close agreement between several proxy indicators increases their reliability. In addition, indirect gradient analysis (PCA, CA, and DCA) can be used to determine whether the inferred environmental variable is related to the main directions of variation in the fossil diatom assemblages (i.e., the sample scores on the first unconstrained axis). The strength of the correlation between the unconstrained axis 1 sample scores and the inferred values indicates the extent to which the principal changes in diatom community structure can be explained by the inferences of the parameter in question.

Implicit in the transfer function approach is that the modern training set samples provide good analogs for the fossil diatom assemblages. This could simply be done by highlighting sediment samples that contain significant numbers of taxa that are absent from the training set. A more robust assessment is to find the closest modern analogs for each fossil sample using MAT (**Fig. 9**). MAT quantifies the dissimilarity between modern and fossil samples using various distance measures. Another way is to assess the goodness-of-fit between modern and fossil species data using constrained canonical correspondence analysis (CCA) with the variable of interest as the sole constraint. A fossil sample with a residual distance equal to or larger than the residual distance of the extreme 10% of the calibration

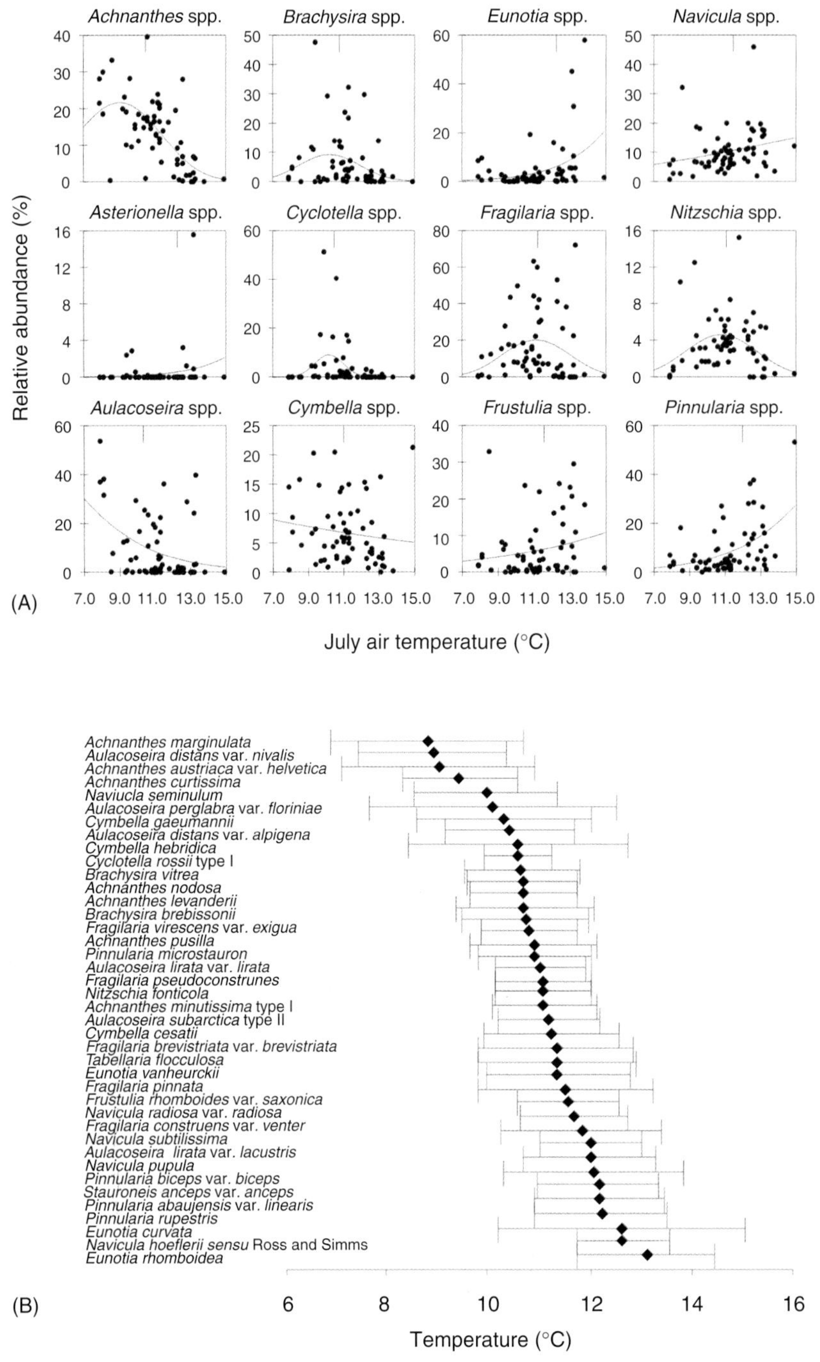

Figure 8 Responses of diatoms to the site-specific air temperature from a training set of 64 lakes in Arctic Lapland. (A) Distribution of the most dominant diatom genera along the temperature gradient. The estimated weighted averaging (WA) temperature optimum for each genus is indicated as a tick. The trend lines are based on a Gaussian logit model. (B) The estimated temperature optima (solid squares) and tolerances (bars) of the 40 most abundant diatom taxa in the data set arranged according to temperature optima. From Weckström and Korkola (2001) Patterns in the distribution, composition and diversity of diatom assemblages iin relation to ecoclimatic factors in Arctic Lapland. *Journal of Biogeography*, **28**, 31–46.

data set is typically considered to have a poor fit to the desired variable (Birks *et al.*, 1990).

Here are two examples in which multiple data manipulation and validation tools have been efficiently used to extract information from fossil diatoms. Wolfe (2003) examined different methods of interpreting diatom stratigraphic data from the late-Holocene sediments of a small Arctic lake on Baffin Island, Nunavut, Canada, with the objective of gaining paleoclimatic insights from fossil assemblages. In this study, diatom relative frequencies and absolute abundances were subjected to several data

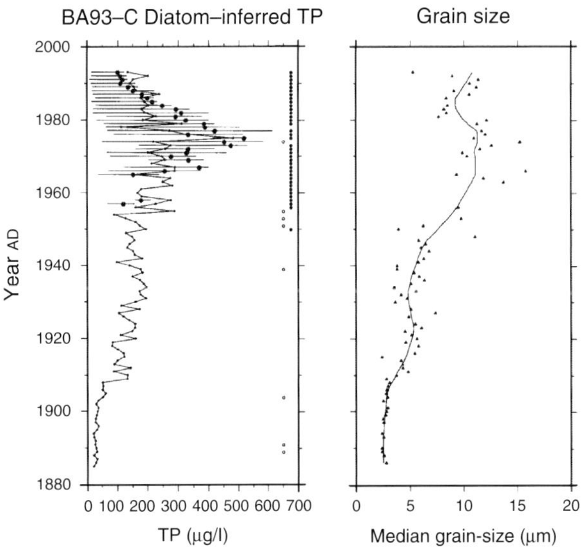

Figure 9 Measured total phosphorus (TP) during spring circulation (filled circles) compared to diatom-inferred TP values and median grain-size distribution in the Baldeggersee annual layers. The horizontal lines represent the annual TP range in the uppermost 15 m of the water column. The dots on the right hand side of the graph represent samples with close (filled dots; 2nd percentile) and good modern analogs (open dots; 5th percentile). Grain-size analyses in annual layers show variations in calcite size (a proxy for PO_4 concentration in the lake water) that parallel the diatom-inferred TP concentrations, hence giving independent support to the reconstruction. From Lotter AF (1998) The recent eutrophication of Badeggersee (Switzerland) as assessed by fossil diatom assemblages. *The Holocene* 8: 395–405.

manipulations, including application of a WA transfer function for summer water temperature, conversion of valve concentrations to cell biovolumes, biostratigraphic rate of change, and taxonomic richness estimated by rarefaction analysis. Hence, from the primary diatom data, a second data set was derived on the basis of the data manipulations, consisting of seven new variables. The major directions of variability within the derived diatom data were then summarized using PCA and the results were considered in relation to independent paleoclimatic proxies to enable an objective assessment of the response of freshwater diatom communities to climatic change.

The second example comes from Laajalahti, a semi-enclosed embayment of the eastern Baltic Sea that has been studied for diatoms to reconstruct its eutrophication history (Weckström *et al.*, 2004). **Figure 10** shows the summary diatom diagram, together with diatom-inferred total dissolved nitrogen (TDN) concentrations, diversity estimates, analog measures, and monitoring data of TN for the last 30 yr (Weckström *et al.*, 2004). The diagram shows the transition from a flora dominated by benthic *Fragilaria* taxa in the late nineteenth century to one dominated by a succession of planktonic *Cyclotella* and *Thalassiosira* species today. All fossil samples in this case have squared chord distances less than 10% of the distribution of all distances among modern samples in the calibration set, as assessed by MAT, suggesting that the samples have good floristic analogs in the eastern Baltic training set. However, one sample had a poor fit to TDN as assessed by the constrained CCA. Comparison between the reconstructed values and monitoring data show that the transfer function is able to reproduce trends and absolute values for the last ca. 10 yr but underestimates concentrations during the most eutrophic period during the 1970s. This underestimation is probably due to the lack of high TN sites in the training set. The down core diversity of diatoms closely follows the pattern of eutrophication deduced from the diatom TDN transfer function. As expected, the diversity decreases with increasing concentrations on TDN, following a negative logarithmic function. These patterns give clear evidence that nutrient enrichment leads to a loss in diatom diversity at this site. There is more variability in the species richness than in the overall species compositions (as deduced on the basis of DCA). These patterns reinforce the general axiom that aquatic biological diversity tends to decrease with increasing nutrient levels.

Conclusions

Stratigraphic diatom data are complex, multivariate, and temporally ordered, most commonly forming irregularly spaced discontinuous time series. Such proxy data thus have special statistical properties and their numerical analysis requires techniques that take particular account of these unique data properties. Much progress has been made in the development of quantitative approaches to infer environmental variables from diatom assemblages. Robust diatom-based quantitative transfer functions have been created for pH, nutrients, salinity, organic carbon, water and air temperature, ice cover duration, lake level and other environmental parameters. However, quantitative inference techniques have their own shortcomings and limitations, so all transfer function approaches have to be assessed critically (Birks, 1998). We need to be sure that our interpretations are not becoming too mechanistic and straightforward, as the transfer functions will always produce results, whether correct or not. Also, it should be remembered that the transfer function approach is not the sole method for gleaning useful information from diatom stratigraphic data. As

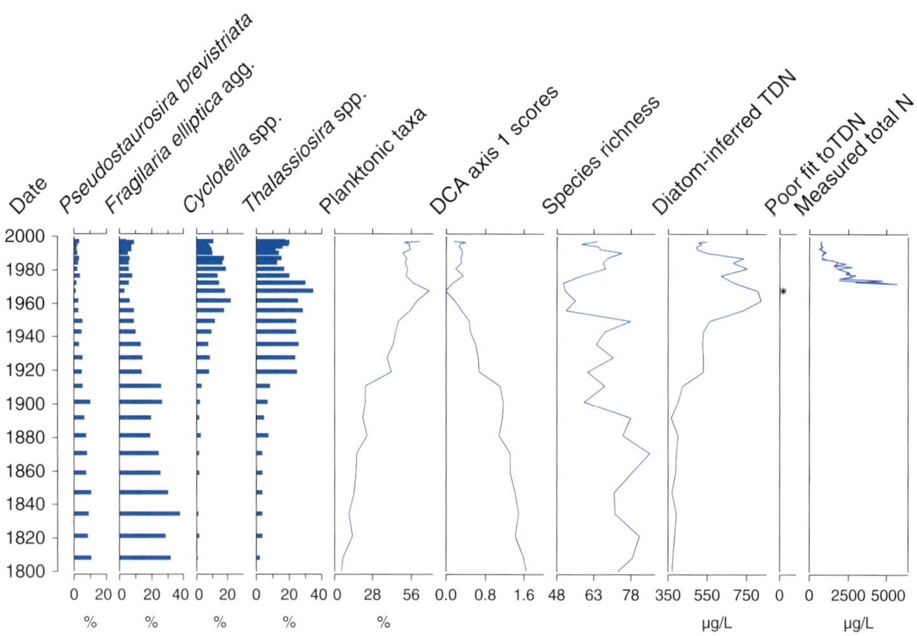

Figure 10 Summary diatom diagram showing major diatom taxa, proportion of planktonic taxa, detrended correspondence analysis (DCA) axis 1 scores, rarefaction species richness, diatom-inferred annual mean TDN and measured TN concentrations for Laajalahti embayment, Helsinki. The inferred TDN values are based on the two-component WAPLS model. The measured TN concentrations are presented as annual, seasonally weighted mean values. The sole sample having a poor CCA fit to TDN is shown by asterisk. Modified from Weckström K, Juggins S, and Korhola A (2004) Quantifying background nutrient concentrations in the coastal zone of the Baltic Sea. *Ambio* 33: 324–327 and Weckström K (in press) Assessing recent eutrophication in coastal waters of the Gulf of Finland (Baltic Sea) using subfossil diatoms. *Journal of Paleolimnology*.

shown by the study of Wolfe (2003), highly valuable environmental information can be extracted from diatom stratigraphic data when subjected to many kinds of creative numerical data manipulations. Other indicators that can be measured in sediments will further strengthen conclusions drawn from the diatom data. In addition, the use of multiple indicators also allows questions of ecological response to environmental forcing to be addressed using diatoms. Perhaps in the future, our research should focus more on using fossil diatom compositions as response variables and early warning indicators rather than using them to reconstruct environmental change *per se*. For this, however, a more comprehensive understanding of the role of diatoms in the functioning of aquatic ecosystems is required.

See also: **Diatom Introduction**. **Diatom Methods**: Quaternary Geological Records. **Quaternary Stratigraphy**: Biostratigraphy.

References

Abrantes, F. (2000). 200,000 yr diatom records from Atlantic upwelling sites reveal maximum productivity during LGM and a shift in phytoplankton community structure at 185 000 yr. *Earth and Planetary Science Letters* **176,** 7–16.

Barker, P., Telford, R., Merdaci, O., *et al.* (2000). The sensitivity of a Tanzanian crater lake to catastrophic tephra input and four millennia of climate change. *The Holocene* **10,** 303–310.

Battarbee, R. W., and Charles, D. F. (1987). The use of diatom assemblages in lake sediment as a means of assessing the timing, trends, and causes of lake acidification. *Progress in Physical Geography* **11,** 552–580.

Battarbee, R. W., Jones, V. J., Flower, R. J., Cameron, N. G., Bennion, H., and Carvalho, L. (2001). Diatoms. In *Tracking Environmental Change Using Lake Sediments. Volume 3: Terrestrial, Algal and Siliceous Indicators* (J. P. Smol, H. J. B. Birks and W. M. Last, Eds.), pp. 155–202. Kluwer, Dordrect.

Bennett, K. (1996). Determination of the number of zones in a biostratigraphical sequence. *New Phytologist* **132,** 155–170.

Birks, H. J. B. (1995). Quantitative palaeoenvironmental reconstructions. In *Statistical Modelling of Quaternary Science Data. Quaternary Research Association Technical Guide 5* (D. Maddy and J. S. Brew, Eds.), pp. 161–254. Quaternary Research Association, Cambridge.

Birks, H. J. B. (1998). Numerical tools in palaeolimnology – Progress, potentialities, and problems. *Journal of Paleolimnology* **20,** 307–332.

Birks, H. J. B., Line, J. M., Juggins, S., Stevenson, A. C., and ter Braak, C. J. F. (1990). Diatoms and pH reconstruction. *Philosophical Transactions of the Royal Society of London B* **327,** 263–278.

Borcard, D., Legendre, P., and Drapeau, P. (1992). Partialling out the spatial component of ecological variation. *Ecology* **73,** 1045–1055.

Davis, R. B. (1987). Paleolimnological diatom studies of acidification of lakes by acid rain: An application to quaternary science. *Quaternary Science Reviews* **6,** 147–163.

Denys, L., and De Wolf, H. (1999). Diatoms as indicators of coastal paleoenvironmental and relative sea-level change. In *The Diatoms: Applications for the Environmental and Earth Sciences* (E. F. Stoermer and J. P. Smol, Eds.), pp. 277–297. Cambridge University Press, Cambridge.
Dixit, S. S., Smol, J. P., Kingston, J. C., and Charles, D. F. (1992). Diatoms: Powerful indicators of environmental change. *Environmental Science and Technology* **26**, 23–33.
Engstrom, D. R., Fritz, S. C., Almendinger, J. E., and Juggins, S. (2000). Chemical and biological trends during lake evolution in recently deglaciated terrain. *Nature* **408**, 161–166.
Fritz, S. C., Cumming, B. F., Gasse, F., and Laird, K. R. (1997). Diatoms as indicators of hydrologic and climatic change in saline lakes. In *The Diatoms: Applications for the Environmental and Earth Sciences* (E. F. Stoermer and J. P. Smol, Eds.), pp. 41–72. Cambridge University Press, Cambridge.
Grimm, E. C. (1988). Data analysis and display. In *Vegetation History* (B. Huntley and T. Webb, Eds.), pp. 43–76. Kluwer, Dordrecht.
Heck, K. L., van Belle, G., and Simberloff, D. S. (1975). Explicit calculation of the rarefaction diversity measurement and the determination of sufficient sample size. *Ecology* **56**, 1459–1461.
Jacobson, G. L., Jr, and Grimm, E. C. (1986). A numerical analysis of Holocene forest and prairie vegetation in central Minnesota. *Ecology* **67**, 958–966.
Korhola, A., Vasko, K., Toivonen, H. T. T., and Olander, H. (2002). Holocene temperature changes in northern Fennoscandia reconstructed from chironomids using Bayesian modelling. *Quaternary Science Reviews* **21**, 1841–1860.
Korhola, A., and Weckström, J. (2004). Paleolimnological studies in Arctic Fennoscandia and the Kola Peninsula (Russia). In *Long-Term Environmental Change in Arctic and Antarctic Lakes* (R. Pienitz, M. S. V. Douglas and J. P. Smol, Eds.), pp. 381–418. Kluwer, Dortrecht.
Kuwae, M., Yoshikawa, S., Tsugeki, N., and Inouchi, Y. (2004). Reconstruction of a climate record for the past 140 kyr based on diatom valve flux data from Lake Biwa, Japan. *Journal of Paleolimnology* **32**, 19–39.
Laird, K. R., Fritz, S. C., Cumming, B. F., and Grimm, E. C. (1998). Early-Holocene limnological and climatic variability in the northern Great Plains. *The Holocene* **8**, 275–285.
Lotter A, F. (1998). The recent eutrophication of Badeggersee (Switzerland) as assessed by fossil diatom assemblages. *The Holocene* **8**, 395–405.
Palmer, A. J. M., and Abbott, W. H. (1986). Diatoms as indicators of sea-level change. In *Sea-Level Research* (O. van de Plassche, Ed.), pp. 457–487. Free University, Amsterdam.
Smol, J. P., Wolfe, A. P., Birks, H. J. B., *et al.* (2005). Climate-driven regime shifts in Arctic lake ecosystems. *Proceedings of the National Academy of Science* **102**, 4397–4402.
Sorvari, S., Korhola, A., and Thompson, R. (2002). Lake diatom response to recent Arctic warming in Finnish Lapland. *Global Change Biology* **8**, 153–163.
Stager, J. C., Cumming, B., and Meeker, L. (1997). A high-resolution 11,400-yr diatom record from Lake Victoria, East Africa. *Quaternary Research* **47**, 81–89.
Stoermer, E. F., and Smol, J. P. (Eds.) (1999). *The Diatoms: Applications for the Environmental and Earth Sciences*, p. 469. Cambridge University Press, Cambridge.
Telford, R. J., and Birks, H. J. B. (2005). The secret assumption of transfer functions: Problems with spatial autocorrelation in evaluating model performance. *Quaternary Science Reviews* **24**, 2173–2179.
ter Braak C. J. F. (1987). *Unimodal Models to Relate Species to Environment*, p. 152. Unpublished PhD Thesis, University of Wageningen.
ter Braak, C. J. F., and Juggins, S. (1993). Weighted averaging partial least squares regression (WA-PLS): An improved method for reconstructing environmental variables from species assemblages. *Hydrobiologia* **269/270**, 485–502.
ter Braak, C. J. F., and van Dam, H. (1989). Inferring pH from diatoms: A comparison of old and new calibration methods. *Hydrobiologia* **178**, 209–223.
Toivonen, H., Mannila, H., Korhola, A., and Olander, H. (2001). Applying Bayesian statistics to organism-based environmental reconstruction. *Ecological Applications* **11**, 618–630.
Vasko, K., Toivonen, H. T. T., and Korhola, A. (2000). A Bayesian multinomial Gaussian response model for organism-based environmental reconstruction. *Journal of Paleolimnology* **24**, 243–250.
Weckström K (2006). Assessing recent eutrophication in coastal waters of the Gulf of Finland (Baltic Sea) using subfossil diatoms. *Journal of Paleolimnology* **35**, 571–592.
Weckström, K., Juggins, S., and Korhola, A. (2004). Quantifying background nutrient concentrations in the coastal zone of the Baltic Sea. *Ambio* **33**, 324–327.
Weckström, J., and Korhola, A. (2001). Patterns in the distribution, composition and diversity of diatom assemblages in relation to ecoclimatic factors in Arctic Lapland. *Journal of Biogeography* **28**, 31–46.
Wolfe, A. (2003). Diatom community responses to late-Holocene climatic variability, Baffin Island, Canada: A comparison of numerical approaches. *The Holocene* **13**, 29–37.

Quaternary Geological Records

R J Flower, University College London, London, UK

© 2007 Elsevier B.V. All rights reserved.

Introduction

Diatoms are ubiquitous microalgae; they grow in both terrestrial and aquatic habitats and occur in most regions of the world. Because these algae possess resistant siliceous cell walls, their remains are often preserved in sedimentary environments and, where they are sufficiently abundant, diatomaceous deposits can accumulate. Such deposits are composed predominantly of diatom valves or frustules (*see* Diatom Introduction) and collectively are referred to as diatomite. Diatomites vary considerably in texture and composition and are known by a variety of names, including kieselguhr, diatomaceous earth, fuller's earth, and tripolite. Despite differences in precise composition that are often attributable to the nature of the diatoms preserved (diatom valves can vary from a few up to around 200 μm in diameter), the word diatomite is generally used to describe any sedimentary deposit where the proportion of diatom remains or of diatom silica is high. When extracted, diatomite can contain a considerable proportion of water as well as non-biogenic clastics and organic material. The term diatomite is often used for a sediment where silica exceeds 50% of the bulk sediment and commercially useful diatomite contains more than 80% silica (Kadey, 1983). Diatomite deposits have a worldwide distribution and can be either freshwater or marine in origin, ancient or modern, and can usually be readily identified in the field as gray sediment (friable when dry) with low specific density. In this account the formation, preservation, distribution, and uses of diatomite are considered.

Diatomite Formation

Whether past or contemporary environments are considered, the process of diatomite formation is essentially the same: prolific diatom growth and subsequent accumulation of their siliceous remains in sedimentary deposits. Key aspects in diatomite formation are high diatom productivity supported by adequate light and dissolved nutrient availability (especially of dissolved silica), good diatom preservation (see below), and perhaps most importantly the lack of any significant terrigenous or other biogenic or clastic materials diluting the diatom component of accumulating sediment (Kadey, 1983). Terrestrial diatomites are all fossiliferous and enjoy a worldwide distribution; both freshwater and marine fossil deposits are common.

The formation of diatom-rich deposits is indicated schematically in **Figure 1**. The figure indicates two extreme processes for diatom deposition. On the left of **Figure 1**, the basic mode of formation of diatomaceous sediment in a deepwater environment is indicated. Here diatom phytoplankton grows in the illuminated upper part of the water column. As the crop develops, diatoms begin to sink and mass sinking events mark the culmination of growth periods as the diatom phytoplankton bloom collapses. Following a period of sinking (that depends on water depth), the dead and senescent diatoms are deposited on to the ocean floor or lake bed. Once deposited on surface sediment, the diatoms are subjected to taphonomic processes (see below) as they are incorporated into the permanent sediment record. In shallow water environments, diatomaceous sediments are formed differently. On the right of **Figure 1** a shallow water environment is indicated and here diatoms grow within a shallow water column as well as in a biofilm on subaqueous sediment and aquatic plant surfaces. Consequently the sinking depth is short or minimal and diatom species are predominantly benthic or tychoplanktonic in habit (*see* Diatom Introduction) rather than planktonic. At the sediment surface, the diatoms are subjected to taphonomic processes and in shallow waters these can be intense. These deposits are typically formed in freshwater environments where the inclusion of aquatic animal and plant macrofossils (such as fish and charophyte remains) often confirm the shallow depth of the system.

In an oceanic environment, diatom-rich sediments are referred to as diatom oozes and they occupy significant areas of the world's ocean sediments, especially in the Southern Hemisphere. These oozes occur most extensively in cold-water regions, notably around the productive Antarctic convergence zone but also in the central eastern Pacific where high diatom productivity is supported by nutrient-rich up welling currents (*see* Pacific). These pelagic sediments are composed of at least 30% biogenic silica and the diatoms are nearly all planktonic. In deep freshwater lakes the formation of diatomaceous sediment is somewhat similar and in Siberian Lake Baikal, for example, a diatom ooze also accumulates. This is especially so in the north basin of this great lake (*see* Large Lakes). In shallow lakes, benthic diatom production requires that the sediment surface lies within the photic zone and that the diatoms are growing in clear water with only small amounts of terrigenous material entering the system. Interestingly, many shallow freshwater diatom deposits are closely associated with geologies that potentially supply abundant dissolved silica for diatom growth. Hence, locations where these diatomites have formed are often associated with

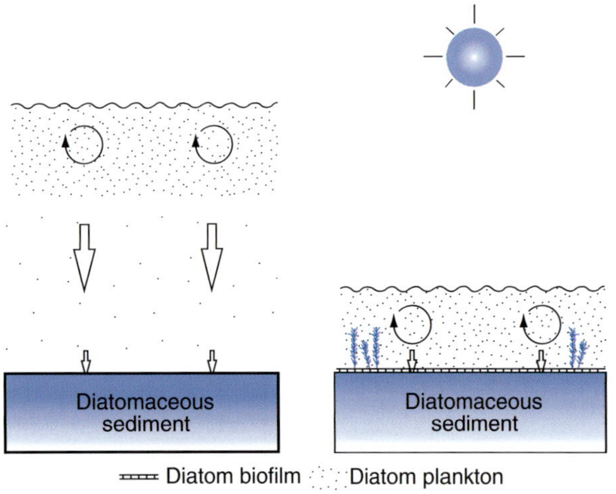

Figure 1 The formation of diatomaceous sediments. *Left* indicates the processes involved in the formation of diatomaceous sediment in a deepwater environment. *Right* indicates the processes involved in the formation of diatomaceous sediment in a shallow-water environment. The essential difference between these environments is light availability; in shallow water light promotes growth of both benthic (as a biofilm on surfaces) and planktonic (free-floating) diatoms but in deepwater diatom plankton growth is restricted to the upper illuminated surface water. Closed arrows indicate water circulation and open arrows indicate sinking diatoms.

characteristic geological formations (basalts, tephras, and volcanic deposits) that are silica-rich. Such associations are found in locations as disparate as Lillicur (Australia), Lomocastra (Costa Rica), Leizhou (China), and Toome Bridge (Northern Ireland).

Irrespective of whether a freshwater or marine environment is considered, diatoms deposited on or growing on subaquatic sediment surfaces are subjected to a variety of sedimentary processes. Those that affect diatom fossilization are referred to as taphonomic processes. These include sediment turbations of various kinds (sediment resuspension and bioturbation, sediment dewatering) that can break or dissolve at least some of the deposited diatoms. Surface sediment taphonomic processes can operate until permanent burial and then slower sedimentary processes (compaction and diagenesis) take over and transform the diatomaceous sediment into diatomite. Most importantly for diatomite formation, the overall sedimentary environment must be conducive to the preservation of diatoms. However, after incorporation into permanent sediments, diatom remains can persist for millions of years.

Preservation of Diatoms

Diatoms are prone to damage and dissolution and these processes can combine to diminish the quantity of diatoms accumulating in sediments. Most processes inimical to the preservation of a high portion of diatom remains in sediments operate within the water column or at the mud–water interface on the sea or lake floor. Diatoms will tend to dissolve whenever the surrounding water is under-saturated regarding dissolved silica (Shemesh *et al.*, 1989) and preservation will be further reduced by breakage often encountered in high energy environments and promoted by grazing (Flower, 1993). Depending on the water column depth, the degree of silicification of individual species of diatom, and a variety of water quality characteristics, diatoms will be preferentially preserved but, even in areas with sediments rich in diatoms, major initial losses by dissolution can occur. In Lake Baikal, for example, over 99% of diatoms produced in the euphotic zone dissolve before being incorporated into sediments (*see* Large Lakes). Some dissolution occurs in the water column but most occurs as a result of surface sediment processes (Ryves *et al.*, 2003). Diatom remains in diatomites can show a range of preservational states from good to very poor and, although dissolution and breakage occurs to various extents during sediment formation, the preservational condition of some terrestrial diatomites is probably decreased during aerial exposure and weathering.

Over geological time, diatom silica changes state and the physical frustular structures disappear as the initial opaline silica of the diatom cell wall looses water, eventually becoming chrystobilite. Despite differing formation processes, we can generalize that diatomites form in shallow or deepwater environments as a result of moderate to high diatom productivity, adequate preservation, and an absence of diluting detrital materials. Because of the instability of opaline silica, diatomite deposits are indicative of the most recent geological period, the Cenozoic.

Diatomite Distributions and Age

Terrestrial diatomite deposits occur on all the continental land masses and, although Quaternary deposits are most frequent, many date from the Miocene Period. The Pre-Quaternary deposits are often substantial, frequently marine in origin, and are particularly common around the margins of the Pacific Ocean (Reading, 1978). A well-known example is the extensive Miocene Monterey Formation in California but other similar examples exist in Japan, South Korea, Eastern Russia, and South America. Although the Monterey Formation is large and variable, the Toro Road exposure (Barron, 1976) is almost pure marine diatoms but preservation is often poor with many valves being broken. Aquitanian-Burdigalian and Messinian diatomites occur in countries bordering the Mediterranean, and Reading (1978) lists Romania, Spain, Sicily, Algeria, and northern Italy as regions with abundant marine Miocene diatomite deposits. The Miocene seems to have been a time of particularly extensive diatomite formation and the worlds largest (marine) diatomite deposit of this period is probably at Lompoc in California (Dolley, 2002). Reading (1978) attributes the prevalence of Miocene diatomites to both tectonic and global climatic changes but Kidder and Gierlowski-Kordesch (2005) suggest that the non-marineMiocene diatomite deposits are associated with the rise of grasslands and changes in the silica cycle.

Freshwater diatomites are common around the Tertiary/Quaternary boundary and those dating across the Plio-Pleistocene periods occur in sections often tens of meters thick around the world (Harwood, 2000). Notable such freshwater deposits occur in Central America (Mathers *et al.*, 1991) and North America (Bradbury and Krebs, 1982), China (Qu and Chen, 2005), and in Africa (Gasse, 1980). See Trauth *et al.* (2005) for a summary of diatomite sequences in East Africa. One Quaternary diatomite exposure (**Fig. 2**) in central Kenya is over 30 m deep; it is near ancient Lake Kariandsi and has been dated to 0.97–0.974 Ma (Trauth *et al.* 2005). Its existence illustrates the development of wet periods in the East

Figure 2 An exposed outcrop of freshwater diatomite from ancient Lake Kariandusi in Kenya. The section dates to 0.97–0.974 Ma and illustrates a former wet period in the East African Rift Valley system (Trauth *et al.* 2005). Note that several caves have been excavated in the basal diatomite section. Photograph: M. Maslin.

African Rift Valley zone at that time (M. Maslin pers. comm.). Elsewhere in Africa, extensive diatomite deposits occur around the Lake Chad area and these extend into the Holocene epoch. Diatomite samples collected as part of the 'ACACIA (Arid Climate, Adaptation and Cultural Innovation in Africa) Project' from the former Lake Megachad area and elsewhere in the region (D. Verschuren, pers. comm.), contain abundant freshwater plankton (*Aulacoseira granulata, Cyclotella stelligera*, and *C. ocellata*) and benthic diatoms (especially *Epithemia, Cocconeis placentula*, and *Denticula*) as well as tephra deposits. The diatom species indicate the former presence of clear dilute alkaline water of varying water depths. The Bodélé depression (northern Chad) contains large areas of exposed diatomite, which are a globally significant source of windblown dust (Giles, 2005).

Another region with extensive Pliocene-Pleistocene diatomites is Central America. Deposits of freshwater mainly planktonic diatoms in Costa Rica can reach 100 m in thickness (Mathers *et al.*, 1991). These also often contain abundant macrofossils of fish and other aquatic organisms. Other very extensive diatomite deposits occur in China and, as in Costa Rica, they are associated with volcanic formations. They date from the Pliocene to Holocene (Shang, 2003). The commercial diatomite deposit in the Leizhou peninsular (North China) is of Middle Pleistocene age (Qu and Chen, 2005) as are those in south-western China, especially Yunnan Province (Shang, 2003).

In the most recent epoch, the Holocene, freshwater diatom deposits occur globally but they tend to be thin, often being only several meters or less deep and localized in distribution. Only a few lakes today seem to be accumulating diatomite quality sediment. Lake Myvatn in Iceland is one lake were diatomaceous sediment is accumulating and where diatomite was extracted commercially until recently (Kjaran *et al.*, 2004). However, many recent diatomite deposits seem to have formed during the early to mid-Holocene. They are regional in distribution but very widespread; Eardley-Wilmot (1928) gives a thorough account and maps of such deposits in Canada (notably in British Columbia and Nova Scotia). In the United Kingdom, Holocene diatomites occur at sites such as Toome Bridge in Northern Ireland (Wilson, 1972), in the Loch Cuithir and other basins on the Isle of Skye in Scotland (Robertson, 1945) and in England, notably at Kentmere (Round, 1957). Today, human activities have accelerated soil erosion and promoted cultural eutrophication of inland water bodies, processes that have tended to diminish the proportion of diatoms in accumulating sediments.

One rather local but environmentally significant Holocene diatomite deposit occurs in the Faiyum Depression (Egypt) to the north of modern Lake Qarun (Flower *et al.*, 2006). This material was explored by Aleem (1958) and it occurs as large expanses of white sediment several meters thick in places and is interspersed with blown sands (**Fig. 3**). Fish remains are abundant and according to Wendorf and Schild (1976), the diatomites were deposited in several stages when the present lake was much larger. However, dates are few and not directly made on the diatomite but carbon-14 measurements on organic matter within diatomite sections indicate

Figure 3 The exposed and wind eroded surface of a Holocene freshwater diatomite deposit near Lake Qarun (Egypt). Note that the desiccation fissures in the exposed diatomite surface are filled with wind blown sand. Photograph: R.J. Flower.

the deposit dates to the early Holocene with some sections post dating 5,500 BP. Other diatomite deposits occur around the northern fringes of the Sahara and especially in Algeria (Baudrimont, 1973), they are often dated to the Neolithic period. Collectively, they evidence a freshwater crisis during the middle Holocene as climate change and aridity impacted much of northern Africa.

The Use of Diatomites

Diatomites have two principal uses, one is commercial and entails extraction and processing into various diatomite products and the other relates to their value as an archive of past environments and of environmental change. Regarding the former, people have used diatomites for several thousands of years but it is only during the last two hundred years that diatomite deposits have been extracted on a commercial scale.

Diatomite extraction

Use of diatomite materials first requires extraction of a suitable deposit followed by fabrication into materials appropriate for distribution. Whether one considers ancient or modern uses, the activities involved are similar and entail both extraction and processing. The ancient methods seemed to have focused on cutting blocks of sediment or manufacturing mud bricks. For the latter, diatomites were mixed (presumably by hand) with water and plant materials (usually straw), moulded and dried in the Sun (Flower, 2006). More recent traditional diatomite extraction in the United Kingdom was also largely done by hand and the techniques depicted by Wilson (1972) for diatomite products from the Toome Bridge area of Northern Ireland are typical. Diatomite bricks were cut by hand, stacked into piles for air drying and then milled to produce a fine dry powder product for bagging and sale. A commercial company (W. Kenyon Ltd.) exploited these deposits from around the outflow of Lough Neagh until recently and the remaining deposits now receive protection by environmental legislation. Small-scale diatomite extraction was also formerly undertaken in Scotland, especially on Skye where eight diatomite sites occur in the Trotternish area. The Loch Cuithir site was worked until 1914 and again after 1937 (see www.bambi.demon.co.uk/skyedata/minerals). Nevertheless, during the 1950s and 1980s, commercial extraction of diatomite effectively ceased in the United Kingdom but the feasibility of small operations is still debated. Elsewhere in the world, diatomite extraction is currently big business and this is particularly so in the United States, China, and the former Soviet Union, countries where advanced plant machinery and high temperature drying of the product are deployed (Kadey, 1983; Harwood, 2000). Other countries with commercially active diatomite extractive industries include Australia, Denmark, Spain, and South and Central America (Eardley-Wilmot, 1928; Dolley, 2002). World production of diatomite averaged about 1.65 million tonnes per year from 1988–1992 (British Geological Survey, 1992).

Ancient uses of diatomites

Archaeology has revealed some interesting uses of diatom-rich materials but the extent of use is unclear. The value of diatomite as a lightweight, easily handled material seems to have attracted interest. Dolley (2002) noted that lightweight building bricks of diatomite were used by the ancient Greeks and that diatomite blocks were used in the construction of the sixth-century Hagia Sophia church, Istanbul. In ancient Egypt, the Ptolemaic inhabitants of the archaeological site of Dimai in the Faiyum (Egypt) also seemed well aware of at least some of the properties of diatomite deposits. Mud bricks at Dimai were fabricated from local diatomite deposits and used in the construction of dwellings, a funerary complex, and official buildings (Flower, 2006). Diatomite can be found frequently exposed near Dimai (**Fig. 4**) but whether it was actively sought for making mud bricks is unknown. It is thought, however, that the insulating properties of diatomite could well have helped to keep buildings internally cool in the hot climate. Also, near Dimai, Caton-Thompson and Gardner (1934) described a region of siliceous (diatomite) deposits that were irrigated

Figure 4 A partly eroded ridge of a Holocene freshwater diatomite near Dimai (Faiyum, Egypt). Note that the exposure is fissured and faulted and is severely wind eroded. Photograph: R. J. Flower.

and used for crop cultivation in Ptolemaic times. Diatoms in building materials are known in archaeological sites elsewhere but they seem to be present as passive tracers of source materials (*see* Use in Archaeology).

Diatomites in modern commercial usage

It is the basic structure of diatomite, it's high porosity, large surface area, abrasive quality, insulation capability, and refractory nature that makes it so useful commercially. The principal commercial uses of processed diatomite are in the food and building industries but there are also applications in the chemical, agricultural, and related industries (Eardley-Wilmot, 1928; Harwood, 2000; Kadey, 1983).

In the food industry, diatomite-derived products (usually powders) are useful as a filtrant or as an absorbent for liquids produced for consumption. There are a variety of liquid treatment aids produced commercially and these are all designed to remove impurities and to clarify liquids. Hence, wine and beers are often treated with processed diatomite as are certain oils and fats. In the chemical industry, diatomite products are used as a filler and as a valuable additive to many paints, rubber, and plastics. Diatomite particles are an effective absorption medium for some macromolecules and can be used to selectively capture dissolved organic compounds. Because of the mildly abrasive nature of diatomite, it is also used in the manufacture of polishes and in some toothpastes. In agriculture, processed diatomite has several uses as a soil additive (e.g. www.maidenwell.com); it improves soil quality (mainly through its ability to take up and retain water) and it can act as a snail and slug repellent. In some agricultural locations, the natural surface soil consists largely of diatomite and the resulting light and freely draining soils are favored in Northern Ireland where they are used for growing root crops (Cruickshank, 1982).

For some of the preceding uses, diatomite is being replaced by modern synthetic materials but it is in the building industry that diatomite has considerable potential. Traditionally diatomite has been the material of choice for the manufacture of fire bricks (Harwood, 2000), where the fine-grained siliceous material confers high insulation value. A range of brick and tile diatomite products has been developed with varying degrees of success (Eardley-Wilmot, 1928). However, more recent research on the addition of diatomite products to building materials, cements, and building bricks, showed that the structural properties of these materials can be improved. Addition of diatomite into fabrication processes of modern building bricks can influence the pozzolanic properties of the product (Wild *et al.*, 1997). The pozzolanic properties of high reactivity silica in diatomite, produce cements and ceramics with increased strength and compressibility resistance (Stamatakis *et al.*, 2003), and similarly for ceramics. The properties of various diatomites are being tested as an additive in modern building materials (Stamatakis *et al.*, 2003), where high strength and other special product features are needed.

Diatomites in environmental science

Diatom-rich sediment sections in marine diatom deposits have long been used for exploring paleodiversity in diatoms. Well-known sites are in New Zealand (the Oamaru Deposit (Grove and Sturt, 1853)) and in California (the Moreno Shales (Hanna, 1927)). Later, diatom-rich marine sediment cores were examined to demark biostratigraphic facies, help define correlations and sedimentary chronozones, and to demonstrate species evolution (Strelikova, 1990). Similarly, terrestrial deposits of freshwater diatoms have been evaluated since the mid-nineteenth century. These were also examined initially from the point of view of diatom diversity and it was not until the 1980s that these deposits were analyzed for purposes of developing a worldwide lacustrine chronobiostratigraphy or inferring climate change. Bradbury and Krebs (1982) studied Miocene to Holocene diatomites in Idaho and Oregon and indicated that at least seven taxa were common to lacustrine deposits at various locations in the Northern Hemisphere. Working in Ethiopia, Gasse (1980) examined the diatom stratigraphy in a 31 m section of diatomite and showed that the diatom record was indicative of fluctuating environmental conditions, especially in water quality and in climate.

On a more recent timescale, Holocene diatomite sections near extant lakes, such as Lough Neagh in Northern Ireland (Plunket *et al.*, 2004) and Lake Qarun in Egypt (Wendorf and Schild, 1976) have been used to document lake level fluctuations and detect high lake stands during the mid-Holocene epoch in response to climate changes. These diatomites have been examined to reveal species changes relative to lake level and the terrestrial deposits can be readily sampled at exposures or cored with hand operated equipment (**Fig. 5**) prior to stratigraphic analysis. Furthermore, the formation of these diatomites is relevant to the presence of archaeologically important human occupation sites (Wendorf and Schild, 1974; Smith and Collins, 1971).

Diatomite products have also been employed as a tracer in freshwater systems (Lund, 1974). Flower (1980) used diatomite dyed with aniline blue

Figure 5 Sampling a Holocene freshwater diatomite deposit in the Toome Bridge area of Northern Ireland. Note the exposed cut face of the diatomite deposit in the foreground and the upper 20–30 cm of soil cover. The semi water-logged deposit is about 1.5 m thick and a Hiller corer is being used to collect a stratigraphically ordered sequence of samples for paleoenvironmental analysis. Photograph: R. J. Flower.

as a stratigraphic marker to estimate sediment accumulation in Lough Neagh (Northern Ireland). The reaction of the silica surface of diatomite to aromatic compounds is the focus of interesting new research and Goa *et al.* (2005) showed that surface modified diatomite can be a very effective trap for aqueous phenol. Hence, treated diatomite could have potential applications for combating water pollution.

Conclusions

Diatomites around the world have provided people with a rich source of material for both commercial exploitation and paleoenvironmental science. Commercial uses of diatomite are changing and some applications are diminishing due to the development of cheaper alternatives. Yet diatomite as an additive to building materials has considerable potential and it also has promising applications in the organic chemistry industry. For paleoenvironmental research, there are rich opportunities for further exploiting the biostratigraphies of diatomite deposits in relation to fossil assemblages as recorders of past climates, ecosystem development, and species distributions through time.

See also: **Diatom Methods**: $\delta^{18}O$ Records; Use in Archaeology. **Diatom Records**: Large Lakes; Pacific.

References

Aleem, A. A. (1958). A taxonomic and palaeoecological investigation of the diatom flora of the extinct Fayoum Lake (Upper Egypt). *University of Alexandria Bulletin* **2**, 17–44.

Barron, J. A. (1976). Marine diatom and silicoflagelate biostratigraphy of the Delmontain Stage and the type *Bolivina obliqua* Zone, *Californian Journal of Research. U.S. Geological Survey* **4**, 339–351.

Baudrimont, R. (1973). Recherches sur la diatomées des eaux continental de l'Algeria: ecology and paleoecology. *Mémoires de la Société d'Historie Naturelle de l'Afrique du Nord* **12**, 1–249.

Bradbury, J. P., and Krebs, W. N. (1982). Neogene and Quaternay lacustrine diatoms of the western Snake River basin Idaho-Oregon, USA. *Acta Geologia Academiae Scientiarum Hungaricae* **25**, 97–122.

British Geological Survey (1992). World Mineral Production 1988–1992. Preliminary statistics. British Geological Survey.

Caton-Thompson, G. W., and Gardner, E. (1934). *The Desert Fayum*. Royal Anthropological Institute, London.

Cruickshank, G. (1982). Soils. In *Environment and Natural Resources* (Cruickshank and D. N. Wilcock, Eds.), pp. 165–184. The Queens University of Belfast, Belfast.

Dolley, T. P. (2002). Diatomite. In: Metals and Minerals. *U.S. Geological Year Book* **1**, 25.1–25.6.

Eardley-Wilmot, V. L. (1928). Diatomite, its Occurrence, Preparation and Uses. Report for the Department of Mines (Mines Branch). Ottawa: F.A. Acland.

Flower, R. J. (1980). Sedimentation in Lough Neagh. Unpublished D.Phil. thesis. Coleraine: University of Ulster.

Flower, R. J. (1993). Diatom preservation: experiments and observations on dissolution and breakage in modern and fossil material. *Hydrobiologia* **269/270**, 473–484.

Flower, R. J. (2006). Diatoms in ancient building materials, application of diatom analysis to Egyptian mud bricks. *Nova Hedwigia*, Beiheft Vol. 130, 245–264.

Flower, R. J., Stickley, C., Rose, N. L., Peglar, S., Fathi, A. A., and Appleby, P. G. (2006). Environmental changes at the desert margin: an assessment of recent paleolimnological records in Lake Qarun, Middle Egypt. *Journal of Paleolimnology* **35**, 1–26.

Gasse, F. (1980). Les diatmeés lacustres Plio-Pleistocene du Gadeb (Ethiopie) Revue Algologique. *Memorie hors-serie* **3**, 1–249.

Giles, J. (2005). The dustiest place on earth. *Nature* **434**, 816–819.

Goa, B. J., Jiang, P. F., An, F. Q., Zhao, S. Y., and Ge, Z. (2005). Studies on the surface modification of diatomite with polyethylenemine and trapping effect of the modified diatomite for phenol. *Applied Surface Science* **250**, 273–279.

Grove, E., and Sturt, G. (1853). On a fossil marine diatom deposit from Oamaru, Otago New Zealand. *Journal of the Quekett Microspocipal Club* **2**, 321–330.

Hanna, D. G. (1927). Cretaceous diatoms from California. Occasional Papers of the California. *Academy of Sciences* **13**, 5–48.

Harwood, D. M. (2000). Diatomite. In *The Diatoms: Applications for the Environmental and Earth Sciences* (E. F. Stoermer and J. P. Smol, Eds.), pp. 436–446. Cambridge University Press, Cambridge.
Kadey, F. L. (1983). Diatomite. In *Industrial Minerals and Rocks* (S. J. Lefond, Ed.), 5th edn, pp. 677–708. Society of Mining Engineers of the American Institute of Mining, Metallurgical and Petroleum Engineers, Inc, New York.
Kidder, D. L., and Gierlowski-Kordesch, E. H. (2005). Impact of grassland radiation on nonmarine silica and Miocene diatomite. *Palaios* **20**, 198–206.
Kjaran, S. P., Holm, S. L., and Myer, E. M. (2004). Lake circulation and sediment transport in Lake Myvatn. *Aquatic Ecology* **38**, 145–162.
Lund, J. W. G. (1974). Phytoplankton as indicators of change in lakes. *Environment and Change* **2**, 273–281.
Mathers, S. J., Chavez, L., Alvarado, L., and Inglethorpe, D. J. (1991). Detailed investigations of selected Costa Rican diatomites. British Geological Survey, Nottingham. Unpublished report to Recope S.A. San Jose. p. 74.
Plunket, G. M., Whitehouse, N. J., Hall, V. A., Brown, D. M., and Baille, M. G. L. (2004). A precisely-dated lake-level rise marked by diatomite formation in northeastern Ireland. *Journal of Quaternary Science* **19**, 3–7.
Qu, R., and Chen, S. (2005). Geological factors controlling ore and mineralization of diatomaceous earth deposits in Leizhou Peninsula. *West China Exploration and Engineering* **107**, 88–89.
Reading, H. G. (Ed.) (1978). *Sedimentary Environments and Facies*. Blackwell Scientific Publications, London.
Robertson, R. H. S. (1945). Diatomite in Scotland. *The Quarry Managers' Journal* April Issue, 451–458.
Round, F. E. (1957). The Late-Glacial and Post-Glacial diatom succession in the Kentmere Valley deposit. *New Phytologist* **56**, 98–126.
Ryves, D. B., Jewson, D. H., Sturm, M., Battarbee, R. W., Flower, R. J., Mackay, A. W., and Granin, N. G. (2003). Quantitative relationships between planktonic diatom communities and diatom assemblages in sedimenting material and surface sediments in Lake Baikal. *Limnology and Oceanography* **48**, 1643–1661.
Shang, Y. (2003). Tengchong diatomite deposit and its genesis. *Yunnan Geology* **22**, 418–425.
Shemesh, A. L., Burkle, L. H., and Froelich, P. N. (1989). Dissolution and preservation of Antarctic diatoms and the effects on sediment thanatocoenoses. *Quaternary Research* **31**, 288–308.
Smith, A. G., and Collins, A. E. P. (1971). The stratigraphy, palynology and archaeology of diatomite deposits at Newferry, Co. Antrim, Northern Ireland. *Ulster Journal of Archaeology* **3**, 1–73.
Stamatakis, M. G., Fragoulis, D., Csirik, G., Bedelean, I., and Pedersen, S. (2003). The influence of biogenic micro-silica-rich rocks on the properties of blended cements. *Cement and Concrete Composites* **25**, 177–184.
Strelikova, N. I. (1990). Evolution of Diatoms during the Cretaceous and Paleogene Periods. In *Proceedings 10th International Diatom Symposium* (H. Simola, Ed.), pp. 195–204. Koeltz, Koenigstein.
Trauth, M. H., Maslin, M. A., Deino, A., and Strecker, M. R. (2005). Late Cenozoic moisture history of East Africa. *Science* **309**, 2051–2053.
Wendorf, F., and Schild, R. (1976). *Prehistory of the Nile Valley*. Academic Press, New York.
Wild, S., Gailius, A., Hansen, H., Pederson, L., and Szwabowski, J. (1997). Pozzolanic properties of a variety of European clay bricks. *Building Research and Information* **25**, 170–175.
Wilson, H. E. (1972). *Regional Geology of Northern Ireland*. Her Majesty's Stationary Office, Belfast.

Salinity Reconstructions from Continental Lakes

S C Fritz, University of Nebraska, NE, USA

© 2007 Elsevier B.V. All rights reserved.

Introduction

Diatoms are widely used to reconstruct past environments, because of their sensitivity to a range of environmental variables. Among the variables that exert a strong influence on diatoms are ionic concentration, expressed as salinity or conductivity, and ionic composition. Because the ionic concentration and composition of inland water bodies are affected by changes in effective moisture (precipitation minus evaporation (P–E)), changes in the composition of diatom species that vary in their salinity tolerances provide a tool for the reconstruction of past climate.

The use of diatoms to infer changes in salinity driven by climate relies on settings where changes in effective moisture are moderate-to-large and where these changes produce relatively large changes in salinity. These conditions are most common in semiarid and arid regions, where evaporative forcing is strong, and in topographically closed lakes where changes in P–E have the potential to concentrate or dilute dissolved salts. These conditions are widespread in the dry interiors of continents, in subtropical deserts, and in dry polar regions. Thus, diatom-inferred changes in salinity can be used to reconstruct effective moisture from lake deposits in many regions of the globe.

In practice, the inference of climate from diatom-inferred salinity is complicated by the diverse controls on the hydrologic budgets of lakes, and occasionally the accurate inference of salinity from diatoms may be obscured by other ecological influences. None-the-less diatom records have played a critical role in developing our understanding of the extent to which twentieth-century climate patterns are representative of long-term climate variation and of the patterns and controls on climate variability at decadal and longer timescales.

Diatoms as Indicators of Salinity and Ion Composition

Salinity Classification of Diatoms

Diatoms have long been recognized as indicators of ionic concentration (salinity). Kolbe (1927) and subsequently Hustedt (1953) developed a system to classify diatoms according to their distribution in waters of varied salinity, the so-called halobion system. This system was based primarily on distributional patterns in sodium chloride (NaCl) dominated marine and coastal environments, which ranged from marshes and estuaries receiving fresh surface runoff to the open ocean where salinity reaches ~35 g L^{-1}. Although this halobion classification was applied to lacustrine diatoms in the mid-1900s, it is not well suited for diatoms in continental lakes, because of their much broader range of salinity and ion composition.

Studies to characterize the salinity ranges of lacustrine diatoms expanded in the 1970s and 1980s, when the potential to reconstruct past climate change from lake sediments became more broadly recognized. During the last few decades, studies of the distribution of diatoms relative to gradients of salinity and ion composition have been carried out on all continents and in lakes from high latitudes to the tropics and have fueled the use of diatoms as paleoclimatic indicators.

Diatoms occur in the full range of lakes from dilute freshwater to hypersaline brines (**Fig. 1**), and classification schemes reflect this diversity of environments. One of the most commonly applied classification systems splits lakes and their associated diatom floras into freshwater (<0.5 g L^{-1}), subsaline (0.5–3.0 g L^{-1}), hyposaline (3.0–20.0 g. L^{-1}), mesosaline (20–50 g. L^{-1}), and hypersaline (>50 g. L^{-1}) waters. Many species have distributional limits near the overlap of the subsaline and hyposaline ranges, and the diversity of taxa is low in mesosaline and hypersaline systems.

Although the distribution patterns of many diatoms are correlated with salinity, the mechanisms for this correlation are not well-known. Salinity imposes an osmotic influence and hence an energetic cost on cells. At high salinity, diatoms manufacture proline or other osmolytes to deal with osmotic gradients; many of these compounds are amino acids and rich in nitrogen. Thus, under high salinity, nutrient requirements may be higher. Other compounds that are correlated with salinity gradients also may affect nutrient uptake by diatoms. For example, high sulfate concentrations may interfere with molybdate uptake, a critical component of enzymes involved in nitrogen use. Many saline lakes in grassland landscapes also have high dissolved organic carbon concentrations (DOC) and low chlorophyll relative to ambient nutrient concentrations. This suggests that DOC may form complexes with a critical nutrient that controls primary production and thereby impact diatom productivity and species composition. Thus, several lines of evidence suggest that one of the major impacts of changing salinity may be via its impact on nutrient availability, uptake, and requirements.

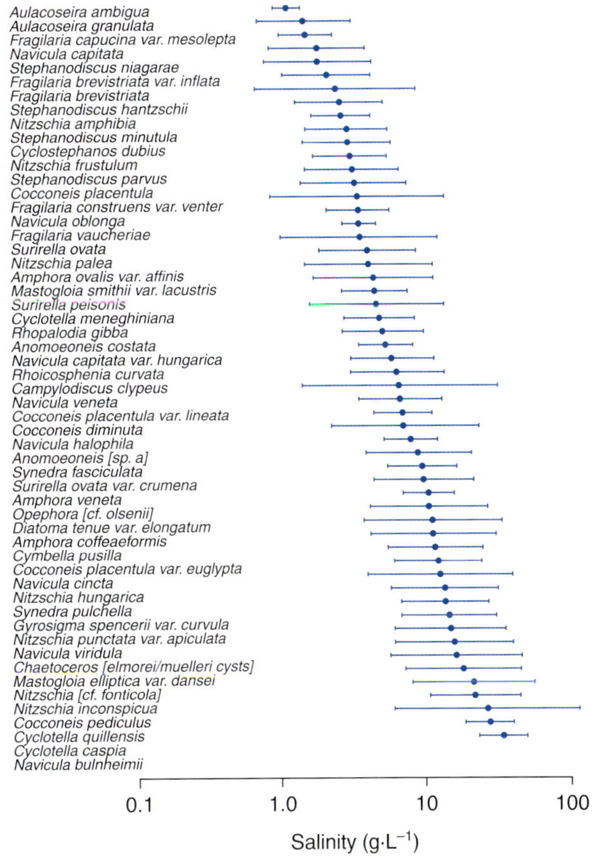

Figure 1 The salinity ranges (horizontal line) and salinity optima (dot) of selected diatom species from the Great Plains region of North America, generated from a survey of water chemistry and diatoms in the surface sediments of 66 regional lakes. Weighted averaging regression was used to calculate ranges and optima (see Korhola). Species are arrayed from top to bottom in order of increasing salinity optima. Reproduced from Fritz *et al.* (1993) with permission.

Diatoms also can be classified according to their distribution in waters of differing anion composition, and some species are characteristic of chloride, sulfate, or carbonate–bicarbonate dominated waters (**Fig. 2**). The reasons for differences in anion association also are unclear, but likely relate, at least in part, to anion impacts on nutrient availability. Shifts in anion composition of lacustrine systems can be indicative of changes in source waters, particularly groundwater, or of progressive changes in brine composition driven by climate.

Quantitative Approaches for Salinity Reconstruction

Since the mid-1980s, transfer functions have been used increasingly in paleoenvironmental studies to quantitatively reconstruct salinity, conductivity, or

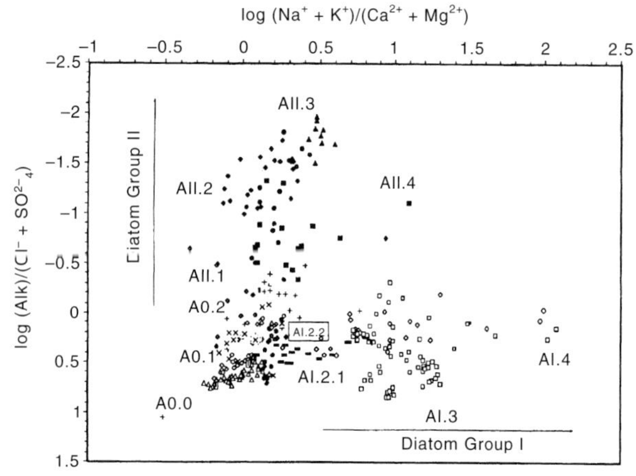

Figure 2 Diatom-inferred cation composition (x axis) and anion composition (y axis) inferred from a series of diatom samples from sediment cores from 12 African lakes. Sediment samples from each of the 12 lakes are represented by a different symbol. A weighted average calibration model developed from over 300 modern lakes in Africa was used to reconstruct cation and anion composition from these sediment samples. The groups (A0, AI, AII) represent different combinations of lake water cation and anion dominance, as indicated in the figure. Subgroups 1–4 represent progressively increasing salinity within a given group. Reproduced from Gasse (2002) with permission.

some other environmental variable of interest from fossil diatom assemblages. These transfer functions are based on modern calibration datasets, which sample a group of contemporary lakes that span a salinity gradient and use these lakes to characterize the salinity range of diatom species and the so-called salinity optimum, the salinity at which each diatom species is most abundant (**Fig. 1**). A variety of statistical techniques can be used to generate transfer functions for the reconstruction of salinity or other environmental variables from modern datasets, including analog matching and several multivariate methods. The transfer function approach does not assume that salinity, or another reconstructed variable, is the proximate or physiological variable impacting diatom distribution, but simply that diatom distribution is highly correlated with salinity and that this relationship is stationary through time. In most cases, due to time and/or resource constraints, a regional transfer function cannot be independently tested by applying the function to a new group of lakes or by comparing a reconstruction from a sediment core with historic measurements of salinity from the same lake. Therefore, statistical techniques for error estimation, such as bootstrapping or jackknifing, are commonly used to evaluate transfer function performance.

Transfer functions for salinity reconstruction are effective in lakes that have undergone moderate-to-large fluctuations in salinity that exceed the standard error of the prediction. Quantitative techniques also work best in systems where other water-chemistry variables that have a strong impact on diatoms, such as pH or nutrients, are less likely to show large changes over time. In theory, statistical approaches, such as variance partitioning, can account for the impact of multiple environmental influences on species distribution. In practice, however, modern data sets are constructed to optimize the environmental gradient of interest, in this case salinity, and other ecologically important variables may not be adequately accounted for. In these cases, the assumption that the relationship between species and salinity does not change with time may be invalid, such as has been demonstrated for diatom-based transfer functions for temperature. Because of these issues, salinity reconstructions are most robust when they are corroborated by inferences from other paleolimnological proxies, such as ostracodes or sediment geochemistry.

Quantitative reconstructions depend on having a modern calibration dataset that accurately defines the environmental ranges and tolerances of diatom species. For most regions, the size of calibration datasets is less than 100 lakes, and although such a sample size can generate a statistically robust transfer function, it may not adequately capture the true optima of all taxa, particularly those that dominate at the ends of the salinity gradient. In some regions, such as the tropical Andes, the density of modern samples currently is insufficient to allow quantitative reconstructions in most settings. In addition, the large number of apparently endemic species makes generalizations from the floras of other regions difficult. Finally, given the large range of environmental change in the past, some regions may no longer contain aquatic environments that are modern analogs for past environmental conditions. Thus, in Africa and Mexico, some of the dominant diatom taxa in fossil records do not occur in the modern flora, even given large spatially distributed modern datasets.

Diatom Preservation in Saline Lakes

Although diatoms are a common component of the algal community in lakes from low-to-high salinity, they are not always well preserved in sedimentary records. Diatom dissolution is highest in NaCl solutions, where high pH (>9) enhances the dissociation of silicic acid, but dissolution also occurs in other salt solutions. Brines can induce the formation of diagenetic silicates (e.g., zeolites and smectite), leading to alteration or dissolution of diatom assemblages. These processes can alter not only total diatom abundance in sediments, but also species composition, because diatom species are differentially sensitive to dissolution and diagenesis. In some cases, dissolution indices can be developed to correct for some of the bias imposed by taphonomic processes.

Lakes, Salinity, and Climate

Climate impacts the water budget of lakes via its influence on precipitation, surface evaporation, and groundwater flow. Shifts in a lake's water balance, in turn, impact lake salinity through concentration or dilution of ions in solution. The magnitude of the climatic imprint depends on the morphometry of the lake and its drainage basin; both lake surface area and volume, for example, affect the degree to which precipitation or evaporation influence lake level and lake salinity. Overall, the best candidates for climate reconstruction from diatom-inferred salinity techniques are subsaline-to-hyposaline lakes in semiarid-to-arid climates, because variations in precipitation and evaporation cause salinity changes that are sufficiently large to produce shifts in diatom-species composition. The rate of lake response to climate is a product of the lake's water residence time and the relative balance of surface flow relative to groundwater. In lakes where groundwater is a large component of the water budget, lake response is governed by the response time of the groundwater system, and short-term climatic change may have very little impact on the water budget. Connection to a groundwater flow system also allows for loss of salts via subsurface outflow, whereas in a terminal basin salts build up over time.

On a global basis, subsaline and saline lakes outnumber freshwater lakes, and systems appropriate for paleoclimatic study are common in most of the regions of the world where water availability is limited. This includes the continental interiors of the Americas, Africa, Australia, and Eurasia, as well as the high-latitude regions of Antarctica and the Arctic. Examples of a range of lake systems where diatom-inferred salinity has been used to infer paleoenvironmental and paleoclimatic history during the Quaternary are given below.

Quaternary Environmental History as Inferred from Diatoms

Quaternary Climate Variation at Orbital and Millennial Timescales

Shifts between freshwater and saline diatom taxa have been used effectively to infer moisture variation on orbital and millennial timescales in extant and former lake basins. In the Owens Lake basin in western North America, peaks in freshwater planktic diatoms and inferred deepwater conditions are broadly correlated with cold periods in the SPECMAP (Spectral Mapping Project; Mapping Species Variability in Global Climate Project) oxygen isotope record after ~400,000 years before present (yr BP; **Fig. 3**), when tectonic processes probably produced an increase in basin depth. Prior to that time, the evidence suggests that the basin was much shallower, and the correlation between intervals when planktic diatoms increased and cold periods is not as strong. In Lake Titicaca in the tropical Andes, the increased abundance of freshwater planktic diatoms (**Fig. 4**) indicates fresh overflowing conditions associated with increased precipitation, whereas benthic diatoms and the saline planktic taxon,

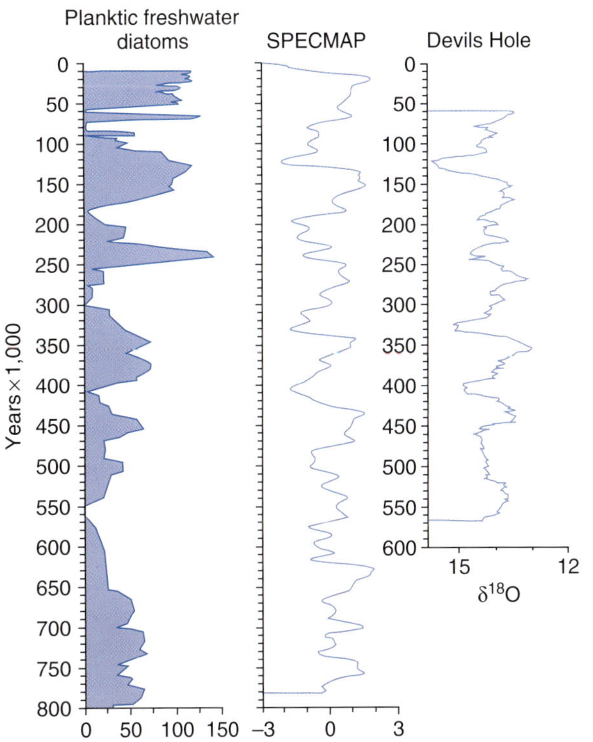

Figure 3 A comparison of the abundance of freshwater planktic diatoms in a drill core from Owens Lake, CA, USA compared with the SPECMAP (Spectral Mapping Project; Mapping Species Variability in Global Climate Project) $\delta^{18}O$ record of global temperature change. High isotopic values represent global cold periods, whereas depleted values are warm periods. The $\delta^{18}O$ record from a calcite vein at Devils Hole, WY, USA also provides a chronology of temperature change; in this case more depleted (smaller) values represent colder periods. The relative proportion of diatoms in each sample was calculated using a standard sample count of 300 diatom valves. In a few samples with very high diatom concentrations, more than 300 valves were counted and thus percentages exceed 100%. Reproduced from Bradbury (1997) with permission.

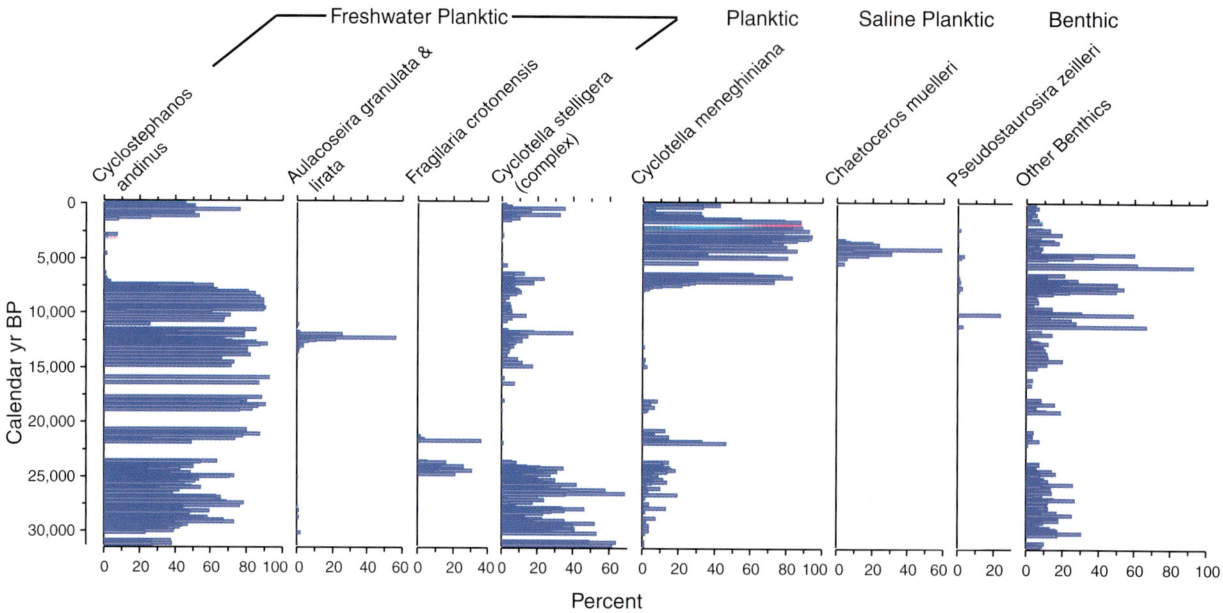

Figure 4 The diatom stratigraphy of the last ~30,000 years from Lake Titicaca, Bolivia, Peru. The dominance of freshwater planktic diatoms are inferred to be times when the lake was fresh and overflowing, whereas benthic (*Pseudostaurosira zelleri* and other benthics) and saline (*Chaetoceros muelleri*) diatoms dominated when the lake level fell below the outlet threshold.

Chaetoceros muelleri, indicate periods when lake-levels fell below the outlet and salinity built up. In this case, the wet intervals were coincident with global cool periods, including the Last Glacial Maximum when sea surface temperatures were cold and the South American summer monsoon was enhanced by high summer insolation. During the contrasting conditions of the early-to-mid-Holocene, lake levels dropped sufficiently to more than double salinity, and seismic evidence suggests that the lake, now 235 m deep, was as much as 85 m lower than today (more discussion of the Lake Titicaca record is given in Diatom Records: Large Lakes). A 50,000 yr diatom record from Tanzania also shows the influence of global temperature change on regional hydrology. Here changes in lake level and lake conductivity show a strong relationship to sea-surface temperature change in the southwestern Indian Ocean (**Fig. 5**), which forces changes in the position of the intertropical convergence zone (ITCZ) and hence influences precipitation in the African tropics.

Late-Glacial and Holocene Climate Change

Changes in salinity inferred from diatom records have played a prominent role in documenting hydrologic changes in the Late-Glacial period through the Holocene, particularly in the Americas and Africa. In northern Mexico and the Great Basin of the southwestern USA, diatom records suggest that the Late-Glacial period was wet as a result of the southerly displacement of the westerly storm tracks, followed by drying in the early-to-mid-Holocene as the storm tracks moved northward. In central Mexico, the opposite climate pattern is evident, with a dry Late-Glacial period, because of a southward displacement of the ITCZ and a reduction in trade-wind sources of moisture. Holocene salinity and moisture patterns across Mexico are spatially complex, as a result of the differential influence of rising sea levels, dating uncertainties, and increasing human impact on the landscape in the late Holocene.

In Africa, diatom analysis has been instrumental in documenting massive precipitation changes during the late Quaternary, including the greening of the Sahara and Sahel in the early-to-mid-Holocene. At Bougdouma in Niger, for example, shifts between alkiliphilous benthic diatoms and freshwater planktic species suggest fluctuating water levels in the deglacial period (**Fig. 6**). Stable wet conditions developed in the early Holocene (~10,300 yr BP), followed by drying after 7,300 yr BP, a trend that is widespread throughout the Sahara and Sahel. The drying trend is interrupted by several wet pulses that produced a rising water table, but conductivity and inferred aridity increased after 3,800 yr BP, with desiccation of the basin during the last 1,000 years. These trends reflect variation in monsoon intensity across northwest Africa, but regional differences in the timing

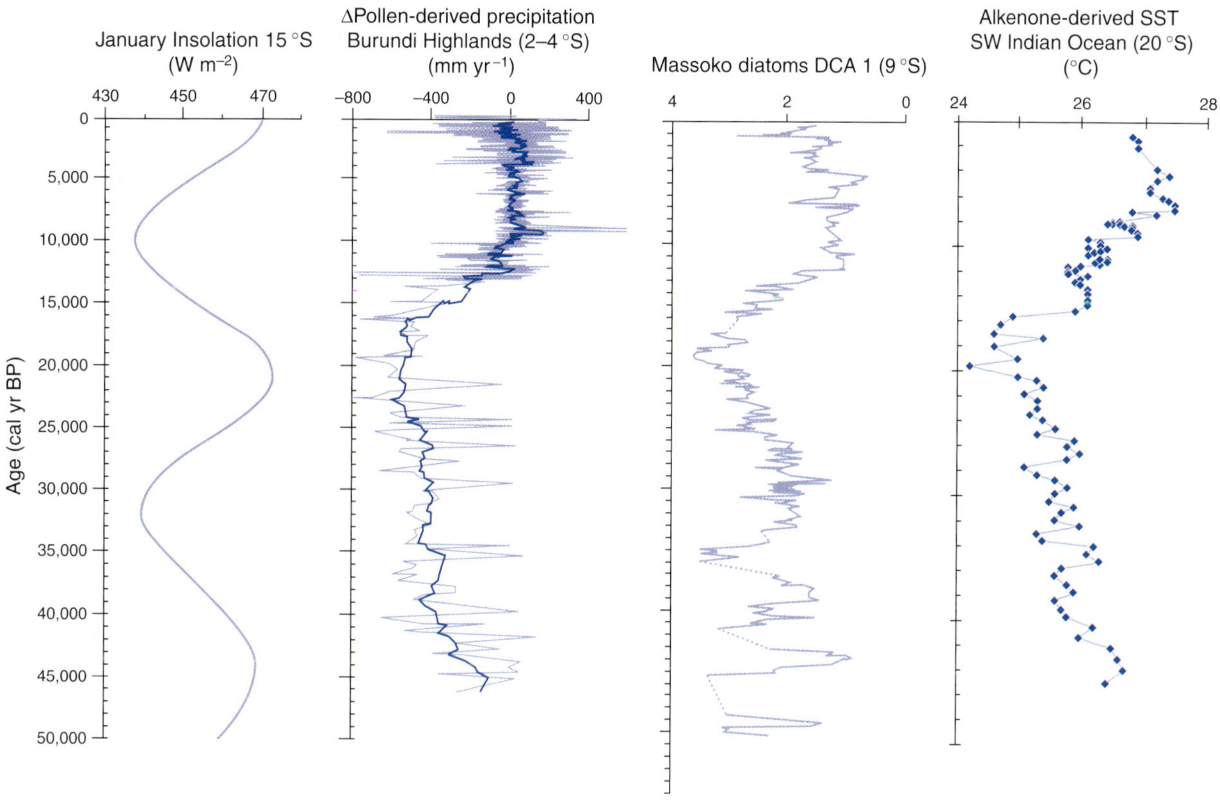

Figure 5 The diatom stratigraphy of Lake Massoko, Tanzania compared with insolation and sea-surface temperature forcing. The detrended correspondence axis (DCA) scores summarize the patterns of change in the diatom species assemblage – samples with high DCA axis 1 scores are samples with high diatom-inferred conductivity and represent times of lowered P–E, an inference that is corroborated by pollen data. Reproduced from Barker *et al.* (2003) with permission.

of lake salinity change reflect differences in the rate of response of different linked aquifer-lake systems.

The impacts of groundwater on lake response to climate have been highlighted in several recent paleolimnological studies. In some lakes of the Ethiopian Rift Valley, diatom-inferred water chemistry does not show patterns of salinity change that would be predicted based on independent evidence for precipitation variation. For example, an interval of high diatom-inferred conductivity from Lake Awassa corresponds with strand lines in an adjacent basin that document a contemporaneous interval of high precipitation (**Fig. 7**). Increased inputs of saline groundwater to one lake but not the other are the likely explanation. Similarly, a karst lake in the Konya Basin of Turkey showed little change in diatom-inferred conductivity during the late Pleistocene, probably because of relatively constant inflow of fresh groundwater, whereas a neighboring lake with limited groundwater flow showed evidence of changes in conductivity and lake levels indicative of changes in effective moisture.

Complex salinity responses to climate change are also evident in intermittently meromictic lakes in West Greenland. Here, the amplitude of salinity variation is apparently affected by the stability of stratification, producing a nonlinear response of lake salinity to climate change. In some intervals, increased diatom-inferred salinity correlates with evidence from paleoshorelines of increased precipitation. Complex groundwater interactions are unlikely in this region, because of permafrost, and the likely explanation is the coalescence of a freshwater lake with a lake of higher salinity during times of increased precipitation.

The use of diatom-inferred salinity to infer changes in effective moisture also can be affected by the sensitivity of diatom assemblages to changes in salinity driven by climate. At high salinity, diatom diversity commonly is quite low, and the ecological ranges of many high-salinity taxa are broad. Thus, species composition may not change as salinity changes. For example, at Moon Lake, ND, USA diatom-inferred salinity was very high during the mid-Holocene (**Fig. 8**), indicative of the dry climate that was characteristic of the mid-latitudes of North America. However, the low variability in mid-Holocene diatom-inferred salinity probably resulted

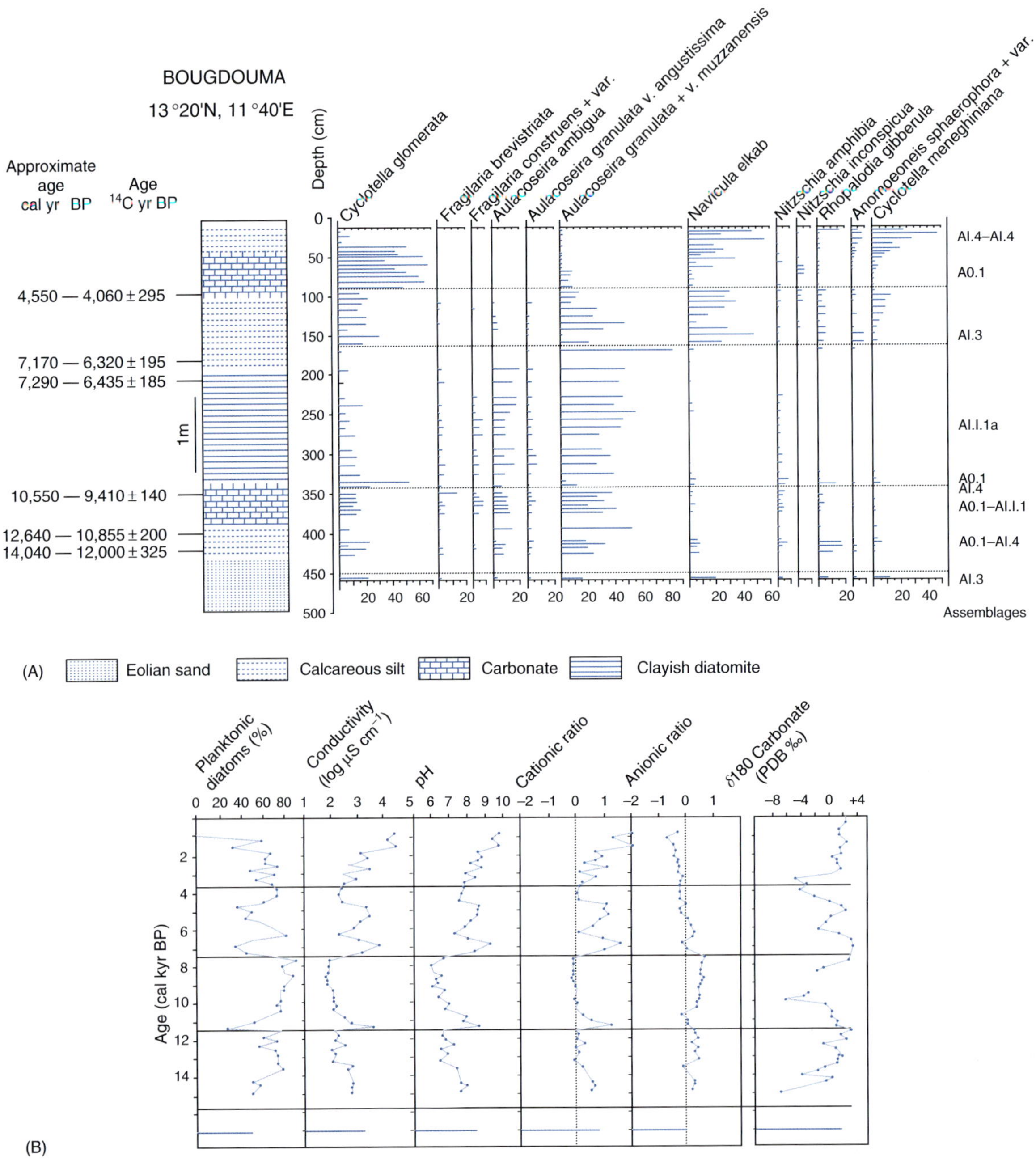

Figure 6 The (A) diatom stratigraphy of Bougdouma in southern Niger and (B) various water-chemistry variables inferred from the diatom stratigraphy using transfer functions (lower figure). The cationic and anionic ratios are defined in **Figure 2**. The assemblages defined on the right-hand panel of the figure (A0, AI, AII) represent the inferred cation and anion water types as defined in **Figure 2**. Reproduced from Gasse (2002) with permission.

because changes in salinity caused by precipitation variation were not sufficiently large to change the composition of the diatom assemblage.

Diatom records from the North American Great Plains also provide an example of the use of diatom-inferred salinity to evaluate natural drought variation and the degree to which the twentieth-century climate patterns are representative of the natural variability of climate. Records from a suite of lakes in the US and Canadian prairies clearly show that droughts equivalent in magnitude to the Dust Bowl droughts of the 1930s and 1940s were recurrent during the last millennium or two. These records also suggest that drought was more prolonged in the recent past than during the twentieth century, with periods of a century or more when drought was persistent. One of the major difficulties with these high-resolution records, however, is

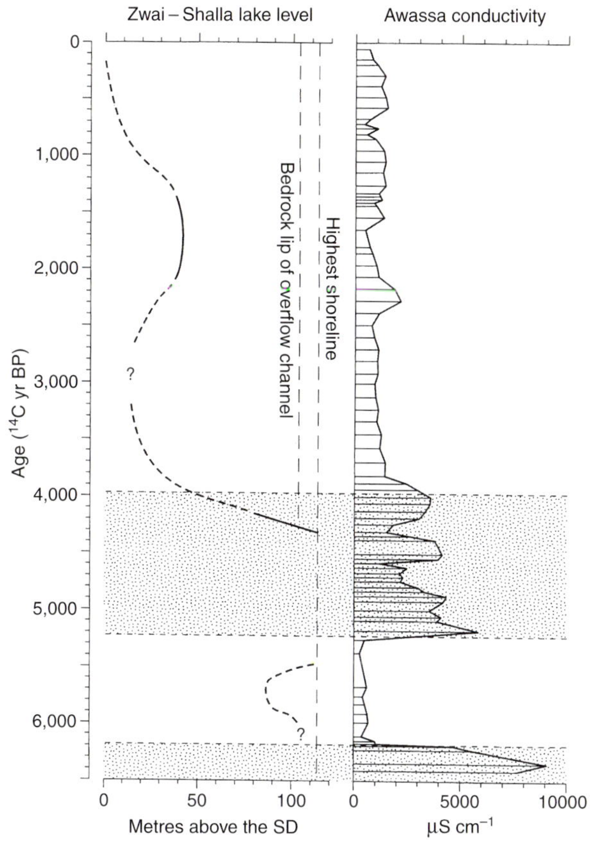

Figure 7 The diatom-inferred conductivity of a lake in Ethiopia compared with shoreline evidence for lake-level change in a nearby basin. Reproduced from Telford *et al.* (1999) with permission.

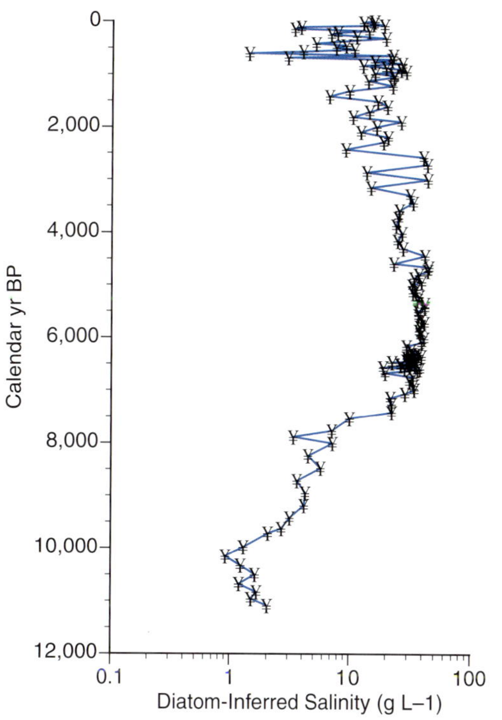

Figure 8 The diatom-inferred salinity of Moon Lake, ND, USA during the last ~12,000 years. Reproduced from Laird *et al.* (1996) with permission.

in the correlation of the pattern of change among multiple sites; the errors associated with a radiocarbon chronology prohibit clear correlation of changes at multidecadal scales. Even directional shifts that are persistent can be difficult to interpret in climatic terms. Thus, a major shift at the onset of Little Ice Age about 800 yr BP is manifest as an increase in salinity at one site and a decrease at another nearby lake (**Fig. 9**), and it is unclear whether the differences reflect a sharp climatic gradient, or whether they are more likely to be due to differing hydrological responses to the same climatic forcing.

Conclusions

The use of diatoms to infer past moisture variation in continental settings has increased greatly during the last two decades, and appropriate basins are widespread in semiarid and arid regions worldwide. In these regions, lacustrine records often provide the best archives of long-term continental climate variation. Although some lakes in extremely dry climates may have discontinuous sedimentary records, in many basins sedimentation rates are sufficiently rapid to allow the sedimentary record to be resolved at subdecadal resolution.

Although more studies are needed to determine how salinity and other ecologically important variables interact to influence diatoms, in many settings our knowledge of diatom ecological tolerances are good enough to enable robust reconstructions of qualitative changes in lake salinity. In some regions, calibration data sets have been developed to allow quantitative salinity reconstruction. These quantitative reconstructions are reasonably good at identifying the direction and relative magnitude of salinity change. In general, our ability to accurately reconstruct patterns of salinity change from diatom assemblages is strong.

The inference of climate from salinity change is considerably more difficult because of the potentially complex series of influences on the hydrologic budget of individual lakes. Although many of the issues involved in disentangling climate from biological and geochemical records are recognized, more frequent application of modeling and empirical studies to quantify their potential impact in individual sites or regions would improve many paleolimnological interpretations. The tandem use of other

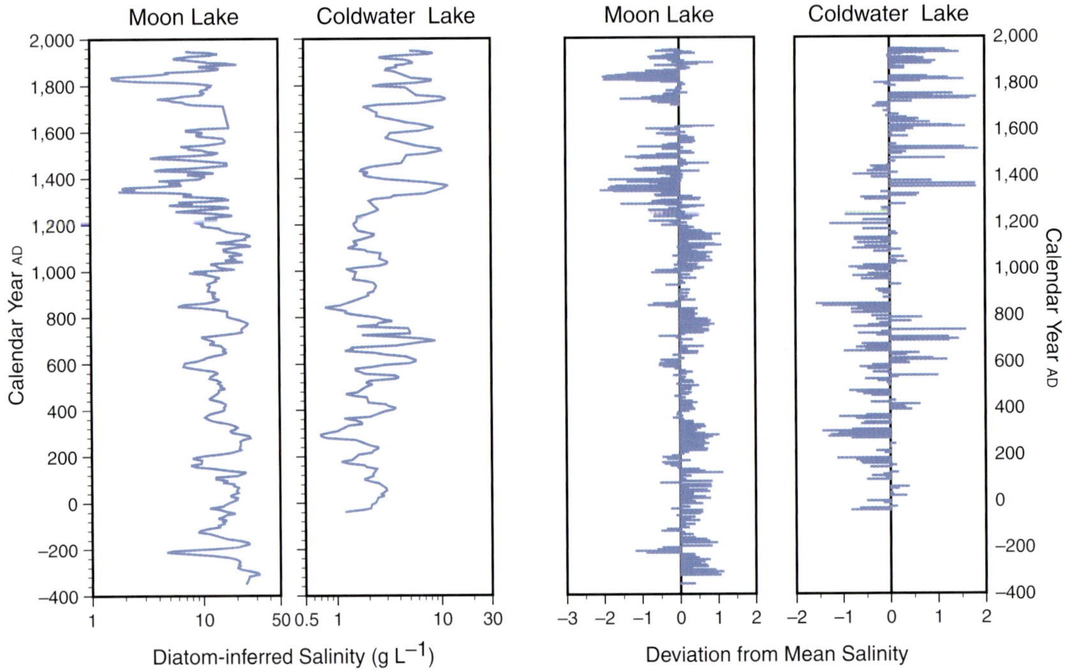

Figure 9 Diatom-inferred salinity spanning the last ~2,000 years of two lakes in ND, USA that are within 200 km of one another. The right hand panel shows the deviation from the long-term mean salinity. Note that Moon Lake became fresher at ~1200 AD, whereas Coldwater Lake becomes saltier at about the same time. Reproduced from Fritz *et al.* (2000) with permission.

proxies can help to constrain environmental interpretation from diatom records, although in some cases, it is not clear how to reconcile apparently divergent inferences. Such cases also point to the need for more frequent integration of paleolimnology with modern process studies and modeling efforts. Overall the number of lacustrine records is still relatively limited, particularly at high temporal resolution and those that span periods much older than the Last Glacial Maximum. Thus, although the diatom paleoclimatology field has expanded in recent years, there is still considerable distance to travel to unravel the patterns of environmental change across space and time and their proximate and ultimate drivers.

Abbreviations

DCA	detrended correspondence axis
DOC	dissolved organic carbon concentration
ITCZ	intertropical convergence zone
NaCl	sodium chloride
P–E	precipitation minus evaporation
SPECMAP	Spectral Mapping Project; Mapping Species Variability in Global Climate Project
yr BP	years before present

See also: **Diatom Introduction**. **Diatom Methods**: Data Interpretation. **Diatom Records**: Large Lakes.

References

Badaut, D., and Risacher, F. (1983). Authigenic smectite on diatom frustules in Bolivian saline lakes. *Geochimica et Cosmochimica Acta* **47,** 363–375.

Baker, P. A., Seltzer, G. O., Fritz, S. C., *et al.* (2001). The history of South American tropical precipitation for the past 25,000 years. *Science* **291,** 640–643.

Barker, P., Fontes, J. C., Gasse, F., and Druart, J. C. (1994). Experimental dissolution of diatom silica in concentrated salt solutions and implications for palaoenvironmental reconstruction. *Limnology and Oceanography* **39,** 99–110.

Barker, P., Williamson, D., Gasse, F., and Gibert, E. (2003). Climatic and volcanic forcing revealed in a 50,000-year diatom record from Lake Massoko, Tanzania. *Quaternary Research* **60,** 368–376.

Bigler, C., Larocque, I., Peglar, S. M., Birks, H. J. B., and Hall, R. I. (2002). Quantitative multiproxy assessment of long-term patterns of Holocene environmental change from a small lake near Abisko, northern Sweden. *The Holocene* **12,** 481–496.

Birks, H. J. B., Frey, D. G., and Deevey, E. S. (1998). Numerical tools in paleolimnology – progress, potentialities, and problems. *Journal of Paleolimnology* **20,** 307–332.

Bradbury, J. P., Grosjean, M., Stine, S., and Sylvestre, F. (2001). Full and Late Glacial lake records along the PEP 1 transect: their role in developing interhemispheric paleoclimate interactions. In *Interhemispheric Climate Linkages* (V. Markgraf, Ed.), pp. 265–292. Academic Press, Boulder, CO, USA.

Bradbury, J. P. (1997). A diatom-based paleohydrologic record of climate change for the past 800 k.y. from Owens Lake, California. *Geological Society of America, Special Paper* **317,** 99–112.

Campbell, C. E., and Prepas, E. E. (1986). Evaluation of factors related to the unusually low chlorophyll levels in prairie saline lakes. *Canadian Journal of Fisheries and Aquatic Sciences* **43**, 846–854.

Ehrlich, A. (1978). The diatoms of the hyperhaline Solar Lake (NE Sinai). *Israel Journal of Botany* **27**, 1–13.

Fritz, S. C. (1990). Twentieth-century salinity and water-level fluctuations in Devils Lake, N. Dakota: a test of a diatom-based transfer function. *Limnology and Oceanography* **35**, 1771–1781.

Fritz, S. C., Cumming, B. F., Gasse, F., and Laird, K. R. (1999). Diatoms as indicators of hydrologic and climatic change in saline lakes. In *The Diatoms: Applications for the Environmental and Earth Sciences* (E. F. Stoermer and J. P. Smol, Eds.), pp. 41–72. Cambridge University Press, Cambridge.

Fritz, S. C., Ito, E., Yu, Z., Laird, K. R., and Engstrom, D. R. (2000). Hydrologic variation in the northern Great Plains during the last two millennia. *Quaternary Research* **53**, 175–184.

Fritz, S. C., Juggins, S., and Battarbee, R. W. (1993). Diatom assemblages and ionic characterization of lakes of the Northern Great Plains, North America: a tool for reconstructing past salinity and climate fluctuations. *Canadian Journal of Fisheries and Aquatic Sciences* **50**, 1844–1856.

Gasse, F. (1977). Evolution of Lake Abhe (Ethiopia and T.F.A.I.) from 70,000 B.P. *Nature* **2**, 42–45.

Gasse, F. (2002). Diatom-infered salinity and carbonate oxygen isotopes in Holocene waterbodies of the western Sahara and Sahel (Africa). *Quaternary Science Reviews* **21**, 737–767.

Hustedt, F. (1953). Die Systematik der Diatomeen in ihren Beziehungen zur Geologie und Ökologie nebst einer Revision des Halobien-systems. *Botanisk Tidskkrift* **47**, 509–519.

Kolbe, R. W. (1927). Zur Okologie, Morphologie und Systematik der Brackwasser-Diatomeen. *Pflanzenforschung* **7**, 1–146.

Laird, K. R., Cumming, F. F., Wunsam, S., *et al.* (2003). Lake sediments record large-scale shifts in moisture regimes across the northern prairies of North America during the past two millennia. *Proceedings National Academy of Sciences* **100**, 2483–2488.

Laird, K. R., Fritz, S. C., Grimm, E. C., and Mueller, P. G. (1996). Century-scale paleoclimatic reconstruction from Moon Lake, a closed-basin lake in the northern Great Plains. *Limnology and Oceanography* **41**, 890–902.

McGowan, S., Ryves, D. B., and Anderson, N. J. (2003). Holocene records of effective precipitation in West Greenland. *The Holocene* **13**, 239–250.

Metcalfe, S., Say, A., Black, S., McCulloch, R., and O'Hara, S. (2002). Wet conditions during the last glaciation in the Chihuahuan Desert, Alta Babicora, Basin, Mexico. *Quaternary Research* **57**, 91–101.

Metcalfe, S. E., O'Hara, S. L., Caballero, M., and Davies, S. J. (2000). Records of Late Pleistocene–Holocene climatic change in Mexico – a review. *Quaternary Science Reviews* **19**, 699–721.

Meybeck, M. (1995). Global distribution of lakes. In *Physics and Chemistry of Lakes* (A. Lerman, D. Imboden, and J. Gat, Eds.), pp. 1–35. Springer-Verlag, Berlin.

Reed, J. M., Roberts, N., and Leng, M. J. (1999). An evaluation of the diatom response to Late Quaternary environmental change in two lakes in the Konya Basin, Turkey, by comparison with stable isotope data. *Quaternary Science Reviews* **18**, 631–646.

Ryves, D. B., Juggins, S., Fritz, S. C., and Battarbee, R. W. (2001). Experimental diatom dissolution and the quantification of microfossil preservation in sediments. *Palaeogeography Palaeoclimatology Palaeoecology* **172**, 99–113.

Sanford, W. E., and Wood, W. W. (1991). Brine evolution and mineral depostion in hydrologically open evaporite basins. *American Journal of Science* **291**, 687–710.

Saros, J. E., and Fritz, S. C. (2000). Changes in the growth rates of saline-lake diatoms in response to variation in salinity, brine type and nitrogen form. *Journal of Plankton Research* **22**, 1071–1083.

Telford, R. J., Lamb, H. F., and Mohammed, M. U. (1999). Diatom-derived palaeoconductivity estimates for Lake Awassa, Ethiopia: evidence for pulsed inflows of saline groundwater? *Journal of Paleolimnology* **21**, 409–421.

Webster, K. E., Soranno, P. A., Baines, S. B., *et al.* (2000). Structuring features of lake districts: landscape controls on lake chemical responses to drought. *Freshwater Biology* **43**, 499–515.

Use in Archaeology

N G Cameron, University College London, London, UK

© 2007 Elsevier B.V. All rights reserved.

Introduction

Diatom analysis has been used in archeological investigations for almost 60 yr (for discussion and references see the review of Battarbee (1988)). Like other Quaternary paleoecological techniques, such as pollen and plant macrofossil analysis, the results of off-site diatom paleoecological studies are often relevant to environmental archeology. However, only diatom research that is integrated with field archeology is considered relevant here. Specific collaboration between field archeologists excavating ancient sites and diatomists interested in site environments has led to the increasing use of the technique in environmental archeology. Unfortunately, the majority of published diatom-archeological studies are of necessity placed in the appendices of archeological site reports, making it difficult to gain an impression of the frequency with which the technique is applied in this context. For example, in much of the routine environmental archeology evaluation work carried out in UK, the aims of a diatom assessment are to determine if diatoms are present and whether there is any potential for their further analysis by percentage counting. Before more detailed analyses are considered, sample characteristics examined include diatom valve concentrations, the quality of diatom preservation (the degree of breakage and dissolution), and the diversity of taxa as well as the types of diatom assemblage and the environmental preferences of those diatoms present (*see* Diatom Introduction). Diatom analysis is now commonly used in archaeology and often, particularly in the prehistoric context, forms part of environmental-led archeological investigations (Nayling and Caseldine, 1997).

Several authors (Battarbee, 1988; Mannion, 1987; Miller and Florin, 1989; Juggins and Cameron, 1999) have reviewed the applications of diatom analysis in archeology. In this article, a number of case studies of on-site diatom-archeological investigations are presented as examples.

Site Environment

As mentioned above, many diatom-paleoecological studies concerned with Holocene environmental reconstruction make reference to catchment archeology (human impact) as a factor in driving environmental change (e.g., de Wolf and Cleveringa 1996) and indeed may discuss the archeological or historical documentary evidence. Diatom environmental studies integrated with archeological excavations are less common. One recent example of such a project is that of Douglas *et al.* (2004) who investigated the impact of prehistoric Inuit whalers upon Arctic freshwater ecosystems. Despite their low population densities and the nomadic nature of Inuit whalers, it was shown through a combination of archeological and paleolimnological data that human settlement markedly changed pond-water ecology from an early date. Winter settlements were constructed from the bones of bowhead whales on Somerset Island in the high Canadian Arctic between about AD 1200 and 1600. Archeological excavation, recovery of artifacts, and radiocarbon dating documented this occupation and allowed estimation of the population size. Diatom analysis (**Fig. 1**) indirectly records marked changes in water chemistry and diatom habitat (*see* Diatom Introduction, Data Interpretation). Although the whalers abandoned the area over 400 years ago, the pond today retains elevated nutrient concentrations and biota atypical of the region.

It is more common to find diatom analysis used in archeological studies of urban contexts in towns in Europe. An example of this kind of analysis is work carried out by Demiddele and Ervynck (1993) using diatoms as ecological indicators in an excavation of Roman and Medieval Oudenberg in the west of Flanders. During the Roman period water-lain sediments in the town showed a high proportion of marine species reaching the town through

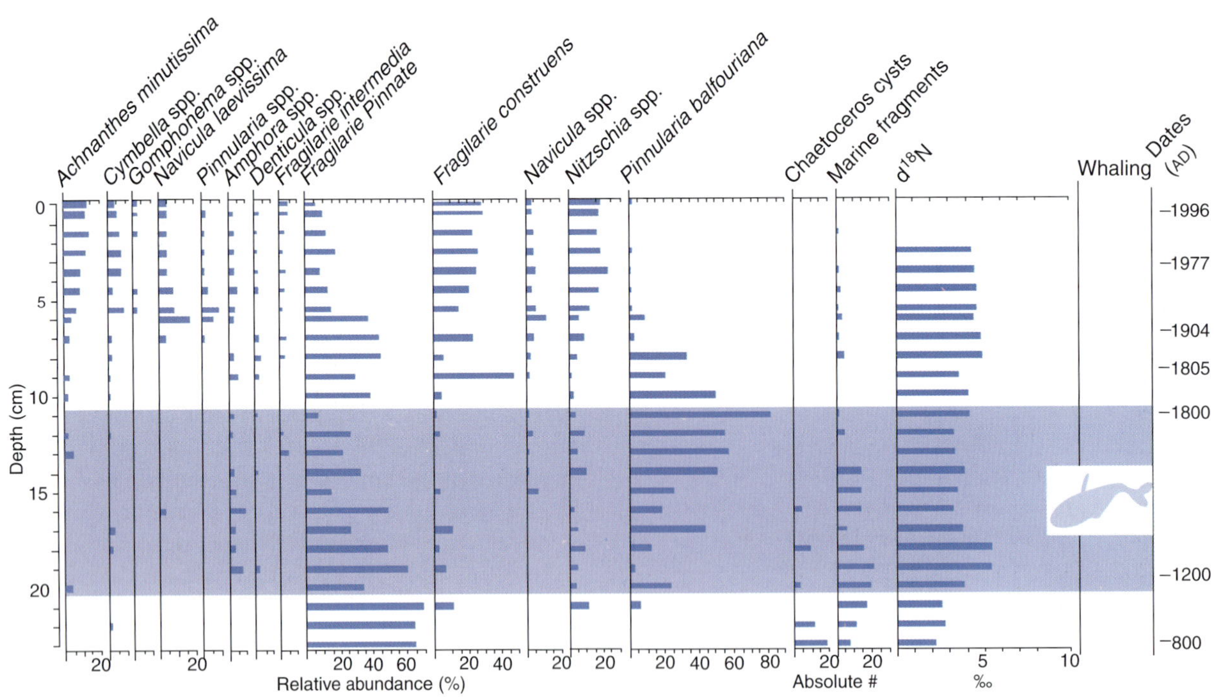

Figure 1 Diatom analysis of lake sediments from the Canadian Arctic showing the impact of Inuit whalers on water chemistry and diatom habitat. From Douglas, M. S. V., Smol, J. P., Savelle, J. M., and Blaise, J. M. (2004). Prehistoric Inuit whalers affected Arctic freshwater ecosystems. *Proceedings of the National Academy of Sciences* **101**, 1613–1617. Copyright (2004) National Academy of Sciences, U.S.A., reproduced with kind permission.

tidal creeks. However, during the Medieval period species of freshwater and low salinity dominate the assemblages indicating a disappearance of marine influence with implications for water transport to the town. In addition, the diatoms from the Medieval contexts corroborate evidence from other sources that there was heavy organic pollution of the water courses. Diatom analysis was therefore of relevance here both in looking at the influence of tidal waters on human activities and also on the impact of humans on site-water quality. Diatoms have also been used for archeological environmental reconstruction over larger areas in landscape archeology projects. This is exemplified by work carried out over several seasons of excavations on wetlands bordering the Severn Estuary, UK where diatom analysis supported evidence for large-scale Roman land reclamation (Rippon, 2000).

Diatom Analysis of Artifacts and Ancient Building Materials

Diatom analysis has been used in the direct study of artifacts particularly in the investigation of pottery clay provenance and pottery typology but also in identifying the sources of materials and manufacturing techniques used in ancient buildings.

Where the firing temperature is low enough to allow the survival of diatom valves, diatom analysis of pottery can assist in the identification of clay sources or the provenance of finished pottery. Where adequate diatom assemblages are preserved, diatom assemblages can also be used to assist in establishing typologies for pottery or in the comparison of typologies with pottery style-based classification schemes (Alhonen *et al.*, 1980). In addition to diatoms derived directly from clay, the mixture of diatoms derived from tempering materials and from water used in the manufacturing process can form part of the species assemblage. The use of diatoms in the analysis of pottery, which, because of the requirement for a low-temperature firing, is usually of prehistoric age, has been reviewed by Battarbee (1988) and by Juggins and Cameron (1999). Stilborg *et al.* 1997 published a microfossil analysis of late Roman Iron Age pottery (AD 200–400) from a region of present-day Denmark. Simple presence and absence of siliceous and calcareous microfossils were used to group and separate pottery sherds. Two groups were established, and the diatomaceous class, known as 'D-ware,' was further subdivided into sherds containing exclusively nonmarine diatoms and a subgroup containing a mixture of marine and nonmarine species. From this classification it was also possible to infer the clay sources used for the pottery. In some of the classes of pottery, it was establi-shed that a mixture of clays was used during manufacture.

In a similar way, diatom analysis has been used to examine ancient building materials (Flower, 2006). Flower examined the diatom content of mud bricks used in ancient Egypt from ca. 5 kyr BP (**Fig. 2**). The bricks were made from a mixture of mud, sand, water, and plant material. Consequently, diatom analysis of the bricks offered the possibility of investigating the provenance and uses of these building materials. Analysis of diatoms from a mud brick from a temple at Dimai showed that the diatom 'mud' component was almost certainly extracted from nearby, now terrestrial, diatomite (*see* Quaternary Geological Records) that accumulated in a former large, shallow lake. In contrast, the few, predominantly planktonic diatoms found in brick from a XIIth Dynasty Pyramid at Hawara were thought to have been introduced from Nile water during manufacture. Flower notes that his results are preliminary, being based on a small number of

Figure 2 Buildings at the ancient Temple at Dimai, Egypt dating from ca. 5 kyr BP and showing mud brick construction. Photograph taken by Dr. Roger Flower.

samples. However, he demonstrates that diatoms in ancient Egyptian (and potentially other mud bricks) can provide important clues to the selection of raw materials, manufacturing processes, and past environments associated with the raw materials. Flower observes that it is unclear whether ancient Egyptian builders exercised any preference for diatom-rich sources for their bricks. However the well-known insulating properties of diatomite would be advantageous in the construction of buildings. He emphasizes that diatom analysis of ancient building materials is also potentially useful not only for mud-brick typification but also for making space–time correlations between mud-brick facies, within and between monuments.

Moats and Medieval Fish Ponds

Man-made water bodies such as moats and fish ponds dating from Roman to Medieval times are associated with sites such as settlements, villas or farmhouses, fortifications and monastic sites, and can provide local and sometimes well-dated diatom-based environmental evidence for the effects of human activities. An example of this is provided by diatom analysis carried out in association with the archeological excavations at the Tower of London moat (Keevill 2004). It should be noted that other types of smaller ancient buildings can have man-made water bodies with valuable diatom environmental records. Such sites include farmhouses, smaller fortifications, and ecclesiastical buildings. At the Tower of London, a number of sources were found to influence water quality over a 600 yr period from 1240 to 1843. The input of estuarine water from the tidal River Thames was evaluated using ecological groups of diatom-salinity preferences. Saline water entered the moat through the Traitor's Gate and the input varied through time. Although there was little archeological evidence for the dumping of refuse directly into the moat, organic pollution was derived from drainage from the Tower, and also from waste disposal into a water course, the City Ditch, which drained from the external city settlement into the moat. Using a diatom–phosphorus transfer function (*see* Data Interpretation), total phosphorus (TP) levels indicated nutrient enrichment (**Fig. 3**) in excess of that found at pre-Medieval sites elsewhere in London that have been analyzed for diatoms. Diatom analysis has been used in a similar context to examine sediments from two profiles taken from a ditch draining the moat surrounding fortifications in Prague, initiated in AD 1230 (Beneš *et al.* 2002). Here the diatom analyses revealed that a relatively clean, fast-flowing stream tended to flush organic pollutants quickly away from the urban environment. Only toward the end of its function as a gutter, when the feature became clogged with sediment, did slowing of the current lead to stagnation. Even then the degree of nutrient enrichment remained surprisingly low, with water quality comparable with 'recreational water supply' according to present-day norms. In Scandinavia diatom analysis of sediments from Medieval ditch and moat-like features was also used by Nunez and Paabo (1990) to ascertain whether these watercourses were connected with the sea. The presence of coastal diatom species in the samples seemed to confirm that this was the case; however, as is common with archeological sediments, the low concentration and fragmentary state of the diatoms from the ditch suggested that in this feature they represented secondary deposition from the ditch bank when the castle island was below sea level. Like many archeological contexts the interpretation of the diatoms at this site was therefore based on the state of diatom preservation in addition to the diatom composition.

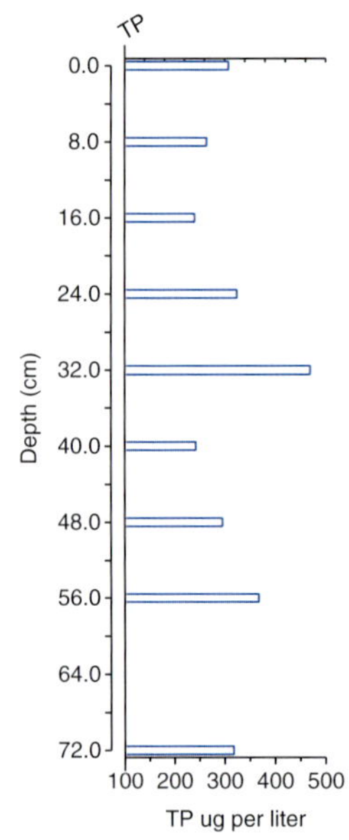

Figure 3 Section through Medieval sediments from the Tower of London moat showing variation in the diatom-based TP reconstruction. From Keevill, G. (2004). *Historic Royal Palaces Monograph No. 1: The Tower of London Moat. Archaeological Excavations 1995–9*, pp. 315. Oxford: Oxford Archaeology with Historic Royal Palaces. Copyright N. G. Cameron.

Waterfronts

The collaboration between diatomists and archeologists concerned with the reconstruction of waterfront environments is exemplified by a series of detailed studies in London, where developer-funded excavations over two decades have stimulated archeological research projects involving diatoms (Milne *et al.*, 1983; Juggins, 1992; Sidell *et al.*, 2000; Sidell, 2003). Previous to these studies, Roman remains lying 4 m below the present high tide level in the outer Thames estuary led to a common assumption that the estuary lay downstream of London during the Roman period. However, within the city the pre-Roman foreshore is buried to the north of the present riverbank. There is a time–space sequence of waterfront structures covering the last 2 kyr and preserved beneath the modern waterfront (**Fig. 4**). The halophilous, estuarine, and allochthonous marine diatoms present in sediments accumulating by well-dated quay structures clearly demonstrated that the river was tidal in the City of London during the Roman period. Juggins (1992) developed models relating diatom assemblages to salinity and taking into account the taphonomy of the diatom assemblages. This allowed the past salinity of the Thames in the city to be better estimated. More recent studies have increased the number of archeological sites with well-dated diatom records and added to the picture of a rapidly changing river (e.g., Sidell *et al.* (2001)). Analyses of sediments, including diatom analysis, show that the upstream progression of the tidal head of the Thames through London began in the Late Bronze Age. However, this rapid transgression is followed by a period of regression and a second transgression. During the Roman period this resulted in continuous rebuilding of the waterfront to maintain the port of London (Sidell, 2003). In addition, signs of possible nutrient enrichment and the effects of catchment deforestation have been detected in sediments from other waterfront and wetland archeological sites in London. The proximity of human activities to accumulating sediments can sometimes be an unwanted complication when interpreting diatom analyses of archeological site sediments. However, the use of archeological artifacts and dendrochronological dating of waterfront structures has the advantage of allowing relatively accurate chronological control for many of the reconstructions.

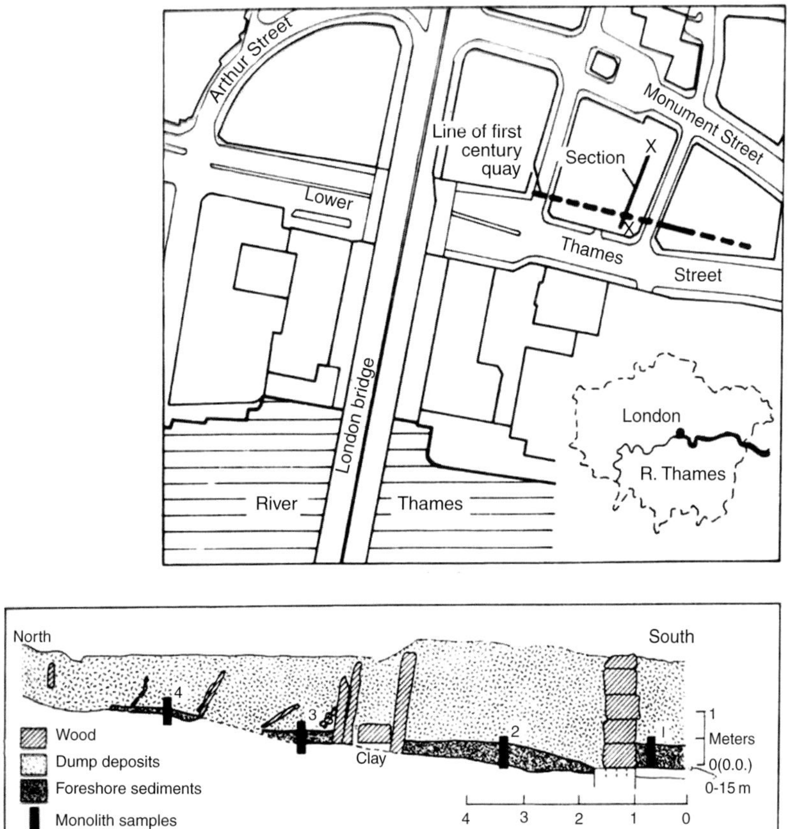

Figure 4 Location and stratigraphy of Roman waterfront sediments from the City of London. Reprinted from Battarbee, R. W. (1988). The use of diatom analysis in archaeology: A review. *Journal of Archaeological Science* 15: 621–644, with permission from Elsevier.

Ancient Boats

The increasing threat to ancient sites from the consequences of sea-level rise and coastal erosion has led to an increasing interest in coastal, estuarine, and intertidal archeology. This has led to the recent discoveries of a number of ancient boats ranging in age from the Bronze Age to the Medieval period (**Fig. 5**). Again dendrochronology has in many cases allowed precise dating of the boat planks (strakes) and other structural wood such as oak frames. There is convincing evidence that several relatively large ancient boat finds represent sea-going craft with implications for long distance transport and trade at an early date in prehistoric Europe. For these finds, diatom analysis has been used to reconstruct the sedimentary environments in which these ancient boats came to lie and in some cases to look at the provenance of materials used in boat construction (Nayling, 1998; Nayling and Caseldine, 1997; Nayling and McGrail, 2004; Clark, 2004). Diatom analysis has also been applied to ancient harbors, exemplified by analysis carried out on samples from a core sequence from Caesarea's ancient harbor with the aim of reconstructing the general salinity conditions of the sedimentary environment (Cameron in Holum *et al.* (1999)).

Figure 5 (A) shows the aft plan of the Newport Medieval (15th Century) Ship found sealed in estuarine alluvium on the bank of the River Usk in the South Wales, UK. (B) shows the Magor Pill wreck (AD 1240), found on the foreshore of the Gwent Levels, South Wales, UK (see Nayling, 1998). Both photographs are by kind permission of Nigel Nayling, Department of Archaeology and Anthropology, University of Wales, Lampeter.

Other Archeological Contexts and Forensic Archeology

Many unusual archeological contexts have been analyzed for diatoms in order to assist in their interpretation. For example, diatoms from Pleistocene postoccupation layers from a cave in the Western Kalahari Desert (Robbins *et al.*, 1996) were found to indicate flowing water. Structures such as sediments in wells (Neely *et al.*, 1995), pits, post holes, and around fish traps (Godbold and Turner, 1993) have been analyzed for diatoms to examine the functions and environments of such features. Salt extraction workings or salterns are sites where salt was produced by the evaporation of salt water in shallow pools, sometimes with the aid of hearths. In Europe, from the Iron Age through to Medieval times, such sites were found near to salt water, in places such as estuaries and salt marshes, and also by inland sources of mineral salt. As a result of their sensitivity to salinity (*see* Data Interpretation) diatoms have been used to assist in the environmental analysis of such sites and in the identification of the functions of specific areas of such sites (e.g., Juggins 1992). Diatom analysis has also been used in experimental archeology, for example, to look at the drinking habits of livestock and the excretion of diatom valves to aid in the soil micromorphological interpretation of habitation floor layers. Like the application of the methods of archeological excavation in forensic science (forensic archeology), diatom analysis has also become more frequently employed as a forensic technique. The collaboration between pathologists and diatom analysts in the diagnosis of drowning is a long-established branch of forensic medicine (Peabody, 1999). The examination of diatoms has also been applied in the archeological investigation of ancient burials (Foged, 1985). However, similar to the fields of archeological stratigraphy and forensic anthropology, diatom analysis has been applied as an environmental technique in the forensic context (Cameron, 2004). This has been particularly useful in the exploration of evidence of association where items such as clothing and footwear, in criminal cases, can be compared with environments at scenes of crime.

Conclusions

As a result of their ubiquity in water, their sensitivity to water quality, an often high species diversity, the habitat specificity of species, and their excellent potential for preservation, diatoms have a wide range of applications in environmental archeology and related fields. Our increasing knowledge of diatom taxonomy and ecology (*see* Diatom Introduction) and the application of numerical techniques developed in the wider field of Holocene paleoecology (*see* Data Interpretation) will improve comparison and analysis of diatom assemblages in the archeological context.

See also: **Diatom Introduction. Diatom Methods**: Data Interpretation.

References

Alhonen, P., Kokkonen, J., Matiskainen, H., and Vuorinen, A. (1980). Applications of AAS and diatom analysis and stylistic studies of Finnish Subneolithic pottery. *Bulletin of the Geological Society of Finland* **52,** 193–206.

Battarbee., R. W. (1988). The use of diatom analysis in archaeology: A review. *Journal of Archaeological Science* **15,** 621–644.

Beneš, J., Kaštovský, J., Kočárová, R., Kočár, P., Kubečková, K., Pokorný, P., and Starec, P. (2002). Archaeobotany of the Old Prague Town defence system, Czech Republic: Archaeology, macro-remains, pollen, and diatoms. Vegetation. *History and Archaeobotony* **11,** 107–119.

Cameron, N. G. (1999). Diatoms from the sediments of the Inner Harbour (Area 114). In *Caesarea Papers 2: Herod's Temple, The Provincial Governor's Praetorium and Granaries, The Later Harbor, A Gold Coin Hoard, and Other Studies Portemouth* (K. G. Holum, A. Raban and J. Patrich, Eds.), [The Journal of Roman Archaeology, Supplement Series Number 35] pp. 343–345. University of Michigan Press, Ann Arbor.

Cameron, N. G. (2004). The use of diatom analysis in forensic geoscience. In *Forensic Geoscience: Principles, Techniques and Applications* (K. Pye and D. J. Croft, Eds.), Vol. 232, pp. 277–280. Geological Society of London, London.

Clark, P. (Ed.) (2004). *The Dover Bronze Age Boat*, p. 340. English Heritage Publications.

Demiddele, H., and Ervynck, A. (1993). Diatomeeën als ecologische indicatoren in de Vlaamse archeologie: Romeins en middeleeuws Oudenburg (prov. West-Vlaanderen). *Archeologie in Vlaanderen III. Instituut voor Het Archeologisch Patrimonium* **3,** 217–231.

De Wolf, H., and Cleveringa, P. (1996). The impact of the beer industry on medieval water quality. In *Proceedings of the 14th Diatom Symposium* (I. Mayama and J. Koizumi, Eds.), pp. 511–522. Koeltz Scientific Books, Koenigstein.

Douglas, M. S. V., Smol, J. P., Savelle, J. M., and Blais, J. M. (2004). Prehistoric Inuit whalers affected Arctic freshwater ecosystems. *Proceedings of the National Academy of Sciences* **101,** 1613–1617.

Flower, R. J. (2006). Diatoms in ancient building materials, application of diatom analysis to Egyptian mud bricks. *Nova Hedwigia Beihefte* **130,** 245–264.

Foged, N. (1985). Diatoms in a tomb from the early Bronze Age. *Nova Hedwigia* **41,** 471–482.

Godbold, S., and Turner, R. C. (1993). *Second Severn Crossing Archaeological Response. Phase 1: The Intertidal Zone in Wales.* Westbury Press, Brentwood.

Juggins, S. (1992). *Diatoms in the Thames Estuary, England: Ecology, Palaeoecology, and Salinity Transfer Function.* p. 216. Cramer, Berlin/Stuttgart.

Juggins, S., and Cameron, N. G. (1999). Diatoms and archaeology. In *The Diatoms: Applications for the Environmental and Earth Sciences* (E. F. Stoermer and J. P. Smol, Eds.), pp. 389–401. Cambridge University Press, Cambridge.

Keevill, G. (2004). *Historic Royal Palaces Monograph No. 1: The Tower of London Moat. Archaeological Excavations 1995–9,* p. 315. Oxford Archaeology with Historic Royal Palaces, Oxford.

Mannion, A. M. (1987). Fossil diatoms and their significance in archaeological research. *Oxford Journal of Archaeology* **6,** 131–147.

Miller, U., and Florin, M.-B. (1989). Diatom analysis. Introduction to methods and applications. In *Geology and Palaeoecology for Archaeologists* (T. Hackens and U. Miller, Eds.), PACT 24, pp. 133–157. Council of Europe, Strasbourg.

Milne, G., Battarbee, R. W., Straker, V., and Yule, B. (1983). The River Thames in London in the mid 1st century AD. *Transactions of the London and Middlesex Archaeological Society* **34,** 19–30.

Nayling, N. (1998). *Council for British Archaeology Research Report 115: The Magor Pill Medieval Wreck*, p. 173. Council for British Archaeology, York.

Nayling, N., and Caseldine, A. (1997). Council for British Archaeology Research Report 108: Excavations at Caldicot, Gwent: Bronze Age Palaeochannels in the Lower Nedern Valley, p. 348. Council for British Archaeology, York.

Nayling, N., and McGrail, S. (2004). *Council for British Archaeology Research Report 115: The Barland's Farm Romano-Celtic Boat*, p. 325. Council for British Archaeology, York.

Neely, J. A., Caran, S. C., Winsborough, B. M., Sorensen, F. R., and Valastro, S., Jr. (1995). An Early Holocene Hand-dug Water Well in the Tehuacan Valley of Puebla, Mexico. *Current Research in the Pleistocene* **12,** 38–40.

Nunez, M., and Paabo, K. (1990). Diatom analysis. *Norwegian Archaeological Review* **23,** 128–130.

Peabody, A. J. (1999). Forensic science and diatoms. In *The Diatoms: Applications for the Environmental and Earth Sciences* (E. F. Stoermer and J. P. Smol, Eds.), pp. 413–418. Cambridge University Press, Cambridge.

Rippon, S. (2000). The Romano-British Exploitation of Coastal Wetlands: Survey and Excavation on the North Somerset Levels, 1993–7. *Britannia* **XXXI,** 69–204.

Robbins, L. H., Murphy, M. L., Stevens, N. J., *et al.* (1996). Palaeoenvironment and archaeology of Drotsky's Cave: Western Kalahari Desert, Botswana. *Journal of Archaeological Science* **23,** 7–22.

Sidell, E. J. (2003) *Holocene Sea Level Change and Archaeology in the Inner Thames Estuary, London, UK.* Unpublished PhD Thesis, p. 614. University of Durham.

Sidell, E. J., Wilkinson, K., Scaife, R., and Cameron, N. (2000). *The Holocene Evolution of the London Thames. Archaeological Excavations (1991–1998) for the London Underground Limited Jubilee Line Extension Project. Museum of London Archaeological Service Monograph 5,* p. 144. Molas, London.

Stilborg, O., Olsson, S., and Håkansson, H. (1997). *Shards of Iron Age Communications Monographs on Ceramics, Keramiska Forskingslaboratoriet (ser.),* p. 326. University of Lund, Sweden.

DIATOM RECORDS

Contents

Antarctic Waters

C E Stickley, Norwegian Polar Institute, Tromsø, Norway
J Pike, Cardiff University, Cardiff, UK
V Jones, University College London, London, UK

© 2007 Elsevier B.V. All rights reserved.

Marine Antarctic Waters: The Southern Ocean Siliceous Ooze Belt

The southern high latitudes play a critical role in the global climate system. Advances in understanding the paleoceanographic and cryospheric evolution of the Southern Ocean and Antarctica are helping us understand global climate change on a number of timescales. Diatoms are arguably the best fossil group for unraveling Antarctic climate change since they are prolific in the Southern Ocean (the area between the coast of Antarctica (~70°S) and the Subtropical Front (~40°S) (**Fig. 1**)) contributing up to 75% of Southern Ocean production, despite this being a high-nitrate, low-chlorophyll (HNLC) region. Such high diatom production in the surface waters combined with a high export flux to the seafloor makes the Southern Ocean the world's main sink for biogenic opal (silica). In particular, the area beneath the Antarctic Circumpolar Current (ACC), a major link in the Thermohaline circulation, comprises a continuous belt of opal completely encircling Antarctica between 45° and 60°S (**Fig. 2**). Diatomaceous sedimentation and preservation has been occurring within this belt more or less continuously since at least Miocene times. Today the nutrient-rich Southern Ocean supports a unique flora comprising at least six planktonic endemic species according to Zielinski and Gersonde (1997): *Eucampia antarctica*, *Fragilariopsis curta*, *F. kerguelensis*, *F. separanda*, *Thalassiosira gracilis* and *T. lentiginosa*. Here, the Polar Frontal Zone (**Fig. 1**) acts as a major geographic barrier to the migration of planktonic diatoms separating cooler-water Antarctic species to the south from warmer-water SubAntarctic species to the north.

The fossil diatom record of the Quaternary obtained from recent sediments of the Southern Ocean is well-known, containing mostly, but not exclusively, extant species in high abundance. For example, Zielinski and Gersonde (1997) report diatom concentrations of $50–200 \times 10^6$ valves/g surface sediment in the Atlantic sector beneath the ACC. The so-called 'siliceous ooze belt of well-preserved diatoms', is bound north and south by sediments containing less well-preserved diatoms (**Fig. 2**); carbonate-rich sediments to the north, and silty diatomaceous clay to the south including the shelf regions. These boundaries, approximating to the modern Polar Front (north) and the mean position of the winter sea-ice edge (south), may have shifted north and south in the past during periods of climate change. Over the Antarctic continental shelf high-biogenic silica deposition also occurred in the Quaternary but records are more patchy, although there are regions where diatom productivity and flux were exceptionally high during the last deglaciation resulting in distinct seasonal layering (*see* Marine Laminated Sequences).

Applications of Quaternary Antarctic Diatoms

Despite preservation issues (*see* Diatom Introduction and Marine Diatoms), marine diatoms are excellent tools for climate reconstruction because they accurately reflect the environment in which they live and are normally the first microplankton group to respond to environmental change. Their use as a paleoenvironmental tool is perhaps of most value for marine sediments recovered from below the calcium carbonate compensation depth (CCD) and in the high-latitude regions such as the Southern Ocean. In both of these environments, siliceous microfossils (diatoms, silicoflagellates, and radiolarians) prevail over calcareous microfossils (nannofossils and

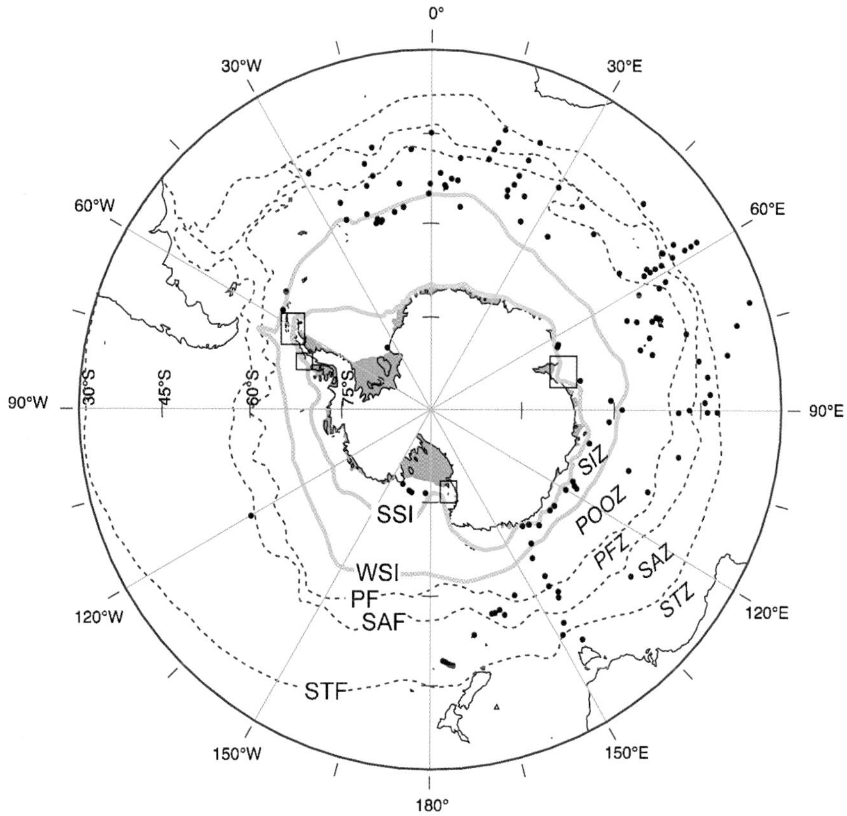

Figure 1 Oceanographic fronts and zones in the modern Southern Ocean. For references to the position of the fronts and zones and of the summer and winter sea-ice edges see abbreviations list. Black dots = the location of the 228 surface samples represented in DD228 of Armand *et al.* (2005), Crosta *et al.* (2005), and Romero *et al.* (2005). Boxed regions represent regions of denser surface sample coverage, which are detailed in Armand *et al.* (2005). Reproduced from Armand *et al.* (2005) *Palaeogeography Palaeoclimatology Palaeoecology* 223: 93–126, with permission from Elsevier BV.

planktonic foraminifera) because carbonate is not preserved below the CCD (preservational advantage), and reproduction of calcareous-walled microplankton is severely inhibited in polar waters (environmental advantage). This means, for example, that Antarctic marine diatoms are arguably the most reliable microfossil group for reconstructing past Southern Ocean environments. Of critical interest are past fluctuations in sea-surface temperature (SST), circum-Antarctic sea-ice patterns, latitudinal shifts in oceanic fronts and their role in carbon dioxide draw-down. To this end, there has been significant recent progress in assessing modern species distribution patterns (biogeography) and in the refinement of Antarctic diatom-based statistical techniques, which has advanced our understanding of the main environmental drivers controlling species abundance. In this way, Antarctic diatoms are the key to elucidating the climatic and oceanographic history of the Southern Ocean. Furthermore, since most Quaternary Antarctic species are extant, ecological-based applications can be applied to the recent past with confidence. Quaternary marine Antarctic diatoms are also widely used in biostratigraphy for both offshore regions and on the Antarctic shelf. Here, they can also help elucidate sea-level changes, past glacial processes, and ice sheet history. North of Antarctica, an interesting but somewhat underdeveloped application, for which they are unequivocal, is in their use for revealing Antarctic-source bottom water in the world's oceans.

The world's richest source of biogenic opal has resulted in a large number of studies involving Quaternary Antarctic diatoms in the Southern Ocean. It is impossible to discuss all of them here or even to acknowledge only the pioneering and best-known works. Instead we summarize those reports which have widespread importance for Antarctic studies and suggest further reading, on these and other topics. From the Southern Ocean we focus on biogeography, SST, and sea-ice reconstructions (including frontal movements), bottom water tracers, and biostratigraphy. Diatoms from lake sediments and ice cores from the Antarctic continent are also briefly discussed.

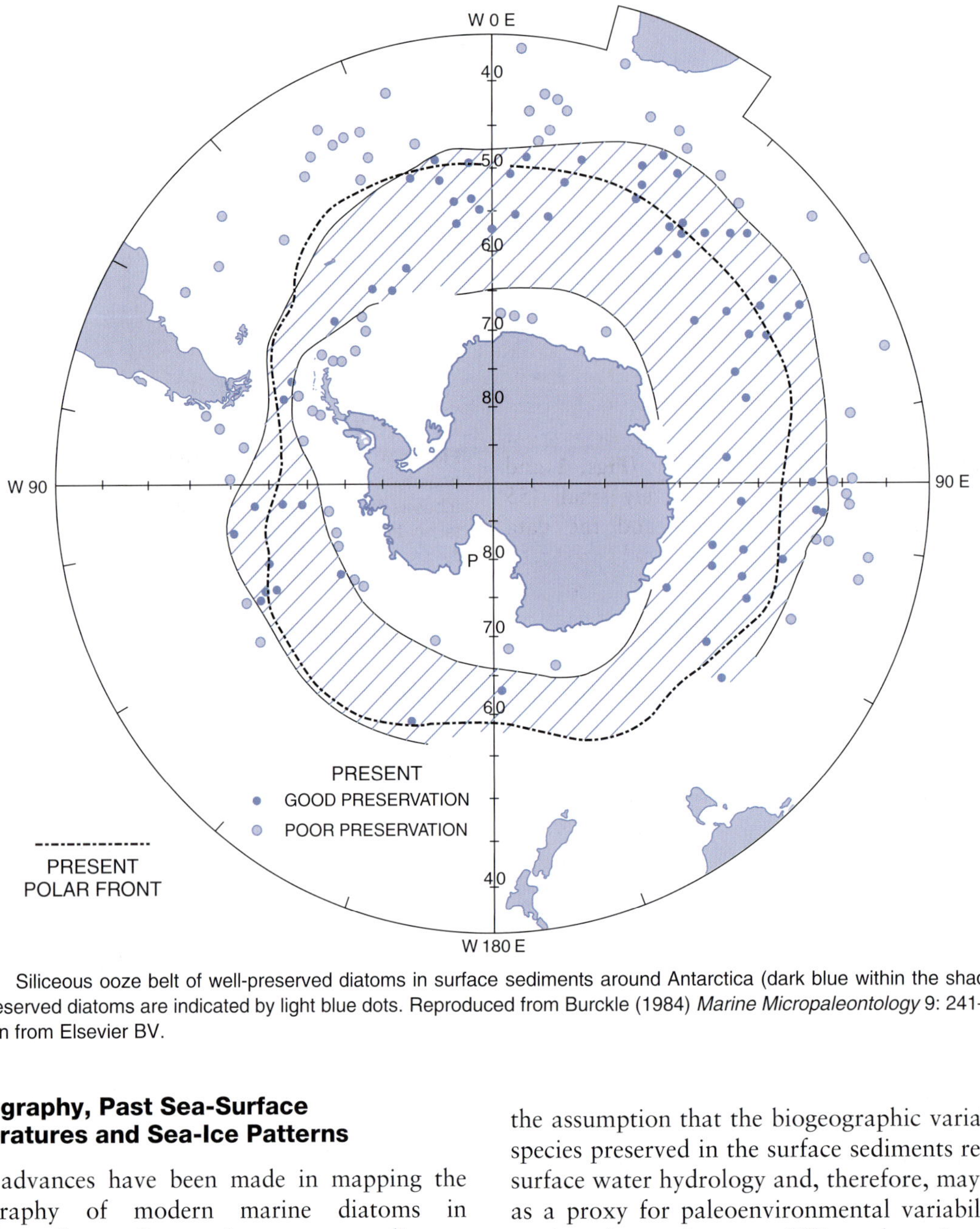

Figure 2 Siliceous ooze belt of well-preserved diatoms in surface sediments around Antarctica (dark blue within the shaded area). Poorly preserved diatoms are indicated by light blue dots. Reproduced from Burckle (1984) *Marine Micropaleontology* 9: 241–261, with permission from Elsevier BV.

Biogeography, Past Sea-Surface Temperatures and Sea-Ice Patterns

Recent advances have been made in mapping the biogeography of modern marine diatoms in Antarctic seafloor sediments (i.e., core-top sediments representing modern or submodern conditions) with respect to environmental controls. This is perhaps one of the most valuable procedures for assessing which hydrographic and cryospheric parameters control the occurrence and abundance of species living in the overlying surface waters. These modern diatom datasets, or training sets, have been analyzed by a variety of statistical techniques and transfer functions have been created. Techniques such as the modern analog technique and the generalized additive model have enabled quantitative reconstructions of SSTs and seasonal sea-ice patterns for the Quaternary. These statistical methods are reproducible and follow the assumption that the biogeographic variability of species preserved in the surface sediments reflect the surface water hydrology and, therefore, may be used as a proxy for paleoenvironmental variability. The merits of assessing past SSTs and sea-ice patterns are clear; the warming and cooling of the ocean surface is a primary gauge for climate change, while sea-ice (coverage, duration, and concentration) is especially important for climate models because of the enormous feedback effects and energy exchanges associated with its growth and decay, as well as its impact on the thermohaline circulation. As most diatom flux to the Southern Ocean sediments occurs during the austral summer, relationships between diatom distributions and environmental parameters are best derived with reference to summer conditions. However, reference to winter conditions is equally valid as it provides a means to contrast seasonal extremes as well as factoring in the controllers for particular species distributions.

Biogeography Until the last decade or so, understanding of the relationship between the distribution of Antarctic diatoms in core-top sediments to surface water hydrology was limited to a few early reports from the 1960s and 1970s. Although pioneering, these reports accounted for either only a limited area of the Southern Ocean or covered a wider area in less detail. One of the more useful later reports is that by Burckle (1984). He mapped species distributions and preservation from core-tops in all sectors of the Southern Ocean, covering its entire latitudinal range from nearshore to just north of the present day position of the Subtropical Front (**Figs. 1** and **2**). Although the dataset was relatively small (55 core-tops; nine significant species) and the data points relatively scattered (**Fig. 2**), Burckle produced one of the first distributional studies to cover the entire circum-Antarctic and he attempted to statistically quantify the relationship between species distribution and environmental parameters. Later, Zielinski and Gersonde (1997) provided the first truly comprehensive modern species-specific distributional dataset for the Atlantic Sector of the Southern Ocean. They clearly showed that the biogeography and abundance of 35 significant diatom species in this region are closely related to summer SSTs and the presence or absence of sea-ice and, therefore concluded that distributional patterns clearly do have paleoenvironmental significance.

A more extensive biogeographical dataset in terms of both coverage and detail is provided in a set of linked research papers by Armand *et al.* (2005), Crosta *et al.* (2005), and Romero *et al.* (2005). Their comprehensive modern database (termed Diatom Database 228 or DD228) incorporates the latest taxonomic concepts and wholly supports the use of fossil diatoms to reconstruct past environments. The application of this dataset will allow a more reliable interpretation of the downcore diatom signal in future Southern Ocean paleoceanographic studies. Appropriately, they broaden the use of the sea-ice control by taking into account its concentration (% water covered) and duration, not merely its presence or absence. This gives a much more complete view of the relationships between diatom distribution and environmental parameters. They covered a much wider area than previous studies, and the biogeography of 32 major diatom species/taxa in 228 core-top sediment samples from all sectors of the Southern Ocean from the coast to just north of the Subtropical Front was mapped (**Fig. 1**). For example, the distribution pattern of the

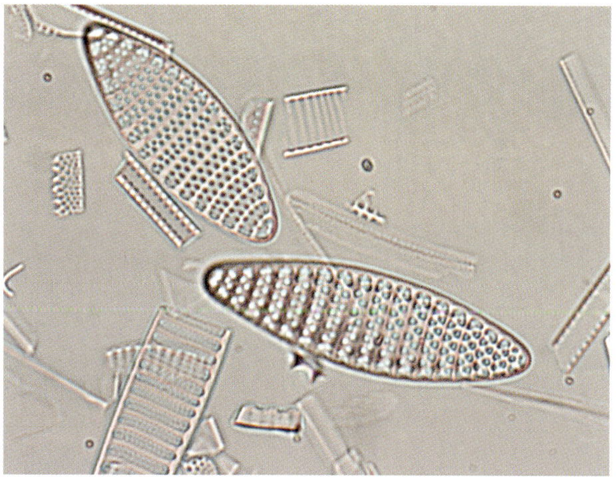

Figure 3 Light microscope image (plane-polarized light) of the endemic diatom *Fragilariopsis kerguelensis*, one of the main contributors to biogenic opal in Southern Ocean sediments. Image taken from late Quaternary laminated sediments of the East Antarctic Margin. The lower specimen is 50 μm long. (Image by EJ Maddison, Cardiff University, UK.)

abundance of the open-ocean endemic diatom *Fragilariopsis kerguelensis* (**Fig. 3**), one of the main contributors to biogenic opal in the Southern Ocean, reveals that it occurs in highest abundance when SSTs are between 1° C and 8° C (**Fig. 4**). In DD228, each diatom species is defined as having most affinity with one of three zonally distinct regions: (1) the sea-ice zone between the maximum average winter and summer sea-ice edges, where sea-ice occurs annually, (2) the open-ocean zone between the maximum average winter sea-ice edge and the SubAntarctic Front, and (3) the tropical/subtropical zone north of the SubAntarctic Front. Although all 32 diatom taxa occur throughout the study region, each to a greater or lesser degree, the correlation of their geographic patterns with sea-surface parameters illustrates the principal environmental control on the occurrence and abundance of each species (see **Table 1**, sea-ice diatoms). For the sea-ice and open-ocean zones, these controls are summer (February) SSTs and sea-ice coverage in terms of (i) its concentration during seasonal extremes (summer and winter) and (ii) its annual duration (months/year). In the tropical/subtropical zone the main controllers are warmer SSTs and minimal influence by sea-ice.

Past Sea-Surface Temperatures and Sea-Ice Patterns Past SSTs and sea-ice patterns should be considered together since they are interrelated, albeit not necessarily linearly, over long timescales. For example, Crosta *et al.* (2004) showed in a quantitative diatom study of the Indian sector of the Southern Ocean that during the transition from interglacials to

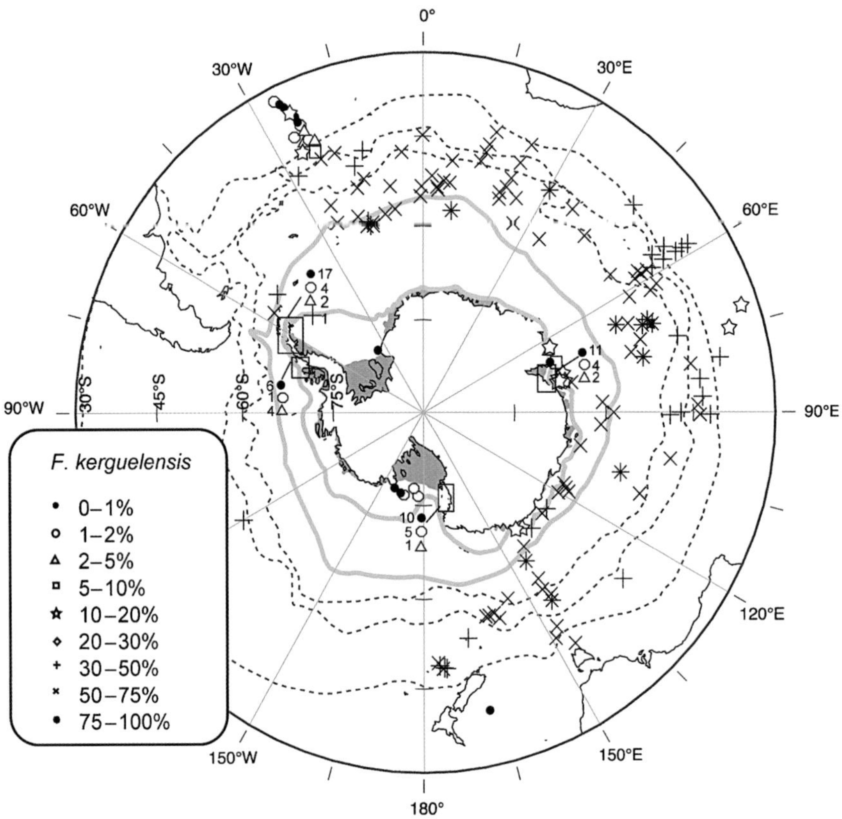

Figure 4 Surface sediment distribution of *Fragilariopsis kerguelensis* relative abundances in DD228. See **Figure 1** for oceanographic fronts and zones. Reproduced from Crosta *et al.* (2005) *Palaeogeography Palaeoclimatology Palaeoecology* 223: 66–92, with permission from Elsevier BV.

glacials, sea-ice expansion lagged the SST decrease by approximately 1 thousand years (ky), while sea-ice retreat was almost synchronous with the SST increase at glacial–interglacial terminations.

Sea-ice (**Fig. 5**) provides an important and vast habitat for Antarctic diatoms, supporting rich assemblages throughout the growing season. At the height of their growth, populations of sea-ice diatoms can become so dense that they color the underside of the sea-ice a yellowy-brown color. In the austral winter sea-ice expands to an area of up to approximately 19×10^6 km^2 (**Fig. 1**) on average, and although inhibiting surface water production and sedimentation, it can stimulate huge blooms of diatoms in the surrounding waters during the spring melt. So-called sea-ice-related taxa, such as the well-known bipolar diatom *Fragilariopsis cylindrus* (**Fig. 6**), have been used to trace winter (August) sea-ice patterns for the Quaternary. Crosta *et al.* (2004), for example, use *F. cylindrus* in their Sea Ice Group to quantitatively reconstruct the sea-ice duration of the Indian sector of the Southern Ocean for the last 220 ky using the modern analog technique. Estimations of past summer (February) sea-ice patterns are more problematic since no microfossils are preserved which allow paleoenvironmental reconstruction. However, occurrences of the cold-water (year-round sea-ice) indicator *F. obliquecostata*, along with a fall in biogenic sedimentation, may indicate the sporadic presence of summer sea-ice.

Although their dataset was never published in detail, Pichon *et al.* (1992) was one of the first studies to develop a transfer function for reconstructing past SSTs from diatoms in core-top material from the Atlantic and Indian sectors of the Southern Ocean and which factored in problems caused by dissolution. They applied their transfer function to two cores in the Indian sector of the Southern Ocean to produce SST estimates for the last ~140 ky. Later, Zielinski *et al.* (1998) developed a more rigorously tested SST transfer function for the same region providing summer SST estimates for a core located in the Atlantic sector over the same time interval.

Other studies have concentrated on reconstructing conditions at particular time-slices such as the Last Glacial Maximum (LGM). For example, Crosta *et al.* (1998), revising the Pichon *et al.* (1992) dataset, were the first to use the modern analog technique to quantitatively reconstruct circum-Antarctic sea-ice patterns from marine diatoms in terms of months

Table 1 Temperature and sea-ice parameter ranges and maximum relative abundances of the 14 sea-ice species and groups in DD228 (adapted from Armand *et al.* (2005) *Palaeogeography Palaeoclimatology Palaeoecology* 223: 93-126, with permission from Elsevier BV)

Sea-ice diatoms (data from Table 1, Armand *et al.,* 2005)	*Max. relative abundance (MRA) (%)*	*Feb. SST at MRA* (°C)	*Feb. SST range* (°C)	*Max. sea ice duration at MRA (months/yr)*	*Max. summer (Feb.) sea-ice concentration at MRA (%)*	*Max. winter (Sept.) sea-ice concentration at MRA (%)*
Actinocyclus actinochilus	2.9	0 to 1	−1.3 to 9	9	2	82
Chaetoceros resting spores	91.8	−0.5 to 1.5	−1.3 to 3.5	7	43	62
Fragilariopsis curta	64.6	0.5 to 1	−1.3 to 2.5	10	20	83
Fragilariopsis cylindrus	2.9	0.5 to 1	−1.3 to 3	8.5	0	87
Fragilariopsis obliquecostata	10.2	−1 to 0	−1.3 to 2	8.5	0	90
Fragilariopsis rhombica	3.8	0.5 to 1	−1.3 to 11.5	9	1	93
Fragilariopsis ritscheri	2.9	0 to 3	−1.3 to 7.5	0 or 9	0 or 28	0 or 72
Fragilariopsis separanda	6.9	−0.5 to 1.5	−1.3 to 1.8	8.5	0	90
Fragilariopsis sublinearis	6.1	0 to 1	−1.3 to 2.5	7.5	24	61
Porosira glacialis	6.4	0 to 0.5	−1.3 to 2	8	0	90
Porosira pseudodenticulata	2.2	0 to 0.5	−1.3 to 2	9.5	28	81
Stellarima microtrias	3.2	−0.5 to 0.5	−1.3 to 3.5	8.5	0	88
Thalassiosira antarctica group	31.8	0 to 0.5	−1.3 to 3.5	8.5	3	92
Thalassiosira tumida	2	−0.5 to 0.5	−1.3 to 7	8.5	0	90

Figure 5 Multiyear sea-ice off the western Antarctic Peninsula in austral summer (December) 2003. (Photograph by CE Stickley taken aboard the *RSS James Clark Ross* of the British Antarctic Survey.)

Figure 6 Secondary electron-scanning microscope image of a colony (eight frustules) of the bipolar diatom *Fragilariopsis cylindrus*, used to trace past winter sea-ice patterns in the circum-Antarctic. Taken from late Quaternary laminated sediments of the East Antarctic Margin. Scale bar = 20 μm. (Figure 4C of Stickley *et al.* (2005) for reference *see* Marine Laminated Sequences.)

per year (duration) at the LGM. In their comprehensive 'state-of-the-art' compilation of published and new datasets, Gersonde *et al.* (2005) provide the most up to date estimation of summer SSTs and sea-ice distribution patterns in the Southern Ocean at the LGM based on quantitative analysis of diatoms and radiolarians. Based on new data and reference datasets of several previously published works Gersonde *et al.* applied three statistical methods to 122 marine cores located in all sectors of the Southern Ocean to quantitatively reconstruct the glacial Southern Ocean at the EPILOG (Environment

Processes of the Ice Age: Land, Ocean, Glaciers) time-slice (19.5–16.0 thousand years ago (ka) or 23–19 calendar years before present (cal ky BP). Their results indicate that at the EPILOG LGM the maximum winter sea-ice extent (at a concentration of >15%) reached as far north as nearly 47°S in the Atlantic and Indian sectors, but only to 57°S in the Pacific (**Fig. 7**). This equates to a total, asymmetric, EPILOG LGM circum-Antarctic winter sea-ice field of approximately 39 × 10^6 km^2, a northward expansion of 100% compared to the present, which agrees with earlier estimations (e.g., CLIMAP (Climate and Environment Monitoring with GPS Atmospheric Profiling)). However, unlike previous studies, they suggest an increased seasonality at the LGM based on the large differences between the extent of the winter and postulated summer sea-ice fields compared with those of today (**Fig. 7**). In addition, their

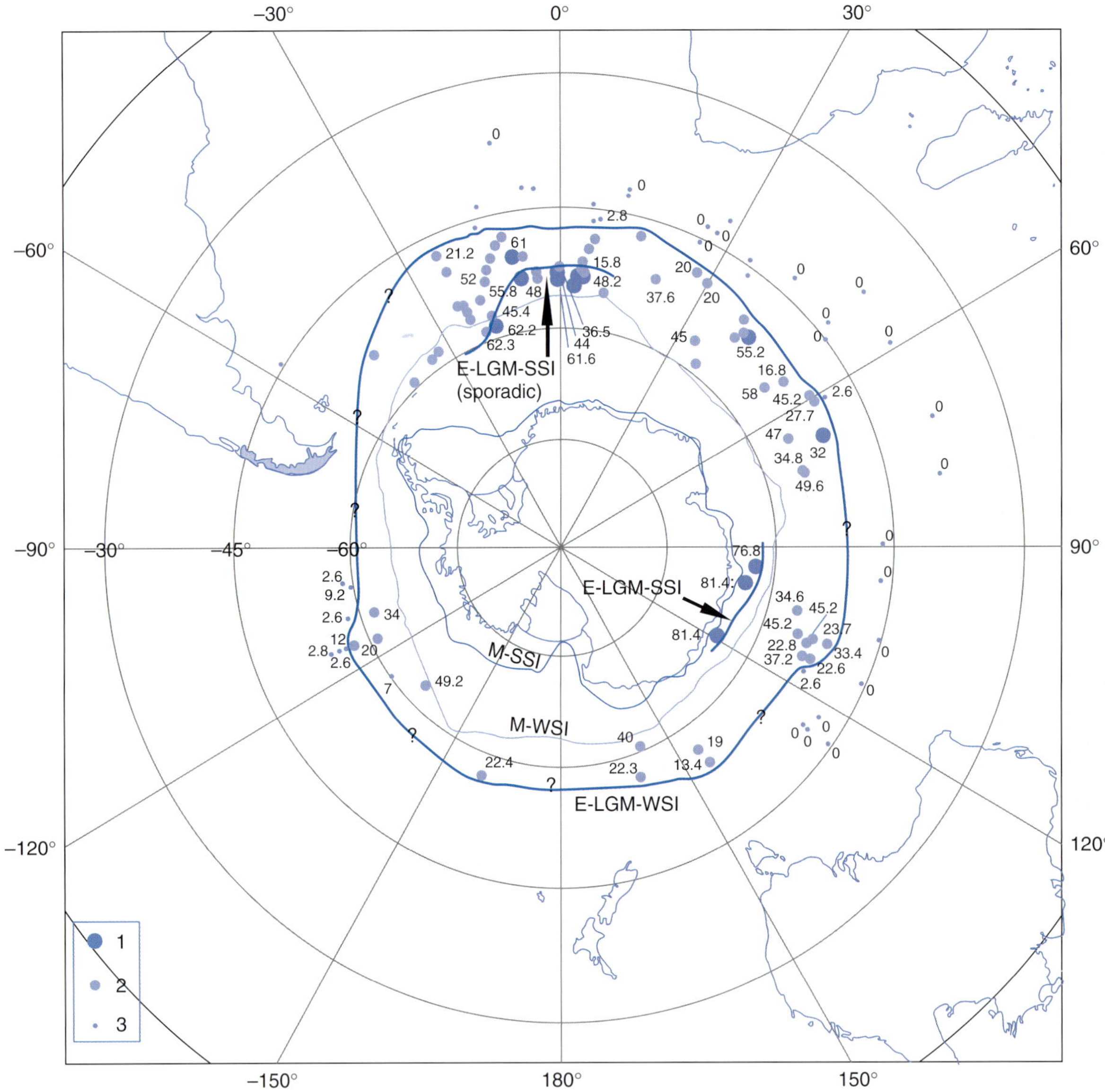

Figure 7 Sea-ice distribution at the EPILOG LGM timeslice compared to modern conditions. EPILOG LGM-WSI, maximum extent of winter sea-ice at the LGM (September, concentration > 15%); M-WSI, maximum extent of modern winter sea-ice (September, concentration > 15%); and M-SSI, maximum extent of modern summer sea-ice. Values indicate estimated winter (September) sea-ice concentration in percent derived with the modern analog technique and the generalized additive model. Large dots, concomitant occurrence of cold-water indicator *Fragilariopsis obliquecostata* (>1% of diatom assemblage) and summer sea-ice (February, concentration > 0%) interpreted to represent sporadic occurrence of EPILOG LGM-SSI (maximum extent of summer sea-ice at the LGM); medium dots, presence of winter sea-ice (September, concentration > 15%, diatom winter sea-ice indicators > 1%); small dots, no winter sea-ice (September, concentration < 15%, diatom winter sea-ice indicators <1%). Reproduced from Gersonde *et al.* (2005) *Quaternary Science Reviews* 24: 869-896, with permission from Elsevier BV.

estimation of EPILOG LGM summer SSTs indicates a northward migration of the LGM Polar Front by 4° and 5–10° latitude in the Atlantic and Indian sectors, respectively, with the strongest cooling (4–6 °C decrease) occurring in these sectors in the modern SubAntarctic Zone (**Fig. 1**). These are larger estimates than previously reported and Gersonde *et al.* (2005) conclude a similar northward shift in the LGM SubAntarctic Front in the Atlantic and Indian sectors but only a relatively minimal northward shift of the LGM Subtropical Front leading to compression of the SubAntarctic Zone and steepened thermal gradients. However, their data from the Pacific sector implies reduced cooling here and therefore possible nonuniform cooling (and asymmetric frontal movement) of the glacial Southern Ocean at the LGM.

Antarctic Diatoms as Tracers of Bottom Water

Marine planktonic diatoms endemic to the Southern Ocean can be transported out of Antarctic waters into the world's oceans via Antarctic Bottom Water (AABW). This is possible where surface-dwelling diatoms are entrained into downwelling AABW during its formation in the Weddell and Ross seas and on the shelves around the Antarctic continent. Entrained diatoms subsequently settling out along the flow path become effective tracers of this deep current. This has been clearly demonstrated for modern times from diatoms in core-top sediments along the AABW flow path in the southwest and central Pacific and the southwest Atlantic. In these locations displaced endemic diatoms can comprise the majority of the sediment assemblage, while they are absent from core-top sediments located in the overlying water mass (e.g., the North Atlantic deep water, NADW). Endemic diatoms displaced by AABW have also been noted in surface and core sediments as far as 60 °N.

Significant abundances of displaced endemic diatoms have been found in late Quaternary (last ~200 ky) core sediments within the AABW flow path in the southwest Pacific and southwest Atlantic oceans. Along with the stable isotopic signal from biogenic carbonate, these assemblages help provide information on AABW production and flow history to these oceanic basins. This is of interest since AABW plays a key role in the climate system, being one of the main transporters of heat and salts around the world's oceans. A higher abundance of displaced diatoms indicates higher AABW flow. For example, Jones and Johnson (1984) found that for the southwest Atlantic, displaced diatoms are more abundant during interglacials (higher AABW flow) than glacial maxima (lower AABW flow). However, unlike stable isotopes, displaced endemic Antarctic diatoms provide an unequivocal tracer for bottom water provenance in the North Atlantic. Furthermore, in the southwest Pacific, downcore variations in the abundance of displaced open-ocean diatoms (e.g., *Fragilariopsis kerguelensis*; **Fig. 3**, *Thalassiosira lentiginosa*) versus sea-ice related diatoms (e.g., *Fragilariopsis curta, F. cylindrus*; **Fig. 6**) may indicate temporal switches in AABW source (open ocean versus shelf, respectively) through the late Quaternary.

Diatom Biostratigraphy

Planktonic marine diatoms make excellent biostratigraphic tools since they are widespread, easily extracted from the sediment, and have undergone several relatively rapid, stepwise, turnover (evolutionary) events during the Cenozoic, driven by climatic and tectonic changes. In the Southern Ocean they are the most useful microfossil group for biostratigraphy. The more recent Southern Ocean zonation scheme(s) provide relative and absolute age information and have been developed over many years from mainly offshore cores drilled by the Deep Sea Drilling Project (DSDP) and the Ocean Drilling Program (ODP). Cenozoic zonation applies well to deep-sea Southern Ocean cores as well as those from the Antarctic shelf. Ramsey and Baldauf (1999) provide a review of the history of the development of the Southern Ocean biochronology.

Although they are most useful for offshore sedimentary sections spanning longer timescales (e.g., Miocene to Recent), there are some well-known diatom events in the Quaternary that can be applied to younger core sediments and complement independent dating methods such as oxygen isotope stratigraphy. Due to latitudinal gradients of surface water parameters, however, the distribution of late Neogene and Quaternary marker diatoms is not the same north and south of the Polar Frontal Zone (**Fig. 1**). This means that the youngest part of the zonation does not work as well in the northern parts of the Southern Ocean as it does for sites south of the Polar Front and on the shelf. For this reason Gersonde and Bárcena (1998) and later Zielinski and Gersonde (2002) modified the late Pliocene and Pleistocene interval of the zonation for sites north and south of the Polar Frontal Zone using cores located mostly in the Atlantic sectors of the Southern Ocean (**Fig. 8**). Their modified northern Southern Ocean zonation critically includes the following biodatums of the subAntarctic (warmer water) marker species *Hemidiscus karstenii*, the abundance of which

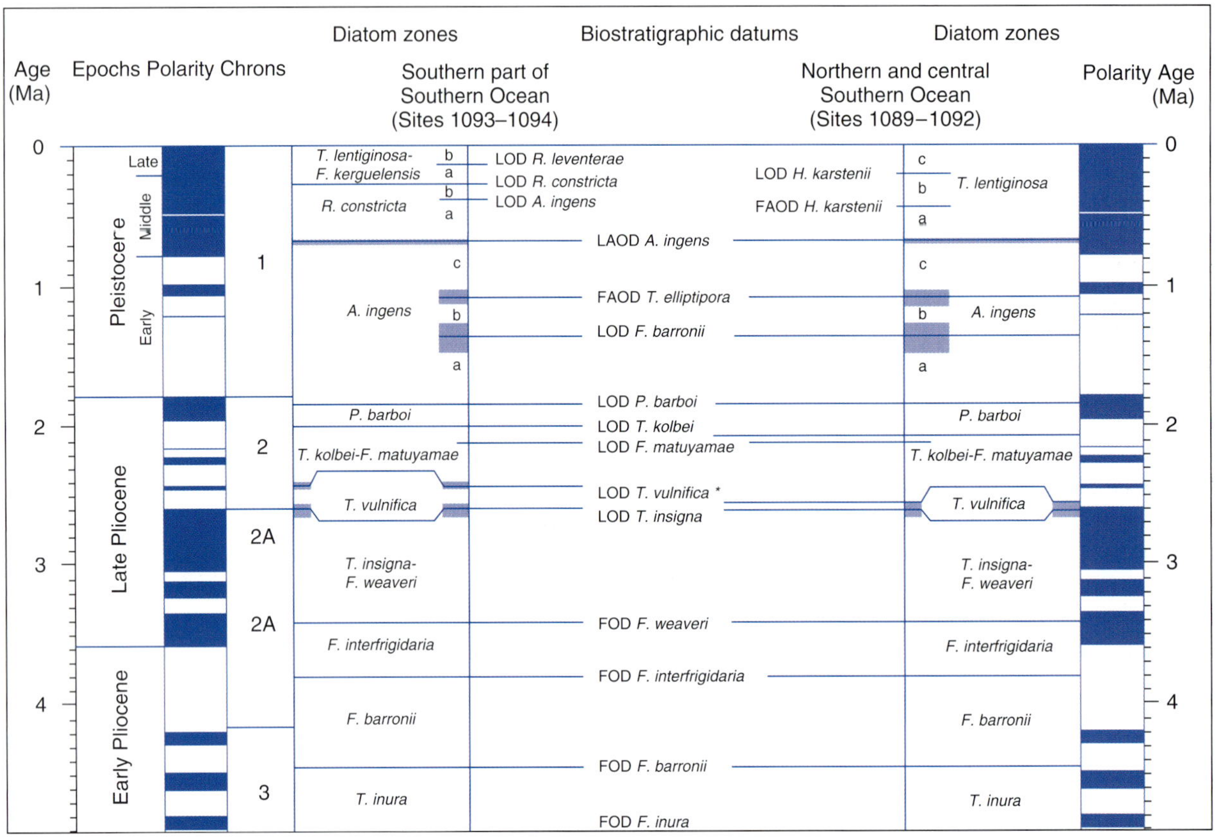

Figure 8 Compilation and correlation of the Plio-Pleistocene diatom zonation for the northern and southern part of the Southern Ocean. Both absolute and relative age information is provided: ages in (Ma); chronostratigraphy (epochs); and the geomagnetic polarity timescale (GPTS, polarity, and chrons). Reproduced from Zielinski and Gersonde (2002) *Marine Micropaleontology* 45: 225–268, with permission from Elsevier BV.

declines south of the Polar Frontal Zone: the First Abundant Occurrence Datum (FAOD) at 420 ky (lower MIS 11), and the Last Occurrence Datum (LOD) at 190 ky (upper marine isotope stage (MIS) 7) (**Fig. 8**). Their southern Southern Ocean zonation (applicable to south of the Polar Front) divides the Pleistocene differently based on biodatums of Antarctic (colder water) markers *Rouxia constricta* and *R. leventerae*: the LOD *R. constricta* at 280 ky (MIS 8), and the LOD *R. leventerae* at 130 ky (MIS 6) (**Fig. 8**). These new biodatums, along with their redefinition of the LOD *Actinocyclus ingens* biodatum improve the temporal resolution for southern Southern Ocean biostratigraphy by up to 100 ky.

Nonmarine Antarctic Waters: Freshwater Diatoms

Less than 1% of the Antarctic continent is ice free and here diatoms can grow in lakes and streams as well as in terrestrial habitats. The majority of the lakes are situated on the coastal oases (e.g., McMurdo Dry Valleys, Vestfold Hills, Larsemann Hills, and Schirmacher Oasis), the ice-free areas of the Antarctic Peninsula (e.g., James Ross Island, King George Island, and Livingston Island), and the Antarctic Islands (e.g., Signy Island). Modern ecological and distributional data have been collected from several areas. The species diversity of nonmarine diatoms in Antarctica is low compared to Arctic regions, probably due to the physical isolation of the continent. The Antarctic Peninsula has a more diverse flora with a general trend of decreasing diversity southwards. Many Antarctic diatoms are classed as aerophilic but are also found in aquatic habitats subject to desiccation or freezing, others are found in association with terrestrial soils and mosses. Some taxa are widespread while others have a regional distribution and some are endemic.

Holocene stratigraphic records have accumulated in many lakes, and due to the low diversity of Antarctic freshwaters and the absence of many other proxies conventionally used in temperate regions (e.g., pollen) diatoms have been extensively used for environmental reconstruction. Hodgson *et al.* (2004) provide a recent review of Antarctic paleolimnology in general and Spaulding and McKnight (1999) review the uses of diatoms in particular. Diatoms are good indicators of salinity, which in some lakes can be a proxy for climate, while in others is related to changes in sealevel. They have also been used as indicators of the changing extent of lake ice cover and lake depth, thus they are good indicators of climate and ecosystem change.

Transfer functions have been developed for a range of environmental variables, for example, Jones and Juggins (1995) created a transfer function for nutrients (chlorophyll a) based on surface sediments obtained from Signy and Livingston Islands. This was applied to Holocene cores from Signy Island, and together with other proxy data showed a Holocene climate optimum of 3,800–1,300 ^{14}C yr BP when the productivity of the lakes was higher. Recent changes in the diatom flora related to nutrient enrichment from birds and seals were also observed.

In the Vestfold and Larsemann Hills, Roberts and McMinn (1996), and Verleyen *et al.* (2003) have developed transfer functions for salinity. The water chemistry of lakes in these areas responds quickly to changes in the moisture balance, with increasing salinity and decreasing lake-water level during dry periods, with the opposite pattern in wet periods. Roberts *et al.* (2001) used a transfer function approach to obtain a high-resolution record of evaporation for the last 650 years from Ace Lake (Vestfold Hills) which was also highly correlated to the seasonal oxygen isotope signal preserved in an ice core from Law Dome, suggesting that the diatom-based reconstructions provide a robust paleoclimate signal.

The McMurdo Dry Valleys located on the western coast of the Ross Sea are among the coldest and driest places on Earth. Lakes here are some of the harshest aquatic environments in the world. The lakes mostly have a perennial ice cover (2.8–6.0 m thick) over liquid water. The volume of these lakes has fluctuated greatly over the last 20,000 years and in recent times lake levels have been rising. Lake Hoare is a perennially ice-covered, closed-basin lake where diatoms form part of the benthic microbial mats, with characteristic species occurring in shallow-water, mid-depth, and deep-water communities. The lake has a unique sedimentary environment where due to lack of water circulation the sediments accumulate *in situ* and so record the history of a particular site rather than integrating events over the whole lake basin. Analysis of the sedimentary diatoms from Lake Hoare indicates that the rapid increase in lake level is a recent event.

Diatoms have also been found in Antarctic ice cores such as Dome C and Taylor Dome; these are a mixture of freshwater and marine taxa which have been blown in by long distance transport. At Taylor Dome, Kellogg and Kellogg (1996a and b) found that low abundances of diatoms occur during glacial intervals (MIS 3–5), and higher abundances occur during interstadials (MIS 5b, 5d, and 6). This is due to the conditions in the source areas. In glacials, the West Antarctic Ice Sheet covers most of the Ross Sea continental shelf and thus the nearest open water source for marine diatoms would be far offshore. The extent of ice-free areas in the dry valleys would also have been limited by lobes of grounded ice pushing into the valleys from the sea and by large ice-covered lakes greatly reducing possible sources for reworked marine diatoms.

Conclusion

In this article we demonstrate how diatoms are a powerful paleoclimate tool in both marine and freshwater systems using fossil record examples from Antarctic waters. The nutrient-rich Southern Ocean is the world's richest source of biogenic opal, resulting in a circum-Antarctic 'siliceous ooze belt'. We summarize the latest information on diatom biogeography in surface sediments used to reconstruct past sea-surface temperatures (SSTs) and sea-ice patterns. We also discuss their use as bottom water tracers and for biostratigraphy. Records of non-marine diatoms have been found in lakes on the Antarctic continent and in ancient ice-cores. We review the uses of these Holocene records for climate and ecosystem change reconstructions.

Abbreviations

AABW	Antarctic Bottom Water
ACC	Antarctic Circumpolar Current
ANDRILL	Antarctic Drilling Program (http://andrill.org/ Also see this website for the history of Antarctic drilling including the Cape Roberts Project (CRP) and the New Zealand-led MSSTS and CIROS drilling projects and for related projects such as Shallow Drilling on the Antarctic Continental Margin (SHALDRILL))
cal ky BP	calendar years before present
CCD	calcium carbonate compensation depth
CLIMAP	CLIMAP (Climate and Environment Monitoring with GPS Atmospheric Profiling
DD228	Diatom Database 228
DSDP	Deep Sea Drilling Project
EPILOG	Environment Processes of the Ice Age: Land, Ocean, Glaciers
FAOD	First Abundant Occurrence Datum
GPTS	geomagnetic polarity timescale

HNLC	high-nitrate, low-chlorophyll
IMAGES	International Marine Past Global Changes Study (http://www.images-pages.org)
IODP	Integrated Ocean Drilling Program
ka	thousand years ago
ky	thousand years
LGM	Last Glacial Maximum
LOD	Last Occurrence Datum
MIS	marine isotope stage
NADW	North Atlantic deep water
Ma	millions of years ago
ODP	Ocean Drilling Program
PF	Antarctic Polar Front
PFZ	Polar Frontal Zone
POOZ	Permanently Open Ocean Zone
SAF	SubAntarctic Front
SAZ	SubAntarctic Zone
SSI	maximum average summer sea-ice edge
SST	sea-surface temperatures
STF	Subtropical Front
STZ	Subtropical Zone
SIZ	Sea Ice Zone
WSI	maximum average winter sea-ice edge

See also: **Diatom Introduction**. **Diatom Methods**: Data Interpretation. **Diatom Records**: Marine Laminated Sequences. **Glacial Climates**: Thermohaline Circulation. **Glaciations**: Late Quaternary of Antarctica. **Paleoceanography, Biological Proxies**: Coccoliths; Marine Diatoms; Planktic Foraminifera; Radiolarians and Silicoflagellates. **Paleoceanography, Physical and Chemical Proxies**: Oxygen Isotope Stratigraphy of the Oceans. **Paleoclimate**: Time Scales of Climate Change. **Quaternary Stratigraphy**: Overview; Biostratigraphy; Chronostratigraphy. **Sea Level Studies**: Microfossil Reconstructions.

References

Armand, L. K., and Leventer, A. (2003). Paleo sea ice distribution—reconstruction and paleoclimatic significance. In *Sea Ice and Introduction to its Physics, Chemistry, Biology and Geology* (D. N. Thomas and G. S. Diekm, Eds.), pp. 333–372. Blackwell, Oxford.

Armand, L. K., Crosta, X., Romero, O. E., and Pichon, J. J. (2005). The biogeography of major diatom taxa in Southern Ocean surface sediments: 1. Sea ice related species. *Palaeogeography Palaeoclimatology Palaeoecology* **223**, 93–126.

Barron, J. A. (2003). Planktonic marine diatom record of the past 18 m.y.: appearances and extinctions in the Pacific and Southern Oceans. *Diatom Research* **18**, 203–224.

Barron, J. A., and Baldauf, J. G. (1989). Tertiary cooling steps and palaeoproductivity as reflected by diatoms and biosiliceous sediments. In *Productivity of the Ocean: Present and Past* (W. H. Berger, V. S. Smetacek and G. Wefer, Eds.), pp. 341–354. John Wiley & Sons, New York.

Barron, J. A., and Baldauf, J. G. (1995). Cenozoic marine diatom biostratigraphy and applications to paleoclimatology and paleoceanography. In *Siliceous Microfossils* (C. Blome, Ed.), pp. 107–108. Paleontological Society, Short Courses in Paleontology, 8. Society of Economic Paleontologists and Mineralogists, Tulsa.

Bohaty, S. M., Scherer, R. P., and Harwood, D. M. (1998). Quaternary diatom biostratigraphy and palaeoenvironments of the CRP-1 Drillcore, Ross Sea, Antarctica. *Terra Antarctica* **5**, 431–453.

Booth, J. D., and Burckle, L. H. (1976). Displaced Antarctic diatoms in the southwestern and central Pacific. *Pacific Geology* **11**, 99–108.

Burckle, L. H. (1984). Diatom distribution and paleoceanographic reconstruction in the Southern Ocean – present and Last Glacial Maximum. *Marine Micropaleontology* **9**, 241–261.

Burckle, L. H., and Cirilli, J. (1987). Origin of diatom ooze belt in the Southern Ocean: implications for late Quaternary paleoceanography. *Micropaleontology* **33**, 82–86.

Crosta, X., Pichon, J. J., and Burckle, L. H. (1998). Application of modern analog technique to marine Antarctic diatoms: Reconstruction of maximum sea-ice extent at the Last Glacial Maximum. *Paleoceanography* **13**, 284–297.

Crosta, X., Sturm, A., Armand, L., and Pichon, J. J. (2004). Late Quaternary sea ice history in the Indian sector of the Southern Ocean as recorded by diatom assemblages. *Marine Micropaleontology* **50**, 209–223.

Crosta, X., Romero, O. E., Armand, L. K., and Pichon, J. J. (2005). The biogeography of major diatom taxa in Southern Ocean surface sediments: 2. Open ocean related species. *Palaeogeography Palaeoclimatology Palaeoecology* **223**, 66–92.

Gersonde, R., and Bárcena, M. A. (1998). Revision of the upper Pliocene-Pleistocene diatom biostratigraphy for the northern belt of the Southern Ocean. *Micropaleontology* **44**, 84–98.

Gersonde, R., Crosta, X., Abelmann, A., and Armand, L. (2005). Sea-surface temperature and sea ice distribution of the Southern Ocean at the EPILOG Last Glacial Maximum—a circum-Antarctic view based on siliceous microfossil records. *Quaternary Science Reviews* **24**, 869–896.

Harwood, D. M., and Maruyama, T. (1992). Middle Eocene to Pleistocene diatom biostratigraphy of Southern Ocean sediments from the Kerguelen Plateau, Leg 120. In *Proceedings of the Ocean Drilling Program, Scientific Results, 120* (S. W. Wise, Jr.,R. Schlich A. Palmer-Julson and E. Thomas, Eds.), pp. 683–734. Texas A&M University, Ocean Drilling Program, College Station, TX, USA.

Hodgson, D. A., Doran, P., Roberts, D., and McMinn, A. (2004). Paleolimnological studies from the Antarctic and subAntarctic islands. In *Long-Term Environmental Change in Arctic and Antarctic Lakes* (R. Pienitz, M. S. V. Douglas and J. P. Smol, Eds.), pp. 419–474. Springer, Dordrecht.

Jones, G. A., and Johnson, D. A. (1984). Displaced Antarctic diatoms in Vema Channel sediments: Late Pleistocene/Holocene fluctuations in AABW flow. *Marine Geology* **58**, 165–186.

Jones, V. J. (1996). The diversity, distribution and ecology of diatoms from Antarctic inland waters. *Biodiversity and Conservation* **5**, 1433–1449.

Jones, V. J., Hodgson, D. A., and Chepstow-Lusty, A. J. (2000). Palaeolimnological evidence for marked Holocene environmental changes on Signy Island, Antarctica. *The Holocene* **10**, 43–60.

Jones, V. J., and Juggins, S. (1995). The construction of a diatom-based chlorophyll a transfer function and its application at three lakes on Signy Island (maritime Antarctic) subject to differing degrees of nutrient enrichment. *Freshwater Biology* **34**, 433–445.

Kellogg, D. E., and Kellogg, T. B. (1996a). Diatoms in South Pole ice: implications for eolian contamination of Sirius Group deposits. *Geology* **24**, 118.

Kellogg, D. E., and Kellogg, T. B. (1996b). Glacial/interglacial variations in the flux of atmospherically transported diatoms in Taylor Dome ice core. *Antarctic Journal of the United States Review*.

Lozano, J. A., and Hays, J. D. (1976). Relationship of radiolarian assemblages to sediment types and physical oceanography in the Atlantic and Western Indian Ocean Sectors of the Antarctic Ocean. In *Investigation of late Quaternary Paleoceanography and Paleoclimatology* (R. M. Cline and J. Hays, Eds.), pp. 303–336. Geological Society of America, Boulder, CO, USA.

Pichon, J. J., Labeyrie, L. D., Bareille, G., *et al.* (1992). Surface water temperature changes in the high latitudes of the Southern Hemisphere over the last glacial–interglacial cycle. *Paleoceanography* **7**, 289–318.

Ramsay, A. T. S., and Baldauf, J. G. (1999). A Reassessment of the Southern Ocean Biochronology, p. 122. Geological Society, London.

Roberts, D., and McMinn, A. (1996). Relationships between surface sediment diatom assemblages and water chemistry gradients in saline lakes of the Vestfold Hills, Antarctica. *Antarctic Science* **8**, 331–341.

Roberts, D., van Ommen, T. D., McMinn, A., Morgan, V., and Roberts, J. L. (2001). Late Holocene East Antarctic climate tends from ice core and lake sediment proxies. *The Holocene* **11**, 117–120.

Romero, O. E., Armand, L. K., Crosta, X., and Pichon, J. J. (2005). The biogeography of major diatom taxa in Southern Ocean surface sediments: 3. Tropical/subtropical species. *Palaeogeography Palaeoclimatology Palaeoecology* **223**, 49–65.

Scherer, R. P. (1991). Quaternary and Tertiary microfossils from beneath Ice Stream B: Evidence for a dynamic West Antarctic Ice Sheet history. *Palaeogeography Palaeoclimatology Palaeoecology* **90**, 395–412.

Sjunneskog, C., and Scherer, R. P. (2005). Mixed diatom assemblages in glacigenic sediment from the Central Ross Sea, Antarctica. *Palaeogeography Palaeoclimatology Palaeoecology* **218**, 287–300.

Spaulding, S. A., and McKnight, D. M. (1999). Diatoms as indicators of environmental change in Antarctic freshwaters. In *The Diatoms: Applications for the Environmental and Earth Sciences* (E. Stoermer and J. P. Smol, Eds.), pp. 245–263. Cambridge University Press, Cambridge.

Spaulding, S. A., McKnight, D. M., Stoermer, E. F., and Doran, P. T. (1997). Diatoms in sediments of perennially ice-covered Lake Hoare, and implications for interpreting lake history in the McMurdo Dry Valleys of Antarctica. *Journal of Paleolimnology* **17**, 403–420.

Stickley, C. E., Carter, L., McCave, I. N., and Weaver, P. P. E. (2001). Lower circumpolar deep water flow through the SW Pacific Gateway for the last 190 ky: evidence from Antarctic diatoms. In *The Oceans and Rapid Climate Change, Past, Present and Future* (B. J. Haupt and M. A. Maslin, Eds.), pp. 101–116. Geophysical Monograph Series, 126, Washington.

Stickley, C. E., Pike, J., Leventer, A., *et al.* (2005). Deglacial ocean and climate seasonality in laminated diatom sediments, Mac. Robertson Shelf, Antarctica. *Palaeogeography Palaeoclimatology, Palaeoecology* **227**, 290–310.

van de Vijver, B., and Beyens, L. (1999). Biogeography and ecology of freshwater diatoms in subAntarctica: a review. *Journal of Biogeography* **26**, 993–100.

Verleyen, E., Hodgson, D. A., Vyverman, W., *et al.* (2003). Modelling diatom responses to climate induced fluctuations in the moisture balance in continental Antarctic lakes. *Journal of Paleolimnology* **30**, 195–215.

Winter, D. M., and Harwood, D. M. (1997). Integrated diatom biostratigraphy of Late Neogene drillholes in Southern Victoria Land and correlation to Southern Ocean records. In *The Antarctic Region: Geological Evolution and Processes* (C. A. Ricci, Ed.), pp. 985–992. Terra Antarctica Publication, Sienna.

Zielinski, U., and Gersonde, R. (1997). Diatom distribution in Southern Ocean surface sediments (Atlantic Sector): implications for paleoenvironmental reconstructions. *Palaeogeography Palaeoclimatology Palaeoecology* **129**, 213–250.

Zielinski, U., and Gersonde, R. (2002). Plio-Pleistocene diatom biostratigraphy from ODP Leg 177, Atlantic Sector of the Southern Ocean. *Marine Micropaleontology* **45**, 225–268.

Zielinski, U., Gersonde, R., Sieger, R., and Fütterer, D. (1998). Quaternary surface water temperature estimations: calibration of a diatom transfer function for the Southern Ocean. *Paleoceanography* **13**, 365–384.

Freshwater Laminated Sequences

H Simola, University of Joensuu, Joensuu, Finland

© 2007 Elsevier B.V. All rights reserved.

Introduction

Varves and Diatoms—Ideal Objects for Paleoecological Studies

Varved or annually laminated sediments contain high-precision records of sedimentation with yearly and sometimes even seasonal resolution (Lamoureux, 2001). Varves, especially in unconsolidated sediments, may be difficult to discern or may easily be disturbed by coring, so they are best sampled by freeze-coring (Glew *et al.*, 2001). Varve formation is a consequence of low-energy sedimentation and an absence of bioturbation, and it can therefore also indicate good preservation of various constituents of the sediment. A particular benefit of varves is that they provide a precise chronology for the sediment sequence. The chronology may be floating or absolute (calendar), depending on whether it is possible to fix the series to the present day or to an historical event, and it may be more or less accurate, depending on the regularity and ease of identification of the annual cycles.

Diatoms are often abundant and are usually well preserved in lake sediments (Battarbee *et al.*, 2001). Sedimentary diatom analysis is a versatile tool for tracking environmental history (Charles *et al.*, 1994; Flower *et al.*, 1997; Smol, 2004). Diatoms in varved sediments often provide environmental records of very high quality, although the relative rarity of varved sediments sets limits to their routine use as environmental archives. In many ways, however, it is possible to improve the interpretation of nonvarved sediment records by comparisons with varved ones.

Diatoms as a Structural Component in Sediment Varves

The siliceous frustules of diatoms are a quantitatively important constituent of many lake sediments. The rapid and direct sedimentation of seasonal blooms of plankton diatoms may result in distinct diatom-rich layers in the sediment. Often yellowish or greenish in color, these may be pure enough to be classified as diatomite. Such layers are often regular and visually significant structural elements in varves (Simola, 1992).

Often, the annual character of a rhythmically laminated sediment can be proved by the microstratigraphy of diatoms reflecting the regular yearly succession and the seasonal blooms of planktonic diatoms (Simola, 1992). In other cases, the absence of such seasonality may indicate a nonannual rhythmite. Occasionally, the annual varve pattern may be verified by diatom stratigraphy even in visually obscure or irregularly laminated sediment sequences (Card, 1997).

Friedrich Nipkow—Splendid Pioneer

Friedrich Nipkow is worth remembering as an early pioneer of paleolimnology, especially for his studies of diatom records in varved sediments. By profession a pharmacist, he studied algae as a hobby in his own laboratory in Zurich, Switzerland. He used a simple tube sampling device to collect profundal sediment cores from the deep lakes Zürichsee and Baldeggersee, which at that time were being strongly eutrophied by untreated sewage effluents.

Nipkow (1927) presented excellent photographs of the varved sediment samples (**Fig. 1**) as well as microphotographs of his diatom slides showing the seasonal assemblages and cell size variation observed within the varves. Varve formation had started in both lakes a few decades earlier as a consequence of hypolimnetic anoxia. Nipkow convincingly verified the laminae as true varves by their seasonal diatom microstratigraphy and by correlating species occurrences in the varves with previous plankton records. Furthermore, he connected several turbidite layers observed in the sediments with historically known landslides of the lake shores.

Nipkow meticulously dissected each varve in his samples into 5–10 seasonal layers and analyzed the diatoms in each in great detail. Besides demonstrating the size reduction–auxosporulation cycle for several species (**Fig. 2**), he also described year-to-year plankton community succession in both lakes and discussed the potential effects of environmental conditions on the diatom development. This study illustrates the advantages provided by the continuous and quantitative sediment records over the laborious long-term monitoring of plankton in water samples.

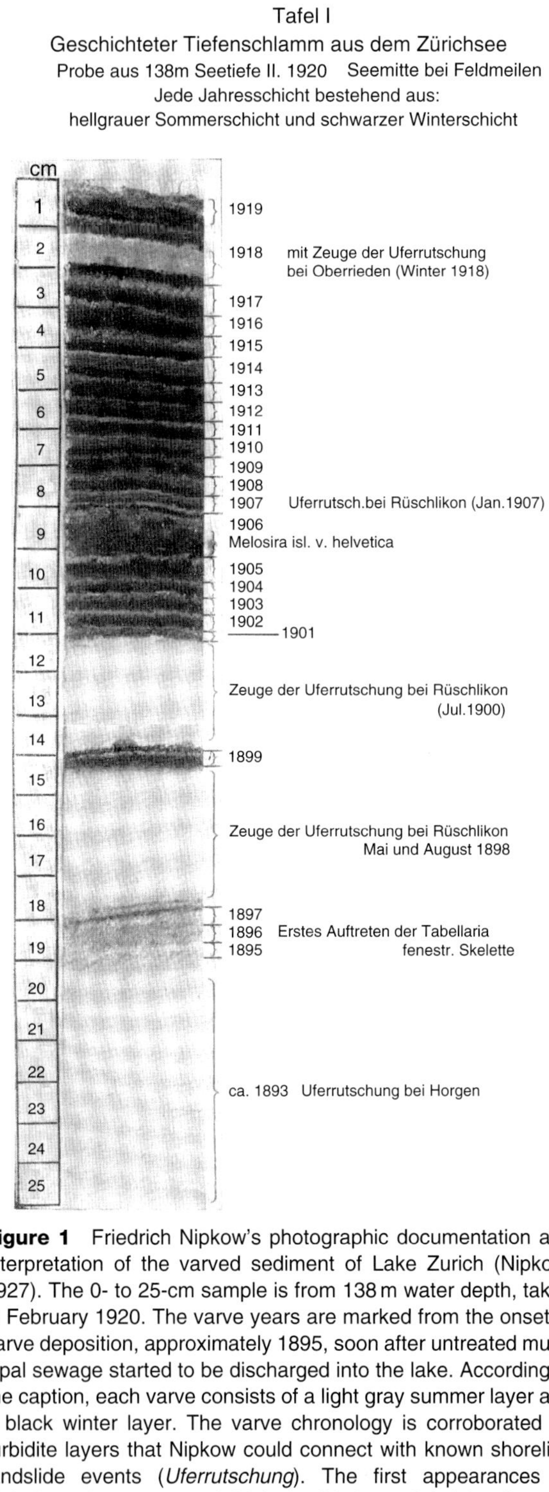

Figure 1 Friedrich Nipkow's photographic documentation and interpretation of the varved sediment of Lake Zurich (Nipkow, 1927). The 0- to 25-cm sample is from 138 m water depth, taken in February 1920. The varve years are marked from the onset of varve deposition, approximately 1895, soon after untreated municipal sewage started to be discharged into the lake. According to the caption, each varve consists of a light gray summer layer and a black winter layer. The varve chronology is corroborated by turbidite layers that Nipkow could connect with known shoreline landslide events (*Uferrutschung*). The first appearances of *Tabellaria fenestrata* and *Melosira* (*Aulacoseira*) *islandica* var. *helvetica* are indicated.

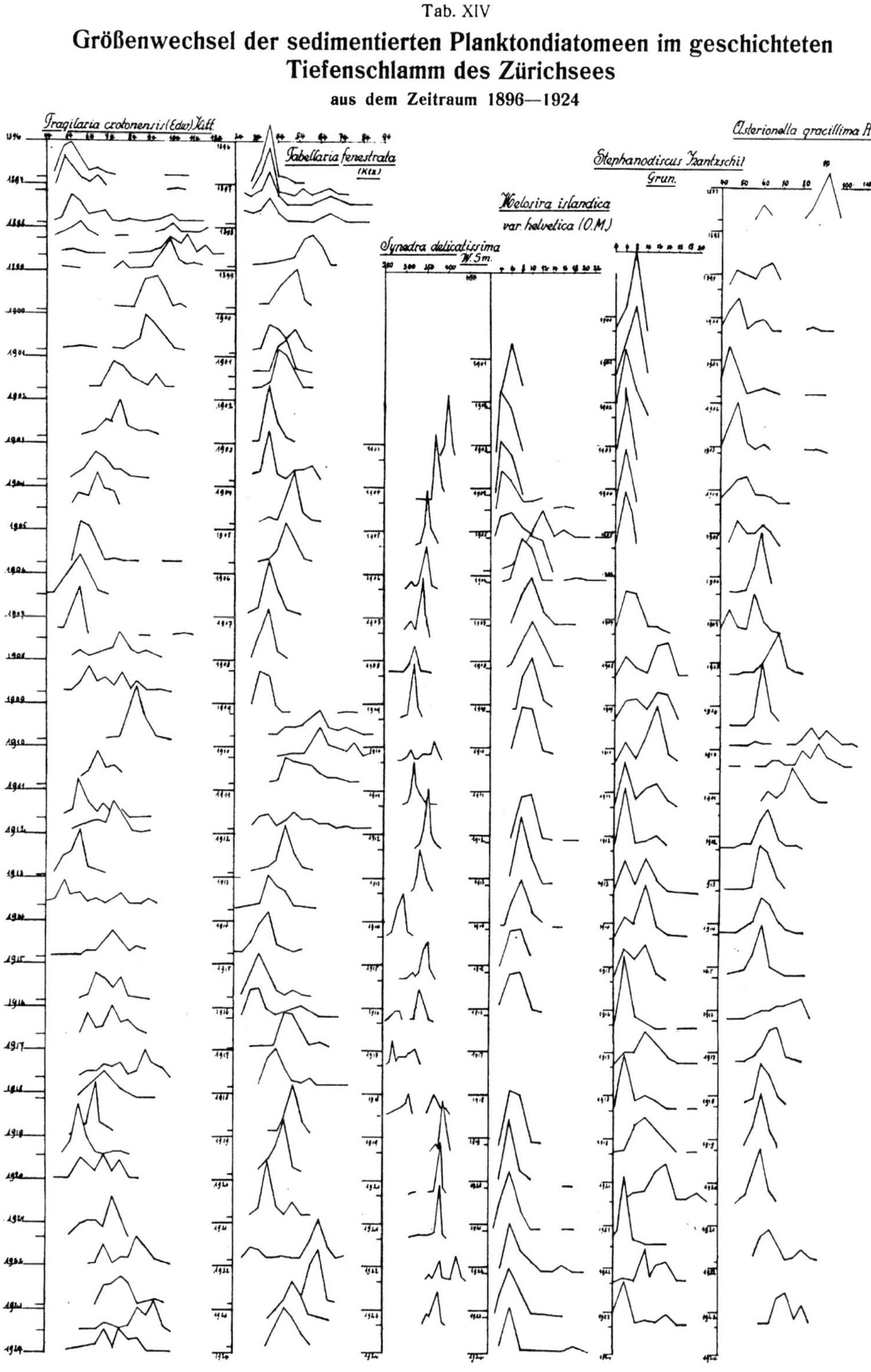

Figure 2 Nipkow's graph of annual, and partly seasonal, size distributions in the populations of major plankton diatoms in the varved sediment of Lake Zurich (Nipkow, 1927). The years from 1896 (top) to 1924 (bottom) are marked on the vertical axes, and short ticks on these indicate from which part of each varve the analyzed sample has been taken. Note that for many taxa, two or three distributions are shown for years within which marked seasonal size distribution changes were observed. The horizontal scales give ranges of apical length for pennate taxa and valve diameter for centric ones as follows: 44–124 μm for *Fragilaria crotonensis* (Edw.) Kitt., 24–94 μm for *Tabellaria fenestrata* (Kütz.), 250–450 μm for *Synedra delicatissima* W. Sm., 4–22 μm for *Melosira islandica* v. *helvetica* (O.M.), 4–20 μm for *Stephanodiscus hantzschii* Grun., and 40–110 μm for *Asterionella gracillima* H. Heib.

Methods of Analyzing Diatom Microstratigraphy

Numerous methods have been developed for studying diatom microstratigraphy, the choice of which depends on the objectives of the particular study. Basically, there are two approaches: (1) dissection of the sediment core into thin horizontal subsamples, in which the constituents of each separated layer can be studied individually, and (2) preparation of vertical sections of the core, which enables the study of the stratigraphic details in a single sample.

Nipkow (1927) was able to dissect varves in his fresh sediment cores into less than 1-mm increments and could estimate even the relative diatom frustule concentrations in these. His samples were fairly compact calcareous sediment, which seems well suited to this type of fresh sediment subsampling. It may not be possible to obtain similar precision with soft organic sediments. Freezing and freeze-drying of the sediment may allow better subsampling precision. Schmidt *et al.* (1995) described a suction sampler for fine-resolution subsampling of varves and their structural components.

An advantage of horizontal slicing is that the material can be properly treated for high-quality diatom slides. Furthermore, sampling of a somewhat extended horizontal area (at least some square millimeters) will help overcome the problem of uneven distribution of diatom frustules in each individual layer. A drawback of this approach, of course, is that it is very laborious, especially when dealing with longer varve sequences.

The tape peeling technique (Simola, 1992) is a simple method for obtaining vertical sediment sections for microscopy. From the cleaned surface of a frozen and slightly freeze-dried sediment core, a thin layer of sediment particles is adhered on strips of adhesive tape. The tape strips can be directly mounted into microscope slides with a suitable liquid mounting medium. Such tape strips are also suitable for direct examination by scanning electron microscopy (Dean *et al.*, 1999). An unmounted tape peel strip is shown in **Figure 3**.

The optical quality of direct tape peel mounts is not always sufficient for high-level diatom identification. Davidson (1988) overcame this by transferring the sediment particles onto a microscope slide, which he then incinerated and mounted with a suitable diatom mountant, thus producing organic-free diatom slides from stratigraphically intact samples.

A number of techniques and embedding materials are available for making permanent thin sections

Figure 3 A tape peel of sediment varves from Lake Laukunlampi, east Finland. The varves are delimited by the dark brown humus-rich winter layers; each varve contains one or more light-colored diatom-rich strata, with the typical spring diatom maximum appearing as an upward fading whitish layer immediately above the dark winter layer. The photograph spans the period 1964–1981; scale in centimeters.

through the replacement of the sediment water with a hardening resin or other suitable embedding medium (Lotter and Lemcke, 1999; Kemp *et al.*, 2001), after which the samples are ground into thin slides, analogous to petrographic thin sections. Being of constant thickness, the thin sections allow quantitative diatom counts. However, diatom analysis in such thin sections may be hampered by the unavoidable breakage of the frustules in the grinding process. This problem is less prominent with the simpler but not fully quantitative tape peeling techniques.

Diatom Assemblages Provide High-Resolution Records of Environmental Changes

Human Impact and Its Relaxation

Since Nipkow's (1927) study, a large number of paleolimnological varve studies have detailed the eutrophication and pollution history of lakes caused by human agency (for a general summary, see Hall and Smol, 1999). In many cases, eutrophication, leading to hypolimnetic anoxia, has actually triggered the onset of varve formation in the sediment.

In some cases, the yearly accumulated numbers of plankton diatom frustules appear to provide a credible record of actual diatom productivity in the lake (Simola *et al.*, 1990; Bradbury *et al.*, 2002). However, because sedimentation conditions vary widely between lakes, it is not possible to directly compare the productivity levels in different lakes by this method. For generally applicable inferences of the lake ecosystem conditions, qualitative characteristics of the diatom assemblages need to be taken into account.

The development of diatom analysis for environmental reconstruction was significantly boosted by acidification research in the 1980s (Battarbee *et al.*, 1986; Charles *et al.*, 1994). The modern approach for diatom-based environment reconstruction typically involves a calibration data set of at least some tens of lakes, from which statistical transfer functions between species assemblages and environmental parameters can be calculated. The transfer functions are then applied for paleoecological inferences. As a result of this development, paleolimnology has matured from descriptive to analytical and predictive science (Birks, 1998).

Bennion *et al.* (1995) were able to verify a diatom-phosphorus transfer function based on a training set of 86 oligo- and mesotrophic European Alpine lakes by applying this transfer function to the microstratigraphically analyzed diatom assemblages of the varved sediment of Lake Mondsee, Austria, and comparing the resulting diatom-inferred total phosphorus (TP) history with the measured annual mean TP concentrations that are available for the lake from 1975 onwards. Overall, the match between the diatom-inferred and measured TP concentrations was very good, with both records tracing the gradual TP decline since the mid-1970s. Especially for the uppermost sediment, the values were in very close agreement. For the lower part of the core, however, the model appeared to underestimate the TP levels, which was attributed to poor diatom assemblage analogs (either a deficiency of the training set or a consequence of diagenetic changes in the sedimentary assemblages). Also, the highest diatom-inferred TP level was recorded in the sediment according to the varve counts somewhat later than the actually observed TP peak year, which could be due to error or uncertainty in the varve chronology.

A cautionary example of TP transfer function application is provided by the long, high-quality varve record in Elk Lake, Minnesota (Bradbury *et al.*, 2002). Within the uppermost century of the analyzed 1,500-year-long sequence, there are clear indications of European settlement and logging activities in the area. These include changes in the diatom species composition (e.g., the advent of thick-walled *Aulacoseira* species implying enhanced turbulence, and somewhat increased overall diatom flux in the sediment), together with increased phytolith concentration as a sign of soil erosion. Contrary to these indications of increasing productivity with increasing land use impact on the lake, a TP reconstruction, based on a Canadian training set, indicates a dramatic decline in the TP level for the past 150 years. Bradbury *et al.* note the strong seasonal variation of TP levels in stratifying eutrophic lakes as a potential source of error in the models. Indeed, diatoms with high TP requirements may bloom in a eutrophic lake during circulation periods when nutrients are abundant, but if the epilimnion is depleted of nutrients during stratification periods, very low TP levels could actually be measured in the epilimnetic water.

The types of uncertainties and error sources demonstrated by the previous examples need to be considered with all environmental inferences based on varve records.

The prospect of testing and improving the reliability of various inference models by sediment varve records is of great practical value for applied water protection work. Active water protection policies (e.g., the European Union Water Framework Directive) will require knowledge of the natural or predisturbance status of impacted water bodies. The baseline conditions need to be known to obtain a realistic goal for the management or restoration measures, and advanced paleolimnology is often the sole provider of this information (Miettinen *et al.*, 2005).

Geological Processes and Catchment Events

In suitable cases, long varve series will allow precise dating and high-resolution diatom studies to assess lake ecosystem dynamics on geological timescales. Long varve series are commonly found in the early postglacial sediments from northern European lakes (Schmidt *et al.*, 1998; Ralska-Jasiewiczowa *et al.*, 1998). Unfortunately, however, these varve records seldom reach to the present day. Typically, they provide floating chronologies for the gradual oligotrophication of the water bodies as a consequence of catchment evolution since the end of the Ice Age through processes of landscape, vegetation, and soil development. Gradual filling in of the basin and declining productivity may have led to weakened stratification and lessened anoxic tendencies, and thus eventual cessation of varve deposition, in the typical development of such lakes.

Coastal meromictic lakes are a special category of basins with a high tendency for varve formation. Especially in areas of isostatic land uplift (e.g., the Baltic coasts in northern Europe and northeastern North America), there are numerous coastal lakes with residual or recurring salt water-induced meromixis. In shallow basins, such meromixis may be a transient postisolation phenomenon, but in deeper ones the condition may persist for millennia. The diatom records in the sediments of such lakes may reveal and date the isolation process with great accuracy, and particularly in Arctic areas, they may serve exceptionally well also as archives of catchment and lake development and climatic events.

This can be demonstrated by the multidisciplinary Taconite Inlet Lakes Project, which studied the limnology and varved sediments of three deep (maximum depth, 51–84 m) coastal meromictic lakes in northern Ellesmere Island, Northwest Territories, Canada (Douglas *et al.*, 1996; Lamoureux and Bradley, 1996). A postisolation varve record of approximately 3,300 years was compiled for one of these lakes (C2) by cross-correlation of several sediment cores taken at different depths along a transect through the oxycline. As a whole, the diatom record indicates rather stable water quality conditions for the lacustrine phase, whereas the meromictic stratification has fluctuated and appears to have weakened during the past centuries. This could be judged by variable preservation of the varve structures in corresponding strata at different depths.

Inferences on Past Climatic Changes

Much research has focused on tracking the climatic signal in sedimentary diatom records. Several studies have indeed shown that diatoms are sensitive indicators of climatic factors (Gasse *et al.*, 1995; Schmidt *et al.*, 1998). However, the indication is often indirect, and it is important to note that in different climatic zones the diatom records may offer quite different types of climate proxies. Ice cover, river discharge, various water quality parameters (e.g., water color, pH, alkalinity, nutrients, conductivity, and salinity), temperature, lake level, and water column stability are climate-related limnological variables that may be inferred from paleolimnological studies (Smol and Cumming, 2000).

Long varved sequences are naturally of great value because they provide good dating and also allow the analysis of year-to-year variation in the records (Lamoureux, 1999). Overall, the main problem is to distinguish the climatic signals from those of catchment processes and especially from human disturbance, which has a long, variable, and often entirely undocumented history on most lake catchments throughout the world (Lotter and Birks, 1997; Schmidt *et al.*, 1998).

The main benefit provided by varved sediments for the development of paleoclimatological diatom inference models is the possibility of testing and calibrating the high-resolution algal record with any existing instrumental climate records for the site. Thus, in the Taconite Inlet Lake C2, comparison of the late 20th-century varves with meteorological records showed that both varve thickness and the proportion of rheophilic (derived from running water) diatoms in the sediment correlate with summer temperature (i.e., meltwater inflow). This allowed Douglas *et al.* (1996) and Lamoureux and Bradley (1996) to make credible long-term inferences of summer temperature fluctuations from the entire varve record of the lake.

In the varved sediment of Elk Lake, Minnesota, limnological monitoring data were used to help interpret the diatom record. Cold spring weather with late ice-out and short spring circulation promotes *Fragilaria crotonensis* as the dominant species of the spring bloom, whereas an early spring and long circulation lead to the dominance of *Stephanodiscus minutulus*, which favors higher nutrient concentrations and tolerates lower light intensity. The relative shifts in the abundances of these two major taxa in a 1500-year sequence appear to have a climatological source, especially as an indicator of climate-related changes in the spring circulation and the development of thermal stratification in the lake (Bradbury *et al.*, 2002). Time series analysis also reveals some regular long-term fluctuations in the diatom record, tentatively assigned by Bradbury *et al.* (2002) to sunspot cycle periodicities.

Avenues for Further Research

Species-Level Studies of Diatoms in Sedimentary Records

The current paleoecological assessments are mainly based on community-level analysis. Inferences are drawn rather robustly of observed correlations between environmental parameters and gross assemblage compositions. No doubt, the ecological inference potential could be further refined by incorporation of species-level ecological information into the models. The specific ecological requirements of diatoms are not fully understood, so basic algological research is still needed, especially concerning planktonic taxa and the factors affecting their population dynamics (i.e., species responses to seasonal limnological events). Actual monitoring of lakes with ongoing varve deposition, combined with follow-up of the sediment record, has great potential for providing this kind of improved ecological understanding.

Detailed studies of single-species dynamics within varved sediments may in turn serve basic algological and ecological research as well. Research questions, partly outlined by Nipkow (1927), include diatom cell cycle dynamics (size reduction–auxosporulation cycles and their timing), within-population morphology variation, and possible genetic studies in sediments with well-preserved diatom cells. Such studies could lead to better understanding of the species concept, especially for common taxa with diffuse and difficult species identification (e.g., *Aulacoseira* and *Stephanodiscus*).

The work of Simola *et al.* (1990) is an example of microstratigraphic diatom analysis conducted with a primary aim to study the population dynamics and species interactions of planktonic taxa. This study of a 418-year-long regularly varved sequence provided data on the seasonality and year-to-year dynamics of the analyzed taxa, but detailed interpretation of the results was hampered by lack of corresponding environmental data (note that the 418-year population histories of some plankton diatoms presented in this work have appeared in a biostatistics textbook as typical examples of 'chaotic' time series).

Revisiting Well-Documented Varves to Gain a Better Understanding of Postdepositional Processes

Taphonomic processes, or the numerous factors and events that tend to alter the representativity of a fossil assemblage in comparison with the original living communities, are a major issue in paleoecological reconstructions. However, this is rarely accounted for in routine analyses, which usually take the analyzed assemblages at their face value.

Postdepositional dissolution is a very significant factor shaping the fossil assemblages or indeed completely destroying diatom frustules in sediment deposits (Flower, 1993; Battarbee *et al.*, 2001). These experimental and empirical studies have demonstrated quite dramatically that the species composition of a fossil assemblage may show very little resemblance to the original one. In some cases, only a few robust forms remain when all the common diatoms have disappeared. In extreme cases, the remaining fossil assemblage may even be dominated by taxa that, due to their initial rarity, would not have been recorded in the analysis of the original sample. As an illustrative empirical example, the common plankton diatoms *Urosolenia* and *Attheya* are rarely included in calibration data sets because their very thin frustules are rarely observed in lake sediments (their cysts may be observed, but this is usually quite rare).

Revisiting well-documented varve deposits would provide very useful information on the potential postdepositional dissolution process, particularly its duration and timing, and how it affects different species in different types of sediments. Perhaps it may be possible to work out species-specific frustule half-life estimates and dissolution rank orders for common diatom taxa in typical conditions, which could be introduced as data correction factors for assemblage-matching analyses. This kind of information could considerably improve the interpretation accuracy and potential in paleoecological inference studies.

See also: **Diatom Introduction**. **Diatom Methods**: Data Interpretation; Quaternary Geological Records. **Diatom Records**: Large Lakes. **Geomagnetic Excursions and Secular Variations**; **Varved Lake Sediments**.

References

Battarbee, R. W., Smol, J. P., and Meriläinen, J. (1986). Diatoms as indicators of pH: A historical review. In *Diatoms and Lake Acidity: The Use of Siliceous Microfossils in the Reconstruction of pH* (J. P. Smol, R. W. Battarbee, R. B. Davis and J. Meriläinen, Eds.), pp. 5–14. Junk, The Hague, The Netherlands.

Battarbee, R., Jones, V., Flower, R., *et al.* (2001). Diatoms. In *Tracking Environmental Change Using Lake Sediments. Vol. 3: Terrestrial, Algal, and Siliceous Indicators* (J. P. Smol, H. J. B. Birks and W. M. Last, Eds.), pp. 155–202. Kluwer, Dordrecht, The Netherlands.

Bennion, H., Wunsam, S., and Schmidt, R. (1995). The validation of diatom-phosphorus transfer functions: An example from Mondsee, Austria. *Freshwater Biology* **34**, 271–283.

Birks, H. J. B. (1998). Numerical tools in paleolimnology—Progress, potentialities, and problems. *Journal of Paleolimnology* **20**, 307–332.

Bradbury, P., Cumming, B., and Laird, K. (2002). A 1500-year record of climatic and environmental change in Elk Lake, Minnesota III: Measures of past primary productivity. *Journal of Paleolimnology* **27**, 321–340.

Card, V. (1997). Varve-counting by the annual pattern of diatoms accumulated in the sediment of Big Watab Lake, Minnesota, AD 1837–1990. *Boreas* **26**, 103–112.

Charles, D., Smol, J. P., and Engstrom, D. (1994). Paleolimnological approaches to biological monitoring. In *Biological Monitoring of Aquatic Systems* (S. L. Loeb and A. Spacie, Eds.), pp. 233–293. CRC Press, Boca Raton, FL.

Davidson, G. A. (1988). A modified tape-peel technique for preparing permanent qualitative microfossil slides. *Journal of Paleolimnology* **1**, 229–234.

Dean, J. M., Kemp, A. E. S., Bull, D., *et al.* (1999). Taking varves to bits: Scanning electron microscopy in the study of laminated sediments and varves. *Journal of Paleolimnology* **22**, 121–136.

Douglas, M., Ludlam, S., and Feeney, S. (1996). Changes in diatom assemblages in Lake C2 (Ellesmere Island, Arctic Canada): Response to basin isolation from the sea and to other environmental changes. *Journal of Paleolimnology* **16**, 217–226.

Flower, R. (1993). Diatom preservation: Experiments and observations on dissolution and breakage in modern and fossil material. *Hydrobiologia* **269/270**, 473–484.

Flower, R. J., Juggins, S., and Battarbee, R. W. (1997). Matching diatom assemblages in lake sediment cores and modern surface sediment samples: The implications for lake conservation and restoration with special reference to acidified systems. *Hydrobiologia* **344**, 27–40.

Gasse, F., Juggins, S., and Ben Khelifa, L. (1995). Diatom-based transfer functions for inferring past hydrochemical characteristics of African lakes. *Paleogeography, Paleoclimatology, Paleoecology* **117**, 31–54.

Glew, J., Smol, J., and Last, W. (2001). Sediment core collection and extrusion. In *Tracking Environmental Change Using Lake Sediments. Vol. 1: Basin Analysis, Coring and Chronological Techniques* (W. M. Last and J. P. Smol, Eds.), pp. 73–105. Kluwer, Dordrecht, The Netherlands.

Hall, R. I., and Smol, J. P. (1999). Diatoms as indicators of lake eutrophication. In *The Diatoms: Applications for the Environment and Earth Sciences* (E. F. Stoermer and J. P. Smol, Eds.), pp. 128–168. Cambridge University Press, Cambridge, UK.

Kemp, A., Dean, J., Pearce, R., and Pike, J. (2001). Recognition and analysis of bedding and sediment fabric features. In *Tracking Environmental Change Using Lake Sediments. Vol. 2: Physical and Geochemical Methods* (W. M. Last and J. P. Smol, Eds.), pp. 7–22. Kluwer, Dordrecht, The Netherlands.

Lamoureux, S. (1999). Spatial and interannual variation in sedimentation patterns recorded in nonglacial varved sediments from the Canadian High Arctic. *Journal of Paleolimnology* **21**, 73–84.

Lamoureux, S. (2001). Varve chronology techniques. In *Tracking Environmental Change Using Lake Sediments. Vol. 1: Basin Analysis, Coring and Chronological Techniques* (W. M. Last and J. P. Smol, Eds.), pp. 247–260. Kluwer, Dordrecht, The Netherlands.

Lamoureux, S., and Bradley, R. (1996). A late Holocene varved sediment record of environmental change from northern Ellesmere Island, Canada. *Journal of Paleolimnology* **16**, 239–255.

Lotter, A., and Birks, H. J. B. (1997). The separation of the influence of nutrients and climate on the varve time-series of Baldeggersee, Switzerland. *Aquatic Sciences* **59**, 362–375.

Lotter, A., and Lemcke, G. (1999). Methods for preparing and counting biochemical varves. *Boreas* **28**, 243–252.

Miettinen, J. O., Kukkonen, M., and Simola, H. (2005). Hindcasting baseline values for water colour and total phosphorus concentration in lakes using sedimentary diatoms—Implications for lake typology in Finland. *Boreal Environment Research* **10**, 31–43.

Nipkow, F. (1927). Über das Verhalten der Skelette planktischer Kieselalgen im geschichteten Tiefenschlamm des Zürich- und Baldeggersees. *Zeitschrift für Hydrologie* **4**, 71–120.

Ralska-Jasiewiczowa, M., Goslar, T., Madeyska, T. and Starkel, L. (Eds.) (1998). *Lake Gościąż, Central Poland, a Monographic Study, Part 1*. W. Szafer Institute of Botany, Polish Academy of Science, Cracow, Poland.

Schmidt, R., Höllerer, H., and Wallner, G. (1995). A vacuum sampler for subsampling freeze-dried laminated sediments with the application to *in situ* frozen varves of Mondsee, Austria. *Journal of Paleolimnology* **14**, 93–96.

Schmidt, R., Wunsam, S., Brosch, U., *et al.* (1998). Late- and post-glacial history of meromictic Längsee (Austria), in respect to climate change and anthropogenic impact. *Aquatic Sciences* **60**, 56–88.

Simola, H. (1992). Structural elements in varved lake sediments. In *INQUA Workshop on Laminated Sediments, Lammi, Finland, June 1990* (M. Saarnisto and A. Kahra, Eds.) Geological Survey of Finland, Espoo, Finland, Special Paper No. 14 pp. 5–9.

Simola, H., Hanski, I., and Liukkonen, M. (1990). Stratigraphy, species richness and seasonal dynamics of plankton diatoms during 418 years in Lake Lovojärvi, South Finland. *Annales Botanici Fennici* **27**, 241–259.

Smol, J. P. (2004). *Pollution of Lakes and Rivers: A Paleoenvironmental Perspective*. Arnold/Oxford University Press, London/Oxford.

Smol, J. P., and Cumming, B. F. (2000). Tracking long-term changes in climate using algal indicators in lake sediments. *Journal of Phycology* **36**, 986–1011.

Large Lakes

A W Mackay, University College London, London, UK

© 2007 Elsevier B.V. All rights reserved.

Introduction

Large lakes are generally defined as those with a surface area of over 500 km^2 (Tilzer and Serruya, 1990). They are irregularly spread out across the globe and contain over 90% of surface freshwater-water resources (Herdendorf, 1982). However, perhaps a more reliable metric of a large lake is the volume of water it contains, as significant changes to surface area can occur if the lake is shallow and hydrological inputs alter. Perhaps the best example of this is the Aral Sea, whose surface area has declined by over 60% in recent decades, and the lake now exists as two separate water bodies. The largest 20 lakes (by volume of water) are summarized in **Table 1** and their geographical position is shown in **Figure 1**. Half of these lakes are tectonic in origin while the other half are formed from glacial activity. These lakes show some specific geographical distributions, which have important implications for Quaternary reconstructions using diatoms and other proxies. For example, all but one of the large, glacial, ice-scour lakes are found in North America north of 40° N latitude between young sedimentary strata and more northerly Precambrian shield rocks (the remaining glacial scour lake is Lake Ladoga in western Russia). Moreover, only one large lake (Hamoun i Helmand) is found between 20° N and 40° N, mainly because of prevailing aridity preventing their development between these latitudes.

By definition, large glacial lakes only contain sedimentary diatom deposits formed since deglaciation, and records therefore most commonly span less than 20 kyr BP. Longer records are therefore best extracted from either extant lakes with deep basins in active rift zones, or paleolakes which have since in-filled, or dried up, but the presence of which, once spanned the Quaternary. Increasingly, the long sedimentary records of tectonic lakes are being exploited. Quaternary diatom studies are most advanced from Lake Baikal, due to the Baikal Drilling Programme (BDP), which began extracting long records in the early 1990s (Williams *et al.*, 2001). However, under the auspices of the Intercontinental Drilling Program (http://www.icdp-online.de), other large lakes in tectonic basins have

Table 1

Volume rank	*Name*	*Volume (km^3)*	*Latitude*	*Longitude*	*Origin*	*Country*
1	Caspian	78,200	42°00′N	50°00′E	Tectonic	Azerbaijan Iran Kazakhstan Russian Federation Turkmenistan
2	Baikal	23,600	54°00′N	109°00′E	Tectonic	Russia
3	Tanganyika	19,000	6°00′S	29°30′E	Tectonic	Burundi Congo (Democratic Republic) Tanzania Zambia
4	Superior	12,100	47°30′N	88°00′W	Glacial	Canada USA
5	Malawi	7,775	12°00′S	34°30′E	Tectonic	Malawi Mozambique Tanzania
6	Michigan	4,920	44°00′N	87°00′W	Glacial	Canada USA
7	Huron	3,540	45°00′N	81°15′W	Glacial	Canada USA
8	Victoria	2,760	01°00′S	33°00′E	Tectonic	Kenya Tanzania Uganda
9	Great Bear	2,292	66°00′N	121°00′W	Glacial	Canada
10	Great Slave	2,088	62°40′N	114°00′W	Glacial	Canada
11	Issyk-Kul	1,738	42°30′N	77°15′E	Tectonic	Kyrgyzstan
12	Ontario	1,640	43°30′N	77°45′W	Glacial	Canada USA
13	Titicaca	932	15°45′S	69°30′W	Tectonic	Bolivia Peru
14	Ladoga	908	61°00′N	31°30′E	Glacial	Russia
15	Hamoun i Helmand	510	31°00′N	61°15′E	Tectonic	Afghanistan Iran
16	Erie	484	42°15′N	81°15′E	Glacial	Canada USA
17	Hovsgol	380	51°00′N	100°30′E	Tectonic	Mongolia
18	Winnipeg	371	52°30′N	97°45′W	Glacial	Canada
19	Kivu	333	2°13′S	29°10′W	Tectonic	Congo (Democratic Republic) Rwanda
20	Nipigon	320	49°50′N	88°31′W	Glacial	Canada

Adapted from Herdendorf (1982) but updated from http://www.worldlakes.org/

Figure 1 Map showing the distribution of the world's 20 most voluminous lakes. The numbers given correspond to the ranking of the lakes, as indicated in **Table 1**.

recently been cored, for example, Lake Titcaca in 2001 and Lake Malawi in 2005. The latter site is notable as sedimentary records were retrieved that extend back over 1.5 Myr, which will allow comparisons with other long Quaternary sequences.

Diatoms as Paleoenvironmental Proxies

Interpretation of the climate record held by diatom proxies in the sediments of large lakes is enhanced in cases where there is knowledge of population growth and succession in the water column, and an understanding of how diatoms are transported to the sediment surface, and finally incorporated into the sedimentary record (Mackay *et al.*, 2003a). In this respect, the quality of diatom preservation can be assessed and key, controlling climatic variables evaluated. Diatom-inferred climatic interpretations therefore require knowledge of taphonomic processes, most notably dissolution (Ryves *et al.*, 2001; *see* Diatom Introduction, Quaternary Geological Records).

The types of diatom communities that can be found in large lakes depend on lake morphology (i.e., is the lake deep or shallow?), age of the lake, and prevailing climate. Large, shallow lakes have a high ratio of benthic to planktonic taxa, because of an extensive littoral region where benthic diatoms can dominate, for example, Lake Victoria. In deep lakes, however, the littoral region is small compared to the pelagial component and open water planktonic taxa go on to predominate in the sedimentary diatom assemblages. In this respect, phytoplankton productivity in large deep lakes is similar to that in the marine system, where the relative contribution of the littoral region to productivity is low (Tilzer, 1990). Planktonic diatoms in large, deep lakes may also be adapted to other hydrological processes that are different from large, shallow lakes such as increased time spent out of the photic zone due to deep-water mixing (e.g., Lake Baikal), or the importance of internal cycling over external inflows on nutrient budgets over short timescales (e.g., Lake Victoria) (Tilzer, 1990). Large tectonic lakes are undoubtedly some of the world's oldest water bodies, and this great age tends to result in planktonic diatom assemblages with low diversity, especially in comparison to large, glacially formed lakes, which can contain species assemblages representing many planktonic genera (Stoermer and Edlund, 1999). This is partly due to processes such as competitive exclusion operating over very long timescales, which can lead to increased levels of endemism within that aquatic ecosystem (e.g., Lake Baikal; Khursevich *et al.* (2000)).

Case Studies

The range of diatom-based paleolimnological studies is vast, although these mostly cover Lateglacial–Holocene periods; there are rather few diatom records from large lakes covering the whole of the Quaternary in comparison to marine records (*see* Marine Laminated Sequences). Three case studies are considered, because they represent ongoing challenges in the field of paleoclimatology/paleolimnology over the Quaternary. Case study 1 is from Lake Baikal in Russia, situated within an active tectonic region of the Eurasian plate (51°28′–55°47′ N and 103°43′–109°58′ E; **Fig. 1**). Quaternary diatom records and associated trends in biogenic silica from this lake have greatly increased our understanding of climate variability in continental regions far from oceanic influences, hence the comprehensive treatment of this lake here. Case study 2 is from Lake Titicaca, which lies between Bolivia and Peru in South America (15°45′ S 69°30′ W; **Fig. 1**). This lake is found to the north of the Altiplano of Peru, the world's second highest plateau (after Tibet), and as such Lake Titcaca is the highest large lake in the world. Paleoclimate records for this site are available for much of the Lateglacial–Holocene period. The third case study uses diatom records from one of the world's largest tropical lakes, Lake Victoria in Africa (01°00′ S 33°00′ E; **Fig. 1**), bordered by Kenya, Tanzania, and Uganda. Lake Victoria is relatively shallow, and observations in the 1960s revealed that the lake had dried out completely toward the end of the last glacial (Kendall, 1969). It was this observation that challenged conventional orthodoxy at the time, which held that tropical regions experienced wetter climates during the Lateglacial. Although there has been much debate since then on the extent and timing of this desiccation event, recent work indicates that it occurred between ca. 15.9 and 14.2 cal kyr BP (Stager *et al.*, 2002).

Lake Baikal

Lake Baikal, situated in southeast Siberia (**Fig. 1**), has long been recognized as one of the world's most unusual freshwater ecosystems, especially in terms of its age (in excess of 20 Myr), depth (ca. 1,642 m) and volume (23,600 km^3), its uninterrupted sedimentary record, and the high degree of endemism that exists within the lake. Another important feature that sets Lake Baikal apart from other large, deep lakes is that its entire water column is saturated throughout with oxygen due to regular renewal of the deep waters every spring and autumn. This results in the oxidation of even the deepest bottom waters, allowing an extensive, and almost wholly, endemic deep-water fauna to survive.

Diatoms are important primary producers and crucial to the ecosystem functioning of Lake Baikal – phytoplankton biomass in spring (which consists mainly of planktonic diatoms) usually determines productivity of the whole year in the pelagic lake (Popovskaya, 2000). The most common planktonic diatoms in Lake Baikal are a mixture of endemic taxa (e.g., *Aulacoseira baicalensis*, *Cyclotella minuta*, and *Stephanodiscus meyerii*) and cosmopolitan taxa (e.g., *Synedra acus* and its varieties and *Nitzschia acicularis*). The distribution of Lake Baikal diatoms exhibits significant spatial variability across the lake, dependent on a number of factors including climate (e.g., ice and snow cover) and hydrological factors such as water depth and fluvial input (Mackay *et al.*, 2003b). As much as 40% of the lake's diatoms may be endemic, although this figure is being revised upward as more comprehensive studies in deeper littoral zones (>20 m) are carried out (Flower, 2005) and centers of hyper-endemicity recognized around the lake. The diatom assemblages in Lake Baikal contain several examples of species flocks, and relict distributions of previously more widespread taxa.

The paleoclimatic archive of Lake Baikal sediments is of global importance, given its position in one of the most continental regions of the world where continuous, high-resolution records spanning the Quaternary are otherwise sparse. Lake Baikal diatoms and associated biogenic silica (BioSi) have proved to be extremely useful paleoclimate proxies, in part because of the intense seasonality of the region, but also the size of the lake, which covers 4 degrees of latitude, across which a distinct climatic gradient exists. Given the depth and volume of Lake Baikal, it is important to assess factors that may affect diatom integrity before their final incorporation in the sedimentary record.

As in the marine environment, the entire water column of Lake Baikal is undersaturated with respect to silica (Falkner *et al.*, 1991), and therefore diatom valve dissolution will always be an important process through glacial and interglacial cycles. In a study comparing recent diatom fluxes through the water column with fluxes being incorporated into the sediments, Ryves *et al.* (2003) demonstrated that in the south basin of Lake Baikal, only ca. 1% of valves are preserved in the sedimentary record. Once incorporated into the sediments, dissolution of diatom silica continues into the pore waters, until saturation of pore waters by Si is reached, thereby buffering against further dissolution. During glacial periods, few diatoms preserve in the sedimentary record, and this is most likely due to more intense relative dissolution processes operating on fewer diatom cells growing during colder climates (Carter and Colman, 1994; Swann and Mackay, 2006).

Although the impact of dissolution on Lake Baikal diatom records has been recognized since some of the earliest studies, only recently have quantitative impacts been determined (Battarbee *et al.*, 2005; Mackay *et al.*, 2005; Ryves *et al.*, 2003). Dissolution is problematic from a paleoenvironmental perspective because partially dissolved valves are often difficult to identify – see **Figure 2**. Furthermore, assemblages may be biased toward more robust, heavily silicified species influencing paleoclimatological reconstructions, or more seriously, taxa may be removed from the record altogether.

Outside of the many mainly taxonomic papers, one of the first studies to relate changes in Lake Baikal diatom composition to changes in orbital forcing was by Grachev *et al.* (1998) who worked on BDP cores extracted from the Academician Ridge. They highlighted that relatively few planktonic taxa tended to dominate the sedimentary assemblages during interglacials. Moreover, they also noted the successionary nature of the changing flora, especially during the last ca. 1 Ma, as diatom abundances suggested progressively longer, more severe glaciations, with different species perhaps dominating the record at different stages for less than 1 kyr. Notably, although many of these taxa are now extinct, several included those commonly found in the lake today, such as *C. minuta*, *A. baicalensis*, and *Aulacoseira skvortzowii*. However, the best and most detailed study on diatom species responses to orbital forcing indicates that over the last 0.8 Ma (since the onset of the mid-Pleistocene revolution), 31 local diatom assemblage zones can be characterized, corresponding to MIS 1–19 (Khursevich *et al.*, 2001) (**Fig. 3**). This study is important because Khursevich *et al.* make the link between diatom evolution and insolation changes, thereby adding to the debate on evolutionary pressures on relatively short, and often abrupt, timescales. More specifically, species responses varied depending on whether insolation amplitude was small (e.g., as occurred between MIS 15a–11, corresponding to the almost continuous presence of *Stephanodiscus flabellatus* var. *distinctus* Khursevich and its varieties) or large (e.g., as occurred between MIS 19–15e, during which time each precessional cycle is dominated by a different species that subsequently became extinct). Furthermore, during the last 0.8 Ma, there are also apparent differential speciation rates between the genera, with 16 new species of *Stephanodiscus* being recorded in comparison to only four for *Cyclotella* (Khursevich *et al.*, 2000, 2001). Importantly, during this time, ice volume increased during each subsequent glaciation, indicative of prolonged glacial periods. Such diatom responses to climate during the Pleistocene have not been witnessed in any other freshwater or marine environment (*see* Freshwater Laminated Sequences; Khursevich *et al.*, 2001). Detailed records also exist for changes in diatom assemblage composition prior to the last 0.8 Ma, with the onset of Northern Hemisphere Glaciation influencing extinctions in at least two genera in the lake toward the beginning of the Quaternary. Following this time, endemic *Cyclotella* species developed and dominated the diatom flora, including *C. praetemperei* till ca. 1.7 Ma, followed by *C. comtaeformica* up to ca. 1.25 Ma (Khursevich *et al.*, 2000).

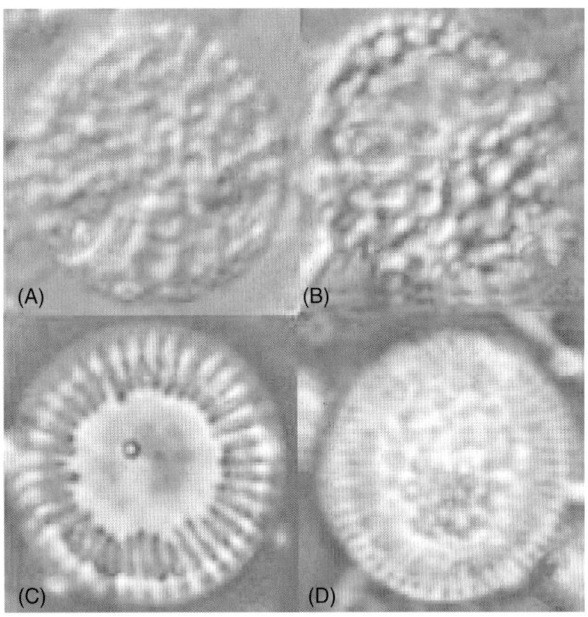

Figure 2 Dissolved and near-pristine images of two endemic, planktonic diatoms found in Lake Baikal sediments. Images (A) and (B) are dissolved valves of *C. gracilis* and *C. minuta*, respectively, found in a marine isotope stage (MIS) 3 sequence extracted from the north basin of Lake Baikal. Images (C) and (D) are of pristine valves of the same species, this time found in the MIS 5e sequence of the same core. Photographs courtesy of George Swann and Patrick Rioual.

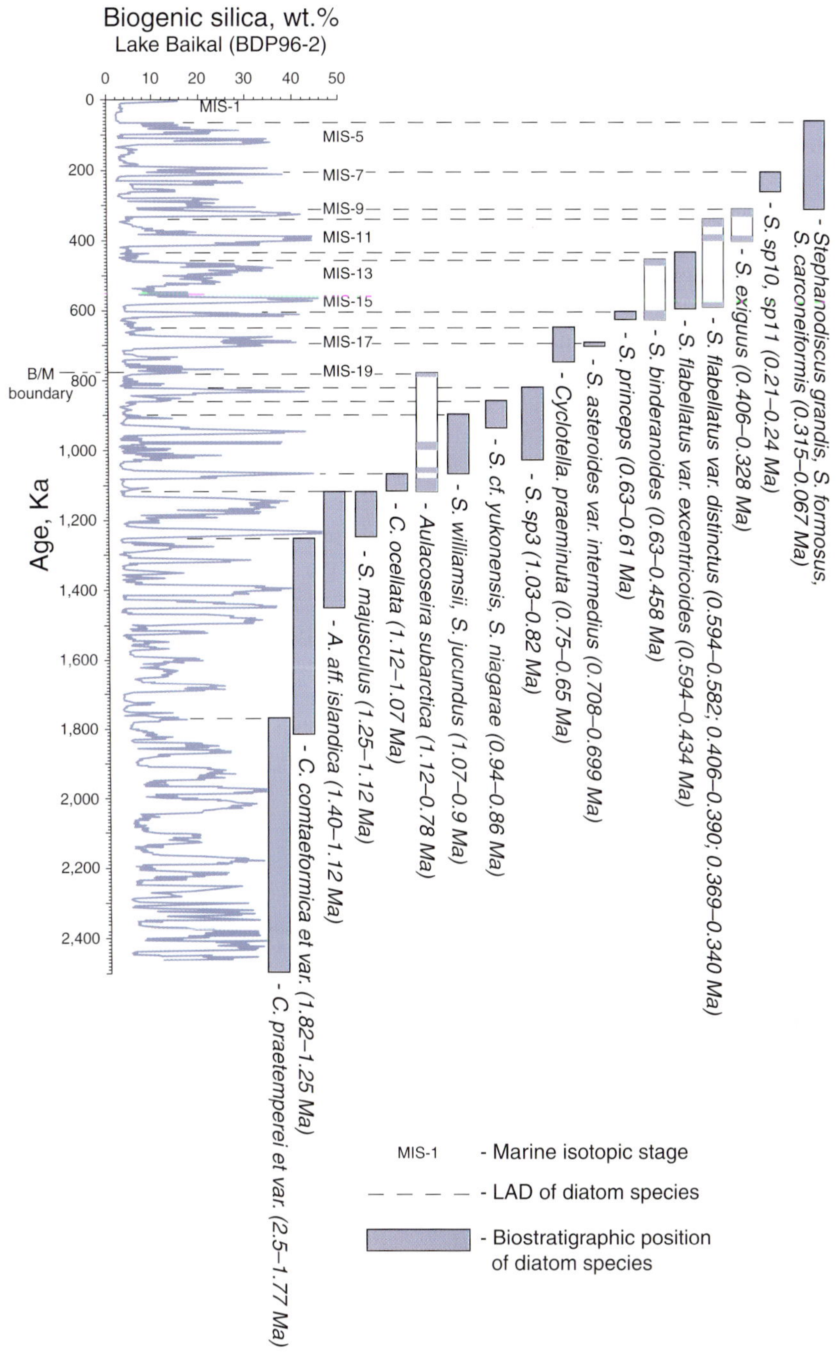

Figure 3 Biostratigraphic position of major diatom species plotted against a biogenic silica record over the entire Quaternary in Lake Baikal. Redrawn from Khursevich, G.K., Karabanov, E.B., Williams, D.F., Kuzmin, M.I., and Prokopenko, A.A (2000) Evolution of freshwater centric diatoms within the Baikal rift zone during the late Cenozoic. In: Minoura K (ed.) *Lake Baikal: A Mirror in Time and Space for Understanding Global Change Processes*, pp. 146–154. Amsterdam: Elsevier.

Lake Titicaca

Lake Titicaca was formed by tectonic activity approximately 3 Ma. This large, monomictic lake is an important site for Quaternary studies because (i) it is the only large, deep, freshwater lake in South America; (ii) the lake itself is sensitive to changes in precipitation (Tapia *et al.*, 2003); and (iii) its sedimentary archive extends back through to at least the last glaciation (>30 ka), and hence offers the possibility of high-resolution paleoclimate reconstructions in this understudied region of the globe (Baker *et al.*, 2001).

Detailed diatom analyses of late Quaternary sediments from Lake Titicaca have contributed significantly to the debate on topical precipitation trends since the last glaciation, especially within the Amazon Basin. In general terms, diatom-based reconstructions have demonstrated that during the Lateglacial period, the climate over the Altiplano was wet, whereas a significant shift occurred to a more prevalent drier climate during the early to middle Holocene (Baker *et al.*, 2001).

Although these scenarios are based on qualitative interpretations of diatom communities present in the lake, they are robust because they are confirmed by other proxies found in the sediment archive (e.g., Cross *et al.* (2000)). Initially, Baker *et al.* (2001) grouped diatoms according to their preferred habitat of living in the planktonic or littoral regions of the lake. High proportions of sedimentary benthic or littoral taxa were therefore attributed to prevailing arid climates resulting in lower lake levels. Conversely, high proportions of planktonic taxa were indicative of higher lake levels (increased lake volume) due to prevailing wetter climates. On this basis, changing diatom assemblages highlight that Lake Titicaca has been sensitive to changes in the North Atlantic thermohaline circulation during both the Younger Dryas and even Bond cycles through the Holocene (Baker *et al.*, 2001).

Tapia *et al.* (2003) considered these diatom responses in more detail. Periods of deep water in the lake, indicative of wetter climate, were characterized by planktonic diatoms, especially *Cyclotella stelligera* complex between 30–24 cal kyr BP and *Cyclotella andina* throughout the Lateglacial period up to ca. 8 cal kyr BP. Two further diatom groups were recognized in the sedimentary profiles: saline indifferent taxa, especially *Cyclotella meneghiniana*, which has a very wide tolerance to varying lake salinities and saline taxa including *Chaetoceros muelleri*, which is indicative of low lake levels and corresponding paleoclimate aridity.

Using these groups, a paleoclimatic synthesis was developed (Tapia *et al.*, 2003) (**Fig. 4**). Between 26–12.5 cal kyr BP, freshwater planktonic taxa (mainly *C. andina* and *C. stelligera*) dominated, suggesting lake levels were high and prevailing climate was therefore wetter, especially during the period coincident with the Younger Dryas in the Northern Hemisphere. High lake levels were interrupted by a period of aridity causing a decline in lake water volume during the transition from the Lateglacial into the early Holocene (as characterized by an increase in benthic taxa). However, this arid period was short-lived, as planktonic diatoms again dominated the lake between ca. 10–8 cal kyr BP, indicating that high lake-level stands can be compared to the wet Coipasa event in the central Altiplano (Sylvestre *et al.*, 1999). The shift to prevailing drier climate during the early to mid-Holocene (as suggested by Baker *et al.* (2001) above) is characterized by marked increases in benthic taxa and *C. meneghiniana*, suggesting that a decline in lake level was accompanied by small increases in salinity. This lake-level decline culminated in the appearance of saline taxa for the first time in Lake Titicaca (*C. muelleri*) from 6–3.5 cal kyr BP. However, low lake-level stands were temporary, and from ca. 4 cal kyr BP freshwater planktonic taxa began to grow and then dominate the flora again.

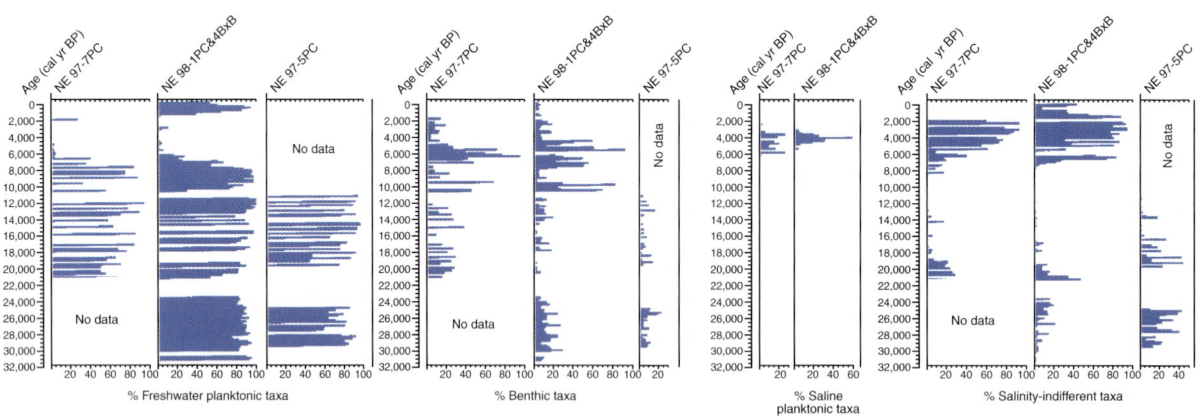

Figure 4 Changes in diatom abundance for the past 31.5 cal kyr BP in Lake Titicaca. Diatom groups represent the four ecological groups highlighted in the text. Gaps in the stratigraphy are zones barren of diatom valves or where diatom concentrations were too low to generate an accurate diatom count. Note that core NE97-7PC does not extend beyond 21 cal kyr BP, while core NE97-5PC does not extend through the Holocene period. Reproduced from Tapia, P.M., Fritz, S.C., Baker, P.A., Seltzer, G.O., and Dunbar, R.B (2003) A late Quaternary diatom record of tropical climatic history from Lake Titicaca (Peru and Bolivia). *Palaeogeography, Palaeoclimatology, Palaeoecology* 194: 139–164.

Lake Victoria

Lake Victoria was also formed by tectonic activity. Although Lake Victoria has an outlet into the Nile River, this accounts for only ca. 10% of water loss from the lake each year; the remaining ca. 90% is lost through evaporative processes. Lake Victoria is relatively shallow, with a mean water depth of ca. 40 m, and consequently small fluctuations in precipitation have marked effects on the lake's hydrology. Lake Victoria is monomictic and overturn occurs during the dry months of June–September, driven by the trade winds. This overturn period determines nutrient availability in the lake and is therefore crucial in controlling algal productivity, allowing diatom populations (including *Aulacoseria* spp. and *Stephanodiscus* spp.) to thrive in the lake. In recent years, however, the period of stratification has been increasing due to eutrophication (Verschuren *et al.*, 2002) and algal production is increasingly determined by other algal groups, including diatoms indicative of more stable water conditions (e.g., *Nitzschia* spp) and cyanobacteria (Hecky, 1993).

Throughout the Lateglacial–Holocene period, Lake Victoria has been shown to be a sensitive recorder of climate variability because its hydrological regime is mainly controlled by precipitation/evaporation. Paleoenvironmental interpretations are based on the knowledge of the ecologies of different diatom taxa. For example, *Aulacoseira* species (e.g., *A. granulata*) are generally heavily silicified, and they therefore require considerable turbulence in the water column to keep their valves suspended in the photic zone. *Nitzschia* species (e.g., *N. acicularis*) on the other hand are often more finely silicified, and have a large surface-area-to-volume ratio, which allows them to 'float' in more stable water conditions. Finally, benthic taxa including *Fragilaria* species are again indicative of shallow water conditions, and when their numbers increase in lake-sediment archives, this is often indicative of declining lake levels. Using these characteristics, diatom-inferred arid phases in this region of tropical Africa are coincident with weakening of Afro-Asian monsoon systems, linked to changes in North Atlantic thermohaline circulation associated with, for example, Heinrich Events and meltwater pulses.

One of the best high-resolution records of Late Glacial diatom-inferred climate changes clearly demonstrates that water levels in Lake Victoria were much lower during the Younger Dryas than at the start of the Holocene (**Fig. 5**) (Stager *et al.*, 2002). Relative lake level is inferred from correspondence analysis axis-2 diatom sample scores from a core taken in the Damba Channel in the north of Lake Victoria. Stager *et al.* (1997) suggests that these benthic assemblages (consisting of taxa such as *Fragilaria* spp.) are indicative of lake depth. Periodicities in the diatom record from Lake Victoria are further interpreted as being indicative of Dansgaard–Oeschger cycles operating on precipitation trends in tropical Africa (Stager *et al.*, 1997).

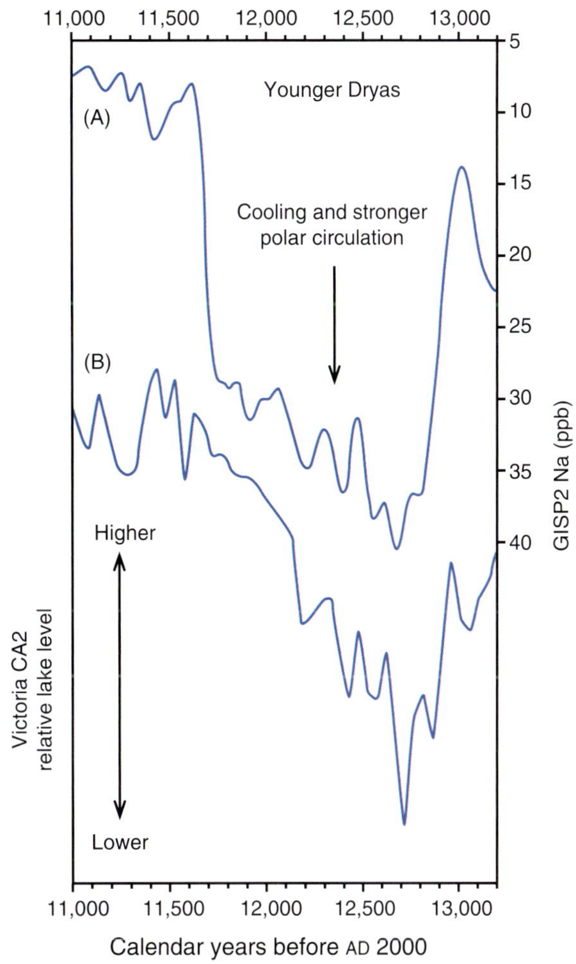

Figure 5 Detail of the Younger Dryas cooling episode registered in the (A), GISP2 Na (ppb) record and (B), Lake Victoria's relative lake level, inferred from correspondence analysis, axis-2 diatom assemblage sample scores. Reproduced from Stager, J.C., Cumming, B., and Meeker, L (1997) A high-resolution 11,400-Yr diatom record from Lake Victoria, East Africa. *Quaternary Research* 47: 81–89.

Further research exploiting Lake Victoria diatoms has reconstructed past changes at very high resolution (1–6 yr) for the last 1 kyr. Diatoms were once again used to infer past changes in lake level, but a diatom-inferred conductivity model was also developed and applied to a 2-m core with a robust chronology (Stager *et al.*, 2005) (**Fig. 6**). Increases in diatom-inferred lake level changes over the last millennium were coincident with declines in solar activity, especially during the period commonly referred to as the Little Ice Age. Although solar fluctuations on these timescales are

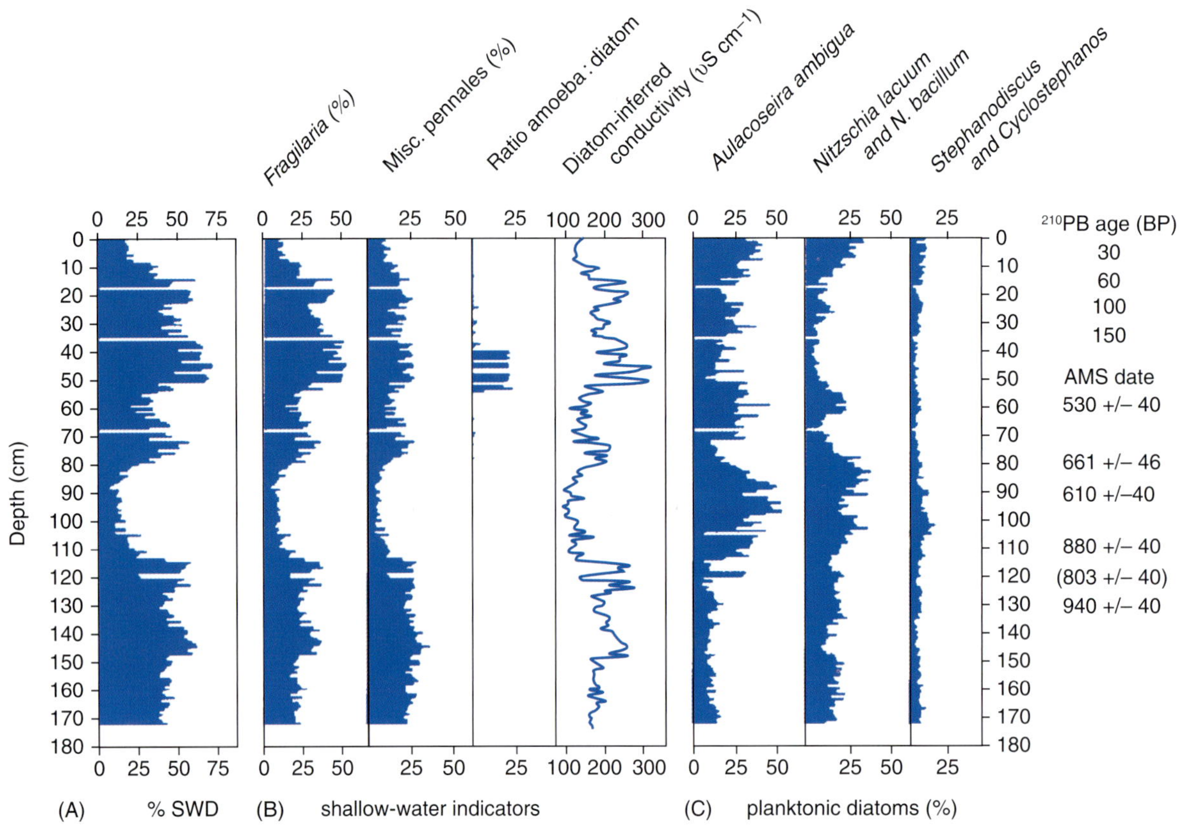

Figure 6 Major diatom taxa in core P2K-1 from Lake Victoria (Stager *et al.*, 2005). (A) Percentages of shallow-water diatoms (SWDs) in fossil assemblages vs. depth; (B) shallow-water indicator taxa and diatom-inferred conductivity; (C) planktonic diatoms indicating greater lake depths and reduced conductivity. To the right are uncorrected AMS dates and selected ^{210}Pb chronology. Reproduced from Stager, J.C., Ryves, D., Cumming, B.F, Meeker, L.D., and Beer, J (2005) Solar variability and the levels of Lake Victoria, East Africa, during the last millennium. *Journal of Paleolimnology* 33: 243–251.

low in amplitude, the authors suggest that when insolation is reduced, cloud cover is increased, resulting in a decrease in evaporation from Lake Victoria. However, in the last 200 yr, higher lake levels are positively associated with increased sunspot numbers, and so interactions are complex and undoubtedly confounded by changes in thermohaline circulation in both the Atlantic and Pacific oceans.

Summary

The case studies highlighted here indicate that diatoms make powerful proxies of past environmental change over the Quaternary period. For large lakes, by far, the majority of the studies are focused on Lateglacial–Holocene changes. This is in part because half of the world's 20 largest lakes are glacial in origin and all but one of these lie in North America. Paleoenvironmental interpretations based on diatoms can be made using both qualitative aspects of diatom autecologies (e.g., habitat preference, morphology) and quantitative inferences, including the use of transfer function methodologies (*see* Data Interpretation). In the field of paleoclimatology, increasingly sophisticated questions are being asked with respect to the magnitude of climate variability over different timescales and across different regions of the world. High-resolution diatom analyses from large lakes are now considered one of the most robust and useful tools in the field of Quaternary research.

See also: **Diatom Introduction**. **Diatom Methods**: Salinity Reconstructions from Continental Lakes. **Diatom Records**: Freshwater Laminated Sequences; Marine Laminated Sequences.

References

Baker, P., Seltzer, G., Fritz, S., *et al.* (2001). The history of South American tropical precipitation for the past 25,000 years. *Science* **291**, 640–643.

Battarbee, R. W., Mackay, A. W., Jewson, D., Ryves, D. B., and Sturm, M. (2005). Differential dissolution of Lake Baikal diatoms: Correction factors and implications for palaeoclimatic reconstruction. *Global and Planetary Change* **46**, 75–86.

Carter, S. J., and Colman, S. M. (1994). Biogenic silica in Lake Baikal sediments: Results from 1990–1992 American cores. *Journal of Great Lakes Research* **20**, 751–760.

Cross, S. L., Baker, P. A., Seltzer, G. O., Fritz, S. C., and Dunbar, R. B. (2000). A new estimate of the Holocene lowstand level of Lake Titicaca, central Andes, and implications for tropical palaeohydrology. *The Holocene* **10**, 21–32.

Falkner, K. K., Measures, C. I., Herbelin, S. E., and Edmond, J. M. (1991). The major and minor element geochemistry of Lake Baikal. *Limnology and Oceanography* **36**, 413–423.

Flower, R. J. (2005). A review of diversification trends in diatom research with special reference to taxonomy and environmental applications using examples from Lake Baikal and elsewhere. *Proceedings of the Californian Academy of Sciences* **56**, 107–128.

Grachev, M. A., Vorobyova, S. S., Likhoshway, Y. V., *et al.* (1998). A high-resolution diatom record of the palaeoclimates of east Siberia for the last 2.5 My from Lake Baikal. *Quaternary Science Reviews* **17**, 1101–1106.

Hecky, R. E. (1993). The eutrophication of Lake Victoria. *Verhandlungen Internationale Vereinigung Limnologie* **25**, 39–48.

Herdendorf, C. E. (1982). Large lakes of the world. *Journal of Great Lakes Research* **8**, 379–412.

Kendall, R. L. (1969). An ecological history of the Lake Victoria basin. *Ecological Monograph* **39**, 121–176.

Khursevich, G. K., Karabanov, E. B., Williams, D. F., Kuzmin, M. I., and Prokopenko, A. A. (2000). Evolution of freshwater centric diatoms within the Baikal rift zone during the late Cenozoic. In *Lake Baikal: A Mirror in Time and Space for Understanding Global Change Processes* (K. Minoura, Ed.), pp. 146–154. Elsevier, Amsterdam.

Khursevich, G. K., Karabanov, E. B., Prokopenko, A. A., *et al.* (2001). Insolation regime in Siberia as a major factor controlling diatom production in Lake Baikal during the past 800,000 years. *Quaternary International* **80–81**, 47–58.

Mackay, A. W., Jones, V. J., and Battarbee, R. W. (2003a). Approaches to Holocene climate reconstruction using diatoms. In *Global Change in the Holocene* (A. W. Mackay, R. W. Battarbee, H. J. B. Birks and F. Oldfield, Eds.), (Chapter 20), pp. 294–309. Arnold.

Mackay, A. W., Battarbee, R. W., Flower, R. J., *et al.* (2003b). Assessing the potential for developing internal diatom-based inference models in Lake Baikal. *Limnology and Oceanography* **48**, 1183–1192.

Mackay, A. W., Ryves, D. B., Battarbee, R. W., *et al.* (2005). 1000 years of climate variability in central Asia: Assessing the evidence using Lake Baikal diatom assemblages and the application of a diatom-inferred model of snow thickness. *Global and Planetary Change* **46**, 281–297.

Popovskaya, G. I. (2000). Ecological monitoring of phytoplankton in Lake Baikal. *Aquatic Ecosystem Health and Management* **3**, 215–225.

Ryves, D. B., Juggins, S., Fritz, S. C., and Battarbee, R. W. (2001). Experimental diatom dissolution and the quantification of microfossil preservation in sediments. *Palaeogeography, Palaeoclimatology, Palaeoecology* **172**, 99–113.

Ryves, D. B., Jewson, D. H., Sturm, M., *et al.* (2003). Quantitative and qualitative relationships between planktonic diatom communities and diatom assemblages in sedimenting material and surface sediments in Lake Baikal, Siberia. *Limnology and Oceanography* **48**, 1643–1661.

Stager, J. C., Cumming, B., and Meeker, L. (1997). A high-resolution 11,400-Yr diatom record from Lake Victoria, East Africa. *Quaternary Research* **47**, 81–89.

Stager, J. C., Mayewski, P. A., and Meeker, L. D. (2002). Cooling cycles, Heinrich event 1, and the desiccation of Lake Victoria. *Palaeogeography, Palaeoclimatology and Palaeoecology* **183**, 169–178.

Stager, J. C., Ryves, D., Cumming, B. F., Meeker, L. D., and Beer, J. (2005). Solar variability and the levels of Lake Victoria, East Africa, during the last millennium. *Journal of Paleolimnology* **33**, 243–251.

Stoermer E, F., and Edlund, M. B. (1999). No paradox in the plankton? – Diatom communities in large lakes. In *Proceedings of the 14th International Diatom Symposium, Tokyo* (Mayama, Idei & Koizumi, Eds.), pp. 49–61. Koeltz Scientific Books, Koenigstein.

Swann, G. E. A., and Mackay, A. W. (2006). Potential limitations of biogenic silica as an indicator or abrupt climate change in Lake Baikal, Russia. *Journal of Paleolimnology* **36**, 81–89.

Sylvestre, F., Servant, M., Servant-Vildary, S., Causse, C., Fournier, M., and Ybert, J.-P. (1999). Lake-level chronology on the Southern Bolivian Altiplano (18°–23°S) during late-Glacial time and the early Holocene. *Quaternary Research* **51**, 54–66.

Tapia, P. M., Fritz, S. C., Baker, P. A., Seltzer, G. O., and Dunbar, R. B. (2003). A late Quaternary diatom record of tropical climatic history from Lake Titicaca (Peru and Bolivia). *Palaeogeography, Palaeoclimatology, Palaeoecology* **194**, 139–164.

Tilzer, M. M. (1990). Environment and physiological control of phytoplankton productivity in large lakes. In *Large Lakes: Ecological Structure and Function* (M. M. Tilzer and C. Serruya, Eds.), pp. 339–367. Springer, Berlin.

Tilzer, M. M., and Serruya, C. (Eds.), (1990). *Large Lakes: Ecological Structure and Function.* Springer, Berlin.

Verschuren, D., Johnson, T. C., Kling, H. J., *et al.* (2002). History and timing of human impact on Lake Victoria, East Africa. *Proceedings of the Royal Society of London B* **269**, 289–294.

Williams, D. F., Kuzmin, M. I., Prokopenko, A. A., Karabanov, E. B., Khursevich, G. K., and Bezrukova, E. V. (2001). The Lake Baikal drilling project in the context of a global lake drilling initiative. *Quaternary International* **80/81**, 3–18.

Relevant Websites

http://www.icdp-online.de – Information Network: International Continental Scientific Drilling Program

http://www.worldlakes.org – LakeNet: World lakes network

Marine Laminated Sequences

J Pike, Cardiff University, Cardiff, UK

C E Stickley, Norwegian Polar Institute, Tromsø, Norway

© 2007 Elsevier B.V. All rights reserved.

Introduction

Historical Perspective

The study of seasonally laminated sediments from the geological record was brought to the fore by Gerard De Geer in the early twentieth century (De Geer, 1912). He introduced the term 'varve' to describe an annually repetitive sediment deposit within Swedish glacimarine clays. These late Quaternary varves comprise a silt-rich lamination, related to spring glacier melt, and a clay-rich lamination, related to lower energy in-wash and biological productivity in the summer. The description of these varves and the seasonal-scale proxy data that they contained, together with the realization that packages of varves could be correlated over tens of kilometers, demonstrated, for the first time, that an annual-scale chronology was possible and led to the first such chronology for the past 12 kyr being erected in Scandinavia (De Geer, 1912). However, after De Geer's work, the study of laminated sediments, varves, and seasonal-scale proxy records was slow to progress. The geological community was skeptical about whether annual resolution was possible from the geological record. In the 1960s, Roger Anderson and his colleagues re-invigorated research into laminated sediment records (e.g., Anderson and Kirkland (1960)) and re-established the credibility of varves as the highest resolution proxy climate archives.

In the marine realm, research was hampered by the technological challenge of recovering long sediment cores from the sea floor. However, early workers first recovered short cores of diatom-rich, marine laminated sediments from the Gulf of California, NW Mexico (**Fig. 1**), and interpreted these as marine varves (Calvert, 1964, 1966; Gilbert and Allen, 1943), relating the alternation between lighter and darker olive-green laminae to seasonally varying diatom flux to the sea floor. High sedimentation rates associated with laminated marine sediments (millimeters to centimeters per year), together with immature coring technologies, meant that the recovery of full Quaternary sequences, or even complete Holocene records, was not possible until the inception of the Deep Sea Drilling Project (DSDP), and the introduction of the hydraulic piston corer (HPC). The HPC was tested during DSDP Leg 64 in 1978, in the Gulf of California, recovering 152 m (~250 ka) of marine laminated diatom ooze, interbedded with homogenous diatom ooze, of late Pleistocene to Holocene age (Schrader *et al.*, 1980, see **Fig. 2A**). DSDP Site 480 represented the first complete Quaternary marine laminated sediment sequence to be recovered. The modification of the HPC into the advanced piston corer (APC) during the Ocean Drilling Program (ODP) led to the recovery of long cores of laminated diatom ooze, including an almost 200 m long, ~160 ka sequence from the Santa Barbara Basin, California margin (Behl and Kennett, 1996) and a 118 m long sequence from Saanich Inlet, British Columbia (**Fig. 1**) with approximately 60 m of Holocene laminated diatom-rich sediments (Blais-Stevens *et al.*, 2001).

Advances in analytical techniques during the 1990s have realized the potential of laminated, diatom-rich marine sediments to provide seasonal-scale proxy ocean and climate records. Almost surgical precision in sampling has been introduced, including small, discrete samples for quantitative diatom abundance work (Barron *et al.*, 2005; Leventer *et al.*, 2002; Sancetta, 1995), and scanning electron microscope

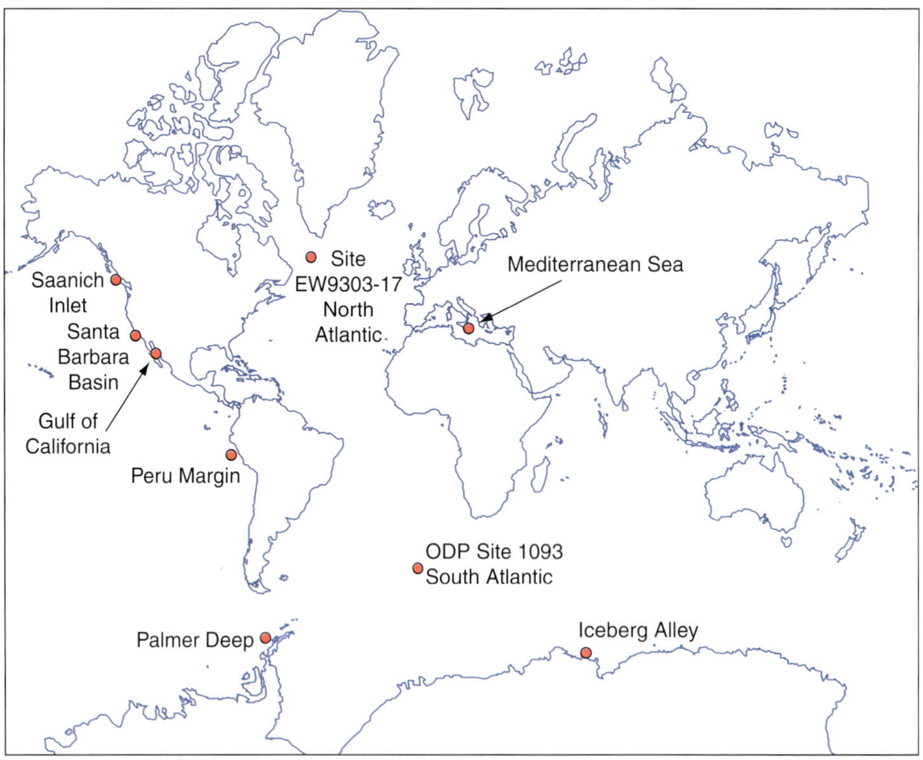

Figure 1 Schematic world map showing the location of sites of accumulation of Quaternary diatom-rich laminated sediments discussed in this chapter.

Figure 2 Examples of Quaternary diatom-rich varved marine sediments. (A) Guaymas Basin, Gulf of California (JPC56, 40-70 cm); (B) Iceberg Alley, MacRobertson shelf, East Antarctica (NBP0101 JPC43B, 23.50–23.80 m).

(SEM) analysis of intact sequences of laminations on the micron scale to investigate diatom species successions through the seasons and relationships between succeeding laminae (Kemp, 1990). The introduction of SEM-scale analysis to the study of laminated, diatom-rich sediments has unlocked Quaternary proxy paleo-ocean and -climate records that can be compared with modern instrumental data retrieved from sediment traps and satellite remote sensing of the sea surface (Thunell *et al.*, 1993). The seasonal variations in diatom flux and terrigenous input into the laminated sediments can often be extrapolated to the interpretation of longer timescale variations through a sequence (Anderson, 1986), and hence laminations can provide analysis of variations from seasonal through to millennial timescales (Kemp, 2003).

Sites of Accumulation of Quaternary Laminated Diatom Ooze

There are two requirements for the formation of laminated diatom ooze in marine sediments: (1) a temporally varying flux of either diatoms or terrigenous grains, or both and (2) a mechanism for preserving the primary sediment fabric. In the Gulf of California (**Fig. 1**), a classic site of laminated diatom-rich sediment deposition, early workers proposed that either diatom flux varied through the year with terrigenous flux remaining constant, or that terrigenous flux varied with biogenic flux remaining constant. The introduction of sediment traps to the study of lakes (e.g., Anderson (1977)), then to the oceans (e.g., Honjo (1982)), made it possible to relate compositionally different fluxes with their seasons of production. This technological advance made it possible to address debates over which flux varied and which remained constant through the year, and, in the case of the Gulf of California, it was discovered that both biogenic and terrigenous fluxes varied, with diatoms having two peak flux events per year in the spring and autumn/early winter (Thunell *et al.*, 1993) related to the 'spring bloom' and 'Fall dump' of diatoms, respectively (Kemp *et al.*, 2000). The application of SEM-scale analytical techniques to Gulf of California sediments further revealed that the fossil diatom assemblage record reflected the seasonal variations in flux through the water column, with up to five discrete diatom productivity events identified in sediments per year (Pike and Kemp, 1997).

Typical sites of accumulation of laminated diatom-rich sediments occur beneath areas of coastal upwelling. Seasonal variations in the strength or direction of alongshore winds influences the influx of nutrients into the euphotic zone from deeper water, hence the temporal abundance and species composition of the diatoms living in the surface waters. Such sites range from continental shelf silled basins (**Fig. 3A**), such as the Santa Barbara Basin (e.g., Schimmelmann and Lange (1996)), to open continental slope sites such as along the California and Peru margins (e.g., Anderson *et al.* (1987), and Brodie and Kemp (1994), see **Figs. 1** and **4B**), to semi-enclosed seas such as the Gulf of California (Pike and Kemp, 1997). Mechanisms for the preservation of laminated sediments that accumulate in shelfal environments are related to levels of benthic dissolved oxygen. Restricted water circulation in silled basins, such as the Santa Barbara Basin (**Fig. 1**), together with high surface water rain of organic material, leads to the rapid consumption of dissolved oxygen below sill depth (concentrations below 0.2 ml l^{-1}), the restriction of an active macrofauna (the source of bioturbation of sediments), and the preservation of the seasonally varying flux as primary laminations (e.g., Reimers *et al.* (1990); Schimmelmann and Lange (1996), see **Fig. 3A**). Along the continental slope, laminated sediments are preserved where the slope intersects the mid-water depth oxygen-minimum

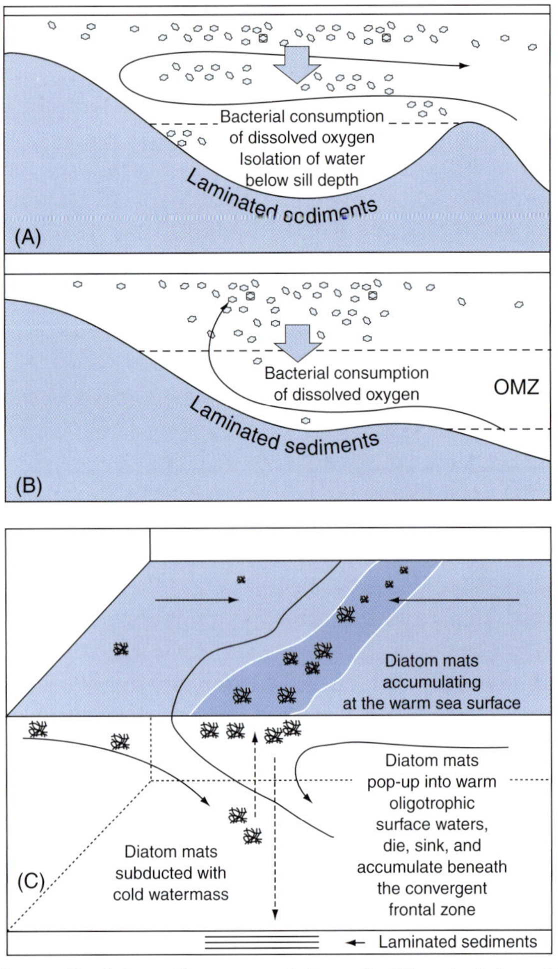

Figure 3 Schematic representation of different diatom-rich laminated sediment accumulation sites. (A) Silled basin, continental shelf; (B) Intersection of the oceanic oxygen minimum zone with the continental shelf; (C) Convergent oceanic frontal zone systems, open ocean settings.

zone (OMZ) (see **Fig. 3B**). The oceanic OMZ is strengthened in regions of coastal upwelling due to bacterial degradation of the high flux of organic debris from the nutrient-rich surface waters. Where this OMZ intersects the continental slope the macrofauna is restricted and variations in flux are preserved in the sediments (e.g., Anderson *et al.* (1987)). In semi-enclosed seas, such as the Gulf of California, the intersection of the basin slopes with the OMZ also leads to the preservation of diatom-rich laminations (Calvert, 1964, see **Fig. 2A**).

During the Quaternary, laminated diatom-rich sediments also have been preserved in the deep-sea beneath areas of open ocean away from the coasts. In this case, a physical mechanism of diatom-rich lamina preservation is invoked as opposed to the chemical mechanism involved with the occurrence of low benthic dissolved oxygen concentrations. Typically, in the deep-sea, levels of benthic dissolved oxygen are high enough to support an active, bioturbating benthic faunal community. When laminated diatom ooze is preserved, a mechanism other than prohibitively low levels of dissolved oxygen is required for the preservation of laminations. In the Quaternary, laminated diatom oozes have been described from the North Atlantic and Southern Ocean (Bodén and Backman 1996; Grigorov *et al.* 2002, see **Fig. 1**). Both of these regions are associated with convergent oceanic frontal systems which concentrate tangled mats of pennate diatoms and lead to a mass flux of preformed diatom mat laminations to the sea floor (**Fig. 3C**), overwhelming the benthic community and preventing bioturbation (Kemp and Baldauf, 1993).

Intra- to Interannual Diatom Productivity Events

Enclosed and Semi-enclosed Seas

Investigation of Holocene laminated diatom ooze from enclosed and semi-enclosed seas, such as the Gulf of California and laminated Pleistocene sapropels from the Mediterranean Sea (**Fig. 1**), highlighted the process of the 'Fall dump' (Kemp *et al.*, 2000) as a major mechanism of diatom flux, silica and carbon export to the sediments. In the Gulf of California, sediment trap (Thunell *et al.*, 1993) and fossil diatom assemblage analysis (Pike and Kemp, 1997) showed that there were two silica flux maxima per year (**Fig. 4A**). In spring, coastal upwelling associated with strong NW winds promotes high primary productivity along the eastern coast of the gulf that results in a layer of *Hyalochaete Chaetoceros* spp. resting spores in the sediments of the deep basins, for example, the Guaymas Basin, where the basin slopes intersect the OMZ. In the fossil record of the coastal oceans, CRS are typically associated with strong spring coastal upwelling. However, sediment traps in the Guaymas Basin showed that the silica flux maxima for the year occurred in late fall/early winter (Thunell *et al.*, 1993), and the diatom fossil record associated with this event, as recorded by SEM-scale sediment fabric analysis, consisted of large centric diatoms such as *Coscinodiscus asteromphalus, C. occulus-iridis, C. granii, Rhizosolenia bergonii,* and *Stephanopyxis palmeriana*. Pike and Kemp (1997) concluded that these diatoms had been growing slowly at depth during the summer when there was a well-stratified water column and low-intensity southerly winds. When the fall/early winter storms mixed the water column as the wind shifted to become more northerly, these diatoms were deposited *en masse*, creating the opal flux maxima record

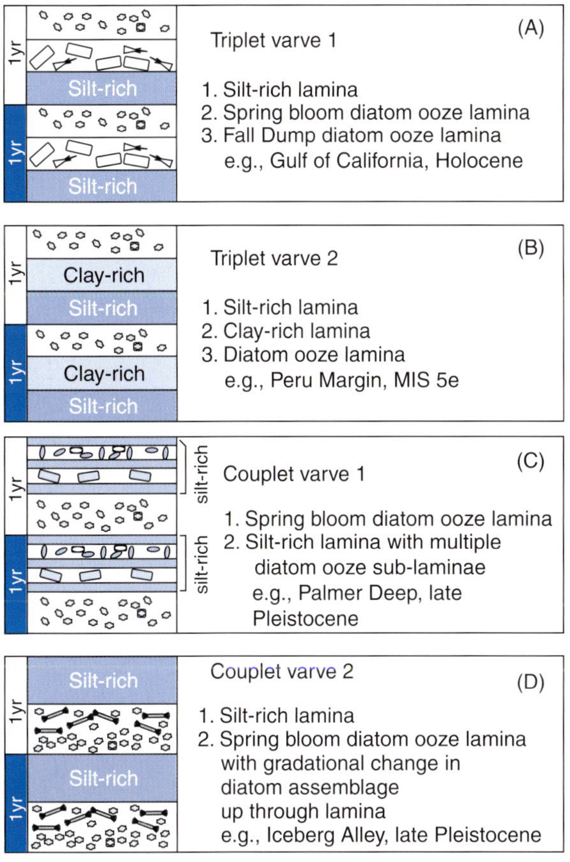

Figure 4 Schematic representation of the different types of marine varves. (A) Triplet varves comprising one terrigenous lamina and two diatom ooze laminae; (B) Triplet varves comprising two terrigenous laminae and one diatom ooze lamina; (C) Couplet varves comprising one diatom ooze lamina and one more terrigenous-rich lamina with the terrigenous lamina containing sublaminae of diatom ooze; (D) Couplet varves comprising one diatom ooze lamina and one terrigenous lamina with the diatom ooze lamina showing a gradational change in diatom assemblage from bottom to top.

in the sediment traps. A similar interpretation was reached for well-preserved Pleistocene sapropels (organic carbon-rich deposits) from the Mediterranean (Kemp *et al.*, 1999). Here, diatoms such as *Rhizosolenia* spp. and *Hemiaulus hauckii*, that form long chains which tangle together to form 'mats,' grow during the summer season of strong water column stratification. These diatoms take advantage of nutrients trapped at depth in the stratified water column and then, by regulating their buoyancy, move up to the surface to photosynthesize (Kemp *et al.*, 2000). Hence, the study of these low-latitude diatom fossil records from the Gulf of California and the Mediterranean raised awareness that high diatom export production need not be closely coupled to coastal or oceanic upwelling, as has been the classic interpretation, but could also be the result of deposition following the breakdown of summer water column stratification and a Fall dump of a diatom population that had been growing slowly over the summer (Kemp *et al.*, 2000).

In the Pleistocene Mediterranean Sea (**Fig. 1**) the primary method of promulgating opal and organic carbon flux to the sediments, to form sapropels, was the deposition of mat-forming diatoms *Rhizosolenia* spp. and *Hemiaulus hauckii*. Although this was not the case for the Holocene Gulf of California laminated diatom-rich sediments, a typical mat-forming diatom was recorded, *Thalassiothix longissima*. Analysis of the position of *T. longissima* diatom mats within the Holocene varves led to the interpretation that *T. longissima* was transported into the gulf with subtropical surface waters in the summer and then was deposited in the fall with the Fall dump flora, as wind speeds increased and wind direction changed to become more northerly (Pike and Kemp, 1997). Today, subtropical Pacific surface waters are advected far into the Gulf of California during El Niño years, as the warm El Niño water mass flows northward along the western edge of continental North America. Frequency analysis of the abundance of *T. longissima* diatom mats within an early Holocene section of laminations revealed three dominant frequencies: ~11 years, ~22–24 years, and ~50 years (Pike and Kemp, 1997). These frequencies match cycles in solar activity and, hence, the diatom record would suggest that in the early Holocene in the eastern Pacific Ocean, the frequency of El Niño, hence the import of *T. longissima* diatom mats into the gulf, was in part regulated by solar activity (Pike and Kemp, 1997).

Continental Shelf Basins

There is an extensive Quaternary laminated sediment fossil diatom record from silled continental shelf basins from both low and high latitudes. A classic site for the accumulation of varved sediments is the Santa Barbara Basin on the California margin (**Fig. 1**), a silled basin situated beneath a region of strong coastal upwelling, hence, a region of seasonally high primary production of diatoms. The Holocene record has been well documented (Schimmelmann and Lange (1996) and references therein); however, the recovery of a long drill core, ODP Site 893, extended the record back into the late Pleistocene, over 160 kyr (e.g., Bull *et al.*, 2000). Laminations are preserved during warm interglacials or interstadials when bottom water ventilation of the basin is weak and bottom water dissolved oxygen concentrations are low (Behl and Kennett, 1996). The laminations form varves which are either couplets of silt-rich and silt-poor laminae, or triplets of silt-rich, silt-poor, and diatom ooze laminae (Bull and Kemp, 1995, see **Fig. 4B**). Lithogenic material is primarily delivered as terrestrial runoff during the winter rainy season whereas diatom flux occurs mainly during the spring and summer, related to the intensity of coastal upwelling over the basin. The diatom ooze laminae range in thickness from < 20 μm to 2 mm, with a mean thickness of 0.25–0.75 mm, and are composed of vegetative remains and resting spores of *Hyalochaete Chaetoceros* species, predominantly *Chaetoceros radicans*. Some thicker laminations show a succession of upward decreasing vegetative remains and increasing resting spores, with the very thickest laminae containing sublaminae of other genera (Bull and Kemp, 1995). The range in thickness of the diatom ooze laminae reflects increased or decreased diatom productivity, hence efficiency of coastal upwelling, whereas the thickness of the terrigenous components is linked to the amount of rainfall and subsequent runoff from the continent. This decoupling of formation processes between diatom and terrigenous laminae means that in the Santa Barbara Basin, the varves can be used to assess the concurrence of changes in marine productivity and terrestrial runoff in relation to oceanographic/climatic events such as El Niño (Bull *et al.*, 2000). Bull *et al.* (2000) investigated the cyclicity and phase relationships between different laminae types from a 160.2 ka (marine isotope stage, MIS 6.2) interval of triplet laminations. They showed that 3.1, 8.4, and 8.8 yr cycles in terrigenous lamina thickness and total varve thickness could be related to similar frequencies in modern El Niño time series. During modern El Niño, there is increased rainfall over the California coast and, hence, increased terrestrial runoff onto the shelf. Cycles of 3.5, 7.3, and 7.6 yr in the diatom laminations also could be related to El Niño, and reflect reduced marine productivity associated with warmer waters expanding northward along the North American coast and reducing the effectiveness of upwelling to bring cold, nutrient-rich deeper waters to the surface (Bull *et al.*, 2000). Further, the changes in thickness of the terrigenous laminae and diatom ooze laminae are antiphased, which is expected because the impact of El Niño is to simultaneously increase runoff from land and decrease marine productivity. This antiphased relationship is strong evidence that variability in the thickness of these late Pleistocene varves is related to the El Niño-Southern Oscillation phenomenon (Bull *et al.*, 2000).

In recent years a number of late Quaternary laminated diatom ooze deposits have been recovered from continental basin and trough sites around Antarctica (Barker *et al.*, 1999; Leventer *et al.*, in review). These basins and troughs are sites of exceptionally high sedimentation, particularly during the rapid warming following the Last Glacial period. For example, Palmer Deep (**Fig. 1**), a funnel-shaped depression on the western Antarctic Peninsula shelf which acts as a large natural sediment trap, had an average radiocarbon-derived linear sedimentation rate of 0.25 cm per year during the last deglaciation (Domack *et al.* (2001); as recalibrated in Maddison *et al.* (2005)). Despite this low radiocarbon-derived sedimentation rate, individual varves can be in excess of 5 cm thick in the early deglaciation (Maddison, personal communication, 2005). Iceberg Alley, a linear trough on the MacRobertson Shelf, East Antarctic margin (**Fig. 1**), had a radiocarbon-derived linear sedimentation rate of 0.5 cm per year during a similar postglacial interval (Stickley *et al.*, 2005), but with individual varves being over 8 cm in thickness (**Fig. 2B**). Compared with average coastal high sedimentation rates measured in cm per kyr, sedimentation rates of cm per yr represent much higher biological productivity in the surface waters, hence flux of diatoms to the sediments. Analysis of the seasonal sequence of diatom assemblages recorded in these sediments reveals evidence of dynamic changes around the coast of Antarctica during deglaciation. In the Palmer Deep postglacial laminated diatom ooze, thick laminations composed of *Hyalochaete Chaetoceros* spp. resting spores (CRS, see **Figs. 5A** and **5B**) are deposited following the spring bloom of *Hyalochaete Chaetoceros* species. In Antarctic coastal waters, as opposed to coastal upwelling at low latitudes, *Hyalochaete Chaetoceros* spp. are typical of spring waters at the sea ice edge. As the sea ice melts, a freshwater cap traps nutrients in the surface waters, leading to a bloom of *Hyalochaete Chaetoceros* spp. diatoms. When the sea ice has completely melted and the nutrients in the water are depleted, CRS are deposited to form a thick lamination in the sediments (Maddison *et al.*, 2005). In the summer, as the water column becomes more oligotrophic, a more typical mixed Antarctic coastal assemblage takes over, comprising diatoms such as *Corethron pennatum*, *Coscinodiscus bouvet*, *Odontella weissflogii*, *Fragilariopsis* spp. and *Thalassiosira antarctica*. During immediate postglacial times, often the summer laminations are characterized by a succession of sublaminae dominated by different species, related to pulses of upwelled circumpolar deep water replenishing nutrients on the shelf, or to advected water masses (**Fig. 4C**). CRS sublaminae occur in earliest summer when the influx of new nutrients restimulates the waning spring *Hyalochaete Chaetoceros* population. *Corethron pennatum* sublaminae occur early in summer, perhaps deposited from a water column that is still partially stratified but with no sea ice remaining. Sublaminae of *Coscindiscus bouvet* are most likely deposited from advected water masses from further north along the peninsula, associated with summer

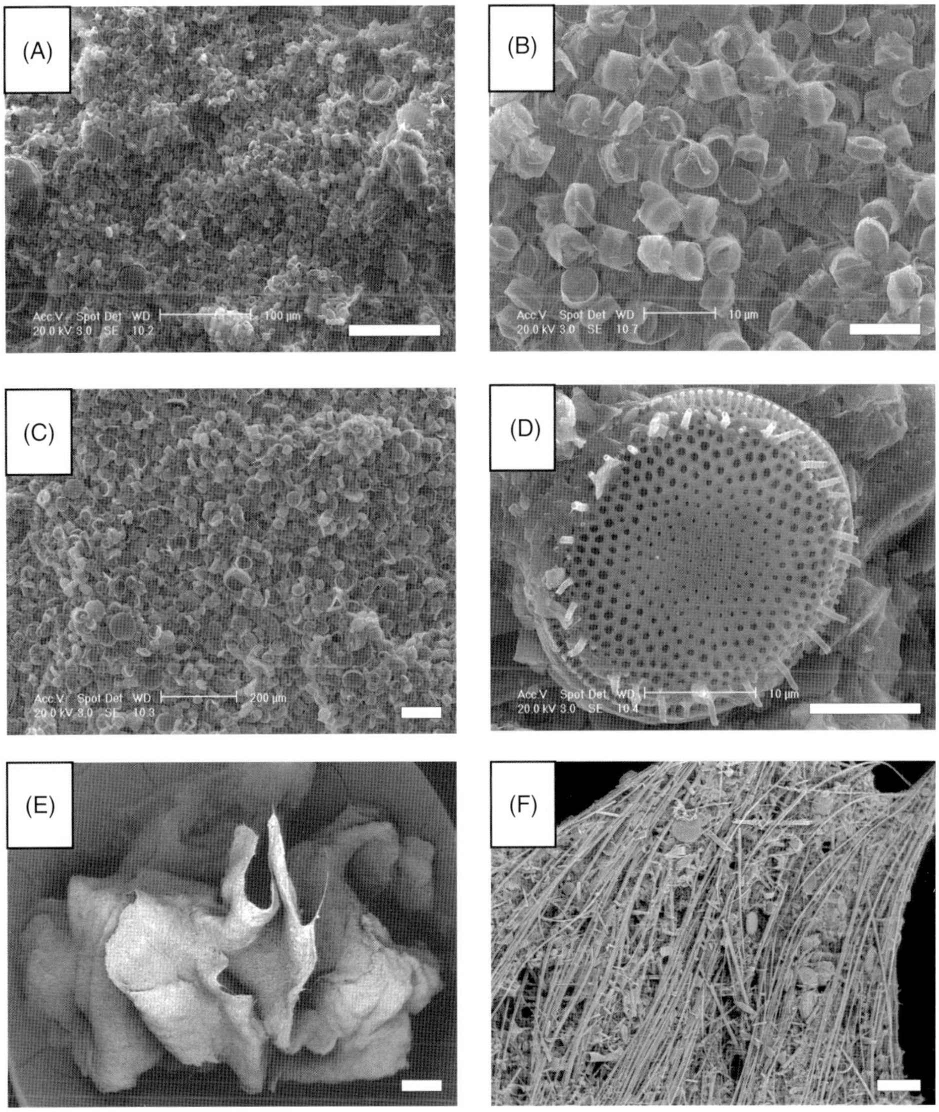

Figure 5 Scanning electron microscope secondary electron images of examples of Quaternary diatom laminations. (A) Low magnification image of an *Hyalochaete Chaetoceros* resting spore lamina-parallel sediment surface (ODP Hole 1098A, 44.1341-44.1261 mcd; scale bar = 100 μm); (B) High magnification image of an *Hyalochaete Chaetoceros* resting spore lamina (scale bar = 10 μm); (C) Low magnification image of a *Thalassiosira antarctica* resting spore lamina-parallel sediment surface (ODP Hole 1098A, 44.3965–44.3735 mcd; scale bar = 100 μm); (D) *Thalassiosira antarctica* resting spore (scale bar = 10 μm); (E) Low magnification image of a pennate diatom mat demonstrating the inherent strength of the mat that enables it to be sculpted (NBP0101 KC09, 255 cm; scale bar = 1 mm); (F) Higher magnification image of a pennate diatom mat showing the long thin frustules of *Thalassiothrix antarctica* that inhibit bioturbation (NBP0101 KC09, 255 cm; scale bar = 100 μm).

storms. Near the top of the summer laminations, *Thalassiosira antarctica* sublaminae often occur as the last bloom of the year before light levels drop and sea ice reforms over the region (Maddison *et al.*, 2005, see **Figs. 5C** and **5D**). It seems from this, and other laminated records recovered from around the Antarctic margin, that *T. antarctica* is a good indicator in the sediment record for Autumn diatom flux.

In Iceberg Alley, MacRobertson Shelf, the postglacial diatom record is similar to that of Palmer Deep; however, here, the seasonal succession of diatoms is expressed within the thick 'spring' laminations, as opposed to within summer laminae in Palmer Deep (**Fig. 4D**). The spring laminae are dominated by CRS as in Palmer Deep; however, near the top of the CRS laminae the concentration of diatoms such as *Corethron pennatum* and *Rhizosolenia* spp. increases (Stickley *et al.*, 2005). The early postglacial environment was characterized by *C. pennatum*, whereas the later postglacial was characterized by increased abundance of *Rhizosolenia* spp. reflecting the gradual move to relative oligotrophy and less persistent sea ice when compared with the early postglacial. This increase in *C. pennatum* and *Rhizosolenia* spp. toward the top of the spring lamination suggests that as the *Hyalochaete Chaetoceros* spp. exhaust the nutrients in the surface waters the other, more oligotrophic, diatoms gradually take over until summer (i.e., the second lamination of the couplet), when the water column stratification eventually breaks down and the typical Antarctic mixed flora of diatoms takes over (Stickley *et al.* 2005). During the postglacial in Iceberg Alley, it is suggested that a calving bay re-entrant existed (a fjord-like indentation in the ice-sheet edge that forms a open water bay surrounded by cliff of ice and, into which, icebergs carve), surrounding the trough and trapping nutrients and flux within the trough (Leventer *et al.*, in review); thus, the extremely high sedimentation rates and diatom flux can be attributed to both high biological productivity resulting from high nutrients in the water column, and sediment focusing within the trough.

High-Latitude Fjords

Thickly laminated diatom-rich sediments also sometimes accumulate in fjords, for example, Saanich Inlet (**Fig. 1**) on the southeastern side of Vancouver Island, British Columbia, Canada (Blais-Stevens *et al.*, 2001). Saanich Inlet is a silled fjord with freshwater input at its head which results in weak flushing of the surface waters and isolation of the waters below sill depth. High surface water productivity leads to severely reduced benthic dissolved oxygen levels and the preservation of sedimentary laminations. ODP drilling in Saanich Inlet recovered ~60 m of Holocene diatom-rich laminated sediments. Centimeter-scale varves consist of couplets of diatom ooze, or diatom mud deposited in spring and summer with silty clay deposited in winter (Dean *et al.*, 2001). Saanich Inlet varves may have up to 19 sublaminae composed of various diatom species, or assemblages. Following from the winter terrigenous lamina, up to two early spring sublaminae characterized by *Thalassiosira pacifica* or *Hyalochaete Chaetoceros* spp. may be deposited as light levels increase and water column stratification begins to be established. In spring, with a strong pycnocline (a zone of rapid increase in density within a layer of the ocean) and high nutrient levels in the surface waters, *Hyalochaete Chaetoceros* spp. bloom and a sublamina is deposited, followed by a sublamina characterized by *Skeletonema costatum*. The summer portion of the varve is characterized by one to many sublaminae of *Hyalochaete Chaetoceros*. The number of summer sublaminae, or blooms in the water column, is related to tidally controlled input of new nutrients and diatom seedstock into the fjord. As the water column stability is broken down in fall, there may be a final bloom and subsequent deposition of a sublamina of *Hyalochaete Chaetoceros* spp. (Dean *et al.*, 2001). Interannual variability in the varved sequence is revealed by the presence or absence of *Thalassiosira pacifica* and/or *Skeletonema costatum* sublaminae. *T. pacifica* is abundant when there is greater input of freshwater into the fjord during El Niño events, whereas *S. costatum* tends to be more abundant during the cooler periods between El Niño events and negative phases of the Pacific-North American index (Dean *et al.*, 2001).

Centennial- to Orbital-Scale Diatom Records

Quaternary laminated sediment sections not only record information about year-on-year variations in diatom assemblages which can be used to reconstruct the annual cycle of hydrographic conditions (Pike and Kemp, 1997), they also record variations over interannual through to millennial timescales. In the Gulf of California, two investigations of diatom species variations revealed long-term variations in the hydrography and climate of the gulf over the last 15 ka (Barron *et al.* 2005; Sancetta 1995). For example, *Azpeita nodulifer*, a tropical species, was abundant in the Guaymas Basin, central Gulf of California (**Fig. 1**), during the Bølling/Allerod interstadial (14.6–12.9 ka) indicating that the seasonal incursion of warm, tropical waters from the Pacific to the south was strong. *A. nodulifer* was still common during the Younger Dryas stadial (12.9–11.6 ka) but is associated with an overall reduction in diatom abundance suggestive of tropical, oligotrophic conditions extending throughout the year (i.e., seasonality was reduced; there were no laminations preserved during the Younger Dryas). Between approximately 11.0 and 6.2 ka, diatom abundance increased but *A. nodulifer* almost disappeared from the records. This indicates intensification in the strength or duration of northwesterly winter winds driving high productivity associated with coastal upwelling and effectively excluding Pacific-derived tropical forms from entering the gulf. The return of *A. nodulifer* to the record during the time period 6.2–5.4 ka indicates the return of southerly sourced tropical water into the gulf and the onset of modern monsoonal conditions with seasonally reversing winds resulting in coastal upwelling in the spring and incursions of warm, tropical waters in the summer. Hence, the diatoms preserved in the Gulf of California laminated sediments can be used to look at long-term, large-scale changes in hydrography and climate of the region (Barron *et al.*, 2005; Sancetta, 1995).

Along the Peru margin, diatom records from laminated sediments have been used to investigate the occurrence of ocean-climate events such as El Niño during earlier Quaternary times (Brodie and Kemp, 1994). Laminated diatom-rich sediments have accumulated intermittently at ODP Site 686, West Pisco Basin (**Fig. 1**), which is a continental shelf basin and part of the rapidly subsiding forearc basin associated with the Peru–Chile Trench. Well-developed triplet laminations (**Fig. 4B**) comprising an alternation of silt-rich and silt-poor terrigenous laminae with occasional near-monospecific *Skeletonema costatum* or, less often, *Hyalochaete Chaetoceros* spp. resting spore diatom ooze laminations, accumulated during the last interglacial, MIS 5e (Brodie and Kemp, 1994). The triplet laminations represent marine varves where the silt-rich and silt-poor terrigenous laminae represent annual variations in continental runoff whereas the intermittent diatom ooze laminations are interpreted as megablooms associated with anti-El Niño conditions, such as La Niña, when upwelling intensity would be increased along the Peru margin. Landward of the West Pisco Basin today, however, there is very little annual rainfall and runoff is intermittent, whereas to the north it is believed that only El Niño is capable of producing any significant surface runoff (Brodie and Kemp (1994) and references therein). The evidence from the diatom-rich triplet laminations and other data such as alkenone sea surface temperature reconstructions (Brodie and Kemp (1994) and references therein) leads to the conclusion that surface waters along the Peru margin were generally warmer during the last interglacial, which would have promoted a more regular rainfall pattern over the continent with more consistent annual runoff – hence the preservation of silt-rich and silt-poor laminations. As stated above, the intermittent *Skeletonema costatum* diatom ooze laminae represent deposition from extremely large productivity events likely to be associated with strong La Niña years (Brodie and Kemp, 1994).

Presence or Absence of Diatom-Rich Laminations: Oceanographic Frontal Zones

Accumulations of Quaternary diatom-rich laminations in the deep-sea, that cannot simply be ascribed a seasonal origin, are associated with the position of major convergent oceanic frontal systems (**Fig. 3C**). The taxa associated with these laminated oozes are mat-forming diatoms from the genus *Thalassiothix* (Bodén and Backman, 1996; Grigorov *et al.*, 2002). The physical oceanographic mechanism responsible for the production of accumulations of laminated, diatom mat-rich sediments was first described for the Neogene sediments of the eastern equatorial Pacific (Kemp and Baldauf, 1993); however, Quaternary occurrences of diatom mat deposits have since been described and ascribed a similar mechanism. In the North Atlantic at Site EW9303-17 on the west flank of the Reykjanes Ridge (**Fig. 1**), a 3-m-thick interval of laminated diatom ooze was described as being monospecific *Thalassiothrix longissima* diatom mats (Bodén and Backman, 1996). Preservation of the diatom mats at the sea floor is not due to low benthic dissolved oxygen levels, as the deep Atlantic is well-oxygenated, but rather to the physical suppression of the benthic faunal burrowers by the impenetrable, mesh-like nature of the diatom mats, themselves. **Figures 5E** and **5F** show examples from the Antarctic that demonstrate the strength of these pennate diatom mats. These mats accumulated during MIS 5e, that is, the last interglacial. The accumulation of these diatom mats at this place and time is related to the position of the sub-arctic convergence where the North Atlantic and Irminger Currents meet cold, less saline Labrador Sea water. *T. longissima* mats (colonies) carried in the colder water get subducted at the frontal zone to then 'pop-up' to the surface on the warmer, more oligotropic side of the front in order to photosynthesize. Here, they exhaust the already low nutrients, the diatoms die and sink to the sea floor to accumulate in a narrow band beneath the front. Oceanic frontal systems are often dynamic and migrate with changing climates so the presence or absence of the laminated *T. longissima* diatom mat deposits can be used to track the frontal systems over time, for example, the sub-Arctic convergence was likely positioned over site EW9303-17 during the last interglacial (Bodén and Backman, 1996) but was elsewhere, perhaps further south, during the ensuing glacial period.

In the South Atlantic, Pleistocene laminated diatom mat deposits have accumulated beneath the Polar Front at approximately 50° S, and were recovered from ODP Site 1093 (**Fig. 1**), among others, during ODP Leg 177 (Grigorov *et al.*, 2002). These Southern Ocean laminated diatom mat sediments are composed of *Thalassiothrix antarctica*. Similar to the North Atlantic, the laminated sediments appear to have been deposited during either interglacials (e.g., MIS 11 and 29) or the transitions into interglacials (e.g., MIS12/11 transition). In contrast to the diatom record from the North Atlantic, the laminated diatom ooze of the South Atlantic appears to contain a regular succession of diatom species in couplets, for example, *T. antarctica* diatom mats followed by a mixed centric/pennate diatom lamina in the MIS 12/11 interval. In an older interglacial (MIS 29), *T. antarctica* is followed by either a mixed centric/pennate diatom lamina or *Fragilariopsis kerguelensis* lamina, and then centric diatom lamina. These successions imply that the laminated diatom mat deposits from the Polar Front region of the Southern Ocean could be recording some seasonal variations as well as the physical position of the Polar Front itself (Grigorov *et al.*, 2002).

Summary

Diatom fossil records from Quaternary laminated marine sediments reveal evidence for climate and ocean variability over subannual, interannual, millennial and orbital timescales. Seasonal diatom species or assemblage successions can reveal information about the nature of the prevailing water column processes producing the silica flux, for example, upwelling or breakdown of water column stratification. Interannual variations in the occurrence of particular species, or in the make-up of the whole diatom assemblage, can provide a valuable indication of the behavior of phenomenon such as El Niño-Southern Oscillation during past climate states. In locations where long, late Quaternary laminated sediment sequences are recovered, the changes in the diatom ooze lamina assemblage over centennial to millennial timescales can reveal changes in local and region-scale ocean circulation and climate. Laminated diatom ooze also accumulates in the open ocean in association with convergent oceanic frontal systems. Here the laminations are composed of tangled diatom mats and are not necessarily seasonally deposited; however, the presence or absence of the laminated diatom mat deposits can be used to track the movement of the oceanic fronts over glacial–interglacial timescales.

See also: **Diatom Introduction**. **Diatom Methods**: Data Interpretation; Quaternary Geological Records. **Diatom Records**: Antarctic Waters. **Paleoceanography, Biological Proxies**: Marine Diatoms.

References

Anderson, R. Y. (1977). Short term sedimentation response in lakes in western United States as measured by automated sampling. *Limnology and Oceanography* **22**, 423–433.

Anderson, R. Y. (1986). The varve microcosm: Propagator of cyclic bedding. *Paleoceanography* **1**, 373–382.

Anderson, R. Y., Hemphill-Haley, E., and Gardner, J. V. (1987). Persistent late Pleistocene-Holocene seasonal upwelling and varves off the coast of California. *Quaternary Research* **28**, 307–313.

Anderson, R. Y., and Kirkland, D. W. (1960). Origin, varves, and cycles of Jurassic Todilto Formation, New Mexico. *Association of American Petroleum Geologists Bulletin* **44**, 37–52.

Barker, P. F., Camerlenghi, A., Acton, G. D. (1999) *Proceedings of ODP, Initial Reports, 178 (CD-ROM)*. Available from: Ocean Drilling Program, Texas A&M University, College Station, TX, USA.

Barron, J. A., Bukry, D., and Dean, W. E. (2005). Paleoceanographic history of the Guaymas Basin, Gulf of California, during the past 15,000 years based on diatoms, silicoflagellates, and biogenic sediments. *Marine Micropaleontology* **56**, 81–102.

Behl, R. J., and Kennett, J. P. (1996). Brief interdstadial events in the Santa Barbara basin, NE Pacific, during the past 60 kyr. *Nature* **379**, 243–246.

Blais-Stevens, A., Bornhold, B. D., Kemp, A. E. S., Dean, J. M., and Vaan, A. A. (2001). Overview of late Quaternary stratigraphy in Saanich inlet, British Columbia: Results of ocean drilling prorgam Leg 169S. *Marine Geology* **174**, 3–26.

Bodén, P., and Backman, J. (1996). A laminated sediment sequence from northern North Atlantic ocean and its climatic record. *Geology* **24**, 507–510.

Brodie, I., and Kemp, A. E. S. (1994). Variation in biogenic and detrital fluxes and formation of laminae in late Quaternary sediments from the Peruvian coastal upwelling zone. *Marine Geology* **116**, 385–398.

Bull, D., Kemp, A. E. S., and Weedon, G. P. (2000). A 160-kyr-old-record of El Niño southern Oscillation in marine production and coastal runoff from Santa Barbara Basin, California, USA. *Geology* **28**, 1007–1010.

Bull, D., and Kemp, A. E. S. (1995). Composition and origins of laminae in late Quaternary and Holocene sediments from the Santa Barbara basin. In *Proceedings of ODP, Scientific Results, 146 (Part 2), Ocean Drilling Program* (J. P. Kennett *et al.*, Eds.), pp. 77–87. (Ocean Drilling Program) College Station, TX.

Calvert, S. E. (1964). Factors affecting distribution of laminated diatomaceous sediments in the Gulf of California. In *Marine Geology of the Gulf of California, Vol: American Association of Petroleum Geologists Memoir 3* (T. H. van Andel and G. G. Shor, Eds.), pp. 311–330. American Association of Petroleum Geologists, Tulsa.

Calvert, S. E. (1966). Origin of diatom-rich varved sediments from the Gulf of California. *Journal of Geology* **74**, 546–565.

Dean, J. M., Kemp, A. E. S., and Pearce, R. B. (2001). Palaeo-flux records from electron microscope studies of Holocene laminated sediments, Saanich inlet, British Columbia. *Marine Geology* **174**, 139–158.

De Geer, G. (1912). A Geochronology of the last 12,000 years. *Comptes Rendus du XI Congres Géologique International (Stockholm 1910)* pp. 241–253.

Domack, E., Leventer, A., Dunbar, R., Taylor, F., and Brachfield, S. (2001). Chronology of the Palmer Deep site, Antarctic Peninsula: A Holocene paleoenvironmental reference for the circum-Antarctic. ODP leg 178 Scientific Party *The Holocene* **11**, 1–9

Gilbert, J. Y., and Allen, W. E. (1943). The phytoplankton of the Gulf of California obtained by the E. W. Scripps in 1939 and 1940. *Journal of Marine Research* **5**, 89–110.

Grigorov, I., Pearce, R. B., and Kemp, A. E. S. (2002). Southern Ocean laminated diatom ooze: Mat deposits and potential for palaeo-flux studies, ODP leg 177, Site 1093. *Deep-Sea Research II* **49**, 3391–3407.

Honjo, S. (1982). Seasonality and interaction of biogenic and lithogenic particulate flux at the Panama basin. *Science* **218**, 883–884.

Kemp, A. E. S. (1990). Sedimentary fabrics and variation in lamination style in Peru continental margin upwelling sediments. In *Proceedings of ODP, Scientific Results,*

112, (E. Suess *et al.*, Eds.), Ocean Drilling Program, College Station, TX, pp. 43–58.

Kemp, A. E. S. (1995). Laminated sediments from coastal and open ocean upwelling systems: what variability do they record? In *Upwelling in the Ocean: Modern Processes and Ancient Records, Vol. ES18: Dahlem Workshop Report* (C. P. Summerhayes *et al.*, Eds.), pp. 239–257. Wiley, Chichester.

Kemp, A. E. S. (2003). Evidence for abrupt climate changes in annually laminated marine sediments. *Philosophical Transactions of the Royal Society, London A* **361**, 1851–1870.

Kemp, A. E. S., Pearce, R. B., Koizumi, I., Pike, J., and Rance, S. J. (1999). The role of mat-forming diatoms in formation of the Mediterranean sparopels. *Nature* **398**, 57–61.

Kemp, A. E. S., Pike, J., Pearce, R. B., and Lange, C. B. (2000). The "Fall dump" – A new perspective on the role of a "shade flora" in the annual cycle of diatom production and export flux. *Deep-Sea Research II* **47**, 2129–2154.

Kemp, A. E. S., and Baldauf, J. G. (1993). Vast Neogene laminated diatom mat deposits from the eastern equatorial Pacific Ocean. *Nature* **362**, 141–144.

Leventer, A., Domack, E., Barkoukis, A., McAndrews, B., and Murray, J. (2002). Laminations from the Palmer Deep: A diatom-based interpretation. *Paleoceanography* **17**(2) 10.1029/2001PA000624.

Leventer A., Domack, E., Dunbar, R., Pike, J., Stickley, C. E., Maddison, E. J., Brachfeld, S., Manley , P., McClennen, C. (2007). Marine sediment record of the deglaciation of the East Antarctic margin, *GSA Today* (in press).

Maddison, E. J., Pike, J., Leventer, A., and Domack, E. W. (2005). Deglacial seasonal and sub-seasonal diatom record from Palmer Deep, Antarctica. *Journal of Quaternary Science* **20**, 435–446.

Pike, J., and Kemp, A. E. S. (1997). Early Holocene decadal-scale ocean variability recorded in Gulf of California laminated sediments. *Paleoceanography* **12**, 227–238.

Reimers, C. E., Lange, C. B., Tabak, M., and Bernhard, J. M. (1990). Seasonal spillover and varve formation in the Santa Barbara basin, California. *Limnology and Oceanography* **35**, 1577–1585.

Sancetta, C. (1995). Diatoms in the Gulf of California: Seasonal flux patterns and the sediment record for the past 15,000 years. *Paleoceanography* **10**, 67–84.

Schimmelmann, A., and Lange, C. B. (1996). Tales of 1001 varves: A review of Santa Barbara basin sediment studies. In *Palaeoclimatology and Palaeoceanography from Laminated Sediments* (A. E. S. Kemp, Ed.), 116, pp. 121–141. Geological Society of London.

Schrader, H., Kelts, K., Curray, J., Moore, D., Aguayo, E., Aubry, H.-P., Einsele, G., Fornari, D., Gieskes, J., Guerrero, J., Kastner, M., Lyle, M., Matoba, Y., Molina-Cruz, A., Niemitz, J., Rueda, J., Saunders, A., Simoneit, B., and Vaquier, V. (1980). Laminated diatomaceous sediments from the Guaymas basin slope (central Gulf of California): 250,000-year climate record. *Science* **207**, 1207–1209.

Stickley, C. E., Pike, J., Leventer, A., Dunbar, R., Domack, E. W., Brachfeld, S., Manley, P., and McClennen, C. (2005). Deglacial ocean and climate seasonality in laminated diatom sediments, MacRobertson Shelf, Antarctica. *Palaeogeography, Palaeoclimatology, Palaeoecology* **227**, 390–1210.

Thunell, R., Pride, C., Tappa, E., and Muller-Karger, F. (1993). Varve formation in the Gulf of California: Insights from time series sediment trap sampling and remote sensing. *Quaternary Science Reviews* **12**, 451–464.

North Atlantic and Arctic

N Koç, Norwegian Polar Institute, Tromso, Norway

© 2007 Elsevier B.V. All rights reserved.

Introduction

Marine planktonic diatoms are single-celled siliceous algae, which are dependent on light for photosynthesis. In the oceans, diatoms are restricted to cold, nutrient-rich regions where silicic acid is not limiting, such as in the polar regions, the coastal and equatorial upwelling systems, and in coastal areas. The most important factors that determine the distribution and abundance of diatoms in surface waters are sea-surface temperatures, sea-ice conditions, macro- and micronutrient levels, stability of the surface water layer, light levels, and grazing. Surface water productivity and dissolution in the water column and at the water–sediment interface are major processes dictating the diatom flux to the seafloor and preservation of diatom valves in sediments. For further details on the ecology of diatoms, *see* Diatom Introduction.

Preservation of diatoms in sediments of the North Atlantic Ocean through geologic time has been highly variable. The North Atlantic Ocean and the Nordic Seas were regions of high biosiliceous productivity until the onset of Northern Hemisphere glaciations during the late Miocene. Diatoms are the most abundant microfossil group preserved in sediments of Paleogene and early to mid-Neogene age. Since diatom species show very rapid evolution throughout the Cenozoic, it has been possible to establish a high-resolution biostratigraphy for the area. Based on Deep Sea Drilling Program (DSDP) Legs 81 and 94, Baldauf (1984, 1987) established a Cenozoic diatom biostratigraphy for the middle- to high-latitude North Atlantic Ocean. Most of the fossil diatom species of the Nordic Seas are, however, endemic to the area. Therefore, a separate diatom biostratigraphy had to be developed for the Nordic Seas. Based on the DSDP Leg 38 material, Schrader and Fenner (1976) and Dzinoridze *et al.* (1978) proposed diatom biostratigraphies for the Nordic Seas. Meanwhile, development of drilling techniques and availability of reliable paleomagnetic stratigraphy enabled the development of a new Neogene (Koç and Scherer, 1996) and late Paleogene (Scherer and Koç, 1996) diatom biostratigraphy for the Nordic Seas based on the Ocean Drilling Program (ODP) Leg 151 material.

Until recently, little information has been available about pre-Pleistocene paleoenvironments in the central Arctic Ocean. Temperate marine conditions existed during the late Cretaceous (Campanian/Maastrichtian) based on evidence provided by silicoflagellates and diatoms from three short T-3 and CESAR cores, all retrieved from the Alpha Ridge in the Amerasian Basin (Clark *et al.*, 1980). One 3.64 m long core (F1-422) containing middle or late Eocene age silicoflagellates and diatoms has also been retrieved from the Alpha Ridge, providing the sole evidence for early Cenozoic marine conditions in the Arctic (Dell'Agnese and Clark, 1994). Thus, existing core material, at best, represented a few percent of the Cenozoic paleoceanographic history of the Arctic Ocean. Recently, the Integrated Ocean Drilling Program's (IODP) 'The Arctic Drilling Expedition 2004 (ACEX)' drilled a 450-m-thick acoustically stratified sediment sequence draping the crest of the Lomonosov Ridge between 87°N and 88°N. A 180-m-long biosiliceous section with abundant diatoms of middle – early Eocene age was recovered. Evidence so far indicates that the central Arctic Ocean was a highly productive biosiliceous ocean until the late middle Eocene. However, since then it has become an ice covered, extremely unproductive ocean as evidenced by the total absence of biosiliceous organisms.

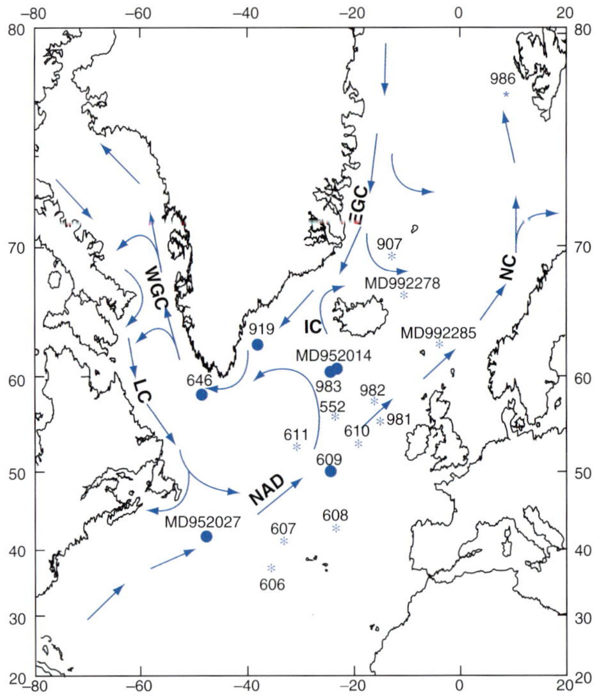

Figure 1 General surface circulation of the northern North Atlantic Ocean and the location of the cores cited for Pleistocene diatom records. In cores marked with dots the last occurrence of *Proboscia curvirostris* is correlated to the oxygen isotope stratigraphy of the core. Stars depict sites where *P. curvirostris* has been observed. MD depicts cores taken by the French RV Marion Dufresne during the IMAGES cruises. All other cores are Deep Sea Drilling (DSDP) or Ocean Drilling Program (ODP) cores. Surface currents are illustrated with arrows as follows: NAD = North Atlantic Drift; IC = Irminger Current; NC = Norwegian Current; EGC = East Greenland Current; WGC = West Greenland Current; LC = Labrador Current. (Koç *et al.*, 2001).

Pleistocene Diatom Record

Biostratigraphy

Pleistocene sediments of the Nordic Seas are mostly barren of diatoms. In contrast to this, sites south of Iceland such as ODP Site 919 from the Irminger Basin and Site 983 from the Gardar Drift are sites with high Pleistocene accumulation rates and well-preserved, abundant diatoms (**Fig. 1**). Based on these sites the resolution of the high-latitude North Atlantic Pleistocene diatom biostratigraphy established by Baldauf (1987) has been refined (Koç and Flower, 1998; Koç *et al.*, 1999, 2001). Eight Pleistocene diatom datum events are identified and, for the first time, tied directly to the oxygen isotope record and paleomagnetic stratigraphy at these sites (**Fig. 2**). These datum events are:

- the last occurrence (LO) of *Proboscia curvirostris* at 0.3 Ma,
- the LO of *Thalassiosira jouseae* at 0.3 Ma,
- the LO of *Nitzschia reinholdii* at 0.6 Ma,
- the LO of *Nitzschia fossilis* at 0.68 Ma,
- the LO of *Neodenticula seminae* at 0.84 Ma,
- the first occurrence (FO) of *Neodenticula seminae* at 1.25 Ma,
- the FO of *Proboscia curvirostris* at 1.53 Ma, and
- the FO of *Pseudoeunotia doliolus* at 1.89 Ma.

Most of these datums are synchronous between the middle-high latitudes of the North Atlantic and the North Pacific. Based on these datums four high-latitude North Atlantic diatom zones have been proposed for the Pleistocene: the *Nitzschia reinholdii* zone, the *Neodenticula seminae* zone, the *Proboscia curvirostris* zone, and the *Thalassiosira oestrupii* zone (Koç *et al.*, 1999).

Paleoceanography

Fluctuating Pleistocene climatic conditions had a major impact on diatom productivity in the high-latitude North Atlantic. Diatom abundance and preservation at Site 919 and 983 covary with the oxygen isotope record of the sites in response to changing oceanographic conditions affecting diatom productivity (Koç and Flower, 1998; Koç *et al.*, 1999; **Fig. 2**). Generally, interglacials are related

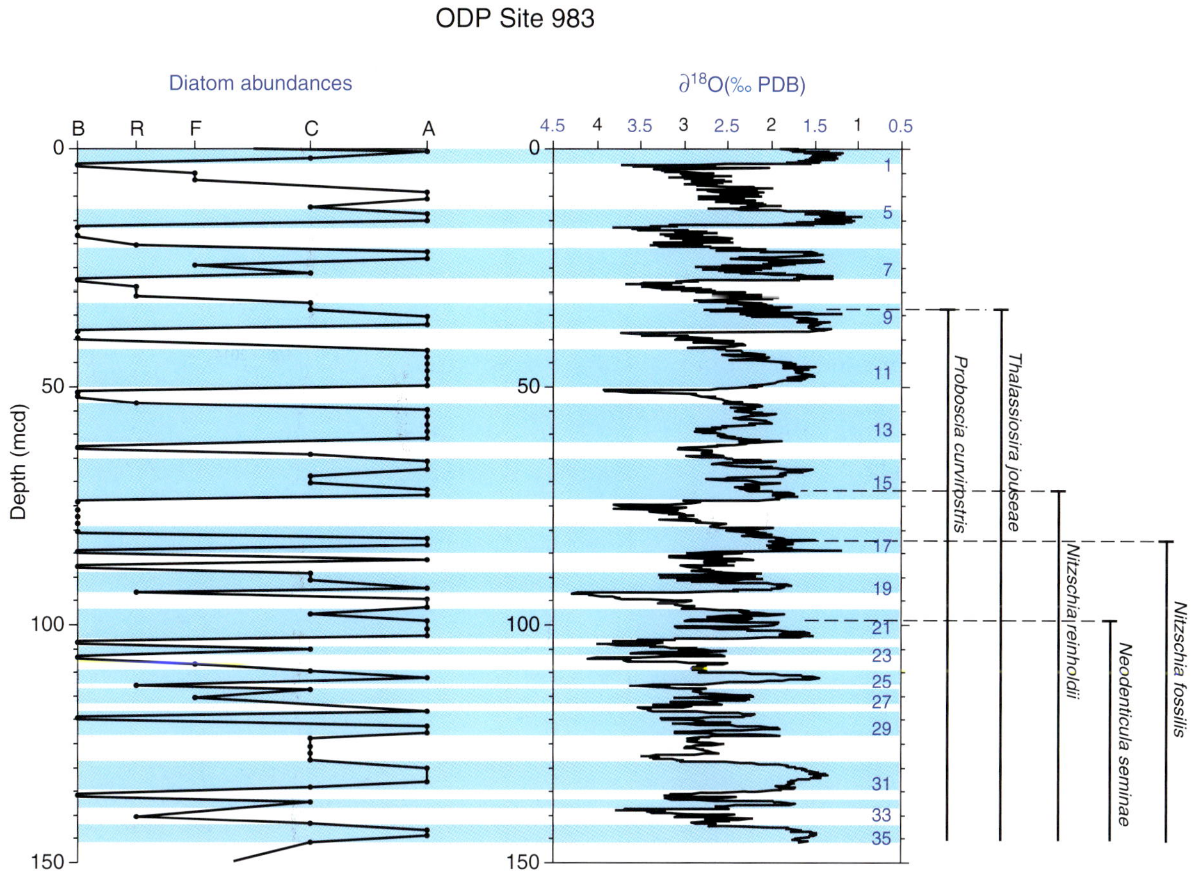

Figure 2 ODP Site 983 relative diatom abundances (B = barren; R = rare; F = few; C = common; A = abundant) plotted together with the planktic δ^{18}O record against composite depth. Interglacial stages are shaded and numbered. Diatom biostratigraphic events and their ranges are indicated, and correlated to the δ^{18}O record (Koç *et al.*, 1999).

to high diatom productivity, whereas glacial stages are often barren of diatoms. However, during glacial stages 18, 20, and 30 there was significant diatom production.

During the late Quaternary glacial periods oceanic polar fronts and sea ice migrated toward the equator, with some of the most dramatic changes occurring in the mid- to high-latitude North Atlantic (CLIMAP Project Members, 1981). Reduced surface water productivity during glacial periods indicates that these were the intervals during which the Polar Front migrated southeastward, thereby covering Sites 919 and 983 with sea ice. Considering the proximity of the sea-ice margin to the area at present, it is highly probable that these diatom-barren glacial intervals reflect periods when these sites were under heavy sea-ice cover. On the other hand, the presence of significant diatom production during glacial stages 18, 20, and 30 indicates open marine conditions at sites south of Iceland during these times. These glacial stages are within the first 100 ka cycles after the mid-Pleistocene transition, and the benthic oxygen isotope records indicate that these glaciations were not as severe as the late Quaternary glacial periods.

Mid-Pleistocene transition *Neodenticula seminae* is part of the modern diatom assemblage of the middle- and high-latitude North Pacific. In the North Atlantic this species is limited to the time interval of 0.84–1.26 Ma (Koç *et al.*, 1999; **Fig. 3**). Baldauf (1986) interpreted the occurrence of *N. seminae* in the North Atlantic as evidence for the presence of cool, low-salinity surface waters in the central North Atlantic during the early Quaternary. The interval of *N. seminae* in the North Atlantic straddles the transition from the dominance of 41 ka cycles in the climate records to the dominance of 100 ka cycles. It is, therefore, possible to take the first occurrence of *N. seminae* in the North Atlantic as a sign for cooling, which started 1.26 Ma, leading to the establishment of the 100 ka cycles with severe glacial periods. The presence of *N. seminae* in the North Atlantic is, therefore, attributable to the unique conditions related to the mid-Pleistocene transition.

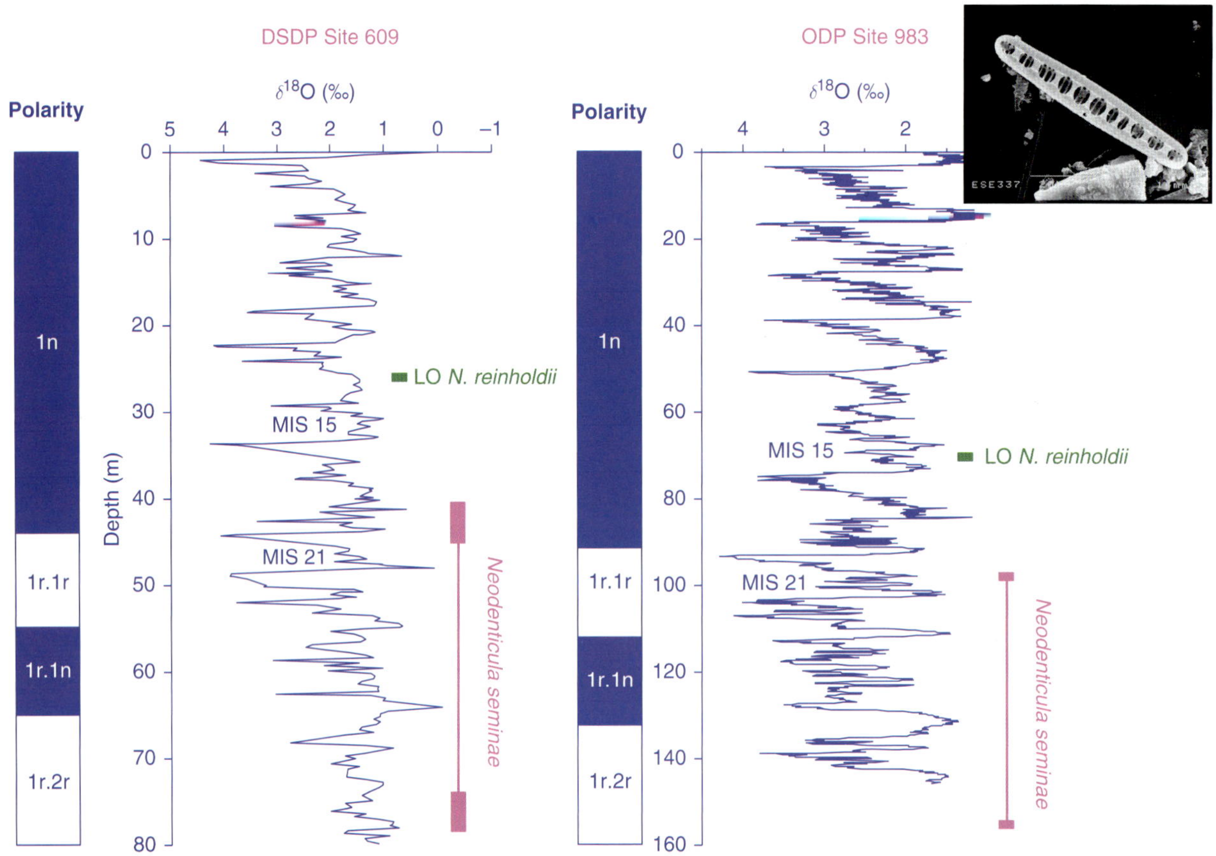

Figure 3 Planktic $\delta^{18}O$ records of DSDP Site 609 (L. Labeyrie, unpublished data) and ODP Site 983 (Hodell and Kleiven, unpublished data) shown together with their respective paleomagnetic stratigraphies (Koç *et al.*, 1999). In order to compare the timing of events between the middle- and high-latitude North Atlantic, the LO of *Nitzschia reinholdii*, and the last and first occurrence of *Neodenticula seminae* are plotted at both sites. MIS = Marine isotope stage. Inset, scanning electron microscope (SEM) picture of *Neodenticula seminae* (Simonsen and Kanaya) Akiba and Yanagisawa from sample 162-983A-13H-7, 49–50 cm (from plate 1 in Koç *et al.* (1999)).

Last occurrence of *Proboscia curvirostris* The youngest diatom event observed in the North Atlantic is the LO of *Proboscia curvirostris* (Jousé) Jordan and Priddle (Koç *et al.*, 2001). *Proboscia curvirostris* is a robust species that can easily be identified in fossil assemblages, and therefore is a practical biostratigraphic tool. Its distribution stretches from 40°N to 80°N in the North Atlantic (**Fig. 1**).

The extinction of *P. curvirostris* is latitudinally diachronous through Marine Isotope Stages (MIS) 9 to 8 within the North Atlantic (**Fig. 4**). This is closely related to the paleoceanographic development of the area. *P. curvirostris* first disappeared within interglacial MIS 9 (324 ka) from the northern areas that are most sensitive to climatic forcing, like the East Greenland current and the sea-ice margin. It survived in the mid-latitude North Atlantic until the conditions of the MIS 8 glaciation became too severe (260 ka). In the North Pacific at ODP Site 883 the LO of *P. curvirostris* also falls within MIS 8. The observed overlap in age between the North Atlantic and the North Pacific strongly suggests that the extinction of *P. curvirostris* is synchronous between these oceans. It is, therefore, potentially a highly valuable datum for correlation between the high-latitude North Atlantic and the North Pacific Oceans.

Last glacial period and Heinrich layers Diatoms are absent in the sediments from the last glacial period in the Nordic Seas. They are also very infrequently found in the sediments of the northern North Atlantic, probably due to extensive sea-ice cover inhibiting diatom production. In contrast, diatom records from the Atlantic upwelling sites reveal maximum productivity during the Last Glacial Maximum (LGM) (Abrantes, 2000).

The last glacial period is perturbed by millennial scale climate variability in the form of Heinrich Events (HE) and Dansgaard–Oeschger cycles (*see* Sub-Milankovitch (DO/Heinrich) Events), and there has been a debate on the degree to which massive

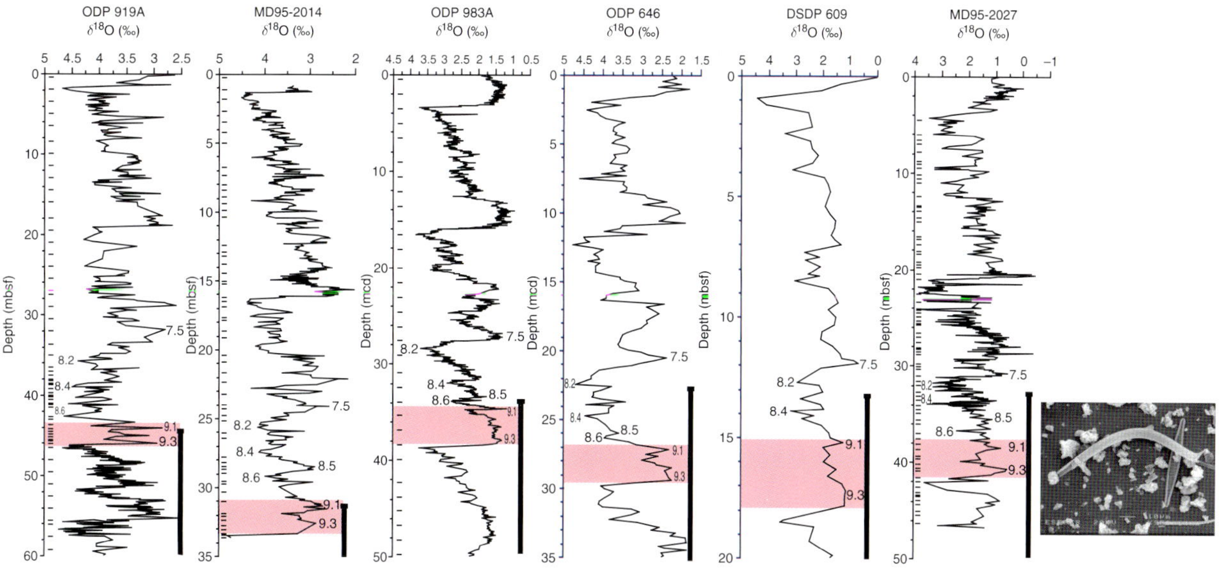

Figure 4 The LO of *P. curvirostris* plotted on the oxygen isotope stratigraphy of each core (Koç *et al.* (2001)). MIS 9 is shaded, and the substages numbered according to Prell *et al.* (1986). Sampling intervals are plotted as bars on the left side of the records. Scanning electron microscope (SEM) photograph of *P. curvirostris* Jordan and Priddle and *Nitzschia fossilis* (Frenguelli) Kanaya on the right side, Sample 162-983A-9H-3, 49–50 cm (from plate 2 in Koç *et al.* (2001)).

iceberg discharges of the HE affected diatom productivity. Sancetta (1992) argued for increased diatom production in the North Atlantic linked to HE suggesting that the presence of numerous icebergs would have supported high production through physical (turbulent mixing) and/or biogeochemical (nutrient input) mechanisms. However, a recent detailed study on HE1 and HE4 shows the opposite effect, namely the reduction of diatom production in relation to massive iceberg discharges. Heinrich layers are deposited mainly between 40° and 55 °N, within the so-called, *Ruddiman belt* (Ruddiman, 1977). North of the Ruddiman belt the HE4 iceberg discharges had a strong impact on productivity resulting in its drastic decrease, whereas during the HE1, productivity showed a progressive increase through the deglaciation. Within the Ruddiman belt massive iceberg discharges had a major effect on productivity during both HE4 and HE1. The possibility that diatoms had been diluted by ice-rafted detritus (IRD) was tested by normalization to a constant ^{230}Th flux, and this showed that the enhancement of particle rain rate does not explain the decrease of diatom flux. The hypothesis that increased nutrient supply from southern source waters could cause an increase in productivity south of the area of maximum iceberg influence was also tested, and no indication for a productivity increase during HE1 and HE4 south of the Ruddiman belt was found. In summary, the massive iceberg discharges associated with HE1 and HE4 had strong impacts on surface ocean productivity, causing its drastic decrease.

Diatoms in Surface Sediments

To gain knowledge about present-day diatom ecology and biogeography, the distribution of diatom species and/or assemblages in the surface sediments of the North Atlantic Ocean and the Arctic Ocean has been mapped, and the resulting information has been applied to fossil records to reconstruct past ocean conditions. The high diversity of diatoms in polar and arctic environments makes them one of the foremost tools available for paleoclimatic reconstructions of the sea-ice margin, oceanic fronts, distribution of water masses and quantitative estimates of, for example, past SSTs. Several types of transfer functions, based on different statistical techniques, have been applied to quantify the relationship between the assemblages and the physical parameters. The most common ones are the Imbrie and Kipp technique (IKM; Imbrie and Kipp, 1971), the Modern Analog Technique (MAT; Hutson, 1980), Weighted averaging partial least square (WA-PLS; ter Braak and Juggins, 1993), Maximum Likelihood (ML); and the Artificial Neural Network approach (ANN; Malmgren and Nordlund, 1997). For a more detailed discussion of diatoms and the transfer function approach, *see* Data Interpretation.

North Atlantic and the Nordic Seas

Diatoms have proved to be an excellent paleoclimatic tool in the Nordic Seas and the northern North Atlantic. Koç and Schrader (1990) mapped the modern distribution of diatoms in this area based on 87 surface sediment samples and demonstrated that diatoms show good analog relations to modern oceanic conditions and are linearly related to sea-surface temperature. Recently, Andersen *et al.* (2004a) expanded this calibration set to 139 surface samples and refined the SST equation. Q-mode factor analysis was applied to this set which defined eight different factors (**Figs. 5A–H**). The Arctic Greenland Assemblage (factor 1) with *Thalassiosira anguste-lineata* and *Thalassiosira trifulta* mirrors Arctic water masses and has its highest factor loadings in the Greenland Sea. The North Atlantic Assemblage (factor 2) with *Thalassiosira oestrupii* reflects the influence of the warm North Atlantic Drift. The sub-Arctic Assemblage (factor 3) with *Rhizosolenia hebetata* f. *semispina* and *Rhizosolenia borealis* has its highest loadings in areas of mixture between Arctic and Atlantic waters, both northeast and southwest of Iceland. The Norwegian Atlantic Current (NwAC) Assemblage (factor 4) with *Thalassionema nitzschioides*, *Proboscia alata*, and *Thalassiosira angulata* can be used as an indicator of the NwAC. The Sea Ice Assemblage (factor 5) with *Fragilariopsis oceanica*, *Fragilariopsis cyclindrus*, and *Thalassiosira hyaline* corresponds to the spring sea-ice limit. The Arctic Assemblage (factor 6) with *Thalassiosira gravida* spores and vegetative cells has its highest loadings in waters between the Polar and Atlantic waters north of Iceland, and in an intermediate area southwest of Iceland. The East and West Greenland Current Assemblage (factor 7) with *Thalassiosira gravida* vegetative cells has its highest factor loadings under the EGC and the West Greenland Current, continuing into the Labrador Sea. A small plume of this assemblage is also seen in close vicinity to the Arctic front, west of the NwAC. The Mixed Water Masses Assemblage (factor 8) with *Rhizosolenia borealis* has the highest loadings in a conjunction area between the Labrador Sea and the North Atlantic. Because of their close affinity to modern hydrographic regimes, these factors provide us with a tool to reconstruct details of past surface current variability.

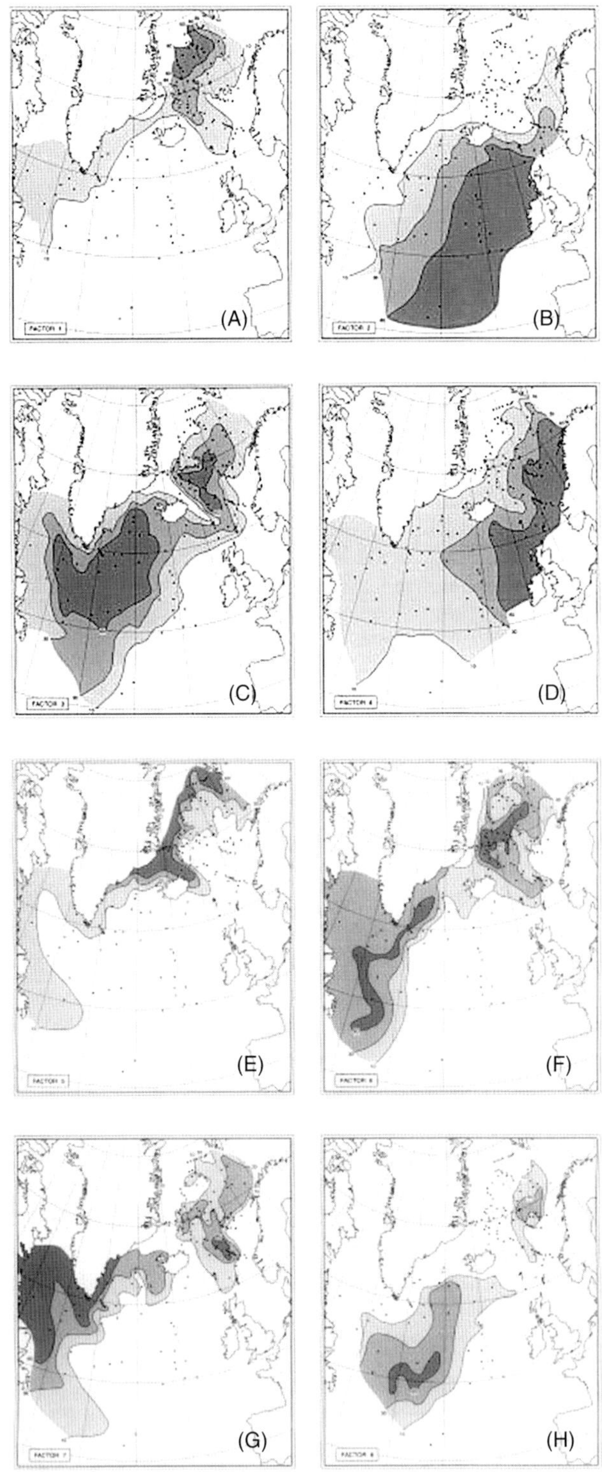

Figure 5 The modern geographic distribution and maximum factor loadings (×100) (shaded for factor loadings 10, 30, and 60) of: (A) the Arctic Greenland Assemblage (factor 1); (B) the North Atlantic Assemblage (factor 2); (C) the sub-Arctic Assemblage (*factor 3*); (D) and the Norwegian Atlantic Current Assemblage (factor 4) Andersen *et al.*, 2004a); (E) the Sea Ice Assemblage (factor 5); (F) the Arctic Assemblage (factor 6); (G) the East and West Greenland Current Assemblage (factor 7); and (H) and the Mixed Water Masses Assemblage (factor 8) (Andersen *et al.*, 2004a).

Portuguese Margin

The entire Portuguese west coast and western Algarve shelf are areas of upwelling. Diatom assemblages of the Portuguese continental margin have been mapped from 169 surface sediment samples (Abrantes, 1988). The results of the cluster and principal component analysis indicate that diatom species and assemblages are good indicators of the changing character of the upwelling system in this area. For example, small species of the genus *Thalassiosira* occur in areas of persistent upwelling.

Baffin Bay and Davis Strait

The Baffin Bay and Davis Strait area is dominated by a counter-clockwise circulation driven by the cold Baffin Current and the West Greenland Current. Baffin Bay is covered by winter sea ice, and pack ice can last as long as August in west central Baffin Bay. Analysis of 74 surface sediment samples resulted in the definition of five diatom assemblages in this area (Williams, 1986). These assemblages are closely related to discrete water masses and ice conditions in the area. The Baffin Current Assemblage (factor 1) is almost totally dominated by spores of *Thalassiosira gravida*. The Summer Pack Ice Assemblage (factor 2) is dominated by *Actinocyclus curvatulus* with a secondary contribution of *Thalassiosira trifulta*. The West Greenland Current Assemblage (factor 3) is dominated by *Thalassiosira hyaline*, and the Fast Ice Assemblage (Factor 4) is characterized by *Porosira* sp. spores.

Laptev and Kara Seas

The modern distribution pattern of diatoms in the Laptev Sea has been mapped from 89 surface sediment samples (Cremer, 1999), and grouped by Q-mode factor analysis into five factors (diatom assemblages): the ice diatom assemblage of the central region of the Laptev Sea, the *Chaetoceros* assemblage of the eastern and southeastern shelf, the *Thalassiosira antarctica* assemblage of the continental slope and deep sea, the freshwater diatom assemblage in the vicinity of river mouths and deltas, and the *Thalassiosira nordenskioeldii* assemblage which shows a patchy occurrence on the central Laptev Sea shelf. The main factors controlling the distribution of diatoms in the Laptev Sea are the riverine freshwater input during the summer, which strongly affects the salinity conditions and the sea-ice extent.

The distribution of diatom assemblages from the outer Ob and Yenisei estuaries and the adjacent inner Kara Sea shelf have been mapped from 62 surface sediment samples (Polyakova, 2003). A total number of 260 diatom species and varieties were identified in these samples and grouped according to their known salinity tolerance and their phytogeographical distribution. The overall distribution pattern of these groups mainly reflects summer surface salinity and the pattern of sea-ice distribution in the area.

SSTs in the North Atlantic and the Nordic Seas

Diatom-based quantitative SSTs have been developed using Imbrie and Kipp's (1971) 'factor analysis regression' and ter Braak and Juggins's (1993) WA-PLS. Both transfer functions have low root mean square error of prediction RMSEP (ca. 1° C), a high coefficient of determination r^2 (ca. 0.9), and a maximum bias of ca. 1° C (Birks and Koç, 2002; Andersen *et al.*, 2004a).

Transfer functions in the high-latitude North Atlantic have primarily been applied to the deglaciation (Koç and Jansen, 1992) and the Holocene periods (Koç *et al.*, 1993; Birks and Koç, 2002; Andersen *et al.*, 2004a, b; Jiang *et al.*, 2005). Due to extensive sea-ice cover during the glacial periods it has been impossible to obtain continuous, long diatom records from the high-latitude North Atlantic.

Deglaciation

Diatoms are absent in the LGM sediments of the northern North Atlantic and the Nordic Seas. Diatoms first occur in the sediments at the start of deglaciation as the sea-ice cover retreated. Reconstruction of SSTs shows that the initial warming of the surface North Atlantic after the LGM happened synchronously between 50°N and 63°N at 13.4 ^{14}C ka, but was delayed by 1 kyr at 72°N (Koç *et al.*, 1996). While a small ice-free corridor opened along the Norwegian coast during the deglaciation, the rest of the Nordic Seas were still covered by sea ice (**Fig. 6**) (Koç *et al.*, 1993). The Bølling–Allerød period was a climatically unstable time interval punctuated by several SST coolings (Koç and Jansen, 1992). The magnitudes of these brief coolings were strongly subdued in the North Atlantic compared to the eastern Norwegian Sea. A strong (9° C) gradient (the Arctic front) between the Norwegian Sea and the North Atlantic was present throughout the Bølling–Allerød period. This gradient became even stronger (15° C) during the Younger Dryas period, and nearly disappeared during the early Holocene. During the Younger Dryas period diatom assemblages off the Norwegian coast are dominated by sea-ice species indicating the proximity of the marginal ice zone to the area.

Holocene

The sea-ice species of the Younger Dryas period were very rapidly substituted by warm Atlantic species at the Holocene–Younger Dryas transition. In the Nordic Seas the early Holocene is dominated by the

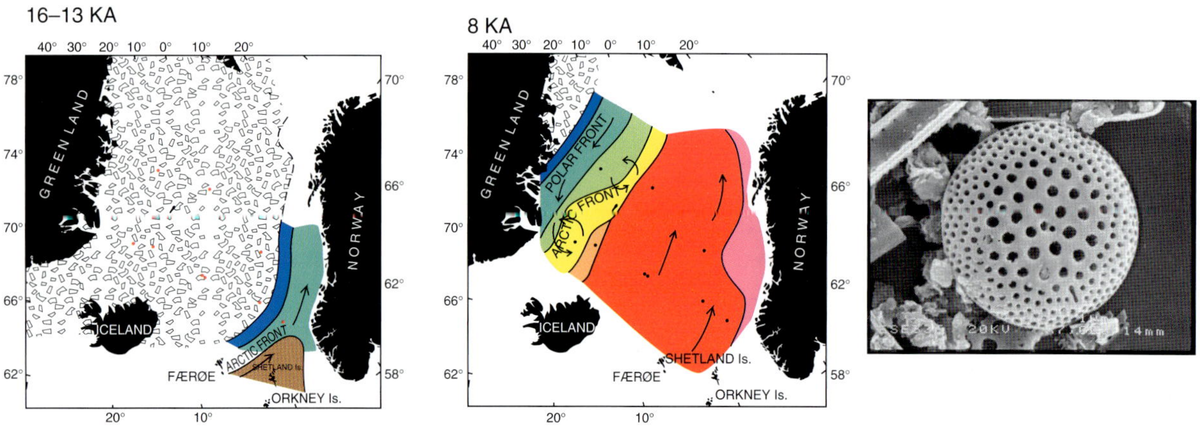

Figure 6 Reconstructions of paleoceanographic conditions of the Nordic Seas during the deglaciation between 16 and 13 ka (calibrated years) and during the Holocene Climate Optimum at 8 ka together with the dominant Atlantic water species *Thalassiosira oestrupii*. Plotted are the sites of the cores used (dots), polar and Arctic fronts with surface ocean circulation (denoted by arrows) and sea-ice cover (broken pattern). Also plotted are the reconstructions of the areal distribution of the various factors described in Koç and Scrader (1990); factor 1 = Norwegian-tlantic Current assemblage (pink); factor 2 = Arctic Waters assemblage (green); factor 3 = sea-ice assemblage (blue); factor 4 = Arctic–Norwegian Waters Mixing assemblage (yellow); factor 5 = Atlantic assemblage (red); factor 6 = Norwegian–Arctic Waters Mixing assemblage (brown). Areas of overlaps are indicated by the mixture of the two overlapping colors (Koç *et al.*, 1993).

North Atlantic Assemblage, with its dominant species *Thalassiosira oestrupii* indicating the invasion of the central and eastern Nordic Seas by warm Atlantic waters (**Fig. 7**) (Koç *et al.*, 1993; Andersen *et al.*, 2004a). After 6.5 cal kyr BP the influence of the North Atlantic Assemblage gradually decreased and arctic species dominated the assemblage. The diatom-based SSTs indicate a division of the Holocene in the Nordic Seas into three periods: the Holocene Climate Optimum (HCO) Period between 9.5 cal kyr BP and 6.5 cal kyr BP with SSTs 4° C warmer than the modern, the Holocene Transition Period (HTP) between 6.5 cal kyr BP and 3 cal kyr BP displaying a gradual SST decrease toward modern values, and the Cool Late Holocene Period (CLPH) between 3 cal kyr BP and 0 cal yr BP with temperatures around the modern SST values (**Fig. 8A**) (Andersen *et al.*, 2004a). The close similarity of the climate development in the Nordic Seas to the high-latitude Northern Hemisphere insolation development indicates that it has mainly been driven by insolation forcing (Koç and Jansen, 1994).

The western sub-Arctic gyre is an area of intense mixing of warm and cold waters, which enhances biological productivity. The Holocene sediments of this area are characterized by abundant diatoms. Several levels within the Holocene are dominated by the species of *Rhizosolenia borealis* and *Thalasiothrix longissima*, often as diatom mats (Andersen *et al.*, 2004a, b).

The development of surface ocean climate in the subpolar North Atlantic deviates from the

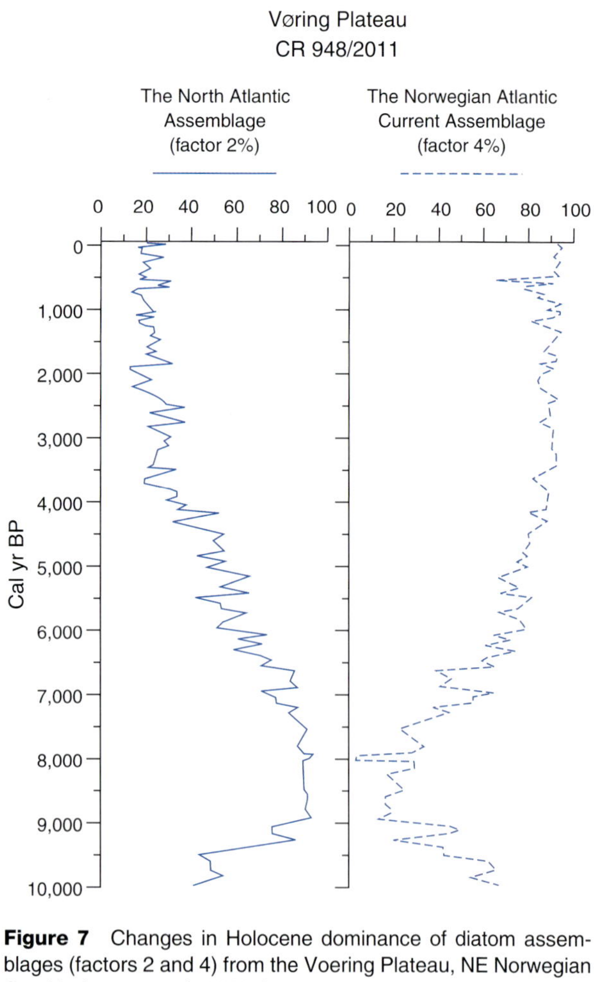

Figure 7 Changes in Holocene dominance of diatom assemblages (factors 2 and 4) from the Voering Plateau, NE Norwegian Sea (Andersen *et al.*, 2004a).

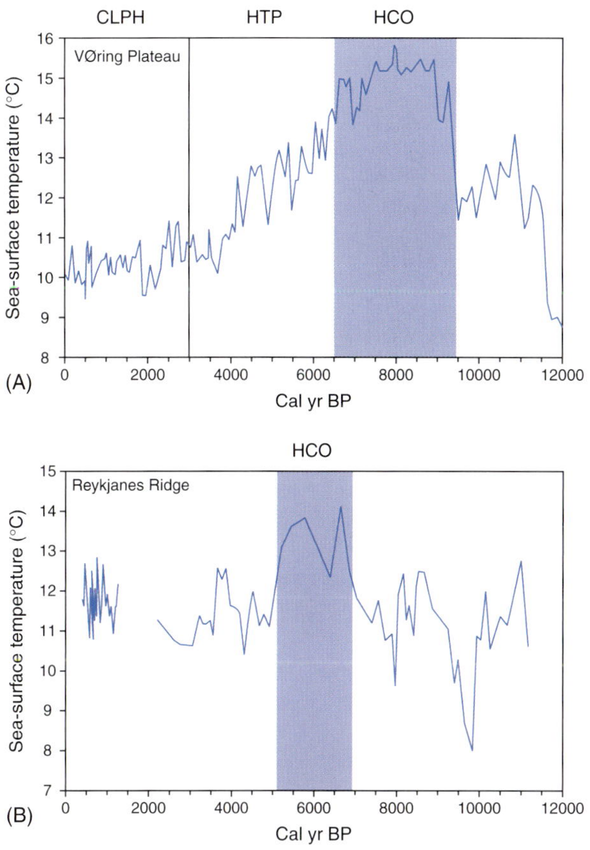

Figure 8 Holocene sea-surface temperature development at A, the Voering Plateau, NE Norwegian Sea and B, the Reykjanes Ridge in the sub-Polar North Atlantic. The Voering Plateau record is divided into the HCO, the HTP, and the CLPH (Andersen *et al.*, 2004a). The shaded area denotes the HCO at both sites. Inset, scanning electron microscope (SEM) picture of the Atlantic water species *Thalassiosira oestrupii*, which shows increased abundances during the HCO.

development in the Nordic Seas, considering both the timing of the Holocene climate optimum and the stability of the SSTs. Over the Reykjanes area the thermal optimum was not reached until 7.5–5 kyr, lagging the insolation maximum by 4 kyr (**Fig. 8B**) (Andersen *et al.*, 2004b). Climate development in the subpolar North Atlantic in the proximity of the Sub-Arctic Front seems to have been mainly affected by (1) incursions of meltwater from the melting remnants of the Laurentide Ice Sheet, (2) increased outflow of polar water from the Canadian Arctic, and (3) variations in the production rate of NADW.

Conclusion

Numerous studies have shown that diatoms, due to their high diversity in polar and arctic environments, are one of the foremost tools available for biostratigraphic and paleoceanographic studies in the mid- to high-latitude North Atlantic. Their high sensitivity to surface ocean conditions have rendered them invaluable especially for reconstructions of the sea-ice margin, oceanic fronts, distribution of surface water masses and quantitative estimates of, for example, past SSTs. They have proved to be especially useful for reconstructing the subtle SST changes of the interglacials (e.g., the Holocene). A great potential exits, therefore, in utilizing diatoms for exploring the natural variability of the mid- to high-latitude North Atlantic with quantitative methods to extract information on the magnitudes and frequency of past ocean changes.

See also: **Diatom Introduction. Diatom Methods**: $\delta^{18}O$ Records; Data Interpretation. **Paleoceanography, Biological Proxies**: Marine Diatoms. **Paleoclimate Reconstruction**: Sub-Milankovitch (DO/Heinrich) Events.

References

Abrantes, F. (1988). Diatom assemblages as upwelling indicators in surface sediments off Portugal. *Marine Geology* **85**, 15–39.

Abrantes, F. (2000). 200,000 yr diatom records from Atlantic upwelling sites reveal maximum productivity during LGM and a shift in phytoplankton community structure at 185,000 yr. *Earth and Planetary Science Letters* **176**, 7–16.

Andersen, C., Koç, N., Jennings, A., and Andrews, J. T. (2004a). Nonuniform response of the major surface currents in the Nordic Seas to insolation forcing: Implications for the Holocene climate variability. *Paleoceanography* **19**, PA2003 (doi: 10.1029/2002PA000873).

Andersen, C., Koç, N., and Moros, M. (2004b). A highly unstable Holocene climate in the subpolar North Atlantic: Evidence from diatoms. *Quaternary Science Reviews* **23**, 2155–2166.

Baldauf, J. G. (1984). Cenozoic diatom biostratigraphy and paleoceanography of the Rockall Plateau region, North Atlantic, Deep Sea Drilling Project, Leg 81. In *Initial Reports of the Deep Sea Drilling Project 81*, pp. 439–478. U S Government Printing Office, Washington DC.

Baldauf, J. G. (1986). Diatom biostratigraphic and paleoceanographic interpretations for the middle to high latitude North Atlantic Ocean. In *Special Publication 21: North Atlantic Paleoceanography* (C. P. Summerhays and N. J. Shackleton, Eds.), pp. 243–252. Geological Society.

Baldauf, J. G. (1987). Diatom biostratigraphy of the middle and high latitude North Atlantic Ocean, Deep Sea Drilling Project, Leg 94. In *Initial Reports of the Deep Sea Drilling Project 94*, pp. 729–762. U S Government Printing Office, Washington DC.

Bauch, H. A., and Polyakova, Y. I. (2003). Diatom-inferred salinity records from the Arctic Siberian Margin: Implications for fluvial runoff patterns during the Holocene. *Paleoceanography* **18**(2), 1027, (doi:10.1029/2002PA000847).

Birks, C. J. A., and Koç, N. (2002). A high-resolution diatom record of late-Quaternary sea-surface temperatures and oceanographic conditions from the eastern Norwegian Sea. *Boreas* **31**(4), 323–344.

Clark, D. L., *et al.* (1980). *GSA Spec. Publ.* **181**, 1–57.

CLIMAP Project Members (1981) Seasonal reconstruction of the earth's surface at the last glacial maximum. *Geological Society of America Map and Chart Series* MC-36.

Cremer, H. (1999). Distribution patterns of diatom surface sediment assemblages in the Laptev Sea (Arctic Ocean). *Marine Micropaleontology* **38**, 39–67.

Dell'agnese, D. J., and Clark, D. L. (1994). Siliceous microfossils from the warm Late Cretaceous and Early Cenozoic Arctic Ocean. *Journal of Paleontology* **68**, 31–47.

Dzinoridze, R. N., Jouse, A. P., Koroleva-Golikova, G. S., *et al.* (1978). Diatom and radiolarian Cenozoic stratigraphy, Norwegian Basin; DSDP Leg 38. In *Initial Reports of Deep Sea Drilling Project* (M. Talwani and G. Udintsev *et al.*, Eds.), Supplement to vols. 38, 39 and 41, pp. 289–428. U.S. Government Printing Office, Washington.

Hutson, W. H. (1980). The Agulhas current during the Late Pleistocene: Analysis of modern faunal analogs. *Science* **207**, 64–66.

Imbrie, J., and Kipp, N. G. (1971). A new micropaleontological method for quantitative micropaleontology: Application to a late Pleistocene Caribbean core. In *Late Cenozoic Glacial Ages* (K. Turekian, Ed.), pp. 71–181. Yale University Press, New Haven.

Jiang, H., Eiriksson, J., Schulz, M., Knudsen, K. L., and Seidenkrantz, M. S. (2005). Evidence for solar forcing of sea-surface temperature on the North Icelandic Shelf during the Late Holocene. *Geology* **33**(1), 73–76.

Koç, K. N., and Jansen, E. (1992). A high resolution diatom record of the last deglaciation from the SE Norwegian Sea: Documentation of rapid climatic changes. *Paleoceanography* **7**, 499–520.

Koç, K. N., and Schrader, H. (1990). Surface sediment diatom distribution and Holocene paleotemperature variations in the Greenland, Iceland and Norwegian Sea. *Paleoceanography* **5**(4), 557–580.

Koç, N., and Flower, B. (1998). High-resolution Pleistocene diatom biostratigraphy and paleoceanography of Site 919 from the Irminger Basin. In *Proceedings of the Ocean Drilling Program: Scientific Results* (A. D. Saunders, H. C. Larsen and S. W. Wise, Jr, Eds.), vol. 152, pp. 209–219. Ocean Drilling Program, College Station.

Koç, N., Hodell, D. A., Kleiven, H., and Labeyrie, L. (1999). High-resolution Pleistocene diatom biostratigraphy of Site 983 and correlations to isotope stratigraphy. In *Proceedings of the Ocean Drilling Program: Scientific Results* (E. Jansen, M. Raymo and P. Blum, Eds.), vol. 162, pp. 51–62. Ocean Drilling Program, College Station.

Koç, N., and Jansen, E. (1994). Response of the high-latitude northern hemisphere to orbital climate forcing: Evidence from the Nordic Seas. *Geology* **22**, 523–526.

Koç, N., Jansen, E., and Haflidason, H. (1993). Paleoceanographic reconstructions of surface ocean conditions in the Greenland, Iceland and Norwegian Seas through the last 14 ka based on diatoms. *Quaternary Science Reviews* **12**, 115–140.

Koç, N., Jansen, E., Hald, M., and Labeyrie, L. (1996). Late glacial-Holocene sea surface temperatures and gradients between the North Atlantic and the Norwegian Seas: Implications for the Nordic heat pump. In *Special Publication No. 111: Late Quaternary Palaeoceanography of the North Atlantic Margins* (J. T. Andrews, W. E. N. Austin, H. Bergsten and A. E. Jennings, Eds.), pp. 117–185. Geological Society.

Koç, N., Labeyrie, L., Manthé, S., Flower, B. P., Hodell, D. A., and Aksu, A. (2001). The last occurrence of *Proboscia curvirostris* in the North Atlantic marine isotope stages 9-8. *Marine Micropaleontology* **41**, 9–23.

Koç, N., and Scherer, R. (1996). Neogene diatom biostratigraphy of the Iceland Sea Site 907. In *Proceedings of the Ocean Drilling Program: Scientific Results* (J. Thiede and A. Myhre, Eds.), vol. 151, pp. 61–74. Ocean Drilling Program, College Station.

Polyakova, Y. I. (2003). Diatom assemblages in surface sediments of the Kara Sea (Siberian Arctic) and their relationship to oceanological conditions. In *Proceedings in Marine Science: Siberian River Run-off in the Kara Sea: Characterization, Quantification, Variability and Environmental Significance* (R. Stein *et al.*, Eds.), pp. 375–399. Springer, New York.

Ruddiman, W. F. (1977). Late Quaternary deposition of ice-rafted sand in the subpolar North Atlantic (lat 40° to 65°N). *Journal of Foraminiferal Research* **19**, 253–267.

Sancetta, C. (1992). Primary production in the glacial North Atlantic and North Pacific Oceans. *Nature* **360**, 249–251.

Schrader, H.-J., and Fenner, J. (1976). Norwegian Sea Cenozoic diatom biostratigraphy and taxonomy. In *Initial Reports of Deep Sea Drilling Project* (M. Talwani and G. Udintsev *et al.*, Eds.), vol. 38, pp. 921–1099. U S Government Printing Office, Washington.

Scherer, R., and Koç, N. (1996). Late Paleogene diatom biostratigraphy of the northern Norwegian-Greenland Sea, ODP Leg 151. In *Proceedings of the Ocean Drilling Program: Scientific Results* (J. Thiede and A. Myhre, Eds.), vol. 151, pp. 75–99. Ocean Drilling Program, College Station.

ter Braak, C. J. F., and Juggins, S. (1993). Weighted averaging partial least squares regression (WA-PLS): An improved method for reconstructing environmental variables from species assemblages. *Hydrobiologia* **269/270**, 485–502.

Williams, K. M. (1986). Recent Arctic marine diatom assemblages from bottom sediments in Baffin Bay and Davis Strait. *Marine Micropaleontology* **10**, 327–341.

Pacific

I Koizumi, Atsubetsu-kita, Sapporo, Japan

© 2007 Elsevier B.V. All rights reserved.

Introduction

The distribution of living diatom species recognized from phytoplankton samples or sediment traps accurately reflects differences in temperature, salinity, nutrients, and other physical-chemical parameters. Many diatom species are endemic to a given region, whereas others have a more cosmopolitan distribution. Therefore, diatoms can be excellent indicators of paleoecology and paleotemperatures (*see* Diatom Introduction, Data Interpretation).

Marine diatoms are well preserved in the siliceous ooze (and diatomaceous rocks) in which diatoms make up more than 30% of the sediment components (**Fig. 1**). Within the Pacific Ocean, siliceous ooze is

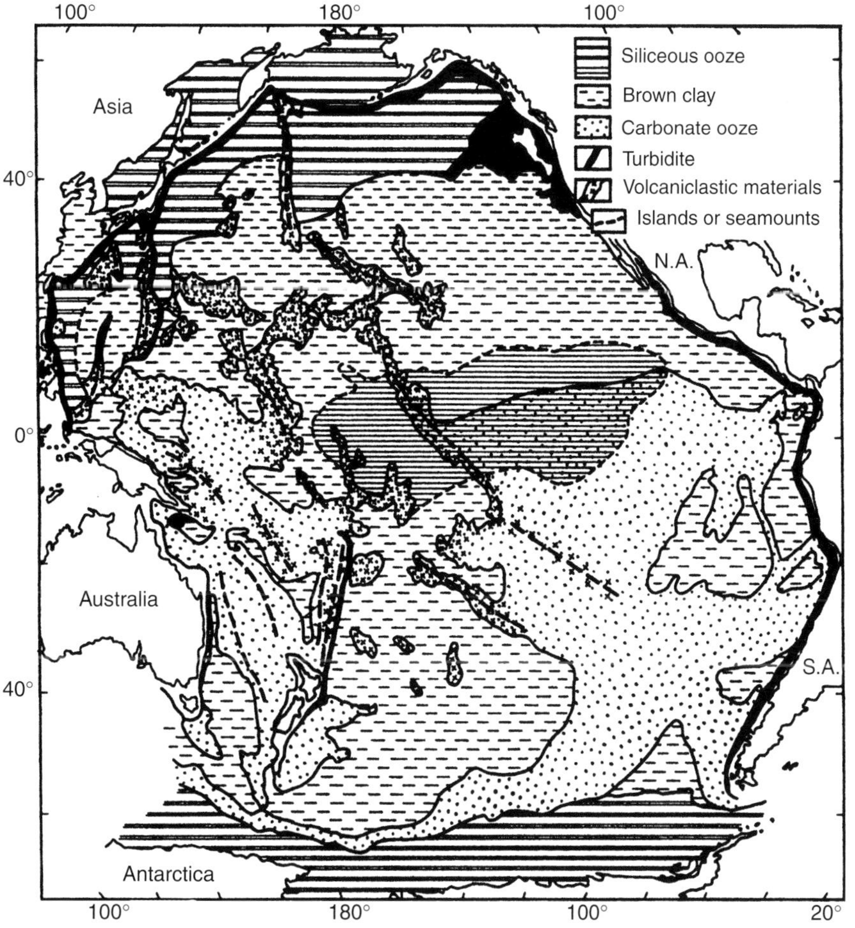

Figure 1 General patterns of recent sediments in the Pacific Ocean (Okada and Kobayash, 1974).

concentrated in (1) the middle to high latitudes of the North Pacific Ocean, including the Bering Sea, Okhotsk Sea, and Japan Sea; (2) the southern belt around Antarctica; and (3) the eastern equatorial Pacific. These areas coincide with regions of increased nutrient levels (nitrogen, phosphorous, silica) and abundant diatoms in the surface wate‑s **Figure 1**. See Stickley, (see Antarctic Waters) this volume, for a discussion of the diatom records from the Southern Ocean.

Kanaya and Koizumi (1966), and Sancetta and Silvestri (1986) demonstrate that diatoms present in the surface sediment of the seafloor in the North Pacific can be successfully used to determine the characteristics of overlying water masses, even though considerable numbers of finely-silicified diatom taxa living in surface waters may not be preserved in deep-sea sediments.

Diatom Habitats

Modern oceanic holoplanktonic and neritic to littoral species ecology and biogeography are important for paleoclimate studies in the North Pacific. Oceanic diatoms live in the open ocean, away from coastal influences, and are moved by wind and current. Neritic diatoms live and reproduce in shallower waters over the continental shelf and are influenced by coastal processes. Littoral diatoms live in regions of reduced salinity resulting from freshwater drainage from land. They are somewhat heterogeneous and their abundance increases under the influences of coastal waters with abundant nutrients (Hendey, 1964). Freshwater diatoms are indicators of continental material (**Fig. 2**).

Oceanic diatoms are holoplanktonic and free-floating. Meroplanktonic diatoms are pelagic only for part of their lives, spending the remainder on the bottom, and they are often found under oceanic conditions. Tychopelagic diatoms spend probably the greater part of their lives as bottom forms lying on mud or sand or attached as ribbon-like bands to a fixed substratum and enter the surface layers of the sea only if they are moved forcibly from their natural habitat by waves or currents.

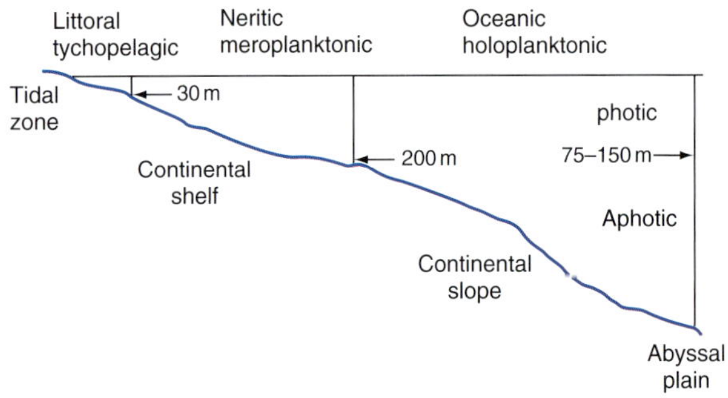

Figure 2 Marine diatom habitats (Koizumi *et al.*, 2004).

Seasonal Fluxes of Diatoms in Time-Series Sediment Traps

The present is a key to resolve the past. One of the best ways to accurately understand paleoenvironments is to adapt present-day data for the high-resolution inference of past environments. Modern ecological and biogeographic information needs to be constructed from data on diatom fluxes, species composition of diatom assemblages, hydrographic data sets, and seafloor-surface sediments representing present conditions. Tanimure (1992) identified and tabulated 136 species or varieties belonging to 55 genera in 39 cup-samples from the sediment trap moorings in the Izu Ogasawara Trench (Tanimura, 1992). Five high flux periods occurred during the spring or autumn months, corresponding to the typical bimodal pattern of diatom production in temperate waters (**Fig. 3**). Four minor peaks reflect high diatom production associated with a nutrient-rich river supply during the early summer months, or with local upwelling in the winter months. The taxonomy and ecology of the *Thalassionema nitzschioides* complex was defined using 65 sediment trap samples and two sediment cores beneath the Kuroshio Current (Tanimura, 1999). *Thalassionema nitzschioides* var.

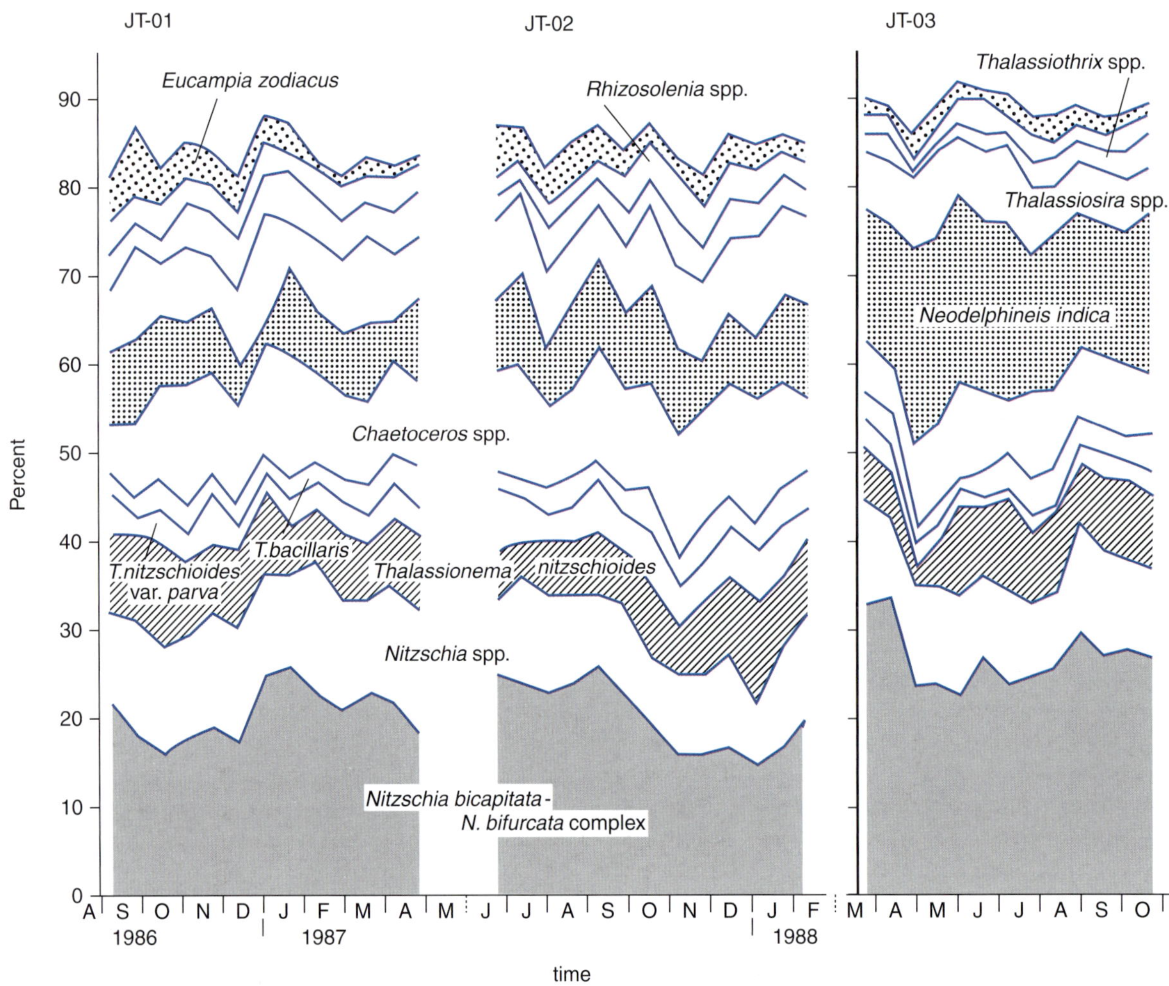

Figure 3 Cumulative percentages of common diatom taxa from the sediment from Station JT-. High abundances of *Nitzschia bicapitata-Nitzschia bifurcata* complex, *Thalassionema nitzschioides*, and *Neodelphineis indica* are all characteristic of assemblages found in sediment traps (Tanimura, 1992).

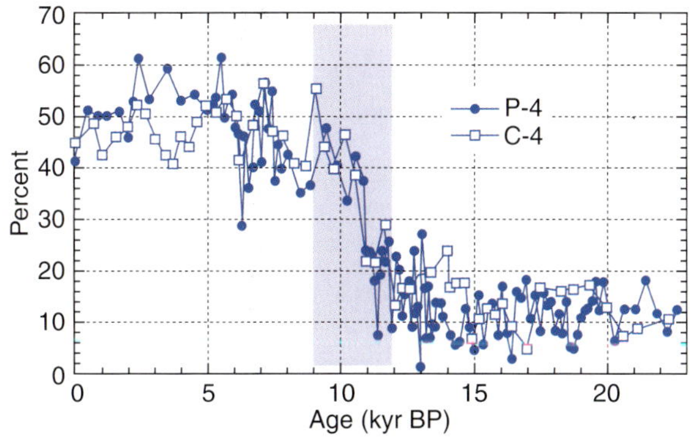

Figure 4 Down-core changes in percentage abundances for three *Thalassionema nitzschioides* varieties (*T. nitzschioides* var. *incurvata*, var. *inflata* and var. *parva*) in the *T. nitzschioides* complex in cores P-4 and C-4. Shaded interval (12 to 9 kyr BP) indicates a transition from the Kuroshio Current taking predominantly offshore paths at 23–12 kyr BP, to the modern current-path after 9 kyr BP (Tanimura, 1999).

incurvata, var. *inflata*, and var. *parva* dominate in warm saline waters in the Subtropical Gyre, while *T. nitzschioides sensu stricto* is abundant in less-saline waters northwest of the Kuroshio Current. Downcore analysis indicated a sharp increase in the abundances of the three *T. nitzschioides* varieties during the period from 12 to 9 kyr BP (**Fig. 4**).

In the equatorial Pacific Ocean, *Ethmodiscus* ooze, which consists mainly of *Ethmodiscus rex*, occurs in distinctive low nutrient regions, where diatoms are otherwise rare in the surface waters (Mikkelsen, 1977; Tanimura, 1981). Specifically, *Ethmodiscus ooze* is formed by differential dissolution and/or winnowing of diatom frustules (Mikkelsen, 1977).

Sea Surface Temperature Reconstruction

The effectiveness of ecological categories of the diatom species, described in **Appendix 1** as paleoceanographic proxies, was extracted using Q-mode factor analysis of the diatom assemblages in core MD01-2421 off central Japan (Koizumi *et al.*, 2004) and five piston cores around the Tsugaru Strait between the Japan Sea and the northwest Pacific (Koizumi *et al.*, 2006). It has been demonstrated that temporal variation in diatom assemblages reflects the variability of physiographical and ecological factors at the given core site (**Fig. 5**).

In order to estimate sea-surface water temperatures, Kanaya and Koizumi (1966) proposed the diatom temperature (*Td*) ratio: $Td = [Xw\ (Xw + Xc)^{-1}] \times 100$, where *Xw* is the frequency of warm-water species and *Xc* is that of cold-water species (**Appendix 1**). *Td* ranges in value from 0 to 100. Going from the sub-Arctic region to the tropical region, *Td* becomes systematically larger, showing a positive correlation with sea-surface temperature over a given geographical site.

Koizumi *et al.* (2004) redefined the *Td* ratio as *Td′* ratio, applied strictly to oceanic holoplanktonic association because *Odontella aurita*, which is a neritic to littoral form and originally defined as a cold-water (*Xc*) species by Kanaya and Koizumi (1966), is not included (**Appendix 1**). Several oceanic holoplanktopnic and neritic to littoral meroplanktonic warm-water (XW) species were also added to the originally defined ones (*Xw*) based on information of Cupp (1943), Hasle (1960, 1976, 2001), Hendey (1964), Jousé *et al.* (1969), Simonsen (1974), Sancetta (1982), Sancetta and Silvestri (1986), Tanimura (1992, 1999), and Shiga and Koizumi (2000). The oceanic holoplanktonic cold-water (XC) species *Fragilariopsis oceanica* was also added (*Xc*). Namely, $Td' = [(Xw + XW)\ (Xw + XW + Xc + XC)^{-1} \times 100$ (**Appendix 1**).

In order to interpret the Pliocene paleoclimatic changing of DSDP Site 580 in the northwestern Pacific, Barron (1992) proposed the Twt ratio: $Twt = (Xw + 0.5Xt)/(Xc + Xt + Xw)$, where Xw equals the distinctly subtropical to tropical taxa (*Azpeitia nodulifera*, *A. tabularis*, *Actinocyclus ellipticus*, *Nitzschia reinholdii-N. marina*, *N. fossilis*, *Nitzschia jouseae* and *Thalassiosira convexa*); *Xt* equals the warm transitional taxa (*Thalassionema nitzschioides*, *Thalassiosira oestrupii* and *Coscinodiscus radiatus*), and *Xc* equals the total percentage of cold-water taxa (*Actinocyclus curvatulus*, *Neodenticula koizumii*, *N. kamtschatica*, *Chaetoceros* spores, *Rhizosolenia barboi*, *Thalassiosira latimarginata* (= *T. trifulta*) and *Coscinodiscus marginatus*).

Statistical analysis of high-resolution time-series records of diatom assemblages in the North Pacific and discussion of their results in terms of regional processes have been encouraged since Sancetta and Silvestri (1986). They performed the cluster analysis of diatom assemblages in the down-core samples together with core top samples in the North Pacific, in order to define the statistically distinct clusters with characteristic taxonomic compositions and oceanographic significance.

The transfer function for sea-surface temperature (SST) and salinity (SSS) estimates was attempted on diatom assemblages in the surface sediment samples and applied to the assemblages spanning 30,000 years in two piston cores from the northwestern Pacific (Koizumi *et al.*, 2001). Six major factors for diatom assemblages were defined as independent variables for summer and winter SST and SSS in regression analysis (**Fig. 6**). These factors account

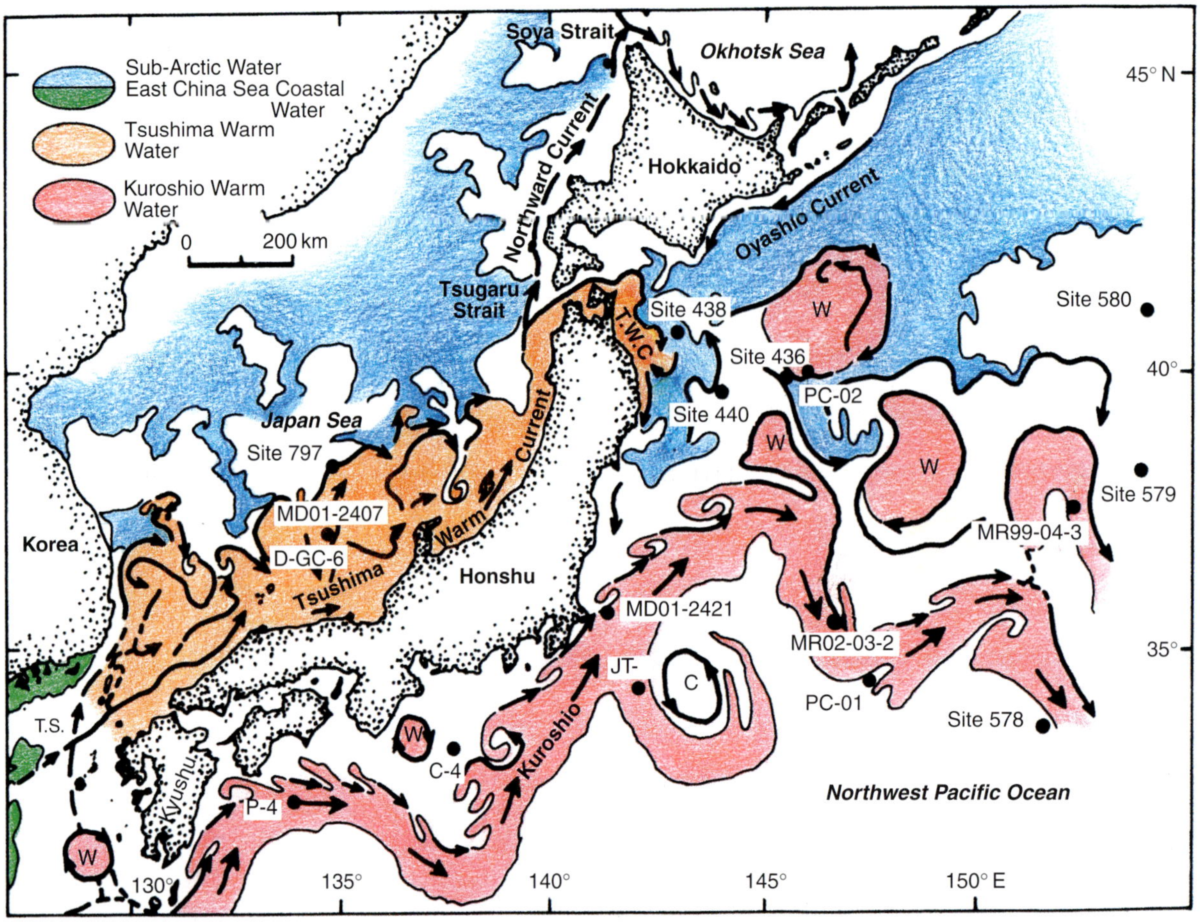

Figure 5 Map showing major current systems, the generalized distribution of surface water masses, and the location sites of cores cited in this paper around the Japanese Islands. Arrows indicate flow of the current. W = warm-water, C = cold-water (Kawai, 1991) T.S. = Tsushima Strait, T.W.C. = Tsugaru Warm Current.

for over 90% of the total variance. As winter and summer SST and SSS remarkably decreased at 24 cm in both PC-01 and PC-02 cores, it is considered that this horizon is dated at about 20,000 years in the last glacial maximum (**Fig. 7**).

Biostratigraphy

Datum Levels, Evolving Lineages, and Evolutionary Processes

Throughout much of the middle to high latitude North Pacific and eastern equatorial Pacific, diatom fossils occur more frequently in sediments than other microfossils, with a greater diversity of species. Since the 1960s, when significant amount of seafloor sediments first began to be obtained by piston coring and later by the Deep Sea Drilling Project (DSDP), diatom biostratigraphy has been well established for the various areas of the Pacific, representing different surface water-masses. Increased coring by the Ocean Drilling Program (ODP) during the past 20 years has extended diatom biostratigraphy further back into the geologic time. Thus, the diatoms whose first appearances or last appearances (extinction) are simultaneous in different areas in the Pacific Ocean have been defined, and consequently accurate datum levels, or ages of evolutionary appearance and extinction, have been established. This build-up of the data has contributed to the refinement of stratigraphy, correlation, and chronology.

Recently, Barron (2003) compiled the ages of appearances and disappearances of diatoms during the past 18 Ma in the equatorial Pacific, North Pacific, and Southern Ocean, and **Table 1** shows the appearance and extinction of diatom species during the Quaternary. The spatial distributions of the appearances and disappearances of diatoms are related to environmental changes and/or to the evolutionary processes. The migration of species from their evolutionary home-area outward to other

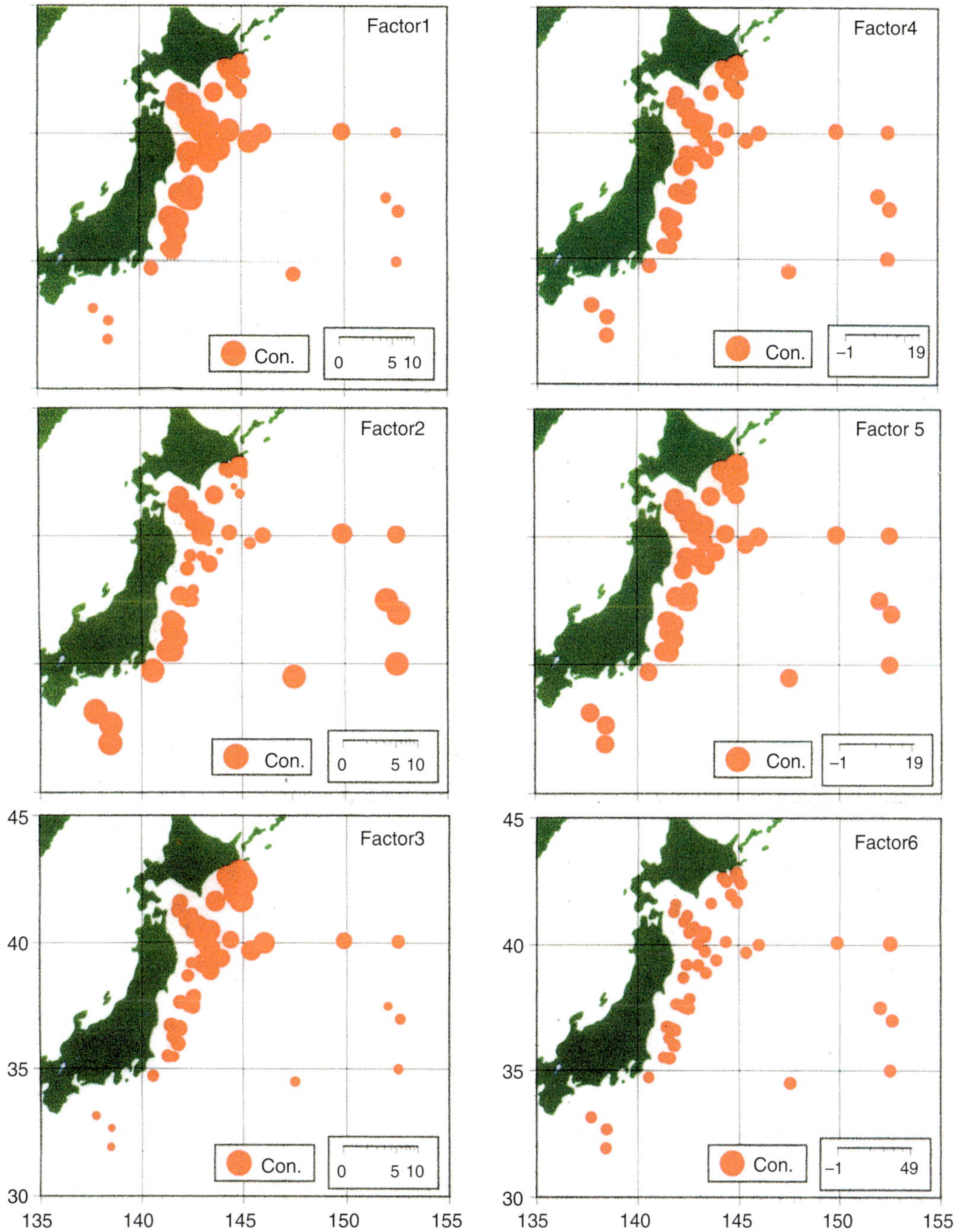

Figure 6 Geographical distribution of the contribution of each factor for diatom fossil assemblages. Factor 1 is composed of littoral tychopelagic species and warm-water species *F. doliolus* which is supposed to have outflowed through the Tsugaru Strait from the Japan Sea. It is distributed in slightly colder mixed water masses. Factor 2 is characterized by the warmer littoral species, and reflects the Kuroshio water mass and the Tsugaru Warm Current. Factor 3 reflects an association with the cold Oyashio Current. Factor 4 is distributed over a large area in the western North Pacific, being composed of a mixed association of cold-water species, freshwater species, and extinct species. Factor 5 is minor but dominant in the northern coastal area with low temperature and low salinity (in summer months, in particular). Factor 6 is dominant in off-coast area and characteristically composed of oceanic and cosmopolitan species (Koizumi *et al.*, 2001).

areas is recognized by delayed first occurrences in those regions and is controlled by the fluctuating frontal boundaries between water-masses. The disappearance of the species in a given region occurs in response to changing surface-water temperatures (and/or salinities), which are beyond the tolerance limitation of the species (Koizumi, 1986b).

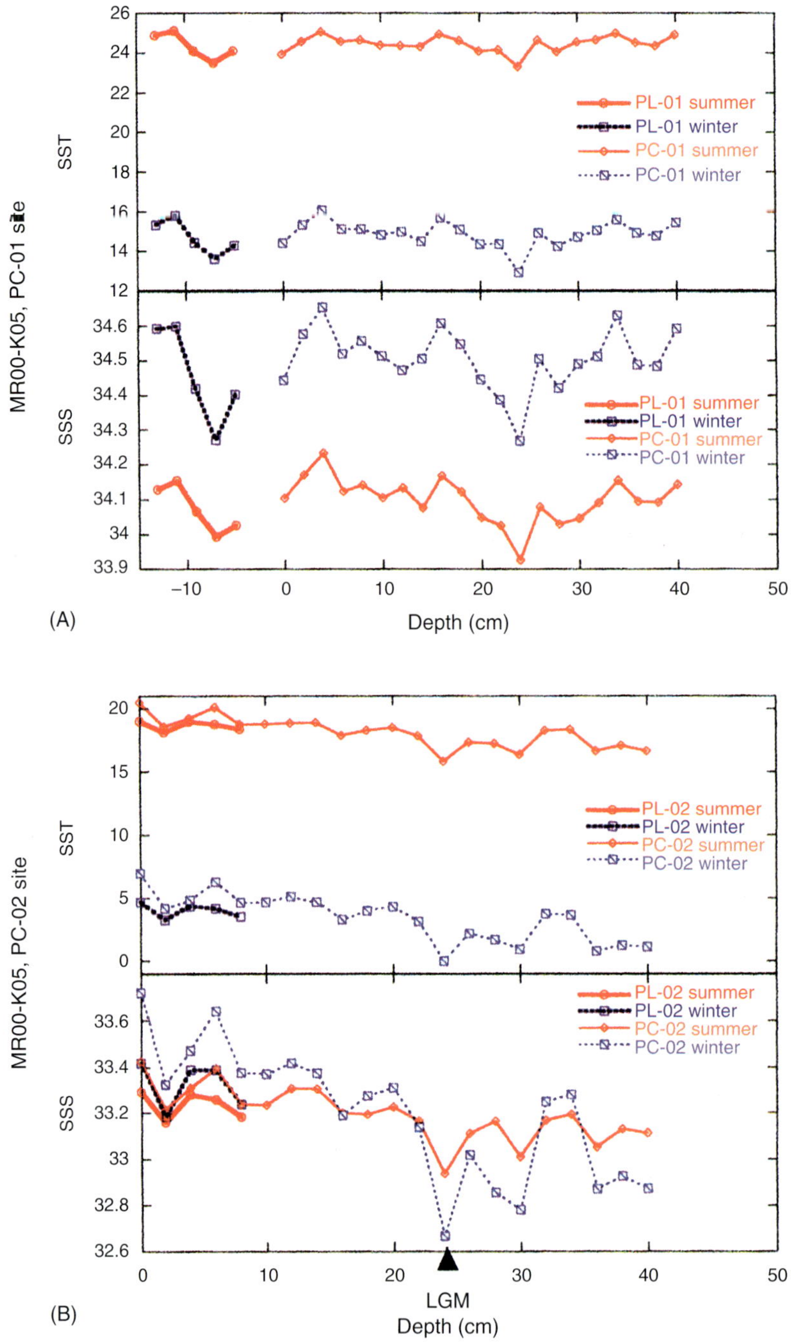

Figure 7 Changes in sea surface temperature and salinity for summer and winter reconstructed by the transfer function method applied to the diatom fossil assemblages from the end of the last glacial to the Present. LGM = Last Glacial Maximum (~20,000 yrs ago BP) (Koizumi *et al.*, 2001).

Datum levels suggest the following the evolving lineages:

(1) from *Neodenticula koizumii* to *Neodenticula seminae*;
(2) from *Rhizosolenia praebergonii* to *Rhizosolenia praebergonii* var. *robusta*;
(3) from *Rhizosolenia bergonii* to *Rhizosolenia matuyamai*;
(4) from *Azpeitia nodulifera* to *Azpeitia barronii*, and
(5) from *Proboscia barboi* to *Proboscia curvirostris*.

Also, the morphological comparison between the ancestor *Nitzschia fossilis* and the descendant *Fragilariopsis doliolus*, based on SEM examination, indicates that *F. doliolus* has a more developed convex dorsal margin than *N. fossilis*, suggesting an evolutionary process (Koizumi and Yanagisawa, 1990). The

Table 1 Ages of appearance and extinction of age diagnostic diatoms in the Pacific Ocean (Barron, 2003)

FA (Ma)	*LA* (Ma)	*Equatorial*	*Middle-to-high latitude*
1.6	0.3		*Proboscia curvirostris*
	0.3–0.4		*Thalassiosira jouseae*
	0.5	*Nitzschia fossilis*	
1.18	1.05	*Rhizosolenia matuyamai*	
	1.5–1.8		*Thalassiosira antiqua*
1.80	living	*Azpeitia barronii*	
2.0	living	*Fragilariopsis doliolus*	
2.5	1.72	*Rhizosolenia praebergonii* v. *robusta*	
2.68	living		*Neodenticula seminae*
3.2	1.9	*Rhizosolenia praebergonii*	
3.9	2.0		*Neodenticula koizumii*
4.0	1.0		*Actinocyclus oculatus*

FA=first appearance, LA=last appearance (extinction).

development of the convex dorsal margin in the valve of *F. doliolus* corresponds with the general decrease of *Td* values observed in the same sedimentary sections. However, no strict correlation can be recognized between the change of the asymmetry index, which defined as a ratio of width in a ventral side and the fluctuations of *Td* values. Therefore, the development of the convex dorsal margin is considered to have been caused by a genetic character rather than the effect of surface-water temperature decline.

Ice Ages

Onset of Glaciation and Climatic Fluctuations

The Quaternary experienced repeated climatic changes, and ice ages are most pronounced after about 800 ka, with alternating cold glacial periods and warm interglacial periods at approximately 100-kyr intervals.

Proxy diatom sea-surface investigations (*Td* values) using DSDP cores from the sites situated on the transitional zone between the modern day sub-Arctic and central water-masses, reconstruct the Pliocene-Pleistocene paleoceanographic history of the northwestern Pacific Ocean (**Fig. 8**; Koizumi, 1986a). These studies have made it clear that the boundary between pre-glacial and glacial occurred at approximately 2.45 Ma at about 40° N latitude, and regional surface-water temperatures were constantly lower than the present-day level. In fact, pronounced drops in surface-water temperature are recognized at 3.15, 2.45, 2.00, 1.60, and 0.90 Ma, with a strong interglacial stage occurring at 0.5 Ma.

Sancetta and Silvestri (1986) defined the relative abundances of diatom taxa in 216 core-tops and compared them with down-core assemblages in six cores beneath different water masses of the entire North Pacific Ocean. The diatom records of high frequency oscillations in glacial–interglacial stages display rapid step-like changes at 2.4, 1.6–1.5, 1.0, 0.8–0.7, and 0.35 Ma. These authors argued that both the high frequency oscillations and the stepwise transitions reflect variations in extent of sub-Arctic water masses, under the influence of the Siberian high pressure and Aleutian lower pressure atmospheric cells.

Stepwise transitions in sub-Arctic water-masses off Sanriku, northeast Japan were recognized in the cold-water species recovered from DSDP Sites 436, 438, and 440 (see **Fig. 5**). Paleo sub-Arctic water masses were replaced by near-modern water masses at approximately 2.3 Ma, whereas the modern cold-water Oyashio Current was established at 1.0 Ma, as its source area started to form sea-ice on a modern scale (Koya, 1995). Also diatom proxy sea-surface temperature values (Twt) show that the warm Kuroshio Current flowed more northerly and in a more coastal position during the late Pliocene, but shifted southward to a near-modern position at 1.8–1.7 Ma (**Fig. 9**). In the Japan Sea, the diatom abundances recognized in ODP Sites 797, 794, and 795 (south-north transect, see **Fig. 5**) decrease remarkably during 2.0–1.6 Ma, across the boundary between Pliocene and Pleistocene. Koya (1998) named this interval, where diatoms are very scarce, the 'Japan Sea Diatom Minimum Interval (JDMI)'. When global sea-level lowered during the glacial stages of the Quaternary, the Japan Sea was under the influence of east-flowing China Sea Coastal Water in the southern area, and diatoms in the northern area are considered to have been reworked, causing the JDMI.

Glacial–Interglacial Cycles

The marked oscillations between warm and cold climates during the Quaternary are demonstrated in the continuous diatom records in the northwestern Pacific Ocean and the Japan Sea. In the area beneath the

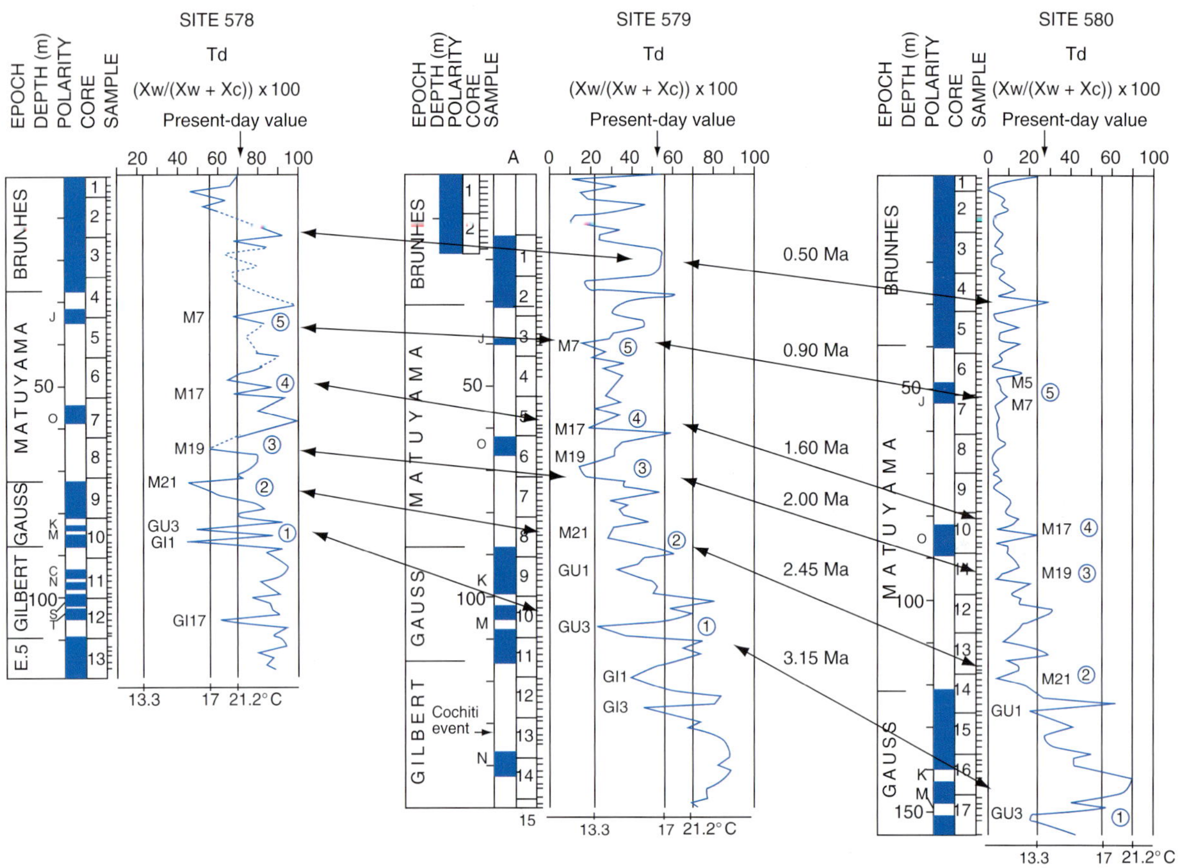

Figure 8 Stratigraphic variations of Td values in DSDP Sites 578, 579 and 580. Circled numbers represent major paleoclimatic events: ① remarkable cooling at ∼3.15 Ma; ② initial of glaciation at ∼2.45 Ma; ③ major glacial event at ∼2.0 Ma; ④ remarkable cooling at ∼1.60 Ma; and ⑤ the beginning of cycles of glacial/interglacial stages at ∼0.90 Ma. Td values during the pre-C1n (Brunhes) Epoch are pronounced carbonate minima in the central North Pacific Ocean. J = Jaramillo, O = Olduvai, K = Kaena, M = Mammoth, C = Coccochiti, S = Sidufjall, T = Thvera (Koizumi, 1986a).

Kuroshio-Kuroshio Extension Current off eastern Honshu, diatom assemblages recognized in three cores reveal two glacial–interglacial cycles during the past 200 kyr (Koizumi *et al.*, 2004; Koizumi and Yamamoto, 2005). Warm water species from the littoral-neritic association are predominant in the near-shore region due to the Kuroshio Current flowing northeast along the coast of southwest Honshu. On the other hand, the cold-water species of the oceanic association are predominant in the Kuroshio Extension Current offshore, where the southward Oyashio Current flows in from the Bering and Okhotsk seas (**Fig. 10**). In the cold belt off the Sanriku coast, diatoms are abundant during the interglacial period but are less abundant and more poorly preserved during the last glacial period (Shimada and Hasegawa, 2001). The predominant occurrence of cold-water species, including sea-ice diatoms during Marine Isotope Stage 3 (MIS-3), implies the inflow of sea-ice from the Okhotsk Sea into the Oyashio Current.

In the Japan Sea, the diatom records in two cores recovered from the small ridge-top basin above the Oki Ridge indicate that the Japan Sea responded to eustatic sea-level fluctuations corresponding with the glacial–interglacial cycles. Oceanic warm-water species, represented by *Fragilariopsis doliolus* and *Thalassiosira oestrupii*, are dominant and are relatively abundant during the last interglacial stage (MIS-5e) and during the Holocene (MIS-1) (**Fig. 11**). This explains the remarkably high *Td′* values occurring at 10-kyr intervals in MIS-5e and MIS-1, and also high values at MIS-5c and 5a. On the other hand, oceanic cold-water species increase during the Late Glacial but decrease during the Younger Dryas cooling event. The shift from an association dominated by cold-water species to an association dominated by warm-water species occurred during the period of 10.0–7.0 ka.

Milankovitch Cyclicity

The diatom records in the Pacific Ocean show correlation with the Earth's orbital parameters during the Quaternary. Results from Fourier analysis on the cold-water diatom *Thalassiosira trifulta*, occurring during the C1n (Brunhes) Paleomagnetic Chron

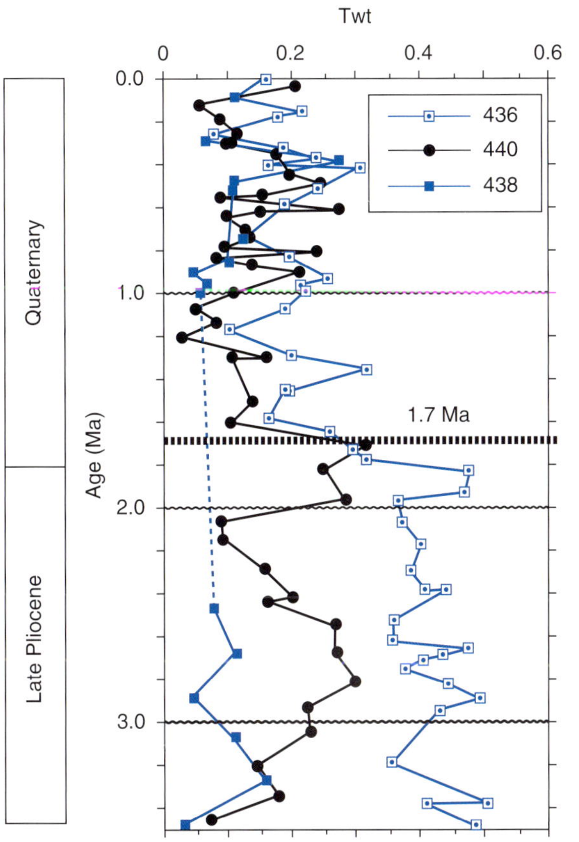

Figure 9 Stratigraphic variations of Twt values (Barron, 1992) at DSDP Sites 436, 438 and 440 (Koya, 1995).

(0–730 ka) at DSDP Site 580, indicate a strong 41-kyr and a weak 23-kyr periodicity (Sancetta, 1985); periods of axial obliquity and of equinoctial precession, respectively. Also, *Td* values calculated for the same site indicate that surface-water temperatures respond to the orbital forcings throughout the Quaternary. The Maximum Entropy Method (MEM) power spectral density (PSD) analyses for *Td* values for the DSDP Sites 579 and 580 show that among the dominant cycles during the C1n Chron are cycles of 411.5-kyr and 126.0-kyr duration for Site 579, and cycles of 467.0-kyr and 136.7-kyr intervals for Site 580. These intervals are interpreted as 413-kyr and 95 to 124-kyr intervals of the orbital eccentricity (Koizumi, 1994b). Minor cycles of 41.2-kyr at Site 579 and 41.7-kyr intervals for Site 580 are near to 41-kyr interval of the obliquity (tilt). During the C1r and C2n (Matuyama) Paleomagnetic Chron, a minor 101.1-kyr cycle and a predominant 398.2-kyr cycle for Site 579 correspond to orbital eccentricity. A dominant cycle of 43.6-kyr interval at Site 580 corresponds to the obliquity.

In the western equatorial Pacific Ocean, changing amounts of dissolved diatom frustules since 400 ka are recognized in ODP Hole 807A on the Ontong Java Plateau, where they closely correspond to climatic precession. This relationship suggests cyclic changes in east-to-west circulation of the water under the influence of the trade winds and subsequent changes in concentration of dissolved silica in the sea-surface water-mass inside the western equatorial region (Ikeda, 1997).

From ODP Site 797 in the Japan Sea, the diatom records indicate three intervals of the Quaternary:

1) the early Pleistocene (1.6-1.1 Ma). with sharp oscillations in abundances;
2) the middle Pleistocene (1.1–0.5 Ma). with moderate abundance but several spike-like intervals; and
3) the late Pleistocene (0.5–0.0 Ma), where the fluctuations in abundances are dominated by the 100-kyr cycles (**Fig. 12**; Koizumi and Ikeda, 1997).

The results of MDM-PSD analysis on diatom abundance for the same site show the predominant cycles of 112.2-kyr and 105.3-kyr intervals since 1.1 Ma, and the cycles of 203.7-kyr and 224.0-kyr intervals are twice the period of eccentricity. The second-order cycles occur within a range from 40.0-kyr to 42.8-kyr. The cycles of 28.6-kyr and 31.0-kyr intervals correspond to 29-kyr intervals in nonlinear interactions of 100- kyr and 41-kyr.

Dansgaard–Oeschger Cycles and Bond Cycles

During the last 200 kyr, spectral analyses for diatom records from the Japan Sea reveal large and abrupt climatic changes in the periodicity of millennial-time scale known as Dansgaard–Oeschger (D–O) cycles and/or Bond cycles. The evaluation of the diatom records at ODP Site 797 was extended to include high resolution spectral analysis for total diatom valve abundance, the abundance of the brackish-water diatom *Paralia sulcata*, and the total abundance of *P. sulcata* and warm-water diatoms. The results of the MEM-PSD analyses are subdivided into five groups of cycle spectra in kyr (**Table 2**). Several periods of millennial timescale from 4 to 7-kyr are present in the high frequency range, as well as cyclic 14.0 to 14.6-kyr units. These cyclic units possibly correlate with the Bond cycles, which comprise the 6 to 16-kyr irregular subsets of progressively colder D–O events. D–O cycles can be correlated with the alternation of dark and light layers in the uppermost sequence in the ODP Site 797. The enhanced inflow of the East China Sea Coastal Water (ECSCW), which caused higher productivity and stronger upwelling during deposition of the dark layers, is considered to be a result of a wetter

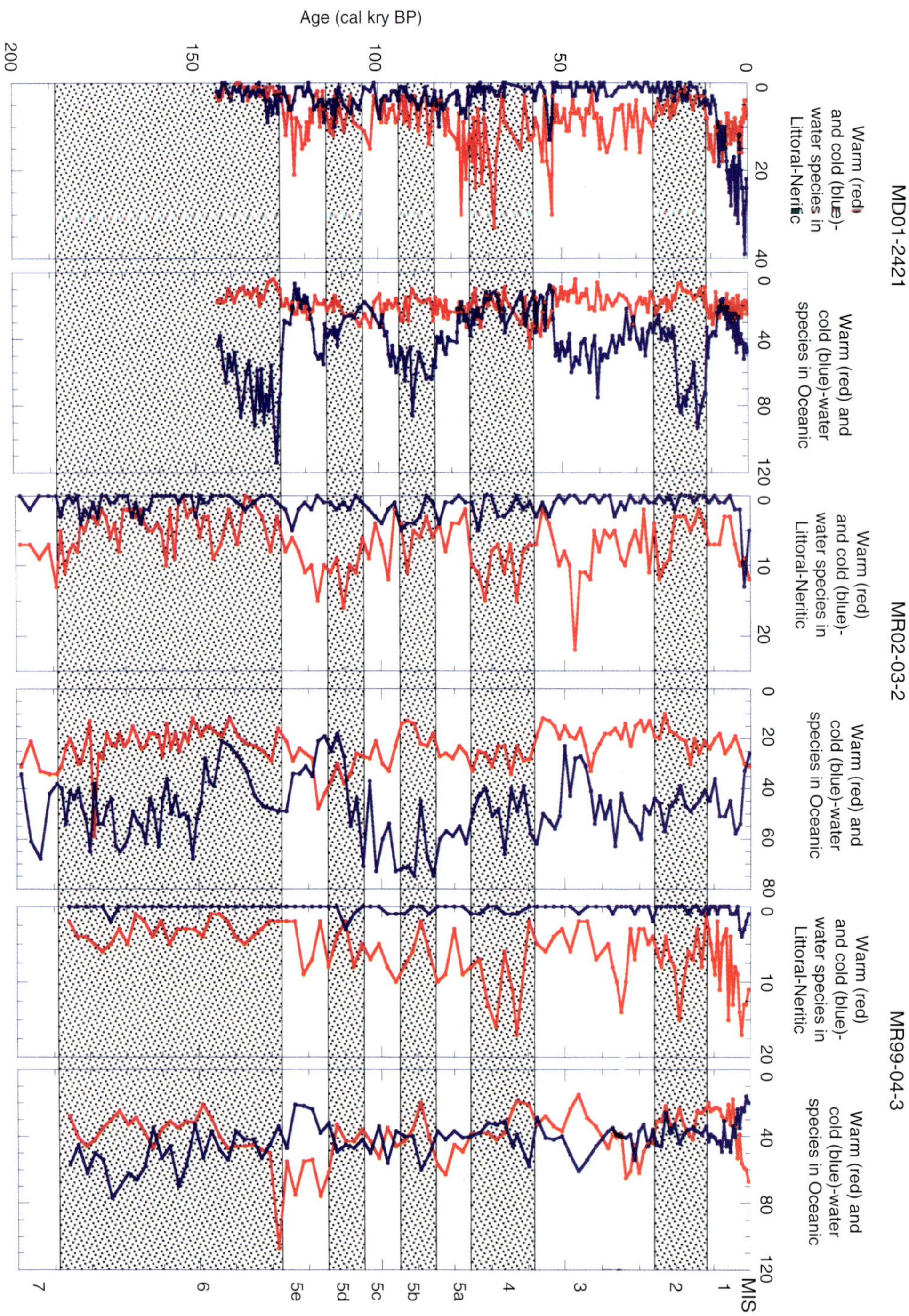

Figure 10 Chronostraigraphic variations of warm- and cold-water diatom species (values/200 valves) in MD01-2421, MR02-03-2 and MR99-04-3 in the northern Pacific Ocean. Shaded interval indicates the glacial phase in marine isotope stage (MIS) (Koizumi and Yamamoto, 2005).

climate in central Asia, judging from the reduced contribution of the Kosa (dust) fraction within the dark layers (Tada *et al.*, 1999). Reduced inflow of the ECSCW, on the other hand, led to deposition of the light layers, which contain exclusively warm-water species.

The Onset of Deglaciation from the Late Glacial to the Holocene

Deglaciation, the warming at the end of the last glaciation, began with a distinct climate change; the Bølling/Allerød warm period and the following Younger Dryas cooling event at 10 ^{14}C ka or 12.9–11.6 cal ka.

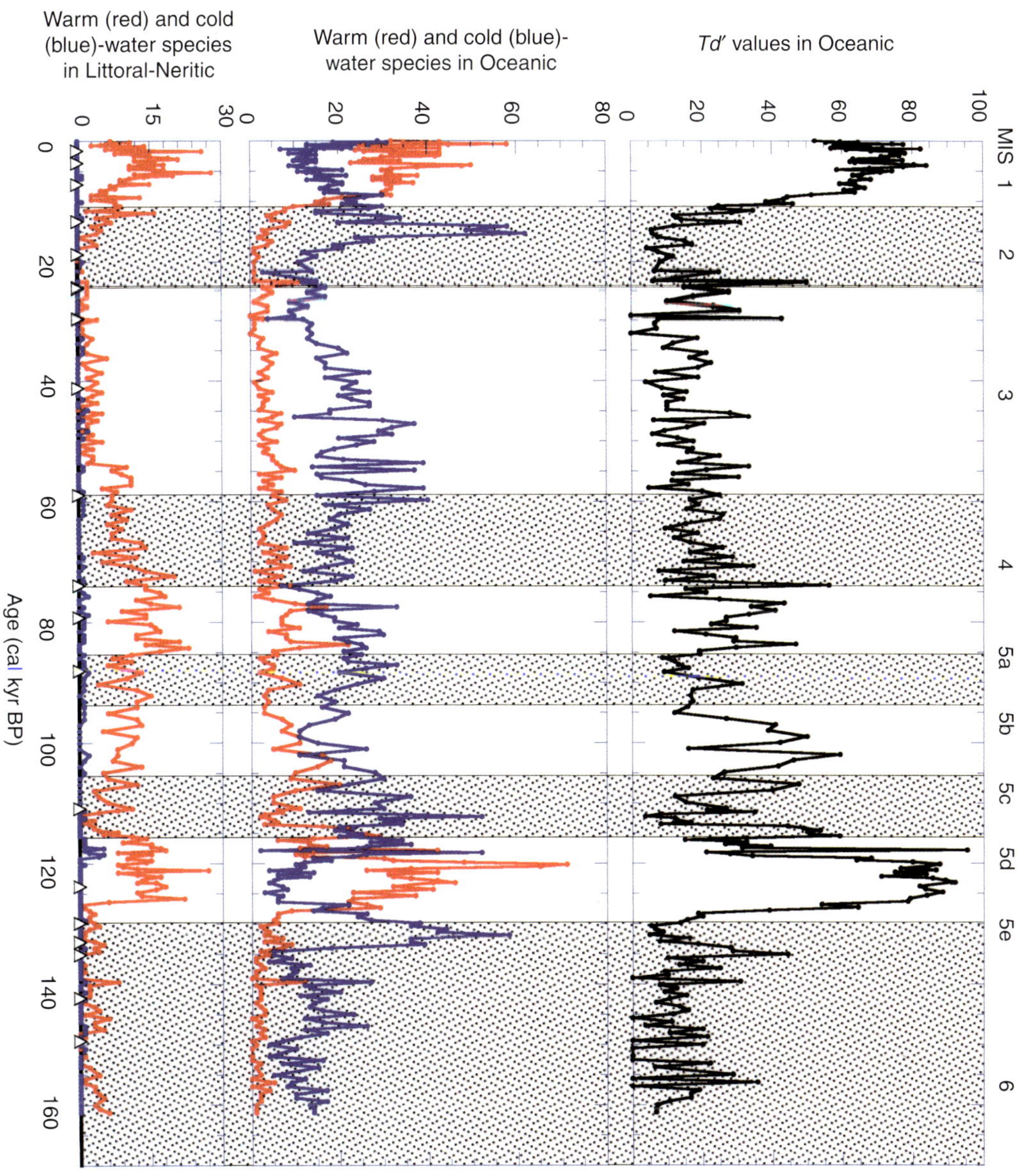

Figure 11 Chronostratigraphic variations of warm- and cold-water diatom species (valves/200 valves), and *Td′* values in MD01-2407 in the Japan Sea. Shaded interval indicates the glacial phase in marine isotope stage (MIS).

The rapidity of change from warm to cool conditions suggests a direct influence on regional climate. In the Bering Sea, the *Thalassiosira trifulta* complex was gradually replaced by *Neodenticula seminae* at the Holocene/Pleistocene boundary (10 ^{14}C ka or 11.6 cal ka), following increasing surface-water temperature and salinity. However, the oxygen isotopes show a much faster change to modern values in the early Holocene (Sancetta, 1985).

In the northeast Pacific off northern California, a rapid increase of *Chaetoceros* spores, a decrease of displaced benthic diatoms, the disappearance of the Pliocene taxa, and the appearance of *Fragilariopsis doliolus* are recognized at 10 ^{14}C ka (Sancetta *et al.*, 1992). Recently the appearance of *F. doliolus* in the same area has been dated more accurately to 10.3 cal ka in the early Holocene (Barron *et al.*, 2003).

In the northwest Pacific off central Japan, deglaciation is recorded by the distinct and rapid decrease of sea-ice species such as *Bacterosira fragilis*, *Fragilariopsis cylindrus*, *Thalassiosira nordenskioldii*, and *Thalassiosira gravida*, and an increase of neritic cold-water species such as *Odontella aurita*, at 12.9–11.6 cal ka (Koizumi *et al.*, 2004).

The deglaciation in the Japan Sea is shown by the increase of oceanic warm-water species *Azpeitia tabularis* and *Thalassiosira oestrupii* and the decrease of cold-water species *Thalassiosira nordenskioldii* and *Thalassiosira trifulta* at 12.9–11.6 cal ka. Also

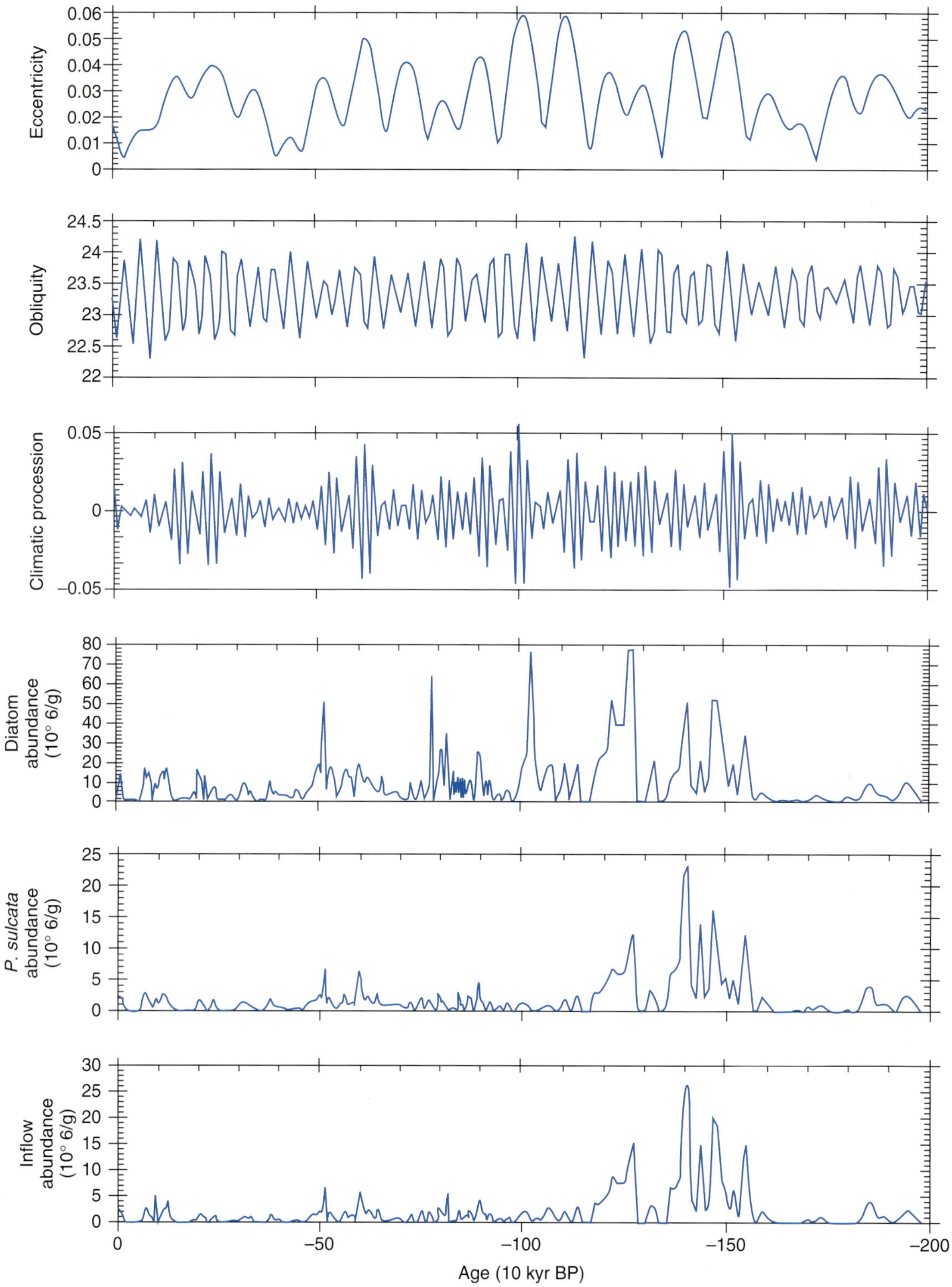

Figure 12 Composition between stratigraphic variations of three kinds of proxy indicators of surface water condition and the Earth's orbital parameters with the insolation at 37° N during the last 2 Ma at ODP Site 797 in the Japan Sea. Inflow = warm-water species and *Paralia sulcata* (Koizumi and Ikeda, 1997).

the decrease of littoral species *Delphineis surirella* and littoral-neritic species *Paralia sulcata* occurs, followed by an increase of *Thalassionema nitzschioides*, suggesting a rise in sea level and salinity, and a transgression from 12.9 ka to 9.8 ka.

Holocene Marine Environments and Climate Changes

In the coastal areas of California and Peru, where seasonal upwelling is driven by eastern boundary

Table 2 The Maximum Entropy Method (MEM)-estimated periods for three kinds of proxy for diatom assemblage during the last 200 kyr at ODP Site 797 in the Japan Sea. Cycles are expressed by kyr and power spectral density in parentheses

Proxy indicators	*Diatom abundance* (10^6/g)	*Paralia sulcata abundance* (10^6/g)	*Warm-water* & *P. sulcata abundance* (10^6/g)	*Orbital forcing and Internal response*
Cycle spectra (kyr)	115.5 (1.000)	120.3 (1.000)	124.4 (1.000)	eccentricity (95–124 kyr)
and	44.5 (0.092)	41.7 (0.027)	50.4 (0.170)	obliquity (41 kyr)
power spectral		30.4 (0.029)	34.1 (0.143)	precession (19–23 kyr)
density			26.4 (0.089)	
	14.0 (0.066)	14.3 (0.037)	14.6 (0.043)	Dansgaard–Oeschger cycles
	11.2 (0.044)			
	6.2 (0.080)	7.0 (0.037)	5.6 (0.064)	Dansgaard–Oeschger cycles
	5.6 (0.051)	6.3 (0.033)	5.1 (0.041)	
		5.5 (0.030)		
		5.1 (0.028)		
		4.3 (0.030)		

currents, diatom-rich sediments are common, although they may be diluted by terrigenous debris from the continents. Sediment trap records from the Gulf of California indicate that early winter diatom blooms are triggered by the onset of strong northwesterly winds, whereas coastal upwelling, leads to late winter–early spring diatom blooms (Sancetta, 1995) (*see* Marine Laminated Sequences). When compared with the sediment trap record, the sediment core record of the Gulf of California indicates that during the early to middle Holocene, very high diatom production was mainly caused by winter northwesterly winds, whereas during the late Holocene, high production also occurred during the spring season due to coastal upwelling. The diatom records in the Peruvian coastal upwelling zone reveals that the increases and decreases in paleoproductivity are not directly related to the glacial–interglacial cycles. The variability in the strength in coastal upwelling is complex, with the largest swings in the magnitude of paleoproductivity across oxygen isotope stage boundaries (Schrader and Sorknes, 1991). During the periods of highest paleoproductivity, *Skeletonema* and *Delphineis* occur abundantly in varved or finely-laminated sediments, while weak paleoproductivity coincides with bioturbated sediments and a high abundance of warm-water taxa such as *Azpeitia nodulifera* and *Fragilariopsis doliolus*.

Along the northern California Margin, dilution of the diatom record is caused by terrigenous debris flowing downslope from the continent. At the coastal ODP Site 1018, such freshwater diatoms as *Aulacosira* spp., *Fragilaria* spp., and *Stephanodiscus* spp. are dominant during periods of low diatom concentration corresponding with the glacial periods. This is apparently due to the lowered sea level during the glacial periods and is a consequence of an increase in terrigenous flux, especially since 780 ka (Ikeda and Koizumi, 2000). At offshore Site 1021, the abundance of freshwater diatoms fluctuates at greater frequency, implying that the freshwater diatoms were transported by seasonal winds and/or there was enhanced precipitation.

Climatic Deterioration and Cultural Changes

The Quaternary is marked with the appearance and expansion of humans and investigations of marine archives have shown that several climatic changes are synchronous with human cultural events.

Diatom assemblages in three cores from off southern Java suggested that modern *Homo sapiens* might have crossed Wallace's line (the boundary between the Oriental region and the Australian region) during the interval from 65 ka to 50 ka (Ikeda *et al.*, 1999). The increase abundance of the coastal species *Actinocyclus octonarius* indicates that during 65–50 ka, the Java Coastal Current became enhanced and started to flow strongly eastward off Java, as the wet northwestern monsoon strengthened. The eastward flowing current could have helped the spread of humans from Sunaland into Sahulland (**Fig. 13**). In fact, the paleolithic sites discovered in Australia are dated far earlier in coastal areas than in inland areas, and one of them dates back 50 ka.

In the Japan Sea and off central Japan during the Holocene period, the *Td′* values show rhythmic fluctuations of the first-order period of ~1.5-kyr and of the second-order period of a century (**Fig. 14**). This indicates a strong and regular inflow of the warm Tsushima Current into the Japan Sea as a branch of the warm Kuroshio Current. The decrease of *Td′* values correspond with the climatic deterioration, poor crops, and starvation when cultural changes occurred in Japan. The paleoclimatic variation and

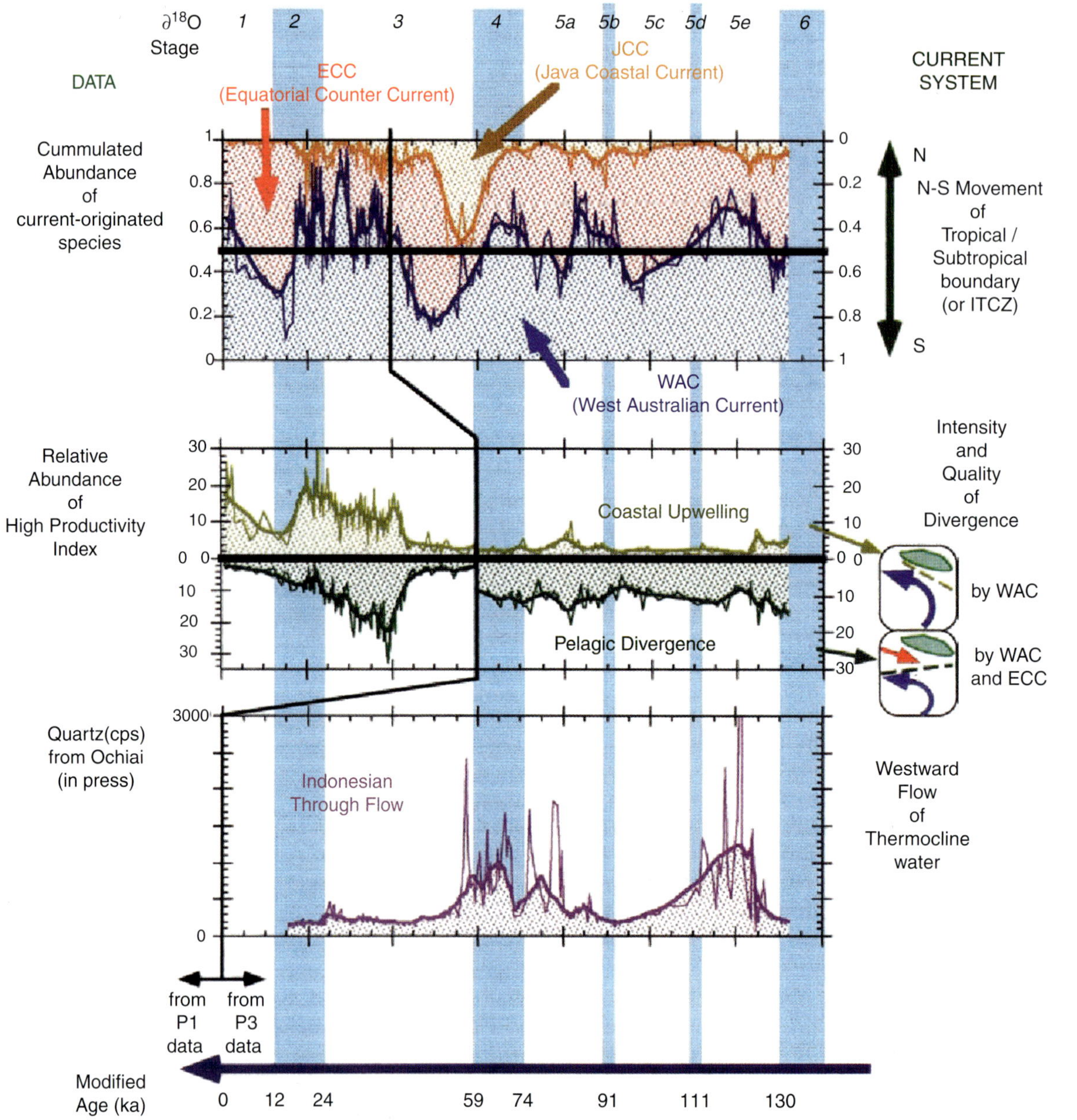

Figure 13 Proposed model for the current-upwelling system in the late Quaternary off southern Java. Upper = proportion of *Thalassiosira oestrupii* representing West Australian Current intensity, *Azpeitia nodulifer* for Equatorial Counter Current intensity, and *Actinoptychus octonarius* for Java Coastal Current intensity, which reflect meridian movement of tropical/subtropical oceanic boundary. Middle = upwelling intensity represented by relative abundance of *Thalassionema nitzschioides* and *Thalassiothrix longissima*. Lower = Quartz content as an indicator of Indonesian through flow (Ikeda *et al.*, 1999).

atmospheric ^{14}C record in the triple events of high ^{14}C denoted by Stuiver *et al.* (1991) are identified in the diatom variations. A similar periodicity is recognized in the large-scale salt transport system operating in today's ocean from the Atlantic to the Pacific Ocean (Broecker and Denton, 1989), and also in lake-level fluctuations in the Jura and northern subAlpine ranges (Magny, 1993, 1995).

The periodicity in the flow of the warm Tsushima Current is also recognized from more northern areas. The modern warm Tsushima Current diverges into two branches; the Tsugaru Warm Current, and the Northward Current, which flows along the west coast of Hokkaido Island and flows out to the Okhotsk Sea through the Soya Strait (**Fig. 3**). Recent investigation has made it clear that the fluctuations at ~1.5-kyr intervals in the abundance of *Fragilariopsis doliolus* correlate with the fluctuations recognized in the Northward Current and the Soya Warm Current (Shimada *et al.*, 2004). The

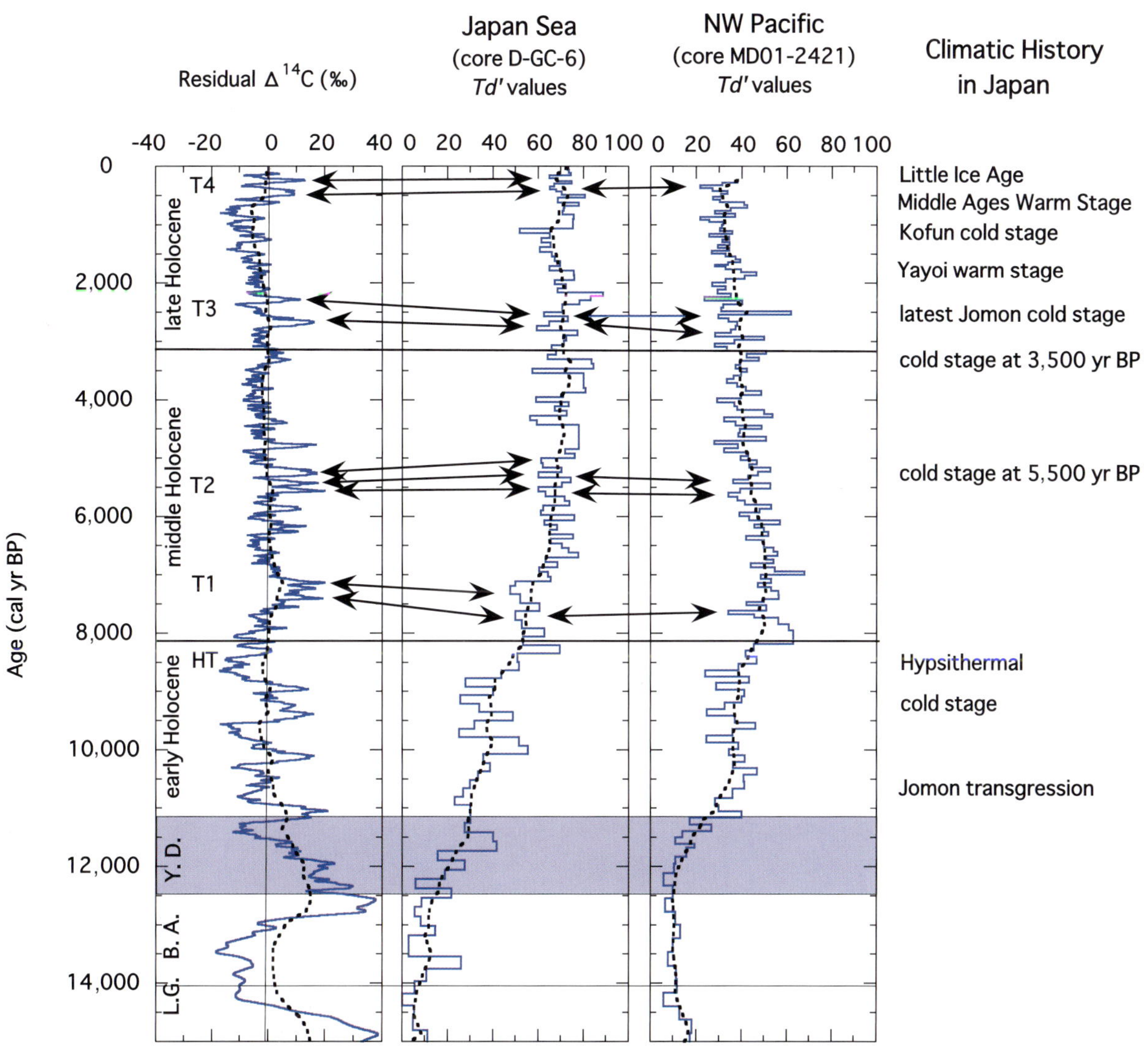

Figure 14 The comparison of the variation of *Td′* values, atmospheric residual ^{14}C record and climatic history in Japan. Y.D. = Younger Dryas (shaded interval), B.A. = Bölling/Alleröd, L.G. = Late Glacial, T1∼T4 = denote triple events of similar nature to Maunder and Sporer type perturbations. The first-order period of ∼2,000 yrs and second-order time scale for *Td′* values correspond with the climatic deterioration and atmospheric residual ^{14}C record.

development of sea-ice cover and/or cold ocean conditions under cooling climate corresponds in turn to development of local cultures around the Okhotsk Sea (**Fig. 15**), with cooling climate correlated with deteriorating changes in cultural environments. On the contrary, there is much evidence to suggest that warm climates allowed local cultures to flourish and expand (Koizumi *et al.*, 2003).

Predicting Climate Changes

Understanding the Quaternary is a key to estimating what the Earth's climate will be like in the future. Such studies demand high resolution analysis of changing processes and the mechanisms that affect the natural environments. In a south-to-north transect of the cores from both the Pacific side and Japan Sea off Honshu, Holocene rhythmic fluctuations in the *Td* values have been recognized. Four warm peaks in the variations of *Td* values are remarkable and can be correlated between the cores, indicating that synchronous climatic events happened around the Japanese Islands. These events are related to the strong and rhythmic flow of the warm Kuroshio Current and inflow of the Tsushima Current into the Japan Sea (Koizumi, 1994a).

A MEM-PSD analysis and a fitting test were performed for the variations of *Td* values during the last 6.5 kyr BP in two of the cores. The modified time series, which standardized to the equal interval between samples, are suitably fitted by five periods. Each of the fitted curves since 6.5 kyr were extended 1.0 kyr into the future (**Fig. 16**). Results lead to the

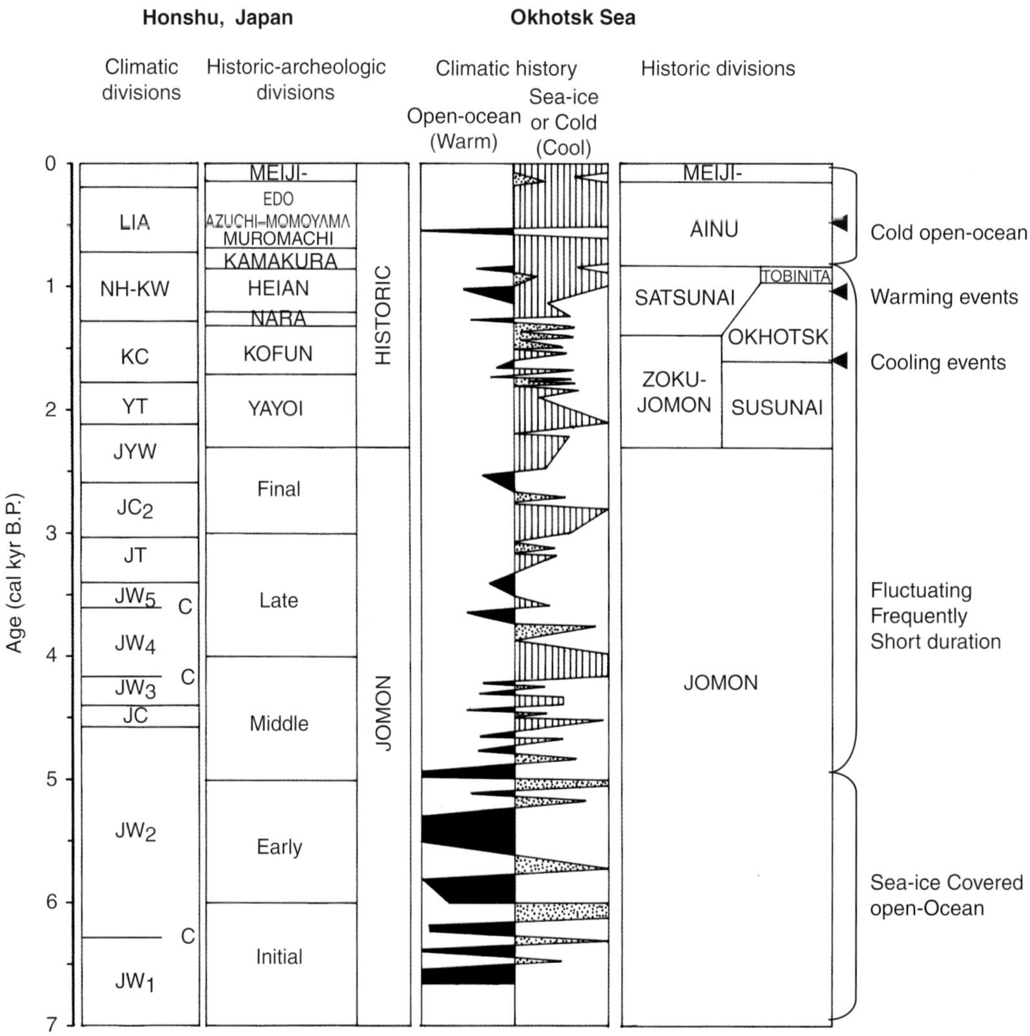

Figure 15 Climatic oscillations reconstructed from diatoms in the Okhotsk Sea. Climatic fluctuation based on pollen records and historical-archeologic divisions in Honshu, Japan are taken from Sakaguchi (1982). Historic divisions in Hokkaido are taken from Ushiro (2000). Warm-open-ocean phases are indicated by black and cold-open with sea-ice phases by dotted area. Cold-open-ocean phases without sea-ice phases are indicated by vertical stripes. JW_1 = first Jomon warm stage, JW_2 = second Jomon warm stage, JC = first Jomon cold stage, JW_3 = third Jomon warm stage, C = cold, JW_4 = fourth Johmon warm stage, JW_5 = fifth Jomon warm stage, JT = Jomon transitional stage, JC_2 = second Jomon cold stage, JYW = Jomon-Yayoi warm stage, YT = Yayoi transitional stage, KC = Kofun cold stage, NH-KW = Nara-Heian-Kamakura warm stage, LIA = Little Ice Age (Koizumi *et al.*, 2003).

estimate that a decrease of *Td* values in core C-4, beginning at 300 years BP and continuing to the present, will continue slowly into the future for a further 500 years. Results from the *Td* values in the Japan Sea L3 core also show the values indicating that the decrease since 250 years BP will continue for the next 350 years. Significant evidence from diatom records from the waters around Japanese Islands, indicate that the Kuroshio Current and Tsushima Warm Current are both weakening at present. This suggests that the present interglacial period is in an unstable condition: this might end and turn to warming within the next 350–500 years.

The mean global temperature has risen 0.5 °C in the last century, caused mainly by increased carbon dioxide in the atmosphere, produced by fossil fuel combustion. Marine phytoplankton, mainly consists of diatoms, produces organic matter by photosynthesis thus contributing to the assimilation of atmospheric carbon. It is estimated that diatoms produce as much organic matter in a year as vegetation on land, this organic matter is oxidized and dissolved in the deep water and seafloor. The resultant carbon dioxide in the deep water, about two times as much quantity in the atmosphere, is segregated and reserved.

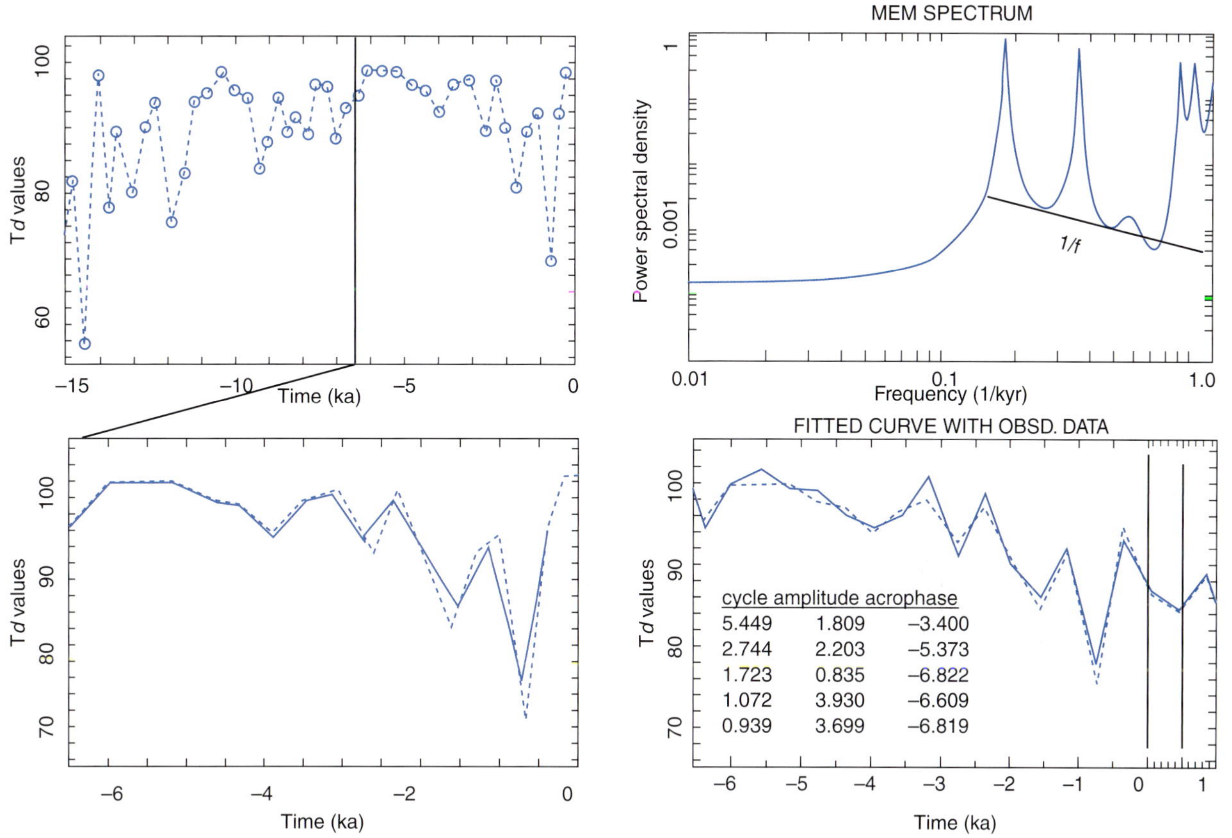

Figure 16 The spectral analysis and fitting test for the *Td* values in C-4. Left-upper = native time-seires during the last 15 kyr, Left-lower = modified time-series since 6.5 ka, right-upper = Maximum Entropy Method (MEM) power spectral density depending on the f^{-1} type, right-lower = fitted curve with modified time-series by five dominant cycles and extended into 1 kyr in future (Koizumi, 1994).

See also: **Diatom Introduction**. **Diatom Methods**: Data Interpretation; Quaternary Geological Records. **Glaciation, Causes**: Milankovitch Theory and Paleoclimate. **Paleoceanography, Biological Proxies**: Marine Diatoms. **Quaternary Stratigraphy**: Biostratigraphy.

References

Barron, J. A. (1992). Pliocene paleoclimatic interpretation of DSDP Site 580 (NW Pacific) using diatoms. *Marine Micropaleontology* **20**, 23–44.

Barron, J. A. (2003). Planktonic marine diatom record of the past 18 my: appearances and extinctions in the Pacific and southern oceans. *Diatom Research* **18**, 203–224.

Barron, J. A., Heusser, L., Herbert, T., and Lyle, M. (2003). High-resolution climatic evolution of coastal north California during the past 16,000 years. *Paleoceanography* **18**, 20-1–20-19.

Ikeda, A. (1997). Cyclic fluctuation in the differential dissolution of late Quaternary diatoms in the western equatorial Pacific Ocean. *The Quaternary Research* **36**, 249–261, (in Japanese with English abstract).

Ikeda, A., Ochiai, K., and Koizumi, I. (1999). Late Quaternary paleoceanographic changes off southern Java. *The Quaternary Research* **38**, 387–399.

Ikeda, A., and Koizumi, I. (2000). Data Report: Diatom flora of the northern California margin since 3Ma. In *Proceedings of the Ocean Drilling Program, Scientific Results 167* (K. Lyle, I. Koisume, C. Richter and R. C. Moore, Jr, Eds.), pp. 119–125. College Station TX (Ocean Drilling Program).

Kanaya, T., and Koizumi, I. (1996). Interpretation of diatom thanatocoenoses from the North Pacific applied to a study of core V20–130 (studies of a deep-se core V20–130, part IV). *Science Report Tohoku University* 2(37), 89–130.

Koizumi, I. (1986a). Late Neogene diatom temperature record in the northwest Pacific Ocean. *Science Report, Co. Gen. Educ. Osaka University* **34**, 145–153.

Koizumi, I. (1986b). Pliocene and Pleistocene diatom datum levels related with paleoceanography in the northwest Pacific. *Marine Micropaleontology* **10**, 309–325.

Koizumi, I. (1994a). The fluctuation and periodicity of marine environments. *Japanese Scientific Monthly* **47**, 29–39 (in Japanese).

Koizumi, I. (1994b). Spectral analysis of the diatom paleotemperature records at DSDP Sites 579 and 580 near the subarctic front in the western North Pacific. *Palaeogeography, Palaeoclimatology, Palaeoecology* **108**, 475–485.

Koizumi, I., and Ikeda, A. (1997). The Plio-Pleistocene diatom record from ODP Site 797 of the Japan Sea. In *Proceedings of the 30th International Geological Congress 11* (W. Naiwen and J. Remane, Eds.), pp. 213–230. VSP, Utrecht.

Koizumi, I., Irino, T., and Oba, T. (2004). Paleoceanography during the last 150 kyr off central Japan based on diatom floras. *Marine Micropaleontology* **53**, 293–365.

Koizumi, I., Irino, T., and Yamamoto, H. (2001). Changes in the warm water Kuroshio Current system from the ending period of the last glacial to the present interglacial based on diatom fossils in the western North Pacific Ocean sampled on the R/V Mirai cruise [MR00-K05]. *JAMSTECR* **44**, 29–40 (in Japanese with English short abstract).

Koizumi, I., Shiga, K., Irino, T., and Ikehara, M. (2003). Diatom record of the late Holocene in the Okhotsk Sea. *Marine Micropaleontology* **49**, 139–156.

Koizumi I, Tada R, Narita H, Irino T, Aramaki T, Oba T, and Yamamoto, H. (2006) Paleoceanographic history around the Tsugaru Strait between the Japan Sea and the northwest Pacific Ocean since 30 cal kyr BP. *Palaeogeography, Palaeoclimatology, Palaeoecology 232*, 36–52.

Koizumi, I., and Yamamoto, H. (2005). Paleohydrography of the Kuroshio-Kuroshio Extension based on fossil diatoms. *JAMSTEC Report of Research and Development* **1**, 57–68.

Koizumi, I., and Yanagisawa, Y. (1990). Evolutionary change in diatom morphology – An example from *Nitzschia fossilis* to *Pseudoeunotia doliolus*. *Trans. Proc. Palaeont. Soc. Japan, N.S* **157**, 347–359.

Koya, K. (1995). Marine environmental changes since late Pliocene, recorded in the diatom fossil assemblage of the DSDP cores off Sanriku, Northeast Japan. *Journal of the Geological Society of Japan* **101**, 148–161 (in Japanese with English abstract).

Koya, K. (1998). Late Pliocene–Pleistocene paleoceanographic study based on diatom assemblage of the Japan Sea cores (ODP Leg 127). Doctoral Thesis, Hokkaido Univ.

Mikkelsen, N. (1977). On the origin of Ethmodiscus ooze. *Marine Micropaleontology* **2**, 35–46.

Sancetta, C. (1885). Wisconsin-Holocene paleoenvironment of the Bering Sea: Evidence from diatoms, pollen, oxygen isotopes and minerals. *Marine Geology* **62**, 55–68.

Sancetta, C. (1995). Diatoms in the Gulf of California: Seasonal flux patterns. *Paleoceanography* **10**, 67–84.

Sancetta, C., Lyle, M., Heusser, L., and Zahn, R. (1992). Late-Glacial to Holocene changes in winds, upwelling, and seasonal production of the northern California current system. *Quaternary Research* **38**, 359–370.

Sancetta, C., and Silvestri, S. M. (1986). Pliocene–Pleistocene evolution of the North Pacific Ocean atmosphere system, interpreted from fossil diatoms. *Paleoceanography* **1**, 163–180.

Schrader, H., and Sorknes, R. (1991). Peruvian coastal upwelling: Late Quaternary productivity changes revealed by diatoms. *Marine Geology* **97**, 233–249.

Shimada, C., Ikehara, K., Tanimura, Y., and Hasegawa, S. (2004). Millennial-scale variability of Holocene hydrography in the southwestern Okhotsk Sea: diatom evidence. *The Holocene* **14**, 641–650.

Tada, R., Irino, T., and Koizumi, I. (1999). Land-ocean linkages in association with Dansgaard-Oeschger cycles recorded in the late Quaternary sediments of the Japan Sea. *Paleoceanography* **14**, 236–247.

Tanimura, Y. (1992). Seasonal changes in flux and species composition of diatoms: sediment trap results from the northwest Pacific, August 1986–November1988. *Bulletin of the National Science Museum, Tokyo* **C18**, 121–154.

Tanimura, Y. (1999). Varieties of a single cosmopolitan diatom species associated with surface water masses in the North Pacific. *Marine Micropaleontology* **37**, 199–218.

Appendix 1 Ecological and biogeographical categories of the important diatom species in the North Pacific Ocean

Taxa	*Categories*	*Paleoecology*	*References*
Actinocyclus elongatus Grunow, in Van Heurck	O-Hp, *w*	tropical-subtropical	5
Alveus marinus (Grunow) Kaczmarska & Fryxell		tropical-subtropical	4, 5, 6
Asterolampra marylandica Ehrenberg		tropical	1, 4, 5
Asteromphalus imbricatus Wallich		tropical-subtropical	5
Azpeitia africana (Janisch) Fryxell & Watkins, in Fryxell, Sims, & Watkins		tropical	4
Azpeitia nodulifera (Schmidt) Fryxell & Sims, in Fryxell, Sims, & Watkins		tropical	4, 6, 8
Fragilariopsis doliolus (Wallich) Medlin & Sims		tropical-subtropical	1, 5, 6, 8
Hemidiscus cuneiformis Wallich		tropical	1, 3, 4, 5, 8
Planktoniella sol (Wallich) Schütt		tropical	1, 3, 4, 5, 6
Rhizosolenia bergonii Peragallo		tropical-subtropical	1, 3, 4, 5
Roperia tessellata (Roper) Grunow		tropical-subtropical	5, 6
Thalassiosira leputopus (Grunow) Hasle & Fryxell		tropical	4
Actinocyclus ellipticus Grunow, in Van Heurck	O-Hp, W	tropical-subtropical	4
Asteromphalus arachne (Brebisson) Ralfs, in Pritchard		tropical-subtropical	5
Asteromphalus flabellatus (Brebisson) Greville		tropical-subtropical	5
Asteromphalus sarcophagus Wallich		tropical	5
Asteromphalus petterssonii (Kolbe) Thrrington-Smith		equatorial	5
Azpeitia tabularia (Grunow) Fryxell & Sims, in Fryxell, Sims, & Watkins		south temperate	8
Nitzschia interruptestriata (Heiden) Simonsen		tropical-subtropical	5
Nitzschia kolaczekii Grunow		warmer	5, 6
Pseudosolenia calcar-avis (Schültze) Sundstrom		tropical-subtropical	1, 3, 5
Rhizosolenia acuminata (Peragallo) Gran		subtropical-temperate	1, 3

(*Continued*)

Appendix 1 (*Continued*)

Taxa	*Categories*	*Paleoecology*	*References*
Rhizosolenia hebetata (Bailey) Gran f. *semispina* (Hersen) Gran		warmer	1, 3
Rhizosolenia imbricata Brightwell		warmer	5
Thalassiosira oestrupii (Ostenfeld) Hasle		subtropical	4, 8
Actinocyclus curvatulus Janisch, in Schmidt	O-Hp, *c*	north temperate-subarctic	3, 4, 7, 8, 10
Actinocyclus ochotensis Jousé		north temperate	4, 7
Asteromphalus hyalinus Karsten		Antarctic-sub-Antarctic	7, 8
Asteromphalus robustus Castracane		north temperate	4, 7
Bacterosira fragilis Gran		Arctic-sub-Arctic (sea ice)	3, 4, 6, 7, 8, 10
Chaetoceros furcellatus Bailey		Arctic-sub-Arctic	3, 4, 7, 8, 10
Coscinodiscus marginatus Ehrenberg		north temperate	3, 4, 7
Coscinodiscus oculus-iridis Ehrenberg		north temperate	7
Fragilariopsis cylindrus (Grunow) Krieger, in Helmck & Krieger		sub-Arctic (sea ice)	4, 6, 7, 10
Neodenticula seminae (Simonsen & Kanaya) Akiba & Yanagisawa		north temperate	4, 6, 7, 8, 10
Porosira glacilis (Grunow) Jorgensen		Arctic-north temperate	3, 4, 6, 7
Rhizosolenia hebetata (Bailey) Gran f. *hiemalis* Gran		north temperate	3, 4, 7, 8, 10
Thalassiosira gravida Cleve		sub-Arctic (sea ice)	3, 4, 6, 7, 10
Thalassiosira hyalina (Grunow) Gran		Arctic-temperate	3, 4, 6, 7
Thalassiosira kryophila (Grunow) Joergensen		Arctic	4
Thalassiosira nordenskioeldii Cleve		Arctic (sea ice)	1, 3, 4, 6, 7, 10
Thalassiosira trifulta Fryxell, in Fryxell & Hasle		sub-Arctic-north temperate	7, 8, 10
Fragilariopsis oceanica (Cleve) Hasle	O-Hp, C	sub-Arctic (sea ice)	4, 6, 7, 10
Asteromphalus Ehrenberg spp.	O-Hp	cosmopolitan	
Coscinodiscus centralis Ehrenberg		cosmopolitan	
Coscinodiscus elegans Greville			
Coscinodiscus obscurus Schmidt			
Coscinodiscus perforatus Ehrenberg		cosmopolitan	
Coscinodiscus radiatus Ehrenberg		cosmopolitan	
Coscinodiscus wailesii Gran & Angust			
Nitzschia bicapitata Cleve, in Kaczmarska & Fryxell		cosmopolitan	
Nitzschia capuluspalae Simonsen			
Nitzschia sicula (Castracanea) Hustedt		cosmopolitan	
Proboscia alata (Brightwell) Sundstrom		cosmopolitan	
Rhizosolenia setigera Bríghtwell		colder	
Rhizosolenia styliformis Bríghtwell		warmer	
Rhizosolenia Bríghtwell spp.			
Stellarima stellaris (Roper) Hasle & Sims			
Thalassiosira decipiens (Grunow) Joergensen		temperate	
Thalassiosira eccentrica (Ehrenberg) Cleve		cosmopolitan	
Thalassiosira ferelineata Hasle & Fryxell			
Thalassiosira gravida Cleve var.			
Thalassiosira lineata Jousé		temperate	
Thalassiosira pacifica Gran & Angst		temperate	
Thalassiosira subtilis (Ostenfeld) Gran		cosmopolitan	
Thalassiosira Cleve 1873			
Thalassiothrix frauenfeldii (Grunow) Hallegraeff			
Thalassiothrix longissima Cleve & Grunow		temperate	
Cyclotella stylorum Brightwell	N~L-Hp~Mp, W	warmer	5
Cymatotheca weissflogii (Grunow) Hendey		tropical	5
Glyphodesmis rhombica (Cleve) Simonsen		warmer	5
Hemiaulus sinensis Greville		subtropical-temperate	1, 5
Lithodesmium undulatum Ehrenberg		warmer	1, 3
Mastogloia Thwaites, in Smith spp.		warm-temperate	5
Odontella longicruris (Greville) Agardh		temperate-subtropical	5
Odontella sinensis (Greville) Grunow		warmer-temperate	3

(*Continued*)

Appendix 1 (*Continued*)

Taxa	*Categories*	*Paleoecology*	*References*
Thalassionema nitzschioides Grunow v. *parva* Heiden, in Heiden & Kolbe		warmer-subtropical	2, 9, 11
Odontella aurita (Lyngbye) Agardh	N~L-Mp, c	north temperate	1, 3, 4, 7
Actinocyclus octonarius Ehrenberg	N~L-Hp~Mp		
Actinoptychus senarius (Ehrenberg) Ehrenberg			
Actinoptychus splendens (Shadbolt) Ralfs, in Pritschard			
Biddulphia pulchella Gray			
Biddulphia tuomeyi (Bailey) Roper			
Odontella reticulata (Roper) De Toni			
Paralia sulcata (Ehrenberg) Cleve			
Skeletonema costatum (Greville) Cleve			
Stephanopyxis (Ehrengerb) Ehrenberg spp.			
Synedra indica Taylor			
Thalassionema frauenfeldii (Grunow) Hallegaeff			
Thalassionema nitzschioides Grunow			
Trigonium alternans (Bailey) Mann			
Achnanthes Bory spp.	L-Mp~Tp		
Amphora Ehrenberg, in Kützing spp.			
Arachnoidiscus ehrenbergi Bailey			
Bacillaria paradoxa Gmelin			
Caloneis Cleve spp.			
Campyloneis grevillei (W. Smith) Grunow			
Campylosira cymbeliformis (A. Schmidt) Grunow, in Van Heurck			
Climacosphenia moniligera Ehrenberg			
Cocconeis Ehrenberg spp.			
Coscinodiscus nitidus Gregory			
Cyclotella striata (Kützing) Grunow			
Cymatosira belgica Grunow			
Cymatosira lorenziana Grunow			
Delphineis margalitalimbata (Mertz) Koizumi			
Delphineis surirella (Ehrenberg) Andrews			
Dimeregramma Rals, in Pritchard spp.			
Diploneis Ehrenberg spp.			
Ditylum brightwellii (West) Grunow			
Grammatophora Ehrenberg spp.			
Gyrosigma Hassall spp.			
Hannaea arcus (Ehrenberg) Patrick, in Patrick & Reimer			
Hantzschia Grunow spp.			
Hemidiscus weissflogii Hustedt			
Hyalodiscus Ehrenberg spp.			
Licmophora Agardh spp.			
Lithodesmium undulatum Ehrenberg			
Melosira nummuloides (Dillwyn) Agardh			
Navicula cancellata Donkin			
Navicula clementis Grunow			
Navicula directa (W. Smith) Ralfs, in Pritchard			
Navicula distans (W. Smith) Schmidt			
Navicula ergadensis (Gregory) Ralfs, in Pritchard			
Navicula finmarchia (Cleve & Grunow) Cleve			
Navicula forcipata Greville			
Navicula granulata Bailey			
Navicula hennedyii W. Smith			
Navicula lyra Ehrenberg			
Navicula marina Ralfs, in Pritchard			
Navicula mutica Kützing			
Navicula peregrina (Ehrenberg) Kützing			
Navicula pygmaea Kützing			

(*Continued*)

Appendix 1 (*Continued*)

Taxa	*Categories*	*Paleoecology*	*References*
Navicula Bory spp.			
Nitzschia compressa (Bailey) Boyer			
Nitzschia constricta (Kützing) Ralfs, in Pritchard			
Nitzschia granulata Grunow, in Cleve & Moller			
Nitzschia lanceolata W. Smith, in Peragallo			
Nitzschia levidensis (W. Smith) Van Heurck, in Van Heurck			
Nitzschia littoralis Grunow, in Cleve & Grunow			
Nitzschia panduriformis Gregory			
Nitzschia trybionella Hantzsch, in Rabenhorst			
Opephora Petit spp.			
Plagiogramma Greville spp.			
Pleurosigma W. Smith 1852			
Rhabdonema arcuatum (Lyngbye) Kützing			
Rhaphoneis amphiceros (Ehrenberg) Ehrenberg			
Rhoicosphenia curvata (Kützing) Grunow			
Rhopalodia O. Müller 1895			
Stauroneis Ehrenberg spp.			
Surirella Turpin spp.			
Synedra tabulata (Agardh) Kützing			
Tabellaria Ehrenberg, in Kützing spp.			
Thalassiosira lacustris (Grunow) Hasle, in Hasle & Fryxell			
Trachyneis antillarum (Cleve & Grunow) Cleve			
Trachyneis aspera (Ehrenberg) Cleve			
Triceratium Ehrenberg spp.			
Aulacosira Thwaites spp.	Freshwater		
Cyclotella (Kützing) Brébisson spp.			
Cymatosira Grunow spp.			
Cymbella Agardh spp.			
Diatomella balfouriana Greville			
Diatoma hiemale (Lyngbye) Heiberg			
Epithemia Kützing spp.			
Eunotia Ehrenberg spp.			
Fragilaria Lyngbye spp.			
Frustulia Rabenhorst spp.			
Gomphonema Agardh spp.			
Meridion circulare (Greville) Agardh			
Navicula americana Ehrenberg			
Neidium iridis (Ehrenberg) Cleve			
Pinnularia Ehrenberg spp.			
Stephanodiscus Ehrenberg spp.			
Synedra ulna (Nitzsch) Ehrenberg			
Synedra vaucheriae Kützing			
Tabellaria fenestrata (Lyngbye) Kützing			
Tabellaria flocculosa (Roth) Kützing			
Tetracyclus lacustris Ralfs			
Actinocyclus ingens Rattray	Extinct	Miocene (18.0–6.5 Ma)	
Actinocyclus oculatus Jousé		Plio-Pleistocene (4.0–1.0 Ma)	
Crucidenticula Akiba & Yanagisawa spp.		Miocene	
Denticulopsis Simonsen spp.		Miocene	
Goniothecium tenue Brun		Miocene	
Kisseleviella carina Sheshukova-Poretzkaya		Miocene (24.0–20.3 Ma)	
Mediaria splendida Sheshukova-Poretzkaya		Miocene (17.8–10.9 Ma)	
Melosira albicans Sheshukova-Poretzkaya		Pliocene?	
Neodenticula Akiba & Yanagisawa spp.		Mio-Pliocene (7.4–2.0 Ma)	
Nitzschia fossilis (Frenguelli) Kanaya, in Schrader		Mio-Pleistocene (8.0–0.6 Ma)	
Nitzschia reunholdii Kanaya, in Barron and Baldauf		Mio-Pleistocene (6.6–0.6 Ma)	
Proboscia barboi (Brun) Jordan & Priddle		Mio-Pliocene (12.3–0.3 Ma)	
Proboscia curvirostris (Jousé) Jordan & Priddle			

(*Continued*)

Appendix 1 (*Continued*)

Taxa	*Categories*	*Paleoecology*	*References*
Pseudopodosira elegans Sheshukova-Poretzkaya		Mio-Pliocene	
Pyxidicula zabelinae (Jousé) Makarova		Pliocene (3.13–2.8 Ma)	
Rhizosolenia matuyamai Burckle, in Burckle, Hammond & Seyb		Pleistocene (1.2–1.0 Ma)	
Rhizosolenia miocenica Schrader		Miocene	
Rouxia Brun & Heribaud, in Heribaud spp.		Miocene (15.0–5.1 Ma)	
Synedra jouseana Sheshukova-Poretzkaya		Miocene	
Thalassionema hirosakiensis (Kanaya) Schrader		Miocene (16.0–7.3 Ma)	
Thalassionema schraderi Akiba		Miocene (8.75–7.0 Ma)	
Thalassiosira antiqua (Grunow) Cleve-Euler		Mio-Pliocene	
Thalassosira convexa Mukhina		Mio-Pliocene (6.6–2.2 Ma)	
Thalassiosira jacksonii Koizumi & Barron, in Koizumi		Pliocene	
Thalassiosira nidulus (Tempère & Brun) Jousé		Pliocene	
Thalassiosira temperi (Brun) Akiba & Yanagisawa		Pliocene	

See Koizumi et al. (2004) for references:
[1] = Cupp (1943),
[2] = Hasle (1960),
[3] = Hendey (1964),
[4] = Jousé et al. (1969),
[5] = Simonsen (1974),
[6] = Hasle (1976),
[7] = Sancetta (1982),
[8] = Sancetta and Silvestri (1986),
[9] = Tanimura (1992, 1999),
[10] = Shiga and Koizumi (2000),
[11] = Hasle (2001).
Ecological and biogeographical categories of diatom species: O-Hp, w=Oceanic holoplanktonic warm-water species defined by Kanaya and Koizumi (1966), O-Hp, W=Oceanic holoplanktonic warm-water species supplimented by Koizumi et al. (2004), O-Hp, c=Oceanic holoplanktonic cold-water species defined by Kanaya and Koizumi (1966), O-Hp, C=Oceanic holoplanktonic cold-water species supplimented by Koizumi et al. (2004), O-Hp=Oceanic holoplanktoic species, N~L-Hp~Mp, W=Neritic~Littoral Holoplanktoic~Meroplanktoic warm-water species, N~L—Mp, c=Neritic~Littoral Meroplanktonic cold-water species defined by Kanaya and Koizumi (1966), N~L-Hp~Mp=Neritic~Littoral Holoplanktonic~Meroplanktonic species.

Diatoms, Marine *see* **Paleoceanography, Biological Proxies**: Marine Diatoms

Dinoflagellates *see* **Paleoceanography, Biological Proxies**: Dinoflagellates

DNA, Fossil Animal *see* **Vertebrate Studies**: Ancient DNA

DNA, Fossil Plants *see* **Paleobotany**: Ancient Plant DNA

Drought History Reconstruction *see* **Paleoclimate Reconstruction**: Paleodroughts and Society

DUNE FIELDS

Contents

High Latitudes

S A Wolfe, Natural Resources Canada, ON, Canada

© 2007 Elsevier B.V. All rights reserved.

Introduction

Quaternary dune fields are widespread throughout high-latitude regions of the Northern Hemisphere including Canada, Alaska, Europe, and Russia. Much of North America north of approximately 45° N latitude, and Europe north of 50° N latitude, was glaciated during the late Quaternary (Wisconsin and Weichselian glacial period, respectively). Most of these areas also presently experience seasonal cold-climate conditions, though there are few large dune fields active today. Cold-climate eolian conditions presently occur in Antarctica, though the few dune fields there are small.

Dune activity is largely controlled by a combination of the three interacting factors of transport capacity (i.e., windiness), sediment supply, and sediment availability. Sediment supply represents the total amount of sandy source material, whereas sediment availability represents the proportion of unprotected, sandy material, i.e., that available for transport, and may be reduced by vegetation cover, moisture, snow, ice, surface lag deposits, and other factors. In high latitudes, sediment supply appears to be the limiting factor on the size of dune fields. Individual dune fields in high latitudes are rarely larger than a few 100 km^2, owing to the relatively small amount of sediment. In comparison, sediment availability, as influenced by the amount of vegetation cover, appears to be a significant control on the degree of dune activity in high latitudes. Transport capacity is also a factor, as high-latitude continental areas may have been windier during late-glacial and early Holocene times than they are today.

Primary Dune Fields

North America Dune fields occur throughout Canada (about 26,000 km^2) and the northern contiguous USA (**Fig. 1**). The majority of dune fields (about 90%) are in central Canada owing to the availability of suitable sandy sediments and strong postglacial winds. The Athabasca Sand Dunes of northern Alberta and Saskatchewan is the largest dune field in Canada (4,800 km^2), and it contains the largest area of active dunes (385 km^2) in the high latitudes (**Fig. 2**). Other large dune fields include the Wood Buffalo Sand Hills, Brandon Sand Hills, Great Sand Hills, and Minot Sand Hills (**Fig. 1**), though they contain only limited areas of active dunes.

Eolian dune fields and sheet sands are also widespread in Alaska (**Fig. 1**), occurring within all major lowlands and alluvial plains beyond the limit of the last glaciation. The most notable dune field is the Great Kobuk Sand Dunes of northwestern Alaska (650 km^2), which contains areas of active dunes (**Fig. 3**).

Europe and Russia Eolian sand deposits form a continuous belt that extend through north-west, central, and east Europe, known as the European sand belt (**Fig. 4**). Stabilized dune fields are found inland or along river systems in the UK, Belgium, The Netherlands, Germany, Poland, Belarus, Ukraine, Lithuania, Latvia, Estonia, and the Russian Plain of European Russia. In addition, relatively small but widespread dune fields are scattered throughout Norway, Sweden, and Finland (Fennoscandia; **Fig. 5**), north of the Arctic Circle.

Other Dunes and Dune Fields

Small active dune fields occur in Antarctica, most notably in the ice-free Victoria Valley of southern Victoria Land. This dune field is about 3.5 km long and less than 1 km wide and consists of sand sheets, and single and compound barchanoid dunes forming transverse ridges between 70 and 600 m long (**Fig. 6**). Other dunes described as 'whalebacks' occur without slipfaces and mantle underlying ground moraine. The dunes contain interbedded snow and sand, and are underlain by permafrost. In addition to inland dunes, active coastal dunes occur in high-latitude, cold-climate, and temperate regions along sandy shorelines in areas of moderate-to-high wind regimes. Active

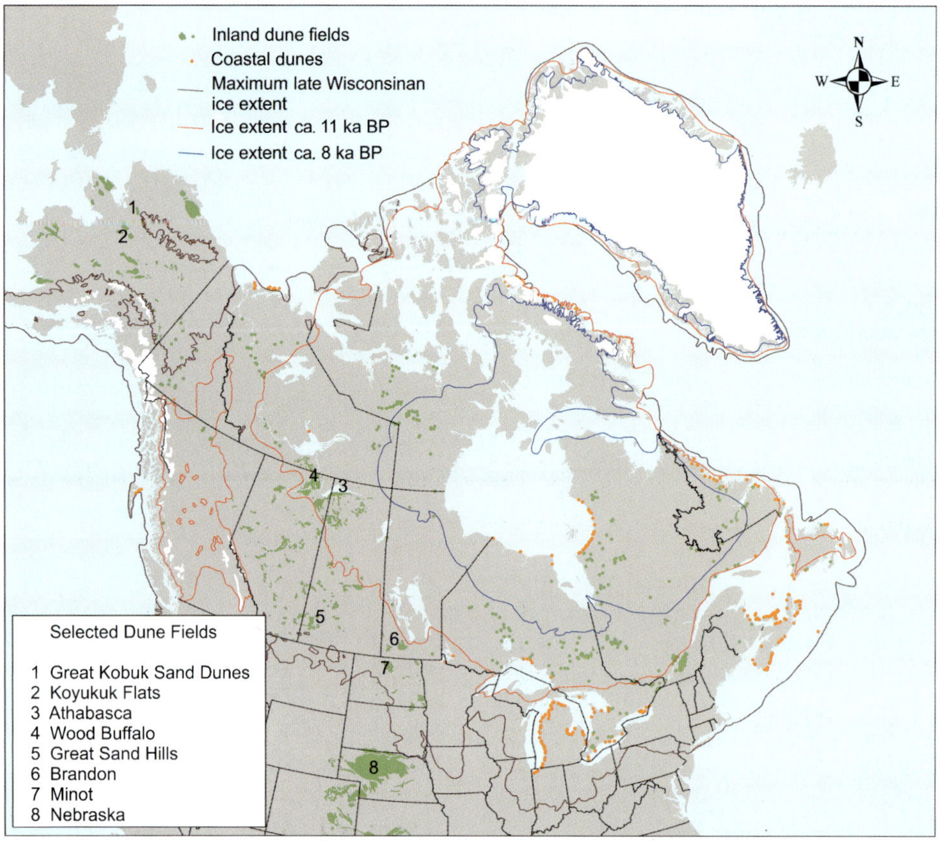

Figure 1 High-latitude dune fields of northern North America. Distribution of dune fields shown in relation to the maximum late Wisconsin glacial ice extent between 18.0 and 15.5 ka BP, and to later ice extents at 11.0 and 8.0 ka BP (uncalibrated radiocarbon years).

Figure 2 Active transverse dunes in the Athabasca sand dunes, northern Saskatchewan, Canada (DG Smith).

high-latitude marine coastal dunes are found in Europe, Alaska, and the Beaufort Sea, Atlantic and Pacific coasts of Canada. Stabilized coastal dunes also occur inland of present-day marine and lake shorelines due to a decline in relative sea level as a result of isostatic uplift or drainage of postglacial lakes during

Figure 3 Large active dune area, approximately 17 km long, of The Great Kobuk Sand Dunes, Alaska (NASA Alaska High Altitude Aerial Photography, courtesy of the Geophysical Institute, University of Alaska Fairbanks).

Figure 4 The European sand belt and selected ice extents (from Zeeberg, 1998).

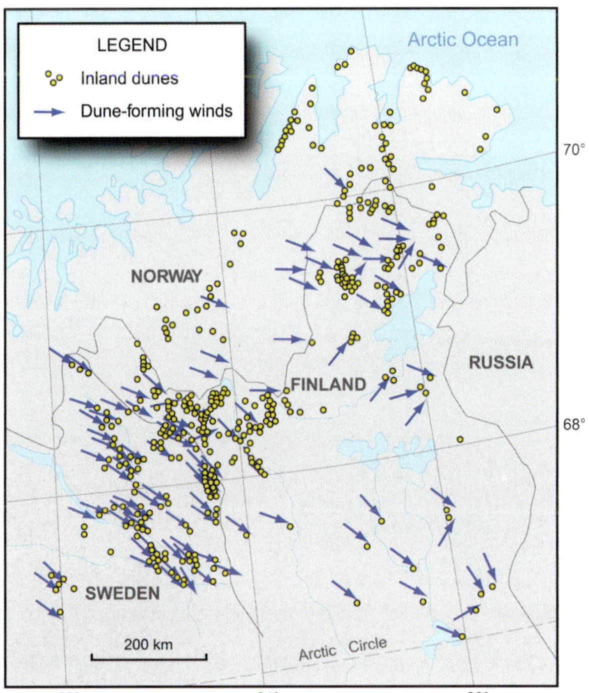

Figure 5 High-latitude dune fields and inferred dune-forming winds in Fennoscandia (from Seppälä, 2004).

the late Quaternary and Holocene. These raised, stabilized coastal dunes may be used to document the timing of former higher lake and sea levels.

Types

The most common dune type in high latitudes is the parabolic dune, both simple and compound forms. Parabolic dunes are indicative of moist sand and vegetation, which modify sediment availability and

Figure 6 Active transverse dunes of the Victoria Valley, Antarctica (N. Lanscaster).

transport. Most high-latitude parabolic dunes typically have partially vegetated heads with lee side slipfaces that are convex downwind in plan view, and trailing arms that are partially or completely vegetated (**Fig. 7**).

Transverse and longitudinal dunes may also occur in high-latitude areas, though some 'longitudinal' dunes actually represent the remnant arms of parabolic dunes. Barchanoid dunes are found in some cold arid regions due to the absence of vegetation and moisture. Blowouts are common where vegetation cover on dune fields is high, and often modify stabilized dunes. Many temperate and cold-climate dune fields of high latitudes are presently stabilized by vegetation, suggesting that they formed under climatic regimes different to those of today.

Although most high-latitude dunes are inactive, they are presently active in some areas of Europe and North America. Small portions of dune fields on the northern Great Plains of North America are active due to

Figure 7 Active parabolic dune in southwestern Saskatchewan, Canada. Dominant transporting winds are from left to right. The head of the dune is partially vegetated whereas the trailing upwind arms are stabilized by vegetation (SA Wolfe).

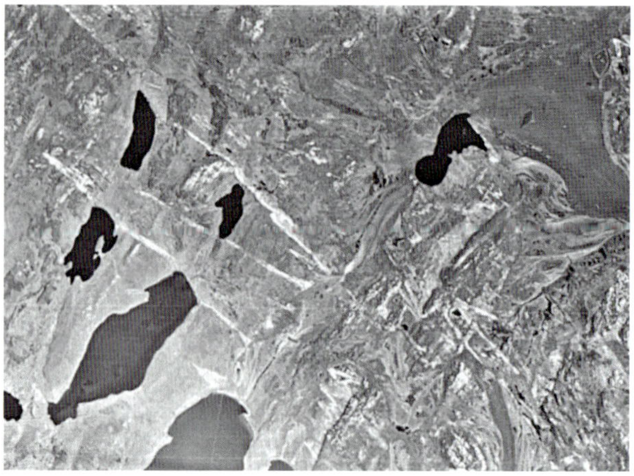

Figure 8 Stabilized elongate parabolic dunes of northern Saskatchewan, Canada. Dune at center left is approximately 3 km long and is oriented towards the northwest. The dunes have migrated over bedrock surfaces with low sediment supplies. Transporting winds were from the southeast, indicating that anticylonic winds associated with the Laurentide Ice Sheet were responsible for their formation. See text and **Figure 12** for details. (National Air Photo Library of Canada photograph A13385-97).

moderate-to-strong wind regimes, and periodic droughts that reduce vegetation cover. More northerly dune fields such as the Kobuk and Athabasca dune fields may be active today due to a combination of factors including moderate present-day wind regimes, moisture stress, and limited growing seasons, and periodic sediment input from rivers, deltas, or lakeshores. Along eastern Hudson Bay, Canada, and in the Lapland region of Finland, small dune areas in subarctic settings are active due to disturbance by fire, and overgrazing and trampling by reindeer, respectively.

North America

Most dune fields in Canada and the northern contiguous USA consist of parabolic dunes and blowouts, with a few dune fields also containing transverse and longitudinal dunes. Various forms of parabolic dunes occur. In northern Canada, very elongate and narrow stabilized parabolic dunes occur on otherwise bare rock of the Canadian Shield (**Fig. 8**). These dunes are typically less than 10 m high, but with trailing arms up to 8 km long. In southwestern Saskatchewan, parabolic dunes tend to be more densely concentrated and less elongate (**Fig. 9**). The Athabasca Sand Dunes contain a wide variety of active dunes, including parabolic, transverse (**Fig. 2**), and 'hybrid' transverse-longitudinal dunes. Well-formed stabilized transverse dunes up to 25 m high also occur in the Holmes Crossing sand hills of central Alberta, Canada (**Fig. 10**). Many dune fields in the northern Great Plains of Canada and the USA are modified by blowouts that have been active at varying times due to drought and disturbance.

Dune fields in Alaska include stabilized parabolic, longitudinal, and compound dunes, ranging from 1.5 to 60 m high. Active dunes in the Kobuk Valley and Koyukuk Flats include transverse and barchanoid dunes (**Fig. 3**). Other areas include partially stabilized parabolic and blowout dunes, and rounded hillocks and hollows or broad undulating ridges and swales with no distinct dune type.

Figure 9 A field of densely concentrated, stabilized circular parabolic dunes, Great Sand Hills, southwestern Saskatchewan, Canada. These dunes were active in the 1800s, possibly as a result of drought in the late 1700s (Wolfe *et al.*, 2001; photograph SA Wolfe).

Europe and Russia

Stabilized dune fields occur on the northern portions of the European sand belt. In the UK, the dunes are typically low hummocks; in The Netherlands, they

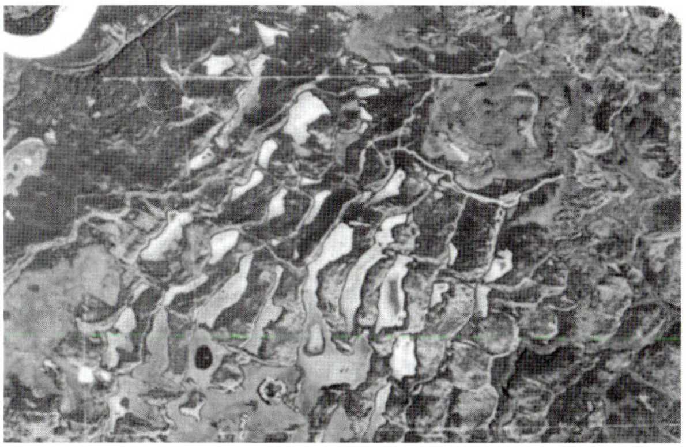

Figure 10 Stabilized transverse dunes, Holmes Crossing, central Alberta, Canada. Optical dating indicates these dunes were last active approximately 15,000 year ago. Transporting winds were from the northwest (upper left), and image view is 7 km wide (National Air Photo Library of Canada photograph A25853-51).

are low ridges of parabolic form; in Germany, parabolic dunes are typically only a few meters high, and up to 12 m; and in Poland, parabolic, longitudinal, transverse, and, rarely, barchanoid dunes occur. Present-day eolian activity in Europe is mostly restricted to areas of human or animal disturbances.

Dune fields on the Russian Plain include individual and compound parabolic dunes, transverse dunes, and blowouts, and may also be associated with sand sheets. Dune heights typically range from 1.5 to 10.0 m, and may reach 20.0 m. Some parabolic dunes are several kilometers long.

Stabilized parabolic dunes are common in Fennoscandia, and may occur in clusters that are also parabolic in form (**Fig. 11**). Individual dunes are typically less than 1,300 m long, up to 150 m wide, and up to 12 m high, but more commonly 4–6 m high. Active parabolic dunes are found in only a few locations, although active blowouts on stabilized dunes are relatively common beyond the treeline.

Origins

The origins of high-latitude dune fields are tied to the interrelated factors of sediment supply, sediment availability, and wind regime. Many dune fields of high latitudes are derived from localized sandy sediments of fluvial, lacustrine, or glacial origin that were deposited by glacial meltwater. Typical source areas include flood plains, glacial outwash plains, braided river valleys, deltas, lake shorelines, eskers, kames, and some till plains and moraines. The size of most high-latitude dune fields is ultimately controlled by the amount of sandy sediment in these deposits. However, the timing of dune activity is more related to periods when these sandy sediments were available for transport by wind. Within or proximal to formerly glaciated areas, initial dune field construction may have occurred shortly after glacial retreat and deposition of the source sediments, when the sandy terrain was sparsely vegetated. It is also hypothesized that winds during these initial periods of dune construction were stronger than today in parts of northern North America and Europe, due to the proximity of continental ice sheets and their effects on pressure gradients and atmospheric circulation patterns.

Figure 11 Active and stabilized parabolic dunes and blowouts at Melajärvi (Paddle Lake), in northern Finland. The source material for the dunes is an esker ridge and surrounding glaciolacustrine/deltaic sediments. The largest dune is about 550 m across and the effective wind direction is from the west-northwest. The area is north of the coniferous treeline, and the surface is vegetated by mountain birch (*Betula pubescens*) dwarf shrubs and lichens (Topographic Division of the Finnish Defense Forces, photograph V9120/111).

Subsequent episodes of dune reactivation during the Holocene may be related to climatic variability, affecting available moisture and vegetation cover, or disturbance by fire, animals, or humans.

North America

Within the glaciated regions of Canada and northern USA, most dune fields are derived from discrete sandy glaciofluvial and glaciolacustrine deposits associated with the retreat of Wisconsin glacial ice. Here, the contrast between relatively small dune fields of formerly glaciated North America and much larger dune fields of unglaciated regions to the south (e.g., the Nebraska Sand Hills of the central Great Plains, USA) is evident (**Fig. 1**). The largest of these northern dune fields, the Athabasca Sand Dunes, was originally supplied by sandy meltwater-derived sediments deposited into glacial Lake Athabasca during the retreat of the Laurentide Ice Sheet about 9,500 calendar years (cal) before present (BP). The dunes have since been supplied by river, delta, and the lakeshore sediments as lake levels declined due, in part, to isostatic rebound.

In the unglaciated lowlands of central and southwestern Alaska, dune fields are derived largely from outwash sediments of major rivers. However, most exposures within these dunes are underlain by subhorizontally stratified sand and subordinate silt, interpreted as eolian sheet sands formed during full-glacial (Wisconsin) conditions. Therefore, the dunes may not have formed from direct reworking of older outwash sediments, but as a result of reworking of eolian sand sheet sediments since the last glacial maximum.

Europe and Russia

Similar to high-latitude dune fields of North America, the European sand belt is closely tied to the extent of glaciation, though the relation is somewhat less distinct. The northerly edges of the sand belt containing most of the dune fields and sand sheets coincides roughly with the maximum position of the late Weichselian ice sheet, whereas finer-grained cover sands and loess approximately coincide with the maximum limit of Pleistocene glaciation (**Fig. 4**). Extensive ice-marginal deposits, such as outwash plains, glaciofluvial, and glaciolacustrine deposits, and abandoned river floodplains and spillways channels supplied sediment for eolian redeposition. Similarly, dune fields on the Russian Plain are located predominantly in areas of outwash plains and meltwater channels, and are derived from glaciofluvial or glaciolacustrine sands. In Fennoscandia, dune fields are evidently derived from glaciofluvial sources, including eskers (**Fig. 5**), kames, deltas, valley trains, ice-dammed lake deposits, and outwash plains.

Ages

High-latitude dune fields throughout much of northern North America and Europe occur within the most recent glacial limits and, therefore, their maximum ages postdate the time of deglaciation. Beyond these limits, dune fields may be older, though it appears that most either originated, or were significantly reworked during the last glacial maximum (late Wisconsin and late Weichselian, respectively). A common problem with determining the age of dune fields is that of partial or complete reworking under conditions of changed climate or disturbance. Thus, reworking may result in superimposed dunes that are potentially misleading indicators of the age and source of the underlying deposits.

North America

The maximum extent of the late Wisconsin glacial ice is the primary factor limiting the age of dune fields in Canada and the northern contiguous USA, as dune fields within this region must postdate deglaciation. **Figure 1** illustrates the maximum extent of the ice sheet between 18.0 and 15.5 thousand years (ka) BP, along with the ice extent at later times during deglaciation, at about 11.0 ka and 8.0 ka BP. This shows that the maximum limiting ages of dune fields within the glacial limit are potentially time transgressive.

In accordance with glacial retreat, the oldest dune fields are expected to be located near the maximum extent of the Wisconsin ice sheet. However, most ages of eolian sands and paleosols in northern Great Plains dune fields along the southerly extents of the Laurentide Ice Sheet relate to late Holocene dune activity that may be associated with episodes of increased aridity in relation to drought. Such a difference in expected ages of eolian sediments highlights the issue of reworking of dune fields in high latitudes.

In contrast, stabilized parabolic dunes have been found in northern Saskatchewan (David, 1981), and southern Quebec, Canada (Filion, 1987), indicating postglacial dune activity initiated at about 10 ka BP. Dune activity is interpreted as having occurred under cool, dry conditions, and with dune-building winds coming from anticylonic air circulation induced by the retreating Laurentide Ice Sheet (**Fig. 12**). Dune activity ceased when circulation patterns similar to present were established following continued retreat of the ice sheet. At both localities, however, activity continued for a short time under modified winds and climate prior to stabilization.

Despite the preceding observations, the oldest recognized late Quaternary dune deposit in Canada is the Kittigazuit Formation of the Tuktoyaktuk Coastlands along the southern Beaufort Sea, Northwest Territories. Radiocarbon dates indicate that these eolian sands were deposited during the mid-Wisconsin, between 37 and 33 ka BP. These sediments were subsequently overridden by glacial ice that advanced over the area in the late Wisconsin (Dallimore *et al.*, 1997).

In Alaska, most dune fields occur in areas that were not glaciated during the late Wisconsin and, therefore, the maximum ages of these dune fields are not constrained by the extent of glacial ice. It is hypothesized

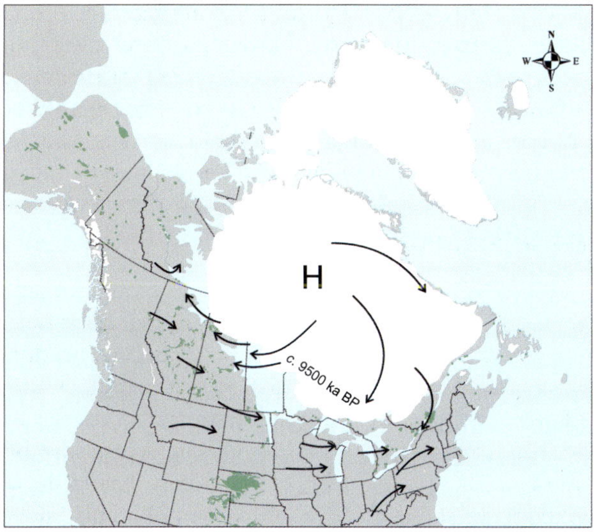

Figure 12 Map of the ice frontal position in North America, ca. 9,500 ka BP, at the time of initial dune formation in northern Saskatchewan and southern Quebec, Canada. Near the end of the Wisconsin glaciation, the retreating continental glacier controlled the distribution of regional air masses by effectively blocking the southward flow of polar air. Instead, it produced its own gravity airflow that descended radially towards the margins of the ice sheet and was gradually deflected clockwise by the Coriolis force. Upon reaching the ice margin, the dry, cold winds were blowing southwestward in the St. Lawrence Lowlands of Quebec in the east and northwestward in northern Saskatchewan and Alberta in the west, producing correspondingly oriented dunes in these areas. Large arrows show principal airflow directions associated with anticyclonic circulation in relation to the high-pressure center (H) over the Laurentide Ice Sheet (after David, 1981).

that many of the subhorizontally stratified deposits represent full-glacial eolian sand sheets; radiocarbon ages indicate that sand sheet aggradation in some lowlands occurred during the last glacial maximum (ca. 25–14 ka BP), whereas many surface dunes may reflect late-glacial and postglacial reworking (ca. 14–8.5 ka BP), and later. The surface morphology and orientations of most dune fields may, therefore, reflect reworking since the last glacial maximum, rather than full-glacial paleowind directions or the age of the underlying deposits. By example, paleosols dating to ca. 24 ka BP within eolian sands in the Kobuk Valley, Alaska, may indicate regional stabilization of the dunes during the mid-Wisconsin, but with dune activity both before and after this period. There have also been repeated episodes of eolian stability and activity of at least some of these dunes in the past 8,000 years.

Europe and Russia

Many detailed chronostratigraphic divisions of European dune sand, cover sand, and loess sequences have been published in the last few decades although a firm chronology of phases of eolian activity for this region is still lacking. Deposition of eolian sediments associated with the European sand belt has probably occurred over at least six Quaternary glacial period. Eolian cover-sand sediments in The Netherlands have been dated by optically stimulated luminescence (OSL) to the last interglacial and the last two glacial stages. Three major phases of eolian deposition have been described within central and northwestern Europe. The first phase occurred during the late Weichselian glacial maximum, ca. 25–15 ka BP, with predominant fluvioeolian deposition occurring during periglacial conditions with widespread permafrost. The second, during the waning stages of glaciation, ca. 15–12 ka BP, representing the major period of sand sheet and low dune deposition, with dominant sand-transporting winds from the northwest and a climate that was not much cooler than today. The last phase occurred between the Younger Dryas and early Holocene, ca. 11–9 ka BP. Dunes formed under west-southwesterly winds, in areas proximal to interfluves and in river valleys, in relation to changing river channel morphology. Dune formation in the earlier phases was probably induced by a combination of factors including an abundance of unconsolidated sediment, absence of topographic obstruction, sparse vegetation cover, and a high wind-energy regime.

On the Russian Plain, radiocarbon ages of paleosols separating dune deposits from underlying alluvial sediments indicate that dune building occurred after about 15.0 to 12.5 ka BP, associated with the cold late glacial stages of the Older and Middle Dryas.

In the UK, late Holocene dune activity has taken place episodically during the last 7,000 yrs in relation to reductions in available moisture and vegetation, and disturbance by humans and livestock.

In Fennoscandia, deglaciation of the ice sheet took place between about 12.0 and 8.5 ka BP. Initial dune building probably began soon after local deglaciation. Stabilization may have occurred in several phases, beginning prior to 7 ka BP with colonization by pine trees, and continuing until about 4 ka BP with increased effective moisture and a regional rise in groundwater. Short-lived late Holocene eolian episodes have occurred due to forest fires. Increased dune activity in the last 800 yrs may be related to changes in climate-vegetation parameters in association with the 'Little Ice Age' and to the effects of trampling by reindeer.

Paleoenvironmental Significance and Future Directions

High-latitude dune fields represent informative sedimentary environments with respect to understanding late-glacial and Holocene paleoclimates. Wind directions can be inferred from dune morphology and orientation, the dip of forest beds, and the relation of dunes to known sediment sources. These provide evidence of past localized wind regimes and regional/continental circulation patterns. Detailed chronostratigraphic sequences of eolian deposits provide information regarding the relative importance of vegetation, niveo-eolian processes, and other cold-climate (i.e., permafrost and periglacial) processes in high latitudes.

In Europe, where the greatest number of chronostratigraphic studies has been conducted, a firm chronology of the phases of dune activity is still lacking. For many other high-latitude dune fields detailed chronostratigraphic studies are still needed. For example, many stabilized dune fields in northern Canada appear to retain morphologies indicating sediment transport directions related to late glacial (anticyclonic) and early Holocene (westerly) wind conditions. However, no detailed chronological studies have been conducted to determine the timing of these events.

Detailed stratigraphic studies provide the means for determining the relative importance of cold-climate eolian processes in dune development in high latitudes. From previous work, it has been suggested that cold (semi-) arid conditions, rather than snow-rich climates favor niveo-eolian transport. However, despite the significance of cold-climate processes in the development of high-latitude dune fields, few analog studies are available for understanding relict eolian features and deposits. Studies are needed that document modern cold-climate eolian processes and sedimentary structures, as well as stratigraphic studies of late Wisconsin eolian deposits in permafrost environments.

In high-latitude areas, it is important to understand how the factors of transport capacity, sediment supply, and sediment availability have affected dune fields through time, and how this compares to mid- to low-latitude dune fields. However, the inaccessibility of many high-latitude dune fields has caused our understanding of them to lag behind that of other areas. Process studies on active transverse and parabolic dunes in high-latitude areas are required in order to understand relative transport rates and sedimentary processes in these environments, and to compare to those in coastal and low-latitude areas.

In the past, dating of eolian deposits has posed a problem due to the reliance on organic materials for radiocarbon dating, which are often scarce in eolian sediments, and which typically provided ages (up to a maximum of about 27 ka BP) that only bracket the time of eolian deposition. The advent of luminescence dating, especially OSL, provides for potentially more detailed chronologies of Mid- to Late Quaternary-aged dune fields based on the eolian sediments themselves.

Coupled with more precise chronology, improved paleoenvironmental reconstructions from high-latitude dune fields are needed by integrating eolian chronostratigraphy with adjacent lake records of biostratigraphy, oxygen-isotopes, and mineralogy.

Finally, studies are needed to determine the relation between eolian sands and various source-sediment deposits. In Alaska, for example, uncertainty still exists about the relation between surficial dune deposits and underlying subhorizontally stratified deposits, variously interpreted as eolian by some workers, and alluvial or lacustrine by others. Further studies are needed here to determine the origin and age of these eolian deposits. In addition, detailed stratigraphic studies of deposits underlying and adjacent to dune deposits, and the relative amount of reworking of source materials are needed here and in other high-latitude dune fields. In this respect, mineralogical and geochemical studies that have provided insight into the origin and evolution of dune fields in mid-to-low latitudes should be applied to high-latitude dune fields.

Abbreviations

BP	before present
cal yr	calendar years
ka	thousand years
OSL	optically stimulated luminescence

Glossary

Cold-climate environment Includes periglacial environments, where the mean annual air temperature is less than +3 °C, as well as regions with snowy climates, where the coldest mean monthly temperature is below −3 °C.

High-latitude dune fields Sand dunes found in more northerly and southerly latitudes and typically associated with seasonal cold-climate conditions.

Parabolic dunes A dune in the shape of a parabola or horseshoe, in plan view, with a concave windward slope and a convex leeward slope. The trailing arms of parabolic dunes are commonly vegetated.

Sand dunes Distinct constructional eolian landforms that generally exceed several meters in height and exhibit slipfaces. They typically comprise stacked cosets of large-scale, high-angle cross strata, representing net deposition from climbing bedforms.

Sand sheets A landform composed of eolian sand of varying thickness but without pronounced relief typically with subhorizontal to low-angle cross-stratification formed by the migration of low-amplitude wind ripples or by sedimentation in the presence of vegetation. The cover sands of the European sand belt are an extensive area of sand sheet deposits.

See also: **Dune Fields**: Mid-Latitudes; Low Latitudes. **Luminescence Dating**: Optically-Stimulated Luminescence.

References

Bateman, M. D., and Godby, S. P. (2004). Late-Holocene inland dune activity in the UK: a case study from Breckland, East Anglia. *The Holocene* **14**, 2004.

Black, R. F. (1951). Eolian deposits in Alaska. *Arctic* **4**, 89–111.

Calkin, P. E., and Rutford, R. H. (1974). The sand dunes of Victoria Valley, Antarctica. *Geographical Review* **64**, 189–216.

Carson, M. A., and McLean, P. A. (1986). Development of hybrid aeolian duines: the William River dune field, northwest Saskatchewan, Canada. *Canadian Journal of Earth Sciences* **23**, 1974–1990.

Dallimore, S. R., Wolfe, S. A., Matthews, J. V., Jr., and Vincent, J. S. (1997). Mid-Wisconsinan eolian deposits of the Kittigazuit Formation, Tuktoyaktuk Coastlands, Northwest Territories, Canada. *Canadian Journal of Earth Sciences* **34**, 1421–1441.

David, P. P. (1981). Stabilized dunes ridges of northern Saskatchewan. *Canadian Journal of Earth Sciences* **18**, 286–310.

David, P. P. (2000). Aeolian landforms. In *The Canadian Encyclopedia* (J. H. Marsh, Ed.), p. 2573. McLelland and Stewart, Toronto, ON, Canada.

Dijkmans, J. W. A. (1990). Niveo-aeolian sedimentation and resulting sedimentary structures; Søndre Strømfjord area, West Greenland and their palaeoenvironmental implications. *Permafrost and Periglacial Processes* **1**, 83–96.

Drenova, A. N., Timireva, S. N., and Chikolini, N. I. (1997). Late glacial dune-building in the Russian Plain. *Quaternary International* **41/42**, 59–66.

Filion, L. (1987). Holocene development of parabolic dunes in the Central St. Lawrence Lowland, Canada. *Quaternary Research* **28**, 196–209.

Filion, L., and Morisset, P. (1983). The late Holocene record of aeolian and fire activity in northern Québec, Canada. *The Holocene* **1**, 201–208.

Hamilton, T. D., Galloway, J. P., and Koster, E. A. (1987). Late Wisconsinan eolian activity and related alluvium, central Kobuk valley. *United States Geological Survey Circular* **1016**, 39–43.

Hansen, E. C., Arbogast, A. F., Packman, S. C., and Hansen, B. (2002). Port-Nipissing origin of a backdune complex along the southeastern shore of Lake Michigan. *Physical Geography* **23**, 233–244.

Kasse, C. (2002). Sandy aeolian deposits and environments and their relation to climate during the Last Glacial Maximim and Lateglacial in northwest and central Europe. *Progress in Physical Geography* **26**, 507–532.

Käyhkö, J. A., Worsley, P., Pye, K., and Clarke, M. L. (1999). A revised chronology for aeolian activity in subarctic Fennoscandia during the Holocene. *The Holocene* **9**, 195–205.

Koster, E. A. (1988). Ancient and modern cold-climate aeolian sand deposition: a review. *Journal of Quaternary Science* **3**, 69–83.

Koster, E. A. (2005). Recent advances in luminescence dating of Late Pleistocene (cold-climate) aeolian sand and loess deposits in western Europe. *Permafrost And Periglacial Processes* **16**, 131–143.

Lea, P. D., and Waythomas C, F. (1990). Late-Pleistocene eolian sand sheets in Alaska. *Quaternary Research* **34**, 269–281.

Mann, D. H., Heiser, P. A., and Finney, B. P. (2002). Holocene history of the Great Kobuk Sand Dunes, northwestern Alaska. *Quaternary Science Reviews* **21**, 709–731.

Muhs, D. R. (2004). Mineralogical maturity in dunefields of North America, Africa, and Australia. *Geomorphology* **50**, 247–269.

Muhs, D. R., and Wolfe, S. A. (1999). Sand dunes of the northern Great Plains of Canada and the United States. In *Holocene Climate and Environmental Change in the Palliser Triangle; a Geoscientific Context for Evaluating the Impacts of Climate Change on the Southern Canadian Prairies* (D. S. Lemmen and R. E. Vance, Eds.), Geological Survey of Canada Bulletin **534**, 293.

Murton, J. B., French, H. M., and Lamothe, M. (1997). Late Wisconsinan erosion and eolian deposition, Summer Island are, Pleistocene Mackenzie Delta, Northwest Territories: optical dating and implications for glacial chronology. *Canadian Journal of Earth Sciences* **34**, 190–199.

Raup, H. M., and Argus, G. W. (1982). The Lake Athabasca Sand Dunes of Northern Saskatchewan and Alberta, Canada I: The Land and Vegetation. *Publications in Botany*, No. 12, National Museums of Canada, p. 96.

Ruz, M. H., and Allard, M. (1994). Foredune development along a subarctic emerging coastline, eastern Hudson Bay. *Marine Geology* **117**, 57–74.

Schokker, J., Cleveringa, P., and Murray, A. S. (2004). Palaeoenvironmental reconstruction and OSL dating of terrestrial Eemian deposits in the southeastern Netherlands. *Journal of Quaternary Science* **19**, 193–202.

Selby, M. J., Rains, R. B., and Palmer, W. P. (1974). Eolian deposits of the ice-free Victoria Valley, southern Victoria Land, Antarctica. *New Zealand Journal of Geology and Geophysics* **17**, 543–562.

Seppälä, M. (2004). *Wind as a Geomorphic Agent in Cold Climates.* Cambridge University Press, Cambridge.

Wolfe, S. A., and Nickling, W. G. (1997). Sensitivity of Eolian Processes to Climate Change in Canada. *Geological Survey of Canada Bulletin* **421**, 30.

Wolfe, S. A., Huntley, D. J., David, P. P., *et al.* (2001). Late 18th century drought-induced sand dune activity, Great Sand Hills, Saskatchewan. *Canadian Journal Earth Sciences* **38**, 105–117.

Wolfe, S. A., Ollerhead, J., and Lian, O. B. (2002). Holocene eolian activity in south-central Saskatchewan and the southern Canadian Prairies. *Géographie physique et Quaternaire* **56**, 215–227.

Zeeberg, J. (1998). The European sand belt in eastern Europe – and comparison of Late Glacial dune orientation with GCM simulation results. *Boreas* **27**, 127–139.

Zieliński, P. (2003). The depositional conditions of longitudinal dunes based on investigations in the western part of the Lubin Upland, SE Poland. *Landform Analysis* **4**, 65–73.

Mid-Latitudes

J Sun, Chinese Academy of Sciences, Beijing, China
D R Muhs, U.S. Geological Survey, CO, USA

© 2007 Elsevier B.V. All rights reserved.

Introduction

Large dune fields, or sand seas, are landscapes often thought to be found only in deserts beneath the great, subtropical high-pressure zones, where subsiding air suppresses rainfall. Dune fields are also quite common in mid-latitude regions, to the north and south of subtropical deserts (**Fig. 1**). Two major characteristics distinguish many mid-latitude dune fields from the sand seas of lower latitudes. One difference is that many of those in mid-latitudes are in semiarid, rather than arid climates, and therefore are not

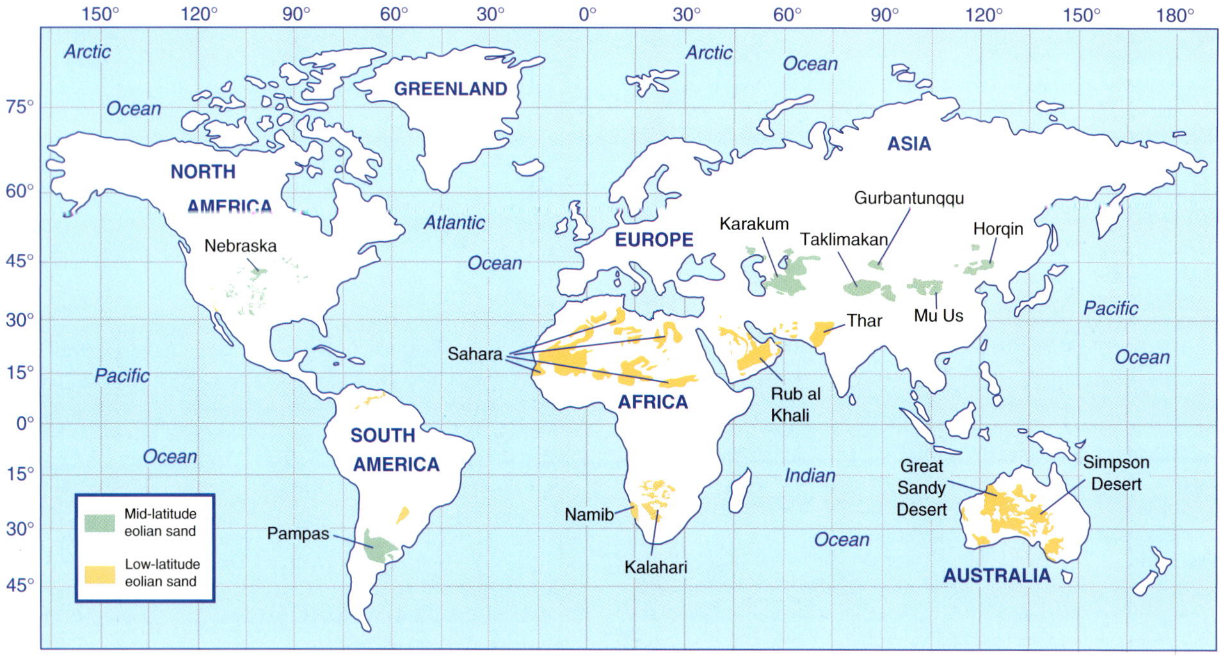

Figure 1 Map showing the global distribution of eolian sand, whether active or stabilized. Sources: North America: Muhs and Zárate (2001), Muhs and Holliday (1995), and Thorp and Smith (1952); South America: Muhs and Zárate (2001) and sources therein; Africa: Wilson (1973), Lancaster (1988), and Thomas and Shaw (1991); Central Asia and India: Wilson (1973); China: Sun and Muhs (this article) and sources herein; Australia: Bowler *et al.* (2001). Mid-latitude dune fields reviewed in this chapter are shown in green; low-latitude dune fields reviewed by N. Lancaster (this volume) are shown in yellow.

presently active. In this article, 'active' refers to eolian sand bodies that are not covered with vegetation, and where particles are currently being transported by the wind. Evidence of active sand transport often takes the form of well-expressed sand ripples on a dune's surface. Where sand is vegetated and particles are not being transported by the wind, a sand body is 'stable.'

A second characteristic of mid-latitude dune fields is that many are situated near mountain ranges or at least areas of relatively high relief. Glaciation in high mountains reduces local bedrock to sand-sized particles, and additional reduction of rock to sand sizes takes place through colluvial and fluvial processes quickly in terrains of steep relief. Through these processes, new sediment becomes available to feed growing dune fields much more rapidly than in areas of low relief, which characterize lower-latitude regions.

Because of their occurrence in semiarid climates, mid-latitude dune fields are particularly sensitive indicators of shifts in the regional moisture balance. Sand dunes move only where there is wind, sand-sized sediment, and a lack of vegetation cover. A shift to drier conditions may result in loss of vegetation cover and activation of sand by wind. A shift back to wetter conditions may stabilize dunes because of enhanced vegetation cover. The geologic record of this alternation of dry and moist climates tends to be preserved in many mid-latitude dune fields as eolian sand–paleosol sequences. In areas downslope of high sediment production, such as mountains, fresh sediment may accumulate in a dune field during periods of eolian activity, bury the pre-existing landscape, and preserve former surface soils as buried soils (or paleosols). In arid, low-latitude environments of low relief, previously stabilized eolian sand may be continually reworked without addition of much new sand, and formerly stable surfaces, which would be marked by paleosols in semiarid climates, may not be preserved.

Throughout this article, we correlate eolian dune-building events, where they can be dated or where the age can be inferred, to the marine oxygen-isotope stratigraphic framework of Martinson *et al.* (1987). Where the abbreviation 'MIS' occurs, this refers to 'marine isotope stage,' using the numbering system and approximate ages in Martinson *et al.* (1987).

Dune Fields in China

Introduction

Dune fields in China form an east–west trending belt in the mid-latitude interior (latitude 35–50° N, longitude 75–125°E) (**Fig. 2**). The total desert area is 1,308,000 km^2, or 13.6% of China's land area

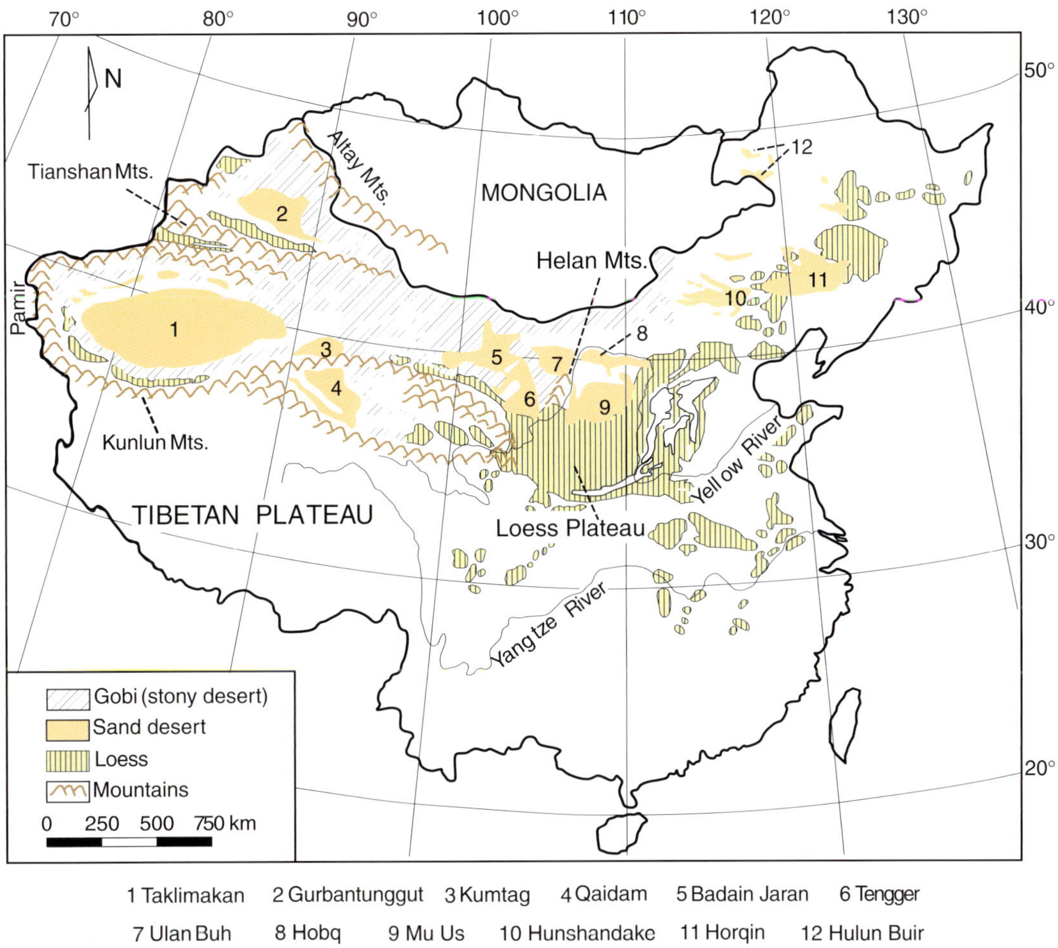

Figure 2 Map showing gobi (stony desert), dune fields, and loess distribution in China.

(Zhu *et al.*, 1980). These dune fields can be divided into two parts by the north–south trending Helan Mountains (**Fig. 2**). Dune fields, west of the Helan Mountains, include the Taklimakan, Gurbantunggut, Kumtag, Qaidam, Badain Jaran, Tengger, and Ulan Buh deserts, which are situated in inland basins or adjacent to high mountains. The formation of these dune fields is a function of their inland geographic location, far from moisture sources. Climate in these deserts is controlled by the westerlies. Dune fields east of the Helan Mountains, situated in the East Asian monsoonal zone, are mainly stabilized and semistabilized at present.

Dune Types in China

The Taklimakan Desert, which is surrounded by the Tianshan and Kunlunshan mountains, is in the Tarim Basin and dominated (85%) by active sand dunes (**Fig. 2**). The climate is extremely arid, with a mean annual precipitation of only 15–40 mm. Local residents call the Taklimakan Desert the 'dead sand sea.' Dune types are quite diverse in this desert (**Fig. 3**), consisting mainly of complex dune chains and transverse dunes. Dunes with heights exceeding 50 m occupy 80% of the area in its interior.

The Gurbantunggut Desert is the second largest dune field in China (48,800 km^2), situated in the Junggar Basin (**Fig. 2**). The mean annual precipitation in the Gurbantunggut Desert is 70–150 mm, which is the highest among the deserts in northwestern China. Stabilized and semistabilized sand dunes occupy 97% of this desert, and linear (or longitudinal) dunes are the dominant type (**Fig. 4**). There are also erosional landforms in the northwestern part of the desert, called 'yardangs' (**Fig. 5**).

The Badain Jaran Desert is the third largest dune field in China (44,300 km^2). The climate is extremely arid, with a mean annual precipitation of less than 50 mm. There are two remarkable properties of the landscape in this desert. First, the desert consists mainly of complex sand hills, dune chains, and pyramidal landforms called 'star' dunes, with an average height of 200–300 m. Second, as many as 144 small lakes are found in the interdune lowlands of the desert (Zhu *et al.*, 1980). The source of these lakes is melting water of the Qilian Mountains, ~500 km to the southwest, which reaches the desert through a subsurface fault system (Chen *et al.*, 2004). The Kumtag Desert also has complex dune chains and star-shaped pyramidal dunes. In this sand sea, heights of the dunes vary between 100 and 200 m (**Fig. 6**).

Figure 3 Sand sea of the Taklimakan Desert. About 80% of this dune field is active at present and precipitation-to-potential-evapotranspiration (P/PE) values are generally less than 0.2 (Wang *et al.*, 2005).

The Tengger Desert is the fourth largest desert in China (42,700 km^2). Sand dunes occupy 71% of the desert, and 93% of the dunes are active. The dominant dune type is transverse dune, with heights usually between 10 and 20 m.

The Qaidam Desert is located in the Qaidam Basin (**Fig. 2**). Yardangs in its interior make up 67% of the desert's area (**Fig. 7**). Sand dunes are mainly distributed in the marginal areas, dominated by transverse dunes.

Transverse dunes dominate other dune fields in China, especially those east of the Helan Mountains. Most dunes are stabilized or semistabilized (**Fig. 8**), and vary in height between 5 and 15 m.

Figure 4 Semistabilized linear dunes in the Gurbantunggut Desert.

Figure 6 Star-shaped pyramidal sand dunes in the Kumtag Desert.

Figure 5 Deflation landforms (yardangs) in the northwestern Gurbantunggut Desert.

Figure 7 Deflation landforms (yardangs) in the interior of the Qaidam Desert.

Figure 8 Stabilized or semistabilized sand dunes are the dominant dune types in the deserts, distributed to the east of the Helan Mountains.

Figure 9 Landsat thematic mapper (TM) image of alluvial fans in the piedmont of the Qilian Mounatins, China. Sediments in these large alluvial fans are important sources of the dune sand and loess in China.

Origin of Chinese Dune Fields

Origins of the dune fields in China can be divided into two categories. Dune fields to the west of the Helan Mountains are situated in the inland basins near the Tianshan, Kunlunshan, and Qilianshan mountains. The mean annual precipitation varies from 20 to 150 mm. Geomorphic processes in the adjacent mountains (especially glacial grinding) have played an important role in producing the vast amounts of sand-sized materials for the formation of these dune fields. Elevations of the Tianshan, Kunlunshan, and Qilianshan mountains range mostly from ~3,000 m to ~5,500 m above sea level. The high elevation gives rise to extremely cold climatic conditions, thus favoring glacial grinding and frost weathering. This leads to the physical weathering of rocks in the surrounding regions, and thus the production of clastic materials (Sun, 2002a, b). The high relief and steep gradients of these mountains give rise to high potential energy of melting water. Thus, the clastic materials can be transported to the piedmont areas, forming large alluvial fans. Examination of multispectral Landsat thematic mapper (TM) imagery indicates that a series of huge alluvial fans have formed in the piedmont of the Qilian Mountains (**Fig. 9**). The zonal distributions of the gobi (stony desert), dune fields, and loess bodies are the result of wind sorting of the piedmont fluvial materials (Sun, 2002a).

Deserts to the east of the Helan Mountains are situated in the high plains of northeastern China. Because this region lies in the East Asian monsoonal zone, the mean annual precipitation ranges from 200 to 450 mm. Most of the dunes are stabilized and semistabilized. However, due to poor land-use practices in historic time, active sand dunes also occur in these deserts, largely due to reworking of last glacial age (MIS 2) sand dunes (Sun, 2000). The provenance of the sediments in these dunes is dominantly from fluvial sediments of several large rivers.

Age of Desert Dunes

One approach for studying long-term dune field history is to examine the chronology of loess deposition, as there are few age estimates of the oldest dunes in China. In the northern piedmont of the Kunlun Mountains, dust derived from the Taklimakan Desert accumulates as loess sediment. Because the Taklimakan Desert lies within the rainshadow of the Tibetan Plateau, which began rising in the late Cenozoic (Gansser, 1964; Molnar and Tapponnier, 1975; Harrison *et al.*, 1992), it is possible that dunes have existed in this basin since the latest Cenozoic.

Recent studies of eolian silt on the Chinese Loess Plateau indicate that the source areas are the gobi (stony) desert in southern Mongolia and the adjoining gobi and sandy deserts (including the Badain Jaran Desert, Tengger Desert, Ulan Buh Desert, Hobq Desert, and Mu Us Desert) in China (**Fig. 10**). These gobi and sand deserts only serve as 'holding areas' of dust and silt, rather than the original source of loess particles. Mountain processes (especially glacial grinding) in the high elevations of Asia have played an important role in producing the vast amounts of loess-sized material for forming the Loess Plateau (Sun, 2002a). Because the basal age of the loess in the Loess Plateau is 2.58 Ma (Liu, 1985; Liu and Ding, 1993), and the loess materials are transported by the near-surface winds from the deserts listed above, these basins have been arid environments for at least 2.58 Ma.

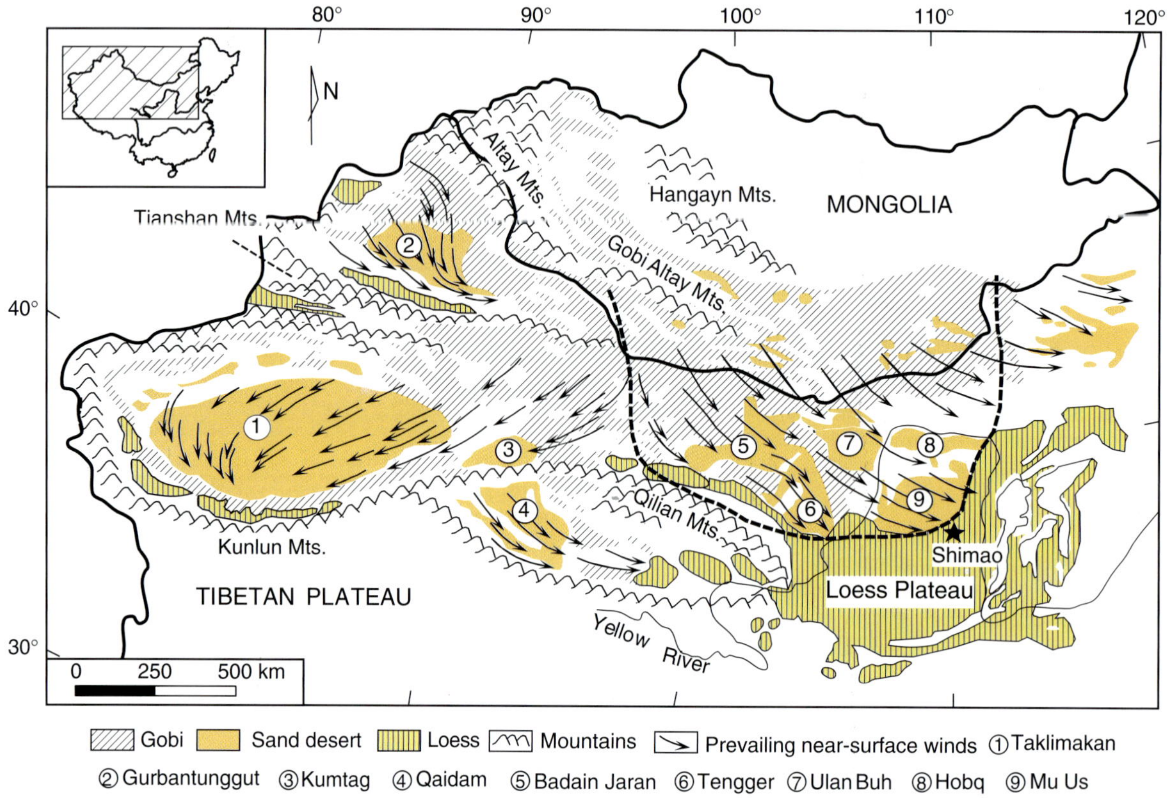

Figure 10 Map showing the near-surface wind directions (arrows) based on dune orientation in the studied regions. Area circled by dashed line indicates the source region of the Loess Plateau (Sun, 2002a).

Relations of Desert Dunes to Loess Deposits

Because Chinese loess is derived from the deserts, desert expansions and contractions, in response to Quaternary glacial–interglacial cycles, may be recorded by eolian deposits at the margins of deserts. Among the 12 deserts in China, only the Mu Us Desert is geographically proximal to the Chinese Loess Plateau. Eolian deposits on the southern margin of the Mu Us Desert are ideal for studying the response of the desert to climatic changes of the Quaternary.

The eolian deposits in the southern marginal Mu Us Desert are characterized by alternations of eolian sand, loess, and paleosols, making it a potential region for recording the ecosystem changes during the Quaternary. Eolian sand accumulated when the Mu Us Desert expanded southeasterly to the Loess Plateau during glacial maxima (Sun *et al.*, 1999).

Paleosols developed under warm and humid conditions of interglacial periods, as a response to the enhanced summer East Asian monsoon winds. The existence of the paleosols in the desert marginal regions indicates that the dunes in the Mu Us Desert were stabilized by vegetation during interglacial maxima.

The loess beds imply intermediate climate conditions, between full glacial and full interglacial climates. The nearly unweathered character of these loess layers suggests that they were deposited under the dominant winter monsoon winds but with greatly reduced wind energy than what is necessary for sand transportation.

The most representative section of these complex eolian deposits is located at Shimao, in the southern marginal Mu Us Desert (Sun *et al.* (1999); see **Fig. 10** for location). The eolian sequences of this section consist of 40 eolian stratigraphic units, of which 13 are sands, 13 are loess, and 14 are paleosols. Based on paleomagnetic polarity studies and a comparison with the composite marine oxygen isotope records (Shackleton and Pisias, 1985; Shackleton *et al.*, 1990), the basal age of this section is about 0.63 Ma (**Fig. 11**). The eolian deposits at Shimao and their correlation with the marine records yield two important implications.

First, during the past 0.63 Ma, there have been at least 13 times of desert expansion in the Mu Us Desert, and the desert expansion occurred not only during glacial maxima, but also during the cold substages of interglacial periods, such as those of MIS 7,

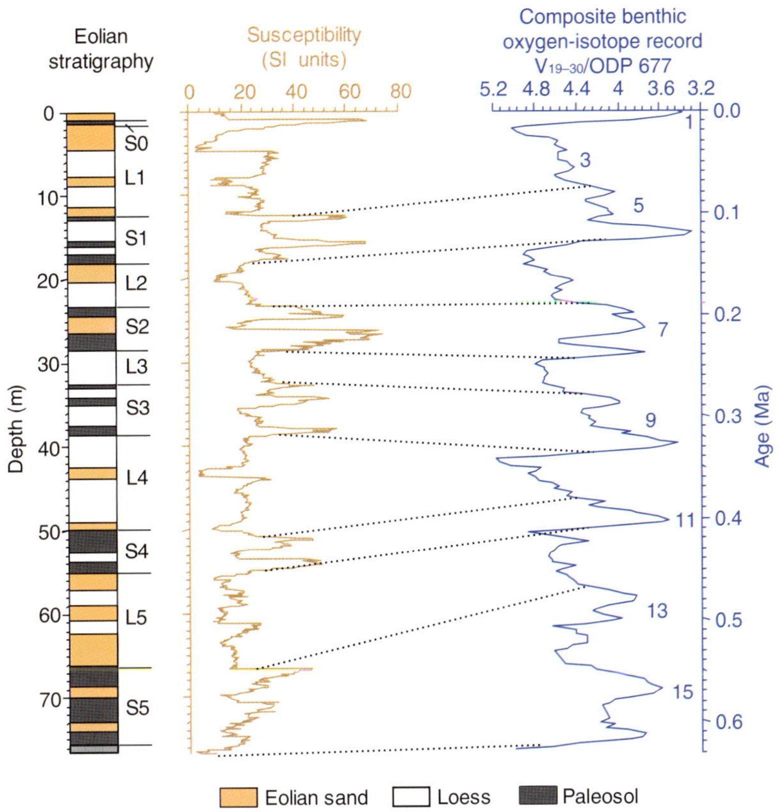

Figure 11 Pedostratigraphy and the magnetic susceptibility curve of the Shimao section and its correlation with the $\delta^{18}O$ records from the Pacific Ocean. The isotope curve is a composite record of V_{19-30} (0–0.34 Myr, Shackleton and Pisias, 1985) and ODP 677 (0.34–0.63 Myr, Shackleton *et al.* (1990)). Modified from Sun, J.M., Ding Z.L., Rokosh, D., and Rutter, N., (1999) 580,000 year environmental reconstruction from eolian deposits at the Mu Us Desert margin, China. *Quaternary Science Reviews* 18: 1351–1364.

13, and 15. Intercalated eolian sand in the interglacial paleosols S2 and S5 support this conclusion (**Fig. 11**).

A second major implication is that the stratigraphy of the section at Shimao shows that except for the Holocene paleosol (S0), all the other paleosols (S1, S2, S3, S4, and S5) are pedocomplexes, with eolian sand or loess beds intercalated within the soil complexes. The sand or loess intercalations indicate that the warm climate during each interglacial time was interrupted by cold–dry episodes.

Desert Distributions during the Holocene Optimum and the Last Glacial Maximum in China

Recent studies indicate that dust from Asia dominates the entire Pacific Ocean north of the intertropical convergence zone (Merrill *et al.*, 1989; Duce *et al.*, 1991). China is considered as one of the two dominant source regions for atmospheric mineral aerosols originating from Asia (Rea, 1994). The waxing and waning of the summer monsoon during the Quaternary must have had a great effect on the expansion and contraction of the deserts in northeastern China, which in turn modulated the flux of dust from Asia to the Pacific Ocean.

Former extent of deserts during the Last Glacial Maximum Variation in global ice volume may have played a key role in modulating and pacing the strength of the East Asian monsoon (Ding *et al.*, 1994). During the Last Glacial Maximum (LGM), the enlarged high-latitude ice sheets of the Northern Hemisphere greatly strengthened the Siberian High, giving rise to an enhanced cold–dry northwest winter monsoon. Under winter monsoonal winds, strong eolian erosion, transportation, and dune building characterized the arid and semiarid regions of northern China.

Since the 1980s, 44 profiles with LGM dune sands and overlying Holocene sandy loam soils have been found in northeastern China, in the monsoonal climatic zone. Thus, during the LGM, active sand dunes occurred not only in the arid inland areas of western

China, but also in the semiarid to subhumid areas, east of the Helan Mountains. The existence of the buried LGM sands implies that the southern limit of the LGM desert extended south to nearly latitude 36°N, and east to longitude 125°E (Sun *et al.* (1998); **Fig. 12A**).

Former extent of deserts during the Holocene optimum During the mid-Holocene climatic optimum, from ~9 ka to ~3 ka, the East Asian summer monsoon was strong. Sandy loam soils on stabilized dunes are widely distributed in desert regions, especially east of the Helan Mountains. Radiocarbon ages and optically stimulated luminescence (OSL) dating indicate that these loamy soils developed mainly between 3.0 and 9.0 ka BP, coincident with the Holocene optimum (Zhou *et al.*, 2001; Li *et al.*, 2002). Pollen analysis of these soils indicates that the vegetation was dominated by steppe taxa, implying that sand dunes in northeastern China were stabilized during the Holocene optimum (Sun *et al.*, 1998).

In contrast to the deserts in northeastern China, the distinctive Holocene sandy loam soil has not been found in the Taklimakan, Qaidam, and Badain Jaran

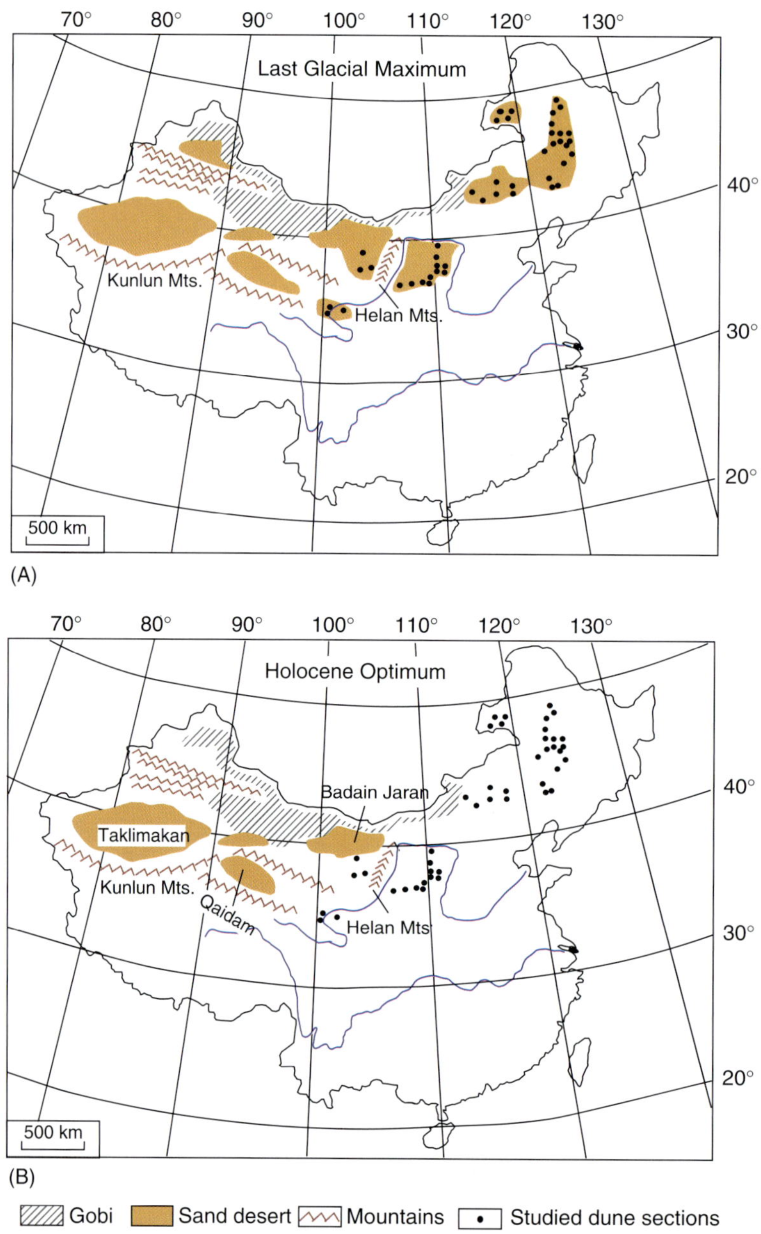

Figure 12 Desert distributions in China during the Last Glacial Maximum (A) and the Holocene Optimum (B), the filled black dots indicating the sites where Holocene loam soils and the underlying sand dunes of the LGM are found (Sun *et al.*, 1998).

deserts. Its absence implies that during the Holocene optimum, the much-enhanced summer monsoon did not penetrate into the interior of the western desert regions, and thus active dune fields existed then (and still exist today). Compared with the desert of the LGM, the extent of desert in the Holocene was greatly reduced, with the eastern limit of desert retreating from longitude 125°E (LGM time, MIS 2) to about 105°E, a distance of well over 1,600 km (**Fig. 12B**).

Dune Fields in Central Asia and South America

Central Asia

To the west of China, dune fields occupy large areas of Central Asia, in the desert basins of Kazakhastan, Uzbekistan, and Turkmenistan. The largest dune fields are situated mostly southeast of the Caspian Sea, Aral Sea, and Lake Balkhash (**Fig. 13**). Wilson (1973) estimates these dune fields to encompass

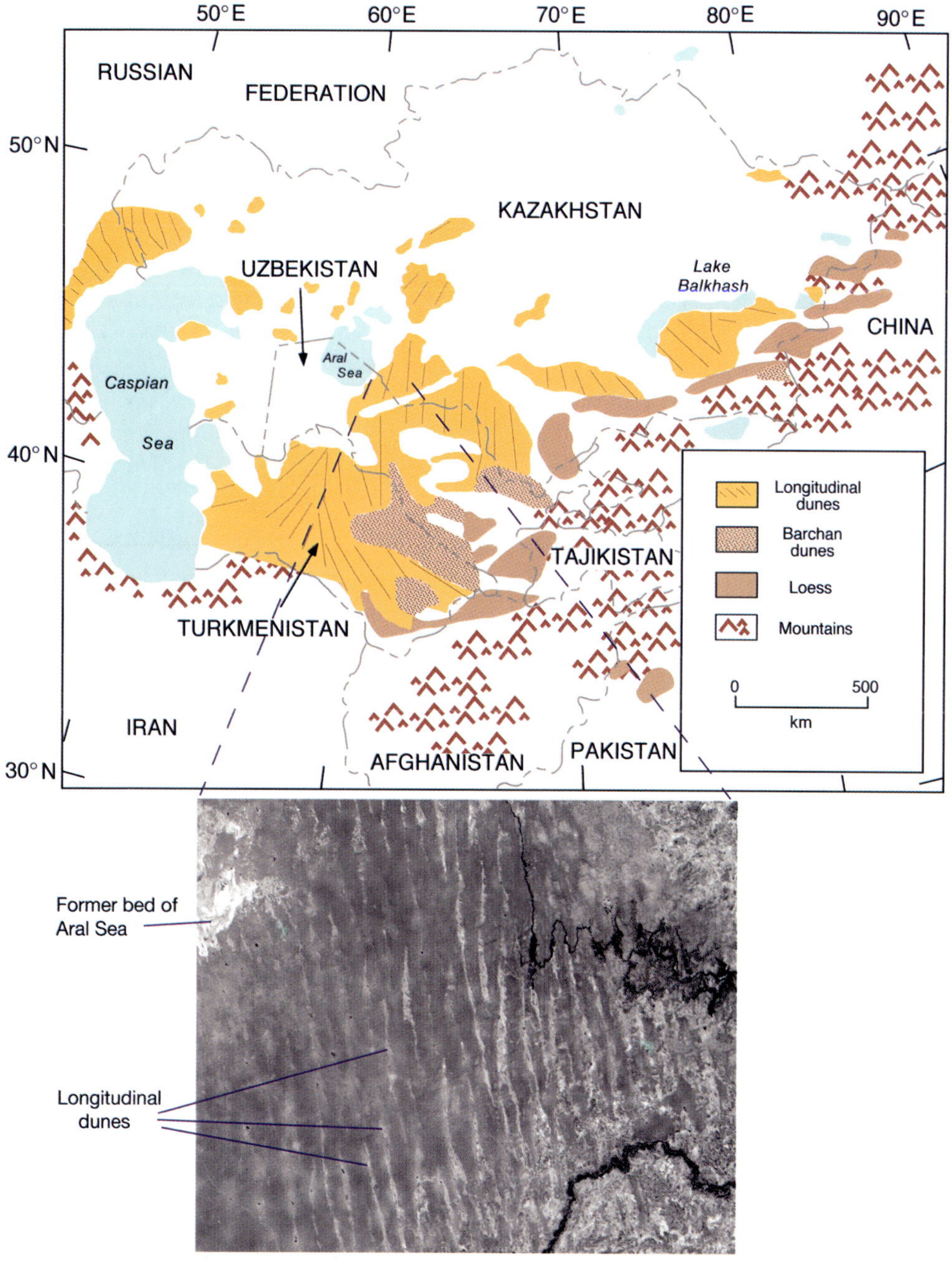

Figure 13 Upper: map showing the distribution of eolian sand and loess in Central Asia (redrawn from data in Lioubimtseva (2002) and sources therein for sand; from Dodonov (1991) for loess). Lower: Landsat TM image (path 159, row 029; band 1, visible blue; image taken 5 August 1987) of longitudinal dunes southeast of the Aral Sea, Kazakhstan and Uzbekistan.

~380,000 km^2 (Karakum sand sea, southeast of the Caspian Sea), ~276,000 km^2 (Kyzylkum sand sea, southeast of the Aral Sea), and ~38,000 km^2 (Myunkum sand sea, southeast of Lake Balkhash). The region is semiarid to arid, and precipitation ranges from ~100 to 400 mm yr^{-1} (Goudie, 2002; Lioubimtseva, 2002).

The main dune forms in the region are longitudinal (linear) dunes, which are easily visible on Landsat imagery (**Fig. 13**). Goudie (2002) reports that the linear dunes are as much as 30 m high and are separated by interdune areas floored with clay-rich sediments. Wilson (1973) indicates that most or all of these dunes are active at present. At the broadest scale, orientations of the dunes (**Fig. 13**) agree fairly closely with modern, dominant wind directions (Letollé and Mainguet, 1993), but in detail, there are some differences, as pointed out by Lioubimtseva (2002). Southeast of the areas of linear dunes, barchans are the dominant landforms, and many of these are also active at present (Goudie, 2002; Lioubimtseva, 2002).

High mountain ranges in Afghanistan, Tajikistan, and China are drained by rivers that flow westward to the basins occupied by the Caspian and Aral Seas. The Volga River, the largest river in Europe, carried sediment from the Fennoscandian ice sheet that covered northern Europe during the LGM (MIS 2). Thus, abundant sediment is delivered to the desert basins of Central Asia at present and it is likely that even more sediment was transported there during glacial periods of the Pleistocene. Because sources of specific dune fields in the region are little studied, it is not known which rivers are most important as dune sand sources.

The paleoclimatic significance of dunes in Central Asia is difficult to assess, as there has been little study of dune stratigraphy, and numerical ages are lacking. However, as with China, some comparisons can be made with chronologies developed for loess deposits that are downwind of the dune fields. Loess deposits are extensive in the southern margins of the desert regions in Central Asia (**Fig. 13**). Dodonov (1991) suggests that this loess is derived from a combination of glaciogenic sources in the neighboring mountains and dust sources in the desert basins. Long-distance transport from desert basin sources is supported by the relatively fine grain size of loess particles in sections exposed in Tadjikistan (Frechen and Dodonov, 1998). Luminescence dating indicates that there was considerable loess deposition during both the early (MIS 4) and later (MIS 2) parts of the last glacial period (Frechen and Dodonov, 1998). These data suggest that winds were strong and desert basin source areas were relatively dry and minimally vegetated during the last glacial period. Nevertheless, during the early part of the last glacial period (MIS 4), the Caspian Sea was connected to the Aral Sea, whereas during the last interglacial period (MIS 5), it was of lesser extent than during the Holocene (Chepalyga, 1984). If the level of the Caspian Sea is at least a partial indicator of the overall moisture balance of the arid basins of the region, these observations suggest that the dune fields may be active primarily during interglacial, rather than glacial periods.

South America

There is a large area of stabilized dunes in the mid-latitudes of South America, primarily in the Pampas region of Argentina. The precipitation gradient in this region is steep, with present mean annual precipitation ranging from less than 500 mm (southwest) to more than 1,000 mm (northeast). In parts of Argentina (particularly in La Pampa and Buenos Aires provinces), there are longitudinal (linear) dunes over 100 km long and 2–5 km wide (**Fig. 14**). Iriondo (1999) refers to this large dune field as the Pampean Sand Sea. The largest dunes in this sand sea are arranged as extensive southwest-to-northeast arch-like forms and are presently stabilized by vegetation. Although the longitudinal dune forms are easily recognizable on Landsat TM imagery (**Fig. 14**), the low relief of these features makes them difficult to identify in the field (Martinez *et al.*, 2001). A large field of parabolic dunes, most of them also stabilized and oriented in a southwest-to-northeast direction, is situated south of the longitudinal dunes (**Fig. 14**). The parabolic dunes may have developed from reworking of pre-existing sands.

Zárate and Blasi (1993) present a model for eolian deposition in Argentina that explains the origin of both dunes and loess. They report sedimentological and mineralogical data that lead them to infer that the main dune sand sources, as well as loess sources, are the Colorado and Negro River floodplains, which are situated to the southwest of the Pampean Sand Sea. The Colorado and Negro rivers drain the Andes, and thus volcaniclastic sediments from these mountains are the ultimate source of much or all of the sediment in the sand sea and loess bodies. Zárate and Blasi (1993) point out that the fluvial regime supplying the particles was dependant, to a large degree, on the extent of glacial cover in the Andes. The Colorado and Negro rivers are presently underfit and have a meandering pattern. In the last glacial period, they apparently had a broad, braided pattern, and during seasonal melting periods, both rivers carried large volumes of meltwater and sediment.

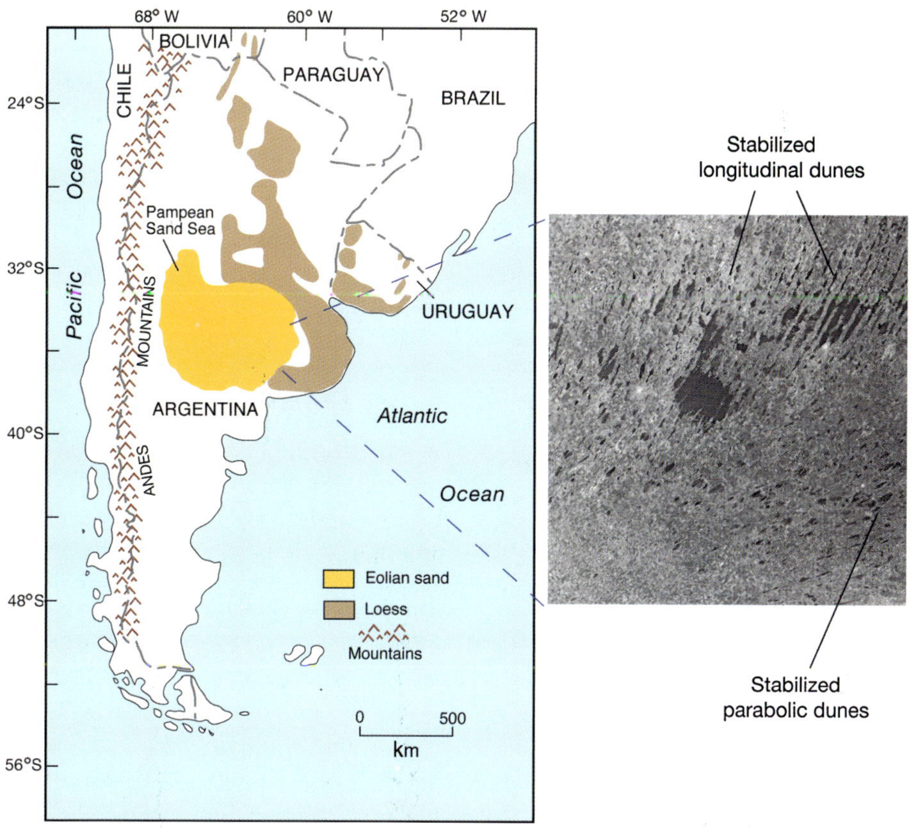

Figure 14 Left: map showing the distribution of eolian sand and loess in southern South America (taken from Zárate (2003), and sources therein). Right: Landsat TM image (path 227, row 085; band 1, visible blue; image taken 17 February 1988) of linear and parabolic dunes in Buenos Aires province, Argentina.

During periods of low flow, sediment availability was large, and provided the source of both sand to build dunes and silt for loess.

There are few data on the ages of eolian sand in Argentina. Iriondo (1999) considers the longitudinal dunes that occur in northwestern Buenos Aires province and surrounding areas to be of late Pleistocene (early last glacial, or MIS 4) age. Thermoluminescence (TL) ages reported by Kröhling (1999) indicate that the earliest part of the Pampean Sand Sea may have formed during the early part of the last glacial period (MIS 4), with reworking of this sand occurring between ~52 and ~36 cal kyr BP. Kröhling (1999) also recognizes an eolian sand facies in a loess-dominated unit called the Tezanos Pinto Formation. Bracketing TL ages of ~32 and ~9 cal kyr BP are reported for the loess facies of the Tezanos Pinto Formation. If these ages are correct, they imply a major period of eolian sand movement during the latter part of the last glacial period (MIS 2). The inference of eolian sand movement occurring during major glacial periods (MIS 4 and 2) is consistent with the sediment origin model of Zárate and Blasi (1993). However, eolian sand movement may also have occurred during the Holocene. For example, Iriondo (1999) considers the parabolic dunes that are found to the southeast of the longitudinal dunes (**Fig. 14**) to be of late Holocene age. Several workers (Zárate and Blasi, 1993; Iriondo, 1999; Kröhling, 1999) use loess and dune formation to infer a relatively dry period for the Pampas region in the late Holocene, although not as dry as during the last glacial period.

Dune Fields in North America

Introduction

Dunes occupy a considerable portion of the mid-latitude regions of North America, particularly in the Great Plains region of the central United States and the Colorado Plateau region of the southwestern United States (**Fig. 15**). The dune fields of the Great Plains have been intensively studied since the 1980s and much is now known about their geographic extent, geomorphic expression, source sediments, chronology, and paleoclimatic significance. The dunes of the Colorado Plateau have been studied less extensively, but work in progress (e.g., Redsteer and Block (2004)) promises important new information about these smaller sand seas.

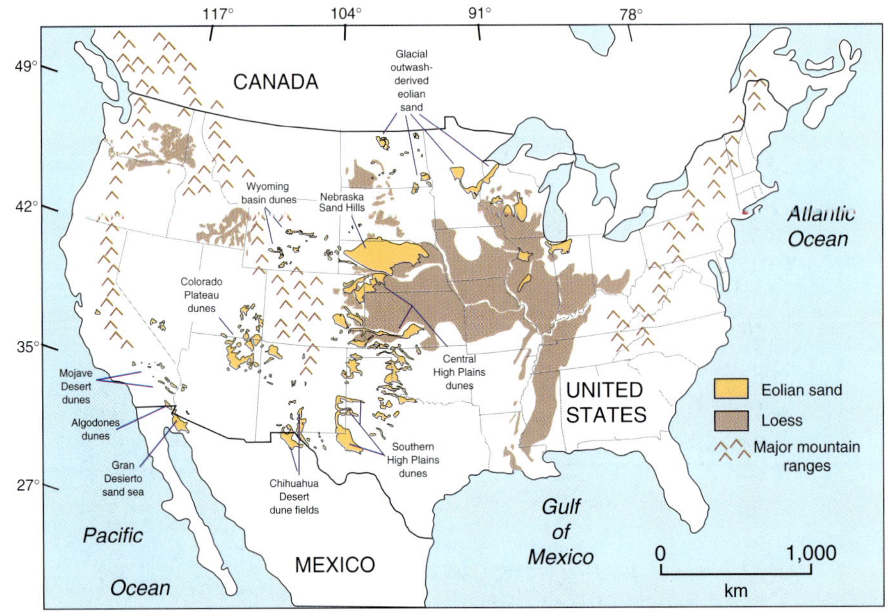

Figure 15 Map showing the distribution of eolian sand and loess in the United States and adjacent parts of Mexico. Redrawn from Thorp and Smith (1952), Muhs and Holliday (1995), Muhs and Zárate (2001), Bettis *et al.* (2003) and sources therein.

Dune Forms in Mid-Latitude Regions of North America

There is a wide range of dune forms in the Great Plains region of North America. The greatest variety of dune forms is found in the Nebraska Sand Hills. In this large (~50,000 km^2) dune field, barchanoid ridges, many 10–20 km long, are present over extensive areas (**Fig. 16**). Most of these barchanoid ridges have smaller, superimposed parabolic dunes, indicating multiple episodes of eolian deposition. In other areas of the Nebraska Sand Hills, barchans are the main dune form (**Fig. 16**), but again superimposed parabolic dunes indicate that the barchan forms were generated during earlier phases of dune formation. Linear (longitudinal) dunes are present over extensive areas of the Nebraska Sand Hills, many of which appear to be superimposed on older dune forms (**Fig. 16**). The Nebraska Sand Hills subregion is semiarid and mean annual precipitation ranges from about 400 mm in the western part to about 600 mm in the eastern part. With the exception of small blowouts on dune crests, all of these dunes are presently stabilized by vegetation (**Fig. 17**).

Dune fields farther south, in Colorado, Kansas, New Mexico, and Texas, are numerous, but are much smaller than the Nebraska Sand Hills. In eastern Colorado, for example, the total area occupied by dunes is ~25,000 km^2 (Madole *et al.*, 2005). As with Nebraska, the region is semiarid and mean annual precipitation ranges from about 300 mm (west) to about 770 mm (east). The most common landforms in these dune fields are stabilized parabolic dunes, often separated by stabilized eolian sand sheets (**Fig. 18**). In many places, compound dunes are common, with small parabolic dunes superimposed on larger, older parabolic dunes.

Still farther south, however, in the Chihuahuan Desert of southern New Mexico, western Texas, and northern Mexico, active barchanoid ridges, barchans, and parabolic dunes are found (**Fig. 18**). Mean annual precipitation in this region is lower than on the Great Plains, and ranges from about 220 to 320 mm. Farther west, on the Colorado Plateau, many different dune forms exist in close proximity to one another. For example, in the Moenkopi Plateau of northeastern Arizona, barchan, linear, and parabolic dunes can all be found within a relatively small area (**Fig. 19**). Mean annual precipitation on this high plateau region is lower still, and ranges from about 150 to 300 mm. As a result, while some dunes are stable, many are active. In places, for example, stabilized eolian sand sheets separate active linear dunes (**Fig. 19**).

Dune Origins in North America

In the past, it was commonly assumed that dunes in most parts of central North America, particularly the Great Plains, were derived from local sandstone bedrock, notably the Tertiary Ogallala Formation. This formation occurs over much of the region, crops out close to the land surface, and is rich in clean sand.

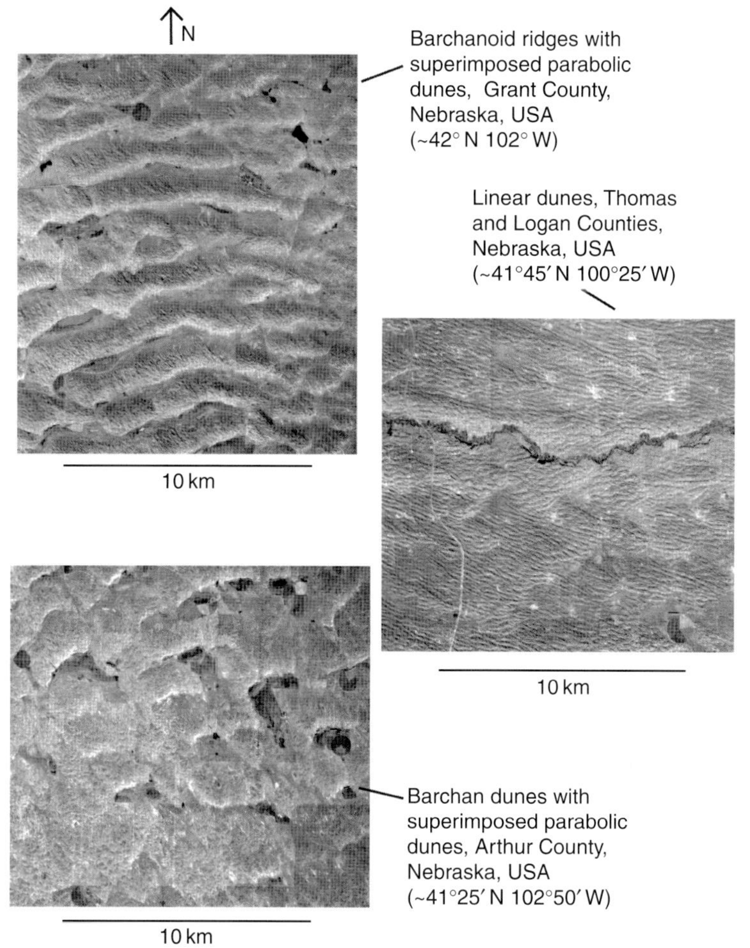

Figure 16 Dune landforms in the Nebraska Sand Hills, USA. All aerial photographs are from the U.S. Geological Survey National High Altitude Aerial Photography (NHAP) program. Frames: Grant County (HAP 182-77, 25 October 1984); Thomas and Logan Counties (HAP 152-52, 31 October 1981); Arthur County (HAP 242-184, 26 September 1985).

It is likely that many early investigators were struck by the close geographic match between the distribution of dune fields and the Ogallala Formation and inferred a genetic link. Nevertheless, much or all of the Ogallala Formation is anchored by a thick, well-cemented calcrete caprock and is not easily eroded.

Later investigators questioned the importance of bedrock as a source sediment for Great Plains dunes. Mineralogical and geochemical studies suggest that in much of the Central High Plains of Colorado and Kansas, dune sediments are derived from major rivers that head in the Rocky Mountains and flow eastward onto the Great Plains. In the Central High Plains, these fluvial sources include the South Platte River, which supplies sediment to dune fields in northeastern Colorado and southwestern Nebraska (not the Nebraska Sand Hills, however), and the Arkansas River, which supplies sediment to dunes in southeastern Colorado and western Kansas (Muhs *et al.*, 1996; Arbogast, 1996). In the Southern High Plains of Texas and New Mexico, dune fields appear to be derived mainly from an older Pleistocene eolian sheet sand called the Blackwater Draw Formation (Muhs and Holliday, 2001), which is, in turn, probably derived from the Pecos River.

One of the greatest unanswered questions about dune fields in North America is the origin of the Nebraska Sand Hills, the largest dune field on the continent. No major rivers are situated upwind (to the northwest) of this dune field. The best available evidence, based on isotopic studies, is that it has multiple sources, most of which are sand-rich Tertiary rocks that underlie, or occur upwind of the dune field (Aleinikoff *et al.*, 1994). Whatever their ultimate sources, dunes of the Nebraska Sand Hills are more mineralogically mature (i.e., quartz rich and depleted in feldspars) than any of their possible source sediments (Muhs *et al.*, 1997; Muhs, 2004). One explanation for this maturity is that most

Western Nebraska Sand Hills, USA
$P = {\sim}490\ \mathrm{mm\,yr^{-1}}$; $P/\mathrm{PE} = 0.76$

Andrews dune field, western Texas, USA
$P = {\sim}350\ \mathrm{mm\,yr^{-1}}$; $P/\mathrm{PE} = 0.36$

Figure 17 Contrasting degrees of dune activity in the Great Plains of the USA as a function of moisture balance (P/PE) in the western Nebraska Sand Hills (upper) and Andrews dune field, Texas (lower). Photographs by D. R. Muhs.

feldspars have been reduced to silt-sized sediments through ballistic impacts, an idea that has support in both theory and laboratory evidence (Dutta *et al.*, 1993). If this explanation is correct, it suggests that the Nebraska Sand Hills dune field has a long history of repeated dune activity, with little input from fresh sand sources.

Dune Ages in North America

For decades, many investigators assumed, with no numerical age control, that dunes in the Great Plains of the United States were last active during full glacial time (MIS 2). More recent studies confirm that the last glacial period was indeed a time of eolian sand activity over much of the region (Swinehart and Diffendal, 1990; Loope *et al.*, 1995; Muhs *et al.*, 1996; Goble *et al.*, 2004; Madole *et al.*, 2005). However, many studies also show that much eolian sand in the Great Plains region is of late Holocene age. In the Great Plains, the areal extent of late Holocene eolian sand is much greater than that of last glacial sand. For example, dunes of the Nebraska Sand Hills cover $\sim$50,000 km^2, yet all have only

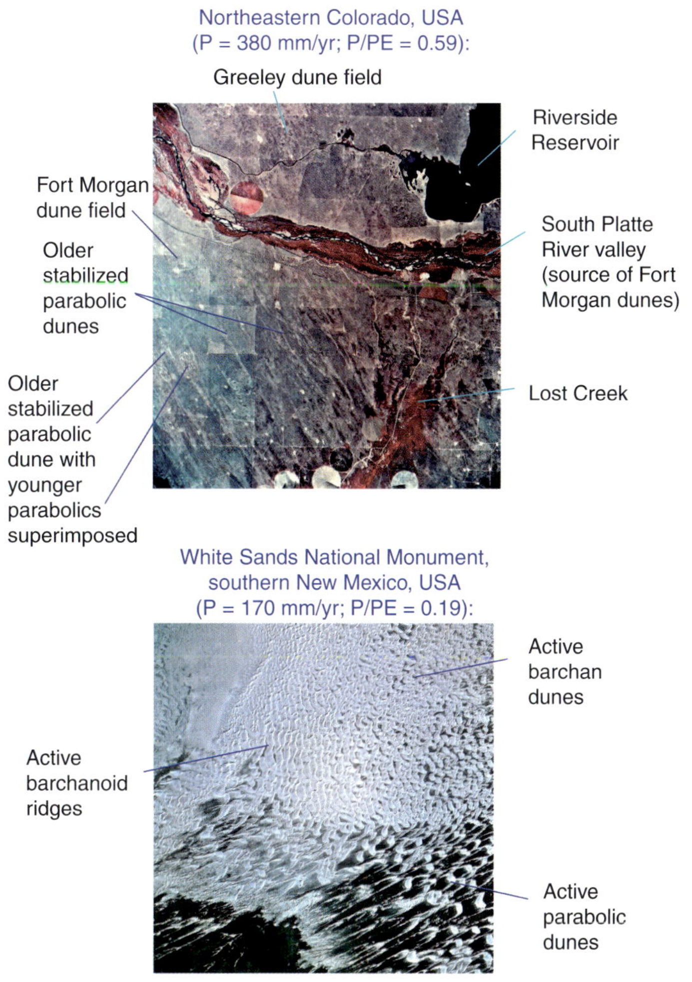

Figure 18 Dune forms in the Central High Plains of Colorado (upper) and northernmost Chihuahuan Desert of New Mexico (lower). Upper photograph is false-colored IR aerial photograph (USGS NHAP 61-107, 22 June 1985); lower photograph is natural-color aerial photograph (USGS NAPP 9524-114, 6 October 1996).

simple A/AC/C soil profiles (Entisols) that develop in a few hundred or few thousand years. This observation, along with late Holocene radiocarbon and optically stimulated luminescence (OSL) ages at numerous localities, suggests that most or all of the Nebraska Sand Hills region has been active at some time in the late Holocene. However, all parts of the dune field were not necessarily active simultaneously.

Late Holocene eolian sand deposition has now been well documented in Nebraska, Colorado, Kansas, Texas, and New Mexico (Swinehart and Diffendal, 1990; Arbogast, 1996; Muhs *et al.*, 1997; Holliday, 2001; Goble *et al.*, 2004; Forman *et al.*, 2005; Madole *et al.*, 2005). Most of these studies have stratigraphic data indicating multiple periods of eolian activity in the late Holocene (**Fig. 20**). The small number of radiocarbon and OSL ages and their large analytical uncertainties preclude testing hypotheses of regionally synchronous activity. However, these observations show that eolian sand in this region has been active under modern climatic regimes. Explorers in the nineteenth century observed active dune sand in many parts of the Great Plains where it is now stable (Muhs and Holliday, 1995).

Paleoclimatic Significance of Dunes in North America

One of the most important paleoclimatic inferences made from stabilized, mid-latitude dune fields is that periods of dune activity indicate drier-than-present climate. Many factors influence the degree of dune activity, including wind strength, sediment availability, and degree of surface disturbance by animals or human activity. Nevertheless, several studies show that in many regions, including southern Africa (Lancaster, 1988; Thomas *et al.*, 2005), China (Wang *et al.*, 2005), and North America (Muhs and Holliday, 1995; Wolfe, 1997), the degree of dune activity is influenced to a great extent by the amount

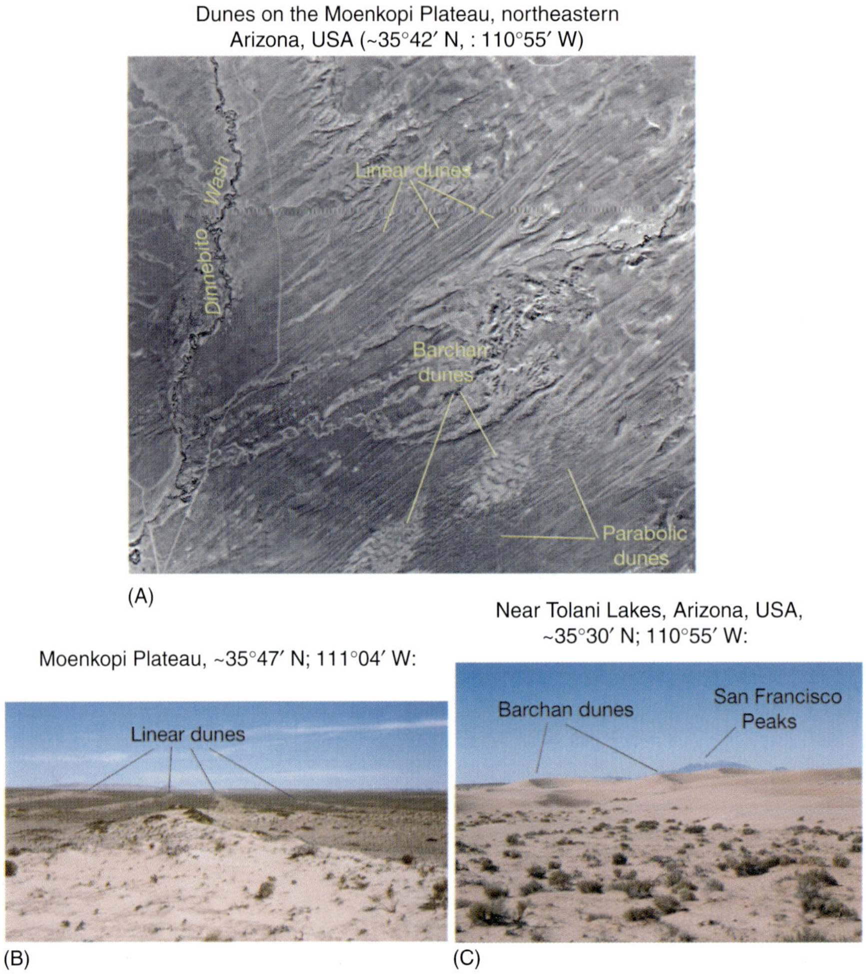

Figure 19 Dune forms on the Colorado Plateau, USA. Upper: aerial photograph showing dune types (USGS NAPP 5240-178, 22 September 1992). Lower: active linear dunes separated by stable eolian sand sheets (left) and active barchan dunes (right). Lower photographs by D.R. Muhs.

of plant cover, which in turn is influenced significantly by moisture balance. Moisture balance can conveniently be calculated, at least to a first approximation, by the ratio of precipitation (P) to potential evapotranspiration (PE). Values less than 1.0 indicate a net moisture deficit and are typical of regions with semiarid, arid, or hyperarid climates. In areas where P/PE is greater than about 0.35–0.40, dunes are stabilized by vegetation. When P/PE is less than about 0.35–0.40, dunes are at least partially active, and as this ratio diminishes, dune activity increases (**Fig. 17**). In desert regions of the southwestern United States and northwestern Mexico, P/PE is less than about 0.25. Thus, dunes in the Chihuahuan Desert, Algodones, and Gran Desierto dune fields (**Fig. 15**) are quite active (**Fig. 18**). In the Taklimakan Desert of China, eolian sand is active over most of the dune field (**Fig. 3**) and P/PE values for stations in and around this sand sea are less than 0.2 (Wang *et al.*, 2005). With these observations in mind, it can be inferred that in areas now characterized by stable dunes, past periods, represented by eolian sand, were more arid than present. In contrast, periods represented by paleosols were times that may have had a moisture balance similar to the present.

One of the remarkable observations of many North American Great Plains dune field records is the presence of multiple paleosols, all dating to the Holocene, intercalated within eolian sand (**Fig. 20**). The evidence for multiple periods of dune activity and stability in the past few thousand years indicates that Great Plains eolian sand is quite sensitive to small changes in the overall moisture balance and degree of vegetation cover. As in the Mu Us Desert in China, the Great Plains region of North America is highly sensitive to fluctuations in the strength of a monsoonal moisture source. In central North America, the main monsoonal flow of moist air is from the Gulf of Mexico in summer. When this flow is diminished, summer precipitation is lower and

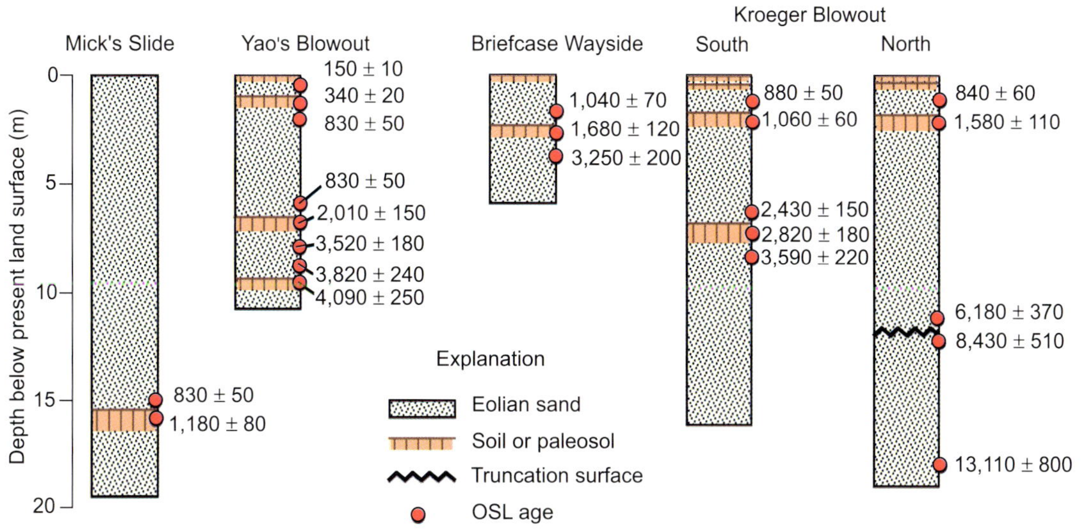

Figure 20 Stratigraphic sections showing optically stimulated luminescence (OSL) ages, in calendar years, of eolian sand in the Nebraska Sand Hills. Redrawn from Goble *et al.* (2004).

vegetation cover is reduced. Dry periods in the Great Plains are characterized not only by decreased summer monsoonal flow from the Gulf of Mexico, but also by an increase in zonal, or westerly flow of Pacific air (typically a winter pattern). Pacific air is usually dry by the time it reaches the Great Plains. Moisture in Pacific air masses is lost on north–south trending mountain ranges between the Pacific Ocean and the Great Plains (**Fig. 15**). Reactivation of stabilized dune sand or mobilization of new sand by wind is enhanced under conditions of reduced vegetation cover during periods of decreased monsoonal flow and increased zonal flow.

Another important application of dunes to paleoclimate studies in mid-latitude North America is wind direction inferred from dune orientation. Most of the late Holocene dunes in the mid-continent of North America are parabolic forms, which are excellent paleowind indicators. Parabolic dunes have arms that point upwind, with noses that face downwind, as seen when they are active in areas such as White Sands, New Mexico (**Fig. 18**). On the Great Plains of North America, orientations of many late Holocene dunes indicate northwesterly winds in Nebraska, and northern and central Colorado (**Fig. 18**), similar to modern, sand-moving wind regimes of fall, winter, and spring (**Fig. 21**). Such orientations are consistent with the scenario of increased zonal or westerly flow discussed above. Late Holocene dunes in southeastern Colorado and Kansas have orientations that indicate southwesterly paleowinds, again similar to present. However, in eastern New Mexico and western Texas, dunes have orientations that indicate northwesterly winds, whereas modern dune-forming winds are dominantly from the southwest. One explanation is that during the periods of late Holocene dune formation, there was less of an influence from southwestern (Gulf of California), monsoonal airflow, and a greater influence from zonal (westerly) circulation.

Summary

Large dune fields, or sand seas, are widespread in the interiors of continental mid-latitudes. Most of them are found in China, Central Asia, Argentina, and the United States. Mid-latitude dune fields in the United States are generally smaller than those of low latitudes, but those in China, Central Asia, and Argentina are quite large. Where climates are arid, such as in western China and Central Asia, active dunes dominate these sand seas. Other dune fields, in the semiarid climates of eastern China, Argentina, and the United States, have mostly stabilized dunes. Mid-latitude dune fields are often near mountains, which supply much of the sediment to the sand seas on a regular basis, particularly during glacial periods. Many different kinds of eolian landforms are found in mid-latitude sand seas, including transverse dunes, longitudinal (linear) dunes, star dunes, parabolic dunes, and eolian sand sheets. Where dune fields of mid-latitudes occur in semiarid climates, they are particularly sensitive indicators of shifts in climate. In China, records of dune field expansion and contraction extend back hundreds of thousands of years.

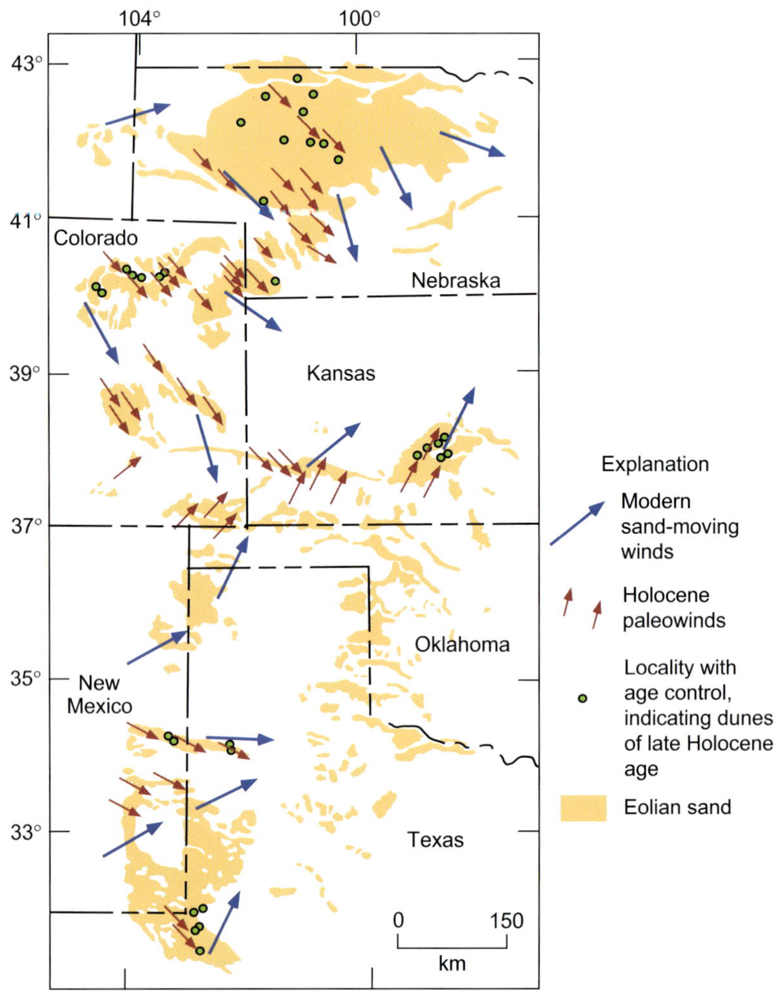

Figure 21 Map showing the distribution of eolian sand in the Great Plains, USA, direction of modern, sand-moving winds, and late Holocene paleowinds based on dune orientations (modified from Muhs and Zárate (2001)). Also shown (green circles) are localities where there are late Holocene radiocarbon or OSL ages for eolian sand. From Swinehart and Diffendal (1990), Arbogast (1996), Muhs *et al.* (1997), Holliday (2001), Goble *et al.* (2004), Forman *et al.* (2005), Madole *et al.* (2005). Modern sand-moving winds constructed using methods of Fryberger and Dean (1979).

See also: **Dune Fields**: High Latitudes; Low Latitudes. **Eolian Records, Deep-Sea Sediments**; **Loess Deposits, Origins and Properties**. **Loess Records**: Central Asia; China; North America; South America. **Luminescence Dating**: Optically-Stimulated Luminescence. **Paleosols and Wind-Blown Sediments**: Nature of Paleosols.

References

Aleinikoff, J. N., Muhs, D. R., and Walter, M. (1994). U–Pb evidence for provenances of Holocene sand dunes, northeastern Colorado and the Nebraska Sand Hills. *US Geological Survey Circular* **1107,** 2.

Arbogast, A. F. (1996). Stratigraphic evidence for late-Holocene eolian sand mobilization and soil formation in south-central Kansas, USA. *Journal of Arid Environments* **34,** 403–414.

Bettis, E. A., III, Muhs, D. R., Roberts, H. M., and Wintle, A. G. (2003). Last glacial loess in the conterminous USA. *Quaternary Science Reviews* **22,** 1907–1946.

Bowler, J. M., Wyrwoll, K.-H., and Lu, Y. (2001). Variations of the northwest Australian summer monsoon over the last 300,000 years: The paleohydrological record of the Gregory (Mulan) Lakes System. *Quaternary International* **83-85,** 63–80.

Chen, J. S., Li, L., Wang, J. Y., *et al.* (2004). Water resources: Groundwater maintains dune landscape. *Nature* **432,** 459.

Chepalyga, A. L. (1984). Inland sea basins. In *Late Quaternary Environments of the Soviet Union* (A. A. Velichko, H. E. Wright, Jr, and C. W. Barnosky, Eds.), pp. 229–247. University of Minnesota Press, Minneapolis.

Ding, Z. L., Yu, Z. W., Rutter, N. W., and Liu, T. S. (1994). Towards an orbital time scale for Chinese loess deposits. *Quaternary Science Reviews* **13,** 39–70.

Dodonov, A. E. (1991). Loess of Central Asia. *GeoJournal* **24**(2), 185–194.

Duce, R. A., Liss, P. S., Merrill, J. T., *et al.* (1991). The atmospheric input of trace species to the world ocean. *Global Biogeochemical Cycles* **5,** 193–259.

Dutta, P. K., Zhou, Z., and dos Santos, P. R. (1993). A theoretical study of mineralogical maturation of eolian sand. *Geological Society of America Special Paper* **284,** 203–209.

Forman, S. L., Marín, L., Pierson, J., Gómez, J., Miller, G. H., and Webb, R. S. (2005). Aeolian sand depositional records from western Nebraska: Landscape response to droughts in the past 1500 years. *The Holocene* **15**, 973–981.

Frechen, M., and Dodonov, A. E. (1998). Loess chronology of the Middle and Upper Pleistocene in Tadjikistan. *Geologische Rundschau* **87**, 2–20.

Fryberger, S. G., and Dean, G. (1979). Dune forms and wind regime. In *US Geological Survey Professional Paper 1052: A Study of Global Sand Seas* (E. D. McKee, Ed.) pp. 137–169 U.S. Government Printing Office, Washington, D.C.

Gansser, A. (1964). *Geology of the Himalayas.* Wiley, London.

Goble, R. J., Mason, J. A., Loope, D. B., and Swinehart, J. B. (2004). Optical and radiocarbon ages of stacked paleosols and dune sands in the Nebraska Sand Hills, USA. *Quaternary Science Reviews* **23**, 1173–1182.

Goudie, A. S. (2002). *Great Warm Deserts of the World.* Oxford University Press, Oxford.

Harrison, T. M., Copeland, P., Kidd, W. S. F., and Yin, A. (1992). Raising Tibet. *Science* **255**, 1663–1670.

Holliday, V. T. (2001). Stratigraphy and geochronology of upper Quaternary eolian sand on the Southern High Plains of Texas and New Mexico, United States. *Geological Society of America Bulletin* **113**, 88–108.

Iriondo, M. (1999). Climatic changes in the South American plains: Records of a continental-scale oscillation. *Quaternary International* **57/58**, 93–112.

Kröhling, D. M. (1999). Upper Quaternary geology of the lower Carcarañá Basin, North Pampa, Argentina. *Quaternary International* **57/58**, 135–148.

Lancaster, N. (1988). Development of linear dunes in the southwestern Kalahari, southern Africa. *Journal of Arid Environments* **14**, 233–244.

Letollé, R., and Mainguet, M. (1993). *Aral.* Springer, Paris.

Li, S. H., Sun, J. M., and Zhao, H. (2002). Optical dating of dune sands in the northeastern deserts of China. *Palaeogeography, Palaeoclimatology, Palaeoecology* **181**, 419–429.

Lioubimtseva, E. (2002). Arid environments. In *The Physical Geography of Northern Eurasia* (M. Shahgedanova, Ed.), pp. 267–283. Oxford University Press, Oxford.

Liu, T. S. (1985). *Loess and the Environment.* China Ocean Press, Beijing.

Liu, T. S., and Ding, Z. L. (1993). Stepwise coupling of monsoon circulation to global ice volume variations during the late Cenozoic. *Global and Planetary Change* **7**, 119–130.

Loope, D. B., Swinehart, J. B., and Mason, J. P. (1995). Dune-dammed paleovalleys of the Nebraska Sand Hills: Intrinsic versus climatic controls on the accumulation of lake and marsh sediments. *Geological Society of America Bulletin* **107**, 396–406.

Madole, R. F., VanSistine, D. P., and Michael, J. A. (2005). Distribution of late Quaternary wind-deposited sand in eastern Colorado. *US Geological Survey Scientific Investigations Map*,2875.

Martinez, G. A., Martinez-Arca, J., Gwyn, Q. H. J., and Bernasconi, M. V. (2001). Combined use of RADARSAT-1 and Landsat TM data for geomorphological applications in lowlands of Buenos aires province, Argentina. *Canadian Journal of Remote Sensing* **27**, 638–642.

Martinson, D. G., Pisias, N. G., Hays, J. D., Imbrie, J., Moore, T. C., Jr, and Shackleton, N. J. (1987). Age dating and the orbital theory of the ice ages: Development of a high-resolution 0 to 300,000-year chronostratigraphy. *Quaternary Research* **27**, 1–29.

Merrill, J. T., Bleck, R., and Uematsu, M. (1989). Meteorological analysis of long-range transport of mineral aerosols over the North Pacific. *Journal of Geophysical Research* **94**, 8584–8598.

Molnar, P., and Tapponnier, P. (1975). Cenozoic tectonics of Asia: Effects of a continental collision. *Science* **189**, 419–426.

Muhs, D. R. (2004). Mineralogical maturity in dune fields of North America, Africa and Australia. *Geomorphology* **59**, 247–269.

Muhs, D. R., and Holliday, V. T. (1995). Evidence for active dune sand on the Great Plains in the 19th century from accounts of early explorers. *Quaternary Research* **43**, 198–208.

Muhs, D. R., and Holliday, V. T. (2001). Origin of late Quaternary dune fields on the Southern High Plains of Texas and New Mexico. *Geological Society of America Bulletin* **113**, 75–87.

Muhs, D. R., and Zárate, M. (2001). Late Quaternary eolian records of the Americas and their paleoclimatic significance. In *Interhemispheric Climate Linkages* (V. Markgraf, Ed.), pp. 183–216. Academic Press, San Diego.

Muhs, D. R., Stafford, T. W., Cowherd, S. D., *et al.* (1996). Origin of the late Quaternary dune fields of northeastern Colorado. *Geomorphology* **17**, 129–149.

Muhs, D. R., Stafford, T. W., Jr, Swinehart, J. B., *et al.* (1997). Late Holocene eolian activity in the mineralogically mature Nebraska Sand Hills. *Quaternary Research* **48**, 162–176.

Muhs, D. R., Reynolds, R., Been, J., and Skipp, G. (2003). Eolian sand transport pathways in the southwestern United States: Importance of the Colorado River and local sources. *Quaternary International* **104**, 3–18.

Rea, D. K. (1994). The paleoclimatic record provided by eolian deposition in the deep sea: The geological history of wind. *Reviews of Geophysics* **32**, 159–195.

Redsteer, M., and Block, D. (2004). Drought conditions accelerate destabilization of sand dunes on the Navajo Nation, southern Colorado Plateau. *Geological Society of America Abstracts with Programs* **36**(5), 171.

Shackleton, N. J., and Pisias, N. G. (1985). Atmospheric carbon dioxide, orbital forcing, and climate. *Geophysical Monograph Series* **32**, 412–417.

Shackleton, N. J., Berger, A., and Peltier, W. R. (1990). An alternative astronomical calibration of the Lower Pleistocene timescale based on ODP Site 677. *Transactions of the Royal Society of Edinburgh: Earth Science* **81**, 251–261.

Sun, J. M. (2000). Origin of eolian sand mobilization during the past 2300 years in the Mu Us Desert, China. *Quaternary Research* **53**, 73–88.

Sun, J. M. (2002a). Provenance of loess material and formation of loess deposits on the Chinese Loess Plateau. *Earth and Planetary Science Letters* **203**, 845–859.

Sun, J. M. (2002b). Source regions and formation of the loess sediments on the high mountain regions of northwestern China. *Quaternary Research* **58**, 341–351.

Sun, J. M., Ding, Z. L., and Liu, T. S. (1998). Desert distribution during the glacial maximum and climatic optimum: Example of China. *Episodes* **21**, 28–31.

Sun, J. M., Ding, Z. L., Rokosh, D., and Rutter, N. (1999). 580,000 year environmental reconstruction from eolian deposits at the Mu Us Desert margin, China. *Quaternary Science Reviews* **18**, 1351–1364.

Swinehart, J. B., and Diffendal, R. F., Jr, (1990). Geology of the pre-dune strata. In *Resource Atlas No. 5a: An Atlas of the Sand Hills* (A. Bleed and C. Flowerday, Eds.), pp. 29–42. University of Nebraska, Lincoln.

Thomas, D. S. G., and Shaw, P. A. (1991). *The Kalahari Environment.* Cambridge University Press, Cambridge.

Thomas, D. S. G., Knight, M., and Wiggs, G. F. S. (2005). Remobilization of southern African desert dune systems by twenty-first century global warming. *Nature* **435**, 1218–1221.

Thorp, J., and Smith, H. T. U. (1952). Pleistocene eolian deposits of the United States, Alaska, and parts of Canada. *National Research Council Committee for the Study of Eolian Deposits, Geological Society of America*, scale **1**: 2,500,000.

Wang, X., Dong, Z., Yan, P., Zhang, J., and Qian, G. (2005). Wind energy environments and dunefield activity in the Chinese deserts. *Geomorphology* **65**, 33–48.
Wilson, I. G. (1973). Ergs. *Sedimentary Geology* **10**, 77–106.
Wolfe, S. A. (1997). Impact of increased aridity on sand dune activity in the Canadian Prairies. *Journal of Arid Environments* **36**, 421–432.
Zárate, M. A. (2003). Loess of southern South America. *Quaternary Science Reviews* **22**, 1987–2006.
Zárate, M. A., and Blasi, A. (1993). Late Pleistocene-Holocene eolian deposits of the southern Buenos Aires Province, Argentina: A preliminary model. *Quaternary International* **17**, 15–20.
Zhou, W. J., Head, M. J., and Deng, L. (2001). Climatic changes in northern China since the late Pleistocene and its response to global change. *Quaternary International* **83/85**, 285–292.
Zhu, Z. D., Wu, Z., Liu, S., and Di, X. M. (1980). *An Outline on Chinese Deserts*. Science Press, Beijing.

Low Latitudes

N Lancaster, Desert Research Institute, NV, USA

© 2007 Elsevier B.V. All rights reserved.

Low-latitude sand seas (also called ergs) are extensive areas of sand dunes located in the tropical and subtropical deserts of the world (**Fig. 1**), where they occupy as much as one-third of the area classified as arid. Major low-latitude sand seas occur in the Sahara and Arabian deserts, the Thar Desert of India, the interior of Australia, and in southern Africa (Namib and Kalahari deserts). Large sand seas are absent in the Americas, where areas of dunes cover less than 1% of the area classified as arid. Important studies and surveys of low-latitude sand seas include those by Wilson (1973) and McKee (1979). Lancaster (1999) provided a review of modern paradigms of sand sea geomorphology. Important summaries of chronologic data include Munyikwa (2005) for Southern Hemisphere sand seas, Glennie and Singhvi (2002) for Arabia, and Swezey (2001) for the Sahara.

Sand seas contain large volumes of sand (tens to hundreds of cubic kilometers) and have accumulated episodically during the Quaternary. Their construction has therefore been determined by climatic, tectonic, and sea-level changes that have affected sand supply, availability, and mobility, as well as the preservation of deposits and landforms from prior episodes of eolian construction. As significant sedimentary deposits in low-latitude desert regions, sand seas also provide an archive of the effects of climate change on these areas.

Sand seas and dune fields typically form part of well-defined regional- and local-scale sediment transport systems in which sand is moved by wind from

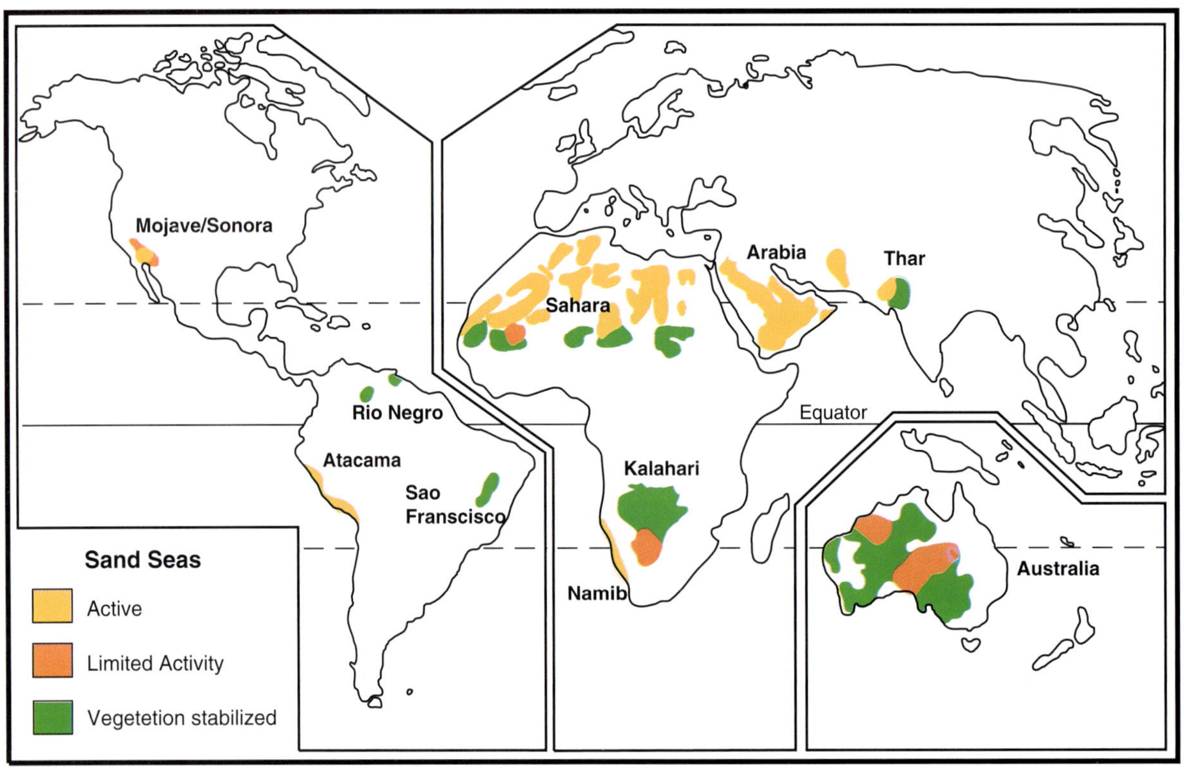

Figure 1 Distribution of major low-latitude sand seas, showing areas of active, partly active, and vegetation-stabilized dunes. For names and details of sand seas, see regional maps.

source areas (e.g., distal fluvial deposits and sandy beaches) via transport pathways to depositional sinks. Sand seas accumulate downwind of source zones wherever sand transport rates decrease as a result of regional and/or local changes in wind speed and directional variability so that the influx of sand exceeds outflux. A conceptual model for the response of low-latitude (and other) sand seas to changes in external forcing factors was provided by Kocurek and Lancaster (1999). In this model, episodes of sand sea construction are viewed as the product of changes in the supply, availability, and mobility of sand sized sediment. Sediment supply is the deposition of sediment that is a source for the eolian system either contemporaneously or at some later time. Availability of sediment for transport by the wind (the probability of entrainment of sand for transport) is strongly influenced by variations in vegetation cover and soil moisture. In addition, changes in sea level and regional hydrology may make sediment stored in shallow marine or lacustrine environments available for transport. The mobility of sediment is determined by the transport capacity of the wind (the potential rate of sediment transport).

Evidence for Paleoenvironments in Low-Latitude Sand Seas

Sand seas provide paleoenvironmental information from three main sources: (a) dune systems, especially those stabilized by vegetation; (b) interdune deposits; and (c) penetration of extradune fluvial deposits into the margins of the sand sea. The quality of paleoclimatic data provided from sand seas is frequently much better than that obtained from adjacent rocky desert regions because sand dune areas tend to accentuate the effects of both dry and wet phases (Rognon, 1982). In periods of increased rainfall, the high infiltration capacity and porosity of dune sands favors the growth and persistence of vegetation, which may lead to the partial or complete stabilization of the dunes and formation of soils on dune sands (**Fig. 2**). If the periods of increased rainfall are of sufficient magnitude and duration, then water tables may rise, leading to the formation of pond and marsh deposits in interdune areas. In contrast, periods of aridity will give rise to a very unfavorable biotic environment in which vegetation cover is reduced as a result of low precipitation and active sand movement. In addition, dune systems provide some of the best indicators of past wind regimes and atmospheric circulation patterns (e.g., Lancaster *et al.*, 2002).

Many sand seas contain a variety of dune types or dunes on different trends, often forming complex patterns. Satellite image analysis, combined with studies of dune geomorphology and sedimentology, can be used to analyze the dune patterns and identify genetically distinct generations of dunes and interpret the conditions in which they formed. Understanding the history of sand seas and dunes has been revolutionized in recent years by the application of luminescence techniques to directly date periods of dune formation and/or reactivation (Stokes, 1997). In addition, large dunes, especially those of linear form, appear to be composite features, reflecting many generations of dune construction and stability. Recent application of ground-penetrating radar techniques has enabled identification of major sedimentary units and bounding surfaces prior to luminescence dating of these units (Bristow *et al.*, 2005).

Sand seas may also contain evidence of periods of climates wetter than present, in which dunes were stabilized by vegetation and soil formation and marshes and shallow lakes formed in interdune areas. Many of these deposits contain fossil biota and/or archeological materials that can be used to reconstruct past hydrology and vegetation communities and provide ages for periods of wetter climates (Gasse, 2002; Teller *et al.*, 1990). Stratigraphic relations between such deposits and those of dunes have traditionally been used as a major source of chronologic information for sand seas in the Sahara (Swezey, 2001) and elsewhere.

In many areas, the evidence for environmental change from the sand seas is corroborated by records of dust flux (indicating arid, windy conditions) from marine sediment cores in adjacent ocean and shelf areas, which frequently provide a more continuous and better dated record than that of the sand seas (deMenocal *et al.*, 2000; Stuut and Lamy, 2004).

Regional Summaries

Sahara Desert

The Sahara Desert contains a significant proportion of the global inventory of wind-blown sand as well as some of the world's largest sand seas, which cover approximately 28% of this desert. There is a large amount of paleoenvironmental information on Saharan sand seas, much of it based on analyses of dated interdune lacustrine deposits and their stratigraphic relations to dune deposits (Swezey, 2001; Gasse and Van Campo, 1994). However, there are few examples of episodes of eolian deposition that have been directly dated by luminescence techniques.

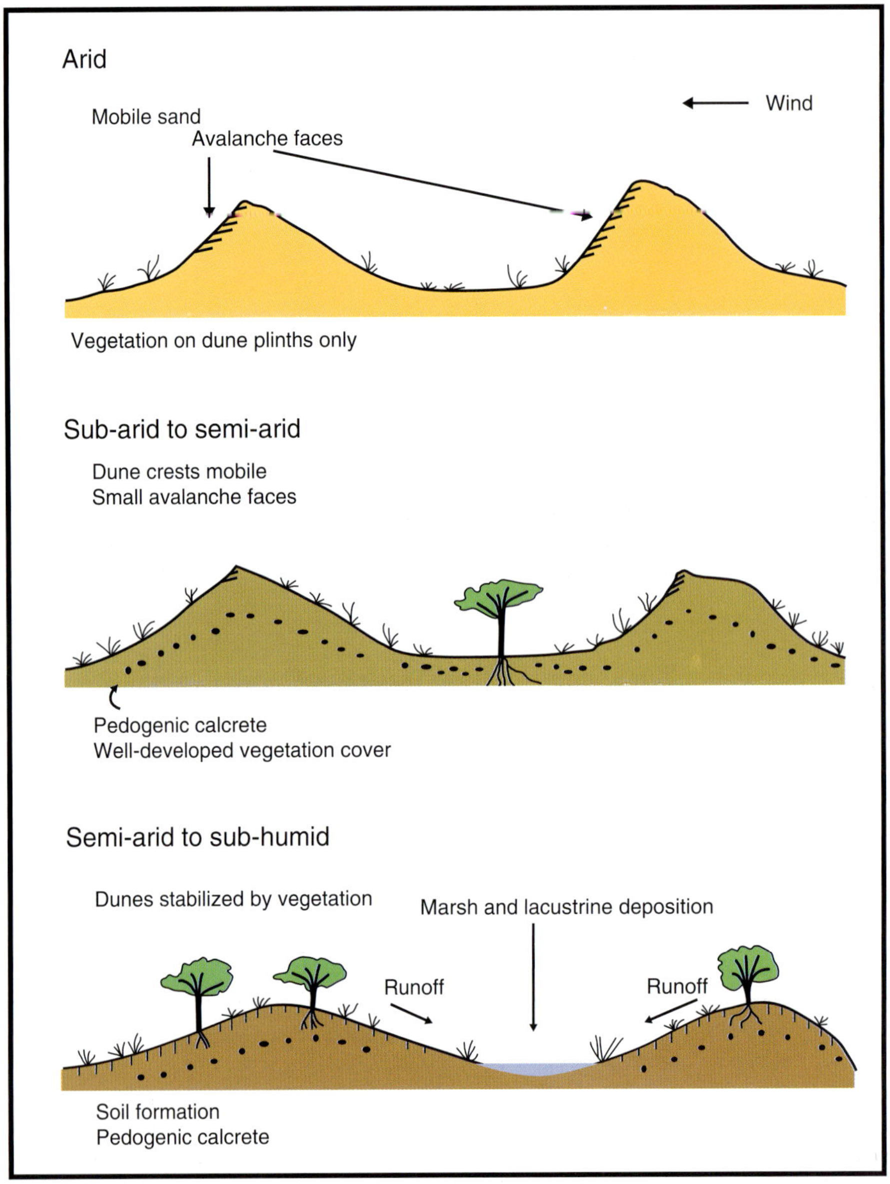

Figure 2 A conceptual model for the response of dune areas to climate change. Note the contrasts between arid and more humid conditions. Based on Rognon P (1982) Pluvial and arid phases in the Sahara: The role of non climatic factors. *Palaeoecology of Africa* 12: 45–62.

The major sand accumulations are situated on each side of the central Saharan uplands, where there is very little sand (**Fig. 3**). In the Sahara, sand is moved long distances between apparent source areas and depositional sinks as well as from one sand sea to another. In many instances, the source of sand is not clear or not known. The high degree of mineralogical maturity of sand from Libyan sand seas (Muhs, 2004) suggests long-distance transport, repeated periods of transport and reworking over short distances, or a source in the mineralogically mature Nubian Sandstone that crops out extensively in the eastern Sahara.

The sand transport pathways, clearly visible on satellite images (Breed *et al.*, 1979; Mainguet, 1984), reflect the dominance of the trade wind circulation that tends to move sand from the sand-poor eastern and central parts of the desert toward thick sand accumulations in the Sahel and from the piedmont of the Atlas Mountains and the Mediterranean coast to the northern and western sand seas (Mainguet and Chemin, 1983; Wilson, 1971). The location of the major Saharan sand seas is controlled both by patterns of winds and by topographic obstacles. The Grand Erg Occidental and the Ubari and Murzuq sand seas of Libya have accumulated against

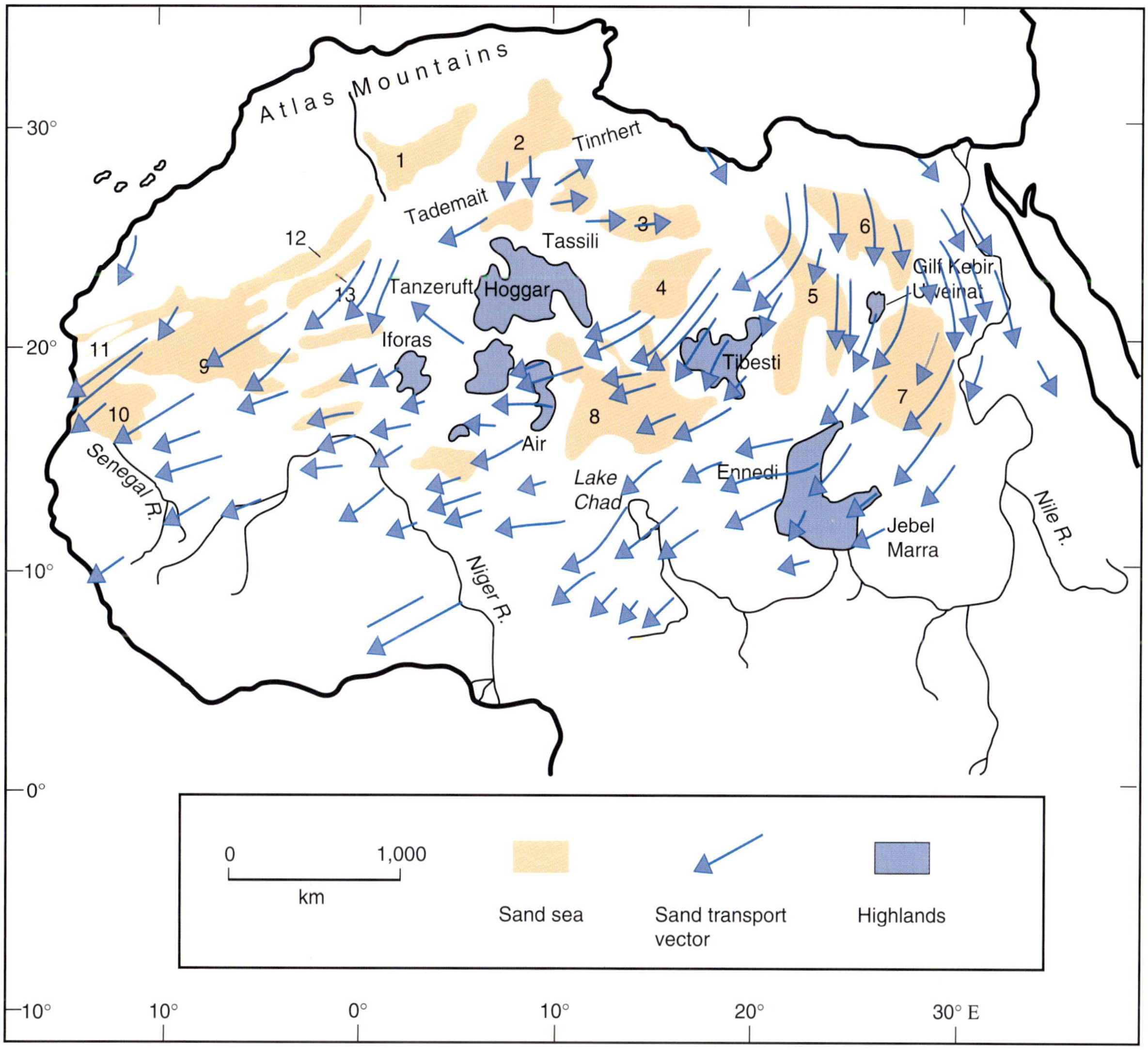

Figure 3 Location of Saharan sand seas and vectors of sand transport. Sand seas are numbered as follows: 1. Grand Erg Occidental, 2. Grand Erg Oriental, 3. Ubari, 4. Murzuq, 5. Calanscio, 6. Great Sand Sea, 7. Selima, 8. Fachi-Bilma and Tenéré, 9. Majabat al Koubra, 10. Aouker, 11. Akchar, 12. Iguidi, 13. Chech. Based on Mainguet, M., and Chemin, M.C., (1983) Sand seas of the Sahara and Sahel: An explanation of their thickness and sand dune type by the sand budget principle. In: Brookfield, M.E, and Ahlbrandt,T.S (Eds.) *Eolian Sediments and Processes. Developments in Sedimentology*, pp. 353–364. Amsterdam: Elsevier; and Mainguet, M (1984) Space observations of Saharan aeolian dynamics. In: El Baz F (Ed.) *Deserts and Arid Lands*, pp. 59–77. The Hague, The Netherlands: Nyhoff.

the northern slopes of the central Saharan uplands (Fryberger and Ahlbrandt, 1979). Sand seas in the Akchar of Mauritania and the Fachi-Bilma region of Tchad lie where winds converge in the lee of escarpments or mountain masses. The thick sand accumulations of the southern Sahara and Sahel have developed because wind energy is reduced on the margins of the trade wind circulation. Many northern Saharan sand seas (e.g., the Grand Erg Occidental and Oriental) occur in areas of complex or multidirectional wind regimes influenced by trade wind and midlatitude cyclonic circulations as well as local topographically induced winds (Wilson, 1971).

Dune patterns in Saharan sand seas are summarized in **Table 1**. The northern Saharan sand seas are characterized by complex crescentic and star dunes developed in a multidirectional wind regime that results from the effects of both trade wind and midlatitude circulations. By contrast, sand seas in the western and southern Sahara are mainly composed of linear dunes, with small areas of crescentic dunes. The dominance of linear dunes in these areas reflects the persistence of trade wind circulations. Eolian accumulations in the eastern Sahara are dominated by extensive sand sheets and areas of gently undulating dunes with a chevron pattern. The formation of extensive sand sheets is

Table 1 Distribution of dunes in selected sand seas

	Percentage of dune area covered								
	Sahara				*Kalahari*	*Namib*	*Arabia*	*Thar*	*Australia*
	South	*North*	*Northeast*	*West*					
Linear dunes	24.08	22.84	17.00	35.50	85.55	32.55	49.81	13.96	95.00
Crescentic dunes	28.37	33.34	14.50	19.20		11.80	14.91	25.61	
Star dunes		7.92	23.90			9.92	5.34		
Sand sheets and streaks	47.54	35.92	39.20	45.30	13.56	45.44	23.24	31.75	5.00
Parabolic dunes					0.60			28.68	
Undifferentiated			4.50				6.71		

Data from Fryberger and Goudie (1981).

attributed to the abundance of coarse sand derived from the Mesozoic Nubian Sandstone that underlies much of the area (Stokes *et al.*, 1998).

In addition to the largely active sand seas of the Sahara, satellite images show the existence of large areas of vegetation-stabilized dunes on the southern margins of the desert, extending south into Sahel and beyond (**Fig. 4**). Extensive systems of mainly linear dunes occur in areas that today receive as much as 1,000 mm of rainfall per year. These dunes, which are today extensively degraded and covered by savanna vegetation, were first recognized on aerial photographs of northern Nigeria and southern Mauritania and subsequently mapped throughout the Sahel (Grove and Warren, 1968). A regionally extensive paleosol is present within these dunes (Mauz and Felix-Henningsen, 2005). This paleosol indicates a past period of stability, similar to the present one. In areas to the north, this paleosol underlies late Holocene to modern dunes, many of them of crescentic form.

The age of the stabilized dunes in the Sahel has been hypothesized to be Last Glacial, spanning the period 20–13 ka (Ogolien period in the western Sahara), mostly by stratigraphic relations to overlying lacustrine deposits dated by radiocarbon methods (Swezey, 2001). Direct age estimates for these dunes using optically stimulated luminscence (OSL) are limited to those in northeastern Nigeria (Stokes and Horrocks, 1998), where maximum rates of construction occurred 18–13 ka, and in the Sahel of Mali, with an age of 12.7 ka.

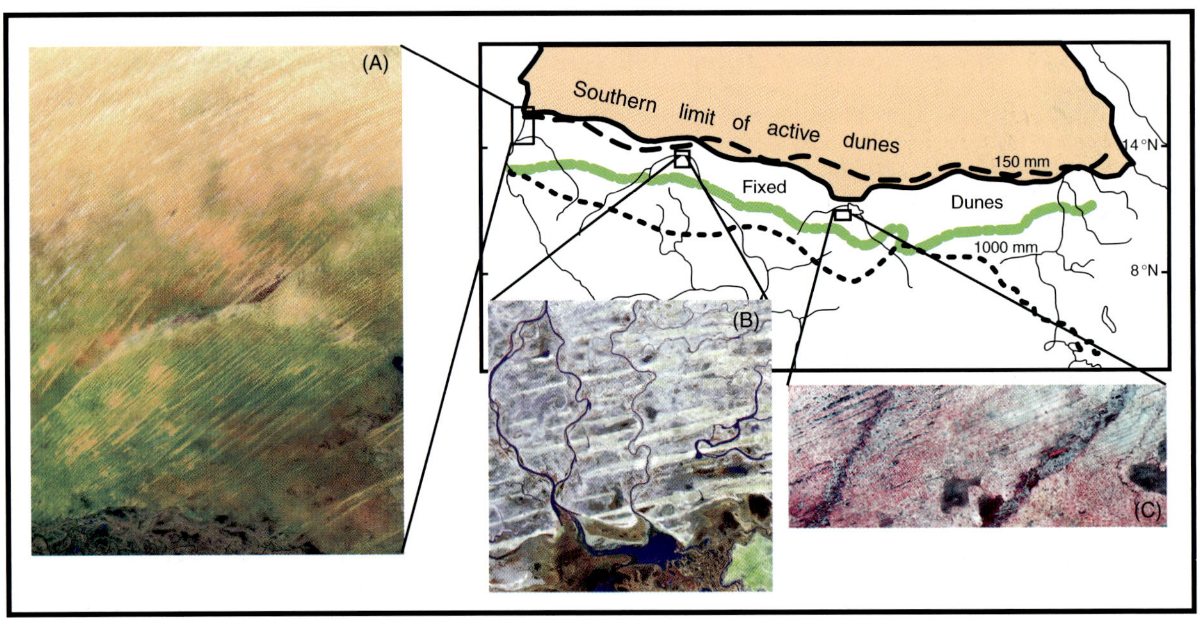

Figure 4 The Sahel and southern Sahara—limits of active and relict vegetation-stabilized dunes. (A) Landsat image of Sahara–Sahel transition in southern Mauritania. Note the change in dune trend and increasing vegetation cover to the south. (B) Aster image of relict dunes and river systems in Mali. (C) Vegetation-stabilized linear dunes west of Lake Chad.

In western Mauritania, where dunes are today largely active, combined geomorphic and OSL dating studies (Lancaster *et al.*, 2002) show that three main generations of linear dunes were formed during the periods 25–15 ka (centered around the Last Glacial Maximum (LGM)), 13–10 ka (spanning the Younger Dryas event), and after 5 ka. The wind regimes that occurred during each of these periods were significantly different, leading to the formation of dunes on three distinct superimposed trends—northeast, north-northeast, and north (**Fig. 5**). The period from approximately 9–5 ka was characterized by formation of shallow lakes in interdune areas and formation of soils on the dunes (Kocurek *et al.*, 1991).

Sedimentary and OSL dating studies of the Selima sand sheet in the eastern Sahara (Stokes *et al.*, 1998) indicate two major episodes of accumulation at 20–15 and 4–3 ka, separated by an early to mid-Holocene pluvial period during which a regionally extensive paleosol formed and local wadi and lacustrine deposits were laid down.

Directly dated information on the response of sand seas on the northern margins of the Sahara to Quaternary climate change is sparse, but a detailed luminescence-dated record from the Chott Rharsa area of southern Tunisia suggests formation of gypsum-rich dunes and low lake levels 12–10, approximately 7.5, and 6.2–5.6 ka (Swezey *et al.*, 1999). In summary, OSL ages, together with stratigraphic relations between eolian sands and ^{14}C dated lacustrine deposits, indicate widespread eolian activity and dune formation in the Sahara from 25–11 ka, with extensive stabilization of dunes in the southern and western Sahara in the period 11–5 ka (the African Humid Period, deMenocal *et al.*, (2000)). The African Humid Period was apparently not uniformly humid throughout this region. Gasse and Van Campo (1994) summarized evidence for a

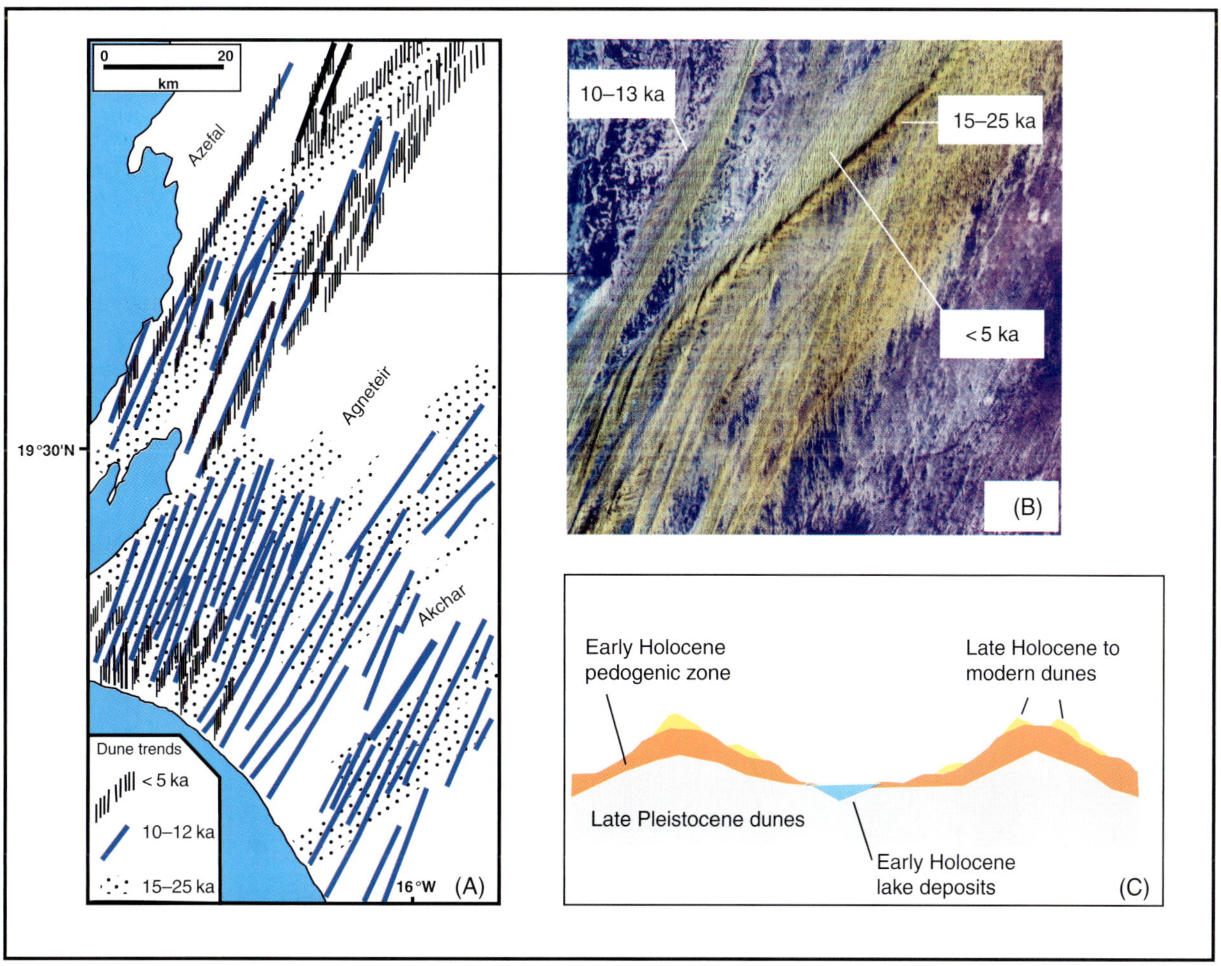

Figure 5 Effects of climate change on Mauritania dune systems. (A) Major dune trends; after Lancaster N *et al.* (2002) Late Pleistocene and Holocene dune activity and wind regimes in the western Sahara of Mauritania. *Geology* 30(11): 991–994. (B) Landsat image of the Azefal, showing three dune generations. (C) Schematic cross section of dunes in this area, showing superimposition of dune genetic units; after Kocurek G, Havholm KG, Deynoux M, and Blakey RC (1991) Amalgamated accumulations resulting from climatic and eustatic changes, Akchar Erg, Mauritania. *Sedimentology* 38(4): 751–772.

widespread period of aridity 8.8–7.8 ka, with some areas experiencing dry conditions at 6.8 ka. The period after 5 ka appears to have been arid, with renewed formation of dunes and/or reworking of existing deposits.

Kalahari Desert

Extensive systems of dunes, mostly of linear form and stabilized by vegetation, occur in the northern and southwestern Kalahari (**Fig. 6**). The dunes can be divided into three groups (Thomas and Shaw, 1991). The eastern and northern groups of dunes are variably degraded and covered by savanna vegetation. They are also truncated by some of the major river systems in the area as well as the Okavango Delta. The southern group of dunes occurs in the southwestern Kalahari in a 100- to 200-km-wide belt that extends from the highlands of Namibia to the Orange River. Dunes consist of 2 to 15 m high simple and compound linear ridges. Vegetation cover on the dunes consists of grasses with scattered trees and shrubs, but the dune crests are commonly unvegetated and active sand movement takes place seasonally.

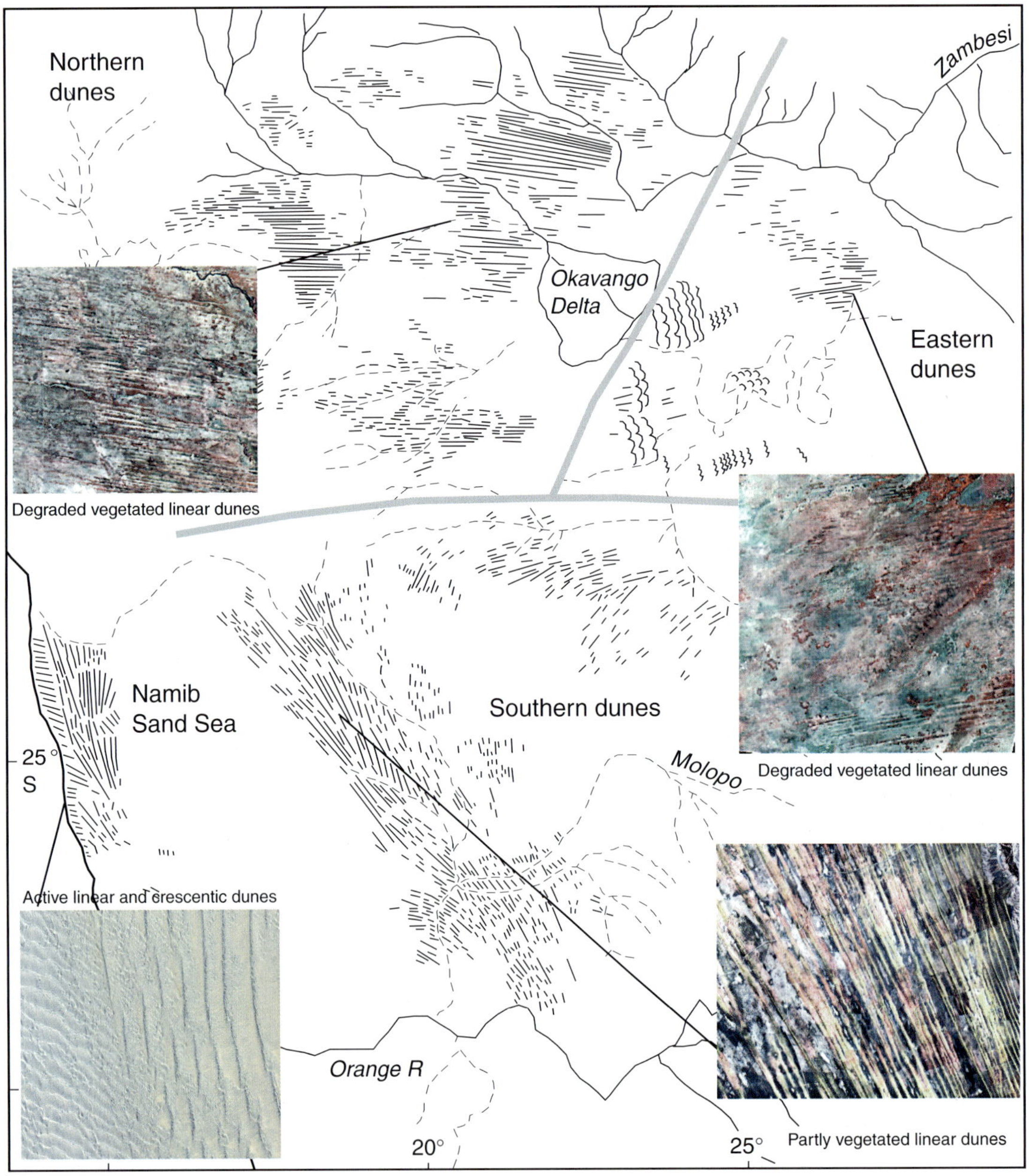

Figure 6 Dune systems in southern Africa, with Landsat images of examples of linear dunes in different areas. Dunes after Thomas DSG and Shaw PA (1991) *The Kalahari Environment*. Cambridge, UK: Cambridge University Press.

Until the recent application of luminescence dating techniques, the age(s) of the Kalahari dune systems was unknown or inferred only by relations to dated lacustrine and fluvial deposits. Luminescence dating and stratigraphic investigations show that the last major period of linear dune development in the southern dune system occurred between 16 and 10 ka and probably involved reworking of sediment deposited in a previous period of dune building from 30–23 ka that may represent the initial deposition of sand in this region (Stokes *et al.*, 1997b). Holocene dune activity in the region was localized in extent and occurred at 6 and –2–1 ka (Thomas *et al.*, 1997), although final stabilization of linear dunes may have occurred as recently as 8–9 ka in drier western areas of the region.

Parallel studies of dune chronology in the northern and eastern dunes indicate that multiple periods of dune building, each spanning several thousand years, occurred during the late Pleistocene and Holocene (Stokes *et al.*, 1997c). OSL ages for these periods are 115–95, 46–41, and 26–20 ka. Short-lived periods of aridity were separated by much longer (40–20 ka) periods of humid conditions. Late Glacial and Holocene eolian activity was restricted to reworking of the crestal areas of the linear dunes in the area of Hwange National Park in the period –16–10 ka and to western Zambia, where linear dune construction from local sediment sources occurred 32–27, 16–13, and 10–8 ka (O'Connor and Thomas, 1999). Dune construction is indicated for the periods 48–41, 36–29, and 23–21 ka in the linear dunes of northern Namibia (Thomas and Shaw, 2002).

Namib Desert

The main Namib sand sea occupies an area of approximately 34,000 km^2 between Luderitz and the Kuiseb River and the coast and the base of the Great Escarpment of Namibia (**Fig. 6**). The sand sea is characterized by crescentic dunes near the coast, with north–south trending linear dunes inland and clusters of star dunes and reticulate or network linear dunes inland (Lancaster, 1989). Despite the existence of multiple generations of dunes in this sand sea, there is little information on their age. OSL dating studies on the northern margin of the sand sea give ages for the base of one linear dune of between 0.34 ± 0.02 and 1.57 ± 0.07 ka, indicating a dune migration rate of approximately 0.12 $m yr^{-1}$ during the past 1,600 years (Bristow *et al.*, 2005). Studies of interdune lacustrine deposits (Teller *et al.*, 1990) indicate widespread conditions favorable for shallow lakes and ponds in the periods 12–8, 24–20, and 32–26 ka, although the state of the adjoining dunes is not clear.

Additional small dune fields dominated by crescentic dunes occur in the northern Namib in the area known as the Skeleton Coast. A Holocene age for the Skeleton Coast dune field is suggested by the small sand thickness represented by the dunes and their simple patterns. Aerial photographs and satellite images indicate that the Engo-Kunene and Baia dos Tigres-Curoca sand seas are characterized by a similar assemblage of dunes, but increased reddening inland suggests a greater chronologic complexity for this area.

Australian Deserts

Dunes cover as much as 40% of the Australian landscape, with the majority in the Simpson Strzlecki and Tirarai deserts (**Fig. 7**). Continent-wide patterns of dunes in Australian sand seas and dune fields have been documented (Wasson *et al.*, 1988). Although linear dunes dominate, short, narrow-crested varieties appear to occur where sand transport is limited by vegetation cover; broad-crested linear dunes are found in the northern parts of the dune system, and narrow-crested varieties are found in the south. Network dunes occur in small basins in the central parts of the continent where the directional variability of the wind regime is high. Mapping of patterns of sediment thickness, together with sedimentological and mineralogical evidence, shows that most sand in the Australian deserts has only moved short distances from its source, although the dune pattern suggests an organized continent-wide sediment transport system (Wasson, 1988; Pell *et al.*, 2000).

A variety of luminescence dating studies indicate discontinuous dune construction in the Simpson–Strzelecki dune field between 35 and 9 ka, with older periods of activity between 65 and 45 ka and at approximately 78 ka (Munyikwa, 2005). Localized periods of eolian activity also occurred during the Holocene. In all instances, periods of eolian construction were coeval with periods of low lake levels in the region (Wasson, 1983).

Arabian Deserts

The Arabian Peninsula is dominated by extensive dune systems, including the world's largest sand sea, the Rub al Khali, the Nafud, and Ad Dahna, all influenced by the northwesterly Shamal wind system (**Fig. 8**). The Wahiba sand sea of Oman comprises a separate system on the coast, influenced by the southwest monsoon circulation.

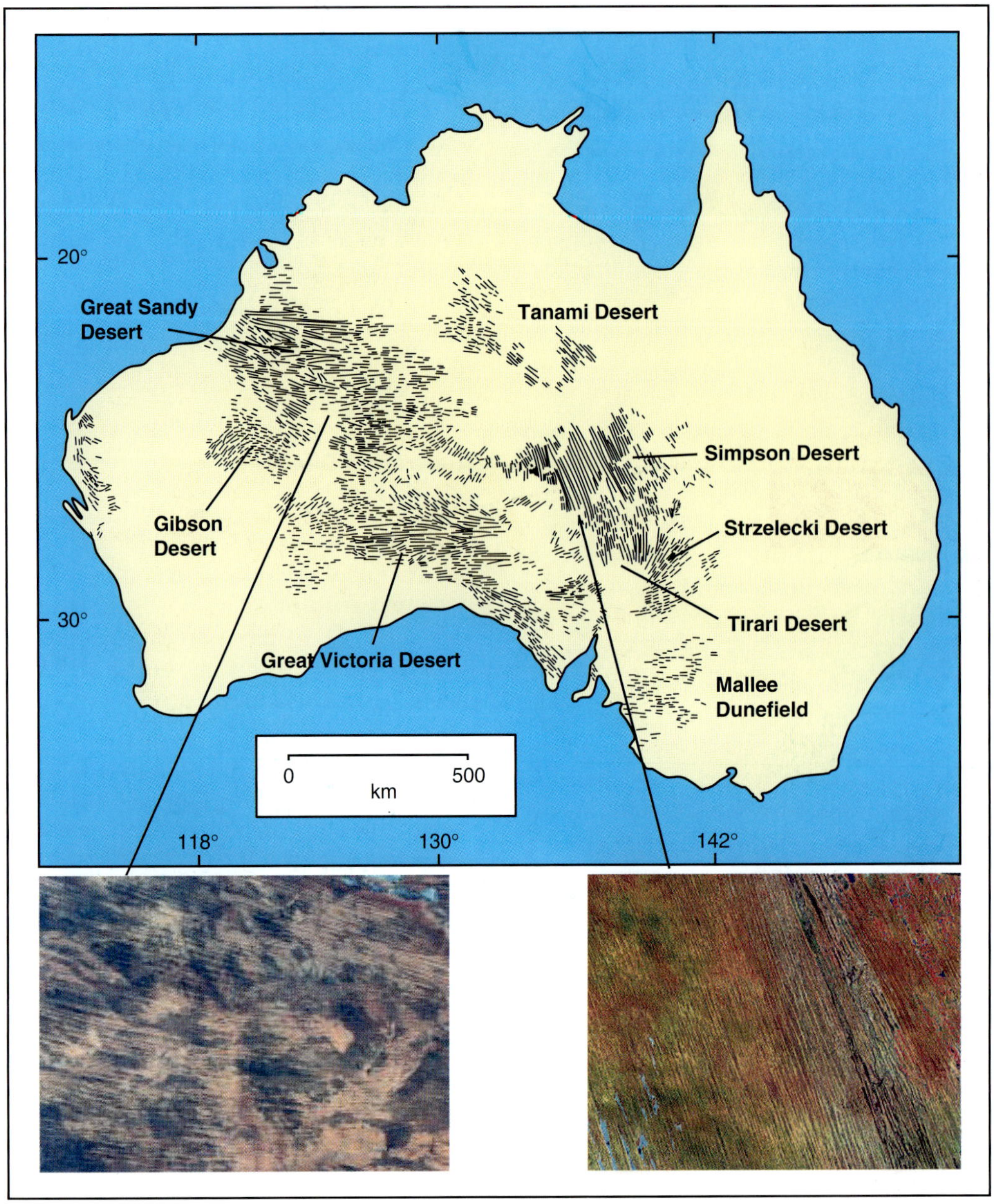

Figure 7 Dune systems in Australia. Landsat images of examples of linear dunes in the Great Sandy and Simpson–Strzelecki deserts. Dune trends and extent after Jennings (1968).

A characteristic feature of eolian sedimentation in the Wahiba sand sea and the United Arab Emirates (UAE) is the existence of thick sequences of carbonate-cemented sand (eolianite) that underlie the more quartz-rich dunes. Limited stratigraphic evidence and clustering of OSL ages on the eolianites in the UAE indicate that they formed during the periods 164–129, 113–99, 64–34, and 22–18 ka from bioclastic sediments deflated from the shelf areas of the Arabian Sea during periods of low sea levels (Glennie and Singhvi, 2002). Wind directions during these periods of dune construction were from the northwest/north-northwest. In the Wahiba, periods of accumulation of eolianites occurred 160–140, 34–24, 16–13, and approximately 9–8 ka (Preusser *et al.*, 2005).

The southern parts of the Rub al Khali are dominated by large complex linear dunes, whereas its northern parts exhibit a complex pattern of dune construction, reactivation, and stabilization in response to changes in sea level and climate. In the UAE and adjacent areas, the largest and most widespread dunes are west-to-east-oriented complex linear dunes, with superimposed crescentic dunes in the

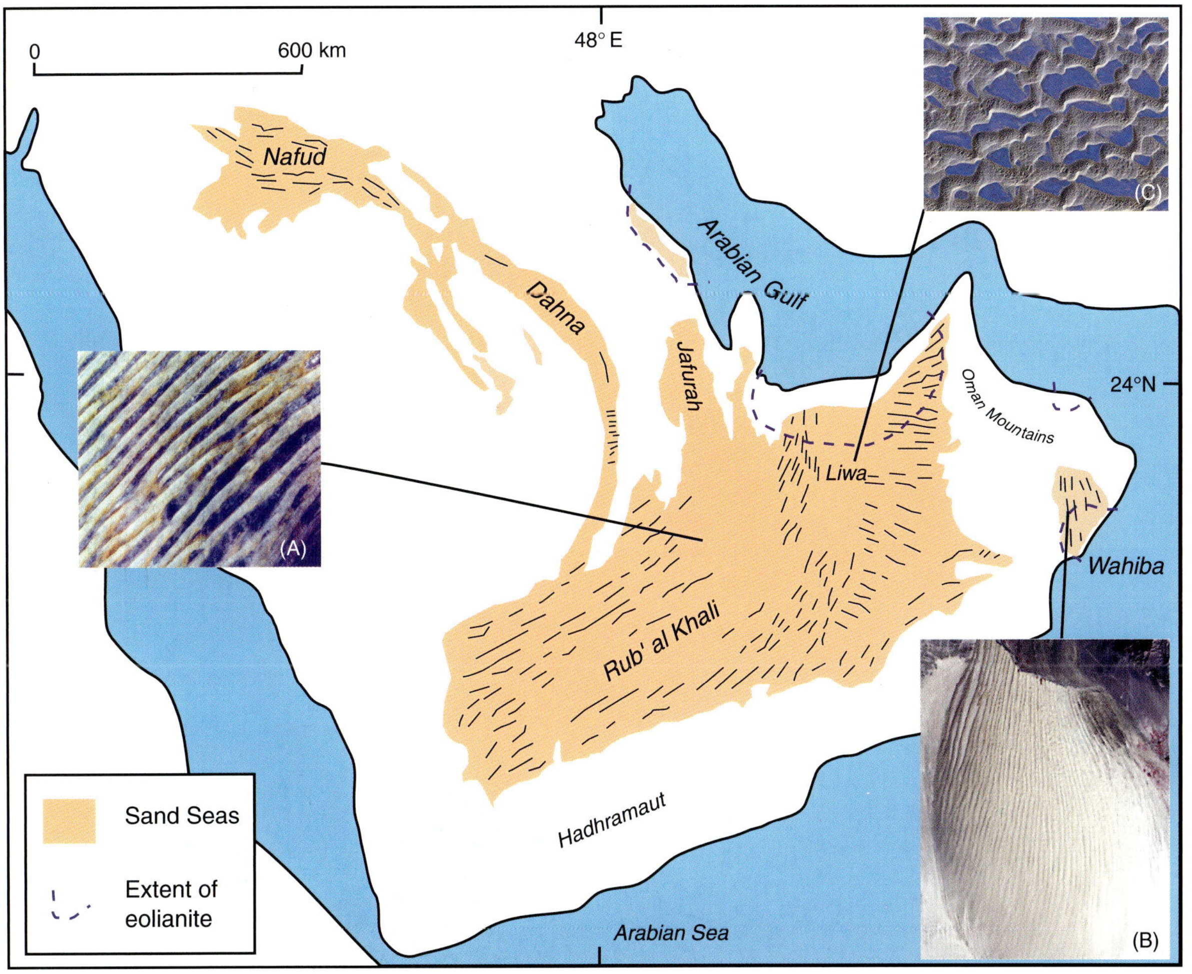

Figure 8 Sand seas and dune fields of the Arabian Peninsula. Insets: (A) aster image of linear dunes in the Rub' al Khali, (B) compound crescentic dunes in the Liwa area, and (C) linear dunes in the Wahiba Sand Sea. Dune extent and schematic dune trends after Glennie KW and Singhvi AK (2002) Event stratigraphy, palaeoenvironment and chronology of SE Arabian deserts. *Quaternary Science Reviews* 21: 853–869.

north and east and star dune peaks in the southeast of the UAE. These dunes are composed mainly of quartz sand and are being reworked in northern areas by northwest winds into crescentic dunes and obliquely oriented small linear dunes that are superimposed on the linear dunes. The linear dunes appear to have been formed (or reformed) during the LGM, as indicated by OSL ages of 42–31, 18–15, and 12– 9 ka in the UAE.

In the Liwa area, large, quartz-rich compound crescentic dunes overlie inland sabkhas. Stratigraphic and geomorphic evidence, together with OSL dating of eolian sands, indicates deposition of eolian sand in the period 130–75 ka (Stokes and Bray, 2005) and after 7.8 ka, peaking at 2.8 ka, but no record from the LGM. The water table in this area was several meters higher than present during the period 41–29 ka (Wood and Imes, 1995).

In the Wahiba sand sea of Oman, a large inland part of the sand sea is characterized by 50- to 100-m-high south–north trending complex linear dunes. In places, the crestal areas of these dunes are overlain by crescentic and network dunes. To the south of the large linear ridges are smaller (10–30 m high) vegetated linear dunes with 300- to 500-m spacing. There are three groups of simple and compound crescentic and network dunes near the coast. These groups of dunes overlie cemented carbonate sands (eolianites), which crop out on the southern and western edges of the sand sea. An extensive program of drilling and luminescence dating of dune sands and eolianites in the central parts of the sand sea indicates that periods of eolian accumulation occurred 24–11, 78–65, 120–100, and 160–140 ka (Preusser *et al.*, 2002; Radies *et al.*, 2004). The main south–north-oriented linear dunes were formed at approximately 20 ka. The coastal part of the Wahiba exhibits a different timing of eolian accumulation, with dune building at 34–24, 16–13, and approximately 9–8 ka (Preusser *et al.*, 2005). In contrast to the main dune area, there was little dune activity at the LGM and during the Younger Dryas. The area experienced wetter conditions with high groundwater levels during the early Holocene.

Thar Desert

The Thar Desert in northwestern India (**Fig. 9**) is dominated by compound parabolic dunes, with smaller areas of linear, crescentic, reticulate, or network dunes and sand sheets (Kar, 1993). The dunes are underlain in part by carbonate-cemented eolianites, which range in age from 21–210 ka. Dune trends indicate their formation by winds

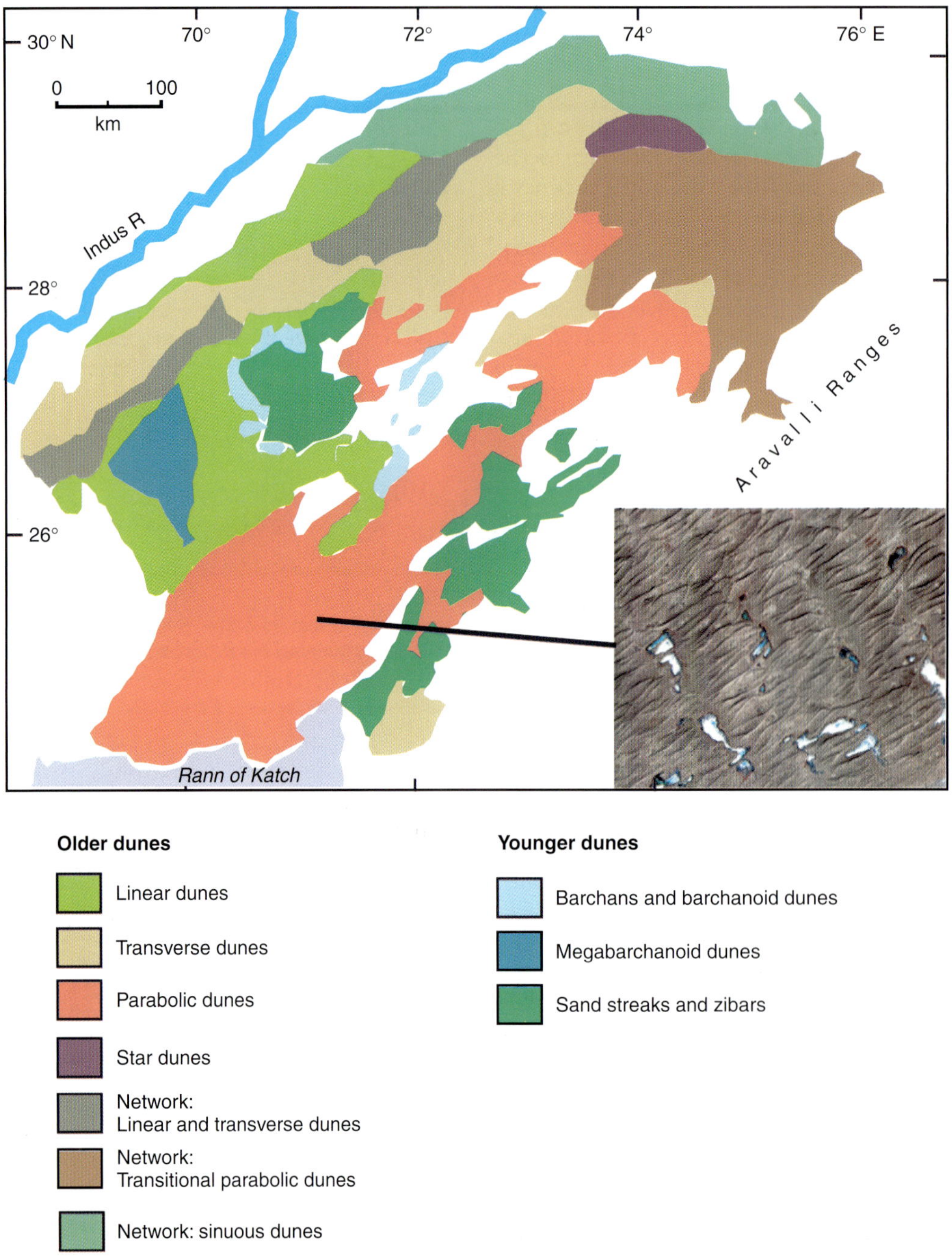

Figure 9 Dunes in the Thar Desert. Insert: aster image of parabolic dunes. After Singhvi AK and Kar A (2004) The aeolian sedimentation record of the Thar Desert. *Proceedings of the Indian Academy of Sciences (Earth Sciences)* 113(3): 371–401.

associated with the southwest monsoon. Many of the dunes are partially or completely stabilized by vegetation, and the limit of stabilized dunes extends into areas where the mean annual rainfall exceeds 500 mm.

The Thar Desert was the site of some of the first luminescence dating studies on eolian deposits (Singhvi *et al.*, 1982), and subsequent studies have resulted in the development of a detailed chronology of dune-building episodes in the region (Singhvi and Kar, 2004). Eolian deposition in the area dates to 150 ka or older; major periods of construction occurred 115–100, ~75, ~55, and between 30 and 25 ka.

The available data suggest that despite an arid climate and an abundant sediment supply from the exposed shelf of the Arabian Sea, late Pleistocene dune building in the Thar Desert did not occur coeval with the LGM but instead took place in a narrow window of time (14–10 ka) as the southwest monsoon was reestablished at the Pleistocene–Holocene transition. Major eolian activity occurred in the Holocene, but with important spatial variability. Dune building ceased in the southern Thar Desert in the early Holocene (10 ka) but continued until 5 ka in the north. In the core arid region of western Rajasthan, dune activity continued after 2 ka to 0.7–0.8 ka.

North American Deserts

Dune fields are limited in extent in the deserts of the southwestern United States and adjacent areas of Mexico (**Fig. 10**). Sand deposits are composed of sand sheets, small dune fields, and sand ramps (amalgamated accumulations of eolian sand, alluvial, and colluvial deposits on the slopes of desert mountains).

Mojave Desert Eolian sediments and landforms in the Mojave Desert of southern California and adjacent areas of Nevada and Arizona are concentrated along well-defined eolian sediment transport corridors that are characterized by areas of active and inactive dunes and sand sheets, together with sand ramps. Geochemical and mineralogical studies indicate several major (e.g., the Mojave River and Colorado River) and many local sources of sediment, but there is a general west-to-east transport of sand.

The history of late Quaternary eolian activity along the major sand transport corridors has been documented using thermoluminescence and infrared-stimulated luminscence (IRSL) dating of dunes and eolian deposits in sand ramps (Clarke and Rendell, 1998; Lancaster and Tchakerian, 2003). These indicate periods of eolian activity 30–20 and 15 or 20 ka–7 ka. Mid- and Late Holocene eolian deposits occur in the Kelso Dunes. Interstratified paleosols in many sand ramps (e.g., near Kelso Dunes and Rice Valley dunefield) suggest one major and a number of minor periods of regional-scale eolian stability 20–15, 14, and 4 ka. Comparison of the record of periods of eolian construction and stability in the Kelso Dunes system with the lacustrine record from Lake Mojave suggests that periods of eolian stability generally coincide with the highest lake levels, whereas periods of eolian construction are linked to periods of fluctuating or desiccating lakes but also overlap with periods of high lake levels. These observations indicate a strong influence of sediment supply on eolian processes in this region.

Multiple, short-duration episodes of late Holocene eolian accumulation are recorded close to modern sand source areas. IRSL ages for sediments in the Kelso Dunes (including the Cronese basins) cluster between 0.20–0.14, 0.5–0.4, 0.85–0.7, 1.7–1.4, 3–2, 4.2–3.5, and 6.7–5.4 ka. Episodes of eolian deposition correlate well with the record of major floods in the Mojave River system and likely reflect the effect of enhanced sediment supply from the Mojave River (Muhs *et al.*, 2003).

Colorado and Sonoran Deserts There are three main areas of eolian deposits in the Colorado and lower Sonoran deserts: the Coachella Valley, the Algodones Dunes, and the Gran Desierto sand sea of northern Mexico (Lancaster, 1995).

The largest dune field in the Sonoran Desert lies in the Gran Desierto of northern Mexico, which has a spatially complex pattern of multiple generations of linear, crescentic, and star dunes, as well as extensive sand sheets (**Fig. 10**), sourced mainly from Colorado River sediments. Recent OSL dating and geomorphic studies show that elements of the complex pattern represent short-lived eolian constructional events since ~25 ka. The simple dune patterns consist of late Pleistocene relict linear dunes formed between 25 and 12 ka, degraded crescentic dunes formed at ~12 ka, eastern crescentic dunes emplaced at ~7 ka, early Holocene western crescentic dunes, and star dunes formed during the past 3 ka (Beveridge, 2004). In the Algodones, Stokes *et al.* (1997a) identified periods of eolian activity >30 ka ~3.1 and <0.4 ka.

South America

Low-latitude dune fields occur in two settings in South America—(a) the largely active dunes of the Atacama Desert on the west coast and (b) the vegetation-

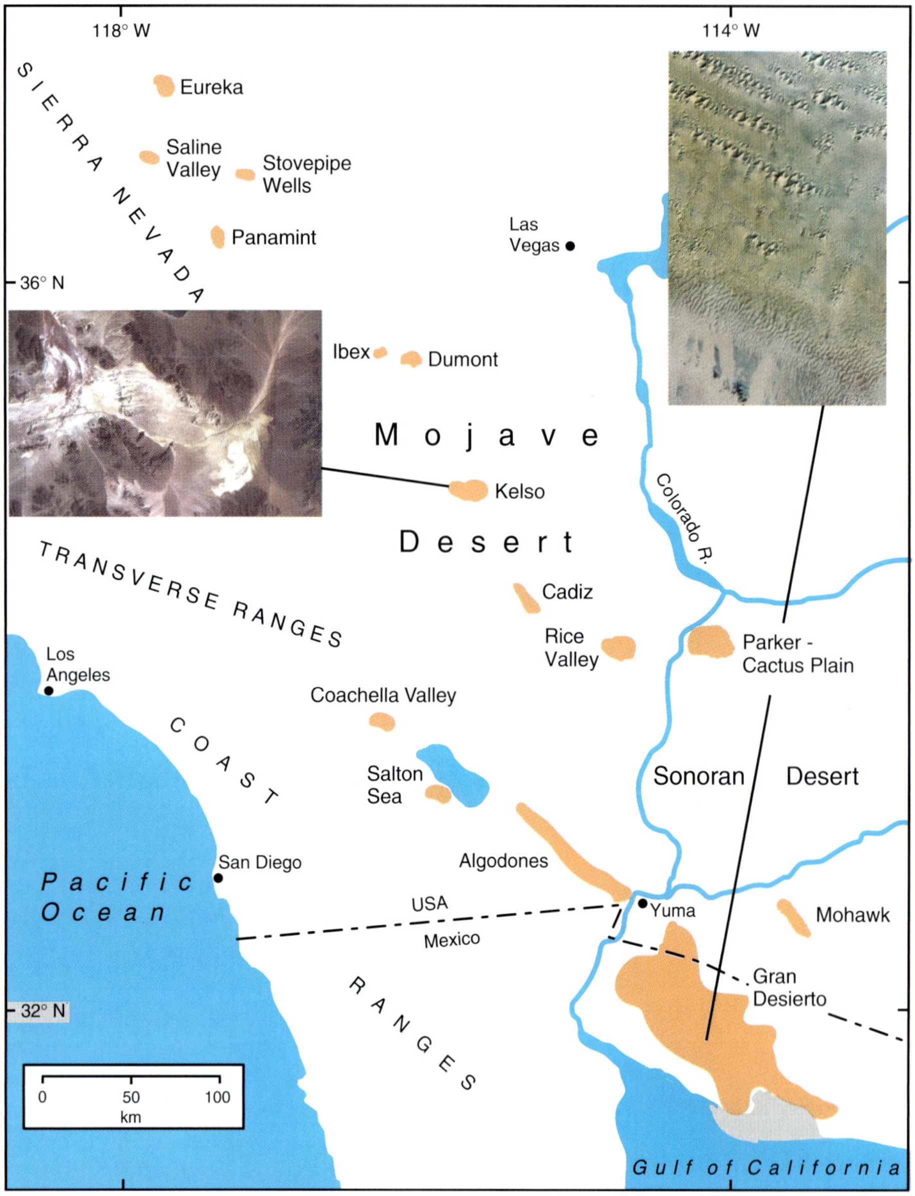

Figure 10 Distribution of dune fields in the Mojave and Sonoran deserts, with Landsat images of selected dune fields. Gran Desierto image shows star dunes in linear chains (26–12 ka) and crescentic dunes formed ~12 ka.

stabilized relict dune fields of the Rio Negro and Sao Francisco areas of Brazil and the northern Pampa in Argentina. Chronologic data for the relict dune fields are summarized by Munyikwa (2005) and suggest eolian activity prior to 69 ka and at 63, 33, 28–27, 23–22, and 17.2–8 ka, with limited Holocene activity.

Conclusions

The available chronologic information shows that most sand seas have experienced episodic accumulation during the past 120 ka, with long periods of stability or nondeposition separated by relatively brief periods in which dune construction occurred (**Fig. 11**). Luminescence dating indicates widespread dune activity in low-latitude deserts during the periods 120–100, 35–10, and since 5 ka. However, subsequent periods of eolian activity have reworked older deposits so that the preserved record in many areas is quite short.

The episodic nature of sand sea construction in low-latitude deserts is a direct consequence of the interplay between sand supply, availability, and mobility (**Fig. 12**). Thus, sediment supplied to source areas during periods of increased fluvial activity is mobilized when it becomes available for transport in

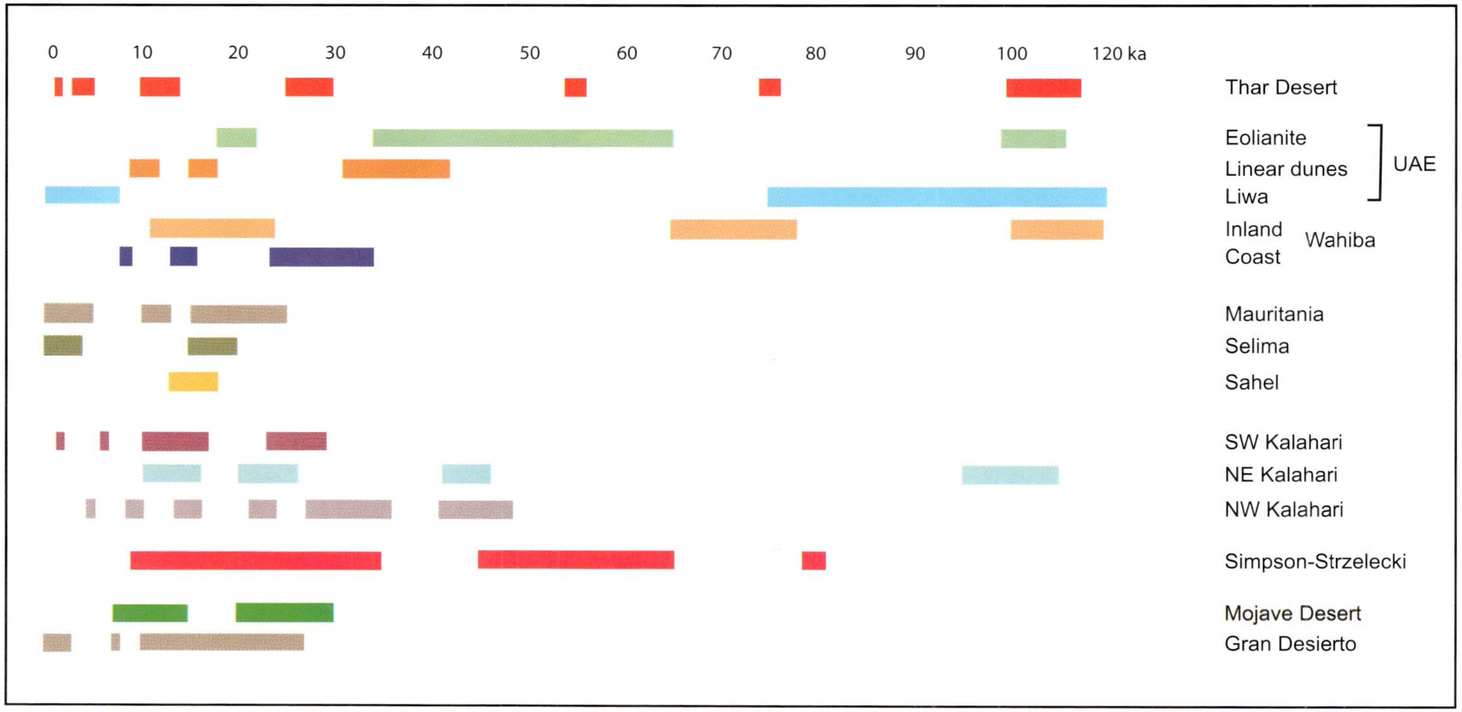

Figure 11 Chronological summary of periods of dune construction and/or activity during the past 120 ka directly dated using luminescence methods. Compiled from a variety of sources referenced in text.

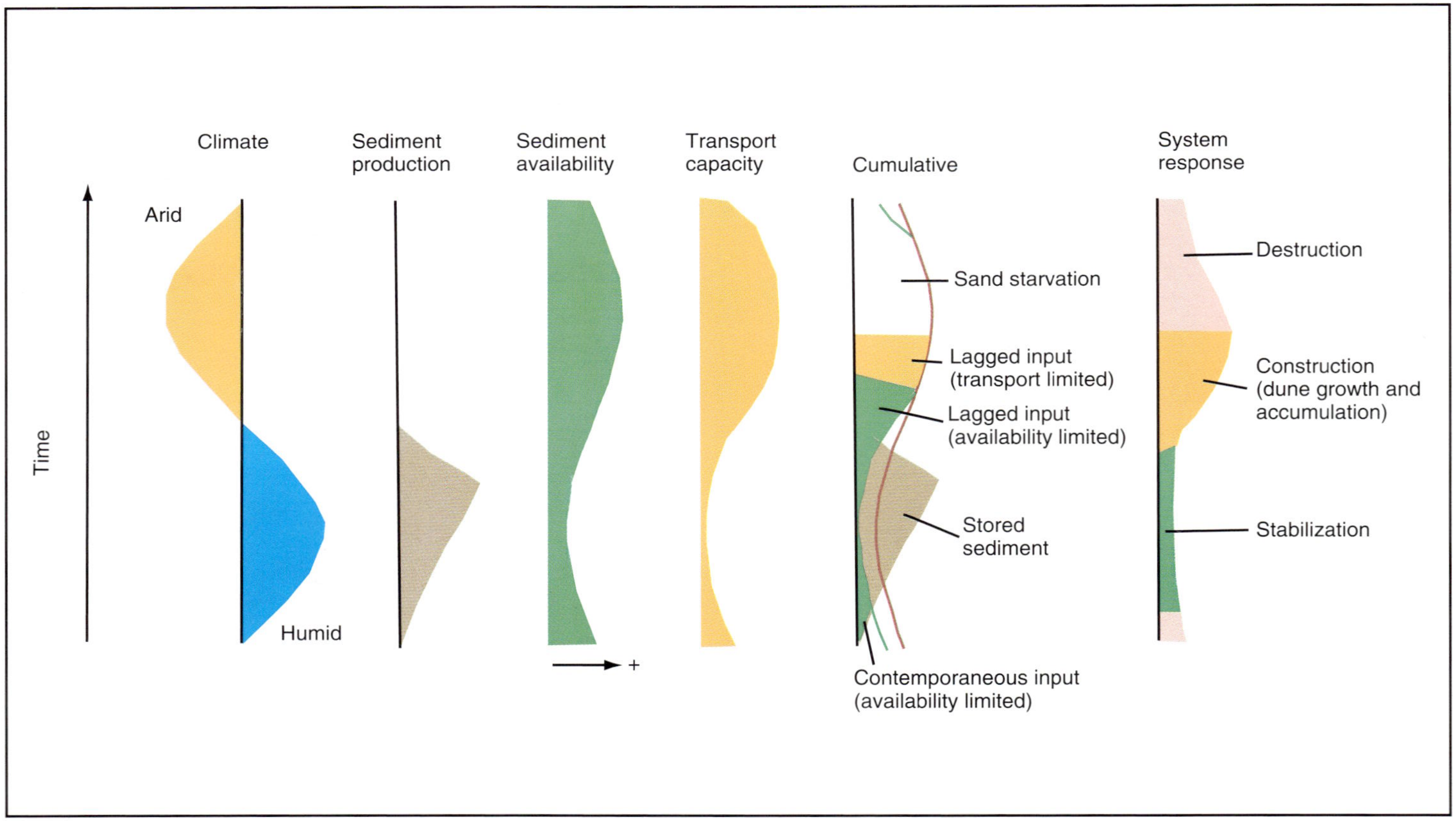

Figure 12 Schematic of the response of sand seas to climate change based on the principles of sediment configuration and sediment state (Kocurek and Lancaster, 1999). Diagram after Kocurek G (1998) Aeolian system response to external forcing factors—A sequence stratigraphic view of the Saharan region. In: Alsharan AS, Glennie KW, Whittle GL, and Kendall CGSC (Eds.), *Quaternary Deserts and Climatic Change*, pp. 327–338. Rotterdam, The Netherlands: Balkema.

subsequent drier episodes, as occurred in the Simpson–Strzlecki deserts in the late Pleistocene, the Mojave Desert in the early Holocene, and throughout the Sahara in the late Holocene. In other areas (e.g., UAE and Wahiba), sediment supplied to shelf zones by biogenic activity became available for transport to dunes as sea level fell. In the case of the Wahiba sand sea and the Thar Desert, this sediment was only mobilized as the strength of the southwest monsoon increased following the LGM.

Sarnthein (1978) summarized the then available evidence for late Quaternary eolian activity in low-latitude arid regions. He concluded that dune fields were much more extensive at the LGM (~20 ka). The recent infusion of chronologic information from luminescence dating studies broadly confirms Sarnthein's hypothesis but suggests that the temporal and spatial patterns of dune construction are more complex than proposed by Sarnthein (**Fig. 11**). For example, there was no dune construction in the northern Kalahari and Thar deserts at this time.

Many areas of dunes, especially those composed of linear dunes, were formed or reformed during the LGM. Arid conditions and eolian activity increased in extent and/or intensity during this period, as corroborated by the record of dust flux in adjacent ocean cores. This was apparently the result of increased intensity and persistence of the trade wind circulation in the Sahara and elsewhere giving rise to increased sediment mobility, coupled with a decrease in the strength of monsoon circulations (and summer rainfall) in Africa and Asia, thereby increasing sediment availability as vegetation cover and soil moisture decreased. In many areas, increased sediment availability as a result of decreased rainfall and enhanced mobility (as a result of increased wind strength) coincided, resulting in widespread dune formation. However, sand seas in which the dominant winds are related to the southwest monsoon (especially Wahiba and Thar sand seas) experienced reduced activity at this time.

The response of low-latitude sand seas to Quaternary climate change is becoming much clearer, in large part as a result of the application of luminescence dating techniques, coupled with new paradigms (especially concepts of dune generations and eolian sediment state) for the geomorphic and sedimentary development of these extensive depositional systems. Future work will be increasingly focused on understanding spatial patterns in the response of sand seas to climate change; the relations between sediment supply, availability, and mobility; and the role of sub-Milankovitch-scale climate change in sand sea evolution.

See also: **Dune Fields**: High Latitudes; Mid-Latitudes. **Luminescence Dating**: Thermoluminescence; Optically-Stimulated Luminescence.

References

Beveridge, C. A. (2004). *The Origin and Evolution of the Gran Desierto Sand Sea, Sonora, Mexico*. Master'sthesis, University of Texas at Austin, Austin.

Breed, C. S., *et al.* (1979). Regional studies of sand seas using LANDSAT (ERTS) imagery. In *A Study of Global Sand Seas* (E. D. McKee, Ed.), pp. 305–398. Washington, DC, U.S. Geological Survey.

Bristow, C. S., Duller, G. A., and Lancaster, N. (2005). Combining ground penetrating radar (GPR) surveys and optical dating to determine dune migration in Namibia. *Journal of the Geological Society of London* **162**(2), 315–322.

Clarke, M. E., and Rendell, H. M. (1998). Climatic change impacts on sand supply and the formation of desert sand dunes in the south-west U.S.A. *Journal of Arid Environments* **39**(3), 517–532.

deMenocal, P., *et al.* (2000). Abrupt onset and termination of the African Humid Period: Rapid climate responses to gradual insolation forcing. *Quaternary Science Reviews* **19**(1–5), 347–361.

Fryberger, S. G., and Ahlbrandt, T. S. (1979). Mechanisms for the formation of aeolian sand seas. *Zeitschrift für Geomorphologie* **23**, 440–460.

Gasse, F. (2002). Diatom-inferred salinity and carbonate oxygen isotopes in Holocene waterbodies of the western Sahara and Sahel (Africa). *Quaternary Science Reviews* **21**, 737–767.

Gasse, F., and Van Campo, E. (1994). Abrupt post-glacial events in West Africa and North Africa monsoon domains. *Earth and Planetary Science Letters* **126**, 435–456.

Glennie, K. W., and Singhvi, A. K. (2002). Event stratigraphy, palaeoenvironment and chronology of SE Arabian deserts. *Quaternary Science Reviews* **21**, 853–869.

Grove, A. T., and Warren, A. (1968). Quaternary landforms and climate on the south side of the Sahara. *Geographical Journal* **134**(Part 2), 189–208.

Jennings, J. N. (1968). A revised map of the desert dunes of Australia. *Australian Geographer* **10**, 408–409.

Kar, A. (1993). Aeolian processes and bedforms in the Thar Desert. *Journal of Arid Environments* **25**, 83–96.

Kocurek, G. (1998). Aeolian system response to external forcing factors—A sequence stratigraphic view of the Saharan region. In *Quaternary Deserts and Climatic Change* (A. S. Alsharan, K. W. Glennie, G. L. Whittle and C. G. S. C. Kendall, Eds.), pp. 327–338. Rotterdam, The Netherlands, Balkema.

Kocurek, G., and Lancaster, N. (1999). Aeolian sediment states: Theory and Mojave Desert kelso dunefield example. *Sedimentology* **46**(3), 505–516.

Kocurek, G., Havholm, K. G., Deynoux, M., and Blakey, R. C. (1991). Amalgamated accumulations resulting from climatic and eustatic changes, Akchar Erg, Mauritania. *Sedimentology* **38**(4), 751–772.

Lancaster, N. (1989). *The Namib Sand Sea: Dune forms, processes, and sediments.* Rotterdam, The Netherlands: Balkema.

Lancaster, N. (1995). Origin of the Gran Desierto Sand Sea: Sonora, Mexico: Evidence from dune morphology and sediments. In *Desert Aeolian Processes* (V. P. Tchakerian, Ed.), pp. 11–36. New York, Chapman & Hall.

Lancaster, N. (1999). Sand seas. In *Aeolian Environments, Sediments, and Landforms* (A. S. Goudie, I. Livingstone and S. Stokes, Eds.), pp. 49–70. New York, Wiley.

Lancaster, N., and Tchakerian, V. P. (2003). Late Quaternary eolian dynamics, Mojave Desert, California. In *Paleoenvironments and Paleohydrology of the Mojave and Southern Great Basin Deserts* (Y. Enzel, S. G. Wells and N. Lancaster, Eds.), pp. 231–249. Boulder, CO, Geological Society of America.

Lancaster, N., *et al.* (2002). Late Pleistocene and Holocene dune activity and wind regimes in the western Sahara of Mauritania. *Geology* **30**(11), 991–994.

Mainguet, M. (1984). Space observations of Saharan aeolian dynamics. In *Deserts and Arid Lands* (F. El Baz, Ed.), pp. 59–77. The Hague, The Netherlands, Nyhoff.

Mainguet, M., and Chemin, M. C. (1983). Sand seas of the Sahara and Sahel: An explanation of their thickness and sand dune type by the sand budget principle. In *Eolian Sediments and Processes. Developments in Sedimentology* (M. E. Brookfield and T. S. Ahlbrandt, Eds.), pp. 353–364. Amsterdam, Elsevier.

Mauz, B., and Felix-Henningsen, P. (2005). Paleosols in Saharan and Sahelian dunes of Chad: Archives of Holocene North African climate changes. *The Holocene* **15**(3), 453–458.

McKee, E. D. (Ed.) (1979). *A Study of Global Sand Seas* Professional paper No. 1052, Washington, DC, U.S. Geological Survey.

Muhs, D. R., Reynolds, R. R., Been, J., and Skipp, G. (2003). Eolian sand transport pathways in the southwestern United States: importance of the Colorado River and local sources. *Quaternary International* **104**, 3–18.

Munyikwa, K. (2005). Synchrony of Southern Hemisphere late Pleistocene arid episodes: A review of luminescence chronologies from arid aeolian landscapes south of the equator. *Quaternary Science Reviews* **24**, 2555–2583.

O'Connor, P. W., and Thomas, D. S. G. (1999). The timing and environmental significance of late Quaternary linear dune development in western Zambia. *Quaternary Research* **52**, 44–55.

Pell, S. D., Chivas, A. R., and Williams, I. S. (2000). The Simpson, Strzelecki and Tirari Deserts: development and sand provenance. *Sedimentary Geology* **130**, 107–130.

Preusser, F., Radies, D., and Matter, A. (2002). A 160,000-year record of dune development and atmospheric circulation in southern Arabia. *Science* **296**, 2018–2020.

Preusser, F., Radies, D., Driehorst, F., and Matter, A. (2005). Late Quaternary history of the coastal Wahiba Sands, Sultanate of Oman. *Journal of Quaternary Science* **20**(4), 395–405.

Radies, D., Preusser, F., Matter, A., and Mange, M. (2004). Eustatic and climatic controls on the development of the Wahiba Sand Sea, Sultanate of Oman. *Sedimentology* **51**(6), 1359–1386.

Rognon, P. (1982). Pluvial and arid phases in the Sahara: The role of non climatic factors. *Palaeoecology of Africa* **12**, 45–62.

Sarnthein, M. (1978). Sand deserts during glacial maximum and climatic optima. *Nature* **272**, 43–46.

Singhvi, A. K., and Kar, A. (2004). The aeolian sedimentation record of the Thar Desert. *Proceedings of the Indian Academy of Sciences (Earth Sciences)* **113**(3), 371–401.

Singhvi, A. R., Sharma, Y. P., and Agrawal, D. P. (1982). Thermoluminescence dating of sand dunes in Rajasthan, India. *Nature* **295**, 313–315.

Stokes, S. (1997). Dating of desert sequences. In *Arid Zone Geomorphology* (D. S. G. Thomas, Ed.), pp. 607–637. New York, Wiley.

Stokes, S., and Bray, H. E. (2005). Late Pleistocene eolian history of the Liwa region, Arabian Peninsula. *Geological Society of America Bulletin* **117**(11–12), 1466–1480.

Stokes, S., and Horrocks, J. (1998). A reconnaissance survey of the linear dunes and loess plains of northwestern Nigeria: Granulometry and geochronology. In *Quaternary Deserts and Climatic Change* (A. S. Alsharan, K. W. Glennie, G. L. Whittle and C. G. S. C. Kendall, Eds.), pp. 165–174. Rotterdam, The Netherlands, Balkema.

Stokes, S., Kocurek, G., Pye, K., and Winspear, N. R. (1997a). New evidence for the timing of aeolian sand supply to the Algodones dunefield and East Mesa area, southeastern California, USA. *Palaeogeography, Palaeoclimatology, Palaeocology* **128**, 63–75.

Stokes, S., Thomas, D. S. G., and Shaw, P. A. (1997b). New chronological evidence for the nature and timing of linear dune development in the southwest Kalahari Desert. *Geomorphology* **20**(1–2), 81–94.

Stokes, S., Thomas, D. S. G., and Washington, R. (1997c). Multiple episodes of aridity in southern Africa since the last interglacial period. *Nature* **388**, 154–158.

Stokes, S., Maxwell, T. A., Haynes, C. V., and Horrocks, J. (1998). Latest Pleistocene and Holocene sand sheet construction in the Selima Sand Sheet, Eastern Sahara. In *Quaternary Deserts and Climatic Change* (A. S. Alsharan, K. W. Glennie, G. L. Whittle and C. G. S. C. Kendall, Eds.), pp. 175–184. Rotterdam, The Netherlands, Balkema.

Stuut, J.-B., and Lamy, F. (2004). Climate variability at the southern boundaries of the Namib (southwestern Africa) and Atacama coastal deserts during the last 120,000 yr. *Quaternary Research* **62**(3), 301–309.

Swezey, C. (2001). Eolian sediment responses to late Quaternary climate changes: Temporal and spatial patterns in the Sahara. *Palaeogeography, Palaeoclimatology, Palaeocology* **167**, 119–155.

Swezey, C., *et al.* (1999). Response of aeolian systems to Holocene climatic and hydrologic changes on the northern margin of the Sahara: A high resolution record from the Chott Rharsa basin, Tunisia. *The Holocene* **9**(2), 141–148.

Teller, J. T., Rutter, N. W., and Lancaster, N. (1990). Sedimentology and palaeohydrology of late Quaternary lake deposits in the northern Namib Sand Sea, Namibia. *Quaternary Science Reviews* **9**, 343–364.

Thomas, D. S. G., and Shaw, P. A. (1991). *The Kalahari Environment.* Cambridge, UK, Cambridge University Press.

Thomas, D. S. G., and Shaw, P. A. (2002). Late Quaternary environmental change in central southern Africa: New data, synthesis, issues and prospects. *Quaternary Science Reviews* **21**, 783–797.

Thomas, D. S. G., Stokes, S., and Shaw, P. A. (1997). Holocene aeolian activity in the southwestern Kalahari Desert, southern Africa: Significance and relationships to late-Pleistocene dune-building events. *The Holocene* 7(3), 273–281.

Wasson, R. J. (1983). The Cainozoic history of the Strzelecki and Simpson dunefields (Australia), and the origin of the desert dunes. *Zeitschrift fur Geomorphologie Supplement* **45**, 85–115.

Wasson, R. J., Fitchett, K., Mackey, B., and Hyde, R. (1988). Large-scale patterns of dune type, spacing, and orientation in the Australian continental dunefield. *Australian Geographer* **19**, 89–104.

Wilson, I. G. (1971). Desert sandflow basins and a model for the development of ergs. *Geographical Journal* **137**(2), 180–199.

Wilson, I. G. (1973). Ergs. *Sedimentary Geology* **10**, 77–106.

Wood, W. W., and Imes, J. L. (1995). How wet is wet? Precipitation constraints on late Quaternary climate in the Arabian Peninsula. *Journal of Hydrology* **164**, 263–268.

E

El Niño Southern Oscillation *see* **Paleoceanography, Records**: Postglacial South Pacific; **Paleoclimate Modeling**: Paleo-ENSO

Electron Spin Resonance Dating *see* **Luminescence Dating**: Electron Spin Resonance Dating

Eolian Sediments *see* **Paleosols and Wind-Blown Sediments**: Overview; Nature of Paleosols; Mineral Magnetic Analysis; Soil Micromorphology; Weathering Profiles; Soil Morphology in Quaternary Studies

EOLIAN RECORDS, DEEP-SEA SEDIMENTS

D K Rea, The University of Michigan, MI, USA

© 2007 Elsevier B.V. All rights reserved.

Introduction

The transport of eolian grains and their deposition in the deep sea has been of interest to scientists for over 150 years, beginning with Darwin's observations made aboard the *Beagle* (Darwin, 1846). Radczewski (1939) examined the reddish quartz grains in the Cape Verde Basin, downwind from the Sahara, and noted the importance of dust transport in bringing those grains to the central Atlantic. Rex and Goldberg (1958) explained the quartz in abyssal sediments of the North Pacific by dust transport from Asia. Since the 1960s, studies of the transport of mineral aerosols have abounded. The reader would find overview papers by Prospero *et al.* (1983, 2002), Rea *et al.* (1985), Duce *et al.* (1991), Rea (1994), and Grousset and Biscaye (2005), and books by Pèwè (1981), Pye (1987), and by Leinen and Sarnthein (1989) to provide useful depictions of the transport and deposition process. More recently, interest in dust transport and deposition has spread to the biological oceanographic community, with the formulation of the 'iron hypothesis', which holds that dust supplies the limiting nutrient iron to otherwise unproductive ocean surface waters (Martin and Fitzwater, 1988; Martin, 1990). Research on various aspects of this topic remains vigorous today (Jickells *et al.*, 2005). Dust is also important in understanding the effects of radiative forcing on the Earth and has received considerable attention from atmospheric scientists (Sokolik, *et al.*, 2001)

The demonstration that availability of dust varies with the climate of the source region (Prospero, *et al.*, 1970; Prospero and Nees, 1977, 1986) caused marine geologists to begin to look down core into the sediments for this useful record of past continental climates. The first such studies were conducted in the North Atlantic, downwind from the Sahara (Parkin and Shackleton, 1973; Parkin, 1974; Parkin and Padgham, 1975). Sarnthein *et al.* (1981, 1982) have studied the same system and provided good overviews of the geologic history of Saharan dust deposition (Ruddiman *et al.*, 1989). Clemens and Prell (1990, 1991), Sirocko and Sarnthein (1989), and Sirocko *et al.* (1991) have studied the monsoon-related dust transport from the deserts of Arabia to the northwestern Indian Ocean. Our work, some of which is described below, has been largely in the Pacific Ocean and stems from the early studies of Heath and Leinen (Heath, 1969; Leinen and Heath, 1981; Leinen, 1989).

Nature of the Data

Arid and semi-arid continental surfaces are the source of dust in the atmosphere and the ocean. Studies of modern climatology show that most of the wind-borne dust transport each year occurs in a few large storms, commonly in the spring or summer. Satellite data are now available that allow us to track a single dust storm from the Sahara to the Caribbean, or the Gobi to California. The dust entrained by these large storms shows a mixture of particle sizes, with the large particles falling out quickly along the transport path, and the fine grains remaining to be transported trans-oceanic distances. These fine grains end up in the ocean sediments and, in the absence of other sedimentary components, form the red or pelagic clay deposits that characterize half of the area of the deep-sea floor.

Much of the work on eolian sediment has focused on the fundamental sedimentology of the dust: the transport and depositional processes and the history of those processes provided by the deep-sea record. More recently, the geochemistry of dust has played an increasing role in the overall research effort. The overarching goal of paleoclimate studies based on the record of dust deposition is to understand the geologic history of continental climates and of the nature and intensity of the zonal wind systems. There are other ways to approach questions of continental climate, but there are no other direct records of the history and patterns of the wind systems (Rea, 1994).

Students of eolian deposition are faced with several problems from the outset. First, dust makes up only 1 or 2% of the 15 to 20×10^{15} grams of terrigenous sediment brought to the oceans every year. This mainly river-borne material becomes the hemipelagic mud (clayey silts that move offshore in suspension) that drapes the continental margins and extends out to sea for unknown distances. Since 1 or 2% of the river loading can overwhelm the eolian signal in a core, one must be certain that sediments used for eolian information are from cores raised far from land or are otherwise shielded from hemipelagic processes. The most useful sediment for paleoceanographic analyses is biogenic ooze, the fossil remains of sea-surface plankton. In such calcite- and/or opal-rich sediments, the dust makes up only a small percent of the total, and so must be isolated from the bulk sediment by a somewhat lengthy laboratory chemical procedure (Rea and Janecek, 1981; Clemens and Prell, 1990; Hovan, 1995). Upon completion of the extraction process, the mineral grains are available for determinations of flux (Mass Accumulation Rate (MAR), commonly given in mg/(cm^2 ky)), particle-size analysis, mineralogy, and geochemistry. It is particularly important to 'clean' the grains prior to geochemical analysis because the oxide coatings on the particles carry their own geochemical signature. As the MAR values are a product of the weight percent of the dust component, the ambient linear sedimentation rate, and the dry bulk density of the sediment, resulting values depend on the quality of all three determinations. In new or well-curated cores, the sedimentation rate is the least well-known of these values; the percent and bulk density values are generally good to $\pm5\%$ or so.

Interpretation of the Data

The mass accumulation rate of eolian dust is a good measure of source area aridity. Humid, vegetated regions put very little dust into the air, while semi-arid and arid regions supply considerable amounts. There are hyper-arid areas, where there is not enough water to break down the continental feldspars into clays, that generate little dust today; such regions have been deflated. We have been able to map the MAR of dust in the deep sea as based on the sedimentary record (Rea, 1994). These maps are similar to those provided by Duce *et al.* (1991) based on modern mineral aerosol transport and show dust flux maxima usually in excess of 1,000 mg/(cm^2 ky) downwind of the main source regions in central and eastern Asia, the Sahara and Sahael, and Arabia and the Horn of Africa. Since most of the land is in the Northern Hemisphere, most of the dust is also. The South Pacific, South Indian, and Southern Ocean are sites of minimal dust input, on the order of 1 mg/(cm^2 ky) or less.

Eolian dust has a grain size distribution that peaks at about 2 microns, with essentially no material coarser than 10 microns (Rea and Hovan, 1995). All the eolian grain-size work has assumed that the grain size is related to the intensity of the transporting wind, but clear evidence of this was not produced until recently when Clemens reported on sediment trap studies in the Indian Ocean, downwind from the Arabian deserts (Clemens, 1998). The size of the dust grains changes very little downwind, after the coarser materials have settled out, such that after perhaps 2,000 km from the original dust storm site, the size is uniform. Maps of eolian grain size show this remarkable uniformity of size across the entire North Pacific Ocean (Rea and Hovan, 1995). Another working hypothesis is that these grains are in equilibrium with the transporting winds, and that the stronger winds will transport a coarser equilibrium grain size of material.

There is one more point to make about these data. At the distal locations sampled in our deep-sea cores, the MAR and grain size of dust vary independently.

The closest analogy is with the competence and capacity of rivers. A large river like the Mississippi might carry a lot of mud, but it will all be fine-grained. Conversely, a mountain stream might move sand and gravel, but not in Mississippi-like quantities. Cross-plots of data from those cores, which have good flux and size information, show no correlation between the two kinds of data (Rea, 1994). This means that the suggestion that more dust and stronger winds are directly related is a misconception.

Spectral analysis of eolian grain size and flux records have demonstrated that both respond to orbital forcing, commonly at the periodicities of approximately 100 kyr (reflecting changing eccentricity of the orbit), 41 kyr (tilt of the spin axis with respect to the orbital plane) and 20 kyr (precession around the orbit), originally calculated by Milankovitch (1941). Milankovitch variability in dust grain size data was shown originally by Janecek and Rea (1984, 1985). The first detailed study was based on a core from the equatorial Pacific and demonstrated that the variability in the eolian grain size data was coherent and in phase with a radiolarian-based proxy for upwelling, quantitatively linking wind intensity and equatorial upwelling in this Quaternary core. The main spectral peak in the grain size and upwelling proxy data occurs at about 30 kyr, a cross-product of eccentricity and tilt periods, rather than one of the primary orbital periods (Pisias and Rea, 1988). Orbital forcing at approximately 30 kyr is a common finding in a variety of spectral-analysis studies of atmospheric and surface ocean proxy records (Rea, 1994).

Studies of a North Pacific core directly downwind from the China Loess Plateau documented orbital variability in both the flux data, which matches loess deposition (see below), and in the dust grain size data (Hovan *et al.*, 1991). The dust accumulation rate record showed spectral power at the primary orbital periods of about 100, 41, and 19 kyr, but eolian grain size data from this same core exhibit spectral peaks at 100, 50, and 25–35 kya (ibid.). Clemens and Prell (1990, 1991) analyzed sediments downwind from the Arabian Desert, to understand the monsoonal influence in the Gulf of Arabia. They show pronounced orbital power at the frequencies associated with tilt and precession, along with noticeable power at about 35 kyr and 54–59 kyr in the grain size record. The MAR of eolian dust varies on longer, 100 kyr and 19–23 kyr periods. This last observation is important because it means that wind intensity is not linked only to the 100-kyr glacial–interglacial cycles of the later Quaternary but varied on shorter periodicities, an observation consistent with the earlier work of Pisias and Rea (1988). The existence of important amounts of variability in the atmospheric transport process at periods other than the primary periodicities determined by Milankovitch (1941) remains a puzzle and is a continuing topic of research.

Much work has concentrated on determining the geochemistry of the dust, as a way to determine both provenance and, if there is a long record, changes in the nature of the eolian source region. Both elemental, largely rare-earth element information (Olivarez *et al.*, 1991; Weber *et al.*, 1996), and isotopic, based on the isotopes of Nd, Sr and Pb (Jones *et al.*, 1994, 2000; Pettke *et al.*, 2000, 2002) data have been presented. The first major finding of this work is that maps of the isotopic values of the mineral component of surficial sediments demonstrate that dust from Asia dominates North Pacific seafloor sediments, from China almost all the way east to North America and south to the Intertropical Convergence Zone (ITCZ) (Jones *et al.*, 1994, 2000). North America supplies surprisingly little dust to the North Pacific (Jones *et al.*, 1994, 2000). South of the ITCZ, dust has the geochemical signature of young andesite, implying a Central or South American source. This transition at the ITCZ is of interest because the first good downcore geochemistry on a red clay core, by Kyte *et al.* (1993), showed a change from a younger Asian source to an older American dust source some time during the late Oligocene. Since the core studied now lies at about 30°N and can be backtracked to about 20°N, this implies that the ITCZ once lay far north of its present location. One of the best examples of using the geochemistry of dust to 'fingerprint' the source regions has been the identification of the source of dust now found in the Antarctic (Argentine Patagonia, Grousset *et al.*, 1992) and Greenland (central Asia, Biscaye *et al.*, 1997) ice cores.

Quaternary records from the North Pacific

A long-standing and continuing goal of paleoclimatologists is to link together records from the land surface, and the ocean, the main engine of climate change. The primary paleoceanographic record of climate change, linked to every paleoceanographic proxy, is the oxygen isotope record. The classic land record of late Cenozoic climate change is the sequence of loess and soil horizons such as those that cover the Loess Plateau in China. The loess horizons form in cold, dusty glacial times, and the soils are the product of warmer and more humid, thus vegetated, interglacials. Loess deposits form a remarkably detailed record of climate in Asia and elsewhere. The only way in which the $\delta^{18}O$ and loess records can be joined is in a marine core that can be tied to a loess sequence and has a good $\delta^{18}O$ record. Hovan *et al.* (1989, 1991) were presented with this circumstance in the form of piston core V21-146

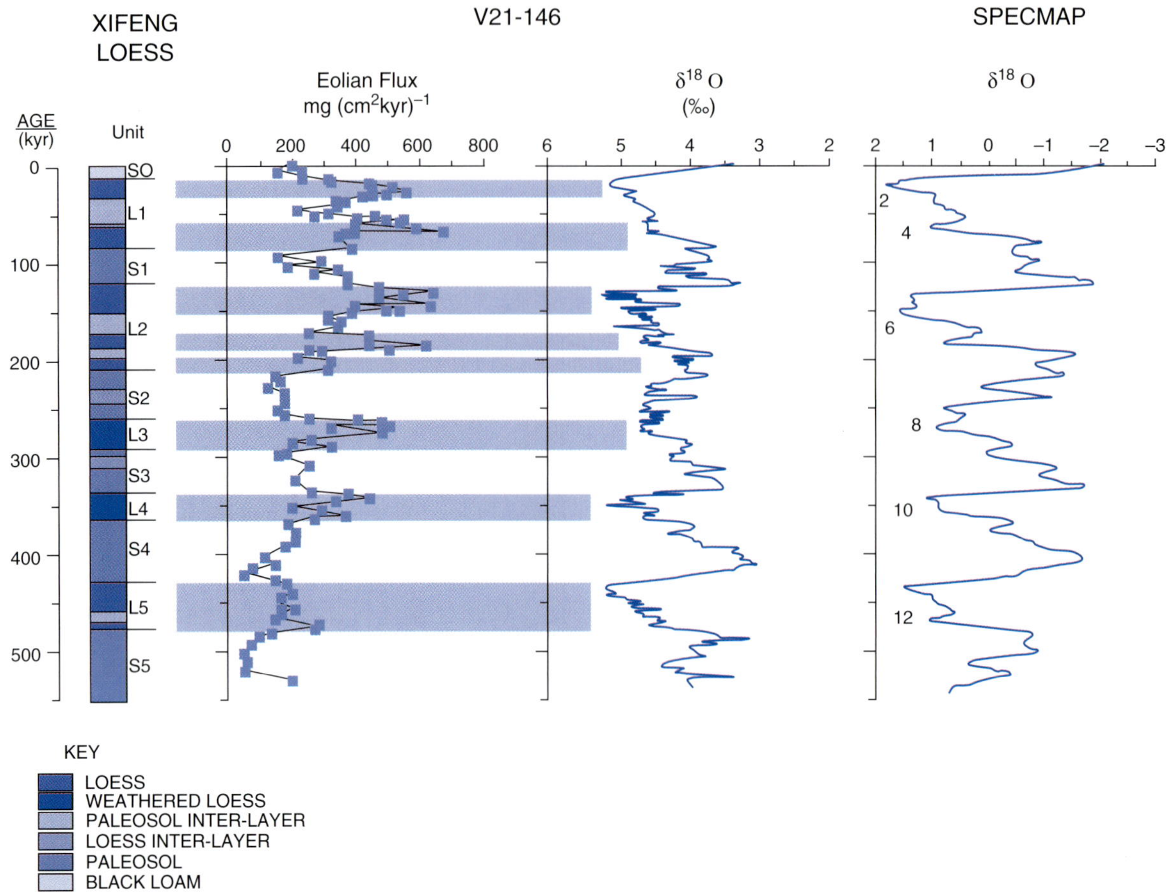

Figure 1 Dust accumulation record spanning the past 500 kyr from core V21-146 in the northwestern Pacific Ocean. Loess (L) and soil (S) horizons are shown at left are from the section at Xifeng. Dust flux maxima faithfully match the loess horizons. $\delta^{18}O$ records from the core, generated originally by Shackleton, show that dust flux maxima occur at glacial maxima as indicated by heavy values on the isotope curve. The global average isotope curve, the SPECMAP stack is shown for comparison. Glacial ages are identified with even numbers on the SPECMAP record. After Hovan *et al.* (1991).

from Shatsky Rise (37°41′N, 163°02′E) in the northwestern Pacific Ocean. The dust MAR record in V21-146 well matches the loess-soil records from China, and enables us to confirm by direct observation that loess maxima occur during the glacial maxima, and that, within the accuracy of the data sets, the $\delta^{18}O$ indication of maximum glaciation matches the loess deposition peak (**Fig. 1**, Hovan *et al.*, 1989, 1991).

Cenozoic records of atmospheric circulation and continental climate

All the good, longer records of dust accumulation are from sediment cores that now underlie the North Pacific prevailing westerlies. The first of these records to have been generated is still the most revealing, that from giant piston core LL44-GPC3 (30°19′N, 157°49′W). Janecek and Rea (1983) determined the 70-million-year-long record of continental aridity and wind intensity from this red clay core. The dust flux data (**Fig. 2**) show little input in Oligocene and older times, a doubling of the dust flux at about 24 Ma, and a large increase associated with late Cenozoic cooling and drying of the Northern Hemisphere. It was the grain-size record from this core that provided the important new information in this study. The middle and late Cenozoic grain size shows a coarsening beginning some time in the late Eocene, and continuing, perhaps in steps, until the present. The surprise in this record is the large change in grain size that occurs lower in the core. This large change in dust size occurs in other North Pacific cores, and has been dated as occurring exactly at the Paleocene–Eocene boundary (see Rea *et al.*, 1990). It records a remarkable reduction in wind intensity from latest Cretaceous and Paleocene winds similar to present, to a regime of very sluggish atmospheric circulation in the Eocene. These data are one of the fundamental pieces of information that underlie the grand quandary of Eocene climate: how, when both atmospheric and oceanic circulation appear sluggish,

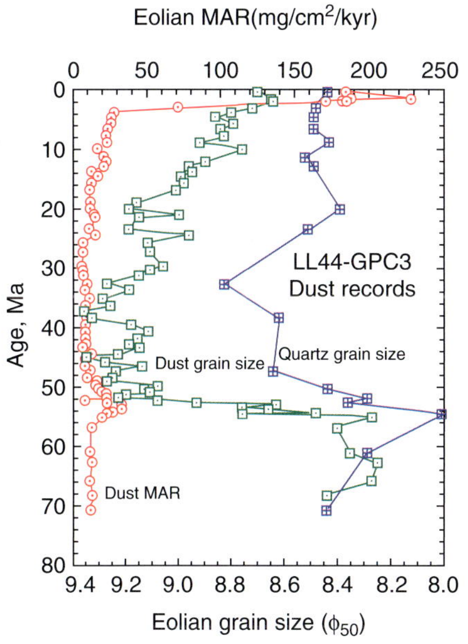

Figure 2 Dust accumulation rate (open circles) and grain size of dust (open squares) and of quartz grains (crossed squares) extracted from piston core LL44-GPC3, raised from the central North Pacific Ocean. Size is calculated as median grain size and shown in phi units (Φ_{50}).

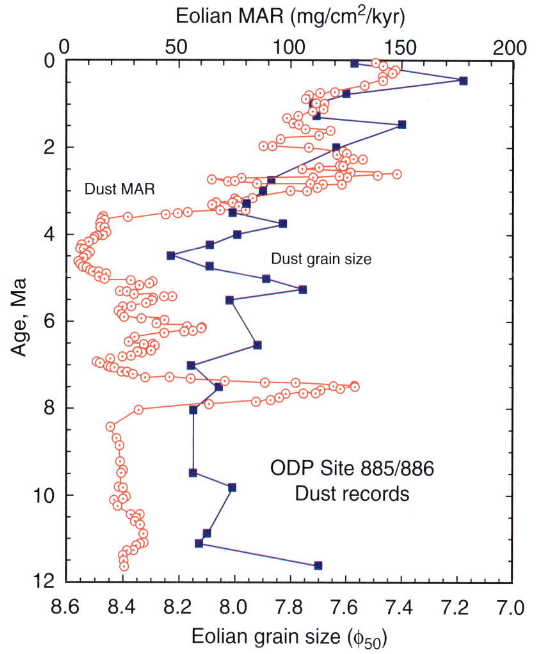

Figure 3 Dust accumulation rate (open circles) and grain size (solid squares) for the past 12 million years from the North Pacific. Size is calculated as median grain size and shown in phi units (Φ_{50}).

is it possible to transport enough heat to keep the polar regions warm enough to support crocodiles and palm trees? The question remains unanswered.

The details of late Cenozoic Asian aridity and wind intensity are best found in sediments recovered at North Pacific ODP Site 885/886 (44°41′N, 168°15′W). The section recovered at Site 885/886 is in biogenic ooze with a good magnetic reversal stratigraphy and so has far better age control than do the red clay cores farther south. Information from this site (**Fig.** 3; Rea *et al.*, 1998) shows a number of important aspects of late Cenozoic climate change. First, the dust flux data show an order of magnitude increase at 3.6 Ma, one million years before the onset of major northern hemisphere glaciation at about 2.6 Ma. This increase reflects the rapid drying of the central Asian basins north of the Tibetan uplift, as these regions were cut off by the rising plateau from their Indian Ocean moisture supply. Other work has shown that conglomeratic sediments sweep north from the north edge of the Tibetan Plateau at this time, a clear indication of rapid uplift then (Zheng *et al.*, 2004; Fang *et al.*, 2005).

The eolian grain size information from Site 885/886 tells a different story. Size and flux data vary independently from each other, as noted above. The record from this site, along with that from GPC3, is a good example of this at longer timescales. The grain-size data show that the increase in late Cenozoic wind intensity, presumably a response to high-latitude cooling, begins to increase at about 4.5 Ma, one million years before the dust flux increased. Together these records show that Northern Hemisphere glaciation did not occur suddenly and without prior indications of change. Rather a series of events, perhaps two or three million years in duration, lead up to the rapid buildup of continental ice sheets at about 2.6 Ma.

Summary

Dust forms only a small percentage of the sediment of the ocean floor, but in broad regions away from continents, the terrigenous component of deep-sea sediment is dominated by dust. These materials provide several kinds of paleoenvironmental information: a history of the aridity of the eolian source region in the MAR data, a history of the strength of the transporting winds in the grain size data, a record of provenance in the dust geochemistry and knowing provenance, the nature of past zonal winds. There is no other kind of direct proxy for the nature of past atmospheric circulation (Rea, 1994). As the atmosphere is responsible for half of the global heat transport, this kind of information is particularly valuable in paleoclimatic reconstructions.

Glossary

Red or Pelagic Clay A common seafloor deposit that consists of mineral dust grains and authigenic oxides and hydroxides. These deposits accumulate very slowly and are devoid of all biogenic material except fossil fish teeth.

Hemipelagic mud Silty sediment delivered to the ocean by rivers and transported off the continental shelf in the water column. The mud grains settle out across the outer continental shelf and the continental slope. This is the sediment that eventually forms shale and mudstone.

Biogenic ooze Sediment composed of the remains of surface-dwelling plankton. These sediments are some combination of calcium carbonate, $CaCO_3$ and biogenic opal, SiO_2nH_2O.

Spectral analysis a mathematical technique that identifies non-random variability in a sample set and discovers the amount and frequency of such fluctuations.

Milankovitch variability Milankovitch hypothesized that variations in the orbit of Earth associated with precession (at periods of 19 and 23 ka), tilt of the spin axis (41 ka), and eccentricity of the orbit (100 and 400 ka) had a pronounced influence on climate.

See also: **Glacial Climates**: Effects of Atmospheric Dust. **Glaciation, Causes**: Milankovitch Theory and Paleoclimate. **Loess Deposits, Origins and Properties**. **Loess Records**: China.

References

Biscaye, P. E., Grousset, F. E., Revel, M., Van der Gaast, S., Zielinsky, G. A., Vaars, A., and Kukla, G. (1997). Asian provenance of last glacial maximum dust in the GISP-2 ice core, Summit, Greenland. *Journal of Geophysical Research* **102**, 26,765–26,781.

Clemens, S. C. (1998). Dust response to seasonal atmospheric forcing: Proxy evaluation and calibration. *Paleoceanography* **13**, 471–490.

Clemens, S. C., and Prell, W. L. (1990). Late Pleistocene variability of Arabian Sea summer monsoon winds and continental aridity: Eolian records from the lithogenic component of deep-sea sediments. *Paleoceanography* **5**, 109–145.

Clemens, S. C., and Prell, W. L. (1991). Late Quaternary forcing of Indian Ocean summer monsoon winds: A comparison of Fourier model and general circulation model results. *Journal of Geophysical Research* **96**, 22,683–22,700.

Darwin, C. (1846). An account of the fine dust which often falls on vessels in the Atlantic Ocean. *Quarterly Journal of the Geological Society of London* **2**, 26–30.

Duce, R. A., Liss, P. S., Merrill, J. T., Atlas, E. L., Buat-Menard, P., Hicks, B. B., Miller, J. M., Prospero, J. M., Arimoto, R., Church, T. M., Ellis, W., Galloway, J. N., Hansen, L., Jickells, T. D., Knap, A. H., Reinhardt, K. H., Schneider, B., Soudine, A., Tokos, J. J., Tsunogai, S., Wollast, and Zhou, M. (1991). The atmospheric input of trace species to the world ocean. *Global Biogeochemical Cycles* **5**, 193–259.

Fang, X., Yan, M., Van der Voo, R., Rea, D. K., Song, C., Pares, J. M., Gao, J., Nie, J., and Dai, S. (2005). Late Cenozoic deformation and uplift of the NE Tibetan Plateau: Evidence from high-resolution magnetostratigraphy of the Guide Basin, Qinghai Province, China. *Geological Society of America Bulletin* **117**, 1208–1225.

Grousset, F. E., and Biscaye, P. E. (2005). Tracing dust sources and transport patterns using Sr, Nd and Pb isotopes. *Chemical Geology* **222**, 149–167.

Grousset, F. E., Biscaye, P. E., Revel, M., Petit, J. R., Pye, K., Joussaume, S., and Jouzel, J. (1992). Antarctic ice core dusts at 18 ky BP, Isotopic constraints on origins and atmospheric circulation. *Earth and Planetary Science Letters* **111**, 175–182.

Heath, G. R. (1969). Mineralogy of Cenozoic deep-sea sediments from the equatorial Pacific Ocean. *Geological Society of America Bulletin* **80**, 1997–2018.

Hovan, S. A. (1995). Late Cenozoic atmospheric circulation intensity and climatic history recorded by eolian deposition in the Eastern Equatorial Pacific, Leg 138. In (N. G. Pisias, L. A. Mayer, T. R. Janecek, A. Palmer Juslon and T. H. van Andel, Eds.) Proceedings of the Ocean Drilling Program, Scientific Results, Vol. 138, pp. 615–625. Ocean Drilling Program, College Station, TX.

Hovan, S. A., Rea, D. K., Pisias, N. G., and Shackleton, N. J. (1989). A direct link between the China loess and marine $\delta^{18}O$ records: aeolian flux to the north Pacific. *Nature* **340**, 296–298.

Hovan, S. A., Rea, D. K., and Pisias, N. G. (1991). Late Pleistocene continental climate and oceanic variability recorded in northwest Pacific sediments. *Paleoceanography* **6**, 349–370.

Janecek, T. R., and Rea, D. K. (1983). Eolian deposition in the northeast Pacific Ocean: Cenozoic history of atmospheric circulation. *Geological Society of America Bulletin* **94**, 730–738.

Janecek, T. R., and Rea, D. K. (1984). Pleistocene fluctuations in Northern Hemisphere tradewinds and westerlies. In *Milankovitch and Climate* (A. Berger, J. Imbrie, J. Hays, G. Kukla and B. Saltzman, Eds.), Part 1, pp. 331–347. D. Reidel Pub. Co, Dordrecht.

Janecek, T. R., and Rea, D. K. (1985). Quaternary fluctuations in the northern hemisphere tradewinds and westerlies. *Quaternary Research* **24**, 150–163.

Jickells, T. D., An, Z. S., Andersen, K. K., Baker, A. R., Bergametti, G., Brooks, N., Cao, J. J., Boyd, P. W., Duce, R. A., Hunter, K. A., Kawahata, H., Kubilay, N., la Roche, J., Liss, P. S., Mahowald, N., Prospero, J. M., Ridgewell, A. J., Tegen, I., and Torres, R. (2005). Global iron connections between desert dust, ocean biogeochemistry, and climate. *Science* **308**, 67–71.

Jones, C. E., Halliday, A. N., Rea, D. K., and Owen, R. M. (1994). Neodymium isotopic variations in North Pacific modern silicate sediment and the insignificance of detrital REE contributions to seawater. *Earth and Planetary Science Letters* **127**, 55–66.

Jones, C. E., Halliday, A. N., Rea, D. K., and Owen, R. M. (2000). Eolian inputs of lead to the North Pacific. *Geochimica et Cosmochimica Acta* **64**, 1405–1416.

Kyte, F. T., Leinen, M., Heath, G. R., and Zhou, L. (1993). Cenozoic sedimentation history of the central North Pacific: Inferences from the elemental geochemistry of core LL44-GPC3. *Geochimica et Cosmochimica Acta* **57**, 1719–1740.

Leinen, M. (1989). The pelagic clay province of the North Pacific Ocean. In (E. L. Winterer, D. M. Hussong and R. W. Drecker, Eds.). *The Eastern Pacific Ocean and Hawaii, Boulder, Colorado, Geological Society of America, The Geology of North America* **N**, 323–335.

Leinen, M., and Heath, G. R. (1981). Sedimentary indicators of atmospheric activity in the northern hemisphere during the Cenozoic. *Palaeogeography, Palaeoclimatology, Palaeoecology* **36**, 1–21.

Leinen, M., and Sarnthein, M. (Eds.) (1989). *Paleoclimatology and Paleometeorology: Modern and Past Patterns of Global Atmospheric Transport*, p. 909. Kluwer Academic Pubs, Dordrecht.

Martin, J. H. (1990). Glacial-interglacial CO_2 change: The iron hypothesis. *Paleoceanography* **5**, 1–13.
Martin, J. H., and Fitzwater, S. E. (1988). Iron deficiency limits phytoplankton growth in the north-east Pacific subarctic. *Nature* **331**, 341–343.
Milankovitch M (1941) *Kanon der Erdbestrahlung und seine Andwendung auf das Eiszeitenproblem. Koninglich Serbische Academie, Beorgrad,* 1–633. (English translation by Israel Program for Scientific Translations, Jerusalem, 1969.)
Olivarez, A. M., Owen, R. M., and Rea, D. K. (1991). Geochemistry of eolian dust in Pacific pelagic sediments: Implications for paleoclimatic interpretations. *Geochimica et Cosmochimica Acta* **55**, 2147–2158.
Parkin, D. W. (1974). Trade-winds during the glacial cycles. *Proceedings of the Royal Society of London, A* **337**, 73–100.
Parkin, D. W., and Padgham, R. C. (1975). Further studies on trade winds during glacial cycles. *Proceedings of the Royal Society of London, A* **346**, 245–260.
Parkin, D. W., and Shackleton, N. J. (1973). Trade wind and temperature correlations down a deep-sea core off the Saharan coast. *Nature* **245**, 455–457.
Pettke, T., Halliday, A. N., Hall, C. M., and Rea, D. K. (2000). Dust production and deposition in Asia and the North Pacific Ocean over the past 12 Myr. *Earth and Planetary Science Letters* **178**, 1–17.
Pettke, T., Halliday, A. N., and Rea, D. K. (2002). Cenozoic evolution of Asian climate and sources of Pacific seawater Pb and Nd derived from eolian dust of sediment core LL44-GPC3. *Paleoceanography* **17**, 3.1–3.13.
Pèwè TL (Ed.) (1981). *Desert Dust: Origin, Characteristics and Effect on Man.* Geological Society of America Special Paper 186, Boulder, Colorado, pp. 303.
Pisias, N. G., and Rea, D. K. (1988). Late Pleistocene paleoclimatology of the central Equatorial Pacific: Sea-surface response to the southeast tradewinds. *Paleoceanography* **3**, 21–37.
Prospero, J. M., and Nees, R. (1977). Dust concentration in the atmosphere of the equatorial North Atlantic: Possible relationship to Sahelian drought. *Science* **196**, 1196–1198.
Prospero, J. M., and Nees, R. T. (1986). Impact of the North African drought and El Niño on mineral dust in the Barbados trade winds. *Nature* **320**, 735–738.
Prospero, J. M., Bonatti, E., Schubert, C., and Carlson, T. N. (1970). Dust in the Caribbean atmosphere traced to an African dust storm. *Earth and Planetary Science Letters* **9**, 287–293.
Prospero, J. M., Charlson, R. J., Mohnen, V., Jaenicke, R., Delany, A. C., Moyers, J., Zoller, W., and Rahn, K. (1983). The atmospheric aerosol system: An overview. *Reviews of Geophysics and Space Physics* **21**, 1607–1629.
Prospero, J. M., Ginoux, P., Torres, O., Nicholson, S. E., and Gill, T. E. (2002). Environmental characterization of global sources of atmospheric soil dust indentified with the Numbus 7 total ozone mapping spectrometer (TOMS) absorbing aerosol product. *Reviews of Geophysics* **40**(1), doi:10.1029/2000RG000095.
Pye, K. (1987). *Aeolian Dust and Dust Deposits*, p. 334. Academic Press, London.
Radczewski, O. E. (1939). Eolian deposits in marine sediments. In *Recent Marine Sediments* (P. D. Trask, Ed.), pp. 496–502. American Association of Petroleum Geologists, Tulsa, OK.
Rea, D. K. (1994). The paleoclimatic record provided by eolian deposition in the deep sea: the geologic history of wind. *Reviews of Geophysics* **32**, 159–195.
Rea, D. K., and Hovan, S. A. (1995). Grain size distribution and depositional processes of the mineral component of abyssal sediments: Lessons from the North Pacific. *Paleoceanography* **10**, 251–258.
Rea, D. K., and Janecek, T. R. (1981). Mass accumulation rates of the non-authigenic inorganic crystalline (eolian) component of deep-sea sediments from the western Mid-Pacific Mountains, Deep Sea Drilling Project Site 463. In (J. Thiede and T. L. Vallier, *et al.*, Eds.) Initial Reports of the Deep Sea Drilling Project, 62, pp. 653–659. U.S. Government Printing Office, Washington, D.C.
Rea, D. K., Leinen, M., and Janecek, T. R. (1985). Geologic approach to the long-term history of atmospheric circulation. *Science* **227**, 721–725.
Rea, D. K., Snoeckx, H., and Joseph, L. H. (1998). Late Cenozoic eolian deposition in the North Pacific: Asian drying, Tibetan uplift, and cooling of the Northern Hemisphere. *Paleoceanography* **13**, 215–224.
Rea, D. K., Zachos, J. C., Owen, R. M., and Gingerich, P. D. (1990). Global change at the Paleocene-Eocene boundary: climatic and evolutionary consequences of tectonic events. *Palaeogeography, Palaeoclimatology, Palaeoecology* **79**, 117–128.
Rex, R. W., and Goldberg, E. D. (1958). Quartz contents of pelagic sediments of the Pacific Ocean. *Tellus* **10**, 153–159.
Ruddiman, W. F., Sarnthein, M., Backman, J., Baldauf, J. G., Curry, W., DuPont, L. M., Janecek, T., Pokras, E. M., Raymo, M. E., Stabell, B., Stein, R., and Tiedemann, R. (1989). Late Miocene to Pleistocene evolution of climate in Africa and the low-latitude Atlantic: Overview of Leg 108 results. In (W. Ruddiman and M. Sarnthein *et al.*, Eds.) Proceedings of the Ocean Drilling Program, Scientific Results, Vol. 108, pp. 463–484. Ocean Drilling Program, College Station, TX.
Sarnthein, M., Tetzlaff, G., Koopmann, B., Wolter, K., and Pflaumann, U. (1981). Glacial and interglacial wind regimes over the eastern Subtropical Atlantic and North-West Africa. *Nature* **293**, 193–196.
Sarnthein, M., Thiede, J., Pflaumann, U., Erlenkeuser, H., Futterer, D., Koopmann, B., Lange, H., and Seibold, E. (1982). Atmospheric and oceanic circulation patterns off Northwest Africa during the past 25 million years. In *Geology of the Northwest African Continental Margin* (U. von Rad, K. Hinz, M. Sarnthein and E. Seibold, Eds.), pp. 545–604. Springer-Verlag, Berlin.
Sirocko, F., and Sarnthein, M. (1989). Wind-borne deposits in the northwestern Indian Ocean: Record of Holocene sediments versus modern satellite data. In *Paleoclimatology and Paleometeorology: Modern and Past Patterns of Global Atmospheric Transport* (M. Leinen and M. Sarnthein, Eds.), pp. 401–433. Kluwer Academic Publishers, Dordrecht.
Sirocko, F., Sarnthein, M., Lange, H., and Erlenkeuser, H. (1991). Atmospheric summer circulation and coastal upwelling in the Arabian Sea during the Holocene and the last glaciation. *Quaternary Research* **36**, 72–93.
Sokolik, I. N., Winker, D. M., Bergametti, G., Gillette, D. A., Carmichael, G. R., Kaufman, Y. J., Gomes, L., Schuetz, L. and Penner, J. E. (Eds.) (2001). *Quantifying the radiative impacts of mineral dust (DUST)*, 106, Journal of Geophysical Research, D, Atmospheres **106**, pp. 18,015–18,493.
Weber, E. T., II, Owen, R. M., Dickens, G. R., Halliday, A. N., Jones, C. E., and Rea, D. K. (1996). Quantitative resolution of eolian continental crustal material and volcanic detritus in North Pacific surface sediment. *Paleoceanography* **11**, 115–127.
Zheng, H., Powell, C., McA, Rea, D. K., Wang, J., and Wang, P. (2004). Late Miocene onset and mid-Pliocene enhancement of the East Asian Monsoon as viewed from land and sea. *Global and Planetary Change* **41**, 147–155.

Equilibrium Line Altitude (ELA) Reconstruction *see* **Glacial Landforms, Ice Sheets**: Paleo ELAs

Erratics, Glacial *see* **Glacial Landforms, Sediments**: Glacial Erratics and Till Dispersal Indicators

Extinctions, Quaternary Vertebrates *see* **Vertebrate Records**: Late Pleistocene Megafaunal Extinctions

F

Felsenmeer (Blockfields) *see* **Periglacial Landforms, Rock Forms**: Block/Rock Streams

FISSION-TRACK DATING

J A Westgate, University of Toronto, ON, Canada
N D Naeser, National Center United States Geological Survey, VA, USA
B Alloway, GNS Science, Lower Hutt, New Zealand

© 2007 Elsevier B.V. All rights reserved.

Introduction

A fission track is a linear region of intense radiation damage formed by the passage of an energetic, charged nuclear particle through an insulating solid. According to a popular hypothesis, known as the 'ion explosion spike' model (Fleischer *et al.*, 1975), the charged particle collides with atoms in its path, displaces them and removes their electrons, producing a narrow zone rich in positive ions, which repulse each other into interstitial positions and form vacancies. An elastic adjustment of the crystal or glass then occurs, straining the undamaged matrix (Fleischer, 2004). Another idea, called the 'thermal spike' model, sees the fast-moving fission fragment producing intense heat along its trajectory. After passage of the charged particle, the track core is quenched by heat loss to the adjacent lattice, leaving it in a disordered state. According to Chadderton (2003), it is likely that both mechanisms are involved in track formation. Maximum density damage exists at the site of fission and decreases to the track termini. This damaged zone persists in the solid after the fission fragment has come to rest and is called a latent track. Its length is equal to the sum of the range of the two charged particles formed by nuclear fission, each moving in opposite directions. ^{238}U fission events form latent tracks that are 6–10 μm wide and 10–20 μm long and account for most tracks in natural, terrestrial materials. Latent tracks can only be observed with a transmission electron microscope but can be made visible by chemical etching (Young, 1958) and readily seen under an optical microscope. Etching takes place by rapid dissolution of the disordered region of the track core which exists in a state of higher free energy than the undamaged material. As can be seen in **Figure 1**, the shape of fission tracks varies in different materials, due mainly to differences in symmetry and magnitude of the difference between the track and bulk etch rates. Fission tracks in glass are relatively wide, conical pits due to its isotropic properties and a track etch rate that is not much greater than the bulk etch rate. Micas have diamond-shaped tracks, whereas tracks in zircon and apatite are needle-like because of the large difference between track and bulk etch rates, the former being much greater.

Fission tracks have varied uses ranging from applications in the fields of engineering, medicine, nuclear physics, and space sciences to the earth sciences (Fleischer, 1998). The most important applications in the earth sciences are thermochronology (Reiners and Ehlers, 2005) and geochronology. Thermochronology – the more active field at present – focuses on understanding the thermal history of rocks and duration and rates of geological processes, such as crustal uplift. Geochronology, on the other hand, is primarily concerned with formation age, the topic of interest to us here.

Zircon and glass are the most suitable phases for dating Quaternary deposits by fission-track methods, although apatite and sphene may also be useful. The most desirable situation for dating Quaternary sediments is where they are associated with silicic tephra beds because not only can they be reliably dated using their zircon and glass shards, but their widespread distribution and common distinctiveness make tephra beds excellent markers, linking the stratigraphies of isolated sites, sometimes over distances exceeding 1,000 km (*see* Tephrochronology).

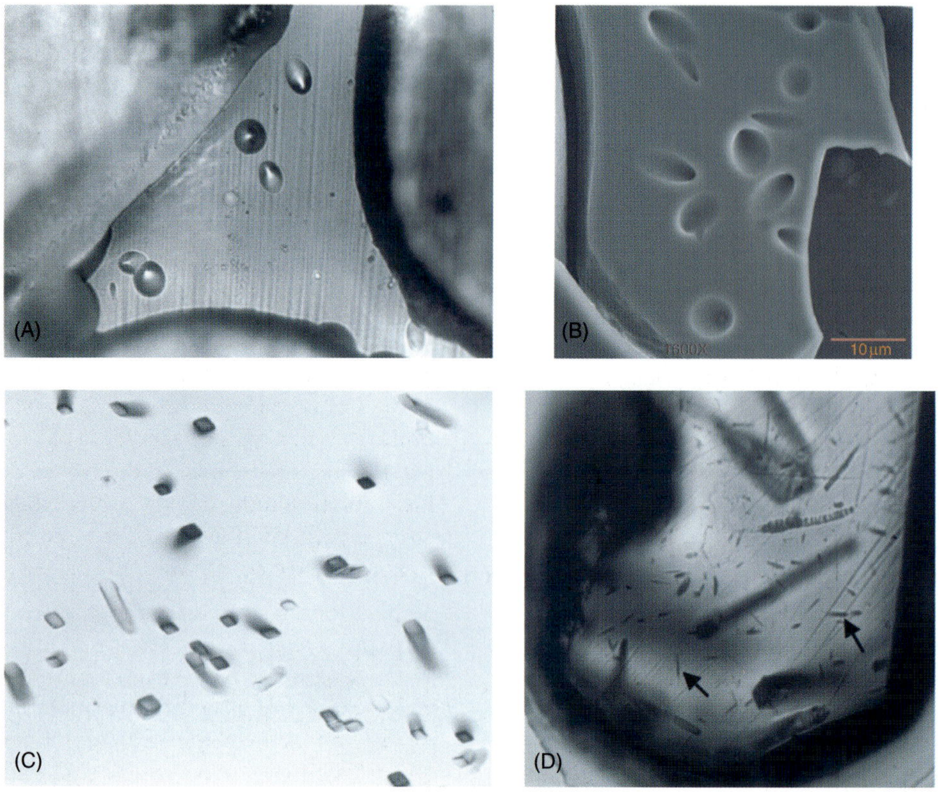

Figure 1 Fission tracks in glass shards, mica, and zircon. (A) Fission tracks (6–8 μm wide) in glass shard as seen using an optical microscope; (B) SEM image of fission tracks in glass shard (darker areas are the deepest part of the conical pits); (C) fission tracks in mica (~5 μm wide); (D) needle-like fission tracks in zircon from the Bishop Tuff, California (crystal is ~0.2 mm wide).

We outline in this article the basic principles of fission-track (FT) dating, detail the different methods used for glass and zircon, describe the correction procedures for partial track fading in glasses, comment on the accuracy and precision of FT dates, and compare the results obtained to those of other dating methods, especially the $^{40}Ar/^{39}Ar$ technique (*see* K/Ar and Ar/Ar Dating). Finally, we illustrate the application of FT dating to the Quaternary sciences by reference to a few case studies.

Age Equation

In principle, FT dating is similar to other radioisotopic dating methods except that the decay product of interest is the spontaneous fission track, not the daughter nuclide. If the rate of decay of ^{238}U by spontaneous fission is known (the decay constant, λ_F), then we need only to know the U concentration and abundance of spontaneous fission tracks, assuming all are preserved, to calculate an age (Price and Walker, 1963). The basic age equation for an FT age, t, given by Wagner and Van den haute (1992) is

$$t = 1/\lambda_D \ \ln\{1 + (\lambda_D/\lambda_F)(\rho_s/\rho_i)QGI\sigma_F\varphi\} \quad [1]$$

where λ_D is the total decay constant for ^{238}U ($1.551 \times 10^{-10}\,yr^{-1}$), λ_F is not well determined (values cluster around $7 \times 10^{-17}\,yr^{-1}$ and $8.5 \times 10^{-17}\,yr^{-1}$ (Bigazzi, 1981)), ρ_s is the areal track density of etched spontaneous tracks, ρ_i is the areal track density of etched induced fission tracks, Q is a procedural factor involving track registration and observation efficiency ($Q = 1$ if track etching and counting procedures are identical), G refers to the initial geometry ratio of the etched surfaces for counting spontaneous and induced tracks ($G = 1$ or 0.5 depending on the dating procedure used), I is the isotopic abundance of $^{235}U/^{238}U$ (7.2527×10^{-3}), σ_F is the cross section for induced nuclear fission of ^{235}U by thermal neutrons ($580.2 \times 10^{-24}\,cm^2$), and φ is the thermal neutron fluence, measured by the induced FT density on a U-doped glass (e.g., NIST 612), ρ_D, irradiated together with the sample, and is equal to $B\rho_D$, where B is a calibration constant that is empirically determined. The irradiation procedure is used to enable the U concentration to be determined through the known relative abundance of ^{235}U to ^{238}U because only ^{235}U fissions when irradiated with thermal neutrons. Hence, given these constants, an FT age is obtained by measuring three areal track densities, ρ_s, ρ_i, and ρ_D. Due to the uncertainties in the values of λ_F and σ_F, and the dependency of ϕ on the nuclear reactor characteristics, especially the degree of thermalization, use of the zeta calibration based on analysis of age standards was recommended by Hurford and Green (1983) and is now widely used. The zeta age calibration factor, ζ, is defined by

$$\zeta = B(I\sigma_F/\lambda_F) \quad [2]$$

Substituting [2] in [1], and considering $G = 1$ and $Q = 1$, gives

$$t = 1/\lambda_D \ \ln\{1 + \lambda_D(\rho_s/\rho_i)\zeta\rho_D\} \quad [3]$$

Rearranging [3] for a standard of known age (t_{std}) gives

$$\zeta = \{\exp(\lambda_D t_{std}) - 1\}/\{\lambda_D(\rho_s/\rho_i)_{std}\rho_D\} \quad [4]$$

The value of ζ can then be used in eqn [3] to give the age of a sample. If the grains within a sample have a common age, the uncertainty on the age is given by (Green, 1981)

$$\sigma(t)/t = \{1/N_s + 1/N_i + 1/N_D + [\sigma(\zeta)/\zeta]^2\}^{1/2} \quad [5]$$

where $\sigma(t)$ is the $\pm 1\sigma$ uncertainty on the age, $\sigma(\zeta)$ is the $\pm 1\sigma$ uncertainty on ζ; N_s, N_i, and N_D are the total number of tracks counted for ρ_s, ρ_i, and ρ_D, respectively, and the area is determined using an eyepiece graticule. If the area is estimated by use of the point-counting technique, commonly the case with glass shards, an additional experimental error is involved. In this case, the relative standard error (se) on the age (t) is given by (Bigazzi and Galbraith, 1999):

$$\begin{aligned} se(t)/t \approx \{&1/Y_s + [1 - (X_s/n_s)]/X_s \\ &+ 1/Y_i + [1 - (X_i/n_i)]/X_i + 1/Y_D\}^{1/2} \end{aligned} \quad [6]$$

where Y_s, Y_i, and Y_D are the total number of spontaneous and induced tracks counted in the sample, and the number of tracks counted in the fluence monitor, respectively. X_s and X_i are the total number of points on glass in determining ρ_s and ρ_i, and n_s and n_i are the corresponding totals for the fields of view.

Prerequisite Conditions for FT Geochronology

The mineral or glass to be dated by the FT method must meet several criteria (Faure, 1977; Dumitru, 2000). The U concentration must be sufficient to give a statistically significant age. Tracks, once formed, must be stable at ambient temperatures and not fully anneal soon after formation. Partial fading of tracks is common in some minerals and glasses, but correction procedures have been developed to account for this condition. Ideally, the material should be free of microlites and other inclusions, defects, and lattice dislocations, which, upon etching, can resemble fission tracks and make track identification and counting difficult (Storzer and Wagner, 1982). The crystal or glass surface area must be large enough to permit an accurate measurement of the areal track densities. For example, the pumiceous grains in **Figure 2A** cannot be dated, but the poorly vesicular, blocky glass shards in **Figure 2B** can be readily dated by fission tracks. The lower limit in grain size is ~80 μm. Ideally, U must be distributed uniformly throughout the material. In some materials, this requirement can now be assessed directly by laser ablation inductively coupled plasma-mass spectrometry (LA-ICP-MS) (Pearce *et al.*, 2004). The external detector method of FT dating, described

Figure 2 (A) Pumiceous grains from the late Pleistocene Sheep Creek-K tephra, Yukon, Canada. (B) Poorly vesicular, blocky glass shards of the middle Pleistocene Gold Run tephra, Yukon, Canada. Sheep Creek-K tephra cannot be dated by glass FT methods because of insufficient glass surface area. Gold Run tephra, on the other hand, is readily dated by fission tracks (**Table 1**).

below, can compensate for the inhomogeneous U distribution typical of minerals such as zircon. The sample must not have suffered any post-emplacement (depositional) heating that would cause resetting of its fission tracks. The nature of the sample's geologic environment would offer insights on the likelihood of such an event, as does the size distribution of its spontaneous tracks. The mineral or glass must be unaltered and form a primary constituent of the rock or sediment being dated. For example, tephra deposits, commonly dated by the FT method, typically contain foreign grains of xenocrystic, lithic, and detrital origin, which must be avoided during the track-counting operation. This requires use of grain-specific methods for FT counting. As with all geochronologic studies, the stratigraphic, and, in some cases, sedimentologic context of the deposit must be well established. Materials that satisfy these criteria are glass (obsidian, tektites, volcanic glass shards, and glasses made by humans), zircon, apatite, and sphene, although the latter two are seldom used to date Quaternary events because of low U and rarity.

Glass FT Methods

General Considerations

One of the earliest successful examples of FT dating involved the use of glass shards to determine the age of Bed 1 at Olduvai Gorge (Fleischer *et al.*, 1965a, b). The FT age of 2.0 ± 0.3 Ma agreed with the K–Ar age of 1.75 ± 0.05 Ma, obtained earlier by Leakey *et al.* (1961). Silicic, hydrated glass shards are an ideal medium for FT dating. They are the most abundant phase of tephra beds; they are mostly homogeneous and seldom contain microlites or inclusions; they are isotropic with uniform etching characteristics; U contents typically range from 2 to 8 ppm for rhyolitic to dacitic glasses, enough to produce a sufficient number of spontaneous tracks for a statistically useful Quaternary age; rapid cooling during the eruption and subsequent storage at near-surface temperatures mean that acceptable formation ages can be obtained; proven correction methods for partial track fading exist; and glass shards as small as $\sim$80 μm can be dated, opening up the geochronology of regions distant from volcanic centers (especially those of calc-alkaline composition where K-rich minerals are rare) because their parent silicic tephra beds are commonly very widespread. It must be admitted, however, that FT dating late Quaternary events by rhyolitic glass shards is a challenge, although possible, if long counting times are accepted (e.g., Kohn (1979); 20.3 ± 7.1 ka), or the glass is unusually rich in U (Alloway *et al.* (2004a); 50 ± 3 ka). Similarly, glass shards of andesitic and basaltic composition commonly have U contents <1 ppm U and cannot be used to date Quaternary events. Obsidians are much more suitable in this respect because of the large surface areas for counting that are available. Bigazzi and Bonadonna (1973) dated Lipari obsidians in the range of 1.4–11.4 ka, and U-rich archeological glasses of late historic age have been dated by fission tracks (Wagner and Van den haute, 1992).

Whereas the problem of contamination in obsidian and tektite glasses is negligible, it is an issue when FT dating glass shards and zircon from tephra beds, which are very susceptible to reworking (**Fig. 3**). Grain-specific methods for compositional characterization and FT dating must be used to neutralize this problem. Tight compositional clustering of glass shards, as seen in Old Crow tephra (**Fig. 4**), shows little if any contamination, but compositional outliers exist in the Dominion Creek tephra and must be addressed before an accurate FT age can be determined (**Fig. 4**). In the latter case, the contaminating glass shards possessed a notably higher track density, allowing them to be recognized and ignored. Samples that show multiple compositional clustering due to magma-mixing or post-depositional reworking cannot be reliably dated by glass-FT methods. Samples with minor contamination are tractable, however, because glass-FT dating methods are grain specific in that the track data are accumulated by examination of individual grains so that any foreign material that is encountered during the counting process can be readily ignored.

A unimodal major-element composition of glass shards in a sample (e.g., Old Crow tephra, **Fig. 4**) argues for U homogeneity. However, a direct measurement of U homogeneity on a shard-to-shard basis can now be made by laser ablation ICP-MS (Pearce *et al.*, 2004), and two examples are shown in **Figure 5**. Uranium contents range from 4 to 12 ppm for both the Lava Creek B and Huckleberry Ridge tephra beds with the average U values obtained by solution ICP-MS and LA-ICP-MS being very similar. In FT dating, U is conventionally measured by irradiating an aliquot of the sample with a specified dose of thermal neutrons that induce fission of a portion of the ^{235}U, which can then be used to determine the amount of ^{238}U in the sample because the ratio of ^{238}U to ^{235}U is a fixed number (I in Eqn. [1]). Thousands of glass shards are scanned to give ρ_i so that a very good average value of the ^{238}U content is obtained, despite the large range in U values seen in **Figure 5**. Further, ρ_s is determined under identical conditions so that ρ_s/ρ_i in age Eqn [1] can be regarded as accurately determined, and proved to be so by the excellent agreement between corrected glass-FT and ^{40}Ar/^{39}Ar dating methods for Huckleberry Ridge

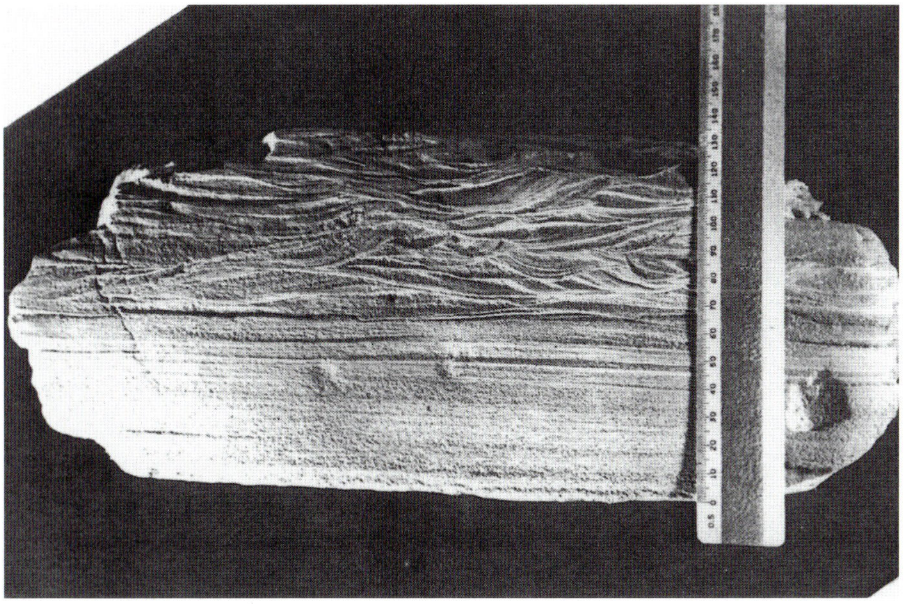

Figure 3 Photograph of Huckleberry Ridge tephra. The lower part of the bed is primary air-fall tephra with horizontal bedding. The upper, cross-bedded part is reworked material washed into the site, and is, therefore, more likely to be contaminated by detrital grains. Scale is mm ruler. Photo by Glen A. Izett.

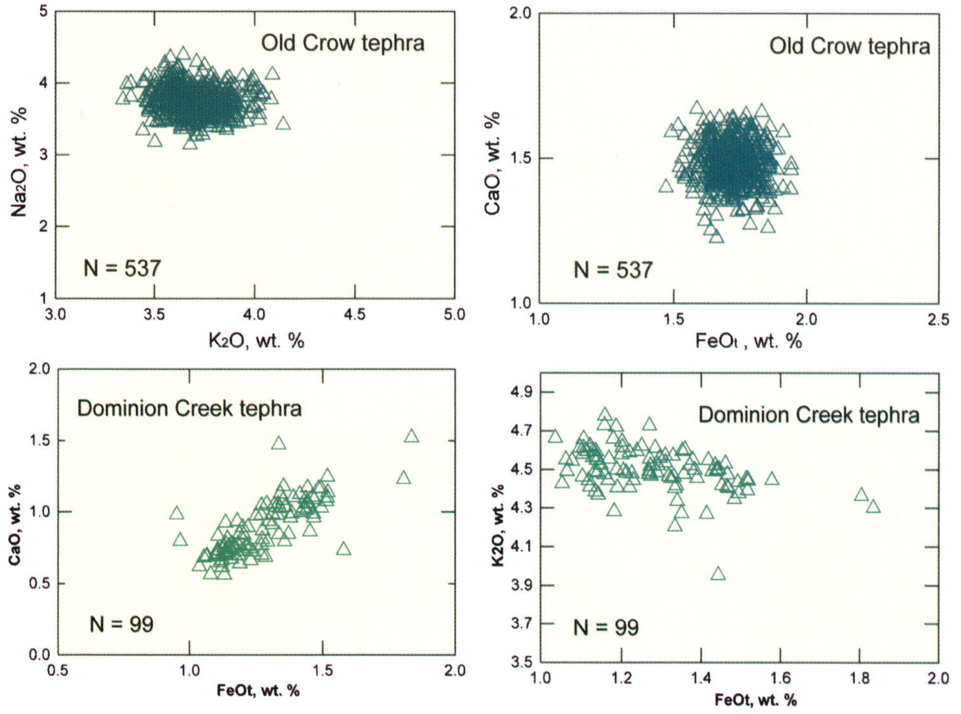

Figure 4 Oxide-variation diagrams showing major-element composition of glass shards from late Pleistocene Old Crow and Dominion Creek tephra beds, Yukon Territory, Canada. Tight clustering in Old Crow tephra demonstrates that contamination is very minor, if present. Several compositional outliers occur in Dominion Creek tephra indicating presence of contamination that must be addressed before an accurate FT age can be obtained. *N* is the number of individual glass-shard microprobe analyses.

tephra (1.97 ± 0.22 and 2.003 ± 0.014 Ma, respectively). In a recent development, FT ages of apatite grains are now being determined using U concentrations determined by LA-ICP-MS (Hasebe *et al.*, 2004). Future investigations should explore the viability of this approach for glass shards, obsidians, and tektites, using both average U values from a large number of LA-ICP-MS analyses as well as solution ICP-MS.

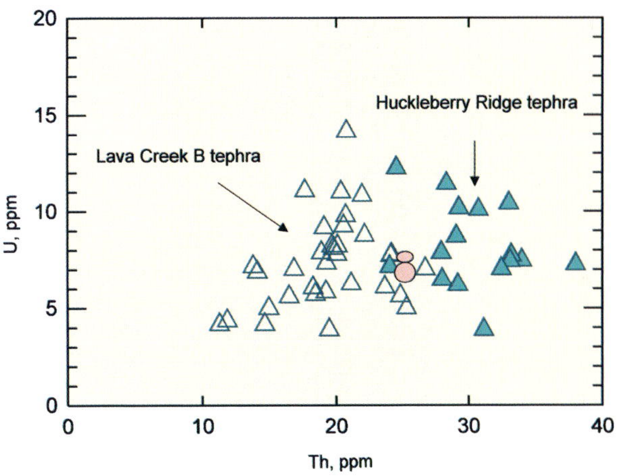

Figure 5 U and Th contents in glass shards from Lava Creek B and Huckleberry Ridge tephra beds, USA, as determined by LA-ICP-MS. Average values for these elements given by solution ICP-MS are shown by pink circle (HRt) and pink oval (LCBt). Average U concentrations by each method are very similar: HRt – 8.2 ppm by LA-ICP-MS, 6.7 ppm by solution ICP-MS; LCBt – corresponding values are 7.4 ppm and 7.5 ppm. Agreement in the average Th values by these two methods is obviously not as good.

There is only one formally accepted glass age standard, Moldavite (Balestrieri *et al.*, 1998). Because the I.U.G.S. Subcommission on Geochronology specified the use of splits from a single Moldavite (Hurford, 1990), and no supplier is known who can provide a sample large enough for distribution to the FT community, in essence, no formal glass age standard presently exists! Other necessary attributes of a formal age standard are that it contains sufficient tracks to give a statistically meaningful age, and that it has not suffered from partial track fading. Whether such a glass exists or not is debatable. The candidate that comes closest to these requirements is the Jankov Moldavite. Other potential age standards recommended by Balestrieri *et al.* (1998) are Roccastrada R2V, Mancusanite, and the Japanese glass JAS-G1. The University of Toronto Laboratory currently uses the Huckleberry Ridge tephra from Meade County, Kansas, USA, to monitor accuracy of its FT age determinations (**Table 1**). This tephra bed has large glass shards with no inclusions, high U content (**Fig. 5**), an $^{40}Ar/^{39}Ar$ age of 2.003 ± 0.014 (2σ) Ma (Gansecki *et al.*, 1998), abundant spontaneous fission tracks, and partial track fading defined by $D_s/D_i = 0.80$ making it a useful informal reference glass.

Population FT Method

The glass-population method can be used for homogeneous samples. In the case of glass shards, the coarsest, least vesicular fraction is obtained using sieves combined with magnetic and heavy-liquid techniques. Obsidian and tektite samples are gently crushed and sieved to concentrate the 0.25–0.125 mm fraction. The sample is split into two aliquots, one of which is heated to remove the spontaneous tracks and irradiated. Both aliquots are then mounted on glass slides with epoxy, polished to expose internal surfaces within the grains, and etched in HF to reveal the tracks. The spontaneous areal track density, ρ_s, is obtained from the unheated shards, and ρ_i from the heated, irradiated shards. Acceptable ages were determined on the Macusanite obsidian using this method (Osorio *et al.*, 2003), the samples being heated for 10 h at 560° C in order to erase the spontaneous tracks. However, for a number of reasons, this method is not recommended for hydrated glass shards. The differential heat treatment means that track-etching efficiencies for ρ_s and ρ_i are different. Heating above 200° C drives off water and produces brittle glass shards, which fracture easily when polished, thereby reducing the glass surface area available for counting because of the larger surface area exposed to chemical attack during etching. Crazed surfaces are also produced and hinder track recognition (Naeser *et al.*, 1980).

Population-Subtraction Method

The important distinction of this method is that the two aliquots are not subjected to any heat treatment prior to irradiation. The induced areal track density, ρ_i, is calculated by determining the areal track density of the irradiated sample and then subtracting ρ_s. An example of an FT age determination using this method is shown in **Table 1**, which also demonstrates the information that should be provided in the documentation of a glass-FT age determination.

Correction Methods for Partial Track Fading in Glasses

Introduction

When a glass or mineral with latent tracks are held at high temperatures for a sufficient period of time, the disordered structure becomes restored, presumably by a diffusion mechanism. A number of factors influence the stability of latent nuclear tracks in solids, but temperature is by far the most important (Fleischer *et al.*, 1965a). As latent tracks are progressively heated, they gradually shorten. For internally distributed tracks, the observed areal track density at a surface depends on the etchable length of the tracks and so this length reduction is accompanied by an areal track density reduction. Thus, if the range of spontaneous fission tracks has been lowered by

Table 1 Glass-fission-track age of Gold Run tephra, Klondike goldfields, Yukon Territory, Canada

Sample number	*Date irradiated*	*Spontaneous track density* (10^2 t cm^{-2})	*Corrected spontaneous track density* (10^2 t cm^{-2})	*Induced track density* (10^5 t cm^{-2})	*Track density on muscovite detector over dosimeter glass* (10^5 t cm^{-2})	*Etching conditions* HF : temp : time (% : C : s)	D_s(μm)	D_i (μm)	D_s/D_i or D_i/D_s^a	*Age* (Ma)
Gold Run tephra										
UT1791	21/01/2002	4.09 ± 0.28 (211)		1.10 ± 0.01 (12,976)	5.15 ± 0.03 (31,171)	24 : 22 : 110	5.39 ± 0.10 [155]	6.24 ± 0.06 [848]	0.86 ± 0.02	0.61 ± 0.06
UT1791[b]	21/01/2002		4.75 ± 0.33 (211)	1.10 ± 0.01 (12,976)	5.15 ± 0.03 (31,171)	24 : 22 : 110	5.39 ± 0.10 [155]	6.24 ± 0.06 [848]	1.16 ± 0.02[a]	0.71 ± 0.07[b]
UT1907[c]	25/06/2004	3.01 ± 0.28 (119)		0.60 ± 0.01 (6,780)	4.90 ± 0.04 (15,640)	24 : 21 : 200	6.46 ± 0.18 [94]	6.81 ± 0.09 [398]	0.95 ± 0.03	0.78 ± 0.09[c]
								Weighted mean age		0.74 ± 0.06
Internal Standard: Huckleberry Ridge tephra										
UT1366[c]	21/01/2002	27.35 ± 1.27 (520)		2.21 ± 0.02 (14,722)	5.01 ± 0.03 (31,171)	24 : 22 : 160	5.89 ± 0.10 [212]	6.26 ± 0.07 [597]	0.94 ± 0.01	1.97 ± 0.22[c]

[a]Indicates application of D_i/D_s in DCFT age determination.

[b]Sample corrected for partial track fading by the track-size (DCFT) method (Sandhu and Westgate 1995); uncorrected age noted simply by sample number.

[c]Samples corrected for partial track fading by thermal pretreatment, the ITPFT method (Westgate 1989); samples heated for 30 days at 150 °C. Ages calculated using the zeta approach and $\lambda_D = 1.551 \times 10^{-10}$ yr^{-1}. Zeta value is 318 ± 3 based on six irradiations at the McMaster Nuclear Reactor, Hamilton, Ontario, using the NIST SRM 612 glass dosimeter and the Moldavite tektite glass (Lhenice locality) with an $^{40}Ar/^{39}Ar$ plateau age of 15.21 ± 0.15 Ma (Staudacher *et al.* 1982). Standard error ($\pm 1\sigma$) on age estimate is calculated according to Bigazzi and Galbraith (1999). Weighted mean age based on UT1791[B] and UT1907[C]. Area estimated using the point-counting method (Naeser *et al.* 1982). D_s = mean spontaneous track diameter, D_i = mean induced track diameter. Number of tracks counted is given in brackets. Number of tracks measured given in square brackets. Analyst, John Westgate. The single-crystal (sanidine) laser-fusion $^{40}Ar/^{39}Ar$ age of Huckleberry Ridge tephra, the internal standard, is 2.003 ± 0.014 Ma (2σ error) (Gansecki *et al.* 1998).
The population-subtraction method was used.

heating, the calculated age will be younger than the true age. Glasses suffer from partial track fading (PTF), which can occur even at ambient surface temperatures (Naeser *et al.*, 1980), so that it is always necessary to apply a correction for this fading effect before a meaningful FT age can be obtained.

Track-Size Correction Methods

The track-size correction method involves measuring the diameters of spontaneous tracks in the natural sample and comparing them to the diameters of induced tracks in the irradiated aliquot. If the mean track diameter of spontaneous tracks, D_s, is less than the mean diameter of induced tracks, D_i , PTF must have affected the sample (**Fig. 6**). A correction for the lowered areal track density is made by reference to a calibration curve, which shows the correlation between reduction in areal track density, ρ/ρ_o, to reduction in mean track diameter, D/D_o, where ρ_o and D_o are the original areal track density and mean track diameter, respectively. The points defining this curve are derived from a series of annealing experiments on the irradiated aliquots (Storzer and Wagner, 1969; Arias *et al.*, 1981). Sandhu and Westgate (1995) developed a modified version of the size correction method in which apparent ages of glass shards from tephra beds are corrected for partial track fading by using a 1 : 1 relationship between ρ/ρ_o and D/D_o – an approach supported by experimental data. If the sample is etched to give D_i in the range of 6–8 μm, the corrected spontaneous areal track density (ρ_{sc}) is given by

$$\rho_{sc} = \rho_s(D_i/D_s) \qquad [7]$$

The near-linear relationship observed by Sandhu and Westgate (1995) contrasts sharply with the distinctly nonlinear correlation found by Storzer and Wagner (1969) (**Fig. 7**). Van den haute's (1985) study explains this difference as due to the different etching conditions used. He theoretically calculated a set of curves showing how ρ/ρ_o varies with D/D_o with progressively longer etch times (i.e., increase in h, the thickness of the glass layer removed by the etchant). At low values of h, the curves are strongly nonlinear, but become near linear with larger values of h (**Fig. 7**). According to these data, the average track diameter in Storzer and Wagner's (1969) study is ~5 μm, whereas Sandhu and Westgate (1995) etched their glasses to give a mean track diameter in the range of 6–8 μm.

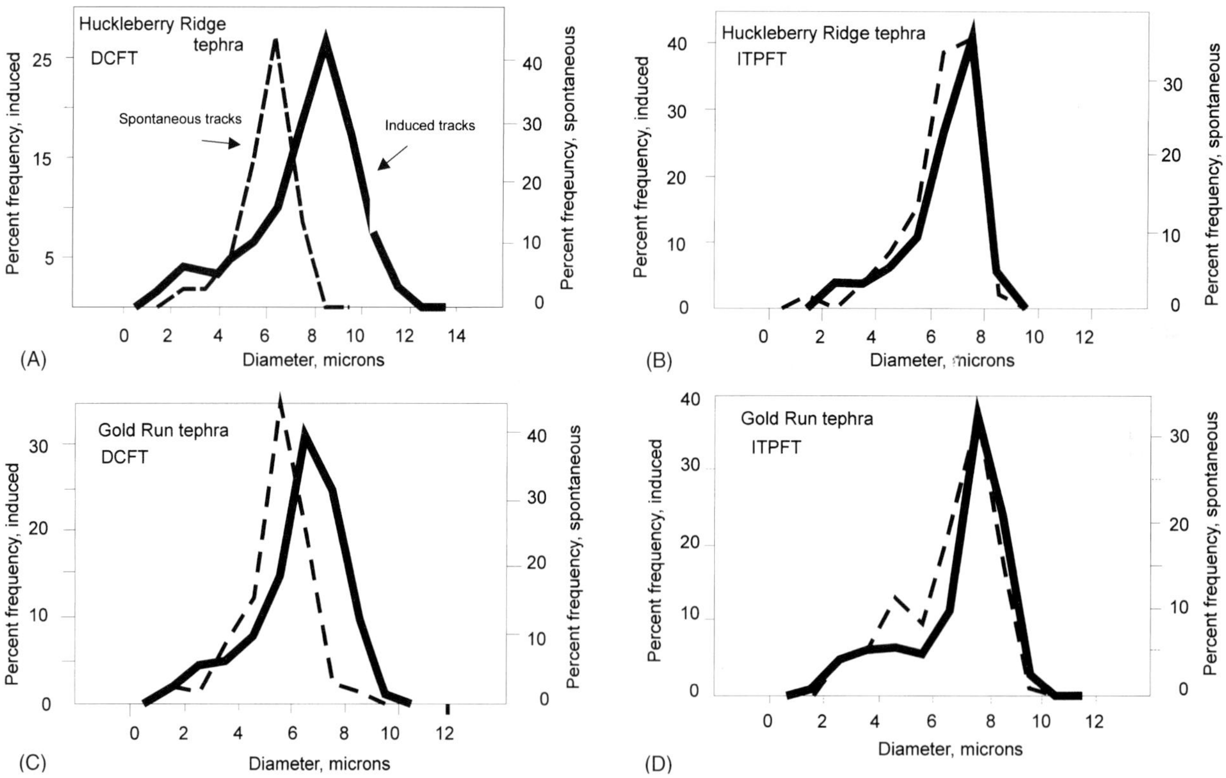

Figure 6 Size frequency curves of FT diameters in glass shards from the Huckleberry Ridge and Gold Run tephra beds showing partial track fading. DCFT samples unheated; ITPFT samples heated 150° C for 30 days. (A) Mean track diameter of spontaneous tracks is 5.98 ± 0.06 μm ($n = 329$); mean diameter of induced tracks is 7.63 ± 0.09 μm ($n = 597$), $D_s/D_i = 0.78 \pm 0.01$. Corresponding values for other samples are: (B) 6.41 ± 0.19 μm ($n = 50$), 6.41 ± 0.11 μm ($n = 199$), 1.00 ± 0.03; (C) 5.39 ± 0.10 μm ($n = 155$), 6.24 ± 0.06 μm ($n = 848$), 0.86 ± 0.02; (D) 6.46 ± 0.18 μm ($n = 94$), 6.81 ± 0.09 μm ($n = 398$), 0.95 ± 0.03.

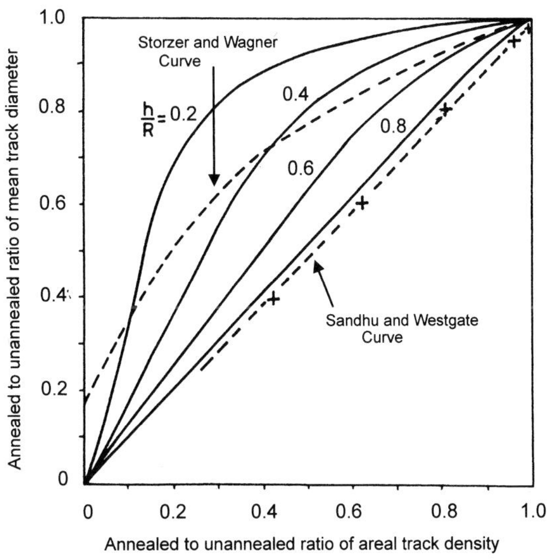

Figure 7 Calculated curves (solid lines) representing the annealed to unannealed ratio of mean track diameter vs. areal track density for different etching conditions in an internal glass surface. The thickness of the glass layer removed by the etching is *h*, and *R* is the etchable range of track of a single fission fragment. Experimental curves of Storzer and Wagner (1969) and Sandhu and Westgate (1995) are indicated by the dashed lines. Modified after Van den haute P (1985) The density and the diameter of fission tracks in glass with respect to age interpretation. *Nuclear Tracks* 10: 335–348.

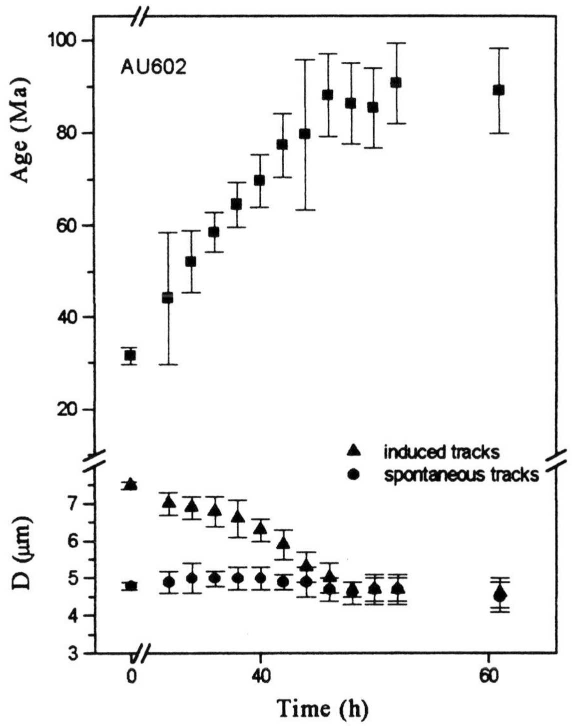

Figure 8 Plateau age for Australian obsidian AU602 obtained by using isothermal treatment at $T = 140°$ C (upper). Induced and spontaneous FT diameters at each step of thermal treatment (lower). Plateau age of 85 Ma is reached at 50 h when $D_s/D_i = 1$. After Bonetti R, Guglielmetti A, Malerba F, Migliorini E, Oddone M, and Bird R (1998) Age determination of obsidian source samples from North Queensland and New South Wales, Australia. In: Van den haute P and De Corte F (eds.) *Advances in Fission-Track Geochronology*, pp. 305–312. Kluwer, Copyright 1998, with kind permission of Springer Science and Business Media.

Plateau Correction Methods

The step-heating plateau method (Storzer and Poupeau, 1973) takes advantage of the higher thermal stability of partially faded spontaneous tracks. The FT age of a sample is proportional to ρ_s/ρ_i. If PTF has occurred, this ratio will be too low. However, when pairs of the natural sample and its irradiated aliquot are heated together for a given time at progressively higher temperatures, ρ_s/ρ_i increases until a plateau level is reached. The value of ρ_s/ρ_i at this plateau level is used to give an FT age corrected for PTF. An example of this approach is given in Arias *et al.* (1981). An alternative method is to keep a constant temperature and change the heating time (Burchart *et al.*, 1975). An excellent example is the work of Bonetti *et al.* (1998) on an Australian obsidian (**Fig. 8**). The steady increase in apparent age with continued heating at 140° C is clearly shown, as is the plateau condition, which is reached after 50 h and indicates an age of 85 Ma. Attainment of the plateau condition is also shown by $D_s/D_i = 1$. A modified version of the plateau technique has been developed by Westgate (1989), who showed that a single heat treatment of 150 C for 30 days is sufficient to correct fully for PTF. Again, the plateau position is recognized by $D_s = D_i$. This procedure has been successfully applied to hydrated, silicic glass shards of many tephra beds, as noted below.

Following up on this work, Kitada and Wadatsumi (1995a) showed that by increasing the temperature to 165° C only a five-day heat treatment was needed to give a plateau age (**Fig. 9**). Other workers have used different time–temperature combinations to reach the plateau (**Table 2**). Plateau annealing conditions should be such as to ensure minimal structural damage to the glass with no deleterious effects on its etching characteristics. Consequently, a low temperature–long time combination is favored, although, in the light of the data shown in **Table 2**, a heating period of over 15 days at 150° C is probably excessive.

Comparison of Glass FT Ages Corrected by the Diameter and Plateau Methods

Glass-FT ages have been determined for many tephra beds at the University of Toronto FT Laboratory using the diameter-correction (glass-DCFT) method of Sandhu and Westgate (1995) and isothermal plateau (glass-ITPFT) method of Westgate (1989) (**Table 3**,

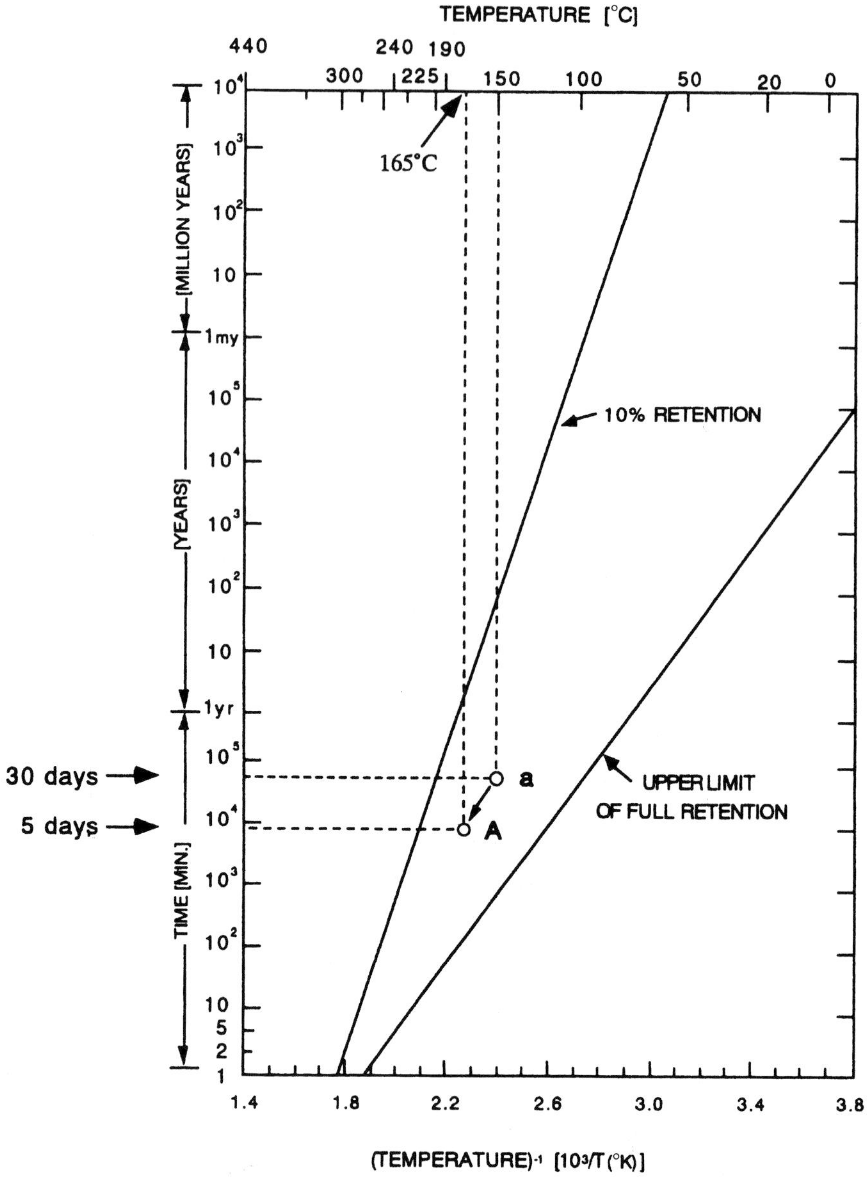

Figure 9 Arrhenius plot of FT annealing experiments for Clearlake glass from Fleischer *et al.* (1968) showing positions of (a) annealing conditions for glass shards used by Westgate (1989), and (A) annealing conditions used by Kitada and Wadatsumi (1995). After Kitada N and Wadatsumi K (1995a) Annealing properties of natural glass fission tracks at the controlled temperature and its implication on the isothermal plateau dating technique. *Journal of Geosciences, Osaka City University* 38: 4.

Fig. 10). Excellent agreement exists, adding confidence to the validity of both methods. Optimal etch times are longer for samples that have been heated, as in the ITPFT method, and limit application of this method to glass shards $\geq$120 μm. On the other hand, glass shards as small as ~80 μm can be dated by the DCFT technique, which requires no heat treatment.

Accuracy and Precision of Glass FT Ages

The glass-FT ages shown in **Table 4** range from 50 ka to ~15 Ma. All are within ~5% of the age determined by other methods, mainly $^{40}Ar/^{39}Ar$, but also K–Ar, thermoluminescence, ^{14}C, zircon-FT, and orbitally tuned ages (**Table 5**). In particular, both the DCFT and ITPFT ages on reference glasses mentioned by Balestrieri *et al.* (1998) agree very well with the $^{40}Ar/^{39}Ar$ results. The glass surface area in all these samples was estimated by the point-counting technique, so that the standard error on each glass-FT age is calculated from Eqn [6] (Bigazzi and Galbraith, 1999). For the most part, the $\pm 1\sigma$ uncertainty is $\leq$10%. An improved precision can be achieved if the sample is dated by several persons, a weighted mean and associated error being determined by the standard inverse variance method. It can be seen, therefore, that glass-FT

Table 2 Plateau annealing conditions for glasses

Material	*Temperature (C)*	*Time*	*Reference*
Obsidian	250	24 h	Osorio *et al.* (2003)
Obsidian	220	4 h	Osorio *et al.* (2003)
Obsidian	250	5 h	Miller and Wagner (1981)
Obsidian	245	24 h	Miller and Wagner (1981)
Obsidian	242	5 h	Miller and Wagner (1981)
Glass shards	165	5–10 d	Kitada and Wadatsumi (1995a)
Obsidian	165	5–10 d	Kitada and Wadatsumi (1995a)
Obsidian	140	50–70 h	Bonetti *et al.* (1998)
Obsidian	205	1–2 h	Dorighel *et al.* (1998)
Glass shards	150	30 d	Westgate (1989)
Glass shards	150	15 d	Sandhu *et al.* (1993)
Glass shards	200	4 h	Bigazzi *et al.* (2000)
Glass shards	210	8 h	Bigazzi *et al.* (2000)
Glass shards Obsidian, Tektite glass	150–350	1 h	Naeser *et al.* (1980)
Chilled rim of basalt flows	200	1 h	Macdougall (1976)
Glass shards	150–220	2–3 h	Arias *et al.* (1981)

In most cases, the plateau condition is recognized by equivalence of the mean spontaneous track diameter (D_s) and mean induced track diameter (D_i).

methods can be successfully applied to volcanic glass shards, obsidians, and tektites, giving accurate and precise ages. Of special note, these grain-specific methods are ideally suited to the dating of distal tephra beds,

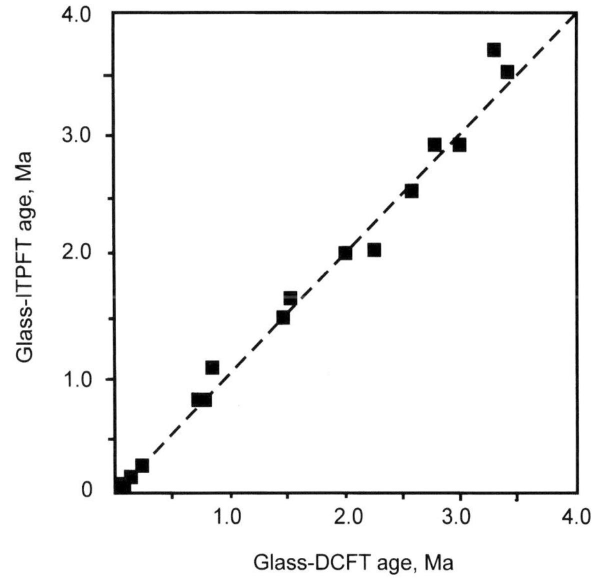

Figure 10 Close correspondence between glass-DCFT and glass-ITPFT ages determined in University of Toronto FT Laboratory. Dashed line is 1 : 1 line. Regression line through data points has $r^2 = 0.989$.

which are typically thin, discontinuous, fine grained, and contaminated by detrital material. They will greatly facilitate the development of detailed chronologies of tephra-bearing sedimentary sequences located far from volcanic centers.

Zircon FT Method

As with glass, zircon FT (ZFT) dating is a grain-specific method in which individual grains are scanned and counted, making it easier to recognize and eliminate contaminating grains (**Fig. 11**). For zircon, an age is

Table 3 A comparison of glass-DCFT and glass-ITPFT ages

Sample name	*glass-DCFT age*	*glass-ITPFT age*	*References*
Gold Run tephra	0.71 ± 0.07	0.78 ± 0.09	This article
Lost Chicken tephra	2.81 ± 0.18	2.90 ± 0.20	Matthews *et al.* (2003)
Quartz Creek tephra	3.00 ± 0.33	2.93 ± 0.36	Westgate *et al.* (2003)
Maninjau Ignimbrite	49.5 ± 3.5 ka	51.0 ± 6.0 ka	Alloway *et al.* (2004a)
Toba tephra	84 ± 17 ka	68 ± 7 ka	Westgate *et al.* (1998)
Fort Selkirk tephra	1.47 ± 0.14	1.48 ± 0.19	Westgate *et al.* (2001)
Huckleberry Ridge tephra	2.00 ± 0.21	2.02 ± 0.12	This article
Jankov moldavite	15.10 ± 0.47	15.38 ± 0.52	This article
Roccastrada R2V glass	2.58 ± 0.09	2.50 ± 0.11	This article
Old Crow tephra	0.14 ± 0.03	0.14 ± 0.01	Sandhu and Westgate (1995)
Mamaku Ignimbrite	0.23 ± 0.02	0.23 ± 0.01	Shane *et al.* (1994)
Ester Ash	0.77 ± 0.12	0.81 ± 0.07	Westgate *et al.* (1990)
Lake Tapps tephra	0.86 ± 0.11	1.06 ± 0.11	Westgate *et al.* (1987)
Pakihikura Pumice	1.53 ± 0.22	1.63 ± 0.15	Alloway *et al.* (1993)
BKT-3 tephra	2.28 ± 0.20	2.05 ± 0.20	Hart *et al.* (1992)
SHT tephra	3.42 ± 0.29	3.53 ± 0.37	Walter and Aronson (1993)
Cindery Tuff	3.32 ± 0.22	3.70 ± 0.14	Westgate (1989)

All ages in Ma unless otherwise noted. Error given is $\pm 1\sigma$. All ages determined at the University of Toronto FT Laboratory.

Table 4 Accuracy and precision of glass-fission-track ages: a comparison with other numerical dating methods

Sample	*Material*	*FT Method*	*Glass-FT age Ma* ($\pm 1\sigma$)	*Age using other methods* (Ma)	*References*
Jankov	Tektite	DCFT	15.10 ± 0.47	Ar–Ar	Staudacher *et al.* (1982)
Moldavite (Czech R.)		ITPFT	15.38 ± 0.52	15.21 ± 0.15	
Roccastrada R2V glass (Italy)	Glass shards	DCFT ITPFT	2.58 ± 0.09 2.50 ± 0.11	Ar–Ar 2.44 ± 0.02	Bigazzi *et al.* (1993)
JAS-G1 glass (Japan)	Obsidian	ITPFT (4)	1.01 ± 0.03	Ar–Ar 0.947 ± 0.005	Kitada and Wadatsumi (1995b)
Huckleberry Ridge tephra (U.S.A.)	Glass shards	DCFT ITPFT (3)	2.00 ± 0.21 2.02 ± 0.12	Ar–Ar 2.003 ± 0.014	Gansecki *et al.* (1998)
Potaka tephra (New Zealand)	Glass shards	ITPFT (6)	1.00 ± 0.03	Ar–Ar 1.04 ± 0.03	Alloway *et al.* (2004b)
Fort Selkirk tephra (Canada)	Glass shards	DCFT ITPFT (2)	1.47 ± 0.14 1.48 ± 0.19	K–Ar >1.47 ± 0.11 <1.60 ± 0.08	Westgate *et al.* (2001)
Quartz Creek tephra (Canada)	Glass shards	DCFT ITPFT	3.00 ± 0.33 2.93 ± 0.36	Ar–Ar >2.71 ≤3.01 2.64 ± 0.24	Westgate *et al.* (2003)
Ongatiti Ignimbrite (New Zealand)	Glass shards	ITPFT	1.21 ± 0.09	Ar–Ar 1.21 ± 0.04	Alloway *et al.* (2004b)
Toba tephra (India)	Glass shards	ITPFT (3)	79 ± 8 ka	Ar–Ar 74 ± 2 ka	Westgate *et al.* (1998) Oppenheimer (2002)
Maninjau Ignimbrite (Sumatra, Indonesia)	Glass shards	ITPFT DCFT (2)	51 ± 6 ka 49.5 ± 3.5 ka	^{14}C 53 ± 2 ka	Alloway *et al.* (2004)
Old Crow tephra (Canada)	Glass shards	ITPFT (7)	140 ± 10 ka	TL >128 ± 22 ka <147 ± 16 ka	Westgate *et al.* (1990) Berger (2003)
Lake Tapps tephra (USA)	Glass shards	ITPFT	1.06 ± 0.11	Ar–Ar 1.15 ± 0.01	Westgate *et al.* (1987) Hildreth (1996)
Bishop tuff (USA)	Glass shards	ITPFT	0.74 ± 0.04	Ar–Ar 0.759 ± 0.002	Sarna-Wojcicki *et al.* (2000)
Cindery Tuff (Ethiopia)	Glass shards	ITPFT	3.70 ± 0.14	Ar–Ar 3.94 ± 0.05	Hall *et al.* (1984)
Rangitawa Tephra (New Zealand)	Glass shards	ITPFT (6)	0.35 ± 0.02	Orbitally tuned 0.35 ± 0.01	Pillans *et al.* (1996)
Mamaku Ignimbrite (New Zealand)	Glass shards	DCFT ITPFT	0.23 ± 0.02 0.23 ± 0.02	Ar–Ar 0.22 ± 0.01	Houghton *et al.* (1995)
SHT tephra (Ethopia)	Glass shards	DCFT ITPFT	3.42 ± 0.29 3.53 ± 0.37	Ar–Ar 3.40 ± 0.03	Walter and Aronson (1993)
Moiti tephra (Ethiopia)	Glass shards	ITPFT	3.89 ± 0.25	K–Ar 4.1	Hart *et al.* (1992)
BKT-3 tephra (Ethiopia)	Glass shards	ITPFT	2.05 ± 0.20	ZFT 2.3	Hart *et al.* (1992)

All FT ages determined at the University of Toronto FT Laboratory. Number in brackets indicates number of samples dated.

obtained on each grain that is counted. If there is a statistically significant difference in age between the primary and contaminating grains, the latter can be identified by their anomalous ages and eliminated from the analysis (e.g., Kowallis (1988), Galbraith (1990), Galbraith and Green (1990), and Seward and Kohn (1997)). Attached glass is a good indicator of a primary zircon (**Fig. 11**), but euhedral zircons are not necessarily primary because they can retain their euhedral form through several erosional cycles.

Zircon has several advantages over glass for FT dating, mainly its significantly higher annealing temperatures. Fission tracks in glass can fade at ambient surface temperatures (Seward, 1979; Naeser *et al.*, 1980; Seward and Kohn, 1997), resulting in an anomalously young age, but zircon does not begin to lose tracks until temperatures of ~160° C or higher, depending on the duration of heating and the α-radiation damage in the zircon (Brandon *et al.*, 1998). Zircon has the added advantage that it can still be dated after glass has altered to clay. On the other hand, it is not present in all tephra beds – silicic tephra beds are more likely to yield usable zircons than mafic tephra deposits. Furthermore, zircons

Table 5 Comparison of glass-ITPFT ages for tephra beds in Wanganui Basin, New Zealand, with paleomagnetic and astronomically calibrated age estimates

Tephra	*Glass-ITPFT age*	*Paleomagnetic age*	*Astronomical age*
Kupe	0.63 ± 0.08	<0.78	0.65
Kaukatea	0.86 ± 0.08	0.78–0.99	0.9
Potaka	1.00 ± 0.03	0.99–1.07	0.99
Rewa	1.20 ± 0.14	*c.* 1.19	1.19
Mangapipi	1.51 ± 0.16	1.19–1.77	1.54
Pakihikura	1.58 ± 0.08	1.19–1.77	1.60
Table Flat	1.71 ± 0.12	1.19–1.77	1.65
Vinegar Hill	1.75 ± 0.20	1.19–1.77	1.75
Waipuru	1.79 ± 0.15	1.77–1.95	1.83

Data from Pillans *et al.* (2005). All ages in Ma with $\pm 1\sigma$.

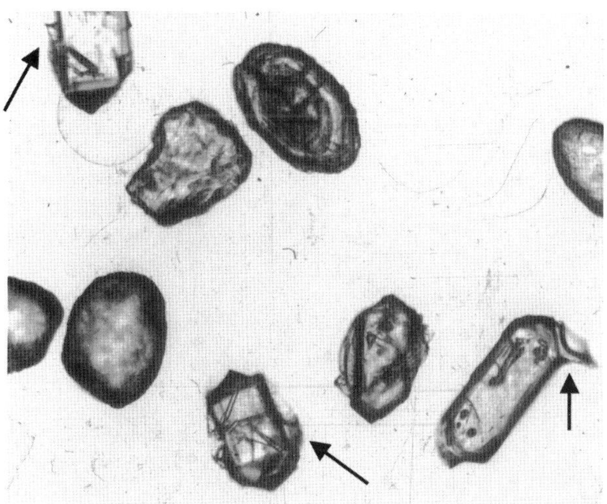

Figure 11 Photomicrograph of zircons separated from a contaminated tephra bed. Primary zircons (arrowed) can be identified by glass adhering to them. Detrital zircons in this tephra are generally darker, more rounded, and lack glass mantles. Zircons are ~150 μm long. Modified from Naeser CW, Briggs ND, Obradovich JD, and Izett GA (1981) Geochronology of Quaternary tephra deposits. In: Self S and Sparks RSJ (eds.) *NATO Advanced Studies Institute Series C: Tephra Studies*, pp. 13–47. Reidel.

may be too small (~80 μm) to be dated by fission tracks, particularly in tephra beds sampled far from their source vents, and, because zircon is so rare in tephra deposits, much larger samples need to be collected compared to glass. If tephra beds are being sampled from drill cores, glass is probably the only phase that can be dated with fission tracks because of the limited amount of material available.

Unlike glass, zircons from a single source tend to have a very inhomogeneous U distribution, both within and between individual crystals, and therefore U content must be determined from exactly the same

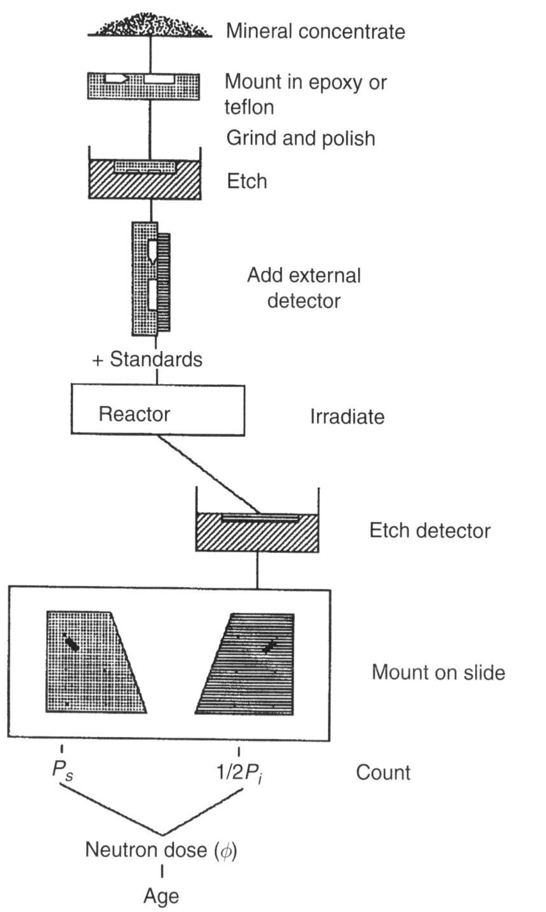

Figure 12 Flow diagram of steps in the external detector method of dating minerals. Details of laboratory procedures are in Naeser (1976). Modified from Naeser CW, Briggs ND, Obradovich JD, and Izett GA (1981) Geochronology of Quaternary tephra deposits. In: Self S and Sparks RSJ (eds.) *NATO Advanced Studies Institute Series C: Tephra Studies*, pp. 13–47. Reidel.

area of a crystal in which spontaneous tracks are counted. This is accomplished using the external detector method (EDM) (Naeser and McKee, 1970) (**Fig. 12**). In the EDM, spontaneous tracks are etched, normally in a eutectic KOH–NaOH melt (Gleadow *et al.*, 1976), and counted in the crystal, whereas induced tracks are counted in the image of the crystal recorded in a plastic or low-U (<10 ppb) muscovite detector that was in direct contact with the crystal during neutron irradiation (**Fig. 13**). An age is obtained for each grain that is analyzed, but experience indicates that a minimum of six grains should be dated to determine a sample age. Seward and Kohn (1997) recommend that for young zircons with low track densities, at least 20 crystals should be dated. An alternative method of determining U content by LA-ICP-MS is being developed in several laboratories (Cox *et al.* 2000; Hasebe *et al.* 2004; Donelick *et al.* 2005). LA-ICP-MS eliminates the need for neutron irradiation, but has thus far been tested mainly on apatite.

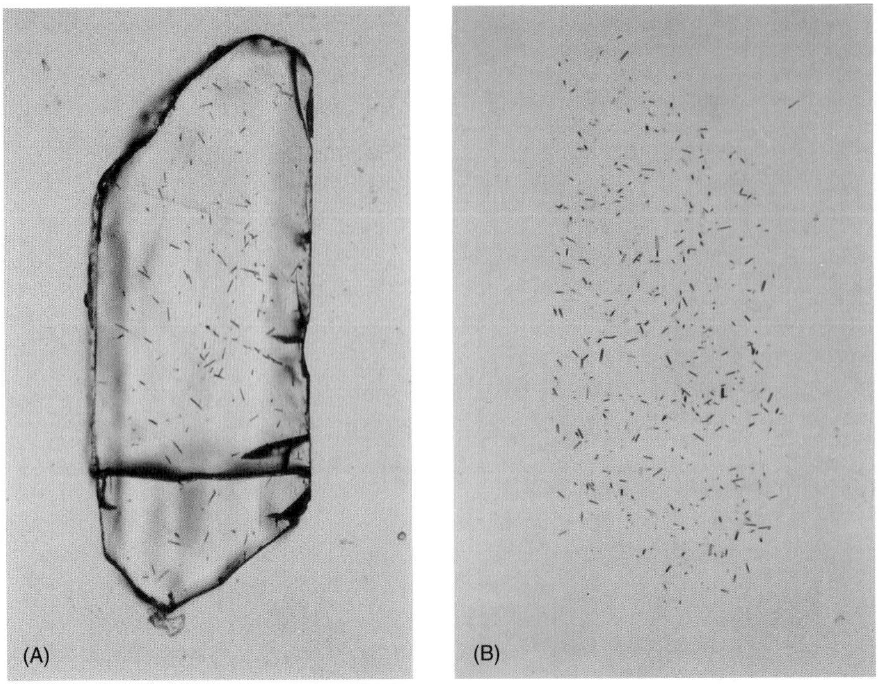

Figure 13 (A) Photomicrograph of crystal (left) and (B) image of the crystal in a muscovite detector (right). In the external detector method, spontaneous fission tracks are counted in the crystal (in this example, an apatite grain), and induced tracks are counted in the detector. From Naeser CW and Naeser ND (1989) Laboratory procedures and techniques. In: Crowley KD, Naeser CW, and Naeser ND (eds.) *Geological Society of America Short Course Notes: Fission-Track Analysis: Theory and Applications*, pp. 3-1–3-17. Boulder: Geological Society of America.

Precautions are necessary when dating zircon, particularly from young tephra beds. Care must be taken to ensure that tracks are fully etched, because even slight underetching can result in ages that are too young (Gleadow, 1980; Naeser *et al.*, 1987; Kohn *et al.*, 1992; Seward and Kohn, 1997). This is a common pitfall in dating young, low-track-density zircons in which the lack of tracks makes it difficult to judge correct etching time. A solution suggested by Gleadow (1980) is irradiating a separate test mount of zircons from the sample with ^{252}Cf tracks and then etching the two mounts together, using the ^{252}Cf tracks to determine correct etching time. Zircon populations in young samples typically include many grains with no spontaneous tracks. It is critical that induced-track counts from these grains be included in the age calculation. Zero-track grains are a real part of low-track-density zircon populations. Failure to include them can result in a calculated age that is significantly older than the true age. A classic illustration is provided by the early attempt to date the ~1.9 Ma KBS Tuff (Hurford *et al.*, 1976; Gleadow, 1980).

The precision of zircon and glass FT ages is similar, and larger than the typical precision on K–Ar and ^{40}Ar/^{39}Ar ages, which can be a few percent or less, depending on the material dated and the age of the sample (McDougall, 1995) (**Tables 6** and **7**). However, one of the advantages of fission tracks is that many tephra beds that do not contain suitable material for K–Ar or ^{40}Ar/^{39}Ar dating can be dated by FT using glass and/or zircon.

Applications

New Zealand

Quaternary sequences in New Zealand and in surrounding oceans up to a distance of 1,000 km contain abundant silicic tephra beds erupted from the Taupo Volcanic Zone (TVZ) in central North Island (e.g., Shane (2000)). The TVZ is considered to be the most active rhyolitic center on Earth and has been the locus of numerous large (>100 km^3) caldera-forming ignimbrites and plinian eruptions. Dispersal of volcaniclastic material around the TVZ is strongly controlled by prevailing westerly winds, position of fluvial drainage systems, and Mesozoic-aged axial ranges forming topographic obstructions (**Fig. 14**). In New Zealand, the predominant tropospheric westerly winds promote dispersal of tephra toward the east coast region of North Island and beyond into the adjacent Pacific Ocean. In sedimentary basins (e.g., Wanganui Basin) located adjacent to the TVZ and outside the main atmospheric dispersal axis, the tephra record is dominated by

Table 6 Comparison of representative glass and zircon fission-track and sanidine K–Ar and $^{40}Ar/^{39}Ar$ ages of the Pearlette family ash beds and correlative Yellowstone Group tuffs, western North America

Tephra bed/tuff	*Pearlette family ash beds*		*Yellowstone Group tuffs*		
	Fission-track, glass age[a] (Ma)	*Fission-track zircon age* (Ma)	*Fission-track zircon age* (Ma)	*K–Ar sanidine age* (Ma)	$^{40}Ar/^{39}Ar$ *sanidine age* (Ma)
Lava Creek B	–	0.7 ± 0.1[b]	0.6 ± 0.2[b]	–	–
	0.60 ± 0.02[c]	0.68 ± 0.15[c]	–	0.616 ± 0.008[d]	–
	0.61 ± 0.12[e]	–	–	0.67 ± 0.01[f]	0.60 ± 0.01[f]
	0.60 ± 0.08[g]	–	–	0.63 ± 0.01[f]	0.61 ± 0.01[f]
	–	–	–	–	0.602 ± 0.002[h]
	–	–	–	–	0.639 ± 0.002[i]
Mesa Falls	–	–	–	1.27 ± 0.10[d]	–
	–	–	–	1.26 ± 0.01[f]	1.29 ± 0.01[f]
	–	–	–	1.55 ± 0.04[f]	1.30 ± 0.01[f]
	–	–	–	–	1.293 ± 0.006[h]
	–	–	–	–	1.285 ± 0.004[i]
Huckleberry Ridge	–	1.9 ± 0.1[b]	1.9 ± 0.2[b]	–	–
	1.39 ± 0.08[j]	1.93 ± 0.16[j]	–	2.02 ± 0.08[d]	–
	–	1.91 ± 0.25[e]	–	2.01 ± 0.16[f]	2.09 ± 0.01[f]
	1.3 ± 0.2[k]	–	–	2.02 ± 0.03[f]	2.04 ± 0.01[f]
	1.21 ± 0.07[l]	–	–	2.05 ± 0.02[f]	2.05 ± 0.01[f]
	2.04 ± 0.16[m]	–	–	–	2.003 ± 0.006[h]
	2.06 ± 0.15[n]	–	–	–	2.059 ± 0.004[i]
	1.97 ± 0.22[o]	–	–	–	–

[a]Ages in italics are corrected ages, determined by either ITPFT or DCFT method (see below, and text).
[b]Naeser *et al.* (1973), Wilcox and Naeser (1992): reported zircon age of Lava Creek B tephra ("Pearlette type O") is an average of four age determinations; age of Huckleberry Ridge tephra ("Pearlette type B") is an average of seven determinations; uncertainty is ±1 standard error of the mean.
[c]Westgate *et al.* (1977): ages determined on correlative Wascana Creek Ash. Glass age is average of five age determinations; uncertainty is ±1 standard error of the mean.
[d]J.D. Obradovich, in Naeser and Naeser (1988).
[e]Naeser *et al.* (1982): glass age of Lava Creek B tephra determined on correlative Wascana Creek Ash; zircon age is on Huckleberry Ridge tephra ("Pearlette type B").
[f]Obradovich and Izett (1991).
[g]Alloway *et al.* (1993): "Lava Creek B ash bed"; ITPFT age.
[h]Gansecki *et al.* 1998.
[i]Lanphere *et al.* (2002): age is weighted mean of total-fusion, plateau, and isochron ages (Lava Creek and Huckleberry Ridge Tuffs), or weighted mean of plateau and isochron ages (Mesa Falls Tuff).
[j]Seward (1979): "Borchers Ash"; reported glass age is an average of four age determinations; uncertainty is ±1 standard error of the mean.
[k]Naeser *et al.* (1980): "Pearlette type B".
[l]Naeser and Naeser (1988): "Huckleberry Ridge ash bed".
[m]Westgate (1989): "Borchers Ash"; ITPFT age.
[n]Sandhu and Westgate (1995): "Borchers Ash", DCFT age.
[o]Table 1.
Analytical uncertainty is ±1 sigma unless otherwise noted.

water-supported mass flow, hyperconcentrated flow, and fluvially transported deposits, which can be considered near-contemporaneous with the proximal eruptive deposits.

In New Zealand, FT dating of tephra and pumice-rich horizons has traditionally played a significant role in advancing the stratigraphy and chronology of Plio-Pleistocene sediments in the southern half of the North Island (e.g., Milne (1973), Seward (1974, 1976), and Boellstorff and Te Punga (1977)). Zircon and glass are the important phases in this respect, although the use of zircon in some cases is thwarted by low abundance and fine-grain size. On the other hand, hydrated glass shards are typically the most abundant component of tephra beds and commonly the only datable phase present.

The glass-ITPFT technique was first applied in New Zealand to three widespread and stratigraphically important rhyolitic horizons from Wanganui Basin (Alloway *et al.*, 1993). In that study, the glass-ITPFT ages, in conjunction with a revised magnetostratigraphy, demonstrated that the then existing Plio-Pleistocene marine chronology required significant age revision and confirmed that the previously

Table 7 Comparison of representative glass and zircon fission-track and sanidine K–Ar and $^{40}Ar/^{39}Ar$ ages of the Bishop Tuff, California

Fission-track, glass age[a] (Ma)	*Fission-track zircon age* (Ma)	*K–Ar sanidine age* (Ma)	$^{40}Ar/^{39}Ar$ *sanidine age* (Ma)
0.56 ± 0.05[b]	0.74 ± 0.03[c]	0.727 ± 0.015[d]	–
0.72 ± 0.05[e]	–	0.741 ± 0.007[f]	0.734 ± 0.012[f]
0.73 ± 0.03[g]	–	0.722 ± 0.007[h]	0.757 ± 0.009[i]
–	–	0.738 ± 0.003[h]	0.764 ± 0.005[i]

[a]Ages in italics are corrected by the ITPFT method (see text).
[b]Naeser and Naeser (1988).
[c]Izett and Naeser (1976).
[d]Dalrymple *et al.* (1965), Mankinen and Dalrymple (1979).
[e]Westgate (1989).
[f]Hurford and Hammerschmidt (1985).
[g]Bishop ash bed, southeast Idaho: weighted mean of two ITPFT ages, 0.72 ± 0.05 Ma (determined by B.V. Alloway) and 0.74 ± 0.04 (J.A. Westgate).
[h]Izett (1982), Izett *et al.* (1988): Izett *et al.* (1988) also report a K–Ar isochron age of 0.74 Ma for sanidine, plagioclase, biotite, and glass from the Bishop Tuff.
[i]Izett and Obradovich (1994).
Analytical uncertainty is ±1 sigma.

published uncorrected glass ages were up to 50% too young (**Fig. 15**). As discussed in Seward and Kohn (1997), the glass ages in Seward (1974, 1979) were determined before the effects of annealing on glass were fully recognized, and they demonstrate the extent to which annealing can lower uncorrected glass-FT ages. This revision had important implications for the evolution of several other sedimentary basins in southern North Island that contained the same tephra beds.

Since that study, the technique has also been used to confirm the age of stratigraphically important rhyolitic horizons (i.e., Potaka Tephra, Shane (1994); Rangitawa Tephra, Pillans *et al.* (1996); Waipuru and Vinegar Hill tephras, Naish *et al.* (1995); Ototoka Tephra, Alloway *et al.* (2004)), and has been pivotal in establishing a robust Plio-Pleistocene tephrostratigraphic framework throughout North Island, New Zealand (Pillans *et al.* (1994, 2005), Naish *et al.* (1998), Shane *et al.* (1996), and Alloway *et al.* (2004b); **Fig. 14**). More recently, ITPFT ages have been determined on glass shards from eight stratigraphically important macroscopic tephra layers from ODP-core 1124C (Carter *et al.*, 2004; Alloway *et al.*, 2005).

In a careful inter-laboratory comparison, Seward and Kohn (1997) reported zircon FT ages for five of the tephra beds in Wanganui Basin, four of which (Pakihikura, Potaka, Vinegar Hill, and Rewa tephras) are in good statistical agreement with previously determined glass-ITPFT ages. Two of the tephras (Rewa and Mangapipi) contain two age populations of zircons, the older interpreted by Seward and Kohn (1997) as detrital contamination, and the younger interpreted as the eruption age. The disagreement in glass and zircon FT ages for the Mangapipi tephra remains to be resolved.

The glass-ITPFT technique is well suited to the dating of distal silicic tephra and has permitted a direct and detailed comparison of stratigraphic and chronologic data in a variety of sedimentary environments in New Zealand. Furthermore, the technique has significantly improved the cross-correlation between tephrochronology, paleomagnetic chronology, and $\delta^{18}O$ stratigraphy in marine cores.

Indonesia

Introduction Indonesia is situated adjacent to the Sunda subduction system, which has produced a 1,600-km-long volcanic arc with as many as 129 historically active volcanoes, extending from Sumatra in the northwest to Sulawesi in the southeast of the Indonesian archipelago. Sumatra, in particular, contains a great variety and number of volcanoes. Many of these volcanoes are dormant but a few have erupted during historic time (i.e., Marapi and Singgalang-Tandikat). To date, much of the volcanological research conducted in Sumatra has focused on the eruptive history of Toba caldera (e.g., Chesner *et al.* (1991)), and the best record of Indonesian arc volcanism is from tephra layers preserved in ODP-sedimentary cores retrieved from the northeastern Indian Ocean adjacent to Sumatra (e.g., Dehn *et al.* 1991 and Lee *et al.* 2004). Many of these tephra layers have yet to be correlated to their onshore eruptive source areas, confirming that there is little known about the eruptive history of many volcanic centers located in Sumatra. In Indonesia, the

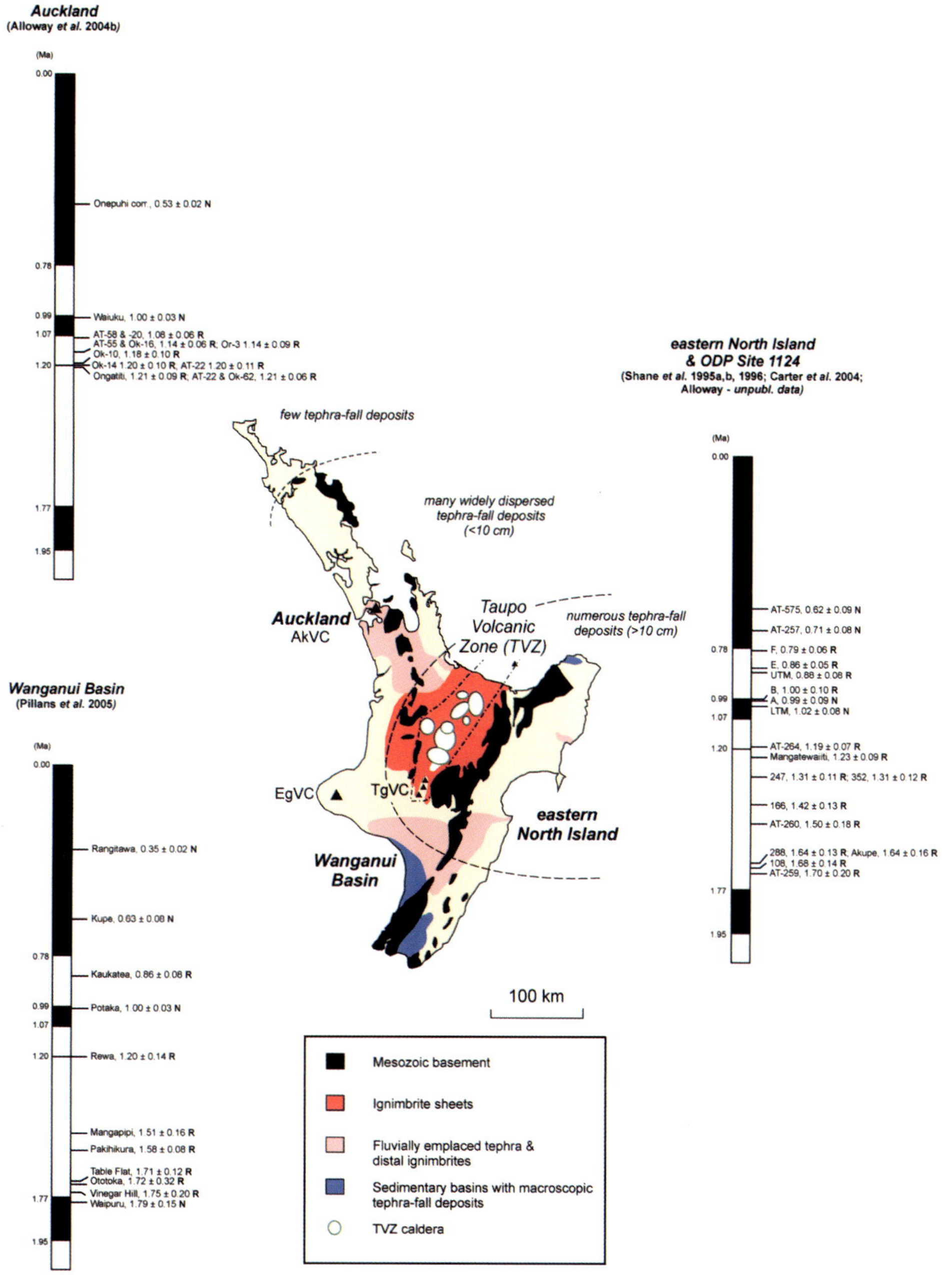

Figure 14 Distribution of Quaternary rhyolitic pyroclastic and associated reworked deposits in North Island, New Zealand and summaries of key Quaternary tephra beds outside of the TVZ that have published glass-ITPFT age control. Paleomagnetic polarity shown as normal (N) or reversed (R) where known. AkVC, Auckland Volcanic Centre; EgVC, Egmont Volcanic Centre; TgVC, Tongariro Volcanic Centre. Modified from Shane PA (2000) Tephrochronology: A New Zealand case study. *Earth-Science Reviews* 49: 223–259.

glass-FT dating technique has been applied to tephra erupted from Toba and Maninjau calderas, situated in northern and west-central Sumatra, respectively (**Fig. 16**).

Toba The Toba caldera complex consists of four overlapping calderas, the youngest of which is the world's largest caldera (100 × 30 km) and envelopes the three older calderas. An estimated 2,500–3,000 km^3 of dense-rock-equivalent pyroclastic material, termed the Youngest Toba Tephra (YTT), was erupted from the youngest caldera during one of the largest single volcanic eruptions in geological history (Rose and Chesner, 1987). Glass-FT ages for YTT have been determined from sites in Malaysia (68 ± 7 ka; Chesner *et al.* (1991)) and distally in northwest India (84 ± 16 ka; Westgate *et al.* (1998)) and are indistinguishable from an $^{40}Ar/^{39}Ar$ age of 73 ± 4 ka (Chesner *et al.*, 1991) and K–Ar ages of 75 ± 12 ka (on biotite) and 74 ± 3 ka (on sanidine)

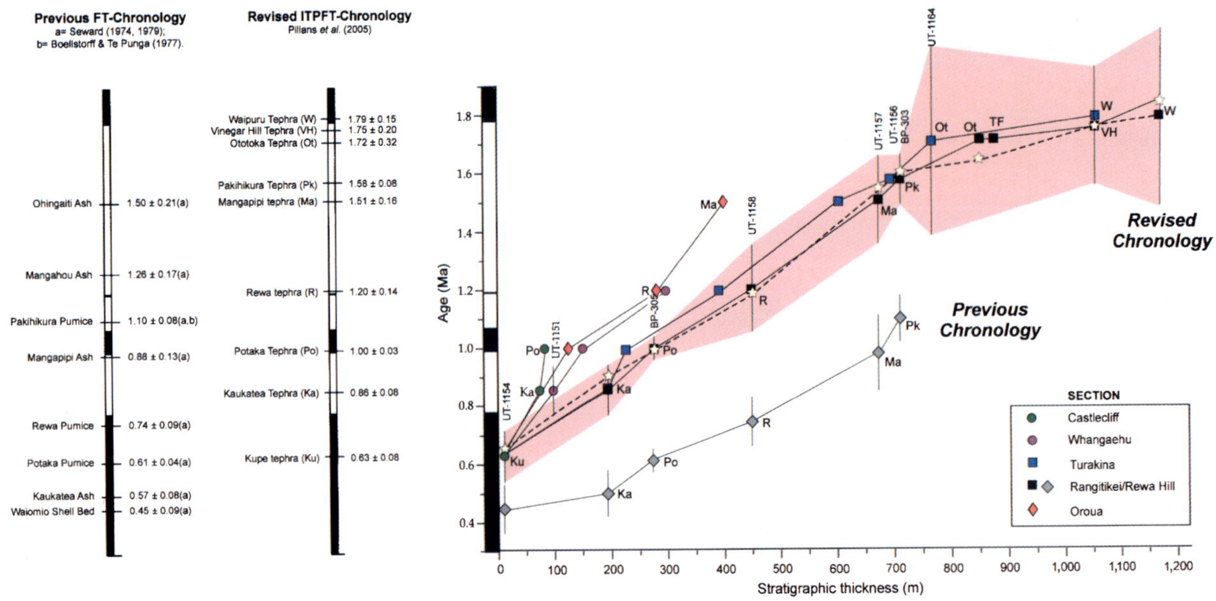

Figure 15 Comparison of previous and revised glass-FT chronologies for middle and lower Pleistocene tephras, Wanganui Basin. Shaded area represents $\pm 1\sigma$ on glass-ITPFT dated tephra beds from Rangitikei/Rewa Hill. Stars connected by dashed lines represent astronomically tuned ages determined for the same tephra beds. From Pillans BJ, Alloway BV, Naish T, Westgate JA, Abbot S, and Palmer AS (2005) Silicic tephras in Pleistocene shallow marine sediments of Wanganui Basin, New Zealand. *New Zealand Journal of the Royal Society of New Zealand* 35: 43–90.

(Ninkovich *et al.*, 1978). Certainly, the very widespread distribution of YTT through southern Southeast Asia and across the Indian Ocean makes it a very useful marker for dating archeological sites on the Indian subcontinent (Westgate *et al.*, 1998).

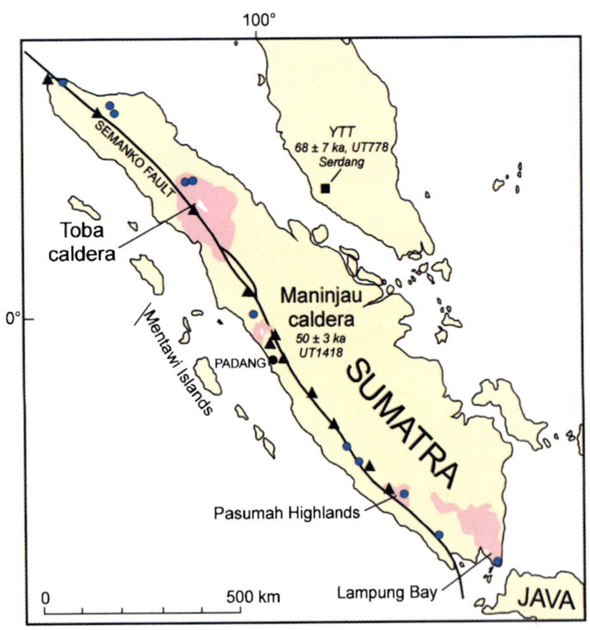

Figure 16 Map of Sumatra showing the position of the Semanko Fault and location of Toba and Maninjau calderas and other volcanic centers. Triangles: volcanoes which have erupted in historic time; circles: volcanoes in fumarolic state. The distribution of silicic volcaniclastic deposits in Sumatra is also shown (pink).

Maninjau Maninjau is a volcanic edifice situated in the Padang Highlands ca. 300 km to the south of Toba in west-central Sumatra (**Fig. 16**). The central part of this edifice is occupied by a collapsed caldera 20 km long and 8 km wide, and is now occupied by a lake with an estimated volume of 100 km^3. The caldera lake parallels the Sumatran volcanic front and is located within the northwest–southeast trending Sumatran fault zone, which runs the entire length of Sumatra. Glass-FT and radiocarbon techniques were used by Alloway *et al.* (2004a) in an attempt to determine an accurate age for the Maninjau eruptive succession. Glass-ITPFT and -DCFT ages give a weighted mean of 50 ± 3 ka ($n = 3$), and, together with concordant ^{14}C ages, give a timing for the latest silicic eruptive activity at 52 ± 3 ka, an age supported by the occurrence of an underlying silicic tephra bed that is geochemically indistinguishable from the ca. 75 ka YTT. This new age for the Maninjau eruptives is significantly different from previous and disparate FT (ca. 70–80 ka, Nishimura *et al.* (1978)) and K–Ar (ca. 0.28 ka, Leo *et al.* (1980)) determinations.

Concluding Statements

We have demonstrated that FT dating methods applied to volcanic glass and zircon can produce accurate and precise ages of Quaternary events in regions affected by Quaternary silicic volcanism. Although precision levels are inferior to those achieved by the $^{40}Ar/^{39}Ar$ method, FT methods are particularly useful for dating K-poor, calc-alkaline volcanics and highly vitric tephra beds. Other favorable attributes include the small sample requirement for glass-FT ages, and the grain-discrete methodologies used, which neutralize the contamination problem. Importantly, setting up an FT dating laboratory is a comparatively inexpensive operation because of its low-technology and low-maintenance requirements. Accordingly, it is very applicable to the lesser developed volcanic regions of the world, where monetary resources are limited.

Acknowledgments

JAW gratefully acknowledges the support of the Natural Sciences and Engineering Research Council of Canada and BVA acknowledges the financial support from the New Zealand Foundation for Research Science and Technology (FRST contract C05X0202). NDN acknowledges support from the USGS National Cooperative Geologic Mapping Program.

Glossary

Fission Track A linear region of intense radiation damage formed by the passage of an energetic, charged nuclear particle through an insulating solid.

See also: **Dating Techniques**. **K/Ar and Ar/Ar Dating**. **Quaternary Stratigraphy**: Tephrochronology. **U-Series Dating**.

References

Alloway, B. V., Pillans, B. J., Carter, L., Naish, T. R., and Westgate, J. A. (2005). Onshore–offshore correlation of Pleistocene rhyolitic eruptions from New Zealand: Implications for TVZ eruptive history and paleoenvironmental construction. *Quaternary Science Reviews* **24**, 1601–1622.

Alloway, B. V., Pillans, B. J., Naish, T. R., and Westgate, J. A. (2004c). Age and correlation of Ototoka tephra. Appendix in "Molluscan biostratigraphy of oxygen isotope stages of the last 2 million years in New Zealand". Part 1: Revised generic positions and recognition of warm- and cool-water migrants. Beu, A. G. *Journal of the Royal Society of New Zealand* **34**, 261–265.

Alloway, B. V., Pillans, B. J., Sandhu, A. S., and Westgate, J. A. (1993). Revision of the marine chronology in the Wanganui Basin, New Zealand, based on the isothermal plateau fission-track dating of tephra horizons. *Sedimentary Geology* **82**, 299–310.

Alloway, B. V., Pribadi, A., Westgate, J. A., *et al.* (2004a). Correspondence between glass-FT and ^{14}C ages of silicic pyroclastic flow deposits sourced from Maninjau caldera, west-central Sumatra. *Earth and Planetary Science Letters* **227**, 121–133.

Alloway, B. V., Westgate, J. A., Pillans, B. J., *et al.* (2004b). Stratigraphy, age and correlation of Middle Pleistocene silicic tephras in the Auckland region, New Zealand: A prolific distal record of Taupo Volcanic Zone volcanism. *New Zealand Journal of Geology and Geophysics* **47**, 447–479.

Arias, C., Bigazzi, G., and Bonadonna, F. P. (1981). Size corrections and plateau age in glass shards. *Nuclear Tracks* **5**, 129–136.

Balestrieri, M. L., Bigazzi, G., Bouska, V., *et al.* (1998). Potential glass age standards for fission-track dating: An overview. In *Advances in Fission-Track Geochronology* (P. Van den haute and F. De Corte, Eds.), pp. 287–304. Kluwer, Dordrecht.

Berger, G. W. (2003). Luminescence chronology of Late Pleistocene loess-paleosol and tephra sequences near Fairbanks, Alaska. *Quaternary Research* **60**, 70–83.

Bigazzi, G. (1981). The problem of the decay constant λ_F of ^{238}U. *Nuclear Tracks* **5**, 35–44.

Bigazzi, G., and Bonadonna, F. (1973). Fission track dating of the obsidian of Lipari Island (Italy). *Nature* **242**, 322–323.

Bigazzi, G., and Galbraith, R. F. (1999). Point-counting technique for fission-track dating of tephra glass shards, and its relative standard error. *Quaternary Research* **51**, 67–73.

Bigazzi, G., Bonadonna, F. P., Centamore, E., *et al.* (2000). New radiometric dating of volcanic ash layers in Periadriatic foredeep basin system, Italy. *Palaeogeography, Palaeoclimatology, Palaeoecology* **155**, 327–340.

Bigazzi, G., Bonadonna, F. P., Laurenzi, M. A., and Tonarini, S. (1993). A test sample for FT dating of glass shards. *Nuclear Tracks and Radiation Measurements* **21**, 489–497.

Boellstorff, J. D., and Te Punga, M. T. (1977). Fission-track ages and correlation of middle and lower Pleistocene sequences from Nebraska and New Zealand. *New Zealand Journal of Geology and Geophysics* **20**, 47–58.

Bonetti, R., Guglielmetti, A., Malerba, F., Migliorini, E., Oddone, M., and Bird, R. (1998). Age determination of obsidian source samples from North Queensland and New South Wales, Australia. In *Advances in Fission-Track Geochronology* (P. Van den haute and F. De Corte, Eds.), pp. 305–312. Kluwer, Dordrecht.

Brandon, M. T., Roden-Tice, M. K., and Garver, J. I. (1998). Late Cenozoic exhumation of the Cascadia accretionary wedge in the Olympic Mountains, northwest Washington State. *Geology* **110**, 985–1009.

Burchart, J., Dakowski, M., and Galazaka, J. (1975). A technique to determine extremely high fission track densities. *Bulletin de l'Academie Polonaise des Sciences, Série des Sciences de la Terre* **23**, 1–7.

Carter, L., Alloway, B., Shane, P., and Westgate, J. A. (2004). Late Cenozoic major rhyolitic eruptions and dispersal – Deep ocean records from off New Zealand. *New Zealand Journal of Geology and Geophysics* **47**, 481–500.

Chadderton, L. T. (2003). Nuclear tracks in solids: Registration physics and the compound spike. *Radiation Measurements* **36**, 13–34.

Chesner, C. A., Rose, W. I., Deino, A., Drake, R., and Westgate, J. A. (1991). Eruptive history of Earth's largest Quaternary caldera (Toba, Indonesia) clarified. *Geology* **19**, 200–203.

Cox, R., Kosler, J., Sylvester, P., and Hodych, J. (2000). Apatite fission-track (FT) dating by LAM-ICO-MS analysis. *Goldschmidt Conference Journal of Conference Abstracts* **5**, 322.

Dalrymple, G. B., Cox, A., and Doell, R. R. (1965). Potassium–Argon age and paleomagnetism of the Bishop tuff, California. *Geological Society of America Bulletin* **76**, 665–674.

Dehn, J., Farrell, J. W., and Schmincke, H.-U. (1991). Neogene tephrochronology from site 758 on northern ninety-east ridge: Indonesia arc volcanism of the past 5 Ma. *Proceedings of the Ocean Drilling Programme Scientific Results* **121**, 273–295.

Donelick, R. A., O'Sullivan, P. B., and Ketcham, R. A. (2005). Apatite fission-track analysis. In *Mineralogical Society of America Reviews in Mineralogy and Geochemistry: Low-Temperature Thermochronology: Techniques, Interpretations, and Applications* (P. W. Reiners and T. A. Ehlers, Eds.)Vol. 58 pp. 49–94. The Mineralogical Society of America, Chantilly, Virginia.

Dorighel, O., Poupeau, G., Bellot-Gurlet, L., and Labrin, E. (1998). Fission track dating and provenience of archaeological obsidian artefacts in Colombia and Ecuador. In *Advances in Fission-Track Geochronology* (P. Van den haute and F. De Corte, Eds.), pp. 313–324. Kluwer, Dordrecht.

Dumitru, T. A. (2000). Fission-track geochronology. In *American Geophysical Union Reference Shelf: Quaternary Geochronology: Methods and Applications* (J. S. Noller, J. M. Sowers and W. R. Lettis, Eds.) Vol. 4 pp. 131–155. American Geophysical Union, Washington, D.C.

Faure, G. (1977). *Principles of Isotope Geology*. Wiley, New York.

Fleischer, R. L. (1998). *Tracks to Innovation: Nuclear Tracks in Science and Technology*. Springer, New York.

Fleischer, R. L. (2004). Fission tracks in solids – Production mechanisms and natural origins. *Journal of Materials Science* **39**, 3901–3911.

Fleischer, R. L., Price, P. B., and Walker, R. M. (1965a). Effects of temperature, pressure and ionization on the formation and stability of fission tracks in minerals and glasses. *Journal of Geophysical Research* **70**, 1497–1502.

Fleischer, R. L., Price, P. B., Walker, R. M., and Leakey, L. S. B. (1965b). Fission track dating of Bed I, Olduvai Gorge. *Science* **149**, 383–393.

Fleischer, R. L., Price, P. B., and Walker, R. M. (1975). *Nuclear Tracks in Solids: Principles and Applications*. University of California Press. Berkeley.

Fleischer, R. L., Viertl, J. R. M., Price, P. B., and Aumento, F. (1968). Potassium–Argon ages and spreading rates on the Mid-Atlantic ridge at 45° north. *Science* **161**, 1338–1342.

Galbraith, R. F. (1990). The radial plot: Graphical assessment of spread in ages. *Nuclear Tracks and Radiation Measurements* **17**, 207–214.

Galbraith, R. F., and Green, P. F. (1990). Estimating the component ages in a finite mixture. *Nuclear Tracks and Radiation Measurements* **17**, 197–206.

Gansecki, C. A., Mahood, G. A., and McWilliams, M. (1998). New ages for the climactic eruptions at Yellowstone: Single-crystal $^{40}Ar/^{39}Ar$ dating identifies contamination. *Geology* **26**, 343–346.

Gleadow, A. J. W. (1980). Fission track age of the KBS tuff and associated hominid remains in northern Kenya. *Nature* **284**, 225–230.

Gleadow, A. J. W., Hurford, A. J., and Quaife, R. D. (1976). Fission track dating of zircon: Improved etching technique. *Earth and Planetary Science Letters* **33**, 273–276.

Green, P. F. (1981). A new look at statistics in fission-track dating. *Nuclear Tracks* **5**, 77–86.

Hall, C. M., Walter, R. C., Westgate, J. A., and York, D. (1984). Geochronology, stratigraphy and geochemistry of Cindery tuff in Pliocene hominid-bearing sediments of the Middle Awash, Ethiopia. *Nature* **308**, 26–31.

Hart, W. K., Walter, R. C., and WoldeGabreil, G. (1992). Tephra sources and correlation in Ethiopia: Application of elemental and neodymium isotope data. *Quaternary International* **13/14**, 77–86.

Hasebe, N., Barbarand, J., Jarvis, K., Carter, A., and Hurford, A. J. (2004). Apatite fission-track chronometry using laser ablation ICP-MS. *Chemical Geology* **207**, 135–145.

Hildreth, W. (1996). Kulshan caldera: A Quaternary subglacial caldera in the North Cascades, Washington. *Geological Society of America Bulletin* **108**, 786–793.

Houghton, B. F., Wilson, C. J. N., McWilliams, M. O., *et al.* (1995). Chronology and dynamics of a large silicic magmatic system: Central Taupo Volcanic Zone, New Zealand. *Geology* **23**, 13–16.

Hurford, A. J. (1990). Standardization of fission track dating calibration: Recommendation by the fission track working group of the IUGS subcommission on geochronology. *Chemical Geology (Isotope Geoscience Section)* **80**, 171–178.

Hurford, A. J., and Green, P. F. (1983). The zeta calibration of fission-track dating. *Isotope Geoscience* **1**, 285–317.

Hurford, A. J., and Hammerschmidt, K. (1985). $^{40}Ar/^{39}Ar$ and K/Ar dating of the Bishop and Fish Canyon tuffs: Calibration ages for fission-track dating standards. *Chemical Geology (Isotope Geoscience Section)* **58**, 23–32.

Hurford, A. J., Gleadow, A. J. W., and Naeser, C. W. (1976). Fission-track dating of pumice from the KBS tuff, East Rudolf, Kenya. *Nature* **263**, 738–744.

Izett, G. A. (1982). The Bishop ash bed and some older compositionally similar ash beds in California, Nevada, and Utah. *United States Geological Survey Open-File Report* **82–582,** 56.

Izett, G. A., and Naeser, C. W. (1976). Age of the Bishop tuff of eastern California as determined by the fission-track method. *Geology* **4**, 587–590.

Izett, G. A., and Obradovich, J. D. (1994). $^{40}Ar/^{39}Ar$ age constraints for the Jaramillo normal subchron and the Matuyama-Bruhnes geomagnetic boundary. *Journal of Geophysical Research* **99**(B2), 2925–2934.

Izett, G. A., Obradovich, J. D., and Mehnert, H. H. (1988). The Bishop ash bed (Middle Pleistocene) and some older (Pliocene and Pleistocene) chemically and mineralogically similar ash beds in California, Nevada, and Utah. *United States Geological Survey Bulletin* **1675**, 37.

Kitada, N., and Wadatsumi, K. (1995a). Annealing properties of natural glass fission tracks at the controlled temperature and its implication on the isothermal plateau dating technique. *Journal of Geosciences, Osaka City University* **38**, 1–12.

Kitada, N., and Wadatsumi, K. (1995b). Proposal for a glass age standard "JAS-G1". *Bulletin of Liaison and Informations, IUGS Subcommission on Geochronology* **13**, 23–27.

Kohn, B. P. (1979). Identification and significance of a Late Pleistocene tephra in Canterbury district, South Island, New Zealand. *Quaternary Research* **11**, 78–92.

Kohn, B. P., Pillans, B., and McGlone, M. S. (1992). Zircon fission track age for Middle Pleistocene Rangitawa Tephra, New Zealand: Stratigraphic and paleoclimatic significance. *Palaeogeography, Palaeoclimatology, Palaeoecology* **95**, 73–94.

Kowallis, B. J. (1988). Fission track dating of contaminated ash beds using age spectra. *Bulletin de Liaison et Informations IGCP Project* **196**(7), 53–57.

Lanphere, M. A., Champion, D. E., Christiansen, R. L., Izett, G. A., and Obradovich, J. D. (2002). Revised ages for tuffs of the Yellowstone Plateau volcanic field: Assignment of the Huckleberry Ridge tuff to a new geomagnetic polarity event. *Geological Society of America Bulletin* **114**, 559–568.

Leakey, L. S. B., Evernden, J. F., and Curtis, G. H. (1961). Age of bed I: Olduvai Gorge, Tanganyika. *Nature* **191**, 478–479.

Lee, M.-Y., Chen, C.-H., Wei, K.-Y., Iizuka, Y., and Carey, S. (2004). First Toba supereruption revival. *Geology* **32**, 61–64.

Leo, G. W., Hedge, C. E., and Marvin, R. F. (1980). Geochemistry, strontium isotope data and potassium–argon ages of the andesite–rhyolite association in the Pandang area, west Sumatra. *Journal of Volcanological and Geothermal Research* **7**, 139–156.

McDougall, I. (1995). Potassium–Argon dating in the Pleistocene. In *Dating Methods for Quaternary Deposits* (N. W. Rutter and N. R. Catto, Eds.), pp. 1–14. Geological Association of Canada, St. John's, NL.

Macdougall, J. D. (1976). Fission track annealing and correction procedures for oceanic basalt glasses. *Earth and Planetary Science Letters* **30**, 19–26.

Mankinen, E. A., and Dalrymple, G. B. (1979). Revised geomagnetic polarity time scale for the interval 0–5 m.y. B.P. *Journal of Geophysical Research* **84**(B2), 615–626.

Matthews, J. V., Jr., Westgate, J. A., Ovenden, L., Carter, D. L., and Fouch, T. (2003). Stratigraphy, fossils, and age of sediments at the upper pit of the Lost Chicken gold mine: New information on the late Pliocene environment of east central Alaska. *Quaternary Research* **60**, 9–18.

Miller, D. S., and Wagner, G. A. (1981). Fission-track ages applied to obsidian artifacts from South America using the plateau-annealing and the track-size age-correction techniques. *Nuclear Tracks* **5**, 147–155.

Milne, J. D. G. (1973). Mt. Curl Tephra, a 230,000-year-old marker bed in New Zealand, and its implications for Quaternary chronology. *New Zealand Journal of Geology and Geophysics* **16**, 519–532.

Naeser, C. W. (1976). Fission track dating. *US Geological Survey Open-File Report* **76–190**, 65.

Naeser, C. W., and McKee, E. H. (1970). Fission-track and K–Ar ages of Tertiary ash-flow tuffs, north-central Nevada. *Geological Society of America Bulletin* **81**, 3375–3384.

Naeser, C. W., and Naeser, N. D. (1988). Fission-track dating of Quaternary events. In *Geological Society of America Special Paper 227: Dating Quaternary Sediments* (D. J. Easterbrook, Ed.), pp. pp. 1–11. Geological Society of America, Boulder.

Naeser, C. W., and Naeser, N. D. (1989). Laboratory procedures and techniques. In *Geological Society of America Short Course Notes: Fission-Track Analysis: Theory and Applications* (K. D. Crowley, C. W. Naeser and N. D. Naeser, Eds.), pp. 3-1–3-17. Geological Society of America, Boulder.

Naeser, C. W., Briggs, N. D., Obradovich, J. D., and Izett, G. A. (1981). Geochronology of Quaternary tephra deposits. In *NATO Advanced Studies Institute Series C: Tephra Studies* (S. Self and R. S. J. Sparks, Eds.), pp. 13–47. Reidel, Dordrecht.

Naeser, C. W., Izett, G. A., and Obradovich, J. D. (1980). Fission-track and K–Ar ages of natural glasses. *United States Geological Survey Bulletin* **1489**, 31.

Naeser, C. W., Izett, G. A., and Wilcox, R. E. (1973). Zircon fission-track ages of Pearlette family ash beds in Meade County, Kansas. *Geology* **1**, 187–189.

Naeser, N. D., Westgate, J. A., Hughes, O., and Péwé, T. L. (1982). Fission-track ages of late Cenozoic distal tephra beds in the Yukon Territory and Alaska. *Canadian Journal of Earth Sciences* **19**, 2167–2178.

Naeser, N. D., Zeitler, P. K., Naeser, C. W., and Cerveny, P. F. (1987). Provenance studies by fission-track dating of zircon – Etching and counting procedures. *Nuclear Tracks and Radiation Measurements* **13**, 121–126.

Naish, T. R., Abbott, S. S., Alloway, B. V., *et al.* (1998). Astronomical calibration of a southern Hemisphere Plio-Pleistocene reference section, Wanganui Basin, New Zealand. *Quaternary Science Reviews* **17**, 695–710.

Naish, T. R., Kamp, P. J. J., Alloway, B. V., Pillans, B. J., Wilson, G. S., and Westgate, J. A. (1995). Tephrostratigraphy and integrated chronology for Pliocene–Pleistocene marine cyclothemic strata, Wanganui Basin: Implications for the Pliocene–Pleistocene boundary in New Zealand. *Quaternary International* **34–36**, 29–49.

Ninkovitch, D., Shackleton, N. J., Abdel-Monem, A. A., Obradovich, J. D., and Izett, G. (1978). K–Ar age of the Late Pleistocene eruption of Toba, north Sumatra. *Nature* **276**, 574–577.

Nishimura, S., Sasajima, S., Hirooka, K., Thio, K. H., and Hehuwat, F., (1978) *Radiometric Ages of Volcanic Products in Sunda Arc*. Parapat, Indonesia: CCOP/SEATAR Workshop.

Obradovich, J. D., and Izett, G. A. (1991). $^{40}Ar/^{39}Ar$ ages of upper Cenozoic Yellowstone group tuffs. *Geological Society of America Abstracts with Programs* **23**(2), 84.

Oppenheimer, C. (2002). Limited global change due to the largest known Quaternary eruption, Toba ~74 kyr BP? *Quaternary Science Reviews* **21**, 1593–1609.

Osorio, A. M., Iunes, P. J., Bigazzi, G., *et al.* (2003). Fission-track dating of Macusanite glasses with plateau and size correction methods. *Radiation Measurements* **36**, 407–412.

Pearce, N. J. G., Westgate, J. A., Perkins, W. T., and Preece, S. J. (2004). The application of ICP-MS methods to tephrochronological problems. *Applied Geochemistry* **19**, 289–322.

Pillans, B. J., Alloway, B. V., Naish, T., Westgate, J. A., Abbot, S., and Palmer, A. S. (2005). Silicic tephras in Pleistocene shallow marine sediments of Wanganui Basin, New Zealand. *New Zealand Journal of the Royal Society of New Zealand* **35**, 43–90.

Pillans, B., Kohn, B. P., Berger, G., *et al.* (1996). Multi-method dating comparison for mid-Pleistocene Rangitawa tephra, New Zealand. *Quaternary Science Reviews* **15**, 641–653.

Pillans, B., Roberts, A. P., Wilson, G. S., and Alloway, B. V. (1994). Magnetostratigraphic, lithostratigraphic and tephrostratigraphic constraints on lower–Middle Pleistocene sea level changes. *Earth and Planetary Science Letters* **121**, 81–98.

Price, P. B., and Walker, R. M. (1963). Fossil tracks of charged particles in mica and the age of minerals. *Journal of Geophysical Research* **68**, 4847–4862.

Reiners, P. W., and Ehlers, T. A. (2005). Low-temperature thermochronology: Techniques, interpretations, and applications. *Mineralogical Society of America, Reviews in Mineralogy and Geochemistry* **58**, 622.

Rose, W. I., and Chesner, C. A. (1987). Dispersal of ash in the great Toba eruption, 75 ka. *Geology* **15**, 913–917.

Sandhu, A. S., and Westgate, J. A. (1995). The correlation between reduction in fission-track diameter and areal track density in volcanic glass shards and its application in dating tephra beds. *Earth and Planetary Science Letters* **131**, 289–299.

Sandhu, A. S., Westgate, J. A., and Alloway, B. V. (1993). Optimizing the isothermal plateau fission-track dating method for volcanic glass shards. *Nuclear Tracks and Radiation Measurements* **21**, 479–488.

Sarna-Wojcicki, A. M., Pringle, M. S., and Wijbrans, J. (2000). New Ar-40/Ar-39 age of the Bishop tuff from multiple sites and sedimentation rate calibration of the Matuyama–Bruhnes boundary. *Journal Geophysical Research – Solid Earth* **105**, 21431–21443.

Seward, D. (1974). Ages of New Zealand Pleistocene substages by fission-track dating of glass shards from tephra horizons. *Earth and Planetary Science Letters* **24**, 242–248.

Seward, D. (1976). Tephrostratigraphy of the marine sediments in the Wanganui Basin, New Zealand. *New Zealand Journal of Geology and Geophysics* **19**, 9–20.

Seward, D. (1979). Comparison of zircon and glass fission-track ages from tephra horizons. *Geology* **7**, 479–482.

Seward, D., and Kohn, B. P. (1997). New zircon fission-track ages from New Zealand Quaternary tephra: An interlaboratory experiment and recommendations for the determination of young ages. *Chemical Geology* **141**, 127–140.

Shane, P. A. (1994). A widespread, Early Pleistocene tephra (Potaka tephra, 1 Ma) in New Zealand: Character, distribution and implications. *New Zealand Journal of Geology and Geophysics* **37**, 25–35.

Shane, P. A. (2000). Tephrochronology: A New Zealand case study. *Earth-Science Reviews* **49**, 223–259.

Shane, P. A., Alloway, B. V., Black, T. S., and Westgate, J. A. (1995a). Isothermal plateau fission-track ages of tephra beds in early–Middle Pleistocene marine and terrestrial sequence, Cape Kidnappers, New Zealand. *Quaternary International* **34–36**, 49–55.

Shane, P. A., Black, T., and Westgate, J. A. (1994). Isothermal plateau fission-track age for a paleomagnetic excursion in the Mamaku Ignimbrite, New Zealand, and implications for late Quaternary stratigraphy. *Geophysical Research Letters* **21**, 1695–1698.

Shane, P. A., Black, T. M., Alloway, B. V., and Westgate, J. A. (1996). Early to Middle Pleistocene tephrochronology of North Island, New Zealand: Implications for volcanism, tectonism and paleoenvironments. *Geological Society of America Bulletin* **108**, 915–925.

Shane, P. A., Froggatt, P., Black, T. S., and Westgate, J. A. (1995b). Chronology of Pliocene and Quaternary bioevents and climatic events from fission-track ages on tephra beds, Wairarapa, New Zealand. *Earth and Planetary Science Letters* **130**, 141–154.

Staudacher, T. H., Jessberger, E. K., Dominik, B., Kirsten, T., and Schaeffer, O. A. (1982). $^{40}Ar/^{39}Ar$ ages of rocks and glasses from the Nordlinger Ries Crater and the temperature history of impact breccias. *Journal of Geophysics* **51**, 1–11.

Storzer, D., and Poupeau, G. (1973). Ages-plateaux de minéraux et verres par la méthode des traces de fission. *Comptes Rendus De L' Academie des Sciences Paris, Série D* **276**, 137–139.

Storzer, D., and Wagner, G. A. (1969). Correction of thermally lowered fission-track ages of tektites. *Earth and Planetary Science Letters* **5**, 463–468.

Storzer, D., and Wagner, G. A. (1982). The application of fission track dating in stratigraphy: A critical review. In *Numerical Dating in Stratigraphy* (G. S. Odin, Ed.), pp. 199–221. Wiley, New Jersey.

Van den haute, P. (1985). The density and the diameter of fission tracks in glass with respect to age interpretation. *Nuclear Tracks* **10**, 335–348.

Wagner, G. A., and Van den haute, P. (1992). *Fission-Track Dating*. Enke, Stuttgart.

Walter, R. C., and Aronson, J. L. (1993). Age and source of the Sidi Hakoma Tuff, Hadar formation, Ethiopia. *Journal of Human Evolution* **25**, 229–240.

Westgate, J. A. (1989). Isothermal plateau fission-track ages of hydrated glass shards from silicic tephra beds. *Earth and Planetary Science Letters* **95**, 226–234.

Westgate, J. A., Christiansen, E. A., and Boellstorff, J. D. (1977). Wascana Creek Ash (Middle Pleistocene) in southern Saskatchewan: Characterization, source, fission-track age, palaeomagnetism and stratigraphic significance. *Canadian Journal of Earth Sciences* **14**, 357–374.

Westgate, J. A., Easterbrook, D. J., Naeser, N. D., and Carson, R. J. (1987). Lake Tapps tephra: An Early Pleistocene stratigraphic marker in the Puget Lowland, Washington. *Quaternary Research* **28**, 340–355.

Westgate, J. A., Preece, S. J., Froese, D. G., Walter, R. C., Sandhu, A. S., and Schweger, C. E. (2001). Dating early and middle (Reid) Pleistocene glaciations in central Yukon by tephrochronology. *Quaternary Research* **56**, 335–348.

Westgate, J. A., Sandhu, A. S., Preece, S. J., and Froese, D. G. (2003). Age of the gold-bearing White Channel gravel, Klondike district, Yukon. In *Yukon Exploration and Geology 2002* (D. S. Emond and L. L. Lewis, Eds.), pp. 241–250. Yukon Geological Survey, Whitehorse.

Westgate, J. A., Shane, P. A., Pearce, N. J. G., *et al.* (1998). All Toba tephra occurrences across peninsula India belong to the 75 ka eruption. *Quaternary Research* **50**, 107–112.

Westgate, J. A., Stemper, B. A., and Péwé, T. L. (1990). A 3 m.y. record of Pliocene–Pleistocene loess in interior Alaska. *Geology* **18**, 858–861.

Wilcox, R. E., and Naeser, C. W. (1992). The Pearlette family ash beds in the Great Plains: Finding their identities and their roots in the Yellowstone Country. In *Quaternary International: Tephrochronology: Stratigraphic Applications of Tephra Beds* (J. A. Westgate, R. C. Walter and N. D. Naeser, Eds.), Vol. 13–14, pp. 9–13. Pergamon Press, Oxford, UK.

Young, D. A. (1958). Etching of radiation damage in lithium fluoride. *Nature* **182**, 375–377.

FLUVIAL ENVIRONMENTS

Contents

Sediments

A Aslan, Mesa State College, CO, USA

© 2007 Elsevier B.V. All rights reserved.

Introduction

Rivers are a dominant feature of most landscapes and thus fluvial sediments are widespread. Fluvial deposits are represented by a continuum of sediment types that range from clay- to gravel-size particles, and include both terrigenous and organic deposits (*see* Terrigenous Sediments). Fluvial sediments provide an important link between weathering and slope processes in source areas as well as deltaic and coastal processes within depositional basins (*see* Deltaic Environments). Investigations of fluvial sediments are legion; recent syntheses include those by Miall (1996), Blum and Törnqvist, (2000), and Bridge (2003). Two additional aspects of fluvial sediments deserve mention. First, these deposits represent economically important reservoirs of oil, natural gas, and water and, include significant coal reserves. Second, fluvial sediments provide critical records of past and present geologic processes and terrestrial environments. This chapter summarizes the major characteristics of fluvial sediments and discusses key differences between river systems of continental interiors and those associated with subsiding continental margins (*see* Responses to Rapid Environmental Change).

Fluvial Sediments

Fluvial sediments consist predominantly of channel and overbank deposits (Allen, 1965). Channel deposits are dominated by sand and gravel and represent the bed load of rivers. Overbank deposits generally consist primarily of clay, silt, and lesser amounts of very fine to fine sand and represent suspended-load sediments that accumulate on floodplains. Organic sediments such as peat, and chemical sediments such as carbonate and evaporite minerals, are also important constituents in some overbank settings.

Channel Deposits

Channel deposits consist of channel-bar and channel-fill sediments (Bridge, 2003). Channel deposits commonly occur as isolated lenticular sand bodies (i.e., ribbon sands) or as thick sheet sands associated with channel belts (**Figs. 1–3**). Channel belts are floodplain areas containing active and abandoned channels and channel bars that record the activity of individual channel systems.

Channel-bar deposits Channel bars typically form in areas where decreasing flow velocities lead to bed-load sedimentation, such as along the inner bend of river meanders (**Fig. 4A**). Channel-bar sediments are stratified sands and gravels, which are

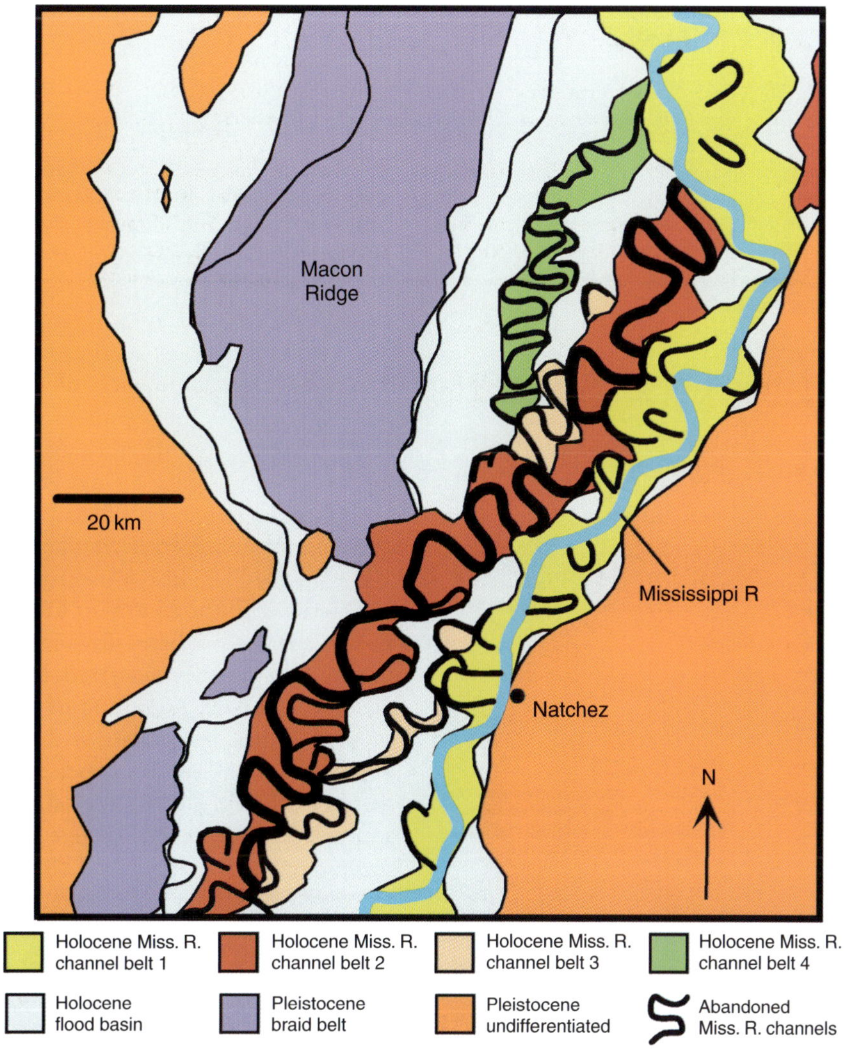

Figure 1 Geologic map of the Mississippi River floodplain near Natchez, Mississippi. Four Holocene Mississippi River channel belts are shown. Modified from Aslan and Autin (1999).

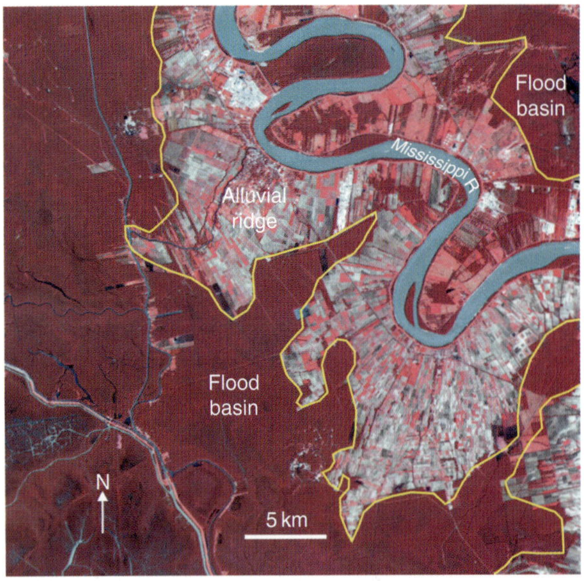

Figure 2 Satellite image of the Mississippi River floodplain south of Baton Rouge, Louisiana. Cultivated areas located adjacent to the river are on natural levees. Flow direction is to the southeast.

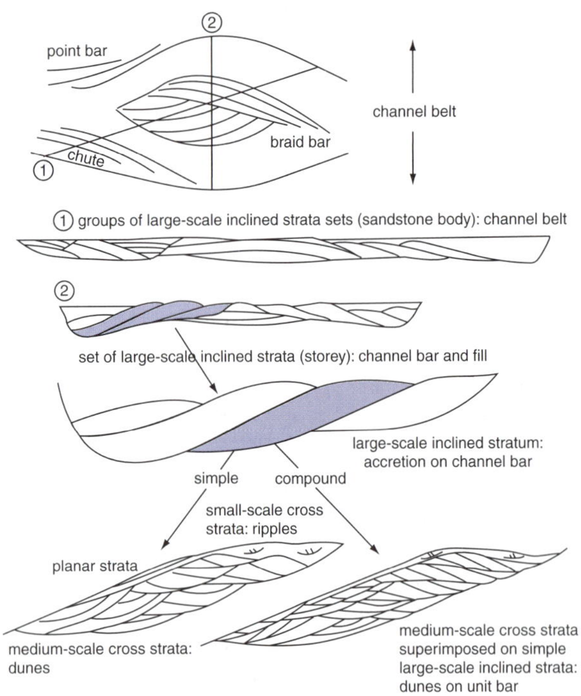

Figure 3 Diagrams showing different scales of fluvial deposits. The upper diagrams show cross-sections through a channel belt composed of multiple depositional units. Large-scale inclined strata (individual examples shown by shaded portions of cross-sections) are the dominant depositional unit of the channel-belt deposits. At the smallest scale, cross strata representing migrating dunes are present within the large-scale inclined strata. From Bridge (2003). Reprinted with permission from Blackwell Publishing.

arranged in sets of large-scale inclined strata (Bridge, 2003) (**Fig. 3**). Large-scale inclined strata consist of gently dipping cross strata that reflect episodic migration of major bar forms (i.e., point bars, side bars) in response to channel translation and/or expansion (**Fig. 4B**). The base of large-scale inclined strata sets is an erosional bounding surface, which often is overlain by rip-up clasts derived from fine-grained bed and bank materials. Within large-scale inclined strata, smaller-scale sets of cross-stratified sand and/or gravel are present (**Figs. 3** and **4C**). These smaller-scale packages of cross-stratified deposits represent bed forms such as dunes and ripples that migrate along the surface of major bars and the bed of channels. Common sedimentary structures associated with these smaller bed forms include small- to medium-scale trough and planar cross-bedding as well as small-scale ripple stratification (Bridge, 2003). Finer-grained sediments such as mud can be present in the upper portion of bar deposits, or occur as mud drapes that separate sets of cross strata.

Channel-fill deposits Channel-fill deposits are highly variable; sediments range from coarse-grained sand and gravel to fine-grained mud and organic sediments. Specific characteristics of channel fills depend on flow conditions at the time of channel abandonment. In vertical profiles through channel-fill deposits, grain size generally decreases upwards. Grain size also tends to decrease in a downstream direction. The dominance of coarser-grained bed-load deposits or finer-grained suspended-load sediments depends on whether or not the channel-fill succession represents an active or passive channel fill. Active fills are typically associated with upstream segments of abandoned channels where flow reduction was gradual. In contrast, passive fills are often associated with the central or lower portions of abandoned channels where flow reduction was severe. Active fills are characterized by bed-load deposits that represent the migration of bar forms and smaller-scale dunes and ripples within the upstream portion of the abandoned channel. As flow wanes, active channel-fill deposits transition upwards to smaller grain sizes and thinner bed sets representing progressively smaller bars and bed forms. Passive fills are represented by clay, silt, and very fine sand deposited from suspension as well as organic sediment such as peat. These deposits are concentrated in the middle and downstream portions of an abandoned channel and are associated with rapid decreases in flow such as occurs during neck cutoff and formation of an oxbow lake.

Figure 4 (A) Mississippi River and major point bar and chute-channel complex located several kilometers northwest of Vicksburg, Mississippi. The river is approximately 1 km wide. View is upstream. (B) Cross-sectional view through an eroding point bar of the Trinity River in Texas showing a set of large-scale inclined strata. The bank is approximately 6 m tall. Flow direction is to the left. (C) Stacked sets of planar cross-bedded sand representing migrating dunes on a point bar of the Trinity River in Texas. Flow is to the right.

Overbank Deposits

Overbank sediments accumulate during floods in natural-levee, crevasse-splay, and flood-basin environments (**Fig. 5**). Sediment grain size and accumulation rate decreases away from the active channel (**Fig. 6A**). Avulsion, the abandonment of all or part of a channel belt in favor of a new course, also plays an important role in the accumulation of overbank deposits (Allen, 1965; Smith *et al.*, 1989). In general, overbank sediments have received considerably less attention than channel deposits (Farrell, 1987). Modern examples of overbank deposits are numerous (Farrell, 1987; Smith *et al.*, 1989; Aslan and Autin, 1999). Classifications of these deposits are presented in Platt and Keller (1992) and Miall (1996). Although overbank deposits are sometimes viewed as monotonous successions of fine-grained alluvium, the recognition of paleosols in ancient overbank sediments has proved useful for subdividing and interpreting these deposits (Kraus, 1999) (*see* Nature of Paleosols). Additionally, there is increasing recognition that fine-grained sediments in both modern and ancient settings that have traditionally been interpreted as products of repeated overbank flooding, may represent episodes of avulsion (Smith *et al.*, 1989; Kraus, 1999; Morozova and Smith, 1999; Stouthamer and Berendsen, 2000; Slingerland and Smith, 2004).

Figure 5 Flooding of the Missouri River in August 1993. Floodwaters have completely inundated the floodplain near Jefferson City, Missouri. Trees on artificial levees outline the sinuous bankfull channel. Areas that are not flooded represent uplands along the margins of the river valley. View is downstream.

Natural-levee sediments Natural levees are asymmetric ridges that slope away from active and abandoned channels towards adjacent flood basins. These features are typically well developed along low-gradient sinuous rivers with large suspended-sediment loads that are subject to deep overbank floods such as the Mississippi (**Fig. 2**). Natural levees can rise up to 5 m above adjacent flood basins and are up to several kilometers wide on floodplains such as the Mississippi (Aslan and Autin, 1999; Törnqvist and Bridge, 2002).

Natural-levee sediments are wedges of sand and silt with minor amounts of clay that typically thin and fine away from channel margins as a result of decelerating overbank flow (**Fig. 6A**). Typical facies are similar to those of upper channel-bar deposits (Bridge, 2003), and consist of horizontal planar sand, small-scale cross-stratified sand, and bioturbated mud (Farrell, 1987). Proximal natural-levee deposits are relatively thick, sandy, and thick bedded. Distal natural-levee

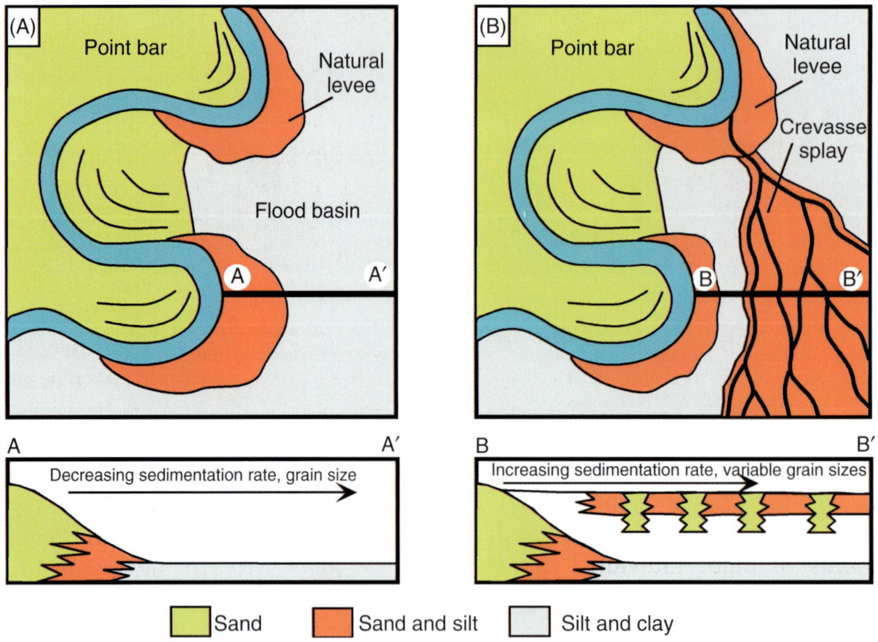

Figure 6 Floodplain maps and cross-sections showing (A) floodplain deposits formed by overbank flooding and sedimentation, and (B) floodplain deposits related to crevassing and avulsion. Note that avulsion deposits thicken away from the active channel belt. Modified from Aslan and Autin (1999).

sediments are thinner, contain a greater percentage of silt and clay, and show more evidence of bioturbation and pedogenesis. Natural-levee facies often occur as vertically stacked meter-thick, fining- or coarsening-upward sequences, which reflect periods of episodic overbank flooding and levee abandonment or progradation, respectively (*see* Sequence Stratigraphy).

Crevasse-channel and crevasse-splay sediments Crevasse channels are breaches in natural levees and transfer water and sediment from mainstem channels to flood basins. Crevasse channels often bifurcate into simple or complex distributary systems that discharge into flood-basin lakes or wetlands (Smith *et al.*, 1989; Smith and Pérez-Arlucea, 1994; Pérez-Arlucea and Smith, 1999). The linear to lobate accumulation of sediment surrounding crevasse channels is referred to as a crevasse splay, and includes the distributary channel network (**Fig. 7A**). Where crevasse channels flow into lakes they form lacustrine deltas. Crevasse channels have natural levees as well as distributary- or terminal-mouth bars (**Fig. 7B**). Crevasse-splay channels can extend up to 10 km into flood basins (Smith, 1986; Smith *et al.*, 1989).

Figure 7 (A) Crevasse splay along Cano Macareo, a major distributary of the Orinoco Delta in Venezuela. Photograph was taken towards the end of the rainy season as water levels began to fall. (B) Sandy distributary mouth-bar deposits at the terminus of one of the crevasse channels shown in (A). The crevasse-splay system discharges into a seasonal wetland.

Crevasse-channel deposits are typically lenticular bodies of sandy channel-bar and sandy to muddy channel-fill sediments. Grain sizes associated with crevasse-channel deposits are similar to those of the bed load fraction of the mainstem channel, and are somewhat coarser than other types of overbank sediments. However, muddy fine-grained crevasse-channel fills do occur (Smith and Pérez-Arlucea, 1994; Pérez-Arlucea and Smith, 1999). Stratification types include small- to medium-scale cross bedding and ripple lamination. Evidence of frequent changes in flow stage such as desiccation cracks and bioturbation features may be present (Bridge, 2003). Crevasse-channel deposit thicknesses will generally be smaller than the thickness of the main-channel sand bodies.

Crevasse-splay deposits range from tabular to wedge-shaped packages of interbedded sand and silt with lesser amounts of mud. Individual crevasse-splay sand bodies are generally 1 to 2 m thick. Strata are commonly bioturbated and ripple laminated (Farrell, 1987). Individual splay packages often show upward-coarsening or upward-fining sequences similar to natural-levee deposits (Pérez-Arlucea and Smith, 1999). Upward-coarsening sequences reflect crevasse-splay progradation whereas deposition of overbank silt and clay during waning flood stages produces upward-fining deposits. Crevasse-splay deposits thin or thicken away from channel margins depending on local floodplain topography (Willis and Behrensmeyer, 1994). Crevasse-splay sedimentation is particularly important within avulsion belts (Smith *et al.*, 1989) and is responsible for rapid fine-grained floodplain aggradation, a key attribute of avulsion.

Flood-basin sediments Flood basins are broad seasonally flooded depressions that separate channel belts (**Fig. 8**). These areas are characterized by low relief, shallow water tables, and complex drainage channel networks. In humid settings, flood basins are vegetated and contain lakes and perennial or seasonal wetlands. In more arid environments, flood basins are sparsely vegetated and can have ephemeral lakes. Eolian processes may also be important in arid settings.

Flood-basin deposits are generally tabular accumulations of clay and silt that bury channel deposits or interfinger with sand bodies representing minor floodplain channels and crevasse-splay deposits (Aslan and Autin, 1999) (**Figs. 9** and **10**). Flood-basin sediments are often cross-cut by channel-belt sand bodies (Stouthamer and Berendsen, 2000). In Quaternary settings, these deposits are typically meters to tens of meters thick, but can be hundreds of meters thick in pre-Quaternary alluvial successions. Slow sediment accumulation rates in flood basins lead to soil development. Bioturbation features such as root mottles and burrows along with soil nodules and mud cracks are common in these deposits.

Flood-basin sediment characteristics largely depend on climatic conditions. In humid areas with a shallow fluctuating water table (i.e., seasonal wetlands), gray sediment colors and yellow-brown to red color mottles reflect episodes of seasonal wetting and drying. Authigenic minerals such as iron and manganese

Figure 8 (A) Seasonal flooding of the Orinoco Delta, Venezuela. Vast tracts of interdistributary flood basins are inundated for several months of each year. During this time, only the crests of natural levees of distributary channels are subaerially exposed. (B) Mud cracks developed in flood-basin silt and clay during the dry season on the Orinoco Delta. This area is completely inundated during the rainy season.

Figure 9 Holocene organic flood-basin mud overlying channel-belt sand on the Mississippi River floodplain near Natchez, Mississippi. Flood-basin mud is approximately 2.5 m thick.

oxides, calcite, and gypsum can be common in these settings (Aslan and Autin, 1999). If smectitic clays are present, seasonal wetting and drying produces pedogenic slickensides and desiccation cracks. In more perennially saturated flood basins, elevated water levels and low clastic input leads to widespread peat accumulation (Flores, 1981). If clastic input is significant, bioturbated flood-basin sediments will often interfinger with laminated lacustrine clay and silt. Lacustrine strata are concentrated in topographic depressions where water levels are perennially elevated and reducing conditions prevail. These deposits often contain reducing minerals such as pyrite, vivianite, and jarosite. In more arid settings, flood-basin sediments and playa-lake deposits contain a greater percentage of mud cracks, bioturbation features, and authigenic minerals such as calcite, gypsum, and other evaporite minerals.

Avulsion deposits Avulsion is common in many low-gradient alluvial river systems and is best known from studies of the Rhine-Meuse Delta, the Saskatchewan River, and the Mississippi River and Delta (Smith *et al.*, 1989; Stouthammer and Berendsen, 2000; Aslan *et al.*, 2005). A significant aspect of avulsion is that this process accounts for large volumes of fine-grained overbank sedimentation on floodplains (Smith *et al.*, 1989; Aslan and Autin, 1999; Pérez-Arlucea and Smith, 1999). During an avulsion, water and sediment escape from the main channel and are transferred to the floodplain through a series of prograding crevasse-splay complexes and/or anastomosing channels (Smith *et al.*, 1989). These channels deposit silt and clay and lesser amounts of sand in local floodplain depressions such as wetlands and lakes. As these topographic depressions fill with sediment, the locus of floodplain deposition shifts downvalley. Because avulsive deposition occurs at all stages of flow and is not restricted to relatively rare large-magnitude floods, this process results in rapid floodplain aggradation. Texturally, avulsive deposits strongly resemble fine-grained sediments that are typically associated with overbank flooding. However, the stratigraphic architecture of avulsive deposits differs from that of typical overbank deposits (**Fig. 6B**). Overbank flooding causes grain size and sediment accumulation rate to decrease away from the active channel. In contrast, avulsion deposit grain sizes do not vary systematically as a function of distance from the active channel. In fact, sedimentation rates can actually increase away from the active channel.

Fluvial Channel Patterns

Fluvial channels have been traditionally subdivided into straight, meandering, and braided (Leopold and Wolman, 1957). More recent classifications recognize anastomosing channels (Miall, 1977, 1996; Rust, 1978) (**Fig. 11A**). These classifications emphasize two factors: channel sinuosity and bifurcation patterns. For example, braided rivers are characterized by channel bifurcation around bars or islands and low channel sinuosities, whereas anastomosing channels are separated by larger-scale flood basins (Schumm, 1977; 1985; Makaske, 2001; Bridge, 2003). Problems with existing channel classification schemes include measurements of the sinuosity of multi-channel rivers, calculations of braiding indices, and delineation of individual channels versus channel belts. Additionally, channel patterns are not mutually exclusive. For example, anastomosing rivers can have braided channel segments (Bridge, 2003). Discrimination of channel patterns using channel characteristics and hydrologic data can be problematic and remains an area of continued research (Lewin and Brewer, 2001; Van den Berg and Bledsoe, 2003). An alternative alluvial river classification scheme proposed by Makaske (2001) subdivides channel systems based on numbers of channel belts, braiding characteristics, and channel sinuosity (**Fig. 11B**). Although floodplain classification has received far less attention than channel classification, Nanson and Croke (1992) provide a recent classification of floodplain systems.

Changes in channel patterns are commonly associated with one or more of the following variables: 1) discharge, 2) sediment load and type, 3) regional gradient, and 4) bed or bank materials (Schumm, 1977; 1985). In general, increases in discharge, bed load, and gradient favor the development of low-sinuosity or braided streams. In contrast, decreases in discharge and gradient and increases in suspended-sediment load favor the development of meandering

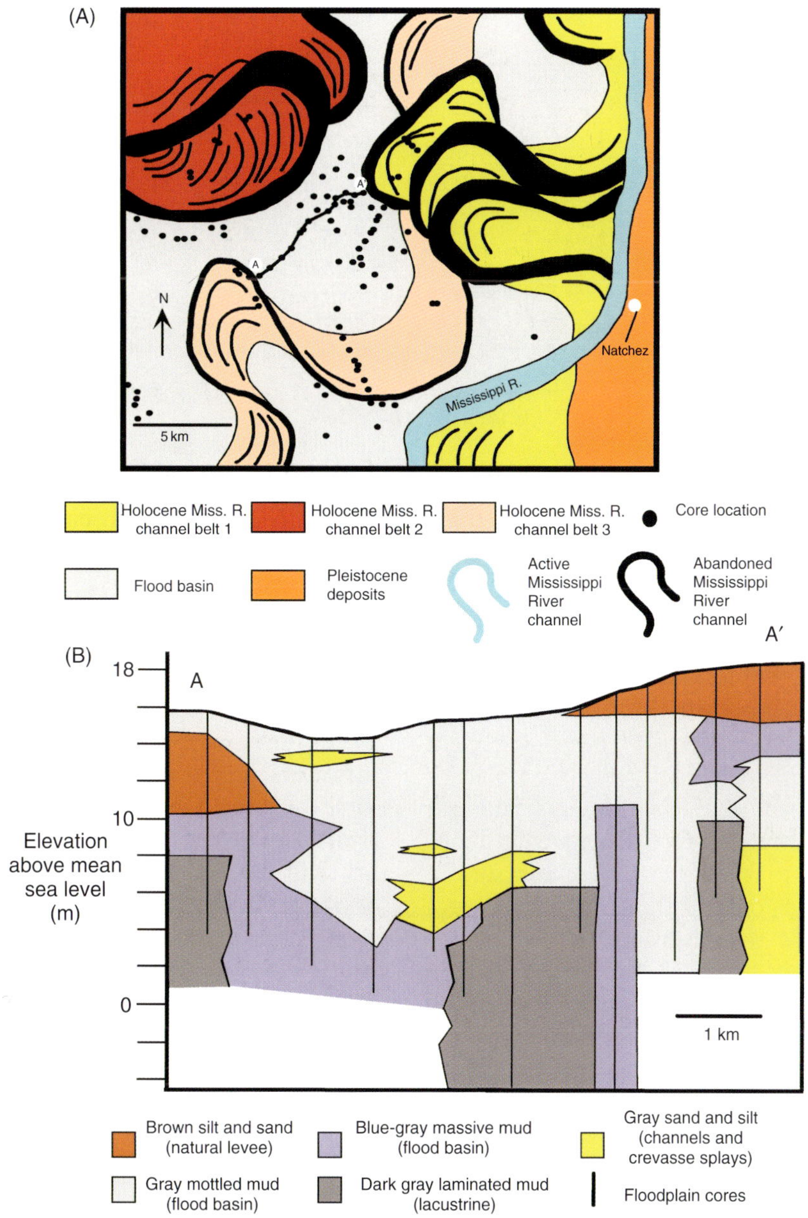

Figure 10 (A) Map of Holocene floodplain deposits of the Mississippi River near Natchez, Mississippi. (B) Cross-section showing typical overbank sediments in the region. Brown silt and sand wedges represent natural-levee deposits. Gray sand and silt represent crevasse-splay sediments (thin sheets to lenticular bodies) whereas the thicker sand accumulations represent floodplain channels. Gray mottled mud and blue-gray massive mud represent flood-basin environments associated with predominantly oxidizing and reducing conditions, respectively. Dark gray laminated mud represents lacustrine strata. From Aslan and Autin (1999).

rivers or anastomosing systems. Anastomosing rivers are especially common where regional rates of aggradation are high and bank materials are cohesive (Smith, 1986; Makaske, 2001).

Sedimentary Records of Quaternary Fluvial Systems

Sedimentary records of Quaternary fluvial systems vary dramatically in terms of their abundance, alluvial architecture, and degree of stratigraphic completeness. These differences are demonstrated by comparing fluvial sediments from eroding continental interiors with those associated with subsiding continental margins.

Fluvial records from eroding continental interiors

Fluvial sediments of continental interiors typically reflect alternating episodes of aggradation, incision,

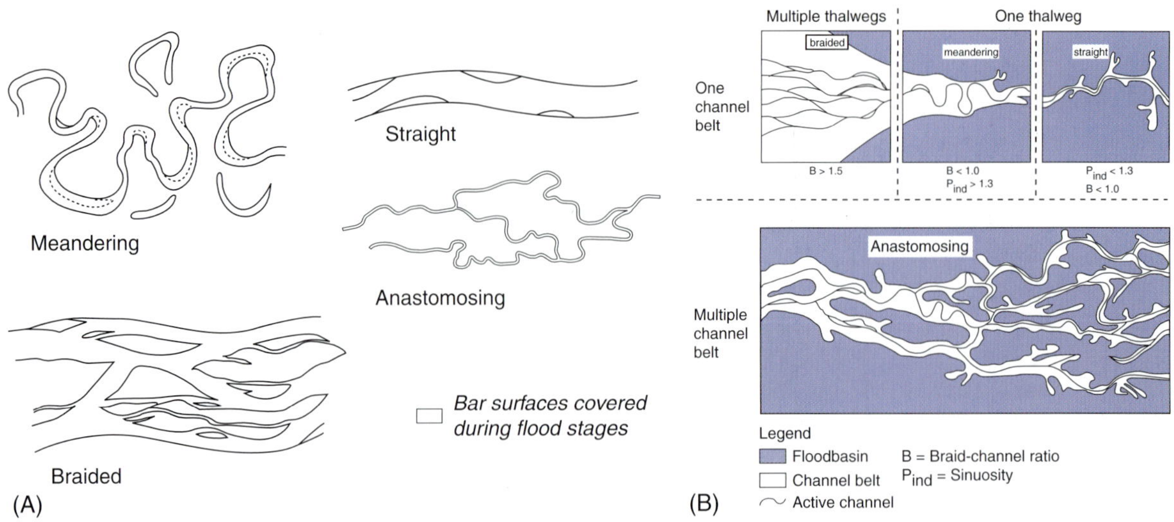

Figure 11 (A) Traditional classification scheme for alluvial channels. From Miall (1977). (B) An alternative classification of alluvial channels based on the number of channel belts present, braiding characteristics, and channel sinuosity. From Makaske (2001). Figures reprinted with permission from Elsevier.

and soil formation, and are located in river valleys. Fluvial sediments in these settings undergo extensive erosion and sediment reworking. Thus they have a low preservation potential and provide incomplete or discontinuous records of fluvial activity. The erosional nature of fluvial records in continental interiors stems largely from long-term rock and surface uplift driven by erosional isostasy or tectonic processes, which leads to fluvial incision (Pederson *et al.*, 2002). Shorter-term episodes of aggradation and incision associated with glacial-interglacial cycles are superimposed on this long-term uplift and incision (*see* Glacial-Interglacial Scale Fluvial Responses) Collectively, these processes lead to the development of flights of down-stepping terraces along many major river valleys of continental interiors (**Fig. 12**).

Sand and gravel are common constituents of fluvial sediments in continental interiors, and are associated with channel bars and bed-load deposition (**Fig. 13A**). Overbank sediments are often represented by thin veneers (**Fig. 13B**). The predominance of coarse-grained sediment reflects the relatively steep slopes of rivers in these settings and low floodplain relief, which combine to inhibit the deposition and preservation of silt and clay. The relative paucity of overbank silt and clay in continental interiors has led some workers to suggest that overbank flooding plays a minor role in the development of floodplains in these settings (Wolman and Leopold, 1957).

Figure 12 (A) River gravels associated with a Quaternary Gunnison River terrace near Grand Junction, Colorado. The terrace strath is cut into Cretaceous sandstone and is approximately 100 m above the modern Gunnison River. This terrace is part of a flight of down-stepping Quaternary terraces in the region, all of which are associated with abundant gravel. View is upstream. (B) Quaternary river gravels overlain by reddish overbank silty sand deposits of a Quaternary Colorado River terrace near Grand Junction, Colorado. The terrace strath is cut into Cretaceous shale and is approximately 30 m above the modern Colorado River. View is upstream.

Figure 13 (A) North Fork of the Gunnison River valley near Hotchkiss, Colorado. This river drains volcanic and sedimentary rocks of the central Rocky Mountains, and the channel and floodplain are represented by abundant gravel and sand. Flat benchlands in the background represent river terraces (arrows). View is upstream. (B) Cutbank along the Colorado River near Grand Junction, Colorado showing channel-bar gravel overlain by overbank silt and sand.

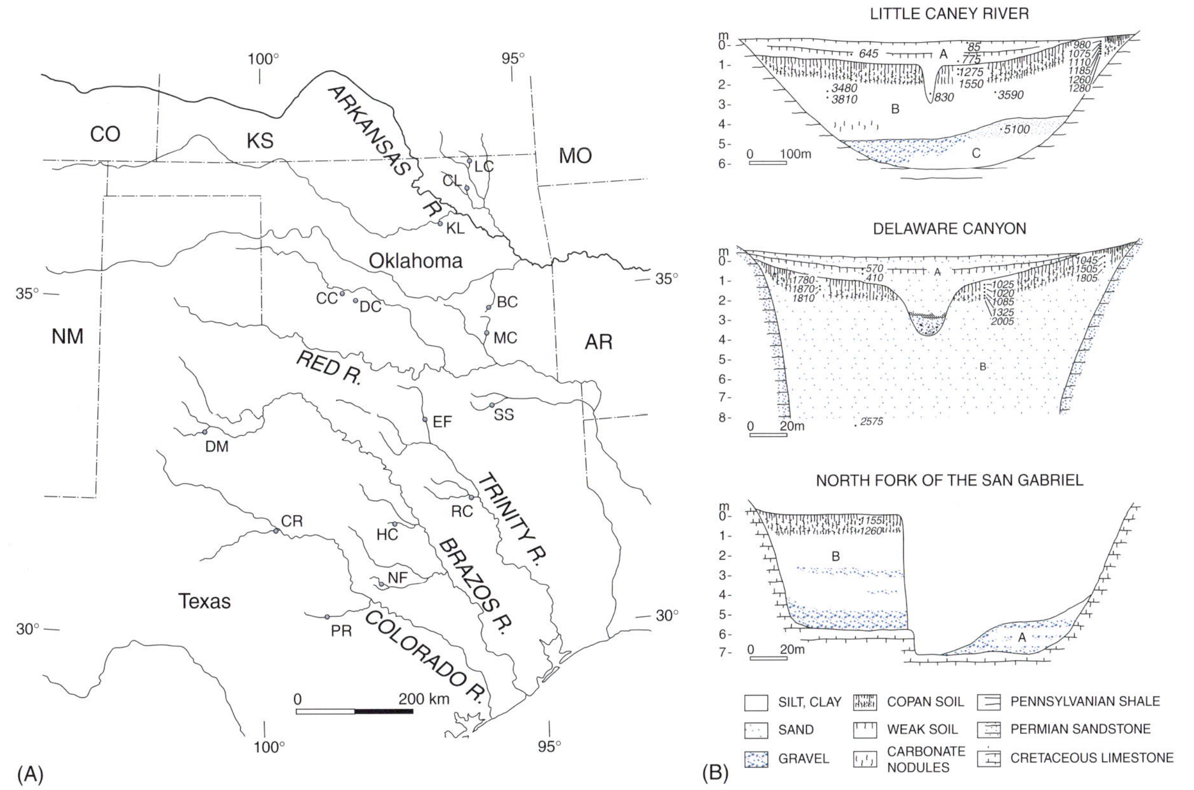

Figure 14 (A) Map showing locations of alluvial rivers that have undergone Holocene filling and entrenchment. BC = Brushy Creek; CC = Carnegie Canyon; CL = Candy Lake; CR = Colorodo River; DC = Deleware Canyon; DM = Double Mountain Fork; HC = Hog Creek; EF = Elm Fork of the Trinity River; KL = Keystone Lake; LC = Little Caney River; MC = McGee Creek; NF = North Fork of San Gabriel River; PR = Pedernales River; RC = Richland Creek; SS = South Sulfur Spring. (B) Cross sections showing the Holocene alluvial stratigraphy at three locations in (A). Sand and gravel deposition and Holocene aggradation commenced ~5 ka in all three locations, which produced depositional unit 'B' in the cross sections. From 2 to 1 ka, sedimentation slowed and a cumulic, organic-rich soil (Copan Soil) developed. At 1 ka, channel trenching occurred throughout the southern Great Plains. Channel trenching was followed by episodic floodplain sedimentation, which produced depositional unit 'A' in the cross sections. Ages are in radiocarbon years before present. Pre-middle Holocene sediments (depositional unit 'C') are rare in the valley fills. These types of valley-fill fluvial deposits have low long-term preservation potential because of the narrow valley widths and the frequency of channel entrenchment and filling. From Hall (1990).

Fluvial sediments of continental interiors (e.g., Holocene arroyo fills) are often associated with complex alluvial stratigraphies that reflect alternating episodes of aggradation, entrenchment, and soil formation (**Fig. 14**). Alluvial successions in the southern Great Plains of the United States provide excellent examples of complex fill histories, and contain regionally correlative stratigraphic units. For example, numerous Holocene valley-fill deposits in this region record aggradation from 5 to 2 ka, soil formation from 2 to 1 ka, and entrenchment since 1 ka followed by episodic floodplain sedimentation (**Fig. 14B**). The correlative nature of the alluvial stratigraphic units in these valleys, and similarities in the timing of the major aggradation and incision events suggest that climatic influences and changes in river hydrology have largely controlled the alluvial stratigraphy of the southern Great Plains (Hall, 1990) as well as elsewhere in the southwestern United States (Walters and Haynes, 2001). Narrow valley widths relative to channel widths cause rivers to rapidly rework older alluvial deposits in these settings. Thus the length of time represented by the fluvial sediments is limited.

Fluvial records from subsiding continental margins

Fluvial sediments of continental margins typically accumulate in subsiding sedimentary basins. For example, the Gulf Coast region of the United States represents a subsiding passive margin that consists of stacked successions of fluvial-deltaic Quaternary units (**Fig. 15**). Because these sediments are accumulating in a subsiding basin, they have a high preservation potential and the sedimentary records of fluvial activity are more complete than those of continental interiors. Rivers in this region occupy broad alluvial valleys separated by low upland divides. Large floodplain widths relative to channel widths, gentle slopes, and depositional relief created by sedimentation along channels (i.e., elevated alluvial ridges) lead to the accumulation and preservation of significant volumes of fine-grained overbank deposits, in addition to coarse-grained channel sediments (**Fig. 16**).

Late Quaternary fluvial sediments of continental margins predominantly occur as valley-fill complexes associated with glacial-interglacial cycles (spans approximately 100 ky) (**Fig. 17**). Packages of fluvial sediments accumulate in response to

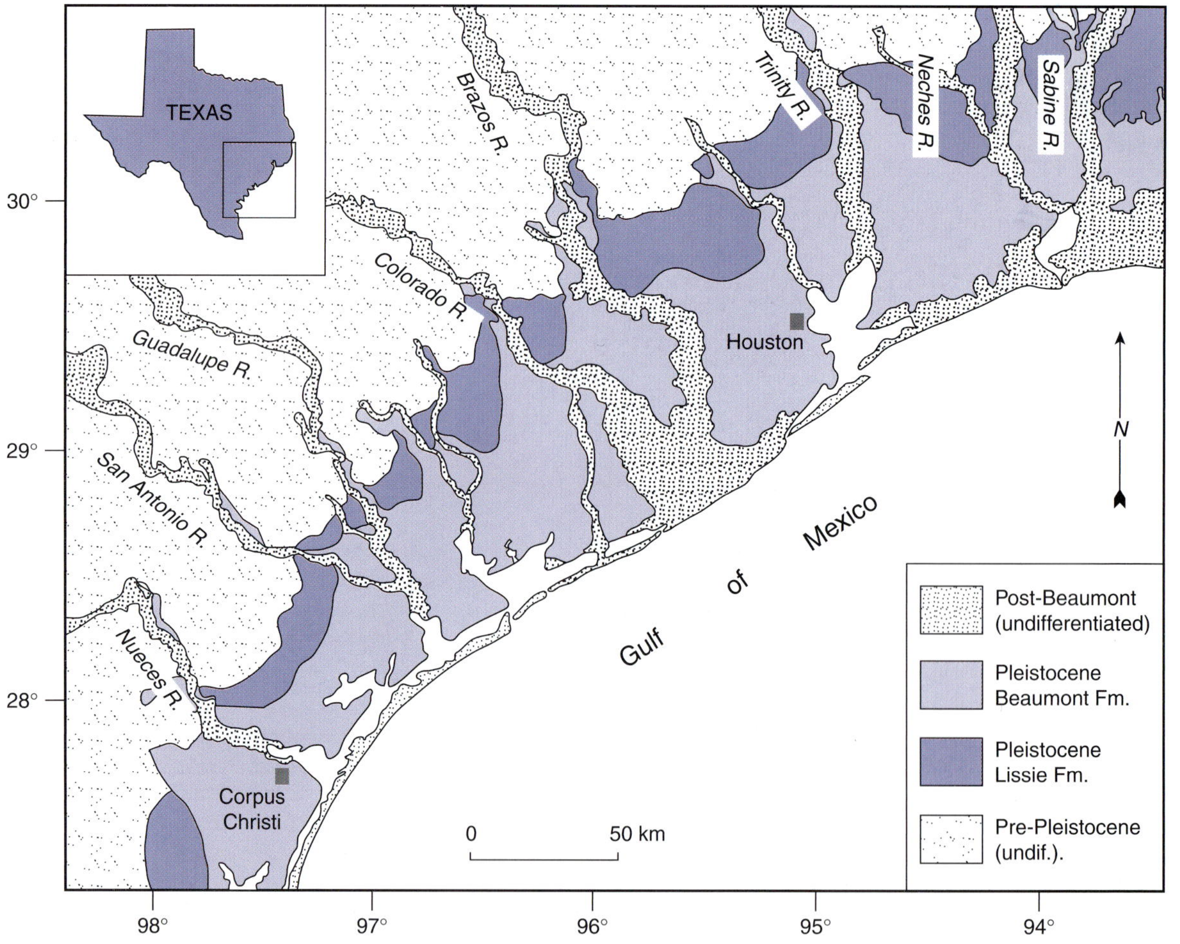

Figure 15 Geologic map of the Texas Gulf Coastal Plain showing the distribution of late Quaternary deposits and major river valleys. Modified from Blum and Price (1998).

Figure 16 Holocene and late Pleistocene fluvial deposits of the Colorado River of south Texas. Holocene overbank deposits veneer a late Pleistocene paleosol that represents the upper boundary of fluvial sediments deposited during the oxygen-isotope-stage (OIS) 5a highstand approximately 70 ka. Most of the sediments below this paleosol accumulated during a period of rising sea level. The light colored zone at the base of the section shows a paleosol that formed during OIS 6, which represented a period of alternating episodes of floodplain construction and valley incision.

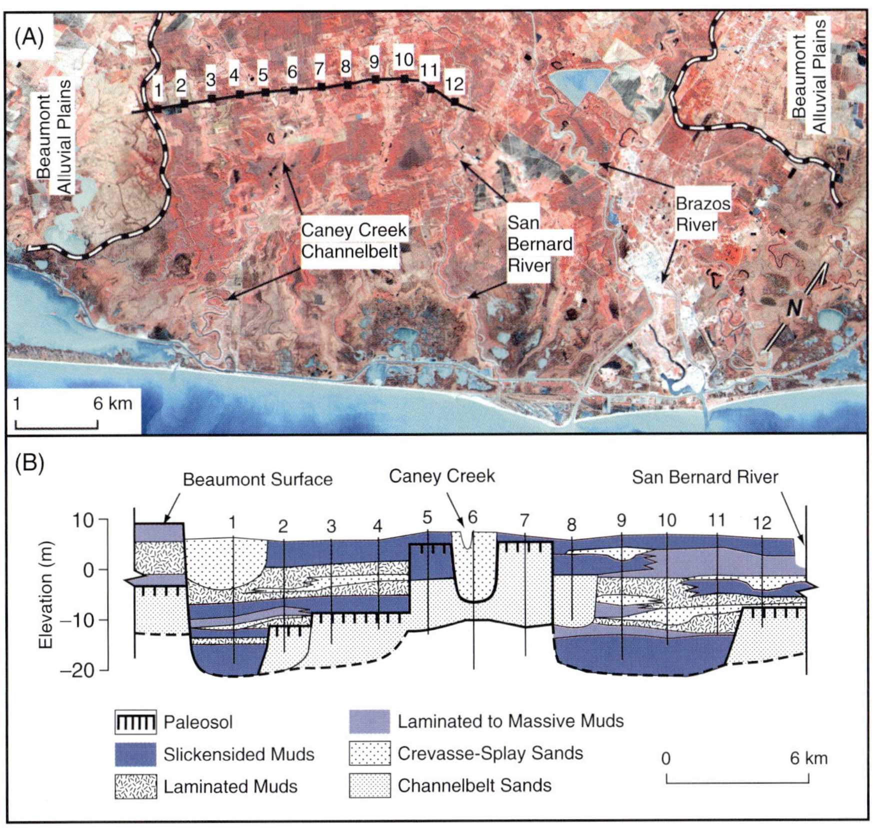

Figure 17 (A) Satellite image of the Lower Colorado River alluvial plain and the transect line and core-sampling sites for the cross section shown in (B). (B) Stratigraphic cross section showing the facies architecture of the late Pleistocene to Holocene Colorado River valley fill complex. The valley fill consists of a series of down-stepping terraces underlain by channel sands and separated from one another by paleosols and erosional unconformities. The sands range in age from approximately 60 to 20 ka. These coarse-grained basal valley-fill deposits are overlain by a mixture of channel and overbank sediments that were deposited during late Pleistocene to Holocene sea-level rise. From Blum and Törnqvist (2000). Reprinted with permission from Blackwell Publishing.

changes in sea level and river hydrology that reflect glacial versus interglacial conditions (Blum and Törnqvist, 2000). In the example of the Colorado River of the Texas Gulf Coast region, the river incised and filled its current valley over the past 70 ka (Blum and Price, 1998). The valley fill consists of a basal Late Pleistocene sand-dominated unit that is overlain by a mixed sand- and mud-rich succession of Holocene age. The basal sand-dominated unit is comprised of a series of down-stepping terraces and well expressed buried soils (**Fig. 17**). During the Holocene, the valley filled with fine-grained muddy overbank deposits that encase channel and minor overbank sand bodies. Similar muddy overbank sediments are thin or absent from the basal Late Pleistocene portion of the valley fill. This major difference in the abundance of overbank sediments in Holocene and Late Pleistocene portion of the valley fill could reflect less frequent episodes of deep overbank flooding during the Late Pleistocene (Blum and Törnqvist, 2000).

Over longer time scales involving multiple glacial-interglacial cycles, avulsion during late stages of valley filling can lead to abandonment of an alluvial valley (Blum and Törnqvist, 2000) (*see* Glacifluvial Landforms of Deposition). Avulsion causes a shift in the location of the valley axis and future valley incision and filling, which causes the ages of valley-fill complexes to vary both laterally and vertically (**Fig. 18**). In areas of active subsidence, alluvial valley-fill complexes locally onlap and cross cut one another. The basal valley-fill unconformities form a composite unconformity that is overlain predominantly by sandy channel-belt deposits with well expressed paleosols. In contrast, the upper portions of the valley fills are a combination of fine-grained overbank and sandy channel deposits.

Summary

Fluvial sediments provide important records of terrestrial environments that are critical for interpreting geologic histories of continental settings. These sediments are also important for evaluating petroleum, coal, and water resources. Fluvial sediments are traditionally subdivided into coarse-grained channel deposits and fine-grained overbank sediments. Channel deposits consist of channel-bar and channel-fill sediments, and accumulate in response to downstream progradation of large-scale bar forms and periodic channel abandonment. Overbank deposits accumulate in natural-levee, crevasse-splay, and flood-basin environments through a combination of overbank flooding and avulsion. Fluvial sediments vary in terms of abundance, architecture, and stratigraphic completeness between continental interiors and margins. In eroding continental interiors,

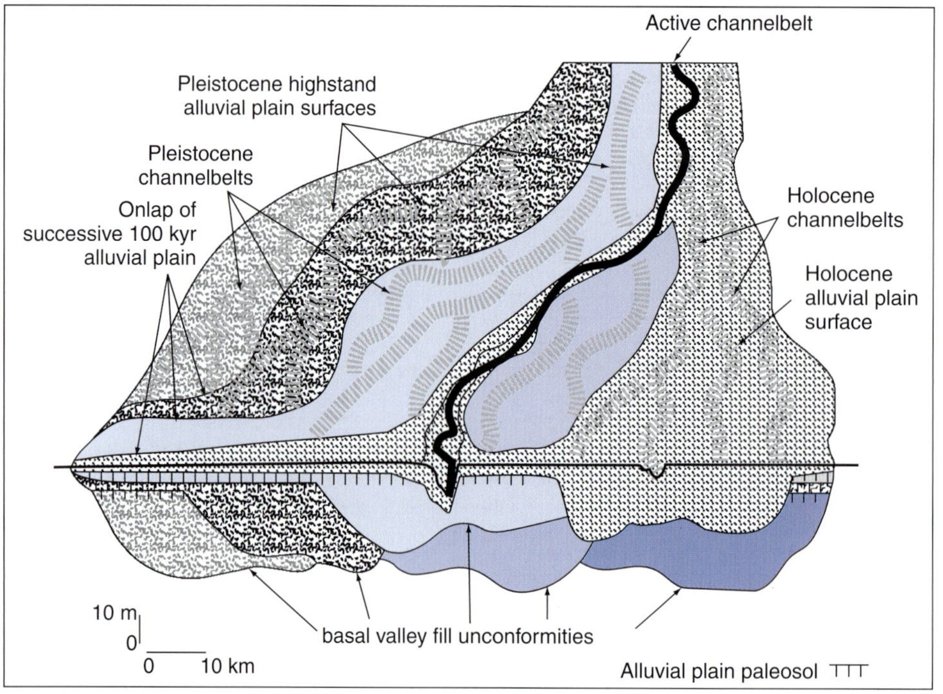

Figure 18 Model showing alluvial-plain construction and the distribution of valley-fill complexes along a subsiding passive margin. From Blum and Törnqvist (2000). Reprinted with permission from Blackwell Publishing.

fluvial sediments are represented by flights of terraces underlain by coarse-grained deposits. Fine-grained overbank sediments are relatively rare. These fluvial sediments reflect long-term incision and common sediment reworking, and thus they preserve relatively incomplete records of fluvial activity. In subsiding continental margins, thick accumulations of both sandy channel and muddy overbank deposits are preserved. These thick fluvial successions provide relatively complete long-term records of fluvial activity.

See also: **Fluvial Environments**: Responses to Rapid Environmental Change; Terrace Sequences; Deltaic Environments. **Glacial Landforms, Sediments**: Glacifluvial Landforms of Deposition. **Glacial-Interglacial Scale Fluvial Responses**. **Paleoceanography, Physical and Chemical Proxies**: Terrigenous Sediments. **Paleosols and Wind-Blown Sediments**: Nature of Paleosols. **Quaternary Stratigraphy**: Lithostratigraphy; Sequence Stratigraphy.

References

Allen, J. R. L. (1965). A review of the origin and characteristics of recent alluvial sediments. *Sedimentology* **5**, 91–191.

Aslan, A., and Autin, W. J. (1999). Evolution of the Holocene Mississippi River floodplain, Ferriday, Louisiana: insights on the origin of fine-grained floodplains. *Journal of Sedimentary Research* **69**, 800–815.

Aslan, A., Autin, W. J., and Blum, M. D. (2005). Causes of river avulsion: insights from the late Holocene avulsion history of the Mississippi River, U.S.A. *Journal of Sedimentological Research* **75**, 648–662.

Blum, M. D., and Price, D. M. (1998). Quaternary alluvial plain construction in response to glaci-eustatic and climatic controls, Texas Gulf Coastal Plain. In *Relative Role of Eustasy, Climate and Tectonism on Continental Rocks* (K. J. Shanley and P. J. McCabe, Eds.) pp. 31–48, SEPM Spec. Pub. 59.

Blum, M. D., and Törnqvist, T. E. (2000). Fluvial responses to climate and sea-level change: a review and look forward. *Sedimentology* **47**, 2–48.

Bridge, J. S. (2003). *Rivers and Floodplains, Forms, Processes, and Sedimentary Record.* Blackwell Science Ltd.

Farrell, K. M. (1987). Sedimentology and facies architecture of overbank deposits of the Mississippi River, False River region, Louisiana. In *Recent Developments in Fluvial Sedimentology* (F. G. Ethridge and R. M. Flores, Eds.) pp. 111–120, SEPM Spec. Pub. 39.

Flores, R. M. (1981). Coal deposition in fluvial paleoenvironments of the Paleocene Tongue River Member of the Fort Union Formation, Powder River area, Powder River Basin, Wyoming and Montana. In *Recent and Ancient Nonmarine Depositional Environments: Models for Exploration* (F. G. Ethridge and R. M. Flores, Eds.) pp. 161–190, SEPM Spec. Pub. 31.

Hall, S. A. (1990). Channel trenching and climatic change in the southern U.S. Great Plains. Geology **18**, 342–345.

Kraus, M. J. (1999). Paleosols in clastic sedimentary rocks: their geologic applications. *Earth Science Reviews* **47**, 41–70.

Leopold, L. B., and Wolman, M. G. (1957). River channel patterns: braided, meandering and straight. *U.S. Geological Survey Professional Paper* **282-B**, 39–85.

Lewin, J., and Brewer, P. A. (2001). Predicting channel patterns. *Geomorphology* **40**, 329–339.

Makaske, B. (2001). Anastomosing rivers; a review of their classification, origin, and sedimentary products. *Earth Science Reviews* **53**, 149–196.

Miall, A. D. (1977). A review of the braided river depositional environment. *Earth Science Reviews* **13**, 1–62.

Miall, A. D. (1996). *The Geology of Fluvial Deposits.* Springer-Verlag.

Morozova, G. S., and Smith, N. D. (1999). Holocene avulsion history of the Lower Saskatchewan fluvial system, Cumberland marshes, Saskatchewan-Manitoba, Canada. In *Fluvial Sedimentology VI* (N. D. Smith and J. Rogers, Eds.) pp. 231–249, Spec. Pub. Int. Ass. Sediment. 28.

Nanson, G. C., and Croke, J. C. (1992). A genetic classification of floodplains. *Geomorphology* **4**, 459–486.

Pederson, J. L., Mackley, R. D., and Eddleman, J. L. (2002). Colorado Plateau uplift and erosion evaluated using GIS. *GSA Today* **12**, 4–10.

Pérez-Arlucea, M., and Smith, N. D. (1999). Depositional patterns following the 1870s avulsion of the Saskatchewan River (Cumberland Marshes, Saskatchewan Canada). *Journal of Sedimentary Research* **69**, 62–73.

Platt, N. H., and Keller, B. (1992). Distal alluvial deposits in a foreland basin setting – the lower Freshwater Molasse (Lower Miocene), Switzerland: sedimentology, architecture and palaeosols. *Sedimentology* **39**, 545–565.

Rust, B. R. (1978). A classification of alluvial rivers. In *Fluvial Sedimentology* (A. D. Miall, Ed.). *Can. Soc. Petrol. Geol. Mem* 5, 187–198.

Schumm, S. A. (1977). *The Fluvial System.* John Wiley and Sons.

Schumm, S. A. (1985). Patterns of alluvial rivers. *Annual Review of Earth and Planetary Sciences* **13**, 5–27.

Slingerland, R., and Smith, N. D. (2004). River avulsions and their deposits. *Annual Review of Earth and Planetary Sciences* **32**, 257–285.

Smith, D. G. (1986). Anastomosing river deposits, sedimentation rates and basin subsidence, Magdalena River, northwestern Colombia, South America. *Sedimentary Geology* **46**, 177–196.

Smith, N. D., and Pérez-Arlucea, M. (1994). Fine-grained splay deposition in the avulsion belt of the lower Saskatchewan River. *Canadian Journal of Sedimentary Research* **B64**, 159–168.

Smith, N. D., Cross, T. A., Difficy, J. P., and Clough, S. R. (1989). Anatomy of an avulsion. *Sedimentary* **36**, 1–23.

Stouthamer, E., and Berendsen, H. J. A. (2000). Factors controlling Holocene avulsion history of the Rhine-Meuse delta (The Netherlands). *Journal of Sedimentary Research* **A70**, 1051–1064.

Törnqvist, T. E., and Bridge, J. S. (2002). Testing a 3-D process-based alluvial architecture model with new field data: the control of spatial variation in floodplain deposition rate on avulsion frequency. *Sedimentology* **49**, 891–905.

Van den Berg, J. H., and Bledsoe, B. P. (2003). Predicting channel patterns; discussion and reply. *Geomorphology* **53**, 333–342.

Walters, M. R., and Haynes, C. V. (2001). Late Quaternary arroyo formation and climate change in the American Southwest. *Geology* **29**, 399–402.

Willis, B. J., and Behrensmeyer, A. K. (1994). Architecture of Miocene overbank deposits in northern Pakistan. *Journal of Sedimentological Research* **B64**, 60–67.

Wolman, M. G., and Leopold, L. B. (1957). River flood plains: some observations on their formation. *U.S. Geological Survey Professional Paper* **282-C**, 87–109.

Responses to Rapid Environmental Change

T E Törnqvist, Tulane University, LA, USA

© 2007 Elsevier B.V. All rights reserved.

Introduction

The recognition of centennial to millennial scale environmental change, notably abrupt climate change, has sparked a great interest into how such rapidly changing boundary conditions influence landscape processes. Rivers are widely considered geomorphic entities that are highly sensitive to environmental change, but it is increasingly recognized that the fluvial response is extremely complex, with considerable spatiotemporal variability (Knox, 1983). This article provides an overview of the variety of fluvial responses to rapid environmental change and is intimately linked to the preceding article that focuses on longer-term (glacial–interglacial scale) fluvial responses (*see* Glacial-Interglacial Scale Fluvial Responses). The article will primarily discuss the effects of rapid climate change on fluvial systems in the alluvial reach (i.e., that portion of the fluvial system that is most likely to preserve – at least in a temporary fashion – a stratigraphic record); bedrock rivers are discussed in more detail elsewhere (*see* Terrace Sequences). The first part consists of a discussion of general concepts and criteria for the recognition of fluvial responses, followed by a selection of case studies that provide examples of the different manifestations of late Quaternary fluvial responses to rapid environmental change.

General Concepts

It is usual to distinguish between climate change, base-level (commonly sea-level) change, and tectonism as the three categories of external (or allogenic) forcing of fluvial systems. In addition, distinction has been made between upstream and downstream controls. While climate change is typically an upstream control (i.e., the forcing propagates downstream through the drainage basin), sea-level change provides a distinct downstream control and can only affect the lowermost reaches of the fluvial system. Considerable debate has centered on the question of how far upstream sea-level control can propagate and it is likely that this varies greatly, depending on variables such as slope and sediment supply (Blum and Törnqvist, 2000). In principle, tectonism can affect any segment of the fluvial system. In addition, the various allogenic controls commonly operate simultaneously.

Modes of Response to Allogenic Forcing

There are different mechanisms of fluvial response to allogenic forcing, including:

1) changes in the discharge regime;
2) changes in channel pattern (fluvial style); and
3) changes in the longitudinal profile by means of aggradation or degradation (incision).

Perhaps the most obvious and direct response to climate forcing is the change of discharge regime and associated changes in hydraulic geometry parameters, such as channel width, depth, and sinuosity. Such adjustments may in some cases lead to full transformations of fluvial style.

Changes in fluvial style have been the subject of research for a long time, commonly focusing on the time period around the Pleistocene–Holocene transition (Kozarski, 1983; and numerous others). Many of these investigations concern the transformation from braided to meandering channel patterns. While there is lively and ongoing debate about the boundary conditions that cause rivers to braid or meander (Van den Berg, 1995; Lewin and Brewer, 2001), it appears likely that braiding is favored by a high bed load and a high stream power (i.e., discharge and valley slope). In addition, several other factors have been mentioned in the literature (e.g., limited bank stability). Rapid climate change can modify boundary conditions that control fluvial style, but changes in fluvial style due to climate forcing can be spatially non-uniform.

Perhaps the most prolific response of the fluvial system to allogenic forcing is the adjustment of the longitudinal profile, either by aggradation or by degradation (incision). The concept of grade (Mackin, 1948) postulates that alluvial rivers strive to attain an equilibrium profile characterized by a more or less balanced sediment budget, typical of stable boundary conditions in terms of base level, discharge, and sediment supply (*see* Terrace Sequences). Disruptions of this delicate balance commonly occur by means of changes in the proportion between water discharge (Q_w) and sediment supply (Q_s). It is important to note that sediment supply in this context refers strictly to bed load, given the fact that rivers are virtually never saturated with respect to the transport of suspended load, while the supply of bed load at times exceeds sediment-transport capacity. **Figure 1** shows how the balance between Q_w and Q_s dictates whether a fluvial reach will incise, aggrade, or maintain equilibrium. In this context, it is important to reiterate that disruptions in the Q_w/Q_s

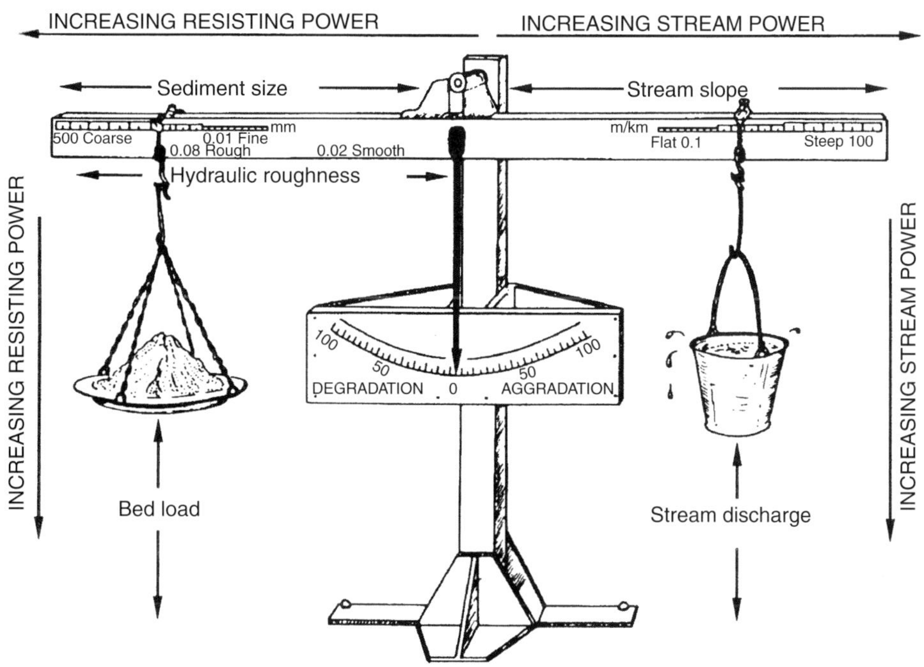

Figure 1 Balance model for aggradation and degradation (incision) in alluvial rivers, as dictated by the relative proportion of water discharge and sediment supply, referred to by Bull (1991) as the balance between stream power and resisting power. Channels aggrade when sediment supply (bed load: amount and grain size) exceeds sediment-transport capacity as determined primarily by discharge and slope, and incise when the reverse is the case. The role of base-level change can also be inferred from this conceptualization. Base-level rise tends to reduce the slope and therefore promote aggradation, whereas base-level fall may increase the slope leading to incision (also depending on the morphology of the surface that is exposed). Originated from an unpublished diagram by W. Borland; after Bull (1991). Reprinted by permission from Oxford University Press.

ratio may be accommodated by other adjustments (such as sinuosity), rather than by lowering or raising the longitudinal profile.

Recognizing Fluvial Incision

Fluvial aggradation and incision not only reflect changes in the sediment budget but also leave the most tangible imprint in the stratigraphic record. In fact, many studies of fluvial archives aim to identify fluvial unconformities resulting from phases of incision with regional significance that may be ascribed to allogenic forcing. However, the proper recognition of regional unconformities representing widespread incision has proven challenging for a number of reasons. This is due to the simple fact that river channels are erosional features by definition. Following Salter (1993) and Best and Ashworth (1997), it is therefore critical that distinction be made between fluvial scour, the autogenic (intrinsic) process that characterizes any alluvial channel (**Fig. 2**) versus allogenic incision. Best and Ashworth (1997), working in the vertically aggrading Ganges-Brahmaputra Delta, showed that channel scour can lead to reworking of the channel belt to depths as much as five times mean channel depth. They go on to show how the resulting stratigraphic phenomena can easily be mistaken for fluvial incision with regional significance. In fact, numerous examples can be gleaned from the literature where channel scour in overall aggradational settings may have been misinterpreted as fluvial incision.

So how can fluvial incision and aggradation be recognized in the stratigraphic record? Given the discussion above, this is a critical, yet by no means a trivial question. Distinction should be made as to whether incision/aggradation is measured relative to the channel bed or the floodplain surface. Incision of the channel bed can be the result of base-level change, vertical crustal movements, or changes in Q_w relative to Q_s (**Fig. 3A**). However, changes in the discharge regime alone (i.e., the 'flashiness' of discharge) can cause floodplain aggradation or degradation, without vertical changes in the channel bed. Floodplains that are frequently inundated by deep overbank floods have the potential to aggrade to levels well above the channel bed (**Fig. 3B**). If lateral channel migration goes along with a reduction of the peak discharge (e.g., as a result of a decrease in the frequency of major storms) a new floodplain will, over time, be established at a lower level, leaving the previous floodplain behind as a terrace. These two mechanisms of fluvial terrace formation are shown schematically in **Figures 4A** and **4B**, respectively, and

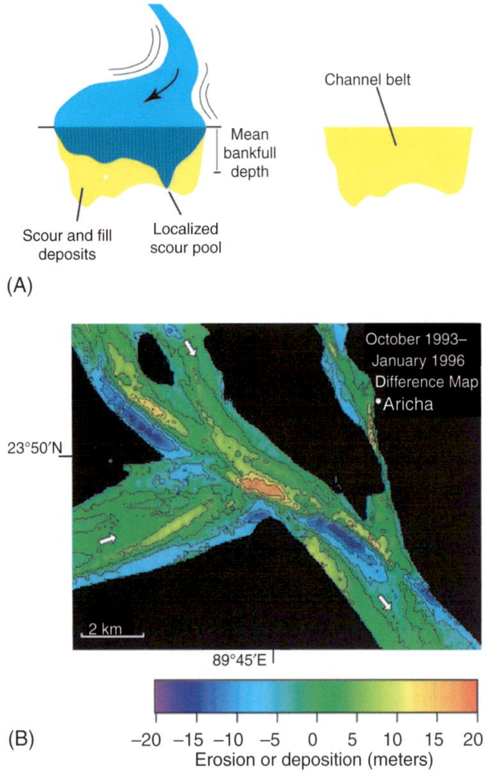

Figure 2 (A) Channel scour and fill in alluvial rivers leads to continuous reworking and deepening of the channel belt, exceeding mean bankfull channel depth (modified after Salter, 1993). (B) Repeated bathymetric surveys of the Ganges-Brahmaputra confluence show the dynamic nature of alluvial channels and their capability to form scour holes about five times the average channel depth that are likely to rework channel belts over annual to decadal timescales (after Best and Ashworth, 1997). Caution must be taken not to confuse such phenomena with fluvial incision and associated floodplain degradation. Reprinted by permission from Macmillan Publishers Ltd.

Figure 3 (A) The Sabarmati River, Gujarat, India, about 100 km upstream from Ahmadabad, bordered by a ~40-m-high terrace. The uppermost terrace deposits date to ~12 ka, whereas the bars are ~4.5 ka in age (Srivastava *et al.*, 2001), indicating spectacular incision that has been attributed to the early Holocene increase in monsoon activity that affected large sections of South Asia during this time (Tandon *et al.*, 2006). Photograph courtesy Martin Gibling. Reprinted by permission from SEPM (Society for Sedimentary Geology). (B) The Amite River in southeast Louisiana after a peak discharge event. Note the high-water mark recognizable by the mud-covered vegetation. The lower arrow shows the approximate thickness of fine-grained overbank deposits overlying sandy point-bar strata. The upper arrow indicates the minimal amount of continued floodplain aggradation that can be expected if the current discharge regime remains unaltered.

are further illustrated by means of animations available from the supplementary materials. Given these different mechanisms of terrace formation, a variety of combinations of processes are capable of producing – at least in theory – a plethora of stratigraphic results that may prove challenging to unravel (Blum and Törnqvist, 2000). For example, it cannot be ruled out that different allogenic forcings operate simultaneously and cause opposite stratigraphic results (Fig. 4C). In this hypothetical example, floodplain aggradation due to an increase in peak discharge would occur simultaneously with channel-bed incision, for instance due to a slight base-level drop, and vice versa. Finally, it should be stressed that incision as defined here is broadly defined so as to encompass floodplain degradation in the absence of channel-bed incision. Obviously, long-term incision over vertical intervals that substantially exceed mean channel depth cannot occur without channel-bed incision.

The overarching conclusion is that inferring fluvial incision and aggradation from the stratigraphic record can be most successfully done when data are available on both floodplain surfaces and channel bases. Unfortunately, this presents the problem that in the naturally incomplete stratigraphic record channel bases tend to have a much higher preservation potential than floodplain surfaces. Therefore, it is absolutely critical that the criteria for recognition of fluvial incision and aggradation be explicitly discussed. Unfortunately, often this is not the case, and the literature is riddled with examples of studies that infer longitudinal-profile adjustments while potentially misinterpreting autogenic scour phenomena as reflecting regional incision

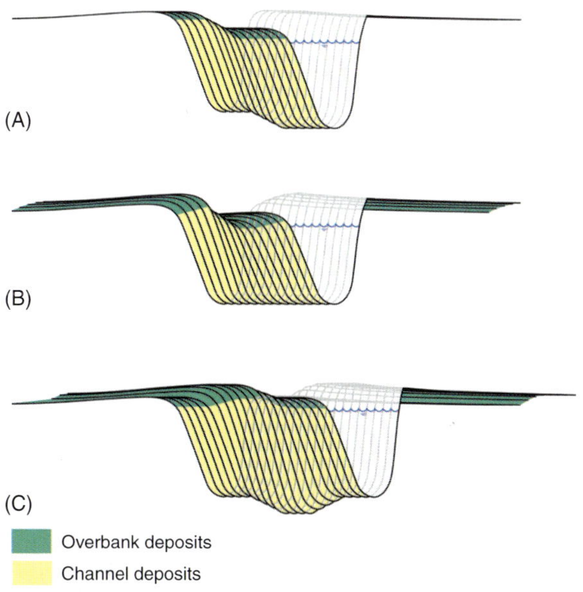

Figure 4 Hypothetical cross-sections illustrating contrasting mechanisms of fluvial incision, floodplain abandonment, and terrace formation (Blum and Törnqvist, 2000). (A) Equilibrium floodplain with lateral migration (from left to right) and minimal vertical aggradation, followed by a short phase of incision characterized by degradation of both the channel bed and the floodplain and a subsequent return to equilibrium conditions. This could be the result of either a reduction in bed load relative to discharge, a drop in base level, or tectonic uplift (or combinations thereof). (B) Fluvial system characterized by lateral migration and vertical aggradation of the floodplain due to frequent overbank flooding. A phase of floodplain degradation due to a reduction in peak discharge is followed by continued conditions of lateral migration and floodplain aggradation at a lower level. Note that the channel bed remains unchanged and that terrace formation is entirely the result of a drastic decrease in peak discharge and resulting floodplain degradation. (C) A combination of the processes operating in the previous two scenarios, where channel-bed degradation occurs simultaneously with floodplain aggradation due to increased overbank flooding. Subsequently, channel-bed aggradation coincides with reduced overbank flooding, resulting in the counterintuitive situation of simultaneous terrace formation and channel-bed aggradation. Animations of all three scenarios are available online as supplementary material.

(e.g., due to a focus exclusively on channel-belt bases rather than floodplain surfaces).

Complex Response and Response Time

A potentially large obstacle in the interpretation of fluvial stratigraphic records is the possibility of the occurrence of nonlinear responses. An experimental study by Schumm and Parker (1973) showed how a simple, rapid drop in base level can lead to alternating fluvial incision and aggradation. The concept of 'complex response' was subsequently discussed at length by Schumm (1977), who advocated that such nonlinearity can be caused by a variety of perturbations (sediment supply, tectonism), both upstream and downstream. The cyclic nature of the Schumm and Parker (1973) experiment was later substantiated by mathematical modeling (Slingerland and Snow, 1988).

Regardless of the possibility of complex response, it has long been recognized that fluvial responses to environmental forcing exhibit time lags of variable duration. For example, Bull (1991) defines the geomorphic response time as the sum of the reaction time (time between the perturbation and the initial adjustment) and the relaxation time (time between the initial adjustment and renewed equilibrium). The spirit of these concepts is not unlike the theoretical work by Paola *et al.* (1992), who introduced the concept of equilibrium time (T_{eq}), defined as

$$T_{eq} = L^2/\upsilon$$

where L = basin (river) length (a proxy for the size of the drainage basin) and υ = diffusivity (efficiency of sediment transport). It follows that small drainage basins that permit rapid sediment transfer are more likely to record abrupt, short-term climate changes than large fluvial systems. Recently, Castelltort and Van Den Driessche (2003) have applied this theory to 93 large rivers worldwide, suggesting that tens to hundreds of thousands of years are needed to fully accommodate changes in sediment supply and restore equilibrium conditions. In addition, these authors suggest that upstream forcing is in many cases accommodated in the fluvial transfer zone, implying that perturbations would not propagate to the river mouth. Nevertheless, as will be exemplified below, several cases have been documented of fairly rapid fluvial responses to environmental forcing, including examples from large drainage basins.

Case Studies

This article focuses on climate controls, because sea-level fluctuations that affect the downstream portions of the longitudinal profile commonly act over longer time intervals (*see* Glacial-Interglacial Scale Fluvial Responses). With the exception of episodic seismic events, tectonism also concerns longer timescales (*see* Terrace Sequences). The last glacial and its transition into the Holocene have witnessed numerous abrupt climate changes that have affected considerable parts of the globe. These changes manifest themselves as drastic fluctuations in temperature and precipitation, particularly well documented in the Greenland ice-core records (*see* Greenland Stable Isotopes). The Dansgaard-Oeschger events serve as some of the most prolific examples that were of specific importance in the North Atlantic region, but lower latitudes have experienced variations in monsoonal activity as well as higher frequency changes such as the

El Niño–Southern Oscillation. In addition, solar forcing has been proposed as another mechanism for short-term climate change.

In many respects, the Pleistocene–Holocene transition can be viewed as one of the most abrupt climate changes of the last glacial–interglacial cycle, in essence corresponding to Termination I. This is particularly the case in temperate regions of the Northern Hemisphere that were strongly influenced, either directly or indirectly, by glaciation. These are also the environments that have the longest tradition of studies of late Quaternary fluvial evolution, so the discussion below will open with a brief overview of some of those findings. The subsequent sections present selected examples of fluvial responses to climate forcing over different timescales, with a premier emphasis on longitudinal-profile adjustments. It should be noted that the subdivision of examples reflects interpretations of the stratigraphic record by the respective authors, based on a variety of observational data. The examples range from monsoonal fluctuations that operate over 10^3–10^4 year timescales, to Dansgaard–Oeschger style abrupt (10^2–10^3 years) climate changes that have been particularly prominent in the North Atlantic region, as well as higher frequency and lower amplitude (10^0–10^3 years) climate fluctuations with variable causes that characterize the Holocene. The examples provided here not only concern different timescales but also contrasting environmental settings, ranging from the tropics to temperate regions (with the latter predominantly subarctic during the last glacial).

Pleistocene–Holocene Transition

A long tradition exists of studies of fluvial channel-pattern changes across the last glacial to interglacial transition, particularly in temperate regions outside of the former centers of glaciation. Such investigations have been numerous in the northwest European lowlands that experienced an overall transformation from periglacial conditions with sparse vegetation cover to temperate, deciduous forests with climate conditions roughly comparable to those of the present. For example, detailed mapping by Berendsen and Stouthamer (2001) of the floodplains and adjacent terraces of the lower Rhine and Meuse Rivers near the Dutch-German border (**Fig. 5**) shows that braided channel patterns dominated during the last glacial, rapidly giving way to meandering planforms in the Holocene. Numerous investigators have argued that braiding was promoted by a high supply of bed load due to a lack of vegetation cover and active periglacial hillslope processes. In contrast, increased stabilization of soils and riverbanks by vegetation during the Holocene favored the transition to meandering patterns. These findings have been mostly substantiated by recent modeling (Bogaart *et al.*, 2003a).

It is interesting to note that strikingly similar changes in fluvial style have been mapped along the Rhine and Meuse (**Fig. 5**), despite the different character of these two rivers around the Pleistocene–Holocene transition, with the Rhine – unlike the Meuse – draining a substantially glaciated hinterland.

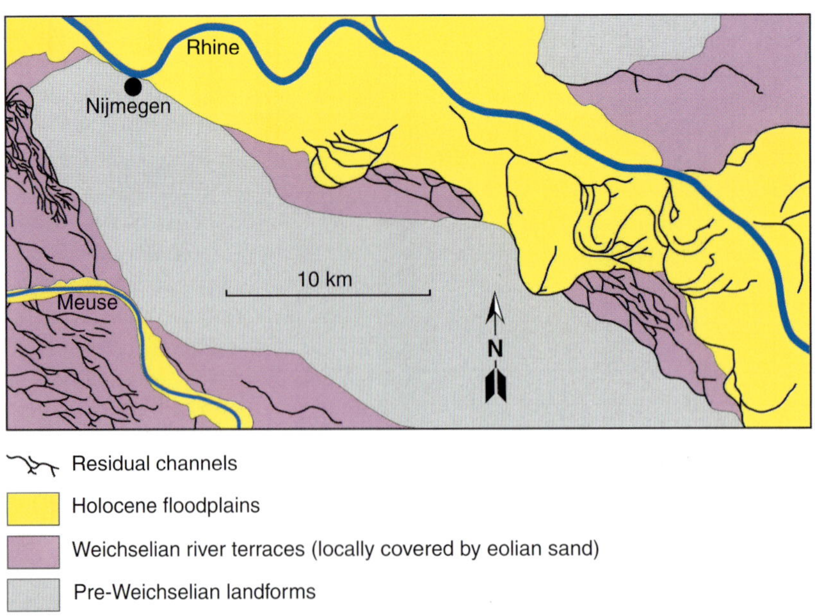

Figure 5 Channel patterns along the Rhine and Meuse Rivers during the late Pleistocene and Holocene near the Dutch-German border (after Berendsen and Stouthamer, 2001). Note the braided channel patterns on the Weichselian (late Pleistocene) terraces, as opposed to the meandering Holocene fluvial style.

Similar observations have been made in North America, with braided channel patterns dominating both glaciated drainage basins such as the Mississippi River (Fisk, 1944; Saucier, 1994) and ice-free settings along the Atlantic Coastal Plain (Leigh *et al.*, 2004) during the last glacial. As will be discussed in more detail below, higher frequency changes in fluvial style and longitudinal-profile adjustments occurred during the Pleistocene–Holocene transition in northwest Europe.

Monsoonal Forcing

Precession-dominated fluctuations in monsoonal activity in low-latitude regions are well documented. A number of recent studies highlight the associated dramatic responses of fluvial systems over millennial or slightly longer timescales, including extensive work in the Himalayan Foreland Basin. Tandon *et al.* (2006) present a large body of data from the drainage basin of the Ganges River, suggesting that widespread fluvial incision (commonly on the order of tens of meters), floodplain abandonment, and terrace formation was especially prominent during time intervals with a strengthened monsoon, particularly the period from about 15 to 5 ka. Similar processes operated farther west in India (**Fig. 3A**). The preceding period around the Last Glacial Maximum, when the southwestern monsoon was suppressed and discharge was substantially lower, was characterized predominantly by fluvial aggradation. Goodbred (2003) pointed out similar relationships and demonstrated the intimate connection between accelerated incision in the drainage basin of the Ganges River during the hypsithermal (notably the early Holocene) and accelerated sedimentation rates in the Ganges-Brahmaputra Delta (*see* Deltaic Environments). The significance of this finding is that this giant sediment-dispersal system was capable of transmitting a climate signal over a very short (millennial) timescale, from source to sink. Studies like the ones discussed here have been possible because of rapid advances in optically stimulated luminescence dating and its application to strata that previously lacked robust age control.

Dansgaard–Oeschger Style Forcing

The North Atlantic region could be viewed as the type area for abrupt climate change, particularly during the last glacial. The Dansgaard–Oeschger events, characterized by sudden warming that lasted for no more than a few thousand years before reversing back to glacial conditions, are likely to have caused major landscape changes. By far the best-documented Dansgaard–Oeschger event is the most recent one (Bølling-Allerød-Younger Dryas), constituting the final episode of the last glacial. Numerous studies of northwest European rivers have revealed at least some commonality in changes of fluvial styles and terracing during this time interval. The Meuse River in The Netherlands has been particularly extensively studied (Huisink, 1997; Tebbens *et al.*, 1999). Although these two investigations use conflicting age models they share a number of findings, including a rapid (Tebbens *et al.*, 1999) or gradual (Huisink, 1997) transition from full-glacial braided channel patterns to a meandering fluvial style during the Allerød, reversing back to braiding during the Younger Dryas. In addition, several phases of incision and aggradation have been inferred, although some of these have been subsequently questioned by Bogaart *et al.* (2003b) who drew the attention to the same 'scour versus incision' problem as discussed above. Nevertheless, Bogaart *et al.* (2003b), using a novel modeling approach aimed explicitly at understanding centennial to millennial scale climatic forcing of fluvial systems, show the plausibility of short-lived incision associated with abrupt climate change, comparable to the pioneering work by Knox (1972) that is discussed below. This finding is consistent with floodplain degradation at the Allerød-Younger Dryas transition along the Meuse, as inferred by Huisink (1997).

One intriguing challenge in studies of fluvial evolution is the possible impact of the numerous earlier Dansgaard–Oeschger events of the last glacial. Studies of fluvial successions from this time interval in The Netherlands (Van Huissteden and Kasse, 2001) have suggested that prior to the last Dansgaard–Oeschger event discussed above, fluvial systems exhibited mostly autogenic behavior with only few episodes of more widespread erosion. They point out that the nature of fluvial responses is dependent on overall environmental boundary conditions, with a particularly prominent role for vegetation and permafrost. Only during the final episode of the last glacial an alternation between open shrub tundra and woodland allowed more dramatic fluvial responses due to the vegetational control on the hydrologic cycle. Nevertheless, there is ample room for more studies of fluvial records of the last glacial, especially in light of the current feasibility of better chronological frameworks.

Holocene Climate Forcing

The potentially nonlinear nature of fluvial response to abrupt environmental change was discussed by Knox (1972), who argued that high-order tributaries in southwest Wisconsin located near a major ecotone responded in a highly sensitive way to Holocene shifts in atmospheric circulation. An abrupt transformation from prairie to hardwood forest ~6,000 ^{14}C years BP was associated with rapid alluviation along these streams. As shown in **Figure 6**, changes in fluvial activity are particularly concentrated in the relatively short interval following an abrupt shift in climate, largely due to the time lag in vegetational response. More specifically, a transition from arid to humid conditions provides a small time window with increased discharge and sparse vegetation cover, allowing accelerated hillslope erosion due to a limited infiltration capacity. This results in increased sediment yield and 'geomorphic work'. Knox (1972) raises the point that the exact nature of the fluvial response, in terms of incision versus aggradation, is highly dependent on location within the drainage network.

Somewhat different conditions prevailed in environments that are characterized by more flashy discharge regimes, with peak discharges as much as two or three orders of magnitude higher than mean annual discharge. This is illustrated by a number of rivers in central Texas, including the Colorado River

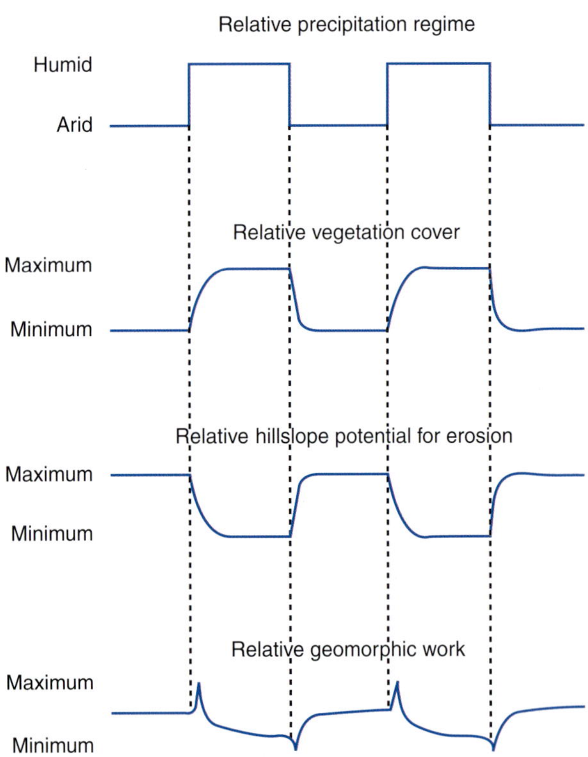

Figure 6 Conceptual model of nonlinear fluvial response to climate forcing (after Knox, 1972). Note that this model is intended to apply specifically to mid-latitude regions with an annual precipitation of 250–1,500 mm. Also note that 'geomorphic work' could refer to both incision and aggradation, depending on location within the drainage basin.

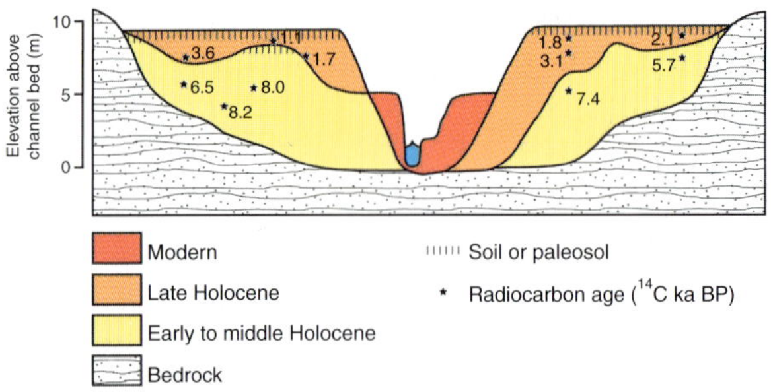

Figure 7 Three Holocene allostratigraphic units along the upper Colorado River, central Texas (after Blum *et al.*, 1994). For explanation see text.

(Blum *et al.*, 1994). **Figure 7** shows a cross-section of this fluvial system featuring three Holocene allostratigraphic units (*see* Morphostratigraphy-Allostratigraphy). While the base of each unit is found at a roughly constant level, their surfaces vary widely, with the early Holocene unit entirely draped by younger strata. Blum *et al.* (1994) explain this by variations in the persistence of deep flooding, which was especially prolific during the relatively humid time period 2,500–1,000 ^{14}C years BP. In contrast, the past millennium was drier and led to the establishment of an inset floodplain at a lower level, along with pedogenesis and terrace formation.

Daniels and Knox (2005) provide an excellent example of a study of century-scale fluvial response, in their case with respect to the drought that characterized the interior of North America during the Medieval Warm Period. They show that tributaries in a small drainage basin in southwest Nebraska incised rapidly around 800–1,100 ^{14}C years BP, roughly synchronous to rivers elsewhere in Nebraska and Kansas. Drought in this semi-arid region is associated with dramatic reductions in vegetation cover and a more variable discharge regime with less frequent, yet more powerful floods. This would enable high erosion rates and the establishment of new floodplain levels several meters lower, within a time period of a few centuries or less. It appears that fluvial incision in this environment is correlated with drought, whereas in examples given above, such as the Meuse River, early Holocene incision was associated with an increase in discharge (Bogaart *et al.*, 2003b). This testifies again to the complexity of the fluvial response and its strong dependence on regionally specific boundary conditions.

Relatively rare are stratigraphic studies that focus on the response of the fluvial sediment transfer system to annual to decadal scale climate forcing. Naturally, it is most practical to study such relationships in the most recent sedimentary records, given the need for exceptionally high temporal resolution. Only a few environments remain that have been essentially unaffected by human intervention over the past decades or century. Recent work in the pristine foreland basin of the Amazon drainage (Aalto *et al.*, 2003) suggests that floodplain aggradation is not a gradual process, but rather consists of rapid pulses of sedimentation separated by multi-year intervals of quiescence. These authors documented a strong correlation with the El Niño–Southern Oscillation, with rapid aggradation in the southern part of the Amazon Basin occurring during rainy La Niña intervals.

Summary

The interest in fluvial responses to rapidly changing climatic boundary conditions is here to stay, and will likely increase given the concerns about the impact of future global change. While there exists a rich literature of investigations of fluvial records for the past ~20 kyr, studies over shorter (decadal or centennial) timescales are less common (Blum and Törnqvist, 2000). One likely reason for this is the formidable task of developing high-resolution geochronological frameworks in fluvial successions, environments that commonly pose serious challenges for numerical dating.

The diverse collection of examples presented herein shows that there is a wide variety of fluvial responses to climate forcing. For example, the impact of increased aridity or humidity can be opposite in different geographic regions, to a large extent depending on initial boundary conditions. As a result, the scientific literature on the topic covered by this article is somewhat overwhelming and confusing, with numerous conflicting data and interpretations. To a considerable extent these conflicts are probably real, reflecting spatial variability, differences in sensitivity, and contrasting response times. However, it is likely that some of these conflicts simply stem from erroneous stratigraphic interpretations.

There are also discrepancies between field evidence and theory, given that large rivers may respond swiftly to environmental forcing, as so eloquently illustrated by the Ganges system. Thus major challenges lie ahead. The emergence of better and more rigorous concepts that are more strongly rooted in theory, as well as better analytical tools (notably dating methods), will likely offer new opportunities to tackle these important issues.

See also: **Dating Techniques**. **Fluvial Environments**: Sediments; Terrace Sequences; Deltaic Environments. **Glacial-Interglacial Scale Fluvial Responses**. **Ice Core Records**: Greenland Stable Isotopes. **Luminescence Dating**: Optically-Stimulated Luminescence. **Paleoclimate Reconstruction**: Sub-Milankovitch (DO/Heinrich) Events; The Last Millennium. **Quaternary Stratigraphy**: Morphostratigraphy-Allostratigraphy.

References

Aalto, R., Maurice-Bourgoin, L., Dunne, T., Montgomery, D. R., Nittrouer, C. A., and Guyot, J.-L. (2003). Episodic sediment accumulation on Amazonian flood plains influenced by El Niño/Southern Oscillation. *Nature* **425**, 493–497.

Berendsen, H. J. A., and Stouthamer, E. (2001). *Palaeogeographic Development of the Rhine-Meuse Delta, The Netherlands.* Koninklijke Van Gorcum, Assen.

Best, J. L., and Ashworth, P. J. (1997). Scour in large braided rivers and the recognition of sequence stratigraphic boundaries. *Nature* **387**, 275–277.

Blum, M. D., and Törnqvist, T. E. (2000). Fluvial responses to climate and sea-level change: a review and look forward. *Sedimentology* **47**(Supplement 1) 2–48.

Blum, M. D., Toomey, R. S., III., and Valastro, S., Jr. (1994). Fluvial response to Late Quaternary climatic and environmental change, Edwards Plateau, Texas. *Palaeogeography, Palaeoclimatology, Palaeoecology* **108**, 1–21.

Bogaart, P. W., Van Balen, R. T., Kasse, C., and Vandenberghe, J. (2003a). Process-based modelling of fluvial system response to rapid climate change – I: model formulation and generic applications. *Quaternary Science Reviews* **22**, 2077–2095.

Bogaart, P. W., Van Balen, R. T., Kasse, C., and Vandenberghe, J. (2003b). Process-based modelling of fluvial system response to rapid climate change II. Application to the River Maas (The Netherlands) during the Last Glacial–Interglacial Transition. *Quaternary Science Reviews* **22**, 2097–2110.

Bull, W. B. (1991). *Geomorphic Responses to Climatic Change.* Oxford University Press, New York.

Castelltort, S., and Van Den Driessche, J. (2003). How plausible are high-frequency sediment supply-driven cycles in the stratigraphic record? *Sedimentary Geology* **157**, 3–13.

Daniels, J. M., and Knox, J. C. (2005). Alluvial stratigraphic evidence for channel incision during the Mediaeval Warm Period on the central Great Plains, USA. *The Holocene* **15**, 736–747.

Fisk, H. N. (1944). *Geological Investigation of the Alluvial Valley of the Lower Mississippi River.* Mississippi River Commission, Vicksburg.

Goodbred, S. L., Jr. (2003). Response of the Ganges dispersal system to climate change: a source-to-sink view since the last interstade. *Sedimentary Geology* **162**, 83–104.

Huisink, M. (1997). Late-glacial sedimentological and morphological changes in a lowland river in response to climatic change: the Maas, southern Netherlands. *Journal of Quaternary Science* **12**, 209–223.

Knox, J. C. (1972). Valley alluviation in southwestern Wisconsin. *Annals of the Association of American Geographers* **62**, 401–410.

Knox, J. C. (1983). Responses of river systems to Holocene climates. In *Late-Quaternary Environments of the United States* (H. E. Wright, Jr., Ed.), Volume 2. The Holocene, p. 26. University of Minnesota Press, Minneapolis.

Kozarski, S. (1983). River channel adjustment to climatic change in west central Poland. In *Background to Palaeohydrology. A Perspective* (K. J. Gregory, Ed.), p. 355. John Wiley, Chichester.

Leigh, D. S., Srivastava, P., and Brook, G. A. (2004). Late Pleistocene braided rivers of the Atlantic Coastal Plain, USA. *Quaternary Science Reviews* **23**, 65–84.

Lewin, J., and Brewer, P. A. (2001). Predicting channel patterns. *Geomorphology* **40**, 329–339.

Mackin, J. H. (1948). Concept of the graded river. *Geological Society of America Bulletin* 59, 463–511.
Paola, C., Heller, P. L., and Angevine, C. L. (1992). The large-scale dynamics of grain-size variation in alluvial basins, 1: Theory. *Basin Research* 4, 73–90.
Salter, T. (1993). Fluvial scour and incision: models for their influence on the development of realistic reservoir geometries. In *Characterization of Fluvial and Aeolian Reservoirs* (C. P. North and D. J. Prosser, Eds.), Vol. 73, pp. 33–51. Geological Society Special Publication.
Saucier, R. T. (1994). *Geomorphology and Quaternary Geologic History of the Lower Mississippi Valley.* Mississippi River Commission, Vicksburg.
Schumm, S. A. (1977). *The Fluvial System.* John Wiley, New York.
Schumm, S. A., and Parker, R. S. (1973). Implications of complex response of drainage systems for Quaternary alluvial stratigraphy. *Nature Physical Science* 243, 99–100.
Slingerland, R. L., and Snow, R. S. (1988). Stability analysis of a rejuvenated fluvial system. *Zeitschrift für Geomorphologie, Neue Folge, Supplementband* 67, 93–102.
Srivastava, P., Juyal, N., Singhvi, A. K., Wasson, R. J., and Bateman, M. D. (2001). Luminescence chronology of river adjustment and incision of Quaternary sediments in the alluvial plain of the Sabarmati River, north Gujarat, India. *Geomorphology* 36, 217–229.
Tandon, S. K., Gibling, M. R., Sinha, R., Singh, V., Ghazanfari, P., Dasgupta, A., Jain, M., and Jain, V. (2006). Alluvial valleys of the Ganga Plains, India: Timing and causes of incision. In *Incised-Valley Systems in Time and Space* (R. W. Dalfymple, D. Leckie and R. W. Tillman, Eds.), SEPM Spec. Pub. in press.
Tebbens, L. A., Veldkamp, A., Westerhoff, W., and Kroonenberg, S. B. (1999). Fluvial incision and channel downcutting as a response to Late-glacial and Early Holocene climate change: the lower reach of the River Meuse (Maas), The Netherlands. *Journal of Quaternary Science* 14, 59–75.
Van den Berg, J. H. (1995). Prediction of alluvial channel pattern of perennial rivers. *Geomorphology* 12, 259–279.
Van Huissteden, J., and Kasse, C. (2001). Detection of rapid climate change in Last Glacial fluvial successions in The Netherlands. *Global and Planetary Change* 28, 319–339.

Terrace Sequences

D J Merritts, Franklin and Marshall College, PA, USA

© 2007 Elsevier B.V. All rights reserved.

Introduction

Fluvial terraces are important chroniclers of (1) climatically driven changes in sediment supply, discharge, or base level that result in channel incision or aggradation, and (2) tectonically driven long-term uplift that results in channel incision and warping or faulting of terrace surfaces. A fluvial terrace, a relatively flat bench occurring along and above a river, is an abandoned floodplain surface that is separated from the modern floodplain (or a lower river terrace) by a steep slope, known as the terrace riser. River valleys often contain flights of multiple terraces that reflect long periods of terrace formation, a history that can be unraveled through mapping, surveying, and dating of fluvial terraces. As such, terraces are useful in studies of Quaternary climate change and on the effects of climate change on landscape processes, such as increased upland erosion and consequent increased sediment supply. In tectonic studies, terraces are useful for assessing both the rate of rock uplift at a given reach and differential rates of deformation along river valleys.

We briefly summarize the causes and processes of terrace formation and then review case studies of terraces in Asia, New Zealand, and North America. As demonstrated by these examples, both climatic and tectonic processes have influenced the formation of fluvial terraces over time periods of thousands to hundreds of thousands of years during the Quaternary period (*see* Glacial-Interglacial Scale Fluvial Responses). The studies cited here used various methods of absolute and relative dating, including cosmogenic nuclide exposure dating, radiocarbon dating, correlation to dated tephra and loess stratigraphies, optically stimulated luminescence dating, and relative age dating from degree of soil development and thickness of weathering rinds on gravel clasts.

Types of Fluvial Terraces and Terrace Terminology

The two basic types of fluvial terraces are aggradational and degradational (**Fig. 1**) (*see* Responses to Rapid Environmental Change). Aggradation, or net deposition of sediment along a river channel and its floodplain over a prolonged period of time, leads to formation of an alluvial fill terrace. At the crossing of a geomorphic threshold from either deposition or stability (i.e., no net deposition) to net degradation, alluvium is incised and a fill terrace is both created and abandoned (Bull, 1991). Radiocarbon dating of organic matter in alluvium from an aggradational terrace along Cajon Creek, a stream that crosses the San Andreas fault in southern California, USA, reveals that it can take as long as 10 kyr for aggradation to form a fill terrace over a distance of ~10 km (Weldon, 1986). Furthermore, the time since the onset of filling, abandonment, and incision decreases upstream along Cajon Creek, indicating a time-transgressive terrace surface.

If incision of an aggradational terrace is punctuated by pauses that allow the stream to cut laterally rather than vertically, then cut terraces are carved into alluvial fill. Cut terraces (sometimes called cut-fill) have a distinctly different origin than the higher fill terrace into which they are carved. It is possible that a subsequent cycle of aggradation and degradation might

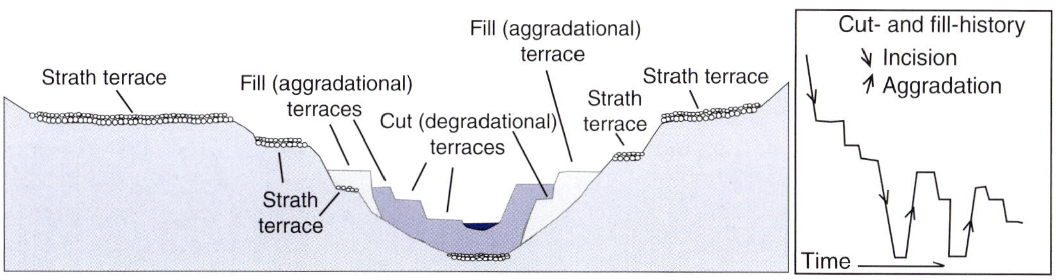

Figure 1 Valley cross-section illustrating a complex sequence of aggradational and degradational (cut and strath) terraces. The history of aggradation and degradation that led to the terrace sequence is outlined in the right-hand box. From Figure 2.11C, Burbank DW and Anderson RS (2001) *Tectonic Geomorphology,* 274p. Blackwell Science.

begin again after some period of vertical downcutting, resulting in an inset fill terrace that is difficult to distinguish from a cut terrace. Only careful stratigraphic work and dating can help to make the distinction between the two possible origins of the terraces.

Lateral cutting of bedrock followed by vertical degradation, or incision, results in formation of a strath terrace. When abandoned, the floodplain associated with the bedrock stream is left behind as well, leaving a veneer of coarse sediment on the strath surface that is proportional to the scour depth and bedload thickness associated with bankfull discharge (e.g., Pazzaglia and Brandon (2001) and Lavé and Avouac (2001)). Whereas the coarse sediment on a strath is generally thought to have been associated with bedload transport, the finer sand, silt, and clay often capping a strath and its gravel veneer probably are deposited by a combination of overbank sedimentation during high-magnitude floods (Merritts *et al.*, 1994) and colluvial deposition from side slopes (Lavé and Avouac, 2000) (*see* Sediments).

Through extensive radiocarbon dating of strath terrace sediments, Merritts *et al.* (1994) have shown that overbank deposition of fine sediment can occur during rare, high-magnitude floods several thousand years after a strath is formed by incision and abandonment of a river's previous channel level. The duration of overbank deposition after strath formation depends on the rate of vertical incision and the stochastic distribution of floods along the river, with low rates of incision and large-magnitude floods favoring the possibility of episodic deposition on the older surface.

Causes and Processes of Fluvial Terrace Formation

Fluvial terraces are formed by one or more changes in three dominant controls on a river's longitudinal profile: base level, water discharge, and sediment supply (*see* Glacial-Interglacial Scale Fluvial Responses and Responses to Rapid Environmental Change). When one or more of these factors changes, the river adjusts its channel to the new conditions and sometimes forms a new channel bed and floodplain at a lower or higher level than the original ones. Changes in the controlling parameters of discharge, sediment supply and caliber, or base level essentially are perturbations that lead to waves of erosion or deposition that can propagate along a stream.

Changes in base level (and local gradient) occur primarily through tectonic processes, as with rock uplift, differential tilting, faulting, and folding. Changes in water discharge and sediment supply occur primarily through changes in prevailing climatic conditions, which affect vegetation type and cover, and rates of soil erosion and runoff of water from upland slopes. For example, when falling sea-level or an uplifting land mass lowers the relative base level of a coastal river and increases the river's gradient near the coast, the river incises and abandons its former floodplain level, leaving a terrace remnant that records the earlier position and gradient of the channel and floodplain. If a cooling climate results in a decrease in vegetation cover on slopes and an increase in sediment supply that exceeds stream-carrying capacity (or perhaps a decrease in effective discharge), then aggradation might occur along the river channel and floodplain. Aggradation continues until the stream gradient is sufficiently steep to transport the prevailing load of incoming sediment. When aggradation ceases, perhaps during subsequent warming and an increase in vegetation cover, the fill surface is abandoned and incised, producing a fill terrace at a higher elevation than the original channel and floodplain.

Previous workers have recognized that a stream can adjust to carry the flux of water and sediment by altering its width, depth, roughness, and/or gradient (Leopold and Bull, 1979). We are most interested in the last of these, adjustments in channel gradient, as these lead to terrace formation. The adjustment of channel long profiles by aggradation or degradation is thought to occur because streams maintain an equilibrium long profile, or graded profile, over sufficiently long time periods (graded timescale of Schumm and Lichty (1965)). A graded stream is one that has sufficient stream power to transport sediment and water supplied from upstream, without aggrading or degrading along its bed (Mackin, 1948; Leopold and Bull, 1979; Bull, 1991; Knox, 1975). In general, long-term degradation of the channel bed and floodplain occurs if stream power is decreased (e.g., decreased discharge or gradient) or if the ability to transport sediment is reduced (e.g., increased bed roughness or grain size).

The graded stream is thought to evolve toward a steady-state profile in which rates of rock uplift and river incision are closely balanced along the length of the long profile of the stream (Bull, 1991; Burbank *et al.*, 1996; Pazzaglia *et al.*, 1998; Pazzaglia and Brandon, 2001). This close connection between uplift and incision implies that the difference in elevation between long profiles of streams and terraces can be used to assess rock uplift, much as marine terraces are used to assess coastal uplift.

Terraces often form during vertical incision by knickpoint, the process by which a steepened reach of channel—the knickpoint—propagates upstream (Gardner, 1983; Zaprowski *et al.*, 2001). As the knickpoint propagates upstream, it leaves an abandoned strath or cut surface in its wake downstream, such that the degradational terrace is graded to the crest of a knickpoint and is time-transgressive in age (Seidl and Dietrich, 1992; Merritts *et al.*, 1994). Experimental work has shown that the rate of knickpoint migration is proportional to discharge, with migration rates diminishing toward headwater reaches. Studies that utilize cosmogenic nuclide exposure dating indicate that knickpoints formed in basalt in Hawaii have migrated upstream at rates of >2 mm yr^{-1} (Seidl *et al.*, 1994), whereas those cutting metamorphic and igneous rocks in the Himalayan Mountains might be as high as 1 m yr^{-1} (Burbank *et al.*, 1996).

Strath terraces in particular have been used as geomorphic markers that record progressive displacement where rivers cross active faults and folds, and in some cases have provided fault slip and fold growth rates (Rockwell *et al.*, 1984; Molnar *et al.*, 1994; Lavé and Avouac, 2001). In the case of faulting that ruptures the surface, each higher terrace records greater displacement than lower terraces (**Fig. 2**).

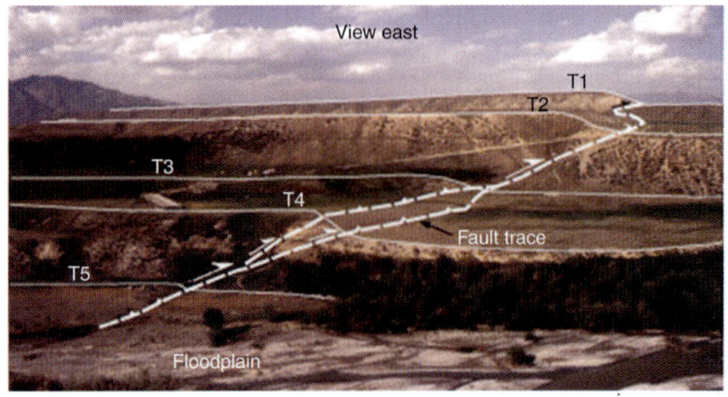

Figure 2 Offset terrace treads (T1 to T5) reveal the trace of a thrust fault (up on the left) and subsidiary splay in the Tien Shan, Kyrgyz Republic. Each higher, older terrace has displaced successively greater amounts. Deformation of such surfaces can be used to estimate rates of faulting and crustal shortening. From Burbank DW and Anderson RS (2001) *Tectonic Geomorphology*, 274p., Figure 9.6. Blackwell Science.

Case Studies of Terrace Formation

Although both climatic and tectonic processes can drive terrace formation simultaneously, the two can often be separated as they operate over different timescales of change. Long-term rock uplift along a tectonically active margin, for example, might operate at roughly uniform rates over a period of hundreds of thousands to millions of years, whereas alternating glacial–interglacial cycles of the Quaternary period have cycles of ~120 kyr. Whereas long-term rock uplift drives long-term river incision, shorter-term variations in climate and associated changes in sea level (i.e., base level) might cause episodes of aggradation or accelerated degradation that produce terraces superimposed on the terrace sequence created by the long-term uplift response. (see Hancock and Anderson (2002), for a modeling approach to fluvial terrace formation in response to changing climatic conditions.)

Tectonically-Driven Terrace Formation

We begin the case studies of tectonically driven terrace formation with four examples of long-term incision and strath terrace formation in response to moderate to rapid rock uplift at two tectonically active plate margins: the Himalayan continent–continent collision zone between the Indian and Eurasian plates, and the ocean–continent Cascadia subduction zone boundary between the Juan de Fuca and North American plates. The fifth case study in this section is an example of strath formation in eastern North America, a low-uplift rate passive plate margin where late Quaternary climate change might be the cause of accelerated incision and strath formation superposed on long-term incision that correlates with post-Miocene falling sea level.

Long-Term Incision and Strath Formation, Tectonically Active Plate Margins

Burbank *et al.* (1996) surveyed and dated strath surfaces along the Indus River near Nanga Parbat in northern Pakistan, where the river crosses the suture zone between the colliding Indian and Eurasian plates in the northwestern Himalayas. Cosmogenic radionuclide exposure ages of seven abandoned strath levels, ranging in height from 65 to 410 m above the Indus River bed (low discharge), span a range in ages from 4 to 68 kyr (**Fig. 3**). The terrace

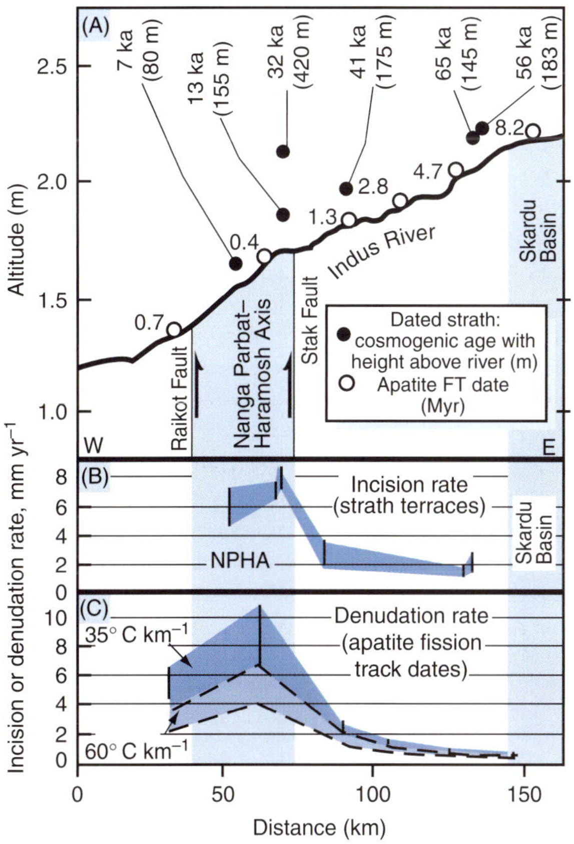

Figure 3 (A) Strath terrace remnants, with their cosmogenic nuclide exposure ages and heights above the modern Indus river level in the Himalayas of Pakistan, are shown with apatite fission-track ages from samples taken along the base of the Indus gorge; (B) Incision rates plotted vs. distance along the Indus river, with incision rates calculated from cosmogenic nuclide exposure ages. Highest rates correspond with the steepest river gradient; (C) Rates of denudation based on apatite fission-track data and two different geothermal gradients (35° C km^{-1} and 60° C km^{-1}. The zone of maximum long-term denudation coincides with the zone of most rapid river incision. From Burbank DW, Leland J, Fielding E, *et al.* (1996) Bedrock incision, rock uplift and threshold hillslopes in the northwestern Himalayas. *Nature* 379: 505–510.

ages and altitudes yield corresponding bedrock incision rates that range from 2 to 12 mm yr^{-1}, some of the highest known rates of bedrock incision in the world. Apatite fission-track ages along the same river reaches generally yield similar denudation rates, indicating that incision and bedrock uplift appear to be balanced.

Early work in Nepal on rivers flowing southward across the Himalayan Mountains documented that terrace profiles on the hanging wall block above the Main Central Thrust, a south-vergent fault, are tilted down to the north and diverge strongly to the south, consistent with terrace formation above a rising block with differential uplift parallel to the river's long profile (Iwata, 1987). Lavé and Avouac (2000) mapped and radiometrically dated four warped and folded Holocene strath terraces along two rivers draining the Siwaliks Hills in central Nepal along the Main Frontal Thrust fault in the foothills of the Himalayas (**Fig. 4**). Combining structural mapping and analysis of active fault bend folding with warped profiles of the dated terraces, they derived rates of bedrock uplift from river incision (as high as 15 mm yr^{-1}) and estimated that much of the shortening across the Himalayas is accommodated by the Main Frontal Thrust fault.

Similar to the work of Personius (1995) in Oregon, USA, and Merritts *et al.* (1994) (discussed below) in northern California, USA, Lavé and Avouac (2000) obtained much younger radiocarbon dates from silty sediments at the tops of terrace deposits than from underlying gravel veneers resting directly on bedrock straths. Corresponding with the above mentioned US research, the authors concluded that radiocarbon dates from the base of the silt or from within the gravel (bedload) itself are better estimates of the timing of strath abandonment, and attribute the overlying silty sediments to postabandonment depositional processes.

Pazzaglia and Brandon (2001) studied six late Quaternary strath terraces along the Clearwater River in northwestern Washington State, USA, in order to compare long-term uplift and incision rates across the Cascadia forearc high formed inboard of the Cascadia subduction zone. Three approaches were used to constrain terrace ages: weathering rind thickness on graywacke gravel clasts; radiocarbon dating; and correlation to dated coastal glacio-fluvial deposits and the global eustatic sea-level curve. Strath terraces formed in the dominantly unglaciated Clearwater drainage while alpine glaciers were advancing in adjacent drainages. During the transition to interglacial times, an increased sediment supply due to local deglaciation, combined with rising base level at the coast, resulted in aggradation along

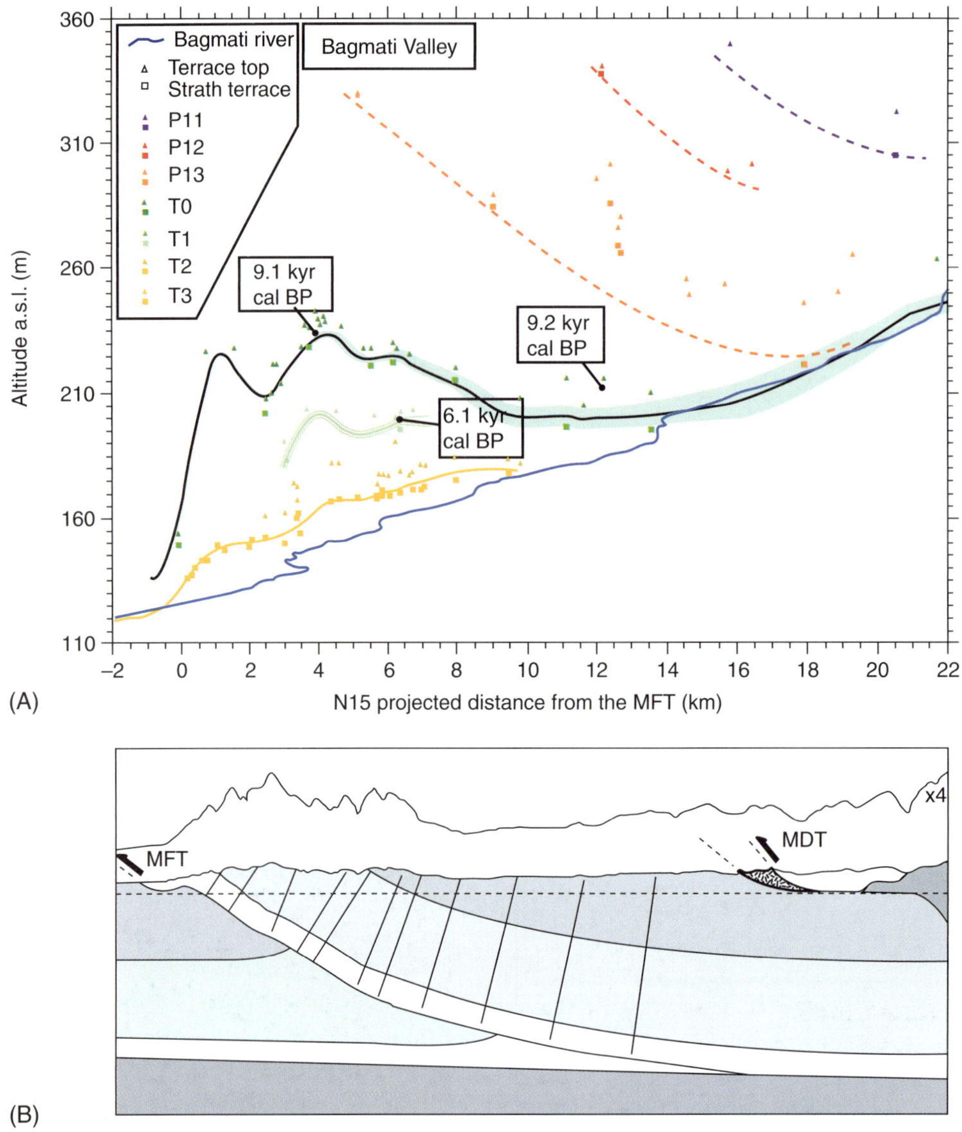

Figure 4 (A) Strath terraces projected along the Bagmati river in the central Himalayas, Nepal. Strath elevations computed from digital elevation model and field measurements (solid thin lines and dashed lines) are shown with individual field measurements. Solid thick lines represent Holocene terraces, and dashed lines represent hypothetical levels of late Pleistocene terraces that have similar soil color and depth of weathering. Note that terraces are folded above a growing thrust fault at 0–14 km; (B) Topographic and structural cross sections, with the same horizontal scale, are shown for comparison with the terrace profiles. The terraces reveal sustained Holocene and late Pleistocene activity of the fold growing above the Main Frontal Thrust fault (MFT). In contrast, the terraces indicate that little offset has occurred across the Main Dun Thrust fault (MDT) during the late Quaternary. From Lavé J and Avouac JP (2000) Active folding of fluvial terraces across the Siwaliks Hills, Himalayas of central Nepal. *Journal of Geophysical Research B* 105(3): 5735–5770, Plate 7A.

the Clearwater River. Nevertheless, the river returned to the same valley profile during each cycle of strath-cutting. Bedrock incision rates increase inland from $<0.1\,\mathrm{m\,kyr^{-1}}$ at the coast to $\sim 0.9\,\mathrm{m\,kyr^{-1}}$ near the headwaters in the Olympic Mountains, and are similar to long-term denudation rates estimated from fission-track cooling ages. As incision keeps pace with rock uplift, terraces diverge upstream. Pazzaglia and Brandon (2001) concluded that the Clearwater River valley has maintained a steady-state profile and a long-term balance between incision and rock uplift over a period of tens of thousands of years.

In a region of relatively high uplift rates at the southernmost tip of the Cascadia subduction zone, where the North American, Pacific, and Juan de Fuca plates join at the Mendocino triple junction, Merritts *et al.* (1994) surveyed and radiometrically dated late Pleistocene and Holocene fill and strath terraces along three rivers in northern California. This work determined that lower reaches of the rivers are dominated by the effects of oscillating sea level, primarily aggradation and formation of fill terraces during sea-level high stands, and deep incision during sea-level low stands (**Fig. 5**). A eustasy-driven depositional wedge that tapers to zero thickness extends tens of kilometers upstream on all three rivers. In contrast, middle to upper reaches of each river are dominated by the effects of long-term uplift, primarily lateral and vertical erosion and formation of steep, unpaired strath terraces exposed only upstream of the depositional wedge. Vertical incision at a rate similar to that of uplift (1–2 m kyr^{-1}) has occurred even during the present sea-level high stand along the two rivers with highest uplift rates. Strath terraces have steeper gradients than the modern channel bed and do not merge with marine terraces at the river mouths.

Long- and Short-Term Incision and Strath Formation, North American Passive Plate Margin

Three ongoing processes have produced a relative, long-term base-level fall at the mouth of the Susquehanna River along the eastern North American passive plate margin: (1) Flexural up-warping of the Appalachian Piedmont in response to denudational unloading in the Susquehanna watershed, (2) downward flexural warping offshore in response to sediment deposition on the continental shelf, and (3) protracted eustatic sea-level fall beginning in the middle Miocene (Pazzaglia and Gardner, 1993; Pazzaglia *et al.*, 1998). The unglaciated, lower Susquehanna River has a convex upward long profile with steep gradients as a result of these processes. Recent cosmogenic exposure dating of three low strath terraces (referred to as levels 1, 2, and 3) within the Holtwood Gorge along the lower

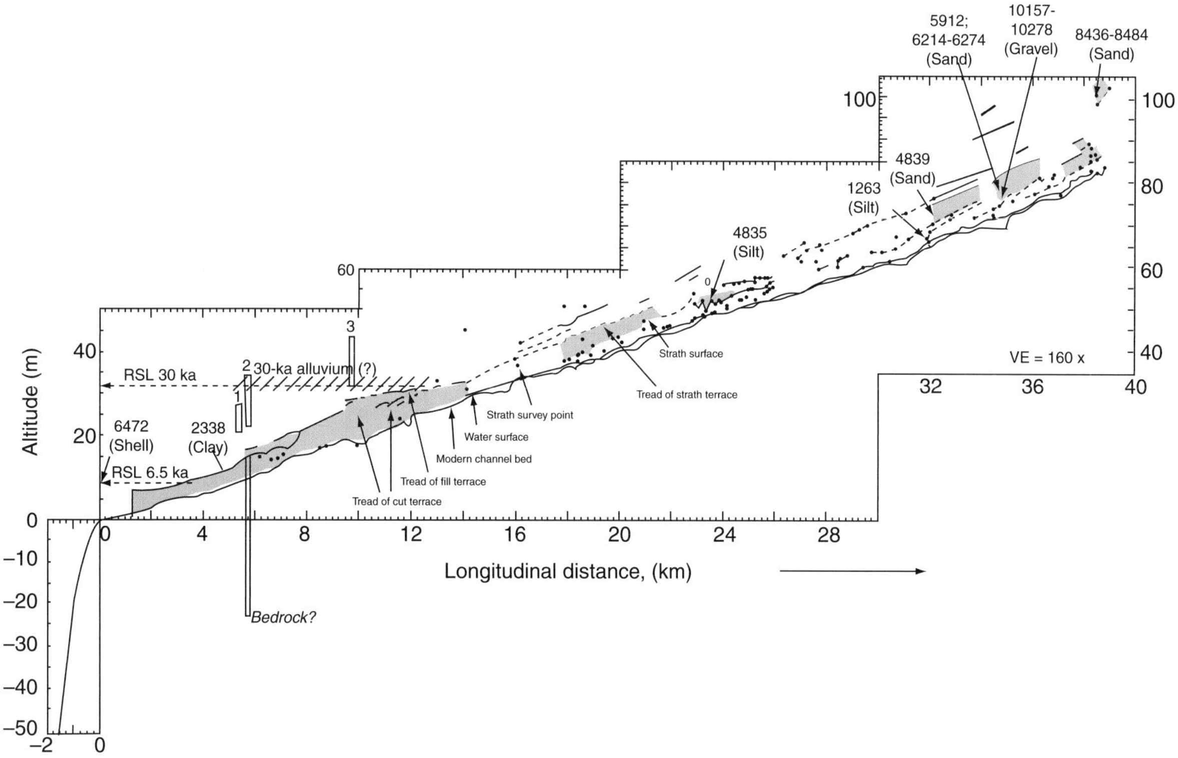

Figure 5 Longitudinal profiles of channel bed, water surface, strath surfaces, and treads of fill terraces from detailed total geodetic station along the Mattole River, northern California. Treads of cut-and-fill terraces are indicated by fine lines (hundreds of survey points not shown), dashed where correlations are tentative, while strath surfaces are indicated by both heavy lines, dashed where correlations are tentative, and survey points (crosses). Stippled pattern downstream of 18 km indicates alluvium where depth can be estimated from well data (drill holes shown as 1, 2, and 3). Multiple unpaired strath terraces occur upstream of 18 km, shown by a bedrock surface (heavy line) and overlying alluvial tread (fine line). Deposits are shown as stippled pattern at several sites to indicate corresponding straths and treads, but not all are included because straths and treads overlap one another. Radiocarbon dates are centroids of calibrated age ranges in years before present. Relative Sea Level (RSL) is determined from both sea-level change and landmass uplift. From Merritts DJ, Vincent KR, and Wohl EE (1994) Long river profiles, tectonism and eustasy: a guide to interpreting fluvial terraces. *Journal of Geophysical Research* 99: 14031–14050.

Susquehanna River indicates that rates of vertical incision dramatically increased between ~35 and ~14 ka (**Fig. 6**). Prior to ~35 ka, maximum vertical incision was ~0.2 m kyr^{-1}, but from ~35 to ~20 ka the rate of incision accelerated to 0.45 m kyr^{-1}. The rate of incision increased again to ~0.52 m kyr^{-1} between ~20 and ~14 ka, when significant incision ceased. An upstream-younging age gradient of ~1.4 kyr km^{-1} on the level 2 terrace is consistent with knickpoint retreat, but no significant time-transgressive relationship exists on either the highest (level 3) or lowest (level 1) terraces, suggesting that these surfaces were abandoned rapidly along the length of the Holtwood Gorge.

Reusser *et al.* (2004) interpret the high rates of late Pleistocene incision to be the result of a prolonged period of climate instability from ~45 to 14 ka, and the cessation of rapid incision to correspond with the transition from Pleistocene full-glacial conditions to the warmer, more climatically stable Holocene Epoch. Using climate data for the North Atlantic region, they note that a marked increase in storminess beginning at ~50 ka and lasting until ~10 ka correlates well with the initiation and cessation of incision within Holtwood Gorge. Furthermore, they suggest that increases in the frequency, magnitude, and duration of flood events that are capable of exceeding critical geomorphic thresholds for detachment of rock might have increased rates of bedrock channel incision during the last glacial cycle. Because of the timing of increased incision, they conclude that climate is likely to be a more important external factor than

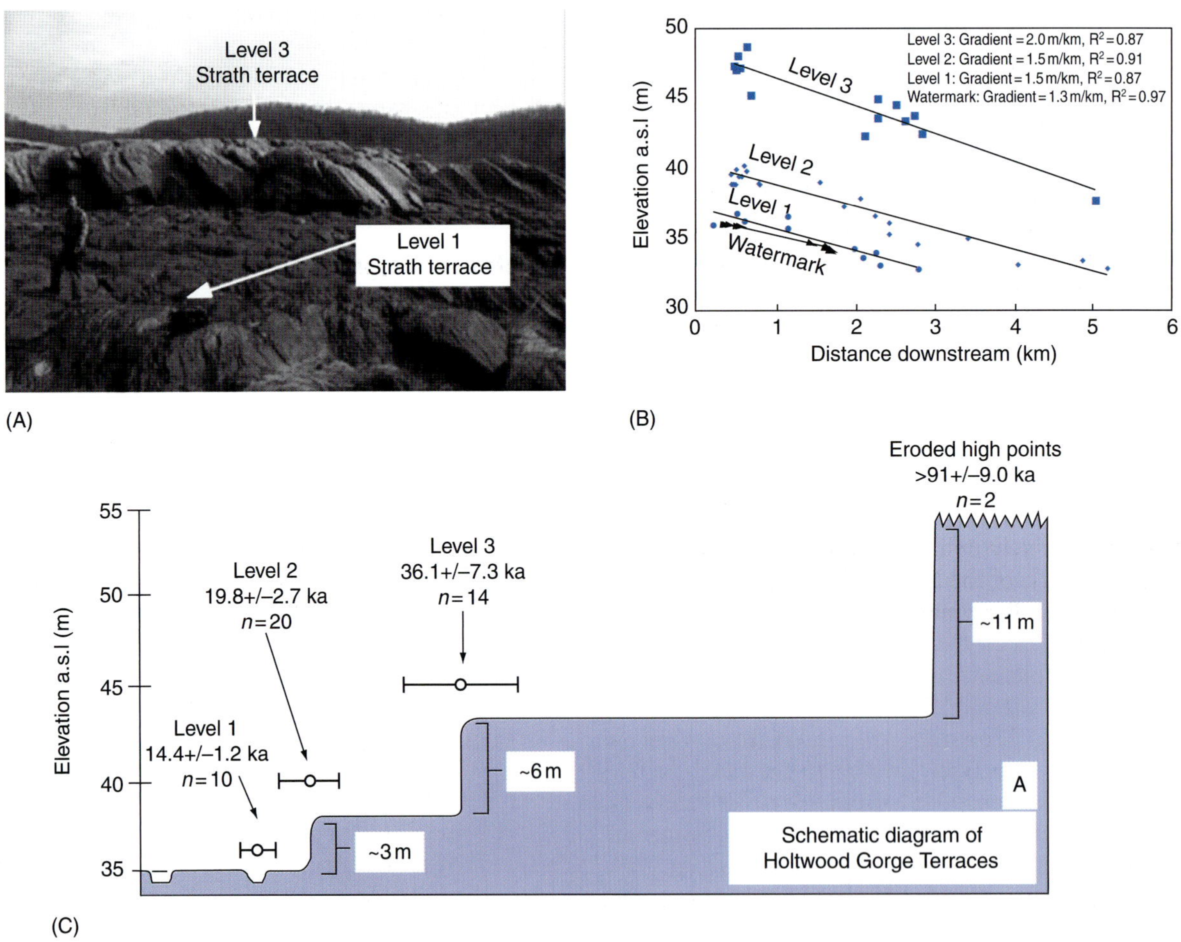

Figure 6 (A) Strath terrace levels 1 and 3 in the Holtwood Gorge along the Susquehanna River at low flow conditions. Level 2 is not preserved at this location. At higher discharges, strath level 1 is submerged. (B) Paleogradients for strath terrace levels 1, 2, and 3 and a prominent watermark in July 2002, were acquired with differential GPS. The watermark is representative of the modern river gradient. (C) The timing of incision and terrace abandonment within Holtwood Gorge is shown diagramatically with respect to time (x-axis) and in relation to documented changes in climate and sea level. From Reusser LJ, Bierman P, Pavich M, Larsen J, and Finkel R (2006) Late Pleistocene bedrock channel incision alongn the U. S. Atlantic passive margin, Holtwood Gorge, Susquehanna River, Pennsyvlania. *American Journal of Science*, Figures 4A, 5, and 17A. Reprinted by permission of the American Journal of Science.

the low sea level and base-level fall associated with full-glacial episodes.

Climatically Driven Terrace Formation

The previous example indicates that incision driven by long-term, slow uplift and relative sea-level fall can be accelerated during cold conditions, as during full-glacial episodes, but many studies in other areas have documented a switch from incision to aggradation during previous glacial periods. Blum and Tornqvist (2000) have reviewed this mismatch between the role of climate and sea level in aggradation and degradation of coastal fluvial systems. Paired aggradational terraces throughout the world have been associated with Quaternary glacial–interglacial cycles, with aggradation commonly thought to have occurred during glacial periods and incision during subsequent warmer periods (cf Penck and Brückner (1909), and Bridgland (2000). In some cases, fill terraces were correlated with glacial moraines and interpreted to be of similar age regionally, so they were correlated with one another among basins, even on unglaciated rivers. Recent work on nonglaciated streams in New Zealand supports the premise of these earlier studies, and is described here as an example of regional aggradational terrace formation driven by climatic change. Other examples of aggradational river terraces documented to have been formed by Quaternary climate change in nonglaciated regions are described in Bull (1991).

Terrace Formation during Quaternary Glacial–Interglacial Cycles

Along the actively uplifting Hikurangi Margin along the eastern side of the North Island in New Zealand, Litchfield and Berryman (2005) and Litchfield and Rieser (2005) extended the results of previous studies to correlate four aggradational terraces among eight nonglaciated drainage basins (**Fig.** 7). The work of Litchfield and Rieser (2005) represents one of the most extensive analyses of the use of optically stimulated luminescence (OSL) dating for river terraces. OSL dating measures the time that has elapsed since sediment was last exposed to light, presumably at the time of deposition. Litchfield and Rieser (2005) used it for both silts and sands near the tops of aggradational terraces, and for loess coverbeds overlying the terraces.

Based on consistent results from tephra chronology, loess stratigraphy, radiocarbon dating, and OSL dating, terrace ages are linked to the following late Pleistocene dry, cool stadial, and/or glacial periods: 15 –30 ka, 31 –40 ka, 55 –70 ka, and 115 ka. These results confirm that aggradation and loess deposition were widespread on the North Island of New Zealand during both full glacial and interstadial periods. As each aggradational terrace formed, the riverbed was the source of loess blown onto higher terraces, so that each higher, older terrace has one more loess cover deposit than the next lower aggradational terrace. Litchfield and Berryman (2005) conclude that limited grass and shrub vegetation cover, combined with periglacial processes, produces an increased sediment supply that led to the widespread aggradation associated with cool periods, whereas re-establishment of forested slopes during postglacial periods led to incision of fill deposits, producing abandoned fill terraces.

Since the formation of the lowest aggradational terrace, T1 at 15 –30 ka, when global eustatic sea level was at its lowest (~ -120 m), North Island streams have been degrading and carving strath terraces into bedrock, despite a rising Holocene sea level. The intriguing conclusion in this region is that climatic controls on sediment and water discharge are more significant to terrace formation than rising or falling base levels.

Cycles of Aggradation and Degradation in Arroyos of the Semiarid American Southwest

Currently incised channel systems in the semiarid, nonglaciated southwestern USA, known as arroyos (Spanish for gulch), have repeatedly aggraded and incised during the Quaternary period, leaving a complex record of cut-and-fill terraces (cf Cooke and Reeves (1976), Bull (1991)). Modern arroyos currently are in a phase of cutting that dates to the nineteenth century, possibly due to grazing and other recent land-use changes, and carry large volumes of sediment during ephemeral to intermittent flows (Cooke and Reeves, 1976). During repeated episodes throughout the Quaternary period, however, arroyos have been unable to transport their sediment load and have aggraded, forming fill terraces.

Decades of investigations of arroyos have led many workers to attribute the cyclicity of arroyo filling and incision to alternations between relatively wet, pluvial climates during glacial periods and drier climates during interglacial and interstadial periods. Wetter conditions are thought to be associated with production of greater sediment yields from upland slopes (e.g., Bull (1991)). Full understanding of the 'arroyo cycle' is complicated, however, by the fact that repeated cycles of cutting and filling have occurred

(A)

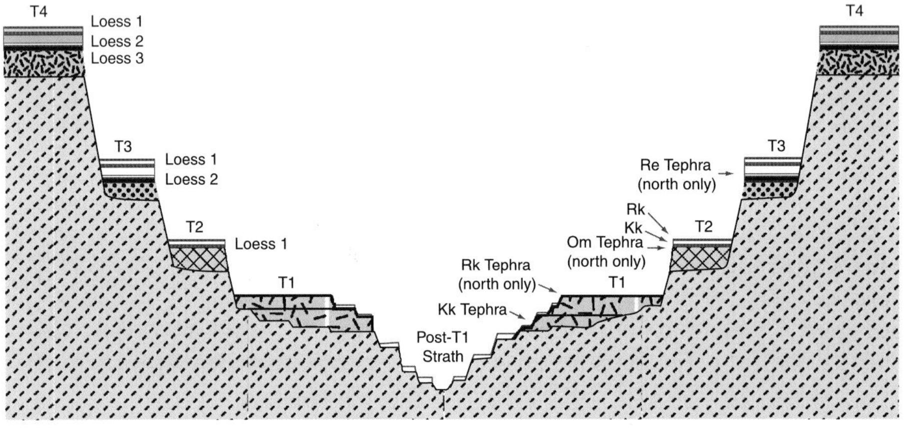

(B)

Figure 7 (A) A suite of fill (T1 and T2) and cut terraces along a river on the North Island of New Zealand. The gravel of T2 is visible above a strath cut into early Pleistocene siltstone bedrock (white). (B) Schematic cross section illustrating the relations among the terraces and overlying loess and tephra coverbeds. From Litchfield NJ and Berryman KR (2005) Correlation of fluvial terraces within the Hikurangi Margin, New Zealand: Implications for climate and base-level controls. *Geomorphology* 68: 291–313, Figure 3. Copyright 2005 with permission from Elsevier.

during the Holocene interglacial warm period (Cooke and Reeves, 1976). Some regional forcing mechanism is thought to be responsible for generally synchronous episodes of cutting and filling that extend over broad areas. Some workers have suggested that complex interactions within arroyo systems, as well as the initial state of the arroyo system at the onset of climatic change, are critical to the response of arroyos to climatic forcing (cf Schumm *et al.* (1984)).

Conclusions

Recent studies of long profiles of rivers and fluvial terraces have illuminated the relation between crustal and surficial processes of uplift and incision, respectively. As described in four examples herein, researchers used altitudinal differences between dated straths and the modern channel to estimate rates of vertical incision at a given stream reach, and then compared these estimates with rates of rock uplift or denudation that were derived independently (Merritts *et al.*, 1994; Burbank *et al.*, 1996; Pazzaglia and Brandon, 2001; Lavé and Avouac, 2001). In all four cases, the rates of uplift and incision are roughly balanced, supporting earlier ideas of steady-state long profiles for graded streams that are adjusted to long-term uplift. Whereas these four examples were in areas of relatively high rates of uplift along tectonically active plate margins, a recent study along the eastern North American passive margin, on the lower Susquehanna River, provides clues to the effect of climate change on river incision. Cosmogenic exposure dating of strath surfaces indicates that relatively short-term climatic change which leads to colder and possibly stormier conditions can result in pulses of accelerated incision, even where long-term uplift rates are low.

Although the Susquehanna river example links glacial climatic conditions to accelerated bedrock incision and strath abandonment, numerous studies in Europe, North America, and New Zealand have documented a switch from incision to aggradation during glacial conditions. Merritts *et al.* (1994) and Pazzaglia and Brandon (2001), in contrast, describe aggradation during late glacial and interglacial times along lower reaches of the coastal rivers that they studied. Blum and Tornqvist (2000) have reviewed this mismatch between the role of climate and sea level (i.e., base-level change) in aggradation and degradation within coastal fluvial systems. One of the best constrained examples, with extensive dating using multiple techniques that include radiocarbon dating, tephra chronology, loess chronology, and OSL dating, is described here for North Island, New Zealand (Litchfield and Berryman, 2005; Litchfield and Rieser, 2005). This work demonstrates conclusively that regional aggradation occurred along coastal rivers during full glacial times, regardless of a markedly lower sea level, and that the Holocene is marked by incision and the formation of cut and strath terraces.

As with the case of the cycles of arroyo cutting and filling described here, it might be that each river system responds differently to climate change, depending on its initial conditions at the onset of external change and on the magnitude of change in sediment supply from upland slopes. Where relief is high and slopes are steep, as in New Zealand, climate cooling and periglacial conditions seem to overwhelm streams with sediment, whereas the low-uplift Susquehanna River basin, with generally low relief and gentle upland slopes, does not generate sufficient excess sediment to cause aggradation along major trunk streams during cold periods. New techniques of dating terraces and terrace deposits, and multiple new sites studied around the world, are providing a growing body of evidence that fluvial responses to climate change might be similar regionally, but have significant differences among regions.

See also: **Fluvial Environments**: Sediments; Responses to Rapid Environmental Change. **Glacial-Interglacial Scale Fluvial Responses**. **Luminescence Dating**: Optically-Stimulated Luminescence.

References

Blum, M. D., and Tornqvist, T. E. (2000). Fluvial responses to climate and sea-level change: A review and look forward. *Sedimentology* 47(Supplement 1), 2–48.

Bridgland, D. R. (2000). River terrace systems in north-west Europe: An archive of environmental change, uplift and early human occupation. *Quaternary Science Reviews* **19**, 1293–1303.

Bull, W. B. (1991). *Geomorphic Responses to Climate Change.* Oxford University Press, New York.

Burbank, D. W., and Anderson, R. S. (2001). *Tectonic Geomorphology*, p.274. Blackwell Science.

Burbank, D. W., Leland, J., Fielding, E., *et al.* (1996). Bedrock incision: Rock uplift and threshold hillslopes in the northwestern Himalayas. *Nature* **379**, 505–510.

Cooke, R. U., and Reeves, R. W. (1976). *Arroyos and Environmental Change in the American South-West.*, p. 213. Clarendon Press, Oxford.

Gardner, T. W. (1983). Experimental study of knickpoint and longitudinal profile evolution in cohesive, homogenous material. *Geological Society of America Bulletin* **94**, 664–672.

Hancock, G. S., and Anderson, R. S. (2002). Numerical modeling of fluvial strath-terrace formation in response to oscillating climate. *Geological Society of America Bulletin* **114**, 1131–1142.

Hartshorn, K., Hovius, N., Dade, W. B., and Slingerland, R. L. (2002). Climate-driven bedrock incision in an active mountain belt. *Science* **297**, 2036–2038.

Iwata, S. (1987). Mode and rate of uplift of the central Nepal Himalaya. *Zeitschrift fur Geomorphologie v. Supp. Bd* **63**, 972–984.

Knox, J. C. (1975). Concept of the graded stream. In *Theories of Landform Development. Proceedings of the 6th Geomorphology Symposium* (W. N. Melhorn and R. C. Flemal, Eds.), pp. 169–198. Publications in Geomorphology, Binghamton:, State University of New York at Binghamton.

Lavé, J., and Avouac, J. P. (2000). Active folding of fluvial terraces across the Siwaliks Hills, Himalayas of central Nepal. *Journal of Geophysical Research B* **105**(3), 5735–5770.

Leopold, L. B., and Bull, W. B. (1979). Base level. aggradation. and grade. *Proceedings of the American Philosophical Society* **123**, 168–202.

Litchfield, N. J., and Berryman, K. R. (2005). Correlation of fluvial terraces within the Hikurangi Margin. New Zealand: Implications for climate and base-level controls. *Geomorphology* **68**, 291–313.

Litchfield, N. J., and Rieser, U. (2005). Optically stimulated luminescence age constraints for fluvial aggradation terraces and loess in the eastern North Island, New Zealand. *New Zealand Journal of Geology and Geophysics* **48**, 581–589.

Mackin, J. H. (1948). Concept of the graded river. *Geological Society of America Bulletin* **59**, 463–512.

Merritts, D. J., Vincent, K. R., and Wohl, E. E. (1994). Long river profiles. tectonism and eustasy: A guide to interpreting fluvial terraces. *Journal of Geophysical Research* **99**, 14031–14050.

Molnar, P., Brown, E. T., Burchfiel, B. C., *et al.* (1994). Quaternary climate change and the formation of river terraces across growing anticlines on the north flank of the Tien Shan, China. *Journal of Geology* **102**, 583–602.

Pazzaglia, F., and Brandon, M. (2001). A fluvial record of long-term steady-state uplift and erosion across the Cascadia forearc high, western Washington state. *American Journal of Science* **301**, 385–431.

Pazzaglia, F., and Gardner, T. (1993). Fluvial terraces of the lower Susquehanna River. *Geomorphology* **8**, 83–113.

Pazzaglia, F. J., Gardner, T. W., and Merritts, D. J. (1998). Bedrock fluvial incision and longitudinal profile development over geologic time scales determined by fluvial terraces. In (K. J. Tinkler and E. E. Wohl, Eds.)Rivers Over Rock: Fluvial Processes in Bedrock Channels, American Geophysical Union Monograph vol. 107, pp. 207–235. American Geophysical Union, Washington DC.

Penck, W., and Brückner, E. (1909). *Die Alpen im Eiszeitalter*, p. 1199. Tauchnitz, Leipzig.

Personius, S. F. (1995). Late Quaternary stream incision and uplift in the forearc of the Cascadia subduction zone, western Oregon. *Journal of Geophysical Research B* **100**, 20193–20210.

Reusser, L. J., Bierman, P., Pavich, M., Larsen, J., and Finkel, R. (2006). An episode of rapid bedrock channel incision during the last glacial cycle, measured with ^{10}Be. *American Journal of Science* **306**, 69–102.

Reusser, L. J., Bierman, P., Pavich, M., Zen, E.-A., Larsen, J., and Finkel, R. (2004). Rapid late Pleistocene incision of Atlantic passive margin river gorges. *Science* **305**, 499–502.

Rockwell, T., Keller, E. A., Clark, M. N., and Johnson, D. L. (1984). Chronology and rates of faulting of Ventura River terraces. California. *Geological Society of America Bulletin* **95**, 1466–1474.

Schumm, S. A., and Lichty, R. W. (1965). Time, space, and causality in geomorphology. *American Journal of Science* **263**, 110–119.

Schumm, S. A., Harvey, M. D., and Watson, C. C. (1984). *Incised Channels: Morphology. Dynamics and Control.*, p. 200. Water Resources Publications, Littleton.

Seidl, M. A., and Dietrich, W. E. (1992). The problem of channel erosion into bedrock. *Catena Supplement* **23**, 101–124.

Seidl, M. A., Dietrich, W. E., and Kirchner, J. W. (1994). Longitudinal profile development into bedrock: An analysis of Hawaiian channels. *Journal of Geology* **102**, 457–474.

Weldon RJ (1986) *Late Cenozoic Geology of Cajon Pass; Implications for Tectonics and Sedimentation Long the San Andreas Fault.* Ph.D. Dissertation. Pasadena: California Institute of Technology.

Zaprowski, B. J., Evenson, E. B., Pazzaglia, F. J., and Epstein, J. B. (2001). Knickzone propagation in the Black Hills and northern High Plains: A different perspective on the late Cenozoic exhumation of the Laramide Rocky Mountains. *Geology* **29**, 547–550.

Deltaic Environments

L Giosan, Woods Hole Oceanographic Institution, MA, USA
S L Goodbred Jr., Vanderbilt University, TN, USA

© 2007 Elsevier B.V. All rights reserved.

Introduction

Definition

Deltas are constructional coastal landforms with both subaerial and subaqueous components that are genetically associated with rivers discharging into a standing body of water such as a lake, estuary, lagoon, sea, or the open-ocean shelf (**Fig. 1**). A delta is usually built by a single river, but exceptions exist (e.g., Ganges–Brahmaputra; Tigris–Euphrates). Reworking of sediments accumulated at the river mouth by basinal processes (e.g., waves, tides, and currents) should be slow enough to allow delta-building to proceed. The river is the main source for sediment delivered to the delta, although in some wave-dominated settings (e.g., Danube, Sao Francisco, and Rio Doce), a significant portion may be transported by wave-driven currents from remote sources (**Fig. 2**).

Importance and Significance of Deltas

Deltas are among the most environmentally and economically important coastal sedimentary environments. Significant oil, gas, and groundwater resources have been exploited from deltaic formations. Historically, modern deltas have been prized by human civilizations for their high natural and agricultural productivity, rich biodiversity, and for the abundance of waterways that provide easy means of transportation. Deltas are ubiquitous: 21 of the world's 25 largest rivers, which deliver 31% of total fluvial sediment reaching the ocean, have formed well-expressed deltas at the coast (Meade, 1996). As a result, ~25% of the world's population lives within deltaic and wetland coastal systems (Syvitski *et al.*, 2005), some in large urban centers such as Shanghai, Bangkok, Dhaka, Saigon, Rangoon, Lagos, Alexandria, Cairo, and New Orleans. River deltas act as filters, repositories, and reactors for a suite of continental materials including sediments, organic carbon, nutrients, and pollutants, significantly affecting both the regional environment at the continent–ocean boundary as well as global biogeochemical cycles (e.g., McKee *et al.* (2004)). However, deltas are fragile geomorphic features that can change dramatically with modest modifications in their boundary conditions. Currently, the largest threat to the world's deltaic systems is the trapping of sediment behind dams erected on delta-building rivers and their tributaries. Deltas also face the threat of accelerated relative sea-level rise due to climate change and fluid extraction.

History of Delta Research

The term 'delta' was introduced by Herodotus (fifth century BC), by analogy to the Greek letter Δ, to describe the shape of the lowlands between the forking distributaries of the river Nile in Egypt (**Fig. 1G**). In 1832, in his *Principles of Geology*, Charles Lyell first defined the modern geological meaning of the term 'delta' as 'an alluvial land, formed by a river at its mouth, without reference to its precise shape.' The first comprehensive review of modern deltas was written by Credner in 1878, where the author correctly identified many of the processes affecting the form, structure, and evolution of deltas. Early landmark papers on deltas also

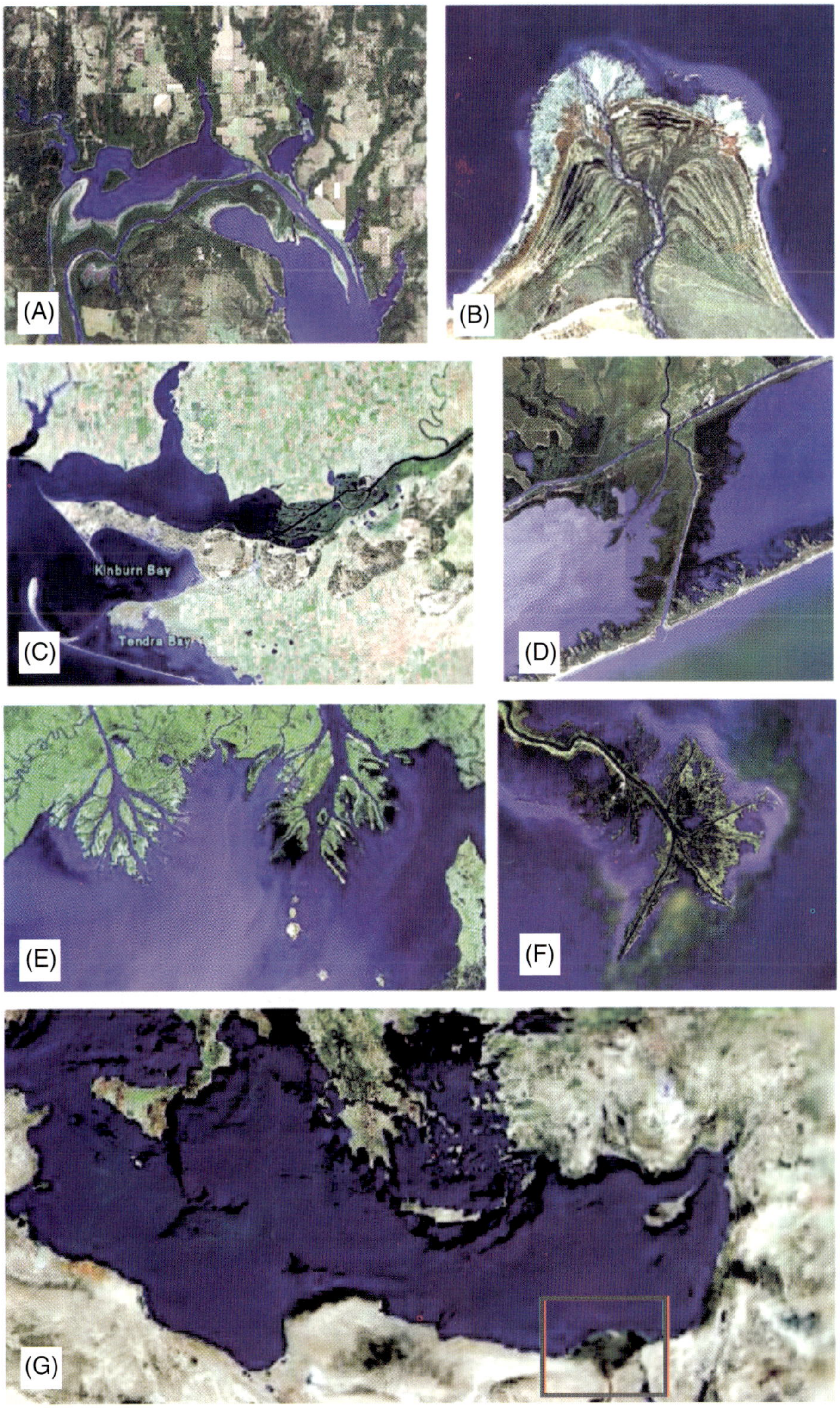

Figure 1 Google Earth satellite images of deltas (http://earth.google.com/). Lacustrine deltas: (A) Red River delta in Lake Texoma (Texas, USA; image ©DigitalGlobe); (B) William River delta in Lake Athabasca (Canada; image ©DigitalGlobe and MDAEarthSat). Deltas in estuaries, lagoons, and bays: (C) bayhead delta in the Southern Bug estuary in the Black Sea (Ukraine; image ©MDAEarthSat and DigitalGlobe); (D) Colorado River delta in the Matagorda lagoon (Texas, USA; image ©DigitalGlobe); (E) Wax Lake and Atchafalaya deltas in the Atchafalaya Bay (Louisiana, USA; image ©MDAEarthSat and DigitalGlobe). The open coast, fluvial-dominated (F) Balize lobe of the Mississippi delta (image ©MDAEarthSat and DigitalGlobe); and (G) the extensive, wave-dominated Nile delta in the eastern Mediterranean Sea (image ©MDAEarthSat).

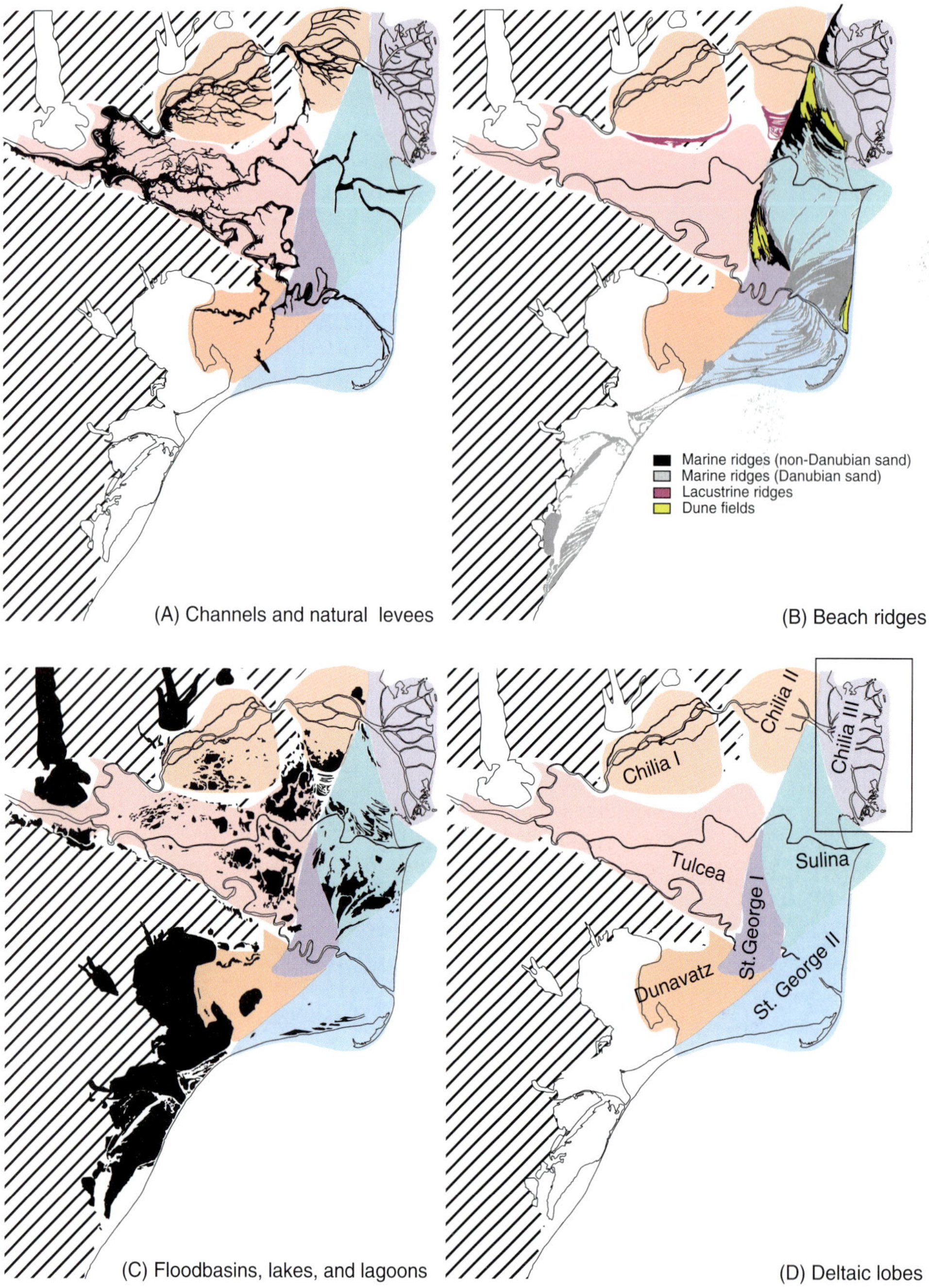

Figure 2 Morphology and depositional environments of the Danube delta in the Black Sea (after Giosan *et al.*, 2005): (A) channels and natural levees; (B) sandy beach ridges; (C) floodbasins, lakes, and lagoons; (D) delta lobes (the box indicates the location of Chilia III lobe described later in **Fig. 5**). The delta started as a bayhead delta (Tulcea lobe), followed by the development of other fluvial-dominated lobes in shallow lakes, and lagoons left unfilled by the Tulcea lobe (Chilia I and II; Dunavatz). Once reaching the open coast, several wave-dominated (St. George I and II and Sulina) or fluvial-dominated (Chilia III) lobes formed. Note that the sand comprising the beach ridge plains built updrift of open coast lobes was delivered via the longshore drift system.

include those of Gilbert (1885), Fisk *et al.* (1954), and Wright and Coleman (1973), among others.

In the 1980s, research on deltas moved from the development of depositional models based on modern highstand deltas, which are ultimately controlled by the delta-building rivers, to sequence stratigraphic interpretations of deltaic evolution through multiple sea-level cycles (e.g., Van Wagoner *et al.* (1988)). More recent research on modern deltas has been spurred by concerns related to human influences on the larger riverine systems, either directly through damming and water consumption, or indirectly via climate changes. Historical stages in deltaic research can be followed in successive comprehensive volumes edited by Morgan (1970), Broussard (1975), Oti and Postma (1995), and Giosan and Bhattacharya (2005), as well as periodic review papers such as Bhattacharya and Walker (1992).

River Deltas

Controls and Processes

The stratigraphic architecture of deltas is the end-result of complex interactions among upstream catchment processes that regulate the location and magnitude of the fluvial sediment discharge, and downstream basinal controls that include the shape and quantity of accommodation space for sediment accumulation and the type and energy of coastal processes that redistribute these sediments. Together, all of these controls determine the mode and degree of partitioning of sediment between the delta and the wider receiving basin (e.g., Wright and Coleman (1973)).

The sediment discharge of a river is determined by (1) climate that regulates the precipitation–evaporation characteristics of the watershed, and thus water flow; (2) the nature and abundance of vegetation that may inhibit erosion; and (3) the tectonic regime and lithology that determines elevation, slope, and degree of erodability of rocks in the watershed. Geology also controls the water gained or lost by a river as groundwater. Accommodation space in the receiving basin is dependent on (1) the relief characteristics of the basin, such as the extent of the shallow sector of the basin (or shelf) and (2) the degree of connection to the open ocean (e.g., a closed lake, indented estuary, or a straight coast). Such basin characteristics may also change drastically as variations in base level expose more or less of the margin. Base-level changes during the Quaternary were predominantly driven by the growth and decay of continental ice sheets, but in the case of enclosed basins such as the Black Sea and Caspian Sea, major base-level changes were forced by climatically driven fluctuations in their water budget. Additional components of relative base-level change are caused by tectonics, compaction, and isostasy.

Sediment in suspension is discharged as a plume at the river mouth, which is the single most important location for partitioning of sediment between the delta and the basin. Some rivers also transport a significant fraction of coarse bedload sands to the coast, where these less mobile sediments are preferentially stored within the delta edifice. As the river plume enters the receiving basin, flow expansion and deceleration exert a strong control on river-mouth sedimentation, which is dependent on the outflow inertia, turbulent bed friction, and plume buoyancy. The latter may be positively, neutrally, or negatively buoyant, depending on the density contrast between the plume and basin waters. Waves and tides are also important in controlling deposition at a river mouth, especially in high-energy coastal settings. Tidal and wind-driven currents, as well as Coriolis rotation, may strongly influence the behavior and path of the plume after exiting the river mouth region. The relative importance of these processes, as well as their history (i.e., frequency of occurrence and intensity), are also important in determining the patterns of sediment deposition across the delta and, ultimately, its stratigraphic architecture.

Delta Sediments, Morphology, and Morphodynamics

Deltas are complex sedimentary systems comprising a multitude of depositional environments. At the apex of the delta plain, which is the subaerial part of a delta, the river usually bifurcates into distributaries that, in many cases, can build their own delta lobes (**Fig. 2**). The number of lobes tends to decrease in wave-dominated settings where the number of distributaries is minimal. Lobes can rarely be distinguished in settings where tides are strong, because tidal currents favor channel stability and suppress avulsions. The entire delta can sometimes change location through large-scale avulsions to build separate subdelta complexes. This is the case for the Huanghe (Yellow) River, which completely switched its course at least twice to build deltas in both the Bohai and South Yellow seas (Saito *et al.*, 2001). The Mississippi River has also built six separate subdeltas comprising 16 individual lobes (see review in Roberts (1997)).

Based on general morphology of the subaerial delta plain, deltas have typically been classified as fluvial, wave, or tide dominated according to the dominant processes affecting sediment delivery, deposition, and dispersal (**Fig. 3**; Galloway (1975); Orton and Reading (1993)). A typical fluvial-dominated delta has an elongate to lobate plan-view morphology with highly crenulated shorelines; mouth bar and channel fill sands are interspersed within fine-grained deposits of the interdistributary region. In contrast, classic wave-dominated deltas assume a cuspate morphology and are mainly composed of beach-ridge sands. Tide-dominated deltas, which appear mostly irregular in plan view, are primarily composed of estuarine and tidal shoals, or islands, flanked by mudflats. However, a continuum of settings with

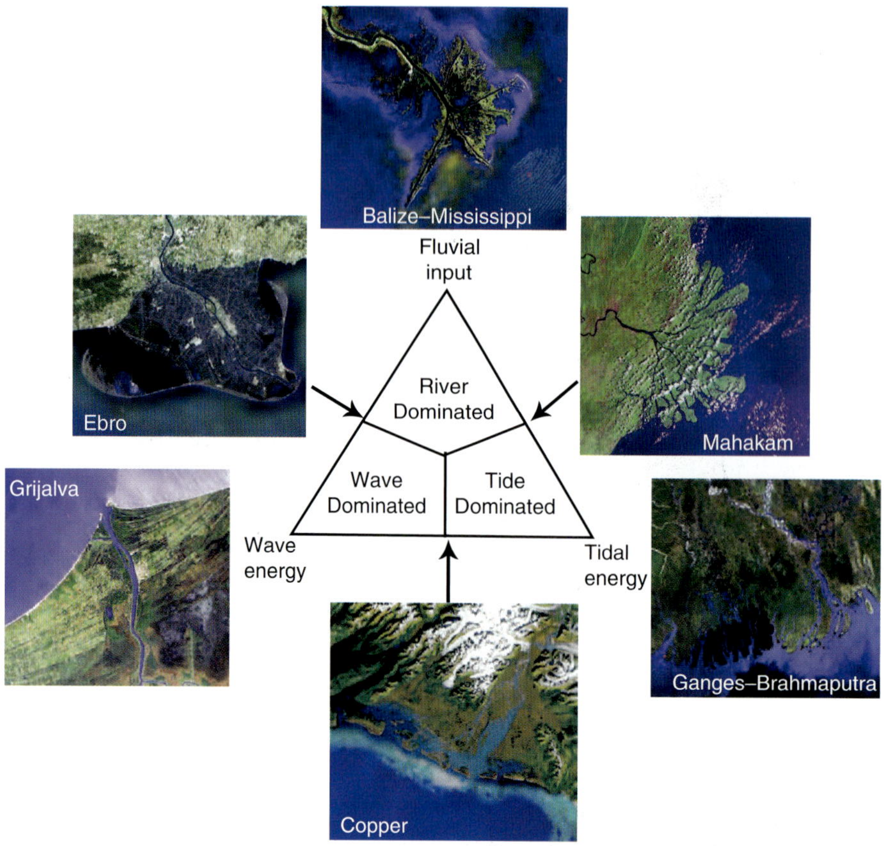

Figure 3 Process-based ternary classification of delta morphology as a function of fluvial discharge and wave and tidal energy. Satellite images of deltas are from Google Earth (http://earth.google.com/; Balize, Ebro, Mahakam, Grijalva, and Copper images © MDAEarthSat and DigitalGlobe; Ganges-Brahmaputra image © MDAEarthSat). After Galloway WE (1975) Process framework for describing the morphologic and stratigraphic evolution of deltaic depositional systems. In: Broussard ML (ed.) *Deltas: Models for Exploration*, Houston Geological Society.

overlapping processes rarely makes this classification operational at a scale larger than an individual delta lobe (Bhattacharya and Giosan, 2003). Even at the lobe scale, an in-depth look at wave influence on deltas shows that morphology and facies distribution is dependent more on the influence of the wave-driven longshore sediment transport relative to fluvial discharge, rather than on wave energy itself (**Fig. 4**). In multilobe deltas, one lobe may assume a vastly different morphology than an adjacent lobe. Of the two youngest lobes of the Danube delta, the Chilia, which is built with ~60% of the river-sediment discharge, is fluvial dominated, whereas the St. George is clearly wave dominated (**Fig. 2**). Temporal variability in sediment discharge or basin energy can also result in drastic morphological changes in a delta/lobe (Giosan *et al.*, 2005). The Po delta lobes had been wave dominated before a drastic increase of sediment discharge during the Little Ice Age (ca. AD 1450–1850) resulted in fluvial-dominated outbuilding phase (Correggiari *et al.*, 2005)). Even without changes in total sediment discharge, the developing Chilia lobe of the Danube delta started to show morphological modifications typical of wave-influenced settings (e.g., straightening of the shoreline, buildup of barrier spits and islands) due to a decrease in sediment delivered by the river per unit shoreline (**Fig. 5**).

Delta morphology is also greatly controlled by geometry of the receiving basin (Wright and Coleman, 1973). Because tides are absent in lakes, lacustrine deltas (**Fig. 1A**) are either fluvially dominated, when they build in shallow, low-energy lakes, or may become wave dominated in larger lakes (**Fig. 1B**), where a longer fetch allows larger waves to form. Bayhead and lagoonal settings are typically low-energy environments that favor fluvial and tidal influences over the effect of waves (**Figs. 1C–1E**). Morphology of deltas building along straight, open coasts is strongly dependent on the shelf configuration (i.e., width, steepness), which changes not only with the type of continental margin (i.e., active vs. passive) but also with sea-level changes that regulate the vertical accommodation space and the degree of coastal indentation. For example, as deltas prograde toward the middle and outer shelf, their morphology

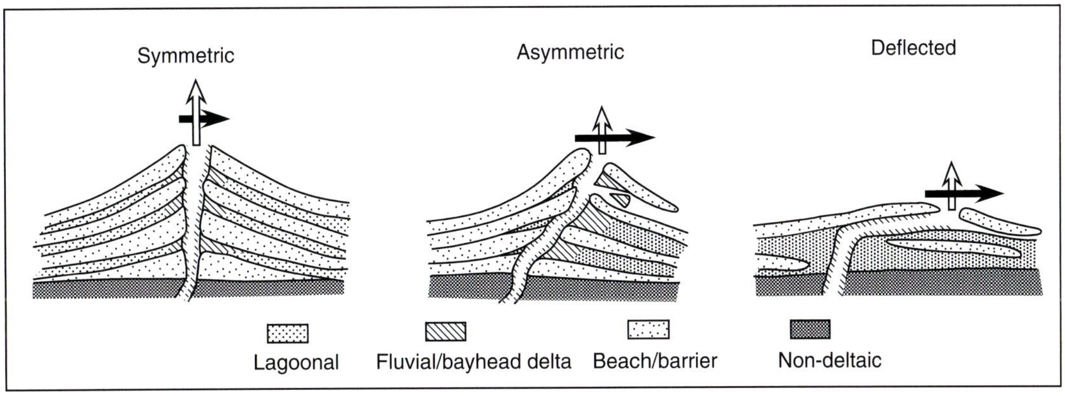

Figure 4 Morphology and depositional environments for mud-rich wave-dominated deltas as a function of the ratio between longshore drift and fluvial discharge (relative magnitudes indicated by black and white-filled arrows respectively). Unlike the classical symmetric wave-dominated deltas, where sediment supplied by the feeding river is reworked on either side by waves, asymmetric deltas develop where the longshore drift becomes important relative to the river discharge and updrift beach ridge plains develop. The downdrift side is a succession of elongate sandy ridges separated by mud-filled troughs. The mouth of the river gets deflected in extreme cases of asymmetry resulting in a series of quasi-parallel sand spits and channel fills. After Bhattacharya JP and Giosan L (2003) Wave-influenced deltas: Geomorphologic implications for facies reconstruction. *Sedimentology* 50: 187–210.

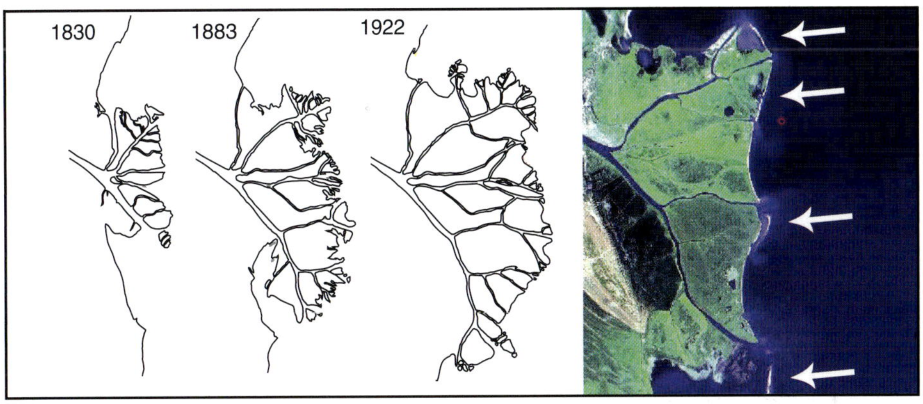

Figure 5 Delta-plain evolution of the Chilia III lobe of the Danube delta. The morphology of the lobe had been fluvial dominated until recently, when the coast begun to straighten and barrier spits and islands started to develop along the coast (indicated by arrows). (Satellite image from Google Earth: http://earth.google.com/; © MDAEarthSat and DigitalGlobe.) After Giosan L, Donnelly JP, Vespremeanu E, and Buonaiuto FS (2005) River delta morphodynamics: Examples from the Danube delta, In: Giosan L and Bhattacharya J (eds.) *River Deltas: Concepts, Models, Examples, SEPM Special Publication* 83, pp. 393–411. SEPM.

may be affected by increasing wave influence associated with deeper nearshore waters. In contrast, a broad, low-gradient inner shelf can attenuate incident waves and amplify tides, thereby favoring more tidally influenced delta systems.

In cross section, the idealized delta morphology and stratigraphy has been described as a prograding clinoform composed of an upward coarsening facies succession that starts with a basal muddy prodelta, progresses into a sandy delta front and is topped by texturally variable delta plain deposits (**Fig. 6**). The subaqueous portion of a delta comprises both the distal prodelta and the steeper, prograding delta front, whereas the delta plain is predominantly subaerial. The muddy prodelta itself often acquires a 'clinoform' geometry, leading to a compound clinoform architecture for the delta that consists of a near-shore sandy clinoform associated with the shoreface and an offshore muddy clinoform (**Fig. 7**) associated with the prodelta (Kuehl *et al.*, 2005). Development of prodelta clinoforms is not universal and appears to be a function of fluvial input and basin hydrodynamics (Swenson *et al.*, 2005). In such settings, gravity-driven flows of suspended mud commonly feed the growth of prodelta clinoforms in the presence of episodic or periodic high-energy processes, such as spring-neap tidal cycles or during storm events.

Morphodynamic response in deltas involves feedbacks between the evolving morphology of a delta and its fluvial and basinal hydrodynamics. Early models identified rotation of the shoreline as responsible for changes in longshore drift along deltaic coasts and affecting the rate of progradation. Asymmetry of the wave climate increases the

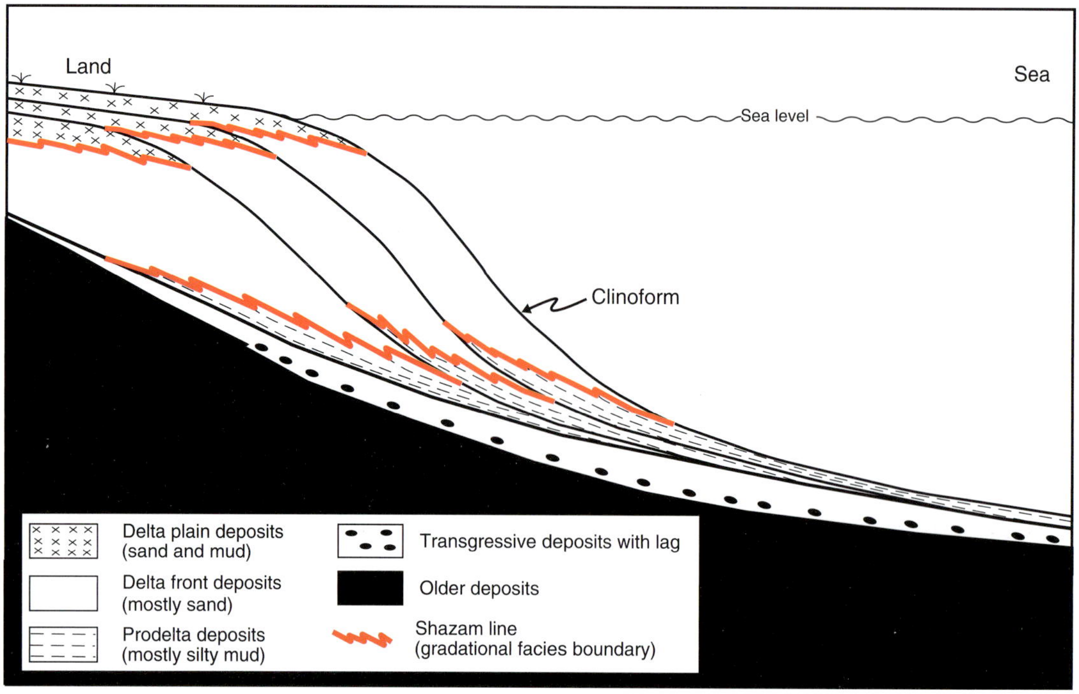

Figure 6 Idealized cross-section showing internal bedding geometry and facies architecture of a prograding deltaic clinoform. After Gani MR and Bhattacharya JP (2005) Lithostratigraphy vs. chronostratigraphy in facies correlations of Quaternary deltas: application of bedding correlations, In: Giosan L and Bhattacharya J (eds.) *River Deltas: Concepts, Models, Examples*, *SEPM Special Publication* 83, pp. 31–48. SEPM.

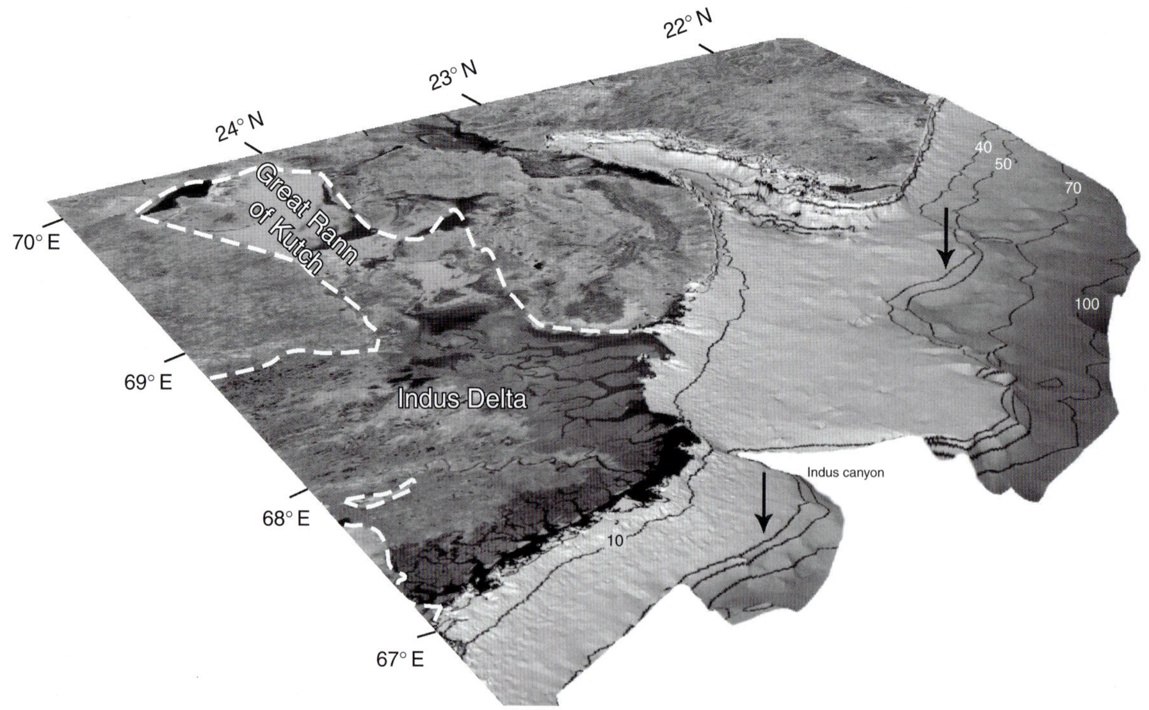

Figure 7 Digital elevation model for the Indus shelf based on the nineteenth century surveys showing the compound clinoform morphology for the subaqueous Indus delta. The near-shore clinoform extends to the ~10 m isobath. Offshore clinoform lobes are indicated by arrows. Inland geography is visualized on a satellite image from 2000 (the limits of the Indus delta and the adjacent mud flats of the Great Rann of Kutch are indicated by the white dashed line).

shoreline instability on the downdrift delta wing, promoting development of spits detached from the main delta coast (Bhattacharya and Giosan, 2003). Hydrodynamic–morphodynamic feedback loops have also been identified in relation to bedload sedimentation at river mouths. The hydraulic groin effect of river plumes is accompanied by the groin effect of the subaqueous delta or of the delta plain, blocking longshore drift at the river mouth (Giosan *et al.*, 2005). On the downdrift wing of asymmetric wave-dominated deltas, periodic emergent barrier islands provide steep, continuous shorefaces for the longshore drift, leading to rapid expansions of the subaqueous delta in the alongshore direction.

Deltas in the Quaternary

Whereas waves and tides are key controls of deltaic systems over decades to centuries, delta evolution at longer timescales is most significantly controlled by (1) sea level, (2) climate, and (3) tectonics. Changes in these first two controls, sea level and climate, are hallmarks of the Quaternary period and its succession of glacial and interglacial phases. As a consequence, deltaic systems have effectively experienced continuous cycles of formation, change, and destruction throughout the Quaternary. Because the major controls of sea level, climate, and tectonics are interrelated, their relative impacts on deltaic systems remain unclear (Goodbred, 2003). Nevertheless, traditional models posit that the development of major deltaic systems is restricted to periods of sea-level highstand, with flooded continental margins and slow to moderate rates of sea-level rise. However, in the past few decades Quaternary deltaic systems have been discovered at a variety of shelf positions and been linked to numerous states of sea-level change. In the past two decades, the advent of multibeam sonar imaging and improved seismic reflection processing has allowed such deltaic shelf deposits to be discovered. A growing knowledge of climate history has in many cases provided explanations (e.g., high sediment supply due to increased precipitation, geomorphic landscape disequilibrium) for delta formation even under rapid rates of sea-level rise.

Influence of Climate on Deltas

Deltas can largely be considered a function of (1) riverine water and sediment discharge to the continental margin and (2) the interaction of these fluxes with coastal and marine processes. In this way, climate plays a major role in both the catchment-based forcing of sediment production and fluvial transport and the modulation of sea level and coastal weather conditions in the receiving basin. As such, there is strong climatic overprinting on deltaic systems, but precise signals are difficult to quantify because of complex feedbacks and secondary influences.

In the terrestrial realm of a fluviodeltaic dispersal system (i.e., the fluvial catchment) perhaps the most prevalent role that climate plays lies in controlling precipitation. With regard to sediment production, precipitation is a primary control on glacial and fluvial incision, weathering, and slope failure. Although highly variable, sediment is readily produced across a range of climates, from arid to humid and polar to tropical. However, one of the main factors that differs among these settings in terms of climate is runoff (= precipitation − evapotranspiration) and thus the ability to transport sediment to the delta. In other words, sediment can be produced in many settings, but is most efficiently transported given sufficient water discharge, which is primarily controlled by climate.

Other important characteristics of precipitation include its phase (liquid vs. frozen), seasonality, and episodicity. First, the phase of precipitation controls the relative role of glacial and fluvial erosion in sediment production, as well as the timing and nature of water discharge (i.e., the hydrograph). Each affects sediment-load grain size and weathering characteristics and, hence, the downstream development of delta stratigraphy. Second, the seasonality of river discharge affects the relative interaction of fluvial and marine processes, whereby strongly seasonal systems are subject to specific periods of river and marine-dominated conditions during the year. Similarly, the episodic discharge typified by storm-dominated or semiarid river systems leaves a strong imprint on deltas, because sediments are delivered in large but infrequent pulses. Finally, precipitation exhibits a strong secondary control on delta systems through the type and density of catchment vegetation, which affects runoff via overland flow and evapotranspiration rates as well as soil erosion rates.

From the marine side of a delta, climate change plays the primary role in modulating eustatic sea level through the orbitally controlled advance and retreat of continental ice sheets. This waxing and waning of ice sheets, a defining characteristic of the Quaternary period, has forced repeated, generally rapid, large-magnitude cycles of sea-level change. For delta systems, being principally coastal features, such glacioeustatic changes in sea level have been a first-order control on their development. In this way, traditional conceptual models suggest that prominent delta systems form only under relatively stable sea-level conditions and more specifically during highstands when fluvial sediment is likely to be trapped on the inner shelf. In contrast, during lowstands of sea level it was considered that most fluvial sediment was discharged beyond the shelf edge, either by off-shelf transport of gravity flows or through bypassing via canyon systems. Intervening periods of sea-level rise were assumed to preclude the development of deltas because of rapid coastal transgression or regression.

Site Cases from the Quaternary

Until recently, few studies have specifically investigated climate signals and impacts in delta systems. The current knowledgebase on this subject comes from several modeling groups, some field studies, and also by revisiting existing data with a better understanding of climate change provided by secular records (e.g., loess deposits, speleothems, and lacustrine sediments). Although the site cases discussed below are not exhaustive, two relevant trends emerge. First, the number of modeling studies equals or exceeds those based on field investigations, and second, the majority of delta systems in which climate has been investigated are located in the subtropics (the Rhine/Meuse a notable exception). Subtropical systems tend to be highly seasonal, receiving nearly all of their precipitation during the wet summer monsoon when the intertropical convergence zone (ITCZ) is most proximal. At orbital timescales, both the position and strength of the ITCZ vary in response to global boundary conditions (i.e., extent of continental ice sheets) and maximum summer insolation, respectively. Thus, such systems have exhibited among the most prominent responses to Quaternary climate variability.

Yellow River delta Extensive databases from a few areas of the world have led to the discovery of significant deltaic deposits on the middle and outer portions of the shelf (see e.g., **Fig. 7**). In the North Yellow Sea, a compilation of seismic data by Liu *et al.* (2004) detailed a massive deltaic clinoform comprising ca. 400 km^3 of sediment from the Yellow (Huanghe) River. The Shandong mud wedge, as the feature is known, is 20–40 m thick and situated on the mid-shelf at water depths of 30 to >50 m. Prominent within the deposit is a marine flooding surface that indicates the delta formed in two separate stages. Dating of the two phases of transgression and deposition suggests that these responses are associated with rapid changes in the rate of sea-level rise during deglaciation, perhaps related to meltwater pulses. In such case, brief but rapid transgressions were followed by progradation of the Yellow River delta, infilling accommodation space generated by the transgression.

This early Holocene pattern of Yellow River delta formation challenges traditional models in several ways. First, the growth occurs during a period of unstable sea level punctuated by sudden changes, yet the location of primary deposition remains largely unchanged during this time. Liu *et al.* (2004) ascribe this in large part to the stability of oceanographic circulation patterns in the North Yellow Sea. Second, size and thickness of the delta far exceed that expected for conditions of rapid Postglacial sea-level rise, when models might suggest formation of a thin transgressive sand sheet. In this case, Postglacial transgression in the early Holocene coincided with intensification of the Asian monsoon, and an apparent increase in fluvial sediment discharge. The notion that climatically induced increases in sediment discharge may be sufficient to force delta progradation during rapid sea level opens many new directions for research on how, where, and when such climate–sea level–delta interactions occur.

Ganges–Brahmaputra river delta The Ganges and Brahmaputra rivers drain the foreslope and backslope of the Himalayan front range, respectively. The two rivers coalesce near the Bengal margin to form the world's largest subaerial delta and deep-sea fan system. The rivers are also strongly driven by the summer SW monsoon, when over 80% of water and 95% of sediments are discharged in the 4–5-month wet season. Given the global significance of the Himalayan/Tibetan uplift and its associated drainage systems, many portions of the Ganges fluvial delta system have been independently investigated in the past 15 yr, revealing important variations in glacial activity, alluvial fan and floodplain development, delta evolution, and regional oceanography. Taken together, the observed patterns show the immense Ganges dispersal system to respond to millennial-scale (<10^4 yr) climate change in a system-wide and largely contemporaneous manner, and that major sedimentary signals are transferred rapidly from source to sink with little apparent attenuation (Goodbred, 2003).

Some of the major signals of climate change recorded in the delta stratigraphy include (1) a thick (20–30 m) coastal mangrove facies that was deposited during rapid sea-level rise in the early Holocene and (2) a massive volume of sediment trapped during Postglacial transgression, reflecting a sediment discharge at least 200% higher than the world's largest modern load of 1 billion tonnes (Goodbred and Kuehl, 2000). This high-discharge period corresponds to the early Holocene hypsithermal, when peak regional insolation supported a stronger than present monsoon and associated precipitation. In contrast, signals recorded in the delta during the

LGM 18–20 ka reflect aridification and very low river discharge. Carbonate ooids, typical of shallow, evaporative tropical waters, formed at this time on the Bengal shelf directly adjacent to the rivermouth. Similarly, reconstructed planktonic foraminifera assemblages from the northern Bay of Bengal (the Ganges receiving basin) indicate peak surface-water salinities and apparently arid conditions at the LGM. These records from the delta and margin correlate well with geomorphic and stratigraphic results from the fluvial catchment, and together reflect the dominance of climate change in controlling this large fluviodeltaic system during the Quaternary (Goodbred, 2003).

Colorado, Brazos, and Trinity deltas (Gulf of Mexico) These three rivers comprise a series of small catchments draining the semiarid to moderately humid hills of central Texas, USA and discharging to the northern Gulf of Mexico (Anderson and Fillon, 2004). Presently, sediment discharge is relatively small and delta progradation is driven by episodic floods associated with humid phases under El Niño conditions. However, during midstands and lowstands of sea level in the late Quaternary, these drainage basins coalesced to form expansive fluviodeltaic sequences on the middle and outer shelf (**Fig. 8**). The size of these shelf deltas and their connection to underfit streams occupying large fluvial valleys reflect formation under higher-than-present discharge, implying significant changes in past climate.

On the Texas shelf, the position and architecture of relict deltas also indicate their formation during a variety of sea-level stages and transitions. Seismic-based reconstructions of the shelf deposits reveal large, forced-regressive sequences formed during the falling stages of sea level at marine isotope stages 5 and 3 (**Fig. 8**; Anderson and Fillon, 2004). During these stages, the relative highstand and midstand of sea level, respectively, placed the deltaic lobes near the modern inner to middle shelf. Overlying these regression-phase deltas are a group of smaller backstepping deltas formed during rapid Postglacial transgression after the LGM. Though smaller than the earlier delta complexes, formation of these systems under rapid rates of sea-level rise was in large part a consequence of a more humid climate and increased sediment discharge in the early Holocene. Together, delta sequences of the Texas shelf have formed under a variety of climatic and sea-level conditions, indicating the critical role that both of these forcings play.

Modeling studies Quaternary climate change has been recognized as an important control on fluviodeltaic systems. However, it has been difficult to confidently link observed stratigraphic and sedimentological signatures to climatic forcings, given the strong overprinting of sea-level and marine processes on delta morphology. Such limits of field investigations are reflected in the relatively limited number of studies, many of which are discussed above. However, the exponential increase in computing power has allowed major advances to be made in the numerical modeling of fluviodeltaic systems. Several current models offer the opportunity to compare the consequences of forcing of fluvial delta systems by any combination of climate, eustasy, and tectonics, thereby helping elucidate the specific character of climate impacts on stratigraphic sequences (Overeem *et al.*, 2005). This is especially relevant given that climate change and glacioeustasy are phased through most of the Quaternary (**Fig. 9**), making it extremely difficult to derive unique interpretations of these forcings from field data alone.

One system that has been a focus of Quaternary response modeling is the Meuse river, which drains central Europe and discharges to the southern North Sea coast. Much of the research on this system during the past decade has had a focus on understanding the signatures of climate and eustasy contained in the late Quaternary stratigraphy and geomorphology (e.g., Veldkamp and Tebbens, 2001). Situated in the periglacial environments of northwest Europe during the Quaternary, climatically driven variations in the magnitude, timing, and nature of discharge appear to have played a role in shaping the fluviodeltaic system upstream of the hinge line. Terrace development in this river-dominated reach of the system is recognized as a primarily climate-controlled process through both modeling and field studies. However, the downstream impacts of such hydrological changes in the Meuse system appear to be insufficient to overcome the dominant influence of glacioeustasy during the Quaternary. Although there are likely to be impacts of climate in the delta, they remain indistinguishable at the resolution of current data and modeling capabilities.

Another system, the Niger delta, has been investigated using an industry-based stratigraphic model (van der Zwan, 2002). This investigation was aimed at testing the impact of presumed Milankovitch-scale variations in sediment discharge over the Neogene. The orbitally linked forcing of regional climate and sediment occurred via global climate, insolation control of monsoon strength, and vegetation. Results suggest that sediment supply does change due to climatic variability at both large-scale (Ma) and Milankovitch-scale timeframes, and that there was a modest impact on stratigraphy. However, under icehouse conditions, fluctuations in climate-forced sediment supply were greatly overshadowed by rapid glacioeustatic sea-level change. These findings echo those of other modeling studies, in that climate is often found to be an

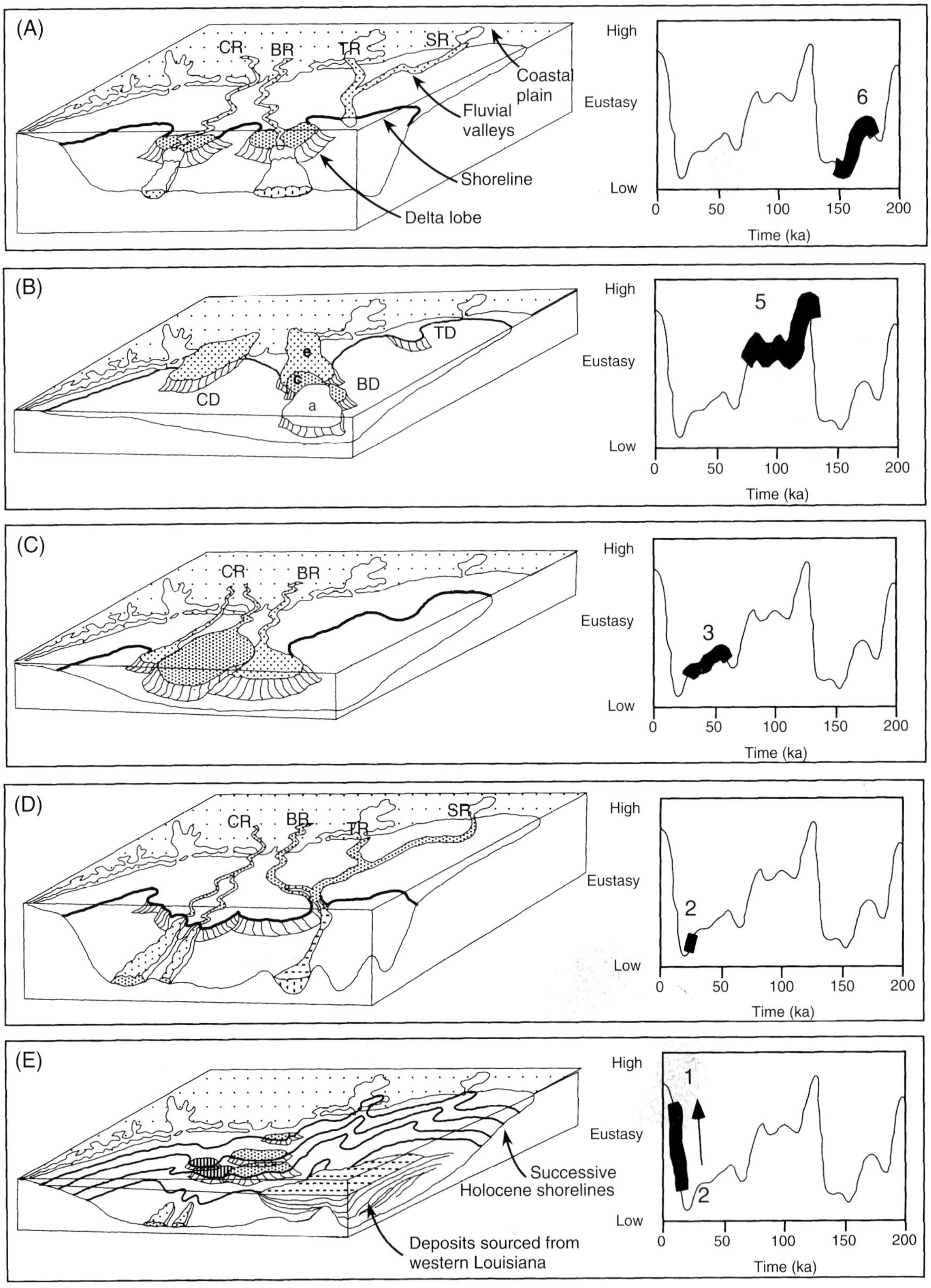

Figure 8 Reconstructed late Quaternary history for the rivers and deltas of the Colorado (CR, CD), Brazos (BR, BD), Trinity (TR, TD), and Sabine (SR) fluviodeltaic systems on the northern Gulf of Mexico shelf. Phases of delta progradation, incision, and backstepping are shown with corresponding phases of sea level, and are also associated with periods of climate change and varying sediment discharge. Note larger-than-present deltaic lobes building on the mid-shelf during MIS-3 interstade, under apparently wetter climatic conditions. Climate and sea level are dominant controls on delta formation and have varied somewhat in-phase during the late Quaternary (see **Fig. 9**), giving rise to complex fluviodeltaic responses. From Anderson JB and Fillon RH (ed.) (2004) *Late Quaternary Stratigraphic Evolution of the Northern Gulf of Mexico Margin. SEPM Special Publication* 79, 314 p. SEPM.

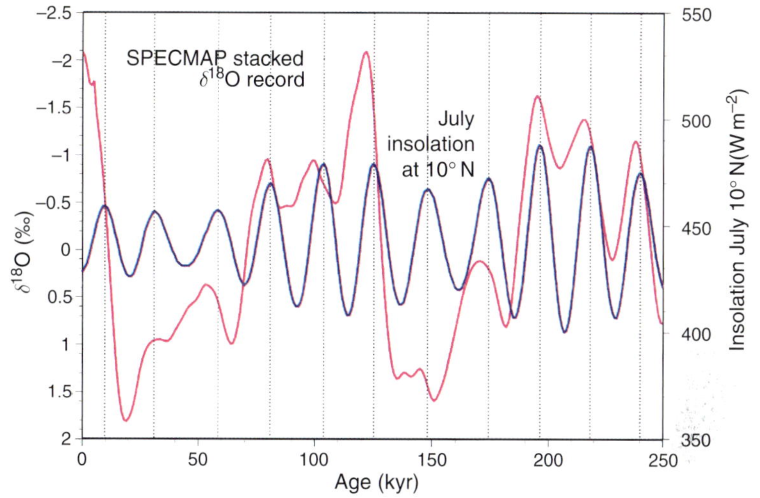

Figure 9 SPECMAP $\delta^{18}O$ and Milankovitch-tuned insolation records for the late Quaternary compiled by Imbrie, and Berger and Loutre, respectively. SPECMAP data represent a proxy for sea-level change and insolation serves as a coarse proxy for wetness in the subtropical (monsoon-influenced) Northern Hemisphere. Intended to show the general cophasing of peak insolation (generally wetter) with sea-level transgression, with the implied consequence of high sediment discharge corresponding with rapid sea-level rise. This pattern has been observed in the Yellow, Yangtze, and Ganges–Brahmaputra delta systems, west African rivers, and most likely in the Texas Gulf of Mexico rivers.

important, but secondary, control on fluviodeltaic sequences. Furthermore, most modeling studies suggest that climatic signals preserved at the continental margin greatly lag the timing of the actual forcing in the fluvial catchment (e.g., Castelltort and Van Den Driessche, 2003). This notion conflicts with recent field evidence from several of the subtropical rivers discussed above. Much of these discrepancies likely arise from our limited knowledge of appropriate input values for pre-Modern climate, riverine, and coastal conditions.

See also: **Fluvial Environments**: Sediments; Responses to Rapid Environmental Change.

References

Anderson, J. B., and Fillon, R. H. (Eds.) (2004). Late Quaternary Stratigraphic Evolution of the Northern Gulf of Mexico Margin. SEPM Special Publication 79, p. 314. SEPM, Tulsa, Oklahoma, USA.

Bhattacharya, J. P., and Giosan, L. (2003). Wave-influenced deltas: Geomorphologic implications for facies reconstruction. *Sedimentology* **50**, 187–210.

Bhattacharya, J. P., and Walker, R. G. (1992). Deltas. In *Facies Models – Response to Sea Level Change* (R. G. Walker and N. P. James, Eds.), pp. 157–177. Geological Association of Canada, St Johns, Newfoundland, Canada.

Broussard, M. L. (1975). *Deltas, Models for Exploration.* Houston Geological Society, Houston, Texas, USA.

Castelltort, S., and Van Den Driessche, J. (2003). How plausible are high-frequency sediment supply-driven cycles in the stratigraphic record? *Sedimentary Geology* **157**, 3–13.

Correggiari, A., Cattaneo, A., and Trincardi, F. (2005). Depositional patterns in the late-Holocene Po delta system. In *River Deltas: Concepts, Models, Examples, SEPM Special Publication* (L. Giosan and J. Bhattacharya, Eds.) 83, pp. 364–392. SEPM, Tulsa, Oklahoma, USA.

Fisk, H. N., McFarlan, E., Kolb, C. R., and Wilbert, L. J. (1954). Sedimentary framework of the modern Mississippi delta. *Journal of Sedimentary Petrology* **24**, 76–99.

Galloway, W. E. (1975). Process framework for describing the morphologic and stratigraphic evolution of deltaic depositional systems. In *Deltas: Models for Exploration* (M. L. Broussard, Ed.). Houston Geological Society, Houston, Texas, USA.

Gani, M. R., and Bhattacharya, J. P. (2005). Lithostratigraphy vs. chronostratigraphy in facies correlations of Quaternary deltas: application of bedding correlations. In (L. Giosan and J. Bhattacharya, Eds.) *River Deltas: Concepts, Models, Examples, SEPM Special Publication* 83, pp. 31–48. SEPM.

Gilbert, G. K. (1885). The topographic features of lake shores. *Annual Report U.S. Geological Survey* **5**, 75–123.

Giosan, L., and Bhattacharya, J. (Eds.) (2005). River Deltas: Concepts, Models, Examples. SEPM Special Publication 83, p. 502. SEPM, Tulsa, Oklahoma, USA.

Giosan, L., Donnelly, J. P., Vespremeanu, E., and Buonaiuto, F. S. (2005). River delta morphodynamics: Examples from the Danube delta. In *River Deltas: Concepts, Models, Examples, SEPM Special Publication* (L. Giosan and J. Bhattacharya, Eds.) 83, pp. 393–411. SEPM, Tulsa, Oklahoma, USA.

Goodbred, S. L. (2003). Response of the Ganges dispersal system to climate change: A source-to-sink view since the last interstade. *Sedimentary Geology* **162**, 83–104.

Goodbred, S. L., and Kuehl, S. A. (2000). The significance of large sediment supply, active tectonism, and eustasy on margin sequence development: Late Quaternary stratigraphy and evolution of the Ganges–Brahmaputra delta. *Sedimentary Geology* **133**, 227–248.

Kuehl, S. A., Allison, M. A., Goodbred, S. L., and Kudrass, H. (2005). The Ganges–Brahmaputra delta. In *River Deltas: Concepts, Models, Examples, SEPM Special Publication* (L. Giosan and J. Bhattacharya, Eds.) 83, pp. 413–434. SEPM, Tulsa, Oklahoma, USA.

Liu, J. P., Milliman, J. D., Gao, S., and Cheng, P. (2004). Holocene development of the Yellow River's subaqueous delta, North Yellow Sea. *Marine Geology* **209**, 45–67.

McKee, B. A., Aller, R. C., Allison, M. A., Bianchi, T. S., and Kineke, G. C. (2004). Transport and transformation of dissolved and particulate materials on continental margins influenced by major rivers: Benthic boundary layer and seabed processes. *Continental Shelf Research* **24**, 899–926.

Meade, R. H. (1996). River-sediment inputs to major deltas. In *Sea Level Rise and Coastal Subsidence* (J. D. Milliman and B. U. Haq, Eds.), pp. 63–85. Kluwer, Dordrecht.

Morgan, J. P. (1970). Deltaic Sedimentation: Modern and Ancient. SEPM Special Publication 15, p. 312. SEPM, Tulsa, Oklahoma, USA.

Orton, G. J., and Reading, H. G. (1993). Variability of deltaic processes in terms of sediment supply, with particular emphasis on grain size. *Sedimentology* **40**, 475–512.

Oti, M. N., and Postma, G. (1995). Geology of Deltas, p. 315. A.A. Balkema, Rotterdam.

Overeem, I., Syvitski, J. P. M., and Hutton, E. W. H. (2005). Three-dimensional numerical modeling of deltas. In River Deltas: Concepts, Models, Examples, SEPM Special Publication (L. Giosan and J. Bhattacharya, Eds.) 83, pp. 13–30. SEPM, Tulsa, Oklahoma, USA.

Roberts, H. H. (1997). Dynamic changes of the Holocene Mississippi River delta plain: The delta cycle. *Journal of Coastal Research* **13**, 605–627.

Saito, Y., Yang, Z. S., and Hori, K. (2001). The Huanghe (Yellow River) and Changjiang (Yangtze River) deltas: A review on their characteristics, evolution and sediment discharge during the Holocene. *Geomorphology* **41**, 219–231.

Syvitski, J. P. M., Harvey, N., Wolanski, E., Burnett, W. C., Perillo, G. M. E., and Gornitz, V. (2005). Dynamics of the Coastal Zone. In *LOICZ: Coastal Change and the Anthropocene* (C. Crossland, Ed.). Springer, Berlin, Germany.

Swenson, J. B., Paola, C., Pratson, L., Voller, V. R., and Murray, A. B. (2005). Fluvial and marine controls on combined subaerial and subaqueous delta progradation: Morphodynamic modeling of compound-clinoform development. *Journal of Geophysical Research* **110**, F02013, (doi:10.1029/2004JF000265).

Van der Zwan, C. J. (2002). The impact of Milankovitch-scale climatic forcing on sediment supply. *Sedimentary Geology* **147**, 271–294.

Van Wagoner, J. C., Posamentier, H. W., Mitchum, R. M., *et al.* (1988). An overview of the fundamentals of sequence stratigraphy and key definitions. In *Sea-Level Changes: An Integrated Approach, SEPM Special Publication* (C. K. Wilgus, B. S. Hastings, C. G. C. Kendall, H. W. Posamentier, C. A. Ross and J. C. Van Wagoner, Eds.) 42, pp. 39–45. SEPM, Tulsa, Oklahoma, USA.

Veldkamp, A., and Tebbens, L. A. (2001). Registration of abrupt climate changes within fluvial systems: insights from numerical modelling experiments. *Global and Planetary Change* **28**, 129–144.

Wright, L. D., and Coleman, J. M. (1973). Variations in morphology of major river deltas as functions of ocean wave and river discharge regimen. *Bulletin of the American Association of Petroleum Geologist* **57**, 370–398.

Relevant Website

http://earth.google.com – Google Earth

Foraminifera Studies *see* **Paleoceanography; Paleoceanography, Biological Proxies**: Benthic Foraminifera; Planktic Foraminifera; **Paleoceanography, Physical and Chemical Proxies**: Oxygen Isotope Stratigraphy of the Oceans; **Paleoceanography, Biological Proxies**: Temperature Proxies, Census Counts

Fossil Mites *see* **Oribatid Mites**

G

GEOMAGNETIC EXCURSIONS AND SECULAR VARIATIONS

R M Twyman, Market Weighton, York, UK

© 2007 Elsevier B.V. All rights reserved.

Introduction

The Earth's magnetic field is approximately dipolar, with one pole in the Northern Hemisphere and the other in the Southern Hemisphere. The magnetic poles are near the geographic north and south poles but are not co-aligned. The angle of this east–west deviation, measured from anywhere on Earth, is called the declination. The degree of declination is not fixed, and the poles may drift, independently of each other, by more than 10 km each year. This independence, and the fact that the two magnetic poles are not directly opposite each other, shows that the similarity between the Earth's magnetic field and that of, for example, a bar magnet, is rather superficial. While the field generated by a bar magnet represents the coordinated movement of electrons in a fixed solid structure, the Earth's magnetic field is much more complex in origin.

The primary source of the Earth's magnetic field is its iron-rich liquid outer core. The movement of this liquid is continuous, driven by convection currents, the planet's rotation (Coriolis effect), and the cycle of melting and solidification that occurs at the boundary between the outer core and the solid inner core. Because the liquid outer core is conductive, its movement relative to the inner core induces an electrical current, which also has a magnetic field. This reinforces the original magnetic field and results in a self-sustaining field-generating system known as the geodynamo (Hollerbach and Jones, 1993; 1995).

The continuous magnetic flux within the Earth's core can be recorded as changes in magnetic field strength and orientation at the surface. The changes are relatively slow, occurring over historical timescales, and this process of gradual change is defined as secular variation (Jackson *et al.*, 2000). Secular variation is punctuated by more dramatic changes, during which the field strength may fall to 10–20% of its normal value within a relatively short time (a few thousand years). Such rapid decreases in strength are known as geomagnetic excursions and the intervals between them are measurable on geological timescales (Gubbins, 1999). In most cases, the field regenerates with the original polarity. Occasionally, however, the excursion is accompanied by a reversal of field polarity, in which case it is known as a geomagnetic reversal. A few such reversals occur every million years on average, but their occurrence does not fit a predictable pattern and some periods of Earth's history are known for the paucity of such events. The most recent was the Brunhes-Matuyama reversal approximately 789,000 years ago. Evidence suggests that since the last reversal there have been at least six global excursions (i.e., large decreases in magnetic field intensity correlated at several different sites) as well as 10–15 further excursions for which evidence exists at a single site only (Champion *et al.*, 1988; Langereis *et al.*, 1997; Lund *et al.*, 1998).

Secular Variation

Secular variation describes continuous drift in the intensity and direction of the Earth's magnetic field (Bloxham and Gubbins, 1985). Such changes are noticeable over relatively short periods of time (tens of years) and thus magnetic charts need to be updated periodically to accommodate them. Humans have observed such changes for around 1000 years, and have collected data from geographically diverse sites for nearly 500 years. Such data show a gradual, monotonic decline in the intensity of the magnetic field and suggest that it has fallen by approximately 20% during that time (Barraclough, 1982; Jackson *et al.*, 2000). This downward trend is supported by archeological evidence, such as mineral-containing Roman artifacts, which indicate that the field intensity at the time of the Roman Empire was twice as high as it is today. If the field strength continues to fall at its current rate, the dipole moment will reach zero by the year 3500. Although the fate of the magnetic field cannot be predicted with any accuracy, this trend has led some to believe that we are experiencing an ongoing excursion.

The orientation of the magnetic field also varies in time. Although the degree of declination differs in different parts of the world, the overall trend is a westward drift at about 0.1° per year together with a decrease in colatitude at a rate of about 0.02° per year. The degree to which the magnetic axis declines from the Earth's axis of rotation is often quoted as 11.3°, but due to the drift described above this is not an accurate statement. Magnetic data collected regularly at fixed points on the Earth's surface, i.e., from ground-based magnetic observatories, demonstrates both the geographical and temporal secular variation in the magnetic field (**Fig. 1**). Geographical maps of declination have been used to generate models showing how the field at the Earth's surface varies in complex patterns in both time and space, with declination contours converging at the geomagnetic poles. The strength of the field at the Earth's surface ranges from less than 30 microteslas (30 μT; 0.3 gauss), for example, over South America, to more than 60 μT (0.6 gauss) at the magnetic poles.

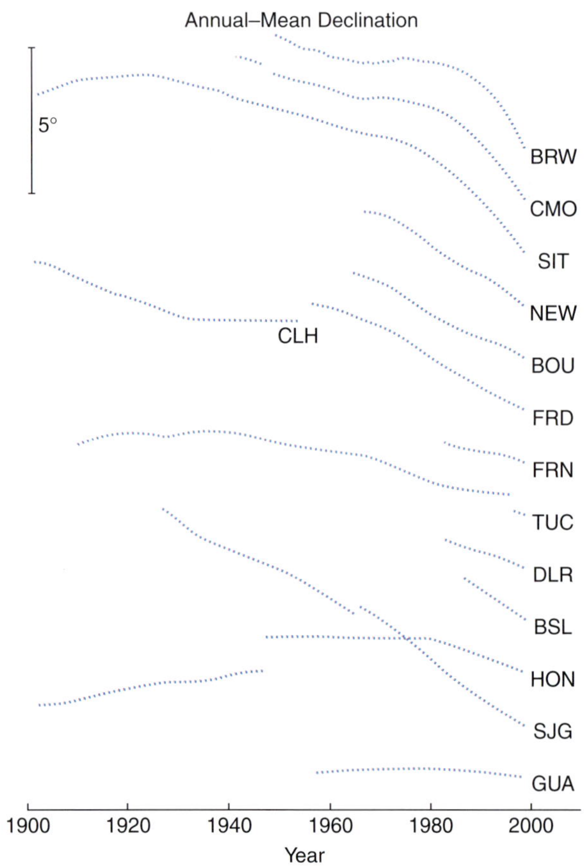

Figure 1 Stackplot of annual-mean declination (D), measured in degrees east, at the United States Geological Survey magnetic observatories over the last century. The field at the surface, and the declination in particular, has a complex form, resulting in equally complex secular variation. Also note the occasional offset, such as in the Honolulu (HON), San Juan (SJG), and Fredericksburg (FRD)/Cheltenham (CLH) annual-mean data, due to change of the absolutes pier or moving of the observatory.

Geomagnetic Excursions

Geomagnetic excursions are radical swings in field direction accompanied by decreases in field strength which take place over a geologically short period of time (thousands of years). The distinction between a geomagnetic excursion and a large secular variation event is the degree of departure between the virtual geomagnetic pole (VGP) and the geographic pole, with the cut-off usually at 45° (Verosub, 1977). Some of the more widely reported geomagnetic excursions since the last reversal event are shown in **Figure 2**.

While the continuous movement of the Earth's molten outer core is sufficient to account for secular variation, it is unclear what causes the much

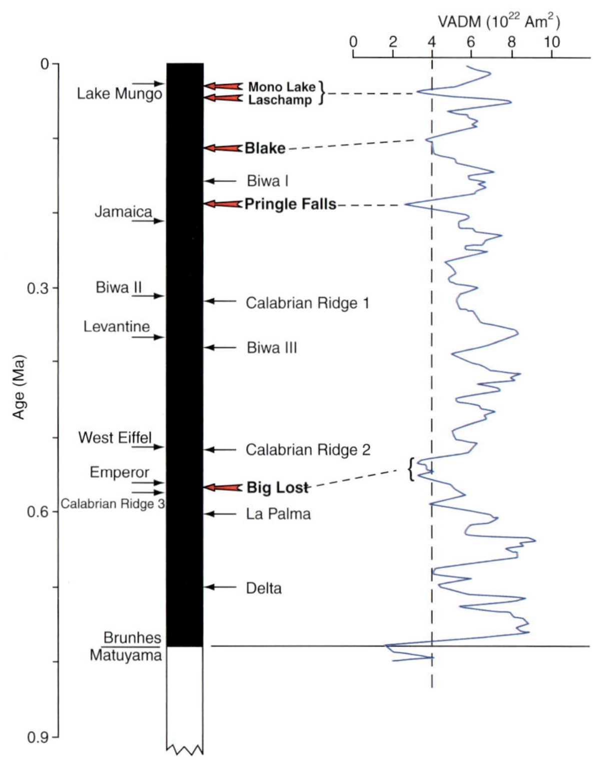

Figure 2 Approximate times of some of the most commonly reported geomagnetic excursions of the Brunhes normal polarity chron. VADM, virtual axial dipole moment.

larger deviations that result in excursions and reversals. One theory is that, occasionally, some aspect of the geodynamo process results in the creation of opposing or competing magnetic fields in the core, such that although the field strength in the core itself is undiminished, the intensity of the field at the Earth's surface falls significantly. This may involve the temporary dissolution of the typical dipole organization of the field and its replacement with multiple north and south poles at different points on the Earth's surface, which would be weaker. Another possibility is that excursions may be initiated by external phenomena such as the movement of continental plates or forces generated by interactions between the core and the mantle. These could disrupt the geodynamo process resulting in a loss of field intensity not only at the Earth's surface, but also within the core itself.

Detecting Geomagnetic Variation

Variations in the magnetic field are measured with sensitive devices called magnetometers. There are two principal types of magnetometer, scalar instruments that measure field intensity and vector instruments that also enable the determination of a directional component. The use of three orthogonal vector magnetometers allows the magnetic field strength, inclination, and declination to be defined precisely.

To recover evidence of geomagnetic events in the distant past, it is necessary to identify materials which preserve the characteristics of the Earth's magnetic field at earlier points in history. Such data are obtained from two major sources: volcanic lava and sedimentary rocks. Both contain iron-rich mineral domains which provide clues to the magnetic field strength and orientation at the time they were deposited. The advantage of sediments is that they are ubiquitous and can yield almost continuous records of magnetic variation, while lavas exist only near volcanic sites and in ocean ridges, and therefore provide a punctuated history of events. Lava data are rarely subject to controversy, whereas sediment data can be unreliable because of chemical and physical changes occurring after deposition. In this context, sediments can become magnetized through a process known as postdepositional remanent magnetization (PDRM), and there is also filtering of the geomagnetic signal (Quidelleur and Valet, 1993).

The discovery of similar patterns of magnetic variations at different locations indicates that both reversals and excursions may be global events. However, reversals are relatively easy to detect using the above methods due to the change in field orientation, whereas excursions may be overlooked because one is essentially searching for the absence of data (i.e., a sharp decline in the magnetic field intensity). Therefore, while we can be fairly certain of the number and nature of geomagnetic reversals at different points during the life of the planet, it is possible that excursions are a much more frequent characteristic of our geomagnetic history than currently accepted.

Using Magnetic Reversals for Dating Within the Quaternary

The regular occurrence of polarity reversals and other events during Earth's history means that such events can be used to date rocks and minerals, a practice known as magnetostratigraphy. Initially, magnetic-polarity intervals were known as epochs, with periods of normal polarity (where a compass needle would point towards the geographical north pole, as it does today) known as intervals of normal polarity, and periods of magnetic reversal (where a compass needle would point towards the geographical south pole) as intervals of reversed polarity. The epochs were named after famous scientists (e.g., Brunhes, Gauss, and Matuyama) with changes of polarity described thereafter as the boundary between two epochs (e.g., the Brunhes-Matuyama reversal; **Fig. 3**). As more events have been discovered, the old nomenclature has been replaced with a simple and universal numbering system, in which epochs have been replaced by 'chrons' and events are now called 'subchrons', each number followed by either n or r to indicate normal or reversed polarity.

Magnetic excursions during the Quaternary have been used to date events and to correlate the timing of events around the world that are otherwise difficult to date, especially events in the early Quaternary. However, the formal status of the Quaternary both in terms of its hierarchical rank on the chronostratigraphic scale and the position of its base have been subject to controversy and revision over the years (Aubrey *et al.*, 2005; Gibbard *et al.*, 2005). In September 2005, a meeting of the International Commission on Stratigraphy recommended that the Quaternary be formally defined as a Subera of the Cenozoic Era, and that its base be defined by the Global Boundary Stratotype Section and Point (GSSP) for the Gelasian Stage of the Upper Pliocene, which has an astronomically calibrated age of 2.588 million years. The Gauss-Matuyama reversal

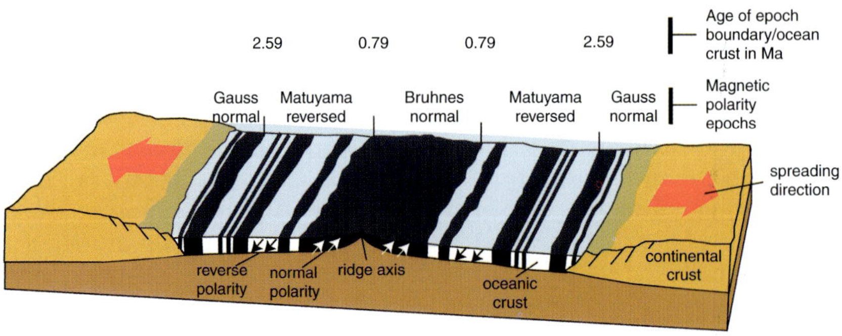

Figure 3 The formation of normal and reversed magnetic polarity patterns in new oceanic crust, on either side of an oceanic ridge axis. Alternating dark (normal) and light (reversed) polarity patterns would normally be recorded by shipborne or satellite surveys. Magnetic reversals are subdivided into major chrons (Bruhnes, normal; Matuyama, reversed; Gauss, normal) peppered with smaller normal and reversed subchrons, which are used for more precise dating.

provides a close approximation for global correlation of the base of the Gelasian, which would be equivalent to the base of propos‘ed Quaternary Subera (Partridge, 1997).

Abbreviations

GSSP	Global Boundary Stratotype Section and Point
PDRM	postdepositional remanent magnetization
USGS	United States Geological Survey
VGP	virtual geomagnetic pole

See also: **Introduction**: History of Dating Methods. **Radiocarbon Dating**: Variations in Atmospheric ^{14}C; Causes of Temporal Variations.

References

Aubry, M. P., Berggren, W. A., van Couvering, J., McGowran, B., Pillans, B., and Hilgen, F. (2005). Quaternary: status, rank, definition, survival. *Episodes* **28**, 118–120.

Barraclough, D. R. (1982). Historical observations of the geomagnetic field. *Philosophical Transactions of the Royal Society of London A* **306**, 71–78.

Bloxham, J., and Gubbins, D. (1985). The secular variation of Earth's magnetic field. *Nature* **317**, 777–781.

Champion, D. E., Lanphere, M. A., and Kuntz, M. A. (1988). Evidence for a new geomagnetic reversal from lava flows in Idaho: discussion of short polarity reversals in the Brunhes and late Matuyama polarity chrons. *Journal of Geophysical Research* **93**, 11,667–11,680.

Clement, B. M., and Constable, C. G. (1991). Polarity transitions, excursions and paleosecular variation of the Earth's magnetic field. *Reviews in Geophysics* **29**, 433–442.

Gibbard, P. L., Smith, A. G., Zalasiewicz, J. A., *et al.* (2005). What status for the Quaternary? *Boreas* **34**, 1–6.

Gubbins, D. (1994). Geomagnetic polarity reversals: a connection with secular variation and core-mantle interaction? *Reviews in Geophysics* **32**, 61–83.

Gubbins, D. (1999). The distinction between geomagnetic excursions and reversals. *Geophysics Journal International* **137**, F1–F3.

Hollerbach, R., and Jones, C. A. (1993). Influence of the Earth's inner core on geomagnetic fluctuations and reversals. *Nature* **365**, 541–543.

Hollerbach, R., and Jones, C. A. (1995). On the magnetically stabilizing influence of the Earth's inner core. *Physics of the Earth and Planetary Interiors* **87**, 171–181.

Jackson, A., Jonkers, A. R. T., and Walker, M. R. (2000). Four centuries of geomagnetic secular variation from historical records. *Philosophical Transactions of the Royal Society of London A* **358**, 957–990.

Jacobs, J. A. (2005). *Reversals of the Earth's Magnetic Field*, Second edition, Cambridge, New York.

Langereis, C. G., Dekkers, M. J., de Lange, G. J., Paterne, M., and van Santvoort, P. J. M. (1997). Magnetostratigraphy and astronomical calibration of the last 1.1 Myr from an eastern Mediterranean piston core and dating of short events in the Brunhes. *Geophysics Journal International* **129**, 75–94.

Love, J. J. (2000). Palaeomagnetic secular variation as a function of intensity. *Philosophical Transactions of the Royal Society of London A* **358**, 1191–1223.

Lund, S. P., Acton, G., Clement, B., Hastedt, M., Okada, M., and Williams, T. (1998). Geomagnetic field excursions occurred often during the last million years. *EOS, Transactions, American Geophysical Union* **79**, 178–179.

Merrill, R. T., and McFadden, P. L. (1990). Paleomagnetism and the nature of the geodynamo. *Science* **248**, 345–350.

Partridge, T. C. (1997). Reassessment of the position of the Plio-Pleistocene boundary: is there a case for lowering it to the Gauss-Matuyama palaeomagnetic reversal? *Quaternary International* **40**, 5–10.

Quidelleur, X., and Valet, J. P. (1994). Paleomagnetic records of excursions and reversals: possible biases caused by magnetization artefacts. *Physics of the Earth and Planetary Interiors* **82**, 27–48.

Robertsa, A. P., and Winklhofer, M. (2004). Why are geomagnetic excursions not always recorded in sediments? Constraints from post-depositional remanent magnetization lock-in modeling. *Earth and Planetary Science Letters* **227**, 345–359.

Valet, J. P. (2003). Time variations in geomagnetic intensity. *Reviews in Geophysics* **41**, 4.1–4.44.

Verosub, K. (1977). Geomagnetic excursions and their paleomagnetic record. *Reviews of Geophysics & Space Physics* **15**, 145–155.

GLACIAL CLIMATES

Contents

Biosphere Feedbacks

V Brovkin, Potsdam Institute for Climate Impact Research (PIK), Potsdam, Germany

© 2007 Elsevier B.V. All rights reserved.

Introduction

A current paradigm in plant ecology is that geographical distribution of dominant land plant forms is mainly controlled by climate (Woodward, 1987). If climate alters, vegetation cover closely follows the change in the climatic patterns. In turn, changes in spatial distribution and composition of terrestrial vegetation (as well as marine biology) alter the climate through modifications of heat and water fluxes, atmospheric gas, and aerosol composition. Biogeochemical and biophysical mechanisms form numerous feedbacks between biosphere and climate that were active during the Earth's geological past and continue to operate at present.

Quaternary changes in terrestrial and marine biology are recorded in sediments around the planet. A picture emerging from pollen and macrofossil reconstructions is that forest cover was greatly reduced at the maxima of Pleistocene glaciations, while during warmest phases of interglacials, expanded forests covered almost a half of the ice-free land surface. These dramatic changes in forest cover had several important implications for climate dynamics. Coldness and aridity of glacial climates were amplified by land cover changes through increase in land-surface albedo (the fraction of solar radiation reflected back into the atmosphere), decrease in transpiration, and enhanced dust formation on exposed bare ground.

Fewer forests and lower productivity during glacial periods should have led to reduced terrestrial biomass and soil carbon storages and consequent release of carbon dioxide (CO_2) to the atmosphere that slightly offset the radiative cooling. This stabilizing effect is invisible in ice-core CO_2 records due to counteracting amplifying feedback between temperature and greenhouse gases that led to lowering of atmospheric CO_2 levels during glacial periods. One of the hypothesized biological mechanisms that receives some support from proxy data is an iron fertilization effect on glacial marine productivity due to increased levels of dust depositions. The role of slow biogeochemical processes such as coral reefs growth and continental weathering in atmospheric CO_2 dynamics is not yet well constrained.

Main Mechanisms of Climate-Biosphere Interaction

Photosynthesis

The process of photosynthesis that has existed on Earth for more than three billion years is a focus point of climate-biosphere interaction. Terrestrial plants capture CO_2 from the atmosphere while marine phytoplankton take up carbon from sea water, which exchanges CO_2 with the atmosphere (House *et al.*, 2005). Respiration of living organisms returns carbon to the atmosphere. In the ocean, a part of dead organic matter (so-called export production) escapes the ocean surface layer and sinks towards the ocean floor. Because of long turnover time of deep ocean waters, the amount of carbon stored in the ocean due to this process (called the biological pump) is comparable to the organic carbon storage on land. Changes in terrestrial biomass and soil organic matter, as well as in volume and composition of marine export production, affect atmospheric CO_2 concentration directly on annual-to-millennial timescales. Altered CO_2 concentration modifies atmospheric radiative properties and, via climatic change and direct CO_2 effect, feeds back on ecosystem production.

Land-Surface Biophysics

In the planetary boundary layer up to 2.5 km above the Earth's surface, vertical profiles of temperature and humidity strongly depend on the partitioning of energy between sensible heat and latent heat. This partitioning is, to a large extent, controlled by terrestrial ecosystems through biophysical mechanisms

(Pitman *et al.*, 2004). Over dry, bare ground, the energy is transported via sensible heat, resulting in relatively high surface-air temperatures. Vegetation canopies transpire water extracted from the root zone, increase latent heat flux upwards, and cool the surface air during the day. Transpiration is tightly coupled to photosynthesis. Both, transpiration and photosynthesis are regulated by the opening of stomata pores in the leaf's surface. Low atmospheric CO_2 levels during glacial periods led to increased stomatal conductance, lower water use efficiency (productivity per water transpired), and reduced levels of photosynthesis.

Increased air humidity due to daily transpiration reduces heat loss in the near-surface atmosphere (greenhouse effect) helping to smooth diurnal fluctuations in temperature. Vegetation canopies also intercept part of the rainfall and effectively reduce runoff, especially during heavy rain events. Moisture that is transported into continental interiors has been transpired and precipitated several times on its way from the ocean; therefore reduction in vegetation cover during glacial periods enhances intracontinental aridity.

Vegetation affects surface albedo and in most cases, it reduces albedo compared to bare ground and snow. Forests are more effective than grasslands in trapping solar radiation. This effect is particularly strong in snow-covered regions where trees extend above the snow pack, while herbaceous vegetation is covered (see **Fig. 1**). The snow-masking effect of tree cover was likely more important in glacial climates when snow cover lasted longer in spring.

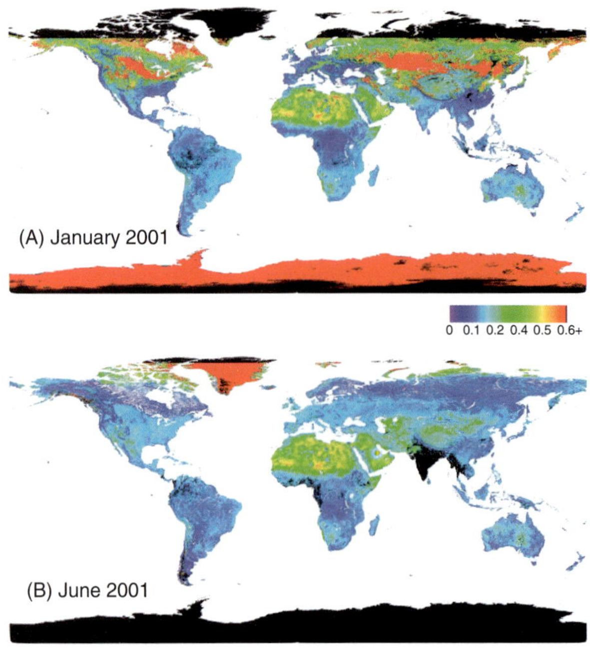

Figure 1 Map of land-surface albedo captured by the MODIS Satellite Instrument (Schaaf *et al.* 2002; Lucht *et al.* 2000). Regions where there were no data available, for example, due to clouds, are indicated in black. (A) January 2001. In the northern areas during the winter season, snow albedo is very high (up to 0.8, red). The boreal forest belt can be clearly seen in blue and green since trees mask snow, reduce albedo, and warm the surface air during the snow season. (B) June 2001. In comparison with January 2001, the northern land areas have a much lower albedo due to the absence of snow. In this map, the area with the highest albedo (up to 0.5, green and yellow) is the Sahara desert. High albedo in this region suppresses rainfall during the summer rainy season. Published from House and Brovkin (2005) with permission.

Long-Term Carbon Cycle Processes

Carbonate sedimentation Marine-calcifying organisms such as corals or coccolithophorids have skeletons or shells of calcium carbonate ($CaCO_3$). After organism death, $CaCO_3$ does not dissolve but forms reefs or drops to the sea floor forming sediments. In the deep seas, carbonates are well preserved above the lysocline (about 3–4 km depth), but they dissolve below this depth (*see* Dissolution of Deep-Sea Carbonates). On timescales of 5,000–10,000 years, carbonate sedimentation in the ocean is constrained by weathering. Since carbonate deposition in continental shelf regions (e.g., coral reef production) changes with sea level during glacial–interglacial cycles, this affects deep-sea sedimentation and leads to changes in the lysocline depth. A temporal imbalance between weathering and sedimentation causes changes in the total inventory of ocean alkalinity. Acidic water dissolves less CO_2 than basic water, and in this way changes in $CaCO_3$ sedimentation affects atmospheric CO_2 concentration. This process is complicated by the fact that export flux contains not only carbonate but also organic matter that becomes mineralized on the seafloor. The later process releases CO_2 making sediment pore water more acidic, leading to carbonate dissolution. Changes in organic-to-carbonate ratio (rain ratio) in export flux affects $CaCO_3$ sedimentation considerably and, through lysocline dynamics, the atmospheric CO_2 level.

Anaerobic decomposition Some dead organic matter is trapped in wetlands or marine sediments in anaerobic conditions. Decomposition of this organic matter by methane (CH_4)-producing bacteria may lead to a release of CH_4 Most of this CH_4 oxidizes in the water column, but a part escapes to the atmosphere. Atmospheric CH_4 concentration is small but important for climate since CH_4 is 20–30 times more potent a greenhouse gas than CO_2. CH_4 sink in the atmosphere is regulated by atmospheric chemistry that in turn is affected by plant emissions of volatile organic compounds (VOCs) such as isoprene.

Numerical Modeling as a Methodology of Feedback Analysis

The climatic effect of planting a single tree is negligibly small. Even for large-scale land cover changes, their climatic consequences are very difficult to detect in observations because several climatic forcings operate simultaneously in the presence of substantial climate variability. In addition, regional climate can be influenced not only by local changes but also by changes in remote regions. Carrying out numerical experiments with climate and ecosystem models is one of the principle methods of testing feedback between climate and biosphere. In the usual design of sensitivity experiments, the simulations are performed twice, once when feedback is neglected and a second time when feedback is included. A difference between these two simulations indicates feedback importance for climate. Biophysical feedbacks studied with global climate models are hereafter referred to as biogeophysical feedbacks to stress their significance for global-scale processes.

Hot Spots of Climate–Biosphere Interaction in Quaternary

The Holocene

'Green' Sahara Multiple proxy data reveal that during the so-called Holocene optimum, ca. 9,000–6,000 years before present (BP), the vegetation cover of the Sahel was greatly extended to the north. During that period, the Sahara's climate was much wetter; many lakes and rivers were present in the region, including the greatly extended Lake Chad (*see* Africa). A conventional explanation of the 'green' Sahara phenomenon in the mid-Holocene, supported by many modeling experiments, is based on the long-term changes in the Earth's orbital parameters. In the early-to-mid-Holocene, the northern hemisphere received considerably more solar radiation during the summer. In the northern subtropical regions, this led to stronger warming over the continent than over the ocean, an increased temperature gradient between land and ocean, and, consequently, intensified monsoon-type circulation in summer which led to increased rainfall over the Sahel/Sahara region.

The North African desert differs from many other subtropical deserts. The radiative balance in the region is negative, i.e., there is a net radiative heat loss over the desert. This loss is particularly due to the high albedo of the desert, as up to 50% of incoming radiation is reflected back into space (**Fig. 1**). The radiative mechanism for the Sahara was first explored by Jule Charney in the 1970s. He pointed out that the heat loss due to high albedo leads to a temperature gradient and induces an atmospheric circulation which maintains the sinking motion of dry atmospheric masses and suppresses rainfall over the region. Low precipitation results in little vegetation cover, and the surface albedo is determined by bare ground with a high albedo. This positive feedback supports a desert that is self-sustaining. Alternatively, if there is more precipitation, there is more vegetation. Vegetation canopy is darker than sand so the albedo is lower, the surface temperature is higher, and the gradient in temperature between land and ocean increases, amplifying monsoon circulation and upward motion over the desert. As a result, the summer rainfall in the region increases. This radiative feedback loop gets additional support through hydrological feedback as vegetation recycles water to the atmosphere, reduces runoff, and enhances rainfall in comparison with bare ground (**Fig. 2**).

Both radiative and hydrological feedbacks are positive, i.e., they amplify changes induced by external forcings or internal variability. Could these positive feedbacks have affected the stability of the regional climate in the past? In the early 1990s, Martin Claussen applied coupled atmospheric-vegetation model ECHAM2-BIOME to investigate an effect of different initial land-surface conditions on the regional climate. In these simulations, vegetation cover responded to climate simulated by the model and vice versa. In the first simulation with present-day boundary conditions (**Fig. 3A**), which started from

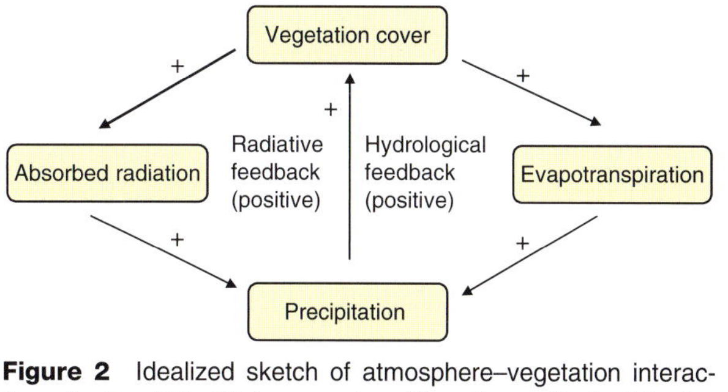

Figure 2 Idealized sketch of atmosphere–vegetation interactions in the Sahel/Sahara region. The signs attached to the arrows indicate the effect of the outgoing box on the incoming box, for example, increased vegetation cover leads to more radiation absorbed at the land surface. Both radiative and hydrological feedback are positive, i.e., they amplify external changes and therefore may cause instabilities. Published from Claussen *et al.* (2004) The global climate. In: Kabat P, Claussen M, Dirmeyer PA, *et al* (eds.) *Vegetation, Water, Humans and the Climate: a New Perspective on an Interactive System.* Global Change: The IGBP Series, pp. 33–57. Berlin, Germany: Springer with permission.

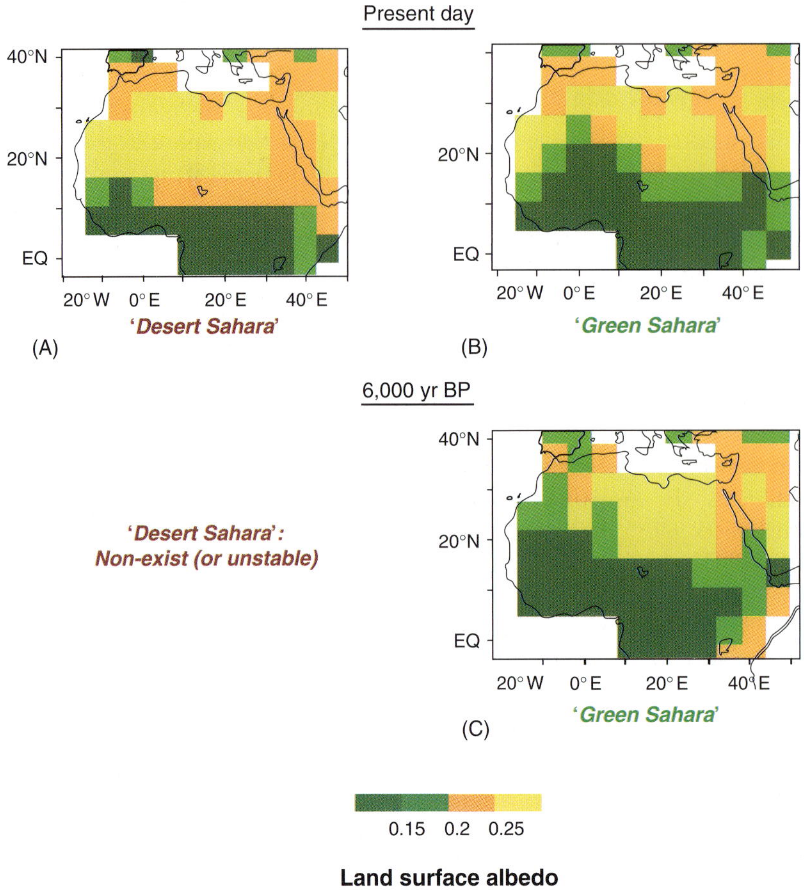

Figure 3 Summary of results obtained with an interactive atmosphere–vegetation ECHAM/BIOME model. For present-day climate, two states are stable in the model: desert Sahara as observed now (A), and 'green Sahara' state with enhanced vegetation cover in western Sahara (B) in accordance with Claussen (1994). (C) For 6,000 yr ago, 'green Sahara' is the only stable state in the model (Claussen and Gayler, 1997). Lower albedo corresponds to a higher fraction of vegetation cover. EQ, equator. Published from Brovkin *et al.* (1998) with permission from American Geophysical Union.

the present-day potential vegetation, the desert in North Africa remained stable. However, when the Sahara was initially covered by forest in the simulation B (**Fig. 3B**), the system converged to another solution, called the 'green' Sahara. The western Sahara is covered by a mixture of shrubs and grass; in addition, vegetation cover in the Sahel region is enhanced. Both desert and green solutions are stable for present-day conditions. At the same time, the coupled model revealed the only steady solution, 'green' Sahara, for the mid-Holocene orbital forcing (**Fig. 3C**). In the dynamical system theory, the fact that the system has one steady state for one parameter value (mid-Holocene insolation) and two states for another value (present-day orbital parameters) means that in between these points the system has undergone a bifurcation that could be associated with abrupt climate changes. Indeed, proxy data suggest abrupt changes of vegetation cover in western Sahara between 5,000 and 6,000 yr BP (deMenocal *et al.*, 2000). Later simulations with another coupled model that included interactive ocean, ECBILT-CLIO-VECODE, supported the fact that the system underwent bifurcation around 4,000–6,000 yr BP as the green state lost its stability (Renssen *et al.*, 2003). Some other coupled models do not show multiple states, but they all reveal a high sensitivity of rainfall in the Sahel/Sahara region to vegetation cover. In summary, strong positive vegetation-atmosphere feedback in North Africa is an important amplifier of externally induced changes in the region during the Quaternary. In particular, this feedback had contributed to abrupt desertification of the region about 5,000–6,000 years ago.

Northern treeline dynamics During the mid-Holocene, the northern tundra–forest boundary was north of its present position in many regions. In particular, in central Siberia the treeline (the forest limit) was shifted to the north by up to 200 km (Bigelow *et al.*, 2003), presumably due to warmer summers, as treeline position corresponds reasonably well with July temperature isotherms of about 10–12°C. In the northern high latitudes, changes in forest cover affect regional climate in several ways. Forests, even deciduous ones, significantly reduce the albedo of snow-covered surfaces. This is the basis for the radiative feedback between forest and surface air temperature. Increased forest cover within large regions leads to decreased surface albedo during the snow season. This results in elevated air temperature and earlier snow melt, and a longer and warmer growing season, which favors further increase in forest area. This feedback (sometimes called taiga–tundra feedback) amplifies the climate system response to the original external forcing (**Fig. 4**). The rate at which forest-cover changes, depends on many local factors such as disturbances (fire), permafrost, soil parent material, and pH. The radiative feedback is dampened during the growing season, when trees have a denser, more productive canopy. Trees also have deeper roots than herbaceous plants and moss and transpire more water. This cools the surface air in forests as compared to tundra during summer. The hydrological feedback, although negative, is secondary to the radiative feedback on the annual average. Besides, enhanced transpiration from deeper soil horizons improves soil drainage and aeration of soil mineral layers that increases viability of tree establishment. In the Arctic environment, the radiative feedback of forests is tightly linked to sea ice–albedo feedback. Elevated surface air temperature due to increased forest cover leads to higher sea-surface temperature (SSTs) and decreased sea-ice area, decreased sea-surface albedo, and, finally, elevated surface-air temperature (**Fig. 4**). Model simulation suggests that taiga–tundra feedback and its amplification by sea ice–albedo feedback led to additional warming of northern high latitudes in response to the changes in orbital forcing during the mid-Holocene (**Fig. 5**).

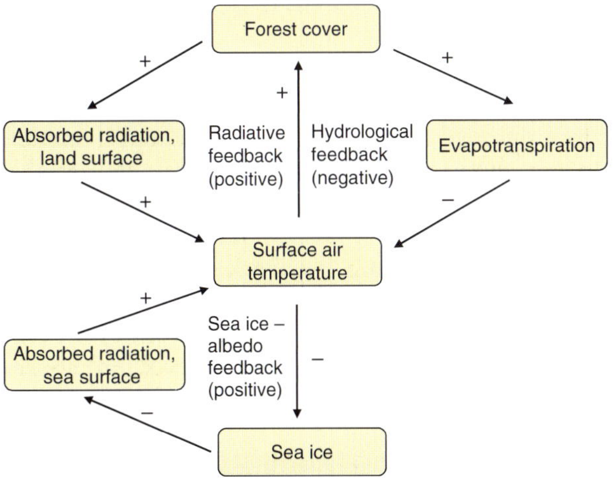

Figure 4 Simplified scheme of climate–vegetation interactions in the northern high latitudes. The signs attached to the arrows indicate the effect of the outgoing box on the incoming box, for example, increased forest (tree) cover leads to more radiation absorbed at the land surface during snow-covered periods. Changes in surface air temperature due to forest cover changes are amplified through sea ice–albedo feedback in the Arctic.

Historical land cover changes At present, about one-third of the Earth's land cover has been modified by agricultural and forestry activities. Human-induced land-cover changes probably began in the early Holocene with development of slash-and-burn agriculture. The magnitude of these changes and their temporal dynamics on large regional scales in the pre-industrial period are mostly unknown. Most models have been used to evaluate the climatic effect of reconstructed potential vegetation versus present-day land cover, and some models have been applied for testing the climatic effect of the last millennium land-cover changes using simplified assumptions about such changes prior to AD 1700. Most model simulations agree that historical deforestation led to some increase in atmospheric CO_2 and consequent warming. At the same time an increase in albedo due to

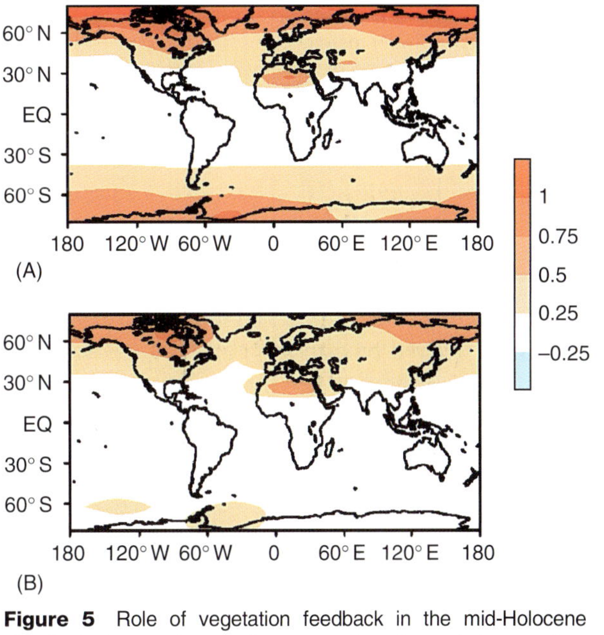

Figure 5 Role of vegetation feedback in the mid-Holocene climate in terms of annually averaged surface air temperature (°C) in CLIMBER-2 model simulations of the mid-Holocene climate (6,000 yr BP). (A) Difference between the mid-Holocene and the present-day simulations. (B) Contribution of the biogeophysical feedback (difference between mid-Holocene simulations with interactive vegetation and vegetation prescribed from present-day simulation). EQ, equator. Adapted from Ganopolski *et al.* (1998) with permission.

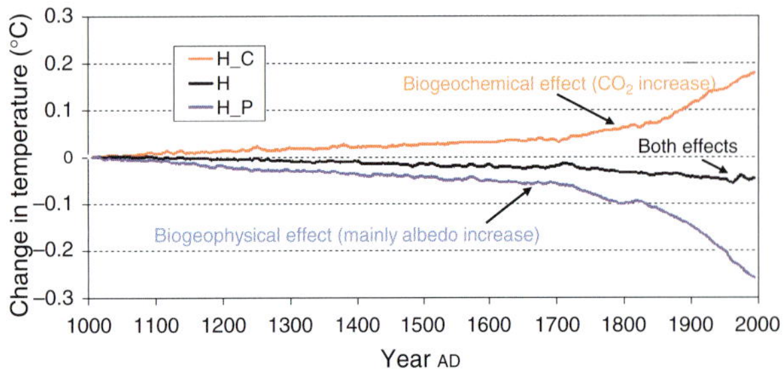

Figure 6 Effects of historical land cover changes on global mean annual temperature (°C) simulated by CLIMBER-2 model. H_P – biogeophysical effect only (mainly due to albedo changes); H_C – biogeochemical effect only in response to atmospheric CO_2 increase; H – both biogeophysical and biogeochemical effects. Shown here are the differences between simulations with and without land cover changes. Published from Brovkin *et al.* (2004) with permission of Blackwell Publishing.

deforestation of regions covered with snow in winter, such as northern Europe, led to a radiative cooling (**Fig. 6**). On a global scale, biogeophysical cooling was likely compensated by biogeochemical warming, but consequences for regional climate could be substantial. Biogeophysical effects due to increasing deforestation rate could have contributed to an observed stabilization of the temperature during the second half of the nineteenth century, and have damped the CO_2 warming in the twentieth century.

Last Glacial Maximum

The last glacial maximum (LGM) was characterized by cold and dry conditions (*see* Last Glacial Maximum GCMs) that led to a great reduction of forest cover in the Northern Hemisphere. Boreal evergreen forests and temperate deciduous forests were fragmented, while European and East Asian steppes were greatly extended (Prentice *et al.*, 2000). Tropical forests in the Americas and Africa were reduced as well. These changes towards herbaceous and sparser vegetation cover had considerable implications for climate.

Biogeophysical feedbacks Model simulations using glacial boundary conditions suggest that reduction in forest cover amplified LGM coldness and aridity. Southward displacement and fragmentation of forest biomes in North America and Eurasia led to increased albedo during the snow period and a regional decrease in temperature by several degrees (**Figure 6**). Global cooling effect of reduced vegetation cover during the LGM is estimated in 0.6°C (Crucifix and Hewitt, 2005; Jahn *et al.*, 2005). Reduction in forest cover led to less transpiration and increased aridity of the continental interiors. Presumably, physiological effects of lowered CO_2 levels during the LGM resulted in increased canopy conductance of C_3 plants and a consequent drop in water-use efficiency (water used per unit of photosynthetic production; **Figure 7**; Harrison and Prentice, 2003).

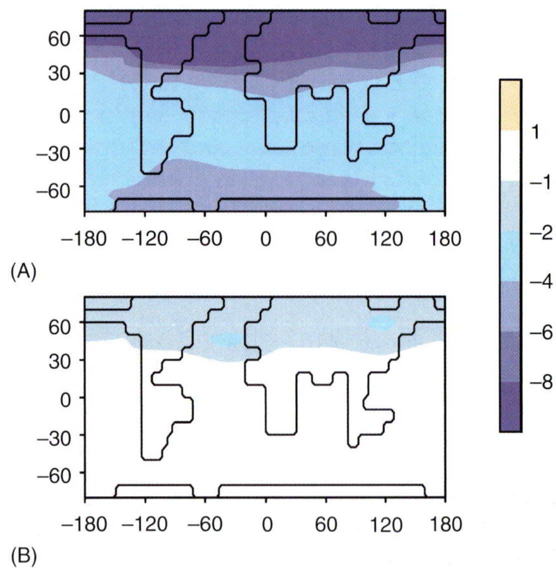

Figure 7 Role of biogeophysical feedback in the LGM climate as simulated by CLIMBER-2 model. Shown are annually averaged surface air temperature differences (°C). (A) A difference between the LGM and the present-day simulations. (B) Change in temperature due to the biogeophysical feedback. Adapted from Jahn *et al.* (2005) with the authors permission.

Terrestrial CO_2 feedback Cold and dry climates as well as low CO_2 levels led to decreased plant productivity and biomass storage during the LGM. Land areas in the Arctic were more extensive due to lowered sea levels but regional ecosystems were not very productive. To some degree, reduction in boreal forests was compensated by the establishment of new forests along the exposed shelf regions in tropical regions such as southeast Asia, but on a global scale, terrestrial carbon storage was drastically reduced. Vegetation models applied to LGM climate-change scenarios reveal a range of decrease in terrestrial storages from several hundred to thousand of petagrams of carbon (PgC; 10^{15}g C). A central estimate of about 600–700 PgC is higher than changes suggested by a decline in deep ocean $\delta^{13}C$ (about 500 PgC). The later proxy should be carefully interpreted since oceanic $\delta^{13}C$ is also affected by changes in oceanic circulation and marine productivity. Decrease in northern peat storage due to expansion of ice sheets and increased aridity also contributed to the release of land carbon to the atmosphere but the magnitude and timing of peat changes are uncertain. A reduction of the terrestrial carbon pool during glacial periods provides a negative (stabilizing) feedback to climate by increasing atmospheric CO_2 concentration. This feedback is rather weak since most of the released terrestrial carbon ends up in the ocean. After carbonate compensation on timescales of 5,000–10,000 years, 500 PgC of land carbon source corresponds to about a 20 parts per million by volume (ppmv) increase in atmospheric CO_2 concentration (Archer *et al.*, 2000).

Dust feedback In semiarid regions, reduced vegetation cover means more bare ground exposed to wind erosion. During glacial periods, these sparsely vegetated regions were a source of mineral aerosols for the atmosphere. Higher eolian dust transport during glacial periods is supported by sediment data (Kohfeld *et al.*, 2005). The mineral aerosols diffuse light and reduce the amount of solar radiation that reached the surface. Radiative transfer models estimate that increased atmospheric dust concentrations during the LGM led to about 2–3 watts per square meter ($W m^{-2}$) reduction in radiative forcing in tropics which roughly corresponds to the radiative effect of lower glacial CO_2 levels in the region (Claquin *et al.*, 2003). As a result, sparser vegetation cover amplified LGM cooling through the dust feedback.

Marine biota feedback Several hypotheses aimed at explaining glacial–interglacial changes in atmospheric CO_2 invoke changes in marine ecosystem structure and export production. SubAntarctic proxy data suggest that export production in the subAntarctic Atlantic was enhanced during the LGM, presumably due to iron fertilization in response to increased eolian dust deposition. Some other regions, such as the Pacific, show few changes, while productivity in Antarctic Ocean was reduced. Iron fertilization could be responsible for some drawdown of atmospheric CO_2 levels at the LGM, but it can not explain the whole range of 80–100 ppmv change (Kohfeld *et al.*, 2005). Increased organic to $CaCO_3$ ratio in export production (e.g., due to silica leakage) could have substantially lowered the atmospheric CO_2 through the sediment response. However, recent data show that the ratio of organic to $CaCO_3$ flux that reached the seafloor is almost constant. Coral-reef production was reduced during the LGM in response to lowered sea levels and surface cooling (Kleypas, 1997). Reduced shallow-water sedimentation could have led to an increase in oceanic total alkalinity, but this effect might also have been offset by decreased continental weathering during the LGM. In summary, the marine biosphere was likely to have provided positive feedback to the atmosphere by amplifying LGM cooling through lowering of atmospheric CO_2, but the magnitude of this feedback and the precise mechanisms are still not well known.

Weathering feedback On a timescale longer than several thousand years, atmospheric CO_2 is strongly affected by changes in carbonate and silicate weathering. Increased weathering lowers atmospheric CO_2 concentration through oceanic carbonate compensation. Vascular plants contribute to the weathering as roots' respiration and litter mineralization lead to elevated soil CO_2 concentrations and the formation of carbonic acid that enhances weathering of parent soil material. Model estimates of changes in weathering rates during the LGM are equivocal as several factors influencing global weathering flux, such as increased ice-sheet areas, shelf exposure, and decreased runoff almost nullify each other (Munhoven, 2002).

Methane and atmospheric chemistry feedback During the LGM, atmospheric CH_4 concentration (about 400 ppbv) was much lower than in the Holocene (600–750 ppbv; see Chapter 338). Decreased concentration of CH_4 led to small but not negligible additional cooling. The plausible explanation of the lowered CH_4 concentration is that terrestrial CH_4 sources (e.g., wetlands and biomass burning) were lower during the LGM. In particular, the extent of wetlands was reduced in the cooler and drier LGM climate, especially in the northern high latitudes, while this reduction might have been offset by establishing of wetlands on exposed tropical shelves. Another important factor could be a change in atmospheric chemistry. Model simulations suggest that emissions of VOC were lower during the LGM due to reduction in forest cover, resulting in increased atmospheric concentration of the OH radical (an atmospheric 'detergent'), and this may have led to an increased sink for CH_4 in the atmosphere (Valdes *et al.*, 2005). In summary, biospheric feedback amplified LGM cooling through lowering of CH_4 concentration, while CH_4-regulating mechanisms involving VOC emissions are not yet fully explored.

The Last Interglacial and Glacial Inception

During the last interglacial, an increase in summer insolation in the northern hemisphere was stronger than during the Holocene. Pollen evidence and model simulations suggest that summer climate was substantially warmer in northern regions than at present and that forest cover was more extensive in the temperate and high northern latitudes (*see* The Last Interglacial). The vegetation feedback amplified the warming in model simulations (**Fig. 8A**). This temperature increase anomaly is mainly due to the northward advance of boreal forest during the warmer climate of the last interglacial. Conversely, the later cooling trend triggered by the decreasing boreal summer insolation was facilitated by the vegetation response (**Fig. 8B**). Most climate-model simulations suggest that a southward shift of northern treeline is crucial for the initiation of glacial inception: if vegetation cover is prescribed to the values simulated for the last interglacial, then cooling in the north due to declining insolation is not sufficient to initiate ice-sheet growth. Treeline retreat amplifies cooling through increased albedo fostering permanent snow cover on land during the glacial inception.

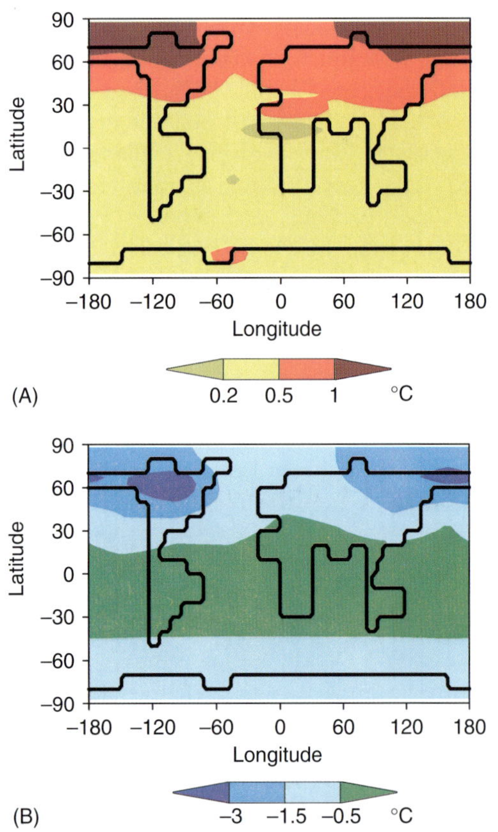

Figure 8 Role of biogeophysical feedback in the climate of last interglacial and glacial inception as simulated by CLIMBER-2 model. Shown are annual temperature differences (°C) between transient experiments with interactive vegetation and vegetation cover prescribed from present-day simulation. Ice sheets are interactive. (A) Warming due to interactive vegetation at 125,000 yr BP. (B) Cooling induced by vegetation changes at 115,000 yr BP. Published from Calov *et al.* (2005) with permission of Springer Science and Business Media.

Conclusions

Analysis of climate-biosphere feedback and their role in climate-system dynamics during the Quaternary is a new, rapidly developing scientific field. Hypotheses about roles of biological mechanisms in past climate dynamics usually emerge from the analysis of geological data, but quantification of feedback is impossible without modeling. Today, the main methodology of feedback analysis is numerical modeling using climate system models of different complexity, from high-resolution general circulation models (GCMs) to simplified conceptual models. Model results are tested against proxy data for biogeochemistry and vegetation cover.

A picture emerging from model simulations and proxy data analysis is that biospheric feedback played a substantial role in the evolution of Quaternary climate, but many details are uncertain. In some cases, such as glacial inception, model simulations agree that the dynamics of northern treeline is an important mechanism amplifying cooling due to summer insolation decrease in the northern high latitudes. Analysis of climate–vegetation interactions in North Africa suggests that biospheric feedback might be strong enough to cause abrupt, irreversible changes in the region. At the same time, the role of biological mechanisms in glacial–interglacial CO_2 changes remains far from clear.

Understanding of the role of biospheric feedback in the past has important implications for the future climate projections. The challenge in Earth system science is to constrain feedback mechanisms within the models based on available – and newly developing – proxies for biological activities in the past.

Abbreviations

BP	before present
$CaCO_3$	calcium carbonate
CH_4	methane
CO_2	carbon dioxide
GCM	general circulation model
LGM	last glacial maximum
PgC	petagrams of carbon
ppmv	parts per million by volume
SST	sea-surface temperature
VOC	volatile organic compound
$W\,m^{-2}$	watts per square meter

See also: **Ice Core Methods**: Methane Studies. **Paleoceanography, Physical and Chemical Proxies**: Dissolution of Deep-Sea Carbonates. **Paleoclimate Modeling**: The Last Interglacial; Last Glacial Maximum GCMs. **Pollen Records, Postglacial**: Africa.

References

Archer, D. E., Winguth, A., Lea, D., and Mahowald, N. (2000). What caused the glacial/interglacial atmospheric pCO_2 cycles? *Reviews of Geophysics* **38**, 159–189.

Bigelow, N. H., Brubaker, L. B., Edwards, M. E., *et al.* (2003). Climate change and Arctic ecosystems I. Vegetation changes north of 55°N between the last glacial maximum, mid-Holocene, and present. *Journal of Geophysical Research–Atmospheres* **108**, 8170.

Bonan, G. (2002). *Ecological Climatology*, p. 690. Cambridge University Press, Cambridge.

Brasseur, G. P., Artaxo, P., Barrie, L. A., *et al.* (2003). An integrated view of the causes and impacts of atmospheric changes. In *Atmospheric Chemistry in a Changing World* (G. P. Brasseur, R. G. Prinn and A. A. P. Pszenny, Eds.), pp. 207–230. Springer-Verlag, Heidelberg, Germany.

Brovkin, V., Claussen, M., Petoukhov, V., and Ganopolski, A. (1998). On the stability of the atmosphere-vegetation system in the Sahara/Sahel region. *Journal of Geophysical Research–Atmospheres* **103**, 31613–31624.

Brovkin, V., Sitch, S., von Bloh, W., *et al.* (2004). Role of land cover changes for atmospheric CO_2 increase and climate change during the last 150 years. *Global Change Biology* **10**, 1–14.

Calov, R., Ganopolski, A., Petoukhov, V., *et al.* (2005). Transient simulation of the last glacial inception. Part II: sensitivity and feedback analysis. *Climate Dynamics* **24**, 563–576.

Claquin, T., Roelandt, C., Kohfeld, K. E., *et al.* (2003). Radiative forcing of climate by ice-age atmospheric dust. *Climate Dynamics* **20**, 193–202.

Claussen, M. (1994). On coupling global biome models with climate models. *Climate Research* **4**, 203–221.

Claussen, M., and Gayler, V. (1997). The greening of Sahara during the mid-Holocene: results of an interactive atmosphere-biome model. *Global Ecology and Biogeography Letters* **6**, 369–377.

Claussen, M., Cox, P. M., Zeng, X., *et al.* (2004). The global climate. In *Vegetation, Water, Humans and the Climate: a New Perspective on an Interactive System. global Change: The IGBP Series* (P. Kabat, M. Claussen, P. A. Dirmeyer *et al.*, Eds.), pp. 33–57. Springer, Berlin, Germany.

Crucifix, M., and Hewitt, C. D. (2005). Impact of vegetation changes on the dynamics of the atmosphere at the Last Glacial Maximum. *Climate Dynamics* **25**, 447–459.

deMenocal, P., Ortiz, J., Guilderson, T., *et al.* (2000). Abrupt onset and termination of the African Humid Period: rapid climate responses to gradual insolation forcing. *Quaternary Science Reviews* **19**, 347–361.

Dickinson, R. E. (1992). Changes in land-use. In *Climate System Modeling* (K. E. Trenberth, Ed.), pp. 689–700. Cambridge University Press, Cambridge.

Ganopolski, A., Kubatzki, C., Claussen, M., Brovkin, V., and Petoukhov, V. (1998). The influence of vegetation-atmosphere-ocean interaction on climate during the mid-Holocene. *Science* **280**, 1916–1919.

Harrison, S., and Prentice, C. I. (2003). Climate and CO_2 controls on global vegetation distribution at the last glacial maximum: analysis based on paleovegetation data, biome modelling and paleoclimate simulations. *Global Change Biology* **9**, 983–1004.

House, J. I., and Brovkin, V. (2005). Climate and air quality. In *Ecosystems and Human Well-Being: Current State and Trends* (R. Hassan, R. Scholes and N. Ash, Eds.), vol. 1, pp. 355–390. Island Press, Washington, DC.

Jahn, A., Claussen, M., Ganopolski, A., and Brovkin, V. (2005). Quantifying the effect of vegetation dynamics on the climate of the last glacial maximum. *Climate of the Past* **1**, 1–7.

Kleypas, J. A. (1997). Modeled estimates of global reef habitat and carbonate production since the last glacial maximum. *Paleoceanography* **12**, 533–545.

Kohfeld, K. E., Le Quéré, K., Harrison, S. P., and Anderson, R. F. (2005). Role of marine biology in glacial–interglacial CO_2 cycles. *Science* **308**, 74–78.

Lucht, W., Schaaf, C. B., and Strahler, A. H. (2000). An algorithm for the retrieval of albedo from space using semiempirical BRDF models. *IEEE Transactions on Geoscience and Remote Sensing* **38**, 977–998.

Munhoven, G. (2002). Glacial–interglacial changes of continental weathering: estimates of the related CO_2 and HCO_3^- flux variations and their uncertainties. *Global and Planetary Change* **33**, 155–176.

Pitman, A. J., Dolman, H., Kruijt, B., Valentini, R., and Baldocchi, D. (2004). The climate near the ground. In *Vegetation, Water, Humans and the Climate: a New Perspective on an Interactive System. Global Change: The IGBP Series* (P. Kabat, M. Claussen, P. A. Dirmeyer *et al.*, Eds.), pp. 9–19. Springer, Berlin.

Prentice, I. C., and Jolly, D. BIOME 6000 Participants (2000). Mid-Holocene and glacial-maximum vegetation geography of the northern continents and Africa. *Journal of Biogeography* **27**, 507–519.

Prentice, I. C., and Raynaud, D. (2001). Paleobiogeochemistry. In *Global Biogeochemical Cycles in the Climate System* (E. D. Schulze, M. Heimann, S. P. Harrison *et al.*, Eds.), pp. 87–93. Academic Press, San Diego, USA.

Renssen, H., Brovkin, V., Fichefet, T., and Goosse, H. (2003). Holocene climate instability during the termination of the African Humid Period. *Geophysical Research Letters* **30**, 1184.

Schaaf, C. B., Gao, F., Strahler, A. H., *et al.* (2002). First operational BRDF, albedo nadir reflectance products from MODIS. *Remote Sensing of Environment* **83**, 135–148.

Valdes, P. J., Beerling, D. J., and Johnson, C. E. (2005). The ice-age methane budget. *Geophysical Research Letters* **32**, L02704.

Vernadsky VI (1926) *Biosfera*, p. 146, Leningrad, Russia: Nauka Citation in accordance with English translation: Vernadsky VI (1998) *The Biosphere*, p. 192, New York: Copernicus Books.

Woodward, F. I. (1987). *Climate and Plant Distribution*, p. 174. Cambridge University Press, New York.

Effects of Atmospheric Dust

I Tegen, Leibniz Institute for Tropospheric Research, Leipzig, Germany

© 2007 Elsevier B.V. All rights reserved.

Micrometer-sized dust particles that are lifted from soil surfaces in arid and semiarid regions by strong surface winds contribute a large fraction to the global atmospheric aerosol mass load and aerosol optical thickness [Intergovernmental Panel on Climate Change (IPCC), 2001]. They influence the Earth's radiation balance, biogeochemical processes, and atmospheric chemistry in several ways, which are only partly understood and remain largely unquantified (**Fig. 1**). Atmospheric dust reflects and to some extent absorbs incoming solar radiation, and it also absorbs and emits outgoing thermal radiation due to

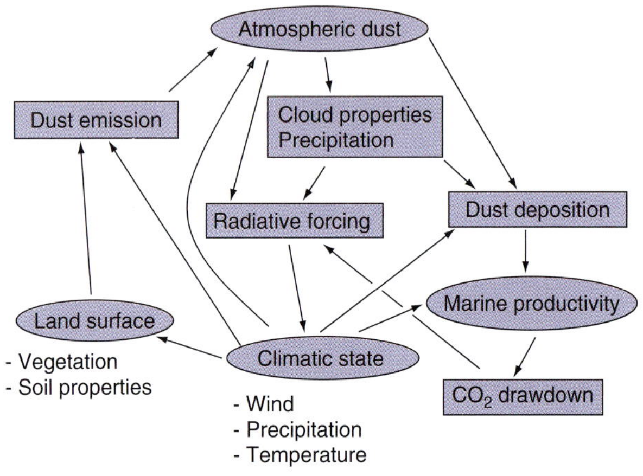

Figure 1 Overview of linkages between atmospheric dust and climate indicating direct and indirect aerosol effects as well as the potentially important role of oceanic dust deposition for marine microorganisms.

its relatively large particle sizes. Thus, it directly influences top-of-the-atmosphere and surface radiation fluxes as well as atmospheric heating rates and stability. Such changes in heating rates also may influence cloud formation. In addition, dust provides cloud condensation nuclei that are particularly effective for ice nucleation and thus potentially influence cloud optical properties, precipitation formation, and cloud lifetimes (indirect aerosol effects). Another potentially important effect of dust is its influence on oceanic export production of marine microorganisms by supplying nutrients into the ocean surface, overcoming limitations of essential nutrients such as iron, or enhancing nitrogen assimilation. Dust fertilizes some terrestrial ecosystems by depositing nutrients into soils, and it influences atmospheric chemistry due to heterogeneous reactions on the particle surfaces. Emission, atmospheric concentration, and deposition of dust are strongly dependent on atmospheric parameters and surface properties, and they are therefore very variable on a wide range of timescales; for example, dust production and dispersal was considerably enhanced in glacial times compared to modern conditions. The strongest emphasis in investigations of the effects of dust aerosol has been on quantification of the direct radiative effect of dust under present-day conditions, whereas only a few studies have examined dust distributions under different climates such as the Last Glacial Maximum (LGM) or with future climate scenarios. Because quantitative data of the atmospheric dust content cannot be obtained from observations alone, such studies must rely on results of large-scale models of dust emission, transport, and deposition.

Atmospheric Dust

Soil dust aerosol particles are produced in arid and semiarid continental regions (**Fig. 2**), including the Sahara and the Sahelian region, the Arabian Peninsula, the Gobi and Taklamakan deserts in Asia, and the Australian and South American deserts

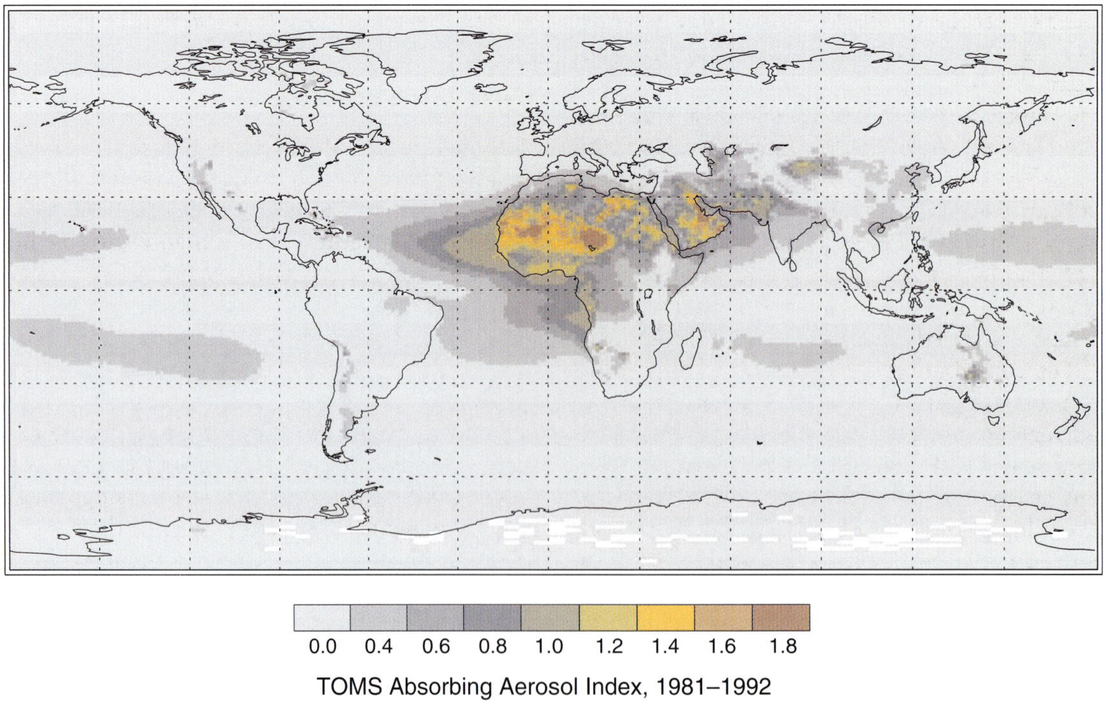

Figure 2 Eleven-year annual average of the Total Ozone Mapping Spectrometer (TOMS) Absorbing Aerosol Index.

(Prospero *et al.*, 2002). Most of the dust contained within the global atmosphere is emitted from the North African (50–70%) and Asian deserts (10–25%). Wind tunnel experiments and theoretical considerations show a dependence of dust emissions to the third or fourth power of the surface wind friction velocity. Dust production is initiated when the surface wind friction velocity in the source region exceeds a threshold value that depends on surface properties such as the size and distribution of roughness elements, the soil texture, and soil moisture (Marticorena and Bergametti, 1995).

Fine soil particles with diameters smaller than 10 μm, which can be transported by winds over large distances, are released from the soil matrix when larger wind-blown sand grains strike on the surface, releasing the smaller particles through the process of saltation. Dust emissions are particularly strong in areas that contain fine and loose sediments. A good agreement between maximal occurrence of dust over land surfaces observed by satellites and the location of enclosed topographic depressions has been observed (Prospero *et al.*, 2002). In such areas, fine sediment may have accumulated during the past. Either lake sediments have formed in such depressions, in case they were filled with water during wetter climate periods, or these areas may have been end points for riverine transport of fine particles. Dust emission strength in a potential source region depends on the aerodynamic surface roughness length because structural elements such as rocks or vegetation increase the transport of wind energy to the surface but also partly absorb this energy, increasing the threshold wind friction velocity required for dust emission (Marticorena and Bergametti, 1995).

The presence of crusts can reduce dust emission from a soil surface. Also, the loss of fine soil material by wind deflation can lead to a decrease of dust production from a specific area over time. On the other hand, surface disturbances as a consequence of cultivation of soils in dry regions may lead to enhanced dust emissions (e.g., as observed during the 'dust bowl' events in the United States in the 1930s and 1950s). The contribution of this enhanced dust emission to the global dust load may be considerable in semiarid regions, but the magnitude is uncertain on the global scale. Although IPCC (2001) estimated a contribution of this enhanced dust of up to 50% to the global dust load, Tegen *et al.* (2004) concluded that this contribution is much smaller, probably less than 10% globally.

Dust injected into high atmospheric levels of up to several kilometers can be transported over horizontal distances of thousands of kilometers by strong wind systems. Remote sensing instruments can clearly distinguish such dust plumes over dark ocean surfaces. Most prominently, dust emitted in the globally strongest dust source, the Sahara desert, can be carried across the North Atlantic to North and Central America (Prospero and Lamb, 2003) and is even found as far downwind as the Amazon basin. Also, large amounts of Asian dust are carried over the North Pacific toward the mid-Pacific islands and North America.

Dust particles are removed from the atmosphere either by wet deposition (i.e., scavenging through precipitation in the water or ice phase) or by dry deposition, which is the gravitational settling in addition to turbulent mixing toward the surface. Close to desert dust source regions, the gravitational sedimentation is responsible for the major part of dust deposition, whereas wet deposition dominates the removal of far-traveled dust over remote ocean areas (Jickells *et al.*, 2005). The efficiency of the dust scavenging from the atmosphere may increase with distance from the source regions, when soluble components coating the dust surfaces make the particles increasingly hygroscopic. Whether dust is removed from the atmosphere by dry or wet processes affects the solubility of any trace elements transported with the dust. The atmospheric lifetime of dust depends on the deposition velocities of the different particle sizes. Atmospheric lifetimes of dust depend on the particle size and range from a few hours for particles larger than 10 μm, which are quickly removed by gravitational settling, to 10–15 days for submicron particles that are mostly removed by wet deposition (Seinfeld and Pandis, 1997).

Dust Measurements

Although *in situ* observations of airborne and deposited dust particles help us to understand atmospheric dust properties, they cannot describe dust distributions globally because dust is highly variable in space and time and the observations can only cover limited regions. *In situ* observations of dust include multiyear concentration measurements from approximately 30 stations of the Miami oceanic network (Prospero and Lamb, 2003), optical thickness measurements by sun photometers (e.g., the AERONET network), and aerosol lidar observations (Ansmann *et al.*, 2005). In contrast, remote sensing using satellite instruments covers large areas of the globe, but it is nearly impossible to accurately quantify dust properties above land surfaces from such retrievals. Over dark ocean surfaces, retrievals of dust optical thickness and angstrom parameters are possible with some confidence (Kaufman *et al.*, 2002), but such measurements only document dust that is far removed from source areas. Multiyear retrievals of the daily distribution of absorbing aerosols with the Total Ozone Mapping Spectrometer instrument (Herman *et al.*, 1997) have provided useful qualitative information of dust plume occurrences over both ocean and land surfaces (**Fig. 3**).

Measurements of deposited dust provide the only method to obtain information on the atmospheric dust content during past climate periods. Such deposition records include terrestrial loess deposits, marine sediments, and dust concentration in ice cores (Kohfeld and Harrison, 2001). Notable examples of loess deposits that were formed during glacial periods are located in North America, the Eurasian Continent, and the South American Pampa region. Deposition fluxes in loess deposits have been as high as 20,000 $g m^{-2}$ per year during glacial periods. Dust particles deposited in the loess regions consists of coarser particles (10–63 μm) that have not traveled far from the source regions. In contrast, marine sediment records are good indicators for intermediate and long-distance transport. Mean grain sizes in marine sediments are typically less than 10 μm, although larger grains are found in sediments close to continental source areas. Accumulation rates in marine sediments are on the order of 1–10 $g m^{-2}$ per year, with maxima as high as 100 $g m^{-2}$ per year. Marine sediments suggest an approximately twofold increase in dust deposition fluxes in the glacial compared to interglacial periods in the tropics and midlatitudes (Kohfeld and Harrison, 2001). Ice core concentration data are records of long-distance transport of dust with particle sizes smaller than 5 μm. Records of increases in the LGM indicate that dust concentrations are highest in the Antarctic ice cores that cover several glacial cycles, with an approximately 10-fold increase in the Vostok and a 40-fold increase in the EPICA core compared to modern concentrations (Delmonte *et al.*, 2002). Although the changes in dust fluxes at these locations are very high, the actual fluxes are only on the order of 0.0001 to 0.01 $g m^{-2}$ per year. The isotopic and mineralogical properties of ice core dust allow the determination of source areas and transport. Current studies indicate that glacial dust arriving at Antarctica originates mostly in southern South America, with minor sources in northern New Zealand, or the Antarctic dry valleys; dust arriving in Greenland originates in the Asian deserts (Biscaye *et al.*, 1997; Basile *et al.*, 1997). Ice core data also show the timing of increase and decrease of dust fluxes compared to changes in other environmental parameters, such as temperature, CO_2, and CH_4 (Petit *et al.*, 1990). Dust concentrations increased after the onset of glacial cooling and lowering of atmospheric CO_2. This timing may lead to explanations of the cause of increased dustiness as well as feedback of dust on climate parameters such as temperature, precipitation, and the radiation balance.

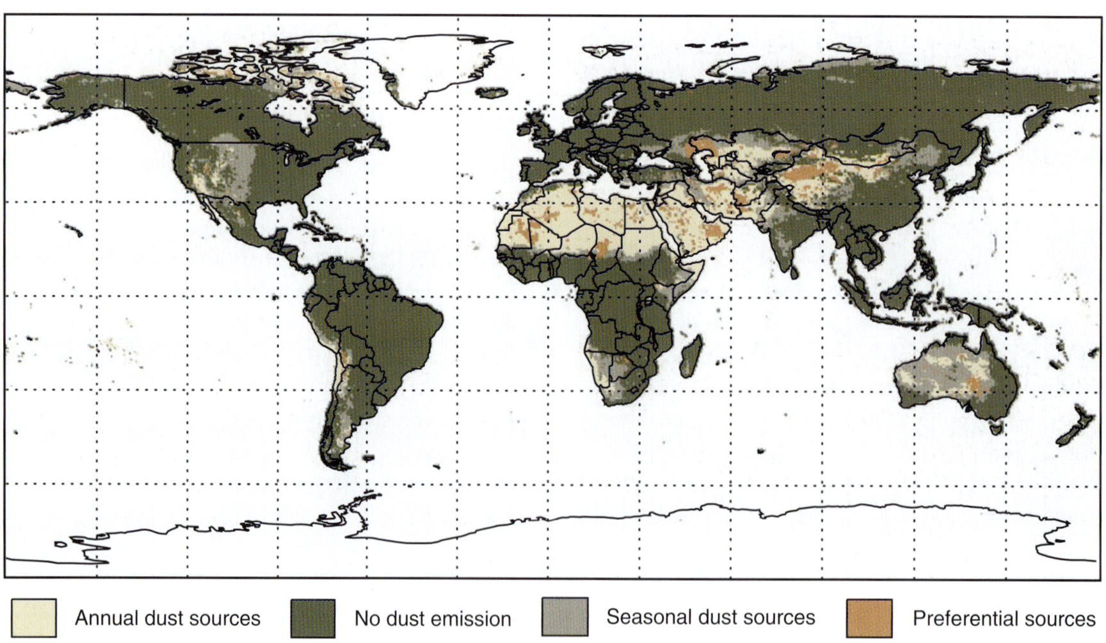

Figure 3 Distribution of global dust source areas, indicating unvegetated desert areas, densely vegetated areas that cannot act as dust sources (dark green), and areas where remote sensing data indicate bare soils for part of the year (light green) and can thus potentially act as seasonal source areas for soil dust emission. Also shown are modeled preferential 'hot spot' dust emission areas (brown). This distribution of source areas was used, for example, in the dust emission model described in Werner *et al.* (2003).

Dust Variability

Soil dust aerosol not only shows distinct spatial patterns but also is temporally very variable at different timescales. Soil dust aerosol distributions respond to changes in global climate conditions, such as the overall intensity of winds that entrain dust into the air, and changes of the hydrological cycle that can affect the extent of source areas as well as the length of time that dust remains in the atmosphere. Within individual source areas, dust emissions mostly follow a seasonal cycle with distinct spring and summer maxima, when strong surface winds occur and the soils are unvegetated (**Fig. 4**). Dust production can also vary from year to year as a consequence of changes in climate conditions. For example, in the Sahara, dust emissions are found to vary interannually in connection with the Sahelian rainfall deficit (Prospero and Lamb, 2003). On decadal timescales, an increase in Saharan dust emissions has been observed since the late 1960s through the 1980s, and the emissions have remained at a high level since then. Long-term records of dust deposition measured in ice cores, marine sediments, and loess deposits show a clear change in deposition fluxes between interglacial and glacial climate periods, when dust was approximately 2–10 times increased globally compared to interglacial periods. This increased dustiness is likely due to a combination of different factors, such as increased aridity, reduced vegetation, increased availability of fine sediment, and stronger surface winds in glacial climates. Higher wind speeds could produce stronger dust emission fluxes from soil surfaces compared to modern conditions; increased aridity expands those areas with drier soils and reduced vegetation where soil deflation by wind is possible. Atmospheric dust loads could also have been enhanced in climate conditions as a consequence of the weakened hydrological cycle with reduced washout of atmospheric dust particles. The best documented period of increased dustiness in the Quaternary is the last glacial period (approximately 17,000–23,000 years ago) (Kohfeld and Harrison, 2001).

Models of the Dust Cycle

Increasingly complex regional and global-scale models of the dust cycle have been developed in order to quantify the atmospheric dust distribution with the aim of supporting investigations of dust effects. The atmospheric dust content computed by these models depends on dust emissions, which are determined by surface wind speeds, and surface properties such as roughness length, vegetation cover, soil moisture, and soil texture. Recently developed models explicitly prescribe seasonal vegetation cover (Werner *et al.*, 2003), masking dust sources, as well as the occurrence of topographic depressions in dry,

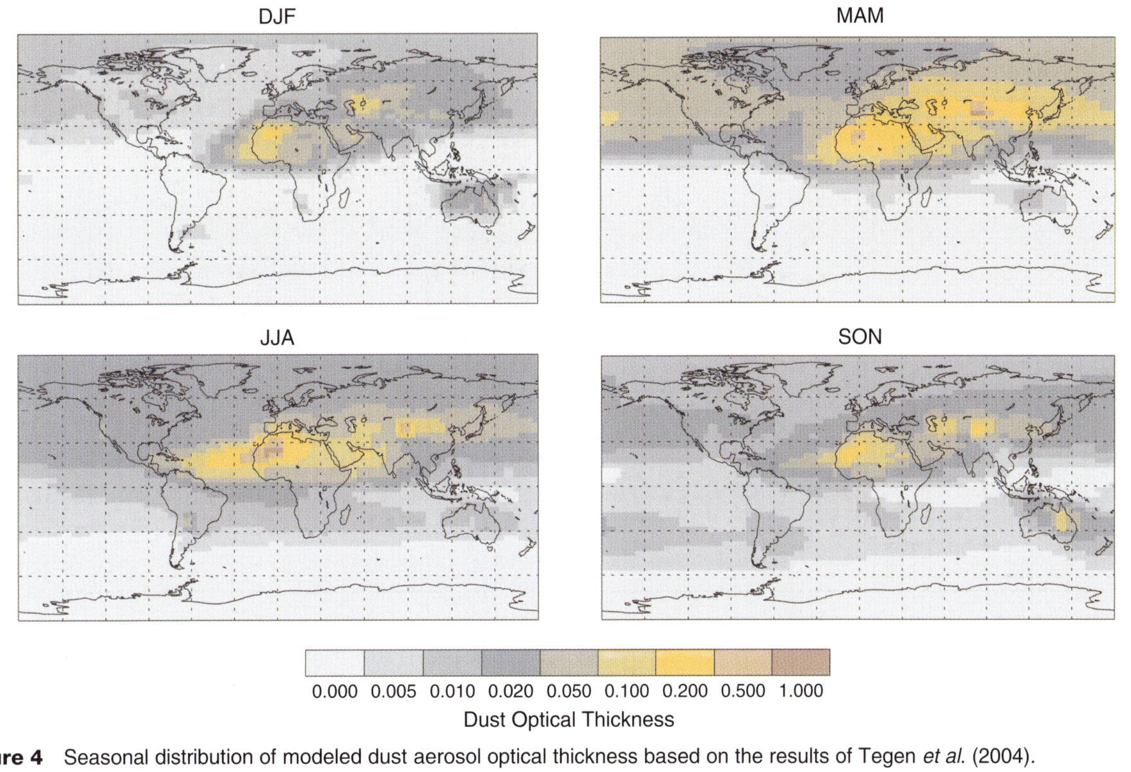

Figure 4 Seasonal distribution of modeled dust aerosol optical thickness based on the results of Tegen *et al.* (2004).

unvegetated areas as indicators of preferential sources containing fine and loose sediment particles that can be easily deflated under strong wind conditions (see **Fig. 2**).

Large-scale data sets of soil properties are difficult to obtain on global scales because the most active dust sources lie in remote areas that cannot easily be accessed. Another major problem in computing dust emissions from wind deflation is insufficient model resolution because peak gusts that are responsible for the strongest dust events remain largely unresolved in large-scale global models. Parameterizations of sub-grid scale variability surface winds (Cakmur *et al.*, 2004) provide only moderate improvements in computations of dust emissions.

Apart from models of present-day dust distributions, the LGM has been a major focus for dust modeling (**Table 1**). During glacial periods, eolian deposition rates were approximately 2–20 times higher than today (Kohfeld and Harrison, 2001) (**Fig. 5**), and changes in boundary conditions such as ice sheet extent and atmospheric trace gas concentrations are reasonably well-known. Models that only consider the influences of increased surface wind speeds and the reduced intensity of the hydrological cycle on dust emission and deposition fluxes at the LGM are unable to reproduce the observed strong increase in dust deposition at high latitudes. The increase in dust source areas caused by exposure of continental shelves is likewise unable to explain the increased high-latitude dust signal. Incorporating increases in dust sources from changes in vegetation cover increases the modeled atmospheric dust loads at high latitudes, consistent with the high dust fluxes observed in polar ice cores (Mahowald *et al.*, 1999; Werner *et al.*, 2003). The potentially large effect of dust aerosol supplied by glacial outwash and the increase in the areal extent of dry lake beds during the LGM have not yet been factored into global dust models.

Table 1 Range of model-based estimates of emission fluxes, atmospheric burden, and atmospheric lifetimes of soil dust aerosols[a]

	Dust emission ($Mt\ yr^{-1}$)	*Atmospheric burden* (Mt)	*Atmospheric lifetime* (days)
Modern	1100–3000	8–36	2.8–7.1
LGM	2400–9000	23–94	3.5–5.5
Ratio LGM/modern	1.4–3.0	1.4–2.8	0.9–1.3

[a]Given are the ranges of minimum and maximum values from model calculations by various authors; summarized in Zender *et al.* (2004) and Harrison *et al.* (2001), including also the results from Werner *et al.* (2003).

Global-scale models are not always useful for investigation of regional-scale dust processes, especially in areas close to source regions, since the horizontal resolution of typically 3–5° in global models and the necessarily simplified emission and deposition parameterizations are too crude to resolve small-scale dust processes and enable comparisons to observations. Regional models of dust emission, transport, and deposition exist for key regions, including the Sahara desert and Asia (Sokolik *et al.*, 2001). Such regional-scale models are well suited for simulation of individual dust storm events as well as for comparisons with *in situ* observations made during field experiments. Although the topography, soil conditions, and small-scale extreme wind events are simulated more realistically in regional compared to global models, uncertainties in the parameterization of key processes and input data on meteorological fields and soil properties remain.

Particle Properties

Dust aerosol particle sizes determine their optical and microphysical properties, as well as atmospheric lifetimes. Close to source regions, the median dust particle diameters are observed to be approximately 60–100 μm; particles in this size range can most effectively be lifted from the ground by surface winds. Because large dust particles settle back to the ground quickly, dust that has been transported thousands of kilometers away from the source has a median size of approximately 2 μm (Seinfeld and Pandis, 1997).

The mineral composition of dust particles reflects the mineralogy of the rocks at the Earth's surface; some main constituents are quartz, feldspars, carbonates, and clays. Dust particles from different source regions can have specific mineralogical composition and, as a consequence, different chemical composition. Claquin *et al.* (1999) summarized available information to compile global maps of mineral composition of soils in dry areas by relating soil mineralogy to soil types. They found that quartz, with approximately 70% of the mass, is the most abundant mineral in the silt size fraction of the soil (particles in the size range of 2–50 μm diameter), and illite is the most abundant clay mineral (contributing up to 50% regionally to the clay fraction). The radiative and chemical properties of dust can change during transport when dust surfaces coat with soluble sulfates and nitrate species and are processed by clouds.

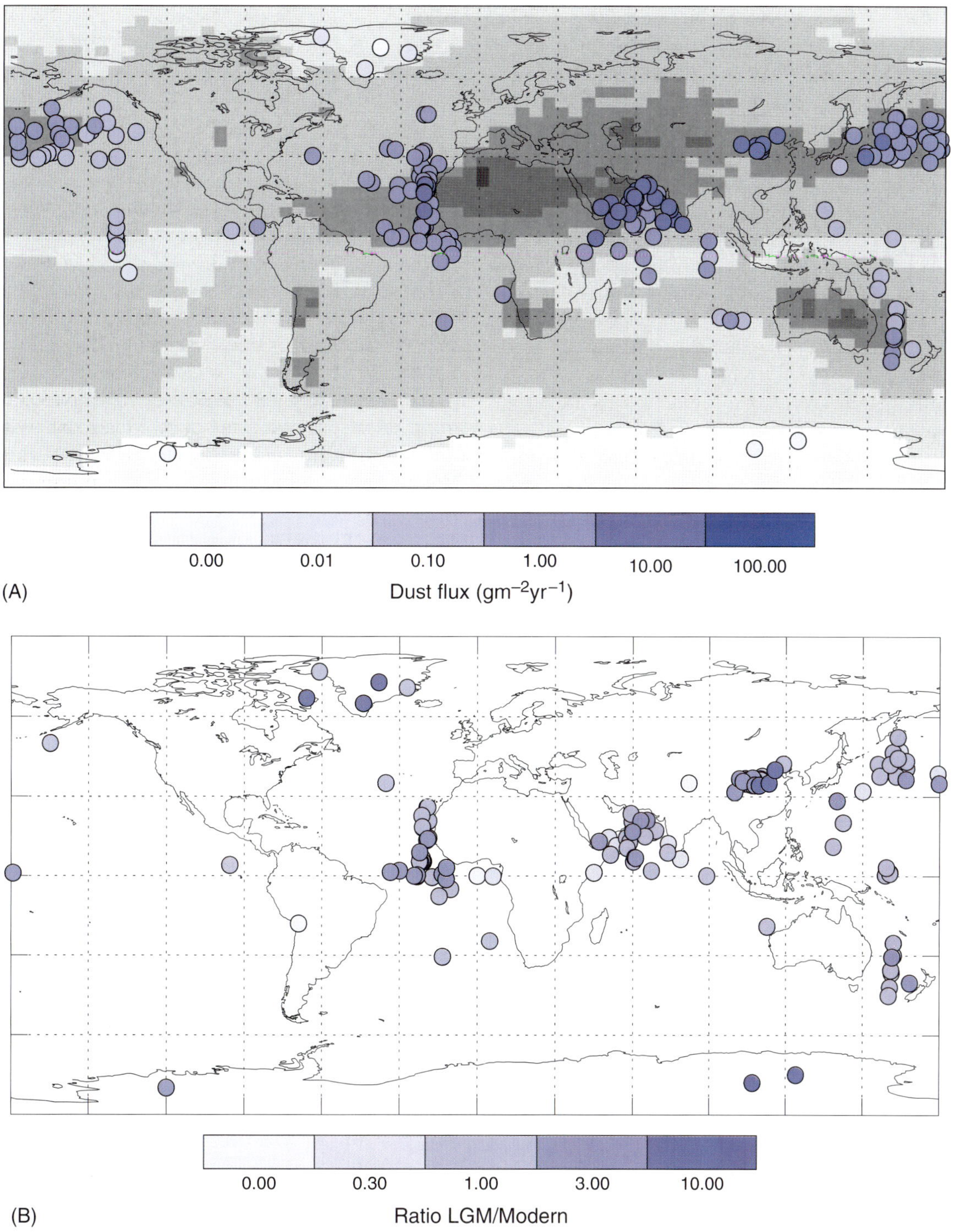

Figure 5 (A) Patterns of modern global dust deposition fluxes comparing a model result (Tegen *et al.*, 2004) with local dust fluxes derived from measurements of marine sediments, sediment traps, terrestrial loess deposits, and ice cores for the late Holocene compiled by Kohfeld and Harrison (2001). (B) To indicate the enhancement of dust deposition fluxes during glacial climate, the ratio of the observed dust fluxes between the Last Glacial Maximum and the late Holocene is shown.

Dust Effects

Direct Effect

Incoming solar radiation is partly reflected and partly absorbed by dust particles in the atmosphere. With mass median diameters of 2 μm or larger, these particles also absorb and emit outgoing terrestrial radiation. The radiative effect of dust particles depends on their optical properties (i.e., their size distribution and complex refractive index), which are either determined by laboratory and field measurements or retrieved by remote sensing (Sokolik *et al.*, 2001). Although direct measurements have a considerable uncertainty range, remote sensing retrievals of dust properties may be influenced by the presence of other aerosol types besides dust or by thin cirrus clouds, or by underestimation of the nonsphericity of dust particles. Dust aerosol radiative properties are strongly influenced by their hematite content, which mostly controls the absorption of incoming solar radiation by the particles. From size and complex refractive indices of the particles, aerosol extinction efficiency (which determines the light attenuation per size fraction), the single scattering albedo (which is the ratio of scattering and total extinction of solar radiation), and the asymmetry parameter (which quantifies the ratio of forward scattered light) can be computed using Mie theory, which is a theory of the scattering of electromagnetic radiation by spherical particles. These parameters are then used to compute the fractions of absorbed and transmitted radiation passing through an aerosol layer. Radiative forcing by dust, defined here as the attenuation of radiation caused by the presence of atmospheric dust, rather than the radiative impact of dust enhancement from human activities, is generally negative over dark ocean surfaces. However, over bright desert surfaces, the net radiative effect of dust depends on their optical properties, and it can be positive for strongly absorbing particles (**Fig. 6**). At the top of the atmosphere, the

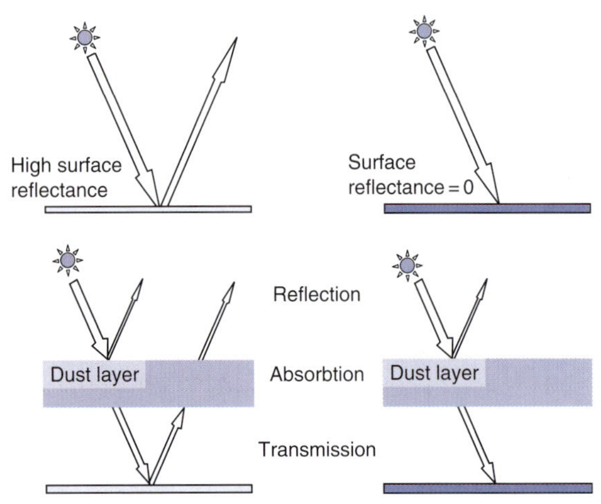

Figure 6 Schematic representation of reflection and absorption of incoming solar radiation in the absence and presence of a layer of dust aerosol, above a bright, highly reflective surface, and a dark absorbing surface. The dust layer partly reflects and absorbs incoming solar radiation; from space the dust appears bright over a dark surface (e.g., the ocean), the dust absorption may be sufficient to appear dark above a bright surface (bright desert or ice cover). Thus, depending on dust optical properties, it may lead to an enhancement or a reduction of energy input into the globe.

negative and radiative forcing by dust can partly cancel in the global mean (IPCC, 2001). Measurements of dust optical properties from remote sensing (Kaufman *et al.*, 2001) have led to the conclusion that the net radiative forcing due to dust is negative in the global mean, indicating that enhanced dust aerosol loads would lead to regional cooling.

The radiative effect of higher dust loads during the LGM has been investigated by Claquin *et al.* (2003). The authors computed the difference in the top-of-the-atmosphere radiation between the LGM and present-day dust loads from a model simulating past dust distribution, which were obtained from results of a simulation taking into account changes in climate parameters and vegetation density for glacial conditions. The estimated change in top-of-the-atmosphere radiative forcing by the enhanced dust load (**Fig. 7**) ranged from -0.3 to $-0.9\,\mathrm{Wm^{-2}}$ at northern high latitudes. This effect is small compared to the radiative effects due to the expansion of ice coverage at the LGM, which results in a radiative forcing of approximately $-19\,\mathrm{Wm^{-2}}$ at latitudes between 45 and 90°N. At low latitudes between 15°S and 15°N, Claquin *et al.* (2003) estimate a decrease in incoming solar radiation fluxes of -1 to $-1.4\,\mathrm{Wm^{-2}}$ caused by the approximately doubled dust load at the LGM, which is of comparable magnitude to the decrease in greenhouse forcing by $-1.8\,\mathrm{Wm^{-2}}$ due to the reduced CO_2 concentration at the LGM. These results indicate that the enhanced atmospheric dust may have contributed to glacial cooling at low latitudes, whereas it played only a small role in the radiation balance at high latitudes.

Semidirect and Indirect Effects

Radiative forcing by dust not only affects the radiation balance of the earth directly but also influences the energy distribution within the atmosphere. At the surface below a dust cloud, incoming solar radiation is reduced, which mainly causes a decrease in latent heating and therefore a weakening of the hydrological cycle (Miller *et al.*, 2004). Together with atmospheric heating in the presence of a dust cloud, this may lead to a reduction in cloud cover, which is called the semidirect aerosol effect. Hansen *et al.* (1997) showed with a simplified global atmospheric model that slightly absorbing aerosols, which have a negative direct radiative effect, may actually cause warming as a consequence of the semidirect effect.

Aerosol particles influence the formation and lifetimes of clouds via the so-called indirect aerosol effect, although the role of soil dust particles in these processes remains unclear. Wurzler *et al.*

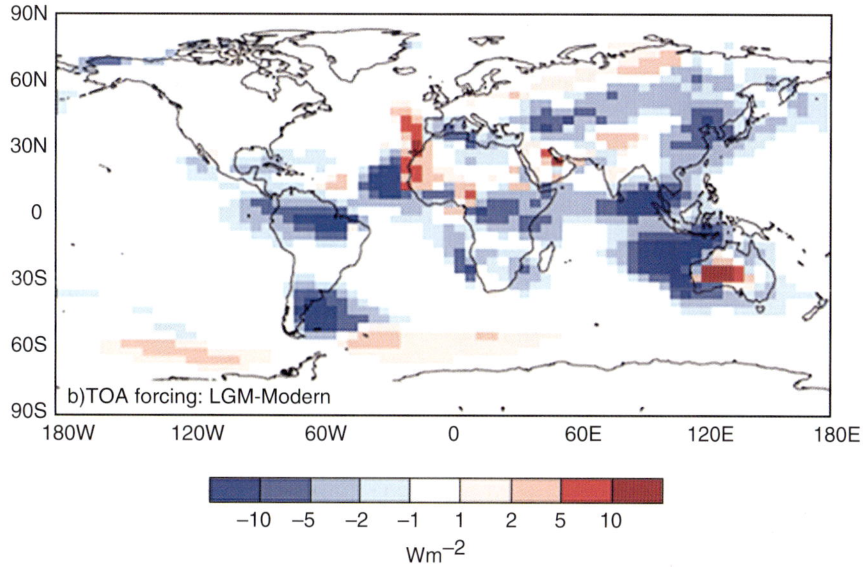

Figure 7 Net changes in solar radiation at the top of the atmosphere by enhanced atmospheric dust loads during the LGM. Shown is the difference in radiation fluxes between LGM and modern conditions, taking into account enhanced dust loads at the LGM. Dust optical properties were based of the assumption of an internal mixture of quartz minerals with 2% hematite. Modified from Claquin, T., Roelandt, C., Kohfeld, K. E., Harrison, S. P., Tegen, I., Prentice, I. C., Balkanski, Y., Bergametti, G., Hansson, M., Mahowald, N., Rodhe, H., and Schulz, M. (2003). Radiative forcing of climate by ice-age atmospheric dust. *Climate Dynamics* **20**, 193–202, Fig. 3b. Copyright Springer-Verlag.

(2000) found that the coating of dust by soluble particles enhances their ability to form cloud condensation nuclei (CCN). In addition, dust particles are effective nuclei for ice clouds; Ansmann *et al.* (2005) found ice formation in an altocumulus cloud in the presence of dust at 10–20 °K higher temperatures than in the absence of insoluble particles, indicating the potentially important role of dust for ice nucleation. This process would not only affect the radiative properties of clouds but also enhance the formation of precipitation. However, the magnitudes of these effects are still unknown.

Effect of Dust on Ocean Biogeochemistry

Deposited soil dust particles are hypothesized to add essential nutrients to terrestrial and marine ecosystems. In particular, the increase in productivity of oceanic microorganisms by addition of iron has received much attention in recent years (Jickells *et al.*, 2005). Microorganisms use iron in a variety of enzyme systems, including those for photosynthesis and nitrogen fixation. Soils contain an average of 3.5% iron. The iron content of dust particles is variable, but this uncertainty is small compared with other uncertainties in the iron cycle. Iron solubility in soil dust particles at source areas is on the order of 1%. Higher solubilities of aerosol iron of locally up to 40% have been reported in atmospheric particles due to photochemical processes (photoreduction of FeIII to FeII) and aerosol cloud processing. In nutrient-rich regions of the ocean, the addition of iron can enhance primary production, species composition, and CO_2 uptake, whereas in oligotrophic regions such as the subtropics, the nutrient supply from dust aerosol can enhance nitrogen fixation. Several field studies show that plankton productivity and CO_2 drawdown are enhanced in nutrient-rich regions when iron is added, but it is unknown if dust is important for oceanic CO_2 uptake at the global scale. Different biogeochemical models have derived different results; one consequence of dust seems to be a change in planktonic species composition (Bopp *et al.*, 2003).

Glacial–interglacial changes in ocean export production are suspected to have contributed to CO_2 drawdown, assuming that stronger dust deposition fluxes (see **Fig. 5**) enhanced primary productivity. The effects of enhanced dust deposition during glacial periods have been investigated by Kohfeld *et al.* (2005), who analyzed sediment records of marine productivity at the peak and the middle of the last glacial cycle and found that iron fertilization and associated mechanisms can be responsible for up to half the observed CO_2 drawdown.

Other Effects

Apart from the direct and indirect effects of soil dust aerosol on the climate, dust also affects atmospheric chemistry by providing reactive surface areas for heterogeneous reactions (Dentener *et al.*, 1996), influencing tropospheric ozone concentrations and the atmospheric sulfate and nitrate cycles. Sulfate reactions occur mainly with calcite in the dust aerosol, which is converted to gypsum. Bonasoni *et al.* (2004) observed that in the Mediterranean region the lowest surface ozone concentration occurred in the presence of coarse dust aerosol, indicating an important role of dust for ozone chemistry.

In addition to these processes, dust storms can adversely affect human quality of life. Dust storms strongly reduce visibility in the atmosphere, causing traffic problems in regions with frequent dust events. Strong exposure to dust has also been related to health problems such as silicosis.

Conclusion

Although the cycle of dust emission, atmospheric transport, and deposition has been well researched and a wide array of data exist on the properties of dust aerosols, their direct and indirect effects on the climate and atmospheric chemistry remain poorly quantified. The strong spatial and temporal variability of the atmospheric content and properties of the dust particles continue to cause challenging problems in quantifying atmospheric dust properties at global scales.

See also: **Paleoclimate Modeling**: Quaternary Environments; Last Glacial Maximum GCMs.

References

Ansmann, A., Mattis, I., Muller, D., *et al.* (2005). Ice formation in Saharan dust over central Europe observed with temperature/humidity/aerosol Raman lidar. *Journal of Geophysical Research* **110**, No. D18S12.

Basile, I., Grousset, F. E., Revel, M., *et al.* (1997). Patagonian origin of glacial dust deposited in East Antarctica (Vostok and Dome C) during glacial stages 2, 4 and 6. *Earth and Planetary Science Letters* **146**, 573–589.

Biscaye, P. E., Grousset, F. E., Revel, M., *et al.* (1997). Asian provenance of glacial dust (stage 2) in the Greenland Ice Sheet Project 2 Ice Core, Summit, Greenland. *Journal of Geophysical Research* **102**, 26,765–26,781.

Bonasoni, P., Cristofanelli, P., Calzolari, F., *et al.* (2004). Aerosol–ozone correlations during dust transport episodes. *Atmospheric Chemistry & Physics* **4**, 1201–1215.

Bopp, L., Kohfeld, K. E., LeQuéré, C., and Aumont, O. (2003). Dust impact on marine biota and atmospheric CO_2 during glacial periods. *Paleoceanography* **18**, Art No. 1046.

Cakmur, R. V., Miller, R. L., and Torres, O. (2004). Incorporating the effect of small-scale circulations upon dust emission in an atmospheric general circulation model. *Journal of Geophysical Research* **109**, No. D07201.

Claquin, T., Schulz, M., and Balkanski, Y. (1999). Modelling the mineralogy of atmospheric dust sources. *Journal of Geophysical Research* **104**, 22,243–22,256.

Claquin, T., Roelandt, C., Kohfeld, K. E., *et al.* (2003). Radiative forcing of climate by ice-age atmospheric dust. *Climate Dynamics* **20**, 193–202.

Delmonte, B., Petit, J. R., and Maggi, V. (2002). Glacial to Holocene implications of the new 27,000-year dust record from the EPICA Dome C (East Antarctica) ice core. *Climate Dynamics* **18**, 647–660.

Dentener, F. J., Carmichael, G. R., Zhang, Y., Lelieveld, J., and Crutzen, P. J. (1996). Role of mineral aerosol as reactive surface in the global troposphere. *Journal of Geophysical Research* **101**, 22,869–22,889.

Hansen, J. E., Sato, M., and Ruedy, R. (1997). Radiative forcing and climate response. *Journal of Geophysical Research* **102**, 6831–6864.

Harrison, S. P., Kohfeld, K. E., Roelandt, C., and Claquin, T. (2001). The role of dust in climate changes today, at the last glacial maximum and in the future. *Earth Science Review* **54**, 43–80.

Herman, J. R., Bhartia, P. K., Torres, O., *et al.* (1997). Global distribution of UV-absorbing aerosols from Nimbus 7/TOMS data. *Journal of Geophysical Research* **102**, 16,911–16,922.

Intergovernmental Panel on Climate Change (2001). *Climate Change 2001: The Scientific Basis* (J. T. Houghton, Y. Ding, D. J. Griggs, M. Noguer, P. J. van der Linden, X. Dai, K. Maskell and C. A. Johnon, Eds.). Cambridge University Press, New York.

Jickells, T. D., An, Z. S., Andersen, K. K., *et al.* (2005). Global iron connections between desert dust, ocean biogeochemistry, and climate. *Science* **308**, 67–71.

Kaufman, Y. J., Tanre, D., *et al.* (2001). Absorption of sunlight by dust as inferred from satellite and ground-based remote sensing. *Geophys. Res. Lett* 1479–1483.

Kohfeld, K. E., and Harrison, S. P. (2001). DIRTMAP: The geological record of dust. *Earth Science Review* **54**, 81–114.

Kohfeld, K. E., LeQuere, C., Harrison, S. P., and Anderson, R. F. (2005). Role of marine biology in glacial–interglacial CO_2 cycles. *Science* **308**, 74–78.

Mahowald, N. M., Kohfeld, K. E., Hansson, M., *et al.* (1999). Dust sources and deposition during the Last Glacial Maximum and current climate: A comparison of model results with paleodata from ice cores and marine sediments. *Journal of Geophysical Research* **104**, 15,895–16,436.

Marticorena, B., and Bergametti, G. (1995). Modeling the atmospheric dust cycle: 1. Design of a soil-derived dust emission scheme. *Journal of Geophysical Research* **100**, 16,415–16,430.

Miller, R. L., Tegen, I., and Perlwitz, J. (2004). Surface radiative forcing by soil dust aerosols and the hydrologic cycle. *Journal of Geophysical Research* **109**, No. D04203.

Petit, J. R., Mounier, L., Jouzel, J., *et al.* (1990). Palaeoclimatological and chronological implications of the Vostok core dust record. *Nature* **343**, 56–58.

Prospero, J. M., and Lamb, P. J. (2003). African droughts and dust transport to the Caribbean: Climate change implications. *Science* **302**, 1024–1027.

Prospero, J. M., Ginoux, P., Torres, O., Nicholson, S. E., and Gill, T. E. (2002). Environmental characterization of global sources of atmospheric soil dust identified with the NIMBUS 7 TOMS absorbing aerosol product. *Reviews of Geophysics* **40**, No. 1002.

Seinfeld, J. H., and Pandis, S. N. (1997). *Atmospheric Chemistry and Physics: From Air Pollution to Climate Change*. Wiley, New York.

Sokolik, I. N., Winker, D. M., Bergametti, G., *et al.* (2001). Introduction to special section: Outstanding problems in quantifying the radiative impacts of mineral dust. *Journal of Geophysical Research* **106**, 18,015–18,027.

Tegen, I., Werner, M., Harrison, S. P., and Kohfeld, K. E. (2004). Relative importance of climate and land use in determining present and future global soil dust emission. *Geophysical Research Letters* **31**, No. L05105.

Werner, M., Tegen, I., Harrison, S. P., *et al.* (2003). Seasonal and interannual variability of the mineral dust cycle under present and glacial climate conditions. *Journal of Geophysical Research* **108** Art. No. 4744.

Wurzler, S., Reisin, T. G., and Levin, Z. (2000). Modification of mineral dust particles by cloud processing and subsequent effects on drop size distributions. *Journal of Geophysical Research* **105**, 4501–4512.

Zender, C., Miller, R. L., and Tegen, I. (2004). Quantifying mineral dust mass budgets: Systematic terminology, constraints, & current estimates. *EOS* **85**, 509–512.

Thermohaline Circulation

S Rahmstorf, Potsdam Institute for Climate Impact Research, Potsdam, Germany

© 2007 Elsevier B.V. All rights reserved.

> It appears to be extremely difficult, if not quite impossible, to account for this degree of cold at the bottom of the sea in the torrid zone, on any other supposition than that of cold currents from the poles; and the utility of these currents in tempering the excessive heats of these climates is too evident to require any illustration.
>
> —Sir Benjamin Thompson (1797)

What is the Thermohaline Circulation?

In 1751, the captain of an English slave-trading ship made the first recorded measurement of deep ocean temperatures—he discovered that the water a mile below his ship was very cold, despite the subtropical location. In 1797, another Englishman, Benjamin Thompson, correctly explained this discovery by cold currents from the poles, as part of what later became known as the thermohaline circulation.

As opposed to wind-driven currents and tides (the latter are due to the gravity of moon and Sun), the thermohaline circulation (THC) is that part of the ocean circulation which is driven by fluxes of heat and freshwater across the sea surface and subsequent interior mixing of heat and salt; hence the name thermohaline. (Geothermal heat sources at the ocean bottom play a minor role.) The term thermohaline circulation thus refers to a particular driving mechanism; it is a physical, not an observational, concept.

The distinction of thermohaline versus wind-driven circulation originates in a nineteenth-century dispute on whether ocean currents are primarily due to the wind pushing along the water or whether they are 'convection currents' due to heating and cooling, or evaporation and precipitation. In 1908, Johan Sandström performed a series of classic tank experiments at Bornö oceanographic station in Sweden to investigate both possibilities (Sandström, 1908), describing the properties of 'wind-driven circulation' and 'thermal circulation.' To include salinity, the latter was later extended to 'thermohaline circulation,' a term that by the 1920s appeared in the classic oceanography textbook by Albert Defant (Defant, 1929).

The ocean's density distribution, which determines pressure gradients and thus circulation, is affected by currents and mixing. Thermohaline and wind-driven currents therefore interact in nonlinear ways and cannot be separated by oceanographic measurements. There are thus two distinct physical forcing mechanisms but not two uniquely separable circulations. Changing the wind stress will alter the thermohaline circulation; altering thermohaline forcing will also change the wind-driven currents.

A related, complementary concept is that of meridional overturning circulation (MOC). This refers to the north–south flow as a function of latitude and depth, often integrated in east–west direction across an ocean basin or the globe and graphically depicted as a stream function. The streamlines typically show a large-scale slow overturning motion of the ocean. The MOC can be easily diagnosed from a model, and in principle it can be measured in the ocean.

Although the terms THC and MOC are often inaccurately used as if synonymous, there strictly is no one-to-one relation between the two. The MOC includes clearly wind-driven parts, namely the Ekman cells consisting of the transport in the near-surface Ekman layer and a return flow below it. A direct contribution of wind-driven currents even to the large scale, deeper overturning is being increasingly discussed. On the other hand, the THC is of course not confined to the meridional direction; rather, it is also associated with zonal overturning cells. Hence, care should be taken with the terminology: The term THC should be reserved for a particular forcing mechanism—for example, when discussing the influence of cooling or freshwater forcing on the ocean circulation. The term MOC should be used when describing a meridional flow field, for example, from a model, which most often will show a mix of both wind-driven and thermohaline flow.

The following are key features of the thermohaline circulation:

- Deep water formation: The sinking of water masses, closely associated with convection, which is a vertical mixing process. Deep water formation takes place in a few localized areas (**Fig. 1**): the Greenland–Norwegian Sea, the Labrador Sea, the Mediterranean Sea, the Weddell Sea, and the Ross Sea.
- Spreading of deep waters (e.g., North Atlantic Deep Water (NADW) and Antarctic Bottom Water (AABW), mainly as deep western boundary currents.
- Upwelling of deep waters: This is not as localized as convection and difficult to observe. It is thought to take place mainly in the Antarctic Circumpolar Current region, possibly aided by the wind (Ekman divergence; **Fig. 2**).
- Near-surface currents: These are required to close the flow. In the Atlantic, the surface currents compensating the outflow of NADW range from the Benguela Current off South Africa via the Gulf Stream and the North Atlantic Current into the Nordic Seas off Scandinavia (**Fig. 3**). It is worth noting that the Gulf Stream is primarily a

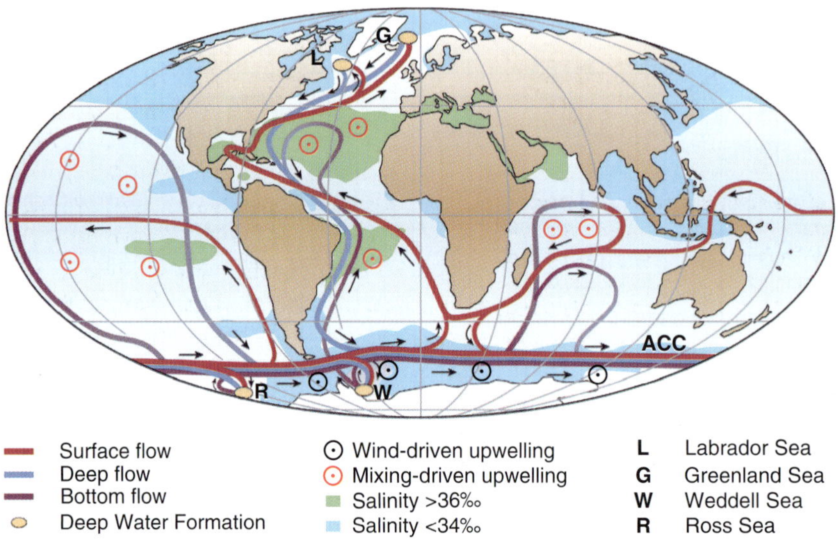

Figure 1 Schematic representation of the global thermohaline circulation. Surface currents are shown in red, deep waters in light blue, and bottom waters in dark blue. The main deep water formation sites are shown in orange. Modified from Broecker, W. S. (1991). The great ocean conveyor. *Oceanography* **4,** 79–89; from Kuhlbrodt, T., *et al.* (2006). On the driving processes of the oceanic meridional overturning circulation. Submitted for publication.

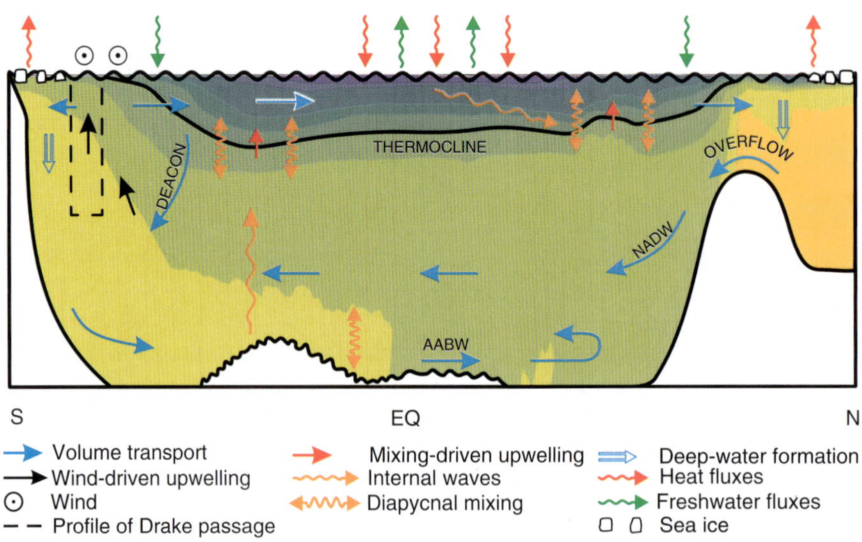

Figure 2 Side view of the circulation in the Atlantic showing various flow components and mechanisms discussed in the text. Color shading shows the observed density stratification, with lightest waters in blue and densest in orange. From Kuhlbrodt, T., *et al.* (2006). On the driving processes of the oceanic meridional overturning circulation. Submitted for publication.

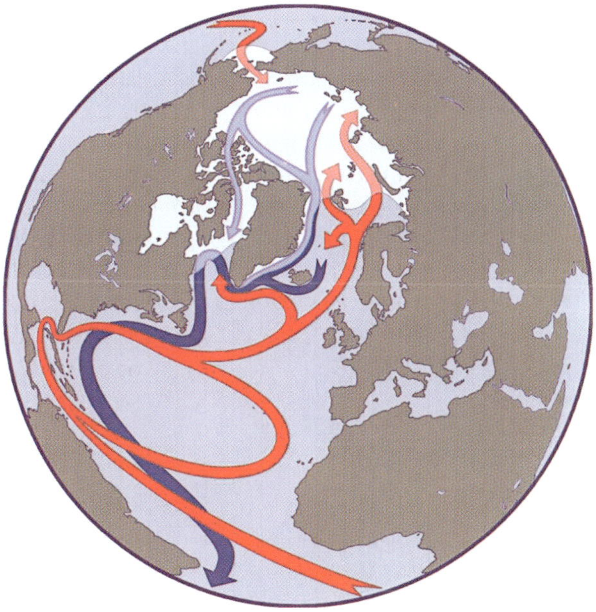

Figure 3 Circulation of the northern Atlantic and Arctic oceans. This simplified figure shows surface currents in red and North Atlantic Deep Water (NADW) in blue. The winter sea ice cover (white) is held back in the Atlantic sector by the warm North Atlantic Current. Figure modified for the Arctic by G. Holloway.

wind-driven current, forming part of the subtropical gyre circulation. The thermohaline circulation—approximated here by the amount of water needed to compensate for the southward flow of NADW—contributes only approximately 20% to the Gulf Stream flow.

Some Observational Data

As explained previously, the THC is not a measurable quantity but, rather, a conceptual idea. However, even many aspects of the MOC are difficult to measure. The vertical motions of sinking and upwelling are too slow to directly capture with current meters. Surface currents are to a large extent part of wind-driven horizontal circulations, so measurements do not easily yield values for the surface component of the MOC. Deep western boundary currents, tracer data and inverse calculations combining various data sources and physical constraints, give the best information on the MOC (**Fig. 4**).

The volume transport of the overturning circulation at 24°N in the Atlantic has been estimated from hydrographic section data (Roemmich and Wunsch, 1985) as 17 Sv ($1\,\text{Sv} = 10^6\,\text{m}^3/\text{s}$) and its heat transport as 1.2 PW ($1\,\text{PW} = 10^{15}\,\text{W}$). Talley *et al.* (2003) estimated 18 ± 5 Sv of NADW formation, and an inverse model (Ganachaud and Wunsch, 2000) yielded 15 ± 2 Sv NADW overturning in the high latitudes.

Narrow channels provide a good opportunity to measure deep water flows. The overflows from the Nordic Seas have been measured as transporting ~3 Sv each between Greenland and Iceland and between Iceland and Scotland (**Fig. 5**), whereas the deep water formation south of these sills, in the Labrador Sea, is estimated as 2–4 Sv. Combined, these numbers do not add up to the total NADW flow estimates given previously because the volume transport of dense water increases through mixing along the

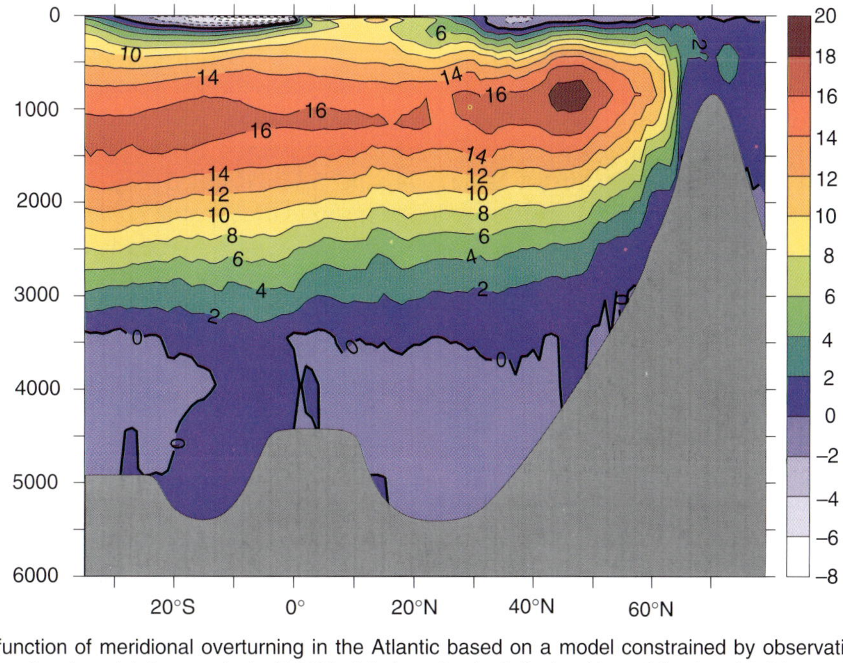

Figure 4 Stream function of meridional overturning in the Atlantic based on a model constrained by observational data (after Köhl *et al.*, 2006). Interannual to decadal changes in the ECCO global synthesis. Submitted for publication. Visible are the Ekman cells in the upper 500 m, the NADW cell down to 3500 m, and the AABW cell near the bottom.

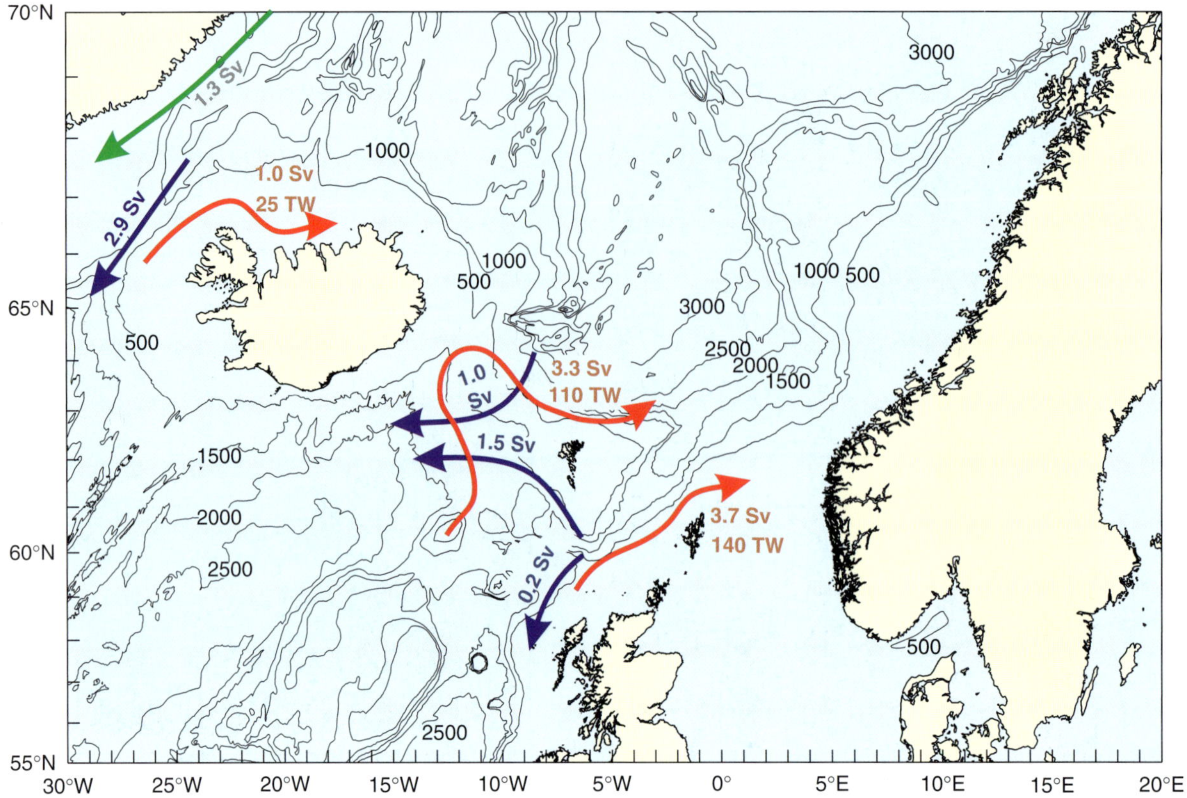

Figure 5 Volume and heat transports across the ridge separating the Nordic Seas from the northern Atlantic. Red marks the surface inflow into the Nordic Seas, blue the deep outflow across the sill, and green the East Greenland Current. From Blindheim, J. (2004). Oceanography and climate. In *The Norwegion Sea Ecosystem* (H. R. Skjoldal, Ed.). Tapir Academic Press, Trondheim, Norway.

way in a process called entrainment (at the expense of being diluted; i.e., the core density decreases).

Less information is available about the second major deep water formation region of the global oceans, namely the Southern Ocean around Antarctica (**Figs. 6** and **7**). Tracer data suggest another ~15 Sv of deep water forming there, bringing the global total to slightly more than 30 Sv. More recent chlorofluorocarbon data suggest only 5 Sv sinking from the surface around Antarctica, which may be reconciled either by a change over time or, more likely, again by entrainment.

What Drives the THC?

The simplest answer to this question would be high-latitude cooling. In cold regions the highest surface water densities are reached; this causes sinking of water, which in turn drives the circulation. This is a thermally dominated circulation in that it is the coldest waters that sink to fill the deep oceans, and for the moment we will ignore the effects of salinity, returning to them in the next section.

Reality is more complex, however. Cooling a previously warmer ocean at the surface will lead to sinking of the coldest, densest water, which will spread out along the bottom causing upwelling elsewhere, according to what is termed 'filling box dynamics.' However, eventually (after ~1,000 years at current deep water formation rates) the box would be full—the ocean basins would be filled with dense water and the 'filling box' circulation would come to a halt.

Maintaining a steady-state circulation thus requires more than cold high latitudes. It requires either the source water to get continually denser to keep replacing the water in the box (which contravenes the idea of a steady state) or a mechanism that continually works to make the water in the box less dense so that the source water remains denser than the water already in the box. This allows a steady state, with newly formed dense waters spreading in the ocean tending to increase the density while the other mechanism tends to reduce the density. The flow field adjusts so that both processes are in equilibrium and the density field remains steady.

This basic constraint was understood by Sandström: He found that heating and cooling his

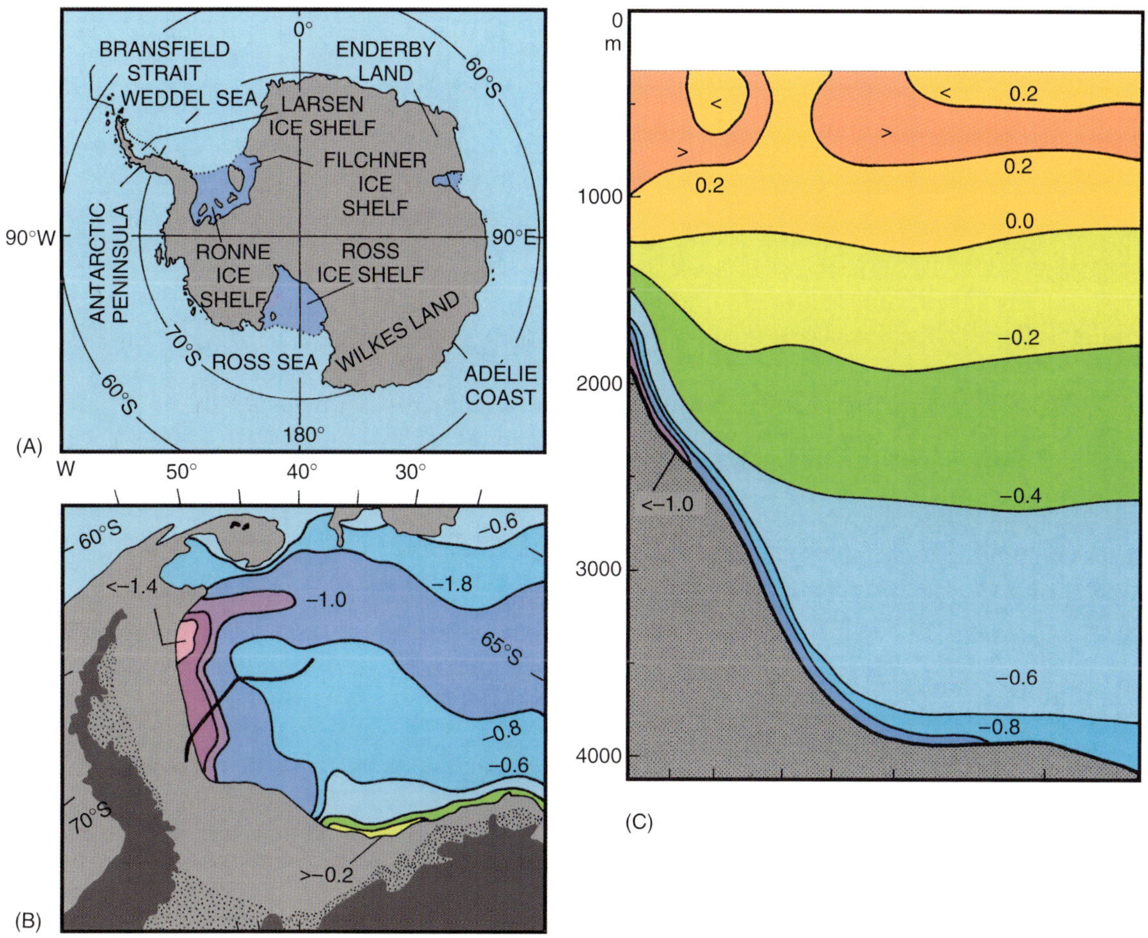

Figure 6 Bottom water formation near Antarctica. (A) Locality map. (B) Bottom potential temperature (°C) in the Wedell Sea; the stippled area indicates ice shelf, and the edge of the shaded region is the approximate 3,000 m contour. (C) A vertical section of potential temperature, along the heavy line in B. Source: Tomczak, M., and Godfrey, J. S. (2003). *Regional Oceanography: An Introduction*, 2nd ed. Available at www.es.flinders.edu.au/~mattom/regoc; after Warren, B. A. (1981). Deep circulation of the world ocean. In *Evolution of Physical Oceanography* (B. A. Warren and C. Wunsch, Eds.), pp. 6–40. MIT Press, Cambridge.

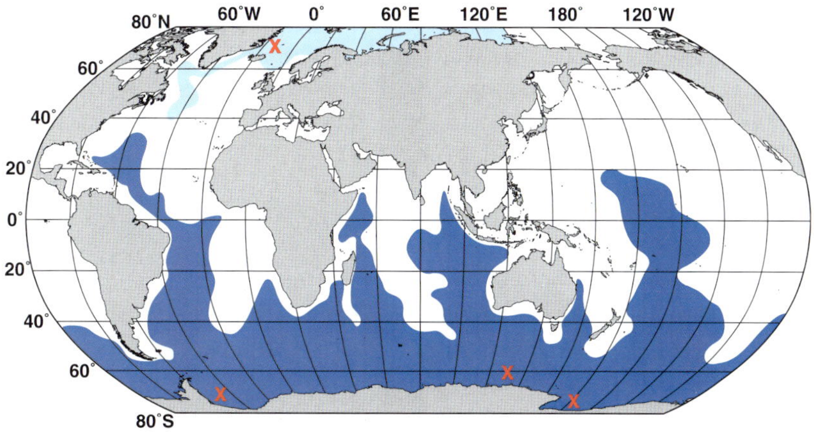

Figure 7 A representation of the spread of Antarctic Bottom Water (AABW, dark blue; based on the $\sigma_4 = 45.92$ kg/m^3 density surface). From Talley, L. D. (1999). Some aspects of ocean heat transport by the shallow, intermediate and deep overturning circulations. In *Mechanisms of Global Climate Change at Millennial Time Scales* (P. U. Clark *et al.*, Eds.), pp. 1–22. American Geophysical Union, Washington, DC.

tank at the surface could not drive a steady-state, deep overturning flow. Rather, heating at greater depth than cooling was required—a conclusion commonly known as Sandström's theorem. This heating at depth is a mechanism that makes the deep waters continually less dense, allowing an ongoing replacement by colder water from above.

What is this mechanism in the real ocean, where heating occurs at (or very near) the surface due to the Sun? As Sandström speculated, it is the downward penetration of heat by mixing. The result is the classic advection–diffusion balance: In steady state, at any given point in the depths of the oceans, the slow diffusion of heat by turbulent mixing is balanced by the equally slow transport of cold waters from the poles.

Downward mixing of heat requires energy to overcome frictional dissipation and to raise the potential energy stored in the water column, as heat penetrating downwards expands the deep water and lifts up the waters above. This is the energy supply of the thermohaline circulation: In an energetic sense, it is driven by turbulent mixing in the ocean's interior. This is why mixing appears in the definition of thermohaline circulation given at the outset of this article (quoted from Rahmstorf, 2002).

The power supply of the turbulence that causes this mixing is both by tidal motions and by the winds. The energy supply needed for generating the observed global overturning motion of ~30 Sv can be estimated from the advection–diffusion balance and the observed density field as ~0.4 TW (Munk and Wunsch, 1998). This can be compared with the power input by winds and tides, estimated as approximately 1 TW each, although these estimates are rather uncertain. With a mixing efficiency of 20%, 0.4 TW would be available for driving the thermohaline circulation (the remaining 80% is dissipated).

Although the energy supply of the thermohaline circulation thus ultimately comes from winds and tides, it is useful to keep the established term 'thermohaline circulation' (as we keep calling 'steam engine' a machine that is powered by coal or wood) in order to distinguish this driving mechanism from a completely different kind: the direct generation of large-scale currents by the frictional action of wind on the water surface. Apart from obviously driving surface currents, this mechanism could also be involved in driving the deep-reaching MOC, in a manner proposed by Toggweiler and Samuels (1995).

This mechanism works as follows: Westerly winds over the Southern Ocean, in conjunction with the Coriolis force, lead to a divergence of surface currents in this region and hence to upwelling. In contrast to the majority of the ocean, where wind-driven upwelling is confined to the upper ocean, the upwelling here comes from deep waters. The reason is a peculiar dynamic constraint on the rotating Earth, termed the 'Drake Passage Effect': Due to the lack of topographic barriers at the latitude band of Drake Passage, no east–west pressure gradient and hence no meridional flow outside frictional boundary layers can be maintained there. This means that the surface water pushed north by the wind (northward Ekman transport resulting from westerly winds) can only be returned to the south at great depth, below 2,500 m, where topographic features exist to support a southward flow. Toggweiler and Samuels (1995) argued that the downward branch required to connect the northward surface flow with this deep southward flow can only occur in the high-latitude north Atlantic because this is the region with the highest surface densities and stable stratification precludes deep water formation elsewhere. In this mechanism, the energy input occurs in the Southern Ocean through the wind creating an Ekman divergence at the surface and convergence at depth, directly 'lifting up' water there, with the MOC looping through the Atlantic to fulfill a continuity requirement.

Which of these two alternative mechanisms actually drives the observed MOC in the world oceans? This discussion is strangely reminiscent of the nineteenth-century debate on wind-driven versus convection currents. Recent literature suggests that both mechanisms play a role (a detailed discussion of the issues raised here and the observational evidence is found in the review article of Kuhlbrodt *et al.* (2006). The MOC would thus be composed of both a thermohaline and a wind-driven component (**Fig. 8**).

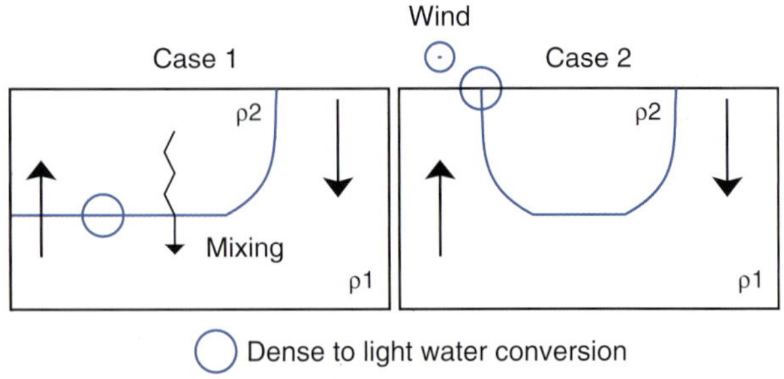

Figure 8 Schematic view of the two driving mechanisms of the MOC discussed in the text. (Left) Thermohaline circulation. (right) Wind-driven circulation through the 'Drake Passage' effect. From Kuhlbrodt, T., *et al.* (2006). On the driving processes of the oceanic meridional overturning circulation. Submitted for publication.

This view is supported by models: If wind forcing at the ocean surface is turned off in a standard climate model, eliminating the second mechanism, the MOC weakens somewhat but qualitatively remains the same. If interior mixing is reduced, weakening the first mechanism, then the MOC also weakens somewhat but then remains unchanged as one goes to extremely small mixing values. Thus, either mechanism is apparently capable of driving a MOC not unlike that observed. Determining which of them dominates in the real ocean requires further study.

It should be noted that these considerations apply to the mechanism for driving a long-term steady flow, not to the transient response to rapid thermohaline forcing such as a big meltwater inflow (discussed later). The mechanism for interrupting the flow by a density perturbation is different from the driving mechanism; the latter will only come into play on longer timescales.

Nonlinear Behavior of the THC

The highest surface densities in the oceans are reached where water is very cold, whereas lower densities are found in the saltier but warmer tropical and subtropical areas. The THC is thus thermally driven. Nevertheless, the influence of salinity is important and gives rise to an interesting nonlinearity of the system.

Defant's 1929 textbook discussed the fact that cooling and freshwater input have an opposing effect on density in the northern Atlantic and, hence, on the THC. This results in a bistability of the system, first described in a classic paper by Stommel (1961) with the help of a simple box model. Salinity is involved in a positive feedback: Higher salinity in the deep water formation area enhances the circulation, and the circulation in turn transports higher salinity waters into the deep water formation regions (because these are regions where precipitation exceeds evaporation, freshwater would accumulate and surface salinity would drop if the circulation stopped). In Stommel's model the high-latitude salinity increases linearly with the flow, and the flow increases linearly with high-latitude salinity, which combined results in a quadratic equation for the flow as function of freshwater input. This leads to two possible equilibrium states: The system is bistable in a certain parameter range.

The situation can be described with a simple stability diagram showing strength of the MOC as a function of the freshwater input into the North Atlantic. **Figure 9** shows the bistable regime and a bifurcation point *S*, where the circulation breaks down. If freshwater input to the northern Atlantic is systematically

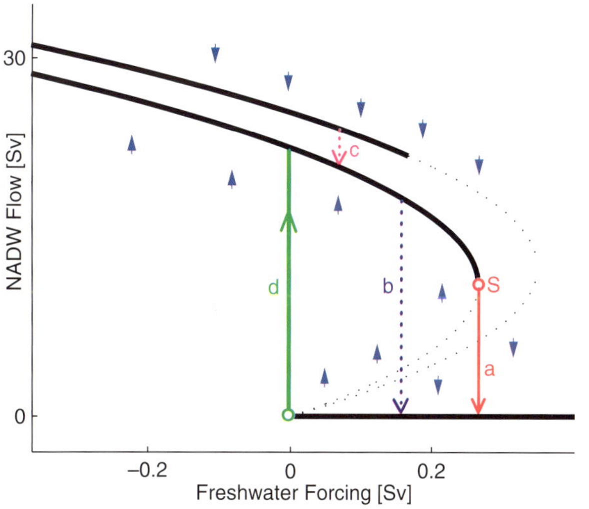

Figure 9 Schematic stability diagram for the Atlantic THC, with solid black lines indicating stable equilibrium states and dotted black lines unstable states. Transitions are indicated by arrows: (a) advective spindown, (b) convective shutdown, (c) transition between different convection patterns, and (d) restart of convection. 'S' marks the Stommel bifurcation beyond which no NADW formation can be sustained. Adapted from Rahmstorf, S. (2000). The thermohaline ocean circulation—A system with dangerous thresholds? *Climate Change* **46**, 247–256.

increased and decreased, a hysteresis loop between transitions *a* and *d* results. This hysteresis behavior appears to be a robust feature that has been found in many different climate models (**Fig. 10**) (Rahmstorf *et al.*, 2005). Although all models tested in this study show qualitatively similar hysteresis behavior, only some have their present-day climate residing in the bistable parameter regime. Whether the real climate system is bistable is unknown.

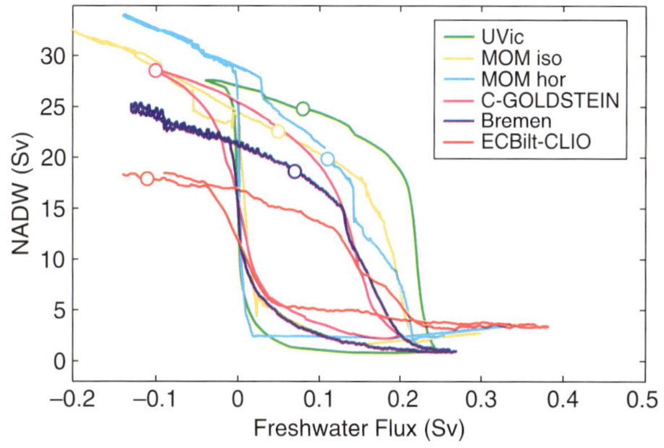

Figure 10 Hysteresis curves from a range of different intermediate-complexity climate models including three-dimensional ocean models. The curves are derived by slowly increasing the freshwater influx to the north Atlantic, then decreasing it again, from a present-day initial state of each model marked by an open circle. From Rahmstorf, S., *et al.* (2005). Thermohaline circulation hysteresis: A model intercomparison. *Geophysical Research Letters* **32**, L23605.

The salt transport feedback is not the only feedback rendering the system nonlinear. The convective mixing process is itself a highly nonlinear, self-sustaining process. In models this leads to multiple stable convection patterns. On the one hand, this can cause artifacts related to the coarse model grid; on the other hand, this may be part of a real mechanism for shifts in convection location, as have apparently occurred during glacial times.

The Circulation's Effect on Climate

The climatic effect of the MOC is due to its large heat transport of up to 1 PW in the North Atlantic. Back-of-the-envelope calculations suggest that this amount of heat transported into the northern North Atlantic (north of 24°N) should warm this region by ~5K. This is indeed approximately the difference between sea surface temperature in the North Atlantic compared to the North Pacific at similar latitudes. An examination of sea ice margins suggests that they are pushed back by the warm surface currents in the Atlantic sector compared to the North Pacific (**Fig. 3**), which in turn leads to reduced reflection of sunlight and thus warming (albedo feedback). Examination of observed surface air temperatures is suggestive (**Fig. 11**): Over the three main deep water formation regions of the global oceans, air temperatures are warmer by up to ~10 °C compared to the latitudinal mean. These observations, however, are no quantitative proof of the climatic effect of the THC, and other explanations can be partly invoked, such as planetary waves in the atmosphere that are locked in place by the geography.

One way to estimate the effect of the THC is to switch off NADW formation in coupled climate models (by adding a large freshwater anomaly to the northern Atlantic). Roughly, this leads to a cooling with a maximum of typically ~10K over the Nordic Seas (**Fig. 12**). The maximum tends to occur near the sea ice margin due to the positive ice albedo feedback, which causes air temperatures to cool much more than sea surface temperatures. The amplitude and spatial extent of the cooling are somewhat model dependent. Most models tend to cool temperatures over land in northwestern Europe (Scandinavia and Britain) by a few degrees, and some show strong cooling further west affecting Canada.

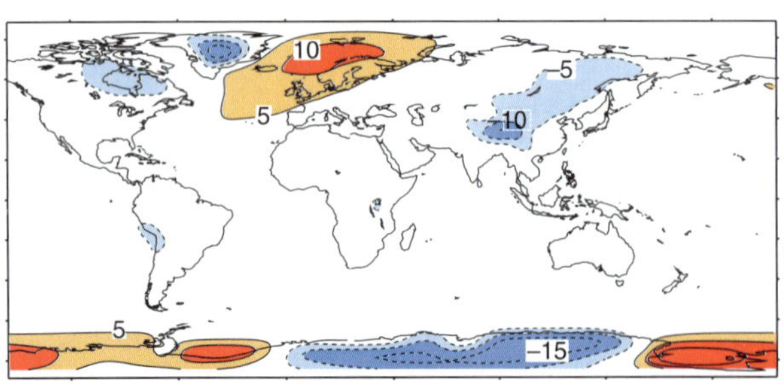

Figure 11 Deviation of surface air temperature from zonal mean (°C). From Rahmstorf, S. (2000). The thermohaline ocean circulation—A system with dangerous thresholds? *Climate Change* **46**, 247–256.

In models the Northern Hemisphere cools and the Southern Hemisphere warms if the THC is brought to a halt because the cross-equatorial heat transport in the ocean is reduced. This change in heat partitioning between the two hemispheres shifts the thermal equator to the south and thus the intertropical convergence zone (ITCZ) and the associated tropical rainfall belts. This is a robust phenomenon seen in models and consistent with paleoclimatic data.

Finally, a change in circulation also leads to a change in sea level. Currently, sea level in the northern Atlantic is almost 1 m lower than in comparable parts of the North Pacific because deep water forms only in the former. This is due to the geostrophic balance of surface currents and sea surface slopes. Any change in surface currents is associated with a rapid dynamic adjustment of the sea surface topography (with no effect on the global average sea level); in case of a shutdown of NADW formation, the sea level around the northern Atlantic would quickly rise by up to 1 m (**Fig. 13**). A further, slower and global sea-level response would result as the deep ocean temperature changes.

The Role of the THC in Quaternary Climate Changes

Sediment data document that the THC has undergone major changes in the Quaternary (Clark *et al.*, 2002; Rahmstorf, 2002). Three major circulation modes were identified: a warm mode similar to the present-day Atlantic, a cold mode with NADW forming south of Iceland in the Irminger Sea, and a switched-off mode without significant formation of NADW (**Fig. 14**). The latter appears to have occurred after major input of freshwater, either from surging glacial ice sheets (Heinrich events) or in the form of meltwater floods (e.g., Younger Dryas event). The most dramatic climate events recorded in Greenland, the Dansgaard–Oeschger (DO) events, could be explained by north–south shifts in convection location (i.e., transitions between warm and cold modes of the Atlantic THC). The coincidence of changes in ocean circulation and in surface climate, as well as the climate effects and stability properties of the THC discussed previously, suggest that ocean circulation changes can play a key dynamical role in

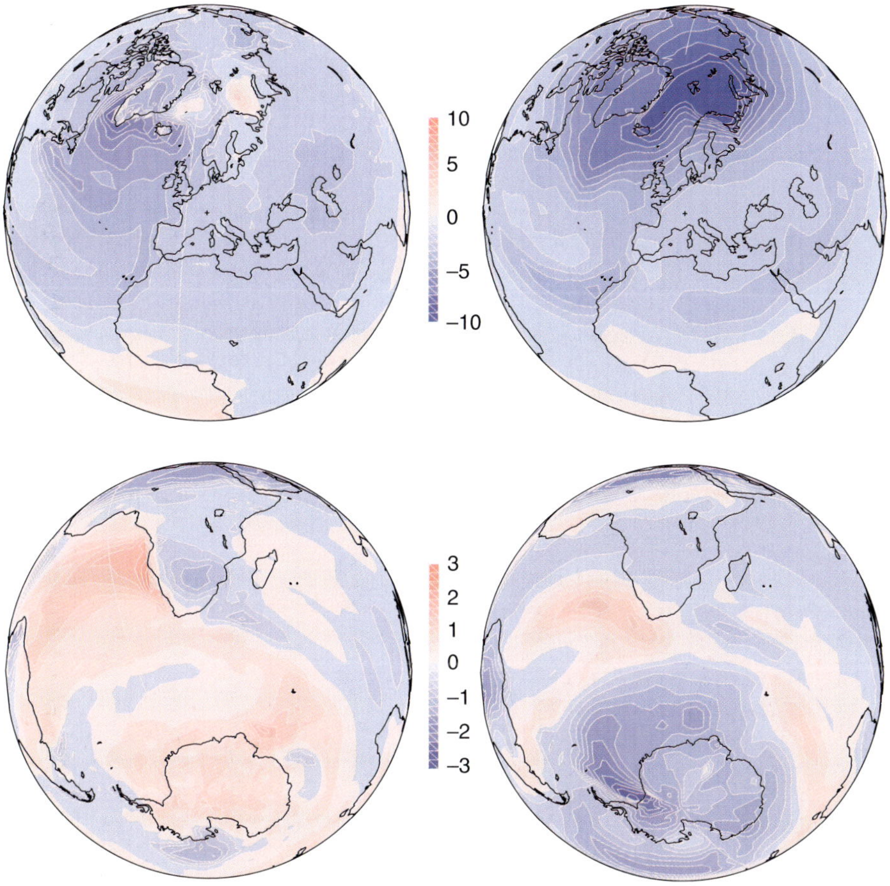

Figure 12 Change in surface air temperature after an induced shutdown of NADW formation in two different climate models. From Stocker, T. F. (2002). North–south connections. *Science* **297,** 1814–1815.

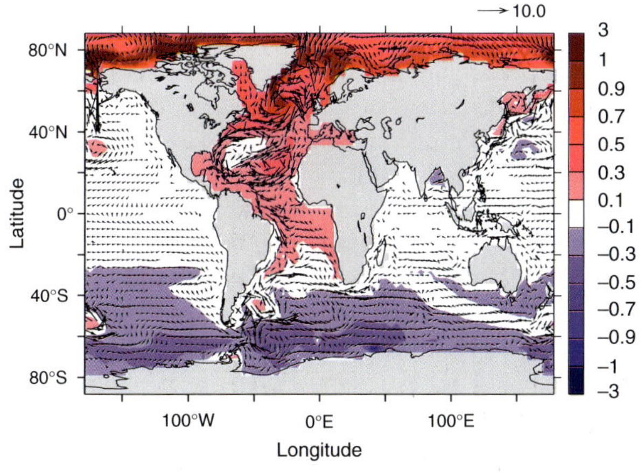

Figure 13 Dynamic sea-level change in a model experiment in which North Atlantic Deep Water formation was shut down. The global mean of this sea-level change is zero; any global mean sea-level changes resulting from the changed climate occur in addition to the shown redistribution of mass. From Levermann, A., *et al.* (2005). Dynamic sea level changes following changes in the thermohaline circulation. *Climate Dynamics* **24,** 347–354.

abrupt glacial climate change. Proxy data show that the South Atlantic cooled when the north warmed and vice versa, a see-saw of Northern and Southern Hemisphere temperatures that is indicative of an ocean heat transport change.

Heinrich events are thought to result from ice sheet instability: An episodic iceberg discharge would have provided a large freshwater addition of the order of 0.1 Sv to the Atlantic for several centuries (Roche *et al.*, 2004). Models suggest this could have been enough to stop all NADW formation due to the reduced density of ocean surface waters; this in turn would explain the cooling found in proxy data, especially around the midlatitude Atlantic. In Greenland, Heinrich events appear to have had little effect, presumably because they occurred during cold phases (cold mode, **Fig. 14**) when the THC already did not reach far enough north to warm Greenland climate.

DO events are dramatic warm events with a large amplitude in Greenland (8–16 °C within approximately a decade; Severinghaus *et al.*, 2003) and

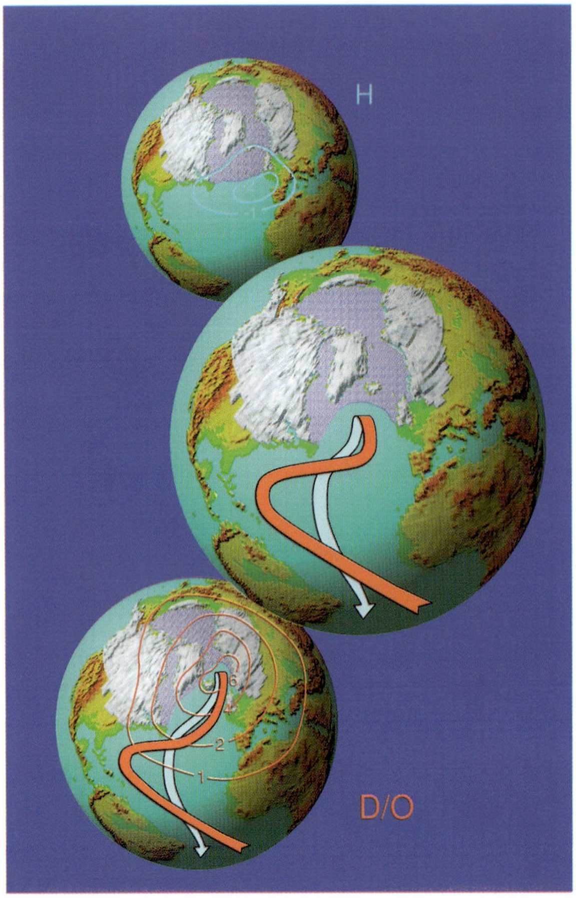

Figure 14 Schematic of three circulation modes of the glacial Atlantic: the prevailing cold mode (middle), the warm mode associated with Dansgaard–Oeschger events (bottom), and the 'off' mode occurring after Heinrich events (top).

areas surrounding the northern Atlantic. A large coinciding increase in salinity in the Irminger Sea suggests a northward push of warm, salty Atlantic waters into the Nordic Seas (Kreveld *et al.*, 2000), which could explain the large warming. What triggered these ocean circulation changes is unknown. Data analysis and modeling have suggested the intriguing possibility that stochastic resonance may have played a role (Alley *et al.*, 2001). This is a physical mechanism that can lead to threshold transitions in systems forced by a weak regular forcing (e.g., a solar cycle) and stochastic variability.

The role of ocean circulation changes on longer timescales (i.e., for the glacial–interglacial cycles) is still under discussion. Reconstructions of the rate of deep water formation and MOC during glacial times are still tentative but are consistent with prevailing roughly similar, perhaps somewhat weaker overturning rates compared to those in modern times (except during events) Models run for glacial conditions also suggest similar or somewhat weaker overturning rates (**Fig. 15**). (although a few indicate moderate strengthening). More important for climate than the rate of overturning is the latitude at which deep water forms. Data and some models indicate a southward shift of the main deep water formation sites from the Nordic Seas to a location south of Iceland, which implies a reduced ocean heat transport into high latitudes and an expansion of sea ice. Models suggest that these factors may have amplified glacial cooling, although ocean circulation changes are probably not a crucial factor for glacial inception.

During deglaciation, melting ice sheets surrounding the North Atlantic have apparently led to episodic influx of meltwater into the ocean. The effect of this meltwater on the THC may explain the oscillations and abrupt cold reversals seen, for example, in Greenland ice cores during this time, including the Younger Dryas and the 8k cold events. This contrasts with the much smoother history of deglaciation found in Antarctic ice cores.

The Future of the THC

Global warming can affect the THC in two ways: surface warming and surface freshening, both reducing the density of high-latitude surface waters and thus inhibiting deep water formation. Most models predict a significant weakening of NADW formation (by 20–50%) in response to anthropogenic global warming during the twenty-first century (Intergovernmental Panel on Climate Change, 2001). Some also find a reduction in AABW formation.

These models did not include meltwater runoff from the Greenland ice sheet, which is difficult to predict. A melting of the ice sheet during 1,000 years would imply an average meltwater influx of ~0.1 Sv, similar to that estimated for Heinrich events. Hence, the future evolution of the Atlantic THC is probably closely tied to the fate of the Greenland ice sheet.

Model simulations—even those that lead to a complete shutdown in the future—find that the influence of anthropogenic warming on the THC until today should be smaller than the natural variability. Therefore, any variations observed to this date are likely related to natural oscillations. However, observations indicate a widespread freshening trend in the northern Atlantic, which, should it continue, could contribute to a future weakening of the THC in the Atlantic (Curry and Mauritzen, 2005).

A major weakening or shutdown of NADW formation could have serious impacts on marine

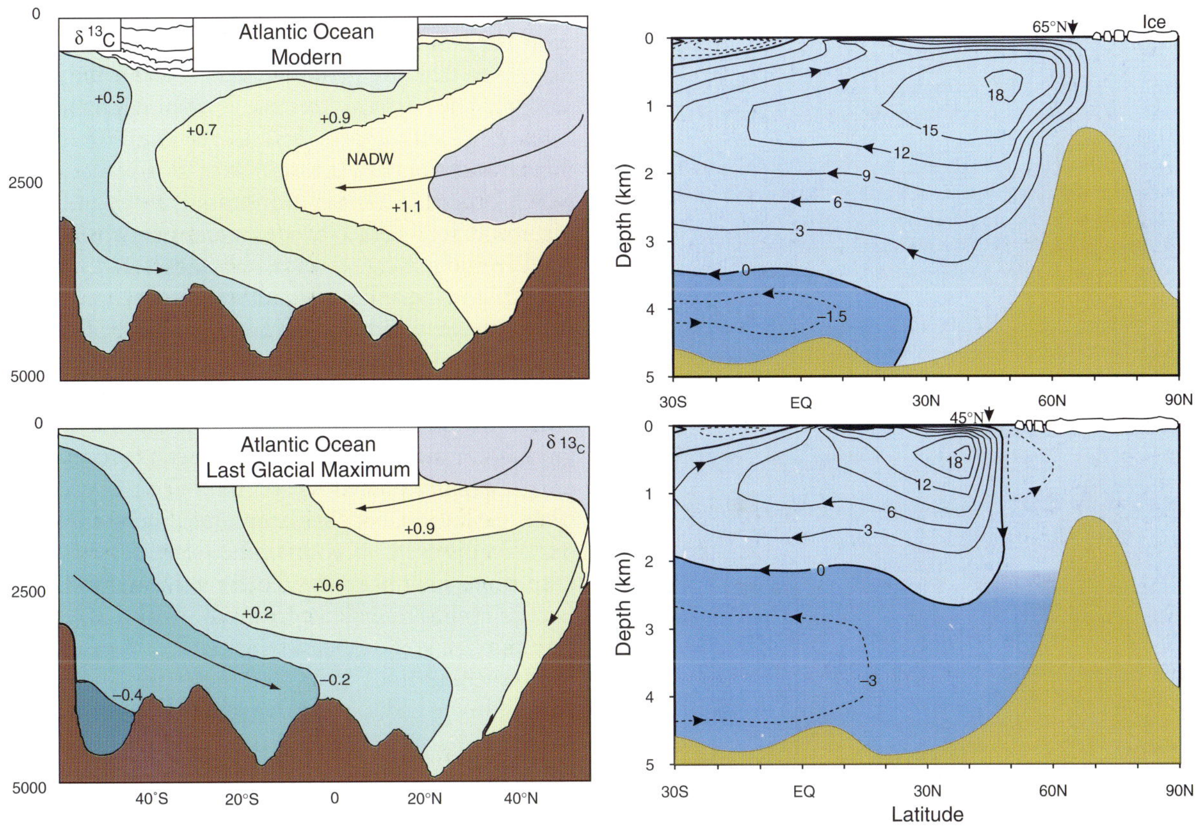

Figure 15 Circulation of modern and glacial Atlantic, as reconstructed by ^{13}C tracer data (left; after Duplessy, J.-C., *et al.* (1988). Deep water source variations during the last climatic cycle and their impact on the global deep water circulation. *Paleoceanography* **3**, 343–360) and from a model simulation (right; after Ganopolski, A., *et al.* (1998). Simulation of modern and glacial climates with a coupled global model of intermediate complexity. *Nature* **391**, 351–356). Note the shallower flow of NADW and the upward and northward expansion of AABW (darker blue) in glacial times, as well as the southward shift of deep water formation in this model.

ecosystems, sea level, and surface climate, including a shift in ITCZ and tropical rainfall belts. A THC collapse is widely discussed as one of a number of 'low probability–high impact' risks associated with global warming.

See also: **Introduction**: History of Recent Major Projects. **Glacial Landforms, Ice Sheets**: Growth and Decay. **Paleoclimate Reconstruction**: Sub-Milankovitch (DO/Heinrich) Events.

References

Alley, R. B., *et al.* (2001). Stochastic resonance in the North Atlantic. *Paleoceanography* **16**, 190–198.

Blindheim, J. (2004). Oceanography and climate. In *The Norwegian Sea Ecosystem* (H. R. Skjoldal, Ed.), Tapir Academic Press, Trondheim, Norway.

Broecker, W. S. (1991). The great ocean conveyor. *Oceanography* **4**, 79–89.

Clark, P. U., *et al.* (2002). The role of the thermohaline circulation in abrupt climate change. *Nature* **415**, 863–869.

Curry, R., and Mauritzen, C. (2005). Dilution of the northern North Atlantic Ocean in recent decades. *Science* **308**, 1772–1774.

Defant, A. (1929). *Dynamische Ozeanographie*. Springer, Berlin.

Duplessy, J.-C., *et al.* (1988). Deep water source variations during the last climatic cycle and their impact on the global deep water circulation. *Paleoceanography* **3**, 343–360.

Ganachaud, A., and Wunsch, C. (2000). Improved estimates of global ocean circulation, heat transport and mixing from hydrographic data. *Nature* **408**, 453–457.

Ganopolski, A., *et al.* (1998). Simulation of modern and glacial climates with a coupled global model of intermediate complexity. *Nature* **391**, 351–356.

Intergovernmental Panel on Climate Change (2001). *Climate Change 2001*. Cambridge University Press, Cambridge, UK.

Köhl, A., *et al.* (2006). Interannual to decadal changes in the ECCO global WOCE synthesis. Submitted for publication.

Kreveld, Sv., *et al.* (2000). Potential links between surging ice sheets, circulation changes, and the Dansgaard–Oeschger cycles in the Irminger Sea, 60–18 kyr. *Paleoceanography* **15**, 425–442.

Kuhlbrodt, T., *et al.* (2006). On the driving processes of the oceanic meridional overturning circulation. Submitted for publication.

Levermann, A., *et al.* (2005). Dynamic sea level changes following changes in the thermohaline circulation. *Climate Dynamics* **24**, 347–354.

Munk, W., and Wunsch, C. (1998). Abyssal recipes II: Energetics of wind and tidal mixing. *Deep-Sea Research* **45**, 1977–2010.

Rahmstorf, S. (2000). The thermohaline ocean circulation—A system with dangerous thresholds? *Climate Change* **46**, 247–256.

Rahmstorf, S. (2002). Ocean circulation and climate during the past 120,000 years. *Nature* **419**, 207–214.

Rahmstorf, S., *et al.* (2005). Thermohaline circulation hysteresis: A model intercomparison. *Geophysical Research Letters* **32,** L23605.
Roche, D., *et al.* (2004). Constraints on the duration and freshwater release of Heinrich event 4 through isotope modelling. *Nature* **432,** 379–382.
Roemmich, D. H., and Wunsch, C. (1985). Two transatlantic sections: Meridional circulation and heat flux in the subtropical North Atlantic Ocean. *Deep-Sea Research* **32,** 619–664.
Sandström, J. W. (1908). Dynamische Versuche mit Meerwasser. *Annalen der Hydrographie und Maritimen Meteorologie* **36,** 6–23.
Severinghaus, J. P., *et al.* (2003). A method for precise measurement of argon 40/36 and krypton/argon ratios in trapped air in polar ice with applications to past firn thickness and abrupt climate change in Greenland and at Siple Dome, Antarctica. *Geochimica et Cosmochimica Acta* **67,** 325–343.
Stocker, T. F. (2002). North–south connections. *Science* **297,** 1814–1815.
Stommel, H. (1961). Thermohaline convection with two stable regimes of flow. *Tellus* **13,** 224–230.
Talley, L. D. (1999). Some aspects of ocean heat transport by the shallow, intermediate and deep overturning circulations. In *Mechanisms of Global Climate Change at Millennial Time Scales* (P. U. Clark *et al.*, Eds.), pp. 1–22. American Geophysical Union, Washington, DC.
Talley, L. D., *et al.* (2003). Data-based meridional overturning stream functions for the global ocean. *Journal of Climatology* **16,** 3213–3226.
Toggweiler, J. R., and Samuels, B. (1995). Effect of Drake Passage on the global thermohaline circulation. *Deep-Sea Research* **42,** 477–500.
Tomczak, M., and Godfrey, J. S. (2003). *Regional Oceanography: An Introduction* 2nd ed. Available at www.es.flinders.edu.au/~mattom/regoc
Warren, B. A. (1981). Deep circulation of the world ocean. In *Evolution of Physical Oceanography* (B. A. Warren and C. Wunsch, Eds.), pp. 6–40. MIT Press, Cambridge.

Volcanic and Solar Forcing

D T Shindell, NASA Goddard Institute for Space Studies and Columbia University, NY, USA

© 2007 Elsevier B.V. All rights reserved.

Solar Variability

Prior to large-scale industrialization, climate forcing was primarily due to natural variations in the Sun's output and in volcanic eruptions for timescales up to several thousand years. The Sun also plays a role in orbital forcing, which is driven by changes in the relative position and orientation of the Earth with respect to the Sun (rather than changes in the output of the Sun). The systematic variations in the Earth's precession, obliquity, and eccentricity create three distinct cycles (Milankovitch cycles) that combine to determine the orbital forcing. Summer insolation reaching northern middle latitudes, which plays a major role in ice sheet growth and retreat, is highly correlated at long timescales (>20,000 years) with long-term climate including glacial–interglacial transitions. This orbital forcing is distinct from solar forcing, which refers instead to variations in the irradiance output from the Sun. Over long timescales for which orbital forcing is substantial, solar and volcanic forcing are thought to be comparatively much smaller and are therefore usually neglected.

Solar forcing takes place because the Sun's output varies on many timescales. Some of the most distinct are the 27-day rotation period of the Sun and the approximately 11-year solar cycle (or Schwabe cycle), also called the sunspot cycle and corresponding to half the 22-year cycle of the Sun's magnetic field. A longer cycle of approximately 80 years, called the Gleissberg cycle, modulates both the amplitude and the length of the sunspot cycle. It is difficult to measure the Sun's output precisely from the Earth's surface due to interference in the atmosphere from clouds and particulate, for example. Precise measurements have been obtained from satellites since 1979, and these clearly show variations up through the ~11-year cycle, over which output varies by approximately 0.1% (Fröhlich and Lean, 2004; Willson and Mordvinov, 2003). However, temporal gaps between instrumental records make estimates of longer term changes difficult.

It is clear from the satellite observations that irradiance variations are not uniform across the spectrum, with variations over the ~11-year solar cycle increasing from slightly less than 0.1% in the visible to 1–10% in the ultraviolet (UV) and shorter wavelengths. Although the Sun's output at these wavelengths is much less than in the visible, the larger percentage changes still imply that a substantial fraction of the integrated energy variations (~15–20%) takes place at these wavelengths. This radiation is largely absorbed by oxygen and ozone before reaching the Earth's surface, or even the troposphere. However, it has the potential to alter the climate and composition of the stratosphere and above. In particular, UV radiation can produce ozone when it is absorbed by an oxygen molecule in the stratosphere so that increased UV irradiance leads to an increase in stratospheric ozone. This phenomenon has been recorded in both satellite and ground-based ozone data. The altered ozone as well as the altered irradiance both affect stratospheric temperatures, primarily in sunlit regions, leading to changes in circulation that could in principle affect surface climate. Thus, it is important to know not only how total solar irradiance has varied but also how that variability has depended on wavelength. Unfortunately, even less information is available about past variations in the solar spectrum than the limited 1979–present total irradiance data set, and proxies offer little promise of reconstructing the spectral variation of past irradiance changes.

Some longer term data that appear to be related to total solar output are available. Sunspots records extend back to the 17th century, and the modern satellite data confirm that sunspots are highly correlated with irradiance over ~11-year solar cycles. The correlation is an inverse one because the dark sunspots are accompanied by bright regions called faculae during periods of high solar activity, and the faculae exert a stronger influence on total solar output. However, exactly how the sunspot number, a value related to both the number of individual spots and the number of sunspot groups, is related to solar output is unclear. Even more confounding is that there is no good evidence as to whether additional changes in solar output that are not reflected in sunspots took place in periods such as the Maunder Minimum from approximately 1650 to 1710, during which very few sunspots were observed. If the Sun was simply in an extended minimum comparable to that seen in the ~11-year solar cycle, then the reduction in irradiance relative to the mean value in typical cycling conditions would be only one-half of one solar cycle's variations, or approximately 0.05%. Many estimates during the past several decades have put the value between several times to more than an order of magnitude greater than this under the assumption that other long-term 'background' changes also took place (Lean, 2000). These were often based on astronomical comparisons with other 'Sun-like' stars, which are now regarded as fairly poor analogs to our own Sun, and recent estimates tend to find substantially reduced long-term solar variations (e.g., Foukal *et al.*, 2004).

Other proxy data have also been used to infer past solar output. The cosmogenic proxies ^{10}Be from ice cores and ^{14}C from tree rings have both been used to reconstruct historical irradiance (Stuiver and Braziunas, 1989). Both isotopes are produced by galactic cosmic rays striking the Earth's atmosphere. The flux of these rays reaching the Earth is modulated by the strength of the solar magnetic field, causing cosmogenic isotope production to vary inversely with solar output. Records of both isotopes show periodicities at many scales, and ^{10}Be, the higher resolution of the two, captures the ~11-year solar cycle reasonably well. Both proxies show periods of extensive lows, of which the Wolf (~1280–1350), Sporer (~1450–1540), and Maunder Minima (~1645–1715) are the most prominent during the past millennium. However, an absolute calibration of these proxies remains elusive because neither the precise relationship between solar output and the solar magnetic field (and hence isotope production) nor the confounding influence of other factors, such as climate change altering the transport of proxies and their deposition onto ice sheets, are well characterized. Nevertheless, the cosmogenic proxies appear to be reasonably good indicators of when solar irradiance varied despite their limited use in determining the magnitude of such variations. It has been suggested that these proxies show that the Sun is now at its brightest level in many millennia, but further research has demonstrated that this inference appears to result from a flawed calibration technique used in the initial analysis. Thus, the various proxy data sets do not provide a reliable estimate of solar variability during the Quaternary. From the solar physics side, although there has been much progress in helioseismometry (the study of the internal oscillations of the Sun) in recent decades, solar models are still in the early stages of development. Although they have made some notable progress (Wang *et al.*, 2005), they are not able to reproduce some of the known features of the solar dynamo or to spontaneously enter a noncycling state such as the Maunder Minimum. Thus, it is not possible to reliably estimate the precise magnitude of past solar irradiance changes, even for the past two decades, although the time dependence of these changes is known fairly accurately.

Volcanism

The other main natural forcing of climate at timescales of thousands of years or fewer is volcanism. As with solar variability, modern measurements of the optical properties of volcanic aerosols injected into the atmosphere are only available for the past several decades, with the 1991 eruptions of Mount Pinatubo in the Philippines being by far the best observed. However, reliable optical data on volcanic aerosols (particulate) injected into the upper atmosphere extend back considerably farther than for solar irradiance (Sato *et al.*, 1993) because the observations are technically much simpler due to a much larger signal-to-noise ratio. Prior to the direct optical measurements, a range of historical, geological, and proxy data are used to estimate explosive volcanism. Documentary evidence records the location of large eruptions for the past several centuries. Geologic evidence allows vulcanologists to estimate the explosivity of eruptions, from which the volume of material ejected into the atmosphere can be estimated. One of the most useful techniques is using particulates found in polar ice cores to reconstruct past volcanism. The technique relies on the presence of ash layers in ice cores from the same years in both Greenland and Antarctica, which then implies an eruption with a worldwide impact (Crowley,

2000). In practice, one is more confident if the ash layer can be related to a large, known eruption so that the coincidental eruption of two high-latitude volcanoes near the ice core locations can be ruled out. Since Greenland receives ash from nearby Iceland quite frequently, such a coincidence is not as unlikely as it might seem. Chemical analysis of the ash could also definitively link the ash layers from the two poles, but it is expensive and difficult. Most important, the thickness of the ash layer must then be used to estimate the global impact of the volcanic aerosols on the Earth's radiation budget (the forcings), an inherently challenging task with substantial uncertainties. Thus, as with solar forcing, the timing of past volcanic forcing is much better known than its magnitude. The records derived from ice cores can be calibrated against modern optical measurements during the period of overlap, however, giving us reasonable confidence in the magnitude of volcanic forcing for at least the past several centuries.

To affect climate, eruptions typically have to inject material into the tropical stratosphere, from which it can be distributed worldwide within a year or two at most. Eruptions that deposit material only in the troposphere usually do not noticeably perturb the global mean climate because the material affects only a portion of the globe and is rapidly removed by the hydrologic cycle. Volcanic aerosols from eruptions that inject material into the stratosphere but occur poleward of the two subtropical jet streams located at approximately 30° latitude, typically spread out toward the closer pole but not into the tropics or the other hemisphere, and thus seem to have much less global impact. Volcanic aerosols are removed from the stratosphere with an exponential decay time of approximately 1½ to 2 years so that a large tropical eruption typically affects climate for 1–3 years. Catalogs of known eruptions now document thousands of events (Simkin and Siebert, 1994). The largest eruption in the past 500 years was that of Tambora in Indonesia during 1815, which caused the famous 'year without a summer' the following year. The largest eruption during the past millennium took place in 1259, but its location has not been determined. In the more distant past, super-eruptions such as the Toba eruption ~75,000 years ago (in Indonesia) may have had a major impact on climate. Proxy evidence for the period following Toba's eruption, for example, suggests that it coincided with a cooling epoch lasting ~1,000 years, although the connection with volcanism is not firmly established. Similarly, extended periods of enhanced volcanic emissions such as those that created the Deccan or Siberian Traps could have induced climate change by increasing tropospheric aerosols for millennia.

Climate Modeling

The climate impact of solar and volcanic forcing can be investigated by a combination of data analysis and modeling studies. Given an estimate of either solar or volcanic forcing—that is, how the radiative balance of the Earth was changed—a climate model can be used to examine the magnitude, pattern, and timing of the climate response and to explore the underlying physical mechanisms that drive it. These models, of varying complexity, include mathematical representation of features of the climate system, such as the land, oceans, atmosphere, and cryosphere, and the physical, chemical, and biological processes in those regions that affect climate. Energy balance models (EBMs) compute the surface temperature response using a one-dimensional column calculation of changes in incoming and outgoing radiation in response to external forcing. Although simple, these models can tell us much about the global mean annual average response (Crowley, 2000). It is clear from these studies, for example, that volcanic eruptions are the primary cause of externally forced interannual variations in global mean annual average surface temperature. To fully understand the response of the climate system, however, requires the use of a three-dimensional general circulation model (GCM). These models generate their own inherent variability due to nonlinear chaotic interactions in the highly coupled climate system, which are referred to as internal or unforced variability. This unforced variability provides a baseline against which the response to forcings from outside the climate system, such as solar variations, volcanic eruptions, or greenhouse gases, can be compared. Investigations with these models provide information on the spatial pattern and timing of the climate response to external forcing as well as its magnitude. Studies with state-of-the-art GCMs are limited by available computer power but have advanced dramatically during recent decades. Simulations with GCMs can be compared with observations or proxy data for a wide range of spatial and temporal scales, and the model response can be analyzed to determine the most important processes at work in a particular case. Such comparisons are discussed next.

Proxy Data

The role of solar and volcanic forcing in climate change during the Holocene has been studied more than any other period during the Quaternary. One reason for this is that climate data covering much of the globe are more widely available, especially for the past 1,000 years. This allows the proxy data to be compared with estimates of past solar and volcanic forcing, and correlations between the proxy climate records and the forcings can then be compared with the results from climate models used to examine the response to these forcings. Proxies with high temporal resolution, typically from decadal to annual or even seasonal resolution, are generally best for comparison with model simulations. Comparisons for the past millennium can be made with reconstructions of surface temperatures based on networks of proxy data. Of these, tree rings are the most abundant proxy. The width and density or tree rings are related to summer temperatures, and these records can be accurately dated to individual years. Ice core oxygen-18 isotopic ratios also serve as a temperature proxy. These extend back to nearly 1 million years ago, making them one of the oldest high-resolution proxies. In sites with high snow accumulation rates, annual temperatures can be retrieved. The limitations of this proxy are primarily that ice cores are restricted largely to the polar regions, and assigning temperatures to calendar years has significant uncertainty. Coral growth rates help fill in the spatial gaps of the ice core records by providing information from the tropics. Documentary evidence is another useful source for the past millennium. Like tree rings, it can be matched exactly with calendar years; however, it is restricted primarily to Europe and East Asia. Annual layers of lake sediments (varves) and annual accumulation in stalagmites (speleotherms) are additional proxies. Finally, deep ocean sediment cores extend farther back in time than any other source, although their interpretation is less straightforward and dating is more uncertain than for the other sources. Additionally, because the ocean integrates transport, temperature, and salinity changes over large spatial and temporal scales, these data typically cannot provide information about the response to short-term perturbations such as volcanism. During the past millennium, data are plentiful enough that multiple data sets can be combined, allowing reconstructions to be developed for this period with confidence in both the global or hemispheric average temperature changes and the spatial patterns (Jones and Mann, 2004). This is not the case further back in time. However, additional proxy records such as pollen can give a qualitative sense of ancient warm or cool periods over middle latitude continental areas that compliments the high-latitude ice core and the oceanic sediment core records.

Model/Data Comparisons

The proxy temperatures clearly show the effects of large eruptions such as Huaynaputina in 1600 or Parker in 1641 (**Fig. 1**), or the later large eruptions of Tambora in 1815 or Krakatua in 1883. In general, these are consistent with the direct observations of the response to recent eruptions such as Pinatubo in 1991. GCM simulations have reproduced the magnitude, spatial pattern, and timing of surface temperature changes in response to the Pinatubo eruption (Robock, 2000; Shindell *et al.*, 2004). Analysis of a large number of eruptions during the past millennium, which improves the statistical reliability of the results, shows a distinct pattern of Northern Hemisphere wintertime warming over the northern continents and cooling over the Labrador Sea region and in the Middle East (Robock, 2000; Shindell *et al.*, 2004). This pattern represents a shift toward the positive phase of the dominant naturally occurring

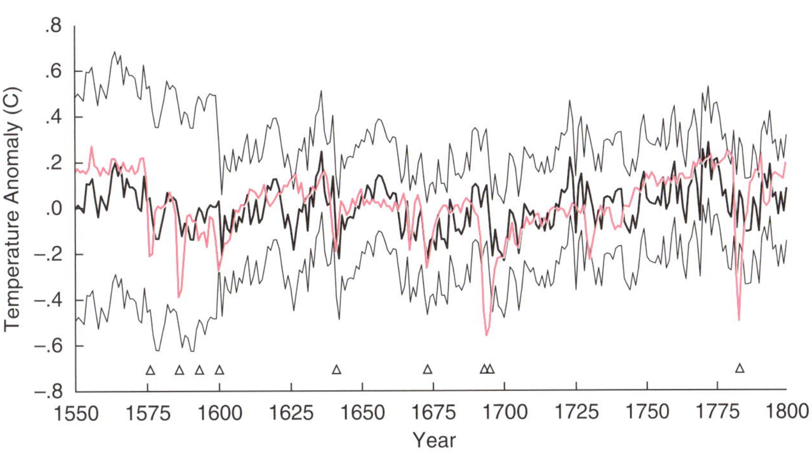

Figure 1 Comparisons between Northern Hemisphere surface temperature anomalies from a proxy reconstruction and a GCM. The proxy reconstruction (Mann *et al.*, 1999) (thick black line) includes an estimate of uncertainty (thin lines bracketing the thick black line), whereas the model results show the mean of six model runs (purple). Large volcanic eruptions included in forcing of the model are marked with triangles, and the response to some of these events is clearly seen in both the model and the proxy data. Adapted from Schmidt *et al.* (2004).

mode of Northern Hemisphere extratropical variability, the large-scale pattern known as either the Northern Annual Mode (NAM) or the Arctic Oscillation (AO). This pattern is closely related to the more regional North Atlantic Oscillation so that in the Atlantic sector they are extremely similar. The hemispheric-scale NAM/AO pattern is visible from the surface up into the stratosphere, however, making it a logical choice for describing the response to solar and volcanic forcing because these have their greatest direct impact on the stratosphere. The relatively good agreement between models and observations in the modulation of the NAM/AO and the resulting Northern Hemisphere continental winter warming following large tropical eruptions is shown in **Figure 2**. The model result shown in the figure is typical of current GCMs that incorporate the stratosphere (where the volcanic aerosols are injected, but a region often left out of climate models used mainly for other types of studies). It shows the best match between the model and the historical data over Eurasia and the Labrador Sea region, both of which are locations where variations in the NAM/AO have a strong influence on surface temperatures. The response over North America is more difficult to simulate or to determine from data because several of the largest recent eruptions coincided with a strong positive phase in the El Niño/Southern Oscillation (ENSO), which has a strong impact in this region. It is a subject of debate as to whether there is a causal link between volcanic eruptions and ENSO events. The winter continental warming response persists with a similar spatial structure into the second winter following large tropical eruptions but has largely faded away by the third.

Proxy data generally show cooler hemispheric-scale temperatures during solar minima, although the magnitude varies between different reconstructions. EBMs can reproduce the general changes, but given that the magnitude of the changes and the solar forcing are both uncertain, this is not a powerful constraint. GCMs face the same uncertainties in matching global or hemispheric changes. There are two primary ways around this problem, namely examining the time dependence of climate change in comparison with forcings or examining the spatial pattern of the climate response. In both cases, the importance of uncertainties in the exact magnitude of forcing and response is minimized.

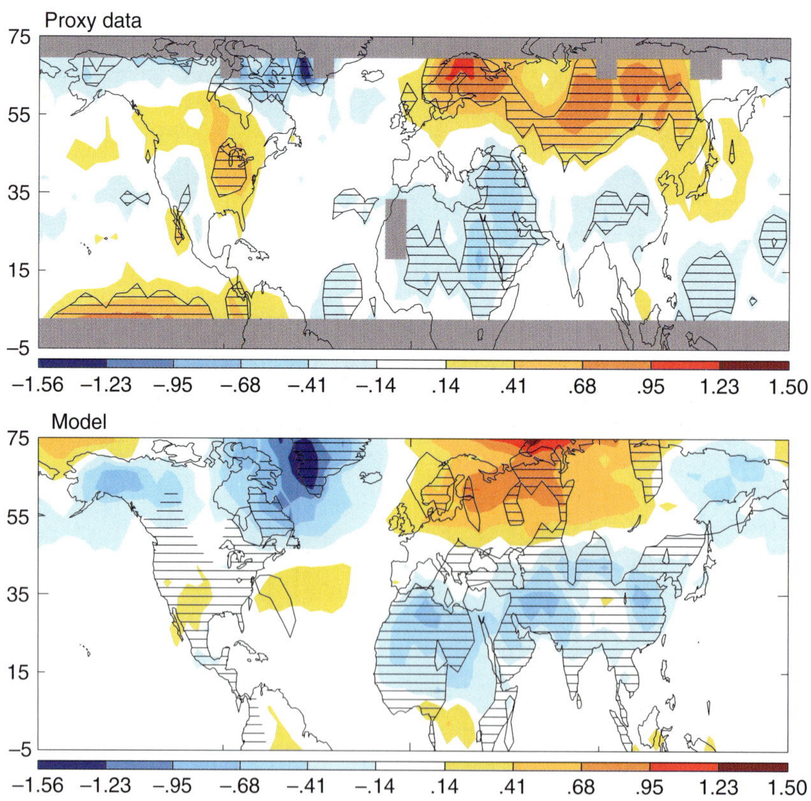

Figure 2 Surface temperature anomalies (°C) during the cold season (October–March) following large tropical volcanic eruptions from proxy reconstructions (top) and in a GCM (bottom). The proxy data are averages of the response to 11 large, well-identified historical eruptions. The model results are the mean cold-season response to 15 simulated eruptions with similar magnitudes to those in the historical analysis. Hatching indicates areas where the response is significant at the 95% confidence level. Modified from Shindell *et al.* (2004).

Climate modeling at the NASA Goddard Institute for Space Studies (GISS) examined the global and regional response to volcanic eruptions and solar variability using the GISS climate model. The results were then compared with reconstructions of surface temperatures from historical times, focusing on the change between the late 17th and 18th centuries. The late 17th century, coinciding with the Maunder Minimum, was one of the coldest parts of the period known as the 'Little Ice Age,' when hemispheric average temperatures dropped by a few tenths of a degree. Both reduced solar output and numerous large volcanic eruptions could have contributed to the hemisphere-wide cooling. Regional changes were much larger, however. Europe and eastern North America, for example, were unusually cold, whereas some other areas appear to have been relatively warm.

In the climate model, both the reduced output from the Sun and the increased amount of the volcanic aerosol particles that reflect sunlight away from the Earth contributed a sizeable amount to the Northern Hemisphere cooling seen in temperature reconstructions (Schmidt *et al.*, 2004; Shindell *et al.*, 2003). Their relative importance depends on uncertain estimates of exactly how much each changed relative to a century later. The global or hemispheric average GCM values are quite similar to those obtained in EBM studies (Crowley, 2000).

Regionally, however, solar forcing appears to have played the dominant role in driving annual average temperature changes over periods longer than a year or two (**Fig. 3**). Multiple factors account for this difference. Volcanoes have a major effect on circulation of the atmosphere for a year or two, as discussed previously, but the aerosols fall fairly rapidly out of the atmosphere, resulting in little long-term effect. Furthermore, the effect during the first and second years following eruptions is to increase westerly airflow around the Northern Hemisphere during winter, when circulation responds most strongly to the latitudinal temperature gradients induced by the volcanic aerosols in the lower stratosphere. This enhanced flow brings relatively warm oceanic air over the continents, leading to the continental winter warming shown in **Figure 2**. In contrast, during the summer, volcanic aerosols lead to cooling via their reflection of sunlight, and this is strongest over the continents, and during summer the volcanic aerosols induce little change in circulation. The regional effects of eruptions during the cold and warm seasons thus tend to cancel one another, leading to a fairly uniform annual average response (**Fig. 3**).

In contrast, the effects of reduced solar output are sustained, lasting for decades, and they also have similar impacts in all seasons. Decreases in solar output lead to reduced heating in the sunlit portion of the stratosphere due to the reduced UV radiation and reduced stratospheric ozone, and ensuing decrease in the meridional gradient of temperature causes decreased wintertime westerly wind flow. This results in less advection of relatively warm maritime air and hence colder Northern Hemisphere continental temperatures (the opposite of the effect following a large tropical eruption) (Shindell *et al.*, 2001). This cooling is reinforced by cooler temperatures everywhere in the summer due to the reduced solar energy, but the land, having a lower heat storage capacity, responds more than the oceans. Thus, during both seasons the land

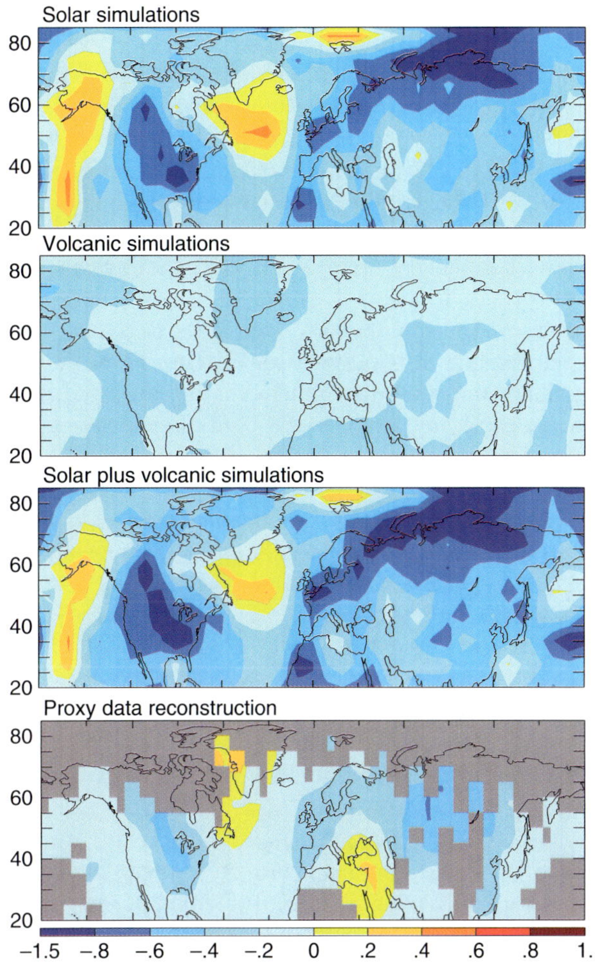

Figure 3 Annual average surface temperature response (°C) to decreased solar irradiance (top), substantial volcanic activity (second row), and both forcings (third row) in a GCM. The solar output reduction used in these simulations with the GISS GCM was approximately $-0.3\,W/m^2$, whereas the mean volcanic forcing was approximately $-0.4\,W/m^2$. Details of the simulations can be found in Shindell *et al.* (2003). For comparison, the difference between proxy temperature reconstructions averaged over 1680–1710 relative to 1780–1800 is also shown (bottom row). The 1680–1710 period, during the late Maunder Minimum, was a time with both reduced solar irradiance and enhanced volcanic activity relative to a century later.

cools in response to reduced solar irradiance, leading to a large regional annual average response.

In the GCM simulations, the annual average response to solar forcing resembles the observations much more closely than the response to volcanic forcing (**Fig. 3**). The response is overall too strong, however, which may indicate an overestimate of how much the Sun dimmed during the late 1600s, an oversensitive model, or both. Nevertheless, it is clear that solar variability seems much more likely to have driven the large regional climate changes seen in historical data. This provides both a useful test of the ability of climate models to simulate patterns of regional changes and important evidence for the importance of solar variability in climate change.

Other modeling studies have demonstrated that the variability in global or hemispheric temperatures induced by volcanic eruptions is fairly well correlated with interannual variations in the proxy reconstructions. However, at timescales greater than a century, volcanic eruptions are only very weakly correlated with the proxy data, whereas solar irradiance variations are highly correlated. This supports the conclusion that solar changes drive much of the multidecadal and longer variations in Quaternary climate, whereas volcanism is most important at shorter timescales. Further evidence for the role of the Sun in pacing millennial timescale variations in climate comes from records of ocean sediments (Bond *et al.*, 2001) and lake sediments (Hu *et al.*, 2003). These show significant correlations between climate proxy indicators and the solar-related cosmogenic proxies. However, the precise relationship between some of these proxies and climate is unclear, and an understanding of the physical mechanisms underlying such millennial scale interactions between the Sun and climate is lacking.

See also: **Glaciation, Causes**: Milankovitch Theory and Paleoclimate. **Paleoclimate Modeling**: Data-Model Comparisons; Last Glacial Maximum GCMs. **Paleoclimate Reconstruction**: The Last Millennium. **Paleoclimate Relevance to Global Warming**. **Paleoclimate**: Introduction.

References

Bond, G., *et al.* (2001). Persistent solar influence on North Atlantic climate during the Holocene. *Science* **294**, 2130–2136.

Crowley, T. J. (2000). Causes of climate change over the past 1000 years. *Science* **289**, 270–277.

Foukal, P., North, G., and Wigley, T. (2004). A stellar view on solar variations and climate. *Science* **306**, 68–69.

Fröhlich, C., and Lean, J. (2004). Solar radiative output and its variability: Evidence and mechanisms. *Astronomy and Astrophysics Review* **12**, 273–320.

Hu, F. S., *et al.* (2003). Cyclic variation and solar forcing of Holocene climate in the Alaskan subarctic. *Science* **301**, 1890–1893.

Jones, P. D., and Mann, M. E. (2004). Climate over past millennia. *Reviews of Geophysics* **42**, RG2002.

Lean, J. (2000). Evolution of the sun's spectral irradiance since the Maunder Minimum. *Geophysical Research Letters* **27**, 2425–2428.

Mann, M. E., Bradley, R. S., and Hughes, M. K. (1999). Northern Hemisphere temperatures during the past millennium: Inferences, uncertainties, and limitations. *Geophysical Research Letters* **26**, 759–762.

Robock, A. (2000). Volcanic eruptions and climate. *Reviews of Geophysics* **38**, 191–219.

Sato, M., Hansen, J. E., McCormick, M. P., and Pollack, J. B. (1993). Stratospheric aerosol optical depths, 1850–1990. *Journal of Geophysical Research* **98**, 22987–22994.

Schmidt, G. A., Shindell, D. T., Miller, R. L., Mann, M. E., and Rind, D. (2004). General circulation modelling of Holocene climate variability. *Quaternary Science Review* **23**, 2167–2181.

Shindell, D. T., Schmidt, G. A., Mann, M. E., Rind, D., and Waple, A. (2001). Solar forcing of regional climate change during the Maunder Minimum. *Science* **294**, 2149–2152.

Shindell, D. T., Schmidt, G. A., Miller, R. L., and Mann, M. E. (2003). Volcanic and solar forcing of climate change during the Preindustrial Era. *Journal of Climatology* **16**, 4094–4107.

Shindell, D. T., Schmidt, G. A., Mann, M. E., and Faluvegi, G. (2004). Dynamic winter climate response to large tropical volcanic eruptions since 1600. *Journal of Geophysical Research* **109**, D05104.

Simkin, T., and Siebert, L. (1994). *Volcanoes of the World: A Regional Directory, Gazetteer, and Chronology of Volcanism during the Last 10,000 Years*, 2nd ed. Geoscience Press, Tuscon, AZ.

Stuiver, M., and Braziunas, T. F. (1989). Atmospheric C-14 and century-scale solar oscillations. *Nature* **338**, 405–408.

Wang, Y. M., Lean, J., and Sheeley, N. R. (2005). Modeling the sun's magnetic field and irradiance since 1713. *Astrophysical Journal* **625**, 522–538.

Willson, R. C., and Mordvinov, A. V. (2003). Secular total solar irradiance trend during solar cycles 21–23. *Geophysical Research Letters* **30**, 3–6.

GLACIAL LANDFORMS

Contents

Introduction

D J A Evans, Durham University, Durham, UK
D I Benn, University of St. Andrews, St. Andrews, UK

© 2007 Elsevier B.V. All rights reserved.

Introduction

The holistic assessment of the character and distribution of glacial sediments and landforms over whole drainage basins or even whole continents facilitates the reconstruction of glacier and ice-sheet dynamics over complete glacial cycles. The accuracy of such paleoglaciological reconstructions relies on analogs with modern examples. Increasing numbers and varieties of modern glacial landsystems are emerging in the research literature based upon intensive studies of glacier processes and forms, and these can be utilized by Quaternary scientists interested in paleoenvironmental reconstruction but not specializing in glacial geology (*see* Glacial Land Systems). Although paleoglaciological reconstructions are often difficult and sometimes controversial, significant developments in interpreting ancient glaciated landscapes have highlighted complex but coherent patterns of glacier activity over much of the last glacial cycle, and this is now being increasingly linked with regional and global climatic and oceanographic changes.

Information on specific glacial processes and their products are reviewed in articles Growth and Decay, Evidence of Glacier and Ice Sheet Extent, Evidence of Glacier Recession, Paleo ELAs, Glacitectonic Structures and Landforms, Glacigenic Lithofacies, Moraine forms and Genesis, Glacifluvial Landforms of Erosion, Glacifluvial Landforms of Erosion, Glacial Erratics and Till Dispersal Indicators, Glacimarine Sediments and Ice-Rafted Debris, Tills, Till Fabric Analysis, Clast Form Analysis, Micromorphology of Glacial Sediments, Glacial Land Systems, Micro to Macro Scale Forms, Major Scale Forms, Trimlines and Palaeonunataks and Quaternary Vulcanism, Subglacial Landforms, but we here focus on basin-wide glacial depositional environments, incorporating interrelated process-form domains in sediment–landform associations that are characteristic of particular topographic and environmental settings or landsystems (*see* Glacial Land Systems).

The Subglacial Environment

Ice-Sheet and Glacier Beds

Ice-sheet and glacier beds, rather than being areas devoid of long-term geomorphic records, are actually rich in evidence of ice buildup as well as recession. Large parts of the areas occupied by the former Pleistocene ice sheets are, in effect, fossil glacier beds, providing a window onto subglacial processes and environments which are only poorly known from modern examples. Regional studies of former glacier beds have shown that the great ice sheets were highly dynamic entities with complex histories of advance and retreat, ice-divide migration, and internal dynamics, and that they shifted their configuration to adapt to climatic change, the position and depth of proglacial water bodies, and basal conditions (e.g., Dyke and Prest, (1987), Boulton and Clark (1990), and Clark (1993, 1999)). Even former ice streams have been identified in the subglacial bedform record (Clark and Stokes, 2003; Stokes *et al.*, 2005).

With respect to the sedimentary record on former glacier beds, large-scale patterns of ice flow and subglacial conditions may be preserved in the form of distinct till units, glacitectonic deformation, and surfaces of erosion or nondeposition. Assemblages of subglacial deposits and landforms created during the advance and retreat of an ice sheet are typically very complex, and often defy analysis by the traditional methods of sedimentary stratigraphy. The products of several glacial advances may be superimposed, and glacitectonic dislocation and subglacial deformation often rework all or part of any pre-existing glacial and nonglacial strata. Due to partial erosion or patchy deposition, tills relating to any one glaciation can be discontinuous so that individual events are only recognizable through the type, style, and orientation of glacitectonic structures.

Theoretical patterns of subglacial deposition and erosion created during an ice-sheet advance and retreat cycle are presented by Boulton (1996a). His model assumes that debris transport below a simple ice sheet occurs entirely by subsole deformation at rates scaled to the balance velocity of the ice sheet. Although this highly simplified model does not deal explicitly with other important subglacial processes such as the erosion of hard substrata, the entrainment of debris into basal ice layers, and deposition by lodgement and meltout, it does provide interesting insights into large-scale time-transgressive patterns of erosion and deposition (**Fig. 1**). The model predicts an outer zone of deposition near the ice-sheet margin and an inner zone of erosion, which gives way to a zone of no erosion below the ice divide. The history of deposition and erosion at any given site reflects the migration of the depositional and erosional zones during ice-sheet advance and retreat. At each site, deposition will occur during ice-sheet advance. The resulting till will be removed from most sites by the subsequent erosion phase, although it will be preserved near the ice-sheet limit. As the ice sheet retreats, the zone of deposition passes over each site in turn, depositing a retreat-phase till. After final deglaciation, the overall pattern of deposition and erosion comprises four major zones: Zone 1 – the ice-divide zone, with slight erosion due to ice-divide migration and a thin till deposited during the retreat phase (sites E and F); Zone 2 – a zone of strong erosion, in which the advance-phase till and part of the preglacial substratum are removed (the

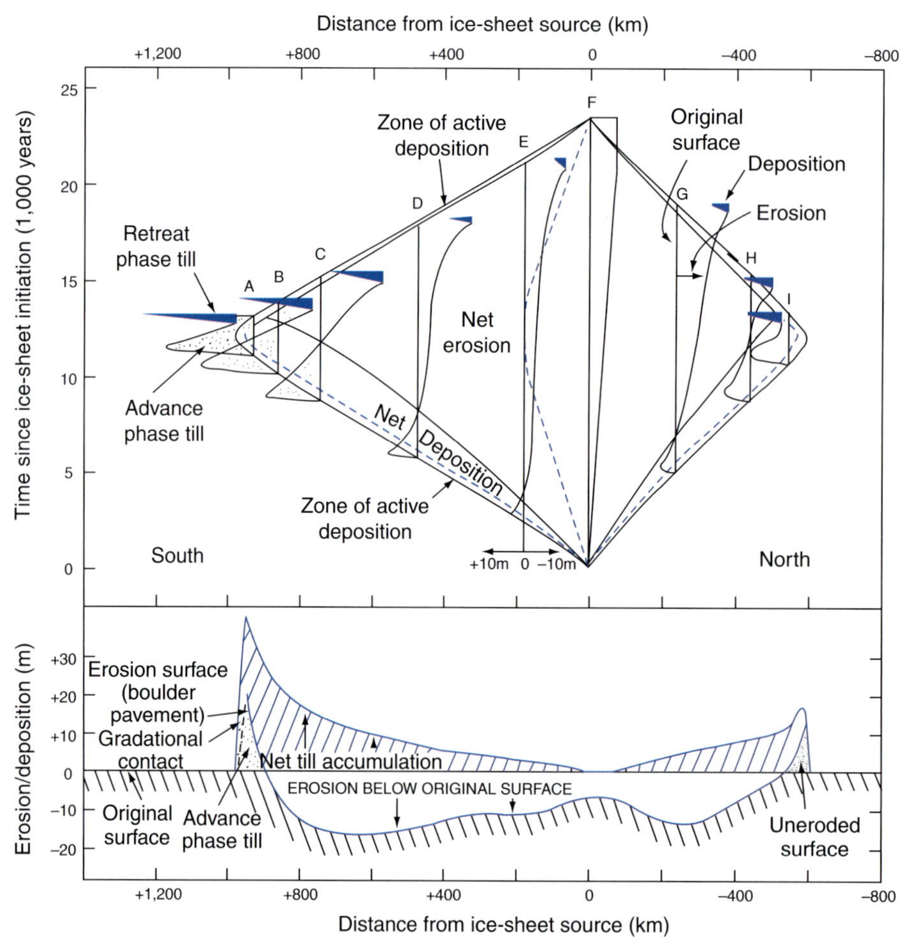

Figure 1 A typical model run from Boulton (1996a), showing subglacial debris transport in a time–space diagram. The ice sheet is initiated at a nucleation point at time $t=0$, and expands to the north and south, reaching its maximum 12 kyr after inception. It then contracts into the area where it first developed, finally vanishing at $t=24$ kyr. The model predicts that there will be an outer zone of deposition near the ice-sheet margin and an inner zone of erosion, although there will be no erosion below the ice divide where the balance velocity is zero. Reprinted from the Journal of Glaciology with permission of the International Glaciological Society.

erosion surface is capped by the retreat-phase till (sites B, C, D, G, and H)); Zone 3 – a zone where both advance- and retreat-phase tills are preserved, but with an intervening erosion surface (some of the advance-phase till has been removed, but erosion was insufficient to bite down into the preglacial surface (sites A and I)); Zone 4 – a zone of continuous till deposition and no erosion (this zone only occurs very close to the ice-sheet limit). More complex stratigraphies will result from oscillations of the ice margin and the migration of different basal thermal regimes (Boulton, 1996a, b).

Subglacial Landform–Sediment Associations

At the scale of continental ice sheets we can identify 'hard' and 'soft' beds. Former hard beds existed in shield terrain, or ancient continental crust, where subglacial deposits tend to be thin and patchy and commonly streamlined at a wide range of scales, forming flutings, megaflutings, megascale lineations, rock-cored drumlins, and leeside cavity fills (*see* Tills). Bedrock surfaces covered in glacially eroded forms, ranging from microscale features (e.g., striae) to macro- and megascale forms (e.g., roches moutonnees and troughs) are common in such areas (*see* Micro to Macro Scale Forms, Major Scale Forms). Subglacially streamlined forms such as megaflutings and drumlins commonly record the last (deglacial) ice-flow directions, although earlier ice-flow directions may be recorded in some areas where older lineations were not obliterated by deglacial flow. Rogen moraine occurs almost exclusively in shield terrains and has been associated with either: (1) local shearing and thickening of debris-rich basal ice (Bouchard, 1989) or (2) partial modification of older streamlined forms following a shift in the ice-divide position (Boulton, 1987; Boulton and Clark, 1990). Long esker networks are also common in areas of hard substrata, where they record late stages of drainage below ablating ice sheets (Shilts *et al.*, 1987; Aylsworth and Shilts, 1989). Clark and Walder (1994) have argued that eskers are characteristic of hard-rock terrains because channelized, conduit flow is possible at the ice–bed interface. At a continental scale, eskers are more widespread on hard substrata, but do occur on soft substrata, in some cases superimposed on drumlins and other bedforms. It is possible that on both hard and soft beds subglacial drainage switched between channelized and distributed systems as ice thickness and gradient and meltwater supply changed over the course of a glacial cycle.

Soft glacier substrata occur in areas underlain by weak sedimentary rocks, or where accumulations of Quaternary sediments mask hard lithologies. Expansive sedimentary bedrock lowlands exist at the margins of the North American and Scandinavian shields, which provided an outer zone of potentially deformable sediment below the great Pleistocene ice sheets. In such areas, the subglacial surface typically consists of fluted and drumlinized terrain underlain by varying thicknesses of subglacial sediments. Late-stage drainage events may be recorded by eskers and/or meltwater channels (Eyles, 1983b; Eyles and Menzies, 1983; Eyles and Eyles, 1992). In many areas occupied by Pleistocene ice sheets, subglacial sedimentary sequences may comprise tens of meters of vertically stacked tills and intervening stratified sediments (**Fig. 2**). Till units may be laterally continuous for tens of kilometers, with remarkably uniform sedimentological characteristics and lithological composition (e.g., Boyce and Eyles 2000).

In many localities, stacked till units record flow interactions between ice lobes (e.g., Eyles *et al.*, 1982 and Hicock, 1992). Near the junction between two adjacent ice-flow units, changes in relative ice discharges can cause gradual or abrupt shifts in the ice-flow direction, as one flow unit becomes dominant over the other. As a result, debris from different source areas can be transported to the site, producing marked compositional stratification in which the lithological composition of till units varies systematically upsection. Such compositional changes are commonly paralleled by shifts in the preferred orientation of particle fabric modes, allowing sequences of ice-flow changes to be reconstructed in some detail (Berthelsen, 1978; Eyles *et al.*, 1982). Vertical compositional changes can also occur in the absence of changes in flow direction. Boulton (1996b) has shown that till composition reflects the duration of glaciation, particularly the length of time available to transport debris from distant locations. Theory

Figure 2 A 30-m thick sequence of multiple tills on the Canadian prairies, Milk River, southern Alberta. Tills can be differentiated by color and bounding surfaces marked by clast pavements.

suggests that till composition should be dominated by local lithologies near the base and shift to predominantly far-traveled lithologies at the top, reflecting the passage of successive waves of debris from further and further upglacier.

Regional Distributions of Subglacial Landforms

Subglacial sediments and landforms exposed by glacier and ice-sheet recession commonly exhibit systematic spatial patterns, in which different types of landforms occur in distinct zones aligned parallel or transverse to former ice flow. Several factors influence the formation of subglacial sediments and landforms, including strength and hydraulic conductivity of the bed, basal thermal regime and the availability of meltwater, ice velocity, shear stress, and effective overburden pressure. Thus, landform zonation can be viewed as a mosaic produced in a variety of subenvironments beneath the parent ice sheet, either at a single moment in time or as a time-transgressive palimpsest resulting from ice advance or retreat. Elongate, flow-parallel zones may mark the former position of flow units such as ice streams and marginal lobes, or meltwater drainage pathways, whereas transverse zones may record downflow variations in basal conditions and their migration through time.

Ice-sheet advance over an area can be recorded by distinctive landform-sediment zones, particularly in regions underlain by weak rocks or unconsolidated sediments. For example, in the northern Great Plains of North America, arcuate belts of hummocky ice-thrust terrain and excavational basins commonly lie at the distal edges of drumlin fields. Some of the thrust belts have a subdued, streamlined appearance, and contain cupola hills and drumlins with low length/width ratios. The excavational basins and thrust ridges document glacitectonic disturbance near the ice margin, probably encouraged by cold-based conditions below the snout (Clayton and Moran, 1974; Bluemle and Clayton, 1984), and the streamlined terrain and drumlin belts record the subsequent advance of wet-based ice over the former marginal zone. In regions where subglacial deformation was intense and prolonged, glacitectonized thrust masses may lie buried, their upper surfaces intensely sheared and incorporated within the subglacial deforming layer. The amount of subglacial modification commonly decreases down-ice, and landform assemblages exhibit a continuum from drumlins with high length/width ratios, through drumlins with low length/width ratios and cupola hills, to largely unmodified thrust masses. Such landform associations apparently represent progressively less effective or short-lived subglacial molding toward the ice limit. Boyce and Eyles (1991) attribute down-ice changes in drumlin morphometry and sedimentology in the Peterborough drumlin field south of the Dummer Moraine, Ontario, to variations in substratum characteristics and duration of glaciation. Specifically, they consider the down-ice variations in drumlin morphometry to be a function of decreasing duration of subglacial deformation toward the ice margin with the northern part of the area having been stripped of its pre-existing sediment because it was located beneath the ice-sheet interior (cf. Boulton (1996a)).

Cross-Cutting Lineations and Ice-Sheet Dynamics

Regional mapping of flow-parallel and transverse landforms can reveal large-scale patterns of ice-sheet flow and their changes through time, yielding valuable information on ice-sheet evolution and dynamics. Reconstructed flow patterns for ice-sheet maxima demonstrate that ice sheets were not simple domes from which ice spread radially, but rather comprised multiple accumulation and dispersal centers (Boulton *et al.*, 1985; Punkari, 1995; Jansson *et al.*, 2003; Stokes *et al.*, 2005). Furthermore, it is possible to reconstruct the migration of ice divides and the switching of ice streams during a glacial cycle by mapping cross-cutting and superimposed subglacial bedforms, assessing till fabrics (*see* Till Fabric Analysis), and tracing indicator erratics (*see* Glacial Erratics and Till Dispersal Indicators). Analysis of Landsat satellite images and aerial photographs of glaciated terrains has enabled glacial geomorphologists to reconstruct changing ice-sheet flow directions and numerous iceflow landform assemblages have been identified, composed of drumlins, megaflutings, and megascale lineations, which are commonly superimposed on or cross-cut each other (**Fig. 3**). Clark (1993) identified two relative age indicators which enable the sequence of ice flows to be reconstructed: (1) 'simple superimposition,' where one set of streamlined lineations is superimposed over another set with a different orientation; and (2) 'pre-existing lineation deformation,' where deformation during a more recent ice flow phase alters the form or continuity of pre-existing lineations. According to Clark, the degree of modification of earlier lineations forms a continuum, ranging from (1) a situation where there is no modification of the pre-existing lineation to (2) superimposition of smaller forms on the surface of older lineations; to (3) substantial breaching or deformation of the pre-existing lineation; to (4) total reorganization of sediment into a new orientation.

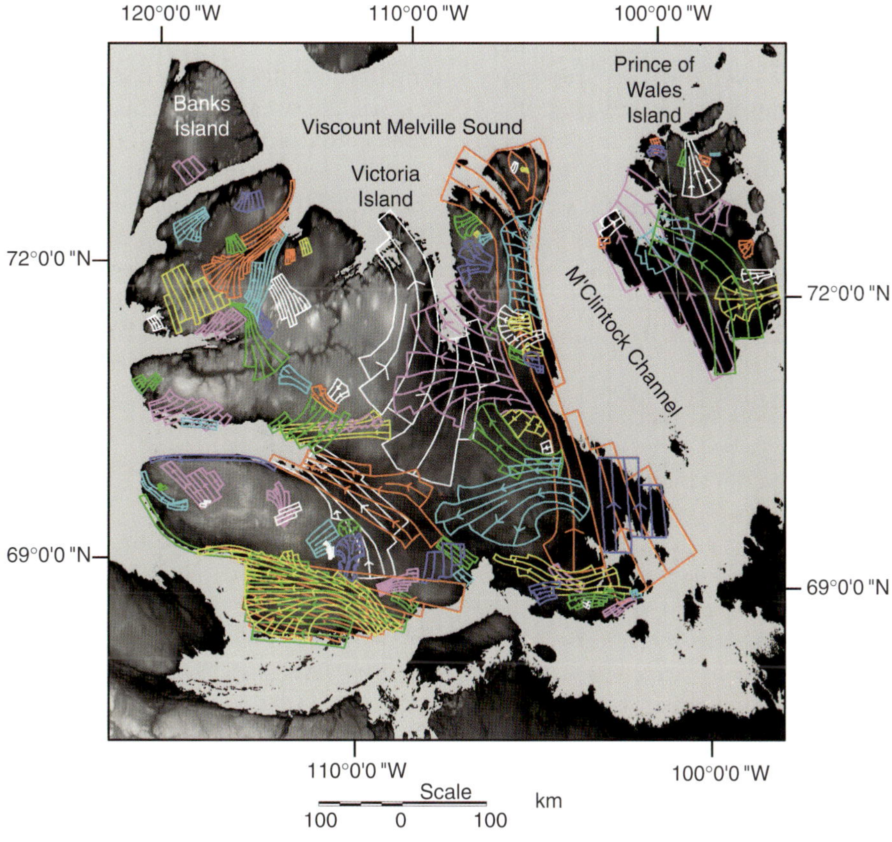

Figure 3 Flow sets comprising coherent assemblages of subglacially streamlined landforms (megalineations) produced during subsequent flow phases at the base of the former Laurentide Ice Sheet over Victoria Island, Arctic Canada. The flow sets record complex cross-cutting by ice streams that changed flow direction due to migrating ice divides. From Stokes CR, Clark CD, Darby DA and Hodgson DA (2005) Late Pleistocene ice export events into the Arctic Ocean from the M'Clure Strait ice stream, Canadian Arctic Archipelago. *Global and Planetary Change* 49: 139–162.

Ice-Sheet Marginal and Supraglacial Landform–Sediment Associations

At the margins of ice sheets and ice caps, supraglacially entrained debris is rare or absent and therefore debris for the construction of ice-marginal landforms must be derived from the glacier foreland or the bed, and elevated to higher levels by either proglacial or englacial glacitectonic processes, or pressurized meltwater flowing in englacial conduits. The processes and patterns of debris entrainment, elevation, and deposition at glacier margins are strongly influenced by glacier thermal regime and ice-margin activity, allowing some of the climatic and dynamic characteristics of former ice masses to be reconstructed from geological evidence. Ice-sheet marginal and supraglacial landform–sediment associations can be differentiated based upon the nature of the glacier body responsible for their emplacement, and Benn and Evans (1998) identify four depositional settings: temperate, subpolar, polar–continental, and surging glacier margins. Temperate margins are mainly wet based but may experience seasonal freezing near the snout during cold conditions in the winter; subpolar margins are characterized by cold-based conditions near the snout and wet-based conditions further upglacier, and polar–continental margins are cold based with insignificant production of meltwater at the surface. Surging glaciers may be of temperate or subpolar type. In reality, there is an unbroken continuum of variation between temperate and polar–continental margins, but it is important to emphasize the distinctiveness of ice-marginal processes and products in different climatic settings.

Temperate Glacier Margins

Temperate glacier margins can be defined as those which are mainly wet based, for at least part of the year, and generally occur in areas of discontinuous or no permafrost. A narrow frozen zone develops below the margin of many temperate glaciers during the winter, due to the penetration of a cold wave from the atmosphere through the thin ice at the edge of the glacier (Krüger, 1994). Additionally, small and discontinuous areas of net freezing can exist below some temperate margins in areas of discontinuous permafrost (e.g., Lawson (1979)), in some instances amounting to thick basal debris-rich ice sequences where supercooling takes place (e.g., Alley *et al.* (1998) and Lawson *et al.* 1998). Deposition at temperate glacier margins typically produces small dump moraines, or push and squeeze moraines derived from sediment exposed on the glacier foreland (*see* Moraine forms and Genesis). Modern examples have been described from Iceland, where thrusting mainly affects ice-contact fans built up against the ice front but can also relocate slabs of subglacial till (Krüger, 1985, 1993, 1994; Humlum, 1985; Boulton, 1986; Evans and Hiemstra, 2005). Glaciers without extensive debris cover respond rapidly to seasonal temperature fluctuations, so glacier margins tend to oscillate on an annual basis. As a result, suites of recessional moraines are commonly formed during deglaciation except where glacier velocities are low and/or ablation continues throughout the year (Sharp, 1984; Boulton, 1986; Krüger 1994). The areas between recessional moraines represent periods of rapid glacier retreat, and commonly exhibit well-preserved subglacial landforms (**Fig. 4**).

Figure 4 Extract from a map of glacial landforms on the foreland of Breidamerkurjokull, Iceland. Tills are orange and their surfaces are fluted (green lines). The flutings are arranged in linear zones separated by push moraines (black line symbol). Overridden push moraines are colored red and eskers are denoted by arrow head symbols. Note that glacifluvial outwash occupies corridors through the subglacial landforms and sediments, where meltwater excavated channels during deglaciation of the foreland. After Evans DJA and Twigg DR (2002) The active temperate glacial landsystem: a model based on Breidamerkurjokull and Fjallsjokull, Iceland. *Quaternary Science Reviews* 21: 2143–2177.

Meltwater can have a profound influence on the landforms and sediments formed at temperate glacier snouts, and subglacial, ice-marginal, and supraglacial depositional assemblages are commonly substantially modified by proglacial glacifluvial deposition and erosion during and following deglaciation (*see* Glacifluvial Landforms of Erosion, Glacifluvial Landforms of Erosion). Landforms such as drumlins, eskers, and moraines may be heavily reworked by meltwater and/or buried by outwash sediments so that they are left as remnants or inliers within sandur plains (**Fig. 4**). Even large ice-marginal moraines can be dissected or partially destroyed by meltstreams directly after emplacement (Price, 1969; Krüger, 1994; Thompson, 1988). In some cases, virtually all of the depositional legacy of a temperate ice lobe takes the form of glacifluvial sediments. Where subglacial and englacial drainage systems are well developed, sands and gravels can accumulate within conduits, and meltstreams emerging at the glacier surface can deposit considerable quantities of glacifluvial sediment in supraglacial positions. Indeed, on some glaciers, glacifluvial sediment constitutes the major part of the englacial and supraglacial sediment load close to the margins (Gustavson and Boothroyd, 1987). During deglaciation, uneven ablation of sediment-covered ice can lead to the development of 'glacier karst' (Clayton, 1964). Sediment deposited within conduits and in supraglacial outwash fans and lakes produces a complex assemblage of landforms including esker systems, kame ridges and plateaus, and pitted outwash (*see* Glacifluvial Landforms of Erosion). Distinctive spatial associations of glacifluvial landforms deposited during ice wastage have been termed morphosequences (Koteff and Pessl, 1981). In such sequences, proglacial outwash passes upvalley into kame and kettle topography, which is inset within ice-marginal kame terraces and moraines. Kame and kettle topography is locally refashioned into suites of river terraces by proglacial and postglacial streams.

Subpolar Glacier Margins

Subpolar margins are characterized by an outer cold-based zone and an upglacier wet-based zone. At present, such glacier margins are found mainly in areas of continuous permafrost (*see* Permafrost). The thermal regime and geomorphic impact of glaciers in permafrost regions is determined by ice thickness and activity (Dyke, 1993). Ice caps growing on thick permafrost must initially be entirely cold-based, but wet-based zones can develop in response to subglacial heating from geothermal heat and the strain heating produced by glacier flow. Thawing of the bed is most likely where glacier velocities are highest (e.g., in the vicinity of the glacier equilibrium line and in areas of flow convergence), and therefore wet-based ice will be most extensive below glaciers with relatively high mass throughput and/or strong converging flow. In the upper parts of glacier accumulation areas, cold-based zones are likely to persist due to low ice velocities and downward advection of cold firn. Cold-based ice occurs beneath the glacier margins where ice is thin, velocities are low, and proglacial permafrost extends below the glacier. Thus in permafrost regions, small low-activity glaciers are likely to be cold based throughout. The presence of wet-based ice upglacier from a frozen margin in subpolar glaciers has a profound impact on processes of debris entrainment, transport, and deposition. First, erosion below wet-based, sliding ice provides a source of debris which can be transported toward the margin in regelation ice or a subglacial deforming layer. As this debris is transported into the frozen zone at the glacier margin, sequences of debris-rich basal ice develop as a result of net adfreezing (Weertman, 1961; Hubbard, 1991). These debris sequences are commonly thickened and elevated toward the glacier surface by strong compressive deformation resulting from ice deceleration at the frozen margin (Goldthwait, 1951; Boulton, 1972; Knight, 1989; Tison *et al.*, 1993; Hambrey and Huddart, 1995; Etzelmüller *et al.*, 1996). Thick basal debris sequences also form where glaciers override buried ice dating from earlier glacial events (Hooke, 1973; Evans, 1989b). The long survival of buried ice in permafrost environments (e.g., Dyke and Evans 2003) means that debris-rich ice may be recycled many times over several glacial cycles as glaciers advance and retreat.

Common landforms at subpolar glacier snouts are composite ridges and thrust-block moraines (*see* Glacitectonic Structures and Landforms), particularly in wide valley bottoms close to sea level where thick sequences of stratified sediment occur (Kalin, 1971; Evans, 1989a, b, 1990; Evans and England, 1991; Hambrey and Huddart, 1995; Etzelmüller *et al.*, 1996; **Fig. 5**). Where proglacial thrusting does not occur, glaciers commonly terminate in steep ice cliffs along which elevated basal debris is exposed and frontal aprons, composed of dry calved ice blocks and fallen or inwashed debris, accumulate (Evans, 1989b). Following ice-margin retreat, belts of hummocky and controlled moraine are deposited upvalley from frontal aprons, as the result of meltout and reworking of englacial debris. If the debris cover is thicker than the depth of summer melting (the active layer thickness), ice cores may survive indefinitely within frontal aprons and

Figure 5 Thrust-block moraine developed in raised marine sediments at the margin of the Eugenie Glacier, Dobbin Bay, Ellesmere Island, Arctic Canada. Note the marine shorelines notched into the front of the moraine attesting to higher sea levels during moraine construction.

associated hummocky and controlled moraine. In cases where bedrock is close to the surface or glacier activity is low, moraines formed at the margins of subpolar glaciers may amount to no more than a line of boulders, attesting to very low turnover of debris. Recessional moraines are uncommon and basal tills are rare or thin and patchy (Evans, 1990). The amount of available debris reaches a minimum where glaciers are predominantly or entirely cold based. Former glacier margins in such areas are most often marked by lateral meltwater channels cut into bedrock (Evans, 1990; Dyke, 1993). These lateral channels form because surface meltwater is prevented from penetrating the glacier by low ice temperatures, so that it must drain over the surface and around the margin of the glacier (*see* Glacifluvial Landforms of Erosion). Where cold glacier tongues descend to the sea or ice-dammed lakes to form glacier ice shelves they can deposit ice-shelf moraines (England *et al.*, 1978; Sugden and Clapperton, 1981; Hodgson and Vincent, 1984; Dyke and Prest, 1987; Hodgson, 1994; *see* Moraine forms and Genesis).

Polar-Continental Glacier Margins

Distinctive glacial landsystems are formed around the margins of glaciers in the polar deserts, for example, fringing parts of the Antarctic Ice Sheet, where the environment is cold, windy, and has a moisture deficit because sublimation exceeds precipitation. Local glaciers are entirely cold based, and are nearly free of debris, although some of the larger terrestrial outlet tongues of the Antarctic Ice Sheet are partly wet based with cold-based margins, and have stacked basal debris sequences similar to those of subpolar glaciers (Shaw, 1977a, b). Where thick basal ice sequences crop out, marginal aprons accumulate by

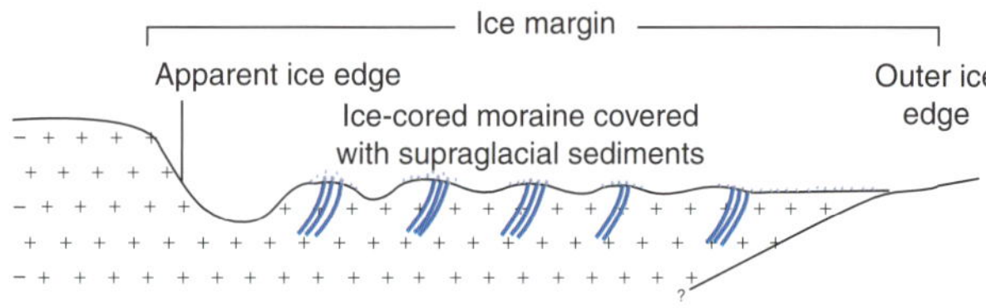

Figure 6 Schematic diagram showing the origin of ice-cored (controlled) moraine ridges at the margin of a polar glacier. From Fitzsimons SJ (2003) Ice-marginal terrestrial landsystems: polar continental glacier margins. In: Evans DJA (ed.) *Glacial Landsystems*, pp. 89–110. London: Arnold. Reproduced with permission.

gravitational processes, but with very limited reworking by meltwater. Within the marginal aprons, belts of hummocky and controlled moraines can develop from stacked and thrust englacial debris sequences (*see* Moraine forms and Genesis). The topography of controlled moraines is inherited from englacial structures (**Fig. 6**), such as debris bands and folds in debris-rich basal ice, and commonly forms ridges transverse to glacier flow (Shaw, 1977a, b). Entirely cold-based glaciers without stacked basal debris sequences leave very little imprint on the landscape, and end moraines commonly consist of little more than low ridges and, unlike subpolar glacier margins, marginal meltwater channels are rare. Thrust-block moraines occurring in front of some polar–continental glaciers (Fitzsimons, 1997) appear to require special conditions, because most of the proglacial sediment is deeply frozen and ice-cemented and has a high shear strength; it is likely that these moraines form by the proglacial thrusting of unfrozen lake-bottom sediments, because lake floors can remain unfrozen below a lake ice cover until glaciers advance into them.

Subaquatic Glacial Depositional Systems

Subaquatic depositional systems form a very important component of the ancient glacial geological record because: (1) large parts of the Pleistocene ice sheets terminated in the sea or large proglacial lakes, depositing huge volumes of sediment in standing water; (2) subaquatic sediments tend to have high preservation potential, and commonly preserve a more complete record of glaciation than terrestrial depositional systems; and (3) the study of subaquatic depositional systems yields important insights into the dynamics of former ice sheets, including the patterns of deglaciation, the location of the grounding lines of former ice streams, and episodes of rapid calving retreat and ice-rafting as recorded for example in Heinrich Events (*see* Glacimarine Sediments and Ice-Rafted Debris).

Stratigraphic Architecture

The large-scale geometry or 'stratigraphic architecture' of subaquatic depositional systems is crucial to understanding environmental change in glacially influenced basins (e.g., Boulton (1990), Eyles and Eyles (1992), and Hambrey (1994)). A large body of information is available on subaqueous depositional systems for hundreds of formerly glaciated sites, and many depositional models have been proposed (e.g., Molnia (1983), Gravenor *et al.* (1984), Eyles *et al.* (1985), Fyfe (1990), Ashley (1995), and Powell and Domack (1995); **Fig. 7**). It has become clear that, although the lateral and vertical relationships between sediment–landform associations such as morainal banks, grounding-line wedges, and distal drapes are very varied, they reflect the interaction of relatively few groups of variables. The most important are (1) the topography and tectonic setting of the basin; (2) the extent, configuration, and dynamics of glacier and floating ice and the position of associated depocenters; and (3) changes in water depth due to eustatic and isostatic cycles and the life cycle of proglacial lakes (Powell, 1984, 1990; Boulton, 1990; Eyles and Eyles, 1992; Martini and Brookfield, 1995; Powell and Domack, 1995).

Subaqueous depositional systems can be classified according to their position of formation relative to the ice front. Powell (1984) defined four zones as subglacial, ice proximal, ice shelf, and ice distal. The subglacial zone lies behind the glacier grounding line, and is reviewed above. Facies deposited in this zone form downglacier-thickening sheets of basal till and associated lenses of water-sorted material. The ice-proximal zone encompasses subaqueous environments adjacent to the glacier-grounding line. The thickness of sediment packages deposited in the ice-proximal zone is related to the glacier snout retreat rate; slow retreat or quasi-stable conditions will encourage the accumulation of large volumes of sediment, whereas during rapid calving retreat a large proportion of debris will be rafted away, restricting the amount of ice-proximal deposition. Sediment associations include push and thrust moraines, grounding-line fans, morainal banks, deltas, DeGeer moraines, and grounding-line fans (*see* Moraine

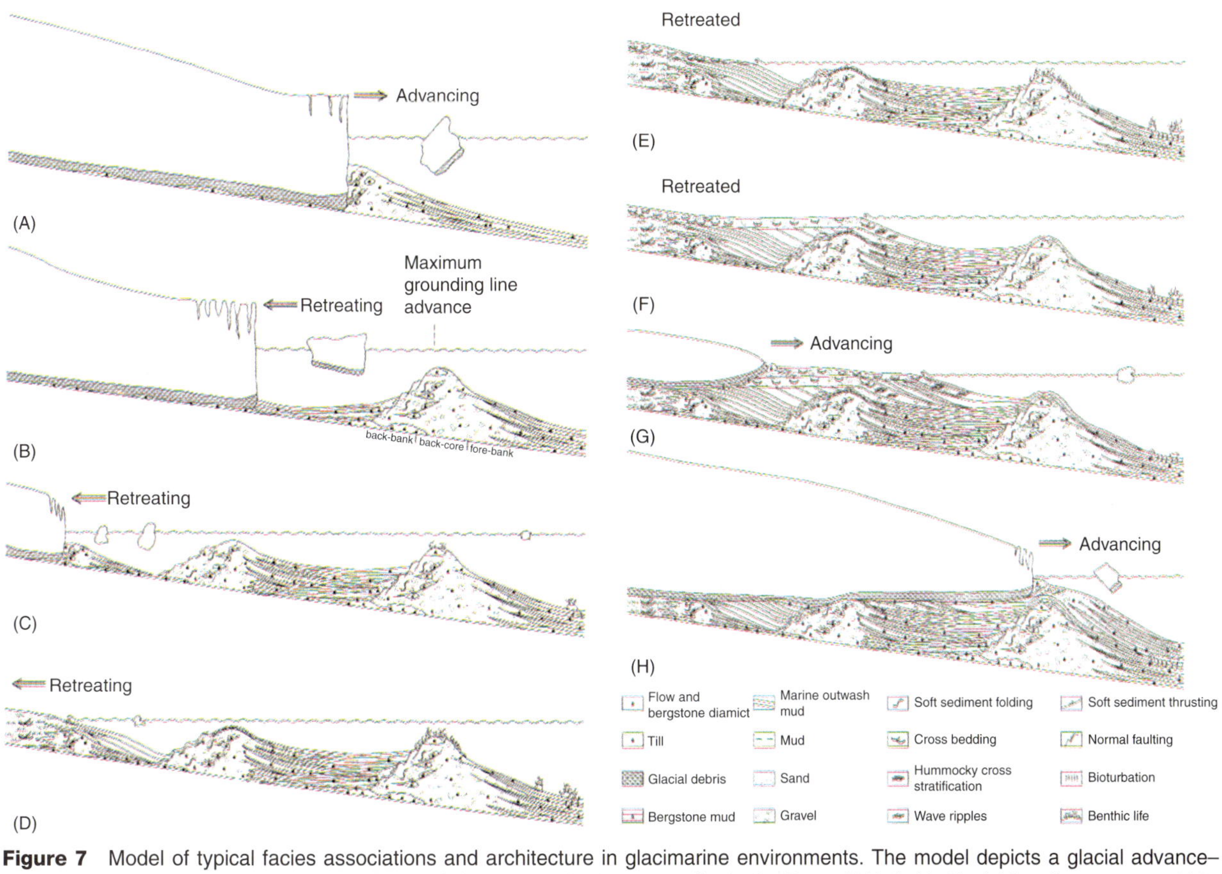

Figure 7 Model of typical facies associations and architecture in glacimarine environments. The model depicts a glacial advance–retreat–readvance cycle. From Powell RD (2003) Subaquatic landsystems: fjords. In: Evans DJA (ed.) *Glacial Landsystems*, pp. 313–347. London: Arnold. Reproduced with permission.

forms and Genesis, Glacimarine Sediments and Ice-Rafted Debris). Sediment gravity flows are common in the ice-proximal zone due to high sedimentation rates, ice push, iceberg calving, storm waves, tidally induced pressures below ice shelves, and morainal bank collapse. Subaqueous outwash and sediment flows both commonly form prograding beds dipping away from the glacier margin, known as clinoforms. The upper parts of such clinoforms may be erosionally truncated during glacier advance, forming a sub-horizontal erosion surface overlain by basal till. The ice-shelf zone occurs in polar climates where ice below the pressure-melting point floats and produces glacier and sea-ice ice shelves. Deposition in this zone is dominated by undermelt and rainout, forming drapes of dropstone diamicton and mud (*see* Glacimarine Sediments and Ice-Rafted Debris). The ice-distal zone is dominated by ice-rafted debris and muds formed by the settling of suspended mineral grains and microfaunal remains. Rainout, current reworking, and gravitational resedimentation produce sequences of massive and stratified sediment tens to hundreds of meters thick, and spanning several glacial cycles. The long-term preservation potential of such sequences is high, and they are well represented in the ancient geological record (Eyles, 1993).

Fjords

Glacial stratigraphic architecture in fjords is strongly influenced by topography. Overdeepened basins act as sediment sinks, whereas bedrock sills provide pinning points where ice-proximal depositional systems can accumulate. Additionally, the presence of steep flanking slopes encourages gravitational reworking of sediment and debris input from ice-free terrain. The sedimentary infill of fjord basins can be usefully divided into three broad units deposited during glacier advance, maximum, and retreat. The advance stage of a glacial cycle is typically represented by a basal till unit or an erosion surface. Till thickness generally increases downfjord toward the glacier limit, reflecting erosion, and downglacier transport of pre-existing sediment and bedrock fragments (Boulton, 1990). Beyond the glacier limit, extensive blankets of mud and diamicton may record distal glacimarine deposition during the advance phase, but are typically buried by younger glacimarine sediments. Sediments relating to glacial maximum conditions are commonly located at pinning points where reduced calving rates encourage ice-margin stability (Warren and Hulton, 1990; Greene, 1992). Typical sediment–landform associations include push or thrust moraines, grounding-line fans, morainal banks, and deltas (Powell, 2003). In low-relief fjord basins, lateral moraines may be preserved on the valley sides, but in high-relief settings such moraines are generally rapidly destroyed by paraglacial reworking after glacier retreat. Lateral moraines with exceptionally low gradients indicate the former extent of floating ice shelves. Beyond the ice limit, blanket-like drapes of fine-grained sediments and diamictic muds record suspension sedimentation from turbid overflows and dumping from icebergs (Powell and Molnia, 1989). During glacier retreat, depositional zones migrate upfjord, and progressively more distal facies are laid down on top of older units. Pauses in glacier retreat may be marked by push or thrust moraines, grounding-line fans, or morainal banks (Boulton, 1986). Substantial ice-marginal accumulations such as large morainal banks and deltas tend to be associated with topographic pinning points. Glacimarine sediments and landforms in fjords may be raised above sea level by isostatic uplift following deglaciation. The emergence of ice-proximal associations above sea level provides a readily accessible source of information on former ice margins in regions where terrestrial ice-marginal accumulations are commonly not well preserved.

Continental Shelves

The glaciation of shelf areas may arise through either the advance of terrestrial margins in a seaward direction, the thickening and grounding of sea ice encouraged by lower sea level, or a combination of these processes. The impact of glaciation on a continental shelf may be direct where the shelf has hosted a grounded glacier mass, or indirect where it merely received distal glacimarine sediments during a glacial cycle. Patterns of sedimentation are dependent on a number of interrelated factors, including the extent and configuration of glacier ice, basin history and tectonic setting, bathymetry, oceanography, glacial thermal regime, and sediment supply (e.g., Powell (1984) and Eyles (1993)). Powell (1984) considered the influence of ice-margin configuration and thermal regime on patterns of sedimentation on continental shelves and identified various glacimarine regimes, based on the nature of the ice source (valley glacier or ice sheet), glacier thermal regime, and glacier front characteristics (tidewater cliff or ice shelf). Tectonic setting also exerts a strong influence on the style and preservation of glacigenic sediments on continental shelves (Eyles, 1993; Hambrey, 1994) and includes passive margins, convergent margins, and rift basins (e.g., Eyles *et al.* (1991), Powell and Domack (1995), and Vorren (2003)).

Large Proglacial Lakes and Epicontinental Seas

During deglaciation of the Pleistocene ice sheets, ice margins were commonly in contact with large water bodies, either proglacial lakes ponded between the ice and topographic barriers, or arms of the sea which flooded isostatically depressed and glacially overdeepened parts of continental interiors (Teller, 2003). The extent and evolution of many of these water bodies has been reconstructed from the sedimentary record, in combination with shoreline evidence, pollen and microfaunal records, and dating programs (e.g., Eronen (1983) and Teller (1995)). These large proglacial water bodies encouraged rapid ice-margin retreat by calving, and may have triggered some episodes of ice advance or surges by encouraging low effective pressures below the glacier margin (Fyfe, 1990). Depositional systems in such settings are similar to those of continental shelves (**Fig.** 7), although there are some important differences. First, fresh lake water and brackish shallow marine conditions reduce the importance of suspended-sediment transport in overflows and interflows, so that underflow deposits are better represented in ice-proximal and distal systems. Second, the seasonal alternation between deposition by underflows in summer and settling of clay particles in winter produces annual varves in lakes (*see* Varved Lake Sediments). Third, proglacial water bodies are prone to sudden changes in water level, due to the opening and closing of outlets by glacier margin fluctuations and periodic jokulhlaups and recharge events. The resulting cyclic water-level changes can be recorded by discrete sediment packages bounded above and below by erosion surfaces, and can add considerable complexity to stratigraphic sequences (Martini and Brookfield, 1995). Fourth, proglacial lakes and epicontinental seas can receive substantial amounts of sediment from recently deglaciated landsurfaces surrounding the basin as well as from the glacier margin (Martini and Brookfield, 1995).

Sediment–Landform Associations of Mountain Glaciation and Glaciated Valleys

The glacial landforms and sediments of mountain environments have been synthesized for a variety of physiographic and latitudinal zones since the concept of a glaciated valley landsystem was introduced by Boulton and Eyles (1979) and Eyles (1983c). A variety of depositional settings has been elucidated by Benn *et al.* (2003) in a model that incorporates the controls of debris supply on landform development (*see* Glacial Land Systems). A plateau icefield landsystem has been proposed by Rea *et al.* (1998) and Rea and Evans (2003) based upon observations on Norwegian and Icelandic ice caps. Glaciated valleys are unique in that they are influenced by valley side debris inputs and the topographic confinement of depositional basins. The influence of these factors varies from valley to valley, and it is useful to subdivide glaciated valley landsystems into low-relief and high-relief types. Low-relief settings include valleys in ancient mountain belts where the vertical distance between valley floor and summit ridges is generally less than 1,000 m (e.g., the glaciated valleys of Scotland, Norway, and Labrador). High-relief settings are characterized by extensive steep valley sides rising thousands of meters above the valley floor. Such environments are commonly associated with young or tectonically active fold mountains such as the European and New Zealand Alps, the High Andes, and the Himalayas, where relative relief may be 3,000 m or more. Features common to both low and high relief mountainous terrains are ice-dammed lakes and paraglacially reworked materials (*see* Paraglacial Geomorphology).

Low-Relief Mountain Environments

Valley glaciers in low-relief mountain terrain carry relatively small amounts of supraglacial debris, so that subglacial landforms commonly escape burial during deglaciation and may be very well exposed at the surface (**Fig.** 8). The upper limits of glacier occupancy in glaciated valleys can be preserved as erosional 'tidemarks,' or trimlines, on valley sides (*see* Trimlines and Palaeonunataks). In recently deglaciated terrain, trimlines are often marked by differences in vegetation cover either side of the former glacier limit, whereas in older glacial landscapes, periglacial trimlines may be preserved where frost-shattered terrain on upper slopes gives way abruptly to ice-scoured bedrock on lower ground (Thorp, 1981, 1986; Ballantyne and McCarroll, 1995). The margins of valley glaciers in low-relief mountain areas are commonly delimited by lateral-frontal dump and push moraines (*see* Moraine forms and Genesis). Moraine characteristics are strongly influenced by catchment lithology, and in valleys underlain by resistant, crystalline rocks, lateral moraines may be little more than ridges of boulders. In contrast, thrust moraines may occur where glaciers come into contact with thick, unconsolidated sediments such as glacimarine clays and silts. Meltwater deposition at some glacier margins is recorded by ice-marginal ramps and fans. Lateral moraines may be much larger on one valley side than the other, a

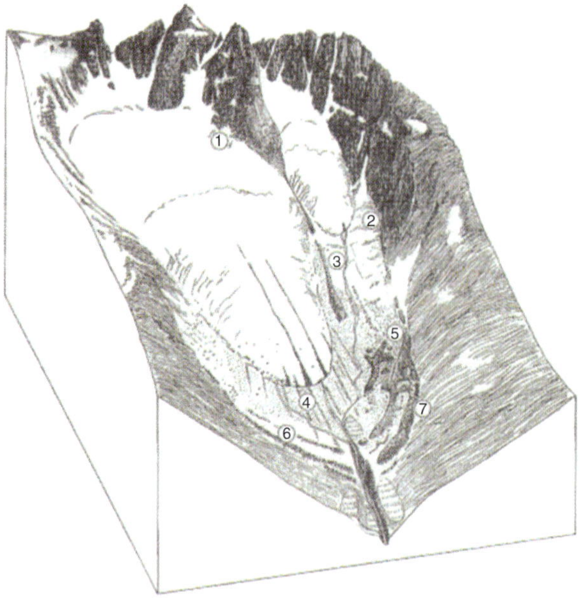

Figure 8 A landsystem model for deposition by valley glaciers with relatively low supraglacial debris concentrations. 1 = supraglacially entrained debris; 2 = periglacial trimline; 3 = medial moraine; 4 = fluted till; 5 = paraglacial reworking; 6 and 7 = lateral moraine displaying within-valley asymmetry. From Benn DI and Evans DJA (1998) *Glaciers and Glaciation*, London: Arnold. Reproduced with permission.

phenomenon known as 'within-valley asymmetry' due to one or a combination of four factors: (1) the differential extent of extensive, debris-supplying rock walls; (2) glacier pushing of pre-existing materials that are unevenly spread over the foreland; (3) cross-valley differences in lithology or structure; and (4) differences in glacier dynamics on either side of a valley (Matthews and Petch, 1982; Benn, 1989; Bennett and Boulton, 1993; Evans, 1999). The combined use of lateral moraines and other ice-marginal landforms and trimlines allows the dimensions of former valley glaciers and their paleo-ELAs (*see* Paleo ELAs) to be reconstructed with considerable accuracy, yielding important input for glaciological models or paleoclimatic studies (Ballantyne, 1989; Thorp, 1991; Nesje and Dahl, 2000). Glacier retreat in low-relief mountain environments may be recorded by recessional moraines and/or linear hummocky moraine superimposed on subglacial sediments and landforms. The size and spacing of such moraines reflects the availability of debris and the frequency and duration of stillstands or readvances of the margin (Eyles, 1979; Benn, 1992; Bennett and Boulton, 1993). Assemblages of chaotic hummocky moraine may occur in pockets within fields of recessional moraines, possibly due to the isolation and stagnation of sediment-covered ice remnants (Benn, 1992). Large areas of hummocky moraine can be produced by incremental marginal stagnation, whereby chaotic moraines are deposited in a 10–500 m wide marginal zone of debris-mantled ice which retreats upvalley during deglaciation (Eyles, 1983a, c). Assemblages indicative of more general stagnation, such as chaotic hummocky moraine inset within 'staircases' of subhorizontal kame terraces, occur on some flat-lying valley floors (Benn, 1992).

High-Relief Mountain Environments

Due to their high sediment yields, high-relief mountain environments contain valley floors that are commonly filled by very large volumes of sediment, particularly in tectonically active ranges such as the Southern Alps of New Zealand, the Himalayas, and the Andes (Owen and Derbyshire, 1989, 1993; Clapperton, 1993; Kirkbride, 1995). Valley glaciers in such settings typically have extensive covers of supraglacial debris, and sedimentation around glacier margins forms large lateral-frontal dump moraines and ice-contact fans and ramps (**Fig. 9**; Humlum (1978), Boulton and Eyles (1979), Small (1983), Owen and Derbyshire (1989, 1993), and Benn *et al.* (2003); *see* Moraine forms and Genesis).

Figure 9 Lateral moraines and ice-contact fans deposited at the margin of a debris-charged valley glacier in the Lahul Himalaya, India.

Debris-mantled glaciers (Nakawo *et al.*, 2000) have relatively muted responses to climatic change, because large lateral–frontal moraines limit the amount of glacier advance, and extensive supraglacial debris cover inhibits ablation of the underlying ice. Therefore, unlike their low-relief counterparts, valley glaciers in high-relief terrain tend not to deposit numerous dump and push moraines but instead remain at the limits imposed by lateral-frontal moraines until advance or retreat is triggered by significant climatic change; so the landform record of a retreating debris-mantled glacier consists of major moraine complexes, representing periods of stability, separated by extensive tracts of hummocky moraine deposited during episodes of ice-margin wastage and stagnation. The vast laterofrontal moraine complexes produced during ice wastage commonly form dams for proglacial and supraglacial lakes, giving rise to periodic glacial lake outburst floods (GLOFs) when dams are breached by lake water (Clague and Evans, 1994; Cenderelli and Wohl, 2003). The thick supraglacial and ice-marginal sediments of valley glaciers in high-relief terrain largely mask subglacial landforms and sediment so that the former beds of such glaciers are rarely preserved in the geomorphological record.

See also: **Glacial Landforms**: Moraine Forms and Genesis; Quaternary Vulcanism, Subglacial Landforms; Evidence of Glacier Recession; Glacial Land Systems; Glacifluvial Landforms of Erosion; Glacitectonic Structures and Landforms. **Glacial Landforms, Erosional Features**: Major Scale Forms; Micro to Macro Scale Forms. **Glacial Landforms, Ice Sheets**: Growth and Decay; Evidence of Glacier and Ice Sheet Extent; Paleo ELAs; Trimlines and Paleonunataks. **Glacial Landforms, Sediments**: Clast Form Analysis; Glacifluvial Landforms of Deposition; Glacigenic Lithofacies; Glacimarine Sediments and Ice-Rafted Debris; Micromorphology of Glacial Sediments; Till Fabric Analysis; Tills; Glacial Erratics and Till Dispersal Indicators.

References

Alley, R. B., Lawson, D. E., Evenson, E. B., Strasser, J. C., and Larson, G. J. (1998). Glaciohydraulic supercooling: A freeze-on mechanism to create stratified, debris-rich basal ice. II: Theory. *Journal of Glaciology* **44**, 563–569.

Ashley, G. M. (1995). Glaciolacustrine environments. In *Glacial Environments, Volume 1: Modern Glacial Environments: Processes, Dynamics and Sediments* (J. Menzies, Ed.), pp. 417–444. Butterworth-Heinemann, Oxford.

Aylsworth, J. M., and Shilts, W. W. (1989). Bedforms of the Keewatin Ice Sheet, Canada. *Sedimentary Geology* **62**, 407–428.

Ballantyne, C. K. (1989). The Loch Lomond Readvance on the Isle of Skye, Scotland: Glacier reconstruction and palaeoclimatic implications. *Journal of Quaternary Science* **4**, 95–108.

Ballantyne, C. K., and McCarroll, D. (1995). The vertical dimensions of Late Devensian glaciation on the mountains of Harris and southeast Lewis, Outer Hebrides, Scotland. *Journal of Quaternary Science* **10**, 211–223.

Benn, D. I. (1989). Debris transport by Loch Lomond Readvance glaciers in northern Scotland, basin form and the within-valley asymmetry of lateral moraines. *Journal of Quaternary Science* **4**, 243–254.

Benn, D. I. (1992). The genesis and significance of "hummocky moraine": Evidence from the Isle of Skye, Scotland. *Quaternary Science Reviews* **11**, 781–799.

Benn, D. I., and Evans, D. J. A. (1998). *Glaciers and Glaciation.* Arnold, London.

Benn, D. I., Kirkbride, M. P., Owen, L. A., and Brazier, V. (2003). Glaciated valley landsystems. In *Glacial Landsystems* (D. J. A. Evans, Ed.), pp. 372–406. Arnold, London.

Bennett, M. R., and Boulton, G. S. (1993). A reinterpretation of Scottish "hummocky moraine" and its significance for the deglaciation of the Scottish highlands during the Younger Dryas or Loch Lomond Stadial. *Geological Magazine* **130**, 301–318.

Berthelsen, A. (1978). The methodology of kineto-stratigraphy as applied to glacial geology. *Bulletin of the Geological Society of Denmark* **27**, 25–38.

Bluemle, J. P., and Clayton, L. (1984). Large-scale glacial thrusting and related processes in North Dakota. *Boreas* **13**, 279–299.

Bouchard, M. A. (1989). Subglacial landforms and deposits in central and northern Quebec, Canada, with emphasis on Rogen moraines. *Sedimentary Geology* **62**, 293–308.

Boulton, G. S. (1972). The role of the thermal regime in glacial sedimentation. In *Polar Geomorphology, Institute of British Geographers, Special Publication* (R. J. Price and D. E. Sugden, Eds.), vol. 4, pp. 1–19. London.

Boulton, G. S. (1986). Push moraines and glacier contact fans in marine and terrestrial environments. *Sedimentology* **33**, 677–698.

Boulton, G. S. (1987). A theory of drumlin formation by subglacial sediment deformation. In *Drumlin Symposium* (J. Menzies and J. Rose, Eds.), pp. 25–80. Balkema, Rotterdam.

Boulton, G. S. (1990). Sedimentary and sea level changes during glacial cycles and their control on glacimarine facies architecture. In *Glacimarine Environments: Processes and Sediments, Geological Society Special Publication* (J. A. Dowdeswell and J. D. Scourse, Eds.), vol. 53, pp. 15–52. London.

Boulton, G. S. (1996a). Theory of glacial erosion, transport and deposition as a consequence of subglacial sediment deformation. *Journal of Glaciology* **42**, 43–62.

Boulton, G. S. (1996b). The origin of till sequences by subglacial sediment deformation beneath mid-latitude ice sheets. *Annals of Glaciology* **22**, 75–84.

Boulton, G. S., and Clark, C. D. (1990). The Laurentide Ice sheet through the last glacial cycle: Drift lineations as a key to the dynamic behaviour of former ice sheets. *Transactions of the Royal Society of Edinburgh–Earth Sciences* **81**, 327–347.

Boulton, G. S., and Eyles, N. (1979). Sedimentation by valley glaciers: a model and genetic Classification. In *Moraines and Varves* (C. Schluchter, Ed.), pp. 11–23. Balkema, Rotterdam.

Boulton, G. S., Smith, G. D., Jones, A. S., and Newsome, J. (1985). Glacial geology and glaciology of the last mid-latitude ice sheets. *Journal of the Geological Society of London* **142**, 447–474.

Boyce, J. I., and Eyles, N. (1991). Drumlins carved by deforming till streams below the Laurentide ice sheet. *Geology* **19**, 787–790.

Boyce, J. I., and Eyles, N. (2000). Architectural element analysis applied to glacial deposits: Internal geometry of a Late Pleistocene till sheet, Ontario, Canada. *Geological Society of America Bulletin* **112**, 98–118.

Cenderelli, D. A., and Wohl, E. E. (2003). Flow hydraulics and geomorphic effects of glacial-lake outburst floods in the Mount Everest region, Nepal. *Earth Surface Processes and Landforms* **28**, 385–407.

Clague, J. J., and Evans, S. G. (1994). Formation and failure of natural dams in the Canadian Cordillera. *Geological Survey of Canada Bulletin* **464**, 35.

Clapperton, C. M. (1993). *Quaternary Geology and Geomorphology of South America*. Elsevier, Amsterdam.

Clark, C. D. (1993). Mega-scale lineations and cross-cutting ice-flow landforms. *Earth Surface Processes and Landforms* **18**, 1–29.

Clark, C. D. (1999). Glaciodynamic context of subglacial bedform generation and preservation. *Annals of Glaciology* **28**, 23–32.

Clark, C. D., and Stokes, C. R. (2003). Palaeo-ice stream landsystem. In *Glacial Landsystems* (D. J. A. Evans, Ed.), pp. 204–227. Arnold, London.

Clark, P. U., and Walder, J. S. (1994). Subglacial drainage, eskers, and deforming beds beneath the Laurentide and Eurasian ice sheets. *Bulletin of the Geological Society of America* **106**, 304–314.

Clayton, L. (1964). Karst topography on stagnant glaciers. *Journal of Glaciology* **5**, 107–112.

Clayton, L., and Moran, S. R. (1974). A glacial process-form model. In *Glacial Geomorphology* (D. R. Coates, Ed.), pp. 89–119. State University of New York, Binghampton.

Dyke, A. S. (1993). Landscapes of cold-centred Late Wisconsinan ice caps, Arctic Canada. *Progress in Physical Geography* **17**, 223–247.

Dyke, A. S., and Evans, D. J. A. (2003). Ice-marginal terrestrial landsystems: Northern Laurentide and Innuitian ice sheet margins. In *Glacial Landsystems* (D. J. A. Evans, Ed.), pp. 143–165. Arnold, London.

Dyke, A. S., and Prest, V. K. (1987). Late Wisconsinan and Holocene history of the Laurentide Ice Sheet. *Geographie Physique et Quaternaire* **XLI**, 237–263.

England, J., Bradley, R. S., and Miller, G. H. (1978). Former ice shelves in the Canadian high Arctic. *Journal of Glaciology* **20**, 393–404.

Eronen, M. (1983). Late Weichselian and Holocene shore displacement in Finland. In *Shorelines and Isostasy* (D. E. Smith and A. G. Dawson, Eds.), pp. 183–208. Academic Press, London.

Etzelmüller, B., Hagen, J. O., Vatne, G., Ødegård, R. S., and Sollid, J. L. (1996). Glacier debris accumulation and sediment deformation influenced by permafrost: Examples from Svalbard. *Annals of Glaciology* **22**, 53–62.

Evans, D. J. A. (1989a). The nature of glacitectonic structures and sediments at subpolar glacier margins, northwest Ellesmere Island, Canada. *Geografiska Annaler* **71A**, 113–123.

Evans, D. J. A. (1989b). Apron entrainment at the margins of subpolar glaciers, northwest Ellesmere Island, Canadian high arctic. *Journal of Glaciology* **35**, 317–324.

Evans, D. J. A. (1990). The effect of glacier morphology on surficial geology and glacial stratigraphy in a high arctic mountainous terrain. *Zeitschrift fur Geomorphologie* **34**, 481–503.

Evans, D. J. A. (1999). Glacial debris transport and moraine deposition: a case study of the Jardalen cirque complex, Sogn-og-Fjordane, western Norway. *Zeitschrift fur Geomorphologie* **43**, 203–234.

Evans, D. J. A., and England, J. (1991). Canadian landform examples. 19: High arctic thrustblock moraines. *Canadian Geographer* **35**, 93–97.

Evans, D. J. A., and Hiemstra, J. F. (2005). Till deposition by glacier submarginal, incremental thickening. *Earth Surface Processes and Landforms* **30**, 1633–1662.

Evans, D. J. A., and Twigg, D. R. (2002). The active temperate glacial landsystem: a model based on Breidamerkurjokull and Fjallsjokull, Iceland. *Quaternary Science Reviews* **21**, 2143–2177.

Eyles, C. H., Eyles, N., and Lagoe, M. B. (1991). The Yakataga Formation: a late Miocene to Pleistocene record of temperate glacial marine sedimentation in the Gulf of Alaska. In *Glacial Marine Sedimentation: Palaeoclimatic Significance, Geological Society of America Special Paper 261* (J. B. Anderson and G. M. Ashley, Eds.), pp. 159–180, Boulder.

Eyles, C. H., Eyles, N., and Miall, A. D. (1985). Models of glaciomarine sedimentation and their application to the interpretation of ancient glacial sequences. *Palaeogeography, Palaeoclimatology, Palaeoecology* **51**, 15–84.

Eyles, N. (1979). Facies of supraglacial sedimentation on Icelandic and alpine temperate glaciers. *Canadian Journal of Earth Sciences* **16**, 1341–1361.

Eyles, N. (1983a). Modern Icelandic glaciers as depositional models for "hummocky moraine" in the Scottish highlands. In *Tills and Related Deposits* (E. B. Evenson, C. Schluchter and J. Rabassa, Eds.), pp. 47–59. Balkema, Rotterdam.

Eyles, N. (1983b). Glacial geology: a landsystems approach. In *Glacial Geology* (N. Eyles, Ed.), pp. 1–18. Pergamon, Oxford.

Eyles, N. (1983c). The glaciated valley landsystem. In *Glacial Geology* (N. Eyles, Ed.), pp. 91–110. Pergamon, Oxford.

Eyles, N. (1993). Earth's glacial record and its tectonic setting. *Earth Science Reviews* **35**, 1–248.

Eyles, N., and Eyles, C. H. (1992). Glacial depositional systems. In *Facies Models: Response to Sea-level Change* (R. G. Walker and N. P. James, Eds.), pp. 73–100. Geological Association of Canada, Toronto.

Eyles, N., and Menzies, J. (1983). The subglacial landsystem. In *Glacial Geology* (N. Eyles, Ed.), pp. 19–70. Pergamon, Oxford.

Eyles, N., Sladen, J. A., and Gilroy, S. (1982). A depositional model for stratigraphic complexes and facies superimposition in lodgement tills. *Boreas* **11**, 317–333.

Fitzsimons, S. J. (1997). Depositional models for moraines in East Antarctic coastal oases. *Journal of Glaciology* **43**, 256–264.

Fitzsimons, S. J. (2003). Ice-marginal terrestrial landsystems: polar continental glacier margins. In *Glacial Landsystems* (D. J. A. Evans, Ed.), pp. 89–110. Arnold, London.

Fyfe, G. J. (1990). The effect of water depth on ice-proximal glaciolacustrine sedimentation: Salpausselka. I: Southern Finland. *Boreas* **19**, 147–164.

Goldthwait, R. P. (1951). Development of end moraines in east-central Baffin Island. *Journal of Geology* **59**, 567–577.

Gravenor, C. P., von Brunn, V., and Dreimanis, A. (1984). Nature and classification of waterlain glaciogenic sediments, exemplified by Pleistocene, late Palaeozoic and late Precambrian deposits. *Earth Science Reviews* **20**, 105–166.

Greene, D. (1992). Topography and former Scottish tidewater glaciers. *Scottish Geographical Magazine* **108**, 164–171.

Gustavson, T. C., and Boothroyd, J. C. (1987). A depositional model for outwash, sediment sources, and hydrologic characteristics, Malaspina Glacier, Alaska: A modern analog of the southeastern margin of the Laurentide Ice Sheet. *Geological Society of America Bulletin* **99**, 187–200.

Hambrey, M. J. (1994). *Glacial Environments*, p. 296. UCL Press: London.

Hambrey, M. J., and Huddart, D. (1995). Englacial and proglacial glaciotectonic processes at the snout of a thermally complex glacier in Svalbard. *Journal of Quaternary Science* **10**, 313–326.

Hicock, S. R. (1992). Lobal interactions and rheologic superposition in subglacial till near Bradtville, Ontario, Canada. *Boreas* **21**, 73–88.

Hodgson, D. A. (1994). Episodic ice streams and ice shelves during retreat of the northwesternmost sector of the late Wisconsinan

Laurentide Ice Sheet over the central Canadian Arctic Archipelago. *Boreas* **23**, 14–28.

Hodgson, D. A., and Vincent, J.-S. (1984). A 10,000yr BP extensive ice shelf over Viscount Melville Sound, Arctic Canada. *Quaternary Research* **22**, 18–30.

Hooke, R..L.eB. (1973). Flow near the margin of the Barnes Ice Cap, and the development of ice-cored moraines. *Geological Society of America Bulletin* **84**, 3929–3948.

Hubbard, B. (1991). Freezing-rate effects on the physical characteristics of basal ice formed by net adfreezing. *Journal of Glaciology* **37**, 339–347.

Humlum, O. (1978). Genesis of layered lateral moraines: Implications for palaeoclimatology and lichenometry. *Geografisk Tidsskrift* 77, 65–72.

Humlum, O. (1985). Genesis of an imbricate push moraine, Höfdabrekkujøkull, Iceland. *Journal of Geology* **93**, 185–195.

Jansson, K. N., Stroeven, A. P., and Kleman, J. (2003). Configuration and timing of Ungava Bay ice streams, Labrador–Ungava, Canada. *Boreas* **32**, 256–262.

Kalin M (1971) The active push moraine of the Thompson Glacier, Axel Heiberg Island, Canadian Arctic Archipelago. McGill University, Axel Heiberg Island, Research Report, Glaciology 4, McGill University, Montreal: p. 68

Kirkbride, M. P. (1995). Processes of transportation. In *Glacial Environments, Vol. 1: Modern Glacial Environments: Processes, Dynamics and Sediments* (J. Menzies, Ed.), pp. 261–292. Butterworth-Heinemann, Oxford.

Knight, P. G. (1989). Stacking, of basal debris layers without bulk freezing on: Isotopic evidence from West Greenland. *Journal of Glaciology* **35**, 214–216.

Koteff C and Pessl F (1981) Systematic ice retreat in New England. United States Geographical Survey Professional Paper 1179: 20.

Krüger, J. (1985). Formation of a push moraine at the margin of Hofdabrekkujökull, south Iceland. *Geografiska Annaler* **67A**, 199–212.

Krüger, J. (1993). Moraine-ridge formation along a stationary ice front in Iceland. *Boreas* **22**, 101–109.

Krüger, J. (1994). Glacial processes, sediments, landforms, and stratigraphy in the terminus region of Myrdalsjökull, Iceland. *Folia Geographica Danica* **21**, 1–233.

Lawson DE (1979) Sedimentological analysis of the western terminus region of the Matanuska Glacier, Alaska. Cold Regions Research and Engineering Laboratory, Report 79-9, Hanover.

Lawson, D. E., Strasser, J. C., Evenson, E. B., Alley, R. B., Larson, G. J., and Arcone, S. A. (1998). Glaciohydraulic supercooling: A freeze-on mechanism to create stratified, debris-rich basal ice. I: Field evidence. *Journal of Glaciology* **44**, 547–562.

Martini, I. P., and Brookfield, M. E. (1995). Sequence analysis of upper Pleistocene (Wisconsinan) glaciolacustrine deposits of the north-shore bluffs of Lake Ontario, Canada. *Journal of Sedimenatary Research* **B65**, 388–400.

Matthews, J. A., and Petch, J. R. (1982). Within-valley asymmetry and related problems of Neoglacial lateral moraine development at certain Jotunheimen glaciers, southern Norway. *Boreas* **11**, 225–247.

Molnia, B. F. (1983). Subarctic glacial-marine sedimentation: a model. In *Glacial-Marine Sedimentation* (B. F. Molnia, Ed.), pp. 95–144. Plenum, New York.

Nakawo, M., Raymond, C. F. and Fountain, A. (Eds.) (2000). *Debris-covered glaciers*. International Association of Hydrological Sciences, Wallingford.

Nesje, A., and Dahl, S.-O. (2000). *Glaciers and Environmental Change*. Arnold, London.

Owen, L. A., and Derbyshire, E. (1989). The Karakoram glacial depositional system. *Zeitschrift fur Geomorphologie Supp.Bd.* **76**, 33–73.

Owen, L. A., and Derbyshire, E. (1993). Quaternary and Holocene intermontane basin sedimentation in the Karakoram Mountains. In *Himalaya to the Sea* (J. F. Shroder, Ed.), pp. 108–131. Routledge, London.

Powell, R. D. (1984). Glacimarine processes and inductive lithofacies modelling of ice shelf and tidewater glacier sediments based on Quaternary examples. *Marine Geology* **57**, 1–52.

Powell, R. D. (1990). Glacimarine processes at grounding-line fans and their growth to ice-contact deltas. In *Glacimarine Enviroments: Processes and Sediments, Geological Society Special Publication* (J. A. Dowdeswell and J. D. Scourse, Eds.), vol. 53, pp. 53–73. London.

Powell, R. D. (2003). Subaquatic landsystems: fjords. In *Glacial Landsystems* (D. J. A. Evans, Ed.), pp. 313–347. Arnold, London.

Powell, R. D., and Domack, E. (1995). Modern glaciomarine environments. In *Glacial Environments, Vol. 1: Modern Glacial Environments: Processes, Dynamics and Sediments* (J. Menzies, Ed.), pp. 445–486. Butterworth-Heinemann, Oxford.

Powell, R. D., and Molnia, B. F. (1989). Glacimarine sedimentary processes, facies and morphology of the south-southeast Alaska shelf and fjords. *Marine Geology* **85**, 359–390.

Price, R. J. (1969). Moraines, sandar, kames and eskers near Breidamerkurjokull, Iceland. *Transactions of the Institute of British Geographers* **46**, 17–43.

Punkari, M. (1995). Glacial flow systems in the zone of confluence between the Scandinavian and Novaya Zemlya ice sheets. *Quaternary Science Reviews* **14**, 589–603.

Rea, B. R., and Evans, D. J. A. (2003). Plateauicefieldlandsystems. In *Glacial Landsystems* (D. J. A. Evans, Ed.), pp. 407–431. Arnold, London.

Rea, B. R., Whalley, W. B., Evans, D. J. A., Gordon, J. E., and McDougall, D. A. (1998). Plateau icefields: Geomorphology and dynamics. *Quaternary Proceedings* **6**, 35–54.

Sharp, M. J. (1984). Annual moraine ridges at Skalafellsjokull, south-east Iceland. *Journal of Glaciology* **30**, 82–93.

Shaw, J. (1977a). Till body morphology and structure related to glacier flow. *Boreas* **6**, 189–201.

Shaw, J. (1977b). Tills deposited in arid polar environments. *Canadian Journal of Earth Sciences* **14**, 1239–1245.

Shilts, W. W., Aylsworth, J. M., Kaszycki, C. A., and Klassen, R. A. (1987). Canadian Shield. In *Geomorphic Systems of North America. Geological Society of America, Centennial Special* (W. L. Graf, Ed.), vol. 2, pp. 119–161.

Small, R. J. (1983). Lateral moraines of Glacier De Tsidjiore Nouve: form, development and implications. *Journal of Glaciology* **29**, 250–259.

Stokes, C. R., Clark, C. D., Darby, D. A., and Hodgson, D. A. (2005). Late Pleistocene ice export events into the Arctic Ocean from the M'Clure Strait ice stream, Canadian Arctic Archipelago. *Global and Planetary Change* **49**, 139–162.

Sugden, D. E., and Clapperton, C. M. (1981). An ice-shelf moraine, George VI sound, Antarctica. *Annals of Glaciology* **2**, 135–141.

Teller, J. T. (1995). History and drainage of large ice-dammed lakes along the Laurentide Ice Sheet. *Quaternary International* **28**, 83–92.

Teller, J. T. (2003). Subaquatic landsystems: Large proglacial lakes. In *Glacial Landsystems* (D. J. A. Evans, Ed.), pp. 348–371. Arnold, London.

Thompson, A. (1988). Historical development of the proglacial landforms of Svinafellsjokull and Skaftafellsjokull, southeast Iceland. *Jokull* **38**, 17–31.

Thorp, P. W. (1981). A trimline method for defining the upper limit of Loch Lomond Readvance glaciers: Examples from the Loch Leven and Glencoe areas. *Scottish Journal of Geology* **17**, 49–64.

Thorp, P. W. (1986). A mountain icefield of Loch Lomond Stadial age, western Grampians, Scotland. *Boreas* **15**, 83–97.

Thorp, P. W. (1991). Surface profiles and basal shear stresses of outlet glaciers from a Lateglacial mountain icefield in western Scotland. *Journal of Glaciology* **37**, 77–89.

Tison, J.-L., Petit, J.-R., Barnola, J.-M., and Mahaney, W. C. (1993). Debris entrainment at the ice-bedrock interface in sub-freezing temperatures, Terre Adélie, Antarctica. *Journal of Glaciology* **39**, 303–315.

Vorren, T. O. (2003). Subaquatic landsystems: Continental margins. In *Glacial Landsystems* (D. J. A. Evans, Ed.), pp. 289–312. Arnold, London.

Warren, C. R., and Hulton, N. R. J. (1990). Topographic and glaciological controls on Holocene ice margin dynamics, central west Greenland. *Annals of Glaciology* **14**, 307–310.

Weertman, J. (1961). Mechanism for the formation of inner moraines found near the edge of cold ice caps and cie sheets. *Journal of Glaciology* **3**, 965–978.

Moraine Forms and Genesis

D J A Evans, Durham University, Durham, UK

© 2007 Elsevier B.V. All rights reserved.

Push and Squeeze Moraines

Push moraines are small moraine ridges, usually less than 10 m in height, produced by minor (often annual) glacier advances in either subaqueous or terrestrial settings (e.g., Rogerson and Batterson, 1982; Sharp, 1984; **Fig. 1**). Other terms include washboard moraines, cross-valley moraines, and corrugated ground moraines. Although the term 'push moraine' has been employed in the past to refer to large-scale glaciotectonized moraine ridges, more specific terms are now available for such features and so they are not included here. Push moraines vary widely in composition, and can consist of subglacial till, mass-movement deposits, water-sorted sediments, or large boulders, depending on the nature of the sediment on the glacier foreland. Where glacier margins are debris covered, push moraines can consist mainly of supraglacial material dumped onto the forefield during the summer and pushed up during winter. Push moraines commonly have asymmetric cross-profiles, with gentle proximal slopes, often covered with small flutings and steep distal slopes. Push moraines are broadly arcuate in planform, but in detail are often irregular and winding, reflecting the morphology of the glacier snout. For example, where the glacier snout is indented by radial crevasses, push moraines have

Figure 1 Contemporary push moraines being constructed at the margin of Flaajokull, Iceland.

sawtooth planforms, with downvalley-pointing teeth and upvalley-pointing notches (Price, 1970; Matthews *et al.*, 1979; Sharp, 1984; Evans and Twigg, 2002; **Fig. 2**). It is common to find younger moraines overlying or completely consuming parts of older moraines wherever summer retreat is small and/or some winter readvances are more substantial than usual (Sharp, 1984; Boulton, 1986; Evans and Twigg, 2002). Where glacier snouts are quasi-stable or have reached a position where they are in equilibrium with their environment, large terminal moraines may accumulate by annual accretion. Similarly, the absence of moraines relating to individual years within a field of annual moraines can be explained by unusually large winter advances or longer term readvances. These can result in ice advance that exceeds several years of annual retreat and so winter push moraines can be destroyed or incorporated into a larger moraine ridge (Rogerson and Batterson, 1982; Boulton, 1986; Krüger, 1993, 1994). The construction of larger push moraine ridges has been explained in some settings as the result of stacking of till slabs due to submarginal freeze-on of subglacial sediment and its release by melt-out over a seasonal cycle (Matthews *et al.*, 1995; Kruger, 1996; Evans and Hiemstra, 2005; **Fig. 3**).

The formation of minor moraine ridges by squeezing of saturated sediment from beneath ice margins has been observed in front of several glaciers whose snouts are predominantly underlain by poorly drained, fine-grained till (Price, 1970; Worsley, 1974; **Fig. 4**). Additionally, subaqueous squeeze moraines have been described by Andrews and Smithson (1966). Newly formed squeeze moraines have steep or vertical sides and are rarely more than 1-m high. Squeeze moraines tend to degrade rapidly, and are commonly subject to reworking by ice push, so that

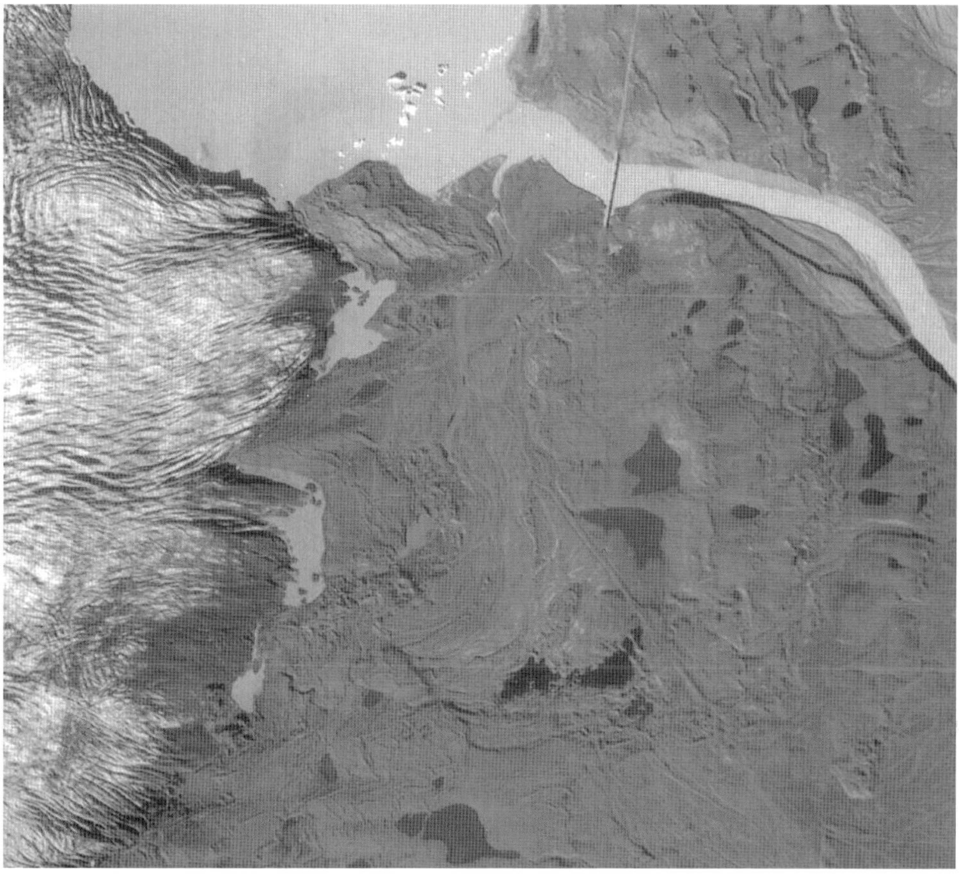

Figure 2 Aerial photograph of the margin of Fjallsjokull, Iceland in 1965, showing sequence of annual push moraines with 'saw-tooth' plan forms (© University of Glasgow and Landmaelingar Islands).

unmodified squeeze moraines are probably rare in the geomorphological record.

Dump Moraines and Ice-marginal Aprons

Material accumulating at the surface of a glacier through the melt-out of debris-rich folia is ultimately subject to remobilization by mass flowage, fall, or fluvial transport. Where such material exists at the glacier margin its remobilization may result in it being dumped onto the adjacent terrain during ice recession. Depending on the rate of debris accumulation and glacier activity, a variety of sediment-landform associations will be created. For retreating glaciers, dumping of supraglacial material onto the former subglacial surface slowly emerging from beneath the ice produces a thin veneer of coarse grained diamicton or perhaps only sporadic collections of boulders (Eyles, 1979). Dump moraines will form where the ice margin remains stationary during debris accumulation (Boulton and Eyles, 1979), although they will be bulldozed into push moraines if the glacier undergoes a subsequent readvance. Dump moraine size is related to supraglacial debris volume and the length of the stillstand period. Small dump moraines will form where glaciers remain stationary during the winter but retreat during the summer, each moraine marking one winter's increment of debris accumulation. The largest and most spectacular dump moraines, however, form at the margins of debris-mantled valley glaciers which occupy similar positions for considerable periods (**Fig. 5**).

The dumping of large quantities of supraglacial debris around glacier margins builds up laterofrontal dump moraines (Boulton and Eyles, 1979; Small, 1983, 1987) through the processes of sliding, rolling, flowage, and fall from the ice surface to produce ice-contact scree. In mountain environments, the debris may be predominantly passively transported rockfall material, consisting of coarse, angular clasts with little matrix. However, actively transported debris can also crop out near valley glacier margins due to the elevation of basal debris septa (linear debris concentrations in glacier ice), thereby introducing matrix-rich debris to some dump moraines. On some laterofrontal moraines, there is a tendency for clast roundness to increase in a down-moraine/

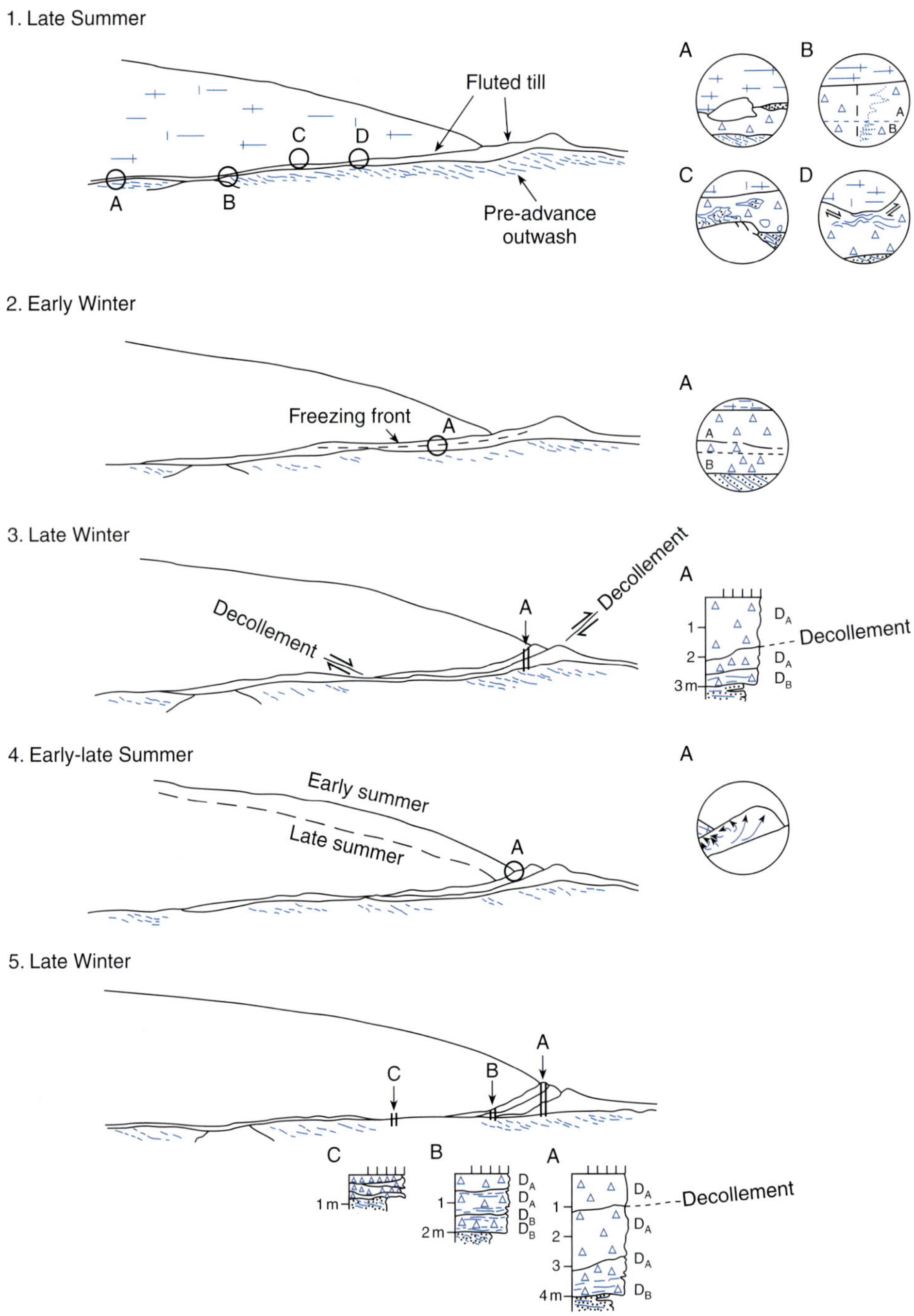

Figure 3 A model of push moraine formation by the incremental stacking of till slabs during an annual cycle of subglacial deformation, freeze-on, and melt-out (from Evans and Hiemstra, 2005). (1) Situation in late summer at typical Icelandic glacier snout where subglacial processes include lodgement and sliding (A), bedrock and sediment plucking (B), subglacial deformation (C), and ice-keel ploughing (D) in a temporally and spatially evolving process mosaic. (2) During early winter, the thin part of the glacier snout freezes onto part of the subglacial till. The till slab that freezes onto the ice sole is likely to be from the more porous A horizon (A). (3) The later winter readvance initiates failure along a decollement plane within the A horizon or at the junction with the more compact B horizon, resulting in the carriage of A horizon till onto the proximal side of the previous year's push moraine. (4) In the early summer, the melt-out of the till slab (A) initiates porewater migration, water escape, and sediment flow (small arrows) and sediment extrusion due to glaciostatic and glaciodynamic stresses. (5) The late summer situation is again followed by winter freeze-on and marginal stacking of subglacial till produced by the reworking of existing subglacial sediments and fresh materials advected to submarginal locations from up-ice. Repeated reworking of the thin end of submarginal till wedges produces overprinted strain signatures and clast pavements.

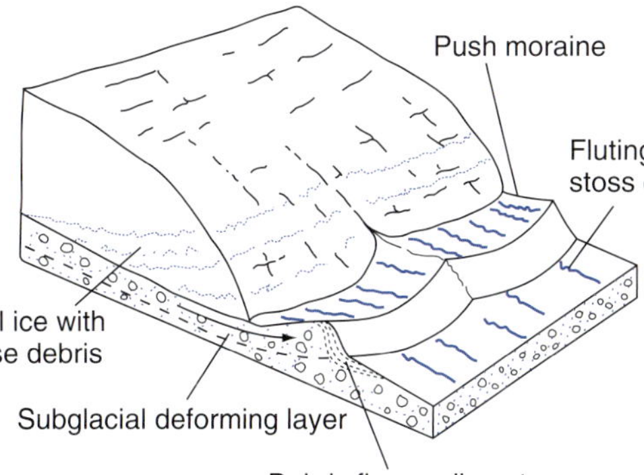

Figure 4 The formation of a push moraine by the combined action of pushing and squeezing from beneath an ice margin (after Price, 1970).

Figure 5 Laterofrontal moraines at the margin of the Hooker Glacier, New Zealand, formed by the accumulation of debris dumped at the highly debris-charged ice margin when it stood at its Little Ice Age limit. Some glacier pushing will also have contributed to moraine construction.

down-valley direction due to the increasing proportion of actively transported debris in the moraines closer to the former glacier center-line (Matthews and Petch, 1982; Benn and Ballantyne, 1994; Evans, 1999). Where dump moraines are deposited on steep valley sides, debris derived directly from the glacier is interdigitated with sediment from the valley walls, producing an internal stratigraphy composed of: (i) stratified debris flows and fall sorted debris avalanche material with beds dipping away from glacier margin; (ii) subhorizontally bedded alluvium deposited as kame terraces along the ice margin; and (iii) debris flows and fall sorted screes from valley-side paraglacial fans. Removal of lateral support by glacier recession results in mass failure of the inner slopes of the moraines. In many cases, such failures clearly extend up into the screes which were deposited against the lateral moraines.

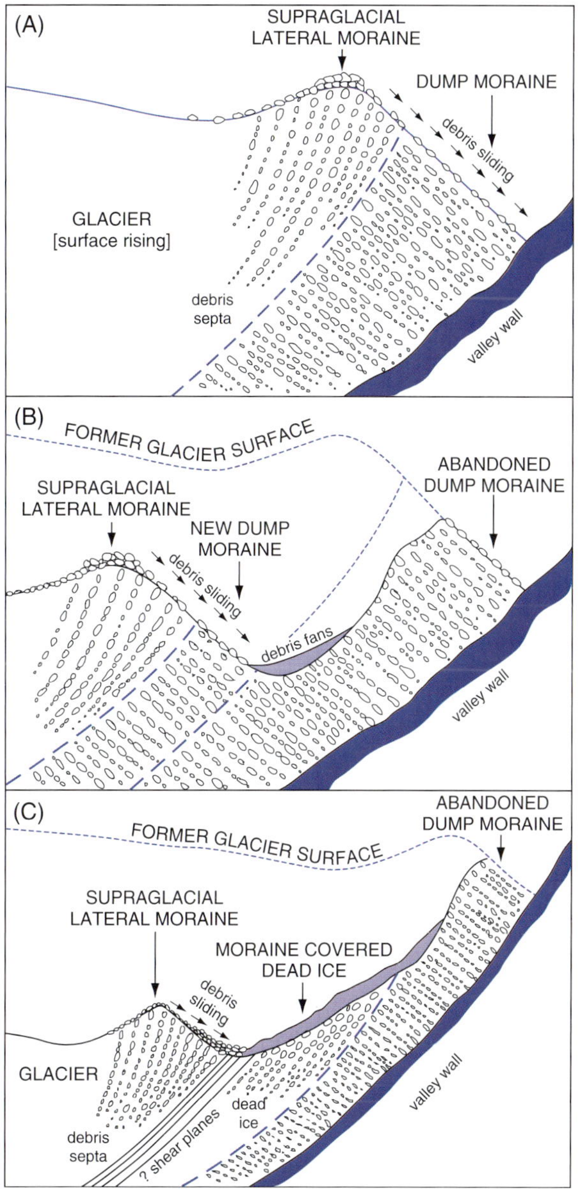

Figure 6 A model of lateral moraine formation by ice-marginal dumping of debris (from Small, 1983).

During glacier thinning, dump moraines are abandoned and their inner faces are subject to collapse and reworking (**Fig. 6**). A new inset dump moraine may be formed within the older one if the glacier restabilizes at a new position, but such moraines will be unstable if they are deposited on top of dead ice masses. Alternatively, renewed thickening of the glacier may dump material on top of the old dump moraine, which may become completely buried. In this case, periods of nondeposition may be recorded by paleosols or even buried trees, which are valuable sources of paleoclimatic data in Alpine environments (Röthlisberger *et al.*, 1980).

Figure 7 Ice-marginal apron produced by dumping of debris and dry calving of ice blocks from a subpolar glacier snout, Ellesmere Island, Arctic Canada.

Distinctive dump moraines known as frontal- or ice-marginal aprons form around the margins of some Arctic and polar glaciers (Shaw, 1977; Evans, 1989; Fitzsimmons, 1997). These moraines form where debris and ice blocks accumulate at the base of steep, terminal ice cliffs by a combination of dry calving and limited melting (**Fig. 7**). When forming, the aprons have steep ice-contact slopes and gentler ice-distal slopes determined by the properties of the debris. However, they will be subject to considerable modification following deglaciation due to the ablation of incorporated ice blocks.

Laterofrontal Fans and Ramps

Deposition by debris flows and glaciofluvial processes around stationary glacier margins can result in the construction of laterofrontal fans and ramps, consisting of coalescent debris fans which descend from the glacier snout (Owen and Derbyshire, 1989; Kuhle, 1990; Owen, 1991; Owen and Derbyshire, 1993; Benn *et al.*, 2003; **Fig. 8**). After glacier recession such landforms are identifiable by their pronounced asymmetric cross profile, comprising a distal slope that is much shallower in gradient than on laterofrontal dump moraines and a steep proximal or ice-contact face. Fans deposited predominantly by glaciofluvial processes grade distally into outwash tracts or sandar, producing 'outwash head' landforms (Kirkbride, 2000; Benn *et al.*, 2003). In some situations where moraines are absent, fans and ramps provide the only evidence of former ice-marginal positions. Ice-contact fan morphology develops over a considerable period, sometimes involving several generations of deposition with intervening periods of abandonment. During fan accumulation, active depo-centers shift position, reflecting changes in the morphology of the ice margin and fan surface. As the fan progrades, surface angles tend to become progressively reduced. Meltwater exiting from englacial (within the ice) positions or draining the glacier surface locally excavates the ice-contact fan/ramp faces to produce inset meltwater fans. In active mountain belts ice-contact fans can reach heights of hundreds of meters, attesting to very high rates of debris transfer. Because debris flow fans and ramps are deposited partly over glacier ice, they may be characterized by pitted or degraded surfaces after ice melt-out. When ice-contact fans are abandoned by glacier retreat, the inner faces are susceptible to collapse and may provide material for lateral moraine development during ice recession.

Figure 8 Debris flow dominated ice-marginal fans marking the former margin of the Batal Glacier in the Lahul Himalaya.

The sedimentology of laterofrontal fans/ramps is dominated by massive, often bouldery debris flow diamictons with beds up to tens of meters thick, interbedded with water-sorted sediments and dipping away from the glacier snout. These sequences comprise coalescent ice-marginal fans. Pockets of glaciolacustrine sediment, which have in some instances been glaciotectonized, attest to the occurrence of short-lived proglacial lakes ponded between the ice margin and the fans during ice retreat phases.

Medial Moraines

Medial moraines are prominent features on the surfaces of many glaciers and are the surface expression of a medial debris septum, which may be a shallow feature or extend to the base of the glacier (**Fig. 9**). Ablation results in the concentration of the debris septum on the glacier surface. Medial moraines on existing glaciers have been classified by Eyles and Rogerson (1978) based on the relationship between debris supply and morphological development. Three broad types were recognized, including (i) ablation-dominant moraines, which emerge at the surface as the result of the melt-out of englacial debris, (ii) ice-stream interaction moraines, which find immediate surface expression downstream from glacier confluences, often by the merging of two supraglacial lateral moraines, and (iii) avalanche-type moraines, which are transient features formed by exceptional rockfall events onto a glacier surface. Medial moraines do not have good preservation potential because they contain relatively little debris and are subject to intense reworking during glacier melt. They sometimes can be traced from modern glacier surfaces onto proglacial surfaces, where they may be recognizable as diffuse spreads of debris with distinctive lithological composition or particle morphologies (Evans and Twigg, 2002). However, such moraines commonly have no marked surface expression, and may be very difficult to recognize once the glacier foreland becomes vegetated. Recognition of ancient medial moraines is easier where they consist of large boulders, which may be traced back to their source outcrops (e.g., Dawson (1979)).

Figure 9 Medial moraines at the margin of Breidamerkurjokull, Iceland. The recession of the glacier snout has resulted in the lowering of the medial moraines on to subglacial landforms to produce a thin rubbly veneer visible in the foreground.

In certain situations ice stream interaction type medial moraines have a high preservation potential and may constitute interlobate moraines. For example, on the west coast of New Zealand very large ice stream interaction medial moraines, which are greater than 100 m high in some areas, have been produced by the coalscence of ancient piedmont glaciers with well-developed lateral moraines. Because these features accumulated essentially as lateral moraines, they contain sediments characteristic of ice-marginal deposition but the moraine cores commonly display glacitectonic structures, and in places, carapaces of subglacial till record glacier coalescence and overriding. Benn and Evans (1998) have suggested that wherever such interlobate features become overridden by glacier ice they may evolve into megaflutings. Some medial/interlobate moraines may comprise large quantities of glaciofluvial sediment, for example where lobate margins of an ice sheet coalesce. Two large examples occur at the southern margins of the former Laurentide Ice Sheet. The Harricana Interlobate Moraine, formerly separating the Hudson Bay and Labrador sectors of the receding ice sheet, stretches for 1,000 km southwards from James Bay to Lake Simcoe, Ontario, Canada (Viellette, 1986, 1988) and comprises a series of ridges up to 10 km wide and 100 m high. The Burntwood-Knife Interlobate Moraine, formerly separating the Hudson Bay and Keewatin sectors of the receding Laurentide Ice Sheet, covers a distance in excess of 500 km on the west side of Hudson Bay and is composed of broad ridges of glaciofluvial sediment and till. Individual segments are up to 12 km long, 4 km wide, and 60 m high (Dredge *et al.*, 1986; Klassen, 1986; Dyke and Dredge, 1989). The large quantities of sands and gravels comprising these moraines were deposited in proglacial lacustrine environments during the final stages of the Laurentide Ice Sheet, attesting to the large-scale reworking of medial moraine debris during glacier recession. Glaciofluvial reworking of medial moraine debris during ice wastage can produce features indicative of former medial moraine locations even though they are strictly different landforms. For example, continuous esker systems may demarcate the former positions of medial moraines in glaciers or ice sheets where debris septum ensures constant sediment supply to englacial drainage networks (e.g., eskers and medial moraines in Breidamerkurjokull, Iceland; Price, 1969, 1973;

Evans and Twigg, 2002). Similarly, a low-relief kame and kettle topography centered on the bottom of the Athabasca Valley near Jasper in Alberta, Canada is interpreted as a former supraglacial medial moraine by Levson and Rutter (1989; **Fig. 10**).

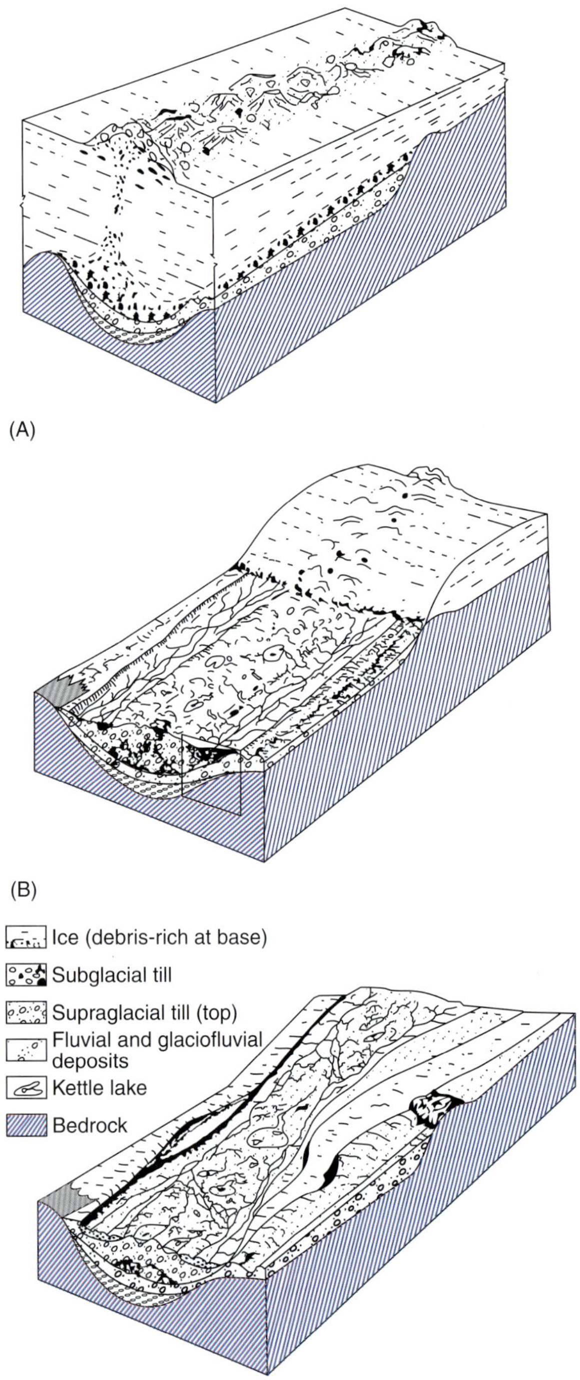

Figure 10 A model of medial moraine formation based on Canadian Rocky Mountain case studies (Levson and Rutter, 1989). Gradual lowering of the medial moraine results in the formation of a linear ice-cored moraine. This develops into a chain of kettle lakes and hummocky terrain comprising rubbly supraglacial debris.

Hummocky Moraine and Controlled Moraine

The term 'hummocky moraine' has been employed in a wide range of senses. As a purely descriptive term, it has been applied to moundy, irregular morainic topography exhibiting varying degrees of order, ranging from entirely chaotic assemblages of mounds to suites of nested transverse ridges (Sissons, 1967; Benn, 1992; Attig and Clayton, 1993; Bennett and Boulton, 1993; **Fig. 11**). Genetically, it has been used to encompass landforms with a variety of origins, including deposition in association with active and stagnant ice, and glacitectonic deformation (e.g., Hoppe, 1952; Gravenor and Kupsch, 1959; Stalker, 1960; Aario, 1977; Bennett and Boulton, 1993; Eyles *et al.*, 1999). The term 'hummocky moraine' is now used in a more restrictive sense (Benn and Evans, 1998) to refer to moraines deposited during the melt-out of debris-mantled glaciers (e.g., Sharp, 1949; Hoppe, 1952, 1959; Sharp, 1985). This genetic definition should only be applied to suites of moraines whose origin has been established by detailed mapping and sedimentological analyses, because generalized descriptions of surface morphology alone are insufficient to determine moraine genesis. For example, many tracts of so-called hummocky moraine in the Highlands of Scotland, originally believed to be chaotic ice-stagnation topography (e.g., Sissons 1967, 1974, 1979), have subsequently been recognized as nested push and dump moraines, or drumlins draped by supraglacial and ice-marginal deposits (Benn, 1992; Bennett and Boulton, 1993; Bennett, 1994). Hummocky moraine deposited from debris-mantled ice can appear chaotic or linear in pattern, depending on the distribution of debris in the parent glacier and the patterns of debris redistribution and reworking during ice wastage. Occurrences with

Figure 11 Hummocky moraine produced by the melting of a debris-charged surge snout, Tungnaarjokull, Iceland.

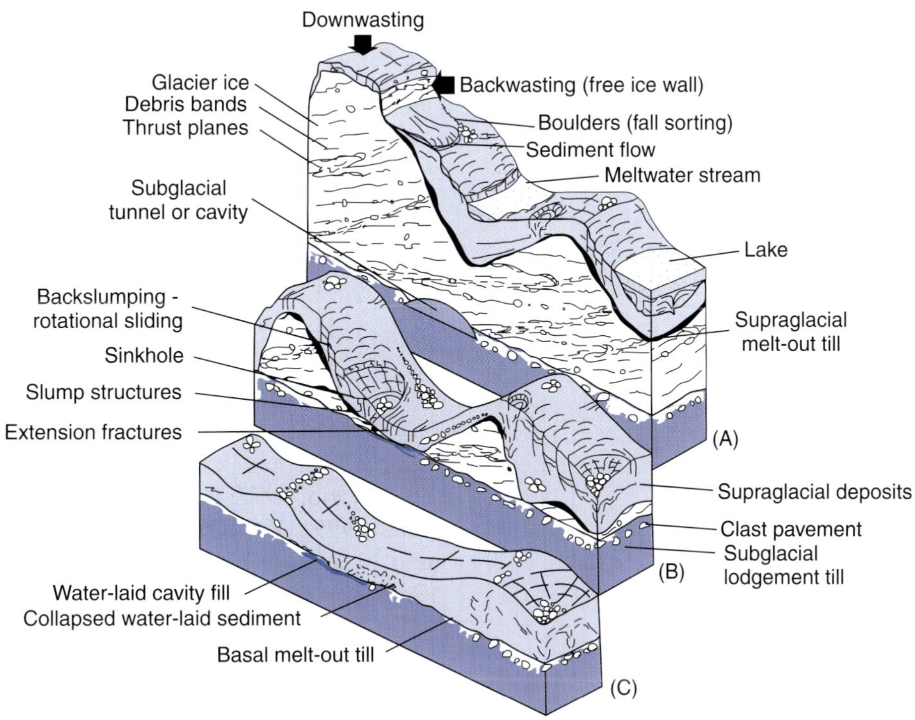

Figure 12 A model of hummocky moraine development due to the melting of a debris-charged glacier snout (from Kjær and Krüger, 2001).

pronounced transverse linear elements are termed controlled moraines (Benn and Evans, 1998).

Hummocky moraine is the end product of topographic inversion cycles during the ablation of debris-mantled ice (**Fig. 12**). During ice ablation, debris is transferred away from topographic highs on the glacier surface by mass movements and meltwater, exposing ice cores to renewed melting and creating new depressions in former high points. Further debris reworking and topographic development is achieved by meltwater streams meandering between dirt cones, the collapse of englacial tunnels, and the enlargement of supraglacial lake basins. Several cycles of debris reworking and topographic inversion may occur before the debris is finally deposited in a series of irregular mounds and ridges.

The distribution of controlled moraines is inherited from debris concentrations or septa within the parent glacier. Transverse debris concentrations close to glacier margins may result from the thusting of debris along shear planes (Goldthwait, 1951; Bishop, 1957; Souchez, 1967), the emergence of basal debris frozen onto the glacier sole (Boulton, 1970), or reworked ice-marginal aprons (Hooke, 1970; Hudleston, 1976; Shaw, 1977; Evans, 1989). Such transverse septa control the pattern of differential ablation on the glacier surface, producing linear ice-cored ridges which eventually become separated from the snout to form ice-cored moraines (Østrem, 1964; Dyke and Evans, 2003). The morphological expression of englacial structures is always most striking while the moraines retain ice cores (**Fig. 13**). However, uneven sediment redistribution during ice-core wastage means that the final deposits tend to consist of discontinuous transverse ridges with intervening hummocks, preserving only a weak impression of the former englacial structure. Reworking by supraglacial meltwater systems also acts to destroy large tracts of controlled moraine before they can be deposited on the ground surface. Patterns of debris reworking and deposition can also be controlled by crevasses or other lines of weakness within the ice, which can guide the final stages of glacier disintegration (Gravenor and Kupsch, 1959).

The sedimentology of hummocky moraine is complex, reflecting multiple cycles of redeposition during its formation (Kjær and Krüger, 2001). Typical facies associations consist of interbedded debris flows and other mass-movement deposits, laminated lacustrine sediments, and glaciofluvial sands and gravels, in varying degrees of disturbance (Eyles, 1979; Paul, 1983). The constituent debris may be actively or passively transported, depending on the position of the moraine relative to debris transport paths in the parent glacier. In some cases, debris from different sources and transport paths may remain segregated within the moraine, giving rise to compositional stratification (Benn, 1992).

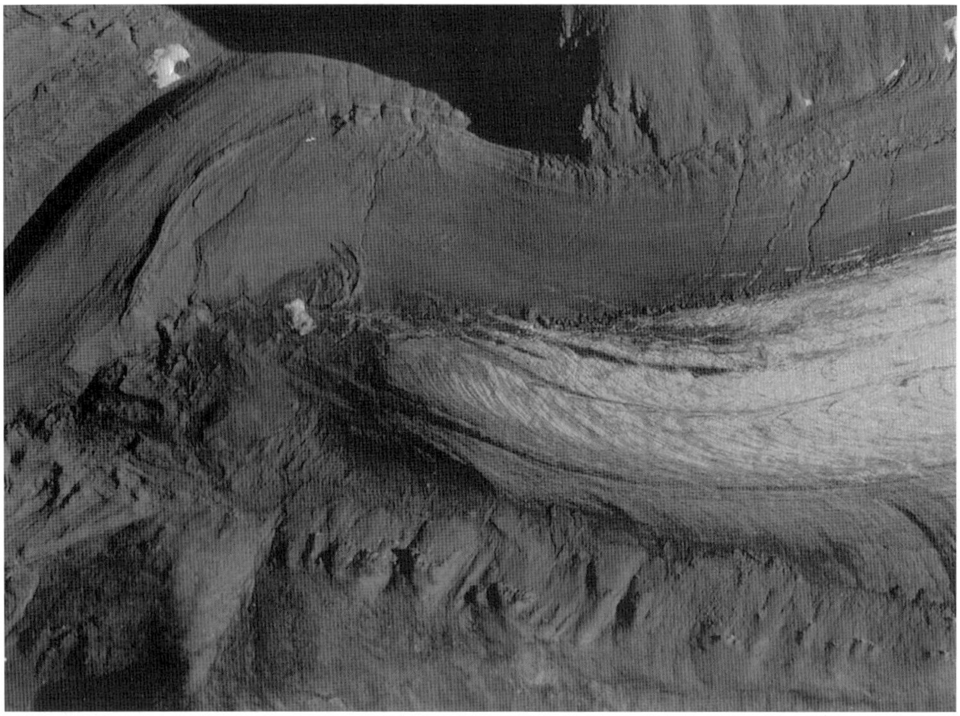

Figure 13 Controlled moraine on the surface of a glacier snout in St. Jonsfjorden, Svalbard. Aerial photograph from the Norsk Polar Institute.

Morainal Banks and Coalescent Subaqueous Fans

Wherever glacier grounding lines remain quasi-stable for long periods of time and/or sediment fluxes to the glacier margin are high, subaqueous moraines will form by the accumulation of sediment along the glacier margin. Debris may be delivered from the ice surface, the melt-out of debris-rich ice and icebergs, the release of unfrozen till from a subglacial deforming layer, or reworking of older deposits (**Fig. 14**). This debris can be reworked in conveyor belt fashion by erosion off the subglacial face and deposition on the subaqueous slope (Powell, 1984). Subaqueous moraines that extend up into shallow water can also be exposed to wave energy, leading to winnowed and sorted gravels and sands in their upper parts (Powell, 1984). Some subaqueous moraines mainly consist of coalescent grounding-line fans, deposited at the same time at the efflux points (exit locations) of adjacent meltwater conduits. For example, the Salpausselka Moraines of Finland consist of

Figure 14 A reconstruction of the depositional environment in which morainal banks accumulate at the glacier grounding line, based on the example at Achnasheen, Scotland (from Benn, 1996). (1) subglacial till; (2) mass flows; (3) mass flow diamictons; (4) subglacial meltwater portal; (5) subaqueous outwash; (6) laminated sediments; and (7) iceberg dump mound.

transverse chains of ice-contact fans and deltas deposited during stillstands of the southern margins of the Scandinavian ice sheet where it contacted deep water (Fyfe, 1990). Similarly, some large-scale moraines, comprising partially glaciotectonized grounding-line fans, occur in eastern Maine, USA (Ashley *et al.*, 1991) where they are arranged laterally along former grounding-line positions and are associated with tunnel mouth and esker deposits. Some subaqueous moraines are formed mainly by dumping and other gravitational processes at the foot of terminal ice cliffs. This is especially the case where subpolar glaciers with thick basal debris sequences contact the sea, releasing large amounts of debris in environments where there is only limited subglacial meltwater (Powell, 1984). At many glacier margins, however, gravitational deposits are interbedded with subaqueous outwash, particularly at the margins of temperate glaciers where meltwater is abundant. For example, McCabe *et al.* (1984, 1987), McCabe (1985, 1986), and Eyles and McCabe (1989b) present Pleistocene examples of interbedded gravitational and outwash deposits from the east-central coast of Ireland in which channelized subaqueous outwash and debris flow facies interdigitate with ice-rafted dropstone muds. These sediments show evidence of glacitectonic deformation and underlie arcuate belts of hummocky topography.

De Geer Moraines

Named after Swedish geologist Gerard De Geer, De Geer moraines commonly occur in fields of closely spaced ridges in association with subaqueous sediments, and mark the intermittent retreat of water terminating glaciers (**Fig. 15**). Many models have

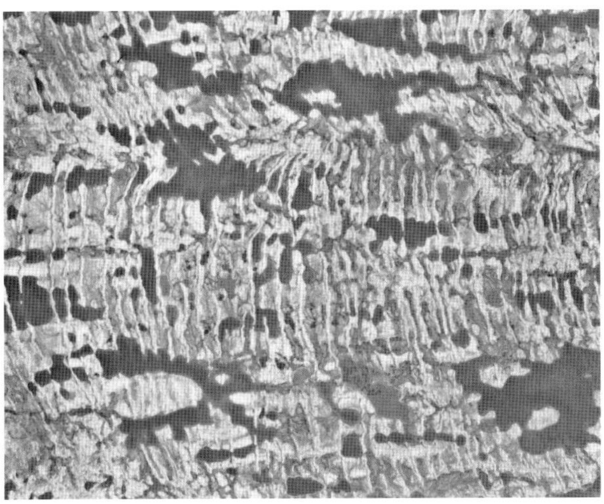

Figure 15 Aerial photograph of De Geer moraines in Quebec, Canada (National Aerial Photograph Library A14882-91).

been proposed to account for De Geer moraine formation but the consensus is that they mark former ice-margin positions, where sediment is deposited or pushed up during brief stillstands or minor readvances. In common with terrestrial push moraines, De Geer moraines are characterized by steeper distal than proximal slopes, lobate forms with acute interlobe angles pointing up glacier, and truncated sections due to partial overriding by later moraines (Boulton, 1986). The formation of subaqueous moraines with steep ice-distal slopes and gentle ice-proximal slopes has also been linked to the calving of floating glacier snouts, perhaps on a yearly basis (Aartolahti, 1972; Holdsworth, 1973b). In contrast, De Geer moraines may result from the squeezing of saturated sediment out from beneath the ice margin (Andrews and Smithson, 1966). The notion that several De Geer moraines may be produced simultaneously has been proposed by Hoppe (1957), Lundqvist (1989), and Zilliacus (1989). This implies formation in basal crevasses a short distance behind glacier margins in a similar fashion to terrestrial crevasse-fills (cf. Solheim, 1986). Specifically, transverse crevasses are formed in areas of extending flow and then widened during partial flotation of the ice margin. Moraines are thought to form when the water level drops and the glacier is lowered onto the substratum, which is squeezed up into the basal crevasses. Tabular icebergs would later break off along the transverse crevasses thus ensuring that the moraines were not later overridden by the glacier. Such an explanation actually implies a crevasse-squeeze origin for the ridges and suggests that the glacier snouts responsible for their construction underwent draw-down and fracturing (Lundqvist, 1989) possibly in response to surging (Zilliacus, 1989). Such an origin would make the ridges directly comparable with terrestrial crevasse-fill ridges on surging glacier forelands (Sharp, 1985; Evans and Rea, 1999, 2003). Similarly, features referred to as 'lift-off moraines' have been identified in seismostratigraphic investigations of the Scotian and mid-Norwegian shelves by King and Fader (1986) and King *et al.* (1991). The production of lift-off moraines is thought to be related to fracture patterns in the base of a retreating glacier snout, involving a genesis similar to that proposed by Lundqvist (1989). It is uncertain whether or not lift-off moraines are produced individually by single events or simultaneously as groups (King *et al.*, 1991).

Ice-shelf Moraines

Ice-shelf moraines are formed where debris is deposited or relocated around the margins of ice

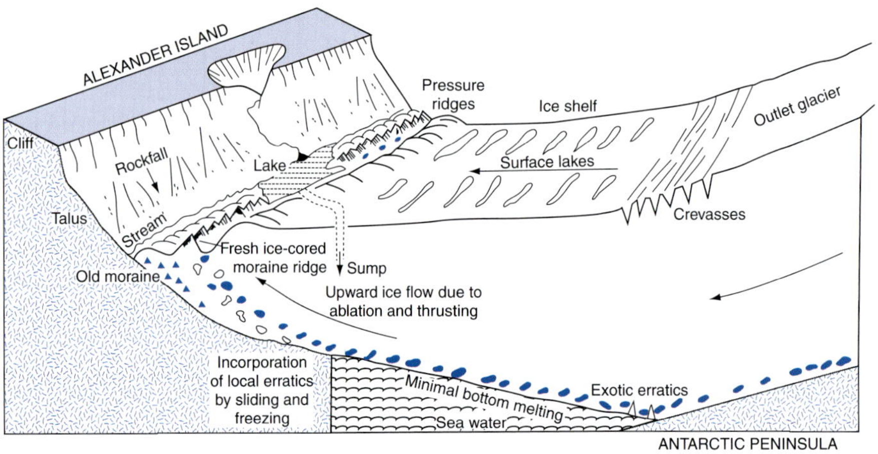

Figure 16 A model for the development of the George VI Sound ice shelf moraine, Antarctica (from Sugden and Clapperton, 1981).

shelves or floating glacier tongues. The crestlines of such moraines tend to be horizontal or have very low gradients, reflecting the low gradient of ice shelves which have zero or very low basal shear stresses. The gradients of ancient ice-shelf moraines, however, may be steeper due to postglacial isostatic tilting, so their gradients should be corrected with reference to the sea-level record. The recent ice-shelf moraine at the margin of the George VI Sound ice shelf, Antarctica, formed where complex ice pressure ridges were induced by flow against Alexander Island. A horizontal moraine ridge consequently formed over a distance of 120 km along the promontories of Alexander Island where basal debris was brought to the glacier surface by ablation and thrusting (**Fig. 16**). Sediments within the moraine include subglacial debris entrained on Alexander Island and the Antarctic Peninsula, glaciofluvial and lacustrine sediments from ice-marginal streams and ponds, and marine sediments and fauna derived from freezing on at the grounding line on Alexander Island (Sugden and Clapperton, 1981). Ancient ice-shelf moraines are rare (e.g., England *et al.*, 1978; Evans *et al.*, 2002) but a spectacular example occurs at the former northwestern margin of the Laurentide Ice Sheet in Viscount Melville Sound, where extensive linear belts of till (Winter Harbour Till), between the altitudes of 60 and 150 m above present sea level, blanket the shores of Byam Martin, Melville, Banks, and Victoria islands and extend eastwards to north Baffin Island (Hodgson and Vincent, 1984; Dyke *et al.*, 1992; Klassen, 1993).

See also: **Glacial Landforms**: Glacitectonic Structures and Landforms. **Glacial Landforms, Sediments**: Glacifluvial Landforms of Deposition; Glacimarine Sediments and Ice-Rafted Debris.

References

Aario, R. (1977). Classification and terminology of moraine landforms in Finland. *Boreas* **6,** 87–100.

Aartolahti, T. (1972). On deglaciation in southern and western Finland. *Fennia* **114,** 1–84.

Andrews, J. T., and Smithson, B. B. (1966). Till fabrics of the cross-valley moraines of north-central Baffin Island, NWT, Canada. *Bulletin of the Geological Society of America* **77,** 271–290.

Ashley, G. M., Boothroyd, J. C., and Borns, H. W. (1991). Sedimentology of Late Pleistocene (Laurentide) deglacial-phase deposits, eastern Maine; An example of a temperate marine grounded ice-sheet margin. In *Geological Society of America Special Paper* (J. B. Anderson and G. M. Ashley, Eds.), Glacial marine sedimentation; paleoclimatic significance, **261**, pp. 107–125.

Attig, J. W., and Clayton, L. (1993). Stratigraphy and origin of an area of hummocky glacial topography, northern Wisconsin, USA. *Quaternary International* **18,** 61–67.

Benn, D. I. (1992). The genesis and significance of 'hummocky moraine': evidence from the Isle of Skye, Scotland. *Quaternary Science Reviews* **11,** 781–799.

Benn, D. I. (1996). Subglacial and subaqueous processes near a glacier grounding line: sedimentological evidence from a former ice-dammed lake, Achnasheen, Scotland. *Boreas* **25,** 23–36.

Benn, D. I., and Ballantyne, C. K. (1994). Reconstructing the transport history of glacigenic sediments: a new approach based on the co-variance of clast form indices. *Sedimentary Geology* **91,** 215–227.

Benn, D. I., and Evans, D. J. A. (1998). *Glaciers and glaciation.* Arnold, London.

Benn, D. I., Kirkbride, M. P., Owen, L. A., and Brazier, V. (2003). Glaciated valley landsystems. In *Glacial landsystems* (D. J. A. Evans, Ed.), pp. 372–406. Arnold, London.

Bennett, M. R. (1994). Morphological evidence as a guide to deglaciation following the Loch Lomond Readvance: a review of research approaches and models. *Scottish Geographical Magazine* **110,** 24–32.

Bennett, M. R., and Boulton, G. S. (1993). A reinterpretation of Scottish 'hummocky moraine' and its significance for the deglaciation of the Scottish highlands during the Younger Dryas or Loch Lomond Stadial. *Geological Magazine* **130,** 301–318.

Bishop, B. C. (1957). Shear moraines in the Thule area, northwest Greenland. US Snow, Ice and Permafrost Research Establishment, Research Report 17.

Boulton, G. S. (1970). On the origin and transport of englacial debris in Svalbard glaciers. *Journal of Glaciology* **9**, 213–229.

Boulton, G. S. (1986). Push moraines and glacier contact fans in marine and terrestrial environments. *Sedimentology* **33**, 677–698.

Boulton, G. S., and Eyles, N. (1979). Sedimentation by valley glaciers: a model and genetic classification. In *Moraines and varves* (C. Schluchter, Ed.), pp. 11–23. Balkema, Rotterdam.

Dawson, A. G. (1979). A Devensian medial moraine in Jura. *Scottish Journal of Geology* **15**, 43–48.

Dredge, L. A., Nixon, F. M., and Richardson, R. J. (1986). Quaternary geology and geomorphology of northwestern Manitoba. *Geological Survey of Canada, Memoir* **418**, 38.

Dyke, A. S., and Dredge, L. A. (1989). Quaternary geology of the northwestern Canadian shield. In *Quaternary geology of Canada and Greenland* (R. J. Fulton, Ed.), 1, pp. 178–214. Geological Survey of Canada, Geology of Canada.

Dyke, A. S., and Evans, D. J. A. (2003). Ice-marginal terrestrial landsystems: Northern Laurentide and Innuition ice sheet margins. In *Glacial Landsystems* (D. J. A. Evans, Ed.), pp. 143–165. Arnold, London.

Dyke, A. S., Morris, T. F., Green, D. E. C., and England, J. (1992). Quaternary geology of Prince of Wales Island, arctic Canada. *Geological Survey of Canada, Memoir* **433**, 142.

England, J., Bradley, R. S., and Miller, G. H. (1978). Former ice shelves in the Canadian high Arctic. *Journal of Glaciology* **20**, 393–404.

Evans, D. J. A. (1989). The nature of glacitectonic structures and sediments at subpolar glacier margins, northwest Ellesmere Island, Canada. *Geografiska Annaler* **71A**, 113–123.

Evans, D. J. A. (1999). Glacial debris transport and moraine deposition: a case study of the Jardalen cirque complex, Sogn-og-Fjordane, western Norway. *Zeitschrift fur Geomorphologie* **43**, 203–234.

Evans, D. J. A., and Hiemstra, J. F. (2005). Till deposition by glacier submarginal, incremental thickening. *Earth Surface Processes and Landforms* **30**, 1633–1662.

Evans, D. J. A., and Rea, B. R. (1999). Geomorphology and sedimentology of surging glaciers: a landsystems approach. *Annals of Glaciology* **28**, 75–82.

Evans, D. J. A., and Rea, B. R. (2003). Surging glacier landsystem. In *Glacial landsystems* (D. J. A. Evans, Ed.), pp. 259–288. Arnold, London.

Evans, D. J. A., and Twigg, D. R. (2002). The active temperate glacial landsystem: a model based on Breidamerkurjokull and Fjallsjokull, Iceland. *Quaternary Science Reviews* **21**, 2143–2177.

Evans, D. J. A., Rea, B. R., Hansom, J. D., and Whalley, W. B. (2002). The geomorphology and style of plateau icefield glaciation in a fjord terrain, Troms-Finnmark, north Norway. *Journal of Quaternary Science* **17**, 221–239.

Eyles, N. (1979). Facies of supraglacial sedimentation on Icelandic and alpine temperate glaciers. *Canadian Journal of Earth Sciences* **16**, 1341–1361.

Eyles, N., Boyce, J. I., and Barendregt, R. W. (1999). Hummocky moraine: sedimentary record of stagnant Laurentide Ice Sheet lobes resting on soft beds. *Sedimentary Geology* **123**, 163–174.

Eyles, N., and McCabe, A. M. (1989). Glaciomarine facies within subglacial tunnel valleys: the sedimentary record of glacio-isostatic downwarping in the Irish Sea Basin. *Sedimentology* **36**, 431–448.

Eyles, N., and Rogerson, R. J. (1978). A framework for the investigation of medial moraine formation: Austerdalsbreen, Norway, and Berendon Glacier, British Columbia, Canada. *Journal of Glaciology* **20**, 99–113.

Fitzsimons, S. J. (1997). Depositional models for moraines in East Antarctic coastal oases. *Journal of Glaciology* **43**, 256–264.

Fyfe, G. J. (1990). The effect of water depth on ice-proximal glaciolacustrine sedimentation: Salpausselka I, southern Finland. *Boreas* **19**, 147–164.

Goldthwait, R. P. (1951). Development of end moraines in east-central Baffin Island. *Journal of Geology* **59**, 567–577.

Gravenor, C. P., and Kupsch, W. O. (1959). Ice disintegration features in western Canada. *Journal of Geology 67, 48–64.*

Hodgson, D. A., and Vincent, J. S. (1984). A 10,000yr BP extensive ice shelf over Viscount Melville Sound, Arctic Canada. *Quaternary Research* **22**, 18–30.

Holdsworth, G. (1973). Ice deformation and moraine formation at the margin of an ice cap adjacent to a proglacial lake. In *Research in polar and alpine geomorphology* (B. D. Fahey and R. D. Thompson, Eds.), pp. 187–199. Geo Abstracts, Norwich.

Hooke, R. LeB. (1970). Morphology of the ice-sheet margin near Thule, Greenland. *Journal of Glaciology* **9**, 303–324.

Hoppe, G. (1952). Hummocky moraine regions, with special reference to the interior of Norrbotten. *Geografiska Annaler* **34**, 1–72.

Hoppe, G. (1957). Problems of glacial morphology and the ice age. *Geografiska Annaler* **39**, 1–18.

Hoppe, G. (1959). Glacial morphology and inland ice recession in northern Sweden. *Geografiska Annaler* **41**, 193–212.

Hudleston, P. J. (1976). Recumbent folding in the base of the Barnes Ice Cap, Baffin Island, Northwest Territories, Canada. *Geological Society of America Bulletin* **87**, 1684–1692.

King, L. H., and Fader, G. (1986). Wisconsinan glaciation of the continental shelf – southeast Atlantic Canada. *Geological Survey of Canada, Bulletin* 363.

King, L. H., Rokoengen, K., Fader, G. B. J., and Gunleiksrud, T. (1991). Till tongue stratigraphy. *Bulletin of the Geological Society of America* **103**, 637–659.

Kirkbride, M. P. (2000). Ice marginal geomorphology and Holocene expansion of debris-covered Tasman Glacier, New Zealand. In *Debris-covered glaciersDebris-covered glaciers* (M. Nakawo, C. Raymond and A. Fountain, Eds.), 264, pp. 211–217. IAHS Publication.

Kjær, K. H., and Krüger, J. (2001). The final phase of dead-ice moraine development: processes and sediment architecture, Kötlujökull, Iceland. *Sedimentology* **48**, 935–952.

Klassen, R. W. (1986). Surficial geology of north-central Manitoba. *Geological Survey of Canada, Memoir* **419**, 57.

Klassen, R. A. (1993). Quaternary geology and glacial history of Bylot Island, Northwest Territories. *Geological Survey of Canada, Memoir* **429**, 93.

Krüger, J. (1993). Moraine-ridge formation along a stationary ice front in Iceland. *Boreas* **22**, 101–109.

Krüger, J. (1994). Glacial processes, sediments, landforms, and stratigraphy in the terminus region of Myrdalsjökull, Iceland. *Folia Geographica Danica* **21**, 1–233.

Kruger, J. (1996). Moraine ridges formed from subglacial frozen on sediment slabs and their differentiation from push moraines. *Boreas* **25**, 57–63.

Kuhle, M. (1990). Ice marginal ramps and alluvial fans in semiarid mountains: convergence and difference. In *Alluvial Fans: a Field Approach* (A. H. Richocki and M. Church, Eds.), pp. 55–68. John Wiley & Sons: Chichester.

Levson, V. M., and Rutter, N. W. (1989). Late Quaternary stratigraphy, sedimentology, and history of the Jasper townsite area, Alberta, Canada. *Canadian Journal of Earth Sciences* **26**, 1325–1342.

Lundqvist, J. (1989a). Rogen (ribbed) moraine – identification and possible origin. *Sedimentary Geology* **62**, 281–292.

Matthews, J. A., and Petch, J. R. (1982). Within-valley asymmetry and related problems of Neoglacial lateral moraine

development at certain Jotunheimen glaciers, southern Norway. *Boreas* **11,** 225–247.
Matthews, J. A., Cornish, R., and Shakesby, R. A. (1979). Saw-tooth' moraines in front of Bodalsbreen, southern Norway. *Journal of Glaciology* **22,** 535–546.
Matthews, J. A., McCarroll, D., and Shakesby, R. A. (1995). Contemporary terminal-moraine ridge formation at a temperate glacier: Styggedalsbreen, Jotunheimen, southern Norway. *Boreas* **24,** 129–139.
McCabe, A. M. (1985). Geomorphology. In *The Quaternary history of Ireland* (K. J. Edwards and W. P. Warren, Eds.), pp. 67–93. Academic Press, London.
McCabe, A. M. (1986). Glaciomarine facies deposited by retreating tidewater glaciers – an example from the Late Pleistocene of Northern Ireland. *Journal of Sedimentary Petrology* **56,** 880–894.
McCabe, A. M., Dardis, G. F., and Hanvey, P. M. (1984). Sedimentology of a Late Pleistocene submarine-moraine complex, County Down, Northern Ireland. *Journal of Sedimentary Petrology* **54,** 716–730.
McCabe, A. M., Dardis, G. F., and Hanvey, P. M. (1987). Sedimentation at the margins of a Late Pleistocene ice lobe terminating in shallow marine environments, Dundalk Bay, eastern Ireland. *Sedimentology* **34,** 473–493.
Østrem, G. (1964). Ice-cored moraines in Scandinavia. *Geografiska Annaler* **46,** 282–337.
Owen, L. A. (1991). Mass movement deposits in the Karakoram Mountains. *Zeitschrift fur Geomorphologie* **35,** 401–424.
Owen, L. A., and Derbyshire, E. (1989). The Karakoram glacial depositional system. *Zeitschrift fur Geomorphologie Supp* **76,** 33–73.
Owen, L. A., and Derbyshire, E. (1993). Quaternary and Holocene intermontane basin sedimentation in the Karakoram Mountains. In *Himalaya to the sea* (J. F. Shroder, Ed.), pp. 108–131. Routledge, London.
Paul, M. A. (1983). The supraglacial landsystem. In *Glacial geology* (N. Eyles, Ed.), pp. 71–90. Pergamon, Oxford.
Powell, R. D. (1984). Glacimarine processes and inductive lithofacies modelling of ice shelf and tidewater glacier sediments based on Quaternary examples. *Marine Geology* **57,** 1–52.
Price, R. J. (1969). Moraines, sandar, kames and eskers near Breidamerkurjokull, Iceland. *Transactions of the Institute of British Geographers* **46,** 17–43.
Price, R. J. (1970). Moraines at Fjallsjokull, Iceland. *Arctic and Alpine Research* **2,** 27–42.
Price, R. J. (1973). *Glacial and fluvioglacial landforms*, p. 242. Oliver and Boyd, Edinburgh.
Rogerson, R. J., and Batterson, M. J. (1982). Contemporary push moraine formation in the Yoho Valley, BC. In *Research in glacial, glacio-fluvial and glacio-lacustrine systems* (R. Davidson-Arnott, W. Nickling and B. D. Fahey, Eds.), pp. 71–90. Geo Books, Norwich.
Röthlisberger, F., Haas, P., Holzhauser, H., Keller, W., Bircher, W., and Renner, F. (1980). Holocene climatic fluctuations – radiocarbon dating of fossil soils (fAh) and woods from moraines and glaciers in the Alps. *Geographica Helvetica* **35,** 21–52.
Sharp, M. J. (1984). Annual moraine ridges at Skalafellsjokull, south-east Iceland. *Journal of Glaciology* **30,** 82–93.
Sharp, M. J. (1985). Sedimentation and stratigraphy at Eyjabakkajokull – an Icelandic surging glacier. *Quaternary Research* **24,** 268–284.
Sharp, R. P. (1949). Studies of superglacial debris on valley glaciers. *American Journal of Science* **247,** 289–315.
Shaw, J. (1977). Till body morphology and structure related to glacier flow. *Boreas* **6,** 189–201.
Sissons, J. B. (1967). *The evolution of Scotland's scenery*. Oliver and Boyd, Edinburgh.
Sissons, J. B. (1974). The Quaternary in Scotland: a review. *Scottish Journal of Geology* **10,** 311–337.
Sissons, J. B. (1979). The Loch Lomond Stadial in the British Isles. *Nature* **280,** 199–203.
Small, R. J. (1983). Lateral moraines of Glacier De Tsidjiore Nouve: form, development and implications. *Journal of Glaciology* **29,** 250–259.
Small, R. J. (1987). Moraine sediment budgets. In *Glacio-fluvial sediment transfer: an Alpine perspective* (A. M. Gurnell and M. J. Clark, Eds.), pp. 165–197. John Wiley & Sons, Chichester.
Solheim, A. (1986). Submarine evidence of glacier surges. *Polar Research* **4,** 91–95.
Souchez, R. A. (1967). The formation of shear moraines: an example from south Victoria Land, Antarctica. *Journal of Glaciology* **6,** 837–843.
Stalker, A. MacS. (1960). Ice-pressed drift forms and associated deposits in Alberta. *Geological Survey of Canada, Bulletin* **57,** 38.
Sugden, D. E., and Clapperton, C. M. (1981). An ice-shelf moraine, George VI Sound, Antarctica. *Annals of Glaciology* **2,** 135–141.
Veillette, J. J. (1986). Former southwesterly ice flows in the Abitibi-Timiskaming region: implications for the configuration of the late Wisconsinan ice sheet. *Canadian Journal of Earth Sciences* **23,** 1724–1741.
Vielette, J. J. (1988). Deglaciation et evolution des lacs proglaciaires Post-Algonquin et Barlow au Temiscamingue, Quebec et Ontario. *Geographie Physique et Quaternaire* **42,** 7–31.
Worsley, P. (1974). Recent 'annual' moraine ridges at Austre Okstindbreen, Okstindan, north Norway. *Journal of Glaciology* **13,** 265–277.
Zilliacus, H. (1989). Genesis of DeGeer moraines in Finland. *Sedimentary Geology* **62,** 309–317.

Quaternary Vulcanism, Subglacial Landforms

J L Smellie, British Antarctic Survey, Cambridge, UK

© 2007 Elsevier B.V. All rights reserved.

Volcano–ice interactions are common on Earth (Smellie and Chapman, 2002). A few eruptions have been observed and there are many extinct edifices of Quaternary age and older. Where ice sheets have shrunk or long since vanished, the subglacial volcanic edifices are frequently highly accessible and permit close-up, detailed, field analysis. They are overwhelmingly the products of basaltic eruptions since basaltic volcanism is the dominant mode of volcanic activity on Earth. Edifices formed by more evolved magmas also occur but are much less common. The internal structure, hydraulic evolution, and eruptive processes have been relatively well investigated (Smellie and Skilling, 1994; Gudmundsson *et al.*, 2004; Smellie, 2000, 2001, 2006a; Tuffen *et al.*, 2002a,b). Most of those studies concentrated on documenting the lithofacies present, which were extremely poorly known prior to the 1990s, and using the sequence architecture to reconstruct the mode(s) of edifice construction and associated eruption dynamics. By contrast, the study of subglacial volcanic landforms has not advanced far and there are very few publications (Hickson, 2000; Smellie, 2006b).

The term 'glaciovolcanic' is used to refer to all types of volcano interactions with ice in all its forms (including snow and firn) and, by implication, any meltwater derived from that ice by volcanic heating (Smellie, 2006a). It includes not only eruptions and their products that took place underneath a glacier or ice sheet (i.e., subglacial) but also subaerial volcanic products (e.g., lavas and pyroclastic flows) that simply ponded and chilled against an ice barrier or else interacted with snow by overriding or entrainment (Thouret, 1990; Lescinsky and Sisson, 1998; Mee *et al.*, 2006). The many different types of glaciovolcanic sequences identified so far can be used to obtain a uniquely wide range of critical parameters of past terrestrial ice sheets, for which other (i.e., non-volcanic) records are typically patchily and poorly preserved and rely on geomorphologic criteria that are highly susceptible to reworking (e.g., moraines and trimlines). They are thus a valuable, but currently underused, proxy for documenting paleoenvironmental changes over time (Edwards and Russell, 2002; Mee *et al*, 2006; Smellie *et al.*, 2006). Reliably identifying and documenting the occurrence of subglacially erupted volcanoes and their distribution over geological time is also central to understanding the inventory and dynamic behavior of past and present water on Earth.

Terrestrial landforms also provide our best empirical basis for interpreting remotely observed features on Earth's nearest planetary neighbors (i.e., the moon, Mars, and Venus) and understanding their geological evolution. Mars, like Earth, also has a substantial inventory of volatile elements, including an unusually thick cryolithosphere and surface ice in the form of small polar ice caps that may have been much more extensive in times past. It is also dominated by basaltic volcanism. The description, quantification, and interpretation of Mars' volcanic landforms are wholly reliant on satellite observations and comparisons with possible terrestrial analogs. The Mars landforms include candidate examples of mafic tuyas (volcanic table mountains) and tephra cones and ridges, which have been erupted subglacially (Chapman and Tanaka, 2001; Head and Platt, 2001; Ghatan and Head, 2002; Chapman and Smellie, 2006). Their identification was used to suggest the former existence of a much expanded south polar ice cap in Hesperian times (ca. 1.8–3.5 Gyr).

This article focuses on volcanic landforms that were erupted and constructed subglacially. There are no known distinctive landforms that were constructed following volcanic interactions with snow or subsurface ice; subaerial volcanic constructs that simply chilled against surface ice are also excluded. The ability to make robust identifications of volcanic landforms and understanding the intrinsic and extrinsic environmental controls on their formation are important research frontiers that will become increasingly important for deciphering ancient planetary glacial records.

Subglacial Volcanic Landforms

Subglacial eruptions on Earth have formed several different types of volcanic landforms (Smellie, 2006b). However, our knowledge of these landforms is so poor that the classification and descriptions of those landforms are almost certainly oversimplified. The descriptions that follow also take little account of possible posteruptive erosional effects. Because of the intimate association with active ice masses, the volcanic outcrops will rapidly become substantially degraded. However, they are described here essentially as primary morphologies. The different volcanic landforms reflect differences in (a) lithofacies and sequence architecture; (b) intrinsic properties of the enclosing ice sheet or glacier (i.e., thickness, thermal regime, and meltwater hydraulics); and (c) magmatic composition, volatile content, rheology, and discharge (Hickson, 2000; Smellie, 2000, 2001, 2006b; Tuffen *et al.*, 2002b). All but two of the landforms identified so far are volcano edifices. The two exceptions (called subglacial sheet-like sequences) are also common subglacial volcanic products, and they form an additional volcanic landform resulting from subglacial eruptions. The six different volcanic landform types are illustrated in **Figure 1** and summarized in **Table 1**.

Practically all of the published examples of subglacial volcanic landforms were constructed during relatively short-lived monogenetic eruptions beneath temperate or wet-based ice. There are far fewer descriptions of long-lived (i.e., millions of years) subglacial stratovolcanoes and there are indications that the morphometry of the larger landforms may be very different from those formed by monogenetic eruptions. Our fundamental knowledge of subglacial volcanic landforms is thus at an early stage, and the following descriptions are heavily biased toward the

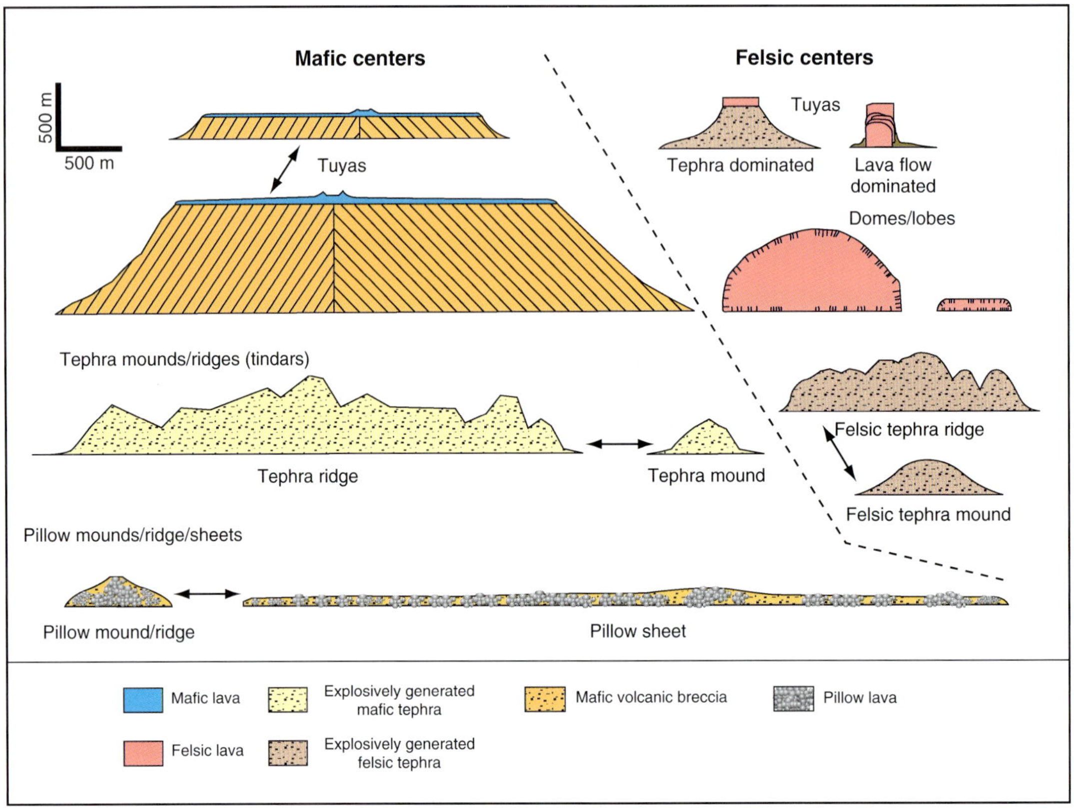

Figure 1 Approximately true-scale vertical profile ketches showing the principal characteristics of the major subglacial volcanic landforms described in this article. Sheet-like sequences are not shown but would resemble pillow sheets in dimensions.

monogenetic examples and should be viewed in that context.

This article is based mainly on a review by Smellie (2006b). The landforms are divided empirically into mafic examples, typically basalt or basaltic andesite in composition, and felsic types, which comprise andesite and more evolved magmas. There is a clear hierarchy of landform types (**Fig. 2**). However, the distinction is unlikely to be so clear-cut in nature and some felsic subglacial volcanoes, particularly those with intermediate compositions, will probably be found that show features more commonly found in mafic systems and vice versa. A preliminary assessment of the landform morphometry is included in each section, using volume and aspect ratio as empirical parameters. Aspect ratio is defined as the ratio of maximum height (above volcano base) to edifice width. For sheet-like sequences, deposit thickness is used rather than height since the coeval edifice dimensions are either unknown or undiagnostic. Aspect ratio is an admittedly crude empirical parameter and it does not closely reproduce many true volcano profiles, but it is still a useful and informative parameter that can enable distinctions to be made. For example, it cannot resolve the distinctive combination of relatively flat tops and steep perimeter cliffs that characterize mafic tuyas, and it also fails to recognize summit craters. However, this is not always a problem. For example, craters are among the first casualties of posteruptive erosion, particularly when constructed of poorly consolidated tephra.

Mafic Tuyas

Mafic tuyas are the most distinctive and frequently described subglacial volcanic edifices. They are also known nongenetically as table mountains or mesas, although not all of those are tuyas. Tuyas are also the most common subglacial volcanic landform to have been identified on Mars. The edifices are typically small and steep-sided, with relatively flat or, more typically, gently domed summit regions with or without a crater, and oval to straight-sided plan shapes (van Bemmelen and Rutten, 1955; Jones, 1969; **Fig. 3**). Although commonly regarded as products of monogenetic eruptions, even some volumetrically quite small examples can be polygenetic (Jones, 1969; Werner and Schmincke, 1999; Smellie, 2001). Published examples of much larger polygenetic mafic tuya volcanoes are scarce, possibly because the edifices would need to be erupted consistently beneath an ice cover that persisted or repeatedly reformed

Table 1 Classification of terrestrial subglacial volcanic landforms and sequences

Landform	*Mafic tuya*	*Felsic tuya (1), tephra dominated*	*Felsic tuya (2), lava flow dominated*	*Felsic domes and lobes*	*Tephra mound/ ridge (tindar)*	*Pillow mound/sheet*	*Sheet-like sequences*
Composition	Basalt	Andesite, dacite, rhyolite, trachyte, phonolite	Andesite, dacite, rhyolite, trachyte, phonolite	Andesite, dacite, rhyolite, trachyte, phonolite	Basalt, andesite, dacite, rhyolite	Basalt	Basalt
Morphology	Table mountain, mesa	Table mountain, mesa	Flat-topped column, bladed ridge	Tall steep-sided domes, lobes, or irregular	Steep cone or ridge	Low oblate smooth mounds, subdued ridges, extensive sheets	Thin sheets, sinuous ribbons and lobes
Characteristic lithologies	Subaerial lava, glassy breccia, tephra	Sheet lava, tephra, pumice, polymict sediments	Sheet lava, tephra, polymict sediments	Lava, glassy breccia	Tephra	Pillow and (probably) sheet lava	Volcanic sand and gravel, sheet and pillow lava, glassy breccia
Architecture	Lava-fed delta on precursor tephra cone (tindar) and/or pillow mound	Lava-capped table mountain; ash core	Lava column	Lava mound (mainly endogenous?)	Cone/ridge	Mound/sheet	Esker-like sediments, sheet lava and tephra
Deposit type	Vent edifice	Vent edifice	Vent edifice	Vent edifice	Vent edifice	Vent edifice	Outflow facies
Eruptive environment	Ice sheet; glacial lake	Ice sheet	Ice sheet	Alpine glacier, mountain ice cap, ice sheet	Ice sheet; glacial vault/lake	Ice sheet; not well-known	Alpine glacier, mountain ice cap; ice sheet
Known distribution	Iceland, Canada, Antarctica	Iceland	Canada	Iceland, Canada	Iceland, Antarctica	Iceland, Canada, Antarctica?	Iceland, Canada, Antarctica

Modified From Smellie (2006b).

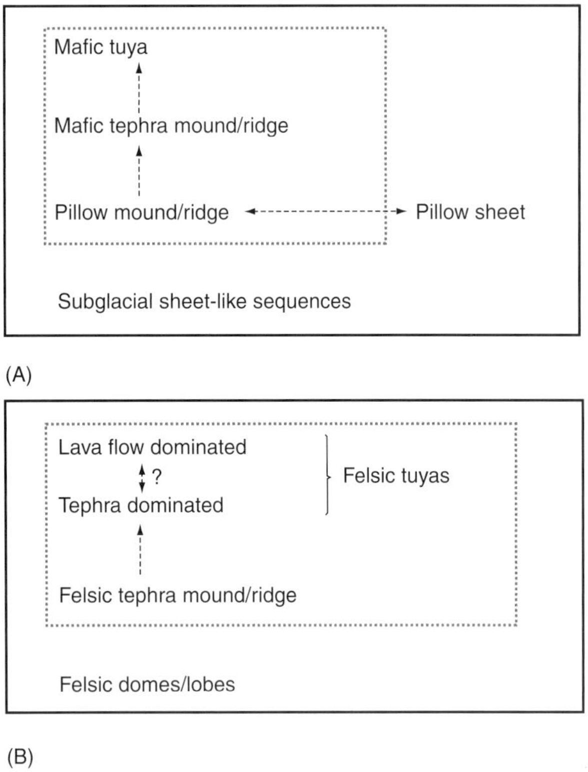

Figure 2 Hierarchy of subglacial volcanic landforms. The dashed arrows indicate possible evolutionary trends between landforms that are commonly related in a single edifice (within dashed-line boxes). With the possible exception of mafic tuyas, each landform can exist without a different precursor landform and some probably always exist in isolation (e.g., felsic domes/lobes).

Figure 3 Examples of mafic tuyas. (A) Herdubreid (Iceland) viewed looking north. It rises approximately 800 m and is probably the best known mafic tuya. (B) Three large mafic tuyas in Iceland (Thorisjökull, Geitlansjökull, and Eiriksjökull, from left to right) exhumed by the postglacial retreat of the Langjökull ice cap. Thorisjökull tuya is approximately 600 m high and 10 km wide. The approximately coincident summit elevations suggest that they erupted through the same ice sheet and thus might be of similar age.

Figure 4 Synthetic view of James Ross Island (Antarctica) viewed looking northwest. This is the world's largest known subglacially erupted basalt stratovolcano, approximately 60 km in diameter and 1.5 km high. It is constructed of multiple superimposed lava-fed deltas. ASTER image draped on a DEM; approximately three times vertical exaggeration.

during the lifetime of the stratovolcano (millions of years). The only documented large polygenetic terrestrial example known is the Mount Haddington volcano on James Ross Island, Antarctica. It rises 1.5 km, has a basal diameter of 60 km, encircling cliffs up to 600 m high, and was constructed in multiple extrusive phases over a period of at least 6.5 million years since the latest Miocene (**Fig. 4**). It has a low surface profile similar to a shield volcano but has the steep margins characteristic of confinement by ice, although modified by numerous posteruptive arcuate collapses.

Mafic tuyas are defined by the presence of lava-fed deltas. Thus, they require a meltwater lake to form in, and if a lava-fed delta is absent, the landform is unlikely to be a mafic tuya (**Fig. 5**). The deltas represent a late subaerial progradational phase of edifice building by relatively rapid lateral extension (Smellie, 2000, 2001), and they have a distinctive sequence architecture comprising a relatively thin caprock formed of flat-lying subaerial lava overlying steep-dipping, homoclinal, glassy volcanic breccias formed by the mechanical disintegration of the caprock lava during quenching where it reached and entered water, in this case a glacial lake (**Fig. 5**). Although tuyas commence construction as a subaqueous tuff

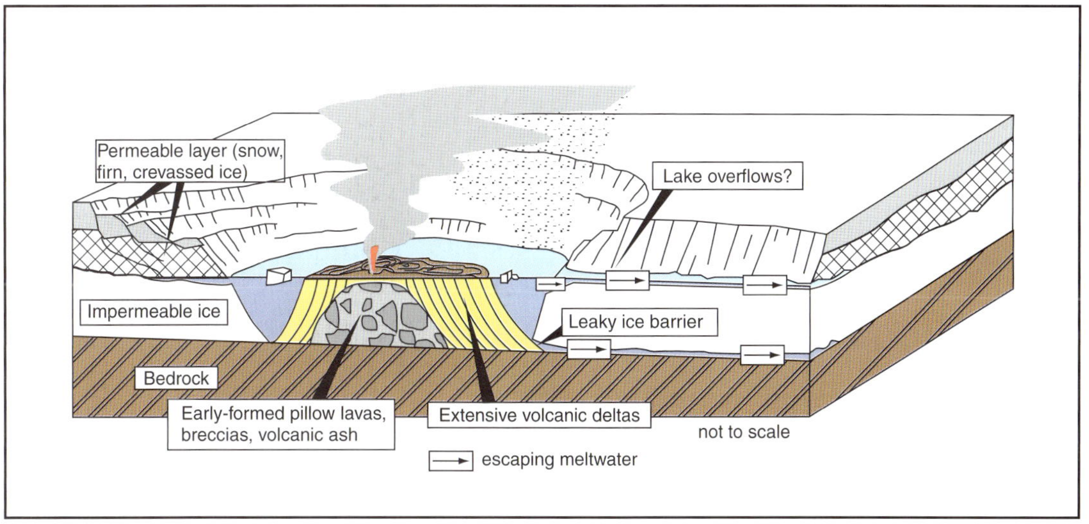

Figure 5 Drawing of a mafic tuya forming in a temperate ice sheet, illustrating the internal composition of the tuya and possible main meltwater discharge routes.

cone and/or pillow mound, those earlier formed constructs do not significantly affect the final shape of the edifice. The flat or gently convex top is a subaerial shield volcano constructed after the ice sheet was melted through completely and it is formed of lavas with a summit crater or small pyroclastic cone of scoria or agglutinate. The principal limitation on the lateral extent of a mafic tuya is the volume of magma discharged over time since deltas will advance as long as space is available and magma is supplied. By contrast, tuyas formed by eruptions within wet-based temperate glaciers might be limited to heights less than 750–1,000 m since eruptions in much thicker ice sheets should always theoretically float the surrounding glacier and a stable meltwater lake (necessary for a lava-fed delta to form) will not occur (Smellie, 2006b). However, this suggestion is speculative, model-based, and not well supported by either theory or observed eruptions, although it remains possible (Smellie, 2006a).

Area is plotted against volume in **Figure 6**. Mafic tuyas have the largest volumes of any subglacial volcanic landform and their areas are also among the largest. The volume:area data for most of the other subglacial volcanic landforms fall within a box bounded by 2.0 and 10 km^2, whereas most of the mafic tuyas fall outside of that box. The mafic tuyas also define a moderately well-defined regression trend that has the highest volume:area ratio for any of the subglacial volcanic landforms apart (possibly) from felsic domes, although the data set is still small (**Fig. 6B**). **Figure 7A** shows aspect ratios using, as the horizontal dimension, the basal length of edifices defined as the maximum horizontal extent. The length is strongly affected by the mode of eruption, whether from fissures or point sources. Fissure length is a tectonically controlled variable and is not an intrinsic volcano-determined dimension. The different types of volcano outcrops generally define discrete fields in the diagram. Tuya landforms in subfield 'a' (**Fig. 7A**) reflect point source eruptions (i.e., relatively symmetrical outcrops), whereas those in subfield 'b' are elongate fissure-erupted tuyas.

In **Figure 7B**, aspect ratio is plotted against width, defined as the minimum horizontal extent. Because the selected widths are perpendicular to fissure length, they reflect volcano construction processes rather than tectonics. The diagram illustrates that there is a clear difference in aspect ratio between mafic and felsic subglacial landforms, with the mafic centers characterized by lower ratios. Moreover, most of the mafic tuyas have a fairly constant aspect ratio of 0.18, comparable to the ratio for point-sourced tuyas in **Figure 7A** (subfield a) that used edifice length as an index. The comparability across both diagrams, using different indices (length versus width), implies that the growth patterns of most mafic tuyas are related to eruptions from central rather than fissure sources. A subordinate subfield of much wider mafic tuyas is also present in **Figure 7B** (subfield b). The simplest explanation for subfield b is that those edifices were constructed during unusually voluminous eruptions, when the associated lava-fed deltas prograded laterally to a greater extent than tuyas in subfield a. If true, both subfields should in reality be continuous. The separation is probably an artifact of the current small data set and the two fields will probably merge as more data become available.

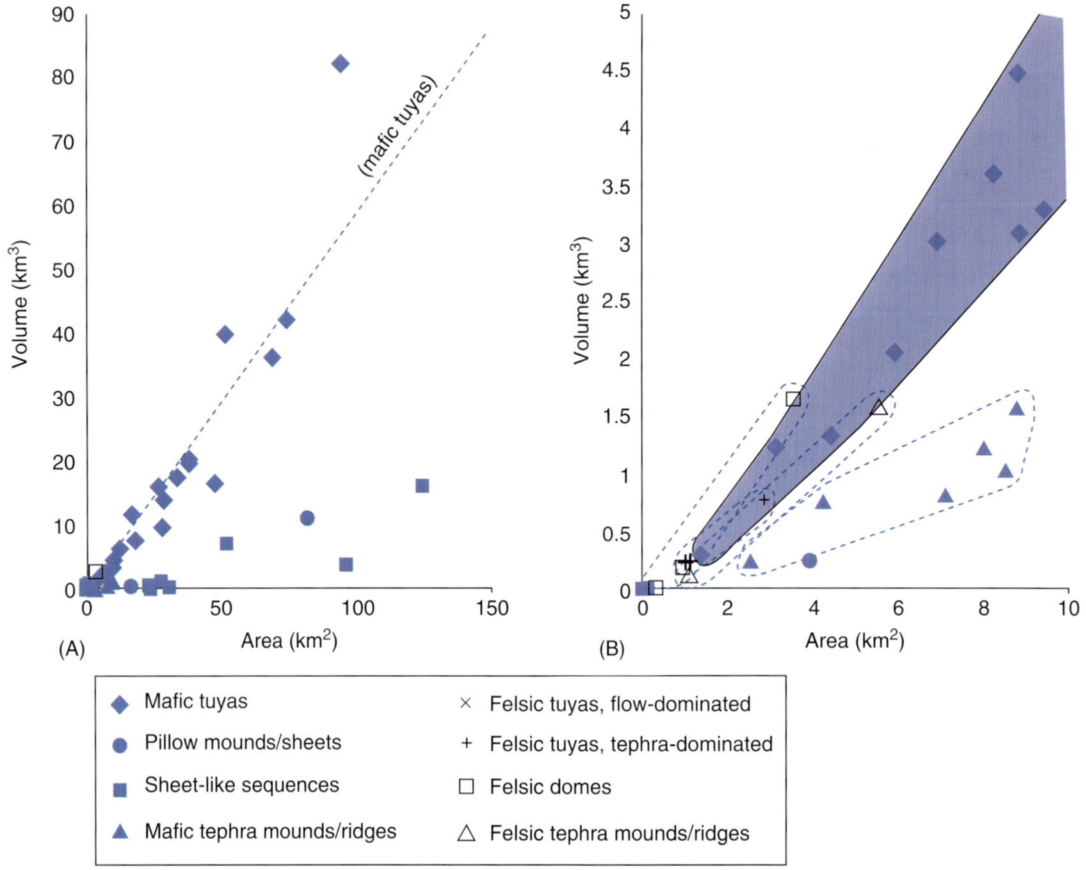

Figure 6 Morphometric data for subglacial volcanic landforms: volume versus area. After Smellie (2006b).

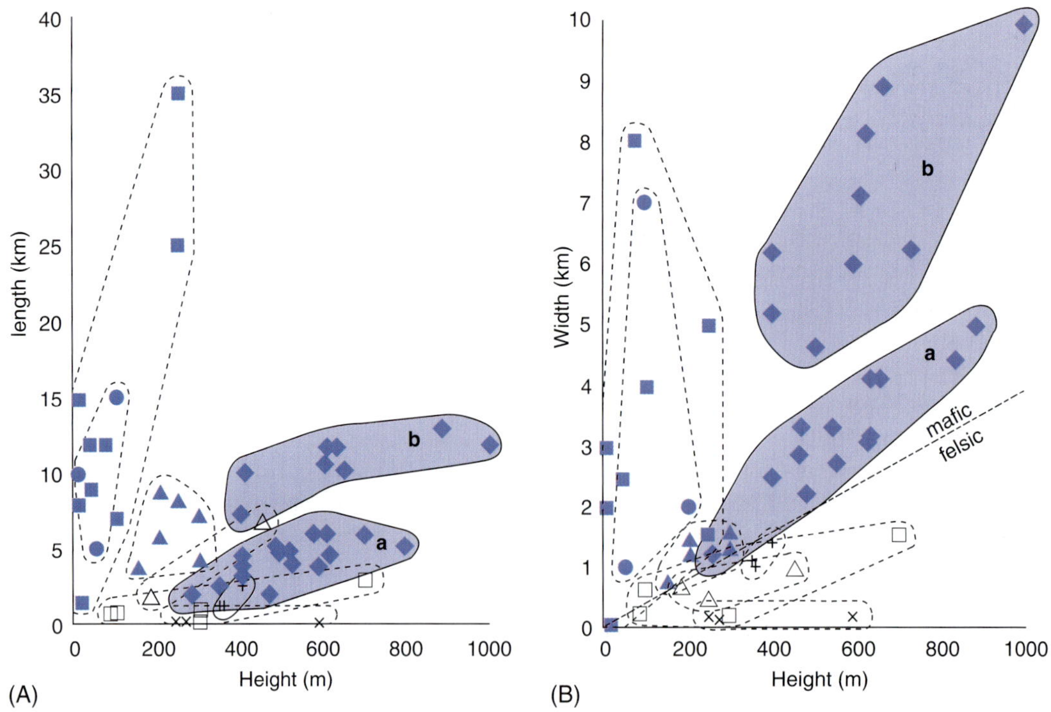

Figure 7 Morphometric data for subglacial volcanic landforms: (A) height versus length and (B) height versus width (aspect ratio). After Smellie (2006b).

The largest monogenetic mafic tuyas measured so far are 800–1,000 m in height, recording eruptions through contemporaneous ice sheets of comparable thickness. Interestingly, those heights are within the range of maximum heights (750–1,000 m) predicted for tuyas erupted in temperate ice, mentioned previously. By contrast, the James Ross Island subglacial polygenetic stratovolcano has considerably larger dimensions (1.5 km high), indicating that the theoretical height limit on mafic tuyas does not apply to polygenetic centers. This is because of the ability of an ice cover on a stratovolcano to reform between each eruptive phase. Thus, each effusive eruption on a stratovolcano only records the thickness of the coeval ice draping the volcano, and it may always be much less than 750–1,000 m. This was the case on James Ross Island, where the ice sheet never exceeded 600 m and was typically much less.

Felsic Tuyas

Some subglacial flat-topped, steep-sided volcanic landforms with felsic compositions have been described as tuyas (Kelman *et al.*, 2002; Tuffen *et al.*, 2002a). Compared to mafic tuyas, very few felsic examples have been described. Their compositions include andesite, dacite, and rhyolite. Apart from a single trachyte lava-fed delta observed by the author on Ross Island, Antarctica, it appears that felsic magmas generally do not form lava-fed deltas (necessary for a mafic tuya). The internal structure of felsic tuyas is also very different from that of mafic tuyas (Tuffen *et al.*, 2002a, **Fig. 3**; Smellie, 2006b).

At least two types of felsic tuyas are identified: tephra dominated and lava flow dominated. In tephra-dominated felsic tuyas, the flat cap is formed of horizontal or gently dipping, relatively thin (typically a few tens of meters) subaerially emplaced felsic lavas, which overlie a much thicker pile (few hundred meters) of fine-grained tephra (**Fig. 8**). The tephra core is usually obscured by lava scree and therefore poorly exposed (Tuffen *et al.*, 2002a). By contrast, the lava flow-dominated felsic tuyas are distinctive flat-topped columns or short blade-like ridges, also flat topped. They are formed almost entirely of subhorizontal ice-impounded subaerial felsic lavas (**Fig. 9**), although some of the lavas may also be plastered onto the subvertical tuya flanks. Only andesite–dacite lava-dominated systems have been described so far, and fragmental deposits are practically absent (Mathews, 1951; Kelman *et al.*, 2002). Like mafic tuyas, the lava caprock in both types of felsic tuyas formed after the ice sheet had melted through completely.

Figure 8 Examples of tephra-dominated felsic tuyas in Iceland. (A) Raudafossafjoll. The summit rises approximately 400 m; (B) Hottur (at right) and unnamed peak, Kerlingarfjoll. Both peaks rise approximately 350 m.

There are very few data for felsic tuyas, so it is impossible to make significant conclusions about their morphometry. The data hint that felsic tuyas may share similar (high) volume:area ratios to the mafic tuyas (**Fig. 6**). In **Figure 7**, it is clear that felsic tuyas have slightly to markedly higher aspect ratios than mafic tuyas. Moreover, despite substantial overlap with data for subglacial felsic domes, there is a strong indication that the lava flow-dominated felsic tuyas have the highest aspect ratios of any subglacial volcanic landform measured to date.

Pillow Mounds, Ridges, and Sheets

Subglacial pillow mounds, ridges, and sheets are formed mainly of pillow lava and less common hyaloclastite breccia. All described examples are basaltic in composition, but there is no intrinsic reason why more andesitic or possibly dacitic magmas should not form pillows. Many of these landforms occur in the cores of basaltic tuyas (and are therefore largely obscured from view), where pillow lava effusion is usually regarded as a ubiquitous early stage of tuya construction (Jones, 1969; Smellie and Hole, 1997; Werner and Schmincke, 1999). The landforms include relatively low oblate mounds with moderately dipping sides and smooth profiles (**Fig. 10**) in examples erupted from central vents, subdued ridges with gentle slopes that were erupted from fissures, and extensive thin sheets (**Fig. 11**). Examples include

Figure 9 Examples of lava flow-dominated felsic tuyas in British Columbia, Canada. (A) The Table (approximately 300 m); (B) Little Ring Mountain (approximately 280 m). Photograph (A) by Cathie Hickson and photograph (B) by Melanie Kelman.

the subglacial volcanoes in British Columbia examined by Dixon *et al.* (2002), which are formed solely of pillow lava extruded under ice sheets that may have been >2 km thick, and pillow-lava centers draped unconformably by agglutinate at Seal Nunataks, Antarctica, which formed within an ice sheet >600 m thick (Smellie and Hole, 1997). By contrast, pillow sheets are only a few tens of meters thick but may extend laterally 10 km or more (Snorrason and Vilmundardottir, 2000). Pillow sheets are only known from Iceland and have not been described in detail. How they form is unclear.

Because they have been so little studied, there are very few morphometric data available for subglacial pillow mounds, ridges, and sheets (**Figs. 6** and **7**). They generally have among the lowest volume:area ratios in the data set, although the data are biased toward the sheet-like examples, which will always have relatively high outcrop areas. Aspect ratios are also very low, comparable only to subglacial sheet-like sequences (**Fig. 7**), subject to the same caveat.

Tephra Mounds and Ridges

Tephra mounds and ridges have been described for subglacial volcanoes of mafic and felsic compositions. They vary from conical edifices or mounds to linear ridges, and mafic examples are by far the most common. They are comparatively steep-sided and, when erupted from fissures, form prominent ridges with multiple summit peaks (**Figs. 10** and **12**). The ridge-like occurrences were called tindars by Jones (1969; also adopted by Hickson, 2000), an etymologically incorrect use of the Icelandic plural word ('tindar') meaning pinnacles or summits (singular 'tindur'). Its usage was extended informally by Smellie (2000) to encompass indiscriminately any subglacially erupted tephra, whether in mounds or in ridges. A 'tindar phase' of subaqueous explosive eruptive activity would thus include those tephra-dominated cores of edifices that subsequently developed into tuyas. It may be preferable to refer to the landforms nongenetically as either subglacial tephra mounds or ridges according to their outcrop morphology, with tindar restricted to occurrences (either ridges only or both ridges and mounds) for which a subglacial eruptive origin can be demonstrated.

The structure and lithofacies of felsic tephra mounds and ridges are not well described, except where they occur below a subaerial lava cap in tephra-dominated felsic (rhyolite) tuyas (Tuffen *et al.*, 2002a). However, they also frequently occur without a lava caprock (e.g., in the Kerlingarfjoll rhyolite complex, central Iceland; **Fig. 13**). They form relatively steep-sided pale-colored successions, commonly with numerous, and sometimes voluminous, craggy syneruptive rhyolite intrusions.

The tephra mounds/ridges have much lower volume:area ratios than the mafic and felsic tuyas. They also appear to show lower aspect ratios and different trajectories to all but a few of the mafic tuyas on the length versus height diagram, which emphasizes the fissure-related aspects of the tephra landforms (**Fig. 7A**). Two subfields are defined compositionally, corresponding to mafic and felsic tephra landforms. The felsic examples have higher aspect ratios than mafic examples, although there are too few data to

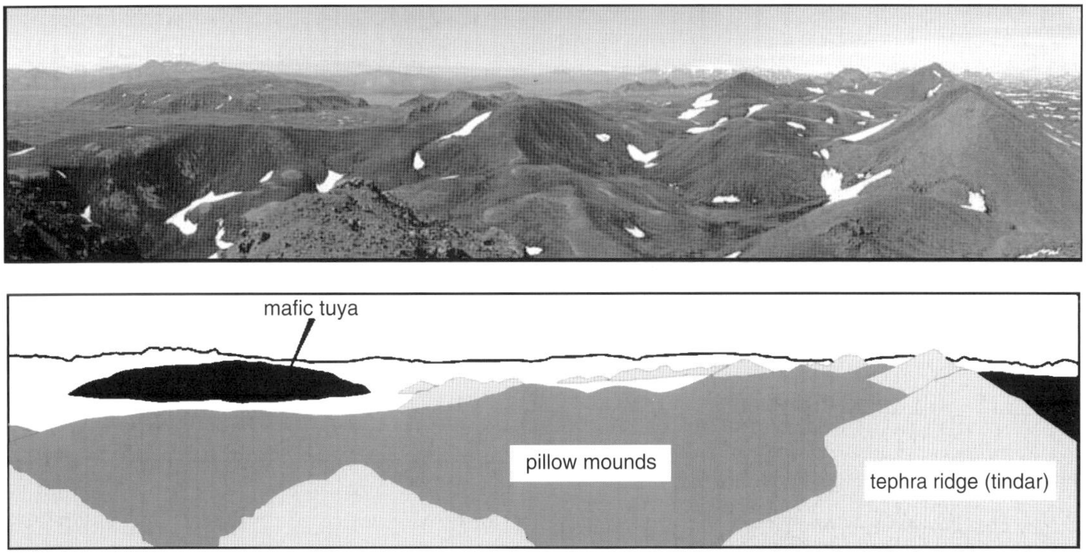

Figure 10 View and explanatory sketch, looking north from Kalfstindar (Iceland) showing contrasting subglacial mafic volcanic landforms, including steep tephra ridges, a tuya, and low-lying smooth-surfaced pillow mounds.

Figure 11 Extensive pillow sheet (dashed line outcrop) at Thorisvatn, Iceland. The prominent ridge in the foreground is a pillow ridge. Photograph by Snorri Snorrason.

Figure 12 View looking south along the pinnacled ridge crest of Kalfstindar, a mafic tephra ridge in Iceland. Note the very steep sides of the ridge, which cut across tephra bedding in the deposit. The ridge rises approximately 250 m above the surrounding plain. The low subdued mounds in the foreground at the right are pillow mounds.

be definitive (**Fig.** 7). In the width versus height diagram (**Fig.** 7B), the data show a good aspect ratio correspondence with compositionally analogous tuyas—that is, mafic tephra mounds/ridges with mafic tuyas, and felsic mounds/ridges with tephra-dominated felsic tuyas.

The distribution of mafic tephra mound data overlaps greatly with that for subaqueous mafic tuff cones erupted in unconfined, nonglacial settings, although displaced to the high aspect ratio side, and they have similar volume:area ratios. Subglacial felsic examples have higher aspect ratios than their subaerial counterparts, but both felsic data sets are very small. The overlapping data for mafic glacial/nonglacial mounds/cones is a surprising result since the subglacial examples are predicted intuitively to be taller as a consequence of a combination of effects (Smellie, 2006b):

1. Buttressing of tephra by the confining ice walls (Smellie, 2000, fig. 5)
2. Rapid early diagenesis and lithification of the warm wet tephra pile, which may be buffered at several tens of degrees Celsius for a substantial period (Gudmundsson *et al.*, 2004), particularly by coeval high-level intrusions
3. Glacial erosion and oversteepening of the indurated volcano flanks once activity has ceased (**Fig. 12**)

Examination of subglacial examples in Iceland seems to support this prediction. The tephra landforms

Figure 13 Examples of felsic tephra mounds/ridges at Kerlingarfjoll, Iceland. (A) Looking west to Ogmundur from Hottur; the low moundy ridges at center-left are tephra mounds located along fissures. Ogmundur, at right, is a lava flow-capped felsic tuya that has been eroded to form the present prominent conical peak. (B) Subdued, fissure-aligned, multiple overlapping felsic tephra mounds, Rodull.

shown in **Figures 10** and **12** are characterized by steep flanks composed of indurated tephra, which contrasts with associated (underlying) coeval pillow mounds/ridges. By contrast, subaqueous tuff cones formed in unconfined marine or (nonglacial) lake settings should have low-gradient flanks because of the long run-out distances possible in unconfined settings (**Fig. 14**). A possible explanation may be that the measured widths of the subglacial examples are incorrect and inadvertently incorporated other landforms. Since most of the examples used in the data set were culled from the literature, this may occur if the tephra mounds/ridges were constructed on pillow lava constructs (with low aspect ratios) if (a) the pillow lava piles (commonly obscured by scree) were incorrectly mapped as tephra or (b) tephra incompletely drapes the underlying, much lower profile, pillow mounds and incorporates their form, thus creating a misleading composite landform, not one constructed solely by tephra. If this is correct, it may provide an explanation for the similar areas and volumes of tephra mounds/ridges in glacial and nonglacial settings.

Felsic Lobes and Domes

Several steep-sided subglacially erupted andesite, dacite, rhyolite, and trachyte–phonolite flows with characteristic fine-scale prismatic jointing have been described from British Columbia and Iceland (Tuffen *et al.*, 2002b; Edwards and Russell, 2002; Kelman *et al.*, 2002). The radiating joint patterns and associated glassy chilled surfaces are interpreted as ice-contact features. Together they define the original landform shapes, which may be conical, lobate, dome-like, or irregular (**Fig. 15**). Associated fragmental volcanic lithofacies are usually uncommon. Meltout tills and reworked volcanic sands and gravels are present in an Icelandic example together with ashy-matrixed breccias.

There are few published morphometric data for felsic lobes and domes. They have high volumes relative to their areas (**Fig. 6B**) and they appear to have higher aspect ratios than almost all of the subglacial volcanic landforms with mafic compositions (**Fig. 7**). Conversely, their aspect ratios are comparable to most other felsic landforms (i.e., tephra-dominated tuyas and tephra mounds). In all but one instance, they have generally lower aspect ratios than the lava flow-dominated felsic tuyas.

Subglacial Sheet-like Sequences

These were also called subglacial 'sheet flow' sequences by Smellie (2001) but were renamed subglacial sheet-like sequences by Smellie (2006b) to incorporate nongenetic terminology. They are relatively thin deposits that, as their name suggests, form either narrow, sinuous, anastomozing low outcrops, which are sometimes esker-like (Mathews, 1958; Smellie and Skilling, 1994), or lobe or sheet-like features. However, the profile, overall shape, and dimensions of an outcrop will depend on the topography of the underlying bedrock (e.g., incised valley), erupted volume of magma, and discharge rate (Loughlin, 2002). Most of the published examples are basaltic, but Edwards and Russell (2002) described a few trachyte and phonolite examples in British Columbia.

At least two different types of sheet-like sequence can be distinguished. Their distinction is based on comparatively subtle, but genetically important, differences in lithofacies and sequence architecture. It is unclear if those differences impact significantly on the landforms, and they are provisionally considered as a single landform group here. The differences do have important implications for interpreting original ice sheet thicknesses, but the origin, formal characterization, and interpretation of the two types are still at an early stage (Smellie, 2006b). Unlike all the other subglacial volcanic landforms described in this article, the sheet-like sequences are strata formed away from the erupting vent(s) (i.e., 'outflow facies')

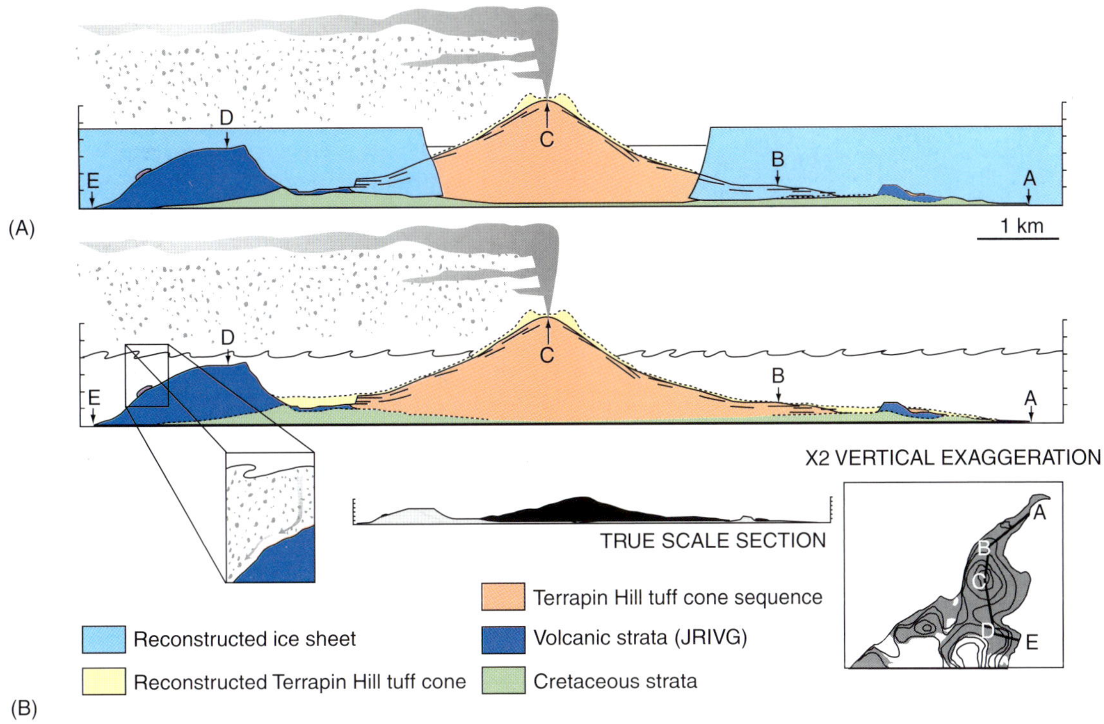

Figure 14 Sketch sections showing the impact of eruptive environment on the morphology of a large tuff cone. (A) An ice sheet is reconstructed. The eruption might have melted a space 3 km in diameter, as sketched. The thermal energy would have been focused in the central parts of the edifice (high heat flow in the vent(s)), and the high thermal inertia of the tephra pile would likely have prevented the melting of a significantly larger space during the eruption. The proximity of an ice sheet would have impinged on the growth of the edifice by buttressing effects of the ice walls (not illustrated), thus leading to an edifice with much steeper flanks, which is not the case. (B) An unconfined marine setting is shown, which is more consistent with the outcrop and can easily explain the presence of isolated outcrops of tephra behind topographical barriers (left) that cannot have formed had an ice sheet been present at the time. Sections based on a 600-m-high, mafic tuff cone on James Ross Island, Antarctica, which erupted in a marine (i.e., nonglacial) setting <880,000 years ago. After Smellie *et al.* (2006b).

rather than volcanic edifices. The few source vents identified so far are simple pyroclastic cones, undiagnostic, in themselves, of a subglacial setting. The full range of source edifices has probably not been identified.

Subglacial sheet-like sequences range from only a few tens of meters thick to a few hundred meters, with widths up to several kilometers. A wide range of examples has been described in Iceland, Antarctica, and British Columbia (Smellie *et al.*, 1993; Kelman *et al.*, 2002; Loughlin, 2002). There are no published whole-outcrop-scale photographs of the sheet-like sequences since the majority described so far form multiple stacked units that together construct large stratovolcanoes (Smellie, 2000; Loughlin, 2002). The morphology is unlikely to be sufficiently distinctive to be identified without detailed lithofacies studies.

Not surprisingly, the sheet-like sequences have the lowest aspect ratios of any subglacial volcanic landform, and they have a distinctive combination of very high area combined with low volume, which overlaps significantly with data for subglacial pillow sheets (**Figs. 6** and **7**).

Influence of Glacier Thermal Regime on Subglacial Volcanic Landforms

All of the foregoing is based on an analysis of subglacial, mainly monogenetic, volcanic landforms that were created by eruptions beneath glaciers and/or ice sheets with temperate (wet-based) thermal regimes. This is important. The thermal regime determines the hydraulics of an ice system, which in turn strongly influences the lithofacies, sequence architecture, and, thus, the landforms ultimately constructed (Björnsson, 1988; Smellie, 2000, 2001, 2006a). Temperate glaciers are dominant on Earth today, except in parts of the Greenland and Antarctic ice sheets, so the results of this landform analysis are likely to have wide application. However, there are different consequences for volcanic landforms constructed within a polar ice sheet or glacier (i.e., one that is frozen to its bed) (Smellie, 2006b). In those, water cannot escape subglacially, either in conduits or in thin films. Any meltwater created by a polar subglacial eruption will therefore be confined in a vault overlying the vent. The system will ultimately

Figure 15 Example of a felsic dome at Ember Ridge northwest, British Columbia, Canada. Despite the prominent posteruptive erosion, the original dome-like morphology can be reconstructed using the orientations of the prominent columnar cooling joints in the outcrop, which grew perpendicular to the original confining ice surface. Photograph by Melanie Kelman.

overflow as volcanic mass (magma) is added and the water is displaced. The colder temperatures of polar glaciers also strongly influence the ice rheology, which is stronger and deforms much more slowly. The energy requirements for melting are also different. Since more energy is needed than in temperate systems (simply to bring the ice temperature up to the melting point), the ice walls of a vault or lake will recede more slowly than in a temperate ice sheet. As a consequence of these combined characteristics, the vault or lake will not enlarge as rapidly as in temperate ice systems. The volcanic products will rapidly fill it and spill out to override the surrounding ice sheet surface. Moreover, because of the high thermal inertia of magma and insulating properties of any underlying autobreccia rubble, any supraglacial lava flows may be capable of traveling considerable distances without melting deeply into the underlying ice.

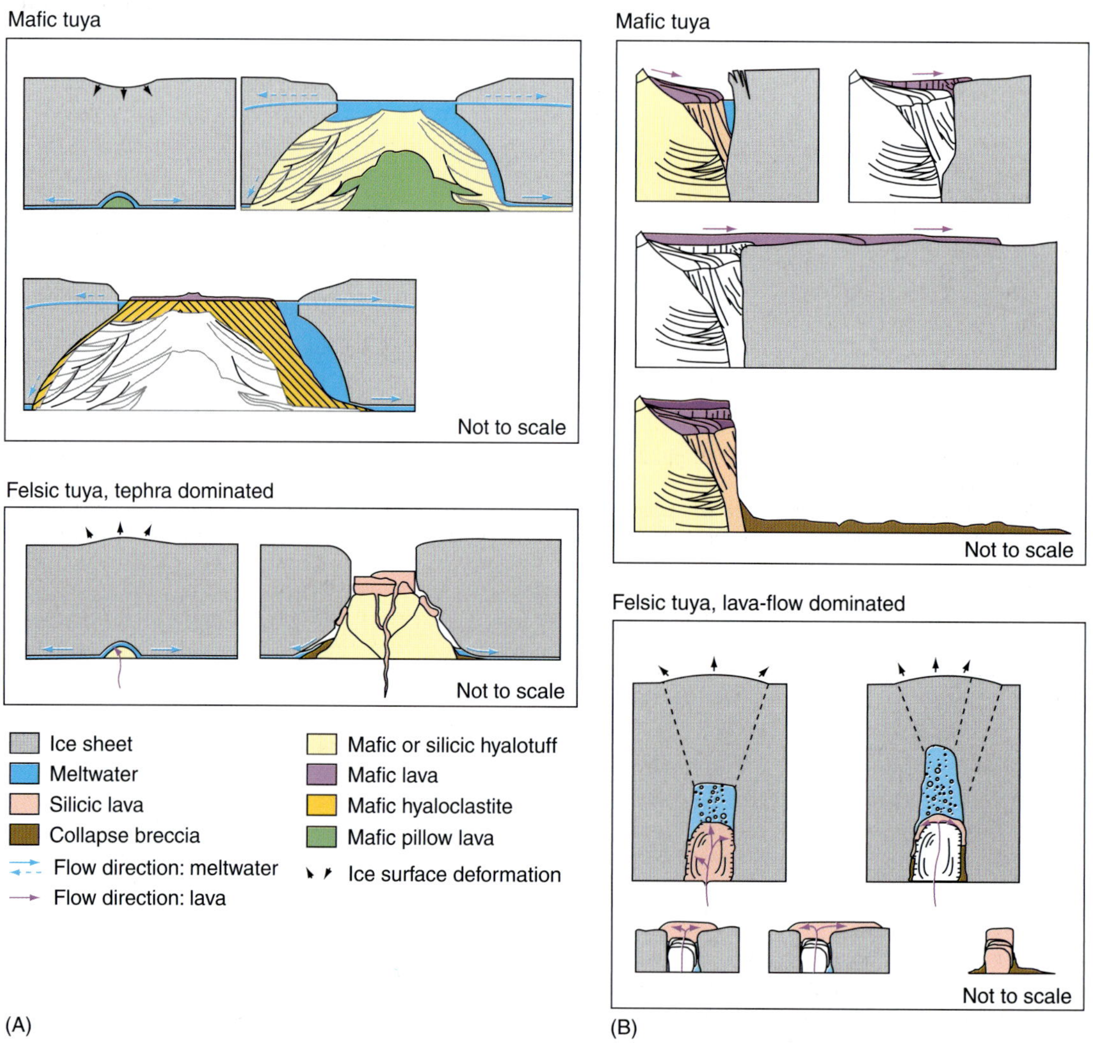

Figure 16 Drawings illustrating the influence of glacier thermal regime on construction and morphology of mafic and felsic tuyas. (A) Temperate glacier regime. (B) Polar glacier regime. The principal differences are shorter lava-fed deltas and taller edifices in polar ice compared with those formed in temperate ice. After Smellie (2006b).

The edifices will also be buttressed by the slowly receding ice walls, presumably leading to even taller constructs than are possible in temperate ice.

The overall effects of eruptions under ice sheets with differing thermal regimes can therefore be broadly predicted, and the main results for mafic and felsic tuyas are shown in **Figure 16**. However, in the absence of any data for volcanic landforms constructed beneath polar ice sheets, it is currently impossible to provide any real morphometric dimensions with which to compare and contrast with analogous landforms associated with temperate ice sheets.

See also: **Glacial Climates**: Volcanic and Solar Forcing. **Glacial Landforms**: Introduction.

References

Björnsson, K. (1988). Hydrology of ice caps in volcanic regions. *Vísindafélag íslendinga. Societas Scientarium Islandica* **45**, 1–139.

Chapman, M. G., and Smellie, J. L. (2006). Mars interior layered deposits and terrestrial sub-ice volcanoes compared: Observations and interpretations of similar geomorphic characteristics. In *Environments on Earth: Clues to the Geology of Mars* (M. G. Chapman and I. P. Skilling, Eds.), Cambridge University Press, Cambridge.

Chapman, M. G., and Tanaka, K. L. (2001). The interior trough deposits on Mars: Sub-ice volcanoes? *Journal of Geophysical Research* **106**, 10,087–10,100.

Dixon, J. E., Filiberto, J. R., Moore, J. G., and Hickson, C. J. (2002). Volatiles in basaltic glasses from a subglacial volcano in northern British Columbia (Canada): Implications for ice sheet thickness and mantle volatiles. In *Volcano–Ice Interaction on Earth and Mars* (J. L. Smellie and M. G. Chapman, Eds.), Special Publications No. 202, pp. 255–271. Geological Society, London.

Edwards, B. R., and Russell, J. K. (2002). Glacial influences on morphology and eruptive products of Hoodoo Mountain volcano, Canada. In *Volcano–Ice Interaction on Earth and Mars* (J. L. Smellie and M. G. Chapman, Eds.), Special Publications No. 202, pp. 179–194. Geological Society, London.

Ghatan, G. J., and Head, J. W. (2002). Candidate subglacial volcanoes in the south polar region of Mars: Morphology, morphometry, and eruption condition. *Journal of Geophysical Research* **107**, E7.

Gudmundsson, M. T., Sigmundsson, F., Björnsson, H., and Högnadóttir, T. (2004). The 1996 eruption at Gjálp, Vatnajökull ice cap, Iceland: Efficiency of heat transfer, ice deformation and subglacial water pressure. *Bulletin of Volcanology* **66**, 46–65.

Head, J. W., and Pratt, S. (2001). Extensive Hesperian-aged south polar ice sheet on Mars: Evidence for massive melting and retreat, and lateral flow and ponding of meltwater. *Journal of Geophysical Research* **106**, 12,275–12,299.

Hickson, C. J. (2000). Physical controls and resulting morphological forms of Quaternary ice-contact volcanoes in western Canada. *Geomorphology* **32**, 239–261.

Jones, J. G. (1969). Intraglacial volcanoes of the Laugarvatn region, south-west Iceland—I. *Quarterly Journal of the Geological Society, London* **124**, 197–211.

Kelman, M. C., Russell, J. K., and Hickson, C. J. (2002). Effusive intermediate glaciovolcanism in the Garibaldi Volcanic Belt, southwestern British Columbia, Canada. In *Volcano–Ice Interaction on Earth and Mars* (J. L. Smellie and M. G. Chapman, Eds.), Special Publications No. 202, pp. 195–211. Geological Society, London.

Lescinsky, D. T., and Sisson, T. W. (1998). Ridge-forming, ice-bounded lava flows at Mount Rainier, Washington. *Geology* **26**, 351–354.

Loughlin, S. C. (2002). Facies analysis of proximal subglacial and proglacial volcaniclastic successions at the Eyjafjallajökull central volcano, southern Iceland. In *Volcano–Ice Interaction on Earth and Mars* (J. L. Smellie and M. G. Chapman, Eds.), Special Publications No. 202, pp. 149–178. Geological Society, London.

Mathews, W. H. (1951). The Table, a flat-topped volcano in southern British Columbia. *American Journal of Science* **249**, 830–841.

Mathews, W. H. (1958). Geology of the Mount Garibaldi area, southwestern British Columbia. Part II. Geomorphology and Quaternary volcanic rocks. *Bulletin of the Geological Society of America* **69**, 161–178.

Mee, K., Tuffen, H., and Gilbert, J. S. (2006). Snow-contact volcanic facies and their use in determining past eruptive environments at Nevados de Chillán volcano, Chile. *Bulletin of Volcanology* **68**, 363–376.

Smellie, J. L. (2000). Subglacial eruptions. In *Encyclopaedia of Volcanoes* (H. Sigurdsson, Ed.), pp. 403–418. Academic Press, San Diego.

Smellie, J. L. (2001). Lithofacies architecture and construction of volcanoes in englacial lakes: Icefall Nunatak, Mount Murphy, eastern Marie Byrd Land, Antarctica. In *Volcaniclastic Sedimentation in Lacustrine Settings* (J. D. L. White and N. R. Riggs, Eds.), Special Publication No. 30, pp. 73–98. International Association of Sedimentologists.

Smellie, J. L. (2006a). The relative importance of supraglacial versus subglacial meltwater escape in basaltic subglacial tuya eruptions: an important unresolved conundrum. *Earth Science Reviews* **74**, 241–268.

Smellie, J. L. (2006b). Terrestrial sub-ice volcanism: landform morphology, sequence characteristics and environmental influences, and implications for candidate Mars examples. In *Preservation of Random Mega-Scale Events on Mars and Earth: Influence on Geologic History* (M. G. Chapman and L. Keszthely, Eds.). Geological Society of America Special Papers.

Smellie, J. L., and Chapman, M. G. (Eds.) (2002). *Volcano–Ice Interaction on Earth and Mars* Special Publications No. 202, Geological Society, London.

Smellie, J. L., and Hole, M. J. (1997). Products and processes in Pliocene—Recent, subaqueous to emergent volcanism in the Antarctic Peninsula: Examples of englacial Surtseyan volcano construction. *Bulletin of Volcanology* **58**, 628–646.

Smellie, J. L., and Skilling, I. P. (1994). Products of subglacial eruptions under different ice thicknesses: Two examples from Antarctica. *Sedimentary Geology* **91**, 115–129.

Smellie, J. L., Hole, M. J., and Nell, P. A. R. (1993). Late Miocene valley-confined subglacial volcanism in northern Alexander Island, Antarctic Peninsula. *Bulletin of Volcanology* **55**, 273–288.

Smellie, J. L., McIntosh, W. C., and Esser, R. (2006a). Eruptive environment of volcanism on Brabant Island: Evidence for thin wet-based ice in northern Antarctic Peninsula during the Late Quaternary. *Palaeogeography, Palaeoclimatology, Palaeoecology* **231**, 233–252.

Smellie, J. L., MrArthur, J. M., McIntosh, W. C., and Esser, R. (2006b). *Late Neogene interglacial events in the James Ross*

Island region, northern Antarctic peninsula, dated by Ar/Ar and Sr-isotope stratigraphy. Palaeoceanography, Palaeoclimatology, Paleoecology.

Snorrason, S. P., and Vilmundardóttir, E. G. (2000). Pillow lava sheets: Origins and flow patterns. In *Volcano/Ice Interaction on Earth and Mars* (V. C. Gulick and M. T. Gudmundsson, Eds.).

Thouret, J.-C. (1990). Effects of the November 13, 1985 eruption on the snow pack and ice cap of Nevado del Ruiz volcano, Colombia. *Journal of Volcanology and Geothermal Research* **41**, 177–201.

Tuffen, H., McGarvie, D. W., Gilbert, J. S., and Pinkerton, H. (2002a). Physical volcanology of a subglacial-to-emergent rhyolite tuya at Raudufossafjöll, Torfajökull, Iceland. In *Volcano–Ice Interaction on Earth and Mars* (J. L. Smellie and M. G. Chapman, Eds.), Special Publications No. 202, pp. 213–236. Geological Society, London.

Tuffen, H., Pinkerton, H., McGarvie, D. W., and Gilbert, J. S. (2002b). Melting of the glacier base during a small-volume subglacial rhyolite eruption: Evidence from Bláhnúkur, Iceland. *Sedimentary Geology* **149**, 183–198.

Van Bemmelen, R. W., and Rutten, M. G. (1955). *Tablemountains of Northern Iceland.* Brill, Leiden, The Netherlands.

Werner, R., and Schmincke, H.-U. (1999). Englacial vs lacustrine origin of volcanic table mountains: Evidence from Iceland. *Bulletin of Volcanology* **60**, 335–354.

Evidence of Glacier Recession

P M Colgan, Grand Valley State University, MI, USA

© 2007 Elsevier B.V. All rights reserved.

Introduction

Glacier recession is the apparent stepwise retreat of the terminus of a glacier or ice sheet towards its source area. This occurs when more ice is melted near the terminus in the ablation zone than can be replaced by ice flowing down glacier from the accumulation zone. Glacier recession usually follows warming air temperatures because an increase in the elevation of the equilibrium line forces the glacier to attempt to reach a new mass balance by decreasing the area of its ablation zone. Glacier recession may also occur because of a decrease in snowfall in the accumulation zone or because of an increase in solar radiation due to decreased cloudiness. This also has the effect of raising the equilibrium line elevation and forcing glacier recession. Ablation rates are commonly an order of magnitude higher than accumulation rates in most glaciers, therefore changes in air temperature usually cause more immediate changes in glacier area than do changes in snowfall. Glaciers commonly retreat in a regressive fashion because glacier ice is always actively flowing toward the terminus to replace what is melting. Because of the relationship between glacier mass balance and climate, both past and present glacier recession is an excellent indicator of climate change within a region. Glaciers may also retreat by widespread down wasting if the terminus is cut off from the accumulation zone as a result of topographic effects or stagnation of ice following a glacier surge (see Paterson (1994) for a detailed explanation about glacier mass balance).

Direct observations, field surveys, photographs, aerial photographs, and remotely sensed data are the primary evidence for historic glacier recession during the late Holocene. A careful comparison of these data gathered in a series of years can provide evidence of both the magnitude and rate of glacier recession. Numerous studies have used these methods to quantify the rate of glacier recession over the last 50–400 years in nearly every region of the world (see Grove, 1988; Dyurgerov and Meier, 2000; Meier *et al.*, 2003).

Geomorphic data in combination with dating methods are used to determine the magnitude and sometimes the rate of Late Pleistocene and Holocene glacier recession. Geomorphic features that can be used to reconstruct the former glacier terminus include moraines, streamlined features, eskers, and outwash plains. There are many different types of moraines, which are ridges of glacial sediment formed along the terminus (end moraine) or sides (lateral moraines) of a glacier. These moraines contain information both of the extent and style of past glaciers.

Subglacial landforms such as drumlins, flutes, and striae are commonly exposed during glacier recession and record ice movement towards the terminus. Sorted stream sediment deposited in front of, or on top of, the ice produces ice-contact features, outwash plains, fans, and channels. Many of these features can be dated with radiocarbon methods or newer methods such as cosmogenic radionuclide exposure dating. Following is a brief summary of some of the historic and geomorphic evidence for glacier recession and the implications for climate and glacier dynamics indicated by that evidence.

Evidence of Twentieth-Century Glacier Recession

Glaciers provide important spiritual, aesthetic, and utilitarian resources for human populations. In many parts of the world glaciers provide economic benefits through water resources, hydroelectric generation, tourism, and wildlife habitat. The current recession of glaciers has important implications for the future of these resources. Glacier recession also contributes significantly to global sea-level rise. Meier and Bahr (1996) estimated that there are over 160,000 mountain glaciers and ice caps with a total area of 680,000 km^2 and a volume of 180,000 km^3. If completely melted these glaciers would cause a rise of approximately 0.5 m in global sea level. Recession of mountain glaciers adds a significant volume of water to oceans and is an important factor in current sea-level rise (Meier *et al.*, 2003). The Greenland and Antarctic ice sheets contain as much as 80 m of equivalent sea-level rise, but less is known about the current mass balance of these ice sheets.

Today, mountain glaciers all over the world are in full retreat threatening natural resources and providing unequivocal evidence of twentieth-century climate change (**Fig. 1**). In a recent compilation of field data collected by the World Glacier Monitoring Service from 88 mountain glaciers from around the world, 82 were reported to have lost mass and are now retreating (data from World Glacier Monitoring Service, 2006). Many glaciers have been monitored continuously for their net mass balance for the last 25–30 years by direct field measurements. Mass balance for a glacier is determined by estimating the total accumulation and ablation in cubic meters of water equivalent. Net mass balance is the change in volume of water equivalent divided by the total area of the glacier. It is commonly reported in mm of water equivalent (Paterson, 1994). The mean net cumulative mass balance for 30 glaciers in nine regions shows a decrease in mass for nearly every balance year since 1980 (data from World Glacier Monitoring Service, 2006).

Many other glaciers have been monitored for terminus position by direct observations, photographs, and with remotely sensed data (**Fig. 2**). Glacier recession has been reported in nearly every region of Earth (see Dyurgerov and Meier, 2000; Meier *et al.*, 2003; Rignot *et al.*, 2003; Cook *et al.*, 2005). Recently, Oerlemans (2005) used 169 glacier retreat records from around the world to reconstruct mean global temperature changes over the last 400 years. This climate record extracted from glacier recession records was very similar to global temperature

Figure 1 Examples of glacier recession as shown in historical photographs. (A) and (B). McCall Glacier in 1958 and 2003. This is a small valley glacier located in the Brooks Range of northern Alaska. Photographs by A. Post (1958) and M. Nolan (2003).

Figure 2 This false-color image shows the Gangotri Glacier, located in the Uttarkashi District of Garhwal Himalaya, India. Gangotri Glacier is one of the largest in the Himalaya and is a source for the Ganges River. The glacier is currently 30.2 km long and 0.5–2.5 km wide. Gangotri Glacier has been receding since 1780, although studies show its retreat accelerated after 1971. (Note that the contour lines drawn here to show the recession of the glacier's terminus over time are approximate.) Over the past 25 years this glacier has retreated more than 850 m, with a recession of 76 m from 1996 to 1999 alone. Source: NASA. (http://earthobservatory.-nasa.gov/Newsroom/NewImages/images.php3?img_id=16584).

reconstructions based on other independent paleoclimate proxy records. Other studies show that glacier recession has accelerated in the last two decades of the twentieth century (Dyurgerov and Meier, 2000; Meier *et al.*, 2003). This most recent global glacier recession began at the end of the Little Ice Age. The Little Ice Age was a period of cooler global temperatures that began in the late Middle Ages and ended in the last half of the nineteenth century (Grove, 1988). Since then, we have seen an acceleration of warming during the late twentieth century and this has led to a quickening in the pace of modern glacier recession.

Global sea-level rise over the last 100 years has been estimated to be 10–30 cm (Warrick and Oerlemans, 1990). This rise is thought to be primarily the result of mountain glacier melting and the rise of sea-surface temperatures leading to expansion of sea water (Meier, 1984). Future sea-level rise over the next 100 years could be higher (up to 70 cm) depending on the rate of global warming, the response of glaciers, and the amount of warming of sea-surface waters (Warrick and Oerlemans, 1990). Future sea-level rise has great implications for societies since a large proportion of the world's population lives close to sea level. Monitoring small mountain glaciers is necessary to help quantify the volume of the water being added to the oceans as well as to monitor regional climate.

Glacier monitoring via remotely sensed data is a commonly used method to examine glacier change (Rignot *et al.*, 2003). The Global Land Ice Measurements from Space (GLIMS) sponsored by the United States Geological Survey, the National Snow and Ice Data Center, and academic researchers is currently developing a database of glacier changes using ASTER satellite images compared to other data sources. Monitoring of glacier recession has and will continue to play an important role in understanding global climate change.

Figure 3 This image shows the Vatnajökull ice cap as it appears in late summer. Vatnajökull covers an area of about 8,100 km^2 and is the largest ice cap in Europe. Note how the glacier's accumulation area is still covered in snow. Along the ice cap margin, are the outlet glaciers that drain the main ice body. Most of the outlet glaciers show up as dark. This is because in the ablation zone all of the previous winter's snow has melted by the end of summer, and bare ice shows up as dark. Medial and lateral moraines on some of the outlet glaciers are apparent. Along the left margin of the photograph are some of the other ice caps in Iceland; Hofsjökull in the upper left and Myrdalsjökull in the lower left. The red areas in this false color image are those areas covered with vegetation. Dark areas are lava fields and water. The large dark area between the ice cap and the coast is the sandur plain of Skeiáðarajökull, one of the largest outlet glaciers that drain Vatnajökull. Some areas along the coast have a bluish tint. This is meltwater carrying silt and discharging into the North Atlantic (Landsat MSS image – USGS).

Modern Glacier Recession and Geomorphology

As glaciers retreat they expose landscapes that were only recently covered by ice. This provides an excellent opportunity for us to explore and understand how glacial landforms and sediments are formed during glacier recession. An excellent example of recent glacier recession is the well-studied Breiðamerkurjökull of southeastern Iceland (Evans and Twigg, 2002). This is a large temperate outlet glacier that drains the ice cap Vatnajökull (**Fig. 3**). Breiðamerkurjökull has retreated between 4 and 5 km since AD 1894 (**Fig. 4**). This works out to an average recession rate of approximately 40–50 m year^{-1} for the last 100 years. While recession has been the overall trend, the glacier did not retreat every year and in fact advanced or remained steady between 1950–1955, 1968–1973, and 1975–1987. A large end moraine was formed during and after Breiðamerkurjökull reached its Little Ice Age maximum sometime between 1886 and 1890. After about 1894 the glacier began to retreat from this Little Ice Age moraine.

The deglaciated foreland in front of Breiðamerkurjökull displays an interesting and varied assemblage of glacial landforms and sediments (**Fig. 5**) that have been mapped in detail by Evans and Twigg (2002) and earlier workers (see references in Evans and Twigg, 2002). The most conspicuous of these landforms are numerous small push moraines (**Figs. 5 and 6**). Push moraines are small ridges that are made up of subglacially deposited till plus some

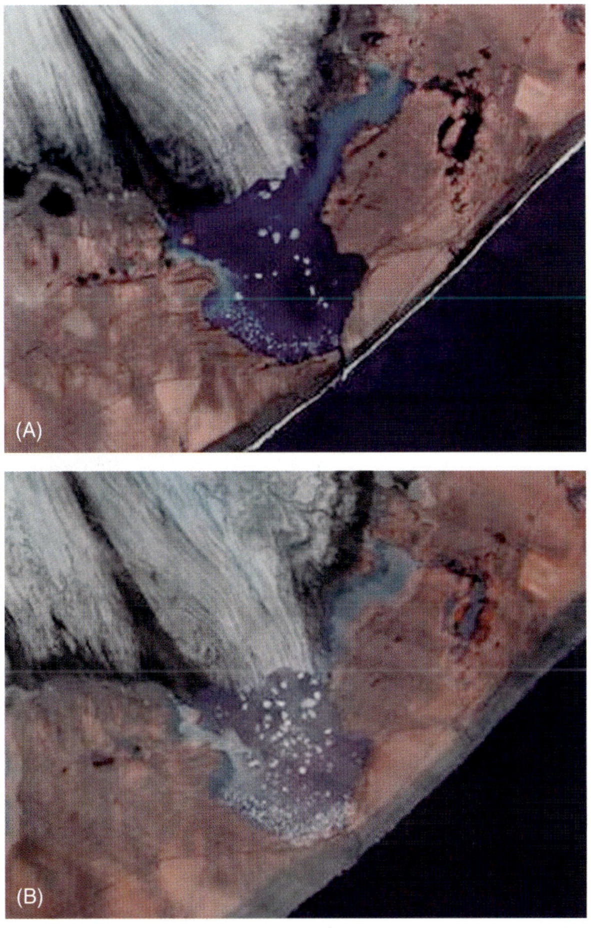

Figure 4 Quicktime movies of the recession of Breiðamerkurjökull 1973–2000. Source: NASA (http://svs.gsfc.nasa.gov/search/Keyword/Breidamerkurjokull.html).

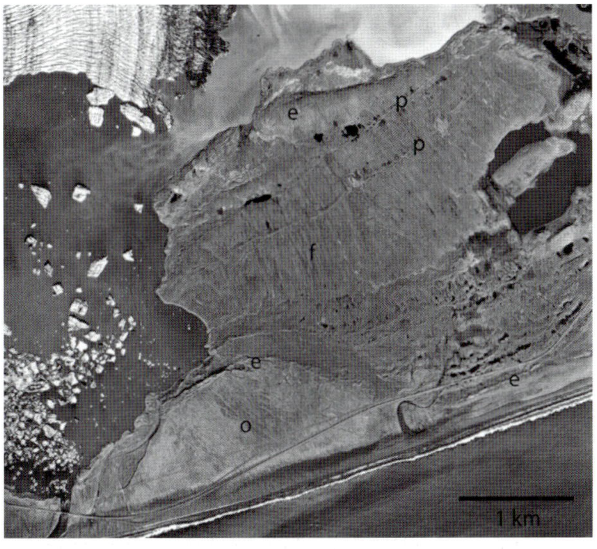

Figure 5 Vertical aerial photograph of the recently deglaciated area in front of the eastern lobe of Breiðamerkurjökull, south Iceland. Features visible include fluted ground moraine (f), small push moraines (p), larger end moraines (e), and outwash plain (o) (Landmaelingar Islands and University of Glasgow, 1998).

outwash deposited in late summer. Ridges are commonly less than 5 m high and 10 m wide (Boulton, 1986). They can be traced in excess of 3 km, and commonly have a saw-tooth pattern that reflects their formation along the crevassed terminus (**Fig. 6**). They tend to be asymmetric in cross-sectional profile with a shallower proximal and steeper distal flank. In late winter (mid-April to mid-May), ablation is very low near the terminus of Breiðamerkurjökull. Because of this the terminus of the glacier advances up to a few meters. This winter readvance pushes up material in front of the ice forming a push ridge. In summer, the terminus retreats because of the high ablation rates in this part of south Iceland ($\sim$10 m year^{-1}). If the overall retreat in summer is greater than the next year's winter readvance, a small push moraine is preserved. Push moraines may coalesce where one year's readvance is more extensive that the previous winter's. Sediment may also be squeezed from underneath the

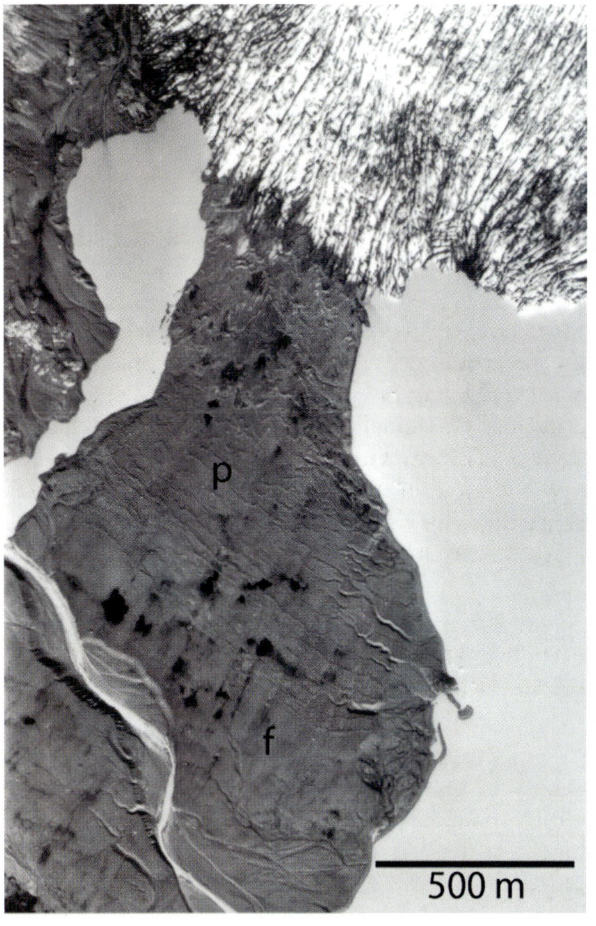

Figure 6 Vertical aerial photograph of the deglaciated surface in front of the western lobe of Breiðamerkurjökull, south Iceland. Feature visible are fluted ground moraine (f), and small push moraines (p) (Landmaelingar Islands and University of Glasgow, 1998).

Figure 7 Small push moraines in front of Breiðamerkurjökull, south Iceland. Photograph by P. M. Colgan (September 2000).

ice margin to form small push ridges. At Breiðamerkurjökull, small push moraines have formed nearly every year since 1965 on the west lobe of the glacier (**Fig. 7**). Because these ridges can be demonstrated to have formed annually or semiannually, they have sometimes been called 'annual moraines'. Similar ridges have been exposed by recent glacier recession in Norway, Svalbard, Baffin Island, the Canadian Rockies, Canadian Arctic, southeastern Alaska, and New Zealand.

Larger ridges of glacial sediment, up to tens of meters high and hundreds of meters wide, formed during more extensive and long-lived readvances or still-stands in the terminus, which commonly results in the stacking and bulldozing of subglacial and proglacial sediments. The larger recessional moraines in front of Breiðamerkurjökull formed during still-stands of the terminus and others are older features that have been overridden by the glacier as it reached its Little Ice Age maximum position. Some of these larger recessional moraines show evidence for the presence of buried ice in the form of hummocky topography. Hummocky topography forms as buried ice within the moraine melts and causes collapse of the overlying sediment.

Flutes and drumlins make up a continuum of streamlined landforms that form under temperate glaciers sliding over a temperate bed (Boulton, 1987). Commonly, obstructions such as boulders or rock outcrops will form cavities under the ice (**Fig. 5**). Subglacial sediment may be deformed into these cavities forming a fluted terrain under the ice margin. Larger obstructions and heterogeneity in the subglacial sediment will lead to the formation of streamlined hills called drumlins. At Breiðamerkurjökull fluted till underlies the small push moraines and larger recessional moraines (**Figs.** 5 and **6**). Small drumlins are also present in the landscape. An important observation is that some of the flutes end in individual push moraines (Evans and Twigg, 2002). This indicates that fluted terrain was being formed during recession and that small push ridges and flutes are genetically related. In contrast, drumlins appear to be related to the position of ice at the Little Ice Age maximum not to recessional positions.

Figure 8 Eskers (k) on the deglaciated foreland in front of Breiðamerkurjökull, south Iceland. Outwash plain (o) (Landmaelingar Islands and University of Glasgow, 1998).

Eskers are another common feature in landscapes exposed by glacier recession. At Breiðamerkurjökull, several small eskers are being exposed as the terminus retreats (**Fig.** 8). Numerous subglacial and englacial ice tunnels drain meltwater in summer from the surface of Breiðamerkurjökull. These tunnels can become filled with sorted stream sediments. Many of these tunnels occur englacially and the sediments filling them are poorly preserved because of reworking during the melting of the underlying ice. Some subglacial tunnels are preserved as esker ridges. Sandar (or outwash plains) have formed at several levels in the proglacial area in front of Breiðamerkurjökull since the glacier began to retreat in 1900. On aerial photographs the abandoned braided stream pattern can be seen on these surfaces (**Figs.** 5 and 8).

Late Pleistocene Glacier Recession

At the end of the Pleistocene Epoch, approximately 10,000 years ago, the great ice sheets that covered North America and northern Europe were rapidly melting. This glacier recession began about 21,000 carbon-14 (^{14}C) years before present (BP) and accelerated after about 14,500 ^{14}C years BP. By 5,000 ^{14}C years BP the great ice sheets were gone (see Dyke and Prest, 1987 for maps of North America). Most of the evidence we have for the timing of this glacier recession comes from radiocarbon dating of wood, peat, shells, and other organic matter that accumulated on top of glacial sediments deposited during recession (see Dyke *et al.*, 2002 for a summary). The timing of major readvances during the deglaciation comes from radiocarbon-dated wood buried under glacial till of the readvance phases. Perhaps the most well-known example is the Two Creeks forest bed in the Great Lakes region of the USA. At this site numerous radiocarbon dates on wood show that ice advance (and the rising lake waters) killed the forest and buried it under a readvancing glacier and its till at about 11,850 ^{14}C years BP (Broecker and Farrand, 1963).

The Laurentide ice sheet, the largest of all the Late Pleistocene ice sheets, produced a truly remarkable record of glacier recession along its southern margin in the northern US. Many of the glacial features we see forming in southeastern Iceland today along the margin of Breiðamerkurjökull were being formed along the southern margin of the Laurentide ice sheet during the Late Pleistocene deglaciation (Evans, 2003; Colgan *et al.*, 2003). By comparing landforms and sediments in Iceland with those that formed during the Late Pleistocene we can better understand how the Late Pleistocene deglaciation progressed and how rapidly the terminus retreated across the landscape.

In central and southern Wisconsin (USA) an area of roughly 40,000 km^2 contains a unique record of glacier recession including landforms such as drumlins and flutes, small push moraines, recessional moraines, eskers, and outwash plains (Colgan and Mickelson, 1997; Colgan, 1999). The distribution of these landforms and their relationships to one another are nearly identical to those forming in southeastern Iceland today. One big difference is that landforms in Wisconsin cover an area hundreds of times larger than that covered in Iceland. Another important difference is the occurrence of features such as hummocky moraines and tunnel channels. These are found along the maximum ice margin and indicate that supraglacial sediment cover and drainage of subglacial meltwater were important processes when ice was at its maximum Late Pleistocene position (see Clayton *et al.*, 1999). In the deglaciated foreland of Breiðamerkurjökull hummocky topography is only a minor landform type and tunnel channels do not appear at all.

Small push moraines are common features in southern Wisconsin in the area once covered by the Green Bay lobe of the southern Laurentide ice sheet (**Fig. 9**). These ridges are usually about 10–50 m wide and usually less than 2 m high. They are traceable over distances of up to 3 km and commonly overlap or cross-cut. Where ridges are well preserved they appear to be spaced 25–100 m apart. The ridges are superimposed on fluted and drumlinlized terrain just as they are in Iceland. If we accept the hypothesis that these ridges formed annually or semiannually as do the small push moraines in Iceland they would indicate a retreat rate of about 50 m year^{-1} for at least hundreds of years between 15,500 ^{14}C years BP and 13,000 ^{14}C years BP. Some especially well-preserved ridges (**Fig. 9**) are also described in northern Wisconsin produced by the Wisconsin Valley sublobe of the Laurentide ice sheet (Ham and Attig, 2001). Estimated retreat rates were 37–82 m year^{-1} when these ridges formed and ablation rates near the margin of the ice sheet must have been 1.7–14.5 m

Figure 9 (A) Small push moraines that may have formed annually during the recession of the Green Bay lobe between 15,500 ^{14}C years BP and 13,000 ^{14}C years BP. (B) Small push moraines created by the Wisconsin Valley lobe in northcentral Wisconsin during deglaciation (Ham and Attig, 2001).

year^{-1}. The common climatic pattern both in Iceland and along the southern margin of the Laurentide ice sheet may have been the very high summer ablation rates followed by a cold winter when ice could advance and push up a small ridge.

Larger recessional moraines are found along the southern margin of the Laurentide ice sheet and can be traced all the way from Ohio to the Dakotas. These moraines have a complex history reflecting both deposition during glacier advance as well as during glacier recession. Many of the end moraines in Illinois, Indiana, southern Michigan, and Ohio are composed of stacked sequences of basal till, outwash, and lake sediment. In Illinois, the very massive moraines are thought to have formed by subglacial transport and deposition of deformable subglacial sediment that was being moved toward the terminus by rapidly flowing outlet glaciers perhaps fed by ice streams (Johnson and Hansel, 1999). In North Dakota, Minnesota, Wisconsin, and Northern Michigan end moraines are very hummocky and indicate that ice near the margin became stagnant and was buried by thick supraglacial debris (Johnson *et al.*, 1995; Ham and Attig, 1996; Johnson and Clayton, 2003). This supraglacial material must have been elevated from the subglacial bed near the terminus because there are no nunataks that could have dropped debris onto this ice surface in the lowland terrain. Processes such as thrusting and shearing of basal debris as has been observed in subpolar glaciers may have been at work. Upon melting, which could have taken several thousand years, the complex topography of kettles, kames, and ice-walled lake plains formed.

Most of the deglaciated surface in Wisconsin is covered by drumlins and flutes. Other regions covered by extensive fields of drumlins include northcentral Minnesota, Michigan, New York, and New England. The subglacial landforms in these areas are covered by landforms formed during glacier recession such as push moraines, and larger recessional moraines.

Ice Marginal Stagnation Along the Southern Margin of the Laurentide Ice Sheet

In many areas along the lowland margins of the great ice sheets in North America and Europe extensive areas of hummocky moraine suggest that vast blocks of glacier ice became stagnant and detached from active ice during recession (Parizek, 1969; Johnson and Clayton, 2003). In the western Great Lakes region of the USA large hummocky end moraines suggest that ice near the margin was covered by thick supraglacial sediment and slowly melted out during deglaciation (see Colgan *et al.*, 2003). The most common landforms produced by this process are hummocks, ring-forms, dump ridges, disintegration ridges, and ice-walled lake plains (Johnson and Clayton, 2003). In the Canadian Prairie region and plains region of the USA there is evidence that large areas of debris-rich stagnant ice covered the landscape during deglaciation in the form of aligned hummocks, ring-forms, and disintegration ridges (see Colgan *et al.*, 2003).

Hummocks are rounded knobs, hillocks, or mounds occurring in widespread tracts (**Figure 10**). Hummocks are usually described as round-to-oval in plan form, 15–400 m in diameter, and are spaced 150–500 m apart. The local relief of hummocks is usually described as between 2 and 70 m with most less than 20 m. Hummocks have been classified into categories based on relief, but a continuum of size and form may also exist. A large area of hummocks is commonly called hummocky topography.

Ring-forms are circular or doughnut shaped forms that commonly form a continuum with hummocks. They can be thought of as small hummocks with a central depression. These forms are thought to have formed when stagnant debris covered ice melts and deposits supraglacial sediment. Ice-walled lake plains form when ice becomes stagnant and meltwater ponds in sinkholes on top of the glacier. The water in the supraglacial lakes melts its way to the bed in areas of supraglacial debris near the margin. Sediment moved by streams and gravity deposit sediment in these lakes. Lacustrine, fluvial, and deltaic sediment may completely fill the sinkhole formed in the ice. Upon melting of surrounding ice the sediment may completely collapse to form hummocky topography or it may form a plateau of lacustrine sediment surround by an ice-contact ridge (**Fig. 11**). Ice-walled lake plains have been described in Canada, North Dakota, Minnesota, Wisconsin, Iowa, and as high kames in southern New England. Ice-walled lake plains are also found in northern Europe in Denmark, Sweden, Germany, Poland, and Russia.

Disintegration ridges are linear ridges 1–10 m high and 10–100 m wide that probably form when sediment is dumped into or squeezed into crevasses in stagnant ice. They commonly show the control of a network of crevasses and may display a box work pattern where the original ice blocks outlined by longitudinal and transverse crevasses may be easily visualized. Disintegration ridges are found both in Canada and the northern US as well as in northern Europe. A dump-ridge is a pile of debris that forms along the margins of stagnant ice. These ridges are found both in front of stagnant ice bodies as well as

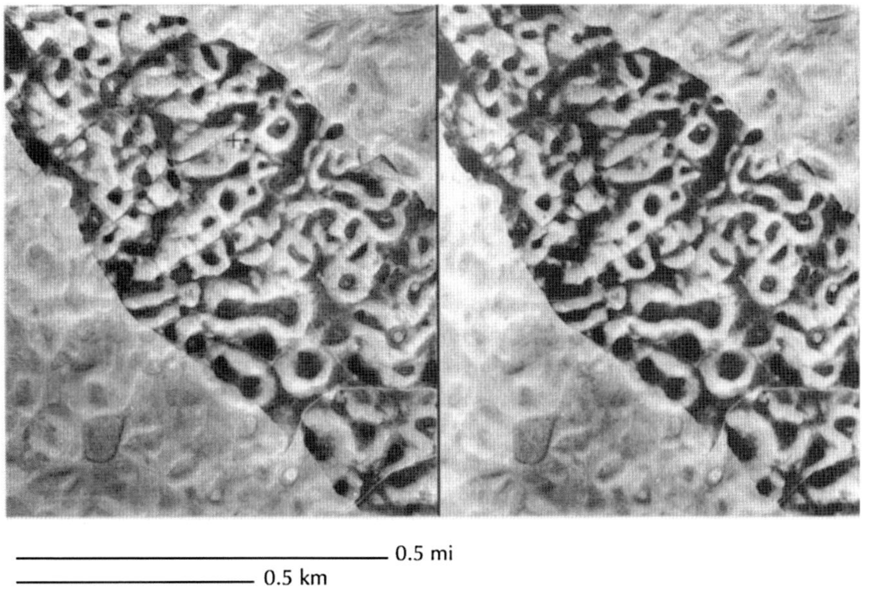

(A)

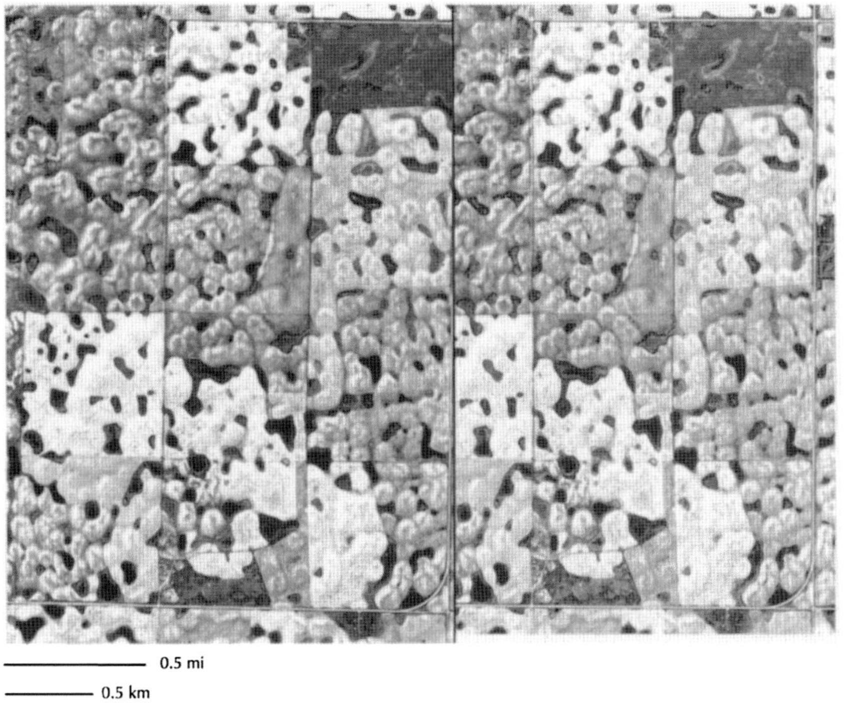

(B)

Figure 10 Stereo aerial photo pairs of hummocks and ring forms deposited along the southwestern margin of the Laurentide ice sheet during deglaciation. (A) Thin lacustrine sediment over clay-rich till southwest of Long Valley Saskatchewan, Canada. Dark areas have been burnt by a prairie fire (A6729-12 and 13, R41-12 and 13). (B) Ring forms near Kenaston, Saskatchewan, Canada (95675-08 68 and 60).

behind large masses of stagnant ice. They have been described in Wisconsin and Minnesota and are probably common wherever stagnant ice was covered by supraglacial debris.

The key environmental factor in developing the features described above is stagnation of supraglacial debris-covered ice near the margin of a glacier during recession. Some have suggested that ice in a zone 2–40 km wide near the ice margin becomes covered by debris and is detached from the actively retreating ice (Johnson *et al.*, 1995; Ham and Attig, 1996; Johnson and Clayton, 2003). This debris-covered ice may then take several thousand years to melt completely. Pleistocene ice-cored moraines in Siberia and in the Mackenzie Delta region of Canada are still melting out and producing topography similar to forms described above thousands of years after glacier recession.

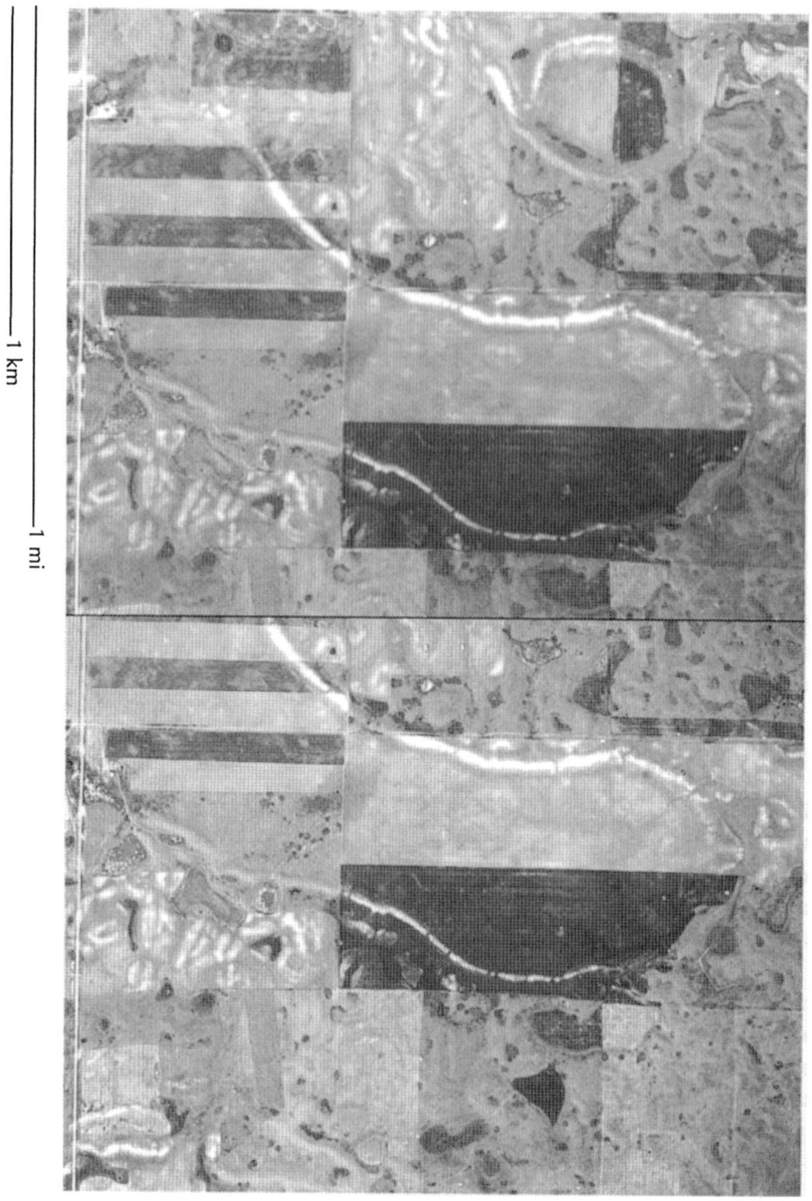

Figure 11 A stereo aerial photo pair of an ice-walled plain located in Montrail County, North Dakota (BAL-4V-9 and 10, USDA). Numerous hummocks and ring forms surround the ice walled lake plain.

Estimated Rates of Glacier Recession During the Late Pleistocene

Many workers have attempted to calculate the rate of Late Pleistocene ice-sheet recession from radiocarbon dating and end moraine mapping. Many of the calculated rates of recession are similar to those seen in modern glaciers today. Kaplan (1999) estimated that the southeastern margin of the Laurentide ice sheet retreated at a rate of about 20 m year^{-1} for the period from 14,000 ^{14}C years BP to 13,000 ^{14}C years BP based on radiocarbon dates and numerous small ridges that formed along a retreating tidewater glacier margin. Gustavson and Boothroyd (1987) compiled a table showing estimates of the glacier recession rates of the southeastern Laurentide ice sheet in southern New England. Most of the rates calculated for the early part of the deglaciation before 13,000 ^{14}C years BP were 50–100 m year^{-1}. Rates of recession calculated from small push moraines in Wisconsin are also in the range of 50–100 m year^{-1}.

Much higher rates of glacier recession have been calculated for areas in which ice may have stagnated during deglaciation. Clayton *et al.* (1985) calculated recession rates as high as 2,000 m year^{-1} on some of the southwestern lobes of the Laurentide ice sheet. They explained these high rates by regional stagnation of ice lobes after a surge.

Implications of Evidence of Glacier Recession

Glacial landforms and sediments produced during glacier recession provide scientists with a unique perspective on reconstructing past glacier recession. As demonstrated above, one way to understand these events is by comparing modern glacier environments and processes to landscapes formed in the past. By these studies and comparisons of past and present events we can interpret the magnitude and sometimes estimate the rate of glacier recession. The biggest limitation in estimating accurate glacier recession rates is the uncertainty in the radiocarbon dating method (500–1,500 years for ages before 10,000 years BP) and the lack of organic material available for radiocarbon dating. A more fundamental problem is that average recession rates do not account for small advances during recession or still-stands. This can be clearly seen in Iceland where an average recession rate of 40–50 m $year^{-1}$ for the period 1894 to 1998 clearly does not include documented still-stands and minor advances. Clearly, evidence for glacier recession can be used both to monitor modern glacier recession as well as reconstruct the style and rates of recession of past ice sheets, but it remains to be seen if we can directly compare modern rates with those of Late Pleistocene glaciers.

Abbreviations

BP	before present
GLIMS	Global Land Ice Measurements from Space

See also: **Glacial Landforms**: Introduction; Moraine forms and Genesis; Glacial Land Systems. **Glacial Landforms, Ice Sheets**: Evidence of Glacier and Ice Sheet Extent. **Glacial Landforms, Sediments**: Glacifluvial Landforms of Deposition. **Glaciations**: Overview; Late Quaternary in North America; Neoglaciation in Europe. **Paleoclimate Reconstruction**: The Last Millennium. **Radiocarbon Dating**: Conventional Method; AMS Method; Plant Macrofossils.

References

Boulton, G. S. (1986). Push moraines and glacier contact fans in marine and terrestrial environments. *Sedimentology* **33**, 677–698.

Boulton, G. S. (1987). A theory of drumlin formation by subglacial sediment deformation. In *Drumlin Symposium* (J. Menzies and J. Rose, Eds.), pp. 25–80. Rotterdam, The Netherlands: Balkema.

Broecker, W. S., and Farrand, W. F. (1963). Radiocarbon age of the Two Creeks forest bed, Wisconsin. *Geological Society of America Bulletin* **74**, 795–802.

Clayton, L., Teller, J. T., and Attig, J. W. (1985). Surging of the southwestern part of the Laurentide ice sheet. *Boreas* **14**, 235–241.

Colgan, P. M. (1999). Reconstruction of the Green Bay lobe, Wisconsin, United States, from 26,000 to 13,000 radiocarbon years BP. In *Glacial Processes Past and Present* (D. M. Mickelson and J. W. Attig, Eds.), pp. 137–150.

Colgan, P. M., and Mickelson, D. M. (1997). Genesis of streamlined landforms and flow history of the Green Bay lobe, Wisconsin, USA. *Sedimentary Geology* **111**, 7–25.

Colgan, P. M., Mickelson, D. M., and Cutler, P. M. (2003). Ice marginal terrestrial landsystems: southern Laurentide ice sheet margin. In *Glacial Landsystems* (D. J. A. Evans, Ed.), pp. 111–142. Arnold, London.

Clayton, L., Attig, J. W., and Mickelson, D. M. (1999). Tunnel channels formed in Wisconsin during the last glaciation. In *Glaciers Past and Present* (D. M. Mickelson and J. W. Attig, Eds.), pp. 69–82. Geological Society of America Special Paper 337.

Cook, A. J., Fox, A. J., Vaughan, D. G., and Ferrigno, J. G. (2005). Retreating glacier fronts on the Antarctic Peninsula over the past half-century. *Science* **308**, 541–544.

Dyke, A. S., and Prest, V. K. (1987). Paleogeography of northern North America, 18,000–5,000 years ago. *Geological Survey of Canada* **1**, Map 1703A, scale 12,500,000,

Dyke, A. S., Andrews, J. T., Clark, P. U., *et al.* (2002). The Laurentide and Innutian ice sheets during the Last Glacial Maximum. *Quaternary Science Reviews* **21**, 9–31.

Dyurgerov, M. B., and Meier, M. F. (2000). Twentieth century climate change: evidence from small glaciers. *Proceedings of the National Academies of Sciences U S A* **97**, 1406–1411.

Evans, D. J. A. (2003). Ice-marginal terrestrial landsystems: active temperate glacier margins. In *Glacial Landsystems* (D. J. A. Evans, Ed.), pp. 12–43. Arnold, London.

Evans, D. J. A., and Twigg, D. R. (2002). The active temperate glacial landsystem: a model based on Breiðmerkurjökull and Fjallsjökull, Iceland. *Quaternary Science Reviews* **21**, 2143–2177.

Global Land Ice Measurements from Space http://www.glims.org/; last accessed 21 June 2006).

Grove, J. M. (1988). *The Little Ice Age*. Routledge, London.

Gustavson, T. C., and Boothroyd, J. C. (1987). A depositional model for outwash, sediment sources, and hydrologic characteristics, Malaspina Glacier, Alaska: a modern anologue for the southeastern margin of the Laurentide ice sheet. *Geological Society of America Bulletin* **99**, 187–200.

Ham, N. R., and Attig, J. W. (1996). Ice wastage and landscape evolution along the southern margin of the Laurentide ice sheet, north-central Wisconsin. *Boreas* **25**, 171–186.

Ham, N. R., and Attig, J. W. (2001). Minor end moraines of the Wisconsin Valley Lobe, north-central Wisconsin, USA. *Boreas* **30**, 31–41.

Johnson, M. D., and Clayton, L. (2003). Supraglacial landsystems in lowland terrain. In *Glacial Landsystems* (D. J. A. Evans, Ed.), pp. 228–258. Arnold, London.

Johnson, M. D., Mickelson, D. M., Clayton, L., and Attig, J. W. (1995). Composition and genesis of glacial hummocks western Wisconsin, USA. *Boreas* **24**, 97–116.

Johnson, W. H., and Hansel, A. K. (1999). Wisconsin Episode glacial landscape of central Illinois: a product of subglacial deforming processes. In *Glacial Processes Past and Present* (D. M. Mickelson and J. W. Attig, Eds.), pp. 122–135.

Kaplan, M. R. (1999). Retreat of a tidewater margin of the Laurentide ice sheet in eastern coastal Maine between ca. 14,000 and 13,000 ^{14}C yr B.P. *Bulletin of the Geological Society of America* **111**, 620–632.

Meier, M. F. (1984). Contribution of small glaciers to global sea level. *Science* **226**, 1418–1421.

Meier, M. F., and Bahr, D. B. (1996). Counting glaciers: use of scaling methods to estimate the number and size distribution of the glaciers of the world. In *Glaciers and Ice Sheets: A Volume Honoring M.F. Meier* (S. D. Colbeck, Ed.), pp. 89–94.

Meier, M. F., Dyurgerov, M. B., and McCabe, G. J. (2003). The health of glaciers: recent changes in glacier regime. *Climate Change* **59**, 123–125.
Oerlemans, J. (2005). Extracting a climate signal from 169 glacier records. *Science* **308**, 675–677.
Parizek, R. R. (1969). Glacial ice-contact rings and ridges. *Geological Society of America Special Paper* **123**, 49–102.
Paterson, W. S. B. (1994). *The Physics of Glaciers*. Pergamon, Oxford.
Rignot, E., Rivera, A., and Casassa, G. (2003). Contribution of the Patagonian Ice Fields of South America to sea level rise. *Science* **302**, 437–438.
Warrick, R., and Oerlemans, J. (1990). Sea level rise. In *Climate Change: The IPCC Scientific Assessment* (J. T. Houghton, G. T. Jenkins and J. J. Ephraums, Eds.), pp. 257–281. Cambridge University Press, Cambridge.
World Glacier Monitoring Service. (2006) (http://www.geo.unizh.ch/wgms/; last accessed 21 June 2006).

Glacial Land Systems

D J A Evans, Durham University, Durham, UK

© 2007 Elsevier B.V. All rights reserved.

The earliest glacial landsystem-type investigations include those of Speight (1963), who mapped and grouped landforms according to common origin and age, and Clayton and Moran (1974), who linked process and form in assessing the spatial distribution of glacial features. Glacial landsystems *per se* were introduced by Fookes *et al.* (1978) to provide engineers with genetic classifications of glacigenic landform–sediment assemblages and included the till plain landsystem, glaciated valley landsystem, and fluvioglacial and ice contact deposit landsystem. The component parts of landsystems are land units (e.g., drumlin or fluting field, Rogen moraine belt, and suite of recessional push moraines) and land elements (e.g., individual drumlins, flutings, eskers, kame terraces, and moraines), with the landsystem thus being a recurrent pattern of genetically linked land units. This was refined by Eyles (1983a,b), Eyles and Menzies (1983), and Paul (1983), who introduced the subglacial, supraglacial, and glaciated valley landsystems. Recent developments of the glacial landsystem concept (Evans, 2003a) have stressed the complexity of glacial depositional environments and the fact that variability in landform–sediment assemblages is dictated not only by the location of deposition but also by the style of glaciation. The combined influence of climate, basement and surficial geology, and topography on glaciation style has necessitated the recognition of a wide range of glacial landsystems for different ice masses and dynamics (Benn and Evans, 1998; Evans, 2003a).

Landsystems as Modern Analogs

Landsystems classifications of modern glacial environments, where landscape evolution can be monitored, provide analogs for the interpretation of ancient glaciated terrain (Price, 1969; Gustavson and Boothroyd, 1987; Kruger, 1994; Dyke and Savelle, 2000; Kjær and Krüger, 2001; Evans and Twigg, 2002). This type of research on modern glaciers has led to the identification of landform–sediment suites indicative of specific styles of glaciation, including plateau icefields (Rea *et al.*, 1998; Rea and Evans, 2003), arid polar glacier margins (Fitzsimons, 2003), and ice dynamics such as surging glaciers (Evans and Rea, 1999, 2003) and ice streams (Stokes and Clark, 1999, 2001; Clark *et al.*, 2003). Once a landform–sediment suite pertaining to a single period of glacier occupancy or activity can be identified, it often becomes possible to differentiate overprinted signatures (Dyke and Morris, 1988; Clark, 1993, 1999; Kruger, 1994) and landscape palimpsests. Examples of some modern analogs are reviewed here, followed by a discussion of their application to ancient glaciated landscapes. The terrestrial examples used here are clearly not exhaustive and the reader is directed to Evans (2003a) for a more comprehensive coverage of the wide range of glacial landsystems that have thus far been identified in both terrestrial and offshore settings.

Active Temperate Glacial Landsystem

Active temperate glacier margins are capable of forward momentum even during overall recession, and this is manifested in the small winter readvances that characterize receding outlet glaciers in places such as Iceland (Sharp, 1984; Boulton, 1986; Kruger, 1995). Details of the glaciology and associated sediment and landform production at the southern Icelandic glacier snouts are presented in numerous publications (Sharp, 1984; Kruger, 1985, 1993, 1994; Boulton, 1986; Boulton and Eyles, 1979; Boulton and Hindmarsh, 1987; Benn, 1995; Evans and Hiemstra, 2005) and it is from this expansive literature that a landsystems model for active temperate glaciers has been compiled (Evans and Twigg, 2002; Evans, 2003b, 2005). This model recognizes three dominant depositional domains with characteristic landform–sediment assemblages (**Fig. 1**). First, the marginal morainic domain consists of areas of extensive low-amplitude marginal dump, push, and squeeze moraines (Price, 1970; Evans *et al.*, 1999; Evans and Twigg, 2000) often recording annual ice recession, although short periods of readvance or stability can produce larger push moraines due to stacking of submarginally frozen sediment slabs (Kruger, 1985, 1993, 1994, 1996; Evans and

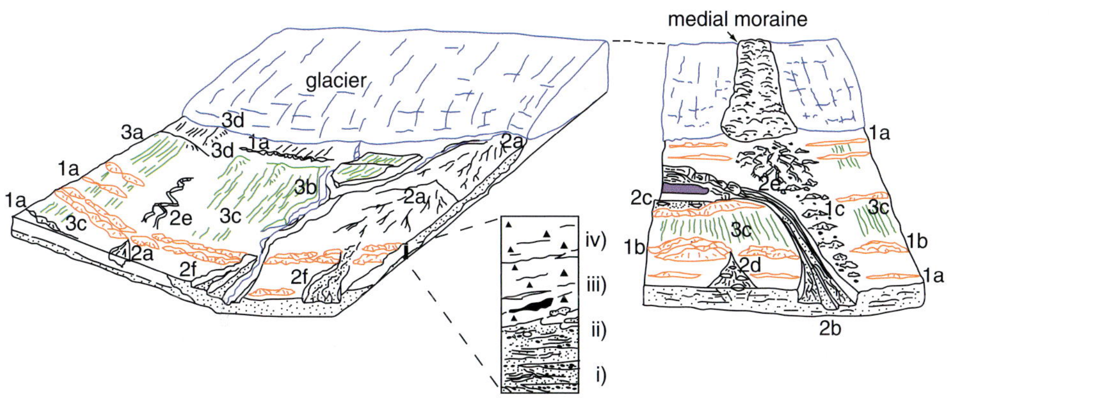

Figure 1 The active temperate glacial landsystem. Landforms are numbered according to their domain (1, morainic domain; 2, glacifluvial domain; and 3, subglacial domain): 1a, small, often annual, push moraines; 1b, superimposed push moraines; 1c, hummocky moraine; 2a, ice-contact sandur fans; 2b, spillway-fed sandur fan; 2c, ice margin-parallel outwash tract/kame terrace; 2d, pitted sandur; 2e, eskers; 2f, entrenched ice-contact outwash fans; 3a, overridden (fluted) push moraines; 3b, overridden, preadvance ice-contact outwash fan; 3c, flutes; and 3d, drumlins. The idealized stratigraphic section log shows a typical depositional sequence recording glacier advance over glacifluvial sediments comprising (i) undeformed outwash; (ii) glacitectonized outwash/glacitctonite; (iii) massive, sheared till with basal inclusions of preadvance peat and glacifluvial sediment; and (iv) massive sheared till with basal erosional contact. After Kruger J (1994) Glacial processes, sediments, landforms and stratigraphy in the terminus region of Myrdalsjökull, Iceland. *Folia Geographica Danica* 21: 1–233; Evans DJA and Twigg DR (2002) The active temperate glacial landsystem: A model based on Breidamerkurjokull and Fjallsjokull, Iceland. *Quaternary Science Reviews* 21: 2143–2177; and Evans, D. J. A. (Ed.) (2003a). *Glacial Landsystem.* Arnold, London.

Hiemstra, 2005), or dump, squeeze, and push mechanisms operating at the same location over several years. Second, the subglacial domain includes assemblages of flutings, drumlins, and overridden push moraines that occur between ice-marginal depocenters and are linked to subglacial deformation and lodgement processes (Boulton, 1987; Boulton and Hindmarsh, 1987; Benn, 1995; Hart, 1995; Benn and Evans, 1996; Evans and Hiemstra, 2005). Third, the glacifluvial and glacilacustrine domain includes (a) the deposits of proglacial outwash streams that drain either away from glacier margins to produce sandur (outwash) fans or parallel to the margin to produce ice margin-parallel outwash tracts and kame terraces, (b) the glacilacustrine depocenters that have developed in large lake-filled depressions uncovered by glacier recession, and (c) simple and complex anabranched eskers that often grade into recessional ice-contact fans and areas of pitted outwash. In addition to the three major depositional domains, anomalies are to be expected due to the signatures of erratic/noncyclic or site-specific processes, including jokulhlaup features (Maizels, 1997; Russell and Knudsen, 1999; Fay, 2002).

Glaciated Valley Landsystem

The glaciated valley landsystem was first compiled by Boulton and Eyles (1979) and Eyles (1983b) in order to encompass the landform–sediment associations produced by valley glaciers. Owen and Derbyshire (1989) highlighted the large volumes of supraglacial debris on Karakoram glaciers and illustrated the fact that valley glaciers do not always deposit clear, inset sequences of proglacially constructed laterofrontal moraines as depicted in the Boulton and Eyles (1979) model. Benn *et al.* (2003) subsequently demonstrated the huge variability in landform and sediment assemblages in highland settings due to their occurrence at every latitude on the Earth's surface, and they proposed a model that comprises a family of landsystems (**Fig. 2**). Recurrent features identified by Benn *et al.* include (a) the strong influence of topography on glacier morphology, sediment transport paths, and depositional basins; (b) the important role of supraglacial debris sources; and (c) the rate of sediment transport from the glacier to the proglacial environment via the glacifluvial system. Their family of landsystems lie on a continuum between the two extremes of 'clean' glaciers, with limited supraglacial debris, and debris-covered glaciers/rock glaciers. Additionally, the relief of upland landscapes often results in temporary lake production during glaciation, either by ice damming of valleys or by moraine damming. The latter are potentially dangerous features in mountain terrain due to their tendency to catastrophically burst through moraine dams and produce glacial lake outburst floods (Clague and Evans, 2000; Cenderelli and Wohl, 2003; Kershaw *et al.*, 2005). The different members of the glaciated valley landsystem family identified by Benn *et al.* (**Fig. 2**) are defined by the efficiency of glacifluvial transport between the glacier and its proglacial area. Efficient transport defines a 'coupled' landsystem association in which moraine development is limited and an outwash head/aggrading sandur

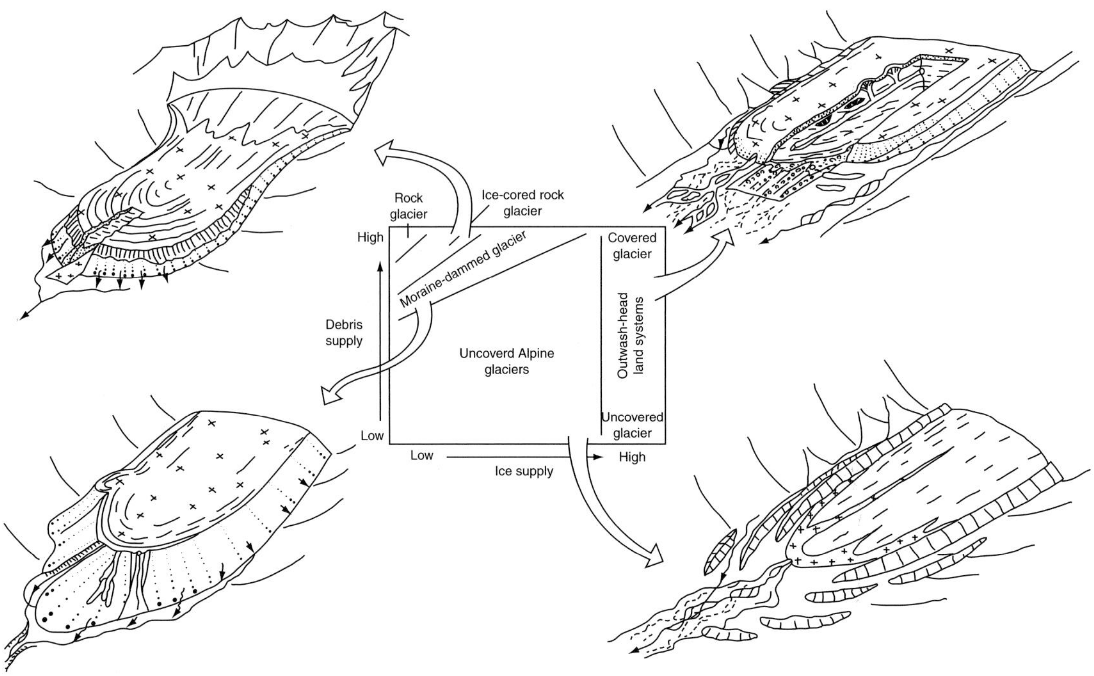

Figure 2 Constraints on landsystems development around valley glaciers, showing development pathways to four landsystem associations. From Benn DI, Kirkbride MP, Owen LA, and Brazier V (2003) Glaciated valley landsystems. In: Evans DJA (ed.) *Glacial Landsystems*, pp. 372–406. London: Arnold. Reproduced with permission.

(Kirkbride, 2000) is common, as typically occurs in the humid mountain ranges of Alaska and New Zealand. Inefficient transport defines a 'decoupled' landsystem in which repeated superposition of large moraines occurs around the margin of a debris-covered glacier, as is common in high-altitude, semiarid to arid mountain ranges such as the central and northern Andes and High Asia.

Plateau Icefield Landsystem

Plateaus are the loci of glacier initiation and expansion during cooling events (Ives *et al.*, 1975) and tend to be a haven for glaciers during interglacials and interstadials, depending on the latitude and altitude of the plateau and the global climate characteristics at the time (Manley, 1955; Rea *et al.*, 1998). Precipitous drops at the margins of plateaus lead to the development of 'reconstituted glaciers' (Benn and Lehmkuhl, 2000) through dry calving at the plateau cliff edge (Gellatly *et al.*, 1986). During the more advanced stages of local-scale glaciation, glaciers develop in the surrounding valley bottoms, and it is during this stage that the most significant landforms are likely to be produced; in most instances, the surrounding valleys contain the bulk of the landform–sediment assemblages used in reconstructing the former glacier cover (Rea and Evans, 2003). On plateau summits, the thermal and dynamic properties of the ice determine the geomorphic signature, which in many instances, especially where ice was cold based, may be very subtle or nonexistent (Dyke, 1993). The assemblage of landforms produced by plateau icefields in the Canadian High Arctic, northern Norway, and Iceland have been used to compile a plateau icefield landsystem (Rea *et al.*, 1998; Rea and Evans, 2003) (**Fig. 3**). Plateau summits may contain nothing more than deglacial meltwater channels cutting through preexisting regolith or blockfield if the former ice cover was cold based. Lateral meltwater channels produced by the retreat of outlet glaciers onto plateaus are characteristic of recession by cold-based ice (Maag, 1969; Dyke, 1993). Evidence for localized erosion by warm-based ice (e.g., striated bedrock and roches moutonnées and the removal of blockfield) occurs in the topographic depressions at plateau edges where ice flow velocities increase due to drawdown into surrounding valleys. Moraines may also be found on plateau edges or leading onto the plateau from the surrounding valley head outlets if warm-based conditions exist locally, but significant till covers do not form due to the commonly short transport distances to plateau outlets and the dynamics of the plateau ice cover (Evans *et al.*, 2006a). If the ice is cold based throughout, lateral and frontal moraines will be found only in the valleys. Valley moraines tend to be dominated by large, cobble- to boulder-sized, angular material even where there is evidence of basal sliding, indicating that the dominant debris source is rock fall. Many laterofrontal moraines in the valley heads that

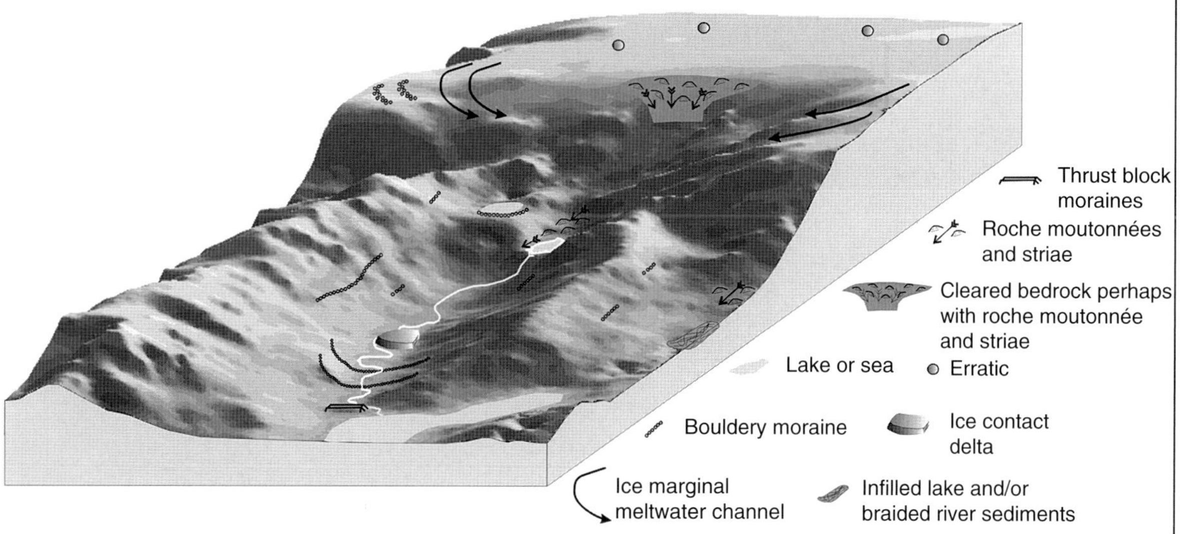

Figure 3 The plateau icefield landsystem. From Rea BR and Evans DJA (2003) Plateau icefield landsystems. In: Evans DJA (ed.) *Glacial Landsystems*, pp. 407–431. London: Arnold. Reproduced with permission.

surround the plateaus are ice cored and display within-valley asymmetry (Benn, 1989). This reflects the distribution of bedrock-free faces and concomitant variability in rock fall/avalanche and debris flow activity. Active free faces may provide sufficient debris to produce supraglacial lateral moraines. In some circumstances, these may develop into rock-glacierized ice-cored moraines that may persist after deglaciation of the valley heads (Evans, 2005). The valley floors are characterized by thin, generally patchy tills, which locally thicken in depressions.

Ice Stream Landsystem

Remotely sensed observations on the processes operating at the beds of ice streams in the Antarctic Ice Sheet (Kamb, 2001) combined with evidence of landform–sediment assemblages on the Antarctic continental shelf, across which ice streams formerly flowed, have informed recent compilations of palaeo-ice stream landsystems (Canals *et al.*, 2000). Ice streams are spatially restricted regions (arteries) of fast-flowing ice in an ice sheet and are readily identifiable by their surface structures and velocities. A landsystem model for ice streams has been developed by Stokes and Clark (1999) and Clark and Stokes (2003) based predominantly on the subglacial bedform evidence for fast-flowing ice and validated by glaciological data on modern ice streams (**Fig.** 4). The bedform evidence comprises megascale glacial lineations (MSGLs), which are elongate streamlined ridges of sediment similar to flutes and drumlins and thought to be indicative of high strain rates and fast ice velocity. The pattern of subglacial bedforms should reflect the dominant flow patterns within an ice stream in that they should converge in an onset zone and lead into a much narrower and well-defined trunk zone where highly attenuated streamlined bedforms (MSGLs) record the rapid velocity of the ice stream trunk. The higher elongation ratios of MSGLs occur in the trunk rather than in the convergent onset zone, and the highest elongation ratios are in the central axis of the trunk. Additionally, elongation ratios in marine-terminating ice streams should steadily increase downstream toward the grounding line, but in terrestrially terminating ice streams they decrease toward the outer margins of the lobate terminus. Ice streams also display abrupt lateral margins where fast ice abuts slower moving ice. This is reflected in the subglacial bedform record, in which a high density of bedforms within the ice stream ceases abruptly. Sediment may concentrate at this boundary, producing an ice stream shear margin moraine (Dyke and Morris, 1988; Dyke *et al.*, 1992; Stokes and Clark, 2002). At the terminus of ice streams, it should be expected that large depocenters will accumulate, such as submarine 'till deltas' (Alley *et al.*, 1989; Vorren and Laberg, 1997) and grounding line wedges (Powell and Domack, 1995; Ó Cofaigh *et al.*, 2003). Because ice streams switch on and off and dominant ice flow trajectories change during the life cycle of an ice sheet, ice stream imprints may cross-cut each other (Dyke and Morris, 1988). Two end member landsystems are envisaged by Clark and Stokes (2003; **Fig.** 5) to account for possible landform records of paleo-ice streams: (a) the 'rubber-stamped' imprint, produced when an ice stream imprint is

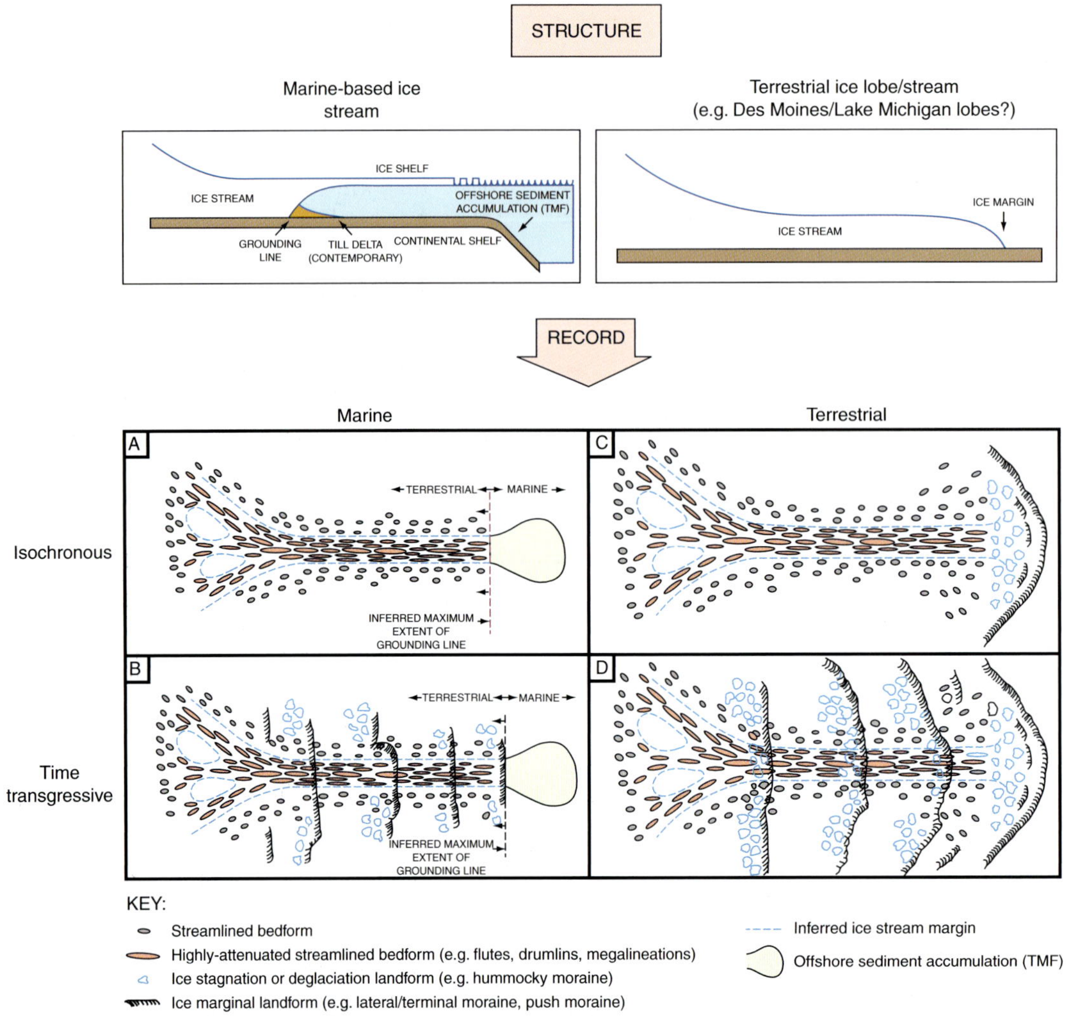

Figure 4 A landsystems model for ice streams showing the four end members of marine-based and terrestrial systems, which are either isochronous/rubber stamped or time transgressive/smudged. From Clark CD and Stokes CR (2003) Palaeo-ice stream landsystem. In: Evans DJA (ed.) *Glacial Landsystems*, pp. 204–227. London: Arnold. Reproduced with permission.

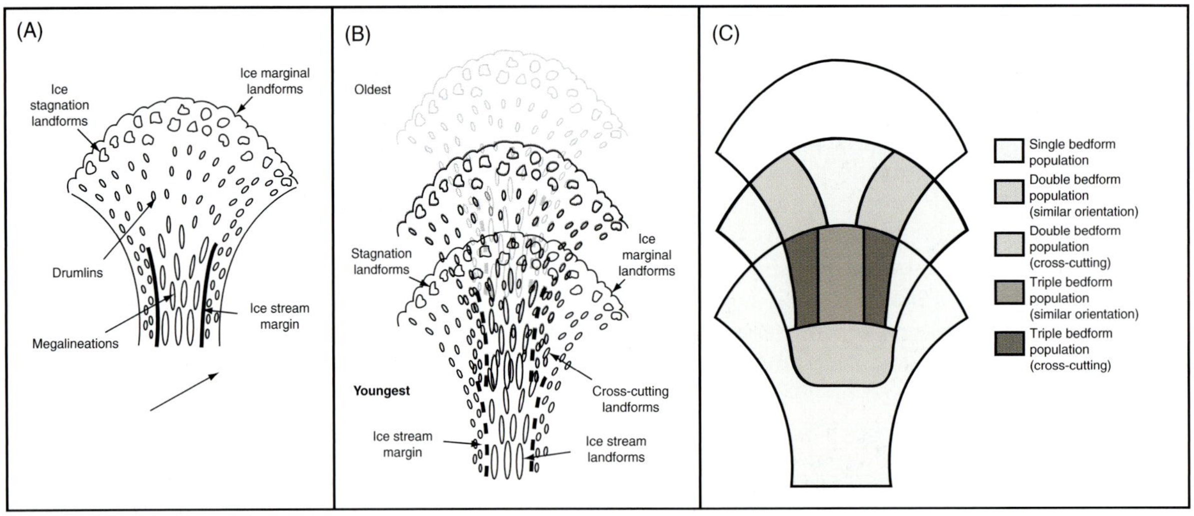

Figure 5 Paleo-ice stream imprints. (A) The imprint of ice stream shut down or the rubber stamped imprint. (B) The overprinted sequence of a continuously operating ice stream during deglaciation or the smudged imprint. (C) Schematic diagram showing the distribution of different cross-cutting bedform imprints associated with a smudged landsystem. From Clark CD and Stokes CR (2003) Palaeo-ice stream landsystem. In: Evans DJA (ed.) *Glacial Landsystems*, pp. 204–227. London: Arnold. Reproduced with permission.

produced at the same time and represents a snapshot of ice stream activity at a point in time when the ice stream switched off, and (b) the 'smudged' or overprinted, time transgressive imprint, produced when an ice stream operates throughout several cycles of advance and retreat or is continuously operating during margin retreat and earlier imprints are modified and overprinted by the younger ice flow patterns.

Applications of Modern Analogs to Quaternary Glacial Landscapes

Active Temperate Glacial Landsystem

Evidence of the recession of a former active temperate glacier lobe at the margin of the southwest Laurentide Ice Sheet has been reported by Evans *et al.* (1999, 2005b) and Benn and Evans (2005). Specifically, the deglaciation of the McGregor Lake area of southern Alberta, Canada, is recorded by numerous closely spaced, low-amplitude moraine ridges that are inset by zones of flutings trending obliquely to individual moraine crests (**Fig. 6**).

Glaciated Valley Landsystem

The concept of a family of glaciated valley landsystems has been applied by Benn *et al.* (2003) to the Ben Ohau Range in New Zealand (Brazier *et al.*, 1998). Holocene climate change in an area with a steep precipitation gradient has resulted in rising equilibrium altitudes and a concomitant response in glacial landsystem

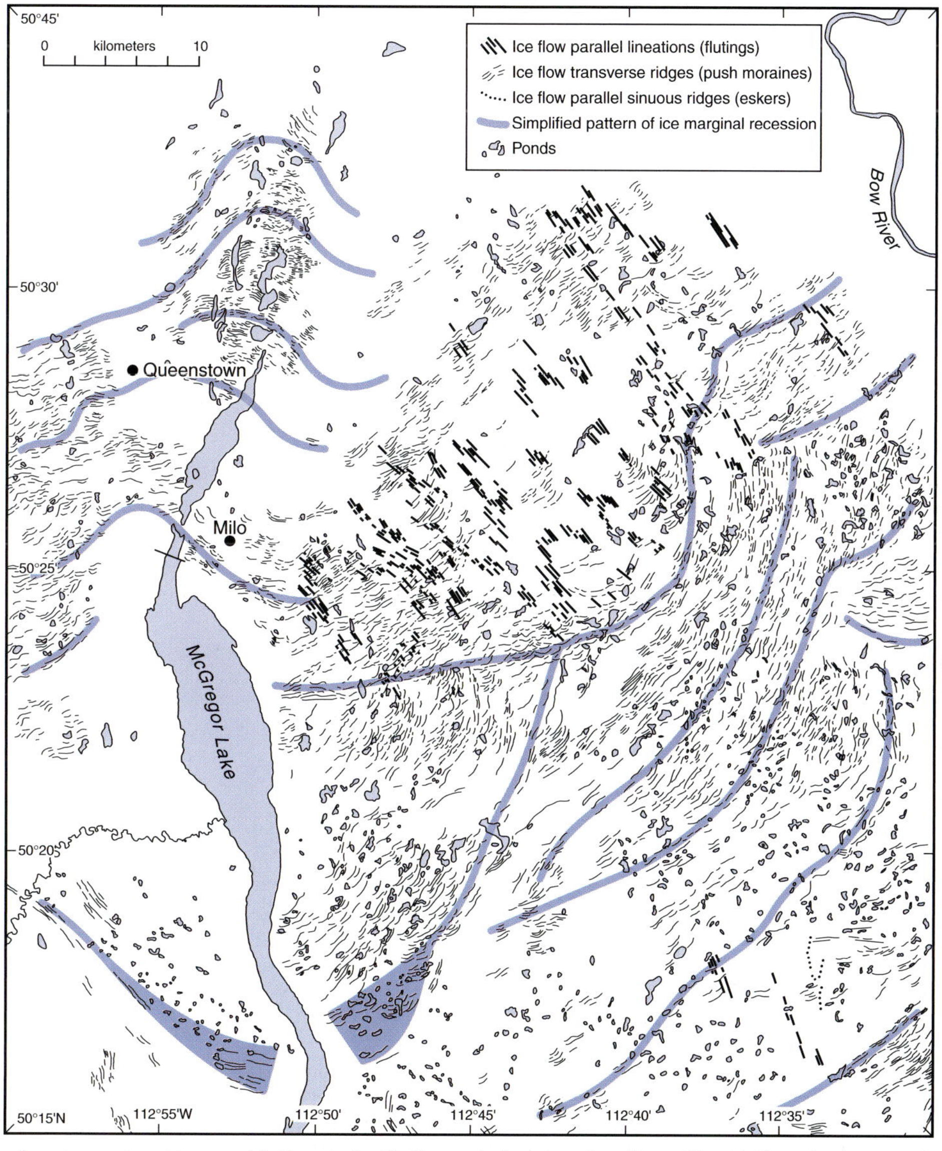

Figure 6 Map of push moraine ridges and flutings in the McGregor Lake area of southern Alberta, Canada, documenting the recession by an active temperate lobe at the southwest margin of the Laurentide Ice Sheet. Light blue shaded lines schematically show the pattern of ice marginal recession and readvance.

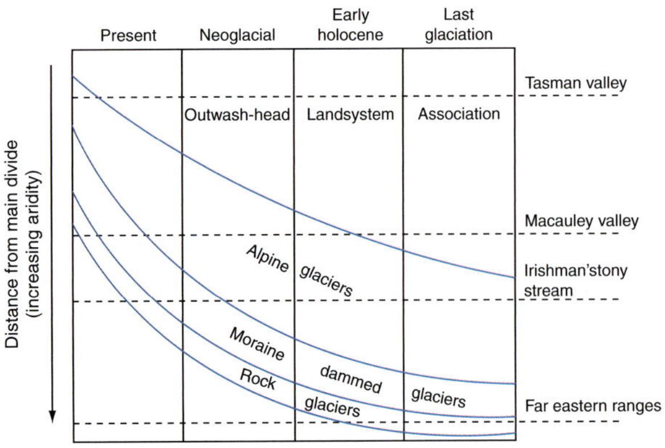

Figure 7 The spatial and temporal occurrence of various members of the glaciated valley landsystem in the Ben Ohau Range, New Zealand, showing the gradual trend from coupled to uncoupled systems. From Benn D. I., Kirkbride, M. P., Owen L. A., and Brazier, V. (2003) Glaciated valley landsystems. In: *Glacial Landsytems* (D. J. A. Evans, Ed.), pp. 372–406. Arnold, London. Reproduced with permission.

development due to altering debris supply and transport. Present-day landsystem distribution demonstrates the crucial role that aridity plays in local factors for landform development. The outwash-dominated landsystems of the major Mount Cook valley glaciers occupy large catchments to the north of the range. Whereas relatively clean Alpine glaciers occupy the uplands of the humid north, debris-covered cirque glaciers with moraine dams occur in central catchments, and active talus rock glaciers dominate toward the arid south. **Figure** 7 shows how process–form domains have shifted in space and time in response to a rise in ELAs since the Last Glacial Maximum, highlighting a shift along the spectrum from coupled to uncoupled landsystems due to reductions in ice-covered area and the importance of glacial transport and runoff.

Plateau Icefield Landsystem

The plateau icefield landsystem concept has been applied by Rea *et al.* (1998) and McDougall (2001) to the reconstruction of Younger Dryas glacier cover in the English Lake District. Initial reconstruction by Sissons (1980) featured extended valley head glaciers that lay below large upland massifs, thereby necessitating the proposal for three separate snowfall intensity zones, in combination with the effects of snowblow and aspect, to account for considerable variations in glacier extent and ELAs (the latter varying by 500 m). Recent mapping of the glacial landforms has revealed that, similar to modern glacierized plateau terrains, the Lake District valley heads contain substantial volumes of glacial sediment in laterofrontal and linear hummocky moraines that extend in many cases onto plateau margins. Evidence of plateau ice is patchy and often restricted to scoured bedrock at plateau edges. Sissons' 'valley' glaciers were therefore actually outlet lobes of a plateau icefield (**Fig. 8**). McDougall's (2001) paleoglacier reconstructions for the English Lake District, incorporating plateau icefields, suggest that ELAs during the Younger Dryas were up to 90 m higher than those calculated by Sissons based on an Alpine style of glaciation with no plateau ice.

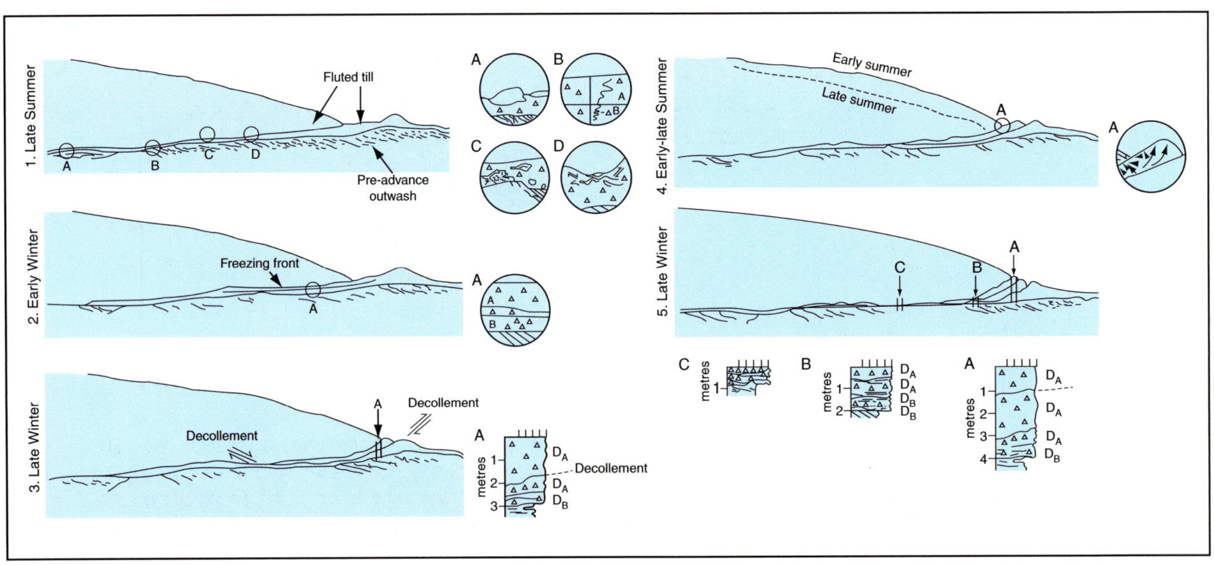

Figure 8 Paleoglaciological reconstruction of Younger Dryas plateau icefield in the English Lake District. From McDougall DA (2001) The geomorphological impact of Loch Lomond (Younger Dryas) Stadial plateau icefields in the central Lake District, northwest England. *Journal of Quaternary Science* 16: 531–543. Reproduced with permission.

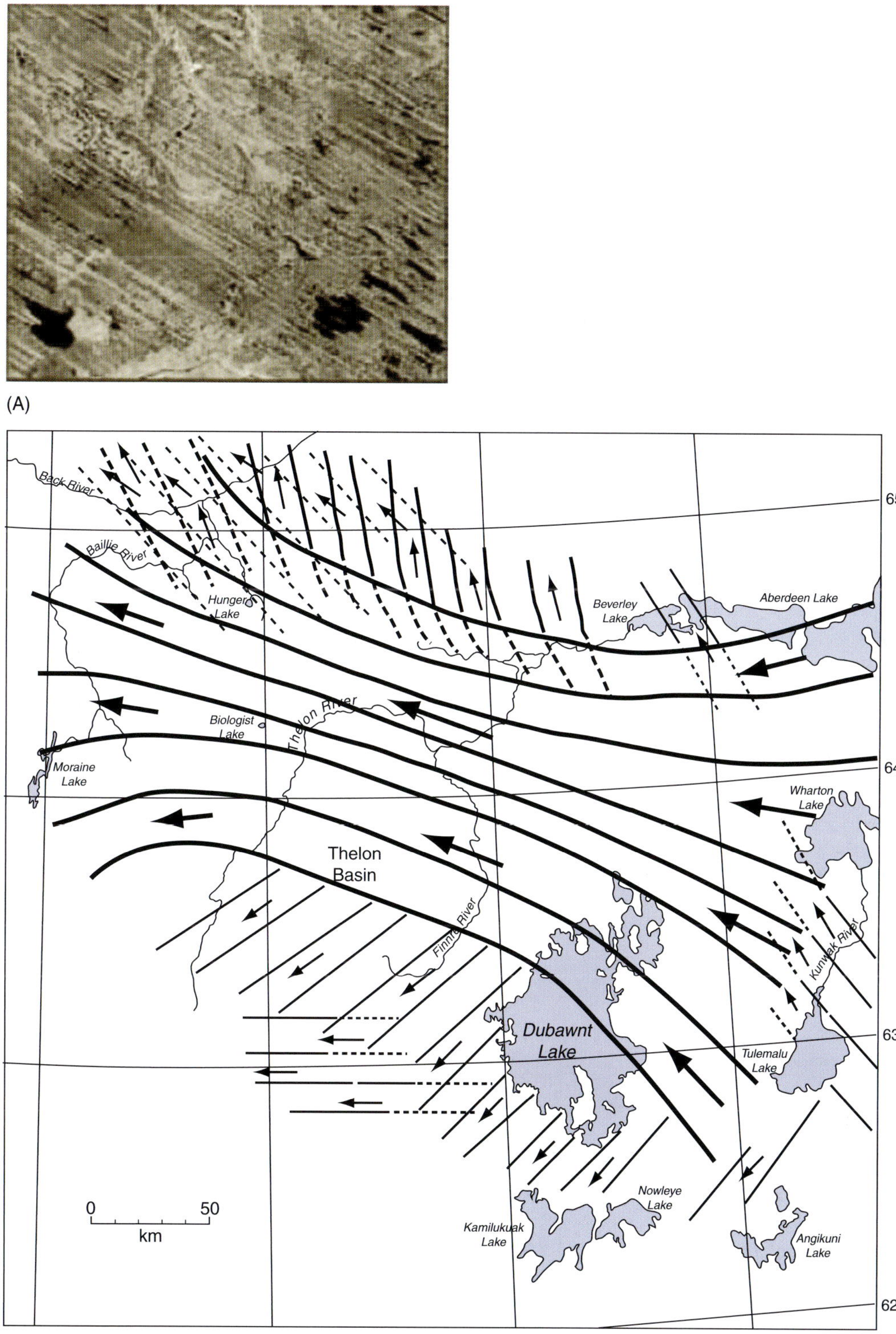

Figure 9 The Dubawnt Lake paleo-ice stream. (A) Satellite image of megascale glacial lineations indicating fast ice flow. (B) Map of the palaeo-ice stream showing the extent of former onset zone, trunk, and terminal divergence overprinting earlier flow patterns. From Stokes CR and Clark CD (2003) The Dubawnt Lake palaeo-ice stream: Evidence for dynamic ice sheet behaviour on the Canadian Shield and insights regarding controls on ice stream location and vigour. *Boreas* 32: 263–279. Reproduced with permission.

Palaeo-ice Stream Landsystem

Evidence for a terrestrially terminating, rubber-stamped, palaeo-ice stream has been identified by Stokes and Clark (2003) in the Dubawnt Lake area of the Northwest Territories of Canada (**Fig. 9**). The ice stream formerly drained the northwestern portion of the Laurentide Ice Sheet and has been recognized by subglacial bedforms that conform to the patterns expected in zones of converging trunk and terminal diverging flow in a terrestrial ice stream. Additionally, the landsystem includes abrupt lateral margins and megascale glacial lineations from which high ice flow velocity is inferred. The clarity of the ice stream bedform signature, with no discordances or cross-cuts, qualified it as a rubber-stamped record of ice stream shutdown.

Conclusion

Paleoglaciological reconstructions, at scales ranging from local to continental, play a significant part in the development of models of global palaeoclimate. However, the great spatial and temporal complexity of glacial processes and forms have necessitated the compilation of process–form models or landsystems that relate to various glaciation styles and systems, which may then be applied to specific physiographic and climatic settings. A number of contemporary glacierized landscapes have been employed as modern analogs for the landform–sediment assemblages produced by different glaciation styles and ice dynamics in differing climatic, geologic, and topographic settings. The identification of modern landsystems (e.g., those pertaining to active temperate glacier snouts, glaciated valleys, plateau icefields, and ice streams) provides glacial geomorphologists with the criteria to decipher paleoglacier dynamics and former glacier–climate relationships in ancient glaciated terrain.

See also: **Glacial Landforms**: Evidence of Glacier Recession. **Glacial Landforms, Erosional Features**: Micro to Macro Scale Forms. **Glacial Landforms, Ice Sheets**: Evidence of Glacier and Ice Sheet Extent. **Glacial Landforms, Sediments**: Glacifluvial Landforms of Deposition; Glacimarine Sediments and Ice-Rafted Debris; Tills.

References

Alley, R. B., Blankenship, D. D., Rooney, S. T., and Bentley, C. R. (1989). Sedimentation beneath ice shelves—The view from ice stream B. *Marine Geology* **85**, 101–120.

Benn, D. I. (1989). Debris transport by Loch Lomond Readvance glaciers in northern Scotland, basin form and the within-valley asymmetry of lateral moraines. *Journal of Quaternary Science* **4**, 243–254.

Benn, D. I. (1995). Fabric signature of subglacial till deformation, Breiðamerkurjökull, Iceland. *Sedimentology* **42**, 735–747.

Benn, D. I., and Evans, D. J. A. (1996). The interpretation and classification of subglacially-deformed materials. *Quaternary Science Reviews* **15**, 23–52.

Benn, D. I., and Evans, D. J. A. (1998). *Glaciers and Glaciation.* Arnold, London.

Benn, D. I., and Evans, D. J. A. (2006). Subglacial floods and the limitations of outrageous hypotheses. In *Glacier Science and Environmental Change* (P. G. Knight, Ed.), Blackwell, London.

Benn, D. I., and Lehmkuhl, F. (2000). Mass balance and equilibrium-line altitudes of glaciers in high mountain environments. *Quaternary International* **65–66**, 15–29.

Benn, D. I., Kirkbride, M. P., Owen, L. A., and Brazier, V. (2003). Glaciated valley landsystems. In *Glacial Landsystems* (D. J. A. Evans, Ed.), pp. 372–406. Arnold, London.

Boulton, G. S. (1986). Push-moraines and glacier-contact fans in marine and terrestrial environments. *Sedimentology* **33**, 677–698.

Boulton, G. S. (1987). A theory of drumlin formation by subglacial deformation. In *Drumlin Symposium* (J. Menzies and J. Rose, Eds.), pp. 25–80. Balkema, Rotterdam.

Boulton, G. S., and Eyles, N. (1979). Sedimentation by valley glaciers: A model and genetic classification. In *Moraines and Varves* (C. Schluchter, Ed.), pp. 11–23. Balkema, Rotterdam.

Boulton, G. S., and Hindmarsh, R. C. A. (1987). Sediment deformation beneath glaciers: Rheology and sedimentological consequences. *Journal of Geophysical Research* **92**, 9059–9082.

Brazier, V., Kirkbride, M. P., and Owens, I. F. (1998). The relationship between climate and rock glacier distribution in the Ben Ohau Range, New Zealand. *Geografiska Annaler* **80A**, 193–207.

Canals, M., Urgeles, R., and Calafat, A. M. (2000). Deep sea-floor evidence of past ice streams off the Antarctic Peninsula. *Geology* **28**, 31–34.

Cenderelli, D. A., and Wohl, E. E. (2003). Flow hydraulics and geomorphic effects of glacial lake outburst floods in the Mount Everest region, Nepal. *Earth Surface Processes and Landforms* **28**, 385–407.

Clague, J. J., and Evans, S. G. (2000). A review of catastrophic drainage of moraine-dammed lakes in British Columbia. *Quaternary Science Reviews* **19**, 1763–1783.

Clark, C. D. (1993). Mega-scale glacial lineations and cross-cutting ice-flow landforms. *Earth Surface Processes and Landforms* **18**, 1–29.

Clark, C. D. (1999). Glaciodynamic context of subglacial bedform generation and preservation. *Annals of Glaciology* **28**, 23–32.

Clark, C. D., and Stokes, C. R. (2003). Palaeo-ice stream landsystem. In *Glacial Landsystems* (D. J. A. Evans, Ed.), pp. 204–227. Arnold, London.

Clark, C. D., Evans, D. J. A., and Piotrowski, J. A., (eds.) (2003). Palaeo-ice streams (Special issue). *Boreas* **32**, 1–280.

Clayton, L., and Moran, S. R. (1974). A glacial process-form model. In *Glacial Geomorphology* (D. R. Coates, Ed.), pp. 89–119. State University of New York, Binghamton.

Dyke, A. S. (1993). Landscapes of cold-centred late Wisconsinan ice caps, Arctic Canada. *Progress in Physical Geography* **17**, 223–247.

Dyke, A. S., and Morris, T. F. (1988). Drumlin fields, dispersal trains and ice streams in Arctic Canada. *Canadian Geographer* **32**, 86–90.

Dyke, A. S., and Savelle, J. M. (2000). Major end moraines of Younger Dryas age on Wollaston Peninsula, Victoria Island, Canadian Arctic: Implications for paleoclimate and for formation of hummocky moraine. *Canadian Journal of Earth Sciences* **37**, 601–619.

Dyke AS, Morris TF, Green DEC, and England J (1992) Quaternary geology of Prince of Wales Island, Arctic Canada. Geological Survey of Canada Memoir 433.

Evans, D. J. A. (Ed.) (2003a). *Glacial Landsystems*. Arnold, London.

Evans, D. J. A. (2003b). Ice-marginal terrestrial landsystems: Active temperate glacier margins. In *Glacial Landsystems* (D. J. A. Evans, Ed.), pp. 12–43. Arnold, London.

Evans, D. J. A. (2005). The glacier-marginal landsystems of Iceland. In *Iceland: Modern Processes and Past Environments* (C. J. Caseldine, A. J. Russell, J. Hardardottir and O. Knudsen, Eds.), pp. 93–126. Elsevier, Amsterdam.

Evans, D. J. A., and Hiemstra, J. F. (2005). Till deposition by glacier submarginal, incremental thickening. *Earth Surface Processes and Landforms* **30**, 1633–1662.

Evans, D. J. A., and Rea, B. R. (1999). Geomorphology and sedimentology of surging glaciers: A landsystems approach. *Annals of Glaciology* **28**, 75–82.

Evans, D. J. A., and Rea, B. R. (2003). Surging glacier landsystem. In *Glacial Landsystems* (D. J. A. Evans, Ed.), pp. 259–288. Arnold, London.

Evans, D. J. A., and Twigg, D. R., (2000) *Breiðamerkurjökull 1998* (1:30,000 scale map). Glasgow/Leicestershire, UK: University of Glasgow/Loughborough University.

Evans, D. J. A., and Twigg, D. R. (2002). The active temperate glacial landsystem: A model based on Breidamerkurjokull and Fjallsjokull, Iceland. *Quaternary Science Reviews* **21**, 2143–2177.

Evans, D. J. A., Lemmen, D. S., and Rea, B. R. (1999). Glacial landsystems of the southwest Laurentide Ice Sheet: Modern Icelandic analogues. *Journal of Quaternary Science* **14**, 673–690.

Evans, D. J. A., Twigg, D. R., and Shand, M. (2006a). Surficial geology and geomorphology of the ßorisjokull plateau icefield, west-central Iceland. *Journal of Maps*.

Evans, D. J. A., Rea, B. R., Hiemstra, J. F., and Ó Cofaigh, C. (2006b). A critical assessment of subglacial mega-floods: A case study of glacial sediments and landforms in south-central Alberta, Canada. *Quaternary Science Reviews*. In press.

Eyles, N. (1983a). Glacial geology: A landsystems approach. In *Glacial Geology* (N. Eyles, Ed.), pp. 1–18. Pergamon, Oxford.

Eyles, N. (1983b). The glaciated valley landsystem. In *Glacial Geology* (N. Eyles, Ed.), pp. 91–110. Pergamon, Oxford.

Eyles, N., and Menzies, J. (1983). The subglacial landsystem. In *Glacial Geology* (N. Eyles, Ed.), pp. 19–70. Pergamon, Oxford.

Fay, H. (2002). Formation of ice block obstacle marks during the November 1996 glacier-outburst flood (jökulhlaup), Skeiðarársandur, southern Iceland. In *Flood and Megaflood Deposits: Recent and Ancient* (I. P. Martini, V. R. Baker and G. Garzon, Eds.), Special Publication No. 32, pp. 85–97. International Association of Sedimentologists.

Fitzsimons, S. J. (2003). Ice-marginal terrestrial landsystems: Polar continental glacier margins. In *Glacial Landsystems* (D. J. A. Evans, Ed.), pp. 89–110. Arnold, London.

Fookes, P. G., Gordon, D. L., and Higginbottom, I. E. (1978). Glacial landforms, their deposits and engineering characteristics. In *The Engineering Behaviour of Glacial Materials*, Proceedings of Symposium, Geoabstracts, pp 18–51. University of Birmingham, Birmingham, UK.

Gellatly, A. F., Whalley, W. B., and Gordon, J. E. (1986). Topographic control over recent glacier changes in Southern Lyngen Peninsula, North Norway. *Norsk Geografisk Tidsskrift* **41**, 211–218.

Gustavson, T. C., and Boothroyd, T. C. (1987). A depositional model for outwash, sediment sources, and hydrologic characteristics, Malaspina Glacier, Alaska: A modern analog of the southeastern margin of the Laurentide Ice Sheet. *Geological Society of America Bulletin* **99**, 187–200.

Hart, J. K. (1995). Recent drumlins, flutes and lineations at Vestari-Hagafellsjökull, Iceland. *Journal of Glaciology* **41**, 596–606.

Ives, J. D., Andrews, J. T., and Barry, R. G. (1975). Growth and decay of the Laurentide Ice Sheet and comparisons with Fenno-Scandinavia. *Naturwissenschaften* **62**, 118–125.

Kamb, B. (2001). Basal zone of the West Antarctic ice streams and its role in lubrication of their rapid motion. In *The West Antarctic Ice Sheet: Behavior and Environment* (R. B. Alley and R. A. Bindschadler, Eds.), Antarctic Research Series No. 77, pp. 157–199. American Geophysical Union, Washington, DC.

Kershaw, J. A., Clague, J. J., and Evans, S. G. (2005). Geomorphic and sedimentological signature of a two-phase outburst flood from moraine-dammed Queen Bess Lake, British Columbia, Canada. *Earth Surface Processes and Landforms* **30**, 1–25.

Kirkbride, M. P. (2000). Ice marginal geomorphology and Holocene expansion of debris-covered Tasman Glacier, New Zealand. In *Debris-Covered Glaciers* (M. Nakawo, C. Raymond and A. Fountain, Eds.), Publication No. 264, pp. 211–217. International Association of Hydrological Sciences, Walling Ford.

Kjær, K. H., and Krüger, J. (2001). The final phase of dead-ice moraine development: Processes and sediment architecture, Kötlujökull, Iceland. *Sedimentology* **48**, 935–952.

Kruger, J. (1985). Formation of a push moraine at the margin of Hofdabrekkujökull, south Iceland. *Geografiska Annaler* **67A**, 199–212.

Kruger, J. (1993). Moraine ridge formation along a stationary ice front in Iceland. *Boreas* **22**, 101–109.

Kruger, J. (1994). Glacial processes, sediments, landforms and stratigraphy in the terminus region of Myrdalsjökull, Iceland. *Folia Geographica Danica* **21**, 1–233.

Kruger, J. (1995). Origin, chronology and climatological significance of annual-moraine ridges at Myrdalsjökull, Iceland. *The Holocene* **5**, 420–427.

Kruger, J. (1996). Moraine ridges formed from subglacial frozen-on sediment slabs and their differentiation from push moraines. *Boreas* **25**, 57–63.

Maag, H. (1969). *Ice-Dammed Lakes and Marginal Glacial Drainage on Axel Heiberg Island*, Axel Heiberg Island Research Report. McGill University, Montreal.

Maizels, J. (1997). Jokulhlaup deposits in proglacial areas. *Quaternary Science Reviews* **16**, 793–819.

Manley, G. (1955). On the occurrence of ice-domes and permanently snow-covered summits. *Journal of Glaciology* **2**, 453–456.

McDougall, D. A. (2001). The geomorphological impact of Loch Lomond (Younger Dryas) Stadial plateau icefields in the central Lake District, northwest England. *Journal of Quaternary Science* **16**, 531–543.

Ó Cofaigh, C., Taylor, J., Dowdeswell, J. A., and Pudsey, C. J. (2003). Palaeo-ice streams, trough mouth fans and high latitude continental slope sedimentation. *Boreas* **32**, 37–55.

Owen, L. A., and Derbyshire, E. (1989). The Karakoram glacial depositional system. *Zeitschrift für Geomorphologie* **76**, 33–73.

Paul, M. A. (1983). The supraglacial landsystem. In *Glacial Geology* (N. Eyles, Ed.), pp. 71–90. Pergamon, Oxford.

Powell, R. D., and Domack, E. W. (1995). Glaciomarine environments. In *Glacial Environments—Processes, Sediments and Landforms* (J. Menzies, Ed.), pp. 445–486. Butterworth-Heinmann, Boston.

Price, R. J. (1969). Moraines, sandar, kames and eskers near Breiðamerkurjökull, Iceland. *Transactions of the Institute of British Geographers* **46**, 17–43.
Price, R. J. (1970). Moraines at Fjallsjökull, Iceland. *Arctic and Alpine Research* **2**, 27–42.
Rea, B. R., and Evans, D. J. A. (2003). Plateau icefield landsystems. In *Glacial Landsystems* (D. J. A. Evans, Ed.), pp. 407–431. Arnold, London.
Rea, B. R., Whalley, W. B., Evans, D. J. A., Gordon, J. E., and McDougall, D. A. (1998). Plateau icefields: Geomorphology and dynamics. *Quaternary Proceedings* **6**, 35–54.
Russell, A. J., and Knudsen, O. (1999). Controls on the sedimentology of the November 1996 jokulhlaup deposits, Skeiðarársandur, Iceland. In *Advances in Fluvial Sedimentology* (N. D. Smith, J. Rogers and A. G. Plint, Eds.), Special Publication No. 28, pp. 315–329. International Association of Sedimentologists.
Sharp, M. J. (1984). Annual moraine ridges at Skalafellsjökull, south-east Iceland. *Journal of Glaciology* **30**, 82–93.
Sissons, J. B. (1980). The Loch Lomond Advance in the Lake District, northern England. *Transactions of the Royal Society of Edinburgh: Earth Sciences* **71**, 13–27.
Speight, J. G. (1963). Late Pleistocene historical geomorphology of the Lake Pukaki area, New Zealand. *New Zealand Journal of Geology and Geophysics* **6**, 160–188.
Stokes, C. R., and Clark, C. D. (1999). Geomorphological criteria for identifying Pleistocene ice streams. *Annals of Glaciology* **28**, 67–74.
Stokes, C. R., and Clark, C. D. (2001). Palaeo-ice streams. *Quaternary Science Reviews* **20**, 1437–1457.
Stokes, C. R., and Clark, C. D. (2002). Ice stream shear margin moraines. *Earth Surface Processes and Landforms* **27**, 547–558.
Stokes, C. R., and Clark, C. D. (2003). The Dubawnt Lake palaeo-ice stream: Evidence for dynamic ice sheet behaviour on the Canadian Shield and insights regarding controls on ice stream location and vigour. *Boreas* **32**, 263–279.
Vorren, T., and Laberg, J. S. (1997). Trough mouth fans—Palaeoclimate and ice sheet monitors. *Quaternary Science Reviews* **16**, 865–881.

Glacifluvial Landforms of Erosion

A E Kehew, Western Michigan University, MI, USA
M L Lord, Western Carolina University, NC, USA
A L Kozlowski, Susquehanna University, Selinsgrove, PA, USA

© 2007 Elsevier B.V. All rights reserved.

Introduction

The vast amount of meltwater released from glaciers has enormous erosional capacity. Temporary impoundments of meltwater occur on top of, within, at the base of, and in front of the ice. The magnitude and intensity of erosion is greatest when these impoundments release sediment-deficient meltwater suddenly or rapidly in the form of outbursts. Erosion is facilitated by high hydraulic potential caused by the cryostatic pressure of the ice or the hydraulic head developed behind a barrier in a proglacial lake. The type and resistance of bed material underlying a region also controls the form and size of erosional landforms produced by these meltwater discharges.

Most glacifluvial landforms of erosion were formed when the great Pleistocene ice sheets of North America and Eurasia produced and ponded vast quantities of meltwater. The erosional landforms included vast channel systems, both subglacial and proglacial, and a variety of forms attributed to subglacial sheet floods. In some cases, subglacial flows switched from an erosional to a depositional regime upon reaching the glacier terminus, typified by aggrading braided streams, sandar, or subaqueous fans of various types.

The discharges that formed large-scale glacifluvial erosional forms had significant, global impacts because they were of such high magnitude. Many large-scale drainage systems in areas affected by glaciation were formed or largely modified by discrete-event, high-discharge meltwater flows. In places these channels constitute important aquifer systems, some of which are buried beneath multiple sheets of glacial drift from earlier glaciations. In addition, numerous studies have argued that the sudden release of vast quantities of cold meltwater correlates with periods of climatic change, possibly by altering oceanic circulation patterns.

Four categories of processes and landforms are considered in this article because they account for the majority of large-scale glacifluvial erosional forms: (1) proglacial lake channel systems formed by glacial lake outbursts, (2) tunnel channels formed by channelized subglacial flows, (3) ice marginal (lateral) meltwater channels associated primarily with cold-based glaciers, and (4) a variety of erosional forms hypothesized to have been formed by massive subglacial sheet floods. This review of erosional landforms and related processes includes ideas that are universally accepted, such as a catastrophic flood origin of the Channeled Scabland in the USA, and a discussion of subglacial landforms attributed to fluvial erosion, which is actively debated and controversial.

Glacial Lake Outbursts

Overview

Glacial lake outbursts, or *jökulhlaups*, are very high magnitude floods caused by the sudden drainage of water from a glacial lake. Glacial lakes form on, in, under, or in front of glaciers and are inherently

Figure 1 Upstream view of *jökulhlaup* on Skeiará river, Iceland. This flood occurred in 1996 and was triggered by a subglacial volcanic eruption. Photograph by Magnus T. Gudmundsson.

unstable. Outbursts occur in many present-day glaciated regions, such as Iceland, Alaska, the Himalayas, and the Andes (**Fig. 1**). Flood discharges typically exceed background rates by orders of magnitude and, during the Pleistocene, carved out spectacular landscapes by flows of unparalleled magnitude. In fact, glacial lake outbursts account for the top seven floods ever known to have occurred in the world (O'Connor and Costa, 2004). Floodwaters from Pleistocene glacial lake outbursts were highly erosive and created a characteristic suite of landforms, now widely accepted to have been formed by catastrophic floods. The geomorphic record of outbursts is most significant from Pleistocene glaciations because of the high magnitude of flows, up to $18 \times 10^6\,m^3\,s^{-1}$ (Baker *et al.*, 1993). The emphasis here, therefore, is on Pleistocene glacial lake outburst erosional landforms, although valuable insights into outburst processes have been gained by observations of modern *jökulhlaups* (Cenderelli and Wohl, 2003).

Recognition of Pleistocene Glacial Lake Outbursts

For more than a century, the incongruity in size between underfit present-day streams and overfit valleys has been apparent. Furthermore, numerous studies suggested that the overfit valleys were paleochannels – not valleys, as evidenced by uniformity in size over long distances and regular meander wavelengths, that conveyed much higher discharges than present during the Pleistocene. Many of these same channels are now widely recognized to have been formed by glacial lake outbursts. This acceptance was not the case when Bretz (1923) first proposed a catastrophic flood origin for the Channeled Scabland: an anastomosing network of channels, covering an area of appoximately 40,000 km^2, in

Figure 2 Waterfalls (about 60 m high) in cataract and gorge eroded by the Spokane Floods of the Channeled Scabland. Eroded basalt and loess landscape is in the background. Photograph by Mark Lord.

the Columbia Plateau of northwestern United States. Bretz, based on years of fieldwork, was compelled to argue for a catastrophic flood, the Spokane Flood, to explain the eroded basalt and loess landscape containing a network of proglacial channels, bedrock cataracts, streamlined islands, dry valleys, and boulder deposits (**Fig. 2**). His hypothesis was widely rejected for some scientific reasons, including the lack of a known water source for the flood, but more importantly because the idea of a catastrophic origin to a landscape was viewed as an assault on the principle of uniformitarianism – the foundation of modern geology. In the 1960s, Bretz's outrageous hypothesis became widely accepted, now well supported by, among other things, a water source (Glacial Lake Missoula), giant current ripples, and high-altitude imagery (**Fig. 3**). This fascinating story

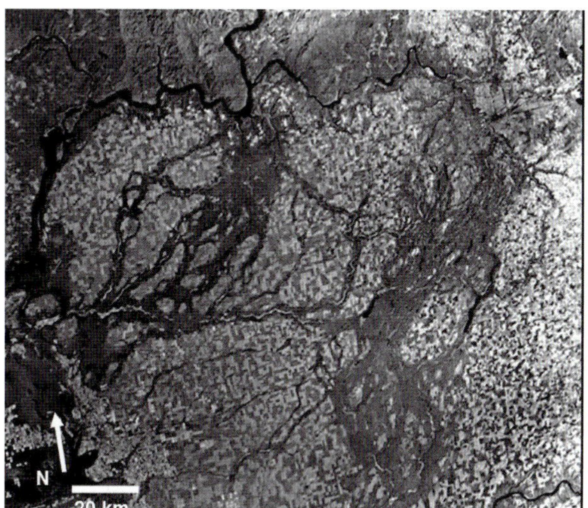

Figure 3 Landsat image of Channeled Scabland, WA, USA. Flood flows were from the northeast toward the southwest.

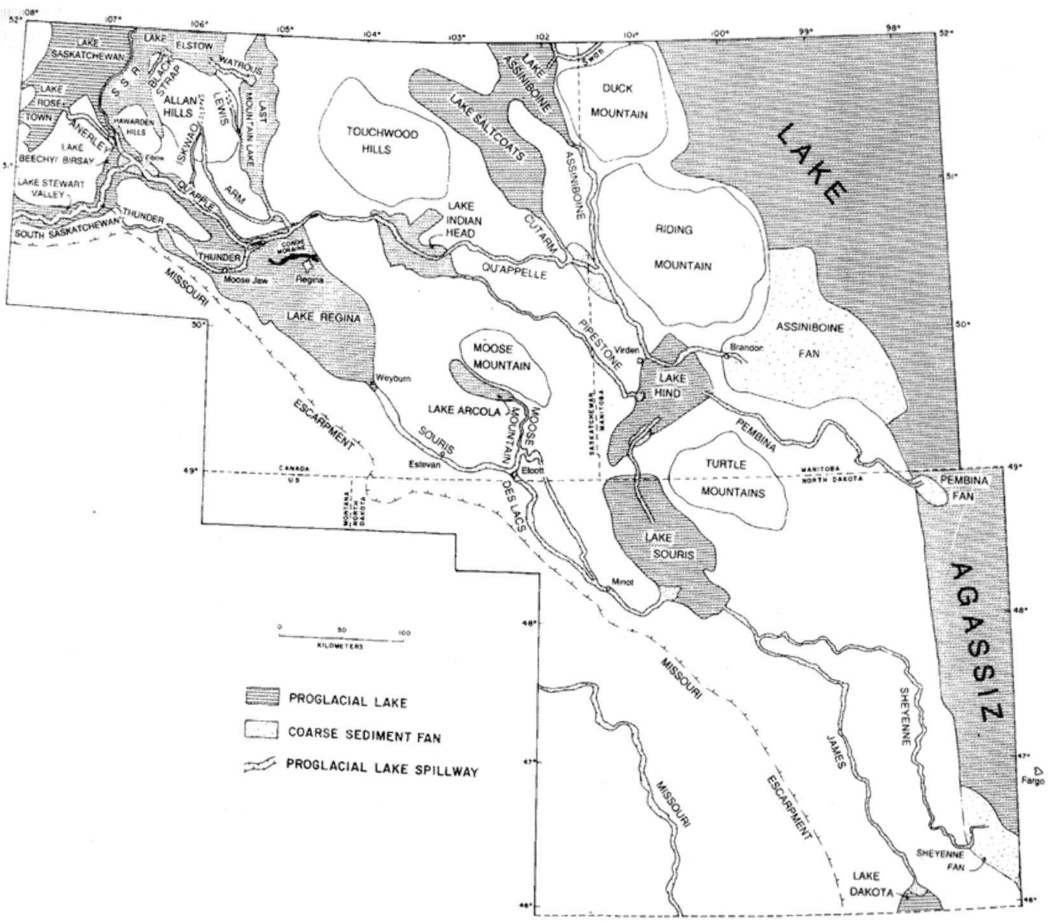

Figure 4 Proglacial lakes and spillways along the southwestern margins of the Laurentide Ice Sheet. From Kehew AE and Teller JT (1994) History of late glacial runoff along the southwestern margin of the Laurentide Ice Sheet. *Quaternary Science Reviews* 13: 859–877.

in the evolution of scientific ideas is well described (e.g., Baker *et al.* (1987)).

Today, dozens of investigations have shown that glacial lake outbursts were not unique to the Channeled Scabland; they were a common, if not inevitable, process during Pleistocene glaciations as the land, ice sheets, water sources, and drainage pathways were in a continuous state of flux. Channel systems, mostly glacial lake spillways, have been attributed to glacial lake outbursts throughout the glaciated interior of North America (**Fig. 4**), Siberia, Scandinavia, and elsewhere.

Glacial Lake Outburst Erosional Landforms

The Channeled Scabland is clearly the most-studied area formed by outburst floods. Many landforms now used as indicators of catastrophic floods were first described in this region (Baker, 1973). Although the landforms of the Scablands are accepted to be formed by very high magnitude flows, several aspects relating to the origin of the Channeled Scabland are actively studied and questioned, such as the number of outbursts, the magnitude of each flood event, the erosional capability of the outbursts, the timing of the floods, and the source of flood waters (Benn and Evans, 1998). Spillway systems that formed by outbursts along the southern margins of the Laurentide Ice Sheet, though not as dramatic as the Scabland landscapes, share many of the erosional landforms and have been well documented (Kehew and Lord, 1987). The most ubiquitous erosional landform created by outburst floods is a trench-shaped channel with steep sides, and a relatively uniform width and depth (**Fig. 5**). The size of the channel varies with flow magnitude, geologic setting, and flow history among other things. Spillways that formed by outbursts in the Great Plains of North America are typically 1–3 km wide and 25–100 m deep (Kehew and Lord, 1987). Common erosional landform types produced by outbursts are summarized in **Table 1**.

General models for the sequence of development of flood-eroded landscapes have been developed based on studies of the Channeled Scabland, outburst spillways in the Great Plains, and experimental flume

Figure 5 Oblique aerial photograph of inner channel and shallow, anastomosing channels on scoured uplands, Souris Spillway, southeastern Saskatchewan. Flow was to the east (top to bottom). Photograph by Mark Lord.

studies (Baker *et al.*, 1987; Kehew and Lord, 1987; and Shepard and Schumm, 1974, respectively). Initial stages of flooding overwhelm the landscape and cause broad scouring, shallow anastomosing channels, and, in some cases, transverse erosional ripples. With increased duration, the flow becomes more organized as the bed deforms, which is reflected in longitudinal grooves and streamlined hills. Continued flow leads to selective deepening to form inner channels, which progressively enlarge by capture of more flow laterally and by upstream knickpoint migration. A spillway formed by a long duration or repeated outburst floods can convey all flow in one channel and may contain only a few very well streamlined hills (**Fig. 6**).

Table 1 Erosional landforms produced by glacial lake outbursts

Landforms	*Characteristics*
Trench-shaped channels (**Fig. 5**)	Commonly parallel to ice margins and connect or bisect glacial lake beds
Anastomosing channel pattern	Network of channels may be from 1 to 100's of kilometers wide
Streamlined erosional residuals (islands) (**Fig. 6**)	Erosional residuals, given long enough flow, will assume a lemniscate form that minimizes fluid drag
Dry valleys (channels)	Variable sizes, commonly on upland areas above deeper inner channels
Perched (hanging) channels	Variable sizes
Longitudinal grooves (**Fig. 5**)	Flow parallel grooves interpreted to be formed by secondary circulation cells; up to 3 m high in Scablands and up to 2 m high in Great Plains
Cataracts (**Fig. 2**)	Dry falls in the Channeled Scabland is 5.5 km wide and 120 m high, probably formed by the coalescence of potholes eroded by vertical vortices (Baker *et al.*, 1987)
Potholes and obstacle scour depressions	Occur at scales of meters to hundreds of meters
Transverse erosional ripples	Scarce landform present in well-preserved parts of Great Plains outburst flow paths (up to 1.2 m high with about 40 m wavelengths)
Erosional scarps	Commonly cut into terraces or valley sides

Compiled from Baker VR (1973) Paleohydrology and sedimentology of the Lake Missoula flooding in eastern Washington. Geological Society of America Special Paper 144. Geological Society of America; Kehew AK and Lord ML (1987) Glacial-lake outburst along the mid-continent margins of the Laurentide Ice Sheet. In: Mayer L and Nash D (eds.) Catastrophic flooding. Proceedings of 18th Annual Binghamton Symposia in Geomorphology, pp. 95–120; Baker VR, Greeley R, Komar PD, Swanson DA, and Waitt RB Jr. (1987) Coloumbia and Snake River Plains. In: Graf WL (ed.) Geomorphic systems of North America, Centennial Special Volume 2, pp. 403–468. Boulder, CO: Geological Society of America.

Hydrologic Characteristics of Pleistocene Glacial Lake Outbursts

The cataclysmic failure of ice dams impounding large glacial lakes released huge volumes of water, more than 2,000 km^3 in the Channeled Scabland, in a matter of days to weeks (O'Connor and Baker, 1992). Outburst flows were highly erosive and, at least in much of the Great Plains spillway systems, became hyperconcentrated as they ripped through generally weak, fine-grained materials that underlie the region (Lord and Kehew, 1987). The sedimentary record of outburst floods is scarce in spillways, is present in some glacial lakes that received floodwaters, but is mostly present in the marine offshore.

The largest documented outburst flood occurred in south-central Siberia in the Altay Mountains, which carved out gorges and produced megaripples in huge gravel bars analogous to those in the Channeled Scabland (Baker *et al.* 1993). Peak discharge is estimated to have been $18 \times 10^6\ m^3\ s^{-1}$ with flows up to 500 m deep and velocities up to 45 $m\,s^{-1}$; the unit stream power generated by this flood is estimated to have been 10^6 W m^{-2}. Paleohydrologic studies of lower magnitude outbursts typically estimate discharges to be between 10^4 and $10^6\ m^3 s^{-1}$ (see Benn and Evans (1998) for summary). The specific characteristics of the erosional landscapes carved by these tremendous flows varied not just with discharge, but with flow mechanics (e.g., flow regime, cavitation, and secondary circulation), flow boundary materials, sediment availability, flow duration, flood hydrograph, and flow history (Baker, 1973; Kehew and Lord, 1987; O'Connor and Baker, 1992; Cenderelli

Figure 6 Oblique aerial photograph of streamlined hill in the Souris-Hind spillway, southern Manitoba; flow was to north (upper right to lower left). Photograph by Mark Lord.

and Wohl, 2003). For example, unit stream power has been used successfully to relate hydraulic conditions to different types of erosional processes and landforms.

Tunnel Channels

Overview

Tunnel channels, or tunnel valleys (also known as tunneldale or rinnentaler), are large meltwater channels eroded into sediment or rock at the base of an ice sheet. Channel dimensions reach >30 m deep, 4 km wide, and more than 100 km long. Tunnel channels have a wide variety of morphologies and associated landforms, and occur in drainage systems covering extensive areas.

The widespread distribution of tunnel channels in the area of the Laurentide and Scandinavian Ice Sheets has important implications for ice-sheet dynamics and glacial meltwater regimes (Clayton *et al.*, 1999; Brennand and Shaw, 1994; Mooers, 1989; Wright, 1973). Although literature on tunnel channels is abundant, controversy exists around the mechanism of tunnel channel genesis and the implications of meltwater volumes associated with channel size. As yet no unified theory on their formation can adequately explain the observed diverse morphologies, landform associations, and substrate lithologies (O'Cofaigh, 1996).

The terms 'tunnel valley' and 'tunnel channel' have often been used interchangeably to denote subglacially eroded drainage courses. However, these terms may also have genetic implications (Clayton *et al.*, 1999; Mooers, 1989). In this discussion the term 'tunnel channel' is used to describe landforms interpreted as large subglacial channels without a specific genetic origin.

Morphology

Despite a general orientation subparallel to longitudinal flow lines of the glacier, tunnel channel drainage patterns are highly variable, including predominantly single isolated straight channels, braided networks, dendritic networks, radial, subparallel channels (**Fig. 7**) and anastomosing and anabranching systems (**Fig. 8**; Sjogren *et al.*, 2002; Clayton *et al.*, 1999; Brennand and Shaw, 1994; Mooers, 1989; Wright, 1973).

Channel morphologies are equally diverse and include straight-walled channels of uniform width (**Figs. 9** and **10**), crenulated channels that expand and contract, and subtle channels only recognizable as linear chains of lakes or depressions (**Fig. 10**). Many tunnel channels display steep walls similar to outburst spillways, but other channels display irregular widths and depths associated with pressurized subglacial meltwater flow and variable degrees of posterosional infill. Longitudinal profiles of flow paths may be undulating and have convex-up or reverse gradients that flow uphill, cross drainage divides, and trend oblique to the regional slope (**Fig. 11**) (Kozlowski *et al.*, 2005; Clayton *et al.*, 1999; Wright, 1973). In other areas, tunnel channels occur as isolated, hanging, channel sections adorning uplands without integration to other channels or modern drainage.

Channel dimensions range from 0.15 to 6 km in width and 2 to 150 km in length, but the majority of channels are 0.5–2.5 km wide and 5–40 km long. Channel depths of 10–50 m are common, with most ranging between 20 and 30 m, although subsurface studies of deeply buried tunnel channels document depths in excess of 100 m (Piotrowski (1993) and references therein). Other subsurface investigations have indicated that the subtle surface expression of some channels is attributed to burial or partial burial by thick sequences of younger sediment that obscures the true depth of the channel bottoms (Clayton *et al.*, 1999), as illustrated in **Figure 12**. Despite a general lack of adequate subsurface information and a scarcity of exposures, available studies suggest that a wide range of sedimentary sequences fills or partially fills tunnel channels (Benn and Evans, 1998; Brennand and Shaw, 1994; Piotrowski, 1993).

Landform Associations

Tunnel channels are often closely associated with drumlins, eskers, moraines, glacifluvial meltwater fans, and hummocky topography. For example, Clayton *et al.* (1999) and Cutler *et al.* (2002) proposed that tunnel channels develop beneath a frozen submarginal zone by the sudden release of meltwater

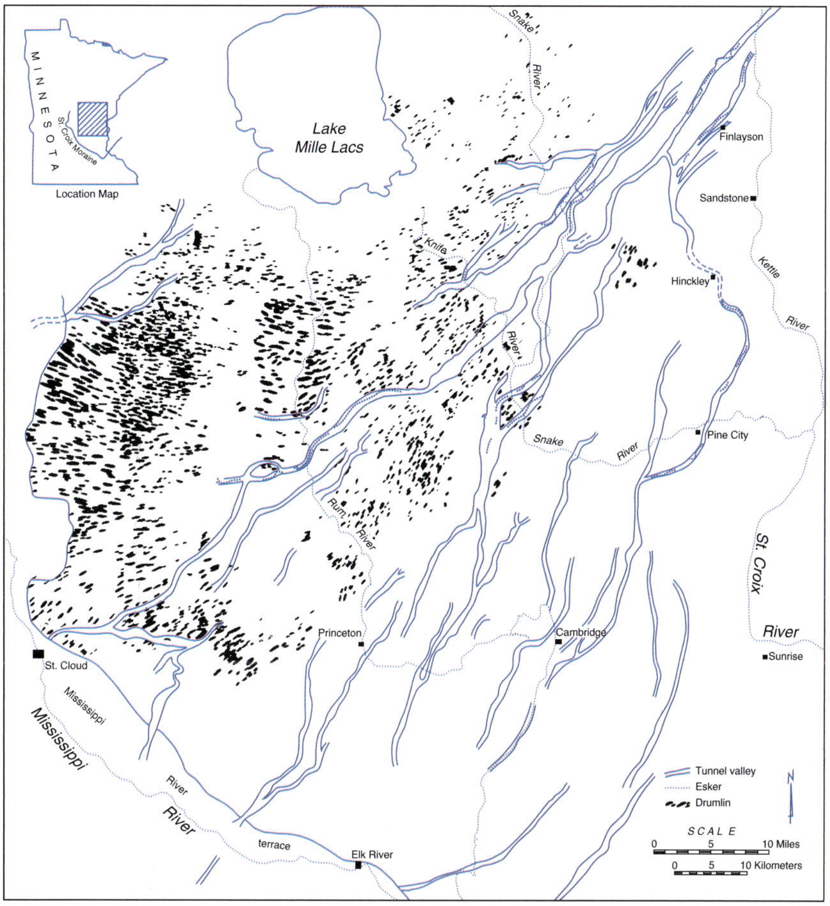

Figure 7 Subparallel tunnel channels and associated landforms of the Superior Lobe, Minnesota, USA. From Wright HE Jr. (1973) Tunnel Valleys, glacial surges, and subglacial hydrology of the Superior Lobe , Minnesota. In: Black R, Goldthwait R, and Willman H (eds.) *The Wisconsinan Stage*, Geological Society of America Memoir 136, pp. 251–276.

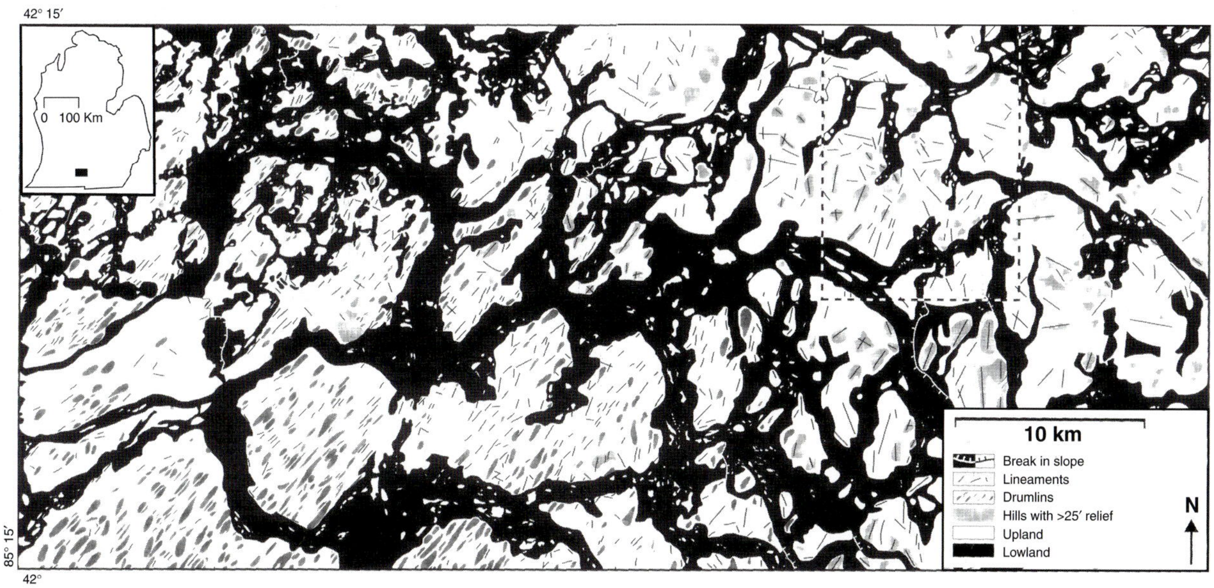

Figure 8 Anabranching system of tunnel channels and associated landforms of the Saginaw Lobe, south-central Michigan, USA. Modified from Sjogren DB, Fisher TG, Taylor LD, Jol HM, and Munro-Stasiuk MJ (2002) Incipient tunnel channels. *Quaternary International* 90: 41–56.

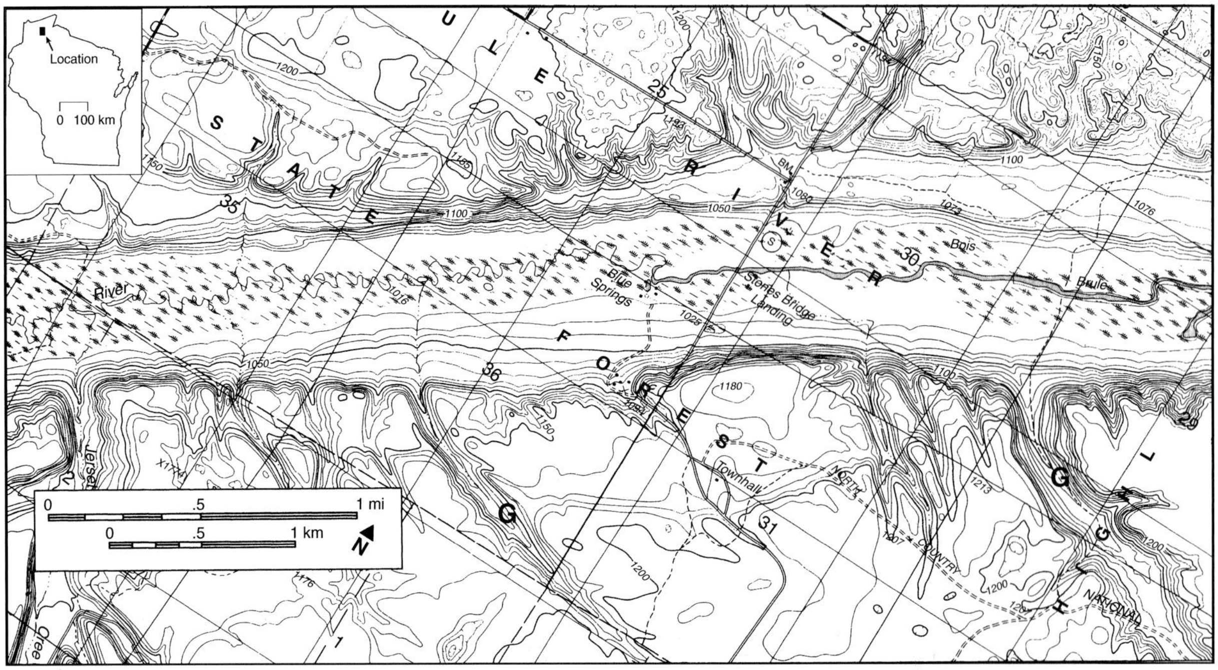

Figure 9 Straight-walled tunnel channel of uniform width in northwestern Wisconsin with post-formational gullies (G). Channel is located in the Lake Minnesuing 7.5 min. quadrangle, contour interval 10 ft (approx. 3 m). Modified from Clayton L, Attig JW, and Mickelson DM (1999) Tunnel channels formed in Wisconsin during the last glaciation. In: Mickelson DM and Attig JW (eds.) *Glacial Processes Past and Present.* Geological Society of America Special Paper 337, 69–82. Geological Society of America.

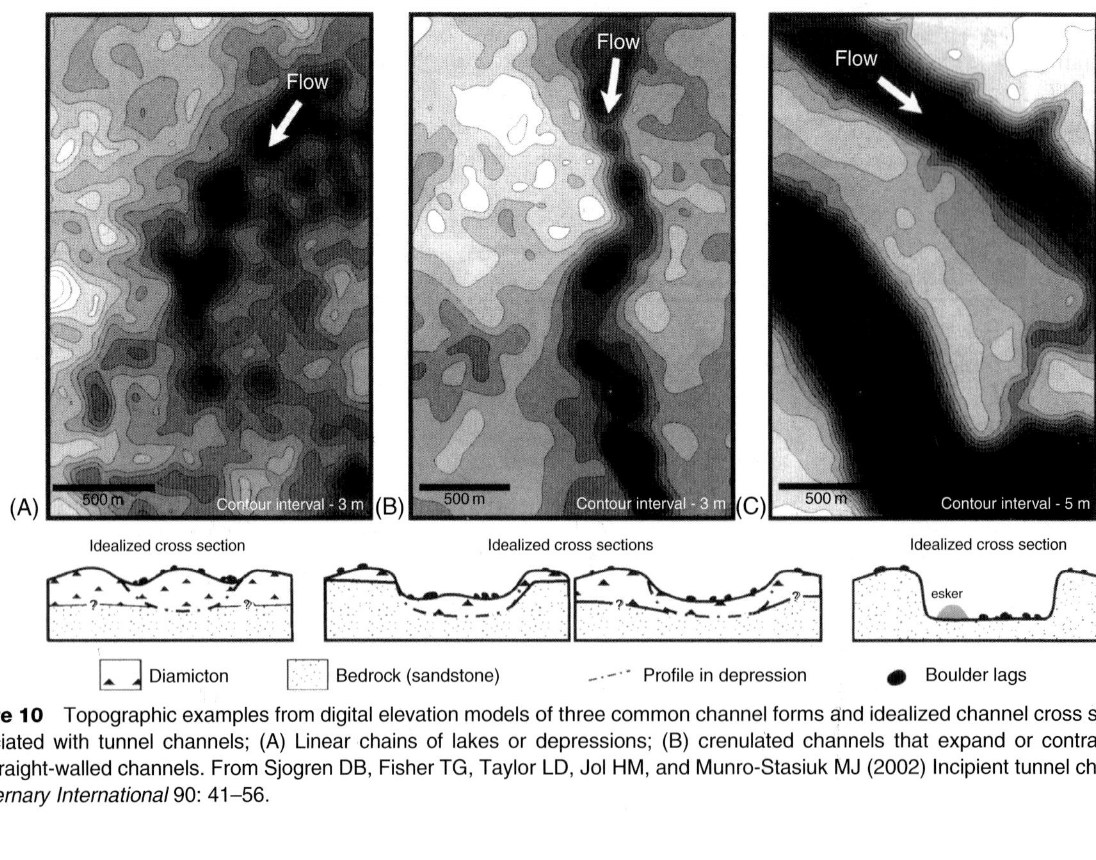

Figure 10 Topographic examples from digital elevation models of three common channel forms and idealized channel cross sections associated with tunnel channels; (A) Linear chains of lakes or depressions; (B) crenulated channels that expand or contract; and (C) straight-walled channels. From Sjogren DB, Fisher TG, Taylor LD, Jol HM, and Munro-Stasiuk MJ (2002) Incipient tunnel channels. *Quaternary International* 90: 41–56.

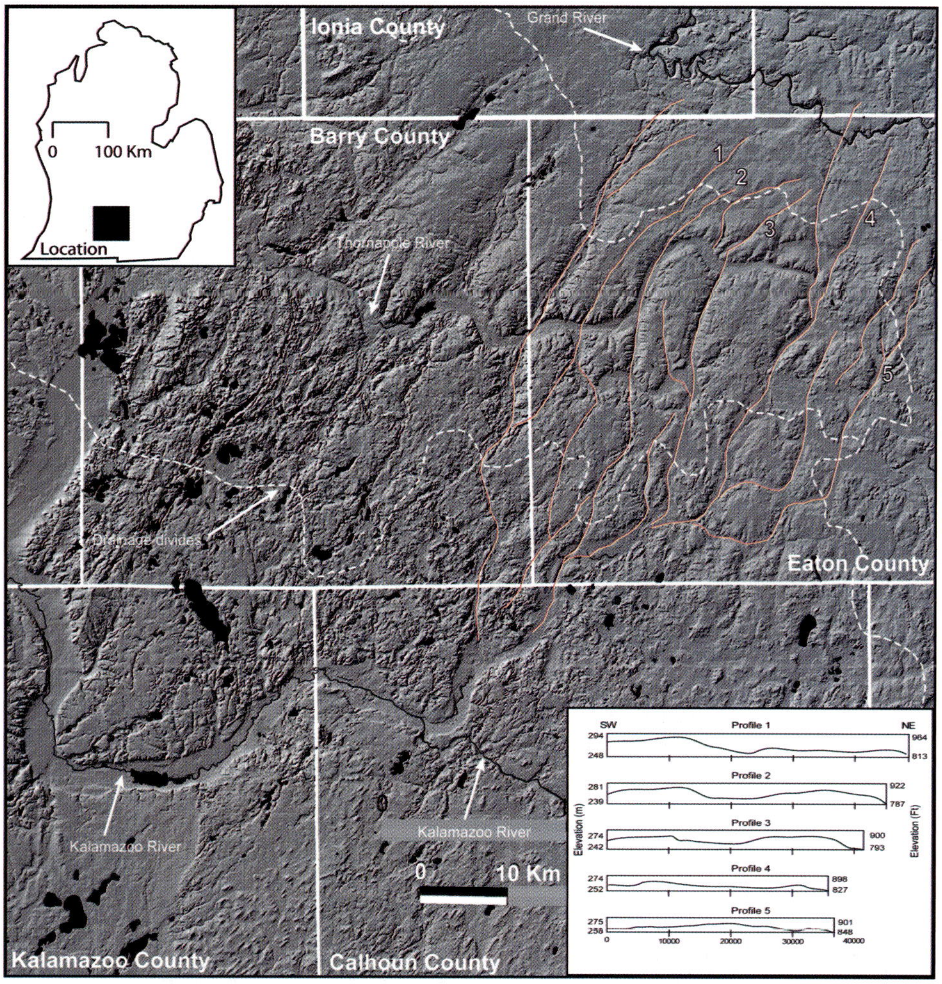

Figure 11 Digital elevation model of south-central Michigan displaying a series of anastomosing Saginaw Lobe tunnel channels (red lines) with segments that flowed uphill and crossed drainage divides (dashed white lines on DEM and inset profiles) to feed a subaerial outburst flood in the Central Kalamazoo River Valley. Modified from Kozlowski AL, Kehew AE, and Bird BC (2005) Outburst flood origin of the Central Kalamazoo River Valley, Michigan, USA. *Quaternary Science Reviews* 24: 2354–2374.

impounded upglacier in the thawed-bed zone where drumlins occur (**Fig. 13**). These channels cut through the terminal moraine, where they end at the apex of large, coarse-grained glacifluvial fans (**Figs. 13** and **14**). Where active ice margins were in close proximity but not necessarily in synchronicity, some tunnel channels display crosscutting relationships with younger landforms and have been interpreted as palimpsest tunnel channels (Kehew *et al.*, 1999). These crosscutting relationships may be used to establish or constrain relative timing of glacial events in complex glaciated regions where the chronology of events is poorly constrained (Kozlowski *et al.*, 2005; Kehew *et al.*, 1999).

Although widely varying in morphology, drainage pattern, and associated landforms, there is a general consensus that tunnel channels:

- are usually straight and lack tributaries except for postglacial gullies;
- are most distinct within several kilometers of the ice margin;
- have undulatory longitudinal profiles with uphill reaches; and
- often breach moraines and terminate at large, coarse-grained glacifluvial fans.

Origin

Most of the controversy surrounding tunnel channels centers on the models proposed to explain their formation (O'Cofaigh, 1996) and relationships to various landform associations (Benn and Evans, 1998). There is a fundamental difference in interpretation between those who use the term 'tunnel channel' to represent bankfull meltwater flow during one or more discharge events (Clayton *et al.*, 1999) and those who use the term 'tunnel valley,' which implies

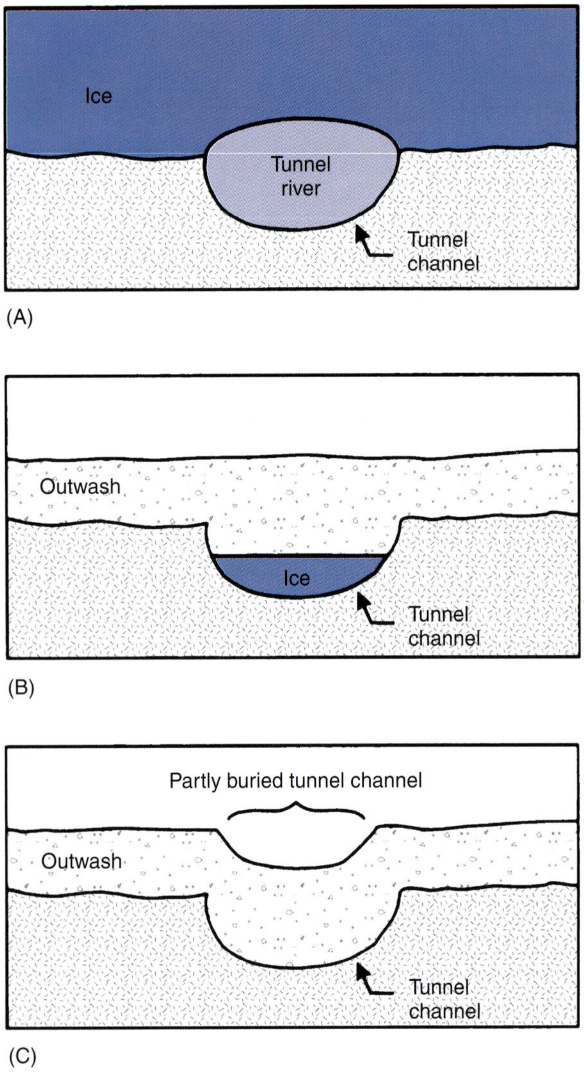

Figure 12 Schematic cross-sectional diagram of: (A) tunnel channel formation; (B) preservation; and (C) modification during and after burial by younger deposits. From Clayton L, Attig JW, and Mickelson DM (1999) Tunnel channels formed in Wisconsin during the last glaciation. In: Mickelson DM and Attig JW (eds.) *Glacial Processes Past and Present* Geological Society of America Special Paper 337, 69–82. Geological Society of America.

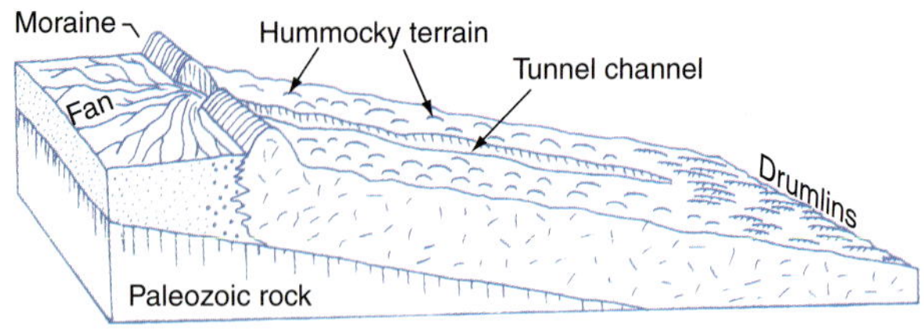

Figure 13 Idealized block model of tunnel channels and spatial relationships of associated landforms observed in Wisconsin, USA. Note uphill flow path and termination of channel at an outwash fan through a breach in a moraine. Modified from Clayton L, Attig JW, and Mickelson DM (1999) Tunnel channels formed in Wisconsin during the last glaciation. In: Mickelson DM and Attig JW (eds.) *Glacial Processes Past and Present*. Geological Society of America Special Paper 337, 69–82. Geological Society of America.

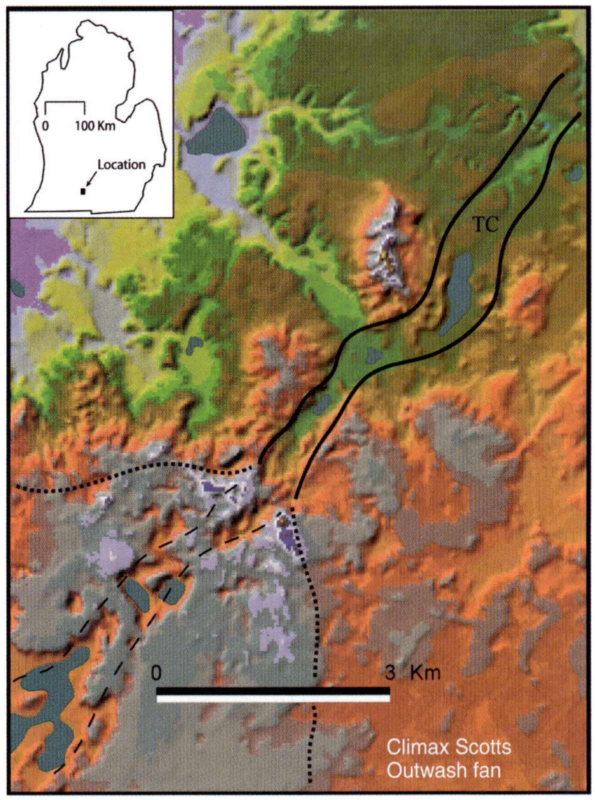

Figure 14 Digital elevation model of a Saginaw Lobe tunnel channel fan in south-central Michigan, USA. Note channel ascends in elevation to the apex of the fan, which is the highest feature on the landscape and delineates a former ice margin. Also, the collapsed appearance of the channel extending through the medial and distal portions of the fan suggests the channel extended beyond this margin and had been occupied prior to burial.

gradual erosion by flows much smaller than the dimensions of the valley (Mooers, 1989; Boulton and Hindmarsh, 1987).

Although multiple models of formation have been proposed, most fall into four categories: (1) catastrophic meltwater floods; (2) time-transgressive erosion; (3) subglacial sediment deformation and piping; and (4) polygenetic origin (**Fig. 15**). Catastrophic meltwater release (**Fig. 15A**) is the most commonly suggested mechanism for tunnel channel formation and, although differences in opinion exist over the exact nature of the flow regime, most authors argue for large-scale transient events. Several researchers have suggested that formation of tunnel channels is largely controlled by glacier-bed conditions and the presence of a permafrost seal at the glacier margin

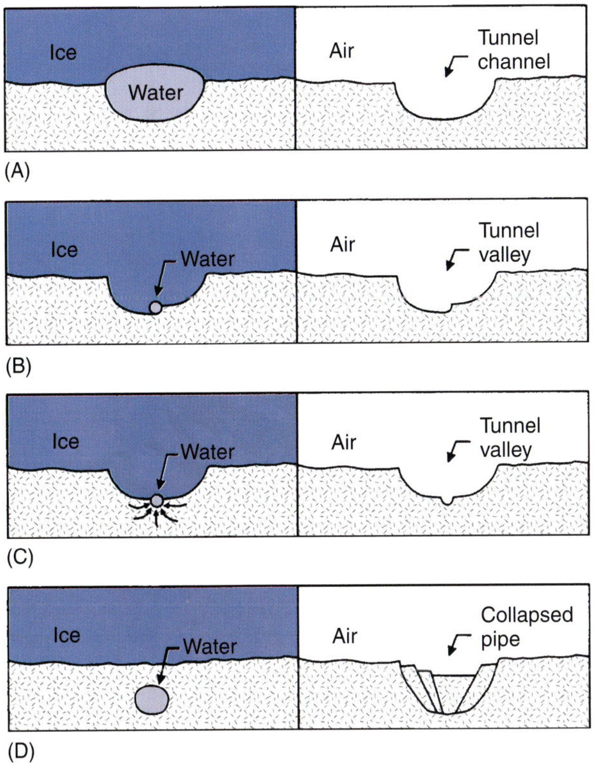

Figure 15 Schematic diagram illustrating proposed models of tunnel channel formation. (A) Catastrophic meltwater floods; (B) time-transgressive erosion; (C) subglacial deformation; and (D) hydraulic sapping or piping and collapse. From Clayton L, Attig JW, and Mickelson DM (1999) Tunnel channels formed in Wisconsin during the last glaciation. In: Mickelson DM and Attig JW (eds.) *Glacial Processes Past and Present.* Geological Society of America Special Paper 337, 69–82. Geological Society of America.

(Hooke and Jennings, in press; Clayton *et al.*, 1999; Wright, 1973). The source of meltwater remains a contentious issue and significant subglacial reservoirs of meltwater are required for rapid erosion of channels in the scale of those observed.

Alternatively, Mooers (1989) suggested that the formation of large erosional troughs in Minnesota resulted from time-transgressive discharge of seasonal supraglacial meltwater that accessed the subglacial system via crevasses and moulins. Valley formation, as opposed to channel formation, resulted incrementally from vertical and lateral planation of subglacial meltwater streams (**Fig. 15B**). This model, which accounts for the association of eskers and meltwater fans sometimes observed with tunnel channels, does not require catastrophic flows, and does not require permafrost to act as a hydraulic seal. In this model, small subglacial flows gradually eroded the observed valleys as the ice margin retreated.

Subglacial deformation in conjunction with piping, or groundwater sapping, is another formational model that has been suggested for the development of tunnel valleys (Boulton and Hindmarsh, 1987). In this model, steady-state meltwater flow over deforming subglacial sediment is unable to infiltrate into subjacent aquifers because of the low transmissivity of the deforming beds. As a result of high fluid pressure, piping is initiated in the substrate to sustain meltwater discharge and maintain glacier equilibrium. Sediment that enters the meltwater conduit by creep or piping is entrained and transported downglacier initiating the formation of a depression or channel (**Fig. 15C**). In time, this process leads to the gradual formation of a valley much larger than the hydraulic conduit.

The diversity in tunnel channel morphology, drainage patterns, and substrate suggests a polygenetic origin for some channels, which may involve preglacial, subglacial, and subaerial erosion (Sjogren *et al.*, 2002; Clayton *et al.*, 1999; Benn and Evans, 1998; O'Cofaigh, 1996; Brennand and Shaw, 1994). Once formed, tunnel channels may be reoccupied during multiple glacial advances or glaciations (Benn and Evans, 1998). Finally, one or more of the basic mechanisms acting in tandem may explain some tunnel channels. Hooke and Jennings (in press), for example, propose the combination of piping and catastrophic flow for the formation of tunnel channels in Minnesota and Wisconsin.

Ice Marginal (Lateral) Meltwater Channels

Overview

Channels eroded along or adjacent to former ice margins are referred to as lateral meltwater channels. These channels often occur in parallel, inset sequences, delimiting the gradual retreat of a former glacier. Although not completely restricted to cold-based glaciers, lateral meltwater channels are best developed where glaciers are frozen to their beds.

Occurrence and Origin

Cold-based glaciers are most common in polar regions such as the Canadian Artic Islands. Dyke (1993) mapped the landforms on several islands that supported ice caps during the last glaciation. The ice caps are now thought to be part of the innuitian Ice Sheet (Dyke *et al.*, 2002), which was confluent with the Greenland and Laurentide Ice Sheets. The thermal regime of these glaciers at the time of their maxium extents consisted of cold-based interiors and warm-based marginal areas. During deglaciation, they became entirely cold-based. Because cold-based glaciers have minimal subglacial meltwater, melting and meltwater production is confined to the ice surface. During the ablation season,

Figure 16 A series of sub-parallel, nested lateral meltwater channels, Melville Peninsula, Nunavut, Arctic Canada. Channels developed adjacent to a cold-based ice margin. (Photograph by Lynda Dredge; Reproduced with the permission of the Minister of Public Works and Government Services Canada, 2006 and Courtesy of Natural Resources Canada, Geological Survey of Canada.)

meltwater from the ice and adjacent areas, if the topography slopes toward the glacier, drains toward the ice margin and then along the margin in the direction of land slope. Channels are eroded by this meltwater flow. As the ice margin retreats, new channels replace those from the last period of meltwater flow, perhaps even forming on an annual basis (Dyke, 1993). In this way, a series of meltwater channels develops as the ice retreats (**Figure 16**).

As a result of the very minor amount of subglacial erosion and deposition performed by cold-based glaciers, lateral meltwater channels represent the most significant landforms produced by these ice masses. The closely spaced, parallel orientation of the channels suggests frequent abandonment in favor of new ice marginal channels during orderly ice margin retreat, and the channels are therefore useful in mapping and defining the pattern of glacial retreat.

Although most common in cold-based glaciers, lateral meltwater channels may also form adjacent to temperate glaciers. Dyke (1993) attributes these occurrences to deflection of surface water to the margin by high subglacial meltwater fluid pressures or to intermittently or permanently frozen basal conditions during ice retreat.

Subglacial Sheetfloods – The Meltwater Hypothesis

Overview

Erosion of glacial substrates by subglacial sheetfloods, also known as the meltwater hypothesis, provides an alternative explanation for the formation of many glacial landforms that are traditionally interpreted as the result of direct action of glaciers upon their beds. A historical summary of the meltwater hypothesis (Shaw, 2002) reviews the major points of the hypothesis and includes a comprehensive citation of the literature.

Bedrock Erosional Forms

Glacier beds composed of hard bedrock exhibit a wide variety of erosional marks and forms. Some, such as striations, were clearly formed by glacial abrasion. Other exposures, in which the bedrock surface is shaped into an assemblage of smoothed, streamlined forms, are more difficult to interpret. The name 'P-forms' (plastically molded) has been applied to these features, with the interpretation that they were produced primarily by the action of plastically deforming ice. Kor *et al.* (1991) concluded instead that these forms were the product of a turbulent, viscous fluid – sediment-rich subglacial meltwater – and suggested the generic term 'S-forms' (sculpted). In their study area on the northeastern side of Georgian Bay, Lake Huron, the forms range in size from centimeters to tens of meters. Their classification of the forms is shown in **Figure 17**. Formation by meltwater, rather than plastically deforming ice, or till, is based on form analogy to marks produced by experimental erosion of cohesive or hard substrates by turbulent fluid flows. The process envisioned for the formation of S-forms by meltwater involves the action of vortices and other flow structures impinging on an irregular bed (Kor *et al.*, 1991). One of these flow structures, the horseshoe vortex (Shaw, 1994), is illustrated in **Figure 18**, and an exposure of sculpted bedrock on Kelley's Island, Lake Erie is shown in **Figure 19**. The Georgian Bay flood was estimated by Kor *et al.* (1991) to have been 70 km wide and 10 m deep, with a velocity of 5–10 $m\,s^{-1}$, producing a discharge of $(0.4–0.7) \times 10^7\,m^3\,s^{-1}$.

Pair (1997), in a study of bedrock erosional forms in northern New York, reached the conclusion that, although many of the forms were indicative of meltwater erosion, they were limited to bedrock valleys in which channelized subglacial drainage would have been concentrated. Pair also points out that plastic deformation of basal ice over an irregular bedrock surface would prevent the kind of broad decoupling of the ice from its bed required by the meltwater hypothesis.

Drumlins, Rogens, Hummocky Terrain, and Megaflutes

Although the immense discharges required by the meltwater hypothesis attract continued criticism, there is general agreement that at least some of the

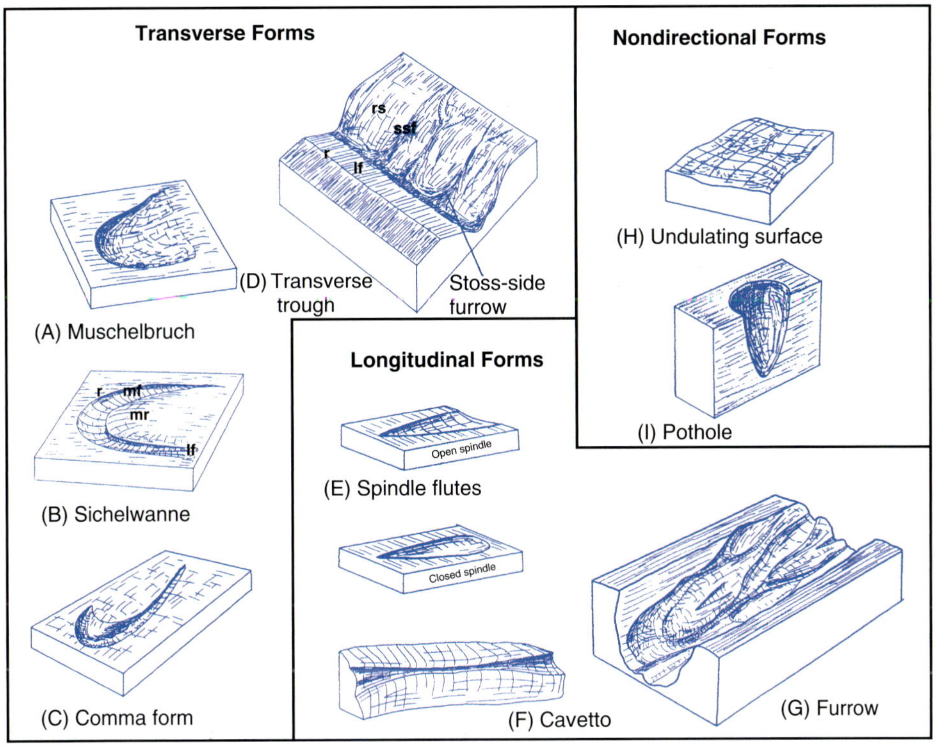

Figure 17 Classification of S-forms, interpreted to be eroded on rock surfaces by subglacial meltwater. From Kor PSG, Shaw J, and Sharpe DR (1991) Erosion of bedrock by subglacial meltwater, Georgian Bay, Ontario: A regional view. *Canadian Journal of Earth Sciences* 28: 623–642.

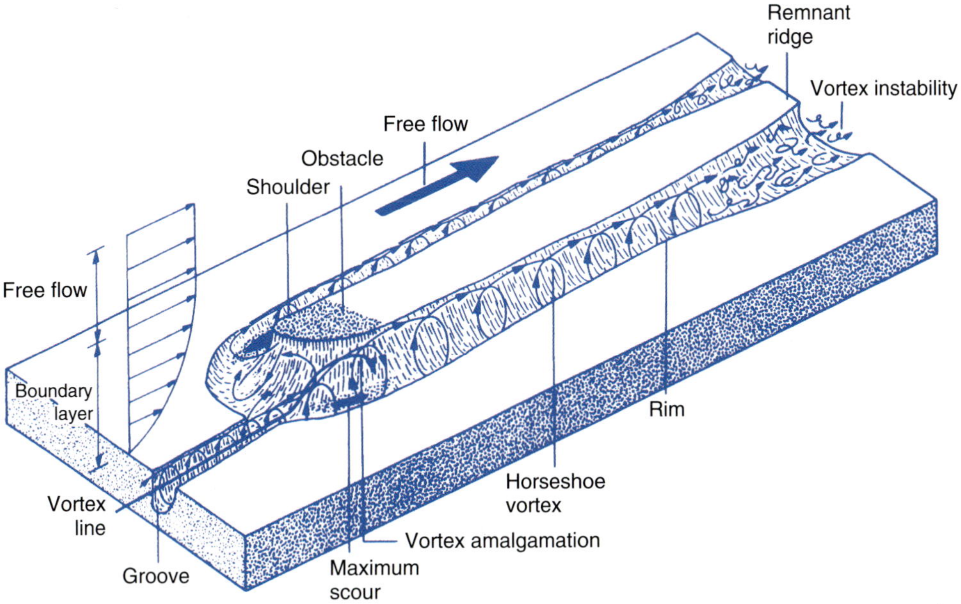

Figure 18 Erosion of bedrock by a horseshoe vortex in subglacial meltwater. From Shaw J (1994) Hairpin erosional marks, horseshoe vortices and subglacial erosion. *Sedimentary Geology* 91: 269–283.

S-forms observed on bedrock surfaces were eroded by turbulent subglacial meltwater. The meltwater hypothesis encounters an even greater degree of controversy, however, in the explanation of drumlins, rogens, and hummocky terrain, landforms occurring in mostly unconsolidated materials and traditionally attributed to direct glacial action as the result of subglacial megafloods. This extension of the hypothesis was initially applied to the Livingstone Lake drumlin field in northern Saskatchewan. Form

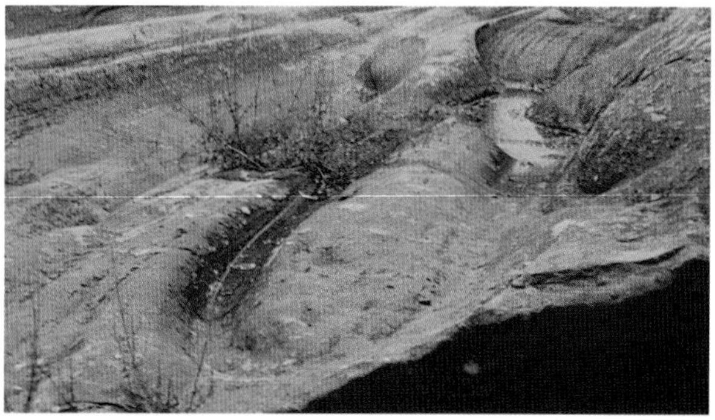

Figure 19 Exposure of a sculpted bedrock surface on Kelley's Island, Lake Erie. Photograph by Alan Kehew.

analogy of the drumlins to the inverted sole of a turbidite bed led to the inference that the drumlins formed by infilling of erosional marks scoured upward into the base of the ice sheet, perhaps during the waning phases of the flood (Shaw and Kvill, 1984). Shaw *et al.* (1989) estimated the total volume of water released as $84 \times 10^3\,km^3$.

The recognition of drumlins with truncated internal structure at land surface led to the broadening of the meltwater hypothesis to include drumlins formed by scour of the subglacial bed by the sheetflood, rather than scour of the base of the ice (Shaw and Sharpe, 1987). Form analogy is also used here as a basis for the process, in this case to the large-scale, streamlined erosional hills of the Channeled Scabland and elsewhere. Interpretation of landforms other than drumlins by similar mechanisms was soon to follow. Rogen moraine was explained by infill of scours in basal ice oriented transverse to the flow direction of the sheetflow (Fisher and Shaw, 1992). Munro and Shaw (1997) added hummocky terrain to the hypothesis. This interpretation was based on landscapes in Alberta where hummocks, covered with boulder lags, have internal structures that are truncated at the land surface. The hummocky terrain occurs adjacent to incised 'flood tracts'; the hummocky terrain itself is attributed to less-severe erosion on the flanks of the main channels or tracts.

Spirited debate continues on the meltwater hypothesis (Evans *et al.*, 2006). Form analogy as the most fundamental evidence for landforms produced by subglacial sheetfloods has been challenged on several grounds. In the case of cavity infill landforms, confirmation of process by form analogy requires extrapolation over orders of magnitude; in addition, Benn and Evans (1998) have pointed out that dissimilar processes can produce similar landforms. The formation of drumlins, rogen moraine, and megaflutes by bed deformation by glacial ice is well established in the literature, although details continue to be controversial. Other problematical aspects of the meltwater hypothesis include the lack of erosional landforms beyond the ice margins from which megafloods emerged, and the lack of evidence for the huge meltwater reservoirs necessary for the generation of megafloods. Clarke *et al.* (2005) contend that both the production of sufficient meltwater and its storage on top of or beneath ice sheets are unlikely without special circumstances.

See also: **Fluvial Environments**: Sediments. **Glacial Landforms**: Introduction; Glacial Land Systems. **Glacial Landforms, Sediments**: Glacifluvial Landforms of Deposition; Glacigenic Lithofacies.

References

Baker V. R. (1973). Paleohydrology and sedimentology of the Lake Missoula flooding in eastern Washington. Geological Society of America Special Paper 144. Geological Society of America.

Baker, V. R., Benito, G., and Rudoy, A. N. (1993). Paleohydrology of Late Pleistocene superflooding, Altay Mountains, Siberia. *Science* **259**, 348–350.

Baker, V. R., Greeley, R., Komar, P. D., Swanson, D. A., and Waitt, R. B., Jr. (1987). Columbia and Snake River Plains. In *Geomorphic Systems of North America* (W. L. Graf, Ed.), Centennial Special Volume 2, pp. 403–468. Geological Society of America, Boulder, CO.

Benn, D. I., and Evans, D. J. A. (1998). *Glaciers and Glaciation*. Arnold (co-published by Oxford University Press), New York.

Brennand, T. A., and Shaw, J. (1994). Tunnel channels and associated landforms, south-central Ontario: Their implications for ice-sheet hydrology. *Canadian Journal of Earth Sciences* **31**, 505–522.

Bretz, J. H. (1923). The Channeled Scabland of the Columbia Plateau. *Journal of Geology* **31**, 617–649.

Cenderelli, D. A., and Wohl, E. A. (2003). Flow hydraulics and geomorphic effects of glacial-lake outburst floods in the Mount Everest region, Nepal. *Earth Surface Processes and Landforms* **28**, 385–407.

Clarke, G. K. C., Levertington, D. W., Teller, J. T., Dyke, A. S., and Marshall, S. J. (2005). Fresh arguments to comments by David Sharpe on "Paleohydraulics of the last outburst flood from glacial Lake Agassiz and 8200 BP cold event." *Quaternary Science Reviews* **24**, 1533–1541.

Clayton, L., Attig, J. W., and Mickelson, D. M. (1999). Tunnel channels formed in Wisconsin during the last glaciation. In *Glacial Processes Past and Present* (D. M. Mickelson and J. W. Attig, Eds.), Geological Society of America Special Paper 337, pp. 69–82. Geological Society of America.

Cutler, P. M., Colgan, P. M., and Mickelson, D. M. (2002). Sedimentologic evidence for outburst floods from the Laurentide Ice Sheet margin in Wisconsin, USA: Implications for tunnel channel formation. *Quaternary International* **90**, 23–40.

Dyke, A. S. (1993). Landscapes of cold-centered Late Wisconsinan ice caps, Arctic Canada. *Progress in Physical Geography* **17**, 223–247.

Dyke, A. S., Andrews, J. T., Clark, P. U., England, J. H., Miller, G. H., Shaw, J., and Veillette, J. J. (2002). The Laurentide and Innuitian ice sheets during the Last Glacial Maxiumum. *Quaternary Science Review* **21**, 9–31.

Evans, D. J. A., Rea, B. R., Hiemstra, J. F., and Cofaigh, C. O. (in press). A critical assessment of subglacial mega-floods: a case

study of glacial sediments and landforms in south-central Alberta, Canada. *Quaternary Science Review*.
Fisher, T. G., and Shaw, J. (1992). A depositional model for rogen moraine, with examples from the Avalon Peninsula, Newfoundland. *Canadian Journal of Earth Sciences* **29**, 669–686.
Hooke R. L., and Jennings C. E. (in press). On the formation of tunnel valleys. *Quaternary Science Reviews* **26**.
Kehew, A. E., Nicks, L. P., and Straw, T. W. (1999). Palimpsest tunnel valleys: Evidence for relative timing of advances in an interlobate area of the Laurentide Ice Sheet. *Annals of Glaciology* **28**, 47–52.
Kehew, A. E., and Teller, J. T. (1994). History of late glacial runoff along the southwestern margin of the Laurentide Ice Sheet. *Quaternary Science Reviews* **13**, 859–877.
Kehew, A. K., and Lord, M. L. (1987). Glacial-lake outburst along the mid-continent margins of the Laurentide Ice Sheet. In *Catastrophic flooding* (L. Mayer and D. Nash, Eds.), Proceedings of 18th Annual Binghamton Symposia in Geomorphology, pp. 95–120.
Kor, P. S. G., Shaw, J., and Sharpe, D. R. (1991). Erosion of bedrock by subglacial meltwater, Georgian Bay, Ontario: A regional view. *Canadian Journal of Earth Sciences* **28**, 623–642.
Kozlowski, A. L., Kehew, A. E., and Bird, B. C. (2005). Outburst flood origin of the Central Kalamazoo River Valley, Michigan, USA. *Quaternary Science Reviews* **24**, 2354–2374.
Lord, M. L., and Kehew, A. K. (1987). Sedimentology and paleohydrology of glacial-lake outbursts in southeastern Saskatchewan and northwestern North Dakota. *Geological Society of America Bulletin* **99**, 663–673.
Mooers, H. D. (1989). On formation of the tunnel valleys of the Superior lobe, central Minnesota. *Quaternary Research* **32**, 24–35.
Munro, M., and Shaw, J. (1997). Erosional origin of hummocky terrain in south-central Alberta. *Geology* **25**, 1027–1030.
O'Cofaigh, C. (1996). Tunnel valley genesis. *Progress in Physical Geography* **20**, 1–19.
O'Connor, J. E., and Baker, V. R. (1992). Magnitudes and implications of peak discharges from Glacial Lake Missoula. *Geological Society of America Bulletin* **104**, 267–279.
O'Connor, J. E., and Costa, J. E. (2004). The world's largest floods, past and present: Their causes and magnitude. *US Geological Survey Circular* **1254**, 13p.
Pair, D. L. (1997). Thin film, channelized drainage, or sheetfloods beneath a portion of the Laurentide Ice Sheet: An examination of glacial erosion forms, northern New York State. *Sedimentary Geology* **111**, 199–216.
Piotrowski, J. A. (1993). Tunnel-valley formation in northwest Germany – geology, mechanisms of formation and subglacial bed conditions for the Bornhoved tunnel valley. *Sedimentary Geology* **89**, 107–141.
Shaw, J. (1994). Hairpin erosional marks, horseshoe vortices and subglacial erosion. *Sedimentary Geology* **91**, 269–283.
Shaw, J. (2002). The meltwater hypothesis for subglacial bedforms. *Quaternary International* **90**, 5–22.
Shaw, J., and Kvill, D. (1984). A glaciofluvial origin for drumlins of the Livingstone Lake Area, Saskatchewan. *Canadian Journal of Earth Sciences* **21**, 1442–1459.
Shaw, J., and Sharpe, D. R. (1987). Drumlin formation by subglacial meltwater erosion. *Canadian Journal of Earth Sciences* **24**, 2316–2322.
Shaw, J., Kvill, D., and Rains, R. B. (1989). Drumlins and catastrophic subglacial floods. *Sedimentary Geology* **62**, 177–202.
Shepard, R. G., and Schumm, S. A. (1974). Experimental study of river incision. *Geological Society of America Bulletin* **85**, 257–268.
Sjogren D, B., Fisher, T. G., Taylor, L. D., Jol, H. M., and Munro-Stasiuk, M. J. (2002). Incipient tunnel channels. *Quaternary International* **90**, 41–56.
Wright, H. E., Jr. (1973). Tunnel valleys, glacial surges, and subglacial hydrology of the Superior Lobe, Minnesota. In *The Wisconsinan Stage*, (R. Black, R. Goldthwait and H. Willman, Eds.), Geological Society of America Memoir136, pp. 251–276.

Glacitectonic Structures and Landforms

D J A Evans, Durham University, Durham, UK

© 2007 Elsevier B.V. All rights reserved.

Introduction

The deformation of bedrock and Quaternary sediment into large moraines has been widely reported from glaciated terrains and is often responsible for some of the most conclusive evidence for former glacier extent in the Quaternary geological record (see Aber *et al.*, 1989; van der Wateren 1995; Benn and Evans, 1998 for reviews). Four types of glacitectonic sediment–landform association are recognized: (1) hill-hole pairs; (2) composite ridges and thrust-block moraines; (3) cupola hills; and (4) megablocks and rafts. Benn and Evans (1998) proposed that glacitectonic moraine ridges are those in which glaciotectonized pre-tectonic and syn-tectonic sediments constitute >25% of the unit area of the moraine. Other types of moraine, such as push moraines, may contain small amounts of glacitectonized sediment (*see* Moraine forms and Genesis). Glacitectonic landforms may be composed of pre-Quaternary bedrock, pre-existing Quaternary sediments, or contemporaneous sediment and where they have been overridden by glacier ice after construction they may be covered by a carapace of glacitectonite or till (see Tills). More prolonged and active glacier overriding will result in the streamlining and molding of glacitectonic features to produce subglacial bedforms.

Glacitectonic Processes

Proglacial glacitectonics involves the large-scale displacement of proglacial materials due to stresses imposed by glacier ice, and involves ductile or brittle deformation or a combination of the two. Ductile deformation involves the production of large, open folds in the sediments or rocks in front of an advancing glacier, which may develop into overfolds or begin to undergo internal thrusting due to continued ice advance. In contrast, brittle deformation involves the thrusting of semi-coherent blocks along discrete planes of failure. In both cases, the lower limit of deformation is usually marked by a basal failure plane or plane of décollement, which often coincides with a sedimentary discontinuity or bedding plane. Although all materials are capable of undergoing both ductile and brittle failure according to variations in the applied stress, temperature, strain rate, and porewater pressure, glacitectonic disruption of frozen sediments appears to be dominated by brittle failure (e.g., Kalin, 1971; Klassen, 1982; Evans and England, 1991). The compression of proglacial materials by glacitectonic deformation is very similar to the processes which occur during continental collision and mountain building (Croot, 1987; Aber, 1988). Indeed, proglacial glacitectonic landforms can be regarded as scale models of mountain chains, and can be studied using well-founded principles of structural geology (see Twiss and Moores, 1992; **Fig. 1**).

The formation of large thrust-moraine complexes standing many tens of meters above the surrounding terrain depends on a number of factors, the most important of which are low-strength proglacial sediments and high, glacially imposed stresses. Sediment strength is dependent upon grain size and sorting, the existence of potential planes of failure, and porewater pressures. High porewater pressures result from subglacial drainage through inefficient distributed systems, such as porewater flow, and can be further encouraged by the existence of proglacial permafrost, which acts to confine pressurized porewater in underlying unfrozen aquifers (Mathews and Mackay, 1960). The presence of weakened sediments, however, is not a sufficient condition for proglacial glacitectonic deformation, which also requires stresses large enough to elevate large masses of sediment above the glacier margin. It is clear that in most situations the shear stresses beneath glacier margins are too low to produce the observed deformation. The solution to this problem was developed by Rotnicki (1976), van der Wateren (1985), and Aber *et al.* (1989), and is known as the 'gravity spreading model'. According to this model, proglacial sediment failure results from the translation of the weight of the glacier into lateral stresses, which push sediment wedges away from the load and then upward from their original position. The stress field producing a thrust block is therefore the product

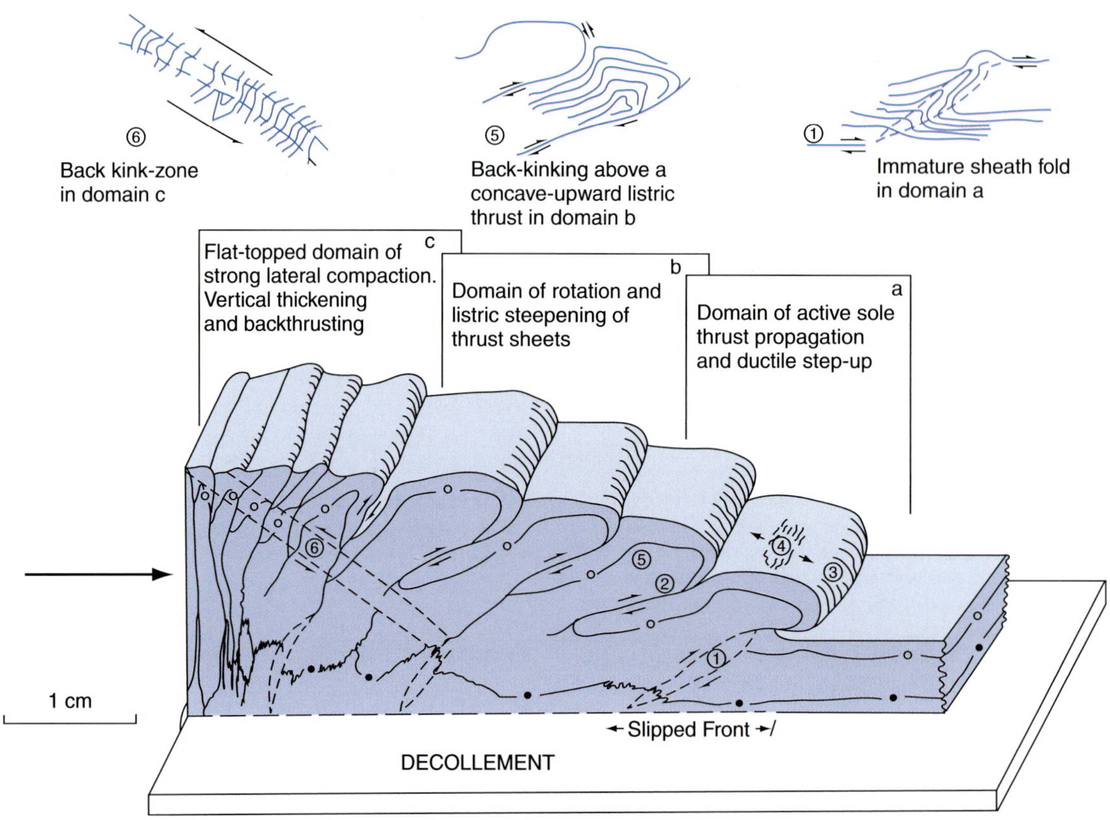

Figure 1 Example of the tectonic structures produced by the thrusting of stratified sand in a squeeze-box. These features are typical of moraines constructed by proglacial glacitectonics: 1) initiation of thrust; 2) thrust fault; 3) slump zone; 4) extension fractures; 5) back kink fold; 6) backthrust zone. From Mulugeta and Koyi (1987).

of the total weight of the ice mass and older thrust blocks. In addition to gravity spreading, two other mechanisms may contribute to the development of thrust systems (Aber *et al.*, 1989; Price and Cosgrove, 1990). First, gravity gliding or gravity sliding involves the deformation of sediment blocks as they move downslope under their own weight. This requires the uplifting of the ice proximal zone by gravity spreading to produce a slope upon which blocks can slide. Second, compression or push-from-behind may occur by the direct shoving of sediment wedges by the forward movement of the glacier. In reality, gravity spreading, gravity gliding, and compression can act in unison in proglacial settings.

The likelihood of glacitectonic disturbance is dictated by various factors (Aber *et al.*, 1989), the most important of which are (i) slope of the proglacial area, whereby ice advance against a topographic obstacle increases the likelihood of proglacial glacitectonics; (ii) the presence of weak layers in the substratum, whereby failure is encouraged by weak layers or potential décollement planes; (iii) the nature of ice/sediment contact in the proglacial area, where partial burial of glacier snouts prior to their advance may lead to the lifting of sedimentary strata; and (iv) subglacial and proglacial drainage, whereby failure is encouraged by high porewater pressures in proglacial and submarginal sediments and rocks.

The shortening of proglacial sediments and rocks during glacitectonic deformation results in complex geological structures analogous in many ways to thrust-fold belts at the collision points of drifting continents, and can be explained using the principles of structural geology (**Fig. 2**). Ductile deformation during glacitectonic compression and shortening results in a wide range of fold types, depending on

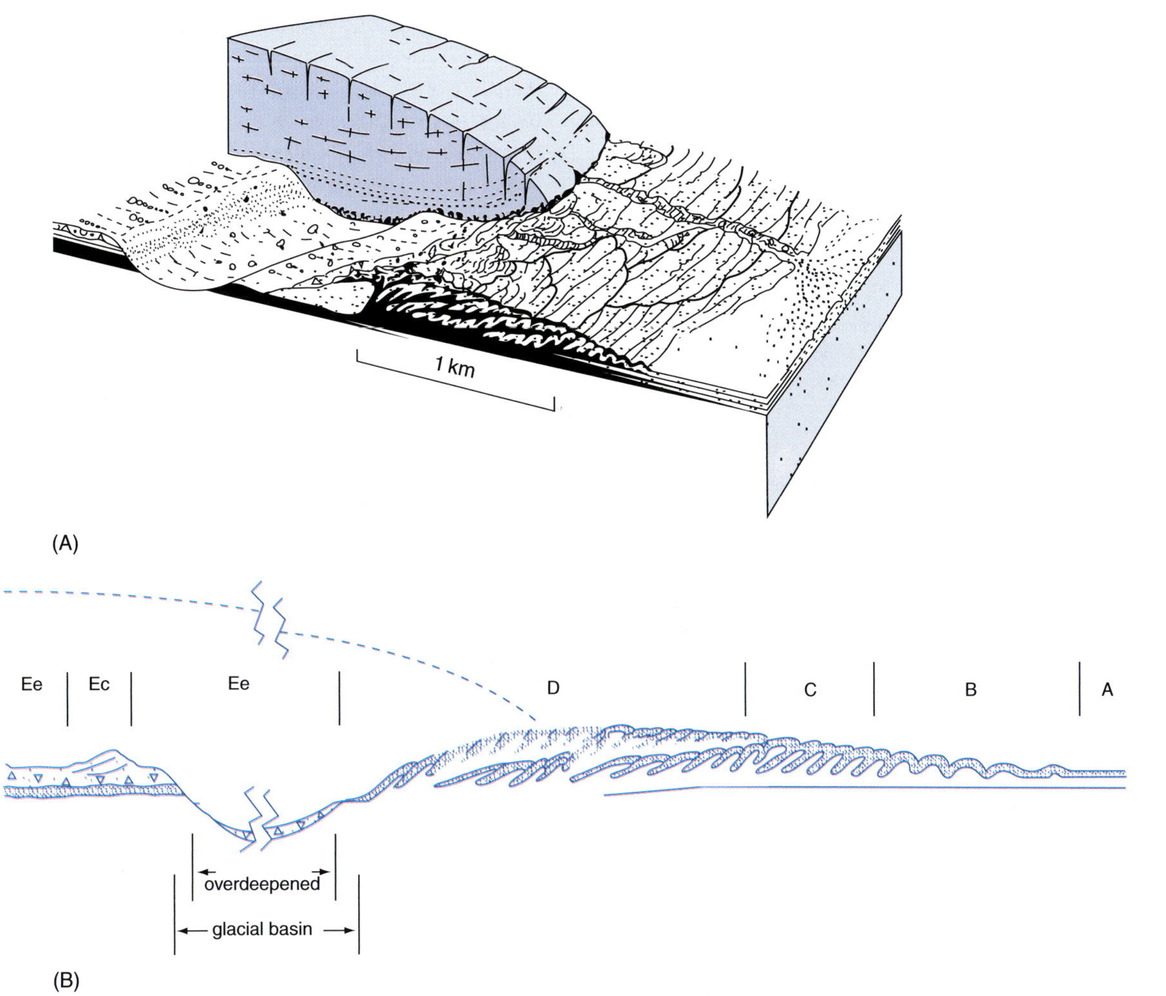

Figure 2 Glacitectonic structures and landforms associated with proglacial thrusting by the snout of Holmstrombreen, Svalbard (from van der Wateren, 1995): (A) block diagram of the thrust and folded sediments that constitute the moraine at the glacier snout; (B) styles of glacitectonic disturbance that can be observed at Holmstrombreen; A = undeformed foreland; B = 'Jura-style' high-angle folding; C = low-angle structures including thrusts and overturned and recumbent folds; D = nappes; E = deformed subglacial sediment/tills where Ec is compressive and Ee is extensional.

the magnitude and orientation of the applied stresses and the strain response of the sediments. Folds range from long wavelength or open folds to isoclinal or recumbent (nappe) folds. Other fold types include chevron, sheath, kink-band, box, and polyclinal folds. Complexities may arise where different sedimentary layers respond differently to stress and produce disharmonic folds, in which the fold wavelengths of some layers are smaller than those of others.

Composite Ridges and Thrust-Block Moraines

The most common and distinctive type of glacitectonic landforms are composite ridges and thrust-block moraines (**Fig. 3A**). Composite ridges are composed of multiple slices of upthrust and contorted bedrock and/or unconsolidated sediments, which are commonly interlayered with and overlain by glacigenic and glaciofluvial sediment (Aber *et al.*, 1989). They are subdivided by Aber *et al.* (1989) into large (>100 m of relief) and small (<100 m of relief) composite ridges. This size differentiation is essentially arbitrary and the processes involved in the production of both large and small composite ridges are the same. However, Aber *et al.* (1989) suggested that large ridges usually contain a considerable volume of pre-Quaternary bedrock, whereas small ridges are composed mostly of unconsolidated Quaternary strata. The term thrust-block moraine was introduced by Evans and England (1991) to describe composite ridges comprising proglacially thrust masses of permafrozen blocks of Quaternary sediments at the margins of Arctic subpolar glaciers. Composite ridges are associated with proglacial or submarginal glacitectonics and they mark the positions of glacier stillstands or readvances.

In planform, composite ridges comprise arcuate suites of subparallel ridges and intervening depressions, arranged in a concave up-glacier pattern and conforming to the general shape of the glacier margin that produced them. Individual ridge crests correspond to the crests of internal folds, the upturned ends or edges of thrust blocks or resistant upturned beds. The depth of disturbance associated with composite ridges is generally a few tens of meters, and may be up to 200 m for the very largest features (Kupsch, 1962; Aber *et al.*, 1989). The internal structures of large composite ridges are very similar to the products of thin-skinned tectonics in thrust and folded mountain belts like the Canadian Rockies, even though composite ridges are at least one order of magnitude smaller than true mountain ranges (Aber *et al.*, 1989). The material in composite ridges commonly forms stacked sheets, with individual thrust slices showing varying degrees of internal deformation. Overturned and rootless folds of bedrock and Quaternary sediments in the composite ridges at Flade Klint, Denmark, have been thrust up to 100 m from their probable source depression (Gry, 1940). Probably the most spectacular exposures through a large composite ridge are located at Møns Klint in Denmark. Several ice advances during the last glaciation have piled up numerous chalk scales with intervening Quaternary sediment to produce a composite ridge characterized by more than 150 m of structural relief (Aber, 1985). The structures within the cliffs at Møns Klint include multiple thrust anticlines and individually folded and stacked chalk floes, which can be traced as a series of ridges on the rugged highland of Høje Møn.

There are numerous examples of small composite ridges, both in front of modern glacier snouts and in the ancient landform record. Spectacular composite

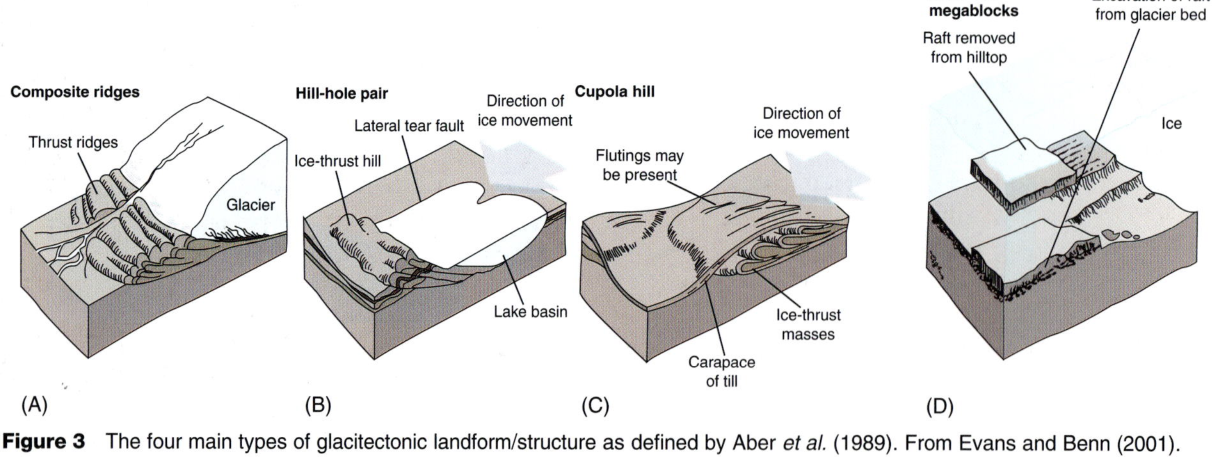

Figure 3 The four main types of glacitectonic landform/structure as defined by Aber *et al.* (1989). From Evans and Benn (2001).

Figure 4 Thrust-block moraine at the margin of a subpolar glacier in the Canadian high Arctic.

ridges composed of outwash and glacimarine sediments form continuous moraines around the snouts of many Spitsbergen glaciers (Gripp and Todtmann, 1925; Gripp, 1929; Croot, 1988). In Spitsbergen and Iceland, composite ridges are strongly associated with surging glaciers, suggesting that conditions during surges are particularly conducive to proglacial tectonics (Sharp, 1985; Croot, 1987; 1988; Evans and Rea, 2003). The most important factors underlying this association are (i) rapid advance of the snout; (ii) extreme compressive deformation at the margin; and (iii) elevated porewater pressures associated with the release of meltwater. The role of porewater pressures is highlighted by the general absence of composite ridges where proglacial areas slope steeply away from the glacier, and water is able to drain away freely. The internal structures of recent composite ridges at Eyjabakkajokull, an Icelandic surging glacier, include low-angle subglacial thrust faults dipping down glacier, over which thrust slices have been transported during overall extension, and proglacial compression structures, such as stacked thrust sheets composed of outwash, tephra, turf, organic soil, and peat beds, which form asymmetric ridges. In the more distal parts of the moraine, smaller asymmetric ridges are the surface expression of anticlines or overturned anticlines (Croot, 1988).

Thrust-block moraines produced by proglacial thrusting of blocks of frozen outwash and raised marine sediments have been documented in the Canadian arctic (Kalin, 1971; Klassen, 1982; Evans, 1989; Evans and England, 1991; **Fig. 4**). In most examples, the bedding within individual blocks and block surfaces dip gently towards the glacier, suggesting that the moraines are composed of either stacked scales or partially rotated deep-seated blocks (Kalin, 1971). In some cases, however, bedding may dip away from the glacier snout suggesting deep-seated wedging of the advancing snout beneath proglacial sediments (Evans and England, 1991). The Canadian Arctic thrust-block moraines occur in areas where permafrost is often in excess of 700 m thick, but they are located in valley bottoms either below the marine limit or below former proglacial lake shorelines relating to the last glaciation. This suggests that permafrost may have been partially degraded below deep-water bodies prior to the Holocene advances responsible for thrusting. Therefore, block failure could have been initiated along an unfrozen *talik* between newly aggraded postglacial surface permafrost and the degraded top of the full glacial permafrost. Thrust-block moraines at the margins of cold-based glaciers in the Dry Valleys area, Antarctica, are interpreted by Fitzsimmons (1996) as the product of thrusting by cold-based glaciers overriding weak, unfrozen lake-floor sediments. Unfrozen areas occur below either saline lakes or wet-based lakes which have a permanent ice cover but do not undergo full-depth freezing. As the glacier enters the lake, sediment blocks become frozen onto the base of the glacier, but can be thrust forward because their deeper layers remain unfrozen. In summary, therefore, the formation of thrust-block moraines in permafrost regions is thought to occur only within areas where unfrozen or only partially frozen sediments occur.

Hill-Hole Pairs

A hill-hole pair is defined as '· · ·*a discrete hill of ice-thrust material, often slightly crumpled, situated a short distance down-glacier from a depression of similar size and shape*' (Bluemle and Clayton, 1984; **Fig. 3B**). Ice-thrust hills may be found at distances of up to 5 km from source depressions, but are usually closer (Bluemle and Clayton, 1984; **Fig. 5**). However, source depressions can become infilled with younger sediment, and are not always visible as surface features. Conversely, glacitectonic depressions are sometimes found without associated hills (Andriashek and Fenton, 1989), perhaps because positive relief features have been destroyed by subglacial erosion.

The typical morphology of a hill-hole pair is as follows (Aber *et al.*, 1989): (i) the hill has an arcuate or crescentic planform, which is concave up-glacier; (ii) the surface of the hill is traversed by a series of transverse, subparallel ridges and depressions; (iii) the hill has an asymmetric cross-profile, with the highest point and steeper slopes on the convex or down-glacier side; and (iv) the topographic depression is approximately the same shape and area as the hill, located on the concave or up-glacier side. These

Figure 5 Bittern Lake, AB, Canada, an example of a hill-hole pair. The glaciotectonically disturbed Cretaceous bedrock that comprises the ice-thrust hill is visible in the road cut.

morphologic features may be less obvious where prolonged subglacial modification has taken place after formation. The largest hill-hole pair so far recognized on the Earth's surface is Herschel Island in Mackenzie Bay, Canada, with an area of more than 100 km^2 (Mackay, 1959; Rampton, 1982) and composed of preglacial Pleistocene sediments thrust up from a now submerged source depression. Its internal structures include low angle thrust faults, synclines, and anticlines, tilted beds which occasionally approach subvertical angles, overturned folds, repeated strata, slickensides, and segregated ice lenses which originally formed horizontally but now parallel the deformed enclosing strata (Mackay, 1959; Mackay *et al.*, 1972).

Cupola Hills

Cupola hill is the English translation of the Danish term *kuppelbakke*, which was introduced by Smed (1962) to describe irregular hills with the general characteristics of glacitectonic landforms but lacking hill-hole pair relationship and/or the transverse ridged morphology of composite ridges (**Fig. 3**). The characteristics of cupola hills can be summarized as: (i) they have a dome-like morphology varying from near circular to oval shapes with lengths of 1–15 km and heights of 20–100 m; (ii) internally, they are composed of detached and deformed floes of Quaternary sediments, older strata, or bedrock; and (iii) they have a carapace of till which truncates the underlying structures. These characteristics indicate that cupola hills are subglacially overridden hill-hole pairs or composite ridges, and represent early stages in the development of drumlins or megaflutings. In cases where the subglacial modification has been slight, transverse ridge morphology may be partially preserved and the cupola form will tend to lie transverse to former ice flow. With increasing subglacial modification ridge forms tend to be replaced by more smoothed, elongated forms. Viewed from the air, cupola hills may appear aligned in arcuate or transverse belts which conform to the former margins of glaciers. Where such belts occur they represent the positions of former large composite ridges that have been modified by subglacial molding.

Examples of cupola hills composed entirely of Quaternary sediments include the island of Ven in Sweden and Ristinge Klint in Denmark (Aber *et al.*, 1989). The island of Ven is composed of gentle synclines disturbed by numerous thrust faults and overturned folds. The glacitectonic disturbance responsible for these structures has produced two major transverse ridges, the surface of which is muted by a cover of discordant till (Adrielsson, 1984). In contrast, the topography of Ristinge Klint is subdued and drumlinized similar to cupola hills on the nearby island of Aero (Smed, 1962). Cliff sections reveal that the area of Ristinge is underlain by more than 30 stacked scales of Quaternary sediments, each scale being up to 20 m thick and containing the same stratigraphic sequence of Saalian, Eemian, and Weichselian strata (Sjorring, 1983). The internal stratification of each scale is little disturbed except for prominent drag folds associated with the thrust faults between scales. Thrusting of the scales is thought to have occurred at the ice margin during glacier advance and deposition of the upper, discordant tills; surface streamlining took place during subsequent ice overriding. Other excellent examples of cupola hills occur in the late Wisconsinan end moraine systems that make up the offshore islands of New England, USA. The islands consist of glaciotectonically disturbed floes or thrust scales of Cretaceous, Tertiary, and Quaternary strata (Oldale and O'Hara, 1984). Sections at Gay Head Cliff, Martha'a Vineyard expose a series of layered scales, each 20–30 m thick and dipping up-glacier, interrupted by overturned folds, low-angle thrust faults and occasional down-dropped blocks or grabens. In places, the strata have been uplifted by over 160 m.

Megablocks and Rafts

Megablocks and rafts are dislocated slabs of rock and unconsolidated strata which have been transported from their original position by glacier action (Ruszcynska-Sjenach, 1987; Stalker, 1973, 1976; Sauer, 1978; **Fig. 3D**). Traditionally, such rafts were assumed to have been frozen on to the base of cold-based ice sheets, although it is now recognized that failure may occur along a basal décollement in a subglacial deforming layer. Megablocks may originate from uprooted cupola hills, although it is often difficult if not impossible to demonstrate detachment from the parent substrate unless there is exceptionally good exposure. Where they are subhorizontal, megablocks and rafts may form flat-topped buttes or plateaus and can be mistaken for bedrock outliers. Similarly, buried megablocks can be mistaken for bedrock if borehole logs or exposures are not available. Although megablocks and rafts contain coherent masses of bedrock or sediment, they are also commonly traversed by shear zones, faults, and brecciated zones, and may be folded.

Rafts of Jurassic clay more than 20 m thick at Lukow, Poland, are derived from Lithuania, more than 300 km to the northeast, and are thought to have been transported in a frozen state (Ruszczynska-Szenajch, 1976). A similar transport distance has been calculated for the Cooking Lake megablock near Edmonton, AB, Canada, which is 10 m thick and covers an area of approximately 10 km^2 (Stalker, 1976). A good example of the subsurface expression of a megablock occurs in the vicinity of the Qu'Appelle Valley, Saskatchewan. This block of Cretaceous shale is approximately 1,000 km^2 in area with a maximum thickness of 100 m, and was discovered by borehole logging and bedrock contour mapping (e.g., Sauer, 1978). A well drilled at the western end of the megablock penetrated 80 m of brecciated and deformed bedrock of the Riding Mountain Formation and then 2 m of till. Steeply dipping rafts are more likely to have some surface expression, although they may be buried and difficult to identify without extensive borehole data. A series of detached chalk rafts occur in till overlying limestone in quarry exposures at Kvarnby in Sweden, suggesting that they were removed from the Baltic Sea bed and transported more than 25 km before being stacked at angles of up to 80° (Ringberg, 1983). These subsurface rafts lie beneath a structurally controlled, low-amplitude hummocky topography which would be difficult to differentiate from a cupola hill without exposures. Similarly, chalk rafts in the Anglian till of Hertfordshire, England, have been identified in borehole records and as pale-toned patches in darker, freshly seeded fields on aerial photographs (Hopson, 1995). Unlike the Swedish examples, the Hertfordshire chalk rafts have been displaced only a short distance (mostly < 1 km) from their origin by the southerly flowing Anglian ice sheet.

See also: **Glacial Landforms**: Moraine forms and Genesis. **Glacial Landforms, Sediments**: Tills.

References

Aber, J. S. (1985). The character of glaciotectonism. *Geologie en Mijnbouw* **64**, 389–395.

Aber, J. S. (1988). *Structural Geology Exercises with Glaciotectonic Examples*, p. 140. Hunter Textbooks Inc., Winston-Salem, NC, USA.

Aber, J. S., Croot, D. G., and Fenton, M. M. (1989). *Glaciotectonic Landforms and Structures*. p. 200. Kluwer, Dordrecht.

Adrielsson L. (1984) *Weichselian Lithostratigraphy and Glacial Environments in the Ven-Gumslov Area, Southern Sweden*, p. 120. Lundqua Thesis 14, Lund University, Sweden.

Andriashek, L. D., and Fenton, M. M. (1989). Quaternary Stratigraphy and Surficial Geology, Sand River Map Sheet 73L, Alberta. *Terrain Sciences Department, Alberta Research Council, Bulletin* 57.

Benn, D. I., and Evans, D. J. A. (1998). *Glaciers and Glaciation*. Arnold, London.

Bluemle, J. P., and Clayton, L. (1984). Large-scale glacial thrusting and related processes in North Dakota. *Boreas* **13**, 279–299.

Croot, D. G. (1987). Glacio-tectonic structures: a mesoscale model of thin-skinned thrust sheets? *Journal of Structural Geology* **9**, 797–808.

Croot, D. G. (1988). Morphological, structural and mechanical analysis of neoglacial ice-pushed ridges in Iceland. In *Glaciotectonics: Forms and Processes* (D. G. Croot, Ed.), pp. 33–47. Balkema, Rotterdam.

Evans, D. J. A. (1989). The nature of glacitectonic structures and sediments at subpolar glacier margins, northwest Ellesmere Island, Canada. *Geografiska Annaler* **71A**, 113–123.

Evans, D. J. A., and Benn, D. I. (2001). Earth's giant bulldozers. *Geography Review* **14**, 29–33.

Evans, D. J. A., and England, J. (1991). Canadian landform examples 19: high arctic thrust block moraines. *Canadian Geographer* **35**, 93–97.

Evans, D. J. A., and Rea, B. R. (2003). Surging glacier landsystem. In *Glacial Landsystems* (D. J. A. Evans, Ed.), Arnold, London.

Fitzsimons, S. J. (1996). Formation of thrust-block moraines at the margins of dry-based glaciers, south Victoria Land, Antarctica. *Annals of Glaciology* **22**, 68–74.

Gripp, K. (1929). Glaciologische und geologische Ergebnisse der Hamburgischen Spitzbergen-Expedition 1927. *Naturwissenschaftlichen Verein in Hamburg Abhandlungen aus dem Gebeite der Naturwissenschaften* **22**, 146–249.

Gripp, K., and Todtmann, E. M. (1925). Die Endmorane des Green-Bay Gletschers auf Spitsbergen, eine studie zum Verstandnis norddeutscher Diluvial-Gebilde. *Mitteilungen aus Geographischen Gesellschaft in Hamburg* **37**, 45–75.

Gry, H. (1940). De istektoniske forhold i moleromraadet. *Meddelelser Dansk Geologisk Forening* **9**, 586–627.

Hopson, P. M. (1995). Chalk rafts in Anglian till in north Hertfordshire. *Proceedings of the Geologists Association* **106**, 151–158.

Kalin, M. (1971). The active push moraine of the Thompson Glacier, Axel Heiberg Island, Canadian Arctic Archipelago.

McGill University, Axel Heiberg Island, Research Report, Glaciology 4, p. 68. Montreal: McGill University.
Klassen, R. A. (1982). Glaciotectonic thrust plates, Bylot Island, District of Franklin. *Geological Survey of Canada, Paper* **82–1A**, 369–373.
Kupsch, W. O. (1962). Ice-thrust ridges in western Canada. *Journal of Geology* **70**, 582–594.
Mackay, J. R. (1959). Glacier ice-thrust features of the Yukon coast. *Geographical Bulletin* **13**, 5–21.
Mackay, J. R., Rampton, V. N., and Fyles, J. G. (1972). Relic Pleistocene permafrost, western Arctic, Canada. *Science* **176**, 1321–1323.
Mathews, W. W., and Mackay, J. R. (1960). Deformation of soils by glacier ice and the influence of pore pressures and permafrost. *Transactions of the Royal Society of Canada, Section 3* **54**, 27–36.
Mulugeta, G., and Koyi, H. (1987). Three-dimensional geometry and kinematics of experimental piggyback thrusting. *Geology* **15**, 1052–1056.
Oldale, R. N., and O'Hara, C. J. (1984). Glaciotectonic origin of the Massachusetts coastal end moraines and a fluctuating late Wisconsinan ice margin. *Bulletin of Geological Society of America* **95**, 61–74.
Price, N. J., and Cosgrove, J. W. (1990). *Analysis of Geological Structures*. p. 502. Cambridge University Press, Cambridge.
Rampton, V. N. (1982). Quaternary geology of the Yukon Coastal Plain. *Geological Survey of Canada, Bulletin* **317**, 49.
Ringberg, B. (1983). Till stratigraphy and glacial rafts of chalk at Kvarnby, southern Sweden. In *Glacial Deposits in North-West Europe* (J. Ehlers, Ed.), pp. 151–154. Balkema, Rotterdam.
Rotnicki, K. (1976). The theoretical basis for and a model of glaciotectonic deformations. *Quaestiones Geographicae* **3**, 103–139.
Ruszczynska-Szenajch, H. (1987). The origin of glacial rafts: detachment, transport, deposition. *Boreas* **16**, 101–112.
Sauer, E. K. (1978). The engineering significance of glacier ice-thrusting. *Canadian Geotechnical Journal* **15**, 457–472.
Sharp, M. J. (1985). Sedimentation and stratigraphy at Eyjabakkajokull – an Icelandic surging glacier. *Quaternary Research* **24**, 268–284.
Sjorring, S. (1983). Ristinge Klint. In *Glacial Deposits in North-West Europe* (J. Ehlers, Ed.), pp. 219–226. Balkema, Rotterdam.
Smed, P. (1962). Studier over den fynske ogruppes glaciale landskabsformer. *Meddelelser Dansk Geologisk Forening* **15**, 1–74.
Stalker, A., and Mac, S. (1973). The large interdrift bedrock blocks of the Canadian Prairies. *Geological Survey of Canada*, Paper **75–1A**, 421–422.
Stalker, A., and Mac, S. (1976). Megablocks, or the enormous erratics of the Albertan Prairies. *Geological Survey of Canada*, Paper **76-1C**, 185–188.
Twiss, R. J., and Moores, E. M. (1992). *Structural Geology*. Freeman and Co., New York.
van der Wateren, D. F. M. (1985). A model of glacial tectonics, applied to the ice-pushed ridges in the central Netherlands. *Bulletin of the Geological Society Denmark* **34**, 55–74.
van der Wateren, D. F. M. (1995). Processes of glaciotectonism. In *Glacial Environments Volume1 – Modern Glacial Environments: Processes, Dynamics and Sediments* (J. Menzies, Ed.), pp. 309–335. Butterworth-Heinemann, Oxford.

GLACIAL LANDFORMS, EROSIONAL FEATURES

Contents

Major Scale Forms

I S Evans, Durham University, Durham, UK

© 2007 Elsevier B.V. All rights reserved.

Introduction

Forms of glacial erosion are both the most obvious of landforms, recognized after a few classes in school, and yet the most difficult to fully explain or to date (Glasser and Bennett, 2004). Medium-sized glacial landforms such as roches moutonnées, whalebacks, and rock drumlins were discussed in the previous chapter (*see* Micro to Macro Scale Forms). Most of these are decametric and hectometric in scale, but some, such as giant roches moutonnées, are kilometric. Here we deal with the largest forms: cirques, troughs, rock basins, and fjords, and with landscapes of glacial erosion. These are on kilometric and larger scales; the smallest glacial cirques are of the order of 200 m long and broad.

Cirques show the former presence of cirque glaciers or the source areas of valley glaciers. Where ice from several cirques converged, forming a valley glacier, a trough was generally eroded. Further glacier growth in areas of high relief produced systems of tributary troughs hanging above major troughs, often with the development of rock basins. Ice cap or ice sheet development led to more complex systems of branching troughs, some formed by ice diffluent across a ridge from a valley with more ice to one with less (Linton, 1963). Ice streams further enlarged major troughs oriented in a direction permitting ice flow to lower or marine areas.

Only a few citations can be made here; further references can be found in Benn and Evans (1998),

Embleton (1972), Evans (1997), and Glasser and Bennett (2004).

Glacial Cirques

Glacial cirques are concave landforms formed at the sources of mountain glaciers. They are doubly concave hollows, open downstream but bounded upstream by the convex crest of a steep headwall (Evans and Cox, 1974). The headwall is arcuate in plan around a more gently sloping floor that should show the effects of glacial erosion, if not buried under moraine and postglacial accumulations. The word is related to 'circus,' but cirques must be open at one side. Their presence shows that a mountain range rose above the snowline long enough for small glaciers to form and to erode their beds cirques vary greatly in their degree of development, and it is useful to recognise five grades (**Fig. 1**).

As rates of glacial erosion are believed to increase with basal shear stress or velocity, and thus with ice thickness and surface slope (Hallet *et al.*, 1996), there may be surprise that such striking forms are produced at glacier sources, where discharge and basal velocity are low. The 'problem of cirque erosion' is to explain the concentration of erosion at the foot of the headwall. This is necessary to convert a nonglacial valley head into a cirque, and to steepen the headwall. Periglacial and ice-marginal processes such as freeze–thaw cannot achieve this; they would cause retreat of the upper part, reducing headwall gradient.

Factors contributing to this explanation include duration of glacier cover, positive feedback on glacier balance, and water-pressure variations. As the snowline falls over a mountain range, the first glaciers will form in high cirques. And as glaciers waste away, those in sheltered, shady cirques are likely to survive longest (**Figure 2**). The more deeply a cirque is eroded, the greater the shelter and (if pole facing) the shade provided, increasing the trapping of wind-blown snow and reducing the rate of melting. The increased mass balance gives a larger, thicker glacier which may erode more effectively, enlarging the cirque further and providing positive feedback. Extra input of snow at the top of a glacier, by wind drifting or avalanching, exaggerates the wedge of net accumulation and is effective in increasing velocity and tendency to rotate.

Cirque glaciers with gradients of roughly 10° to 30° tend to flow in a rotational manner, as a wedge of net accumulation in the upper half flows to replace a wedge of net ablation in the lower. In warm-based ice, this maximizes basal sliding and concentrates erosion at the foot of the headwall, deepening the cirque floor and helping develop a closed basin (Lewis, 1960). Many cirque glaciers are polythermal, and the boundary between ice frozen to the bed and that which is sliding gives possible instabilities that facilitate quarrying (Richardson and Holmlund, 1996). This is most effective in sedimentary or well-jointed rock, in which thrust sheets may develop (Bennett *et al.*, 1999). Continued erosion where ice is thickest, while peripheral ice is frozen to the headwall, may be a further factor undercutting the headwall. It is, however, limited to colder climates (continental and polar), as cold ice is not found near equilibrium line altitude (ELA) in snowy, maritime climates (see below).

Finally, the bergschrund and other crevasses formed as ice moves away from the headwall permit the periodic entry of meltwater. This causes variation in water pressure in subglacial pockets, facilitating the extension and opening of joints and the quarrying (plucking) of blocks of rock. This water-pressure mechanism (Hooke, 1991; Iverson, 1991) has replaced Lewis's freeze–thaw and meltwater hypotheses, as temperatures beneath warm ice vary very little. Although abrasion is sensitive to basal velocity, quarrying is not so directly affected and is important on lee slopes under shallow ice flowing.

Rapid glacial erosion near the sources of mountain glaciers is thus explicable, and has the further effect of steepening the headwall. This causes a threshold of stability to be exceeded, and the headwall retreats by episodic collapse in rock avalanches (Evans, 1997; McSaveney, 2002). They may occur when the glacier is active, but by removing support ('debuttressing' the slope) glacier recession makes them more likely. Again we note that frost action is not necessary; this is simply a slope stability problem, with slope gradient and height a function of rock mass strength (Augustinus, 1992). **Figure 3** shows a cirque floor covered by two rock avalanche deposits, the results of headwall collapse in ultrabasic rocks. Headwall collapse is a normal part of cirque erosion, as basal glacial erosion steepens headwalls beyond their maximum stable angle and height.

As many mountain areas have undergone tens of major glacial episodes in the Quaternary (and late Pliocene), considerable time is available for cirques to develop. Larsen and Mangerud (1981) studied deposits from a small cirque glacier on resistant mica-rich gneiss in western Norway. It eroded 118,000 m^3 during 700 yr of the Younger Dryas, equivalent to vertical lowering of the 0.21 km^2 cirque by 0.8 mm a^{-1}. At this modest rate, 200 m would be eroded in 250 kyr of glaciation, which is sufficient to account for development of most cirques. Glaciation must, however, be of the appropriate style; ice should be wet based, at least in part, and with steep surface gradients in cirques. The rapidity of climatic changes

Figure 1 Cirques of different grades. (A) Bowscale Tarn, a classic (grade 1) cirque in the northern Lake District, England. The headwall consists of slates with varying degrees of metamorphism: the lake is up to 16.5 m deep. The unusually large moraine (bottom right) was produced by a Loch Lomond Stadial (Younger Dryas) glacier with an ELA around 536 m and a surface slope of 17°; it was up to 95 m thick. (B) Craig Trum y Ddysgl, the west side of a well-defined (grade 2) cirque in tuffaceous Ordovician sandstones above the Nantlle, North Wales. (C) Lonscale Cove, a definite (grade 3) cirque in slates east of Skiddaw in the northern Lake District, England. (D) Browncove Crags, a poor (grade 4) cirque in Ordovician volcanic rocks on the northwest slopes of Helvellyn, eastern Lake District, England (E) Craig Rhiw-erch, a cirque in tuffaceous Ordovician siltstones, on Maesglase in the upper Dyfi basin, Mid Wales. It is marginal (grade 5) because of its steep (although drift-filled) floor. (*Photographs; I.S. Evans.*)

Figure 2 Coalescent north-facing cirques at the head of Truax Creek, Bendor Range, British Columbia Coast Mountains, with remnants of LIA cirque glaciers. The three summits are at 2,466, 2,517 and 2,589 m altitude: the rocks are thermally metamorphosed mudrocks. (*Photograph; I.S. Evans.*)

Figure 3 Rock avalanche covering a cirque floor in the Shulaps Range, British Columbia Coast Mountains. The outer deposit (left) is covered by the 2,500 B.P. Bridge River Ash; the inner deposit is younger. The rock is serpentinised peridotite (harzburgite). (*Photograph; I.S. Evans.*)

suggests that a prolonged phase of snowpatch activity (nivation) is unlikely before glaciers form. The nivation process-complex erodes more slowly than do warm-based glaciers; its importance has often been exaggerated, as has freeze–thaw action.

Spatially, glacial ELAs currently vary by 6,000 m globally, by 600–2,000 m between wet and dry sides of major mountain ranges, and by several hundred meters locally (Evans, 2006a). Glacial cirques likewise occur at lower altitudes on the snowy, windward sides of mid-latitude mountains. Globally, cirque floors are found down to an altitude below, and roughly paralleling, present ELA. As the zero (or other) isotherms follow a different pattern, both these observations relate cirques to glacial erosion rather than to frost action or any other factor directly related to atmospheric temperature. On a local scale, cirque glaciers and thus cirques favor shady and leeward slopes. In most areas they face poleward (**Fig. 4**), and in the west-wind belt this is deflected to the east, for example, in Europe, Tasmania, the Falklands, and the American Rockies.

Techniques for defining and measuring cirques are presented by Evans and Cox (1995) and, using geographic information system (GIS), by Federici and Spagnolo (2004). Where cirques are eroded into a plateau, their outlines are broadly rounded and the headwall crest approaches semicircular. Classically, simple forms are found in horizontally bedded rocks such as basalts (**Fig. 5**) or well-bedded sandstones. In sharp-crested mountains, coalescence of cirques gives more angular outlines (**Fig. 2**). Cirques are scale-specific forms, commonly between 200 m and 4 km broad and long with the upper limit dictated by valley spacing. They are rarely over 2 km in Britain; larger cirques found in high mountain areas tend to be complex, with smaller cirques nested within. Cirques around Mount Everest reach widths of 4 km, and larger features in the Antarctic (Alpine valley heads, some over 8 km wide and long; Haynes (1998)) reflect much longer periods of Alpine glaciation during the last 36 Myr. After deep glacial erosion, high-Alpine cirque forms develop around trough heads (**Fig. 6**; see also figure 2 of Evans (1997)).

Cirque development involves flattening of the floor (eventually developing a reversed slope), steepening of the headwall, increased closure in plan as the headwall recedes more deeply into the mountain mass, and increased length, width, and depth. As cirques develop, as expressed by their overall size, it appears that they increase in length faster than in width, and in depth least of all (Gordon, 1977; Federici and Spagnolo, 2004; Evans, 2006b) (**Fig. 7**). Cirque development is thus allometric rather than isometric; shape changes with size. For part of the Tien Shan, Oskin and Burbank (2005) have invoked cirque headwall retreat two to three times faster than downward erosion. On the other hand, in the Ben Ohau Range, New Zealand, cirques seem to lengthen and deepen faster than they widen (Brook *et al.*, 2006). Haynes (1998) invoked long-term branching and coalescence keeping width and length in balance, so that alpine valley heads attain an equilibrium shape in plan: continual addition of new small cirques maintains positively skewed frequency distributions as older cirques widen and lengthen. Attainment of a steady state, however, requires downward erosion to match headwall retreat not only in cirques but also in their surroundings, including the glacial troughs below. This is possible only after cirque coalescence has eliminated gently sloping

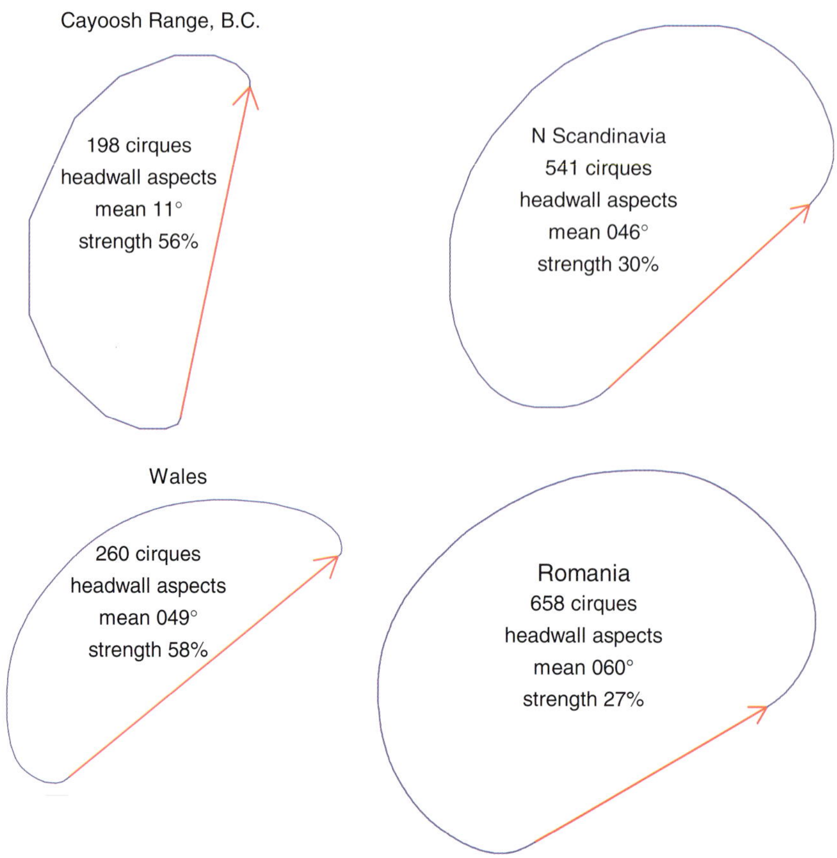

Figure 4 Vector plots of cirque aspect. The resultant vector is in red, the individual aspects in blue, plotted in cumulated vector form; scales vary. The Cayoosh Range, with a resultant a little east of north, is typical of fairly dry mountains of high relief, as in Central Asia and parts of British Columbia. The other three plots are all from the west-wind belt in Europe and show eastward tendencies as strong as northward, and stronger in Romania. Data for northern Scandinavia kindly supplied by S. Hassinen; for Romania, by M. Mindrescu. (*Graphs; I.S. Evans.*)

higher surfaces, producing a landscape of arêtes and horns. Arêtes are sharp ridges produced initially between cirques side-by-side on the glacially favored side of a ridge. Those between cirques back-to-back are more likely to produce deep cols, but this requires glaciation that is more symmetrical between slopes of different aspects. Locally symmetrical glaciation (Evans, 2006a) is found where either ELA is well below ridge altitudes or the climate is neither sunny nor very windy.

Figure 5 A trough-side cirque eroded into the basalt plateau above Isafjordur, northwest Iceland (Photo supplied by D. J. A. Evans).

Figure 6 High-Alpine cirques around a trough-head south of Tzoonie Mountain, 66 km NNW of Vancouver, British Columbia. (*Photograph; I.S. Evans.*)

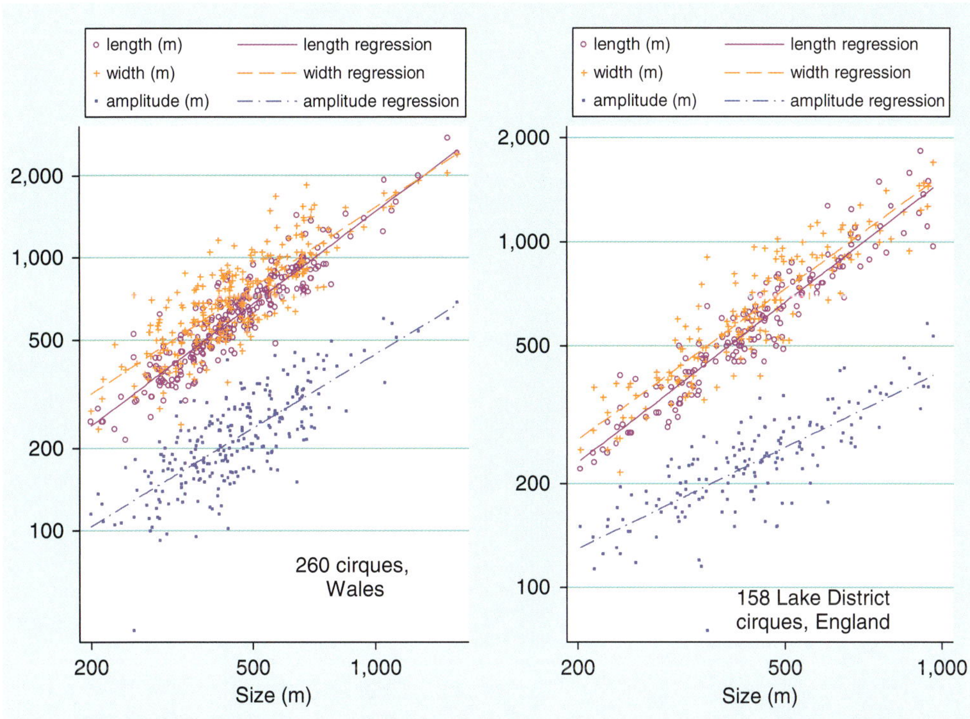

Figure 7 Allometric plots of length, width and height (amplitude) against overall size, for cirques in Wales and the English Lake District (scales differ and are logarithmic). In both regions the gradient for amplitude is significantly less than those for length and width. (*Graphs and data; I.S. Evans.*)

When the vertical dimension is considered, however, the form of cirques clearly changes (flattens) as they enlarge: they are not equilibrium forms. The factors discussed above permit quite rapid cirque erosion both downward and headward, and this is an important part of the 'glacial buzzsaw' (Mitchell and Montgomery, 2006) which gives a climatic limit to the altitude of all but the most rapidly rising mountains. Thus small glaciers are found in many mountain ranges, but the erosional effects of warm-based mountain glaciation mean that extensive ice fields are found outside polar regions only where uplift rates are very high: in southeast Alaska, Patagonia, and around the Tibetan Plateau.

Glacial Troughs

Calibration

Popularly termed 'U-shaped valleys,' glacial troughs are the channels of present or former glaciers or ice streams. Like rivers, glaciers obey the necessary geometric relationship:

$$Q = Av \cong wdv$$

Discharge equals mean velocity times cross-sectional area; the latter is approximately width times depth.

Ice discharge increases downstream to the equilibrium line (this is the firnline or end-of-summer snowline on glaciers outside polar regions) and declines thereafter through the ablation zone. Both A and v vary to accommodate discharge, so the largest troughs (glacier channels) are required where discharge is greatest, around the equilibrium line or where major glacier channels join below this. Erosion of troughs in proportion to ice discharge is known as calibration (Evans, 1997). A varies also with rock erodibility, and v varies with gradient, implying that narrow sections of troughs had steeper ice-surface gradients.

Given the great variation in discharge through the Quaternary, the question of the timing of formative discharge is important. Often this seems to relate to average Quaternary conditions, but toward glacier margins the late Quaternary glacial maxima seem more relevant. Diminished glaciers require smaller A and v and are expected to be much less effective unless there is a change of thermal state or basal hydrology permitting more rapid sliding. Increased gradient might also make smaller glaciers more effective.

Cross-Profiles

Channel cross-sections are roughly parabolic, 'U'-shaped or catenary; this may develop through erosion

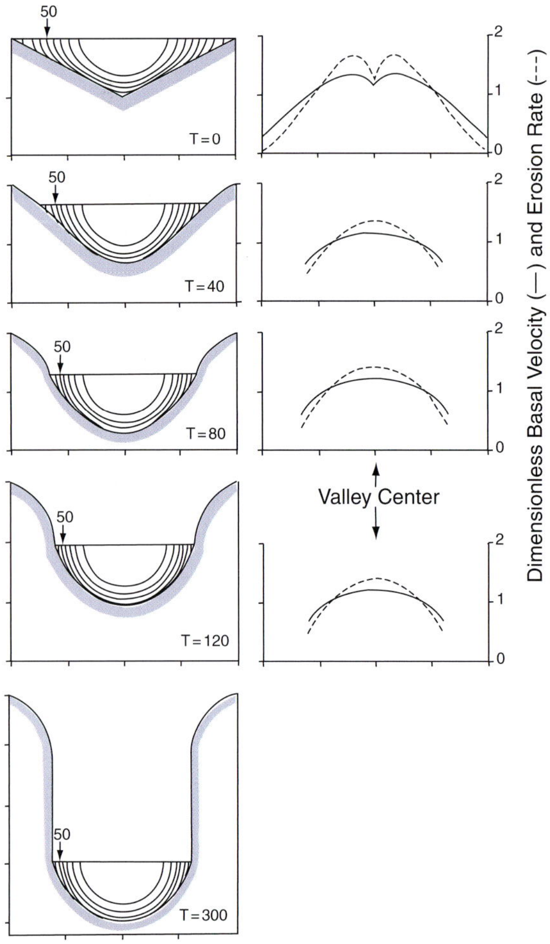

Figure 8 Simulation of 'U' cross profile development with erosion proportional to basal velocity squared. If the time-step, T, is in units of 1,000a, an erosion rate of 1 mm a-1 is represented. Glacier velocity contours (e.g. 50) are shown at intervals of 10% of maximum velocity in that profile. (*From Harbor 1992: 1368, Fig.5.*)

proportional to basal ice velocity or a power thereof. Harbor (1992) showed how the reduced shear stress and velocity at the apex of a V-shaped valley leads to erosion maxima at either side, until a 'U' cross-profile is formed (**Fig. 8**). Subglacial water erosion is a complicating factor, and even some deeply eroded glacial troughs remain closer to a V cross-profile in their lower reaches. Complex 'trough-in-trough' forms are unlikely, unless subglacial or interglacial water erosion incises a trough floor.

The depth/width ratio varies greatly: glacier paleochannels (troughs) are narrower and steeper-sided on rocks with greater rock mass strength (RMS), which most often arises from joints being widely spaced; rock sample strength (e.g., abrasion resistance) seems to have much less effect (Augustinus, 1992). Because of the importance of basal ice velocity, high discharges of warm-based ice are most effective in eroding troughs. Many major troughs relate to ice streams in former ice sheets, where basal sliding rates were high.

Trough deepening causes stress concentration, producing rock deformation or extra jointing, and thus aiding continued deepening: stress concentration on valley floors and shoulders, and at the bares of valley sides, is probably more effective than traditional 'pressure release' (Evans, 1997). Especially on weak rocks, with closer, wider, or more continuous joints or partings, the width of troughs comes largely from collapse of their sides on deglaciation, when glacier support is removed. Bovis (1990) has described the effects of 'debuttressing' over the last two centuries, opening up fissures that cut recent alluvium and moraines (**Fig. 9**). Many shoulders above troughs are cut by slope trenches or antislope scarps, which show arrested failures. Collapse is most likely during deglaciation after further glacial undercutting. This widening of troughs makes the deep narrow troughs of Harbor's long-term simulations unlikely except on the most massive rocks such as the atypical troughs illustrated in textbooks: Yosemite and Lauterbrunnen. Vertical incision is also limited by the need to maintain continuous slope downstream on the glacier surface profile. **Figure 10** shows a more typical, but still well-developed, glacial trough in the British Columbia Coast Mountains.

Trough cross-profiles can be described by the equation $y = a\,x^b$, where the origin is in the middle of the valley floor. b values are commonly between 1.5 and 2.0; the latter gives a true parabola. When fitting to observed heights y as a function of distance x, only points on bedrock should be included: points on alluvium, talus, or till are one-sided constraints – the bedrock surface is somewhere below. To specify trough size and define the scope of the equation, trough width and depth are also required. Alternative equations, such as quadratic $y = a + bx + cx^2$, have been found more flexible (**Fig. 11**).

Long Profiles

Glaciated valleys (troughs) are more deeply concave in long profile (i.e., lower near their sources) than fluvial valleys, because of the effectiveness of glacial erosion around the equilibrium line (**Fig. 12**). They are also stepped, influenced by geology (e.g., joint spacing) and by the initial, preglacial profile. Initial variations in gradient are exaggerated by a positive feedback, basins are excavated in flatter sections, and the height of steps is increased. Steps are exaggerated because fluctuations in water pressure on lee slopes

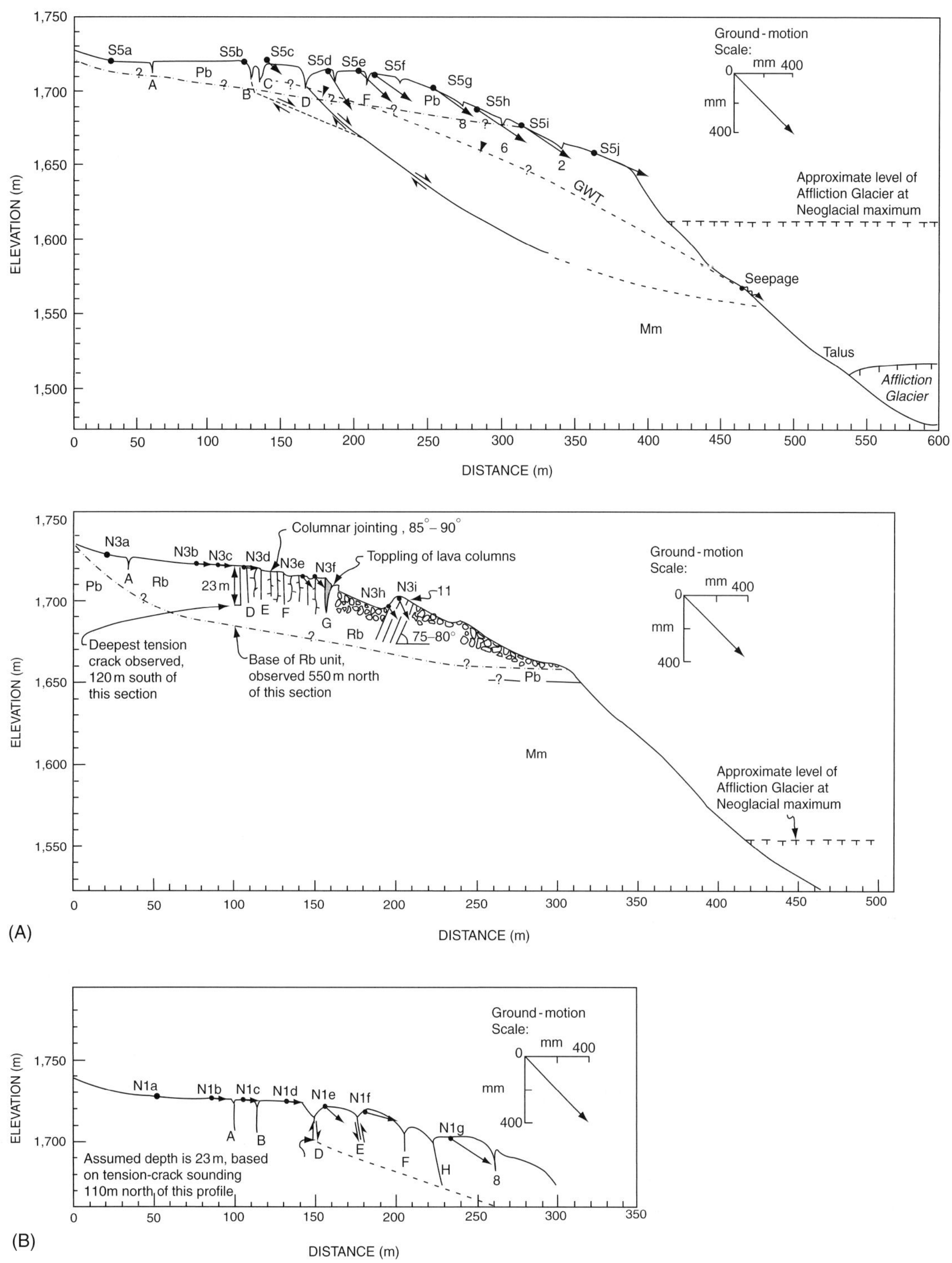

Figure 9 Debuttressing of a trough side in the Mount Meager massif by post-LIA glacier recession. Arrows are ground-motion vectors between 1982 and 1986. S5, N3 and N1 are three survey lines. Mm = Miocene monzonite; Pb = Pleistocene basalt; Rb = Recent basalt. (*From Bovis 1990. Canadian J. Earth Sciences. 27: 251, Fig.7 and 253, Fig. 9.*)

aid quarrying (*see* Micro to Macro Scale Forms). Many steps (riegels, verrous) relate to zones of more resistant (usually less jointed) rock, but others relate to position within the system of glacial channels; as troughs tend to be calibrated to ice discharge, steps are also expected at the confluences of major tributaries. Where the troughs are glacially calibrated, the degree of "hang" relates to the difference in ice

Figure 10 (A, B) Two views of the trough of Keary Creek, Bendor Range, British Columbia Coast Mountains. The foreground (of both photos) is on quartz-diorite; beyond the lake the slopes are on thermally metamorphosed mudrocks. The upper slopes are gullied; talus covers those below a mid-slope trimline. These are typical trough-sides: Keary Lake is a ribbon lake in a rock basin. (*Photographs; I.S. Evans.*)

discharge: large tributaries hang less than small (**Fig. 13**). This development of 'hanging valleys' has been simulated by Macgregor *et al.* (2000) (**Fig. 14**). When simulating long profiles, erosion is proportional either to basal velocity or to its rate of change, depending on whether erosion is 'production limited' or 'transport limited.'

Rock Basins and Fjords

The glacier bed can be irregular, with reversed slopes enclosing rock basins. The glacier surface must slope continuously down-valley to provide the necessary potential energy for flow (via shear stress). Of course, basins are also found in river beds, but on a much smaller scale. Glacier flow is approximately laminar whereas most rivers are turbulent; glacier velocities are lower than river velocities, but pressures and stresses are higher. Glaciers are able to transport large blocks of rock up reversed bed slopes and are better placed than rivers to excavate fractured material from their beds: there is no limit to their competence. Glaciers carry material upward either by thrusting or by emergent flow in the ablation zone (to replace surface ice lost by melting) (**Fig. 13**). Most rock basins thus relate to lithology or structure, but some are related to position, for example, in the lower ablation zone where ice flow has an upward component to balance surface melt. Around the Alps, it is notable how many deep lake basins lie just within the terminal moraines of major Pleistocene glaciers. In Scandinavia, British Columbia, Patagonia, and New Zealand the west coast fjords are almost matched by long ribbon lakes, east of the mountains.

Modern seismic work on sediments in large lakes (mainly valley rock basins) has shown great thicknesses of glacial, glaciofluvial, and postglacial sediment. The rock basins are even deeper than formerly thought (often twice as deep as the water: **Fig. 16**), and they are often dammed by massive rock thresholds, strengthening the case for glacial excavation. However, some of the deeper sediment is pre-last-glaciation, showing that deepest excavation was by larger, earlier glaciers. Rock basins are larger and deeper in the west of Scotland than in east, because of greater ice discharge due to heavier snowfall.

Fjords (**Figs. 5** and **17**) are spectacular troughs, often with deep rock basins (Holtedahl, 1967), and shallow water near their mouths (**Fig. 18**). Their thresholds are often where ice flow can diverge, at the edge of the mountains. Flotation and calving of ice (reducing discharge more abruptly than does surface melting) provide an extra explanation for the reversed slope. Never the less ribbon lakes which are almost as spectacular of all ranges with fjords, formed largely under ablation zone, are found on

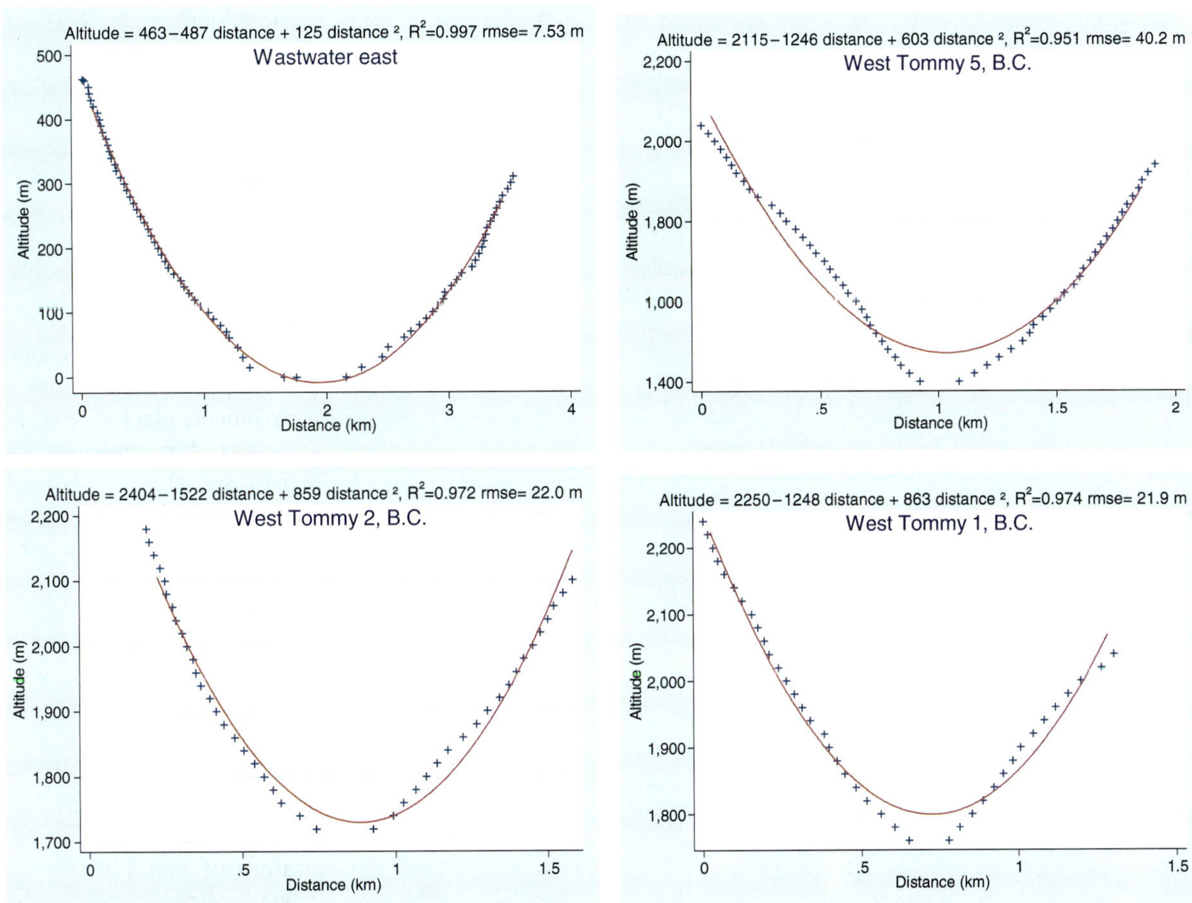

Figure 11 Fitting quadratic curves to a glacial trough cross-section. Each point plots the intersection of a contour with the profile. Note that an R^2 of 0.95 is obtained for a poor fit (top right), and even an R^2 of 0.97 (bottom two) does not guarantee a good fit, though 0.997 (top left) does. Quadratic curves are more robust than power curves (for which a complex iterative procedure is recommended). (*Graphs; I.S. Evans.*)

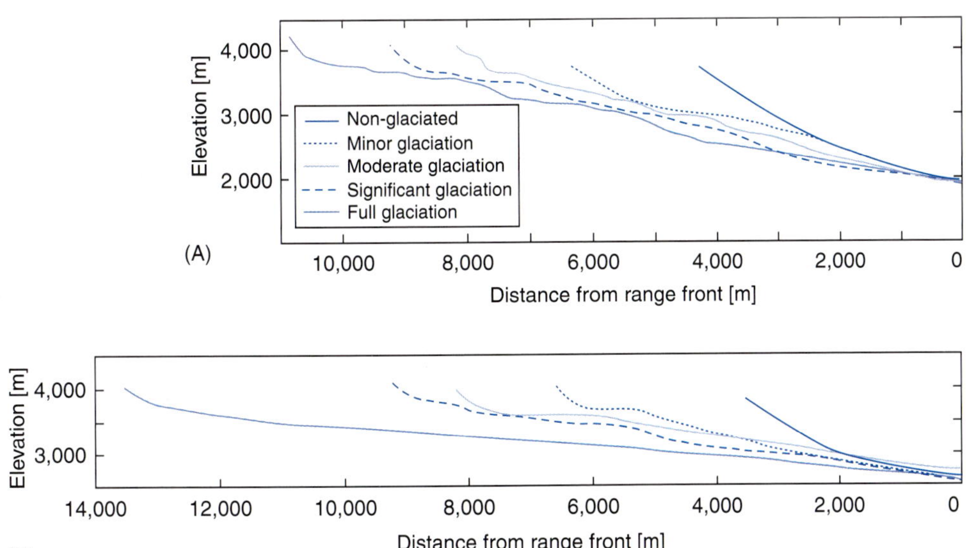

Figure 12 Long profiles of streams on the east slope of (a) the Sierra Nevada, California, and (b) the Sangre de Cristo Mountains, Colorado: no vertical exaggeration. These show the effects of glacial erosion in deepening and, especially, extending valleys, in areas of marginal glaciation (valley glaciers). (*From Brocklehurst and Whipple 2006: 287, Fig.2.*)

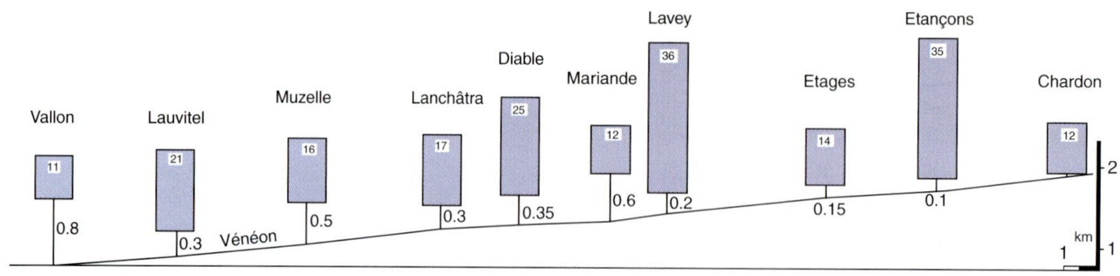

Figure 13 On uniform granitic rocks, large tributaries of the Véněon (Oisans, French Alps) hang less than small. Shaded bars are proportional to tributary catchment areas in km^2; numbers below give step height in km. Step height also increases downstream as the trunk catchment area increases. (*From G. Monjuvent 1974: 486, Fig.11.*)

landward slopes. Upward flow of ice encourages erosion of rock basins (**Fig. 15**).

Fjords sometimes follow geological structures, but often cut across contrasting rock types. They are, of course, structural landforms due to erosion exploiting fracture systems (**Fig. 19**) (Holtedahl, 1967), not tectonic landforms due to earth movements. Classic fjords are transverse to coasts, but spectacular longitudinal fjords are found in Alaska, British Columbia, and Chile. These reflect diffluent flow of ice, probably eroding lines of structural weakness.

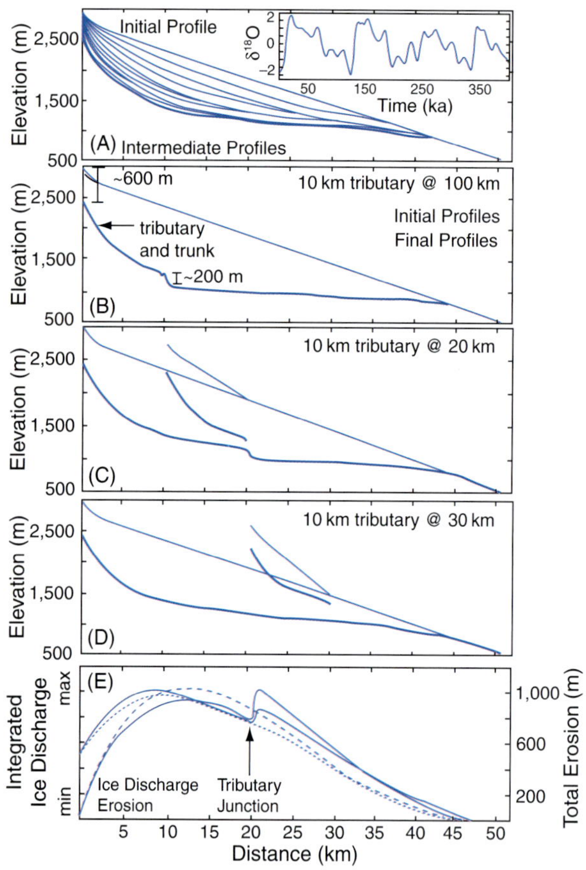

Figure 14 Simulation of long profiles, with glaciers varying over the last 400,000 years of environmental conditions and erosion proportional to basal sliding velocity. (A) Single glacier with no tributary; profiles at 40ka intervals. INSET marine $\delta^{18}O$ curve used to simulate temperature. (B) Initial and final profiles with tributary joining at 10 km from sources, i.e. equal to trunk; both 'hang' by 200 m. (C) Tributary at 20 km; a 100 m step is generated, with the tributary hanging by a further 100 m, and a shallow rock basin below. (D) Tributary at 30 km hangs by 200 m and trunk profile is little affected. (E) Long-term ice discharge (black) and total erosion (grey) in trunk valley, for tributary at 20 km (solid lines) and for no tributary (dashed lines). (*From MacGregor et al. 2000: 1033, Fig. 3.*)

Dating and Rates

It is difficult to date erosional forms; they relate mainly to periods of greatest basal ice velocity, allowing for duration. Some troughs and basins, for example, around the periphery of the Alps, developed only around times of glacial maxima. Others relate to intermediate, 'average Quaternary' conditions of much-repeated valley glaciation in mountains. Most troughs developed by the deepening and widening of preglacial valleys, but some appear to have developed solely by glacial erosion.

In Norway and British Columbia, great contrasts in precipitation between coastal and inland areas have affected the degree of glacial activity. Rudberg (1997) made estimates of the volume of rock removed to produce various landforms in central Norway, comparing the 15,100 km^2 coastal province of Møre og Romsdal with an adjacent comparable area inland:

Volume in km^3	*Coastal*	*Interior*
V-shaped fluvial valleys	15	22
Cirques	60	4
Trough valleys	2,150	330
Fjords/lakes	1,290	4
Sounds and submarine valleys	265	
Strandflat	270	

Although very rough, these estimates clearly show the much more intense glacial transformation of the land surface in the coastal area, of the order of tenfold.

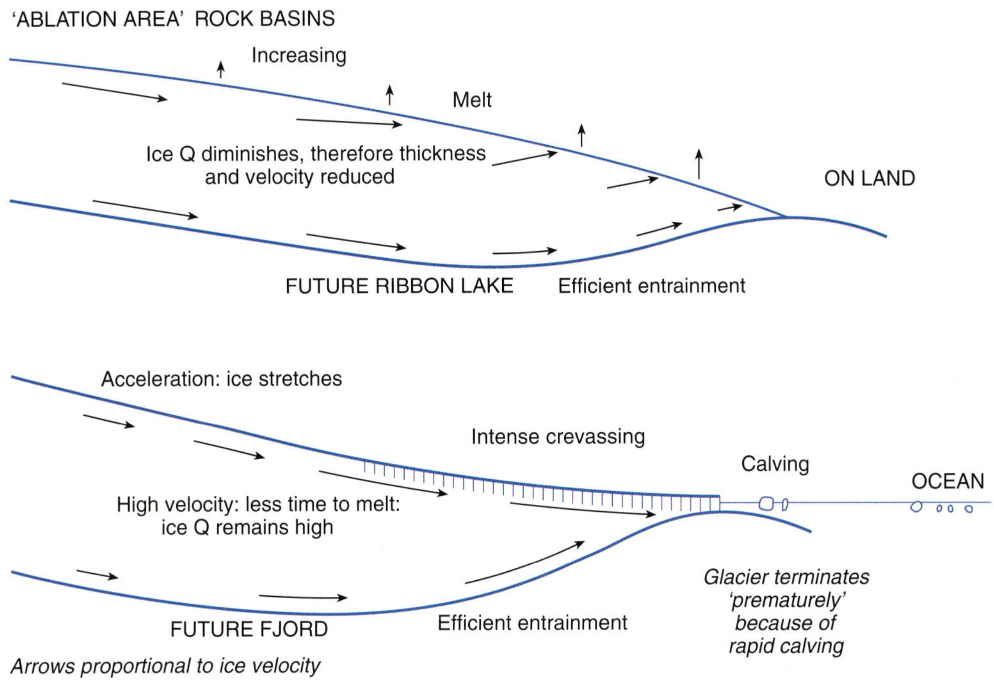

Figure 15 Erosion of 'ablation area' rock basins; similarities and differences between land- and water-terminating glaciers, producing ribbon lakes and fjords respectively. The upward flow tendency of a normal ablation area aids entrainment and excavation of a rock basin as the glacier thins toward the snout. This is enhanced in a tidewater glacier as it accelerates toward its calving front. (*Sketch I.S. Evans.*)

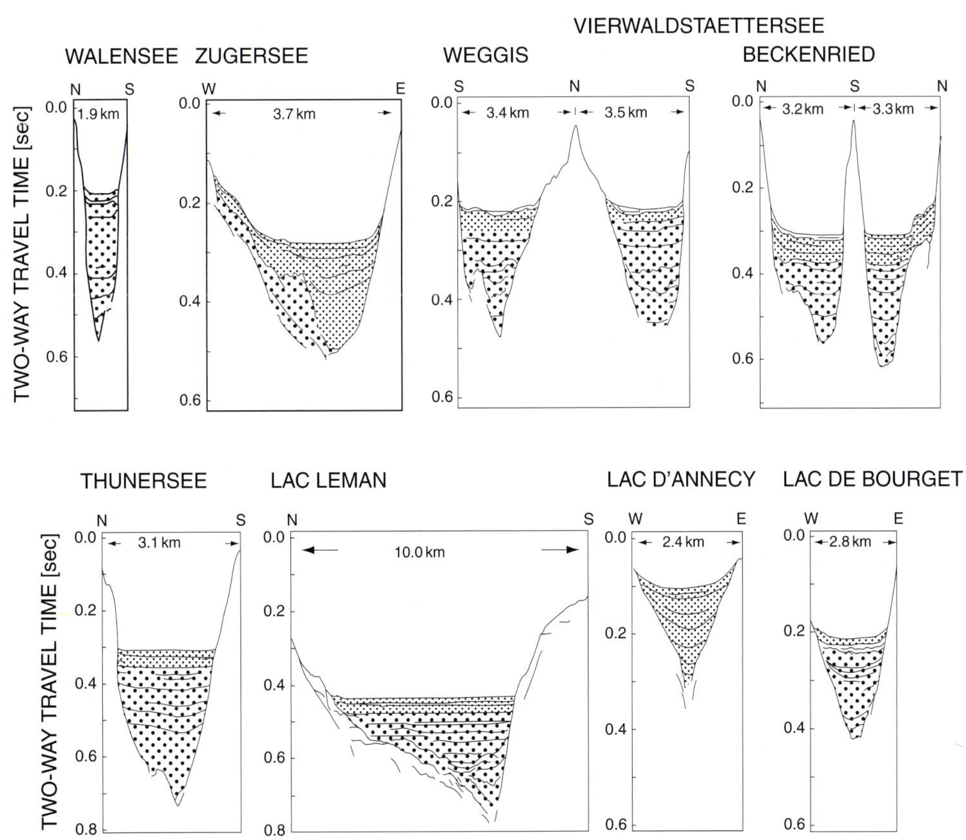

Figure 16 Cross-profiles of lake basins and sediments in the northwestern Alps (Switzerland and France). Basin depth is often twice depth, as rapid sedimentation during deglaciation tends to fill the sediment trap formed by a lake. (*From Finckh, P., Kelts, K. and Lambert, A. 1984: 1121-22, Fig.2.*)

Figure 17 A British Columbian Fjord: Jervis Inlet. (A) Queen's Reach, looking north: altitudes rise to 1982 m (left) and 1753 m (right); (B) Princess Royal Reach, looking southwest: altitudes rise to 2282 m on the right and around 2600 m in the distance. (*Photographs: I.S. Evans*)

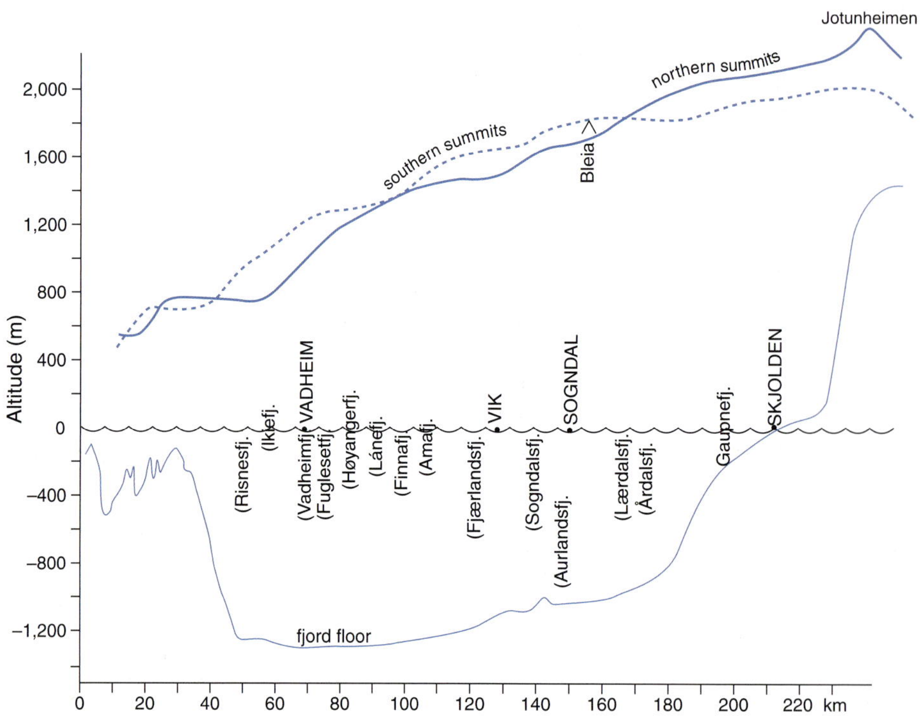

Figure 18 Long profile of Sognefjord (with its Lustrafjord source), western Norway. The level of hanging tributary fjord floors is shown by (Name. Profiles of summit levels on both sides are shown. The fjord floor is covered by an average thickness of 200 m of sediment. (*From Nesje and Whillans 1994: 35, Fig.2.*)

Full trough development requires erosion of 50–1,000 m, which can be achieved at 0.1–2 mm a^{-1} for several hundred thousand years. These are the rates observed from small Norwegian glaciers in recent decades, and greater rates are found for larger glaciers especially where basal velocities are high (Hallet *et al.*, 1996). Recent studies using terrestrial cosmogenic nuclide concentrations confirm the selectivity of glacial erosion. For example, from striated surfaces near the downstream ends of two glacial troughs, some 2.5 m has been eroded during the last major glaciation near the trough center, declining to lesser depths near the margins (Fabel *et al.*, 2004), as predicted by Harbor's model. This study was based on abraded areas, not quarried. Further upstream, the depth of scour was greater, with complete removal of inherited nuclides throughout the cross-section.

If troughs are calibrated, are they equilibrium forms unlike most cirques? Once a well-calibrated trough has developed, erosion may slow down and sediments may protect bedrock until greater ice

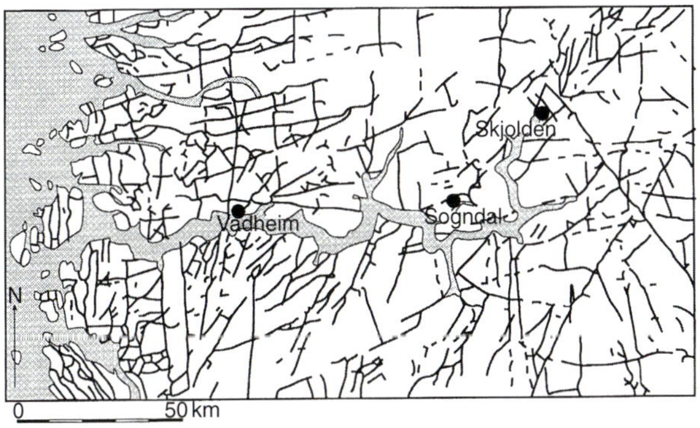

Figure 19 Map showing the relation of fjords to the fracture system around Sognefjord, western Norway. (*From Nesje and Whillans 1994: 39, Fig.6.*)

discharges or velocities are achieved. Smaller, slower glaciers are likely to erode less bedrock, but abrading and quarrying (as well as slope collapse) do continue during deglaciation. Trough widening is a one-way process, unlikely to be matched by downward erosion in a steady state; the positive feedbacks involved make attainment of equilibrium unlikely.

Landscapes of Glacial Erosion

Sugden (1974) noted the importance of relief and basal temperature in developing different landscapes of glacial erosion. Where ice had remained frozen to its bed, there was hardly any erosion unless meltwater from upstream froze onto the glacier base. Where deeper valleys had permitted basal melting and thus sliding, they were deepened further while intervening uplands were unaffected (**Fig. 20**). Valleys focusing ice flow and facilitating evacuation to areas of net ablation permitted faster flow and generation of more heat and more basal melting; positive feedback enhanced the landscape contrasts. Where a whole landscape was covered by warm-based ice, it was scoured into rocky hills and lake basins. These three types were labeled, respectively, landscapes of no erosion, of selective linear erosion, and of areal scouring. They contrast also with a fourth, alpine landscape, of cirques and troughs. Kleman and Stroeven (1997) have demonstrated the selectivity of glacial erosion in the Swedish mountains.

Landscapes of areal scouring might be expected to show streamlining and thus lineation. This is common on weak rocks, for example, around the Forth in central Scotland (Linton, 1963). On resistant metamorphic and igneous rocks of the Canadian and Fennoscandian shields, there is strong structural control of glacial erosion with lake basins along fault and major joint lines, and on weaker rocks (**Fig. 21**). This

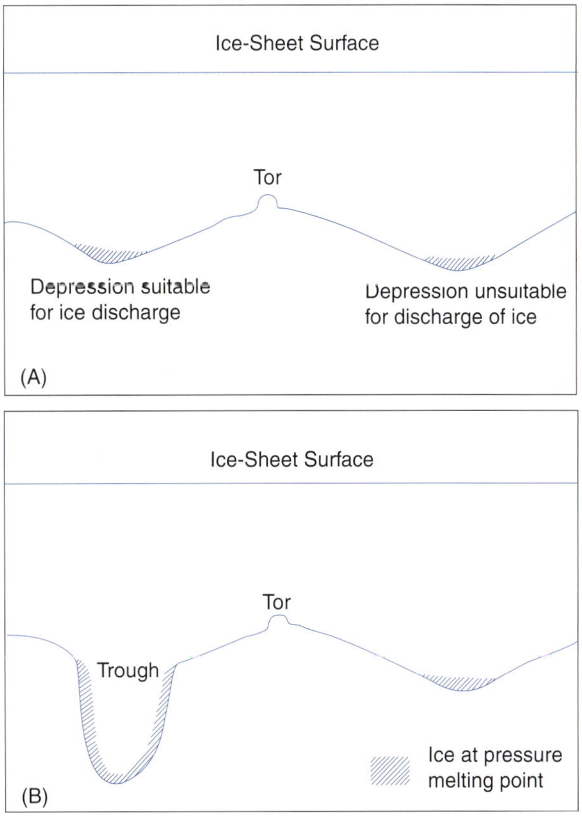

Figure 20 Selective linear erosion of troughs requires basal ice at pressure melting point, in a depression that can channel ice flow. (*From Sugden, D.E. 1974: 185, Fig. 8.*)

Figure 21 NASA Landsat image of Canadian Shield around the eastern arm of Great Slave Lake, Canada. Cuestas in the Lake show differential glacial erosion among bedded volcanic rocks. Elsewhere the pattern of small lakes shows erosion of thousands of rock basins in Precambrian (Archaean) igneous and metamorphic rocks by the Laurentide Ice Sheet.

leads some to infer that glaciation has simply removed a mantle of weathered rocks; this is clearly the case in some marginal areas of decelerating flow. In the absence of deeply weathered material, erosion of bedrock may have been considerable, but is difficult to measure.

See also: **Glacial Landforms**: Glacifluvial Landforms of Erosion. **Glacial Landforms, Erosional Features**: Micro to Macro Scale Forms.

References

Augustinus, P. C. (1992). The influence of rock mass strength on glacial valley cross-profile morphometry: N.Z. *Earth Surface Processes and Landforms* **17**, 39–51.

Benn, D. I., and Evans, D. J. A. (1998). *Glaciers and Glaciation*, pp. 178–196. Arnold, London.

Bennett, M. R., Huddart, D., and Glasser, N. F. (1999). Large-scale bedrock displacement by cirque glaciers. *Arctic, Antarctic and Alpine Research* **31**, 99–107.

Bovis, M. J. (1990). Rock-slope deformation at Affliction Creek, southern Coast Mountains, British Columbia. *Canadian Journal of Earth Sciences* **27**, 243–254, (postscript, 2000; 1998 collapse; *Canadian Journal of Earth Sciences* **37**: 1321–34).

Brocklehurst, S. H., and Whipple, K. X. (2006). Assessing the relative efficiency of fluvial and glacial erosion through simulation of fluvial landscapes. *Geomorphology* **75**, 283–299.

Brook, M. S., Kirkbride, M. P., and Brock, B. W. (2006). Cirque development in a steadily uplifting range: Rates of erosion and long-term morphometric change in alpine cirques in the Ben Ohau Range, New Zealand. *Earth Surface Processes and Landforms* 10.1002/esp.1327, available on-line 5 April 2006; not yet paginated.

Embleton, C. E. (Ed.) (1972). *Glaciers and Glacial Erosion* (readings, translations), Macmillan, London.

Evans, I. S. (1997). Process and form in the erosion of glaciated mountains. In *Process and Form in Geomorphology* (D. R. Stoddart, Ed.), pp. 145–174. Routledge, London.

Evans, I. S. (2006a). Local aspect asymmetry of mountain glaciation: A global survey of consistency of favoured directions for glacier numbers and altitudes. *Geomorphology* **73**(1–2), 166–184.

Evans, I. S. (2006b). Allometric development of glacial cirque form: Geological, relief and regional effects on the cirques of Wales. *Geomorphology* 10.1016/j.geomorph.2006.02.013, available on-line 20 March 2006; not yet paginated.

Evans, I. S., and Cox, N. J. (1974). Geomorphometry and the operational definition of cirques. *Area* **6**, 150–153, (IBG, London).

Evans, I. S., and Cox, N. J. (1995). The form of glacial cirques in the English Lake District, Cumbria. *Zeitschrift für Geomorphologie, N.F.* **39**, 175–202.

Fabel, D., Harbor, J., Dahms, D., *et al.* (2004). Spatial patterns of glacial erosion at a valley scale derived from terrestrial cosmogenic ^{10}Be and ^{26}Al concentrations in rock. *Annals of the Association of American Geographers* **94**, 241–255.

Federici, P. R., and Spagnolo, M. (2004). Morphometric analysis on the size, shape and areal distribution of glacial cirques in the Maritime Alps (Western French-Italian Alps). *Geografiska Annaler* **86A**, 235–248.

Finckh, P., Kelts, K., and Lambert, A. (1984). Seismic stratigraphy and bedrock forms in perialpine lakes. *Bulletin of the Geological Society of America* **95**, 1118–28.

Glasser, N. F., and Bennett, M. R. (2004). Glacial erosional landforms: Origins and significance for palaeoglaciology. *Progress in Physical Geography* **28**, 43–75.

Gordon, J. (1977). Morphometry of cirques in the Kintail-Affric-Cannich area of NW Scotland. *Geografiska Annaler* **59A**, 177–194.

Hallet, B., Hunter, L., and Bogen, J. (1996). Rates of glacial erosion and sediment evacuation by glaciers: A review of field data and their implications. *Global and Planetary Change* **12**, 213–235.

Harbor, J. M. (1992). Numerical modeling of the development of U-shaped valleys by glacial erosion. *Bulletin of the Geological Society of America* **104**, 1364–1375.

Haynes, V. M. (1998). The morphological development of alpine valley heads in the Antarctic Peninsula. *Earth Surface Processes and Landforms* **23**, 53–68.

Holtedahl, H. (1967). Notes on the formation of fjords and fjord-valleys. *Geografiska Annaler* **49A**, 188–203.

Hooke, R. Le B. (1991). Positive feedbacks associated with erosion of glacial cirques and overdeepenings. *Bulletin of the Geological Society of America* **103**, 1104–1108.

Iverson, N. R. (1991). Potential effects of subglacial water-pressure fluctuations on quarrying. *Journal of Glaciology* **37**(125), 27–36.

Kleman, J., and Stroeven, A. P. (1997). Preglacial surface remnants and Quaternary glacial regimes in northwestern Sweden. *Geomorphology* **19**, 35–54.

Larson, E., and Mangerud, J. (1981). Erosion rate of a Younger Dryas cirque glacier at Kråkenes, western Norway. *Annals of Glaciology* **2**, 153–158.

Lewis, W. V. (Ed.) (1960). *Norwegian Cirque Glaciers, R.G.S. Research Series, 4, 104p.* Royal Geographical Society, London.

Linton, D. L. (1963). The forms of glacial erosion. *Transactions of the Institute of British Geographers* **33**, 1–28.

Macgregor, K. R., Anderson, R. S., Anderson, S. P., and Waddington, E. D. (2000). Numerical simulations of glacial-valley longitudinal profile evolution. *Geology* **28**, 1031–1034.

McSaveney, M. J. (2002). Recent rockfalls and rock avalanches in Mount Cook National Park, New Zealand. In *Catastrophic Landslides: Effects, Occurrence, and Mechanisms* (S. G. Evans and J. V. DeGraff, Eds.), (Reviews in Engineering Geology XV)., pp. 35–70. Geological Society of America, Boulder, CO.

Mitchell, S. G., and Montgomery, D. R. (2006). Influence of a glacial buzzsaw on the height and morphology of the Cascade Range in central Washington State, USA. *Quaternary Research* **65**, 96–107.

Monjuvent, G. (1974). Considérations sur le relief glaciaire a propose des Alpes de Dauphiné. *Revue de Géographie Physique et de Géologie dynamique*, série **2**, 16, 466–502.

Nesje, A. and Whillans, I. M. (1994). Erosion of Sognefjord, Norway. *Geomorphology* **9**, 33–45.

Oskin, M., and Burbank, D. W. (2005). Alpine landscape evolution dominated by cirque retreat. *Geology* **33**, 933–936.

Richardson, C., and Holmlund, P. (1996). Glacial cirque formation in northern Scandinavia. *Annals of Glaciology* **22**, 102–106.

Rudberg, S. (1997). Glacial and interglacial erosion in Scandinavian mountains in a west–east comparison including an approach to a quantitative calculation. *Zeitschrift fur Geomorphologie* **41**, 183–204.

Sugden, D. E. (1974). Landscapes of glacial erosion in Greenland and their relationship to ice, topographic and bedrock conditions. In *Progress in Geomorphology* (E. H. Brown and R. S. Waters, Eds.), 7, pp. 177–195. Institute of British Geographers Special Publication. IBG, London.

Set
ISBN-13: 978-0-444-51919-1
ISBN-10: 0-444-51919-X

Volume 1 of 4 Volumes
ISBN-13: 978-0-444-51920-7
ISBN-10: 0-444-51920-3